ADVANCES IN MARITIME TECHNOLOGY AND ENGINEERING

This book comprises a collection of the papers presented at the 7th International Conference on Maritime Technology and Engineering (MARTECH 2024) held in Lisbon, Portugal, on 14-16 May 2024. This Conference has evolved from the series of biannual national conferences in Portugal, which have become an international event, reflecting the internationalization of the maritime sector and its activities. MARTECH 2024 is the seventh of this new series of biannual conferences. This book comprises 142 contributions that were reviewed by an International Scientific Committee.

This book is dedicated to maritime transportation and ports as well as maritime traffic modelling and maritime safety and reliability. It further comprises sections dedicated to structural analysis and design, such as ultimate strength and composites, as well as maintenance of marine structures.

The "Proceedings in Marine Technology and Ocean Engineering" series is dedicated to the publication of proceedings of peer-reviewed international conferences dealing with various aspects of "Marine Technology and Ocean Engineering". The series includes the proceedings of the following conferences: the International Maritime Association of the Mediterranean (IMAM) conferences, the Marine Structures (MARSTRUCT) conferences, the Renewable Energies Offshore (RENEW) conferences and the Maritime Technology (MARTECH) conferences. The "Marine Technology and Ocean Engineering" series is also open to new conferences that cover topics on the sustainable exploration of marine resources in various fields, such as maritime transport and ports, usage of the ocean including coastal areas, nautical activities, the exploration and exploitation of mineral resources, the protection of the marine environment and is resources, and risk analysis, safety and reliability. The aim of the series is to stimulate advanced education and training through the wide dissemination of the results of scientific research.

Proceedings in Marine Technology and Ocean Engineering

ABOUT THE SERIES

The 'Proceedings in Marine Technology and Ocean Engineering' series is devoted to the publication of proceedings of peer-reviewed international conferences dealing with various aspects of 'Marine Technology and Ocean Engineering'. The Series includes the proceedings of the following conferences: the International Maritime Association of the Mediterranean (IMAM) Conferences, the Marine Structures (MARSTRUCT) Conferences, the Renewable Energies Offshore (RENEW) Conferences and the Maritime Technology and Engineering (MARTECH) Conferences. The 'Marine Technology and Ocean Engineering' series is also open to new conferences that cover topics on the sustainable exploration and exploitation of marine resources in various fields, such as maritime transport and ports, usage of the ocean including coastal areas, nautical activities, the exploration and exploitation of mineral resources, the protection of the marine environment and its resources, and risk analysis, safety and reliability. The aim of the series is to stimulate advanced education and training through the wide dissemination of the results of scientific research.

BOOKS IN THE SERIES

Volume 1: Advances in Renewable Energies Offshore, 2019, C. Guedes Soares (Ed.).
Volume 2: Trends in the Analysis and Design of Marine Structures, 2019, J. Parunov and C. Guedes Soares (Eds.).
Volume 3: Sustainable Development and Innovations in Marine Technologies, 2020, P. Georgiev and C. Guedes Soares (Eds.).
Volume 4: Developments in the Collision and Grounding of Ships and Offshore Structures, 2020, C. Guedes Soares (Ed.).
Volume 5: Developments in Renewable Energies Offshore, 2021, C. Guedes Soares (Ed.).
Volume 6: Developments in Maritime Technology and Engineering, 2021, C. Guedes Soares and T.A. Santos (Eds.)
Volume 7: Developments in the Analysis and Design of Marine Structures, 2021, J. Amdahl and C. Guedes Soares (Eds.)
Volume 8: Trends in Maritime Technology and Engineering, 2022, C. Guedes Soares and T.A. Santos (Eds.)
Volume 9: Sustainable Development and Innovations in Marine Technologies, 2023, Selma Ergin and C. Guedes Soares (Eds.)
Volume 10: Trends in Renewable Energies Offshore, 2023, C. Guedes Soares (Ed.)
Volume 11: Advances in the Analysis and Design of Marine Structures, 2023, J.W. Ringsberg and C. Guedes Soares (Eds.)
Volume 12: Advances in the Collision and Grounding of Ships and Offshore Structures, 2023, H. Le Sourne and C. Guedes Soares (Eds.)
Volume 13: Advances in Maritime Technology and Engineering, 2024, C. Guedes Soares and T.A. Santos (Eds.)

Proceedings in Marine Technology and Ocean Engineering (Print): ISSN: 2638-647X
Proceedings in Marine Technology and Ocean Engineering (Online): eISSN: 2638-6461

PROCEEDINGS OF THE 7TH INTERNATIONAL CONFERENCE ON MARITIME TECHNOLOGY AND ENGINEERING (MARTECH 2024), 14–16 MAY 2023, LISBON, PORTUGAL

Advances in Maritime Technology and Engineering

Celebrating 30 years of the Centre for Marine Technology and Ocean Engineering (CENTEC)

Volume 1

Edited by

C. Guedes Soares

Centre for Marine Technology and Ocean Engineering (CENTEC), Técnico Lisboa, Universidade de Lisboa, Portugal

Tiago A. Santos

Ordem dos Engenheiros, Portugal

CRC Press
Taylor & Francis Group
Boca Raton London New York

CRC Press is an imprint of the
Taylor & Francis Group, an **informa** business
A BALKEMA BOOK

Designed cover image: Authors

First published 2024
by CRC Press/Balkema
4 Park Square, Milton Park, Abingdon, Oxon, OX14 4RN

and by CRC Press/Balkema
2385 NW Executive Center Drive, Suite 320, Boca Raton FL 33431

CRC Press/Balkema is an imprint of the Taylor & Francis Group, an informa business

Typeset by Integra Software Services Pvt. Ltd., Pondicherry, India

British Library Cataloguing-in-Publication Data
A catalogue record for this book is available from the British Library

Library of Congress Cataloging-in-Publication Data
A catalog record has been requested for this book

Set:
ISBN: 978-1-032-83327-9 (hbk)
ISBN: 978-1-032-83328-6 (pbk)

Volume 1:
ISBN: 978-1-032-83099-5 (hbk)
ISBN: 978-1-032-83310-1 (pbk)
ISBN: 978-1-003-50876-2 (ebk)
DOI: 10.1201/9781003508762

Volume 2:
ISBN: 978-1-032-83107-7 (hbk)
ISBN: 978-1-032-83316-3 (pbk)
ISBN: 978-1-003-50877-9 (ebk)
DOI: 10.1201/ 9781003508779

Advances in Maritime Technology and Engineering – Guedes Soares & Santos (Eds)
© 2024 The Editor(s), ISBN 978-1-032-83099-5

Table of contents

Preface

Since 1987, the Naval Architecture and Marine Engineering branch of the Portuguese Association of Engineers (Ordem dos Engenheiros) and the Centre for Marine Technology and Ocean Engineering (CENTEC) of the Instituto Superior Técnico (IST), University of Lisbon, (formerly Technical University of Lisbon) have been organizing national conferences on Naval Architecture and Marine Engineering. Initially, they were organised annually and later became biannual events.

These conferences had the objective of bringing together Portuguese professionals giving them an opportunity to present and discuss the ongoing technical activities. The conferences have been typically attended by 150 to 200 participants and the number of papers presented at each one was in the order of 30 in the beginning and 140 in the latest events.

At the same time as the conferences have matured, the international contacts have also increased and the industry became more international in such a way that the fact that the conference was in Portuguese started to hinder its further development with wider participation. Therefore, a decision was made to experiment with having also papers in English, mixed with the usual papers in Portuguese. This was first implemented in the First International Conference of Maritime Technology and Engineering (MARTECH 2011), but, subsequently, five more MARTECH conferences have been organized in 2014, 2016, 2018, 2020 and 2022.

This year, the MARTECH 2024 Conference also celebrates the 30th anniversary of the foundation of the Centre for Marine Technology and Ocean Engineering (CENTEC) in 1994, which has been one of the main contributors to the organization of this Conference representing the contribution of Técnico Lisboa (IST).

For the Seventh International Conference of Maritime Technology and Engineering (MARTECH 2024), a total of around 200 abstracts have been received and 142 full papers were finally accepted. Also included in the book are abstracts of the four Keynote Lectures.

The Scientific Committee had a major role in the review process of the papers although several other anonymous reviewers have also contributed and are deserving of our thanks for the detailed comments provided to the authors allowing them to improve their papers. The participation is coming from research and industry from almost every continent, which is also a demonstration of the wide geographical reach of the conference.

The contents of the present books are organized in the main subject areas corresponding to the sessions in the Conference and within each group the papers are listed by the alphabetic order of the authors.

We want to thank all contributors for their efforts and the sponsoring Institutions for their support, hoping that this Conference will be continued and improved in the future.

C. Guedes Soares & T.A. Santos

Organizing Committee

Conference Chairmen
Carlos Guedes Soares, Técnico Lisboa, Universidade de Lisboa, Portugal
Dina Dimas, Ordem dos Engenheiros, Portugal

Organizing Committee
Tiago A. Santos, Ordem dos Engenheiros & Técnico Lisboa, Portugal
Ângelo Teixeira, Técnico Lisboa, Universidade de Lisboa, Portugal
José Manuel Cruz, Ordem dos Engenheiros, Portugal
Manuel Ventura, Técnico Lisboa, Universidade de Lisboa, Portugal
Francisco C. Salvado, Ordem dos Engenheiros, Portugal
José Gordo, Técnico Lisboa, Universidade de Lisboa, Portugal
Raúl Caria, Ordem dos Engenheiros, Portugal
José Gaspar, Técnico Lisboa, Universidade de Lisboa, Portugal
Pedro Pereira da Silva, Ordem dos Engenheiros, Portugal

Scientific Programme Committee
Ahmed Dursun Alkan, Yildiz Techn. University, Turkey
Ole Andreas Alsos, NTNU, Norway
Felice Arena, Mediterranea Univ of Reggio Calabria, Italy
Ermina Begovic, University of Naples, Italy
Kostas Belibassakis, NTUA, Greece
Franz von Bock und Polach, TUHH, Germany
Maria Boile, University of Piraeus, Greece
Evangelos Boulougouris, Univ. of Strathclyde, UK
Nian Zhong Chen, Tianjin University, China
Ranadev Datta, IIT Kharagpur, India
Nastia Degiuli, University of Zagreb, Croatia
Vicente Díaz-Casás, Universidad A Coruña, Spain
Leonard Domnisoru, Univ. "Dunarea de Jos, Galati, Romania
Bettar Ould el Moctar, Univ. of Duisburg-Essen, Germany
Segen F. Estefen, UFRJ, Brazil
Massimo Figari, University of Genova, Italy
C.J. Fortes, LNEC, Portugal
Yordan Garbatov, Técnico Lisboa, Portugal
Spyros Hirdaris, Aalto University, Finland
Zhiqiang Hu, Newcastle University, UK

Xiaoli Jiang, TU Delft, The Netherlands
Debabrata Karmakar, National Inst. Techn. Karnataka, India
Iraklis Lazakis, Univ. of Strathclyde, UK
Bernt Leira, NTNU, Norway
Rui Li, Dalian University of Technology, China
Xavier Martínez-Garcia, Uni. Politécnica de Catalunya, Spain
Alba Martínez-López, Univ. Las Palmas Gran Canária, Spain
Jakub Montewka, Gdansk University of Techn., Poland
Lorenzo Moro, Memorial Univ. Newfoundland, Canada
Joško Parunov, University of Zagreb, Croatia
L. Prasad Perera, Arctic University of Norway, Norway
Jasna Prpić-Oršić, University of Rijeka, Croatia
Harilaos N. Psaraftis, DTU, Denmark
Franco Quaranta, University of Naples, Italy
Suresh Rajendran, IIT Madras, India
Jonas W. Ringsberg, Chalmers Univ. of Technology, Sweden
Cesare Rizzo, University of Genova, Italy
Jani Romanoff, Aalto University, Finland
Liliana Rusu, University "Dunarea de Jos" Galati, Romania
Herve Le Sourne, ICAM, Univ. of Nantes, France
António Souto-Iglesias, Univ. Politécnica de Madrid, Spain
Francisco Taveira-Pinto, FEUP, Portugal
Lourdes Trujillo, Univ. of Las Palmas, Gran Canaria, Spain
Nicholas Tsouvalis, NTUA, Greece
Michele Viviani, University of Genova, Italy
Deyu Wang, Shanghai Jiao Tong University, China
Jin Wang, Liverpool John Moores Univ., UK
Xin Wang, Dalian Maritime Academy, China
Wei Guo Wu, Wuhan University of Technology, China
X.B. Xiang, Huazhong Univ. of Science & Technology, China
Xinping Yan, Wuhan University of Technology, China
Xueqian Zhou, Harbin Engineering University, China

Technical Programme & Conference Secretariat

Maria de Fátima Pina, Técnico Lisboa, Universidade de Lisboa, Portugal
Sandra Ponce, Técnico Lisboa, Universidade de Lisboa, Portugal
Sónia Vicente, Técnico Lisboa, Universidade de Lisboa, Portugal
Mariana Mateus, Técnico Lisboa, Universidade de Lisboa, Portugal

 ISBN 978-1-032-83099-5

Sponsors

Sponsored by

GRUPO ETE

DNV

Portline Bulk International S.A.

BUREAU VERITAS 1828

LISNAVE ESTALEIROS NAVAIS, S.A.

GRUPO SOUSA

WestSEA VIANA SHIPYARD

PORTO DE SINES

Supported by

TÉCNICO LISBOA

Keynote lectures

Advances in Maritime Technology and Engineering – Guedes Soares & Santos (Eds)

Roadmap for the European research in waterborne transportation

M. Boile
Quantitative Analysis in Shipping Laboratory Department of Maritime Studies, University of Piraeus, Piraeus, Greece

ABSTRACT: Energy transition in the waterborne transport sector will translate into compliance costs and will have an impact on the market. The extent of this impact and the level of the compliance cost will depend on many factors, including the pace of the technological development, the availability and the cost of alternative fuels, market adoption, and, of course, on the level of regulation and its effectiveness. Additional work is needed to ensure that transition to a low carbon (or zero carbon) waterborne transport industry is done in a fair and equitable manner. Research, development and innovation play a major role in this transition. The Research and Innovation Agenda of the waterborne transport sector, as this has been developed within the context of the Waterborne Technology Platform activities and especially the Zero Emission Waterborne Transport Partnership, will be discussed.

SHORT BIOGRAPHY: Maria Boile is Director of the MSc in Shipping and Director of the Quantitative Analysis in Shipping Laboratory at the University of Piraeus, Department of Maritime Studies. She is Head of the Transport Economics and Environment, Maritime and Air Transport Sector, at the Hellenic Institute of Transport (HIT), Centre for Research and Technology Hellas (CERTH). She is Affiliated Faculty with the Center for Advanced Infrastructure and Transportation, at Rutgers, the State University of New Jersey, and Coordinator of the European Waterborne Technology Platform.

Professor Boile has served in senior academic and research positions in Europe and the U.S. She holds a diploma in Civil (Transportation) Engineering from the National Technical University of Athens, Greece, M.Sc. in Civil and Environmental Engineering from Rutgers University, US, and Ph.D. in Transportation Engineering from the New Jersey Institute of Technology, US. She has participated in over 80 sponsored research projects, as a principal investigator in over half of them. She has authored and co-authored over 170 technical articles, 12 book chapters and two books.

DOI: 10.1201/9781003508762-1

Advances in Maritime Technology and Engineering – Guedes Soares & Santos (Eds)

Challenges for future maritime technology and engineering

B. Buchner
MARIN, Wageningen, The Netherlands

ABSTRACT: In a time of large transitions and societal challenges, the role of Maritime Technology and Engineering is changing. The energy and digital transition on one hand, combined with the challenges of safe, secure and free seas on the other, require us to rethink this role. This analysis shows that traditional fields of knowledge, such as hydrodynamics, structures and marine power systems, cannot provide us the required solutions at their own. An integrated approach is needed, coupling these technologies and combining the traditional model (or physics) based approach with data driven methods and Artificial Intelligence. It is also important to combine these methods with disruptive innovation to bring the maritime sector forward. Based on MARIN's 'Better Ships, Blue Oceans' strategy, this presentation demonstrates on how research institutes and companies can potentially deal with these challenges.

DOI: 10.1201/9781003508762-2

Advances in Maritime Technology and Engineering – Guedes Soares & Santos (Eds)

Harnessing energy from the oceans

A. Incecik
University of Strathclyde, Glasgow, Scotland, UK

ABSTRACT: The demand for energy will increase as the world population and the prosperity in the world increase. By 2050, it is expected that the global population will increase by about 20% to 9.6 billion. This presentation will focus on how the world's energy demand can be met by offshore oil and gas and renewable sources in the 'energy transition' era.

Today about 28% of oil and 38% of gas are produced from the offshore locations. However, only 14 percent of all energy production comes from renewable sources. Renewable energy is estimated to make up 77% of the world's primary energy supply by 2050. Today 16% of electricity output produced from offshore wind. For the near future (2030), it is expected that global installed capacity of offshore wind will reach 130–140 GW, compared to less than 20 GW installed at the end of 2018. By 2050, the capacity in Europe is expected to reach 400 GW and globally well above 500 GW.

The presentation will start by describing the primary energy consumption by fuel types as function of three different scenarios: "Accelerated" and "Net Zero" when the world collectively takes actions for CO2-equivalent emissions to fall by around 75% by 2050 (relative to 2019 levels) in "Accelerated" and 95% in "Net Zero". In the 'New Momentum' scenario CO2 emissions peak in the 2020s and are around 30% below 2019 levels by 2050.

The presentation will continue by illustrating the locations of the global deep-water oil and gas reserves and different types of oil and gas exploration and production platform. The design principles in selecting different types of oil and gas production platform and analysis methods will be discussed.

The presentation will be concluded by describing the offshore renewable energy devices, life cycle assessment of these devices, and some recent floating offshore wind turbine installations in Scotland.

DOI: 10.1201/9781003508762-3

Advances in Maritime Technology and Engineering – Guedes Soares & Santos (Eds)
© 2024 The Author(s), ISBN 978-1-032-83099-5

The practice and outlook of intelligent navigation of ships in China

M.X. Li, J.F. Zhang, Y. Ma, F. Ma, J.L. Liu, C.G. Liu & X.P. Yan
State Key Laboratory of Maritime Technology and Safety, Wuhan University of Technology, Wuhan, P.R. China
National Engineering Research Center for Water Transport Safety, Wuhan University of Technology, Wuhan, P.R. China
Intelligent Transportation Systems Research Center, Wuhan University of Technology, Wuhan, P.R. China
Wuhan University of Technology, Wuhan, P.R. China

ABSTRACT: Intelligent ship technology is expected to significantly improve the safety and green of shipping systems, and gradually reduce or replace seafarers. It fundamentally propels the transformation of shipping technologies and has become one of the focal points in shipping industry. This paper first analyzes the latest progress of key intelligent navigation technologies such as intelligent perception, intelligent decision-making, intelligent control, and communication and network security. Then, it comprehensively introduces typical application cases carried out in China in assisted navigation, remote ship control, autonomous navigation, and formation navigation. Finally, the key development directions of ship-shore collaboration technology and ship-port intelligent equipment under the new generation of waterborne transportation system architecture are discussed. This paper can provide theoretical support for the research of intelligent navigation technology.

1 INTRODUCTION

With the deepening application of big data, artificial intelligence, the Internet, and the Internet of Things in different fields, ship navigation technology is developing towards digitization and intelligence. The International Maritime Organization (IMO) divides the autonomy level of ship into four levels: (1) L1, Ship with automated processes and decision support; (2) L2, Remotely controlled ship with seafarers on board; (3) L3, Remotely controlled ship without seafarers on board; (4) L4, Fully autonomous ship. Significant progress has been made in intelligent ships (L1), and the development of remote ship control (L2 and L3) is a global consensus and a hot interdisciplinary topic. To promote the development of intelligent ships, IMO has introduced the Maritime Autonomous Surface Ship (MASS) Code, with non-mandatory implementation set for 2024 and mandatory implementation by 2028.

In the gradual development of intelligent ship technology, Europe, America, Japan, and South Korea have successively initiated critical technology research on intelligent ships, remote ship control, and others (see Figure 1). In 2012, the European Union launched the Maritime Unmanned Navigation through Intelligence in Networks (MUNIN) project, researching intelligent navigation for large vessels (Retschmann, 2015). In 2015, Finland initiated the Advanced Autonomous Waterborne Applications (AAWA) project aimed at achieving remote navigation and unmanned ship transportation (Jokioinen, 2016). In 2020, Japan launched the fully autonomous ship program MEGURI2040, subsequently completing the development and accurate ship testing of intelligent navigation systems for six types of vessels (Hirata, 2022). In 2022, South Korea's large liquefied natural gas carrier, the PrismCourage, completed open-sea intelligent navigation experiments (Zhang, 2022). Contemporarily, IBM's three-body electric vessel Mayflower accomplished the first fully autonomous unmanned transatlantic navigation (Rivkin, 2021).

China has also launched projects related to intelligent ships. Intelligent Ship Development Action Plan (2019-2021), Intelligent Ship Regulations (2020), and the Implementation Opinions on Accelerating the Green and Intelligent Development of Inland River Ships have successively proposed to research a new mode of ship navigation, focusing on "shore-based control", supplemented by "ship-based monitoring and watch-keeping". This new mode follows the path of "assisted navigation-remote ship control-autonomous navigation", focusing on overcoming key technical challenges in the intelligent navigation of ships (Li, 2021). The "Intelligent Ship 1.0" project, which was launched in 2016, focused on the key technology of ship-assisted decision-making. In 2017, the "Da Zhi" intelligent-assisted decision-making bulk carrier was delivered. In 2018, the large ore carrier "Mingyuan" was delivered, equipped with assisted autonomous navigation functions. In the same year, Wuhan University of Technology developed a ferry assisted navigation system that was installed on the

DOI: 10.1201/9781003508762-4

ferry "Banxin 2", with functions such as intelligent obstacle recognition and optimal route recommendation (Yan, 2019). In 2019, the "Zhiteng" completed its sea trial and had functions such as autonomous collision avoidance and autonomous navigation. In 2020, "Hai Dian Yuan Intelligent 1" completed the sea trial and has autonomous collision avoidance function. In 2021, the container ship "Zhifei" also successfully conducted the sea trials, equipped with functions such as unmanned ship navigation, remote ship control, autonomous navigation, and automatic berthing and unberthing. In 2022, the "Zhuhaiyun" was successfully launched and can achieve autonomous navigation and remote ship control.

To promote the further development of intelligent ship navigation technology, this paper delves into the key technologies of intelligent navigation, aiming to provide safer and more reliable intelligent navigation solutions for the industry. At the same time, combined with typical application cases of intelligent navigation in China in recent years, the performance of intelligent navigation technology in practical scenarios and their substantial impact on the industry were fully analyzed. Finally, the future development direction of intelligent navigation technology was discussed, providing new theoretical and technical support for research and practice in this field.

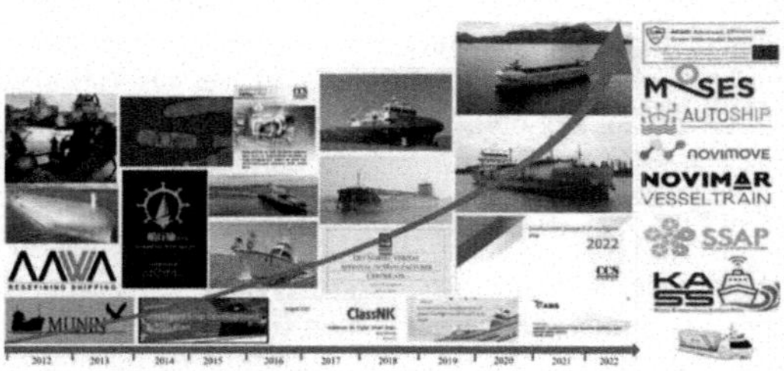

Figure 1. The global development status of intelligent ships.

2 KEY TECHNOLOGIES FOR INTELLIGENT SHIP

The primary objective of intelligent navigation is to gradually reduce and replace the tasks of seafarers through the adoption of intelligent means, achieving a new navigation mode for ships where "shore-based control", supplemented by "ship-based monitoring and watch-keeping" (Ma, 2023). Under this new mode, the shore-based support system utilizes the ship-shore communication link to monitor the real-time ship navigation situation. Simultaneously, AI-assisted decision-making subsystems intervene in the vessel's navigation, enabling remote control. Seafarers are liberated from traditional watch-keeping duties, with most of whose work being replaced by AI or completed by shore-based operators. Seafarers only need to intervene in emergency.

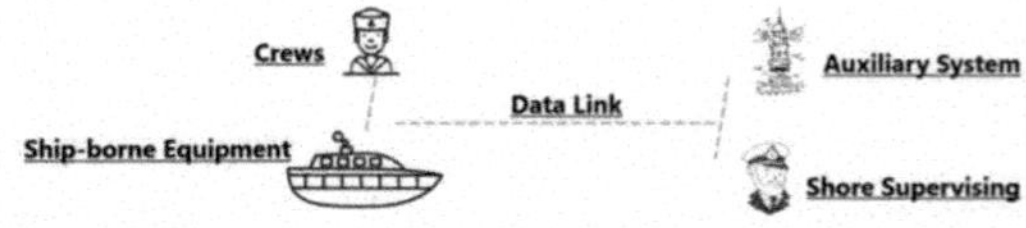

Figure 2. A new ship navigation mode of "shore-based control", supplemented by "ship-based monitoring and watch-keeping".

To achieve this goal, the conceptual design of the "Navigation Brain System" (NBS) (YAN, 2017) draws inspiration from the working mechanism of the human brain. NBS is divided into interconnected and mutually cooperative modules that collectively perform complex tasks. The main functional modules include perception, decision-making, control, and communication network security (Figure 3).

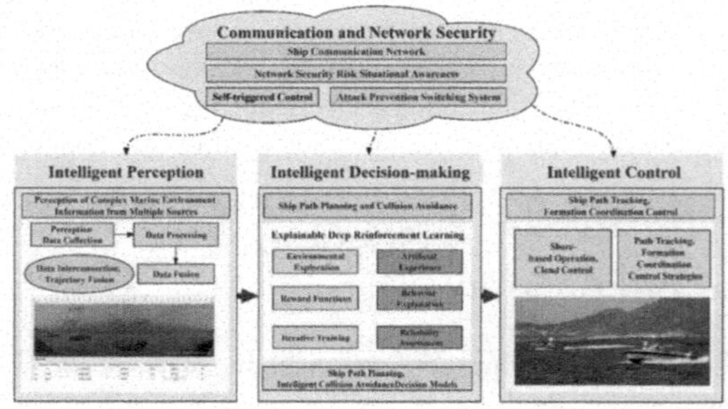

Figure 3. The core functional modules of the "Navigation Brain system".

2.1 *Intelligent perception technology*

Intelligent perception technology is the foundation of ship intelligent navigation, which determines the level of intelligent navigation for ships to a large degree (Ma, 2023). Intelligent perception utilizes onboard sensor equipment and information processing technology to acquire real-time status information of the ship's own status and various information around the ship, providing more reliable data support for key technologies such as intelligent decision-making and control (Figure 4). Intelligent perception is primarily from ship-related data and environmental data during navigating. In detail, one vessel requires real-time data of equipment, such as the main engine, steering gear, hydraulic systems, propellers, and electrical systems, monitoring the ship's behavior and having the ability to decide on the basis of existing information. The external effects including hydro-meteorological conditions and wind-wave-current status are considered as well.

The complexity and diversity of perception devices lead to untimely and imprecise perceptions of navigational environmental information and ship operating conditions. Moreover, data heterogeneity, diverse sources, high redundancy, and intense noise make information fusion and processing challenging. To address these challenges, there is an urgent need to explore advanced sensor technologies and multimodal information fusion techniques and improve the coordination rules and integrated standards for onboard sensor deployment. Through fast, accurate, and integrated multi-source heterogeneous data, effective real-time navigation environment reconstruction techniques can be implemented in aspects such as navigational conditions, hydrometeorology, target identification, and obstacle recognition. This technology contributes to further authenticating the perception of ship operation status, accurately monitoring navigational environments, and achieving awareness and multi-target identification of obstacles.

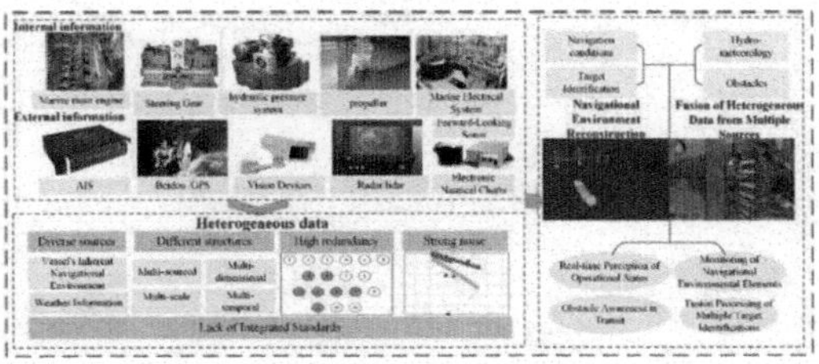

Figure 4. Intelligent perception technology for ship.

2.2 *Intelligent decision-making technology*

Intelligent decision-making is the core technology of ship intelligent navigation, with the fundamental approach being the dynamic optimization of navigation decisions and route planning based on the analysis of navigational environment, traffic, routes, navigation tasks, and situational assessment (Ma, 2023). Through the extension and enhancement of human cognition, decision-making, planning, and execution capabilities, intelligent decision-making can accomplish specific tasks that humans may find challenging or achieve certain tasks more efficiently than humans (Chai, 2022), achieving the transformation of the control subject from the seafarers to the human-machine cooperation mode.

As shown in Figure 5, intelligent decision-making technologies encompass situational awareness, intelligent route planning, and intelligent collision avoidance. Situational awareness requires an initial analysis of factors such as the ship's navigation status and environmental conditions, extracting Risk Influence Factors (RIFs). Subsequently, visual analysis, risk assessment, prediction, and other techniques are employed to evaluate and forecast the safety situation of ship navigation, providing information for decision-making. The primary step in intelligent route planning involves analyzing Electronic Chart Display and Information System (ECDIS) and extracting ship navigable waters. Based on environmental modeling, intelligent route planning is eventually achieved under multiple objectives and constraints. Intelligent collision avoidance can integrate the International Maritime Collision Avoidance Rules (COLREGs) into collision avoidance decision-making (Johansen, 2016). Firstly, by analyzing the requirements in COLREGs, a collision avoidance decision-making knowledge base is established. Based on the knowledge of reserved rules, identify ship encounter situations, evaluate collision risks, and ultimately make navigation decisions to achieve intelligent collision avoidance. The perceptual information collected during ship navigation is highly uncertain (Thombre, 2022), and intelligent decision-making technology needs to fully understand and master information from different sources to solve key problems such as inaccurate target intention recognition and difficulty in predicting nonlinear operations. Correctly conducting situational assessment and decision-making on ship course and speed, planning routes, and intelligent collision avoidance are key technologies that need to be addressed in future intelligent decision-making. In addition, it is necessary to comprehensively apply ECDIS, COLREGs, expert systems, and other technologies to the route planning and collision avoidance decision-making subsystems. Quantitative analysis on COLREGs and good seamanship can improve the safety and reliability of decision-making. Regarding to engineering applications, introducing natural language models and intelligent interaction technology can facilitate human-machine interaction, improving information communication efficiency between crews and ships.

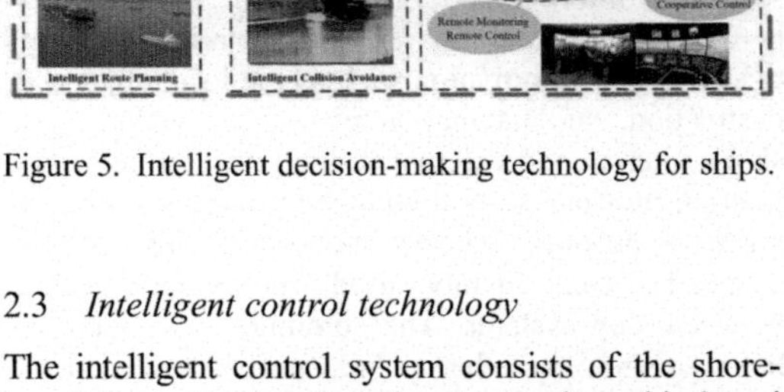

Figure 5. Intelligent decision-making technology for ships.

2.3 *Intelligent control technology*

The intelligent control system consists of the shore-based remote-control subsystem, the ship-based motion control subsystem, and the cloud-based collaborative control subsystem. The shore-based remote-control system, with the shore-based navigation control center and remote navigator as its core, features new functions such as remote navigation control, situational monitoring, and human-machine interaction. The relevant control instructions are transmitted through the ship-shore communication module. Upon receiving control instructions, the ship-based motion control system intelligently controls the ship's speed/heading to achieve navigation control. The cloud-based collaborative control system distributes navigation task information and backs up navigation data based on the information from the shore-based human-machine interaction module and the ship- based environmental perception module. During the navigation process, ships face complex environmental conditions such as sea conditions, water depth, wind, and atmospheric pressure, all of which can impact the ship's control system. Therefore, control algorithms need to have high adaptability, flexibility, and control accuracy. The controller needs to have strong robustness against these disturbances and be able to accurately track the desired route (Islam, 2021). In future engineering applications, ships are expected to move towards remote and autonomous navigation. Hence, there is an urgent need to establish an integrated "ship-shore-cloud" control center (Ma, 2023) to achieve the convergence of intelligent control technologies (Figure 6).

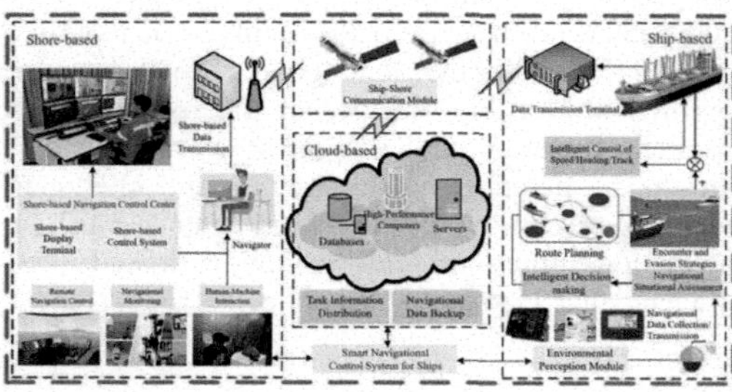

Figure 6. "Ship-Shore-Cloud" integrated control.

2.4 *Communication and network security technology*

Communication and network security technology safeguard information exchange between ships and between ships and shores. It primarily transmits essential ship information such as the location, speed, destination, and navigation plan. This includes the ship's internal and external communication networks. Internal network system equipment includes wireless switches, network switches, wired terminals, wireless terminals, etc., mainly used for communication between ship systems. The communication facilities between ships and other include various base stations, VTS (Vessel Traffic Service, VTS) centers, remote control centers, etc. Ships can use these facilities to exchange data through satellite communication, wireless local area networks (Wi-Fi), LTE/4G, 5G, and wired communication (Figure 7).

To meet the development needs of intelligent ships and achieve large-scale parallel ship-to-ship and ship-to-shore comprehensive business communication. On the one hand, it is urgent to improve communication bandwidth and speed. On the other hand, ship clusters can be divided into multiple decentralized groups, each with the same communication module. When a communication network node fails, communication network restructuring can be achieved to reduce communication costs and ensure communication quality (Ma, 2023). Additionally, the adoption of the software-defined networking paradigm is recommended to counter emerging network threats and evolving methods of network attacks. This paradigm provides dynamic and automated mitigation mechanisms for attack traffic in ship communication networks, enhancing resilience against network attacks (Sahay, 2019).

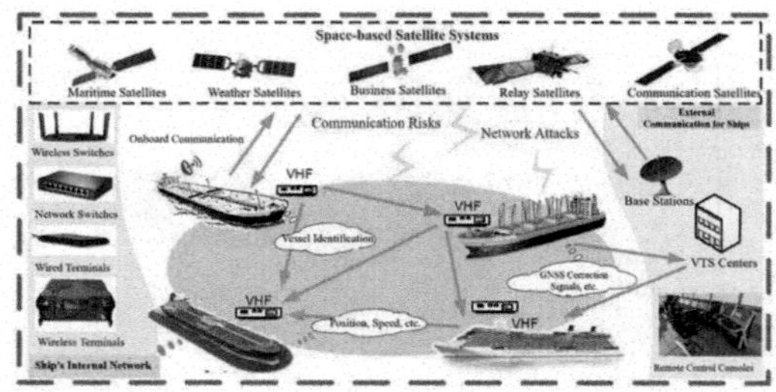

Figure 7. The structure of maritime communication networks.

3 INTELLIGENT NAVIGATION APPLICATION

3.1 *Auxiliary navigation*

In July 2018, Wuhan University of Technology developed first assisted intelligent navigation system for ferry vessel in China, which was applied to the inland ferry vessel "Banxin 2" on the Nanjing section of the Yangtze River (Figure 8). The "Banxin 2" ferry is also the first ferry vessel in China that involve artificial intelligence technology and put into application. The intelligent assisted navigation system of the ferry can provide seafarers with intelligent collision avoidance, recommended routes, operation in different weather conditions, three-dimensional and real-time management and other functions. At the perception level, the system integrates sensors such as radar, cameras, laser scanners, etc., which can obtain various data of the ship's surrounding environment, including distance, speed, position, obstacles, etc. By integrating data from multiple sensors, the system can obtain more comprehensive and accurate environmental perception information. The data processing and perception enhancement system, through deep improvements such as ViT (Vision Transformer, ViT), achieves an effective tracking rate of over 95% for ship targets in congested waters, and reduces typical errors such as target exchange by 20%. Based on methods such as open learning, the vision capture rate has reached 98%. We have focused on optimizing the autonomous capture performance in foggy and nighttime conditions. At the decision-making level, the system integrates three key technologies: fusion and reconstruction, recognition and expression, and auxiliary decision-making, solving the problems of fusion and reconstruction, recognition and expression of the complex navigation environment and ship status of inland ferries, and establishing an intelligent safety auxiliary driving control model for under-actuated ships.

Figure 8. “Banxin 2” ferry vessel and assisted navigation system.

3.2 *Remote navigation*

The “Zhechangxinghuo 7308” ship is a typical case of remote navigation control technology. In December 2021, the ship conducted a 10-day trial of autonomous navigation and remote navigation in the Huzhou navigation section of Zhejiang Province. The ship is based on technologies such as scene cognitive interaction, deep reinforcement learning, and adaptive control, and has established control intelligence similar to that of seafarer (Figure 9). During autonomous navigation, a total of 33 ship encounter situation, and the frequency of encounter situations reached a maximum of 5 times per 3 minutes. The error of ship route tracking is always within the range of ± 20 meters, with a maximum of 19.8 meters and an average of 5.6 meters. Through actual ship testing, the intelligent navigation system can autonomously identify target ships and make autonomous collision avoidance decisions.

Figure 9. Remote driving platform of “Zhechangxinghuo 7308” vessel.

To address the lack of electrified propeller and rudder systems on inland vessels, the platform developed three types of mechanical arms: “grasping”, “steering wheel”, and “slant lever”, aimed at mainstream steering wheel and main engine control methods. These mechanical arms imitate human hands to achieve control of traditional inland vessels. These improvements have endowed the remote navigation platform with stronger intelligence and autonomy capabilities.

Figure 10. “Zhechangxinghuo 7308” vessel and control console.

3.3 *Autonomous navigation*

In April 2021, a 300-TEU container ship, named “Zhifei”, was officially delivered and put into operation at Qingdao Port (Figure 11). The ship has a displacement of approximately 8,000 tons, a capacity of 310 TEUs, a designed speed of 12 knots, and a range of 4,500 nautical miles. “Zhifei” is first autonomous container cargo vessel in China. It is integrated with multiple advanced technologies and is primarily used for domestic coastal container shipping and intelligent system testing.

Figure 11. Container cargo vessel “Zhifei”.

The “Zhifei” can collaborate with shore-based facilities such as ports, shipping, maritime, and others through multi-network and multi-mode communication systems such as 5G and satellite communication. “Zhifei” has overcome three technical difficulties, namely intelligent perception, intelligent decision-making, and intelligent control technology. It can achieve intelligent perception, autonomous tracking, autonomous route planning, intelligent collision avoidance, automatic berthing and departure, and remote navigation control. In addition, the ship is equipped with intelligent navigation, intelligent operation and maintenance, intelligent energy efficiency management, intelligent hull, information integration and other equipment, which can also be used to cultivate intelligent shipping talents.

3.4 *Formation navigation*

In January and April 2023, Wuhan University of Technology completed the trial of formation navigation involving three inland vessels in the Guangyangba and Fengjie waters of Chongqing, respectively (Figure 12). Vessel formation can make vessel operation safer, more efficient, and beneficial, especially in locks or canals. The hardware of the ship formation navigation system includes millimeter wave radar, differential GNSS, gyro-compass, industrial control computer, and servo potentiometer. The software mainly includes ship formation navigation control algorithms, control software design, and shore-based system software design, enhancing coordination and communication between ships and shore. In the future, this system is expected to be applied in a wider range of scenarios, providing support for fleet navigation, and further improving the safety and economy of ships.

Figure 12. Formation control testing scene in the trial navigation area.

4 FUTURE PROSPECTS

Intelligent ships do not operate in isolation, where their development relies on the collaborative progress between ship and shore facilities. To achieve intellectually, environmentally-friendly, and automatically, each stage of navigation needs to be supported by multi-aspect technologies. Consequently, New generation waterborne transportation systems (NEW-WTS) is introduced to meet the comprehensive demands of future maritime navigation. The following will emphatically focus on the architecture of this system, the ship-shore collaboration, and the intelligent equipment on ships and ports.

4.1 *The architecture and operational mode transformation*

By utilizing advanced technologies, NEW-WTS can realize a novel complex where the elements of shipping are interconnected and self-consistent. This comprises physical-dispersed but network-connected units such as ships, ports, waterways, and cargo/passengers, as shown in Figure 13. The remote-control system in NEW-WTS aims to establish a subsystem that the shore control station is central, modern information and communication technologies are the key approaches, and the ultimate goal is to achieve remote monitoring, command, and management of ships with high reliability. This subsystem should be equipped with high-bandwidth, low-latency, and highly reliable communication interaction between ships and shores, along with shore functions such as water-traffic control, ship monitoring, and capacity scheduling. China has already achieved full coverage of the Vessel Traffic System (VTS) and Automatic Identification System (AIS) for waterborne vessel traffic (Ministry of Transport of the People's Republic of China, 2021), providing a foundational guarantee for the development of NEW-WTS.

The goal of the new operational mode based on the remote-control system is to eliminate or minimize crew requirements. The shore control center utilizes advanced human-machine interfaces and information systems to achieve remote control over ship navigation, steering, and mechanical operations (WANG, 2021). This new mode will significantly improve the efficiency of the shipping system and reduce labor costs. Simultaneously, ship watch-keeping is an auxiliary safeguard, ensuring the safety of ships during berthing, departure, emergencies, or extreme situations. Different from traditional navigation and control methods, the future will witness a new paradigm of "shore-based control", supplemented by "ship-based monitoring and watch-keeping" (YAN, 2021). This represents a key transformation in the ship operational mode under NEW-WTS.

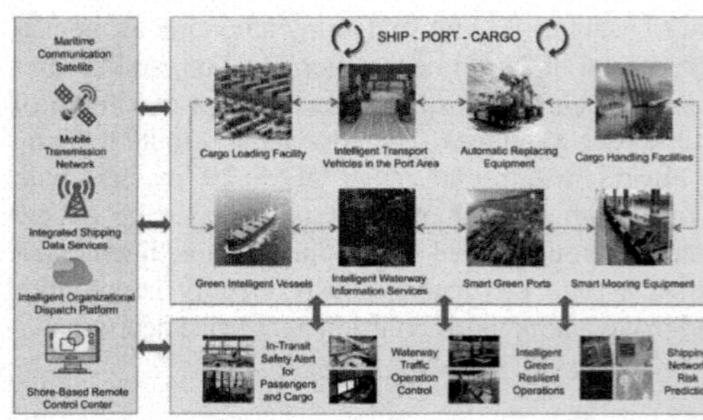

Figure 13. Ship-Port-Cargo complex.

4.2 *Ship-shore collaboration technology*

Under the architecture of NEW-WTS, shore and ship collaboration are mutually complemented. The core of the shore control is the shore system platform, which can extract the useful information and features and reconstruct the traffic scene of ships at a specific time and space based on different data source, as shown in Figure 15. The data from AIS, radar, satellite are collected from each ship control stations in order to analyze ocean and weather information, nautical charts, navigation logs, VTS records, and so on. The obtained information undergoes corresponding interaction, including data fusion, feature analysis, behavior pattern assessment, etc., to acquire a more comprehensive and accurate perspective. Navigation behavioral logic, potential risks, and tendency should be further conducted or invertibly retrieved by information analysis. By regulating and optimizing models and algorithms, navigation safety can be enhanced, ship traffic management been optimized, and improved channel utilization efficiency. Meanwhile, the ship-shore support system will cover the components of various operational information and scheduling service, providing fleets with real-time monitoring, tracking, managing, organizing, and dispatching. In case of emergencies, it provides necessary information and decision support to enhance rescue effectiveness.

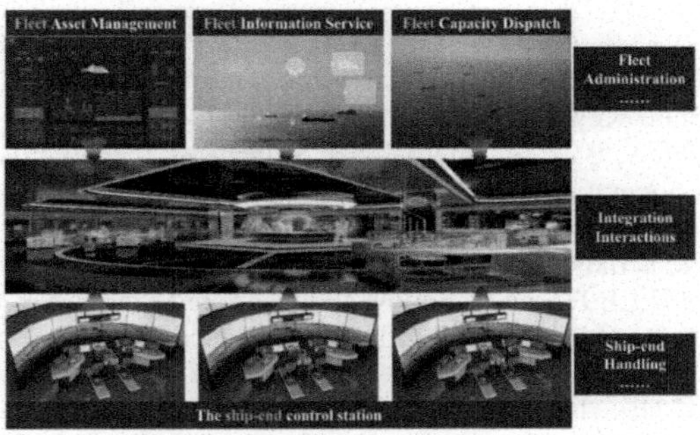

Figure 14. Shore system conception.

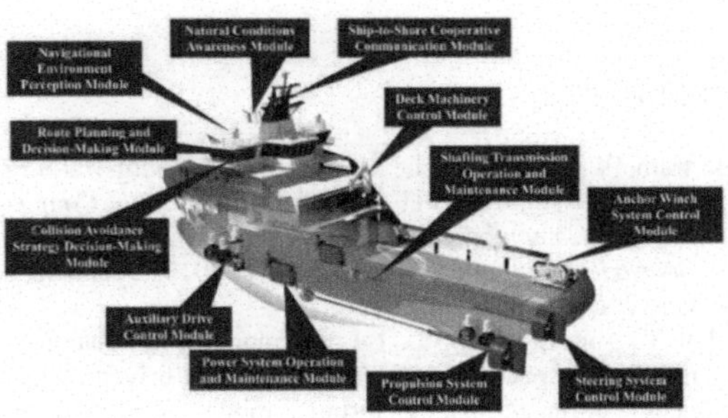

Figure 15. Ship system conception.

The ship-borne system will provide intelligent and digital support for ship navigation, maneuvering, communication, monitoring, management, etc. The specific modules are shown in Figure 16. Firstly, the modules of navigation environment perception and natural conditions sensation are integrated by meteorological monitoring system, GPS, ECDIS, and other subsystems. By transmitting the data, the required information can be conveyed, which reduces the adverse effects of environmental factors on navigation and guarantees the efficiency and safety. Secondly, the ship-shore cooperative communication module needs to include satellite communication system, maritime satellite communication (such as International Maritime Mobile Satellite System, Inmarsat), Very High Frequency (VHF) radio, etc., to provide support for communication effectiveness and reliability among the ship-shore facilities and emergency service organizations. Then, according to the acquired ship information and environmental factors, the route planning and collision avoidance decision-making module needs to formulate optimal routes to avoid collision, and take necessary measures timely in the face of special circumstances. Finally, the power system operation and maintenance module and other closely-relevant modules conformably control the ships under various environmental and operational conditions. Advanced simulating and computing capabilities are often required to automatically maintain the ships' positions and heading by using computer-controlled propulsion and maneuvering systems. It can lead to safely and efficiently perform every task, such as maintaining routes, stably maneuvering, and stably mooring.

4.3 *Intelligent equipment and systems in ship and port*

Under NEW-WTS, it is significant to develop ship and port intelligent equipment. By improving the automation, intelligence and informatization level of waterway transportation system, it can drive the technological progress of the industry on related ship equipment, promote the deep integration of information technology and maritime transportation, and establish a new infrastructure system in the future. This will enhance the economic and social benefits of waterway transportation, as well as realizing more efficient loading operations, navigation scheduling, and energy management. It is hopeful to cultivate the collection of transportation equipment industries. The development of intelligent ship equipment can be carried out from the transformation of existed ship equipment and the establishment of new types of shipping equipment.

For the existing ship equipment, it can add the intelligent perception component, automatic control element and others, which improves the safety, economicality and efficiency of operation. In order to achieve these goals, it is necessary to conduct a comprehensive assessment of existing marine equipment, understanding their performance, technical level and maintenance status, and identifying needful-upgraded or urgently-replaced systems. According to the assessment results, it follows taking appropriate measures such as increasing sensors, updating communication modules, selecting adaptable data analysis software, upgrading control systems, adding human-machine interaction interfaces, and so on. Thus, the intelligentization and humanization of the ship-borne equipment can be improved, and the operation can be better coordinated. At the same time, it is necessary to increase the automatic fault diagnosis and alarm function to ensure the timely maintenance and repairing. In the process of intelligent transformation, it may require the crew having new operating skills and related knowledge, so that necessary training and technical support should be provided to ensure the transformation forms a closed loop.

The development of new ship intelligent equipment is also oriented to the needs of future intelligent ships under NEW-WTS, where it can design novel, highly intelligent and flexible shipborne equipment and other key equipment, to meet the different scenarios. Examples include various ship-borne robots, autopilot systems, remote control and monitoring equipment. Specifically, cabin operation and maintenance robots (MENG, 2021) need to be equipped with intelligent functions such as perception, learning, decision-making, control, and human-machine interaction, so that they can carry out inspections, cleaning, maintenance, and other work in the cabin, and replace the personnel to enter the hazardous environment such as high temperature, high pressure, high noise and so on. Cabin-cleaning robots (Li, 2022) can perform tasks such as cargo cleaning in the corners of containers and other places that are not easily accessible, significantly reducing the labor intensity of the crew. The development of new types of marine intelligent equipment requires increasing independent research for the new types of core sensors, actuators, key components, etc. Establishing engineering centers and industrial alliances for new types of marine intelligent equipment, will realize breakthroughs in key technologies and transform scientific research results into products, incubating and cultivating industrial ecology, and encouraging enterprises to develop innovative marine intelligent equipment to promote the transformation and upgrading of the ship industry.

5 CONCLUSIONS

With the worldwide emphasis on intelligent ships and the increasing requirements of intelligence and greening, intelligent ships will have increasing application scenarios and face new historical missions and development opportunities. Discussions will be carried out around the key technologies of intelligent navigation and their main applications to explore the future development trend of intelligent ships, including the new generation of shipping system control system, ship-shore terminal synergy mechanism, and ship-port intelligent equipment and systems, so as to improve the operation efficiency and safety of waterway transportation system. In the future, a new type of transportation system will be realized with the integration of all elements and self-coordination and sharing of "shore-based control", supplemented by "ship-based monitoring and watch-keeping", which will realize the interconnection and sharing of information among ships, ports, shipping, passengers and cargoes, and realize the comfortable carriage and personalized, efficient and precise displacement of passengers and cargoes, which is the direction for the development of the field of intelligent ships and the direction of efforts. Eventually, a series of autonomous transportation systems and equipment will arise to build an intelligent waterway transportation system.

ACKNOWLEDGEMENTS

The research of this paper was supported by National Natural Science Foundation of China (519201 05014).

SHORT BIOGRAPHY OF THE KEYNOTE LECTURER

Dr. Xinping Yan is Chair Professor and Director of State Key Laboratory of Maritime Technology and Safety, Base for International Science & Technology Cooperation on Smart Shipping and Maritime Safety (MoST), Wuhan University of Technology, China. He is the Chair of Intelligent Water Transport Safety of IEEE ITS and a Fellow and Chartered Marine Scientist of IMarEST. In 2019, he was elected Academician of the Chinese Academy of Engineering (CAE). His research interests include intelligent transport system key technologies, intelligent fault diagnosis of marine engines, renewable energy, energy efficiency management of the vessel, navigation systems for vessels, amongst others.

REFERENCES

Bertram, V., & Technische Universität Hamburg-Harburg (Eds.). (2019). COMPIT '19: *18th International Conference on Computer and IT Applications in the Maritime Industries: Tullamore*, 25-27 March 2019. Technische Universität Hamburg-Harburg.

Chai T., Liu Q., Ding J., Lu S., Song Y., & Zhang Y. (2022). Perspectives on industrial-internet-driven intelligent optimizedmanufacturing mode for process industries. *SCIENTIA SINICA Technologica*, 52(1), 14–25.

China Classification Society. (2020). Code for Intelligent Ships (2020) (in Chinese). *China Classification Society.*

Hirata, E., Watanabe, D., & Lambrou, M. (2022). Shipping Digitalization and Automation for the Smart Port. In T. Bányai, Á. Bányai, & I. Kaczmar (Eds.), *Supply Chain-Recent Advances and New Perspectives in the Industry 4.0 Era. IntechOpen.*

Islam, M. M., Siffat, S. A., Ahmad, I., & Liaquat, M. (2021). Robust integral backstepping and terminal synergetic control of course keeping for ships. *Ocean Engineering*, 221, 108532.

JIANG Zhonglian, CHU Xiumin, & YAN Xinping. (2015). Developments and Prospects of Intelligent Water Transport:A Review of the Symposium on Intelligent Water Transport, in Conjunction with the 10th Annual Meeting of China ITS. *Journal of Transport Information and Safety*, 6, 1–8.

Johansen, T. A., Perez, T., & Cristofaro, A. (2016). Ship Collision Avoidance and COLREGS Compliance Using Simulation-Based Control Behavior Selection With

Predictive Hazard Assessment. *IEEE Transactions on Intelligent Transportation Systems*, 17(12), 3407–3422.
Jokioinen E, Poikonen J, & Hyvonen M. (2016). Remote and autonomous ships-the next steps [AAWA Position Paper]. *Rolls Royce PLC*.
Li, P., Liu, W., & Zheng, Y. (2022). An automatic warehouse clearance scheme for bulk cargo unloading in ports. *2022 2nd Asia-Pacific Conference on Communications Technology and Computer Science (ACCTCS)*, 296–299.
Li Y J, Zhang R, & Wei M H. (2021). State-of-the-art research and prospects of key technologies for ship autonomous navigation (in Chinese). *China Journal Ship Research*, 16, 32–44.
Ma Y., Wang W., & Yan X. (2023). Research progress of vessel intelligent navigation technology for the new generation of waterborne transportation system. *Scientia Sinica Technologica*, 53(11), 1795–1806.
Meng J., Wang S.T., Li G., Jiang L.Q., Zhang X.L., & Liu C. (2021). Iterative-Learning Error Compensation for Autonomous Parking of Mobile Manipulator in Harsh Industrial Environment. *Robotics and Computer-Integrated Manufacturing*, 68, 102077.
Ministry of Transport of the People's Republic of China. (2021, December 16). Special press conference on the construction of the 'four modernizations' of the maritime team [Personal communication].
Retschmann L, Rodseth A, & Tjora B. (2015). Maritime unmanned navigation through intelligence in networks-qualitative assessment [Technical Report]. *MUNIN project report*.
Sahay, R., Meng, W., Estay, D. A. S., Jensen, C. D., & Barfod, M. B. (2019). CyberShip-IoT: A dynamic and adaptive SDN-based security policy enforcement framework for ships. *Future Generation Computer Systems*, 100, 736–750.
Thombre, S., Zhao, Z., Ramm-Schmidt, H., Vallet Garcia, J. M., Malkamaki, T., Nikolskiy, S., Hammarberg, T., Nuortie, H., H. Bhuiyan, M. Z., Sarkka, S., & Lehtola, V. V. (2022). Sensors and AI Techniques for Situational Awareness in Autonomous Ships: A Review. *IEEE Transactions on Intelligent Transportation Systems*, 23(1), 64–83.
Xinping Yan, Feng Ma, Jialun Liu, & Xuming Wang. (2019). Applying the Navigation Brain System to Inland Ferries.
Wang Y.Y., Liu J.L., Ma F., Wang X.P., & Yan X.P. (2021). Review and prospect of remote control intelligent ships. *Chinese Journal of Ship Research*, 16(1), 18–31.
Yan Xin-ping, Li Chen, Liu Jia-lun, You Xu, Wang Shu-wu, & MA Feng. (2021). Architecture and Key Technologies for New Generation of Waterborne Transportation System. *Journal of Transportation Systems Engineering and Information Technology*, 21(5), 22–29.
Yan Xinping & LIU Chenguang. (2016). Review and prospect for intelligent waterway transportation system. *CAAI Transactions on Intelligent Systems*, 6, 807–817.
Yan Xinping, Wu Chao, & Ma Feng. (2017). Conceptual Design of Navigation Brain System for Intelligent Cargo Ship. Navigation of China, 40(04), 95-98+136.
Zhang B C, Geng X F, & Li Y B. (2022). Development status and trend of intelligent navigation technology (in Chinese). *Science Technology Revision*, 40, 51–56.

Maritime transportation

Advances in Maritime Technology and Engineering – Guedes Soares & Santos (Eds)

Economic feasibility of a short sea shipping route in Southern Africa

P.M. Batista Santos & T.A. Santos
Centre for Marine Technology and Ocean Engineering, Instituto Superior Técnico, Universidade de Lisboa, Portugal

ABSTRACT: The Southern Africa Development Community (SADC) has recognized the pivotal role of enhancing its regional freight transport network in reaching its development objectives. To this purpose, the adoption of Short Sea Shipping (SSS) has emerged as a feasible approach. This paper is dedicated to investigating and implementing a methodology for the systematic planning of a regular liner route connecting a network of ports, while taking into account the efficient distribution of cargo within the hinterlands of these ports via the road infrastructure to their ultimate destinations. This study leverages a sophisticated numerical tool for the economic evaluation of a SSS based transport chain. It builds upon a model of the existing transport infrastructure, encompassing roads, ports, and maritime routes. A case study is presented, focusing on a route that links ports within Southern Africa. The findings support the idea that, given the current logistics landscape in the region, implementing a SSS route would be a profitable investment. If freight rates were more aligned with international levels, the route would not be feasible, partly also due to the elevated port costs in the region.

1 INTRODUCTION

The SADC comprises 16 countries across the Southern African region, presenting a vast and diverse landscape. Despite its abundant resources and economic potential, the region grapples with transportation and logistics challenges, leading to disruptions in efficient trade flows. The current state of the transportation infrastructure in the region faces a spectrum of issues. These include inadequate road networks, limited rail connectivity, and suboptimal port operations. What exacerbates the situation is the lack of synchronized planning and coordination among various modes of transport, impeding the development of consistent liner routes. This lack of cohesion results in delays, increased costs, and heightened complexities for businesses engaged in regional trade. The absence of well-defined routes and standardized procedures makes it challenging for liner operators to optimize their services, negatively impacting the reliability and efficiency of supply chains.

SSS holds great promise for improving regional connectivity and streamlining maritime transportation in Southern Africa, ultimately enhancing the overall transport landscape. Eurostat (2023) defines SSS as “the maritime transport of goods over relatively short distances, as opposed to the intercontinental cross-ocean deep-sea shipping” and this definition applies well also along this extensive African coastline. The SADC is thus well positioned to harness the potential of SSS as a viable transportation mode. This form of shipping offers an alternative to difficult and lengthy road journeys, in addition to alleviating road congestion, thereby promoting trade and increasing economic growth. SSS could also effectively move goods and passengers between countries in the region, fostering regional integration and collaboration. Furthermore, it provides a cost-effective and environmentally conscious alternative to land-based transportation, leading to a reduction in greenhouse gas emissions and air pollution, thereby mitigating other adverse environmental impacts.

This paper aims to employ numerical tools dedicated to the analysis of intermodal transport chains within the context of Southern Africa. The central focus lies in the analysis and planning of a liner route while accounting for door-to-door transport chains, with the ultimate goal of facilitating and promoting trade within the region. The establishment of consistent connectivity and efficient logistics infrastructure holds the potential to advance regional integration, broaden market access, and invigorate cross-border investments.

The paper is organized in the following manner. Section 2 comprises a literature review of the relevant topics for the paper. Section 3 provides an overview of transportation issues in Southern Africa. Section 4 presents the methodology adopted to study transport operation in the region. Section 5 shows how freight flows were estimated and describes the transport network model built to support the transport chain’s assessment. Section 6 provides the numerical results regarding individual transport

DOI: 10.1201/9781003508762-5

chains, modal splits, aggregate freight flows and economic feasibility of the SSS service. Finally, section 7 presents the main conclusions of the study and recommendations for further work.

2 LITERATURE REVIEW

The case study addressed in this paper encompasses several distinct yet interconnected sub-problems. Firstly, the need arises to estimate cargo flows between the different regions within each country, to allow the disaggregation of trade flows at the country level. Subsequently, intermodal chains that connect pairs of regions are assessed, considering factors such as cost and transit time. Lastly, drawing upon these findings, a discrete choice methodology is applied to determine modal splits. These three sub-problems provide the motivation for the following sub-sections.

2.1 *Gravitational models for estimating cargo flows*

The assessment of cargo flows between specific origin and destination pairs (O/D) is essential since accurate statistics on cargo flows are typically lacking in underdeveloped countries. In such cases, gravity models are commonly employed to estimate cargo flows, as they are considered to be proportionate to trade between countries or regions.

Gravity models, initially developed by drawing parallels with Newton's Universal Law of Gravitation, utilize variables that influence bilateral trade flows, as discussed by Baier and Standaert (2020). This model has proven to be a successful method for predicting trade flows between two countries across various types of goods. Notably, it can be applied to any trade scenario where trade diminishes as the distance between countries increases.

Nevertheless, the application of the gravity model does present challenges, with the primary one being the wide array of models available for use. Selecting the most appropriate model can be a complex task. Sinaga *et al.* (2019) compare Newton's law of gravitation with trade flows, using equations for both the gravity force, as presented below (1).

$$F_{ij} = G\frac{(M_i M_j)}{(D_{ij}^{\ 2})} \tag{1}$$

where F_{ij} is the force between and i and j, with G being the gravitational constant, M being the mass and D representing the distance between i and j.

2.2 *Numerical models for intermodal transport Chains*

Intermodal freight transport, as defined by UNECE (2001), involves the movement of goods using multiple modes of transportation within a single loading unit or vehicle, with no direct handling of the cargo during mode changes. As highlighted by Rodrigue (2020), the development of intermodal transport has been significantly driven by the standardization of containerized freight, leading to a substantial increase in container volume over recent decades.

There are several advantages to intermodal transport, as noted by Crainic and Bektaş (2007). These benefits include the flexibility to choose carriers at each stage of the shipment, potential cost reduction, and the ease of transitioning containers between different modes of transport. However, there are also drawbacks, such as potential delays due to cargo handling during intermodal transitions, and the vulnerability of the entire operation to service disruptions since it relies on various transport modes. Reducing transport costs primarily involves competitive freight rates, maintaining adequate capacity on trade routes, and ensuring suitable transport times. A well-designed liner shipping network is crucial for achieving this, as emphasized in Krogsgaard *et al.* (2018). The size of the fleet and appropriate ship sizing is also vital, as seen in Fagerholt (2004).

Some researchers focus on evaluating the economic feasibility of liner routes. Santos *et al.* (2021) present a four-step process that begins with assessing the demand for freight transport between origin-destination pairs, integrating data from maritime and global trade. Using a logistic regression model that accounts for both cost and transit time, the projected demand is allocated among different transport chains. The assessment of these transport chains is supported by a transport network model. Finally, a link-by-link aggregation of demand is performed to estimate the required ship size to meet the demand in maritime links. While most studies on SSS and intermodality have historically concentrated on routes within the European Union, recent research has explored other geographic regions. For example, Konstantinus *et al.* (2019a) analyze the potential for the development of SSS in the SADC.

2.3 *Discrete choice models in intermodal transport*

Discrete choice models are employed to elucidate or predict decisions when faced with multiple discrete options, such as the selection of transport modes. These models operate on the assumption that when presented with a discrete set of choices, individuals opt for the one that maximizes their utility or benefit. The utility function is statistically inferred for each specific case study. For instance, Ben-Akiva *et al.* (2008) present models of this kind. One of these models begins by establishing a total logistics cost model based on revealed preference data and then incorporates service quality into the model. To bridge the gap between service perceptions and perceptual indicators, a factor analytic approach is adopted, as service perceptions are not directly observable. After extracting these indicators, the model is estimated.

In a related topic, Konstantinus *et al.* (2020) conduct a stated preference experiment in three inter-urban freight corridors within the SADC. Their paper introduces a discrete choice model aimed at understanding the conditions under which SSS could serve as an alternative to road transport. To assess the influence of random heterogeneity in freight mode choice, they subsequently estimate a mixed multinomial logit model and a latent class choice model based on supply chain data. This study utilizes certain transport data presented in the work by Konstantinus *et al.* (2020).

3 CHARACTERIZATION OF TRANSPORT IN SOUTHWESTERN AFRICA

3.1 *Geographical definition of region under study*

This paper's geographical focus covers the western coast of SADC, stretching from Cape Town to Luanda. This coast represents a coherent maritime expanse where road connections are challenging, and potential transport enhancements through SSS between South Africa, Namibia, and Angola could be of substantial significance. The transport network examined within this paper encompasses Angola, Namibia, and South Africa's Western Cape province, all of which host ports equipped with container terminals along the coast. The regions and ports covered in this coastal area are illustrated in Figure 1. For the purposes of this study, this shaded region will be referred to as "Southwestern Africa."

Figure 1. Geographical region under study: SADC countries and Southwestern Africa.

SADC has recognized the imperative of improving its regional freight transport network to achieve its development objectives. The adoption of SSS has been put forth as a means to enhance transportation conditions within the region, as mentioned in Konstantinus *et al.* (2019a). This necessity stems from the underdeveloped transport infrastructure in the region and the prevalent reliance on road transport. Shipping along the SADC coastlines faces two major challenges: extended turnaround times in ports and exceedingly high port charges.

3.2 *Characterization of maritime transport in Southwestern Africa*

A survey of SSS services in operation along the western coast of the SADC was conducted. It was found that Naiber, an Angolan shipping company, operates small routes between Luanda, Lobito, and Namibe. However, these routes primarily involve domestic transport (cabotage) and handle a relatively modest volume of cargo. Another company, Safmarine, provides services for freight transportation utilizing cargo liners and container ships in the same region. Currently, Maersk operates a few routes passing through the area, but it doesn't run any intra-Africa routes that qualify as SSS in this specific region. This aligns with the conclusion drawn by Konstantinus (2021), stating that, both at that time and in the present, there are no SSS routes operating in the region.

The success of SSS is contingent on various factors, including freight volumes, port competitiveness, geographical characteristics, and the regulatory environment in which it operates. Given the extensive geographic area, projected freight volumes, and the customs and trade agreements that SADC is actively pursuing, SSS theoretically holds the potential for success in this region. Moreover, SSS offers numerous advantages when compared to road or rail transport, notably its reduced environmental impact, leading to fewer external costs associated with transportation.

3.3 *Land-based transport in Southwestern Africa*

Trade between African countries is heavily reliant on limited road networks, as rail systems are often inadequate or non-existent. Therefore, the integration of SSS within intermodal chains, whenever feasible, would significantly facilitate international (and even national) trade. To model these intermodal chains, a road network model is essential for assessing the functionality of complete intermodal transport chains and for comparing the competitiveness of SSS with road-only transportation.

The condition of the road network varies within this region. Angola's road network, and to a lesser extent Namibia's, is predominantly in poor condition, while South Africa generally maintains its roads well preserved. However, South Africa experiences congestion issues in and around major cities. To model the road network and its condition, maps provided by the Angola Ministry of Transports (2019), Roads Authority of Namibia (2010), and the Western Cape Government (2023) were employed for Angola, Namibia, and South Africa, respectively. The road conditions were modeled by adjusting the average speed attributed to each road segment (motorways: 80 km/h; good roads: 60 km/h; poor roads: 40 km/h; dirt roads: 30 km/h).

Border controls and transit time at border crossings are recognized as significant challenges in this region's land transport. Reports indicate that customs delays constitute approximately 75 percent of the time consumed during cross-border trips. As observed in Konstantinus *et al.* (2019a), several border stations in the SADC induce prolonged delays due to traffic, with queues sometimes extending up to 60 kilometers. Predominantly, pre-clearance and border processing delays are the primary contributors to these delays. The total time spent at border stations is incorporated into the model as a prescribed delay at the node representing the border station.

3.4 *Characterization of country level cargo flows in Southwestern Africa*

To comprehend the general cargo and containerized cargo flows among the countries in the region, data was gathered from official government reports, specifically the Namibian Government (2013) and the Road Transport Agency (2022), for Namibia and South Africa, respectively. In 2010, South Africa dominated the seaborne cargo volume in the SADC, accounting for 85% of the total (250 million tons of cargo). Angola followed with 5%, and Namibia with 2%.

Regarding cargo flow between SADC countries, Konstantinus (2019b) offers insights into projected intra-regional trade volumes within SADC, determined through a gravitational model. These projections indicate a positive outlook for the region, as most countries are expected to witness a substantial increase in trade by the year 2044. These results underscore the significance of gravitational models in predicting bilateral trade between two countries or regions.

It's worth noting that Angola and Namibia are not major exporters of containerized goods, while South Africa serves as the primary exporting country in the region, handling a significant portion of the containerized trade in the area. These figures will be further examined and tailored to the requirements of this study in section 5.

4 METHODOLOGY

4.1 *General*

The flowchart depicted in Figure 2 outlines the various steps of the methodology, as required for assessing the economic feasibility of a SSS service. It is important to emphasize that this methodology is of a general nature, making it applicable for use in different geographic regions. In this paper, it will be applied in a case study in the southwestern coast of Africa. The subsequent sub-sections will provide a comprehensive breakdown of each step in this assessment process.

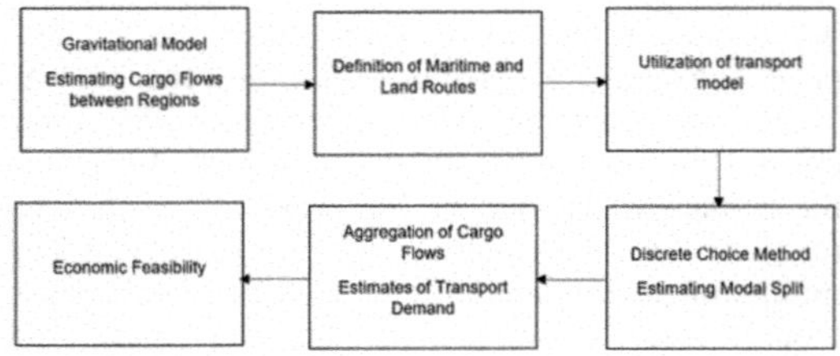

Figure 2. Methodology for assessing the economic feasibility of liner services in Southwestern Africa.

4.2 *Gravitational model to estimate cargo flows between regions*

The initial step involves estimating cargo flows between regions using a gravitational model. Two distinct approaches can be employed within the gravitational model to make the necessary estimations. The most common approach is to use Gross Domestic Product (GDP) as the economic mass, while the alternative approach involves using population data. This model enables the estimation of cargo flows between regions with only the available information on total trade between countries. The first stage, therefore, entails determining the most appropriate variable, based on available data, to serve as the economic mass within a gravitational model such as this:

$$X_{ij} = \frac{(Y_i Y_j)}{D_{ij}^k} \tag{2}$$

where the X_{ij} represents the trade flows between i and j, Y represents the economic mass (GDP or population) and D_{ij} is the distance between i and j, and k being a user specified exponent. To obtain the distance between i and j the shortest distance land route is identified using the transport network model, which includes real road distances for each road segment between consecutive nodes. The percentage of cargo flow that each route (corresponding to a connection between region i and j) will represent in the overall trade between two countries is given by:

$$P_{ij} = \frac{X_{ij}}{TCF_{ij}} \tag{3}$$

TCF_{ij} is the total representative value of cargo flow between two countries and NR the number of routes between countries:

$$TCF_{ij} = \sum_{i=0}^{NR} X_{ij} \tag{4}$$

The percentage of cargo flow corresponding to the pair i and j is then multiplied by the total cargo flow between the two countries, allowing an estimate of the real cargo flow between regions (*CFR*):

$$CFR_{ij} = P_{ij} \times CFC_{ij} \quad (5)$$

4.3 *Transport analysis model*

The following step involves the use of a transport analysis tool designed to compute key parameters that influence transport decision-making, such as cost and transit time, for any pair O/D in the network. In this study, the numerical tool for transport analysis is a software called Intermodal Analyst (IA version 2.0). This software has been developed at the Centre for Marine Technology and Ocean Engineering (CENTEC) of the University of Lisbon since 2016. Its main objective is to calculate the time and cost associated with transportation between any pair O/D. The software performs this assessment using a model of the region's transport network. Routes are modelled as paths, p, created by an ordered set of links so that an O/D pair is interconnected. A set k is used to group paths that have the same O/D. Each link, designated by the letter i in this instance, is given attributes that help describe it, such as the length (distance) Dij, the mode of transportation, and the average speed of operation Sij. The software is able to handle intermodal paths (routes) with different transport modes articulated through port terminals. The software is used to compare the feasibility of SSS with road transport only.

4.4 *Discrete choice method to estimate modal shift*

The subsequent step involves the use of the discrete choice method to estimate modal shares. When a new transport alternative is introduced, the preferences of decision-makers may shift, leading to what is known as a modal shift. In this paper, the comparison is made between SSS (including pre-carriage and on-carriage) and road-only transportation.

To assess the preferences of transport decision-makers for each mode of transportation under different configurations of SSS routes, a binary logit model is employed. The binary logit model operates on the assumption that individuals choose the option that offers them the greatest benefit, as defined by a utility function that is contingent on various route transport parameters, including cost and transit time.

To estimate the extent of modal shift, a comparison is drawn between the modal shares predicted by the binary logit model before and after the implementation of different SSS routes or changes in the model's parameters. This helps gauge how the introduction of SSS affects the choices made by transportation decision-makers.

4.5 *Aggregation of cargo flows and estimates of transport demand*

In order to aggregate cargo flows, data on the quantity of containers being transported between various origins and destinations must be collected. This data can be derived from the numerical results generated by the software, which involves multiplying modal shares by the total number of containers that need to be transported annually between specific origin and destination pairs. The process of aggregating cargo flows encompasses several stages.

The first step involves gathering information on the movement of containers between different O/D pairs. This information should provide details about the cargo's O/D, the number of containers used, and the mode of transportation employed. The collected data can then be categorized based on the mode of transportation and the locations involved. By combining the data in this manner, it becomes possible to assess the overall demand for transportation services, including SSS. Additionally, this approach allows for the estimation of cargo flows within specific segments of the road system.

4.6 *Economic feasibility of liner services*

The final step in the process involves the evaluation of the costs and revenues associated with liner services. Various factors are taken into consideration, including operating costs, capital costs, demand for cargo transport, and the availability of resources such as ships, ports, and labor. The results of this assessment play a crucial role in informing decisions related to route planning.

The sections that follow will delve into the analysis of the factors mentioned above, comparing the two modes of transportation, SSS and road-only transportation. This comparative analysis will allow for the assessment of the economic feasibility of SSS liner services in comparison to road transport. The methodology applied to gauge the economic feasibility will entail conducting a financial analysis for the SSS route, encompassing a time span ranging from 20 to 30 years.

5 MODEL OF CARGO FLOWS AND TRANSPORT NETWORKS

5.1 *Estimating regional cargo flow using a gravitational model*

5.1.1 *Estimating the GDP of the regions under study*

To identify the cargo flows per O/D pair, it is necessary to refer to existing literature or, alternatively, to estimate them using a gravitational model. The O/D pairs in this context correspond to large regions within the countries under study, specifically Angola, Namibia, and South Africa (Western Cape). A total of 274 O/D pairs need to be considered and will be incorporated into the transport model. These pairs are based on four origins (regions) located in South Africa, five in Namibia, and thirteen in Angola, representing the major regions and population centres.

Given that GDP per capita values for the provinces in Namibia and Angola were not readily available, similar indicators or proxies for GDP were identified for these provinces to apply in the gravitational model. The selected indicators were per capita consumption for Namibia and per capita average income for Angola. These indicators were then converted in order to estimate the GDP per capita. After this conversion, the final step involves utilizing the gravitational model to estimate the percentage of cargo flows for each of the routes in the context of total trade between the countries served by the respective route. It is worth noting that GDP per region had to be estimated prior to applying the gravitational model using certain parameters. Finally, being GDPR the GDP per Region, the PR the population of the region, and GDPCR the gross domestic product per capita of the region.

$$GDPR = GDPCR \times PR \tag{6}$$

5.1.2 *Cargo flow per route*

The equation for the gravitational model used in this study is the following one:

$$X_{ij} = \frac{(Y_i\ Y_j)}{(D_{ij}^2)} \tag{7}$$

where the X_{ij} is a representative value of the trade flows between i and j, Y represents the GDP per region for node i and j and D_{ij} is the distance between i and j. In order to use this expression, the initial step involved estimating the volume of cargo flows between countries. Subsequently, to derive an estimate of the cargos exported and imported by each SADC country, data from the Observatory of Economic Complexity (OEC) was used. It is important to note that this data is expressed in monetary value rather than in tons. Unfortunately, disaggregated data for cargo flows in tons between Angola and other countries was not available and approximations were used. Table 1 presents a summary of the results obtained through this process.

Table 1. Estimate of imports and exports in equivalent TEUs (14 tons per TEU) for 2020 between countries under study.

	TEUs/year traded between countries		
Origin/Destination	Angola	Namibia	South Africa
Angola	-	357	1, 115
Namibia	6, 968	-	35, 450
South Africa	48,840	144, 723	-

It is worth acknowledging that the volumes presented in Table 1 are not particularly high. The highest values are observed in the exports of South Africa to both Namibia and Angola, which is in line with expectations. In opposition, the exports in the southbound direction are notably small, reflecting the relatively limited containerized goods exports from Angola and Namibia.

Nevertheless, it is highly plausible that a new SSS service along this coast could potentially capture a portion of the substantial cargo volume originating from China, transhipping it to Namibia and Angola. To estimate this potential cargo volume, the value of Chinese imports into Angola and Namibia was first identified using data from the Observatory of Economic Complexity (OEC) for the year 2020. By assuming that the import profile remains consistent, and that value and weight maintain a similar proportion, it was possible to extrapolate the number of equivalent twenty-foot equivalent units (TEUs) arriving from China into Angola and Namibia.

Lastly, it is assumed that the new SSS service could capture a moderate 20% of these volume for its operations. This presents a potential opportunity for the SSS service to handle a portion of the significant freight flow between China and the coastal regions of Namibia and Angola. Table 2 shows the corrected estimates of imports and exports considering this additional volume.

Table 2. Estimate of imports and exports in equivalent TEUs (14 tons per TEU) for 2020 between countries under study including China captured TEUs.

	TEUs/year traded between countries		
Origin/Destination	Angola	Namibia	South Africa
Angola	-	357	1,115
Namibia	6,968	-	35,450
South Africa	80,798	150,257	-

The next step is to multiply the percentage of cargo flow that each route (pair origin/destination) represents by the TEUs handled per year between the countries which the route connects, in order to obtain the TEUs for each route which will be an input in the software:

$$TEUER = TEUCR \times PCFR \tag{8}$$

Being *PCFR* the percentage of cargo flow for the route, and *TEUER* the TEUs for each route, and *TEUCR* the TEUs between countries of said route.

Table 3. Container terminals fees along with call and dwell time.

Container Terminal	Handle Fee for Full containers (USD/move for TEU)	Handle Fee for Empty containers (USD/move for TEU)	Storage fee (USD/day for TEU)	Call Time	Dwell Time	Free Time
Luanda	130 USD	70 USD	60 USD	12 hours	12 days	5 days
Walvis Bay	415 USD	415 USD	9 USD	12 hours	9.5 days	5 days
Cape Town	111.8 USD	96 USD	140 USD	12 hours	5-6 days	3 days

5.2 *Transport network model for Southwestern Africa*

5.2.1 *Road network model*

The primary function of the software is to calculate the time and cost involved in travelling between a pair O/D. The transport network is stored in a database that includes the main roads within the geographical area under study, with a focus on connecting major cities in each region. This network enables the connection of main pairs O/D in the three countries under examination. The roads are defined in such a way that border control posts, intersections, and port terminals constitute nodes in the network.

For nodes within the network that represent port terminals, specific attributes are included. These attributes encompass the average call time, handling charges, dwell fees, and free time available at each container terminal. The information included in the model, which is stored in the database, can be found in Table 3. This data was sourced from DP World MPL (2021), Deutsche Afrika-Linien (2023), and Transnet (2023) for the ports of Luanda, Walvis Bay, and Cape Town, respectively.

Another important aspect of modelling the transport network is to characterize the time spent in border control posts and if possible, the administrative (crossing) fees applied in such posts, so information regarding the border control posts for the border between Namibia and South Africa and between Namibia and Angola was identified in PSCG (2023).

Table 4. Time spent in border posts and cross border fees.

Border	Border Post	Time spent in border in hours	Crossing Fee [USD]
Namibia-South Africa	Noordoewer/ Vioolsdrif BCP	2	92
Namibia-South Africa	Alenxader Bay	2	92
Namibia-Angola	Ruacana	89.5	92
Namibia-Angola	Omahenene	89.5	92
Namibia-Angola	Katwitwi	89.5	92
Namibia-Angola	Sarasungu	89.5	92

5.2.2 *Short sea shipping route*

In addition to the road-based transport network briefly described above, this paper proposes the addition of a SSS service covering the western coast of Southern Africa. The route being considered is shown in Figure 3. The numbers in the figure indicate the identification of the links composing the route and the nodes through which the route pass. It is then included in the transport network database. The route calls in one port per country, namely Cape Town, Walvis Bay and Luanda. The ship completes a complete symmetric round voyage.

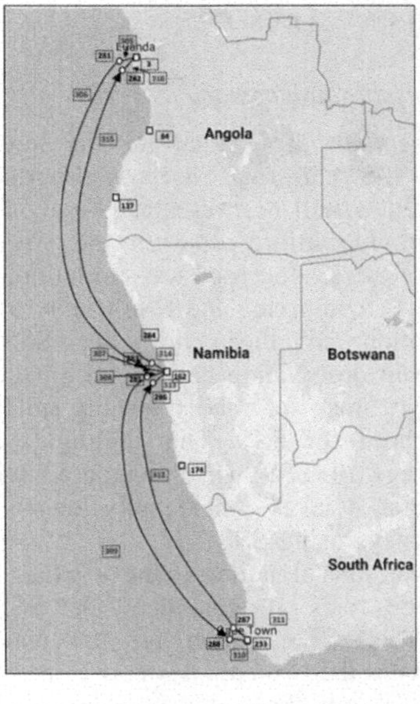

Figure 3. Proposal of a short sea shipping route.

5.2.3 *Cost levels for transport modes*

The transport analysis software also requires the specification of cost functions, that is cost of freight for different transportation distances and modes of transport. These functions needed to be adapted to the local conditions of the SADC region and this was carried out using values provided in Konstantinus (2019b) for different routes (land and/or maritime).

5.2.4 *Logistic performance of ports and border posts*

Finally, the model of the transport network also includes values for dwell time and time spent in border control, which were taken from Pushak and Foster (2011), African Development Bank (2020) and TRANSNET (2023), respectively for the ports of Angola, Namibia and South Africa. The cross-border times were retrieved from PSCG (2023). Table 5 shows the values adopted.

Table 5. Time spent in port terminals and border posts.

Port	Dwell Time in days
Cape Town	5.5
Walvis Bay	9.5
Luanda	12
Border Posts	**Time in Border in days**
Angola-Namibia	3.7
South Africa-Namibia	0.1

6 NUMERICAL RESULTS

6.1 *Transportation demand*

In this section, the results for the modal split between SSS and road-based transportation for various routes will be presented. Firstly, Tables 6 and 7 provide information on the voyage cost and time for SSS and road transportation, respectively. When comparing the results for road-based transportation to those for the SSS route, a significant observation is that, for all the routes, the voyage times for road transport are less than half of those for SSS. This highlights a major disadvantage of SSS when compared with road transport, as it takes considerably longer to complete journeys using SSS.

However, the situation changes when comparing voyage costs. It is evident that SSS is the more cost-effective option for the majority of routes, with the exception of two paths: Cape Town to Keetmanshoop and Otjiwarongo to Uige. This deviation occurs because the distances between the nodes are very short in these cases, and the distance travelled by road with SSS is nearly the same as in road-only transport. For the remaining paths, SSS is approximately 500 to 1,000 USD cheaper than road transport.

Figure 4 shows the modal split for paths with origin in Cape Town, the main south African city considered in thus study, and George with

Table 6. Voyage time and costs for SSS.

Paths being analyzed	Voyage time for SSS in days	Voyage cost for SSS in USD
George-Menongue	25.1	8,137
Cape Town-Keetmanshoop	18.4	5,109
Otjiwarongo-CapeTown	17.5	4,383
Keepmanshoop-Luanda	25.4	5,632
Otjiwarongo-Uige	24.8	6,544

Table 7. Voyage time and costs for road only transportation.

Paths being analyzed	Voyage time for Road in days	Voyage cost for Road in USD
George-Menongue	8.5	8,237
Cape Town-Keetmanshoop	1.3	2,884
Otjiwarongo-CapeTown	3.0	4,635
Keepmanshoop-Luanda	7.9	6,157
Otjiwarongo-Uige	7.5	4,850

all the possible destinations in Angola and Namibia. In each destination a pie chart indicates the SSS modal share in blue, with the remainder corresponding to road transportation. Figure 5 displays the same characteristics but for paths with origin in Vredenburg and Stellenbosch.

For the paths ending in Namibia, the results are favourable to road transportation. As seen in Figures 4 and 5, only the paths ending in Walvis Bay show a slight increase in preference for SSS, however road transportation was still the favourite for these paths. Comparing the results for the different origins it is possible to conclude that the further away the origin is from the container terminal in Cape Town, the lower the competitiveness of SSS. This was to be expected since the distance travelled by sea in these paths is not large enough to compensate when compared to SSS.

As regards Angola, it can also be seen that for the destination closer to the border with Namibia, road transportation is still favoured. The same conclusion can be taken for the paths originating in other origins in South Africa. However, for the paths ending closer to the container terminal in Luanda, and to the paths ending in regions in the interior of Angola it is possible to see that SSS became favoured over road transportation. This leads to the conclusion that SSS benefits with proximity of the destination to container terminals along with the increase in distance.

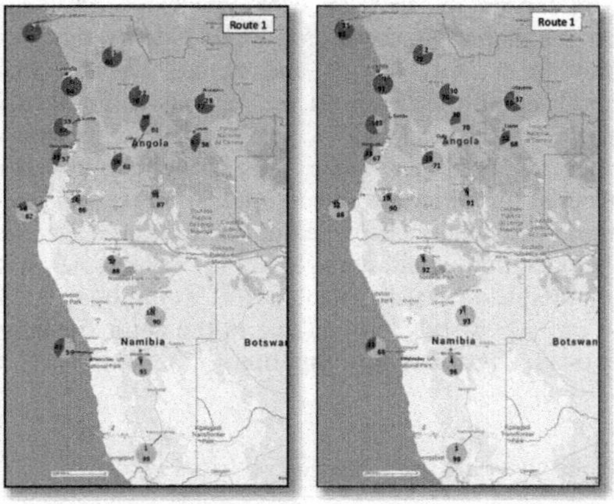

Figure 4. Modal splits in routes between cape town (left), george (right) and regions of Angola and Namibia.

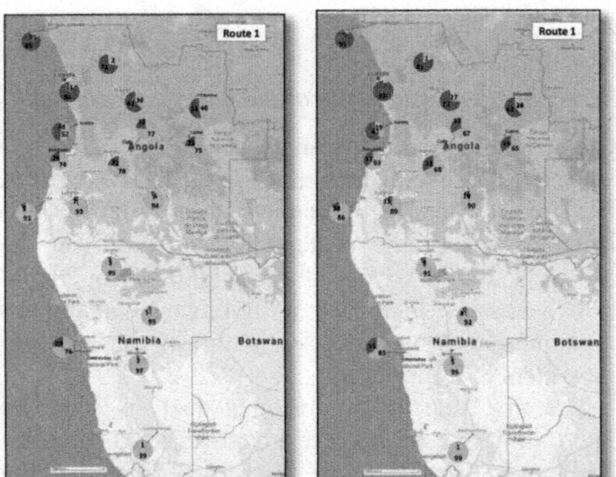

Figure 5. Modal splits in routes between Vredenburg (left), Stellenbosch (right) and regions of Angola and Namibia.

Table 8. Transport demand for the different legs of the SSS operation.

TEUs/year handled in each leg	
SSS Route	TEUs/year
Cape Town-Walvis Bay	42,291
Walvis Bay-Luanda	30,550
Luanda-Walvis Bay	686
Walvis Bay-Cape Town	3,686

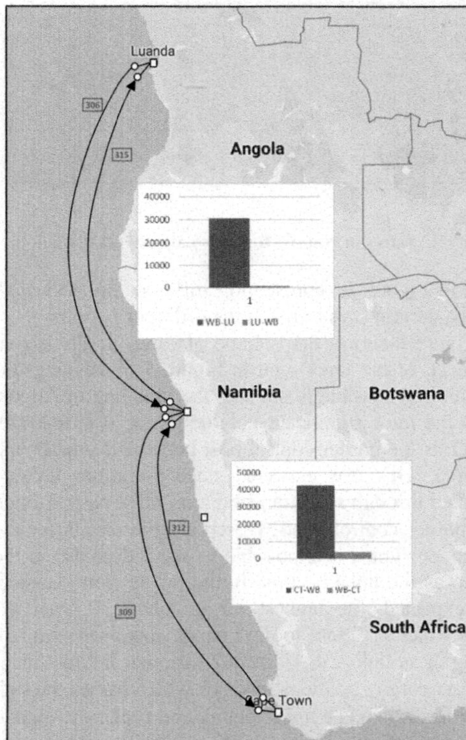

Figure 6. Volume of cargo for each maritime link.

Table 8 shows the total aggregated volumes of TEUs (transport demand) expected per year in different legs of the SSS route. The numbers are the summation of all TEUs expected to use the container ship in a particular leg of its voyage, being that these containers will have many different final destinations. A rational assumption has been made by considering that each container will be loaded/unloaded in the port located closer to its origin/destination.

Figure 6 shows the comparison between the volumes of cargo of the SSS route comparing the northbound and southbound legs of the route. Analyzing this figure, it is possible to see there is a large difference in the cargo volumes between the northbound and southbound parts of the route, with the cargo volume going north being much larger than comes south. This was expected since the major exporting country of containerized goods in the region is South Africa and as seen in Table 2, the most significant cargo flows are between South Africa and the two other countries, in addition to the fact that Namibia exports some cargo to South Africa. In contrast to that, Angola is not an exporting country of containerized goods so there is only a small amount of cargo coming south from Luanda.

Figure 7 displays the links existing in the road network of Angola, this will help to understand how the cargo disperses over this country. The links analyzed in this study are highlighted by the orange arrows along with the box containing the numbering of the

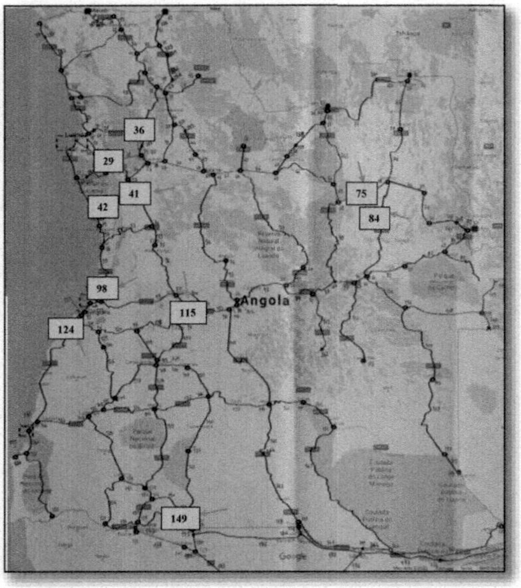

Figure 7. Links in Angola for the road network model.

link in Figure 7, representing different links through which cargo flows to reach the different regions.

Figure 8 shows the volume of cargo in TEUs/year for each of the links seen in Figure 7. Analyzing this figure, it is possible to see that the road link in Angola with the most significant volume of cargo is link 149 which is the border control post between Namibia and Angola. This was expected since the cargo volume handled through road transportation must pass through the border control posts. Observing the results for the remainder links, it is possible to see a decrease in the cargo as the links become further away from Luanda. As expected, the highest cargo volume is seen the links which are close to the Luanda container terminal, being these links 29, 41 and 42. Another link with significant cargo volume is link 115 which passes through Huambo which is a major population center in Angola. The rest of the links which are in the interior of Angola see a decrease in the cargo volume as expected since there is less population in these areas.

To determine the required vessel size for the SSS operation serving the Southwestern coast of Africa,

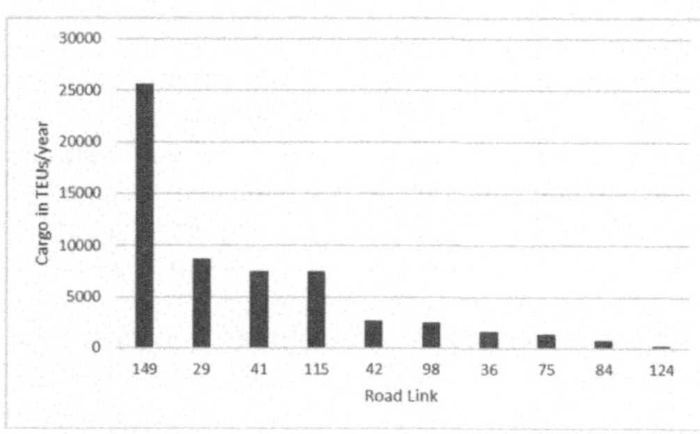

Figure 8. Cargo volume for road links.

several factors were considered. These factors included the transport demand (as shown in Table 8), the round voyage time (approximately two weeks) and the need for weekly departures. The goal was to calculate the average cargo volume that the vessel would need to carry on each voyage, taking into account the requirement for weekly departures. To provide a safety margin and ensure that the SSS routes could reliably meet the region's cargo needs, approximately 200 TEUs were added to the minimum required vessel size. This safety margin helps account for variations in cargo volume between voyages. The results, displayed in Table 9, indicate that the operation would require a 1,000 TEU vessel. It's important to note that two vessels would be needed to provide weekly service due to the approximately two-week round voyage duration.

Table 9. Vessel size estimation.

	Required size of vessel	Size of vessel with safety margin
SSS operation	813	1,000

6.2 *Economic feasibility of the novel service*

The economic feasibility analysis of the SSS route is impacted by several factors. One of these factors is the need for empty containers to return to Cape Town to ensure the route's feasibility. Given the two-week round voyage duration and the operation of two vessels, there was a significant demand for containers, making it necessary to bring back empty containers to Cape Town.

The first step in the financial analysis involved determining the initial investment required to establish the SSS route. This investment included the acquisition of container ships and, in addition, the cost of the fleet of containers necessary for the operation. Typically, it is considered necessary to have a fleet of containers amounting to the double of the capacity of the ships, and since the operation would involve two vessels, 4,000 containers need to be purchased. These initial investment costs are essential to understand the financial requirements for setting up the SSS route and ensuring its operational capacity.

Table 10. Initial investment of the SSS operation.

Initial Investment in USD	
Two container vessels of 1,000 TEUs along with 4,000 containers	65,200,000 USD

The analysis further extended to financial considerations, such as operating costs, voyage costs, revenue, and profitability, to assess the route's overall economic feasibility. The operating costs of the SSS route were calculated taking into account various factors and components, mainly the technical characteristics of the vessels being used in this operation. Fuel costs were also calculated considering the propulsion power of the vessels and the duration of the voyages on both routes. After that the port-related costs were calculated and found to be a significant part of the expenses. This included port fees and cargo handling costs. Notably, the handling charges for empty containers had a substantial impact on the overall port costs.

Figure 9 illustrates the distribution of yearly costs. The majority of the costs were attributed to port costs, accounting for 55% of the total costs. This high proportion was primarily due to the significant handling charges applied to empty containers, particularly at the Walvis Bay container terminal. Fuel costs represented 24% of the total costs, while capital costs and operating costs accounted for 14% and 7%, respectively. Understanding this breakdown of operational costs is essential for assessing the financial feasibility of the SSS route, as it allows for the identification of cost drivers and areas where potential cost reductions or optimizations may be feasible.

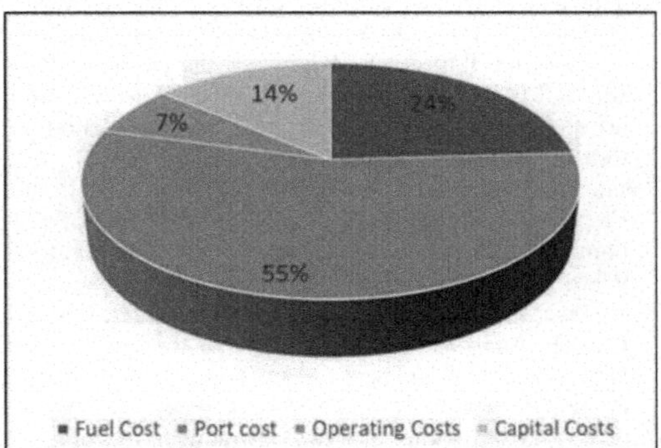

Figure 9. Division of voyage costs for the operation.

Revenue was calculated from the freight rate to be applied in each voyage (leg or legs) carried out by the container. The method chosen to estimate the revenue was to multiply the number of containers that pass through each maritime link by the transport cost for SSS provided by the software for the respective link. After having obtained the revenue, the cashflows were calculated, discounted at a rate of 8%, and the Net Present Value (NPV) and the payback period were obtained. The final results are displayed in Table 11.

Analysing Table 11 some conclusions can be taken regarding the economic feasibility of the SSS operation in the Southwestern Africa region.

Table 11. Net present value and payback period.

Route	Net Present Value (USD)	Payback period (years)
Cape Town-Walvis Bay-Luanda	242,870,448	2

Observing the results, the NPV is around 242 million USD. The payback period is considered to be very low with 2 years. With these results it can be stated that, for the freight rates under consideration, the SSS operation would be economically feasible since the payback period is very short and the NPV is almost four times the initial investment.

The high economic feasibility of the operation is mainly driven by the high freight rates indicated by Konstantinus (2019b). These rates were found to be considerably higher than those mentioned in UNCTAD (2018). For voyages with 1,000 kms length, in the region of study, the unit cost is 0.9 USD/TEU.km while the unit cost, for similar voyages in other regions of the world, is only 0.15 USD/TEU.km, which is a significant difference.

Another observation taken after analysing the results was that the financial analysis was done with the same cargo volume every year, which is unrealistic, particularly as the SADC region is expected to see significant economic growth in the future as mentioned in Chapter 3.4. Therefore, additional calculations were run for a unit cost of 0.15 USD/TEU.km, and with an increase of cargo volume of 2.5% every year to consider the economic performance estimated for the region.

Table 12 displays the results for these lower freight rates. Comparing the results with those in Table 11, it can be seen that the NPV is now -42 million USD, implying a significant loss. This result was expected since the 2.5% yearly increase in the freight volume was not enough to compensate the 78% decrease in the freight rate. This significantly lowered the revenue stream of the operation leading to its unprofitability. Consequently, no pay-back period is obtained.

Table 12. Net present value with increased cargo volume and freight rates in line with international values.

Route	Net Present Value (USD)
Cape Town-Walvis Bay-Luanda	-42,144,568

7 CONCLUSIONS

This paper has provided a brief review of the existing literature on the definition and planning of intermodal transport chains, including SSS, as well as the use of the discrete choice method for evaluating modal shares and gravity models for disaggregating transport statistics. These methods have formed the basis for the development of an integrated methodology for assessing SSS routes within the broader context of intermodal transport chains. The methodology was applied in a case study dedicated to transport in the southwestern coastline of the African continent, an area that has not been extensively explored in previous literature. Specifically, the case study focused on the economic feasibility of a SSS route operating in the southwestern coast of Africa, connecting South Africa, Namibia, and Angola. This case study builds upon previous research conducted by Konstantinus (2019b).

The application of this methodology revealed the challenges associated with studying remote geographical areas that encompass countries with limited statistical data. Nonetheless, even within this region, notable disparities in the availability and depth of data were observed between South Africa and the other countries. In any case, the gravity model proved to be a suitable tool for estimating cargo flows between regions in these countries. Additionally, it was evident that trade between South Africa and its two northern neighbors is significantly unbalanced, with minimal cargo moving southbound. Therefore, it became necessary to assume that some additional cargo transshipped in Cape Town from container ships arriving from Asia would contribute to the new SSS service, helping to achieve the required critical mass.

The study evaluated the transport chains (comprising fully road-based and SSS-based routes) between 274 origin-destination pairs. The evaluation took into account road and sea distances, road conditions, and the performance of port and border crossings. One of the key findings was that road transportation is consistently faster than SSS, as expected, due to the slower travel speeds of ships and the current relative performance of ports and border crossings. In contrast, for the majority of O/D pairs, SSS was found to be a more cost-effective option compared to road transportation, specifically for O/D pairs closer to container terminals.

Modal splits for each origin-destination pair were estimated using a discrete choice model. SSS was favored over road transportation, particularly for routes ending in Angola. The aggregation of cargo volumes from all 274 O/D pairs allowed for the estimation of total demand for SSS and the determination of the required ship size for this operation.

Finally, the paper explored the economic feasibility of the SSS service through an investment analysis. The results indicated that the SSS operation is economically feasible, as it demonstrated a high NPV with a return that exceeded four times the initial investment. Additionally, it had a relatively short payback period. It must be pointed out that there is a severe cargo imbalance in the route and that it was assumed that some cargo coming from the Far East could be captured for SSS in the port of Cape Town. Furthermore, additional analysis revealed that when applying freight rates more in line with those common in other regions of the world, for routes with similar characteristics, the NPV became significantly negative.

Future research in this area could benefit from engaging with local authorities to obtain more information on cargo flows and logistic performance in the various countries. Additionally, considering the existing railway lines in the region as a means to distribute containerized cargo within the interior of Angola, particularly along the Lobito Corridor, could be an interesting topic of study, potentially mitigating the challenges posed by poorly maintained roads.

ACKOWLEDGEMENTS

This research contributes to the Strategic Research Plan of the Centre for Marine Technology and Ocean Engineering (CENTEC), which is financed by FCT under contract UIDB/UIDP/00134/2020.

REFERENCES

African Development Bank (2020). The new Walvis Bay container terminal, https://projectsportal.afdb.org/dataportal/VProject/show/P-NA-DD0-002 (retrieved on 2 February 2023).

Angola Ministry of Transports (2019). National Master Plan for the Transport Sector & Preliminary Feasibility Study of the Railway Link between the Benguela Railway (CFB) and Zambia (in portuguese) Retrieved from https://adfersit.pt/uploads/article/27/angola-planodirectornacionaldosectordostransportes.pdf

Baier, S., & Standaert, S. (2020). Gravity Models and Empirical Trade. Oxford Research Encyclopedia of Economics and Finance. https://doi.org/10.1093/acrefore/9780190625979.013.327

Ben-Akiva, M., Bolduc, D., & Park, J. Q. (2008). Discrete Choice Analysis of Shippers' Preferences. Recent Developments in Transport Modelling, 135–155. https://doi.org/10.1108/9781786359537-008

Crainic, T. G., & Bektaş, T. (2007). Brief overview of Intermodal transportation. In CRC Press eBooks (pp. 28–16). https://doi.org/10.1201/9781420004588.ch28

DEUTSCHE AFRIKA-LINIEN (2023). Walvis Bay Container Terminal Tariff Book, Retrieved from https://rantzaugrshot.blob.core.windows.net/public/Uploadedpdfs/20200512113115_200401_DAL_Agency_Tariff_for_Namibia.pdf

DP World MPL (2021). Tariff Book for Luanda Container Terminal, https://www.dpworld.com/luanda/-/media/project/dpwg/dpwg-tenant/africa/luanda/media-files/tariff-and-charges/tariff-and-charges-eng.pdf?rev=2119e206d08a4ebfb7590f5f2fde7e05

Eurostat (2023). https://ec.europa.eu/eurostat/statisticsexplained/index.php?title=GDP_per_capita,_consumption_per_capita_and_price_level_indices#Relative_volumes_of_consumption_per_capita

Fagerholt, K. (2004). Designing optimal routes in a liner Shipping problem. Maritime Policy & Management, 31 (4), 259–268. https://doi.org/10.1080/0308883042000259819

Konstantinus, A., Zuidgeest, M., Christodoulou, A., Raza, Z., & Woxenius, J. (2019a). Barriers and Enablers for Short Sea Shipping in the Southern African Development Community. Sustainability, 11(6), 1532. https://doi.org/10.3390/su11061532

Konstantinus, A. (2019b). Opportunities for Short-Sea Shipping in the Southern African Development Community (SADC) Region: Evidence based on Discrete Choice Modelling, PhD thesis, Department of Civil Engineering, Faculty of Engineering, University of Cape Town, Cape Town, South Africa.

Konstantinus, A., & Zuidgeest, M. (2019c). An investigation into the factors influencing inter-urban freight mode choice decisions in the Southern African Development Community region. Journal of Transport and Supply Chain Management, 13. https://doi.org/10.4102/jtscm.v13i0.463

Konstantinus, A., Zuidgeest, M., Hess, S., & de Jong, G. (2020). Assessing inter-urban freight mode choice preference for short-sea shipping in the Southern African Development Community region. Journal of Transport Geography, 88, 102816. https://doi.org/10.1016/j.jtrangeo.2020.102816

Konstantinus, A. (2021). Short sea shipping: Stated intentions of shipowners and operators in the Southern Africa Development Community Region. Maritime Transport Research, 2, 100015. https://doi.org/10.1016/j.martra.2021.100015

Krogsgaard, A., Pisinger, D., & Thorsen, J. (2018). A flow-first route-next heuristic for liner shipping network design. Networks, 72(3), 358–381. https://doi.org/10.1002/net.21819

Namibian Government (2013). Master Plan for Development of an International Logistics Hub for SADC Countries in the Republic of Namibia, Logistics Network in SADC and Namibia, Retrieved from https://d3rp5jatom3eyn.cloudfront.net/cms/assets/documents/NHIES_2015-16.pdf on 5th of March of 2023

OEC (2023). data regarding cargo flows, Retrieved from https://oec.world/en/profile/bilateral-country/chn/partner/ago

Paixão, A. C., & Marlow, P. B. (2002). Strengths and weaknesses of short sea shipping. Marine Policy, 26(3), 167–178. https://doi.org/10.1016/s0308-597x(01)00047-

Pushak N., & Foster V. (2011). Angola's Infrastructure: A Continental Perspective, https://web.worldbank.org/archive/website01327/WEB/IMAGES/AICD_ANG.pdf

PSCG (2023). World Customs Organization, East & Southern Africa Region, Regional Private Sector Group, Weekly Cross-Border Report, Week ending 5 March 2023.

Roads Authority (2010). Geographic Information System (GIS) and Road Referencing System (RRS), Retrieved from https://www.ra.org.na/Pages/network.aspx

Road Transport Agency (2022). Statistics on Trade Volumes and Value Flowing Through South African Commercial Border Posts and Destination Countries, Retrieved from, https://www.cbrta.co.za/uploads/files/Assessment-of-Trade-Volumes-and-Values-Passing-Through-RSA-Border-Posts-Report.pdf, on 7th of March of 2023

Rodrigue, J.-P. (2020). The Geography of Transport Systems (5th ed.). Routledge. https://doi.org/10.4324/9780429346323

Santos, T. A., dos Santos, G. L., Martins, P., & Guedes Soares, C. (2021). A methodology for short-sea-shipping service design within intermodal transport chains. Maritime Economics & Logistics, 24(1), 138–167. https://doi.org/10.1057/s41278-021-00193-8

Sinaga, A. M., Masyhuri, Darwanto, D. H., & Widodo, S. (2019). Employing Gravity Model to Measure International Trade Potential. IOP Conference Series: Materials Science and Engineering, 546(5), 052072. https://doi.org/10.1088/1757-899x/546/5/052072

TRANSNET (2023). Tariff Book for Cape Town Container Terminal, Retrieved from https://www.transnetportterminals.net/Customer/Operating%20Procedures/TPT%20Tariff%20Book%201%20April%202023v2.pdf

UNECE (2001). Terminology on Combined Transport. Genua: The United Nations Economic Commission for Europe, New York.

UNCTAD (2018). Review of Maritime Transportation, United Nations, Geneva, Switzerland.

Western Cape Government (2023). Road Network Information System, Retrieved from https://rnis.westerncape.gov.za/rnis/rnis_web_reports.main

Advances in Maritime Technology and Engineering – Guedes Soares & Santos (Eds)

Platform-based marketplaces for sustainable logistic solutions in port ecosystems

O. Dinu, E. Rosca, A. Rusca, A. Ilie & V. Radu
National University for Science and Technology Politehnica Bucharest, Bucharest, Romania

ABSTRACT: Well-integrated logistic networks are crucial for safe, resilient transport and smart mobility services for goods and can also create room for finding efficient ways to store or transport materials and products. To make possible the integration, harmonization, and orchestration of independent logistics operations in ports' ecosystems, smart platforms that can effectively connect the providers of sustainable transport solutions and those who need them are needed. Platform-based marketplaces will help to operate in ports and other logistics hubs by simplifying cooperation between the actors. In response to this opening, the paper proposes to gain a clear insight into the main components and the key findings of previous projects conducted in the field, to identify which are the factors that can have a hindering effect and to analyze sustainability issues in the form of emissions caused by subcontracting (Scope 3), in order to drive logistics to climate neutrality. The second step proposes the following stages of evolution in developing digital platforms for ports along multiple dimensions, starting from the most commonly identified solution employed presently.

1 INTRODUCTION

In general terms, an ecosystem is a complex network or an interconnected system. The concept of ecosystem has its origins in biological sciences. Many definitions of the term ecosystem can be identified in the literature, one of the first notables being Tansley's 1935, which emphasizes the continuous interaction in which living organisms are engaged with all the elements of the environment in which they coexist. Only at the end of the 20th century (Moore 1996) was a notable contribution made to the definition by introducing the notion of a community of living organisms that constantly interact with the environment, generating internal energy in the system. He also transitions to business ecosystems, which he regards as economic communities of interacting organizations and individuals, guided by commercial interests, whose roles are to provide services and products for users who are also part of these ecosystems, generating added value. Over time, the community members align, co-evolve, share visions, and mutually support themselves.

A port community comprises a set of organizations in which different actors are interested in creating value for the final customer. Various roles are present in the port community: authorities, members, system operators, supply chain partners, and managers (Chandra & Hillegersberg 2017). It is necessary to manage the relationships between the identified distinct actors interacting with the port, considering common goals (Henesey et al. 2003). A port community can also be seen as an organization that manages the links between maritime and land transport systems. A seaport's existing infrastructure and software are essential to its success.

Platforms are critical components of modern ecosystems. The primary condition for a platform's existence is that the ecosystem's elements depend upon common standards and interfaces (Robertson & Ulrich 1998). Through the platforms, matches are made between users and providers in a multi-sided model, resulting in integration and network effects (Sanchez-Cartas & Leon 2021). They use digitalization to optimise the use of resources and reduce the costs of all participants in the logistics chain, creating scenarios in which any actor can play one or multiple roles (Derave et al. 2021). They are similar to marketplaces, bringing together the main port actors and creating an ecosystem of partners. However, they present many differences compared to traditional fixed and linear value chains that companies use in the conventional supply chain. The critical functions of platforms, generating added value, are facilitating interaction, collaboration and transaction of the services offered on the platform, coordinating operations and building trust. The major challenge for any such platform is to build trust and to summon a critical mass of transport service providers to make such platforms a reality. Due to network effects, the value of such platforms depends on

DOI: 10.1201/9781003508762-6

and increases with the number of participants (Meyer et al. 2018).

Since the beginning of 2023, incoming EU regulations regarding minimum standards for EU Climate Transition Benchmarks and EU Paris-aligned Benchmarks have demanded large companies to report their scope 3 emissions -indirect value chain emissions. Scope 3 includes all indirect emissions that occur in the value chain of a reporting company. The Environmental Protection Agency (EPA) describes scope 3 emissions as "the result of activities from assets not owned or controlled by the reporting organization, but that the organization indirectly impacts in its value chain." Even though these emissions are out of the reporting company's control, they can represent the most significant portion of its GHG emissions inventory. Scope 3 emissions can be divided into upstream and downstream emissions and classified into categories. In the context of the present paper, the relevant categories are the upstream transportation and distribution and the downstream transportation and distribution. The ISO 14083 standard came into effect, providing a globally aligned approach to quantifying and reporting greenhouse gas emissions in multimodal and global transportation chains. The new standard is the latest consensus for calculating and reporting logistics emissions, providing a reliable and consistent approach for all stakeholders involved. The reporting of emissions is also possible on a clearly defined aggregation level. Aggregation, in this case, means that transport legs and hub operations with similar characteristics are grouped, and the aggregated emission intensity can be applied.

It is essential to identify how different sustainability dimensions are integrated into the port ecosystem, the scope of actions, and the type of stakeholder involvement to better analyze their interests and expected contributions. In response to this opening, the article discusses how our findings relate to past experiences from previous and ongoing European projects and proposes a more integrative and holistic approach to exploring new business models.

The paper proposes to gain a clear insight into the main components and the key findings of previous projects conducted in the field and identify which factors can have a hindering effect. The second step facilitates the identification of the elements and competing options in platform-based marketplaces along multiple dimensions and the analysis of sustainability issues in the form of emissions caused by subcontracting (Scope 3) to drive logistics to climate neutrality. The second step proposes the following stages of evolution in developing digital platforms for ports along multiple dimensions, starting from the most commonly identified solution employed presently.

2 EXAMINING PREVIOUS AND ONGOING PROJECTS

2.1 *Studies integrating scientific and practitioner solutions*

The knowledge of the system is typically assessed to obtain a general view of the key outputs of previous and ongoing projects. The paper aims to draw on a wealth of experience from previous results, seeking a high association level with already proposed models. The main research questions to be answered are:

What are the main components and the key findings of this study?

Which are the factors that can have a hindering effect?

Our considered data corpus consists of a broad sample of studies integrating scientific and practitioner literature. The most representative and essential projects considering digitalization, integrated logistic networks, ecosystems, regulation issues, and climate neutrality mainly address port solutions from the significant directions of research and industrial technology transfer, EU.1.3. - EXCELLENT SCIENCE, EU.3.4. - SOCIETAL CHALLENGES and EU.2.1.5. - INDUSTRIAL LEADERSHIP, financed through EU H2020, CEF Transport and FP7 – Transport programmes, were selected and presented briefly:

ADMIRAL (Programme: H2020 HORIZON-CL5 -2022). *Advanced Multimodal Marketplace for Low Emission and Energy Transportation* is an ongoing project aiming to transform supply chain management in freight transportation by developing a cutting-edge digital marketplace for multimodal logistics. It seeks to shift the focus on indirect emissions, reduce overall emissions in logistics and transportation and enhance transparency throughout the supply chain.

AEOLIX (Programme: H2020-EU.3.4. - SOCIETAL CHALLENGES). *Architecture for EurOpean Logistics Information eXchange* project envisions implementing collaborative IT infrastructure for operational connection of logistics information systems. It also utilizes 3rd party logistics services and a mutual exchange of information between the stakeholders. Develops AEOLIX platform and AEOLIX toolkit, covering core logistics services to support and implement the business needs of end-users.

CLOSER FP7 (Programme: FP7-TRANSPORT). *Connecting Long and Short-distance Networks for Efficient Transport, the* CLOSER project focused on several themes, including emerging mobility schemes, interchanges between short- and long-distance transport for passengers and freight, and the regulatory environment. With the help of 30 different indications, the project analysed various methods for linking short- and long-distance modalities.

CLUSTERS 2.0 (Programme: H2020-EU.3.4. - SOCIETAL CHALLENGES). The CLUSTERS 2.0 project developed low-capital, cost-efficient and investment-intensive enhanced goods handling and transhipment solutions. Its hyper-connected open network of logistics clusters and hubs functions "in the frame of Ten-T". CLUSTERS 2.0 laid the groundwork for developing ways to link freight hubs through the physical internet.

COFRET FP7 (Programme: FP7-TRANSPORT). *The carbon footprint of freight transport reviewed existing methodologies for calculating carbon footprint and greenhouse gas emissions of freight transport and logistics in the supply chain context. It evaluated their compatibility with the European standard EN 16258, suggesting the* next steps needed to achieve a global alignment to support global standardization.

COG-LO (Programme: H2020-EU.3.4. - SOCIETAL CHALLENGES). COG-LO offers a framework and tools that add cognition and collaboration features to logistics processes. Operational and logistics frameworks that enabled ad-hoc cross-border logistics operations were proposed. These frameworks can be beneficial for further development of generic cross-logistics operations.

CO3 (Programme: FP7-TRANSPORT) *Collaboration Concepts for Comodality* project aimed to develop, professionalize, and disseminate information on the business strategy of logistics collaboration in Europe. The project's goal was to deliver a concrete contribution to increasing load factors, reducing empty movements and stimulating comodality through collaboration between industry partners, thereby reducing transport externalities such as greenhouse gas emissions and costs.

FEDeRATED (Programme: Connecting Europe Facility – CEF Transport). FEDeRATED *Network of platforms* grew out of the Digital Transport and Logistics Forum (DTLF) and aims to create an EU network of platforms for logistics. The initiative focuses on developing data standards that can be used to exchange data between various stakeholders in the logistics ecosystems. The initiative is concerned with data security, ensuring operationally and commercially sensitive data is protected while providing the infrastructure for deep data sharing and process integration.

FENIX (Programme: H2020-EU.2.1.5. - INDUSTRIAL LEADERSHIP). *European Federated Network of Information eXchange in LogistiX* is aiming to support the development, validation, and deployment of the digital information systems along the EU transport Core Network, based on DTLF's suggestions. It develops a European federated architecture for data sharing and digital corridor information systems for the European logistics community.

GECKO (Programme: H2020-EU.1.3. - EXCELLENT SCIENCE). *Governance principles and methods enabling decision makers to manage and regulate the changing mobility systems* project aims to support authorities in developing the most appropriate regulatory framework and governance model to transition to a new mobility era of cooperative, inclusive, competitive, sustainable, and interconnected mobility.

ILIAD (Programme: H2020-EU.2.1.1. - INDUSTRIAL LEADERSHIP). *Intra-Logistics with Integrated Automatic Deployment: Safe and scalable fleets in shared spaces* project will enable the transition to automation of intra-logistic services with key stakeholders. Robotic solutions that integrate with current warehouse facilities will be developed, extending state-of-the-art to achieve self-deploying fleets of heterogeneous robots in multiple-actor systems, life-long self-optimization, efficient and safe operation in environments shared with humans, and efficient fleet management.

ISEA-NET (Programme INTERREG IIIA) aims to improve the skills of the maritime community in three ports in southeast Ireland and three ports in south Wales. The design and development of a Maritime Community Portal to deliver training services and support. It will also bring about an awareness of how an Electronic Community Information Service can achieve practical business benefits.

KnowlEdge (Programme: H2020-EU.2.1.1. - INDUSTRIAL LEADERSHIP). The project addresses the need for new AI solutions by developing a platform that ensures secure distributed data management and facilitates knowledge exchange. As an outcome, barriers towards using AI are removed, and product and service quality and business sustainability are achieved.

LOGISTAR (Programme: H2020-EU.3.4. - SOCIETAL CHALLENGES). *Enhanced data management techniques for real-time logistics planning and scheduling projects propose the intensive use of the* Internet of Things, open data, artificial intelligence, optimization techniques, and other ICT advances to plan and optimise transport in the logistics sector effectively. The relevant stakeholders will test the project results in the Living Labs: Fast Mover Consumer Goods, Logistic Transport operators and 4-party logistics.

LSW (Programme COMPETE2020). *Logistic Single Window* is a collaborative platform comprising the entire transport chain (private and public players), aiming to achieve a far more agile and efficient logistic chain.

MOVE21(Programme: H2020-EU.3.4. - SOCIETAL CHALLENGES). *Multimodal and interconnected hubs for transport contribute to a zero-emission 21st-century* project envisioned to strengthen the effect of multimodal transportation systems in terms of efficiency, cost and emissions. Additionally, it explores potential barriers and impacts, collaboration models, and policy integration.

MULTIRELOAD (Programme: H2020 HORIZON-CL5-2021). *Port solutions for sustainable mobility* project consortium comprises of ports, highly innovative technology, logistics and service providers, leading European research institutions, and well-connected networks will cooperate intensely to facilitate the significant modal shift to sustainable modes of transport, radical efficiency increase of supply chains and capacity-sharing (data, infrastructure, loading space, etc.) for all operators within the critical Trans-European Transport Networks (TEN-T) Rhine-Alpine and Rhine-Danube Corridors.

NEXTRUST (Programme: H2020-EU.3.4. - SOCIETAL CHALLENGES). *Building sustainable logistics through trusted collaborative networks across the entire supply chain* project's general objective was to increase efficiency and sustainability in logistics by developing an innovative business model with interconnected, trusted, collaborative networks along the entire supply chain. The NexTrust project focuses on the basis for a formal integrated framework for logistics ecosystems by using digital technologies. However, despite these advances, these solutions have not been widely spread in pan-European logistics because of the lack of incorporation of identification and control mechanisms.

NEXUS is a project of the Port Administration of Sines (located in the southwest of Portugal) to promote high-performance operations for the supply chains crossing the seaport of Sines. The project aims to foster the synchronization of processes between all actors, making use of total visibility over operations and traceability of goods. This project involves 28 companies (national and international organizations). All participants in this network, including the authorities, will gradually share information in real-time and align their processes.

NextLogic (https://www.nextlogic.nl/). The Port of Rotterdam introduced a parallel initiative–NextLogic–established for and by the market and has a bottom-up approach from the beginning; the platform is aimed at providing solutions to various optimization problems faced by logistics companies operating at the Port of Rotterdam. NextLogic is a data integration platform that addresses the systems of container transport and logistics companies (for example, barge operators, land terminals, port terminals and warehouses, shipowners, and freight forwarders at the port of Rotterdam). NextLogic focuses on dealing more efficiently with inland container shipping, providing a platform on which the entire container transport chain can work together. The benefits for supply chain stakeholders that adhere to NextLogic include more reliable planning and predictable response time and optimising the use of port terminals, cranes, and barges with connections to the port.

NxtPort: The NxtPort data usage platform was created to facilitate data-sharing practices among users in the Port of Antwerp. NxtPort addresses the specific type of data transfer that other community systems have not covered. A private company created it, and its main objective is to integrate data from terminal operators. In addition, the platform aims to create added value for data owners and users, allowing market applications to be built with existing data.

NOVELOG (Programme: H2020-EU.3.4. - SOCIETAL CHALLENGES). *New cooperative business models and guidance for sustainable logistics* projects propose new cooperative business models and guidance for sustainable city logistics in the context of stakeholder collaboration and the development, field testing and transfer of best governance and business models.

PORTFOREWARD (Programme: H2020 - MG-7-3-2017). *Towards a green and sustainable ecosystem for the EU Port of the Future* project deals with the port of the future - must be oriented to port community and have an operative strategic capability to work, in line with European purposes, on smart, through ICT solutions, because it is essential to improvement exchange of information flows between port and port community; interconnected with the use of a combination of different modes of transport and the integration of different technologies, because it is essential to achieve better monitoring and controlling of the freight flows.

SELIS (Programme: H2020-EU.3.4. - SOCIETAL CHALLENGES). *Towards a Shared European Logistics Intelligent Information Space* project proposes a Shared European Logistics Intelligent Information Space, SELIS, connecting nodes, providing a distributed standard communication and navigation platform for Pan-European logistics applications.

SYNCHRO-NET (Programme: H2020-EU.3.4. - SOCIETAL CHALLENGES). *Synchro-modal Supply Chain Eco-Net* considers synchro-modal supply chain operations, contributing towards cost-effective solutions that destress the supply chain and reduce emissions and costs for logistics operations. Additionally, it supports integrated optimizations and synchro-modal risk/benefit analysis. Net tool for the optimization of the transport chain and governance recommendations

Transforming Transport Project (Programme: H2020-EU.2.1.1. - INDUSTRIAL LEADERSHIP) demonstrated the transformations that Big Data will bring to the mobility and logistics market in a measurable and replicable way.

WINN (Programme: FP7-TRANSPORT). European Platform Driving Knowledge to Innovations in Freight Logistics project aimed to increase collaboration and consensus building of the different stakeholders dealing with sustainability in the freight transport logistics and intermodality, establish a broad collaboration framework built upon main stakeholders operating at the European level (CO-Tree, EIRAC heritage) and national triple helix networks that include public bodies, companies, and excellence research centers in logistics operating in different countries in Europe.

2.2 *Mapping factors that may determine a hindering effect*

After examining the study sample in the previous sub-section, we identified six main components/ key findings, summarized in Table 1.

It is clear from the assessment of the previous projects that the first steps providing the basis for a formal integrated framework for port logistics ecosystems by using digital technologies were carried out. For example, NxtPort is an integration platform that aims at development opportunities after integrating information from different sources. NextLogic project has a bottom-up approach; the platform is aimed at providing solutions to various optimization problems faced by logistics companies operating in the port and is a data integration platform that addresses the systems of container transport and logistics companies (for example, barge operators, land terminals, port terminals and warehouses, shipowners, and freight forwarders at the port of Rotterdam).

Despite these advances, a minor replication at the European level was identified mainly because of the difficulty of absorbing the technologies and the actors' resistance. The political and economic changes in Europe and the associated complexity of logistics processes regarding information exchange are some factors to be considered.

Some of the projects proposed connected information systems of supply chain members over a single platform, automated negotiation and planning optimization in freight transport and virtual logistics networks. The port community involves various public and private actors, usually operating the port business fragmentedly. Although those projects rely on multiagents' reactiveness and communication characteristics, the sequence of necessary logistics actions still lacks clarity.

An analysis using PESTLE (Cadle et al. 2010) was carried out, summarizing the political (P), economic (E), social (S), technological (T), legal (L) and environmental (E) frequently identified barriers and their anticipated temporal effect used for mapping *diverse factors that may determine a hindering effect.*

(P) From the political point of view, two main directions were identified: the first and one of the most frequently identified barriers is the freight transport sector's hesitancy to implement stricter carbon neutrality regulations. The second identified factor is that platforms need standards, interaction norms, and proper governance, and this barrier's effect is expected to decrease over time.

(E) From the economic point of view (E), huge investments are required to build port infrastructure across the EU to facilitate the application of innovations. This barrier's effect over time is expected to be continuous.

(S) From the social point of view, one of the most critical issues facing the logistics industry is workforce shortages - an ageing logistics sector workforce unable to use new digital services. This is

Table 1. The main components and the key findings of the relevant studies.

PROJECT ACRONYM/Link	Digitalization	Collaboration/Integration	Climate neutrality	Ecosystem	Platforms/Business models	Regulation
ADMIRAL	✓	✓	✓	✓	✓	
AEOLIX	✓	✓	✓	✓	✓	
CLOSER FP7		✓		✓		✓
CLUSTERS 2.0	✓	✓			✓	
CO3	✓	✓		✓	✓	
COFRET FP7			✓			✓
COG-LO		✓			✓	
FEDeRATED	✓	✓		✓		✓
FENIX	✓	✓		✓		
ILIAD-RIA	✓	✓			✓	
ISEA-NET	✓	✓		✓		
knowlEdge					✓	✓
LOGISTAR-RIA		✓			✓	✓
MOVE21	✓	✓			✓	
MULTIRELOAD	✓	✓			✓	
NEXTRUST		✓	✓	✓	✓	
NOVELOG		✓	✓		✓	
PORTFOREWARD		✓	✓		✓	
SELIS	✓	✓	✓			
SYNCHRO-NET	✓	✓			✓	
Transforming	✓					
WINN		✓	✓		✓	

a noticeable trend with the expected exacerbation of the barrier's effect over time. While automation and robotics are expanding, simplifying complex tasks, and removing manual paperwork — the need for skilled personnel will never be replaced.

(T) The transport analysis showed that the significant barriers are the failure to standardize data format to describe cargo and logistics services, the lack of uptake by Logistic Service Providers, Logistics Service Buyers, and Transport Hubs and the lack of applicability of technology to specific needs of intermodal transport, and corridor-based transport services. However, there is an expected significant reduction over time in this field.

(L) From the legal point of view, the lack of a legal framework that deals with data generated from new and innovative sources seems to be the most frequently identified barrier, with an expected reduction in the barrier's effect over time. Ports are de low trust environments. The lack of transparency has been one of the key commercial drivers of trade, so the appetite for data sharing to improve overall visibility and, therefore, productivity is understandably low. So, for any port digital platform to succeed, a key priority is a neutral data exchange with solid data governance whereby data providers ultimately control what happens to their data. To mitigate this barrier, a neutral trustee party must take responsibility for the collaboration's legal foundations, ensuring that the shared data remains confidential (Cruijssen 2012).

(E) Finally, from the environmental point of view, the alignment with the goals for emission reduction is still missing. Since the beginning of 2023, incoming EU regulations regarding minimum standards for EU Climate Transition Benchmarks and EU Paris-aligned Benchmarks entered into force, and The ISO 14083 standard came into effect, providing a globally aligned approach to quantifying and reporting greenhouse gas emissions in multimodal and global transportation chains. It is expected that this barrier's effect will be reduced over time.

3 PORT STAKEHOLDERS AND THEIR ROLES

As identified in the previous chapter, the port community involves various public and private actors, usually operating the port business in a fragmented way. The assessment of the critical categories of companies involved in the port ecosystem led to identifying the main stakeholders: the individual companies that comprise port communities, the government in the form of public authorities, regulators and standardization agencies, the research and development institutions, and the civil society.

Port authorities, as public authorities, regulators, and standardization agencies, are government bodies that exercise control over certain activities in the port's transport and logistics sector, including standardization, social, safety, security, and public goods-related issues.

Innovative solutions providers are the companies and technological institutes that develop and provide innovative technologies and solutions.

Shipping lines are the companies that own and operate ships in the port.

Shippers are the owners of the cargo that needs to be shipped. They can be found, in terminology, as material suppliers, manufacturers, vendors, and distributors.

Logistics service providers (LSPs)/ (Multimodal) transport operators represent the companies that organize and manage all aspects of cargo shipment. There are various terms to refer to types of Logistics Service Providers: transport service providers, carriers, freight forwarders, warehouse service providers, multimodal transport operators, and LSPs.

Haulage contractors are responsible for transporting a freight consignment via land, water or air.

Freight forwarders provide services for the international transport of goods, are involved in all types of international transport operations, including maritime or inland waterways, air, rail, and road, and ensure continuity by using different means of transport. It also coordinates the administrative operations related to international transport; this includes customs formalities, financial management and insurance contracts.

Freight hub and terminal operators can be found in different forms (e.g. maritime, inland port terminals, railway terminals, road terminals) conditioned so freight can be transferred from one mode of transport to another.

The key organization categories identified by (Sweeney & Evangelista 2005) are presented in Figure 1 and consist of port authorities, shipping lines, shipping agents, logistics service providers (LSPs), customs clearance, haulage contractors, freight forwarders, and stevedores. The findings from assessed projects align with the scientific literature. However, new categories of actors are identified, e.g. Innovative solutions providers in the form of technology providers, data analytics companies, and platform developers.

The identified actors in the port ecosystem must help fulfil the primary function and feature of ports: sea and land access, providing infrastructure for ships berthing, inland waterways, road and rail link, industrial areas, control of vehicles during entering and leaving port, environmental control, control of damages of cargo, safety and security within the port vicinity, commercial papers control, pilotage, tugging and mooring arrangement.

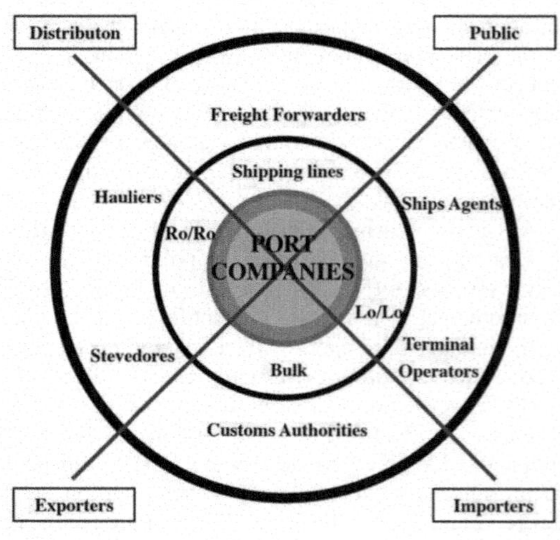

Figure 1. Typical Port Community Participants (Sweeney & Evangelista 2005).

The roles of the identified stakeholders cover a wide range of activities. When the platform-based solutions are employed, all categories of stakeholders will be involved and will benefit. The end-users, including operators, cargo owners, freight forwarders, port authorities, port operators, and logistics companies, will benefit from a maximum throughput, reduced waiting times, optimized processes and exact tracking/ tracing of emissions throughout the transport chain (incl. subcontractors), ultimately leading to lower costs (and meeting (future) regulations. The developer community, including software developers, solution providers, service providers, and start-ups, will integrate their tools and developments for sustainable transport into the platform, allowing data sharing with a broader ecosystem.

4 PORT PLATFORM-BASED MARKETPLACES

Port platform-based marketplaces can facilitate integration at two levels (Sweeney & Evangelista 2005):

- Between companies that comprise port communities (port community integration) and
- Between the port community and external organizations.

There is a clear interaction between port ecosystems and digital platforms. (Täuscher &Laudien 2018). The main identified elements of the platform are the participants, the technology employed, the data availability, the applications (Apps) involved, and, last but not least, the interactions between them.

On the other hand, the traditional ecosystem of port actors such as port authorities, port operators, cargo owners, and ship operators is increasing and opening up to data partners, developers, and technology partners. This is the area where the platforms and the ecosystem connect even more. The analysis of past and ongoing projects shows that various digital platform solutions are developed at the port level. The main elements and characteristics of each proposed platform development are presented in Figure 2.

Most of them are network-type platform solutions. The paper proposes the following stages of evolution in developing digital platforms for ports, starting with the network-type platform, the data aggregator-type platform, and the final version of the marketplace-type platform.

The proposed platform solution will evolve into a modern market network ecosystem with multiple-sided marketplace components. The participants interact in a network-like frame to perform complex transactions like port calls (ship coming to port, using services, and leaving a port) or newer transactions like remote operations. The "Single Window" philosophy implemented for the management of port information has been achieving very positive results; as far as maritime transport is concerned (a sector with particular characteristics), being imperious to apply this philosophy to all the means of transport and port logistics (Simão & da Anunciação 2017). The final goal is to coordinate the whole supply chain in an optimized and sustainable manner.

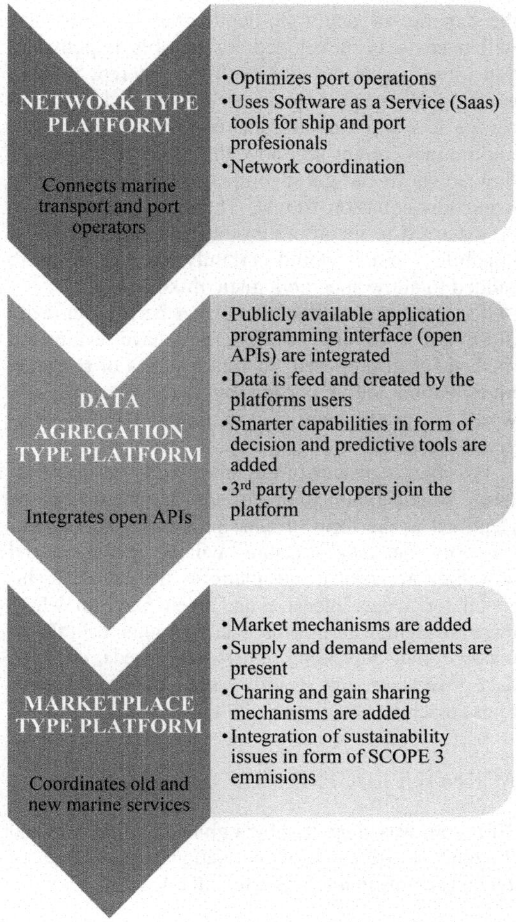

Figure 2. The proposed model for the evolution of the port platform-based solution.

5 CONCLUSIONS

The goal was to identify the appropriate sample of studies integrating scientific and practitioner literature. The results of the analysis of the most representative and essential projects considering digitalization, integrated logistic networks, ecosystems, regulation issues, and climate neutrality, mainly addressing port solutions, from the significant directions of research and industrial technology transfer, led to contributions to the literature by clearly identifying key output areas and by analyzing and mapping various factors that may determine a hindering effect. This research only provides a starting point in exploring platform-based marketplaces for sustainable logistic solutions in port ecosystems, and there are plenty of limitations and possible extensions of this study to overcome with further research.

The proposed platform evolution framework applies to a broad form of platform-based business models for logistics. However, this scale comes at the expense of depth: In-depth granular constructs will need to be developed for testable hypotheses, and these constructs may well be different for and across different port solutions. This includes how to source low-emission freight, consolidate, and offer multimodal freight and how the information should flow so cargo can get a competitive advantage from using low-emission freight. The study may have to be extended to incorporate other dimensions of sustainability, and it would certainly need to be more refined to allow specific logistic mixes.

The proposed model has not yet been empirically illustrated with case studies, nor have evaluation methods been proposed. At future stages of research, investigation and reporting by way of case study would be pretty helpful as a foundation for 'integrated' research.

Possible extensions of this study in further research, when sustainability solutions are integrated, can be identified in the form of conceptual data model solutions of what data structures will be processed and how they are related and platform framework architected for a data-intensive and open platform where there are significant incoming and outgoing data flows, massive data processing and storage needs, complex data governance and security issues and requirements to use machine learning capabilities.

ACKNOWLEDGEMENTS

This work was supported by a grant from the National Program for Research of the National Association of Technical Universities - GNAC ARUT 2023.

REFERENCES

Cadle, J., Paul, D., Turner, P. 2010. *Business analysis techniques: 72 essential tools for success*. London: BCS, The Chartered Institute.

Chandra, D.R., van Hillegersberg, J. 2017. Governance lifecycles of inter-organizational collaboration: A case study of the port of Rotterdam. *Procedia Computer Science* 121: 656–663.

Cruijssen, F. 2012, *CO^3 Position Paper: Framework for Collaboration*.

Derave, T.; Prince Sales, T.; Gailly, F.; Poels, G. 2021. Comparing digital platform types in the platform economy. In Proceedings of the International *Conference on Advanced Information Systems Engineering*, pp. 417–431, Cham: Springer International Publishing

Henesey, E.L., Notteboom, E.T., Davidson, P. 2003. Agent-based simulation of stakeholders relations: An approach to sustainable port terminal management. *Proceedings of the International Association of Maritime Economists Annual Conference*, Busan, Korea.

Meyer, N., Horvat, D., Hitzler, M., Doll, C. 2018. Business Models for Freight and Logistics Services. *Working Paper Sustainability and Innovation*, No. S 08/ 2018.

Moore, J.F. 1996. *The Death of Competition: Leadership & Strategy in the Age of Business Ecosystems*. New York: Harper Collins

Robertson, D., Ulrich, K. 1998. Planning for Product Platforms. *Sloan Management Review* 39(4): 19–31.

Sanchez-Cartas, J. M.; Leon, G. 2021. Multisided platforms and markets: a literature review. Journal of Economic Survey, 35(2), pp. 452–487.

Sweeney, E., Evangelista, P. 2005. Supply chain learning needs: Towards a port community perspective. *Annals of Maritime Studies* 42(1): 93–108.

Tansley, A.G. 1935. The Use and Abuse of Vegetational Concepts and Terms. *Ecology* 16 (3): 284–307.

Täuscher, K.; Laudien, S. M. 2018. Understanding platform business models: A mixed methods study of marketplaces. *European Management Journal*, 36(3), pp. 319–329.

Simão, J.C. D.; da Anunciação P.F. 2017. Port Single Window and Logistics Single Window: Two Competitiveness Proposals in the Port Value Chain, in *Handbook of Research on Information Management for Effective Logistics and Supply Chains*

Advances in Maritime Technology and Engineering – Guedes Soares & Santos (Eds)
© 2024 The Author(s), ISBN 978-1-032-83099-5

Uncertainty analysis of EEDI and bulk carrier conceptual design

Y. Garbatov
Centre for Marine Technology and Ocean Engineering (CENTEC), Instituto Superior Técnico, Universidade de Lisboa, Lisbon, Portugal

P. Georgiev
Technical University of Varna, Varna, Bulgaria

ABSTRACT: The present study analyses uncertainties and their impact on ship design. A Sensitivity Analysis (SA) is used to understand the uncertainty related to different stages of ship design, which will permit an uncertainty-based decision-making approach to be employed. It brings the SA into the uncertainty analysis of the Energy Efficiency Design Index (EEDI). A general workflow is introduced in implementing SA as a best practice through a multi-objective reliability-based robust design optimisation of handy-size bulk carriers. Additionally, it verifies the application of SA, focusing on the most widely used algorithms that can be used in ship design. The SA models are initially used to evaluate the EEDI and later to ship design. Reliability-based robust design optimisation has been applied as a part of a multi-objective concept design optimisation framework.

1 INTRODUCTION

International Maritime Organisation (IMO) has been working on reducing greenhouse gas emissions from shipping for over ten years, adopting a revised strategy recently (IMO, 2023). The revised strategy aims to reach the peak of the Green House Gases (GHG) emissions from international shipping as soon as possible and attempt to achieve a near net-zero GHG emissions balance by 2050.

The start date of Phase 3 for implementation of the reduction factors for the EEDI concerning the EEDI reference line has been revised for gas carriers, containerships, general cargo carriers, LNG carriers, and passenger ships. The new effective date for these vessels is April 1, 2022. For all other ships, Phase 3 will commence on January 1, 2025 (MEPC, 2020). So, the reduction factor for bulk about reference line for bulk carriers with DW>20,000 tonnes will be 30%.

Uncertainty-based design optimisation (UDO) for ships has gained attention for reflecting ship performance under uncertainties accurately (Wei et al., 2019), where it can lead to high dimensionality problems and inefficient optimisation. Diez & Peri (2010) presented two-stage stochastic programming for robust design optimisation. They are minimising the EEDI of a passenger ship shown by increasing the value at increasing the number of uncertain factors while decreasing the energy-saving effect (Wang et al., 2021). Utilisation of surrogate models for robust ship design optimisation, taking into account uncertainty and multiple objectives, is presented by Priftis et al. (2020).

The safety requirements introduced by IMO impact the development of the risk-based ship design formulation (Papanikolaou et al., 2009, Papadimitriou et al., 2018).

A study developing a risk-based conceptual ship design method has been presented in (Garbatov & Georgiev, 2021) using long-time experience and statistics, which are used as input variables in a risk-based analysis. The needs of cargo transportation demand as part of a multimodal transportation frame were identified, estimating the potential ship fleet of multipurpose ships (Georgiev & Garbatov, 2021). The optimal design solution was based on capital, operational expenditure, and energy efficiency. A risk-based framework was developed for ship hull accounting for maintenance planning (Garbatov et al., 2018), where the analysis included various operational scenarios. The risk-centered maintenance methodology was applied for maintenance planning.

The present study deals with uncertainty assessment of EEDI and bulk carrier designs. Accounting for the uncertainties in the design solutions, different sensitivities of the decision variables impacting the capital and operational expenditures are considered. In this regard, the ship design optimisation problem is solved from a probabilistic point of view. The statistical descriptors of the objective functions are used as objectives of the reliability-based robust optimisation problem, accounting for the propagation of the uncertainties involved. To solve this problem, a series of polynomial equations are used as a mathematical model and constraints of ship design. The second-order Taylor series expansion derives

DOI: 10.1201/9781003508762-7

the stochastic variables' mean values and standard deviation.

This study introduces a new probabilistic approach for multi-objective design optimisation, which considers both reliability and robustness in the concept design optimal solution of a bulk carrier, demonstrating how uncertainty can be accounted for in a very efficient and fast manner using modern state-of-the-art approaches.

2 SENSITIVITY APPROACHES

Analysing uncertainty is essential in ship design, manufacturing, and operation. Ships are affected by numerous uncertainty sources, including human, weather, operation and cost-dependent conditions, load behaviour, incomplete information, ageing, etc. SA (Ju et al., 2011) is vital in making informed decisions with considerable uncertainty. The study first analyses the variations in the inputs on the uncertainties of the EEDI estimation using different approaches based on the most suitable ones. Secondly, the ship's reliability-based robust optimal design will be defined as a numerical procedure for identifying optimal design solutions by satisfying the probabilistic constraints in determining the optimal robust object functions.

The employed design model comprises a series of parametric equations representing the simplified ship's physical behaviour and uncertainty. The results identify the best design solution using the model decision variables and statistical descriptors by minimising or maximising the objective functions and satisfying the boundary conditions.

SA is combined with the Uncertainty Analysis (UA) to investigate the effect of uncertainty on the model output. It characterises the model output uncertainty due to the uncertainty in the inputs. SA shows how the input uncertainties propagated in the output uncertainty model.

Figure 1 shows the UA-SA framework, where the model inputs are first defined, and their uncertainty is characterised. Then, the uncertainty of the model inputs propagates through the model, and the resulting uncertainty is estimated with UA by determining the statistical descriptors, such as the mean value and variance. UA ends with estimating output uncertainty and whether the model will function with some limitations when the model inputs are severely affected by uncertainty. SA identifies which modelling assumptions or decision variables mainly induce uncertainties in the objective functions and quantifies the relative contribution of the input variables to the model output uncertainty by computing the sensitivity factors.

SA can be identified as a "perturb-and-observe" approach, where the model input values are repeatedly changed, and the impact on the model output is evaluated. SA methods are also classified into "local" or "global", considering the input space as "One-At-a-Time" (OAT) or "All-At-a-Time" (AAT).

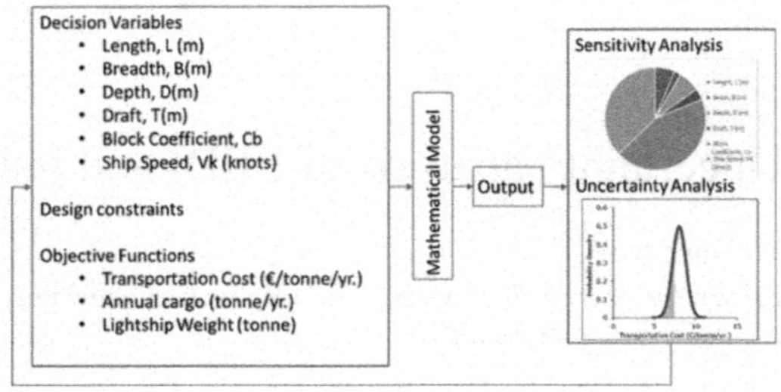

Figure 1. High-level UA-SA framework.

In the local SA, the input variability is analysed around a reference point $x°$, varied "locally", which is the operational/nominal point or the "appropriate" or "best known" configuration. Consequently, local sensitivity measures such as $\frac{\partial y(x)}{\partial x_i}$, i.e., the model output partial derivative concerning each model input variable x_i depends on the specific input space location x_i^0 at which they are computed.

The global SA approaches aim to explore the input space, investigating the model input variability, including the neighbourhood of x_i^0. The local SA should be applied only for slight variations in the neighbourhood of x_i^0 unless the model is proven linear in all its inputs, whereby $\frac{\partial y(x)}{\partial x_i}$ remains constant for any x_i^0. When the model is non-linear, considerable uncertainty affects the model inputs; local SA provides only an incomplete view of the model sensitivity behaviour. The global SA allows for a complete exploration of the input space.

2.1 *EEDI model*

Regarding air pollution, EEDI (gCO_2/tonne-nmile) can be measured by the EEDI-reference (MEPC, 2011, 2018b) in the case of a bulk carrier of 20,000 DW and above, Phase 2, January 1 2020, to December 31 2024. For the direct-drive diesel engine propulsion system with only propulsion power, EEDI can be expressed as follows (Ren et al., 2019):

$$EEDI = \frac{P_{ME}\, C_{FME}\, SFC_{ME} + P_{AE}\, C_{FAE}\, SFC_{AE}}{f_{iCSR} \cdot Capacity \cdot V_{ref}} \quad (1)$$

where: P_{ME} is 75% of the rated installed power (MCR_{ME}) for the main engine and P_{AE} is the auxiliary engine power, and for a specific ship it is as follows (MEPC, 2018a):

$$P_{AE} = \begin{cases} 0.025 MCR_{ME} + 250 & MCR_{ME} \geq 10,000\ kW \\ 0.05 MCR_{ME} & MCR_{ME} < 10,000\ kW \end{cases} \quad (2)$$

The engine brake power (P_B) is the power developed by the engine and the shaft power (P_S) is the power available at the propeller shaft for propulsion. It accounts for losses due to the presence of a gearbox. The efficiency losses in one-step reduction gearboxes for medium-speed diesel setups are relatively modest, typically ranging from 1% to 2% (Woud & Stapersma, 2002). However, in more intricate gearboxes featuring two or three reduction stages, as commonly used in multi-engine configurations, losses can be significantly higher, ranging from 3% to 5%. In the study, a gearbox efficiency $\eta_{GB} = \frac{P_s}{P_B}$ of 98% is accepted.

In conceptual ship design, the shaft power P_S (kW) is estimated by:

$$P_s = \frac{\Delta^{2/3}\, V k^3}{C_{ADM}} \tag{3}$$

where: Δ is the weight displacement, V_k is the ship speed (kn), C_{ADM} is the Admiralty coefficient. C_{ADM} ranges from 400 to 600. The higher the value, the more economical the vessel is (Wärtsilä, 2023).

Considering Eqn (2), the EEDI from Eqn (1) is simplified to:

$$EEDI = \frac{0.75\left(\frac{P}{0.98}\right)(3.114)190 + 0.05\left(\frac{P}{0.98}\right)(3.206)215}{1.02\, DW\, V_{ref}} \tag{4}$$

Finally, the relation for EEDI is presented as:

$$Y = g(x) = \frac{478.4 x_1}{x_2 x_3} \tag{5}$$

where Y is the scalar model output and $\boldsymbol{x} = [x_1, x_2, x_3]$ is the vector of the model inputs, x_1 is the power of the ship (kW), x_2 is the deadweight (tonne), x_3 is the ship speed, (kn)

The EEDI model is assumed to be deterministic when the inputs take some initial value x°, the resulting model output always takes the same value $Y_0 = g(x^0)$. Despite its simplicity, the Eqn (5) model is characterised by several terms representing quadratic relations, including a negative and an interactive relation. This model is fundamentally non-linear (due to x_2, x_3) and non-additive due to x_2, x_3.

The three model inputs are considered uncertain. It is assumed that the true values of x_1, x_2, x_3 are not precisely known, and the ranges of variation are defined as $x_i \in \left[x_i^{min}, x_i^{max}\right]$. Considering the minimum x_i^{min} and maximum x_i^{max} for each model, input variables are sufficient for performing a local SA since they are performed under a deterministic framework. Any value between x_i^{min} and x_i^{max} is a possible value of x_i, without a particular probability information associated with it.

2.2 *Tornado diagram*

The simplest SA method is the Tornado diagram (Eschenbach, 1992) based on OAT. Each model input generally takes three values: min, max and a reference point. The reference point is identified as x_i^0.

Given the reference case x_i^0 and the point $\left(x_i^+, x_{\sim_i}^0\right)$, it is possible to calculate the sensitivity measure $\Delta_i^+ y$, i.e., the variation Δ of the model output y due to changing alone of the model input x_i from x_i^0 to $\left(x_i^+, x_{\sim_i}^0\right)$:

$$\Delta_i^+ y = g\left(x_i^+, x_{\sim_i}^0\right) - g\left(x_i^0\right) \tag{6}$$

The analysis goes by moving each input OAT from the reference case x_i^0 to its lower limit x_i^- and then the following equation is used:

$$\Delta_i^- y = g\left(x_i^-, x_{\sim i}^0\right) - g\left(x_i^0\right) \tag{7}$$

For the EEDI case, assuming that the reference case is $x^0 = \left[x_1^0, x_2^0, x_3^0\right]$, where the two endpoints are $x^+ = \left[x_1^+, x_2^+, x_3^+\right]$ and $x^- = \left[x_1^-, x_2^-, x_3^-\right]$.

The two sets of OAT sensitivity measures $\Delta_i^- y$ and $\Delta_i^+ y$ are plotted in the Tornado diagram, as shown in Figure 2.

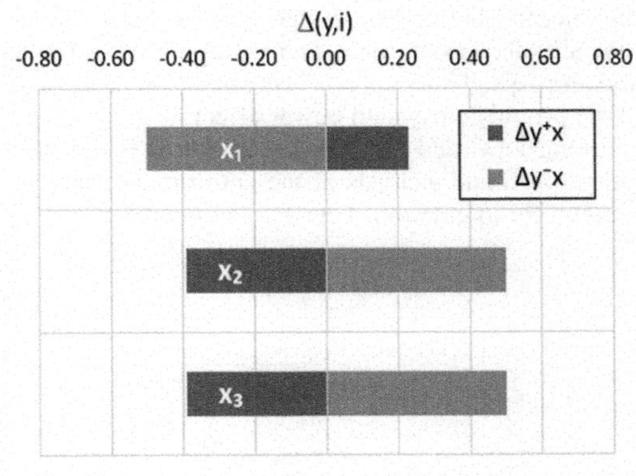

Figure 2. Tornado diagram for EEDI when the model inputs are subjected to a change of 50% concerning the reference case.

The information provided by the Tornado diagram gives the direction of the output change due to a variation in each model input alone. According to Figure 2, a 50% increase of x_1, x_2, x_3 concerning their reference cases causes a positive $(x_1^0 \rightarrow x_1^+)$, and negative $(x_2^0 \rightarrow x_2^+)$ and $(x_3^0 \rightarrow x_3^+)$ effect on the model output y, respectively. Equally, a 50% decrease in each of the three model inputs produces a negative $(x_1^0 \rightarrow x_1^-)$, and positive for $(x_2^0 \rightarrow x_2^-)$, $(x_3^0 \rightarrow x_3^-)$ effect, respectively.

It is sufficient to recall the EEDI function to realise that the interactive term involving x_2, x_3, the Tornado diagram does not capture Type II errors. Since no interactive effects are caught due to the OAT design, a Tornado diagram might risk the decision-making, neglecting the interactions between the model's inputs.

2.3 *One-way sensitivity analysis*

The model output is analysed at more than three points when each input individually varies inside its acceptable range of variation. In this case, the so-called one-way SA is used, which is seen as a generalisation of the sensitivity measures provided by the Tornado diagram.

A reference case x^0 and variation for the inputs, $x^+ = [x_1^+, x_2^+, \ldots, x_k^+]$ and $x^- = [x_1^-, x_2^-, \ldots, x_k^-]$, are defined in the one-way SA.

The inputs are set at their reference case values, except for the input x_i, which can vary between its range extremes. The one-way sensitivity function $h(x_i)$ is defined as a function of the input x_i as:

$$h(x_i) = g(x_i, x^0_{\sim i}) \tag{8}$$

Evaluating $h(x_i)$ at the N different x_i values (with the other inputs held constant at $x^0_{\sim i}$), the model output is plotted against x_i. The input in the EEDI model varies to define $N = 10$ points and the one-way sensitivity function $h(x_i)$ calculated at any point, leading to a sequence of 10 model output values and are plotted against x_i in Figure 3 left. The plots are then normalised and show that y is quadratic monotonically decreasing for x_2 and x_3, representing DWT and ship speed and linearly increasing for x_1 related to the propulsion system power effect.

Being local and OAT, one-way SA does not provide a profound analysis of the entire input space or account for interactions between the inputs.

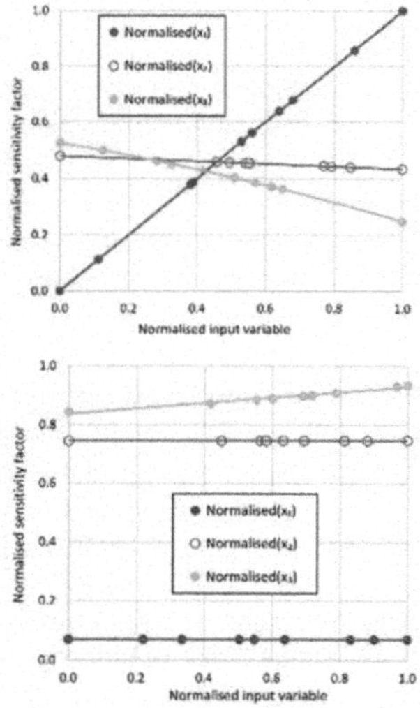

Figure 3. One-way SA (left) and differential SA (right).

2.4 *Differential sensitivity analysis*

Differential SA (Helton, 1993) is a type of local sensitivity estimation based on the partial derivative of the model output. The differential SA provides a straightforward translation of the "sensitivity" notion and explores the model function's Taylor expansion at the reference case x^0. It is applicable only in the vicinity of the investigated input space portion around x^0, and unless the model is proven linear, it might be misleading if it is used to interpret the model behaviour further away from the reference case.

The basic procedure for numerically estimating S_i^∂ is using an arbitrarily small size of the finite change x_i, which is generally the same for all model inputs. The absolute values of S_i^∂ cannot be compared, and a normalisation actor for S_i^∂ is introduced. Frequently adopted weights are the input-output reference case values, i.e., $\frac{x_i^0}{g(x_1^0)}$, or the input-output standard deviation, $\frac{\sigma_{xi}}{\sigma_y}$. In this latter case, the associated differential becomes:

$$S_i^\sigma = \frac{\partial g(x)}{\partial x_i}\bigg|_{x^0} \frac{\sigma_{xi}}{\sigma_y}, i = 1, 2, \ldots, n \tag{9}$$

where σ_{xi} reflects the standard deviation of the input x_i and σ_y is the standard deviation of the output respectively. S_i^∂ estimate the impact on the model output of perturbing x_i by a fraction of the standard deviation of x_i. $S_i^\partial$and S_i^σ (or other normalised versions of S_i^∂) are seen as "absolute" and "relative" sensitivity, respectively. It is noteworthy that the adoption of S_i^σ as a sensitivity measure instead of S_i^∂ requires to assume the model input variation range for deriving the standard deviation values σ_{xi} and σ_y. Normalised differential sensitivity measures such as S_i^σ are preferred to S_i^∂ for analysing the model sensitivity at a global level by considering information about the input variability (see Figure 3 right).

Differential SA is a hybrid local–global sensitivity measure with a more consistent model input ranking. However, its effectiveness reduces with the increase of the model nonlinearity.

In contrast to local SA, the global SA (Saltelli et al., 2007) is performed by accounting for the probabilistic description of the model inputs, which transforms them into random variables X_i.

Different possible global SA approaches are derivative-based global sensitivity methods, density-based methods, entropy-based measures, and variogram-based methods.

The non-linear differentiation SA is used due to the simplicity in the implementation when a parametric model is present and can efficiently deal with the imposed objectives.

The second-order approximation of the variance is based on the Taylor series expansion. The uncertainty related to a descriptor $Y(x_i)$ when the input parameters x_i are in a non-linear relationship, the higher-order terms of the Taylor series expansion are included, and the second-order approximation of variance is given (Harr, 1987) as:

$$\mu[Y] = Y\ [\mu(x_i)] + \frac{1}{2}\sum_{i=1}^{n}\left(\frac{\partial^2 Y}{\partial x_i^2}\right)\sigma_{x_i}^2 \quad (10)$$

$$\sigma_Y^2 = \sum_{i=1}^{n}\left(\frac{\partial Y}{\partial x_i}\right)^2 \sigma_{x_i}^2 + \frac{1}{2}\sum_{i=1}^{n}\left(\frac{\partial^2 Y}{\partial x_i^2}\right)^2 \sigma_{x_i}^4 \quad (11)$$

where $\frac{\partial Y}{\partial x_i}$ and $\frac{\partial^2 Y}{\partial x_i^2}$ are the first and second-order sensitivities of the input parameters x_i.

The model responses are generated by evaluating the model for each combination of the input parameters. Exploring the input stochastic variables, the uncertainty is propagated forward through the model to the output. The uncertainty assessment may characterise the variability associated with the model output by extracting the output distribution and statistical descriptors.

The conclusions drawn after the SA may be used as feedback to progressively refine the uncertainty of a specific input to guide removing non-influential ones. This is a more accurate modelling or selective investigation of different model features.

3 UNCERTAINTY-BASED CONCEPT SHIP DESIGN OPTIMISATION

A reliability-based robust optimisation of concept ship design is performed using the global SA to study the decision variables' sensitivity inputs' impact on the degrees of uncertainty of the objective functions design solution. During the design process, a series of relationships are employed related to steel weight, outfit weight, displacement, propulsion power, machinery weight, deadweight, daily fuel consumption, energy efficiency, sea days, fuel carried, crew, stores, and water, cargo deadweight, port stay days, round trips per year, ship cost, capital and operational cost, voyage cost and annual cost.

The design solution satisfies the design constraints related to the initial stability, and dimensional ratios *L/B, L/D,* and *L/T* are within the limits (Lyon & Mistree, 1985). Multiple objective functions and constraints define the design to find a suitable solution (Rao, 2009). Not all encountered are optimal for any set of objective functions, and a trade-off between the objective functions needs to be defined (Keane et al., 1991).

In this regard, the Pareto Frontier optimality solution may be employed (Komuro et al., 2006) to determine the optimal solutions in the design space. The Pareto optimal solution is where any improvement in one objective worsens at least one other purpose (Messac & Mullur, 2007).

3.1 *Design formulation*

According to different design requirements, uncertainty-based design optimisation can be defined as robust design optimisation (RDO), reliability-based design optimisation (RBDO), and reliability-based robust design optimisation (RBRDO). Compared with traditional design optimisation, RDO considers the interference of external factors on outputs. The robust optimal solution is a relatively stable optimal solution that improves the objective performance and reduces sensitivity to uncertainty (Kovach & Cho, 2008, Liang & Mourelatos, 2008).

RBRDO takes the influence of the uncertainty on the objective functions and the constraints into account simultaneously. Its mathematical expression is as follows:

$$\begin{aligned} & min[R(F_l)],\ l = 1,2,\ \ldots,\ o \\ & \beta_i[P(g_i \leq 0)] \geq r,\ i = 1,2,\ \ldots,\ n \\ & \mu(h_j) = 0,\ j = 1,2,\ \ldots, m \end{aligned} \quad (12)$$

The objective function $R(F_l)$ is using a weighted-sum approach (Delpiano & Sepúlveda, 2006, Beyer & Sendhoff, 2007):

$$R(F_l) = \alpha\mu(F_l) + (1-\alpha)\sigma(F_l) \quad (13)$$

where $\mu(F_l)$ and $\sigma(F_l)$ are the mean and standard deviation of the response F_i at the design point, where α is a weighting parameter varying as $0 \leq \alpha \leq 1$, representing the importance of the mean and standard deviation for any object function. $\beta_i[P(g_i \leq 0)] \geq r$ denotes the beta reliability index, which is the probability that the ith constraint is not violated, r indicates the beta target reliability level that the ith constraint is required not to cross, and $\mu(h_j)$ is the mean of the jth equality constraint.

To compute the mean and variance of the stochastic variables, representing the first-order moment and variance representing the second-order moment, Eqns (10) and (11) are used, assuming that they follow the Normal probability density function (Ang & Tang, 1975). The Mean Value First-Order Second Reliability Method (MVFOSRM) is used to estimate the uncertainty in the constraints, where the reliability is defined as the probability that constraints are not going to be violated under a given set of conditions as

$$R_l = 1 - P_{f,l} \quad (14)$$

The constraints g_l of the design process will be considered violated when the imposed limit g_l^C is overcrossed. The probability of failure $P_{f,l}$ of the constraint can be defined as:

$$P_{f,l} = P\left(g_l^C \leq g_l\right) = P\left(g_l^C - g_l^E \leq 0\right) \quad (15)$$

$$P_{f,l} = P\left[g_l\left(g_l^C, g_l^E\right) \leq 0\right] \quad (16)$$

where $g_l()$ is defined as a limit state function, defining the probability the limit state (constraint) is violated. The limit state function that represents the design constraints can also be described as a safety margin $M_l = g_l^C - g_l^E$, where g_l^C and g_l^E are independent normally distributed variables, and the probability of failure is expressed as:

$$P_{f,l} = (M_l \leq 0) \quad (17)$$

The safety margin is also a function of the random variables, which is random, too. The two first moments of the safety margin M_i are determined by:

$$\mu(M_l) = \mu(g_l^C) - \mu(g_l^E) \quad (18)$$

$$\sigma(M_l) = \sqrt{\sigma^2(g_l^C) - \sigma^2(g_l^E)} \quad (19)$$

Since M_l is a linear function of g_l^C and g_l^E and $(\mu(M_l))/\sigma(M_l)$ is the unit standard normal, then the probability of failure is described by:

$$P_{f,l} = \Phi\left(\frac{\mu(M_l)}{\sigma(M_l)}\right) \quad (20)$$

3.2 *Mathematical model*

The ship design optimisation analysis is based on polynomial solutions of the sets of equations representing ship behaviour. The framework used in the study starts with defining a mathematical model for design optimisation, determining the decision variables, their ranges, and objective functions, and setting up suitable constraints for the designed objective functions. Next is to carry out multi-objective optimisation, which results in a set of Pareto-optimal solutions—establishing metamodels using the response surface methodology (RSM), identifying the optimal solution, and conducting sensitivity analysis to analyse the effects of decision variables on design objectives and establishing sensitivity charts, showing the behaviour of all objective criteria in all optimal solutions. The decision variables and their range of the bulk carriers analysed here can be seen in Table 1.

Table 1. Decision variable.

Variable	Min	Max
Length, L (m)	120.00	192.07
Beam, B (m)	15.00	31.06
Depth, D (m)	12.00	14.71
Draft, T (m)	9.00	10.94
Block Coefficient, C_B	0.63	0.83
Ship Speed, V_k^* (knots)	8.00	10.32

In the ship operation, due to unavoidable disturbances such as the waves, wind and currents, the service/operational speed of the ship varies and can be modelled as an uncertain variable. Additionally, capital and operational descriptors are considered uncertainty parameters, too. All uncertainty parameters propagate through the mathematical model and impact the boundary conditions and objective functions.

The magnitude of perturbation uncertainty is relatively small in most cases, and its coupling effects with other parameters will transmit a significant deviation in the optimisation design process. Therefore, in general, the perturbed uncertain variables are modelled as:

$$X_i = x_i \in_i \quad (21)$$

where x_i is the actual parameter estimate, X_i is the stochastic variable, $\in_i$ is the perturbation factor, which follows the normal probability density function, where the mean value is 1, and the standard deviation is assumed as $0.1x_i$. The design solution must comply with constraints, as shown in Table 2.

The ship design variability sources contribute to the design state uncertainty, and SA ranks inputs to their relative contribution to the global uncertainty. The ship design model used in the optimisation comprises a series of parametric equations (Sen & Yang, 1998, Hannapel & Vlahopoulos, 2010).

Table 2. Design constraints.

Relations	Limit
$L/B >$	5.00
$L/D <$	15.00
$L/T <$	19.00
$F_n <$	0.32
EEDI=478.4 $P/DW/V_k <$	$f(DW)$
c=T-0.45 $DW^{0.31} <$	0.00
d=T-0.7 D-0.7 <	0.00
f=0.07 B-0.53 T-((0.085 C_B-0.002) B^2) <	0.00

The reliability-based robust design formulation integrates algorithms for multi-objective design, applied to the preliminary design of a bulk carrier. Transportation cost (TC), Lightweight (LW), and Annual cargo (AC) are chosen as objective functions to be optimised. The goal is to minimise TC and LW while maximising AC.

3.3 *Design optimisation*

The reliability levels of 85%, 90% and 95% are prescribed and analysed for the probabilistic constraints. The mean μ and variance σ of the object functions are evaluated at the design point (L, B, D, T, CB, V_k). The reliability-based robust optimisation uses the weighting factor of α= 0.6 to combine the objective functions' statistical descriptors. The NSGA II approach is employed for multi-objective optimisation (Deb, 2001, Wong et al., 2016). The analysis is carried out by setting several parameters of the genetic algorithm. The number of generations is 100, the population is 50, the SBX crossover probability for uniform crossover p_c=0.9, the mutation probability 0.1, the random seed 0.5, and the time taken is 134 sec.

3.4 *Optimal solution*

The number of admissible optimal ship design solutions was arbitrarily fixed to 50. Some solutions are very similar. To simplify the subsequent analysis of solutions, their number is reduced by applying the Gray Relational Analysis (Wang & Rangaiah, 2017), where the optimal solution is identified from the set of nondominated solutions from the Pareto-optimal front. Different studies indicated that Gray Relational Analysis and simple additive weighting are better for choosing one of the Pareto-optimal solutions.

The Gray Relational Analysis (GRA) describes the similarity between each objective value of each optimal solution by choosing the best value of each objective. GRA is implemented in three steps: data normalisation defining the ideal sequence (Martınez-Morales et al., 2010). This method does not require weights.

Figure 4 shows all six decision variables, *L, B, D, T,* C_b and V_k, as a function of the LW, AC and TC for the reliability-based robust design optimisation and respective optimal design solutions.

Figure 5 shows the Pareto front for the conflicting objective functions, where each represents one specific ship design solution, marking a trade-off between the objective functions.

Figure 5 also shows the effect of the reliability level of the constraints in the obtained Pareto frontier. With an increase in reliability, the Pareto front worsens. A more reliable design demands a higher TC for an identical AC capacity.

3.5 *Response surface*

A response function approximating the relation between the objective functions and decision variables is established near the optimum design solution (Montgomery, 1997). A linear relationship has resulted in 3 response equations for the three objective functions. The response *Y* as a function of six decision variables $x_1, x_2, \cdots, x_6$ representing *L, B, D, T,* C_b and V_k is defined as:

$$Y = f(x_1, x_2, \cdots, x_6) + \varepsilon \tag{22}$$

where ε represents the error due to some unknown or uncontrollable variables.

The first-order response model is defined as:

$$\eta = \beta_0 + \sum_{i=1}^{6} \beta_i \xi_i \tag{23}$$

where β_0 is a constant and β_i are the six first-order coefficients for the model written in terms of normalised decision parameters, ξ_i. The developed response

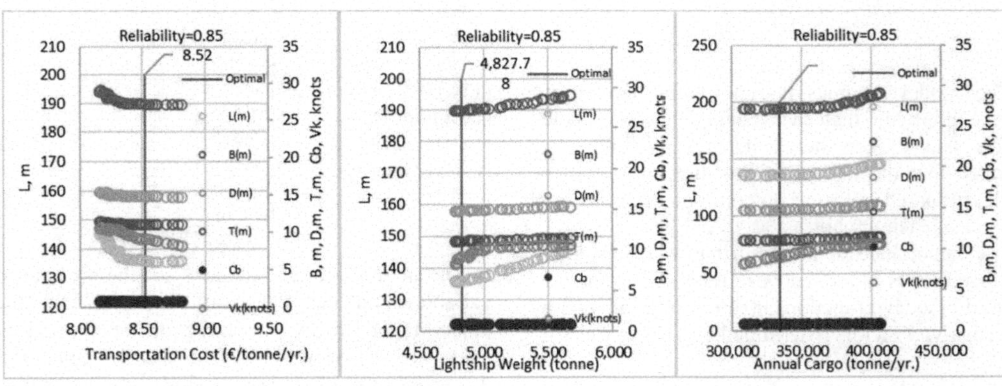

Figure 4. Variation of the decision variables as a function of TC (left), LW (central) and AC (right).

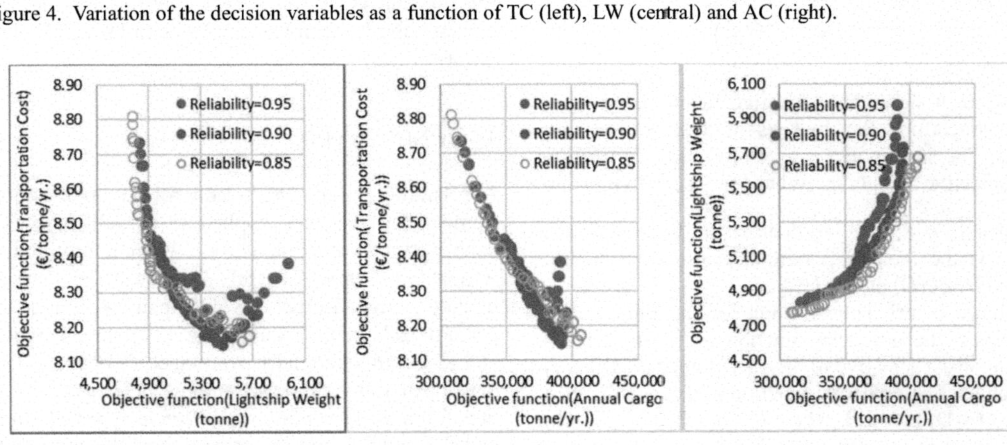

Figure 5. Pareto-optimal solutions for bi-objective ship design solutions.

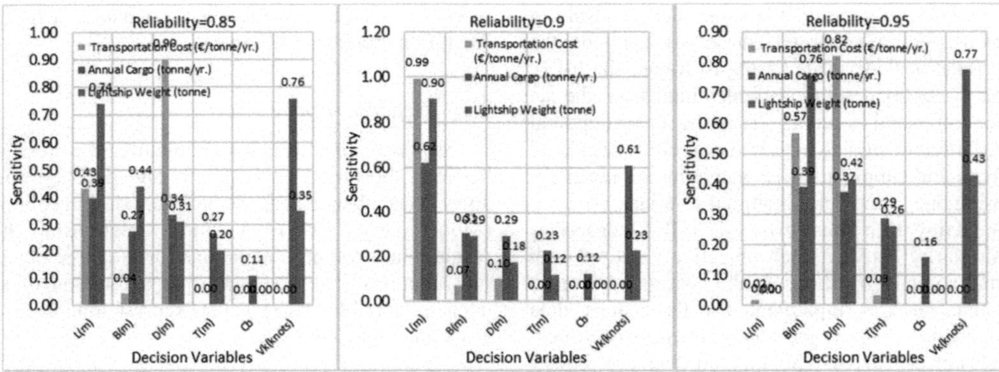

Figure 6. Sensitivity analysis.

surface function is used to define the design sensitivity.

3.6 *Design sensitivity*

The sensitivity of the normalised objective function to any of the decision variables is estimated as the first partial derivation concerning the decision variable, ξ_i under consideration:

$$\frac{\partial \eta}{\partial \xi_i} = \frac{\partial}{\partial \xi_i}\left[\sum_{i=1}^{6} \beta_i \xi_i\right] \tag{24}$$

The model assumes no correlation among the effects, implying that each change has cumulative effects and is considered dependent. Still, it is assumed that the impact of the interaction will be small compared with the outcome of the individual changes of each variable.

The importance of the contribution of any decision parameters to the response functions or the normalised objective functions $\eta(\xi)$ can be evaluated by the sensitivity factors determined by:

$$\alpha_i(\xi) = \frac{1}{\sqrt{\sum_{i=1}^{6}\left(\frac{\partial \eta(\xi)}{\partial \xi_i}\right)^2}} \frac{\partial \eta(\xi)}{\partial \xi_i} \tag{25}$$

Figure 6 shows the objective functions' sensitivities concerning the changes in the decision variables. The sensitivities vary from objective function to objective function. When the sensitivities are averaged, $\bar{\alpha}_i = \frac{\sum_{j=1}^{6} \alpha_j(\xi)}{6}$ then, it can be noticed that $\xi_1(L)$ is the most important decision variable, followed by ξ_3 (D) and ξ_6 (V_k).

Figure 6 also shows which of the three design solutions, the objective function is the most stable and the least sensitive to the variation of the decision variables. The objective function of TC is only influenced by D, L and B, where Cb and D are less influential decision variables. L, D and Vk are the most influential decision variables

Table 3. Descriptors of ship design optimisation.

Reliability	0.85	0.90	0.95
Scenario No	26.00	19.00	50.00
L (m)	135.79	137.94	138.39
B (m)	27.09	27.27	27.04
D (m)	14.83	15.06	14.82
T (m)	11.02	11.10	11.04
C_b	0.85	0.84	0.84
V_k (knots)	9.16	9.89	9.61
μ(LW) (tonne)	4,828	5,019	4,964
μ(AC) (tonne/yr.)	334,235	359,047	348,382
μ(TC) (€/tonne/yr.)	8.52	8.38	8.45
σ(LW) (tonne)	401.18	416.74	412.61
σ(AC) (tonne/yr.)	34,993	36,603	35,925
σ(TC) (€/tonne/yr.)	0.97	0.95	0.96

The optimal design solutions for the six decision variables and the statistical descriptors of three object functions for the three different reliability levels are shown in Table 3. It must be pointed out that higher reliability is identified with intermediate LW, AC and TC for the analysed design solutions. In contrast, the lower reliability leads to lower LW and AC and higher TC.

4 CONCLUSIONS

A reliability-based design and robust optimisation framework have been applied to a multi-objective concept design of a bulk carrier, where the following framework uses several advanced numerical methods. Second-order Taylor expansion is used to perform SA, demonstrating an efficient, accurate, and fast computational performance (134 sec.). The developed reliability-based robust design optimisation framework can analyse different types of ships and shipbuilding and operational conditions accounting for the propagating uncertainties.

ACKNOWLEDGEMENTS

The first author has been supported by the Strategic Research Plan of the Centre for Marine Technology and Ocean Engineering, financed by the Portuguese Foundation for Science and Technology (Fundação para a Ciência e Tecnologia - FCT) under contract UIDB/UIDP/00134/2020. This work was performed within the Technical University of Varna Research Plan, funded by the State Budget under the contract NP13 for 2023.

REFERENCES

Ang, A. H. S. & Tang, W. H. 1975. Probability Concepts in Engineering Planning and Design, Vol. 1—Basic Principles, New York, John Wily & Sons.

Beyer, H. G. & Sendhoff, B. 2007. Robust Optimization – A Comprehensive Survey. Computer Methods in Applied Mechanics and Engineering, 196, 3190–3218.

Deb, K. 2001. Multi-objective Objective Optimization using Evolutionary Algorithms, Wiley and Sons Ltd.

Delpiano, J. & Sepúlveda, M. 2006. Combining Iterative Heuristic Optimization and Uncertainty Analysis Methods for Robust Parameter Design. Engineering Optimization, 38, 821–831.

Diez, M. & Peri, D. 2010. Two-stage Stochastic Programming Formulation for Ship Design Optimisation under Uncertainty. Ship Technology Research, Schiffstechnik, VOL. 57, 172–181.

Eschenbach, T. G. 1992. Spider Plots versus Tornado Diagrams for Sensitivity Analysis. J. Appl. Anal., 22, 40–46.

Garbatov, Y. & Georgiev, P. 2021. Risk-based conceptual ship design of a bulk carrier accounting for energy efficiency design index (EEDI). Transactions of The Royal Institution of Naval Architects - International Journal of Maritime Engineering, 163, A 51–A 62.

Garbatov, Y., Sisci, F. & Ventura, M. 2018. Risk-based framework for ship and structural design accounting for maintenance planning. Ocean Engineering, 166, 12–25.

Georgiev, P. & Garbatov, Y. 2021. Multipurpose vessel fleet for short black sea shipping through multimodal transport corridors. Brodogradnja, 72, 79–101.

Hannapel, S. & Vlahopoulos, N. Robust and Reliable Multidiscipline Ship Design. 13th AIAA/ISSMO Multidisciplinary Analysis Optimization Conference, AIAA 2010-9394 2010 Fort Worth, Texas.

Harr, M. E. 1987. Reliability-Based Design in Civil Engineering, Dover Publications Inc.

Helton, J. C. 1993. Uncertainty and sensitivity analysis techniques for use in performance assessment for radioactive waste disposal. Reliability Engineering & System Safety, 42, 327–367.

IMO. 2023. IMO's work to cut GHG emissions from ships [Online]. Available: https://www.imo.org/en/MediaCentre/HotTopics/Pages/Cutting-GHG-emissions.aspx [Accessed October 10 2023].

Ju, P., Qin, C., Wu, F., Xie, H. & Ning, Y. 2011. Load Modeling for Wide Area Power System. Int. J. Electr. Power Energy Syst, 33, 909–917.

Keane, A. J., Price, W. G. & Schachter, R. D. 1991. Optimization Techniques In Ship Concept Design. Transaction of the Royal Institute of Naval Architecture, A.133, 123–129.

Komuro, R., Ford, E. D. & Reynolds, J. H. 2006. The use of multi-criteria assessment in developing a process model. Ecological Modelling, 197, 320–330.

Kovach, J. & Cho, B. R. 2008. Development of a Multidisciplinary-Multiresponse Robust Design Optimization Model. Engineering Optimization, 40, 805–819.

Liang, J. & Mourelatos, Z. P. 2008. A Single-Loop Method for Reliability-Based Design Optimization. International Journal of Product Development, 5, 76–92.

Lyon, T. & Mistree, F. 1985. A Computer-based Method for the Preliminary Design of Ships. Journal of Ship Research, 29, 251–269.

Martınez-Morales, J. D., Pineda-Rico, U. & Stevens-Navarro, E. Performance comparison between madm algorithms for vertical handoff in 4g networks. 7th International Conference on Electrical Engineering Computing Science and Automatic Control, 2010 Tuxtla Gutierrez, Mexico. 309–314.

MEPC 2011. Amendments to the annex of the protocol of 1997 to amend the International Convention for the Prevention of Pollution from Ships, 1973, as modified by the protocol of 1978 relating thereto MEPC 203(62). The Marine Environment Protection Committee (MEPC).

MEPC 2018a. MEPC.308(73). 2018 Guidelines on the method of calculation of the attained Energy Efficiency Design Index (EEDI) for new ships. London: IMO.

MEPC 2018b. Resolution MEPC.301(72) - Amendments to MARPOL Annex VI (ECAs and required EEDI for ro-ro cargo ships and ro-ro passenger ships. The Marine Environment Protection Committee (MEPC), MEPC.308(73)-19/Add.1.

MEPC 2020. MEPC.324(75). Amendments to the annex of the protocol of 1997 to amend the International Convention for the Prevention of Pollution from Ships, 1973, as modified by the protocol of 1978 relating thereto. IMO.

Messac, A. & Mullur, A. A. 2007. Multiobjective optimization: Concepts and methods. In: Arora, J. (ed.) Optimization of structural and mechanical systems. World Scientific, 121–147.

Montgomery, D. 1997. Design and Analysis of Experiments, John Wiley & Sons Inc.

Papadimitriou, S., Koliousis, I. G., Sdoukopoulos, E., Lyridis, D. V., Tsioumas, V. & Stavroulakis, P. J. 2018. The Dynamics of Short Sea Shipping. New Practices and Trends, Cham, Switzerland, Palgrave Macmillan.

Papanikolaou, A. D., Guedes Soares, C., Jasionowski, A., Jensen, J. J., McGeorge, D., Poylio, E. & Vassalos, D. 2009. Risk-Based Ship Design, Berlin/Heidelberg, Germany, Springer.

Priftis, A., Boulougouris, E., Turan, O. & Atzampos, G. 2020. Multi-objective robust early-stage ship design optimisation under uncertainty utilising surrogate models. Ocean Engineering, 197.

Rao, S. S. 2009. Engineering optimization: theory and practice, John Wiley & Sons.

Ren, H., Ding, Y. & Sui, C. 2019. Influence of EEDI (Energy Efficiency Design Index) on Ship–Engine–Propeller Matching. Journal of Marine Science and Engineering, 7, 425.

Saltelli, A., Ratto, M., Andres, T., Campolongo, F., Cariboni, J., Gatelli, D., Saisana, M. & Tarantola, S. 2007. Global Sensitivity Analysis: The Primer, Chichester, UK, John Wiley & Sons, Ltd.

Sen, P. & Yang, J. 1998. Multiple Criteria Decision Support in Engineering Design, London, Springer.

Wang, H., Hou, Y., Xiong, Y. & Liang, X. 2021. Research on multi-interval coupling optimization of ship main dimensions for minimum EEDI. Ocean Engineering, 237.

Wang, Z. & Rangaiah, G. P. 2017. Application and Analysis of Methods for Selecting an Optimal Solution from the Pareto-Optimal Front obtained by Multiobjective Optimization. Ind. Eng. Chem. Res., 56, 560–574.

Wärtsilä. 2023. Encyclopedia of Marine and Energy Technology [Online]. Available: https://www.wartsila.com/encyclopedia/term/admiralty-coefficient-admiralty-constant [Accessed September 15, 2023].

Wei, X., Chang, H., Feng, B. & Liu, Z. 2019. Sensitivity Analysis Based on Polynomial Chaos Expansions and Its Application in Ship Uncertainty-Based Design Optimization. Mathematical Problems in Engineering, 2019, 7498526.

Wong, J. Y. Q., Sharma, S. & Rangaiah, G. P. 2016. Design of Shell-and-Tube Heat Exchangers for Multiple Objectives Using Elitist Non-dominated Sorting Genetic Algorithm with Termination Criteria. Applied Thermal Engineering, 93, 888–899.

Woud, H. K. & Stapersma, D. 2002. Design of Propulsion and |Electric Power Generation Systems., London, Institute of Marine Engineers.

Advances in Maritime Technology and Engineering – Guedes Soares & Santos (Eds)
© 2024 The Author(s), ISBN 978-1-032-83099-5

A novel ship berthing decision-making modelling based on random forest method

D. Han, Y. Wang & B. Wu
State Key Laboratory of Maritime Technology and Safety (SKL MTS), Wuhan University of Technology, Wuhan, P.R. China
Intelligent Transport Systems Research Centre (ITSC), Wuhan University of Technology, Wuhan, P.R. China
National Engineering Research Centre for Water Transport Safety (WTSC), Wuhan University of Technology, Wuhan, P.R. China

R. Xiong
Shenzhen Maritime Safety Administration of the People's Republic of China, Shenzhen, P.R. China

ABSTRACT: With the growing development of large-scale ships and autonomous ships, ship berthing maneuvers faces many challenges in complex port waters. Currently, most modelling methods of ship berthing decisions based on the simulation data from ship-handling simulator, which lacks the validation of a real scenarios. This paper aims to fill the gap in this problem by utilizing the real pilotage data to establish a decision-making model for ship berthing maneuvers. Consequently, the pilotage database was collected, and random forest method is employed to implement the decision-making modelling of ship berthing. The evaluation results indicate that the accuracy of the constructed model reaches over 90% (i.e., accuracy score is 0.9369, precision score is 0.9140, recall score is 0.9369, and F1 score is 0.9243). The experimental results provide a novel perspective, practical and useful guidance for piloting behavior, port security management, and the development of automatic berthing.

1 INTRODUCTION

Ship berthing is a dynamic process of maneuvering a ship to berth, which is a challenging task even for experienced shipmasters and pilots, and it is easy to fall into an urgent situation and even cause marine accidents when encounter unexpected events suddenly (e.g., gale, strong current, misjudgment, and underestimation) according to navigation experience. At present, ship berthing operation is an indispensable step in maritime transportation, there are many researches focus on this topic from different perspectives. To prevent berthing accidents (e.g., collision, fire, oil spill, and grounding) in harbor area, it is critically important to identify the influence factors of ship berthing operations from the perspective of risk analysis or assessment (Khan et al, 2021; Kuzu et al, 2019; Sokukcu & Sakar, 2022). Similarly, the ship handling simulator was used to train seamen for safe berthing maneuvering (Benedict et al, 2018), and tugboats were allocated to assist ship berthing (Bui et al, 2011; Liu et al, 2013). Moreover, the numerical simulation of hydrodynamic response during ship berthing operations (Xiao et al, 2021) is also the favorite research directions.

With the rapid development of Maritime Autonomous Surface Ships (MASS), achieving the overall goal of automatic berthing has been paid more attentions, which is one of the intelligent techniques of marine ships. Because of the low-velocity motion deteriorates the maneuverability of the marine ships, the underactuated characteristics of berthing ship become the main challenge for automatic berthing control (Liu et al, 2022). Meanwhile, the artificial neural network (Nguyen, 2020; Tran et al, 2012), Proportion Integral Differential (PID) control (Yang et al, 2017), and adaptive control (Xiong et al, 2021) are the most used research methods in automatic berthing control.

There are a great of researches on ship berthing provided many valuable references, especially the development of automatic berthing technology. To best of our knowledge, there is no maneuvering decision-making in relation to ship berthing have been conducted, and only two journal papers (Xue et al, 2019; Xue et al, 2019) have focused on this topic. Moreover, the dataset sample used in both of the above studies is the simulation data from ship-handling simulator, which lacks of validation in real-world scenarios. However, the dataset used in this paper is the real ship navigation data, and we consider the situation of "Stop engine" and "Midships", which was not present in these two previous studies.

The remainder of this paper is structured as follows: In Section 2, the research methodology of this study is proposed, the process of data collection and preprocessing, and the preliminary knowledge related to the random forests are introduced. In Section 3, the experimental parameters are set to the optimal value,

DOI: 10.1201/9781003508762-8

then the correlation analysis of variables and feature importance measures are carried out. Finally, the summaries and conclusions of this paper are presented in Section 4.

2 THE PROPOSED METHODOLOGY FOR SHIP BERTHING DECISION-MAKING

In order to accurately predict ship berthing maneuver decisions (i.e., telegraphy orders and rudder orders used by shipmasters or pilots under current environmental conditions), a framework based on random forest is developed, which is shown in Figure 1. First, the dataset is collected from the SEAiq Pilot software and data processing is carried out. Second, the random forest algorithm can be used to train dataset sample, the evaluation matrix and feature importance are also given at the same time. Finally, the optimal decision tree is visualized by calculating prediction accuracy of every tree.

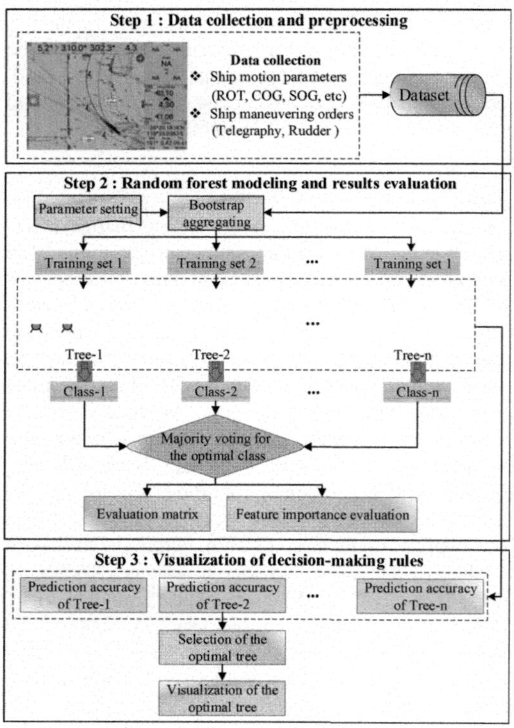

Figure 1. The proposed random forest model for ship berthing decision-making.

2.1 *Data collection and preprocessing*

We collect the data from the SEAiq Pilot software, which is widely used in pilot stations worldwide, can be utilized to assist pilots during ship berthing by calculating and displaying distance, bearing, and speed. This software can collect pilotage data in real-time, including information on location, ship handling, and so on (Han et al., 2023). The data collected from this software is the real-life scenarios, which is different from the simulation data from ship-handling simulator. When pilots conduct piloting operations, they will input the related parameter information of the ship into this software, so as to make the appropriate berthing maneuvering decisions. The interface of SEAiq Pilot software is shown in Figure 2.

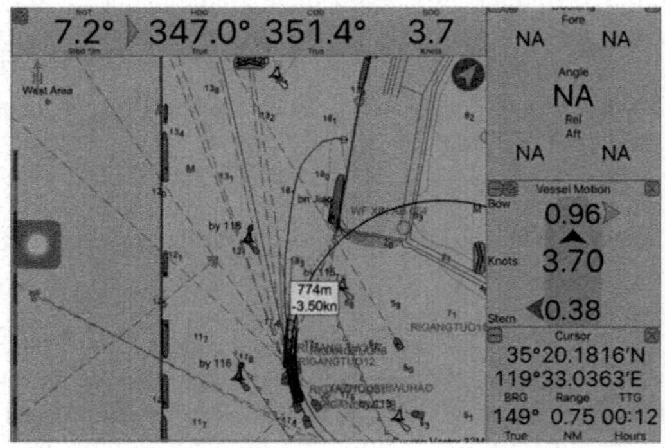

Figure 2. Interface of SEAiq Pilot software.

As shown in Table 1, there are eight input variables are extracted, including ship motion parameters (e.g., ROT, SOG, and SSM) and location (e.g., DBB, and HDG). It is should be noted that, in our experimental scenario, we did not consider the influence of environmental factors owe to as follows, (1) our experiment is carried out at Rizhao Port, the ships in this port are less affected by current and waves due to the breakwaters, (2) The deadweight of the ship we tested is 250877 tons, belongings to the super large ships, which makes the influence of wind is slight.

Table 1. Meaning of data variables (input).

Variables	Meaning	Units
DBB	Distance from bow to berth	Meters
ROT	Rate of turning	Degrees/min
HDG	Heading	Degrees
COG	Course over ground	Degrees
SOG	Speed over ground	Knots
SSM	Speed of ship motion	Knots
LSBM	Lateral speed of bow motion	Knots
LSSM	Lateral speed of stern motion	Knots

To best of our knowledge, the shipmasters or pilots maneuver the ship by operating different telegraph and rudder orders to change ship's speed and bearing. However, because of the differences in operating habits and experience, every operator may have different operating orders. To solve this problem, we quantified the combining telegraph and rudder orders to control the ship speed and course, the maneuvering decision-making orders is shown in Table 2. Moreover, it should be noted that our

Table 2. Maneuvering decision-making orders (output).

Speed control				Course control			
Telegraph orders	Symbols	Telegraph orders	Symbols	Rudder orders	Symbols	Rudder orders	Symbols
Full ahead	F1	Full astern	F2	Port five	Pf1	Starboard five	Sf1
Half ahead	H1	Half astern	H2	Port ten	Pt1	Starboard ten	St1
Slow ahead	S1	Slow astern	S2	Port fifteen	Pf2	Starboard fifteen	Sf2
Dead slow ahead	D1	Dead slow astern	D2	Port twenty	Pt2	Starboard twenty	St2
Stop engine	S	–	–	Midships	M	–	–
Decisions	Symbols	Decisions	Symbols	Decisions	Symbols	Decisions	Symbols
F1Pf1	Y1	S1Pt2	Y22	F2Sf2	Y43	D2Pf1	Y64
F1Pt1	Y2	S1Sf1	Y23	F2St2	Y44	D2Pt1	Y65
F1Pf2	Y3	S1St1	Y24	F2M	Y45	D2Pf2	Y66
F1Pt2	Y4	S1Sf2	Y25	H2Pf1	Y46	D2Pt2	Y67
F1Sf1	Y5	S1St2	Y26	H2Pt1	Y47	D2Sf1	Y68
F1St1	Y6	S1M	Y27	H2Pf2	Y48	D2St1	Y69
F1Sf2	Y7	D1Pf1	Y28	H2Pt2	Y49	D2Sf2	Y70
F1St2	Y8	D1Pt1	Y29	H2Sf1	Y50	D2St2	Y71
F1M	Y9	D1Pf2	Y30	H2St1	Y51	D2M	Y72
H1Pf1	Y10	D1Pt2	Y31	H2Sf2	Y52	SPf1	Y73
H1Pt1	Y11	D1Sf1	Y32	H2St2	Y53	SPt1	Y74
H1Pf2	Y12	D1St1	Y33	H2M	Y54	SPf2	Y75
H1Pt2	Y13	D1Sf2	Y34	S2Pf1	Y55	SPt2	Y76
H1Sf1	Y14	D1St2	Y35	S2Pt1	Y56	SSf1	Y77
H1St1	Y15	D1M	Y36	S2Pf2	Y57	SSt1	Y78
H1Sf2	Y16	F2Pf1	Y37	S2Pt2	Y58	SSf2	Y79
H1St2	Y17	F2Pt1	Y38	S2Sf1	Y59	SSt2	Y80
H1M	Y18	F2Pf2	Y39	S2St1	Y60	SM	Y81
S1Pf1	Y19	F2Pt2	Y40	S2Sf2	Y61	–	–
S1Pt1	Y20	F2Sf1	Y41	S2St2	Y62	–	–
S1Pf2	Y21	F2St1	Y42	S2M	Y63	–	–

experiment is the actual scenario at sea, that is why we consider the situation of "Stop engine" and "Midships", which was not present in some previous studies (Xue et al, 2019). However, we did not consider the process of tugboat assisted berthing due to the complexity of tugboat-ship collaborative control. According to the input and output variables explained above, the training samples for the proposed RF model (partially) is listed in Table 3.

2.2 *Random forest modeling and results evaluation*

Random forest (RF) is an ensemble learning method that combines multiple decision trees for classification. It's theoretical foundation lies in decision trees and randomness. The basic principle of RF is to make predictions by constructing multiple decision trees, each trained on randomly selected feature subsets and sample subsets. During prediction, RF aggregates the predictions of each decision tree through voting or averaging to obtain the final prediction. RF is applicable to various classification problems, particularly suitable for handling high-dimensional and large-scale datasets, it can handle data with complex and nonlinear relationships, and also exhibits good robustness to missing values and outliers. Furthermore, RF can assess the importance of features by calculating the frequency of feature usage in decision trees and its impact on prediction results, this helps with feature selection and data analysis.

2.2.1 *Random forest training and prediction*

The RF training model is established by taking the data variables as the input feature, and taking the maneuvering decision-making orders as the prediction results. In this process, the bootstrap aggregating method is introduced to divide the dataset into a training set and a test set, owe to it is able to improve accuracy and stability while avoiding overfitting by reducing the variance of the results.

Table 3. Training samples for the proposed RF model (partially).

No.	DBB	ROT	HDG	COG	SOG	SSM	LSBM	LSSM	Decisions
1	1850	-0.1	296.00	296.90	5.10	5.10	0.07	0.09	Y19
2	1848	-0.1	296.00	296.90	5.10	5.10	0.07	0.09	Y19
3	1845	-0.1	296.00	296.80	5.10	5.10	0.06	0.08	Y19
...	...	...	...	...	...	...	...	...	...
282	1070	8.3	328.00	322.30	4.00	3.98	0.38	-1.17	Y35
283	1069	8.1	328.00	322.60	3.90	3.85	0.38	-1.12	Y35
284	1068	7.8	328.00	322.50	3.90	3.85	0.36	-1.10	Y35
...	...	...	...	...	...	...	...	...	...
468	492	4.0	353.00	1.10	3.60	3.60	0.89	0.13	Y33
469	486	3.6	353.00	0.40	3.60	3.60	0.80	0.13	Y33
470	482	3.2	353.00	359.70	3.60	3.60	0.72	0.13	Y33
...	...	...	...	...	...	...	...	...	...

2.2.2 *Random forest training and prediction*

The RF training model is established by taking the data variables as the input feature, and taking the maneuvering decision-making orders as the prediction results. In this process, the bootstrap aggregating method is introduced to divide the dataset into a training set and a test set, owe to it is able to improve accuracy and stability while avoiding overfitting by reducing the variance of the results.

2.2.3 *Results evaluation*

One of the most significant advantages of RF is the measurement of feature importance, which can be determined through the calculation of the Gini coefficient for features (Lu et al, 2022). We use *VIM*to represent feature importance measurement and *GI* to represent the Gini index. Assuming there are m features $X_1, X_2, \cdots, X_m$, now we need to calculate the Gini index score $VIM_j^{(Gini)}$ for each feature X_j, i.e., the average change in the impure node segmentation of the j-th feature in all decision trees of a random forest. The formula for calculating the Gini index is:

$$GI_m = 1 - \sum_{k=1}^{|K|} p_{mk}^2 \tag{1}$$

where $|K|$ represents the number of categories, and p_{mk} represents the proportion of categories in node m. It refers to the probability of randomly selecting two samples from node m with inconsistent category labels.

The importance of feature X_j on node m, i.e., the change in Gini index before and after the branch of node m is:

$$VIM_{jm}^{(Gini)} = GI_m - GI_l - GI_r \tag{2}$$

where GI_l and GI_r respectively represent the Gini indices of the two new nodes after the branch.

If the node where feature X_j appears in decision tree i is set M, then the importance of X_j in the i-th tree is:

$$VIM_{ij}^{(Gini)} = \sum_{m \in M} VIM_{jm}^{(Gini)} \tag{3}$$

Assuming that there are n trees in a random forest, then the Gini index score of feature X_j is:

$$VIM_j^{(Gini)} = \sum_{i=1}^{n} VIM_{ij}^{(Gini)} \tag{4}$$

Finally, normalize all obtained importance scores:

$$VIM_j = VIM_j / \sum_{i=1}^{n} VIM_i \tag{5}$$

The approach of feature importance measures is used to quantify the overall contribution of each feature in the decision-making process across all trees, which can help visualize the most influential features, detect potential biases in the model, and provide a more intuitive understanding of the decision-making process.

2.3 *Visualization of decision-making rules*

We usually treat the RF as a black box, input data to obtain prediction results, and do not care about how the model calculates, leads to the RF training process cannot be visualized. The performance and prediction accuracy of this black box are determined by parameters. However, we can visualize the process of each decision tree by calculating the prediction accuracy of each decision tree, and the optimal decision tree can be determined by comparison. Then, the we can visualize the optimal decision tree, and the decision-making rules of the optimal decision tree can represent the whole forest because of the parameters of max_features is set to None (i.e., each tree in the forest has all features of the sample data).

The evaluation of prediction results of RF model is an indispensable process. As shown in Table 5, The evaluation results of the model indicate that the accuracy of the random forest model

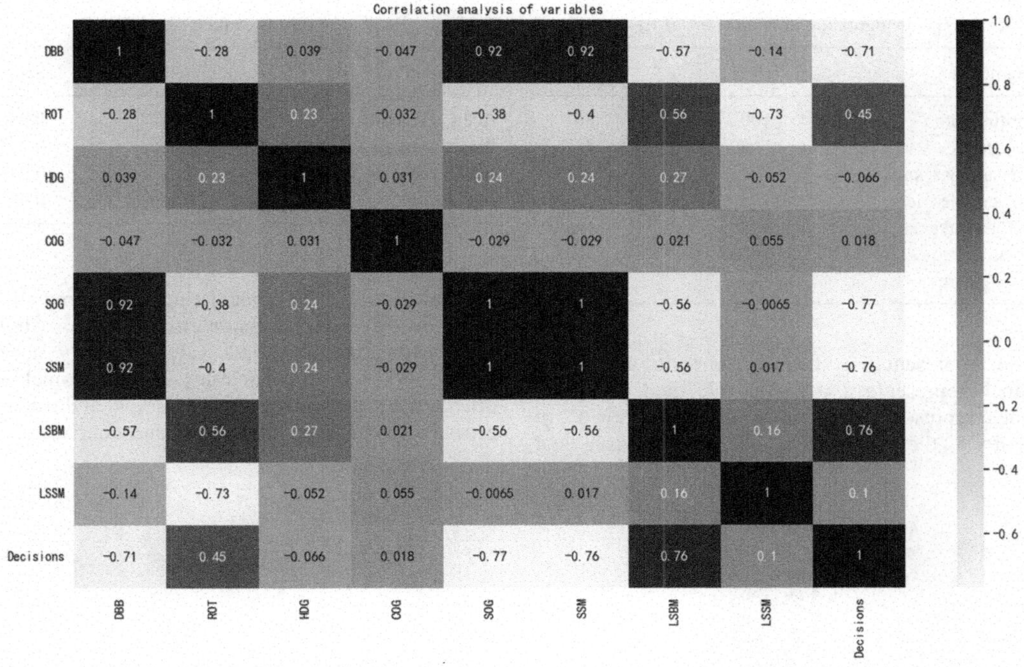

Figure 3. Correlation analysis of variables.

constructed in this article reaches over 90% (i.e., accuracy score is 0.9369, precision score is 0.9140, recall score is 0.9369, and F1 score is 0.9243).

3 EXPERIMENTS

3.1 *Parameters setting*

In our experiments, all processed dataset have been used as the experimental sample, and the RF algorithms have been implemented with Scikit-Learn (Pedregosa et al, 2011). It is the most significant to set the experimental parameters of RF algorithms before data training. Compared to other algorithms, the RF algorithm has more parameters (i.e., the parameters related to bootstrap aggregating method and the parameters related to each sub decision tree of the random forest). However, based on experience, we only need to set several parameters that have a significant impact, while selecting default values for the remaining parameters. By this way, the optimization of relevant parameters are carried out, as shown in Table 4. Note that bootstrap = Ture indicates the bootstrap samples are used when building trees, and max_features = None indicates all sample features are applied to each tree.

3.2 *Correlation analysis of variables*

The correlation analysis of variables in this article is shown in Figure 3. It illustrates that DBB has a strong correlation with SSM and SOG. When the DBB gradually decreases, it indicates that the ship is gradually approaching the berth, in order to ensure the safety of the port water area, the shipmaster or pilot must reduce the ship speed, which is consistent with sailing habits and experience. Moreover, Decisions as the output variables, it can be seen that it is highly correlated with most input variables (e.g., SOG, SSM, and LSBM).

3.3 *Feature importance measures*

Feature importance measures provide insights into the data used for training the model, highlighting which features have the most significant impact on the model's predictions. Moreover, it can help in understanding the model's behavior, improving its performance, and simplifying it by removing irrelevant features. it can also be used for feature selection, model interpretation, and understanding the relationships between different features. Feature importance measures are particularly useful in complex models where it is challenging to understand the relationship between the input features and the output prediction. As shown in Figure 4, it illustrates that ROT and DBB are the most influential features for ship berthing maneuvering decision-making, and LSSM, SSM

3.4 *Visualization of the optimal decision tree*

The visualization of the optimal decision tree is a graphical representation that helps in understanding the decision-making model. It is a flowchart-like structure in which each internal

Table 4. The configuration of parameters of RF algorithm.

Parameters	Default	Setting
n_estimators	100	80
max_depth	None	4
min_samples_split	2	2
min_samples_leaf	1	3
max_features	sqrt	None
bootstrap	True	True
max_features	None	None

node represents a feature or attribute, each branch represents a decision rule, and each leaf node represents an outcome. As shown in Figure 5, the optimal decision tree is visualized

Table 5. Prediction results evaluation of RF.

Evaluation methods	Score
Accuracy score	0.9369
Precision score	0.9140
Recall score	0.9369
F1 score	0.9243

by calculating the calculating the prediction accuracy of each decision tree in the whole forest. Note that the class value in every nodes represents the maneuvering decision-making orders (i.e., the combining telegraph and rudder orders to control the ship speed and course).

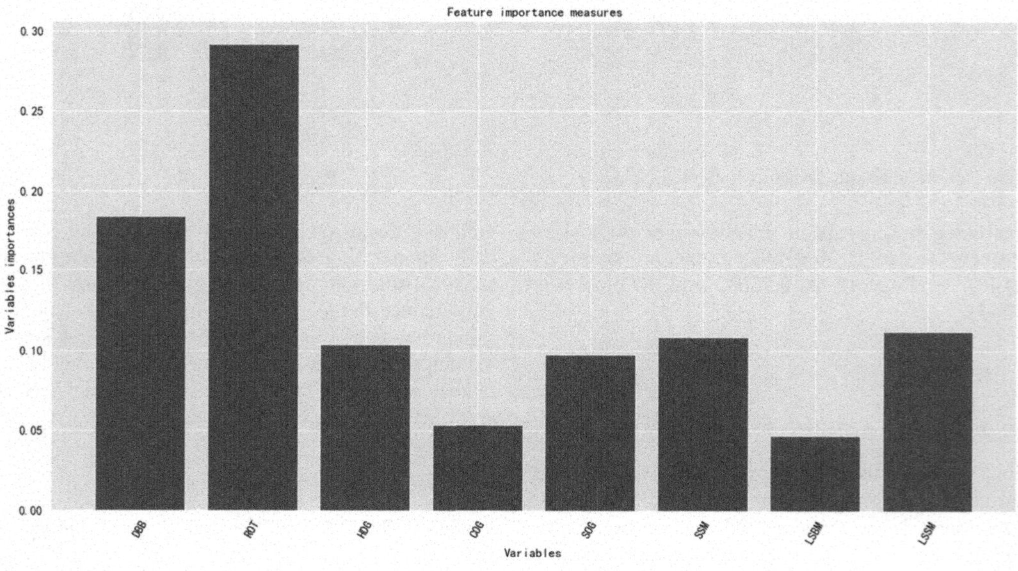

Figure 4. Feature importance measures.

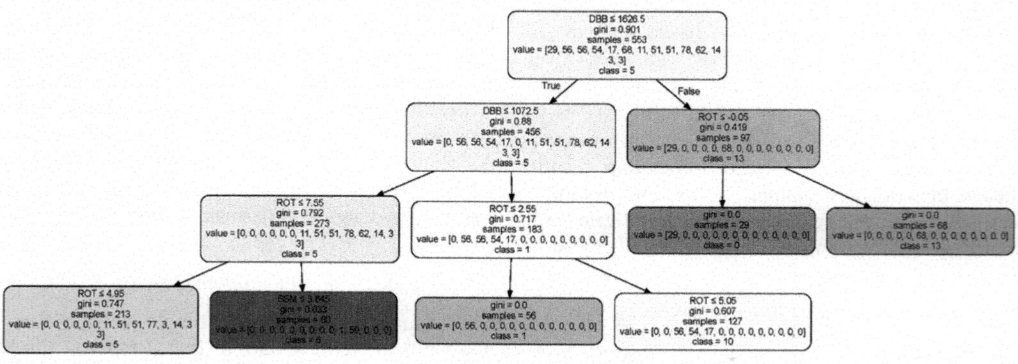

Figure 5. Visualization of the optimal decision tree.

3.5 *Feature importance measures*

Feature importance measures provide insights into the data used for training the model, highlighting which features have the most significant impact on the model's predictions. Moreover, it can help in understanding the model's behavior, improving its performance, and simplifying it by removing irrelevant features. it can also be used for feature selection, model interpretation, and understanding the relationships between different features. Feature importance measures are particularly useful in complex models where it is challenging to understand the relationship between the input features and the output prediction. As shown in Figure 4, it illustrates that ROT and DBB are the most influential features for ship berthing maneuvering decision-making, and LSSM, SSM followed.

3.6 *Visualization of the optimal decision tree*

The visualization of the optimal decision tree is a graphical representation that helps in understanding the decision-making model. It is a flowchart-like structure in which each internal node represents a feature or attribute, each branch represents a decision rule, and each leaf node represents an outcome. As shown in Figure 5, the optimal decision tree is visualized by calculating the calculating the prediction accuracy of each decision tree in the whole forest. Note that the class value in every nodes represents the maneuvering decision-making orders (i.e., the combining telegraph and rudder orders to control the ship speed and course).

4 CONCLUSIONS

This study proposes a methodology for ship berthing decision-making based on RF method, which aims to fill the gap in this problem by utilizing the real pilotage data to establish a decision-making model for ship berthing maneuvers. The ship motion parameters (e.g., ROT, SOG, and SSM) and location parameters (e.g., DBB, and HDG), etc., are collected and analyzed quantitatively to train the proposed model. The experimental results provide a novel perspective, practical and useful guidance for piloting behavior, port security management, and the development of automatic berthing. Nevertheless, the proposed RF model still has some shortcomings (e.g., we need to collect more data to further train the machine learning model and improve recognition accuracy, and the used data should be processed as the fuzzy variables due to the different dimensionality of data), which need to be improved in further research.

ACKNOWLEDGEMENTS

This work supported by National Natural Science Foundation of China (Grant No. 51920105014, 52071248, and 52272422), Hubei Natural Science Foundation (Grant No. 2021CFB312).

REFERENCES

Benedict, K., Gluch, M., Kirchhoff, M., Schaub, M., Tuschling, G., Baldauf, M., Gehrke, M. En-hanced Fast-Time-Simulation Features to support Ship-Handling Simulator Training. 19th Annual General Assembly (AGA) of the International-Association-of-Maritime-Universities (IAMU), Barcelona, Spain. 2018, pp. 94–112.

Bui, V.P., Kawai, H., Kim, Y.B., Lee, K.S. A ship berthing system design with four tug boats. Journal of Mechanical Science and Technology. 2011, 25(5), 1257–1264.

Han, D., Chen, H.Z., Song, G., Yip, T.L., & Wu, B. Mapping the landscape of ship berthing research: A bibliometric analysis and literature review over two decades. Ocean & Coastal Management. 2023, 242, 106730.

Khan, R.U., Yin, J.B., Mustafa, F.S., Anning, N. Risk assessment for berthing of hazardous cargo vessels using Bayesian networks. Ocean & Coastal Management. 2021, 210, 105673.

Kuzu, A.C., Akyuz, E., Arslan, O. Application of Fuzzy Fault Tree Analysis (FFTA) to maritime industry: A risk analysing of ship mooring opera-tion. Ocean Engineering. 2019, 179, 128–134.

Liu, X., Ren, W., Zhang, Q., Zhang, S.H. Re-search of Marine Simulator Cables' Mathematical Models. 4th International Conference on Ad-vances in Materials and Manufacturing (ICAMMP 2013), Kunming, China. 2013, pp. 551–554.

Liu, Y., Im, N.K., Zhang, Q., Zhu, G.B. Adap-tive Auto-Berthing Control of Underactuated Vessel Based on Barrier Lyapunov Function. Journal of Marine Science and Engineering. 2022, 10(2), 279.

Lu, J., Su, W., Jiang, M., & Ji, Y. Severity prediction and risk assessment for non-traditional safety events in sea lanes based on a random forest approach. Ocean & Coastal Management. 2022, 225, 106202.

Nguyen, V.S. Investigation of a Multitasking System for Automatic Ship Berthing in Marine Practice Based on an Integrated Neural Control-ler. Mathematics. 2020, 8(7), 1167.

Pedregosa et al. Scikit-learn: Machine Learning in Python, JMLR 12, 2011, pp. 2825–2830.

Sokukcu, M., Sakar, C. Risk analysis of collision accidents during underway STS berthing maneu-ver through integrating fault tree analysis (FTA) into Bayesian network (BN). Applied Ocean Re-search. 2022, 126, 103290.

Tran, V.L., Im, N. A study on ship automatic berthing with assistance of auxiliary devices. In-ternational Journal of Naval Architecture and Ocean Engineering. 2012, 4(3), 199–210.

Xiao, G.Q., Tong, C., Wang, Y., Guan, S.S., Hong, X.B., Shang, B. CFD Simulation of the Safety of Unmanned Ship Berthing under the In-fluence of Various Factors. Applied Sciences-Basel. 2021, 11(15), 7102.

Xiong, Y., Yu, J.J., Tu, Y.J., Pan, L., Zhu, Q.G., Mou, J.M. Research on data driven adaptive berthing method and technology. Ocean Engineering. 2021, 222, 108620.

Xue, J., Wu, C., Chen, Z., Gelder, P.H., & Yan, X. Modeling human-like decision-making for inbound smart ships based on fuzzy decision trees. Expert systems with applications, 2019a, 115, 172–188.

Xue, J., Chen, Z., Papadimitriou, E., Wu, C., & Gelder, P. A. Influence of environmental factors on human-like decision-making for intelligent ship. Ocean Engineering. 2019b, 186, 106060.

Yang, K.U., Hur, J.G., Choi, M.S., Yeo, D.J., Byun J.H. Study on ship automatic berthing system with mooring lines. China Ocean Engineering. 2017, 31(1), 19–29.

Advances in Maritime Technology and Engineering – Guedes Soares & Santos (Eds)

Development of ICT networks in maritime transport applications

T. Neumann
Gdynia Maritime University, Gdynia, Poland

ABSTRACT: In recent years, interest in digital technologies to automate industry has been increasing dramatically, modern technologies in shipping promise many benefits for growth. The concept of Industry 4.0 includes a range of solutions, using ICT (Information and Communication Technologies), for a variety of industries offering increased productivity and industrial functioning. In this text, the author attempts to characterize the directions of development in which ICT networks and their applications in maritime transport are heading, and describes the most important sectors in which the introduction of modern technologies may have the greatest impact on the operation and efficiency of systems. The text also contains an analysis of the security of maritime transport in cyberspace, the author lists the most important threats, actors and motivations behind the attacks as well as the weaknesses of currently used systems. Using the example of the 2017 NotPetya attack, the author seeks to examine the vulnerabilities and make recommendations to reduce the threat of the attack.

1 INTRODUCTION

Maritime transport is an important industry for the functioning of the global economy, goods transported by sea constitute over 80% of all goods transported in the world and, according to reports, their number is expected to increase in the coming years. Sea vessels are considered the most efficient means of transport, their importance for the operation of the global economy has influenced the interest in the use of modern Information and Communication Technologies (ICT).

The use of modern solutions is undoubtedly driven by the desire to improve the most important parameters of maritime transport. The use of ICT technologies opens the way to many areas that can significantly affect the efficiency, safety and reliability of maritime transport.

In this work, the author, based on the analysis of current literature, tries to examine the directions in which the broadly understood use of ICT networks in maritime transport is heading. The text presents the application and argumentation behind the use of ICT solutions as a development of existing factors, as well as innovative ideas to increase automation in the maritime industry.

This work is divided into three sections, in the first section the author presents the history of communication in maritime transport, currently used solutions and characterizes the concept of Industry 4.0 as a new industry model emphasizing the automation and optimization of processes using ICT networks.

In the next part of the work, the author characterizes the current applications of ICT networks in maritime transport and describes their use in currently used systems. The following subsections contain a description of the areas of greatest interest in the use of ICT technologies in particular fields of maritime transport, together with the identification of the logic of introducing these technologies into use as well as the obstacles standing in the way of the development of these concepts.

The last section of the work presents cybersecurity of maritime transport, the text lists the weaknesses of systems used in maritime transport, the main threats and actors posing a threat. The author of the text also describes the most devastating cyber attack in the history of maritime transport, and proposes measures to be taken to protect the systems (Cichocki, 2023).

The author's motivation for choosing the topic is the ongoing technological development and the desire to examine the direction in which maritime transport is heading using modern technologies while paying attention to the cybersecurity aspect of maritime transport.

2 ICT NETWORKS IN MARITIME TRANSPORT

2.1 *The concept of an IT network in maritime transport*

Communication plays an important role in maritime transport, facilitating navigation, increasing shipping safety and influencing the efficiency of trade. The history of maritime transport communication dates back to the 5th century BC, when ancient Greek

DOI: 10.1201/9781003508762-9

ships communicated with each other using trumpets to increase the intensity and direction of the speaker's voice. For many centuries, the transfer of information from ship to shore or as inter-ship communication was carried out physically by means of pigeons or small boats. In the 18th century, semaphore flags and the semaphore alphabet became the main means of communication between ships. In this system, the manipulation of flags, shields or pennants corresponds to letters of the alphabet and numbers. During night or poor visibility conditions, flags were replaced with torches. Nowadays, flags are still used in maritime communications in the form of International Code of Signals flags, where each flag stands for a letter, a number or a given signal. Maxwell's development of the theory of electromagnetism and Marconi's invention of the telephone in the 19th century allowed for wireless communication using Morse code. A device allowing wireless transmission of messages in Morse code is called a telegraph, this device was operated by radio operators receiving and transmitting up to 200 words per minute. A revolutionary achievement for maritime communications was the introduction of radiotelephones. They converted sound waves into radio waves using a transmitter and intensity modulation, and then the receivers converted the received radio signals into sound waves using frequencies in the range of 2 to 23 MHz. In 1950, VHF communications became the standard for the United States Navy. Another revolution in maritime communications was initiated by the introduction of the first telecommunications satellites into orbit in 1962. Many international organizations such as IALA (International Association of Lighthouse Authorities), ITU (International Telecommunications Union) and IMO (International Maritime Organization) have realized awareness of the benefits of wireless information exchange in maritime transport. Nowadays, a wide range of devices and systems, including: satellites, UAV (Unmanned Aerial Vehicle), HAP (High altitude Platform), operating on different frequencies to provide coverage.

Humanity has experienced three industrial revolutions in its history and is on the verge of another - the fourth. The first industrial revolution, industry 1.0, or as it is colloquially called the "age of steam", began in England at the end of the 18th century. Revolution 1.0 was characterized by the mechanization of production using the steam engine. In the early years of the 20th century, the US began Industrial Revolution 2.0, the era of electricity, introducing mass production using electrically powered devices. Then, in the mid-20th century, the information age began. The last of the revolutions led to the appearance of computers and the development of information and communication technologies for the purposes of production automation. Currently, humanity is experiencing the fourth industrial revolution, industry 4.0, the era of cyber physical systems (Cps). Maritime transport accounts for over 80% of global transport operations. In 2021, the total weight of goods transported by ships was almost 11 billion tons (UNCTAD, 2022). The achievements of the Fourth Industrial Revolution are expected to have a significant impact on the maritime transport ecosystem, covering a wide range of activities, including core maritime activities (e.g. fishing, shipbuilding, shipping, port operations and offshore energy) and maritime related activities (e.g. manufacturing industries and logistics services) (Aiello, et al, 2020). The latest report "Oceans Economy 2030" issued by the OECD (Organization for Economic Co-operation and Development) estimates that the maritime industry may double its contribution to global value creation by 2030 with increased demand for shipping, construction of new ships, equipment marine and related services. These forecasts depend largely on the technological innovations introduced by Industry 4.0, which will be the basis of the future maritime ecosystem in which specific cyber-physical systems "smart ships", characterized by new operational and design requirements, will replace traditional ships in the fully connected maritime ecosystem of Shipping 4.0 with increased efficiency and sustainable development (OECD, 2016).

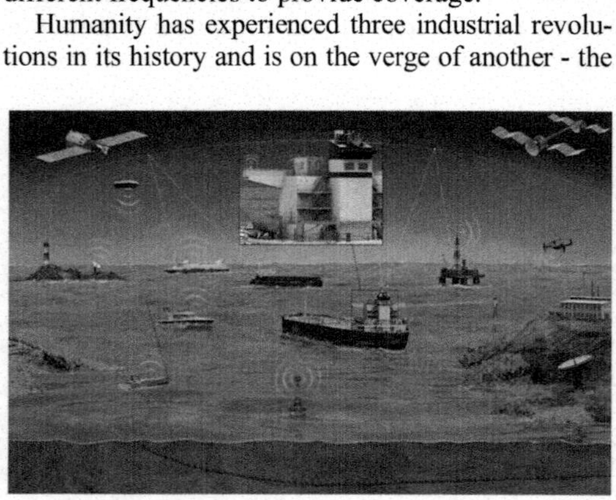

Figure 1. Communication in maritime transport. Source: (Alqurashi et al., 2023).

The history of ICT began in the 1950s with the development of the first computers, which were later connected to each other in a network, and thus the global Internet network was born. Information and Communication Technology (ICT) is one of the most important branches of IT (Information Technology) dealing with the production and use of telecommunications and IT devices and services. ICT is related to a range of products and services that produce, process, store and distribute information. The concept of ICT includes media for recording and transmitting information, technologies enabling the transmission of information, networks, satellite systems with applications and services related to them, all communication devices and applications, radio, TV and mobile phones (Zych, 2018).

A network is a structure connecting devices and elements that allow two or more computers or other devices to communicate. The classification of networks depends on the range of their operation, the method of communication and their distribution in a given area. Due to the range of operation, there are two basic types of networks: LAN (Local Area Network) and WAN (Wide Area Network). Networks can operate in wired and wireless ways (Onishchenko et al., 2022).

A LAN is the most common type of network. Its scope usually covers the area of one building or institution. This network may include up to several hundred computers, connected with each other by coaxial cable, twisted pair or optical fiber.

WAN is a network that covers a large geographical area, such as a city or region. It even consists of several thousand computers.

WLAN (Wireless Local Area Network) is a local wireless network in which radio waves are the information carrier. The WLAN network allows the user to move around the facility without losing coverage. The advantages of this structure include: mobility, range and flexibility. There are three variants of wireless networks: IBSS, BSS and ESS.

IBSS (Independent Basic Service Set) is a network without an access point. Users communicate directly. Stations should be located no more than 60 meters apart.

BSS (Basic Service Set) is a network operating on the principle of an access point. Network users can only transmit data within the range of one AP (Access Point). In theory, the number of clients is unlimited, but it depends on the network bandwidth.

ESS (Extended Service Set) connects at least two BSS networks into a backbone network. This solution allows the user to move between BSS networks without signal loss (Hudański, 2015).

On seagoing ships, the most common type of network is a LAN connecting ship systems with the main computer collecting data from on-board devices. An example of such a network is the use of the NMEA 2000 standard or its predecessor NMEA0183, which allows for the connection of sensors and ship devices in a bus topology. This means that all components are connected to one bus, most often to a coaxial cable. The NMEA system is a source of data such as geographical position, engine parameters, weather outside the ship, and conditions inside the ship. Both versions of the NMEA system can be installed on a ship, version 0183 is compatible with older types of equipment and devices available on seagoing ships, while standard 2000 is a more advanced system and is compatible with modern devices. In the case of the latter, you just need to add a branch and use it to connect additional devices. Both systems can be connected to one unit after using an appropriate adapter (Melnyk et al., 2022).

2.2 *Maritime transport and Industry 4.0 technologies*

Maritime transport is one of the oldest industries in the world and has played a significant role in the evolution of humanity and the economic development of many countries. The first transport ships were small and could be used to transport small amounts of cargo, but now maritime transport is a highly advanced activity. Shipping was one of the few industries that operated around the clock to ensure a continuous and stable flow of goods from one country to another. The maritime industry is the backbone of the world economy. Thanks to it, it is possible to import and export goods over long distances. Currently, most goods are produced in places rich in cheap raw materials, and then the goods are transported all over the world. Statistical data indicate a continuous increase in the value of maritime transport, which will significantly impact the burden on ports as well as people finding employment in the maritime industry. Maritime transport is a labor-intensive industry that requires niche capabilities to operate specialized equipment and devices. To make the entire import and export process run quickly and efficiently, a technological impulse and transformation of the entire industry are needed. The fourth industrial revolution, industry 4.0 or industrial revolution 4.0, began and has been ongoing since the beginning of the 21st century, based on the achievements of the third industrial revolution, which introduced computerization and automation of industry (Donepudi, 2014 and Neumann, 2018).

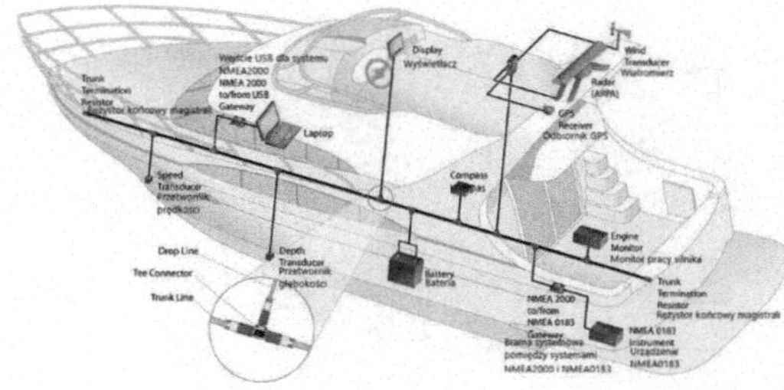

Figure 2. Sample NMEA2000 system architecture. Source: (Anon, 2023).

Industry 4.0 is a breakthrough process aimed at reducing human participation in ordinary and innovative activities in favor of intelligent machines. These devices produce data, analyze it, learn from it and make decisions based on it. The digitization of maritime transport and the adaptation of the solutions of the fourth industrial revolution are not as dynamic as in the case of other industries. Maritime transport uses many types of machines and devices, the multitude of these devices guarantees a database that can be collected, processed, analyzed and used in decision-making processes, with as little human intervention as possible. By using the solutions of the fourth industrial revolution, the efficiency of maritime transport and port operations would increase many times (Erstad et al., 2023).

The figure above is a collection of technology introduced by the fourth industrial revolution. It includes the following elements:

- Internet of Things (IoT) - Internet of Things refers to a type of network connecting to the

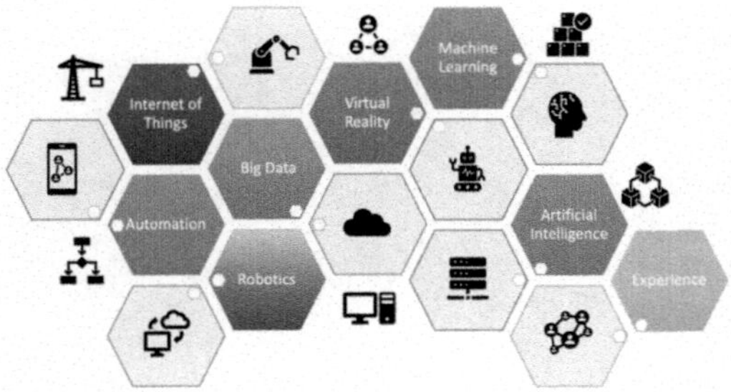

Figure 3. Industry 4.0 technologies. Source: (Filippopoulos and Stamoulis 2017).

Internet or cloud computing, based on specific protocols, devices that produce data in order to exchange information, intelligent recognition, positioning, tracking, monitoring and management. IoT-related objects become recognizable and gain data on the basis of which they conduct decision-making processes and enable context-related decisions by sharing data about themselves. Devices may have access to data collected and generated by other entities in the network or are part of complex services (Patel, et al, 2016).

- Automation - the process of performing a predetermined sequence of operations with little or no human work, using specialized equipment and devices that use and control manufacturing processes.
- Robotics - a field of science dealing with the construction, programming and control of robots. A robot is a computer-controlled machine that has been programmed to move, move objects and interact with the environment.
- Big Data Analytics - the term big data was given to data sets that were too large to be analyzed in the past due to technological limitations. With the development of devices and the increase in the computing power of computers, the issue of large files also refers to data analysis, allowing the interpretation of the content contained in the files, which can be used in decision-making processes (Mirović, et al, 2018).
- Virtual reality – computer simulation of a three-dimensional, interactive space in which objects have the character of spatial presence (Bryson, 2013).
- Machine learning - an area of artificial intelligence dealing with algorithms that improve their performance based on experience gained while performing tasks (Mitchell, 1997).
- Artificial Intelligence – simulation of human intelligence using devices, mainly computers (Laskowski, 2023).
- Experience.

Industry 4.0 has ushered in digital transformation with the increase in access to the Internet, smart devices, robotics, interconnection of devices, automatic collection of large amounts of data, machine learning, artificial intelligence, enhanced and virtual reality, and experience exchange.

The maritime industry is going through a series of changes and transformations in all aspects of operations, including business drivers, compliance with legal and environmental regulations, the implementation of new technologies, macroeconomic trends and employee expectations (Donepudi, 2014).

3 SAFETY OF MARITIME TRANSPORT IN CYBERSPACE

Recent years have brought considerable interest in the use of digital technologies, sophisticated IT and ICT solutions in maritime transport. This process is desirable because it is a factor driving the development of transport and increasing its efficiency. Despite many benefits resulting from the development and use of modern technologies in maritime transport, the number of threats to the emerging cyberactive environment is also growing. Maritime transport is a highly competitive industry, commercial data generated by companies is often the subject of attacks aimed at stealing and selling data to competitors. Ships are equipped with many devices and systems sensitive to attacks, an example of such a device is GPS, which is critical to the safety of the ship, the signal of which may be distorted or falsified, which translates into the accuracy of the received position, as a result, the ECDIS system does not receive the exact position, and the ship may encounter a navigational obstacle (Oruc, 2022; DNV, 2024; Lloyd's Register, 2024).

The maritime industry has special characteristics that make the cyber threat model significantly different from the model represented by land-based systems. Ships are sets of complex systems that create a "floating village" equipped with many ICT technologies, such as office systems, life support systems, engine automation and navigation systems. The life cycle of ships is on average 25-35 years, the software of the devices is updated at different time intervals, the result is a very diverse set of equipment between ships for administrative use and for information technology (IT) and operational technology (OT). Maritime transport is very cost-sensitive and high competitiveness results in insufficient attention being paid to the cybersecurity aspect.

3.1 *Types of threats to maritime transport cybersecurity*

Maritime transport is increasingly becoming dependent on the use of digital technologies, and its sensitivity to threats in the digital domain is increasing. Currently, all elements of maritime transport are exposed to threats related to cybercrime. Maritime vessels, port infrastructure, cargo, shipowners' offices, crew personal devices, drilling platforms, terminals and many other components of this industry are increasingly dependent on digital technologies. There are many motivations aimed at compromising the operational safety of transport, the most important of which are (Androjna & Twrdy, 2020):

- Cyber Abuse – low-level criminal activity including vandalism, disruption of systems, damage to websites or unauthorized access to systems.
- Activists' activities – In foreign literature they are referred to as "hacktivists". The group's activities are most often intended to seek publicity or to exert pressure to achieve a goal or support a cause, for example to prevent the handling of certain cargo or to disrupt the operation of a ship. The target of the attack may be the ship itself, the persons responsible for the ship or cargo, and the cargo itself.
- Espionage – unauthorized access to private information (intellectual property, commercial data, financial strategies, personal data). Espionage is most often commercial in nature and involves stealing or falsifying data for profit or on behalf of states.
- Organized crime – operating largely with a view to financial gain, this activity includes criminal damage, cargo theft, smuggling of goods and people, tax and excise tax evasion.
- Terrorism – unlawful action aimed at forcing the authorities to take or refrain from taking some action.
- Warfare – during a war, countries try to disrupt the opponent's logistics chains or neutralize specific ships, such as tankers and product carriers, supplying fuel to the country.
- Accidental/unconscious actions – actions of an employee who unknowingly affects the cybersecurity exposure of data or systems.

Threat actors can be divided into categories: individual actors, activist groups, competitors, cybercriminals, terrorists and state-sponsored entities. Each entity poses a real threat to remotely controlled ship systems, services provided by third parties, ship information collected on external servers and to the ship's supply chain.

Threats to maritime transport cybersecurity are not always planned and aimed at a specific system, device or person, and cybersecurity breaches may also occur through the actions of an employee, crew member or third party. Cyber threats can be planned and targeted at a given system or untargeted. Untargeted attacks use means most often available on the Internet to find the weak points of the system(s) to which they want to gain unauthorized access, while targeted attacks are more sophisticated and use "tools" specifically prepared to attack a specific ship or company. Untargeted attacks include (BIMCO, 2023):

- Malware – programmed to gain access to or harm a computer without the owner's knowledge. There are many types of malware, the most popular ones include: viruses, Trojans, ransomware (ransomware), spyware and worms. Malware can exploit deficiencies and problems in insecure or outdated software, such deficiencies or problems may constitute code errors, system weaknesses, improper design, hardware faults and/or protocol implementation errors. Malware can exploit weaknesses in the system and hardware remotely or be launched by the system user, e.g. malicious code may be placed in an email attachment or on a website.
- Water holing – creating a fake website or compromising the security of a real website in order to compromise the security of unsuspecting users.
- Scanning – searching through large portions of Internet information to find weaknesses that can be exploited.
- Typosquatting (URL Hijacking) – This method takes advantage of typos made by users when entering a website address into a search engine. If the user enters an incorrect address, he or she may be redirected to an alternative website, most often infected with malware.

Whereas targeted attacks represent:

- Social engineering - a non-technical type of attack aimed at breaking security procedures by manipulating people with access to confidential information. This attack may or may not be carried out via social media.
- Brute force – An attack aimed at trying as many passwords as possible with the intention of getting the right one. The attacker systematically checks all passwords until he guesses the right one.
- Credential stuffing – the use of previously obtained credentials or specific common passwords to gain unauthorized access to systems or applications.
- Denial of service DoS - prevents legal and authorized users from accessing information, most often by "clogging" the network with data. Distributed Denial of Service (DDoS) most often targets several computers or systems to launch a DoS attack.
- Phishing – an attack aimed at sending a large number of e-mail messages containing requests for individual pieces of information or confidential data. Emails may also contain malicious attachments or a request for the recipient to use a link in the message.
- Spear-Phishing – similarly to phishing, but the target is individual people who receive personalized e-mails containing attachments or links downloading malware. In some cases, Inmarsat C messages were used to reduce recipient suspicion.
- Supply chain subversion – attacking a company or ship by compromising the security of equipment, systems or ancillary services provided on the ship or to the company.

Attacks on maritime transport can be divided into those carried out directly on the ship and on land infrastructure (ports, companies), the difference is in the use of means to achieve the intended effect; in the case of ships, hackers aim to disrupt the operation of the ship or its systems, while in the case of land infrastructure they pay more attention to obtaining sensitive data or disrupting the operation of the entire port system in order to obtain a ransom in exchange for unlocking the devices.

Seaports are a frequent target of attacks because many operations are related to the processing of proprietary data (commercial data of companies). The factors that influence the vulnerability of ports to attacks include (Androjna & Twrdy, 2020):

- Low awareness and skills in relation to information and cybersecurity,
- Technical complexity of the port ecosystem,
- Lack of balance between business performance and cybersecurity,
- Using outdated and sensitive hardware and software,
- Lack of framework regulating the implementation of cybersecurity,
- Combination of information and operational technologies,
- Security risks in the supply chain (lack of certificate verification, remote access by suppliers),
- Heterogeneous networks/systems,
- Involvement of all stakeholders in security matters,
- Cybersecurity does not "keep up" with technological development, the emergence of new technologies influencing the increased digitization of the port.

The European Commission defines critical infrastructure as a resource or system that is necessary to maintain key social functions. Any disruption to the operation of critical infrastructure may have a direct impact on the security of the European Union or the well-being of its inhabitants. Ports are considered to be critical infrastructure, which in many cases are places where many modes of transport connect, as well as being a place of supply for the functioning of a large number of industries.

Cybersecurity is not entirely dependent on the technological level and development of devices and systems, in many cases the human factor has a significant impact on the effectiveness of the attack. The role of humans in the cyberspace zone is very large, because while working, they may unknowingly allow the system or device to be infected with malware or share company commercial data or other private information. Attack methods are becoming more and more sophisticated and are starting to be based on manipulation and empathy, which allows hackers to achieve their goals more easily. Risk factors related to people and their activities can be divided into user-related or management-related, user-related factors include: lack of motivation, beliefs, risky behavior, inappropriate use of technology, while the second group includes factors related to inappropriate staff or heavy workload (Kadena & Gupi, 2021). Based on (Meland et al., 2021), it can be concluded that at least 25% of recorded attacks in 2011-2020 are carried out using malware, viruses, ransomware or the newer ransomtheft, which, in addition to blocking the system, also steals data that is attached to e-mail messages as attachments. or links to infected websites. Most dangerous situations concern information technology (IT), but there are cases of violation of operational technology (OT) security, which occurs when an infected medium is connected to ship or office equipment or a computer is unknowingly connected to the wrong network.

Security in the cyberspace zone is a complex problem and requires constant attention due to new threats related to constant development.

3.2 *Security analysis of the maritime industry based on the example of the NotPetya attack*

Cyberspace attacks on maritime transport take various forms and have a greater or lesser impact on the functioning of the ecosystem, but there are cases in which a serious incident occurs and the value of millions of dollars or euros is lost. The most devastating cyberattack in the history of maritime transport was an attack using malware that pretended to be ransomware, but turned out to be wiper (from wipe - to wipe). Unlike ransomware, wiper is not programmed to decrypt the data it has encrypted. The software used for the attack was called NotPetya and was used for this purpose on June 27, 2017. The Maersk shipowner is responsible for approximately 20% of world transport, on the day of the attack it had a fleet of 800 sea vessels transporting millions of tons of cargo annually, and the company was also responsible for 76 ports around the world. NotPetya encoded system files and displayed a message requiring a deposit of $300 in Bitcoin to unlock the files.

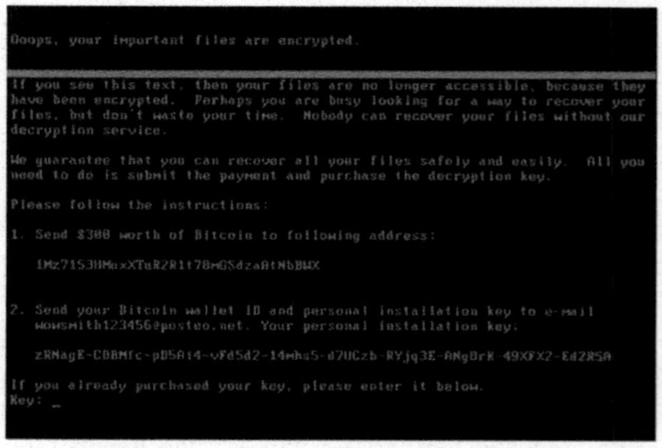

Figure 4. A message displayed on an infected computer. Source: website (Editor 2020).

The NotPetya malware initially seemed to experts to be closely related to previous types of ransomware. It combined the very strong encryption methods of previous "Petya" attacks with the very high infectiousness achieved by using the once-secret government-developed "Eternalblue" tool, which just a month before the NotPteya attack allowed the WannaCry malware to contaminate 200,000 computers in 150 countries around the world. Eternalblue is an exploit malware (a program designed to take advantage of bugs in the operating system). This malware exploited a vulnerability in the Microsoft Server Message Block 1.0 protocol and thus gained the ability to remotely execute the supplied code on the victim's computer. In 2016, Microsoft released a patch allowing protection against this software, but not all business owners had the latest versions of their systems, and in extreme cases their equipment was equipped with software that had already reached the end of its useful life (Walton, 2023). Additionally, the NotPetya malware used opensource software, i.e. software in which the source code is released under a license and allows the user to modify, research and distribute the software under a free software license. This software is called Mimikatz and is used to obtain passwords and PIN codes, it is a tool used when the device has already been attacked by a hacker. The combination of Eternalblue and Mimikatz allowed the bypass of a patch released by Microsoft, in a computer that did not have the patch, Mimikatz searched for passwords to gain access to computers that are protected.

The cyberattack using the NotPetya malware did not target Maersk, in fact the attack aimed to block Ukrainian computers running MEDoc, a program approved by the Ukrainian government for submitting reports to regulatory authorities and exchanging electronic documents between contractors. Hackers broke into the MEDoc application servers and forced an update of the application software on June 27, 2017, but instead of update files it contained malware. NotPetya began to spread quickly across Ukrainian computers, mainly banks.

Maersk's systems were accidentally infected because Maersk's Odessa branch was using MEDoc software, and within seven minutes the virus spread throughout the company's IT system and compromised all key business areas. The malware activity led to the destruction of 49,000 laptops and 1,000 applications, including the central website, and the immediate loss of access to 1,200 applications. In order to minimize the effects of the attack, the systems were disconnected from the global network and returned to manual systems, which resulted in a 20% reduction in turnover (Pownall, 2019). As a result of disconnection from the global network, port terminals stopped serving cars to collect cargo from ships because there was no information about the cargo transported in containers. The company regained full IT functionality only four weeks after the incident; financial losses as a result of the attack amounted to approximately $300 million. The attack did not damage any of the ship's systems, but due to terminal failures it was not possible to unload the ships.

The example of Maersk and the NotPetya attack shows that an attack can be very devastating even if a given entity is not the target of the attack. The relatively low losses are the result of quick action by those responsible for the systems, disconnecting them from the network and immediate efforts to restore the systems. Nowadays, there is no industry immune to threats in cyberspace, there is always a risk of security breaches, whether conscious or unconscious. Developing technology gives rise to new threats, the Internet of Things or Internet of Vessels gives rise to many opportunities, but by connecting many devices with various protocols and other security methods, it may be difficult to properly protect them against unauthorized access. The highest levels of autonomous ships (functioning without human intervention) also require the best security measures to avoid hackers taking over propulsion, control or navigation systems, the improper operation of which may lead to a threat to the environment, goods and even life or health.

In order to reduce the risk of serious consequences of an attack, it is important to train employees and increase their awareness of cybersecurity, because a human is responsible for a cybersecurity breach in 95% of cases (Worth, 2023; Bacasdoon & Bolmsten 2022). Unfortunately, many attack methods use manipulation techniques, appealing to empathy or authorities, and thus a person may unknowingly expose the entire system to attack. An additional problem is the low level of security - passwords are often visible in public places, in the case of ships they are printed and laminated and are located near computers. Despite the nature of the bridge as a place reserved for unauthorized persons, access to it is not protected and security rules may be broken and thus ship systems may be infected or data may be stolen.

4 CONCLUSIONS

This work discusses the issue of the development of ICT networks in maritime transport applications; the examples cited in the text above indicate great interest in the development of ICT solutions in maritime transport.

The literature analysis showed specific plans for implementing solutions using the latest technologies, as well as the benefits resulting from their use.

Without a doubt, digitization and digitalization are approaching maritime transport, but the dynamics with which it is progressing is not as great as in other industries. Maritime transport is a sensitive industry using many different systems and it is not possible to introduce modern solutions in all areas at once. The benefits of using technologies based on ICT solutions and the assumptions of the Industry 4.0 concept presented in the above work are promising for the maritime industry, which is still based on the use of outdated equipment and work methods with a low degree of automation.

Many of the modern areas of interest in the use of ICT networks in maritime transport are based on the use of reliable telecommunications systems capable of transmitting data at high speed and over the greatest possible distances. The telecommunications systems described in this work do not meet all the requirements for communication, the main assumptions of an ideal system are global coverage, low price, ease of use and high transmission speed, many solutions related to the use of ICT networks assume the use of systems that meet all the above-mentioned requirements, which in nowadays it is not possible. The development of autonomous ships or the Internet of Maritime Things, which require constant connection and contact with interested persons and entities, depends on the further development of the systems described.

With the increasing importance of digital solutions in maritime transport, its sensitivity to threats in the cyberspace zone is also increasing. Systems in maritime transport are not properly protected and their operation can be easily disrupted, which is unacceptable in the case of systems ensuring safety for health, life and the environment. It is necessary to update software and security systems for the systems used by companies in the maritime industry, as well as education to increase employee awareness to minimize the risk.

Considerable interest in increasing the share of digital technologies in maritime transport will not be met with greater interest in the issue of cybersecurity, which is a critical issue in the modern digital ecosystem. Using the example of the devastating NotPetya attack on Maersk, the author presented the paralyzing effects of security breaches.

The use of ICT networks and related solutions in maritime transport is not a novel thing; despite the increasing interest in this topic in recent years, the dynamics of their implementation is effectively stopped by the lack of legal bases containing requirements for systems and modern technologies, as well as technological and social barriers. ICT networks offer great benefits from their use in maritime transport, but they also come with risks related to cyberspace, attacks, excessive reliance on technology and inappropriate use of the system. There is still a long way to go before digitization and the introduction of modern technologies into maritime transport, legal, technological and social problems slow down this process, but the use of these solutions is inevitable.

ACKNOWLEDGEMENTS

This study was funded by the Gdynia Maritime University, under research project WN/2024/PZ/07.

REFERENCES

Aiello, G., A. Giallanza, and G. Mascarella. 2020. Towards Shipping 4.0. A Preliminary Gap Analysis. *International Conference on Industry 4.0 and Smart Manufacturing (ISM 2019)* 42 (January 1): 24–29.

Alqurashi, F.S., A. Trichili, N. Saeed, B.S. Ooi, and M.-S. Alouini. 2023. Maritime Communications: A Survey on Enabling Technologies, Opportunities, and Challenges. *IEEE Internet of Things Journal* 10, no. 4 (February 15): 3525–3547.

Androjna, A., and E. Twrdy. 2020. Cyber Threats to Maritime Critical Infrastructure. *Cyber Terrorism and Extremism as Threat to Critical Infrastructure Protection.*

Anon. 2023. Maretron. *Maretron*. https://www.maretron.com/.

Bacasdoon, J., and J. Bolmsten. 2022. A Multiple Case Study of METI Cybersecurity Education and Training: A Basis for the Development of a Guiding Framework for Educational Approaches. *TransNav, the International Journal on Marine Navigation and Safety of Sea Transportation* 16, no. 2: 319–334.

BIMCO. 2023. The Guidelines on Cyber Security Onboard Ships. https://www.bimco.org/about-us-and-our-members/publications/the-guidelines-on-cyber-security-onboard-ships.

Bryson, S. 2013. Virtual Reality: A Definition History - A Personal Essay. *CoRR* abs/1312.4322. http://arxiv.org/abs/1312.4322.

Cichocki, R. 2023. State-Sponsored and Organized Crime Threats to Maritime Transportation Systems in the Context of the Attack on Ukraine. *TransNav, the International Journal on Marine Navigation and Safety of Sea Transportation* 17, no. 3: 717–721.

DNV. 2024. Cyber Secure Class Notation. Accessed January 29. https://www.dnv.com/services/cyber-secure-class notation-124600.

Donepudi, P.K. 2014. Technology Growth in Shipping Industry: An Overview. *American Journal of Trade and Policy* 1, no. 3 (December): 137–142.

Editor, S. 2020. Artificial Intelligence and the Era of Autonomous Shipping - Sea News. https://seanews.co.uk/shipping-news/artificial-intelligence-and-the-era-of-autonomous-shipping/.

Erstad, E., R. Hopcraft, J.D. Palbar, and K. Tam. 2023. CERP: A Maritime Cyber Risk Decision Making Tool. *TransNav, the International Journal on Marine Navigation and Safety of Sea Transportation* 17, no. 2: 269–279.

Filippopoulos, I., and G. Stamoulis. 2017. Collecting and Using Vessel's Live Data from on Board Equipment Using "Internet of Vessels (IoV) Platform" (May 2017). In *2017 South Eastern European Design Automation, Computer Engineering, Computer Networks and Social Media Conference (SEEDA-CECNSM)*, 1–6.

Hudański, P. 2015. Projektowanie sieci teleinformatycznych. *Problemy Nauk Stosowanych* 3: 33–42.

Kadena, E., and M. Gupi. 2021. Human Factors in Cybersecurity: Risks and Impacts. *Security Science Journal* 2, no. 2 (December 19): 51–64.

Laskowski, N. 2023. What Is Artificial Intelligence and How Does AI Work? | Definition from TechTarget. *Enterprise AI*. https://www.techtarget.com/searchenterpriseai/definition/AI-Artificial-Intelligence.

Lloyd's Register. 2024. Cyber Safe for Marine. Accessed January 29. https://www.lr.org/en/services/classification-certification/cyber-resilience/cyber-safe-for-marine/.

Meland, P.H., K. Bernsmed, E. Wille, Ø.J. Rødseth, and D. A. Nesheim. 2021. A Retrospective Analysis of Maritime Cyber Security Incidents. *TransNav, the International Journal on Marine Navigation and Safety of Sea Transportation* 15, no. 3: 519–530.

Melnyk, O., S. Onyshchenko, O. Onishchenko, O. Shumylo, A. Voloshyn, Y. Koskina, and Y. Volianska. 2022. Review of Ship Information Security Risks and Safety of Maritime Transportation Issues. *TransNav, the International Journal on Marine Navigation and Safety of Sea Transportation* 16, no. 4: 717–722.

Mirović, M., M. Miličević, and I. Obradović. 2018. Big Data in the Maritime Industry. *Naše More*: 56–62.

Mitchell, T.M. 1997. *Machine Learning*. 1st edition. New York: McGraw-Hill Education.

Neumann, T. 2018. The Importance of Telematics in the Transport System. *TransNav, the International Journal on Marine Navigation and Safety of Sea Transportation* 12, no. 3: 617–623.

OECD. 2016. *The Ocean Economy in 2030*. https://www.oecd-ilibrary.org/content/publication/9789264251724-en.

Onishchenko, O., K. Shumilova, S. Volyanskyy, Y. Volyanskaya, and Y. Volianskyi. 2022. Ensuring Cyber Resilience of Ship Information Systems. *TransNav, the International Journal on Marine Navigation and Safety of Sea Transportation* 16, no. 1: 43–50.

Oruc, A. 2022. Ethical Considerations in Maritime Cybersecurity Research. *TransNav, the International Journal on Marine Navigation and Safety of Sea Transportation* 16, no. 2: 309–318.

Patel, K.K., S.M. Patel, and P. Scholar. 2016. Internet of Things-IOT: Definition, Characteristics, Architecture, Enabling Technologies, Application & Future Challenges. *International Journal of Engineering Science and Computing*: 6122–3131.

Pownall, C. 2019. *The Context and Impact of Maerk's NotPetya Cyber Attack*.

UNCTAD. 2022. Review of Maritime Transport. *Geneva: United Nations*.

Walton, H. 2023. The Maersk Cyber Attack - How Malware Can Hit Companies of All Sizes. https://www.kordia.co.nz/news-and-views/the-maersk-cyber-attack.

Worth. 2023. Cybersecurity Statistics & Cybercrime Facts. https://www.worthinsurance.com/post/cybersecurity-statistics.

Zych, J. 2018. *Teleinformatyka dla bezpieczeństwa*. Fundacja na rzecz czystej energii. https://politechnik.poznan.pl/ksiegarnia/teleinformatyka-dla-bezpieczenstwa/.

Advances in Maritime Technology and Engineering – Guedes Soares & Santos (Eds)

Characterizing offshore supply operations in the North Sea

J.F. Oliveira & T.A. Santos
Centre for Marine Technology and Ocean Engineering (CENTEC), Instituto Superior Técnico, Universidade de Lisboa, Lisbon, Portugal

R.C. Botter
University of São Paulo, São Paulo, Brazil

ABSTRACT: This paper presents a survey of offshore supply operations in the oil and gas industry in the North Sea, aiming to improve Brazilian operations in the same sector. It comprises an assessment of the physical locations and characteristics of supply bases in the Norwegian, Dutch and United Kingdom sectors of the North Sea and Norwegian Sea. The main characteristics of the supply bases are identified and the driving parameters in planning their location are identified. The fleet of supply vessels operating in the various sectors of this area is characterized as to type, main dimensions and cargo capacity. The patterns of operation of the vessels are identified, characterized and grouped in typical profiles. Finally, the operations of specific vessels are analyzed in detail for a period of a month. Conclusions are drawn regarding the bases, the fleet and its operations, including a comparison between the different sectors.

1 INTRODUCTION

The initiation of a study on offshore support vessel operations in Brazil stems from the need to enhance the efficiency of such activities. Despite Brazil's oil production being roughly twice that of the United States and Norway, the country still concentrates its operations in a single base in the city of Macaé, Rio de Janeiro. Brazil faces challenges such as limited pipeline infrastructure for oil and gas transport, allocating two vessels for cargo logistics to each platform, that is four times more than Norway.

To improve the efficiency of Brazilian operations, it is crucial to understand how these activities are conducted in other exploration regions, such as the North Sea and the Gulf of Mexico, known for their operational effectiveness. This paper will focus mainly on the study of North Sea operations.

The North Sea is a sea within the Atlantic Ocean situated in the northern part of the European continent, bordered by the coasts of Norway and Denmark to the east and the British Isles, France, Germany, the Netherlands and Belgium to the south. Among the primary economic activities in this sea, the extraction of oil and natural gas stands out. Fishing and maritime transport have always been crucial activities in the region and, more recently, offshore wind turbine farms have been installed.

In this area, the oil and natural gas extraction industry operates not far from the coast, utilizing platforms anchored to the seabed, floating platforms or FPSOs (Floating Production, Storage, and Offloading vessels). Furthermore, unlike Brazil, where wells can reach water depths exceeding 2,000 meters, the majority of wells in the North Sea are situated in waters with depths of up to 200 meters.

The oil and gas industry holds significant economic importance for Brazil, the United Kingdom, Norway, the Netherlands, and Denmark. To grasp the relevance of this industry in each country, Figure 1 offers a comprehensive overview of the oil and gas landscape among these nations.

	Proved oil reserves (billion barrels)	Extracted oil (billion barrels per year)	Proved gas reserves (tera cubic feet)	Extracted gas (tera cubic feet per year)
United Kingdom	2.50	0.40	6.40	1.40
Norway	8.10	0.73	54.50	3.90
Netherlands	0.14	0.03	4.70	1.10
Denmark	0.44	0.03	1.00	0.05
Brazil	12.70	1.39	12.80	0.89

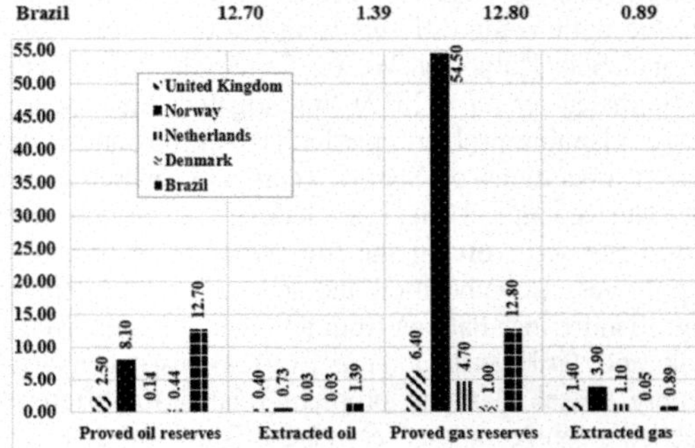

Figure 1. Proved reserves and extraction data in 2021. (Source: AENERT (2023)).

DOI: 10.1201/9781003508762-10

Figure 1 illustrates that Brazil and Norway possess the most extensive oil and gas reserves among the featured nations. Regarding production, the United Kingdom, Norway and Brazil are actively extracting substantial quantities of oil and gas, while the Netherlands and Denmark exhibit more modest reserves and production in comparison to their counterparts. Additionally, the United Kingdom displays relatively high gas production concerning its reserves; conversely, Norway maintains a more favorable balance between gas reserves and production. Brazil's oil production is remarkable, even though its gas production is relatively smaller in comparison to its substantial reserves. Considering the potential of Brazil in the oil and gas industry it is important indeed to improve the efficiency of its supply operations, and this is the motivation for this paper.

The remainder of this paper is organized as follows. Section 2 provides a literature review of studies and papers covering offshore supply operations, the infrastructure, types of ships and operational patterns. Section 3 shows the offshore supply bases identified across different European countries surrounding the North Sea and their technical characteristics. Section 4 contains an analysis of a representative sub-set of the offshore supply vessel fleet used in the North Sea. Section 5 shows the different types of operation regarding supply of offshore platforms, proposing a categorization of these operations in relevant types. Section 6 examines in detail the operations of a number of supply vessels throughout a period of time in the North Sea. Finally, section 7 summarizes the main conclusions of this study.

2 LITERATURE REVIEW

The literature on describing offshore supply operations conducted in the North Sea is not very extensive and the majority of it is focused on enhancing supply vessel planning. Fagerholt and Lindstadb (1999) studied the optimization of supply operations in the Norwegian Sea. Halvorsen-Weare, Fagerholt, Nonås and Asbjørnslett *et al.* (2012) investigated the supply vessel planning problem, particularly in the context of energy operations of Statoil - company operating in the Norwegian oil and gas industry exploring the North Sea -, involving the improvement of offshore supply vessel composition and scheduling for supplying offshore installations from an onshore depot. Maisiuk and Gribkovskaia *et al.* (2014) also analyzed supply vessel planning problem related to different fleet size configurations servicing offshore oil and gas installations. Esteves and Branski (2021) had the objective of mapping logistical practices – regarding supply bases, vessel fleets and patterns of operation - in offshore oil exploration in Brazil, Norway and the United States in the Gulf of Mexico, analyzing the main operating companies in each region.

Esteves and Branski (2021) state that Norwegian exploration relies on the geographic proximity of support bases that encourage collaboration between companies, favoring the exchange of information, economies of scale and innovation. Such operations involve sending materials and equipment from platforms to bases on land, with the aim of keeping these resources at the bases for as long as possible before being sent to platforms when necessary (just in time). Additionally, in Norway, Statoil oversees maritime activities, including routing (used to rationalize and optimize platform supply), loading and unloading of ships. They also argue that, in Brazil, Petrobrás uses 190 Platform Supply Vessels (PSVs) to transport cargo, transporting approximately 210 thousand tons of cargo per month, equivalent to 42 thousand items. The PSV journey begins and ends at the Macaé supply base, there are fixed departure times every day for a specific period, the maximum planning period is one week, each departure is assigned a ship and only one platform can be visited during the trip.

According to Schipper (2017), in the Netherlands, the Peterson Offshore Group acts as a logistics facilitator for the Southern North Sea Pool (SNS Pool), which means strategic management of the supply chain. This pool presents impressive data: with approximately 40 offshore locations visited as part of operations, 5 to 7 vessels are employed, which are dedicated to supporting production. Furthermore, in these operations, a notable number of 290 loading/unloading movements (lifting to the dock) are carried out daily, reflecting the intense movement of materials and equipment. Furthermore, vessels are generally assigned to a set of platforms, round-trip journeys lasting a maximum of 4 days – because of the fresh food – and depending on requests for this purpose. day and location of each of them. For this planning, Peterson developed the Lighthouse software, which centralizes the logistics operation. It is through this software that operators place orders, determine the position of cargo on the ship, how the vessel is loading and unloading, among other activities. Lighthouse allows Peterson and its customers to communicate effectively, ensuring everyone stays up to date on operational developments in real time.

Figure 2 offers in-sights into the frequency of trips taken within a three-month duration through the SNS Pool. This chart pro-vides a snapshot of the transportation dynamics within this specific timeframe, shedding light on the patterns and distribution of trips facilitated by the SNS Pool.

3 SUPPLY BASES IN OFFSHORE OIL AND GAS LOGISTICS IN THE NORTH SEA

In the North Sea, the oil and gas industry of Norway and the United Kingdom takes center stage, although other countries such as the Netherlands and Denmark also have a presence. Concerning supply bases, their location adheres to a criterion of proximity to platforms and FPSOs, aiming to minimize the distances covered by logistics support vessels and, consequently, travel costs.

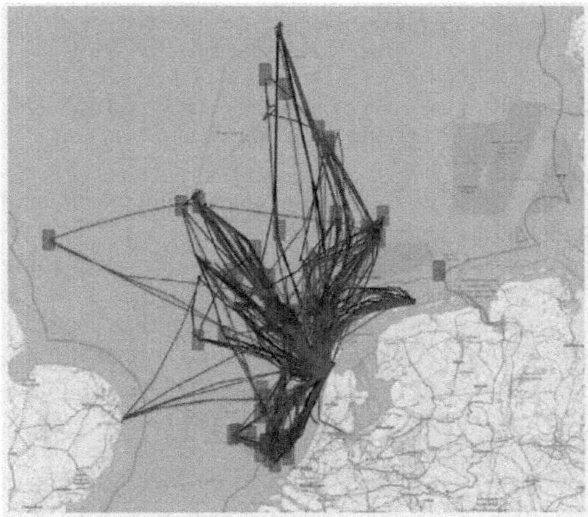

Figure 2. Trips taken within a 90-day period via the SNS Pool (Source: Schipper (2017)).

The case study of Norway and its three exploration zones (North Sea, Norwegian Sea and Barents Sea) reveals that the distances from supply bases to platforms are generally around 250 km. In addition, the Norwegian coastline from Farsund (southern extremity) to Sandnessjoen (northern extremity of the Norwegian Sea sector) spans approximately 1080 km with a total of 14 supply bases, depicted in purple in Figure 3.

The UK industry concentrates the majority of support services for oil and gas extraction wells around the cities of Lerwick, Peterhead, Aberdeen and Great Yarmouth, illustrated with a blue mark

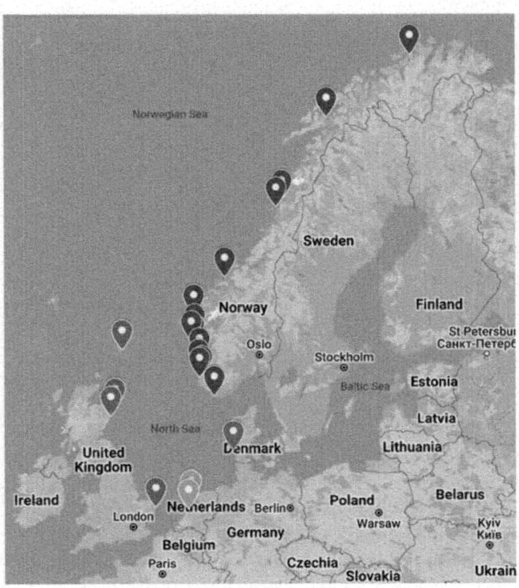

Figure 3. Supply bases in the North Sea.

in Figure 3. The supply bases clustered around Lerwick and Peterhead have a distance to the blocks of less than 250 km; The bases around Great Yarmouth are closer, less than 150 km away.

The oil and gas industry in the Netherlands is concentrated around the cities of Den Helder, Ijmuiden and Rotterdam, not far from natural gas (and some oil) extraction wells. These cities and bases contain support services and related industries. The distance between supply bases on the Dutch coast and most gas fields is mostly less than 150 km.

Denmark's oil and gas industry's main support point is the base located in Esbjerg. The distance between the supply base on the Danish coast and the blocks is approximately 300 km. Therefore, Denmark is among the countries that explore the North Sea that have greater distances between the coast and blocks.

In contrast, in Brazil, as shown in Figure 4, the Santos Basin exhibits distances between the coast and platforms reaching up to 350 km, while the Campos Basin is closer to the coastline. In all sectors of the North Sea, support bases are strategically situated on the coasts closest to the oil fields, with no trend toward centralizing operations in one large base. Conversely, along the Brazilian coast, the distance between the northern border of Paraná to the southern border of Espírito Santo is approximately 950 km, a mere 130 km difference from Norway. However, in Brazil, logistics support activities along this coast seem concentrated in a few ports.

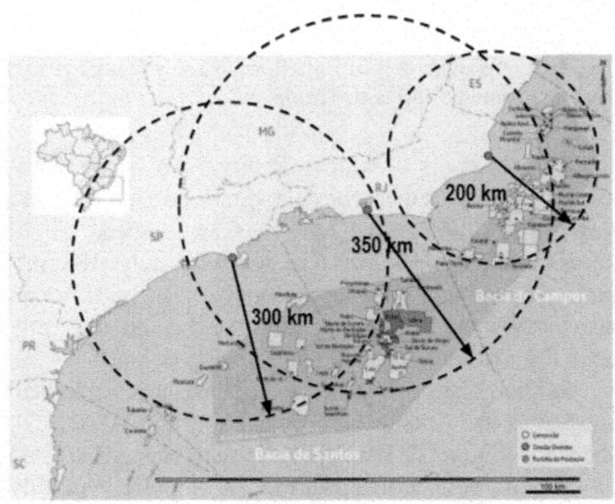

Figure 4. Map of the Campos and Santos basin and distances to the Brazilian coast.

4 SUPPLY VESSEL FLEET IN THE NORTH SEA

Offshore exploration and production operations are supported by a logistics and service system that requires a wide variety of specialized vessels, helicopters, ports, airports, warehouses, trucks and many other components. Regarding specialized vessels, the following stand out as the most used for logistics activities: Platform Supply Vessels (PSVs) and Utility Vessels (UTs).

However, Standby Safety and Multi-Purpose Offshore Vessels are also widely used. The first comprises vessels designed to provide emergency support and response to critical situations; they can be equipped with firefighting and rescue equipment and trained personnel to deal with emergencies. On the other hand, Multi-Purpose Offshore Vessels are versatile vessels that can perform various functions in the offshore industry, such as transporting general cargo, transporting bulk liquids, support underwater dives, and even firefighting operations. The ability to perform multiple functions makes these vessels adaptable to a wide range of offshore operations, providing flexibility and efficiency in operations.

Platform Supply Vessels are specially designed for loading and unloading operations at sea and can transport deck cargo, water, diesel oil, fluids and solid bulk. Furthermore, PSVs – which were represented in Figure 5 – can also be used as rescue and stand by vessels, the service speed is around 10 knots, the deck area can exceed 1000 m^2 and they have excellent maneuverability.

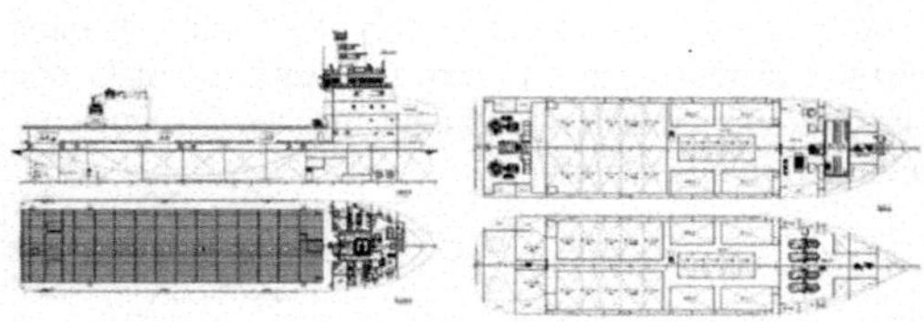

Figure 5. Arrangement of a Platform Supply Vessel (PSV) (Source: Damen Shipyards Group).

Utility Vessels – shown in Figure 6 – are commonly used for emergency service loads, as they are faster than PSVs, they have a service speed of 20 knots and the deck area is approximately 183 m^2, much smaller than Platform Supply Vessels. UTs can also carry water and diesel, although they are currently used for smaller and emergency transport loads.

Another type of vessel is the offshore logistical support is the Anchor Handling Tug Supply (AHTS), shown in Figure 7. This operates with towage, related to anchors and transportation of supplies, mainly pipes, fresh water, oil, mud, brine, cement and parts. AHTS can also be adapted to transport crew or even

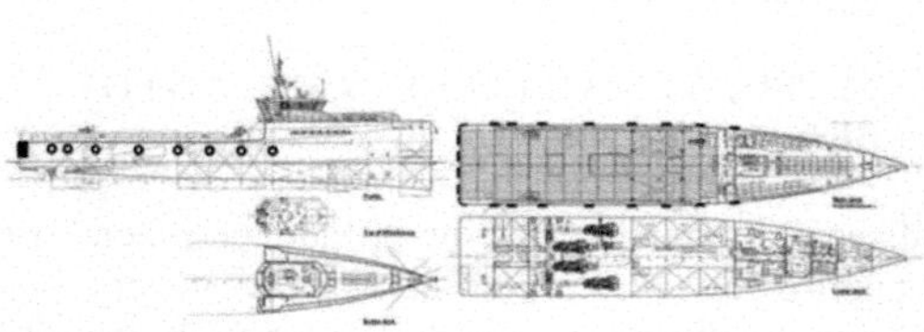

Figure 6. Arrangement of a Utility Vessel (UT) (Source: Damen Shipyards Group).

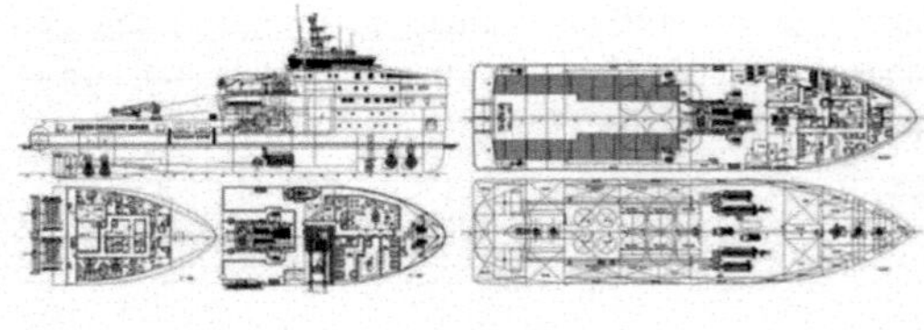

Figure 7. Arrangement of an Anchor Handling Tug Supply (AHTS) (Source: Damen Shipyards Group).

equipped with first-class firefighting, oil recovery and sound positioning facilities.

To understand how the fleet of ships participating in offshore logistics operations in the North Sea is composed, ships – from the categories: Standby Safety Vessel, Offshore Supply Ship, Anchor Handling Vessel, Multi-Purpose Offshore Vessel, and Utility Vessel – that docked or undocked at the studied terminals in Norway, Netherlands, Denmark and the United Kingdom were recorded on March 6, 8, and 10, 2023. In this record, 125 vessels were listed and distributed according to Table 1 by country and type. The characteristics of the vessels were recorded in a database.

It is important to highlight that the term "Off-shore Supply Vessel" was used in this analysis due to the nomenclature used on the Marine Traffic website, source of information. This term covers several vessels, including Platform Supply Vessels (PSVs), which provide supplies and services to offshore installations.

Table 1. Number of ships by country and type.

Vessel type	Denmark	United Kingdom	Netherlands	Norway
Standby Safety Vessel	1	9	5	0
Offshore Supply Vessel	7	40	12	22
Anchor Handling Vessel	0	1	1	7
Utility Vessel	0	1	2	0
Multi-Purpose Offshore Vessel	0	8	1	8
Total	8	59	21	37

In order to characterize this fleet of vessels, graphs of beam, summer deadweight, gross tonnage, year of construction and deck area, all in function of the overall length, were plotted, as shown in Figures 8 to 12.

In Figure 8, it can be observed that the beam mostly exhibits linear growth as the overall length of the ship increases. About the summer deadweight (Figure 9),

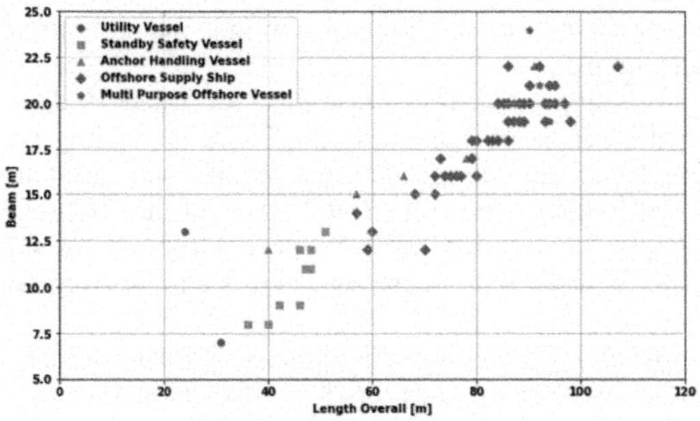

Figure 8. Beam vs. Overall Length of the ships in the database.

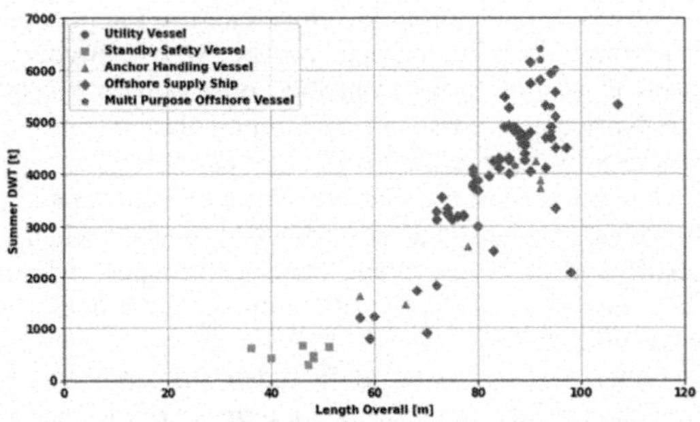

Figure 9. Summer Deadweight vs. Overall Length of the ships in the database.

the data concentration occurs between 70 and 100 m in length, which corresponds to the range of 3 to 6 thousand tons. Analyzing figures above, Multi-Purpose Offshore Vessel and Offshore Supply Ship categories have the largest dimensions in this sample.

Regarding the year of built (Figure 10), Offshore Supply Ships are the ones with the most recent year of construction (2014), while the Utility Vessel has the lowest value (1968). Figure 11 relates gross tonnage and overall length and is better represented by an exponential trend line. Additionally, in all diagrams, there is an accumulation of points between 80 and 100 meters in length, which corresponds to the range of 15 to 25 meters in beam and 3 to 6 thousand.

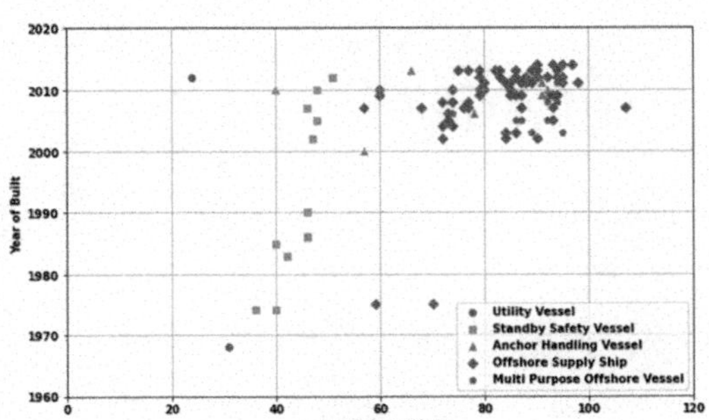

Figure 10. Year of Built vs. Overall Length of the ships in the database.

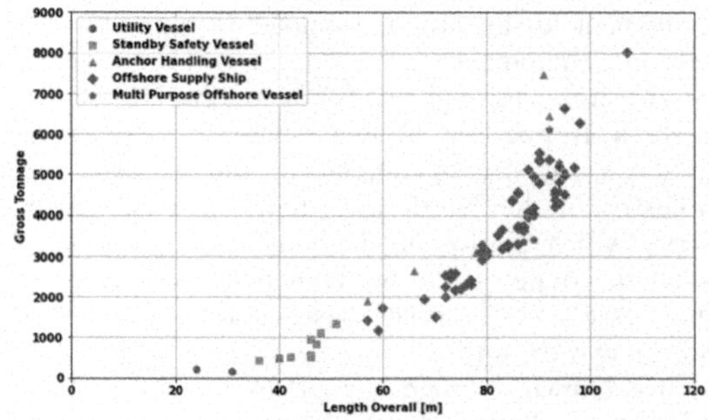

Figure 11. Gross Tonnage vs. Overall Length of the ships in the database.

The last graph, in Figure 12, shows the deck area of the different ships under investigation, and it is important to highlight that Standby Safety Vessels were not included in this analysis as the deck area is less important in this type of ship. In Figure 12, it is notable that Multi-Purpose Offshore Vessel and Utility Vessel stand out respectively as the largest and smallest average deck areas. Most ships have a deck area between 600 and 1100 m^2.

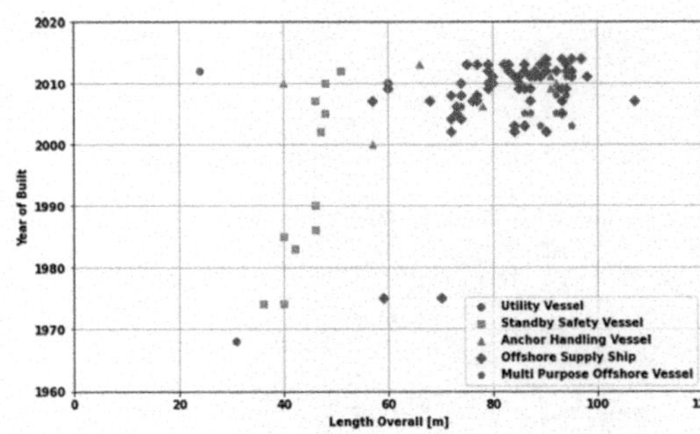

Figure 12. Year of Built vs. Overall Length of the ships in the database.

5 TYPICAL PATTERNS OF OPERATION IN OFFSHORE OIL AND GAS LOGISTICS

The pattern of offshore support operations in the oil and gas industry in the North Sea is characterized by a highly specialized and efficient approach. In this region, where there is intense oil and gas exploration and production activity, offshore logistics plays a crucial role in supporting operations.

To achieve this, the North Sea has a well-developed support infrastructure, as shown in section 3. There is a network of ports strategically located to meet the logistical needs of the offshore industry. These ports have modern facilities and specialized equipment for loading and unloading supplies, vessels, and equipment.

Therefore, through the study of the Norwegian case, it is possible to understand the pattern of offshore support operations in the oil and gas industry in the North Sea, that is: which bases serve which platforms, the most common routes, what the route is like between platforms, whether the vessels go to other bases and whether they are known or not.

It is known, through a report by the Government of Norway (2000) which indicates the allocation of several oil fields to the various logistical support bases in Norway, that Tananger supplies the platforms in the southernmost, Dusavik supplies the fields in front from Stavanger. Mongstad, Agotnes and Floro supply the fields north of Bergen and Kristiansund and Sandnessjoen the Norwegian Sea fields. Figure 13 confirms an essentially geographic allocation criterion for determining the location of the bases.

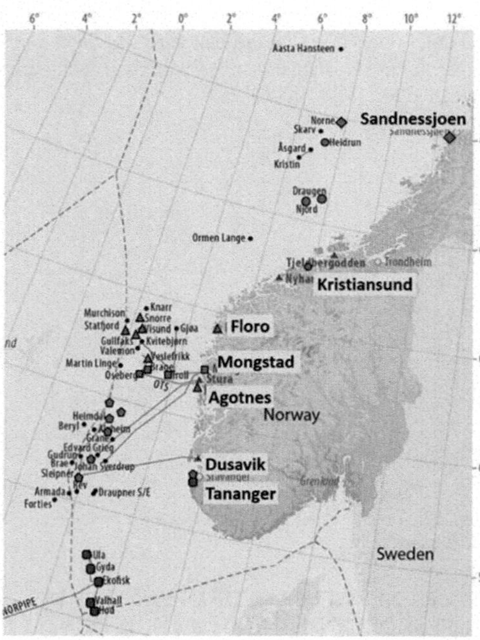

Figure 13. Allocation of oil fields to Norwegian logistical support bases (Source: Government of Norway (2000)).

In this context, Dusavik, Tananger, Floro, Agotnes and Mongstad were the bases analyzed in the North Sea. These ports are in southwestern Norway, which is an important center for offshore activities in the country. For the Norwegian Sea, the port of Kristiansund, located on the west coast of the country, was chosen. Finally, for the Barents Sea, in northern Norway, the Hammerfest base was chosen, located on the island of Kvaløya and known for being an important shipping hub and for its natural gas industry.

It is important to mention that, for 7 and 10 days in the period of June and July 2023, depending on the base, the Marine Traffic website was used to identify the routes of the last 30 days of the Offshore Supply Vessel and Multi-Purpose Offshore Vessel type vessels that departed from such Norwegian ports.

To understand the predominance of operating patterns in Norway, categories were established that delineate the identified route types. These categories cover different aspects, providing a comprehensive view of the logistics involved. Regarding route types, three main classifications stand out. Linear routes, characterized by ships that do not visit other terminals or that stop in locations close to the port of departure. Multidirectional routes, which involve stops at terminals far from the home port or along the Norwegian coast. Finally, international routes, where ships visit terminals in other countries.

The position of the platforms was also considered, being categorized as dispersed, when they are widely distributed in the coverage zone, or grouped, when they are close to each other and concentrated in a specific region.

Other factors analyzed included the number of bases visited, which reflects the number of points on the coast visited by the ship, also considering the point of origin. Furthermore, the number of bases used to support the platforms was examined, highlighting the offshore support terminals among the scaled bases.

Figures 14 and 15 depict two distinct cases. The first one depicts the *Eldborg* vessel from Dusavik sailing in a multidirectional route over the North Sea, stopping at distributed platforms and using 5 supply bases. In the second case, *North Purpose* departs from Tananger follows a linear route across the North Sea that is distinguished by grouped platforms with only one base utilized for support.

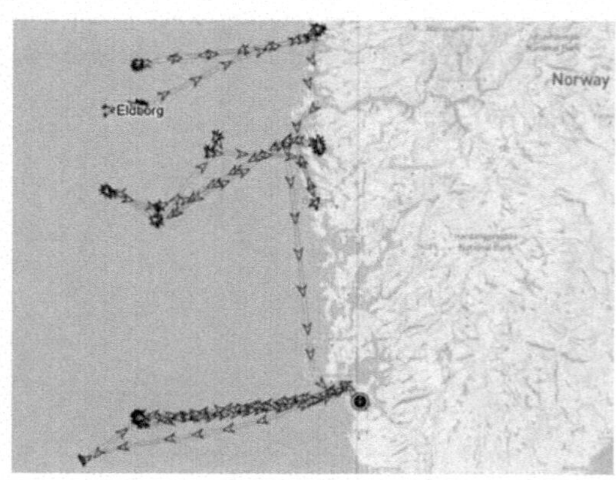

Figure 14. Multidirectional route of the Eldborg ship visiting scattered platforms in the North Sea and departing from Dusavik.

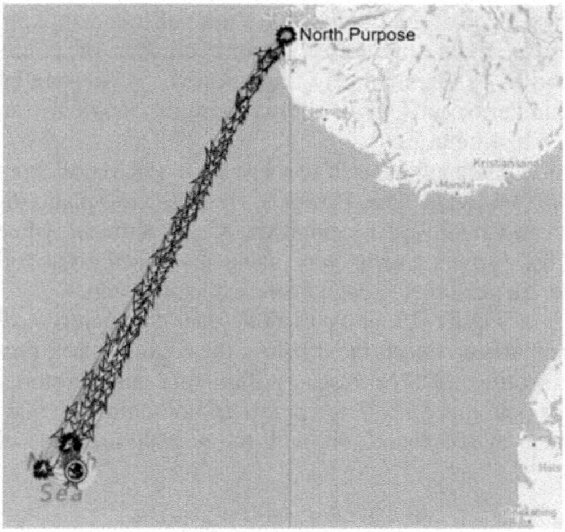

Figure 15. Linear route of the North Purpose vessel visiting grouped platforms in the North Sea and departing from Tananger.

Table 2 presents data on the distribution of ships across various origin bases, categorized by the type of route they follow. In the overall context, the dataset comprises a total of 65 ships. The majority of them are engaged in multidirectional and linear routes, with 31 vessels each. The predominant involvement in multidirectional routes suggests a diversified approach to destination selection among the ships. Notably, Tananger stands out for its substantial number of ships on linear routes.

Table 2. Segmentation of the routes of ships departing from Norwegian bases.

Origin base	Total number of ships	Type of route: Multidir-ectional	Linear	International
Dusavik	15	7	8	-
Tananger	16	3	12	1
Floro	4	3	1	-
Agotnes	9	6	2	1
Mongstad	13	6	6	1
Kristiansund	7	6	1	-
Hammerfest	1	-	1	-
Total	65	31	31	3

Table 3 presents the position of associated platforms, which the majority are scattered (43). The prevalence of scattered platforms indicates a preference for dispersing activities across various location. Tananger also stands out for a more concentrated distribution of position of platforms. Besides that, Table 3 indicates that Agotnes demonstrates a more dispersed operational footprint (with 3.4 and 2.9 as respectively the average number of bases called and utilized for supporting platforms across different origin bases), while Tananger and Hammerfest exhibit a more focused operational strategy.

Table 3. Segmentation of platforms visited by ships departing from Norwegian bases.

Origin base	Position of platforms: Scatterd	Grouped	Average number of bases scaled	Average number of bases used to support platforms
Dusavik	9	6	2,6	2,1
Tananger	6	10	1,7	1,1
Floro	4	0	2,3	2,0
Agotnes	9	0	3,4	2,9
Mongstad	9	4	2,5	1,4
Kristiansund	6	1	2,1	1,6
Hammerfest	0	1	1,0	1,0
Total	43	22	-	-

In addition, Table 4 offers a overview of coverage areas. It is worth mentioning that the North Sea stands out as the area that concentrates the largest number of routes, highlighting its importance in this operational context.

Table 4. Classification of the areas covered by ships departing from Norwegian bases.

Origin base	Areas covered: North Sea	Norwegian Sea	Barents Sea
Dusavik	15	2	-
Tananger	15	1	-
Floro	4	1	-
Agotnes	9	6	-
Mongstad	13	3	-
Kristiansund	5	7	-
Hammerfest	-	-	1

6 CASE STUDY: FEATURES OF SPECIFIC SUPPLY VESSEL OPERATIONS

A more detailed analysis of the routes of a number of ships was carried out for four ships using data from Marine Traffic. The ship's operations were recorded for a period of 30 days. The ships are the Norwegian sector's *North Purpose* and *Eldborg* (Figures 14 and 15 respectively) and Dutch's sector *Vos Base* and *Pool Express* (Figure 16) vessels. The location of the bases, arrival and departure times, time on platform and average round trip speeds were analyzed.

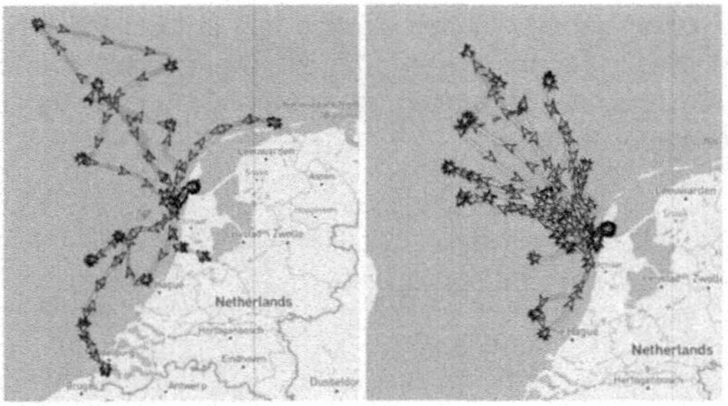

Figure 16. Routes of the vessels Pool Express (left) and Vos Base (right) departing from Den Helder.

Firstly, the Norwegian sector ships operational patterns will be studied. *North Purpose* begins its journey in the port of Tananger and follows a linear route, that is, this ship does not visit any terminal other than the port of origin; on the other hand, Eldborg departs from the port of Dusavik and follows a multidirectional route, the vessel makes stops at terminals far from the port of origin along the Norwegian coast.

Figure 17 shows the presence of 2 groups: one above and one below 2.5h, with 65% of the platforms having a time of less than 2.5h. Furthermore, a platform stands out for having the longest call time, 4h 55min. In Figure 18, the Eldborg vessel appears to follow the pattern seen previously. In this case, 60% of the points have a time of less than 2.5h. Furthermore, there is a greater number of points above 4 hours of stay. Additionally, a substantial number of calls are of shorter duration, ranging from 0.5 to 1 hour.

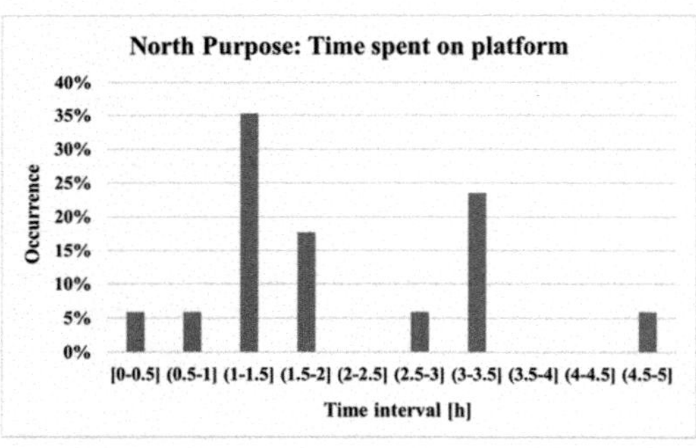

Figure 17. Histogram of the time spent on the platforms of the North Purpose vessel.

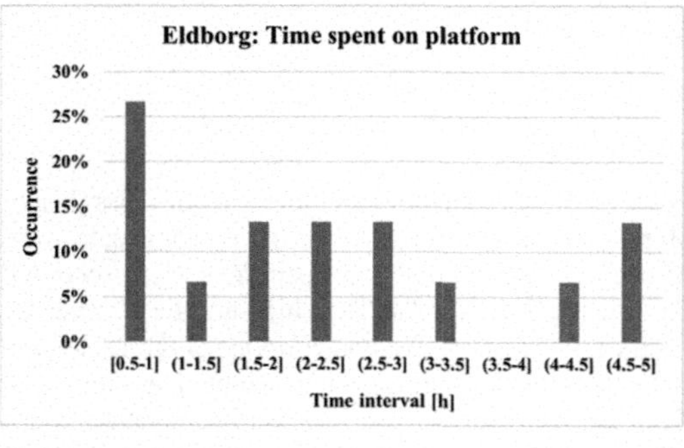

Figure 18. Histogram of the time spent on the platforms of the Eldborg vessel.

Regarding the Dutch sector vessels, both depart from the Den Helder base. However, *Vos Base* visits platforms located northwest to southwest of this terminal, while *Pool Express* makes stops along the Dutch coast and serves platforms throughout the exploration zone.

In Figure 19, it is evident that the majority of points are concentrated below the 6-hour mark, representing 90% of cases. Within this concentration, we note that 56% of the points exhibit a time on platform of less than 2.5h, with the notable presence of

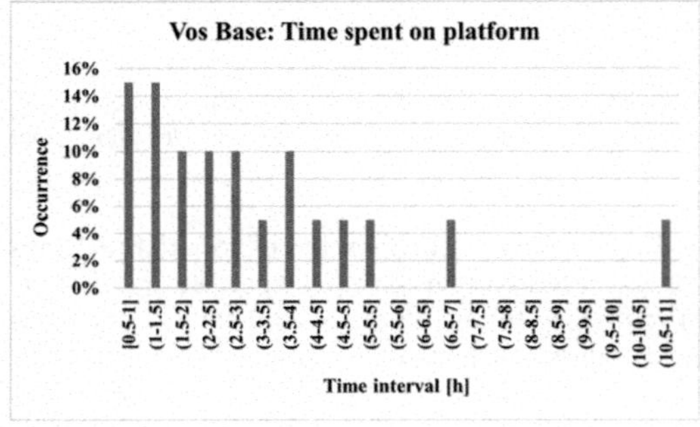

Figure 19. Histogram of the time spent on the platform of the Vos Base vessel.

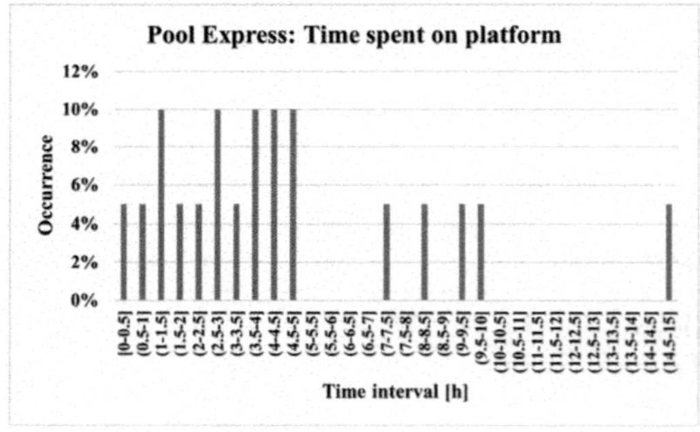

Figure 20. Histogram of the time spent on the platform of the Pool Express vessel.

a clear outlier, which records an exceptional time of 10 h and 44 min. Finally, when analyzing Figure 20, this is the longest length of stay and most points are below the 6-hour mark, covering 75% of cases. Within this set, only 30% of the points exhibit a stay of less than 2.5 h and the presence of an atypical value of 14 h and 58 min stands out.

Such graphs above play a fundamental role in understanding maritime operations throughout the study period in each case, which provides essential data to evaluate the efficiency and logistics of vessels, as well as to identify patterns of behavior on their

Table 5. Statistical analysis of vessel trajectories.

		Norway		Netherlands	
Time on the platform		North Purpose	Eldborg	Pool Express	Vos Base
Average		01h 43m	02h 16m	03h 46m	02h 33m
Standard deviation (h)		0.051	0.061	0.152	0.103
Maximum		04h 55m	04h 54m	14h 58m	10h 44m
Minimum		00h 29m	00h 28m	00h 21m	00h 37m
Percentage of calls that take	[0h, 2h]	65%	47%	25%	40%
	[2h, 4h]	29%	33%	30%	35%
	[4h, 6h]	6%	20%	20%	15%

journeys. In this way, Table 5 was prepared, with some statistical analyzes of the information collected in order to compare them with each other.

Table 5 provides information about the time spent on the platform for different ships (*North Purpose* and *Eldborg* in Norway, *Pool Express* and *Vos Base* in the Netherlands). The average time spent on the platform varies across locations, with the shortest average time in Norway at *North Purpose* (01h 43min) and the longest in the Netherlands at *Pool Express* (03h 46min). The standard deviations are small, which suggests that the majority of values are close to the average, that is, there is low dispersal. *Eldborg* in Norway has the smallest standard deviation (0.061), indicating relatively consistent time spent, and *Pool Express* in the Netherlands has the largest standard deviation (0.152), suggesting more variability. The maximum time spent on the platform is highest at *Pool Express* (14h 58min), indicating a potential outlier or longer duration. The minimum time spent is shortest at *Eldborg* (00h 28min).

In comparison to the other time intervals, North Purpose exhibits a larger percentage of scales in the [0, 2h] interval and the percentage declines with increasing time interval, reaching its lowest point in the (4h, 6h] interval. Eldborg exhibits a notable decline in percentage in the (4h, 6h] interval, while it has the maximum percentage of scales in the [0, 2h] interval, much like North Purpose. In terms of Netherlands, Pool Express shows a comparatively constant proportion throughout the three time intervals and has the lowest percentage of scales in the [0, 2h] interval as compared to other vessels. Finally, Vos Base concentrates 75% of stops on platforms lasting up to 4 hours.

Therefore, it is interesting to highlight that the Norwegian operation appears to be more efficient, since, represented by North Purpose and Eldborg, Norway has a significantly shorter average time on the platform compared to ships in the Netherlands (Pool Express and Vos Base), which may indicate faster and more efficient processes. Eldborg presents the lowest standard variation, indicating a relatively high consistency in the time spent on the platform, this suggests a more predictable and efficient operation in terms of time management. Furthermore, the two Norwegian ships, unlike the Dutch, have 100% of their vessels analyzed with stops of less than 6h, which indicates that the stops have shorter time intervals, especially [0, 2h].

In summary, Table 4 provides insights into the average, variability, and extreme values of the time spent on the platform for different locations. Further investigation into the factors influencing these times, especially outliers, could provide more context to the observed variations. There are also outliers, whose analysis is important to understand extreme events or errors in the data. There is a high maximum stay time on the platform, because, in some cases, there are closing times, which means the ship must wait all night to load and/or unload. No events of this type were observed on the two Norwegian ships. It should be noted that this study focused on 4 ships and 2 areas of operation, thus not allowing these conclusions to be extrapolated to the set of fleets used in these two areas of operation.

7 CONCLUSIONS

The aim of this study was to carry out an analysis of offshore supply operations in the North Sea, with a more detailed focus on the Norwegian and Dutch sectors. Understanding the operational difference between these nations and Brazil is crucial to improving the efficiency of Brazilian operations in the oil and gas industries. Diverging from operations in the North Sea, activities in Brazil are centralized at a single base in Macaé, Rio de Janeiro, presenting challenges associated with numerous cargo-focused vessels and a limited pipeline system.

The North Sea provided a standard for best practices because of its reputation for operational efficiency. The study highlighted the value of offshore logistics in assisting with oil and gas operations while offering insights into the geographical and operational features of the North Sea. The area's closeness to the coast, use of different kinds of platforms, and shallower sea depths than Brazil are important aspects of operations.

Strategically placed supply bases in the North Sea followed a minimum voyage distance requirement by placing themselves in close geographic vicinity to platforms and FPSOs. Bases in Norway, the UK, the Netherlands, and Denmark are distributed in order to accommodate the scattered character of exploration zones.

Moreover, the study examined the specialized boats used in offshore logistics, such as the Anchor Handling Tug Supply (AHTS), Platform Supply (PSV), and Utility (UT) vessels. Details on the vessels, including their beam, summer deadweight, gross tonnage, year of construction, and deck area, revealed information about the fleet makeup of supply vessels.

A highly specialized and effective approach was the hallmark of North Sea offshore oil and gas logistics operational patterns. Modern facilities for loading and unloading commodities were installed in strategically placed ports as part of the region's well-developed support infrastructure. Insights into efficiency and dynamism were obtained by classifying Norway's operational patterns into three categories: international, multidirectional, and linear.

A thorough examination of the amount of time spent on platforms was part of a case study pertaining to particular supply vessel operations in the Netherlands and Norway. Operational efficiency varied, with Norway displaying consistent time frames and the Netherlands displaying more variability, according to statistical comparisons.

In conclusion, this study provides insightful information about offshore supply operations in the North Sea by providing a detailed understanding of operational, efficiency, and geographic factors. Further research and inquiry in this crucial area can be built upon the foundation provided by the detailed examination of supply vessel operations and the comparison analysis with the Brazilian setting.

ACKNOWLEDGEMENTS

This work was performed within the scope of the research project "Integrated System for Strategic and Operational Planning of Offshore Oil & Gas Support Activities on the Brazilian Coast," funded by the National Council for Scientific and Technological Development (CNPq) through the CNPq/MCTI/FNDCT Call No. 14/2022 - Development of Science, Technology, and Innovation for the Maritime Transport and Shipbuilding Sectors. The project is under the auspices of the University of São Paulo.

This work contributes to the Strategic Research Plan of the Center for Marine Technology and Ocean Engineering (CENTEC), which is financed by FCT under contract UIDB/UIDP/00134/2020.

REFERENCES

Aas, B., Halskau Sr, Ø. & Wallace, S. 2009. The role of supply vessels in offshore logistics, *Maritime Economics & Logistics*, 11, 302–325.

AENERT (2023). *Energy Industry*, https://aenert.com/countries/europe/ (accessed on 6/11/2023).

Esteves, R.L., Branski, R. (2021). *Logística na Cadeia de Petróleo offshore: Brasil, Noruega e Golfo do México.* In: Campello, Marcos Luiz Costa (org.). Logística: contribuições para melhorias na produção e nos resultados. Editora Científica, p. 29–45.

Fagerholt, K., Lindstadb, H. (1999). *Optimal policies for maintaining a supply service in the Norwegian Sea.* Omega – The International Journal of Management Science.

Friedberg, D.Ø., Uglane, V.T. (2013). *Routing and Scheduling of Platform Supply Vessels: Case from the Brazilian Petroleum Industry.* Industrial Economics and Technology Management, Department of Industrial Economics and Technology Management, Norwegian University of Science and Technology (NTNU).

Halvorsen-Weare, E. E., Fagerholt, K., Nonås, L. M., & Asbjørnslett, B. E. (2012). *Optimal fleet composition and periodic routing of offshore supply vessels.* European Journal of Operational Research, 10.

Kaiser, M.J. 2015. *Offshore service industry and logistics modelling in the Gulf of Mexico*, Springer International, Switzerland.

Maisiuk, Y., Gribkovskaia, I. (2014). *Fleet Sizing for Offshore Supply Vessels with Stochastic Sailing and Service Times.* In Proceedings of the 2nd International Conference on Information Technology and Quantitative Management, ITQM 2014 (pp. 10). Molde University College, Specialized University in Logistics.

Schipper, J. (2017). A new Model for Energy Logistics. Learning from collaboration in the North Sea.

Advances in Maritime Technology and Engineering – Guedes Soares & Santos (Eds)

Evaluating the impact of using smart equipment for preselecting cargo vehicles in the maritime port access area

A. Rusca, O. Dinu, F. Rusca, E. Rosca, M. Rosca & A. Ilie
Faculty of Transports, National University for Science and Technology Politehnica Bucharest, Romania

ABSTRACT: Access within seaports can represent a weak link for logistics chain that include a maritime component. If the number of arriving trucks exceeds the processing capacity of the maritime terminals, cargo vehicles are blocked at the port access gates. The long queues of cargo vehicles that appear more and more often at the access gates area usually disrupt the activity of maritime terminals. Thus, there is a need to implement new intelligent technologies for traffic monitoring and control. The use of smart equipment helps the port administration in the classification and ranking of vehicles in relation to the term at which are accepted for processing by the maritime terminals. However, the implementation of these new technologies is difficult without a previous evaluation of the effects induced by them. The impact assessment induced by use of smart equipment was studied by the authors. For the specific case of a port in Romania, a discrete simulation model was developed. Based on it, a series of possible scenarios were studied. The obtained results can help any port administration in taking the decision to implement such smart equipment.

1 INTRODUCTION

Access gates in a seaport play an important role in ensuring the security, efficiency, and in the smooth operation of the port terminals. Access gates act as a primary layer of security by controlling entry and exit points to the port facility. By restricting unauthorized access, seaports can prevent potential security threats such as terrorism, smuggling, theft, and sabotage. In the access area are often installed advanced security measures like biometric identification, security checks, and surveillance systems. Seaports are busy environments with a constant flow of vehicles, including trucks, containers, and other heavy equipment (Dinu et al. 2017). Access gates with a suitable control can help seaport administration to manage this traffic efficiently by controlling the entry and exit points, reducing congestion, and ensuring a smooth flow of traffic entities like trucks. In the same terms the access gates are often equipped with technology that allows the collection of data related to entry and exit times, vehicle information, and personnel movements. This data can be used in collaboration with smart equipment for a better management of traffic to avoid the jams on the access roads to seaports.

The emergence of disruptive events, such as the war in Ukraine, led to the disruption of logistics chain and a rerouting of grain flows to neighboring ports. Such a situation occurs in the area of access gates in seaport of Constanta in Romania, where the flow of grain from Ukraine has overloaded the processing and storage capacities of the maritime terminals. The effects were visible especially in the area of the access gates to the port by the appearance of queues with grain trucks with waiting times for many days.

The appearance of this phenomenon forces the port administration to implement control optimization solutions within the port in relation to the processing capacities of the grain terminals. At this moment, the control is carried out manually by the port staff. The objective of the research is to present the generic structure of the smart equipment that can be used and the evaluation of the impact over the waiting times in the queue that the introduction of these technologies has. A discrete simulation model was created and tested for different traffic management scenarios.

2 LITERATURE REVIEW

According to Steenken et al. (2005) are defined three decision levels (terminal design, operative planning, real time control) in maritime activity when the port authorities or maritime terminal owner can intervene on the technologies and equipment used. The occurrence of an unforeseen event, however, reduces this intervention at current time horizon without the port authorities being able to make a complex analysis. In these conditions, the use of discrete modeling in the impact assessment is a powerful tool (Rusca et al. 2016, Rusca et al. 2018, Mathias et al. 2018, Coman et al. 2022, da Silva & Botter, 2009)

DOI: 10.1201/9781003508762-11

Some events are generated by connections with land networks. It is important to evaluate processes and identify the causes that lead to the system blocking (Guedes Soares & Teixeira 2001, Abdirassilov & Sladkowski. 2018, Antognoli et al. 2018, Belcore et al. 2023).

The evolution of technology allowed the implementation of new smart equipment for monitoring and controlling the activity in seaports. These are used for the optimization of handling and storage activities (Belcore et al. 2023). Smart equipment is also used for monitoring and controlling vehicles moving to/from the port. The main objective is to avoid blockages in port access areas with negative effects on the activity.

3 THE STRUCTURE OF SMART EQUIPMENT

The technological evolution of smart equipment imposes them as a robust and credible tool in the traffic management process. For the specific case of the access area in seaports, this equipment must have the following components:

- Sensors: are used to gather data about cargo vehicles flows incoming or outgoing from maritime port area. It might include cameras to recognize the type of vehicles and read the license plates, RFID readers to track cargo containers and other shipments, and various other sensors to measure factors such as temperature, humidity and air quality;
- Data aggregation: The data from the various sensors would be collected and aggregated at local level or in a central database. This real-time analysis can provide for users the traffic patterns or trends, as well as can detect anomalies and potential security threats;
- Analytics and decision-making: Advanced analytics and machine learning algorithms can be used to process the data and identify patterns and trends. The system can be set to alert port authorities about any unusual or potentially dangerous activity, allowing them to take appropriate actions;
- GUI: A system developed on the base of smart equipment must have a user-friendly interface that would allow port authorities and other authorized users to monitor traffic in real-time and take decisions based on the data.

The generic structure of smart equipment aggregated is depicted in Figure 1. Depending on the particularities of the access gates area or on the requirements of port authorities, more specific modules can be added. But the main objective of smart equipment in the case of our research is to monitor and control the access in seaport.

The logical process in this case supposes the read of license plate for cargo truck, to compare the readed value with the recordings of data provided by the terminals and to allow the entrance of the trucks inside the seaport or to redirect these ones to a parking area until the terminal will give the accept of entrance.

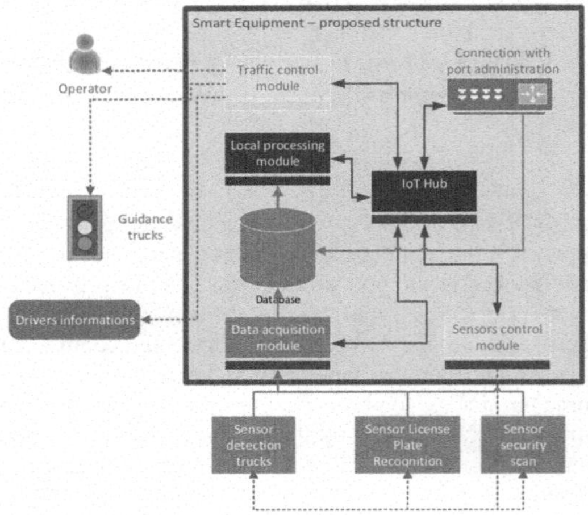

Figure 1. Smart equipment general structure.

4 THE MODEL USED TO EVALUATE THE IMPACT

The assessment of the impact caused by the usage of smart equipment for preselecting cargo vehicles in the maritime port access process is made using a discrete simulation model. The access process inside the seaport and the waiting process outside the port in parking areas are modelled considering the queuing systems theory. The tool used is Arena Rockwell (Kelton et al. 1998). The main attributes for a simulation model are:

- Event (defined as change in queuing system status): represented by the arriving moment of trucks at the access gate of port area or by the moment when the truck receives the acceptance from maritime terminal to start the logistics activity.
- Activity (defined as status of an object over a specified time interval between two successive events): represented by access gate busy status or no free place in the parking area.
- Process (defined as Sequence of the states of an object): represented by access inside the seaport or in the parking area.
- Plan (defined as orderly set of chronological conditions): represented by truck arrivals (scheduled or random).

The main blocks of the discrete simulation model developed in ARENA are depicted in Figures 2-4. In Figure 2 is shown the logical block used to generate the simulation entities associated to trucks arrived in the access gate area.

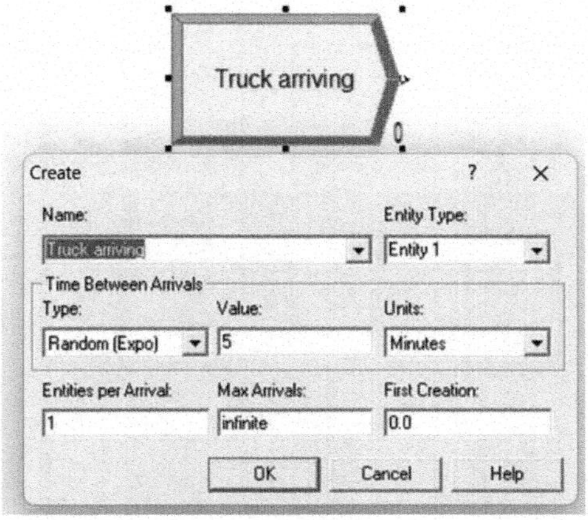

Figure 2. Generating simulation entities block.

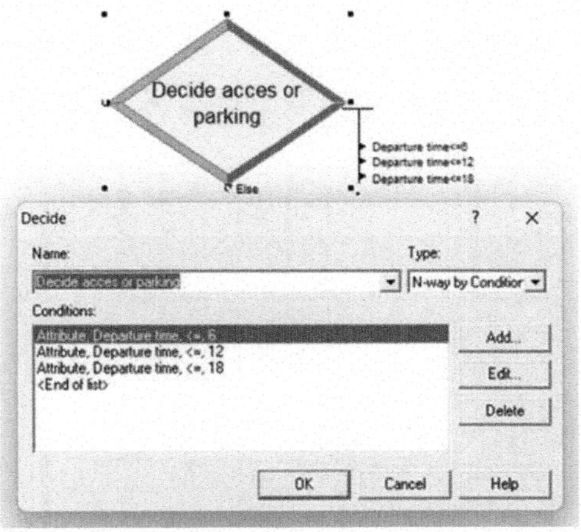

Figure 3. Time attribute initialization block.

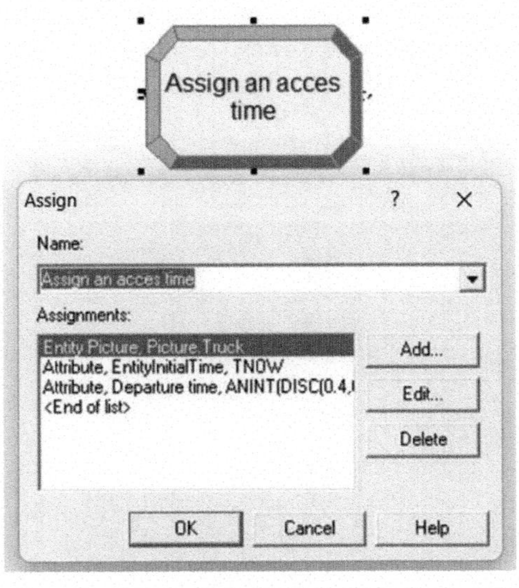

Figure 4. Time attribute initialization block.

In Figure 3 are set time attributes like *Departure_time* to represent the time interval between truck arrival moment (*EntityInitialTime*) and the forecasted time of access to the seaport. In Figure 4 is represented the decisional block. The trucks accepted by the maritime terminal are guided to the access gates while the others are guided to parking areas in accordance with the waiting time stored in the *Departure_Time* attribute.

5 THE STUDY CASE. SIMULATION SCENARIOS AND RESULTS

The largest maritime port of the European Union on the Black Sea is the Port of Constanta. The main flows handled in maritime terminals of it are grain flows. These come from the hinterland of the port predominantly located on the territory of Romania. The geopolitical evolution of the area, the beginning of the conflict in Ukraine in 2022, led to the emergence of new grain flows directed to the port of Constanta. The behavior of drivers from Ukraine can be assimilated to that of small transport companies from Romania. They have a low level of communication with the port authority in contradiction with a desire to unload the grain as quickly as possible in the terminal. Also, these carriers will use recently developed grain terminals in a short period of time that do not yet have well-implemented grain handling and storage technologies. However, the main problem is the large number of trucks that arrive in the access area within the port. With the overcrowding of the storage areas and handling equipment in the maritime terminals, queues appeared in this new situation in the access gate area. Thus, waiting times have increased to values of 3-5 days in peak periods. The solution from the port authority was to find a series of free spaces where the trucks can wait before getting access to the port. However, the trucks are checked and directed manually by the staff at the access gates. The purpose of the research is to evaluate through a discrete simulation model the impact of the introduction of smart equipment in guiding trucks to the parking area. In the development of the model, the hypothesis of three parking areas (P1, P2 and P3) was proposed to which trucks should be guided in relation with the waiting time until they are accepted for unloading in the maritime terminals (more than 6 hours but less than 12 hours-P1, up to 18 hours-P2 and up to 24 hours-P3). For trucks with longer waiting times, it is considered that these ones can wait in the parking lots located on the highway bordering the port area. The proposed parking areas and the layout of the access network to the main gate used by cereals flows is presented in Figure 5.

We assumed in our model the next considerations:

- The simulation entities are the representation of trucks arrived at the port access gates;
- The trucks arrival time is generated in accordance with a repartition function,

- An attribute *Departure_Time* is associated for every simulation entity. It is set with a value between 1 to 24 and represents the moment for entrance in the port area,
- After 6, 12 and 18 hours the *Departure_Time* for entities from every specific store is set to 6 hours and the entity is guided to an access gate;
- A Process block is used to simulate the transit through the access gate area;
- If an entity is guided to a parking area P1, P2 or P3 (with a waiting time of 6, 12 or 18 hours) and this parking area is full a supplementary waiting time appears until a place in the store become free.

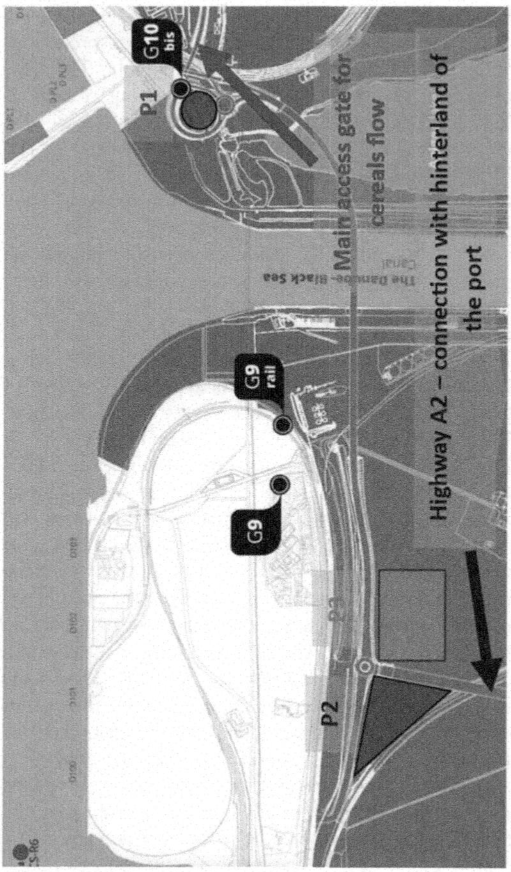

Figure 5. Time attribute initialization block.

The location of parking areas were proposed considering that it must be outside the port, near the access gates and on the transport corridor, in our case the A2 highway, which connects the hinterland of the port and the border areas with the neighboring country Ukraine.

The simulation model is depicted in Figure 6.

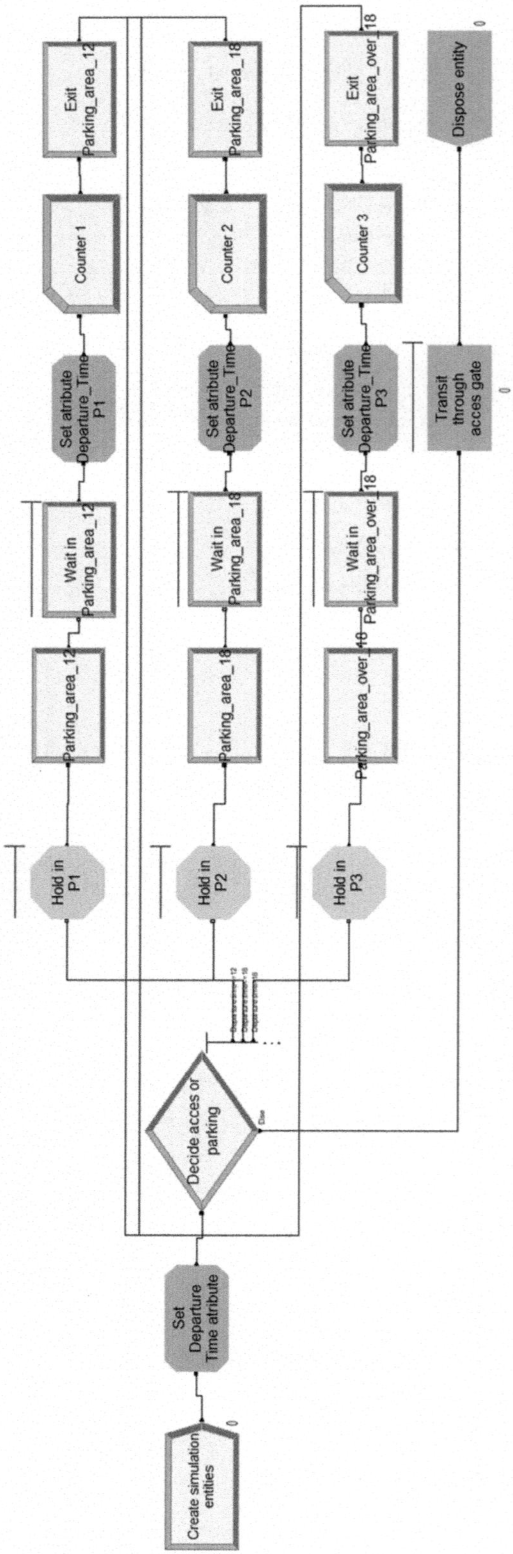

Figure 6. The developed simulation model.

To evaluate the impact of using smart equipment for preselecting cargo vehicles in the maritime port access area a set of 20 scenarios are created. Is evaluated the impact when the distribution of trucks arrival flow is increased with 10%, 20%, 30% and 40% compared to a base scenario. Another modified attribute of the simulation is the percentage of trucks waiting over 6, 12 and 18 hours. For reasons of mathematical formalization, it is considered that trucks with waiting times longer than 24 hours will spend this time interval in the service spaces on the A2 highway that connects the port to the hinterland. The input parameters are presented in Table 1.

The base scenario is built using real data. The distribution of trucks arrival from the highway network of Romania is assimilated with the negative exponential distribution as the authors determined in the paper Rusca et al. (2022). The repartition parameter λ is set at a value of 5 minutes using data collected by the authors. The parking areas are considered with a value (number) of parking slots as follows: 20 for P1, 40 for P2 and 60 for P3.

Table 1. The input parameters.

		The ratio of trucks [%]			
Scenario	λ [minutes]	Gates	P1	P2	P3
S0*	5	40	30	20	10
S01	5	30	40	20	10
S02	5	20	30	30	20
S03	5	25	25	25	25
S10	4.54	40	30	20	10
S11	4.54	30	40	20	10
S12	4.54	20	30	30	20
S13	4.54	25	25	25	25
S20	4.16	40	30	20	10
S21	4.16	30	40	20	10
S22	4.16	20	30	30	20
S23	4.16	25	25	25	25
S30	3.84	40	30	20	10
S31	3.84	30	40	20	10
S32	3.84	20	30	30	20
S33	3.84	25	25	25	25
S40	3.57	40	30	20	10
S41	3.57	30	40	20	10
S42	3.57	20	30	30	20
S43	3.57	25	25	25	25

* Base scenario

The results obtained are presented in Table 2 and 3. In the simulation process, a warm period of 2 days was taken into account in which no values were re-corded in order to avoid the transient status of the system. Also, for each scenario, the simulation had 5 replicants to obtain average values. The simulation period was 30 days considering a 24-hour work schedule. The number of access gates in the port was considered to be 2, as is the real case with access point no.10Bis in the port of Constanta dedicated to grain terminals.

Table 2. The waiting time to enter the seaport or in a parking area [hours].

Scenario	Waiting time at gates	Waiting time P1	Waiting time P2	Waiting time P3
S0	0.02	0.59	0.01	0.00
S01	0.02	1.87	0.01	0.00
S02	0.02	0.76	1.52	0.00
S03	0.02	0.42	0.44	0.31
S10	0.03	1.21	0.21	0.00
S11	0.03	2.26	0.08	0.00
S12	0.03	1.01	2.32	0.01
S13	0.02	0.43	1.01	0.92
S20	0.03	1.51	0.25	0.00
S21	0.03	2.58	0.16	0.00
S22	0.04	1.44	2.80	0.11
S23	0.03	0.64	1.48	1.57
S30	0.04	1.82	0.65	0.00
S31	0.04	2.83	0.69	0.00
S32	0.04	1.73	3.55	0.10
S33	0.04	0.94	2.04	2.33
S40	0.04	2.05	0.90	0.00
S41	0.04	2.98	0.76	0.00
S42	0.05	1.99	4.49	1.38
S43	0.05	1.29	2.89	3.46

Table 3. The average number of trucks in the queue to enter the seaport or in a parking area.

Scenario	Queue for gates	Queue for P1	Queue for P2	Queue for P3
S0	0.28	2.05	0.03	0.00
S01	0.30	8.98	0.01	0.00
S02	0.29	2.69	5.64	0.00
S03	0.34	1.26	1.39	0.93
S10	0.45	4.91	0.58	0.00
S11	0.42	12.03	0.21	0.00
S12	0.41	3.89	9.58	0.01
S13	0.38	1.39	3.58	3.14
S20	0.49	6.61	0.74	0.00
S21	0.48	14.91	0.49	0.00
S22	0.58	6.2	12.20	0.35
S23	0.44	2.28	5.43	5.57
S30	0.63	8.64	2.08	0.00
S31	0.65	17.64	2.21	0.00
S32	0.68	8.01	16.80	0.31
S33	0.75	3.57	8.14	8.88
S40	0.81	10.38	3.12	0.00
S41	0.80	19.55	2.57	0.00
S42	0.89	9.82	24.07	4.73
S43	0.89	5.39	12.73	14.42

The simulation results show the influence of the value of λ on the service parameters of the systems. The increase of the value of this parameter leads to the increase of the value of the waiting times. Values of over 4 hours are thus noted, such as the waiting time to enter P2 in the case of the S42 scenario. The behavior of truck drivers is also important. In the case of scenarios S02, S12, S22, S32 and S42, the percentage of 30% of drivers who arrive and must wait 12 hours before being accepted in the port area has the effect of an overload of the corresponding parking zones. When a percentage of 40% of drivers who will be accepted over 6 hours at the port arrives earlier, in the case of scenarios S11, S21, S31 and S41, for parking area P1, average waiting times in the access queue can be found with values higher than 2 hours. However, as a positive conclusion to this research, the queue at the access gates to the seaport is much reduced with the help of a smart equipment for preselecting cargo vehicles in the maritime port access area, thus avoiding the blockages that appear now in the case of the port of Constanta.

6 CONCLUSIONS

Sea ports are important components in logistics chain. The activity in the maritime terminals has strict rules in relation to the storage capacity and the movement of maritime vessels. The arrival of an oversized flow of vehicles from the hinterland on the road network can lead to the blocking of the activity. The solution is the usage of smart equipment to perform a pre-selection of these vehicles in the area of the access points and guide them to external parking lots until the vehicles can be received inside the maritime terminals. The research aims to develop a discrete simulation model to evaluate the impact of the usage of smart equipment. The obtained results are useful for the port administration in the feasibility studies to justify the purchase and implementation of these smart equipment.

Problems encountered in carrying out this research were: identifying the input data in the simulation model and understanding of the technological processes specific to port activity. In future research, the development of an evolutionary algorithm is considered to manage scenarios specific to this field, such as: the group arrival of trucks for a certain terminal, non-compliance of system indications and priority access of certain trucks.

ACKNOWLEDGEMENTS

This work was supported by a grant from the National Program for Research of the National Association of Technical Universities - GNAC ARUT 2023.

REFERENCES

Abdirassilov, Z.& Sladkowski, A. 2018 Application of artificial neural network for short-term prediction of container train flows in direction of China-Europe via Kazakhstan. *Transport Problems* 4(4):103–113.

Antognoli M.& Capodilupo L.& Marinacci C.& Ricci, S. 2018. Present and Future Operation of Rail Freight Terminals – Studies in Systems. In Sładkowski, A. (eds) *Transport Systems and Delivery of Cargo on East–West Routes* 15:233–273. Springer, Cham

Belcore, O.M.& Di Gangi, M.& Polimeni, A. 2023. Connected Vehicles and Digital Infrastructures: A Framework for Assessing the Port Efficiency. *Sustainability* 15(10):8168

Coman, C.& Rusca, A.& Rosca, E.& Burciu, S.& Rusca, F. 2022. The influence of smart equipment Smart Hub over highway traffic in the Romanian border points, *Transportation Research Procedia*. 64(C): 157–165.

Dinu, O. & Roşca, E. & Popa, M. & Roşca, M. A. & Rusca, A. 2017. Assessing materials handling and storage capacities in port terminals. In *IOP Conference Series: Materials Science and Engineering* 227(1): 012039. IOP Publishing.

Guedes Soares, C. & Teixeira, A.P. 2001. Risk assessment in maritime transportation, In *Reliability Engineering & System Safety*, 74(3):299–309.

Kelton, W. D., Sadowski, R. P., & Sadowski, D. A. 1998. *Simulation with Arena*, New York, Editorial Mc Graw Hill.

Mathias, N.A.S.& Santos, T.A.& Guedes Soares, C. 2018. Analysis of a new container terminal using a simulation approach. Progress In *Maritime Technology and Engineering 4th International Conference on Maritime Technology and Engineering (MARTECH) May 07–09, 2018 Lisbon, Portugal*:43–52. Rotterdam: Balkema.

Rusca, F.& Popa, M.& Rosca, E.& Rusca, A. 2016 Capacity analysis of storage area in a maritime container terminal. In *Maritime Technology and Engineering III: Proceedings of the 3rd International Conference on Maritime Technology and Engineering (MARTECH 2016, Lisbon, Portugal, 4–6 July 2016)*: 92–99. Rotterdam: Balkema.

Rusca, F.& Popa, M.& Rosca. E.& Rusca, A. 2018. Simulation model for maritime container terminal. *Transport problems* 13(4): 47–54.

Rusca, F.& Rusca, A. & Rosca, E. & Coman, C. & Burciu, S. & Oprea, C. 2022. Evaluating the Influence of Data Entropy in the Use of a Smart Equipment for Traffic Management at Border Check Point. *Machines* 10(10): 937

da Silva, A. & Botter, R. 2009. Method for assessing and se-lecting discrete event simulation software applied to the analysis of logistic systems In *Journal of Simulation* 3: 95

Steenken, D., Voß, S. & Stahlbock R. 2005. Container terminal operation and operations research - a classification and literature review. In Günther, H.-O., Kim K.H. (Eds.), *Container Terminals and Automated Transport Systems*, Berlin, Springer, 3–49.

Advances in Maritime Technology and Engineering – Guedes Soares & Santos (Eds)
© 2024 The Author(s), ISBN 978-1-032-83099-5

Safety and economic optimization of offshore production systems

L.M.R. Silva & C. Guedes Soares
Centre for Marine Technology and Ocean Engineering (CENTEC), Instituto Superior Técnico, Universidade de Lisboa, Lisbon, Portugal

ABSTRACT: A robust optimization model is proposed aiming to improve the search for an optimal configuration of offshore production systems. The new mathematical model integrates safety and economic contraints by looking for the minimum infrastructure cost and, simultaneously, the minimum expected loss of pipelines. The present approach enhances the literature and decision-making process by allowing the model to consider the structural material of the pipeline as rigid and flexible getting closer to real projects. The worst-case scenario is adopted to represent the uncertainty of the pipeline failure, which also enables a tractable formulation to solve the problem. The efficiency of the proposed model is demonstrated using a case study. The subsea system layout elements may be identified by the model considering the safety and economic concerns, and the initial production system costs are also defined.

1 INTRODUCTION

The design optimization ensuring an efficient and cost-benefit operation of offshore production systems are essential concern to consider during the oilfield development phase (Sales et al., 2023). An efficient system guarantees that the equipment and components perform adequately over the lifetime and under specific conditions. Thus, the cost-benefit aspect ensures that the project designed fulfils not only the technical requirements but also maximizes their economic potential.

An oilfield project starts with a conceptual design phase gathering some general information about the technology, and hazards and risks are analyzed (Redutskiy et al., 2021). The gathered information allows the identification of different project alternatives and the selection of the concept that best fits all the project requirements. It has driven the development of rapid screening tools attempting to help the decision-making process. Nevertheless, most of the research works developed a cost-effective or uncertain decision tool (Silva & Guedes Soares, 2023).

Crawley & Grant, (1997) developed a concept risk assessment tool, which emphasizes the technical uncertainty in the decision process of an oilfield development. The risk assessment is based on pre-processed risk building blocks for each item of hazardous equipment. Gu & Gudmestad, (2011) presented a risk assessment approach that analyses the uncertainties, risks, and opportunities during the development of an oilfield. Silva et al., (2019) established an oilfield concept selection tool based on the risk assessment of the pipeline system alternatives. Bhardwaj et al., (2021) adopted the finite element analysis to predict the assessment of the effect of corrosion defects on the burst strength of subsea pipeline systems. Bhardwaj et al., (2022) presented a Bayesian network to analyze the reliability of the subsea production system. However, the researchers cited above considered only the risk aspect of an oilfield development.

Silva & Guedes Soares, (2019) developed a mathematical programming method to optimize the oilfield development helping decision-makers to select the production system that minimizes the project cost. Silva & Guedes Soares, (2021) integrated the economic aspect with the reservoir uncertainty presenting a discrete model that runs different production scenarios to analyze the maximization of the project revenue. Silva & Guedes Soares, (2022) extended the economic and reservoir uncertain analyses inserting the production scenarios into a stochastic programming model. However, only the economic aspect is analyzed by the described papers, considering or not the reservoir uncertain.

Bhattacharyya & Cheliyan, (2019) developed a method for optimizing the cost and reliability of a subsea production system. Sales et al., (2023) proposed an optimization approach for a subsea processing system, which considers both economic and safety aspects. Silva & Guedes Soares, (2023) presented a robust optimization model for production system cost and pipeline risk. These models improve the decision-making process at an early stage of the oilfield development allowing to maximize the

DOI: 10.1201/9781003508762-12

economic gains in the project lifetime and preventing disastrous consequences.

The present approach extends the optimization model described in Silva & Guedes Soares, (2023) by enlarging the safety aspect inserting the risk of failure of rigid and flexible pipelines. The main idea is to allow the model to design the most optimal production system taking into consideration not only the economic aspect but also the structural material of the pipeline that minimizes the risk of failure.

The remainder of this paper is organized as follows. Section 2 provides the pipeline risk assessment methodology; Section 3 contains the production system optimization approach; Section 4 describes the case study, results, and discussions; and Section 5 presents the final conclusions.

2 PIPELINE RISK ASSESSMENT

Focusing on the pipeline system, which is a fundamental component in the subsea production system, the assessment of the risk of oil leakage allows the decision-makers to identify where actions could be taken to reduce the risk and improve safety in the offshore industry (Silva & Guedes Soares, 2016).

The DNV (2017) model to estimate the pipeline failure frequency is adopted. As discussed in Silva et al., (2019) the approach is in line with the failure rates available in the literature. The model considers four main factors influencing the failure frequencies: age, size, length, and location. During the operational phase of the pipeline system, the age and size, in association with site-specific conditions, may affect the subsea pipeline system causing accidental events with extreme consequences (Alencar et al., 2023). Nonetheless, Silva & Guedes Soares, (2023) concluded that the pipeline length is the most relevant issue due to smaller pipes presenting smaller frequencies and vice-versa. Thus, the combination of a greater length and higher production flow rate can cause a catastrophic incident.

Accordingly, the DNV (2017) model classifies the annual frequency of failure per length unit considering the pipe size (small or subsea and larger diameter, and riser) and structural material (rigid and flexible).

Because only the gathering system is considered in this paper, all the pipelines are considered as small diameter subsea pipelines. Thus, for pipelines considering the structural material as rigid, the report presented a specific equation for assessing the annual frequency of failure per kilometer, following:

$$FF = (f_{km} \text{ x Pipeline Length}) + f_{pipeline} + f_{\text{Dragged Anchor}} \quad (1)$$

where f_{km} is the failure factor proportional to pipeline length (equals 4.4×10^{-4} per km-year); $f_{pipeline}$ is the failure factor related to failures that are not dependent on pipeline length (equals 2.3×10^{-3} per km-year); $f_{DraggedAnchor}$ factor reflects the failures related to dragged anchors from ships underway (equals 1.9×10^{-7} per km-year).

Besides Equation (1), the DNV, (2017) model recommends failure rates based on data sources, such as PARLOC, CONCAWE and MCS Advanced Subsea Engineering, which are shown in Table 1.

Silva et al., (2019) adopted the DNV (2017) model and developed a methodology that can estimate the risk index of each pipeline segment of the pipeline system. The authors specified that the pipeline risk index can be expressed as the equivalent expected loss (EEL), which is a function of the financial loss of the production flow rate through the pipelines by the likelihood of the pipeline failure per year, following:

$$\text{Risk} = \text{EEL} = \text{Severity} \times \text{Frequency of Failure} \ (\text{US\$/year}) \quad (2)$$

The production flow through each pipeline segment is directly related to the severity criterion. The greater the amount of hydrocarbon the greater the probability of an accidental catastrophe. In accordance with each national authority, the production loss may impose indemnities to be paid. Each country has its own regulations for the damage caused to the environment. These regulations need to be studied by the oil producer before system implementation.

3 PRODUCTION SYSTEM OPTIMIZATION

The production system optimization integrates the safety and economic concerns of a subsea production system. A robust optimization model is developed aiming at describing the relationship between uncertain factors in the decision environment and the set of input parameters. The robust framework may assume a set of scenarios to represent the uncertain data. However, the model complexity directly increases with the number of scenarios. Moreover, the present robust model is also considered as a MILP model, due to the adoption of binary and discrete variables.

Table 1. Recommended failure frequencies.

Structural Material	Description	Failure Rate	Unit
Rigid Pipeline	Small Diameter	2.5×10^{-4}	per km-year
Flexible Pipeline	Subsea	2.3×10^{-3}	per km-year

3.1 *Model assumptions*

A set of simplifications regarding how the system works is necessary to reduce the size of the combinatorial problem. The main production system assumptions are summarized below:

1. The number, flow rate, and location of the wells in x-y axis are pre-defined data. Thus, the maximum number of wells that can be connected to each manifold is pre-defined data.
2. The standardized maximum capacity cannot exceed the total oil production.
3. The installation operation is not impacted by topographic limitations. The pipeline routes are considered as straight lines.

   The main robust approach assumptions are:

4. Attempting to reduce the model size, the present optimization model assumes a worst-case scenario.
5. The robust criterion is considered as absolute robustness aiming at minimizing the maximum expected value among the feasible decisions.

A detailed list of assumptions is described in Silva & Guedes Soares, (2023).

3.2 *Conceptual model*

The problem of optimizing the offshore production system is formulated in the next sub-section, but, schematically, it can be stated as follows bellow:

- *Given* the number and location of the wells; the reservoir volume of oil-in-place; the minimum flowrate that each manifold can receive; the constant price of barrel price; the length of the risers; the manifold slot sizes; the platform capacities; the cost of pipeline and risers considered as rigid and flexible; the cost of each size of the platform and manifold slots.
- *Obtain* the distance between the touch-down point of the platforms and the wells; the distance between the manifolds and the wells; the distance between the touch-down point of the platforms and the manifolds; and the probability of occurrence of each production scenario. Thus, the frequency of failure of the pipeline segment that connects the wells and the platforms considering the structural material as flexible and rigid; the frequency of failure of the pipeline segment that connects the wells and the manifolds considering the structural material as flexible and rigid; the frequency of failure of the pipeline segment that connects the manifolds and the platforms considering the structural material as flexible and rigid.
- *Subject to* well assignment constraints; manifold installation and assignment constraints; flowrate conservation constraints; pipeline connection constraints and expected loss constraints.

3.3 *Mathematical model*

All the necessary notation of the model index, sets, parameters, and variables are presented in Tables 2–4. The objective function aims at minimizing the expected loss of the pipeline system and the total cost of the production system infrastructure, following (Equation (3)-(12)):

$$\text{minimize}\,\{ELSystem + TotalCost\} \qquad (3)$$

Table 2. Sets index.

Index	Description
$w \in W$	Set of wells
$m \in M$	Set of possible manifolds
$p \in P$	Set of possible platforms
$s \in S$	Set of size of platforms
$r \in R$	Set of reservoirs

Table 3. Parameters.

Parameters	Description
RISERLENGTH	length of risers;
MAXWELLS	maximum number of wells;
$OHIP_r$	volume of original hydrocarbon-in-place of each reservoir;
QMAX	maximum flowrate that manifolds can receive from the wells;
BP	a constant price of oil barrel;
FLEXIBLECOST	cost of flexible pipelines;
RIGIDCOST	cost of rigid pipelines;
$PLATCOST_{ps}$	cost of platform *p* of size *s*;
$MCOST_m$	cost of manifold *m* according to the slot capacity;
RISERCOST	cost of riser;
$DIST_{pw}$	distance between the touch-down point of platform *p* and well *w*;
$DIST_{mw}$	distance between manifold *m* and well *w*;
$DIST_{pm}$	distance between the touch-down point of the platform *p* and manifold *m*;
$SLOT_m$	manifold slot size;
$CAPACITY_{ps}$	maximum platform capacity size;
FFF_{pw}	Failure frequency of flexible pipeline connecting the well *w* and platform *p*;
FFF_{mw}	Failure frequency of flexible pipeline connecting the well *w* and manifold *m*;
FFF_{pm}	Failure frequency of flexible pipeline connecting the manifold *m* and platform *p*;
FFR_{pw}	Failure frequency of rigid pipeline connecting the well *w* and platform *p*;
FFR_{mw}	Failure frequency of rigid pipeline connecting the well *w* and manifold *m*;
FFR_{pm}	Failure frequency of rigid pipeline connecting the manifold *m* and platform *p*;
DY	number of days per year

$$\text{ELSystem} = \{ELPW + ELMW + ELPM\} \quad (4)$$

$$\text{Total Cost} = \{Cost\ Of\ Platform + Cost\ Of\ Pipelines + Cost\ Of\ Risers + Cost\ Of\ Manifolds\} \quad (5)$$

Being,

$$ELPW = \sum_{p \in P} \sum_{w \in W} el_{pw} \quad (6)$$

$$ELMW = \sum_{m \in M} \sum_{w \in W} el_{mw} \quad (7)$$

$$ELPM = \sum_{p \in P} \sum_{m \in M} el_{pm} \quad (8)$$

$$Cost\ Of\ Platform = \sum_{p \in P} \sum_{s \in S} y_{ps} \times PLATCOST_{ps} \quad (9)$$

Thus,

$$Cost\ of\ Pipelines = \left(\sum_{p \in P} \sum_{w \in W} DIST_{mw} \times f_{pw} \times FLEXIBLECOST + \left(\sum_{m \in M} \sum_{w \in W} DIST_{mw} \times f_{mw} + \sum_{p \in P} \sum_{m \in M} DIST_{pm} \times f_{pm} \right) \times FLEXIBLECOST) + \left(\sum_{p \in P} \sum_{w \in W} DIST_{mw} \times r_{pw} \times RIGIDCOST + \left(\sum_{m \in M} \sum_{w \in W} DIST_{mw} \times r_{mw} + \sum_{p \in P} \sum_{m \in M} DIST_{pm} \times r_{pm} \right) \times RIGIDCOST \right) \quad (10)$$

$$Cost\ Of\ Risers = \sum_{p \in P} \sum_{w \in w} RISERLENGTH \times x_{PW} \times RISERCOST + \sum_{p \in P} \sum_{m \in M} RISERLENGTH \times x_{Pm} \times RISERCOST \quad (11)$$

$$Cost\ Of\ Manifolds = \sum_{p \in P} \sum_{m \in M} y_m \times MCOST_m \quad (12)$$

The mathematical model is presented as follows:

$$\sum_{p \in P} x_{pw} \leq 1 \forall w \in W \quad (13)$$

$$\sum_{m \in M} x_{mw} \leq 1 \forall w \in W \quad (14)$$

$$\sum_{p \in P} x_{pw} + \sum_{m \in M} x_{mw} = 1 \forall w \in W \quad (15)$$

$$\sum_{p \in P} \sum_{w \in W} x_{pw} + \sum_{m \in M} \sum_{w \in W} x_{mw} = MAXWELLS \quad (16)$$

$$\sum_{m \in M} y_m \leq 1 \quad (17)$$

$$\sum_{p \in P} x_{pm} \leq 1 \forall m \in M \quad (18)$$

$$y_m \leq \sum_{p \in P} x_{pm} \forall m \in M \quad (19)$$

$$\sum_{w \in W} x_{mw} \leq SLOT_m \times y_m \in M \quad (20)$$

$$\sum_{w \in W} x_{mw} \leq (SLOT_m - 1) \times y_m \in M \quad (21)$$

$$\sum_{s \in S} y_{ps} \leq 1 \forall p \in P \quad (22)$$

$$\sum_{w \in W} x_{pw} + \sum_{m \in M} x_{pm} \leq MAXWELLS \times \sum_{s \in S} y_{ps} \forall p \in P \quad (23)$$

$$\sum_{m \in M} z_{pm} + q_p \leq \sum_{s} CAPACITY_{ps} \forall p \in P \quad (24)$$

$$\sum_{p \in P} \sum_{w \in W} x_{pw} \times Q_w + \sum_{m \in M} \sum_{w \in W} x_{mw} \times Q_w \leq \sum_{r \in R} OOIP_r \quad (25)$$

$$q_p = \sum_{w \in W} x_{pw} \times Q_w \forall p \in P \quad (26)$$

$$q_m = \sum_{w \in W} x_{mw} \times Q_w \forall m \in M \quad (27)$$

$$-q_m + pr_{pm} \leq 0 \forall m \in M, p \in P \quad (28)$$

$$-(x_{pm} \times QMAX) + pr_{pm} \leq 0 \forall m \in M, p \in P \quad (29)$$

$$q_m + (x_{pm} \times QMAX) - pr_{pm} \leq QMAX \forall m \in M, p \in P \quad (30)$$

$$-q_m + pf_{pm} \leq 0 \forall m \in M, p \in P \quad (31)$$

$$-\left(x_{pm} \times QMAX\right) + pf_{pm} \leq 0 \forall \mathrm{m} \in \mathrm{M}, \mathrm{p} \in \mathrm{P} \quad (32)$$

$$q_m + \left(x_{pm} \times QMAX\right) - pf_{pm} \leq QMAX \forall \mathrm{m} \in \mathrm{M}, \mathrm{p} \in \mathrm{P} \quad (33)$$

$$f_{pw} + r_{pw} \geq x_{pw} \forall \mathrm{w} \in \mathrm{W}, \mathrm{p} \in \mathrm{P} \quad (34)$$

$$f_{pm} + r_{pm} \geq x_{pm} \forall \mathrm{m} \in \mathrm{M}, \mathrm{p} \in \mathrm{P} \quad (35)$$

$$f_{mw} + r_{mw} \geq x_{mw} \forall \mathrm{w} \in \mathrm{W}, \mathrm{m} \in \mathrm{M} \quad (36)$$

$$el_{pw} = \left(Q_w \times BP \times DY \times FFF_{pw} \times f_{pw}\right) + \left(Q_w \times BP \times DY \times FFR_{pw} \times r_{pw}\right) \quad (37)$$
$$\forall p \in P,\ w \in W$$

$$el_{mw} = \left(Q_w \times BP \times DY \times FFF_{mw} \times f_{pm}\right) + \left(Q_w \times BP \times DY \times FFR_{mw} \times r_{pm}\right) \quad (38)$$
$$\forall m \in M,\ w \in W$$

$$el_{pm} = \left(pf_{pm} \times BP \times DY \times FFF_{pm}\right) + \left(pr_{pm} \times BP \times DY \times FFR_{pm}\right) \quad (39)$$
$$\forall p \in P,\ m \in M$$

The wells assignment constraints, Equation (13)-(16), ensure a proper pipeline connection to the other infrastructures, to the manifold or directly to the platform.

Table 4. Variables.

Variables	Description
$y_{ps} \in \{0,1\}$	assumes the value 1 if the platform *p* of size *s* is installed;
$x_{pw} \in \{0,1\}$	assumes the value 1 if the well *w* is assigned directly to the touch-down point of the platform *p*;
$x_{mw} \in \{0,1\}$	assumes the value 1 if the well *w* is assigned to the manifold *m*;
$x_{pm} \in \{0,1\}$	assumes the value 1 if the manifold *m* is assigned to the touch-down point of the platform *p*;
$y_m \in \{0,1\}$	assumes the value 1 if the manifold *m* is installed;
$q_m \in IR^+$	oil flowrate received at the manifold *m*;
$q_p \in IR^+$	oil flowrate from well *w* received at the platform *p*;
$pr_{pm} \in IR^+$	auxiliary variable flow to linearize rigid pipe production from the manifold *m* to the platform *p* if the manifold *m* is assigned to the platform *p*;
$pf_{pm} \in IR^+$	auxiliary variable flow to linearize flexible pipe production from the manifold *m* to the platform *p* if the manifold *m* is assigned to the platform *p*;

(*Continued*)

Table 4. (*Continued*)

Variables	Description
Uncertain Variables	
$el_{pw} \in IR^+$	expected loss of pipeline connection from well *w* to platform *p*;
$el_{mw} \in IR^+$	expected loss of pipeline connection from well *w* to manifold *m*;
$el_{pm} \in IR^+$	expected loss of pipeline connection from manifold *m* to platform *p*;
$f_{pw} \in \{0,1\}$	assumes the value 1 if the pipeline material is flexible connecting the platform *p* to well *w*;
$f_{mw} \in \{0,1\}$	assumes the value 1 if the pipeline material is flexible connecting the manifold *m* to well *w*;
$f_{pm} \in \{0,1\}$	assumes the value 1 if the pipeline material is flexible connecting the platform *p* to manifold *m*;
$r_{pw} \in \{0,1\}$	assumes the value 1 if the pipeline material is rigid connecting the platform *p* to well *w*;
$r_{mw} \in \{0,1\}$	assumes the value 1 if the pipeline material is rigid connecting the manifold *m* to well *w*;
$r_{pm} \in \{0,1\}$	assumes the value 1 if the pipeline material is rigid connecting the platform *p* to manifold *m*;

Equation (17)-(21) represent the manifold installation and assignment constraints and guarantees that if a manifold is installed it is connected to a platform. As there is a physical constraint on the number of wells assigned to a manifold, Equation (20) ensures that the number of wells cannot be exceeded. Attempting to provide a model that takes into consideration the system flexibility, Equation (21) designs an early production system that considers a future tie-back development to the chosen manifold.

Platform installation and received flow rate are represented by Equation (22)-(24). These constraints guarantee that if a platform is installed, it receives a certain amount of production, from wells and/or from manifolds.

Equation (25)-(33) are the flow rate conservation constraints. These represent a set of important constraints, which ensures that the production flow from the wells may be delivered directly to a chosen platform or joined into a manifold, and, if it is the case, from the installed manifold it is sent to the platform.

The pipeline connection constraints ensure that, if there is a pipeline connection between infrastructures, the structural material of the pipeline is rigid or flexible, Equation (34)-(36).

The expected loss constraints, Equation (37)-(39), evaluate the risk level of each pipeline segment, assessing the severity and failure frequency, for flexible or rigid structural material.

4 CASE STUDY

A case scenario is described in the section. A new reservoir discovery, in which the estimated volume of the original oil-in-place is equal to 180,000,000 *bbl* and a pre-defined number of wells is 5. The well clustering division is evaluated for manifolds with 4 slots, and only one possible platform location, with two possible platform capacities, 200,000 and 110,000 *bbl/d*. The mean water depth of the oilfield location is 155m and the respective riser length is 186m (1.2 times the value of the mean water depth was used for the safety of all risers).

The gross cost estimations of the main production system infrastructures are presented in Table 5. The distance between the main infrastructures and the respective failure frequencies of each possible pipeline segment is shown in Table 6. Moreover, the wells, manifold and platform locations are presented in Figure 1.

Table 5. Cost estimates for the production system components.

Production System Components	Costs	Unit
Platform Cost (200,000 bbl/d)	$80 \times 10^{+07}$	US$ per unit
Platform Cost (110,000 bbl/d)	$65 \times 10^{+07}$	US$ per unit
Manifold Cost (4 Slots)	4,000.00	US$ per unit
Rigid Pipeline Cost	270.00	US$ per meter
Flexible Pipeline Cost	350.00	US$ per meter

Attempting to investigate the solution provided by the model, a second run is proposed changing the input data of the pipeline costs. For the second instance, the cost of the rigid pipes increases to 370US$/meter, and the cost of the flexible pipes decreases to 150 US$/meter.

Table 6. Distance between infrastructure and the respective failure frequencies of the pipeline segments, as rigid and flexible material.

Distances	W1	W2	W3	W4	W5	M1
P1	26	18	14	28	15	26
M1	3	8.67	12.67	3	11.67	-
Rigid Pipeline - Failure Frequencies						
P1	5.4×10^{-02}	3.7×10^{-02}	2.9×10^{-02}	5.8×10^{-02}	3.1×10^{-02}	5.4×10^{-02}
M1	6.3×10^{-03}	1.8×10^{-02}	2.6×10^{-02}	6.3×10^{-03}	2.4×10^{-02}	-
Flexible Pipeline - Failure Frequencies						
P1	1.3×10^{-02}	1.0×10^{-02}	8.4×10^{-03}	1.4×10^{-02}	8.9×10^{-03}	1.3×10^{-02}
M1	3.6×10^{-03}	6.1×10^{-03}	7.8×10^{-03}	3.6×10^{-03}	7.4×10^{-03}	-

4.1 *Results and discussions*

The problem and mathematical model are implemented at the IBM ILOG CPLEX Studio Optimization environment and solved through the dynamic search technique on a computer with Intel(R) Core (TM) i7 of 2.30 GHz and 16GB RAM. The main features of the solutions provided by the model are summarized in Table 7. Thus, Figure 2 and Figure 3 present the subsea layout and pipeline system.

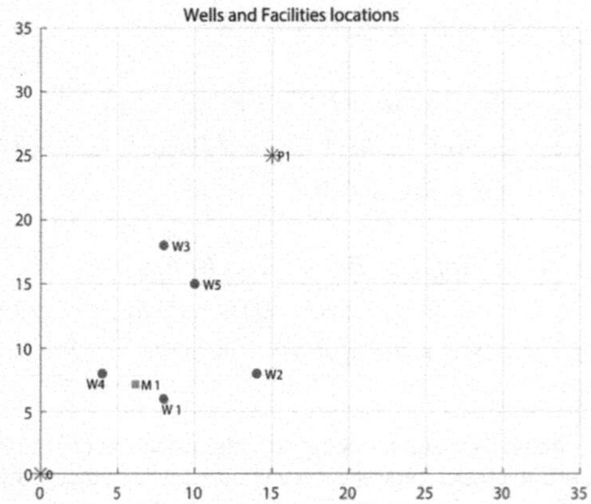

Figure 1. Production system infrastructure's locations.

At first glance, it is possible to note that, as discussed by Silva & Guedes Soares, (2023), the new model maintains the search for the most optimal tradeoff between the total production system cost and the total expected loss of the system, because it proposes that for the wells that are closer to the platform,

Table 7. Main features of the model solutions.

Case Study	Optimal Solutions	
	Solution 1	Solution 2
CPU time (s)	0.01	0.21
GAP (%)	0.00	0.00
Number of binary variables	36	36
Number of other variables	17	17
Number of non-zeros coefficients	165	165
Number of constraints	57	57
Obj. function	$6.82\times10^{+08}$	$6.88\times10^{+08}$
Total Cost (US$)	$6.70\times10^{+08}$	$6.74\times10^{+08}$
Expected Loss (US$)	$1.11\times10^{+07}$	$1.38\times10^{+07}$
Cost of Platform (US$)	$6.50\times10^{+08}$	$6.50\times10^{+08}$
Cost of Pipeline (US$)	$1.88\times10^{+07}$	$2.25\times10^{+07}$
Cost of Manifold (US$)	$4.00\times10^{+03}$	$4.00\times10^{+03}$
Cost of Riser (US$)	$1.95\times10^{+06}$	$1.95\times10^{+06}$

a direct path reduces the severity criterion and, respectively, reduces the expected loss. Thus, for the wells that are closer to the manifold, the model suggests the cluster satellite layout because it reduces the pipeline costs.

The improvement of the new model solution is that the model also considers the difference between the price and the frequency of failure of the pipeline structural material. The first solution assumes all the pipelines as rigid pipes. In this instance, even with a higher frequency of failure in the pipeline segments that connect the wells to the manifold, rigid pipelines are cheaper than flexible ones. For the second solution, because the rigid pipes are more costly than flexible, the pipeline segments of the cluster system are all flexible, showing that the system balances the risk level of the pipeline system and the total cost of infrastructures.

Moreover, the present case scenario is a small case study due to the reduced number of wells, which directly impacts on the complexity of the

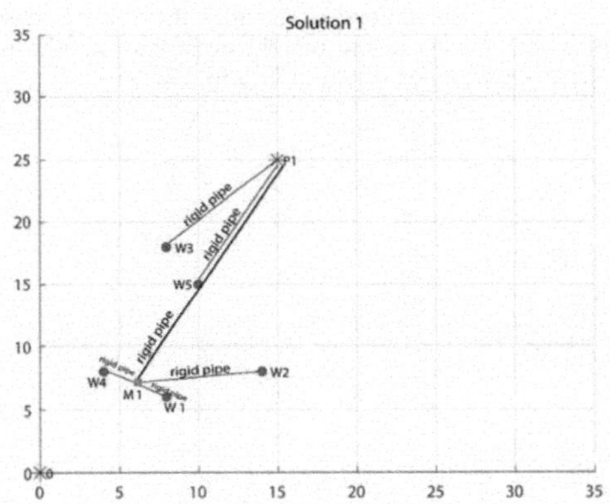

Figure 2. Subsea layout and pipeline system solution one.

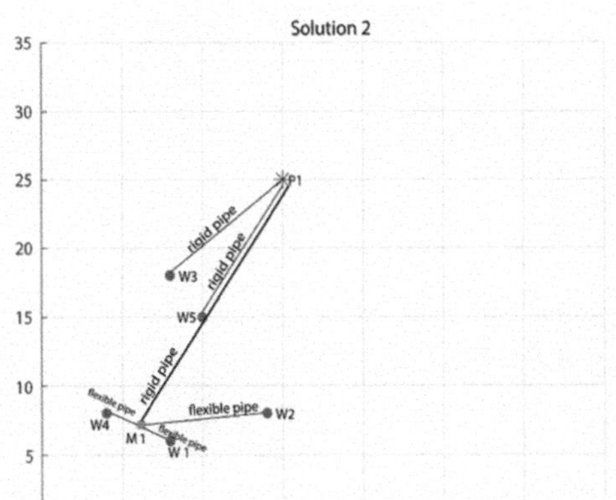

Figure 3. Subsea layout and pipeline system solution two.

model. This optimization problem is considered as an NP-hard, which means that the growth in the computational time can rapidly increase making it difficult to solve the problem using the branch-and-bound algorithm. To find the maximum polynomial time for the proposed model to find the exact solutions it is necessary to increase the problem size and test different instances of input data.

5 CONCLUSIONS

The paper presents a robust optimization model that integrates the structural material of the pipeline into the search for the optimal production system that minimizes the total capital expenditures of the system infrastructures and, simultaneously, minimizes the total risk level of the pipeline system. The main contribution of the present model is the refinement of the search not only for the most cost-benefit but also for combining the safety-economic search. This search refinement is based on the distinction of the pipeline structural material that changes the price and the failure frequencies.

A case study is provided to demonstrate the efficiency of the proposed robust mixed integer linear programming model. In conclusion, the robust model not only defines all the production system elements, as well as the proper system layout concept, the technology selection, location, number, and capacity of the production unit and/or size of the manifold, the proper assignment of the wells, and, the initial capital expenditures of main facilities and equipment but also, screens out the expected risk for each pipeline connection and defines the structural material of each segment. The results show that the model is a useful tool during the concept selection of an oilfield development helping the decision makers to visualize whether the risk is acceptable or not. In the cases that the risk is unacceptable, many precautions and barriers can be established, focusing on the minimization of the hazardous consequences of an unexpected event.

ACKNOWLEDGEMENTS

This work was performed within the scope of the Strategic Research Plan of the Centre for Marine Technology and Ocean Engineering (CENTEC), which is financed by the Portuguese Foundation for Science and Technology (Fundação para a Ciência e Tecnologia - FCT) under contract UIDB/UIDP/00134/2020.

REFERENCES

Alencar, M. H., Silva, L. M. R., Bhardwaj, U., & Guedes Soares, C. (2023). Risk Identification and Bowtie Analysis for Risk Management of Subsea Pipelines. *33rd European Safety and Reliability Conference*, 2107–2114.

Bhardwaj, U., Teixeira, A. P., & Guedes Soares, C. (2021). Burst strength assessment of X100 to X120 ultra-high strength corroded pipes. *Ocean Engineering, 241* 110004.

Bhardwaj, U., Teixeira, A. P., & Guedes Soares, C. (2022). Bayesian framework for reliability prediction of subsea processing systems accounting for influencing factors uncertainty. *Reliability Engineering & System Safety, 218*(PA), 108143.

Bhattacharyya, S. K., & Cheliyan, A. S. (2019). Optimization of a subsea production system for cost and reliability using its fault tree model. *Reliability Engineering and System Safety, 185* 213–219. https://doi.org/10.1016/j.ress.2018.12.030

Crawley, F. K., & Grant, M. M. (1997). Concept risk assessment of offshore hydrocarbon production installations. *Process Safety and Environmental Protection, 75*(3), 157–163.

DNV (2017). *Recommended Failure Rates for Pipelines*.

Gu, M., & Gudmestad, O. T. (2011). Uncertainties, risks, and opportunities in development of hydrocarbon fields. *ASME. International Conference on Ocean, Offshore and Arctic Engineering, 1*, 149–158.

Redutskiy, Y., Camitz-Leidland, C. M., Vysochyna, A., Anderson, K. T., & Balycheva, M. (2021). Safety systems for the oil and gas industrial facilities: Design, maintenance policy choice, and crew scheduling. *Reliability Engineering and System Safety, 210*, 107545.

Sales, L., Stolpnes, T., Stanko, M., & Faanes, A. (2023). Subsea Processing Optimization Considering Reliability and Maintenance. *Proceedings of the International Conference on Offshore Mechanics and Arctic Engineering - OMAE, 1*, 1–9.

Silva, L. M. R., & Guedes Soares, C. (2021). Oilfield development system optimization under reservoir production uncertainty. *Ocean Engineering, 225*, 108758.

Silva, L. M. R., & Guedes Soares, C. (2022). A stochastic programming model for designing offshore production systems. In Guedes Soares, C. & Santos T. A., (Eds.). *Trends in Maritime Technology and Engineering Vol.2*, 545–552.

Silva, L. M. R., & Guedes Soares, C. (2023). Robust optimization model of an offshore oil production system for cost and pipeline risk of failure. *Reliability Engineering and System Safety, 232*, 109052.

Silva, L. M. R., & Guedes Soares, C. (2016). Study of the risk to export crude oil in pipeline systems. In C. Guedes Soares & T. A. Santos (Eds.), *Maritime Technology and Engineering* 3 Taylor & Francis Group. 1013–1018.

Silva, L. M. R., & Guedes Soares, C. (2019). An integrated optimization of the floating and subsea layouts. *Ocean Engineering, 191*.

Silva, L. M. R., Teixeira, A. P., & Guedes Soares, C. (2019). A methodology to quantify the risk of subsea pipeline systems at the oilfield development selection phase. *Ocean Engineering*, 179, 213–225.

Advances in Maritime Technology and Engineering – Guedes Soares & Santos (Eds)

Coordinated scheduling of multiple resources in multi-functional seaports

X.Y. Zhang & W.Q. Guo
College of Navigation, Dalian Maritime University, Dalian, P.R. China

J.J. Li
School of Electromechanical and Automotive Engineering, Yantai University, YanTai, P.R. China

J.T. Wang
Yingkou AtoN division, Northern Navigation Service Center, Yingkou, P.R. China

J. Lin
College of Navigation, Longyan University, Longyan, P.R. China

ABSTRACT: In the evolution of ports towards multifunctionality, the collaborative scheduling of multi-resources plays a crucial role in increasing ports' operational efficiency. Coordinated scheduling of multiple resources aims at fully utilizing the port's anchorages, channels (one-way, two-way, compound, and restricted), and berths resources to provide optimal scheduling solutions for ships entering and leaving the port. This paper primarily analyzed the characteristics of the different types of channels and assessed the effects of anchorage, channel, and berth resources to improve the vessel scheduling efficiency using a multi-objective vessel scheduling–optimization model. Taking the minimum total scheduling and waiting times of vessels, the model was established based on these channels by considering safety, efficiency, and fairness. To solve the proposed model, a multi-objective genetic algorithm (MOGA) was used in this paper. Compared with the first-come, first-served (FCFS) scheduling method, the capacity-conversion times of vessels entering and leaving a port and the total scheduling time are decreased efficiently. These works improve the safety and optimization of ship scheduling catering to the future trend of seaport multifunctionality and intelligent maritime services.

1 INTRODUCTION

With the development of global trade globalization, ports have evolved from singular load and unload points to multifaceted centers for logistics, supply chains, and commercial activities. This multifunctional development demands a more efficient and cohesive utilization of port resources to adapt to ever-changing needs. However, the current utilization of port resources such as anchorages, channels, and berths does not consistently achieve optimal states. At times, resources may remain idle, while improper scheduling leads to traffic congestion at others. In this context, the imperative to explore coordinated scheduling of multiple resources in seaports becomes evident. This exploration stands as a cornerstone for bolstering the utilization of port resources and enhancing the traffic safety and operational efficiency of vessels.

Therefore, this paper first analyzes the ship scheduling characteristics in unidirectional, bidirectional, compound, and restricted channels from the perspective of coordinated multi-resource scheduling in ports. Subsequently, it constructs multi-objective scheduling optimization models based on these characteristics to evaluate the impact of resources such as anchorages, channels, berths, and terminal handling equipment on enhancing vessel scheduling efficiency. Finally, a multi-objective genetic algorithm is employed to attain the approximately optimal vessel scheduling plan within the multi-functional seaport.

The structure of the article is organized as follows: Section 2 of this paper conducts a comprehensive review of relevant literature. Section 3 delineates the challenges encountered in vessel scheduling within different types of channels. Simulation experiments were conducted in Section 4. Finally, the concluding Section 5 provides a comprehensive summary of the entire study and offers prospects.

2 LITERATURE REVIEW

For vessel traffic organization service, Byeong-Deok (2009) states that the different communication protocols and the data-exchange format of VTS can cause some problems in the overlap area. Therefore, they created a standard communication protocol and some supplementary information. Seung-hee and Daehee

DOI: 10.1201/9781003508762-13

(2015) concluded that e-navigation focuses on information exchange and the marine information service, not the connection and coordination of the security issues between vessels. Thus, they believe that e-navigation should have more emphasis on security. They researched the security-communication protocol and added the security information service into the marine service portfolios (MSPs). Byunggil Lee and Namje Park (2012) researched the security architecture of the inter-VTS exchange-format protocol. They designed a secure inter-VTS network security structure and security protocol based on the sharing of sensitive maritime data. Zhang (2013) focused on e-navigation dynamic information services such as the dynamic changes of basic geographic information, hydrological and meteorological information, water-transportation environmental information, and ship dynamic information. Huang and Chang (2013) believe that maritime-safety information (MSI) with navigation display is a high-priority user need in the e-navigation survey, and they built a software module to send MSI messages through the ship's navigation sensor. Through the literature review, we found that one research hot point concerns how to use pertinent information more efficiently. Therefore, this paper reports our studies under the MSPs framework to explore how to best use maritime traffic information by the VTS.

For the vessel traffic organization approach, previous studies have been conducted to carry out relevant research based on different channel types. Zhang et al. (2016, 2019) studied the integration problem of vessel scheduling and berth allocation in a one-way channel. They first studied the problem in a fixed planning period for a discrete berth bulk cargo port with a single harbor basin, then studied the problem in a multi-harbor basin. Zhang et al. (2019) expanded the mathematical model to incorporate vessel traffic conflict within a compound channel, aiming to minimize both the total waiting time of vessels and the channel occupancy time. Zhang et al. (2017) proposed a ship scheduling model for restricted two-way waterways, minimizing the weighted average of mean and maximum waiting times. The model prohibits two-way transportation in the presence of large ships, utilizing a sequential scheduling algorithm that divides ships into rounds. Experiments show the proposed algorithm significantly reduces waiting times, especially with a higher proportion of large ships, compared to the first-come-first-serve model. Zhang et al. (2020) studied the optimization model and algorithm of vessel traffic scheduling in a restricted two-way channel in Huanghua Port, taking the channel as the main object. Li et al. (2021) investigated the traffic scheduling problem of vessels entering and leaving restricted channels in a multi-harbor basin and generated an optimal traffic scheduling scheme for each vessel to ensure the safety and efficiency of vessel navigation.

In addition, many other scholars have researched vessel traffic scheduling, considering different resources involved in the port production process. Jia et al. (2019, 2020) considered the characteristics of different channel types and took the vessel entering and leaving port process as the critical factor to study the traffic scheduling problem under the combination of vessel and berth. Niu et al. (2020) studied the coordination problem of anchorage allocation and vessel traffic scheduling in Shanghai Yangshan deep-water port. Abou Kasm et al. (2021) studied the mathematical model of the vessel scheduling problem with tug and pilotage constraints and channel restrictions; then, they designed an exact solution method based on constraint separation. Liu et al. (2022) studied the seaport berth and channel planning problem, aiming to minimize the expected total weighted completion times of ships. However, most of the aforementioned studies tend to introduce port resources as one or a few constraints into vessel scheduling models, without considering the efficiency of utilizing multiple resources and their impact on the overall vessel scheduling plan.

3 SCHEDULING MODELING OF VESSELS IN DIFFERENT TYPES OF CHANNELS

The concise procedural framework of vessel ingress and egress from ports is outlined as follows: Departing from the originating port towards the destined harbor, vessels transmit anticipated arrival reports to the designated harbor scheduling department. Based on the received vessel reports, the harbor scheduling authority formulates berth plans, subsequently dispatching these plans to the Vessel Traffic Service (VTS) center. Upon reaching the port's boundary, vessels relay arrival notifications to the VTS center. Concurrently, vessels engaged in port-side docking operations forward departure reports to the VTS center upon their exit. The VTS center orchestrates vessel ingress and egress following the current navigational conditions, exemplified in Figure 1. The diversity in vessel count, tonnage, and distinct requirements pose challenges for decision-makers to manually devise optimal scheduling schemes. Hence, there arises a necessity to establish corresponding supportive decision models tailored to real-world scenarios.

3.1 *Mathematical model of vessel-scheduling optimization in a one-way channel*

The scheduling of vessels within one-way channels revolves around effectively coordinating navigational pathways and berth resources to establish an optimal scheduling sequence. The optimization objective is to minimize the total waiting time and total scheduling time of the ships. The total scheduling time represents the interest of the Port Authority and is calculated as the sum of the difference between the scheduling completion time and the scheduling start time for all vessels within the scheduling plan. The total waiting time represents the benefit to the ship operator and is calculated as the sum of the difference between the time of the start of the schedule (the time the ship anchors) and the time actual arrival time at the port for all ships within the scheduling plan. The queuing time is included in the waiting time.

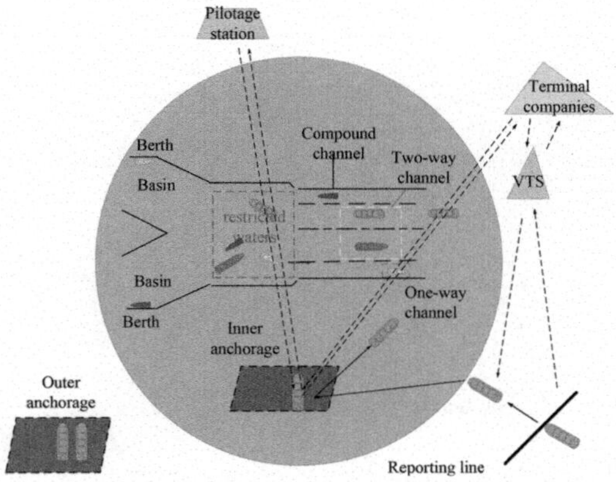

Figure 1. Vessel-operation flow in a seaport.

3.1.1 *Notation*

In this section, the parameters and decision variables utilized to build the MILP are defined.

F	the total time required by all vessels to complete the inbound and outbound processes;
W	the total waiting time of all vessels to be inbound or outbound after correction;
b_i	the starting time of vessel i to inbound or outbound;
f_i	the ending time of vessel i to inbound or outbound;
b_o	the starting time of the last vessel in the previous stage to inbound or outbound;
f_o	the ending time of the last vessel in the previous stage to inbound or outbound;
p_i	the theoretical time needed for a vessel to pass the fairway;
s_i	the theoretical distance of vessel i from the entrance of the fairway to the berth;
v_i	the average speed of vessel i from the entrance of the fairway to the berth;
$t0_i$	the time needed by vessel i to cross the report line;
IO_i	a binary that = 1 only if the ship enters the port; otherwise, it = 0;
R_{ij}	a binary that = 1 only if ship j is later than ship i inbound or outbound; otherwise, it = 0;
B_{ij}	a binary that = 1 only if the ship moors the near distance berth when ship i is scheduled earlier than vessel j; otherwise, it = 0;
$Berth_{im}$	a binary that = 1 if berth m is occupied; otherwise, it = 0
s	the distance from the report line to the entrance of the fairway;
h	the safety time interval of vessels in the same direction;
$\bar{h}$	the safety time interval of vessels in different directions;
M	a fixed constant number to ensure that the inequality is valid in the case where the conditions meet the requirement.

3.1.2 *Mathematical model*

Given the input parameters and decision variables, formulate the MILP model is formulated as follows:

$$\min[F, W] \tag{1}$$

$$F = \sum_{i=1}^{n}(f_i - f_{i-1}) + f_1 - b_0 - h_0 \tag{2}$$

$$W = \sum_{i=1}^{n}(b_i - t0_i - \frac{s+s_i}{v_i}) + \max(b_i - t0_i) - \min(b_i - t0_i) \tag{3}$$

$$b - t0 - \frac{s}{v_i} > 0 \tag{4}$$

$$(b_i - b_0 - h - B_{ij}h_1) \times (1 - (IO_i - IO_0)^2) \geq 0 \tag{5}$$

$$(f_i - f_0 - h - B_{ij}h_1) \times (1 - (IO_i - IO_0)^2) \geq 0 \tag{6}$$

$$(b_i - f_0 - \bar{h}) \times (IO_i - IO_0)^2 \geq 0 \tag{7}$$

$$f_i - b_i - p_i \geq 0 \tag{8}$$

$$p - \frac{s_i}{v_i} \geq 0 \tag{9}$$

$$(b_i - b_j + M \times R_{ij} - p_j - \bar{h}) \times (IO_i - IO_j)^2 \geq 0 \tag{10}$$

$$(-b_i + b_j + M \times (1 - R_{ij}) - p_j - \bar{h}) \times (IO_i - IO_j)^2 \geq 0 \tag{11}$$

$$(b_i - b_j + M \times R_{ij} - h - B_{ij}h_1) \times (1 - (IO_i - IO_j)^2) \geq 0 \tag{12}$$

$$(-b_i - b_j + M \times (1 - R_{ij}) - h - B_{ij}h_1) \times (1 - (IO_i - IO_j)^2) \geq 0 \tag{13}$$

$$(f_i - f_j + M \times R_{ij} - h - B_{ij}h_1) \times (1 - (IO_i - IO_j)^2) \geq 0 \tag{14}$$

$$(-f_i - f_j + M \times (1 - R_{ij}) - h - B_{ij}h_1) \times (1 - (IO_i - IO_j)^2) \geq 0 \tag{15}$$

$$(1 - Berth_{in}) + (1 - IO_I) \times M > 0, t = b_i \tag{16}$$

The objective functions (1) - (3) represent the minimization of total scheduling time and total waiting time. Constraint (4) indicates that the start time of a vessel's scheduling must occur after it arrives at the channel entrance. Constraints (5) to (7) ensure a safety time interval between vessels. Constraints (8) and (9) specify that the actual time taken by a vessel to traverse the channel should not be less than the theoretical passage time. Constraints (10) and (11) limit the time vessels spend on the channel to prevent conflicts at the entrance for vessels navigating in opposite directions. Constraints (12) - (15) ensure a safe time interval among vessels navigating in the same direction. Constraint (16) guarantees berth availability when a vessel enters the channel for docking.

3.2 *Vessel scheduling in a two-way channel*

Navigational management of vessels within two-way channels shares similarities with that of one-way channels. However, the primary distinction lies in two-way channels necessitating a transition to one-way usage under specific circumstances, such as impaired visibility, adverse wind conditions, or discrepancies in vessel capabilities. To accommodate these influential factors, scheduling within two-way channels requires the incorporation of a model for two-way-to-one-way conversion. The following excerpt outlines the criteria for mandating one-way navigation within a specific two-way port.

(1) Visibility less than VR1 meters.

(2) Wind force equal to or greater than WS1 level.

(3) Single vessel width equal to or greater than B1 meters.

(4) Single vessel width equal to or greater than B_1' meters, carrying hazardous cargo.

(5) The sum of the widths of two vessels intended for a passing maneuver is equal to or greater than B2.

When any of the aforementioned conditions are met, two-way channels must universally transition into one-way usage. This represents a distinctive attribute and fundamental requirement distinguishing two-way channels from their one-way counterparts. Hence, the regulations governing both one-way and two-way navigation are prescribed as per Eq. 17. Additionally, within two-way channels, when vessels are navigating in opposite directions and intend to pass each other, they are required to maintain a specific lateral separation.

Due to the restriction of two-way channels for one-way vessel traffic under certain circumstances, optimizing vessel scheduling in two-way channels necessitates an additional incorporation of a model for two-way-to-one-way traffic conversion. As depicted in Figure 2, this model encompasses four types of navigation mode transitions. Apart from the transitions between one-way-to-one-way and two-way-to-two-way traffic, two other mode conversions, namely one-way-to-two-way and two-way-to-one-way traffic shifts, need consideration. Moreover, the model should also account for transitions in traffic flow of vessels entering or exiting the channel, as well as the proximity of berths to the terminal ends of the channel.

$$\begin{cases} \textit{if } V \le V_1, \text{then } H_i = 1; \text{else } H_i = 0 \\ \textit{if } W \ge W_1, \text{then } H_i = 1; \text{else } H_i = 0 \\ \textit{if } B_i \ge Y_1, \text{then } H_i = 1; \text{else } H_i = 0 \\ \textit{if } B_i + B_j \ge Y_2, \text{then } H_i = 1; \text{else } H_i = 0 \\ \textit{if } B_i \ge Y_1' \text{ and the transportation of hazardous} \\ \quad \text{goods then } H_i = 1; \text{else } H_i = 0 \end{cases} \tag{17}$$

3.3 *Vessel scheduling in a compound channel*

On January 1, 2014, Tianjin Port, the largest port in northern China, inaugurated a compound channel system. Prior to this, access to Tianjin Port was limited to a single two-way main channel, which gradually became inadequate to accommodate the increasing volume of vessel traffic. Furthermore, the presence of numerous smaller vessel types such as patrol boats, port operation vessels, and fishing boats entering and exiting the main channel significantly impeded the navigation of larger vessels. Consequently, in both the northern and southern sections of the main channel, additional channels were excavated and dredged to create two smaller vessel channels, permitting unidirectional passage exclusively for small-sized vessels. The introduction of the compound channel system aimed primarily to cater to the diverse needs of vessels with varying sizes and types entering and leaving the port, mitigating the impact of smaller vessel types on medium to large vessels. This entailed dredging multiple channels of varying depths within the same cross-section, depicted in a simplified compound channel model as illustrated in Figure 3.

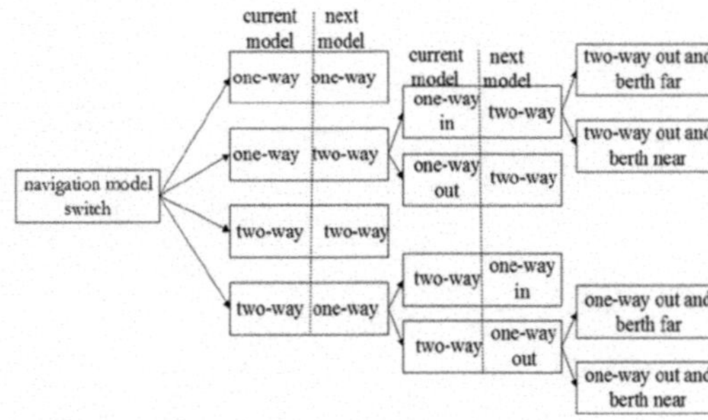

Figure 2. Navigational-model convert.

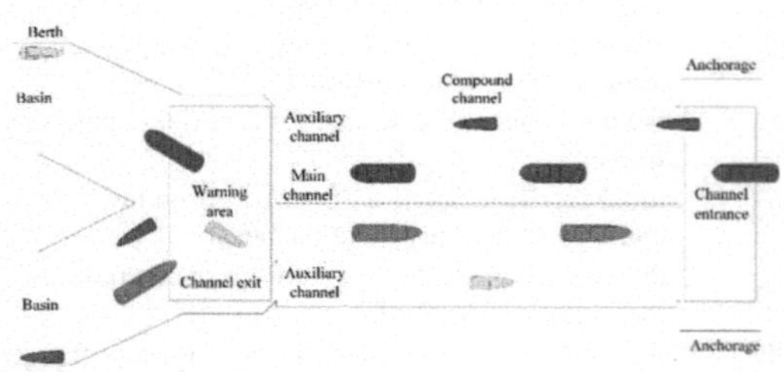

Figure 3. Compound channel.

Compared to two-way channels, the scheduling complexity within a compound channel system is heightened due to the diversity of vessel types and destinations, potentially accommodating multiple streams of vessel traffic simultaneously. Moreover, the presence of numerous intersecting waterways within the actual channels further complicates scheduling within the compound channel setup beyond that of one-way or two-way channels. As illustrated in Figure 4, considering inbound vessels, the red arrows indicate larger vessels entering the harbor via the deep-water channel, while the blue arrows represent smaller vessels entering through the shallower northern channel. In the left diagram of Figure 4, the

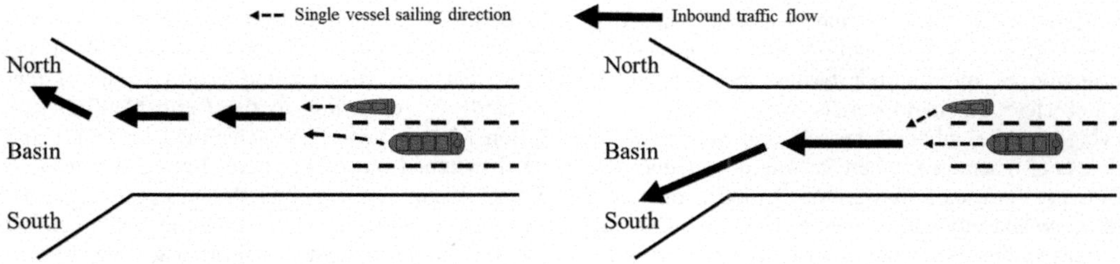

Figure 4. Inbound traffic flow within a compound channel.

overall traffic flow directs vessels toward the northern basin, while the right diagram indicates a southward flow toward the southern basin. When vessels from different channels converge upon reaching their respective terminals, variations in berth distribution lead to the intersection of vessels from different channels at the channel terminus.

The crux of scheduling complexities within a compound channel system lies in the strategic and rational management of traffic flow to minimize instances of vessel intersections and crossings. This constitutes a focal point of ongoing research within our research group

3.4 *Vessel scheduling in a restricted channel*

Through the analysis of the channel of Hamburg Port and Huanghua Coal Port, we design a layout of a restricted channel in multi-harbor basins shown in Figure 5. The two ends of the restricted channel are the multi-harbor basin entrance and the restricted channel entrance. There is an area connecting the one-way segment and a two-way segment in the restricted channel, which is called the one-way and two-way connection area.

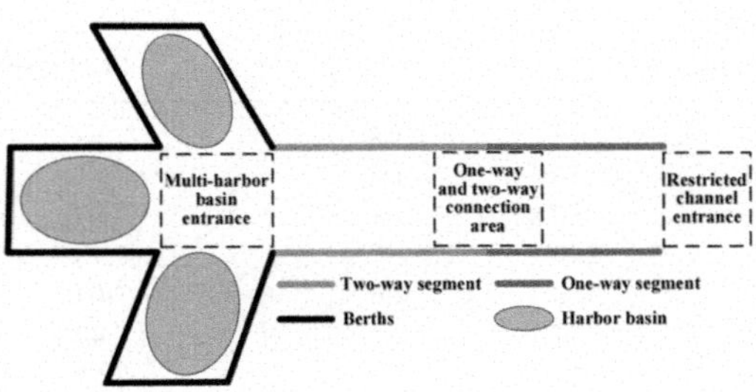

Figure 5. Physical layout of the restricted channel.

There is a one-way segment and a two-way segment in the restricted channel. Therefore, there are two navigational modes for vessels in a restricted channel: one is one-way and two-way mixed navigational mode, and the other is one-way navigational mode. One-way and two-way mixed navigational mode means that vessels need to change the navigational mode when sailing in one-way and two-way connection areas in a restricted channel. One-way navigational mode means that vessels sail in one direction in the whole restricted channel. Therefore, there are two types of navigational mode conversions in the restricted channel. In this section, we analyze the characteristics of vessel traffic conflicts according to the vessel traffic flow in the restricted channel.

The characteristics of vessel traffic conflict are mainly reflected in three situations: overtaking, head-on situation, and crossing situation. Figure 6 shows that when vessel traffic flow in the restricted channel is the single out or double-single out, the inbound and outbound flow with a head-on situation often occurs at the restricted channel entrance near the one-way segment. Similarly, when vessel traffic flow in the restricted channel is single-double in and double-single out, the vessels with a head-on situation usu-

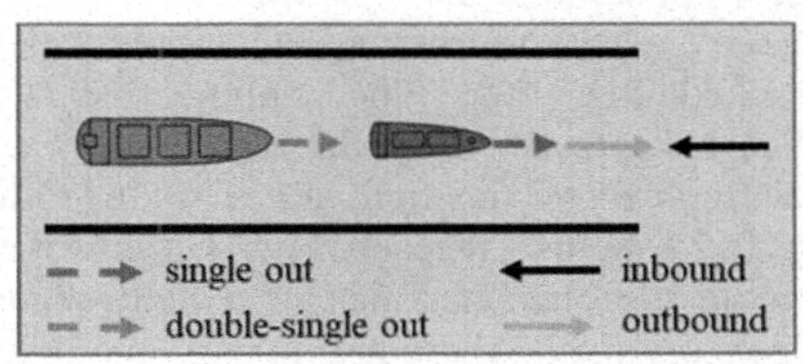

Figure 6. Vessel traffic conflict at the restricted channel entrance.

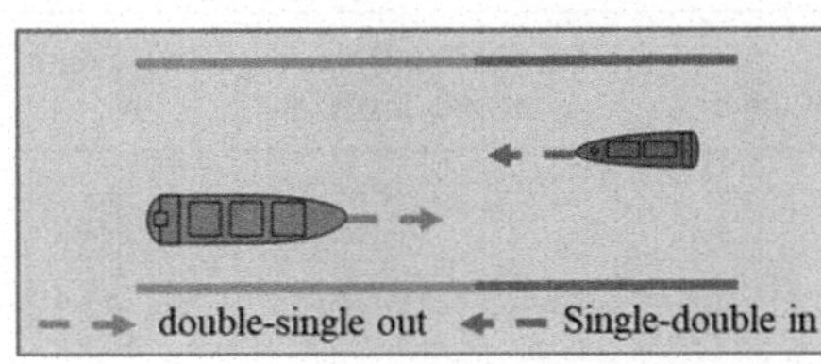

Figure 7. Vessel traffic conflict at the one-way and two-way connection area.

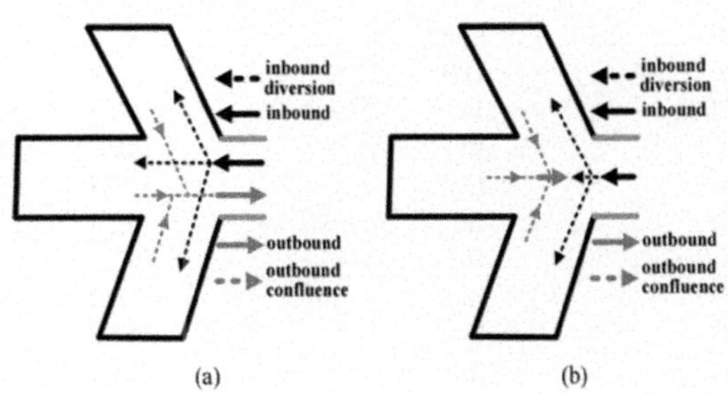

Figure 8. Vessel traffic conflict at the multi-harbor basins entrance. (a) One-way and two-way mixed navigational mode. (b) One-way navigational mode.

ally occur at the one-way and two-way connection area in Figure 7. A multi-harbor basin entrance is the intersection of multi-harbor basins and a restricted channel. Here, outbound vessels from different basins converge, while inbound vessels sail separately to each basin. Therefore, when vessels are in one-way and two-way mixed navigational mode, outbound confluence and inbound diversion often have crossing situations in this area (Figure 8(a)). When vessels are in one-way navigational mode, outbound confluence and inbound diversion often have head-on situations in this area (Figure 8(b)).

4 SIMULATION EXPERIMENTS FOR MULTI-RESOURCE COORDINATION SCHEDULING

The vessel data from the Vessel Traffic Service (VTS) center is presented in Table 1. Following computational modeling, eight Pareto optimal solutions are derived, as shown in Table 2. Among these sets of solutions, considering the minimization of total waiting time as the primary perspective, the optimal scheduling sequence is represented by the first entry in the table, with objective function values of 2.6186 hours and 24.7768 hours, respectively. Conversely, when prioritizing the minimization of total scheduling time, the optimal scheduling sequence is represented by the eighth entry in the table, with objective function values of 2.6985 hours and 23.0124 hours, respectively. These solutions conform to the characteristics of a multi-objective optimization model, where optimizing one objective value in a solution may lead to a deterioration in the outcome of another objective value. Table 1 indicates that both Vessel 1 and Vessel 19 are assigned to Berth 3. According to the Berth Conflict Resolution Model, Vessel 19, scheduled to depart, must be dispatched out of the berth before the arrival of Vessel 1. Among the eight Pareto optimal solutions depicted in Table 2, after resolving berth conflicts, the scheduling schemes prioritize the departure of vessels assigned to the same berth before the arrival of incoming vessels. This validation confirms the model's rationality and adherence to berth assignment constraints within vessel fleets exhibiting similar berth requirements.

In terms of optimization efficiency, vessels navigating within unidirectional channels are unable to encounter vessels traveling in the opposite direction. Hence, during the transition between inbound and outbound scheduling within these channels, it's imperative to ensure an absence of vessels navigating in the opposite direction. Consequently, each transition between inbound and outbound traffic necessitates a considerable amount of time to clear the channel. Addressing this issue theoretically suggests that minimizing the frequency of transitions between inbound and outbound traffic within unidirectional channels correlates with a reduction in overall scheduling time. A thorough examination of the scheduling sequences represented by Pareto 1 through Pareto 8 in Table 2 reveals that the scheduling sequences derived from the model significantly reduce the number of transitions between inbound and outbound traffic flows within the channel. Consequently, this substantially enhances the efficiency of vessel scheduling by minimizing the

Table 1. The information on vessels to be scheduled.

Ship number	Inbound and outbound directions	Length m	Berth number	Speed kn	Berth distance n mile	Start time
1	In	148	3	8	1.8	7:08
2	In	142.7	6	14	1.8	7:01
3	In	84.4	1	8.8	1.3	7:07
4	Out	144.8	5	9	1.7	7:18
5	Out	278.7	14	8.3	2.6	7:04
6	In	241.3	12	7.1	2.4	7:40
7	In	115.8	9	11.2	2.1	7:56
8	Out	97.2	13	9.5	2.5	7:49
9	Out	150.4	6	15.5	1.8	7:36
10	In	88.8	13	7.7	2.5	7:23
11	In	130	17	10.7	2.9	7:24
12	Out	108.4	10	11.7	2.2	8:29
13	In	142.7	5	11.5	1.7	7:12
14	In	72	2	9.6	1.4	7:47
15	In	166.2	7	12.3	1.9	7:38
16	In	147.5	11	10.9	*23*	7:09
17	In	334.1	14	12.9	2.6	7:03
18	In	85.5	10	8.6	2.2	7:01
19	Out	145.1	3	16.5	1.5	7:11
20	In	95.2	4	7.8	1.6	7:00

Table 2. Approximate optimal scheduling sequence.

No.	Target value 1	Target value 2	Scheduling plan
1	2.6186	24.7767	5 4 13 15 6 17 11 19 9 8 12 2 7 18 10 3 14 1 20 16
2	2.6398	23.8463	5 4 13 15 16 17 11 3 19 9 8 12 2 6 10 14 1 20 7 18
3	2.6589	23.5049	5 4 13 15 16 17 11 7 19 9 8 12 2 18 10 3 14 1 20 6
4	2.6349	23.9836	5 4 13 15 16 17 11 20 19 9 8 12 2 6 10 3 14 1 7 18
5	2.6675	23.1771	5 4 13 15 16 17 11 20 19 9 8 12 2 18 10 3 14 1 7 6
6	2.6765	23.1661	5 4 19 15 16 17 11 14 1 9 8 12 2 18 10 3 20 13 7 6
7	2.6455	23.8244	5 4 19 15 16 17 11 14 13 9 8 12 2 6 10 3 1 20 7 18
8	2.6985	23.0124	5 4 19 15 16 17 11 14 20 9 8 12 2 18 10 3 1 13 7 6

Table 3. Schedules of the movement plan for two vessels.

Pareto 1				Pareto 8			
Scheduling plan	Inbound and outbound directions	Start time	End time	Scheduling plan	Inbound and outbound directions	Start time	End time
5	Out	7:07	7:30	5	Out	7:07	7:30
4	Out	7:19	7:34	4	Out	7:19	7:34
13	In	7:34	7:46	19	Out	7:25	7:38
15	In	7:39	7:49	15	In	7:40	7:49
6	In	7:43	8:03	16	In	7:43	7:55
17	In	7:52	8:14	17	In	7:48	8:30
11	In	7:56	8:20	11	In	7:51	8:07
19	Out	8:22	8:27	14	In	8:07	8:16
9	Out	8:24	8:31	20	In	8:09	8:21
8	Out	8:25	8:41	9	Out	8:23	8:30
12	Out	8:29	8:43	8	Out	8:24	8:40
2	In	8:45	8:53	12	Out	8:29	8:42
7	In	8:46	8:58	2	In	8:44	8:52
18	In	8:48	9:03	18	In	8:46	9:01
10	In	8:50	9:09	10	In	8:48	9:07
3	In	9:07	9:17	3	In	9:05	9:15
14	In	9:09	9:20	1	In	9:09	9:20
1	In	9:13	9:24	13	In	9:12	9:25
20	In	9:15	9:27	7	In	9:15	9:31
16	In	9:19	9:37	6	In	9:21	9:41

frequency of transitions between inbound and outbound traffic within unidirectional channels.

Table 3 provides detailed insights into two optimal scheduling sequences. Examining Pareto 1 in Table 3 reveals that the departure time for Vessel 4 is at 7:34 AM, while the arrival time for Vessel 13 commences at 7:36 AM, ensuring a safe transition between inbound and outbound vessel traffic by accounting for time intervals between their movements. Additionally, adhering to a first-come-first-serve rule, scheduling the aforementioned 20 vessels results in a total scheduling time of 4.967 hours and a total waiting time of 49.5 hours. Comparing the model-derived eight optimal scheduling sequences against the first-come-first-serve method, Pareto 1 demonstrates a 47.2% reduction in total scheduling time, while Pareto 2 showcases a 45.6% reduction in total waiting time. In practical application, if the port prioritizes minimizing overall scheduling time, the optimal choice would be Pareto 1, which exhibits the lowest value for the first objective. Conversely, if minimizing vessel waiting time is paramount, selecting Pareto 8, with the lowest value for the second objective, would be more favorable. Based on this analysis, decision-makers can opt for an appropriate solution based on the prevailing requirements at the time

5 CONCLUSIONS AND FUTURE RESEARCH

The article analyzes the challenges in scheduling within one-way, two-way, compound, restrictive channels, and multi-resource coordinated scheduling. It emphasizes the establishment of an optimization model tailored to vessel scheduling, capitalizing on the characteristics of vessel movement in and out of these channels. The model, solved through genetic algorithms, translates the plethora of information associated with vessel movements in and out of these channels into streamlined and viable scheduling solutions. The complexities of vessel traffic within two-way, compound, and restrictive channels are addressed by augmenting the foundational one-way channel model with additional safety constraint models, such as two-way traffic conversion models for two-way channels and traffic conflict resolution models for compound channels. Simulation experiments validate the channel vessel scheduling optimization model, meeting the expected criteria.

Nevertheless, given the extensive nodes within vessel traffic organization, diverse vessel types, and intricate relational constraints, our model needs to be further refined to better capture the characteristics of the vessel scheduling problem. Moreover, with the escalation in vessel numbers, the scale of the scheduling conundrum burgeons, demands iterative algorithmic enhancements for more efficacious problem resolution.

ACKNOWLEDGEMENTS

The study is supported by the National Natural Science Foundation of China (Grant No. 52371359) and Dalian Science and Technology Innovation Fund (Grant No. 2022JJ12GX015).

REFERENCES

Abou Kasm, O., Diabat, A., Bierlaire, M., 2021. Vessel scheduling with pilotage and tugging considerations. Transp. Res. Part E Logist. Transp. Rev. 148. 10.1016/j.tre.2021.102231

Byeong-Deok YEA., 2009. Introduction of the Standardization in VTS data exchange. In: Proceedings of Asia navigation conference.

Huang, C. H., & Chang, S. J., 2012. Onboard integration of maritime safety information for navigation display and route safety. In 2012 12th International Conference on ITS Telecommunications (pp. 461–465). IEEE.

Jia, S., Li, C.L., Xu, Z., 2019. Managing Navigation Channel Traffic and Anchorage Area Utilization of a Container Port. Transp. Sci. 53, 728–745. https://doi.org/10.1287/trsc.2018.0879

Jia, S., Wu, L.X., Meng, Q., 2020. Joint Scheduling of Vessel Traffic and Pilots in Seaport Waters. Transp. Sci. 54, 1495–1515. https://doi.org/10.1287/trsc.2020.0990

Lee, B., & Park, N., 2012. Security architecture of inter VTS exchange format protocol for secure u-Navigation. In Embedded and Multimedia Computing Technology and Service: EMC 2012 (pp. 229–236). Springer Netherlands.

Li, J.J., Zhang, X.Y., Yang, B.D., Wang, N.N., 2021. Vessel traffic scheduling optimization for restricted channel in ports. Comput. Ind. Eng. 152. https://doi.org/10.1016/j.cie.2020. 107014

Liu, B.L., Li, Z. C., Wang, Y.D., 2022. A two-stage stochastic programming model for seaport berth and channel planning with uncertainties in ship arrival and handling times. Transp. Res. Part E Logist. Transp. Rev. 167, 102919. https://doi.org/10.1016/j.tre.2022.102919

Niu, M., Wang, Z.Y., Li, Y.J., Zheng H.X., 2020. Berth allocation and ship scheduling integrated optimization considering the priority of berth in use. Chin. High Technol. Letters 30(09), 972–981. https://doi.org/10.3772/j.issn.1002-0470.2020.09.012

Oh S, Seo D, Lee B., 2015. S3(Secure Ship to Ship) Information sharing scheme using ship authentication in the e-navigation. Int J Secur Appl 9:97–110.

Zhang A M., 2013. Research on some key techniques of dynamic information services in e-Navigation. Wuhan

Zhang, J., Santos, T.A., Guedes Soares, C., Yan, X., 2017, Sequential ship traffic scheduling model for restricted two-way waterway transportation, Journal of Engineering for the Maritime Environment, Vol. 231, Issue 1, pp. 86–97. https://doi.org/10.1177/1475090215621580

Zhang, X.Y., Lin, J., Guo, Z.J., Liu, T.S., 2016. Vessel transportation scheduling optimization based on channel-berth coordination. Ocean Eng. 112, 145–152. https://doi.org/10.1016/j.oceaneng.2015.12.011

Zhang, X.Y., Li, R.J., Chen, X., Li, J.J., Wang, C.B., 2019. Multi-object-based Vessel Traffic Scheduling Optimisation in a Compound Waterway of a Large Harbour. J. Navig. 72, 609–627. https://doi.org/10.1017/S0373463318000863

Zhang, X.Y., Wang, Z.Q., Deng, Z.P.,2020. Ship Traffic Organization Optimization of Constrained Two-Way Channel in Huanghua Port. Navig. China 43(04):1–7. https://doi.org/10.3969/j.issn.1000-4653.2020.04.001

Advances in Maritime Technology and Engineering – Guedes Soares & Santos (Eds)
 ISBN 978-1-032-83099-5

A bibliometric analysis on ship scheduling from 2000 to 2023

H.T. Zhu, Z.Y. Li & B. Wu
State Key Laboratory of Maritime Technology and Safety (SKL MTS), Wuhan University of Technology, Wuhan, P.R. China
Intelligent Transport Systems Research Centre (ITSC), Wuhan University of Technology, Wuhan, P.R. China
National Engineering Research Centre for Water Transport Safety (WTSC), Wuhan University of Technology, Wuhan, P.R. China

ABSTRACT: The ship scheduling problem has received significant research attention. So far, a bibliometric analysis that offers a summary review of this topic is yet to be conducted, this paper presents a literature review on ship scheduling. 381 publications from 2000 to 2023 were collected from the core collection database of Web of Science, to analyze co-authorship network, co-authorship network, keyword co-occurrence network and citation network by using VOSviewer for visual analysis. The research hotspots, publications quantity trends, and future research directions on this topic are indicated. The visual analysis illustrates the limitations of current research and future research hotspots, and also provides guidance and help to promote the development of ship scheduling.

1 INTRODUCTION

Many processes of ship navigation require scheduling management, so ship scheduling is very important in ocean transportation. Ship scheduling is the optimization and scheduling of ship types on a shipping route based on the actual freight situation of the shipping company when determining the route. The model is complex, difficult to solve, and has limited practicality. Subsequently, more and more scholars have studied the model and solution algorithm of ship allocation scheduling on their basis. In recent years, research has also been strongly related to the following keywords, such as waterways, berths, docks, ports, locks, etc. Ship scheduling has strong research value in these keywords.

Many research on ship scheduling has provided valuable references, but currently there has been no bibliometric analysis on ship scheduling. Based on this, this article aims to provide an academic achievement to fill the gap in visual analysis of ship scheduling overview.

Bibliometrics quantitatively analyzes knowledge systems, by using mathematical and statistical techniques, and forms a scientific methodology for this purpose. It is employed for the analysis, comparison, and quantification of literature to comprehend the development and trends of scientific research (Jiang et al, 2023).

In the early days, bibliometrics was mainly applied in fields such as informatics, mathematics, and medicine. Later, due to the maturity of this method, it was also applied in the field of shipping industry (Yang et al., 2023).

Therefore, the motivation of this article is to provide an academic achievement to fill the cross-research gap between bibliometric overview and visual review analysis in the field of ship scheduling. Therefore, 388 publications are collected from the Web of Science (WOS) core database as a valid dataset. On this basis, VOSviewer is used to perform mapping analysis on the collaborative author network (i.e. country, organization, and author), author co-citation network, keyword co-occurrence network, and citation network of the paper. In addition, the growth trend of paper publications and the contribution of scientific papers (i.e. number of citations, number of first-author papers, and journals) are analyzed, and Origin is used to visualize the analysis results. Finally, based on the analysis results of keyword co-occurrence networks, the main research hotspots are summarized and reviewed, and suggestions for future work are proposed. Therefore, the overall goal of this article is to: (1) Analyzing the growth trend of ship scheduling publications; (2) Identifying the most influential articles, journals, and authors; (3) Identifying the most productive authors, countries, organizations, and their collaborative relationships; (4) Looking ahead to future research hotspots.

The remaining part of this article is structured as follows: In Section 2, the research methodology of this study is described. Section 3 combines bibliometrics and statistical theory to analyze the literature dataset. Section 4 looks forward to future research

DOI: 10.1201/9781003508762-14

hotspots. Finally, the fifth section is the summary and conclusion of this article.

2 METHODOLOGY

This study chose the Web of Science Core Collection for research, using advanced retrieval with the search formula set to "ship scheduling" OR "vessel scheduling" and the search content set to "topic". A total of 4215 results were obtained from the search, but the content was too much and complicated. The search content was changed to "title", resulting in 388 articles that were exported in plain text file format. Subsequently, we used the extraction function of WOS and set the extraction content to "Full Record and Cited References", including full title, author, affiliation, citation frequency, keywords, and other information. Based on the exported information, conduct subsequent bibliometric analysis. And, after checking the exported information and removing 5 articles containing invalid fields, a total of 381 articles were obtained as the research dataset for this article.

Next, VOSviewer and Origin are used to visually analyze the literature review of the dataset in this article. The VOSviewer software is a powerful analytical tool that offers visualization capabilities for collaborative relationships among authors, countries, and organizations, as well as keyword co-occurrence networks, citation networks, and co-citation networks. By employing VOSviewer, researchers can easily interpret and represent knowledge from their datasets (Van Eck and Waltman, 2010). In this study, VOSviewer is utilized to visualize network nodes related to sources, countries, organizations, authors, and keywords. When constructing and visualizing bibliometric networks using bibliographic data, VOSviewer offers two counting methods: full count and fractional count (Van Eck and Waltman, 2010). However, the full count method is commonly preferred over fractional counting. Therefore, this research employs the full count method for analysis.

3 ANALYSIS OF LITERATURE DATASET

3.1 *Publications growth tendency*

Figure 3 displays the distribution of the 381 selected papers according to their publication year. The figure demonstrates a consistent upward trend in the number of papers published annually. It is noteworthy that two distinct developmental stages can be observed in the field of ship berthing. From 2011 to 2023, there has been a significant surge in published papers, culminating in a peak of 39 papers in 2020. Importantly, the future growth trajectory of publications can be predicted using the S-curve logic model proposed by Fisher and Pry in 1971 (Fisher and Pry, 1971). According to this model, the number of publications on ship scheduling is expected to continue increasing over an extended period.

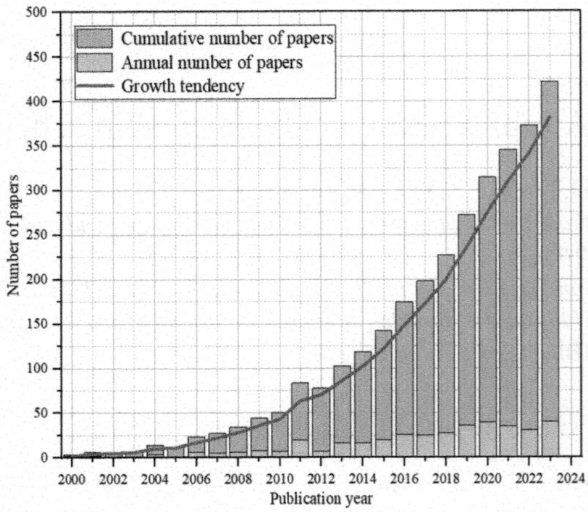

Figure 1. The distribution of published papers by publication year.

3.2 *Contribution of scientific papers*

3.2.1 *The cited times of papers*

Figure 2 presents the information on papers with more than 95 citations, including their publication year, cited authors, and connections represented by the same color. There is a total of 15 articles in this category.

It is important to note that only the first author is considered in this analysis. The most frequently cited paper, authored by Christiansen Marielle and published in 2004, has been cited 465 times. The second-ranked paper, also authored by Christiansen Marielle and published in 2013, has received 366 citations.

Furthermore, in both 2013 and 2014, two articles each achieved high citation rankings. The two articles from 2013 received 366 and 118 citations respectively, while the two articles from 2014 received 303 and 103 citations respectively. The ranking combination for these two years included one article with a high ranking and another with a lower ranking. In 2012, three articles simultaneously made it to the high citation list, with citation counts of 183, 132, and 108 respectively, positioning them in the middle of the list.

Upon examining the raw data, it is evident that Meng Qiang and Wang Shuian have a collaborative relationship in the three articles within this ranking, with a cooperation rate of 20%. This finding suggests that the high degree of collaboration between these two authors contributes to the recognition of their resulting articles within the academic community. Additionally, the articles ranked 9th and 14th are both single-author articles.

The number of citations serves as an indicator of an author's contribution to some extent. Therefore, in the context of ship scheduling development, Christiansen Marielle from the Norwegian University of Science and Technology has made the most significant contributions according to the citation analysis.

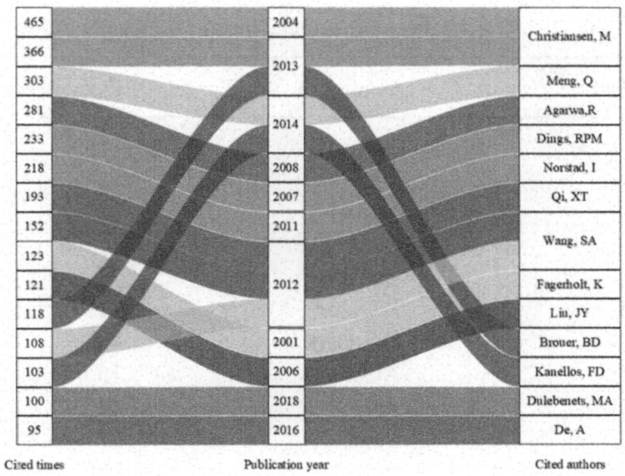

Figure 2. Papers with high citations by the publication year and cited authors.

Table 1. The significant information of the authors with at least three papers.

The first author	Countries	Organizations	Number of publications
Dulebenets, MA	USA	State University System of Florida	6
Fagerholt, K	Norway	Norwegian University of Science and Technology	6
Christiansen, M	Norway	Norwegian University of Science and Technology	5
Jiang,X	China	Southeast University	4
Zhang, XY	China	Dalian Maritime University	4
Bronmo, G	Norway	Norwegian University of Science and Technology	3
Cheraghchi, F	Canada	University of Ottawa	3
De, A	India	Indian Institute of Technology	3
Jia, S	China	Hong Kong Polytechnic University	3
Korsvik, JE	Norway	Norwegian University of Science and Technology	3

3.2.2 *Number of papers as the first author*

Table 1 presents the key information regarding authors who have published at least three papers. The table includes 10 authors, along with their respective countries, organizations, and the number of published papers. Notably, two authors stand out with the highest publication volume: Maxim A. Dulebenets and Fagerholt Kjetil, both of whom have published six papers. Dulebenets is affiliated with the State University System of Florida, while Fagerholt is associated with the Norwegian University of Science and Technology. Following closely is Christian Marielle, who has published five articles, and Jiang Xi and Zhang Xinyu, both with a publication volume of four articles. The remaining five authors have each published three articles, collectively accounting for 50% of the authors listed in this ranking.

Figure 3 illustrates the contributions made by authors who have published at least three papers as first authors. As one of the most prolific authors, Maxim A. Dulebenets published three papers in 2018, and one paper each in 2017, 2021, and 2022, respectively. Fagerholt Kjetil, another author with a large number of publication records, published two papers in 2013 and one paper each in 2000, 2001, 2004, and 2009. Christian Marielle contributed one paper each in 2002, 2004, 2013, and 2015. Additionally, it is worth noting that Jiang Xi began to publish papers in the past three years, which indicates his recent entry into this field.

3.2.3 *Journal of publications*

This article analyzes at least five journals that have published papers, most of which have been published in the fields of operations research optimization, marine science and engineering, and computer science and technology.

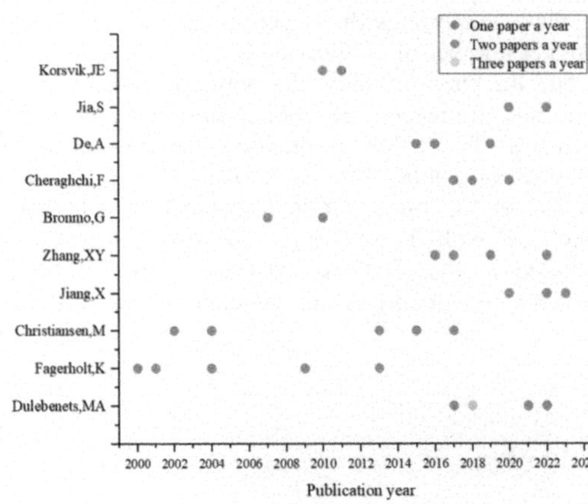

Figure 3. Production authors as the first author by year.

The number of articles published in journals that reached five or more papers per year is shown in Figure 4. According to Figure 4, at least five papers have been published in 15 journals. Specifically, 132 papers were published in these journals, accounting for 34.5% of the total number of papers in the dataset (381). Among them, the top three journals with the highest number of publications are Computers & Industrial Engineering (CIE) with 16 articles, European Journal of Operational Research (EJOR) with 15 articles, and Transportation Research Part E-Logistics and Transportation Review with 14 articles. These three journals focus on operations optimization and computer science and technology. The remaining 12 journals published fewer papers than the above three journals. Among them, the fourth-ranked journal, Journal of Marine Science and Engineering (JMSE), has published ten papers, while the fifth-ranked journal, Computers & Operations Research (COR), has published nine papers. The former journal's research field is marine science and engineering, and the latter journal's research field is the intersection of operations research optimization and computer science and technology.

3.3 *Analysis of co-authorship*

Within this subsection, an analysis of co-authorship among authors, organizations, and countries is conducted. To present the cooperative relationships clearly, a threshold is established. The full counting method is employed, treating each paper with equal importance and assigning equal weight to each co-authorship link. The purpose of this co-authorship analysis is to identify countries, organizations, and authors that demonstrate high productivity in the field of ship scheduling.

3.3.1 *Co-authorship network analysis of countries*

Figure 5 shows the network of cooperative relationships in the field of ship scheduling. The size of each node corresponds to the number of published papers in that country, with larger nodes representing a higher number of publications.

The thickness of the links between nodes represents the frequency of cooperation between two countries. The color of the nodes corresponds to the year of publication, as indicated by the color bar below the image. It is important to note that countries with a minimum of two papers are included in the analysis. Based on this criterion, a total of 31 countries are selected as samples for analyzing the co-author network. The color clustering in the image reveals that these 31 countries are grouped into seven clusters, with a total of 72 connected links.

Table 2. Top-10 productive countries on ship scheduling.

Rank	Countries	Documents
A	China	158
B	USA	51
C	Norway	38
D	Singapore	36
E	Canada	22
F	England	21
G	Denmark	18
H	Australia	15
I	Japan	14
J	India	11

Table 2 highlights the top 10 countries with the highest productivity in ship scheduling fields. The rankings are determined based on the number of published papers. Notably, Japan, Taiwan, China, and Norway have been consistently publishing papers on ship scheduling since 2014. Among them, Norway has the highest number of published papers (38), followed by Japan with 14 papers. In 2016, the United States, Australia, and Mexico also began to contribute papers to the field. China and Singapore started showing interest in ship scheduling around 2018.

From Figure 5, it is evident that there is close cooperation between China and Singapore, the United States and South Korea, as well as Norway and Canada. China leads the pack with 158 papers,

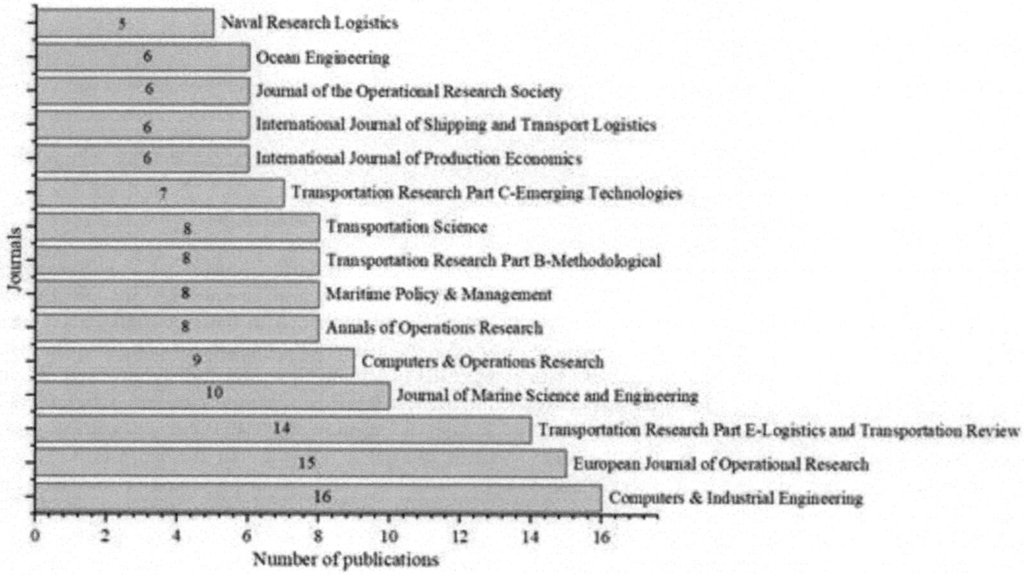

Figure 4. Number of publications by journals.

claiming the top spot in terms of paper productivity. The United States follows with 51 papers, and Norway with 38 papers, accounting for 41.15%, 13.28%, and 9.89% respectively. Based on this analysis, it can be concluded that China, the United States, and Norway exhibit the highest productivity not only in terms of publication volume but also in terms of their close collaboration with other countries. These three countries play an essential role in the development of ship scheduling.

3.3.2 *Co-authorship network analysis organizations*
To analyze the co-authorship network between the organizations, the minimum article publishing threshold for an organization is set to 4.

There are a total of 28 organizations in the available analysis chart, which are connected. The collaborative network of organizations is shown in Figure 6, where the size of nodes represents the number of papers published by these organizations, the thickness of links between each node reflects the weight of cooperation, and the color of the organization represents different clusters. There are seven clusters of different colors.

Norwegian University of Science and Technology (Norway) is more willing to cooperate with Shanghai Maritime University (China) and the National University of Singapore (Singapore), while Wuhan University of Technology (China) is more willing to cooperate with Hong Kong Polytechnic University (Hong Kong, China), Hong Kong University of Science and Technology (Hong Kong, China), and Shanghai Jiao Tong University (China). In addition, Dalian Maritime University has a more significant cooperative relationship with Nanyang Technological University, Ningbo University, and Huazhong University of Science and Technology. Tsinghua University has also formed a strong cooperative network with Hong Kong Polytechnic University (Hong Kong, China) and Hong Kong University of Science and Technology (Hong Kong, China). In terms of European schools, the cooperative relationship between Delft University of Technology and the University of Aalborg is closer.

Table 3 summarizes the seven most prolific institutions with over eight published papers, ranked based on the number of papers published by the institutions. It can be seen that the highest output of articles is 35 papers, corresponding to Dalian Maritime University, followed by the second-ranked Norwegian University of Science and Technology and the National University of Singapore, with 22 and 19 papers respectively. It is worth noting that although there are research organizations such as the Chinese Ministry of Education involved in Figure 6, in the list in Table 3, these organizations are all universities, indicating that universities, as the most important research institutions, play an important role in ship scheduling research.

3.3.3 *Co-authorship network analysis of authors*
Set the minimum publication number for a single author to 2, and connect the authors with other authors. The set of connected authors is shown in Figure 7, where the color reflects the cluster and the thickness of the links represents the strength of cooperation. Please note that the same author refers to the publication name of the paper (including first author, corresponding author, etc.). The collaborative networks of these authors can be divided into seven clusters of different colors, namely the blue cluster, orange cluster, green cluster, red cluster, blue cluster, purple cluster, and yellow cluster. Among them, it

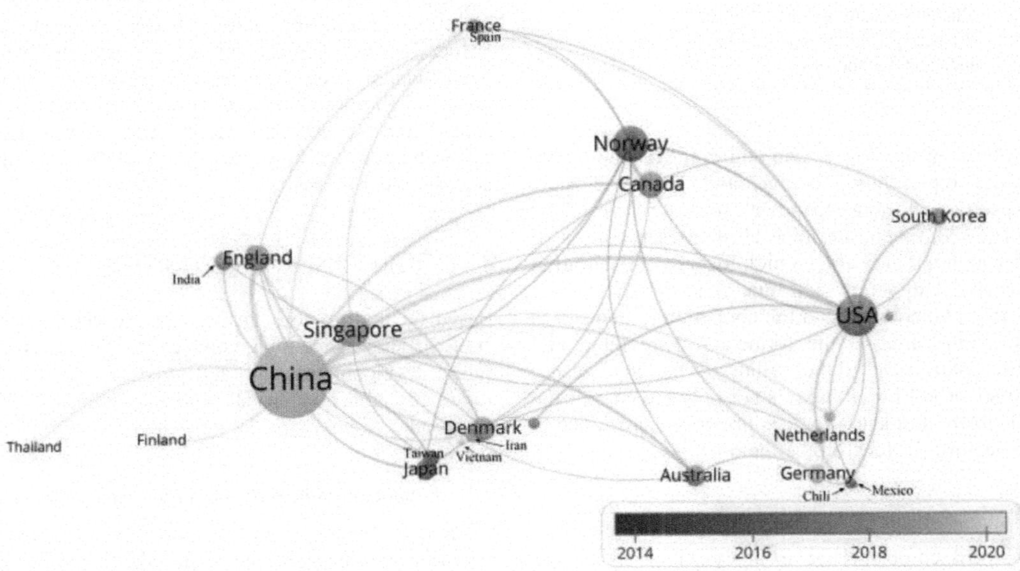

Figure 5. Visualization of co-authorship network between countries.

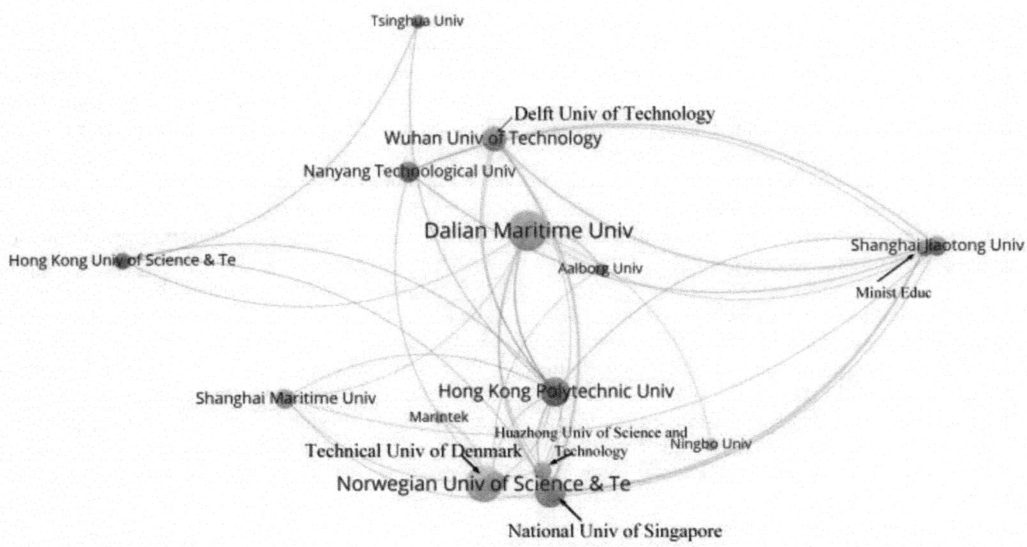

Figure 6. The mapping of organizations co-authorship network.

Table 3. Top-7 productive organizations on ship scheduling.

Rank	Organizations	Countries	Documents
A	Dalian Maritime Univ	China	35
B	Norwegian Univ of Science and Technology	Norway	31
C	National Univ of Singapore	Singapore	22
D	Hong Kong Polytechnic Univ	Hong Kong, China	19
E	Wuhan Univ of Technology	China	14
F	Shanghai Jiaotong Univ	China	10
G	Huazhong Univ of Science and Technology	China	8

can be seen that Yang Xiao and Wang Yadong from China, representing orange and purple clusters respectively, form a separate cluster.

The red, green, and blue clusters each consist of three authors, indicating a high level of collaboration intensity. The yellow and blue clusters each consist of only two authors, and the cooperation intensity is lower compared to the former. Among them, the most noteworthy is the author of the red cluster, Christiansen Mariell, who is not only the core of the red cluster, but also provides important assistance in connecting clusters of other colors.

3.4 *Analysis of keywords co-occurrence*

Keyword co-occurrence is analyzed based on the number of identical keywords in published papers. The co-occurrence network of keywords divided by publication year is shown in Figure 8, where the minimum occurrence threshold for one keyword is set to 15, and 27 keywords meet this threshold. The size of the nodes in the graph represents the frequency of each keyword appearing, and the thickness of the links between each node represents the co-occurrence frequency of two keywords. In order to fully display the co-occurrence network of these keywords, some synonyms of certain keywords are not merged (such as "ship scheduling" and "vessel scheduling").

Based on the visualization results, it can be found that the hot topics of ship scheduling represent the research hotspots of the past and the future research directions. From the graph, it can be seen that "Model", "Algorithm", and "Liner shipping" have been widely used keywords, resulting in larger nodes. In addition, "Time windows", "Genetic Algorithm", and "speed optimization" have been popular research keywords in recent years, and the nodes are correspondingly smaller.

4 FUTURE RESEARCH HOTSPOTS

Based on the previous analysis of the current situation of ship scheduling, this section discusses the future development trend, and puts forward several future research hotspots of ship scheduling in four directions.

4.1 *Optimization models and algorithms*

Continuous research is essential for developing advanced optimization models and algorithms in ship scheduling. Although significant progress has been made in research in this field, there are still issues of

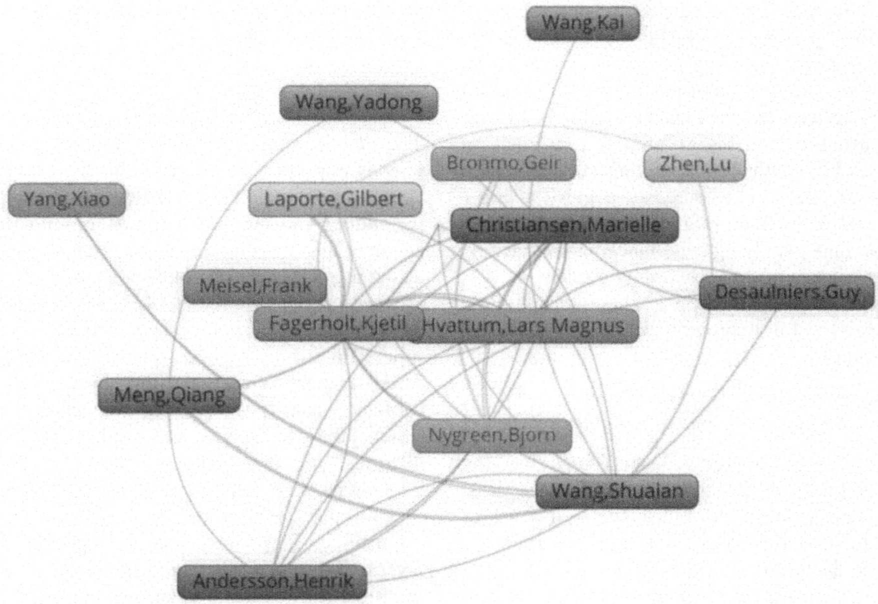

Figure 7. Co-authorship network analysis of authors.

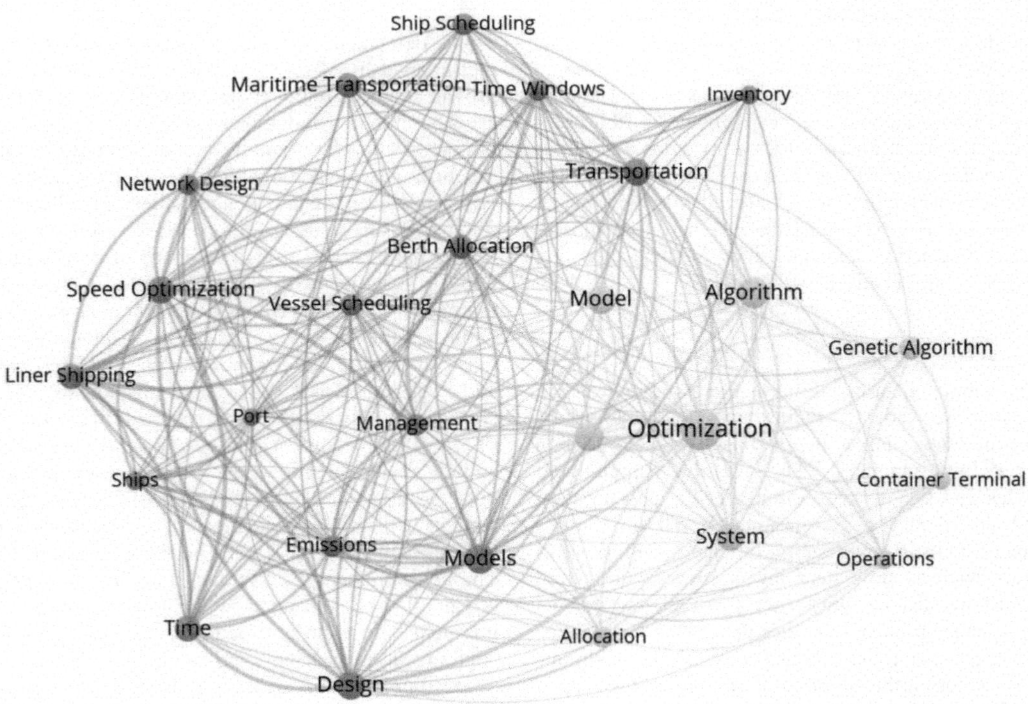

Figure 8. The visualization of keywords co-occurrence network.

complexity and uncertainty that need to be addressed. Future research should focus on exploring new methods that combine metaheuristic and mathematical programming to achieve better scheduling solutions. Among them, in terms of optimization methods, more attention needs to be paid to optimization mathematical models under uncertain scenarios. The model parameters in the marine field are also plagued by uncertainty. Ksciuk has conducted a comprehensive study on the optimization of ship scheduling under uncertain scenarios, including the travel time of ships, loading and unloading time of ports, available demand of ports, fuel prices, etc. (Ksciuk J et al,2023). In addition, by using hybrid algorithms, factors such as dynamic weather conditions, port congestion, and ship characteristics can be considered to achieve a larger scale and more efficient solution process (Al et al, 2024). In addition, studying the application of machine learning techniques in ship scheduling models, such as reinforcement learning or deep learning, can provide valuable insights for improving scheduling accuracy and adaptability.

4.2 *Integration of emerging technologies*

The integration of emerging technologies has the potential to promote the ship scheduling practices. Based on Tang's paper, the block chain, Internet of Things (IoT), and cloud computing offer opportunities for enhanced data sharing, collaboration, and decision-making in the maritime industry (Tang Wenzhi et al,2023). Future research should delve into how these technologies can be effectively integrated into ship scheduling systems. For example, blockchain can facilitate secure and transparent transactions, improving trust and accountability in supply chain operations. The IoT can enable real-time monitoring of vessel conditions, cargo status, and port congestion, leading to more accurate and faster responsive scheduling decisions. Cloud computing can provide scalable computing resources for complex optimization algorithms and enable efficient communication and coordination among stakeholders in a distributed environment. Exploring the synergistic effects of these technologies and their practical implementation in ship scheduling will be helpful in advancing the field.

4.3 *Sustainable and green ship scheduling*

Environmental sustainability is a pressing concern in the maritime industry. Future research should prioritize the integration of green considerations into ship scheduling practices. This involves optimizing routes and schedules to minimize fuel consumption and greenhouse gas emissions. Based on Meng's paper, investigating energy-efficient routing strategies, such as slow steaming or route optimization algorithms that account for weather patterns and currents, can significantly reduce the carbon footprint of maritime operations (Meng L et al,2023). Additionally, exploring the adoption of alternative fuels, such as liquefied natural gas (LNG) or hydrogen, and evaluating their impact on scheduling and operational efficiency is essential. Furthermore, incorporating emission reduction strategies, such as shore power utilization in ports or exhaust gas cleaning systems, can contribute to sustainable ship scheduling practices. By aligning ship scheduling with environmental objectives, the industry can make significant strides towards a greener and more sustainable future.

4.4 *Real-time decision support systems*

The development of real-time decision support systems is important for efficient ship scheduling. Leveraging big data analytics, predictive modeling, and AI techniques can provide participants with timely and accurate information. Future research should focus on designing robust and scalable systems that can handle the dynamic and unpredictable characteristics of maritime operations. These systems should integrate real-time data from multiple sources, including weather forecasts, port congestion information, and vessel performance data. By employing machine learning algorithms, these systems can continuously learn from historical data and make accurate predictions, enabling proactive scheduling adjustments and improving overall operational efficiency. In addition, incorporating artificial loops into decision support systems can leverage the expertise and domain knowledge of maritime professionals to ensure the practicality and effectiveness of scheduling decisions.

5 CONCLUSION

This article aims to provide a comprehensive bibliometric analysis of ship scheduling by analyzing the growth trend of publications, identifying influential articles, journals, and authors, and exploring collaborative relationships among authors, countries, and organizations. Additionally, the study identified future research hotspots in ship scheduling. The key findings of this study are summarized as follows.

The publications growth tendency on ship scheduling is steadily increasing, and the most impactful journals are CIE and EJOR. The most cited paper is published by Christiansen Marielle in 2004. (2) The most productive authors are Maxim A. Dulebenets and Fagerholt Kjetil from USA and Norway. The Norwegian University of Science and Technology (Norway), Shanghai Maritime University (China), and the National University of Singapore (Singapore) are identified as the most collaborative organizations. China, USA, and Norway are recognized as the key contributors to ship berthing research. (3) The representative cited authors on ship scheduling is Christiansen Mariell, and the future research in this field is expected to focus on key areas such as “optimization”, “design”, and “models”.

In summary, this article provides a visual overview and analysis of the current status of ship scheduling research, and identifies future research directions. It is hoped that this article will inspire further research and contribute to the continued evolution of ship scheduling practices. Furthermore, VOSviewer can be used for other types of literature analysis to provide more comprehensive bibliometric analysis in the future.

ACKNOWLEDGEMENTS

This work is supported by National Natural Science Foundation of China (Grant No. 51920105014, 52071248, and 52272422), Fundamental Research Funds for the Central Universities (WUT: 2023IVB079), Fundamental Research Funds for the Central Universities (WUT: 2023IVB079).

REFERENCES

Ronen D. Cargo ships routing and scheduling: Survey of models and problems. European Journal of Operational Research. 1983, 12(2), 119–126.

Jiang S, Ma D, Tan X, et al. Bibliometric analysis of the current status and trends on medical hyperspectral imaging. Frontiers in Medicine. 2023, 10, 1235955.

Yang Y, Gai T, Cao M, et al. Application of group decision making in shipping industry 4.0: Bibliometric Analysis, Trends, and Future Directions. Systems. 2023, 11(2), 69.

Van Eck, N.J., Waltman, L, 2010. Software survey: VOSviewer, a computer program for bibliometric mapping. Scientometrics, 84 (2), 523–538.

Petropoulos F, Apiletti D, Assimakopoulos V, et al. Forecasting: theory and practice. International Journal of Forecasting. 2022, 38(3), 705–871.

Ksciuk J, Kuhlemann S, Tierney K, et al. Uncertainty in maritime ship routing and scheduling: a literature review. European Journal of Operational Research. 2023, 308(2), 499–524.

Al Samrout M, Sbihi A, Yassine A. An improved genetic algorithm for the berth scheduling with ship-to-ship transshipment operations integrated model. Computers & Operations Research. 2024, 161, 106409.

Tang Wenzhi, Yin Zhengxing, Wang Liping, et al. Planning and Key Technologies of Smart Ports for Large Bulk Cargo in Inland Rivers. Logistics Technology. 2023,42 (09), 121419.

Meng L, Wang X, Jin J, et al. Optimization model for container liner ship scheduling considering disruption risks and carbon emission reduction. Journal of Marine Science and Engineering. 2023, 11(7), 1449.

Advances in Maritime Technology and Engineering – Guedes Soares & Santos (Eds)
© 2024 The Author(s), ISBN 978-1-032-83099-5

The potential of offshore charging in the Greek coastal shipping network – A GIS approach

A. Ziakas
University of Piraeus, Piraeus, Greece

M. Boile
University of Piraeus, Piraeus, Greece
Hellenic Institute of Transport, Athens, Greece

ABSTRACT: International and European regulations have been introduced with the aim to address the increasing concerns regarding the decarbonization of the maritime sector. Electrification presents a strong potential in the reduction of greenhouse gas emissions and externalities, if electricity is generated by renewable energy sources. A possible solution to address the limited sailing range of electric ferries is the electricity supply to electric ships along their journey. Offshore charging for electric maritime vessels has been tested in a simulated environment with promising results. In this paper, the concept of electric propulsion in the ferry sector is reviewed, focusing on the electricity supply from both onshore (ports) and offshore installations. This paper serves to demonstrate the potential of offshore charging in extending the use of electric vessels and increasing their coverage of ferry routes at a network level.

1 INTRODUCTION

The shipping sector, which represents 13.5% (European Environmental Agency) of the GHGs related to transport, aims at a carbon neutral future. Several measures aiming to reduce emissions have been presented. This paper focuses on electric propulsion and its potential to provide zero-emission transportation of passengers, vehicles, and goods. Electric propulsion in the ferry sector has already been implemented in Baltic and Scandinavian countries since the previous decade (Gagatsi et al., 2016). In Greece, the ferry sector is responsible for providing mobility and accessibility to the island regions and supporting territorial cohesion, connecting the islands among them and with the mainland. The frequency of port calls, however, and the time spent in the port lead to significant emissions in the port area, making ferries a sector with strong potential in adopting electric propulsion. In this context, the paper explores the potential of electric propulsion in the ferry sector, focusing on the electricity supply from both onshore and offshore installations.

The rest of the paper is organized as follows. Section 2 gives the methodology used in this paper; Section 3 presents the state-of-the-art in electricity supply technologies and compatibility standards; Section 4 describes the Greek Coastal Shipping Network using QGIS software; Section 5 presents the most suitable locations for Offshore Wind Farms (OWFs) in Greece along with the criteria considered for their siting; Section 6 concludes with the findings of this review and the benefits of developing offshore charging stations in the selected regions, while future research topics are indicated.

2 METHODOLOGY

This study has been built in three thematic layers and presents the technical examination for the installation of offshore chargers for electric ships in the areas that are characterized as "suitable" for the development of offshore wind farms. First, the study presents the state-of-the-art in onshore and offshore power supply, along with the compatibility standards. Second, the Greek Coastal Shipping Network is being described in order to provide a better understanding of the area under consideration. Third, the most "suitable" locations for the development of OWFs along with the criteria considered for their siting are presented based on the literature to date. For the detailed documentation of the aforementioned, it was crucial to examine the pertinent literature, with special attention being given to those that refer to applications in Greece. For this purpose, the authors used a range of keywords to identify the papers in Google Scholar, Scopus, Web of Science, ResearchGate, Zenodo, Academia, and BASE Search. In addition, an artificial intelligence tool was used to cross-check the references on a literature map and find similar ones if those were missed during the first stage of papers' collection. A number of reports from either industry or research

DOI: 10.1201/9781003508762-15

institutions has also been included in the review. The keywords "offshore charging", "onshore power supply", "shoreside electricity", "electric ships charging", "electricity supply standards", "electric propulsion" "electricity supply standards", "offshore wind farms development" were selected. 85 papers relevant to the keywords and topic of research have been selected. Initially, duplicates were erased following an abstract review. The articles extracted from this process were classified into thematic folders, studied, and reviewed. The authors conclude with 62 scientific papers, reports, and white papers which cover the range of the necessary information for this research. Subsequently, in order to describe and depict the Greek Coastal Shipping Network, the QGIS software was used. The coastal shipping lines visualized in the selected maps have been drawn according to data collected from the Ministry of Maritime Affairs and Insular Policy (2021), and Marine Traffic (2021). It is important to mention that most of the shipping lines in the Greek Coastal Shipping Network are established and only a few changes are observed in November of each year, when the subsidised lines are being decided.

3 ELECTRICITY SUPPLY TO SHIPS

3.1 *Onshore power supply*

Electric energy supply will be key service for ports as more and more authorities are striving towards the greening of their operations (Krämer & Czermański, 2020). Onshore power supply (OPS) may be a viable option to reduce emissions in maritime industry (Zis, 2019). OPS systems may be fixed or mobile (Krämer & Czermański, 2020). In addition, waterborne vessels can either use cold-ironing (CI) from ship-to-ship (military operations) or from shore-to-ship (typical operation). Full compliance with CI can lead to reductions of in-port CO_2, SO_2, NO_X and BC (black carbon) emissions by 48–70 %, 3–60 %, 40–60 % and 57–70 %, respectively (Zis et al., 2014). A vessel connected to the shire grid is able to continue its operations such as cargo handling, lighting, heating, hot water, and ventilation with all of its engines shut down. An OPS system consists of the following components (Wang et al., 2021):

- a substation to receive power from the local distribution grid at a higher voltage level;
- one or more transformers and frequency converters to convert the magnitude and frequency of the voltage to be compatible with the vessel's electrical specification;
- the cables management system;
- and switchgear, circuit breakers, shore-to-ship cables, plugs, receptacles, converters, safety interlocks, and other power and communication equipment.

The implementation of OPS systems, however, is challenged by the lack of compatibility between ship and shore, the high installation cost that mitigates the adoption, and the so-called *"chicken-egg problem"*-port authorities need ships to use the OPS system and ships need OPS infrastructures in ports (Zis, 2019). One way to break this dilemma is to require from ports to have *"limited"* capability for *"shorepower"* providing (as in Europe by 31st December 2025 through Directive 2014/94/EU).

However, the *"chicken-egg problem"* is currently being addressed by the Alternative Fuels Infrastructure Regulation (AFIR) and the Fuel EU Maritime which work hand-in-hand. The key dates for the AFIR are January 2025, which is the deadline for the EU Member States to submit national policy frameworks, and the period 2025-2030, when potential mandatory onshore power supply (OPS) for inland and seaports may take place.

For a ship up to 5,000 GT such an investments costs between $50,000 and $350,000, while for ships larger than 100,000 GT the investment may require up to $750,000 (Zis, 2019). According to the same study, the annual cost for a port authority depends on the elements presented below:

- the terminal type and layout;
- the number of berths able to supply ships with electricity;
- the annual operating costs;
- the number of ships using it at their calls;
- the price per kWh sold by the grid;
- the price per kWh sold to the calling ship;
- monetary incentives provided to a ship to influence a decision to use OPS;
- potential subsidies provided to a ship towards retrofitting costs.

Regarding the types of chargers, ferries and passenger boats use combined charging system combo 2 and Charge de Move for charging operations (Mehammer et al., 2023).

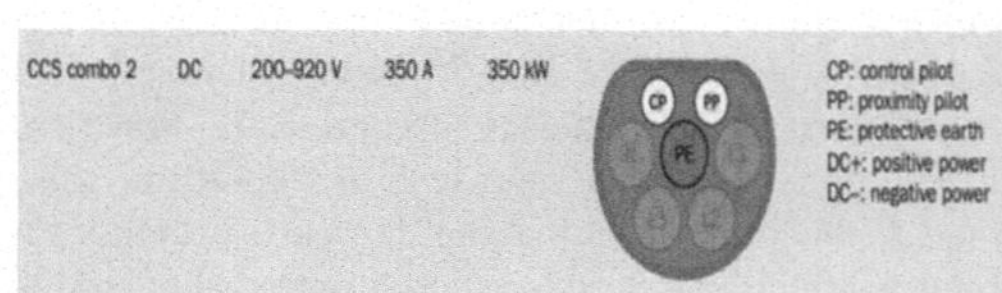

Figure 1. The CCS combo 2 charger.

CCS combo 2 (Figure 1) (Source: Mliu92) consists of a type 2 connector with the addition of two pins for DC voltages up to 920 V, with the power rating being specified as 350 kW using DC (Mehammer et al., 2023). The Mine Smart Ferry, in Thailand, uses 26 CCS combo 2 connectors for charging (Mehammer et al., 2023).

Charge de Move (CHAdeMO) (Figure 2) (Source: Mliu92) is a DC rapid charging system with a connector developed by Japanese automotive manufacturers and Tokyo Electric Power. The connector supports DC only, has a rated charging power of 400 kW, and supports high-level communication (Mehammer et al., 2023). The cost is similar to CCS combo 2 and has been used on the Raicho 110-passenger vessel in Japan, in 2011 (Mehammer et al., 2023).

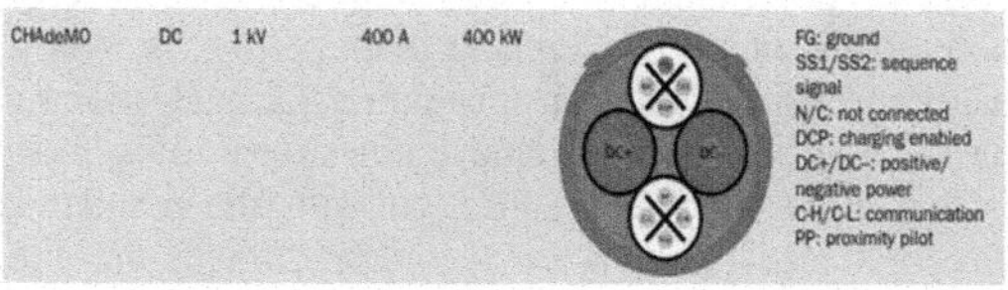

Figure 2. The CHAdeMO charger.

3.2 *Offshore charging*

Despite the notable process of electric propulsion in the ferry sector, the range of an electric waterborne vessel remains limited (Yang et al., 2022). In order to expand its sailing distance, more energy is required. A charging station located between the ports of departure and arrival may offer a viable solution to address such challenge. Only a few studies have been conducted related to offshore charging for electric waterborne vessels, describing its potential and operation (Yang et al., 2022; Sruthy et al., 2021). An electric tanker designed and developed in Japan, named “e5” was compared with a conventional one (Yang et al., 2022). The study indicated that with today’s battery technology, electrifying the maritime sector with FNPPs as the ultimate energy source can already achieve cost-competitiveness against the use of conventional bunker fuels, even in the absence of a carbon tax.

Marinized charging stations can radically become game changers in ocean shipping leading to the energy transition of the sector (Sruthy et al., 2021). The charging of an electric ship in an offshore charging station was referred to as an important issue that need to be solved (Sruthy et al., 2021). The direct access of large vessels to the charging station may potentially cause stability issues to the platform structure, over which the station is built. Secondly, the station must be close to the route, as, in most cases, ships are not able to deviate from it. In order to access the offshore floating station, two solutions were proposed; a remotely controlled charge carrier vehicle, appropriate also for emergency situations; and a floating pole shaped charging point for the station, which minimizes stability issues. The presence of multiple floating charging stations along the route increases the reliability of the service, and reduces the need for battery storage onboard.

The charging of electric ships using an offshore station connected with a floating solar PV system was also investigated (Sruthy et al., 2019). More specifically, the study examined the charging of the ship during its journey without diverting to a port. For this purpose, a floating charging station (FCS) connected with a hybrid energy storage system was designed and verified using a MATLAB/Simulink model. The concept of offshore charging saves space from the port. The FCS uses renewable energy sources and is totally independent from the grid. By constructing it close to the shipping route, additional expenses may be avoided. The FCS consists of a solar PV, the primary energy source generator which is connected to the DC bus through a converter. The load and storage devices are connected to the same DC bus. The hybrid storage devices are connected by using a bidirectional converter. The construction of several offshore charging stations will significantly reduce ship weight and cost, by reducing the equipment for on board storage and renewable energy sources (Sruthy et al., 2019).

Offshore charging may also facilitate other offshore operations such as the crew transfer vehicles, which can charge in a station connected with a wind turbine (Brindley et al., 2022). The study pointed out that developing a method to charge batteries in the field would be a key enabler for the large-scale deployment of electric marine vehicles towards carbon emission reductions.

In the most recent study examining the concept of offshore charging, Sruthy & Preetha (2023) presented a detailed study considering the area of North Sea for the development of offshore chargers in order to promote sustainable marine transportation. The study examined an energy management strategy to fulfil real-time energy demands with balanced energy flows, and ensure minimal dependency on backup energy sources. The operation of offshore charger aims to minimize carbon footprint (Sruthy & Preetha, 2023). The vessel used to model the energy consumption and needs has an average battery capacity of 2500 kWh. The study aimed to achieve the following objectives: satisfy electric ship’s energy demands; maximize employment of RESs for charging E-vessel; recharge of the depleted energy level of energy storage systems (ESS); assurance of balanced energy flow in FCS. The study points out that offshore chargers facilitate the recharging of electric ships and can be placed on high seas or closer to land on floating infrastructures. The concept of offshore chargers lies on the concepts of RES and ESS (Sruthy & Preetha, 2023). The ESS act as back-up power in case of a potential RES shortage. From a technical perspective, such an infrastructure requires electronic power converters and controllers for modulations of the power flow, charging of E-vessel batteries, and charging & discharging scenarios in energy storage systems (Sruthy and Preetha, 2023). The offshore charging station receives power from a different combinations of energy sources: RES alone, RES and ESS, ESS alone, RES and ESS and BES, ESS and BES, and BES alone. The energy management system allocates the power required by the vessel based on the energy availability of each source (Sruthy & Preetha, 2023).

However, such an innovative project needs to emphasize in addressing the challenges that are created. Apart from the financial obstacles that may be met for the construction of the station, a study by CENEX (2022) indicated that the weather conditions influence the adaptability of the vessel to the charging infrastructure. In fixed infrastructure, the cable must be long enough considering the tidal range. If the charging infrastructure is constructed on a floating platform, the simultaneous movement of the vessel and the platform must be considered to avoid the hyperextension of the cable. Finally, such a novel concept generates plethora of questions such as the stability issue of the vessel during offshore charging. Further research is required to address the abovementioned challenges.

It can be deduced that offshore charging constitutes a promising option for waterborne vessels that operate in short routes, despite the fact that it has only been tested in a simulated environment.

4 THE GREEK COASTAL SHIPPING NETWORK

According to the XRTC Annual Reports on the Greek Ferry Market for 2022 and 2023, the Greek Coastal Shipping Network reached the record number of 19 million passengers being transported in 2019 (XRTC, 2022). The COVID-19 pandemic slowed the pace. However, year 2021 could be characterized as a - recovery year, as passenger traffic increased by 47% compared to 2020. In 2021, inbound tourism doubled as compared to the lowest tourist number of the last decade, which was 7.4 million visitors in 2020. In 2021 14.7, million tourists visited the country, a figure which is still far from the 31 million tourists in 2019. Passenger traffic in the Greek coastal shipping increased by 36% in 2022, compared to 2021, reaching the number of 17.5 million, while the car movement increased by 12% reaching the 3.7 million. The forecast is that in 2023 the number will be equal or more to that of 2019, before the onset of the pandemic.

From a fleet perspective, the Greek ferry companies don't have new orders placed in shipyards for the renewal of their vessels. The only exception is the Attica Group, which completed the construction of three Aero Catamaran ships which are now launched on the Argosaronicos line (XRTC, 2022). The biggest challenge this sector faces is the renewal of the fleet as around 86% of the ships is above 20 years of age and 41% of the vessels operating today will be over 40 years old in 2035. The main reasons of the aforementioned challenge are the lack of capital and the fact that there is no clear target as well as a developed low/zero carbon fuel market, making the shipowners' options risky (XRTC, 2022). The same report proposed that investments in the renewal of low-carbon and zero-emission fleets can be provided by subsidies for green investments. Coastal shipping presents strong potential in adopting innovative solutions, as it is characterized as the "early adopter" due to the lower costs in comparison with ocean going vessels.

Considering the operational characteristics of the Greek Coastal Shipping Network, there are 33 companies which operate 163 ships that can transport 132,546 passengers and 30,938 vehicles. The economic and social development of the Greek islands is based exclusively on coastal shipping. Coastal shipping has a great impact on island connectivity as in most islands it is the only connection with the mainland. Greece's insular character is visible through the 115 inhabited islands, which represent 14.6% of the country's population. Only 25, from the 95 islands served by ferries have an airport, while 85% of the supplies needed to be transported to and from the islands are covered by coastal shipping. Over the past 14 years, especially since the beginning of the Greek debt crisis, the market's greatest weakness has been the lack of a business model that brings new capital to the markets that have to be profitable in order to justify new investments. Overviewing past opportunities the report highlights the adaption of the Greek coastal shipping to a competitive environment, the entrance to the Eurozone. Nowadays, the condition of the market requires the renewal of the fleet towards decarbonization, raising capital from EU's green funds (XRTC, 2023). After conducting a SWOT analysis of the market, it is certified that Greek coastal shipping is currently undergoing a new development cycle and opportunities in a transparent environment. For instance, the banks are seeking their exit from the coastal shipping market, making room for new investors.

The major characteristic of the Greek Coastal Shipping Network is the seasonality in passenger and vehicle flows. The main driver for this is the large touristic flows from early April until late October. Figure 3 depicts the seasonality in demand for car transport of the Festos Palace vessel which provides connection between Piraeus and Crete (the biggest Greek island). Similarly, Figure 4 presents the seasonality in passenger flows considering the whole Greek coastal shipping fleet (data of 2021).

For the purpose of this study the Greek Coastal Shipping Network is visualized using Geographical Information System (GIS). The aim is to present its complexity and density and focus in specific areas that have been characterized as "suitable" for the development of offshore wind farms. The whole Greek Coastal Shipping Network is presented in Figure 5.

5 OFFSHORE WIND FARMS SITTING

5.1 *Offshore wind farm location selection criteria*

A study on the sitting of offshore wind farms in Greece (Vagiona & Karanikolas 2012) indicates that the location is the most important parameter for environmental spatial analysis and that the location

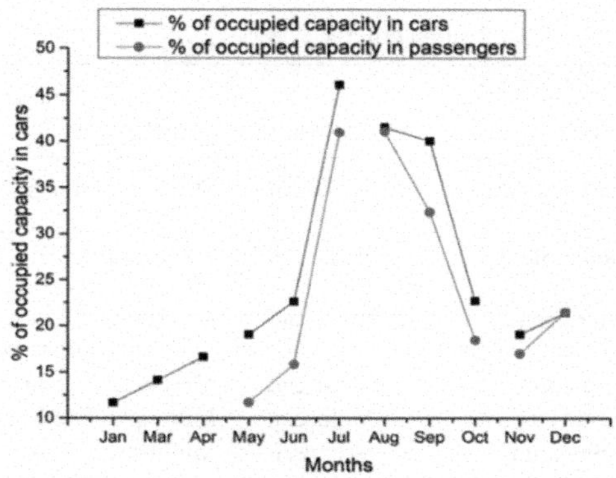

Figure 3. Seasonality phenomenon (Festos Palace vessel).

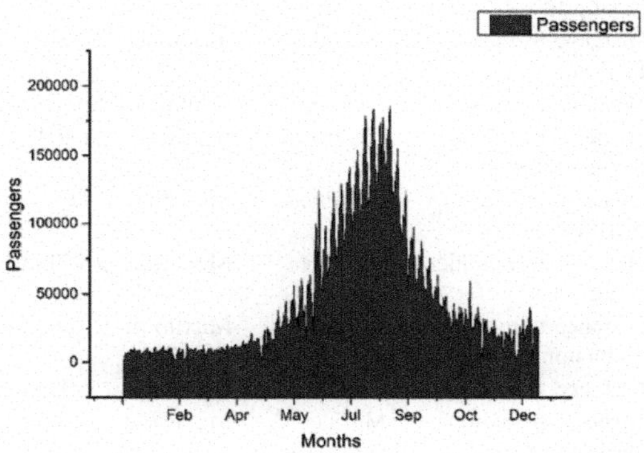

Figure 4. Seasonality phenomenon (whole fleet).

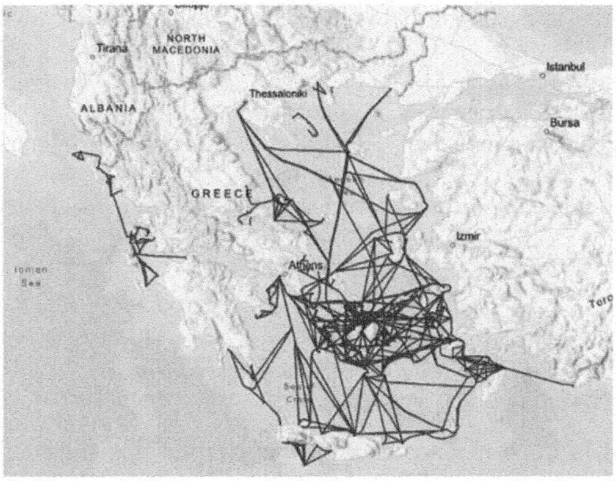

Figure 5. GIS visualization of the Greek Coastal Shipping Network.

can be punctually determined by using GIS. The study considers three criteria of exclusion, which are the following:

- areas that don't fulfil the minimum wind velocity (6 m/s) for the operation of wind turbines;
- areas that are characterized as protected areas either by National or European legislation;
- and areas with sea depth exceeding 30 meters.

Subsequently, a range of evaluation criteria were applied, in order to evaluate the candidate locations. These criteria include average wind velocity, distance to protected areas, distance from shipping routes, distance from shore, and connection to electricity network.

Since then, many studies have been conducted in order to provide the optimal siting framework for OWFs considering an extended number of criteria. The study of Díaz & Guedes Soares, (2020) is one of the most comprehensive in the literature to date. In the first stage, data collection is conducted, in order to build GIS layers. Secondly, the unsuitable locations are excluded, and finally, the remaining sites are evaluated by applying a set of criteria. A recent study aimed to present a framework for the optimal siting of OWFs considering the cost of investment (Spyridonidou & Vagiona, 2020a). The research was conducted in four stages: identification of proper sites; determination of technical specifications of the wind turbines; estimation of total investment cost; and prioritization of the most suitable sites. Kim et al., (2013) also took into account a wide range of factors that affect the development of OWFs. In that study (Kim et al., 2013), target sites were limited to below 20 m in depth and to over 10 km in distance from the coastline. Vagiona & Kamilakis, (2018) conducted an analysis for the site selection of OWFs. The methodological framework included three tools; GIS; AHP; and TOPSIS. First, areas unsuitable for the siting of OWFs were identified. Subsequently, thematic maps were created and a set of evaluation criteria were applied determining the most appropriate sites for OWFs development in the South Aegean. Similar to the abovementioned, another study (Spyridonidou and Vagiona, 2019) examined the optimal location selection for offshore wind farms in shallow waters (water depth ≤ 50m), as it facilitates the installation in terms of both technological and economical perspectives. The study follows a similar approach with others found in the literature: exclusion and evaluation of the potential sites, and, subsequently, integration to GIS. In order to weight the criteria, a multicriteria analysis method (TOPSIS) was implemented. Similarly, another study (Caceoğlu et al., 2022) combined Analytical Hierarchy Process and GIS to properly site potential OWFs in the Northwest Turkey. The criteria were classified as decision and exclusion criteria. The factors considered in the study were the wind potential, population to be served, energy self-sufficiency on a regional scale, commercial fishing activity, distance to shipping routes, distance to existing large ports, social acceptance, visual impact, tourism sites, water depth, distance to bird migration routes, distance to environmentally protected areas and key biodiversity areas, soil conditions and earthquake hazard, electricity grid connection, military zones, ship wrecks, civil aviation, existing pipelines in the seabed, offshore natural gas/petroleum wells, length of onshore and offshore cables. More recently, a study (Gkeka-Serpetsidaki & Tsoutsos, 2022) recognized sixteen evaluation criteria based on the available literature and by using AHP, pairwise comparisons were created among the evaluation criteria. As most of the reviewed research works, the study of Spyridonidou & Vagiona, (2020b) also conducted an AHP and integrated the weighted criteria to the GIS software in order to present the most suitable sites for OWFs development. Finally, Vasileiou et al., (2017) presented a study for the OWFs siting in Greece considering its strong potential.

Table 1. Exclusion criteria.

Criteria	Unsuitable areas
Military areas	All
Hydrocarbons and minerals	All
Sand and Gravel	All
Aquaculture and fishing	All
Marine renewable energies pilot zones	All
Environmental protected areas	All
Underwater lines and pipelines	<500m
Maritime traffic	<500m
Heritage areas	All
Wind velocity	<4m/s
Water depth	<50 and >1000m
Distance from shore	>200m

Tables 1 and 2 (Díaz & Guedes Soares, (2020) based on Christoforaki and Tsoutsos (2017), Vasileiou et al., (2017), Ziemba et al., (2017), Fetanat and Khorasaninejad (2015), Bagocius et al., (2014), Sourianos et al., (2017), Wu et al., (2016)) and 2 (Díaz & Guedes Soares, (2020) based on Kim et al., (2016), Mekonnen & Gorsevski (2015), Schillings et al., (2012), Moller (2011), Vagiona & Kamilakis (2018), Stefanakou et al., (2019), Gaveriaux et al., (2019), Van Haaren & Fthenakis (2011), Fetanat and Khorasaninejad (2015), Yue and Yang (2009), Peters et al., (2020), Huffmeier & Goldberg (2019), Dinh and McKeogh (2018), Castro-Santos et al., (2020a), Castro-Santos et al., (2020b), Castro-Santos et al., (2018), Argin et al., (2019), Uzunoglu and Guedes Soares (2020), Costa et al., (2010), Kausche et al., (2018), Guedes Soares et al., (2014), Silva et al., (2015), Costa et al., (2020), Beiter et al., (2015), Yu et al., (2020), UNCLOS (1995), Thomsen et al., (2006), Energitenet dk (2015), Pacheco et al., (2017), Ministerio de la presidencia (2007), The Ministry of Infrastructure and the Environment (Holland) (2014), de la Vega et al., (2013)) depict the exclusion and evaluation criteria, respectively, that were used in the majority of the studies considering the proper siting of OWFs.

Table 2. Evaluation criteria.

Criteria	Criteria type	Objective	Criteria Value
Wind velocity	Metocean	Maximize	4 m/s - 25 m/s
Wind potential	Metocean	Maximize	>2700 h/year
Water depth	Metocean	Minimize	50 m–1000 m
Wave conditions	Metocean	Minimize	No limit
Marine currents	Metocean	Minimize	No limit
Temperature	Metocean	Minimize	—20 °C - 40 °C
Technical feasibility	Viability	Maximize	No limit
Sufficient study times	Viability	Maximize	No limit
Distance to local electrical grid	Logistics	Minimize	<200 km
Distance from coastal facilities	Logistics	Minimize	No limit
Distance from shore	Facilities	Minimize	Specific
Distance from residential areas	Facilities	Maximize	Specific
Distance from the maritime routes	Facilities	Maximize	>500 m
Distance from underwater lines	Facilities	Maximize	>750 m
Distance to marine recreational activities	Facilities	Maximize	>1000 m
Distance from airport	Facilities	Maximize	>500 m
Distance from protected areas	Marine environment	Maximize	>1000 m
Proximity to migratory birds' paths	Marine environment	Minimize	>1000 m
Proximity to migratory marine life paths	Marine environment	Minimize	>1000 m
Area of the territory	Techno-economic	Maximize	>11 km^2
Proximity to the area of electric demand	Techno-economic	Maximize	No limit
Population served	Techno-economic	Maximize	No limit
Multiple resources	Techno-economic	Minimize	No limit

5.2 *Suitable sites for offshore wind farm development in Greece*

It is important to mention that despite that most of the criteria in the OWF location selection studies are the same, results vary. The draft National Programme for the development of offshore wind farms (OWF) in Greece includes 25 areas, covering a total area of 2,712 km^2 with an estimated minimum capacity of 12,4 GW. Based on this roadmap, Greece gains a competitive advantage in introducing new technologies, acquire know-how and develop the necessary supply chain. According to the report (HELEMA, 2023), the deployment of OWFs can bring economic prosperity boosting GDP of up to

€1.9 billion per year on average over the period 2024-2050 and in government revenues of up to €440 million per year, while 44,400 jobs may be supported every year. The plan indicates ten eligible areas for development by 2030-2032, with a total capacity of approximately 4.9 GW, mainly for floating projects. The above areas do not include the marine area between Evros-Samothraki, which is defined as an area for the development of pilot OWF projects, according to Law 4964/2022. The following areas are included:

- Eastern Crete, where it is estimated that projects with a total capacity of 800 MW will be developed.
- Southern Rhodes, with a maximum installed capacity of between 300 MW and 550 MW.
- Central Aegean, with a maximum installed capacity of between 200 MW and 450 MW.
- The Evia-Chios axis, with a maximum installed capacity of 300 MW.
- Ionian Sea, with a maximum installed capacity of 450 MW.

The study (Vasileiou et al., 2017) identified twelve potential locations (Figure 6 (Vasileiou et al., 2017)) with the most promising for developing an offshore wind farm being:

- East of Crete
- South-East of Mykonos
- North-West of Crete

The rest are:

- East of Mykonos
- South-East of Mykonos
- South-West of Rhodes
- North-East of Skyros
- East of Skyros
- South-East of Astypalaia
- South-South-East of Skyros
- South-East of Skyros
- East-North-East of Skyros

Under the same scope, the study of Spyridonidou et al., (2020) identified sixteen potential sites (depicted in Figure 7 (Spyridonidou et al., 2020)):

1. Southeast of Euboea
2. Southwest of Thasos
3. Southeast of Euboea/North of Andros
4. Southeast of Thasos/Northwest of Samothrace
5. Southeast of Euboea
6. South of Hydra
7. East of Hydra
8. Southeast of Euboea
9. Northeast of Volos
10. Northwest of Lefkada/West of Preveza

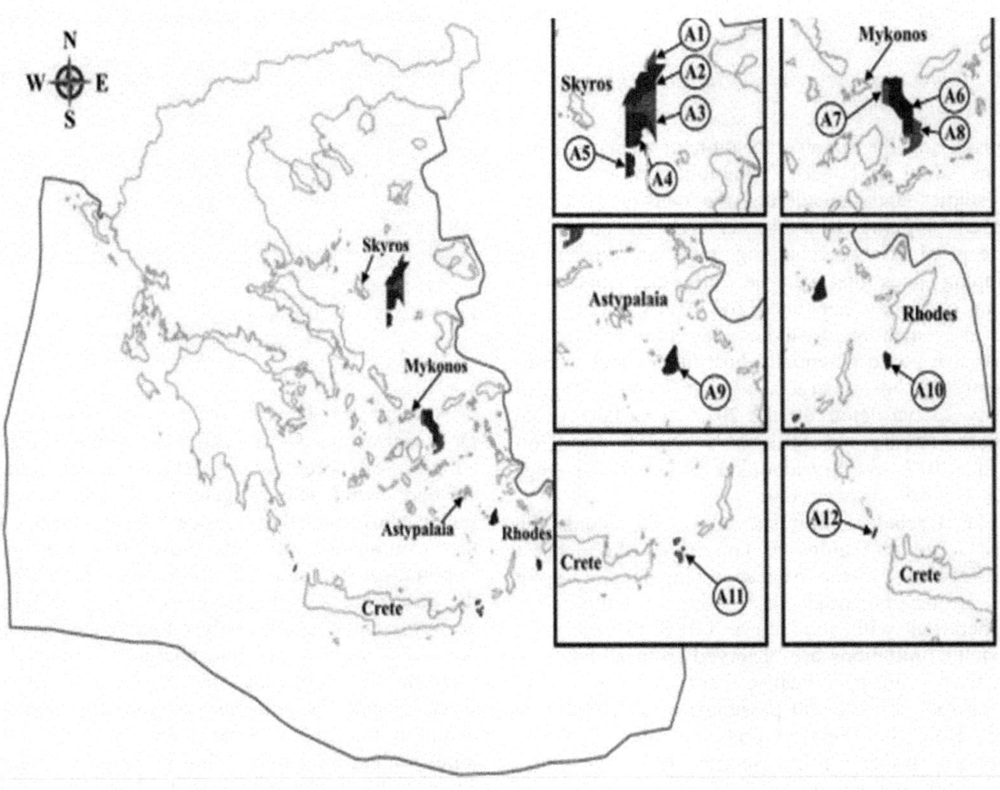

Figure 6. Potential/Eligible areas for the siting of offshore wind farms in Greece.

11. Southeast of Euboea/North of Andros
12. Northwest of Skyros
13. Northeast of Euboea/South of Skopelos
14. North of Volos
15. East of Thasos
16. East of Poros

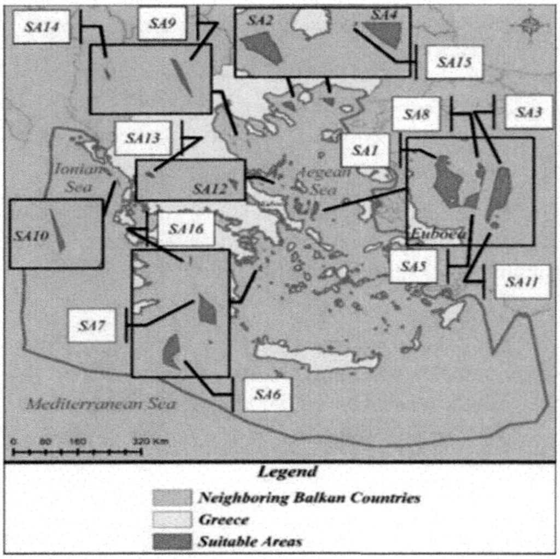

Figure 7. Suitable areas for offshore wind farms siting in Greece.

5.3 *Combining the novelty of offshore charging with OWFs*

Considering the suitable location of offshore wind farms siting in Greece (combining the results of all studies and considering the density of Greek coastal shipping network) as well as the novel concept of offshore charging, there are clusters of shipping lines that can be fully electrified with clean electricity generated by OWFs. Combining the abovementioned studies, the two highlighted areas that could potentially host OWFs and, subsequently offshore chargers, are the Central Aegean (study of Vasileiou et al., 2017) and East Crete regions (studies of HELEMA (2023), Vasileiou et al., 2017 and Sryridonidou et al., 2020) (Figures 8 and 9, respectively).

The Cyclades complex, which is among the most touristic regions in Greece, is the densest "subnetwork" in the Aegean archipelago. Despite the strong potential of offshore charging in cooperation with the future OWF in the area, capacity limitations are observed from a ship perspective. More specifically, there are vessels able to transport over 2,000 passengers and 1,000 cars (e.g., Blue Star Ferries) that serve the Cyclades complex. Today's fully electric ships can transport 200 passengers and 30 cars (reference vessel: E-ferry Ellen). It can reasonably be deduced that in order to match the demand for transportation, more electric ships will be needed, creating cost inefficiencies. The research, development and innovation in shipbuilding may contribute to the re-examination of the Central Aegean subnetwork, as bigger, more energy- and cost-efficient ships will be built.

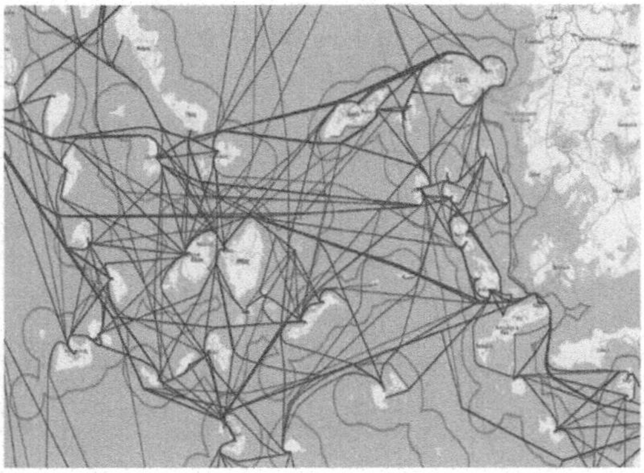

Figure 8. GIS visualization of the Cyclades.

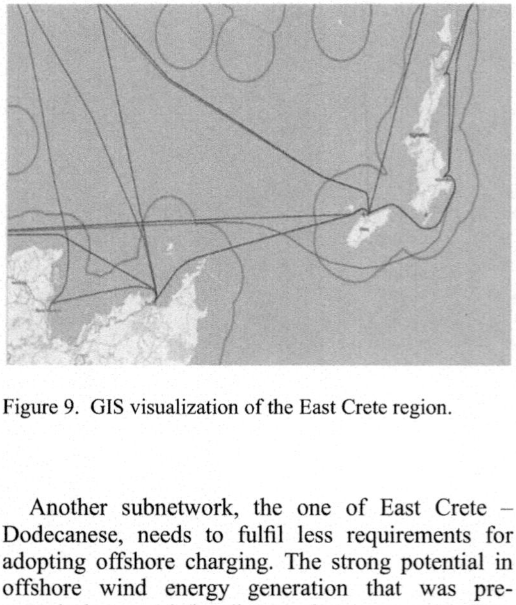

Figure 9. GIS visualization of the East Crete region.

Another subnetwork, the one of East Crete – Dodecanese, needs to fulfil less requirements for adopting offshore charging. The strong potential in offshore wind energy generation that was presented above and the distance between ports allow the examination of this novel concept. It is important to mention that Siteia was considered as the base port, and the routes presented in Table 3 are representative. The only criterion that has been set is that only 1 offshore charger is allowed per transport leg (ensuring the passengers' level of service), and the only assumption that has been made is that the electric ferry can sail for 50 miles without recharge. Table 3 gives the total distance of the routes while Table 4 depicts the distances per transport leg.

Table 3. Route distances.

Route	Distance (miles)
Siteia-Agios Nikolaos-Thira	101
Siteia-Thira	87
Siteia-Kasos	48
Siteia-Kasos-Karpathos	72
Siteia-Kasos-Karpathos-Diafani-Chalki-Rhodes	156
Siteia-Kasos-Karpathos-Kasos-Rhodes	206
Siteia-Kasos-Karpathos-Rhodes	162

Table 4. Transport leg distances.

Port A	Port B	Distance (miles)
Siteia	Agios Nikolaos	23
Siteia	Kasos	48
Agios Nikolaos	Oia (Thira)	78
Kasos	Karpathos	24
Karpathos	Diafani	17
Diafani	Chalki	33
Chalki	Rhodes	34
Rhodes	Chalki	34
Kasos	Rhodes	110
Karpathos	Rhodes	90

In 2021, only five vessels were operating in the presented routes: CHAMPION JET 1, PAROS JET, SUPPERRUNNER JET, HELLENIC HIGHSPEED, and POWER JET. The vessels' age is between 25 and 28 years, and they are all fossil-fueled. The vessels capacity, in terms of passengers, vary between 727 and 1200, and in terms of cars between 140 and 200. The vessels' average occupied capacity has a range between 4.61% and 10.07% for passengers, and 2.44% and 8.75% for cars. The low average utilization of available capacity indicates that electric ferries with today's capacities may serve efficiently the selected subnetwork. Electric ships will recharge their batteries: (a) at port during call; (b) and at offshore chargers, directly connected with the OWF, where necessary.

The e-ferry Ellen, one of the most advanced electric ferries can sail for 22 nautical miles without recharging. Taking into account the assumption that future electric ferries would be able to sail for 50 nautical miles, it is observed that only the transport leg of Kasos-Rhodes is facing recharging issues (more than one offshore charger is required). On the other hand, the transport legs of Agios Nikolaos-Thira and Karpathos-Rhodes constitute those with strong potential for hosting a floating offshore charger. Clusters of shipping lines, and even whole networks can benefit from the presence and operation of this novel, although yet immature, technology, contributing to the elimination of emissions by 2050 and "greening" both shipping and ports. Similarly, such concept may be applicable to every archipelagic Polynesian, and/or insular maritime region of the world. With today's uncertainty around zero-emission fuel selection, electrification provides a clear pathway towards decarbonization (considering clean energy production).

6 CONCLUSIONS

In this paper a number of studies have been reviewed considering the concepts of onshore power supply, offshore charging, and OWFs siting using GIS. The state-of-the-art in electricity supply technologies and standards was presented, followed by the exclusion and evaluation criteria used for the proper siting of OWFs. Considering the Greek Coastal Shipping Network, one of the most complex and dense networks in the world, this paper aimed to suggest the coexistence of these two promising technologies. Having determined the suitable locations for OWFs, clusters of coastal shipping lines may be benefited if ships have the opportunity to charge in the open sea. The location selection of offshore chargers, however, may differ from this of OWFs. Subsequently, further research is required in order to identify the constraints of this specific location selection problem. In addition, the vessel routing need to be determined assuming that the fleet is fully electric and the vessels are capacitated in terms of power storage. Moreover, the electricity supply both onshore and offshore engages "players" from different sections. Potential synergies and conflicts, and the equilibrium point may be identified by conducting a game theoretical analysis.

ACKNOWLEDGEMENTS

This work has been partly supported by the University of Piraeus Research Center.

REFERENCES

Argin, M., Yerci, V., Erdogan, N., Kucuksari, S., & Cali, U. (2019). Exploring the offshore wind energy potential of Turkey based on multi-criteria site selection. Energy Strategy Reviews, 23, 33–46.

Bagočius, V., Zavadskas, E. K., & Turskis, Z. (2014). Sequence determining of construction of the offshore wind farm construction applying permutation method. E&M Economics and management, 17(3), 50–61.

Beiter, P., Musial, W., Smith, A., Kilcher, L., Damiani, R., Maness, M., … & Scott, G. (2016). A spatial-economic cost-reduction pathway analysis for US offshore wind energy development from 2015–2030 (No. NREL/TP-6A20-66579). National Renewable Energy Lab. (NREL), Golden, CO (United States).

Brindley, W. (2022), Offshore wind turbine vessel charging, Catapult Offshore Renewable Energy, PN000510-RPT-003 – Rev. 1.

Caceoğlu, E., Yildiz, H. K., Oğuz, E., Huvaj, N., & Guerrero, J. M. (2022). Offshore wind power plant site selection using Analytical Hierarchy Process for Northwest Turkey. Ocean Engineering, 252, 111178.

Castro-Santos, L., Bento, A. R., Silva, D., Salvação, N., & Guedes Soares, C. (2020b). Economic feasibility of floating offshore wind farms in the north of Spain. Journal of Marine Science and Engineering, 8(1), 58.

Castro-Santos, L., Filgueira-Vizoso, A., Álvarez-Feal, C., & Carral, L. (2018). Influence of size on the economic feasibility of floating offshore wind farms. Sustainability, 10(12), 4484.

Castro-Santos, L., Silva, D., Bento, A. R., Salvação, N., & Guedes Soares, C. (2020a). Economic feasibility of floating offshore wind farms in Portugal. Ocean Engineering, 207, 107393.

CENEX, 2022, Charging Infrastructure for Near-Shore Electric Vehicles, Part 2: The infrastrucrure considerations for the UK's near-shore maritime sector.

Christoforaki, M., & Tsoutsos, T. (2017). Sustainable siting of an offshore wind park a case in Chania, Crete. Renewable Energy, 109, 624–633.

Costa, P. S., Simões, T., & Estanqueiro, A. (2010). Sustainable offshore wind potential in continental Portugal. In Workshop on Oceans as a Source of Energy.

Costa, T. C., Pereira, L. T., Marta-Almeida, M., & Guedes Soares, C. (2020). Mapping of currents off the northwestern Iberian coast with the Regional Ocean Modelling System. Journal of Operational Oceanography, 13(2), 71–83.

De la Vega, D., Matthews, J. C., Norin, L., & Angulo, I. (2013). Mitigation techniques to reduce the impact of wind turbines on radar services. Energies, 6(6), 2859–2873.

Díaz, H., & Guedes Soares, C. (2020). An integrated GIS approach for site selection of floating offshore wind farms in the Atlantic Continental European coastline. Renewable and Sustainable Energy Reviews, 134, 110328.

Dinh, V. N., & McKeogh, E. (2019). Offshore wind energy: technology opportunities and challenges. In Proceedings of the 1st Vietnam Symposium on Advances in Offshore Engineering: Energy and Geotechnics (pp. 3–22). Springer Singapore.

Energinet dk. Technical project description for offshore wind farms (200 MW). 2015

Fetanat, A., & Khorasaninejad, E. (2015). A novel hybrid MCDM approach for offshore wind farm site selection: A case study of Iran. Ocean & Coastal Management, 109, 17–28.

Gagatsi, E., Estrup, T., & Halatsis, A. (2016). Exploring the potentials of electrical waterborne transport in Europe: The E-ferry concept. *Transportation Research Procedia*, *14*, 1571–1580.

Gavériaux, L., Laverrière, G., Wang, T., Maslov, N., & Claramunt, C. (2019). GIS-based multi-criteria analysis for offshore wind turbine deployment in Hong Kong. Annals of GIS, 25(3), 207–218.

Gkeka-Serpetsidaki, P., & Tsoutsos, T. (2022). A methodological framework for optimal siting of offshore wind farms: A case study on the island of Crete. Energy, 239, 122296.

Guedes Soares, C., Bento, A. R., Gonçalves, M., Silva, D., & Martinho, P. (2014). Numerical evaluation of the wave energy resource along the Atlantic European coast. Computers & Geosciences, 71, 37–49.

HEREMA-The Draft National Programme for Offshore Wind Energy, unlocking a natural wealth for clean energy and billions of euros' investments

Hüffmeier, J., & Goldberg, M. (2019). Offshore wind and grid in the Baltic sea—status and outlook until 2050. In Baltic Lines-Coherent Linear Infrastructures in Baltic Maritime Spatial Plans. Swedish Agency for Marine and Water Management.

Kausche, M., Adam, F., Dahlhaus, F., & Großmann, J. (2018). Floating offshore wind-Economic and ecological challenges of a TLP solution. Renewable Energy, 126, 270–280.

Kim, J. Y., Oh, K. Y., Kang, K. S., & Lee, J. S. (2013). Site selection of offshore wind farms around the Korean Peninsula through economic evaluation. Renewable Energy, 54, 189–195.

Kim, T., Park, J. I., & Maeng, J. (2016). Offshore wind farm site selection study around Jeju Island, South Korea. Renewable energy, 94, 619–628.

Krämer, I., & Czermański, E. (2020). Onshore power one option to reduce air emissions in ports. In Sustainability Management Forum| NachhaltigkeitsManagementForum (Vol. 28, No. 1-2, pp. 13–20). Berlin/Heidelberg: Springer Berlin Heidelberg.

Mehammer, E. B., Strand, H., Magnusson, N., Thinn, K. S., & Eberg, E. (2023). How to Plug In the Fishing Fleet: Connectors in charging infrastructure for small fishing boats. IEEE Electrification Magazine, 11 (1), 73–82.

Mekonnen, A. D., & Gorsevski, P. V. (2015). A web-based participatory GIS (PGIS) for offshore wind farm suitability within Lake Erie, Ohio. Renewable and Sustainable Energy Reviews, 41, 162–177.

Ministerio de la presidencia. Real Decreto 1028/2007. BOE 2007; 183:33171–9.

Möller, B. (2011). Continuous spatial modelling to analyse planning and economic consequences of offshore wind energy. Energy Policy, 39(2), 511–517.

Pacheco, A., Gorbeña, E., Sequeira, C., & Jerez, S. (2017). An evaluation of offshore wind power production by floatable systems: A case study from SW Portugal. Energy, 131, 239–250.

Peters, J. L., Remmers, T., Wheeler, A. J., Murphy, J., & Cummins, V. (2020). A systematic review and meta-analysis of GIS use to reveal trends in offshore wind energy research and offer insights on best practices. Renewable and Sustainable Energy Reviews, 128, 109916.

Schillings, C., Wanderer, T., Cameron, L., van der Wal, J. T., Jacquemin, J., & Veum, K. (2012). A decision support system for assessing offshore wind energy potential in the North Sea. Energy Policy, 49, 541–551.

Silva, D., Bento, A. R., Martinho, P., & Guedes Soares, C. (2015). High resolution local wave energy modelling in the Iberian Peninsula. Energy, 91, 1099–1112.

Sourianos, E., Kyriakou, K., & Hatiris, G. A. (2017). GIS-based spatial decision support system for the optimum siting of offshore windfarms. Eur. Water, 58, 337–343.

Spyridonidou, S., & Vagiona, D. (2019). Offshore Windfarm Siting in Greece using GIS and TOPSIS. In 16th International conference on environmental science and technology Rhodes, Greece.

Spyridonidou, S., & Vagiona, D. G. (2020a). Spatial energy planning of offshore wind farms in Greece using GIS and a hybrid MCDM methodological approach. Euro-Mediterranean Journal for Environmental Integration, 5, 1–13.

Spyridonidou, S., & Vagiona, D. G. (2020b). Systematic review of site-selection processes in onshore and offshore wind energy research. Energies, 13(22), 5906.

Spyridonidou, S., Vagiona, D. G., & Loukogeorgaki, E. (2020). Strategic planning of offshore wind farms in Greece. Sustainability, 12(3), 905.

Sruthy, V., & Preetha, P. K. (2023). Implementation and operational feasibility of an offshore floating charging station for sustainable marine transportation. Environment, Development and Sustainability, 1–32.

Sruthy, V., Raj, B., & Preetha, P. K. (2021). An offshore floating charging station for electric ships: accessibility enhancement schemes for recharging. Ships and Offshore Structures, 16(10), 1143–1150.

Sruthy, V., Raj, B., Preetha, P. K., & Ilango, K. (2019, April). SPV based floating charging station with hybrid energy storage. In 2019 IEEE International Conference on Intelligent Techniques in Control, Optimization and Signal Processing (INCOS) (pp. 1–6). IEEE.

Stefanakou, A. A., Nikitakos, N., Lilas, T., & Pavlogeorgatos, G. (2019). A GIS-based decision support model for offshore floating wind turbine installation. International Journal of Sustainable Energy, 38(7), 673–691.

The Ministry of Infrastructure and the Environment (Holland). White paper on offshore wind energy. 2014.

Thomsen, F., Lüdemann, K., Kafemann, R., & Piper, W. (2006). Effects of offshore wind farm noise on marine mammals and fish. Biola, Hamburg, Germany on behalf of COWRIE Ltd, 62, 1–62.

UNCLOS. Ocean. Development and International Law 1995;26.

Uzunoglu, E., & Guedes Soares, C. (2020). Hydrodynamic design of a free-float capable tension leg platform for a 10 MW wind turbine. Ocean Engineering, 197, 106888.

Vagiona, D. G., & Kamilakis, M. (2018). Sustainable site selection for offshore wind farms in the South Aegean —Greece. Sustainability, 10(3), 749.

Vagiona, D. G., & Karanikolas, N. M. (2012). A multicriteria approach to evaluate offshore wind farms siting in Greece. Global NEST Journal, 14(2), 235–243.

Van Haaren, R., & Fthenakis, V. (2011). GIS-based wind farm site selection using spatial multi-criteria analysis (SMCA): Evaluating the case for New York State. Renewable and sustainable energy reviews, 15(7), 3332–3340.

Vasileiou, M., Loukogeorgaki, E., & Vagiona, D. G. (2017). GIS-based multi-criteria decision analysis for site selection of hybrid offshore wind and wave energy systems in Greece. *Renewable and sustainable energy reviews*, *73*, 745–757.

Wang, L., Liang, C., Shi, J., Molavi, A., Lim, G., & Zhang, Y. (2021). A bilevel hybrid economic approach for optimal deployment of onshore power supply in maritime ports. Applied Energy, 292, 116892.

Wu, Y., Zhang, J., Yuan, J., Geng, S., & Zhang, H. (2016). Study of decision framework of offshore wind power station site selection based on ELECTRE-III under intuitionistic fuzzy environment: A case of China. Energy Conversion and Management, 113, 66–81.

XRTC Annual Report on Greek Ferry Market, 2022

XRTC Annual Report on Greek Ferry Market, 2023

Yang, S., Yuan, J., Nian, V., Li, L., & Li, H. (2022). Economics of marinised offshore charging stations for electrifying the maritime sector. Applied Energy, 322, 119389.

Yu, Q., Liu, K., Teixeira, A. P., & Guedes Soares, C. (2020). Assessment of the influence of offshore wind farms on ship traffic flow based on AIS data. The Journal of Navigation, 73(1), 131–148.

Yue, C. D., & Yang, M. H. (2009). Exploring the potential of wind energy for a coastal state. Energy Policy, 37 (10), 3925–3940.

Ziemba, P., Wątróbski, J., Ziolo, M., & Karczmarczyk, A. (2017). Using the PROSA method in offshore wind farm location problems. Energies, 10(11), 1755.

Zis, T. P. (2019). Prospects of cold ironing as an emissions reduction option. Transportation Research Part A: Policy and Practice, 119, 82–95.

Zis, T., North, R. J., Angeloudis, P., Ochieng, W. Y., & Harrison Bell, M. G. (2014). Evaluation of cold ironing and speed reduction policies to reduce ship emissions near and at ports. Maritime Economics & Logistics, 16, 371–398.

Maritime traffic modelling

Advances in Maritime Technology and Engineering – Guedes Soares & Santos (Eds)

Detection and analysis of ship domain parameters in confined waters

D.P. Liu
Centre for Marine Technology and Ocean Engineering (CENTEC), Instituto Superior Técnico, Universidade de Lisboa, Lisbon, Portugal

C.-L. Siow & H.-S. Kang
Marine Technology Centre, Institute for Vehicle System & Engineering, Faculty of Mechanical Engineering, Universiti Teknologi Malaysia, Johor Bahru, Malaysia

C. Guedes Soares
Centre for Marine Technology and Ocean Engineering (CENTEC), Instituto Superior Técnico, Universidade de Lisboa, Lisbon, Portugal

ABSTRACT: A novel concept of a multi-observational eccentric elliptical ship domain is introduced, based on AIS data extraction. By considering varying data cumulative densities and characteristics of different directions, the method adapts the ship domain to reflect the characteristics of local traffic patterns. In the mean while, the study analyzes the relationship between the ship domain and speed across a selected ship type. Subsequently, the tanker dada from historical AIS data received in Malacca Strait is analyzed, which constituted the highest proportion in the dataset. The established multi-observational eccentric elliptical ship domain model is applied to identify near miss scenarios in the study area, offering reference insights for maritime traffic supervision.

1 INTRODUCTION

Maritime transportation involves risks of incidents such as fires, groundings, and collisions (Guedes Soares & Teixeira, 2001), with areas of high traffic density and specific geographical constraints often presenting heightened risks. Consequently, it is crucial to study maritime transportation safety in specific traffic regions. As a significant traffic characteristic in restricted waterways such as ports, channels, and straits, the concept of ship domains has been extensively applied in various studies, including collision situation detection (Szłapczyński & Niksa-Rynkiewicz, 2018; Breithaupt et al., 2022), ship collision avoidance analysis (Goerlandt & Montewka, 2015; Szlapczynski & Szlapczynska, 2016), maritime traffic pattern extraction (Liu et al., 2023; Rong et al., 2021), and maritime traffic complexity analysis (Liu et al., 2019; Xin et al., 2023).

Since the concept of ship domain was first defined by Fujii & Tanaka (1971), extensive studies have been conducted on various vessels in different regions.

In the context of empirical methods, based on traffic patterns in the North Sea, Goodwin (1975) defined the boundaries of the classical ship domain model using the distance at the highest ship density. The ship domain was segmented into various sectors around the vessel. Based on the passage distances between ships in the Great Belt and the Drogden Strait, Hansen et al. (2013) represented the ship domain parameters with the Length Over All (LOA). The length of the elliptical domain was defined as 8 LOA, and the width as 3.2 LOA. Wang & Chin (2016) calibrated ship domains using Automatic Identification System (AIS) data from the Singapore port area and proposed a polygonal domain model that functionally relates to ship length and speed.

Analytical methods have also found application in related research. To improve the reliability and flexibility of the ship domain concept, Wang (2010) proposed modelling ship domains with fuzzy quaternions. This model is utilized for assessing maritime spatial collision risks considering ambiguity and uncertainty. Dinh & Im (2016) modelled target ship domains with 'blocking areas' and 'action areas' polygons, employing statistical methods to determine the ship dynamic parameters. This approach establishes the boundaries of ship domains and is applied to guide ship collision avoidance manoeuvres.

Additionally, researchers have employed knowledge-based approaches to conduct studies on ship domains. Zhu et al. (2001) discussed the concept of the subjective domain and established the relevant model with a neural network approach. This model analyzed factors influencing ship length, ship manoeuvrability, and traffic visibility. Remzi et al. (2020) developed an asymmetric polygonal ship domain with a model based on fuzzy inference

DOI: 10.1201/9781003508762-16

systems and discussed the proposed domain under different ship encounter scenarios.

Recently, the research on ship domains has focused on extracting domain characteristics for different vessel categories from AIS data. Hörteborn et al. (2019) analyzed ship interactions and the domain characteristics influenced by waterway geographical patterns from AIS data, particularly the relative parameters explanation in narrow channels and high-traffic areas. Zhang and Meng (2019) utilized AIS data-driven probabilistic domains to characterize ship domain boundaries as fuzzy values and employed this probabilistic approach to assess the risk of ship collisions in the Strait of Singapore. Silveira et al. (2022) utilized AIS data to model an empirical quaternion ship domain model, applying modified quaternion ship domain models with different shape parameters in each quadrant. This approach better fits the general empirical shape of ship domains and was further used to assess collision risks in Portugal's coastal water. Kundakçı et al. (2023) employed AIS data and Kernel Density Estimation (KDE) methods to construct ship domain boundaries in Turkish coastal waters. This was achieved by using linear and nonlinear polynomial regression functions based on the probability boundary intersections in different directions from the central ship.

This study aims to extract ship domain characteristics corresponding to different ship characteristics from AIS data. It involves extraction of ship domain patterns, analyzing, and related parameters for target ship types in the study area. Initially, the research details the eccentric ellipse ship domain method for fitting historical data's cumulative probability distribution characteristics, along with the improved Ship Domain Overlap Index (SDOI) parameters and near-collision scenario classification based on this domain mining method. Subsequently, a case study was conducted using historical AIS data from selected navigational lanes in the Strait of Malacca. This study modelled and analyzed ship domains based on selected ship types' length and speed characteristics. Furthermore, the case analysis results facilitate extracting and categorizing near-collision situations in the study area, followed by visualization and analysis of the corresponding traffic patterns.

2 THE ASSESSMENT OF NEAR MISS SCENARIOS

This study employs a hybrid approach based on AIS data mining and probability distributions, utilizing ship position data within the given distance from historical records to calculate relevant parameters of the target ship domain and the corresponding Ship Domain Overlap Index (SDOI). Those indicators facilitate identifying and categorizing near miss scenarios for selected ship types. Given the flexibility, practicality, and computational advantages of using ellipse as the standard for the target domain fitting, the ellipse-based fitting approach is adopted in the proposed research.

The filtered AIS data is segmented into distinct trajectory sets based on Maritime Mobile Service Identity (MMSI) statistics and corresponding time intervals. Within each trajectory, the latitude, longitude, speed, and heading information from AIS records are interpolated at 60-second intervals, facilitating the analysis of the target domain at corresponding time intervals. Moreover, when the difference in heading values exceeds 180°, the linear interpolation is employed as given by:

$$\theta i = \theta i + 1 + 2\pi sign(\theta i + \pi - \theta i + 1) \tag{1}$$

where θ is the course data, this study employs a probabilistic-based approach to explore the ship domain. The research data is constructed based on the vessel distance recorded in AIS data.

The ship coordinate records in the AIS data, formatted in World Geodetic System 1984 (WGS 1984), should be transformed before the domain centre determination. The conversion of ship position coordinates from latitude and longitude to a Cartesian coordinate system is given by:

$$\begin{cases} x = lon \cdot \dfrac{C}{2\pi} \\ y = \log\left(\tan\left(\dfrac{(90 + lat) \cdot \pi}{360}\right)\right) \cdot \dfrac{C}{2\pi} \end{cases} \tag{2}$$

where *lon* is the given longitude, *lat* is the given latitude, *C* is half the circumference of the Earth at the equator (20,037,508.34 m). The Euclidean measurement defines the distance between the own ship and the target ship:

$$d = \sqrt{(x_o - x_t)^2 + (y_o - y_t)^2} \tag{3}$$

where the coordinate parameters of the own ship and the target ship are donated by (x_o, y_o) and (x_t, y_t).

In the scatter plot of the target ships extracted from historical data, the Probability Density Function (PDF) of data points within a given distance towards the direction of true north, with the geometric centre of the target domain as the origin, can be represented as follows:

$$\int_0^d f(r)dr = 1, \omega = 0 \tag{4}$$

where *d* is the given distance, *r* is the distance from the target domain centre falls in [0, *d*], ω represents the angle of deviation measured clockwise from true north. Consequently, the PDF for data points within the exemplar sector of the target domain can be articulated as follows:

$$\int_{\theta_1}^{\theta_2} \int_0^d f(r, \omega) jdrd\omega = 1 \tag{5}$$

where θ_1 and θ_2 are the angles of deviation for the sector's edges measured clockwise from true north,

j is the Jacobian determinant necessary for the transformation from Cartesian to polar coordinates. Eq. 5 represents the PDF expression for all data within the target domain when θ_1=0 and θ_2=2π.

This study employs an ellipse as a standard for ship domain determination to fit the target area. The boundary of the target domain can be fitted with the given probability values in the Cumulative Distribution Function (CDF) of the target domain around the central ship (x, y), as Figure 1 shows. The geometric parameters of the fitted ellipse can be determined by the vertical boundary limit $[y_d, y_u]$ and horizontal boundary limit $[x_l, x_r]$ given by the CDF:

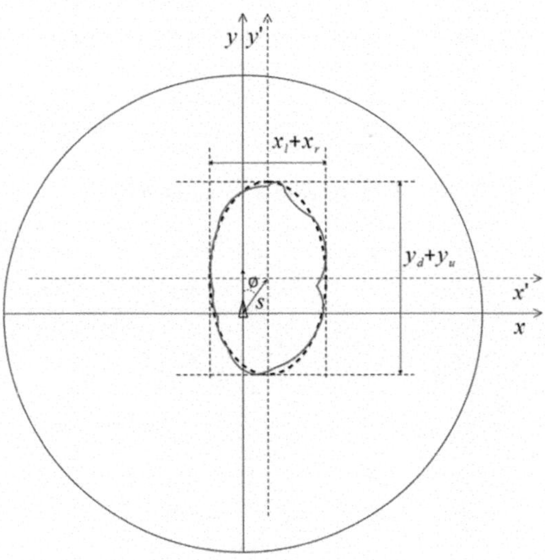

Figure 1. Ship domain in the proposed research.

$$\begin{cases} x' = \dfrac{x_l + x_r}{2} \\ y' = \dfrac{y_d + y_u}{2} \end{cases} \tag{6}$$

$$\begin{cases} a = \dfrac{\sqrt{(x_l - x_r)^2}}{2} \\ b = \dfrac{\sqrt{(y_d - y_u)^2}}{2} \end{cases} \tag{7}$$

$$\begin{cases} s = \sqrt{x'^2 + y'^2} \\ \phi = \dfrac{\pi}{2} - \arctan 2(y', x') \end{cases} \tag{8}$$

where (x, y) is the geometric centre of the target domain, a is the minor axis parameter, b is the major axis parameter of the target domain. ϕ is the clockwise angle of deviation from true north centred on the target domain, and s is the distance from the actual ship position to the centre of the target domain.

The SDOI is a coefficient inversely proportional to the separation distance between the domains of two vessels (Liu et al., 2019). When multiplied by the SDOI, the domains of the two ships are brought into a tangential state. As a practical maritime traffic safety metric, the SDOI has been employed as a critical parameter in formulating collision risk assessment criteria for vessels in open waters.

This study introduces a probabilistic ship domain model based on historical AIS data. The proposed computational approach involves SDOI determination by calculating the geographical coordinates of both the own ship and the target ship, as well as the projection lengths of the ship domains in various directions. The statistical values of the horizontal and vertical projection lengths of the ship domains for both the own ship and the target ship can be determined through projection calculations, with the formula as follows:

$$P_\eta = 2(a_o \sin(\theta_o) + a_t \sin(\theta_t) + b_o \cos(\theta_o) + b_t \cos(\theta_t)) \tag{9}$$

$$P_\xi = 2(a_o \cos(\theta_o) + a_t \cos(\theta_t) + b_o \sin(\theta_o) + b_t \sin(\theta_t)) \tag{10}$$

where a_o, b_o are the minor axis parameter and the major axis parameter of the own ship domain, θ_o is the own ship course; a_t, b_t are the minor axis parameter, major axis parameter of the target ship domain, and θ_t is the target ship course.

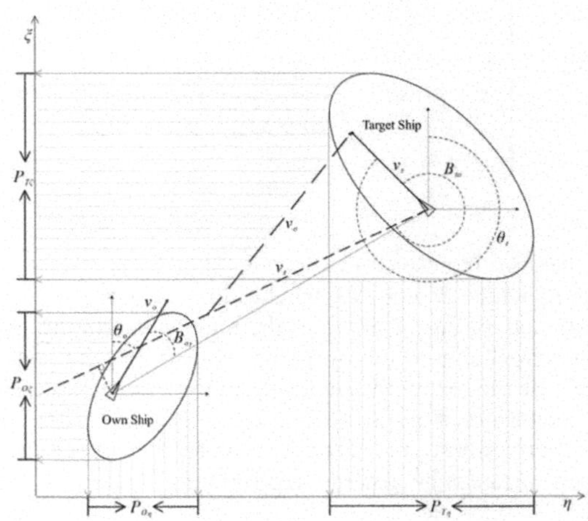

Figure 2. Ship motion parameters.

Subsequently, the distances between the centres of the own ship $(\eta o, \xi o)$ and the target ship $(\eta t, \xi t)$ in both horizontal and vertical projection directions can be calculated as follows:

$$SDOI\eta = \frac{2|\eta o - \eta t|}{P\eta} \tag{11}$$

$$SDOI\xi = \frac{2|\xi o - \xi t|}{P\xi} \tag{12}$$

$$SDOI = \max\{SDOI\eta, SDOI\xi\} \tag{13}$$

The horizontal projection of SDOI can be represented by $SDOI\eta$, while the vertical projection is represented as $SDOI\xi$. The revised SDOI value is denoted by the maximum of $SDOI\eta$ and $SDOI\xi$. When the SDOI values of two ships exceed 1, their ship domains only be expanded to overlap, indicating a lower risk of collision. An SDOI value equal to 1 implies that the ship domains are tangential. For SDOI values greater than 1, the risk of collision increases. Therefore, smaller SDOI values suggest overlap in the domains. As the SDOI value decreases, the overlapping area of the ship domains increases, correspondingly elevating the risk of collision.

As illustrated in Figure 2, the relative bearings can be calculated with the respective coordinates of the own ship (xo, yo) and the target ship (xt, yt), which can be given by:

$$B_{ot} = \begin{cases} \arcsin\dfrac{x_t - x_0}{\sqrt{(xt-xo)^2+(yt-yo)^2}} - \theta o, \\ if\ y_t < y_0 \\ \pi - \arcsin(\dfrac{x_t - x_0}{\sqrt{(xt-xo)^2+(yt-yo)^2}}) - \theta o, \\ if\ y_t \geq y_0 \end{cases} \tag{14}$$

$$B_{to} = \begin{cases} \arcsin\dfrac{xo - xt}{\sqrt{(xo-xt)^2+(yo-yt)^2}} - \theta t, \\ if\ y_t \geq y_0 \\ \pi - \arcsin(\dfrac{xo - xt}{\sqrt{(xo-xt)^2+(yo-yt)^2}}) - \theta t, \\ if\ y_t < y_0 \end{cases} \tag{15}$$

where θo represents the heading statistic of the own ship, while θt denotes the heading statistic of the target ship. Eq. 14 provides the relative bearing from the target ship to the own ship; conversely, the relative bearing from the own ship to the target ship can be determined by Eq. 15.

Furthermore, the relative velocity v_r between the two ships can be generated as follows:

$$v_r = \sqrt{v_0^2 + v_t^2 - 2v_0 v_t \cos(\theta_t - \theta_0)} \tag{16}$$

where v_o is the ship speed of the own ship, v_t is the velocity of the target ship. Consequently, when an SDOI value less than 1 is utilized to identify near-collision scenarios, situations where the SDOI value is less than 1 are marked as near misses.

Subsequently, based on the v_r, B_{ot}, B_{to} of the two ships, the detected near miss scenarios can be categorized into head-on, crossing, and overtaking scenarios. Figure 3 delineates the criteria adopted for determining different scenarios based on the proposed ship domain and the dynamic parameters of the two ships.

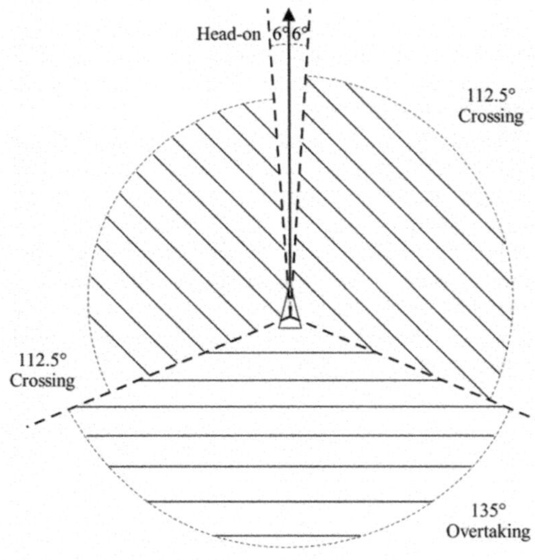

Figure 3. Near miss scenarios determination plot.

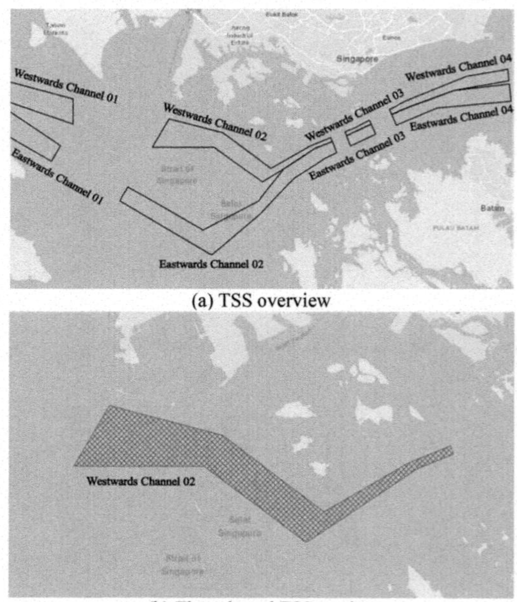

(a) TSS overview

(b) The selected TSS section

Figure 4. Research area plot.

3 CASE STUDY

3.1 *Data selection*

A study is conducted to validate the established theoretical framework in the Strait of Malacca, one of the busiest waterways in Malaysia, Indonesia, and Singapore. The Strait of Malacca plays a significant role in global maritime trade as a primary transportation channel between the Indian and Pacific oceans. As depicted in Figure 4, the research data was sourced from a section of the Traffic Separation

Scheme (TSS) within the selected area, collected from July 3 to August 4, 2022.

3.2 *Research data analysis*

3.2.1 *AIS data processing*

Recovering and accurately extracting ship movement patterns from historical data is crucial for AIS data mining tasks. Initially, information from the maritime traffic website is utilized to supply and correct col details in the collected AIS data. Data containing wrong speed values, incorrect MMSI information, and overlaps with land are removed. To minimize the impact of unsuitable data on the fitting results, the data with ship types labelled as 'Pilot vessel,' 'Towing,' and 'Tug' are also deleted from the database.

3.2.2 *Research ship type determination*

Subsequently, the filtered AIS data was used to compile statistics on traffic patterns and ship types within the study area, as shown in Table 1 and Figure 5.

Table 1. Near miss scenarios classification criteria.

Traffic statistic	
Average daily passenger ship number	2
Average passenger ship speed ($v \geq 1$ knot)	16.24
Average passenger ship LOA	325.49
Average daily cargo ship number	46
Average cargo ship speed ($v \geq 1$ knot)	11.60
Average cargo ship LOA	228.19
Average daily tanker number	60
Average tanker speed ($v \geq 1$ knot)	9.92
Average tanker LOA	227.56
Average daily container number	7
Average container speed ($v \geq 1$ knot)	12.21
Average container LOA	260.59
Average fishing vessel number	3
Average fishing vessel speed ($v \geq 1$ knot)	10.13
Average fishing vessel LOA	89.77
Average daily tug number	2
Average tug speed ($v \geq 1$ knot)	4.92
Average tug LOA	27.22

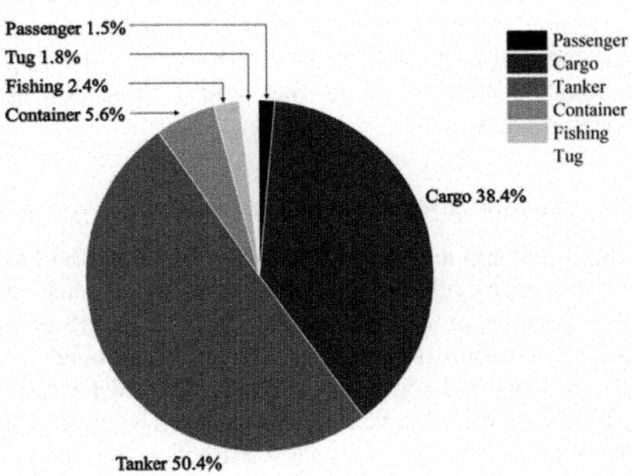

Figure 5. Ship category statistics plot.

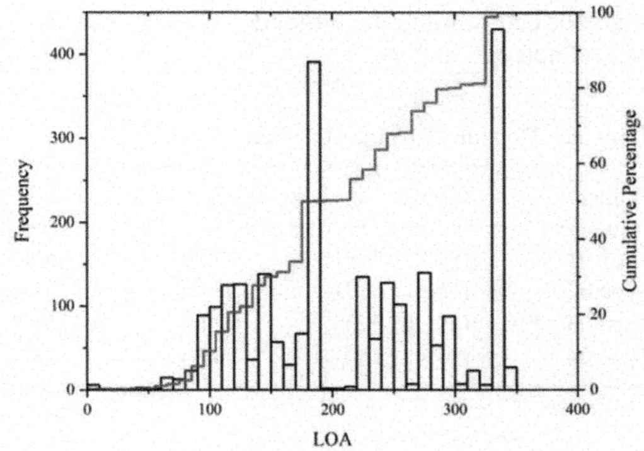

(a) LOA statistics

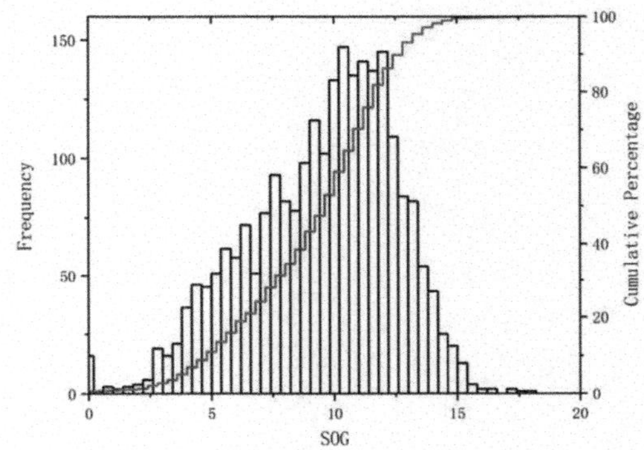

(b) SOG statistics

Figure 6. Histogram and CDF plots of tanker statistic.

According to Figure 5, further data analysis is conducted on tankers, representing the highest data proportion in the study area.

Figures 6(a) and (b) display tankers' length and speed statistics within this area. Consequently, in subsequent ship domain studies, tankers in the study area will be categorized into four types based on varying lengths and speeds, and each category will be discussed separately as Table 2 displays.

Table 2. Tanker type partition for target domain study.

Tanker	LOA (m)	SOG (knot)
Type 01	(0, 200]	(0, 10]
Type 02	(0, 200]	(10, 20]
Type 03	(200, 400]	(0, 10]
Type 04	(200, 400]	(10, 20]

3.3 *Ship domain construction*

The boundary construction of the target domains with 5% CDF value of data points is calculated from the central ship. The horizontal and vertical projections of the areas formed by the CDF contour line are marked (as shown in Table 3) and used as the basis for determining the target domains' true centres and related parameters.

Table 3. Domain statistics of tanker.

Tanker	Centre	a	b
Type 01	(93.0, -29.9)	799.8	1035.2
Type 02	(66.5, -72.7)	658.4	1131.2
Type 03	(39.7, 27.8)	724.6	1057.4
Type 04	(97.4, 23.6)	652.4	1178.4

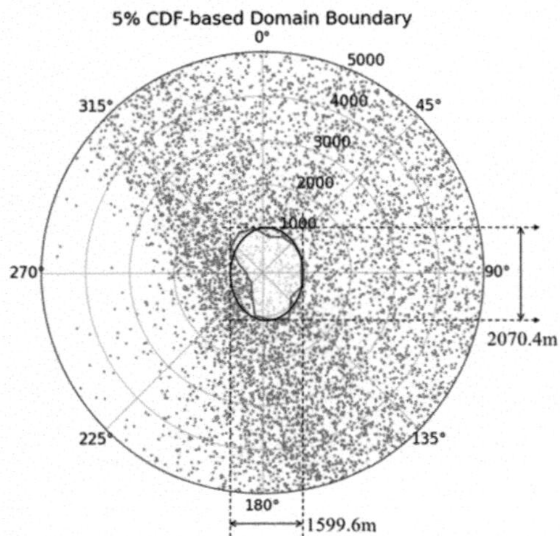

(a) LOA (0, 200], SOG (0, 10]

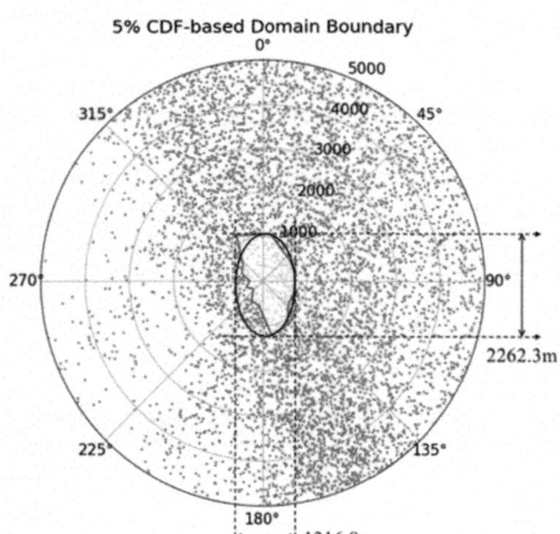

(b) LOA (0, 200], SOG (10, 20]

Figure 7. Domain plot of tanker in types 01~02.

Figures 7(a) and (b) illustrate the fitting results of ship domains for tankers on Type 1 and Type 2 in the study area, using AIS data from July 3 to August 4, 2022. In Figure 7, the data points are displayed in bow-up orientation. For tankers with lengths in the (0, 200] interval, the width and length parameters of the ship domain decrease as speed increases.

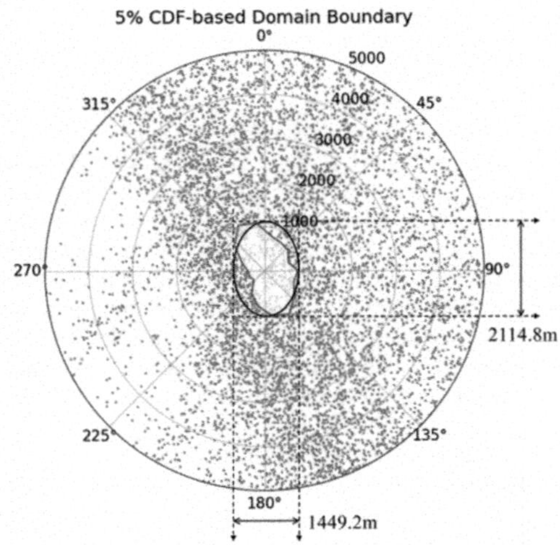

(a) LOA (200, 400], SOG (0, 10]

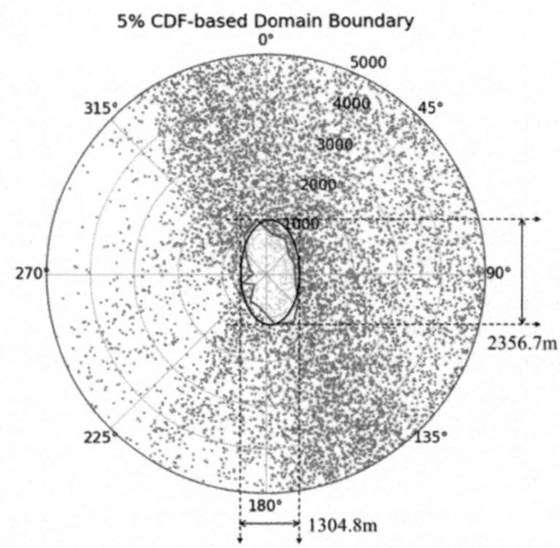

(b) LOA (200, 400], SOG (10, 20]

Figure 8. Domains plot of tankers in types 03~04.

Figure 8 demonstrates that for tankers with lengths in the (0, 200] interval, the width of the ship domain decreases while the length increases as speed increases.

3.4 *Near miss scenarios detection*

This study applied the corresponding SDOI method to identify near-collision scenarios with the established ship domains in the selected study area. The AIS data collected within the selected area on September 22, 2022, is applied for scenario extraction and analysis with the extracted tanker ship domains. The approach from the near-collision scenario extraction framework developed by Liu et al. is employed to process the data under other vessel types.

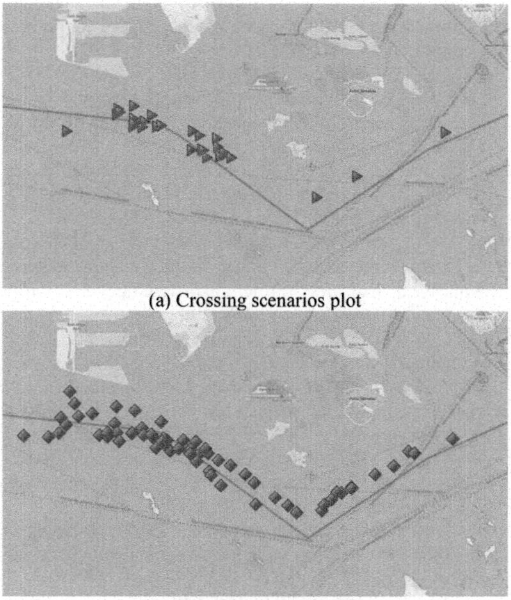

(a) Crossing scenarios plot

(b) Overtaking scenarios plot

Figure 9. Detected near miss situations.

Consequently, the identified near-collision scenarios are categorized and depicted in Figure 9. Figure 9(a) shows the crossing scenarios recognized in the study area, and the geographic distribution of overtaking scenarios is marked as green squares in Figure 9(b).

4 DISCUSSION

This research is carried out in a section of the westward TTS of Malacca Strait, near the Singapore port area.

Figure 10. Scenarios heat plot.

The modelling results of ship domains for tankers indicate that there is a high-density data band on the left side when ship speeds are below 10 knots (Figures 7(a) and 8(a)). This reflects a traffic pattern of slower tankers being frequently overtaken within the lane. Additionally, within a 5-kilometer radius of the central ship, a sparse data area on the left side is consistently observed in the images (Figures 7(a), (b) and 8(a), (b)). This corresponds to the distribution of anchorage areas on the left side of the westwards channel.

Historical AIS data from the selected area on September 22, 2022, revealed 91 near miss scenarios. 63 overtaking scenarios accounted for 69.2% of all situations, aligning with the high-density data band observed in Figures 7(a) and 8(a). The crossing scenarios mainly occurred in the junction regions of Singapore port's entry and exit waters with the research channel. Figure 10 displays the overall distribution of near miss scenarios, with the highest-density areas in red gradually transitioning to areas with fewer near miss scenarios.

5 CONCLUSIONS

The constructed tanker domains from AIS data and CDF reveal that tankers with varying static and dynamic characteristics have distinct ship domain features. For tankers with lengths in the (0, 200] interval, the actual domain centre is located towards the rear-right of the vessel, and both width and length parameters decrease with increasing speed. In contrast, for tankers with lengths in the (200, 400] interval, the domain centre is positioned towards the front-right of the vessel, with the width decreasing and length increasing as speed increases.

The detected near-collision scenarios with the constructed domains and corresponding SDOI show that overtaking scenarios are distributed relatively evenly in the research area. However, crossing scenarios are concentrated in the port entry/exit water and channel junction area. The junction area is a hotspot for near miss scenarios with crossing and overtaking.

ACKNOWLEDGEMENTS

This work of the first author has been supported by the China Scholarship Council under Grant: 202106570012. The visits to Lisbon of the second and third authors have been made under the by project RESET (REliability and Safety Engineering and Technology for large maritime engineering systems), which is partially financed by the European Union Horizon 2020 research and innovation program, under the Marie Skodowska-Curie grant agreement No. 73088 RESET. This work contributes to the Strategic Research Plan of the Centre for Marine Technology and Ocean Engineering, which is financed by the Portuguese Foundation for Science and Technology (Fundação para a Ciência e Tecnologia-FCT) under contract UIDB/UIDP/00134/2020.

REFERENCES

Breithaupt, S. A., Bensi, M., & Copping, A. (2022). AIS-based characterization of navigation conflicts along the US Atlantic Coast prior to development of wind energy. *Ocean Engineering*, 264.

Dinh, G. H., & Im, N. (2016). The combination of analytical and statistical method to define polygonal ship domain and reflect human experiences in estimating dangerous area. *International Journal of E-Navigation and Maritime Economy*, 4, 97–108.

Fiskin, R., Nasiboglu, E., & Yardimci, M. O. (2020). A knowledge-based framework for two-dimensional (2D) asymmetrical polygonal ship domain. *Ocean Engineering*, 202.

Fujii, Y., Tanaka, K., 1971. Traffic capacity. *Journal of Navigation*, 24, 543–552.

Goerlandt, F., & Montewka, J. (2015). Maritime transportation risk analysis: Review and analysis in light of some foundational issues. *Reliability Engineering and System Safety*, 138, 115–134.

Goodwin, E.M., 1975. A statistical study of ship domains. *Journal of Navigation*, 28, 328–344.

Guedes Soares, C., Teixeira, A.P., 2001. Risk assessment in maritime transportation. *Reliability Engineering and System Safety*, 74, 299–309.

Hansen, M. G., Jensen, T. K., Lehn-Schioler, T., Melchild, K., Rasmussen, F. M., & Ennemark, F. (2013). Empirical ship domain based on AIS data. *Journal of Navigation*, 66(6), 931–940.

Hörteborn, A., Ringsberg, J. W., Svanberg, M., & Holm, H. (2019). A Revisit of the Definition of the Ship Domain based on AIS Analysis. *Journal of Navigation*, 72(3), 777–794.

Kundakçı, B., Nas, S., & Gucma, L. (2023). Prediction of ship domain on coastal waters by using AIS data. *Ocean Engineering*, 273.

Liu, D., Siow, C. L., Kang, H. S., & Guedes Soares, C. (2023). Detection and analysis of near collision scenarios in the Malacca Strait. I*n Advances in the Collision and Grounding of Ships and Offshore Structures*, H. Le Sourne & C. Guedes Soares, (Eds.), Taylor & Francis Group, London, UK, pp. 75–81.

Liu, Z., Li, Y., Zhang, Z., Yu, W., Han, B., & Du, Y. G. (2023). Spatial modeling and analysis approach for ship-encounters dynamic spatial–temporal domain. *Ocean Engineering*, 278.

Liu, Z., Wu, Z., & Zheng, Z. (2019). A novel framework for regional collision risk identification based on AIS data. *Applied Ocean Research*, 89, 261–272.

Rong, H., Teixeira, A. P., & Guedes Soares, C. (2021). Spatial correlation analysis of near ship collision hotspots with local maritime traffic characteristics. *Reliability Engineering and System Safety*, 209.

Silveira, P., Teixeira, A. P., & Guedes Soares, C. (2022). A method to extract the Quaternion Ship Domain parameters from AIS data. *Ocean Engineering*, 257.

Szłapczyński, R., & Niksa-Rynkiewicz, T. (2018). A Framework of A Ship Domain-Based Near-Miss Detection Method Using Mamdani Neuro-Fuzzy Classification. *Polish Maritime Research*, 25(s1), 14–21.

Szlapczynski, R., & Szlapczynska, J. (2016). An analysis of domain-based ship collision risk parameters. *Ocean Engineering*, 126, 47–56.

Wang, N. (2010). An intelligent spatial collision risk based on the quaternion ship domain. *Journal of Navigation*, 63(4), 733–749.

Wang, Y., & Chin, H. C. (2016). An Empirically-Calibrated Ship Domain as a Safety Criterion for Navigation in Confined Waters. *Journal of Navigation*, 69(2), 257–276.

Xin, X., Liu, K., Loughney, S., Wang, J., & Yang, Z. (2023). Maritime traffic clustering to capture high-risk multi-ship encounters in complex waters. *Reliability Engineering and System Safety*, 230, 36

Zhang, L., & Meng, Q. (2019). Probabilistic ship domain with applications to ship collision risk assessment. *Ocean Engineering*, 186.

Zhu, X., Xu, H., Lin, J. (2001). Domain and its model based on neural networks. *Journal of Navigation*, 54, 97–103.

Advances in Maritime Technology and Engineering – Guedes Soares & Santos (Eds)

Collision risk assessment in ship encounter scenarios using AIS trajectory data

E. Lotovskyi, H. Rong & A.P. Teixeira
Centre for Marine Technology and Ocean Engineering (CENTEC), Instituto Superior Técnico, Universidade de Lisboa, Lisbon, Portugal

ABSTRACT: This paper presents an approach to collision risk assessment in maritime traffic, utilising AIS trajectory data and manoeuvrability characteristics of the vessels. The proposed model evaluates the change in collision risk during ship evasive manoeuvres considering the vessels' manoeuvrability, ensuring a realistic representation. The methodology introduces two parameters to assess and manage the risk of collision: the probability of collision and the safety window of rudder input. A case study explores near-collision scenarios. Results are discussed, encompassing trajectories, collision probabilities and safety window assessments, providing insights into the collision risk during ship evasive manoeuvres. The study can be further developed for collision avoidance applications.

1 INTRODUCTION

Ship collisions are navigational accidents that account for 13% of all maritime casualties in the EU (EMSA, 2021). Between 2014 and 2020, these collisions were responsible for 32% of fatalities in maritime accidents. On a global scale, collisions rank as the second most frequent cause of incidents, following machinery damage/failure (Allianz, 2022). Therefore, the assessment of the risk of collision remains an important research topic, as evidenced by many recent studies (e.g. Antão et al., 2023; Rong et al., 2022; Silveira et al., 2022).

Assessment of collision risk aboard vessels is typically done by navigation watch officers using radar and Automatic Radar Plotting Aid (ARPA) technology (Krata & Montewka, 2015). This technology monitors nearby vessels, displaying the Closest Point of Approach (CPA) and Time to Closest Point of Approach (TCPA). However, in urgent or high-velocity scenarios, CPA and TCPA may be less effective (Silveira et al., 2016).

In collision situations, mariners follow the International Regulations for Preventing Collision at Sea (COLREG). However, the correct following of COLREG rules does not guarantee collision avoidance due to varying risk interpretations (Silveira et al., 2021). The definition of collision risk in COLREG is subjective, relying on mariner experience and judgment (Silveira et al., 2016). Different interpretations can lead to close-quarter encounters even with the same set of rules (Kuwata et al., 2014).

Ambiguity in COLREG arises when a ship must take collision avoidance measures (Silveira et al., 2021). For example, it is unclear when a ship initially not obligated to act should do so if the other vessel fails to respond adequately.

Approximately 50% of ship accidents result from insufficient vigilance of navigation officers (Kuwahara et al., 2016). Even in good weather conditions, 60% of collisions occur due to late detection.

The risk of collision can be assessed using different methodologies depending on available data and research aims. Collision diameter (Fujii et al., 1970) and ship domain (Kuwahara et al., 2016) are frequently employed concepts for studying collision risk. However, the collision diameter approach only detects situations with ships in physical contact. Thus, if the distance between vessels exceeds the collision diameter, Fujii's approach may incorrectly suggest collision avoidance (Rong et al., 2022). On the other hand, a ship domain refers to a ship's comfort zone rather than its vital region (Krata & Montewka, 2015).

Collision risk assessment often relies on statistical analysis. However, it is crucial to acknowledge that real maritime accident data is rarely sufficient due to its infrequent occurrence (Rong et al., 2023). Instead, collision risk is commonly assessed using non-accidental events data, such as traffic conflicts or near collisions, derived from Automatic Identification System (AIS) trajectory data. Rong et al. (2019) used AIS data to evaluate collision risk and generate spatial density maps employing the Kernel Density Estimation (KDE) method. Rong et al. (2022) developed a method using an improved Sliding Window Algorithm to accurately detect ship evasive behaviour from AIS trajectory data. Rong et al. (2023) explored the relationship between selected collision risk indicators and decision-making in ship evasive manoeuvres.

DOI: 10.1201/9781003508762-17

In collision probability studies seeking to enhance risk assessment, manoeuvrability plays a crucial role. Krata & Montewka (2015) applied a quasi-linear modular hydrodynamic model to estimate the critical area for a ship involved in a collision encounter. Silveira et al. (2016) analysed ship evasive manoeuvres in near-collision scenarios using AIS data and the first-order Nomoto model. Lotovskyi & Teixeira (2023) assessed the minimum safe distance in different close-quarter situations, considering vessel manoeuvring capabilities through the use of Nomoto's first-order equation.

The objective of this paper is to suggest a collision risk assessment approach using AIS trajectory input data that can be further developed for collision avoidance applications. The model assesses the change in collision risk during a generic ship collision avoidance manoeuvre obtained from the methodology outlined by Rong et al. (2022). The proposed model considers the vessels' manoeuvrability characteristics, modelled using Nomoto's first-order model. The subsequent sections of this paper are structured as follows. Section 2 presents the methodology of this study. Section 3 provides a case study description. Section 4 discusses results, encompassing trajectories, probability of collision and safety window assessments, contributing to the understanding of collision risk during ship evasive manoeuvres.

2 METHODOLOGY

2.1 *Collision risk assessment*

This study aims to evaluate the variation in collision risk during ship evasive manoeuvres adopted by vessels in crossing situations. The flowchart of the procedure employed in this study is shown in Figure 1.

The near-collision situation is randomly selected from the database of ship evasive manoeuvres registered in a specific area. The chosen case study data is pre-processed and divided into two scenarios. The first scenario represents the actual behaviour of both vessels (i.e. collision avoidance), while the second scenario assumes that neither of the vessels takes appropriate action to avoid a near-collision situation, maintaining the same heading (i.e. no action).

The collision risk resulting from the adoption of evasive manoeuvres is assessed using two parameters, namely the probability of collision and the safety window of rudder input. Both parameters are applied to both scenarios, and the obtained results provide the information on which a preliminary judgment regarding the variation of collision risk can be based.

2.2 *Ship evasive manoeuvre database*

To accurately detect the ship evasive behaviour from AIS trajectory data, Rong et al. (2022) proposed a novel approach to automatically identify the ship evasive behaviour from ship trajectories based on an improved Sliding Window Algorithm. The proposed

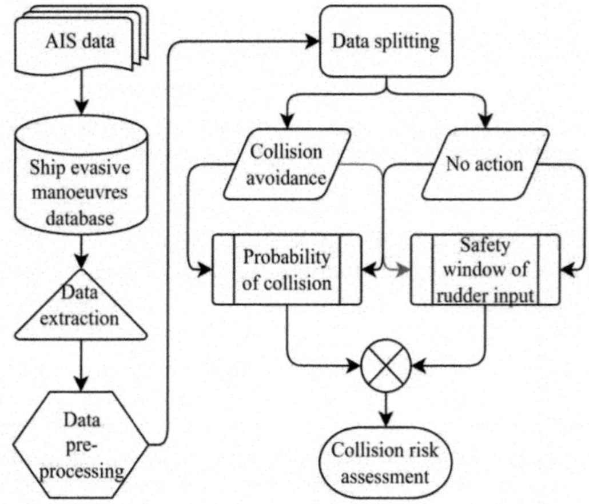

Figure 1. Simplified flowchart of the implemented model.

method precisely determines the time at which a ship takes significant action to avoid a collision with a target ship. For this purpose, improved Sliding Window algorithm scans ship trajectory involved in encounter scenarios considering the relative distance between the involved ships, the rate of turn (ROT) and the derivative of the ROT. The method is applied to the maritime traffic off the coast of Portugal. The resulted database contains the ship trajectories represented by the data points in chronological order as shown in Equation 1.

$$Traj_s = \left\{p_1^s,\, p_2^s, \ldots,\, p_{n_s}^s\right\}, \quad \text{where } p_i^s = \{long_i, lat_i, v_i, \psi_i, t_i\} \tag{1}$$

where $Traj_s$ = registered trajectory of ship s; n_s = number of data points in a trajectory; p_i^s = data point i of the trajectory expressed as a quintuple vector; $long_i$ = longitude of the point i; v_i = speed over ground (SOG); ψ_i = heading; t_i = timestamp.

2.3 *Data pre-processing*

The trajectories obtained from the database only contain the points that correspond to the ship evasive manoeuvre. The pre-processing of the data starts by the conversion of geographic coordinates to Cartesian coordinates. This conversion is done with the support of haversine formula (Bowditch, 2019). The haversine formula (see Equation 2) calculates the distance between two points on the surface of a sphere (e.g., the Earth) given their latitude and longitude.

$$\begin{aligned} d &= 2 \times R \times atan2\left(\sqrt{a},\, \sqrt{1-a}\right) \\ \text{where } a &= \sin^2\left(\frac{\phi_2 - \phi_1}{2}\right) + \\ &+ \cos(\phi_1) \times \cos(\phi_2) \times \sin^2\left(\frac{\lambda_2 - \lambda_1}{2}\right) \end{aligned} \tag{2}$$

where d = distance between two points; R = radius of the Earth; ϕ_1, ϕ_2 = latitude of the point 1 and point 2, respectively; λ_1, λ_2 = longitude of the point 1 and point 2, respectively.

For the case of head-on encounter, the geographic coordinates of the first point of the southern vessel correspond to the origin of the reference, i.e. $(x_1^S,\ y_1^S) = (0,\ 0)$. Thus, the cartesian coordinates of the first point of the northern vessel, i.e. $(x_1^N,\ y_1^N)$, can be determined by Equation 3.

$$\begin{cases} x_1^N = x_1^S - d \times \cos(90^\circ - \psi_1^N) \\ y_1^N = y_1^S - d \times \sin(90^\circ - \psi_1^N) \end{cases} \quad (3)$$

where ψ_1^N = heading angle of the first point of the northern vessel.

2.4 *Data splitting*

To assess the behaviour of vessels during ship evasive manoeuvres and evaluate collision risk, the case study data is divided into two scenarios: collision avoidance and no action.

In the collision avoidance scenario, the vessel's behaviour is assessed as it occurs. The probability of collision and the safety window of rudder input are applied from the moment both ships apply COLREG rules until the moment of the closest point of approach that corresponds to the higher risk of collision. The procedure will be applied every 2 minutes until the time of the closest point of approach reaches zero minutes. The objective is to evaluate how these parameters vary at each point along the real trajectory of both vessels.

In the no action scenario, neither vessel takes appropriate action to avoid a dangerous situation, maintaining the same course and speed. In this scenario, the values of speed over ground and heading before starting the avoiding manoeuvre are registered and considered constant throughout the assessment. Consequently, the trajectories of both vessels are modified. Using the new positions of both vessels, the probability of collision and the safety window of rudder input are applied every 2 minutes to the new trajectories of the vessels.

The collision risk is assessed by comparing the results obtained from both scenarios.

2.5 *Probability of collision*

When two vessels approach each other, involving a risk of collision, both ships adhere to the COLREG rules to avoid a close-quarter situation. However, even if both vessels follow the rules, the probability does not reduce to zero.

In this study, the probability of collision is determined every 2 minutes until the vessels pass each other (i.e. TCPA is zero). At each point of the trajectories, it is considered that both vessels are "blind", and they can apply any rudder angle for a certain period of time at that instant. This results in different hypothetical trajectories that the vessel can follow.

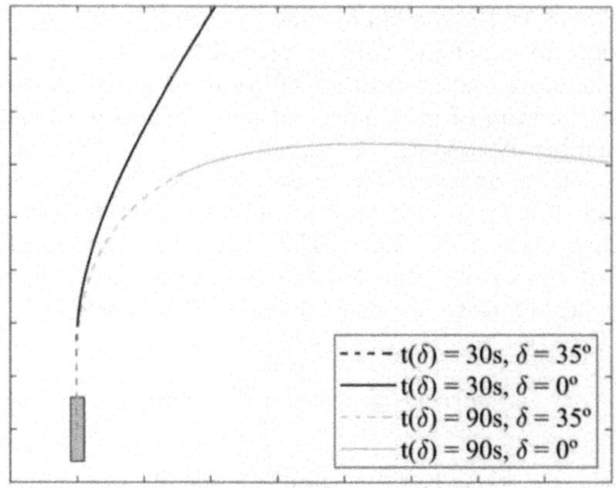

Figure 2. Influence of the time of rudder application on the trajectory of the ship.

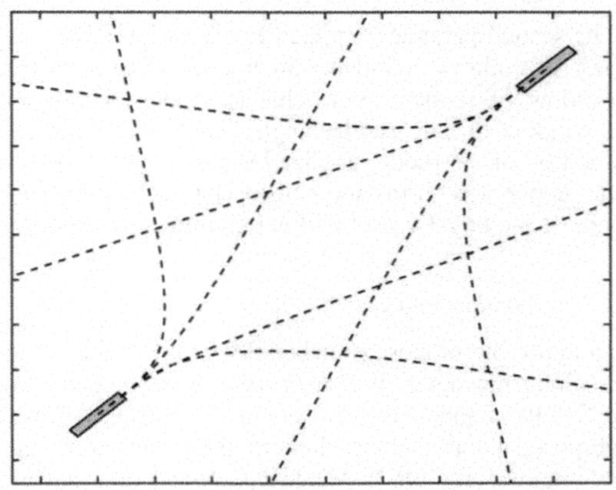

Figure 3. Different collision scenarios simulated by the MCS.

Figure 2 illustrates a generic example of the possible trajectories that a vessel can have depending on the time the rudder is applied, considering an initially applied rudder angle of 35º in both cases. The dotted line corresponds to the moment when the rudder is applied, and the thick line corresponds to the trajectory of the vessel when the rudder is put to 0º.

The blindness of the ships may arise due to various reasons, such as the "blind" vessel misinterpreting information from another ship, the "blind" vessel identifying the encounter situation too late, communication errors between bridges, or a misinterpretation of each other's actions.

The maximum duration of rudder application is determined based on the parameters: advance and transfer, that can be estimated from the turning circle manoeuvre (see Figure 5).

A sample of size n representing various trajectories that each vessel can follow is obtained by the Monte Carlo simulation (MCS) technique (see

Figure 3). By combining these trajectories, it is possible to count the number of collisions that occur. Therefore, the probability of collision is calculated as the ratio of the number of collisions to the total number of samples.

The accuracy of the results provided by MCS is influenced by the number of simulation histories (Lotovskyi et al., 2022). The appropriate number of histories can be estimated by setting a maximum error related to the Confidence Interval (CI) estimated by:

$$Error_{CI\ 95\%} = \frac{1.96\frac{\sigma}{\sqrt{n}}}{\mu} \times 100 \quad (4)$$

where σ = standard deviation; n = sample size; and μ = sample mean.

2.6 *Safety window of rudder input*

The second parameter applied to assess the effectiveness of collision avoidance manoeuvres is the safety window of rudder input. This parameter represents a window of safe rudder angles combined with the duration of the rudder application that the own vessel can apply at each instant, considering that the behaviour of the target vessel will not significantly change.

2.7 *Nomoto's model*

To apply the proposed methodology and evaluate the resulting outcomes, it is necessary to mathematically define the manoeuvring behaviour of a vessel. For this purpose, the first-order Nomoto model is employed (Nomoto et al., 1957). While this model may not be the most precise method to predict the vessel's trajectory during manoeuvring (Sutulo & Guedes Soares, 2011), it is a straightforward model to implement and does not require any specialised information, such as data from hydrodynamic experiments or zigzag sea trial tests in open waters. The easily accessible data (such as AIS data) can be used to identify the main dimensions and vessel type, which are sufficient for implementing the manoeuvre in this case study.

The first order Nomoto model in time domain can be expressed by Equation 5 (Nomoto et al., 1957).

$$T\dot{r} + r = K\delta \quad (5)$$

where K = static yaw rate gain; T = governance time constant; r = ROT; and δ = rudder angle.

Besides, the model can be defined by Equation 6.

$$T\ddot{\psi} + \dot{\psi} = K\delta, \dot{\psi} = r \quad (6)$$

where, ψ = heading angle of the ship that can be defined by Equation 7.

$$\psi_{i+1} = \psi_i + r_{i+1}(t_{i+1} - t_i), i \in N \quad (7)$$

From Equation 5, ROT is given by:

$$r = \frac{r_0}{e^{t/T}} + \delta K\left(1 - \frac{1}{e^{t/T}}\right) \quad (8)$$

where r_0 = initial ROT.

An initial approximation of the parameter T can be obtained by considering the vessel's length and type. For this study, the triangular probability distribution suggested by Silveira et al. (2016) is adopted to model the T parameter.

Static yaw rate gain, K, can be estimated by Equation 9.

$$K = \frac{V_{STD}^{(35)}}{\delta R}, \ where\ R = 2 \times Lpp \quad (9)$$

where R = steady turning radius (SNAME, 1989); and $V_{STD}^{(35)}$ = steady turn speed for the maximum rudder angle defined in Equation 10.

$$V_{STD}^{(35)} = V_0 \times \frac{V_{ST}}{V_0}, \quad where\ \frac{V_{ST}}{V_0} \sim U\ (0.6,\ 0.7) \quad (10)$$

where V_0 = initial speed; and V_{ST}/V_0 = ratio sampled from a uniform probability distribution for turning circles executed using maximum rudder angle.

The speed of the vessel, $V^{(\delta)}$, at instant t and rudder angle δ is modelled in Equation 11.

$$V^{(\delta)} = V_0\left(1 - C_{Vred}^{(\delta)} \times \frac{tg^2(\beta)}{tg^2(\beta_{STD})}\right), \quad where\ \beta_{STD} \sim U\left(18\frac{Lpp}{R}, 22.5\frac{Lpp}{R} + 1.45\right) \quad (11)$$

where $C_{Vred}^{(\delta)}$ = speed reduction coefficient for rudder angle δ (see Equations 12 and 13); β = drift angle; and β_{STD} = steady turning drift angle defined probabilistically by a uniform distribution (SNAME, 1989).

$$C_{Vred}^{(\delta)} = \frac{\delta}{35^\circ} \times C_{Vred}^{(35^\circ)} \quad (12)$$

$$C_{Vred}^{(35^\circ)} = 1 - \frac{V_{ST}}{V_0} \quad (13)$$

The effect of drift angle, β, on the turning circle manoeuvre can be estimated by Equation 14.

$$T\dot{\beta} + \beta = K_\beta \delta \quad (14)$$

Considering:

$$\beta = \beta_{STD} \Leftrightarrow \dot{\beta} = 0 \quad (15)$$

$$K_\beta = \frac{\beta_{STD}}{\delta} \quad (16)$$

From Equation 15, the drift angle at instant t can be estimated by Equation 17.

$$\beta = \frac{\beta_0}{e^{t/T}} + \delta K_\beta\left(1 - \frac{1}{e^{t/T}}\right) \quad (17)$$

The position change of pivot point is described by Equation 18.

$$\begin{cases} x_{i+1} = x_i + \Delta d \times \cos(COG) \\ y_{i+1} = y_i + \Delta d \times \sin(COG) \end{cases}, \quad (18)$$
$$where\ COG = \psi - \beta$$

where Δd = sailed distance between t_i and t_{i+1}.

3 CASE STUDY DESCRIPTION

The database of ship evasive behaviours is derived from the maritime traffic off the continental coast of Portugal provided by a three-month AIS dataset, from 1st October to 31st December 2015. The database contains 2846 evasive manoeuvres from which there are 1328, 1011 and 507 identified scenarios for overtaking, crossing and head-on encounters, respectively.

From the database, an encounter scenario is randomly selected. The data of the vessels engaged in the selected case study is summarised in Table 1. To classify the scenario by the type of encounter, the relative heading between the vessels is analysed using the classification proposed by Silveira et al. (2021). As indicated in Table 1, the relative heading angle before the vessels execute evasive manoeuvres is less than 0°. Thus, the selected scenario corresponds to a head-on encounter situation involving two similar cargo ships.

Table 1. Initial input data considerations.

Variables	Southern ship (vessel 1)	Northern ship (vessel 2)
Type of the vessel	Cargo ship	Cargo ship
Lpp (m)	120	127
B (m)	20	20
V_0 (knots)	13.5	15.4
V_{max} (knots)	16.5	16.5
ψ_0 (°)	45.4	225.6
β_0 (°)	0	0
ROT (0°/s)	0	0
δ (°)	U (−35, 35)	U (−35, 35)
T parameters:		
- Minimum	70	70
- Most likely	100	100
- Maximum	130	130

It is important to note that when vessels are approaching each other head-on, Rule 14 of COLREG dictates: "*when two power-driven vessels are meeting on reciprocal or nearly reciprocal courses so as to involve risk of collision each shall alter her course to starboard so that each shall pass on the port side of the other*" (Cockcroft & Lameijer, 2012).

The probability of collision and the safety window of rudder input are assessed through trajectory simulation of two ships modelled as rectangles, with their movement estimated based on available data. The model assumes a fixed pivot point located at the midship position. The simulation considers deep water and good visibility conditions, excluding weather conditions from the analysis. The initial heading, ψ_0, is set equal to the initial COG, the initial drift angle, β_0, and the initial rate of turn are held at zero.

Both the blind rudder angle behaviour and the random duration of rudder application follow a uniform distribution.

4 DISCUSSION AND RESULTS

4.1 *Trajectories of the ships*

Figure 4 shows the trajectories of both vessels in two case scenarios: collision avoidance and no action. The trajectories are depicted from the moment when both vessels have applied to the COLREG rules. The solid line represents the actual trajectories of the vessels, while the dashed line depicts their initial trajectories. As evident from the results, both vessels followed COLREG rules and altered their course to starboard. Additionally, the trajectories reveal that vessel 2 altered her course before vessel 1.

4.2 *Turning circle manoeuvre*

Figure 5 shows a generic trajectory of the midship point of the vessel during a turning manoeuvre, including the main measures obtained based on the turning manoeuvre.

Using Nomoto equations, the main dimensions of the manoeuvre can be determined. In this analysis, it is assumed that the initial speed of both vessels is 16.5 knots, and the rudder angle is 35°. Under these conditions, the resulting transfer, advance, tactical diameter and turning radius are summarised in Table 2. The obtained results are then compared with the recommended values in the literature (Cockcroft & Lameijer, 2012; Sutulo, 2013). As observed, all the obtained results fall within the recommended intervals.

Based on the turning circle manoeuvre, it is possible to determine the time the vessel takes to reach the advance and transfer for 90°. In the case of vessel 1, this time is equal to 102 s, while for vessel 2, it is equal to 104 s. Thus, the random duration of rudder application for both vessels is given by:

$$\Delta t_\delta \sim U\ (1,\ 90) \quad (19)$$

Please note that the maximum duration of rudder application is intentionally set slightly lower than

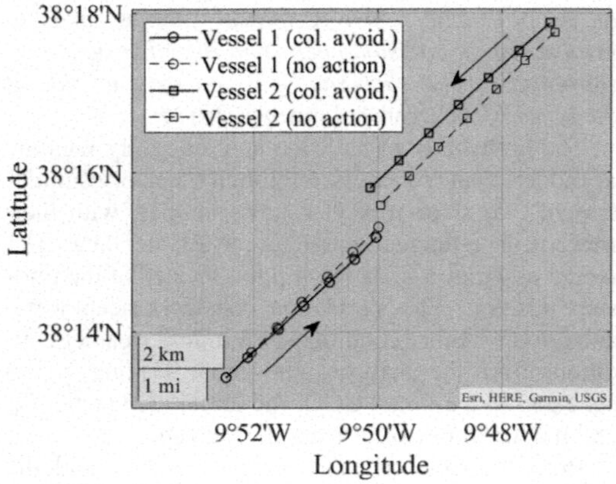

Figure 4. Trajectories of vessels in collision avoidance and no action case scenarios.

Table 2. Turning circle results of the case study vessels.

Main measures	Vessel 1	Vessel 2
Transfer (1.5 *Lpp* – 3.5 *Lpp*)	280 m	296 m
Advance (3 *Lpp* – 5 *Lpp*)	599 m	586 m
Tactical diameter (3 *Lpp* – 7 *Lpp*)	586 m	597 m
Diameter of turn (3 *Lpp* – 7 *Lpp*)	488 m	513 m

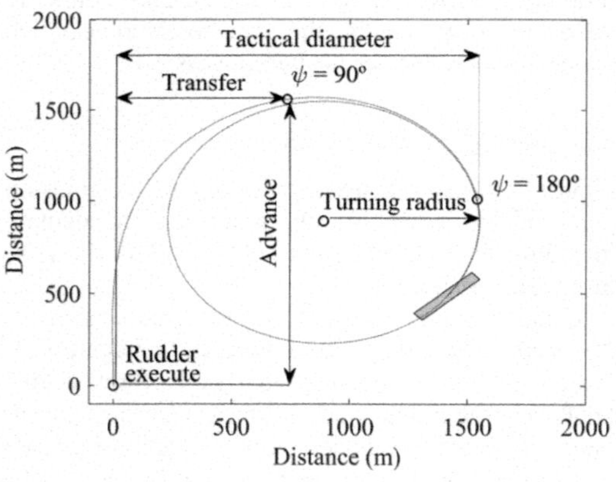

Figure 5. Path of midship point in turning circle and main dimensions of manoeuvre.

predicted by the turning circle model. This allows to account for uncertainties associated with the parameters of the manoeuvrability model.

4.3 *Number of simulation histories*

Any simulation analysis starts by determining the optimal number of simulated scenarios (i.e. histories) for a given level of accuracy. Increasing the number of simulations yields more precise results. To estimate the necessary number of simulations, the error associated with the 95% Confidence Interval of the probability of collision is considered. Two distinct models are utilised for this purpose: one for the collision avoidance scenario and another for the no action scenario.

To make this estimation, the probability of collision is calculated as the ratio of the number of collisions that occur to the total number of simulations conducted for various rudder angles and time of rudder application. In each simulation, the initial distance between vessels is kept constant and the uncertainties associated with the parameters of the manoeuvrability model are considered.

The optimal number of simulation histories is determined individually for the collision avoidance and no action scenarios. Figure 6 displays the obtained results. Due to limitations in computer processing power and the errors associated with the 95% CI, 20,000 simulation histories are chosen for both models. The resulting error for the avoidance collision scenario is 5.82%, while for the no action scenario it is 5.17%.

4.4 *Probability of collision*

Considering the manoeuvring characteristics of the vessels and applying the optimum number of simulation histories, this section presents a preliminary risk assessment based on the probability of collision.

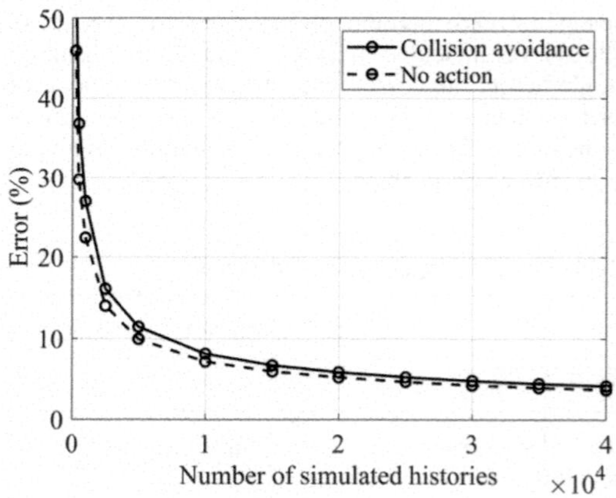

Figure 6. 95% CI error of the probability of collision for different number of simulated histories.

Figure 7 shows the variation in the probability of collision in both case scenarios along the decreasing time to the closest point of approach (TCPA). By observing the results, four levels of risk described in Cockcroft & Lameijer (2012) can be identified. When TCPA is 10-12 minutes, the probability of collision is close to zero. The relative distance between the vessels in this case is more than 5 nm, which, according to the level of risk described in COLREG, corresponds to no risk of collision. As TCPA decreases to 6-8 minutes, there is a noticeable increase in the probability of collision, reflecting a relative distance between the vessels of more than 3 nm, indicating a risk of collision stage. The probability of collision significantly rises when TCPA is 2-4 minutes, representing a close-quarter situation with a relative distance between the vessels less than 3 nm. When TCPA reaches zero, the probability of collision reaches the highest value, corresponding to an immediate danger stage.

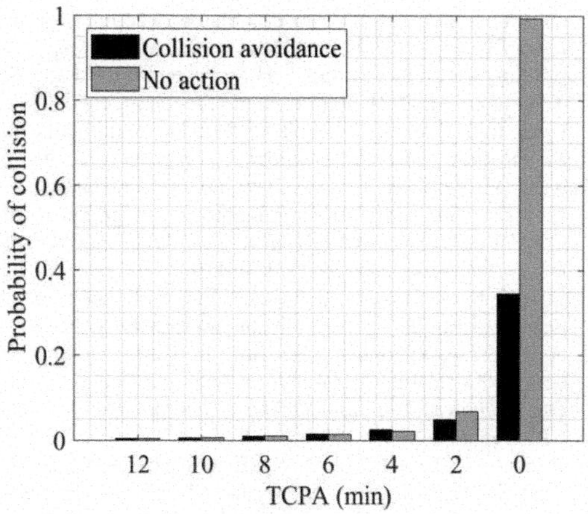

Figure 7. Probability of collision in two case scenarios for different TCPA.

By comparing the variation in the probability of collision between both scenarios, it is evident that the probability of collision remains almost the same in both scenarios until TCPA = 2 minutes. From this moment onward, it becomes apparent how the probability of collision increases if no action is taken by the bridge. The obtained results show that by adhering to COLREG rules, the vessels decrease the probability of collision by 65%. Moreover, it can be affirmed that the effect of vessels manoeuvring to starboard on the safety level assessment is significant only when the relative distance between the vessels is less than 3 nm.

4.5 *Safety window of rudder input*

Considering the manoeuvring characteristics of the vessels, this section presents a preliminary collision risk assessment based on the safety window of rudder input.

Considering that the northern vessel is a target vessel, and the southern vessel is an own vessel, it becomes possible to estimate what combination of rudder angle and time of rudder application is safe to avoid a collision. In this study, the safety window of rudder input approach is first proposed. This concept is especially relevant in the presence of more than two ships involved in the scenario. It is necessary to determine what kind of manoeuvrers are safe to avoid a collision with more than one ship.

Figure 8 illustrates an example of the safety window for the own ship in the collision avoidance scenario when the TCPA is zero. The black squares represent the values of the rudder's angle and time of rudder application that can lead to a collision with the target vessel, while the white squares represent the safe combinations. This window assumes that the behaviour of the target vessel is expected, and this parameter can be extended to more than one target vessel.

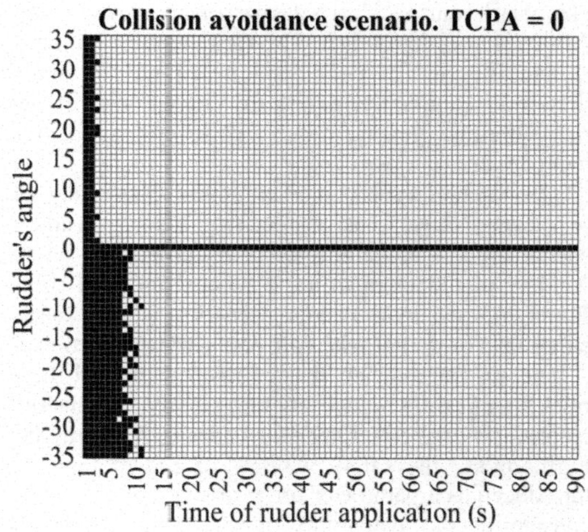

Figure 8. Safety window of rudder input for the vessel 1.

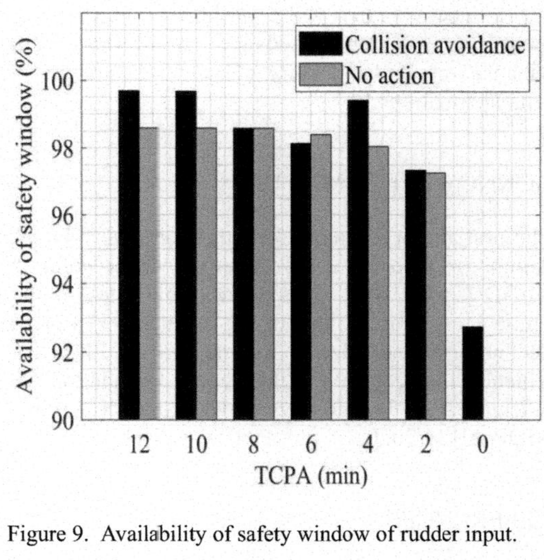

Figure 9. Availability of safety window of rudder input.

By calculating the percentage of white squares, it is possible to determine the availability of the safety window. This availability corresponds to the percentage of safe combinations of rudder angle and time of rudder application that can be adopted by the bridge. Figure 9 summarises the availabilities for both scenarios. As observed, in the collision avoidance scenario, the availability is consistently higher than in the no action scenario. When TCPA is zero, it is apparent that, due to the fact that the own vessel turned to starboard, at the most critical point, she still has an availability of a safety window of 92%, while in the no action scenario, this availability is approximately equal to 0%, corresponding to an imminent collision.

5 CONCLUSIONS

This paper presents a preliminary assessment of collision risk during ship evasive manoeuvres, using AIS trajectory input data. The study employs a manoeuvrability model developed based on Nomoto's first-order equation to assess the change in collision risk during generic ship collision avoidance manoeuvres.

The study focuses on a particular case study of two cargo ships evasive behaviours in a head-on encounter. The probability of collision and the safety window of rudder input are evaluated for both collision avoidance and no action scenarios, where vessels either take or do not take collision avoidance actions.

The results indicate that adherence to COLREG rules significantly reduces the probability of collision, emphasising the effectiveness of evasive manoeuvres. Furthermore, the safety window of rudder input is introduced as a parameter for assessing collision risk, considering the availability of safe combinations of rudder angle and time of rudder application.

Overall, the analysis provides valuable insights into the dynamics of collision risk during ship evasive manoeuvres that will contribute to the development of collision avoidance applications and enhance the understanding of vessel behaviour in critical situations.

ACKNOWLEDGEMENTS

The first author is financed by the Portuguese Foundation for Science and Technology under contract UI/BD/154389/2022. This work contributes to the Strategic Research Plan of the Centre for Marine Technology and Ocean Engineering (CENTEC), which is financed by the Portuguese Foundation for Science and Technology under contract UIDB/UIDP/00134/2020.

REFERENCES

Allianz. 2022. *Safety and shipping review 2022. An annual review of trends and developments in shipping losses and safety*. Official website of Allianz SE. https://commercial.allianz.com/news-and-insights/reports/shipping-safety.html#download. Date accessed: 2024-02-28.

Antão, P., Sun, S., Teixeira, A.P., & Guedes Soares, C. 2023. Quantitative assessment of ship collision risk influencing factors from worldwide accident and fleet data. *Reliability Engineering and System Safety, 234*: 109166.

Bowditch, N. 2019. *American Practical Navigator. An Epitome of Navigation. Useful tables. Calculations. Glossary of marine navigation: Vol. II*. National Geospatial Intelligence Agency.

Cockcroft, A.N., & Lameijer, J.N.F. 2012. *A guide to the collision avoidance rules: international regulations for preventing collisions at sea*. Elsevier.

EMSA. 2021. *Annual overview of marine casualties and incidents 2021*. European Maritime Safety Agency.

Fujii, Y., Yamanouchi, H., & Mizuki, N. 1970. *On the fundamentals of marine traffic control. Part 1: Probabilities of collision and evasive actions*. Electronic Navigation Research Institute.

Krata, P., & Montewka, J. 2015. Assessment of a critical area for a give-way ship in a collision encounter. *Archives of Transport, 34*(2): 51–60.

Kuwahara, S., Nishimura, H., Nakagawa, K., & Yoshinaga, M. 2021. Research and Development of Collision Risk Decision Method for Safe Navigation and Its Verification. *ClassNK Technical Journal, 3*(1): 13–40.

Kuwata, Y., Wolf, M.T., Zarzhitsky, D., & Huntsberger, T. L. 2014. Safe maritime autonomous navigation with COLREGS, using velocity obstacles. *IEEE Journal of Oceanic Engineering, 39*(1): 110–119.

Lotovskyi, E., & Teixeira, A.P. 2023. Effect of timely manoeuvre execution on the collision probability in head-on and crossing encounter scenarios. In H. Le Sourne & C. Guedes Soares (Eds.), *Advances in the Collision and Grounding of Ships and Offshore Structures* (113–120). CRC Press.

Lotovskyi, E., Teixeira, A.P., & Guedes Soares, C. 2022. Availability Analysis of an Offshore Wind Turbine Subjected to Age-Based Preventive Maintenance by Petri Nets. *Marine Science and Engineering, 10*(7): 1000.

Nomoto, K., Taguchi, T., Honda, K., & Hirano, S. 1957. On the steering qualities of ships. *International Shipbuilding Progress, ISP, 4*(35): 354–370.

Rong, H., Teixeira, A.P., & Guedes Soares, C. 2019. Ship trajectory uncertainty prediction based on a Gaussian Process model. *Ocean Engineering, 182*: 499–511.

Rong, H., Teixeira, A.P., & Guedes Soares, C. 2022. Ship collision avoidance behaviour recognition and analysis based on AIS data. *Ocean Engineering, 245*: 110479.

Rong, H., Teixeira, A.P., & Guedes Soares, C. 2023. Statistical analysis of collision risk indicators in ship evasive manoeuvres. In H. Le Sourne & C. Guedes Soares (Eds.), *Advances in the Collision and Grounding of Ships and Offshore Structures* (121–128). CRC Press.

Silveira, P., Teixeira, A.P., & Guedes Soares, C. 2016. Probabilistic modelling of evasive manoeuvring actions to avoid collisions. In C. Guedes Soares & T. Santos (Eds.), *Maritime Technology and Engineering III* (887–893). CRC Press.

Silveira, P., Teixeira, A.P., Figueira, J.R., & Guedes Soares, C. 2021. A multicriteria outranking approach for ship collision risk assessment. *Reliability Engineering and System Safety, 214*: 107789.

Silveira, P., Teixeira, A.P., & Guedes Soares, C. 2022. A method to extract the Quaternion Ship Domain parameters from AIS data. *Ocean Engineering, 257*: 111568.

SNAME. 1989. *Principles of Naval Architecture. Motions in Waves and Controllability: Vol. III* (E. V. Lewis, Ed.; 2nd ed.). The Society of Naval Architects and Marine Engineers.

Sutulo, S. 2013. *Basics of Ship Manoeuvrability*. Class notes. Instituto Superior Técnico (IST).

Sutulo, S., & Guedes Soares, C. 2011. Mathematical models for simulation of manoeuvring performance of ships. In C. Guedes Soares, Y. Garbatov, N. Fonseca, & A. P. Teixeira (Eds.), *Marine Technology and Engineering* (661–698). Taylor & Francis Group.

Advances in Maritime Technology and Engineering – Guedes Soares & Santos (Eds)
© 2024 The Author(s), ISBN 978-1-032-83099-5

Analysis of collision risk indicators in ship collision avoidance behaviour using logistic regression

H. Rong, A.P. Teixeira & C. Guedes Soares
Centre for Marine Technology and Ocean Engineering (CENTEC), Instituto Superior Técnico, Universidade de Lisboa, Lisbon, Portugal

ABSTRACT: The aim of the present study is to analyse the collision risk indicators that influence the collision avoidance behaviour. First, Risk Influencing Factors, including the Distance to Closest Point of Approach, Time to Closest Point of Approach, relative speed, relative bearing, approaching angle and ship domain, are calculated at the time the evasive action is taken and at the closest point in near collision scenarios. Subsequently, a logistic regression model is developed to investigate whether there are significant changes in the Risk Influencing Factors taken their values at the closest point as baseline categories. The results of this evaluation contribute to a deeper understanding of collision risks and avoidance behaviour in encounter scenarios, facilitating improved safety measures and informed decision-making for ship operators and maritime authorities.

1 INTRODUCTION

Maritime accident statistics reveal that ship-ship collisions rank among the most common ship accident types, especially in waterways linking large ports and important maritime corridors that are characterized by high traffic density and complex navigational conditions (EMSA, 2021; Guedes Soares & Teixeira, 2001). However, applying statistical analysis methods for collision risk assessment based on accident data is challenging in regions with infrequent maritime accidents. In such cases, insights into specific sea areas and local waterways can be derived from non-accidental events (namely traffic conflicts or near collisions) identified from Automatic Identification System (AIS) trajectory data (Du et al., 2020; Goerlandt et al., 2012; Qu et al., 2011; Rong et al., 2021, 2020; Yoo, 2018).

Numerous studies have proposed methodologies for detecting and assessing the risk of ship-ship collisions based on various Risk Influencing Factors (RIFs) (Chen et al., 2019; Huang et al., 2020). Montewka et al. (2010) developed an approach replacing geometric collision diameter with Minimum Distance To Collision (MDTC), a critical distance beyond which collision avoidance actions may fail to prevent a collision. Using the MDTC model, the probability of maritime accidents was estimated (Montewka et al., 2012). Zhang et al. (2015) introduced the Vessel Conflict Ranking Operator (VCRO) model, which assesses the risk level of ship encounter situations based on expert judgments. This model considers ship interactions, including relative speed, angle, and distance, and for further refinement, Zhang et al. (2016) incorporated ship size and the concept of Minimum Distance to Collision (MDTC), reflecting the urgency of evasive manoeuvres. Wang et al. (2010) characterized a vessel conflict as a critical situation wherein a vessel is anticipated to enter another vessel's ship domain in a certain time, and afterwards estimated the vessel collision frequency in the Singapore Strait based on the analysis of vessel conflicts. Rong et al. (2020) utilized collision risk indicators, such as Distance at the Closest Point of Approach (DCPA), Time to Closest Point of Approach (TCPA), and relative distance between ships, to evaluate collision risk in near-collision scenarios off the coast of Portugal and estimated spatial density maps of collision risk based on the Kernel Density Estimation (KDE) method. Subsequently, Rong et al. (2021) employed a spatial autocorrelation method to explore the co-occurrence of near-collision hotspots and local maritime traffic characteristics, considering average speed, speed dispersion, degree of speed acceleration, and ship route overlaps. In the domain of ship collision risk assessment, various Risk Influence Factors (RIFs) have been adopted to evaluate the potential for collisions at sea. However, the use of these RIFs lacks experimental support in many cases, relying instead on subjective assessments by experts.

When there is a collision risk, the officer of the watch (OOW) should take measures to increase the expected passing distance in a timely manner to avoid a potential collision (Perera & Guedes Soares, 2015). The time at which ships employ collision avoidance strategies effectively reflect the instances

DOI: 10.1201/9781003508762-18

when navigators perceive the collision risk. This collision avoidance behaviour serves as an indication of the navigators' proactive measures to mitigate potential collisions.

Rong et al., (2022) introduced a novel approach for automatically identifying ship evasive behaviour from ship trajectories based on an improved Sliding Window Algorithm. The effects of these evasive manoeuvres may not be immediately apparent in terms of the relative distance between the two ships involved in near-collision scenarios. Following the execution of the evasive manoeuvre, the relative distance may continue to decrease for a certain duration.

Although the relative distance is still decreasing, the risk of collision is now considerably lower than it was at the time of evasive manoeuvre initiation.

The evasive manoeuvre serves its primary purpose of altering the vessels' trajectories to create a safer separation, and the subsequent decrease in relative distance should be interpreted in the context of the reduced collision risk achieved through these intentional navigational measures. Therefore, comparing the changes in Risk Influencing Factors at the time the evasive manoeuvre is taken and at the closest point can help in understanding the contribution of certain RIFs to collision risks in real navigation scenarios.

This paper presents an analysis of collision risk indicators in ship collision avoidance behaviour using the Logistic Regression. The remainder of this paper is organized as follows. In Section 2, the research methodology is illustrated, including the RIFs adopted in this study and the Logistic Regression model. Then, the RIFs at the time the collision evasive manoeuvre is taken and at the closest point in near collision scenario are statistically characterised in Section 3. Finally, conclusions are drawn in Section 4.

2 METHODOLOGY

In the evaluation of maritime safety during near-collision scenarios, the analysis of RIFs plays a pivotal role. These indicators serve as quantitative measures that capture various aspects contributing to collision risk perception by a ship.

During the critical phases of an encounter, particularly between the initiation of evasive manoeuvres and the closest point of approach, the ship's crew or automated system keenly monitor changes in RIFs. The fundamental principle guiding this analysis is that significant changes in RIFs correspond to tangible alterations in the factors influencing the collision risk.

Through an exploration of patterns in RIF changes, maritime operators can gain insights into which risk-related parameters significantly influence the collision risk perceived by a ship. This study employs a Logistic Regression model to examine whether there are substantial changes in RIFs at the time of evasive manoeuvres compared to those at the closest point. The considered RIFs in this study include the relative speed, relative bearing, approaching angle, Distance at the Closest Point of Approach (DCPA), Time to Closest Point of Approach (TCPA) and ship domain.

2.1 *Collision risk indicators*

Relative speed: it refers to the rate of change of distance between the two ships, and it is calculated by the vector sum of the speeds of the two vessels. Relative speed indicates the relative motion in relation to each other, and higher relative speeds amplify the potential severity of a collision and reduce the time available for evasive actions. In the context of navigation, navigators closely monitor relative speed to make informed decisions and adhere to the Convention on the International Regulations for Preventing Collisions at Sea (COLREGs).

Relative bearing: refers to the angle between the ship's heading and the location of another object. Relative bearing is a parameter in identifying encounter situations and assessing the risk of collision in maritime navigation. The "Rule of Constant Bearing" is a fundamental principle in collision avoidance. It states that if the relative bearing of an approaching vessel remains constant, a collision is likely. Navigators use this rule to assess the risk of collision and take evasive actions if needed.

Approaching angle: refers to the angle between relative speed and relative distance. In a maritime context, it can provide insights into the dynamics of the encounter between two vessels. It indicates whether the vessels are moving directly toward or away from each other, or if there is a lateral component to their movement.

A small approaching angle (0 to 90 degrees) suggests that the vessels have a component of relative motion toward each other. It might indicate a closing situation where vessels are approaching each other. A large approaching angle (90 to 180 degrees) suggests a component of relative motion away from each other. It might indicate a situation where vessels are moving apart.

Understanding the angle between these vectors can aid in assessing the relative motion and potential collision risk. For instance, a decreasing angle might signal a closing situation, while an increasing angle might indicate a situation where vessels are moving apart. This information is valuable for mariners in making informed decisions for collision avoidance manoeuvres based on the evolving dynamics of the encounter.

Closest Point of Approach (CPA): In the process of vessel collision avoidance, DCPA (Distance of Closest Point of Approach) and TCPA (Time to Closest Point of Approach) are typically regarded as the most effective indicators for measuring the level of collision risk. These values are used to determine the spatial and temporal proximity of two ships involved in encounter scenarios, and can be calculated from the ships' position, speed and heading data.

Let data points $p_o = (lng_o, lat_o, v_o, \psi_o)$ and $p_t = (lng_t, lat_t, v_t, \psi_t)$ denote the position of the own ship and target ship, respectively. The relative distance and relative speed of own ship to target ship denoted by v_r, can be calculated by:

$$D_r = 2R_e sin^{-1}\left(\sqrt{\begin{array}{c} sin^2\left(\frac{lat_i - lat_j}{2}\right) + \\ cos(lat_i)cos(lat_j)sin^2\left(\frac{lng_i - lng_j}{2}\right) \end{array}}\right) \tag{1}$$

$$v_r = \sqrt{\left(v_i cos\psi_i - v_j cos\psi_j\right)^2 + \left(v_i sin\psi_i - v_j sin\psi_j\right)^2} \tag{2}$$

where D_r represents the relative distance between the own and target ships, v_r denotes the relative velocity, R_e is the Earth's radius.

According to Figure 1, given the own ship (o) and target (t) ship coordinates, headings and velocities, respectively, the DCPA and TCPA can be obtained by:

$$\psi_r = cos^{-1}\left(\frac{v_0 - v_t \times cos(\psi_o - \psi_t)}{v_r}\right) \tag{3}$$

$$d_{cpa} = D_r \times sin(\psi_R - \alpha_t - \pi) \tag{4}$$

$$t_{cpa} = \frac{D_r \times cos(\psi_r - \alpha_t - \pi)}{v_r} \tag{5}$$

where φ_r denotes the relative course and α_t denotes the azimuth of the target ship.

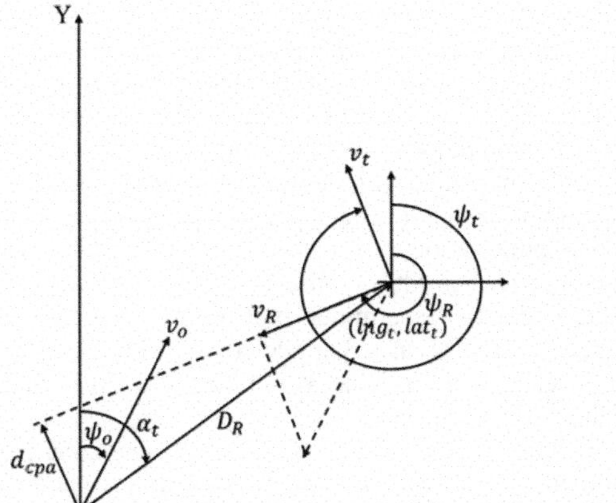

Figure 1. Illustration of DCPA and TCPA calculation.

Ship domain: the concept of ship domain has been widely applied in navigational safety studies. Originating from the work of Fujii & Tanaka (1971), the ship safety domain represents a generalized protective space around a ship, which is an essential risk indicator for collision avoidance with other ships. Numerous researchers have introduced variations in the dimensions and shapes of ship domains. A comprehensive review by Szlapczynski & Szlapczynska (2017) over the past decades reveals the introduction of diverse ship domains with varying dimensions and shapes. This study aims to examine the impact of ship domain on ship collision avoidance behaviour. The ship domain proposed by Fujii & Tanaka (1971) has been widely recognized and applied in previous near collision analysis, providing a well-established foundation for assessing collision risk in maritime scenarios. Therefore, in this paper this ship domain is adopted, which consists of an ellipse centred at the position of the ship with semi-major and semi-minor equal to 4 and 1.6 times the ship length (L), respectively.

2.2 *Classification of encounter scenarios by COLREGs*

Measures and actions required to avoid collision accidents are described in the International Regulations for Preventing Collisions at Sea (COLREGs) adopted by the International Maritime Organization (IMO) in 1972 and still valid, with several minor amendments since then. In this study, three types of encounters defined in COLREGs are considered, namely head-on, overtaking and crossing situations:

- Overtaking: a situation in which two vessels are proceeding on the same route, lying on almost parallel courses, with their own vessel's relative bearing from112.5° to 247.5°.
- Head-on: a situation in which vessels are lying on almost reciprocal courses and the course difference falls in the range ±5°.
- Crossing: covers the area of own vessel's relative bearing from 5° to112.5° or from 247.5° to 355°.

2.3 *Hypothesis of the study*

In the context of maritime navigation, understanding the factors influencing collision avoidance manoeuvres is of great importance for ensuring safety. The objective of this study is to investigate the importance of Risk Influencing Factors by examining their changes between two pivotal points of the encounter scenarios, which are the evasive manoeuvre initiation point and the closest point. This study focuses on six key RIFs that have been widely applied in ship collision risk assessment studies, which are the relative speed, relative bearing, approaching angle, DCPA, TCPA and ship domain. The aim is to test whether these variables exhibit significant variations as ships move from the evasive manoeuvre initiation to the closest point.

A regression model is useful in this study to quantitatively assess the significance of RIFs by providing a statistical framework to analyse the relationship and variations in critical variables. The regression analysis allows for the identification of the individual contributions of these factors and helps determine their statistical significance in influencing collision avoidance manoeuvres. To structure the model in this study, two sets of hypotheses are proposed. Firstly, the overall model significance hypothesis assesses whether the combined effect of the RIFs is statistically significant in explaining variations in critical variables during collision avoidance manoeuvres. Secondly, individual variable significance hypotheses examine whether each specific RIF, such as relative speed, relative bearing, approaching angle, DCPA, TCPA, and ship domain, has a significant contribution to explaining the observed changes during collision avoidance manoeuvres.

1) Overall model significance:
   Null hypothesis (H0): The combined effect of Risk Influencing Factors (RIFs) in the regression model is not statistically significant, indicating that these factors do not collectively explain variations in critical variables during collision avoidance manoeuvres.
   Alternative Hypothesis (H1): The combined effect of RIFs in the regression model is statistically significant, suggesting that these factors collectively contribute to explaining variations in critical variables during collision avoidance manoeuvres.
2) Individual Variable Significance:
   For each specific RIF (e.g., relative speed, relative bearing, approaching angle, DCPA, TCPA, and ship domain).
   Null Hypothesis (H0): The coefficient of RIF is equal to zero, indicating that this specific factor does not have a significant individual contribution to explaining variations in critical variables during collision avoidance manoeuvres.
   Alternative Hypothesis (H1): The coefficient of RIF is not equal to zero, suggesting that this specific factor has a significant individual contribution to explaining variations in critical variables during collision avoidance manoeuvres.

2.4 *Logistic regression*

Logistic regression is a type of linear regression used to analyse dichotomous dependent variables. It is based on the construction of a statistical model that describes the relationship between observed dependent and independent variables, also called response and explanatory variables. In this study, binomial logistic regression is applied to test for an association between the dependent variable and the independent variables. In the logistic regression model, the dependent variable is coded as 1 for the collision avoidance behaviour initiation and 0 for the closest point. The independent variables consist of relevant RIFs measured at both the time at which the evasive manoeuvre is taken and the closest point, allowing for an analysis of their impact on the likelihood of successful collision avoidance behaviour.

By applying the logistic function in,

$$NP = \frac{e^{(\alpha+\beta_1 x_1+\ldots+\beta_n x_n)}}{1+e^{(\alpha+\beta_1 x_1+\ldots+\beta_n x_n)}} \tag{6}$$

the probability of collision avoidance behaviour becomes a value between 0 and 1, regardless of the size of the independent variables.

For estimating the model coefficients ofx_i, a logit transformation function similar to a linear regression model is used, shown in Eq. 7:

$$logit(NP) = ln\left(\frac{NP}{1-NP}\right) = \alpha + \beta_1 x_1 + \ldots + \beta_n x_n \tag{7}$$

The regression equation coefficients (β) are estimated using the Maximum Likelihood method proposed by Cox & Snell (1989).

After training the logistic regression model, the next step involves testing for significance. This entails assessing the statistical significance of the model and individual coefficients through methods such as likelihood ratio tests, Wald tests, or score tests. The significance tests provide insights into whether the selected independent variables have a significant impact on predicting the outcome of collision avoidance behaviour. For the study, the hypotheses for the test of overall significance follow:

$$H_0 : \beta_1 = \beta_2 = \ldots = \beta_n = 0$$
$$H_1 : \textit{one of the parameters is not equal to zero}$$

The test for overall significance is based on the value of a χ^2 test statistic. If the null hypothesis is true, the sampling distribution of χ^2 follows a chi-square distribution with degrees of freedom equal to the number of independent variables in the model. The p-value associated with this test statistic indicates the probability of observing the calculated χ^2 value under the assumption that the null hypothesis is true. Therefore, at any predetermined level of significance, a small p-value would lead to the rejection of the null hypothesis.

If the χ^2 test shows an overall significance, another χ^2 test can be used to determine whether each of the individual independent variables is has a significant contribution to the overall model. For the independent variables x_i, the hypotheses are:

$$H_0 : \beta_i = 0$$
$$H_1 : \beta_i \neq 0$$

The test of significance for an independent variable is also based on the value of a χ^2 test statistic. If the null hypothesis is true, the sampling distribution of χ^2 follows a chi-square distribution with one degree of freedom. A low p-value for an individual variable indicates that there is evidence to reject the null hypothesis for that variable. Therefore, in testing the significance of individual independent variables, a small p-value would suggest that the specific variable has a significant contribution to predicting the outcome of collision avoidance behaviour.

3 CASE STUDY

The ship evasive behaviour recognition method is now applied to near-collision scenarios identified in the maritime traffic off the continental coast of Portugal provided by a three-month AIS dataset, from 1st October to 31st December 2015.

3.1 *Near collision scenarios*

Figure 2 shows a total number of 2846 near-collision scenarios identified off the coast of Portugal. There are 1328, 1011 and 507 near collisions identified for overtaking, crossing and head-on near encounters, respectively. It should be mentioned that encounters involving tugs and pilot vessels as well as in port areas have been excluded in this study. Overtaking near-collision locations are represented by red circles, crossing by blue triangles and head-on near-collision locations by yellow squares. Due to the high volume of maritime traffic guided and organized by the Traffic Separation Schemes (TSSs) off Cape Roca and off Cape St. Vincent, most overtaking cases are evenly distributed within the ship traffic lanes formed by the two TSSs. The geographic distributions of crossing and head-on near-collision scenarios are correlated to port areas and junction areas where the traffic flow diverges from TSSs to the ports of Portugal.

Figure 3 shows the relative distance distributions of the target ship when the own ship takes evasive manoeuvres in overtaking, crossing and head-on scenarios. Around 90% of the relative distances are in the ranges [1nm, 4.5nm], [2nm, 9nm] and [3nm, 10nm] for overtaking, crossing and head-on scenarios, respectively. According to the relative distance distributions, it is found that the vessels involved in head-on scenarios tend to make evasive manoeuvre at a larger distance compared with those in overtaking and crossing scenarios.

Figure 4 shows the relative distance distributions of two ships at the closest point in near collision scenarios. It is seen that there are no significant differences were observed in the distribution of

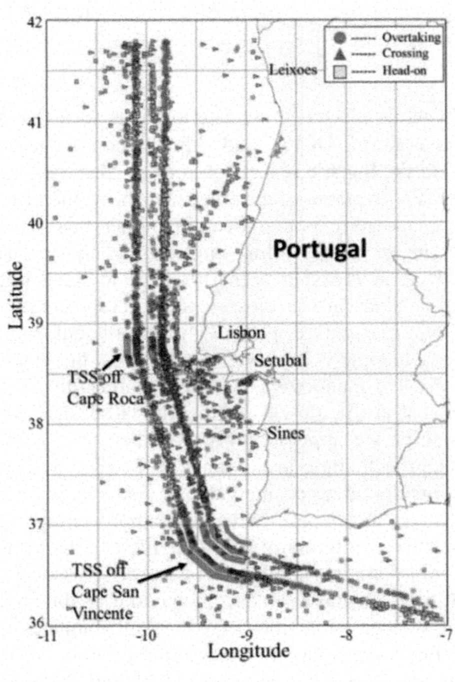

Figure. 2. Near collision scenarios detected off the continental coast of Portugal.

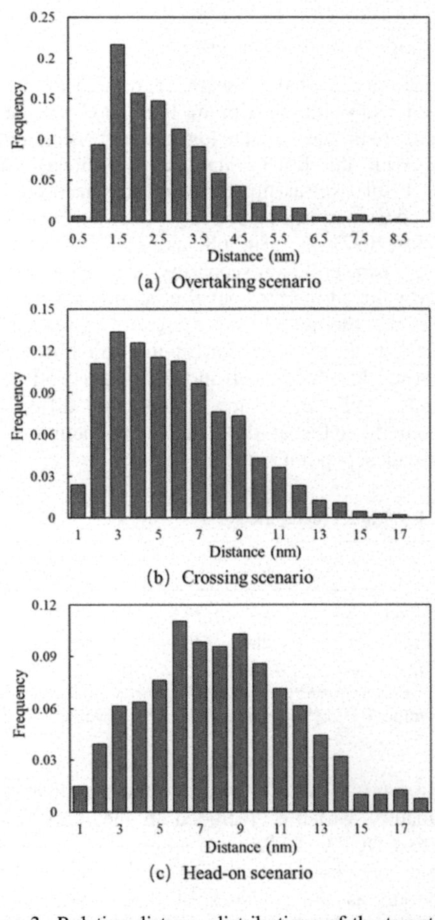

Figure 3. Relative distance distributions of the target ship when the own ship takes evasive manoeuvre.

distances between ships in overtaking, crossing and head on near collision scenarios. The distances between vessels were consistently confined within the range of 0.5 nautical miles to 1.5 nautical miles. It should be mentioned that the distance distributions in all three encounter types exhibited a similar and normal distribution pattern, suggesting similarities in the proximity dynamics during the near-collision scenarios.

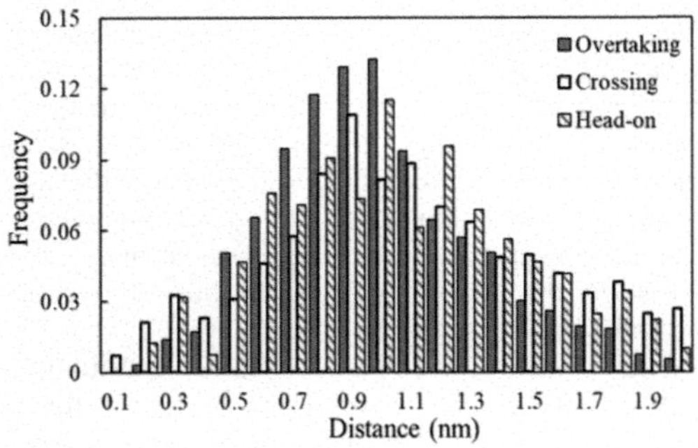

Figure 4. Relative distance distributions at the closest point.

3.2 *Logistic regression result*

The parameters of the logistic regression model are derived using the maximum likelihood estimation method. Estimates of the logistic regression model along with the fitness statistics are presented in Table 1 for overtaking, crossing and head-on near collision scenarios, respectively. The likelihood ratio statistics, 1977.46, 2842.68 and 1008.09 for three different near collision scenarios, are well above the critical value at the 99% level of significance, which implies that the model has a reasonably good fit. The results lead to the rejection of the null hypothesis, indicating that the overall model is statistically significant, and there are significant differences between the RIFs at the evasive manoeuvre point and the closest point.

Table 1. Model fit summary.

	chi2	Prob>chi2
Overtaking	1977.46***	0.0
Crossing	2842.68***	0.0
Head on	1008.09***	0.0

* : p-value<0.05; **: p-value<0.01; ***: p-value<0.001

The z-test is used to examine the significance of explanatory variables included in the model. It is seen that DCPA, TCPA, and ship domain have significant associations with the collision avoidance behaviour, as the p-values of these variables are smaller than 0.001. In terms of DCPA and TCPA, the increasing values are found to have a negative influence on collision avoidance behaviour (as the coefficients are -0.893 and -0.024, respectively). This finding is consistent with other studies such as (Mehdi et al., 2017; Xu & Wang, 2014) who reported that decreasing the value of DCPA and TCPA increases the probability of collision. However, it is found that the ship domain violation has a negative association with the collision avoidance behaviour as the coefficient is negative (-1.097). This result indicates that the ship domain violation has a higher probability of occurring at the closest point rather than at the evasive manoeuvre point. The variable "approaching angle" exhibits a significant negative association (9.72) with the collision avoidance behaviour, suggesting that following the collision avoidance behaviour the approaching angle tends to decrease.

Table 2. Logistic regression model for overtaking scenarios.

Explanatory Variable	Coefficient	S.E.	z	P>|z|
Relative bearing	0.001	0.001	0.53	0.595
Relative speed	-0.107	0.069	-1.55	0.121
DCPA	-0.893***	0.001	5.31	0.000
TCPA	-0.025*	0.001	-2.46	0.014
Approaching angle	9.720***	0.547	-17.78	0.000
Ship domain	-1.090***	0.341	-3.22	0.001
Cons	-4.560	0.567	-8.04	0.000

* : p-value<0.05; **: p-value<0.01; ***: p-value<0.001

In terms of crossing scenarios, it is seen that the relative speed, DCPA, and ship domain are significant in the logistic regression model. The coefficient of DCPA is positive, meaning that the value at evasive manoeuvre is significantly higher than that at the closest point. This implies that the vessels involved in crossing scenarios tend to have larger DCPA values at the closest point compared to the evasive manoeuvre point. The coefficient of the approach angle is positive, meaning that the value at the evasive manoeuvre point is significantly higher than that at the closest point. This is because when two ships are approaching each other, the value of the approach angle in the range of $[-\pi/2, \pi/2]$ and the corresponding cosine value is positive, but when two ships leave apart the value of approaching angle is negative. In terms of ship domain, it is found that the ship domain violation has a negative association with the collision avoidance behaviour as the coefficient is negative (-1.867). This result indicates that the ship domain violation has a higher probability of occurring at the closest point rather than at the evasive manoeuvre point.

Table 3. Logistic regression model for crossing scenarios.

Explanatory Variable	Coefficient	S.E.	z	P>|z|
Relative bearing	-0.001	0.002	-0.71	0.481
Relative speed	-0.149***	0.042	-3.59	0.000
DCPA	-0.630***	0.001	3.19	0.001
TCPA	-0.097	0.012	1.29	0.196
Approaching angle	10.054***	1.007	-9.98	0.000
Ship domain	-1.867*	0.758	-2.46	0.014
Cons	-4.058	0.771	-5.27	0.000

* : p-value<0.05; **: p-value<0.01; ***: p-value<0.001

In terms of head-on scenarios, the logistic regression analysis revealed that among the examined collision avoidance indicators, only the approaching angle (10.054) demonstrates a significant positive association with collision avoidance behaviour. Interestingly, other collision avoidance indicators included in the model were found to be non-significant. This implies that, in the specific condition of head-on scenarios, variables such as relative speed, DCPA, and ship domain violation do not significantly contribute to collision avoidance behaviour.

Table 4. Logistic regression model for head-on scenarios.

Explanatory Variable	Coefficient	S.E.	z	P>|z|
Relative bearing	-.032	0.061	-0.52	0.600
Relative speed	0.018	0.207	0.09	0.932
DCPA	0.657	0.001	0.65	0.515
TCPA	0.007	0.012	0.01	0.992
Approaching angle	24.733*	11.473	-2.16	0.031
Ship domain	-0.347	0.085	-0.63	0.524
cons	-4.325	11.003	-0.39	-25.891

* : p-value<0.05; **: p-value<0.01; ***: p-value<0.001

4 CONCLUSIONS

This paper employs a Logistic Regression model to systematically analyse the contribution of Risk Influencing Factors (RIFs), including relative speed, relative bearing, Distance at the Closest Point of Approach (DCPA), Time to Closest Point of Approach (TCPA), and ship domain, to collision avoidance behaviour. The study not only contributes to the understanding of maritime collision dynamics but also provides practical implications for optimizing collision risk assessment measures in real-world navigation scenarios.

The analysis of 2846 near-collision scenarios identified in a three-month AIS dataset shows that the relative distance between ships during evasive manoeuvres in overtaking, crossing and head-on scenarios reveals distinct patterns. The ships involved in head-on scenarios tend to initiate evasive manoeuvres at a larger distance compared to those in overtaking and crossing scenarios.

The logistic regression model, derived by the maximum likelihood estimation, provides a good fit for overtaking, crossing, and head-on near-collision scenarios, as evidenced by the high likelihood ratio statistics well exceeding the critical values at a 99% significance level. In addition, the rejection of the null hypothesis indicates the statistical significance of the model, indicating notable differences in Risk Influencing Factors (RIFs) at the time the evasive manoeuvre is taken and at the closest point in each scenario.

The z-test analysis reveals the significance of the explanatory variables in the logistic regression model for both overtaking and crossing scenarios. DCPA and TCPA exhibit a significant relationship with the collision avoidance behaviour. These findings align with existing studies, demonstrating that increasing the values of DCPA and TCPA decreases the probability of collision, while ship domain violations are more likely at the closest point than during the evasive manoeuvres.

ACKNOWLEDGEMENTS

The paper contributes to the project "Monitoring and Surveillance of Maritime Traffic off the continental coast of Portugal (MoniTraffic)", which has been co-funded by the "Fundo Azul" programme of the Portuguese Directorate-General for Maritime Policy, under contract no. FA-04-2017-005. This work also contributes to the Strategic Research Plan of the Centre for Marine Technology and Ocean Engineering (CENTEC), which is financed by the Portuguese Foundation for Science and Technology under contract UIDB/UIDP/00134/2020.

REFERENCES

Chen, P., Huang, Y., Mou, J., van Gelder, P.H.A.J.M., 2019. Probabilistic risk analysis for ship-ship collision: State-of-the-art. Safety Science 117, 108–122. doi:10.1016/j.ssci.2019.04.014

Cox, D.R., Snell, E.J., 1989. The analysis of binary data. Chapman and Hall/CRC, London, UK.

Du, L., Goerlandt, F., Kujala, P., 2020. Review and analysis of methods for assessing maritime waterway risk based on non-accident critical events detected from AIS data. Reliability Engineering and System Safety 200, 106933. doi:10.1016/j.ress.2020.106933

EMSA, 2021. Annual Overview of Marine Casualties and Incidents 2021.

Fujii, Y., Tanaka, K., 1971. Traffic Capacity. Journal of Navigation 24, 543–552.

Goerlandt, F., Montewka, J., Lammi, H., Kujala, P., 2012. Analysis of near collisions in the Gulf of Finland. Advances in Safety, Reliability and Risk Management - Proceedings of the European Safety and Reliability Conference, ESREL 2011 2880–2886. doi:10.1201/b11433-409

Guedes Soares, C., Teixeira, A.P., 2001. Risk assessment in maritime transportation. Reliability Engineering & System Safety 74, 299–309.

Huang, Y., Chen, L., Chen, P., Negenborn, R.R., van Gelder, P.H.A.J.M., 2020. Ship collision avoidance methods: State-of-the-art. Safety Science 121, 451–473. doi:10.1016/j.ssci.2019.09.018

Mehdi, R., Gluch, M., Fischer, S., Baldauf, M., 2017. A perfect warning to avoid collisions at sea? Scientific Journals of the Maritime University of Szczecin 49, 53–64. doi:10.17402/245

Montewka, J., Goerlandt, F., Kujala, P., 2012. Determination of collision criteria and causation factors appropriate to a model for estimating the probability of maritime accidents. Ocean Engineering 40, 50–61. doi:10.1016/j.oceaneng.2011.12.006

Montewka, J., Hinz, T., Kujala, P., Matusiak, J., 2010. Probability modelling of vessel collisions. Reliability Engineering and System Safety 95, 573–589. doi:10.1016/j.ress.2010.01.009

Perera, L.P., Guedes Soares, C., 2015. Collision risk detection and quantification in ship navigation with integrated bridge systems. Ocean Engineering 109, 344–354. doi:10.1016/j.oceaneng.2015.08.016

Qu, X., Meng, Q., Suyi, L., 2011. Ship collision risk assessment for the Singapore Strait. Accident Analysis and Prevention 43, 2030–2036. doi:10.1016/j.aap.2011.05.022

Rong, H., Teixeira, A.P., Guedes Soares, C., 2022. Ship collision avoidance behaviour recognition and analysis based on AIS data. Ocean Engineering 245, 110479. doi:10.1016/j.oceaneng.2021.110479

Rong, H., Teixeira, A.P., Guedes Soares, C., 2021. Spatial correlation analysis of near ship collision hotspots with local maritime traffic characteristics. Reliability Engineering and System Safety 209, 107463. doi:10.1016/j.ress.2021.107463

Rong, H., Teixeira, A.P., Guedes Soares, C., 2020. Risk of Ship Near Collision Scenarios off the Coast of Portugal, in: Proceedings of the 29th European Safety and Reliability Conference, ESREL 2019. pp. 3660–3666. doi:10.3850/978-981-11-2724-3 0625–cd

Szlapczynski, R., Szlapczynska, J., 2017. Review of ship safety domains: Models and applications. Ocean Engineering 145, 277–289. doi:10.1016/j.oceaneng.2017.09.020

Wang, N., 2010. An Intelligent Spatial Collision Risk Based on the Quaternion Ship Domain. Journal of Navigation 63, 733–749. doi:10.1017/S0373463310000202

Xu, Q., Wang, N., 2014. A survey on ship collision risk evaluation. Traffic Management 26, 475–486.

Yoo, S.L., 2018. Near-miss density map for safe navigation of ships. Ocean Engineering 163, 15–21. doi:10.1016/j.oceaneng.2018.05.065

Zhang, W., Goerlandt, F., Kujala, P., Wang, Y., 2016. An advanced method for detecting possible near miss ship collisions from AIS data. Ocean Engineering 124, 141–156. doi:10.1016/j.oceaneng.2016.07.059

Zhang, W., Goerlandt, F., Montewka, J., Kujala, P., 2015. A method for detecting possible near miss ship collisions from AIS data. Ocean Engineering 107, 60–69. doi:10.1016/j.oceaneng.2015.07.046

Advances in Maritime Technology and Engineering – Guedes Soares & Santos (Eds)

Assessing deep learning methods for sea-surface multi object tracking from visible light video

Z.Y. Shao
Navigation College, Dalian Maritime University, P.R. China
Centre for Marine Technology and Ocean Engineering (CENTEC), Instituto Superior Tecnico, Universidade de Lisboa, Lisbon, Portugal

Y. Yin, H.G. Lyu & S. Sun
Navigation College, Dalian Maritime University, P.R. China

C. Guedes Soares
Centre for Marine Technology and Ocean Engineering (CENTEC), Instituto Superior Tecnico, Universidade de Lisboa, Lisbon, Portugal

ABSTRACT: This study assesses the effectiveness of several deep learning approaches for multi-object tracking on the sea surface, employing visible light videos captured by electro-optical sensors on unmanned surface vehicles. Utilizing the Jary Maritime Tracking Dataset, six leading deep learning methods including SORT, DeepSORT, DeepMOT, UAVMOT, ByteTrack and BoT-SORT are quantitatively compared. Focusing on seven types of common of sea-surface objects as main interest categories for navigation, performance is quantified by calculating the cardinality of true and false positives and negatives as well as ID Switches (IDs) in the test set. Employing Multiple Object Tracking Accuracy (MOTA), Identity F1 (IDF1) score, IDs and Frames Per Second (FPS) as performance metrics, noticeable differences are found that substantial variations in the efficacy of diverse multi-object tracking methods. The results identify BoT-SORT as the most effective method regarding accuracy and stability for sea-surface object tracking, highlighting the need for refined association strategies in visual tracking to enhance maritime navigation.

1 INTRODUCTION

Stable tracking of sea-surface objects is a crucial step for autonomous ships to achieve situational awareness and autonomous navigation (Thombre et al., 2022). Relevant objects include large ships such as cargo ships and ferries, as well as small objects with weak radar signals such as small fishing boats, kayaks, and pontoons. The comprehensive acquisition of information on these objects is essential to ensure navigation safety (Lyu et al., 2023).

During the execution of assisted or fully automated navigation, it is necessary to identify and track objects within a specific range around the ship (Perera et al., 2012). Marine radar and Automatic Identification System (AIS) are now widely available in present-day navigation systems onboard. However, they may have difficulty detecting some small objects, such as wooden fishing boats and buoys. In search and rescue operations (Vettor & Guedes Soares, 2015) the targets are also very small.

Utilizing electro-optical (EO) sensors for object recognition and tracking has the potential to effectively address this issue (Gao et al., 2023; Shao et al., 2022; Sun et al., 2023). Consequently, it is necessary to utilize efficient visual multi-object tracking methods for more efficient maritime applications.

Tracking sea-surface objects using visible light videos acquired by EO sensors is challenging (Shan et al., 2021). The maritime environment, compared to other scenes, presents high dynamics and uncertainty, primarily reflected in several aspects: (1) Complex environmental changes on the sea surface, such as sea-surface glare, and sea fog; (2) Ship-borne cameras follow the wave motion, such as rocking and shaking, can make target association more difficult. (3) The presence of dense targets near ports and significant occlusions; (4) The prevalence of many small objects, leading to substantial loss of object feature information.

In response to these challenges, various methods have been proposed in the literature. Bloisi et al. (2011) employed Haar saliency features for detecting and tracking different maritime vessels. Prasad et al. (2017) achieved object tracking in maritime videos

DOI: 10.1201/9781003508762-19

through horizon detection, video frame registration, dynamic background removal and object tracking. To address multiple foreground objects against complex ocean backgrounds, compared to using single and shallow features, methods utilizing multi-view learning can more effectively handle significant target scale variations by integrating descriptors extracted from diverse feature sets. The study by Chen et al. (2019) investigates various ship feature sets, including local binary pattern (LBP) descriptors, Gabor filters, histograms of oriented gradients (HOG), and Canny descriptors, to represent the target's geometry, texture, and contour information. Utilizing multi-view learning in conjunction with a sparse representation framework enables more stable tracking in complex maritime scenarios. However, traditional maritime object tracking techniques, based on manually crafted features, face issues of extensive computation and limited generalizability.

In recent years, deep learning methods have excelled in various multi-object tracking tasks using visible light videos, particularly in pedestrian and vehicle tracking (Tian et al., 2020). These methods effectively learn complex feature representations from vast video datasets through advanced convolutional neural network (CNN) architectures. Multiobject tracking methods based on deep learning can be categorized into two types depending on their framework: detection-based tracking (DBT) and joint detection and tracking (JDT). With the rapid evolution of object detectors, especially the continuous development of the YOLO series (Bochkovskiy et al., 2020; Redmon et al., 2016; Redmon and Farhadi, 2018; Wang et al., 2023), DBT has emerged as the predominant approach in multi-object tracking.

However, comparative studies and guidance on method selection for multi-object tracking tasks in maritime autonomous navigation scenarios are scarce. This paper therefore focuses on the application of these methods and evaluates their performance. 6 advanced deep learning-based multi-object tracking methods are utilized: SORT ((Bewley et al., 2016)), DeepSORT (Wojke et al., 2017), DeepMOT (Xu et al., 2020), UAVMOT (Liu et al., 2022), ByteTrack (Zhang et al., 2022) and BoT-SORT (Aharon et al., 2022).

This study concentrates on shipboard visible light video sequences. The dataset utilized is the Jary Maritime Tracking Dataset (JMTD) collected from unmanned surface vehicles (USVs). Objects are categorized into 7 types: sailing boats, fishing boats, floaters, passenger ships, speedboats, cargo ships, and special ships. To enable a fair comparison of multiobject tracking performance in autonomous navigation scenarios on the sea surface, each method employs YOLOv7 as the detector (Wang et al., 2023).

Performance indicators from different test sequences demonstrate that deep learning methods are effective for maritime visual multi-object tracking tasks. Furthermore, this study highlights the challenges associated with using visual tracking to assist navigation.

The remainder of this paper is structured as follows. Section 2 introduces the training and test data, along with details on the various multi-object tracking methods compared in this study. Section 3 experiments to verify the performance of different multi-object tracking methods. Section 4 discusses the experimental results and navigation perspectives, and Section 5 presents the conclusions and future prospects.

2 METHODS

This section describes the datasets employed for training and testing and introduces 6 classics multi-object tracking methods selected for this study.

2.1 *Dataset*

Video sequences totaling 38 in number (comprising 29,776 images) from the JMTD are utilized to evaluate multi-object tracking methods on the sea surface. Of these, 8 videos are randomly selected for testing, with video serial numbers being clip-01, clip-20, clip-38, clip-44, clip-52, clip-68, clip-70, clip-77.

Video images are collected using the EO sensors mounted on the USVs, capturing 7 types of sea-surface objects: sailing boats, fishing boats, floaters, passenger ships, speedboats, cargo and special ships. The dataset includes five types of image resolutions: 1920×1080, 1920×960, 1920×720, 1920×680, and 1920×640.

The videos in the dataset are segmented into frames at equal intervals, with each video having a frame rate of either 1 frame/s or 5 frames/s to simulate the complex and dynamic real sea environment. Each frame is meticulously annotated following the MOT16 standard format. Additionally, the JMTD dataset encompasses four typical real sea scenes, as depicted in Figure 1: harbour, departure, open sea and entering scenes.

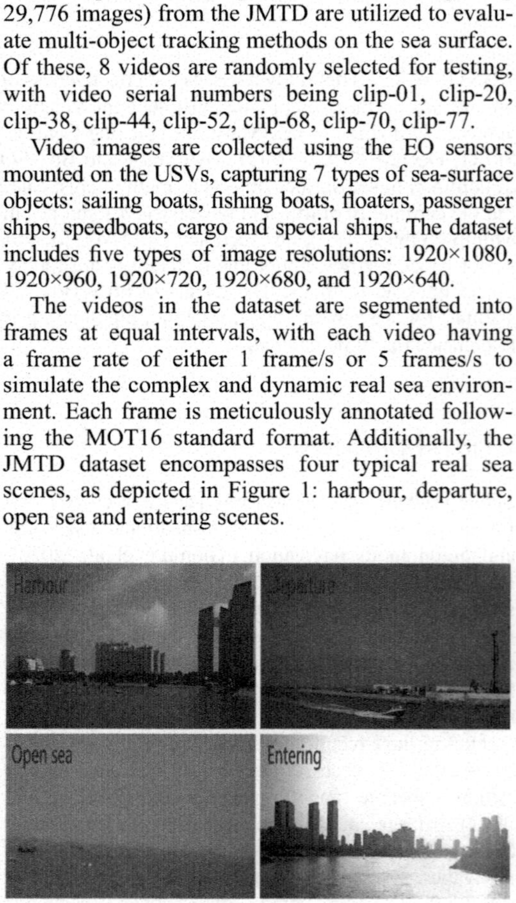

Figure 1. Different real sea scenes in the JMTD dataset.

Among these challenges, dense occlusions between sea-surface objects, the shake of shipboard cameras, and the complex conditions of the sea surface, including strong reflection and low illumination, all present difficulties in achieving stable tracking of sea-surface objects.

2.2 *Multi-object tracking methods*

This section evaluates 6 advanced deep learning-based multi-object tracking methods; SORT, DeepSORT, DeepMOT, UAVMOT, ByteTrack and BoT-SORT. Below the architecture of each method is briefly introduced, as well as its advantages and limitations.

The selection of these 6 methods is partly due to their demonstrated effectiveness on the MOT16, VisDrone, and UAVDT datasets, as detailed in the referenced literature (Liu et al., 2022; Milan et al., 2016). Additionally, all 6 multi-object tracking methods are based on TBD. This framework enables the employment of a unified detector to more accurately assess the association performance of the various tracking methods.

The common point of using multi-object tracking methods based on TBD is that the detector needs to be trained first. During the training process, the loss function is utilized to optimize the performance of the detector, ensuring its ability to accurately identify and locate multiple targets within an image. For this study, the efficient YOLOv7 is employed as the detector. The loss functions involved include object confidence loss, category loss, and bounding box regression loss.

The confidence loss is used to assess the model's accuracy in predicting the presence or absence of a target. The Binary Cross-Entropy Loss, commonly used for this purpose, is detailed in

$$\mathcal{L}_{obj} = -\sum_{i=0}^{N}[p_i \cdot \log(\hat{p}_i) + (1 - p_i) \cdot \log(1 - \hat{p}_i)] \tag{1}$$

In this context, p_i denotes the ground truth for the presence of the target in the ith prediction (where 1 indicates the target's existence and 0 its absence), while $\hat{p}_i$ represents the model's confidence in its prediction.

Category loss is used to evaluate the model's prediction accuracy for categories, typically employing multi-category cross-entropy loss, as detailed in

$$\mathcal{L}_{cls} = -\sum_{i=0}^{N}\sum_{c=1}^{C} y_{i,c} \cdot \log\left(\hat{y}_{i,c}\right) \tag{2}$$

where $y_{i,c}$ is the true label of ith prediction for category c, and $\hat{y}_{i,c}$ is the probability of the model prediction.

Regarding bounding box loss, a variant of Intersection over Union (IoU) is commonly used, as illustrated in

$$\mathcal{L}_{\text{box}} = 1 - \text{IoU} \tag{3}$$

SORT SORT is an efficient and simple multi-object tracking method. It employs a fast association matching strategy to link detected objects with existing trajectories. Primarily, SORT utilizes the Kalman filter for predicting target locations in videos and the Hungarian algorithm for data association. This method is especially suitable for scenarios requiring high real-time responsiveness. However, its limitation lies in comparatively lower tracking accuracy.

DeepSORT Buiding on SORT, DeepSORT adds the depth appearance features of the target as the association cost to cope with changes in the target appearance. Furthermore, DeepSORT employs more intricate motion models and state estimation, along with cascade matching within a brief time window, to minimize the likelihood of ID switching. Compared to SORT, DeepSORT excels in addressing short-term target occlusion issues. However, in scenarios where appearance feature information is less distinct, such as environments with densely packed small objects or poor lighting conditions, the algorithm may struggle to learn effective feature representations to handle these situations.

DeepMOT DeepMOT integrates deep learning methods with traditional motion models to enhance tracking accuracy. Similar to DeepSORT, DeepMOT employs a deep neural network for extracting targe features in images. However, it further incorporates a target interaction model to more effectively manage the interrelationships and dynamic changes among different targets. The primary advantages of DeepMOT include proficient handling of interaction dynamics. Yet, this approach demands greater computational resources and imposes more stringent requirements on training data.

UAVMOT UAVMOT is a method specially designed for multi-object tracking tasks in unmanned aerial vehicles (UAVs) videos. It enhances object tracking accuracy and robustness by introducing an ID feature update module, an adaptive motion filter module, and a gradient-balanced focus loss, tailored to the distinctive dynamic environment and perspective changes characteristic of UAVs.

ByteTrack ByteTrack merges the Kalman filter with a straightforward association strategy. Its key strength lies in handling low-confidence results effectively, enhancing the tracking of occluded objects in complex environments. By integrating detectors and simple correlation strategies, ByteTrack achieves efficient tracking and exhibits strong robustness against occlusions and low-confidence objects. Nonetheless, it may face challenges in dynamic camera scenes and with prolonged occlusion situations.

BoT-SORT As a TBD method, BoT-SORT constructs a cost matrix by adapting Kalman filtering, using global motion estimation technology from OpenCV, and calculating IoU distance and cosine similarity. This approach is particularly adept at addressing tracking challenges in dynamic camera environments. Its primary advantage lies in maintaining target IDs more stably while ensuring high tracking accuracy. However, the method's limitation is its significant computational complexity.

3 EXPERIMENTS

To ensure a fair comparison of these different multi-object tracking methods, all experiments are conducted under identical environmental conditions. Regarding hardware, the configuration used is Inter (R) Core (TM) i7-10700KF CPU; 32.0GB memory; NVIDIA GeForce RTX 3090 GPU. The software environment was Python3.8, Pytorch1.8.1, CUDA11.6. For the detector component, the standard YOLOv7 (with both depth multiple and width multiple set to 1) was utilized, employing the official pre-trained weights provided by the developers. The initial learning rate was set at 0.01, the batch size at 24, the number of iterations at 200, and the input model's image size was uniformly resized to 640×640.

For evaluation metrics, the performance of the tracking method is comprehensively evaluated by calculating Multiple Object Tracking Accuracy (MOTA), IDF1, ID switch (IDs) and Frames Per Second (FPS). MOTA is a comprehensive performance index to evaluate the accuracy of multi-object tracking, taking into account false detection FP, missed detection FN and IDs. It is achieved by using 1 minus the proportion of FP, FN and IDs to the total number of real targets, which can be expressed by

$$\text{MOTA} = 1 - \frac{\sum_t (FN_t + FP_t + IDs_t)}{\sum_t GT_t} \quad (4)$$

Among them, FN_t is the number of missed detections at frame t, FP_t is the number of false detections at frame t, IDs is the number of ID switches at frame t, and GT is the total number of real targets at frame t. The value range of MOTA is from negative infinity to 100. The higher the value, the better the tracking accuracy performance.

IDF1 is a metric based on the F1 score, which combines the two indicators IDP and IDR. It can be formulated as follows:

$$\text{IDP} = \frac{IDTP}{IDTP + IDFP} \quad (5)$$

$$\text{IDR} = \frac{IDTP}{IDTP + IDFN} \quad (6)$$

$$\text{IDF1} = \frac{2IDTP}{2IDTP + IDFP + IDFN} \quad (7)$$

Among them, IDTP is the number of correctly matched identities, IDFP denotes the number of incorrect matches, and IDFN represents the number of missing identities. The higher the IDF1 value, the better its performance in correctly matching the target identity.

IDs are used to measure the number of target identity changes during the tracking process, which directly reflects the stability and reliability of the tracking process. In ideal multi-object tracking, the values of IDs should be as low as possible.

Table 1. MOTA, IDF1, IDs results of multi-object tracking for each video test clip.

Video	Methods	MOTA↑	IDF1↑	IDs↓
Clip-01	SORT	39.1	20.4	378
	DeepSORT	25.7	27.6	457
	DeepMOT	43.7	57.6	47
	UAVMOT	44.7	37.5	156
	ByteTrack	71.1	49.5	178
	BoT-SORT	**86.8**	**82.0**	**19**
Clip-20	SORT	51.7	29.4	539
	DeepSORT	27.6	24.0	664
	DeepMOT	55.7	38.9	171
	UAVMOT	55.8	44.0	219
	ByteTrack	73.6	50.4	236
	BoT-SORT	**80.6**	**62.4**	**66**
Clip-38	SORT	35.6	16.3	760
	DeepSORT	18.4	17.9	986
	DeepMOT	50.0	46.4	139
	UAVMOT	48.7	42.9	241
	ByteTrack	73.4	59.1	181
	BoT-SORT	**82.3**	**63.8**	**72**
Clip-44	SORT	41.7	37.6	286
	DeepSORT	27.8	26.9	506
	DeepMOT	30.5	46.1	92
	UAVMOT	46.9	50.3	183
	ByteTrack	70.7	66.6	121
	BoT-SORT	**91.2**	**93.9**	**4**
Clip-52	SORT	73.1	71.9	122
	DeepSORT	55.5	58.8	143
	DeepMOT	69.7	76.6	30
	UAVMOT	68.1	73.5	28
	ByteTrack	84.7	78.4	53
	BoT-SORT	**89.2**	**82.2**	**22**
Clip-68	SORT	40.6	30.0	1298
	DeepSORT	29.7	30.6	1300
	DeepMOT	34.2	26.6	562
	UAVMOT	42.7	40.2	644
	ByteTrack	63.0	46.9	609
	BoT-SORT	**74.5**	**57.3**	**242**
Clip-70	SORT	32.9	3.3	705
	DeepSORT	20.2	11.7	1091
	DeepMOT	39.2	**57.0**	**2**
	UAVMOT	34.0	7.1	570
	ByteTrack	56.8	36.9	463
	BoT-SORT	**67.6**	26.5	289
Clip-77	SORT	17.9	2.8	332
	DeepSORT	12.2	2.6	555
	DeepMOT	18.8	26.6	**5**
	UAVMOT	21.7	5.0	169
	ByteTrack	25.5	11.8	91
	BoT-SORT	**30.5**	**35.3**	29

For assessing real-time performance, FPS (Frames Per Second) is utilized to gauge the speed of the multi-object tracking method. FPS indicates the number of video frames the tracking method can process per second. A higher FPS signifies a faster analysis of video data by the tracking methods.

Table 1 provides a detailed evaluation of 6 different deep learning-based multi-object tracking methods on 8 test sequences. Among them, Clip-01, Clip-20, Clip-38, and Clip-52 are harbour scenes, Clip-44 is an entry scene, Clip-68 is a departure scene, and Clip-70 and Clip-77 are open sea scenes. Based on the analysis of MOTA and IDF1 metrics, BoT-SORT demonstrates significant potential in multi-object tracking tasks on the sea surface. Notably, in Clip-44 and Clip-52, tracking sea-surface objects with BoT-SORT achieves exceptionally high MOTA scores (91.2 and 89.2) and IDF1 scores (93.9 and 82.2). The ByteTrack method also shows high performance, especially in Clip-52. SORT performs poorly in most scenarios, especially in open sea scenarios (Clip-70 and Clip-77). DeepSORT adds appearance features to SORT, but it is not suitable for autonomous navigation scenarios and performs poorly in various accuracy. This may be attributed to the limited feature information of small sea-surface objects. DeepMOT shows good performance in some specific scenarios, especially on the IDF1 metric. This indicates its relative robustness in maintaining the continuity of target identity. In the harbour scene Clip-52, its IDF1 score can reach 76.6.

However, in more challenging scenes, such as open sea scenes (Clip-70 and Clip-77), DeepMOT's performance drops significantly. This reflects the limitations of this approach in dealing with a complex dynamic background. UAVMOT exhibits some volatility across different scenarios, and its performance in terms of MOTA and IDF1 is not as stable as BoT-SORT and ByteTrack.

Similar to DeepMOT, UAVMOT also faces challenges in open-sea scenarios. Based on the analysis of IDs indicators, BoT-SORT maintains a low number of ID switching in most scenarios, demonstrating its advantage in the stability of tracking sea-surface objects. For example, in Clip-44 and Clip-52, the BoT-SORT method resulted in only 4 and 22 ID switches, respectively. DeepMOT exhibits superior IDs performance in open sea scenes. However, in scenes with dense occlusions, its stability does not match that of BoT-SORT.

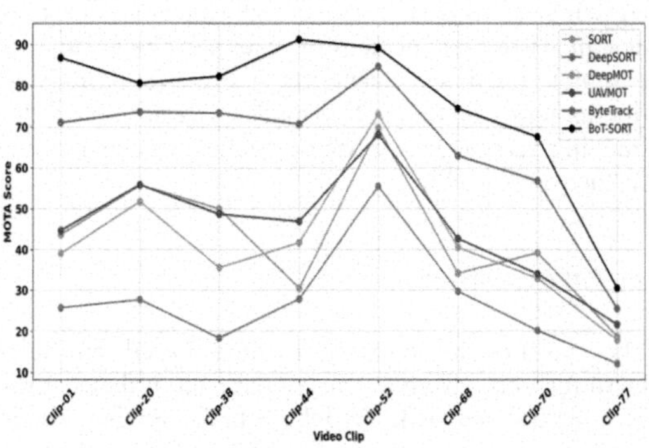

Figure 2. MOTA performance of different multi tracking methods by each video clip.

ByteTrack also maintains a relatively low number of ID switches across various scenes, particularly in harbour scenes, indicating better tracking continuity.

Table 2. Overall performance metrics of multi object tracking methods across all test videos.

Methods	MOTA↑	IDF1↑	IDs↓	FPS↑
SORT	39.5	25.0	4420	**34.4**
DeepSORT	25.8	24.5	5702	14.9
DeepMOT	40.3	41.9	1048	16.0
UAVMOT	43.4	36.3	2210	19.0
ByteTrack	63.6	48.3	1932	30.4
BoT-SORT	**72.9**	**59.3**	**743**	11.7

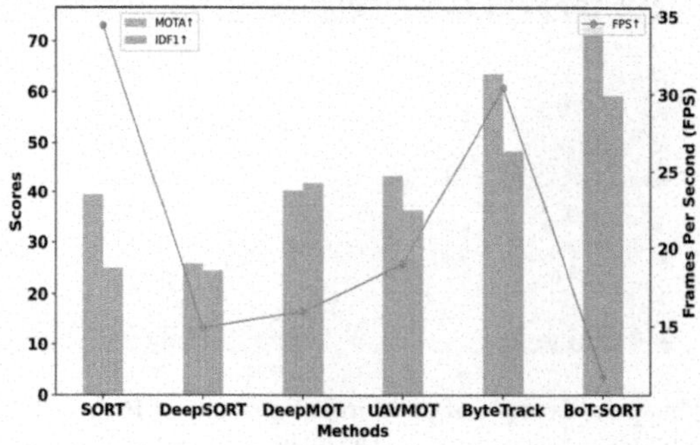

Figure 3. Comparison of multi object tracking methods on MOTA, IDF1 and FPS indicators.

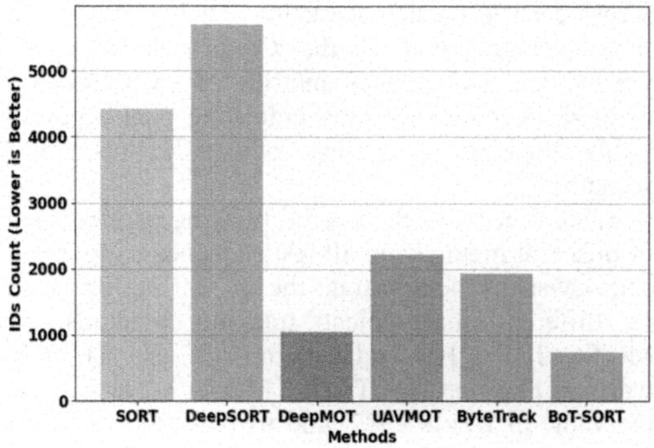

Figure 4. Comparison of IDs counts across multi object tracking methods.

However, in dense or dynamic scenes, the frequency of ID switches tends to increase. The two methods, SORT and DeepSORT, exhibit higher IDs values in multiple test sequences, particularly in the open sea and dense scenes. This indicates their limitations in handling complex environments and maintaining the continuity of target identity.

Figure 5. Example of multi-tracking results of BOT-SORT on some test sets of the JMTD. From top to bottom, each row represents a sequence clip. The image has been enlarged somewhat for better viewing.

Figure 2 visually shows the tracking accuracy metrics for various multi-object tracking methods across each test video sequence. Analyzing specific scenes reveals that in the Clip-70 and Clip-77 videos, the performance metrics of each tracking method are relatively low, indicating challenges in stably tracking sea-surface objects in open sea scenarios.

Table 2 displays the overall tracking performance of different methods on all test sequences. Figures 3 and 4 visually demonstrate the overall performance of different multi-object tracking methods on MOTA, IDF1, IDs and FPS on the test set. It is worth noting that ByteTrack's MOTA score is 63.6, the value of IDs is 1932, and its real-time performance can reach 30.4FPS, making it a choice with good overall performance. BoT-SORT performs best in terms of trackingaccuracy indicators, with a MOTA score of 72.9, IDF1 of 59.3, and IDs of 743, but the real-time performance is relatively low, only 11.7FPS.

Figure 5 shows some visualization results of the BoT-SORT method on the test set. The different coloured boxes marked in the image represent the results of multi-object tracking. Each box represents a tracked object. The text label in it indicates the category of the target, and the number next to it represents the target's identity so that it distinguishes each object in a video sequence. The size and shape of these boxes provide a visual estimate of the size and location of the target, while the numerical IDs corresponding to these boxes remain consistent between consecutive frames of the sequence, demonstrating the ability of this tracking method to maintain tracking coherence between different frames.

BoT-SORT effectively identifies and tracks multiple objects across various shipboard video sequences, resulting in a favourable overall visualization effect due to its camera motion compensation design. However, in dense scenes near ports and in complex open sea scenarios, there are noticeable fluctuations in the IDs values. Furthermore, the tracking accuracy for extremely small objects require further optimization.

4 DISCUSSION

The multi-object tracking method based on deep learning learns from extensive data and utilizes complex neural network models, which can be effectively applied to maritime autonomous navigation scenarios. The application of this technology significantly improves the intelligence level of autonomous navigation systems, allowing them to understand and adapt to the changing marine environment.

However, despite significant progress, several technical challenges remain when applying these object-tracking methods to associating sea-surface objects in autonomous navigation scenarios.

These challenges can be summarized as follows:

- **Shaking of ship-borne cameras:** Ship-borne cameras continuously move with the sea waves, and a stabilization algorithm needs to be designed to maintain accurate tracking.
- **Nonlinear movement of targets:** The nonlinear and unpredictable movement of sea-surface objects demands tracking algorithms to possess high adaptability and predictive capabilities.
- **Target occlusion:** In areas near ports with dense targets, algorithms must accurately distinguish and continuously track each independent target amidst occlusions.
- **Small object tracking capability:** With the continuous development of small object detection technology, the performance of multi-object tracking on the sea surface can be further promoted.
- **Influence of the complex sea surface environment:** Reflections on the sea surface or low illumination may interfere with visual imaging, resulting in the loss of sea-surface object feature information and posing challenges for the design of target similarity measurement methods.
- **Real-time performance of the tracker:** In maritime navigation scenarios, the tracking system must process input data in real time and rapidly update target location information to keep pace with environmental changes.
- **Class imbalance in maritime dataset:** The maritime dataset exhibits a class imbalance issue. The volume of learning data for small objects, like fishing boats and floaters, is notably limited. Consequently, further exploration into data augmentation techniques, such as employing Generative Adversarial Networks (GANs) to enlarge the dataset of small samples, is essential.

In evaluating tracker performance, BoT-SORT takes into account the jitter of cameras and exhibits commendable tracking accuracy and stability. This makes it particularly valuable in complex autonomous maritime navigation scenarios. However, there is still room for enhancing the tracker's performance, especially its real-time capabilities, to ensure efficient operation on resource-limited onboard platforms.

5 CONCLUSIONS

In this study, the application of deep learning-based multi-object tracking technology in maritime autonomous navigation scenarios is thoroughly investigated. A qualitative and quantitative analysis of 6 advanced multi-object tracking methods is conducted. Particular focus is placed on the performance of these technologies in processing maritime visible light videos, with an emphasis on multi-object tracking tasks. The findings reveal that these deep learning-based methods are apt at effectively identifying and tracking diverse sea-surface objects. Among all evaluated methods, BoT-SORT stands out in terms of accuracy and stability for multi-object tracking, underscoring its suitability for autonomous navigation tasks on the sea surface.

However, it is also evident that applying existing deep learning-based multi-object tracking technology, including BoT-SORT, directly to autonomous maritime navigation presents certain challenges. Future developments should aim to further integrate shipborne video characteristics to develop a more robust and efficient maritime multi-object tracker.

ACKNOWLEDGEMENTS

This work was supported by the National Key R&D Program of China (2022YFB4300803, 2022YFB4301402); in part by the National Natural Science Foundation of China (No. 52071049); in part by the Liaoning Provincial Science and Technology Plan (Key) project (No. 2022JH1/10800096); in part by the international cooperation training program for innovative talents of Chinese Scholarships Council (Grant No. CSC [2022] 2260). The authors acknowledge Jiangsu Automation Research Institute for the dataset. This work contributes to the Strategic Research Plan of the Centre for Marine Technology and Ocean Engineering (CENTEC), which is financed by the Portuguese Foundation for Science and Technology (Fundação para a Ciência e Tecnologia - FCT) under contract UIDB/UIDP/00134/2020.

REFERENCES

Aharon N, Orfaig R, Bobrovsky BZ., 2022. BoT-SORT: Robust associations multi-pedestrian tracking. ArXiv:2206.14651.

Bewley, A., Ge, Z., Ott, L., Ramos, F., Upcroft, B., 2016. Simple Online and Realtime Tracking, Proceedings of the IEEE International Conference on Image Processing (ICIP). pp. 3464–3468.

Bloisi D, Iocchi L, Fiorini M, Graziano G., 2011. Automatic maritime surveillance with visual target detection. Proceedings of the International Defense and Homeland Security Simulation Workshop (DHSS). pp. 141–145.

Bochkovskiy A, Wang C Y, Liao H Y M, 2020. YOLOv4: Optimal Speed and Accuracy of Object Detection. ArXiv: 2004.10934.

Chen, X., Wang, S., Shi, C., Wu, H., Zhao, J., Fu, J., 2019. Robust Ship Tracking via Multi-view Learning and Sparse Representation. The Journal of Navigation 72, 176–192.

Gao, D., Zhu, Y., Guedes Soares, C., 2023. Uncertainty modelling and dynamic risk assessment for long-sequence AIS trajectory based on multivariate Gaussian Process. Reliability Engineering & System Safety 230, 108963.

Liu S, Li X, Lu H, et al. 2022. Multi-object tracking meets moving UAV. Proceedings of the IEEE/CVF Conference on Computer Vision and Pattern Recognition. pp. 8876–8885.

Lyu, H., Shao, Z., Cheng, T., Yin, Y., Gao, X., 2023. Sea-Surface Object Detection Based on Electro-Optical Sensors: A Review. IEEE Intelligent Transportation Systems Magazine. 15, 190–216.

Milan A, Leal-Taixé L, Reid I, Roth S, Schindler K., 2016. MOT16: A benchmark for multi-object tracking. ArXiv:1603.00831.

Perera, L.P., Oliveira, P., Guedes Soares, C., 2012. Maritime Traffic Monitoring Based on Vessel Detection, Tracking, State Estimation, and Trajectory Prediction. IEEE Transactions on Intelligent Transportation Systems. 13, 1188–1200.

Prasad, D.K., Rajan, D., Rachmawati, L., Rajabally, E., Quek, C., 2017. Video Processing from Electro-Optical Sensors for Object Detection and Tracking in a Maritime Environment: A Survey. IEEE Transactions on Intelligent Transportation Systems. 18, 1993–2016.

Redmon, J., Divvala, S., Girshick, R., Farhadi, A., 2016. You Only Look Once: Unified, Real-Time Object Detection, Proceedings of the IEEE Conference on Computer Vision and Pattern Recognition (CVPR). pp. 779–788.

Redmon, J., Farhadi, A., 2018. YOLOv3: An Incremental Improvement. ArXiv:1804.02767.

Shan, Y., Zhou, X., Liu, S., Zhang, Y., Huang, K., 2021. SiamFPN: A Deep Learning Method for Accurate and Real-Time Maritime Ship Tracking. IEEE Transactions on Circuits and Systems for Video Technology. 31, 315–325.

Shao, Z., Lyu, H., Yin, Y., Cheng, T., Gao, X., Zhang, W., Jing, Q., Zhao, Y., Zhang, L., 2022. Multi-Scale Object Detection Model for Autonomous Ship Navigation in Maritime Environment. Journal of Marine Science and Engineering. 10, 1783.

Sun, S., Lyu, H., Dong, C., 2023. AIS aided marine radar target tracking in a detection occluded environment. Ocean Engineering 288, 116133.

Thombre, S., Zhao, Z., Ramm-Schmidt, H., Vallet Garcia, J.M., Malkamaki, T., Nikolskiy, S., Hammarberg, T., Nuortie, H., H. Bhuiyan, M.Z., Sarkka, S., Lehtola, V.V., 2022. Sensors and AI Techniques for Situational Awareness in Autonomous Ships: A Review. IEEE Transactions on Intelligent Transportation Systems. 23, 64–83.

Tian, W., Lauer, M., Chen, L., 2020. Online Multi-Object Tracking Using Joint Domain Information in Traffic Scenarios. IEEE Transactions on Intelligent Transportation Systems. 21, 374–384.

Vettor, R. and Guedes Soares, C. 2015. Computational system for planning search and rescue operations at sea. Procedia Computer Science. 51:2848–2853.

Wang, C.-Y., Bochkovskiy, A., Liao, H.-Y.M., 2023. YOLOv7: Trainable bag-of-freebies sets new state-of-the-art for real-time object detectors. Proceedings of the IEEE/CVF Conference on Computer Vision and Pattern Recognition. pp. 7464–7475.

Wojke, N., Bewley, A., Paulus, D., 2017. Simple Online and Realtime Tracking with a Deep Association Metric. Proceedings of the IEEE international conference on image processing (ICIP). pp.36453649.

Xu, Y., Osep, A., Ban, Y., Horaud, R., Leal-Taixe, L., Alameda-Pineda, X., 2020. How to Train Your Deep Multi-Object Tracker. Proceedings of the IEEE/CVF conference on computer vision and pattern recognition. pp. 6787–6796.

Zhang, Y., Sun, P., Jiang, Y., Yu, D., Weng, F., Yuan, Z., Luo, P., Liu, W., Wang, X., 2022. ByteTrack: Multi-Object Tracking by Associating Every Detection Box. European Conference on Computer Vision. Cham: Springer Nature Switzerland, pp. 1–21.

Advances in Maritime Technology and Engineering – Guedes Soares & Santos (Eds)

Maritime traffic complexity evaluation in the Yangtze River using AIS data

X. Xu & X. Yan
State Key Laboratory of Maritime Technology and Safety, Wuhan University of Technology, Wuhan, P.R. China
Centre for Marine Technology and Ocean Engineering (CENTEC), Instituto Superior Técnico, Universidade de Lisboa, Lisbon, Portugal
Intelligent Transportation System Research Center, Wuhan University of Technology, Wuhan, P.R. China

B. Wu
State Key Laboratory of Maritime Technology and Safety, Wuhan University of Technology, Wuhan, P.R. China
Intelligent Transportation System Research Center, Wuhan University of Technology, Wuhan, P.R. China
Inland Port and Shipping Industry Research Co., Ltd. of Guangdong Province, Shaoguan, P.R. China

A.P. Teixeira & C. Guedes Soares
Centre for Marine Technology and Ocean Engineering (CENTEC), Instituto Superior Técnico, Universidade de Lisboa, Lisbon, Portugal

ABSTRACT: Traffic complexity evaluation is an element of marine situation awareness that plays a critical role in ensuring navigation safety. This paper proposes a method for evaluating marine traffic flow complexity using AIS data that can be used by marine supervisors in vessel traffic service (VTS) centers to enhance the understanding of the marine traffic condition. The traffic flow complexity is evaluated in three steps, including data preprocessing, weight determination and traffic complexity modelling. First, abnormal AIS data are deleted and then the remaining data are synchronized. Second, the distance between all ship pairs is calculated using AIS data and an exponential function of the distances is adopted as weight to express the relationship among the ships. Third, the graph theory is introduced to describe marine traffic conditions, and the indicators edge density and strength are used to evaluate the complexity. The experimental results in a selected section of the Yangtze River show that the proposed method can quantitatively evaluate the marine traffic complexity and effectively strengthen the marine traffic condition situation assessment by the VTS operators.

1 INTRODUCTION

The transportation in Yangtze River accounts for over 80% of inland shipping and is the busiest inland waterway in China. The Gezhouba waterway is one of the segments with the higher traffic volume, resulting in a complicated traffic situation (Wu et al., 2020).

Ship navigation safety is an active field for inland waterway transportation in the Yangtze River. To guarantee ship navigation safety, the marine traffic situation is perceived by maritime supervisors using various ways, which support traffic control decisions. An important step is to evaluate maritime traffic complexity using situation information, including the number of ships, ship location, and ship behaviour. In practice, a difficult question to address is how to effectively assess the maritime traffic situation for ship navigation safety.

The Automatic Identification System (AIS) is a maritime communication system between ships and ships, and shore-based stations, that transmits ship static and dynamic information (Xu et al., 2023). Static information, such as maritime mobile service identity (MMSI), ship type and size are affected by the ship's attributes. Dynamic information, such as ship longitude and latitude, speed, and course are affected by the real-time ship navigation status (Liu et al., 2023). The neighbour ships and shore-based supervision centres can be aware of the ship navigation situation by AIS data. The transmission interval of AIS data is usually between 6s-10s, depending on the ship navigation status (Wu et al., 2022). Owing to this mechanism, the transmission interval for every ship is different, which causes the asynchrony of information. Moreover, extreme weather conditions, terrain, and distance also cause information and packet loss. Therefore, before using AIS data, the quality of AIS data is usually improved by data cleaning and interpolation (Guo et al., 2021).

Some studies have focused on maritime traffic complexity evaluation using AIS because of the comprehensive information provided by AIS data. Statistical

DOI: 10.1201/9781003508762-20

analysis is a common method to draw the traffic situation and to derive the traffic characteristics of some special waterway segments. Rong et al. (2021) analysed the correlation between near-collision hotspots and the maritime traffic characteristics off the continental coast of Portugal. The proposed approach enabled a quantitative assessment of the geo-characteristics of maritime traffic and their relationship with hotspots of high collision risk, which can help authorities implement schemes and efficient preventive and mitigation strategies to improve maritime traffic safety in some locations. Huang et al. (2019) represented the ship traffic conditions using a maritime time-space diagram and discussed the relationship between flow rate, traffic density and local speed. Besides, Wen et al. (2015) introduced a traffic complexity evaluation model for quantitative analysis of the crowding degree and navigation risk. This model is constructed by using the relative distance, relative speed, and trajectory between ship pairs.

Recently, some methods, including complex networks (Sui et al., 2021; Sui et al., 2022), radial distribution functions (Liu et al., 2020), and the technique for order preference by similarity to an ideal solution (TOPSIS) (Zhang et al., 2022) have been used to assess maritime traffic complexity. Specifically, Sui et al. (2021) proposed a maritime traffic complexity evaluation method that combines the Voronoi diagram and complex networks. A waterway segment is divided into some cells and the safety distances for avoiding collision are set between ships. In subsequent work, the node importance of the marine traffic situation complex network (MTSCN) is evaluated, which reflects the ship navigation risk and the traffic network complexity (Sui et al., 2022). The traffic condition of every ship is seen as a molecule system, and the neighbouring ships are treated as other atoms around this ship. The maritime traffic complexity model established includes three sub-models with radial distribution functions. Finally, the overall traffic complexity is identified by integrating three results of sub-models (Liu et al., 2020). Moreover, the speed, course and position of the ships are used to construct each sub-model (Liu et al., 2021). The concept of maritime traffic complexity is transformed as the irregularity and unpredictability of ship travel time sequences. The higher irregularity of ship travel time sequences indicates a higher traffic complexity (Zhang et al., 2022).

From the above analysis, some studies assess maritime traffic complexity from macroscopic characteristics from AIS data, which reflects the navigation rule of the whole waterway over a long period, and it can be treated as a characteristic of this waterway. However, the real-time maritime traffic complexity evaluation is helpful for maritime supervision and ship navigation safety. The objective of this paper is to construct a real-time traffic complexity evaluation method, to help maritime supervisors understand the traffic conditions.

The remainder of this paper is organized as follows. Section 2 describes the details of the proposed maritime traffic complexity evaluation method. Section 3 describes the application scenario in the Yangtze River, and conclusions are drawn in Section 4.

2 METHODOLOGY

2.1 *Framework of maritime traffic complexity evaluation*

The framework of the proposed method is presented in this section. Specifically, the traffic complexity evaluation of inland waterways is carried out in three steps. The first step is to preprocess AIS data. The quality of AIS data is improved by removing the abnormal data, and the remaining AIS data is interpolated as a continuous trajectory with a 1s-interval. The second step is to calculate the distance of all ship pairs, and the weight between ships is determined based on the distance distribution. The third step is to construct the maritime traffic structure graph, and two indicators are introduced to evaluate the inland waterway traffic complexity. The framework of the proposed method is shown in Figure 1.

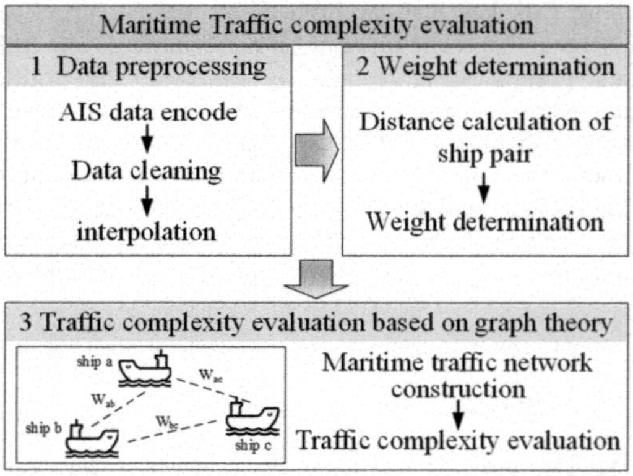

Figure 1. Framework of the proposed Maritime Traffic Complexity Evaluation Method.

2.2 *AIS data pre-processing*

Due to extreme weather conditions, obstacle occlusion, and packet loss in transmission, the original AIS datasets have some erroneous data. Therefore, a data cleaning process is used to improve the quality of AIS data. Specifically, the ship's longitude and latitude not located within the range of the waterway, are regarded as abnormal data. Also, the position data is removed when the distance between the adjacent positions is larger than a threshold. The distance between adjacent positions is calculated as:

$$\mathrm{D} = Geo\big((lon_i, lat_i), (lon_j, lat_j)\big) \qquad (1)$$

where *Geo* denotes the distance between two points on the earth's surface, and the (lon_i, lat_i), (lon_j, lat_j)

are the longitude and longitude of the two points, respectively.

Considering that AIS data is not transmitted at the same time, interpolation is carried out to ensure the synchronism of the ship navigation status before the evaluation of the maritime traffic complexity. In this paper, a cubic spline interpolation with 1s-interval is used to process the ship's longitude and latitude, speed, and course, which contributes to the real-time traffic complexity evaluation (Zaman et al., 2023).

A continuous ship trajectory is expressed as $[(lon_1, lat_1), \ (lon_2, lat_2), \ldots, (lon_t, lat_t)]$. In every interval, a set of cubic polynomials $S_n(lon)$ are constructed to supplement the ship trajectory under 1 second. The cubic polynomials are shown in Equation (2), and the cubic polynomial must conform to the requirement, as given in Equation (3). (Gao et al., 2021).

$$S_n(lon) = a_n + b_n(lon - lon_n) + c_n(lon - lon_n)^2 + d_n(lon - lon_n)^3 \tag{2}$$

$$S_n(lon_n) = lat_n \tag{3}$$

where a_n, b_n, c_n, and d_n are undetermined coefficients, which can be solved by expressing a system of linear equations for adjacent trajectories. After obtaining these coefficients, the interpolated trajectory points at 1s intervals are determined.

2.3 *Maritime traffic complexity evaluation*

The marine traffic condition is described as an undirected graph g(P, S), where P is a set of ships at a certain time, and S is a set of edges, which denotes the relationship between two ships (Isufaj et al., 2021). This graph describes the marine traffic condition per time step (e.g., 1s). Moreover, an undirected weighted graph is constructed to distinguish the degree of influence between different ship pairs in this paper. The weight is assigned to these edges, and the range of weight is 0 to 1. The weight of each vertex is given in

$$W(x) = \sum\nolimits_{y=1}^{M} \omega_{x,y} \tag{4}$$

and the weight is determined by the distance between two ships, as shown in

$$\omega_{x,y} = \mathcal{L}(\mathrm{D}) = e^{-\beta\frac{\|\vec{D}\|}{\rho_{x,y}}} \tag{5}$$

where $\omega_{x,y}$ denotes the weight between vertex x and vertex y, and $W(x)$ denotes the sum of the weights of vertex x and other connected vertices. Note that $\omega_{x,y} = \omega_{y,x}$. β is a correction parameter related to the environment, defined as 1.81 (Wen et al., 2015). As a traffic separation scheme is implemented in the Three Gorges Waterway, the ship domain is introduced to determine the safety distance between two ships. The ships navigating in the Three Gorges Waterway have a length of around 100 m, and the safety distance is determined as 4 times the ship length, which is defined as 0.22 nautical miles.

To evaluate the maritime traffic complexity, two indicators are introduced, including edge density and strength. Edge density (ED) is the ratio of the sum of weight to the number of unweighted edges in the graph. Compared with the ship density, the influence between ship pairs is integrated into edge density, which is given in Equation (6), and the number of edges in a fully connected graph with the same size is given in Equation (7). The strength of every ship is defined by the sum of the connected weight, which measures the importance of the ship in marine traffic, as shown in Equation (8). The strength of the graph is the average of all ships.

$$\mathrm{ED}(\mathrm{g}, \mathrm{t}) = \frac{\sum_{(x,y)} \omega_{x,y}(t)}{A(P_t)} \tag{6}$$

$$A(P_t) = \frac{m(m-1)}{2} \tag{7}$$

$$W(x, t) = \sum\nolimits_{y=1}^{M} \omega_{x,y}(t) \tag{8}$$

where t is the ship navigation time, m is the number of vertices in the constructed graph, and M is the number of edges connected by every vertex.

3 APPLICATION IN THE YANGTZE RIVER

3.1 *Data description*

The Three Gorges Waterway shown in Figure 2 is selected to apply the proposed method. AIS data of this segment was collected on August 22, 2019. Note that the selected AIS data only includes three hours (from 10:00 to 13:00) in this paper, the reason is that the proposed method focuses on real-time maritime traffic complexity evaluation. The original AIS data of different ships are asynchronous, which is influenced by the ship's dynamic status. It cannot be directly used to evaluate traffic complexity in real time. Therefore, the original AIS data are synchronized by interpolation. The resulting interpolated AIS trajectories with 1s-interval are shown in Figure 3.

3.2 *Results of maritime traffic complexity evaluation*

After obtaining the ship's longitude and latitude information with 1s interval from the interpolated AIS data, the distance between each ship pair for each second is calculated using Equation (1). The distance distribution of two different areas within three hours is shown in Figure 4. It can be seen that the

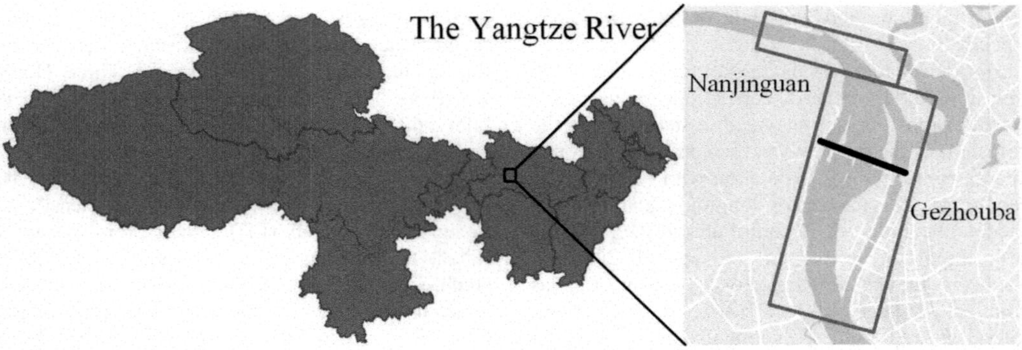

Figure 2. Case-study scenario in the Yangtze River.

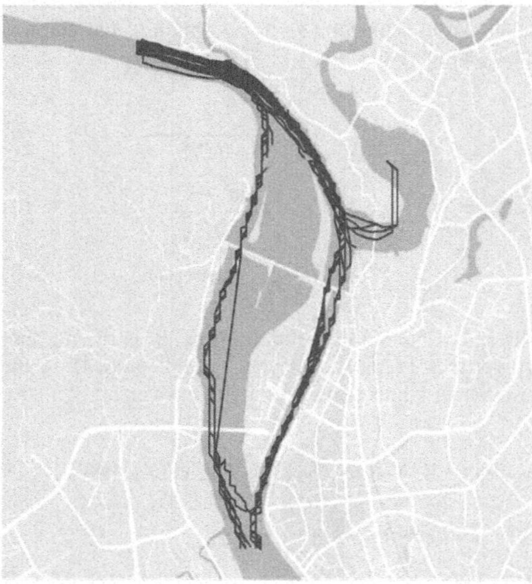

Figure 3. Interpreted AIS data mapping in the Yangtze River.

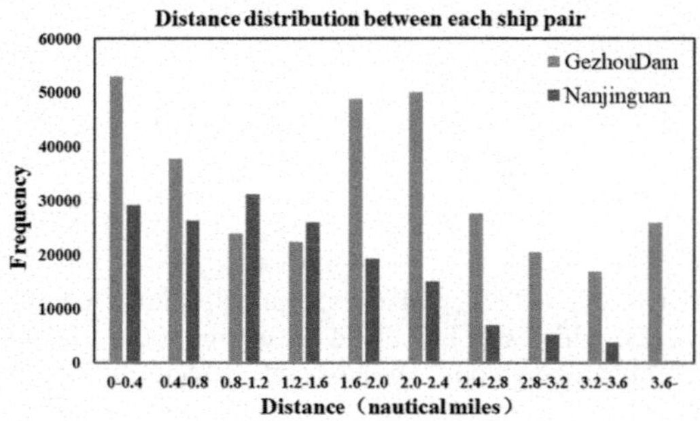

Figure 4. Distance distribution for two selected areas within three hours.

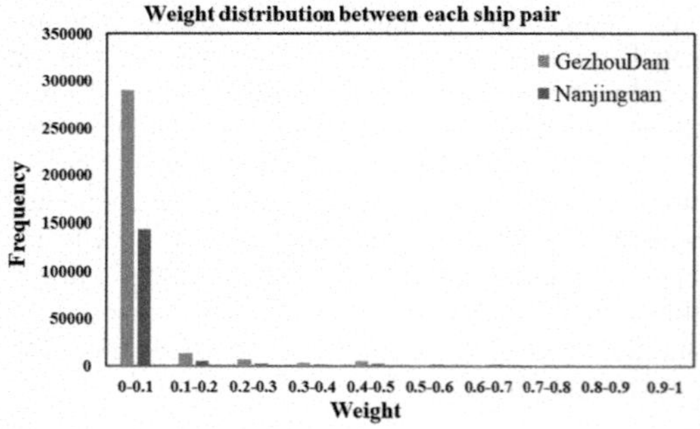

Figure 5. Weight distribution for two selected areas within three hours.

distance of ships is mainly in the range of 0-4 nautical miles in the Gezhouba area, and the number of ships in this area is higher than in the Nanjinguan area.

The weight of a ship pair is determined by the distance between two ships, which reflects the degree of influence. The weight distribution of two areas within three hours is shown in Figure 5. The majority of the weights are less than 0.1 in the Gezhouba area. The reason is that the upward and downward ships can navigate in two separate channels in this area. Moreover, ships need to wait for the ship in front to pass through the lock, and the distance between ship pairs must be maintained.

After calculating the weights of ships' pairs, the marine traffic structure is constructed using the ship's longitude and latitude per 1s and the weights. Note that the weights of ship pairs are only determined in one area, and the relationship of two ships in different areas is not considered. The reason is that the maritime traffic complexity of every segment is separately assessed by supervisors in the Three Gorges Waterway. The maritime traffic structure at 10:00:13 is described in Figure 6. The red dot represents the ship's position, and the black line represents the interaction of two ships, the degree of influence is determined by the weight.

By Equation (5)-(8), the edge density and strength of the constructed marine traffic condition are calculated, which are shown in Figures 7 and 8, respectively. It can be seen that the edge density of the Gezhouba area is lower than the Nanjinguan area. The reason is that the upward and downward channels are separated in this area, and ships maintain

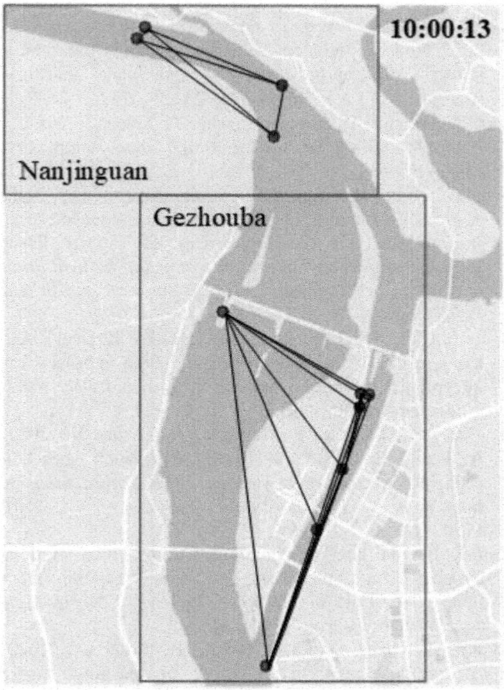

Figure 6. The marine traffic condition at 10:00:13 of two selected areas.

a large distance from nearby ships in preparation for passing through the lock. The edge density of the Nanjinguan area increased to 0.35 at about 11:50, which is caused by the change in the distance between the ships in the encountering.

As there are many ships at anchor in the Gezhouba area, and ships are far from other ships, the weights between ships are small. Therefore, the strength of the Gezhouba area is close to 0, as can be seen in Figure 8.

The average Edge Density (ED) and Strength of the Gezhouba and Nanjinguan areas are derived, as shown in Table 1. The ED is 0.039 and 0.067 in the two areas, respectively. From the above analysis, the traffic conditions in the Nanjinguan area are more complex than in the Gezhouba area.

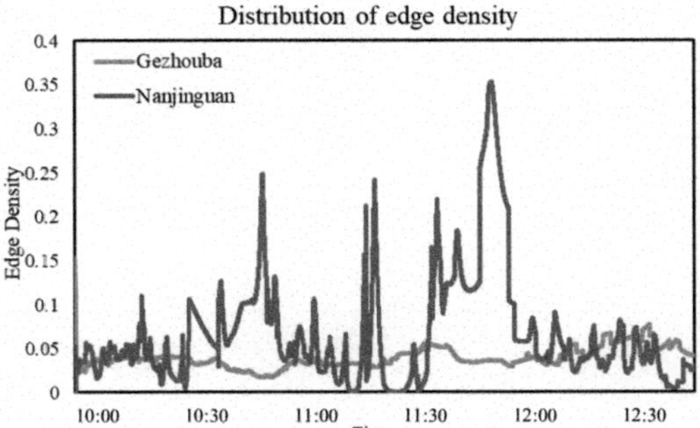

Figure 7. The distribution of edge density of the constructed marine traffic condition.

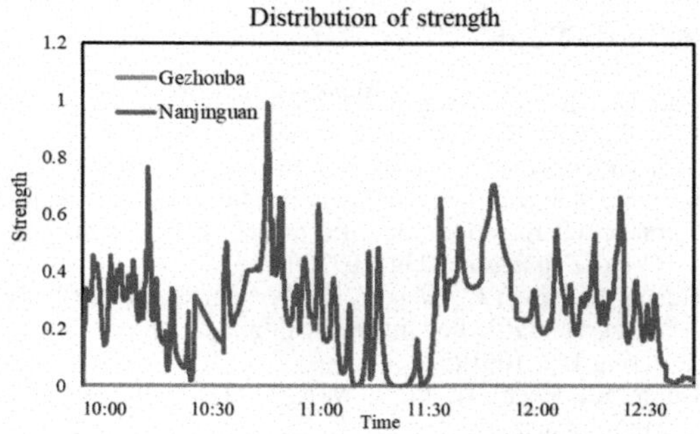

Figure 8. Distribution of strength of the constructed marine traffic condition.

Table 1. Average ED and strength of the different areas.

Average	ED	Strength
Gezhouba	0.039	0.001
Nanjinguan	0.067	0.271

4 CONCLUSIONS

This paper proposes a method for maritime traffic complexity evaluation from the micro perspective, which constructs the maritime traffic graph using AIS data. The complexity is influenced by the ship interaction, and it is considered in the constructed traffic graph through the distance between ship pairs. The proposed method can evaluate the traffic complexity in real time, and it contributes to traffic supervisors of VTS. Specifically, the abnormal values in AIS data are removed, and a cubic spline interpolation with 1s-interval is introduced to maintain the synchronism of AIS data. Afterwards, the weight based on the distance between ship pairs is calculated. Finally, the marine traffic condition is constructed using graph theory. The feasibility of the proposed method is applied to AIS data in two areas of the Three Gorges Waterway.

ACKNOWLEDGEMENTS

This work was supported in part by the International Cooperation and Exchange of the National Natural Science Foundation of China under Grant 51920105014, in part by the National Natural Science Foundation of China under Grant 52272422 and Grant 52071248, and in part by the China Scholarship Council under Grant 202206950046. This work contributes to the Strategic Research Plan of the Centre for Marine Technology and Ocean Engineering (CENTEC), which is financed by the Portuguese Foundation for Science and Technology (Fundação para a Ciência e Tecnologia - FCT) under contract UIDB/UIDP/00134/2020.

REFERENCES

Gao, D., Zhu, Y., Zhang, J, He, Y., Yan, K., Yan, B., 2021. A novel MP-LSTM method for ship trajectory prediction based on AIS data. *Ocean Engineering* 228, 108956.

Guo, S., Mou, J., Chen, L., Chen, P., 2021. Improved kinematic interpolation for AIS trajectory reconstruction. *Ocean Engineering* 234, 109256.

Huang, Y., Yip, T.L., Wen, Y., 2019. Comparative analysis of marine traffic flow in classical models. *Ocean Engineering* 187, 106195.

Isufaj, R., Koca, T., Piera, M.A., 2021. Spatiotemporal graph indicators for air traffic complexity analysis. *Aerospace* 8 (12), 364.

Liu, D., Rong, H., Guedes Soares, C., 2023. Shipping route modelling of AIS maritime traffic data at the approach to ports. *Ocean Engineering* 289, 115868.

Liu, Z., Wu, Z., Zheng, Z., 2020. Modelling ship density using a molecular dynamics approach. *The Journal of Navigation* 73 (3), 628–645.

Liu, Z., Wu, Z., Zheng, Z., Wang, X., Guedes Soares, C., 2021. Modelling dynamic maritime traffic complexity with radial distribution functions. *Ocean Engineering* 241, 109990.

Rong, H., Teixeira, A.P., Guedes Soares, C., 2021. Spatial correlation analysis of near ship collision hotspots with local maritime traffic characteristics. *Reliability Engineering & System Safety* 209, 107463.

Sui, Z., Huang, Y., Wen, Y., Zhou, C., Huang, X., 2021. Marine traffic profile for enhancing situational awareness based on complex network theory. *Ocean Engineering* 241, 110049.

Sui, Z., Wen, Y., Huang, Y., Zhou, C., Du, L., Piera, M.A., 2022. Node importance evaluation in marine traffic situation complex network for intelligent maritime supervision. *Ocean Engineering* 247, 110742.

Wen, Y., Huang, Y., Zhou, C., Yang, J., Xiao, C., Wu, X., 2015. Modelling of marine traffic flow complexity. *Ocean Engineering* 104, 500–510.

Wu, B., Tian, H., Yan, X., Guedes Soares, C., 2020. A probabilistic consequence estimation model for collision accidents in the downstream of Yangtze River using Bayesian Networks. *Proceedings of the Institution of Mechanical Engineers, Part O: Journal of Risk and Reliability* 234 (2), 422–436.

Wu, Y., Chu, X., Deng, L., Lei, J., He, W., Królczyk, G., Li, Z., 2022. A new multi-sensor fusion approach for integrated ship motion perception in inland waterways. *Measurement* 200, 111630.

Xu, X., Wu, B., Xie, L., Teixeira, A.P., Yan, X., 2023. A Novel Ship Speed and Heading Estimation Approach Using Radar Sequential Images. *IEEE Transactions on Intelligent Transportation Systems* 24(10), 11107–11120.

Zaman, B., Marijan, D., Kholodna, T., 2023. Interpolation-Based Inference of Vessel Trajectory Waypoints from Sparse AIS Data in Maritime. *Journal of Marine Science and Engineering* 11 (3), 615.

Zhang, M., Zhang, D., Fu, S., Kujala, P., Hirdaris, S., 2022. A predictive analytics method for maritime traffic flow complexity estimation in inland waterways. *Reliability Engineering & System Safety* 220, 108317.

Advances in Maritime Technology and Engineering – Guedes Soares & Santos (Eds)

A novel Kino-dynamic RRT path planning algorithm considering ship maneuverability restrictions

H. Zhang, J.F. Zhang & J.J. Liu
State Key Laboratory of Maritime Technology and Safety (Wuhan University of Technology), National Engineering Research Center for Water Transport Safety, Intelligent Transportation Systems Research Center, Wuhan University of Technology, Wuhan, China

C. Guedes Soares
Centre for Marine Technology and Engineering (CENTEC), Instituto Superior Técnico, Universidade de Lisboa, Portugal

ABSTRACT: Path planning and collision avoidance is one of the most important ways to navigation safety and efficiency. The trajectory smoothness is a key factor to adjust to ship maneuverability, which can also reduce operation frequency and reduce fuel consumption. In this paper, an improved dynamic Rapidly-Exploring Random Tree (RRT) algorithm, called Kino-dynamic RRT Algorithm, is proposed under multiple ship encounter situations in open sea with four-dimensional linear double integrator. The results show that the proposed method can generate a smoothing trajectory both considering vessel maneuverability and speed change, while keeping clearance of other obstacles under different encounter situations. The research can support ship path planning and making collision avoidance decisions in a systematic way.

1 INTRODUCTION

Maritime transportation is the backbone of international trade and the global economy. Due to the high volume with low cost of global trade, shipping transportation has been attracting steady attention (Yuan et al., 2022). During the shipping technologies development, navigational safety has always been the top priority of research. Historical data from 2014 to 2019 shows that 44% of accidents are caused by collision, contact and grounding (ESMA, 2022). The rate can be even higher for bulk carriers, but lower for containerships (Guedes Soares & Teixeira, 2001).

Guarantees of navigational safety could be divided into multiple categories, and automatic ship path planning is one of the most effective technologies in reducing accidents and enhancing intelligence in maritime transportation studies. There are many different path-planning algorithms, which mainly include graph-based planning algorithms, sampling-based algorithms and intelligent planning algorithms.

The graph-based algorithms can be categorized into two types, which are Dijkstra (Johnson., 1973) and A* algorithm (Hart et al., 1973). Graph-based planning algorithms are characterized by transforming the entire map into grids and finding the best path by computing the information contained in all the grids. For instance, the Dijkstra algorithm is more focused on breadth searching and the A* algorithm is more focused on depth searching. Dijkstra aims to find the shortest path to all nodes, whereas A* finds the best path from start to goal with limited search space. Based on the two algorithms, there are more algorithms with further modifications. The D* algorithm (Stentz, 1997) can calculate the shortest paths considering both static and dynamic environments in a short time. The Hybrid A* algorithm (Dolgov et al., 2010; Montemerlo et al, 2008) can generate nodes in any position of a grid and use a curve to connect nodes to get a smooth path.

In recent years, intelligent methods have developed rapidly and extensively. These methods have also been modified and applied to path planning. These algorithms can find a feasible path in a complex environment and a real-time optimal path, but they need many calculation resources and may fall into a local optimum. Some studies improve the calculation efficiency or compensate for the disadvantages through modifications (Dong et al., 2022; Güven et al, 2022; Zhou et al, 2022).

The ant colony algorithm has high calculated efficiency with distributed computing capabilities, but can easily fall in a local optimum. Wang et al. (2019) proposed a modified version of the ant colony algorithm which combined the A* algorithm to avoid falling in a local optimum. The neural network algorithm is a method that emulates the neural

DOI: 10.1201/9781003508762-21

work in the human brain. Sung et al. (2021) combined neural network and path planning for autonomous driving with satisfactory performance. However, the algorithm needs massive data for training to get a feasible path compared to other algorithms.

The Rapidly-Exploring Random Tree (RRT) is a sampling-based algorithm. The main idea is to perform a random search with fixed steps in the search space, until reaching the destination. The search procedure usually makes a trade-off between efficiency and randomness. RRT was first introduced to robot path planning and works well in many applications (Li et al., 2020). In recent years, some researchers extended its application to ship path planning. A COLREG-RRT algorithm (Chiang et al., 2018) was proposed for surface vehicle navigation, and it was realized by constructing virtual obstacles for dynamic obstacles in the tree growth procedure to make the path comply with COLREG and using conducting joint forward simulations for all vessels to maintain a high success rate for navigation. However, few researchers have focused on bridging the gap between path planning and collision avoidance, which makes the planned trajectory not smooth enough.

To fix the gap between path planning and collision avoidance for ships, this study proposed a novel Kino-dynamic RRT path planning algorithm considering ship manoeuvrability restrictions. The rest of the paper is organized as follows: A literature review on ship path planning and collision avoidance algorithms is discussed in Section 2. In section 3, the dynamic path planning methodology is formulated, including dynamic collision detection, Quaternion Ship Domain (QSD) model, dynamic path planning algorithms and Kino-dynamic RRT* algorithm. A comprehensive simulation of the path planning under multiple ship encounter situations is performed and discussed in section 4. Finally, conclusions are summarized in section 5.

2 LITERATURE REVIEW

2.1 *Local path planning*

Local path planning is a method focusing on a small scope with both static and dynamic environments in real-time, and it is realized by dynamic collision detection and route re-planning. Many deterministic and heuristic algorithms have been combined in local path planning. Lyu et al. (2018) proposed a modified Artificial Potential Field (APF) that includes a new repulsion potential field function and the corresponding virtual forces were added functional and safety requirements to realize collision avoidance with dynamic targets and static obstacles while complying with COLREGs.

Wang et al. (2019) proposed an improved ant colony optimization algorithm to enhance the real-time performance of path planning. To realize this, a dynamic viewable method was first used to identify the local environment. A reverse eccentric expansion was designed to deal with dynamic obstacles based on the motion velocity and COLREGs, and an improved pseudo-random proportional rule was used to select ant state transition for algorithm convergence acceleration. Meantime, the wolf pack allocation principle and the maximum-minimum ant system were added to prevent the ant colony algorithm from falling into the local optimum. Fiskin et al. (2021) proposed a decision support system which is based on the genetic algorithm and fuzzy logic for vessel operators under real-time encounter situations and tested in a real environment with autonomous surface vehicles.

2.2 *RRT algorithm*

RRT is a sampling-based algorithm that has been widely used in path planning (Li et al., 2020; Kang et al, 2021; Sun et al, 2018). Its main idea is uniform random sampling in the whole map and finding a feasible path through all random nodes by limited steps. The algorithm has advantages in terms of fast planning and complete solutions, but its disadvantages are that it is time-consuming or even impossible for the algorithm to find a way out when getting through a narrow area. Moreover, due to random sampling, the path generated by the algorithm may not be optimum. Many studies have tried to overcome such problems from different perspectives.

Karaman et al. (2011) proposed the RRT* algorithm to make the path asymptotic optimality by adding a rewrite procedure after a new node is added to the tree. However, the calculation efficiency was reduced to a large degree. According to path length and positions of two origin nodes, Jonathan et al. (2011) proposed inform RRT* algorithm to narrow down the sampling space into an ellipse space, and the ellipse space was iterated relative to the best path length. This method increased the convergence speed of the RRT* algorithm.

After the series of improved RRT algorithms, it was introduced into ship path planning and support decision-making (Enevoldsen et al., 2021; Gu et al, 2023; Huang et al, 2018). However, to adapt to the ship's navigation environment, a dynamic detection model is necessary to detect other obstacles in a dynamic environment. Chiang et al. (2018) introduced a COLREG-RRT algorithm for ship navigation to realize dynamic collision detection by constructing virtual obstacles. However, the algorithm cannot ensure finding a feasible trajectory.

Enevoldsen et al. (2021) proposed a Grounding-aware RRT* algorithm to address the risk of grounding during evasive manoeuvres to avoid collision. A probabilistic model was built with a multivariate kernel density estimation method based on historical AIS data and encoded into RRT* cost function to penalize path deviations, which led the ship to sail in shallow waters. It considered ship manoeuvrability and the planning path, which complies with the ordinary practice of seamanship. Dusting et al. (2021) proposed a kino-dynamic RRT* algorithm, which was constructed by a controllable linear system and weighted controllability

Gramian equations, and they were solved by the closed-form method or the 4-th order Runge-Kutta method and encoded in the sampling method of RRT* algorithm. The algorithm was verified to find a good path with high efficiency by kino-dynamic model.

To ensure navigation safety, most researchers perform well in specific situations. However, further modification or combination is still necessary for more general situations. Meantime, efficiency is equally important, so the procedure needs a trade-off between efficiency and information complexity.

3 METHODOLOGY

In this section, a modified RRT algorithm, called Kino-dynamic RRT* algorithm is proposed for local path planning with a dynamic collision detection based on the Quarterion Ship Domain (QSD) model (Silveira et al., 2022). Before local path planning, the algorithm needs to get a reference path considering the static obstacles in the map. The flow chart is shown in Figure 1.

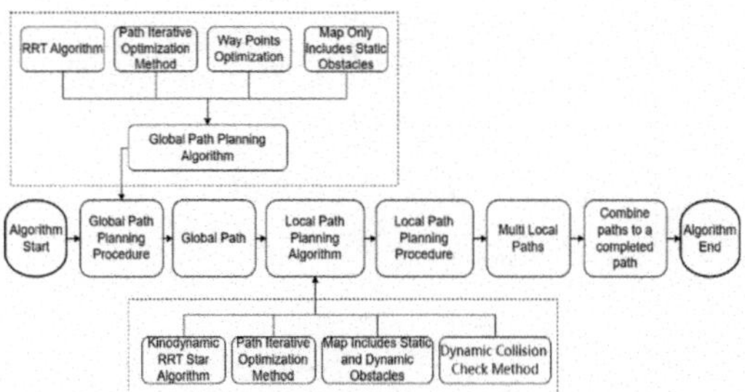

Figure 1. Flow chart of the path planning algorithm.

The whole procedure of the global path planning algorithm includes two main phases and three sub-phases. Two main phases are global path planning and local path planning. Global path planning includes the core of the traditional RRT algorithm additional path iterative optimization method and waypoint optimization method. It only considers the static obstacles in the map. Local path planning includes a modified version of Kino-dynamic RRT* algorithm with a sequence sampling method additional path iterative optimization method and dynamic collision detection method. It considers both static and dynamic obstacles in the map. The sub-phases are global path data processing, local paths data processing and local paths assembling.

3.1 *Global path planning procedure*

The RRT algorithm is used for global path planning, and its flowchart is shown in Figure 2. Firstly, an initial path is generated by the RRT algorithm, and the sampling method with the target as the new node is used to improve the convergence rate. A path is generated only considering the static environment. However, the path has many nodes making the algorithm converging much slower.

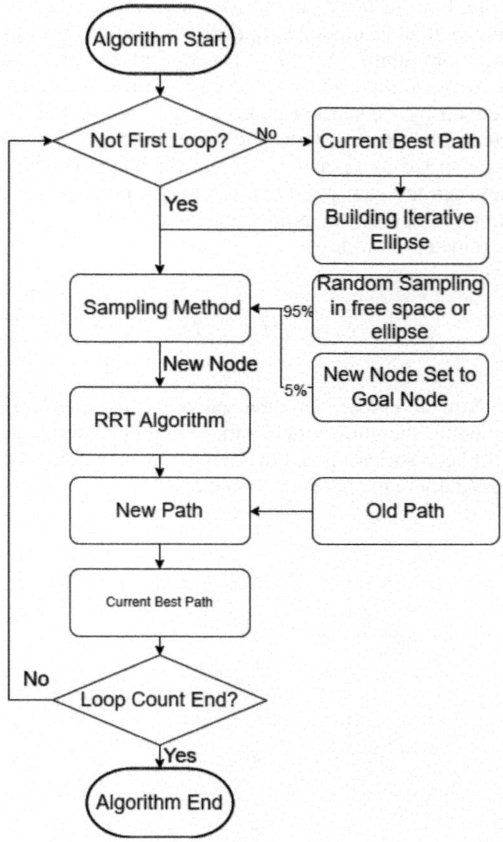

Figure 2. Flowchart of global path planning based on RRT algorithm.

A waypoint optimization method should be applied to optimize the initial trajectory, which is shown in Figure 3. Finally, the path is optimized by using an iterative ellipse sampling space with the RRT algorithm. The sampling space is been used for both global and local path planning.

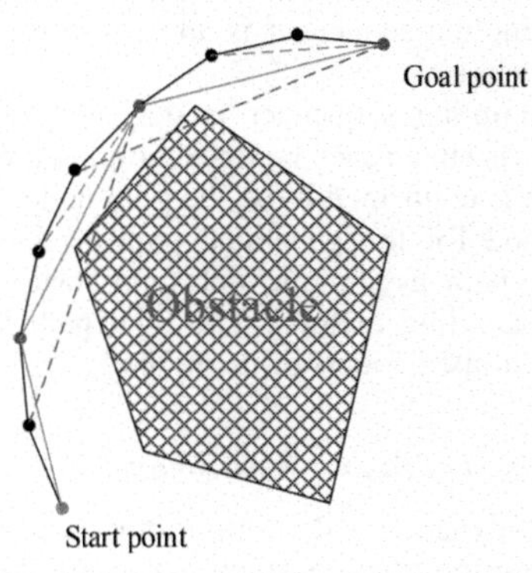

Figure 3. Diagram of waypoints optimization.

Another aspect is that the path generated by the traditional RRT algorithm is non-asymptotic optimality because of free-space random sampling. Therefore, in the dynamic RRT algorithm, an iterative method is applied to optimise the path to asymptotic optimum. The core concept of the method revolves around an iterative ellipse sampling space defined by the starting point, the goal point, and the current best path length. The current best path length is recorded as L_p and L_{se} is the Euclidean distance between the start point and the goal point. Then the minor axis of the ellipse L_d is defined by both parameters as follows:

$$L_d = \sqrt{{L_p}^2 - {L_{se}}^2} \tag{1}$$

Now an ellipse sampling space is constructed. By updating the current best path length L_p, the ellipse will keep shrinking and make the new path converge gradually to the optimum trajectory.

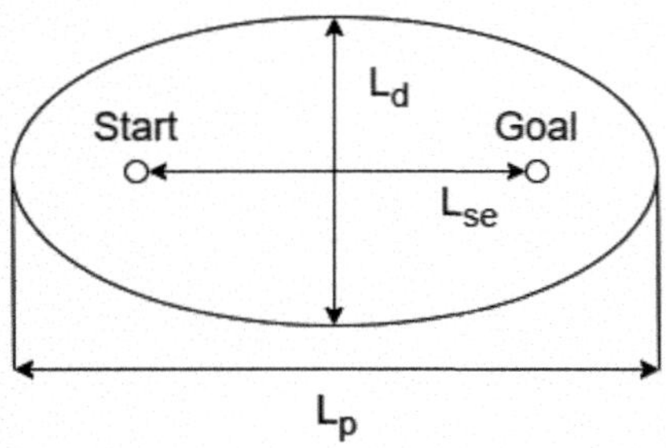

Figure 4. Diagram of ellipse sampling space construction.

3.2 *Kino-dynamic RRT* algorithm*

In this section, the Kino-dynamic RRT* algorithm is proposed for local path planning, and the flow chart is shown in Figure 5. Firstly, a four-dimension linear model is applied to calculate the path from node to node. Since the Kino-dynamic RRT* is connecting nodes by equations and the step length can be uncertain, a sampling sequencer is applied in the random sampling phrase.

The sampling sequencer is designed to set the new node as state space random sampling. When the new node is confirmed to be the goal node, an iterative method for the generated path is activated to make the path asymptotic optimum instead of the RRT* resampling method. The final path is generated with limited counts of iterations.

3.3 *Solution of Kino-dynamic model*

The Kino-dynamic RRT* algorithm is a modified RRT* algorithm that considers the controllability of the planned target. First, let $\chi \in R^n$ and $U \in R^m$, be

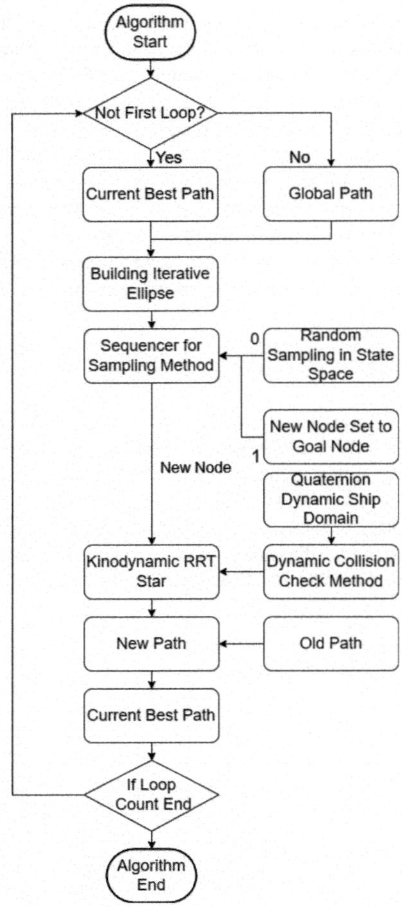

Figure 5. Flow chart of Kino-dynamic RRT* algorithm.

the state space and control input space. Then define a linear system which requires to be formally controllable as follows:

$$\dot{x}[t] = Ax[t] + Bu[t] + c \tag{2}$$

where $x[t] \in \chi$ is the target state and $u[t] \in U$ is the target control input. The rest variables, $A \in R^{n\times n}$, $B \in R^n$, $c \in R^n$, are constant and available.

Use tuple $\pi = (x[], u[], \tau)$ to define the target path, where τ is the arrival time or duration of the path, $u : [0, \tau] \rightarrow U$ defines the control input and $x : [0, \tau] \rightarrow \chi$ defines the states along the path. The path cost function $c[\pi]$ could be defined as follows:

$$c[\pi] = \int_0^\tau \left(1 + u[t]^T Ru[t]\right) dt \tag{3}$$

where $R \in R^m$ is positive-definite, constant and given parameter, which weights the cost of the control inputs relative to each other and to the duration of the trajectory. Let $G[t]$ be weighted controllability Gramian given by:

$$G[t] = \int_{0}^{t} \exp[A(t-t')]BR^{-1}B^{T}\exp\left[A^{T}(t-t')\right]dt' \tag{4}$$

Further, let $\bar{x}[t]$ describe what the state x would be at t if no control input were applied:

$$\bar{x}[t] = \exp[At]x_0 + \int_{0}^{t} \exp[A(t-t')]c\, dt' \tag{5}$$

Then, the optimum control policy for the fixed final state, fixed final time optimal control problem is given by:

$$u[t] = R^{-1}B^{T}\exp[A^{T}(\tau-t)]G[\tau]^{-1}(x_1-\bar{x}[\tau]) \tag{6}$$

So, the cost function $c[\pi]$ can be formulated as:

$$c[\tau] = \tau + (x_1-\bar{x}[\tau])^{T}G[\tau]^{-1}(x_1-\bar{x}[\tau]) \tag{7}$$

The optimal arrival time τ^* can be calculated by taking the derivative of $c[\tau]$ with respect to τ and solving $\dot{c}[\tau]=0$ for τ. The derivative is given by:

$$\dot{c}[\tau] = 1 - 2(Ax_1+c)^{T}d[\tau] - d[\tau]^{T}BR^{-1}B^{T}d[\tau] \tag{8}$$

where $d[\tau]$ is defined as:

$$d[\tau] = G[\tau]^{-1}(x_1-\bar{x}[\tau]) \tag{9}$$

Now the optimal arrival time τ^* has been given. According to the optimal control policy, $x[t]$ has solution with $u[t]$ for all $0<t<\tau^*$ which completely determines $\pi^*[x_0,x_1]$:

$$\begin{bmatrix} x[t] \\ y[t] \end{bmatrix} = exp\left[\begin{bmatrix} A & BR^{-1}B^{T} \\ 0 & -A^{T} \end{bmatrix}(t-\tau^*)\right]\begin{bmatrix} x_1 \\ d[\tau^*] \end{bmatrix} + \int_{\tau^*}^{t} exp\left[\begin{bmatrix} A & BR^{-1}B^{T} \\ 0 & -A^{T} \end{bmatrix}(t-t')\right]\begin{bmatrix} c \\ 0 \end{bmatrix}dt' \tag{10}$$

where $y[t]$ is defined as:

$$y[t] = \exp[A^{T}(\tau^*-t)]d[\tau^*] \tag{11}$$

3.4 *Dynamical collision risk detection based on ship domain*

In this study, the Quaternion Ship Domain (QSD) model is chosen to construct a dynamic collision detection method. The QSD model is built for detecting collision risk with other vessels and it brings the COLREGs consideration into its definition. In the dynamic collision detection method, a collision is deemed to have occurred if a vessel enters the ship domain area of other vessels. The QSD model is shown in Figure 6, and the parameters are as follows:

$$\begin{cases} R_{fore} = \left(1+1.34\sqrt{k_{AD}{}^2+\left(\frac{k_{DT}}{2}\right)^2}\right)L \\ R_{aft} = \left(1+0.67\sqrt{k_{AD}{}^2+\left(\frac{k_{DT}}{2}\right)^2}\right)L \\ R_{starb} = (0.2+k_{AD})L \\ R_{port} = (0.2+0.75k_{AD})L \end{cases} \tag{12}$$

The parameters of the ellipse are related to vessel dynamic parameters. k_{AD} and k_{DT} represent gains of the advance and tactical diameter. Both of them are defined by the sailing speed of vessels. k_{AD} and k_{DT} can be calculated as follows:

$$\begin{cases} k_{AD} = 10^{0.3591\,\lg(v)+0.0952} \\ k_{DT} = 10^{0.5441\,\lg(v)-0.0795} \end{cases} \tag{13}$$

The dynamic collision detection model is built based on the QSD model. As depicted in Figure 7, an additional parameter 't' is incorporated into the tree vertices, representing the time taken to reach the corresponding vertices. Considering the new sampling node as $V_{n+1} = \{x_{n+1}, y_{n+1}, t_{n+1}\}$, the navigation time from V_n to V_{n+1} can be broken down into several segments, each viewed as successive moments. This allows for the prediction of positions, speeds, and courses of both OS and TS and consequently facilitates collision risk detection.

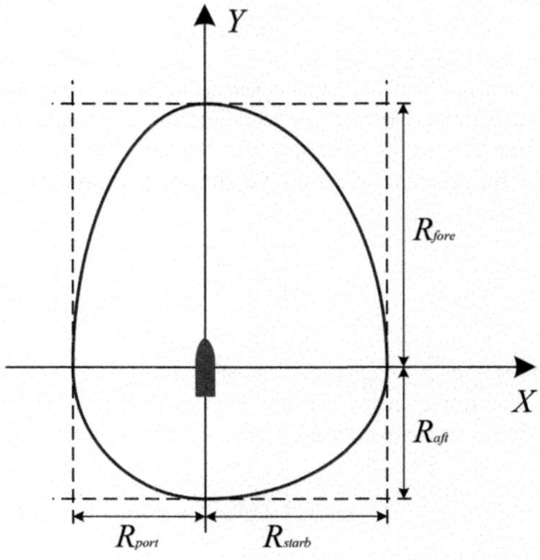

Figure 6. The diagram of the Quarternion ship domain (QSD) model.

Should any domain violation appear (i.e., the interval node V_n' lands within the TS's domain), it is assumed that the path from V_n to V_{n+1} is obstructed. If there is no domain violation, the new trajectory is deemed to be free of collision and can be added to the tree. It is important to note that even though it presumes TS maintaining its course and speed, the model remains applicable as long as TS's motion state can be anticipated.

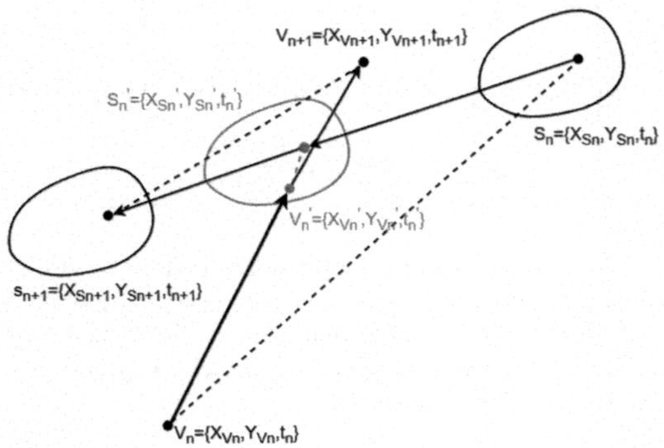

Figure 7. The illustrative demonstration of dynamic collision risk detection.

The dynamic collision risk detection method makes the algorithm generate a path to avoid moving ships in the planning map.

3.5 *Local path planning procedure*

3.5.1 *Four-dimension model linear system*

In this part, a four-dimension controllable linear system is introduced into the algorithm. First, the four-dimension state space $x[t] \in R^{4\times1}$ is defined as follows:

$$x[t] = [Px, Py, Vx, Vy]^{\mathrm{T}} \tag{14}$$

where Px and Py represent the vessel position x and y, Vx and Vy represent the vessel velocities in x and y directions. According to equation (2), once the parameters of the state space is defined, then the control input space $u[t]$ can be uniquely identified as follow~s:

$$u[t] = [ax, ay]^{\mathrm{T}} \tag{15}$$

where ax, ay represent accelerations in x and y directions, respectively. Then the whole linear system can be expressed by the following equation:

$$\dot{x}[t] = Ax[t] + Bu[t] + c \tag{16}$$

where $\dot{x}[t]$ stands for the derivative of $x[t]$. It can be conveyed that the matrixes A, B and c is fixed as:

$$A = \begin{bmatrix} 0 & 0 & 1 & 0 \\ 0 & 0 & 0 & 1 \\ 0 & 0 & 0 & 0 \\ 0 & 0 & 0 & 0 \end{bmatrix}, \quad B = \begin{bmatrix} 0 & 0 \\ 0 & 0 \\ 1 & 0 \\ 0 & 1 \end{bmatrix}, \quad c = \begin{bmatrix} 0 & 0 \\ 0 & 0 \\ 0 & 0 \\ 0 & 0 \end{bmatrix} \tag{17}$$

It is easy to verify the system is controllable. If $A \in R^{n\times n}$ is nilpotent, which is defined as $A^n = 0$. Therefore, $\exp[At]$ has a closed-form expression in the form of an $(n-1)$-degree matrix polynomial in t. In summary, if matrix $A \in R^{n\times n}$ is nilpotent, the system is controllable and can be solved. Now a four-dimention linear system model has been built, and the algorithm will input start state and goal state. By using the solution method of Kino-dynamic model and get a solution which can describe a path from start point to goal point.

3.5.2 *Sequencer for new node generation*

In Kino-dynamic RRT* algorithm, the expansion is based on the equation of two state nodes, and the classical Euclidean distance step length expansion method is not suitable, which can use the time step instead of the distance step (Tyler et al., 2018). In this case, the concept of expand step is removed and connects new sample nodes and parent nodes through functions.

The sequencer is designed to choose state space random sampling and the goal node as a new node. If a random sampling node is not added to the tree, the sequencer will keep random sampling until a new node is added. When a target node has been the new one, it only tries to connect to the last node. Its advantages are reducing the complexity of calculation and avoiding the missing distance step expansion function.

3.5.3 *Sampling space selection considering COLREG restrictions*

The COLREG is a crucial consideration. If two vessels encounter collision risk, according to COLREGs the give-way vessel must initiate actions to prevent a collision. Rules 13 to 15 of COLREGs are concerned with overtaking, head-on situations, and cross paths. The encounter is defined by both vessels' relative bearing and courses. Concurrently, it is important to stipulate the distance at which preventive measures against collision should be employed in every situation. Customarily, sailors start their collision evasion manoeuvres at a span of 6 nautical miles. A schematic diagram outlining the direction of ship collision avoidance for each case can be seen in Figure 7.

According to Figure 8, the give-way ship can choose to navigate towards starboard or port when an overtaking situation arises. However, it is suggested that the give-way ship moves towards starboard during head-on and crossing situations. Thus, depending on each scenario, the planned path should align with COLREGs by confining the sampling space for positions x and y. In the case of overtaking, the sampling space remains unrestricted, whereas in head-on and crossing situations, the sampling space is confined to the half ellipse.

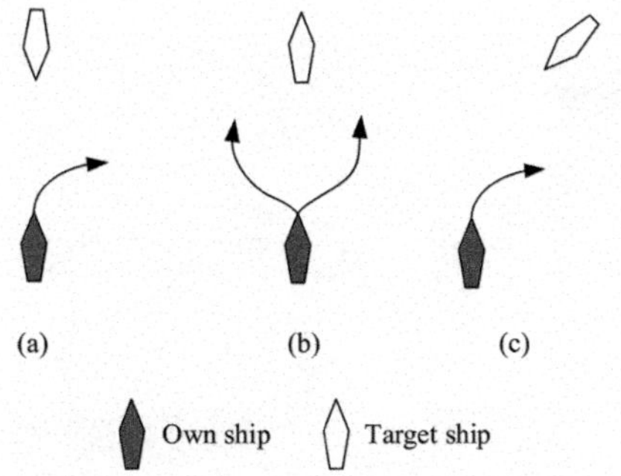

Figure 8. Ship collision avoidance direction in each situation. (a) Head-on. (b) Overtaking. (c) Crossing.

Table 1. The pseudo-code of local path planning.

Algorithm 1: Kino-dynamic RRT Based Local Planning
Input: $\mathbf{P}_{start}$, $\mathbf{P}_{goal}$, $\mathbf{O}_{sta}$, $\mathbf{O}_{dyn}$, P_{global}, $\mathbf{L}_{statespace}$

Output: *Tree* τ
$\tau \leftarrow \{\mathbf{P}_{start}\}$;
$\mathbf{E}_{halfspace} \leftarrow$ **EllipseBuilding**$(\mathbf{P}_{start}, \mathbf{P}_{goal}, P_{global})$;
$\mathbf{P}_{new} \leftarrow$ **Sequence**(0 : **EllipseSampling** $(\mathbf{E}_{halfspace}, \mathbf{L}_{statespace})$, 1 : $\mathbf{P}_{goal}$);
$\mathbf{P}_{nearest} \leftarrow$ **GetNearestPoint**(τ);
$x[t] \leftarrow$ **KinodynamicCalculation**$(\mathbf{P}_{new}, \mathbf{P}_{nearest})$
if CoChek$(x[t], \mathbf{O}_{sta}, \mathbf{O}_{dyn})$ **then**
SetParent$(\mathbf{P}_{nearest}, \mathbf{P}_{new})$;
$\tau \cup \{\mathbf{P}_{new}\}$;
if $\mathbf{P}_{new} = \mathbf{P}_{goal}$ **then**
Continue;
for $i = 1, 2, \ldots, n$ **do**
$\tau_{new} \leftarrow$ **PathOptimizing** $(\mathbf{P}_{start}, \mathbf{P}_{goal}, \mathbf{O}_{sta}, \mathbf{O}_{dyn}, \mathbf{L}_{best})$;
$\mathbf{L}_{new} \leftarrow$ **GetPathLength**(τ_{new});
if $\mathbf{L}_{new} < \mathbf{L}_{best}$ **then**
$\mathbf{L}_{best} \leftarrow \mathbf{L}_{new}$;
$\tau_{best} \leftarrow \tau_{new}$;
return τ_{best};

4 RESULTS

In this section, three different ship encounter situations, including overtaking, crossing and head-on, are formulated to verify the effectiveness of the algorithm. The global path is generated by the first stage of the dynamic RRT* algorithm, and the local path planning is processed by the second phase of the dynamic RRT* algorithm.

4.1 *Initiative of local path planning*

To further validate the effectiveness of local path planning in ship encounter scenarios based on the global path planning results. The other dynamic ships are placed on the map and forming overtaking, crossing, and head-on situations with the OS respectively. The parameters related to ship encounters are detailed in Table 2.

The own ship start point is (0, 0) and the goal point is (20, 20). From start to goal, three encounters will happen between the own ship and target ships. So, there will be three local paths to re-plan.

Table 2. Parameters and encounter situations between OS and TSs.

Own Ship (OS)		Target ship (TSs)		
Position/nm	Heading/°	Position/nm	Heading/°	Situation
(0.0, 0.0)	52.4	(1.43, 1.10)	52.4	Overtaking
(6.46, 4.98)	48.0	(14.45, 4.56)	318.0	Crossing
(16.05, 13.60)	31.7	(20.0, 20.0)	211.7	Head-on

4.2 *Overtaking scenario*

In an overtaking scenario, the OS, acting as the give-way ship, overtakes the TS. The initial gap between them is 1.80 nautical miles, less than the standard distance for collision avoidance in such a situation. Figure 8 illustrates the planned paths of the vessels on the navigable map, where the magenta colour indicates the OS black denotes the TS (the overtaken vessel), and identical colours indicate synchronized ship positions. The dashed domain around each ship corresponds to its nearest point. The dotted line shows the optimized non-optimal paths including all figures follow.

As shown in Figure 9, the own ship steers to starboard to keep clearance of the target ship. According to the COLREGs, both port and starboard turns are compliant in an overtaking situation. However, at the initial moment, the relative bearing of the target ship to the own ship is zero, and in the QSD model, the starboard side is larger than the port side. Hence, a turn to starboard allows the ship to

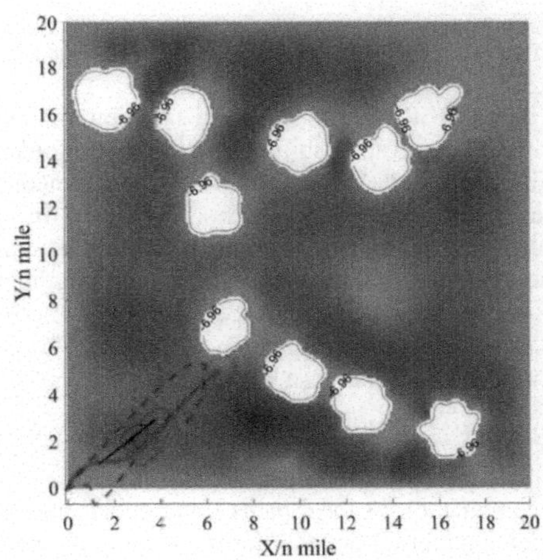

Figure 9. The result of local path planning for the overtaking scenario.

safely bypass the TS with a lower collision avoidance magnitude.

4.3 *Crossing scenario*

In this simulation, a crossing scenario is established between the own ship and the target ship. It is important to note that the own ship is on the port side of the target ship, acting as the give-way ship while the TS is the stand-on vessel. The outcome of local path planning can be viewed in Figure 10.

The lines and colours in Figure 10 share the same interpretation as those in Figure 9. As revealed in Figure 8, the own ship steers starboard to avoid the target ship, adhering to the stipulations of COLREGs for a crossing situation. At their closest proximity, the ships did not intrude into each other's domains, which validates that the algorithm is capable of identifying a safe navigational route on the map.

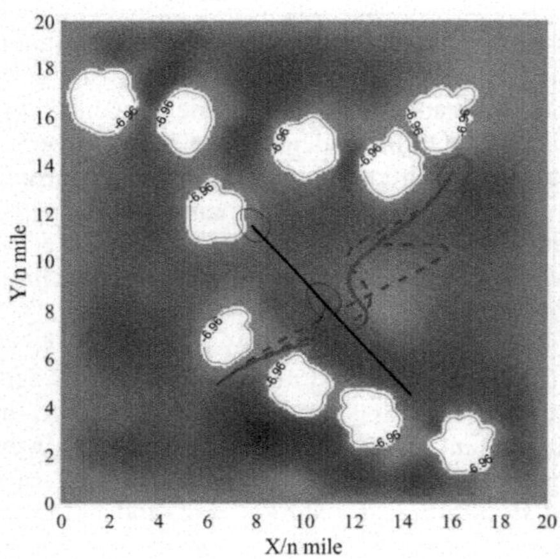

Figure 10. The result of local path planning for the crossing scenario.

4.4 *Head-On scenario*

Based on COLREGs, in a head-on situation, both vessels share the responsibility of collision avoidance and are expected to turn starboard simultaneously, passing each other's port sides. However, in this simulation, obstacles are present on the starboard side of the target ship's intended path, making coordinated manoeuvres impossible. Consequently, only the own ship turns starboard to maintain clearance. The findings are depicted in Figure 11.

4.5 *Whole scene*

The combination of local paths (blue dotted line) and the global path (red dotted line) is shown in Figure 12. It's shows that the local path makes own vessel avoid collision and along the global path afterwards.

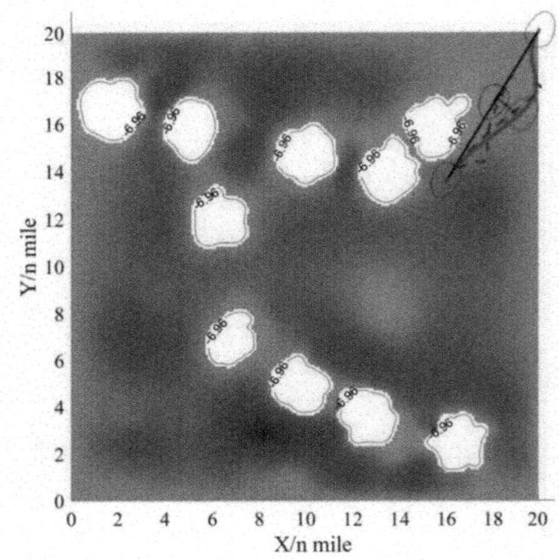

Figure 11. The result of local path planning for the head-on scenario.

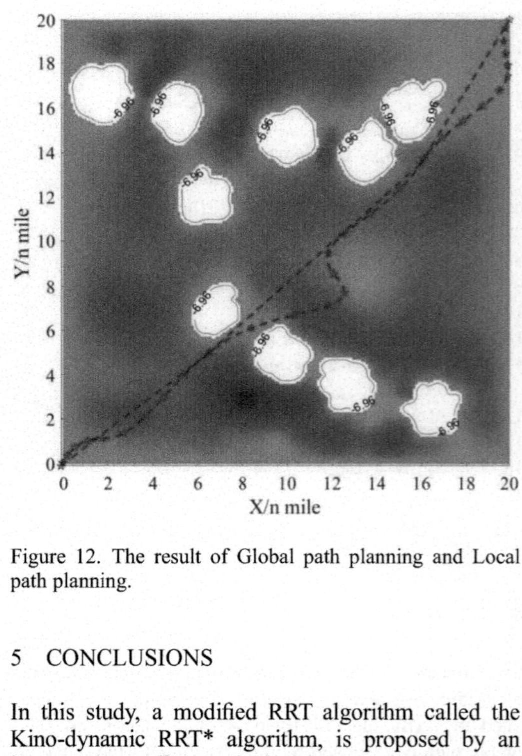

Figure 12. The result of Global path planning and Local path planning.

5 CONCLUSIONS

In this study, a modified RRT algorithm called the Kino-dynamic RRT* algorithm, is proposed by an improved sampling method and tree expanding to the origin methods. The algorithm considers a simplification linear controllable system to connect two state spaces and generate a smooth curve. The results show that the proposed path planning method can find a satisfactory path with fewer iterations than the origin two-stage dynamic RRT* algorithm, and can also keep clear from static obstacles and dynamic ships with a non-interpolated smoothing curve. The research can help make planning routes before sailing and decision-making during ship encounters.

In this study, certain simplifications have been made. Future studies can explore these aspects in more detail. For example, in the local path planning, the kino-dynamic model is simplified as a four-dimensional model, and a suitable ship kinetic and dynamic model should be involved in future research.

ACKNOWLEDGEMENTS

The research of this paper was supported by National Key Technologies Research and Development Program (2023YFB4302303), National Natural Science Foundation of China (52071247; 51920105014), Key Technologies Research and Development Program of Guangxi Zhuang Autonomous Region (2022AB 04010) and Innovation and entrepreneurship team import project of Shaoguan city (201208176230693).

This work contributes to the Strategic Research Plan of the Centre for Marine Technology and Ocean Engineering, which is financed by the Portuguese Foundation for Science and Technology (Fundação para a Ciência e Tecnologia - FCT) under contract UIDB/UIDP/00134/2020.

REFERENCES

Chiang, H. & Tapia, L. 2018. COLREG-RRT: an RRT-based COLREGS-compliant motion planner for surface vehicle navigation. *IEEE Robotics and Automation Letters*. 3(3): 2024–2031.

Dong, Z.R. & Bian, X.Y. & Zhao, S. 2022. Ship pipe route design using improved multi-objective ant colony optimization. *Ocean Engineering*. 258, 111789.

Dolgov, D. & Thrun, S. & Montemerlo, M. & Diebel, J. 2010. Path Planning for Autonomous Vehicles in Unknown Semi-structured Environments. *The International Journal of Robotics Research*. 29(5):485–501

Dustin, J.W. & Jur, V.D.B. 2012. Kinodynamic RRT*: Optimal Motion Planning for Systems with Linear Differential Constraints. *Computer Science*. arXiv:1205.5088

Enevoldsen, T. & Reinartz, C. & Galeazzi, R. 2021. COLREGs-Informed RRT* for Collision Avoidance of Marine Crafts. *In Proceedings of the 2021 IEEE International Conference on Robotics and Automation (ICRA)*. pp. 8083–8089.

Enevoldsen, T. & Galeazzi, R. 2021. Grounding-aware RRT* for Path Planning and Safe Navigation of Marine Crafts in Confined Waters. *IFAC-PapersOnLine*. 54, 195–201.

Fiskin, R. & Atik, O. & Kisi, H. & Nasibov, E. & Johansen, T.A. 2021. Fuzzy domain and meta-heuristic algorithm-based collision avoidance control for ships: Experimental validation in virtual and real environment. *Ocean Engineering*. 220.

Gu, Q.Y. & Zhen, R. & Liu, J.L. & Li, C. 2023. An improved RRT algorithm based on prior AIS information and DP compression for ship path planning. *Ocean Engineering*. 279(114595).

Güven, A.F. & Yörükeren, N. & Samy, M.M. 2022. Design optimization of a stand-alone green energy system of university campus based on Jaya-Harmony Search and Ant Colony Optimization algorithms approaches. *Energy*. 253, 124089.

Guedes Soares, C. & Teixeira, A.P. 2001. Risk assessment in maritime transportation. *Reliability Engineering and System Safety*. 74, 299–309.

Hart, P.E. & Nilsson, N.J. & Raphael, B.A. 1973. Formal Basis for the Heuristic Determination of Minimum Cost Paths. *IEEE Transactions on Systems, Man, and Cybernetics: Systems*. 4, 28–29.

Huang, D.S. & Bevilacqua, V. & Premaratne, P. & Gupta, P. 2018. Unmanned Ship Path Planning Based on RRT. *14th International Conference on Intelligent Computing*. 10594.

Jonathan, D.G. & Siddhartha, S.S. & Timothy D.B. 2014. Informed RRT*: Optimal Sampling-based Path Planning Focused via Direct Sampling of an Admissible Ellipsoidal Heuristic. *International Conference on Intelligent Robots and Systems*. arXiv:1404.2334

Johnson, D.B. 1973. A note on Dijkstra's shortest path algorithm. *Journal of the ACM*. 20, 385–388.

Kang, J.G. & Lim, D.W. & Choi, Y.S. & Jang, W.J., Jung, J.W. 2021. Improved RRT-Connect Algorithm Based on Triangular Inequality for Robot Path Planning. *Sensors*. 21(2).

Karaman, S. & Frazzoli, E. 2011. Sampling-based algorithms for optimal motion planning. *The International Journal of Robotics Research*. 30(7). 846–894.

Li, Y.J. & Wei, W. & Gao, Y. & Wang, D.L. & Fan, Z. 2020. PQ-RRT*: An improved path planning algorithm for mobile robots. *Expert Systems with Applications*. 152: 113425.

Lyu, H.G. & Yin, Y. 2018. COLREGS-Constrained Real-time Path Planning for Autonomous Ships Using Modified Artificial Potential Fields. *The Journal of Navigation*. 72(3).

Montemerlo, M. & Diebel, J. & Dolgov, D. & Thrun, S. 2008. Practical Search Techniques in Path Planning for Autonomous Driving. *Computer Science*.

Silveira, P. A. M.; Teixeira, A. P., & Guedes Soares, C. 2022. A method to extract the Quaternion Ship Domain parameters from AIS data. *Ocean Engineering*. 257:111568.

Sung, I. & Choi, B. & Nielsen, P. 2021. On the training of a neural network for online path planning with offline path planning algorithms. *International Journal of Information Management*. 57, 102142.

Sun, Q.P. & Li, M. & Wang, T.H. & Zhao, C.P. 2018. UAV path planning based on Improved Rapidly -exploring Random Tree. *Proceedings of the 30th Chinese Control and Decision*. 6420–6424

Stentz, A. 1997. Optimal and efficient path planning for partially known environments. *In Intelligent Unmanned Ground Vehicles*. pp. 203–220.

Tyler, S.2018. Distributionally Robust Sampling-Based Motion Planning Under Uncertainty. *International Conference on Intelligent Robots and Systems*. pp. 6518–6523.

Wang, H.J. & Guo, F. & Yao, HF. & He, S.S. & Xu, X. 2019. Collision Avoidance Planning Method of USV Based on Improved Ant Colony Optimization Algorithm. *IEEE Access*. 7, 52964–52975.

Yuan, J.F. & Wang, J. & Zhang, X., Xu, Y. & Zeng, Y. & Ren, Y.J. 2022. A second-order dynamic and static ship path planning model based on reinforcement learning and heuristic search algorithms. *EURASIP Journal on Wireless Communications and Networking*. 128.

Zhou, Y. & Li, W.D. & Wang, X.M. & Qiu, Y.M. & Shen, W. M. 2022. Adaptive gradient descent enabled ant colony optimization for routing problems. *Swarm and Evolutionary Computation*. 70, 101046.

Advances in Maritime Technology and Engineering – Guedes Soares & Santos (Eds)

Identifying collision avoidance behaviour in AIS data from a heavy traffic area

L. Zhang
School of Navigation, Wuhan University of Technology, Wuhan, P.R. China
Centre for Marine Technology and Ocean Engineering (CENTEC), Instituto Superior Técnico, Universidade de Lisboa, Lisboa, Portugal

P.F. Chen, Y. Luo & J.M. Mou
School of Navigation, Wuhan University of Technology, Wuhan, P.R. China

C. Guedes Soares
Centre for Marine Technology and Ocean Engineering (CENTEC), Instituto Superior Técnico, Universidade de Lisboa, Lisboa, Portugal

ABSTRACT: This paper introduces a method for identifying collision avoidance behaviour using Automatic Identification System data from a heavy traffic area. The method encompasses two primary stages: (1) The method clusters the ship trajectory using the Ordering Points to Identify the Clustering Structure algorithm and then extracts the route boundary by the Kernel Density Estimation method. (2) It extracts trajectory pairs involving collision risks by the Velocity Obstacle method, followed by the extraction of avoidance behaviour based on heading changes. To demonstrate the effectiveness of the proposed method, the method is applied to a heavy traffic area in Shanghai, China, where collision avoidance behaviours were extracted from one week's worth of data.

1 INTRODUCTION

With the vigorous development of the shipping industry in recent years, the number of maritime traffic accidents continues high, among which ship collision accidents have always been one of the most important types of shipping accidents (Guedes Soares and Teixeira, 2001; Zhang et al., 2021). Furthermore, ships are gradually developing on a large-scale, and the harm and loss caused by ship collision accidents are increasing. Therefore, the avoidance of ship collisions has always been an important topic in shipping safety research.

In recent years, ship trajectory data have become more accessible because of the convenience of data acquisition and storage with the widespread adoption of Automatic Identification System (AIS) equipment. The AIS data contains dynamic and static information of ship navigation, which can intuitively show the process of ship navigation, and plays an important role in enhancing the situation awareness of ship navigation. Based on AIS data, a lot of research on ship collision avoidance has been done, among them, the extraction of ship collision avoidance behaviour can guide the formulation of collision avoidance strategies, which is of great significance to the safety of ship navigation.

However, research on extracting avoidance behaviour is limited, with most studies extracting near-misses from AIS data by establishing of the collision risk model. Chen et al., (2018) proposes a Time Discrete Non-linear Velocity Obstacle (TD-NLVO) method for detecting near-misses. (Zhang et al., 2015) developed an innovative approach to identify near-misses using AIS data, incorporating factors such as minimum distance, relative velocity, and headings. In a subsequent study, Zhang et al., (2016) refined this model by integrating the principle of minimum distance to collision, thereby improving the precision and dependability in detecting near-misses. Yoo, (2018) proposed a near-miss identification and density description method, and verified the validity of the model by describing the distribution of actual collision accident locations off the south coast of South Korea. Du et al. (2021) considered a variety of factors to link ship behaviour with collision risk to detect near miss, and applied it to the detection of near miss in the North Baltic Sea to identify highly dangerous waters.

Due to the subjectivity in defining collision risk, different collision risk methods may extract varying near-miss databases from the same AIS data. It means that a near-miss identified by one method might not be classified as such by another. Data-driven approaches are also highly sensitive to the accuracy of the dataset,

DOI: 10.1201/9781003508762-22

which could affect their reliability and accuracy. When seamen perceive that their ship is in a near-miss situation, they will manoeuvre the ship based on the current encounter scenario and their navigational experience to avoid a collision. Therefore, the ship's avoidance behaviour can be considered a relatively objective indicator for identifying near-miss.

Mestl et al., (2016) developed a novel approach for identifying the avoidance behaviour utilizing high-resolution AIS data and emphasizing the rate of turn. Rong et al., (2022) proposed an innovative method for the automatic detection of ship collision avoidance behaviour using AIS data. This method, which is grounded in the principles of COLREG, employs an enhanced sliding window algorithm, and considers the rate of turn.

Research on extracting ship avoidance behaviour is currently limited. To address this gap, this paper proposes a novel method for extracting ship avoidance behaviour from AIS data, which also contributes to the development of a high quality near-miss database.

2 METHODOLOGY

In this research, a novel collision avoidance behaviour extracting method is proposed. This section briefly introduces the main steps of the method, whose framework is illustrated in Figure 1. The first step involves clustering the ship trajectory using the Ordering Points to Identify the Clustering Structure (OPTICS) algorithm and extracting the route boundary with the Kernel Density Estimation (KDE) method. The objective is to identify the navigational characteristics of the selected area. The second step employs the Velocity Obstacle (VO) method to extract trajectory pairs involving collision risk, followed by the extraction of avoidance behaviour based on heading changes. It also involves analyzing the spatial distribution of the extracted avoidance behaviors and their correlation with navigational characteristics. In Section 2.1 and 2.2, the detailed strategies of the methods used in the identification of the navigation trajectory and the extraction of the collision avoidance behaviour are described, respectively.

2.1 *Navigation route recognition*

2.1.1 *Trajectory clustering*

Grouping ship trajectories with similar behaviour characteristics into the same cluster facilitates the extraction of ship behaviour patterns in specific areas. In this paper, the OPTICS clustering algorithm is used for trajectory clustering. It can cluster irregular samples and is insensitive to noise data, effectively solving the problem that the initial AIS data is messy and contains a lot of noise. It is widely utilized in ship trajectory clustering research.

The OPTICS algorithm does not explicitly generate clustering results. Instead, it performs a ranking of objects within the dataset and generates a reachability distance distribution plot known as the reachability plot. The reachability plot provides insights into the density-based structure of the data, allowing for more in-formed decisions in selecting suitable parameters for the clustering. As shown in Figure 2, by selecting different threshold values of reachability distance, the samples can be divided into different clusters.

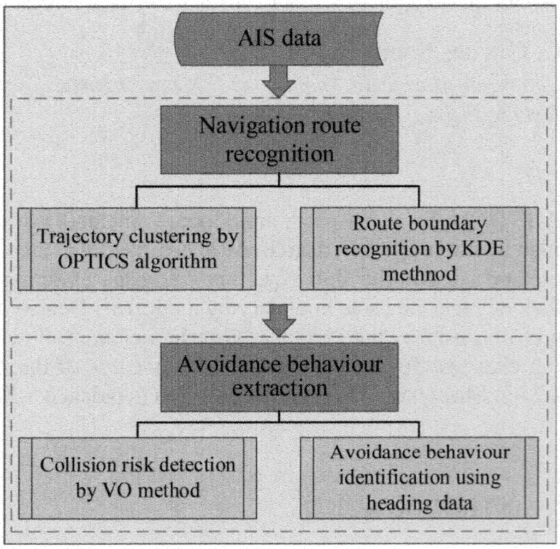

Figure 1. The overall flowchart of the research.

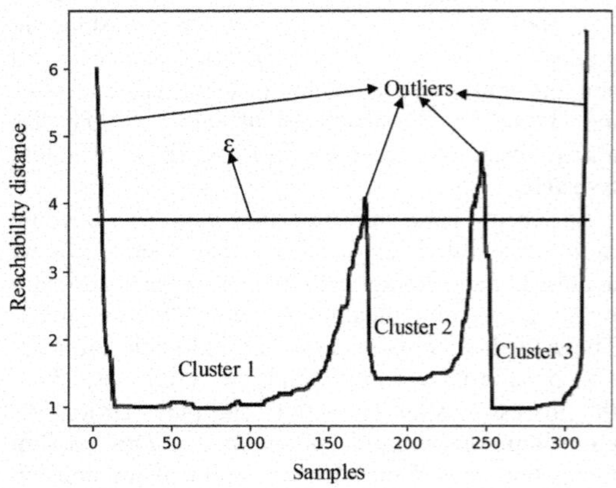

Figure 2. The diagram of the reachability plot.

The core objects MinPts and reachability distance ε are parameters of the OPTICS algorithm. As shown in Figure3, MinPts means the number of samples inside the circle with the radius of ε.

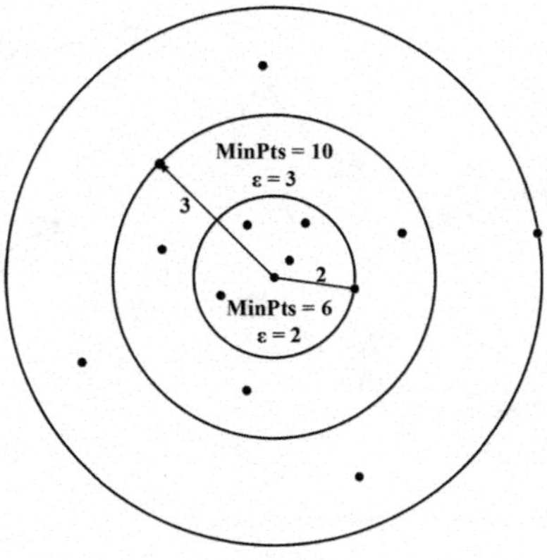

Figure 3. The diagram of the OPTICS algorithm.

The core idea of OPTICS is to determine the minimum radius ε, named the reachability distance, that can include a given number of samples MinPts. The expression of the reachability distance for a given MinPts can be described as follows:

$$\begin{cases} N_\varepsilon(p) = \{p_i \in D | dis(p_i, p_i) \le \varepsilon\} \\ |N_\varepsilon(p)| = MinPts \end{cases} \quad (1)$$

where p_i means one of the samples D; $|N_\varepsilon(p)|$ means the number of p that satisfy the upper equation; ε is the reachability distance of sample p when the core objects is *MinPts*.

In this part, the Hausdorff distance (Sutulo & Guedes Soares, 2014) is used to measure the similarity of trajectories, that is, the distance between two sets. The equation is as follows:

$$dhau(T_a, T_b) = max\begin{Bmatrix} \sup\limits_{a\in M}\left[\inf\limits_{b\in N}(dis(ab))\right], \\ \sup\limits_{b\in N}\left[\inf\limits_{a\in M}(dis(ba))\right] \end{Bmatrix} \quad (2)$$

where $\sup\limits_{a\in M}\left[\inf\limits_{b\in N}(dis(ab))\right]$ is one-way distance set from trajectory $T_a$to trajectoryT_b, $\sup\limits_{b\in N}\left[\inf\limits_{a\in M}(dis(ba))\right]$ is one-way distance set from trajectory T_b to T_a, $dis(ab)$ and$dis(ba)$ are the Euclidean distance paradigms between record positions in trajectory T_a and T_b.

2.1.2 *Recognition of route boundary*

Generally, the buoys are arranged along the channel to mark the boundaries, and to guide ships to pass safely. However, due to the limitations of various conditions (e.g. wind, current water depth. etc.), the buoys in many waterways are not fully equipped and the position changes with the change of the flow. Therefore, some great challenges and uncertainty rely on buoys to identify the channel boundary. Recently, the rapid development of AIS has improved the convenience of trajectory data acquisition. The AIS data recorded the re-al trajectory of ships passing through the waterways. To sum up, it is a more accurate and reliable way to identify the navigation route boundary through historical AIS data.

In many studies, the identification of channel boundary is carried out by analysing the distribution of buoys. However, due to various restrictions in the actual process of buoy layout, many buoys in navigation waterways are not fully equipped, and the position changes with the change of flow, which brings great challenges and uncertainties to the identification of channel boundaries. In recent years, the rapid development of AIS has provided a new strategy for the method of channel identification. By analysing the historical AIS data of ships sailing in the waters, the channel boundary can be extracted accurately, reliably, and efficiently.

Since the AIS data only contains the relevant information of ship navigation, it is difficult to directly extract the channel boundary from the preprocessed AIS data, and the ship trajectory density can well reflect the position and boundary of the channel. In this paper, the trajectory density map is obtained by meshing the trajectory of the ship 's navigation area to extract the channel boundary. Assume that the range of the study water is (x_{min}, y_{min}) to(x_{max}, y_{max}) and is divided into M×N small grids. The index of the grid about each trajectory point can be calculated as follows:

$$\begin{cases} W_m = \frac{M\cdot(x_j - x_{min})}{x_{max} - x_{min}} \\ H_n = N - \frac{N\cdot(y_j - y_{min})}{y_{max} - y_{min}} \end{cases} \quad (3)$$

where W_n and H_n denote the index of the grid where the trajectory point $P = (x, y)$ is located.

After obtaining the trajectory density map, the KDE method is used to extract the channel boundary from the map. The KDE method is a non-parametric method that applies kernel smoothing to probability density estimation, that is, the kernel is used as a weight to estimate the probability density function of random variables. The spatial distribution characteristics can be studied according to the location information of the ship 's navigation trajectory points. The equation of KDE can be expressed as:

$$\hat{f}_h(x) = \frac{1}{n}\sum\nolimits_{i=1}^{n} K_h(x - x_i) = \frac{1}{nh}\sum\nolimits_{i=1}^{n} K\left(\frac{x - x_i}{h}\right) \quad (4)$$

where x_i represents independently distributed trajectory points; h is a smoothing parameter called bandwidth; $K(x)$ is the kernel function, including Uniform, Triangular, Gaussian, and Normal, etc. Gaussian distribution is the most widely used distribution in statistics and many statistical tests, has great influence in many aspects of statistics. Furthermore, there are many random variables obeying or approximately obeying the Gaussian distribution in real life. In this paper, Gaussian kernel is also selected as the kernel function of KDE method. The Gaussian kernel is as follows:

$$K(x) = \frac{1}{\sqrt{2\pi}} e^{\left(-\frac{x^2}{2}\right)} \tag{5}$$

In the trajectory density map, the number of trajectory points in each small grid is the grey value of the grid. As shown in Figure 4, select the column i of small grids, and then use the KDE method to calculate the probability density curve of the grey values for selected grids. The fitted probability density curve has a maximum point p^i_{max}, which represents the centre point of the route corresponding to column i. Subsequently, set a boundary probability threshold p^i_θ to extract the two boundary points p^i_b. After iteratively calculating the probability density curve of each column by the KDE method, two boundary lines and one centre line are identified.

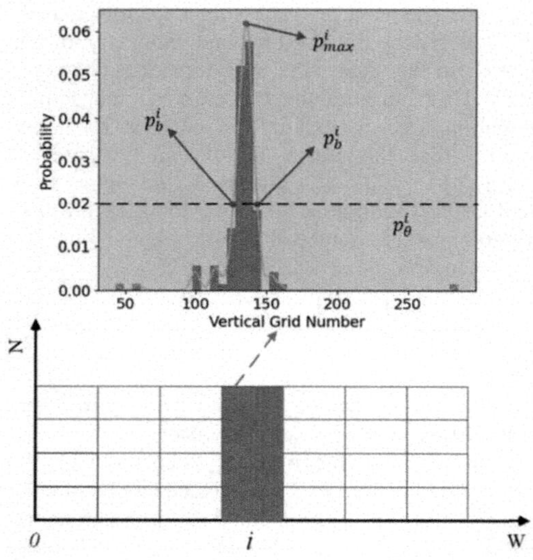

Figure 4. Fitting of the trajectory density map by KDE.

2.2 *The extraction of the avoidance behaviour*

2.2.1 *Risk detection*

In this section, the VO method is used to extract trajectory pairs involving collision risk, by randomly selecting and retaining the time overlap of two ship trajectories, as depicted in Figure 5. Consider one sub-trajectory as the trajectory of the Own Ship (OS), which can be expressed as $traj_{os}^{[t_i,t_j]} = \left\{p_{os}^{t_i}, \ldots, p_{os}^{t_j}\right\}$, where $p_{os}^{t_f}$ represent positions of OS at the time t_f. Consider another sub-trajectory $traj_{ts}^{[t_i,t_j]} = \left\{p_{ts}^{t_i}, \ldots, p_{ts}^{t_j}\right\}$ as the trajectory of the Target Ship (TS).

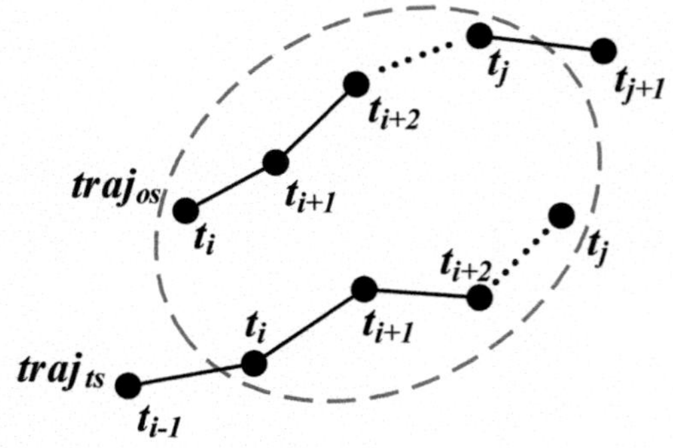

Figure 5. Trajectory pair extraction.

Furthermore, the velocity obstacle method is employed to assess collision risk for pairs of trajectory points$\left(p^t_{os}, p^t_{ts}\right)$ with the same time sequence t. The fundamental equation for the velocity obstacle method is as follows:

$$p_{os}^{t_f} \in p_{ts}^{t_f} \oplus ConfP(O) \tag{6}$$

where $p_{os}^{t_f}$ and $p_{ts}^{t_f}$ represent positions of OS and TS at the time t_f, respectively; $ConfP(O)$ is an abbreviation of "Conflict Position"; O is the origin of $ConfP(O)$.

Assuming that OS and TS sail in a straight line at constant speed $\overrightarrow{V_{os}}$ and $\overrightarrow{V_{ts}}$, Eq. (6) can be substituted with Eq. (7):

$$\overrightarrow{V_{os}} \in \left(\frac{p^t_{ts} - p^t_{os}}{t_f - t_0} + \overrightarrow{V_{ts}}\right) \oplus \frac{ConfP(O)}{t_f - t_0} \tag{7}$$

where p^t_{os} and p^t_{ts} represent the detected point pair at time t.

The diagram illustrating the velocity method is presented in Figure 8, When Eq. (7) is satisfied, this signifies a potential collision risk between OS and TS. As demonstrated in the Figure 7, the trajectory pair, denoted as $traj_{pair} = \left\{traj_{os}^{[t_i,t_j]}, traj_{ts}^{[t_i,t_j]}\right\}$ is extracted if any trajectory point pair $\left(p^t_{os}, p^t_{ts}\right)$ between the time intervals t_i to t_j is identified to have a potential collision risk.

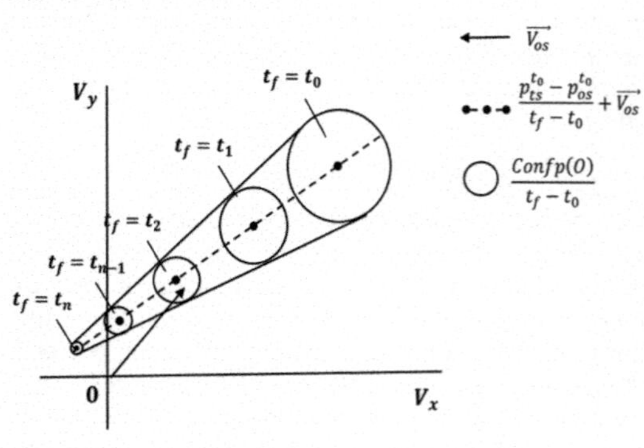

Figure 6. The diagram of the velocity obstacle method.

2.2.2 *Avoidance behaviour identification*

After extracting trajectory pairs with potential collision risk by the VO method, the avoidance behaviour of ship navigation is subsequently extracted based on heading changes within a fixed time window.

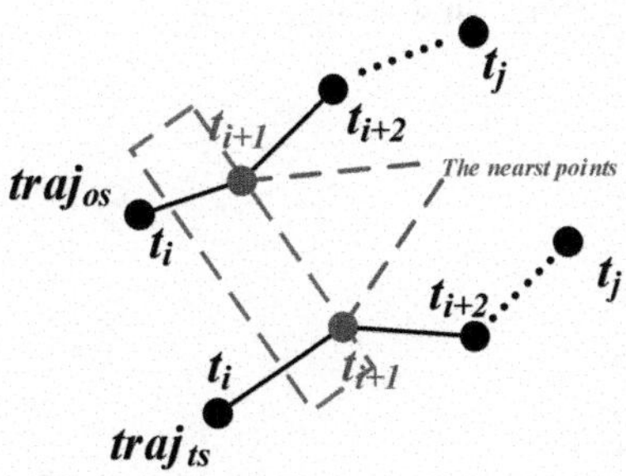

Figure 7. The extraction of avoidance behaviour.

As illustrated in Figure 7, the points $(p_{OS}^{t_{min}}, p_{TS}^{t_{min}})$ with the closest distance can be calculated using the following equation:

$$t_{min} = argmin\ D_R(t),\ t \in [t_i, t_j] \tag{8}$$

where t_{min} denotes the time when the relative distance between the two ships reaches the minimum. The $D_R(t)$ can be expressed as $D_R = \| p_{ts}^t - p_{os}^t \|$.

Furthermore, the time window $(t_s = t_{min} - \Delta t, t_e = t_{min})$ during which the avoidance behaviour might occur is achieved by applying a fixed time window threshold Δt.

To identify the avoidance behaviour, the heading data is differentiated to obtain the Rate of Turn (ROT). As shown in Figure 8, each extreme point such as p^{t_e} corresponds to two points $p_{t_e}^{t_L}$ and $p_{t_e}^{t_R}$ where the ROT equals 0. The areas formed by these three points can be considered as a single behaviour or noise caused by environmental factors.

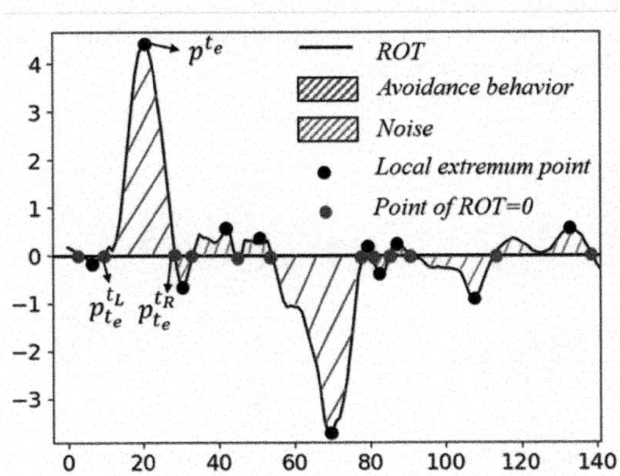

Figure 8. The ROT of OS.

According to Rule 8(b) of the COLREGS: "*a succession of small alterations of course and/or speed should be avoided if the circumstances of the case admit*". Therefore, the relatively small heading change can be regarded as non-collision avoidance behaviour, and the ROT changes can be used to distinguish the avoidance behaviours and noise caused by environmental factors. The ROT changes of the areas formed by $p_{t_e}^{t_L}$ to $p_{t_e}^{t_R}$ can be expressed as the following equation:

$$\Delta\theta(p_{t_e}) = p_{OS}^{t_R}.\theta - p_{OS}^{t_L}.\theta \tag{9}$$

where t_L andt_R represent the times of point $p_{t_e}^{t_L}$ and $p_{t_e}^{t_R}$; $p_{OS}^{t_L}.\theta$ and $p_{OS}^{t_R}.\theta$ represents the heading of OS at the time t_L and t_R, respectively.

When $\Delta\theta(p_{t_e})$ exceeds the defined turning threshold τ_θ, the changes in the area are recognized as stemming from avoidance behaviour. The specific value of the turning threshold τ_θ can be set based on the characteristics of the research waterways. It is considered that there is avoidance behaviour in the time window (t_s, t_e) when $\Delta\theta(p_{t_e})$ within the time window is greater than τ_θ.

3 CASE STUDY

3.1 *Case design*

To verify the feasibility and effectiveness of the method proposed in this paper, a waterway in Shanghai, China, is chosen as the experimental waterway. The waterway extends over a range of longitudes from 122.4° to 122.6° and latitudes from 30.9° to 31.2°, as shown in Figure 9. The experimental waterway consists of six main navigation routes, including two inbound routes, two outbound routes, and two outside routes. The two outside routes intersect with the other four inbound and outbound routes, creating two areas of route intersection. This uses one week of AIS data from the waterways, from April 18 to April 25, 2019, for navigation route recognition and ship avoidance behaviour extraction.

The OPTICS algorithm is employed to cluster the collected AIS data. Subsequently, the KDE method is used to recognize the route boundaries of each trajectory cluster and compare them with actual routes, demonstrating the effectiveness of the proposed route recognition method. Following this, the VO method is utilized to extract trajectory pairs with collision risks, serving as the dataset for avoidance behaviours extraction. Furthermore, based on the extracted dataset, this study identifies ship avoidance behaviours by analysing the heading data of the ship. Finally, the spatial distribution of the identified avoidance behaviour in the waterway is analysed to validate the feasibility and effectiveness of the avoidance behaviour ex-traction method proposed in this paper.

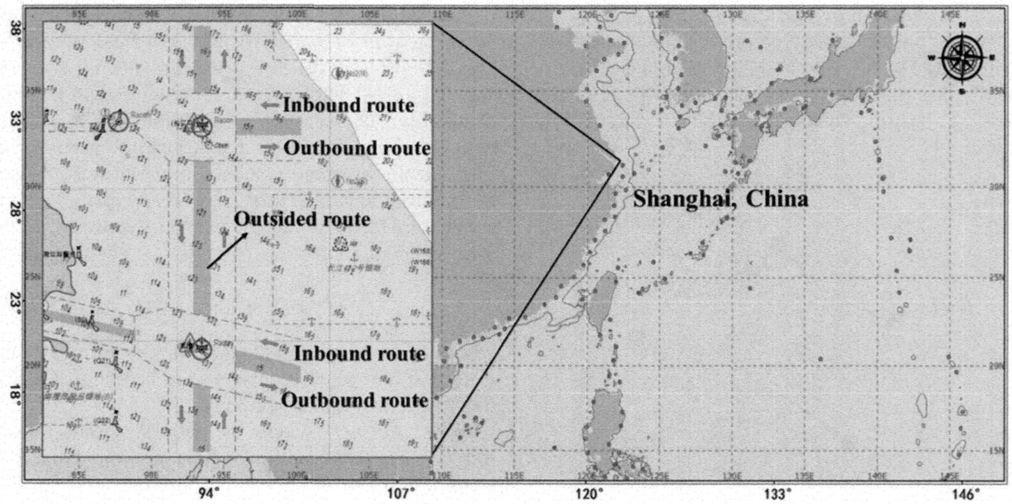

Figure 9. The experimental waterway.

3.2 *The result of navigation route recognition*

After pre-processing, the AIS data collected for this study are shown in Figure 10(a). The AIS data reflects some characteristics of the shipping routes, but they are not clearly defined, therefore, the OPTICS clustering algorithm is applied to the clustering This study categorizes ship trajectories into two groups based on average speeds: greater than 0.5 knots and less than 0.5 knots, as shown in Figure 10(b) and Figure 10(c). Trajectories with speeds greater than 0.5 knots are used for clustering.

Density-based clustering algorithms require setting the number of samples MinPts and the minimum radius ε. However, the OPTICS algorithm only requires setting the value of the parameter MinPts and then the reachability plot generated by the algorithm can be used to determine the value of parameter ε. In this case, the MinPts is set to 16, and the corresponding reachability plot is shown in Figure 11(a). The chosen parameter ε equals 250,000, dividing the reachability plot into different parts. Twelve of these parts correspond to reachability values less than ε are considered trajectory clusters, while the parts with values greater than ε are considered outliers and are removed. Therefore, the results obtained through OPTICS clustering are as shown in Figure 11(b). By comparing the clustering result in Figure 11(b) with Figure 9, it can be seen that the clustered trajectory clusters correspond to the actual routes.

After clustering the AIS data using the OPTICS algorithm, this paper employs the KDE algorithm to determine the boundaries of each trajectory cluster. Figure 12(a) shows one of the twelve trajectory clusters. Firstly, the selected waterway is gridded into a 500x500 grid according to Eq. (3). Then, for each column of the grid, the density probability distribution is calculated using Eq. (4), as shown in Figure 12(b). Furthermore, by setting a probability density threshold p two row numbers N_1 and N_2 corresponding to the threshold p are obtained, as indicated by the two red points in Figure 12(b). These points determine two boundary points for the column. Repeating this process for each column yields the navigational boundaries as shown in Figure 12(c). After applying KDE to all trajectory clusters, the navigational route boundaries of the research waterway are obtained as shown in Figure 12(d).

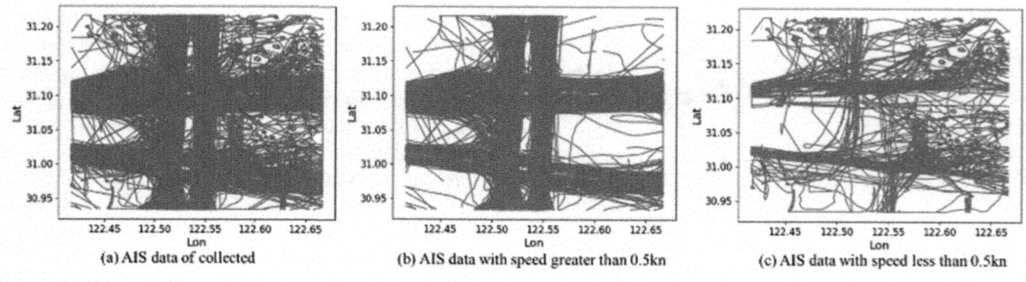

Figure 10. The AIS data of research waterway.

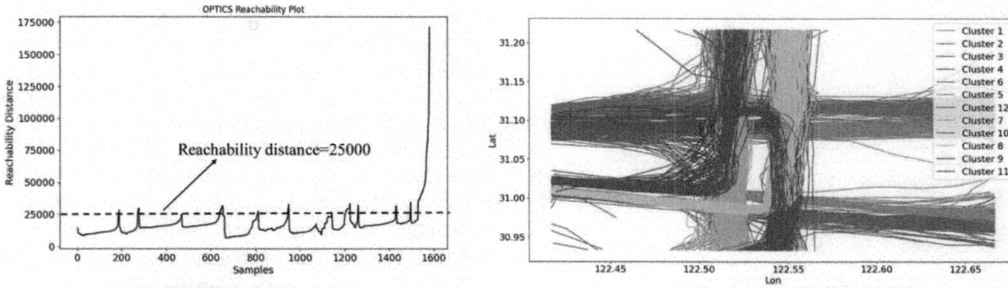

(a) The OPTICS reachability plot

(b) The clusters by OPTICS algorithm

Figure 11. The results of clustering by OPTICS algorithm.

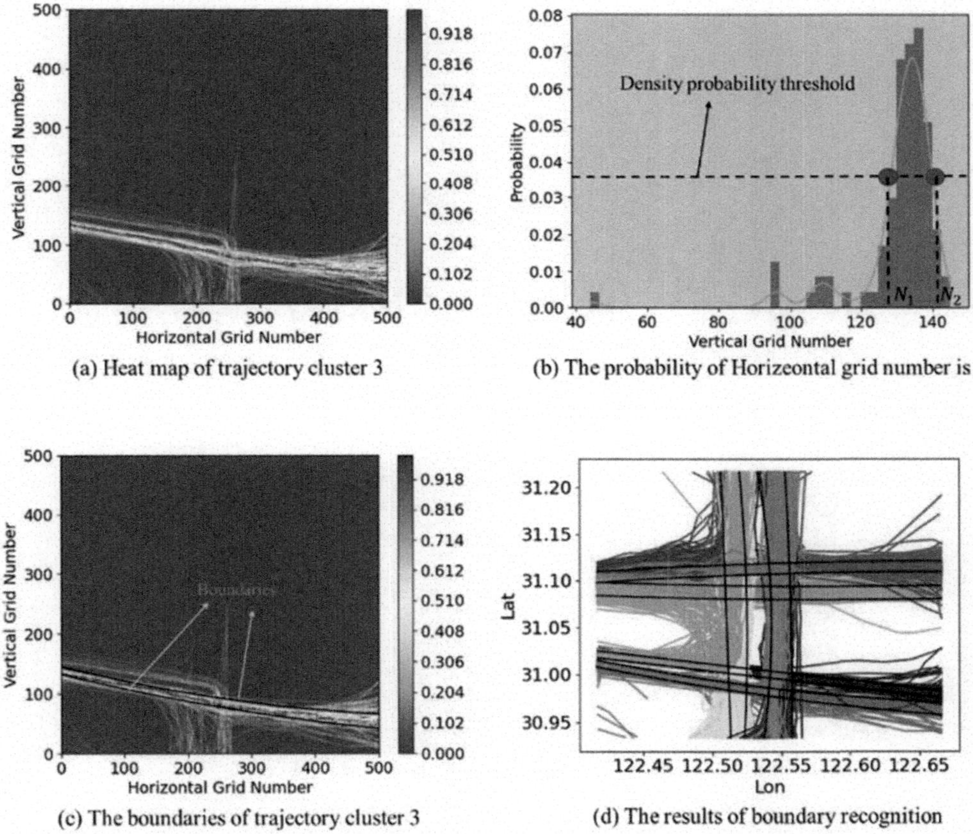

(a) Heat map of trajectory cluster 3

(b) The probability of Horizeontal grid number is 1

(c) The boundaries of trajectory cluster 3

(d) The results of boundary recognition

Figure 12. The results of route boundary recognition.

3.3 *The result of avoidance behaviour extraction*

By applying the VO algorithm to the collected AIS data for collision risk detection, a total of 6526 risky trajectory pairs are identified. Subsequently, using the ship heading data, different turning thresholds τ_θ are set for extracting ship avoidance behaviour. As shown in Figure 13, the turning thresholds τ_θ are negatively correlated with the number of extracted avoidance behaviours. In this paper, a τ_θ of 5 is adopted, resulting in the extraction of 1200 avoidance behaviours.

Figure 14 illustrates two of these extracted avoidance behaviours, where the blue lines represent the trajectory of own ship, and the black lines represent the trajectory of target ship. The blue and red dots indicate the start and end points of the avoidance behaviour, respectively. As shown in Figure 14(a), to

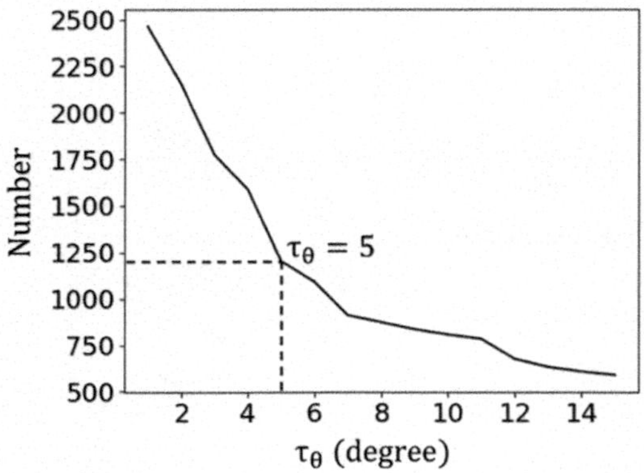

Figure 13. The number of avoidance behaviours under different turning thresholds τ_θ.

avoid a collision, the own ship executed a right-turn avoidance behaviour and success-fully passed behind the target ship. In Figure 14(b), the own ship also performed a right-turn avoidance behaviour, but in this case, it successfully passed in front of the target ship.

To conduct further analysis of the extracted avoidance behaviours, this section employs the method proposed by (He et al., 2017; Liu et al., 2022) to categorize the avoidance behaviours into situations of head-on, crossing, and overtaking according to the relative bearing and relative speed. In this experiment, a total of 1200 avoidance behaviours are extracted from the one week of AIS data collected. As illustrated in Figure 15(a), the 1200 extracted avoidance behaviours have been categorized into three encounter situations, of which 886 are crossing situations, 144 are head-on situations, and 53 are overtaking situations.

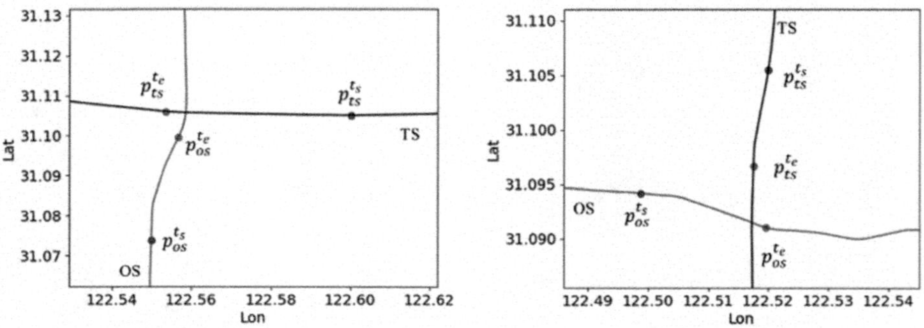

(a) Avoidance behaviour 1

(b) Avoidance behaviour 2

Figure 14. The examples of extracted avoidance behaviour.

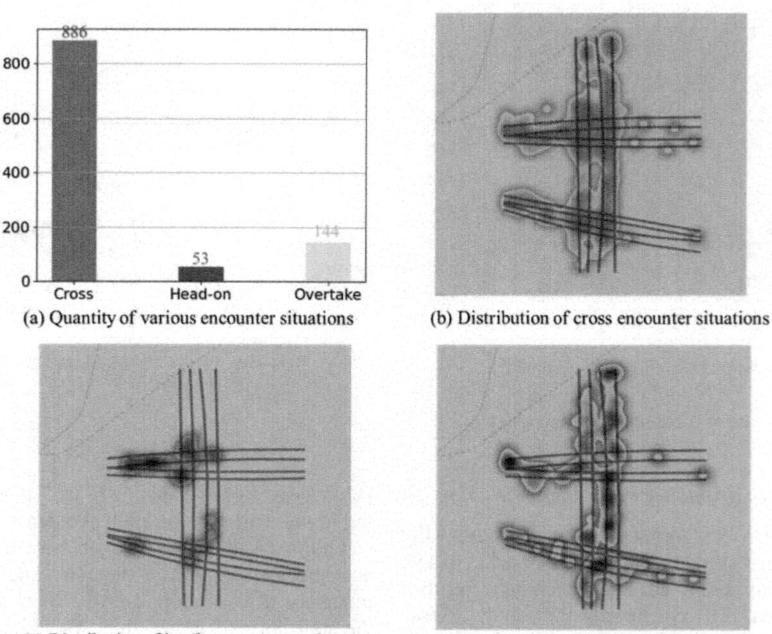

(a) Quantity of various encounter situations

(b) Distribution of cross encounter situations

(c) Distribution of head-on encounter situations

(d) Distribution of overtake encounter situations

Figure 15. Distribution of encounter situations of research waterways.

The distributions of crossing, head-on, and overtaking encounter situations are shown in Figure 15(b) to Figure 15(d), respectively. From the perspective of crossing scenarios, they are primarily concentrated in the two route intersection areas, with some also present on the left side of the outside routes, and only a few on the right side of the outside routes. The fewer head-on encounter situations might be due to the traffic separation schemes implemented in the experimental area. As for overtaking encounter situations, they are primarily found on the two outside routes, particularly on the right one. These distributions align with the characteristics of the experimental waterway, thus demonstrating the effectiveness of the ship avoidance behaviour extraction method proposed in this paper.

4 CONCLUSION

The research on ship collision avoidance behaviour extraction is of great significance to ensuring the safety of ship navigation. To further explore the valuable information in AIS data for the research of ship collision avoidance, a method to identify collision avoidance behaviour in AIS data from a heavy traffic area is pro-posed in this paper. This study employs the OPTICS algorithm to cluster ship trajectories and the KDE method to extract navigational route boundaries. The VO method is applied to extract trajectory pairs involving collision risk, followed by the extraction of avoidance behaviour based on heading changes within a fixed time window.

To validate the feasibility and effectiveness of the method proposed in this study, a week's worth of AIS data are utilized to identify navigation route boundaries and extract avoidance behaviours in the case study. This experiment successfully identified the boundaries of six navigation routes and extracted 1200 avoidance behaviours from the AIS data. Furthermore, the encounter situation features of these 1200 avoidance behaviours were found to be consistent with the features of the research waterways.

The method proposed in this study still has some limitations. In the recognition of navigation routes, the recognized route boundaries exhibit significant discrepancies with the actual boundaries. In the process of ship navigation avoidance behaviour extraction, this paper regards the change of ship heading as the result of ship collision avoidance behaviour. However, the change of ship heading will be affected by many factors, including channel direction and route planning, which are not taken into account in this paper. Therefore, the above factors will be taken into consideration in future research.

ACKNOWLEDGEMENTS

The work presented in this study is financially supported by the China Scholarship Council under grants 202206950031, the National Key Research and Development Program of China (2023YFE0203700) and the National Natural Science Foundation of China under grant 52271367. This work contributes to the Strategic Research Plan of the Centre for Marine Technology and Ocean Engineering (CENTEC), which is financed by the Portuguese Foundation for Science and Technology (Fundação para a Ciência e Tecnologia - FCT) under contract UIDB/UIDP/00134/2020

REFERENCES

Chen, P., Huang, Y., Mou, J., van Gelder, P.H.A.J.M., 2018a. Ship collision candidate detection method: A velocity obstacle approach. Ocean Engineering 170, 186–198. https://doi.org/10.1016/j.oceaneng.2018.10.023

Du, L., Banda, O.A.V., Huang, Y., Goerlandt, F., Kujala, P., Zhang, W., 2021. An empirical ship domain based on evasive maneuver and perceived collision risk. Reliability Engineering & System Safety 213, 107752. https://doi.org/10.1016/j.ress.2021.107752

Guedes Soares, C., Teixeira, A.P., 2001. Risk assessment in maritime transportation. Reliability Engineering & System Safety 74, 299–309. https://doi.org/10.1016/S0951-8320(01)00104-1

He, Y., Jin, Y., Huang, L., Xiong, Y., Chen, P., Mou, J., 2017. Quantitative analysis of COLREG rules and seamanship for autonomous collision avoidance at open sea. Ocean Engineering 140, 281–291. https://doi.org/10.1016/j.oceaneng.2017.05.029

Liu, J., Zhang, J., Yan, X., Guedes Soares, C., 2022. Multi-ship collision avoidance decision-making and coordination mechanism in Mixed Navigation Scenarios. Ocean Engineering 257, 111666. https://doi.org/10.1016/j.oceaneng.2022.111666

Mestl, T., Tallakstad, K.T., Castberg, R., 2016. Identifying and Analyzing Safety Critical Maneuvers from High Resolution AIS Data. TransNav 10, 69–77. https://doi.org/10.12716/1001.10.01.07

Rong, H., Teixeira, A.P., Guedes Soares, C., 2022. Ship collision avoidance behaviour recognition and analysis based on AIS data. Ocean Engineering 245, 110479. https://doi.org/10.1016/j.oceaneng.2021.110479

Sutulo, S. and Guedes Soares, C. 2014; An Algorithm for Offline Identification of Ship Manoeuvring Mathematical Models after Free-Running Tests. Ocean Engineering. 79:10–25. http://dx.doi.org/10.1016/j.oceaneng.2014.01.007

Yoo, S.-L., 2018. Near-miss density map for safe navigation of ships. Ocean Engineering 163, 15–21. https://doi.org/10.1016/j.oceaneng.2018.05.065

Zhang, W., Goerlandt, F., Kujala, P., Wang, Y., 2016. An advanced method for detecting possible near miss ship collisions from AIS data. Ocean Engineering 124, 141–156. https://doi.org/10.1016/j.oceaneng.2016.07.059

Zhang, W., Goerlandt, F., Montewka, J., Kujala, P., 2015. A method for detecting possible near miss ship collisions from AIS data. Ocean Engineering 107, 60–69. https://doi.org/10.1016/j.oceaneng.2015.07.046

Zhang, Y., Sun, X., Chen, J., Cheng, C., 2021. Spatial patterns and characteristics of global maritime accidents. Reliability Engineering & System Safety 206, 107310. https://doi.org/10.1016/j.ress.2020.107310.

Advances in Maritime Technology and Engineering – Guedes Soares & Santos (Eds)

An integrated ship collision avoidance strategy based on deep reinforcement learning and model predictive control

C.X. Zhao, X. Wang & H.B. Li
College of Navigation, Dalian Maritime University, Dalian, China

C.W. Zhang
School of Information Science and Technology, Dalian Maritime University, Dalian, China

ABSTRACT: In this paper, an integrated ship collision avoidance strategy is proposed based on deep reinforcement learning with model predictive control. According to COLREGs, the reward function is designed by combining sparse reward and dense reward, and proximal policy optimization (PPO) algorithm is applied to generate collision avoidance path planning. Then, an improved model predictive control (MPC) based on maneuvering modeling group (MMG) model is applied for ship path following. Following this approach, i.e. by incorporating the PPO into MPC, the computational burden for large-scale is decreased and a more stable performance is achieved. The experimental results show that OPP-MPC model has shorter training time and better convergence effect than only OPP for path planning. The environment simulation verifies that the path planning method has the obstacle avoidance ability in accordance with COLREGs and can make the ship back on course in a shorter time.

1 INTRODUCTION

Intelligent ships have a clear competitive advantage over conventional ships in terms of energy efficiency, environmental protection, safety, and economics, owing to their cutting-edge cognition, decision-making ability, and superior control technology.

The sequence of points or curves connecting the starting position and the end position is called the path, and the strategy forming the path is called the path planning, which has a wide range of applications in many fields, mainly used in intelligent robot autonomous collision avoidance, UAV obstacle avoidance and unmanned driving. Ship path planning involves finding an optimal and collision-free route from a starting point to a destination within a specified area. The primary challenge in ship path planning is to ensure that the ship can avoid potential collisions with other ships or obstacles while reaching its intended destination efficiently. A*, Artificial Potential Field (APF), Simulated Annealing (SA) and other algorithms have been studied and developed for many years. Compared with the path planning in the field of robots and manipulators on lands, there are only a limited number of studies on the path planning of ships. They have a poor perception of input received from a multi-dimensional environment and is primarily concerned with distances, smoothness and accessibility of paths without considering the dynamic characteristics of the corresponding agent.

In the past decade of development, intelligent prediction technology based on machine learning has developed rapidly. Reinforcement learning algorithms can be divided into value-based, policy-based, and value and policy based methods.

The value-based method defines the value function and selects the action according to the size of the value function. The value-based methods mainly include TD, Q-Learning, SARSA, Dyna and so on. The strategy based on approach parameterizes the strategy and maximizes the cumulative return by optimizing the parameters. The value-based approach is mainly applicable to discrete action Spaces, where the goal is to get the optimal strategy by maximizing the value function of each state. The value function is used to measure the degree of the robot's choice strategy in the current state. Yoo & Kim (2016) used the Q learning algorithm based on temporal-difference learning (TD learning) for marine vehicles path optimisation in ocean environments without considering obstacles. Cheng & Zhang (2018) realised the obstacle avoidance for unmanned ships based on DQN, and the vessel had no knowledge of the disturbance of the environment. Zhou et al. (2019) proposed a route planning method for unmanned ship cooperation based on DQN. By defining 13 action choices, ships can choose the best action in the current state. However, the action selection space of this method is discrete and cannot be applied to continuous action environment. Shen et al. (2019) applied the Dueling-DQN algorithm to local route planning. The algorithm selects the sample with the largest absolute value in the experience buffer pool to accelerate the learning speed of Q function, but the programming

DOI: 10.1201/9781003508762-23

speed sometimes fails to meet the timeliness requirement of local programming.

The policy-based approach obtains the optimal strategy by directly optimizing the strategy. Policy-based methods mainly include strategy gradient, imitation learning and so on. Pfeiffer et al. (2018) combined imitation learning and reinforcement learning for model training, and through goal-driven demonstration of imitation learning. This mothed can significantly improve the exploration ability in reinforcement learning process, and the obstacle avoidance performance is also better. In 1983, Grondman et al. (2012) proposed Actor-Critic algorithm, which combines the advantages of value-based algorithm and policy-based algorithm. Compared with traditional strategy gradient method, it is more efficient and a reinforcement learning algorithm with better performance. Actor-Critic framework is widely used, and on this basis, Asynchronous Advantage Actor-Critic(A3C), Trust Region Policy Optimization (TRPO), Proximal Policy Optimization(PPO), Deep Deterministic Policy Gradient(DDPG), Soft Actor-Critic(SAC) and other deep reinforcement learning algorithms have been developed. Cao et al. (2019) proposed an underwater vehicle target search method based on A3C algorithm. By designing an asynchronous superiority evaluation network structure, the method enables the underwater vehicle to learn from its own experience, thus generating search strategies, and applying the algorithm to underwater vehicle navigation to further optimize search routes. Xie et al. (2020) proposed a composite learning method based on A3C algorithm, Long Short Term Memory (LSTM) and Q-learning, and verified its advantages through multi-ship collision avoidance simulation. Yang & Han (2023) optimized the parameter update and exploration strategy of Actor-Critic algorithm, which can solve the collision avoidance problem under complex meteorological conditions. Chun et al. (2021) designed a ship path planning algorithm based on PPO algorithm based on ship collision avoidance risk, and compared it with A* algorithm to verify the reliability and flexibility of PPO algorithm. Zheng et al. (2023) proposed a new algorithm under intensive reward based on PPO algorithm, which can significantly improve training efficiency and perform better than traditional methods in complex and uncertain environments. Wu et al. (2023) used dynamic window method combined with PPO algorithm to realize multi-ship autonomous collision avoidance path planning in complex environment. However, the path planning algorithm based on deep reinforcement learning algorithm usually realizes ship path planning by interacting with the environment to predict the feasible path, which has some problems such as poor generalization ability, low security guarantee, difficulty in practical application, and limitations of different reinforcement learning algorithms.

Due to the characteristics of large inertia and complex maneuvering, collision avoidance has always been the key to safe navigation of ships. At present, most intelligent algorithms do not consider the maneuverability, but some scholars have already proved the importance of ship maneuverability. Wang et al. (2017) confirmed the significance and necessity of ship maneuverability in the collision avoidance by consisting two ships encounter scenarios in close-quarters situation. On this basis, some collision avoidance decision-making studies considering ship maneuverability have achieved good results (Wang et al. 2021, Li et al. 2023a, b). Hence, it is necessary to consider the influence of ship maneuverability in collision avoidance path planning.

The objective of this paper is to settle the elimination collision path planning problem of the Intelligent ships with known local dynamic environmental information, and an integrated ship collision avoidance strategy based on PPO and MPC is proposed. This method combines the characteristics of PPO and MPC. The PPO action-state model under the constraint of COLREGS is established, and the improved model predictive control (MPC) based on maneuvering modeling group (MMG) model is applied for ship path following to realize Intelligent ships path planning and collision avoidance in a dynamic environment. The main contributions of this study are summarized as follows:

(1) A PPO method is designed the action space and reward function base on COLREGS to collision avoidance path planning, which can ensure that each action of the Intelligent ships is the optimal solution in the current state.
(2) An improved model predictive control (MPC) based on maneuvering modeling group (MMG) model is applied for ship path following control.
(3) Compared with the control effect of PPO path planning, the superiority of PPO-MPC algorithm is verified.

The rest of the paper is organized as follows. Section 2 introduces the PPO-MPC algorithm structure of path planning. Path planning and COLREGS collision avoidance problems are described in Section 3. Section 4 introduces the path tracking control of ships based on MPC. The simulation experimental design and experimental result analysis are presented in Section 5. Conclusions and future work are presented in Section 6.

2 PPO-MPC ALGORITHM STRUCTURE OF PATH PLANNING

A simple framework of the proposed method is shown in Figure 1 for brief explanation. This method divides collision avoidance planning control into path planning and trajectory tracking. The PPO algorithm was used in the path planning process, and the trained data and model were input into the MPC controller for trajectory tracking control to complete the collision avoidance behavior.

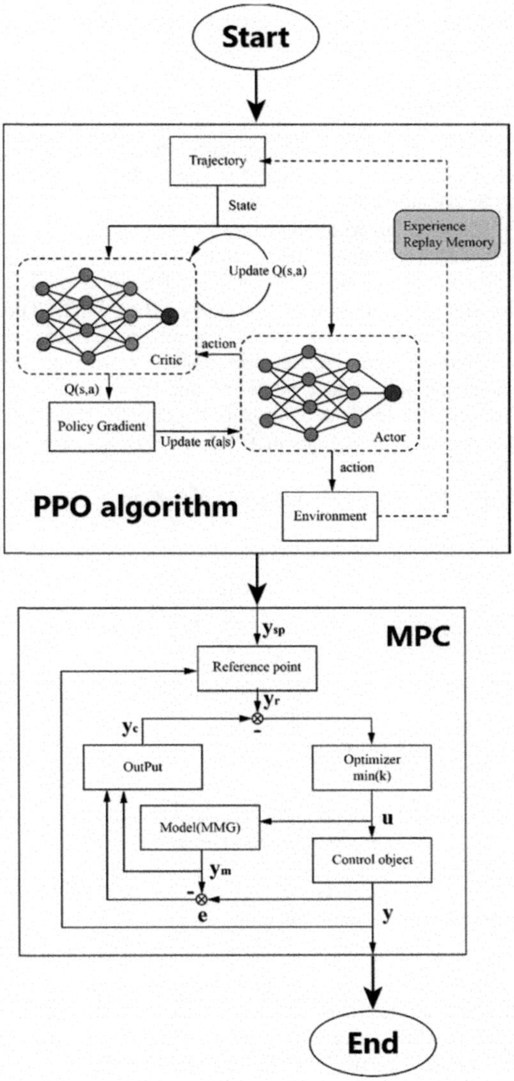

Figure 1. PPO-MPC algorithm structure of path planning.

3 COLLISION AVOIDANCE OF SHIPS BASED ON PPO

Strategy Gradient algorithm is a reinforcement learning algorithm based on strategy iteration, which can obtain the expectation of maximum reward directly by sampling state, action and reward. proximal policy optimization (PPO) (Schulman et al. 2017) is an improved algorithm of policy gradient. There are two forms of PPO, one is PPO-Penalty, and the other is PPO-Clip. Many experiments have proved that PPO-Clip performs better than PPO-Penalty. Ship collision avoidance is a typical MDP, and the current decision is only related to the current navigational state and not the past state. Ship collision avoidance is a process that requires a continuous series of actions and belongs to the sequential decision-making problem. In the training phase, the PPO was trained during the entire navigation process. That is, based on the PPO algorithm, the ship must safely and economically avoid obstacles and reach the target point to speed up the convergence rate of the algorithm.

The core principle of deep learning is to combine low-level features by stacking multi-layer network structures and nonlinear transformation methods to realize the hierarchical expression of input data. In contrast to deep learning, RL lacks direct supervision signals to steer the behavior of the agent, and the agent must interact with the environment through a trial-and-error mechanism to achieve the maximum reward value. DRL applied to ship collision avoidance can construct an end-to-end intelligent collision avoidance decision-making system by inputting raw data and outputting action control by combining the ability of deep learning to perceive data and the decision-making ability of RL. We adopted the PPO algorithm based on the actor–critic framework to parameterize, and directly learn the strategy by interacting with the environment to realize the control of the ship in a continuous action space. The PPO algorithm demonstrates a high learning performance and stable learning convergence for the control problem of a given target, such as the collision avoidance method.

In this section, we discuss the design of the state space, action space, reward function, and the PPO model.

The navigational information and dynamic changes that the agent can sense are represented in the state space of DRL. The design of the state space directly affects the convergence, convergence speed, and final performance of an algorithm. Additionally, it is vital to stop relying overly on end-to-end feature learning and extreme feature engineering while constructing the state space. The performance of the algorithm is affected if the state space is extremely complicated and redundant. In this study, the state is divided into OS state and TS state. And all states must be observable. The total state composed of 9 elements is defined by combining these two states. The states of the OS and TS can be expressed as:

$$State = [\psi_{OS}, x_{TS}, y_{TS}, x_{OS}, y_{OS}, DCPA, TCPA, X_{oS}, Y_{oS}] \tag{1}$$

To avoid collisions with other ships, the navigator needs to adjust the direction of the OS to ensure the safety of the ship in a complicated path, and it is necessary to make appropriate decisions based on long experience. The collision avoidance method of an autonomous ship can be created through a sufficient learning process by simulating the appropriate decision-making skills that the navigator acquires over a long period of experience. In this study, the rudder angle is defined as the action selected by the policy of the agent, and this angle could change the direction and path of the OS. The direction of the ship could be maintained, changed to the left, or to the right. Therefore, in this study, the action that can be selected by the OS is defined as

$$action = [-\delta_{max}, \delta_{max}], \delta_{max} = 35^\circ \tag{2}$$

The DRL process proceeds in the direction to maximize the value function for each state when the agent selects actions for a given scenario. The process of the entire DRL proceeds in the direction in which the value of the loss function is minimized, and the total reward is maximized.

In this study, the rewards obtained for each time step are divided into two types. First, rewards for collision avoidance (RCA) reflect whether the OS is avoiding collisions with TS. The RCA is subdivided into collision, warning and COLREGs. In addition, rewards for path following (RPF) reflect whether the OS reaches the destination along the reference path. The RPF is subdivided into time, goal and route. To normalize the effect of each dense reward, the value of each reward is adjusted to be within − 1 and 1. and to improve convergence speed, the value of each sparse reward is adjusted to be − 1000 and 1000. A detailed description of each reward is provided below.

$$R_{dcpa} = \begin{cases} 1, \text{DCPA} > 0 \\ -1, \text{others} \end{cases} \tag{3}$$

$$R_{route} = 2 * e^{(-d_O * 2)} - 1, d_O = \sqrt{(x_{oS} - X_{oS})^2 + (y_{oS} - Y_{oS})^2} \tag{4}$$

$$R_{goal} = \begin{cases} 2 * e^{(-L*2)} - 1, \text{other} \\ 1000, L \leq 0.1\text{n mile} \end{cases} L = \sqrt{(x_{oS} - X)^2 + (y_{oS} - Y)^2} \tag{5}$$

$$R_{collision} = -1000, \sqrt{(x_{oS} - x_{TS})^2 + (y_{oS} - y_{TS})^2} < 0.1\text{nmile} \tag{6}$$

$$R_{COLREGs} = -1000, \text{if satisfies COLREGs} \tag{7}$$

For the OS to arrive at a destination from a starting point, a calculation of a number of time steps is required. One iteration is performed after the number of time steps reaches the batch size, and the DRL process continues until the Max iteration. The reward summed in each iteration is called total reward. As learning progresses and iterations increase, the collision avoidance performance of the OS can be determined by examining the change in the total reward.

$$R_{time} = -1 \tag{8}$$

$$reward = R_{time} + R_{dcpa} + R_{route} + R_{goal} + R_{collision} + R_{COLREGs} \tag{9}$$

4 PATH TRACKING CONTROL OF SHIPS BASED ON MPC

4.1 *Maneuvering Modeling Group (MMG) model*

MPC is a model-based control strategy, and the choice of a predictive model is one of the core issues to be considered.

The maneuvering modeling group (MMG) (Yasukawa & Yoshimura, 2015) model is one of the well-known and high-fidelity 6 DOF mathematical models for ship manoeuvring motion proposed by JTTC. The primary features of basic MMG are the division of all hydrodynamic forces and moments working on the ship's hull, rudder, propeller, and their interaction. To simplify the model, in this paper only considers the forces of 3 degrees of freedom (3-DOF), namely the ship's surge, sway, and yaw, and ignores the heave, roll, and pitch 3 degrees of freedom.

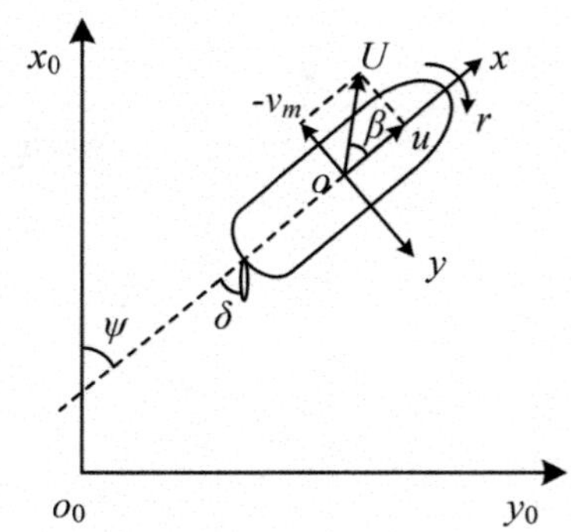

Figure 2. Coordinate systems.

In Figure 2, ψ, u, r and δ represent the ship's heading angle, forward speed, head-turning angular speed and rudder angle respectively, v_m is the ship's transverse speed, then the ship's center of gravity transverse speed v, drift angle β and ship's combined speed U are:

$$v = v_m + x_G r \tag{10}$$

$$\beta = \arctan\left(-\frac{v_m}{u}\right) \tag{11}$$

$$U = \sqrt{u^2 + v_m^2} \tag{12}$$

where x_G is the position of the ship's center of gravity.

The standardized 3-DOF MMG model is:

$$\begin{cases} (m+m_x)\dot{u} - (m+m_y)v_m r - x_G m r^2 = X_H + X_R + X_P \\ (m+m_y)\dot{v} + (m+m_x)ur + x_G m\dot{r} = Y_H + Y_R \\ (I_{zG} + x_G^2 m + J_z)\dot{r} + x_G m(\dot{v}_m + ur) = N_H + N_R \end{cases} \tag{13}$$

where: m is the quality of the ship; m_x and m_y are the additional mass of the ship longitudinally and laterally;

X, Y and N represent the longitudinal force, transverse force and turning moment in the ship, respectively. The subscripts H, R, and P represent the hydrodynamic forces acting on the hull, propeller, and rudder, respectively.

Hull hydrodynamic X_H, Y_H and N_H are:

$$\begin{cases} X_H = \left(\frac{1}{2}\right)\rho L_{pp} d U^2 X'_H\left(v'_m, r'\right) \\ Y_H = \left(\frac{1}{2}\right)\rho L_{pp} d U^2 Y'_H\left(v'_m, r'\right) \\ N_H = \left(\frac{1}{2}\right)\rho L_{pp}^2 d U^2 N'_H\left(v'_m, r'\right) \end{cases} \tag{14}$$

where: ρ is the density of water; L_{pp} is the length between the pillars of a ship; d is the ship's draft; v'_m and r' are dimensionless values.

Propeller propulsion force X_p is:

$$\begin{cases} X_P = \left(1 - t_p\right)\rho n_p^2 D_p^4\left(k_{t2} J_P^2 + k_{t1} J_P + k_{t0}\right) \\ J_P = \dfrac{u\left[1 - w_{p0}\exp\left(C_0 \beta_p^2\right)\right]}{n_p D_p} \\ \beta_p = \beta - x'_p r' \end{cases} \tag{15}$$

where: t_p is thrust reduction factor; n_p is propeller speed; D_p is the propeller diameter; k_{t0}, k_{t1} and k_{t2} are propeller open water test coefficients; J_P is the speed factor; w_{p0} is the wake coefficient of propeller when the ship is sailing directly. β_p is the inflow Angle of the propeller during the control movement. x'_p is the dimensionless value of the propeller position.

The rudder forces X_R, Y_R and N_R during steering are:

$$\begin{cases} X_R = -\left(1 - t_R\right) F_N \sin\delta \\ Y_R = -\left(1 + a_H\right) F_N \cos\delta \\ N_R = -\left(x_R + a_H x_H\right) F_N \cos\delta \end{cases} \tag{16}$$

where: t_R is the drag reduction coefficient caused by the rudder Angle; a_H and x_H are the coefficients expressing the mutual interference between hull and rudder. F_N is the normal force of the rudder.

4.2 *Basic principles of MPC*

According to MMG mathematical model, the ship's dynamic model is established, and then the MPC algorithm is used to design the controller. The basic principle of MPC is as follows: First, the output and parameters are measured according to the current time and the reference trajectory. Then the optimal predictive control input sequence within the predictive step size is solved to minimize the objective function defined by the predictive output and the reference trajectory. The first control input in the optimal predictive control input sequence is used as the system control input at time after it is input into the system, the calculation at time is repeated at time, to obtain the optimal system input at time, this rolling optimization strategy can guarantee every step. The input of the system is based on the optimal value calculated based on the current state, which is the difference between MPC and optimal control. The control performance of MPC is related to modeling accuracy, prediction step size, and optimal solution efficiency of the objective function. In the control process, appropriate simplified models, appropriate control parameters, and objective functions should be selected to meet different control systems.

Firstly, the motion state of the ship is combined with the unified prediction model of the system. In the ship path tracking process, the motion state of the ship is combined with the system unified forecasting model, control input at the previous moment, unknown input sequence to calculate future predicted output trajectory sequence. Then, the predicted output trajectory sequence, reference trajectory and unknown input sequence are substituted into the objective function. The optimal input sequence which makes the objective function value optimal under the constraint is obtained. Finally, the first input of the optimal input sequence acts on the ship and maintains a control cycle. Due to the existence of external interference, the real state of the ship is different from the predicted state, so you need to obtain the true movement status of the ship through the sensor, and use this status to carry out the next optimal input calculation (Ye et al. 2019.). The principle of ship path tracking control based on MPC is shown in Figure 3: y_{sp} represents the set output of the system, y_r represents the reference trajectory, u is the actual output value, y_m is the model output, and y_c is the predicted output.

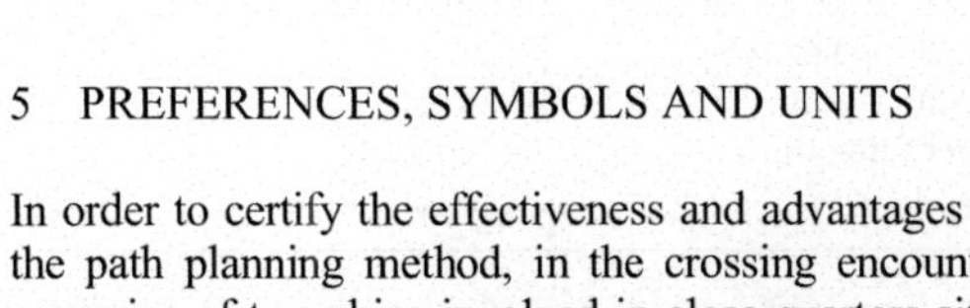

Figure 3. Path tracking control block diagram of MPC.

5 PREFERENCES, SYMBOLS AND UNITS

In order to certify the effectiveness and advantages of the path planning method, in the crossing encounter scenarios of two ships involved in close-quarters situation are provided for simulation. The own ship and the target ship have the same principal particulars and maneuverability characteristics, as shown in Table 1.

In the simulation, the encounter situation similar to the cross-encounter situation in the literature [8] is selected. $\psi_o = 30°$, $\psi_t = 290°$, $r_o = r_t = 0°$ /s, $\delta_o = \delta_t = 0°$, $u_o = u_t = 15.5$ knots, $v_{mo} = v_{mt} = 0$ knots, $R_O = 1$ nautical mile, $\alpha_{r0} = 40°$, O_O is at the origin (0, 0). It was clear that the two ships were in a close crossing situation, and according to the International Collision

Table 1. Principal particulars of ship.

Lpp (m)	Breadth (m)	Draught (m)	Displacement (m^3)	Barycentric coordinates (m)
320.0	58.0	20.8	312,600	11.2
Square coefficient	Propeller diameter (m)	Rudder height (m)	Rudder area (m^2)	
0.81	9.86	15.8	112.5	

Prevention at Sea Regulations 1972, the ship was a yielding ship and should have turned to the right and avoided crossing in front of it. The TensorFlow framework and Python are used to build the algorithm model and use to create a visual simulation environment that can observe the training and verification process in real time. The visual simulation environment and algorithm model, as well as the analysis of experimental results are introduced in this chapter.

5.1 *PPO-MPC training result*

The neural network hyperparameters of PPO-MPC are designed through experiments. The hyperparameters of the neural network are displayed in Table 2.

The model learns the action strategy by continuously interacting with the environment, and the learning effect is represented by the cumulative reward value of each training episode. The maximum number of action of the agent in a single training session is 800.

Table 2. Principal particulars of ship.

Hyper parameter	Symbol	Value
Actor learning rate	actor_lr	1e-4
Critic learning rate	critic_lr	5e-3
number episodes	num_episodes	10000
num of hidden layer units	hidden_dim	128
Reward decay rate	γ	0.922
lmbda	λ	0.9
num of iterations	epochs	10
episode	eps	0.15

The total reward for each training episode is displayed in Figure 4. The experimental results show that the model is trying random exploratory actions and triggers the training termination condition. The agent reaches the goal through some action strategies, and the training failure constraint is triggered by the remaining action strategies. After approximately 8000 training episodes, the reward value curve fluctuates to approximately -100, and the

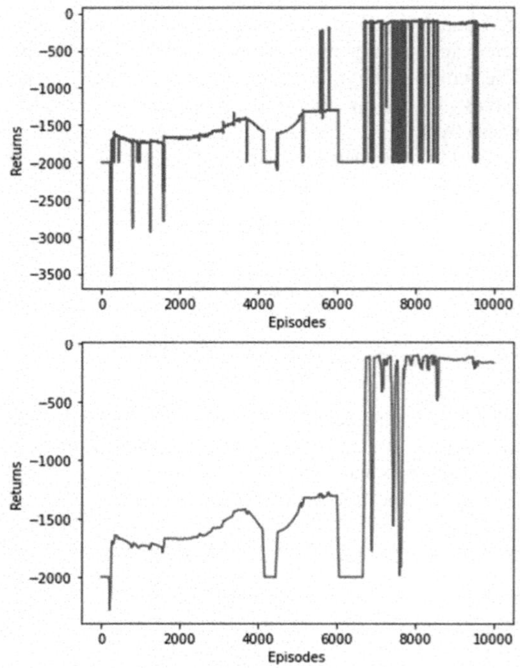

Figure 4. Total reward for each episode of PPO-MPC.

result shows that the agent reaches the goal and triggers the goal reward mechanism.

The collision avoidance experiments were designed to verify whether PPO-MPC can achieve COLREGS-compliant collision avoidance while implementing path planning. The experimental results are illustrated in Figure 5 and Figure 6.

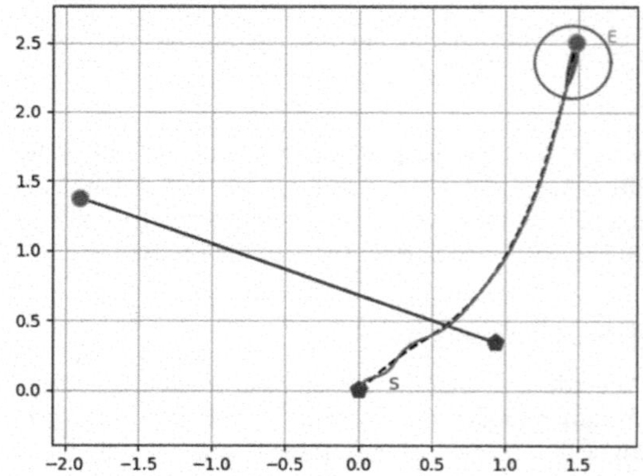

Figure 5. Trajectory of two meeting ships of PPO-MPC.

Figure 7 shows the real-time change curves of rudder Angle, forward speed, lateral speed and combined speed during the maneuvering movement of the ship. It can be seen that the maneuvering performance of the ship is good and all data are reasonable.

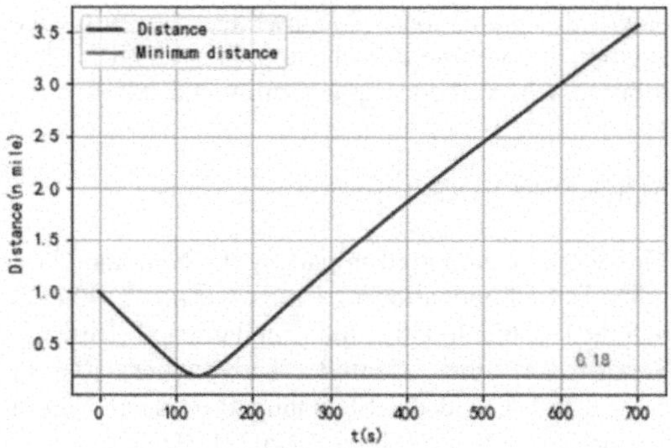

Figure 6. Distance of two meeting ships of PPO-MPC.

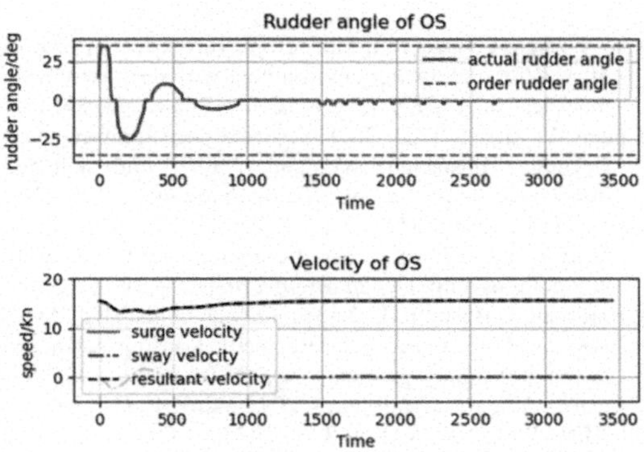

Figure 7. Rudder angle and velocity of two meeting ships of PPO-MPC.

5.2 *Compare PPO-MPC and PPO*

The neural network hyperparameters of only PPO are designed through experiments. The hyperparameters of the neural network are displayed in Table 3.

Table 3. Hyper parameters of the only PPO training algorithm.

Hyper parameter	Symbol	Value
Actor learning rate	actor_lr_o	1e-4
Critic learning rate	critic_lr_o	5e-3
number episodes	num_episodes_o	15000
num of hidden layer units	hidden_dim_o	128
Reward decay rate	γ_o	0.943
lmbda	λ_o	0.9
num of iterations	epochs_o	10
episode	eps_o	0.2

The model learns the action strategy by continuously interacting with the environment, and the learning effect is represented by the cumulative reward value of each training episode. The maximum number of action of the agent in a single training session is 900.

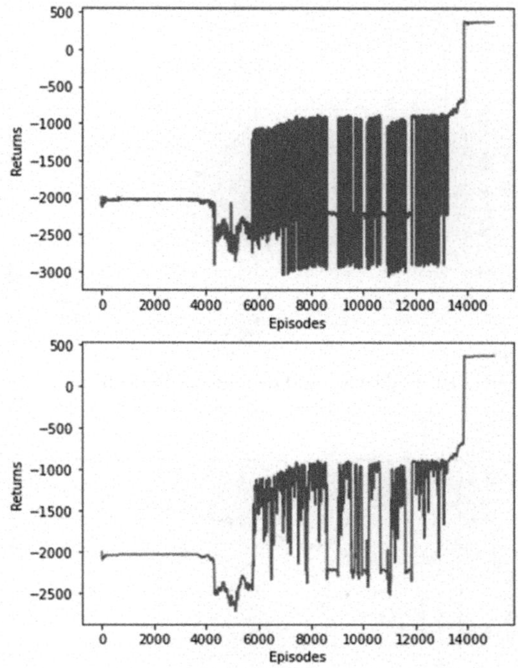

Figure 8. Total reward for each episode of only PPO.

The total reward of only OPP for path planning is displayed in Figure 8. After approximately 14000 training episodes, the reward value curve tends to stabilize, and the result shows that the agent reaches the goal and triggers the goal reward mechanism.

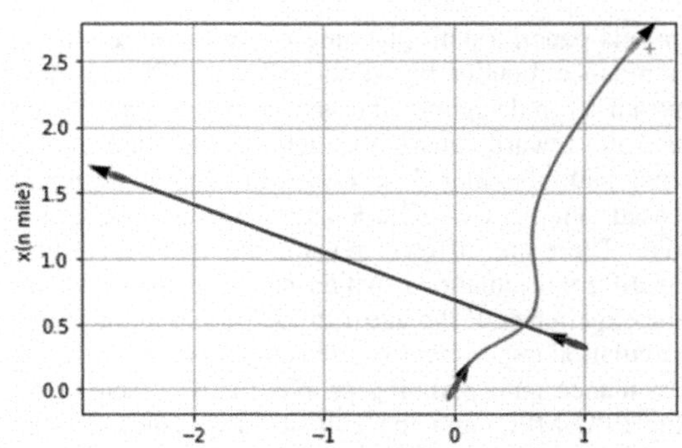

Figure 9. Trajectory of two meeting ships of only PPO.

The experimental results of only PPO are illustrated in Figure 9-11. Training time of PPO-MPC for path planning is 14 minutes, but PPO-MPC requires 89 minutes. The experimental results show that only PPO training time and movement time are longer than OPP-MPC for path planning and less accurate.

In other words, OPP-MPC model has shorter training time and better convergence effect than only OPP for path planning.

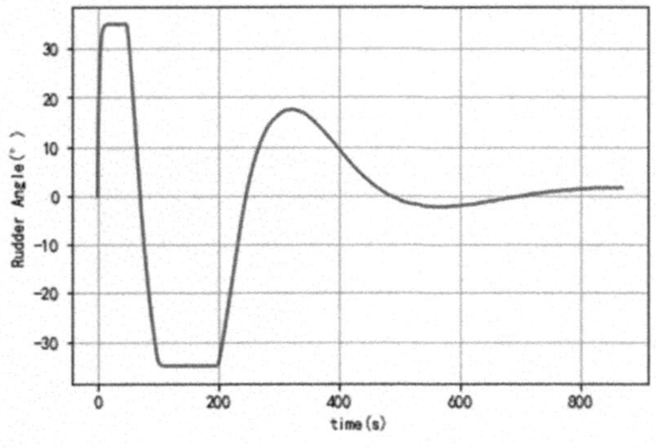

Figure 10. Rudder angle of own ship of PPO-MPC.

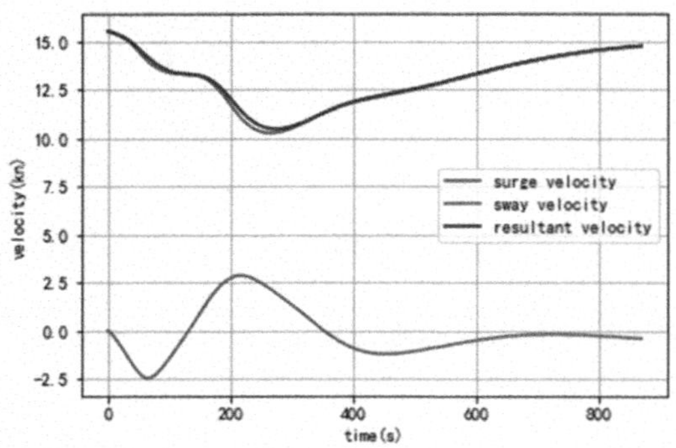

Figure 11. Velocity of own ship of only PPO.

6 CONCLUSION

In this paper, a path planning method with collision avoidance function based on PPO and MPC was proposed by redesigning the action space, state space and the reward function to achieve collision avoidance path planning. The reward function is designed based on the COLREGS and Collision avoidance risk. There are different reward functions according to different situations, which can effectively reflect the experience of the agent in various situations. The simulation experiment results show that the collision avoidance path planning problem can be settled by the PPO-MPC, and the collision avoidance actions conform to COLREGS.

However, the proposed method also has some limitations. As a model-free DRL, the PPO algorithm does not incorporate perceptual functions, so the output of the algorithm model only represents an action strategy, which will be affected by the physical factors of the ship in practical applications. The deep neural network is trained in the simulated environment, so the trained PPO-MPC, will be affected by the ship model and environmental uncertainty in practical applications.

For future work, we will try to consider the role of uncertain environmental factors in collision avoidance decision-making and add to perceptive function to improve the reliability of the proposed method in practical applications. Meanwhile, we will also attempt to combine it with model-based DRL algorithm to achieve better applicability and stability.

ACKNOWLEDGEMENTS

This work is supported in part by the National Natural Science Foundation of China (Grant. 51909022, 61976033, 62376048), the Fundamental Research Funds for the Central Universities (Grant. 3132023623), and the Shandong Provincial Natural Science Foundation (Grant ZR2022LZH002).

REFERENCES

Cao, X. Sun, C.Y. & Yan, M.Z. 2019. Target search control of AUV in underwater environment with deep reinforcement learning. *IEEE Access*, 2019 7, 96549–96559.

Cheng, Y. Zhang, W.D. 2018. Concise deep reinforcement learning obstacle avoidance for underactuated unmanned marine vessels. *Neurocomputing*, 272:63–73.

Chun, D.H. Roh, M.I. Lee, H.W. et al. 2021. Deep reinforcement learning-based collision avoidance for an autonomous ship. *Ocean Engineering*, 234:109216.

Grondman, I. Busoniu, L. Lopes, G.A.D. et al. 2012. Asurvey of Actor-Critic reinforcement learning: standard andnatural policy gradients. *IEEE Transactions on Systems, Man, and Cybernetics*, 2012, 42(6):1291–1307.

Li, H.B. Wang, X. & Wu, T.H. 2023a. Model Predictive Control and Artificial Potential Field based collision avoidance path planning method with ship maneuverability//*The 9th International Conference on Collision and Grounding of Ships and Offshore Structures (ICCGS 2023)*, 11-13 September, Nantes, France (p. 145–151).

Li, H.B. Wang, X. Wu, T.H. & Ni, S.K. 2023b. A COLREGs-Compliant Ship Collision Avoidance Decision-Making Support Scheme Based on Improved APF and NMPC. *Journal of Marine Science and Engineering*, 2023, 11(7), 1408.

Pfeiffer, M. Shukla, S. Turchetta, M. et al. 2018. Reinforced imitation: sample efficient deep reinforcement learning for Mapless navigation by leveraging priordemonstrations. *IEEE Robotics and Automation Letters*, 2018, 3(4):4423–4430.

Schulman, J. Wolski, F. Dhariwal, P. et al. 2017. Proximal policy optimization algorithms. *Machine Learning*, 1707.06347

Shen, H.Q. Hashimoto, H. Matsuda, A. et al. 2019. Automatic collision avoidance of multiple ships based on deep Q-learning. *Applied Ocean Research*, 2019, 86: 268–288.

Wang, X. Liu, Z.J. & Cai, Y. 2017. The ship maneu-verability based collision avoidance dynamic sup-port system in close-quarters situation. *Ocean Engineering*, 146:486–497.

Wang, X. Liu, Z.J. & Li, T.S. 2021. Collision avoidance dynamic support model for ships meeting at a close range. *Journal of Harbin Engineering University*, 2021, 42:1256–1261.

Wu, C. Yu, W. Li, G. & Liao, W. 2023. Deep reinforcement learning with dynamic window approach based collision avoidance path planning for maritime autonomous surface ships. Ocean Engineering, 284:115208

Xie, S. Chu, X.M. Zheng, M. et al. 2020. A composite learning method for multi-ship collision avoidance based on reinforcement learning and inverse control. *Neurocomputing*, 411: 375–392

Yang, X. & Han, Q.L. 2023. Improved reinforcement learning for collision-free local path planning of dynamic obstacle. *Ocean Engineering*, 283:115040

Yasukawa, H. & Yoshimura, Y. 2015. Introduction of MMG standard method for ship maneuvering predictions. *Journal of marine science and technology*, 2015, 20: 37–52.

Ye, H. Jiang, H.B. Shidian, M. et al. 2019. "Linear model predictive control of automatic parking path tracking with soft constraints." *International Journal of Advanced Robotic Systems*.

Yoo, B. & Kim. J. 2016. Path optimization for marine vehicles in ocean currents using reinforcement learning. *Journal of Marine Science and Technology*, 2016, 21: 334–343.

Zheng, K.J. Zhang, X.Y. Wang, C.B. et al. 2023. A partially observable multi-ship collision avoidance decision-making model based on deep reinforcement learning. *Ocean & Coastal Management*, 242:106689

Zhou, X.Y. Wu, P. Zhang, H.F. et al. 2019. Learn to navigate: cooperative path planning for unmanned surface vehicles using deep reinforcement learning. *IEEE Access*, 2019 7, 165262–165278.

Maritime safety and reliability

Advances in Maritime Technology and Engineering – Guedes Soares & Santos (Eds)
© 2024 The Author(s), ISBN 978-1-032-83099-5

Safety barriers for preventing leaks due to failure of maintenance operations on-board LNG floaters

M. Abdelmalek & C. Guedes Soares
Centre for Marine Technology and Ocean Engineering (CENTEC), Instituto Superior Tecnico, Universidade de Lisboa, Lisbon, Portugal

ABSTRACT: This paper establishes safety barriers for preventing leaks from failures of maintenance operations on-board LNG floaters. The establishment of safety barriers is conducted according to the developed guidelines for safety barrier management for the oil and gas industry of the Norwegian Continental Shelf. The background on the concepts of safety barriers and safety barrier management is presented. The recently developed literature for investigating failure of maintenance operations in the process industry is consulted for providing the input for this study regarding the process of maintenance operation and its failure-related aspects. Finally, the study is concluded by proposing safety barriers for preventing leaks from failure of maintenance operations, in addition to, discussing the benefits of utilising the safety barrier management concept for improving the operational safety and the possibility of its integration in modelling the risk dynamically.

1 INTRODUCTION

Since 2010 the maritime industry has witnessed an exponential growth in the number of ships that carry Liquified Natural Gas (LNG) as a cargo or as a ship fuel (Abdelmalek & Guedes Soares 2022). The growth in this segment of ships is attributed to i) the need for reducing global warming and climate change (UNEP 2007), ii) fulfilling the International Maritime Organisation's (IMO) requirements and ambitions for reducing Nitrogen Oxides (NOx), Sulphur Oxides (SOx), and greenhouse gas (GHG) emissions from ships (IMO 2018, 2019a,b), in addition to iii) the economic competitiveness of LNG as a ship fuel (WARTSILA 2017, Wang *et al.* 2021), and iv) as a domestic energy source (Jaramillo 2007, GIIGNL 2021) during the current energy transition period, when compared with the other traditional alternatives.

A simultaneous increase in the related publications on the safety and risk management of LNG floaters has been observed since the same date (Abdelmalek & Guedes Soares 2023). The increasing interest in this research domain is primarily motivated by the need to assess the risks of the newly introduced ship concepts and operational activities to the maritime LNG sector. Examples of the developed studies are fire and explosion risk analysis of bunkering operations (Wu *et al.* 2021), determination of safety exclusion zone during bunkering operations (Jeong *et al.* 2020), and risk-based layout evaluation of LNG fuel gas supply systems (FGSS) (Nubli & Sohn 2022) of LNG fuelled ships (LNGFSs), as well as collision risk analysis between LNG carriers (LNGCs) and LNG Floating, production, Storage, and offloading Units (FPSOs) (Antao & Guedes Soares 2012, Hu *et al.* 2017, Abdussamie *et al.* 2018), and fire and explosion risk analysis of cargo transfer operation between LNGCs and LNG Floating Storage and Regasification Units (FSRUs) (Martins *et al.* 2016).

Furthermore, the lack of specific data regarding human error probabilities (HEPs) for the critical operations of LNGFSs, such as LNG bunkering, has been also a motivator to develop studies for filling this gap of information (e.g. Ahn *et al.* 2022, Fan *et al.* 2022; Uflaz *et al.* 2022). Although the research domain of risk analysis and safety management of LNG floaters has been witnessing rapid development recently, failure analysis and the management of human and organisational factors in operational activities other than external LNG transfers (i.e. bunkering and cargo transfer) were not sufficiently investigated in the literature.

To contribute to filling the identified research gap above, the developed guidelines for safety barrier management (SBM) in the Norwegian Continental Shelf (NCS) (PSA 2013, DNV 2014, Hauge & Øien 2016) is adopted in this study for establishing safety barrier elements for preventing leakage on-board LNG floaters. The adopted approach is applied to the failure of maintenance operations. Failure of maintenance operations is selected as a case study for this research due to the significant contribution of the maintenance-related aspects in realising the historical major accidents in the process industry (Okoh & Haugen 2013a, 2014a), as well as for contributing to

DOI: 10.1201/9781003508762-24

improving the management of human and organisational factors which are the primary elements that influence the safe performance of maintenance operations on-board LNG floaters.

The present research is divided into five sections including an introductory and concluding sections. Sections two and three introduce background on LNG hazards, and the process of safety barrier manage-ment, respectively. In section four, the relevant maintenance-related aspects of LNG leaks in addition to the case study and the final discussion are presented. However, the presentation of the related definitions of safety barriers concept in the text is restricted only to the adopted definitions in this study, unless otherwise mentioned.

2 LNG HAZARDS

LNG is a cryogenic liquid consisting primarily of methane that liquifies at the atmospheric pressure at -162°C (Nerheim *et al.* 2021). The flammability range of methane is between 4.5% and 16.5%, and its minimum ignition energy is 0.28 mJ, at atmospheric pressure and 20°C (Vandebroek & Berghmans 2012). In addition, LNG is colourless, odourless, non-toxic, and non-corrosive. Accidental release of LNG can result in severe outcomes to the atmosphere, human health, and physical assets in case of ignition or non-ignition (Abdelmalek & Guedes Soares 2023). The various hazards of LNG leaks are outlined in Table 1.

3 SAFETY BARRIERS AND SAFETY BARRIER MANAGEMENT

3.1 *Background*

The interest in the principle of safety barriers has noticeably increased since the introduction of the ARAMIS (Accidental Risk Assessment Methodology for Industries in the context of the SEVESO II Directive) project (2002-2005) (Andersen *et al.* 2004), which among its objectives are the classification of safety barriers and the definition of the related terminologies to safety barriers. Furthermore, Sklet (2006b) has introduced an extensive review that presents a broad diversity of the presented concepts, definitions, and classifications of safety barriers in the literature. Further classification of safety barriers was presented by Hollnagel (2008), which also introduced a comparison between the presented categories of safety barriers (physical, functional, symbolic, and immaterial barriers) against a set of benefit (e.g. efficiency and availability) and cost (e.g. resource needs and evaluation ability) qualities.

Other than proposing definitions and classifications for safety barriers, the focus on establishing a holistic approach for SBM has started since the publication of the Norwegian Petroleum Safety Authority's (PSA) framework for SBM for the oil and gas industry in 2013 (PSA 2013). Later, on behalf of the Norwegian Shipowners' Association, DNV published a guidance note for SBM in the operations phase for rig owners (DNV 2014). The period between 2012 and 2015 also witnessed the development of the PETROMAKS project which was developed to present tools and guidelines for improving barrier management in the petroleum industry (Øien *et al.* 2015, Hauge & Øien 2016).

A safety barrier can be defined as "*physical and/or non-physical means planned to prevent, control, or mitigate undesired events or accidents*" (Sklet 2006b, mentioned in Rausand 2011). However, as indicated in the outlined definitions below, safety barriers are also arranged, hierarchically as barrier function, barrier system, and barrier element, in addition to the performance influencing factors (PIFs). The main idea of distinguishing between the term safety barrier and the above-mentioned terminologies is to make a distinction between what a barrier is and what a barrier does (Hollnagel 2008).

- Barrier function: "*A function planned to prevent, control, or mitigate undesired events or accidents*" (Sklet 2006b, mentioned in Rausand 2011).

Table 1. Hazards of LNG leaks.

Non-ignited LNG leaks' hazards
• Frostbites to humans, will often develop to amputation of the affected organ (Sever *et al.* 2010). • Hypothermia (SEPS 2018, Mayo Clinic 2022), and/or oxygen deficiency asphyxiation to humans (SEPS 2018). • Methane is a precursor to the tropospheric (ground level) ozone (Royal Society 2008, Mohajan 2012). • Methane consumes hydroxyl radicals (OH) which is a natural cleaner of the troposphere (Sanderson 2008). • A physical explosion is known as a rapid phase transition (RPT) (Alderman 2005). • Brittle fractures to the exposed non-cryogenic structures to the cold LNG temperature (Alderman 2005).
Ignited LNG leaks' hazards
• Immediate ignition of LNG will lead to pool fire (Betteridge *et al.* 2014), or jet fire of pressurised leaks (Zhang *et al.* 2019). • Pool and jet fires can lead to boiling liquid expanding vapour explosion (BLEVE) (Pitblado 2007) to the engulfed LNG equipment in their thermal radiation zone. • Delayed ignition of LNG vapour will lead to flash fires (Margolin 2013), but when deflagration to detonation transition (DDT) occurs vapour cloud explosion (VCE) (ABS 2004, Oran *et al.* 2020), will occur leading to significant consequences.

- Barrier system: "*A system that has been designed and implemented to perform one or more barrier functions*" (Sklet 2006b, mentioned in Rausand 2011).
- Barrier element: "*Engineered system, structures, or other design features which realize one or several barrier functions*" (DNV 2014).
- Performance influencing factors: "*Conditions which are significant for the ability of barrier functions and elements to perform as intended*" (PSA 2013, mentioned in Hauge & Øien 2016).

Furthermore, as indicated below, other definitions are presented for distinguishing between technical, operational, and organisational barrier elements:

- Technical barrier elements: *"Engineered systems, structures, or other design features which realize one or several barrier functions"* (DNV 2014).
- An operational barrier element: *"A task performed by an operator, or team of operators, which realizes one or several barrier functions"* (DNV 2014).
- Organisational barrier elements: *"Personnel responsible for, and directly involved in, realizing one or several barrier functions"* (DNV 2014).

In addition to presenting definitions of the terminologies of safety barriers, safety barriers are also classified with respect to their functions (de Dianous & Fiévez 2006, Rausand 2011), their systems (Rausand 2011), and their nature or type (Hollnagel 2008). For example, Rasuand (2011) has classified safety barriers based on their functions as preventive and mitigation barriers, whereas based on their system as active and passive barriers.

3.2 *Safety barrier management process*

In general, barrier management is defined in PSA (2013) as "*coordinated activities to establish and maintain barriers so that they maintain their function at all times*". The definition above is adopted in the proposed frameworks for barrier management by DNV (2014), and Hauge & Øien (2016).

In the present research, the proposed approach for SBM by Hauge & Øien (2016) is adopted. As illustrated in Figure 1, the process of barrier management in the design phase consists of the briefly described seven activities below (Hauge & Øien 2016):

1. Facility description and area division (Step 1)
   In addition to presenting a brief description of the facility for which barrier management is conducted, area division should also be carried out. Area division is necessary since barrier strategy, which is a main deliverable of the SBM process, is developed for each area to indicate the main hazards and the established barrier functions and elements in that area.
2. Defined situations for hazards and accidents (DSHAs) (Step 2.a)
   The main objective of this activity is the determination of the events that could lead to major accident hazards. The DSHAs are usually described in the risk picture of the facility. According to Rausand (2011), in the risk picture, detailed information about the hazardous events in addition to their frequencies, as well as the consequences and the associated frequencies of the resulting accident scenarios from each event are presented.

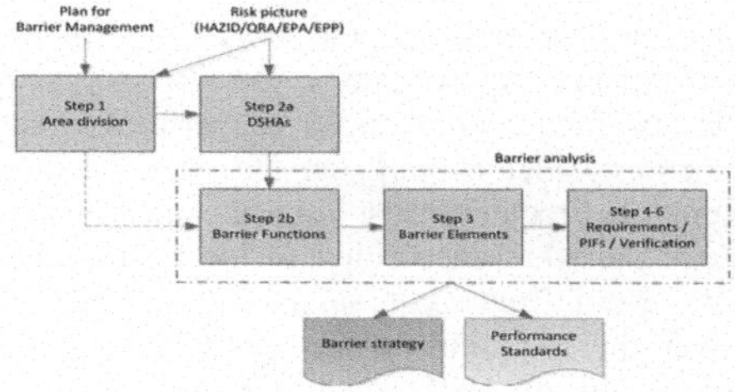

Figure 1. The process of safety barrier management (Hauge & Øien 2016, p23).

3. Defining barrier functions (Step 2.b.)
   The definition of the required proactive and reactive barrier functions of each DSHA is the first step in the process of barrier analysis. In this step, questions should be asked on how a particular DSHA can be realised, accordingly, barrier functions are determined for impeding the progression of the triggering and escalation events of that DSHA. In addition, as indicated in Sklet (2006b), the barrier function should commence with a verb to indicate the intended purpose of that function such as prevent, limit, or mitigate.
4. Barrier systems and elements in each area (Step 3).
   In this step, the breakdown structure of barrier functions is developed. In this structure, the relationship between barrier functions and their underlying barrier elements is presented. One or more levels of barrier sub-functions can be used whenever needed between the technical, operational, and organisational (TO&O) barrier elements and their corresponding barrier functions.
5. Performance requirements (Step 4)
   Performance requirements must be identified for the established TO&O barrier elements. The word performance is defined in PSA (2013) as "*the properties which a barrier element must possess in order to ensure that individual barrier and its function will be effective*". Hauge & Øien (2016) represented that performance requirements for technical barrier elements are classified as functional requirements (capacity and effectiveness), integrity requirements (reliability, availability, and integrity), and survivability requirements (load resistance and robustness).
   As represented in Sklet (2006b), functional requirements of technical barrier elements can be obtained from design codes and standards, regulation, or through risk-based design. On the other hand, the stated safety integrity levels (SIL) of the safety instrumented systems in standards (e.g. Norwegian Oil and Gas 2018) are used as the

input for determining the reliability/availability performance requirements of the SISs as probabilities of failure on demand to perform a safety function (Sklet 2006b). On the other hand, performance requirements for operational barrier elements, which according to the adopted definition herein represent safety-critical tasks, will cover the requirements for conducting the task safely and efficiently such as response time and number of the required personnel. Therefore, performance requirements for organisational barrier elements will cover only the requirements needed for the individual to perform his role efficiently, such as training and competency requirements.

6. Performance influencing factors (Step 5)
   Identification of PIFs that influence the performance of barrier elements is necessary since such factors are used for monitoring the performance of barrier elements in operations. PIFs can also be among the performance requirements of barrier elements such as training and level of experience in relation to organisational barrier elements. It is also advisable that the number of the selected PIFs for each element should not be extensive for the sake of simplification of the management of such PIFs in operations (Hauge & Øien 2016).
7. Verification of performance requirements (Step 6): Verification activities are performed to ensure that barrier elements are performing as stated in the performance standards. Although such activities are conducted only in operations, verification activities should be planned in the design phase during the development of the barrier management process.
   Furthermore, as shown in Figure 1, the process of SBM should result in two deliverables which are barrier strategy and performance standards (PSA 2013). Barrier strategy, according to DNV (2014), should be area-specific and indicate information on the implemented safety barriers to confront the identified hazards in that area. On the other hand, performance standards should be developed to document the relevant performance requirements to the established safety barriers on the system or function level (PSA 2013).

4 APPLICATION

4.1 *Failure of maintenance operations*

Maintenance is paramount for the safe and efficient operation of LNG floaters. In addition, maintenance is one of the most significant PIFs to the performance of technical barrier elements. However, despite the significant role of maintenance in preventing major accidents, maintenance may also result in critical events such as process leaks (Vinnem *et al.* 2016). Okoh & Haugen (2013a) represented that 30-40% of accidents and precursor events in the chemical process industry have resulted from failures of the maintenance-related aspects.

Furthermore, in the review of Okoh & Haugen (2014a) it is indicated that maintenance failures were responsible for 53% of the major accidents with completed investigations (i.e. 64 out of 74 accidents) between the years 2000 and 2011 in the process industry of the United States, whereas 38% of the occurred major accidents in the same interval in Europe are maintenance related. Also, maintenance failures were among the responsible factors for the occurrence of some of the well-known disasters in the history of the process industry such as Piper Alpha (1988) and Bhopal (1984) disasters, and others as represented in Okoh & Haugen (2013b, 2014b).

Moreover, maintenance-related causes of major accidents are classified in relation to the accident process and work process in Okoh & Haugen (2014a). For example, in the accident process classification, Okoh & Haugen (2014a) have adopted the presented definitions of active and latent failures in Reason (1997) for classifying the maintenance-related causes of major accidents. The term active failures indicates failures that could directly result in accidents, while latent failures refer to the dormant factors that have indirect effects on realising accidents (Okoh & Haugen 2013b). Further information regarding active and latent failures and the distinction between them is introduced by Cohen (2017).

Okoh & Haugen (2013b, 2014b), have also classified maintenance-related aspects of major accidents with respect to the accident process (same as the active failure category of Okoh & Haugen 2014a), and maintenance management cycle. The classified maintenance failures under the accident process and maintenance management cycle categories are presented in Table 2. In addition, the presented failure aspects of maintenance in Table 2 are adopted herein since they are more relevant to the scope of the present research. However, the relevant maintenance failure aspects to the scope of the study are marked in Table 2.

4.2 *Case study*

4.2.1 *Maintenance of LNG equipment*

In general, there is a wide diversity of LNG systems with various capacities installed on-board LNG floaters such as LNG regasification and LNG reliquefaction systems (Abdelmalek & Guedes Soares 2022). Maintenance operations that require the dismantling of such equipment and its fittings should be performed with extreme caution to avoid the release of LNG to the external environment.

The starting point for identifying the required TO&O barrier elements for preventing leaks from maintenance of the LNG-containing equipment onboard LNG floaters is the identification of the related tasks to that operation. For this purpose, the presented phases of preventive maintenance of process equipment in Vinnem *et al.* (2016, p350) and the introduced elements of the maintenance work process in Okoh & Haugen (2015, p217) were utilised

Table 2. Maintenance-related aspects of major accidents' classification (Okoh & Haugen 2013b, 2014b).

The accident process classification		
Type	Description	Relevant
Lack of barrier maintenance	Barrier maintenance is not sufficient so the degradation (e.g. increased corrosion) of barriers is not rectified.	No
Barrier maintenance error	Erroneous performance of maintenance has introduced hidden failures (e.g. wrong calibration of sensors or improper assembly of equipment). i.e. task performed incorrectly.	Yes
New hazard	Maintenance presents new hazards such as ignition sources.	No
Initiating event	Maintenance operation initiate a hazardous event (e.g. process leak).	Yes
The maintenance management cycle classification		
Type	Description	Relevant
Lack of maintainability	Maintenance could not restore the system to perform the required function. i.e. a task was not conducted.	Yes
Deficient fault diagnosis	The process of detecting faults or failure causes is deficient.	No
Deficient planning	Cover aspects related to maintenance planning errors such as improper documentation, resources, and time scale for performing maintenance. (In the present research represents a failure of pre-job planning).	Yes
Deficient scheduling	Improper scheduling of frequency of maintenance of equipment.	No
Deficient execution	Same as barrier maintenance error (from the scope of this study).	Yes
Deficient checking	Verification or supervision tasks are deficient.	Yes

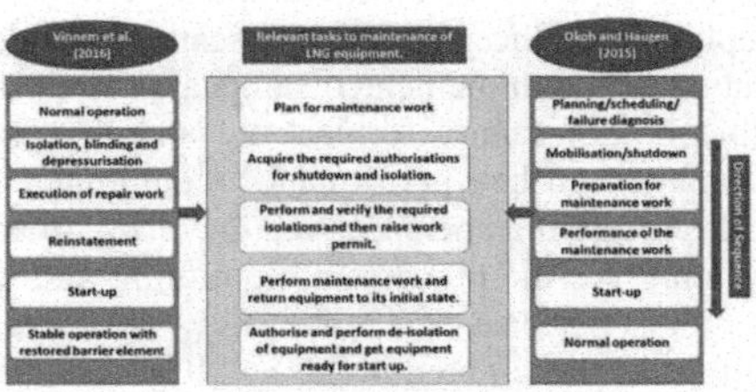

Figure 2. Steps of maintenance operation of LNG equipment.

for presenting the steps of the analysed maintenance operation as illustrated in Figure 2.

4.2.2 *Establishment of barrier elements*

Maintenance failure is one of the foreseeable causes of leaks in the process industry in general. Accordingly, a specified 2nd level barrier sub-function for preventing leaks from maintenance operations should be established under the common barrier function of preventing LNG leaks. As illustrated in Figure 3, two 3rd-level sub-functions are developed for representing the established TO&O barrier elements in this study. The first is "*prevent leak due to poor maintenance planning*" which represents the responsible barrier elements for the proper planning of maintenance work. The second is "*prevent leak from poor maintenance execution*" which is under which the required barrier elements for the successful execution of isolation and maintenance work are classified. The determination of barrier elements for the indicated 3rd level sub-functions in Figure 3 has been conducted based on the presented information in the related literature to the context of this research. For example, the determination of the corresponding tasks to maintenance work in the process industry was based on Okoh & Haugen (2015), and Vinnem *et al.* (2016), for the process of LNG removal, purging and gas freeing tasks the presented information in SIGTTO (n.d.) was utilised, and Sklet (2006a) and Hauge & Øien (2016) were considered for the possible leak scenarios due to maintenance operations failure.

In addition, since the scope of the present study is on the resulting leaks from the maintenance task only, barriers against the resulting hazards from maintenance such as being a source of ignition or dropped objects should be addressed in other barrier functions. Similarly, poor scheduling or lack of barrier maintenance could be addressed under the prevent leak due to equipment/barrier degradation sub-function. On the other hand, poor quality of maintenance instructions and the required documentation for describing maintenance and isolation, as well as the quality of the control of work system are considered as PIFs to the operational barrier elements. However, Table 3 presents the established barrier elements for the presented two 3rd-level barrier sub-functions in Figure 3.

In the provided barrier elements in Table 3, it is assumed that organisational barrier elements involve three levels of organisations which are facility management (e.g. Captain), technical management (e.g.

chief engineer), and maintenance team (from ship crew, 3rd party, or combination of both).

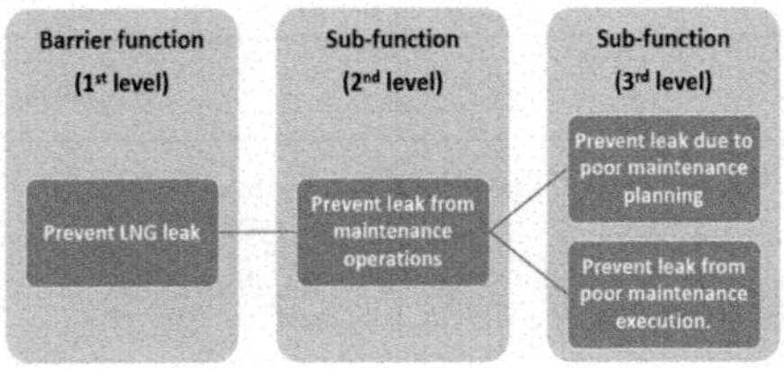

Figure 3. Barrier function breakdown.

4.2.3 *Performance requirements*

Performance requirements shall be identified for the established TO&O barrier elements to indicate the properties that a barrier element must possess to ensure the effectiveness of that barrier and its function (PSA 2013). In the present case, the gas sensor is the only technical barrier element in the proposed barrier sub-functions. Usually, gas sensors are calibrated regularly on-board the ship, and maintained and tested periodically by the authorised entity by the manufacturer to fulfil the integrity and efficiency requirements i.e. being as good as new after maintenance, as well as indicating acceptable results upon the pre-use calibration.

Furthermore, performance requirements of operational barrier elements (i.e. safety-critical tasks) are in the form of written procedures indicating how the task must be performed, for example, in the form of the required actions and their sequence, as well as the completion criteria of and the required time (e.g. response time for manual emergency shutdown (ESD)) for achieving each action, and the required resources for completing the task(s).

As indicated in DNV (2014), task analysis techniques, such as safety critical task analysis (SCTA) and human reliability analysis (HRA), are required for the accurate determination of performance requirements of operational barrier elements. However, such analysis is needed for critical and/or complex tasks (DNV 2014, p48). With regards to the present case study, maintenance, isolation, gas freeing and permit-to-work procedures are the reference for performance requirements in the presented barrier elements in Table 3. Therefore, such procedures must be consulted and communicated during job planning and followed during the work execution. An example of the components of the permit-to-work system is presented in section 9 of OCIMF & CCNR (2023).

Performance requirements of organisational barrier elements according to the adopted definition of these elements in this study will represent the requirements that an individual should possess for performing the safety-critical task efficiently. Therefore, performance requirements in this case represent the related requirements to competence, experience, or training courses. However, the determination of the performance requirements of the various organisations is performed according to the regulatory requirements, in addition to, the company-specific requirements. For instance, the minimum level of training requirements and competency level of each rank on-board LNGFSs are determined by the IMO in the resolution MCS.397(95) (IMO 2015).

4.2.4 *Performance influencing factors*

Maintaining the proper health of barrier elements depends on the condition of their PIFs. In addition, PIFs are the factors that can be monitored in operations to indicate the performance of safety barriers (Abdelmalek & Guedes Soares 2021). However, Table 4 presents a set of relevant PIFs of the presented TO&O barrier elements in Table 3. The selected PIFs in Table 4 are based on the presented information in DNV (2014) and Hauge & Øien (2016).

4.3 *Discussion*

Failure of maintenance operations of LNG equipment on-board LNG floaters can result in accidental leaks during the performance of the task due to deficient planning or execution of isolation, gas freeing, and equipment de-isolation and restoration, or after the completion of the task due to the erroneous planning and execution of the repair tasks. Therefore, safety barriers against leak-related failures should be adequately established and managed in operations to ensure the proper planning and safe completion of maintenance work on-board LNG floaters.

From one perspective, the adoption of the SBM concept ensures the proper identification and documentation of the implemented safeguards against the possible failure causes for leakage and the proper determination of their performance requirements and PIFs. From another perspective, the SBM methodology improves the process of monitoring and managing the performance of safety barriers in operations as represented in DNV (2014) and Hauge & Øien (2016) which also allows the identification of the status of the individual barrier as well as the status of the safety level on-board the ship.

Furthermore, human, and organisational factors which pose a primary challenge to the operational safety of some types of LNG floaters (Abdelmalek & Guedes Soares 2022) are properly addressed through the process of SBM. Also, the development of leading indicators to the PIFs of the TO&O barrier elements (Abdelmalek & Guedes Soares 2021) will improve the prediction of early warning signals of failure (or weak signals of disturbance (Hollnagel 2022)).

5 CONCLUSIONS

In this research, the proposed SBM framework in Hauge & Øien (2016) is adopted for establishing safety barrier elements for preventing leaks from failures of maintenance operations of LNG equipment on-board LNG floaters. For this purpose, background on the causes and the possible scenarios of LNG leaks due to failure of maintenance operations were presented, in addition to proposing a set of TO&O barrier elements based on the

Table 3. The established TO&O barrier elements.

Barrier element	Brief description of the objective of the barrier element	Failure aspect
Organisational barrier elements (relevant to both 3rd level barrier sub-functions)		
Performance of facility management	Performance of the person in charge of the ship who oversees and coordinates the conducted activities on-board (e.g. Captain/offshore installation manager (OIM), or his/her deputy).	Deficient planning, and checking.
Performance of technical management	Represent the performance of the responsible person on the technical team on- board the ship (e.g. chief engineer/maintenance supervisor, or his/her deputy).	Deficient planning and checking.
Performance of maintenance team members	The performance of personnel who are conducting the work (isolation, gas freeing, and repair) physically.	Deficient execution and maintainability.
Sub-function (3rd level). Prevent leaks due to failure of maintenance planning		
Operational barrier elements (Detection and prevention safety-critical tasks (DNV 2014)*).*		
Identify the required resources.	Technical management is responsible for determining the correct resources (e.g. manpower and the correct spare parts) for the maintenance work according to procedures, instruction manual, job safety analysis etc.	Deficient planning.
Plan for isolation work.	Technical management is responsible for preparing the proper plan of isolation to ensure that all sources of LNG and other flammable substances to the target equipment are isolated.	Deficient planning.
Conduct pre-job meeting	Pre-job meetings should be conducted by technical management with the involved maintenance team in the work to discuss the isolation and maintenance plan. This should be repeated during the handover with the maintenance team that coming on shift.	Deficient planning and checking.
Sub-function (3rd level). Prevent leaks due to failure of isolation and maintenance execution		
Technical barrier elements		
Gas concentration measuring sensor	Gas detectors that used for measuring the flammable gas concentration prior to dismantling the equipment.	Initiating event
Operational barrier elements (Detection and prevention safety-critical tasks (DNV 2014)*).*		
Acquire authorisation of isolation work.	The isolation certificate must be approved before the performance of the isolation.	Deficient checking.
Perform isolation work.	Perform isolation work according to plan including lockout/tagout (LOTO), system draining, and nitrogen purging as indicated, for example, in SIGTTO (n.d.).	Deficient execution, and initiating event.
Verification of isolation.	Technical management must verify the proper conduction of isolation as well as ensure that the equipment to be dismantled is purged sufficiently and depressurised before dismantling. As mentioned in SIGTTO (n.d.) gas concentration inside equipment should be less than 2% by volume before dismantling.	Deficient checking.
Issue permit to work	Permit to work must be issued and authorised by the facility and technical management before starting the work.	Deficient checking.
Conduct maintenance according to plan.	Maintenance work should be carried out according to the plan to avoid the introduction of hidden failures that can lead to LNG leaks during testing or normal operation. Also, segment isolation should be maintained till the completion of the maintenance work.	Deficient execution, initiating events, and lack of maintainability.

(*Continued*)

Table 3. (*Continued*)

Barrier element	Brief description of the objective of the barrier element	Failure aspect
Verify proper completion of maintenance.	Technical management must verify that maintenance work has been conducted according to plan, and equipment is properly installed.	Deficient checking.
Obtain the required authorisation for de-isolation.	Authorisation must be obtained from the facility and technical management before commencing the de-isolation of equipment.	Deficient checking.
Conduct de-isolation to restore equipment as normal according to plan.	The maintenance team should perform de-isolation of equipment according to the agreed plan during the pre-job meeting.	Deficient execution, initiating events, and lack of maintainability.
Verify the proper restoration of equipment.	Technical management must verify that the de-isolation of equipment has been conducted properly so that accidental leaks will not occur during testing or starting up of equipment.	Deficient checking.

Table 4. Proposed PIFs of the selected TO&O barrier elements.

Barrier elements *Technical*	*Operational*	*Organisational*
• Quality of maintenance management systems. • Quality of third-party inspection and maintenance. • Quality of maintenance and inspection actions and verification.	• Quality of procedures. • Quality of safety management system. • Ergonomics. • Complexity. • Working conditions (environmental).	• Generic competence. • Training and exercises. • Experience. • Individual workload. • Communications.

relevant literature to the scope of this study. However, in real-case situations, other aspects must be considered for establishing and verifying the adequacy of the ship-specific safety barriers, as well as for the determination of their performance requirements. Examples of those aspects are ship-specific data, regulatory and regional requirements that apply to ship, industrial and/or company-specific requirements etc.

Based on the conducted work herein it can be concluded that the adoption of the SBM concept is beneficial for improving the management of the operational safety on the ship, as well as on the fleet level through the continuous monitoring of PIFs that influence the performance of safety barriers through the identification of early warning signals of failure. In addition, performance indicators of safety barriers can present the required input for operationalising the concept of dynamic risk analysis (DRA) for critical marine LNG units.

ACKNOWLEDGEMENTS

This work was performed within the scope of the Strategic Research Plan of the Centre for Marine Technology and Ocean Engineering (CENTEC), which is financed by the Portuguese Foundation for Science and Technology (Fundação para a Ciência e Tecnologia) under contract UIDB/UIDP/00134/2020.

REFERENCES

Abdelmalek, M., & Guedes Soares, C. 2021. Performance-based leading risk indicators of safety barriers on liquefied natural gas carriers. in: Guedes Soares, C. & Santos T. A., *Developments in Maritime Technology and Engineering* 5 Vol. 1, London, UK: Taylor and Francis Group" 211–220.

Abdelmalek, M., & Guedes Soares, C. 2022. A review of failure causes and critical factors of maritime LNG leaks. in: Guedes Soares, C. & Santos T. A., *Trends in Maritime Technology and Engineering*, London, UK: Taylor and Francis Group; Vol. 2, 173–183.

Abdelmalek, M,, & Guedes Soares, C. 2023. Review of Risk Analysis Studies in the Maritime LNG Sector. *Journal of Marine Science and Application*, 22(4): 693–715. https://doi.org/10.1007/s11804-023-00376-0

Abdussamie, N., Daboos, M., Elferjani, I., Shuhong, C., & Alaktiwi, A. 2018. Risk assessment of LNG and FLNG vessels during manoeuvring in open sea. *Journal of Ocean Engineering and Science*, *3*(1): 56–66. https://doi.org/10.1016/j.joes.2017.12.002

ABS 2004. Consequence assessment methods for incidents involving releases from Liquefied Natural Gas carriers.

Ahn, S. I., Kurt, R. E., & Turan, O. 2022. The hybrid method combined STPA and SLIM to assess the reliability of the human interaction system to the emergency shutdown system of LNG ship-to-ship bunkering. *Ocean Engineering*, *265*: 112643.

Alderman, J. A. 2005. Introduction to LNG safety. In *Process Safety Progress* Vol. 24: 144–151. https://doi.org/10.1002/prs.10085

Andersen, H., Casal, J., Dandrieux, A., Debray, B., de Dianous, V., Duijm, N. J., … Tixier, J. 2004. Accidental risk assessment methodology for industries in the context of the Seveso II directive, The risk severity index. EVG1-CT-2001-00036.

Antao, P., & Guedes Soares, C. 2012. Risk assessment to the approach and berth of LNG vessels at the FLNG systems. In Guedes Soares, C., Garbatov Y. Sutulo S. & Santos T. A., (ed. 1). *Maritime Engineering and Technology.* Taylor & Francis Group: 425–432.

Betteridge, S., Hoyes, J. R., Gant, S. E., & Ivings, M. J. 2014. Consequence modelling of large LNG pool fires on water. In *Institution of Chemical Engineers Symposium Series* (Vol. 24:1–11. Edinburgh, UK.

Cohen, T. N. 2017. A human factors approach for identifying latent failures in healthcare settings. Embry-Riddle Aeronautical University, PhD Theses.

de Dianous, V., & Fiévez, C. 2006. ARAMIS project: A more explicit demonstration of risk control through the use of bow–tie diagrams and the evaluation of safety barrier performance. *Journal of Hazardous Materials*, *130*(3): 220–233.

DNV. 2014. Barrier management in operation for the rig industry - Good Practices. Høvik.

Fan, H., Enshaei, H., & Jayasinghe, S. G. 2022. Human error probability assessment for LNG bunkering based on fuzzy Bayesian network-CREAM model. *Journal of Marine Science and Engineering*, *10*(3): 333. https://doi.org/10.3390/jmse10030333

GIIGNL 2021. GIIGNL Annual Report 2021. *GIIGNL*. Paris.

Hauge, S., & Øien, K. 2016. Guidance for barrier management in the petroleum industry. *SINTEF report A* (Vol. 27623). Trondheim.

Hollnagel, E. 2008. Risk + barriers = safety? *Safety Science*, *46*(2): 221–229.

Hollnagel, E. 2022. Systemic potentials for resilient performance. In *Resilience in a Digital Age*: 7–17. Springer.

Hu, Z-q., Zhang, D-w., Zhao, D-y., & Chen, G. 2017. Structural safety assessment for FLNG-LNGC system during offloading operation scenario. *China Ocean Engineering*, *31*(2): 192–201. https://doi.org/10.1007/s13344-017-0023-4

IMO. 2015. Resolution MSC.397(95) (adopted on 11 June 2015) Amendments to Part A of the Seafarers' Training, Certification and Watchkeeping (STCW) Code, *397*(June). London.

IMO. 2018. Adoption of The Initial IMO Strategy on Reduction of GHG Emissions Recalling Article 38 (e) of the Convention on the International Maritime Organization (the Organization) concerning the functions of the Marine Environment Protection Committee. London.

IMO. 2019a. IMO 2020 – cutting sulphur oxide emissions. Retrieved December 13, 2023, from https://www.imo.org/en/MediaCentre/HotTopics/Pages/Sulphur-2020.aspx#:~:text=Known as "IMO 2020"%2C,were already stricter (0.10%25).

IMO. 2019b. Nitrogen Oxides (NOX) - Regulation 13. Retrieved December 13, 2023 from https://www.imo.org/en/OurWork/Environment/Pages/Nitrogen-oxides-(NOx)-–-Regulation-13.aspx

Jaramillo, P., Griffin, W. M., & Matthews, H. S. 2007. Comparative life-cycle air emissions of coal, domestic natural gas, LNG, and SNG for electricity generation. *Environmental Science and Technology*, *41*(17): 6290–6296. https://doi.org/10.1021/es063031o

Jeong, B., Park, S., Ha, S., & Lee, J-u. 2020. Safety evaluation on LNG bunkering: To enhance practical establishment of safety zone. *Ocean Engineering*, *216*: 107804. https://doi.org/10.1016/j.oceaneng.2020.107804

Margolin, S. 2013. *Flash Fire (duration and heat flux). Westex*. https://westex.com/webfoo/wp-content/uploads/Flash_Firepart11.pdf.

Martins, M. R., Pestana, M. A., Souza, G. F. M., & Schleder, A. M. 2016. Quantitative risk analysis of loading and offloading liquefied natural gas (LNG) on a floating storage and regasification unit (FSRU). *Journal of Loss Prevention in the Process Industries*, *43*: 629–653. https://doi.org/10.1016/j.jlp.2016.08.001

Mayo Clinic. 2022. Hypothermia. Retrieved December 14, 2023, from https://www.mayoclinic.org/diseases-conditions/hypothermia/symptoms-causes/syc-20352682

Mohajan, H. K. 2012. Munich Personal RePEc Archive Dangerous effects of methane gas in atmosphere. *International Journal of Economic and Political Integration*, *1*(2): 3–10. Retrieved from https://mpra.ub.uni-muenchen.de/id/eprint/50844

Nerheim, A. R., Æsøy, V., & Holmeset, F. T. 2021. Hydrogen as a maritime fuel–can experiences with LNG be transferred to hydrogen systems? *Journal of Marine Science and Engineering*, *9(7)*: 743. https://doi.org/10.3390/jmse9070743

Norwegian Oil and Gas. 2018. Application of IEC 61508 and IEC 61511 in the Norwegian petroleum industry (Recommended SIL requirements).

Nubli, H., & Sohn, J. M. 2022. CFD-based simulation of accidental fuel release from LNG-fuelled ships. *Ships and Offshore Structures*, *17*(2): 339–358. https://doi.org/10.1080/17445302.2020.1827806

OCIMF & CCNR. 2023. International safety guide for inland navigation tank-barges and terminal (ISGINTT) (2nd Editio).

Øien, K., Hauge, S., Størseth, F., & Tinmannsvik, R. K. 2015. Towards a holistic approach for barrier management in the petroleum industry. Trondheim: SINTEF.

Okoh, P., & Haugen, S. 2013a. Maintenance-related major accidents: classification of causes and case study. *Journal of Loss Prevention in the Process Industries*, *26* (6): 1060–1070.

Okoh, P., & Haugen, S. 2013b. The influence of maintenance on some selected major accidents. *CEt Chemical Engineering Transactions*, *(31)*.

Okoh, P., & Haugen, S. 2014a. A study of maintenance-related major accident cases in the 21st century. *Process Safety and Environmental Protection*, *92*(4): 346–356.

Okoh, P., & Haugen, S. 2014b. The implication of maintenance in major accident causation. *Loss Prevention Bulletin (236)*.

Okoh, P., & Haugen, S. 2015. Improving the robustness and resilience properties of maintenance. *Process Safety and Environmental Protection*, *94*: 212–226.

Oran, E. S., Chamberlain, G., & Pekalski, A. 2020. Mechanisms and occurrence of detonations in vapor cloud explosions. *Progress in Energy and Combustion Science*, *77*: 100804. https://doi.org/10.1016/j.pecs.2019.100804

Pitblado, R. 2007. Potential for BLEVE associated with marine LNG vessel fires. *Journal of Hazardous Materials*, *140*(3): 527–534. https://doi.org/10.1016/j.jhazmat.2006.10.021

PSA. 2013. Principles for barrier management in the petroleum industry.PSA,http://www.havtil.no/contentassets/11851dc03a84473e8299a2d80e656356/principles-for-barrier-management-in-the-petroleum-industry_2013.pdf

Rausand, M. (ed. 1) 2011. *Risk assessment: theory, methods, and applications.*Hobokin, New Jersey: John Wiley & Sons.
Reason, J. 1997. *Managing the risks of organizational accidents*. Aldershot: Ashgate.
Royal Society. 2008. Ground-level ozone in the 21st century: future trends, impacts and policy implications. Edinburgh, UK. Retrieved from http://www.royalsociety.org/Ground-level-ozone-in-the-21st-century-future-trends-impacts-and-policy-implications-/
Sanderson, K. 2008. How pollution can help to clean the air. *Nature*: 1–6.
SEPS. 2018. Cryogenic Liquid/Solid Safety. Glasgow.
Sever, C., Kulahcı, Y., Acar, A., & Karabacak, E. 2010. Unusual hand frostbite caused by refrigerant liquids and gases. *Ulusal Travma ve Acil Cerrahi Dergisi = Turkish Journal of Trauma & Emergency Surgery : TJTES*, *16* (5): 433–438. Retrieved from http://www.ncbi.nlm.nih.gov/pubmed/21038121
SIGTTO. n.d. LNG marine loading arms and manifold draining, purging and disconnection procedure. London.
Sklet, S. 2006a. Hydrocarbon releases on oil and gas production platforms: Release scenarios and safety barriers. *Journal of Loss Prevention in the Process Industries*, *19* (5): 481–493. https://doi.org/10.1016/j.jlp.2005.12.003
Sklet, S. 2006b. Safety barriers: Definition, classification, and performance. *Journal of Loss Prevention in the Process Industries*, *19*(5): 494–506.
Uflaz, E., Sezer, S. I., Akyuz, E., Arslan, O., & Kurt, R. E. 2022. A human reliability analysis for ship to ship LNG bunkering process under DS evidence fusion HEART approach. *Journal of Loss Prevention in the Process Industries*, *80*: 104887.
UNEP. 2007. How to cut greenhouse gas emissions and minimize global warming. Geneva.
Vandebroek, L., & Berghmans, J. 2012. Safety aspects of the use of LNG for marine propulsion. *Procedia Engineering*, *45*: 21–26. https://doi.org/10.1016/j.proeng.2012.08.114
Vinnem, J. E., Haugen, S., & Okoh, P. 2016. Maintenance of petroleum process plant systems as a source of major accidents? *Journal of Loss Prevention in the Process Industries*, *40*: 348–356.
Wang, Y., Wright, L., & Zhang, P. 2021. Economic Feasibility of LNG Fuel for Ocean-going Ships: A Case Study of Container Vessels. *Maritime Technology and Research*, *3*(2): 202–222. https://doi.org/10.33175/mtr.2021.248055
Wartsila. 2017. LNG as a marine fuel boosts profitability while ensuring compliance Market trends favour the greener choice. *Wärtsilä Services business white paper.*
Wu, J., Bai, Y., Zhao, H., Hu, X., & Cozzani, V. 2021. A quantitative LNG risk assessment model based on integrated Bayesian-Catastrophe-EPE method. *Safety Science*, *137*: 105184. https://doi.org/10.1016/j.ssci.2021.105184
Zhang, Q. xi, Liang, D., & Wen, J. 2019. Experimental study of flashing LNG jet fires following horizontal releases. *Journal of Loss Prevention in the Process Industries*, *57*: 245–253. https://doi.org/10.1016/j.jlp.2018.12.007

Advances in Maritime Technology and Engineering – Guedes Soares & Santos (Eds)
© 2024 The Author(s), ISBN 978-1-032-83099-5

Ship safety guard – safety analysis of buoy anchoring in bad weather

C.J. Bai, J.S. Chen, J.H. Tang, H.B. Sui & M. Yang
Maritime College of Dalian Maritime University, Dalian, Liaoning, P.R. China

ABSTRACT: Navigation aids are artificial signs that help guide ships to navigate, locate and mark obstacles and give warnings. They are called "water traffic lights". If navigation aids are displaced or lost in bad weather, it will bring great hidden dangers to water traffic safety, so the configuration of navigation aids anchorage is very important. By using the parameters related to typical navigation aids in a wind farm, and by counting the data of wind, wave and current in the navigation aids' layout area, the characteristics of different anchor types and mooring methods are compared and selected. By using the methods of modeling and simulation, the mooring safety of navigation aids in bad weather is comprehensively analyzed, also the influence of various factors on the mooring safety of navigation aids is evaluated. Then, the mooring configuration of navigation aids is determined. In the meantime, a complete mooring layout scheme is summarized. Finally, corresponding improvement measures and suggestions are put forward according to the experimental results.

1 INTRODUCTION

In the waterway safety system, the navigation mark is very important, and it plays an irreplaceable role in the field of marine environment monitoring, early warning and forecasting, disaster prevention and mitigation, resource development and maritime traffic (Wang et al. 2019).

At present, the cause of navigation aids in China is full of vitality, and there are more and more researches on navigation aids. However, it is still found that many external factors cause navigation aids to be abnormal, such as steep rise of water level, abnormal weather, ship collision, man-made destruction, etc (Yan et al. 2011), resulting in serious consequences, such as navigation aids failure, damage, loss and so on.

Nowadays, there are few research results on how to prevent the adverse effects of sea weather and bad sea conditions on navigation aids. Experts and scholars generally take ships as the research object, and summarize the related factors that affect the anchoring force, such as anchoring system and mooring mode (Gu et al. 1996), and discuss their respective characteristics and adaptation scenarios. The formulas of wind and water flow force and mooring force (Wang et al. 1985, Xia et al. 2010) are given, which are used to calculate the external load on the ship and deduce the anchor grasping force and anchor chain friction force (Lu et al. 2015). In the field of navigation mark safety, the existing papers are similar to those of ship anchoring safety, and the configuration mode of navigation mark anchorage system is also based on this, but there are still some defects in related research, such as not considering the characteristics of navigation marks, resulting in unscientific mooring mode; Without using long-term meteorological statistical data, it is easy to cause the anchorage distribution to be unsuitable for the local actual situation.

In addition, because the theoretical research is insufficient, it has not been fully applied to practice. In most projects, the layout of navigation aids still relies on the experience of engineers, blindly ensuring the safety of navigation aids, and adopting biased anchors and longer anchor chains, which is a great waste of materials.

Taking the navigation mark layout of a wind farm project as an example, according to the obtained local hydrometeorological data, this paper scientifically designs the navigation mark, determines the anchor type and anchor chain size, and enhances its anchoring safety. Finally, it summarizes the anchoring process, which provides reference for the anchoring design of related navigation marks.

2 HYDROMETEOROLOGICAL AND GEOLOGICAL CONDITIONS

This wind farm is located in the coastal area, near the traditional shipping channel, and various types of ships come and go frequently. In order to ensure the safety of ships and wind farms, provide real-time hydrometeorological information for wind power construction and ensure the smooth construction of the project, it is necessary to arrange navigation aids. The weather in

DOI: 10.1201/9781003508762-25

the area where the wind farm is located is complicated, and it is easily affected by tropical cyclones, cold waves, cold air and temperate cyclones all year round, and it is often affected by severe weather and sea conditions such as offshore squalls, storm surges and huge waves. In order to ensure the safety of navigation mark anchorage, this paper collects the long-term hydrometeorological data observed by local stations and designs the navigation mark anchorage.

2.1 *Wind*

Located in the west coast of Guangdong, this wind power project is one of the main areas where typhoons and tropical storms move and land in the northwest Pacific and South China Sea, and suffers from extreme windy weather caused by tropical cyclones all the year round. It belongs to the subtropical monsoon climate, with obvious monsoon activity. In winter, it is controlled by the continental cold high pressure, and the northeast wind prevails. In summer, it is in front of the southwest low pressure trough, and it is influenced by the subtropical high pressure, and the southeast wind blows more. The seasonal variation of wind speed is also obvious. Generally, the wind speed is higher in winter. Although windy weather sometimes occurs in summer due to the influence of tropical cyclones, the wind speed in winter is still higher than that in summer according to the monthly average wind speed.

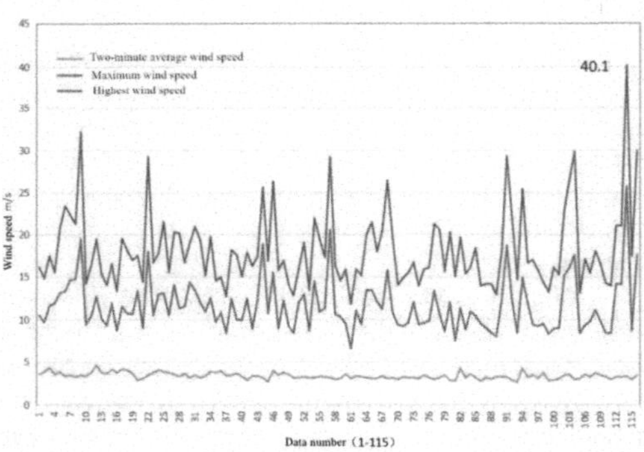

Figure 1. Wind speed statistics for recent 10 years.

In this paper, the wind speed in this area from 2013 to 2023 was obtained from the National Meteorological Center-China Meteorological Data Network, and the data mainly includes:

Maximum wind speed: the maximum instantaneous wind speed in a given period of time. The line chart shows that its maximum value is 40.1 m/s.

Maximum wind speed: the maximum value of the 10-minute average wind speed in a given period. As shown in Figure 1, its maximum value is 25.7 m/s.

Two-minute average wind speed: As can be seen from Figure 1, its maximum value is within 5 m/s.

According to the classification of wind speed by Frincis Beanfort, the maximum wind speed in this area can reach 13 (above 12 is a hurricane). The data show that the 13-level wind can generate more than 1000 kilograms of pressure per square meter, which is quite powerful. If it acts on the navigation mark, it will easily make it out of the safe design position and reasonable swing range.

2.2 *Current*

Usually, the current in the area where the navigation mark is laid tends to relatively stable, so the measured current data can be directly used. In winter, the measured maximum velocity is 0.58m/s and the minimum velocity is 0.20 m/s. The tidal current flow direction is concentrated between 230°and 290°; The tidal current direction varies widely, ranging from 50°to 140°in high tide, 70°to 150°in middle tide and 160°to 260°in low tide. In summer, the measured maximum flow velocity in the sea area where the navigation mark is located is 1.23 m/s. The tidal current direction is concentrated between 240°and 320°, and the ebb current direction changes between 30° and 110°. The duration of ebb tide in winter is longer than that of flood tide, and the average velocity of ebb tide is less than that of flood tide. In summer, the duration of flood tide is slightly longer than that of ebb tide, and the average velocity of flood tide is greater than that of ebb tide.

In the sea area where the project is located, the current basically flows along the coastline or seabed topography. In the deep water area, the current direction is generally SE ~ NW, and in the shallow water area (within 20 m), the current direction is roughly the same as the coastline direction. In addition, it also shows the vertical distribution characteristics of decreasing velocity from spring tide to neap tide and decreasing velocity from surface to bottom.

2.3 *Geological conditions*

The anchoring force of navigation aids mainly comes from the grasping force of anchor embedded in the bottom material and the friction between the seabed and the anchor chain. Table 1 mainly refers to the research of previous scholars (Chen et al., 2015) and shows that the composition of bottom

Table 1. The effect of buoy anchor system in different bottom material.

Bottom material	Ooze	Hard mud	Sandy mud	Sand	Sand shell	Gravel	Scree
Anchoring force coefficient	10	9	8	7	7	6	5
Friction coefficient of chain	3	2	2	2	2	1.5	1.5

material in different sea areas has great influence on the anchoring of navigation aids, and usually the bottom material is too hard or too soft and too loose to provide too much grasping force.

The wind farm project area is located offshore at a distance of 25 to 47 km, with a water depth of 29 to 37 m, and the sediments are mostly sand, clay and mud. The sea area in the site is wide, and no islands or reefs are found. According to the classification of wind farm topography in Code for Engineering Geological Survey of Land and Offshore Wind Farms (NB/T 31030-2012), the site belongs to offshore landform.

3 BUOY PARAMETERS

The buoy used this time is a large buoy with a diameter of 10 m, which has many advantages, such as strong wind and waves resistance, high performance, good maintainability, low comprehensive cost, stability, reliability and safety.

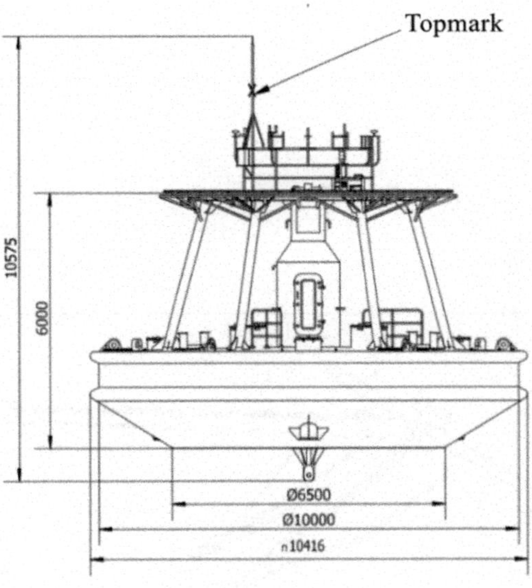

Figure 2. Structure and size of the buoy.

This buoy is developed and improved according to the application requirements of this project, based on the mainstream buoy product in China, the 10m-diameter large-scale ocean buoy system. The structural dimensions are shown in Figure 2, and the buoy body parameter information is shown in Table 2.

Table 2. Main parameters of buoy.

Buoy diameter	Pre-tender diameter	Moulded depth	Drinking water	Water discharge
10 m	6.5 m	2.4 m	1.1 m	57 t

3.1 *Buoy anchor system*

Anchorage equipment is the fixed equipment of floating navigation mark. Anchorage equipment consists of standard cable and anchorage. A beacon is a cable that binds the beacon body and the anchorage. Commonly used standard cables are anchor chain, synthetic fiber cable and steel wire rope. Anchorage is a device that fixes the target body to the bottom of the water by means of a standard cable. Commonly used anchors are rod anchors and sinking stones.

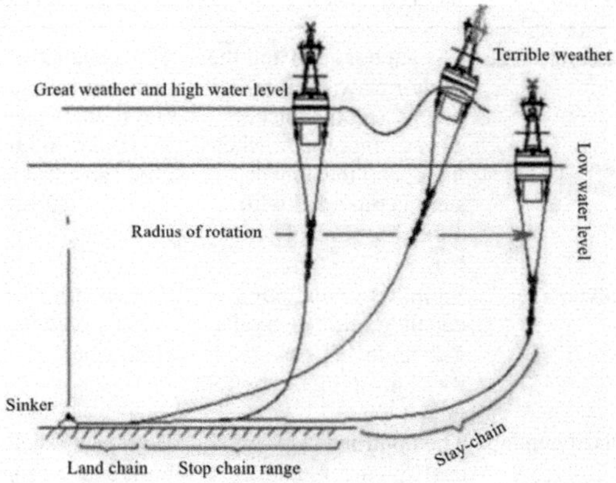

Figure 3. Sketch of buoy anchorage system.

This buoy mooring system is mainly composed of anchor, anchor chain, anchor chain swivel and connector, which is an important part of buoy monitoring system and an important part to ensure the normal work of the system, as shown in Figure 3. Its function is to provide a stable mooring force, and the balance buoy is subjected to the wind force above the water surface and the seawater flow force of the submerged part, so that the system can be moored and positioned for a long time in the harsh marine environment, without anchor and cable. The design of buoy anchor system is the focus of this paper, and then the characteristics of different anchor types and different mooring methods are compared.

3.2 *Selection of anchor type*

Different types of anchors can usually be divided into rod anchors and rodless anchors according to their structures, and can be divided into large-grip anchors and special anchors according to their functions, with different application scenarios, as shown in Table 3.

The structures of different anchors are shown in Figure 4-7.

In the navigation mark anchoring system, in order to ensure the anchoring safety, a large-grip anchor is usually used, and its gripping force is at least twice that of an ordinary rodless anchor with the same weight. Common ones are Ac-14 anchor, Danford anchor, Bohr anchor and Steven anchor. From

Table 4, it can be seen that the anchor holding force of different anchor types is not constant and has high dispersion under different bottom material.

According to the related research, Danforth anchor is a kind of anchor with big grip, with wide crown, thick claws and longitudinal edges, which can adapt to a variety of sediments and has good stability, and the grip-to-weight ratio is about 6. In addition, from the comparison of various types of anchors in Table 4, it can be seen that even in the case of anchoring, Danforth anchor can still provide a good grip, which is very conducive to coping with extreme weather such as strong wind.

After comprehensive consideration, Danforth anchor is selected for this buoy anchor system.

Table 3. Characteristics and analysis of various anchors.

Classify	Stocked anchor	Stockless anchor	Large grip anchor	Special anchor
Characteristic	The anchor stem and the anchor claw are integrated, and the anchor claw cannot rotate when fixed, and the anchor stem is provided with a fixed or foldable cross bar.	The anchor shaft and claw are cast separately, and there is no cross bar. The fluke and the anchor pipe can rotate around the pin shaft.	It mainly provides a large gripping force for mooring objects, with wide and long claws, deep soil biting and good stability, and can be divided into large gripping anchors with rods and large gripping anchors without rods.	Used for special purposes, such as permanent mooring of buoys, buoys, light boats, floating docks and floating docks.
Advantage	Simple structure, large catching ratio, generally 4-8, up to 12; The stability of grasping the bottom is good	Simple structure and convenient operation and collection.		
Disadvantage	The upturned claw is easy to entangle the anchor chain, and the shallow anchorage is easy to scratch the bottom of the ship; It is not convenient to anchor and put away.	Both paws shall be buried at the same time, and the middle grasping ratio shall be 2.5-4, not exceeding 8. The grip force is small, and it is easy to rake loose the soil during diversion, causing anchor walking.		

Table 4. Anchoring power of various anchors in different bottom material.

Anchor type	Hall anchor	AC-14 anchor	Danforth anchor
Sand bottom	3.5	7	5
Mud bottom	3.0	10	6
Dragging anchor	1.5	2	5.5

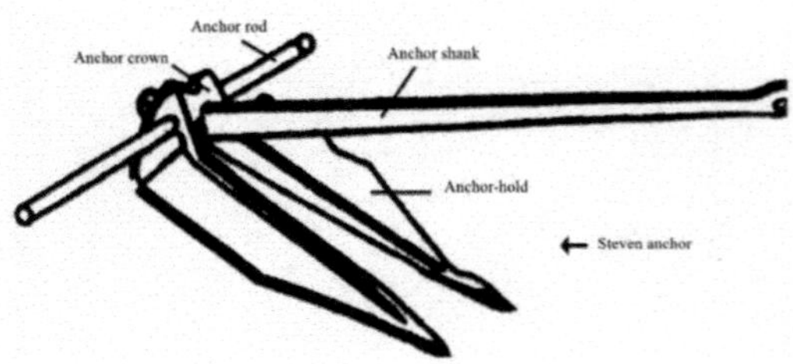

Figure 5. Anchors without rod.

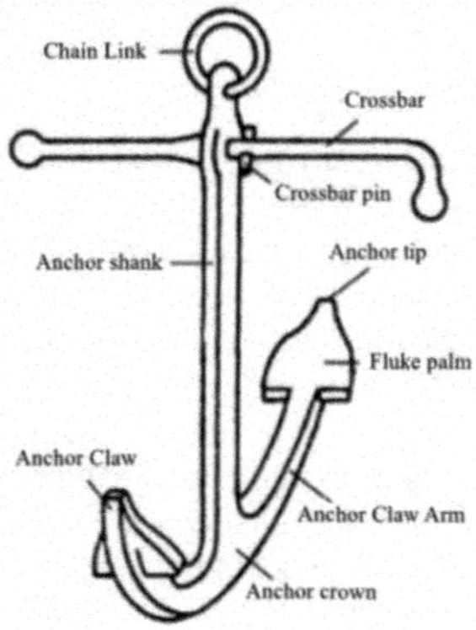

Figure 4. Anchors with rod.

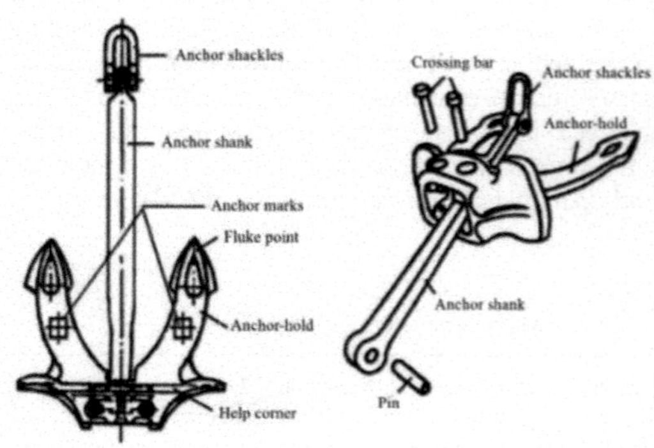

Figure 6. Big grip anchors.

Figure 7. Special Anchors (umbrella anchor, spiral anchor and single claw anchor from left to right).

3.3 *Mooring mode*

Anchor chain fixing system is an important part of anchoring at sea. According to the number of anchor points, the mooring system of ocean buoys can be divided into single-point mooring system and multi-point mooring system. Single-point mooring is a commonly used wind-proof mooring method for navigation aids in sheltered anchorage. Its most obvious force and motion characteristics are that the moored navigation aids can rotate around a single point by 360°, that is, the weathervane effect, which rotates in the direction that produces the minimum external force according to the wind, waves and currents, and the mooring force acting on a single point can always be kept to a minimum. Multi-point mooring is an improvement of the former, which is suitable for areas with very bad anchoring environment. The working procedure is complex but the safety factor is reliable. See Table 5 for the characteristics and application scope of different buoy anchoring methods.

Because the waters near the buoy deployment test point and the official point are relatively open, this project will choose a simple single-point mooring method to reduce the construction difficulty and deployment time and maximize the efficiency.

3.4 *Design parameters*

According to the buoy profile structure in this project, combined with the experience of coastal navigation mark engineering layout, it is planned to adopt the anchor chain mode with mature technology and high safety factor, which is suitable for the actual sea area environmental conditions of the buoy to be placed. As the main research part, the mooring parameters of buoys are designed, as shown in Table 6.

4 ANCHORAGE CONFIGURATION

The seabed bottom material at the planned launching point of the navigation mark is muddy and sandy, and the selection of floating body has also been determined. The force of the main anchor system of the navigation mark comes from the water surface target, which is mainly the wind, wave and current forces in different horizontal directions on the floating body itself. At the same time, the mooring system is subjected to the action of current under water (too small to be considered in this paper), and the combined force of them finally acts on the mooring system of the buoy. Therefore, the design basis of buoy anchor chain diameter and connection mode is the force of buoy under the worst sea conditions, that is, it is assumed that the wind, wave and current are in the same direction.

The research object of this part is to determine the appropriate anchor weight and anchor chain length, to ensure that the generated anchor force and anchor chain grasping force meet the requirements of safe anchoring of navigation aids, and at the same the time, to save anchor system materials as much as possible.

Table 5. Selection of buoy anchoring modes.

Anchoring mode	Characteristic	Area of application
Single point mooring mode	The moving range of the buoy body is large, and the arrangement is simple.	The waters are wide and calm, the wind and current directions are relatively consistent, or the water surface is open, and the requirements for the fixed position of buoys are relatively low, so it is suitable for deep-sea waters.
Multi point mooring mode	It can reduce the influence of ocean currents and waves on the horizontal movement of the surface buoy, ensure the buoy will not break anchor or run off the mark. Keep the moving range of the mark within a small range.	It is suitable for the offshore shallow sea (10-50 m), narrow waterway, anchorage place with poor sediment and high flow velocity in the sea area where the drift and deflection of floating body are small and the moving range of standard body is relatively fixed.

Table 6. Design parameters of buoy mooring system.

Anchor type:	Danforth anchor
Anchor weight	It is assumed to be *M* tons (to be calculated later)
Anchoring force coefficient	According to the change diagram of sediment and anchorage force, 6 is selected in this paper.
Anchor chain specification	Φ41
Full link length	27.5 meters/knot
Unit anchor chain weight	0.0353 tons per meter (in air)
Anchor chain material	CM490 grade anchor chain steel
Anchor chain length	Generally, the preset water depth is 2 to 5 times, and it is assumed to be *L* meters (to be calculated later).

4.1 *Wind load*

4.1.1 *Wind pressure*

At present, the empirical formula of wind pressure is:

$$P = \frac{1}{2}\delta_a \rho_a k_0 v^2 \quad (1)$$

Among them, δ_a is the air resistance coefficient, which is slightly different for different navigation aids, and the common range is 0.4 to 0.9; The air density ρ_a is 1.25 kg/m^3; The wind direction influence coefficient k_0 is lower than 1.2;V is the corrected wind speed (the wind speed in the same area varies with height, unit: m/s).

In this paper, in order to simplify the calculation, the Code for Classification of Offshore Floating Units (2014) is directly adopted, and the relationship between wind speed and wind pressure is as follows:

$$P = 0.613v^2 \quad (2)$$

4.1.2 *Wind power*

$$F_{wind} = C_s C_h S P \quad (3)$$

Where, C_s is the shape coefficient (dimensionless) of the wind-affected member; C_h is the height coefficient of the wind-affected member (dimensionless);S is the wind area (unit:m^2).

The vertical distribution formula of wind speed is:

$$\frac{v_z}{v_{10}} = \frac{\lg\left(\frac{z}{z_0}\right)}{\lg\left(\frac{10}{z_0}\right)} \quad (4)$$

v_z represents the wind speed at a height of z meters; v_{10} represents the wind speed at the standard height of 10 m, the paper selects the maximum wind speed of 40.1 m/s in Figure 1 for calculation; z_0 is the roughness of the ground, taking 0.003 m.

Corrected wind speed:

$$v_{0.625} = \frac{\lg\frac{0.625}{0.003}}{\lg\frac{10}{0.003}} * v_{10} = 26.39\text{m/s}$$

$$v_{3.87} = \frac{\lg\frac{3.87}{0.003}}{\lg\frac{10}{0.003}} * v_{10} = 35.41\text{m/s}$$

According to the structural diagram of the buoy, determine the wind area of the buoy:

S_1 is the wind area of the lower part of the buoy (13.34 m^2), and the centroid of the area is 0.625 meters from the ground;

S_2 is the windward area of the upper buoy part (11.14 m^2), and the centroid of the area is 3.87 meters above the ground.

Table 7. Shape coefficient C_s.

Component shape	Form factor
Sphericity	0.4
Cylinder-shaped	0.5
(Hull, Deck Room, Smooth Deck Lower Surface)	1.0
Deck room group or similar structure	1.1
Cable wire	1.2

Table 8. Height coefficient C_h.

Height above sea level (unit: m)	Height coefficient
0-15.3	1.00
15.3-30.5	1.10
30.5-46.0	1.20
46.0-61.0	1.30
91.5-106.5	1.48
106.5-122.0	1.52
122.0-137.0	1.56
137.0-152.5	1.60
152.5-167.5	1.63
167.5-183.0	1.67
183.0-198.0	1.70
198.0-213.5	1.72
213.5-228.5	1.75
228.5-244.0	1.77
244.0-259.0	1.79
Above 259	1.80

Calculate in combination with Table 7 and Table 8:

$$F_{w1} = C_s C_h S_1^* 0.613^* v_{0.625}^2 = 5695\text{N} = 609\text{kg}$$

$$F_{w2} = C_s C_h S_1^* 0.613^* v_{3.87}^2 = 11131\text{N} = 1136\text{kg}$$

4.2 *Flow load*

According to the structural diagram of the buoy, get the force of the water on the buoy is:

$$F_{current} = \frac{1}{2} C\rho A V^2 = 6877\text{N} = 702\text{kg} \quad (5)$$

Where C is the resistance coefficient of the buoy in seawater, which is related to the shape of the buoy, and takes 1 (dimensionless); The density of seawater $\rho$is 1025 kg/m; The current receiving area of the buoy A is 8.87m^2; $V$is the seawater velocity. According to the local current conditions, the maximum velocity in winter is 1.23 m/s.

4.3 *Resultant force of wind and current*

Summarizing the calculation formulas of wind load and current load, it is found that they are directly proportional to the square of wind speed and the square of seawater velocity, and the image presents a quadratic curve, as shown in Figure 8.

Obviously, the velocity of seawater has a greater influence on the external force generated by the buoy. The seawater velocity of 3 m/s can generate a flow load of more than 4000 kg, and the wind speed needs to reach 60 m/s.

4.4 *Anchoring force*

The anchoring force of an anchor, also known as the mooring force, refers to the binding force on the navigation mark when it is anchored. Under the action of gravity, the chain length of the anchored ship is divided into two parts, the part hanging in the water is called catenary length, and the part lying on the seabed is called undercover chain length, as shown in the figure. The contact between undercover chain length and geology produces friction and forms anchor chain grip, which is often ignored in previous studies. This paper takes it into account in the calculation process.

Statistics of the weights of anchor chains with different chain diameters in water and air are shown in Figure 9, and the following relationships are summarized:

$$W_c = 0.87W_a \tag{6}$$

W_a is the weight of the anchor chain in the air; W_c is the weight of the same length of anchor chain in water.

The anchoring force of an anchor, also known as the mooring force, refers to the binding force on the navigation mark when it is anchored. Under the action of gravity, the chain length of the anchored ship is divided into two parts, the part hanging in the water is called catenary length, and the part lying on the seabed is called undercover chain length, as shown in Figure 10. The contact between undercover chain length and geology produces friction and forms anchor chain grip, which is often ignored in previous studies. This paper takes it into account in the calculation process.

The anchoring force is divided into two parts: the grasping force of anchor and the grasping force of anchor chain:

$$F_{\text{anchor}} = F_a + F_c = 6000M + 21.8 * (L - 85.25) \tag{7}$$

F_a is the anchoring force (the grasping force of the anchor embedded in the ground); Grasping force F_c is for anchor chain (friction between anchor chain and seabed); λ_a is the anchor grip coefficient; M is the weight of the anchor in the air (t); λ_c is the grip coefficient of anchor chain, and the mud bottom is 0.6; W_c (0.0353 t/m)is for the mass of anchor chain in air; l is the undercover chain length (m).

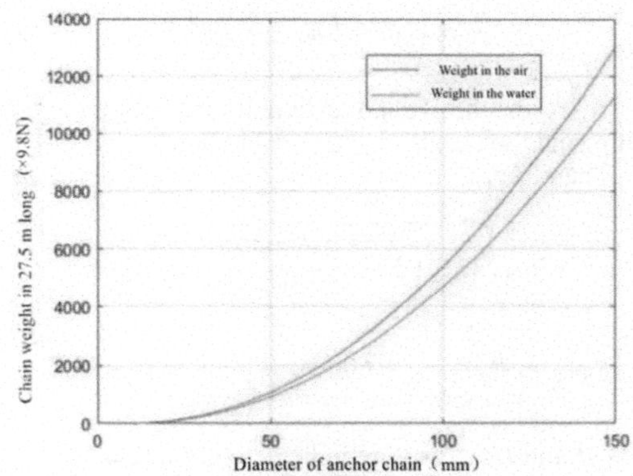

Figure 9. Weight of anchors with different chain diameters in water and air.

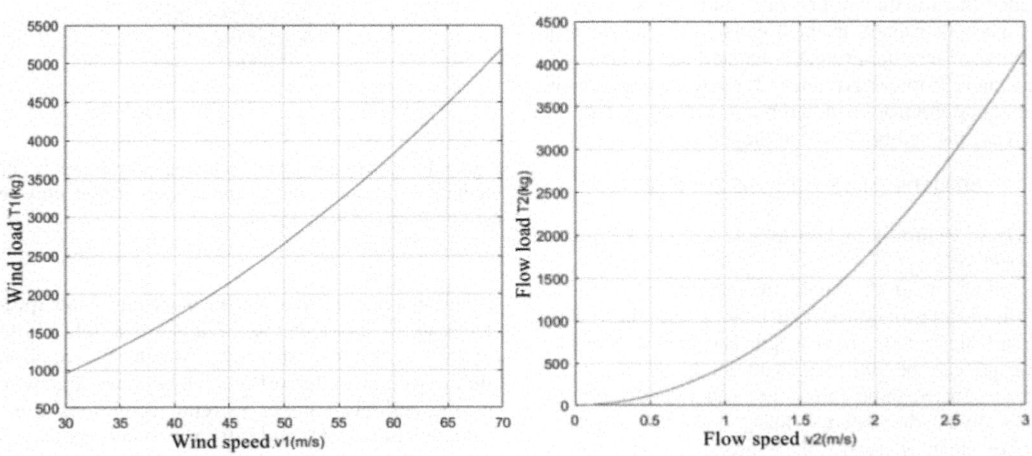

Figure 8. Relationship between wind speed (current speed) and wind load (current load).

The process of calculating the anchor chain grasping force F_c is as follows:

$$F = 1000 * 6 * M = 6000M(kg)$$

First, find the length of undercover chain = the length of outgoing chain-the length of catenary:

$$l = L - s \tag{8}$$

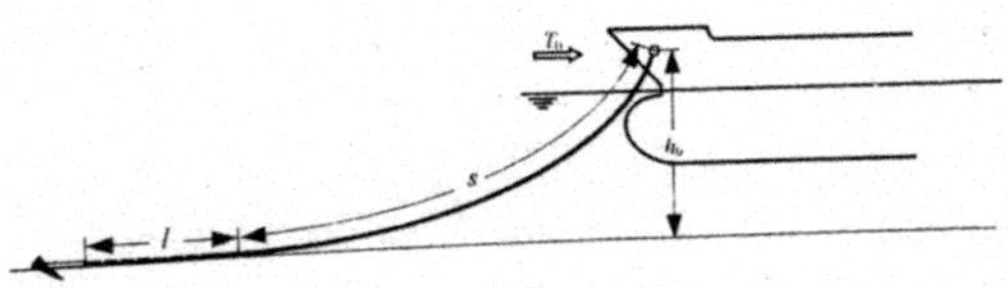

Figure 10. Grip schematic of chain.

From mechanical knowledge and linear integral, we can get:

$$s = \sqrt{h_0 * \left(h_0 + \frac{2 * T_0}{W_c}\right)} \tag{9}$$

T_0 is the horizontal external force (kg) on the ship, and the size is $F_{\text{external}} = 2447\text{kg}$.
Therefore,

$$s = \sqrt{37 * \left(37 + \frac{2 * 2447}{30.7}\right)} = 85.25\text{m}$$

$$F_{\text{anchor}} = F_a + F_c = 6000M + 21.8 * (L - 85.25)$$

(M and L are waiting to be determined).

4.5 *Analysis of results*

According to the above calculation, it can be concluded that the maximum horizontal external load on the navigation mark in the area is 2447. Because the sea area where the project is located has large swells and many tropical cyclones, the external load of the horizontal environment can be multiplied in order to increase the safety factor, so there are:

$$6000M + 21.18 * (L - 85.25) > 2447n$$

Draw Figure 11 when taking n at 1, 1.5 and 2 respectively.

Among them, L should be between 85.25m and 185m (5 times of water depth), and a feasible solution of suitable anchor chain length and Danforth anchor tonnage can be obtained accordingly. In particular, when L takes three times the water depth (111 m) under twice the external load, since the length of anchor chain is usually three times the water depth, M values 0.72t.

Because the calculation result of anchor chain gripping force is small, in the case of only considering anchor gripping force, Figure 12 of minimum anchor weight corresponding to different seawater velocity and wind speed is drawn for reference of engineering layout.

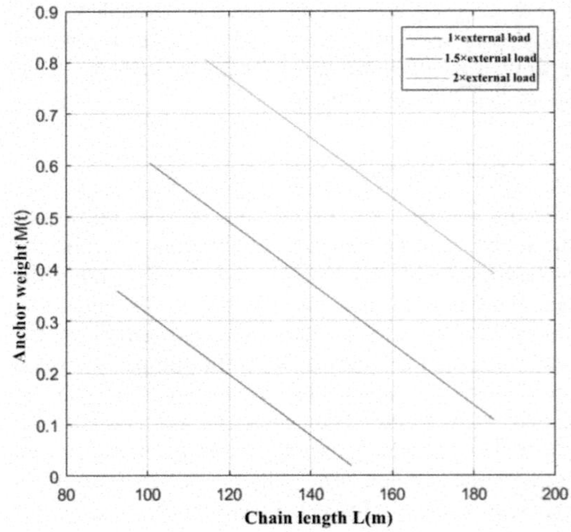

Figure 11. Relation between anchor weight (M) and chain length (L) under different load conditions.

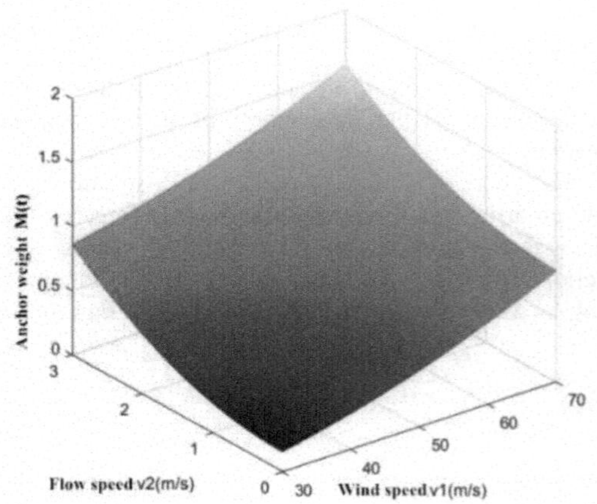

Figure 12. Considering the anchoring power merely, the minimum anchoring weight corresponds to the different current speed and wind speed.

The increase of seawater velocity and wind speed can lead to the increase of anchor weight. From Figure 12, we can directly estimate the required anchor weight under different flow rates and wind speeds, and even under extreme conditions (3m/s seawater flow rate and 70 m/s wind speed), 3t anchor (the anchor weight used in most projects) is enough to ensure the safety of navigation aids.

Table 9. Grip provided by unit weight of anchor and each knot of chain in different bottom material.

Bottom material	Anchoring force coefficient	Friction coefficient of chain	Anchoring force provided by 1-ton anchor	Anchoring force provided by 27.5-meter anchor chain
Ooze	10	2	10000	1165
Hard mud	9	1.5	9000	874
Sandy mud	8	1.2	8000	699
Sand	7	1.2	7000	699
Sand shell	7	1	7000	583
Gravel	6	1	6000	583
Scree	5	1	5000	583

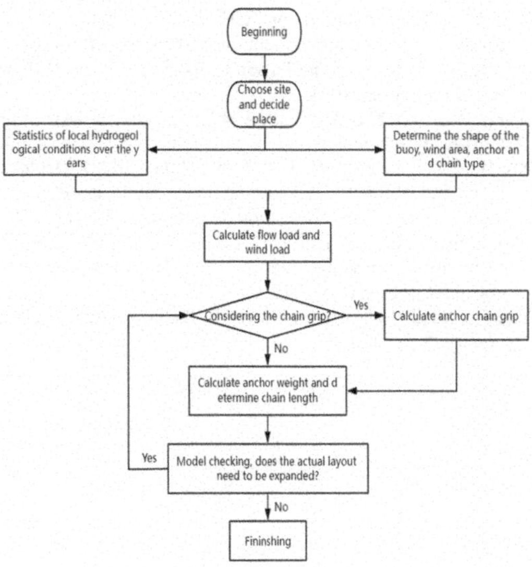

Figure 13. Anchor and chain of navigation aids arrangement diagram based on local hydrology and meteorology.

5 DISCUSSION

According to the above calculation process, the flow chart of anchor weight and anchor chain distribution based on local hydrometeorology can be obtained, as shown in Figure 13.

In addition, due to the great influence of different sediments on the anchoring safety of navigation aids, this paper calculates the unit anchor weight and the grasping force provided by each anchor chain under different sediments for reference, as shown in Table 9.

Through the in-depth analysis of the anchoring safety of navigation aids in bad weather, it is found that the external load on navigation aids is closely related to the shape of navigation aids, the area affected by wind and current, and the height of centroid, and is approximately proportional to the square of local seawater velocity and the square of wind speed. However, the anchoring force of navigation aids mainly comes from anchor grasping force and anchor chain grasping force, and is influenced by different geological conditions, anchor types, anchor weights, mooring methods and anchor chain length.

6 CONCLUSION

Meteorological element is an important factor affecting the safety of navigation mark anchoring. Proper selection of the strength and length of the anchor chain or rope, and proper anchoring position can help to cope with the specific wind direction. Waves and tidal currents also have an important influence on the anchorage of navigation marks. Choosing a stable anchorage can reduce the impact of waves and tidal currents on anchorage and ensure the stability of navigation marks. The state and reliability of mooring equipment are very important for the safety of navigation mark mooring. Regularly check and maintain the anchor chain, anchor rope, anchor rod, slip ring and other equipment to ensure their normal work and prevent damage or fracture. Anchoring plan and emergency response measures are the key to ensure the safety of navigation mark anchoring. Make a detailed anchoring plan, including the selection of anchoring position, the length of anchor chain or rope, the weight of anchor itself, pay close attention to weather forecast and alarm, and adjust the anchoring plan in time to deal with sudden bad weather.

This paper is based on the existing data and model analysis, and the actual situation may be different due to various factors. Therefore, in actual operation, it is necessary to adjust and optimize according to the specific situation, and strengthen the ability to deal with emergencies. Finally, in view of the importance of anchoring safety of navigation aids in bad weather, future research can further explore new methods and technologies to improve anchoring safety and cope with changing bad weather conditions.

REFERENCES

C Bai, C.J., Lu, Y. & Sun, G. (2017). Study on the safety chain scope of LNG carriers anchoring in designated anchorage. Cybernetics and Computational Social Systems (ICCSS): 40–43.

Chen, Z.B. & Chen, W.W. (2018). Analysis on the configuration scheme of anchorage equipment for floating signs in special environment. Pearl River Waterway (24): 38–39.

Chen, C.Y., Dai, R., Jiang C., Wang H.M. & Shou Z.Y. (2015). Anchorage safety of Dalian LNG anchorage. China Science and Technology Information (17): 23–25.

Dai, H.L., Mou, N.X., Wang, C.Y. & Tian, M.Y. (2014). Current situation and trend of China's marine buoys. Meteorological and Hydrological Marine Instruments 31 (02): 118–121.

Gu, Y.J. & Yu, X.L. (1996). Estimation of anchoring force and its change. World Shipping (04).

Jin, Y.X. (2016). Analysis and discussion on the displacement and drift of light buoys. Pearl River Waterway 418 (19): 54–55.

Lu, Y. & Bai, C.J. (2015). Dragging Anchor Event and Theoretical Verification of Single Mooring Ship. 2015 Ninth International Conference on Frontier of Computer Science and Technology (FCST): 209–213.

Lu, K., Rao, X., Wang H.M., Zhang, Q. (2023). Influence of different mooring systems on hydrodynamic performance of marine data buoys. Journal of Military Engineering: 1–11.

Ni, H.J., Jiang, Z.L., Chu, X.M. & Nie, Z.Y. (2019). Analysis of the drift characteristics of navigation AIDS and research on anti-collision strategies. China Intelligent Transportation Association: 10.

Sang, L.Z. & Hong, S.X. (2020). Perfection and Optimization of Navigation Mark Telemetry and Remote Control System in South China Sea. China Navigation 43(02): 35–40.

Wang, J.C. (2022). A new generation of ocean monitoring technology-integrated intelligent observation buoy. Journal of Intelligent Systems 17(03): 447.

Wang, J.C., Li, Y.Z., Yang, Y.D., Liu S.X., K,Q.L., Zheng L. (2013). Research status and development trend of attitude information measurement technology of marine data buoys. Acta Automatica Sinica 39(6): 723–729.

Wang, L., Wang, Z.Q., Han Y., Yang, T.X., Song, S.Z. (2022). Study on temporary emergency anchorage of LNG based on gyration radius. China Waterway 22(05): 71–73.

Wang, J.C. & Li, Y.Z. (2019). Development and application of marine data buoy technology in China. Shandong Science 32(05): 1–20.

Wang, M.Q. (1985). Determination of ship's release length at anchor and its practical chart. China Shipbuilding (02).

Xia, Y.Q., Tang, X.N., Jiang, K.H. & Wang, L.Q. (2010). Study on theoretical calculation formula of mooring force of windproof single-point mooring system. Waterway Engineering (04).

Wang, Y.Z. (2021). Navigation science. Dalian Maritime University Press: 9.

Xu, H., Liu, Y., Zhou, W.Q. & Zhang, Y.Q. (2022). Deployment and recovery of mooring data buoys in deep-sea thermohaline chain. Navigation Technology 253(01): 4–7.

Yan, G.Q. & Pang, R.Y. (2011). Cause analysis and control of navigation aids disorder. Western Transportation Science and Technology (02): 90–92.

Yang, L.X. (2016). Navigation mark. Dalian Maritime University Press: 10.

Advances in Maritime Technology and Engineering – Guedes Soares & Santos (Eds)

Weighted grey relational analysis of main engine faults on ships in port areas

C.J. Bai, H.B. Sui, Z.L. Yang, J.S. Chen & M. Yang
Maritime College of Dalian Maritime University, Dalian, Liaoning, P.R. China

ABSTRACT: In this paper, we will make statistics on the data of the main engine failure of a large domestic port area, and analyze the condition of the ship where the main engine failure occurs. The weighted grey correlation analysis method is used to weight the factors affecting the failure of the main engine of the ship, and the correlation between the failure of the main engine and the characteristics of the ship is studied, and the correlation degree between the failure of the main engine and the failure event of the main engine is calculated. The study found that the age of the ship, the main location of the failure, and the deadweight of the ship were the main factors. Finally, based on the analysis results, this paper puts forward suggestions for preventing the occurrence of host failure, which can provide reference for port management departments.

1 INTRODUCTION

The design and manufacture of modern ships are developing in the direction of gradually larger hulls (Li 2023), and such ships are bound to have complex structures. As the power unit of the whole ship, the ship main engine is the heart of the ship, and its safety and reliability are very important. If the ship's main engine can't work normally, it is difficult to ensure the safety of ship navigation. In the same way, in the harbor area, the failure of ship main engine will lead to the already very busy waterway to become more congested, in the limited navigation resources ship main engine failure will lead to more tense navigation resources, the probability of ship accidents will also increase. In addition, the failure of the main engine of the ship in the harbor area will also occupy and consume the re-sources of VTS, tugboats, rescue vessels, etc., and consume a lot of manpower and material resources of the relevant units.

As the heart of the ship, the technical performance and working condition of the main engine have great influence on the navigation safety, especially when the load of the ship changes frequently (Luo & Yu 2023), such as entering and leaving the harbor, berthing at the pier, and inclement weather, etc. The duration of the main engine failure is short. The duration of ship main engine failure is as short as tens of minutes, or as long as a few days, although most of them will not cause ship collision, channel closure, grounding and other serious accidents, but it will adversely affect the safety of navigation of the surrounding ships and the efficiency of port navigation, and become an important factor of the risk of waterborne traffic safety. According to "Hein's Law", behind every serious accident, there must be 29 minor accidents and 300 attempted precursors and 1,000 accidents hazards. And ship failure ac-cording to the port data, the number of monthly fluctuations in the number of failures, but the fault repair time is generally long, and even more than 24 hours; once the failure of the ship tonnage is too large, the need for port departments to assist, a large amount of consumption of manpower and material resources, occupying the waterway access.

Gray correlation analysis on the number of data and information accuracy requirements are not high and do not require the sample data with a typical probability distribution, the calculation of this analysis method is simple and very consistent with the characteristics of the analysis of various types of ship accidents, so gray correlation analysis can be used as a reference to analyze the causes of ship marine accidents. However, the traditional gray correlation analysis method has certain limitations, such as the lack of the curve and reference to the ontological trend of the grasp of the various types of factors in the accident itself and the weight of the proportion of the accident itself, so this paper adopts the weighted gray correlation analysis method for the analysis of main engine failures of ships in the port area, which can be more intuitive under-standing of the factors related to the occurrence of main engine accidents of the ships in the port area. In this paper, the age of the ship that has main engine failure, the location of the failure and the type of the ship are taken as the factors, and the correlation matrix is established on the basis of the statistics of the number of ship failures per month, the ship's load, and the length of time for repairing the failures, etc. Afterwards, the weighted gray correlation analysis is conducted to derive the correlation degree after the weighting, and the result is

DOI: 10.1201/9781003508762-26

ranked to find out the most strongly correlated factors. The results of this paper can provide a reference for the statistical analysis of ship main engine accident data in harbor area.

2 WEIGHTED GRAY CORRELATION THEORY

2.1 *Introduction to weighted gray correlation analysis*

The advantage of gray correlation analysis (Sun 2007) is that it can deal with small samples, incomplete data and uncertainty problems, and it is also well adapted to the case of uneven data distribution.

The weighted gray correlation analysis method is based on the gray system theory, which is a system analysis method based on incomplete information, uncertainty and stochasticity, and is suitable for dealing with systems with small amount of data, insufficient information, and nonlinear relationships. Gray correlation analysis requires the se-lection of a reference series and a comparison series, which represent the research object and the reference object, respectively. The reference and comparison series are standardized so that their values are between 0 and 1, which is convenient for comparison and calculation. Calculate the difference sequence between the reference and comparison series, accumulate and process the difference sequence to obtain a correlation coefficient, which is used to indicate the degree of correlation between the two series. By comparing the correlation coefficients between different series, the degree of correlation and mutual influence between them can be analyzed. At the same time, according to the size of the correlation coefficient, it can also be ranked and sorted to select the most relevant data series. The calculation uses Dunn's correlation, which, as the earliest gray correlation model, focuses on the influence of the distance between the data on the correlation, and then weights it to synthesize the key correlation factors in the model.

2.2 *Weighted gray correlation analysis theoretical model*

In order to improve the accuracy of the analysis of the factors related to the failure of the main engine of the ships in the port area, the weighted gray correlation analysis model was introduced (Liu et al. 2022,Wu, Liu & Fang 2023), and its basic principles are as follows:

Assume X= $\{x_i|i=0,1,2,, m\}$ is the set of sequence correlators, X_0 is the reference factor (parent factor), X_i is a comparative factor (sub-factor), X_i (k) is the value at point k, where k=1, 2, 3, …, n.

Let X'=(X'(1),X'(2),X'(3),, X'(n)) be the cumulant generating sequence of the sequence X'(1) = X (1),X'(k)=X(k)- X(k-1), k=2,3,4, …, n.

For X_0 and X_i, first secondly the sequence of variables is homogenized.

The matrix formed after homogenization becomes: $(X_0^*, X_1^*, X_2^*, X_3', \ldots, X_n^*)$

For after that the sequence of features is derived:

$$x_i' = \frac{x_i}{x_i^*} \cdot i = 0, 1, 2, 3, \ldots, n \quad (1)$$

Afterwards find the sequence of differences, the maximum and minimum differences

$$\Delta_i = |x'_0 - x'_i|, i = 1, 2, 3, \ldots, n \quad (2)$$

Form the matrix of absolute differences:

$$(\Delta_1, \Delta_2, \Delta_3, \ldots, \Delta_n)$$

Also find the largest number (maximum difference) in the absolute difference matrix and denote it respectively as Δ_{max}.

The cumulative generating sequence is also performed after the same steps.

Calculate the weighted gray correlation coefficient

$$\xi_i(k) = \frac{\varepsilon * \Delta_{max}}{\lambda_1 * \Delta_i + \lambda_2 * \Delta_i' + \varepsilon * \Delta_{max}} i = 1, 2, 3, \ldots, n \quad (3)$$

Where: λ_1 for the displacement weighting coefficient; λ_2 for the rate of change weighting coefficient are generally 0.5; ε for the resolution coefficient ∈ (0,1), the value can't be too small too small may lead to changes in the degree of correlation; the value can't be too large too large may lead to the correlation of the resolution of small. After reviewing the relevant literature to take ε = 0.5 or 0.6 is more appropriate at this time on the basis of which the resolution can be improved, for the convenience of calculation, this paper takes ε = 0.5.

Gray weighted correlation

$$\gamma_i = \gamma(X_0, X_i) = \sum_{k=1}^{n} \beta_k \gamma[X_0(k), X_i(k)] \quad (4)$$

Where: β_k is the weight coefficient of factor k normalized.

In this paper, the whitening weight function method is used to determine the weight coefficients of each factor, and the whitening weight function is set as

$$f(x) = xe^{1-x} + (1-x)e^x - 1 \quad (5)$$

Known sequences:

$x_i(k)=(x_i(1), x_i(2), x_i(3), \ldots, x_i(k), \ldots, x_i(n))$, i=1,2, 3, …, m; k=1, 2, 3, …, n.

First sum the attribute factors in each sequence:

$$D_k = \sum_{i=1}^{m} x_k(i) \quad (6)$$

After that the factor entropy is derived:

$$e_k = \frac{1}{n}\sum_{i=1}^{m} f\left(\frac{x_i(k)}{D_k}\right) \tag{7}$$

Find the entropy sum:

$$E = \sum_{k=1}^{n} e_k \tag{8}$$

Find the relative weights:

$$\lambda_k = \frac{1 - e_k}{m - E} \tag{9}$$

Finally, the weights of the factors are derived using the regularization method:

$$\beta_k = \frac{\lambda_k}{\sum_{k=1}^{n} \lambda_k} \tag{10}$$

3 ANALYSIS OF RESULTS

3.1 *Mainframe failure statistics*

Statistics on the data of main engine failure of ships in a port area from January to September found that most of the main engine failures occurred in the port area did not cause accidents of general grade and above such as collision, fire, grounding, etc., but they affected the safety of navigation of the surrounding ships and the navigational efficiency of the port, and became a risk factor for the safety of waterborne traffic in the port area, and the following results were obtained by organizing the obtained data:

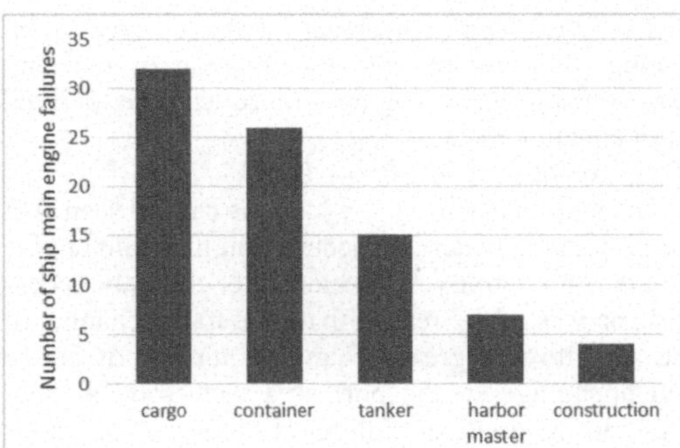

Figure 1. Number of types of vessels.

In terms of vessel types, the types of vessels that have experienced main engine failures in the port area include bulk carriers, container vessels, harbor craft, general cargo vessels, oil tankers, and engineering vessels, etc. Most of these vessels are commercial vessels such as bulk carriers, container vessels, and oil tankers. Most of them are mainly commercial vessels such as bulk carriers, container ships and oil

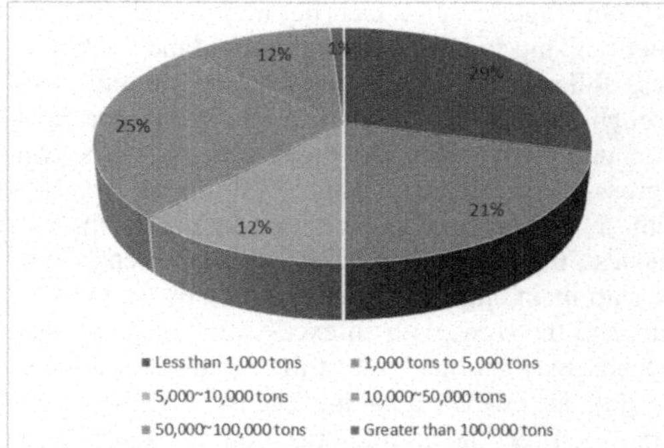

Figure 2. Distribution of Ship's deadweight tonnage.

tankers. As the number of engineering ships and harbor crafts in the harbor area does not change much, the number of main engine failures of these two types of ships in the harbor area is not large.

As shown in Figure 2, from the point of view of the total tons of ships in danger, the ships less than 50,000 tons accounted for the majority of ship main engine failure, after statistics and combined with the actual analysis can be judged to be due to the majority of oil tankers and Hong Kong as well as other port area work ships will not be more than 1,000 tons of total tonnage, of which 1,000 to 5,000 tons of the ship type is the most common, more than 5,000 tons of the general for the bulk carriers and container ships.

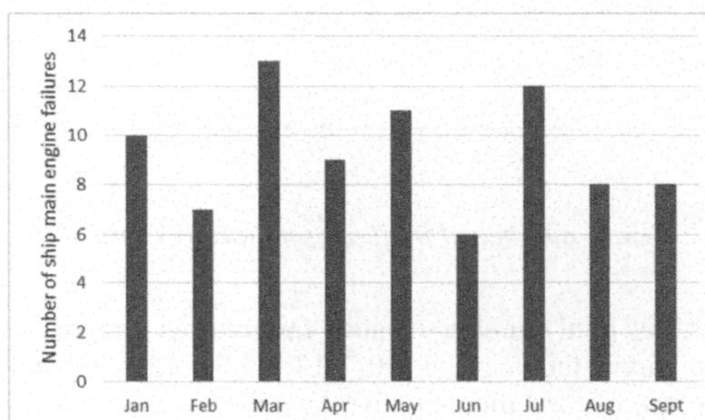

Figure3. Distribution of Ship's fault repair duration.

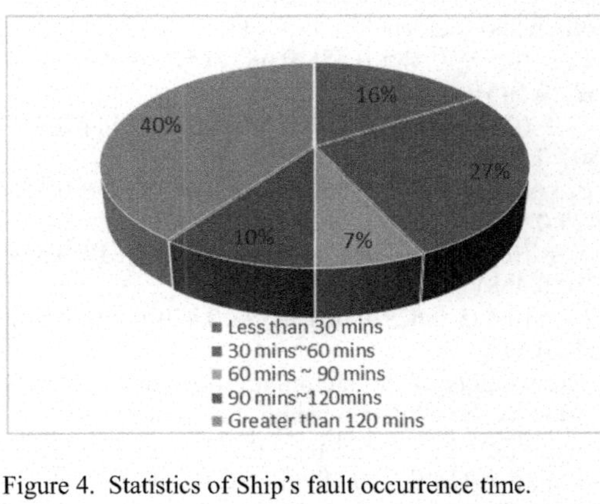

Figure 4. Statistics of Ship's fault occurrence time.

From Figure 3, it can be understood that the length of ship failure is generally more than 120 minutes, followed by 30 minutes to 60 minutes, and through the statistics of ship failure in the port area, it is understood that the proportion of ships with a gross tonnage greater than 5,000 among the ships with a failure length of more than 120 minutes reaches 50%, which can be seen that the repair time for ship main engine failures is generally on the long side, and the ships with an excessively large carrying capacity may seriously affect the normal operation of the port area when a failure occurs. Port area may seriously affect the normal operation of the port area.

Combined with the data obtained in the port area, the statistics are plotted in Figure 4. From the perspective of main engine failure time, the number of ship main engine failures occurred most frequently in the first quarter, followed by the third quarter, with little fluctuation on the whole.

Table 1. Number of Vessel Failures in the Port Area by Age Category and Month.

Age of vessel (years)	Jan	Feb	Mar	Apr	May	Jun	Jul	Aug	Sept
1~5	2	0	0	1	1	0	2	1	1
6~10	2	4	7	2	8	1	8	3	2
11~15	1	1	4	5	1	1	1	3	5
16~20	2	0	0	1	1	1	0	1	0
>20	3	2	2	0	0	3	1	0	0
Total faults	10	7	13	9	11	6	12	8	8

3.2 *Total number of main engine failures and ship age*

Let the total number of ship failures per month be the parent factor X_0 = (10 7 13 9 11 6 12 8 8), and the age of the failed ship be the sub-factor X_i(i=1, 2, 3, 4, 5), the weighted correlation coefficients of the total number of accidents and age of the ship can be obtained as the sequence of weighted correlation coefficients calculated by Eqs. (1), (2), and (3) as:

ξ_1 = (0.595 0.485 0.558 0.683 0.529 0.596 0.574 0.623 0.546)

ξ_2 = (0.549 0.832 0.862 0.595 0.715 0.644 0.745 0.807 0.585)

ξ_3 = (0.550 0.696 0.550 0.537 0.442 0.859 0.466 0.424 0.587)

ξ_4 = (0.460 0.462 0.528 0.585 0.492 0.618 0.544 0.750 0.468)

ξ_5 = (0.495 0.629 0.624 0.555 0.479 0.422 0.647 0.545 0.572)

The weights of the factors are obtained after calculation by the whitened weight function method:

$\beta_1 = 0.092$ $\beta_2 = 0.123$ $\beta_3 = 0.111$ $\beta_4 = 0.109$ $\beta_5 = 0.110$ $\beta_6 = 0.114$ $\beta_7 = 0.109$ $\beta_8 = 0.112$ $\beta_9 = 0.120$

The weighted correlation between the total number of accidents and the age of the ship can be derived from equation (4) as:

$\gamma_1 = 0.571$ $\gamma_2 = 0.707$ $\gamma_3 = 0.575$

$\gamma_4 = 0.545$ $\gamma_5 = 0.553$

After sorting out $\gamma_2 > \gamma_3 > \gamma_1 > \gamma_5 > \gamma_4$, that is, the ship age in 6~10 years ship is most likely to occur mainframe failure followed by 11~15 years ship, which is consistent with the intuitive analysis, and the combination of the previous article can be speculated that the ship mainframe failure and the ship mainframe wear parts are not timely maintenance and replacement of the ship mainframe failure and can be seen that the number of mainframe failures in the first three months is the most, according to the port location can be judged to be high probability of the mainframe failure due to low temperatures and windy and high waves, such as the port environmental factors, which leads to the probability of mainframe failure increases.

3.3 *Location of ship's main engine failure and duration of incident*

Let the total number of failures be the parent factor X_0 = (13 23 6 8 34) and the locations of failures be the sub-factors X_i(i=1, 2, 3). The weighted correlation coefficients for the locations of main engine failures on a ship can be derived from the calculations in equations (1), (2), and (3):

ξ_1 = (0.388 0.306 0.447 0.337 0.285)

ξ_2 = (0.675 0.486 0.559 0.516 0.324)

ξ_3 = (0.685 0.636 0.723 0.669 0.865)

The weights of the factors are obtained after calculation by the whitened weight function method:

$\beta_1 = 0.197$ $\beta_2 = 0.162$ $\beta_3 = 0.275$

$\beta_4 = 0.238$ $\beta_5 = 0.128$

The weighted correlation between the total number of accidents and the location of the ship experiencing main engine failure can be derived from equation (4) as:

$\gamma_1 = 0.366$ $\gamma_2 = 0.530$ $\gamma_3 = 0.707$

Sorting results in $\gamma_3 > \gamma_2 > \gamma_1$,it can be seen that the ship in the waterway occurred in the main engine accident probability is greater, after the data access and analysis of the reason that: due to the channel of the ship flow is greater than the number of berths and anchorage so the port area waterway is more prone to main engine failure, at the same time after

Table 2. Locations and repair durations of Ship's faults in port area.

Failure duration (min)	<30	30~60	60~90	90~120	>120
berths	2	2	1	2	3
anchorage	3	4	0	1	8
fairway	8	7	5	5	23
Total faults	13	23	6	8	34

the statistics can be seen, the ship main engine accidents occur in the hours of greater than 120 minutes, followed by 30 ~ 60 minutes.

Combined with the statistical data, it can be seen that the ships with failure time of 30~60 minutes are mainly port work ships, and a small portion of them are cargo ships, which do not have much impact on the navigation of the harbor channel; however, most of the ships with failure time of more than 120 minutes are the cargo ships with tonnage of more than 10,000 tons and most of them need tugboats to assist them, which have a greater impact on the navigation of the harbor channel.

Table 3. Types of Ship's faults and tonnage of faulty ships in port area.

Tonnage of Ship (ton)	<1000	1000~ 5000	5000~ 10000	10000~ 50000	50000~ 100000
cargo ship	4	7	8	9	2
container ship	1	7	2	8	8
Oil tanker	13	1	0	1	0
other	6	2	0	3	0
Total faults	24	17	10	21	10

3.4 *Types and tonnage of ships with main engine failures*

Let the total number of failures be the parent factor X_0 = (13 23 6 8 34), and the types of ships with main engine failures be the sub-factors $X_i(i=1,2,3,4)$. The weighted correlation coefficients between ship types and main engine failures can be derived from equations (1), (2), and (3):

ξ_1 = (0.595 0.485 0.558 0.683 0.529 0.596 0.574 0.623 0.546)

ξ_2 = (0.549 0.832 0.862 0.595 0.715 0.644 0.745 0.807 0.585)

ξ_3 = (0.550 0.696 0.550 0.537 0.442 0.859 0.466 0.424 0.587)

ξ_4 = (0.460 0.462 0.528 0.585 0.492 0.618 0.544 0.750 0.468)

The weights of the factors are obtained after calculation by the whitened weight function method:

$\beta_1 = 0.179$ $\beta_2 = 0.184$ $\beta_3 = 0.236$

$\beta_4 = 0.167$ $\beta_5 = 0.234$

The weighted correlation between the total number of accidents and ship types can be derived from equation (4) as:

$\gamma_1 = 0.762$ $\gamma_2 = 0.682$ $\gamma_3 = 0.657$ $\gamma_4 = 0.707$

Sorting results in $\gamma_1 > \gamma_4 > \gamma_2 > \gamma_3$,After analysis, it can be seen that the probability of failure of cargo ships in the port area is greater, combined with statistical point of view, in the port area of the failure of the main engine of the cargo ship of the ship's tonnage is generally more than 5,000 tons, most of the international voyage of the cargo ship, followed by other types of ships, because there are many types of ships, most of them for the port for the ship is not a big ship type, generally in the 1,000 tons or less, so the ship's tonnage in less than 1,000 tons of the number of failures in the bias of the number of failures, and the same time, in the failure of the ship in the tonnage of oil tankers is not high, but once a failure of the main engine in the port is still the larger hidden danger.

3.5 *Discussions*

Through several weighted gray correlation analysis, it can be concluded that in the first quarter of the main engine failure of ships in the port area is mainly in the age of 6~10 years, after analysis, it is believed that the main engine failure is more likely to occur from January to March due to the temperature, wind and waves and other environmental reasons; the probability of ships in the port area in the waterway is more likely to occur, and most of the failure repair time is longer than 120 minutes, and from the acquired data of main engine failure of ships in the port area, it can be concluded that the longest time to repair the main engine failure of ships is more than 24 hours. Ship main engine failure data can be statistically obtained from the ship main engine failure repair the longest time more than 24 hours, of which 50% of the ship's total deadweight tonnage greater than 5000 tons, this kind of larger tonnage of ships in the waterway main engine failure and repair time is too long will be on the normal navigation in the port area to cause the impact of the large number of ships in the port area navigation, the probability of main engine failure of the cargo ships this kind of commercial ships is higher and the general deadweight is more than 5000 tons, the main engine failure is more than 5000 tons, the general load is more than 5000 tons. The probability of main engine failure is higher in cargo ships, and the general deadweight is more than 5000 tons, and it can be seen from the table, there are also more than 100,000 tons of super-large ships occurring main engine failure, which causes great security risks to the port area, and the probability of main engine failure of the Hong Kong ship is second only to the cargo ship, which affects the daily maintenance of the port area.

After analyzing the statistical data of ship mainframe failure in the harbor area (Alarcin et al. 2014, Balin et al. 2021, Bian et al. 2012), combined with the conclusion to reduce the ship mainframe failure in the harbor area, we propose and analyze the following recommendations:

First of all, after analyzing the results of the ship main engine failure occurs mainly in the age of 6 ~ 10 years, combined with the analysis of the previous article can be speculated, mainly in the ship main engine part of the wear parts may reach the life expectancy, need to be replaced in a timely manner. At the same time, it is also possible that some of the ships have just finished the maintenance from the dockyard, and the various parts of the ship are not in place, leading to the failure of the ship's main engine. And combined with the port area in the first quarter of the ship main engine failure in the number of more, combined with the port area is located in the geographic location, after deducing that the possible reasons are, the port area where the temperature is not high, the ship's main engine fuel atomization is not good, resulting in insufficient combustion, and other reasons, and the time period in the port area of the wind and waves is larger, the rudder steering of the ship sailing in the port area is frequent, the main engine shift times, resulting in the failure of the main engine Chances of increasing, so it can be seen that most of the ship mainframe failure of the ship's own poor condition, poor maintenance in the maintenance, especially part of the old ship in the maintenance of the old ship and the newly built part of the ship, the probability of these failures is higher; Secondly, most of the ship mainframe failure occurs in the port area of the ship repair time more than 120 minutes, combined with the location of the ship mainframe failure occurs in the port area can be seen that most of the ship mainframe failure occurs in operation. Ship main engine is in operation failure, and ship main engine composition is complex, many parts, it is difficult to determine the source of failure and repair in time, which leads to failure repair time is higher, the navigation of the waterway caused by a greater impact, so it should be produced as early as possible to optimize a more accurate failure source determination program, as early as possible to determine the source of the failure, as early as possible to repair, can be restored as early as possible to reduce the loss of all parties.

At the same time the shipping company should be in accordance with the maintenance of equipment on the maintenance cycle, the various equipment for regular inspection, testing, dismantling and maintenance operations, especially the ship's main engine, such as this, a long period of time, belonging to the ship's core equipment, these key equipment, should be done, early inspection, early maintenance, so that the ship's equipment throughout the good operating conditions, to ensure that the key equipment is reliable, reduce the probability of failure. The ship should be inspected and maintained as early as possible, so that the ship's equipment is in good working condition throughout the whole journey. Even if there is an unavoidable failure of the ship under perfect maintenance, we should not panic, the company and the captain should make a plan for the situation early, actively and effectively communicate with the relevant departments or other ships to seek effective help, and also can solve the accident as early as possible.

Finally, through the analysis of the results can be seen, most of the failure of the ship is a cargo ship of this kind of commercial ships, most of them have a large load, its voyage to the equipment maintenance is more difficult, and the voyage period is more compact so this kind of ship in the port area after the ship docking, the company and the regulatory authorities should pay more attention to the ship, especially the large tonnage, especially the age of the ship in the 6 ~ 10 years of the condition of the ship for a careful inspection, on the Defects found, damage should be repaired in time, and adequate maintenance and repair, the ship's inspection agency is the most authoritative authority on the quality of the ship after the ship sailing, in order to ensure the safety of the port, the ship's inspection agency should be the focus of the inspection of the ship, the shortcomings found must be required to correct the other party immediately, in the shipbuilding and repair of the ship should be more stringent management at the time of the source of the accident may be prevented, the ship's inspection and maintenance of the ship near the overhaul and maintenance of the ship. For the approaching maintenance and repair of the ship is more should focus on inspection, to prevent the ship failure because of fluke.

4 CONCLUSION

With the construction of modern ports, the number of ships and cargo throughput in the port area will increase more and more, avoiding accidents in the port area is the basis for maintaining the smooth operation of the port area, this paper combines the data obtained on ship mainframe failures in the port area with the knowledge of ship mainframe failures and the study of weighted grey correlation analysis (Mihara et al. 1989), using the weighted grey correlation analysis method to summarize the factors associated with the occurrence of ship mainframe failures in the port area and put forward the proposed measures to reduce mainframe failures. The main research results of this paper are:

1. Ships aged 6~10 years are more likely to have main engine failures and statistically, the number of main engine failures of ships in the harbor area in the first 3 months is significantly higher;
2. Most of the main engine failures in harbor navigation occurred in the channel and the repair time was generally more than 120 minutes;
3. Most of the ships in the port area are cargo ships, which have the highest probability of main engine failure and have a deadweight of more than 5000 tons.

The sample data analyzed in this paper is not detailed enough, and there is no detailed statistics for more factors, such as: specific to each ship's registry, ship's maintenance records, ship main engine type and so on. Ship host device structure and operating state is complex, the working environment is harsh, at the same time, it is not easy to real-time detection of the internal value of the ship host work, for the host failure causes of the analysis of the host has brought greater difficulties, to achieve a more comprehensive analysis of the situation, need further exploration and research.

REFERENCES

Alarcin, F., Balin, A, & Demirel, H (2014). Fuzzy AHP and Fuzzy TOPSIS integrated hybrid method for auxiliary systems of ship main engines. *Journal of marine engineering & technology, 13(1)*, 3-11.

Balin,A., Demirel, H., Celik, E, & Alarçin, F(2021). A fuzzy dematel model proposal for the cause and effect of the fault occurring in the auxiliary systems of the ships' main engine. *International Journal of Maritime Engineering*.

Bian, H.T., Liu, Q., Sun, J.M. & Wang, C (2012). Application of weighted gray correlation analysis in ship accident analysis. *China Water Transportation (Next Half Month) (11)*, 42-44.

Burak Göksu & Kadir Emrah Erginer (2020).Prediction of Ship Main Engine Failures by Artificial Neural Networks. *Journal of ETA Maritime Science*.

Feng, S.H., Li, Z & He, Y.H (2017). Statistical analysis and insights of aircraft fire accidents based on gray correlation method. *Safety and Environmental Engineering (03)*, 138-143+149.

Hu, Y.H., Hu, X.F & Xie, S. W (2000). Hierarchical diagnosis of ship main engine faults. *Diesel Engine (03)*, 16-20.

Jiang, B.C., Tasi, S.L & Wang, C.C (2002). Machine vision-based gray relational theory applied to IC marking inspection. *IEEE Transactions on Semiconductor Manufacturing, 15(4)*, 531-539.

Li G. (2023). Emergency maneuvering of main engine failure in and out of Ganyu Port and its revelation. *China Water Transportation (03)*, 129-131.

Liu, S. F., Tang, W., Xie, N. M. & Tao, L. Y (2022). The creation, development and international dissemination of China's original gray system theory. *Journal of Nanjing University of Aeronautics and Astronautics (Social Science Edition) (04)*, 1-10.

Luo, L.K. & Yu, L. (2023). Emergency Response to Ship Loss of Control in Harbor. *Marine Technology (03)*, 10-13.

Mihara, Y., Murakami, F. & Iwasaki, H (1989). Approach to marine engine fault diagnosis method. *Journal of the Marine Engineering Society in Japan, 24(2)*, 160-166.

Sun, Y.G. (2007). Research on grey correlation analysis and its application. *Nanjing University of Aeronautics and Astronautics*.

Wu, H.H., Liu, S.F. & Fang, Z.K. (2023). A general gray number type panel data correlation analysis model and applications. *Systems Engineering and Electronics (03)*, 755-761.

Xu, H., Sun, L.Z., Gu, P& Wang, B.C (2023). The Influence Factors of Incremental Forming Geometric Accuracy based on the Weighted Gray Correlation Analysis Method. *Journal of Physics: Conference Series, 2587.*

Xue, J., Van Gelder, P. H. A. J. M., Reniers, G., Papadimitriou, E., & Wu, C (2019). Multi-attribute decision-making method for prioritizing maritime traffic safety influencing factors of autonomous ships' maneuvering decisions using grey and fuzzy theories. *Safety Science, 120*, 323-340.

Yang, X.D (2017). Introduction to ship main engine management and maintenance. *China New Technology and New Products (18)*, 52-53.

Yang, Y.C., & Shen, K. Y (2013). Comparison of the operating performance of automated and traditional container terminals. *International journal of logistics research and applications, 16(2)*, 158-173.

Advances in Maritime Technology and Engineering – Guedes Soares & Santos (Eds)
© 2024 The Author(s), ISBN 978-1-032-83099-5

Fatigue reliability assessment of an additive manufacturing material

U. Bhardwaj, A.P. Teixeira & C. Guedes Soares
Centre for Marine Technology and Ocean Engineering (CENTEC), Instituto Superior Técnico, Universidade de Lisboa, Lisbon, Portugal

S. Kamil & A.K. Ariffin
Universiti Kebangsaan, Malaysia

ABSTRACT: This paper assesses the fatigue reliability of the additive manufacturing material of a compact tension specimen. The limit state function is constituted using an analytical formulation, model uncertainty factor, and fracture toughness at the crack initiation. A case study is developed comprising a compact tension specimen with appropriate statistical models in load, material properties and crack geometry. The First Order Reliability Method is used to predict the probabilty of fracture under plane stress and plane strain conditions. The relative importance of the parameters is also evaluated using sensitivity analysis. The results explain the reliability reduction with the increase in loading and uncertainties in the crack length.

1 INTRODUCTION

Additive Manufacturing (AM) is gaining attention for its potential application and benefits in the aerospace, automotive, and medical sectors (Azri et al., 2018; Yim et al., 2023). Recent advances in rapid prototyping technology benefits from additive manufacturing rather than conventional subtractive manufacturing (Gong et al., 2022). This process comprises a three-dimensional object constructed by adding consecutive layers of AM material. Recent developments have made it possible to manufacture metallic objects in addition to plastic or other materials by AM (Gong et al., 2022). The direct metal laser sintering method (DMLS) is widely used for manufacturing metal products using AM (Uzan et al., 2018). Yet, the scientific information lacks to explicitly determine AM materials' mechanical properties and failure mechanism. In particular, fatigue crack analysis is the most critical issue for AM materials, which is responsible for its structural integrity and reliability (Nadot et al., 2020). The presence of cracks in a stressed structure can significantly affect material strength and design life (Feng et al., 2012).

Fatigue analysis is essential for new materials, such as AM materials, since different manufacturing processes may introduce an entirely new fatigue crack growth process (Nadot et al., 2020). Due to the variability of loading, environment, and the location of crack initiation parameters, the evaluation of fatigue stress and crack growth is based on assumptions or limited experience. Therefore, further research is required to assess material characterization at the micro and macro level for AM material's characteristics (Ciampaglia et al., 2023).

The structural integrity and, hence, the reliability of AM materials based on the metal powder bed laser fused method can be questioned, as it is a relatively novel concept and limited data are available in the public domain (Hou et al., 2022). The real failure data and integrity issues may be obtained from AM consumers, which is an arduous process. Furthermore, the safety and reliability of AM end products is a critical issue, especially when the application is among engineering structures with catastrophic failures (Aboulkhair et al., 2016; Maskery et al., 2016). For such cases, fatigue reliability becomes a primary concern.

This study strives to fill this gap by assessing the fatigue reliability of AlSi10Mg, an additive manufacturing material. The mechanical characteristics of this material are produced by DMLS (Azri et al., 2018).

The need for probabilistic fracture mechanics involving statistical uncertainties in the design parameters has been recognized for a long time (Anderson, 2017). Fracture mechanics theory consists of relationships between the maximum permissible load on a structure and crack characteristics in that structure (Anderson, 2017). If the statistical uncertainties of the variables are adequately modelled, probabilistic approaches can account for the mechanic and statistical aspects of fracture problems with high confidence (Provan, 2013).

The fracture analysis can be based on linear or non-linear elastic models. It is highly material-specific as to which model is more appropriate to model the mechanical response when a crack grows. In fact, the mode-I (tensile opening) loading system is the most evaluated mode for expressing the mechanical behavior of a structure having specific geometry under particular loading.

DOI: 10.1201/9781003508762-27

Due to various approximations and limitations in analytical formulations, Finite Element Method (FEM) based approaches are used in the probabilistic analysis (Reddy & Rao, 2008; Huang et al. 2013).

This paper focuses on the probabilistic safety assessment of an AM material with side edge crack geometry in the compact tension specimen under tensile load. An analytical model for mode-I loading is adopted, while the model uncertainty factor is applied to account for modelling uncertainty. A case study is developed with adequate statistical modelling of limit state random variables. First Order Reliability Method (FORM) is employed to estimate the structural reliability index and sensitivity factors.

2 FATIGUE OF ALSI10MG MATERIAL FABRICATED FROM DMLS PROCESS

The AM technology has a history of 4 decades when a 3D prototyping technology was developed for plastic prototypes using photo-hardening polymer (Kodama, 1981). Later, this technology has seen augmentation along with the introduction of stereolithography (SLA) technology and fused deposition modelling (FDM) (Lonjon, 2017).

In recent times, the worldwide application of AM has increased significantly. Figure 1 shows the distribution of the consumer sectors of AM (Wohlers & Campbell, 2017). The application of AM to current and novel consumers is also expected to grow with time. Furthermore, AM technology is gaining wide popularity in the industrial and engineering sectors. Some distinguishable advantages of AM include building materials with complex geometry, manufacturing products layer by layer, creating lightweight metallic structures, and shortening the supply chain.

Studies conducted for the identification of mechanical properties of AlSi10Mg material reveal its brittle behaviour. This fact is corroborated in a number of studies based on strain values and compression testing (Aboulkhair et al., 2016; Maskery et al., 2016).

The fatigue crack failure for brittle material can be assumed following the linear elastic fracture mechanics. Furthermore, crack propagation is more significant in ductile materials than brittle materials in practice (Anderson, 2017).

A critical parameter for fatigue strength is fracture toughness. The common practice for its calculation is conducting experiments with various crack geometries using fracture mechanics concepts.

In the literature, an experiment has been conducted on AlSi10Mg material to evaluate the stress intensity factor (Shaari, 2018). Figure 2 presents a schematic of the standard specimen used in the fatigue test and the basic dimensions of the parameters.

It should be noted that when the size of test specimens satisfies the requirements of linear elastic fracture mechanics, the critical stress intensity factor (fracture toughness K_{IC}) can be calculated. This experiment evaluates the critical stress intensity factor

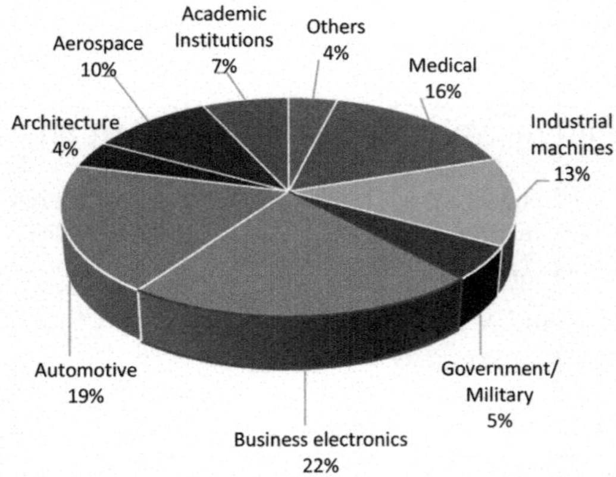

Figure 1. Distribution of worldwide consumer sectors of AM.

(K_{IC}) as 15.54 MPa$\sqrt{m}$ (Shaari, 2018). These results are also corroborated by finite element analysis.

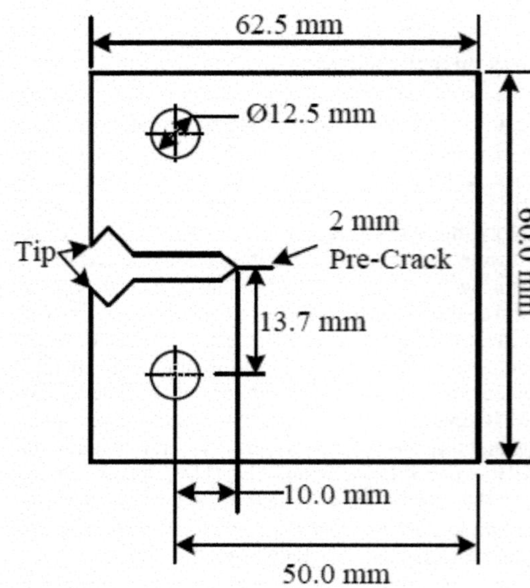

Figure 2. Compact tension specimen used in the fatigue test (Shaari, 2018).

In fracture mechanics, the crack driving force is usually described as J-integral, which represents the crack tip stress and strain field. The critical stress intensity factor (K_{IC}), the primary crack driving force, can be used to derive the J-integral (J_{IC}).

3 FATIGUE RELIABILITY OF THE ALSI10MG MATERIAL

As discussed in the previous section, the AlSi10Mg material is a linear-elastic structure. Thus, mode-I loading can be considered to characterize the fracture failure. It is well-understood that other modes and mixed mode failures of any structure under fatigue failure exist. Later study will address more complex mixed-mode failures,as it requires comprehensive numerical simulations to solve such failure scenarios under uncertain mechanical and geometric characteristics and the specimen is subjected to random loads.

Let **X** be a vector of the random variable with $\mathbf{X_1}$, $\mathbf{X_2}$, and... .$\mathbf{X_N}$ characterizing uncertainties in the load, crack geometry and material properties. In the present case, crack size a, elastic modulus E, Poisson's ratio v, far-field applied stress magnitude σ^∞ and mode-I fracture toughness at the crack initiation J_{IC} are the random variables, then $\mathbf{X} = \{a, E, v, \sigma^\infty, J_{IC}\}$.

Under pure mode – I, loading condition, the J-integral (J) is the crack driving force that can be obtained as

$$J = \frac{K_I^2}{E'},\ E' = \begin{cases} E, & \text{Plane stress} \\ \frac{E}{1-v^2} & \text{Plane strain} \end{cases} \tag{1}$$

The structure is supposed to fail when $J > J_{IC}$, consequently, a limit state $g(\mathbf{X})$ function or performance function can be defined as

$$g(\mathbf{X}) = JIC(\mathbf{x}) - M.J(\mathbf{x}) \tag{2}$$

where M is the model uncertainty factor that accounts for the uncertainty associated with the mathematical modelling.

Failure occurs when $g(\mathbf{X})$ is less or equal than zero. This region is called a failure region, whereas $g(\mathbf{X}) > 0$ represents the safe region.

Based on the generalized form $g(\mathbf{X})$ of the limit state function formulated in Eq. 2, the probability of a limit state violation, typically referred to as the probability of failure (P_f), can be calculated as:

$$P_f = \int_{g(\mathbf{X})\le 0} f_{\mathbf{x}}(\mathbf{x})d\mathbf{x} \tag{3}$$

where, $f_{\mathbf{x}}(\mathbf{x})$ is the joint probability density function of the vector **X**.

The probability of failure in Eq. 3 typically presents the probability of crack growth initiation and provides a conservative estimate of structural performance. A better failure criterion could be based on crack instability; however, it requires crack growth simulation in some structural analysis software.

4 RELIABILITY ANALYSIS BY FORM

A Multidimensional integration is required to solve Eq. 3. The direct calculation of P_f from the integral is rather complicated and computational techniques are frequently unfeasible. Therefore, approximate reliability methods such as the Monte Carlo Simulation (MCS) or First-/Second-Order Reliability Methods (FORM/ SORM) are frequently used.

First Order Reliability Methods (FORM) involve a first-order Taylor series expansion (linearisation in a standard normal space) of the limit state function around a point known as the most probable failure point or design point (u*). The first step in this method consists of transforming non-normal variables to standard normal variables (see e.g. (Ditlevsen & Madsen, 2005) for more details on the transformation methods). The design point is obtained by solving a constrained optimization problem corresponding to finding the point located on the limit-state surface, which has the minimum distance from the origin in the standard normal space. The shortest distance between the origin and this particular point on the limit state surface is known as the reliability index (β).

The reliability index (β) is related with the probability of failure by:

$$\beta^{MC} = -\phi^{-1}(P_f) \tag{4}$$

where ϕ^{-1} is the inverse normal distribution function with zero mean and unit standard deviation.

Further, through a sensitivity analysis, this paper examines the importance of the basic limit state variables on the reliability indices. The sensitivity measure or factor (α_i) can be calculated using the FORM approach (Ang & Tang, 2006), as:

$$\alpha_i = \frac{\left(\frac{\partial g(\mathrm{u})}{\partial u_i}\right)_{\mathrm{u}=\mathrm{u}^*}}{\sqrt{\sum_{i=1}^{n}\left(\frac{\partial g(\mathrm{u})}{\partial u_i}\right)^2_{\mathrm{u}=\mathrm{u}^*}}} \tag{5}$$

where i = 1,2,3……n is the index associated with each random variable considered in reliability analysis. The partial derivatives are calculated at the FORM design point u*, representing the most probable failure point of the design random variables in the standard Gaussian space. The positive α_i value indicates that the random variable (i) contributes positively (i.e. increases safety level or β when it increases) and vice versa.

5 CASE STUDY

The fatigue reliability of the AM material fabricated from DMLS method is investigated in this study. A reference case is developed comprising a crack geometry of a compact tension specimen subjected to load P.

Figure 3 presents the main parameters of the compact tension specimen adopted in the present study. The main dimensions are the same as seen in Figure 2. In the development of linear elastic and elastic-plastic fracture mechanics, significant progress has been perceived in deterministic modelling. For the successful application of linear elastic fracture mechanics to fatigue analysis, stress concentration factor equation corresponding to a particular specimen is required.

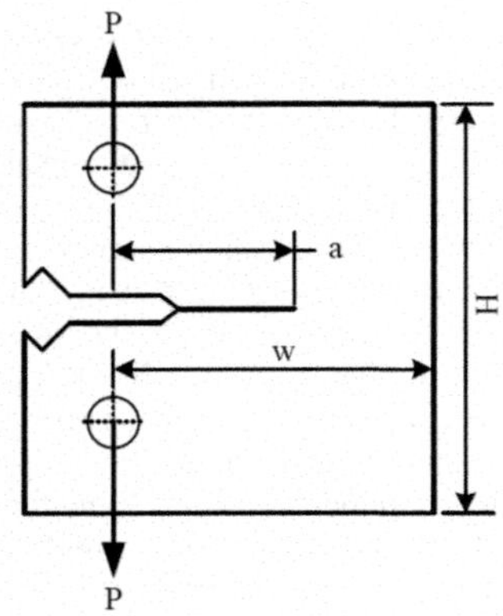

Figure 3. Parameters of the compact tension specimen.

In the present context, there are some approximations of mode-I stress concentration factor (K_I) developed analytically such for compact tension specimens (Figure 3). Considering a far-field tensile stress (σ^{∞}) in the fracture analysis is more practical and can be easily derived. According to (Perez, 2016), the equation for K_I can be written as

$$K_I = Y\sigma^{\infty}\sqrt{\pi a} \tag{6}$$

where Y is the geometry correction factor, a function of the specimen's normalized crack length (*a/W*). For the present loading case and for $0.45 < a/W < 0.55$, *Y* is given as

$$Y = \frac{2+\frac{a}{W}}{\left(1-\frac{a}{W}\right)^{\frac{3}{2}}}\left(0.886 + 4.64\frac{a}{W} - 13.32\frac{a}{W}^2 + 14.72\frac{a}{W}^3 - 5.60\frac{a}{W}^4\right) \tag{7}$$

Table 1. Statistical properties of random input for mode-I problems.

Random variable	Mean	COV	PD
Normalized crack length (*a/W*)	0.5	Var	LN
Elastic modulus (*E*)	$62^{a}/74^{b}$ GPa	0.05	N
Poisson's ratio (*v*)	0.3	0.05	N
Model uncertainty factor (*M*)	1	0.05	GB
Initiation fracture toughness (J_{IC})	3895.03 kJ/m^2	0.47	LN
Far-field tensile stress (σ^{∞})	Var	0.1	N

COV–Coefficient Of Variation, PD – Probability Distribution, LN – Log Normal, N – Normal, GB – Gumbel, a – As-built, b – Heat-treated, Var-Variable.

The case under study assumes all limit state variables as statistically independent random variables. Table 1 presents the mean value, coefficient of variation, and probability distribution of load, crack size, and material properties. The uncertainty in Initiation fracture toughness is adopted from (Rahman et al., 1995), while all other COVs are adopted from the research by (Reddy & Rao, 2008). The fracture toughness for the specimen is calculated from the K_{IC} given by (Shaari, 2018).. The elastic modulus is assumed for two cases where the structure is as-built and heat-treated (Azri et al., 2018). Model uncertainty factor is assumed with unit mean and minimum uncertainty among the random variable. The distribution of *M* is assumed to be Gumbel distributed.

6 RESULTS

FORM is employed to conduct probabilistic fracture analysis under linear elastic mode-I loading. Various probabilistic analyses are performed to estimate the reliability index (β) as a function of the mean of the far field tensile stress E[σ^{∞}], here E[] is the expectation operator. Both plane stress and plane strain conditions are considered. These two material types are also analyzed as the specimen is finally used as - built or heat-treated. Thus, four scenarios are obtained as assessed in the following subsection.

6.1 *Plane stress & as-built specimen (A)*

Figure 4 shows the reliability results (β) vs E[σ^{∞}] for the present case under plane stress conditions with a variation of COV of the normalized crack length - COV(*a/W*). Both deterministic COV = 0 and random values (COV 5%, 10% and 15%) are incorporated in the analysis. The reliability of the specimen decreases non-linearly with an increase in far-field tensile stress. Expectedly, the results indicate that the reliability decreases with increasing uncertainty in normalized crack length. Interestingly, at higher stress, the impact of COV of *a/W* becomes less significant as the reliabilities calculated at different $COVs_{(a/W)}$ are about to converge.

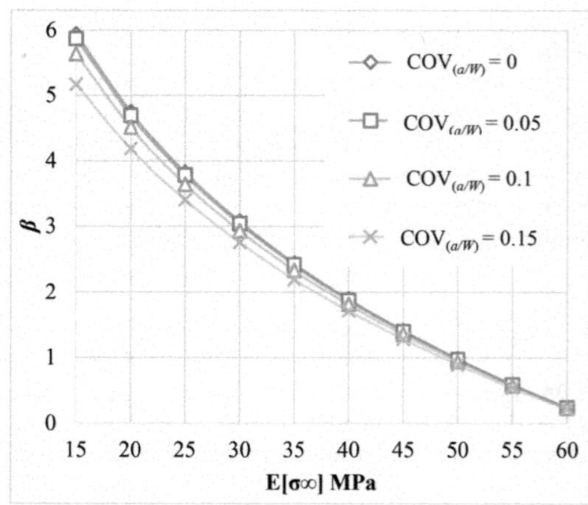

Figure 4. Reliability as a function of far-field tensile stress (plane stress & As-built).

6.2 *Plane strain & as-built specimen (B)*

Now, the scenario is created using plane strain criteria with the specimen as-built. Reliability variations with the far-field stress are shown in Figure 5. Furthermore, the plots with a deviation of COV of the normalized crack length are illustrated. The reliability estimates are slightly higher than in the previous case, while similar trends are observed in Figure 4.

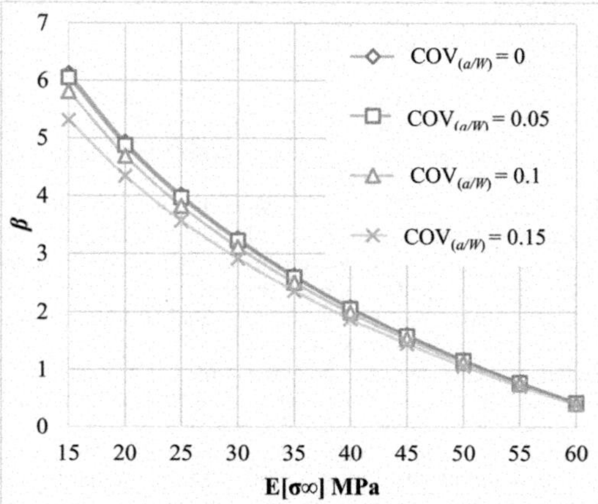

Figure 5. Reliability as a function of far-field tensile stress (plane strain & As-built).

6.3 *Plane stress & heat-treated specimen (C)*

If the specimen is heat-treated, higher material strength is apparent. Probabilistic fatigue failure is investigated in this section first concerning plane stress conditions. Figure 6 shows the reliability as a function of the far-field tensile stress. Compared to previous scenarios (Scenario A – Plane stress), a slight increase in reliability vs. far-field stress is evident. Reliability is much lower when the uncertainty in the crack size is high.

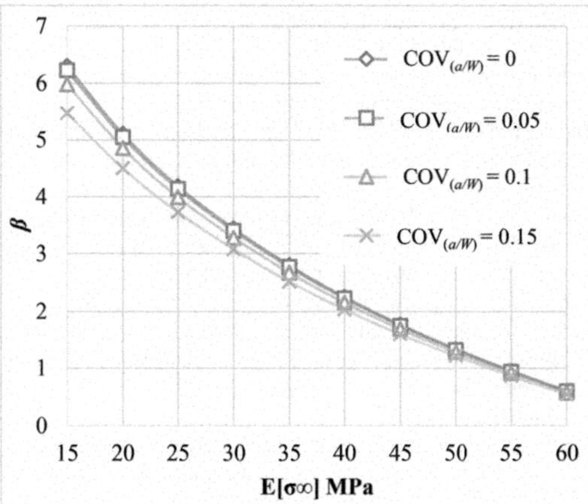

Figure 6. Reliability as a function of far-field tensile stress (Plane stress & Heat-treated).

6.4 *Plane stress & heat-treated specimen (D)*

Finally, the plane strain condition is analyzed in the present case for the heat-treated specimen. Figure 7 shows the trends of β vs $E[\sigma^{\infty}]$ obtained using FORM with varying $COV_{(a/W)}$. As expected, the trend of reduction in reliability with an increase in far-field stress is perceived. The reliability of in-plane strain is slightly higher than in-plane stress regardless of the load intensity since the J-integral for plane stress is $(1\text{-}\nu^2)$ times larger than that for plane strain.

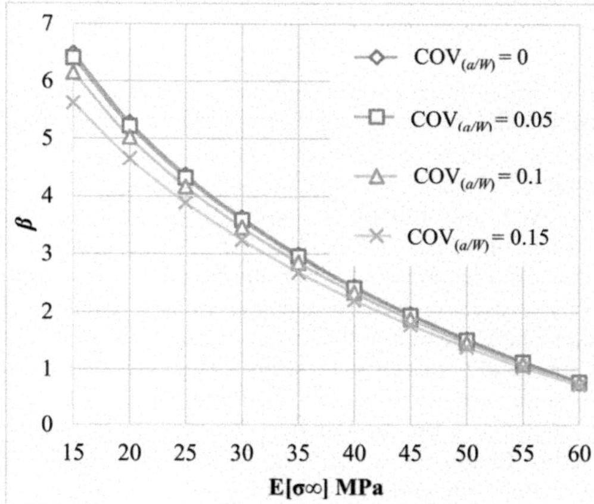

Figure 7. Reliability as a function of far-field tensile stress (Plane strain & Heat-treated).

6.5 *Sensitivity analysis*

As discussed in Section 4, FORM can be used to evaluate sensitivity factors (α) at the design points. As in the present analyses some of the design variables are considered to vary, It is considered $E[\sigma^{\infty}] = 20.4$ MPa and $COV_{(a/W)} = 0.1$ for the sensitivity analysis. The four scenarios discussed earlier under the headings A, B, C and D are studied. Table 2 presents the estimated sensitivity factors (α) in four scenarios. Similar α – values are obtained in all four scenarios for the random variables. J_{IC} is the most significant variable for reliability, followed by crack geometry. Obviously, the variables associated with positive α factors increase reliability when increased and vice versa is valid for the variables with negative sensitivity factors. Model uncertainty factor and poisson's ratio (C and D scenarios) have a marginal impact on reliability.

7 CONCLUSIONS

This paper assesses the reliability of AlSi10Mg, an additive manufacturing material produced by direct metal laser sintering. The basic mechanical properties of this material are obtained from the literature. Linear elastic mode-I fracture failure is adopted as failure scenario.

A case study of a compact tension specimenunder tensile loading is adopted. Limit state is formulated

Table 2. α – values of variables in the four scenarios.

	Scenario			
Variables	A	B	C	D
J_{IC}	0.778	0.871	0.87	0.869
σ^{∞}	-0.346	-0.336	-0.337	-0.334
E	0.087	0.1	0.1	0.1
a/W	-0.511	-0.335	-0.332	-0.337
M	-0.08	-0.079	-0.079	-0.079
ν			0.061	0.06

using initiation fracture toughness, model uncertainties and the crack driving force. Adequate uncertainties are applied to limit state variables based on the literature. First Order Reliability Method is then employed and reliability indices are calculated as a function of far-field stress. Trends of reliability reduction are obtained when applied load or uncertainty in crack geometry increase. Four scenarios are established regarding the material (as-build or heat-treated) and loading (plane strain and stress). Sensitivity factors are calculated based on the FORM sensitivity analysis of the mode-I fracture analysis. Initiation fracture toughness is the most significant variable in reliability, followed by the normalized crack length.

Further study will focus on fracture reliability of double-edge tension and middle tension specimens. Due to space limitations it is not viable to present a comparison of fatigue behaviour of additive manufacturing material to that of traditional materials (steel or composite). However, the same will also be included in further extension of present research work.

ACKNOWLEDGEMENTS

This work was performed within the scope of the Strategic Research Plan of the Centre for Marine Technology and Ocean Engineering (CENTEC), which is financed by the Portuguese Foundation for Science and Technology (Fundação para a Ciência e Tecnologia - FCT) under contract UIDB/UIDP/00134/2020, and in the scope of the Marie Skodowska-Curie grant agreement No. 73088 RESET financed by the European Union Horizon 2020 research and innovation.

REFERENCES

Aboulkhair, N.T., Maskery, I., Tuck, C., Ashcroft, I., Everitt, N.M., 2016. The microstructure and mechanical properties of selectively laser melted AlSi10Mg: The effect of a conventional T6-like heat treatment. Mater. Sci. Eng. A 667, 139–146.

Anderson, T.L., 2017. Fracture Mechanics: Fundamentals and Applications, Fourth Edi. ed, Taylor and Francis.

Ang, A.H.S., Tang, W.H., 2006. Probability concepts in engineering: Emphasis on applications to civil and environmental engineering. John Wiley & Sons Inc.

Azri, M.A., Shaari, S., Kamal Ariffin, A., Abdullah, S., 2018. Microstructure, Mechanical Properties and Fatigue Behavior of AlSi10Mg: an Additive Manufacturing Material. Int. J. Eng. Technol. 7, 186–190.

Ciampaglia, A., Tridello, A., Paolino, D.S., Berto, F., 2023. Data driven method for predicting the effect of process parameters on the fatigue response of additive manufactured AlSi10Mg parts. Int. J. Fatigue 170, 107500.

Ditlevsen, O., Madsen, H.O., 2005. Structural Reliability Methods. John Wiley & Sons Ltd, Denmark.

Feng, G.Q., Garbatov, Y., Guedes Soares C., 2012. Fatigue reliability of a stiffened panel subjected to correlated crack growth. Struct. Saf. 36–37, 39–46.

Gong, J., Wei, K., Liu, M., Song, W., Li, X., Zeng, X., 2022. Microstructure and mechanical properties of AlSi10Mg alloy built by laser powder bed fusion/direct energy deposition hybrid laser additive manufacturing. Addit. Manuf. 59, 103160.

Hou, Y., Li, Y., Cai, X., Pan, C., Wang, J., Zhang, W., Xu, P., Fan, Z., Gao, Y., Li, Z., Li, J., Sun, H., Huang, W., Yang, W., 2022. Mechanical response and response mechanism of AlSi10Mg porous structures manufactured by laser powder bed fusion: Experimental, theoretical and numerical studies. Mater. Sci. Eng. A 849, 143381.

Huang, W., Garbatov, Y., Guedes Soares C., 2013. Fatigue reliability of a stiffened panel subjected to correlated crack growth. Engg. Struct. 56, 868–879.

Kodama, H., 1981. Automatic method for fabricating a three-dimensional plastic model with photo-hardening polymer. Rev. Sci. Instrum. 52, 1770–1773.

Lonjon, C., 2017. The history of 3d printer: from rapid prototyping to additive fabrication [WWW Document]. sculpteo. URL https://www.sculpteo.com/blog/2017/03/01/whos-behind-the-three-main-3d-printing-technologies/ (accessed 5.24.23).

Maskery, I., Aboulkhair, N.T., Aremu, A.O., Tuck, C.J., Ashcroft, I.A., Wildman, R.D., Hague, R.J.M., 2016. A mechanical property evaluation of graded density Al-Si10-Mg lattice structures manufactured by selective laser melting. Mater. Sci. Eng. A 670, 264–274.

Nadot, Y., Nadot-Martin, C., Kan, W.H., Boufadene, S., Foley, M., Cairney, J., Proust, G., Ridosz, L., 2020. Predicting the fatigue life of an AlSi10Mg alloy manufactured via laser powder bed fusion by using data from computed tomography. Addit. Manuf. 32, 100899.

Perez, N., 2016. Fracture Mechanics, Second. ed. Springer Cham, Switzerland.

Provan, J.W., 2013. Probabilistic fracture mechanics and reliability. Springer Dordrecht.

Rahman, S., Ghadiali, N., Paul, D., Wilkowski, G., 1995. Probabilistic pipe fracture evaluations for leak-rate-detection applications. Washington, DC.

Reddy, R.M., Rao, B.N., 2008. Stochastic fracture mechanics by fractal finite element method. Comput. Methods Appl. Mech. Eng. 198, 459–474.

Shaari, M.S. Bin, 2018. Fatigue crack growth behaviour for additive manufacturing material using finite element computation. Universiti Kebangsaan Malaysia.

Uzan, N.E., Shneck, R., Yeheskel, O., Frage, N., 2018. High-temperature mechanical properties of AlSi10Mg specimens fabricated by additive manufacturing using selective laser melting technologies (AM-SLM). Addit. Manuf. 24, 257–263. h

Wohlers, T., Campbell, R.I., 2017. Wohlers Report 2017.

Yim, S., Bian, H., Aoyagi, K., Yamanaka, K., Chiba, A., 2023. Effect of powder morphology on flowability and spreading behavior in powder bed fusion additive manufacturing process: A particle-scale modeling study. Addit. Manuf. 72, 103612.

Advances in Maritime Technology and Engineering – Guedes Soares & Santos (Eds)
© 2024 The Author(s), ISBN 978-1-032-83099-5

The underwater venting of methanol vapour

C. De Boom, A. Twerda, A.W. Vredeveldt & M. Hoogeland
Netherlands Organisation for Applied Scientific Research TNO, Den Haag, The Netherlands

ABSTRACT: Sailing on methanol supports the decarbonisation of shipping. Venting under water is a solution to avoid dangerous zones, normally created by venting toxic methanol above deck. An Eulerian based Computational Fluid Dynamics (CFD) model and a simple integral model are used to predict the methanol concentration above the waterline after underwater venting of methanol vapour. Both models are successfully validated against experimental data from a methane release at the bottom of a water basin. The CFD model showed superiority over the integral model. The numerical models are applied on a mixture of methanol and nitrogen being vented during bunkering. The gas dissolution depends strongly on the bubble diameter and venting depth. The CFD model showed in the most critical case the gas dissolution being sufficient and no methanol vapour reaches the deck of the ship. The underwater venting of methanol-nitrogen vapour proved to be a safe alternative compared to the venting above deck.

1 INTRODUCTION

To meet the shipping industry's emissions reduction goals, it is imperative to explore and adopt alternative marine fuels. Methanol (or methyl alcohol) is expected to play a large role in the future of decarbonization of shipping. However, current regulations limit the attractiveness of methanol as marine fuel for some ship types due to the need of a (high) venting mast to avoid large unusable space around a venting point. Hazardous area zones are needed around fuel tank vapour outlets due to the flammability and toxicity of methanol vapour. To avoid risk for humans, the outlets shall be positioned such that the distance to crewed areas is sufficiently large. Consequently, for smaller ships, the outlets need a large mast to create these distances which is impractical feature, also in view of air draught, or deck spaces are closed off for daily operations.

Usually, a critical concentration at 30 minutes' exposure of 6000ppm (0.6 vol%) is used which corresponds to the Immediate Danger to Life and Health of methanol (IDLH) (National Institute for Occupational Safety and Health, 1994).

An alternative venting arrangement is proposed by creating an underwater vent. Methanol dissolves well in water and is considered not toxic for aquatic life.

This novel solution has not been incorporated in ship design due to a lack of available supporting calculations and experiments. In this paper numerical supporting information is provided.

Very little research has been conducted on the underwater venting of vapour. Nevertheless, information can be obtained from similar scenarios, such as gas release of subsea pipeline ruptures such as performed in the Rotvoll experiment (Ross, 2009, Ahmed et al, 2021). From this research different zones were chosen to represent the governing physics. After a subsea gas release, the bubbles will rise through these different zones in the water column. These zones (flow establishment, established flow, surface and atmosphere zone) are depicted in Figure 1. The dominating physics will change in each zone. An important driving force is the buoyancy which depends on the density ratio of the released gas and the displaced liquid. Other mechanisms governing the bubble plume are drag, turbulence and gas dissolution.

This paper investigates the venting of the fuel tank vapour below the waterline instead on deck to be able to limit/ eliminate measures such as large vent musts or unusable deck space. Numerical models are applied to compute the concentration on deck. These are described in Section 2. The cases are detailed in Section 3 and their results respectively in Section 4. Finally, conclusions will be presented in Section 5.

2 NUMERICAL INVESTIGATION

2.1 *Integral method*

Integral models assume that plume velocities and gas fractions can be represented by certain mathematical profiles, mostly Gaussian profiles. These models are called "integral" because the governing set of equations is based on an integral over the width of the profiles (Olsen and Skjetne, 2015). Integral models use a constant angle to describe the widening of the plume. See Figure 1. The applied integral model is based on the widely used so-called Friedl model (Friedl and

DOI: 10.1201/9781003508762-28

Fanneløp, 2000) and the model built by Loes and Fanneløp, (1989) for the plume above the water surface. The Friedl model is a theoretical model which mainly focuses on the fountain, which is the elevation of the water surface due to a subsea gas release.

2.2 *Computational fluid dynamics*

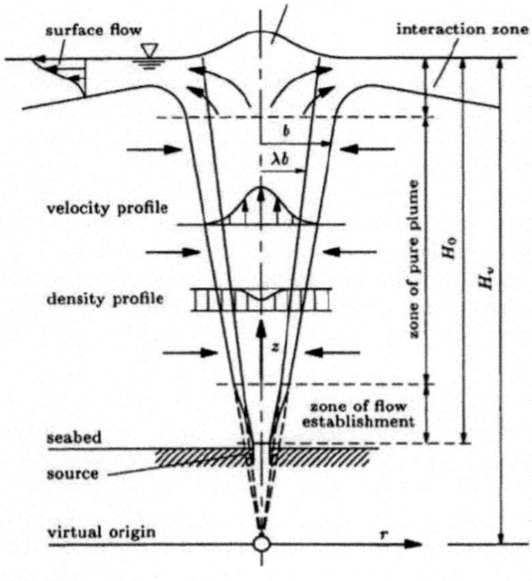

Figure 1. Sketch of time-averaged bubble plume used in the integral method taken from Friedl and Fanneløp (2000).

In CFD models the transport equations describing the conservation of mass, momentum, energy, and species are solved. The CFD simulation is set up in ANSYS-Fluent (version 22.2). Buoyancy and gravity are included in the model and the density is calculated using the incompressible ideal gas model. The Reynolds Averaged Navier Stokes (RANS) model k-ε is used as turbulence model.

Two types of models can solve particle transport (gas through water), namely Euler–Euler and Euler–Lagrange models. In the Euler-Euler model, the particles are considered as a continuum phase, just like the water. The particles are a virtual fluid described by momentum transport equations similar to the Navier-Stokes equations. In the Euler-Lagrange model, the particles are followed individually and not as a fluid. The description for the movement of a particle is described by Newton's second law. The Lagrangian model (in comparison to the Eulerian model) can calculate the particle size distribution with a lower computational cost (Wojciech and Adamczyk, 2014). The Eulerian-Lagrangian Discrete Phase Model (DPM) is applied for the simulation of the underwater gas release. This model will track a parcel of bubbles in the Lagrangian frame and will monitor the change in properties in the Eulerian frame. The drag force of the bubble is taken from Xia et.al. (2001) which according to Cloete et al (2009) represents the behaviour of a bubble plume. The ambient fluid is treated as a continuum by solving the Navier-Stokes equations. The surface (interface between seawater and air) is tracked by a scheme in the Volume of Fluid (VOF) model. This method uses explicit discretization and represents the interface between fluids using a piecewise-linear approach.

2.2.1 *Dissolution*

Because methanol can dissolve easily in water, this effect must be incorporated in the model. The mass transfer rate through the surface of the bubble with diameter (d_b) can be calculated as follows (Skjetne and Olsen, 2012):

$$\dot{m} = \pi d_b^2 k_i \rho_l n_i^{sol} \frac{M_i}{M_l} \tag{1}$$

Where ρ_l is the liquid density, M is the molar weight and n is the solubility of the compound in the surrounding liquid. The concentration of the dissolved gas in the surrounding liquid is assumed to be negligible. k is the mass transfer coefficient and the subscript i represents a certain chemical compound (methane or methanol). For the mass transfer coefficient, the following Sherwood relation as function of Reynolds (Re) and Schmidt (Sc) number is used.

$$Sh = \frac{k}{D/d_b} = 2 + 0.552 Re^{1/2} Sc^{1/3} \tag{2}$$

With D the molecular diffusivity. The solubility of gases in water can be calculated by the well-known Henry's law (Herman, 2017).

2.2.2 *Bubble size*

The drag coefficient is dependent on the bubble size. In addition, the gas dissolution or mass transfer from gas bubbles to the surrounding liquid is dependent on the bubble surface area. A correct bubble size distribution is therefore required to achieve accurate simulation results. The instantaneous local mean bubble diameter (d_b) is described by the balance between the local time derivative of the local bubble diameter and an equilibrium term which accounts for breakup and coalescence (Cloete et al, 2009):

$$\frac{\partial d_b}{\partial t} = \frac{\left(d_b^{eq} - d_b\right)}{\tau_{rel}} \tag{3}$$

τ_{rel} is the relaxation time, which is the time needed for the bubble to return to the bubble equilibrium diameter. The relaxation time is determined by the speed of breakup or coalescence and is modelled as a characteristic timescale for both processes according to (Pan, 2014) and (Laux and Johansen, 1999). A mathematical expression for the equilibrium diameter (d_b^{eq}), derived from experiments in a stirred tank, is given by Calderbank (1958).

$$d_b^{eq} = C_1 \phi_b^{0.5} \left(\frac{\sigma}{\rho}\right)^{0.5} \left(\frac{\mu_b}{\mu}\right)^{0.25} \varepsilon^{-0.5} + C_2 \quad (4)$$

Where ϵ is the dissipation of turbulent kinetic energy, μ_b is the viscosity of the bubble phase and σ is the surface tension between bubbles and the surrounding fluid. ϕ_b is the bubble void fraction which is the volume fraction that bubbles occupy in the water. This bubble fraction is calculated by the gas concentration (in kg/m3) divided by the gas density which is computed by means of the ideal gas law. C_1 is a dimensionless constant and is equal to 4.0 and C_2 is the minimum bubble size and is equal to 0.1mm (Pan, 2014)

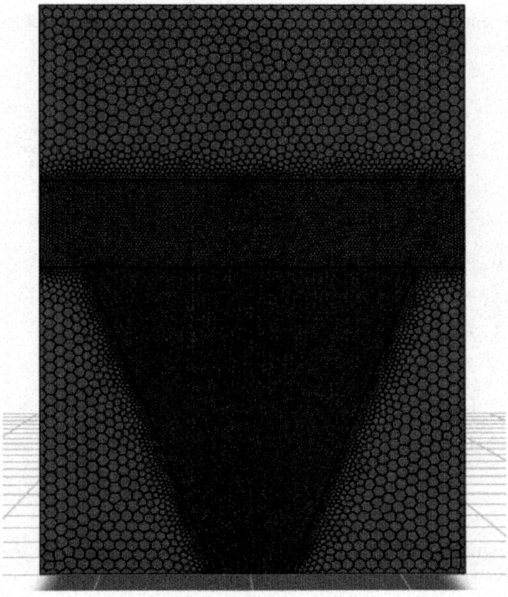

Figure 2. Polyhedral mesh with conical regional refinements.

3 CASES

3.1 *Rotvoll experiment*

The numerical models, described in Section 2, should have a solid basis. Therefore, first a model validation should be performed where the measured gas concentrations from an underwater gas release are compared to the predicted concentrations from the numerical model. This model will be afterwards adapted to the methanol venting experiment. In the past, multiple experiments of subsea gas releases have been carried out. In 1997 a series of experiments were conducted in a 7-meter-deep freshwater basin with a surface area of 6x9m at Statoil's (now Equinor) Research center in Trondheim. A mixture of helium and air was used, to obtain the same density as natural gas, to monitor the gas concentration above the surface. To measure gas concentration and velocities below the surface, air was injected. The gas was released at different flow rates from a circular opening of 0.17m diameter at 6.9m from the water surface. The plume characteristics were captured by a high-speed camera. Table 1 shows the parameters and results of the experiment.

3.1.1 *CFD case set-up*

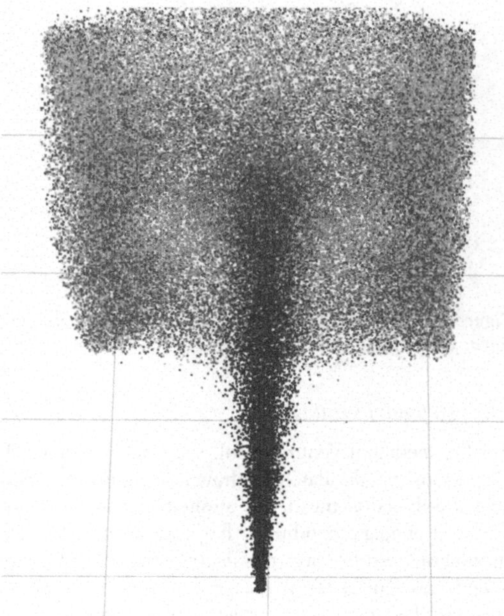

Figure 3. Particle tracks of Rotvoll experiment (0.20 kg/s release rate), seen from below.

For the CFD a polyhedral mesh was mode with local grid refinement in the cone and near the air-water surface as can be seen in Figure 2. All settings mentioned in Section 2.2 are applied.

Boundary conditions are set to be able to solve momentum and continuity equations. Except for the

Table 1. Parameters and results of Rotvoll experiment (Sjoen et al. 1997).

Flow rates (m3/s)/ (Nm3/s)	Mass flow rates (kg/s) methane/air	Inlet Velocity (m/s)	Rise time (s)	Initial fountain height (m)	Maximum fountain height (m)
0.05/0.083	0.03/0.10	2.2	6.0	-	-
0.1/0.17	0.06/0.21	4.4	4.8	0.3	0.65
0.45/0.75	0.3/0.92	19.6	3.1	0.45	1.24

top surface of the mesh, the boundary type is designated as "wall" with a no-slip condition. Since bubbles will reflect upon hitting a wall, it is important to position these surfaces in such a way that they do not affect the results. The initial bubble size is constant and taken from the equilibrium diameter formula given in Equation 4. A typical result of the flow of the particles is depicted in Figure 3, where the colouring is done with the residence time of the particle.

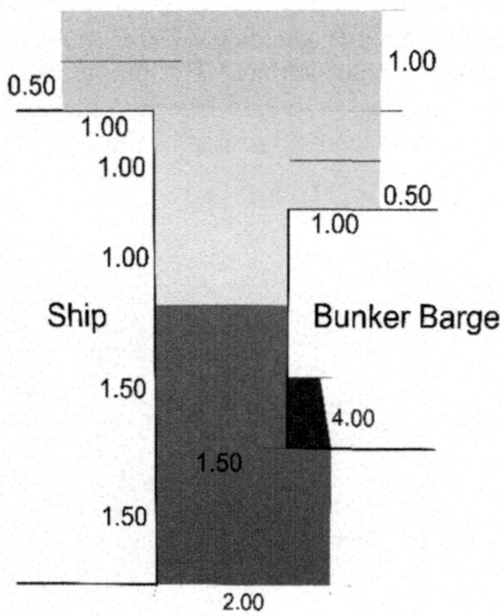

Figure 4. Geometry full scale methanol-nitrogen venting (blue: water, grey: air).

3.2 *Methanol venting*

For the methanol venting a full-scale setup was used. The geometry simulated is shown in Figure 4. Also, here a polyhedral mesh was applied. A careful selection of boundary conditions helps to ensure that the simulation results are physically meaningful. The boundaries (surfaces) of the geometry (see Figure 4) are described here. The surfaces of the ship and the bunker barge are set as walls. The interface between the water and air surface is an internal boundary type. The particles can leave the domain at the bottom surface of the geometry, below the (keel of the) bunker barge, 2m to the aft or front of the release point, above the deck of the bunker barge and above the deck of the ship. These surfaces below the waterline are walls but with the DPM escape setting and the surfaces above the waterline are pressure outlets.

A pressure relief valve (PRV) is used to build up pressure in the bunker tank to overcome the hydrostatic back pressure before venting. For relief pressure 14 kPa is used and the closing pressure is 9 kPa.

These values are characteristic for a PRV on a vessel. The mass release rate also depends on the volume of the bunker tank. As the pressure will decrease when the valve opens, the mass flow will decrease. This can be seen in Figure 5 where the released mass flow rate is plotted versus time without closing. Note that the closing pressure of the

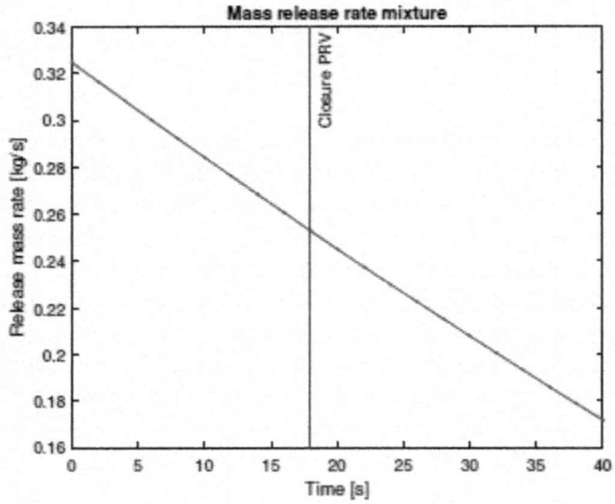

Figure 5. Released mass in function of time without closing. The valve.

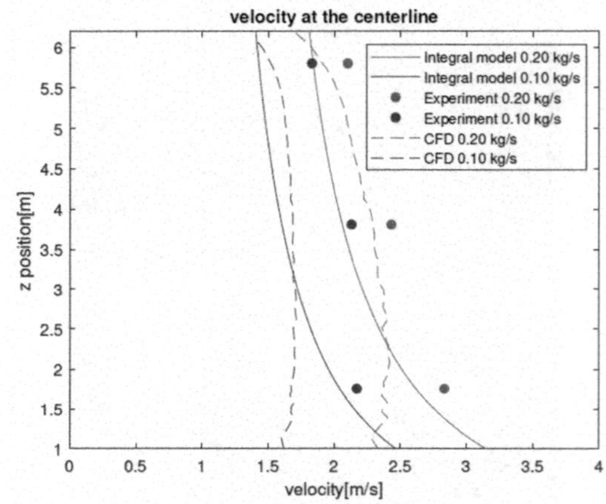

Figure 6. Comparison velocity at centerline between the Rotvoll experiment, integral model and CFD.

valve is reached after approx. 18s. After this time the pressure rises again until the opening pressure is reached, and the release process is repeated.

4 RESULTS AND DISCUSSION

4.1 *Rotvoll experiment*

Figure 6 shows the velocity at the centerline in function of the height, where 0 is the bottom of the basin and 7m is at the water surface. At low heights there is a large difference between the CFD model and the integral model. This is due to zero initial velocity, of the liquid, which is not taken into account in the integral model. After the release, the velocity increases due to the buoyancy and afterwards it decreases due to the drag force. The integral model shows a larger discrepancy near the surface due to the radial movement of the water.

This can be explained due to lacking surface interaction in the model. When the gas reaches the surface there is an increasing number of turbulent eddies, leading to an erroneous velocity profile close to the surface. Both the CFD and the integral deviate more from the experiments in the case with a release rate of 0.10 kg/s. This is mainly due to the use of parameters which are fitted to the 0.20 kg/s release rate case. The underestimation of the velocity in the CFD model is probably due to incorrect initial bubble size. When a larger bubble is injected, the buoyancy is larger and consequently also the velocity.

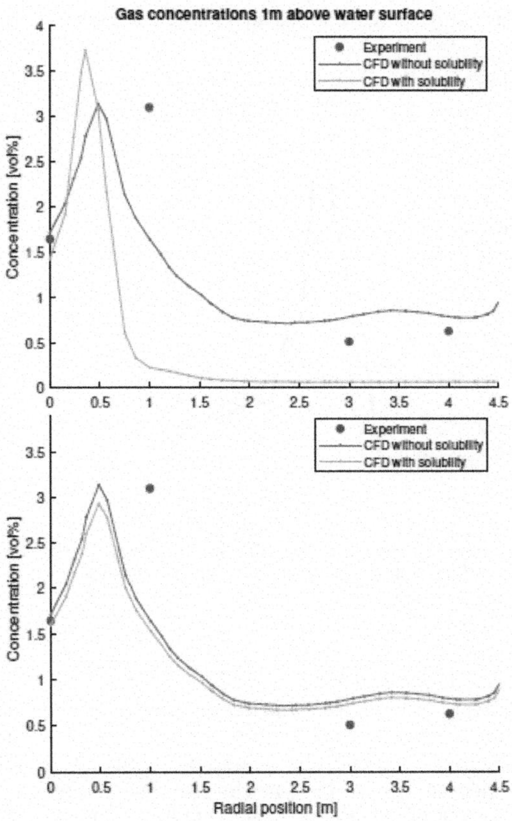

Figure 7. Comparison methane concentrations between the experiment and CFD for 1m above the water surface. Top without alteration below with alteration of the mass transfer for the dissolution.

Figure 7 shows the concentration of methane 1m above the water surface. In the top figure the original implementation is shown with and without the solubility considered (time averaged over 180s). The highest concentration is not measured directly above the release point (at the center of the plume) but at 1m away from the center. This is just outside of the fountain region (where the water is pushed upwards). There seems to be an overestimation of gas dissolution of bubbles further away from the centerline. Due to the radial flow below the waterline these bubbles have a longer residence time in the water and therefore more time to dissolve. A reduction factor in the mass transfer from the gas to the water could be used to avoid this discrepancy. An alternative computation method to take a lower gas dissolution into account is used and can be seen in bottom plot of Figure 7. This matches the experimental results better.

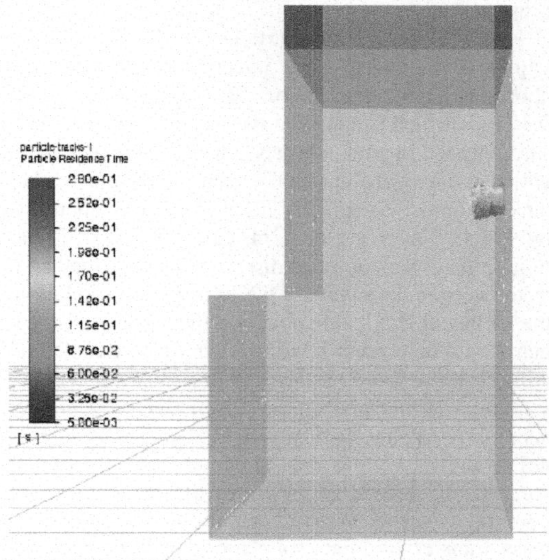

Figure 8. Methanol particle tracks for the full-scale setup from venting at 0.9m depth.

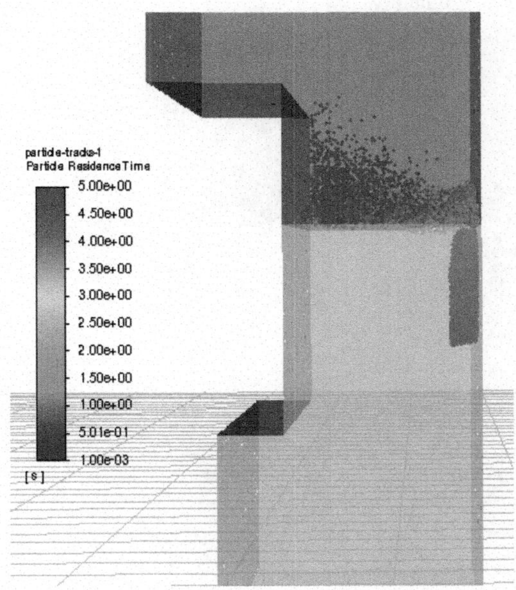

Figure 9. Methanol particle tracks for the full-scale setup (initial bubble size: 0.08m) from venting at 0.9m depth.

4.2 *Methanol venting*

In the simulations of the Rotvoll experiment, detailed in the previous section, it was shown that the gas dissolution close to the surface was overestimated by CFD see Figure 7). Therefore, the mass transfer in Equation 2 is multiplied with 0.5 to be on the conservative side. As the closing pressure of the pressure relief valve is 9 kPa, the maximal venting depth should be 0.9m to overcome the hydrostatic pressure.

Figure 8 shows the particle tracks of methanol vapour at the end of a gas injection. The injection is chosen to have a duration of 18s with a release rate of 0.30 kg/s which is conservative compared to Figure 4. The released vapour consists out of 100% methanol, which is an overestimation except when the bunker tank is almost completely filled. The with an initial bubble size diameter 0.002m, calculated with Equation 4, there is a complete dissolution of the methanol in the surrounding water. Due to this gas dissolution, the methanol is not able to rise to the surface. If nitrogen would be present it will remain in the bubble and will reach the water surface. As previously stated, the saturation of methanol in the surrounding water is not taken into account. If this were the case, more methanol would be able to rise. However, due to external currents and waves, there will be always a supply of new/ under saturated water leading to an increase of mass transfer of methanol vapour to the water. To quantify that is very difficult. Figure 8 only shows methanol. If the released vapour would consist of a mixture of methanol and nitrogen, there will still be nitrogen vapour above the water surface. Nitrogen has a low solubility.

Figure 10. Methanol particle tracks for the full-scale setup (initial bubble size: 0.08m) from venting at 0.5m depth.

4.2.1 *Bubble size*

Due to the shallow injection depth, the bubbles have very limited time to reach the equilibrium diameter. Figure 9 shows the methanol particle tracks if the departure bubble size diameter would be equal to the diameter of the release opening (0.08m). This is most likely to represent the reality, as in the jetting regime the departure bubble size will lay close to the orifice diameter. In this case, methanol vapour can rise above the water surface at a release depth of 0.5m. As an indication, the methanol peak concentration at 0.25m above the water surface is 800ppm. Higher above the water surface, the concentration decreases rapidly and at 0.5m above the water surface only sporadic a methanol particle is found.

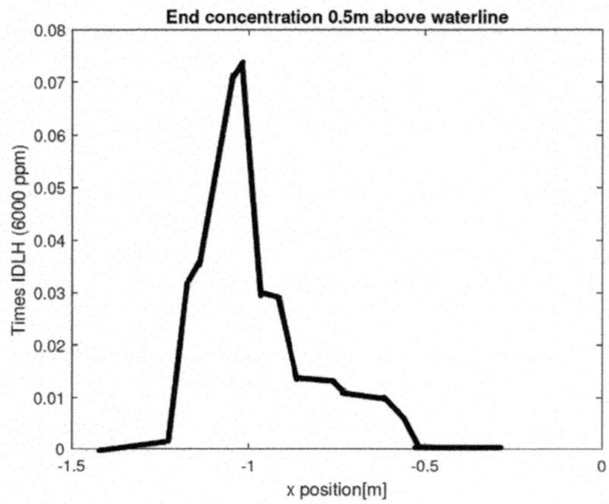

Figure 11. Methanol concentration at 0.5m above the waterline from the venting at 0.5m (x=0: ship, x=-1.5: barge).

4.2.2 *Depth of 0.5m*

Vessels with a smaller draught will vent the methanol at a smaller distance from the waterline (venting depth). The smaller the venting depth, the lower the residence time of methanol in the water leading to a lower gas dissolution rate. Figure 10 shows the result at the end of the venting at 0.5m below the waterline (with an initial bubble size equal to the opening diameter of 0.08m). Even at this lower venting depth no methanol vapour can reach the deck of the barge (or ship). The application of a reduction factor (of 0.5) has little influence due to the short residence time of the methanol vapour in the water. In Figure 11 the methanol concentration at 0.5m above the waterline is depicted. The highest methanol concentration is found close to halfway the bunker barge and the ship. The concentrations are well below the IDLH of 6000ppm.

5 CONCLUSIONS

The validity of the integral model and CFD model have been tested against an experiment (carried out in 1997) where Helium/ air mixture is injected in a water tank. Results indicated that both models performed reasonably well in predicting velocity and void fraction measurements below the waterline. However, the integral model was found to be less reliable in predicting gas concentrations above the waterline due to several limitations. The model's lack of consideration for solubility, coupled with the difficulty in estimating coefficients, restricted its general applicability. In contrast, the CFD model's predictions were in good agreement with the gas measurements obtained in the experiment. Although the CFD model over-predicted gas dissolution, this was accounted for by introducing a mass reduction factor. Further investigation in this mass reduction factor or solubility model is required. Looking at the full-scale model the methanol concentration above the waterline is strongly dependent on the gas release rate, venting depth, and the initial bubble size of the released gas. When the initial bubble size is small (e.g., 1cm), the methanol vapour will quickly dissolve after the release. For larger bubble sizes, the venting depth plays an important role. In the most conservative case (8cm initial bubble size and venting depth of 0.5m), the methanol does not fully dissolve in the water, but no methanol vapour is found above 0.5mm above the waterline. Further research to the relation of the orifice diameter (diameter of sea outlet for gas release) and initial bubble diameter will be needed.

The comparison between the methanol concentration on deck due to venting above the waterline, as dictated by the IEC-code, and the methanol concentration on deck due to venting below the waterline, as predicted by the CFD model, has yielded some interesting results. A better validation of the CFD model with methanol would be a recommendation for further research to determine the effect of some model parameters. Furthermore, environmental conditions like wind, water movements, temperature differences are not included and will have an impact on the surfacing plume. These factors might influence the results, but generally will increase the dissolution of methanol in water.

Based on the model assumptions taken, this investigation shows that venting below the waterline has potential to be a valid option to prevent any or strongly reduce methanol vapour on deck. This could be a technique to reduce or even eliminate any potential hazards resulting from toxic or explosive methanol vapour. Hence, it can be stated that venting below the waterline could be used to significantly decrease the hazardous areas and safety distances on deck. This reduction in hazardous areas and safety distances can also result in improved operational efficiency, as it allows for greater flexibility in the placement and operation of equipment and machinery on board the vessel.

ACKNOWLEDGEMENTS

This work was performed within the MENENS project, a consortium of 21 partners (companies and knowledge institutes) covering the entire Dutch maritime sector, sponsored by RVO (Rijksdienst voor Ondernemend Nederland).

REFERENCES

Adamczyk, W. P. et al. 2014. Comparison of the standard Euler–Euler and hybrid Euler–Lagrange approaches for modeling particle transport in a pilot-scale circulating fluidized bed In: Particuology 15.

Calderbank, P.H. 1958. Physical Rate Processes in Industrial Fermentation, Part I. In: Trans. Inst. Chem. Engrs. (London) 36, p. 443.

Cloete, S., Olsen, J. E & Skjetne. P. 2009. CFD modeling of plume and free surface behavior resulting from a sub-sea gas release. In: Applied Ocean Research 31.3, pp. 220–225.

Ellethy, A. M. et al. 2021. Modelling and Assessment of Accidental Gas Release from Damaged Subsea Pipelines. In: International Journal of Environmental Science and Development 12.6, pp. 162–168

Friedl, M. J. & Fanneløp, T. K. 2000. Bubble plumes and their interaction with the water surface. In: Applied Ocean Research 22.2, pp. 119–128

Herman, S. 2017. Chapter 18 - Fragrance. In: Cosmetic Science and Technology. Ed. by Kazutami Sakamoto et al. Amsterdam: Elsevier

Laux, H. & Johansen, S.T. 1999. A CFD analysis of the air entrainment rate due to a plunging steel jet combining mathematical models for dispersed and separated multiphase flows. In: Proceedings of the Annual TMS Meeting on Fluid Flow Phenomena in Metal Processing, San Diego

Loes, M. & Fanneløp, T. K. 1989. Concentration Measurements Above an Underwater Release of Natural Gas. In: SPE Drilling Engineering 4.02, pp. 171–178

National Institute for Occupational Safety and Health. 1994. Methyl alcohol: Immediately dangerous to life or health concentrations.

Olsen, J. E. & Skjetne. P. 2015. Current understanding of subsea gas release: A review. In: The Canadian Journal of Chemical Engineering 94.2, pp. 209–219

Pan, Q. 2014. Modelling of turbulent flows with strong dispersed phase-continuous fluid interactions. PhD Thesis NTNU.

Ross, S. L., 2009. Assessing Risk and Modeling a Sudden Gas Release Due to Gas Pipeline Ruptures. SINTEF and Wellflow Dynamics.

Sjoen, K., Engebresten, T., Northug, T. & Fanneløp, T. K. 1997. Surface Flow and GasDispersion from a Subsea Release of Natural Gas. In: Proceedings of the 7th International Offshore & Polar Engineering Conference. International Society of Offshore & Polar Eng

Skjetne, P. & Olsen, J. E. 2012. A parcel based modelling concept for studying subsea gas release and the effect of gas dissolution. In: Progress in Computational Fluid Dynamics, An International Journal 12.2/3, p. 187.

Xia, J.L., Ahokainen, T. & Holappa, L. 2001. Analysis of flows in a ladle with gas-stirred melt". In: Scandinavian Journal of Metallurgy 30.2 (Apr. 2001), pp. 69–76.

Advances in Maritime Technology and Engineering – Guedes Soares & Santos (Eds)
 ISBN 978-1-032-83099-5

Heterogeneity analysis of risk factors of marine accidents in different ship types

W.J. Cao, X.J. Wang, Y.W. Feng, H.X. Wang & Z.J. Liu
Navigation College, Dalian Maritime University, Dalian, P.R. China

Q. Yu
Navigation College, Jimei University, Xiamen, P.R. China

ABSTRACT: There are significant differences in risk influential factors (RIFs) among different ship types of marine accidents. This study aims to quantitatively analyse the characteristics of different types of marine accidents using advanced analytical techniques, identify similarities and differences among RIFs. Firstly, the potential relationships of RIFs of marine accidents of four ship types (bulk carriers, container ships, fishing vessels, and oil tankers) are mined through the association rules. Secondly, a nonlinear causal relationship model of accident RIFs, the Weighted Influence Nonlinear Measurement System (WINGS) model is developed, and used to analyse the causal relationships of RIFs from the perspective of dynamic information transmission. Finally, the results of the analysis are used to further identify the key RIFs of different ship types and reveal the differences in the mechanisms of the evolution of marine accidents of different ship types. This study can enhance the safety management of vessels.

1 INTRODUCTION

The ongoing growth and expansion of global trade have intensified maritime transportation, resulting in a heightened density of ships and consequently elevating the risk of marine accidents (Cao et al., 2023). Simultaneously, the escalating dimensions of ships compound the risk of marine accidents (Weng and Yang, 2015). The incidence of marine accidents not only poses a substantial threat to personal safety and the marine environment but also results in economic losses for both national and regional economies. Consequently, there is paramount importance in systematically analysing the risk influential factors (RIFs) of marine accidents to enhance preventive measures.

A ship constitutes a complex system, and the incidence of marine accidents can be influenced by a combination of RIFs. Numerous researchers worldwide have extensively studied the RIFs contributing to marine accidents, Bye and Aalberg (2018) conducted statistical analysis on marine accident data and Automatic Identification System (AIS) data in Norwegian waters. The study findings suggested that certain ship types contributed to a higher occurrence of both ship collisions and groundings. Uğurlu et al. (2016) conducted a statistical analysis of 850 significant marine accidents in the Turkish Straits spanning from 2001 to 2010, employing the hierarchical analysis method (AHP). The findings of the study underscored the significance of ship type as one of the most RIFs in marine accidents. Utilizing GIS technology and establishing a comprehensive database, Sigua and Aguilar (2003) conducted a systematic examination of marine accidents in Philippine waters. The findings revealed that fishing vessels and sailing vessels constituted the most prevalent types of vessels involved in accidents. Utilizing the association rule approach, Weng and Li (2019) conducted a comprehensive analysis of 10 years' worth of marine accident data in Fujian waters. The study resulted in the identification of 66 association rules for non-severe accidents and 228 association rules for severe accidents. The findings indicated that ship type, weather conditions, season, and offshore distance collectively served as determinants influencing the severity of marine accidents.

Previous studies have revealed notable variations in the factors influencing marine accidents across different ship types (Li et al., 2021). However, the current literatures tend to overlook this gap, resulting in deficiencies in risk prevention and control strategies. This study develops a Weighted Influence Nonlinear Measurement System (WINGS) model, utilizing association rules to analyse the dynamic causality of information transfer among RIFs. The aim of this study is to reveal variations in the evolutionary mechanisms of marine accidents across different ship types. The findings of this study can assist organizations, including maritime authorities and shipping companies, in devising tailored prevention measures for diverse types of marine accidents.

DOI: 10.1201/9781003508762-29

2 MATERIALS AND METHODOLOGY

2.1 *Data source and pre-processing*

For this study, accident investigation reports spanning 2000-2019 from the databases of seven global maritime agencies—such as the China Maritime Safety Administration (China MSA), Federal Bureau of Maritime Casualty Investigation (BSU), National Transportation Safety Board (NTSB), Japan Transportation Safety Board (JTSB), Australian Transport Safety Board (ATSB), Canadian Transportation Safety Board(TSB), and the Marine Accident Investigation Branch (MAIB) have been chosen as the primary data source.

Analysis of the investigation reports in the mentioned database exposed inaccuracies and/or incompleteness in certain data, attributed to variations in the level of detail and different countries uploading the accident records. Consequently, accident investigation reports lacking essential data, such as environmental conditions, must be excluded to uphold the principles of data completeness. Firstly, after filtering and eliminating duplicate accident reports, a total of 1294 unique accident investigation reports were obtained. Secondly, according to relevant literature, it is known that bulk carriers, container ships, fishing vessels and oil tankers are several types of ships with the most frequent marine accidents (Jin, 2014, Uğurlu, 2016). Therefore, this study focuses on the heterogeneity analysis of factors affecting ship accidents of these four types of ships. Thirdly, the study further screened 1294 accident investigation data. Details of the screening process for this accident investigation report can be found in previous studies (Cao et al., 2023). Finally, a comprehensive database is established, comprising 910 marine accident records spanning from 2000 to 2019.

This study developed a first-level index system for RIFs encompassing accident, human, vessel, environment, management, along with 36 second-level RIFs, adopting the perspective of system safety engineering through literature review and expert input. As illustrated in Figure 1, the distribution of RIFs for accidents is presented, with a focus on accident information due to space limitations in this paper. Table 1 provides details, including categories, serial numbers, and descriptions of the RIFs explored in this study.

2.2 *Association rule mining*

Association rule mining is a method intended for uncovering frequent patterns, correlations, associations, or causal structures within datasets across diverse databases and repositories (Zhou et al., 2019). In essence, association rule mining represents a straightforward and practical data mining technique employed to identify frequent item sets amid uncertainties, generating robust association rules from extensive datasets. Notably, association rules

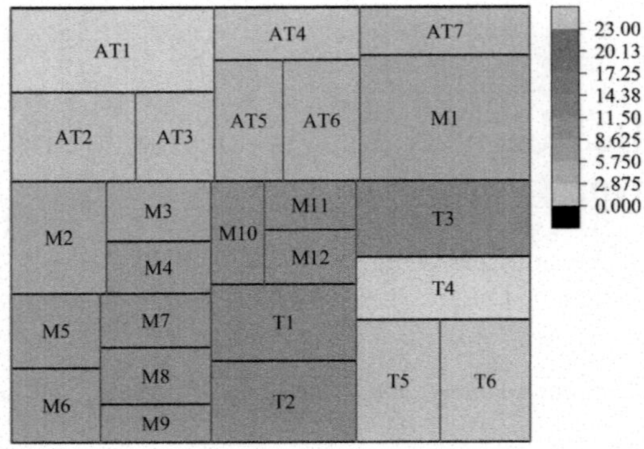

Figure 1. Tree map of accident information.

prioritize the exploration and discovery of patterns in data over hypothesis confirmation and remain unaffected by missing data (Montella et al., 2021). The equations for the association rules are depicted in Equation (1) to (3).

$$Sup(E) = \frac{LE}{L} \quad (1)$$

$$Con(E \Rightarrow F) = \frac{Sup(EF)}{Sup(F)} \quad (2)$$

$$Lift(E \Rightarrow F) = \frac{Con(E \Rightarrow F)}{Supt(F)} \quad (3)$$

Where *Sup*(*E*) indicates the degree of support for Matter *E*; L_E indicates the number of occurrences of Matter *E* in the data; *L* indicates the number of all the data; $Con(E \Rightarrow F)$ indicates the degree of confidence from Matter *E* to Matter *F*; and $Lift(E \Rightarrow F)$ indicates the degree of enhancement from Matter *E* to Matter *F*.

2.3 *WINGS*

Originally designed for addressing complex and interrelated groups of components or systems, decision-making and trial evaluation laboratory (DEMATEL) is a systematic structural modelling approach. It proves particularly effective in establishing and evaluating causal relationships (dependencies) between various components within a system (Si et al., 2018). WINGS, a structural model that extending the capabilities of DEMATEL, serves as a tool for analysing systems with diverse interrelated components. As an extension of DEMATEL, WINGS inherits all the advantages of its predecessor, effectively managing the complexity of reaction factors through clear and simple mathematical operations. Moreover, in comparison to DEMATEL, WINGS possesses unique features allowing for the assessment of both the strength of acting factors and the strength of RIFs. To scrutinize the causal relationship among RIFs, this study employs WINGS for analysing the characteristics

Table 1. Marine accidents effect factors database.

Level I	Level II	N	Value/definition	Cv
Accident	Accident type	AT	collision, stranding/grounding, fire/explosion, contact, capsize/sinking, hull/machinery damage, other	1,2,3,4,5,6,7
	Month	M	January, February, March, April, May, June, July, August, September, October, November, December	1,2,3,4,5,6,7,8,9,10,11,12
	Time	T	0000-0400, 0400–0800, 0800–1200, 1200–1600, 1600–2000, 2000-2400	1,2,3,4,5,6
Human	Physical & Psychological state	PPS	poor, good	1,2
	Education background	EB	poor, good	1,2
	Time at sea	TAS	<5 years, $5 \leq$ time <10 years, ≥ 10 years	1,2,3
	Time in present rank	TIR	<1 year, $1 \leq$ time <5 years, ≥ 5 years	1,2,3
	Communication problem	C	yes, no	1,2
	Operational error	OE	yes, no, unknown	1,2,3
	Violation operation	VO	yes, no, unknown	1,2,3
Ship	Age	SA	0–10 years, 10–20 years, 20–30 years, ≥ 30 years	1,2,3,4
	Gross tonnage	GT	<500 t, 500–3000 t, ≥ 3000t	1,2,3
	Engine power	EP	<750 KW, 750–3000 KW, ≥ 3000 KW	1,2,3
	Flag state	FS	Flag of convenience, Not flag of convenience	1,2
	Ship's certificates	SC	complete and valid, incomplete or invalid	1,2
	Ship manning	SM	adequate, inadequate	1,2
	Seafarers' certificates	SFC	complete and valid, incomplete or invalid	1,2
	Seaworthiness	SEA	yes, no	1,2
	PSC/FSC inspection	PF	unsure, sure	1,2
Environment	Location	L	Inland waters, Port, Coastal waters, Open Sea	1,2,3,4
	Visibility	V	very poor - Vis <0.5 nm, Poor - $0.5 \leq$ Vis <2 nm, Moderate - $2 \leq$Vis <5 nm, Good and very good - Vis $\geq$5 nm	1,2,3,4
	Wind force	WF	0-5, 6–7, 8–9, 10-12	1,2,3,4
	Sea state	SS	0-3, 4–5, 6–7, 8-9	1,2,3,4
	Current speed	CS	<2kn, 2-4kn, ≥ 4kn	1,2,3
	Traffic density	TD	low, high	1,2
	Fairway width/ ship length	WL	$w/l < 1$, $1 \leq w/l < 2$, $w/l \geq 2$	1,2,3
	Depth-draft ratio (h/ d)	DDR	$h/d < 1.2$, $1.2 \leq h/d < 1.5$, $1.5 \leq h/d < 3$, $h/d \geq 3$	1,2,3,4
Management	Regulation	R	inadequate, adequate	1,2
	Supervision	SUP	inadequate, adequate	1,2
	Violation in supervision	VIS	yes, no	1,2
	Safety management system	SFS	defective, non-defective	1,2
	Safety management	SFM	inadequate, adequate	1,2
	Rectification of problems	ROP	unresponsive, responsive	1,2
	Company safety culture	CC	poor,good	1,2
	Training	TRA	inadequate, adequate	1,2
	Drill	D	off schedule, stick to the schedule	1,2

Note: N indicates notation; Cv indicates corresponding values.

of marine accident RIFs across different ship types. The analysis unfolds in several steps. Firstly, association rules among RIFs are mined to establish a confidence matrix. Secondly, as the confidence value in association rule mining signifies the correlation magnitude between factors, it aligns with the value in the direct influence matrix of the WINGS method, representing the degree of correlation between factors (Cui, 2021). Consequently, the confidence matrix from association rule mining can be directly transformed into the direct influence matrix. Thirdly, the comprehensive influence matrix is calculated. Finally, centrality degree, reason degree, weights, and other properties of the RIFs on accidents are calculated. The specific analytical steps are outlined as follows.

Step 1: Constructing a direct impact matrix H based on association rules. The equations are shown in Equations (4) to (6).

$$H = \begin{matrix} & x_1 & x_2 & \cdots & x_n \\ x_1 \\ x_2 \\ \vdots \\ xn \end{matrix} \begin{bmatrix} h_{11} & h_{12} & \cdots & h_{1n} \\ h_{21} & h_{22} & \cdots & h2n \\ \vdots & \vdots & & \vdots \\ h_{n1} & h_{n2} & \cdots & h_{nn} \end{bmatrix} \tag{4}$$

$$h_{ij} = f_{ij} \times Con(i \Rightarrow j) \tag{5}$$

$$f_{ij} = \begin{cases} 1, \ when\ an\ association\ rule\ exists\ from\ i\ to\ j \\ 0,\ else \end{cases} \tag{6}$$

Where x_i indicates an RIF analysed in this study; h_{ij} indicates the degree of influence of factor i on factor j, when $i = j$, $h_{ij} = 0$; f_{ij} indicates whether there is a significant influence of factor i on factor j, if there is a significant influence of 1, and vice versa for 0.

Step 2: The direct influence matrix is normalized to obtain the normalized influence matrix $StaH$, which is calculated as shown in Equation (7) to (10).

$$Nopa = \frac{1}{\left(\max(c_1, c_2, \cdots, c_n)^2 + \max(g_1, g_2, \cdots, g_n)^2\right)^{1/2}} \tag{7}$$

$$c_i = \sum_j h_{ij} \tag{8}$$

$$g_i = \sum_j h_{ji} \tag{9}$$

$$StaH = Nopa \times H \tag{10}$$

Where $Nopa$ is the normalization parameter; c_i represents the sum of row i of the direct influence matrix; and g_i represents the sum of column i of the direct influence matrix.

Step 3: Using the theory of transmissibility of influence, a comprehensive influence matrix TH is generated, calculated as in Equation (11).

$$TH = \sum_{k=1}^{\infty} StaH^k = StaH \cdot (Eh - StaH)^{-1} = (t_{ij})_{e \times e} \tag{11}$$

Where denotes the inner product of the matrix; E_h denotes the unit matrix; and t_{ij} denotes the combined effect of factor i on factor j in the combined effects matrix.

Step 4: The comprehensive influence matrix was used to calculate the influence degree, influenced degree, centrality degree, and reason degree of the RIFs and the weights of each RIF, and the calculation equations are shown in Equations (12) to (16).

$$l_i = \sum_j t_{ij} \tag{12}$$

$$m_i = \sum_j t_{ji} \tag{13}$$

$$CT_i = l_i + m_i \tag{14}$$

$$RT_i = l_i - m_i \tag{15}$$

$$W_i = \frac{(CT_i^2 + RT_i^2)^{1/2}}{\sum_i (CT_i^2 + RT_i^2)^{1/2}} \tag{16}$$

Where l_i denotes the influence of factor i; m_i denotes the influence of factor i; CT_i denotes the centrality of factor i; RT_i denotes the cause of factor i; and W_i denotes the weight of factor i.

3 RESULTS AND ANALYSIS

3.1 *Association rules for RIFs*

In order to investigate potential correlations among RIFs, this study necessitates the application of association rule mining to the pertinent dataset. To establish suitable thresholds, this study conducted a series of experiments and ultimately determined the minimum support threshold to be 0.3 and the minimum confidence threshold to be 0.5 (Ashrafi et al., 2005). Subsequently, association rules are extracted for accident data pertaining to four distinct ship types: bulk carriers, container ships, fishing vessels, and oil tankers.

Figure 2 illustrates the distribution of association rules concerning RIFs influencing marine accidents in bulk carriers. For space considerations, this paper exclusively displays the distribution of RIFs impacting marine accidents on bulk carriers. An analysis of Figure 2 reveals that the association rules are predominantly situated within the support range of 0.3-0.9 and the confidence range of 0.5-1.0. Simultaneously, association rules exhibiting higher lift values are primarily located in comparatively sparser regions.

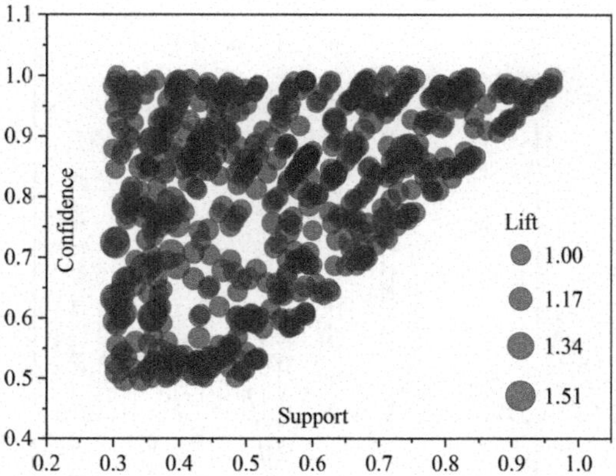

Figure 2. Distribution of effect factors association rules for marine accidents.

3.2 *Causality analysis based on WINGS inter impacts*

This section examines the causal relationships among RIFs utilizing centrality degree, reason degree, and weights. Figure 3 illustrates the causal distribution of factors impacting bulk carriers. Due to space constraints, this paper exclusively presents the causal distribution related to bulk carriers. The centrality degree (CT) signifies the extent of influence a factor holds over the entire complex system, with larger values indicating a more pronounced impact on marine accidents. The foremost factors significantly impacting marine accidents on bulk carriers include SC1 (Ship's certificates complete and valid), SM1 (Ship manning adequate), and EB2 (Good educational background of the crew). The centrality scores for these factors were 0.0851, 0.0845, and 0.0833, respectively. Likewise, the most RIFs affecting marine accidents involving container ships comprise CS1 (Current speed < 2 kn), VIS2 (No violation in supervision), and EB2 (Good educational background of the crew), with centrality scores of 0.0783, 0.0780, and 0.0767, respectively. The primary factors exerting significant influence on marine accidents involving fishing vessels include SM1 (Ship manning adequate), VIS2 (No violation in supervision), and CC2 (Good company safety culture). These factors possess centrality scores of 0.0763, 0.0762, and 0.0750, respectively. Similarly, the foremost factors influencing marine accidents involving oil tankers encompass SFS2 (Shipping company's non-defective Safety management system), EB2 (Good educational background of the crew), and VIS2 (No violation in supervision). The respective centrality scores for these factors are 0.0885, 0.0858, and 0.0814. Through an analysis of these centrality values, it becomes evident that the educational background of the crew significantly influences marine accidents across all four types of vessels. This commonality suggests that the level of education among the crew has a notable impact on maritime safety, irrespective of the specific vessel type—be it a container ship, fishing vessel, or oil tanker. Recognizing and addressing crew education emerges as a crucial factor in enhancing safety measures and mitigating the risk of marine accidents within diverse maritime sectors.

The reason degree (RT) reflects the extent to which factors hold a causal position in a complex system. $RT > 0$, indicating that the factor is a causal element influencing other factors, a higher value of RT implies a greater impact of this factor on other elements within the complex system. Conversely, $RT < 0$, a smaller value of RT indicates that the element is a resultant factor susceptible to the influence of other factors in the complex system. In the context of marine accidents involving bulk carriers, the three most prominent causal factors are OE2 (No human operational error), TD1 (Traffic density low), and WL2 (1≤Restricted fairway/channel < 2). Regarding the consequential factors in marine accidents of bulk carrier, the three most notable RIFs are C2 (No communication problems), SM1 (Ship manning adequate), and SEA1 (Seaworthiness). In the context of marine accidents involving container ships, the three most prominent causal RIFs are V4 (Visibility ≥ 5 nm), DDR3 (1.5 ≤ Depth-draft ratio < 3), and SA1 (Ship's age 0-10 years). Regarding consequential RIFs in container ship marine accidents, the three most notable RIFs are SM1 (Ship manning adequate), SC1 (Ship's certificates complete and valid), and SFC1 (Seafarers' certificates are complete and valid). In the context of marine accidents involving fishing vessels, the three most prominent causal RIFs are DDR4 (Depth-draft ratio≥3), TD1 (Traffic density low), and DDR3 (1.5 ≤ Depth-draft ratio < 3). Regarding consequential factors in marine accidents of fishing vessel, the three most notable RIFs are SEA1 (Seaworthiness), C2 (No communication problems), and SC1 (Ship's certificates complete and valid). In the context of marine accidents involving oil tankers, the three most prominent causal RIFs are OE2 (No human operational error), SFM2 (Shipping company's adequate safety management), and V4 (Visibility ≥ 5 nm). Regarding consequential factors in oil tanker marine accidents, the three most notable RIFs are SEA1 (Seaworthiness), R2 (Adequacy of regulators' regulations), and SFC1 (Seafarers' certificates are complete and valid).

By employing causality values analysis to examine the differences and similarities in marine accidents among four types of vessels, it becomes apparent that the causation degree of environmental factors in the navigational areas of ships is relatively high across all four vessel types, indicating a significant impact of environmental factors on other RIFs. Conversely, issues related to communication among personnel and the seaworthiness of vessels exhibit comparatively lower causation degrees in marine accidents involving the four vessel types, suggesting their susceptibility to influence from other RIFs. Concerning bulk carriers, particular attention should be given to the impact of human

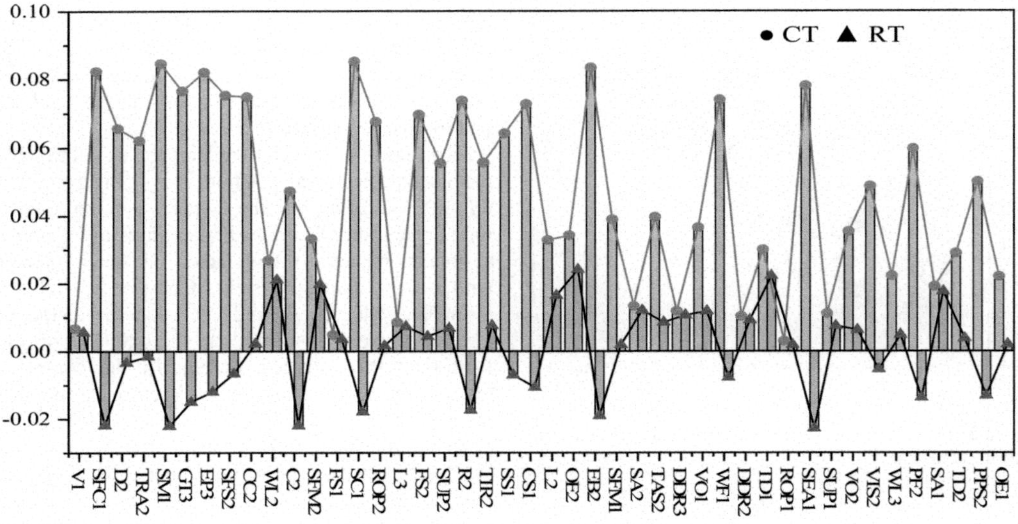

Figure 3. Causal distribution of RIFs.

errors in vessel manipulation and navigational density. In contrast, for container ships, fishing vessels, and oil tankers, additional consideration is necessary regarding the influence of management factors such as company safety management systems.

Considering both centrality degree and reason degree, for bulk carriers, the navigational environmental factors and ship company management play a significant role in influencing accidents. In the case of container ships, the impact of human factors, specifically the educational background of crew members, takes precedence. For fishing vessels, the significant factors influencing accidents are ship company management and navigational environmental conditions. In the case of oil tankers, the adequacy and effectiveness of crew certificates, along with regulatory authorities' management factors, have a significant impact. In summary, the RIFs in marine accidents vary among the four types of vessels, necessitating a more meticulous approach to risk management for each specific type of ship.

In the complex system of marine accident evolution, the weight distribution of RIFs is illustrated in Figure 4. Due to spatial constraints, this paper exclusively presents the weight distribution of container ships. In the complex system of interacting factors in marine accidents of bulk carrier, the key RIFs with the top three weight rankings are SM1 (Ship manning adequate), SC1 (Complete and valid ship's certificate), and EB2 (Good educational background of the crew). In the complex system of interacting factors in container ship marine accidents, the key RIFs with the top three weight rankings are CS1 (Current speed < 2 kn), EB2 (Good educational background of the crew), and VIS2 (No violation in supervision). In the complex system of interacting factors in marine accidents of fishing vessel, the key RIFs with the top three weight rankings are PF2 (NO port State and flag State inspections), SM1 (Ship manning adequate), and VIS2 (No violation in supervision). In the complex system of interacting factors in marine accidents of oil tanker, the key RIFs with the top three weight rankings are SFS2 (Shipping company's non-defective Safety management system), EB2 (Good educational background of the crew), and VIS2 (No violation in supervision). These weights comprehensively consider the centrality and causation degree attributes, mitigating the influence of positive and negative values in causation degree.

From a weight perspective, a common characteristic among marine accidents involving the four types of vessels is the significant influence of crew educational background in their respective accident systems. Simultaneously, there are variations in the weight distribution of RIFs in marine accidents among the four types of vessels: for bulk carriers, the weight of ship personnel allocation is substantial. For container ships, the weight of water flow speed is significant. In the case of fishing vessels, the weight of regulatory authorities' management factors is substantial. For oil tankers, the weight of shipping company management factors is substantial. Therefore, for each specific vessel type, these RIFs should be carefully addressed and controlled accordingly.

4 CONCLUSION

4.1 *Conclusions*

The results of this study indicate that in the interactive system of RIFs influencing marine accidents across the four types of ships, a commonality is observed in the significant role played by the human factor. Consequently, reducing the occurrence of marine accidents necessitates an enhancement of

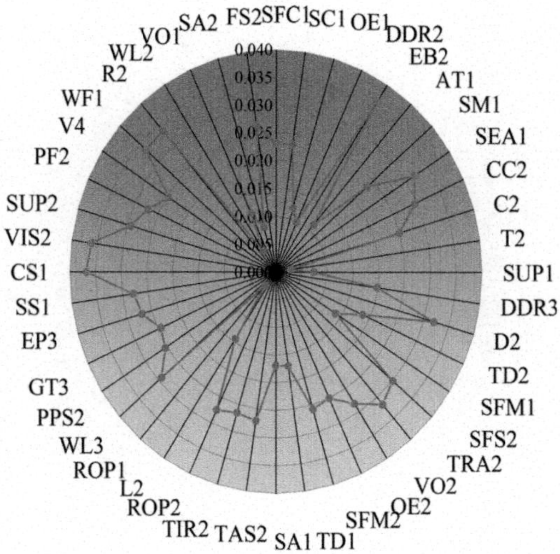

Figure 4. Weight distribution of RIFs.

safety management, along with the improvement of safety awareness and ship-handling skills among crew members. The distinction lies in the fact that bulk carriers should specifically address the impact of inherent ship-related factors, emphasizing the need for robust self-maintenance and judicious control of the vessel's equipment. Container ships, on the other hand, must consider navigational environmental factors and enhance preparedness for extreme weather, adverse sea conditions, and complex waters. Fishing vessels should be attentive to the influence of maritime surveillance within the management factors, necessitating improvements in regulations governing safe ship operations, coupled with heightened seafarer training and surveillance. Oil tankers should focus on the impact of shipping company within the management factors, necessitating the reinforcement of a robust safety culture in the company's management practices.

4.2 *Limitations*

Through an analysis of the RIFs influencing marine accidents across various types of ships, this study can assist stakeholders, including maritime authorities and shipping companies, in formulating analytical measures for preventing marine accidents specific to each ship type. The study holds reference value in the development of a marine accident RIFs identification system and the enhancement of ship safety management. Nonetheless, it is important to acknowledge certain limitations in this study.

Firstly, this study focused its analysis on the RIFs of marine accidents within only four ship types that experienced a higher incidence of accidents, thus limiting the range of analysed ship types. Secondly, the RIFs database utilized in this study is extracted manually, this processing procedure inevitably brings uncertainty. Finally, this study concentrates on analysing the RIFs influencing marine accidents across various ship types. However, it does not delve deeply into the interplay among factors affecting marine accidents within the same type of ship. These limitations will be further explored in future studies.

REFERENCES

Ashrafi, M. Z., Taniar, D. & Smith, K. 2005. Redundant association rules reduction techniques. *In*: Zhang, S. & Jarvis, R. (eds.) *AI 2005: ADVANCES IN ARTIFICIAL INTELLIGENCE*.

Bye, R. J. & Aalberg, A. L. 2018. Maritime navigation accidents and risk indicators: An exploratory statistical analysis using AIS data and accident reports. *Reliability Engineering & System Safety*, 176, 174–186.

Cao, Y. H., Wang, X. J., Wang, Y. H., Fan, S. Q., Wang, H. X., Yang, Z. L., Liu, Z. J., Wang, J. & Shi, R. J. 2023. Analysis of factors affecting the severity of marine accidents using a data-driven Bayesian network. *OCEAN ENGINEERING*, 269.

Cui, Y. M. 2021. *DevelopmentResearch on the Barriers Based on ARM—DEMATEL Smart Port*.

Jin, D. 2014. The determinants of fishing vessel accident severity. *Accident Analysis & Prevention*, 66, 1–7.

Li, B. D., Lu, J. & Li, J. 2021. Investigation of accident severity in sea lanes from an emergency response perspective based on data mining technology. *OCEAN ENGINEERING*, 239.

Montella, A., Mauriello, F., Pernetti, M. & Rella riccardi, M. 2021. Rule discovery to identify patterns contributing to overrepresentation and severity of run-off-the-road crashes. *Accident Analysis & Prevention*, 155, 106119.

Si, S.-L., You, X.-Y., Liu, H.-C. & Zhang, P. 2018. DEMATEL Technique: A Systematic Review of the State-of-the-Art Literature on Methodologies and Applications. *Mathematical Problems in Engineering*, 2018, 3696457.

Sigua & Aguilar. Maritime Incident Analysis using GIS. 5th EASTS Conference, 2003.

Uğurlu, Ö. 2016. Analysis of fire and explosion accidents occurring in tankers transporting hazardous cargoes. *International Journal of Industrial Ergonomics*, 55, 1–11.

Uğurlu, Ö., Erol, S. & Başar, E. 2016. The analysis of life safety and economic loss in marine accidents occurring in the Turkish Straits. *Maritime Policy & Management*, 43, 356–370.

Weng, J. & Li, G. 2019. Exploring shipping accident contributory factors using association rules. *Journal of Transportation Safety & Security*, 11, 36–57.

Weng, J. & Yang, D. 2015. Investigation of shipping accident injury severity and mortality. *Accident Analysis & Prevention*, 76, 92–101.

Zhou, Y., Li, C., Ding, L., Sekula, P., Love, P. E. D. & Zhou, C. 2019. Combining association rules mining with complex networks to monitor coupled risks. *Reliability Engineering & System Safety*, 186, 194–208.

Advances in Maritime Technology and Engineering – Guedes Soares & Santos (Eds)

Learning operational envelopes

N. Clemett & M. Collette
Department of Naval Architecture and Marine Engineering, University of Michigan, USA

ABSTRACT: For Autonomous Surface Vehicles (ASVs), rapid adaptation to on-water data post-deployment is crucial. Delayed adaptation may lead ASVs into hazardous situations, mistakenly identified as safe by pre-deployment models. This work demonstrates the effectiveness of meta-learning, specifically Model Agnostic Meta-Learning, to accurately predict vertical bending moment Root-Mean Square (RMS) values from weather data and ship characteristics. Emphasis is placed on few-shot learning, where limited data is available for training. National Oceanographic and Atmospheric Administration (NOAA) hindcast weather data from the North Pacific is used to train the models with real world conditions. After initial training on a data set generated from closed-form solutions, the models are retrained and evaluated for accuracy using a second, 'contaminated' data-set representing nonlinear behaviors of ships in open water.

1 INTRODUCTION

Crewed vessels have traditionally depended on human expertise to navigate between safe and hazardous weather conditions. This reliance poses a significant challenge for autonomous ships which, in the absence of onboard human decision-making, risk venturing into weather that could lead to severe vessel damage. Additionally, autonomous ships are faced with overcoming their theoretical initial loading predictions presumed at deployment, to quickly learn from real world data. Faced with on-water data after initial deployment, ASV's must quickly adapt their models to account for on-water measurements.

The rapid adaptation of ASVs to on-water conditions is not just a technical challenge but also a pivotal aspect of mission planning. Understanding the volume of real-world data required to update the ASVs' models for making precise predictions based on environmental inputs is crucial. This presents a solution for the process of refining an initial engineering-based model with real-world data.

Utilizing Model Agnostic Meta-Learning (MAML), an initial model is developed based on closed-form solutions that estimate vertical bending moments. Due to the lack of human input in decision making on safe conditions, structural loads have been a concern for ASVs. Because of the nonlinearity of ocean waves, predictions of individual bending moments are often computationally expensive and inaccurate. To combat this, the root mean square of vertical bending moments was used and represents the typical level of bending in prevailing conditions. Despite the vertical bending moment being used for this work, this approach is valid for any linear Response Amplitude Operators (pitch, heave, etc.). In the absence of available real-world data, simulated sensor data is created. The initial model is then loaded and evaluated on its performance after few-shot learning from a different and scaled data set representing on-water measurements.

2 BACKGROUND

2.1 *Model Agnostic Meta-Learning (MAML)*

Meta-Learning, often termed as ``learning to learn", is a paradigm designed to enhance a learning algorithm's capacity to adapt to new tasks. This approach yields highly adaptable models that are well suited to be trained with limited data. Model Agnostic Meta-Learning (MAML), is such an algorithm, and specializes in its flexibility. MAML can be integrated with any model trained by gradient descent, and is viable for classification, regression, and reinforcement learning.

The core of MAML and similar Meta-Learning algorithms lies in their task-oriented structure. These tasks are defined following the framework proposed (Finn et al. 2017) ($\mathcal{T}$, Equation 1). Each task ($\mathcal{T}$) encompasses several components: a loss function (L), an initial distribution over states ($q(x_1)$), a transition distribution that defines the probability of moving to the next state ($q(x_{t+1}|x_t,a_t)$), and the episode length (H). This framework allows MAML to efficiently train models by rapidly adjusting to new and varied tasks, a feature that is valuable in scenarios where data is scarce or rapidly changing.

DOI: 10.1201/9781003508762-30

$$\mathcal{T} = \{L(x_1, a_1, \ldots, x_H, a_H), q(x_1), q(x_{t+1}|x_t, a_t), H\} \tag{1}$$

One feature of MAML is its ability to retain knowledge from previous tasks and use this accumulated experience to learn highly responsive parameters (θ). These parameters are specifically tuned to adapt rapidly to new tasks, even when only a minimal amount of task-specific data is available. Retention of past information makes MAML a suitable candidate for structural predictions as memory of past large scale events is of utmost importance.

When MAML is implemented in regression tasks, the loss function uses the mean-squared error (MSE) of the networks prediction and the actual value (Equation 2). This assists in the learning process by refining predictions based on the accuracy of previous outputs.

$$\mathcal{L}_{\mathcal{T}_)}(f_\phi) = \sum_{x^{(j)}, y^{(j)} \sim \mathcal{T}_i} \|f_\phi(x^{(j)}) - y^{(j)}\|_2^2 \tag{2}$$

The MAML algorithm operates through two distinct loops, each with its respective hyperparameters and application of gradient descent. Before the loops begin, the initial model parameters, θ, are randomly initialized. The inner loop, or adaptation loop, focuses on individual tasks, each with k-shot sampled data points. In this loop, the model uses its current parameters θ to predict outputs for each of the inputs. The predictions are then evaluated against the actual outputs using the loss function (as outlined in Equation 2). This loss is scaled from the hyperparameter α and Stochastic Gradient Descent (SGD) is performed to generate the adapted model parameters, denoted as $\{\theta_i'$ for each task (Equation 3). The ability of the model to adjust its parameters (θ to θ_i) effectively based on immediate feedback from the loss function is key to its performance in new and varied tasks.

$$\theta_i' = \theta - \alpha \nabla_\theta \mathcal{L}_{\mathcal{T}_)}(f_\theta) \tag{3}$$

After the tasks have been iterated through in the inner loop, a meta-update is performed (Equation 4). This step consolidates the learning from individual tasks to improve the overall model. The initial parameters θ are updated through SGD performed on the task-specific losses with respect to the adapted parameters $\{\theta_i'$. This process ensures that the model parameters (θ) are not tailored to any single task but rather optimized in a way that balances and incorporates learning from all encountered tasks. The complete MAML process is detailed in Algorithm 1.

$$\theta \leftarrow \theta - \beta \nabla_\theta \sum_{\mathcal{T}_i \sim p(\mathcal{T})} \mathcal{L}_{\mathcal{T}_i}(f_{\theta'_i}) \tag{4}$$

Algorithm 1 MAML for Few-Shot Supervised Learning (Finn, Abbeel, & Levine 2017)

Require: $p(\mathcal{T})p(\mathcal{T})$: distribution over tasks
Require: α, β: step size hyperparameters
randomly initialize θ
while not done **do**
  Sample batch of tasks $\mathcal{T}_i \sim p(\mathcal{T})$
  for all $\mathcal{T}_i$ **do**
    Sample K datapoints $D = \{(x^{(j)}, y^{(j)})$ from $\mathcal{T}_i$
    Evaluate $\nabla_\theta \mathcal{L}_{\mathcal{T}_)}(f_\theta)$ using D and $\mathcal{L}_{\mathcal{T}_)}$ in Equation 2
    Compute adapted parameters with gradient descent:

$$\theta_i' = \theta - \alpha \nabla_\theta \mathcal{L}_{\mathcal{T}_i}(f_\theta)$$

    Sample datapoints $D_i' = \{(x^{(j)}, y^{(j)})\}$ from $\mathcal{T}_i$ for the meta-update
  end for
  Update $\theta \leftarrow \theta - \beta \nabla_\theta \sum_{\mathcal{T}_i \sim \mathbf{p}(\mathcal{T})} \mathcal{L}_{\mathcal{T}_i}(f_{\theta'_i})$ using each D_i' and $L_{\mathcal{T}_i}$ in Equation 2
end while

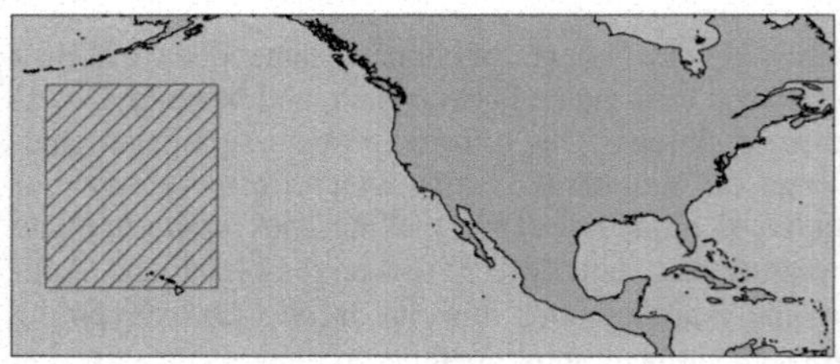

Figure 1. Hindcast weather data collection zone.

3 METHODOLOGY

3.1 *Weather data*

For this study, a comprehensive dataset of hindcast weather data was obtained from the National Oceanic and Atmospheric Administration (NOAA) for the entire year of 2018. The data collected includes significant wave height H_s, peak period T_p, prominent wave direction β, wind direction, and wind speed. The geographic focus of the data collection was a rectangular area in the North Pacific, delineated by the coordinates [(20.0,-175.0),(50.0, -150.0)] (Figure 1).

3.2 *Initial design-stage VBM estimate*

Estimation of the wave-induced vertical bending moments (VBM) in mono-hull ships, as used in this study, is based on the closed-form solutions developed by Jensen & Mansour (2002). These solutions start with using strip-theory to analytically derive the vertical bending moment of a homogeneously loaded box-shaped vessel. Correction factors are then introduced to modify the initial box-shaped vessel model to better represent the characteristics and dynamics of an actual ship. The vertical bending moment (VBM) as a function of frequency is found through the closed-form expression in Equation 5.

$$\frac{VBM(\omega)}{\rho g B_0 L^2} = [1 - cos(\frac{k_e L}{2}) - \frac{k_e L}{4} sin(\frac{k_e L}{2})] * \kappa * \frac{1-kT}{(k_e L)^2} * F_V(Fn) F_C(C_b) \times \sqrt[3]{|cos\mu|} \quad (5)$$

Where κ is the approximation of the Smith correction factor defined by Equation 6 (Jensen 2001). It is dependent on the effective wave number (Equation 7) and the peak period and accounts for dynamic pressure decay as distance increases with vertical distance away from the free surface.

$$\kappa = exp(-k_e T_p) \quad (6)$$

$$k_e = |kcos\ \mu| \quad (7)$$

$F_V(Fn)$ represents the speed correction factor from a homogeneously loaded box-shaped vessel to a range of ships. It accounts for speed through the incorporation of the Froude number (valid for $Fn < 0.3$) (Equation 8).

$$F_V(Fn) = 1 + 3Fn^2; \qquad Fn = \frac{V}{gL} \quad (8)$$

Due to the smaller volume of the foreward part of the vessel, ships with lower block coefficients have lower bending moments. The block coefficient C_b is corrected through the piece-wise Equation 9.

$$F_C(C_b) = [(1-\varphi)^2 + 0.6\alpha(2-\varphi)]$$
$$\varphi = 2.5(1 - C_b); \quad C_b = \max(0.6, C_b) \quad (9)$$

3.3 *Simulated at-sea data*

The response spectrum of the vertical bending moment is determined by applying the transfer function to the wave energy density spectrum and the VBM spectrum, as outlined in Equation 12. A Bretschneider wave energy density spectrum was used in this work (Equation 11). The Bretschneider wave energy density spectrum is generated for the prevailing weather and ship conditions. To account for the frequency change due to vessel heading and speed, the frequency of encounter is used in the Bretschneider spectrum. It is dependent on the ship velocity V, and relative heading between the ship and prominent wave direction μ (180°representing head seas) (Equation 10).

$$\omega_e = \omega - kVcos\mu \equiv \alpha\omega \quad (10)$$

$$S_\zeta(\omega) = \frac{173H_s^2}{T_p^4}\omega_e^{-5} exp\{\frac{-692}{T_p^4}\omega_e^{-4}\} \quad (11)$$

Table 1. Main particulars of vessel.

Characteristic	Value
Length	270.0 m
Beam	32.2 m
Draft	10.85 m

$$S_{VBM}(\omega) = |VBM(\omega)|^2 \cdot S_\zeta(\omega) \quad (12)$$

The area under the VBM response curve is the variance (Equation 13).

$$m_0 = \int_0^\infty S_{VBM}(\omega)d\omega \quad (13)$$

Lastly, the Root-Mean Square of the VBM is calculated through the square root of the VBM response's variance (Equation 14).

$$VBM_{RMS} = \sqrt{m_0} \quad (14)$$

To generate an initial operational envelope, MAML was used to train a model using 5000 samples. Each sample used a randomly selected weather sample from the NOAA North Pacific hindcast data. Ship velocity and heading were then randomly selected. Due to limitations of the closed-form solutions, headings within +/- 40° of beam seas (90, 270) were not selected. After weather, heading, and speed were selected, closed-form solutions were used to generate the VBM RMS values. The network was trained using this data set ($task_1$) and saved as the $task_1$ model. This model is representative of predeployment training, where models and operational envelopes are developed based fully on simulations or closed-form solutions.

To simulate post-deployment data, a second data set was then generated. In the absence of on water data, a data set was generated in the same way as $task_1$. However, to highlight discrepancies between predictions and real-world measurements, the RMS value was scaled depending on the significant wave height (Equation 15).

$$f(H_s) = \begin{cases} RMS, & \text{if } H_s \leq 3.0m, \\ RMS \cdot (\frac{0.1H_s}{3} + 1), & \text{if } H_s > 3.0m \end{cases}$$
$$f(H_s) = RMS_{\text{task2}} \quad (15)$$

After the data set was collected, the saved model from $task_1$ was loaded and trained with a varying amount of shots from $task_2$. This problem design was constructed to evaluate a state-of-the-art machine learning technique for an autonomous vessel deployment and study its ability to quickly and accurately adapt its model to match real-world measurements.

3.4 *Machine learning*

The initial network used was a five layered fully-connected neural network with 32 neurons per layer. The dropout rate between layers was set to 0.1 and used to combat over-fitting. The MAML algorithm used this network to predict VBM RMS values from inputs of significant wave height H_s, wave period T_p, ship speed V, and the difference in heading between the ship and prominent wave direction μ. After experimentation, the MAML hyperparameters used were an adaptation learning rate of 0.02 and meta learning rate of 0.001.

Models were trained for up to 500 epochs per task, but given a early stopping patience condition of 20 epochs. If the validation loss continued to increase for 20 epochs, training would terminate early to avoid over-fitting. The model with the lowest meta-loss was then selected to be used for predictions.

4 RESULTS

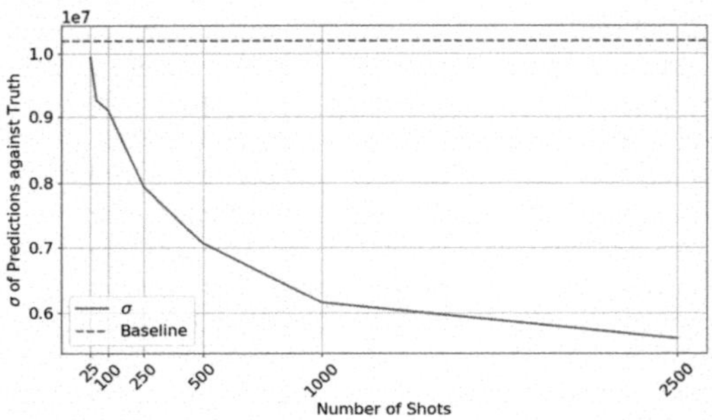

Figure 2. Performance increases rapidly before tapering off as number of shots is increased.

Models trained with a varying amount of shots per task were evaluated against a second unseen scaled-RMS data-set. To assess performance, the standard deviation of the models predictions minus the ground-truth was used, with results in Table 2. Lower shot models have the possibility to contain scarce wave conditions over 3.0 meters which results in scaled VBM RMS values. Because of this, 100 models were trained for each shot condition and their predictions against the evaluation data-set were averaged.

The baseline condition involves the original trained model from $task_1$ and is evaluated against the $task_2$ data-set. The $task_2$ 5000 condition is unique in that it does not start from the trained $task_1$ model. By being trained only with the scaled data-set from $task_2$, this model does not retain the prior information from the previous task. To keep the total amount of data the same between the models, this unique condition is trained with 5000 shots as opposed to 2500.

Table 2. Effect of numbers of shots on prediction accuracy.

Shot number	σ of VBM RMS predictions (N)
Baseline	10198270
25	9944380
50	9264317
100	9100765
250	7934630
500	7065943
1000	6159750
2500	5611591
$task_2$ 5000	5526475

As the number of shots increases, prediction performance rapidly improves before beginning to taper off at the 1000 shot mark (Figure 2). The performance improvement of the 1500 shot increase from 1000 shots to 2500 is 8.89%. In contrast, the marginal 225 shot increase from 25 shots to 250 results in an improvement of 20.2%.

Because MAML remembers prior experiences, the model tends to under-fit large RMS values. This is due to being first trained on the pre-deployment unscaled data from $task_1$. This is exemplified in Figure 3 where the $task_2$ 5000 condition which does not use the pre-deployment data is the only model to over-estimate a large RMS value.

5 CONCLUSIONS

This work shows that MAML is an adequate choice in machine learning to use when data is scarce. With 1000 samples, MAML successfully adapted to a new data-set and halved the standard deviation of its predictions against ground truth from the baseline model. By focusing on improving the model parameters, it is able to successfully adapt to new tasks with small amounts of data. This attribute, along with the algorithm's inherent flexibility, make it suitable for use with autonomous vessels, where quick adaptation and remembrance of prior events is essential.

Through accurate predictions of VBM RMS values, autonomous vessels can use weather forecasts and mission routing to predict structural loads during deployment. If predicted loads exceed the safety threshold, mission replanning can take place using the model's load predictions to plot a safe route. As the vessel gains experience in untested conditions, the operational envelope can be refined to ensure safe operation of the vessel.

5.1 *Future work*

To further explore MAML's ability to adapt to the inherent variability of ocean environments, the next step is to use multi-directional wave data in the model.

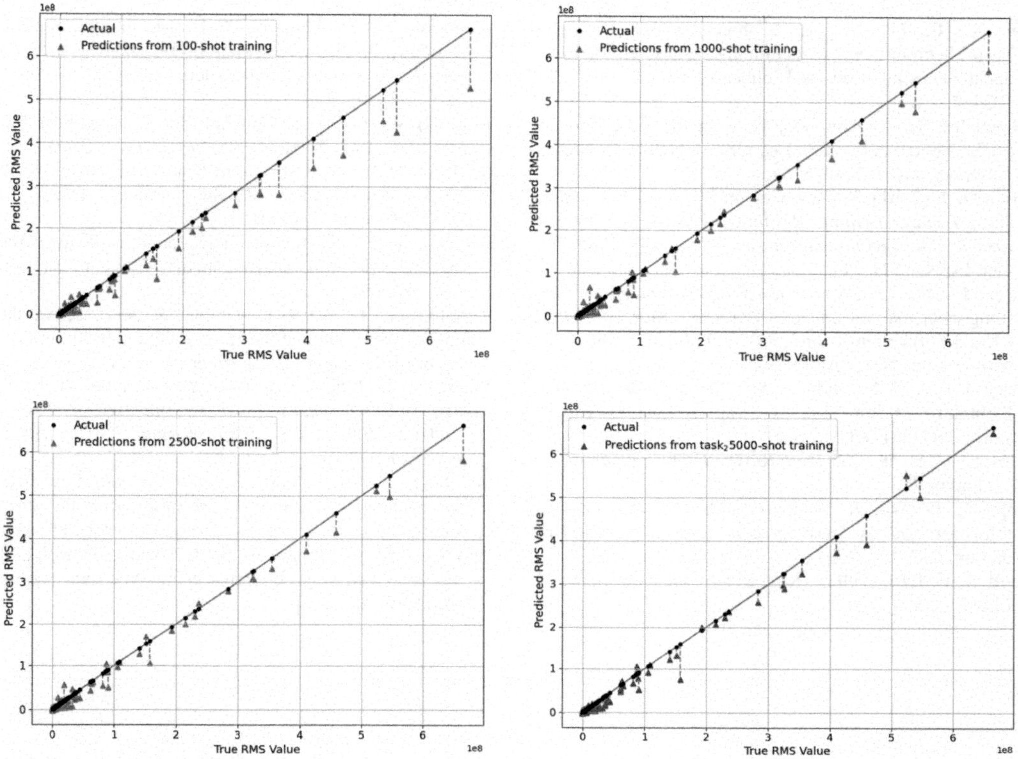

Figure 3. Prediction accuracy improving as number of shots increases.

Despite using a more complex data-set, as MAML is presented with new tasks or missions, accuracy of predictions will increase, and initial pre-deployment training will be phased out. With the addition of new and diverse data, MAML parameters will benefit and become increasingly sensitive to new conditions.

The first time an ASV encounters large scale wave events, it needs to quickly adapt its model to learn safe/unsafe conditions and update its operational envelope as needed. To test this, initial training will be done on a data-set where large wave conditions are absent. The second task will explore few-shot learning when the large events are introduced, and use MAML to quickly adapt to unseen and potentially dangerous conditions.

Accelerometers are widely accessible and inexpensive instruments. By using closed-form expressions for vertical motions, simulated accelerometer data can be generated. Future work will use MAML and a Long Short-Term Memory network (LSTM) to use a time sequence of sensor data and predict VBM RMS values.

ACKNOWLEDGEMENTS

The authors would like to thank Dr. Jessica Dibelka, ONR code 331 for supporting work to date under grant Number N00014-20-1-2044.

REFERENCES

Cetin, A., V. R. Solum, & C. M. Evans (2022, June). Machine Learning Based Prediction of Wave-Induced Vessel Response. In *Volume 5A: Ocean Engineering*, Hamburg, Germany, pp. V05AT06A034. American Society of Mechanical Engineers.

Ciaburro, G. & G. Iannace (2021, May). Machine Learning-Based Algorithms to Knowledge Extraction from Time Series Data: A Review. *6*(6), 55.

Collette, M., R. Bielski, P. Rohrer, A. Magistro, B. Sulkowski, & J. Van Houten (2022, August). Needs Exploration for Long-Term Autonomous Marine Systems: Working Report. Publisher: My University.

Decò, A. & D. M. Frangopol (2015, March). Real-time risk of ship structures integrating structural health monitoring data: Application to multi-objective optimal ship routing. *96*, 312–329.

Entezami, A. & H. Shariatmadar (2018, March). An unsupervised learning approach by novel damage indices in structural health monitoring for damage localization and quantification. *17*(2), 325–345.

Finn, C., P. Abbeel, & S. Levine (2017, July). Model-Agnostic Meta-Learning for Fast Adaptation of Deep Networks. arXiv:1703.03400 [cs].

Fjørtoft, K. E. & J. Rødseth (2020). Using the Operational Envelope to Make Autonomous Ships Safer. In *Proceedings of the 30th European Safety and Reliability Conference and 15th Probabilistic Safety Assessment and Management Conference*, pp. 76–83. Research Publishing Services.

Han, X., B. J. Leira, & S. Sævik (2021, January). Vessel hydrodynamic model tuning by Discrete Bayesian updating using simulated onboard sensor data. *220*, 108407.

Hospedales, T., A. Antoniou, P. Micaelli, & A. Storkey (2022, September). Meta-Learning in Neural Networks: A Survey. *44*(9), 5149–5169.

Jensen, J. J. (2001). Wave Loads on Ships. In J. J. Jensen (Ed.), *Load and Global Response of Ships*, Volume 4 of *Elsevier Ocean Engineering Series*, pp. 123–178. Elsevier. ISSN: 1571-9952.

Jensen, J. J. & A. E. Mansour (2002). Estimation of Ship Long-term Wave-induced Bending Moment using Closed-Form Expressions., 41–55. Publisher: The Royal Institution of Naval Architects.

Jensen, J. J., A. E. Mansour, & A. S. Olsen (2004, January). Estimation of ship motions using closed-form expressions. *31*(1), 61–85.

Journee, J. & W. Massie (2001). Delft University of Technology.

Lang, X., D. Wu, & W. Mao (2024, March). Physics-informed machine learning models for ship speed prediction. *238*, 121877.

Lloyd, A. R. J. M. (1998). (Rev. ed ed.). Gosport: A.R.J.M. Lloyd.

Mondoro, A., M. Soliman, & D. M. Frangopol (2016, October). Prediction of structural response of naval vessels based on available structural health monitoring data. *125*, 295–307.

Nugroho, W. H. & A. S. Mujahid (2015, November). Head Sea Slamming Pressures Prediction on a Frigate Ship Hull (A Numerical Study). In *Day 1 Wed*, November 04, 2015, Providence, Rhode Island, USA, pp. D011S003R011. SNAME.

Perera, L. P. & C. G. Soares (2017, January). Weather routing and safe ship handling in the future of shipping. *130*, 684–695.

Pezeshki, H., H. Adeli, D. Pavlou, & S. C. Siriwardane (2023, April). State of the art in structural health monitoring of offshore and marine structures. *176*(2), 89–108.

Raghu, A., M. Raghu, S. Bengio, & O. Vinyals (2020, February). Rapid Learning or Feature Reuse? Towards Understanding the Effectiveness of MAML. arXiv:1909.09157 [cs, stat].

Wu, R.-T. & M. R. Jahanshahi (2020, March). Data fusion approaches for structural health monitoring and system identification: Past, present, and future. *19*(2), 552–586.

Zhu, B. & D. M. Frangopol (2013, November). Reliability assessment of ship structures using Bayesian updating. *56*, 1836–1847.

Advances in Maritime Technology and Engineering – Guedes Soares & Santos (Eds)

Research on the characteristics of collision accidents between merchant ships and fishing vessels within the waters of China

C. Dong, X. Wang & Y. Gong
Dalian Maritime University, Dalian, China

ABSTRACT: Commercial fishing is a traditional risk industry, and the safety of fishing vessels and fishermen has always been a concern. Collision accidents between fishing vessels and merchant ships have received increasing attention due to the consequences of the accidents, which usually result in human casualties. Based on detailed information collected from the China Maritime Safety Administration on 129 collision accidents between merchant ships and fishing vessels that occurred in the waters of China from 2014 to 2023, the characteristics of the accidents are analyzed, including the time of the accidents; the nationality, type, and speed of the merchant ships; the length, hull material used and speed of the fishing vessels; the weather, navigation environment at the time of the accident; and the reasons for the accidents are discussed. Finally, collision prevention measures are proposed to provide a reference for reducing collisions between fishing vessels and merchant ships.

1 INTRODUCTION

Fishing is an important industry that directly or indirectly affects the lives and income of many people. According to the Food and Agriculture Organization of the United Nations, there are about 4.6 million fishing vessels in operation around the world, these fishing vessels supply the world's growing needs for fish and fish products. As carriers of fishing activities, the safety of fishing vessels has become a major concern today (Obeng et al, 2022). Generally, compared to other industries, fishing is considered one of the most dangerous activities globally, and the number of fishing vessels and fishermen lost due to fishing vessel accidents is much higher than other marine industries. On average, 24,000 fishermen die each year in this profession as a result of accidents (Ugurlu et al., 2020). According to the European Maritime Safety Agency, the total number of reported fishing vessel casualties and incidents over the period from 2014 to 2022 was 4440, and fishing vessels have a higher proportion of very serious marine casualties than in other types of ships. Nowadays, fishing has always been listed as one of the most dangerous professions in many countries, such as the United States, United Kingdom, and Canada (Jin et al., 2001; Turan et al., 2003; Wu et al., 2005; Yoo, 2019). Commercial fishing is a traditional risk industry, and the safety of fishing vessels and fishermen has always been a concern.

There are various forms of accidents involving fishing vessels, including collisions, capsizing, fires, etc. Commonly, the collision accident of fishing vessels often leads to other forms of accidents, such as explosions, flooding or even sinking, which can have fatal consequences on the lifespan of crew members and vessels (Pérez-Labajos et al., 2005). According to incomplete statistics from 1999 to 2008, the main threats to the safety of fishing vessels in the waters of China include collision, damage, grounding, and fire. The number of collision accidents accounts for almost 60% of the total accidents (Zhang, 2012). Fishing vessel collision accidents account for a large proportion of the total number of fishing vessel accidents. Wang et al. (2005) found that the majority of the collision accidents involved a fishing vessel and a merchant ship in the UK. Collision accidents between merchant ships and fishing vessels often occur in the waters of China. According to the accident report from the China Maritime Safety Administration, collisions between merchant ships and fishing vessels account for nearly half of the collision accidents (Zhuang et al., 2021). The safety of fishing vessels has always been a concern for the Chinese government. Recently, collision risk zones of merchant ships and fishing vessels have been established along the coast of China to alert passing merchant ships and fishing vessels.

However, the causes of collision accidents between merchant ships and fishing vessels are very complex. Due to different fishing methods, natural environments, and encounter scenarios between merchant ships and fishing vessels, there are significant differences in the causes of collision accidents. Generally, the fishing vessel has a higher risk of vessel sinking and casualties in collision accidents. Compared to

DOI: 10.1201/9781003508762-31

other types of ships, fishing vessels have their characteristics such as weak structural strength, incomplete life-saving equipment, and low professional quality. Therefore, after a collision accident between fishing vessels and merchant ships, the mortality rate of personnel is higher. Therefore, to ensure the safety of fishing vessels' crews and merchant ships' crews, it is very important to research the collision problem between fishing vessels and merchant ships.

The present paper mainly analyzes the characteristics of collision accidents between merchant ships and fishing vessels along the coast of China, finds out the patterns of collision accidents, provides certain assistance for avoiding collision accidents, and offers effective information for maritime search and rescue work.

2 DATA ON COLLISION ACCIDENTS

The present paper mainly analyzes 129 cases of collisions between fishing vessels and merchant ships in the waters of China. The collision data between fishing vessels and merchant ships were collected from accident reports of the China Maritime Safety Administration, that occurred during the 10 years 2014–2023. These data do not include the collision accidents that have not been reported. The data mainly includes visibility, accident location, accident time, accident level, the hull material used by fishing vessels, fishing vessel speed and length, merchant vessel type, merchant vessel nationality, merchant vessel speed and length, accident casualties, etc.

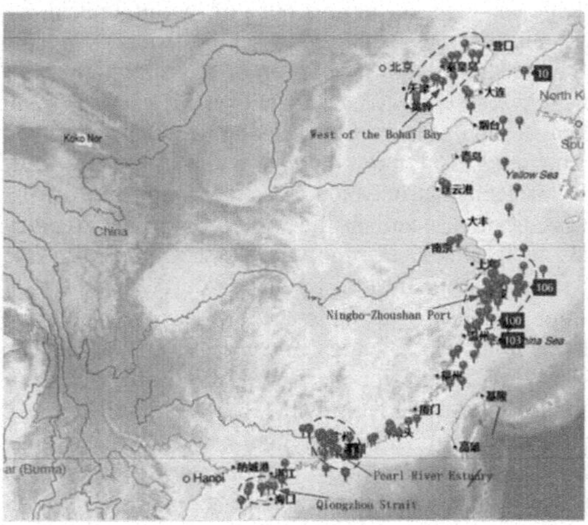

Figure 1. Location of collision accidents between merchant ships and fishing vessels.

3 RESULTS AND ANALYSIS

3.1 *Accident area*

Through data analysis, it is found that the collision accidents between merchant ships and fishing vessels are mainly located in the waters closer to the coastline. The most frequent areas of collision accidents between merchant ships and fishing vessels are the waters of Ningbo Zhoushan Harbour, the waters of the Pearl River Estuary, the west side of Bohai Bay, and the waters near the Qiongzhou Strait, as shown in Figure 1. These intersecting water areas are near busy waterways with intensive ship traffic flows, and ships navigating in these areas are prone to collision accidents.

3.2 *Time and period of accident*

As shown in Figure 2, the monthly distribution of collision accidents shows that the largest number of collisions between merchant ships and fishing vessels occurred in January, with 18 accidents. More than 10 accidents occurred from September to December, March and April. Fewer accidents occurred from May to July, with the number of accidents below 5. The low number of accidents from May to July may be because the waters of China are closed to fishing during this period, and fishing vessels seldom carry out fishing activities off the coast of China.

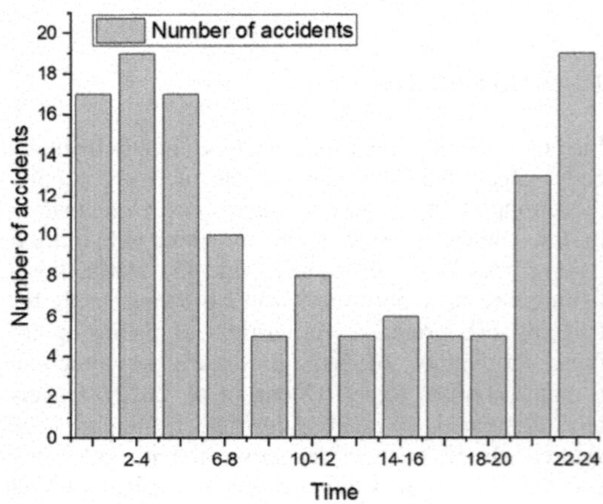

Figure 2. The monthly distribution of collision accidents between merchant ships and fishing vessels.

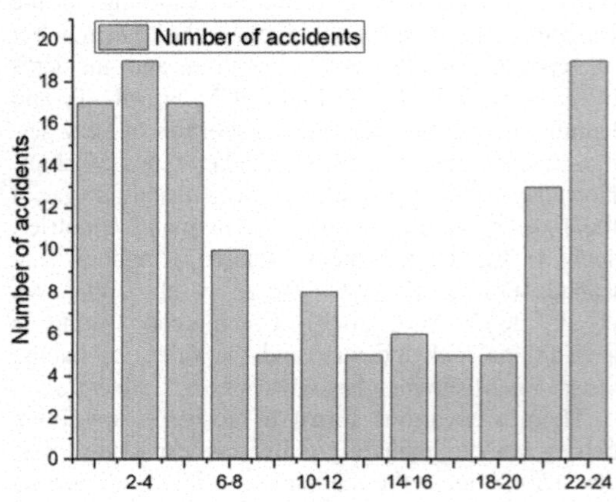

Figure 3. The hourly distribution of collision accidents between merchant ships and fishing vessels.

From the hourly distribution of collision accidents in Figure 3, it can be seen that most collision accidents between merchant ships and fishing vessels occur at night, with the high incidence of accidents occurring between 2200 hours and 0600 hours on the following day, and the number of collision accidents occurring at hourly intervals of these periods being more than 17. During these periods, the crews of merchant ships and fishing vessels are generally more tired and prone to negligence in keeping watch, which makes them more prone to accidents.

3.3 *Characteristics of the fishing vessels*

As can be seen from the statistical data on the hull materials used by fishing vessels involved in collision accidents in Figure 4, the hull materials used by fishing vessels are mostly steel, accounting for 62% of the total. Among them, wooden vessels accounted for 32.6%, and glass fiber and cement materials accounted for 3.9% and 1.6%, respectively.

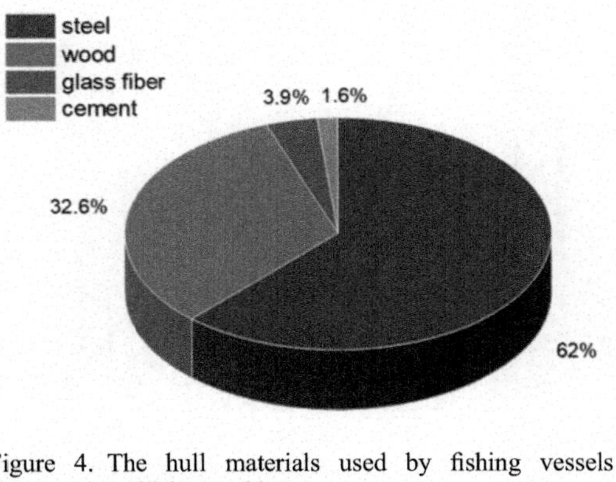

Figure 4. The hull materials used by fishing vessels involved in collision accidents.

As can be seen from the statistical data on the length of fishing vessels involved in collision accidents in Figure 5, most of the fishing vessels involved in collision accidents were less than 12 m in length, with the number of vessels being 89; among them, the number of vessels of 12-24 m and those of more than 24 m were comparable, with 20 vessels each.

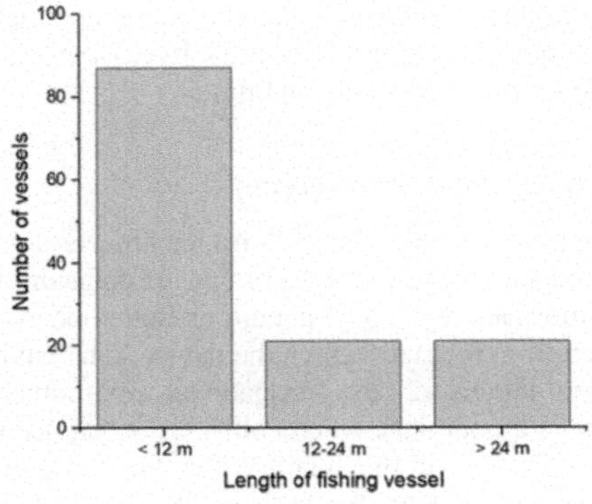

Figure 5. The length of fishing vessels involved in collision accidents.

3.4 *Characteristics of the merchant ships*

As can be seen from the statistical data on the lengths of merchant ships involved in collision accidents in Figure 6, the lengths of merchant ships involved in accidents are mostly concentrated in the range of 20-300 m, with 58 merchant ships in the range of 20-100 m, 62 merchant ships in the range of 100-300 m, and only 7 merchant ships larger than 300 m.

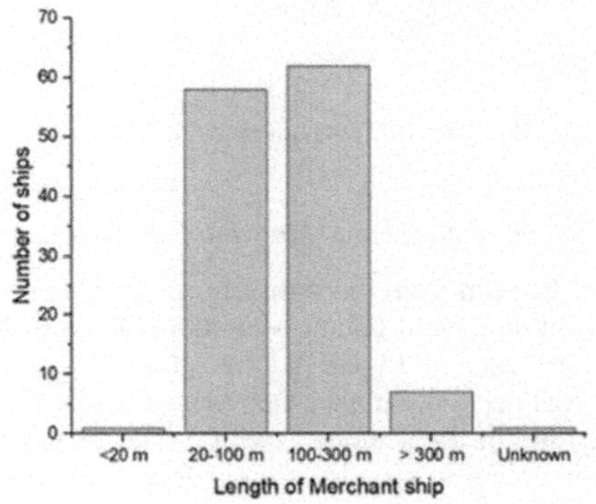

Figure 6. The length of merchant ships involved in collision accidents.

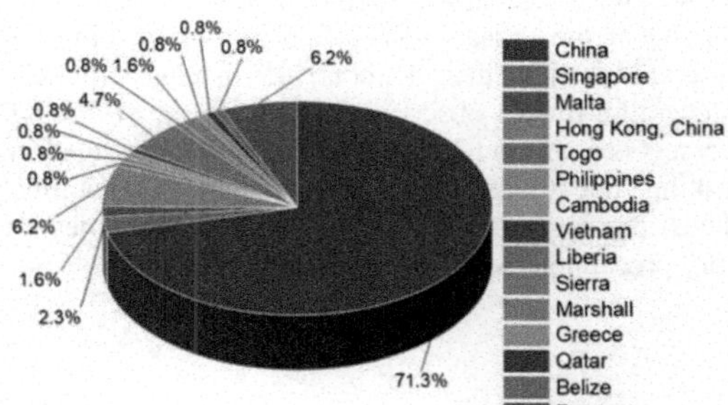

Figure 7. The nationalities of merchant ships involved in collision accidents.

As can be seen from the statistics on the nationalities of merchant ships involved in collision accidents in Figure 7, the majority of merchant ships involved in accidents are of Chinese nationality, accounting for 71.3%. Other nationalities of merchant ships involved in collision accidents include Panama, Liberia, Hong Kong of China, Singapore, and the Marshall Islands.

As can be seen from the statistics on the types of merchant ships involved in collision accidents in Figure 8, most merchant ships involved in accidents with fishing vessels were bulk carriers, which accounted for 58.9% of the total. In addition, among the merchant ships involved in collision accidents, general cargo ships accounted for 14.7%, container ships for 11.6%, oil tankers for 7%, roll-on/roll-off vessels for 3.1%, and multipurpose vessels for 2.3 %.

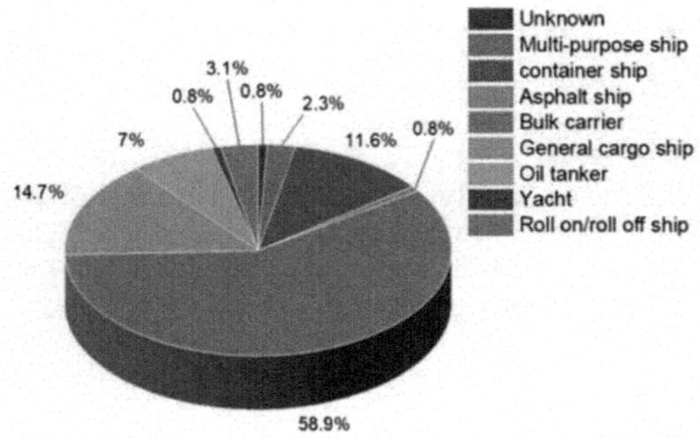

Figure 8. The types of merchant ships involved in collision accidents.

3.5 *Speed of merchant ships and fishing vessels*

As can be seen from the statistics on vessel speeds of merchant ships and fishing vessels involved in collision accidents in Figure 9, the speed of merchant ships was generally higher than that of fishing vessels at the time of the accidents, with 54.3% of merchant ships traveling at speeds of more than 8 kn, and the highest speed of merchant ships involved in the collision was 23.9 kn, while three merchant ships were at anchor at the time of the collision accidents. At the time of the collision accidents, only 31.8% of fishing vessels were traveling at speeds greater than 8 kn, and 41.1% of the vessels were at anchor or traveling at low speeds (less than 4 knots) for fishing activities. Generally, if it is observed that the speed of fishing vessels is at around 3 knots, they are often engaged in fishing operations with restricted maneuverability. Thus, passing merchant ships should take the necessary precautions to avoid collision accidents.

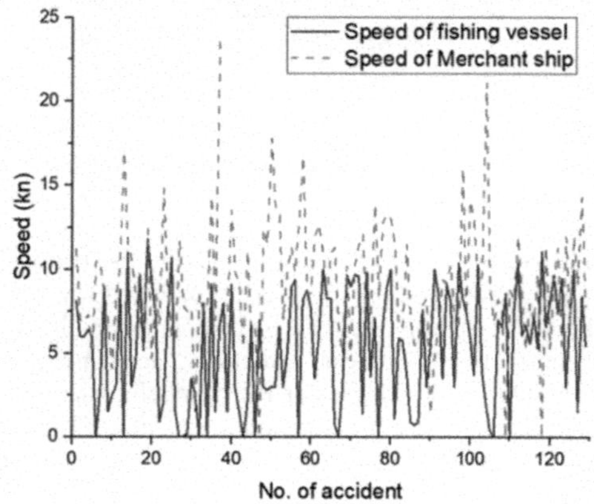

Figure 9. The speed of merchant ships and fishing vessels involved in collision accidents.

3.6 *Level of collision accidents*

According to China's classification standards for ship accidents, ship accidents are classified into four levels, namely, particularly major accidents, major accidents, large accidents, and general accidents. As can be seen from the statistics on collision accidents between merchant ships and fishing vessels in Figure 9, most of the collision accidents that occurred were general accidents, accounting for 61.2% of the total, followed by larger accidents, accounting for 31%, and major accidents, accounting for 7.8%; there were no collision accidents of particularly major accident.

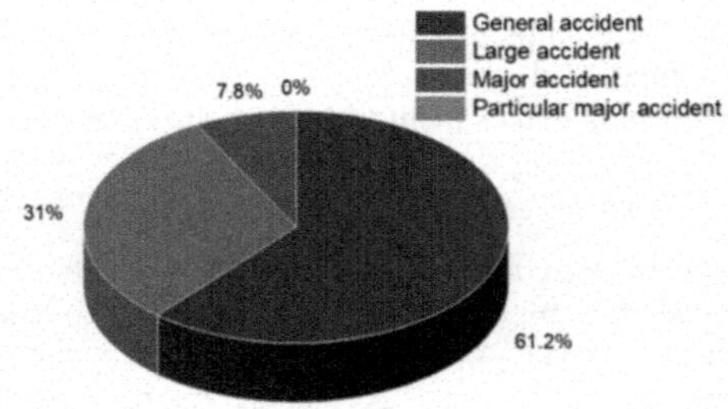

Figure 10. The level of collision accidents between merchant ships and fishing vessels.

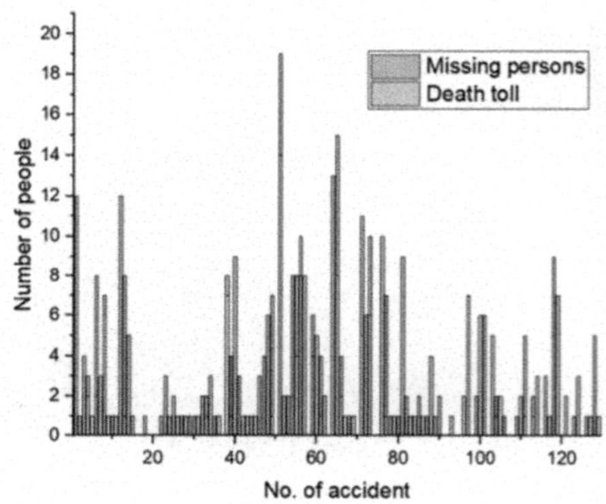

Figure 11. The number of persons killed and missing in collision accidents.

As shown in Figure 11, most collision accidents resulted in deaths and missing persons, with 28 accidents with more than 6 deaths and missing persons and the largest number of deaths and missing persons at 19. There were only 25 accidents that did not result in death and missing persons. Therefore, collision accidents between merchant ships and fishing vessels pose a serious threat to the lives of crews and should be given adequate attention.

3.7 *Navigational environment*

In the present study, navigational conditions are categorized into three types according to collision accident information, namely, good, fair, and poor. As can be seen from the statistics on the navigational environment in Figure 12, the navigational environment is mostly in a poor state when collisions occur, accounting for 70.5% of the total. When the navigational environment is poor, the passing ship traffic flow is denser and the navigational waterway conditions are restricted, and the probability of collision between merchant ships and fishing vessels will increase significantly. The proportions of fair and poor navigational conditions were 7% and 22%, respectively.

In the present paper, visibility levels are categorized as good, fair, and poor, where good visibility refers to a visibility distance greater than 6 nautical miles, fair visibility is between 3 and 6 nautical miles, and poor visibility is a visibility distance of less than 3 nautical miles. As shown in Figure 11, at the time of the collision accidents, visibility conditions were mostly in good visibility, accounting for 68.2% of the total, with fair visibility accounting for 25.5% and poor visibility accounting for only 6.2%. Therefore, under good visibility conditions, collision accidents are more likely to occur if the ship crews are negligent. On the contrary, when visibility is poor, collision accidents can be avoided by adopting a regular lookout and steering the vessel.

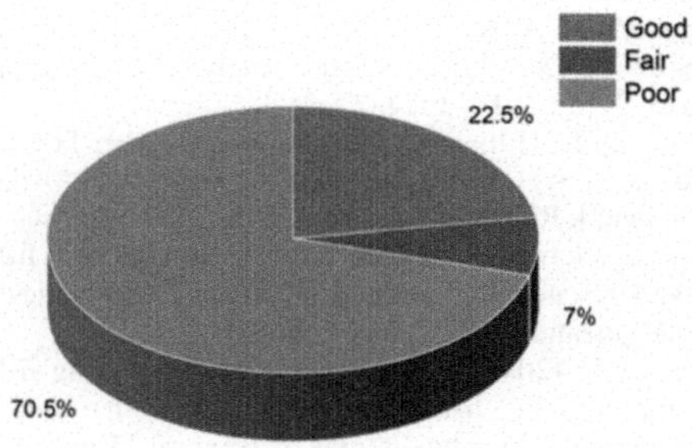

Figure 12. The navigational conditions of collision accidents.

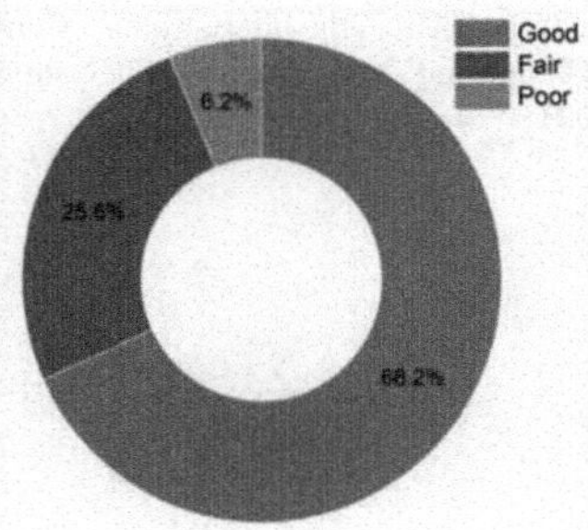

Figure 13. The visibility levels of collision accidents.

4 DISCUSSIONS

Collision accidents between merchant ships and fishing vessels tend to occur in water areas that are rich in fishery resources and densely passing merchant ships, and the number of encounters between merchant ships and fishing vessels is more frequent, which increases the probability of collision accidents. The government of China has designated high-risk areas for collision accidents between merchant ships and fishing vessels to warn passing merchant ships. However, as the crews of fishing vessels seldom receive training and are seldom aware of the need to comply with maritime traffic regulations, there is a great deal of arbitrariness in the navigation of fishing vessels, all of which creates a great deal of trouble for the Maritime Safety Administration (MSA) in the work of maritime traffic organization. It is therefore recommended that the training of fishing vessel crews should be strengthened and that in-depth cooperation between the fisheries and maritime authorities in law enforcement should be promoted to reduce the incidence of ship collisions.

Collision accidents between fishing vessels and merchant ships mostly occur during the fishing season, when the crews of fishing vessels neglect the rules of maritime traffic to obtain a bigger harvest, thus leading to accidents. In addition, during the duty period 22:00-05:00, when the crews of merchant ships are more tired, the crews often neglect to look out and fail to detect the surrounding activities of fishing vessels, which in turn leads to the occurrence of collision accidents. It is recommended to strengthen the safety education of the crew of fishing vessels and to restrain their maritime traffic behavior, such as not anchoring in the waterway and not crossing the bows of merchant ships at will. In the dense area of fishing vessels, it is recommended that a captain experienced in maneuvering be on the bridge to supervise and guide vessel maneuvering, pay attention to and maintain a formal lookout, and avoid collision with the surrounding fishing vessels.

The communication equipment of fishing vessels is relatively weak, and many collision accidents are caused by the lack of effective communication between merchant ships and fishing vessels. It is recommended that fishing vessels be mandatorily equipped with some essential communication equipment, to provide a communication channel with the surrounding vessels to avoid collision, thus reducing the occurrence of collision accidents between merchant ships and fishing vessels caused by poor communication.

5 CONCLUSIONS

With the development of science and technology and the improvement of management standards, the frequency of collisions between merchant ships and fishing vessels is gradually decreasing. To reduce the occurrence of collision accidents between merchant ships and fishing vessels, many aspects need to be strengthened and improved.

(1) Pay attention to the organization of vessel traffic in important waters and accident-prone periods. Most collision accidents between merchant ships and fishing vessels occur between 2300 and 0400. When the summer fishing ban is lifted, the density of fishing vessels will normally increase to the highest. Thus, crews and

supervisors need to focus on these times and periods.

(2) Focus on the training and education of ship pilots of merchant ships and fishing vessels to make them comply with collision avoidance rules and adopt good seamanship, with particular attention to lookout on the bridge. According to Morel and Chauvin (2006), the main cause of collisions is to be found in a poor lookout performance, both on the merchant ships and the fishing vessels. Crews of the watch should adapt the level of watchkeeping based on the density of fishing vessels and workload.

(3) Strengthen the sorting out of ship traffic flow of merchant ships and fishing vessels. It has done a good job in delineating the boundaries of coastal merchant shipping routes and fishing zones, remediating and removing obstruction of navigation facilities, continuing to promote the broadcasting of maritime safety information, and realizing online updating of electronic charts for fishing vessels.

(4) Government departments should increase the frequency of joint enforcement activities. The Maritime Safety Administration and the Fisheries Administration have strengthened cooperation to achieve multi-point coverage of merchant and fishing vessels, seafarers, fishermen, shipping enterprises, fishing vessel owners, etc., and the ability to jointly prevent and control the risk of collision between merchant and fishing vessels has been steadily improved.

REFERENCES

Jin D, Kite-Powell H, Talley W. The safety of commercial fishing: determinants of vessel total losses and injuries[J]. Journal of Safety Research, 2001, 32(2): 209–228.

Morel G, Chauvin C. A socio-technical approach of risk management applied to collisions involving fishing vessels[J]. Safety science, 2006, 44(7): 599–619.

Obeng F, Domeh V, Khan F, et al. Analyzing operational risk for small fishing vessels considering crew effectiveness[J]. Ocean Engineering, 2022, 249: 110512.

Pérez-Labajos C A, Azofra M, Blanco B, et al. Collision of fishing vessels: Lorenz curves and GINI indices[J]. Journal of maritime research, 2005, 2(3): 97–109.

Turan O, Öiçer A İ, Martin P L. Risk assessment of loss of life for fishing vessels in fuzzy environment[C]//Safety and reliability. Taylor & Francis, 2003, 23(2): 19–38.

Uğurlu F, Yıldız S, Boran M, et al. Analysis of fishing vessel accidents with Bayesian network and Chi-square methods[J]. Ocean Engineering, 2020, 198: 106956.

Wang J, Pillay A, Kwon Y S, et al. An analysis of fishing vessel accidents[J]. Accident Analysis & Prevention, 2005, 37(6): 1019–1024.

Wu Y, Pelot R, Hilliard C. The effect of weather factors on the severity of fishing boat accidents in Atlantic Canada[J]. Risk Management, 2005, 7: 21–40.

Yoo S L. Network analysis by fishing type for fishing vessel rescue[J]. Physica A: Statistical Mechanics and Its Applications, 2019, 514: 892–901.

Zhang, Zh.L. (2012). Analysis of accidents during fishing operations in fishing vessels in China and its countermeasures,The Journal of Chinese Fishery Quality and Standards, Shanghai, China.

Zhuang R, Wu Y, Chen W, et al. Design of Intelligent Collision Avoidance System for Fishing Vessels Based on AIS[C]//2021 IEEE 7th International Conference on Cloud Computing and Intelligent Systems (CCIS). IEEE, 2021: 334–337.

Advances in Maritime Technology and Engineering – Guedes Soares & Santos (Eds)

Evaluation of technical and safety aspects of nitrogen use onboard tanker ships

G. Elidolu
Maritime Faculty, Istanbul Technical University, Tuzla, Istanbul, Turkey
Centre for Marine Technology and Ocean Engineering (CENTEC), Instituto Superior Técnico, Universidade de Lisboa, Lisbon, Portugal

Y. Arslanoglu
Maritime Faculty, Istanbul Technical University, Tuzla, Istanbul, Turkey

A.P. Teixeira
Centre for Marine Technology and Ocean Engineering (CENTEC), Instituto Superior Técnico, Universidade de Lisboa, Lisbon, Portugal

ABSTRACT: Nitrogen is used onboard tanker ships to safeguard cargo during cargo handling operations and ship voyages. Although the use of nitrogen minimises the formation of combustible or dangerous gas mixtures, its application on board ships can also involve dangerous aspects that should be of concern to the tanker industry and require careful analysis. This study evaluates, from a technical and safety point of view, the use of nitrogen in critical cargo processes onboard tankers. The paper presents the processes of nitrogen utilisation onboard tanker ships with regard to undesirable situations that can occur during cargo operations and also discusses proper safety measures.

1 INTRODUCTION

Maritime transportation is the most frequently used transportation mode for global trade of different types of cargo (UNCTAD, 2023). In particular, liquid cargoes such as crude oil, petroleum products, chemicals, and liquefied natural gas (LNG) are typically handled and shipped via tanker vessels in a variety of tonnage options. The increase in the volume of liquid cargo trade via seaways, particularly oil and derivatives, makes tanker ships indispensable for the energy sector and chemical industries (UNCTAD, 2023).

Tanker vessels are equipped with safety and navigation systems to mitigate the risks associated with the transport of hazardous materials across the oceans. Stringent international regulations and standards brought by the industry members govern their construction and operation to ensure environmental protection and prevent accidents (IMO, 2014; IMO, 2020; IMO, 2022; OCIMF, 2019). However, accidents still occur on tanker ships (OCIMF, 2017). Fire and explosion are among the serious accident types since they could be responsible for human injury, loss of lives, loss of cargo, and property and environmental damages. Mostly, these accidents occur due to hot work operations, pressure increase in tanks, and failures during the tank cleaning process (Uğurlu, 2016). Table 1 illustrates some of the explosion accidents that have occurred on tanker ships since 2010, with remarks on the accident causes (MAIB, 2019; MSIU 2011; MSIU 2013; MSIU, 2015; MSIU, 2021; JTSB 2013a; JTSB, 2018; UEIM, 2014; UEIM, 2017; UEIM, 2019).

Several studies about risks and technical innovations have been conducted by the maritime industry to address the safety issues of tanker ships, particularly to reduce the likelihood of accidents, enhance reliability and mitigate their consequences (Bubbico et al., 2009; Gao, 2022; Psarros et al., 2011; Yu et al., 2021). For instance, double-hull designs, tank radar systems, and inert gas systems play key roles in ship cargo operations (Yip et al., 2011). In particular, inert gas systems have become one of the crucial features, to prevent explosions, as revealed in a study that examined explosion and fire incidents in cargo areas of tankers between 2004 and 2015. This study indicates that 15 accidents occurred due to a lack of inerting processes, resulting in 50 fatalities (OCIMF, 2017). The use of inerting systems contributes to safety against accidents, however, their operation may also involve significant aspects that need to be evaluated from a technical and safety point of view. Therefore, this study aims to assess the use of nitrogen, an inerting element, onboard tanker vessels.

The remainder of this study is as follows: Section 2 explains the nitrogen characteristics and its

DOI: 10.1201/9781003508762-32

onboard implications. Section 3 evaluates the use of nitrogen from both technical and safety perspectives. It also presents useful practices for safe and efficient nitrogen operations. Finally, the conclusions of the study are presented in Section 4.

Table 1. Explosion accidents onboard tanker ships.

Ship	Year	Location	Remarks on accidents
Mar Cristina	2011	Norway	Air compression in blocked line.
Umar 1	2013	France	High pressure in tanks when loading methyl tert-butyl ether.
Tibil	2014	Türkiye	Hot work during tank cleaning after acrylonitrile.
Bomar Mercury	2015	Netherlands	Tank cleaning after aniline.
Eiwa Maru 3	2016	Japan	Welding during the tank cleaning after base oil cargo.
Habaş	2017	Türkiye	Line clearing after discharge.
Golden Sunny Hana	2018	Japan	Steam application in the tank cleaning after gasoline.
Syn Zania	2019	Türkiye	Pressure increase during the loading of propylene.
Stolt Groenland	2019	South Korea	Heating of cargo.
Seascout	2021	Oman	Spark due to hot work on deck when loaded with hydrocarbon

2 ONBOARD NITROGEN OPERATIONS

Tanker ships carry various liquids, including petroleum products and chemicals, which may be sensitive to contamination and reaction, besides having dangerous characteristics such as flammability and explosiveness. The storage in ships' tanks and the transfer of such cargo pose significant hazards for the crew, the cargo itself, property, and the marine environment.

Considering the confined tank atmosphere, an inert gas process should be applied for safety by replacing the air in the tanks with nitrogen.

Nitrogen is a non-metal element and constitutes about 78% of the earth's atmosphere by volume in N_2 form. It is a pure, colourless, odorless and non-flammable gas, and can provide an unreactive atmosphere in confined spaces (IMO, 2011). It is used onboard tankers mainly for preventing explosion and fire, avoiding contamination and collecting moisture on the cargo (ICS et al., 2020; IMO, 2020).

Onboard operations of nitrogen are presented as follows, which are schematised in Figure 1.

2.1 *Inerting*

Prior to handling flammable liquids, tanker ships are to apply inert gas operation. This aims to prevent the forming of a fire triangle inside the cargo tank, which includes flammable matter, oxygen, and the ignition factor.

In crude oil tankers, the inert gas could be flue gas, which is produced by fuel burning in ship's boilers. The oxygen is targeted to be lowered by volume in the tank, mostly about 8% (ICS et al., 2020). However, in oil chemical tankers, N_2 is suitable for the same purpose since it is pure and does not react with the cargo.

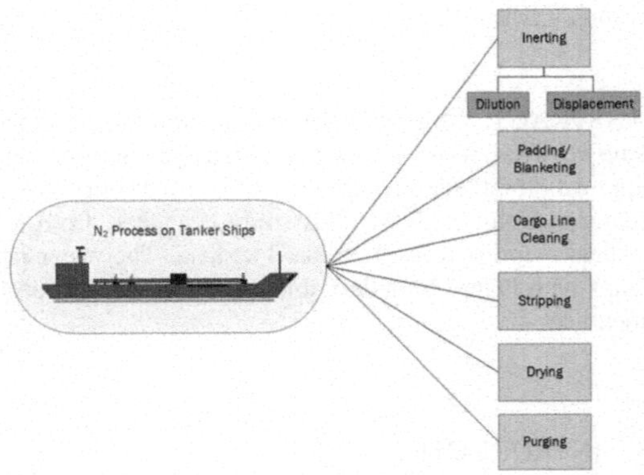

Figure 1. Gas inerting processes onboard tanker ships.

2.2 *Padding/Blanketing*

Padding is performed by filling the empty tank atmosphere with N_2. It covers the top of the cargo as a dense blanket. By replacing the air on the cargo with N_2, the atmosphere is kept safe, and out of flammable range.

2.3 *Cargo line clearing*

After the loading process is completed, cargo lines including the hoses connected between the ship and the terminal could be blown through the tanks to prevent any spillage. While air could be suitable for non-flammable cargoes, N_2 could be preferred for hazardous liquids to avoid unwanted reactions (ICS et al., 2020).

2.4 *Stripping*

Towards the end of the cargo discharge operation, pumping out the liquid could be difficult and the pumping efficiency of the main line could be decreased. To perform final discharge from the ship to an extent the amount of the ship and terminal agreement, independent and thinner lines connected to cargo pumps could be used for transfer. This approach is encouraged in particular when discharging cargoes that harm to environment, since the remaining liquids in the tanks are washed and could be discharged into the sea, subject to the conditions stated in international regulations (IMO, 2022). The stripping could be made by using N_2 if the cargo is sensitive to contamination or has hazardous characteristics (ICS et al., 2020).

2.5 *Drying*

After tank cleaning, there could be moisture in cargo tanks, which may affect the next cargo such as contamination. N_2 could be used for drying the cargo tank since its dew point is -40 Celsius (IMO, 2011).

2.6 *Purging*

Purging is the process of applying N_2 into a cargo tank before starting gas freeing operation. It aims to prevent forming an explosive atmosphere in tanks due to remaining gas from hazardous cargoes.

3 EVALUATION OF TECHNICAL AND SAFETY ASPECTS OF NITROGEN USE

Nitrogen use onboard tanker ships is encouraged in the maritime industry in order to prevent explosions. However, its application should be carefully considered for the safety of life, property and the marine environment. In normal circumstances, nitrogen is considered a non-poisonous gas, yet its use onboard could be dangerous for humans since it depletes the oxygen in the atmosphere of cargo tanks. Besides, producing and monitoring the nitrogen via the nitrogen generator, which is a common and efficient way of providing N_2 to tankers, requires some procedures to maintain proper operations.

3.1 *Technical aspects*

The use of Nitrogen on chemical tankers involves the application of nitrogen gas in various processes to maintain safe and smooth cargo operations. Inerting, padding, stripping, drying and other processes require a functional nitrogen-generating source and also planning and managing properly the technical aspects of the N_2 use.

3.1.1 *Replacing tank atmosphere*

To prevent a flammable atmosphere in tanks, the oxygen level, which is 21% in normal air conditions, should be lowered to 8% levels (ICS et al., 2020). Achieving the desired oxygen level demands a timely process. Therefore, nitrogen could be introduced into the cargo tanks by dilution and displacement methods. In the dilution method, the air inside the tank is diluted with N_2. The velocity of the N_2 supplied to the tank is significant for this process since it is expected to force the air out of the tank. Therefore, the dilution method could not be an efficient option if there are several tanks to be inertised.

The displacement method is performed by servicing the N_2 from the loading line of the tank, which is connected to the manifold. It is supplied to the tank from the bottom of the cargo and the air is displaced by N_2. Unlike the dilution method, the velocity of the N_2 should be slow to prevent its loss before the release of air.

3.1.2 *Tank atmosphere monitoring*

When using nitrogen, the tank atmosphere should be monitored constantly. This could be done by using tank radar systems and portable gas detection equipment. Since the targeted oxygen content is about 8%, a nitrogen generator system should be arranged to provide a safe atmosphere in the tank against explosion and fire.

3.1.3 *Pressure control*

The pressure inside the tanks should be controlled when filling them with nitrogen. It is recommended to maintain the tanks with positive pressure in an inert condition to avoid air ingress (ICS et al., 2020). Moreover, the crew should be aware that environmental conditions could also affect the pressure of cargo tanks. In particular, when navigating in warm climates, the pressure of cargo vapours may tend to increase. The conditions of pressure and vacuum valves should be periodically checked and blanketing should be applied to secure the inert condition of the void spaces. Additionally, when supplying the N_2 from the shore terminal, the rate of the N_2 could be high levels. The ship's crew should take precautions against over pressurization of tanks with nitrogen since the tank structure may not resist high pressure and may rupture (ICS, 2021; NTSB, 2020).

3.2 *Safety aspects*

The use of nitrogen requires the ship's crew to take several safety issues into account. The integrity of cargo should be maintained throughout the voyage, as well as taking safety measures for the ship's tanks, lines and cargo systems. Furthermore, the crew should consider the health hazards of using nitrogen when performing all stages of cargo operations.

3.2.1 *Cargo safety*

Cargoes with hazardous properties, such as flammability and explosiveness, should be handled under safe atmospheric conditions (ICS et al., 2020). Nitrogen contributes to protecting the cargo's purity while providing a non-flammable environment in tanks. However, some of the cargoes may require special treatment such as inhibition. According to the type of inhibitor, the oxygen content inside the tank could be important for its effectiveness (IMO, 2020). Therefore, the nitrogen should be serviced into tanks by observing the necessary oxygen level. Likewise, some compounds can react with moisture in the tank. Before starting the loading, the crew should ensure that the tank is free of moisture by applying nitrogen to avoid off-specification concerns.

3.2.2 *Static electricity-related issues*

Static electricity is a phenomenon that can pose significant risks onboard tanker ships. It can be built up due to several reasons such as loading dangerous cargoes at high rates, contact of metals to ship's tanks, and using static accumulator materials on deck (Elidolu et al., 2022; ICS et al., 2020). These causes could be present when handling flammable cargoes under normal air conditions. Accordingly, explosions and fires may occur onboard ships. To prevent such accidents, the needs of cargoes should be known by the ship's crew to apply nitrogen for the inerting and padding process for safety.

3.2.3 *Exposure to nitrogen*

Nitrogen has an asphyxiant effect on humans when the oxygen level in the atmosphere is below 21%. The respiratory system works by considering the carbon dioxide (CO_2) level in the lungs; hence, the rate of respiration rises if CO_2 is increasing. However, when a person inhales N_2 in large concentrations, the respiratory system could collapse. Ship's crew must be vigilant against the health effects of nitrogen onboard tanker vessels (DSB, 2023; IMO, 2011; JTSB, 2013b).

Exposure to nitrogen could occur when entering into tanks, as well as when working on the deck area. The former should not be performed if the atmosphere in tanks has no standard breathing air levels, in other words, tanks must be in a gas-free state. The condition to enter into a tank should be that the oxygen level is 21% and the lower flammable limit (LFL) is less than 1%. Before entry, safety controls and emergency preparations, including organising oxygen resuscitation kits should be made by the crew. The atmosphere in the tank should be checked using portable gas detection equipment, in particular those that can check toxic gas contents. That equipment should also be type-approved and calibrated before use. On the other hand, exposure to nitrogen may occur in an open area. N_2 may accumulate on the deck region due to its movement affected by deck structural properties. Special attention should be given before working on the deck. As a precaution, tanks engaged in nitrogen operations should be marked for warning. A portable oxygen analyzer is necessary for constant measuring of the atmosphere (IMO, 2011).

4 CONCLUSIONS

Tankers are essential for the transport of liquid cargo worldwide. Due to the dangerous characteristics of cargoes, operating these vessels involves risks to human health, the ship, the integrity of cargo, and the marine environment. Rules and regulations brought by IMO in the last decade contribute to preventing and mitigating a significant part of those risks. However, accidents still occur, resulting in loss of lives and environmental damage. To prevent such accidents, nitrogen is the gas most commonly used for inerting. It is supplied to cargo tanks to create a non-flammable atmosphere. However, its operation can be complex and may also involve serious risks for the ship's crew. In this study, the nitrogen processes on tanker vessels are explained by including the application methods such as inerting, padding, drying, stripping and line clearing. During these stages, significant aspects of the operation are identified in order to maintain safety. Besides, the health hazards associated with nitrogen use are also discussed since onboard personnel be could exposed to N_2 both when entering into tanks or working on the deck area. The study aims to provide appropriate measures to users of nitrogen onboard ships by addressing risks at different stages of its use. Future research could focus on evaluating the risks of N_2 operation using suitable methodologies for probability prediction, accident analysis or failure assessment.

ACKNOWLEDGEMENTS

This work was supported by the Scientific Research Projects Department of Istanbul Technical University (project number 43690). In addition, this study was supported by funding from The Scientific and Technological Research Council of Türkiye (grant n. 1059B142201561).

This work contributes to the Strategic Research Plan of the Centre for Marine Technology and Ocean Engineering (CENTEC), which is financed by the Portuguese Foundation for Science and Technology (Fundação para a Ciência e Tecnologia - FCT) under the contract UIDB/UIDP/00134/2020.

REFERENCES

Bubbico, R., Di Cave, S. Mazzarotta, B. 2009. Preliminary risk analysis for LNG tankers approaching a maritime terminal, *Journal of Loss Prevention in the Process Industries*, 22(5), 634–638.

DSB. 2023. Fatal accident in cargo tank of chemical tanker. Lessons learned from the occurrence on board the NCC SAFA. *Dutch Safety Board*. Hague: Netherlands.

Elidolu, G., Akyuz, E., Arslan, O., Arslanoğlu, Y. 2022. Quantitative failure analysis for static electricity-related explosion and fire accidents on tanker vessels under fuzzy bow-tie CREAM approach, *Eng. Fail. Anal.*, 131, 105917.

Gao, F. 2022. An integrated risk analysis method for tanker cargo handling operation using the cloud model and DEMATEL method, *Ocean Engineering*, 266(4), 113021.

ICS. 2021. Tanker Safety Guide (Chemicals). Fifth edition. *International Chamber of Shipping*.

ICS, OCIMF, IAPH. 2020. ISGOTT. International Safety Guide for Oil Tankers and Terminals. Sixth edition.

International Chamber of Shipping, Oil Companies International Marine Forum, International Association of Ports and Harbors.

IMO. 2011. Guidelines on Tank Entry for Tankers Using Nitrogen as an Inerting Medium. MSC.1/Circ.1401. *International Maritime Organization.* London: United Kingdom.

IMO. 2014. ISM Code: International safety management code with guidelines for its implementation. 2014 edition. *International Maritime Organization.* London: United Kingdom.

IMO. 2020. IBC Code: International Code for the Construction and Equipment of Ships Carrying Dangerous Chemicals in Bulk. 2020 edition. *International Maritime Organization.* London: United Kingdom.

IMO. 2022. International Convention for the Prevention of Pollution from Ships (MARPOL). Consolidated Edition. *International Maritime Organization.* London: United Kingdom.

JTSB. 2013a. Marine Accident Investigation Report for Oil/Chemical Tanker EIWA MARU 3. MA2017-10. *Japan Transport Safety Board.*

JTSB. 2013b. Marine Accident Investigation Report for Chemical tanker KYOKUHO MARU NO.2. MA2013-4. *Japan Transport Safety Board.*

JTSB. 2018. Marine Accident Investigation Report for Chemical Tanker GOLDEN SUNNY HANA. MA2019-3. *Japan Transport Safety Board.*

MAIB. 2019. Report on the investigation of the cargo tank explosion and fire on board the chemical tanker Stolt Groenland. 9/2021. Ulsan, Republic of Korea 28 September 2019. *Marine Accident Investigation Branch.*

MSIU. 2011. Safety investigation into the explosion and subsequent structural damage on board the oil/chemical tanker MAR CRISTINA in the inland port of Sandefjord, Norway on 30 August 2011 201108/033. 13/2012. *Marine Safety Investigation Unit.* Malta.

MSIU. 2013. Safety investigation into the over-pressurisation of a cargo tank on board the Maltese registered chemical tanker UMAR 1 in Fos sur Mer, France on 19 October 2013 201310/016. 27/2014. *Marine Safety Investigation Unit.* Malta.

MSIU. 2015. Safety investigation into the gas inhalation by seven crew members on board the Maltese registered oil/chemical tanker BOMAR MERCURY in the port of Rotterdam, The Netherlands on 21 July 2015. 201507/017. 15/2016. *Marine Safety Investigation Unit.* Malta.

MSIU. 2021. MT Seascout Explosion leading to equipment and structural damages in position 21° 07.1′ N 059° 31.9′ E 18 March 2021. 05/2022. *Marine Safety Investigation Unit.* Malta.

NTSB. 2020. Overpressurization and Rupture of Cargo Tank on Cargo Vessel Fairchem Filly. NTSB/MAB-20/14, DCA19FM037. *National Transportation Safety Board.* United States. Available at: https://www.ntsb.gov/investigations/AccidentReports/Reports/MAB2014.pdf)

OCIMF. 2017. Inert Gas Systems. The use of inert gas for the carriage of flammable cargoes. First edition. The Oil Companies International Marine Forum. London: United Kingdom.

OCIMF. 2019. Ship Inspection Report (SIRE) Programme. Vessel Inspection Questionnaires for Oil Tankers, Combination Carriers, Shuttle Tankers, Chemical Tankers and Gas Tankers, Seventh Edition.

Psarros, G. Skjong, R. Vanem, E. 2011. Risk acceptance criterion for tanker oil spill risk reduction measures, *Marine Pollution Bulletin*, 62(1), 116–127.

UNCTAD. 2023. United Nations Conference on Trade and Development, Review of Maritime Transport 2023. Available at: https://unctad.org/system/files/official-document/rmt2023_en.pdf

UEİM. 2014. Marine Casualty Investigation Report for Tibil (No.01/2014). Ulaşım Emniyeti İnceleme Merkezi, Türkiye. Available at: https://ulasimemniyeti.uab.gov.tr/uploads/pages/deniz/tibil.pdf

UEİM. 2017. Marine Casualty Investigation Report for Habaş (DNZ-04/2018). Ulaşım Emniyeti İnceleme Merkezi, Türkiye. Available at: https://ulasimemniyeti.uab.gov.tr/uploads/pages/deniz/habas-5d974d598549e.pdf

UEİM. 2019. Marine Casualty Investigation Report for Syn Zania (41 /DNZ-07 /2020). Ulaşım Emniyeti İnceleme Merkezi, Türkiye. Available at: https://ulasimemniyeti.uab.gov.tr/uploads/pages/deniz/kaza-inceleme-raporu-syn-zania.pdf

Uğurlu, Ö. 2016. Analysis of fire and explosion accidents occurring in tankers transporting hazardous cargoes, *International Journal of Industrial Ergonomics*, 55, 1–11.

Yip, T.L., Talley, W.K., Jin, D. 2011. The effectiveness of double hulls in reducing vessel-accident oil spillage, *Marine Pollution Bulletin*, 62(11), 2427–2432.

Yu, Q., Teixeira, A. P., Liu, K., Rong, H., Guedes Soares, C. 2021. An integrated dynamic ship risk model based on Bayesian Networks and Evidential Reasoning, *Reliability Engineering & System Safety*, 216, 107993.

Advances in Maritime Technology and Engineering – Guedes Soares & Santos (Eds)
© 2024 The Author(s), ISBN 978-1-032-83099-5

A study on the human evacuation on heeling ships considering the utilization rate of facilities

S.M. Fang, Z.J. Liu, X.J. Wang & J. Ning
Navigation College, Dalian Maritime University, Dalian, P.R. China

ABSTRACT: In the event of an emergency onboard, human evacuation is a crucial measure to ensure safety. This study proposes a method to assess the utilization rate of ship facilities, aiming to quantify the human evacuation process and efficiency onboard. The proposed method involves calculating the utilization time of available stairs and assessing the efficiency of using all stairs on each deck. The simulation focuses on a passenger ship accommodating 236 individuals, exploring different heeling angles in evacuation scenarios. The research results indicate that as the heeling angle increases, evacuation time rises, and the utilization of ship stairs becomes more uneven. To optimize the human evacuation process, the study suggests selecting the evacuation path with the best overall performance instead of the shortest path. The optimized path reduces evacuation time by 6.5%-34.3% across different heeling scenarios, while increasing facility utilization rates by 6.3%-37.0%.

1 INTRODUCTION

As the primary mode of passenger transportation, the safety of passenger ships remains a paramount concern for society. When an accident occurred, particularly during sea navigation, can have catastrophic consequences. The emergency scenarios, often involving heeling and trimming of the ship's hull due to collisions or water ingress, significantly impact human movement (Fang et al., 2022a, Kang et al., 2019). Consequently, human evacuation onboard emerges as a critical component directly tied to the safety of lives and property. The effectiveness and efficiency of evacuation procedures play a pivotal role in minimizing casualties and losses resulting from accidents (Wang et al., 2023, Liu et al., 2022). In light of this, researchers are diligently exploring avenues to enhance evacuation efficiency, aiming to better ensure the safety of both passengers and crew in diverse maritime situations.

During ship evacuations, people typically choose the most familiar and nearest evacuation routes, often resulting in heavier evacuation tasks on stairs near cabins. This concentration of people leads to significant congestion, with some stairs being underutilized. Such imbalances in the overall evacuation process highlight the need for well-defined evacuation strategies to expedite the process and enhance efficiency (Fang et al., 2023, Liao et al., 2016). Especially in the case of ship heeling, people may exhibit heterogeneous behaviors, including panic, thereby necessitating more sophisticated evacuation management strategies (Bartolucci et al., 2021).

However, accurately assessing the effects of evacuation has always been challenging due to the complexity and uncertainty of the evacuation process. While previous studies focused primarily on evaluating overall evacuation results, there has been a relative lack of precise quantitative evaluations of efficiency within the evacuation process (Wang et al., 2021, Sun et al., 2018a). To fill the research gap, a new method is introduced to assess occupant pass situation for ship (OPSS) in this study. This method quantifies the temporal utilization rate of facilities by calculating the use time of each evacuation channel, considering the efficiency of different evacuation channels. It offers a more comprehensive understanding of facility utilization in the evacuation process. To better reflect the actual situation, the use order of facilities at different decks is taken into account, integrating the calculated facility utilization rate of each deck to obtain the total OPSS. This innovative evaluation method provides practical and operational data support for enhancing passenger ship safety management.

2 PARAMETERS AND EVALUATION INDICATORS

2.1 *Parameters setting*

Previous studies have made it clear that heeling is prone to occur when a ship is involved in a maritime accident because the length is larger than the width. The ship heeling phenomenon has a negative effect on the human movement, which has been further

DOI: 10.1201/9781003508762-33

confirmed by the current experimental and simulation studies (Bles et al., 2001, Kim et al., 2004, Sun et al., 2018b). Although the speed attenuation degree under different heeling angles has been initially understood, the detailed influence of heeling angles on large-scale human evacuation, especially the evacuation efficiency of details, is still not clear (Arshad et al., 2022).

To address this concern, a series of scenarios with heeling angles ranging from 0° to 25° are established in this study. Specific speed reduction parameters are assigned for each scenario, as detailed in Table 1 (Fang et al., 2022a, Fang et al., 2022b). The chosen range aims to provide a comprehensive understanding of the influence of heeling angles on the evacuation process. This includes assessing whether there is a noticeable decline in evacuation efficiency as the heeling angle progressively increases and determining the extent to which it significantly impacts overall evacuation speed and results. By pursuing this objective, it is expected that the influence mechanism of heeling ships on human evacuation can be investigated more comprehensively and deeply.

Table 1. Speed reduction factors at different heeling angles.

Heeling angle	0	5	10	15	20	25
Speed reduction factor	1	0.978	0.923	0.860	0.728	0.394

2.2 *Evaluation indicators*

(i) Total evacuation time

The total evacuation time encompasses the duration from when all people receive the evacuation signal to the moment the last person exits the ship. Given the primary focus on evacuation efficiency in human movement for this study, the following assumptions are applied:

a) Evacuation is deemed complete when all people arrive at the assembly station.
b) Evacuation begins immediately upon receiving the evacuation signal, disregarding any preparation stage.
c) The assembly station serves as the final destination, prohibiting the use of elevators during evacuation, and people do not return for activities such as wearing life jackets.

(ii) Utilization rate of stairs

In land-based evacuation scenarios, assessing overall evacuation efficiency typically involves quantifying the utilization rate of exits (Ping et al., 2018). However, directly applying this method to human evacuation on ships faces challenges due to the proximity of assembly station locations on ships, and the overall evacuation efficiency cannot be truly assessed by quantifying the exit utilization rate alone. To address this, the concept of OPSS, an evaluation indicator tailored for ships are introduced in this study. OPSS focuses on evaluating the utilization rate of stairs since stairs serve as the sole passages connecting each deck, and people's choices of stairs on each floor directly impact the overall evacuation efficiency.

It is essential to consider that the assembly station deck is typically not on the same level as the living cabins, leading to longer travel times for stairs closer to the assembly station deck. Therefore, it becomes necessary to quantitatively assess the use time of stairs on each deck separately, as expressed in Equations 1-3:

$$T_D = T_{last\_out} - T_{first\_in} \quad (1)$$

$$OPSS_i = \frac{\sum_{m=1}^{S_m} TD^i max - TD_m}{(S_m - 1) \times TD^i max} \quad (2)$$

$$\text{Total OPSS} = \sum_{i=1}^{n} OPSS_i \quad (3)$$

where, T_D is the travel time of stair, T_{last_out} and T_{first_in} represent the time when the first people enter the stair and the time when the last people come out, respectively. $OPSS_i$ is the value of the OPSS on i^{th} deck, $TD^i{}_{max}$ is the maximum use time of all stairs on i^{th} deck, S_m is the number of stairs on i^{th} deck, and the comprehensive OPSS value of human evacuation on board (Total OPSS) is calculated accordingly. This detailed evaluation method is crucial for understanding people's stair-selection behavior during the evacuation process on ships, and then improve the accurate evaluation of overall evacuation efficiency. This calculation aims to delve into the relationship between the assembly station and stair usage, providing a scientific basis for improving the efficiency of ship personnel evacuation.

In comparison to the exit utilization rate in traditional land-based evacuations, the proposed stair utilization rate offers several advancements:

a) It replaces the evaluation of exit utilization with the quantification of stair utilization. This is achieved by normalizing the duration of stair usage on each deck, mitigating assessment distortions resulting from the concentration of ship exits.
b) Unlike the evaluation of exit utilization, stair usage is sequential, the assessment parameter for ship stair utilization rate is optimized to account for the duration of usage for each stair.
c) The evaluation method of stair utilization rate consolidates the efficiency of stair usage on each deck, preventing inaccuracies arising from variations in the duration of usage across different decks.

3 SIMULATION AND RESULTS

3.1 *Evacuation scenarios*

To systematically evaluate the impact of various heeling angles on evacuation particulars and overall results, the Chinese training ship, MV Yukun, is selected as the subject of the case study. MV Yukun has an overall length of 116.0 m and a molded breadth of 18.0 m with a maximum passenger capacity of 236 people, as shown in Figure 1(a). Cabins are predominantly situated on the Tween deck and the Main deck, while the assembly station is positioned on the Boat deck, as illustrated in Figure 1 (b).

At the onset of evacuation, all people must navigate from their respective cabins to the assembly station until the last person reaches this designated area, signifying the completion of evacuation. To replicate real-world conditions, the population distribution and initial speeds adhere to recommendations by the International Maritime Organization (IMO), as shown in Table 2 (IMO, 2016). In the benchmark scenario, everyone evacuates based on the principle of the nearest route, wherein people choose the path and stairs closest to their cabins to reach the assembly station.

Given that the initial distribution of individuals is confined to the deck, the speed parameter predominantly signifies the standard range of walking speeds on level ground, adhering to a normal distribution. The speed parameter put into the simulation software limit the maximum speed during the simulation, representing the free walking speed when individuals are unaffected by external factors. In reality, individuals must adjust their walking speed to the maximum allowable speed to expedite evacuation. However, during emergency evacuation, influential factors such as congestion, stair ascent or descent, and other impediments often prevent individuals from attaining the ideal speed. In the simulation process, real-time speed on stairs is determined by several parameters, including collision avoidance algorithms, slope angle and dimensions of the stairs. Consequently, this study specifically references the initial speed on flat terrain, integrating it with the rules of the multi-agent model to calculate real-time speeds under different situations.

3.2 *Results of benchmark scenario*

In the benchmark scenario, the evacuation process is simulated under six different heeling angles. Analyzing the curve depicting the number of evacuees over time in Figure 2, it is evident that in the 0° scenario, 236 evacuees complete the evacuation in 380.5 s. Conversely, as the heeling angle escalates to 25°, the overall evacuation time surges to 1244.5 s. Additionally, the curve illustrating the number of evacuees changing over time exhibits a more gradual trend in the 25° scenario compared

a) b)

Figure 1. Training ship MY "Yunkun" and its simulation schematic diagram.

Table 2. Speed ranges for different population groups.

Population groups	Maximum speed (m/s)	Minimum speed (m/s)
Females younger than 30 years	1.55	0.93
Females 30-50 years old	1.19	0.71
Females older than 50 years	0.94	0.56
Females older than 50, mobility impaired (1)	0.71	0.43
Females older than 50, mobility impaired (2)	0.61	0.37
Males younger than 30 years	1.85	1.11
Males 30-50 years old	1.62	0.97
Males older than 50 years	1.4	0.84
Males older than 50, mobility impaired (1)	1.06	0.64
Males older than 50, mobility impaired (2)	0.91	0.55

to other scenarios, indicating reduced evacuation efficiency in this specific condition.

It is essential to highlight that the relationship between heeling angle and evacuation time is not linear. For instance, when the heeling angle increases from 0° to 5°, the evacuation time only experiences a modest increase of 7.2 s. In contrast, elevating the heeling angle from 20° to 25° leads to a significant increase of 671.7 s, marking a 117% surge compared to the total evacuation time of 572.8 s at 20°. It is indicated that the difficulty in people's movement intensifies at larger heeling angles, resulting in a notable slowdown of the overall evacuation process (Walter et al., 2019, Ma et al., 2020).

To provide a more detailed assessment of stair usage during evacuation in the benchmark scenario, the black line chart in Figure 3 is employed to illustrate the total utilization rate of stairs under various heeling angles. It's evident that the heeling angle not only significantly impacts the overall evacuation time but also increases the total OPSS (the sum of stair utilization) as the heeling angle increases. This increase is attributed to the slower moving speeds caused by the elevated heeling angle, prolonging the use of stairs and resulting in a higher Total OPSS.

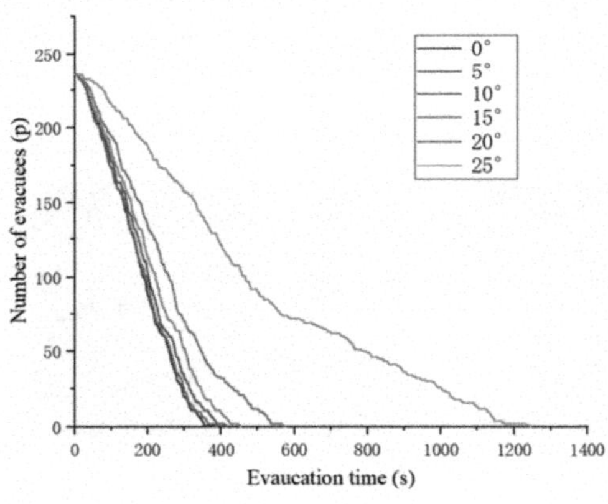

Figure 2. The change curve of the number of evacuees at different heeling angles.

For a more granular presentation of OPSS on single deck, bar charts describe the OPSS of the Tween deck, Main deck, and Promenade deck are included in Figure 3. Notably, the OPSS value of the Tween deck consistently ranks the highest, staying around 0.6. As decks get closer to the Boat deck, the OPSS value diminishes, but with increasing heeling angles, the OPSS differences between various decks gradually decrease. This phenomenon is rooted in the fact that people on the Tween deck must travel the Main deck and the Promenade deck to reach the Boat deck. Consequently, stairs on decks closer to the assembly station tend to endure longer usage, leading to smaller disparities in the utilization of different stairs.

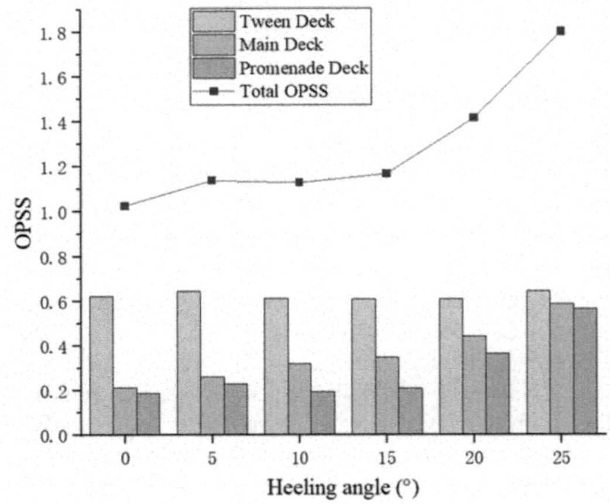

Figure 3. OPSS values for different decks and Total OPSS at different heeling angles.

3.3 *Optimized evacuation strategy*

In the benchmark scenario, the focus was solely on studying the shortest path selection, overlooking the potential impact of route choices on overall evacuation efficiency. In real emergency evacuations, crew members often assume leadership roles, enhancing efficiency by strategically organizing crowd movement, including the selection of more optimized routes based on people positions. To directly compare evacuation times between the shortest path and the optimized path, Figure 4 illustrates the evacuation time reduction rate resulting from path optimization under varying heeling angles. The reduction rate ηT is calculated using Equation 4:

$$\eta_{\mathrm{T}} = \frac{T_S - T_O}{T_S} \tag{4}$$

where, T_S and T_O are the total evacuation time of the shortest path and the optimized path, respectively. It becomes evident that at 5° and 0° heeling angles, the evacuation time for the optimized path exhibits a modest improvement in Figure 4, with reductions of only 6.7% and 6.5%, respectively. However, as the heeling angle increases further, η_{T} shows a consistent rise, signifying that the optimized evacuation strategy can markedly enhance human evacuation efficiency during ship heeling. At a 25° heeling angle, the overall evacuation time experiences a substantial 34.3% reduction, with the optimized evacuation time reduced to 817.3 s compared to 1244.5 s in the benchmark scenario.

In Figure 5, the Total OPSS at different heeling angles is presented to facilitate a comprehensive comparison between the optimized and benchmark evacuation processes. Notably, the optimized OPSS values consistently remain below 1.0, reaching only 1.07 and 1.13 at 5° and 25° heeling angles, respectively. At a 5° heeling angle, where the reduction in

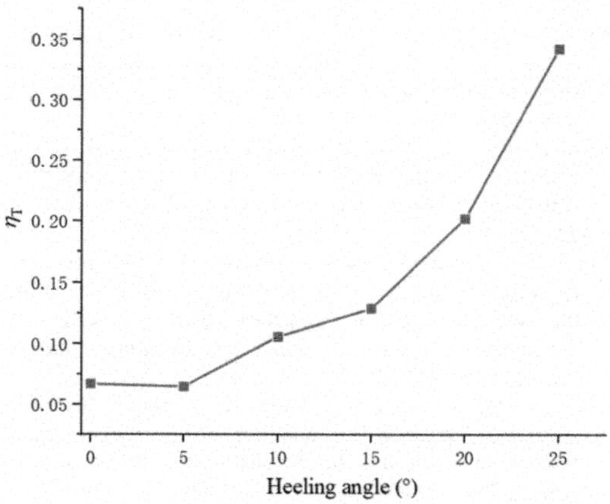

Figure 4. The time reduction rate obtained by optimized strategy.

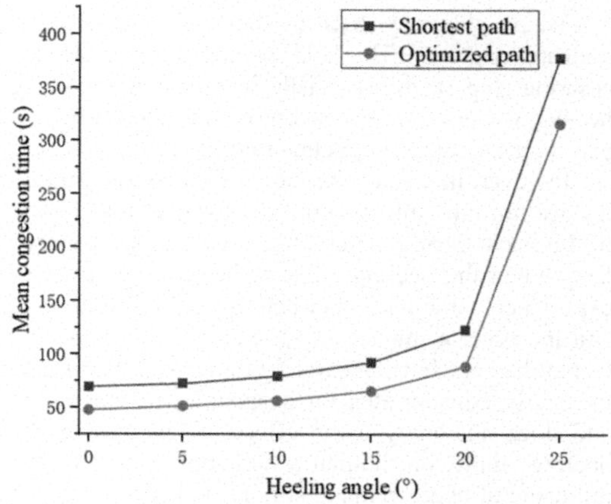

Figure 6. Comparison of mean congestion time of the optimized path and the shortest path.

human moving speed is not substantial, the psychological urgency to complete evacuation promptly intensifies due to ship heeling. Consequently, a more pronounced gathering of people occurs in bottleneck areas like stairs, resulting in similar OPSS values before and after optimization. This finding aligns with the marginal improvement of only 6.5% in total evacuation time at 5° heeling angle. However, at higher heeling angles, particularly 20° and 25°, the optimized OPSS values exhibit significant reductions of 33.5% and 37.0%, respectively. This emphasizes that judicious route selection can effectively enhance the balance of stair utilization, expediting the overall evacuation process (Fang et al., 2023, Yang et al., 2020). The optimized evacuation strategy demonstrates superior efficiency in the context of ship heeling, offering practical operational guidance for managing ship emergencies.

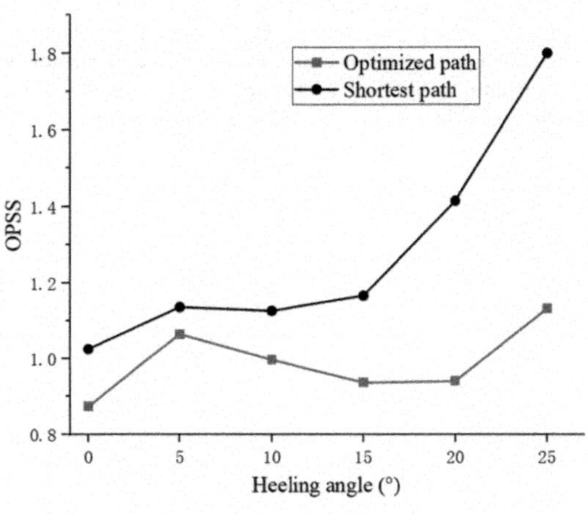

Figure 5. Comparison of OPSS values of the optimized path and the shortest path.

Figure 6 illustrates the quantification of mean congestion time at both optimized and benchmark scenarios along the evacuation path. As the heeling angle increases, there is an exponential rise in mean congestion time in both scenarios. Particularly, at a 25° heeling angle, the benchmark scenario records a mean congestion time of 378.1 s, indicative of a markedly slow evacuation process with significant congestion issues affecting most individuals. Notably, all heeling scenarios demonstrate lower mean congestion times compared to the benchmark scenario, highlighting the positive impact of judiciously optimizing evacuation paths. For instance, at a 25° heeling angle, the optimized mean congestion time is reduced to 316.3 s, reflecting a notable 16.3% improvement. This underscores that effective route optimization can substantially mitigate congestion, enhance the efficiency of human evacuation, and offer a more dependable solution for ensuring ship safety in emergency evacuation scenarios.

4 CONCLUSION

Given the current lack of effective quantification for onboard evacuation efficiency, a meticulous evaluation of the evacuation process under varying heeling angles is undertaken, focusing on total evacuation time, OPSS, and mean congestion time. The analysis reveals a substantial impact of heeling ship on overall evacuation efficiency, with increased heeling angles leading to reduced movement speeds and prolonged evacuation. Particularly at significant heeling angles, efficiency experiences a notable decline, elongating the evacuation duration. In response, optimization of the evacuation path proves effective in enhancing efficiency. Optimized strategies not only reduce evacuation time but also diminish OPSS and alleviate channel congestion. This underscores the practical significance of rational guidance and organization by leaders in ship evacuation management. Such measures can effectively enhance evacuation efficiency, balance ship facility usage, and mitigate congestion. This holds particular importance for ensuring ship safety, especially in scenarios involving heeling, where effective congestion mitigation strategies play a crucial role in promoting safety.

However, this study has some limitations. Firstly, it does not take into account the psychological state of humans during emergency evacuation, particularly when the heeling angle is larger. Crowds may experience panic and other heterogeneous behaviors, and the precise impact on evacuation time and efficiency has not been accurately quantified. Secondly, the study concentrates on a specific heeling angle and does not analyze the effect of changes in the heeling angle on human evacuation. Finally, the influence of pedestrian aids on individual's choice of path and movement is not considered. These limitations will be further explored in future studies.

REFERENCES

Arshad, H., Emblemsvåg, J., Li, G. & Ostnes, R. 2022. Determinants, methods, and solutions of evacuation models for passenger ships: A systematic literature review. *Ocean Engineering*, 263, 112371.

Bartolucci, A., Casareale, C. & Drury, J. 2021. Cooperative and competitive behaviour among passengers during the costa concordia disaster. *Safety Science*, 134, 105055.

Bles, W., Nooy, S. A. E. & Boer, L. C. Influence of ship listing and ship motion on walking speed. 2001 2001. Springer, 437.

Fang, S., Liu, Z., Wang, X., Wang, J. & Yang, Z. 2022a. Simulation of evacuation in an inclined passenger vessel based on an improved social force model. *Safety Science*, 148, 105675.

Fang, S., Liu, Z., Yang, X., Wang, X., Wang, J. & Yang, Z. 2023. A quantitative study of the factors influencing human evacuation from ships. *Ocean Engineering*, 285, 115156.

Fang, S., Liu, Z., Zhang, S., Wang, X., Wang, Y. & Ni, S. 2022b. Evacuation simulation of an Ro-Ro passenger ship considering the effects of inclination and crew's guidance. *Proceedings of the Institution of Mechanical Engineers, Part M: Journal of Engineering for the Maritime Environment*, 237, 192–205.

IMO 2016. Revised Guidelines for Evacuation Analysis for New and Existing Passenger Ships. IMO MSC.1/Circ 1533, 6 June 2016.

Kang, Z., Zhang, L. & Li, K. 2019. An improved social force model for pedestrian dynamics in shipwrecks. *Applied Mathematics and Computation*, 348, 355–362.

Kim, H.-T., Lee, D.-K., Park, J.-H. & Hong, S.-K. 2004. The effect on the mobility of evacuating passengers in ship with regard to list and motion. *IE interfaces*, 17, 22–32.

Liao, W., Tordeux, A., Seyfried, A., Chraibi, M., Drzycimski, K., Zheng, X. & Zhao, Y. 2016. Measuring the steady state of pedestrian flow in bottleneck experiments. *Physica A: Statistical Mechanics and its Applications*, 461, 248–261.

Liu, K., Ma, Y., Chen, M., Wang, K. & Zheng, K. 2022. A survey of crowd evacuation on passenger ships: Recent advances and future challenges. *Ocean Engineering*, 263, 112403.

Ma, Y., Liu, K., Chen, M., Ma, J., Zeng, X., Wang, K. & Liu, C. 2020. ANT: Deadline-Aware Adaptive Emergency Navigation Strategy for Dynamic Hazardous Ship Evacuation With Wireless Sensor Networks. *IEEE Access*, 8, 135758–135769.

Ping, P., Wang, K. & Kong, D. 2018. Analysis of emergency evacuation in an offshore platform using evacuation simulation modeling. *Physica A: Statistical Mechanics and its Applications*, 505, 601–612.

Sun, J., Guo, Y., Li, C., Lo, S. & Lu, S. 2018a. An experimental study on individual walking speed during ship evacuation with the combined effect of heeling and trim. *Ocean Engineering*, 166, 396–403.

Sun, J., Lu, S., Lo, S., Ma, J. & Xie, Q. 2018b. Moving characteristics of single file passengers considering the effect of ship trim and heeling. *Physica A: Statistical Mechanics and its Applications*, 490, 476–487.

Walter, H. J., Li, R., Wagman, J. B. & Stoffregen, T. A. 2019. Adaptive perception of changes in affordances for walking on a ship at sea. *Hum Mov Sci*, 64, 28–37.

Wang, X., Liu, Z., Wang, J., Loughney, S., Yang, Z. & Gao, X. 2021. Experimental study on individual walking speed during emergency evacuation with the influence of ship motion. *Physica A: Statistical Mechanics and its Applications*, 562, 125369.

Wang, X., Xia, G., Zhao, J., Wang, J., Yang, Z., Loughney, S., Fang, S., Zhang, S., Xing, Y. & Liu, Z. 2023. A novel method for the risk assessment of human evacuation from cruise ships in maritime transportation. *Reliability Engineering & System Safety*, 230, 108887.

Yang, X., Yang, X., Wang, Q., Kang, Y. & Pan, F. 2020. Guide optimization in pedestrian emergency evacuation. *Applied Mathematics and Computation*, 365, 124711.

Advances in Maritime Technology and Engineering – Guedes Soares & Santos (Eds)

A fire risk assessment method for offshore platforms: Applications of 24Model with BN

M.Y. Guo & M. Chen
College of Shipbuilding Engineering, Harbin Engineering University, Harbin, Heilongjiang, P.R. China

ABSTRACT: This paper integrates the 24Model with a Bayesian network (BN) and proposes a new fire risk assessment method for offshore platforms. Triangular fuzzy membership function fuzzy set theory, coefficient of variation method, and the combination of node state distance function and fuzzy inference are used to address the subjectivity and uncertainty due to insufficient objective data and conflicting opinions of multiple experts. Based on the proposed method, the human, organizational, unsafe conditions, and technical factors of fire safety failure in the living quarters of an offshore platform are comprehensively analyzed from the perspective of the system. Based on diagnostic inference and sensitivity analysis, the main affecting factors and causal path of the risk scenario are determined. The results show that the proposed method could help in identifying the root causes of the failures, provide new insights into the probabilities and relationships among different factors, and enable effective risk management.

1 INTRODUCTION

Despite the continuous improvement of safety management technology for offshore platforms, fires still occur from time to time, resulting in significant damage to human life, assets, and the environment (Shi et al. 2021). The statistical report on offshore platform accidents issued by the Bureau of Safety and Environmental Enforcement (BSEE) highlights that fires and explosion incidents caused by fires pose a major threat to offshore exploration. Specifically, fire accidents impacting living quarters often lead to severe consequences and can escalate into catastrophic incidents. Risk analysis is essential to effectively reduce fire risk on offshore platforms.

According to statistics, human and organizational are among the main factors that cause fire accidents on offshore platforms (Wrobel 2021). The European Maritime Safety Agency (EMSA) released a study of 4104 fire accidents on offshore platforms from 2011 to 2018 in 2019, which pointed out that 65.8% of the accidents were caused by human behavior. In-depth investigations into major offshore platform fires by the Norwegian Petroleum Security Agency (PSA) have shown that misconduct by operators is the leading cause of risk accidents, directly or indirectly (Robb and Miller 2012). Hence, the impact of human and organizational factors should be considered in the effective quantitative risk analysis (QRA) of offshore platforms (Steijn et al. 2020; Ade and Peres 2022).

At present, the research on fire accident risk assessment of offshore platforms mainly focuses on the reliability of equipment technology. The most existing risk assessment methods do not fully consider human and organizational factors (HOF), and the coupling with technical and human factors is insufficient (Skogdalen and Vinnem 2011; Torres et al. 2016; Steijn et al. 2020). There are differences in the definition of HOF; misunderstanding the research object and insufficient risk identification will bring uncertainty to risk modeling. And due to the lack of data related to human and organizational factors, quantifying the impact of HOF is usually estimated based on expert judgment or seemingly arbitrary (Pan et al. 2020; Steijn et al. 2020). Coupled with the complex socio-technical nature of offshore platforms, the effective integration of HOF in QRA remains a challenge. Ideally, QRA encompass the integration of technological, human, and organizational factors into a unified framework. (Steijn et al. 2020).

The accident causation model is an expression of all the causes of accidents within an organization and their interrelationships (Fu et al. 2017). By using the accident causation model for accident analysis, all the causes leading to the occurrence of accidents can be identified clearly, along with the paths through which accidents unfold (Xie et al. 2022; Fu et al. 2020b). The initial accident causation models was a simple linear model based on causality. With the continuous advancement of research and analysis of accidents and disasters, the accident causation models have evolved from simple linear models to complex linear models based on epidemiology (Fu et al. 2020; Zhang et al. 2020). Reason (1990) proposed the Swiss Cheese Model (SCM), a conceptual model that describes how failures at the individual

DOI: 10.1201/9781003508762-34

and organizational levels can lead to system failure, emphasizing the direct linear relationship between accident causes and outcomes. HFACS, 24Model provide specific hierarchical categorization and catalogs of accident causes, demonstrating good scalability and applicability (Wu et al. 2023; Kaptan et al. 2021). In response to complex social systems, scholars have subsequently proposed the FRAM model, Social Technical Systems Model, Accimap method, STAMP Model, and others (Hollnagel 2002; Zhang et al. 2020; Wu et al. 2023).

HFACS is one of the most famous models for analyzing human and organizational factors. The 24Model was proposed in 2005 by Fu (Fu et al. 2016; Fu et al. 2019). 24Model and HFACS do share several similarities. They both consider organizational factors at the organizational level, such as safety management systems and safety culture (Fu et al. 2017; Wu et al. 2023). They provide explicit definitions of the constituent factors in accident models and offer systematic frameworks for accident analysis. HFACS has been widely applied in fields such as aviation, healthcare, and the military (Wiegmann and Shappell 2001; Chen et al. 2013; Lyu et al. 2019). It places human factors at the core of accident analysis, facilitating a deeper understanding and resolution of accidents caused by human errors and behaviors. However, it should be noted that the factors included in HFACS may exhibit some overlap (Fu et al. 2017; Fu et al. 2020a; Fu et al. 2020b). In strict terms, HFACS could be considered a model that closely aligns with human factors in accident analysis. The 24Model takes into account a comprehensive range of factors, including human factors, technological factors, organizational factors, and external unsafe factors. By incorporating a broader spectrum of causal elements, it provides a more holistic perspective during risk analysis. This allows for a more comprehensive understanding and assessment of risks when conducting a risk analysis, making it particularly suitable for evaluating risks in complex systems. That's why we chose the 24Model instead of HFACS. However, 24Model only has a qualitative function. That is, it can only identify and classify risk factors in accidents. It needs to be combined with other probability analysis methods to quantify the impact of risk factors.

Bayesian network (BN) approaches is widely used probability analysis methods (Ghasemi et al. 2018; Vojkovic et al. 2021). The BN method utilizes conditional probability to depict the relationships between different information elements and enables inference in situations where information is limited, incomplete, and uncertain (Steijn et al. 2020; Ghasemi et al. 2018). Therefore, the BN method has been widely used in offshore platform fire risk assessment in the past decades. For example, Wang used a BN to conduct a probability analysis of offshore platform fire and explosion accidents and established the probability prediction model of offshore platform fires (Wang et al. 2011; Wang et al. 2015; Wang et al. 2017). Roed (2018 used a BN to build a comprehensive probability model for fires and explosions on offshore platforms.

One major characteristic of modern safety management is the shift from reactive to proactive safety management. Based on the logical relationships in accident causation models and combined with probability analysis methods, it is possible to obtain a comprehensive understanding of accident causes and their occurrence probabilities. This approach provides assistance in achieving precise accident prevention.

Based on the above research, this paper uses a method that integrates 24Model and a BN to construct a new fire risk assessment method for offshore platforms. The BN enables a more realistic representation of probabilistic causality and effectively incorporates and combines information from multiple sources. The integration of 24Model and a BN can deeply explore the impact of potential human and organizational factors and quantify the interrelationships among the risk sources and their total effects on accidents. However, it is worth noting that due to the lack of human statistics in the maritime field, the data available for model quantification is insufficient, and the conditional probability table (CPT) of a BN is usually derived by expert judgment. The scale of the CPT increases exponentially with the number of parent nodes. When the network scale of the BN is large, it is difficult for experts to provide accurate probability values for it, and it is also challenging to maintain consistency when feeding a large number of conditional probability values (Wang et al. 2011; Wang et al.,2017). Fuzzy inference utilizes fuzzy sets, membership functions, and fuzzy operation rules to address the fuzziness and uncertainty in fuzzy decision-making (Wang 2009; Yazdi and Kabir 2020). It allows for the avoidance of the requirement for experts to provide precise probability values, thus reducing the subjectivity of expert evaluation. Therefore, this paper employs the node state distance function and fuzzy inference to derive the conditional probability table of a BN.

The rest of this article is organized as follows. Section 2 introduces the 24Model and Bayesian network. Section 3 takes the fire safety of the N500 deep-water semi-submersible support platform living quarters as an example to construct a risk assessment example, shows how to use the risk scenario modeling and quantify the probability for various types of nodes, and presents analyze results. Section 4 discusses the characteristics and limitations of the method proposed in this paper.

2 24MODEL AND BAYESIAN NETWORK

2.1 *24Model*

The 24Model has undergone five versions of evolution (Fu et al. 2016; Fu et al. 2019; Fu et al. 2020a; Fu et al. 2020b; Wu et al. 2023). Figure 1 shows the details of the latest 24Model framework. According to the 24Model, the causation of accidents is attributed to hazards, which encompass various factors including human factors, technical factors,

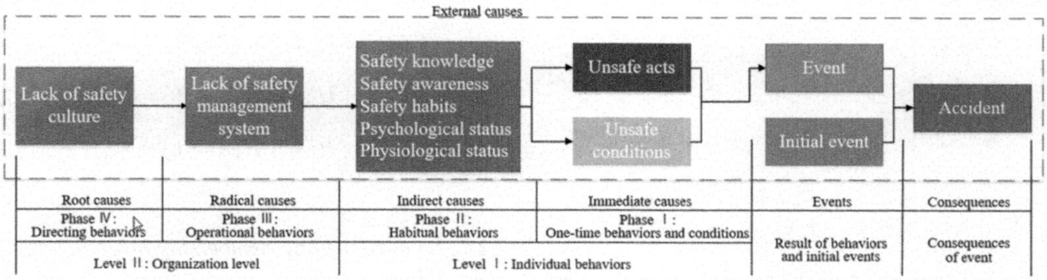

Figure 1. Overview of 24Model framework (Fu et al. 2017).

organizational factors, and external unsafe factors. Human factors encompass unsafe actions (by individuals at different levels of the organization, including managers, individual factors (insufficient safety knowledge, low safety awareness, poor safety habits, negative safety attitudes, and compromised physical fitness for safety). Unsafe conditions include unsafe conditions, environment, and external states. Organizational factors encompass deficiencies in safety management systems and safety culture. External factors include higher-level organizations, government agencies, suppliers and their products/services, the members' families, genetics, upbringing and natural factors, market competition, and socio-political, economic, legal, and cultural factors that influence the occurrence of accidents within the organization (Wu et al. 2023; Li et al. 2021; Suo et al. 2017).

2.2 *Quantitative method of Bayesian network*

2.2.1 *Conditional probability table*

The BN has a causal relationship between the parent node and the child node. That is, the parent node's state affects the child node. As shown in Figure 2, the BN assumes that nodes A and B have three states of "High," "Medium," and "Low," and node C may have states of "High," "Medium," and "Low."

1) Calculation of parent nodes' weight values

The number of parent nodes of BN child nodes is set to be m to form set $p = \{p_1, p_2, \ldots, p_m\}$, and there are n experts participating in the evaluation to create set $e=\{e_1,e_2,\ldots,e_n\}$. For the parent node $p_i(i = 1, 2, \ldots, m)$, the weight estimation interval corresponding to expert opinions is $[u_{1i}, v_{1i}]$,$[u_{2i}, v_{2i}]$, …,$[u_{ni}, v_{ni}]$. The sum of the weights of the parent node is 100. *Kr* is the expert weight, this paper uses the coefficient of variation method to determine the weight of experts, described in detail in section 3.2 For the parent node p_i, the relative weight ω formula is as follows:

$$\omega_i = \frac{0.5 \times \sum_{r=1}^{n} Kr[v_{ri}^2 - u_{ri}^2]}{\sum_{r=1}^{n} Kr[v_{ri} - u_{ri}]} \quad i = 1, 2, \mathrm{K}\, m \tag{1}$$

The normalized weight value is

$$k_i = \frac{\omega_i}{\sum_{i=1}^{m} \omega_i} \tag{2}$$

The weight values $k = (k_1, k_2, \ldots, k_m)$ of all parent nodes can be obtained.

2) Calculation of nodes state distances

The node state distance is the degree of similarity and difference between the child node and parent node state combination:

$$Z_j = \sum_{k=1}^{n} |Z_{ij}| \cdot k_i \tag{3}$$

where there are *n* parent nodes, the node state distance is Z_j, *i* is the i-th parent node, *j* is the state of the child node, ω_i is the weight value of the parent node *i*, and the node state distance Z_j is the difference between the state of the parent node *i* and the state of the child node in state *j*. The absolute value represents the distance, whether the positive or negative parent node has the same significance degree as the child node. The value range of Z_j is [0,2]. When the states of the child node and its parent node are not different, $Z_j = 0$; when the states of the child node are completely opposite to its parent node, $Z_j = 2$.

3) Fuzzy inference

The node state distance is the degree of similarity and difference between the combined state *j* of the Z_j child node and the current parent node, and the value of the node state distance Z_j determines the value of the conditional probability of the child node. The variable Q_j is the size of the conditional probability of the child node in the state *j* under the current parent node state combination. The mapping relationship between the node state distance and the conditional probability is constructed by a fuzzy inference system. Through the fuzzy inference system, given the state distance between the child node and the current parent node in state *j*, the conditional probability in *j* state can be calculated.

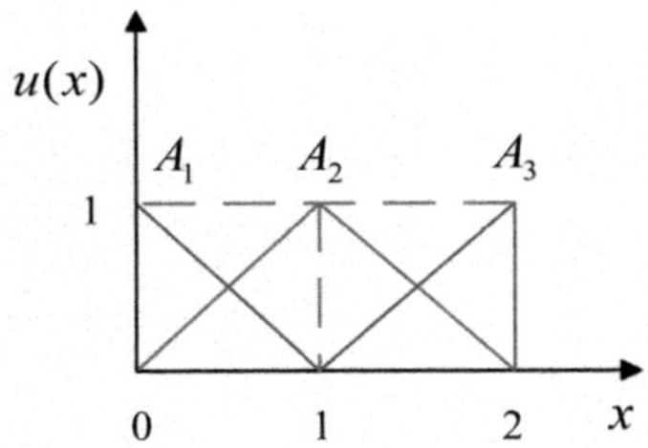

Figure 2. Fuzzy membership function of node state distance function.

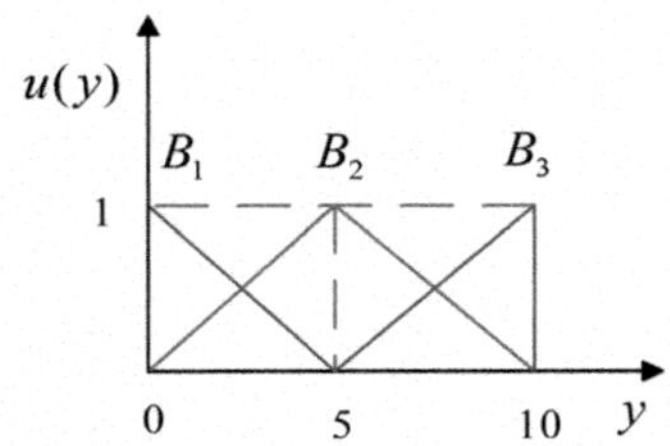

Figure 3. Fuzzy membership function of consequence probability tendency.

The graphical solution of the fuzzy membership function of the node state distance function and the fuzzy membership function of the consequence probability tendency are shown in Figures 2 and 3, respectively. $u_{A_1}(x)$, $u_{A_2}(x)$, and $u_{A_3}(x)$ are membership functions for near, medium, and far distances, respectively, and $u_{B_1}(y)$, $u_{B_1}(y)$, and $u_{B_3}(y)$are membership functions with low, medium, and high probability, respectively.

The antecedent of the fuzzy rule is the set of node state distances $Z_j\{near, medium, far\}$, and the consequence is the set of probability tendencies $Q_j\{low\,probability;\ medium\,probabbility;\ high\,probabbility\}$. Then, the fuzzy reasoning rules are as follows:

R1: IF node state distance Z_j is near, THEN consequence probability tendency Q_j is high probability;
R2: IF node state distance Z_j is medium, THEN consequence probability tendency Q_j is medium probability;
R3: IF node state distance Z_j is far, THEN consequence probability tendency Q_j is low probability.

4) Calculation of conditional probability of child nodes

The conditional probability value Q_j of each state can be obtained by fuzzy reasoning, and the conditional probability value of the child node can be obtained by normalizing the Q_j of each state. The calculation is as follows:

$$\begin{cases} P(C=H|A=H,B=L)=\frac{Q_H}{\sum_{j=H}^{L}Q_j} \\ P(C=H|A=H,B=L)=\frac{Q_M}{\sum_{j=H}^{L}Q_j} \\ P(C=H|A=H,B=L)=\frac{Q_L}{\sum_{j=H}^{L}Q_j} \end{cases} \tag{4}$$

2.2.2 *Triangular fuzzy membership function fuzzy set theory*

For the acquisition of the prior probability of the root node, this paper uses triangular fuzzy membership function fuzzy set theory to deal with the uncertainty and inaccuracy of the expert knowledge. A fuzzy number $g=(P_L,P_M,P_U)$ is called a triangular fuzzy number, where P_L, P_M, and P_U are the lower boundary, the most likely boundary, and the upper boundary, respectively. The membership function is given as follows:

$$f_g(x)=\begin{cases} 0 & x<P_L \\ \frac{x-P_L}{P_M-P_L} & P_L\le x\le P_M \\ \frac{P_U-x}{P_U-P_M} & P_M\le x\le P_U \\ 0 & x>P_U \end{cases} \tag{5}$$

The confidence membership of the uncertainty estimation is calculated by using the defuzzification method of the regional center to obtain the possible Ps after transformation, as follows:

$$Ps=\frac{\int f_i(x)xdx}{\int f_i(x)dx} \tag{6}$$

The occurrence probability P_r is obtained by the following formula (Yazdi and Kabir 2020; Hu et al. 2014):

$$P_r=\begin{cases} \frac{1}{10^K} & P_s\neq 0 \\ 0 & P_s=0 \end{cases} \tag{7}$$

3 CASE ANALYSIS

As an important place to support oil and gas exploitation, the N500 deep-water semi-submersible support platform can accommodate 500 people. It has large numbers of cabins and staff members and dense fire safety hazards. Its living quarters is where staff undergo a series of living activities, such as rest, catering, entertainment, and sanitation, with a large number of people (Zhang et al. 2021; Lotsberg et al. 2004; Veritas 2007; Scandpower 2013). Once a fire accident occurs, it causes great harm to people on the platform. Therefore, the fire safety risk assessment of the living quarters is very important for the safe operation and maintenance of the N500 deep-water semi-submersible support platform.

3.1 *Risk factors analysis with 24Model*

The living quarters of the N500 deep-water semi-submersible support platform is equipped with measures to prevent fire spread, fire extinguishing system, and emergency evacuation system to deal with fire evacuation risks.

In this case, to determine risk factors, we invited relevant experts. Experts selected and modified risk factors we gleaned from accident investigations, supplemented by some factors that were difficult to document in accident analysis. All nodes were classified according to the 24Model framework, as shown in Table 1.

Based on the above analysis, the risk factors contributing to the fire safety failure of the N500 deep-water semi-submersible support platform living quarters were identified, as shown in Figure 4.

3.2 *Quantifying risk factor nodes of BN*

In this case, the opinions of four marine engineering experts were collected and sorted to assign the status of intermediate nodes in the BN. To objectively describe the weights of the expert opinions, according to three factors, the expert's engineering level, professional working years, and number of engineering projects participated in, the variance method was introduced to determine the importance of the three factors. After, the opinion weights *Kr* of the four experts were determined according to the importance of the three factors. There are three steps in this method:

1) Raw data collation

Assuming that there are *n* experts to be evaluated and *P* evaluation indicators, the original numerical index data matrix of the j-th evaluation indicator of the i-th sample is as follows:

$$X = \begin{pmatrix} x_{11} & \cdots & x_{1p} \\ \vdots & \ddots & \vdots \\ x_{n1} & \cdots & x_{np} \end{pmatrix} \tag{8}$$

2) Calculation of standard deviation S_j and coefficient of variation v_j for item *j*

The standard deviation and coefficient of variation are defined as follows:

$$\begin{cases} \bar{x}_j = \frac{1}{n}\sum_{i=1}^{n} x_{ij} \\ S_j = \sqrt{\frac{\sum_{i=1}^{n}(x_{ij}-\bar{x}_j)^2}{n-1}} \end{cases} \tag{9}$$

$$v_j = \frac{S_j}{\bar{x}_j} \tag{10}$$

3) Calculation of normalized coefficient of variation *Kr*

The normalized coefficient of variation is as follows:

$$Kr = \frac{v_j}{\sum_{j=1}^{p} v_j} \tag{11}$$

Four normalized weights of the expert opinions were calculated according to the weights of the variation coefficient of each index. The calculated weights of the expert opinions were $Kr = \{Kr_1 = 0.4954, Kr_2 = 0.204, Kr_3 = 0.205, Kr_4 = 0.096\}$.

According to the calculation method of the Bayesian conditional probability table introduced in Section 2, the weights of the parent nodes of the BN were obtained by calculating the expert-estimated weight interval of each BN node. According to the estimated weight value, the CPT allocation algorithm was used to complete the entire BN's conditional probability allocation combined with the root node's prior probability. Taking the HF8 node as an example, the CPT calculation result is shown in 1.

Table 1. HF8 node conditional probability table.

		HF8		
HF4	HF5	High	Medium	Low
High	High	0.56	0.33	0.11
	Medium	0.35	0.37	0.28
	Low	0.33	0.33	0.34
Medium	High	0.37	0.36	0.27
	Medium	0.27	0.46	0.27
	Low	0.28	0.36	0.36
Low	High	0.33	0.34	0.33
	Medium	0.28	0.36	0.36
	Low	0.11	0.33	0.56

3.3 *Sensitivity analysis*

Information entropy defines the uncertainty of variables in the model and is a statistic of quantitative dispersion. The information entropy *H(Y)* of the BN is calculated by the following formula:

$$H(Y) = -\sum_{t} P(y) \log P(y) \tag{12}$$

where $P(y)$ is the probability distribution of variable *Y*.

The importance of basic events to top events can be evaluated by mutual information. Mutual information can quantitatively calculate the impact of basic events on top events. Based on the order of importance, we can pay more attention to which basic events and security measures should be taken. Mutual information between variables is calculated by the following formula:

$$H(X:Y) = \sum_{x \in X} \sum_{y \in Y} P(x,y) \log_2 \left(\frac{P(x,y)}{p(x) \cdot p(y)} \right) \quad (13)$$

Where $H(X:Y)$ is the mutual information between variables X and Y, $P(x,y)$ is the joint probability distribution function of X and Y, and $P(x)$ and $P(y)$ are the edge probability distribution functions of X and Y, respectively.

Sensitivity analysis of the BN was carried out, and the results of mutual information of events are shown in Figures 5-8. By analyzing the results in these tables and Figures, the factors contributing more to the probability of the final accident were obtained. Controlling these critical causal events would greatly reduce the likelihood of an eventual accident.

From 5, the mutual information risk factor "Not handled in time as required" is the highest followed by "Detector failure", "Communication malfunction", "Aging of facilities". The individual and organizational factor node "Ignoring alarm signals" is the highest followed by "Procedure violation", "Intentionally violating the existing regulations", "Lack of personnel training", "Bad organizational culture".

From Figure 6, for failure of fire extinguishing system, the risk factors "Failure of broadcasting system" is the highest followed by "Line fault", "Equipment damage", "Damage to central control room", "Failure of Smoke detector". The individual and organizational factor node "Ignoring alarm signals" is the highest followed by "Using inappropriate tools", "Judgment/decision error", "Poor visibility".

From Figure 7, for failure of personnel emergency evacuation, the risk factor "Evacuation route selection error" is the highest followed by "Failure of ventilation system", "Failure of escape guidance device". The individual and organizational factor node "Judgment/decision error" is the highest followed by "Risk-taking behavior", "Poor safety knowledge", "Poor safety habits", "Poor security awareness".

3.4 *Causal path analysis and probabilistic inference*

Causal path analysis aims to identify the different stages and factors that contribute to the occurrence of accidents to improve safety management and prevent similar incidents. By analyzing the process and path leading to an accident, potential risks and weak links can be identified, enabling the implementation of appropriate measures to reduce the probability of accidents.

As depicted in Figure 9, within the 24Model framework, there are interrelated and influential relationships among the modules of accident causation. There are five influence paths that exist between the different modules.

For failure of fire extinguishing system, the main unsafe paths are: ① Bad organizational culture → Insufficient safety supervision → poor safety knowledge → Procedure violation → Not handled in time

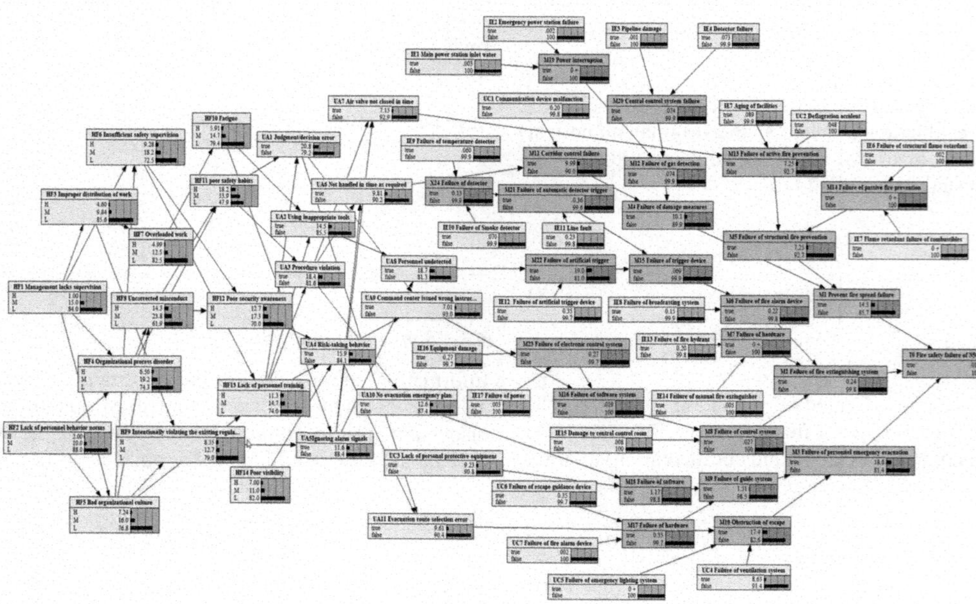

Figure 4. BN structure diagram of fire safety risk assessment for N500 deep-water semi-submersible support platform living quarter.

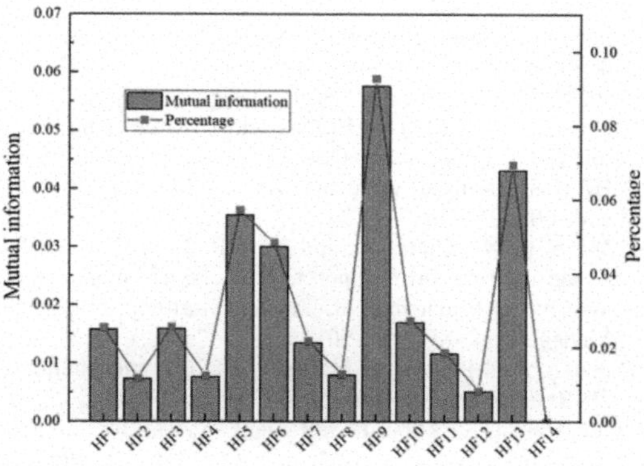

Figure 5. Sensitivity analysis of "Prevent fire spread failure".

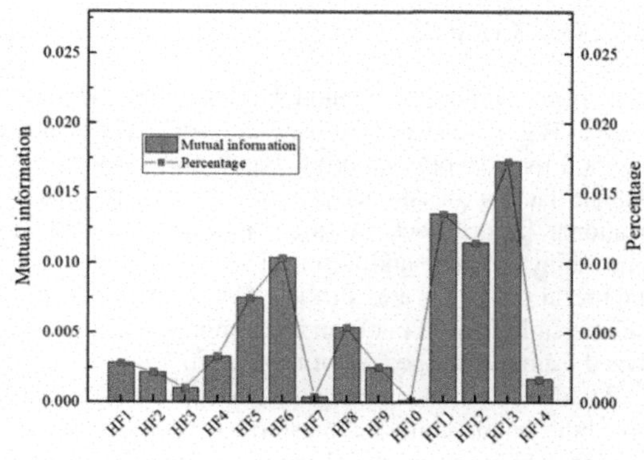

Figure 7. Sensitivity analysis of "Failure of personnel emergency evacuation".

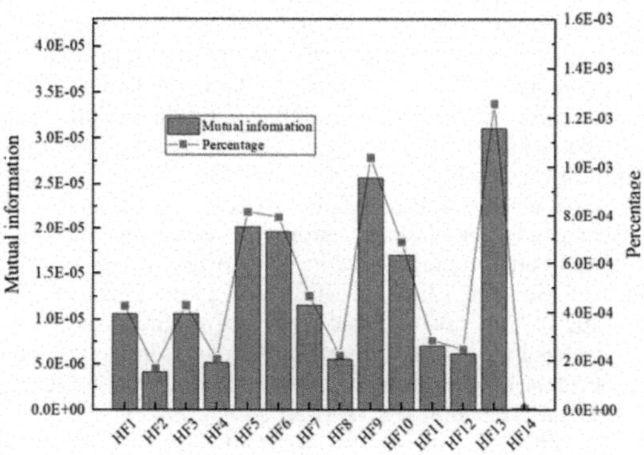

Figure 6. Sensitivity analysis of "Failure of fire extinguishing system".

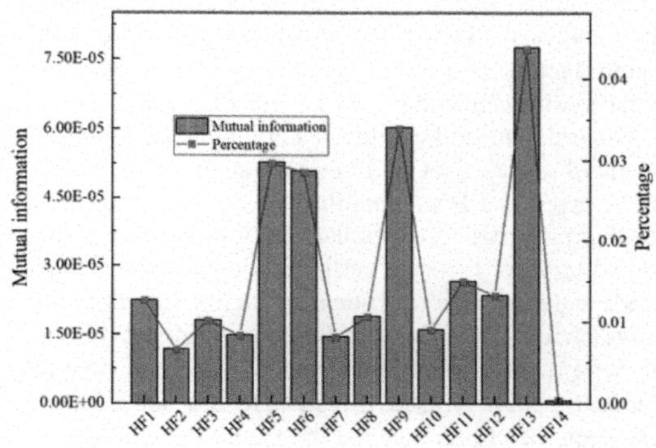

Figure 8. Sensitivity analysis of "Fire safety failure of N500 deep-water semi-submersible support platform living quarters".

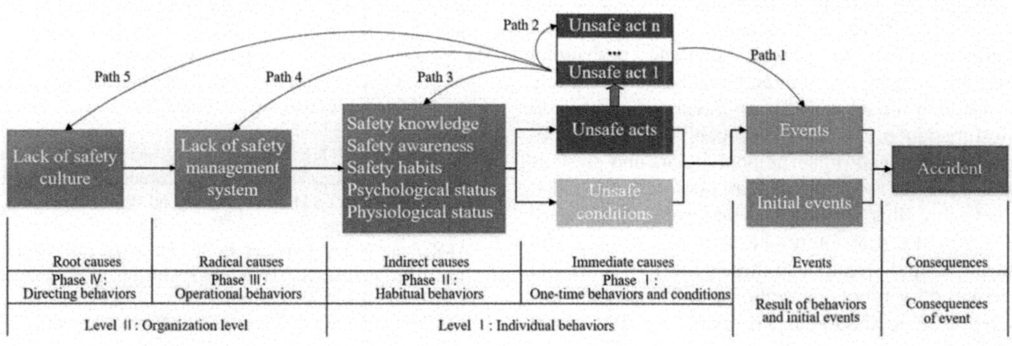

Figure 9. The influence path in 24Model.

as required → Corridor control failure → Failure of damage measures → Prevent fire spread failure; ② Intentionally violating the existing regulations → Ignoring alarm signals → Air valve not closed in time → Failure of active fire prevention → Failure of structural fire prevention → Prevent fire spread failure.

For failure of fire extinguishing system, the main unsafe path is: ① Failure of broadcasting system → Failure of fire alarm device → Failure of fire extinguishing system.

For failure of personnel emergency evacuation, the main unsafe paths are: ① Failure of ventilation system → Obstruction of escape → Failure of personnel emergency evacuation; ② Poor safety habits → Judgment/decision error → Evacuation route selection error → Obstruction of escape → Failure of personnel emergency evacuation.

4 CONCLUSION

This paper proposed a method that integrates 24Model and a BN to construct a new fire risk assessment method for offshore platforms. Based on 24Model, the human factors (unsafe acts), unsafe conditions, organizational factors (lack of safety management system and safety culture), and technical factors of fire safety failure in the living area of the N500 deepwater semi-submersible support platform are comprehensively analyzed from the perspective of the system.

In addition, by integrating the BN approach, qualitative information could be transformed into quantitative information in a structured manner. The BN method allows for the modeling of probabilistic relationships between variables and facilitates the assessment of risk and the identification of critical factors affecting fire safety in the platform's living quarters. This quantitative analysis enhances the understanding of the underlying factors and provides a more robust basis for decision-making and preventive measures.

In order to address the uncertainty due to insufficient objective data and conflicting opinions of multiple experts, a BN quantification method is designed in this paper using triangular fuzzy membership function fuzzy set theory, coefficient of variation method, node state distance functions, and fuzzy inference. Triangular fuzzy membership function fuzzy set theory is used to quantify root factor nodes. The node state distance function is used to calculate the node distance between the child node and the parent node. The fuzzy inference system can derive the CPT based on the node distance. The method of combining the node state distance function with fuzzy inference could reason according to expert opinions. In the combination process, the weights of experts were taken into account by using the coefficient of variation method. By not requiring experts to provide accurate probability values, the subjectivity of expert evaluation is reduced. The resulting model provides an integrated and effective estimation of the failure probabilities for technological, human, and organizational factors and presents new insights for reducing offshore platforms' fire risks.

Inevitably, this method also has some limitations. The occurrence probability of a node may change with time and deeper research, so these general data need to be replaced by process specific data. The input information of the method is expert opinions and historical data. In theory, the evaluation results' accuracy could be improved with the accumulation of data.

REFERENCES

Ade, N. and Peres, S. C. (2022). A review of human reliability assessment methods for proposed application in quantitative risk analysis of offshore industries. International Journal of Industrial Ergonomics 87.

Bird F E, G. G. L. (1966). Damage Control:A New Horizon in Accident Prevention and Cost Improvement. New York, American Management Association.BSEE, http://www.bsee.gov/Inspection-and-Enforcement/Accidents-and-Incidents/Fires-Explosions/.

Chen, S.-T., Wall, A., Davies, P., Yang, Z., Wang, J. and Chou, Y.-H. (2013). A Human and Organisational Factors (HOFs) analysis method for marine casualties using HFACS-Maritime Accidents (HFACS-MA). Safety Science 60: 105–114.

de Maya, B. N., Khalid, H. and Kurt, R. E. (2021). Application of card-sorting approach to classify human factors of past maritime accidents. Maritime Policy & Management 48(1): 75–90.

EMSA Annual Review of Marine Casualties and Incidents 2019. European Maritime Safety Agency.

Fu, G., Cao, J.-L., Zhou, L. and Xiang, Y.-C. (2017). Comparative study of HFACS and the 24Model accident causation models. Petroleum Science 14(3): 570–578.

Fu, G., Wang, J. and Yan, M. (2016). Anatomy of Tianjin Port Fire and Explosion: Process and Causes. Process Safety Progress 35(3): 216–220.

Fu, G., Xie, X., Jia, Q., Li, Z., Chen, P. and Ge, Y. (2020). The development history of accident causation models in the past 100 years: 24Model, a more modern accident causation model. Process Safety and Environmental Protection 134: 47–82.

Fu, G., Xie, X., Jia, Q., Tong, W. and Ge, Y. (2020). Accidents analysis and prevention of coal and gas outburst: Understanding human errors in accidents. Process Safety and Environmental Protection 134: 1–23.

Fu, G., Xie, X. C., Jia, Q. S., Li, Z. H., Chen, P. and Ge, Y. (2020). The development history of accident causation models in the past 100 years: 24Model, a more modern accident causation model. Process Safety and Environmental Protection 134: 47–82.

Fu, G., Zhao, Z., Hao, C. and Wu, Q. (2019). The Accident Path of Coal Mine Gas Explosion Based on 24Model: A Case Study of the Ruizhiyuan Gas Explosion Accident. Processes 7(2).

Ghasemi, F., Sari, M. H. M., Yousefi, V., Falsafi, R. and Tamosaitiene, J. (2018). Project Portfolio Risk Identification and Analysis, Considering Project Risk Interactions and Using Bayesian Networks. Sustainability 10(5).

He, X. and Oyadiji, S. O. (2001). Application of coefficient of variation in reliability-based mechanical design and manufacture. Journal of Materials Processing Technology 119(1-3): 374–378.

Hollnagel, E. (2002). Understanding accidents-from root causes to performance variability. Proceedings of the IEEE 7th Conference on Human Factors and Power Plants,1-1

Hu, C. F., Adivar, M. and Fang, S. C. (2014). Non-L-R Type Fuzzy Parameters in Mathematical Programming Problems. Ieee Transactions on Fuzzy Systems 22 (5): 1062–1073.

IMO (2008). Casualty Investigation Code. London, UK, IMO Publishing.

Jia, Q., Fu, G., Xie, X. and Hu, S. (2022). Method Construction and Evaluation: A More Comprehensive and Reliable Classification of Coal Mine Gas Explosion Causes. Energies 15(22).

Jiang, W. and Han, W. (2018). Analysis of "2.28" KEEPER Chemical Industries Hazardous Chemical Explosion Accident Based on FTA and HFACS. International Journal of Environmental Research and Public Health 15(10).

Kaptan, M., Sarialioglu, S., Ugurlu, O. and Wang, J. (2021). The evolution of the HFACS method used in analysis of marine accidents: A review. International Journal of Industrial Ergonomics 86.

Kurtz, R. S. (2013). Oil Spill Causation and the Deepwater Horizon Spill. Review of Policy Research 30(4): 366–380.

Li, J., Qin, Y. R., Wang, Z. and Xin, Y. L. (2021). How to analyse the injury based on 24Model: a case study of coal mine gas explosion injury. Injury Prevention 27(6): 542–553.

Li, Y., Cheng, Z., Yip, T. L., Fan, X. and Wu, B. (2022). Use of HFACS and Bayesian network for human and organizational factors analysis of ship collision accidents in the Yangtze River. Maritime Policy & Management 49(8): 1169–1183.

Lotsberg, I., Olufsen, O., Solland, G., Dalane, J. I. and Haver, S. (2004). Risk assessment of loss of structural integrity of a floating production platform due to gross errors. Marine Structures 17(7): 551–573.

Lyu, T., Song, W. and Du, K. (2019). Human Factors Analysis of Air Traffic Safety Based on HFACS-BN Model. Applied Sciences-Basel 9(23).

Pan, Y., Zhang, L., Li, Z. and Ding, L. (2020). Improved Fuzzy Bayesian Network-Based Risk Analysis With Interval-Valued Fuzzy Sets and D-S Evidence Theory. Ieee Transactions on Fuzzy Systems 28(9): 2063–2077.

Ping, P., Wang, K., Kong, D. and Chen, G. (2018). Estimating probability of success of escape, evacuation, and rescue (EER) on the offshore platform by integrating Bayesian Network and Fuzzy AHP. Journal of Loss Prevention in the Process Industries 54: 57–68.

Reason, J. Human Error. Cambridge University Press, New York, USA, ISBN0521314194.

Robb, M. and Miller, G. (2012). Human factors engineering in oil and gas - a review of industry guidance. Work-a Journal of Prevention Assessment & Rehabilitation 41: 752–762.

Roed, W. (2018). Unforeseen events with a major accident potential-a study of some examples from the Norwegian oil and gas industry. 28th Annual International European Safety and Reliability Conference (ESREL), Trondheim, NORWAY,1841–1846

Scandpower (2013). QRA for N500 Accommodation Unit.

Shi, X., Zhuang, H. and Xu, D. (2021). Structured survey of human factor-related maritime accident research. Ocean Engineering 237.

Skogdalen, J. E. and Vinnem, J. E. (2011). Quantitative risk analysis Offshore-Human and organizational factors. Reliability Engineering & System Safety 96(4): 468–479.

Steijn, W. M. P., Van Kampen, J. N., Van der Beek, D., Groeneweg, J. and Van Gelder, P. H. A. J. M. (2020). An integration of human factors into quantitative risk analysis using Bayesian Belief Networks towards developing a 'QRA plus '. Safety Science 122.

Suo, X., Fu, G., Wang, C. X. and Jia, Q. S. (2017). AN APPLICATION OF 24MODEL TO ANALYSE CAPSIZING OF THE EASTERN STAR FERRY. Polish Maritime Research 24: 116–122.

Torres, L., Yadav, O. P. and Khan, E. (2016). A review on risk assessment techniques for hydraulic fracturing water and produced water management implemented in onshore unconventional oil and gas production. Science of the Total Environment 539: 478–493.

Toth-Laufer, E., Rudas, I. J. and Takacs, M. (2014). Operator Dependent Variations of the Mamdani-Type Inference System Model to Reduce the Computational Needs in Real-Time Evaluation. International Journal of Fuzzy Systems 16(1): 57–72.

Ugurlu, O., Yildiz, S., Loughney, S., Wang, J., Kuntchulia, S. and Sharabidze, I. (2020). Analyzing Collision, Grounding, and Sinking Accidents Occurring in the Black Sea Utilizing HFACS and Bayesian Networks. Risk Analysis 40(12): 2610–2638.

Vederhus, L., Odegard, A., Nistad, S. and Havold, J. I. (2018). Perceptions of demanding work in maritime operations. Safety Science 110: 72–82.

Veritas, D. N. (2007). Accident statistics for Fixed Offshore Units on the UK Continental Shelf 1980-2005. Statistics Reports. H. Statistics Reports. Norway. Norway, HSE.

Vinnem, J. E. (2011). Evaluation of offshore emergency preparedness in view of rare accidents. Safety Science 49(2): 178–191.

Vojkovic, L., Kuzmanic Skelin, A., Mohovic, D. and Zec, D. (2021). The Development of a Bayesian Network Framework with Model Validation for Maritime Accident Risk Factor Assessment. Applied Sciences-Basel 11(22).

Wang, Y. (2009). Fuzzy Inferences Methodologies for Cognitive Informatics and Computational Intelligence. 8th IEEE International Conference on Cognitive Informatics (ICCI 2009), Hong Kong Polytechn Univ, Hong Kong, PEOPLES R CHINA,241–248

Wang, Y. F., Li, Y. L., Zhang, B., Yan, P. N. and Zhang, L. (2015). Quantitative Risk Analysis of Offshore Fire and Explosion Based on the Analysis of Human and Organizational Factors. Mathematical Problems in Engineering 2015.

Wang, Y. F., Qin, T., Li, B., Sun, X. F. and Li, Y. L. (2017). Fire probability prediction of offshore platform based on Dynamic Bayesian Network. Ocean Engineering 145: 112–123.

Wang, Y. F., Xie, M., Ng, K. M. and Habibullah, M. S. (2011). Probability analysis of offshore fire by incorporating human and organizational factor. Ocean Engineering 38(17-18): 2042–2055.

Wiegmann, D. A. and Shappell, S. A. (2001). Human error analysis of commercial aviation accidents: Application of the human factors analysis and classification system (HFACS). Aviation Space and Environmental Medicine 72(11): 1006–1016.

Wrobel, K. (2021). Searching for the origins of the myth: 80% human error impact on maritime safety. Reliability Engineering & System Safety 216.

Wu, Y., Fu, G., Wu, Z., Wang, Y., Xie, X., Han, M. and Lyu, Q. (2023). A popular systemic accident model in China: Theory and applications of 24Model. Safety Science 159.

Xie, X., Shen, S., Fu, G., Shu, X., Hu, J., Jia, Q. and Shi, Z. (2022). Accident case data-accident cause model hybrid-driven coal and gas outburst accident analysis: Evidence from 84 accidents in China during 2008-2018. Process Safety and Environmental Protection 164: 67–90.

Yazdi, M. and Kabir, S. (2020). Fuzzy evidence theory and Bayesian networks for process systems risk analysis. Human and Ecological Risk Assessment 26(1): 57–86.

Zhang, H., Zhou, Z., Chen, M. and Yi, D. (2021). Fire Risk Quantitative Assessment of Offshore Supported Platform. Journal of Marine Science and Application 20(2): 381–392.

Zhang, J., Fu, J., Hao, H., Fu, G., Nie, F. and Zhang, W. (2020). Root causes of coal mine accidents: Characteristics of safety culture deficiencies based on accident statistics. Process Safety and Environmental Protection 136: 78–91.

Zhang, J., Fu, J., Hao, H., Fu, G., Nie, F. and Zhang, W. (2020). Root causes of coal mine accidents: Characteristics of safety culture deficiencies based on accident statistics. Process Safety and Environmental Protection 136: 78–91.

Advances in Maritime Technology and Engineering – Guedes Soares & Santos (Eds)

Establishing safety zones for ammonia bunkering operations: A quantitative risk assessment

I. Hwang, P. Zhou, B. Jeong, H. Wang, H. Jang & M.P. Palliparambil
Department of Naval Architecture, Ocean and Marine Engineering, University of Strathclyde, Glasgow, Scotland, UK

ABSTRACT: To respond to climate change, the maritime sector is actively investigating the use of alternative fuels. Ammonia is recognized as a promising option due to its potential for sustainable energy. Nonetheless, the deployment of ammonia as a marine fuel must contend with significant safety hazards, particularly its high toxicity. This research evaluates the consequences of potential ammonia spills during bunkering operations and determines appropriate safety zones. Employing a holistic risk assessment methodology, the study examines various environmental conditions, the duration of ammonia emissions, and the establishment of specific bunkering accident scenarios. It conducts a thorough frequency analysis of such incidents and employs accident impact analysis modelling. The findings of this research are of critical importance to vessels utilizing ammonia as a fuel source and extend to safety management professionals within the maritime sector. The insights gleaned here are poised to enhance ammonia bunkering activities' readiness and response measures.

1 INTRODUCTION

Ammonia represents a carbon-neutral fuel with considerable traction in the maritime domain, eliciting substantial interest as a substitute for fossil fuels. Its ability to maintain a liquid state at temperatures below -33°C under atmospheric pressure significantly facilitates onboard storage and utilization, thereby offering operational efficiencies for marine vessels. Moreover, the favorable economic considerations associated with ammonia, relative to other zero-carbon alternatives, incentivize decision-makers to explore its applications within the marine industry. However, the adoption of ammonia as a marine fuel is not without its challenges, encompassing technical, safety, and regulatory dimensions (Lin et al. 2023). Of particular concern are the toxic and corrosive properties inherent to ammonia, which pose significant hazards during handling and potential leakage scenarios. Notably, the toxicity of ammonia presents acute risks to human health and safety, with the potential for fatal consequences even upon minimal exposure (Bureau Veritas Marine & Offshore 2022). Addressing these challenges necessitates comprehensive safety protocols and regulatory frameworks, which, while currently evolving, remain nascent within the maritime regulatory landscape. Bunkering, the pivotal process of fuel supply to ships, assumes critical importance in maritime operations, requiring meticulous attention to ensure seamless vessel functionality. While existing research predominantly focuses on the storage, utilization, and onboard handling of ammonia, safety considerations pertaining to bunkering operations warrant commensurate scrutiny. Bunkering operations entail multifaceted interactions involving diverse stakeholders, including external fuel suppliers, port authorities, and regulatory agencies, occurring concurrently with other cargo operations or crew embarkation/disembarkation processes. Given the propensity for bunkering incidents to yield far-reaching consequences, particularly within port environments, the potential ramifications of ammonia leaks extend beyond onboard personnel to encompass broader externalities. Consequently, research efforts should encompass a comprehensive assessment of the scope and mitigation strategies for ammonia-related incidents during bunkering operations, reflecting a proactive approach towards enhancing maritime safety standards and regulatory frameworks (Duong et al. 2023).

2 BACKGROUND

The existing regulatory landscape regarding Ammonia bunkering lacks definitive guidelines, with pertinent information dispersed within the bunkering sections of the Guidelines on Ammonia Fuels issued by respective Classification societies. However, these references offer scant elaboration, typically comprising only one or two pages, and largely mirror provisions outlined in the International Code of Safety for Ship Using Gases or Other Low-flashpoint Fuels (IGF Code) (Organization and Morska 2016), which governs liquefied natural gas (LNG) operations. While both LNG and ammonia

DOI: 10.1201/9781003508762-35

share characteristics as low-flashpoint fuels, it is imperative to distinguish between the two, as the primary risks associated with LNG—namely, fire and explosion—are mitigated by its non-toxic nature. Conversely, ammonia poses significant toxicity risks, necessitating a nuanced approach to risk assessment and management to mitigate potential human casualties stemming from leaks. The maritime sector stands at a pivotal juncture, marked by evolving paradigms and a burgeoning need for robust risk assessment methodologies. Qualitative risk assessment methodologies heavily rely on the collective expertise of subject matter experts, necessitating the selection of individuals possessing practical experience and comprehensive understanding of relevant systems (Cheliotis et al. 2021). However, the nascent adoption of ammonia as a marine fuel translates into a dearth of experienced users or experts, complicating efforts to conduct accurate risk assessments specific to ammonia-fueled systems. Consequently, the reliance on existing systems for risk evaluation introduces inherent limitations in comprehensively identifying and analyzing potential risks inherent to novel concepts (Trivyza et al. 2021). Quantitative risk assessment offers an alternative avenue for safety evaluation, leveraging data-driven calculations or simulations to delineate areas overlooked by qualitative assessments. However, the scarcity of operational data pertaining to ammonia-fueled vessels impedes the establishment of comprehensive databases essential for reliable quantitative analyses. Inadequate fundamental data reliability engenders skepticism surrounding the accuracy and robustness of calculations or simulations predicated on such datasets. In light of these constraints, the scope of accident analysis is inherently circumscribed. Despite employing both qualitative and quantitative risk assessment methodologies to identify plausible risks, inherent limitations persist. While efforts to mitigate accident causation or probability remain pivotal, the efficacy of such measures is inherently constrained. Consequently, emphasis shifts towards researching strategies to mitigate the impact post-accident occurrence. This study endeavors to identify factors capable of mitigating the impact and scope of ammonia bunkering fuel spill accidents. By analyzing variations in parameters influencing accident outcomes, pivotal factors with maximal impact are discerned, facilitating the presentation of control measures aimed at minimizing post-accident repercussions to the greatest extent possible.

Figure 1 presents the methodology of this study endeavors to identify the paramount parameters governing the severity and consequences of ammonia bunkering spills through a parametric analysis approach. To this end, a systematic examination of factors influencing spill outcomes was conducted, involving iterative adjustments to individual parameters to discern their relative impacts. Initially, identify pertinent parameters via system identification. These parameters are identified as variables anticipated to influence the probability or consequences of an accident. A parametric analysis was employed to ascertain the impact of variations in these parameters on the outcomes. The outcomes are articulated through both Consequence Analysis and Frequency Analysis. Presenting the results in a trend format enables the identification of parameters that have a significant influence on both the consequence and frequency of incidents. By constraining or managing the factors deemed most critical, the most effective strategies can be implemented to diminish the likelihood of an accident or lessen its repercussions. The DNV Phast software served as the primary analytical tool for assessing accident repercussions. Leveraging this software, Consequence and Frequency analyses were conducted to visualize the spatial distribution of spill consequences and the probability-severity spectrum of accidents, respectively. Consequence analysis delineated the spatial extent of spill impacts, considering factors such as the potential for ammonia-induced fire and explosion, as well as toxic effects at concentrations exceeding 30 ppm. Meanwhile, Frequency analysis categorized the likelihood of accidents based on their severity into distinct color-coded ranges, facilitating a comprehensive understanding of accident risk spatially. Notably, while exacerbating individual factors in worst-case scenarios may attenuate accident likelihood, it concurrently amplifies the spatial reach and severity of resultant consequences.

3 CASE STUDY

3.1 *Case ship and system*

This case study forms an integral component of the NH3CRAFT project, an initiative aimed at showcasing ammonia-fueled ships, funded by the European Union. The project entails the utilization of 770 m3 of ammonia as onboard fuel, replenished with 750 m3 of ammonia every 5 days. While deliberations concerning propulsion systems are ongoing, the focus remains on systems utilizing liquid ammonia. The scenario of fuel leakage during bunkering was delineated, envisioning liquid ammonia seepage from various components—namely pipelines, valves, and pumps—during the transfer process from the supply-side facility's storage tank to the target vessel's storage tank. This study postulates that fuel leakage ensues from punctures occurring independently in each element, with three distinct hole sizes (5 mm, 10 mm, and 100 mm) considered, capable of manifesting singularly or of Hamburg.tif of Rotterdam.tif.

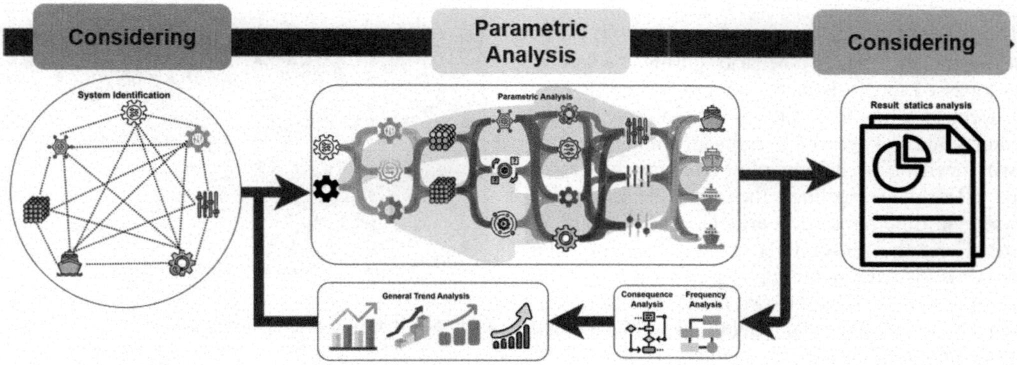

Figure 1. Methodology.

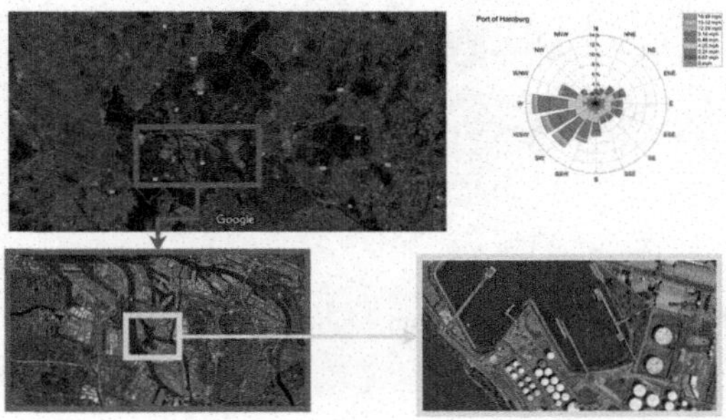

Figure 2. Port of Hamburg (Maps 2023a, meteoblue 2023a).

3.2 *Potential bunkering location*

Two potential ammonia supply sites have been identified: the Port of Hamburg and the Port of Rotterdam, anticipated to emerge as significant ammonia provisioning centers for vessels throughout Europe in the near future. These ports are currently in the process of constructing pertinent facilities in accordance with their respective development plans, encompassing all requisite aspects for ammonia fuel utilization, including transportation, storage, and supply infrastructure (Pathways 2023).

While existing literature indicates the absence of operational ammonia bunkering facilities, several ongoing ammonia bunkering projects are underway worldwide. Notably, Mabanaft and Hapag-Lloyd are actively engaged in ammonia bunkering initiatives in and around the Port of Hamburg, Germany in Figure 2. This endeavor is evidenced by their collaboration and the signing of a memorandum of understanding (MoU) aimed at evaluating ammonia supply options. The proposed import terminal, slated to commence hydrogen provision to Germany by 2026, is set to be established at Mabanaft's existing Blumensand port terminal (Pathways 2023).

Similarly as presented in Figure 3, Cepsa and ACE Terminal have entered into an MoU to facilitate the transoceanic transport of green ammonia, its redistribution to hinterland end markets, and its conversion into green hydrogen for utilization by end customers in Northwest Europe. ACE Terminal's strategic location within the Port of Rotterdam offers direct connectivity to the region's industrial base and the planned national hydrogen network, presenting a robust logistical infrastructure poised for integration into Northwest Europe's broader energy landscape by 2027 []. The comparison of two port locations serves as a means to discern differential environmental considerations. By scrutinizing accident outcomes under varying conditions, such as annual wind speed and direction, and diverse population densities, insights into the nuanced impact of environmental factors on accident scenarios can be gleaned.

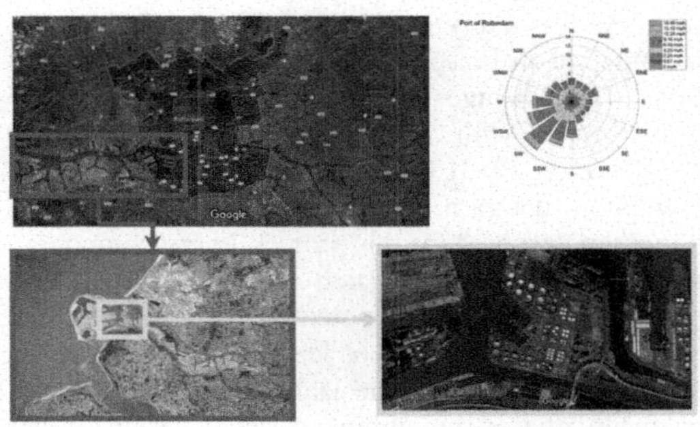

Figure 3. Port of Rotterdam, (Maps 2023b, meteoblue 2023b).

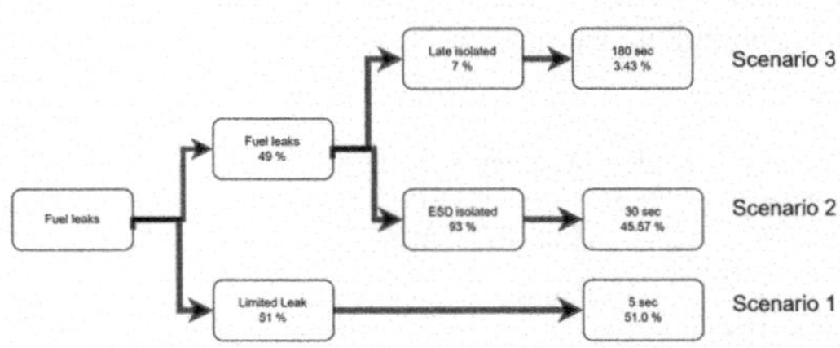

Figure 4. Accidental scenarios,(DNV 2013).

3.3 *Leakage scenarios*

The ammonia leakage scenario during bunkering encompassed three distinct situations in Figure 4. In the event of a fuel leak, the sensor promptly detects the ammonia leakage and triggers an automatic shutdown mechanism, halting all bunkering processes and effectively sealing the fuel leak within a 5-second timeframe (Scenario 1). Alternatively, in cases where the sensor fails to detect an ammonia leak but is visually identified by an operator or manager, an Emergency Shut Down (ESD) device is activated to cease bunkering operations. This manual intervention requires approximately 30 seconds to effectively mitigate the ammonia leak (Scenario 2). However, in instances where the ESD device malfunctions, necessitating manual intervention, the operator must manually cease pump operations and close all valves. This manual shutdown process is more time-consuming, taking up to 180 seconds to complete, and represents the worst-case scenario (Scenario 3). The probability of the above scenarios was determined by referring to DNV's Failure frequency analysis (DNV 2013), and the researcher estimated the time for each scenario. Furthermore, the analysis considers the occurrence of fuel leakage through three different hole sizes in the event of an accident. These hole sizes—5 mm, 10 mm, and 100 mm—possess independent probabilities of occurrence. Fuel leakage may ensue from a single 5 mm hole, or alternatively, leakage may manifest from three holes occurring simultaneously. The quantity of fuel leaked varies based on the hole size, with corresponding assessments conducted to ascertain the extent of influence in the event of an accident. As this study is a part of the project, there is a limitation in presenting all datas to open publication. The leakage frequency was calculated by a researcher with failure freqnency guidance by DNV (DNV 2013) based on the piping and instrumentation diagram (PID) designed for ammonia operation designed for this project. Every component in the piping diagram, such as pipes, valves, and pumps, is considered a potential part of a leakage accident. The frequency of leakage per each size is shown in Table 1. As the leakage frequency is not calculated based on ammonia use, there is doubt about the frequency of ammonia leakage. The shortage of reliable data for ammonia leakage will remain until sufficient operation data is collected. For this reason, this study focuses on the mitigation of the consequences of accidents, not reducing accident frequency.

Table 1. Leakae frequency per hole.

	Hole size		
	5 mm	10 mm	100 mm
Frequency	8.25E-04	3.36E-04	9.99E-05

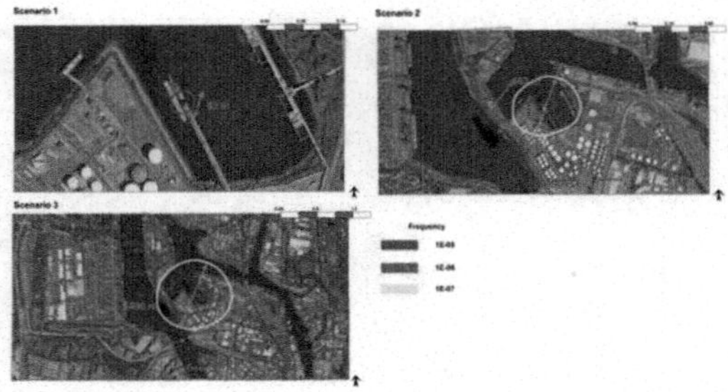

Figure 5. Port of Hamburg, low frequency.

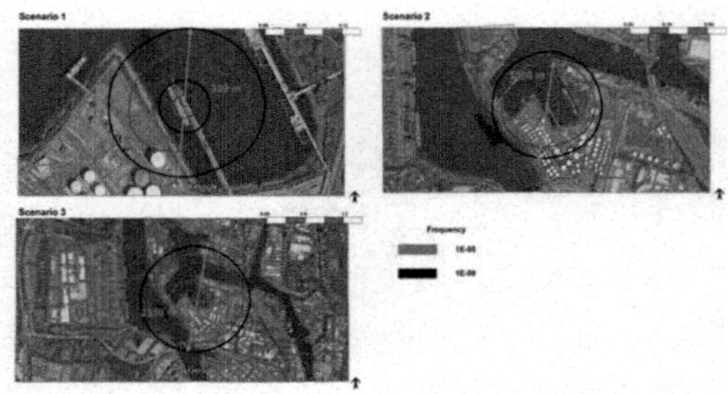

Figure 6. Port of Rotterdam, high frequency.

3.4 *Simulation results*

The simulation results are derived from both Frequency analysis and Consequence analysis, each represented by different colored lines delineating the frequency and severity of potential accidents across various scenarios. Instances of higher frequency typically correspond to less severe accidents, while a confluence of diverse conditions suggests the likelihood of more severe accidents occurring within shorter intervals. Generally, accident cycles ranging from 1e-05 to 1e-07 are deemed probable considering the operational volume of ships, with cycles exceeding 1e-08 deemed increasingly likely. Accidents occurring at lower frequencies, below 1e-08, are considered sufficiently rare as to be improbable during routine ship or port operations. However, the unpractical frequencies considered in this research can differ from the determination of port authority. Each port sets its own acceptable accidental frequencies for setting the bunkering safety zone. As it is completely up to the decision of the port authority, there is a limitation to setting a realistic acceptable accidental frequency by the researcher.

Consequence analysis is depicted as a range delineated on the map, with each accident cycle associated with a distinct impact range reflective of potential repercussions. These impact ranges encompass the spatial extent of potential fire, explosion, and toxicity resulting from fuel spills, with individuals within this range typically deemed at risk of

Table 2. Simulation results.

	Port of Hamburg			Port of Rotterdam		
Accident Frequency	Scenario 1	Scenario 2	Scenario 3	Scenario 1	Scenario 2	Scenario 3
1E-05	14m	-	-	10m	-	-
1E-06	50m	295m	1060m	51m	349m	1034m
1E-07	52m	563m	1490m	51m	610m	1567m
1E-08	265m	842m	1955m	253m	904m	2022m
1E-09	330m	1050m	2350m	345m	1070m	2360m

fatality. Designated as safety zones, these areas should restrict access solely to essential personnel during bunkering operations to mitigate the risk of fatalities, with non-essential activities suspended to ensure safety.

3.4.1 *Port of Hamburg*

Upon examining the simulation outcomes concerning the Port of Hamburg in Figure 5, it becomes evident that in scenario 1, characterized by a minor fuel spill, the impact range spans approximately 50 meters. However, should the fuel leakage persist unabated, the influence area escalates significantly. Under scenario 2, the impact area exceeds 550 meters, ultimately extending to approximately 1,500 meters.

While accidents at lower cycles are generally deemed improbable, this study accounts for worst-case scenarios from a safety standpoint as shown in Figure 6, necessitating the presentation of simulation results even for lower cycles. Consequently, such scenarios entail considerable severity. In scenario 1, the impact area expands to 330 meters, roughly six times larger, while scenario 3 manifests potential ramifications over a considerably broader expanse, spanning up to 2,300 meters. Although the likelihood of such accidents is deemed low, their potential consequences are deemed profoundly serious.

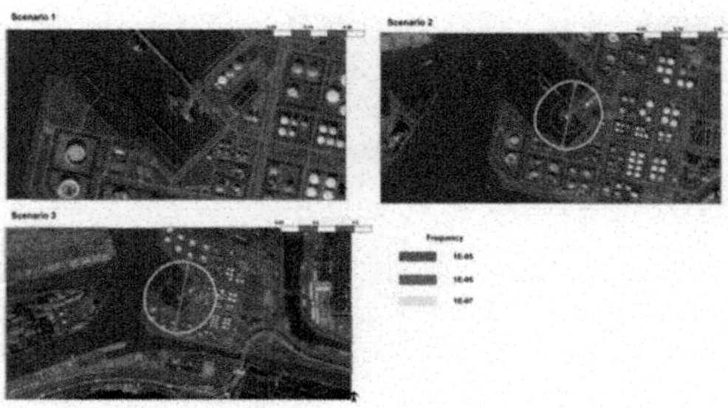

Figure 7. Port of Rotterdam, high frequency.

3.4.2 *Port of Rotterdam*

The simulation outcomes at the Port of Rotterdam exhibit a comparable spectrum of accident consequences to those observed at the Port of Hamburg. Despite variations in annual wind speeds and directions between the two ports, the Port of Rotterdam demonstrates a generally broader accident impact area, albeit without significant disparities.

As shown in Figure 7, in scenario 1, the swift activation of the sensor resulting in an initial fuel cut manifests a relatively modest accident impact range of 51 meters. However, under scenario 2, this range expands to 610 meters, and further escalates to over 1500 meters under scenario 3, surpassing the extent observed at the Port of Hamburg.

Even in scenarios involving lower frequency accidents as presented in Figure 8, the Port of Rotterdam displays a wider range of accident impacts compared to the Port of Hamburg. Scenario 1 encompasses an impact range exceeding 340 meters, while scenario 2 yields an accident impact area exceeding 1,000 meters, culminating in a notably expansive range of 2,300 meters.

Table 2 presents the organization of the outcomes. The variation in port conditions, encompassing factors like wind direction and velocity, does not significantly influence the findings. Significant disparities in the outcomes are delineated, contingent upon the scenarios outlined previously.

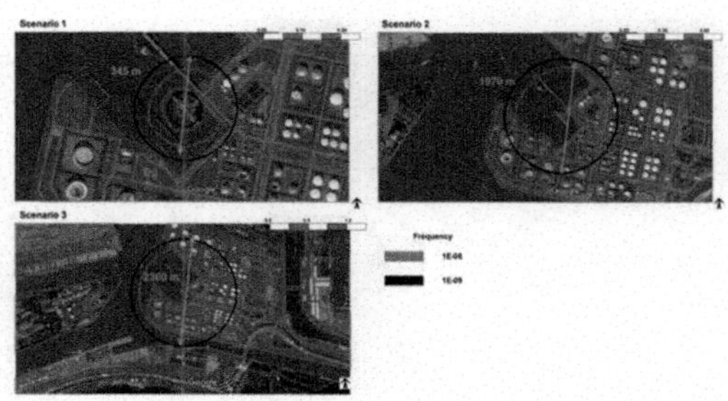

Figure 8. Port of Rotterdam, low frequency.

4 CONCLUSIONS

The simulation yields several significant findings. While the outcome of accidents is influenced by prevailing wind speed and direction, their impact is not notably dramatic. The paramount determinant of accident severity is the quantity of leaked fuel and the duration of the leakage. Consequently, promptly shutting off the fuel emerges as the most effective strategy for mitigating accident severity.

In light of these findings, two safety recommendations are proposed for consideration during the design phase. Firstly, enhancing the precision of sensors is paramount. Rapid sensor-triggered shut-offs result in accident ranges of approximately 560 meters, underscoring the efficacy of swift sensor-based responses. However, detecting ammonia leaks amid open bunkering spaces poses challenges, as ammonia disperses rapidly in open air. Despite this, given ammonia's heightened toxicity—even minute concentrations as low as 25 ppm pose significant health risks—additional sensors capable of detecting sudden changes in pressure or flow within the ammonia supply line could be instrumental in automatic shut-off mechanisms.

Secondly, a paradigm shift in bunkering practices is advocated. Historically, the gravest accidents during bunkering involved fossil fuel leaks leading to fires or explosions, necessitating reliance on natural wind to disperse fuel gas concentrations. However, in the case of ammonia, toxicity outweighs fire or explosion risks. The simulation, which assumed direct ammonia leakage into the atmosphere without safety barriers, underscores the potential for extensive fatalities across broad ranges. Implementing secondary barriers to prevent direct ammonia leakage into the atmosphere, such as enclosing the bunker station where supply lines converge, coupled with heightened ammonia concentration detection capabilities, presents a viable strategy to enhance precision and avert catastrophic outcomes.

REFERENCES

Association, A. E. (2023, Feb). Cepsa: Renewable ammonia in spain.

Bureau Veritas Marine & Offshore (2022, Jun). Bureau veritas completes new study on ammonia fuel. https://marine-offshore.bureauveritas.com/newsroom/bureau-veritas-completes-new-study-ammonia-fuel.

Cheliotis, M., E. Boulougouris, N. L. Trivyza, G. Theotokatos, G. Livanos, G. Mantalos, A. Stubos, E. Stamatakis, & A. Venetsanos (2021). Review on the safe use of ammonia fuel cells in the maritime industry. *Energies 14*(11).

DNV (2013, Mar). *Failure Frequency Guidance*.

Duong, P. A., B. R. Ryu, M. K. Song, H. V. Nguyen, D. Nam, & H. Kang (2023). Safety assessment of the ammonia bunkering process in the maritime sector: A review. *Energies 16*(10).

Lin, C.-Y., P.-C. Wu, & H. Yang (2023). Alternative gaseous fuels for marine vessels towards zero-carbon emissions. *Gases 3*(4), 158–164.

Maps, G. (2023a, Oct). Port of hamburg. Google Maps.

Maps, G. (2023b, Nov). port of rotterdam. Google Maps.

meteoblue (2023a, Oct). Weather hamburg. https://www.meteoblue.com/en/weather/historyclimate/climatemodelled/hamburg_*germany*_2911298.

meteoblue (2023b, Nov). Weather rotterdam. https://www.meteoblue.com/en/weather/historyclimate/climatemodelled/rotterdam_*netherlands*_2747891.

Organization, I. M. & M. O. Morska (2016). *IGF Code: International Code of Safety for Ships Using Gases Or Low Flashpoint Fuels*. IMO publication. International Maritime Organization.

Pathways, G. (2023, Jan). Mabanaft, hapag-lloyd plan ammonia bunkering projects. https://gaspathways.com/mabanaft-hapag-lloyd-plan-ammonia-bunkering-projects-1674.

Trivyza, N. L., M. Cheliotis, E. Boulougouris, & G. Theotokatos (2021). Safety and reliability analysis of an ammonia-powered fuel-cell system. *Safety 7*(4).

Advances in Maritime Technology and Engineering – Guedes Soares & Santos (Eds)
© 2024 The Author(s), ISBN 978-1-032-83099-5

A data-driven Bayesian Network for risk modeling and causal analysis of global maritime accidents

H.Y. Jiang, J.F. Zhang & C.P. Wan
State Key Laboratory of Maritime Technology and Safety (Wuhan University of Technology), Wuhan, P.R. China
National Engineering Research Center for Water Transport Safety, Wuhan, P.R. China
Intelligent Transportation Systems Research Center, Wuhan, P.R. China
Wuhan University of Technology, Wuhan, P.R.China

M.Y. Zhang
Department of Mechanical Engineering, School of Engineering, Aalto University, Espoo, Finland

C. Guedes Soares
Centre for Marine Technology and Engineering (CENTEC), Instituto Superior Técnico, Universidade de Lisboa, Lisbon, Portugal

ABSTRACT: Multiple factors may cause maritime accidents. Investigating the impact of risk factors on maritime accidents is imperative. This paper employs a data-driven Bayesian network approach to explore the impact of risk factors on maritime safety using a large dataset of maritime accidents. The interdependencies among risk influencing factors are modeled using a Tree Augmented Network, followed by the sensitivity analysis and model validation. The results indicate that the key risk influencing factors influencing maritime accidents mainly include ship location, type, age, gross tonnage, and deadweight tonnage. This study contributes to the prevention of specific types of maritime accidents.

1 INTRODUCTION

Most maritime accidents exhibit the characteristic of low probability-high consequences, in which catastrophic maritime disasters can result in significant loss of life, and substantial loss of property, instigate societal panic, and disruption of marine ecosystems. Relevant studies in existing literature indicate that maritime transportation safety is influenced by a combination of multiple factors, including ship conditions, environmental conditions, human errors, management issues, and dynamic factors that change over time.

Analyzing maritime accidents has become an effective approach to mitigating risks in maritime transportation.it is essential to collect and investigate the most recent data on maritime accidents, analyze the causes of accidents, identify the key Risk Influencing Factors (RIFs) in different scenarios, and subsequently predict the associated risks. Thus, maritime regulations can review and revise regulations, standards, and management practices.

2 LITERATURE REVIEW

2.1 *Maritime risk analysis research*

Maritime risk research involves both qualitative and quantitative assessments. The former usually provides detailed observational insights through the analysis of small samples and commonly used qualitative assessment methods include the Functional Resonance Analysis Method (FRAM), Root Cause Analysis (RCA), and risk rating scales. In particular, qualitative methods such as Human Factor Analysis and Classification System (HFACS) and Accident Analysis Map (AcciMap) are used to analyze the human factors in Marine accidents.

To list some, Kristiansen et al. (1999) have developed in the European project CASMET a method for marine casualty analysis accounting for human and organisational factors, which was later adopted by the European Maririme Safety Agency. Chen et al. (2013) developed a HOFs analysis method for marine casualties using HFACS for Maritime Accidents (HFACS-MA), and used a case to demonstrate the proposed method and illustrated several advantages of the framework in accident analysis. Salihoglu & Besikçi (2021) conducted a qualitative assessment of operational risks in shipping using the FRAM. Ma et al. (2022) integrated HFACS, DEMATEL (Decision-Making Trial and Evaluation Laboratory), and FCM (Fuzzy Cognitive Map) to measure human factors that cause maritime accidents. However, qualitative assessment methods are unable to quantify risks, leading to questions about their reliability and effectiveness.

DOI: 10.1201/9781003508762-36

To solve these problems, Quantitative Risk Assessment (QRA) methodologies have been proposed to measure the causal relationship between maritime disasters and relevant influencing factors. These mainly include Fuzzy Logic (FL), Evidence Reasoning (ER), Event Tree Analysis (ETA), Fault Tree Analysis (FTA), and Bayesian networks (BN), among others.

Sur & Kim (2020) used a fuzzy comprehensive evaluation method to evaluate comprehensive risk and risk level of fishing ship accidents for nine accident types from 1988 to 2016. Weng & Yang (2015) developed a binary logistic regression model and a zero-truncated binomial regression model to predict the probability of fatal shipping accidents and corresponding mortalities. Wang et al. (2016) used the zero-inflated negative binomial regression model, Pearson correlation analysis, and ZINB regression model to acquire and evaluate the influencing factors of casualties in maritime accidents, respectively. Xu and Liu (2013) used the FTA method to establish a fault tree model of water traffic accidents in the Yangtze River. The results indicate that fatigue driving, poor safety awareness, poor driving skills, heavy weather, and inadequate outlook are significant causes. Yu et al. (2022) proposed an integrated multi-criteria framework employing an ER approach to calculate collision risk in the coastal waters of mainland Portugal. Comparative analysis were conducted against linear regression and fuzzy regression models to evaluate the rationality of the proposed framework.

Among all the QRA methods that have been applied to maritime transportation, BN has an advantage in terms of powerful modelling capabilities, especially in handling data tolerance and bidirectional risk diagnosis and predictive analysis. Compared to methods like FL and ER, BN demonstrates an advantage in modelling the causal relationships between influencing factors. In contrast to FTA, BN can handle multiple state variables and multiple outputs. Moreover, in comparison to other QRA methods, BN can jointly model Human and Organizational Factors (HOFs) along with other RIFs. These advantages make BN a suitable method for maritime accident modelling.

2.2 *BN in maritime risk analysis*

The advantages of BN in risk modelling have led to extensive research and application in the field of maritime safety. Li et al. (2023) used BN and integrated maritime accident data to quantify the influence of 23 critical RIFs on different types of maritime accidents. The six RIFs are ship type, ship operation, voyage segment, deadweight, length, and power. Fan et al. (2020) proposed an advanced method of human factor analysis and maritime accident prevention by incorporating BN and Technique for Order of Preference by Similarity to Ideal Solution (TOPSIS) into the Multi Criteria Decision Making (MCDM) system. Wu et al. (2020) utilized BNs to model the consequences of collisions in the downstream of the Yangtze River. The introduced consequence estimation model provided a detailed description of the development of collision accidents with a high level of accuracy in the estimation results. Antão & Guedes Soares (2019) developed a Bayesian belief network model to assess the impact of human error in maritime accidents under different weather conditions. The results indicate a significantly high-risk acceptance level among fishing vessel crews, while recreational boaters exhibit a noticeable low-risk perception. Zhang et al. (2018) proposed a quantitative maritime risk assessment method utilizing Bayesian rules and the least squares estimation approach to identify the dominant factors leading to collision accidents. The results indicate that small vessels with a length less than 100 m have a significantly higher collision risk compared to larger vessels. The collision probability for vessels without pilots is approximately nine times higher than that for vessels with pilots. Antão et al. (2023) integrating Bayesian rule with the least-squares method, utilized a sample of 936 global collision events from the period 2005 to 2017 to assess the impact of risk factors on collision probability. The findings indicate that the prominent risk-influencing factors for ship collisions are ship type and geographical region.

When applying BN to maritime accident analysis, the first step is to establish the BN structure. In general, literature reviews, expert knowledge, and data learning are the primary methods for BN structure learning. Expert knowledge serves as a crucial source of data for maritime accident modelling when important data from relevant investigations is unavailable or incomplete (Zhang et al. 2016). Silveira et al. (2021) introduced a judgmental expert-based method for assessing collision risk in shipping using the ELECTRE Tri-nC multi criteria outranking method. The results were compared with expert assessments of collision risk in same scenarios. The model is applicable for real-time collision risk assessment and collision avoidance in both vessels and VTS (Vessel Traffic Service) centers. However, expert knowledge is considered subjective and uncertain (Zhang & Thai 2016).

To avoid subjectivity and incompleteness introduced by subjective data in BN modelling, many scholars have developed data-driven BNs for maritime risk analysis, such as Naive Bayesian Networks (NBN), Augmented Naive Bayesian Networks (ABN), the K2 algorithm, and Tree Augmented Naive Bayes (TAN). Fan et al. (2020) applied NBN to construct a network model that captured interdependencies between RIFs and quantified the influence of different factors on various types of maritime accidents. However, the NBN model assumes that the influencing factors are independent, which does not align with the actual scenarios of maritime accidents. Wang & Yang (2018), in a comparative study with the NBN model, introduced the ABN model to analyze the key RIFs affecting the severity of channel accidents. Nonetheless, the network structure is relatively complex and requires manual adjustments to address unreasonable causal relationships among RIFs.

Friedman et al. (1997) pointed out the superiority of TAN over NBN. TAN not only maintains the robustness and computational simplicity of NBN but also enhances the accuracy of results. Fan et al. (2020) constructed a risk-based TAN model to analyze RIFs containing human factors in maritime accidents and conducted a sensitivity analysis, scenario analysis, and Most Probable Explanations (MPE). The key RIFs affecting maritime accident types are the ship age, ship operation, voyage segment, information, and ship condition. Therefore, this study employs the TAN method to construct a maritime accident BN model, enhancing risk prediction accuracy.

Data-driven BNs require a substantial amount of historical accident data. In this paper, we used an extensive commercial accident database with a wide time span and global accident coverage. The abundance of accident cases addressed the limitations in previous research regarding the quantity and regional diversity of accident data. The sea area and flag of convenience were added to the model as new RIF, and the BN structure was established by TAN method. Sensitivity analysis and model verification were carried out to study the joint influence of 6 kinds of RIFs on the types of maritime accidents. RIF analysis was conducted according to the results, and some enlightenment was obtained.

3 DATA COLLECTION AND PROCESSING

The data in this paper was obtained from Lloyd's (https://lloydslist.maritimeintelligence.informa.com/sectors/casualty), spanning a period of twenty years from 2002 to 2022, and encompassing 55,469 maritime accident cases worldwide.

Seven types of accident information are selected from the dataset for modeling. However, some attributes such as accident location and ship type had an excessive number of states, which required reclassification before modeling. The reclassified RIFs information is presented in Table 1.

4 ACCIDENT MODELING AND RESULTS ANALYSIS

4.1 *Tan*

The BN structure learning aims to specify the dependencies between variables to construct a Directed Acyclic Graph (DAG) mechanism. In this study, the maritime accident dataset consists of 55,469 records, making it suitable for data-driven risk analysis. The TAN method is adopted here to develop the BN structure. It eliminates the attribute independence assumption of NBN to accommodate dependencies between attributes.

Table 1. RIFs from historical accident data.

RIFs	States
Accident type	Hull damage, Machinery damage, Grounding, Fire/Explosion, Contact, Foundered, Collision, Other
Location	North Atlantic zone, Mediterranean zone, Indian Ocean zone, Pacific zone, Polar zone, South Atlantic zone, Unknown zone
Ship type	Oil and chemical tankers, Personnel and passenger ships, Cargo ships, Fishery ships, Engineering and maintenance ships, Specialized cargo ships and others
Gross tonnage (GT)	(0, 3000], (3000, 10000], (10000, 30000], >30000
Deadweight tonnage (DWT)	(0, 3000], (3000, 10000], (10000, 30000], >30000
Age	(0, 5], [6, 10], [11, 15], [16, 20], >20, Unknown
Flag	Non-convenient, Convenient

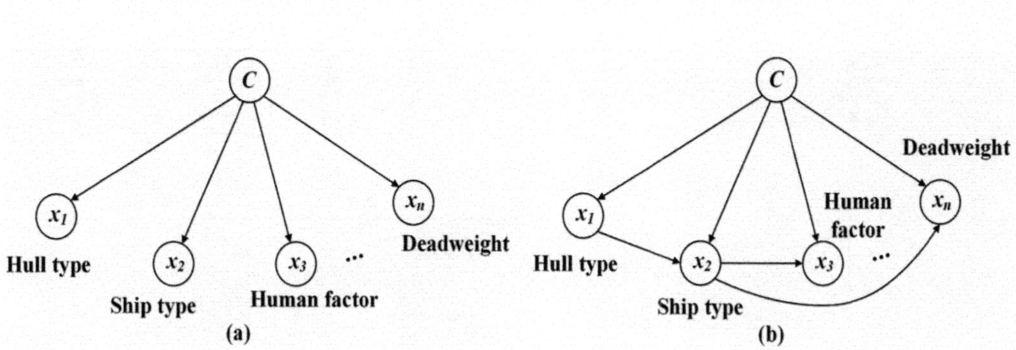

Figure 1. The structure of (a) NB network and (b) TAN network.

As a result, TAN not only retains the robustness of NBN but also makes the network structure more realistic. In a TAN network, each attribute depends on the class variable and another attribute. The structures of NBN and TAN models are depicted in Figure 1.

A BN encodes a joint probability distribution over a set of random variables U, which is an annotated DAG. Let $U = \{A_1, \ldots, A_n, C\}$ where n stands for the number of RIFs, the variables $A_1, \ldots, A_n$ are the RIFs and C is the class variable (accident types). Consider a graph structure where the class variable is the root, that is, $\prod C = \emptyset$ ($\prod C$ denotes the set of parents of C in U), and each RIF has the class variable as its unique parent, *i.e.*$\prod A_i = \{C\}$ for$1 \leq i \leq n$. A BN defines a unique joint probability distribution over U given by

$$P(A_1, \ldots A_n, C) = P(C)\ \Pi_{i=1}^{n} P(A_i|C) \quad (1)$$

The DAG$\{A_1, \ldots, A_n\}$ is a tree if $\prod A_i$ contains only one parent for allA_i, except for one variable without parents (referred to as the root). There is a function π that can define a tree over $A_1, \ldots, A_n$, if there is exactly one i such that $\pi(i) = 0$ (*i.e.* the root of the tree), and there is no sequence$i_1, \ldots, i_k$ such that $\pi(ij) = i_{j+1}$ for $i \leq j < k$ and $\pi(i_k) = i_1$ (*i.e.* no cycles). Such a function defines a tree network where $\prod A_i = \{C, \ldots, A_{\pi(i)}\}$ if $\pi(i) > 0$, and $\prod A_i = \{C\}$ if $\pi(i) = 0$.

Learning a TAN structure is an optimization problem. Solving this problem follows the general procedure proposed by Chow & Liu (1968), who used conditional mutual information between attributes. The function can be defined as

$$I_P(A_i, A_j|C) = \sum_{a_{ii}, a_{ji}, c_i} P(a_{ii}, a_{ji}, c_i) \log \frac{P(a_{ii}, a_{ji}|c_i)}{P(a_{ii}|c_i) P(a_{ji}|c_i)} \quad (2)$$

where I_P represents the conditional mutual information, a_{ii} is the ith state of RIF A_i, a_{ji} is the ith state of RIF A_j, c_i is the ith state of "accident type". The optimisation problem, *i.e.* learning a TAN structure, is to find a tree definition function π over$A_1, \ldots, A_n$ such that the log-likelihood is maximized.

To generate the BN model, six RIFs are considered to illustrate their relationships with the dependent variable (*i.e.* accident types). Netica software is utilized to assist in the calculations. It features a "Learning network" function, which is used to develop the TAN network based on Equation (2). By importing the accident database, the resulting BN structure is shown in Figure 2.

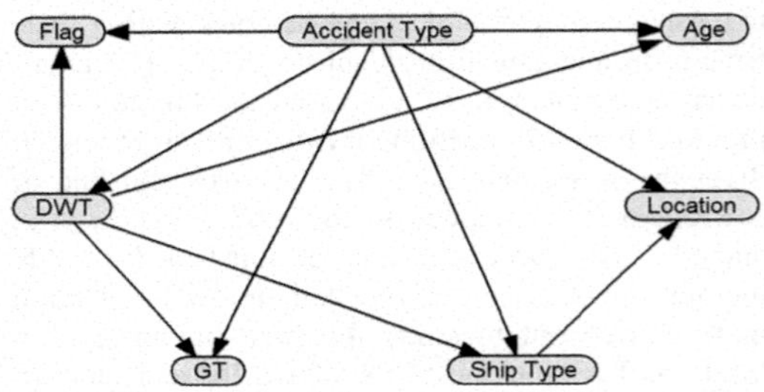

Figure 2. The resulting BN structure.

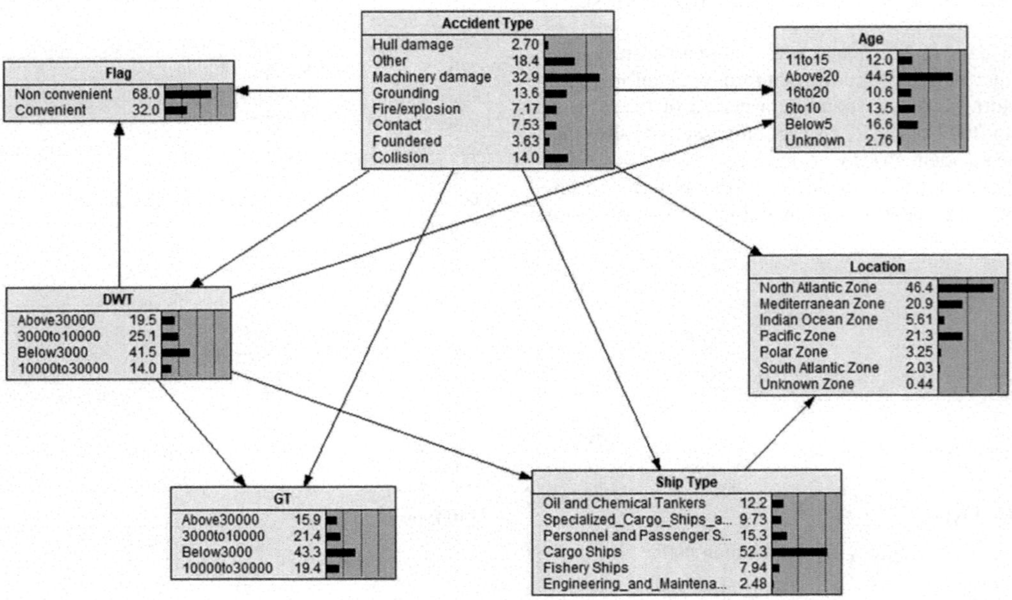

Figure 3. The final TAN-based BN.

Using the TAN model and Netica software, the parameter learning of conditional probability tables (CPTs) for the cases was conducted with the counting learning algorithm. Once the structure is constructed and CPTs are obtained, posterior probabilities for the various variables can be calculated accordingly. Figure 3 presents the TAN results containing all six retained RIFs.

4.2 *Sensitivity analysis*

Sensitivity analysis is a commonly used method for uncertainty analysis. In BN-based maritime risk analysis, its essence lies in identifying the RIFs that have a significant impact on the target variable to implement cost-effective measures to reduce risk for key factors (Yang et al. 2018). To perform a comprehensive evaluation, sensitivity analysis is carried out using mutual information, joint probability, and TRI (True Risk Influence). Specifically, mutual information values are used to assess the importance and priority of RIFs in influencing the target node "accident types". A higher mutual information value indicates a stronger correlation between variables. Subsequently, joint probability and TRI are employed to explore the detailed effects and interactions of these RIFs. A joint probability is utilized to measure the impact of each state of important RIFs on different types of accidents. TRI is an effective approach for testing multivariable sensitivity. Furthermore, ranking the RIFs for the eight accident types based on TRI values can effectively determine the extent of the influence of significant RIFs on each accident type.

4.2.1 *Mutual information*

Mutual information represents the dependence between two variables in probability theory (Trucco et al. 2008). Stemming from information theory, mutual information is described as a measure of uncertainty in the dataset and is interpreted as a reduction in entropy. Mutual information elucidates how strong the relationship is between RIFs and "accident type".

This study aims to analyze the impact of RIFs on different types of maritime accidents. Therefore, by treating "accident types" as a fixed variable, the mutual information values between RIFs and "accident types" are defined as follows:

$$I(C, X_{ij}) = -\sum_{c,i} P(C, X_{ij}) \log_b \frac{P(C, X_{ij})}{P(C)P(X_{ij})} \quad (3)$$

where C is the accident type, X_i indicates the ith RIF, X_{ij} expresses the jth state of the ith RIF, and $I(C, X_i)$ represents the mutual information value between the accident type and the ith RIF. A higher mutual information value indicates a stronger relationship between RIFs and "accident types". By computing mutual information values, it is possible to rank RIFs based on their impact on the target node "accident types". Subsequently, less relevant RIFs with relatively lower correlations to "accident types" can be eliminated. The remaining RIFs are then considered significant variables in the model when selecting accident types.

The mutual information values, the percentage reduction in entropy, and the variance of beliefs between the target node "accident type" and RIFs are computed and presented in Table 2. When "accident type" is the parent node, the mutual information value for "ship type" is 0.05062, followed by "GT" at 0.0491. This implies that "accident type" and "GT" have a substantial influence on "accident types".

Furthermore, Table 2 reveals that the mutual information value for "flag" is below 0.02. In this study, a threshold of 0.02 is chosen as the criterion for selecting key factors for further discussion. Variables "location", "ship type", "GT", "DWT" and "age" are selected for the subsequent factor analysis.

Table 2. Mutual information between "Accident Type" and RIFs.

Node	Mutual Information	Percent	Variance of beliefs
Accident Type	2.63418	100	0.6613808
Ship Type	0.05062	1.92	0.0017009
GT	0.0491	1.86	0.001689
Location	0.0488	1.85	0.0033952
DWT	0.04469	1.7	0.0014774
Age	0.02962	1.12	0.0008108
Flag	0.01536	0.583	0.0004168

4.2.2 *Joint probability*

For the most important variables identified through mutual information computations, joint probability calculation is employed to explore the detailed impact of these RIFs on "accident type". The calculation of joint probability involves when the state of one child node (e.g. ship type) is assigned a different value and locks the state of the other child nodes, calculating the value of the target node "accident type" (Wang & Yang 2018).

For example, there are only two variables "accident type" and "ship type", and "accident type" is the parent node of "ship type", Set "accident type" as M, "ship type" as N, "$M = M_i$" means the ship is at its ith "accident type" state, and the same goes to "$N = N_j$". According to Baye's rules, the joint probability can be calculated as (Fan et al. 2020):

$$P(M = M_i,\ N = N_j) = P(M = M_i) \times P(N = N_j | M = M_i) \quad (4)$$

where $P(M = M_i, N = N_j)$ refers to the joint probability that events "$M = M_i$" and "$N = N_j$" both occur, $P(M = M_i)$ is the prior probability of the ith "accident

Table 3. Joint probability (%).

	Hull damage	Other	Machinery Damage	Grounding	Fire/ Explosion	Contact	Foundered	Collision
Initial	2.70	18.39	32.93	13.63	7.17	7.53	3.63	14.01
Location								
North Atlantic	3.08	16.3	36.4	14	6.94	9.46	2.72	11
Mediterranean	2.07	14.5	42	12.5	6.16	6.9	2.16	13.7
Indian Ocean	3.02	36.4	17	9.65	7.47	3.2	6.6	16.6
Pacific	2.14	21.9	21.7	13.2	7.98	5.71	6.06	21.3
Polar	4.54	14.4	38.3	17.3	9.57	4.92	4.05	6.94
South Atlantic	2.33	17.3	17.4	26.8	10.2	5.35	5.41	15.3
Unknown	3.15	50.2	13	7.66	5.22	8.02	3.95	8.88
Ship Type								
Oil and chemical tankers	1.77	29.6	25.5	12.4	6.64	5.73	1.05	17.3
Specialized cargo ships and others	2.68	20.9	26	12.7	9.13	8.01	6.37	14.2
Personnel and passenger ships	2.15	15.7	40.4	10.1	8.59	11.6	2.23	9.23
Cargo ships	2.62	17.2	34	15.4	5.08	7.67	2.69	15.4
Fishery ships	5.44	9.2	35.3	13.1	14.9	1.7	12.2	8.16
Engineering and maintenance ships	3.44	25.1	21	9.86	12.7	5.64	6.88	15.4
GT								
(0,3000]	3.14	12.3	33.1	17	7.61	6.68	7.18	12.9
(3000,10000]	2.15	16.8	38	11.5	6.76	8.1	1.77	15
(10000,30000]	2.69	23.2	32	10.9	6.44	9.46	0.5	14.9
>30000	2.23	31.2	26.9	10.5	7.42	6.78	0.29	14.6
DWT								
(0,3000]	3.16	12.6	33.1	15.6	8.5	7.6	7.22	12.2
(3000,10000]	2.07	16.9	37.7	13	6.3	7.36	1.95	14.7
(10000,30000]	2.7	22.9	31.5	10.5	7.04	9.35	0.64	15.4
>30000	2.5	29.4	27.4	12.4	5.57	6.33	0.31	16
Age								
(0,5]	1.87	24.7	28.1	12.8	5.51	8.84	1.36	16.8
(5,10]	1.68	22.9	32.9	11.7	5.77	7.68	1.53	15.9
(10,15]	1.89	21.2	35.9	10.6	6.09	7.62	1.48	15.3
(15,20]	2.31	17.7	36.1	11.8	6.79	7.54	2.21	15.5
>20	3.61	14.5	33.8	15.6	8.5	6.95	5.35	11.7
Unknown	2.97	11.6	23.9	16.1	8.85	7.94	14.9	13.8

type", $P(N = N_j \mid M = M_i)$ denotes the conditional probability of the occurrence of ith "accident type" state given that j^{th} "ship type" state occurs.

In summary, to obtain the joint probability of "accident type = collision" and "ship type = cargo ship" in Netica, we simply need to set the state of the child node "ship type" to "100% cargo ships" without manipulating other variables. The value corresponding to the "collision" state in the brief bar of the target node "accident type" is the joint probability of "accident type = collision" and "ship type = cargo ships".

For the most important variables screened from mutual information calculations, additional sensitivity methods are used to explore the detailed effects of these RIFs on 'accident type'. The probability of each state of each variable is sequentially increased to 100% to obtain the joint probabilities (Wang & Yang 2018), as shown in Table 3, which provides the joint probability values for each variable and "accident type", comparing them to the original probabilities. It shows how the probability of each accident type changes when the selected variable is set in a specific state.

Based on Table 3, the states of each variable that have the greatest and least impact on different accident types are highlighted in bold. For instance, when ships are sailing in the "pacific zone", the probability of "collision" is the highest (21.3%), whereas when ships are in the "polar zone", the probability of "collision" accidents is the lowest (6.94%). Regarding ship gross tonnage, when it is below 3000 tons, the probability of "grounding" accidents is the highest (17%), but the probability of "contact" accidents is the lowest (6.68%)

4.2.3 *True Risk Influence*

The True Risk Influence (TRI) method, proposed by Alyami et al. (2019), is an effective approach for testing multivariable sensitivity. When analyzing the impact of selected variables, this method elevates the probability of the states that have the most significant impact on a specific accident type (*i.e.* collision) to 100%, resulting in High-Risk Inference (HRI). Then, the probability of the states that have the least impact on the same accident type is increased to 100%, yielding Low-Risk Inference (LRI). By calculating the average of HRI and LRI, the TRI for the variable concerning collisions is obtained. This process is applied to other accident types and variables in the same way, resulting in the corresponding TRI values.

A higher TRI value indicates a greater impact of the variable on accident types. Table 4 lists the TRI values for all accident types. To assess the impact level of these variables within accident types, the variable importance is ranked from 1 (most important) to 5 (least important) based on TRI values of the 8 accident types, as shown in Table 5. In general, the most important variables for "accident type" are as follows:

Location> Ship type > Age > GT > DWT

Table 4. TRI of RIFs for all accident types.

	Ship type	GT	Location	DWT	Age
Hull damage	1.84	0.50	1.24	0.55	0.97
Other	10.20	9.45	17.90	8.40	6.55
Machinery damage	9.70	5.55	14.50	5.15	6.10
Grounding	2.77	3.25	9.57	2.55	2.75
Fire/Explosion	4.91	0.59	2.49	1.47	1.67
Contact	4.95	1.39	3.13	1.51	0.95
Foundered	5.58	3.45	2.22	3.46	6.77
Collision	4.57	1.05	7.18	1.90	2.55
Average	5.56	3.15	7.28	3.12	3.54

Table 5. The importance degree of RIFs.

	Ship type	GT	Location	DWT	Age
Hull damage	1	5	2	4	3
Other	2	3	1	4	5
Machinery damage	2	4	1	5	3
Grounding	3	2	1	5	4
Fire/Explosion	1	5	2	4	3
Contact	1	4	2	3	5
Foundered	2	4	5	3	1
Collision	2	5	1	4	3
Average	2	4	1	5	3

Different accident types are related to different variable priorities. For example, the most important RIF for "fire/explosion" accidents is "ship type", while the most important RIF for "foundered" accidents is "age". Conversely, "age" is the least important RIF for "contact", and "location" is the least important RIF for "foundered" accidents. The influence of "GT" on "grounding" accidents is greater than its impact on other types of accidents.

4.3 *Model validation*

To verify the comprehensive impact of multiple RIFs on "accident type" and validate the correctness of the BN model during the inference process in the sensitivity analysis (Zhang et al. 2013; Yang et al. 2009), at least two axioms need to be satisfied:

Axiom 1: A slight increase or decrease in the prior probabilities of each RIF, should contribute to the correspondence increase or decrease in the posterior probability of the target node (*i.e.* accident type).

Axiom 2: The total influence of the integration of the probability variations of x parameters should be no smaller than the one from the set of y ($y \in x$) RIFs.

To validate the model, we examined the comprehensive impact of multiple RIFs on accident types. The start is with "location" as the first node and increase the probability of the state that had the highest change value for generating "hull damage" in "accident type" to 100%, denoted as "100%" in Table 6. Then, the same method is applied to the next RIF, obtaining and updating the cumulative change value, repeating this process for all RIFs. Similarly, the steps are applied to the other accident types. The results are presented in Table 6. The first column shows the original probability values for each accident type, while the subsequent columns represent the cumulative change values for the results. By comparing the changes in probability values, the comprehensive impact of these RIFs on "accident type" can be assessed.

According to Table 6, the increase or decrease in the prior probabilities of variable nodes leads to an increase or decrease in the posterior probabilities of the target node, thereby verifying Axiom 1. As shown in Table 6, as the RIFs are continuously updated, the probability values of the target node gradually increase, thus verifying Axiom 2.

4.4 *Implications*

By calculating the joint probabilities and TRI values using the BN model, the following conclusions can be drawn:

(1) Regarding different locations, the North Atlantic zone exhibits the highest risk of contacts, the Mediterranean zone has the highest risk of machinery damage, the Indian Ocean zone carries the highest risk of foundering, the Pacific

Table 6. Change in probability.

Location	/	100%	100%	100%	100%	100%	100%
Ship Type	/	/	100%	100%	100%	100%	100%
GT	/	/	/	100%	100%	100%	100%
DWT	/	/	/	/	100%	100%	100%
Age	/	/	/	/	/	100%	100%
Flag	/	/	/	/	/	/	100%
Hull damage	2.70	4.54	5.29	5.53	5.56	6	6.09
Other	18.4	50.2	60.8	74.8	75.1	78.1	79.2
Machinery damage	32.9	42.0	45.4	47.4	47.7	48.7	48.8
Grounding	13.6	26.8	41.1	53.2	53.6	57.7	59.5
Fire/ Explosion	7.17	10.2	19.1	21.4	31.2	32.7	35.3
Contact	7.53	9.46	10.6	13.5	14.5	18.7	19.9
Foundered	3.63	6.6	15.3	17.3	17.4	18.8	19.6
Collision	14.0	21.3	26.7	29.7	30.6	34.9	36.7

zone poses the highest risk of collisions, the Polar zone experiences the highest risk of hull damage, and the South Atlantic zone is associated with the highest risk of grounding and fire/explosion accidents.

(2) Concerning ship types, oil and chemical tankers present the highest risk of collisions, personnel and passenger ships have the highest risk of machinery damage and contacts, cargo ships are at the highest risk of grounding, and Fishery ships exhibit the highest risk of hull damage, fire/explosion, and foundered accidents.

(3) In terms of ship gross tonnage, ships with a tonnage of below 3000 tons are associated with the highest risks of hull damage, grounding, fire/explosion, and foundered. Ships with tonnages between 3000 and 10,000 tons exhibit the highest risks of machinery damage and collisions, while ships in the range of 10,000 to 30,000 tons are at the highest risk of contacts.

(4) Considering the deadweight tonnage of ships, those with less than 3000 tons are associated with the highest risks of hull damage, grounding, fire/explosion, and foundered. Ships in the 3000 to 10,000-ton range exhibit the highest risk of machinery damage, while those in the 10,000 to 30,000-ton category are at the highest risk of contact. Ships exceeding 30,000 tons are associated with the highest risk of collisions.

(5) Concerning ship age, ships aged below 5 years are at the highest risk of collisions and contacts. Ships aged 16 to 20 years are at the highest risk of machinery damage, and those with an age exceeding 20 years have the highest risk of hull damage.

(6) The five most crucial RIFs affecting maritime accident types are location, ship type, ship age, gross tonnage, and deadweight tonnage. Therefore, decision-makers in the maritime industry can derive valuable insights for accident prevention, route planning, and risk reduction. Additionally, the variability in accident types across different regions emphasizes the importance of collaboration with other stakeholders in the maritime sector to develop maritime safety policies and regulations, particularly focusing on collision, foundered, and fire/explosion incidents.

5 CONCLUSIONS

Using a comprehensive global maritime accident database spanning the past 20 years, this study constructs a risk analysis model based on data-driven BN. It quantifies the influence of six RIFs on different types of maritime accidents. Through sensitivity analysis and model validation, the study identifies the most crucial influencing factors. The results of this research provide valuable insights for accident prevention.

1. By calculating mutual information, the study ranked key RIFs for different accident types. The results indicate that the top five RIFs with significant influence on maritime accident types are location, ship type, ship age, gross tonnage, and deadweight tonnage.
2. This study enhances the construction method of data-driven BN risk models and benefits from a large database, surpassing the sample size of previous research, which reduces prediction errors and enables accurate risk prediction.

Furthermore, while the overall number of cases is sufficient, the information in individual case is not comprehensive. There are only six accident-related variables. To present more detailed results, continuous data collection is needed to support model development. In future work, accident reports will be searched based on the accident ship's IMO number and the accident date to obtain more comprehensive accident information, exploring the impact of more RIFs on maritime accidents.

ACKNOWLEDGEMENTS

The research of this paper was supported by National Natural Science Foundation of China (52071247; 51920105014), Key Technologies Research and Development Program of Guangxi Zhuang Autonomous Region (2022AB04010) and Innovation and entrepreneurship team import project of Shaoguan city (201208176230693). This work contributes to the Strategic Research Plan of the Centre for Marine Technology and Ocean Engineering (CENTEC), which is financed by the Portuguese Foundation for Science and Technology (Fundação para a Ciência e Tecnologia) under contract UIDB/UIDP/00134/2020.

REFERENCES

Alyami, H., Yang, Z., Riahi, R. et al. 2019. Advanced uncertainty modelling for container port risk analysis. *Accident Analysis & Prevention* 123: 411–421.

Antão, P. & Guedes Soares, C. 2019. Analysis of the Influence of Human Errors on the Occurrence of Coastal Ship Accidents in Different Wave Conditions using Bayesian Belief Networks. *Accident Analysis and Prevention* 133: 105262.

Antão, P., Sun, S., Teixeira, A.P. et al. 2023. Quantitative assessment of ship collision risk influencing factors from worldwide accident and fleet data. *Reliability Engineering and System Safety* 234: 109166.

Chen, S.-T., Alan, W., Philip, D. et al. 2013. A Human and Organisational Factors (HOFs) Analysis Method for Marine Casualties Using HFACS-Maritime Accidents (HFACS-MA). *Safety Science* 60: 105–114.

Chow, C. & Liu, C. 1968. Approximating discrete probability distributions with dependence trees. *IEEE Transactions on Information Theory* 14(3): 462–467.

Fan, S.Q., Yang, Z., Blanco-Davis, E. et al. 2020. Analysis of maritime transport accidents using Bayesian networks. *Proceedings of the Institution of Mechanical Engineers, Part O: Journal of Risk and Reliability* 234 (3): 439–454.

Fan, S.Q., Blanco-Davis, E., Yang, Z. et al. 2020. Incorporation of human factors into maritime accident analysis using a data-driven Bayesian network. *Reliability Engineering & System Safety* 203: 107070.

Fan, S.Q., Zhang, J.F., Blanco-Davis, E. et al. 2020. Maritime Accident Prevention Strategy Formulation from a Human Factor Perspective Using Bayesian Networks and TOPSIS. *Ocean Engineering* 210: 107544.

Friedman, N., Geiger, D. & Goldszmidt, M. 1997. Bayesian network classifiers. *Machine Learning* 29: 131–163.

Kristiansen, S.; Koster, E.; Schmidt, W. F.; Olofsson, M.; Guedes Soares, C., and Caridis, P. A 1999. New Methodology for Marine Casualty Analysis Accounting for Human and Organisational Factors. *Proc. Int. Conf. on Learning from Marine Incidents*; London. Royal Institution of Naval Architects; Paper No.14.

Li, H.H., Ren, X. & Yang, Z. 2023. Data-Driven Bayesian Network for Risk Analysis of Global Maritime Accidents. *Reliability Engineering & System Safety* 230: 108938.

Ma, L.H., Ma, X.X., Lan, H. et al. 2022. A Data-Driven Method for Modeling Human Factors in Maritime Accidents by Integrating DEMATEL and FCM Based on HFACS: A Case of Ship Collisions. *Ocean Engineering* 266: 112699.

Salihoglu, E. & Besikçi, E.B. 2021. The use of functional resonance analysis method (FRAM) in a maritime accident: a case study of prestige. *Ocean Engineering* 219: 108223.

Silveira, P.A.M., Teixeira, A.P., Figueira, J.R. et al. 2021. A multicriteria outranking approach for ship collision risk assessment. *Reliability Engineering and System Safety* 214: 107789.

Sur, J.M. & Kim, D.J. 2020. Comprehensive Risk Estimation of Maritime Accident Using Fuzzy Evaluation Method - Focusing on Fishing Vessel Accident in Korean Waters. *The Asian Journal of Shipping and Logistics* 36(3): 127–135.

Trucco, P., Cagno, E., Ruggeri, F. et al. 2008. A Bayesian Belief Network modelling of organisational factors in risk analysis: a case study in maritime transportation. *Reliability Engineering & System Safety* 93(6): 845–856.

Wang, H., Yang, Y.D. & Ma, Y. 2016. Research on the Yangtze River Accident Casualties Using Zero-Inflated Negative Binomial Regression Technique. *In 2016 IEEE International Conference on Intelligent Transportation Engineering (ICITE)*: 72–75.

Wang, L.K. & Yang, Z.L. 2018. Bayesian network modelling and analysis of accident severity in waterborne transportation: a case study in China. *Reliability Engineering & System Safety* 180: 277–289.

Weng, J.X. & Yang, D. 2015. Investigation of Shipping Accident Injury Severity and Mortality. *Accident Analysis & Prevention* 76: 92–101.

Wu, B., Tian, H.B., Yan, X.P. et al. 2020. A probabilistic consequence estimation model for collision accidents in the downstream of Yangtze River using Bayesian Networks. *Journal of Risk and Reliability* 234(2): 422–436.

Xu, L. & Liu, Q. 2013. Human Factors Analysis to Yangtze River Water Traffic Accidents Based on Fault Tree. *2013 6th International Conference on Information Management, Innovation Management and Industrial Engineering*: 23–25.

Yang, Z.L., Wang, J., Bonsall, S. et al. 2009. Use of fuzzy evidential reasoning in maritime security assessment. *Risk Analysis* 29: 95–120.

Yang, Z.S., Yang, Z.L. & Yin, J.B. 2018. Realising advanced risk-based port state control inspection using data-driven Bayesian networks. *Transportation Research Part A: Policy and Practice* 110: 38–56.

Yu, Q., Teixeira, A.P., Liu, K. et al. 2022. Framework and application of multi-criteria ship collision risk assessment. *Ocean Engineering* 250: 111006.

Zhang, D., Yan, X.P., Yang, Z.L. et al. 2013. Incorporation of formal safety assessment and Bayesian network in navigational risk estimation of the Yangtze River. *Reliability Engineering & System Safety* 118: 93–105.

Zhang, G.Z. & Thai, V.V. 2016. Expert elicitation and Bayesian Network modeling for shipping accidents: a literature review. *Safety Science* 87: 53–62.

Zhang, J.F., Teixeira, A.P., Guedes Soares, C. et al. 2016. Maritime transportation risk assessment of Tianjin port with Bayesian belief networks. *Risk Analysis* 36(6): 1171–1187.

Zhang, J.F., Teixeira, A.P., Guedes Soares, C. et al. 2018. Quantitative assessment of collision risk influence factors in the Tianjin port. *Safety Science* 110: 363–371.

Advances in Maritime Technology and Engineering – Guedes Soares & Santos (Eds)

Structural reliability assessment of secondary hull component subjected to crack growth due to local vibration

S.K. Kleivane & B.J. Leira
Department of Marine Technology, Norwegian University of Science and Technology, Trondheim, Norway

ABSTRACT: Ship hull vibration occurs at both global and local levels, and local resonance is frequently a major contributor regarding fatigue loading on hull components. A local finite element model of the pipe stack connected to a cargo pump, which constitutes a structural system that is located inside a cargo tank, is developed. The secondary components under investigation are the supports attaching this pipe stack to the tank wall. Modal analysis is conducted to establish frequencies and mode shapes of the structural system. Engine excitation and hydrodynamic loading (in the form of springing and whipping wave-induced vertical bending moment) are applied as loading based on statistical models describing the midship bending moment. This represents a stochastic model for vibration response, which is used for a probabilistic formulation of crack propagation. An initial crack is assumed to be present in the structure, and different reliability methods are used in failure probability calculations.

1 INTRODUCTION

Structural vibrations continue to be a major concern in the design and operation of ocean-going vessels. Excessive vibration may lead to the malfunction of machinery and equipment or fatigue failure of local structural members. The main engine, the propeller and the hydrodynamic loading are identified as the main sources of vibration. Vibrations are observed at both global and local levels and constitute a complex vibration problem within the ship hull. Hydrodynamic loading is considered to result in both local and global wave-induced vibration, generally described as springing and whipping (Storhaug 2007). Vibration from the main engine is transmitted into the ship hull by dynamic forces from the engine foundation and can cause vibration of both the hull girder and local structure. Wave-induced vibration, along with main engine vibration, is dominant in the amidship region, while the propeller excitation is predominantly located in the aft end of the vessel, especially in the hull area above the propellers. The local structural system investigated in this work is a cargo pump pipe stack system, which is a structural part located inside a cargo tank. The cargo tanks are located in the midship region or the vicinity, and the focus of this work will be on vibration caused by the main engine and wave-induced loading.

Structural vibration analysis based on finite element methods (FEM) is frequently used in the design stage of ship structures for calculations and estimations of natural frequencies and mode shapes (free vibration analysis), and loads and responses of the structure (forced vibration analysis) to avoid the occurrence of excessive vibrations in the structure. However, due to the many modes of vibration which develop within the ship response, the occurrence of resonance conditions is practically impossible to avoid. The main goal for vibration analysis is thus to identify resonance conditions and apply measures to mitigate the dynamic response of the structure (ABS 2022).

Structural reliability analysis in engineering has become an essential step in the design of ship structures, especially concerning fatigue loading and conducting inspection, maintenance and repair. This is due to the large uncertainty and the complexity associated with fatigue assessment, and also concerning other aspects such as corrosion which may accelerate fatigue cracking. Widely used methods for reliability calculations are the Monte Carlo simulation methods and asymptotic methods such as the first-order and second-order reliability methods (FORM/SORM).

Some previous research has looked into the investigation of details in the ship hull, mostly concerning panel stiffeners in the ship side and bottom. Some investigations have also been done on the boundaries of ballast and cargo tanks. Many studies look at fatigue crack growth using experiments and finite element analysis (FEA) based on linear elastic fracture mechanics assumption (Leheta et al. 2016, Božić et al. 2011), and incorporating reliability assessment for ship hulls under fatigue cracking (Tamimi et al. 2023, Guedes Soares and Garbatov 1999). However, considerations on secondary components or equipment installed inside the tanks, such as the cargo pump pipe stack, have not been considered to the same extent. This paper aims to present such an

DOI: 10.1201/9781003508762-37

investigation and develop a reliability model for failure probability estimations related to crack propagation of secondary structural components.

2 VIBRATION SOURCES AND STOCHASTIC MODEL FOR VIBRATION RESPONSE

The major vibration sources on the ship hull are generally considered as the main engine vibration, the wave-induced vibration and vibrations from the propulsion system and propellers. In the midship region, the main engine and wave-induced loading are considered the most important regarding the contribution to the vibration of structural systems and components located in this region. The propeller excitation is more dominant in the aft end of the ship, especially on the hull surface located above the propellers which is frequently exposed to pressure impact pulses due to the operation of the propellers in the water (Lewis & SNAME 1988, ABS 2023).

The purpose of vibration analysis is to estimate the response of a vessel to different excitation sources and assist in the design phase to eliminate significant vibration problems. The vibration analyses are done by FEA using Ansys Inc. software. The establishment of the stochastic model for vibration response due to main engine vibration and wave-induced loading is based on the previous research by the authors for the global cargo hull model of a generic coastal tanker (Kleivane et al. 2023b). This model is extended and transferred to the local analysis to investigate local excitation and fatigue crack propagation of the pipe stack supports.

2.1 *Main engine*

The preferred choice for ocean-going merchant ships has been the two-stroke low-speed marine diesel engine, where tankers typically have engines with six cylinders or more (MAN Energy Solutions 2023, Bukovac et al. 2015). These engines are large in size and mass and are often installed directly onto the ship structure, causing their vibrations to be fully transmitted directly into the hull structure. By this, the main engine can be a significant source of vibration excitation of the ship hull girder and local structures and equipment. Of specific concern in the assessment of the ship hull is the magnitude of the vibration-induced forces and moments, the location of the engine in the ship structure, and the possibility of correlation of engine vibration with the lower natural frequencies of the ship hull girder (Noonan 1990).

Two main types of excitations generated by the engine can be defined. The rotating and reciprocating action of the engine parts causes mass inertia forces and moments, where the 2nd order external vertical inertia moment is the most critical. These moments are typically mitigated by moment compensators installed in the ship hull, which then makes the so-called guide force couples the major source of engine vibration. The guide force couples are forces and moments caused by the combustion process in the engine and generate two main modes of vibration, as seen in Figure 1.

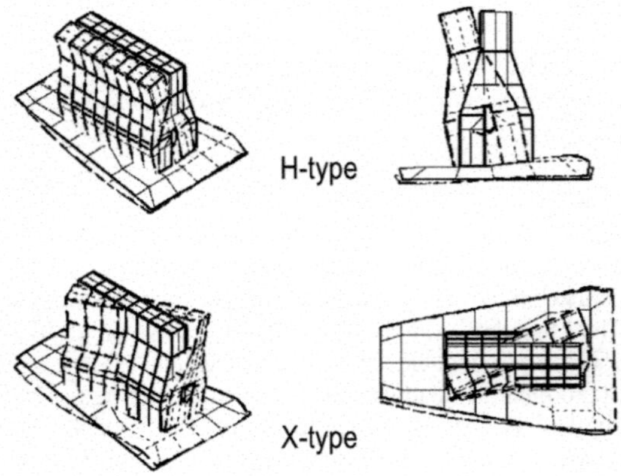

Figure 1. X-type and H-type vibration modes of the engine block (Asmussen, Menzel, & Mumm 2001).

The X-mode causes a twisting of the engine block and the H-mode causes a rocking of the engine block. The magnitude of the forces and moments generated by the engine is provided by the engine manufacturer, typically tabulated in the engine specification for individual engine ratings.

In the current study, the engine vibration is represented by the engine operational speed in rpms, since the magnitude of the forces and moment emanating from the engine is influenced by the engine speed. Engine rpm is then defined as a stochastic variable in the reliability model presented and is assumed to be normally distributed between an operational speed range of 80 rpm to 90 rpm. Further elaboration on the stochastic model and its variables is given in Section 2.3.

2.2 *Wave-induced loading*

Wave-induced vibration caused by springing and whipping can produce a high-frequency elastic response, which may contribute to fatigue loading on the hull and structural components (Storhaug 2007). Of particular concern is the vertical bending moments, which result from the change of distribution of the buoyancy forces along the length of the ship combined with the hydrodynamic and internal forces associated with the wave-induced ship motions (Gaspar et al. 2016).

Generally, springing is defined as the stationary resonance vibration of the ship's two-node vertical mode when the wave encounter frequency corresponds with the structure's natural frequency. On the other hand, whipping is generally defined as the transient hull girder vibration caused by an impact, typically occurring from slamming. Both vibrations may contribute to fatigue loading, and the loading can be simplified as a superposition of high-frequency and low-frequency load components, which can give rise to coupled damage effects due to the interaction between different frequencies and components (Gan et al. 2021, Gan et al. 2022).

2.3 *Stochastic model for vibration response*

The stochastic model for vibration response is defined with the two stochastic variables as described above; the engine speed and the wave-induced vertical bending moment. In addition, the tank filling level is defined as a stochastic variable, since the tank filling level will influence the natural frequency of the structure and thus the vibration response. The development of the stochastic model with three stochastic parameters, defined as X, Y and Z, is described in previous studies by the authors (Kleivane et al. 2023a, Kleivane et al. 2023b) and adopted for the current analysis of the local pipe stack model. A brief description is given in the following.

The engine speed is described by a normal distribution with a mean of 86.2 rpm and a standard deviation of 1.1, based on a study which investigated fuel consumption speed curves for ships operating at sea. (Adland et al. 2020). A simplified approach for representing the probabilistic modelling of vertical wave-induced bending moment (VBM) is adopted in the current study. The VBM is defined as a stochastic variable and the long-term distribution of extreme values can be formulated based on the linear prediction model by the IACS-CSR formulation. Generally, a two-parameter Weibull distribution function can be applied to describe the long-term probability distribution function of the maximum VBM (Guedes Soares and Moan 1991, Gaspar et al. 2016). The applied Weibull distribution function has scale parameter w = 52.2 MNm and shape parameter $k = 1$, and the vertical bending moment is defined for a range of 1.5 - 105 MNm. The filling level is given as the percentage of the tank filled and is assumed to have a uniform distribution with a range from 0 - 100 %.

The frequency range for the engine speed implemented in the numerical analysis spans from 2 - 20 Hz, while the frequency range for the vertical wave-induced bending moment is somewhat lower, in the range of 5 - 10 Hz. This can be evaluated against the natural frequencies of the local model to assess the possible occurrence of resonance conditions.

3 FINITE ELEMENT MODEL OF CARGO PUMP PIPE STACK

The local model of the cargo pump pipe stack and its supports is established based on a generic cargo hold model of a common oil tanker design with a double hull and a centreline bulkhead, and the size corresponds to a coastal or medium-size tanker. The model of the cargo pump pipe stack is a simplified model consisting of two pipes and a pump head at the bottom (circular cylinder component), shown in Figure 2. The cargo pipe, which is the largest pipe seen in Figure 2, is used to pump the cargo up to the deck, and the hydraulic pipe is connected to the pump head where the engine operating the cargo pump is located. The model consists of solid elements, specifically, SOLID187 is used in Ansys for the structural elements. For the pipe stack supports, which are the secondary components under consideration, the welds are included in the model, as seen in Figure 3.

The material properties of the structural elements correspond to those of steel given in Ansys. Furthermore, the pump head is given a lower density of 625 kg/m^3 to give a weight of 200 kg for this part. This pump head itself is not in contact with the tank bottom, but it is installed in the tank with a ring support bolted to the tank bottom to restrict movement of the pump. Moreover, a distributed mass of 130kg is applied to the hydraulic pipe to simulate the hydraulic oil contained within the pipe. These assignments are established through communication with an industry supplier of similar cargo pump systems as to the simplified model presented herein.

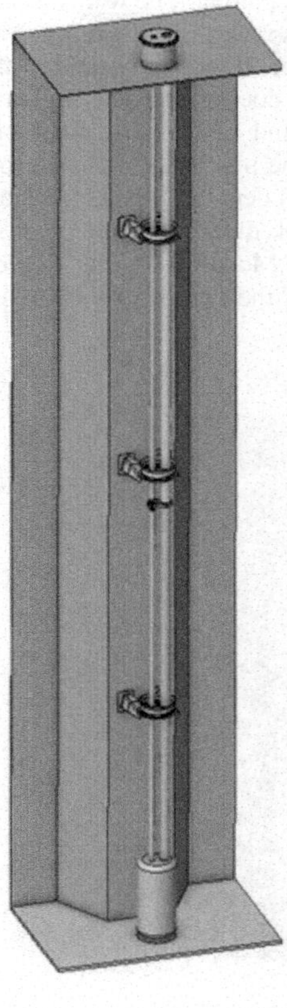

Figure 2. Local finite element model of the cargo pump pipe stack.

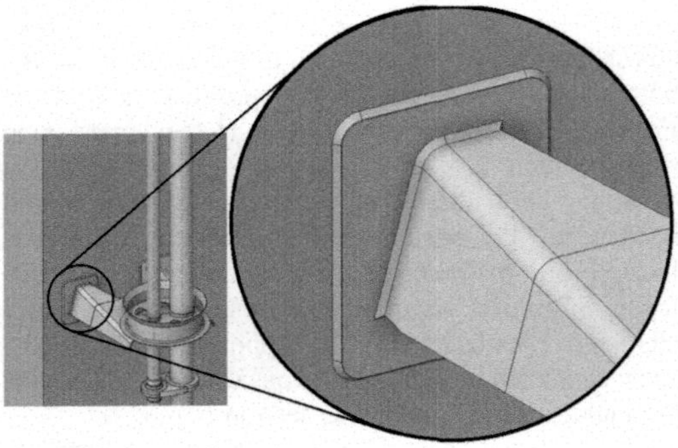

Figure 3. Pipe stack support with the welds included in the model.

4 MODAL ANALYSIS OF LOCAL MODEL

Modal analysis (free vibration analysis) of the local model of the pipe stack is performed using Ansys Workbench to establish the natural frequencies and corresponding mode shapes. The combined model of the pipe stack connected to the tank wall is analysed, and also, separate models of only the tank wall and the pipe stack are analysed to identify the local modes of the pipe stack and of the tank wall, and how they may influence each other. The tank wall is fixed at its top and bottom plating. The wall sides are constrained with frictionless support, which constrains translation movement in the direction normal to the surface it is applied on and does not restrict in-plane motion.

Figures 4 and 5 show the first eight modes of vibration for the pipe stack attached to the tank wall. This analysis is denoted Modal 1. Figure 6 shows the first four local modes of the pipe stack structure alone, denoted Modal 2. Figure 7 gives the first two local modes of the tank wall section, denoted Modal 3. The natural frequencies for the eight first modes for the three analysis cases Modal 1, Modal 2 and Modal 3 are given in Table 1.

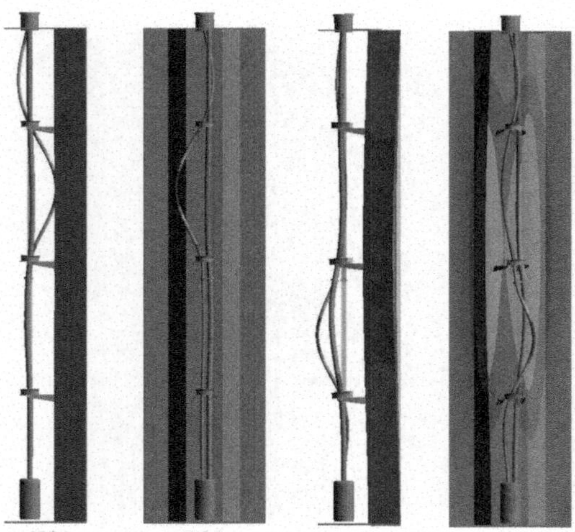

Figure 4. The first four modes of vibration for the local model.

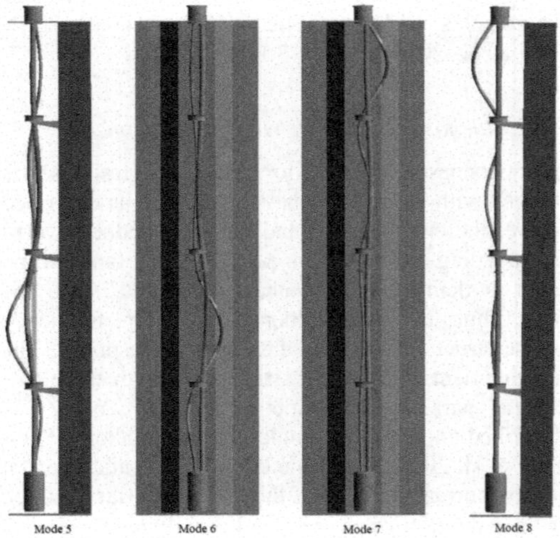

Figure 5. The next four modes of vibration for the local model.

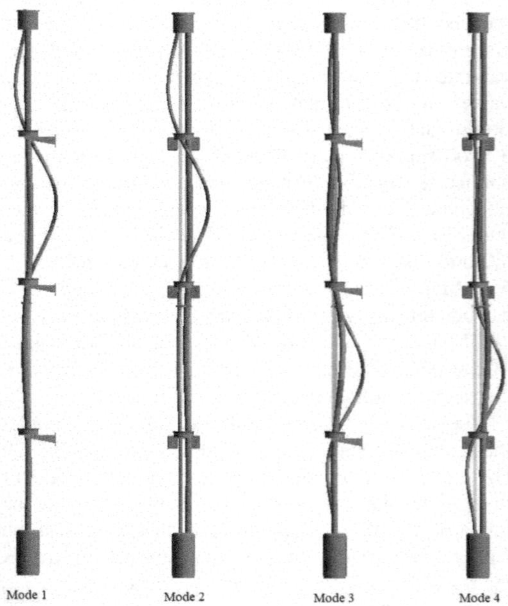

Figure 6. The first four local modes of vibration for the pipe stack structure only.

Comparing Modal 1 and Modal 3, it appears that the local mode 1 of the tank wall interacts with mode 4 of the pipes stack connected to the tank wall. In addition, it is seen that the tank wall influences the direction of the mode shape of the pipe stack support.

Furthermore, from Table 1 it is noted that the modes come as approximate pairs with comparable frequencies of the two following modes. The lowest natural frequencies of the local model are within the frequency range of the engine vibration, but not within the range of the wave-induced bending moment.

5 RELIABILITY MODEL OF FATIGUE CRACK GROWTH

Fatigue crack growth is caused by cyclic stresses at the crack tip. A common formulation of fatigue crack propagation is the Paris-Erdogan relation, also called the Paris law. This formulation of crack propagation is based on the assumption that linear elastic fracture mechanics (LEFM) is applicable and that the crack growth stage (stage II) of the fatigue history is dominant. For welded joints, this is typically the case, and welded components are especially susceptible to fatigue cracking due to the impact of the welding procedure on the components and surrounding material. The Paris crack growth law is given as:

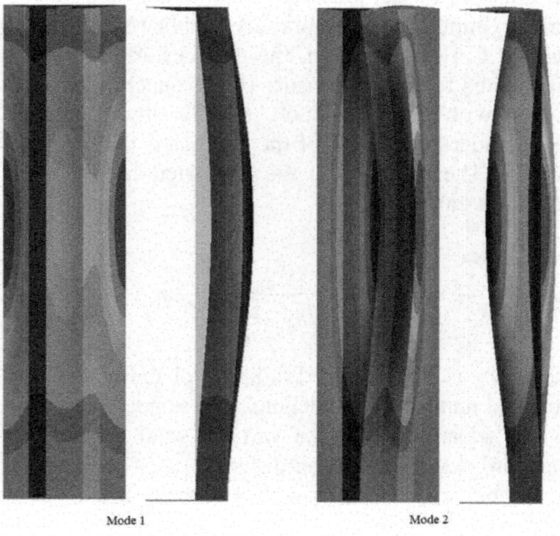

Figure 7. The first two modes of vibration for the tank wall section only.

Table 1. Natural frequencies for local model.

Mode	Modal 1 Frequency [Hz]	Modal 2 Frequency [Hz]	Modal 3 Frequency [Hz]
1	17.78	17.87	21.75
2	17.85	17.99	21.85
3	21.12	23.90	35.12
4	21.79	24.85	34.75
5	24.06	29.36	36.68
6	24.71	29.75	36.91
7	29.11	38.40	38.43
8	29.64	40.21	38.87

$$\frac{da}{dN} = C \cdot (\Delta K)^m \tag{1}$$

where C and m are material parameters, and ΔK is the stress intensity factor range, given as:

$$\Delta K = \Delta S \sqrt{\pi a} F(a) \tag{2}$$

where ΔS is the stress range, a is the crack size, and $F(a)$ is called the geometry function, which in general depends on the size of the initial crack, the geometry of the crack and the configuration of the loading.

For fatigue loading on a structural component, a stress limit exists for which an initial crack will not start to propagate if the stresses occurring at the location of the crack are below this limit. The stress limit can be defined as the maximum allowable stress before crack propagation occurs. For a given initial crack size, a_i, and a given threshold stress intensity factor (SIF), ΔK_{th}, it is possible to find the limiting stress. The threshold SIF is a material parameter for a given load ratio and environmental conditions, and different recommendations for its value are found in the literature (e.g., British Standard 2005, Lindley & Richards 1982). Based on this, a limit state formulation is defined for the possible occurrence of crack propagation of a crack in the pipe stack support. The initial crack is assumed to be a surface crack with a semi-elliptical shape. Assuming the threshold conditions for the initial crack are known, the stress limit may be calculated by:

$$\Delta S_l = \frac{\Delta K_{th}}{\sqrt{\pi a_i} F(a_i)} \tag{3}$$

It is to be noted that this formulation has a limitation in that the threshold parameter ΔK_{th} is obtained by testing physically 'long' cracks, and LEFM is only valid under such conditions. For short cracks, the plastic zone at the crack tip becomes large compared to the crack length, and LEFM is no longer valid (Berge & Ås 2017). Under such circumstances, elastic-plastic fracture mechanics (EPFM) is applied to describe the behaviour of the crack. However, for the current investigations, it is assumed that LEFM is applicable.

The geometry function is determined based on the initial crack size and empirical formulations of SIF calculations (Newman & Raju 1981). For the failure probability estimation, different initial crack sizes are investigated. When it comes to the threshold SIF, different values are given in the literature depending on the type of metal and the environment to which the metal is exposed (corrosive or non-corrosive). A standard providing methods for assessing flaws in metallic structures gives a fatigue crack growth threshold for welded joints of 2 MPa$\sqrt{m}$ (British Standard 2005), while in another study it was shown experimentally that the threshold SIF for stainless steel is about 10 MPa$\sqrt{m}$ (Singh et al. 2003).

Looking at Eq. (3) it is seen that for a larger threshold SIF, given the same crack size, the fatigue limit becomes larger. Therefore, only the lowest threshold SIF of 2 MPa$\sqrt{m}$ is implemented in the failure probability calculation, as this will be the critical case of the two. Table 2 gives the calculated fatigue limit for five different initial crack sizes.

Table 2. Threshold stress limit for different initial crack sizes.

Case	a_i [mm]	F [-]	S_l [MPa]
1	0.1	1	125
2	0.2	1	91
3	0.3	1	82
4	0.4	1	68
5	0.5	1	60

In reliability analysis, failure criteria or acceptance criteria need to be defined for the evaluation of the significance of the estimated failure probabilities. However, this represents one of the major challenges with probability estimation, as these acceptance criteria are difficult to define. The conditions for defining criteria are generally only known on an approximate level, for instance, based on experimental data or operational experience. For the current investigations in this study, an acceptance criterion is set for which it can be argued that crack propagation is not likely to occur at the supports. A probability of failure between 1-5 % is established as the acceptance criteria.

6 FAILURE PROBABILITY FOR CRACK PROPAGATION

The failure probabilities are calculated using Proban, where different reliability methods are applied to evaluate the integrity of the calculated failure probabilities. Proban is a general-purpose probabilistic calculation tool which contains a variety of failure probability simulation and approximation methods, making it suitable for structural reliability calculations. The results from the failure probability calculations are presented in Table 3, as percentages (%).

Table 3. Estimated failure probabilities for the occurrence of crack propagation, given in percentage (%).

Case	S_l[MPa]	MC	DSPS	FORM	SORM
1	125	0	0	0	0
2	91	0.021	0.022	0.050	0.020
3	82	0.204	0.212	0.449	0.190
4	68	2.512	2.459	4.458	2.336
5	60	6.520	6.520	9.831	6.756

The MC stands for Monte Carlo simulation, DSPS stands for design point simulation, also called importance sampling, which is an improved Monte Carlo method, and FORM and SORM are the first-order and second-order reliability methods, respectively.

A calculated failure probability between 1 - 5 % is defined as the acceptance criterion, meaning for the current circumstances investigated, crack propagation is not likely to occur. From Table 3, it is seen that for a larger initial crack size, there is a larger possibility of crack propagation to occur. For the largest initial crack size analysed (case 5), the acceptance criterion is exceeded, and there is a relative possibility that cracks of size 0.6 mm and larger will start to propagate. Furthermore, the results from the Monte Carlo simulation methods coincide well with the SORM approximation, while the first-order approximation deviates from the others. This may indicate that, for the investigations presented in this study, a first-order approximation is not an adequate approach, which may be explained by the failure surface displaying nonlinear characteristics, and a second-order approximation seems more accurate.

To give a measure of uncertainty associated with the failure probability calculation using the two Monte Carlo methods, error estimates and uncertainty importance factors are evaluated. For the Monte Carlo simulation, the error estimates can be calculated to give a measure of the uncertainty associated with the calculation, specifically calculating the standard deviation of the estimated failure probability. The variance of the estimated failure probability is calculated by:

$$\mathrm{Var}\left[\hat{p}_f\right] = \frac{1}{N-1}\hat{p}_f(1-\hat{p}_f) \qquad (4)$$

where $\hat{p}_f$ is the estimated failure probability and N is the total number of simulations. The standard deviation is the square root of the variance, and the average standard deviation for the different cases investigated is in the order of 10^{-4} with a sample size of N = 500000. The standard deviation is relatively small, which may be due to the large number of sample simulations. Proban is fast with respect to calculation of failure probability, and the relatively large sample sizes in the present analyses are processed in just a few seconds.

For the importance sampling simulation, calculating the standard deviation is more complicated and requires further information that is not explicitly provided in the calculations using Proban. Therefore, for the current analysis, the uncertainty importance factors are evaluated at the design point for the DSPS procedure. The uncertainty importance factors quantify the relative contribution of the various uncertainties (i.e., random variables) to the calculated failure probability. This essentially quantifies the influence of each random variable on the results, and the importance factors for the DSPS analysis as averaged values for the different cases are given in Table 4.

Table 4. Uncertainty importance factors for DSPS failure probability estimations (N=100000).

Variable	Importance factors
Engine rpm	50.7
Filling level	39.2
VBM	10.1

The engine speed is the most influential variable and the filling level also has a significant impact on the results, whereas the wave-induced VBM has a minor role compared to the two other stochastic variables.

6.1 *Preliminary investigation of crack propagation*

A preliminary estimation of fatigue crack growth over the lifetime of the structure is conducted, where it is assumed that the structure experiences 10^7 cycles during its lifetime of, say, 25-30 years. The limit state is formulated as the crack propagation reaching a critical crack size, typically taken as the thickness of the plate where the crack is growing or some percentage of this thickness.

The thickness of the pipe stack support where the crack is assumed to be present is 8 mm, which is taken as the critical crack size. Two cases are investigated; the probability that the crack grows to 50 % of the critical crack size during the lifetime of the structure, and the probability that the crack grows through the thickness. The initial crack size is assumed to be exponentially distributed with a mean of 1.0, a standard deviation of 1.0 and a lower value of 0.1 mm. The parameters in the Pairs law are taken from a simplified fatigue crack growth law recommended by British Standard 2005, as $C = 5.21\ 10^{-13}$ and $m = 3$. The estimated failure probability for crack propagation is given in Table 5.

Table 5. Estimated failure probabilities for crack propagation during the lifetime, given in percentage (%).

Critical crack length [mm]	MC	DSPS	FORM	SORM
4	1.307	1.311	1.312	1.312
8	0.015	0.015	0.015	0.015

Two cases are investigated; the probability that the crack grows to 50 % of the critical crack size during the lifetime of the structure, and the probability that the crack grows through the thickness. The thickness of the pipe stack support is 8 mm, and this is then taken as the critical crack size. The initial crack size is assumed to be exponentially distributed with a mean of 1.0, a standard deviation of 1.0 and a lower value of 0.1 mm. The parameters in the Pairs law are taken as $C = 5.21\ 10^{-13}$ and $m = 3$ according to recommended values for simplified fatigue crack growth law given by (British Standard 2005). The estimated failure probability for crack propagation is given in Table 5.

The failure probability that a crack grows to 50 % of the critical crack size, that is, grows to 4 mm, during the lifetime of the structure is 1.3 %. The failure probability that the crack grows through the thickness of the pipe stack support is 0.015 %. Note that considerations of inspection and maintenance operations are not taken into account here. Typically, inspections are done every 5 years, sometimes with intermediate inspections, say, every 2.5 years. Further investigation of the progression of crack propagation and considerations such as time until failure are needed for the implementation of fatigue crack growth considerations in the reliability model. Moreover, this is only a preliminary analysis, and the calculation of crack growth depends on several parameters that are based on experimental data or measurements, which gives large uncertainties associated with these parameters (for instance the material parameters C and m). A more comprehensive investigation of this is also needed.

7 DISCUSSION AND CONCLUSION

The vibration response assessment includes the main engine and wave-induced loading, in addition to the effect of the filling level in the cargo tanks. Other sources of vibration which may contribute to fatigue loading on the secondary hull component should be implemented in further development of the reliability model. Such contributors to vibration may be sloshing, which is significant for components installed in cargo tanks where they can be exposed to liquid sloshing, and propeller excitation for secondary structural components located in the vicinity of the propellers. The excitation frequencies of the vibration sources should also be studied more closely and compared with the natural frequencies of the local structure to further investigate vibration response for different load combinations and operational conditions for the possibility of resonant conditions to occur.

The estimated failure probabilities shown in Table 3 indicate that the probability of the occurrence of crack propagation increases with increasing crack size. For the smallest crack sizes, cases 1-5, crack propagation is not likely to occur since the probability of failure for these cases is within the acceptance criteria currently defined. For case 6, with an initial crack size of 0.6 mm, the estimated failure probability exceeds the established criteria, and thus, crack propagation will in this case occur under the defined circumstances. Therefore, it is likely that cracks of sizes larger than 0.5 mm (as this is the last case investigated within the defined criteria) will start to propagate under the investigated loading conditions.

In fatigue loading conditions, stresses are generally well below yield and LEFM may be applied, but the initial size of the crack may be argued as too small for application of LEFM. This should be investigated further, as it may be necessary to incorporate elastic-plastic fracture mechanics to properly represent the short crack and its behaviour at the crack tip. Furthermore, a stress limit was defined for which it is assumed that crack propagation does not occur if stresses at the component are below this limit. However, it is to be noted that ships in operation are exposed to variable amplitude loading, and cracks may propagate at stresses below this limiting stress.

Preliminary analyses of fatigue crack propagation during the lifetime of the ship structure are conducted. The results presented in Table 5 show a relatively small probability for an initial crack to reach a critical size during the lifetime of the ship. However, as stated above, LEFM is assumed here, and the probability distribution for the initial crack size is applied with a lower crack size value of 0.1 mm, which may be considered a short crack. There are also large uncertainties associated with the different parameters used to calculate the crack propagation and this need to be investigated further. Moreover, no consideration of inspection and maintenance is taken into account. Inspection and maintenance are conducted for all ships regularly and it will influence the fatigue cracking of components in the structure. Therefore, it should be considered to include an evaluation of the influence of inspection and maintenance on the fatigue cracking of ships in service. Hence, the formulation of crack propagation should be developed further to investigate the propagation of cracks during the ship's or the component's lifetime, and also investigate the time until failure for different loading conditions. This might be helpful in the fatigue design of such secondary components as investigated in this paper.

REFERENCES

ABS (2022). Insights into ship vibration analysis. Technical report, American Bureau of Shipping.

ABS (2023). Guidance note on ship vibration. Technical report, American Bureau of Shipping.

Adland, R., P. Cariou, & F.-C. Wolff (2020). Optimal ship speed and the cubic law revisited: Empirical evidence from an oil tanker fleet. *Transportation Research Part E: Logistics and Transportation Review 140*, 101972.

Asmussen, I., W. Menzel, & H. Mumm (2001). *Ship Vibration*. Hamburg, Germany: Germanischer Lloyd. GL Technology.

Berge, S. & S. K. Ås (2017). Fatigue and fracture design of marine structures. In *(Compendium)*. Norwegian University of Science and Technology. Trondheim, Norway.

Božić, Z., M. Mlikota, & S. Schmauder (2011). Application of the k, j and ctod parameters in fatigue crack growth modelling. *Tehnicki Vjesnik 18*, 459–466.

British Standard (2005). Guide to methods for assessing the acceptability of flaws in metallic structures. (BS 7910:2005).

Bukovac, O., V. Medica, & V. Mrzljak (2015). Steady state performances analysis of modern marine two-stroke low speed diesel engine using mlp neural network model. *Brodogradnja 66*, 57–70.

Gan, J., X. Liu, H. Liu, Z. Wang, & W. Wu (2022). Effect of high-low frequency superimposed loading on the fatigue crack propagation of longi-web connection joint. *International Journal of Fatigue 163*, 107043.

Gan, J., X. Liu, Z. Wang, & W. Wu (2021). Experimental study on the fatigue damage of designed t-type specimen with high-low frequency superimposed loading. *International Journal of Fatigue 143*, 105985.

Gaspar, B., A. P. Teixeira, & C. Guedes Soares (2016). Effect of the nonlinear vertical wave-induced bending moments on the ship hull girder reliability. *Ocean Engineering 119*, 193–207.

Guedes Soares, C. & Y. Garbatov (1999). Reliability of maintained ship hulls subjected to corrosion and fatigue under combined loading. *Journal of Constructional Steel Research 52*(1), 93–115.

Guedes Soares, C. & T. Moan (1991). Model uncertainty in the long-term distribution of wave-induced bending moments for fatigue design of ship structures. *Marine Structures 4*(4), 295–315.

Kleivane, S. K., B. J. Leira, & S. Steen (2023a). Development of a reliability model for crack growth occurrence for a secondary hull component. *Brodogradnja 74*(1), 99–115.

Kleivane, S. K., B. J. Leira, & S. Steen (2023b). Reliability Analysis of Crack Growth Occurrence for a Secondary Structural Component due to Vibration Excitation. Volume Volume 2: Structures, Safety, and Reliability of *International Conference on Offshore Mechanics and Arctic Engineering*, pp. V002T02A042.

Leheta, H., A. Elhewy, & H. Younes (2016). Analysis of fatigue crack growth in ship structural details. *Polish Maritime Research* 23.

Lewis, E. V. & SNAME (1988). *Principles of naval architecture: 2: Resistance, propulsion and vibration* (2nd revision. ed.), Volume 2. Jersey City, N.J: Society of Naval Architects and Marine Engineers.

Lindley, T. C. & C. E. Richards (1982). Near-threshold fatigue crack growth in materials used in the electricity supply industry. Proc. Second Int. Conf. on Fatigue and Fatigue Threshold, Engineering Materials Advisory Services Ltd. England.

MAN Energy Solutions (2023). MAN BW Project Guidelines. https://www.man-es.com/marine/products/planning-tools-and-downloads/project-guides. MAN Marine Energy Systems.

Newman, J. C. & I. S. Raju (1981). An empirical stress-intensity factor equation for the surface crack. *Engineering Fracture Mechanics 15*(1), 185–192.

Noonan, E. F. (1990). Ship vibration design guide. http://www.shipstructure.org/. Ship Structure Committee (SSC-350).

Singh, P., B. Guha, & D. R. G. Achar (2003). Fatigue life prediction for stainless steel welded plate cct geometry based on lawrence's local-stress approach. *Engineering Failure Analysis 10*(6), 655–665.

Storhaug, G. (2007). *Experimental investigation of wave induced vibrations and their effect on the fatigue loading of ships*. Ph. D. thesis, Norwegian University of Science and Technology (NTNU), Department of Marine Technology, Faculty of Engineering Science. Oslo.

Tamimi, M. F., M. Soliman, & O. Khandel (2023). A comprehensive approach for quantifying the reliability of ship hulls under propagating fatigue cracks. *Ocean Engineering 279*, 114488.

Advances in Maritime Technology and Engineering – Guedes Soares & Santos (Eds)
© 2024 The Author(s), ISBN 978-1-032-83099-5

Numerical investigation of fire risk in ammonia tank leak incidents: Implications for ammonia-fueled vessels

M.P. Mujeeb-Ahmed, B. Jeong, P. Zhou, H. Jang, I. Hwang & H. Wang
Department of Naval Architecture, Ocean and Marine Engineering, University of Strathclyde, Glasgow, UK

ABSTRACT: The potential of ammonia as a sustainable marine fuel is accompanied by significant safety concerns. This study delves into the aftermath of ammonia tank leakage incidents, with a primary focus on evaluating the associated fire risks. Employing a general cargo ammonia-retrofit ship comprising of several high-pressure ammonia storage tanks, we conduct an in-depth analysis utilizing CFD modelling of the multifaceted factors that contribute to the severity of jet fire scenarios, encompassing different ammonia release rates, and leak directions (horizontal and vertical). The results show that in the event of fire, it poses significant safety concerns to the storage tanks, especially composite tanks, and therefore passive fire protection measure including the installation of the removable fire barrier is suggested as the design recommendation, apart from the active CO_2 measures. The knowledge garnered from this research will empower stakeholders to refine their preparedness and response strategies in the event of fire accidents.

1 INTRODUCTION

The global pursuit of sustainable and environmentally friendly energy sources has led to a renewed interest in ammonia as a potential carbon-neutral fuel. Ammonia, with its high energy density and zero carbon emissions during combustion, has emerged as a promising candidate for alternative marine fuel. However, the utilization of ammonia as a fuel comes with inherent challenges, particularly regarding safety considerations in the event of accidental releases (Cheliotis et al. 2021, Jang et al. 2023).

One critical aspect of ensuring the safe deployment of ammonia-fueled vessels is the assessment and mitigation of fire risks associated with ammonia tank leak incidents. The consequences of such incidents extend beyond the immediate threat to personnel and infrastructure, encompassing broader implications for the viability and acceptance of ammonia as a mainstream fuel option.

The likelihood of a fire resulting from ammonia leakage is generally considered to be a secondary hazard, given the thermophysical properties of ammonia, as detailed in Table 1. This is attributed to its high autoignition temperature and a limited flammability range, among other factors.

Although toxicity is the main hazard associated with ammonia, fire and usually combined with explosion cannot be neglected and should be considered a critical hazard for the following reasons:

Table 1. Characteristics of ammonia fuel (The Engineering ToolBox, 2008).

Sl No.	Properties (unit)	Value
1	Flashpoint (°C)	132
2	Auto-ignition temperature (°C)	630
3	Boiling point (°C)	-33
4	Low heating value (LHV) (MJ/kg)	18.6
5	High heating value (HHV) (MJ/kg)	22.5
6	Flammability range [% volume in air, LFL (lower flammability limit) – UFL (upper flammability limit)]	15-28
7	Density, gas (kg/m^3)	0.69
8	Density, liquid at -33.3°C, 1 atm (kg/m^3)	696
9	Energy density (GJ/m^3)	11.4
10	Volume per unit energy (m^3/GJ)	3.14

1. Flammability: While not flammable in itself, ammonia can support combustion when mixed with air within a specific concentration range (16% to 25% by volume). If a leak occurs and ammonia vapour mixes with air in this range, it can ignite and cause a fire.
2. Reactivity: Ammonia is highly reactive with certain materials, such as certain metals and oxidizing agents. Contact with incompatible materials can lead to chemical reactions that generate heat, sparks, or even fire.
3. Corrosivity: Ammonia is corrosive to many materials commonly used in marine equipment and infrastructure. Corrosion can weaken structural integrity and increase the risk of leaks,

DOI: 10.1201/9781003508762-38

which could lead to fire hazards if ammonia contacts reactive surfaces or materials.

4. Toxicity: Ammonia is toxic when inhaled at high concentrations. In the event of a fire involving ammonia as a marine fuel, the combustion products can include toxic gases such as nitrogen oxides and hydrogen cyanide, which pose additional risks to human health and safety.

The flammable gas concentrations are defined between the Lower Explosive Limit (LEL) of 15% (150,000 ppm) and the Upper Explosive Limit (UEL) of 28% (280,000 ppm). This is particularly important in an enclosed area (such as ammonia tanks stored in the cargo hold space considered in this study) (DNV GL, 2020) and the presence of oil will increase the fire hazard or lower the LEL to as low as 8% (80,000 ppm). Further, ammonia stored in pressurized tanks (at 10 bar) can rupture when overexposed to heat.

The adoption of ammonia as a fuel, especially in the marine field, is still in the early stages of development and testing. Further, the likelihood and severity of fire accidents depend on various factors, including fuel properties, ship design, operational procedures, and safety measures in place.

Research on hydrogen jet fire (Liu et al. 2024, Rengel et al. 2018 & Yuan et al. 2021), and LNG leaked fire (Chu & Chang, 2017, Lee et al. 2015) are available. Recently, (Pomonis et al. 2022) performed an ammonia fire simulation in the engine room. In conclusion, no specific studies focus on the ammonia fire from the high-pressure storage tank installed in the ammonia-powered ship.

In summary, there has been a lack of research on the fire generated from the leaked ammonia stored in high-pressure tanks for ammonia-powered ships.

This paper presents a comprehensive numerical investigation aimed at better understanding the fire risk dynamics in high-pressure ammonia storage tank leak scenarios. By employing advanced computational models and simulations, we delve into the complex interplay of factors influencing jet fire propagation, and severity in the context of varying ammonia release rates. Further, passive fire protection measures such as the use of fire barriers have been considered as the mitigation action used to study fire behaviour.

2 CASE STUDY

This study uses a real full-scale 31,000 dwt general cargo ship retrofit with ammonia storage tanks and its auxiliaries, see Figure 1. Ammonia with low energy content, requiring approximately 1.5 times more space than LNG (Machaj et al. 2022) and 2.4 times more tank volume than heavy fuel oil (HFO) (ABS, 2020), prompted the exploration of various potential locations within the ship to store ammonia in tanks with sufficient capacity to meet a single ship voyage. Figure 2 depicts the conceptual layout arrangement of different storage tanks, featuring two large metallic tanks positioned on the upper deck. Further, within cargo hold no. 5 space, a large metallic tank with a capacity of 155 m^3 is situated, along with two columns comprising metallic and composite tanks of 5 TEUs each, considering the scalability and modularity of the tanks. Each TEU of the composite tank comprises 8 small tanks, each with a capacity of approximately 55 m^3, while each TEU of the metallic tank contains a standalone tank with a capacity of approximately 100 m^3. Table 2 summarizes the tank arrangements in the cargo hold space where a total of 46 different tanks have been installed. Table 3 provides the design material properties of the storage tanks. This study considers fire scenarios arising from tanks located within the enclosed cargo hold space.

Figure 1. 31,000 dwt general cargo ship for ammonia retrofitting.

Figure 2. 3D model of tanks in No.5 hold within the cloud (blue-metallic tanks, yellow-composite tanks).

3 FIRE SCENARIO

The fire scenarios in the cargo hold space are evaluated based on the worst-case scenario involving

Table 2. Summary of proposed tanks and their arrangements in the No.5 cargo hold space.

	Metallic tanks	Composite tanks	Total number of tanks	Total ammonia capacity
Concept layout design	i. 1 × 155 m^3 ii. 5 × TEU (~100 m^3) *(~5 stand-alone tanks)*	i. 5 × TEU (~55 m^3) *(~40 stand-alone tanks)*	46	~ 300 m^3

Table 3. Proposed design material properties of ammonia storage tanks

Sl No.	Properties	Description/ value	Unit
1	Tank type	Type C	-
2	Material	P-355 NL1	-
3	Yield strength	345	MPa
4	Thickness	16-40	mm
5	Design internal pressure	19	barg
6	Design internal temperature	45	°C
7	Corrosion allowance	1.5	mm
8	Filling limit	95%	
9	Max. temperature resistance of metallic tank material (metallic)	500-600	°C
10	Max temperature resistance of composite tank material	100	°C

potential leakage from a large metallic tank (capacity of 155 m^3) and its impact on adjacent tanks placed in open containers. With ammonia stored at a high pressure of 10 bar and pressure relief valves (PRVs) set at 20 bar, any leakage above this pressure could lead to a jet leak and subsequent jet fire upon rapid ignition. Thus, only jet fires are considered, and both horizontal and vertical jet fires are assessed.

The analysis of fire scenarios focuses on single leaks, considering three different sizes—5 mm, 10 mm, and 100 mm—to encompass small, medium, and large leakage situations, respectively for horizontal jet fires. For comparison purposes, we have included an additional scenario involving a 50 mm sized leak for the vertical jet fire scenarios. This decision was made due to the absence of fire occurrences in the cases of 5 mm and 10 mm leak sizes (see, Section 5.1.2). The corresponding mass flow rates are calculated based on pressure, temperature, volume, and leak sizes, resulting in values of 0.062, 0.25, and 24.8 kg/s, respectively.

Regarding the impact of horizontal jet fire on adjacent tanks, three fire scenarios were considered:

Scenario 1: No fire barrier, evaluating the impact on open containers.

Scenario 2: A fire barrier is placed in front of the composite tanks, while metallic tanks in the containers are unprotected.

Scenario 3: A fire barrier is placed covering both metallic and composite containers, offering complete protection against the jet fire.

Illustrated in Figure 3, at present, a distance of 1.2 m separates a large metallic tank and containerized tanks, and therefore, a fire barrier is strategically positioned 700 mm away from the containerized tanks in accordance with SOLAS Reg.II-2/13 regulations and fire safety systems Chapter 13 (IMO, 2007), for easy inspection and access. The dimensions of the fire barrier, including its height and width, have been chosen to fully cover the containerized tanks in the specified scenarios. In this study, the fire barrier has been modelled as an obstruction, assuming an ideal case of an adiabatic surface with no exchange of heat across the fire barrier surface.

4 FIRE MODELLING SETUP

In conducting jet fire simulations, we employed the FLACS-Fire module to execute and analyse various fire scenarios. This software effectively considers and models fire scenarios, having undergone validation with experimental and real fire situations

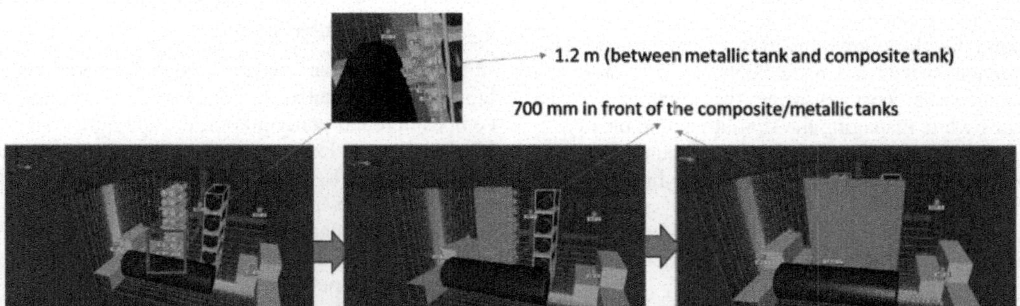

Figure 3. Three jet fire scenarios considered for the fire simulation: (a) existing arrangement, (b) Fire barrier installed for the composite tanks, (c) Fire barrier covering both metallic and composite tanks.

(Hansen et al. 2010). For a comprehensive understanding and detailed information, please refer to the FLACS user manual (FLACS, 2022).

The initial step in our simulation process, based on the general arrangement drawing, involves creating a 3D model of cargo hold space No. 5. This model accurately captures the key tank locations, arrangements, and structure of cargo hold walls and ceilings. Subsequently, the developed model is imported into FLACS, where necessary geometric adjustments are made to eliminate potential errors in the model (see Figure 4 (a)).

4.1 *Grid modelling*

In FLACS, the total simulation volume (12.8 m × 25.4 m × 13.4 m) of cargo hold space is partitioned into grid cells or control volumes using three sets of grid planes, each oriented along one of the three axes (*x*, *y*, *z*). To effectively represent simulation results, FLACS employs three grid formats: core domain, stretched domain, and refined leak domain.

The core domain encompasses the central area of interest with cells of uniform size, ensuring an accurate representation of the key simulation parameters. The stretched region accounts for potential fire expansion beyond the core domain, while the refined leak domain focuses on precisely capturing the leak area to reduce unnecessary refinement in the remaining space.

In Figure 4, the FLACS model and grid details for the fire study are illustrated, including different monitor points and the position of the leak for horizontal jet fires. This strategic grid division allows for a comprehensive and efficient analysis of the simulated fire scenarios.

At first, grid sensitivity in the fire simulation, specifically regarding temperature outputs, was assessed. The findings indicated that the results remain relatively stable even with mesh refinement around the leak area. Consequently, to optimize computational efficiency, a coarse mesh is employed. Accordingly, for the core and stretched domains, a total of 37,444 cells are utilized, with a stretching factor of 1.25. Additionally, 45,584 cells are implemented, accounting for the refined mesh specifically around the leak area in all fire simulations. This approach strikes a balance between accuracy and computational cost, ensuring the reliability of the results while managing the simulation complexity.

Boundary condition Nozzle-type boundary conditions have been considered for all the directions. No wind condition is considered as the space in the cargo hold No. 5 is assumed to be sealed during the fire and no air flow takes place to restrict the oxygen inflow during the ammonia fire.

4.2 *Fire models employed*

The combustion and radiation models employed are the Eddy Dissipation Concept (EDC) and Discrete Transfer method, both well-validated by FLACS

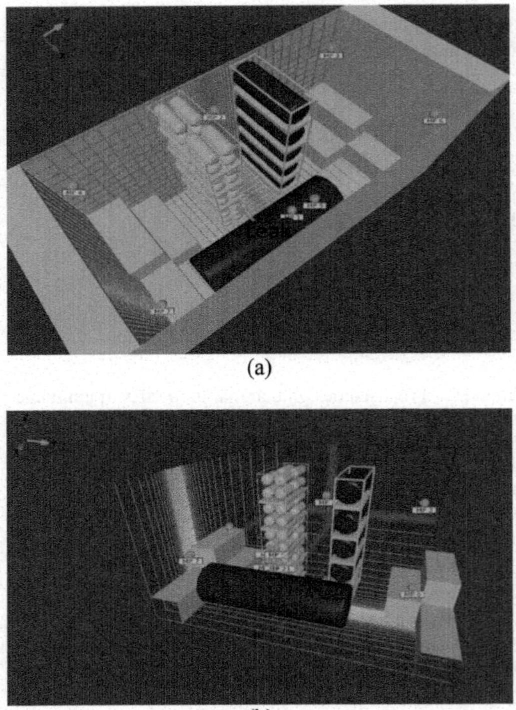
(a)

(b)

Figure 4. (a) FLACS jet fire model setup in the cargo hold space; (b) grid details and position of monitoring points (MP's) for horizontal jet fires.

through experimental fire case studies and recommended for the fire scenarios. No conduction models are used to measure the thermal activity of walls, as their effectiveness is yet to be validated, and FLACS currently discourages their use. Soot formation is not considered, assuming a carbon-free chemical combustion reaction during the combustion of ammonia gas. Lastly, no fire extinction has been activated during the simulation, again, the functionality is yet to be validated.

4.3 *Fire simulation controls*

In FLACS-Fire, the consideration of the potential fire ignitions is computed initially based on the defined composition and concentration of ammonia, flammability range, release properties, etc. An ignition region, shaped as a cube, encompasses the entire cargo hold space, and the fire ignites where the gas jet falls within the flammable limit. The ignition time is set to 0.5 s to allow sufficient ammonia leakage and the potential formation of a flammable cloud.

Furthermore, a 10-s ignition duration is set with a total simulation time of 100 s employing a using a transient solver, to enable the fire to reach a steady state. Crew members' presence in the cargo hold space is deemed of very low probability, resulting in an assumption of no fatalities due to the fire. The discharge coefficient is set at 0.85, and a 100% composition of ammonia is assumed. Ambient temperature and pressure are defined as 25°C and 1 bara, respectively. Given the absence of expected fatalities, toxicity is disregarded in the simulation.

5 RESULTS AND DISCUSSIONS

This section provides the findings from the different fire scenarios considered.

5.1 *No fire barrier*

5.1.1 *Horizontal jet fire scenario*

Figure 5 displays the temperature-time plots for horizontal jet fire scenarios with leak sizes of 100 mm and 10 mm. Notably, no fire ignition occurred in the scenario with a 5 mm leak size. Additionally, a high initial turbulence was observed at the outset for 100 mm leak scenario, and a steady state temperature of 1500°C and 1850°C is observed for leak sizes of 100 mm and 10 mm, respectively, suggesting that the impact of fire, in terms of the maximum measured temperature, is independent of the leak sizes in these scenarios. This observation may be attributed to the relatively small distance (1.2 m) between the large metallic tanks and the containerized tanks, resulting in a direct impact of the fire on the tanks.

Additionally, the plots reveal high initial turbulence followed by the establishment of a steady flame temperature. The maximum temperature reached in the steady state is identified as 1500°C (approx.). This information provides valuable insights into the fire dynamics and temperature profiles associated with different leak sizes in the specified scenarios.

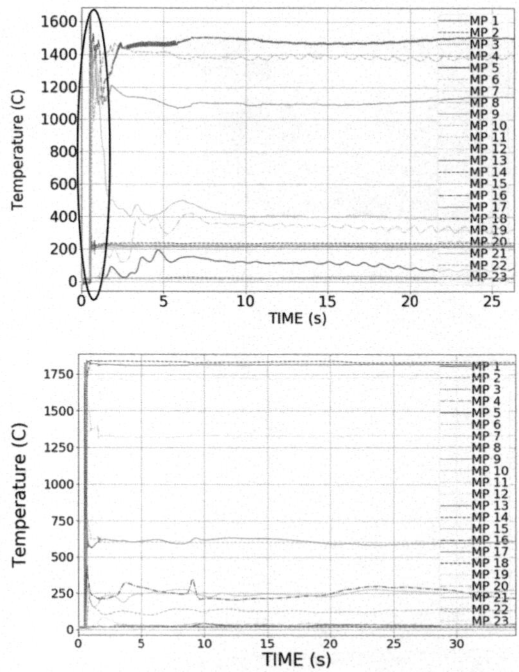

Figure 5. Temperature-time plot for the horizontal jet fire jet scenarios – 100 mm (top) and 10 mm (bottom).

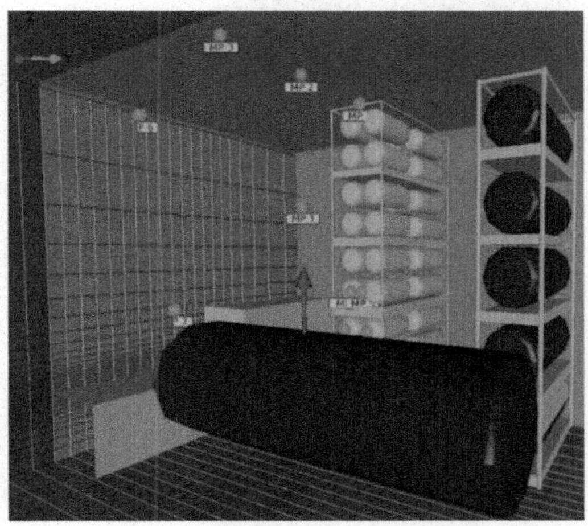

Figure 6. Location of monitor points for vertical jet fire scenario.

5.1.2 *Vertical jet fire scenario*

Figure 7 shows the results of the temperature measured at different monitor points (shown in Figure 6) for the vertical jet fire scenarios. No fire ignition happened for the scenario with a leak size of 5 mm, and 10 mm and the impact of fire in terms of steady-state temperature is found to be similar for 50 mm and 100 m leaked jet fires. Further, the impact of vertical jet fire on the storage tanks is less compared to the cargo hold ceilings itself. As a risk reduction measure, it is imperative to have active fire protection measures, including the CO_2 fire extinguishing system, aimed at swiftly extinguishing the fire and protect the cargo hold ceilings.

5.2 *Impact of installing fire barrier*

To actively mitigate horizontal jet fires impacting composite and metallic tanks within containers, the implementation of a fire barrier in front of the large metallic tank is considered a passive fire preventive measure. Figure 8-10 depict temperature-time plots obtained from strategically positioned monitor points critical for measuring temperature increases on the tanks and walls, offering valuable insights into fire behaviour.

Referring back to Table 3, it's crucial to emphasize that the maximum external fire temperatures for which metallic and composite tank material could withstand are 500-600°C and 100°C, respectively. Exposure to external fire temperatures exceeding the

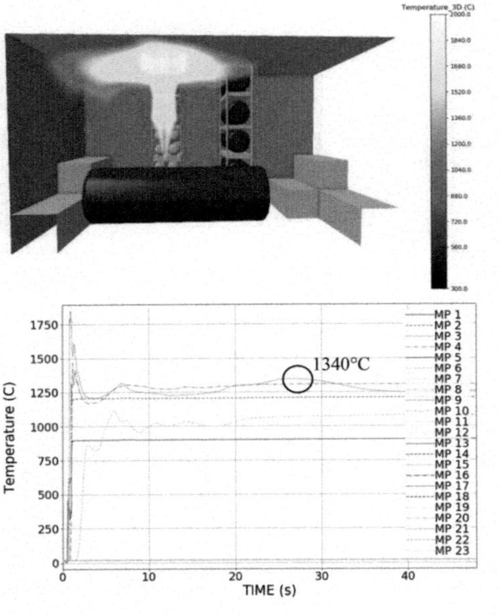

(a)

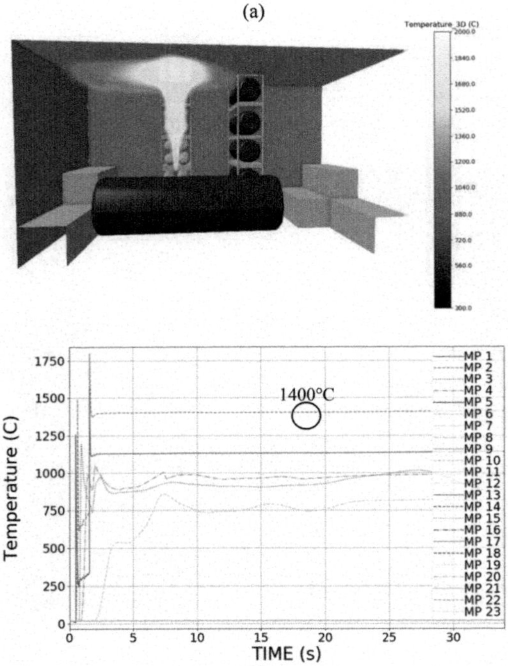

(b)

Figure 7. Temperature measured against time during vertical jet fires – (a)100 mm (top) and (b) 50 mm (bottom) leakage sizes.

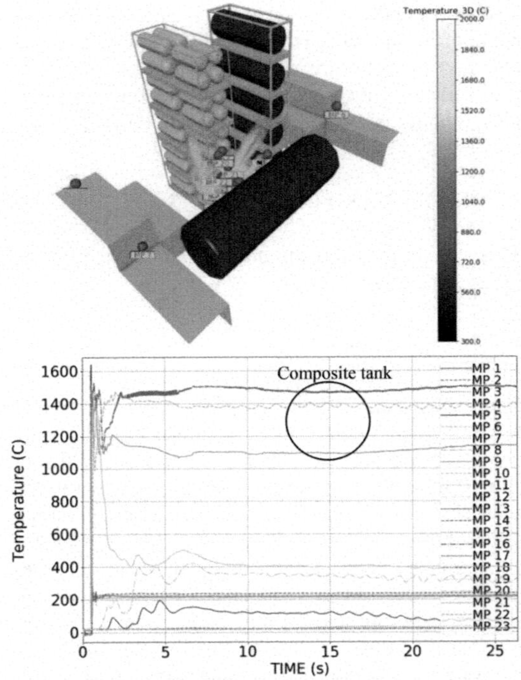

Figure 8. Fire scenario for the existing tank arrangements.

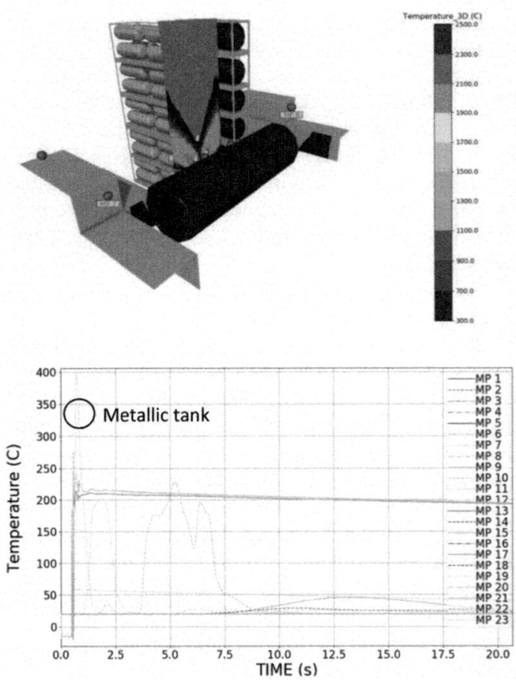

Figure 9. Fire scenario for the proposed tank arrangement scenario with fire barrier for composite tanks.

design limits can lead to the loss of tank integrity, potentially resulting in tank rupture.

In the absence of fireproofing materials or insulation, there is a risk of increased overpressure, contingent upon the material, whether metallic or composite. Such scenarios heighten the overall risk level, as outlined in Table 4, which summarizes key findings from horizontal jet fire scenarios, showcasing the maximum temperatures measured for tanks stored in containerized solutions. Importantly, the

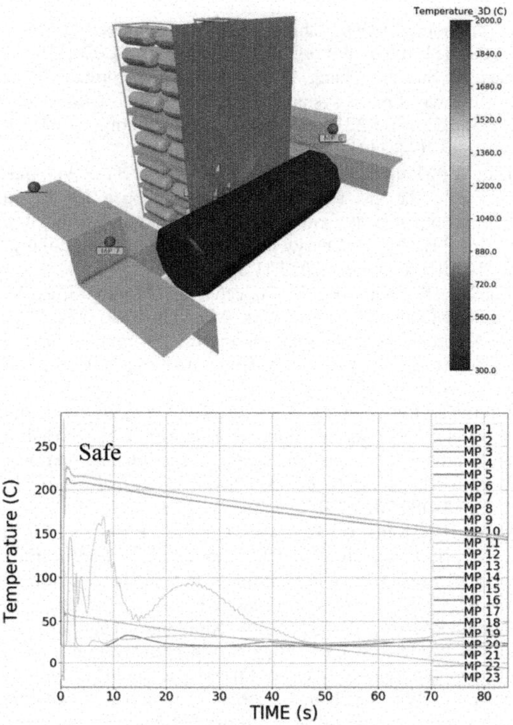

Figure 10. Fire scenario for the proposed tank arrangement scenario with fire barrier for both composite and metallic tanks.

Table 4. Summary of fire scenarios.

Fire Scenarios	Leak scenario	Max. temperature (°C) (M-->metallic, C-->composite)	Risk level
A – jet fire impact assessment (no barrier)	Horizontal jet fire, leak size=5 mm (0.006 kg/s)	No fire	Loaw
	Horizontal jet fire, leak size=10 mm (0.24 kg/s)	1500 °C	High
	Horizontal jet fire, leak size=100 mm (24.4 kg/s)	1500 °C	High
Impact of installing a barrier for composite tanks only	Horizontal jet fire, leak size=5 mm	No fire	Low
	Horizontal jet fire, leak size=10 mm	25 °C (C) 395 °C (M)	Medium
	Horizontal jet fire, leak size=100 mm	25 °C (C) 395 °C (M)	Medium
Impact of installing a barrier for both composite and metallic tanks	Horizontal jet fire, leak size=5 mm	No fire	Low
	Horizontal jet fire, leak size=10 mm	25 °C (C) 25 °C (M)	Low
	Horizontal jet fire, leak size=100 mm	25 °C (C) 25 °C (M)	Low

installation of a fire barrier significantly reduces the risk level and offers complete protection for the composite tanks (for barrier placed in front of composite tanks) and both composite and metallic tanks (for barrier covering metallic tanks as well). This underscores the effectiveness of employing fire safety measures to mitigate potential risks associated with horizontal jet fires.

6 CONCLUSIONS AND RECOMMENDATIONS

Following are the main conclusions and recommendations derived from the research findings:

- Though employing open-type containers of ammonia storage tanks has the advantage of easy access and maintenance, however, this could potentially lead to direct exposure to the horizontal jet fire from the adjacent large tank and impair structural integrity.
- By incorporating passive fire protection measures, for example, by placing a removable fire barrier of suitable height and width between the large metallic tank and adjacent containers, it was found that the facility enhances its overall safety posture against a potential jet fire and demonstrates a commitment to mitigating potential fire-related risks. This recommendation aligns with industry standards and emphasizes a proactive approach to fire safety in ammonia storage facilities.
- A removable fire barrier could be considered considering access points in the fire barrier to allow for emergency response personnel and equipment and regular inspection and maintenance schedule for the fire barrier to ensure tank structural integrity and effectiveness over time.
- Given the presence of active fire extinguishing measures like CO_2 supply in cargo hold No. 5 space, it is advisable to implement a fire barrier with an A-15 or A-30 rating. However, this recommendation can be validated through CFD simulations of fire extinction using a CO_2 extinguishing system.

The outcomes of this research are anticipated to play a pivotal role in shaping fire safety standards and fostering confidence in the storage and utilization of ammonia as a sustainable and secure energy solution in the shipping industry.

ACKNOWLEDGMENTS

The work presented in this paper has been carried out within the UKRI HORIZON-CL5-2021-D5-01 NH_3 CRAFT project, co-funded by the European Commission (Project Number: 101056831 (HE), 10105683 (IUK)). The authors would like to express sincere gratitude to all the partners in the project including designers, class societies, industries, and academia for their critical review and constructive feedback during the course of this study. The views and opinions presented in this paper are those of the authors alone.

REFERENCES

ABS, 2020. Ammonia as marine fuel. Sustainability whitepaper.

Cheliotis, M., Boulougouris, E., Trivyza, N.L., Theotokatos, G., Livanos, G., Mantalos, G., Stubos, A., Stamatakis, E., Venetsanos, A., 2021. Review on the safe use of ammonia fuel cells in the maritime industry. Energies (Basel) 14, 1–20. https://doi.org/10.3390/en14113023

Chu, B., Chang, D., 2017. Effect of full-bore natural gas release on fire and individual risks: A case study for an LNG-Fueled ship. J Nat Gas Sci Eng 37, 234–247. https://doi.org/10.1016/j.jngse.2016.11.043

DNV GL, 2020. Ammonia as a marine fuel safety handbook.

FLACS, 2022. FLACS-CFD v22.1r2 User's Manual.

Hansen, O.R., Gavelli, F., Ichard, M., Davis, S.G., 2010. Validation of FLACS against experimental data sets from the model evaluation database for LNG vapor dispersion. J Loss Prev Process Ind 23, 857–877. https://doi.org/10.1016/J.JLP.2010.08.005

IMO, 2007. FSS code: International code for fire safety systems. Resolution MSC. 98(73) 1–190.

Jang, H., Mujeeb-Ahmed, M.P., Wang, H., Park, C., Hwang, I., Jeong, B., Zhou, P., Mickeviciene, R., 2023. Regulatory gap analysis for risk assessment of ammonia-fuelled ships. Ocean Engineering 287, 115751. https://doi.org/10.1016/j.oceaneng.2023.115751

Lee, S., Seo, S., Chang, D., 2015. Fire risk comparison of fuel gas supply systems for LNG fuelled ships. J Nat Gas Sci Eng 27, 1788–1795. https://doi.org/10.1016/j.jngse.2015.11.003

Liu, W., Markert, F., Giuliani, L., Hostikka, S., 2024. Analyzing the gas temperature of a hydrogen jet fire in a compartment with the Fire Dynamics Simulator. Int J Hydrogen Energy 53, 1097–1106. https://doi.org/10.1016/j.ijhydene.2023.11.306

Machaj, K., Kupecki, J., Malecha, Z., Morawski, A.W., Skrzypkiewicz, M., Stanclik, M., Chorowski, M., 2022. Ammonia as a potential marine fuel: A review. Energy Strategy Reviews. https://doi.org/10.1016/j.esr.2022.100926

Pomonis, T., Jeong, B., Kuo, C., 2022. Engine room fire safety evaluation of ammonia as marine fuel. Journal of International Maritime Safety, Environmental Affairs, and Shipping 6, 67–90. https://doi.org/10.1080/25725084.2021.2015867

Rengel, B., Mata, C., Pastor, E., Casal, J., Planas, E., 2018. A priori validation of CFD modelling of hydrocarbon pool fires. J Loss Prev Process Ind 56, 18–31. https://doi.org/10.1016/j.jlp.2018.08.002

The Engineering ToolBox, 2008. Ammonia - Thermophysical Properties. URL https://www.engineeringtoolbox.com/ammonia-d_1413.html (accessed 3.4.24).

Yuan, Y., Wu, S., Shen, B., 2021. A numerical simulation of the suppression of hydrogen jet fires on hydrogen fuel cell ships using a fine water mist. Int J Hydrogen Energy 46, 13353–13364. https://doi.org/10.1016/j.ijhydene.2021.01.130

Advances in Maritime Technology and Engineering – Guedes Soares & Santos (Eds)

A numerical study of the behaviour of Hydrogen in ventilated spaces

A. Twerda, B. van de Krol, A.W. Vredeveldt & C. Lombardi
Netherlands Organisation for Applied Scientific Research TNO, Den Haag, The Netherlands

ABSTRACT: This paper discusses several numerical tools to model scenarios of a hydrogen leak. The first one is a CFD model and the second is HYRAM$^+$, an engineering tool based on publicly available data and models. The CFD model was validated using experimental data and benchmark results of similar models and used to investigate the effect of leak rates and forced ventilation in a 40-foot container. The outcome was compared with HYRAM$^+$ results and differences in layer height and H_2 concentration were found, showing that HYRAM$^+$ underestimates both. One of the CFD dispersion results was used as initial situation to simulate the ignition of a Hydrogen layer, resulting in a fast propagation of the flame front and a significant increase in pressure. The results show the applicability of CFD and other numerical tools for the early risk assessment of hydrogen-fuelled ships and, the outcome can be used designing the fuel-cell ventilation systems and performing explosion risk assessments.

1 INTRODUCTION

To reduce the emission of greenhouse gasses the maritime transport sector is looking to alternatives for fossil fuels. One of these alternatives is hydrogen which, after combustion, has only water as "waste" product. Besides this positive property, the use of Hydrogen as a fuel also has challenges. Next to storage and transport of the fluid in either liquid or gaseous phase, the main challenge is related to safety. The reason for this is that H2 has, compared to diesel or Natural Gas, a (much) larger flammability range, a lower autoignition temperature and a lower activation energy (Miao et al. 2011, Hankinson et al. 2009). In this paper we will evaluate two tools to determine the hydrogen dispersion in a ventilated space after a leak. In Section 2 details of both modelling approaches will be presented. Results of a test case will be given in Section 3. Finally, some discussions and conclusions are detailed in Section 4.

2 MODELLING

Two kinds of modelling approaches are being pursued. The first one is based on Computational Fluid Dynamics (CFD), where the transport equations for the conservation of mass, momentum, energy, and species are solved. The second is the "Accumulation in Confined Areas model" included in the HYRAM+ toolkit and based on physics models, which allows rapid computational time.

2.1 *Computational Fluid Dynamics*

The CFD simulation is set up in ANSYS-Fluent. Next to solving for the flow and pressure, species transport for 2 species, i.e., Air and H_2 respectively, is computed. Each specie has constant properties, while for the mixture properties mass-averaging is applied. Buoyancy and gravity are included in the model and the density is calculated using the incompressible ideal gas model. The RANS model realizable k-ε is used as turbulence model, with full buoyancy effects incorporated and standard wall functions applied in the regions near solid walls (Nieuwstadt et al. 2016).

2.1.1 *Validation*

The CFD model is validated by performing simulations according to a case from literature described in Giannissi et al. (2015) This case compares computational results to experimental data from Cariteau et al. (2013). The case concerns the release of helium in a passively ventilated space with one vent. The experimental results are compared to results from different methods of numerical simulation, provided by the Joint Research Centre (JRC), Notional Centre for Scientific Research Demokritos (NCSRD) and the University of Ulster (UU). A typical result is depicted in Figure 1, where a contour plot of the concentration of Helium is shown. A comparison of Helium concentrations at different locations can be seen in Figure 2, where results from our simulations are compared to experimental results and results from another simulation which uses Large Eddy Simulation (LES).

2.2 *Hyram$^+$*

Hyram$^+$ (Hyram+, Growth et al. 2017) is an engineering tool designed by Sandia energy for quick evaluation of behavior, hazards, and consequences of different release scenarios of H_2 or (blends of) some other species. Amongst others, there are

DOI: 10.1201/9781003508762-39

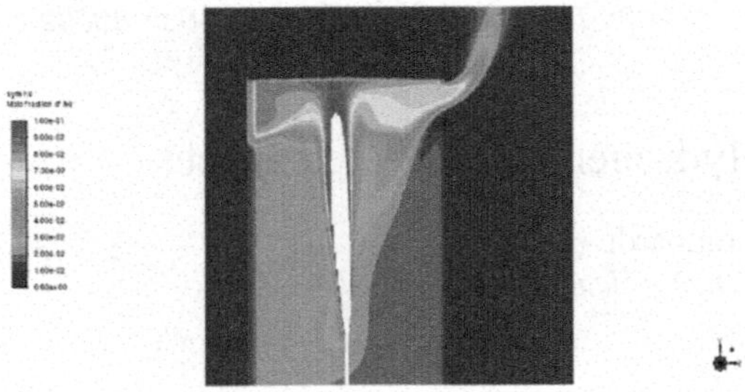

Figure 1. Contour plot of the Helium concentration at the centre plane after 400s. Also, the locations for the detailed comparison displayed in Figure 2 are depicted in this plot.

options for a release in a vented enclosure or in open space.

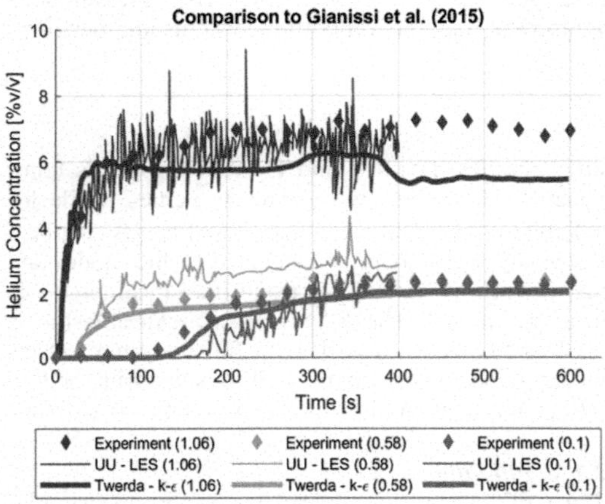

Figure 2. Comparison of our numerical results obtained by RANS k-ε simulation in Ansys Fluent to experimental and LES results for the reference case described in Gianissi et al.

This work focuses on the release in a vented enclosure model of Hyram$^+$, which takes inputs such as leak pressure, leak diameter, enclosure geometry, ventilation geometry and ventilation flow. It then assumes isentropic flow through a circular orifice as a leak.

Next, the notional nozzle model according to Birch et al. (1987) is used to expand the flow to ambient conditions.

After this, an area of development takes place which links the plug flow from the notional nozzle model to a Gaussian plume model. Lastly, this Gaussian plume model is used as input for the confined accumulation model according to Lowesmith et al. (2009).

This accumulation model assumes a stratified layer with a certain depth and uniform concentration of the fuel released in the enclosure.

The depth and concentration are computed over time using conservation of mass and species in the upper buoyant layer.

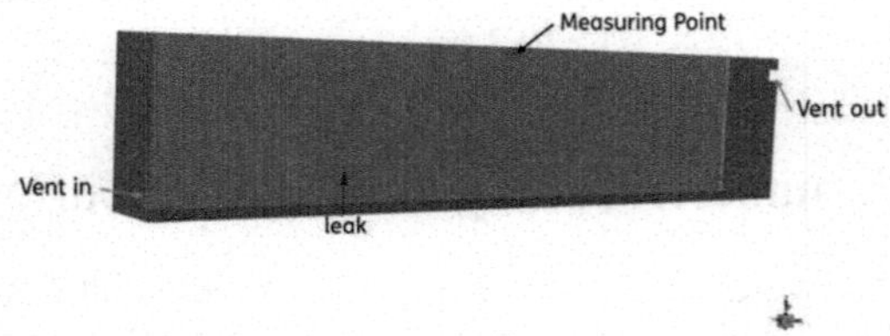

Figure 3. Geometry of the simulated case, indication the leak ventilation position and measurement point.

3 RESULTS

3.1 *Case description*

The selected case consists of a 40ft container 12m x 2.5m x 2.5m (LxBxH) with a ventilation inlet and outlet. A set of different leak sizes, pressures, ventilation rates and dimensions are calculated with Hyram$^+$ tool which are listed in Table 1. The chosen values are based either on conditions commonly found in Hydrogen Fuel Systems or are in the same range as one would expect on board ships of lengths-over-all between 20-120m. The positions of the vent in and vent out are kept constant.

Table 1. Hyram$^+$ simulation matrix. Bold numbers are also computed with the CFD model.

Variable [#]	Values	Unit
Leak pressure [2]	**10, 350**	Bar
Leak diameter [6]	0.05, 0.1, 0.2, 0.158, **0.316**, 0.632	mm
Ventilation size [3]	200x100, **400x200**, 800x400	mm
Number of air changes [5]	1, **2, 4**, 8, 16, 30	Per hour

A selection of those, denoted in bold in Table 1, are also calculated using CFD. The 30 changes per hour is according to the IGF code for an Emergency Shut Down (ESD) designated space (IMO Gitushi et al. 2022). In Figure 3, a picture of the geometry is shown. Only half the geometry is simulated using CFD, assuming symmetry in the y-plane.

3.2 *Hyram$^+$ results*

The equilibrium concentrations in the upper buoyant layer for the different parameter combinations in the Hyram$^+$ simulations are displayed in Figure 4. In this figure each combination of leak pressure and diameter is indicated by the leak mass flow. Also, only the results for the smallest and largest ventilation areas are displayed for brevity. Based on Figure 4 it can be concluded that the ventilation area has a significant impact on the Hyram$^+$ results, especially the effectiveness of active ventilation. This huge effect of the ventilation area on the Hyram$^+$

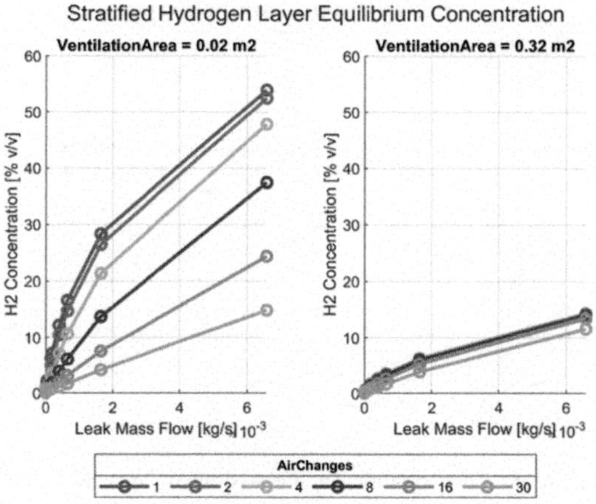

Figure 4. Comparison of the equilibrium hydrogen concentrations in the upper buoyant layer for the different combinations of parameters in the Hyram$^+$ simulations.

results can be explained by the way active ventilation is implemented. The original expression for the flow into the enclosure comes from Warren et al. (1980) and is displayed in Equation 1.

$$Q_{in}=\sqrt{Q_b^2+Q_v^2} \tag{1}$$

In Equation 1, Q_b is the buoyancy driven flow according to Equation 2 and Q_v is the ventilation driven flow according to Equation 3.

$$Q_b = \frac{C_d A_v}{\sqrt{2}}\sqrt{2g\frac{\rho_{air}-\rho_{layer}}{\rho_{air}}H_1} \tag{2}$$

$$Q_v = \frac{C_d Q_{vent}}{\sqrt{2}} \tag{3}$$

In these equations, C_d is a discharge coefficient, A_v is the ventilation area and H_1 is the distance between the bottom of the layer and the middle of the top vent.

The idea behind this implementation is that fuel leaves the enclosure due to buoyancy, leading to under-pressure and the drawing in of fresh air.

However, this leads to discrepancies with the CFD simulations, where the flow through the active ventilation is specified. Therefore, the Hyram$^+$ simulations are compared to Hyram$^+$ simulations with only ventilation driven flow by setting the ventilation area to 0 and multiplying the ventilation flow by $\sqrt{2}$. This comparison is displayed in Figure 5 where it is shown that just the ventilation driven flow leads to higher concentrations in the upper buoyant layer when compared to the implementation of both ventilation and buoyancy driven flow.

3.3 *CFD results*

Additional to the setup described in Section 2.2.1 some calculation details of the CFD model are shown in Table 2. As can be seen in Table 1, 4 cases are calculated, *i.e.,* two type of leak (high and low pressure) and two types of air changes (2 and 4 air changes per hour). For the case with the high-pressure leak and 2 air changes per hour, two snapshots of the H_2 concentration in the symmetry plane are shown in Figure 6. As can be seen a layer of Hydrogen is formed at the top of the container as assumed in the Hyram$^+$ model.

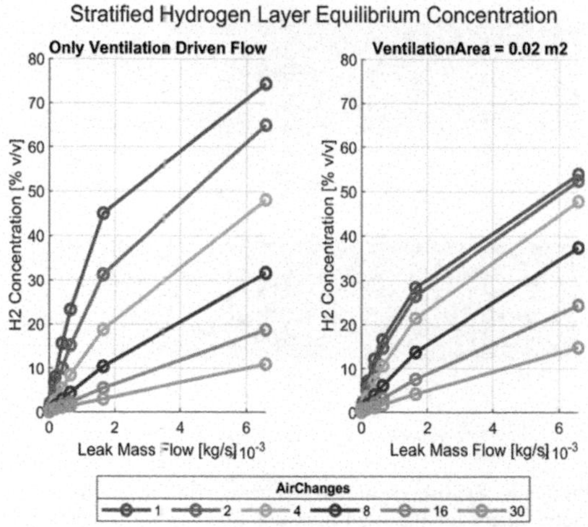

Figure 5. Comparison of the equilibrium hydrogen concentrations in the upper buoyant layer of Hyram$^+$ for the different parameters with only ventilation driven flow and with both ventilation and buoyancy driven flow.

Table 2. CFD simulation details.

Case	# cells	H_2 Mass flow
Low leak pressure	2.5 M	4.89E-5 kg/s
High leak pressure	1.4 M	1.65E-3 kg/s
Setting		**Value**
Spatial discretisation scheme		2nd Order
Time discretisation scheme		1st Order
Time step		0.01s

3.4 *Comparison*

Figure 7 shows the results of Hyram$^+$ and CFD for the high leak pressure case with 2 and 4 air changes per hour. Several observations can be made. Firstly, for the Hyram$^+$ cases with buoyancy and ventilation present, there is practically no differences between the results for different amounts of ventilation. Furthermore, this is not the case for the CFD results where the concentration of H_2 is almost halved when doubling the ventilation rate. When eliminating the buoyancy term in the Hyram$^+$ computations both models almost predict the same value. Hence, the buoyancy overestimates the effective ventilation and under predicts the H_2 concentration as already observed in Section 3.2.

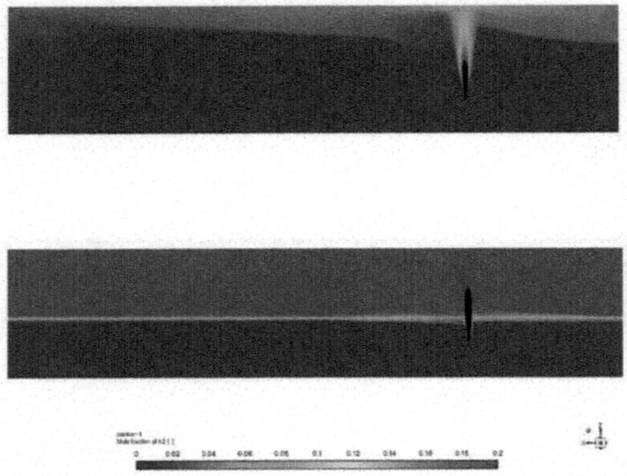

Figure 6. Snapshots of the h2 concentration at the symmetry plane for the high-pressure case with 2 air changes per hour.

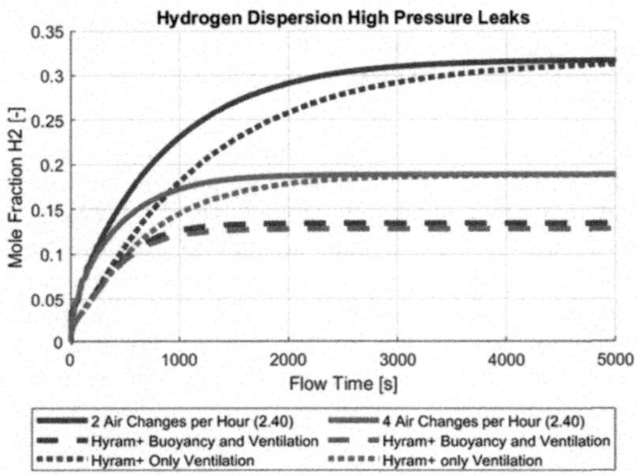

Figure 7. Concentration of the hydrogen layer with the high-pressure leak for 2 and 4 air changes per hour. Solid lines: CFD; dashed lines: Hyram^{+}; dotted lines: Hyram^{+} without the buoyancy term.

Similar results are obtained for the low-pressure release which is shown in Figure 8. However, the difference is less drastic than the high-pressure case because the buoyancy term has less influence here due to a lower hydrogen concentration.

3.5 *Ignition*

The case with the high-pressure leak and lower ventilation rate of 2 air-changes per hour is used as initial condition to simulate an ignited hydrogen layer. The situation after 5000s is used and the mole fraction of H_2 in the layer is 32%, which is much higher than the stochiometric value. The combustion model used is the partially pre-mix model (Bray et al. 2005). In this model two non-dimensionless scalars, both with values between 0 and 1, determine the composition in a cell. The first is the mean mixture fraction (f) which determines the mixing of the 2 streams *i.e.,* in pure H_2 (f=1) and in pure air (f=0). The second variable is the progress variable (c) which determines if the mixture is non-burned (c=0) or burned (c=1).

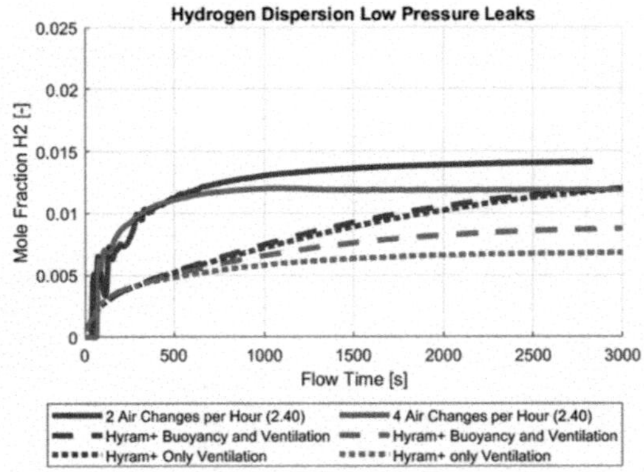

Figure 8. Concentration of the hydrogen layer with the low-pressure leak for 2 and 4 air changes per hour. Solid lines: CFD; dashed lines: Hyram^{+}; dotted lines: Hyram^{+} without the buoyancy term.

The ignition was modelled by setting the progress variable to 1 in very small region containing only 2 grid cells. Figure 9 shows the flame front at 0.02s after ignition. The pressure build-up in the container is shown in Figure 10. The maximum pressure is achieved at ~0.2s and is almost 5 bar. After that, most of the oxygen in the container is used in the combustion.

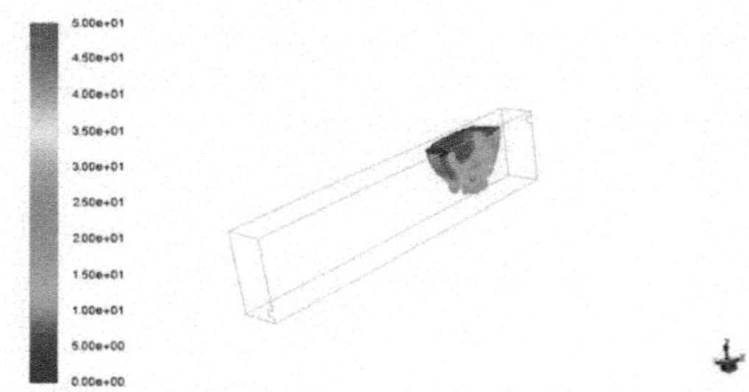

Figure 9. Flame front at 0.02 s after ignition. Colouring is done by turbulent flame speed in m/s.

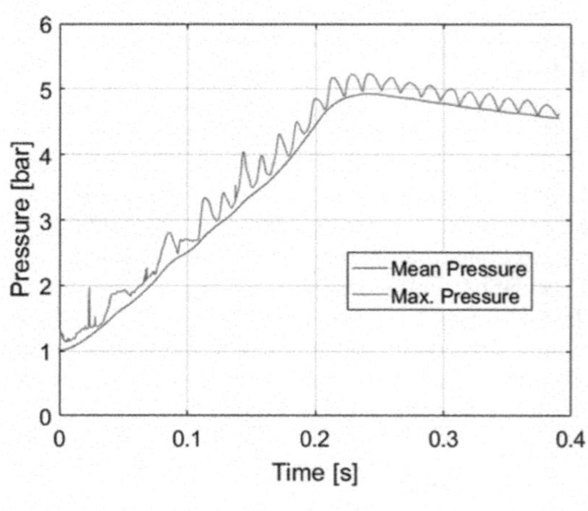

Figure 10. Pressure build-up in the container after ignition. Mean and maximum pressure in the domain is plotted.

4 DISCUSSION & CONCLUSION

Both methods applied have their pros and cons. Hyram$^+$ is and easy to use tool which gives reasonable answers with in second or minutes. CFD on the other hand is a more complex tool where sufficient knowledge of the underlaying models and methods is necessary. Validation of the results and methods is recommended to ensure proper use. As the applied CFD model is validated against experiments for hydrogen dispersion, for the explosion part this is still work to be performed in more detail in the future. Next to that, the simulation times of a CFD model can run up to several days or weeks depending on the number of cells and resources available. The ventilation calculated with buoyancy term in Hyram$^+$ clearly overpredicts the total ventilation flow resulting in a low estimate of the Hydrogen layer in the domain.

Also, if the ventilation is enlarged to 30 air changes per hour, there will not always be a layer H_2 formed, and the assumptions of the tool are compromised. This can be seen in Figure 11 where the high-pressure case is simulated with 30 air changes per hour. Here, there are still regions where Hydrogen concentrations are higher that he LEL, while a major part of the room is properly ventilated and all the H_2 is removed. This particulate case could certainly not be predicted with Hyram$^+$.

Figure 11. Concentration of H_2 after 520s with a ventilation rate of 30 air changes per hour.

Further study should include the internal arrangement of the container to identify scenario where hydrogen could be locally trapped.

The internal arrangement is critical for the risk assessment for a concept design of a hydrogen fueled ship. The effect of which is difficult to capture with Hyram+.

ACKNOWLEDGEMENTS

This work was performed within $SH_2IPDRIVE$, a Dutch consortium of 25 partners, sponsored by RVO (Rijksdienst voor Ondernemend Nederland).

REFERENCES

ANSYS Fluent, www.ansys.com.

Birch, A. D., Hughes, D. J., & F. Swaffield, 1987, Velocity decay of high-pressure jets, In: Combustion Science and Technology, vol. 52, no. 1-3, pp. 161–171.

Bray K., Domingo P. & Vervisch. L., 2005, Role of the Progress Variable in Models for Partially Premixed Turbulent Combustion, Combustion and Flame. 141. 431–437.

Cariteau, B. & Tkatschenko, I., 2013, Experimental study of the effects of vent geometry on the dispersion of a buoyant gas in a small enclosure. In: Int J Hydrogen Energy. 38 (19).

Giannissi, S., Shentsov, V., Melideo, D., Cariteau, B., Baraldi, D., Venetsanos, A.G., et al., 2015, CFD benchmark on hydrogen release and dispersion in confined, naturally ventilated space with one vent. In: Int J. Hydrogen Energy.

Gitushi, K.M., Blaylock, M.L. & Klebanoff, L.E., 2022, Hydrogen gas dispersion studies for hydrogen fuel cell vessels II: Fuel cell room releases and the influence of ventilation, In: International Journal of Hydrogen Energy, vol. 47, pp. 21492–21505.

Growth, K. M., & Hecht, E. S. 2017. HyRAM: A methodology and toolkit for Quantitative Risk Assessment of hydrogen systems, International Journal of Hydrogen Energy 42.

Hankinson G., Mathurkar H., Lowensmith B. J., 2009, Ignition energy and ignition probability of methane-hydrogen-air mixtures, 3rd International Conference on Hydrogen Safety, Ajaccio, France.

Hyram$^+$, HyRAM$^+$. Sandia Energy.

International Code of Safety for Ship Using Gases or Other Low-flashpoint Fuels (IGF Code)". International Maritime Organization. www.imo.org

Lowesmith, B., Hankinson, G., Spataru, C., & Stobbart, M., 2009, Gas build-up in a domestic property following releases of methane/hydrogen mixtures, In: International Journal of Hydrogen Energy, vol. 34, pp. 5932–5939.

Nieuwstadt, F.T.M, Westerweel, J., & Boersma, B.J., 2016, Turbulence: Introduction to Theory and Applications of Turbulent Flows. 1st ed.

Miao, H., Lu, L., Huang., 2011, Flammability limits of hydrogen-enriched natural gas, International Journal of Hydrogen Energy, Volume 36, Issue 11.

Warren, P.R. & Webb, B.C., 1980, The relationship between tracer gas and pressurisation techniques in dwellings. In: Proc. 1st AIC Conf. Instrumentation and Measurement Techniques. Windsor, UK; pp. 245–276.

Advances in Maritime Technology and Engineering – Guedes Soares & Santos (Eds)

A complex network-based method for modelling and analyzing risk factors influencing maritime transport safety

C.P. Wan & B. Wu
State Key Laboratory of Maritime Technology and Safety, Wuhan University of Technology, Wuhan, P.R. China
Intelligent Transportation Systems Research Center, Wuhan University of Technology, P.R. China

ABSTRACT: This study collected 694 reports on maritime accident investigations from 2011 to 2021. The text mining approach is adopted to identify the risk factors and the complex network analysis theory is integrated to construct the maritime accident risk evolution network model. The constructed network is utilized to ascertain the key contributory factors of maritime accidents during different stages of its evolution. The research findings reveal that ship collisions are often caused by improper crew operation and emergency response. Furthermore, these accidents show distinct topologic features such as short transmission paths, high levels of randomness in the networks, and the presence of interdependent risk factors. This study introduced an innovative approach to examine the mechanisms behind maritime accidents, which can serve as a theoretical foundation for maritime authorities and other stakeholders in managing and ensuring the safety of maritime traffic.

1 INTRODUCTION

Maritime transportation is the main mode of international trade, accounting for more than 2/3 of the global trade volume, and 80%~90% of the world's import and export freight volume. While, the number of ships increases with the development of international shipping industry, and the navigation environment becomes more and more complex. In busy sea areas, increased vessel traffic may lead to congestion, and crew members may face greater work pressure and emergency situations. Intensive shipping activities inevitably increase the probability of ship accidents, including collision, contact, grounding, fire or explosion. Once a ship accident occurs, it often leads to serious consequences, such as environmental pollution, property loss and even casualties. According to the 2022 Annual Overview of Maritime Casualties and Casualties, released by the European Maritime Safety Agency (EMSA), 59.6% of maritime accidents from January 1, 2014 to December 31, 2021, were caused by human factors. 68.3% of the factors were related to human factors. Taken together, human factors were associated with 81.1% of marine casualties and accidents.

At present, scholars have carried out several studies in the risk assessment domain. Parts of them identified the risk influencing factor (RIF) of maritime accidents from different angles, and parts of them carried out a series of research on risk evolution, risk warning, accident prevention and so on. Different methods including system simulation techniques, machine learning methods have been used to infer and evaluate the risk of maritime accidents, which has promoted the development of the risk analysis technology of maritime traffic accidents and enriched the theory of maritime traffic safety supervision. However, the current research mainly aims to identify the importance of RIF, ignore their interaction and the accident evolution mechanisms. Due to the complexity of traffic accident evolution, the set of RIFs is expanding from a small number of single-level factors to multi-type, multi-dimensional and multi-level factors.

2 LITERATURE REVIEW

2.1 *Risk factors influencing maritime safety*

In the analysis of maritime accidents, identifying the causes of accidents and exploring the correlation between multiple factors of accidents is crucial for risk prevention and control of maritime accidents. Previous studies have been conducted form different aspects such as human error, environment, vessel, and management.

Chen et al. (2020) proposed a Reason SHEL model to establish a human causal index system and utilized multidimensional association rule algorithms to mine human causal accident chains, which can provide new theories and methods for extracting accident causal chains and reducing maritime accidents. Lan et al. (2023) studied all types of accidents involving total loss of ships through association rule algorithms and found that the main types of accidents involving total

DOI: 10.1201/9781003508762-40

loss of ships are hull/mechanical damage and sinking. Uyanik et al. (2021) studied that meteorological knowledge is one of the important factors affecting safe navigation and used machine learning methods to estimate the visibility during the passage of the strait based on various meteorological indicators. Wu et al. (2015) proposed an improved data envelopment analysis model to analyze the effectiveness of maritime safety control from the perspective of the safety level along the Yangtze River, with a focus on studying the impact of different navigation environments on safety levels. In terms of "management" factors, the maritime management agency mainly establishes and implements a safety management system for shipping enterprises (Chang, 2009), which should reflect the requirements of the International Ship Safety Operation and Pollution Prevention Management Regulations and the People's Republic of China Ship Safety Operation and Pollution Prevention Management Regulations.

2.2 *Risk evolution analysis of maritime accidents*

To control and reduce water transportation risks, it is necessary to study the evolution trend of maritime accidents. Analyzing the temporal and spatial distribution characteristics of accidents from different perspectives, further extracting accident mechanisms and models from a large number of typical accident cause analysis, reflecting the regularity of accident occurrence, providing a scientific and complete theoretical basis for qualitative and quantitative analysis of accident causes, accident prevention, etc.

Ren et al. (2022) constructed a safety risk evolution network for the Yangtze River Estuary waters using complex network accident chain analysis technology, conducted vulnerability simulation evaluation, and proposed a key vulnerability factor identification method. Xu et al. (2017) utilized multi resolution modeling of complex systems and Bayesian networks for risk evolution analysis, avoiding risk evolution networks determined solely through subjective judgment. Fu et al. (2022) developed an accident map Bayesian network model and proposed a probability model to evaluate the probability and severity of grounding accidents, quantitatively analyze the evolution law of grounding accidents, and better identify key RIF. By using methods such as complex networks, system dynamics, and Bayesian network models to analyze the causal logic relationship between accident causes, revealing the risk occurrence patterns and changing trends in different traffic water areas, new risk control strategies for complex systems can be provided.

3 METHODOLOGY

3.1 *Modelling of maritime accident chains*

An accident causation chain shows the cause, origin and end process of the accident and the consequence of the accident, to analyze the occurrence and development of the accident phenomenon (Barrt et al., 2004). The analysis of the accident chain of navigation risk is based on the method of accident chain modeling, which adopts the method of accident tree and event tree to model the accident chain. The usual steps include determining the analysis object, collecting and analyzing the accident report, extracting the RIF node and constructing the accident chain.

STEP 1 Identification of the nodes of the accident evolution chain

STEP 2 Identification of directed edges of accident evolution chain

STEP 3 Analysis of accident evolution chain

3.2 *Complex network analysis*

3.2.1 *Evaluation of accident risk evolution network node*

In the complex network analysis, the causative factors in the accident risk evolution network are analyzed by node strength, weighted closeness centrality and weighted betweenness centrality.

(1) degree

In the accident risk evolution model, the node is the extracted RIF, and the edge represents the logical relationship between the RIF. The larger the degree value, the higher the connectivity between RIF. For a directed network, the degree k_i of node *i* includes an out-degree k_i^{out} and an in-degree k_i^{in} (Rentzeperis & Van, 2021). Then:

$$k_i = k_i^{out} + k_i^{in} = \sum_{j=1}^{m} l_{ij} + \sum_{j=1}^{m} l_{ji} \quad (1)$$

where, l_{ij} indicates that a directed connecting edge of the node i pointing to the node j exists between the node i and the node j; l_{ji} indicates that the node j points to the directed connected edge of the node i between the node j and the node i.

(2) Node strength

Node strength is defined as the sum of the weights of all edges connected to the node. The greater the strength of the node, the higher the risk of the factor in an accident, which are,

$$S_i = \sum_{j=N_i} W_{ij} \quad (2)$$

where, S_i is the node strength; N_i is the set of nodes directly connected to node i; W_{ij} is the edge weight value.

(3) Weighted closeness centrality

Weighted closeness centrality is used to measure the reachability and connectivity efficiency of the node with other nodes. The smaller the sum of the weighted shortest distances between a node and all other nodes, the stronger the weighted closeness centrality (He et al., 2020) and the greater the importance to the network.

$$WCC_i = [\frac{1}{|N_i|}\sum_{j=1,j\neq i}^{|N_i|} d_{ij}^{w}]^{-1}, d_{ij}^{w} = \min\left\{\frac{1}{w_{ih}}+\ldots+\frac{1}{w_{hj}}\right\} \quad (3)$$

where, d_{ij}^{w} is the weighted shortest path length between node i and node j. w_{ih} represents the weight of the joint edge between node i and node h, w_{hj} represents the weight value of the joint edge between node h and node j, and w represents the weighted value.

(4) Weighted betweenness centrality

Weighted centrality reflects the importance of the node to the overall connectivity of the network, that is, the node's control over the network flow along the shortest path. The more the weighted shortest path passes through the node, the greater the centrality of the weighted betweenness, and the closer the relationship between the RIF. The calculation formula is as follows.

$$WBC_i = \frac{\sum_{k\neq i\neq j}\frac{n_{kj}^{(i)}}{n_{kj}}}{\sum_{i=1}^{n}\sum_{k\neq i\neq j}\frac{n_{kj}^{(i)}}{n_{kj}}} \quad (4)$$

where, n_{kj} is the sum of the weighted shortest paths between node k and node j; $n_{kj}^{(i)}$ is the number of times the weighted shortest path between node k and node j passes through node i.

3.2.2 *Topological characteristics analysis of accident risk evolution network*

In the accident evolution network, the topology characteristics of a complex network are analyzed by using indicators such as the average weight degree, average clustering coefficient, average path length, and network diameter. The related indicators include:

(1) Average weighting degree

The average weighting degree refers to the average value of the weight sum between the node and other nodes. In this study, this index represents the average occurrence frequency of each RIF in the accident risk evolution network, so the average weighted degree is the measurement of the risk occurrence rate of the whole accident evolution network, as shown in the following formula:

$$K_i = \frac{1}{N}\sum_{i} k_i \quad (5)$$

where, K_i is the average weighted degree of the whole accident evolution network; N is the number of all nodes in the accident evolution network; K_i is the degree of the accident node i.

(2) Average clustering coefficient

A parameter that reflects the degree of convergence of nodes in the entire network. In the accident risk evolution network, the average clustering coefficient is large, indicating that the single RIF will not cause too much influence on the network, and it has certain fault tolerance. The formula for calculating the average aggregation coefficient Ci is as follows.

$$C_i = \frac{1}{N}\sum_{i\in N}\frac{2E_i}{k_i(k_i-1)} \quad (6)$$

where, there are E_i actual edges between k_i accident nodes with E_i.

(3) Network diameter and average path length

The maximum value of the shortest path length between RIF nodes is called the network diameter[36]. The larger the network diameter, the stronger the correlation between RIF and the faster the propagation efficiency. The average path length refers to the average shortest path length of all nodes in a complex network. The shorter the average path, the better the location conditions of the accident nodes in the accident evolution network, and the more compact the network is:

$$L = \frac{1}{N(N-1)}\sum_{i\neq j} d_{ij} \quad (7)$$

where, L represents the average path length; d_{ij} represents the distance from node i to node j.

3.3 *Risk evolution of maritime accidents*

The process of analyzing the evolution process of maritime accidents contains the following steps (Fu et al., 2022).

Step 1. Risk incubation stage. Risk incubation stage refers to the possibility of risk occurrence in various signs, such as some potential threats, some nodes fail, and some nodes are in high-risk state, although the risk has not yet appeared.

Step 2. Risk diffusion stage. The risk diffusion stage can be understood as the risk is spreading in the system, from one or more potential risk events to a certain degree of high-risk events. Although the risk accident in the network has not occurred at this time, the whole system has a high-risk level.

Step 3. Risk occurrence stage. The occurrence of accidents follows four steps: potential hidden dangers, high-risk scenarios, accident occurrence, and accident consequences. In the risk occurrence stage, risks have arrived, and the loss caused by risks is unpredictable, and different consequences may still spread and intensify.

4 CASE STUDY

4.1 *Data collection*

From the official website of the Maritime Safety Administration of the People's Republic of China, 694 open maritime accident investigation reports were collected from 2011-2021, mainly including ship collision, grounding, overturning, oil spill, sinking and other types of accidents. The accident investigation reports include information such as accident profile, investigation and evidence collection, and accident process and causes, which are used to construct the dataset for maritime risk evolution analysis. According to the statistical analysis, the types of accidents involved in China's coastal waters include collision accidents, rock-hit accidents, damage accidents, stranding accidents, self-sink accidents, fire (or explosion) accidents, capsize accidents, operational pollution accidents, wind accidents and other accidents.

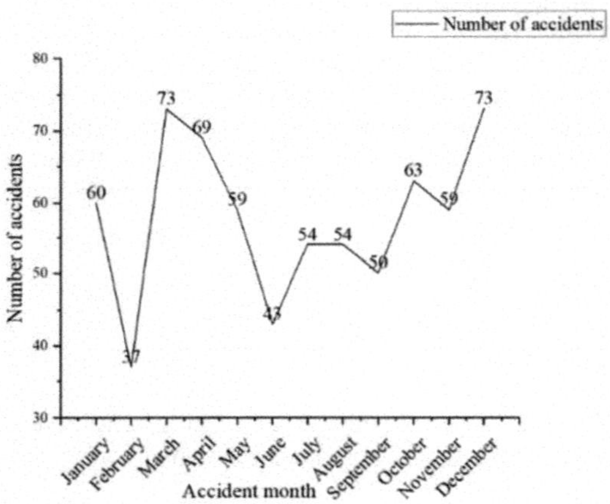

Figure 1. Statistics of accident occurrence by month.

From the trend of accident occurrence months in Figure 1, the number of accidents in spring is higher than that in other seasons, while the number of accidents in summer is the lowest. This may be because spring is usually the season with strong winds and waves in China's coastal areas, and the meteorological conditions at sea are relatively harsh. This unfavorable weather condition can easily lead to difficulties in ship maneuvering and reduced visibility, increasing the probability of accidents occurring. Compared to spring, the wind and waves in summer are usually smaller, and the meteorological conditions are relatively good, which reduces the instability of ships at sea and reduces the risk of accidents.

4.2 *Network modelling of maritime accidents*

4.2.1 *Investigation of accident causes*

This study involved 694 investigation reports of maritime traffic accidents. To make the segmentation result more ideal, stop word and professional dictionaries should be customized before segmentation, and irrelevant words should be deleted appropriately. The values of each feature are calculated as feature weights. The higher the weight, the more important or frequent the term is in the text. The segmentation results of the selected parts (ranking the top 50 in importance) are shown in the Table 1.

Table 1. Statistical results of partial word frequency.

Characteristic term	Importance	Characteristic term	Importance
collision	0.84	safety awareness	0.06
Improper	0.54	rules	0.06
emergency	0.45	Unseaworthiness	0.06
response	0.41	seamanship	0.06
underestimating	0.26	use	0.05
lack of watchkeeping	0.21	spilling	0.05
safety	0.19	driving	0.04
In place	0.19	navigation aids	0.04
crew	0.17	care	0.04
risk	0.15	bridge	0.03
sinking	0.15	certificate	0.03
manning	0.12	traffic flow	0.03
Management	0.10	maintenance	0.03
speed	0.10	navigational	0.03
discomfort	0.10	visibility	0.03
fire	0.09	environment	0.03
Wind and waves	0.08	stowage	0.03
Grounding	0.08	current	0.03
training	0.07	good	0.03
duties	0.07	failure	0.03
comply	0.07	Unfamiliar	0.02
navigation	0.07	cargo	0.03
effectively	0.07	shore-based	0.03
operation	0.06	poor	0.03
explosion	0.06	overload	0.03

This study outputs the feature items and their corresponding TF-IDF values using the Wordcloud module in Python language as a word cloud graph, displaying visual statistical results. Words such as "collision", "emergency", and "response" are of high importance in the word cloud diagram, and this reason has received high attention in maritime accident investigation reports.

4.2.2 *Identification of contributing factors of maritime accidents*

The high-frequency words selected through Python text mining technology only contain key words related to the causes of maritime accidents, which cannot be fully expressed. It is also necessary to combine the original accident corpus and manually analyze the accident causes of maritime accident reports to establish associations between these words to reveal potential accident cause patterns. Text mining is an auxiliary method that still requires the combination of professional knowledge and experience for in-depth analysis and interpretation of accidents. Finally, the word cloud results are described based on factors obtained from human analysis to extract the causes of maritime accidents.

According to Text mining technology and artificial analysis of the original accident corpus, the RIF extracted by this study, including four categories of personnel risk, ship risk, environmental risk and management risk (Chen et al., 2020), can be subdivided into 39 risk causal factors, as shown in Table 2.

Table 2. Causes of maritime accidents.

Category	No.	Causative factors	Interpretation
Human factors	A1	Lack of maintenance	Neglect inspection, non-standard daily maintenance and maintenance work
	A2	Lack of safety awareness	Ignoring security risks and Lack of safety awareness
	A3	Improper operation	Reliability of operation behavior
	A4	Lack of watchkeeping	Failure to maintain a regular lookout through monitoring equipment such as visual or radar
	A5	Underestimating the risk	Underestimation of the urgent situation of risks such as collision and stranding in complex environments such as wind and current
	A6	The crew did not perform their duties effectively	Failure to effectively fulfill responsibilities due to lack of sense of responsibility and experience
	A7	Failure to comply with navigation rules	Failure to abide by navigation regulations, collision avoidance rules and other relevant rules and regulations
	A8	Carelessness of crew	negligence and carelessness of personnel

(*Continued*)

Table 2. (*Continued*)

Category	No.	Causative factors	Interpretation
	A9	Not following good seamanship	Reasonable maneuvering of the ship without using good operating skills
	A10	Not navigating with care	not cautiously manoeuvring the ship
	A11	Improper emergency response	The emergency measures are not correct, timely, or in place.
	A12	Unfamiliar with the navigational environment	Unfamiliar with the navigational environment of waterways and port waters
	A13	fatigue	Drivers are physically and mentally tired, which affects driving and on-duty.
	A14	Insufficient communication	Communication between personnel, inter-ship coordination, avoiding and dynamic untimely and inadequately
	A15	Failure to keep a safe speed	Failure to adopt a safe speed consistent with the circumstances and urgency of the time
ship factor	B1	structural defect	Ship structure material is unqualified, and structure has hidden safety hazard
	B2	Overloaded	Heavy overload of cargo, damage to the strength of the vessel and exceeding the authorized full-load displacement
	B3	Steering gear failure	The ship lost control due to the sudden failure of the rudder.
	B4	Hull aging	Bottom rust and water penetration, corrosion and damage of ship structure
	B5	Severe circuit aging	Circuit equipment and line aging
	B6	Failure of navigation aids	Radar, AIS and other navigational aid equipment, main engine, electrical equipment and other navigational equipment failure.

(*Continued*)

Table 2. (*Continued*)

Category	No.	Causative factors	Interpretation
	B7	Insufficient under-keel clearance	Over draft, insufficient under-keel clearance, etc., cannot guarantee the safe navigation of the ship.
	B8	Lack of effective weathertightness of ship cargo hold	Due to the lack of tight hatch covers and the failure to effectively seal the cargo hold, etc., the ship enters water during strong storms and waves.
	B9	Carriage of dangerous goods	Transport of dangerous goods such as inflammable and explosive
	B10	Improper stowage of goods	Risks such as improper ship stowage and improper cargo tie-down
	B11	Insufficient light and sound equipment	The ship fails to set up lights and sound signals in accordance with the provisions during navigation.
	B12	Inadequate nautical information or equipment	Inadequate nautical data or equipment such as charts and ECDIS
	B13	Legal channel crossings	Blindly chase through narrow waterways and cross other waterways.
Environmental factors	C1	Rapid current	rapid change such as ebb tide and rise tide
	C2	Nighttime	Nighttime
	C3	Strong wind and heavy waves	Winds greater than Pu's six, waves higher than two meters.
	C4	Dense traffic flow	Ship navigation dense area, complex navigation environment
	C5	Poor visibility	Poor visibility due to bad weather such as rain and smog
Administrative factors	D1	Poor Security Management	The safety supervision and administration system of shipping companies is not implemented in place, and the management personnel fail to fulfill the main responsibility of safety.
	D2	insufficiency of crew	Failure to provide sufficient qualified personnel as required
	D3	Failure to provide effective shore-based support	Insufficient daily or emergency support for ships by shipping companies
	D4	Inadequate bridge watch arrangement	Lack of bridge resources, neglect of duty, and failure to detect the danger in time
	D5	Defects in the safety management system	The safety management system of shipping companies is not established or has defects, resulting in problems in safety supervision and administration.
	D6	Insufficient training of crew	Inadequate education and training on safety knowledge and operation skills of shipping companies
	D7	Invalid ship certificate	Failure to apply for ship survey as required, relevant certificates have expired
	D8	Crew incompetence	Whether the crew members have the certificate of competence, the corresponding knowledge and skills
	D9	Unseaworthiness	The ship does not meet the seaworthiness conditions due to structural modification, manning and unqualified certificate.

4.2.3 *Extraction of accident chains*

Taking the investigation report of collision between "10.19" Fumingcheng" and "Jileyu 07079" in Tangshan as an example, the construction process of accident chain is introduced. In the accident, there are five causing chains of ship accidents, including:

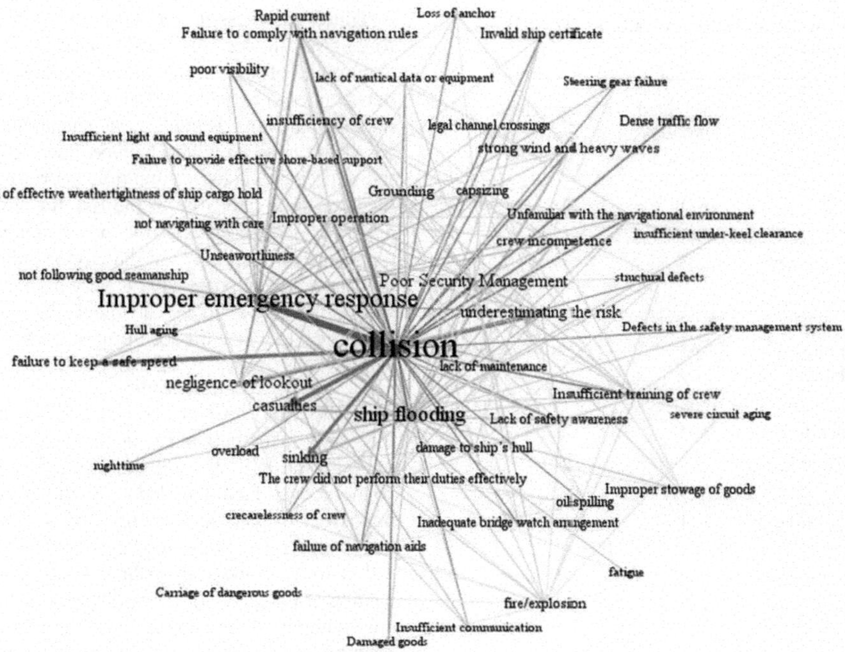

Figure 2. Maritime accident risk evolution network model.

1) lack of watchkeeping → underestimating the risk → Improper emergency response → collision → (damage to ship's hull, sinking, casualties, damage to cargo);
2) insufficiency of crew → Unseaworthiness → collision → (damage to ship's hull, sinking, casualties, damage to cargo);
3) not following good seamanship → Improper emergency response → collision → (damage to ship's hull, sinking, casualties, damage to cargo);
4) failure to keep a safe speed → collision →. (damage to ship's hull, sinking, casualties, damage to cargo);
5) crecarelessness of crew → Improper emergency response →collision → (damage to ship's hull, sinking, casualties, damage to cargo). The combined effect of the above five accident chains led to the final occurrence of the accident.

Similarly, the other 693 accident investigation reports can be analyzed to extract the accident cause chain in all accident cases.

4.2.4 *Construction of accident risk evolution model*

Then, the accident chain is transformed into adjacency matrix, and then the RIF are taken as nodes in the complex network, and the directed connection relationship between RIF is taken as the connecting edge, and the accident risk evolution model is constructed. The nodes of the risk evolution network model include four main RIF (human, ship, environment and management), accident types and accident consequences. By combining all the cause chains of 694 accidents in China's coastal waters, a model of the evolution network of maritime accident risk can be constructed, as shown in Figure 2. The size of a node reflects the degree value of the node. The larger the degree value, the more connections between RIF. The darker the color of the connecting edges of the node, the thicker the line of the edges, the greater the weight between RIFs.

4.3 *Analysis of risk evolution network of maritime accidents*

This section will conduct evolution analysis from the latent stage, diffusion stage and occurrence stage of accidents, and reveal the evolution law of maritime traffic safety risk from four aspects: man, ship, environment and management, and identify the key RIF of maritime accidents.

4.3.1 *Analysis of topology characteristics of risk evolution network*

The average weighting degree of the accident risk evolution network in the area is 57.55, that is, the average frequency of risk factors in the network is about 58 times; The average clustering coefficient is 0.29, indicating that the accident factors on the network have a large clustering characteristic. Generally, the higher the aggregate characteristics of one kind of factors, the more likely the abnormality of the factors will lead to the exposure of the connected factors to the risk state, and the "domino effect" will cause ship accidents. The diameter of accident risk evolution network is 6.00, indicating that the correlation between RIF is strong and the ship efficiency is faster.

Table 3. Network characteristic values of key RIF for maritime accidents.

factors	weighted closeness centrality	weighted betweenness centrality
Poor Security Management	0.71	0
Failure of navigation aids	0.68	39.67
Lack of maintenance	0.65	30.09
The crew did not perform their duties effectively	0.65	50.44
Improper operation	0.64	23.84
Strong wind and heavy waves	0.64	0
Insufficient training of crew	0.61	15.56
underestimating the risk	0.58	26.48
Improper emergency response	0.54	34.99

The average path length of the network is 2.05. That is, one factor on the network can affect another factor in only two steps on average. This indicates that RIF are closely related, and chain reactions are likely to occur.

Table 3 shows that the weighted approach centrality of "inadequate safety management" in the accident risk evolution network is the largest. The greater the weighted closeness to centrality, the higher the driving force, will induce other RIF, which will lead to the occurrence of ship accidents. In addition, the weighted betweenness centrality of "crews fail to perform their duties effectively" is the largest. The greater the centrality of the node weighted betweenness, it indicates that the node is traversed by multiple shortest paths, which plays an important role in the risk transmission of the network. Among the RIF, the crew's failure to perform their duties effectively will directly or indirectly lead to the ship accident probability is very high. Effective control of these RIF nodes can block the spread of risks and prevent the occurrence of maritime accidents.

4.3.2 *Risk incubation stage*

The RIFs from different aspects can be analyzed. And this study takes human factors as an example. Among the potential RIF affecting the navigation of ships, personnel-related factors include: insufficient safety awareness (A2), neglectful lookout (A4), failure to observe navigation rules (A7), negligence of personnel (A8), failure to use good seamanship (A9), failure to exercise caution Driving (A10), unfamiliarity with the navigation environment (A12), fatigue (A13), insufficient communication (A14), failure to maintain safe speed (A15).

4.3.3 *Risk diffusion stage*

The following high-risk events may be caused under the individual or interaction of potential hazards, including lack of maintenance, improper operation, underestimating the risk, the crew did not perform their duties effectively, improper emergency treatment, failure of navigation aids, and Insufficient training of crew. The conjugation relationship between these nodes is shown in Figure 3.

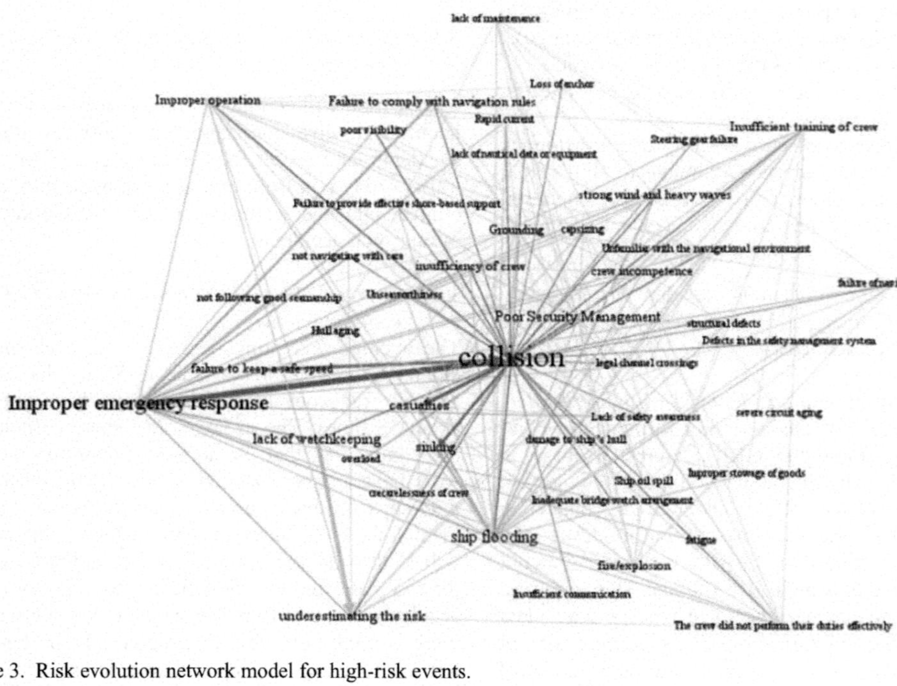

Figure 3. Risk evolution network model for high-risk events.

7 most important RIFs in the risk diffusion stage are calculated, which are lack of maintenance, improper operation, underestimation of hazard, the crew did not perform their duties effectively, improper emergency response, failure of navigation aids, and insufficient training of crew. Taking the lack of maintenance as an example, the node strength was 49, which was associated with 11 threat factors, including failure of navigation aids, hull aging, etc. The weighted closeness centrality and betweenness centrality of the node are relatively high, which are the key causes of accident risk. In all high-risk events, this node mainly threatens the ship's condition, resulting in equipment failure, hull rust and water penetration, resulting in ship sinking and other serious consequences.

Table 4. Fault type node information.

Type of accident	In-degree	Out-degree	Node strength	Weighted closeness centrality	Weighted betweenness centrality
Grounding	19	1	111	0.31	7.69
collision	35	8	1273	0.68	216.63
sinking	4	1	200	0.24	15.67
oil spill	6	2	45	0.30	0.42
fire/ explosion	14	4	106	0.34	53.51

4.3.4 *Risk occurrence stage*

The main types of accidents in the risk occurrence stage are grounding (TOC1), collision (TOC2), sinking (TOC3), oil spill (TOC4), fire/explosion (TOC5). The risk evolution network at this stage is shown in Figure 4.

Figure 4 Shows the risk evolution model centered on the type of accidents in the occurrence stage. Ship collision is the most common type of accidents in maritime accidents. According to the model, grounding (TOC1), collision (TOC2), sinking (TOC3), oil spill (TOC4) and fire/explosion (TOC5) are the five most likely accidents in ship navigation. Among them, the grounding is related to 19 RIF, the collision is related to 26 RIF, the capsizing is related to 9 RIF, the oil spill is related to 14 RIF, and the fire/explosion is related to 9 RIF, as shown in Table 4.

Among the five types of accidents, the weighted closeness centrality and the weighted betweenness centrality of "collision" are the highest, which are 0.68 and 216.63, respectively. The results show that "collision" is the central hub of the whole network, and the possible consequences of accidents include ship capsize, ship damage, sinking, ship flooding and casualties. The weighted media centrality of "sinking" and "fire/explosion" is second only to "collision", indicating that the preconditions of these two accidents are also relatively high, and the potential threat of accidents is second only to "collision".

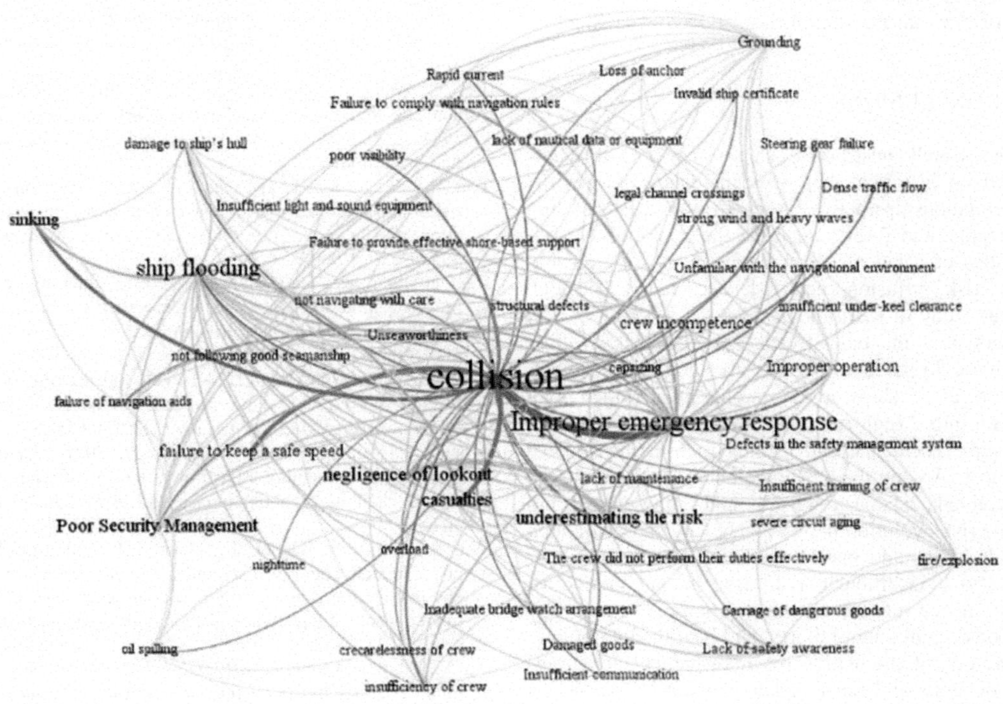

Figure 4. Evolution model of ship navigation accident type in coastal waters of China.

5 DISCUSSIONS

The research shows that ship accidents are the result of the superimposed action of many RIF, which have different causal chain effects. From the network analysis results, the key RIF include lack of maintenance, improper operation, underestimating the risk, the crew did not perform their duties effectively, improper emergency response, failure of navigation aids and insufficient training of crew.

Taking the analysis of the lack of maintenance as an example, the accident chain includes "The crew did not perform their duties effectively → lack of maintenance → the consequences of accidents", "Poor Security Management → lack of maintenance → the consequences of accidents", to name but a few. As can be seen from the figure, "lack of maintenance" is the result of cascade failure of factors such as "Poor Security Management" and "The crew did not perform their duties effectively". Among the 11 factors of lack of maintenance, 9 RIF will occur, including failure of navigation aids, hull aging, severe circuit aging and other threat factors. In all high-risk events, lack of maintenance will threaten the condition of the ship, resulting in serious aging of the circuit system, failure of navigation aids, rust and water penetration of the hull, resulting in fire, explosion, sinking of the ship and other serious consequences. Therefore, it is suggested to strictly standardize the daily maintenance system of shipping companies, implement multi-person supervision and multiple inspections to ensure the crew's duties in place, curb the primary hidden dangers of lack of maintenance, cut off the transmission of risks and avoid accidents.

6 CONCLUSIONS

This research constructs a chain of maritime accidents based on historical ship accident data, analyzes the interrelationships between the causes and RIF of ship accidents, and further analyzes the evolution characteristics of accidents from three stages: risk incubation, risk diffusion, and risk occurrence. Through a case study of maritime accidents in China's coastal waters over the past decade, the results show that there are 36 potential hazards involved in the navigation risks of ships in this water area, which may spread into 7 high-risk accidents, forming 5 types of maritime accidents, and leading to 6 serious consequences. Among them, ship collision is the main type of accident, related to 43 RIF, and is the central hub in the risk evolution network. The key causes of such accidents include inadequate maintenance, improper operation, insufficient risk assessment, ineffective performance of crew duties, improper emergency response, equipment failures, and insufficient training. Therefore, strengthening the management and training of ships and personnel to contain initial hidden dangers and prevent the occurrence of high-risk events can help reduce the occurrence of accidents. In future study, AIS data can be combined to explore the impact of the complexity of ship traffic flow on maritime accidents

ACKNOWLEDGEMENTS

This work is supported by the National Natural Science Foundation of China (51920105014; 51909202), and the Young Elite Scientist Sponsorship Program by CAST (2021QNRC001).

REFERENCES

Barrat, A., Barthelemy, M., Pastor-Satorras, R. and Vespignani, A., 2004. The architecture of complex weighted networks. *Proceedings of the national academy of sciences*, *101*(11), 3747–3752.

Chang, I.S., 2009. A study on the effective safety management measures for the prevention of marine accidents. *Journal of the Korean Society of Marine Environment & Safety*, *15*(1), 33–39.

Chen, D., Pei, Y. and Xia, Q., 2020. Research on human factors cause chain of ship accidents based on multidimensional association rules. *Ocean Engineering*, *218*, 107717.

Chen, X., Li, X. R., and Sheng, Y. 2020. Characteristics of Aviation Network in Yangtze River Delta Based on Complex Network. *Journal of Transport Information and Safety*, 38(04): 139–146.

Feng, R., and Li, S. 2019. Application of FDA Accident Causation Model in Safety Management of Marine Ship Collision Accident. *Journal of Safety Science and Technology*, 15(07): 179–185.

Fu, S., Yu, Y., and Chen, J. 2022. A framework for quantitative analysis of the causation of grounding accidents in arctic shipping. *Reliability Engineering & System Safety*, 226: 108706.

He, X., Wang, L., Zhu, H. and Liu, Z., 2021. Statistical analysis of complex weighted network for seismicity. *Physica A: Statistical Mechanics and its Applications*, *563*, p.125468.

Lan, H., Ma, X., Ma, L. and Qiao, W., 2023. Pattern investigation of total loss maritime accidents based on association rule mining. *Reliability Engineering & System Safety*, *229*, 108893.

Rentzeperis, I. and Van Leeuwen, C., 2021. Adaptive rewiring in weighted networks shows specificity, robustness, and flexibility. *Frontiers in Systems Neuroscience*, *15*, 580569.

Ren, Z. Z., Chen, W. J., and Li. X. 2022. Vulnerability analysis of safety risk evolution in traffic-intensive waters. *Journal of Safety and Environment*, 22(02): 615–622.

Uyanık, T., Karatuğ, Ç. and Arslanoğlu, Y., 2021. Machine learning based visibility estimation to ensure safer navigation in strait of Istanbul. *Applied Ocean Research*, *112*, 102693.

Wu, B., Wang, Y., Zhang, J., Savan, E.E. and Yan, X., 2015. Effectiveness of maritime safety control in different navigation zones using a spatial sequential DEA model: Yangtze River case. *Accident analysis & prevention*, *81*, 232–242.

Xu, Y., Lv, J., and Yang, J. 2017. Complex system risk evolution analysis based on multi-resolution modeling and Bayesian networks. In *2017 Second International Conference on Reliability Systems Engineering (ICRSE)*, IEEE.

Structures

Advances in Maritime Technology and Engineering – Guedes Soares & Santos (Eds)
© 2024 The Author(s), ISBN 978-1-032-83099-5

Effect of mooring lines to hydroelastic response of floating flexible circular structure based on analytical approach

P. Amouzadrad, S.C. Mohapatra & C. Guedes Soares
Centre for Marine Technology and Ocean Engineering (CENTEC), Instituto Superior Técnico, Universidade de Lisboa, Lisbon, Portugal

ABSTRACT: Under linearized small amplitude water wave theory and structural response, an analytical model associated with effect of a moored floating flexible circular structure under Timoshenko-Mindlin beam theory is developed. The theoretical solution is obtained by matched eigenfunction expansion method along with orthogonal relation and Bessel's functions. The accuracy of the present solution is checked with existing published results, the roots of the dispersion relations are demonstrated via contour plots, and several results of bending moment, displacements, and reflection and transmission coefficients for different structural and physical parameters are analysed. It is observed that the displacement of the circular structure decreases with increase in the values of mooring stiffness, thickness, and elastic modulus of the structure. The bending moment becomes lower as the mooring stiffness and structural thickness increase whilst, the reflection coefficient increases and this effect is reverse to that of transmission coefficient.

1 INTRODUCTION

Due to the growing interest in the applications of floating horizontal flexible circular structure on a water-free surface on the hydroelastic problem that has been investigated by many researchers based on the different approaches (Michele et al. 2023, Meylan, 2021, Mohapatra & Guedes Soares, 2023,). In wave scattering, the understanding of the hydroelastic response of a floating flexible circular structure is relevant to many physical problems arising in ocean engineering, covering from offshore wave energy converters and solar wind devices to large floating structures and ice sheets (Michele et al, 2023). Therefore, for the practical application of floating flexible circular structures in engineering design, the hydroelastic response to the effects of structural properties combined with mooring lines is essential for achieving the stability of the flexible circular structure in the marine environment.

Therefore, for engineering analysis, in the present paper, an anlaytical model of wave interaction with a moored floating flexible circular structure based on the Timoshenko Mindlin beam theory is developed. For instance, Li et al. (2022) studied the hydroelastic response of a very large floating laminated disk under wave action in water of finite depth. Andrianov & Hermans (2005) investigated on analytical and numerical study of the plate hydroelastic behavior between a floating elastic circular plate and incident surface waves for finite and infinite water depth.

Recently, experimental data on floating flexible disk dynamics in water waves is compared with potential flow theory in Michele et al. (2023). The research results of hydroelastic responses of Very Large Floating Structure (VLFS) in complex environments have been systematically summarized in Ding et al. (2021). The solution for incident waves on a circular elastic plate in three-dimensions was given in finite depth and in shallow water by Sturova (2003). Khabakhpasheva & Korobkin (2002) analysed several two-dimensional problems of hydroelastic behaviour of floating plates under the influence of periodic surface water waves. Montiel et al. (2013) reported results from a series of wave basin experiments and a linear numerical model on the flexural response of a single or pair of circular floating thin elastic plate(s) subject to monochromatic gravity waves. Peter et al. (2004) presented a closed form solution for a circular thin plate of shallow draft floating on the water surface subject to linear wave forcing. Zilman & Miloh (2000) presented a three-dimensional closed–form solution for water wave interaction with a floating circular thin elastic plate in shallow water.

The hydroelastic problem of a circular VLFS subjected to a regular wave is analysed based on the Mindlin plate theory by Watanabe et al. (2006). Shen et al. (2023) studied the deformation and hydrodynamic forces of single net cage under currents and verified against the previous physical model tests. The solution for incident waves on an elastic floating plate in two-dimensions was given in finite depth using the Wiener-Hopf technique by Tkacheva (2004). The solution for wave scattering by multiple floating elastic plates with spring connectors or hinges at the plate edges was addressed using matched eigenfunction expansion and Green's function method in Kohout & Meylan (2009). An experimental validation of theoretical models of regular-water-wave transmission through arrays of floating discs in a wave basin based on combined potential-flow and thin-plate theories under the

DOI: 10.1201/9781003508762-41

assumption of linear motions is presented in Bennetts & Williams (2015).

Damaren (2010) presented the added mass, damping, and diffraction force coefficients over a range of wave numbers for the optimal plate as well as circular and square plates. The hydroelastic problem of a circular VLFS with a submerged horizontal annular plate attached to its perimeter is analyzed and the coupled fluid–structure interaction problem is solved by using the modal expansion method in the frequency domain by Pham et al. (2008). Wang et al. (2004) presented exact vibration mode shapes and modal stress-resultants for freely vibrating, circular Mindlin plates with free edges which are important in the hydroelastic analysis of very large circular floating structures (VLFS).

The previous works of the author are relevant to the present paper in terms of the study on the hydro-elastic response of the floating flexible structure (see for more detail Amouzadrad et al. 2023a, 2023b, 2023c). On the other hand, recently Mohapatra and Guedes Soares (2023) developed a Boundary Integral Equation Method code for the model of wave interaction with a floating flexible circular structure and behaviour of the structural properties are analysed.

From the above literature it is noticed that there is no study related with mooring lines with a floating flexible circular structure based on Timoshenko-Mindlin beam theory under analytical approach is reported till date. Therefore, in the present paper, an analytical model under linearized small amplitude water wave theory and structural response is developed. By applying matching technique along with Bessel's function and orthogonal conditions in open water region, the analytical solution is derived. The obtained analytical result of bending moment is verified against other calculations existing in the literature. The real and complex roots of the dispersion relations are demonstrated in contour plots. The numerical results for the structural displacements, bending moments, and reflection and transmission coefficients are presented and the detailed analysis on the mooring stiffness associated with structural parameters are also provided.

2 MODEL DEFINITION

The mathematical formulation of the physical prob-lem is based on the cylindrical polar coordinate system (r, θ, z). The circular floating flexible plate is modelled based on the Timoshenko-Mindlin beam theory. The circular flexible plate is assumed to be floating in a finite water depth. For simplicity, it is assumed that the system is symmetric in terms of rotation about the vertical axis z. Therefore, the axi-symmetric coordinate system (r, z) is chosen that is independent of the rotational angle θ. The origin of the coordinate system is assumed to be at the centre of the circular plate of radius a and small thickness d and the z - axis being vertically downward positive (see Figure 1). The bottom topography is assumed to be uniform at $z = h$. The domain of consideration is divided into two regions: region 1 is the undisturbed free surface region $r\ a,\ 0<z<h$, and region 2 is the plate covered region $r\ a,\ 0<z<h$ as shown in Figure 1.

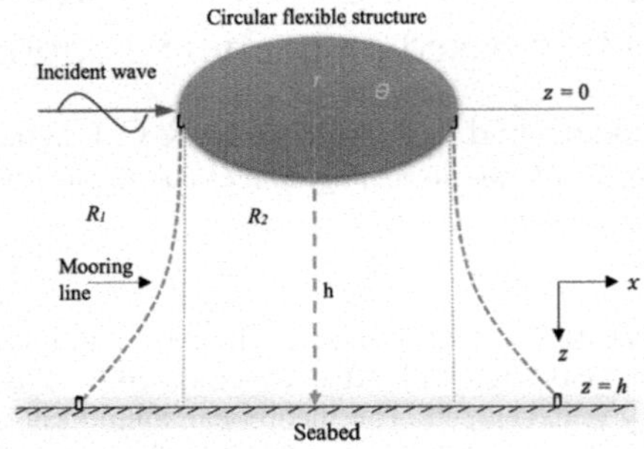

Figure 1. Geometry of a moored circular structure and coordinate system.

Under the assumption of the linearized water wave theory, the fluid is considered to be incompressible, inviscid and the motion is irrotational and time harmonic with angular frequency ω. Therefore, there exists a velocity potential $\Phi\ (r,\ z,\ t)$ of the form $\Phi\ (r,\ z,\ t) = Re[\phi\ (r,\ z)e^{-i\omega t}]$, where the spatial velocity potential $\phi\ (r,\ z)$ satisfies the two-dimensional Laplace's equation as (as in Mohapatra and Guedes Soares, 2024):

$$\nabla^2\phi = 0 \text{ in the fluid domain} \tag{1}$$

where

$$\nabla^2 = \frac{1}{r}\frac{\partial}{\partial r}\left(r\frac{\partial}{\partial r}\right) + \left(\frac{\partial^2}{\partial z^2}\right)$$

The no flow condition at the rigid bottom yields

$$\frac{\partial\phi}{\partial z} = 0 \text{ on } r \geq 0,\ z = h \tag{2}$$

It is assumed that on the upper surface, the draft of the floating flexible circular plate to be very small compared to the radius of the plate. Using the Timoshenko-Mindlin beam theory, the boundary condition on the circular plate covered surface is given by

$$\left\{EI\frac{\partial^4}{\partial r^4}\frac{1}{cos^4\theta} - \left(\frac{EI\rho_b}{\mu G} + \frac{\rho_b d^3}{12}\right)\frac{\partial^2}{\partial t^2}\frac{\partial^2}{\partial r^2}\frac{1}{cos^2\theta}\right.$$

$$+\frac{\partial_b{}^2 d^3}{12\mu G}\frac{\partial^4}{\partial t^4} + \rho_b d\frac{\partial^2}{\partial t^2}$$

$$\left.+\rho g\left(1 - \frac{EI}{\mu G d}\frac{\partial^2}{\partial r^2}\frac{1}{cos^2\theta} + \frac{\rho_b d^2}{12\mu G}\frac{\partial^2}{\partial t^2}\right)\right\}$$

$$\frac{\partial\phi_1}{\partial z} = \left(\frac{EI}{\mu G d}\frac{\partial}{\partial r^2}\frac{1}{cos^2\theta} - \frac{\rho_b d^2}{12\mu G}\frac{\partial^2}{\partial t^2} - 1\right)\rho\left(\frac{\partial\phi_2}{\partial t}\right) \tag{3}$$

where ρ_b, μ, G = $E/2(1+\mu)$, d, EI, and $\rho_b d = m_s$ are the density, transverse deformation, shear modulus, thickness, rigidity, and mass of the plate, respectively. Further, Φ_1 denotes the velocity potential for the region 1 (as defined in Figure 1).

In region 1, the free boundary condition yields

$$\left(\frac{\partial}{\partial t}+\frac{\partial}{\partial r}\right)^2\phi+ g\frac{\partial\phi}{\partial z}=0,\ r>a, z=0 \quad (4)$$

where g is the gravitational constant.

The continuity of velocity and pressure across the plate – water vertical interface at $r = a$ and $0 < z < h$ yields

$$\frac{\partial\phi_1}{\partial r}=\frac{\partial\phi_2}{\partial r},\ \phi_1=\phi_2,\ r=a,\ 0<z<h \quad (5)$$

where Φ_2 denotes the velocity potential for the region 2. As the floating flexible circular plate is connected with mooring lines to the bottom with stiffness q at $r=a$, the shear force and bending moment yield

$$\frac{\partial^4\phi_2}{\partial z^3\partial r}=0,\ r\rightarrow a,\ z=0 \quad (6)$$

$$\frac{\partial^3\phi_2}{\partial z^3}+\left(\frac{1-\nu}{r}\right)\left\{\frac{m_s\omega^2(I_r+S_d)}{EI}\right\}\frac{\partial^2\phi_2}{\partial r\partial z}=q\frac{\partial\phi_2}{\partial z},\quad r\rightarrow a,\ z=0 \quad (7)$$

The far-field condition is assumed to be of the form

$$\phi_1(r,z)=\left\{H_0^{(2)}(k_0r)+B_0H_0^{(1)}(k_0r)\right\}\times\left\{\frac{cosh(k_0(h-z))}{cosh\ k_0h}\right\}as\ r\rightarrow\infty \quad (8)$$

where k_0 is the corresponding wave number and B_0 is the constant associated with the amplitude of the progressive wave. Further, $H_0^{(1)}(k_0r)$ and $H_0^{(2)}(k_0r)$ are the Hankel functions.

3 MATCHING TECHNIQUE AND ORTHOGONAL CONDITION

Using the Fourier expansion formulae and the method of the separation of variables, the velocity potentials ϕ_1 and ϕ_2 satisfying the Laplace Equation (1) and boundary conditions (2-4) are expanded as:

$$\phi_1(r,z)=\{H_0^{(2)}(k_0r)+B_0H_0^{(1)}(k_0r)\}g_0(z)+\sum_{n=1}^{\infty}B_nK_0(k_nr)g_n(z) \quad (9)$$

$$\phi_2(r,z)=A_0J_0(p_0r)f_0(z)+\sum_{n=I,II,1}^{\infty}A_nI_0(p_nr)f_n(z) \quad (10)$$

where J_0 is the Bessel function of first kind of order zero, I_0 is the modified Bessel function of first kind of order zero, and K_0 is the modified Bessel function of the second kind of order zero. Further, $H_0^{(1)}$ and $H_0^{(2)}$ are the Hankel functions of order zero. The eigenfuncions $f_n(z)$ and $g_n(z)$ in Eqs. (9-10) are given by

$$g_n(z)=\frac{\cosh k_n(h-z)}{\cosh k_nh},\ n=0;\quad g_n(z)=\frac{\cos k_n(h-z)}{\cos k_nh},\ n=1,2\ldots, \quad (11)$$

$$f_n(z)=\frac{\cosh p_n(h-z)}{\cosh p_nh},\ n=0;\quad f_n(z)=\frac{\cos p_n(h-z)}{\cos p_nh},\ n=I,II,1,2,\ldots, \quad (12)$$

with the eigenvalues k_n, n = 0, 1, 2, …satisfying the dispersion relation

$$\left.\begin{array}{l}\omega^2-gk_0tanh(k_0h)=0,\ \text{for } n=0,\\ \omega^2+gk_ntan(k_nh)=0,\ \text{for } n=1,\ 2,\ldots\end{array}\right\} \quad (13)$$

and p_n, n = 0, I,II,1,2, satisfying the dispersion relation

$$\{EIp_0^4-(m_s\omega^2(I_r+S_d)-S_d\rho g)p_0^2-\left(\frac{\chi(\rho g-m_s\omega^2)}{EI}\right)\}p_0\ tanh\ p_0h-\left(\rho S_d+\rho\left(1-\frac{\chi}{EI}\right)p_0^2\right)\omega^2=0,\ \text{for } n=0$$

$$\{EIp_n^4-(m_s\omega^2(I_r+S_d)-S_d\rho g)p_n^2-\left(\frac{\chi(\rho g-m_s\omega^2)}{EI}\right)\}p_ntan\ p_nh-\left(\rho S_d+\rho\left(1-\frac{\chi}{EI}\right)\rho_n^2\right)\omega^2=0, n=1,11,1,2,\ldots\} \quad (14)$$

where $\chi=(m_s\omega^2I_rS_d)$.

Further, the eigenfunctions g_n and f_n satisfy the orthogonal relations

In region 1

$$\langle g_m,g_n\rangle=\left\{\begin{array}{ll}0, & m\neq n,\\ E_n, & m=n,\end{array}\right. \quad (15)$$

with

$$E_n=\left\{\begin{array}{ll}\frac{2k_nh+sinh\ 2k_nh}{4k_n\ cosh^2\ k_nh}, & n=0,\\ \frac{2k_nh+\sin 2k_nh}{4k_n\ cos^2\ k_nh}, & n=1,2,\ldots,\end{array}\right.$$

In region 2

$$F_{ORT} = f_m, f_n = \begin{cases} 0, & m \neq n, \\ Q_n, & m = n, \end{cases} \quad (16)$$

With

$$Q_n = \frac{\{\Psi_1 + \Psi_2 + \Psi_3\}}{4p_n cosh^2(p_n h)\Upsilon(p_n)\Lambda(p_n)}, n = 0$$

$$Q_n = \frac{\{\Psi_1^* + \Psi_2^* + \Psi_3^*\}}{4p_n cos^2(p_n h)\Upsilon(p_n)\Lambda(p_n)}, n = 1, 11, 1, 2, \ldots,$$

where

$$\Psi_1 = \Upsilon(p_n)\Lambda(p_n)(2p_n h + sinh(2p_n h)),$$

$$\Psi_2 = \left(8\aleph p_n^3 - 4\beta p_n\right)\Lambda(p_n)p_n^2 tanh^2(p_n h)cosh^2(p_n h),$$

$$\Psi_3 = 4\tau_2 \Upsilon(p_n)p_n cosh^2(p_n h),$$

$$\Psi_1^* = \Upsilon(p_n)\Lambda(p_n)(2p_n h + sin(2p_n h)),$$

$$\Psi_2^* = \left(8\aleph p_n^3 - 4\beta p_n\right)\Lambda(p_n)p_n^2 tan^2(p_n h)\cos^2(p_n h),$$

$$\Psi_3^* = 4\tau_2 \Upsilon(p_n)p_n cos^2(p_n h).$$

$$\begin{aligned} \langle f_m, f_n\rangle &= \int_0^h f_m(z)f_n(z) \\ &+ \tfrac{\aleph}{\Upsilon(P_n)}[f'''_n(z).f'_n(z) + f'_n(z).f'''_n(z)]_{z=0} \\ &- \tfrac{\beta}{\Upsilon(P_n)}[f'_n(z).f'_n(z)]_{z=0} + \tfrac{\tau_2}{\Lambda(P_n)}[f_n(z)f_n(z)]_{z=0} \end{aligned}$$

with $f_n'''(0) = -P_n^3 tanP_n h, f'_n(0) = -P_n tanP_n h,$

$$\beta = \left\{m_s\omega^2(I_r + S_d) - S_d\rho g\right\}$$

$$\aleph = \left\{\left(\omega^4 m_s^2 I_r S_d - \rho g m_s \omega^2 I_r S_d\right)/EI_r\right\} + \kappa,$$

$$\tau_1 = \rho\omega^2 S_d,\ \tau_2 = \rho\omega^2\left\{1 - \left(m_s\omega^2 I_r S_d\right)/EI_r\right\},$$

$$\Upsilon(P_n) = \left(EIp_n^4 - \beta p_n^2 + \aleph\right),\ \Lambda(p_n) = \left(\tau_1 + \tau_2 p_n^2\right).$$

where the orthogonal mode-coupling relations g_m, g_n and f_m, f_n are defined as

$$\langle g_m, g_n\rangle = \int_0^h g_m g_n dz, \quad (17)$$

$$\begin{aligned} \langle f_m, f_n\rangle &= \int_0^h f_m f_n dz \quad (18) \\ &+ \tfrac{\aleph}{\Upsilon(P_n)}\left[f_m'''(z).f_n'(z) + f_m'(z).f_n'''(z)\right]_{z=0} \\ &- \tfrac{\beta}{\Upsilon(P_n)}\left[f_m'(z).f_n'(z)\right]_{z=0} \\ &+ \tfrac{\tau_2}{\Lambda(P_n)}\left[f_m(z)f_n(z)\right]_{z=0}. \end{aligned}$$

Applying the mode coupling relation as in Eq. (18) on ϕ_2 and f_n and using the continuity condition (5) along with the orthogonal relation as in Eq. (16), the expansion formula yields

$$A_n p_n J'_0(p_n a)Q_n = \left\{H_0{}^{(2)'}(k_0 a) + B_0 H_0{}^{(1)'}(k_0 a)\right\} (k_0 G_{0n}) + \textstyle\sum_{n=1}^{\infty} B_n k_n K'_0(k_n a)G_{mn} - \frac{D_n - D_n^*}{\Upsilon(p_n)\Lambda(p_n)}, n = 0 \quad (19)$$

$$A_n p_n I'_0(p_n a)Q_n = \left\{H_0{}^{(2)'}(k_0 a) + B_0 H_0{}^{(1)'}(k_0 a)\right\} (k_0 G_{0n}) + \textstyle\sum_{n=1}^{\infty} B_n k_n K'_0(k_n a)G_{mn} - \frac{C_n - C_n^*}{\Upsilon(p_n)\Lambda(p_n)}, \quad n = 1, ll, 1, 2, \ldots \quad (20)$$

where

$$G_{mn} = \int_0^h g_m f_n dz, n = 0, 1, 11, 1, 2, \ldots m = 0, 1, 2, \ldots,$$

$$D_n = \Lambda(P_n)\aleph(2p_n^4 tanh^2 p_n h),$$

$$D_n^* = \Lambda(P_n)\beta(p_n^2 tanh^2 p_n h) + \Upsilon(P_n)\tau_2,$$

$$C_n = \Lambda(P_n)\aleph(2p_n^4 tan^2 p_n h),$$

$C_n^* = \Lambda(P_n)\beta(p_n^2 tan^2 p_n h) + \Upsilon(P_n)\tau_2,$ and the value of Q_n being the same as in Eq. (16).

Applying the orthogonal relation (15) along with mode coupling relation (17) on ϕ_1 and g_n and the continuity of pressure (5) yields

$$\left\{H_0^{(2)}(k_0 a) + B_0 H_0^{(1)}(k_0 a)\right\}E_n = A_0 J_0(p_0 a)G_{0n} + \textstyle\sum_{n=l,ll,1}^{\infty} A_n I_0(p_n a)G_{mn}, \ n = 0 \quad (21)$$

$$B_n K_0(k_n a)E_n = A_0 J_0(p_0 a)G_{0n} + \textstyle\sum_{n=l,ll,1}^{\infty} A_n I_0(p_n a)G_{mn}, \ n = 1, 2, \ldots \quad (22)$$

The moored edge conditions as in (6)(7) yields the final equation as follows

$$A_0 p_0 J'_0(p_0 a)\left(-p_0{}^3 tanhp_0 h\right) + \textstyle\sum_{n=l,ll,1}^{\infty} A_n p_n I'_0(p_n a)\left(-p_n^3 tanp_n h\right) = 0 \quad (23)$$

$$A_0\left[J_0(p_0 a)p_0{}^2 + A\{p_0 J'_0(p_0 a)\}\right] + A_n\left[A\left\{\textstyle\sum_{n=l,ll,1}^{\infty} p_n I'_0(p_n a)\right\} + \textstyle\sum_{n=l,ll,1}^{\infty} I_0(p_n a)p_n{}^2\right] - q = 0 \quad (24)$$

With $A = ((1-\nu)/r)\left\{\left(m_s\ \omega^2\ (I_r + S_d\ )\right)/EI\right\}$

The vertical deflection of the floating flexible circular structure with rotary inertia and shear deformation can be obtained from the following relation

$$\frac{\partial \eta}{\partial t} = \frac{\partial \phi_2}{\partial z} \text{ gives}$$

$$\eta(r,z) = \frac{-1}{2i\omega}\Bigg[\{A_0 J_0(p_0 r)\}p_0 \; tanh \; p_0 h + \sum_{n=1,\text{II},1}^{\infty} \{A_n I_0(p_n r)\}p_n \; tan \; p_n h\Bigg] \quad (25)$$

The bending moment and shear force can be computed by the following formulae

$$M_B = \frac{\partial^3 \phi_2}{\partial z^3} + \left(\frac{1-\nu}{r}\right)\left\{\frac{m_s \omega^2 (I_r + S_d)}{EI}\right\}\frac{\partial^2 \phi_2}{\partial r \partial z} - q\frac{\partial \phi_2}{\partial z} \quad (26)$$

$$F_S = \frac{\partial^4 \phi_2}{\partial z^3 \partial r} \quad (27)$$

NB: It may be noted that in the case of shallow water waves, one can derive the shallow water dispersion relation by approximating *tanhkh≈kh* in relations (13-14).

Finally, the reflection and transmission coefficients can be defined as: the reflection coefficient C_r is the ratio of amplitude of the reflected and the incident progressive waves in region 1 whilst, the transmission coefficient C_t is the ratio of transmitted flexural gravity wave amplitude in region 2 to the incident wave amplitude in region 1. The formulae for the the C_r and C_t can be read as

$$C_r = \frac{B_0}{A_0} \quad (28)$$

$$C_t = \frac{A_0 p_0 \; tanh(p_0 h)}{k_0 \; tanh(k_0 h)} \quad (29)$$

In the next section, the numerical results based on the analytical solution for the displacement, bending moment, and wave quantities (reflection and transmission coefficients) are presented for the hydroelastic response of the circular structure on different design parameters for the engineering practice cases.

To solve the system of linear equations (19-24), the infinite series sums present in the series expansion are truncated up to a finite number of terms N (say) to perform the numerical calculation by developing MATLAB codes to present numerical results based on the analytical solution. The MATLAB codes are developed by incorporating Bessel functions, Hankel functions, and considering12 terms in the series solution to compute the displacements, bending moments, and wave quantities.

In the numerical simulation, all computations are performed by considering water density $\rho = 1025\text{kgm}^{-3}$ acceleration due to gravity $g = 9.8\text{ms}^{-1}$, structural thickness $d = 0.29$ m, water depth $h = 15$ m, unless otherwise stated and Elastic modulus $E = 8$ GPa in which for the practical case the wood type material could be used for the floating circular structure.

For wave data, the wave period and wave amplitude are 2 s and 1 m respectively considered for the numerical calculation.

The numerical results are presented based on the analytical solution by using MATLAB R2023b, 64-bit (win64) to perform calculations in a desktop machine with Intel ® core i7-4790 CPU with 3.60GHz processor and 16 GB of ram memory, 3601 MHz, 4 Core(s), and 8 Logical Processor(s). The data was written on an SSD disk and each case took roughly 5-8 minutes to finish.

4 RESULTS AND DISCUSSIONS

4.1 *Determination and logical numerical scheme for roots of the dispersion relations*

The nature and location of the roots of the dispersion relation (13) and (14) are shown in Figures (2, 3a-3c). It may be mentioned that the dispersion relation (14) contains four complex roots p_n of the form $\pm a \pm ib$ (see the contour plots in Figures 3b,3c), two real roots of the form $\pm p_0$ and an infinite number of purely imaginary roots $p_n = ip_n$ (see the contour plot in Figure 3a). Out of two real roots, only the positive real root along with four complex roots as well as imaginary roots are considered for numerical computation in the series solution. On the other hand, the roots of the water dispersion relation (13) are also obtained using the iterative numerical scheme Newton-Raphson method for real root where relation (13) contains two real roots ($+k_0$ and $-k_0$) and an infinitely imaginary roots which are shown in the contour plot (Figure 2).

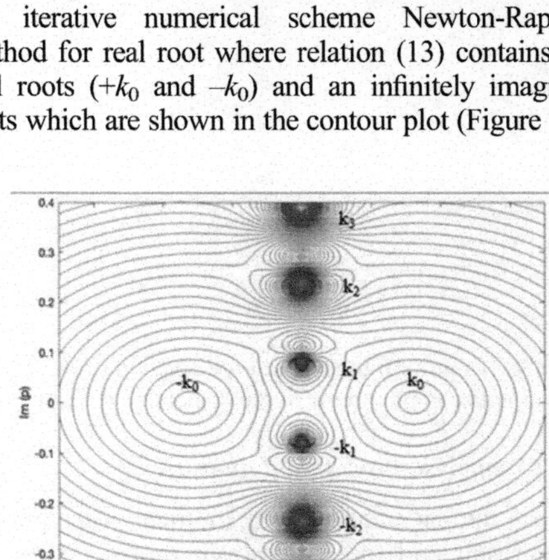

Figure 2. Contour plot showing the position of the real and imaginary roots of the water dispersion relation (13).

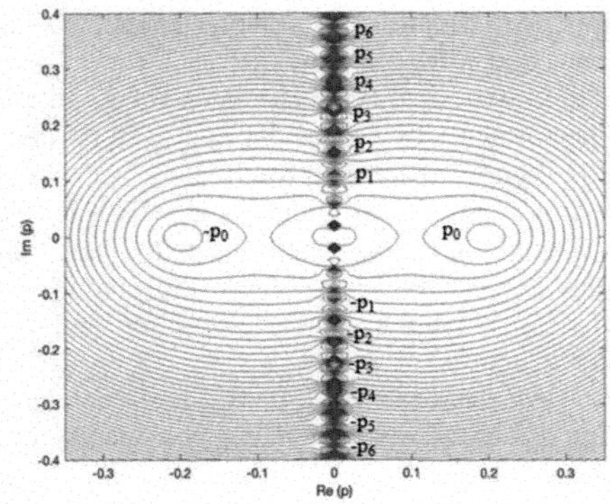

Figure 3a. Contour plot showing the position of the real and imaginary roots of the dispersion relation (14).

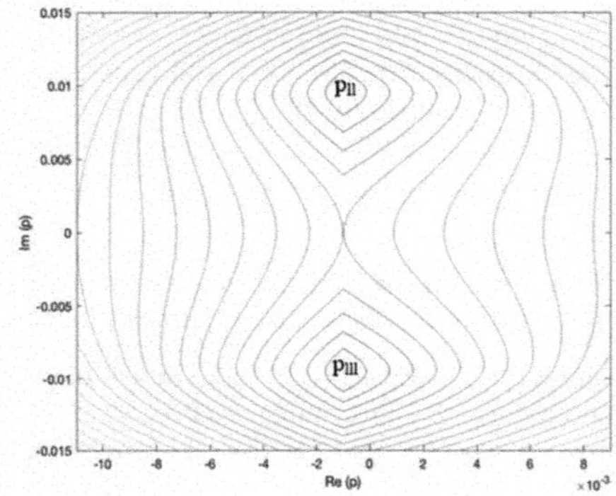

Figure 3b. Contour plot showing the position of the 2nd and 3rd quadrants complex roots of the dispersion relation (14).

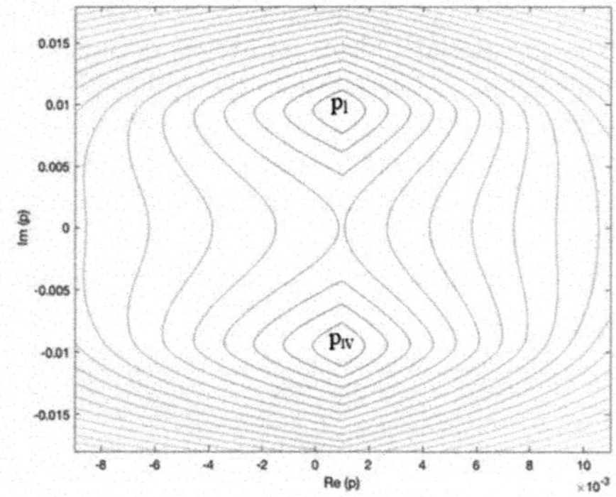

Figure 3c. Contour plot showing the position of the 1st and 4th quadrants complex roots of the dispersion relation (14).

4.2 *Comparison result and analysis*

Figure 4 compares the bending moment of the present model result and the result from model (Damaren, 2010) versus wavenumber *kr* with mooring stiffness $q = 10^{2.5}$ Nm^{-1}. To attend a similar trend of the result between the present and the model result (Damaren, 2010), a smaller value of mooring stiffness and thinner structure were chosen to a similar trend.

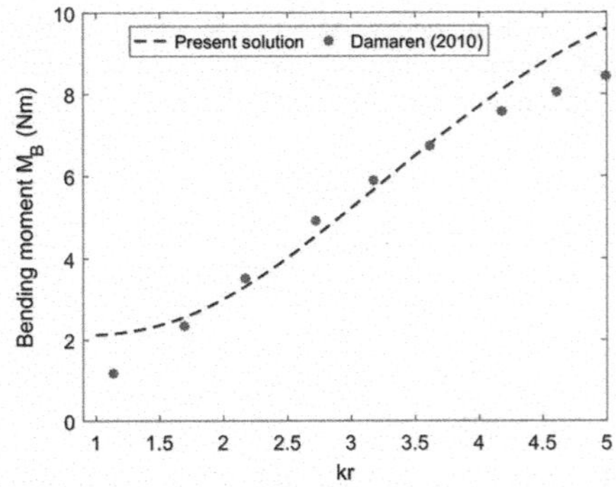

Figure 4. Comparison of bending moment between the present and the model Damaren (2010).

The comparison indicates that the present model solution supported by the model exists in the literature. However, the differences can be explained by the formulation under the assumption of Timoshenko-Mindlin beam theory of the present model and thin plate theory in Damaren (2010). Stiffness properties of the model data from the reference versus the present model shown in Table 1

Table 1. Stiffness properties of the circular structure.

Present model	0.0148
Reference model of Damaren (2010)	0.01

4.3 *Hydroelastic response of circular structure via structural displacements and bending moments*

Figure 5 simulates the effect of displacements on different mooring stiffness versus non-dimensional wavenumber *kr*. It can be easily seen that by increasing the mooring stiffness, the displacement becomes lower, which indicates that higher stiffness prevents the displacement that leads to more stability in the structure.

Figure 6 shows the structural displacements on different values of the elastic modulus E versus non-dimensional wavenumber *kr* with mooring stiffness $q = 10^3$ Nm^{-1}. With an increase in the values of E, the displacement decreases, which indicates that increasing the elastic modulus can effectively decrease the displacement of the floating flexible circular structure.

As can be seen from Figure 7, when the circular structure is thicker, the displacement goes on decreasing with occurring resonating pattern. However, for thinner structure, the displacement becomes higher with attending a very peak value for higher wavenumber. It can be concluded that a thicker structure is more stable than a thinner which is due to the increase its rigidity.

It may be mentioned that, in Figures 5-7, the general trend is that the displacements decrease with increase in the values of mooring stiffness, elastic modulus, and thickness of the circular structure. Overall, the results and observations are similar as in

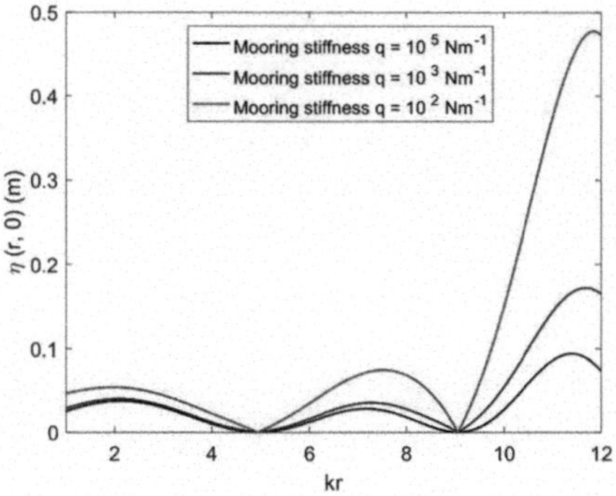

Figure 5. Influence of mooring stiffness q on the displacement.

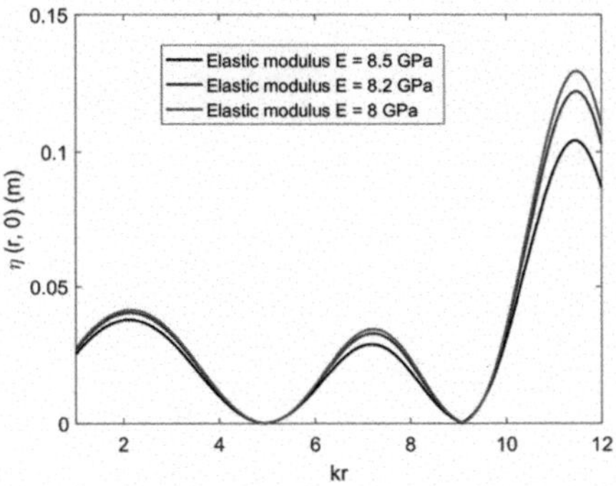

Figure 6. Influence of elastic modulus E on the displacement with mooring stiffness $q = 10^3$ Nm^{-1}.

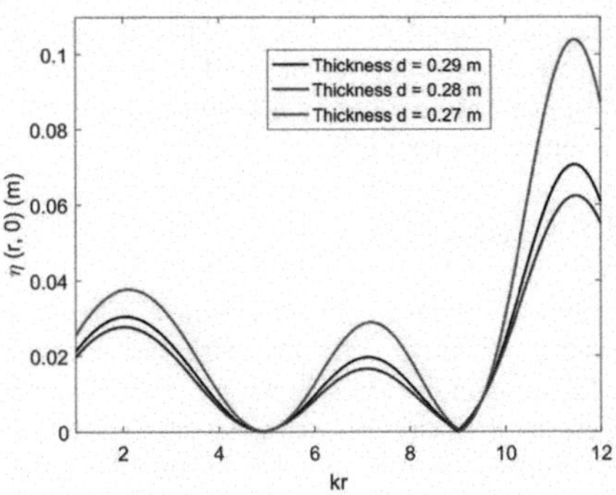

Figure 7. Effect of thickness on the displacement with $q = 10^3$ Nm^{-1}.

Mohapatra and Guedes Soares (2023) obtained from numerical Boundary Integral Equation Method.

Figure 8 presents the effects of mooring stiffness q on the bending moment versus kr. It is seen that the bending moment becomes higher for lower values of mooring stiffness, however, the variations of bending moment for different mooring stiffness are lessen for very smaller values of wavenumber.

Figure 9 gives the effects of the bending moment for different E versus kr with mooring stiffness $q = 10^3$ Nm^{-1}. As it is observed that, the bending moment decreases for higher values of E, however, the peak is occurred in the resonating pattern of bending moment for lower value of E.

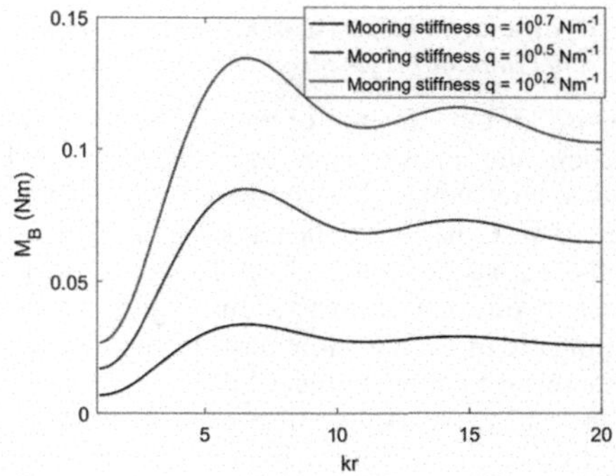

Figure 8. Effect of mooring stiffness on the bending moment.

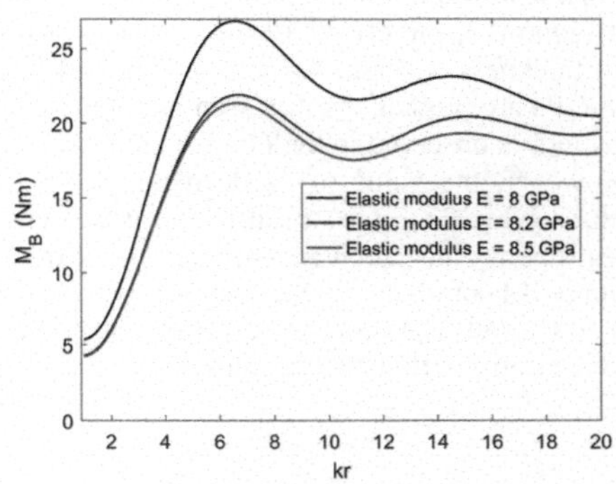

Figure 9. Effect of E on the bending moment with mooring stiffness $q = 10^3$ Nm^{-1}.

In Figure 10, the bending moment for different thickness d versus kr with mooring stiffness $q = 10^3$ Nm^{-1} are plotted. The structural thickness seems to be important for the bending moment as for thicker structure the bending moment becomes lower, whilst, a peak is occurred for smaller values of wavenumber in the pattern of bending moment.

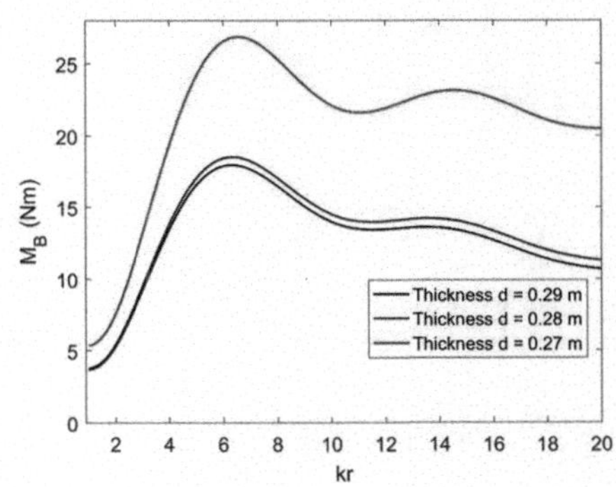

Figure 10. Effect of thickness on the bending moment with $q = 10^3$ Nm^{-1}.

4.4 *Hydroelastic response of circular structure via reflection and transmission coefficients*

In Figure 11, the influence of mooring stiffness on the reflection and transmission coefficients C_r and C_t versus *kr* are plotted. It is observed that for smaller values of *kr*, C_t has a significant variation for different *q* whilst, C_r has negligible effect. However, the transmission coefficient decreases for higher values of *q* but in case of C_r, the variation is lessening and provides a reverse pattern to that of C_t.

Figure 12 presents the effect of different *E* on the C_r and C_t versus *kr* with a certain value of mooring stiffness $q = 10^3 \text{Nm}^{-1}$ and thickness $d = 0.29$ m. In the case of C_r, minor variation is seen due to mainly increase in the values of rigidity with mooring stiffness that leads high transmission.

In Figure 13, the effects of the C_r and C_t for different thickness *d* versus *kr* with $q = 10^3$ Nm^{-1} and $E = 8$ GPa are plotted. As it is seen that, the resonating pattern is observed in the C_t for smaller values of *kr*. However, the value of C_t decreases for thicker structure which is due to the less wave energy passes below the circular structure as expected.

Figure 14 simulates the effects of water depths *h* on te C_r and C_t versus *kr*. Figure 14 reveals that the reflection coefficient becomes lower for deeper water whilst, the transmission of wave energy decreases for shallower water. This is because of more wave energy passes below the structure as the depth of water increases as a result the transmission coefficient becomes higher.

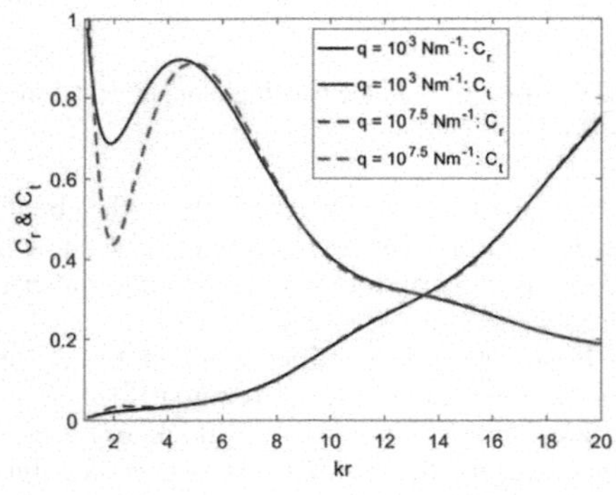

Figure 11. Influence of mooring stiffness on the C_r and C_t with elastic modulus $E = 8$ GPa and water depth $h = 15$ m.

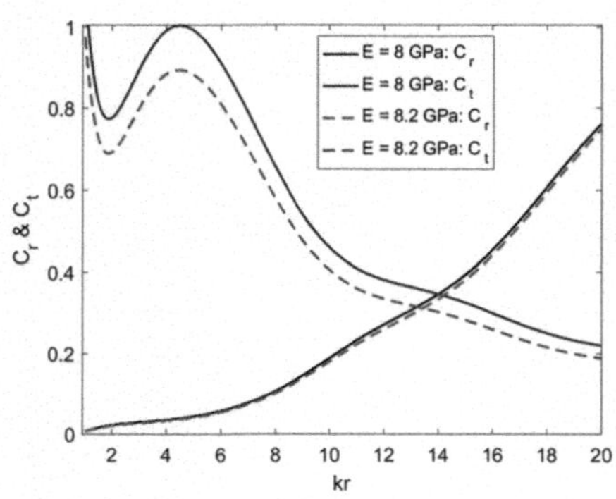

Figure 12. Effect of elastic modulus on the C_r and C_t with $q = 10^3$ Nm^{-1}.

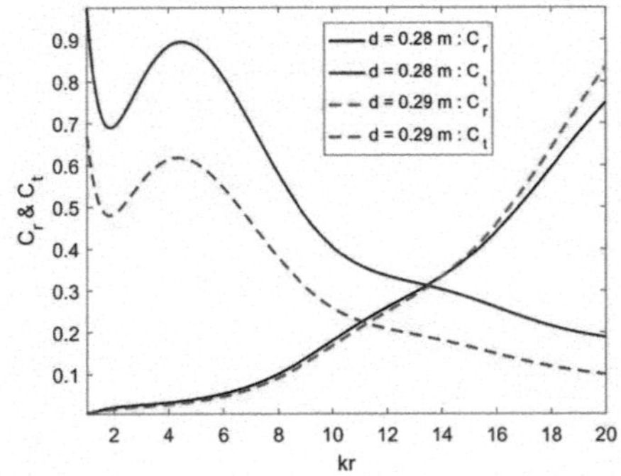

Figure 13. Effect of structural thickness on the C_r and C_t with mooring stiffness $q = 10^3$ Nm^{-1}.

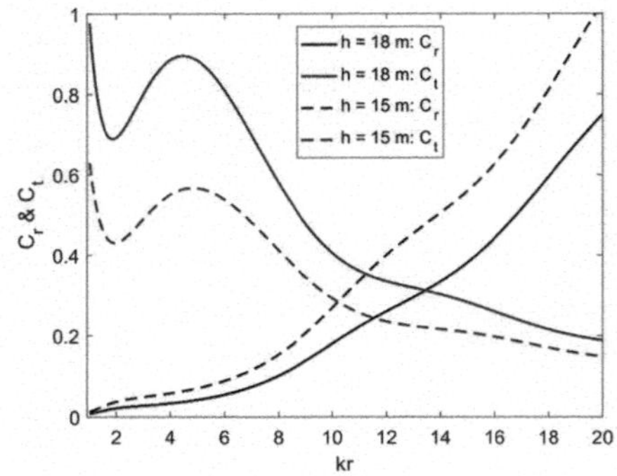

Figure 14. Effect of water depth on the C_r and C_t with mooring stiffness $q = 10^3$ Nm^{-1}.

5 CONCLUSIONS

An analytical model for the effect of mooring lines on a floating flexible circular structure based on Timoshenko-Mindlin beam theory is developed to study the hydroelastic response via structural displacements, bending moment, and wave quantities. The theoretical solution is derived based on the matching technique and Bessel's function along with orthogonal relation. The level of accuracy of the analytical result is checked by comparing with existing published literature and it has a good level of agreement. In addition, effect of different structural parameters on the structural displacements, bending moment, and wave quantities are analysed in different cases in detail, the following important concluding remarks are obtained as:

1. The present analytical result is supported by the model result existing published literature.
2. The contour plots show that the correctness of the real and complex roots are considered in the series solution.

3. The displacement result indicates that for higher values of mooring stiffness, the stability of the structure becomes higher.
4. It is noticed that as the mooring stiffness and structural thickness increases, the bending moment becomes lower whilst, the bending moment depends significantly on the elastic modulus of the circular structure.
5. It is noticed that the transmission coefficient decreases for higher values of elastic modulus, mooring stiffness, and thicker structure with occurring resonating pattern for smaller values of wavenumber whilst, there is a negligible variation in the reflection coefficients and the results become reverse in nature to that of transmission coefficients.
6. The present formulation will be helpful to model a moored floating flexible structure under wave-current interactions to analyze the performance for the floating platform via hydroelastic response.

ACKNOWLEDGEMENTS

This work contributes to the Strategic Research Plan of the Centre for Marine Technology and Ocean Engineering (CENTEC), which is financed by FCT under contract UIDB/UIDP/00134/2020. The first author has been funded by the Portuguese Foundation for Science and Technology (Fundação para a Ciência e a Tecnologia-FCT), through a doctoral fellowship under the Contracts no. UI/BD/154592/2023. The second author has been contracted as a Researcher by FCT, through Scientific Employment Stimulus, Individual support under Contract No. CEECIND/04879/2017.

REFERENCES

Amouzadrad, P., Mohapatra, S.C. & Guedes Soares, C. 2023a. Hydroelastic response of a moored floating flexible offshore structure based on Timoshenko-Mindlin-Beam theory. *Trends in Renewable Energies Offshore*, Guedes Soares, C. (Ed.), Taylor and Francis, London, UK, pp. 879–888.

Amouzadrad, P., Mohapatra, S.C. & Guedes Soares, C. 2023b. Hydroelastic response of a moored floating flexible structure based on Timoshenko-Mindlin beam theory. *Advances in the Analysis and Design of Marine Structures*, Ringsberg, J.W. & Guedes Soares C., (Eds.), Taylor & Francis, London, UK.

Amouzadrad, P., Mohapatra, S.C. & Guedes Soares, C. 2023c. Hydroelastic Response to the Effect of Current Loads on Floating Flexible Offshore Platform. *Journal of Marine Science and Engineering* 11(2): 437.

Andrianov, A.I. & Hermans, A.J. 2005. Hydroelasticity of a circular plate on water of finite or infinite depth. *Journal of Fluids and Structures* 20 (5): 719–733.

Bennetts, L.G. & Williams, T.D. 2015. Water wave transmission by an array of floating discs. *Proceedings of the Royal Society A Mathematics, Physics, and Engineering Science* 471(2173): 20140698.

Damaren, C.J. 2010. Hydroelastic analysis of the floating plate optimized for maximum radiation damping. *Ocean Engineering* 37 (2-3): 252–261.

Ding, J., Wu, Y., Xie, Z., Yang, W., Wang, S., Yu, J. & Yu, T. 2021. Overview: Research on hydroelastic responses of VLFS in complex environments, *Marine Structures* 78: 102978.

Khabakhpasheva, T. & Korobkin, A. 2002. Hydroelastic behaviour of compound floating plate in waves. *Journal of Engineering Mathematics* 44: 21–40.

Kohout, A.L. & Meylan, M.H. 2009. Wave Scattering by Multiple Floating Elastic Plates with Spring or Hinged Boundary Conditions. *Marine Structures* 22: 712–729.

Li, A.J., Fang, H. & Liu, Y. 2022. Hydroelastic analysis of interaction between water waves and a floating laminated disk. *Physics of Fluids* 34: 047121.

Meylan, M.H. 2021. Time-Dependent Motion of a Floating Circular Elastic Plate. *Fluids* 6(1): 29.

Michele, S., Zheng, S., Buriani, F., Borthwick, A. & Greaves, D. 2023. Floating hydroelastic circular plate in regular and irregular waves. *European Journal of Mechanics - B/Fluids* 99: 148–162.

Mohapatra, S.C. & Guedes Soares, C. 2023. Hydroelastic Response of a Moored Floating Flexible Circular Structure Applying BIEM. *Journal of Marine Science and Engineering* 11(12): 2322.

Mohapatra, S.C. & Guedes Soares, C. 2024. A semi-analytical model of an array of moored floating offshore net cages under current loads. *Ocean Engineering* 291: 116309.

Montiel, F., Bennetts, L. G., Squire, V. A., Bonnefoy, F. & Ferrant, P. 2013. Hydroelastic response of floating elastic discs to regular waves. part 2. modal analysis. *Journal of Fluid Mechanics* 723: 629–652.

Peter, M. A., Meylan, M. H. & Chung, H. 2004. Wave scattering by a circular elastic plate in water of finite depth: a closed form solution. *International Journal of Offshore and Polar Engineering* 14: 81–85.

Pham, D.C., Wang, C.M. &Utsunomiya, T. 2008. Hydroelastic analysis of pontoon-type circular VLFS with an attached submerged plate. *Applied Ocean Research* 30 (4): 287–296.

Shen, H., Zhao, Y.P., Bi, C.W. & Xu, Z. 2023. Nonlinear dynamics of an aquaculture cage array induced by wave-structure interactions. *Ocean Engineering* 269: 113711.

Sturova, I.V. 2003. The action of an unsteady external load on a circular elastic plate floating on shallow water. *Journal of Applied Mathematics and Mechanics* 67: 407–416.

Tkacheva, L. 2004. Plane problem of vibrations of an elastic floating plate under periodic external loading. *Journal of Applied Mechanics and Technical Physics* 45: 420–427.

Wang, C., Xiang, Y., Watanabe, E. & Utsunomiya, T. 2004. Mode shapes and stress-resultants of circular Mindlin plates with free edges. *Journal of Sound and Vibration* 276: 511–525.

Watanabe, E., Utsunomiya, T., Wang, C.M. & Hang, L.T.T. 2006. Benchmark hydroelastic responses of a circular VLFS under wave action. *Engineering Structures* 28: 423–430.

Watanabe, E., Utsunomiya, T., Wang, C.M. & Xiang, Y. 2003. Hydroelastic analysis of pontoon-type circular VLFS. *Proceedings of 13th International Offshore & Polar Engineering Conference*, Vol. 1, Honolulu, USA, 93–99.

Zilman, G. & Miloh, T. 2000. Hydroelastic buoyant circular plate in shallow water: a closed form solution. *Applied Ocean Research* 22: 191–198.

Advances in Maritime Technology and Engineering – Guedes Soares & Santos (Eds)
© 2024 The Author(s), ISBN 978-1-032-83099-5

Numerical analysis of the effect of current and wind on the dynamics of large floating flexible platform

P. Amouzadrad, S.C. Mohapatra & C. Guedes Soares
Centre for Marine Technology and Ocean Engineering (CENTEC), Instituto Superior Técnico, Universidade de Lisboa, Lisbon, Portugal

ABSTRACT: Drawing upon linear potential flow theory, a numerical model of wave interaction with a moored articulated large floating flexible structure is presented. The floating flexible structure is combined by six plates with hinges and connected with mooring lines at its edges. The mathematical equations associated with the numerical model formulations and multi-body hydrodynamics in frequency and time domain are provided. The numerical result of vertical displacement is compared with existing published experimental datasets and other numerical model results as well as comparisons of without current and two different wave models are also presented. The effects of current and wind on the dynamics of the articulated floating flexible structure are studied by analysing several numerical results of structural displacements for different current and wind speeds with constant values of stiffness of mooring and hinges. Further, the transmission wave amplitudes around the structure are analysed via contour plots, showing that, as the current speed increases the structural displacement increases with a constant value of wind speed, whilst, no effect of wind on the displacement is observed.

1 INTRODUCTION

Due to the growing interest in the applications of very large floating flexible horizontal structures (VLFSs) for utilization of ocean space and marine resources (Jiang et al., 2023), the dynamics response via hydroelasitc has becoming a significant analysis in the area of hydroelasticity in the wave-structure interaction problems (see Yang et al., 2019). However, it is difficult to construct a single continuum structure because of the large size which can occupy a large area of the sea and cause massive loads in structures.

So, it is necessary to develop modern and sophisticated multi-module floating interconnected structures that can address new trends, needs, and satisfy new requirements for different applications in marine environments, especially considering floating emergency bases, floating islands for tourism (see Wang et al., 2024 for Figure 1(a)), or the production of energy such as hydrogen, logistics, living (see Jiang et al., 2023 for Figures 1(b-d) and Flikkema et al., 2021 for Figure 1(e)).

Therefore, it is necessary to model a moored large floating interconnected flexible structure under different environmental conditions; current, winds, and waves. The multi-module or interconnected or hinged structure brings several advantages, including; flexibility for expansion of the several modules, increasing reliability of the interconnected as the system has higher redundancy, lower costs for manufacturing and deployment owing to the modularity in design, and the whole system be stable due to the connection of mooring lines (Bispo et al., 2022a).

The study of different aspects of VLFSs combined with several interconnected structures based on numerical approaches has been added to the literature. The three-dimensional hydroelastic responses of a multiply-connected tourism platform based on numerical and experimental were carried out by Wang et al. (2024). A simplified method was established by Ding et al. (2019) to estimate the hydroelastic responses of VLFS using 3D hydroelasticity theory and inhomogeneous wave theory. Based on the Boussinesq equation and Cummins' theory, a time-domain hydroelastic method for an eight-module semi-submersible VLFS was studied for numerical assessment by Yang et al. (2019). A numerical frequency-domain model was developed to study the hydrodynamic performance of a hybrid floating platform (Zhao et al., 2023). Ding et al. (2020) investigated experimentally and numerically the connector loads of a 3-module VLFS in variable seabed bathymetry. Using the FEM, Shi et al. (2022) proposed a new type of face-contact connector to reduce the local stress level at connection areas. Cheng et al. (2016) investigated the hydroelastic analysis of a pontoon-type floating structure edged with dual horizontal/inclined perforated plated based on the Boundary Element Method (BEM) in the time domain and the finite element method was adopted to present the

DOI: 10.1201/9781003508762-42

deflection of the floating structure. Tay (2019) studied the energy extraction from an articulated plate anti-motion device that was connected to the VLFS by hinges by implementing a hybrid boundary element-finite element method. Zhang, et al. (2018) studied numerically the dynamic response and power capture performance of a VLFS composed of two hinged modules using the discrete-module-beam-bending based hydroelasticity method in the frequency domain. Chen et al. (2023) developed a coupled hydrodynamic–structural model to investigate the dynamic responses of a multi-module floating system with flexible connectors. The hydrodynamic experimental and numerical model of a double floating body connected by a hinged connecter was conducted by Song et al. (2023).

Yoon et al. (2014) proposed a numerical procedure to analyze floating plate structures with multiple hinge connections in regular waves to investigate the bending moment and deflection of the plate structures based on the BEM for the fluid portion and a Finite Element Method (FEM) for the structural part. Zhao et al. (2015) studied the hydroelastic response of a 2D floating plate with hinged conditions under the action of monochromatic incident waves based on the Mindlin plate theory and Wiener-Hopf technique. Ding et al. (2020) investigated the hydroelastic response of an 8-module VLFS based on a numerical and experimental study using the Boussinesq equation and Rankine source method. Wu et al. (2018) established a direct-coupled method to analyze the response of a VLFS consisting of three-module floating body, based on the Rankine source method.

Recently, Ding et al. (2021) and Bispo et al. (2021) reviewed the numerical approaches in the hydroelastic response of VLFSs different geometrics with continuous floating structure or interconnected structures. Bispo et al. (2022a) employed the numerical approach using ANSYS® AQWA to simulate a moored articulated floating structure composed of a set of hinged plates The analysed the structural displacements and wave quantities for different structural parameters along with mooring stiffness and rotational stiffness. Further, a WEC model of a moored submerged horizontal set of hinged plates attached with a vertical plate at one end of its edge based on the ANSYS® AQWA was developed by Bispo et al. (2022b).

On the other hand, the previous works of Amouzadrad et al. (2023a, 2023b, 2023c) are relevant to the present paper in terms of the hydroelastic response of the continuous floating flexible structures. In parallel, Mohapatra & Guedes Soares (2021, 2023) developed a Boundary Integral Equation Method (BIEM) code associated with wave interaction with a floating flexible rectangular and circular flexible structure and analysed the hydroelastic response for different design and structural parameters.

(a) Floating islands for tourism adopted from Wang et al. (2024)

(b) Logistic (Jiang et al. 2023)

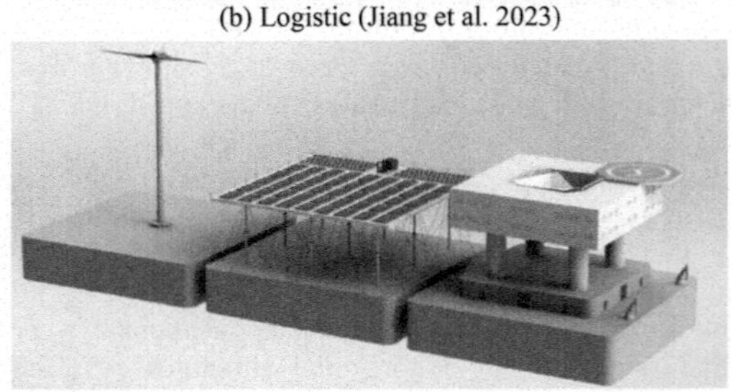

(c)Energy platform adopted from Jiang et al. (2023)

(d) Living adopted from Jiang et al. (2023)

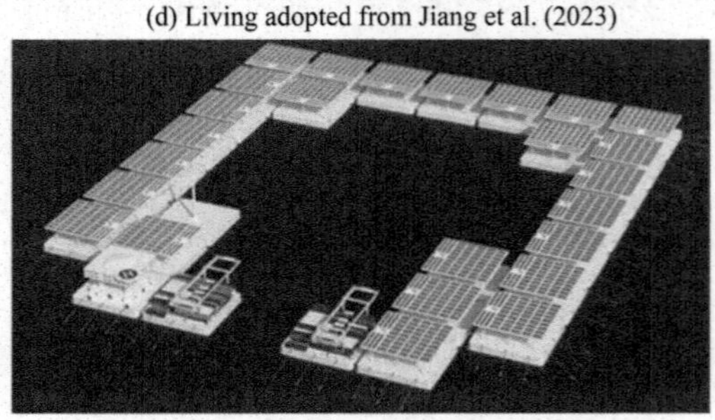

(e) Multi-use floating island (Flikkema et al. 2021)

Figure 1. Various VLFS (articulated floating structure of different geometrics) applications.

From the above literature, it is confirmed that there is no numerical analysis on the effect of current and wind performed on the large floating articulated flexible structures, based on the BEM code (ANSYS® AQWA).

Therefore, here, a three-dimensional numerical model of a moored articulated floating horizontal structure is presented based on BEM code by employing numerical ANSYS® AQWA. The paper is organized as follows: Section 2 presents the numerical model details and, briefly, the mathematical description of the problems being solved in frequency and time domains. In sub-section 3.1, the present numerical results of vertical displacement are compared with existing published experimental data sets and other numerical model results. Further, details of the implemented cases and the respective results on the effect of current and wind on the dynamics of flaoting articulated structure are presented for different structural and physical parameters and discussed in sub-section 3.2. Further, the contour plots on the transmission wave amplitudes around the structures are presented and the wave characteristics are analysed for different wave periods in sub-section 3.3. Finally, in Section 4, conclusions are drawn based on the present analysis and the findings are highlighted in bullet points, followed by the next steps to be developed in the present research.

2 NUMERICAL MODEL FORMULATION AND DESCRIPTION

2.1 *Description of numerical model*

The numerical ANSYS AQWA based on BEM code under linear potential-flow theory is adopted to model an articulated moored floating structure composed of six hinged plates under regular wave action. AQWA Hydrodynamic Diffraction provides an integrated environment for developing the principal hydrodynamic parameters called for complex motions and response analyses based on the BEM in the frequency domain. AQWA Hydrodynamic Response provides dynamic analysis capabilities in the time domain for undertaking global performance assessment of floating structures by obtaining the impulsive response from the diffraction module and solving the equation of motion through the convolution integration method. A wide range of physical connections, such as mooring lines, and joints, are provided to model the restraining conditions on the floating structure.

The numerical AQWA model is an articulated floating structure, moored to the seabed with linear elastic cables and composed of six sets of plates joined by hinges. The joints are located 0.5m below the calm waterline, with only one degree of freedom which rotates around the local x-axis parallel to the global coordinate y. In the numerical model, the wave range is from -180º to -180º with an interval of 90º, the sea bottom is assumed to be flat, and the hydrodynamic coefficients are defined for the frequencies corresponding to a wave period range from T=2s to T=60s. The general design parameters of the present model are provided in Table 1.

Table 1. Design parameters of the articulated floating structure.

Parameter (unit)	symbol	Value
Length (m)	L	195
Width (m)	b	45
Thickness (m)	d	4
Draft (m)	D	2.5
Individual length (m)	L_p	30
Gap between plates (m)	G_p	3
Water depth (m)	h	60
Mass (tonne)	m	19500
Current speed (m/s)	V	0.5-2.5
Wind speed (m/s)	U	10-20
Mooring stiffness (N/m)	q	6.00E+06
Rotational stiffness (Nm/rad)	k_s	4.00E+06

2.2 *Multi-body hydrodynamics – Frequency and time domain*

The mathematical model acquired in this numerical implementation is like the one presented by Lu et al. (2016) and extended (see Jin et al., 2020; Xu et al., 2017).

It is also assumed that there is a uniform current flowing with a constant speed $V = V.cos\theta$ along the direction of wave propagation in the positive direction of the x-axis. Considering the fluid as ideal and assuming the linear theory of waves, the total velocity potential with time-dependent can be written as $\Phi = \varphi\ e^{-i\omega t}$ where φ is the total time-independent velocity potential, ω is the angular frequency rad/sec, and t (s) is the time. The following Boundary Value Problem (BVP) can be individually satisfied by the radiated and diffracted potentials:

$$\nabla^2\varphi = 0, \text{in}\ \Omega \tag{1}$$

$$g\frac{\partial\varphi}{\partial z} - \omega^2\varphi = 0, \text{on}\ S_F \tag{2}$$

$$\frac{\partial\varphi}{\partial n_k} = \vec{V}_{S_k}.\ \vec{n}_k, \text{on}\ S_k \tag{3}$$

$$\frac{\partial\varphi}{\partial z} = 0, \text{on}\ S_B \tag{4}$$

$$\lim_{r\to\infty}\sqrt{r}\left(\frac{\partial\varphi_{R_k}}{\partial r} - \frac{i\omega^2}{g}\varphi_{R_k}\right) = 0, \text{on}\ S_\infty \tag{5}$$

where, the fluid domain is represented by Ω, S_F is the free surface, S_B is the bottom surface, S_∞ represents the boundary at infinity, and the wetted surface of the body is given by S_k, with k from 1 to 6, representing each plate module, as it can be seen in Figure 2. For the kth plate module, $\vec{n}_k$ is the unit vector normal to the wetted surface, pointing outwards, and $\vec{v}_{s_k}$ is the fluid velocity on the wetted surface of each plate.

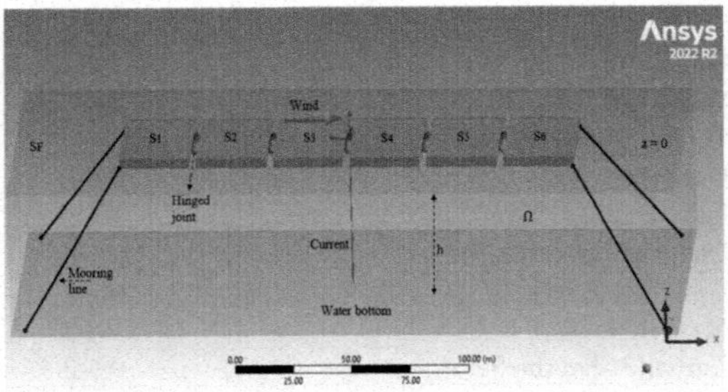

Figure 2. The geometry of the structure and hinges emulates a long flexible floating structure.

The BVP can be individually satisfied by the radiated potentials φ_R and diffracted potentials φ_D. The velocity potential can be decomposed as

$$\varphi = \varphi_{\mathrm{I}} + \varphi_{\mathrm{D}} + \varphi_{\mathrm{R}}, \tag{6}$$

where

$$\varphi\mathrm{I} = \sum_{m=1}^{\infty} \frac{ig\, I_{m0}}{\omega} \frac{\cosh \lambda_0 (h - z)}{\cos h\lambda_0 h} \cos \lambda_0 z\, e^{-ip_{m0}x},$$

where $i=\sqrt{-1}$, λ_0 is the wavenumber associated with the gravity wave dispersion relation. Further, and p_{m0} and I_{m0} are the wavenumber and incident wave amplitude associated with structure covered region, respectively (as in Bispo et al., 2022a; Mohapatra & Guedes Soares, 2022 in case of three dimensions).

The total velocity potential satisfies the Laplace equation (1), free surface boundary condition (2), bottom boundary condition (4), and along with the following conditions

$$\frac{\partial \varphi}{\partial \eta_n} = \vec{V}_{\mathrm{Sn}} \cdot \vec{\eta}\mathrm{n}, \text{ on the wetted surface.} \tag{7}$$

$$\lim_{r\to\infty} \sqrt{r}\left(\frac{\partial \phi}{\partial r} - \mathrm{i}\frac{2\pi}{\lambda}\phi\right) = 0, \text{ on } \mathrm{S}\infty, \tag{8}$$

where λ is the wavelength, r is the radial distance from the origin and $\vec{\eta}_n$ is the inward unit vector at the n-th body surface. For the n-th plate module, $\vec{\eta}_n$ is the unit vector normal to the wetted surface, pointing outwards, and $\vec{V}_{\mathrm{Sn}}$ is the fluid velocity on the wetted surface of each plate where n=6. Once the velocity potential, φ, is obtained, the added mass and the radiation damping can be computed. Therefore, the excitation forces $F_{w_k}^{(n)}$, can be related to the incident and diffraction potentials for the k-th degree of freedom and the n-th body by:

$$F_{w_k}^{(n)} = \rho i\omega \iint_{\bar{S}_n} (\varphi\mathrm{I} + \varphi\mathrm{D}) \cdot \vec{\eta}_n \mathrm{d}S, \tag{9}$$

in which ρ is the fluid density, ω is the angular frequency, $\bar{S}_n$ is the mean wetted surface of each plate module n, and i is the complex unit. The generalized added mass and radiation damping are given by:

$$[\mathrm{a}] + \frac{i}{\omega}[\mathrm{b}] = \rho \iint_{\bar{S}_n} \varphi_R \cdot \vec{\eta}_n \mathrm{d}S, \tag{10}$$

where [a]=$A_{kn}^{(rs)}$ and [b]=$B_{kn}^{(rs)}$ are the added mass and the radiation damping coefficients of the n-th mode of the r-th plate induced by the n-th mode of the s-th plate. For six degrees of freedom, proceeding as in [26], the equation of motion of the free multi-body floating structure is defined in the frequency domain and for unitary wave amplitude as:

$$[\mathrm{M} + \mathrm{A}(\omega)]\,\ddot{\mathrm{X}} + \mathrm{B}(\omega)\,\dot{\mathrm{X}} + \mathrm{KX} = \mathrm{F_W}(\omega), \tag{11}$$

where M is a matrix of masses for all bodies and its diagonal is composed by 6 × 6 matrices of masses for each plate n. The matrix K is the hydrostatic stiffness matrix and it is determined similarly to the mass matrix, for each body and 6 DOF. A(ω) and B(ω) are the added mass and radiation damping matrices respectively that are composed of k × k matrices of 6 × 6 size of added masses and radiation damping. The vector K denotes the displacement, from which follows that $\dot{X}$ and $\ddot{X}$ are its first- and second-time derivatives, respectively. Lastly, F_{W}(ω) is the vector of wave excitation forces for all n bodies. Once the hydrodynamic coefficients (wave excitation force, added mass and radiation damping) are acquired in the frequency domain by AQWA to solve the previously presented equations, by assuming the linearity of wave theory, the system can also be simulated in the time domain. The system of differential equations applied in the time domain can be summarized in an equation (see Xu et al. 2017)

$$[(\mathrm{M} + \mathrm{A}_\infty)\ddot{x}(t) + (\mathrm{K}+\mathrm{K_E})x(t) = f_{\mathrm{W}}(t) + f_E(t) + f_C(t), \tag{12}$$

where the structural mass matrix is represented by M and the fluid added mass matrix at infinite frequency is represented by A_∞. This equation includes the additional stiffness matrix, K_E, due to the connections of the bodies. Displacement as a function of time is denoted by x(t), and likewise in the frequency domain, its first- and second-time derivatives are represented by one and two upper dots. Forces actuating are the wave excitation force, $f_W(t)\ f_W\ (t)$, external forces such as the ones provided by the mooring, f_E (t) and the so-called convolution forces, f_C (t) (see Jin et al. 2020 for more details).

This system of equations is integrated in time by AQWA using a two-stage predictor-corrector numerical scheme and the details of its implementation can be found in its Theory Manual (ANSYS, 2020).

3 RESULTS AND DISCUSSIONS

To check the level of accuracy and analyse the dynamics of the large floating articulated flexible structure, the results of comparison and several numerical results on the effect of current and wind on the vertical displacements and transmission wave amplitudes on design parameters are analysed.

The numerical simulations using ANSYS were performed in an 8-core Intel® i7 CPU @ 3.6GHz, with 16GB of RAM. For this setup, it takes roughly 2 hours to acquire the hydrodynamic coefficients by solving the problem in the frequency domain and an additional 1 hours to perform the simulations for the considered six wave frequencies in the time domain, totalizing approximately 3 hours of simulation for each parameter combination setup for hydrodynamic response.

3.1 *Comparison results and analysis*

In Figure 3, the present numerical result of vertical displacements of floating structure with a current speed of 0.5 m/s and wavelength of 60 m is compared with existing published experimental datasets in the literature (Yago & Endo, 1996) (Model without current speed). In Figure 3, experiment values (Yago & Endo, 1996) are shown as 'Experiment', and the amplitude of the heave motion of the centre of gravity along the present model for 6 interconnected plates is plotted. The number of modules is related to the modal order of the structure vibration. Furthermore, the result from the experimental model for the first plate is 0.18 m, while the present one is 0.24 m, which is 6% larger. However, in the present model in the case of the last three modules (4 to 6) compared to the experiment, the results are slightly smaller, 6.9%, 16% and 22% respectively.

Figure 4 compares the present result of vertical displacement against other numerical model of five bodies existing in the literature (Lu et al. 2016). It is observed that for lower and higher structural length, there is in good level of agreement. However, the difference in the intermediate part of the structural length is noticed which may be explained due to the presence of current 2 m/s in the present model and mooring lines of 6 bodies and these parameters were not considered in the model Lu et al. (2016) (without current, free beam model with 5 bodies).

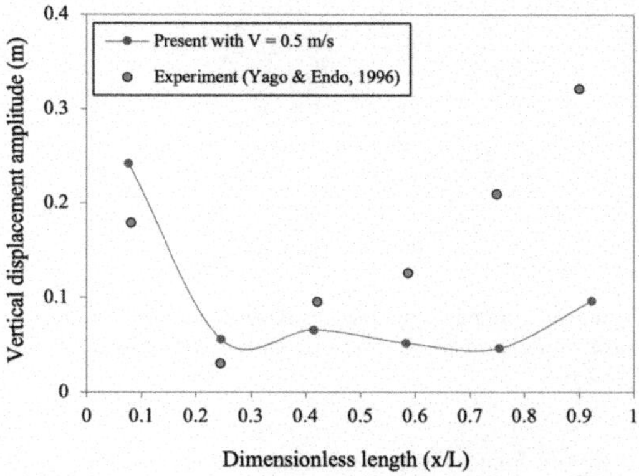

Figure 3. Comparison of vertical displacements between the present and experimental datasets.

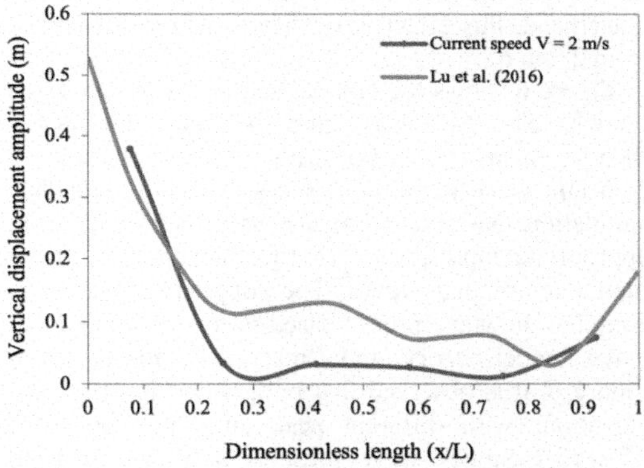

Figure 4. Comparison of vertical displacements between the present and numerical model of 5 bodies (Lu et al., 2016).

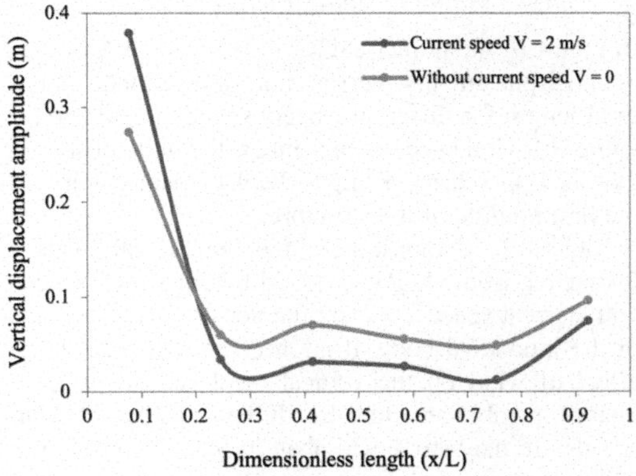

Figure 5. Comparison: displacements between without and with current speed V.

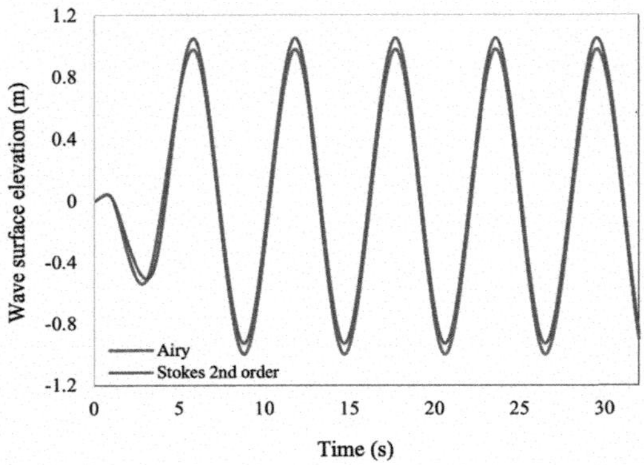

Figure 6. Comparison between Airy and 2nd order Stokes wave models on free surface elevation with current speed $V = 2$ m/s.

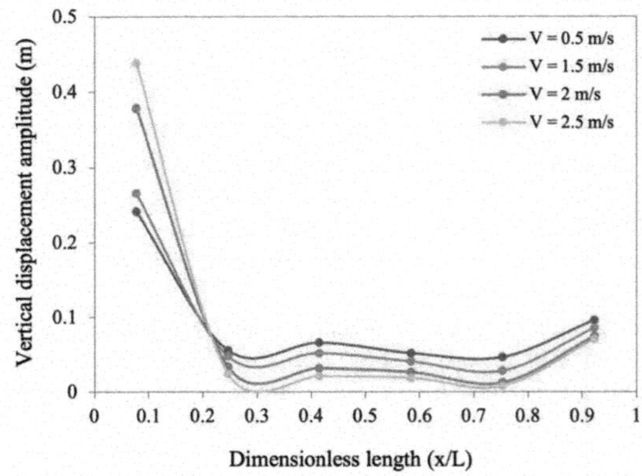

Figure 7. Effect of different current speed on the vertical displacement with certain value of wind speed U= 10 m/s.

Figure 5 shows the comparison of vertical displacement between cases $V = 0$ and $V = 2$ m/s versus non-dimensional structural length (x/L) with other parametric values are provided in Table 1. It is seen that displacement without the current speed becomes lower in 1st plate than that of displacements with V. On the other hand, this effect is reverse in nature for bodies from 2nd to 6th.

Figure 6 shows the comparison between two wave models: The Airy wave and the 2nd-order Stokes theory versus time (s) with wave amplitude = 1 m and wave period $T = 6.2$ s. The time response simulations based on BEM are performed by defining appropriate inputs, such as the mathematical wave model for regular waves. The adopted software can compute the wave profile based on two mathematical wave models: Airy and 2nd order Stokes. From Figure 6, it is observed that both waves have similar elevations with different peak amplitude however, 2nd order Stokes wave presents a higher peak in crests, but they behave like a sinusoidal harmonic wave due to the symmetrical shape.

3.2 *Effect of current and wind on the dynamics of large floating flexible structure*

In Figure 7, the numerical result from AQWA of vertical displacements versus non-dimensional structural length for different current speeds V and certain values of wind speed is presented. It can be observed that as V increases, a larger displacement occurs on the first module of the structure.

Table 2 demonstrates the numerical values (extracted from AQWA) for different wind speeds with current speed 2 m/s on the vertical displacements of the individual body. It is observed that there is no effect of wind on the vertical displacement of large floating flexible structure for different values of wind speeds. It may be noted that in AQWA (ANSYS, 2020), the wind drag force is calculated using the wind speed at reference height 10 m above water line. In addition, the wind speed near the still water line (SWL) is small and thus the wind loads can be neglected for the floating foundation near SWL (as in Chu et al. 2022).

Table 2. Max displacement (Global Z) with different wind speed for a constant current speed 2 m/s.

Wind speed (U)	η_1	η_2	η_3	η_4	η_5	η_6
10 m/s	0.3786	0.0334	0.0316	0.0263	0.0123	0.0745
15 m/s	0.3786	0.0334	0.0316	0.0263	0.0123	0.0745
20 m/s	0.3786	0.0334	0.0316	0.0263	0.0123	0.0745

It may be highlighted that to avoid the repetition of the analysis on mooring stiffness and rotational stiffness for the vertical displacements of the articulated floating flexible structure as in Bispo et al. (2022a), here, the numerical results on displacements for different design parameters are deferred.

3.3 *Wave characteristics around the structure via contour plots based on AQWA*

Figure 8 shows the contours of the wave elevation amplitude normalized with respect to the incident wave amplitude (1 m) under regular waves for different wave periods T. The contours of the normalized wave elevation amplitude behind the structure reflect the transmission waves. It can be seen that the transmission coefficients generally increase as wave period increases from 5.4 s to 10 s.

On the other hand, it is worth mentioning that the results from the reflection coefficients decrease as wave period increases in front of the structure. It is also observed that the variations of the transmission coefficient may increase as the wavelength λ increases.

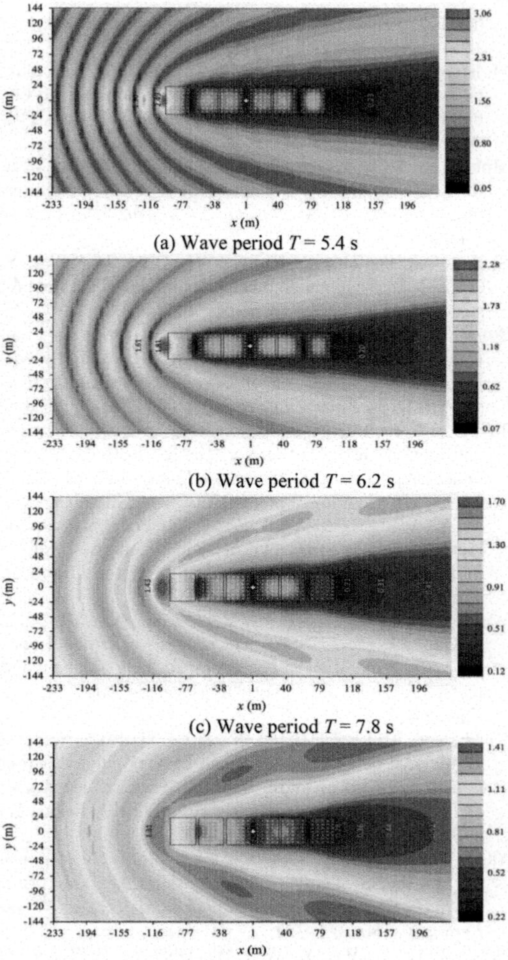

(a) Wave period $T = 5.4$ s

(b) Wave period $T = 6.2$ s

(c) Wave period $T = 7.8$ s

(d) Wave period $T = 10$ s

Figure 8. Contours of normalized wave amplitude around the floating flexible structure under $\theta = 0°$ for different T.

4 CONCLUSIONS

In the present paper, under potential linear flow theory, a numerical model based on a moored large floating flexible structure to analyse the dynamics of the floating structure by considering current and wind loads. The result of vertical displacement is compared with existing published experimental datasets. Further, numerical results on comparisons between with and without current speeds and two different wave models are presented. In addition, the results of vertical displacements for different current and wind speeds and contour plots for transmission wave amplitudes are presented and discussion is provided. From the present numerical analysis, the following significant concluding remarks can be obtained as:

1. The numerical simulation of vertical displacement is supported by the existing published experimental datasets and other numerical model results.
2. The selection of proper wave model is of great importance when dealing with large floating flexible structure in nearshore region.
3. It is observed that the vertical displacement becomes higher for larger values of current speeds on the 1st plate, however, this effect becomes opposite as the wave go on propagating from 2nd to six plates.
4. No wind effect on the vertical displacement is observed due to the fact that the wind drag force is calculated using the wind speed at reference height 10 m above water line.
5. Contour results reveal that the transmission wave amplitudes become higher for larger value of wave period.
6. The present analysis helps to extend to very large floating flexible structures by considering a finite number of interconnected plates connected with mooring lines under current and by neglecting wind loads.

ACKNOWLEDGEMENTS

This work contributes to the Strategic Research Plan of the Centre for Marine Technology and Ocean Engineering (CENTEC), which is financed by FCT under contract UIDB/UIDP/00134/2020. The first author has been funded by the Portuguese Foundation for Science and Technology (Fundação para a Ciência e a Tecnologia-FCT), through a doctoral fellowship under the Contracts no. UI/BD/154592/2023. The second author has been contracted as a Researcher by FCT, through Scientific Employment Stimulus, Individual support under Contract No. CEECIND/04879/2017.

REFERENCES

Amouzadrad, P., Mohapatra, S. C., & Guedes Soares, C. 2023a. Hydroelastic response of a moored floating flexible offshore structure based on Timoshenko-Mindlin-Beam theory. *Trends in Renewable Energies Offshore*, Guedes Soares, C. (ed.), Taylor and Francis, London, UK, pp. 879–888.

Amouzadrad, P., Mohapatra, S. C., & Guedes Soares, C. 2023b. Hydroelastic response of a moored floating flexible structure based on Timoshenko-Mindlin beam theory. *Advances in the Analysis and Design of Marine Structures*, Ringsberg, J.W. & Guedes Soares C., (eds.), Taylor & Francis, London, UK.

Amouzadrad, P., Mohapatra, S. C., & Guedes Soares, C. 2023c. Hydroelastic Response to the Effect of Current Loads on Floating Flexible Offshore Platform. *Journal of Marine Science and Engineering* 11(2): 437.

AQWA theory manual, 2020. ANSYS, Inc. and ANSYS, Europe. USA.

Bispo, I.B.S., Mohapatra, S.C., & Guedes Soares, C. 2021. A review on numerical approaches in the hydroelastic responses of very large floating elastic structures. In:

Guedes Soares, C. & Santos, T. A. (eds.), *Developments in Maritime Technology and Engineering*, Taylor & Francis Group, London, Vol 1, pp. 425–436.

Bispo, I.B.S., Mohapatra, S.C., & Guedes Soares, C. 2022a. Numerical analysis of a moored very large floating structure composed by a set of hinged plates. *Ocean Engineering* 253: 110785.

Bispo, I.B.S., Mohapatra, S.C., & Guedes Soares, C. 2022b. Numerical model of a WEC-type attachment of a moored submerged horizontal set of articulated plates. *Trends in Maritime Technology and Engineering*, Guedes Soares, C. & Santos T.A. (eds.), Taylor and Francis, London, UK, Vol. 2, pp. 335–344.

Chen, M., Ouyang, M., Guo, H., Zou, M., & Zhang, C. A. 2023. Coupled Hydrodynamic–Structural Model for Flexible Interconnected Multiple Floating Bodies. *Journal of Marine Science and Engineering* 11(4): 813.

Cheng, Y., Ji, C., Zhai, G., & Gaidai, O. 2016. Hydroelastic analysis of oblique irregular waves with a pontoon-type VLFS edged with dual inclined perforated plates. *Marine Structures* 49: 31–57.

Chu, Y.I., Wang, C.M., & Zhang, H. 2022. A frequency domain approach for analyzing motion responses of integrated offshore fish cage and wind turbine under wind and wave actions. *Aquacultural Engineering* 97: 102241.

Ding, J., Tian, C., Wu, Y., Wang, X., Liu, X., & Zhang, K. 2019. A simplified method to estimate the hydroelastic responses of VLFS in the inhomogeneous waves. *Ocean Engineering* 172: 434–445.

Ding, J., Wu, Y., Xie, Z., Yang, W., Wang, S., Yu, J., & Yu, T. 2021. Overview: Research on hydroelastic responses of VLFS in complex environments. *Marine Structures* 78: 102978.

Ding, J., Wu, Y., Zhou, Y., Ma, X., Ling, H., & Xie, Z. 2020. Investigation of connector loads of a 3-module VLFS using experimental and numerical methods. *Ocean Engineering* 195: 106684.

Ding, J., Xie, Z., Wu, Y., Xu, S., Qiu, G., Wang, Y., & Wang, Q. 2020. Numerical and experimental investigation on hydroelastic responses of an 8-module VLFS near a typical island. *Ocean Engineering* 214: 107841.

Flikkema, M., Breuls, M., Jak, R., de Ruijter, R., Drummen, I., Jordaens, A., Adam, F., Czapiewska, K., Lin, F.Y., Schott, D., et al. 2021. Floating Island Development and Deployment Roadmap; *Technical Report*; Space@Sea: Wageningen, The Netherlands.

Jiang, C., Xu, P., Bai, X., Zhao, Z., Moctar, O., & Zhang, G. 2023. A review of advances in modeling hydrodynamics and hydroelasticity for very large floating structures. *Ocean Engineering* 285: 115319.

Jin, C., Bakti, F.P., & Kim, M.H. 2020. Multi-floater-mooring coupled time-domain hydro-elastic analysis in regular and irregular waves. *Applied Ocean Research* 101: 102276.

Lu, D., Fu, S., Zhang, X., Guo, F., & Gao, Y. 2016. A method to estimate the hydroelastic behaviour of VLFS based on multi-rigid-body dynamics and beam bending. *Ships and Offshore Structures* 14 (4): 354–362.

Mohapatra, S.C., & Guedes Soares, C. 2021. Effect of mooring lines on the hydroelastic response of a floating flexible plate using the BIEM approach. *Journal of Marine Science and Engineering* 9 (9): 941.

Mohapatra, S.C., & Guedes Soares, C. 2022. 3D hydroelastic modelling of fluid-structure interactions of porous flexible structures. *Journal of Fluids and Structures* 112: 103588.

Mohapatra, S.C., & Guedes Soares, C. 2023. Hydroelastic Response of a Moored Floating Flexible Circular Structure Applying BIEM. *Journal of Marine Science and Engineering* 11(12): 2322.

Shi, Q., Xu, D., Zhang, H., Zhao, H., Ding, J., & Wu, Y. 2022. A face-contact connector for modularized floating structures. *Marine Structures* 82: 103149.

Song, X., Liu, W., & Wu, H. 2023. Investigation on Load Characteristics of Hinged Connector for a Large Floating Structure Model under Wave Actions. *Journal of Marine Science and Engineering* 11(4): 786.

Tay, Z.Y. 2019. Energy extraction from an articulated plate anti-motion device of a very large floating structure under irregular waves. *Renewable Energy* 130: 206–222.

Wang, Q., He, C., Ding, J., Geng, Y., Cheng, X., Zhang, X., Ni, X., Wang, J., & Wu, Y. 2024. Numerical and experimental studies on hydroelastic responses of a multiply-connected domain tourism platform in head waves. *Ocean Engineering* 294: 116795.

Wu, Y.S., Ding, J., Tian, C., Li, Z.W., Ling, H.J., Ma, X.Z., & Gao, J.L. 2018. Numerical analysis and model tests of a three-module VLFS deployed near islands and reefs. *Journal of Ocean Engineering and Marine Energy* 4: 111–122.

Xu, J., Sun, Y., Li, Z., Zhang, X., & Lu, D. 2017. Analysis of the hydroelastic performance of very large floating structures based on multi modules beam theory. *Mathematical Problems in Engineering* vol. 2017, Article ID 6482527: 14 pages.

Yago, K., & Endo, H. 1996. On the hydroelastic response of box-shaped floating structure with shallow draft. *Journal of Japan Society of Naval Architects and Ocean Engineers* 180: 341–352.

Yang, P., Li, Z., Wu, Y., Wen, W., Ding, J., & Zhang, Z. 2019. Boussinesq-Hydroelasticity coupled model to investigate hydroelastic responses and connector loads of an eight-module VLFS near islands in time domain. *Ocean Engineering* 190: 106418.

Yoon, J., Cho, S., Jiwinangun, R.G., & Lee, P. 2014. Hydroelastic analysis of floating plates with multiple hinge connections in regular waves. *Marine Structures* 36: 65–87.

Zhang, X., Lu, Da., Gao, Y., & Chen, L. 2018. A time domain discrete-module-beam-bending-based hydroelasticity methd for the transient response of very large floating structures under unsteady external loads. *Ocean Engineering* 164: 332–49.

Zhao, C., Hao, X., Liang, R., & Lu, J. 2015. Influence of hinged conditions on the hydroelastic response of compound floating structures. *Ocean Engineering* 101: 12–24.

Zhao, X., Xue, F., Chen, L., Göteman, M., Han, D., Geng, J., & Sun, S. 2023. Hydrodynamic analysis of a floating platform coupled with an array of oscillating bodies. *Ocean Engineering* 287: 115439.

Advances in Maritime Technology and Engineering – Guedes Soares & Santos (Eds)
© 2024 The Author(s), ISBN 978-1-032-83099-5

Structural behaviour of a windsurfer fin

G. Bandeira & L.S. Sutherland
Centre for Marine Technology and Ocean Engineering (CENTEC), Instituto Superior Técnico, Universidade de Lisboa, Lisbon, Portugal

P.H. Miller
Department of Naval Architecture and Marine Engineering, U.S. Coast Guard Academy, USA

ABSTRACT: A numerical Finite Element Analysis model of a production E-glass and carbon fibre hand-laminated composite material slalom windsurfer fin has been successfully validated in terms of both longitudinal bending and lateral twist via experimental mechanical cantilever tests utilising video-extensometry techniques. The effects on fin bending and twist (instrumental in determining sailing performance) of hydro-elastic tailoring of the internal layup was explored via rotations firstly of the internal 'wing spar' UD 0° carbon plies, and secondly of the external 'wing skin' woven ±45° carbon plies. Also, the effects of substituting expensive carbon with more economical E-glass internal UD plies was investigated. Various methods of increasing and/or decreasing fin twist and/or bending stiffness were found, and the information presented on this will aid the designer to achieve the specific structural and hence sailing behaviour desired.

1 INTRODUCTION

Despite its small size, the fin is at least as important as the other two main components of a windsurfer; the rig (sail, boom and mast) and board. The performance (in terms of speed, 'lift' upwind, early planning and control) of the fin is highly sensitive to even very small changes to its longitudinal bending and lateral torsional twisting response. There is significant unexploited potential to use the internal lay-up of these composite fibre reinforced plastic laminated fins to optimize this behaviour, but this must be done without compromising fin strength.

Since the fin is a small part, and the industry itself a small one, optimised designs have been arrived at by an 'evolutionary' 'try it and see' process whereby different potential solutions have been manufactured and evaluated on the water (often in competitions) and only 'the fittest survived'. This approach has been very successful - with some notable extinctions along the way, including acrylic (with 1/20 the stiffness of current materials) 'fenced' and 'football' fins – but comes at the expense of validated knowledge of how the fin actually achieves this performance. Although the professionals involved definitely 'know what works' the anecdotal explanations are not always scientifically 'waterproof'.

Scientific research of windsurfing fins was undertaken mostly in the 1990's (Broers *et al.*, 1992; Chiu *et al.*, 1992, 1993; Sutherland, 1993; Sutherland & Wilson, 1994; Fagg & Velay, 1996b, 1996a; Fagg, 1997b, 1997a; Kunoth et al., 2007) and work has continued on explanations of fin performance in terms of the findings of scientific studies (Nascimento et al., 2018; Saldanha, 2019; Cardoso de Brito, 2021; Cardoso de Brito *et al.*, 2022; Sutherland *et al.*, 2022) .

This has been achieved via the modelling of a 37 cm production glass and carbon fibre/epoxy laminated composite slalom fin (Figure 1) designed and fabricated by Steve Cook of F-Hot fins Ltd (Cook, 2003), which is also the subject of this present paper.

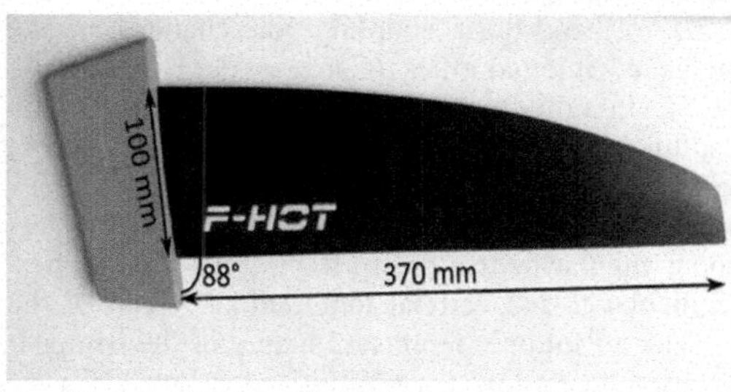

Figure 1. 37 cm Slalom fin.

Figure 2. Carbon and E-glass internal plies (one side).

DOI: 10.1201/9781003508762-43

The (commercially sensitive) internal structure of the fin (Figure 2) consists of 10 central Uni-Directional (UD) E-glass plies surrounded on either side by 11 span-wise UD carbon plies. These internal carbon and glass plies consist of 50 or 70 mm wide tapes which together act as a central, stiffening 'wing spar'. On the outside of this 'wing spar', on either side, are five 'full face' carbon woven and UD plies which cover the whole surface of the "foil" (hydro-foil).

The three main stages of the previous work were:

1. Computer Finite Element Analyses (FEA) structural analysis including bending validation via mechanical testing,
2. Computational Fluid Dynamics (CFD) numerical modelling,
3. Fluid-Structure Interaction (FSI) modelling combing 1 & 3 above, and including validation via water tunnel testing and on-the-water testing by professional windsurfers.

The present study aims to further experimentally validate the ABAQUS FEA structural model in terms of fin twist (about its longitudinal axis), which is crucial in terms of fin performance for upwind 'lift' ('lift' in the classical aeronautical sense with respect to the geometry of the fin, but in a horizontal, upwind sense with respect to the windsurfer) and control in wind gusts and waves. It also aims to use and adapt the FEA model to investigate the effects on fin structural behaviour of various changes to the internal ply layup which are of interest to the manufacturer, and then to validate that fin strength has not been compromised by these changes.

Previous work has explored the use of hydro-elastic tailoring of the laminate lay-up schedule, in terms of bend-twist coupling via changes in the angles of internal plies (Cardoso de Brito *et al.*, 2022). Longitudinal cantilever flexure of the fin resulting in lateral (when affixed into the board) deflections of the fin is referred to here as 'bending', and as mentioned above, 'twist' refers to the torsional angular deflections of the fin in the plane perpendicular to the vertical longitudinal 'span' of the fin. The anisotropic laminated nature of the composites used here allow a 'coupling', or linking, of these bend and twist deflections whereby bending results in twist (and vice versa, although this is not the active mechanism here).

Hydro-elastic tailoring for bend-twist coupling has also been researched for other marine applications, for example marine propeller blades (Young & Liu, 2007; Lin et al., 2010) and hydrofoils (Young *et al.*, 2018), and for the wind energy sector, for example (Cox & Echtermeyer, 2013). Here, the relative effects of rotating the inner 'wing spar' or the outer 'full face' plies on this fin twist behaviour will be investigated here.

Also, the effects of substituting some of the internal 'wing spar' carbon plies for glass ones, both as a potential cost-cutting exercise and to research methods of altering the longitudinal 'bending' stiffness of the fin, was investigated.

2 EXPERIMENTAL FIN TWIST TESTS

As described above, two structural responses can be used to characterise (and hence modify) the sailing performance of a windsurfer fin; longitudinal 'bending' and lateral 'twist'. In previous work (Cardoso de Brito *et al.*, 2022) the FEA model developed and used in this paper was successfully validated in terms of fin *bending* via a series of mechanical cantilever tests. Here, the aim is to further validate this numerical model, in terms of twist via modified and enhanced versions of these point loaded cantilever experimental tests. The current experimental setup is shown in Figure 3.

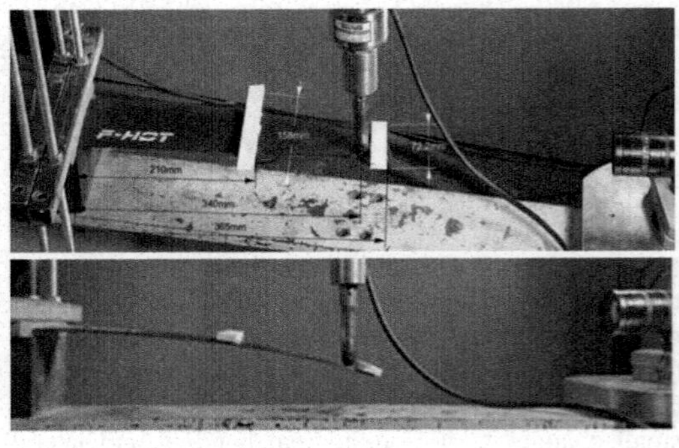

Figure 3. Cantilever test setup (camera at right).

The challenge here is that twist angles were thought to be very small (but still significant with respect to the low leeway angles, i.e. angles of attack, whilst sailing (Sutherland, 1993), requiring very accurate and robust measurement methods. Mechanical displacement transducers would not provide sufficient accuracy, and the use of strain gauges requires calculations to convert readings into angles of twist, into which the complex and locally variable internal layup would introduce large errors.

The solution successfully applied here, was that of video-extensometry. With the software and camera used (FLIR BFS-U3-51S5M-C camera, Fujinon HF25SA-1 lens, MatchID software for data acquisition), simple 'bullseye' self-adhesive paper targets are locked onto by the software, tracked and their exact position recorded with time.

Hence, by fixing a balsa wood beam with a target at each end across the fin chord (with stiff moulding putty) at a given span, and knowing the relative vertical movement of each target and the distance between them, the fin chord 'angle of attack' and hence fin twist at that span of the fin can be calculated via simple trigonometry. The camera was hence positioned as shown on the right-hand side of Figure 3, giving the view of the targets shown in Figure 4.

As can be seen in these figures, in fact two separate beams were used to give independent measurements of twist at two separate spans, 210 (57 % span) and 365 mm (as close to the fin tip as practically possible) from the fin base (with lateral target separations of 158 and 75.2 mm, respectively).

Figure 4. Video-extensometry camera view of targets.

To give significant bending, and hence bend-twist induced twist, the load was applied at a point 340 mm (90% span) from the base as the maximum distance practically possible. Since the centre of pressure of a symmetrical hydrofoil is known to be at approximately ¼ chord length from the leading edge, (Molland and Turnock, 2021) the load was applied at this position (corresponding to a point 34 mm from the trailing edge). Since the shear centre of the fin was not known (due to complex geometry and internal lay-up material and ply architecture variations along the fin span), tests were also carried out at a loading point at the same span but much closer to the trailing edge (at a distance of 10 mm) with the aim of inducing more fin twist. Using two loading cases also provide for a more rigorous experimental twist validation of the FEA model.

The cantilever tests were carried out on a calibrated servo-hydraulic test machine (Sutherland, 2024) at a loading rate of 2 mm/s up to a deflection at the loading point of 40 mm (to ensure that the fin was not broken and that the targets remained within the field of vision of the camera). Another challenge was that the test machine force-displacement data acquisition system was not synchronised with that of the video-extensometry equipment. Hence, the tests were carried out in increasing (and then decreasing) displacement steps (e.g. Figure 5), interspaced with easily identified (on both measurement systems) stationary periods, giving a series of corresponding (at equal displacements) load-displacement and target positon points (and hence twist angles) that could then be plotted against each other (thus side-stepping the need for any problematic and error-inducing measurement systems synchronisation).

The test machine records raw force-displacement data without any filtering (in contrast to that from most commercially available equipment), allowing full control and knowledge of, and hence confidence in, the subsequent filtering process. In this case a low noise to signal ratio with respect to a very high (50Hz) sampling rate means that the use of a simple moving average filter gave very good results (Figure 6), and exhibited an extremely linear response.

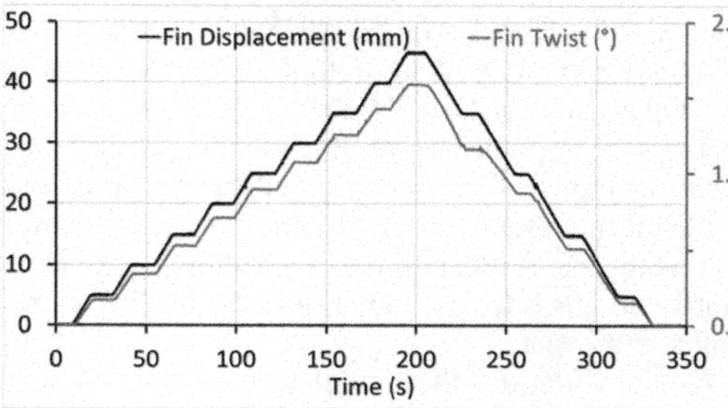

Figure 5. Incremental displacement steps.

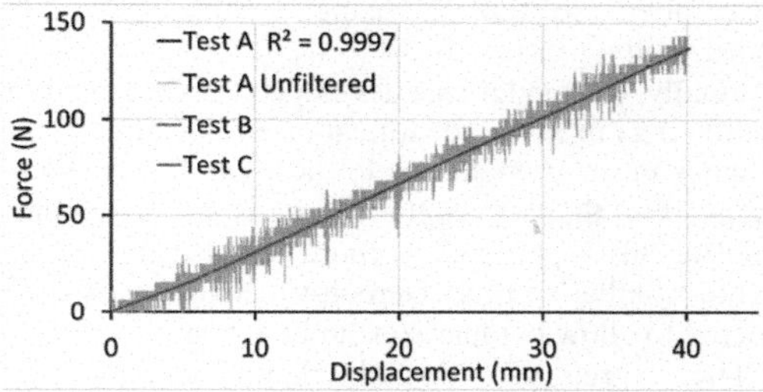

Figure 6. ¼ Chord loading force-displacement.

Three test replicates were made at each loading point (¼ chord and 10 mm from trailing edge), each group of three showing extremely good repeatability for both bending displacement (Figure 6) and twist (Figure 7).

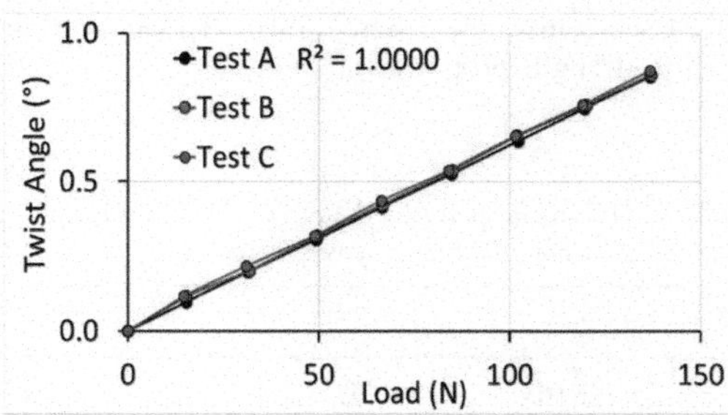

Figure 7. ¼ Chord loading twist angle –force.

Twist angles were considered positive for 'wash-out' - when the trailing edge deflects further than the leading edge (corresponding to a reduced angle of attack in service), and negative for the opposite 'wash-in'. 'Wash-out' was observed in all experimental tests here.

The results of these tests are presented and used to validate the numerical model in section 3.1.

3 NUMERICAL STRUCTURAL MODELLING

The ABAQUS (ABAQUS, 2018) FEA numerical structural model used as a tool in this work was developed in (Cardoso de Brito, 2021) where full details may be found. Here, this model was first validated in terms of fin bending and twist against the results of the tests described in section 2 (see section 3.1) using the two experimental loading points at ¼ chord and close to (10 mm from) the trailing edge, both at 90% span.

After validation, the model was used to numerically investigate the effects on fin bending and twist of different modifications to the internal ply lay-ups to investigate hydro-elastic tailoring (section 4) and then the substitution of carbon for glass plies (section 5), using a 90% span - ¼ chord loading case.

Finally, the model was developed (as described in section 3.2) to allow identification of any significant changes in fin strength due to these ply lay-up alterations. For these strength comparisons, a loading point at 80% span and ¼ chord as used in (Nascimento, Sutherland & Garbatov, 2018) was considered to allow comparisons to be made.

The degree of longitudinal fin bending was described using the vertical displacement of the applied loading point in the current cantilever setup. Fin twist was obtained from the results of the FEA model in the same way that the experimental twist was trigonometrically calculated from the positions of the two targets on the balsa beams (section 2), but here the positions of two nodes, one at the leading, and one at the trailing edge of the model at the appropriate fin span (210 or 365mm) were used. The elastic material properties developed in (Cardoso de Brito *et al.*, 2022) were also used here (Table 1, see section 3.2 for strength properties).

Table 1. Elastic properties: (1) Epoxy UD E-Glass; (2) Epoxy UD carbon; (3) Epoxy woven carbon.

	(1)	(2)	(3)	Units
ρ	1.72E+03	1.40E+03	1.37E+03	kg/m^3
E_x	2.23E+10	5.66E+10	2.84E+10	Pa
E_y	7.90E+09	5.11E+09	2.84E+10	Pa
E_z	7.90E+09	5.11E+09	5.54E+09	Pa
v_{xy}	0.29	0.31	0.04	
v_{yz}	0.40	0.42	0.30	
v_{xz}	0.29	0.31	0.30	
G_{xy}	4.12E+09	3.26E+09	3.30E+09	Pa
G_{yz}	3.50E+09	3.08E+09	2.70E+09	Pa
G_{xz}	4.12E+09	3.26E+09	2.70E+09	Pa

3.1 *Model validation*

The excellent repeatability between tests (Figure 6 and Figure 7) allowed average values of the three test results for each loading case to be used.

As can be seen from Figure 8, the FEA model is again validated by the current experimental fin bending displacement data. As expected, the lateral chord-wise position of the loading point does not affect the fin bending displacement to a large degree.

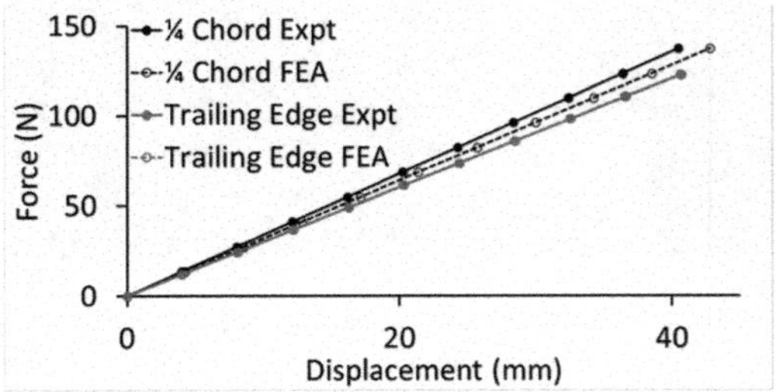

Figure 8. FEA model bending validation.

Successful validation of the FEA model by the experimental twist angle data is shown in Figure 9 and Figure 10, where the same axes scales have been used to enable easy comparison between the two figures.

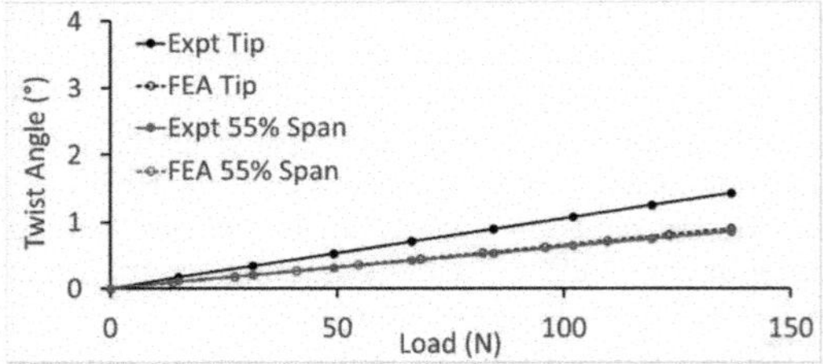

Figure 9. FEA model twist validation (¼ chord load). FEA tip coincident with both 55% span plots).

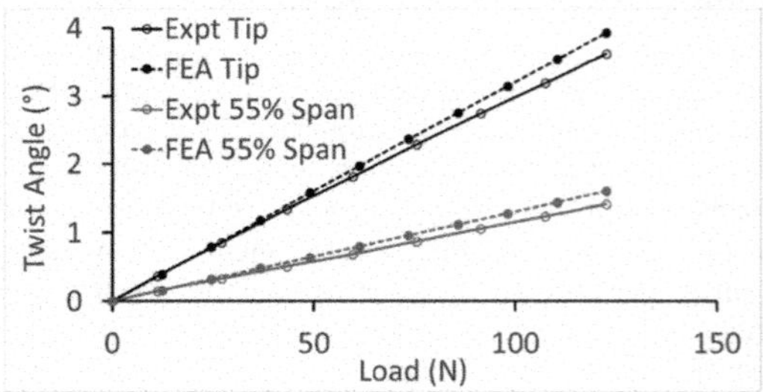

Figure 10. FEA model twist validation (trailing edge load).

Excellent agreement between FEA and experiment is seen for all cases except for the ¼ chord loaded tip twist case, where the experimental tip twist is significantly (50%) higher than the corresponding FEA predicted values, although to put this into context, this is only an error of 0.5° and both the trend and twist direction are correctly predicted.

However, this discrepancy highlights the inherent difficulties of numerically modelling such foils, especially when they are hand-made and with extremely thin trailing edges at the tip (thought to be vital for good sailing performance). It is thought, due to difficulties in, or even the impossibility of, ensuring that all relevant plies extend right to the trailing edge at the tip, that the layup in a real fin is not identical to that specified by the laminate schedule in this zone, and hence that modelled by the FEA. These aspects would manifest themselves most clearly in the tip deflection where twist due to bend twist coupling is likely to be dominant over load-shear centre offset moments (when loading occurs at ¼ chord), as has been observed. The exact mechanism(s) for this discrepancy is(are), nevertheless, not currently verifiable.

3.2 *Fin strength*

The same methodology used in (Cardoso de Brito *et al.*, 2022) to obtain the elastic properties (Table 1) was followed here to give estimates of the ply strength properties (Table 2). Experimentally based values for 0.6 fibre volume fraction (FVF) plies from both the ANSYS material library ('ANSYS© 2022′, 2022) and (Daniel & Ishai, 2006) were 'adjusted' to correspond to the (lower) FVF of the fin using rule of mixtures assumptions.

Table 2. Strength properties: (1) Epoxy UD E-Glass; (2) Epoxy UD carbon; (3) Epoxy woven carbon.

	(1)	(2)	(3)	Units
$\sigma_{x\,(T)}$	6.56E+08	1.00E+09	3.39E+08	Pa
$\sigma_{y\,(T)}$	3.49E+07	4.22E+07	3.39E+08	Pa
$\sigma_{z\,(T)}$	3.49E+07	4.22E+07	3.39E+08	Pa
$\sigma_{x\,(C)}$	-4.10E+08	-4.46E+08	-2.92E+08	Pa
$\sigma_{y\,(C)}$	-9.25E+07	-6.27E+07	-2.92E+08	Pa
$\sigma_{z\,(C)}$	-9.25E+07	-6.27E+07	-2.92E+08	Pa
τ_{xy}	5.87E+07	6.89E+07	9.42E+07	Pa

Although the rule of mixtures is not known to be particularly accurate for strength calculations, considering that it was only used to modify experimentally obtained values, its use is justified here. Further, the strength estimates here are only made in a comparative sense, and *it is not intended that they should be used to give accurate failure load estimates*. The aim here is to compare the strength of the fin with the modifications to the internal layup (as described in sections 4 and 5) with that of the 'nominal', as-fabricated, fin *using the same ply strength estimates*, simply to identify any possible resulting *significant reductions* in strength.

The Tsai-Wu failure criteria (Tsai & Wu, 1972) (which may be thought of as an orthotropic 'equivalent' to the von Mises criteria for isotropic materials) safety factor output from ABAQUS was used to estimate the fin strength. Figure 11 shows the Tsai-Wu failure criteria safety factor contour plot for the as-fabricated fin, where the region highlighted in red indicates a safety factor of less than 1, i.e. failure. This was verified to occur first on the compression side of the fin, as expected due to the lower longitudinal compression strengths of these fibre reinforced polymer plies.

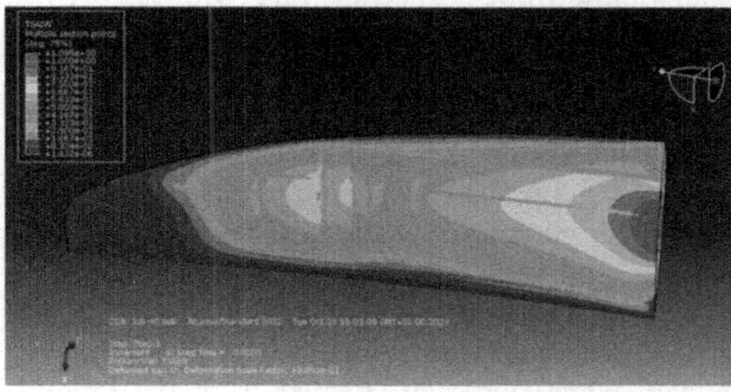

Figure 11. Tsai-Wu safety factor.

An element at the first appearance of this zone of first failure was identified and its safety factor plotted against the load applied to the fin, allowing the first failure load to be obtained when this safety factor reached unity. This approach gave an estimated initial failure load of approximately 680 N. This is only 75% of the 910 N obtained using the Tsai-Wu criteria by the ANSYS model of (Nascimento, Sutherland and Garbatov, 2018), but material properties used in the previous case were not as developed as those used here; for example, the present value of UD carbon compressive strength is only 72% of that used in the previous ANSYS model (those of woven carbon and UD glass are 57% and 60% of those previously used, respectively).

The Tsai-Wu failure criterion cannot identify the mode of this initial failure, be it fibre failure, matrix failure, or fibre-matrix interface failure, but it is probable that it is due to matrix cracking and not fibre failure, and almost certainly would not immediately give catastrophic failure of the fin (which would follow a progressive, multi-mode damage path). However, as previously discussed, these estimates of initial failure load are of interest in a comparative sense.

4 HYDO-ELASTIC TAILORING

The use of bend-twist coupling to customise the structural and hence hydrodynamic performance of the fin is explored here using the FEA model previously validated in both bend and twist. This information can then potentially help to decouple the bend-twist behaviour from the fin's geometric shape, enabling new designs that are at the same time highly efficient hydro-dynamically (fast) *and* bend and twist suitably under hydrodynamic loading (enabling early planing and good control in gusts and waves).

The aim here is not, and cannot be, an optimisation exercise since there is no one 'optimum' solution – this will depend on many interacting factors such as wind and water conditions, board and sail type, sailor weight and strength and even style, and sailing 'discipline' (recreational, wave, slalom, freestyle, etc.).

Hence, since the various aspects of fin performance can be deduced from its bend and twist characteristics, how these responses vary with changes to the internal layup giving different degrees of bend-twist coupling is 'mapped' here to give valuable design data and trends. It is important to remember that the fin is not a simple rectangular bar with a uniform layup; its complex internal ply architecture and its geometric shape will both modify any ply-level bend-twist coupling effects.

As verified in (Cardoso de Brito *et al.*, 2022), it is known that maximum bend-twist coupling is achieved when fibres are rotated from the longitudinal axis by approximately 30°, and hence this was the degree of ply rotation used here. It was also confirmed that rotating the plies in the fin to rake aft (defined as a positive rotation) will give more wash-in (increased angle of attack, defined as negative twist), and vice versa - see Figure 22 in (Cardoso de Brito *et al.*, 2022) and Figure 61 in (Talay, 1975).

4.1 *Central UD carbon ply rotations*

Firstly, the effects of rotating the longitudinally-aligned (0°) carbon UD 'tape' plies in the central 'wing spar' volume of the fin laminate were explored. This was done incrementally, rotating first the outer most UD ply on both sides of the fin (both in the same direction - either aft or forward, NOT clockwise or anti-clockwise when viewing from the appropriate side), then sequentially rotating more plies (on both sides) at a time. Loading was applied at 90% span and ¼ chord (the latter coinciding the hydrodynamic centre of pressure). The effects on bending deflection (at the 90% span ¼ chord loading point) and (tip) twist are shown in Figure 12, which shows that:

- Zero rotated plies correspond to the 'as fabricated' nominal fin with a 52.8 mm bending deflection and a tip twist of +0.9° (wash-out).
- Both forward and aft rotations (-30° and + 30°, respectively) give an increase in bending deflection (up to 21.5 mm, 41% and 16.8 mm, 32%, respectively).
- Aft rotations (+30°) give an increase in wash-in (an increase in angle of attack, of up to 2.4°, 267%). This switches the overall twist of the fin from wash-out to wash-in.
- Forward rotations (-30°) give an increase in wash-out (a decrease in angle of attack, of up to 2.1°, 233%).

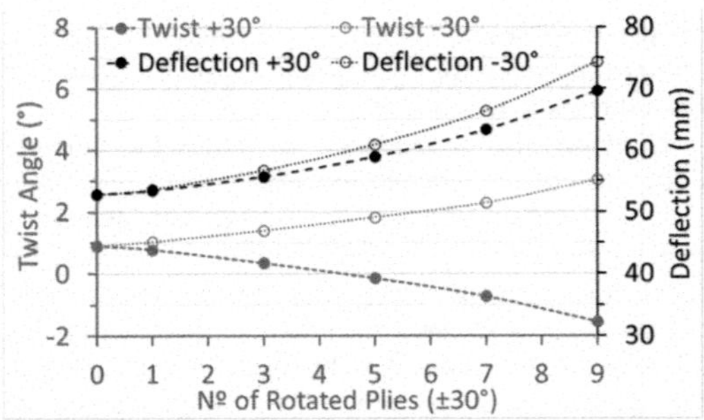

Figure 12. Bend-twist with 30° inner carbon UD rotations.

Hence, the fin tip can be induced to wash-in (or wash-out) far further as it is loaded up by rotating the inner UD carbon plies aft (or forward). However, rotations in either direction will lead to an increase in longitudinal flexibility, which will be significant if all plies are rotated. It is up to the designer and sailor to decide which specific behaviour (if any of the above, since foil flexibility is not always advantageous) is desirable for the specific case at hand.

4.2 *Outer woven carbon ply rotations*

Secondly, the effects of rotating the three ±45° aligned woven carbon 'full face' plies of the outer 'wing skin' area of the fin laminate were then explored. Again, this was done incrementally, rotating first the outermost ply on both sides of the fin (again, both in the same direction - either aft or forward, NOT clockwise or anti-clockwise when viewing from the appropriate side), then sequentially more plies (on both sides) at a time. Since these woven plies were initially at ±45°, +30° and -30° rotations led to +75°/-15° and 15°/-75° layups, respectively. Loading was again applied at 90% span and ¼ chord. The effects on bending deflection (at the loading point) and (tip) twist are shown in Figure 13, which shows that:

- No rotated plies again correspond to the 'as fabricated' nominal fin with a 52.8 mm bending deflection and a tip twist of +0.9° (wash-out).
- Both forward and aft rotations (-30° and + 30°, respectively) give a decrease in bending deflection (of up to 11 mm, 21% and 3.2 mm, 6%, respectively).
- Aft rotations (+30°) give a large increase in wash-out (a decrease in angle of attack, of up to 5.8°, 644%).
- Forward rotations (-30°) give an increase in wash-in (an increase in angle of attack, of up to 1.7°, 189%). This switches the overall twist of the fin from wash-out to wash-in.

The final two points made above appear to contradict the statement previously made (in the fourth paragraph of section 4) that a positive ply rotation should give more wash-in (defined as negative twist), and vice versa. However, this statement was made in the context of 0° UD reinforcements, and these plies are ±45°. In this case a positive 30° rotation brings the -45° fibres into a -15° alignment which induces more wash-out due to bend-twist coupling. Equally, but conversely, a negative 30° rotation brings the +45° fibres into a +15° alignment which induces more wash-in due to bend-twist coupling. The fact that these fibres are closer to 0° in both cases explains why the fin becomes stiffer in longitudinal bending in both cases, despite the fact that the orthogonal fibres become aligned further away from 0° at +75° or -75° (for positive and negative 30° rotations, respectively).

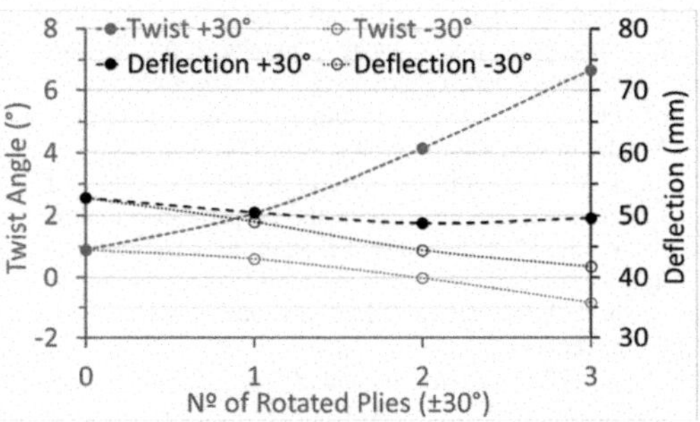

Figure 13. Bend-twist with 30° outer woven carbon rotations.

Hence, the fin tip can be induced to wash-out significantly (or wash-in) as it is loaded up by rotating the outer woven carbon plies aft (or forward). In this case however, rotations in either direction will lead to an increase in longitudinal stiffness. Again, it is up to the designer and sailor to decide which specific behaviour (if any of the above, since foil flexibility is sometimes advantageous) is desirable for the specific case at hand.

The Tsai-Wu criteria estimate of the initial failure load of the fin with all three outer woven carbon plies rotated by 30° was 675 N – a reduction of less than 1% from the original fin. Although this is reassuring, further work would have to be carried out to be sure that the final catastrophic failure load is also not compromised. Given the complexity of, and hence resources and time required for, numerically based progressive damage modelling, and since there would be no confidence in this approach anyway without experimental strength validation, this would best be evaluated by simply mechanically testing both original and modified design fins to destruction.

5 REPLACING UD CARBON WITH GLASS

The high cost of carbon fibre is a considerable restriction, especially for small manufacturers such as F-Hot Ltd. Hence, the FEA model is now used to explore the implications of substituting some of this expensive carbon with more economical E-glass fibres. The 'wing spar' assembly of rectangular UD tapes in the 'as manufactured' fin consists of 10 innermost E-glass plies surrounded by 11 carbon UD plies on each side. Two cases are studied here:

(i) 4 UD carbon plies substituted for 2 E-glass UD
(ii) 10 UD carbon plies substituted for 6 E-glass UD

More carbon plies are removed than glass plies substituted in each case since the heavier (in terms of both fibre density and gm^{-2}) glass UD plies are 1.8 times thicker than the lighter carbon UD plies. The effects of these two cases on fin tip twist and deflection (at the 90% span ¼ chord loading point) are shown in Figure 14.

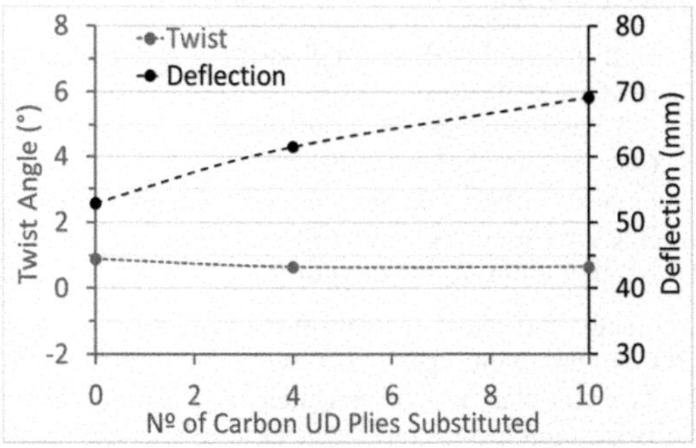

Figure 14. Substitution of carbon UD with E-glass.

Figure 14 shows that twist is barely affected by these changes (by 0.2° in both cases). However, the fin bending stiffness is significantly reduced, with increases in deflection of 8.7 mm (16%) and 16.4 mm (31%) for substitution of 4 and 10 plies (out of the original 38 carbon 'wing spar' UD plies), respectively.

The Tsai-Wu criteria estimate of the initial failure load of the fin for case (ii) with 10 carbon UD inner plies substituted by E-glass UD was 555 N – a concerning reduction of nearly 20% from that of the original fin.

Hence, costs may be reduced in this way without affecting the twist response if a more flexible fin is acceptable, but the associated weakening of the fin is of concern and must be further investigated (again, experimentally would be the best method here) to evaluate if this is tolerable *and* worth the amount of cost savings made.

6 CONCLUSION

A numerical Finite Element Analysis model of a production E-glass and carbon fibre hand-laminated composite material slalom windsurfer fin has been successfully validated in terms of both longitudinal bending and lateral twist via experimental mechanical cantilever tests utilising video-extensometry techniques.

Hydro-elastic tailoring of the fin internal ply layup was completed to explore its effect on the fin bending and twist responses (which characterise and control sailing performance). Firstly, the internal 'wing spar' UD 0° carbon plies were rotated forward and then aft:

- Both forward and aft rotations gave an increase in bending deflection.
- Aft rotations gave an increase in wash-in (increase in angle of attack).
- Forward rotations gave an increase in wash-out (decrease in angle of attack).

Secondly, increasing numbers of the external 'wing skin' woven ±45° carbon plies were rotated forward and aft:

- Both forward and aft rotations gave a decrease in bending deflection.
- Aft rotations gave a large increase in wash-out (decrease in angle of attack).
- Forward rotations gave an increase in wash-in (increase in angle of attack).
- No reduction in fin strength was seen.

Finally, the effects of substitution of some internal UD carbon 'wing spar' plies with more economical E-glass UD plies was investigated. This did not affect fin twist significantly but did lead to a less stiff fin in bending and, concerningly, weakened the fin by 20%.

Hence, the fin tip can be induced to wash-in or wash-out, and the bending stiffness varied, by extremely simple to implement ply rotations during manufacture. The information presented will aid the designer to achieve the specific behaviour they decide desirable for the specific windsurfing discipline/wind and water conditions/sailor etc. permutation of interest.

ACKNOWLEDGEMENTS

Video-extensometry equipment and technical expertise was kindly provided by our colleague Dr Mário Garrido of CERIS, Instituto Superior Técnico.

L.S. Sutherland was funded by the Portuguese Foundation for Science and Technology (Fundação para a Ciência e Tecnologia - FCT), under the Transitional Standard with contract IST-ID/164/2018. This work contributes to the Strategic Research Plan of the Centre for Marine Technology and Ocean Engineering (CENTEC), which is financed by FCT under contract UIDB/UIDP/00134/2020.

REFERENCES

ABAQUS (2018) 'Abaqus Unified FEA-3DEXPERIENCE R2018′. Rhode Island: Dassault Systèmes.

'ANSYS© 2022′ (2022). Cannonsburg, PA, USA: ANSYS Inc.

Broers, C. *et al.* (1992) 'Effects of Fin Geometry and Surface Finish on Sailboard Performance and Manoeuvrability', in P.A. Wilson (ed.) *Manoeuvering and control of marine craft, MCMC 92. 2nd International conference, Manoeuvring and control of marine craft*, Southampton, UK: Computational Mechanics Publications, pp. 275–289.

Cardoso de Brito, M. (2021) *An integrated FSI analysis of a windsurfer fin*. MSc Thesis. Instituto Superior Técnico.

Cardoso de Brito, M. *et al.* (2022) 'Fluid-Structure Interaction Analyses for Hydro-Elastic Tailoring of a Windsurfer Fin', *Journal of Marine Science and Engineering*, 10(10), p. 1371.

Chiu, T.W. *et al.* (1992) 'The Effects of Tip Flexibility on the Performance of a Blade-Type Windsurfer Fin', in *Manoeuvering and control of marine craft, MCMC 92. 2nd International conference, Manoeuvring and control of marine craft*, Southampton, UK: Computational Mechanics Publications, pp. 261–274.

Chiu, T.W. *et al.* (1993) 'An experimental study of the effects of deformable tip on the performance of fins and finite wings', in 23rd Fluid Dynamics, Plasmadynamics, and Lasers Conference. American Institute of Aeronautics and Astronautics (Fluid Dynamics and Co-located Conferences).

Cook, S. (2003) *F-Hot Windsurf Fins, F-hot Mouldings*. Available at: http://www.f-hot.com/fins (Accessed: 21 December 2023).

Cox, K. and Echtermeyer, A. (2013) 'Geometric Scaling Effects of Bend-twist Coupling in Rotor Blades', *Energy Procedia*, 35, pp. 2–11.

Fagg, S. (1997a) 'Reversible and finitely variable cambered lifting section'.

Fagg, S. (1997b) *The development of a reversible and finitely variable camber windsurf fin*. PhD. Bournemouth University.

Fagg, S. and Velay, X. (1996a) 'Simulating the operation of a novel variable camber hydrofoil', in *1996 IEEE Aerospace Applications Conference. Proceedings*, pp. 261–271 vol.3.

Fagg, S. and Velay, X. (1996b) 'The development of a novel composite material hydrofoil through advanced computational techniques', *WIT Transactions on Engineering Sciences*, 10.

Kunoth, A., Schlichtenmayer, M. and Schneider, C. (2007) 'Speed windsurfing: Modeling and numerics', *International Journal of Numerical Analysis and Modeling*, 4 (3–4), pp. 548–558.

Lin, H.J., Lai, W.M. and Kuo, Y.M. (2010) 'Effects of Stacking Sequence on Nonlinear Hydroelastic Behavior of Composite Propeller Blade', *Journal of Mechanics*, 26(03), pp. 293–298.

Molland, A.F. and Turnock, S.R. (2021) Marine Rudders, Hydrofoils and Control Surfaces: Principles, Data, Design and Applications. Butterworth-Heinemann.

Nascimento, F., Sutherland, L.S. and Garbatov, Y. (2018) 'Experimental and numerical structural analysis of a windsurf fin', in C. Guedes Soares and T. A. Santos (eds) *Progress in Maritime Technology and Engineering: Proceedings of the 4th International Conference on Maritime Technology and Engineering (MARTECH 2018), May 7–9, 2018, Lisbon, Portugal*. London, UK: Taylor & Francis Group, pp. 387–394.

Saldanha, A.R.Q. (2019) *Hydrodynamic and Fluid-Structure Interaction analysis of a Windsurf Fin*. MSc. Instituto Superior Técnico.

Sutherland, L.S. (1993) *Windsurfer Fin Hydrodynamics*. MSc. University of Southampton.

Sutherland, L.S. *et al.* (2022) 'Fluid-structure interaction analyses of a composite windsurf fin', in *Trends in Maritime Technology and Engineering Volume 1*. London, UK: CRC Press, Taylor and Francis Group, UK, p. 9.

Sutherland, L.S. (2024) *Creation of Experimental Facilities*. Available at: https://leighsutherland.com/design-construction-of-experimental-facilities-equipment/ (Accessed: 15 January 2024).

Sutherland, L.S. and Wilson, P.A. (1994) 'Windsurfer Fin Hydrodynamics', in *Marine, Offshore and Ice Technology*. Computational Mechanics Publications, pp. 51–58.

Talay, T.A. (1975) *Introduction to the aerodynamics of flight*. SP-367. Langley Research Center, Hampton Virginia, USA: NASA.

Tsai, S.W. and Wu, E. (1972) *A general theory of strength for anisotropic materials*. Technical Report AFML-TR -71-12. Wright-Patterson Air Force base, Ohio, USA: Air Force Materials Laboratory.

Young, Y.L. *et al.* (2018) 'Load-dependent bend-twist coupling effects on the steady-state hydroelastic response of composite hydrofoils', *Composite Structures*, 189, pp. 398–418.

Young, Y.L. and Liu, Z. (2007) 'Hydroelastic Tailoring of Composite Naval Propulsors', in Volume 4: Materials Technology; Ocean Engineering. ASME 2007 26th International Conference on Offshore Mechanics and Arctic Engineering, San Diego, California, USA: ASME, pp. 777–787.

Advances in Maritime Technology and Engineering – Guedes Soares & Santos (Eds)

Influence of the laminate stiffness on the slamming response of composite wedges

M. Calvário, S. Wang & C. Guedes Soares
Centre for Marine Technology and Ocean Engineering (CENTEC), Instituto Superior Técnico, Universidade de Lisboa, Lisbon, Portugal

ABSTRACT: This paper assesses the impact response of a panel made of composite materials, with different couplings on the laminate stiffness. The analysis is undertaken considering quasi-isotropic, symmetric and cross-ply laminates based on a glass fibre reinforced polymer. The wedge geometry is derived from a case study presented in the literature and two impact velocities corresponding to free-fall and constant drop. The comparison between laminates takes into consideration the results of the slamming force, pressure and strain in sensors, in which the fluid-structure interaction is obtained with an explicit finite element approach. According to the results, significant differences are found between the velocity scenarios and high deformations are reached for the cross ply-laminate in the constant velocity scenario.

1 INTRODUCTION

The wave slamming analysis is an important issue to guarantee the integrity of ship structures. The large relative motions between ships and waves, create impulsive pressures which might damage locally and globally the hull structure. Furthermore, the wave slamming analysis is also critical for the design of other marine structures such as the Wave Energy Converters (WECs).

Different studies have been presented over the years to assess the impact between bodies and water, which have been initiated by van Karman, whose theory was developed for a two-dimensional wedge impact on calm water (Wang & Guedes Soares 2017a). Further contributions were given by Wagner, whose approach took into consideration the piled-up water on the wedge (Wang & Guedes Soares 2020) and Debrovol'skaya (1969), who proposed a similarity solution for a wedge with constant velocity.

With regards to experimental research, Zhao & Faltisen (1996) studied the drop test of a wedge with an impact angle of 30º and a bow-flare section, stressing that maximum pressure occurred before flow separation. Other important experimental research has been conducted by Luo et al. (2012) which studied numerically a V-shaped wedge (with stiffened panel construction) where a good agreement between numerical and experimental was obtained.

In the last decades, numerical approaches such as the explicit finite element method (FEM) have been developed to study the wave slamming. The work conducted by Aquelet et al. (2006) using the Euler-Lagrange coupling, highlighted the fluid-structure interaction in particular stressing the role of the penalty coupling method. The role of this penalty-coupling method was also investigated on the drop of a three-dimensional body (Wang & Guedes Soares 2014a) and on the drop of a composite wedge (Calvário et al. 2023).

The Arbitrary Euler-Lagrange coupling is suitable to analyse the wave slamming problem and was implemented in different studies on the impact of elastic wedges (Wang & Guedes Soares 2014b) and plates (Wang et al 2016, Wang & Guedes Soares 2016, 2017b) revealing the importance of structure thickness, length and boundary conditions on the hydroelasticity. A review of different slamming studies on ships can be found in Wang & Guedes Soares (2017a).

Furthermore, the role of wave slamming might be even more critical when the structure is made of composite materials since these structures can be more flexible than steel (Qin & Batra 2009, Wang et al. 2022).

The research work carried out by Hasson et al. (2017) analysed the behaviour of composite wedges under slamming impact with the presence of damage where a good agreement was found between numerical and experimental approaches.

Composite materials due to their advantages in the harsh marine environment such as lightness and corrosion resistance, have been implemented in different sectors of marine structures such as in the hulls of different types of ships (Mouritz et al. 2001) and the structures of wave and tidal energy devices (Calvário et al. 2018).

DOI: 10.1201/9781003508762-44

The laminate configurations influence the coupling between forces and moments and so the mechanical performance. This paper addresses the slamming response of a composite wedge using different laminate stiffnesses.

The same glass fibre-reinforced polymer is used although three different laminate cases are studied: quasi-isotropic, symmetric and cross-ply. The objective is to verify differences in the slamming response (slamming force, pressure and deformations) under two impact velocities.

The paper is structured in five sections. Section 2 presents a brief theoretical description of wave slamming, while the case study is presented in Section 3. The results are discussed in Section 4 and in Section 5 the conclusions are presented.

2 THEORETICAL FORMULATION

The fluid-structure interaction is a complex phenomenon in which different theories have been proposed. Figure 1 presents the water entry of a two-dimensional wedge. A cartesian coordinate system is defined, where the entry velocity is defined as *dz/dt*, the vertical distance from the lowest point to the calm water Z_d, a deadrise angle β and half-width of the horizontal wetted surface $L(t)$ (Wang & Guedes Soares 2012).

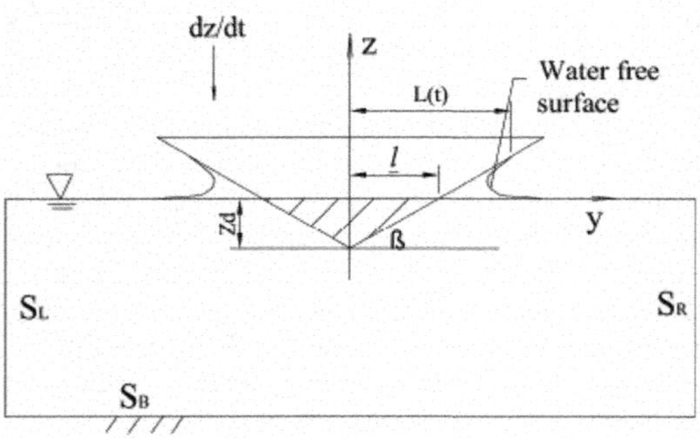

Figure 1. Coordinate system for a water entry of two-dimensional wedge. Retrieved from Wang & Guedes Soares (2012).

In this paper, the ALE approach is followed to account for body elasticity. The total time derivative of a variable *f*, relative to a reference coordinate is given by (Aquelet et al. 2006):

$$\frac{\partial f(\vec{X},t)}{\partial t} = \frac{\partial f(\vec{x},t)}{\partial t} + (\vec{v} - \vec{w}) \cdot \overrightarrow{gradf}(\vec{x},t) \quad (1)$$

where $\vec{X}$ = Lagrangian coordinate; $\vec{x}$ = ALE coordinate; $\vec{v}$ = particle velocity and $\vec{w}$ = velocity of the reference coordinate.

The fluid-structure interaction is modelled with a penalty coupling, following the approaches of Wang & Guedes Soares (2014a) and Calvário et al. (2023).

3 CASE STUDY

Taking into consideration the angle, material, ply thickness and symmetry of a laminate, it is possible to zero out some elements of the stiffness matrices of composite materials. This is particularly important because it can reduce or even eliminate the coupling of forces and bending moments, normal and shear forces or bending and twisting moments which may produce a desired mechanical performance (Kaw 2006).

In this paper three cases of laminates (quasi-isotropic, symmetric and cross-ply) are considered to assess their resistance against wave slamming.

The three laminates use the same glass fibre-reinforced polymer material (Table 1), each ply has 0.5 (mm) thickness and the overall laminate thickness is 5 (mm). The lay-up is presented in Table 2.

Table 1. Material properties of skin and core. Adapted from Hasson et al. (2017).

Parameter	Skin	Core	Units
ρ	1960	80	kg m^{-3}
t	7	20	mm
E_{11}	48.16	0.077	GPa
E_{22}	11.21	0.077	GPa
E_{33}	11.21	0.077	GPa
G_{12}	4.42	0.029	GPa
G_{13}	4.42	0.029	GPa
G_{23}	9	0.029	GPa
v_{12}	0.274	0.3	—
v_{13}	0.274	0.3	—
v_{23}	0.096	0.3	—

Table 2. Composite laminate.

Definition	Lay-up
Quasi-isotropic	[18/36/54/72/90/108/126/144/162/180]
Symmetric	[0/15/30/45/60/60/45/30/15/0]
Cross-ply	[0/90/0/90/0/90/0/90/0/90]

The research presented by Hasson et al. (2017) is used to validate the scenarios of constant velocity and free-fall. This reference analysed the drop of a wedge with a constant velocity of 4 (m/s) made-off sandwich panels (500x250x27 mm) with polymeric skins and PVC foam (Table 1) is derived. The equipment used in the test is illustrated in Figure 2.

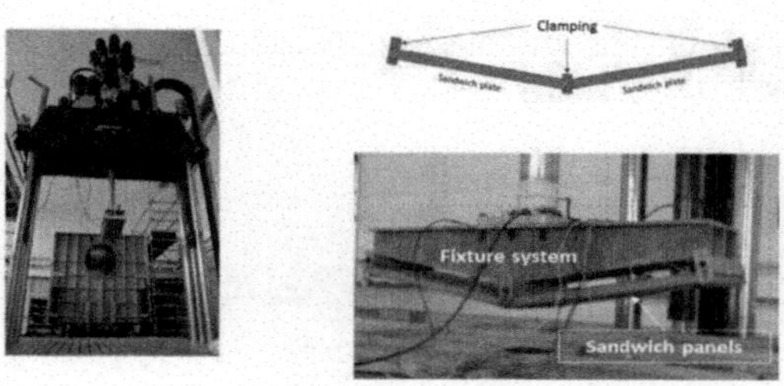

Figure 2. Impact test equipment. Adapted from Hasson et al. (2017).

The deadrise angle was 10°, and the strain gauges (*A*, *C* and *E*) were glued to the panel's surface to measure the strain in the *x* direction (Figure 3). In addition to the experimental test, a numerical simulation using a Coupled Eulerian-Lagrangian model was undertaken using Abaqus software, and generally, the results of numerical simulations were in agreement with experimental ones.

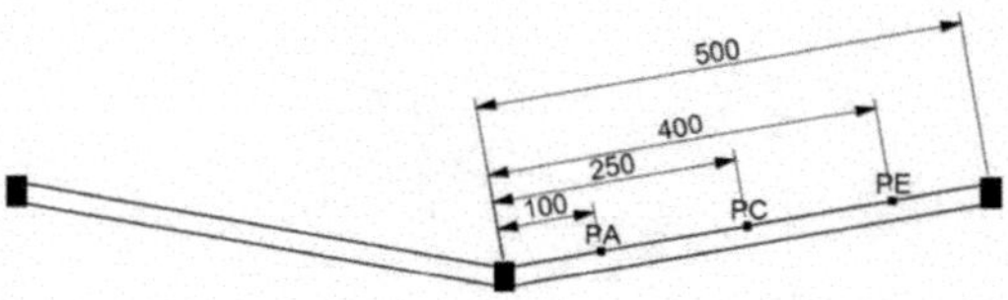

Figure 3. Illustration of the strain gauges positions on the panels. Retrieved from Silva et al. (2020).

In the presented paper, the LS-Dyna software is used to model the fluid-structure interaction with the ALE solver. The fluid is modelled with solid finite elements while the structure with shell elements uses the MAT 054 option. The main input parameters are presented in Table 3, while the mesh is illustrated in Figure 4.

Table 3. Input parameters for the numerical simulation.

Parameter	Value	Units
Domain width	2.5	m
Air height	0.2	m
Water depth	0.6	m
Velocity model	Free-fall and constant	—
Initial velocity	2	m/s
Penalty factor	0.1	—
Simulation time	0.05	s
Mesh size	2.5	mm
N. Elements (fluid)	30630	—
N: Elements (structure)	200	—

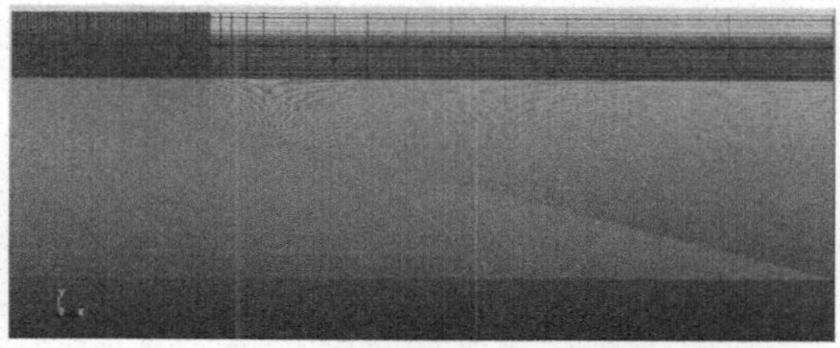

Figure 4. Mesh illustration used for the numerical simulation.

4 RESULTS

This section presents the simulation results. In Section 4.1 the validation is undertaken, where results are compared with published data. In Section 4.2 the case of free fall is considered while in Section 4.3 deals with the case of constant impact velocity.

4.1 *Validation*

The comparison between the numerical and experimental results presented by Hasson et al. (2017) takes into consideration the slamming force and strains at the sensor. In the case of the slamming force, F_S (Figure 5), there is a difference between the peak force simulated in LS-Dyna (18.5 kN) relative to the experiments (approximately 10.4 kN). Nevertheless, the peak force is relatively close to the one obtained with the Abaqus simulation (16. 1 kN), with a similar Lagrangian-Eulerian formulation. The discrepancies between experimental and numerical results can be derived by misalignments in the experiments, numerical errors in the data processing and for instance a not well-defined mesh in finite element model.

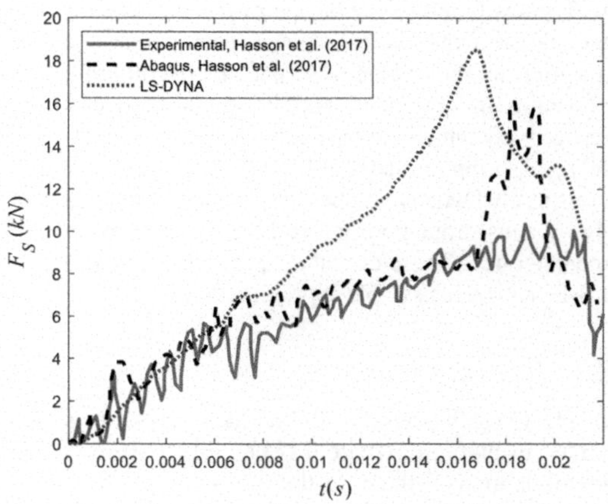

Figure 5. Slamming force for numerical and experimental models.

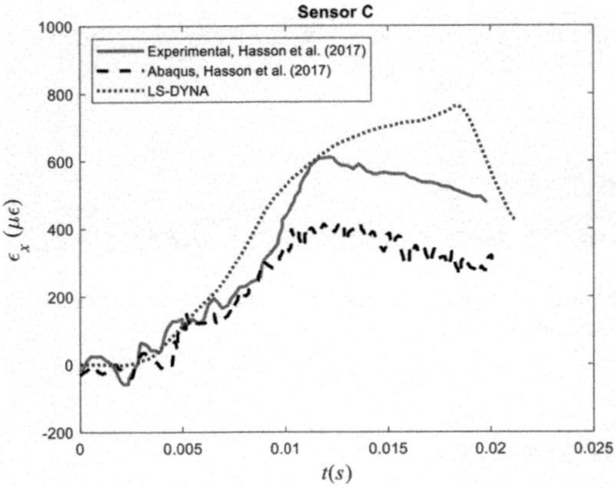

Figure 6. Strains in sensor C for numerical and experimental models.

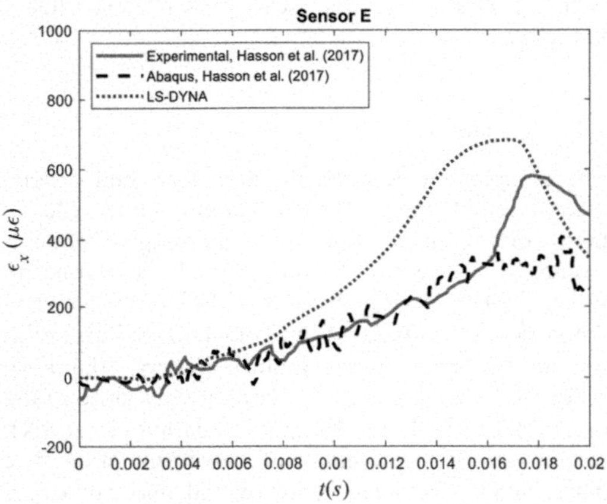

Figure 7. Strains in sensor E for numerical and experimental models.

With regards to the strains in sensors C and E (Figures 6 and 7), the shape of the plot is similar to the experiments where after a peak value there is a significant drop value of strain. The maximum values of strains in sensors C, and E are 763 and 683 (με) respectively which are relatively close to the experimental results (approximately 612 and 582 με). The simulations in LS-Dyna are higher than the ones obtained with the Abaqus simulations but still are closer to the experimental ones. Therefore, the approach carried out in LS-Dyna is a reasonable approximation.

4.2 *Free fall scenario*

In the free-fall scenario, the surface elevation is presented in Figure 8, where it is possible to observe that in all laminates there is the restoring of the initial shape, during the entrance (approximately at 0.04 s), which is an indication that the simulation is in the elastic regime. The results of F_S are presented in Figure 9, while the pressures in sensors A and C in Figures 10 and 11 respectively.

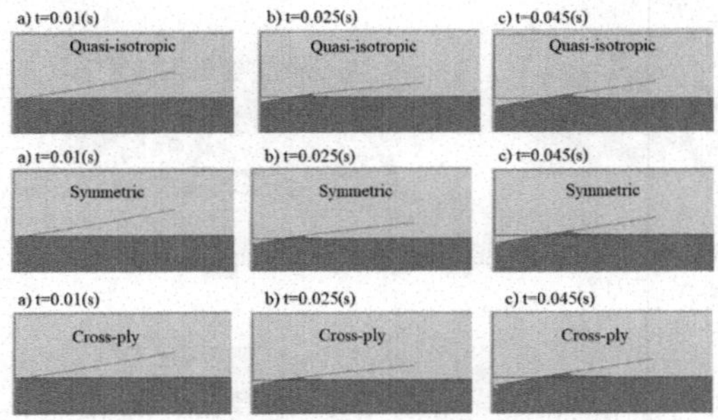

Figure 8. Surface elevation at 0.01, 0.025 and 0.045 (s) for the three laminates in the free-fall scenario.

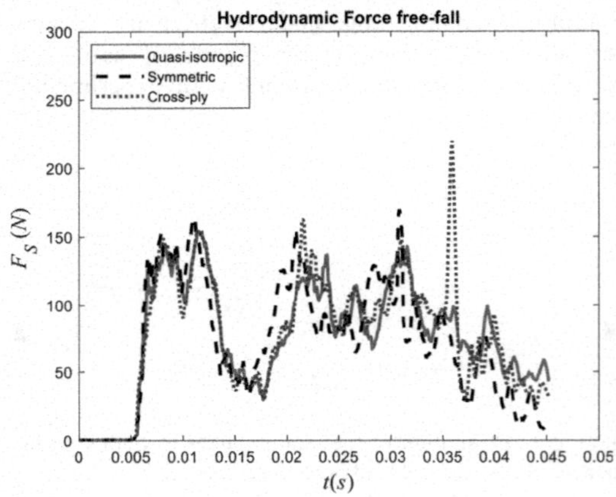

Figure 9. Slamming force for the three laminates in the free-fall scenario.

In the case of F_S, the plot shape is similar among the laminates, where quasi-isotropic and symmetric have similar results while a larger peak (220 N) is obtained for the cross-ply laminate. Moreover, this behaviour is also observed in the pressure in sensor A where the peak value of the cross-ply laminate is 35 (kPa) compared to 14 and 19 (kPa) for quasi-isotropic and symmetric respectively. In the case of sensor C, a large reduction in pressure is observed (maximum peak of 2.2 kPa) where the difference between laminates is not meaningful and no relevant pressure values are obtained in sensor E for all laminates. This is a result of the recovery of the wedge deformation.

With regard to the deformations in sensors A, C and E (Figures 12-14), results also reveal that the cross-ply laminate reaches the highest values (739, 472 and 132 (με) in sensors A, C, and E respectively) with similar peaks between quasi-isotropic and symmetric laminates. The magnitude of peak deformation is maximum in sensor A and decreases in sensor C and especially in sensor E (with a small value in this sensor), due to the recovery of the wedge deformation.

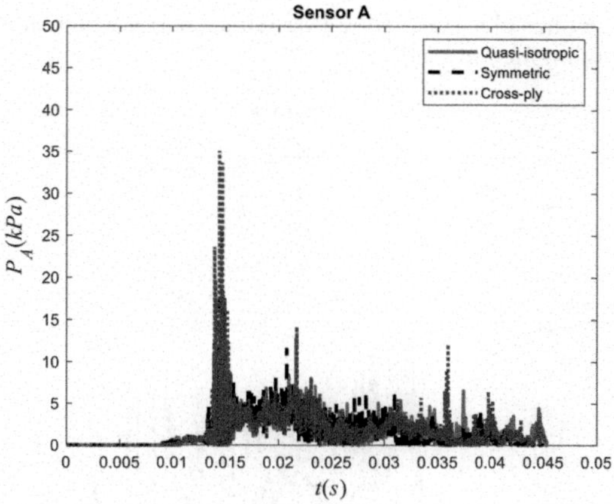

Figure 10. Pressure in sensor *A* for the three laminates in the free-fall scenario.

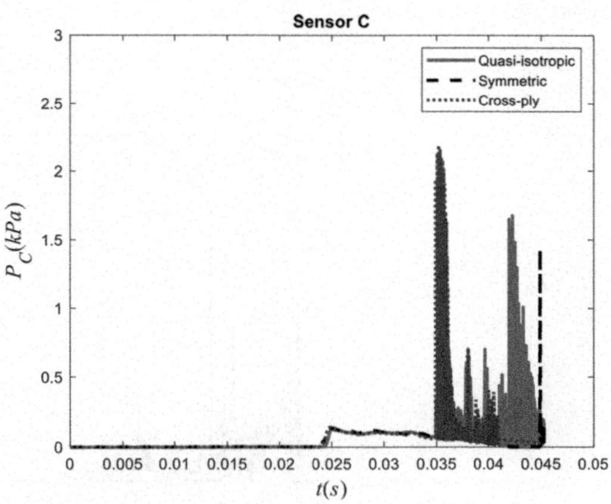

Figure 11. Pressure in sensor *C* for the three laminates in the free-fall scenario.

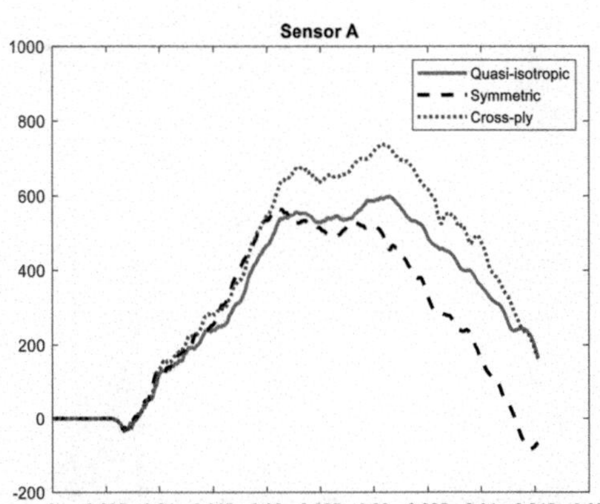

Figure 12. Strain in sensor *A* for the three laminates in the free-fall scenario.

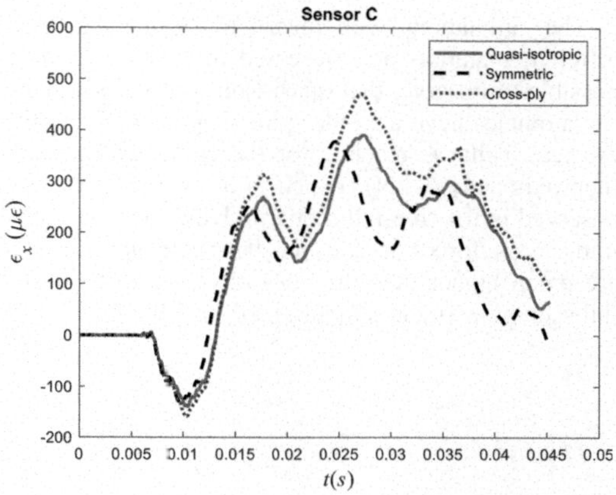

Figure 13. Strain in sensor *C* for the three laminates in the free-fall scenario.

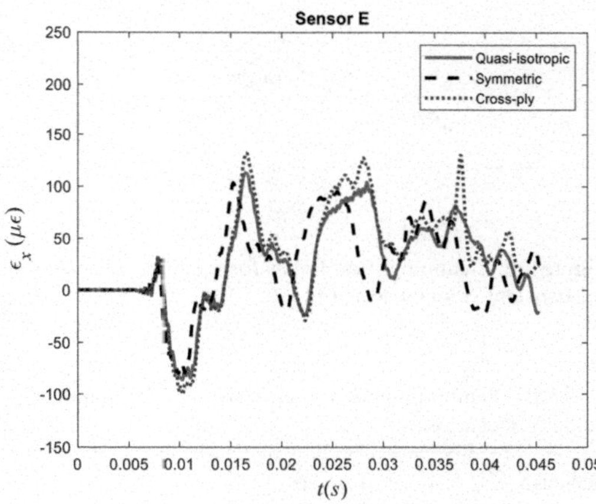

Figure 14. Strain in sensor *E* for the three laminates in the free-fall scenario.

4.3 *Constant velocity*

In the constant velocity scenario of 2 (m/s), the surface elevation is presented in Figure 15 where it is possible to observe that in all materials a larger deformation compared to the free-fall scenario.

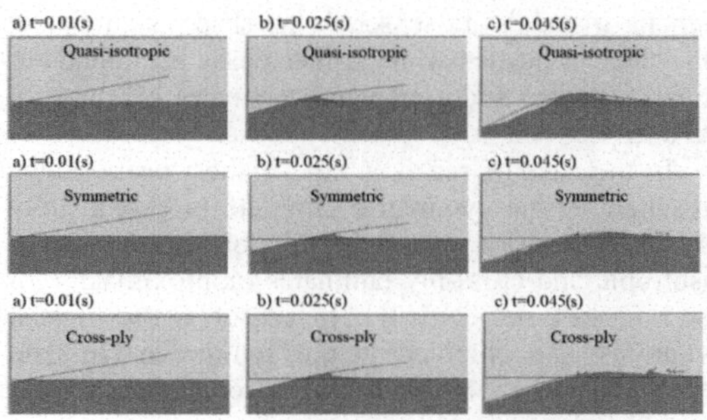

Figure 15. Surface elevation at 0.01, 0.025 and 0.045 (s) for the three laminates in a constant velocity scenario.

The simulations with regard to F_S for the three different materials are presented in Figure 16. It is possible to observe, that quasi-isotropic and symmetric laminates have a similar plot shape, and the highest peak value is reached for the cross-ply laminate (approximately 3 kN), which is a similar behaviour observed in the free-fall scenario. However, the slamming peak forces in the constant velocity scenario are much higher than the ones observed in the free-fall scenario, which is highlighted in Table 4.

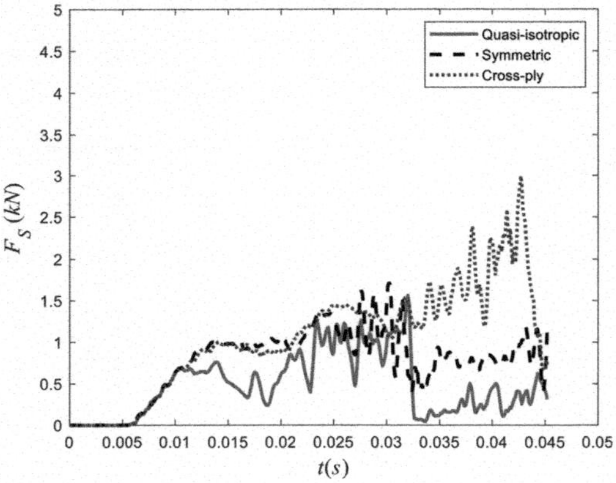

Figure 16. Slamming peak forces for the three laminates in the constant velocity scenario.

Table 4. Slamming peak values in free-fall and constant velocity scenarios.

Scenario	Lay-up	*F* (*N*)
	Quasi-isotropic	154
Free fall	Symmetric	170
	Cross-ply	220
	Quasi-isotropic	1563
Constant	Symmetric	1715
velocity	Cross-ply	2999

In the case of pressure in the sensors (Figure 17-19), important differences are revealed among materials. In sensor A the shape is similar in the three laminates while higher peaks are found for quasi-isotropic and symmetric laminates (57 and 78 kPa respectively).

In the case of sensor *C*, the highest peak is also reached by the symmetric laminate (approximately 141 kPa), which is roughly double of the quasi-isotropic and cross-ply laminates (approximately 73 and 77 kPa respectively). In sensor *E* the highest value is also obtained in the symmetric laminate (168 kPa), where a high peak is also obtained for the cross-ply laminate (123 kPa).

Furthermore, the pressure peak values are also much higher in the constant velocity scenario than the ones observed in the free-fall scenario, which are highlighted in Table 5.

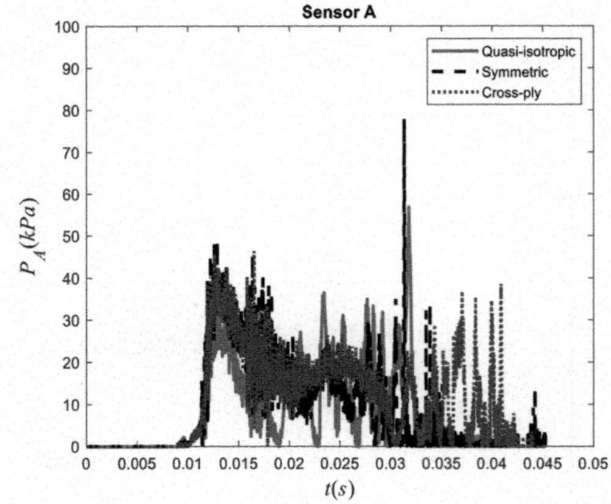

Figure 17. Pressure in sensor *A* for the three laminates in a constant velocity scenario.

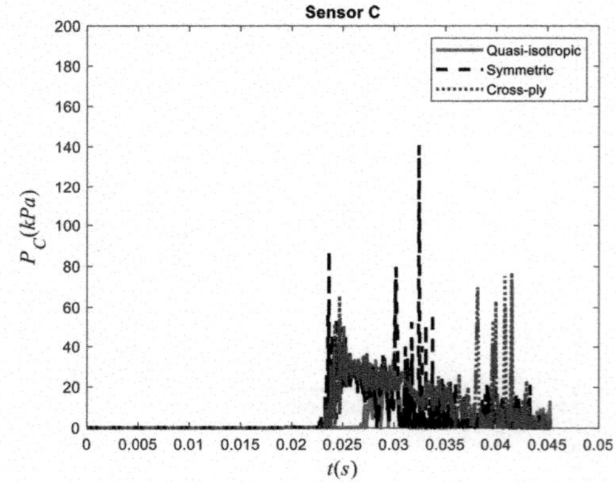

Figure 18. Pressure in sensor *C* for the three laminates in a constant velocity scenario.

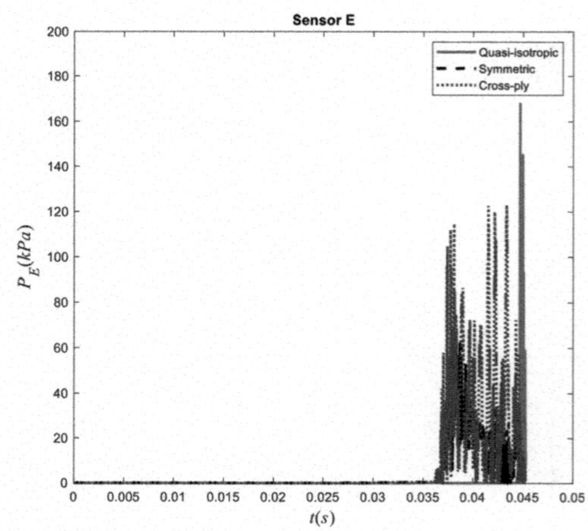

Figure 19. Pressure in sensor *E* for the three laminates in a constant velocity scenario.

Table 5. Pressure at sensors for both free-fall and constant velocity scenarios.

Scenario	Lay-up	P_A (kPa)	P_C (kPa)	P_E (kPa)
	Quasi-isotropic	14	1.7	—
Free fall	Symmetric	19	1.4	—
	Cross-ply	35	2.2	—
	Quasi-isotropic	57	73	168
Constant	Symmetric	78	141	74
velocity	Cross-ply	47	77	123

With regards to the strain results (Figures 20-22), there are also differences among laminates. In sensor *A*, similar peak values are reached among materials, while the highest value is obtained by the cross-ply laminate (1115 με) and also a relatively large value in compression (623 με) is observed.

In sensor *C* important differences are obtained, where the deformation in the cross-ply laminate reached 2427(με) which is much higher than the other two laminates. The same behaviour is observed in sensor *E*, where the peak deformation of 3067 (με) which is a very large deformation.

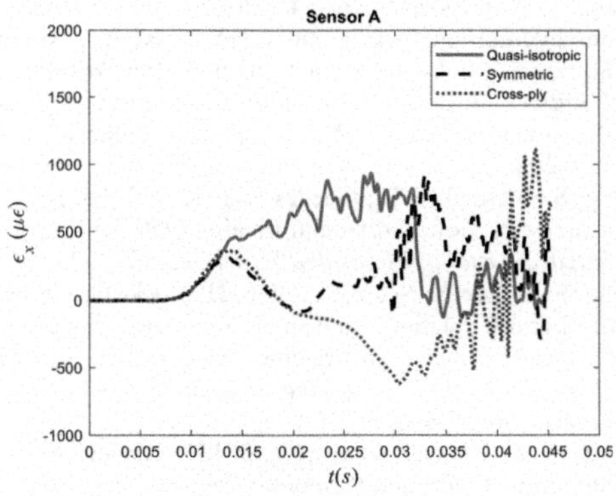

Figure 20. Strain in sensor *A* for the three laminates in a constant velocity scenario.

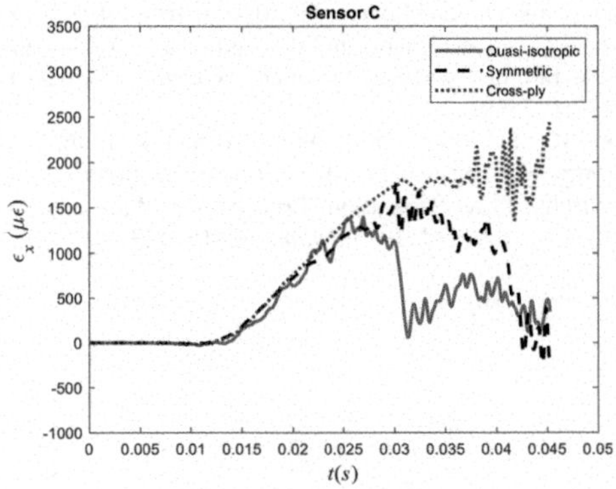

Figure 21. Strain in sensor *C* for the three laminates in a constant velocity scenario.

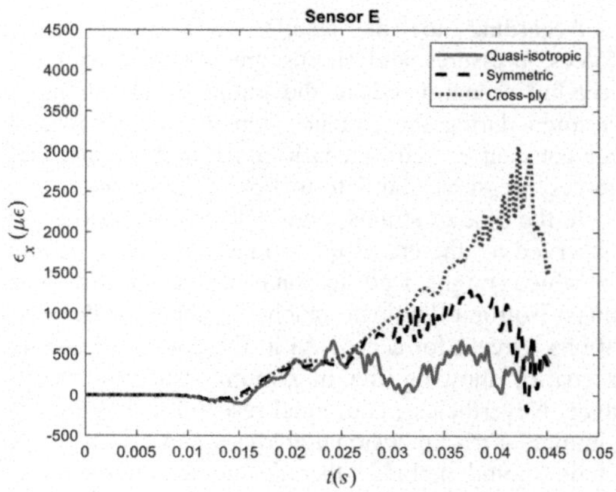

Figure 22. Strain in sensor E for the three laminates in a constant velocity scenario.

The strain peak values are also much higher in the constant velocity scenario than the ones observed in the free-fall scenario (Table 6). In sensor *A* the ratio between results is roughly 50 %, increasing considerably in sensor *C* and the ratio is maximum in sensor *E*, particularly for the cross-ply laminate. In all sensors, the cross-ply laminate reached the highest deformations, being particularly high in sensors *C* and *E* which might lead to the failure of the laminate.

The quasi-isotropic and symmetric laminate reached similar results in the free-fall scenario while in the constant velocity case, in sensor *E* the quasi-isotropic laminate reached a much lower deformation, showing a larger structural resistance in the most critical scenario.

Table 6. Strains at sensors for both free-fall and constant velocity scenarios.

Scenario	Laminate	ε_{xA} ($\mu\varepsilon$)	ε_{xC} ($\mu\varepsilon$)	ε_{xE} ($\mu\varepsilon$)
	Quasi-isotropic	598	390	113
Free fall	Symmetric	566	378	104
	Cross-ply	739	472	132
	Quasi-isotropic	936	1396	663
Constant	Symmetric	905	1787	1287
velocity	Cross-ply	1115	2427	3067

5 CONCLUSIONS

A study with regard to the impact response of a composite wedge made of different laminates is conducted. The study considers both free-fall and constant velocity scenarios (initial velocity of about 2 m/s). The simulations are undertaken with LS-Dyna software using the ALE approach and a comparison with published data was undertaken, being the validation satisfactory.

According to the results, significantly lower forces, pressures and strains are obtained with the free-fall scenario, where the initial wedge shape is restored during the entrance. On the other hand, in the constant velocity scenario much higher slamming forces, pressures and deformation are observed.

In the case of strains, high deformation values are observed for the cross-ply laminate in sensors C and E which might lead to catastrophic failure. The quasi-isotropic laminate reached the lowest deformations (except for sensor A) in the constant velocity scenario, showing the maximum structural resistance. Nevertheless, additional research is required to simulate a tri-dimensional analysis and also a future study should include other design features such as the implementation of transverse stiffeners.

ACKNOWLEDGEMENTS

The first author has been financed by the Portuguese Foundation for Science and Technology (Fundação para a Ciência e Tecnologia - FCT) under the grant SFRH/BD/137429/2018. This work was performed within the scope of the Strategic Research Plan of the Centre for Marine Technology and Ocean Engineering (CENTEC), which is financed by the Portuguese Foundation for Science and Technology (Fundação para a Ciência e Tecnologia - FCT) under contract UIDB/UIDP/00134/2020.

REFERENCES

Aquelet, N., Souli, M., Olovsson, L. 2006. Euler-Lagrande coupling with damping effects: Application to slamming problems. *Computer methods in applied mechanics and engineering* 195: 110–132.

Calvário, M., Sutherland, L. S., Guedes Soares, C. 2018. A review of the applications composite materials in wave and tidal energy devices. Guedes Soares, C. & Teixeira A. P., (Eds.). *Maritime Transportation and Harvesting of Sea Resources*: 695–701. London, UK: Taylor & Francis Group.

Calvário, M., Wang, S., Guedes Soares, C., 2023. Analysis of slamming loads on a sandwich composite wedge. C. Guedes Soares (Ed.). *Trends in Renewable Energies Offshore*: London, UK: Taylor & Francis Group. 889–898

Dobrovol'skaya Z.N. 1969. On some problems of similarity flow of fluids with a free surface. *Journal of Fluid Mechanics* 36(4): 805–829.

Hasson, O.H., Tarfaoui, M., El Malki Alaoui, A., El Moumen, A. 2017. Experimental and numerical investigation on the dynamic response of sandwich composite panels under hydrodynamic slamming loads. *Composite Structures* 178: 297–307.

Kaw, A. (2006). *Mechanics of Composite Materials*. Boca Raton: CRC Press, Taylor & Francis Group.

Luo, HB.; Wang, H., Guedes Soares, C. 2012. Numerical and experimental study of hydrodynamic impact and elastic response for one free-drop wedge with stiffened panels. *Ocean Engineering* 40:1–14.

Mouritz, A.P, Gellert, E., Burchill, P., Challis, K. 2001. Review of advanced composites structures for naval ships and submarines. *Composite Structures* 53: 21–41.

Qin, Z., Batra R.C., 2009. Local slamming impact of sandwich composite hulls. *International Journal of Solids and Structures*, 46(10): 2011–2035.

Silva, N., Wang, S., Guedes Soares, C. 2020. ALE and finite element investigation of water impact one composite panels. In: C. Guedes Soares (Ed.), *Developments in the Collision and Grounding of Ships and Offshore Structures*: UK: Taylor & Francis Group. 227–236.

Wang, S. & Guedes Soares, C. 2012. Analysis of the water impact of symmetric wedges with a multi-material Eulerian formulation. *Trans RINA* 154 (4): 191–206.

Wang, S. & Guedes Soares, C. 2014a. Numerical study on the water impact of 3D bodies by an explicit finite element method. *Ocean Engineering* 78: 73–88.

Wang, S. & Guedes Soares, C. 2014b. Numerical study on hydroelastic water entry of a wedge. Guedes Soares, C. & Lopez Pena F., (Eds.). *Developments in Maritime Transportation and Exploitation of Sea Resources*: London, UK: Taylor and Francis Group. 199–205.

Wang, S.; Karmakar, D., Guedes Soares, C. 2016. Hydroelastic impact of a horizontal floating plate with forward speed. *Journal of Fluids and Structures*: 97–113.

Wang, S. & Guedes Soares, C.2016. Hydroelastic vibration of bottom plating subjected to wave impact. In C. Guedes Soares and Santos T. A. (Eds.), *Maritime Technology and Engineering 3*: London, UK Taylor & Francis Group. 401:407.

Wang, S. & Guedes Soares, C. 2017a. Review of Ship Slamming Loads and Responses. *Journal of Marine and Science Application* 16: 427–445.

Wang, S. & Guedes Soares, C. 2017b. Hydroelastic analysis of a rectangular plate subjected to slamming loads. *Journal of Marine Science and Application*, 16: 405–416.

Wang, S. & Guedes Soares, C. 2020. Effects of compressibility, three-dimensionality and air cavity on a free-falling wedge cylinder. *Ocean Engineering*. 217:107589.

Zhao, R., Faltinsen, O.M. and Aarsnes, J. 1996. Water Entry of Arbitrary Two-Dimensional Sections with and Without Flow Separation. *Proceedings of the 21st Symposium on Naval Hydrodynamics*. 408-423.

Advances in Maritime Technology and Engineering – Guedes Soares & Santos (Eds)
© 2024 The Author(s), ISBN 978-1-032-83099-5

Preliminary validation of an isogeometric MatLab code for cruise ship glazed openings

C. Chianese & F. Marmo
Department of Structures for Engineering and Architecture, University of Naples Federico II, Naples, Italy

M. Acanfora
Department of Industrial Engineering, University of Naples Federico II, Naples, Italy

ABSTRACT: A general tendency in the design of modern cruise ships concerns the massive use of lightweight materials combined with large glazed openings in the vessel superstructure. Owing to their general validity, current regulatory prescriptions tend to fail in delivering accurate thickness estimations due to the inadequacy to keep up with modern glass technology or the effects of ongoing climate change. With growing proportions in number and dimensions, the problem of improving the design of window panes onboard cruise ships urges for the development of reliable numerical tools able to predict the mechanical behaviour of glass units. In this context, an isogeometric structural code for 3-D linearly elastic continua is being implemented in MatLab. At this embryonic stage, this work discusses the effectiveness of the present code in terms of accuracy and computational cost, relative to standard finite element analysis, with reference to benchmark cases of in-plane loaded 2-D glass plates.

1 INTRODUCTION

In recent years, it has become prominent in cruise ship design to include expansive glazed openings in the superstructure. Owing to its transparency and lightness, glass is now a recurring option for designers to conform to current standards of comfort and sustainability onboard without compromising on stability and weight-saving criteria (Heiskari et al. 2023). Use of glass enables passengers to preserve connection with the outside while bringing natural daylight to interior living spaces (Vergassola & Boote 2020). It also improves ship's appeal, lowering its centre of gravity at the same time (Wium et al. 2022).

However, although a firm scientific foundation has been established for the use of glass in land-based structures, the design of glass elements for ships has not come without efforts due to the many challenges posed by the marine environment (Moupagitsoglou 2020). To mention but a few, glass panes at sea are frequently exposed to relevant accelerations due to the ship's motions in addition to basic wind pressures. Further, high brittleness and a significant spread on glass strength (Haldimann et al. 2008) makes it not a natural candidate as a load-carrying material, especially because watertightness is a crucial prerequisite both for integrity and safety of the vessel. Added to this is the inadequacy of glazing to deform plastically, which leaves it particularly susceptible to stress concentrations induced to accommodate hull girder global deformations (Boote & Vergassola 2016).

In most cases, relevant guidelines of Classification Societies (DNV 2022) have provided valuable assistance to engineers in design pressure and scantlings determination. Yet, the general nature of those rules sometimes tends to fail in delivering accurate thickness estimations as a result of the inability to either fully capture the stress mechanisms of the specific case or input proper design pressures. For instance, the regulatory approach is found to deliver unnecessarily heavy insulating glass units in that it ignores the load sharing effect between outer and inner panes (Heiskari et al. 2023). While conservative estimates may bring about weight and stability issues, insufficient scantlings may pose serious threats to the safety of passengers. This is the case of glazing on high-level superstructures, whose design is mainly based on modest regulatory pressures given their limited wave exposure. However, ongoing climate change is making extreme wave events even more likely, thereby leaving superstructures vulnerable to rogue wave impacts.

With increasing proportions in number and dimensions, the problem of improving the design of glass elements in cruise ship superstructures is of great concern given its influence on multiple aspects of the ship's design process. In this regard, numerical investigations have proven to be a viable approach

DOI: 10.1201/9781003508762-45

to gain insight in the mechanical behaviour of glass units (Gerlach & Fricke 2016).

In this wake, the scope of our contribution is to provide the Designer with an efficient numerical tool, suited to reliably predict the structural response of a glass unit, so as to inform the design of glazed openings in cruise ship superstructures. Glass is an amorphous medium with perfectly elastic isotropic properties. As such, a multi-patch isogeometric structural code for 3-D linearly elastic isotropic continua is being developed in MatLab. Given the current level of implementation, the present code is able to deal with in-plane loaded 2-D solids modelled as a single NURBS patch.

Although glass panes are designed to mainly sustain out-of-plane loads locally exerted by the external environment, the current trend of increasing the extension of glazed surfaces is making glass progressively involved in the global strength of the vessel. In this context, plane stress states may arise from hull girder deformations or accelerations due to the ship's translational oscillations. For instance glass elements, located on a ship's side, may experience in-plane normal stresses when the vessel is vertically bent. Similarly, in-plane shear stresses may be induced in glass units of transversal extent due to ship's racking.

Within the scope of validation, this paper illustrates the effectiveness of the proposed isogeometric approach in terms of accuracy and computational effort, relative to standard finite element analysis, with reference to benchmark cases of in-plane loaded glass plates.

2 MATHEMATICAL BACKGROUND

2.1 *Isogeometric analysis*

Initiated by Hughes et al. (2005), the core idea of isogeometric analysis (IGA) is to enhance interoperability between design and analysis by using smooth, high-order spline basis functions both for geometry representation and approximation of physical fields. Compared to standard FEM, the method delivers key advantages, including reduced pre-processing time and improved accuracy as a result of a simplification of the time-intensive mesh generation step and the possibility to perform analysis on exact geometry regardless of mesh refinement level. Besides, high-order interelement continuity allows for a smoother solution space, thereby eliminating the need for stress recovery procedures during post-processing.

Before outlining the details of the IGA computational model, the plane stress problem of a linearly elastic isotropic plate is briefly presented in the following for the reader's convenience.

2.2 *The boundary-value problem (BVP)*

Let Ω be a thin plate bounded by prescribed displacement Γ_u and traction Γ_t boundaries, whose union defines a partition of the domain frontier Γ (Figure 1).

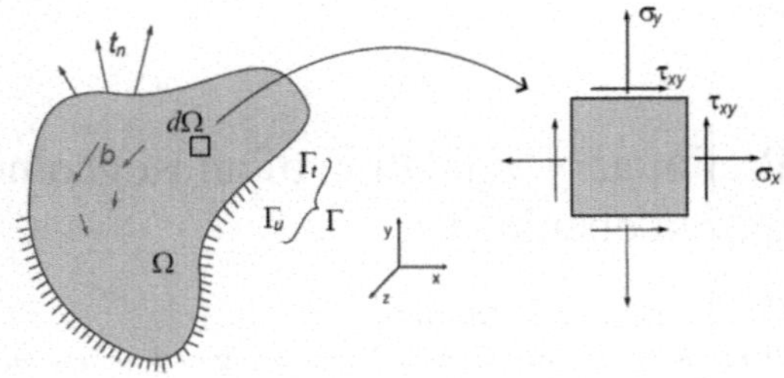

Figure 1. Schematic of a plane stress problem.

If the plate is acted upon by constraints and loads applied parallel to its plane and distributed uniformly over thickness, the state of stress is then specified by σ_x, σ_y and τ_{xy} only, and structural analysis can be limited to the xy-plane. As customary, the displacement formulation of the above plane stress problem can be cast in the form (Timoshenko & Goodier 1951):

$$\begin{aligned}&\text{Seek } \mathbf{u}(\mathrm{x},\mathrm{y}) \text{ such that}\\ &\mathcal{D}^{\mathrm{T}}\mathcal{E}(\mathcal{D}\mathbf{u}) + \mathbf{b} = \mathbf{0} \text{ on } \Omega\\ &\mathbf{u} = \bar{\mathbf{u}} \text{ on } \Gamma_{\mathrm{u}}\\ &\mathbf{t}^{n} = \bar{\mathbf{t}} \text{ on } \Gamma_{\mathrm{t}}\end{aligned} \tag{1}$$

where the Navier field equation is obtained by replacing the constitutive law in the equilibrium equations and then expliciting the strains in terms of displacements using the compatibility relations. In case of isotropic elasticity, the elastic tensor reads

$$\mathcal{E} = \frac{\mathrm{E}}{\infty - \nu^{\in}}\begin{bmatrix}\infty & \nu & / \\ \nu & \infty & / \\ / & / & (\infty - \nu)/\in\end{bmatrix} \tag{2}$$

being E the Young modulus and υ the Poisson ratio, while the definition of the differential operator

$$\mathcal{D} = \begin{bmatrix}\partial/\partial\S & / & \partial/\partial\dagger \\ / & \partial/\partial\dagger & \partial/\partial\S\end{bmatrix}^{\mathcal{T}} \tag{3}$$

reflects the use of Voight notation for strain and stress tensors.

The strong form (1) can be solved analytically only in exceptionally simple cases. To find a numerical solution via isogeometric analysis, it is necessary to shift to the variational formulation of the BVP:

$$\begin{aligned}&\text{Seek } \mathbf{u}(\mathrm{x},\mathrm{y}) \in \mathrm{U} \text{ such that}\\ &\iint_{\Omega} \mathcal{E}(\mathcal{D}\mathbf{u}) \cdot (\mathcal{D}\mathbf{v}) d\Omega = \int_{\Gamma_t} \mathbf{t}^{n} \cdot \mathbf{v} d\Gamma_{\mathrm{t}} + \iint_{\Omega} \mathbf{b} \cdot \mathbf{v} d\Omega\\ &\text{for all } \mathbf{v}(x,y) \in V\end{aligned} \tag{4}$$

being U the space of trial solutions satisfying the Dirichlet boundary conditions of $(1)_2$ and V the set of arbitrary weight functions complying with the homogeneous counterpart of those essential constraints. Eq. (1) is perfectly equivalent to the strong form. Yet, it brings the advantage of weakening the strong continuity requirements on the displacement field **u**, delivering an integral relation satisfied on average throughout the domain, rather than at each point of it.

2.3 *The IGA computational model*

In practice, a numerical solution to the weak form is sought by considering a finite-dimensional approximation of the functional spaces, i.e. $U^h \subset U$ and $V^h \subset V$, so as to convert Eq. (1) into a finite system of linear algebraic equations.

To this purpose, the physical domain Ω is discretized into N^e elements of a single NURBS patch, identified by pairs of knot spans in the parametric directions ξ and η. The support of the NURBS basis is sistematically localised so that basis functions taking on nonzero values in an element are known apriori, depending on the domain parameterization.

Therefore, using Galerkin approach, the solution field within an element Ω^e is computed as a linear combination of local NURBS basis functions and displacement values at associated control points:

$$\mathbf{u}(x,y) = \boldsymbol{S}^e\mathbf{u}^e = \sum_{i=1}^{n^e_{cp}} \begin{bmatrix} R_i(\xi,\eta) & 0 \\ 0 & R_i(\xi,\eta) \end{bmatrix} \begin{bmatrix} \mathrm{u}_{xi} \\ \mathrm{u}_{yi} \end{bmatrix} \tag{5}$$

$$\mathbf{v}(x,y) = \boldsymbol{S}^e\mathbf{v}^e = \sum_{i=1}^{n^e_{cp}} \begin{bmatrix} R_i(\xi,\eta) & 0 \\ 0 & R_i(\xi,\eta) \end{bmatrix} \begin{bmatrix} \mathrm{v}_{xi} \\ \mathrm{v}_{yi} \end{bmatrix} \tag{6}$$

In the above relations, it is highlighted the fact that the element displacement operator $\mathbf{S}^e$ can be decomposed into a set of 2 x 2 sub-matrices, each depending on the basis function related to the current control point.

Similarly, strain components in an element Ω^e are obtained by approximating control point displacements via the relevant derivatives of the corresponding local basis function:

$$\mathbf{e}(x,y) = \mathcal{B}^{e}\mathbf{u}^e = \sum_{i=1}^{n^e_{cp}} \begin{bmatrix} R_{i,x}(\xi,\eta) & 0 \\ 0 & R_{i,y}(\xi,\eta) \\ R_{i,y}(\xi,\eta) & R_{i,x}(\xi,\eta) \end{bmatrix} \begin{bmatrix} \mathrm{u}_{xi} \\ \mathrm{u}_{yi} \end{bmatrix} \tag{7}$$

In the IGA framework, the discretized weak form reads:

$$\sum_{i=1}^{n^e_{cp}} \left[\left(\iint_{\Omega^e} \mathcal{B}^{e^T} \mathcal{E} \mathcal{B}^e \mathrm{d}\Omega^e \right) \mathbf{u}^e \right] \cdot \mathbf{v}^e = \sum_{i=1}^{n^e_{cp}} \left(\int_{\Gamma^e_t} \boldsymbol{S}^{e^T} \mathbf{t}^n \mathrm{d}\Gamma^e_t + \iint_{\Omega^e} \boldsymbol{S}^{e^T} \mathbf{b}\, \mathrm{d}\Omega^e \right) \cdot \mathbf{v}^e \ \forall \mathbf{v} \in V^h \tag{8}$$

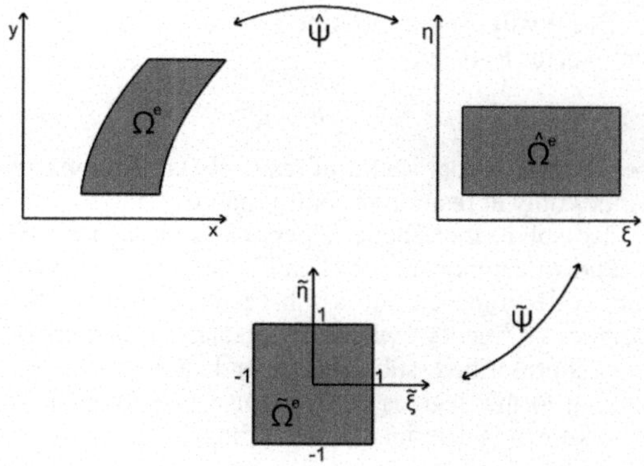

Figure 2. Mapping of an element from physical to parent domain in IGA.

In this work, Gauss-Legendre numerical quadrature is adopted for numerical integrations in Eq. (8). Because NURBS basis functions are defined in the parametric space, an additional mapping from the Cartesian to the parametric domain is introduced, compared to standard FEM: $\hat{\Psi} : \Omega^e \rightarrow \hat{\Omega}^e$. For numerical integrations, elements are then mapped to a regularly shaped parent element: $\tilde{\Psi} : \hat{\Omega}^e \rightarrow \tilde{\Omega}^e$ (see Figure 2).

In particular, the inverse mapping $\tilde{\Psi}^{-1}$ to a child element $\hat{\Omega}^e = \left[\xi_i, \xi_{i+1}\right] \times \left[\eta_j, \eta_{j+1}\right]$ is regulated by:

$$\begin{aligned} \xi &= \frac{1}{2}\left[(\xi_{i+1} - \xi_i)\,\tilde{\xi} + (\xi_{i+1} + \xi_i)\right] \\ \eta &= \frac{1}{2}\left[\left(\eta_{j+1} - \eta_j\right)\tilde{\eta} + \left(\eta_{j+1} + \eta_j\right)\right] \end{aligned} \tag{9}$$

where $\tilde{\xi}$ and $\tilde{\eta}$ denote the coordinates of Gaussian points in the parent one.

Thus, the numerical integration of a generic integrand $f(x,y)$ over a physical element is performed as:

$$\iint_{\Omega^e} (x,y)\mathrm{d}\Omega^e = \sum_{q=1}^{n_{gp}} f(\tilde{\xi}_q, \tilde{\eta}_q) w_{gp_q} |\boldsymbol{J}_{\hat{\Psi}}||\mathbf{J}_{\tilde{\Psi}}| \tag{10}$$

where n_{gp} and w_{gp} indicate the number and weights of Gaussian points in turn, while the Jacobian determinants related to the two mappings are defined as:

$$\begin{aligned} |\boldsymbol{J}_{\hat{\Psi}}| &= \begin{vmatrix} \partial x/\partial\xi & \partial x/\partial\eta \\ \partial y/\partial\xi & \partial y/\partial\eta \end{vmatrix} \\ |J_{\tilde{\Psi}}| &= \left|\frac{\partial\xi}{\partial\tilde{\xi}}\frac{\partial\eta}{\partial\tilde{\eta}}\right| \end{aligned} \tag{11}$$

By assembling over elements, Eq. (8) can be cast into a global standard form:

$$(\mathcal{K}\mathbf{U}) \cdot \mathbf{V} = \mathbf{Q} \cdot \mathbf{V} \ \forall \mathbf{V} \in V^h \tag{12}$$

where $\boldsymbol{\kappa}$ is the plate stiffness matrix and **Q** is the vector of control point forces equivalent to the edge $\mathbf{t}^n$ and volume **b** loads pertaining to the whole patch.

Eventually, the arbitrariness of the weight function vector leads to:

$$\mathcal{K}\mathbf{U} = \mathbf{F} + \mathbf{Q} \tag{13}$$

being **F** the global vector of residuals taking nonzero values only at restrained control points.

To solve the above algebraic system for the unknown components of **U** and **F**, each of those vectors is decomposed into a sub-vector relevant to the degrees of freedom where the displacement component is prescribed (subscript P), and another one pertaining to the degrees of freedom whose kinematics is unknown (subscript U). Accordingly:

$$\begin{bmatrix} \mathcal{K}_{PP} & \mathcal{K}_{PU} \\ \mathcal{K}_{UP} & \mathcal{K}_{UU} \end{bmatrix} \begin{bmatrix} \mathbf{U}_P \\ \mathbf{U}_U \end{bmatrix} = \begin{bmatrix} \mathbf{F}_P \\ \mathbf{F}_U \end{bmatrix} + \begin{bmatrix} \mathbf{Q}_P \\ \mathbf{Q}_U \end{bmatrix} \tag{14}$$

so that the unknown components of the displacement field result from:

$$\mathbf{U}_U = \mathcal{K}_{UU}{}^{-1}[\mathbf{F}_U + \mathbf{Q}_U - \mathcal{K}_{UP}\mathbf{U}_P] \tag{15}$$

and constraint reactions at plate edge supports are:

$$\mathbf{F}_P = \mathcal{K}_{PP}\mathbf{U}_P + \mathcal{K}_{PU}\mathbf{U}_U - \mathbf{Q}_P \tag{16}$$

Distinctly from standard FEM, no recovery procedure is required for stress components, provided that a C^1 interelement continuity is ensured by the selected NURBS parameterization, as they directly follow from discrete Hooke's law:

$$\mathbf{s}(x,y) = \mathcal{E}(\mathcal{B}\mathbf{U})\ \forall(x,y) \in \Omega \tag{17}$$

being **ß** the patch strain operator.

3 RESULTS

3.1 *Infinite plate with a circular hole*

In this subsection, the IGA computational model is validated in the MatLab environment against a benchmark case having an exact solution: an infinite plate with a circular hole (Kirsch 1989). In light of symmetry and a plane state of stress, the solid is modelled as a finite quarter plate of zero thickness subject to uniform uniaxial tension on the upper edge (Figure 3). The contribution of the plate thickness t to its stiffness is still accounted for as a multiplier of the Young modulus during analysis.

The domain parameterization is defined by a quadratic NURBS basis having support on an open uniform knot vector with internal knots of single multiplicity in both circumferential (ξ) and radial (η) parametric directions. This choice has been opted for as it delivers accurate geometry at all levels of discretization as well as a C^1- continuous stress field across element boundaries. Yet, a discontinuity in the inverse of the geometrical mapping is subtly introduced along the segment joining the plate upper right corner and the midpoint of the circular side due to a repeated control point in that corner. However this issue, inevitable when using a single patch, does not cause problems in the simulations since Gauss points never fall at the position of that singularity.

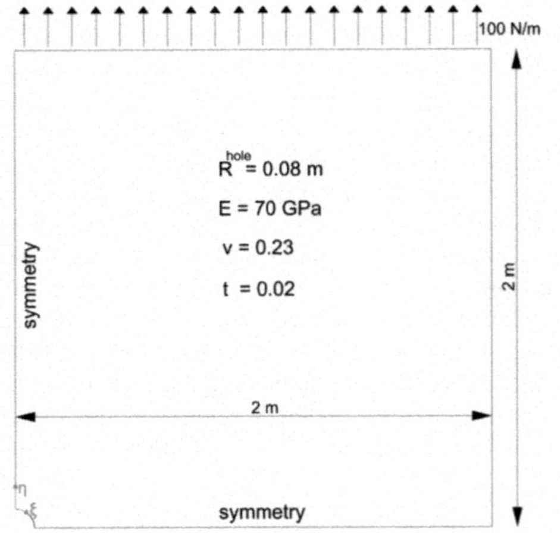

Figure 3. Plate with circular hole under uniaxial loading.

Within the scope of validation, a family of mesh grids has been generated by knot insertion while keeping the order of the NURBS basis constant. For each run, the geometry has been exported directly from a CAD software without further discretization.

To put the advantages of isogeometric analysis in perspective, the numerical solution is compared with the output from a standard FEM code based on quadrilateral nine-node elements. Details of the tested FEM and IGA mesh grids are available in Table 1 and Table 2, respectively. There, the mesh parameter, h_{MAX}, coincides with the length of the largest element side in metres in the physical space. Figure 4 shows the outcome of the structural analysis performed on the finest FEM and IGA mesh grids. Figure 5 compares the absolute error measured on the hoop stress profiles along the two axes of symmetry of the plate across all the main mesh grids tested relative to the analytical solution in the region of stress gradients.

Table 1. Computational details of the tested FEM mesh grids.

Mesh ID	h_{max}	N^{nodes}	$N^{elements}$	max $\|\|\sigma\|\|_2$
FEM mesh 1	0.314	231	50	0.704
FEM mesh 2	0.157	861	200	0.549
FEM mesh 3	0.105	2251	538	0.321
FEM mesh 4	0.079	4041	977	0.297
FEM mesh 5	0.063	6351	1546	0.285
FEM mesh 6	0.052	9184	2245	0.241
FEM mesh 7	0.045	12091	2965	0.223
FEM mesh 8	0.039	15837	3908	0.185
FEM mesh 9	0.035	20224	4981	0.179

Table 2. Computational details of the tested IGA mesh grids.

Mesh ID	h_{max}	N^{cp}	$N^{elements}$	max $\|\|\sigma\|\|_2$
IGA mesh 1	2.748	12	2	0.947
IGA mesh 2	1.647	24	8	0.645
IGA mesh 3	0.988	60	32	0.458
IGA mesh 4	0.551	180	128	0.102
IGA mesh 5	0.374	612	512	0.062
IGA mesh 6	0.153	6351	2048	0.043

For quantitative indications on the maximum value of the L^2-norm of the relative error in the stress field observed in each run, the reader is referred to Table 1 and Table 2.

As shown, as the level of refinement is increased, the accuracy of the present isogeometric code improves until the structural response is found to be in good agreement with the reference solution. Interestingly, IGA is found to deliver a better overall accuracy with respect to standard FEM at the expense of a significantly reduced computational cost. Such a superior convergence velocity can be ascribed to the optimal geometry accuracy and higher smoothness of the basis enabled by the NURBS surface technology.

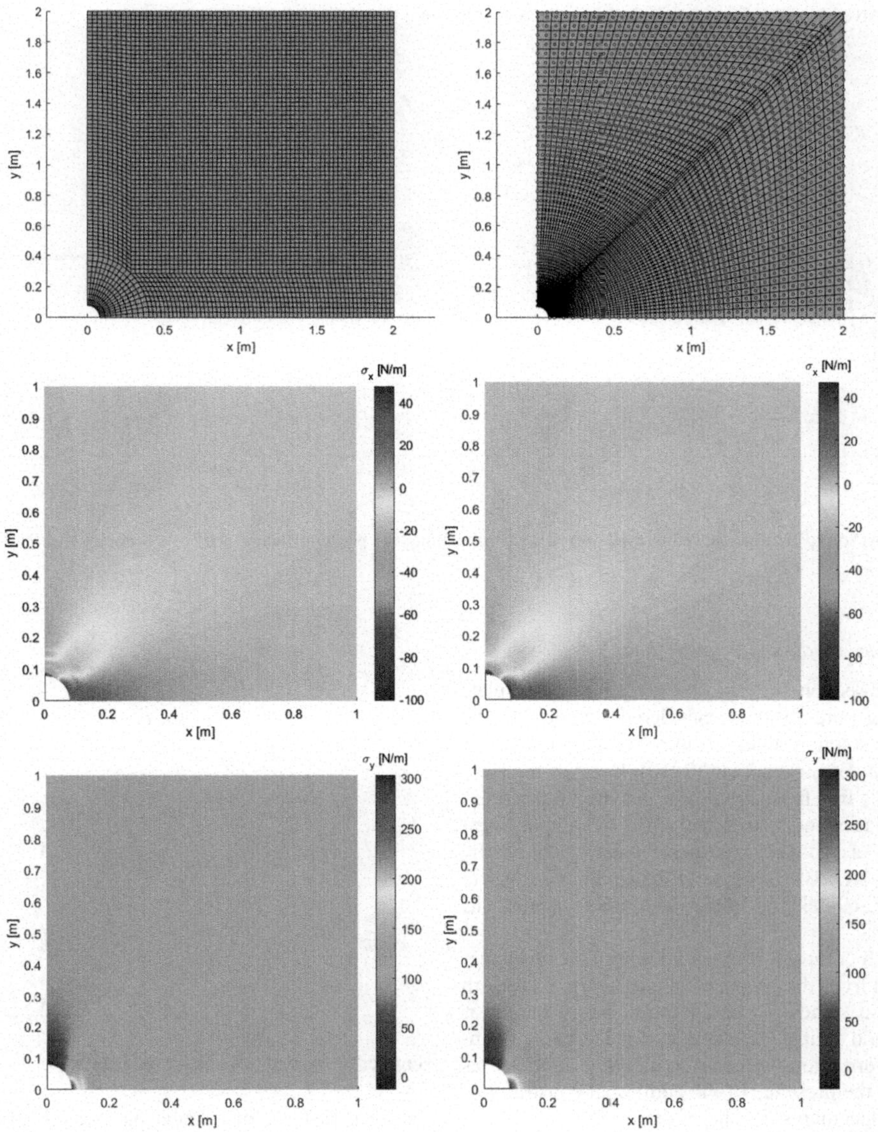

Figure 4. Finest FEM (left) and IGA (right) mesh grids and corresponding plate structural response in enhanced view of region of stress gradients. Nine quadrature points per element have been used in each simulation.

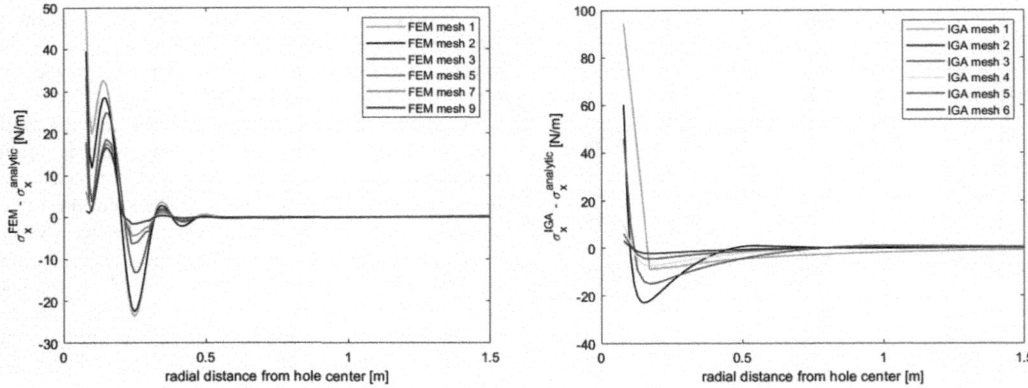

Figure 5. Absolute error measured in the hoop stress profile along the plate vertical axis of symmetry.

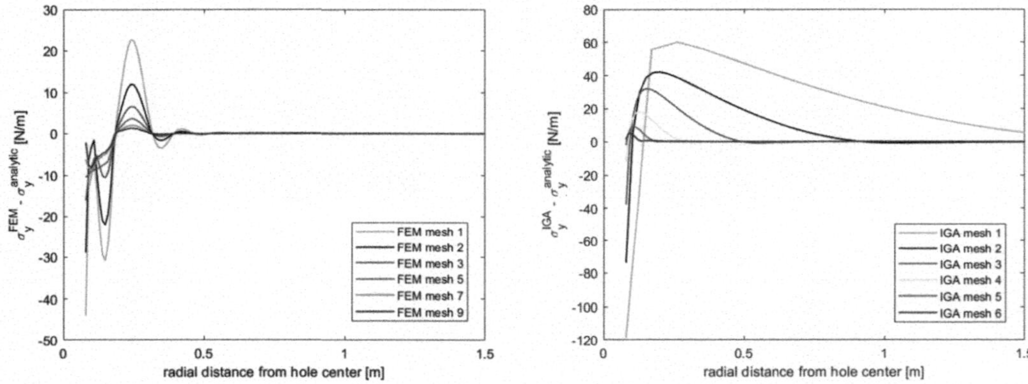

Figure 6. Absolute error measured in the hoop stress profile along the plate horizontal axis of symmetry.

3.2 *Squared plate under pure shear*

In this subsection the case of a squared glass plate undergoing pure shear is considered (Figure 7). The analytical solution field of this benchmark case is available in Blaauwendraad (2010). Because the plate geometry is free from curvilinear details and the reference displacement field is described by simple linear functions, a domain parameterisation defined by a bilinear NURBS basis is sufficient in this case to accurately model the plate and fully capture its kinematics.

Figure 8 compares the magnitude of the displacement field from the proposed isogeometric code with that from a standard FEM solution based on linear quadrilateral four-node elements for the cases of in-plane deformation without rigid body motions as well as in the presence of the additional constraint of zero displacements in the y-direction. As shown, a perfect agreement with the reference solution is found on both FEM and IGA mesh grids when discretizing the plate into 16 elements. In particular, it is noted that use of a linear parameterization allows for a direct correspondence between the position of control points in the IGA model with that of nodes in its FEM counterpart.

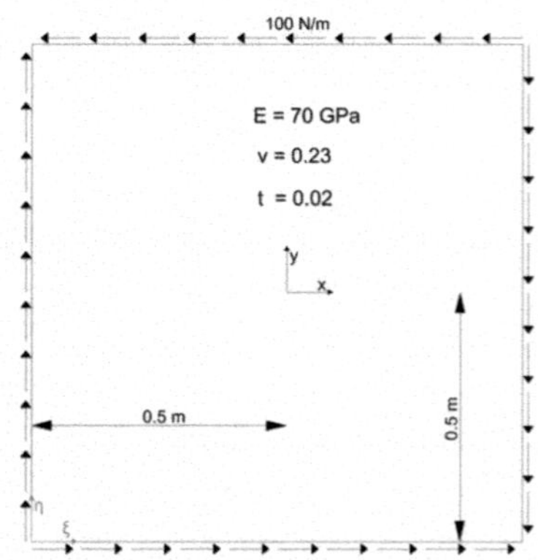

Figure 7. Squared plate under pure shear.

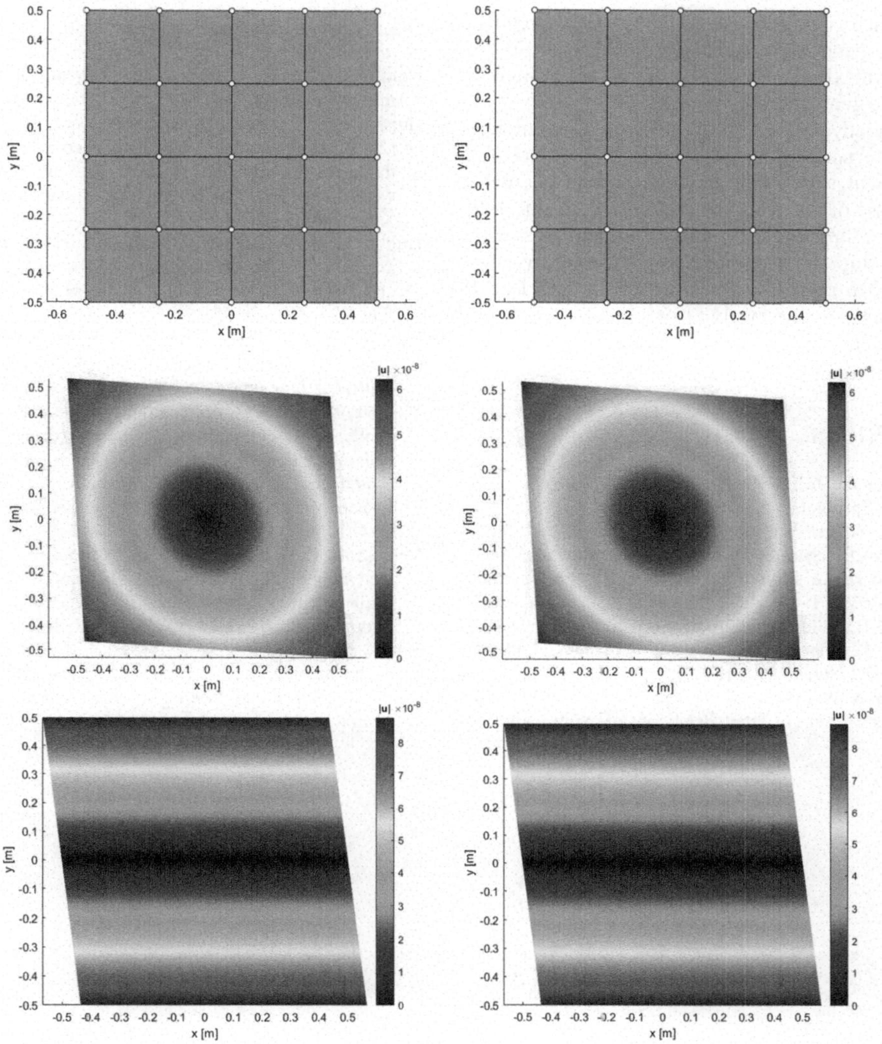

Figure 8. Finest FEM (left) and IGA (right) mesh grids and corresponding plate deformed configuration in the absence of rigid body motions (middle row) and the additional constraint of zero vertical displacements (bottom row). Four quadrature points per element have been used in each simulation.

4 CONCLUSIONS

In this paper, an isogeometric structural code for isotropic linearly elastic continua has been presented. At the current state of implementation, the code can be used for the analysis of in-plane loaded glass plates modelled as a single 2-D NURBS patch. The validation procedure, performed against classical benchmark cases, confirms the reliability of the structural response predicted by the present code and unveils the potential of using isogeometric analysis. More explicitly, IGA is found to be advantageous compared to traditional finite element method in that a higher overall accuracy can be achieved at the expense of a reduced computational effort. This particularly holds when geometries with complex details are addressed.

The key feature of isogeometric analysis is to employ the same high-order, smooth NURBS basis functions for modelling as well as structural assessment. On one hand, this considerably simplifies pre-processing by making time-intensive mesh generation and defeaturing pointless. On the other hand, use of NURBS technology completely eliminates the geometric error associated to discretization as isogeometric elements still constitute a perfect partition of the initial domain at all levels of grid refinement.

The 3-D multi-patch extension of the proposed code is currently under development. This advancement will allow us to address more realistic cases of generally-loaded glass elements in the marine context while avoiding the onset of singularities in the geometrical mapping.

Taking advantage of the ability of isogeometric analysis to perform simulations in a more accurate and efficient way while preserving exact geometry, a future use of the proposed code is foreseen to fully capture the structural response of modern glass units in cruise ship superstructures so as to improve their design when regulatory prescriptions would lead to unacceptably conservative or unsafe thickness determination.

REFERENCES

Blaauwendraad, J. (2010). *Plates and FEM: surprises and pitfalls*. Springer.

Boote, D. & G. Vergassola (2016). Local and global analysis of large motoryacht superstructure. *24th International Hiswa Symposium on Yacht Design and Yacht Construction*, 1–16.

DNV (2022). Rules for classification: Ships, part 3 hull, chapter 12 openings and closing appliances. Technical report, Det Norske Veritas.

Gerlach, B. & W. Fricke (2016). Experimental and numericalinvestigation of the behavior of ship windows subjected toquasi-static pressure loads. *Marine Structures* 46, 255–272.

Haldimann, M., A. Luible, & M. Overend (2008). Structural use of glass. IABSE.

Heiskari, J., J. Romanoff, A. Laakso, & J. W. Ringsberg (2023). Influence of the design constraints on the thickness optimization of glass panes to achieve lightweight insulating glass units in cruise ships. *Marine Structures* 89.

Hughes, T., J. Cottrell, & Y. Bazilevs (2005). Isogeometric analysis: Cad, finite elements, nurbs, exact geometry and mesh refinement. *Comp. Meth. in Appl. Mech. and Eng.* 194, 4135–4195.

Kirsch, E. (1989). Die theorie der elastizität und die bedürfnisse der festigkeitslehre. *Zeitschrift des Vereines deutscher Ingenieure* 42, 797–807.

Moupagitsoglou, K. (2020). Structural glass in superyacht applications: overcoming challenges, design standards and analysis methods. *Conference on Architectural and Structural Applications of Glass, Belis, Bos Louter Eds* .

Timoshenko, S. & J. Goodier (1951). *Theory of Elasticity*. McGraw-Hill Book Company.

Vergassola, G. & D. Boote (2020). A simplifed approach to thedynamic efective thickness of laminated glass for ships andpassenger yachts. *International Journal on Interactive Design and Manufacturing* 14, 123–135.

Wium, D., E. Lataire, & J. Belis (2022). Considerations forthe integration of glass in superyacht structures. *Challenging Glass Conference Proceedings, Belis, Bos Louter Eds* .

Advances in Maritime Technology and Engineering – Guedes Soares & Santos (Eds)

Durability of flax and glass fibre reinforced epoxy laminates for marine applications

J. Domingues & L.S. Sutherland
Centre for Marine Technology and Ocean Engineering (CENTEC), Instituto Superior Técnico, Universidade de Lisboa, Lisboa, Portugal

M. Garrido
Civil Engineering Research and Innovation for Sustainability (CERIS), Instituto Superior Técnico, Universidade de Lisboa, Lisboa, Portugal

ABSTRACT: The durability of an infused flax fibre reinforced plastic (FFRP) marine composite laminate has been extensively investigated via hygrothermal ageing and impact, quasi-static (QS) point loading, flexural and interlaminar shear testing. Comparisons were made with an equivalent thickness 'traditional' glass reinforced plastic (GRP) marine laminate. FFRP impact and QS indentation performance improved with ageing. That of GRP deteriorated significantly but only when aged at 60 °C did it drop to that of the much lower performing FFRP. In flexure GRP performed far better than FFRP, and both laminates deteriorated significantly with ageing. Loss of FFRP shear stiffness after ageing was substantial. Plasticisation of the resin and flax fibres is thought to be important. Sustainability-driven design decisions concerning flax composites in a marine environment should consider that they appear to be heavier, less stiff, far less robust and very susceptible to the effects of moisture. However, under severe ageing conditions, plasticisation may temper FFRP's disadvantages under impact and point (but not flexure) loadings.

1 INTRODUCTION

Composite materials are extensively used in marine applications due to such advantages as low weight, corrosion resistance, low maintenance, and the ability to create complex, seamless forms (Rubino et al., 2020). 'Traditional' marine composites use glass and/or carbon fibres, but carbon fibres are resource-intensive and energy-consuming, leading to substantial CO_2 emissions (Navaratnam et al., 2023). Synthetic fibres such as aramid also pose environmental concerns due to poor degradability. Recycling fibre-reinforced polymers (FRPs) remains a concern (De Fazio et al., 2023).

Flax is a plant fibre that offers low density, ease of handling and non-abrasiveness, making it a possible alternative to synthetic fibres (Yan et al., 2014). Practical applications, such as a racing sailboat built with flax composites have demonstrate the feasibility, at least in theory, of high-performance structures (Castegnaro et al., 2017). However, the adoption of flax fibres in the naval industry is limited due to higher costs and a lack of long-term performance data in wet, aggressive marine environments.

Plant fibres in composite structures pose challenges due to their hydrophilic nature, causing sensitivity to moisture and poor compatibility with hydrophobic polymers (Scida et al., 2013). When exposed to water, composites of these fibres absorb moisture, increasing weight, giving swelling and reducing mechanical properties, particularly when concerning flax fibre reinforced polymers (FFRPs) in marine use (Li and Xue, 2016).

While FFRPs with bio-epoxy can offer some sustainability advantages (Duflou et al., 2014), understanding their durability in marine environments and addressing hydrophilic challenges is crucial to ensure that the overall balance is not negative. In fact, Duflou et al. (2014) determined a lower environmental impact for equivalent strength GRP when compared to FFRPs.

Currently, there is a lack of experimental data on marine FFRP durability, particularly compared to GRP (since the latter is the established practice which would be substituted in the marine industry). The very rapid and sudden recent increase in the use of flax fibres and bio-based epoxy resin in the marine sector underscores the need for more

DOI: 10.1201/9781003508762-46

experimental data. This study aims to help fill this knowledge gap.

Due to the nature of the materials used, marine composite structural design is often flexure controlled, and hence such tests are a very relevant means to quantify their durability. Similarly, interlaminar shear failure, delamination, is a common failure mode in such applications (and is especially important for point and impact loadings) and hence short beam tests were also used as a durability measure. Susceptibility to impact is a known potential fragility of marine composites (Sutherland, 2018) and the only study found (Squires, 2022) indicates a very worrying fragility of flax composites in this respect. Further, durability of these materials in this respect does not appear to have been studied to date. Quasi-static (QS) indentation testing allows both the response to point loads to be evaluated, and also provides valuable information about the impact response (Garrido et al., 2021; Sutherland and Guedes Soares, 2012).

Hence, the approach taken here is to evaluate, and compare with GRP, the durability of FFRP with a bio-based epoxy matrix before, during and after hygrothermal ageing at a range of temperatures via short- and long-beam flexural, and QS indentation and dynamic impact testing.

2 EXPERIMENTAL PROCEDURE

2.1 *Materials*

The FFRP and GRP laminates were manufactured using biaxial fabrics and a blended bio-epoxy resin:

Glass fibre: 840 g/m^2 Non-crimp 0°/90° biaxial (*Rebelco*, Portugal).

Flax fibre: A*mpliTex*™ *5008*, 350 g/m^2 Non-crimp ± 45° biaxial (*Bcomp,* Switzerland).

Epoxy: *Greenpoxy 56* resin, *SD 7561* hardener (*Sicomin,* France).

Equivalent thickness was selected here as the laminate equivalence criteria between GRP and FFRP to give both comparable moisture diffusion and specimen dimensions in the subsequent durability assessment. A thickness of approximately 4.1 mm was deduced with forethought for the mechanical tests, which required 6 plies in each material (since this aspect had been taken into account in the selection of the fibres).

Vacuum assisted resin transfer moulding (VARTM), or 'infusion' was used to produce the laminates since it both reduces variability and may eliminate any problems with wetting out and compacting of the flax fibres. Before infusing, glass and flax fibres were dried for 3 and 5 hours, respectively, under vacuum below 20 mbar to ensure the hydrophilic nature of the flax fibres did not lead to variations in initial moisture content. After infusion was completed the laminated plates were left to cure at ambient temperature (for a minimum of one week), after which a post-cure cycle of 10 h at 80 °C was performed. According to the resin's technical datasheet, this should result in a glass transition temperature (T_g) of 78 °C (SICOMIN Composites, 2018). This post cure ensured that the T_g was approximately 20 °C above the highest hygrothermal ageing temperature to be used (detailed in section 2.2), thus avoiding any state transition effects.

After production, all laminated plates were cut into 340 mm by 264 mm rectangular plates, to give equivalent surface areas for moisture absorption and allowing conservatively for all the planned test specimens to be cut out at the various following stages of hygrothermal ageing. FFRP plates were cut out at an angle of ± 45° to give the same 0°/90° fibre orientation as in the GRP plates.

Ignition loss tests as per the *ASTM D2584-02* standard were conducted on 6 dry GRP specimens giving an estimated fibre volume fraction (FVF) of 0.52. An alternative, method (*ASTM D3171-22*) had to be applied for the heat-degradable flax fibres, measuring the weight and exact dimensions of nominally 70 mm by 70 mm specimens and considering the densities and areal weights given in the datasheets:

Epoxy resin density: 1.198 g/cm^3, Flax density: 1.35 g/cm^3, Glass density: 2.54 g/cm^3.

In this way, both matrix and fibre volume fractions may be independently calculated. This *ASTM D3171* method was also applied to the GRP laminate (12 specimens for both laminates) giving the following results:

FFRP FVF 0.37

GRP FVF 0.50

Thus giving confidence in the results due to the practically equal values obtained for both the GRP FVF using the different methods.

To give an additional comparison of the unaged materials, ISO 527-1,4 (2012) tensile tests along the longitudinal direction were also conducted:

GRP (7 specimens):

Strength 437 ± 30 MPa, Modulus 25.2 ± 1.1 GPa

FFRP (5 specimens):

Strength 112 ± 4 MPa, Modulus 8.5 ± 0.3 GPa.

2.2 *Hygrothermal ageing*

Three temperatures were selected for the hygrothermal ageing in order to investigate distinct ageing environments up to easily conceivable summer or tropical deck temperatures, but below the cured resin glass transition temperature:

20 ± 2 °C, 40 ± 2 °C, and 60 ± 2 °C.

The 340 mm by 264 mm plates were immersed inside plastic boxes containing water. The use of sea water was considered, but was discounted at this stage in order to decouple any durability effects of water absorption from the inevitable and difficult to control fluctuations in the complex composition of seawater under hygrothermal ageing conditions. Further, this aspect is not at all straightforward, for example concentrations of natural salts and other chemicals (from the composite laminate itself, paints, fuel etc) will vary wildly between, and within, cases (for example at the outer hull and in the bilges).

The currently reported phase of the work fitted into the timescale of an MSc thesis and hence withdrawal of specimens for durability testing was carried out after 1 and 3 months' ageing. However, enough additional laminated material was produced and immersed to allow removal and durability testing after 6, 12 and, depending on results, 24 months).

Specimens were tested 'wet'; defined here as immediate immersion in water at room temperature on removal of the aged plates from the hygrothermal ageing bath, being only removed from this water for CNC router cutting into individual specimens (taking a maximum of 1 hour) and for removal for testing (after wiping dry with a cloth) – which occurred at the latest on the morning after specimen cutting if this cutting occurred in the afternoon.

To try to avoid wicking along the plate edges (which should not, in theory anyway, occur in service) and hence to promote equivalent (for both materials) unidirectional diffusion through the panel faces and thence the thickness of the laminate, the cut edges of the immersed plates were coated with a two-part epoxy adhesive (*Icosit K-101 N)*.

To measure the rate of this moisture diffusion into the laminates, a set of 12 (70 mm by 70 mm epoxy adhesive edge-sealed) control specimens for both materials was also placed in the heated ageing water baths (4 of each at each of the three temperatures for each material). These specimens were accurately weighed before immersion and then daily for the initial period, and then progressively less often as their water mass uptake stabilised for a period up to 3720 h. These measurements are ongoing and will continue up to the end of the planned future work. The relative mass uptake, M_t (%), was defined and calculated via Equation 1 (Scida et al., 2013),

$$M_t = \frac{w_t - w_0}{w_0} \times 100 \tag{1}$$

where w_t is the weight of the "wet" specimen at time t (g); w_0 is the weight of the unaged specimen (g).

2.3 *Durability testing*

2.3.1 *Flexural properties*

Flexural stiffness and strength were determined via tests based on the *ASTM D7264-07* standard. Seven specimens were tested in each case with a *Instron 5982-100kN* universal testing machine (UTM) using a three-point bending setup with a span of 132 mm, loading nose and support roller radii of 5 and 3 mm, respectively, and a loading rate of 20 mm/min.

The stress and strain at the outer surface at mid-span, σ (MPa) and ε (mm/mm), respectively, were determined as per the equations found in the *ASTM D7264-07* standard. The flexural chord modulus of elasticity, E_f^{chord} (MPa), was estimated between strains of 0.003 and 0.005.

2.3.2 *Short-beam strength*

The short beam tests were based on the *ASTM D2344-00* standard. Seven specimens were tested in each case with the same UTM using a three-point bending setup with a span of 16.5 mm, loading nose and support roller radii of 5 and 2 mm, respectively, and a loading rate of 2 mm/min.

2.3.3 *Quasi-static indentation*

Quasi-static (QS) indentation tests were based on standards *ASTM D7136/D7136M-15* and *ASTM D6264-12,* and were performed on a calibrated computer controlled servo-hydraulic mechanical test machine.

70 mm by 70 mm test specimens were impacted on a 50 mm internal diameter circular support – simply supported to avoid any disparities in the achievable level of clamping between static and dynamic tests that would occur in a clamped configuration. In all cases, five specimens were tested for both QS indentation and low-velocity impact. A cylindrical indenter with a 10 mm diameter hemispherical tip was used, both to try to ensure that the damage area did not propagate to the specimen edges, and to remain true to typical marine impact scenarios where impacts with flat objects are rare. A constant rate of 0.5 mm/s was used and absorbed energy, *AE* (J), was determined by integrating the force-displacement data.

Perforation is both an important response and yet is surprisingly difficult to define, the most obvious definitions – the creation of a 'hole', or the complete passing of the hemispherical part of the indenter, through the laminate are actually practically impossible to define, identify and measure. Hence, an easily and consistently identifiable point in the force-displacement data corresponding to a repeatable degree of 'perforation' was used as a definition here, as was done in previous work (Sutherland et al., 2017). For these materials, a clear peak in the force-displacement plot is followed by a significant drop in force until this levels out into a 'plateau' when the indenter's hemispherical tip has passed through the specimen (see Figure 4). The start of this plateau, however is not well defined and its exact form a source of variability. Hence, in order to accurately and repeatedly define this onset of 'perforation', a given and relevant indenter displacement is taken. This displacement is constant for a given material, but to define the same degree of perforation one needs to take on different values for different materials (and ageing periods). This approach makes sense practically since 'perforation' will occur for a given material at a given indenter displacement. Values of 13 and 15 mm for glass and flax fibre laminates (the higher latter value due to the lower stiffness of flax laminates) were found to give equivalent points in the force–displacement plots for both laminates and this consistent approach minimized variations in absorbed energy measured at perforation.

2.3.4 *Low-velocity impact*

The low-velocity impact tests were based on the standard *ASTM D7136/D7136M-15* and were performed on a Rosand IFW5 Instrumented Drop Weight machine (see Sutherland and Guedes Soares (2012) for more details). To ensure valid comparisons between QS indentation and these low-velocity impact tests, identical test setups were employed by using the same specimen dimensions and physical support and indenter (in this case 'impactor').

The QS AE results obtained here were used to estimate the impact incident kinetic energies (IKE's) required for the same degrees of damage as seen in the QS tests. This was achieved using the method proposed in Sutherland and Guedes Soares (2012) which showed that the first damage (usually start of delamination and/or contact damage) occurs at the same force/displacement/AE, but that the AE to perforation energy for impact and QS tests are related via a simple proportional ratio, or 'dynamic scaling factor', and as outlined in Castilho et al (2015), Sutherland et al (2017) & Garrido et al (2021). This factor for GRP was estimated in this previous work at between 1.5 and 2, and hence a value of 1.6 was taken here initially and then refined using the first impact results obtained.

Since GRP impacted specimens exhibited very similar behaviour (in terms of displacement) up to 'perforation' (as defined above) as was seen for the equivalent QS tests, the same displacement value of 13 mm was defined as the onset of impact perforation. However, for FFRP the displacements required to achieve this level of perforation for impact were seen to be higher than the equivalent QS values, and hence a value of 20 mm (versus 15 mm for QS) was defined for the impact of the flax laminates.

3 RESULTS AND DISCUSSION

3.1 *Mass uptake*

The mass increase was significantly higher for FFRP than for GRP as shown in Figure 1. This is mainly due to the far more hydrophilic nature of the flax fibres. Very significant mass uptake occurred for FFRP during the first hours of immersion (2.5% in the first 96 h at 20 °C, versus 0.1% for the GFRP), which was accompanied by a noticeable thickness increases (~7%, versus <1% for the GRP). During the 3720 hours of mass measurements only the FFRP aged at 60 °C had apparently reached saturation at a mass uptake of about 12%; the mass of the laminates aged in the other environments were absorbing water. It should also be noted that at the end of the immersion period considered here the water uptake for FFRP laminates even at 20 °C was approximately 75% of that at 60 °C whilst for GRP this figure was only around 20%. Hence, water absorption is severe for FFRP even at relatively low temperatures, whereas for GRP water uptake exhibited a much more pronounced temperature dependence.

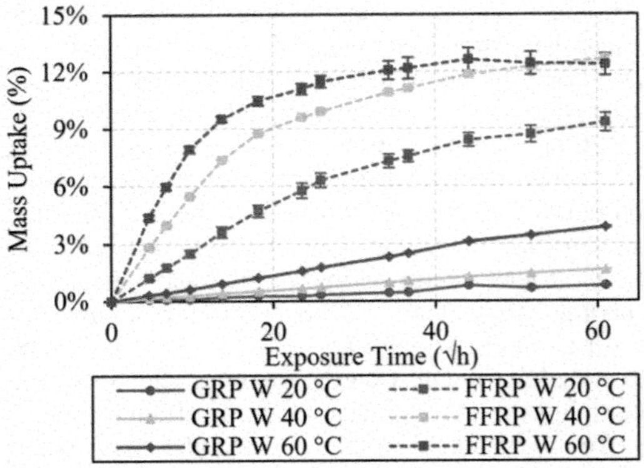

Figure 1. Mass uptake for FFRP and GRP laminates.

GRP mass uptake was approximately linear with respect to the square root of time, following the Fickian diffusion model. However, since after the 3720 hours immersion saturation had not yet been achieved these results have not yet been fitted to this model. For the FFRP an initial higher mass uptake was followed by a gradual reduction in the rate of mass increase. The first stage could well be dominated by water absorption into the fibres, which progressively (but relatively quickly) saturate leading to the second stage, where absorption in the matrix becomes more dominant, but further work would be required to confirm this or otherwise.

3.2 *Flexural behaviour*

GRP exhibited consistent failure modes in flexure for unaged and aged specimens across all ageing conditions. Failure firstly occurred in compression of the top ply, as shown in Figure 2(a), followed by tensile failure of the lower ply. For FFRP the unaged specimens failed in tension at the lower ply at large deflections (around 34 mm) as shown in Figure 2(b) and aged FFRP suffered even larger deflections and rupture was not achieved.

The flexural stress-strain curves shown in Figure 3(a) show these different behaviours. The GRP response was approximately linear up to the first ply failure, which was immediately preceded by a slight stiffness reduction and followed by a clear drop in force. The aged GRP suffered reductions in stiffness and maximum stress with increasing ageing temperature, as illustrated in Figure 3(b) by representative stress-strain curves for specimens aged for 3 months. The aged FFRP exhibited severe reductions in flexural chord modulus of elasticity. These likely occurred mostly in the initial days of

(a) (b) (c) (d)

Figure 2. Typical failure modes for the flexural tests on (a) GRP and (b) FFRP, and short-beam tests on (c) GRP and (d) FFRP.

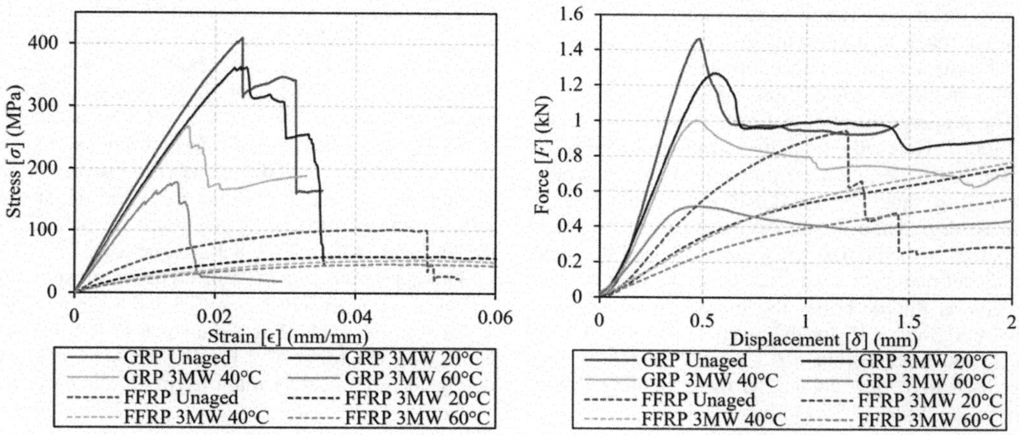

(a) (b)

Figure 3. GRP and FFRP aged 3 months: (a) flexural stress-strain curves and (b) short-beam force-displacement curves.

hygrothermal ageing, with similar retention between 1 and 3 months for all ageing temperatures.

The average flexural chord modulus of elasticity and maximum flexural stress were 20.0 ± 0.7 GPa and 382.2 ± 33.5 MPa for the unaged GRP and 6.1 ± 0.2 GPa and 103.8 ± 2.8 MPa for the unaged FFRP, respectively. The property retention values after ageing are presented in Table 1.

3.3 *Short-beam strength*

Valid failure modes for interlaminar shear strength (ILSS) were only obtained for the GRP (Figure 2(c)), with the unaged FFRP exhibiting tensile failure and the aged FFRP experiencing large deflections followed by crushing at the supports (invalid failures), as shown in Figure 2(d). For this reason, it was only possible to obtain ILSS estimates for the GRP.

The typical force-displacement curves for the GRP and FFRP are displayed in Figure 3(b) where representative curves for the 3-month ageing results are displayed for clarity (curves for other ageing periods exhibited similar behaviour). Even though the FFRP did not suffer valid failure modes, a clear and significant decrease in the apparent stiffness of the material was seen with hygrothermal ageing. This trend was also observed by Cuinat-Guerraz et al. (2016), where FFRP laminates with a moisture content higher than 3% did not present clear failure in short-beam tests due to excessive deflections.

The ILSS value obtained for the unaged GRP was 33.7 ± 0.7 MPa and the corresponding ILSS retention is shown in Table 1. A severe loss in ILSS retention was seen, especially for the GRP aged at 60 °C. This was likely caused by the significant deterioration of the matrix's mechanical properties due to severe plasticisation. Similar results have been reported in Ray (2006), where the hygrothermal ageing of glass/epoxy laminates led to a reduction of 40% in the ILSS for laminates aged for 150 h at 70 °C at a relative humidity of 95%.

3.4 *Quasi-static indentation tests*

The QS indentation tests force-displacement curves are shown in Figure 4 and the respective failure modes in Figure 5. As expected, the first GRP damage was delamination (Point A in Figure 4), reflected by the sudden force-displacement stiffness reduction accompanied and a visible whitening in the area around the indenter contact point. After this point, stiffness increased again for the unaged GRP, but not for the aged laminates, as shown in Figure 4. For the FFRP laminates delamination was seen in the specimens but its effects were not apparent in the force-displacement plots.

At higher displacements both materials experienced initial fibre failure, evidenced by a small force drops against a still ascending trend (Point B). Generally, this point was close to the maximum force for the GRP. Fibre damage was more evident on the back face of specimens, as shown in Figure 5.

Both delamination and fibre failure were more significant in unaged specimens or in those aged at lower temperatures. The final damage stage, perforation, was, as previously mentioned, defined here at indenter displacements of 13 mm for the GRP and 15 mm for the FFRP, and corresponded to an asymptotic approach to a constant low force as the cylindrical indenter passed through a hole of its own making in the specimen (Point D).

The property retentions of each material are given in Table 1. Generally, property retentions were higher for the FFRP, despite these properties being generally lower than those of the GRP (even with ageing induced degradation), with the exception of the longest period at the highest temperature. One of the most significant parameters was the *AE* at the defined perforation as this describes the impact absorbing capacity. Perforation *AE* for the unaged GRP was 47.2 ± 1.1 J, and after ageing for 3 months at 60 °C this decreased to 19.9 ± 0.4 J. The corresponding values for FFRP were 23.9 ± 0.7 J and 24.5 ± 0.4 J, with this increase due to the effects of the higher forces seen with ageing just exceeding those of the reduced stiffness exhibited. Both effects are most likely due to fibre and/or matrix plasticisation reducing the degree of damage. This could be due to arresting resin cracking (delamination) and/or reducing brittle fibre breakage, but although these results provide valuable clues, it is impossible to explain this behaviour in terms of damage modes with the available data.

For force and *AE* at initial fibre failure, the unaged GRP achieved values of 6.8 ± 0.2 kN and 15.4 ± 1.3 J, whereas for FFRP these values were 1.9 ± 0.1 kN and 2.5 ± 0.2 J. These values both actually increased with ageing for FFRP; for three months at 60 °C retentions reached 126% (2.4 ± 0.1 kN) and 254% (6.4 ± 0.7 J) for force and *AE* at fibre failure, respectively. Conversely, the GRP experienced severe deterioration with ageing in both the force and *AE* at fibre failure, with corresponding retentions of 34% (2.3 ± 0.2 kN) and 28% (4.2 ± 0.8 J), respectively.

The maximum force (Point C) and corresponding *AE* for the unaged GRP were 7.3 ± 0.4 kN and 20.0 ± 2.6 J, respectively. For the unaged FFRP, an average maximum force of 2.1 ± 0.1 kN was obtained; due to the significant variability in the displacement

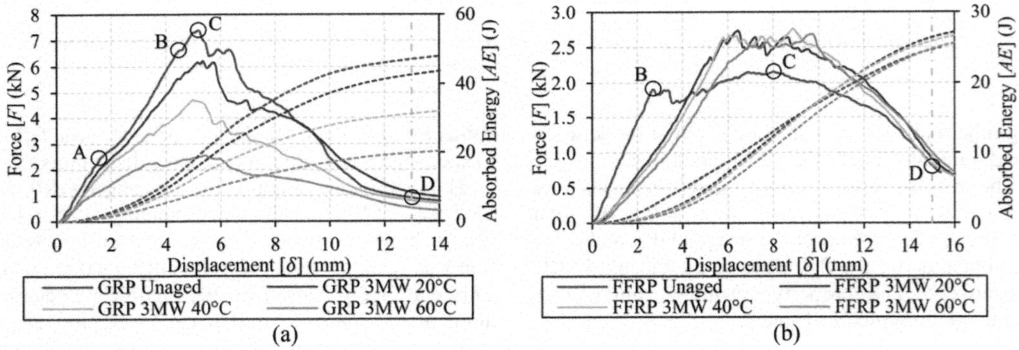

Figure 4. Force-displacement curves (solid lines) and *AE*-displacement curves (dashed lines) from QS indentation tests on (a) GRP and (b) FFRP aged 3 months and tested "wet"; perforation is marked with a yellow dot-dash line.

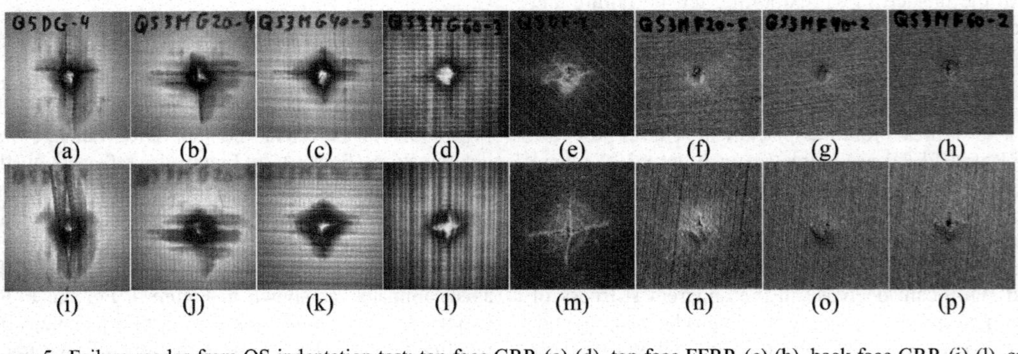

Figure 5. Failure modes from QS indentation test: top face GRP, (a)-(d), top face FFRP, (e)-(h), back face GRP, (i)-(l), and top face FFRP, (m)-(p), for unaged specimens and aged at 20 °C, 40 °C, and 60 °C, respectively.

at which the maximum load occurred in the fibre failure plateau for FFRP, *AE* estimates were not taken from the obtained results. With ageing, similar trends as those described above for force at initial fibre failure were found for these parameters.

3.5 *Low-velocity impact tests*

In general, and as expected, the low-velocity impact force-displacement curves were similar in form to those of the QS indentation tests and are presented in Figure 6. Impact failure modes are illustrated in Figure 7 and property retentions with ageing are summarised in Table 1.

Impact *AE* at perforation (as defined previously) was 68.9 ± 5.7 J for the GRP, decreasing to 30.7 ± 3.4 J after ageing for 3 months at 60 °C. The corresponding unaged and aged FFRP values were 23.8 ± 2.0 J and 24.8 ± 1.3 J, respectively. The plasticising effect of the flax fibres and matrix can explain the increment seen for FFRP. The 'dynamic scaling factor' ratio between impact and QS *AE* at perforation values were very different for the two materials: approximately 1.5 for the GRP (agreeing with previous work) yet only 1.0 for the FFRP.

The unaged GRP maximum impact force and corresponding *AE* at this force were 11.3 ± 0.6 kN and 36.2 ± 5.6 J, respectively. For unaged FFRP the maximum force was 2.3 ± 0.1 kN, again very similar to that for the equivalent QS indentation tests (consistent values of the *AE* at maximum force for FFRP were unavailable, as described previously).

Hence, a rate effect for maximum force was seen for GRP but not for FFRP, for both maximum force and for perforation. The hypothesis that some sort of flax fibre 'plasticization' occurs with aging (and these fibres are far less brittle in their natural very hydrated state) and the fact that no rate effect was seen for FFRP in the onset of fibre controlled penetration (around maximum force) and perforation is consistent with the hypothesis made in Sutherland, L.S., Guedes Soares, C., 2012 that this rate effect in GRP is connected to the speed of propagation of the (brittle) glass fibre failure modes when compared to the loading rate - i.e. at impact loading rates the brittle fibre (progressive bundle) failure simply does not have enough time to propagate as far as it can at QS loading rates.

As expected, force oscillations induced by the test machine and/or vibrations around the material contact stiffness seen in the low-velocity impact test results made identification (or even confirmation) of features of the force-displacement plots corresponding to initial matrix delamination and initial fibre failure (identifiable from the corresponding QS curves) impossible.

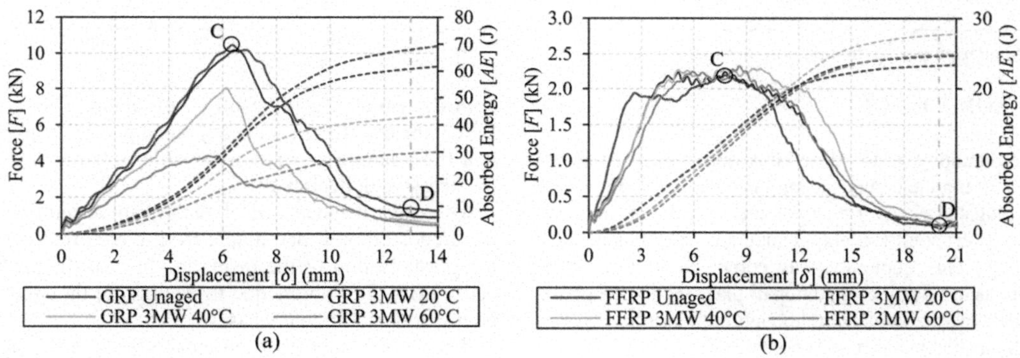

Figure 6. Force-displacement curves (solid lines) and *AE*-displacement curves (dashed lines) from low-velocity impact tests on (a) GRP and (b) FFRP aged 3 months and tested "wet"; perforation is marked with a yellow dot-dash line.

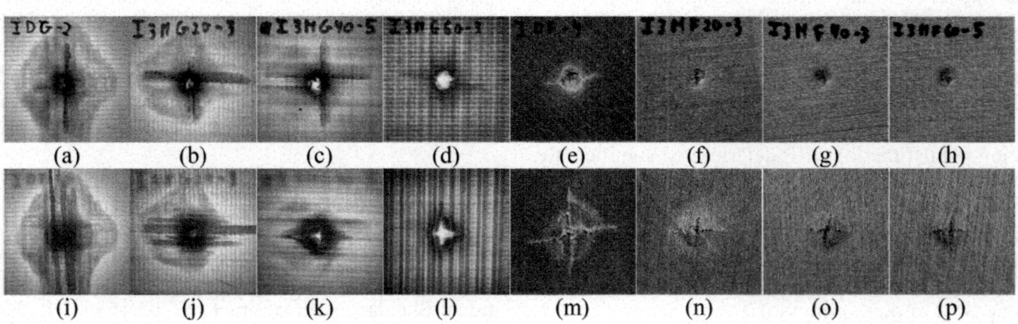

Figure 7. Failure modes from low-velocity impact test: top face GRP, (a)-(d), top face FFRP, (e)-(h), back face GRP, (i)-(l), and top face FFRP, (m)-(p), for unaged specimens and aged at 20 °C, 40 °C, and 60 °C, respectively.

Table 1. Property retention (%) after ageing for GRP and FFRP specimens.

		Aged at 20 °C		Aged at 40 °C		Aged at 60 °C	
Material	Variable	1M	3M	1M	3M	1M	3M
GRP	Flexural Maximum Stress	85%	99%	72%	66%	51%	42%
	Flexural Chord Modulus of Elasticity	100%	96%	93%	88%	86%	73%
	ILSS	97%	91%	81%	68%	48%	32%
	QS indentation AE at Perforation	90%	88%	77%	60%	47%	42%
	QS indentation Maximum Force	89%	85%	79%	59%	43%	35%
	QS indentation AE at Maximum Force	100%	89%	75%	57%	43%	39%
	QS indentation Initial Fibre Damage Force	92%	86%	83%	50%	40%	34%
	QS indentation AE at Initial Fibre Damage	91%	86%	88%	45%	32%	28%
	QS indentation AE at Initial Matrix Damage	83%	73%	75%	65%	37%	26%
	QS indentation Initial Matrix Damage Force	91%	84%	85%	73%	55%	38%
	Low-velocity impact AE at Perforation	94%	89%	83%	63%	46%	45%
	Low-velocity impact Maximum Force	94%	90%	86%	67%	43%	39%
	Low-velocity impact AE at Maximum Force	93%	88%	88%	66%	39%	39%
FFRP	Flexural Maximum Stress	59%	56%	50%	52%	45%	44%
	Flexural Chord Modulus of Elasticity	52%	50%	37%	38%	38%	36%
	QS indentation AE at Perforation	104%	107%	107%	107%	102%	102%
	QS indentation Maximum Force	122%	123%	130%	127%	134%	125%
	QS indentation Initial Fibre Damage Force	131%	123%	134%	133%	137%	126%
	QS indentation AE at Initial Fibre Damage	205%	208%	263%	240%	260%	254%
	Low-velocity impact AE at Perforation	117%	107%	115%	111%	112%	104%
	Low-velocity impact Maximum Force	102%	102%	104%	106%	105%	100%

NB better property retention does not mean better properties

4 CONCLUSION

The durability of an infused flax fibre reinforced plastic (FFRP) marine composite laminate has been extensively investigated via hygrothermal ageing and impact, quasi-static (QS) point loading, flexural and interlaminar shear testing. This durability was compared to that of an equivalent thickness 'traditional' glass reinforced plastic (GRP) marine laminate (although both used a commercially available bio-based epoxy resin).

FFRP resin content was 25% higher than that of GRP, and FFRP initial mechanical properties were much lower. Hence, for equivalent structural designs FFP laminates would need to be significantly thicker and heavier.

Hydrothermal ageing produced complex trends dependent on which property/response was considered, but overall the performance of FFRP under impact and QS indentation actually improved with ageing, whilst that of GRP deteriorated significantly. However, it is imperative to note that the impact and indentation performance of the unaged FFRP was substantially worse than that of the unaged GRP, and that only under the most severe 60 °C ageing conditions did the GRP performance drop to the much lower level of that of the FFRP.

In terms of flexure, FFRP exhibited far lower properties compared to GRP, and the performance of both deteriorated significantly with ageing. Loss of FFRP shear stiffness with any ageing was so great that ILSS failure was not achievable in the tests due to excessive deflections.

The exact mechanisms of each of these multifarious trends are not identifiable from the current data, but plasticisation of the resin, and also of the artificially dried flax fibres, due to ageing are thought to be instrumental in many cases.

The results show that sustainability-driven design decisions concerning flax composites in a marine environment must take into account the fact that they do not perform at all well under flexure, impact and point loadings. This may well potentially more than negate any apparent sustainability advantages of flax, for example where product resistance and durability is considered. However, ageing under severe conditions may temper FFRP's disadvantages under impact and point loadings, if not under flexure.

ACKNOWLEDGEMENTS

L.S. Sutherland was funded by Fundação para a Ciência e a Tecnologia (FCT), under the Transitional Standard with contract IST-ID/164/2018. This work contributes to the Strategic Research Plan of the Centre for Marine Technology and Ocean

Engineering (CENTEC), which is financed by FCT under contract UIDB/UIDP/00134/2020.

The support from FCT through project PTDC/ECI-EGC/29597/2017 (EcoComposite) and through funding of the research unit CERIS (UIDB/04625/2020) is gratefully acknowledged.

REFERENCES

Castegnaro, S., Gomiero, C., Battisti, C., Poli, M., Basile, M., Barucco, P., Pizzarello, U., Quaresimin, M., Lazzaretto, A., 2017. A bio-composite racing sailboat: Materials selection, design, manufacturing and sailing. Ocean Eng. 133, 142–150.

Castilho, T., Sutherland, L.S., Guedes Soares, C., 2015. Impact resistance of marine sandwich composites, in: Guedes Soares, C., Santos, T.A. (Eds.), Maritime Technology and Engineering. Taylor & Francis Group, London, UK, pp. 607–618.

Cuinat-Guerraz, N., Dumont, M.-J., Hubert, P., 2016. Environmental resistance of flax/bio-based epoxy and flax/polyurethane composites manufactured by resin transfer moulding. Compos. Part Appl. Sci. Manuf. 88, 140–147.

De Fazio, D., Boccarusso, L., Formisano, A., Viscusi, A., Durante, M., 2023. A Review on the Recycling Technologies of Fibre-Reinforced Plastic (FRP) Materials Used in Industrial Fields. J. Mar. Sci. Eng. 11, 851.

Duflou, J.R., Yelin, D., Van Acker, K., Dewulf, W., 2014. Comparative impact assessment for flax fibre versus conventional glass fibre reinforced composites: Are bio-based reinforcement materials the way to go?. CIRP Ann. 63, 45–48.

Espert, A., Vilaplana, F., Karlsson, S., 2004. Comparison of water absorption in natural cellulosic fibres from wood and one-year crops in polypropylene composites and its influence on their mechanical properties. Compos. Part Appl. Sci. Manuf. 35, 1267–1276.

Garrido, M., Teixeira, R., Correia, J., Sutherland, L., 2021. Quasi-static indentation and impact in glass-fibre reinforced polymer sandwich panels for civil and ocean engineering applications. J. Sandw. Struct. Mater. 23, 194–221. https://doi.org/10.1177/1099636219830134

Li, Y., Xue, B., 2016. Hydrothermal ageing mechanisms of unidirectional flax fabric reinforced epoxy composites. Polym. Degrad. Stab. 126, 144–158.

Navaratnam, S., Selvaranjan, K., Jayasooriya, D., Rajeev, P., Sanjayan, J., 2023. Applications of natural and synthetic fiber reinforced polymer in infrastructure: A suitability assessment. J. Build. Eng. 66, 105835.

Ray, B.C., 2006. Temperature effect during humid ageing on interfaces of glass and carbon fibers reinforced epoxy composites. J. Colloid Interface Sci. 298, 111–117.

Rubino, F., Nisticò, A., Tucci, F., Carlone, P., 2020. Marine Application of Fiber Reinforced Composites: A Review. J. Mar. Sci. Eng. 8, 26.

Scida, D., Assarar, M., Poilâne, C., Ayad, R., 2013. Influence of hygrothermal ageing on the damage mechanisms of flax-fibre reinforced epoxy composite. Compos. Part B Eng. 48, 51–58.

Springer, G.S., Shen, C.-H., 1976. Moisture Absorption and Desorption of Composite Materials.

Squires, Q.M., 2022. Manufacturing a RHIB hull out of sustainable materials. Focusing on damage tolerance of flax fibre composites. MSc Thesis, Ship Science, University of Southampton, UK.

Sutherland, L.S., Guedes Soares, C., 2012. The use of quasi-static testing to obtain the low-velocity impact damage resistance of marine GRP laminates. Compos. Part B Eng. 43, 1459–1467.

Sutherland, L.S., Sá, M.F., Correia, J.R., Guedes Soares, C., Gomes, A., Silvestre, N., 2017. Impact response of pedestrian bridge multicellular pultruded GFRP deck panels. Compos. Struct. 171, 473–485.

Yan, L., Chouw, N., Jayaraman, K., 2014. Flax fibre and its composites – A review. Compos. Part B Eng. 56, 296–317.

Advances in Maritime Technology and Engineering – Guedes Soares & Santos (Eds)

Impact of uniform and grooving corrosion on hot-spot stresses of a T-shaped tubular joint

Y. Dong
Yantai Research Institute of Harbin Engineering University, Yantai, P.R. China
College of Shipbuilding Engineering, Harbin Engineering University, Harbin, P.R. China
HEU-UL International Joint Laboratory of Naval Architecture and Offshore Technology, Harbin, P.R. China

L.G. Liu, H.K. Yang & X. Liu
Yantai Research Institute of Harbin Engineering University, Yantai, P.R. China

Y. Garbatov
Centre for Marine Technology and Ocean Engineering (CENTEC), Instituto Superior Técnico, Universidade de Lisboa, Lisbon, Portugal
HEU-UL International Joint Laboratory of Naval Architecture and Offshore Technology, Harbin, P.R. China

ABSTRACT: Tubular joints are widely used in offshore structures and are prone to fatigue due to the high stress concentration. The fatigue life may be significantly reduced after the corrosion protection fails for tubular joints in a corrosive environment. The study uses the finite element method to investigate the effect of uniform and grooving corrosion on hot-spot stresses of a T-shaped tubular joint. The weld geometry is considered in the model, and the grooving corrosion is modelled, assuming that the corrosion rate of weld metal is 1.4 times that of the base metal. The results show that the Stress Concentration Factors (SCFs) increase with the corrosion depth for the cases of uniform corrosion. At the same time, the SCFs decrease with the increase of the corrosion depth for the instances of grooving corrosion. The complex stress distribution around the grooving corrosion area may invalidate the surface extrapolation method for determining hot-spot stresses.

1 INTRODUCTION

During their service life, ships and offshore structures are exposed to cyclic loading in aggressive environments. Fatigue and corrosion are the two most significant strength degradation phenomena that can reduce structural safety (Guedes Soares & Garbatov, 1999, Guedes Soares et al., 2006). Many studies have investigated the effect of both corrosion and fatigue.

The strength degradation process due to fatigue and corrosion is extraordinarily complex. Different forms of corrosion, e.g., general, pitting, grooving, and edge corrosion, occur after corrosion protective measures such as paint coatings or cathodic protection become ineffective (DNV, 2014). Fatigue cracks can start from the corrosion damage because the local stress is improved by the corrosion damage (Shojai et al., 2022). The thickness reduction and surface irregularities caused by corrosion can reduce the fatigue crack initiation life due to the increased local stress. The crack propagation life is also reduced due to the increased local stresses and the altered material's fatigue crack propagation properties (Moan & Ayala-Uraga, 2008, Dong et al., 2012, Han et al., 2019). In addition, the applied stress can increase the corrosion rate, and the high stress region around the weld leads to non-uniform corrosion degradation of welded joints, which reduces the fatigue life (Yang et al., 2016).

Design standards consider the effect of corrosion on fatigue strength. When calculating the local stress for fatigue strength assessment, the finite element model considers a certain amount of thickness reduction caused by corrosion. In the case of ships (DNV, 2021), it is assumed that the dominating part of the design fatigue life is associated with the in-air environment because of the effectiveness of the corrosion protection measures. In contrast, a minor part is considered corrosive. The in-air S-N curve can be used for all the environments, but the fatigue damage in a corrosive environment should be multiplied by a factor of 2. For offshore structures (DNVGL, 2019), the S-N curves in air, seawater with cathodic protection, and seawater with free corrosion are provided. The S-N curves for cathodic protection and free corrosion are derived from those in-air by reducing the S-N curves with the so-called environment

DOI: 10.1201/9781003508762-47

reduction factor, which is 3 for free corrosion. These values are based on a few fatigue tests performed on welded joints.

The approach specified in design standards relies on local stresses not altered by corrosion. At the same time, some studies modelled the effect of corrosion in more detail. Garbatov et al. (2002) examined the influence of uniform corrosion wastage on the stress concentration factors (SCFs) of typical ship structural details. They compared the fatigue damage for different corrosion models. Nguyen et al. (2013) performed spectrum fatigue analyses of a double-hull tanker structural detail considering corrosion wastage over time. The corrosion effect on the fatigue damage was explicitly considered through the time-varying hull girder section modulus to the deck. Yang et al. (2016) investigated the accelerated corrosion of highly stressed steel, and the time-dependent SCF as a function of corrosion deterioration was analysed considering the non-uniform corrosion of welded joints.

Moan and Ayala-Uraga (2008) established a reliability-based model for assessing deteriorating ships subjected to multiple environmental conditions. The model was based on a fracture mechanics formulation, and the corrosion-induced increased crack growth rate was considered in two aspects: (1) the increased stress range produced by the plate thinning (wastage) effect and (2) corrosion fatigue itself. Following the same strategy, Dong et al. (2012) performed the fatigue reliability assessment of welded multi-planar tubular joints of the support structure of a fixed jacket offshore wind turbine, considering the effect of corrosion and inspection. Han et al. (2019) also considered the two aspects in developing a probabilistic life-cycle management framework for ship structures subjected to coupled corrosion–fatigue deterioration processes. Based on limited experimental data, Garbatov and Guedes Soares (2017) performed a fatigue analysis of dented pipelines employing different failure criteria accounting for the dent size. The pipeline was assumed to be a series of segments modelled as randomly correlated elements.

Tubular joints are widely used in offshore structures and are prone to fatigue due to the high stress concentration. The SCF quantifies stress concertation as the ratio of the hot-spot stress of a point of the weld toe to the nominal stress, which depends on the joint type, geometric dimensions, and loading type. Extensive studies have been performed to develop parametric equations for calculating the SCFs of tubular joints. The Efthymiou equations are widely adopted in practice and incorporated in design codes (DNVGL, 2019). Reviews on the SCFs of tubular joints can be found in (Ahmadi et al., 2012, Hectors & De Waele, 2020, 2021).

The grooving corrosion is referred to as preferential weld line corrosion, with a localised line of material deterioration typically adjacent to welded joints long abutting stiffeners and at stiffener or plate butts or seams (DNV, 2014). Incidents probably caused by grooving corrosion in a bulk carrier were reported (DNV, 2013). Some studies have been performed to evaluate the effect of grooving corrosion on the ultimate strength of stiffened plate structures (Wang et al., 2015, Zhu et al., 2021). The impact of grooving corrosion on the load-bearing capacity of welded hollow spherical joints was investigated using the finite element method (Chen et al., 2020, 2021). The grooving corrosion may also happen around the brace-to-chord weld of tubular joints, which can affect the SCFs. However, there are no detailed studies on the modelling of the grooving corrosion in the SCF analyses of tubular joints and the effect of grooving corrosion on the SCF.

In the present study, the effect of corrosion on the SCFs of a T-shaped tubular joint is investigated using the finite element (FE) method. In addition to the uniform corrosion, the grooving corrosion around the weld is also considered. The grooving corrosion in the tubular joint for the calculation of SCFs is modelled, and the effect of grooving corrosion on the SCFs is quantified.

2 FE MODEL TO DETERMINE SCF

This section introduces the FE model of a T-shaped tubular joint for the SCF calculation. The main feature of the model is that the weld geometry is constructed, which is believed to be more realistic (Hectors & De Waele, 2021). The uniform and grooving corrosion will be introduced in the next section.

The geometry of the T-shaped tubular joint is shown in Figure 1. The geometric ratios are within the validity range of the Efthymiou equations recommended (DNVGL, 2019).

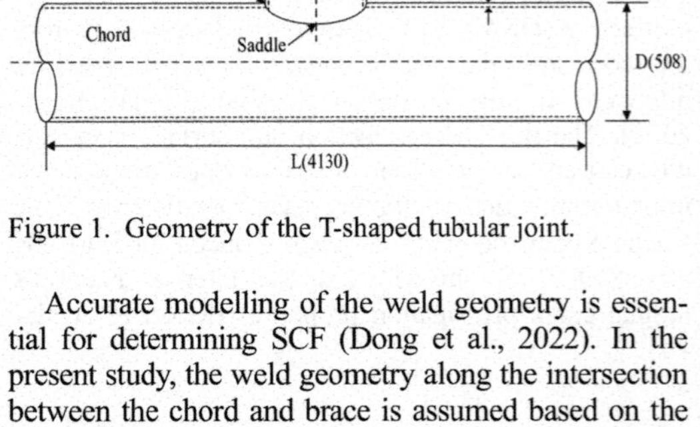

Figure 1. Geometry of the T-shaped tubular joint.

Accurate modelling of the weld geometry is essential for determining SCF (Dong et al., 2022). In the present study, the weld geometry along the intersection between the chord and brace is assumed based on the specifications as stipulated in (AWS, 2020). The dihedral angle ψ is the angle between the chord and brace surface, which varies along the intersection curve. For the T-shaped tubular joint, ψ goes from 90° at the crown to $\pi - \cos^{-1}(\beta) \approx 143°$ at the saddle (Ahmadi et al., 2012). The weld geometry specified by AWS is closely related to ψ. Prequalified joint dimensions and groove angle for the tubular T joint are shown in Table 1. The typical details are shown in Figure 2.

Table 1. Prequalified joint dimensions and groove angle for the tubular T joint with 90°≤ψ≤143°.

Parameter	Range
End preparation angle, ω	From 10° (or 45° for ψ>105°) to 90°
Root opening, R	From 2 mm to 6 mm
Root face, c	From 0 to 2 mm
Joint included angle, φ	From 37.5° to 60° for ψ≤105° Not less than 37.5° for ψ>105°
Weld leg size, F	From 0 to t/2 as ψ varies from 135° to 90°
Weld thickness, t_w	Not less than t

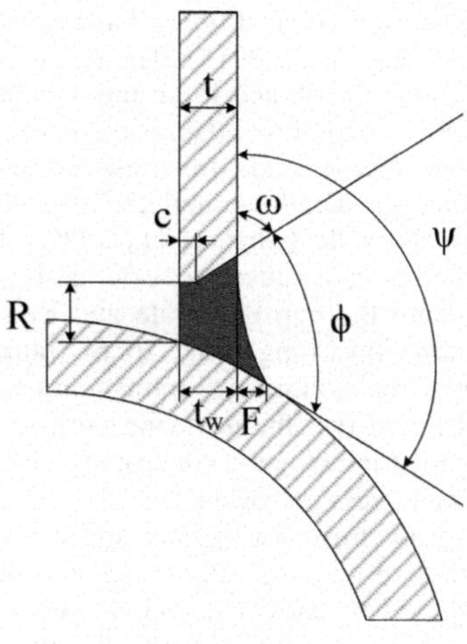

Figure 2. Joint details of the T-shaped tubular joint.

According to the requirement shown in Table 1, the following joint dimensions are assumed: ω=45°, R=2 mm, c=0 mm, F=t/2, and t_w=t. The joint included angle is the difference between ψ and ω, which satisfies the requirement. The weld leg size can be less than t/2 for ψ>90°. A conservative assumption that F has a constant value equal to t/2 along the intersection curve is adopted to simplify the modelling.

The FE analyses are performed using ANSYS (2012). The element type SOLID186 is used. The mesh condition is shown in Figure 3. The FE model is divided into several zones. Coarse mesh is used in the zone away from the weld, and dense mesh is employed in the zone, including the weld. A mesh convergent study is conducted to justify the mesh condition. The number of elements for the thickness of the chord and brace is 5 and 6, respectively. Eight hundred eighty finite elements are used along the intersection curve of the weld toe. The approximate mesh size around the weld toe is 3 mm. The Young's modulus and Poisson's ratio were 206 GPa and 0.3, respectively. Both chord ends are assumed to be fixed, and the three load cases acting on the brace are considered: axial loading, in-plane bending and out-of-plane bending.

The SCF is calculated as the ratio between the hot-spot stress at the weld toe and the nominal brace stress. The surface stress extrapolation method is generally used to determine the hot-spot stress, which requires the stress at two or three reference points whose distance from the weld toe is specified by standards (Hobbacher, 2015, DNVGL, 2019). In the present study, the reference points set by IIW (Hobbacher, 2015) are used to determine the hot-spot stress located at 0.4T and T from the weld toe. The mesh is designed so that the nodes are located on the reference points. For some exceptional cases, if the nodes are not found on the reference points, the stress of the reference point is obtained by linear interpolating stresses of the two closest nodes. The stress component perpendicular to the weld toe is obtained by defining a local coordinate system. The hot-spot stress is determined by linear extrapolation of the stresses of the two reference points.

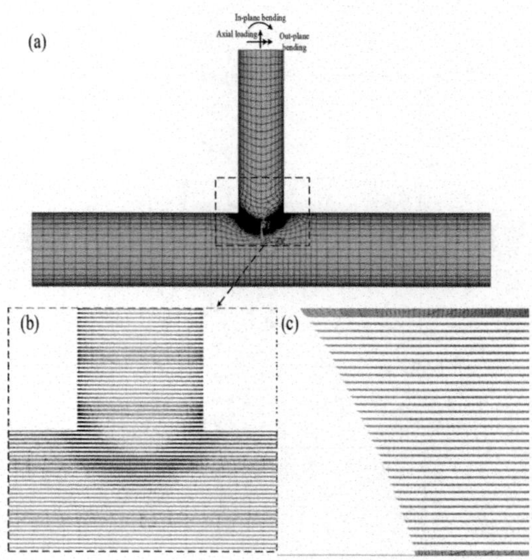

Figure 3. Mesh condition of the Tubular joint: (a) global view; (b) around the intersection; (c) around the weld toe.

To verify the FEM estimates, the SCF results are compared with the experimental results provided by Yeoh et al. (1995) and Soh (1997) and with the results of the Efthymiou equations (DNVGL, 2019). The comparisons are shown in Figure 4. The errors of the FEM results concerning the experimental results and Efthymiou equations results at the critical points are listed in Table 1. It can be seen that the results obtained using the FE method agree with those from experiments. The shape of the SCF distributions originating from the experimental measurements is captured very well. Discrepancies can be seen around the crown under axial loading and in-plane bending cases. The discrepancies are expected because the location of strain gauges concerning the weld toe and the weld geometry were not provided, and several inaccurate measurements were reported (Yeoh et al., 1995). Compared with the results from Efthymiou equations, the SCFs from the FE method are underestimated for the in-plane bending and out-plane bending cases. The underestimation may be attributed to modelling the weld geometry in the FE method, which is not considered in developing the Efthymiou equations.

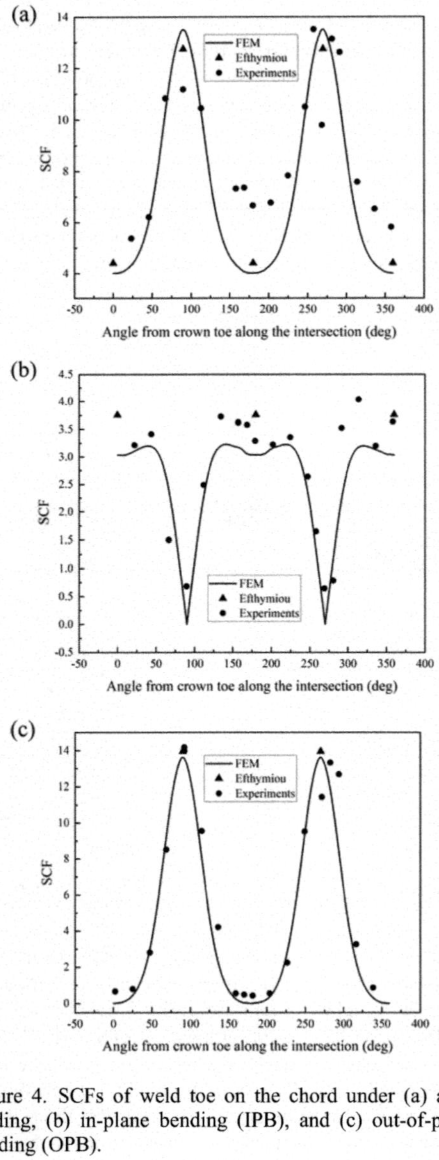

Figure 4. SCFs of weld toe on the chord under (a) axial loading, (b) in-plane bending (IPB), and (c) out-of-plane bending (OPB).

Table 2. Errors of FEM results at critical points.

Loading	Axial loading		IPB	OPB
Critical point	Saddle	Crown	Crown	Saddle
Errors to Experimental results (%)	16.00	27.03	2.76	3.31
Errors to Efthymiou equations (%)	5.88	9.13	19.33	2.44

3 CORROSION MODELLING

The grooving corrosion mechanism was investigated by Kato et al. (1978). Rapid heating and cooling of welding can produce a sulphide-enriched portion surrounding the MnS inclusions, which dissolves anodically concerning MnS inclusions and the base metal. Corrosion pits were initiated immediately on the MnS inclusions. They were then developed into grooving corrosion with the aid of the macro-cell between the anodic weld and cathodic base metal. In addition, the protective coating around the weld may not be as good as that on the base metal because of weld defects, which can lead to localised grooving corrosion (Wang et al., 2015). The grooving corrosion can also be attributed to an unfavourable local corrosive environment and the residual stress caused by welding, which leads to stress corrosion and greater corrosion depth around the weld than that away from the weld (Chen et al., 2020). The high stress at the hot spot caused by external loading can also accelerate the corrosion rate and lead to non-uniform corrosion (Yang et al., 2016). Tanaka et al. (2007) conducted continuous immersion tests in seawater and found that the grooving corrosion occurring on welded joints is the consequence of galvanic corrosion and stress corrosion.

The hot spot stress distributions and stress concentration factors of longitudinally stiffened welded deck joints subjected to corrosion and fatigue were analysed by Chakarov et al. (2008), Charakov et al. (2008). The stress concentration factors were defined alongside the transverse welds, accounting for locally non-uniform distributed corrosion degradation. Several stages of structural deterioration were evaluated, and the correlation between the status of degradation and stress concentration factor was established.

Only FEM results are used for the comparison between the SCF values affected by the two corrosion degradation types. The uniform and grooving corrosion are modelled, as shown in Figure 5. The outer surface of the tubular joint is uniformly removed due to uniform corrosion, whose corrosion depth is d. The modelling of grooving corrosion is mainly based on the work of Tanaka et al. (2007). It is assumed that the weld metal and base metal are subjected to uniform corrosion, but their corrosion rates are different; the corrosion rate of weld metal is

1.4 times that of the base metal, i.e. the corrosion depth of weld metal d_g equals 1.4d. The above assumptions are consistent with the observations (Tanaka et al., 2007).

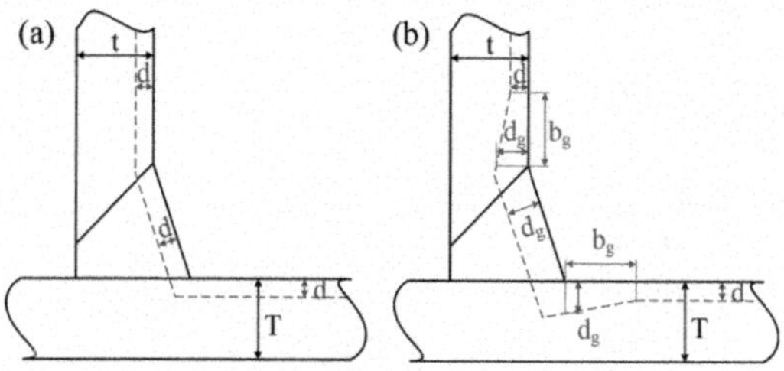

Figure 5. Modelling of (a) uniform and (b) grooving corrosion.

The corrosion rate of the affected zone is assumed to be linearly varied from the weld metal to the base metal, and the width of the heat-affected zone b_g is considered to be 3 mm. It has been shown that plastic deformation is the main reason for stress corrosion (Tanaka et al., 2007). Because plastic deformation is limited in practice, the contribution of stress corrosion to grooving corrosion is ignored. The reference points for determining the hot-spot stress concern the new weld toe and plate thickness due to corrosion. The FE models of uniform and grooving corrosion are shown in Figure 6.

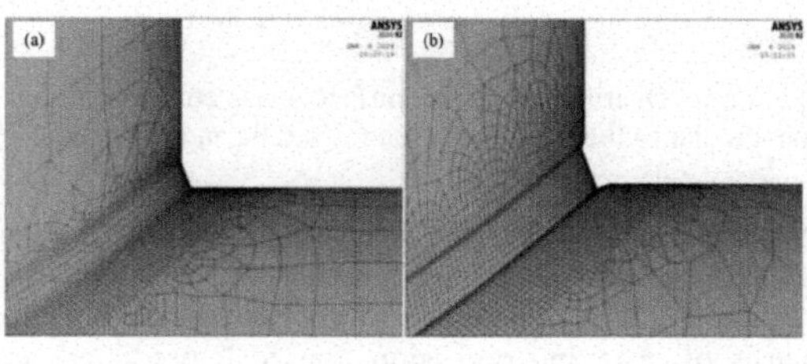

Figure 6. FE model of (a) uniform and (b) grooving corrosion.

4 RESULTS

The effect of uniform corrosion on the SCFs is investigated. The corrosion depth d ranges from 0 to 2.5 mm. The maximum uniform corrosion depth is determined based on the corrosion rate of 0.1 mm/year (Tanaka et al., 2007) and the service life of 25 years. The SCFs of the critical points as a function of uniform corrosion depth are shown in Figure 7. The SCFs are determined by the FE method and the Efthymiou equations, respectively.

The FE results show that the SCFs increase with d, and the effects of d on the SCFs are linear. The impact of uniform corrosion on the SCFs cannot be neglected. The effects of d on the SCFs depend on the loading condition and the location. For the case of axial loading, the increase of the SCF of the saddle is about 18% when d varies from 0 to 2.5 mm. For the case of IPB, d has a minor effect on the SCF of the crown.

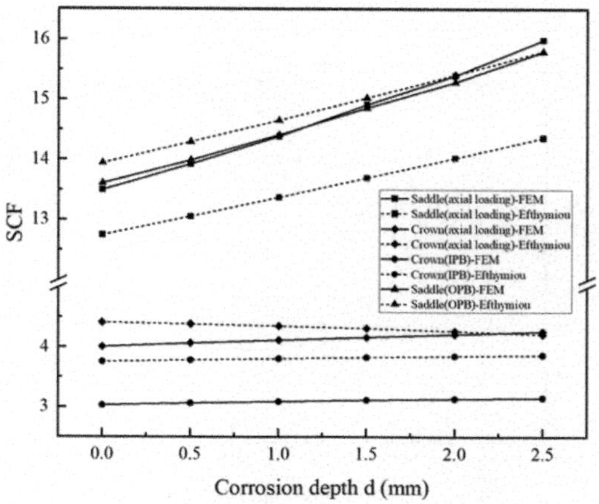

Figure 7. SCFs as a function of uniform corrosion depth.

Although the results of the Efthymiou equations deviate from those of the FE method, similar trends with the corrosion depth are observed. For the case of axial loading, the SCF of the crown slightly decreases with the increase of d, which is inconsistent with the FE results. The other SCFs are all increased with d. Compared with the FE results, the Efthymiou equations underestimate the increase of the SCFs for the saddle under axial loading and OPB when d becomes large. The difference in the FE and Efthymiou equations results may be partly attributed to the different thinning strategies of the two methods. In the FE method, it is assumed that the corrosion does not happen in the interior region, and the material inside the brace, including the part material located at the outer surface of the chord, remains unchanged. It is assumed that the thicknesses of the chord and brace are both reduced uniformly when using the Efthymiou equations.

For the grooving corroded tubular joint, only the loading condition of the axial loading is investigated. The SCFs as a function of corrosion depth are shown in Figure 8. The results of the grooving corrosion are compared with those of uniform corrosion. In contrast to the increase of SCF with d for uniform corrosion cases, the SCF decreases with the growth of d for grooving corrosion cases. The saddle's SCF decrease is about 60% when d varies from 0 to 2.5 mm. The SCFs are not continuously decreased with the increase of d. The decrease in the SCF for grooving corrosion cases is expected because the transition around the weld toe of grooving corroded welded joints is observed to be smoother than the original condition (Yang et al., 2016).

The distributions of the surface stress component perpendicular to the weld toe at the saddle and crown are shown in Figure 9. The stresses are normalised by dividing the nominal stress. In many previous studies of hot spot stress analyses, the stress usually increases when approaching the weld toe, and the stress of the first reference point (0.4T) is higher than the second reference

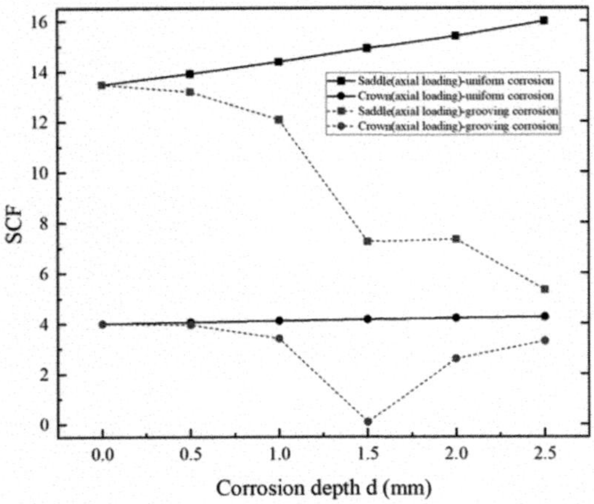

Figure 8. Comparison of SCFs for grooving corrosion and uniform corrosion.

point (1.4T). However, there is a decrease and subsequent increase of the stress when approaching the weld toe for grooving corrosion cases, as shown in Figure 9. Stress variation occurs around the grooving corroded area. As the corrosion depth increases, the first reference point is located in the groove corroded area, and the second reference point is located outside the groove. As a result, the stress of the first reference point is lower than that of the second reference point. Low SCFs are obtained for some grooving corrosion cases using surface extrapolation to determine the hot spot stress (Hobbacher, 2015).

The surface stress extrapolation method is applied, assuming that at a distance larger than 0.4T from the weld toe on the plate surface, the non-linear component has practically vanished, and the stress and strain increase almost linearly when approaching the weld toe (Hobbacher, 2015). It has been shown from Figure 9 that the stress distribution is complex and inconsistent with the assumption for grooving corrosion cases, which makes using the surface extrapolation method questionable. Further analyses may be conducted using the through-thickness linearisation method.

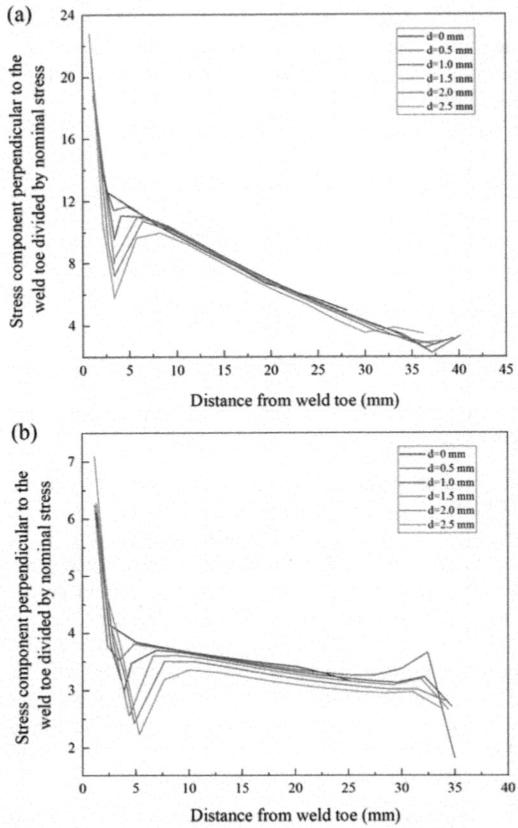

Figure 9. Distribution of the surface stress component perpendicular to the weld toe at the (a) saddle and (b) crown.

5 CONCLUSIONS

The effect of uniform and grooving corrosion on the SCFs of a T-shaped tubular joint is investigated using the FE method. The following conclusions can be established.

The FE results of the uncorroded tubular joint generally agree with the experimental results and results from empirical equations. The uniform corrosion increases the SCFs, and the influence depends on the loading condition and the location. The SCF variations due to uniform corrosion predicted by the FE method differ slightly from those indicated by empirical equations. The grooving corrosion is modelled such that the corrosion rate of weld metal is 1.4 times that of the base metal, and the corrosion rate of the heat-affected zone is assumed to be linearly varied from the weld metal to the base metal. For the cases of grooving corrosion, the SCF predicted by the surface extrapolation method decreases with the increase of the corrosion depth. The increasing surface stress suffers a decrease and subsequent increase when approaching the weld toe around the area of grooving corrosion, which may make the use of the surface extrapolation method questionable.

ACKNOWLEDGEMENTS

The present work is supported by the National Natural Science Foundation of China (Grant No. 52101350).

REFERENCES

Ahmadi, H., Lotfollahi-Yaghin, M. A., Yong-Bo, S. & Aminfar, M. H. 2012. Parametric Study and Formulation of Outer-Brace Geometric Stress Concentration Factors in Internally Ring-Stiffened Tubular Kt-Joints of Offshore Structures. *Applied Ocean Research*, 38, 74–91.

ANSYS 2012. Online Manuals.

AWS 2020. Structural Welding Code: Aws D1.1. Miami, US.

Chakarov, K., Garbatov, Y. & Guedes Soares, C. 2008. Hot Spot Stress and Stress Concentration Factors Due to Different Fabrication Imperfections in Deck Structures. *International Shipbuilding Progress*, 55, 47–62.

Charakov, K., Garbatov, Y. & Guedes Soares, C. 2008. Hot-Spot Stress Analysis of Deck Structures Subjected to Corrosion and Fatigue. *In*: Guedes Soares, C. & Kolev, P. (eds.) *Maritime Industry, Ocean Engineering and Coastal Resources*. London, UK: Taylor & Francis Group, 153-160.

Chen, H., Liu, H. & Chen, Z. 2020. Compressive Strength of Corroded Special-Shaped Welded Hollow Spherical Joints Based on Numerical Simulation. *Thin-Walled Structures*, 149, 106531.

Chen, H., Liu, H. & Chen, Z. 2021. Failure Pressure of Welded Hollow Spherical Joints Containing Grooving Corrosion Defects and Wall Reduction. *International Journal of Steel Structures*, 21, 35–51.

DNV. 2013. *Incident Information on Grooving Corrosion on Ship's Side* [Online]. Available: https://officerofthewatch.com/2013/11/29/incident-information-on-grooving-corrosion-on-ships-side/.

DNV 2014. Allowable Thickness Diminution for Hull Structure. Hovik: Det Norske Verit.

DNV 2021. Fatigue Assessment of Ship Structures. *Class Guidline DNV-CG-0129*. Hovik: Det Norske Veritas.

DNVGL 2019. Fatigue Design of Offshore Steel Structures. *Recommended Practice DNVGL-RP-C203*. Hovik: Det Norske Veritas.

Dong, W., Moan, T. & Gao, Z. 2012. Fatigue Reliability Analysis of the Jacket Support Structure for Offshore Wind Turbine Considering the Effect of Corrosion and Inspection. *Reliability Engineering & System Safety*, 106, 11–27.

Dong, Y., Garbatov, Y. & Guedes Soares, C. 2022. Recent Developments in Fatigue Assessment of Ships and Offshore Structures. *Journal of Marine Science and Application*, 21, 3–25.

Garbatov, Y. & Guedes Soares, C. 2017. Fatigue Reliability of Dented Pipeline Based on Limited Experimental Data. *International Journal of Pressure Vessels and Piping*, 155, 15–26.

Garbatov, Y., Rudan, S. & Guedes Soares, C. 2002. Fatigue Damage of Structural Joints Accounting for Non-linear Corrosion. *Journal of Ship Research*, 46, 289–298.

Guedes Soares, C. & Garbatov, Y. 1999. Reliability of Maintained Ship Hulls Subjected to Corrosion and Fatigue under Combined Loading. *Journal of Constructional Steel Research*, 52, 93–115.

Guedes Soares, C., Garbatov, Y., Zayed, A. & Wang, G. 2006. Non-Linear Corrosion Model for Immersed Steel Plates Accounting for Environmental Factors. *2005 SNAME Maritime Technology Conference and Expo and Ship Production Symposium*.

Han, X., Yang, D. Y. & Frangopol, D. M. 2019. Probabilistic Life-Cycle Management Framework for Ship Structures Subjected to Coupled Corrosion–Fatigue Deterioration Processes. *Journal of Structural Engineering*, 145, 04019116.

Hectors, K. & De Waele, W. 2020. A Numerical Framework for Determination of Stress Concentration Factor Distributions in Tubular Joints. *International Journal of Mechanical Sciences*, 174, 105511.

Hectors, K. & De Waele, W. 2021. Influence of Weld Geometry on Stress Concentration Factor Distributions in Tubular Joints. *Journal of constructional steel research*, 176, 106376.

Hobbacher, A. 2015. *Recommendations for Fatigue Design of Welded Joints and Components*, Springer.

Kato, C., Otoguro, Y., Kado, S. & Hisamatsu, Y. 1978. Grooving Corrosion in Electric Resistance Welded Steel Pipe in Sea Water. *Corrosion Science*, 18, 61–74.

Moan, T. & Ayala-Uraga, E. 2008. Reliability-Based Assessment of Deteriorating Ship Structures Operating in Multiple Sea Loading Climates. *Reliability Engineering & System Safety*, 93, 433–446.

Nguyen, K. T., Garbatov, Y. & Guedes Soares, C. 2013. Spectral Fatigue Damage Assessment of Tanker Deck Structural Detail Subjected to Time-Dependent Corrosion. *International Journal of Fatigue*, 48, 147–155.

Shojai, S., Schaumann, P., Braun, M. & Ehlers, S. 2022. Influence of Pitting Corrosion on the Fatigue Strength of Offshore Steel Structures Based on 3d Surface Scans. *International Journal of Fatigue*, 164, 107128.

Soh, A.-K. 1997. An Improved Procedure for the Determination of Hot Spot Stresses in Tubular Joints. *Fatigue & Fracture of Engineering Materials & Structures*, 20, 1709–1718.

Tanaka, Y., Nakai, T., Matsushita, H. & Niwa, T. 2007. Experimental Study on Grooving Corrosion Occurring on Welded Joints for Ship Structure. *Journal of the Japan Society of Naval Architects and Ocean Engineers*, 5, 261–268.

Wang, Y., Wharton, J. A. & Shenoi, R. A. 2015. Ultimate Strength Assessment of Steel Stiffened Plate Structures with Grooving Corrosion Damage. *Engineering Structures*, 94, 29–42.

Yang, S., Yang, H., Liu, G., Huang, Y. & Wang, L. 2016. Approach for Fatigue Damage Assessment of Welded Structure Considering Coupling Effect between Stress and Corrosion. *International Journal of Fatigue*, 88, 88–95.

Yeoh, S.-K., Soh, A.-K. & Soh, C.-K. 1995. Behaviour of Tubular T-Joints Subjected to Combined Loadings. *Journal of constructional steel research*, 32, 259–280.

Zhu, Y., Zhang, Y. & Du, F. 2021. Ultimate Strength of Hull Structural Stiffened Plate with Grooving Corrosion Damage under Uniaxial Compression. *Journal of Ship Research*, 65, 309–319.

Advances in Maritime Technology and Engineering – Guedes Soares & Santos (Eds)

Ultimate strength of corrugated-core sandwich plates under in-plane compression

M. Elsaka & C. Guedes Soares
Centre for Marine Technology and Ocean Engineering (CENTEC), Instituto Superior Técnico, Universidade de Lisboa, Lisbon, Portugal

ABSTRACT: To achieve a lightweight hull design, the conventional stiffened plate structure is replaced with a metal sandwich structure, reducing the weight of the hull while maintaining the original structure's load-bearing capabilities. The study aims to evaluate the load-bearing capabilities of I-type, V-type and U-type metal corrugated sandwich panels when subjected to uniaxial compression. This evaluation involves the implementation of buckling analysis and nonlinear finite element analysis using ANSYS software. The assessed the load-bearing capabilities of both conventional and corrugated-core sandwich panels under identical weights are compared. A significant increase in the hull bearing capacity based on corrugated-core sandwich panel is observed. The discernible differences in lightweight design and weight reduction effectiveness among various structural types are apparent. A comprehensive analysis of the fundamental principles governing lightweight characteristics in metal sandwich panels can serve as an informative guide for their incorporation into the design of hull structures.

1 INTRODUCTION

In recent years, there has been a growing adoption of lightweight sandwich structures across diverse industrial sectors, encompassing aerospace, automotive, and shipbuilding applications (D'Alessandro et al. 2012). Sandwich constructions can be classified as an all-metal, hybrid metal, and composite materials (Kujala et al. 2003, Kujala & Klanac 2005, SANDCORe 2013).

The metal sandwich panel structure is a type of metallic sandwich construction that consists of upper and lower metal panels connected through specific technologies to various structural elements such as corrugations, honeycombs, or trusses, forming a unified sandwich structure (Zenkert 1997, Carlsson & Kardomateas 2011).

The metal sandwich panel structure offers distinctive advantages compared to conventional steel buildings, encompassing efficient design optimization, weight reduction, robust load-bearing capacity, and impact resistance, (Metschkow 2006, Kozak 2009, Elsaka et al. 2019, Elsaka & Guedes Soares 2023).

Hence, there is a need to investigate the in-plane load-bearing capacity of metal sandwich plate structures. (Jelovica & Romanoff 2020) conducted analytical and numerical examinations to study the bifurcation buckling behaviour of sandwich panels incorporating unidirectional stiffeners. (Zaid et al. 2016) improved the vertical deformation resistance may be provided by a corrugated core and high strength-to-weight ratio. (Zhengjie et al. 2020) conducted buckling and non-linear finite element analyses on metal sandwich panels subjected to ultimate unidirectional compression, evaluating V-type and I-type metal corrugated structures in comparison with conventional stiffened plate structures, revealing that I-type metal corrugated panels achieve greater weight reduction than both V-type panels and stiffened panels.

In this study, novel steel sandwich panel stiffener cross-sections are utilized to improve the ultimate strength-to-weight ratio of ship structures. An examination is conducted to assess the uniaxial compression ultimate bearing characteristics of various metal corrugated sandwich panels - specifically, those of I-type, V-type, and U-type. The assessment of ultimate strength and maximum load-carrying capacity under in-plane compression is conducted through non-linear finite element analysis (NLFEA), utilizing the ANSYS software (version 2022 R2). Comparative evaluations are performed against conventional stiffened panels. The primary objective is to provide scientifically derived insights that serve as a reference for applying metal sandwich plates in ship hull structures.

2 CORE GEOMETRY

Corrugated cores combined with metal sandwich panels serve as a compelling industrial choice for

DOI: 10.1201/9781003508762-48

structural components due to their notable stiffness-to-mass ratio. Nevertheless, a typical practice involves simplifying the intricate structure of the core to an equivalent homogeneous material (Bartolozzi et al. 2013).

Over the course of recent decades, there has been an increase of proposed corrugated core configurations that draw inspiration from natural forms. These configurations include geometric shapes such as rectangles, triangles, and trapezoids, as well as mathematical functions like arc-tangent and sinusoidal patterns (Buannic et al. 2003, Metschkow 2006, Zhang et al. 2019).

This study chose three distinct core shapes for investigation, selected based on their facile production through laser welding. The unidirectional corrugated core configurations are Straight core, trapezoidal core, and triangular core as shown in Figure 1.

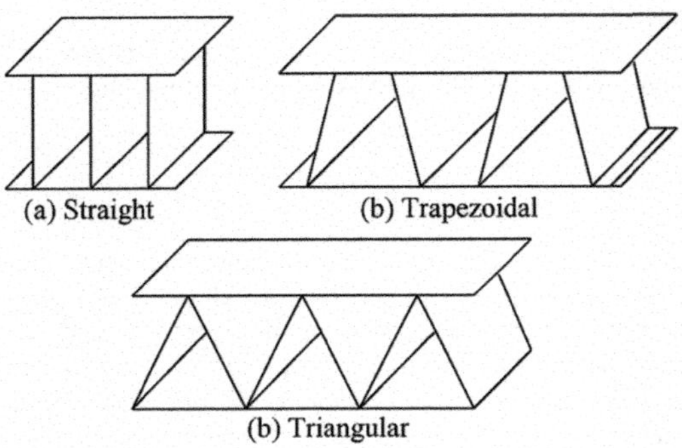

Figure 1. Various corrugated sandwich cores: (a) Straight, (b) trapezoidal, (c) triangular.

2.1 *Straight core – I core*

The strength and properties of corrugated cores with a vertical web cores have been a subject of discussion by various authors. For example, (Zhong et al. 2022) investigated the ultimate strength and collapse behaviour of the hull girder, considering the existence of an upper sandwich deck and its susceptibility to vertical bending, horizontal bending, torsion, and combinations of these loads. (Elsaka & Guedes Soares 2023) investigated the ultimate compressive capacity of vertical web-core steel sandwich panels and concurrently assessed the weight reduction achieved for an equivalent ultimate load when compared to conventional stiffened panels.

Figure 2 presents a schematic illustration delineating the architectural configuration of an I-type metal corrugated sandwich panel. Key dimensions include the length (L) and width (B) of the plate, with uniform thickness for both upper and lower plates (t_f). Further specifications involve the core thickness (t_c), core layer height (h_c), and the unit spacing between sandwich layers denoted as (S).

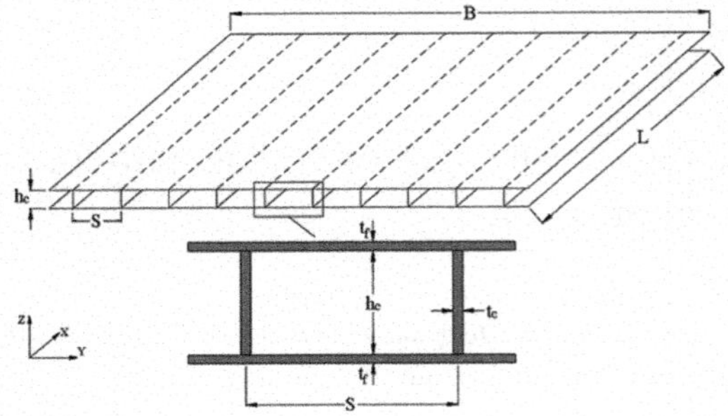

Figure 2. Visual representation outlining the design of a steel corrugated sandwich panel with a I-type configuration.

2.2 *Trapezoidal core – U core*

(Liu et al. 2022) investigated the structural deformation mechanisms exhibited by the U-type corrugated core sandwich panel used in ship structures when subjected to lateral quasi-static compression loads. The examination employs a comprehensive approach, incorporating experimental, numerical, and analytical methods. (Knox et al. 1998) developed a sandwich design that incorporated an adhesively bonded steel U-core, designed for use in marine structures. Their study delved into the performance characteristics and failure modes, comparing them with alternative manufacturing methods.

Figure 3 displays trapezoidal corrugated cores characterized by a trapezium shape with a distinct angle.

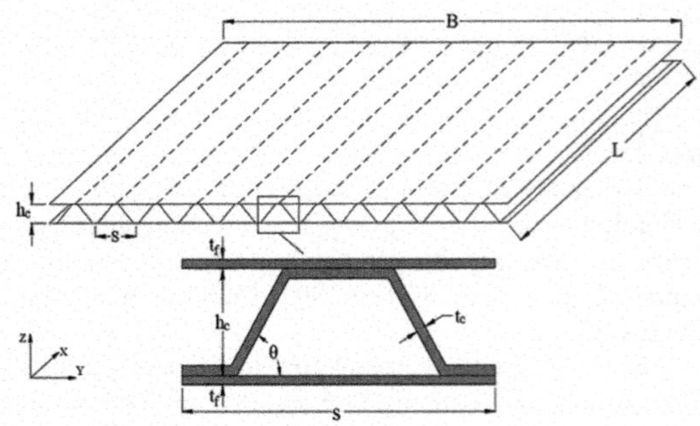

Figure 3. Visual representation outlining the design of a steel corrugated sandwich panel with a U-type configuration.

2.3 *Triangular core – V core*

(Rejab & Cantwell 2013) conducted a sequence of experimental inquiries and presented numerical analyses on the compression response in V-core sandwich panels made from an aluminum alloy.

Schematic representation elucidating the structure of a V-type metal corrugated sandwich panel, as depicted in Figure 4.

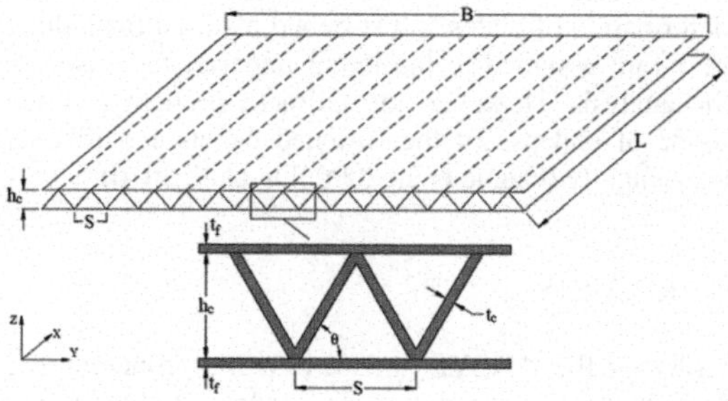

Figure 4. Visual representation outlining the design of a steel corrugated sandwich panel with a V-type configuration.

3 FINITE ELEMENT MODELLING

Determining the ultimate strength of structural elements and systems is of paramount importance, signifying their maximum load-bearing capacity, as emphasized by (Tekgoz et al. 2014). In contemporary structural research, numerical techniques have become indispensable, and the finite element method is a crucial tool for analysing thin-walled structures. The assessment of ultimate strength through NLFEA follows established procedures, involving the evaluation of buckling and ultimate strength capacities in stiffened panels. This methodology, validated by (Xu et al. 2013), has gained widespread acceptance, particularly in the study of conventional stiffened panels.

The current investigation employs NLFEA to explore the uniaxial compressive capabilities of sandwich panels in contrast to conventionally stiffened panels. A linear eigenvalue analysis is conducted to establish initial global imperfections, subsequently utilized in the nonlinear buckling analysis of the stiffened plate. To derive load end shortening curves, a forced displacement is applied to one edge of the model, and the corresponding reaction force on the opposite edge is computed. The magnitude of the applied displacement must be sufficiently large to induce significant deformation beyond the ultimate buckling capacity of the panel.

3.1 *Numerical Models Characteristics*

To facilitate a comparative analysis of the load-bearing capabilities between the metal sandwich panels and a stiffened panel, a structure with identical length, width, and weight was devised, as depicted in Figure 5. This reinforced panel features a strip thickness of 8 mm and incorporates four evenly distributed reinforcing L-stiffeners. The exact cross-sectional area of the stiffened panel is 24064mm², and it corresponds to a mass of 340kg.

The innovative corrugated sandwich panels proposed in this research aim to achieve an equivalent weight to that of conventional panels. The process of parameter identification plays a crucial role in understanding the pivotal input variables influencing the model's output parameters. Significantly, this study focuses exclusively on the variation of the core configuration thickness, maintaining a fixed set of parameters including face thicknesses, spacing between sandwich unit cores, and core height. By isolating the core thickness as the variable of interest, the study seeks to elucidate its specific impact on the overall performance of the corrugated sandwich panels, providing valuable insights into the intricate interplay between design parameters and structural behaviour. Table 1 provides a comprehensive overview of the geometrical parameters pertaining to I-type, U-type, and V-type metal corrugated sandwich panels. It is noteworthy that the cross-sectional areas of both variants demonstrate a close correspondence.

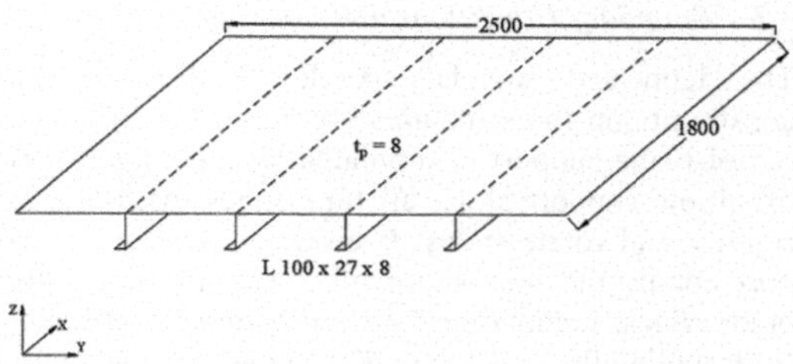

Figure 5. Schematic diagram of conventional stiffened panel (all measurements in mm).

Table 1. Dimensional parameters of sandwich panel.

Sandwich Panel	L mm	B mm	tf mm	tc mm	hc mm	S mm	Weight kg
I-core	2500	1800	3	6	60	100	339.12
V-core	2500	1800	3	2.3	60	100	338.86
U-core	2500	1800	3	2	60	100	335.00

3.2 *Material characteristics*

The same steel grade is used for both conventional and sandwich panels as indicated in Table 2.

Table 2. Main material characteristics.

Density	Young's Modulus	Poisson's Ratio	Yield Stress	Strain Hardening
kg/m3	E (MPa)	ν	σy (MPa)	Et/E
7850	210000	0.3	235	0

Assuming an idealized stress–strain curve, the material is considered to exhibit elastic, perfectly plastic behavior, with no consideration for material hardening effects as shown in Figure 6.

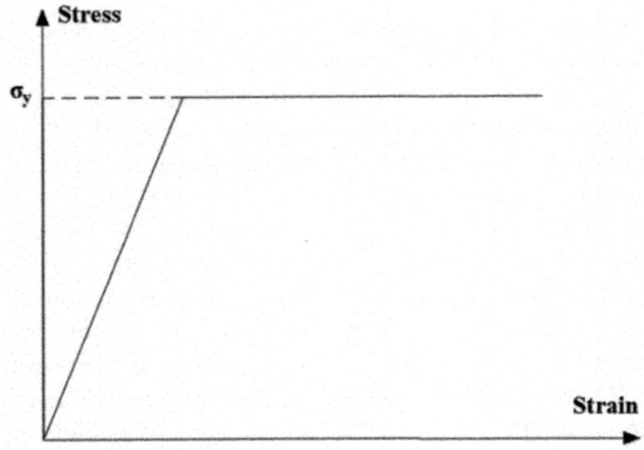

Figure 6. Idealized elastic perfectly plastic stress–strain curve.

3.3 *Boundary Conditions and Loading*

The delineated boundary conditions dictating the constraints on the structures are defined as follows: out-of-plane support is stipulated by a configuration of simple support along all four edges, with lateral rigidity and rotate freely. Conversely, in-plane support entails the imposition of constraints upon the outer edges, mandating their retention of rectilinear form while allowing for in-plane mobility. This particular constraint mechanism assumes a pivotal role in shaping the post-buckling behavior, thereby intricately impacting the ultimate structural capacity.

The implementation of the appropriate boundary conditions for a load case is shown in Figure 7. The boundary conditions are established to involve displacement constraints along the X, Y, and Z axes. These constraints dictate how the structure is fixed or restrained in these particular directions. The same boundary conditions are applied for all models.

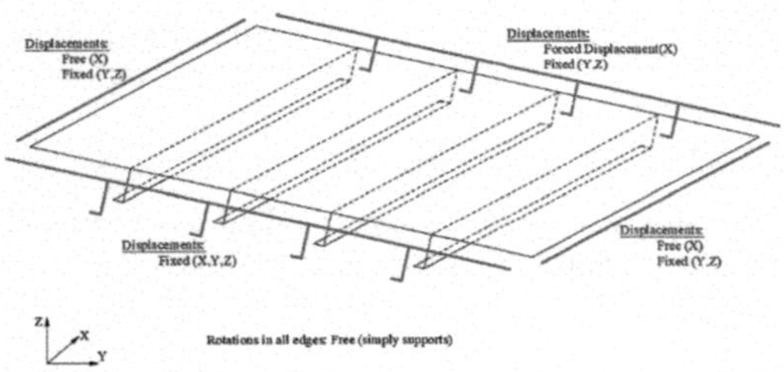

Figure 7. Boundary conditions and loading.

3.4 *Initial imperfections*

The initial deflections have a significant impact on the ultimate strength analysis of ship hull structures, and should be accounted for in the ultimate strength assessment (Paik & Seo 2009, Saad-Eldeen et al. 2012). The shape of the first eigenmode can be introduced as an initial imperfection (Jelovica & Romanoff 2013).

In the present study, the first mode of linear eigenvalue buckling analysis is chosen to simulate the initial deformation of plates and stiffened plates, a magnitude of 1 mm assigned to the initial imperfections used in the analysis. However, the ultimate strength and the mode of collapse of the stiffened panels are affected only slightly (Ozguc et al. 2006, Du et al. 2016).

4 RESULTS

Utilizing the ANSYS software, a unit concentrated load is applied to the leading edge of the model, causing it to withstand a unidirectional axial compressive load. Subsequently, a buckling analysis is performed on the four structural models, with Figure 8 to Figure 11 illustrating the first-order buckling eigenvalues of these models.

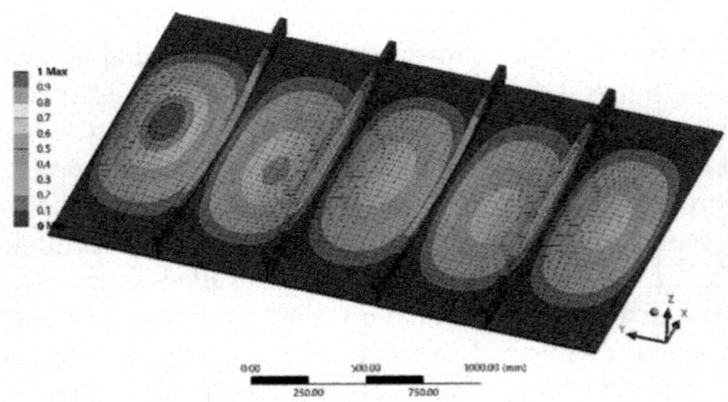

Figure 8. Eigenvalue Buckling Mode of the Conventional Stiffened Panel.

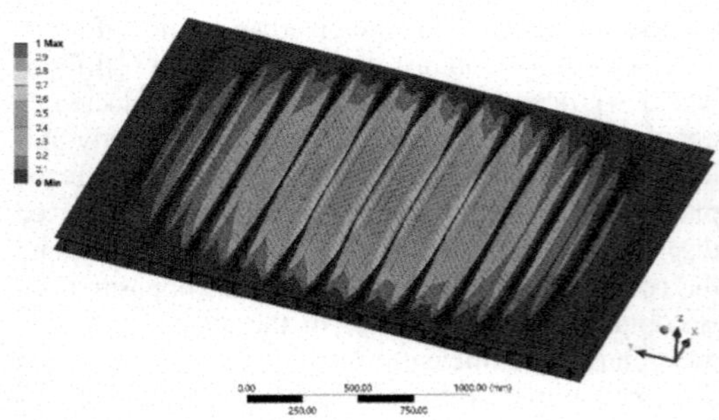

Figure 9. Eigenvalue Buckling Mode of the I-core sandwich Panel.

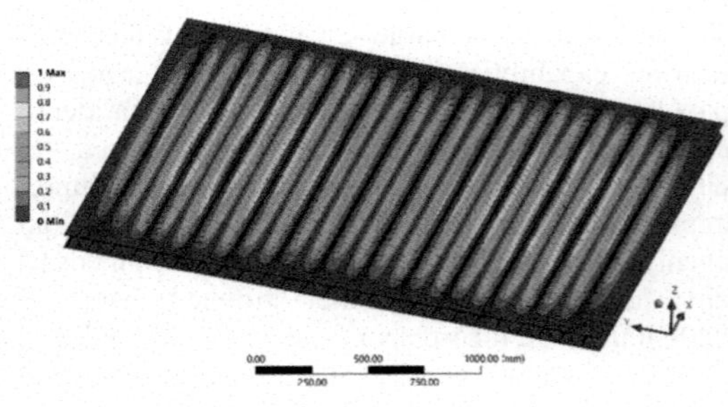

Figure 10. Eigenvalue Buckling Mode of the U-core sandwich Panel.

Figure 11. Eigenvalue Buckling Mode of the V-core sandwich Panel.

A forced displacement is implemented at the leading edge of the model, subjecting it to a unidirectional axial compressive load. Subsequent to this, an analysis of the structural models is conducted to determine their ultimate bearing capacity and failure mode. Figures 13 to 16 present the Mises stress distribution of the four structural models under the limit state, each respectively.

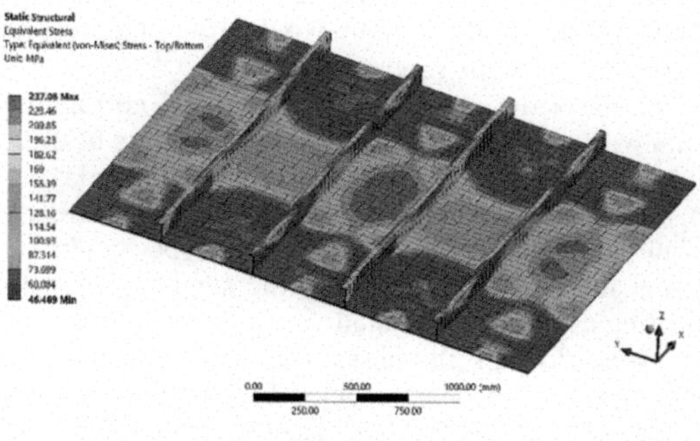

Figure 12. Von-Mises stress distribution at the ultimate strength limit of the Conventional Stiffened Panel.

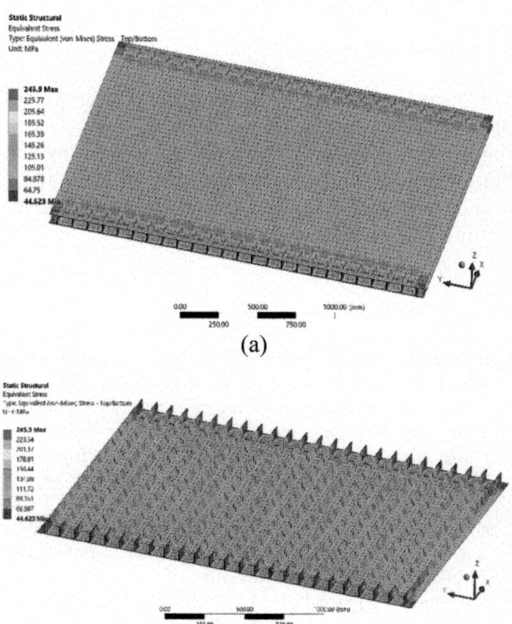

(a)

(b)

Figure 13. Von-Mises stress distribution at the ultimate strength limit of the(a) I-core sandwich Panel, (b) After removing the top covering face.

In Figure 16, the load-displacement curves depict the outcomes of NLFEA conducted in this study. Notably, during the linear phase of the load-displacement curve, the slopes are relatively similar, indicating comparable stiffness among the four structural models in the elastic stage. However, as the plastic stage is reached, distinctive failure modes emerge, resulting in more pronounced differences in the nonlinear phase of the curves. The extreme point on the load-displacement curve signifies the ultimate bearing capacity of each model. In Table 3, the data showcases the ultimate load capacities and their corresponding percentage difference in comparison to the reference conventional stiffened panel. Notably, the I-type metal corrugated sandwich plate emerges as the preeminent structure, boasting the highest ultimate bearing capacity and a remarkable addition percentage of 11.64%. Following closely is the V-type, which exhibits a comparable performance to the conventional stiffened plate structure. This indicates the V-type's efficacy in maintaining competitive load-bearing capabilities. Conversely, the U-type sandwich panel lags behind, demonstrating the smallest ultimate bearing capacity in this study, falling short by 3.81% compared to the conventional stiffened panel.

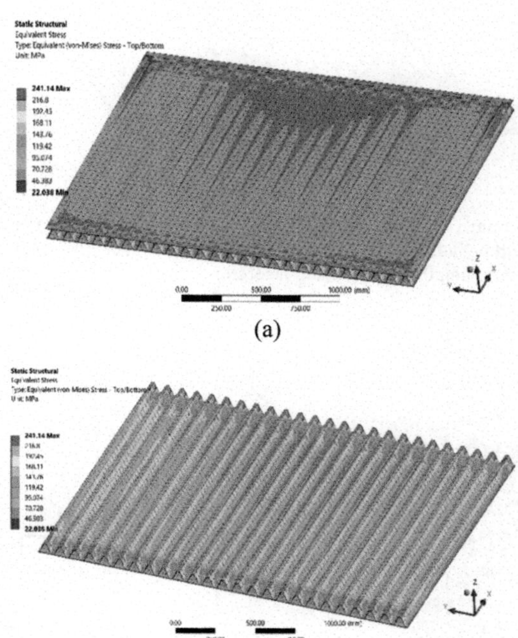

(a)

(b)

Figure 14. Von-Mises stress distribution at the ultimate strength limit of the (a) U-core sandwich Panel, (b) After removing the top covering face.

Table 3. Comparison of ultimate Compressive Load Capacities through NLFEA.

Structural model	Weight (kg)		Ultimate strength (N)	
	Calculated	Error %	Calculated	difference %
Stiffened panel	340	0.00	4.73E+06	0.00
U-core panel	338.86	-0.34	4.55E+06	-3.81
V-core panel	335	-1.47	4.76E+06	0.61
I-core panel	339.12	-0.26	5.28E+06	11.64

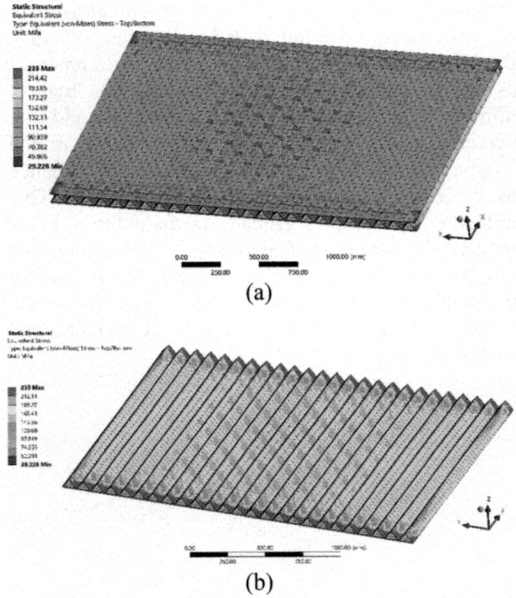

Figure 15. Von-Mises stress distribution at the ultimate strength limit of the (a)V-core sandwich Panel, (b) After removing the top covering face.

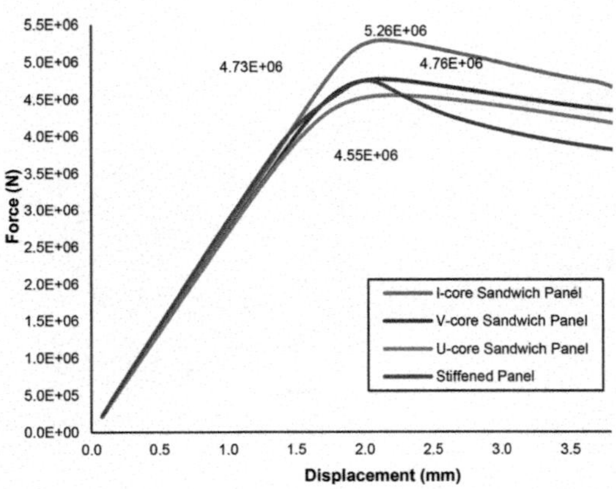

Figure 16. Load-displacement curves of the four structure models.

5 CONCLUSIONS

The primary focus of this paper is to conduct an evaluation of the potential enhancement in load capacity attainable through the utilization of metallic corrugated sandwich panels. Specifically targeting the in-plane unidirectional compression ultimate bearing performance of such panels, our investigation encompasses the I-type, U-core, and V-type metal corrugated sandwich panel structures, in comparison with a conventional stiffened plate structure.

The design of the sandwich panels is mainly influenced by one key factor: the thickness of the core. The other factors like face thickness, core height, and the space between core units kept within specific limits. By focusing on adjusting only the core thickness, the aim is to make the design process more straightforward. This approach allows to explore how changes in the core thickness affect the overall performance of the sandwich panels.

At the same weight the I-core metal corrugated sandwich panel exhibits the greatest ultimate bearing capacity. The V-type closely follows, displaying a performance similar to the conventional stiffened plate structure. In contrast, the U-type sandwich panel falls behind, displaying the smallest ultimate bearing capacity in this study.

These findings illuminate the varying degrees of structural enhancement achievable through distinct corrugated sandwich plate designs, with the I-type standing out as particularly promising for applications where superior load-bearing capacity is paramount.

ACKNOWLEDGEMENTS

This work contributes to the Strategic Research Plan of the Centre for Marine Technology and Ocean Engineering (CENTEC), which is financed by the Portuguese Foundation for Science and Technology (Fundação para a Ciência e Tecnologia - FCT) under contract UIDB/UIDP/00134/2020.

REFERENCES

Bartolozzi, G., Pierini, M., Orrenius, U. & Baldanzini, N. (2013). "An Equivalent Material Formulation for Sinusoidal Corrugated Cores of Structural Sandwich Panels." *Composite Structures 100*: 173–185.

Buannic, N., Cartraud, P. & Quesnel, T. (2003). "Homogenization of Corrugated Core Sandwich Panels." *Composite Structures 59*(3): 299–312.

Carlsson, L. A. & Kardomateas, G. A. (2011). Structural and Failure Mechanics of Sandwich Composites, Springer Science & Business Media.

D'Alessandro, V., Amabili, M., De Rosa, S. & Franco, F. (2012). Preliminary Identifications of Sandwich Panels. *INTER-NOISE and NOISE-CON Congress and Conference Proceedings*, Institute of Noise Control Engineering.

Du, J., Yang, P., Cui, C. & Xia, T. (2016). Ultimate Strength of Steel Panels and Stiffened Plates with Longitudinal through-Thickness Cracks under Compression. *4th International Conference on Sustainable Energy and Environmental Engineering*, Atlantis Press.

Elsaka, M. & Guedes Soares, C. (2023). "Ultimate Strength Behaviour and Optimization of Laser-Welded Web-Core Sandwich Panels under in-Plane Compression." *Journal of Marine Science Engineering 11*(11): 2200.

Elsaka, M., Leheta, H., Zayed, A. & Badran, S. (2019). Strength and Weight Characteristics of a Self-Propelled Barge Based on Sandwich Panel System Construction. *Sustainable Development and Innovations in Marine Technologies: Proceedings of the 18th International Congress of the Maritme Association of the Mediterranean (IMAM 2019)*, September 9-11, 2019, *Varna, Bulgaria*, CRC Press.

Jelovica, J. & Romanoff, J. (2013). "Load-Carrying Behaviour of Web-Core Sandwich Plates in Compression." *Thin-Walled Structures 73*: 264–272.

Jelovica, J. & Romanoff, J. (2020). "Buckling of Sandwich Panels with Transversely Flexible Core: Correction of the Equivalent Single-Layer Model Using Thick-Faces Effect." *Journal of Sandwich Structures Materials 22*(5): 1612–1634.

Knox, E., Cowling, M. & Winkle, I. (1998). "Adhesively Bonded Steel Corrugated Core Sandwich Construction for Marine Applications." *Marine structures 11*(4-5): 185–204.

Kozak, J. (2009). "Selected Problems on Application of Steel Sandwich Panels to Marine Structures." *Polish Maritime Research 16*(4): 9–15.

Kujala, P. & Klanac, A. (2005). "Steel Sandwich Panels in Marine Applications." *Brodogradnja: Teorija i praksa brodogradnje i pomorske tehnike 56*(4): 305–314.

Kujala, P., Romanoff, J., Salminen, A., Varis, J. & Vilpas, M. (2003). "All Steel Sandwich Panels." *Finnish Federation of Metallic Industry, MET.*

Liu, K., Zong, S., Li, Y., Wang, Z., Hu, Z. & Wang, Z. (2022). "Structural Response of the U-Type Corrugated Core Sandwich Panel Used in Ship Structures under the Lateral Quasi-Static Compression Load." *Marine structures 84*: 103198.

Metschkow, B. (2006). "Sandwich Panels in Shipbuilding." *Polish Maritime Research 13*(special issue 1): 5–8.

Ozguc, O., Das, P. & Barltrop, N. (2006). "A Proposed Method to Evaluate Hull Girder Ultimate Strength." *Ships Offshore Structures 1*(4): 335–345.

Paik, J. K. & Seo, J. K. (2009). "Nonlinear Finite Element Method Models for Ultimate Strength Analysis of Steel Stiffened-Plate Structures under Combined Biaxial Compression and Lateral Pressure Actions—Part I: Plate Elements." *Thin-Walled Structures 47*(8-9): 1008–1017.

Rejab, M. & Cantwell, W. (2013). "The Mechanical Behaviour of Corrugated-Core Sandwich Panels." *Composites Part B: Engineering 47*: 267–277.

Saad-Eldeen, S., Garbatov, Y. & Guedes Soares, C. (2012). "Analysis of Plate Deflections During Ultimate Strength Experiments of Corroded Box Girders." *Thin-Walled Structures 54*: 164–176.

SANDCORe (2013). Best Practice Guide for Sandwich Structures in Marine Applications. *Under European Commission Contract No. FP6-506330, SAND. CORe.*

Tekgoz, M., Garbatov, Y. & Guedes Soares, C. (2014). Finite Element Modelling of the Ultimate Strength of Stiffened Plates with Residual Stresses. Analysis Design of Marine Structures. C. S. R. A. Guedes Soares, (Eds.), Taylor & Francis Group: 309–317.

Xu, M. C., Yanagihara, D., Fujikubo, M. & Guedes Soares, C. (2013). "Influence of Boundary Conditions on the Collapse Behaviour of Stiffened Panels under Combined Loads." *Marine structures 34*: 205–225.

Zaid, N., Rejab, M. & Mohamed, N. (2016). Sandwich Structure Based on Corrugated-Core: A Review. *MATEC web of conferences*, EDP Sciences.

Zenkert, D. (1997). The Handbook of Sandwich Construction, Engineering Materials Advisory Services: Solihull, UK.

Zhang, Z., Lei, H., Xu, M., Hua, J., Li, C. & Fang, D. (2019). "Out-of-Plane Compressive Performance and Energy Absorption of Multi-Layer Graded Sinusoidal Corrugated Sandwich Panels." *Materials Design 178*: 107858.

Zhengjie, L., Lu, H., Nan, Z., Junjie, L., Jiajun, H. & Enrong, Q. (2020). "Ultimate Bearing Capacity for Steel Sandwich Panels under Uniaxial Compression (in Chinese)." *Chinese Ship Research 15*(4): 53–58.

Zhong, Q., Wu, G., Han, Z. & Wang, D. (2022). "Comparative Investigation on Ultimate Strength of Hull Girder with Laser-Welded Web-Core Sandwich Deck." *Ocean Engineering 264*: 112483.

Advances in Maritime Technology and Engineering – Guedes Soares & Santos (Eds)
© 2024 The Author(s), ISBN 978-1-032-83099-5

Ultimate strength tests on the stiffened plate panels

S.S. Guo
Key Laboratory of High Performance Ship Technology (Wuhan University of Technology), Ministry of Education, Wuhan, PR China

L. Zhu
School of Naval Architecture, Ocean and Energy Power Engineering, Wuhan University of Technology, Wuhan, PR China

S.M. Zhang
Lloyd's Register EMEA, Global Technology Centre, Southampton, UK

ABSTRACT: Ultimate strength of stiffened plate panels is important to the safety of ship structures. Experimental research on the stiffened panels helps to understand the ultimate strength characteristics and failure mechanism of the structure. In order to investigate the ultimate buckling capacity of a stiffener stiffened panel, uniaxial compression tests were carried out. Test apparatus were designed to accomplish the simply supported boundary condition at four edges of the model. The initial imperfections of the model were measured before the test, and the strains at critical parts of the model were measured in the test. Eventually, the failure modes and collapse characteristics of stiffener stiffened panels were analyzed. The research reported in this paper can provide a good reference for verifying the relevant results of theoretical studies, numerical methods and empirical formulas.

1 INTRODUCTION

Ships are subject to a variety of external loads in the process of the navigation, which may lead to deformations or even destruction of the ship structure, and then affect the safety of the ship. As an important structure of the ship, the stiffened plate panels can resist the deformations of ship structure due to the external loads. And the stiffened plate panels are mainly subjected to the axial compression loads caused by longitudinal bending of the hull girder. In order to obtain the realistic data on the collapse failure of stiffened plate panels under axial compressive loads, scholars have carried out a large number of ultimate strength test studies.

Smith (1975) conducted the axial compressive ultimate strength tests on 11 sets of stiffened plates and studied the effects of failure mode, initial deflections and lateral pressure on the ultimate strength of the stiffened plates. Horne et al. (1976) carried out 34 axial compression tests on the stiffened plates, and analyzed the effects of width to thickness ratio of the plate, initial deflections, weld residual stresses on the ultimate strength of the stiffened plates. Faulkner (1977) used an experimental method to investigate the effect of large initial deflections on the ultimate strength of the stiffened plates with initial eccentricity. Tanaka et al. (1992) conducted axial compression tests on 3 sets of stiffened plates and studied the lateral inclination of the web and the failure mode of local buckling. Aalberg et al. (2001) carried out longitudinal axial compression tests on the aluminium alloy stiffened plates with simply supported boundary conditions at four edges. Fujikubo et al (2005) made studies of ultimate strength of continuous stiffened panel under combined transverse thrust and lateral pressure. Paik et al. (2008) carried out axial compressive ultimate strength tests of aluminum alloy stiffened panels, and established a statistical database of initial deflections in welding manufacturing. The ultimate strength of aluminium alloy stiffened plates under in-plane compressive loading was investigated experimentally by simulating simply supported boundary conditions on all sides by Rønning et al. (2010), and the initial imperfections of the models were measured before the test, the failure modes and post-buckling characteristics of the models with different spans and stiffeners were comparatively analyzed. Xu et al. (2013) studied the collapse strength of wide stiffened panels under simply supported boundary conditions by using axial compression tests respectively, in which measured the initial imperfections on the plates and the distributions of the strains. Li et al. (2018) simulated the manufacturing imperfections of extrusion moulding aluminium alloy stiffened plates, and studied the sensitivity of ultimate strength to geometrical parameters, initial imperfections and material properties. The initial

DOI: 10.1201/9781003508762-49

imperfections of the stiffeners were scanned and introduced the initial defects into the finite element models for calculation by Yu et al. (2019). Sun et al. (2021a) conducted the axial compression tests and designed the corresponding simple support apparatus to study ultimate strength and collapse behavior of the novel aluminum stiffened panel with floating girder and fixed girder. Then, Sun et al. (2021b) studied the residual strength of stiffened panels with dent damages under uniaxial compression conditions.

Many researches have shown the importance of experimental studies on the ultimate strength of the stiffened plate panels under axial compressive loads. The ultimate strength tests were mostly on the longitudinally-framed stiffened panels, and the loading method was mainly uniaxial compression. Many scholars have set up simply-supported boundary conditions for axial compression tests, and a few scholars have measured the distributions of the strains of the models.

In this paper, an ultimate strength test under the simply supported conditions at four edges was conducted to investigate the buckling behaviour and ultimate buckling capacity of a stiffener stiffened panel. The test model is approximately $1/5^{th}$ of large container ship bottom structures. The initial imperfections of the plate and the stiffener web tilt angles were measured before the test, and the strains at critical parts of the model were recorded in the test. The collapse mode and stress distributions of the stiffener stiffened panel were analyzed.

2 TEST MODEL

2.1 *Definitions of the test model*

The stiffened panel model with overall dimensions 3000 mm by 688 mm, the plate thickness is 4 mm and its geometry is shown in Figure 1. The plate is stiffened with longitudinals and transversals with frame spacings of 172 mm and 1000 mm, respectively.

Continuous welding is required for all welding with a weld throat thickness not less than t/3 (t is the plate thickness). The model is manufactured by a qualified shipyard who follows Lloyd's Register's welding procedures and standards. End plates are welded to the two ends of the model, and the stiffening brackets are welded at the junction of the stiffened panel and the two end plates to prevent premature edge collapse.

2.2 *Property of material*

The stiffened panel model is manufactured by 355 high tensile steel. In order to obtain the realistic material properties, the quasi-static tensile tests were carried out by using a universal testing machine. The specimens for the tensile tests should be obtained from the base material to ensure that the properties of specimens are consistent with that of the stiffened panel. And the sampling directions were divided into transverse and longitudinal (rolling), three samples were made for each direction for each thickness, for a total of 12 specimens. The test results are presented in Table 1.

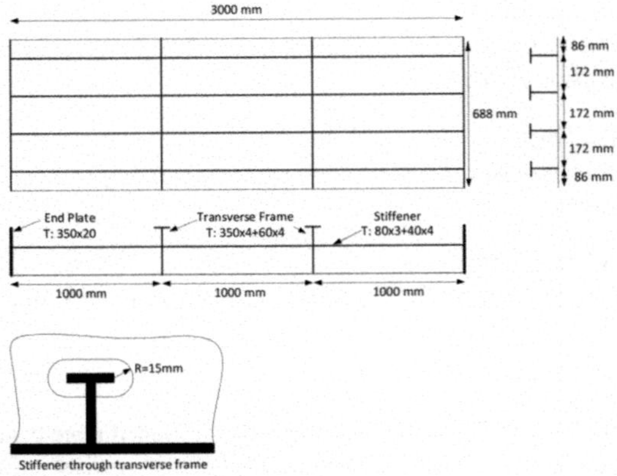

Figure 1. Dimensions of the stiffened panel.

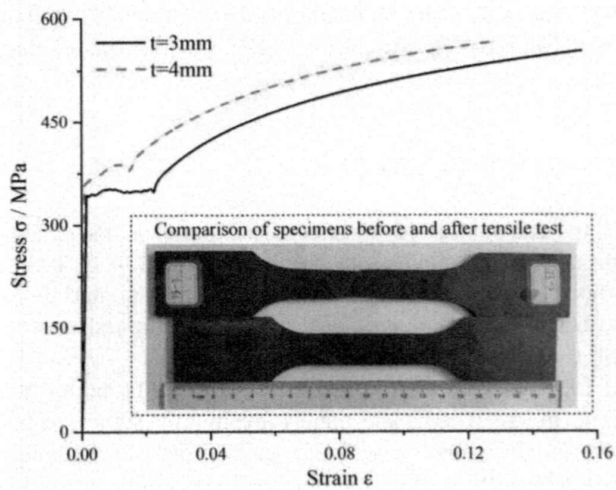

Figure 2. Comparison of specimens before and after tensile test.

Table 1. Material properties of stiffened panel.

Specification	Webs of stiffener	Plate Flanges of stiffener Transverse frame	Unit
Nominal thickness	3	4	mm
Measured thickness	2.94	3.76	mm
Young's modulus	206	206	GPa
Poisson's ratio	0.3	0.3	/
Yield stress	341	360	MPa
Ultimate tensile stress	555	576	MPa

It can be seen from Figure 2 that the specimens produced longitudinal elongation and transverse neck contraction after tensile, the fracture occurred in the middle part of the parallel length. The fracture was approximately 45° with the axis of the specimen, which was caused by the material slipping along the maximum shear stress surface during the tensile process. The true stress-strain curves of two thicknesses of the plate material obtained by the tensile tests are shown in Figure 3. According to the results of the material tensile tests, the average value of specimens at each thickness is considered to represent the material tensile properties of the corresponding plate. The average yield strengths of steel plate with thickness of 3 mm and 4 mm are 341 MPa and 360 MPa respectively.

2.3 *Initial imperfections*

In the model tests, the initial imperfections of the stiffened panel model were scanned by a 3D scanner before starting the test.

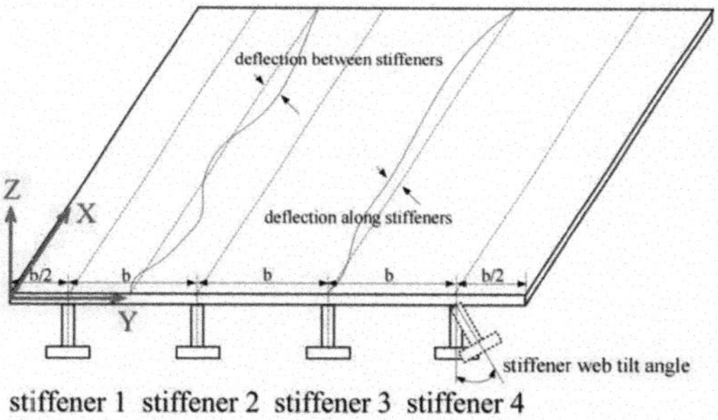

Figure 3. Illustration of initial imperfections.

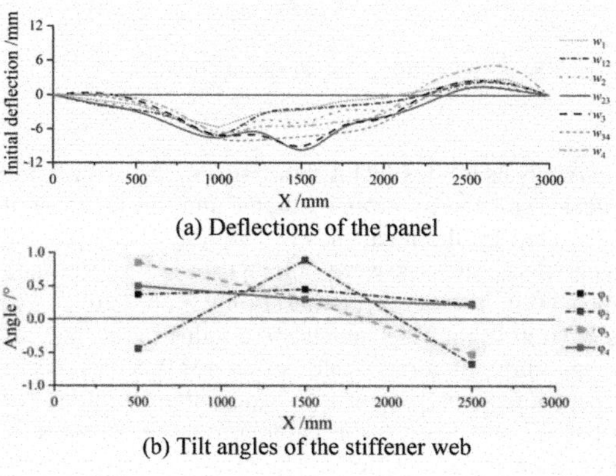

Figure 4. The measured initial imperfections.

The X-axis is set along the longitudinal direction of the stiffened panels, and in line with the direction of loading force as shown in Figure 3. Initial imperfections are measured which include the deflections of the stiffener (wi, along the line at plate and stiffener intersections), deflections between stiffeners (wi,i+1) and the stiffener web tilt angle (φ_i). The forward deflections along the Z-axis and the angles of counterclockwise rotation around the X-axis are set to be positive values. The measured results are presented in Figure 4. It can be seen from Figure 4 that the initial deflections of the panel were asymmetrical. The maximum deflection was located in the middle area of the stiffened panel, and there were small tilt angles of the stiffeners.

3 EXPERIMENT SET-UP

3.1 *Test apparatus*

Generally, the boundary conditions of the stiffened panels in the ship are between simply supported and clamped. In this paper, the boundary conditions were simply supported at four edges of the stiffened panel. Special apparatuses were designed to simulate the simply supported conditions in the test.

A transfer beam was designed to ensure the loading force of MTS machine apply to the stiffened panel uniformly. Cylinders were fixed on two end plates separately along the neutral axis of the stiffened panel which are accommodated by suitable grooves connected to the loading head and the pedestal base. The top groove with a boundary plate was connected to the transfer beam by bolts while the pedestal with the groove is fixed on the ground. In this case, the loaded edges could rotate as the model buckled and deflected, but there was no displacement movement.

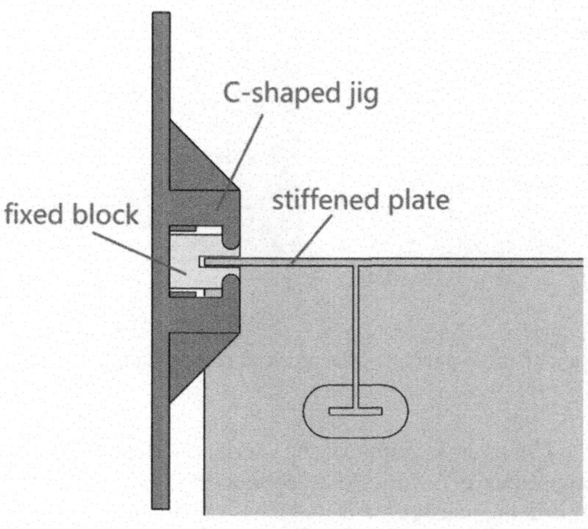

Figure 5. Assembly sketch of side edges.

The side edges of the model were simply supported by clamping devices with C-shaped jigs, which is shown in Figure 5. Side clamping devices were sufficiently rigid to withstand the lateral deformation of the stiffened panel. The C-shaped jig could constrain the lateral displacement of the side edges but except for rotation. In addition, fixed blocks were designed at the transverse frame to ensure the displacement can only occur along the length of the stiffener.

3.2 *Test details*

Strain gauges (Figure 6) were arranged at critical parts of the model to monitor the strain data during the test. A 500T multi-channel electro-hydraulic servo loading system (MTS testing machine) was used to apply axial compression load to the test model as shown in Figure 7. The stiffened panel was loaded by the combination of force-control loading and displacement-control loading in the test. The MTS testing machine is equipped with displacement and pressure sensors, thus the force and displacement data could be measured directly through the machine.

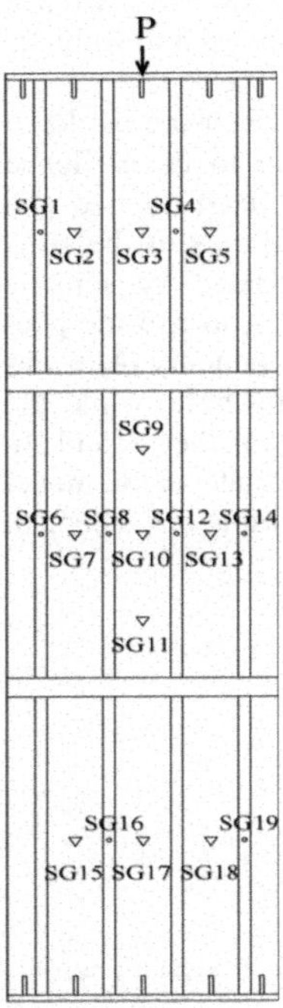

Figure 6. The location of strain gauges.

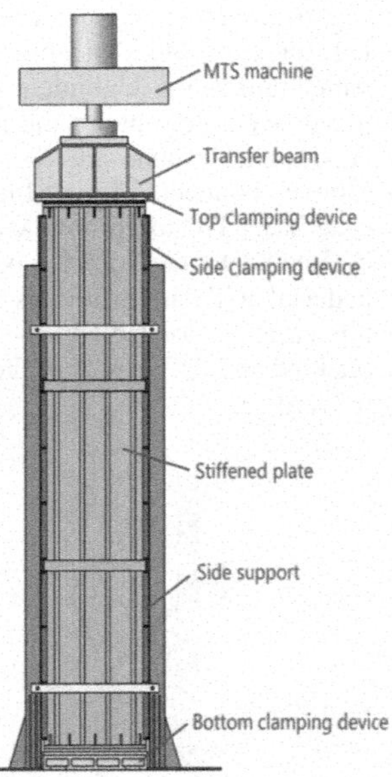

Figure 7. Experimental set-up of the uniaxial compression test.

The overall height of the model and test clamping apparatus is 3.5 m. After connecting and debugging the data measurement systems and loading system, a preloading test of the stiffened panel was conducted first. The preloading is to test the loading equipment and measuring instruments. Next, the uniaxial compression test was carried out. It was necessary to monitor the deformation of the stiffened panel and the load display in the test. When the structure collapsed, the testing machine is stopped.

4 RESULTS AND DISCUSSION

In the first stage of the test, the machine was loaded smoothly by force-control loading, the strain rate changed slowly. At this time, the loading force and displacement were roughly correlated linearly, and the deformation of the structure was within the elastic range. When the load reached 400 kN, switched the loading into displacement-control loading. Displacement-control loading was slightly faster but easier to control than the force-control loading. When the loading force was over 1000 kN, it could be seen that the middle of the stiffened panel began to bend slightly, and the stiffener webs in the section 2 began to expand outwards slowly. As the loading continued, the deformation of the stiffened panel increased and the loading force-displacement curve is given in Figure 8. The loading force reached its peak value at 1119.98 KN. The loading force-displacement curve plummeted as the test machine continued to move downward, indicating that the structure had collapsed. At this moment, the out-of-plane deformation of plate panels are seen most clearly. After unloading the deformation of the stiffened panel was less obvious.

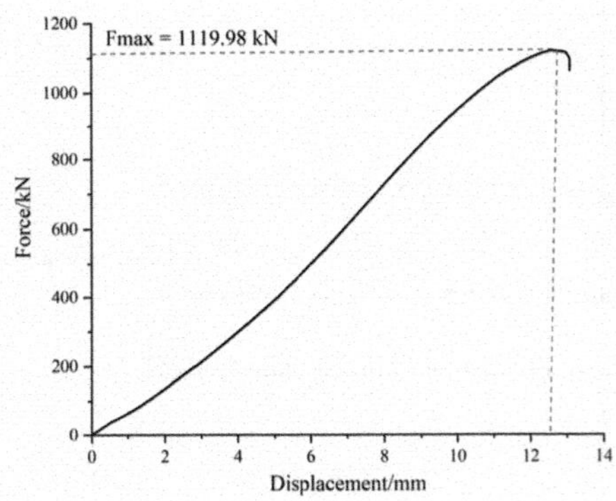

Figure 8. Loading force-displacement curve.

Analysis of test data were made first using the measurements of strain gauges and these are not restricted by the local plastic yielding of the panels. Von-Mises stresses were obtained based on the data measured by the three-direction strain gauges, as shown in Figure 9. The analyzed value takes the corresponding material yield stress (360 MPa) when exceeding this magnitude, since the formulae of material mechanics for calculating the stresses is based on linear elasticity. The stresses were linearly related to the displacement of the MTS loading machine at the early stage of the test. And the stress growth rate of some measured points became faster in the later period of the test, indicating that is a possibility of local buckling around the point. The stress fluctuated at a few measured points. The increase of stress at SG19 was the fastest, indicating that the stress was concentrated on the stiffener 4 in section 3. The material at SG7 reached the yield limit first, shows that the middle plate of the stiffened panel was the first part to reach yielding.

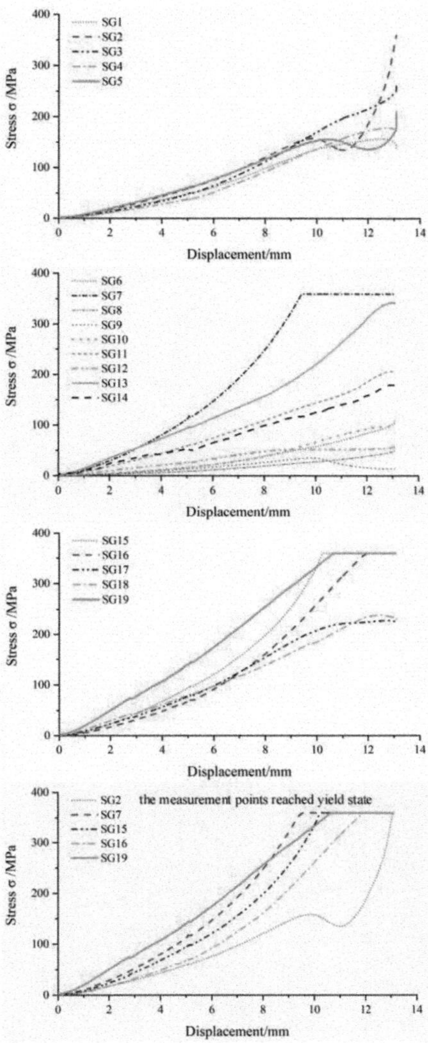

Figure 9. The measured points at critical parts.

Subsequently, the steel material yielded at SG15, SG19 and SG16, which indicate that the plate and stiffeners in section 3 reached the yield state successively. While the test machine dropped 12.62 mm, the loading force reached its maximum value, and the material yielded at SG2 at this time. The stiffened panel could not bear further increased load.

5 CONCLUSIONS

In this paper, the test investigation on the ultimate strength of a stiffener stiffened panel was reported. According to the results of the test, the following conclusions can be drawn.

The ultimate buckling capacity of the stiffener stiffened panel was 1119.98 kN. The materials at particular locations had yielded while the structure collapsed. Stress distributions of the stiffened panel were asymmetric due to the weld residual stresses and initial imperfections in the manufacturing process. The failure modes of the stiffened panel were reflected in the buckling of the plate and the deformation of the local stiffeners. The test results can be used to validate numerical calculation procedures and to verify design rules formulations.

REFERENCES

Aalberg, A. Langseth, M. & Larsen, P.K. 2011. Stiffened aluminum panels subjected to axial compression. *Thin-Walled Struct* 39(10): 861–885.

Faulkner, D. 1977. Compression tests on welded eccentrically stiffened plate panels. *Steel Plated Struct* p.130–139.

Fujikubo, M. Yao, T. & Khedmati M.R. 2005. Estimation of ultimate strength of continuous stiffened panel under combined transverse thrust and lateral pressure Part 1: Continuous plate. *Marine Struct* 18(5):411–427.

Horne, M.R. & Narayanan, R. 1976. Ultimate capacity of stiffened plates used in box girders. *Proc Instn Civ Engrs* 61: 841–844.

Li, C.F. Ren, H.L. et al. 2018. Numerical investigation on the ultimate strength of aluminium integrally stiffened panels subjected to uniaxial compressive load. *Thin-walled Struct* 127: 221–234.

Paik, J.K. Andrieu, C. & Cojeen, H.P. 2018. Mechanical collapse testing on aluminum stiffened plate structures for marine applications. *Marine technology* 45(4): 228–240.

Rønning, L. Aalberg, A. & Larsen, P.K. 2010. An experimental study of ultimate compressive strength of transversely stiffened aluminum panels. *Thin-Walled Struct* 48(6): 357–372.

Smith, C.S. 1975. Compressive strength of welded steel ship grillages. *Trans. RINA* 117: 325–359.

Sun, K. Zhang, G.L. Zhu, L. et al. 2021a. Experimental study on ultimate strength of novel aluminum alloy stiffened panel with floating girder. *OMAE* 2021–63761.

Sun, K. Zhu, L. Xu, L. et al. 2021b. Study on residual strength of stiffened panels with dent damages. *Developments in Maritime Technology and Eng* p.591–598.

Tanaka, Y. & Endo, H. 1992. Ultimate strength of stiffened plates with their stiffeners locally buckled in compression (2nd report) - effects of initial imperfections of stiffeners due to welding. *Society of Naval Architects of Japan & Ocean Eng* p.391–401.

Xu, M.C. & Guedes Soares, C. 2013. Experimental study on the collapse strength of wide stiffened panels. *Marine Struct* 30: 33–62.

Yu, Y.Z. Feng, G.Q. et al. 2019. Experimental and Numerical Investigation on the Ultimate Strength of Stiffened panels with Scanned Initial Geometrical Imperfection. *China Ocean Eng* 33(4): 446–458.

Advances in Maritime Technology and Engineering – Guedes Soares & Santos (Eds)
© 2024 The Author(s), ISBN 978-1-032-83099-5

Recent developments in ship structural components subjected to repeated impact loadings

X. He & C. Guedes Soares
Centre for Marine Technology and Ocean Engineering (CENTEC), Instituto Superior Técnico, Universidade de Lisboa, Lisbon, Portugal

ABSTRACT: This paper aims to enhance the understanding of dynamic response of ship structural components subjected to repeated impacts and consequently to offer some useful guidance for anti-repeated impact design. Repeated impacts are one of the most common loading scenarios in marine structures, such as ice floe impacts, slamming, and minor collisions between supply vessels and offshore platforms. Repeated impacts are prone to accumulate plastic deformation on marine structures, even leading to structural failure. The characteristics of dynamic response of structural components under repeated impacts are demonstrated based on numerical and experimental studies. Special attention is given to pseudo-shakedown analysis under repeated impacts. The advantages of pseudo-shakedown phenomenon in withstanding repeated impacts are discussed.

1 INTRODUCTION

The Arctic sea ice extent is declining at a dramatic rate due to global climate change (Borgerson, 2008), which has led to increased interest in Arctic shipping, as shipping costs are greatly saved due to reduced fuel consumption and increased voyage frequency (Lasserre, 2014; Zhang et al., 2016). This trend also brings more opportunities but also challenges to ship building industry. On one hand, more ice breakers and ice region ships are in high demand. On the other hand, the safety of ship structures is the most important concern. One of the biggest threats for ships sailing in ice covered region is the impacts from ice floe.

The impacts from ice floes are usually repeated impacts (Daley et al., 1986; Jordaan et al., 1993; Ralph & Jordaan, 2013), which are able to deform the ship hull cumulatively and even cause a fracture on ship panel. From the perspective of ice ship design, it is necessary to design reinforced ship structures to withstand the ice floe impacts. However, before the design phase, upstanding clearly how ship structures behave under these repeated impacts is the first step.

Besides this important engineering problem, there are other practical situations that ships and offshore structures suffer repeated impacts, for example, slamming on offshore platforms or on the bow and stern of ships, collision between supply vessels and platforms, sloshing and jet flow in LNG tanks, green water loading on ship decks and drop objects, among others. Slamming refers to the impact of water on a significant area, leading to ship whipping (Wang & Guedes Soares, 2017) and occasionally causing localized damage to the hull (Liu et al., 2020). Therefore, the structural design of ship and marine structure against repeated impacts is also an important task.

Considering these engineering situations, this article reviews studies on the dynamic response of structures subjected to repeated impacts, aiming at providing a comprehensive understanding of how structures behave dynamically under such conditions. Section 2 reviews the investigation works on repeated impacts. Section 3 reviews the progress in the analyses of pseudo-shakedown, which is a specific phenomenon that can occur during repeated impacts. Section 4 closes this article with some conclusions and remarks.

2 STUDIES ON REPEATED IMPACTS

Zhu (1990) conducted a systematic theoretical, experimental, and numerical study on the rectangular plates subjected to repeated impacts from rigid indenter, considering the engineering practice that the platforms are repeatedly struck by supply vessels. The mechanical mechanisms of the dynamic response of rectangular plates under repeated impacts were reasonably illustrated. Later, Zhu & Faulkner (1996) reported experimental results of repeated impact tests on mild steel and aluminium alloy rectangular plates. The experimental variations of the displacement of plate and the rebound velocity of striker vs. impact number were obtained. They also presented a rigid, perfectly plastic analytical

DOI: 10.1201/9781003508762-50

solution to predict the maximum deflection, impact duration and acceleration. The comparison between the analytical results and experimental results were made. It shows that reasonable agreement was achieved under the first several impacts for the displacement. While the discrepancy becomes larger with an increase of impact number, mainly due to neglecting the elasticity in the rigid, perfectly plastic analysis. Applying the analytical solution to engineering practice of ship plate design, Zhu et al. (2015) proposed some design formulae about the design thickness when the limited maximum displacement is required, which are straightforward for engineering design.

Furthermore, Zhu et al. (2018a) conducted repeated impacts on stiffened rectangular plates. Similarly, a rigid, perfectly plastic analysis for stiffened plates was proposed and compared with the experimental results. The rigid, perfectly plastic solution satisfactorily predicted the displacement of stiffened plates under the early repeated impacts. However, obvious discrepancy appears when the impact number is relatively large, due to neglecting the elastic effect of the material, as shown in Figure 1.

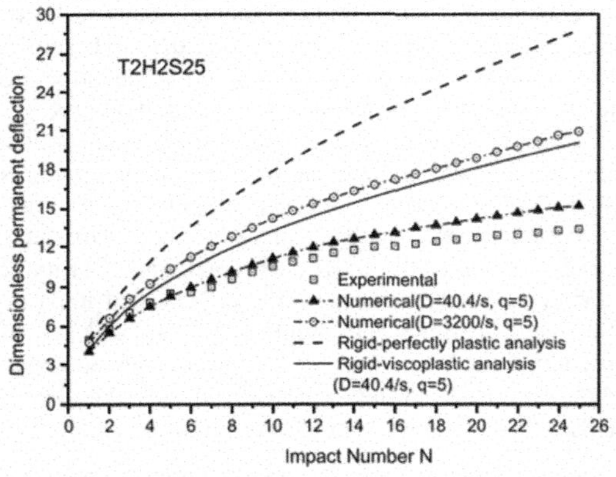

(b) Dimensionless permanent deflections δ_{cp} in the case of T2H2S25

Figure 1. Dimensionless permanent deflections at the center of the dented line vs. impact number. (Figure 6 in Zhu et al., 2018a).

Ghajar et al. (2015) conducted experimental and numerical investigations of repeated impacts on rectangular aluminium plates using drop weights impacting from either constant or variable heights. The deformation processes of the plates from plastic deformation, strength enhancement, crack initiation and propagation to penetration were observed. Variations of contact force, displacement for different impacts were analysed and compared. Numerical simulations of repeated impacts were also conducted using continuum damage mechanics by the FE code ABAQUS. The damage model was written as a user-defined material model implemented in user subroutine VUMAT.

Shin et al. (2018) performed repeated wet free drop tests on steel rectangular plates, where the plate thicknesses are 3 mm, 5 mm, 8 mm and the drop heights are 1 m and 2 m for each plate thickness. Impact pressure, deflection of the plate and strain time histories at some pints of the plates were measured. The main findings of the tests include following aspects. In wet free drop tests in the small restricted water tank, several pressure peaks are measured during a very short time immediately after the water entry. The 1st peak is due to slamming and the 2nd peak is due to the interaction between trapped air. The thin plate with thickness of 3 mm shows the "M" deformed shape after the 1st drop, which is the 3rd mode observed in the plate free vibration. For the tests on 3 mm thick plate under 2 m drop height, the maximum accumulative displacement reaches 20 times of the plate thickness. The tests gave an insight into the dynamic response of plates under repeated pressure loadings and also provide some useful experimental data, which might be helpful for developing numerical models.

Kim et al. (2019) investigated experimentally the dynamic behaviour of corrugated steel sheets under repeated impacts to simulate the performance of LNG cargo containment system under repeated jet flow impacts. Two structural types of specimen were fabricated. The repeated impacts were conducted under different energy and a different number of impacts. The effect of energy and the number of impacts were discussed. Zeng et al. (2022) conducted repeated impact tests on corrugated-core sandwich structures, whose panels and alloy corrugated core are both made from aluminium alloy. The effects of different impact locations and different impact energies were investigated.

Cai et al. (2022) conducted repeated impact tests of ice on steel plates, analysing the collision forces and plastic deformations, and performed numerical simulations to compare with experimental results for a group of tests.

Recently, He & Guedes Soares (2021a) performed repeated drop-weight impact tests on mild steel and aluminium alloy beams in a fully instrumented Rosand IFW5 drop-weight machine. Further, the repeated impact tests on mild steel rectangular plates were also conducted (He & Guedes Soares, 2023, 2024a). The experimental results related to impact force, the velocity of the indenter, displacement, contact duration and absorbed energy process were analysed in detail. Subsequently, the dynamic response characteristics of the beams and plates under repeated impacts were clearly uncovered, which can be summarized as follows.

The maximum accumulated permanent displacement of the structure, the peak contact force during an impact and the rebound velocity of the impactor generally keep increasing with the impact number but along with a decreasing growth rate. This trend is due to the strengthening effects of geometry changes and strain hardening. If the structure eventually suffers a failure, the peak contact force and rebound velocity experience a drop after the fracture initiates. These general characteristics are shown in Figure 2 for an example. In contrast, the contact duration during an impact and the dissipated energy generally decreases with the impact number before failure, and along with a reduced decrement rate. The contact duration and dissipated energy then start to increase after the fracture initiates, as shown in Figure 3. Besides, He & Guedes Soares (2024b) reported some experimental results of different shaped impactors repeatedly impacting the rectangular plates. The results showed that the impactor shape influences significant the dynamic response of rectangular plates subjected to repeated impacts.

Considering ships and marine structures may suffer repeated impacts under extreme environments, for example, in ice covered regions where the temperature is quite low, some works have been carried out to explore the effects of low temperature on repeated impacts. Cho et al. (2014) and Truong et al. (2018a, 2018b) investigated experimentally the repeated impacts of "T" shaped steel beams (Cho et al., 2014; Truong et al., 2018a) and steel grillages (Truong et al., 2018b) at room and sub-zero (-50°C) temperatures. The permanent deflections were measured after each impact test. The experimental results show that the permanent displacement decreases under sub-zero temperature compared to that under room temperature, as shown in Figure 4, because the material gets strengthened under low temperature. Numerical simulations were also performed by using an FE code and compared to the experimental results. Besides, Truong et al. (2018b) proposed a semi-analytical rigid-plastic solution, which incorporates the rebound velocity of the indenter, strain hardening and strain rate sensitivity, to predict the displacement for the structure.

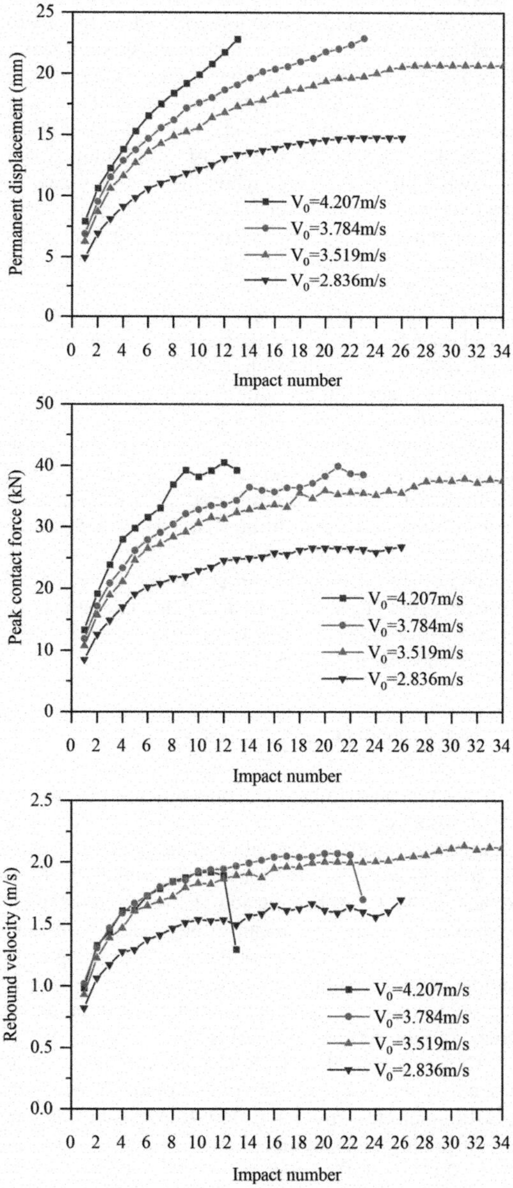

Figure 2. Accumulated permanent displacement, peak contact force, rebound velocity vs. impact number.

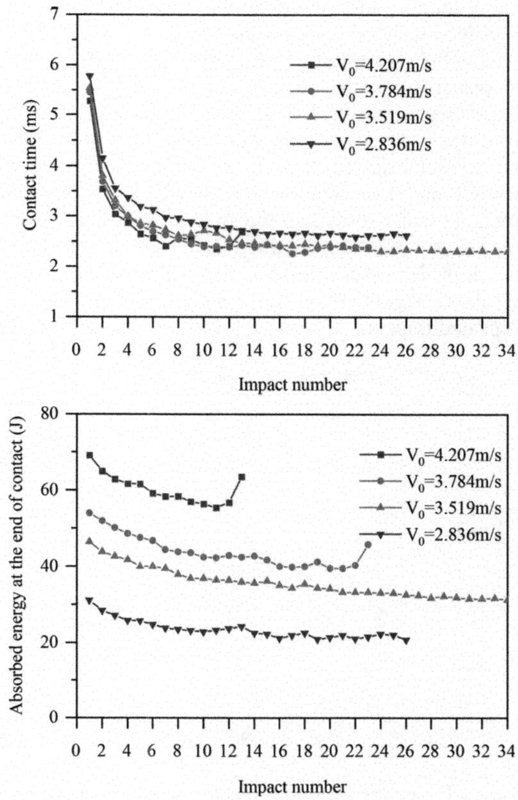

Figure 3. Contact time, dissipated energy vs. impact number.

Considering some engineering structures working under repeated impacts are prone to initial imperfections, the effects of initial imperfection on the repeated impact behaviour are also pursued. Li & Sun (2016) investigated experimentally and numerically the crack behaviour and life prediction of notched and unnotched beams made of 18Cr2Ni4WA steel subjected to repeated impact loadings. Numerical simulations were carried out by an FE code. They observed that the failure mode for notched specimens under repeated impacts is a progressive failure with crack growth differing from the single impact failure. The repeated impacts are extremely sensitive to notch due to stress concentration on the crack tip. An energy-based life prediction equation, i.e., the relationship of number of impacts to failure vs. dissipated energy, for repeated impacts was established empirically and validated against experiments.

Duan et al. (2019) investigated experimentally the dynamic behaviour of circular aluminium plates under repeated impacts. Three series of tests were conducted, i.e. plates without crack, having surface cracks with varying length, and having cracks with varying depths. The response and failure modes are obtained. Plates with larger cracks carry smaller impact forces and assume larger deformations. With the increase in the number of impacts, the effects of crack lengths and depths on the dynamic behaviour become much more significant.

Finally, rigid-plastic analytical predictions of the displacement were conducted and compared with experimental results. Zeng et al. (2020) conducted repeated impact tests on circular mild steel plates with prefabricated cracks at room temperature (20°C) and low temperature (-40°C). The experimental results show that the strength of the material increases while the ductility decreases as the temperature decreases. Besides, the prefabricated crack weakens significantly the resistance of plates to repeated impacts.

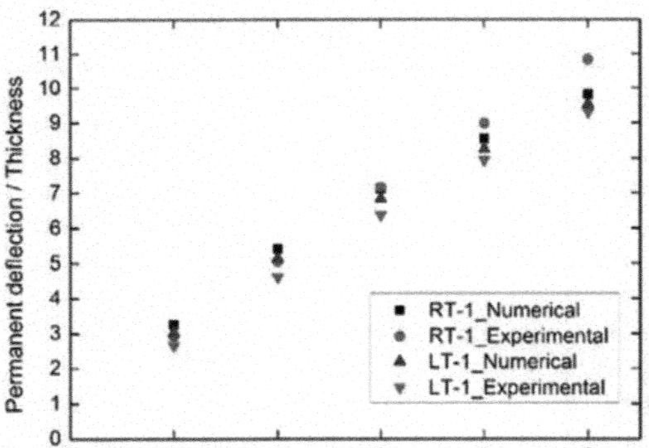

Figure 4. Comparison of the dimensionless permanent displacement vs. impact number between under room temperature and low temperature. (Fig. 19 in Cho et al., 2014).

Nowadays, the application of either composite structure or composite material for ship and marine structure manufacturing has been more and more extensive. The temperature effects on the repeated impact behaviour of composite structures have also been investigated. Zhu et al. (2018b) and Guo et al. (2018) conducted repeated impacts on aluminium foam sandwich plates at room temperature (20°C) and low temperature (-60°C) by using an INSTRON 9350 Drop Tower. The experimental results show that the deflections and absorbed energies of aluminium foam sandwich plates at low temperature are smaller than those at room temperature. With the increase of impact number, the effect of low temperature on dynamic behaviour of aluminium foam sandwich plates becomes much more significant. Besides, as the impact number increases but before fracture occurs, the growth rate of displacement decreases; the stiffness in the loading stage increases; the elastic energy of aluminium foam sandwich plates increases while the absorbed energy decreases. These characteristics are generally consistent with the dynamic response of simple structures under repeated impacts, as for example, the mild steel beams and plates of He & Guedes Soares (2021a, 2023, 2024a). Furthermore, Zhang et al. (2021) performed elastoplastic FE simulations on aluminium foam sandwich panels under repeated impacts by using an FE code. The accumulation processes of plastic deformation, the impact force-displacement curves as well as the energy absorption performances were numerically calculated. The numerical results were compared with the repeated impact tests conducted in the INSTRON 9350 Drop Tower.

There are also some other works on repeated impacts, but they are involved in the study of pseudo-shakedown. Therefore, these works are reviewed separately in next section.

3 PSEUDO-SHAKDOWN ANALYSIS

During repeated impacts, a phenomenon known as pseudo-shakedown can occur, wherein a structure develops a plastic deformation pattern that prevents a further increase in plastic deformation from repeated loads. When structures reach a pseudo-shakedown state, it means they can withstand repeated loads over time without additional plastic deformation or failure, as long as fatigue failure is avoided. Thus, pseudo-shakedown has the potential to extend the service life of ships.

The concept of pseudo-shakedown was introduced by Jones (1973) for plates made from rigid-plastic material and subjected to repeated impact loads having a triangular-shaped pressure-time history, based on the observation of repeated wave impacts on ship bows. When a plate is subjected to a pressure pulse with a short duration, under certain circumstances (the duration is shorter than a critical duration), the transverse displacement of the plate can be smaller than that predicted by an analysis for the same maximum pressure applied statically.

If the identical pressure pulses are repeated, the transverse displacement increases and tends towards the associated permanent transverse displacement predicted by a static analysis when the influence of finite-displacements (geometry changes, membrane effects) are retained in the governing equations. This behaviour is termed pseudo-shakedown. Once a pseudo-shakedown state has been achieved, then no further external work is done by the repeated identical pressure pulses, since a rigid, perfectly plastic structure remains rigid, and thereby, no energy can be absorbed by the structure. The pseudo-shakedown phenomenon was first observed and reported in the water drop tests on mild steel panels by Yuhara (1975).

Shen & Jones (1992) proved a conjecture for the pseudo-shakedown phenomenon for rigid, perfectly plastic beams and plates subjected to repeated dynamic pressure pulses. They proved that the sufficient and necessary condition for the pseudo-shakedown phenomenon is that the transverse displacement produced by a dynamic pressure pulse should be no more than that predicted by a static analysis using the peak value of the pressure pulses. Some comments were made on the number of repeated dynamic loadings necessary to reach a pseudo-shakedown state for a fully clamped beam.

Jones (2014) extended the study of the pseudo-shakedown phenomenon to circular and rectangular plates under identical repeated mass impacts which allow the accumulation of masses from earlier impacts on the plate surface as, for example, in the loading of cargo in a hold of a ship, using a rigid-plastic analysis. Previous rigid, perfectly plastic analysis (Zhu & Faulkner, 1996) explored the effect of only the current impact mass and ignored the accumulation of any masses from earlier impacts on a plate. This led to a steady growth of the transverse displacements without the achievement of pseudo-shakedown. However, the new expressions for predicted displacement which account for the accumulated mass have asymptotic values, which reveal that a pseudo-shakedown state is achieved for this case and has an associated modest value of the maximum permanent transverse displacement. The equations are useful design tools for an estimate of the pseudo-shakedown behaviour of plating, which is loaded repeatedly with cargo and other masses in various practical situations.

The pseudo-shakedown presented by Jones (1973, 2014) and Shen & Jones (1992), is obtained based on rigid, perfectly plastic analyses, where the elastic and strain hardening effect of the realistic ductile material are not considered.

The pseudo-shakedown phenomenon can also occur in elastoplastic structures subjected to repeated impacts. Huang et al. (2000) carried out quasi-static cyclic loading tests to draw force-deflection hysteretic curves. By integrating the force-deflection hysteretic curves, the total strain energy-deflection curves and elastic strain energy-deflection curves were drawn. By analysing the energy-deflection curves, an energy criterion to judge the pseudo-shakedown state of the structure was proposed. It suggests that the structure will reach a pseudo-shakedown state when the maximum elastic strain energy to be absorbed in the deformed structure is equal to or larger than the external work of the dynamic load in magnitude. This postulation was confirmed by their repeated drop mass impact results. It should be noted that the analysis method proposed in this article is suitable only for the particular cases of heavy-striker, low-speed impacts. For these cases, the transverse displacement profile for dynamic loads remains identical to the corresponding one for static and quasi-static loads. Hence the quasi-static cyclic deformation curves of the structures can be used for the response analysis of repeated mass impacts.

Stoffel (2006a) conducted impact tests on steel and copper circular plates subjected to repeated shock wave loadings. The experimental results indicated three types of magnitude relationship between the displacement produced by dynamic impulse and that by a quasi-statically loaded pressure equal to the peak pressure of the impulsive loading. (1) After the first impulse, the displacement caused by the dynamic pressure pulse is larger than the quasi-statically obtained deflection and tends towards a limit state after repeated shock wave-loadings. (2) The displacement caused by the dynamic pressure pulse exceeds the quasi-statically obtained deflection and does not increase further after repeated impulsive loadings. (3) The displacement caused by the dynamic pressure pulse is smaller than the quasi-statically obtained deflection and tends towards the quasi-statically obtained displacement after repeated shock wave loadings. The third relationship confirms to the pseudo-shakedown phenomenon developed by Jones et al. (Jones, 1973; Shen & Jones, 1992). The second relationship suggests the pseudo-shakedown doesn't occur. While the first relationship, which is different from the third one, also indicates the occurrence of pseudo-shakedown. Stoffel (2006b) clarified the pseudo-shakedown for the first scenario is due to the elastic effect by using a simplified model assuming elastic-plastic material behaviour.

Recently, He & Guedes Soares (2021a) experimentally examined the occurrence of the pseudo-shakedown in the repeated drop-weight impact tests on mild steel and aluminium alloy beams. Through analysing the experimental results in detail, including the impact force, the velocity of the indenter, displacement, contact duration and absorbed energy process, the dynamic behaviour of beams under a pseudo-shakedown state was demonstrated. Furthermore, He & Guedes Soares (2021b) verified the pseudo-shakedown phenomenon on the tested mild steel and aluminium alloy beams through numerical simulations, providing further evidence for their findings. Detailed numerical analyses effectively illustrated the dynamic behaviour of beam structures subjected to repeated impacts.

Expanding their research, He & Guedes Soares (2022) investigated the pseudo-shakedown behaviour of mild steel rectangular plates under repeated impacts by performing numerical simulations. This study revised the conclusion made by Huang et al. (2000) regarding the cause of pseudo-shakedown. It has revealed that the pseudo-shakedown phenomenon occurs when the maximum internal energy absorbed by the specimen in a single impact is less than the elastic energy storage capacity of the specimen, rather than the impact energy being lower than the elastic energy capacity. This is because not all impact energy is dissipated as internal energy within the structure. The study also highlighted the significant influence of the material's necking effect on pseudo-shakedown, which was not considered in Huang et al. (2000).

Based on these findings, new criteria based on critical displacements were established, introducing the concepts of two critical displacements, namely the 'necking displacement' and 'failure displacement', which correspond to the points when the maximum contact force is reached and the first element fails, respectively. The 'necking displacement' and 'failure displacement' are marked in the curve of elastic energy vs. displacement in Figure 5 for an example. The new criteria are more applicable in practical engineering scenarios for predicting the occurrence of pseudo-shakedown and failure.

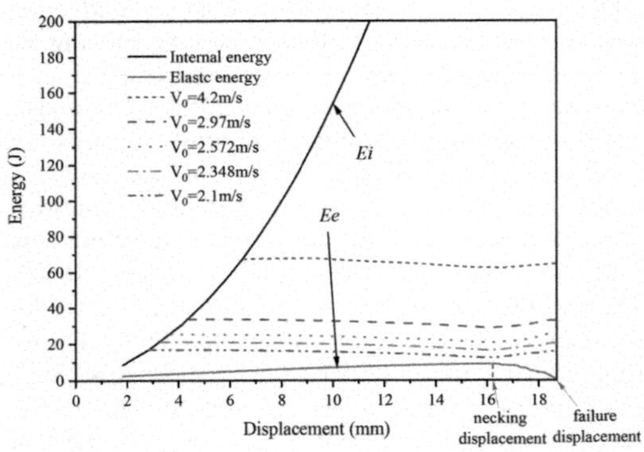

Figure 5. The 'necking displacement' and 'failure displacement' in the curve of elastic energy vs. displacement. (Fig. 15 in He & Guedes Soares, 2022a).

He et al. (2022) further examined the displacement-based criteria for the pseudo-shakedown and the failure for mild steel and aluminium alloy rectangular plates subjected to low velocity repeated impacts under a sphere indenter. The validity of the criteria is based on the conjecture that the modal deformation profiles are the same for a plate under various low-velocity impacts. This conjecture was verified by the uniformity of the variation of internal energy vs. displacement and the identical deformation profiles under repeated impacts, single impact and quasi-static punching, as shown in Figure 6. Based on this mechanical characteristic, the 'necking displacement' and 'failure displacement' under repeated impacts, can be accurately predicted by either a single impact or quasi-static punching. These proposed critical displacements are very useful in predicting the occurrence of pseudo-shakedown and failure.

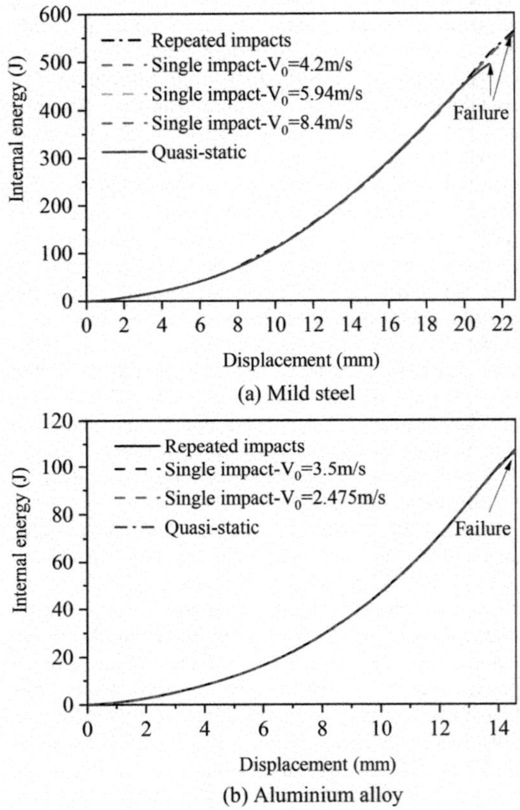

Figure 6. Contact force-time history of a rectangular plate subjected to repeated impacts and achieving pseudo-shakedown finally. (Fig. 12 in He et al., 2022b).

More recently, He & Guedes Soares (2024) performed repeated impact tests on mild steel rectangular plates. It was observed the pseudo-shakedown phenomenon also occurred for the plates. The pseudo-shakedown behaviour of plates subjected to repeated impacts was given a clear insight. When a pseudo-shakedown state is reached, the plate exhibits consistent behaviour under subsequent identical impacts, with the contact force time history remaining the same, as shown in Figure 7 for an example. The peak contact force, maximum displacement, rebound velocity of the impactor, and contact duration all stabilize at a constant value, and the permanent

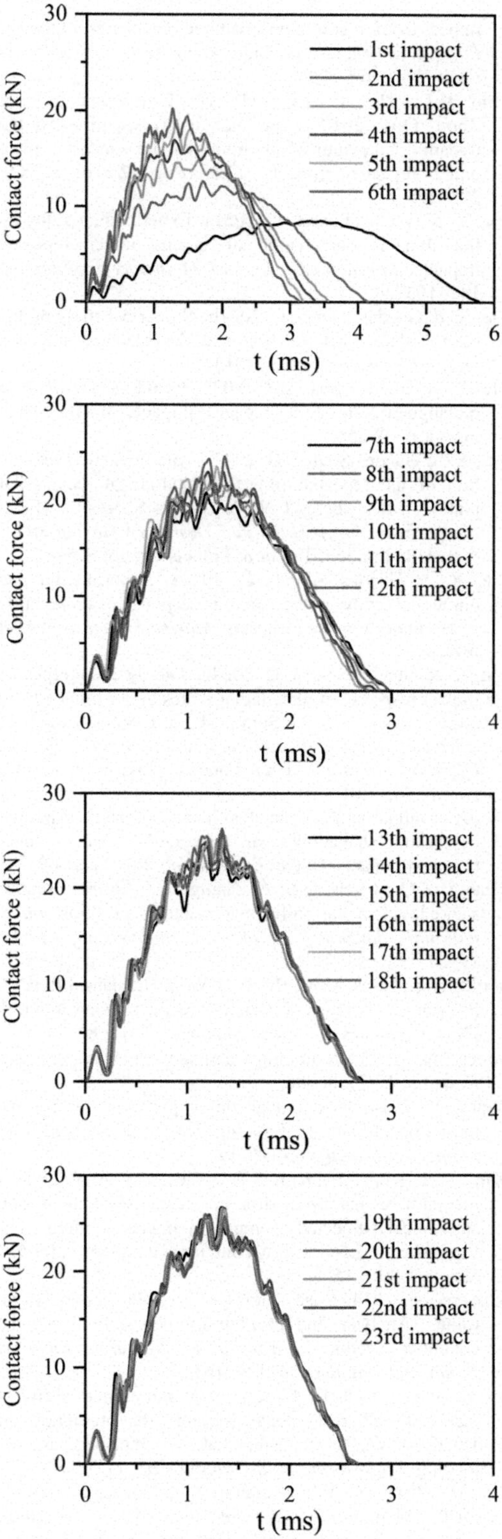

Figure 7. Contact force-time history of a rectangular plate subjected to repeated impacts and achieving pseudo-shakedown finally.

displacement of the plate does not increase further, as shown by the curves for repeated impact velocities of 3.519 m/s and 2.836 m/s in Figures 2 and 3.

It is noted that the pseudo-shakedown phenomenon was also observed under the repeated impacts on a corrugated-core sandwich structure with relatively low impact energy (Zeng et al., 2022). It should be also mentioned that in the repeated impact experiments (Zhu and Faulkner, 1996; Zhu et al., 2018a, 2018b; Guo et al., 2018; Zhang et al., 2021; Cho et al., 2014; Truong et al., 2018a, 2018b; Ghajar et al., 2015; Li and Sun, 2016; Duan et al., 2019; Kim et al., 2019), the pseudo-shakedown phenomenon was not observed, because the impact energies in their test exceed the limit energy for pseudo-shakedown.

4 CONCLUSIONS

In this paper, the state of art on structures subjected to repeated impacts and the associated pseudo-shakedown behaviour, are reviewed. Through the experimental findings, some common characteristics of the dynamic response of structures subjected to repeated impacts can be concluded. The structures strengthen with the geometry changes and strain hardening as the impact number increases. Before reaching initial failure or the pseudo-shakedown state, the accumulated permanent displacement, peak contact force, and rebound velocity generally increase with the number of impacts, but at a decreasing growth rate. On the other hand, the contact duration and dissipated energy ratio decrease as the number of impacts increases. In cases where a plate experiences fracture, the peak contact force and rebound velocity decrease, while the contact duration and dissipated energy increase.

When a pseudo-shakedown state is reached, the plate exhibits consistent behaviour under subsequent identical impacts, with the contact force time history remaining the same. The peak contact force, maximum displacement, rebound velocity of the impactor, and contact duration all stabilize at a constant value, and the permanent displacement of the plate does not increase further.

The primary cause for pseudo-shakedown under repeated mass impacts is that the maximum internal energy absorbed during an impact is no more than the elastic energy that the structure can store.

Under low velocity repeated impacts, the proposed concepts of 'necking displacement' and 'failure displacement' can be simple but useful tools predict the occurrence of pseudo-shakedown and failure, which are helpful for engineering practice.

From the perspective of engineering design of ship structures withstanding repeated impacts, the concept of pseudo-shakedown should be taken into consideration. For example, if the operating conditions of a ship are well-known, the ship plate thickness can be precisely determined based on the loading condition to achieve a pseudo-shakedown for the ship panel. As a result, this allows for a reduction in material consumption during design and accordingly contributes to a lightweight structure, which will potentially improve the ship's performance and efficiency.

On the other hand, developing numerical model to predict the damage from repeated impact loadings is the most economical and also the most widely adoptive way in the primary design stage. But there are still some challenges on how to create precise numerical model to predict the repeated impacts. The dynamic response under repeated impacts is sensitive to the material strain rate sensitivity effects, since repeated impacts are usually involved in a wide strain range. Besides, an accurate failure criterion is crucial to predict the occurrence of pseudo-shakedown. They are parts of the works worth our efforts in the future study on repeated impacts.

ACKNOWLEDGEMENTS

The authors are grateful to Prof Norman Jones for his suggestions and encouragement to pursue the study of pseudo-shakedown. The first author is grateful for the support of the Scholarship from the China Scholarship Council (CSC) under Grant No. 201706950008. This work contributes to the Strategic Research Plan of the Centre for Marine Technology and Ocean Engineering (CENTEC), which is financed by the Portuguese Foundation for Science and Technology (Fundação para a Ciência e Tecnologia - FCT) under contract UIDB/UIDP/00134/2020.

REFERENCES

Borgerson, S.G. 2008. Arctic meltdown-The economic and security implications of global warming. *Foreign Aff.*, 87: 63.

Cai, W., Zhu, L. & Qian, X.D., 2022. Dynamic responses of steel plates under repeated ice impacts. *International Journal of Impact Engineering* 162: 104129.

Cho, S.R., Truong, D.D. & Shin, H.K. 2014. Repeated lateral impacts on steel beams at room and sub-zero temperatures. *International Journal of Impact Engineering* 72: 75–84.

Daley, C.G., St. John, J.W., Brown, R. & Glen, I. 1986. Consolidation of local ice impact pressures measured aboard USCGC Polar Sea (1982-1984), Report No. TP 8533E, submitted to Transportation Development Centre, Arctec Canada Limited, Kanata, Ontario, 214.

Duan, F., Liu, J., Wang, G. & Yu, Z.L. 2019. Dynamic behaviour of aluminium alloy plates with surface cracks subjected to repeated impacts. *Ships and Offshore Structures 14*(5): 478–491.

Ghajar, R., Khalili, S.M.R., Yarmohammad Tooski, M. & Alderliesten, R.C. 2015. Investigation of the response of an aluminium plate subjected to repeated low velocity impact using a continuum damage mechanics approach. *Fatigue & Fracture of Engineering Materials & Structures 38*(4): 475–488.

Guo, K.L., Zhu, L., Li, Y.G., Yu, T.X., Shenoi, A. & Zhou, Q.W. 2018. Experimental investigation on the dynamic behaviour of aluminium foam sandwich plate under repeated impacts. *Composite Structures* 200: 298–305.

He, X. & Guedes Soares, C. 2021a. Experimental study on the dynamic behaviour of beams under repeated impacts. *International Journal of Impact Engineering* 147: 103724.

He, X. & Guedes Soares, C. 2021b. Numerical study on the pseudo-shakedown of beams under repeated impacts. *Ocean Engineering* 242: 110137.

He, X. & Guedes Soares, C. 2022. Pseudo-shakedown of rectangular plates under repeated impacts. *Marine Structures* 85: 103258.

He, X. & Guedes Soares, C. 2023. Experimental investigation of the repeated impact behaviour of rectangular plates. *In* Ringsberg, J. W. & Guedes Soares C., (Eds.) *Advances in the Analysis and Design of Marine Structures* UK: London, Taylor & Francis Group; 615–620.

He, X. & Guedes Soares, C. 2024a. Experimental and numerical study on the dynamic response of rectangular plates under repeated impacts. *Marine Structures*, under review.

He, X. & Guedes Soares, C. 2024b. The dynamic repeated impact response of rectangular plates under rounded and blunt impactors. *In* Le Sourne, H. & Guedes Soares, C. *Advances in the Collision and Grounding of Ships and Offshore Structures* UK: London, Taylor & Francis Group; 295–301

He, X., Garbatov, Y. & Guedes Soares, C. 2022. Analysis of pseudo-shakedown of rectangular plates under repeated impacts. *Ocean Engineering* 265: 1122609.

Huang, Z.Q., Chen, Q.S. & Zhang, W.T. 2000. Pseudo-shakedown in the collision mechanics of ships. *International Journal of Impact Engineering* 24(1): 19–31.

Jordaan, I.J., Maes, M.A., Brown, P.W. & Hermans, I.P. 1993. Probabilistic analysis of local ice pressures. *Journal of Offshore Mechanics and Arctic Engineering* 115: 83–89.

Jones, N. 1973. Slamming damage. *Journal of Ship Research* 17(2): 80–86.

Jones, N. 2014. Pseudo-shakedown phenomenon for the mass impact loading of plating. *International Journal of Impact Engineering* 65: 33–39.

Kim, M.S., Kim, J.H., Kim, S.K. & Lee, J.M. 2019. Experimental investigation of structural response of corrugated steel sheet subjected to repeated impact loading: performance of LNG cargo containment system. *Applied Sciences 9*(8): 1558.

Lasserre, F. 2014. Case studies of shipping along Arctic routes. Analysis and profitability perspectives for the container sector. *Transportation Research Part A: Policy and Practice*, *66*: 144–161.

Li, L., & Sun, L. 2016. Experimental and numerical investigations of crack behavior and life prediction of 18Cr2Ni4WA steel subjected to repeated impact loading. *Engineering failure analysis 65*: 11–25.

Liu, B., Wang, S., Villavicencio, R. & Guedes Soares, C. 2020. Slamming load and hydroelastic structural response of bow flare areas of aluminium fast displacement crafts. *Ocean Engineering* 218: 108207.

Ralph, F. & Jordaan, I. 2013. Probabilistic methodology for design of arctic ships. *In International Conference on*

Offshore Mechanics and Arctic Engineering 55409: V006T07A010. American Society of Mechanical Engineers.

Shen, W.Q. & Jones, N. 1992. The pseudo-shakedown of beams and plates when subjected to repeated dynamic loads. *Journal of Applied Mechanics* 59(1): 168–175.

Shin, H., Seo, B. & Cho, S.R. 2018. Experimental investigation of slamming impact acted on flat bottom bodies and cumulative damage. *International Journal of Naval Architecture and Ocean Engineering* 10(3): 294–306.

Stoffel, M. 2006a. Limit states of elastic–viscoplastic plate deformations caused by repeated shock wave-loadings. Part 1: Experimental observation. *Mechanics Research Communications 33*(6): 771–774.

Stoffel, M. 2006b. Limit states of elastic–viscoplastic plate deformations caused by repeated shock wave-loadings. Part 2: Theoretical modelling. *Mechanics Research Communications 33*(6): 775–779.

Truong, D.D., Jung, H.J., Shin, H.K. & Cho, S.R. 2018a. Response of low-temperature steel beams subjected to single and repeated lateral impacts. *International Journal of Naval Architecture and Ocean Engineering* 10(6): 670–682.

Truong, D.D., Shin, H.K. & Cho, S.R. 2018b. Repeated lateral impacts on steel grillage structures at room and sub-zero temperatures. *International Journal of Impact Engineering* 113: 40–53.

Wang, S. & Guedes Soares, C. 2017. Review of ship slamming loads and responses. *Journal of Marine Science and Application* 16: 427–445.

Yuhara, T. 1975. Fundamental Study of Wave Impact Loads on Ship Bow (3rd Report) Simulation of Bow Damage. *Journal of the Society of Naval Architects of Japan, 1975*(137): 240–245.

Zeng, Y., Chen, H., Yu, R., Shen, Z., Yu, Z. & Liu, J. 2020. Experimental research on dynamic behavior of circular mild steel plates with surface cracks subjected to repeated impacts in low temperature. *Shock and Vibration* 2020: 3713709.

Zeng, Y., Liu, J., & Zhao, Y. 2022. Experimental Research on the Repeated Impacts Behavior of Aluminum Corrugated-Core Sandwich Structures. *Shock and Vibration* 2022: 4644029.

Zhang, Y., Li, Y.G., Guo, K.L. & Zhu, L. 2021. Dynamic mechanical behaviour and energy absorption of aluminium honeycomb sandwich panels under repeated impact loads. *Ocean Engineering* 219: 108344.

Zhang, Y., Meng, Q. & Ng, S.H. 2016. Shipping efficiency comparison between Northern Sea Route and the conventional Asia-Europe shipping route via Suez Canal. *Journal of Transport Geography 57*: 241–249.

Zhu, L. 1990. Dynamic inelastic behaviour of ship plates in collision (*PhD thesis*). Glasgow: University of Glasgow.

Zhu, L. & Faulkner, D. 1996. Damage estimate for plating of ships and platforms under repeated impacts. *Marine Structures* 9(7): 697–720.

Zhu L., Guo K.L., Li Y.G., Yu, T.X. & Zhou, Q.W. 2018b. Experimental study on the dynamic behaviour of aluminium foam sandwich plates under single and repeated impacts at low temperature. *International Journal of Impact Engineering* 114: 123–132.

Zhu, L., Shi, S.Y. & Jones, N. 2018a. Dynamic response of stiffened plates under repeated impacts. *International Journal of Impact Engineering* 117: 113–122.

Zhu, L., Shi, S.Y. & Yu, T.X. 2015. A new ice load-response model for structural design of ice classed ships. In *ISOPE International Ocean and Polar Engineering Conference* (pp. ISOPE–I). ISOPE.

Advances in Maritime Technology and Engineering – Guedes Soares & Santos (Eds)

Deep active learning with KD-Tree based greedy sampling in structural simulation

C. Jiang
School of Civil Engineering, Tianjin University, Tianjin, P.R. China

N.Z. Chen
State Key Laboratory of Hydraulic Engineering Intelligent Construction and Operation, Tianjin University, Tianjin, P.R. China
School of Civil Engineering, Tianjin University, Tianjin, P.R. China

ABSTRACT: Deep learning has found extensive application in the realm of maritime technology and engineering, such as the development of surrogate model for structural simulation. However, it is expensive to generate training examples, particularly in structural simulation. To address the need for reducing the expensive process of generating training examples, a deep active learning (DAL) with KD-Tree based greedy sampling is proposed. Within this active learning method, a predefined acquisition function is used to select samples for querying from unlabeled sample pool. The acquisition function is based on the greedy sampling approach where the diversity in input and output spaces is increased by selecting new samples from the unlabeled sample pool. The data in the unlabeled sample pool is structured for KD-Tree construction, resulting in a significant reduction in the time complexity of the DAL method. A comparative study with random sampling is made to demonstrate the effectiveness and advantage of the proposed DAL method. A significantly lower test loss and variance is achieved by the proposed method compared to the random sampling approach under conditions of a small sample size. This approach can be applied more broadly to enhance other deep learning models used in the maritime technology and engineering domain, as it provides a computationally efficient way to achieve additional data.

1 INTRODUCTION

The Finite Element Method (FEM) is commonly used in structural analysis. However, with the increasing scale and complexity of engineering structures, the growing degrees of freedom (DOF) make solving FEM extremely time-consuming. Recently, concepts like Simulation Intelligence (SI) (Lavin et al. 2021) and Digital Twins (DT) (Booyse et al. 2020) have gained popularity, emphasizing real-time simulation. To address the challenge of time efficiency, numerous surrogate models, or big models are developed, enabling accelerated FEM computation. Once these surrogate models are trained, real-time access to structural responses becomes feasible by inputting parameters of the engineering structure.

In recent years, significant progress is achieved in the development of deep learning (DL), allowing its widespread application in engineering and scientific fields (Lorente et al. 2017; Haderbache et al. 2021; Perera et al. 2022; Zhao & Chen 2023a; Zhao & Chen 2023b). Due to its powerful generalization and learning capability, DL has become the predominant technology for developing surrogate models (Dalton et al. 2022; Jiang & Chen 2023; Pfaff et al. 2020; Black & Najafi 2022). However, a fundamental challenge in deep learning is obtaining large number of labeled training data. In other words, the FEM needs to be solved numerous times under various parameters, including different boundary conditions and loads. In contrast to the fields of computer vision (CV) and natural language processing (NLP), training datasets are often numbered in millions. This scale of data required in the structural domain is challenging to envision. Thus, learning from a small amount of data is essential for the use of DL in structural simulation.

Several methods are developed to reduce the number of labeled data for DL, such as transfer learning (Torrey & Shavlik 2010) and few-shot learning (Snell et al., 2017). Transfer learning is the reuse of a pre-trained model on a new problem and it is helpful in settings where there is limited data on the task of interest but abundant data on a related problem. Few-shot learning is another technique through which the goal of training models on a small number of samples is achieved by learning the similarity between samples. It is worth noting that transfer learning requires abundant data on a related problem, while few-shot learning is more suitable for classification tasks, whereas structural simulation often involves regression problems.

To reduce the cost while maximizing the learning performance, a deep active learning (DAL) with KD-

DOI: 10.1201/9781003508762-51

Tree based greedy sampling is proposed in this paper. In most machine learning research, the learners (or models) are treated as a passive recipient of data (Cohn et al. 1996). However, data collected in this "passive" way is not always effective. Therefore, it is desirable to only use examples with maximal information about the function. Methods where examples pointed out by the learners are called active learning (AL) (Cohn et al. 1994), a concept upon which our proposed method is based. More specifically, AL is a method that the most useful samples are selected from the unlabeled dataset, and they are delivered to the oracle (e.g., FEM solver) for labeling. The objective is to minimize the cost of labeling as much as possible while still maintaining performance. AL approaches can be divided into membership query synthesis (Schumann & Rehbein 2019), stream-based selective sampling (Loy et al. 2012) and pool-based selective sampling (Wu 2018). Pool-based sampling strategy is used in this paper because it is more suitable for most DL problem (Ren et al. 2021). The framework diagram of the pool-based AL cycle is illustrated in Figure 1. In the initial stage, a small number of samples is randomly selected from the unlabeled dataset. These selected samples are submitted to the oracle for labeling and are subsequently incorporated into the labeled training dataset. Following this, the DL model is trained based on the updated labeled dataset. In the subsequent stages, a new sample is strategically chosen from the unlabeled dataset based on the knowledge gained from the previous round of training. The newly selected sample is then forwarded to the oracle for labeling, and the training process is repeated. This iterative cycle continues until either the label budget is exhausted or the predetermined termination conditions are met.

The most important part of AL learning is selecting sample from unlabeled pool, also known as the acquisition function. The proposed DAL model is inspired by the greedy sampling (GS) approach (Yu & Kim 2010; Wu et al. 2019). The most diverse samples from the labeled training dataset are selected by GS and submitted to the oracle for labeling. Three different active learning (AL) approaches are employed in our study: greedy sampling on the input space (GSx), greedy sampling on the output space (GSy), and greedy sampling on both input and output space (iGS). Although these approaches are easy to understand, implement, and have proven useful in various tasks, a major challenge is their high time complexity. In the proposed DAL model, K-Dimensional Tree (KD-Tree) (Bentley 1975) is utilized to significantly reduce the time complexity of GS-based AL approaches.

For evaluation, the deep learning (DL) model based on the graph neural network (GNN) proposed by Jiang and Chen (2023) is employed to examine our proposed DAL model. The DL model from their study is specifically designed to accelerate FEM. The comparative study with randomly sampling is made to demonstrate the capability of our proposed DAL model. As a result, high accuracy is observed with the same number of samples under the DAL method.

In Section 2, the fundamentals of our proposed model are described. In Section 3, numerical cases designed to test the capability of the proposed DAL model are presented. Finally, in Section 4, our conclusions are drawn.

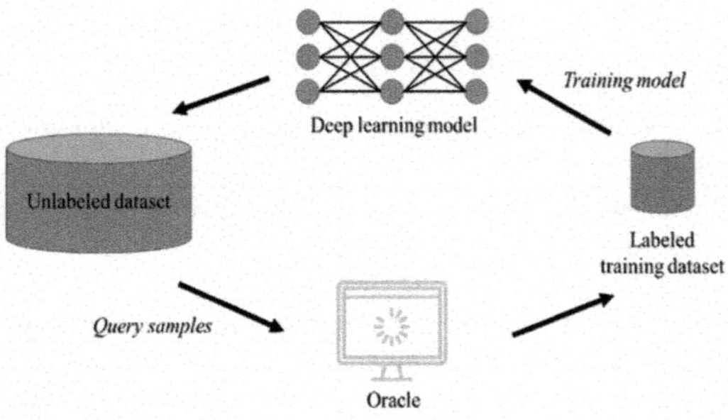

Figure 1. The pool-based AL cycle.

2 METHODOLOGY

In this section, the existing GS approaches in the literature and the KD-tree based GS approach proposed in this paper are introduced.

2.1 *Greedy sampling*

In most AL approaches, DL model is updated in each iteration and the predictions are computed for unlabeled samples in the acquisition function. In GSx approach, which is a passive sampling method, samples are selected based entirely on its location in the feature space. Therefore, it is dependent on the DL model and characterized by low computational cost.

Assuming the pool Z consists of N samples $\{\mathbf{x}_n\}_{n=1}^{N}$, initially, a sample is randomly selected from it and submitted to oracle for labelling. The objective of this approach is to subsequently select $K-1$ samples from Z, label them, and add them to the training dataset S for training. For ease of exposition, assuming the first k samples have already been selected. For each of the remaining $N-k$ unlabeled samples $\{\mathbf{x}_n\}_{n=k+1}^{N}$, the distance $d_{nm}^{\mathbf{x}}$ to each of the k labeled samples in training dataset S is computed by GSx:

$$d_{nm}^{\mathbf{x}} = ||\mathbf{x}_n - \mathbf{x}_m||, m = 1, \ldots, k; n = k+1, \ldots, N \tag{1}$$

where $\mathbf{x}_n$ is the input feature of sample n and $\mathbf{x}_m$ is the input feature of sample m. Then, the shortest distance from $\mathbf{x}_n$ to all samples in dataset S is:

$$d_n^{\mathbf{x}} = \min_n d_{nm}^{\mathbf{x}}, n = k+1, \ldots, N_m \tag{2}$$

Finally, the sample with the maximum $d_n^{\mathbf{x}}$ is selected to label and added to dataset S:

$$\mathbf{x}_s = \underset{n}{argmax}\ d_n^{\mathbf{x}}, n = k+1, \ldots, N \tag{3}$$

In summary, the sample farthest away from the training dataset S in the input space is selected to enhance diversity within the training dataset in each iteration. The algorithm of GSx is provided in Table 1.

The AL is achieved by increasing the diversity of input space of training dataset in GSx approach. The GSy aims to enhance the diversity of output space of dataset.

Assuming the pool Z consists of N samples $\{\mathbf{x}_n\}_{n=1}^N$, initially, few (K_0) samples are selected according to the GSx approach from it and submitted to oracle for labelling. The objective of this approach is to select one sample from Z, label it, and add it to the training dataset S for training. This process is repeated for $K - K_0$ iterations.

For simplicity, assuming the first $k(k > K_0)$ samples have been labeled and added to training dataset S, and the DL model $f(x)$ is constructed based on $S = \{\mathbf{x}_n\}_{n=1}^k$. For each of the remaining $N - k$ unlabeled samples $\{\mathbf{x}_n\}_{n=k+1}^N$, the distance $d_{nm}^{\mathbf{y}}$ to each of the output of the k labeled samples in training dataset is computed by GSy:

$$d_{nm}^{\mathbf{y}} = ||f(\mathbf{x}_n) - \mathbf{y}_m||, m = 1, \ldots, k; n = k+1, \ldots, N \tag{4}$$

where $f(\mathbf{x}_n)$ is the prediction of $\mathbf{x}_n$, $\mathbf{y}_m$ is the label of $\mathbf{x}_m$. Then, the shortest distance from $f(\mathbf{x}_n)$ to the labels of all training dataset S is:

$$d_n^{\mathbf{y}} = \underset{n}{min}\ d_{nm}^{\mathbf{y}}, n = k+1, \ldots, N_m \tag{5}$$

Eventually, the samples with maximum $d_n^{\mathbf{y}}$ is selected to label and added to training dataset S:

$$\mathbf{x}_s = \underset{n}{argmax}\ d_n^{\mathbf{y}}, n = k+1, \ldots, N \tag{6}$$

In summary, an initial DL model is constructed according to GSx method, and then the sample farthest from the training dataset S in the output space is selected to enhance diversity among these samples in training dataset in each iteration. The algorithm of GSy is provided in Table 2.

The diversity in the input space is considered in the GSx approach. However, not all features are deemed useful, and their importance varies. In other words, the weights of features are not taken into account. On the other hand, the weights of features are implicitly considered in the GSy method because the diversity in the output space is taken into account (Wu et al. 2019). Nevertheless, the DL model $f(x)$, used as a sensitive predictor in GSy, is constructed based on a very small number of samples and may not be highly accurate. In other words, feature selection/weighting in GSy may not be considered reliable.

Table 1. The algorithm of GSx.

The algorithm of GSx.
Set $Z = \{\mathbf{x}_n\}_{n=1}^N$, and $S = \emptyset$;
Randomly sample $\mathbf{x}_r$ from Z;
Label $\mathbf{x}_r$;
Move $\mathbf{x}_r$ from Z to S;
for $k = 1, \ldots, K-1$
  for $n = k+1, \ldots, N$
    compute $d_n^{\mathbf{x}}$ in (2);
  end
  Select $\mathbf{x}_s$ in (3);
  Label $\mathbf{x}_s$;
  Move $\mathbf{x}_s$ from Z to S;
end
Train DL model $f(x)$ based on $S = \{\mathbf{x}_n\}_{n=1}^K$

Table 2. The algorithm of GSy.

The algorithm of GSy.
Set $Z = \{\mathbf{x}_n\}_{n=1}^N$, and $S = \emptyset$;
Randomly sample $\mathbf{x}_r$ from Z;
Label $\mathbf{x}_r$;
Move $\mathbf{x}_r$ from Z to S;
for $k = 1, \ldots, K_0 - 1$
  for $n = k+1, \ldots, N$
    compute $d_n^{\mathbf{x}}$ in (2);
  end
  Select $\mathbf{x}_s$ in (3);
  Label $\mathbf{x}_s$;
  Move $\mathbf{x}_s$ from Z to S;
end
Train DL model $f(x)$ based on $S = \{\mathbf{x}_n\}_{n=1}^{K_0}$.
for $k = K_0, \ldots, K-1$
  for $n = k+1, \ldots, N$
    compute $d_n^{\mathbf{y}}$ in (5);
  end
  Select $\mathbf{x}_s$ in (6);
  Label $\mathbf{x}_s$;
  Move $\mathbf{x}_s$ from Z to S;
  Train DL model $f(x)$ based on $S = \{\mathbf{x}_n\}_{n=1}^k$.
end

Another approach, iGS, combines GSx and GSy at the same time. Like GSy, assuming the pool Z consists of N samples $\{\mathbf{x}_n\}_{n=1}^N$, initially, few (K_0) samples are selected according to the GSx approach from it and submitted to oracle for labelling. The objective of this approach is to select one sample from Z, label it, and add it to the training dataset S for training. This process is repeated for $K - K_0$ iterations.

For simplicity, assuming the first $k(k > K_0)$ samples have been labeled and added to training dataset S, and the DL model $f(x)$ is constructed based on $S = \{\mathbf{x}_n\}_{n=1}^k$. For each of the remaining $N - k$ unlabeled samples $\{\mathbf{x}_n\}_{n=k+1}^N$, the distance $d_{nm}^{\mathbf{xy}}$ to each of the output and input of the k labeled samples in training dataset is computed by iGS:

$$d_n^{\mathbf{xy}} = \min_m d_{nm}^{\mathbf{y}}, d_{nm}^{\mathbf{x}}, n = k+1, \ldots, N \quad (7)$$

Then, the samples with maximum $d_n^{\mathbf{xy}}$ is selected to label and added to training dataset S:

$$\mathbf{x}_s = \underset{n}{argmax}\ d_n^{\mathbf{xy}}, n = k+1, \ldots, N \quad (8)$$

In summary, the sample farthest away from the training dataset S in the input space and output space is selected to enhance diversity within the training dataset in each iteration. The algorithm of iGS is provided in Table 3.

Table 3. The algorithm of iGS.

```
The algorithm of iGS.
Set Z = {x_n}_{n=1}^N, and S = ∅;
Randomly sample x_r from Z;
Label x_r;
Move x_r from Z to S;
for k = 1,...,K_0 − 1
   for n = k + 1,...,N
      compute d_n^x in (2);
   end
   Select x_s in (3);
   Label x_s;
   Move x_s from Z to S;
end
Train DL model f(x) based on S = {x_n}_{n=1}^{K_0}.
for k = K_0,...,K − 1
   for n = k + 1,...,N
      compute d_n^xy in (7);
   end
   Select x_s in (8);
   Label x_s;
   Move x_s from Z to S;
   Train DL model f(x) based on S = {x_n}_{n=1}^k
end
```

Although these GS approaches are useful in many machine learning tasks (Wu et al., 2019), their time complexity is high. Assuming that the forward time of the DL model $f(x)$ is very small and training time in GSy and iGS is not considered, the time complexity of one epoch of these GS algorithms is $O(N\ k)$. Therefore, KD-Tree is used to decrease the time complexity of GS algorithms in our study.

2.2 *KD-Tree based greedy sampling*

In data science, KD Tree is a popular data structure used to divide points in K-dimensional space, where K is usually a large number. Due to its structure, it is highly efficient in the search for points in multidimensional space, including nearest-neighbor searches and range searches (Bentley 1990). A KD Tree is a binary tree in which each k-dimensional point is represented by each node. Every non-leaf node in the tree is served as a hyperplane, dividing the space into two partitions. This hyperplane is perpendicular to the chosen axis, which is associated with one of the K dimensions. There are different strategies for choosing an axis when dividing, but the most common one is to cycle through each of the K dimensions repeatedly and select a midpoint along it to divide the space.

In the construction of KD Tree, points are recursively partitioned in the space. First, the process is started by selecting an axis. The median point is then selected along the axis and the space is split. The remaining points are lied into two subsets based on their position relative to the splitting hyperplane. The left and right children of the root are then created with the points in the subset that are lied to each side of the hyperplane. This process is repeated recursively for each child node, where a new axis is selected and the points are split based on their position relative to the new hyperplane.

Nearest-neighbor searches (NNS) are efficiently facilitated by KD Trees, which are recursively traversed based on axis-aligned hyperplanes. The closest point to a query point is identified, and the "current best" distance is updated at each step. The search is optimized through the use of a hypersphere concept, determining whether the other tree branch should be explored. This approach, involving the comparison of absolute distances and the elimination of unnecessary branches, results in an effective and streamlined nearest-neighbor search process.

This efficient search process is applied in the GS approaches, significantly enhancing the efficiency of the proposed DAL model. In GSx, assuming the pool Z consists of N samples $\{\mathbf{x}_n\}_{n=1}^N$, initially, a sample is randomly selected from it and submitted to oracle for labelling. The objective of this approach is to subsequently select $K-1$ samples from Z, label them, and add them to the training dataset S for training. For ease of exposition, assuming the first k samples have already been selected. A KD Tree is then built based on the training dataset $S = \{\mathbf{x}_n\}_{n=1}^k$. For each unlabeled sample in pool Z, the minimum distance $d_n^{\mathbf{x}}$ to dataset S is computed by using NNS based on KD Tree. Finally, the sample $\mathbf{x}_s$ with the maximum $d_n^{\mathbf{x}}$ is selected for labelling and added to dataset S according to Equation (3). The time complexity for each iteration of the existing GSx approach is approximately $O(N\ k)$, while the KD-Tree GSx approach is $O(N\ logk)$ under balanced conditions. In the case of a highly unbalanced tree, the time complexity reverts to $O(N\ k)$.

Furthermore, the KD Tree can also be applied to GSy and iGS approaches. The minimum distances, $d_n^{\mathbf{y}}$ in Equation (5) and $d_n^{\mathbf{xy}}$ in Equation (7), are computed using NNS based on the KD Tree structure, reducing the complexity to $O(N\ logk)$ at most.

In summary, a KD Tree is built as the data structure in training dataset, expediting the process of find minimum distance from each sample in pool Z to dataset S. The reduction in time complexity makes the proposed DAL model suitable for decreasing the required labeling times in the DL model for structural analysis.

3 NUMERICAL EXAMPLES

Numerical experiments on FEM datasets and GNN model from the work of Jiang and Chen (2023) are conducted in this section to demonstrate the performance of the proposed DAL model.

3.1 *Datasets and GNN model*

Case 1 and case 2 in the work of Jiang and Chen (2023) are used as datasets in this paper. The graph data structure is used to represent the FEM mesh in their model. The information of coordinate, boundary and external force of each FEM element are embedded in the graph. The von-Mises stress of each FEM element is the output of neural network. In case 1, the input feature is the magnitude of external force, while the output is the von-Mises stress of the structure. In case 2, the input features include the magnitude and the point of application of external force, while the output is von-Mises stress. In each dataset, a total of 136 samples are present, with 128 designated for training and 8 for validation. Their summary is given in Table 4. The diagram is illustrated in Figure 2.

Table 4. Summary of datasets.

Dataset	No. of samples	Dimension of input of a node	Dimension of output of a node
Case 1	136	6	1
Case 2	136	8	1

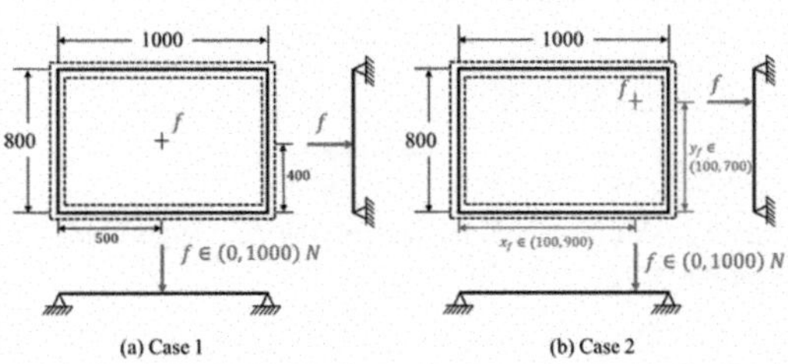

Figure 2. Illustration of the variables (marked in red) in two cases. (a) Case 1; (b) Case 2 (Jiang and Chen (2023)).

The DL model is based on GNN. This DL model can be trained on non-Euclidean data, which is well-suited for structure analysis, especially in cases where the FEM mesh represents non-Euclidean data. Through the utilization of message passing techniques (Gilmer et al. 2017) and neural networks, FEM meshes can be effectively trained on by GNN. The optimizer in our experiments is stochastic gradient descent (SGD) (Robbins & Monro 1951). The initial learning rate is 0.01 and the Cosine Annealing Warm Restarts (CAWR) is used as the scheduler to adjust the learning rate during training (Loshchilov & Hutter 2016). The number of iterations for the first restart is 10, while the factor increases the iterations is 2 in CAWR scheduler. The number of training epochs is set as 5110 in all experiments. Further details of architecture of neural networks can be referenced in the work of Jiang and Chen (2023).

3.2 *Algorithms*

The performance of the following algorithms is compared in this section.

1. Random sampling (RS), which involves the random selection of all K samples.
2. GSx, which has been introduced in Section 2.1.
3. GSy, which has been introduced in Section 2.1.
4. iGS, which has been introduced in Section 2.1.

All four algorithms are implemented on the datasets and GNN model mentioned in Section 3.1, and they minimize the following loss function:

$$\mathcal{L} = \sqrt{\frac{1}{n}\sum_{i=1}^{n}\left(\log(\hat{\sigma}_i + 1) - \log(\sigma_i + 1)\right)^2} \quad (9)$$

where σ_i is the stress at node i generated by the FEM solver, $\hat{\sigma}_i$ is the prediction at node i generated by GNN-based DL model.

Moreover, mean absolute percentage error (MAPE) is chosen as the error function to represent the accuracy of the proposed model, which can be formulated as:

$$\varepsilon = \frac{100}{n}\sum_{i=1}^{n}\left|\frac{\sigma_i - \hat{\sigma}_i}{\sigma_i + 0.001}\right| \quad (10)$$

3.3 *Results and discussion*

128 samples are selected as unlabeled training dataset Z and 8 samples are chosen for test dataset from both case 1 and case 2. The MAPEs and their variance for the four algorithms on the test dataset of case 1, averaged over 5 runs, are shown in Figure 3 and Figure 4, respectively.

When the number of training samples is only 9, GSx, GSy and iGS yield similar results as all training samples are selected by GSx. The test loss is only 7% for all of these three GS methods while it is 53% for random sampling. With the increase in the number of samples, the test loss is gradually decreased. When the number of samples reaches 20 to 23, it is suddenly declined to around 0.9%, whereas the average loss of random sampling remains at 7%.

On the other hand, the loss range of GS methods is lower than that of random sampling. For example, at the point where GS methods start to decline (number of samples is 23), the maximum loss of iGS is 0.87%, while the minimum loss is 0.53%. Meanwhile, for random sampling, the maximum loss is 19.20%, and the minimum loss is 1.57%. It is noteworthy that, in certain instances (e.g., when the number of samples is 25), the minimum loss of random sampling may be smaller than that of GS methods. However, it is crucial to emphasize that their maximum loss substantially exceeds that of GS methods. It can be observed that the GS approaches are stable, as they train a predictor with small variance.

In case 2, the input feature is more complex than that of case 1 as the point of application of external force is variable. The MAPEs and their variance for the four algorithms on the test dataset of case 2, averaged over 5 runs, are shown in Figure 5 and Figure 6, respectively.

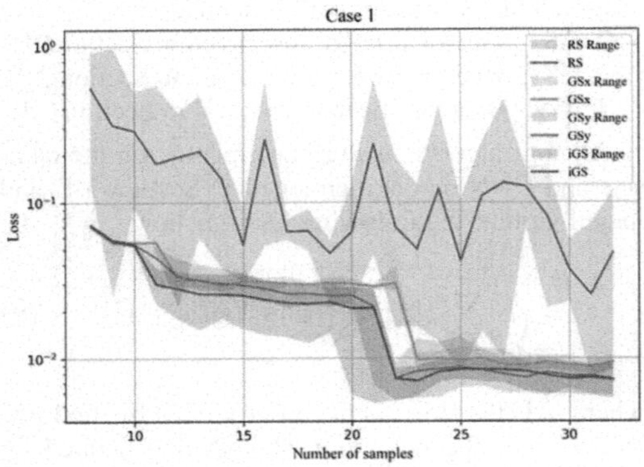

Figure 3. Performances of the four algorithms on case 1, averaged over 5 runs.

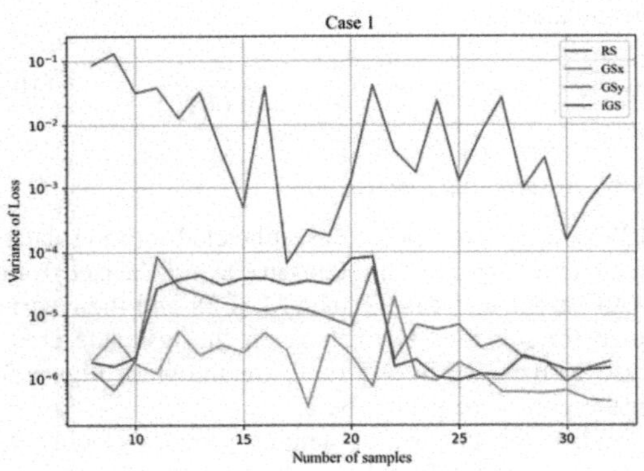

Figure 4. Variance of the loss of the four algorithms on case 1, averaged over 5 runs.

In alignment with case 1, comparable results are yielded by GSx, GSy, and iGS when the number of samples is 9. With an increase in the number of samples, a sharp decrease in average loss is observed for GSx, GSy, and iGS at counts of 12, 19, and 13, respectively, reaching 14.6%, 5.86%, and 14.2%, respectively. In contrast, notably higher losses are observed for random sampling at counts of 12, 19, and 13, standing at 46.0%, 19.1%, and 36.1%, respectively. Upon reaching 32 samples, the test loss for random

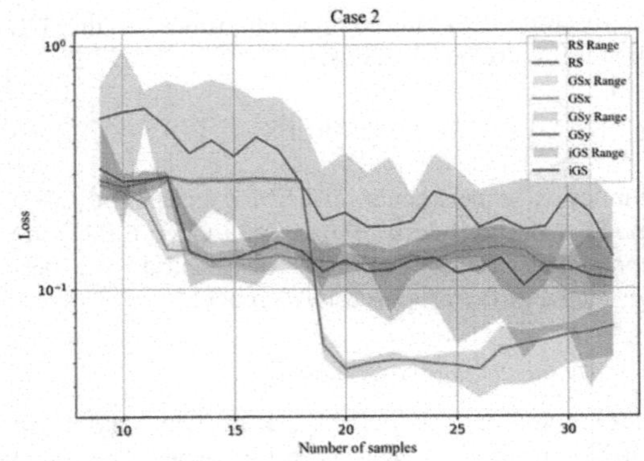

Figure 5. Performances of the four algorithms on case 2, averaged over 5 runs.

sampling is 13.6%, while for GS approaches, it is 11.6%, 6.99%, and 10.9%, respectively, with GSy significantly outperforming random sampling. Similarly, the loss range of GS methods in case 2 is smaller than that of random sampling. The minimum mean, maximum, and minimum test losses of four sampling methods in Case 1 and Case 2 for a range of samples from 9 to 32, along with the variance at the sample count of 32 are presented in Table 5.

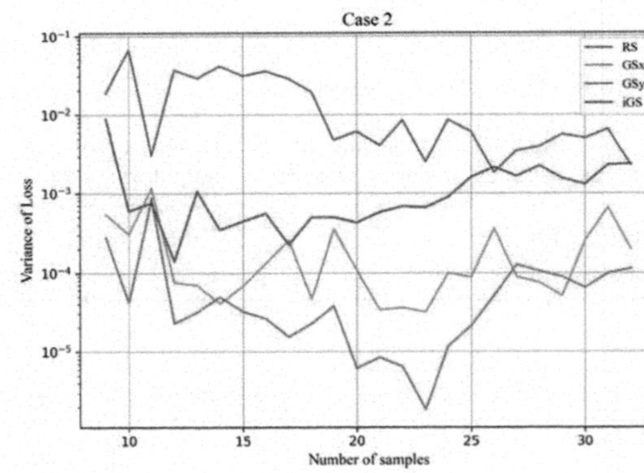

Figure 6. Variance of the loss of the four algorithms on case 2, averaged over 5 runs.

Table 5. Detailed results for Case 1 and Case 2.

		Mean(%)	Max(%)	Min(%)	Variance
Case 1	RS	2.26	5.74	0.76	1.52e-3
	GSx	0.88	0.98	0.78	4.50e-7
	GSy	0.73	0.89	0.54	1.89e-6
	iGS	0.71	0.87	0.52	1.49e-6
Case 2	RS	13.61	23.01	7.84	2.31e-3
	GSx	11.64	13.22	9.16	1.97e-4
	GSy	4.66	5.03	3.56	8.75e-5
	iGS	10.30	14.51	3.93	1.38e-5

4 CONCLUSION

A novel approach was introduced to reduce the demand for labeled samples in deep learning (DL) models employed in structural analysis—deep active learning (DAL) model. In the proposed DAL model, the samples farthest away from the training dataset in the input space and/or output space was selected in the unlabeled pool by using greedy sampling (GS) approaches to enhance diversity within the training dataset in each iteration. To address potential concerns about the time complexity associated with GS approaches, a KD-Tree structure was employed for the training dataset, resulting in a significant reduction in time complexity from $O(N\ k)$ to at most $O(N\ logk)$. A numerical experiment was conducted to investigate the performance of the proposed DAL model and the random sampling method. The results demonstrate that a smaller test loss and a smaller variance are achieved by the proposed DAL model, attesting to the reliability and robustness of the model. The potential of DAL with KD-Tree based greedy sampling is underscored in this study as a promising approach for efficient DL model training in structural analysis. The proposed method will be employed in handling more complex datasets and integrating with deep learning models in future research.

REFERENCES

Battaglia, P. W., Hamrick, J. B., Bapst, V., Sanchez-Gonzalez, A., Zambaldi, V., Malinowski, M., … & Pascanu, R. 2018. Relational inductive biases, deep learning, and graph networks. arXiv preprint arXiv:1806.01261.

Bentley, J. L. 1975. Multidimensional binary search trees used for associative searching. Communications of the ACM, 18(9), 509–517.

Bentley, J. L. 1990, May. K-d trees for semidynamic point sets. In Proceedings of the sixth annual symposium on Computational geometry (pp. 187–197).

Black, N., & Najafi, A. R. 2022. Learning finite element convergence with the multi-fidelity graph neural network. Computer Methods in Applied Mechanics and Engineering, 397, 115120.

Booyse, W., Wilke, D. N., & Heyns, S. 2020. Deep digital twins for detection, diagnostics and prognostics. Mechanical Systems and Signal Processing, 140, 106612.

Cohn, D., Atlas, L., & Ladner, R. 1994. Improving generalization with active learning. Machine learning, 15, 201–221.

Cohn, D. A., Ghahramani, Z., & Jordan, M. I. 1996. Active learning with statistical models. Journal of artificial intelligence research, 4, 129–145.

Dalton, D., Gao, H., & Husmeier, D. 2022. Emulation of cardiac mechanics using Graph Neural Networks. Computer Methods in Applied Mechanics and Engineering, 401, 115645.

Gilmer, J., Schoenholz, S. S., Riley, P. F., Vinyals, O., & Dahl, G. E. 2017, July. Neural message passing for quantum chemistry. In International conference on machine learning (pp. 1263–1272). PMLR.

Haderbache, A., Shirahata, K., Tabaru, T., & Okuda, H. 2021. Surrogate Model for Structural Analysis Simulation using Graph Convolutional Network (Unrefereed Workshop Manuscript).

Jiang, C., & Chen, N. Z. 2023. Graph Neural Networks (GNNs) based accelerated numerical simulation. Engineering Applications of Artificial Intelligence, 123, 106370.

Lavin, A., Krakauer, D., Zenil, H., Gottschlich, J., Mattson, T., Brehmer, J., … & Pfeffer, A. 2021. Simulation intelligence: Towards a new generation of scientific methods. arXiv preprint arXiv:2112.03235.

Lorente, D., Martínez-Martínez, F., Rupérez, M. J., Lago, M. A., Martínez-Sober, M., Escandell-Montero, P., … & Martín-Guerrero, J. D. 2017. A framework for modelling the biomechanical behaviour of the human liver during breathing in real time using machine learning. Expert Systems with Applications, 71, 342–357.

Loy, C. C., Hospedales, T. M., Xiang, T., & Gong, S. 2012, June. Stream-based joint exploration-exploitation active learning. In 2012 IEEE Conference on Computer Vision and Pattern Recognition (pp. 1560–1567). IEEE.

Loshchilov, I., & Hutter, F. 2016. Sgdr: Stochastic gradient descent with warm restarts. arXiv preprint arXiv:1608.03983.

Perera, R., Guzzetti, D., & Agrawal, V. 2022. Graph neural networks for simulating crack coalescence and propagation in brittle materials. Computer Methods in Applied Mechanics and Engineering, 395, 115021.

Pfaff, T., Fortunato, M., Sanchez-Gonzalez, A., & Battaglia, P. W. 2020. Learning mesh-based simulation with graph networks. arXiv preprint arXiv:2010.03409.

Ren, P., Xiao, Y., Chang, X., Huang, P. Y., Li, Z., Gupta, B. B., … & Wang, X. 2021. A survey of deep active learning. ACM computing surveys (CSUR), 54(9), 1–40.

Robbins, H., & Monro, S. 1951. A stochastic approximation method. The annals of mathematical statistics, 400–407.

Schumann, R., & Rehbein, I. 2019, November. Active learning via membership query synthesis for semi-supervised sentence classification. In Proceedings of the 23rd conference on computational natural language learning (CoNLL) (pp. 472–481).

Snell, J., Swersky, K., & Zemel, R. 2017. Prototypical networks for few-shot learning. Advances in neural information processing systems, 30.

Torrey, L., & Shavlik, J. 2010. Transfer learning. In Handbook of research on machine learning applications and trends: algorithms, methods, and techniques (pp. 242–264). IGI global.

Wu, D. 2018. Pool-based sequential active learning for regression. IEEE transactions on neural networks and learning systems, 30(5), 1348–1359.

Wu, D., Lin, C. T., & Huang, J. 2019. Active learning for regression using greedy sampling. Information Sciences, 474, 90–105.

Yu, H., & Kim, S. 2010, December. Passive sampling for regression. In 2010 IEEE International Conference on Data Mining (pp. 1151–1156). IEEE.

Zhao, Z., & Chen, N. Z. 2023a. Spatial-temporal graph convolutional networks (STGCN) based method for localizing acoustic emission sources in composite panels. Composite Structures, 323, 117496.

Zhao, Z., & Chen, N. Z. 2023b. Acoustic Emission Based Damage Source Localization for Heterogeneous Structure of Wind Turbine Blades Using Long Short-Term Memory Neural Networks. In International Conference on Offshore Mechanics and Arctic Engineering (Vol. 86847, p. V002T02A009). American Society of Mechanical Engineers.

Advances in Maritime Technology and Engineering – Guedes Soares & Santos (Eds)
© 2024 The Author(s), ISBN 978-1-032-83099-5

Uncertainty analysis of ultimate strength models of stiffened plates

A. Kakaie, Y. Garbatov & C. Guedes Soares
Centre for Marine Technology and Ocean Engineering (CENTEC), Instituto Superior Técnico, Universidade de Lisboa, Lisbon, Portugal

ABSTRACT: This study analyses the uncertainties embedded in different ultimate strength models of stiffened plates accounting for model, material properties and geometrical parameter variability. The model uncertainty is estimated based on the available experimental data for two different collapse modes, and the model's statistical uncertainty is propagated based on the material properties and geometrical variability. The accuracy of the developed probabilistic models is investigated based on the available experimental data, including the plate and column slenderness. The output of the present analysis is fundamental for any further structural reliability analysis of marine structures built of stiffened plates.

1 INTRODUCTION

Stiffened plates are fundamental for building ship and offshore structures, satisfying the requirements for the ultimate limit state according to the Classification Society Rules. The ship hull consists of stiffened panels that define the deck, bottom, side shell, and bulkhead structures. The stiffened panels are composed of plates reinforced by transverse frames and longitudinal stiffeners.

Estimating the ultimate strength of stiffened plates is challenging due to the embedded uncertainties due to the combined effect of geometries, nonlinear materials, initial imperfections, residual stresses and load combination that may trigger a variety of collapse modes (Guedes Soares & Søreide, 1983; Ringsberg *et al.*, 2021). Therefore, analysing the stiffened plate strength is essential.

Uncertainties are divided into aleatory ones, representing the natural variability of the physical process, and epistemic uncertainty due to the lack of knowledge, and they can be reduced by collecting more information (Guedes Soares, 1997). A typical example of aleatory uncertainty in ship design is the variability of wave-induced loads (Parunov *et al.*, 2022). Epistemic uncertainty can be classified into (1) measurement uncertainty due to the inaccuracies in the measuring instruments, variability in environmental conditions and human factors, (2) model uncertainty due to simplifications, assumptions and idealisations in the mathematical model formulations of physical processes, (3) statistical uncertainty, which is due to limited information such as a limited number of observations (Nowak & Collins, 2012).

The structural response of the stiffened panels subjected to compressive loads is essential due to the inherent instability associated with potential failures, particularly on the deck and bottom ship hull structures, since the buckling collapse of stiffened plates governs the ship's hull ultimate strength under vertical bending (Guedes Soares & Gordo, 1997; Gaspar et al., 2011;). Compressive collapse failure mode occurs in an unstable mode, posing considerable safety risks and harmful consequences (Garbatov *et al.*, 2011).

Several experimental tests have been performed to predict the ultimate strength of stiffened plates subjected to uniaxial compressive forces and their combination with lateral loads. Dowling *et al.* (1973) and Smith (1975) presented a series of tests on full-scale steel grillages representing typical warship decks and single-bottom structures under compressive load combined, in some cases with lateral pressure. Faulkner (1977) and Horne *et al.* (1976). analysed the ultimate strength of stiffened panels subjected to longitudinal compressive load. Tanaka & Endo (1988) conducted experimental investigations on longitudinally stiffened panels, analysing the ultimate compressive strength.

Chen (2003) tested several stiffened panels under uniaxial compressive, purely or in combination with lateral loads. In addition, several experimental studies have been conducted on stiffened panels with associated plates of very high tensile, considering different combinations of mechanical material properties and geometric configurations for the stiffeners Gordo & Guedes Soares, 2008, 2009, 2011, 2012). Following these studies, several experimental tests have been performed to investigate the impact of various boundary conditions and spans on the ultimate strength of the stiffened panels and different boundary conditions and spans (Xu & Guedes Soares, 2013a, 2013b, 2021).

Several studies have investigated the behaviour and design of stiffened panels under compressive loads based on analytical and numerical approaches

DOI: 10.1201/9781003508762-52

(e.g. Guedes Soares & Søreide, 1983; Guedes Soares, 1988a, Paik & Kim, 2002; Paik & Thayamballi, 2003). In addition, empirical formulations based on experimental or numerical data were proposed to predict a stiffened panel's ultimate compressive strength. Lin (1985) proposed an empirical relation for stiffened panels with large dimensions of stiffeners, which has an acceptable accuracy for a lower slenderness ratio (λ). Based on the experimental data, Paik & Thayamballi (1997) modified Lin's formula for a broader range of stiffened panels.

With the rapid development in computing technology, numerical simulation-based approaches such as finite element (FE) approaches have been widely applied in many engineering fields and uncertainty quantifications have been developed for those approaches (Mathies et al. 1997). Empirical and semi-empirical formulations based on linear and non-linear FE-based numerical simulations have been proposed in several studies in the last decades.

Zhang & Khan (2009) presented extensive non-linear FE analyses and developed a formulation for the ultimate compressive strength of plates and stiffened panels with $\lambda \leq \sqrt{2}$ for ship structures. The proposed formulation method was verified against a large number of published analyses and was also compared to experimental results. Kim *et al.* (2017) proposed an empirical formula based on non-linear FE analyses, which covers larger column slenderness ($\lambda \leq 5$). Considering the effect of longitudinal compressive load with lateral pressure, Xu *et al.* (2018) proposed an empirical formulation with ten coefficients by conducting 1296 cases of numerical simulations.

Although using empirical formulations is one of the possible ways to predict ultimate strength, their applicability depends on their accuracy. The proposed empirical models are calibrated based on the limited experimental or numerical data. Therefore, the estimation of the ultimate strength of stiffened panels involves uncertainties arising from several sources: material properties, such as yield strength and module of elasticity, which can be influenced by factors like manufacturing processes, material composition, and environmental conditions; fabrication, such as initial imperfections and irregularities in structural profiles and loadings, including dynamic forces and varying operational conditions.

In addition, an incomplete understanding of the behaviour of stiffened plates, especially in complex loading scenarios, and the challenges associated with accurately modelling and estimating their response adds another layer of uncertainty, which can be considered model uncertainty.

The main objective of this work is to analyse the uncertainties associated with the available empirical and semi-empirical models for predicting the ultimate strength of stiffened plates subjected to uniaxial compressive load, expanding earlier work that addressed plate elements (Guedes Soares, 1988b).

The uncertainties related to the predicted ultimate strength of the stiffened plates are estimated based on the model uncertainty and uncertainties related to the model parameters. Model uncertainty is calculated based on the available experimental data for two different ranges of the slenderness ratio. The uncertainty associated with the model parameter is obtained based on the Monte-Carlo simulation method regarding the variability of material properties and dimensions.

The remainder of the paper is structured as follows. The available mathematical models and collected experimental data are described in Section 2. The uncertainty analysis of the ultimate strength models of stiffened plates considering various sources of uncertainty is discussed in Section 3. Concluding remarks are provided in the last section.

2 ULTIMATE STRENGTH FORMULATIONS AND EXPERIMENTAL DATA

Over the last decades, several ultimate strength empirical formulations have been proposed based on experimental and numerical data, which are analysed in this section. The basic format of most of these empirical models can be expressed as:

$$R = \frac{\sigma_u}{\sigma_{Yeq}} = f(\lambda, \beta) \tag{1}$$

where σ_u and σ_{Yeq} are the ultimate strength and equivalent yield strength of the stiffened plates subjected to uniaxial compressive load, and λ and β are the column and plate slenderness ratio, respectively. In this study, four empirical formulations proposed by Lin (1985), Paik & Thayamballi (1997), Zhang & Khan (2009) and *Kim et al.* (2017) are investigated.

Lin (1985) proposed an empirical formulation to estimate the ultimate strength of stiffened plates:

$$R_L = \left[0.960 + 0.765\lambda^2 + 0.176\beta^2 + 0.131\lambda^2\beta^2 + 1.064\lambda^4\right]^{-\frac{1}{2}} \tag{2}$$

where

$$\lambda = \frac{l}{\pi r}\sqrt{\frac{\sigma_{Yeq}}{E}} \tag{3}$$

$$\beta = \frac{b}{t}\sqrt{\frac{\sigma_{Yeq}}{E}} \tag{4}$$

Paik & Thayamballi (1997) accepted the structure of the Lin's formulation and modified the coefficient based on the collected experimental data:

$$R_{PT} = \left[0.995 + 0.936\lambda^2 + 0.170\beta^2 + 0.88\lambda^2\beta^2 - 0.067\lambda^4\right]^{-\frac{1}{2}} \leq \frac{1}{\lambda^2} \tag{5}$$

Zhang & Khan (2009) proposed an empirical formulation based on the 61 numerical simulations obtained from the non-linear FE method analyses.

They also compared the formulation with a limited number of experimental data.

$$R_{ZK} = \frac{1}{\beta^{0.28}\sqrt{1+\lambda^{3.2}}} \qquad for\ \lambda \leq \sqrt{2} \qquad (6)$$

Recently, Kim *et al.* (2017) proposed a generalised empirical formulation based on the numerical analysis of 124 stiffened panels with Tr and Flat-bar stiffeners.

$$R_K = \frac{1}{0.8884 + e^{\lambda^2}} + \frac{1}{0.4112 + e^{\beta^{0.5}}} \qquad (7)$$

In addition to the empirical formulations, several design formulations, such as Euler, Johnson-Robertson and Perry-Robertson formulations, have been proposed to predict the ultimate strength of stiffened plates. However, numerical studies indicated that these design formulations tend to overestimate the ultimate strength of the stiffened plates subjected to uniaxial loads (Kim *et al.*, 2017, 2018).

Several test programs have been made on the stiffened plates under compressive load. In this study, 87 model tests were collected from (Dowling *et al.*, 1973; Smith, 1975; Horne *et al.*, 1976; Faulkner, 1977; Tanaka & Endo, 1988; Chen, 2003; Gordo & Guedes Soares, 2008, 2011, 2012; Xu & Guedes Soares, 2013a, 2013b, 2021). The histogram of the analysed data with the mean and standard deviation values and 95% confidence interval (CI) are presented in Figure 1.

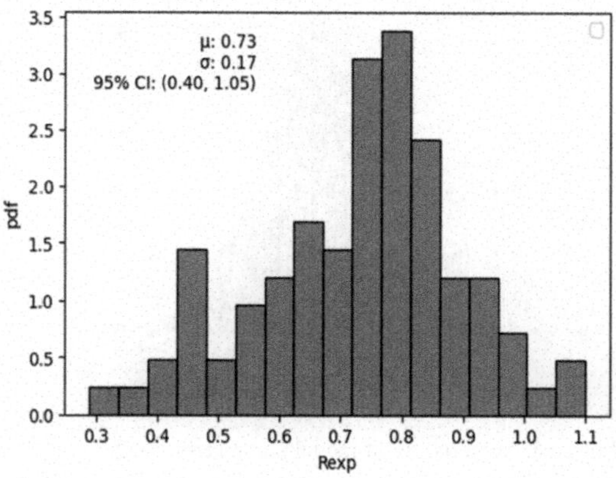

Figure 1. Histogram of collected experimental test data.

Ozdemir *et al.* (2018) proposed three collapse modes of stiffened plates based on the different ranges of column slenderness:

$$\begin{cases} Plate & for\ 0 < \lambda \leq 0.6 \\ Column & for\ 0.6 < \lambda \leq 1.3 \\ Overall & for\ 1.3 < \lambda \end{cases} \qquad (8)$$

and the available experimental data is classified based on this classification. The scatter diagram of collected experimental data found on the column slenderness is depicted in Figure 2. It can be seen that the available experimental data is in the range of 0<λ≤0.6 and 0.6<λ≤1.3, which is indicated by blue and red colours, respectively.

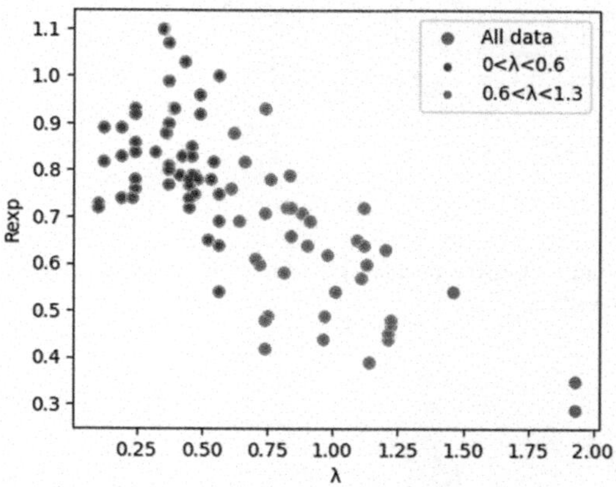

Figure 2. Experimental test results based on λ.

3 UNCERTAINTY ANALYSIS

Regarding the prediction of the ultimate strength of stiffened plates, uncertainty arises due to the uncertainty in the experimental measurements ε_l, statistical uncertainty of model parameters λ, β and model uncertainty ε_m due to simplifications and idealisations made in the mathematical models. Mathematically, the probabilistic model of ultimate strength can be expressed as:

$$R = R_m\left(\lambda, \beta | \sigma_{Yeq}, E, l, b, t\right)\varepsilon_m\varepsilon_l \qquad (9)$$

The model uncertainty of the investigated empirical formulations for the ultimate strength of stiffened panels is based on experimental data. Also, the Monte Carlo simulation estimates the statistical uncertainty of model parameters accounting for the variability in material properties and dimensions. This study does not investigate the measurement uncertainty since the information related to the measuring instruments used, including precision, accuracy, resolution, and calibration uncertainties, is unavailable.

The comparison between the results of experimental tests and those of empirical models is shown in Figure 3. It can be observed that Lin (1985), Paik and Thayamballi (1997) and Kim *et al.* (2017) models predict the lower bound for ultimate strength compared with experimental tests. However, Zhang and Khan (2009) models are closer to the experimental results. In addition, the results show higher dispersion in the lower ultimate strength (*R*<0.7) for all investigated models, implying a higher uncertainty in predicting the ultimate strength.

The distributions of $\varepsilon_m = R_{exp}/R_m$ for different empirical models are depicted in Figures 4 and 5 for

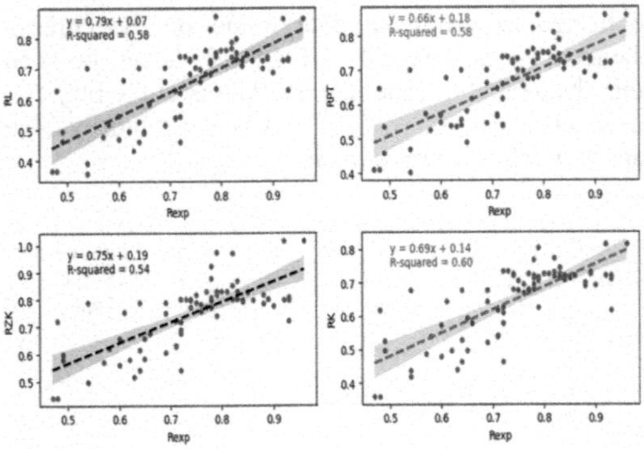

Figure 3. Experimental test data and empirical model results.

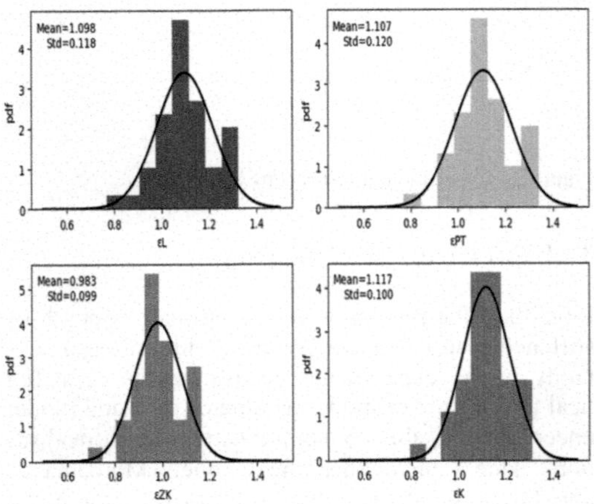

Figure 4. Model uncertainty of different empirical models, $0<\lambda\leq0.6$.

$0<\lambda\leq0.6$ and $0.6<\lambda\leq1.3$, respectively. These two different ranges of column slenderness are separated due to the different collapse types. Two different model uncertainties are introduced due to the different behaviours of stiffened plates when changing the column slenderness. It can be observed that the mean values of the model uncertainty in Lin (1985), Paik and Thayamballi (1997) and Kim *et al.* (2017) models are bigger than 1, which is due to the underestimation of ultimate strength compared to the experimental results for all λ. The model uncertainty increases with the increase of λ. For instance, the standard deviation of ε_L, ε_{PT}, ε_{ZK} and ε_K are 0.12, 0.12, 0.1 and 0.1, respectively, for $0<\lambda\leq0.6$, while these values are 0.17, 0.13. 0.14 and 0.17 for $0.6<\lambda\leq1.3$.

The model parameters λ and β are functions of material properties and thickness of plates and stiffeners. The yield strength and the elasticity modulus E are assumed to follow Log-normal distribution with the mean values equal to nominal values and coefficient of variations of 0.05, 0.08 and 0.03, respectively (Gaspar & Guedes Soares, 2013). It is assumed that the radius of gyration r has similar statistical parameters of the thickness, and other dimensions parameters such as b and l are deterministic. Ten thousand random numbers are generated for each mentioned random variable, and λ and β.

The histogram of the generated data and fitted log-normal distribution are shown in Figure 6. Plate and column slenderness have the same distribution with a mean value equal to nominal values and a coefficient of variation of 0.07.

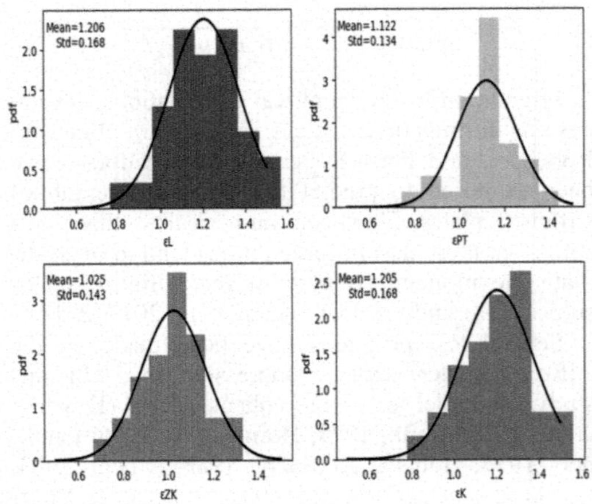

Figure 5. Model uncertainty of different empirical models, $0.6<\lambda\leq1.3$.

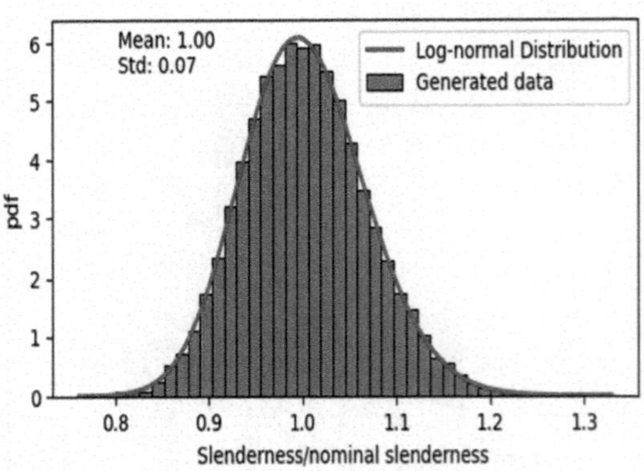

Figure 6. Uncertainty of model parameters λ and β.

Regarding the obtained model uncertainty and the uncertainty of model parameters, the results of the ultimate strength of the stiffened plates for different stiffness, λ are calculated based on the Monte-Carlo simulation. Figure 7 shows the results of each model's mean values and 0.95 CI regarding β=2.25 and compared with the experimental data with $2<\beta<2.5$. It can be observed that the proposed probabilistic modelling covers the experimental data. Also, the predicted ultimate strength results obtained from the models with β=2.25 are compared to the experimental data with $1.6<\beta<2$. Regarding the variability of the β and measurement uncertainty in empirical tests, the overall results of the proposed probabilistic models are reasonable.

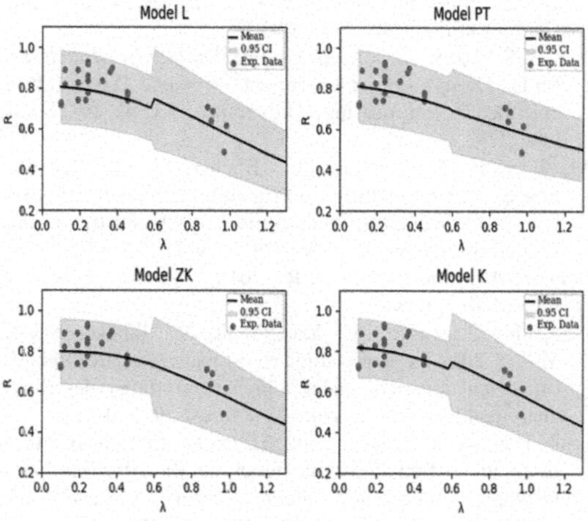

Figure 7. Probabilistic modelling of ultimate strength with 0.95 CI regarding β_{models}=2.25.

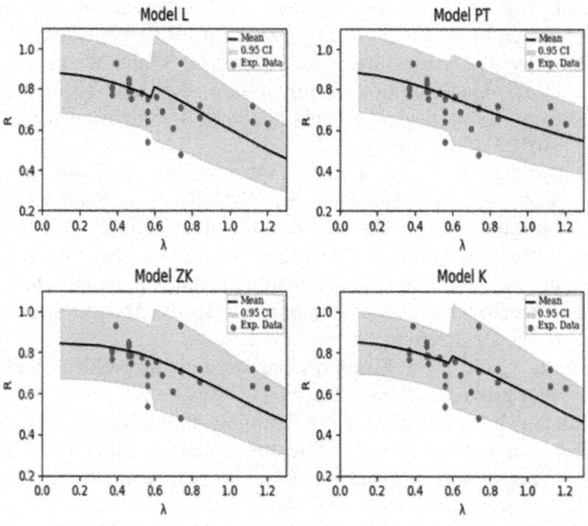

Figure 8. Probabilistic modelling of ultimate strength with 0.95 CI regarding β_{models}=1.75.

It should be noted that the results of this study should be understood within the context of the limited amount of available experimental data. Undoubtedly, in the presence of a larger dataset, the accuracy of the proposed probabilistic models would increase.

Another approach that can be used for uncertainty analysis of stiffened plates is the Bayesian approach. The Bayesian approach is beneficial when dealing with limited data because it allows for incorporating prior knowledge or beliefs about the system in addition to the observed data. This approach has been applied for uncertainty and reliability analysis of different types of marine structures (Garbatov & Guedes Soares, 2002; Kakaie *et al.*, 2023; Li *et al.*, 2021) and can be an alternative way for the probabilistic modelling of ultimate strength of stiffened plates which is left for consideration in future studies.

4 CONCLUSIONS

Uncertainty analysis of the ultimate strength models of the stiffened plates is carried out considering different empirical formulations. The model uncertainty is developed based on available experimental data exploring the Normal distribution with statistical descriptors as a function of stiffness for two collapse modes: local plate buckling type where $0<\lambda\leq0.6$ and column type with $0.6<\lambda\leq1.3$. The results indicates higher variability of the model uncertainty factor in all investigated models for $0.6<\lambda\leq1.3$, implying higher uncertainty in predicting the ultimate strength in that range. Also, the statistical uncertainty of the model's parameters propagates, accounting for the variability of the material properties and dimensions. The log-normal distribution with mean values equal to nominal values and a coefficient of variation of 0.07 was proposed for plate and column slenderness. Finally, the developed probabilistic uncertainty models of the ultimate strength formulations of stiffened panels were compared to the available experimental data considering the plate and column slenderness. The developed probabilistic models for investigated empirical formulations of ultimate strength can be used in the reliability assessment of structures composed of stiffened plates. It is also part of a framework to estimate the model uncertainty of the empirical formulation of the ultimate strength of stiffened plates considering material and structural behaviour descriptors.

ACKNOWLEDGEMENTS

The first author has been funded by the Portuguese Foundation for Science and Technology (Fundação para a Ciência e a Tecnologia-FCT) through a doctoral fellowship under Contract no. 2023.03056.BD. This work contributes to the Strategic Research Plan of the Centre for Marine Technology and Ocean Engineering, which is financed by the Portuguese Foundation for Science and Technology (Fundação para a Ciência e Tecnologia - FCT) under contract UIDB/UIDP/00134/2020.

REFERENCES

Chen, Y. 2003. *Ultimate strength analysis of stiffened panels using a beam-column method.* Ph.D. Dissertation. Virginia Polytechnic Institute and State University, Blacksburg, VA, USA

Dowling, P. J., Chatterjee, S., Frieze, P. A., & Moolani, F. M. 1973. Experimental and predicted collapse behaviour of rectangular steel box girders. In *Proceeding of the International Conference on Steel Box*

Girder Bridges. Institute of Civil Engineers, London, UK, 77–94.
Faulkner, D. 1977. Compression tests on welded eccentrically stiffened plate panels. *Steel Plated Structures*, 130–139.
Garbatov, Y., & Guedes Soares, C. 2002. Bayesian updating in the reliability assessment of maintained floating structures. *J. Offshore Mech. Arct. Eng.*, *124*(3), 139–145.
Garbatov, Y., Tekgoz, M., & Guedes Soares, C. 2011. Uncertainty assessment of the ultimate strength of a stiffened panel. *Advances in Marine Structures. Guedes Soares, C. & Fricke, W. (Eds.), Taylor & Francis Group, London, UK*, 659–668.
Gaspar, B., & Guedes Soares, C. 2013. Hull girder reliability using a Monte Carlo-based simulation method. *Probabilistic Engineering Mechanics*, *31*: 65–75.
Gaspar, B., Teixeira, A. P., Guedes Soares, C., & Wang, G. 2011. Assessment of IACS-CSR implicit safety levels for buckling strength of stiffened panels for double hull tankers. *Marine Structures*, *24*(4): 478–502.
Gordo, J. M., & Guedes Soares, C. 2008. Compressive tests on short continuous panels. *Marine Structures*, *21* (2–3): 113–137.
Gordo, J. M., & Guedes Soares, C. 2009. Tests on ultimate strength of hull box girders made of high tensile steel. *Marine Structures*, *22*(4): 770–790.
Gordo, J. M., & Guedes Soares, C. 2011. Compressive tests on stiffened panels of intermediate slenderness. *Thin-Walled Structures*, *49*(6): 782–794.
Gordo, J. M., & Guedes Soares, C. 2012. Compressive tests on long continuous panels. *Journal of Offshore Mechanics and Arctic Engineering*. 134(2):021403
Guedes Soares, C. 1988a. Design equation for the compressive strength of unstiffened plate elements with initial imperfections. *Journal of Constructional Steel Research*, *9*(4): 287–310.
Guedes Soares, C. 1988b; Uncertainty Modelling in Plate Buckling. *Structural Safety*. 5(1):17–34
Guedes Soares, C. 1997. Quantification of Model Uncertainty in Structural Reliability. Guedes Soares, C. (Ed.) *Probabilistic Methods for Structural Design*. Kluwer Academic Publishers; pp. 17–38.
Guedes Soares, C., & Gordo, J. M. 1997. Design methods for stiffened plates under predominantly uniaxial compression. *Marine Structures*, *10*(6): 465–497.
Guedes Soares, C., & Søreide, T. H. 1983. Behaviour and design of stiffened plates under predominantly compressive loads. *International Shipbuilding Progress*, *30* (341): 13–27.
Horne, M. R., Narayanan, R., & Merrison. 1976. Ultimate capacity of longitudinally stiffened plate used in the box girder. *Proceedings of the Institution of Civil Engineers*, *61*(2): 253–280.
Kakaie, A., Guedes Soares, C., Ariffin, A. K., & Punurai, W. 2023. Fatigue Reliability Analysis of Submarine Pipelines Using the Bayesian Approach. *Journal of Marine Science and Engineering*, *11*(3): 580.
Kim, D. K., Lim, H. L., Kim, M. S., Hwang, O. J., & Park, K. S. 2017. An empirical formulation for predicting the ultimate strength of stiffened panels subjected to longitudinal compression. *Ocean Engineering*, *140*: 270–280.
Kim, D. K., Lim, H. L., & Yu, S. Y. 2018. A technical review on ultimate strength prediction of stiffened panels in axial compression. *Ocean Engineering*, *170*: 392–406. https://doi.org/10.1016/j.oceaneng.2018.10.022
Li, X., Zhang, Y., Abbassi, R., Khan, F., & Chen, G. 2021. Probabilistic fatigue failure assessment of free spanning subsea pipeline using dynamic Bayesian network. *Ocean Engineering*, *234*: 109323.
Lin, Y.T., 1985. *Structural Longitudinal Ship Modelling*. Ph.D. Dissertation. Department of Naval Architecture and Ocean Engineering, University of Glasgow, Scotland, UK.
Matthies, H. G.; Brenner, C.; Bucher, C., and Guedes Soares, C. Uncertainties in Probabilistic Numerical Analysis of Structures and Solids - Stochastic Finite Elements. *Structural Safety*. 1997; 19(3):283–336
Nowak, A.S. and Collins, K.R., 2012. *Reliability of structures*. CRC press.
Ozdemir, M., Ergin, A., Yanagihara, D., Tanaka, S., & Yao, T. 2018. A new method to estimate ultimate strength of stiffened panels under longitudinal thrust based on analytical formulas. *Marine Structures*, *59*: 510–535.
Paik, J. K., & Kim, B. J. 2002. Ultimate strength formulations for stiffened panels under combined axial load, in-plane bending and lateral pressure: a benchmark study. *Thin-Walled Structures*, *40*(1): 45–83.
Paik, J. K., & Thayamballi, A. K. 1997. An empirical formulation for predicting the ultimate compressive strength of stiffened panels. *ISOPE International Ocean and Polar Engineering Conference*, ISOPE-I.
Paik, J. K., & Thayamballi, A. K. 2003. *Ultimate limit state design of steel-plated structures*. John Wiley & Sons.
Parunov, J., Guedes Soares, C., Hirdaris, S. and Wang, X., 2022. Uncertainties in modelling the low-frequency wave-induced global loads in ships. *Marine Structures*, 86:103307.
Ringsberg, J. W., Darie, I., Nahshon, K., Shilling, G., Vaz, M. A., Benson, S., Brubak, L., Feng, G., Fujikubo, M., & Gaiotti, M. 2021. The ISSC 2022 committee III. 1-Ultimate strength benchmark study on the ultimate limit state analysis of a stiffened plate structure subjected to uniaxial compressive loads. *Marine Structures*, *79*, 103026.
Smith, C. S. 1975. Compressive strength of welded steel ship grillages. *Trans. RINA*, *117*, 325–359.
Tanaka, Y., & Endo, H. 1988. Ultimate strength of stiffened plates with their stiffeners locally buckled in compression. *Journal of the Society of Naval Architects of Japan*, *1988*(164): 456–467.
Xu, M. C., & Guedes Soares, C. 2013a. Experimental study on the collapse strength of narrow stiffened panels. *Journal of Offshore Mechanics and Arctic Engineering*, *135*(2): 21402.
Xu, M. C., & Guedes Soares, C. 2013b. Experimental study on the collapse strength of wide stiffened panels. *Marine Structures*, *30*: 33–62.
Xu, M. C., & Guedes Soares, C. 2021. Experimental evaluation of the ultimate strength of stiffened panels under longitudinal compression. *Ocean Engineering*, *220*: 108496.
Xu, M. C., Song, Z. J., Zhang, B. W., & Pan, J. 2018. Empirical formula for predicting ultimate strength of stiffened panel of ship structure under combined longitudinal compression and lateral loads. *Ocean Engineering*, *162*: 161–175.
Zhang, S., & Khan, I. 2009. Buckling and ultimate capability of plates and stiffened panels in axial compression. *Marine Structures*, *22*(4): 791–808.

Advances in Maritime Technology and Engineering – Guedes Soares & Santos (Eds)
© 2024 The Author(s), ISBN 978-1-032-83099-5

Research on technologies of block lifting sequence optimization based on improved genetic algorithm

R. Li, Z.J. Xiao, Q.Y. Zhang & X. Chen
School of Naval Architecture and Ocean Engineering, Dalian University of Technology, Dalian, P.R. China

ABSTRACT: Dock block lifting is an important step in ship construction, in which the gantry cranes, the dock and other facilities are important construction resources. Creating a reasonable and efficient lifting plan can improve the efficiency of the gantry cranes and the dock. Based on the case of two gantry cranes performing lifting in a shipyard to build a container ship, a mathematical model was established in this study, considering the factors of the speed of lifting, such as the unloaded traveling distance of the cranes, the number of wire and shackle replacements, the avoidance of the cranes, etc. The adaptive genetic algorithm with parallel migration strategy is adopted to solve the model, based on its characteristics. Finally, a software system of dock block quick lifting planning was developed, and the validity of the system was verified.

1 INTRODUCTION

Dock block lifting accounts for approximately 45% of the entire shipbuilding process. The quality of the block-lifting plan will not only affect the reasonable use of both building dock and the equipment, also determine the completion time of both the block construction plan and the zone outfitting plan. Therefore, it is of vital significance to study the quick block lifting technology for shipbuilding in order to shorten the ship construction cycle. Dock block lifting refers to the lifting of the completed blocks from stockyard to dock for welding, using lifting equipment in accordance with block lifting sequence in the lifting plan, to finally complete the construction of the entire ship. Block lifting process is shown in Figure 1, in which block initial location refers to the specific location of the block in the block stockyard, and block target location refers to the specific location of the block in dock.

Before block lifting, a block lifting plan will be formulated. The main component of the plan is the block lifting sequence, which will affect the replacement times of wire and shackle and the unloaded traveling distance of the crane during lifting, and thus increase the lifting cycle. Therefore, the optimization of the block lifting sequence is necessary in developing a reasonable lifting plan and improving the efficiency of the lifting process.

As shown in Table 1, many researchers at home and abroad have conducted many research on the quick block lifting, in which modles are generally established for block lifting process and then solved, but the emphases of their models are different, and thus the selected solution algorithms are different accordingly

Xu et al. (2006a) solved the reasonable lifting sequence of each block using genetic algorithm, in order to achieve effective control of the number of block required. Xu et al. (2006b) also considered block construction constraints and the block manufacturing ability of the shop during modeling. Meng et al. (2019) adopted the hybrid heuristic algorithm combining both genetic algorithms and tabu search

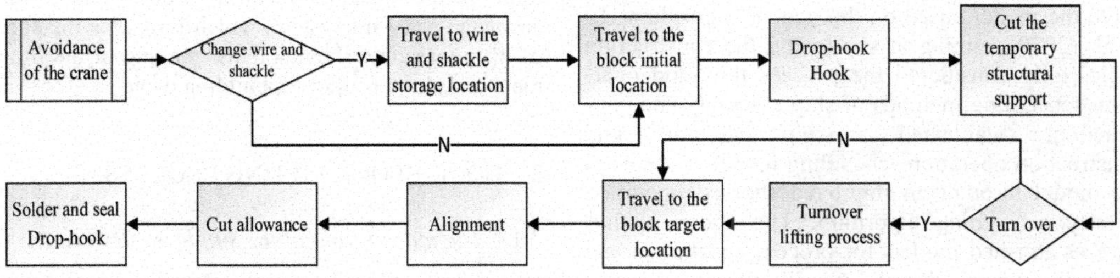

Figure 1. Block lifting process.

DOI: 10.1201/9781003508762-53

Table 1. Comparison of related studies.

Related studies	Xu Aimin (2006)	Wang Quanhua (2015)	Roh(2010)	Yilang (2014)	Main content of the study
Target of optimization	Block lifting sequence	Block lifting sequence	Block lifting sequence	Block lifting sequence	Block lifting sequence
Focus of constraint condition	Process constraint	Equipment resources	Unloaded traveling distance of the crane and replacements of wire and shackle	Public resources and non-public resources	Process constraint, crane unloaded traveling distance, replacements of wire and shackle, and avoidance of the crane
Optimization algorithm	Immune genetic algorithm	Genetic algorithm	Genetic algorithm	Improved Ant Colony optimization algorithm	Improved genetic algorithm
Lifting model	Single gantry crane lifting model	Single gantry crane lifting model	Single gantry crane lifting model	Single gantry crane lifting model	Double gantry crane lifting model

methods to optimize the sequence of block lifting, with the objective to minimize the total number of obstructive blocks. Zhang et al. (2015) used the improved genetic algorithm to solve the optimal plan for blocks storage in stockyard and establish heuristic rules to obtain the optimal entry and exit routes of the block. Wang et al. (2019) proposed a virtual flow block building mode which combines fixed position sectional production and flow mode of operation and created both a worker scheduling algorithm and a spatial arrangement scheduling algorithm based on the virtual assembly mode. Wang et al. (2017) studied the schedule for the transportation by the block transporter, constructed an optimization model aiming at fewer block transporters and shorter operating time, and proposed two heuristic algorithms based on the genetic algorithm to solve the model. Roh et al. (2010) used the genetic algorithm to achieve the optimal block lifting sequence, considering the wire and shackle replacement nodes as the optimization points during block lifting. Li et al. (2022) used a method of multi-workshop collaborative scheduling for ship block construction, established an optimization model targeting site utilization and segment completion rates as decision variables, and finally proposed a two-stage solution method for the optimization model to solve the model based on the genetic algorithms. Li et al. (2020). aiming at shortening the construction cycle, fully considered the process flow and operation scheduling methods of ship assembly and construction, established a ship assembly and construction operation scheduling model, and solved the model through an improved shortest operation priority scheduling algorithm. The above studies have established models for process constraints and lifting processes during the lifting process and solved them with algorithms, optimized the lifting sequence and improved lifting efficiency.

Wang et al. (2015) used AOE (Directed Acyclic Network) network planning technology and directed graph to construct a lifting network optimization model and solved it based on the improved genetic algorithm. They obtained the shortest lifting cycle and conducted simulations. Yi(2014) built an information data structure for block lifting during lifting process based on an improved Ant Colony Optimization. He also considered shipyard resource constraint and lifting process constraint and created the lifting network graph for shipbuilding. In summary, given that lifting network graph is also an effective way to describe the lifting plan, and that the main body of the graph is the block lifting sequence, we can see the importance of optimizing the lifting sequence.

Lu et al.(2016) proposed a scheduling technique that comprehensively arranges lifting tasks and coordinates two gantry cranes during block production, which forms a scheduling priority determination rule and a crane task allocation mechanism, and realizes the coordination of double gantry cranes under various constraints. This collaborative scheduling method of double gantry cranes is worth learning from, though the research objective of this study is not block lifting.

This study analyzes and establishes a mathematical model for the block lifting process of double gantry crane dock, solves it through genetic algorithm to optimize the lifting plan, and develops a block lifting sequence software system, which improves the speed of lifting, realizes the automatic making of the lifting plan, and shortens the shipbuilding cycle.

2 DOCK BLOCK LIFTING PROCESS

2.1 *The quick lifting of the single gantry crane*

Generally, in the block lifting by one gantry crane, the speed is mainly reduced by the unreasonable block lifting sequence, which could cause an increase in unloaded traveling distance of the crane and also in the number of wire and shackle replacements.

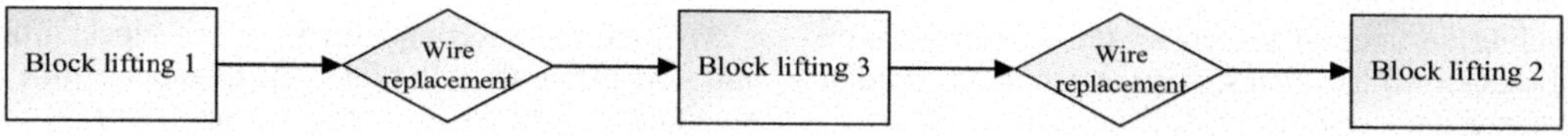

Lifting sequence of block 1, block 3, block 2

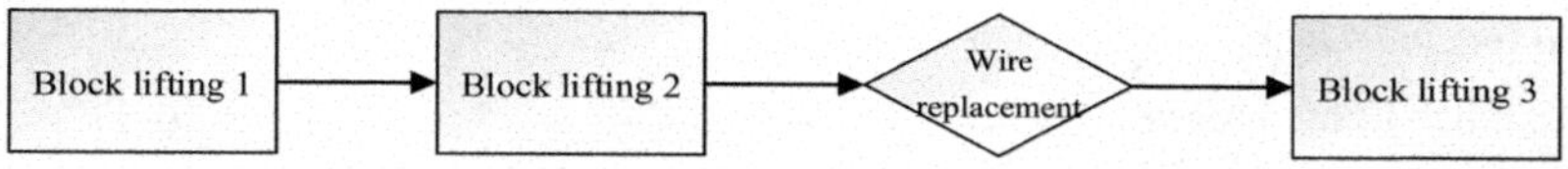

Lifting sequence block 1, block 2, block 3

Figure 2. Number of wire and shackle replacements.

To be specific, firstly, the different lifting sequences among blocks will lead to non-unique moving path of the crane. Take block 1 and block 2 as examples. Suppose the moving path of the crane, which lifts block 1 first and then block 2, is shorter than the that of the crane, which lifts block 2 first and then block 1, then in the latter situation the speed of lifting will inevitably be reduced. Moreover, when lifting different blocks, if block weights differ greatly, it is necessary to replace wire and shackle, such as hooks, wires, etc. Assuming that the same wire and shackle are used for both block 1 and block 2, while different wire and shackle are used for block 3, the relationship between the order of lifting and the number of wire and shackle replacement is as shown in Figure 2. The increase in the number of wire and shackle replacement will not only increase the time length for which the crane stays at the wire and shackle replacement site, but also increase the moving distance of the crane.

2.2 *The quick lifting of double gantry cranes*

According to the shipyard survey, in order to improve the lifting efficiency, shipyard usually uses two gantry cranes to carry out lifting work at the same time. It has been proved from the actual work experience that the block lifting by double gantry cranes can indeed greatly improve the lifting speed, but problems still exist such as unloaded traveling distance of the crane and the number of wire and shackle replacements, Besides, moving path interference could also occur when two gantry cranes perform lifting on the same track, when one crane has to avoid the other crane, and such avoidance may occur twice during a lifting process.

As shown in Figure3, two gantry cranes A and B perform block lifting at the same time. When they are moving from block stockyard to dock, if the preparation time of the blocks lifted by Crane B is longer than that by Crane A in the stockyard, Crane A needs to wait until Crane B completes its work, and the two cranes could then move to dock.

When the cranes moves from dock to block stockyard, as shown in Figure 4, if the positioning, soldering and sealing of the blocks lifted by Crane A at dock consumes longer time than those by Crane B, Crane B needs to wait until Crane A completes its work, and the two cranes could then move to block stockyard.

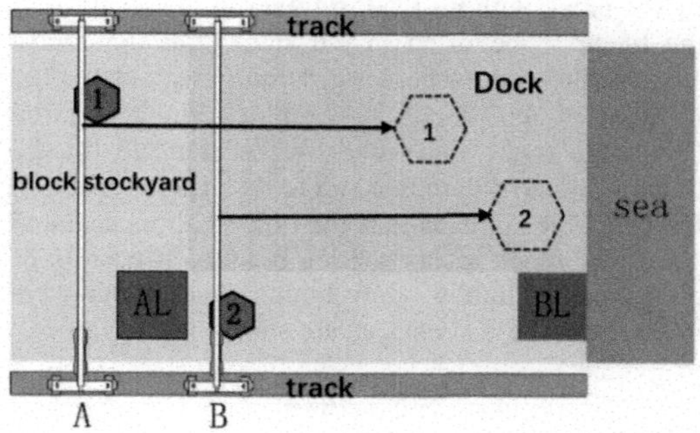

Figure 3. Avoidance by Crane A.

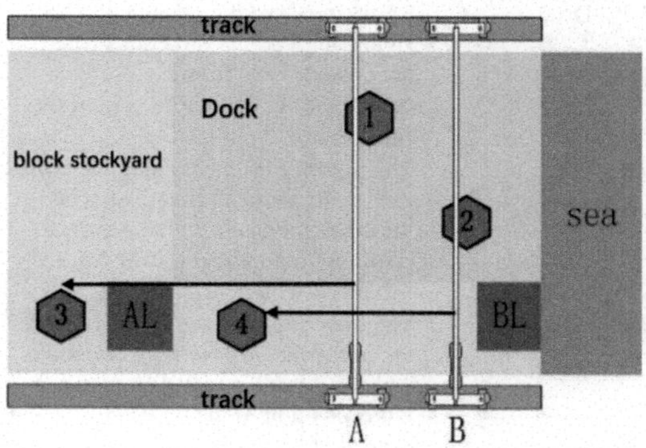

Figure 4. Avoidance by Crane B.

Avoidance time, as ineffective working time of the gantry crane. could be effectively reduced by optimizing block lifting sequence. However, the time spent on hooking, cutting the temporary structural support, and cutting allowances depends on the munual proficiency of the workers, and thus cannot be improved by optimizing block lifting sequence. Therefore, this study focuses on the factors that affect the lifting speed, such as the unloaded traveling distance of the crane, the number of wire and shackle replacements, and the avoidance time of the crane, and performs analysis and optimization.

3 DOCK LIFTING MODELING AND SOLUTION

3.1 *Dock lifting modeling*

According to block lifting process, this study divides the time spent by the gantry crane in block lifting into six stages: the first stage is the avoidance time of the gantry crane; the second stage is the time used by the gantry crane with no load to travle to the wire and shackle storage site and the time for wire and shackle replacements; the third stage is the time used by the crane with no load to travel to the location of the block to be lifted in the stockyard; the fourth stage is the time for hooking, turning over, etc. at the location of the blocks in the block stockyard; the fifth stage is the time used by the crane to lift the block from stockyard and run to its target location in the dock; the sixth stage is the time to align, seal and welde the block after reaching the target location of dock, until finally drop-hook. The parameters involved in these six stages are shown in the Table 2.

Table 2. Table of parameters.

Symbol	Meaning
T^i_{A1}, T^i_{A2}, T^i_{A3}, T^i_{A4}, T^i_{A5}, T^i_{A6}	The time taken by Crane A to lift the i-th block for six stages
T^i_{B1}, T^i_{B2}, T^i_{B2}, T^i_{B4}, T^i_{B5}, T^i_{B6}	The time taken by Crane B to lift the i-th block for six stages
Di^i_x, Di^i_y	The distance of the crane in the horizontal and vertical directions when the crane travels from the previous position to the initial position of the i-th block without load
Dl^i_x, Dl^i_y	The distance of the crane in the horizontal and vertical directions when the crane travels from the previous position to the initial position of the i-th block with load
Dw^i_x, Dw^i_y	The traveling distance of the crane in the horizontal and vertical directions when the crane travels from the previous position to the wire and shackle stockyyard without load
N_A, N_B	The number of blocks lifted by Crane A and Crane B
N^i_{lug}	Number of lifting lugs of the i-th block lifted
N^i_{wire}	Number of wires and shackles of the i-th block lifted
V_{it}, V_{il}	Horizontal movement speed of the crane with no load Vertical movement speed of the crane without load
V_{lt}, V_{ll}	Horizontal movement speed of the crane with load Vertical movement speed of the crane with load

When Crane A completes its i-1-th block lifting and start lifting the i-th block, the avoidance time of the crane at the first stage can be shown as Equation (1), where r^i_{A1} equals to 0, and the crane does not need to avoid at this stage.

$$T^i_{A1}=r^i_{A1}\times[T^{i-1}_B+T^i_{B2}+T^i_{B3}+T^i_{B4}-(T^{i-1}_A+T^i_{A2}+T^i_{A3}+T^i_{A4})] \quad (1)$$

$$r^i_{A1}\in\{0,1\}$$

After Crane B completes its i-1-th block lifting, and starts lifting the i-th block, its time at the first stage can be shown in Equation (2), where r^i_{B1} equals to 0, and the crane does not need to avoid at this stage.

$$T^i_{B1}=r^i_{B1}\times[T^{i-1}_A-(T^{i-1}_B+T^i_{B2})] \quad (2)$$

$$r^i_{B1}\in\{0,1\}$$

At the second stage, the time for replacement of the wire and shackle is shown as Equation (3) for Crane A, and Equation (4) for Crane B, respectively, where T_{change} is the time to replace a single wire and shackle; and when r^i_{A2} and r^i_{B2} equal to 0, there is no need to replace the wire and shackle at this stage.

$$T^i_{A2}=r^i_{A2}\times(DW^i_X/V_{it}+DW^i_Y/V_{il}+T_{change}+N^i_{wire}) \quad (3)$$

$$r^i_{A2}\in\{0,1\}$$

$$T^i_{B2}=r^i_{B2}\times(DW^i_X/V_{it}+DW^i_Y/V_{il}+T_{change}+N^i_{wire}) \quad (4)$$

$$r^i_{B2}\in\{0,1\}$$

At the third stage, the time of movement without load could be shown as Equation (5) for Crane A, and Equation (6) for Crane B, respectively.

$$T^i_{A3}=DI^i_X/V_{lt}+DI^i_Y/V_{ll} \quad (5)$$

$$T^i_{B3}=DI^i_X/V_{lt}+DI^i_Y/V_{ll} \quad (6)$$

At the fourth stage, the time to prepare for block lifting can be shown in Equation (7) for Crane A, and Equation (8) for Crane B, respectively, where T_{put} is the time for hooking or drop-hook at one lifting lug, V_{dl} is the speed of the lifting up or lifting down, T_{turn} is the time for block turningover, H_{C1} and H_{C2} are the distances between the crane hook and jig before and after block lifting, and when r^i_{A4} and r^i_{B4} equal to 0, the block is under forward construction and does not require turnover lifting.

$$T^i_{A4}=H_{C1}/V_{dl}+T_{put}\times N^i_{lug}+H_{C2}/V_{dl}+r^i_{A4}\times T_{turn} \quad (7)$$

$$r^i_{A4}\in\{0,1\}$$

$$T^i_{B4}=H_{C1}/V_{dl}+T_{put}\times N^i_{lug}+H_{C2}/V_{dl}+r^i_{B4}\times T_{turn} \quad (8)$$

$$r^i_{B4}\in\{0,1\}$$

At the fifth stage, the time for the cranes to run with load is shown in Equation (9) for Crane A, and Equation (10) for Crane B, respectively.

$$T_{A5}^{i} = DL_X^i / V_{lt} + DL_Y^i / V_{ll} + H_{C1} / V_{dl} \quad (9)$$

$$T_{B5}^{i} = DL_X^i / V_{lt} + DL_Y^i / V_{ll} + H_{C1} / V_{dl} \quad (10)$$

At the sixth stage, the time for soldering and sealing is shown in Equation (11) for Crane A, and Equation (12) for Crane B, where T_P is the time for block alignment, and T_{weld} is the time for soldering and sealing.

$$T_{A6}^{i} = T_p + T_{weld} + T_{put} \times N_{lug}^i + H_{C2} / V_{dl} \quad (11)$$

$$T_{B6}^{i} = T_p + T_{weld} + T_{put} \times N_{lug}^i + H_{C2} / V_{dl} \quad (12)$$

To sum up, the total time required to complete the lifting operations for all blocks by Crane A and Crane B respectively are:

$$T_A = \sum_{i=1}^{N_A} (T_{A1}^i + T_{A2}^i + T_{A3}^i + T_{A4}^i + T_{A5}^i + T_{A6}^i) \quad (13)$$

$$T_B = \sum_{i=1}^{N_A} (T_{B1}^i + T_{B2}^i + T_{B3}^i + T_{B4}^i + T_{B5}^i + T_{B6}^i) \quad (14)$$

For the entire lifting plan, Crane A and Crane B perform lifting operations at the same time, and the larger value between T_A and T_B becomes the time spent on the entire lifting plan.

3.2 *Improved genetic algorithm*

In genetic algorithm, crossover rate Pc and mutation rate Pm are important parameters which could determine the behavior and performance of the genetic algorithm. This study adopts Srinvivas' adaptive genetic algorithm, in which Pc and Pm can automatically adjust themselves according to individual fitness. When each population individual fitness tends to be consistent or converges to the local optimal solution, Pc and Pm increase; on the contrary, when the population fitness is relatively dispersed, Pc and Pm decrease. In this way, the convergence of the genetic algorithm is maintained with the diversity of population ensured. The calculations of Pc and Pm are shown as follows:

$$P_c = \begin{cases} P_{C1} - \frac{(P_{C1}-P_{C2})(f'-f_{avg})}{f_{max}-f_{avg}}, f' \geq f_{avg} \\ P_{C1}, f' < f_{avg} \end{cases} \quad (15)$$

$$P_m = \begin{cases} P_{m1} - \frac{(P_{m1}-P_{m2})(f_{max}-f)}{f_{max}-f_{avg}}, f \geq f_{avg} \\ P_{m1}, f < f_{avg} \end{cases} \quad (16)$$

In the above equations, P_{c1}=0.9, P_{c2}=0.6, P_{m1} =0.1, P_{m2}=0.001, f_{avg} is the average fitness of each population generation, f_{max} is the maximum fitness of the population, f' is the larger fitness value in the two individuals of the crossover, and f is the fitness value of the individual to mutate.

The traditional genetic algorithm usually uses single population, which could solve the problem. However, algorithms with multiple populations are believed to obtain better results. The parallel migration strategy[9] with multiple populations indicates that when each sub population performs independent evolution and calculation based on the genetic algorithm, there is also individual exchange among sub populations. This multiple population genetic algorithm is closer to the evolution of natural populations, thus having more self-learning and self-adapting capabilities. This study chooses a parallel genetic algorithm with a ring topology form to optimize and solve the problem of quick lifting, in whcih the configuration of migration is shown in Figure 5.

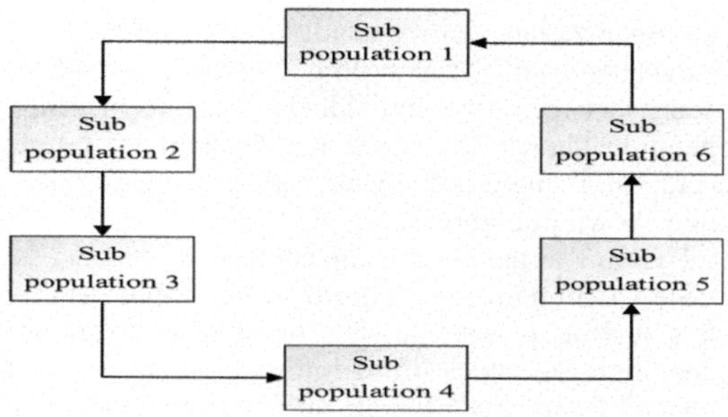

Figure 5. Configuration of migration.

3.3 *Genetic algorithm solution*

The genetic algorithm is an intelligent algorithm to solve multi-objective optimization problems, which has excellent parallelism, robustness and global efficiency. It consists of several important components: the fitness function, the population, the crossover and the mutation.

(1) The fitness function uses the calculation of the total time by two gantry cranes to lift a certain number of blocks as its algorithm:

$$F = \max\{T_A, T_B\} \quad (17)$$

(2) Population, as the feasible solutions of several fitness functions, is the sequence of block lifting in block lifting plan. Assume that there are 10 blocks waiting to be lifted; population size is set to 7; and each block is represented by a positive integer from 1 to 10, then the feasible solution population matrix by genetic algorithm is randomly generated and shown in Equation (18), where each row is a population individual, that is, the block lifting sequence in block lifting plan.

$$\begin{pmatrix} 3 & 4 & 6 & 7 & 9 & 1 & 10 & 2 & 5 & 8 \\ 8 & 5 & 2 & 10 & 9 & 6 & 7 & 3 & 4 & 1 \\ 1 & 3 & 6 & 7 & 5 & 10 & 9 & 8 & 4 & 2 \\ 2 & 1 & 5 & 6 & 7 & 9 & 3 & 8 & 10 & 4 \\ 4 & 2 & 5 & 7 & 6 & 1 & 10 & 9 & 8 & 3 \\ 5 & 6 & 8 & 10 & 3 & 4 & 1 & 2 & 7 & 9 \\ 1 & 2 & 4 & 5 & 6 & 7 & 10 & 3 & 8 & 9 \end{pmatrix} \quad (18)$$

(3) Crossover and mutation are important genetic operations in the genetic algorithm, which reflect the

self-organization, self-adaptation, and self-learning characteristics of the algorithm. Crossover operation occurs between two individuals, with its specific operation shown in Figure 6. Mutation operation occurs to a single individual, with its specific operation shown in Figure 7.

The flow of the genetic algorithm:

Step 1: Initialize the algorithm, and input parameters such as population size, number of iterations, generation gap, block lifting information etc.

Step2 Randomly generate an initial population of solutions;

Step3 Calculate the fitness of each individual in all population groups;

Step 4: Perform selection, crossover, and mutation on the previous population to obtain a new generation population;

Step5 Determine whether to perform population migration. If so, implement the migration strategy among populations; otherwise, proceed to the next step;

Step6 Determine whether the maximum number of iterations has been reached. If so, the algorithm stops and the individual with the highest fitness in the group is output as the result; otherwise, return to Step3.

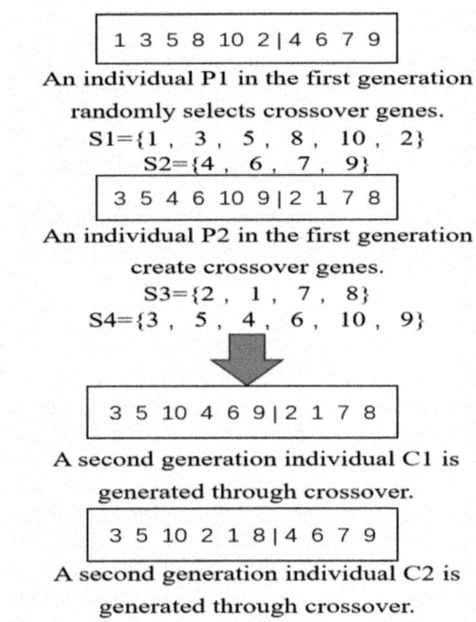

Figure 6. Crossover operation.

3.4 *Calculation and analysis of the actual case*

Take block lifting container ship as an example. Crane A and Crane B each lift 10 blocks, and the traditional genetic algorithm and the improved genetic algorithm are used to calculate 10 times respectively. During the calculation, the parameters of the genetic algorithm are shown in Table 3, and some of the block lifting information are shown in Tables 4 and 5.

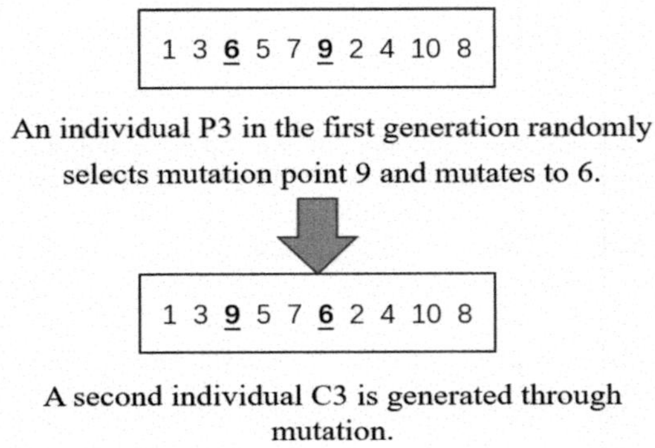

Figure 7. Mutation operation.

Upon the calculation completion, the results are summarized in Table 6, from which it can be seen that the best results could be calculated both before and after the improvement, but the traditional genetic algorithm cannot stably calculate the optimal results. With the comparison of the average calculation time of the two algorithms, it can be established that the calculation time of the improved algorithm is about 50% longer than that of the traditional genetic algorithm. As shown in Figure 8, during calculation, the traditional genetic algorithm will converge earlier than the improved genetic algorithm, resulting in unsatisfactory calculation results.

In summary, it is proved that the improved genetic algorithm can be used to stably solve the optimal solution or approximate solution of the block lifting mathematical model. Since the improved genetic algorithm is structurally more complex than the traditional one, the inevitable increase in calculation time is acceptable. Therefore, the improved genetic algorithm is feasible in solving the problems investigated in this study.

Table 3. The calculation parameters of the actual case.

Parameter symbols	Numerical value	Meaning
l	10	Number of blocks in the lifting plan
NIND	30	Population size
MAXGEN	200	Number of iterations in algorithm calculation
GGAP	0.8	Generation gap
SUBPOP	6	Population number
MIGR	0.2	Population migration rate
MIGGEN	12	Intervals for individual migration (in generations)

4 THE DEVELOPMENT OF DOCK LIFTING PLANNING SOFTWARE SYSTEM

4.1 *System framework design*

The system optimizes the block lifting sequence by analyzing dock resource information, dock gantry

Table 4. Block lifting information for Crane A.

Block name	Algorithmic coding	Block weight /t	Immediate predecessor activity	Immediate successor activity	Lifting equipment	Number of lifting lugs
207P	1	625	-	-	600T	10
207S	2	320	1	-	600T	6
205P	3	663	1	-	600T	10
205S	4	338	2,3	-	600T	6
608P	5	468	7	6	600T	10
608S	6	468	7	5	600T	8
408C	7	380	4	-	600T	10
406P	8	340	4	-	600T	10
606P	9	456	8	10	600T	10
606S	10	456	8	9	600T	10

Table 5. Block lifting information for Crane B.

Block name	Algorithmic coding	Block weight /t	Immediate predecessor activity	Immediate successor activity	Lifting equipment	Number of lifting lugs
209P	1	663	-	-	900T	10
209S	2	338	1	-	900T	6
409P	3	340	7	-	900T	10
610P	4	456	8	5	900T	10
610S	5	456	8	4	900T	10
211P	6	643	1	-	900T	10
211S	7	323	6,2	-	900T	6
410C	8	380	7	-	900T	10
411P	9	340	-	-	900T	10
213P	10	685	6,2	-	900T	10

Table 6. Calculation results before and after algorithm improvement.

Algorithm / Calculation Times	Traditional genetic algorithm calculation results/min	Traditional genetic algorithm calculation time/min	Improved genetic algorithm calculation results/min	Improved genetic algorithm calculation time/min
1	3966	10.6	3966	23.5
2	3966	10.8	3966	24.3
3	3966	11	3966	25
4	3966	10.8	3966	24.4
5	4125	9.8	3966	23.8
6	3966	10.6	3966	23.6
7	4125	9.6	3966	24.1
8	3966	10.5	3966	23.9
9	4125	10	3966	24.7
10	3966	10.8	3966	24
Average value	4013.7	10.45	3966	24.13

crane position information, hull block and other information during dock block lifting, and finally creates a lifting plan report. The system is mainly divided into four parts: the user interface that provides interaction between the system and the user, the data layer that stores basic block information, the algorithm supporting layer that forms an improved genetic algorithm to optimize quick lifting calculations, and the system layer that includes the operating system and database management system. The system layer structure is shown in Figure 9.

4.2 *System function modules*

The system function module includes dock block lifting data management module, optimization

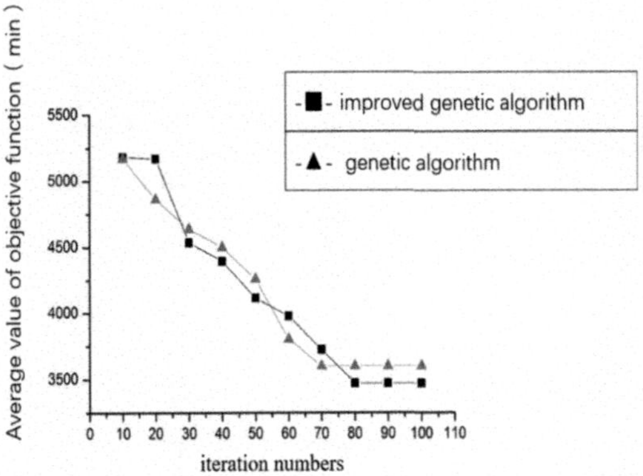

Figure 8. The iterative convergence curve of algorithm.

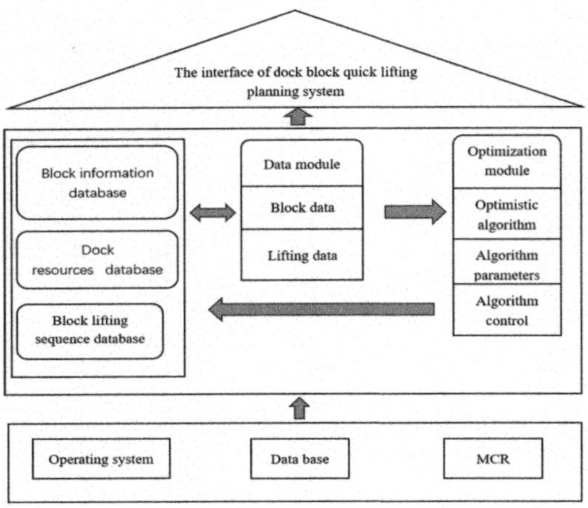

Figure 9. System architecture framework.

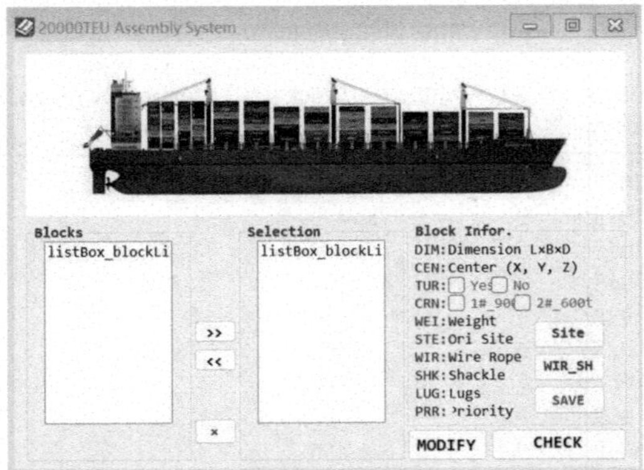

Figure 10. Lifting plan information editing interface 1.

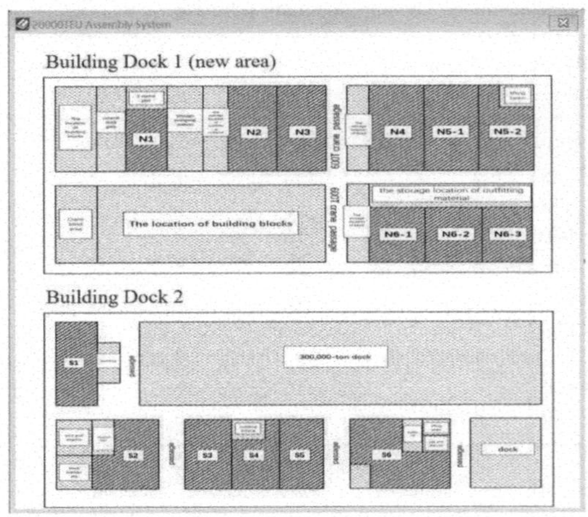

Figure 11. Lifting plan information editing interface 2.

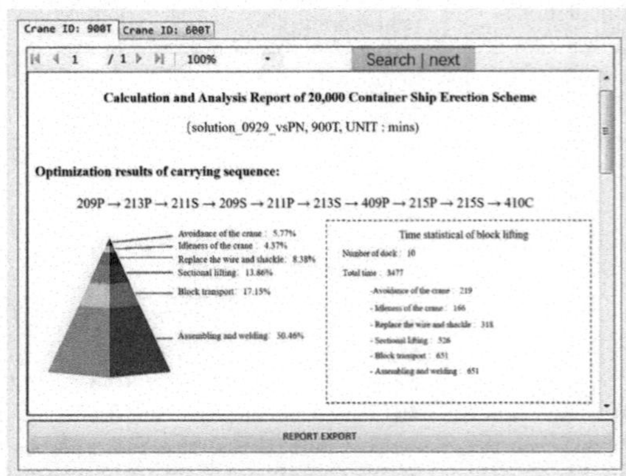

Figure 12. Sample report.

calculation module and reporting module. The dock block lifting data management module is responsible for the maintenance of various databases in the calculation of dock block quick lifting, and can query, modify, add and delete data in the databases. The databases are divided into two groups, including a database that stores all block information of the entire ship and a database that stores the information and calculation results of the blocks to be lifted. Through the user interface shown in Figure 10, the user selects the blocks, of which the lifting sequence needs to be optimized. From the user interface, the user could enter the interface shown by Figure 11 to select the locations of the blocks, and then edit the number of lifting lugs and other information for blocks through the lifting plan information editing interface. Through these operations, all information can be stored in the block lifting information data table and directly provide data for calculations.

The optimization calculation moudule calculates the block lifting sequences of the 600T and 900T cranes and the timetables of the cranes at each stage of lifting, based on all the input information of the blocks to be optimized and through the genetic algorithm calculation.

The reporting module collects the result data of the optimization calculation and displays them visually. The output report includes the lifting plans, lifting sequences, time required for each lifting stage and the total planned time of the 600T and 900T cranes, thus enabling users to easily analyze and summarize the entire lifting plan. The report is shown in Figure 12.

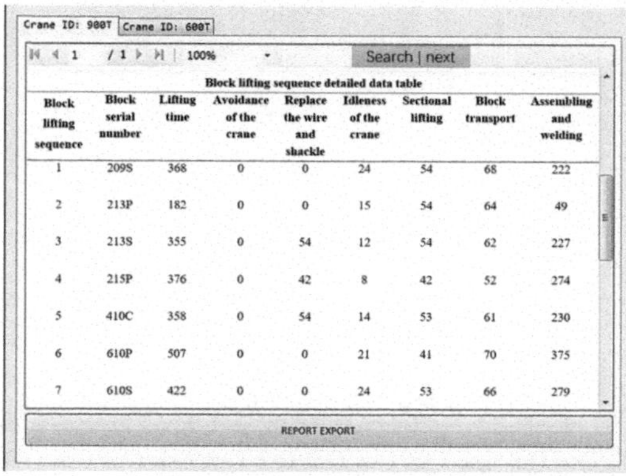

Block lifting sequence detailed data table

Block lifting sequence	Block serial number	Lifting time	Avoidance of the crane	Replace the wire and shackle	Idleness of the crane	Sectional lifting	Block transport	Assembling and welding
1	209S	368	0	0	24	54	68	222
2	213P	182	0	0	15	54	64	49
3	213S	355	0	54	12	54	62	227
4	215P	376	0	42	8	42	52	274
5	410C	358	0	54	14	53	61	230
6	610P	507	0	0	21	41	70	375
7	610S	422	0	0	24	53	66	279

Figure 13. Sample report.

5 SUMMARY

This study has investigated the current dock block lifting process requirements and block lifting sequence, analyzed the main factors affecting the speed and progress of dock block lifting, quantified the influence of the factors on the dock block lifting working hours, lifting times, etc. in view of the standard process plan for dock block lifting, established a mathematical model and an optimization algorithm for dock block lifting sequence based on the improved genetic algorithm, solved it, and verified the feasibility of the algorithm and the reliability of lifting model. To summarize, the study has completed the design and development of dock block quick lifting planning system, which realizes the automatic planning of block lifting, the maintenance of block information, as well as the display of the results of quick block lifting optimization.

REFERENCES

Li, L. et al. (2020). Digital Simulation and Optimization of the Job Scheduling Process of Ship Curved Block [J] Mechanical design & manufacture,2020(02):141–144.

Li, Z. et al. (2022). Method of Multi-Factory Collaborative Scheduling for Ship Block Construction Plan [J]. Ship Engineering,44(08):115–121.

Lu, Z.B. & Wang, A.M. (2016). Double yard cane scheduling for curved block in shipbuilding [J]. Computer Integrated Manufacturing Systems, 22(09):105–116.

Meng, L.T. et al. (2019). Combined Assembly Block Scheduling in Storage Yard of Shipbuilding [J]. Journal of Shanghai Jiao Tong university,50(07):780–788.

Roh, M., Lee, K. (2010). Optimal scheduling of block lifting in consideration of the minimization of traveling distance while unloaded and wire and shackle replacement of a gantry crane [J]. Journal of Marine Science and Technology,15(2):190–200.

Wang, C. et al. (2017). Ship Block Transportation Scheduling Approach Based on Genetic Algorithm [J] Journal of Shanghai Jiao Tong university, 2017(03):338–343.

Wang, H. et al. (2019). Ship Block Dynamic Scheduling under Multi-Constraints Based on Virtual Flow [J]. Ship Engineering,41(12):128–137.

Wang, Q.H. & Tang, J. (2015). Research on optimization and simulation of berth hoisting network [J]. Journal of Wuhan university of technology, 2015(02):218–222.

Wang, X.P. & Cao, L.M. (2002). Genetic Algorithm - Theory, Application and Software Implementation [M]. Xi'an: Xi'an Jiaotong University Press, 2002.

Xu, A.M. & Jin, Y. (2006a). Hull lifting sequence optimization of large bulk carrier [J]. Journal of Donghua university (Natural Science Edition), 2006(05):1–4.

Xu, A.M. & Jin, Y. (2006b). Optimization of network of block erection on dock using modified immune genetic algorithm [J]. Computer Integrated Manufacturing Systems, 2006(05):715–720.

Ying, L. (2014). Research on Technologies of Block Erection Sequence Optimization on the berth based on Improved Ant Colony Algorithm [D]. Dalian University of Technology, 2014.

Zhang, Z.Y. et al. (2015). Block stockyard scheduling approach based on an improved genetic algorithm [J]. Journal of Harbin Engineering University, 2015 (08):1103–1108.

Advances in Maritime Technology and Engineering – Guedes Soares & Santos (Eds)
© 2024 The Author(s), ISBN 978-1-032-83099-5

Optimization design and stability analysis of vehicle deck

Z.Q. Li & J.X. Yue
School of Naval Architecture, Ocean and Energy Power Engineering, Wuhan University of Technology, Wuhan, P.R. China

S.Q. Feng
Marine Design and Research Institute of China, Shanghai, P.R. China

ABSTRACT: Fewer transverse structures and longer spans inside the vehicle decks of Ro-Ro ships make the structure less stable. Meanwhile, the mass of the vehicle deck accounts for a relatively large proportion of the ship's total weight. Therefore, a lightweight design method and an optimal design scheme for the vehicle deck of Ro-Ro ships are proposed in this paper. Firstly, parametric modeling and direct FEA were used. As a result, the critical loading conditions of the deck under vehicle load were determined. Secondly, based on the Mixed Integer Sequential Quadratic Programming algorithm, a lightweight design procedure for vehicle decks with qualified strength was proposed. Finally, the stability study of the optimized vehicle deck was carried out. The results indicate that stress levels can be significantly reduced by a reasonable arrangement of vehicles. Optimizing deck plate thickness is the effective solution to reduce the mass of vehicle decks.

1 INTRODUCTION

For Ro-Ro ships, vehicle decks bear various loadings from different kinds of vehicle layouts and wheel prints. So vehicle decks have a higher demand on local strength. However, the structural mass of vehicle decks accounts for a considerable proportion of the ship's total mass. Therefore, this paper aims to determine the critical loading conditions for the vehicle deck and implement lightweight optimization design for vehicle deck structures, with the contribution to the structural safety of ships and the alleviation of structural weight (Jia, 2006).

Theoretical, numerical, and experimental studies have been conducted on vehicle decks' strength analysis and structural optimization for lightweight. The relationship between stress and deformation under the wheel print load on the deck has been analyzed by (Romanoff, 2007). By combining theoretical analysis with the finite element method. The distribution pattern of the wheel print load and the corresponding deck design methods have been studied. By simulating the actual loading conditions of the stiffened plate structure through the application of localized load simulation techniques and real tire loading, as conducted by (Jackson, 1981). The variation pattern of the stiffened plate structure's strength under the wheel print load has been investigated through the strength tests of stiffened plates conducted by (Zhu, 2020). Meanwhile, a design method for the stiffened plates has been proposed.

The optimization design of the ship's midship section structure has been proposed by (Jang, 1996) through the optimized design of local structures. These structures include reinforced panels, complex beams, frames, and other structures. (Sekulski, 2009, Sekulski, 2010)used a genetic algorithm to perform single-objective and multi-objective optimization designs on the topology layout of the deck. This method has been proven to be effective.

Research on the ultimate strength of large openings and long-span structures on cruise decks has been conducted using a combination of experimental and nonlinear finite element methods by (Liu, 2021), and the factors affecting longitudinal stability have been analyzed. (Paik, 2008) discusses the effect of initial defects, type of loading, and other factors on ultimate strength by analyzing the longitudinal bearing capacity of the hull structure using the nonlinear finite element method.

Above all, research on vehicle decks mainly focused on structural strength analysis. However, structural optimization designs based on strength analyses are not widely accepted. Therefore in this paper, the lightweight design of the vehicle deck of Ro-Ro ships is implemented by optimization algorithms combined with direct structural analysis. The critical loading condition caused by vehicle weights is determined by parametric modeling in finite element analysis (FEA). The size and number of structural components are optimized under critical

DOI: 10.1201/9781003508762-54

loading conditions. The longitudinal stability of the optimized vehicle deck is checked by nonlinear FEA. The above procedures can be used for the lightweight design of vehicle decks for Ro-Ro ships.

The paper is organized as follows. Section 2 discusses the critical loading conditions of the vehicle deck based on parametric modeling. Section 3 proposes an optimized design of the vehicle deck based on the critical loading condition. In Section 4, we analyze the stability of the optimized vehicle deck using the nonlinear finite element method. Finally, some conclusions and open questions are presented in Section 5.

2 ANALYSIS OF CRITICAL LOADING CONDITIONS ON VEHICLE DECK

For the vehicle deck of a Ro-Ro ship, the wheels are concentrated on the deck plating, with a relatively small area of contact (Cai, 2021). Therefore, within the local range at the position where the load is applied, the deck is supported by significant stress and deformation by the vehicles. It is necessary to discuss the effect of wheel print loads at different positions on the local vehicle deck. Vehicle deck condition analysis entails that, with the geometric component parameters of the vehicle deck being established. The stress values at different positions on the deck due to wheel print loads are obtained through parametric modeling. Meanwhile, the critical loading condition is determined.

2.1 *Vehicle load*

Many classification societies (CCS, 2022, DNV, 2022) have formulated that the vehicle load is simplified to a uniform load acting on a rectangular area. It is also ruled that vehicles are to be parked along the breadth or length direction of the ship. Figure 1 gives a schematic diagram of a set of wheel prints on the rear wheels of the vehicle, where '*a*' is the longer side of the print, '*b*' is the shorter side, and '*η*' is the wheelbase.

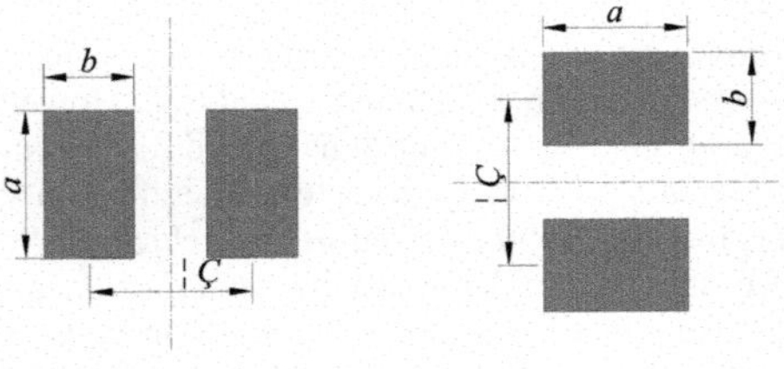

Figure 1. Vehicle wheel print diagram.

This section refers to the (CCS, 2022) specification and calculates the vehicle load under the navigation state of the ship. The state of the ship's navigation refers to the condition of the ship sailing on a dynamic sea-surface. At this point, it is necessary to consider the influence of acceleration a_v caused by the ship's motion.

The vehicle load P_h is shown as:

$$P_h = \frac{Q(g + 0.5a_v)}{n_o ab} \tag{1}$$

where Q is the vehicle design mass. g is the gravitational acceleration. n_0 is the number of wheels. a is the wheel print length. b is the wheel print width. a_v is the vertical acceleration shown as:

$$a_v = \frac{7a_0}{\sqrt{C_b}} \tag{2}$$

where C_b is the block coefficient. a_0 is the acceleration factor shown as:

$$a_0 = \frac{3C}{L_{pp}} + \frac{V}{50} \tag{3}$$

where V is the maximum service speed. L_{pp} is the captain. C is a factor shown as:

$$C = 10.75 - \left[\frac{(300 - L_{pp})}{100}\right]^{\frac{3}{2}} \tag{4}$$

In addition, to consider the structural strength of the vehicle deck after the ship's hull is breached, an additional water head pressure of 0.5 meters must be applied to the vehicle deck (CCS, 2022).

2.2 *Parametric modeling of the vehicle deck*

In order to study the effect of vehicle loads on the vehicle deck and the dimensions of each component on the structural strength. This section uses parameterized calculation methods to calculate the vehicle deck and vehicle position. The purpose is to determine critical loading conditions.

The typical plate armor length L and breadth B of the vehicle deck researched in this paper are determined. These parameters include the deck thickness D_T, T-profile type S_T, deck longitudinal type S_L, deck longitudinal spacing s, deck longitudinal span l, and the equivalent vehicle load position, etc. In this case, the parameterization of the equivalent load position is achieved by changing the values of the parameters m and n. Parameter m is the distance of the ship's longitudinal direction wheel print from the edge of the vehicle's deck, and n is the distance of the ship's breadth direction wheel print from the edge of the vehicle's deck. In the process of parameterization modeling, it is necessary to consider both the parking of vehicles along the longitudinal and breadth directions of the ships. Figures 3 and 4 give schematic diagrams of the vehicle deck arrangement.

Based on the vehicle deck parametric calculation method, the strength of the deck under different vehicle arrangement schemes can be determined.

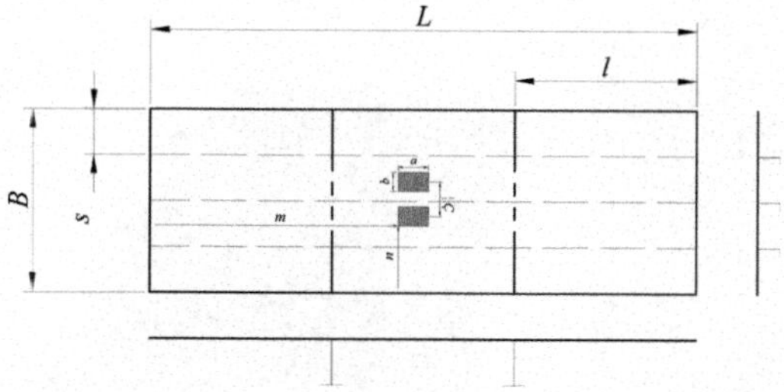

Figure 2. The equivalent load position along the longitudinal direction of the ship.

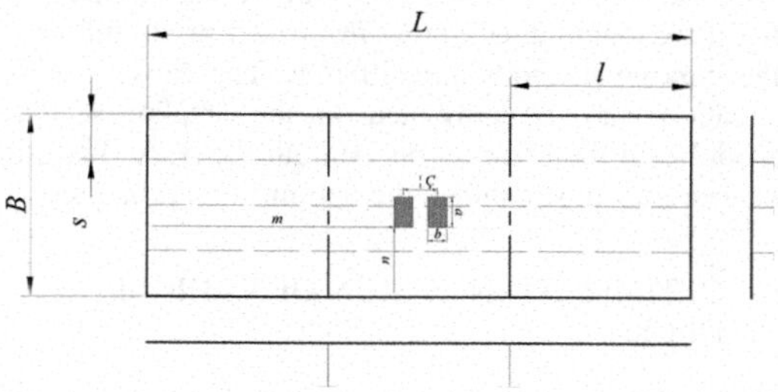

Figure 3. The equivalent load position along the breadth direction of the ship.

Meanwhile, it is considered that the loading condition corresponding to the maximum Mises stress on the vehicle deck is the critical loading condition.

2.3 *Critical loading conditions analysis*

Based on the above method, a typical plate frame with vehicles on the vehicle deck of a Ro-Ro ship is selected to be analyzed in this paper. The vehicle and deck parameters are shown in Table 1.

Table 1. Vehicle and deck parameters.

Parameter	Value
design mass Q	19t
print length a	0.4m
print width b	0.25m
number of wheels n_0	6
block coefficient C_b	0.683
captain L_{pp}	136m
deck length L	7.2m
deck breadth B	2.4m
deck thickness D_T	18mm
deck longitudinal spacing s	0.6m
deck longitudinal span l	2.4m

According to the data in Table 1 and equations (1)-(4) can be used to derive the vehicle loads P_h, which equals to 384.52kN/m^2. The region of the equivalent load is then determined based on the location of the wheel prints in Figures 2 and 3. In order to find the critical loading conditions of this typical deck, the parameters 'm' and 'n' need to be selected. The selection range of parameters 'm' and 'n' is shown in Table 2.

Table 2. The range of m and n values and loading conditions.

Condition	Breadth direction	Longitudinal direction
Range of m	The range of values is 0.25m-6.25mm, and the increase is 1m	The range of values is 0.4m-6.4m, and the increase is 1m
Range of n	The range of values is 0.2m-1.8m, and the increase is 0.4m	The range of values is 0.25m-1.45m, and the increase is 0.3m
Number of conditions	35	35

This simulation uses commercial finite element software ABAQUS. During the finite element analysis, a uniformly distributed vehicle load is applied to the wheel print area. Since the selected vehicle deck is surrounded by a rigid structure, the boundary conditions are set to be rigidly fixed around the perimeter to simulate real working conditions. The material selected for the vehicle deck is EH36 steel with an elastic modulus of E=206GPa, Poisson's ratio υ=0.33, and density ρ=7850kg/m^3. The overall model is modeled using shell elements, with a size of 25mm.

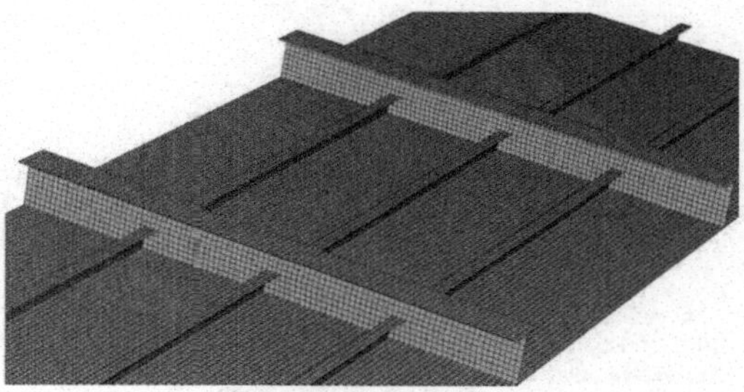

Figure 4. Vehicle deck local mesh.

Figures 5 and 6 respectively give the surface plots of the variation of Mises stress with parameters 'm' and 'n' for parking along the ship's breadth and the ship's length directions. It can be seen from the figure that the stress for the layout along the ship's breadth is generally less than that for the layout along the ship's length. Meanwhile, the stress is greatest when the wheel print is located at the center of the vehicle deck. Among all the stress conditions mentioned above, the maximum value of Mises stress is 162.3 MPa. The loading condition corresponding to this value is that the vehicle is laid out at the center of the deck in the direction of the ship's length directions. Therefore, this loading condition is defined as a critical loading condition.

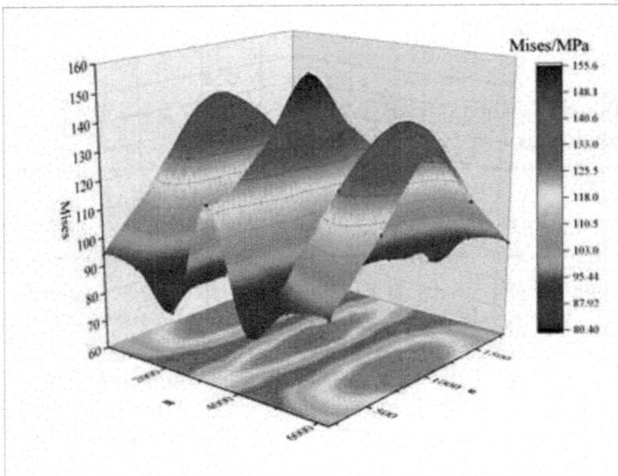

Figure 5. Variation of stress along the breadth direction of the ship concerning for to m, n.

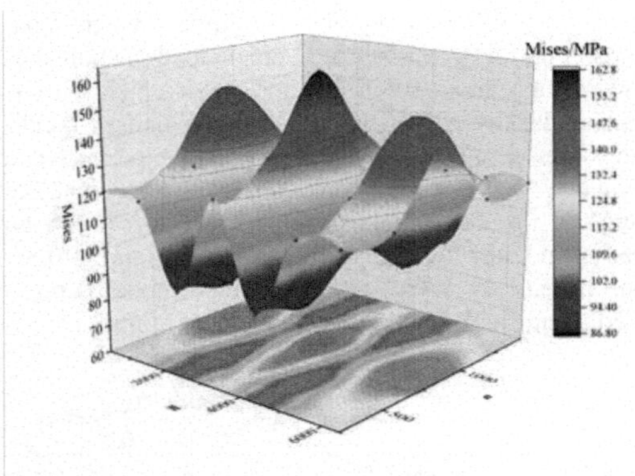

Figure 6. Variation of stress along the length direction concerning for to m, n.

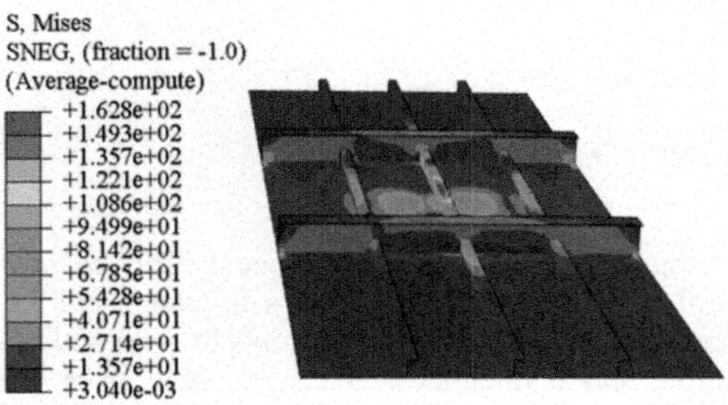

Figure 7. Stress cloud map for parking along the width direction of the ship.

The cloud diagrams of the Mises stress maximum loading condition for the two different layout methods are given in Figures 7 and 8. The figures show that the

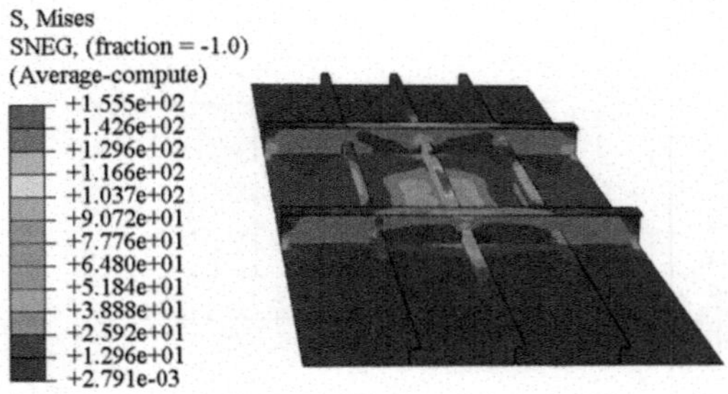

Figure 8. Stress cloud map for parking along the captain's direction.

areas of maximum Mises stress are all concentrated at the deck longitudinal below the wheel print, although the parking methods are different. The cause of this situation may be attributed to the smaller section modulus of the components beneath the deck. Thus, it is necessary to strengthen the optimization process.

3 OPTIMIZATION DESIGN OF VEHICLE DECK

Once the critical loading of the vehicle deck has been determined, the size and mass of the vehicle deck components that meet the constraint conditions are altered to achieve the purpose of reducing the structural weight of the vehicle deck. Based on the results obtained from the analysis of the critical condition of the vehicle deck in Section 2.3, this section carries out an optimization design for the critical condition. The optimization design process for the vehicle deck is shown in Figure 9.

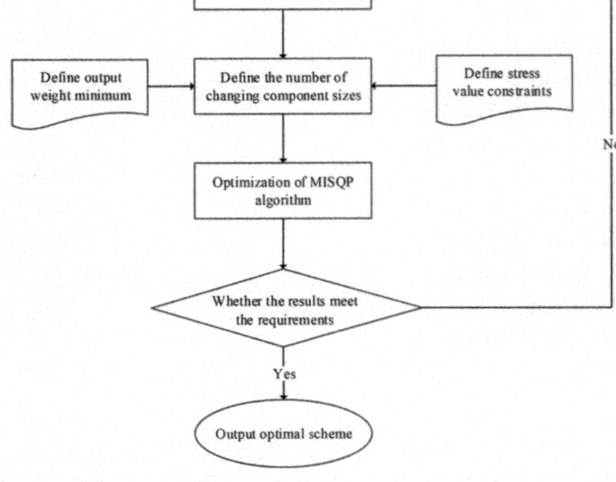

Figure 9. Vehicle deck optimization process.

3.1 *Optimization algorithm*

The Mixed Integer Sequential Quadratic Programming algorithm (MISQP) is an efficient gradient optimization algorithm. The algorithm utilizes the characteristics of the function's gradient and derivatives to determine the iteration step size and iteration direction. When the objective function exhibits continuous and unimodal characteristics within the design space, the algorithm is capable of obtaining the optimal solution of the problem along the direction of the steepest descent (Exler, 2012). The MISQP algorithm is employed for optimization because it supports the single-objective, multi-constraint problem proposed in this paper.

In order to describe the fundamental concept of the algorithm, the optimization problem of the model is first transformed into the problem of finding the optimal solution in mathematics, which is described as follows.

$$\begin{gathered} \min f(x) \\ g_j(x) \leq C, j = 1, \cdots, m \\ x \in R \end{gathered} \tag{5}$$

where $f(x)$ is the objective function. $g_j(x)$ is the constraint condition, m is the number of inequality constraints, C is constant, and they are continuously differentiable in the domain of definition.

The basic idea of the confidence domain approach is to perform a new iteration through a second-order model and approximation. The idea restricts the step size by trusting the radius Δ_k and k denotes the kth iteration. The quality function is then determined by constraints to obtain a rate of change r_k that is predictable and can be improved. And the trust radius increases or decreases depending on the deviation of r_k from the ideal value $r_k=1$. If r_k is less than 1, then Δ_k decreases, conversely Δ_k increases. If r_k is approximately equal to 1, Δ_k is considered to remain constant.

Lagrange's equation for the $f(x)$ objective function is shown as:

$$L(x,u) = f(x) - \sum\nolimits_{j=1}^{m} u_j g_j(x) \tag{6}$$

where u is the multiplier vector.

The precise definition of the penalty function is shown as:

$$p_\sigma(x) = f(x) + \delta \left\| g_j(x) \right\| \tag{7}$$

where δ is the penalty parameter, which should be greater than zero. Here all the constraints are combined into a vector that $g(x)=(g_1(x),\ldots,g_m(x))^T$.

The penalty function can transform a constrained objective function into an unconstrained one, which can enhance convergence. Hence equation (7) is also known as the quality function. In order to determine the method of quadratic programming of sequences, it is necessary to perform a discriminant to the next iteration which can be realized through equation (8).

$$\begin{gathered} \min \frac{1}{2} d^T B_K d + \nabla f(x_k)^T d + \sigma_k \left\| \left(\nabla g_j(x_k)^T d + g_j(x_k) \right) \right\|_\infty \\ \left\| d \right\|_\infty \leq \Delta_k \\ d \in R \end{gathered} \tag{8}$$

where d is the solution for each iteration. d_k is the optimal solution. σ_k and Δ_k are optimal parameters. B_K is the symmetric positive definite matrix.

The second key component of the confidence domain approach is to predict a new radius for the next iteration. This idea is to test the quotient of the actual value of the quality function and the value of the predictable improvement. It can be calculated by equation (9).

$$r_k = \frac{P_k(x_k) - P_k(x_k + d_k)}{\varphi_k(0) - \varphi_k(d_k)} \tag{9}$$

where $P_k(x)=P_{\sigma k}(x)$, it denotes the penalty function (7) used to record actual improvements. φ_k is shown in equation (10). This equation is used to estimate predictable improvements.

$$\varphi_k(d) = \nabla f(x_k)^T d + \frac{1}{2} d^T B_k d + \sigma_k \left\| \left(g(x_k) + \nabla g(x_k)^T d \right) \right\| \tag{10}$$

Through the above mathematical model, the optimal solution that satisfies the conditions can be determined.

3.2 *Optimization model*

(1) Design variable

The design parameters of the vehicle deck are considered as the design variables for the purpose of structural optimization design. These parameters include the thickness of the deck plate (D_T), the cross-sectional parameters of the deck transverses (S_T), the cross-sectional parameters of the deck longitudinal of the deck (S_L), the number of deck transverses (N_B), and the number of deck longitudinals (N_L) on the deck.

For the thickness of the deck plate, the optimization range is set to be between 12-18mm, with the thickness being an integer value. The optimization of the number of the deck transverses and deck longitudinals require that their spacing be selected as an integer value. Consequently, the number of deck transverses selected for optimization includes 2, 3, 4, 5, 7, 8, and 9, and the number of deck longitudinals chosen for optimization includes 3, 4, 5, and 7.

For the cross-sectional parameters of the deck transverses and deck longitudinals, dimensional optimization was carried out concerning the table of steel profiles. Figure 10 gives a schematic diagram of the T-profile, where 'w' is the web height, 'W' is the wing width, 't_1' is the web thickness, and 't_2' is the wing thickness. Figure 11 gives a schematic diagram of the L-profile, where 'H' is the long side width, 'h' is the short side width, and 'd' is the side thickness. The selection of numerical schemes for dimensional optimization of T-profiles and L-profiles are given in Table 3 and Table 4, respectively.

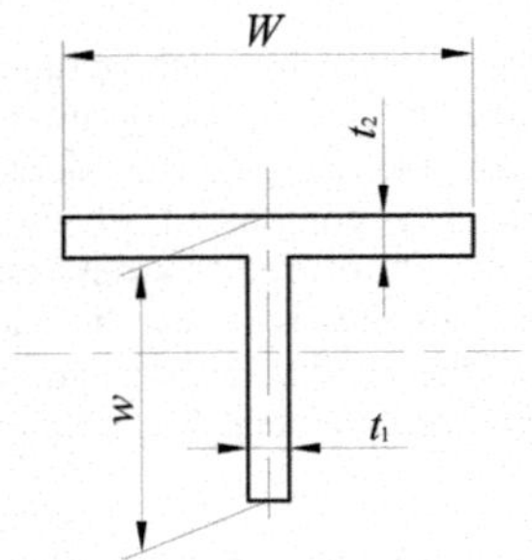

Figure 10. T-profile section schematic diagram.

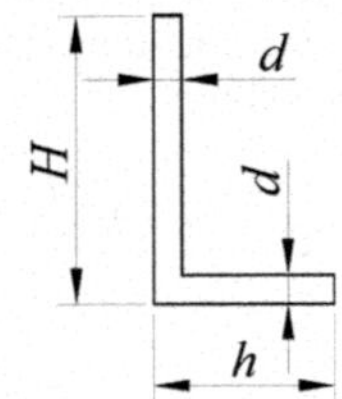

Figure 11. L-profile section schematic diagram.

Table 3. Range of T-profile dimensions for the deck transverses.

T-Profile No. S_T	Web height /w	Wing width /W	Web thickness /t_1	Wing thickness /t_2
	mm	mm	mm	mm
1	175	175	7	11
2	170	250	9	14
3	200	200	8	13
4	150	305	15	15
5	225	200	9	14
6	220	300	11	18
7	250	200	10	16
8	244	300	11	18
9	275	200	10	16
10	303	201	12	20

(2) Constraint condition

The constraint condition for this structural optimization design is the allowable stress of the material used for the vehicle deck, whose expression is shown in equation (11).

$$\sigma_{\max} \leq [\sigma], \tau_{\max} \leq [\tau] \quad (11)$$

Table 4. Range of unequal angle sizes for the deck longitudinals.

L-profile No. S_L	Long side width /H	Short side width /h	Side thickness /d
	mm	mm	mm
1	75	50	10
2	80	50	8
3	90	56	8
4	100	63	10
5	110	70	8
6	125	80	8

where $\sigma_{\max}$ is the maximum normal stress. $\tau_{\max}$ is the maximum shearing stress. $[\sigma]$ is the normal allowable stress. $[\tau]$ is the shearing allowable stress.

The materials selected for this study have an allowable normal stress and allowable shearing stress of 252.31MPa and 119.51 MPa, respectively.

(3) Objective function

The objective function for the optimization is defined as the mass of the vehicle deck (M), to minimize the structural mass of the vehicle deck. The expression is shown in equation (12).

$$M = \min f(x) \quad (12)$$

where $f(x)$ is the mass of the vehicle deck for each iteration.

3.3 *Optimization result*

Based on the MISQP algorithm, the optimization of the vehicle deck was iterated 27 times, resulting in the final vehicle deck mass becoming stable and the stress values all reaching the convergence range. Figure 12 gives the iteration curve for the mass of the vehicle deck.

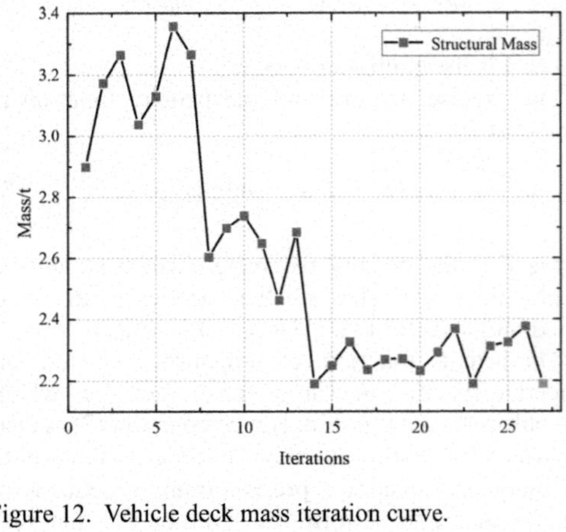

Figure 12. Vehicle deck mass iteration curve.

It can be seen from Figure 10 that the mass of the vehicle deck gradually decreases and converges with the increase in the number of iterations, reaching an optimal solution, which reflects the rationality of the algorithm.

Table 5. The optimized scheme and the initial plan.

Optimization parameters	Initial scheme	Optimization scheme	percentage
D_T /mm	18	12	-33.33%
N_B	2	4	50.00%
N_L	3	4	25.00%
No. S_T	5	1	—
No. S_L	4	5	—
σ_{max} /MPa	160.5	108.8	-32.22%
τ_{max} /MPa	29.37	32.5	10.65%
M /t	2.90	2.19	-24.48%

The comparison of the optimized scheme with the initial scheme for each parameter is given in Table 5. From the table, it can be seen that the normal stress in the optimized scheme has decreased compared to before optimization, while the shearing stress has increased. The structural mass of the vehicle deck after optimization is observed to have reduced by 24.48% compared to the initial scheme.

The normal stress clouds before and after optimization are given in Figures 13 and 14, respectively. It can be seen from the figure that the maximum stress is concentrated in the longitudinal girder area beneath the wheel imprint. The shearing stress clouds before and after optimization are given in Figures 15 and 16, respectively. It is known from the figure that the maximum shearing stress is concentrated in the area of the web of the deck transverse.

Figure 17 gives the global impact factor of the design variables given by this optimization. The global impact factors for all the design variables and the objective function are calculated through each iteration step of the optimization platform, as represented by this influence factor. The impact factor allows for the determination of the proportion of influence between the design variables and the objective function.

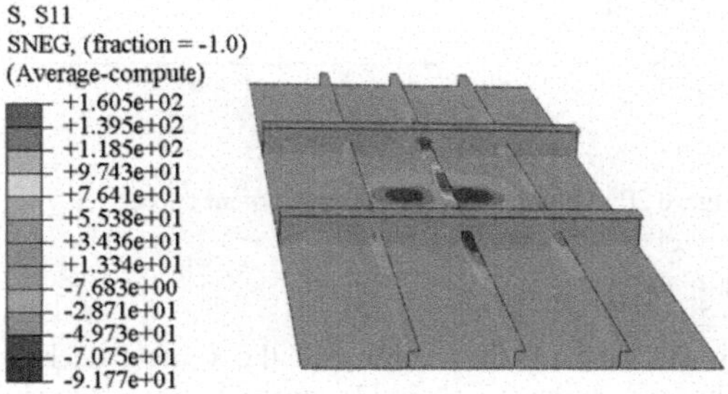

Figure 13. Normal stress cloud before optimization.

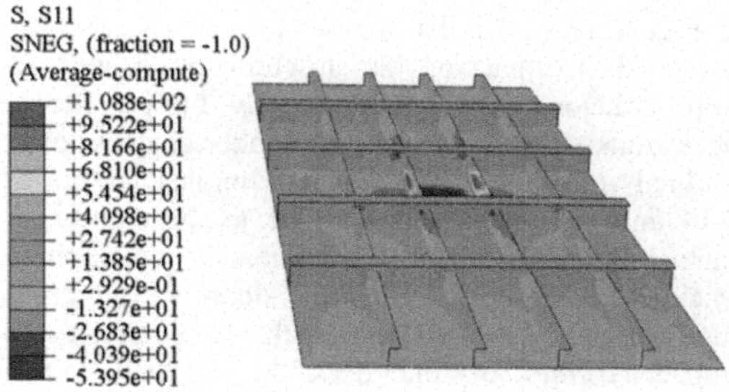

Figure 14. Normal stress cloud after optimization.

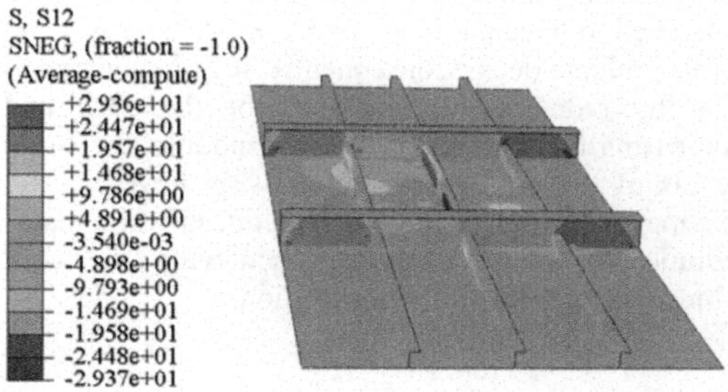

Figure 15. Shearing stress cloud before optimization.

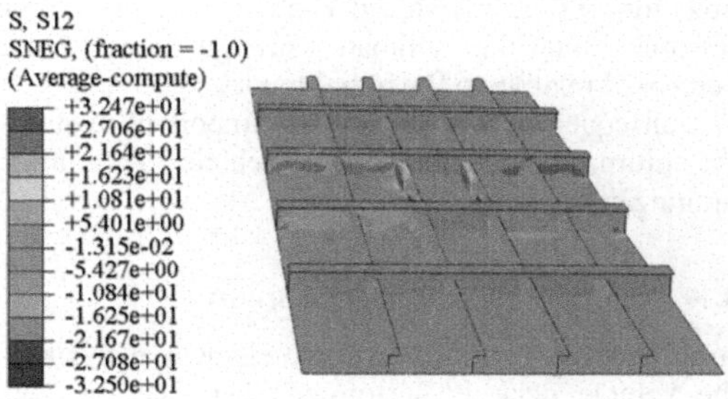

Figure 16. Shearing stress cloud after optimization.

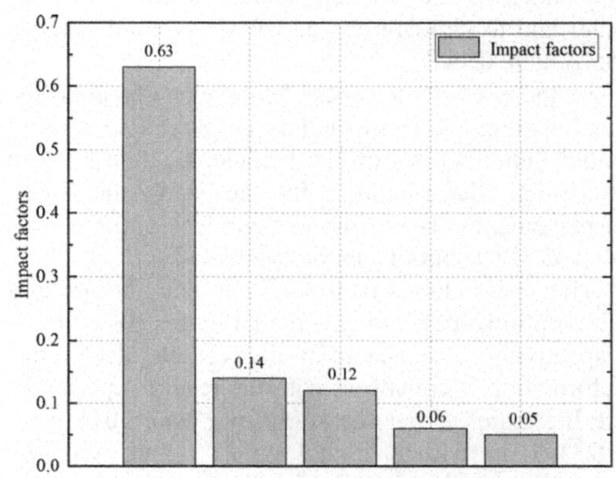

Figure 17. Design variables and output mass impact factors.

Figure 17 shows that the deck thickness (D_T) has the greatest impact on the structural mass, with an impact factor of 0.63. Subsequently, the number of deck transverses (N_B) and the number of deck longitudinals (N_L) are shown to have impact factors of 0.14 and 0.12, respectively. Due to the significant impact factor of the deck thickness, a reduction in thickness is observed to rapidly decrease the structural mass of the vehicle deck. In contrast, the impact factors of the deck longitudinal cross-sectional parameter (S_L) and the deck transverses cross-sectional parameter (S_T) on the structural mass are observed to be minimal. Altering the cross-sectional parameters of the deck components is observed to have a minor impact on the overall mass of the vehicle deck. Consequently, it is demonstrated that by reducing the thickness of the deck and increasing the components profile module, the stress levels of the deck can be effectively lowered, and the mass of the structure can be reduced while maintaining its integrity, thereby achieving the dual objectives of structural optimization.

4 VEHICLE DECK STABILITY

Considering the lightweight design of the vehicle deck, it is acknowledged that the mass after optimization has been reduced compared to the initial scheme, and the ultimate strength may have decreased relative to the initial model.

Consequently, it is particularly important to study the ultimate strength of the vehicle deck after optimization.

4.1 *Nonlinear finite element analysis*

In this section, a nonlinear finite element analysis of the vehicle deck is performed. The material and mesh are the same as those described in Section 2.3, with boundary conditions being rigidly fixed at one end and a forced displacement load applied at the other end to simulate the actual stress conditions of the vehicle deck.

In the present nonlinear finite element analysis, the buckling mode method is employed to impose initial imperfections on the vehicle deck, simulating its initial deformation. In the end, the load-displacement curve of the vehicle deck and the state of model distribution can be calculated.

The stress clouds of the vehicle deck before and after optimization are given in Figures 18 and 19, respectively. The shape of the vehicle deck after deformation is consistent with the results reported in the literature, as per the study by (Wan, 2011, Cui, 2023). It is evident from Figure 19 that, with the continuous increase in load, deformation begins to occur in the vehicle deck, predominantly concentrated in the upper section. The deformation is primarily manifested as buckling of the plate grid and distortion of the deck longitudinals.

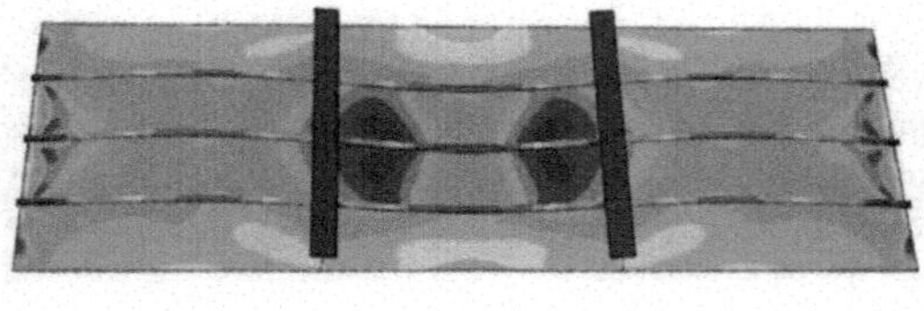

Figure 18. Vehicle deck deformation before optimization.

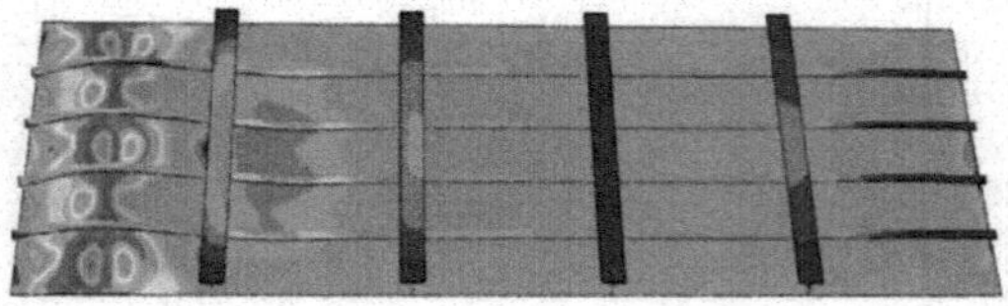

Figure 19. Vehicle deck deformation after optimization.

The load-displacement curves obtained from the nonlinear buckling analysis are given in Figure 20, from which the ultimate load-carrying capacity of the vehicle deck can be derived.

From Figure 20, it is observed that the ultimate strength of the vehicle deck before optimization was 12265kN, corresponding to an ultimate displacement of 16.21 mm. After optimization, the ultimate strength of the vehicle deck increased to 14294kN, with a corresponding ultimate displacement of 15.74 mm. Compared to the pre-optimization, the ultimate strength has increased by 16.54%. The increase in the ultimate strength of the overall structure of the vehicle deck is attributed to the augmented number of the deck transverse and deck longitudinals.

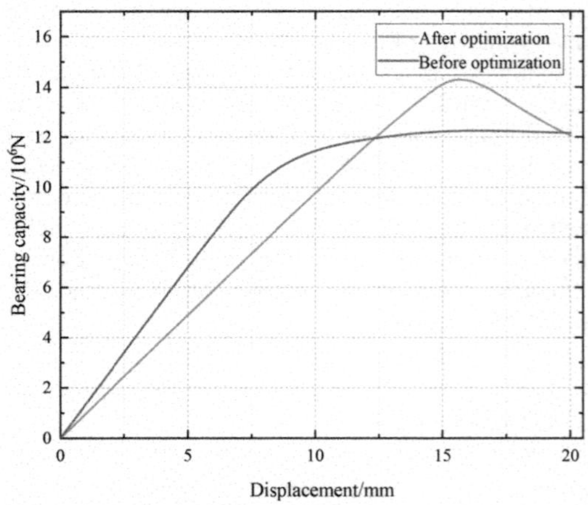

Figure 20. Vehicle deck load-displacement curve.

4.2 *Stability check*

In this section, the stability of the vehicle deck is checked by the (CCS, 2022) specification. For buckling calculations of the deck, the minimum buckling safety factor $\lambda > 0.8$ is specified to satisfy the buckling strength scale.

The minimum buckling safety factor λ is shown as:

$$\lambda = \frac{\sigma_{cr}}{\sigma_A} \tag{13}$$

where σ_A is the working stress, obtained from the finite element analysis is considered here. σ_{cr} is the critical buckling stress of the plate grid.

Critical buckling stress of plate grid σ_{cr} is shown as:

$$\sigma_{cr} = kD\left[\frac{\pi^2 E}{12(1-\nu^2)}\right]\left(\frac{t}{s}\right)^2 \tag{14}$$

where k is the bending factor, take 4 here. D is the constraint factor, take 1.1 here. t is the deck thickness. s is the longitudinal bone spacing. E is the modulus of elasticity.

The critical buckling stress σ_{cr} is calculated to be 512.01N/mm^2 using Equation (13). The stress determined from the finite element analysis is 562.50N/mm^2. Consequently, the buckling safety factor for the plate grid is calculated to be 0.91, which exceeds the value of 0.8 and thus meets the standard requirements. It also indicates that the optimization carried out is meaningful.

5 CONCLUSIONS

In this paper, an optimization design method for the vehicle deck of Ro-Ro ships is proposed based on The Mixed Integer Sequential Quadratic Programming algorithm (MISQP) and finite element direct calculation under critical loading conditions. The main research conclusions are as follows:

(1) An analysis procedure based on parametric modeling was proposed to determine the most critical loading condition of the vehicle deck. Under this loading condition, vehicles' layout and wheel print loads would cause the maximum local stress in the vehicle deck.
(2) The structural optimization design was implemented by using the MISQP algorithm for the vehicle deck and resulted in a 24.48% weight reduction compared to the initial design. The deck thickness had the greatest impact on the overall mass of the vehicle deck.
(3) The stability of the optimized vehicle deck satisfied the requirement of classification rules. Ultimate strength analysis was carried out on the optimized vehicle deck, and the ultimate bearing capacity was improved by 16.54%. The buckling of the vehicle deck was mainly reflected in the deformation of plate grids and longitudinals.

REFERENCES

Cai, W. & Zhu, L. 2021. Plastic deformation of ship plate subjected to repeated patch loads at different locations - Elastoplastic numerical analysis and design equation. *Marine Structures* 76: 102901.

CCS 2022. Steel Seagoing Ships Classification Code. *China Classification Society*. Beijing: People's Transportation Press.

Cui, H.X. & Wan, Q. 2023. Stability Study of Long-span Double-deck Under Axial Compression Load. *Journal of Wuhan University of Technology(Transportation Science & Engineering)* 23(6):1–14.

DNV 2022. Rules For Classification High speed and light craft Part 5 Chapter 2. *Det Norske Veritas*. Norway: DET NORSKE VERITAS

Exler, O. & Lehman, T. 2012. MISQP: A Fortran subroutine of a trust region SQP algorithm for mixed-integer nonlinear programming-user's guide. *Report, Department of Computer Science, University of Bayreuth*.

Jackson, R.J.& Frieze, P. 1981. Design of deck structures under wheel loads. *The Royal Institution of Naval Architects* 3: 119–144.

Jang, C.D. & Seo, S. 1996. A Study on the Optimum Structural Design of Surface Effect Ships. *Marine Structures* 9(5): 519–544.

Jia, J. & Ulfvarson, A. 2006. Dynamic analysis of vehicle–deck interactions. *Ocean Engineering* 33(13): 1765–1795.

Liu, B. & Yao, X.N. 2021. Experimental and numerical analysis of ultimate compressive strength of long-span stiffened panels. *Ocean Engineering* 237: 109633.

Paik, J.K. & Kim, B.J. 2008. Methods for ultimate limit state assessment of ships and ship-shaped offshore structures. *Ocean Engineering* 35(2): 261–270.

Romanoff, J. & Varsta, P. 2007. Laser-welded web-core sandwich plates under patch loading. *Marine Structures* 20(1-2): 25–48.

Sekulski, Z. 2009. Least-weight topology and size optimization of high speed vehicle-passenger catamaran structure by genetic algorithm. *Marine Structures* 22(4): 691–711.

Sekulski, Z. 2010. Multi-objective topology and size optimization of high-speed vehicle-passenger catamaran structure by genetic algorithm. *Marine Structures* 23(4): 405–433.

Wan, Q. & Wang, F.H. 2011. Longitudinal Stability Analysis and Optimum Design of Supportless Long-span Deck Structure. *Shipbuilding of China* 52(1): 17–25.

Zhu, L. & Cai, W. 2020. Design method for steel deck plates under quasi-static patch loads with allowable plastic deformations. *Marine Structures* 71: 102702.

Advances in Maritime Technology and Engineering – Guedes Soares & Santos (Eds)
© 2024 The Author(s), ISBN 978-1-032-83099-5

Ultimate strength assessment of panels with thin-plates and small-sized stiffeners under combined biaxial loads

M. Ozdemir, S. Zhang, L. Yu & H. Ocakli
Lloyd's Register EMEA, Southampton, UK

ABSTRACT: Panels composed of thin plates and small-sized stiffeners built in high-tensile strength steels are commonly found in cruise and naval ships. Due to their reduced thickness, these panels are more susceptible to elastic instability. In this paper, investigations on the influence of boundary conditions (BCs), initial imperfections and the number of stiffeners in nonlinear finite element analyses (FEAs) on the prediction of the ultimate strength of panels consisting of thin plates and small-sized stiffeners were carried out. As a benchmark case, the BCs and initial imperfections of the LR nonlinear FEA guidance notes were followed. Subsequently, different BCs for the eigenvalue analysis and initial imperfections were employed. Various load combinations were systematically examined and ultimate strength interaction envelopes were generated. Additionally, elastic large deflection analyses were conducted to determine whether secondary global buckling of panels could occur solely due to elastic instabilities. The results were then compared with the eigenvalue analysis results.

1 INTRODUCTION

The buckling and ultimate strength assessment of plates and stiffened panels is a design process that demands careful attention, particularly in the ultimate strength assessment using nonlinear Finite Element Analysis (FEA). This is primarily due to the uncertainties inherent in nonlinear collapse analysis, which can be categorized as aleatory and epistemic (ISSC 2022). Aleatory uncertainties are essentially stochastic in nature, involving factors like material properties and scantlings. In contrast, epistemic uncertainties arise from a lack of knowledge or data regarding the actual response of the structure or system, with numerical modelling techniques falling under this category. The aleatory uncertainties can be quantified and eliminated by additional testing or introducing stochastic models in the calculations, while the epistemic uncertainties cannot be easily quantified (ISSC 2022).

The buckling collapse behaviour of the plates and stiffened panels are more sensitive to the initial imperfections including the initial deflections and welding residual stresses. In this regard, there have been significant research works dedicated to investigating the influence of those parameters.

Yao & Ueda (1985) carried out measurements on the stiffened plates of the actual ships and formulized the complex initial deflections in terms of combined harmonic functions. The initial imperfections in this complex shape are so called "thin-horse mode" of initial deflections. In a nonlinear collapse analysis, the deflection, representing the lowest energy level will be triggered and lead to the failure of the panel while the other deflection components may remain constant or decrease (Yao & Fujikubo 2016).

Gordo (2008) investigated the effect of the shape and amplitude of distortions in restrained unstiffened plates under longitudinal compression. It was concluded that the shape of initial deflections is more influential than the amplitude of the imperfections.

The ultimate strength committee of ISSC (2009) presented an extensive benchmark study on nonlinear FEA of stiffened panels under combined biaxial thrust. The investigation covered various factors, including boundary conditions along the longitudinal edges, lateral pressure, mesh density, material modelling, residual stresses, and geometric initial deflections. The study systematically examined the influence of initial deflections by removing each deflection component one by one from the base condition, involving all initial deflection components. The findings highlighted that local plate deflections significantly impact ultimate strength under longitudinal thrust conditions, while the influence of initial deflection was less pronounced in cases with predominant transverse thrust loads.

Tekgoz et al. (2013a) and (2013b) studied the influence of welding residual stresses on the ultimate strength of stiffened panels using the idealized welding residual stress distribution and modified stress-strain curves for welding residual stress fields.

Recently, the ultimate strength committee of ISSC (2022) conducted an extensive benchmark study to systematically determine uncertainties using nonlinear FEAs and fast-running codes. The numerical

DOI: 10.1201/9781003508762-55

simulations were compared with a reference experiment, revealing that neither of the numerical models could capture the failure mode and location observed in the experiment. However, the predicted ultimate strength values generally showed good agreement with the experimental results.

The initial geometric deflections are applied usually by the combination of eigenmodes in the nonlinear collapse analyses. As it is known that the shape of the initial deflections is more influential on the ultimate strength than the magnitude of them, we will particularly focus on the derivation of the eigenmodes to be employed in the nonlinear FEAs.

The nonlinear FEA guidance of LR (2016) will be followed first to derive the eigenmodes of the stiffened panels, then the constraint conditions of the stiffeners will be modified to observe the possible changes of eigenmodes as well as their impacts on the collapse mode and load of the panels.

2 FINITE ELEMENT MODELLING

The stiffened panels were modelled following the nonlinear FEA guidelines of LR (2016). The modelling extent of the stiffened panel was considered as 1/2 + 2 + 1/2 frame spacing in the longitudinal direction. This approach aims to minimize the impact of boundary conditions along the transverse edges on the failure of panels. A study by Xu et al. (2013) clearly demonstrated the significant influence of boundary conditions on the ultimate strength assessment of stiffened panels. However, ISSC (2009) ultimate strength committee reported that the influence of boundary conditions along the longitudinal edges is less pronounced.

The modelling extent of a panel with 4 stiffeners is indicated in Figure 1. In the given figure, the spacing between the frames is a = 4250 mm, the stiffener spacing is taken as b = 750 mm. The pate thickness is considered to be t_p = 6 mm.

The stiffener has angle bar cross-section with the dimensions: h_w = 100, t_w = 7, b_f = 24.6 and t_f = 8.9 mm. The stiffener cross-section is illustrated in Figure 2.

The number of stiffeners was taken as n_s = 4 and 10 so as to capture influence of possible global buckling collapse. The model discretization was carried out according to the minimum requirement of LR (2016) using shell elements as follows.

- Element size on the plating is $b/8$,
- Stiffener web is divided into three elements,
- Flange is modelled with one element.

The material is typical high-tensile strength steel with the properties as follows.

- Elastic modulus: $E = 206{,}000$ MPa
- Yield stress: $\sigma_Y = 355$ MPa
- Poisson's ratio: $v = 0.3$
- Strain hardening: $H = 0$

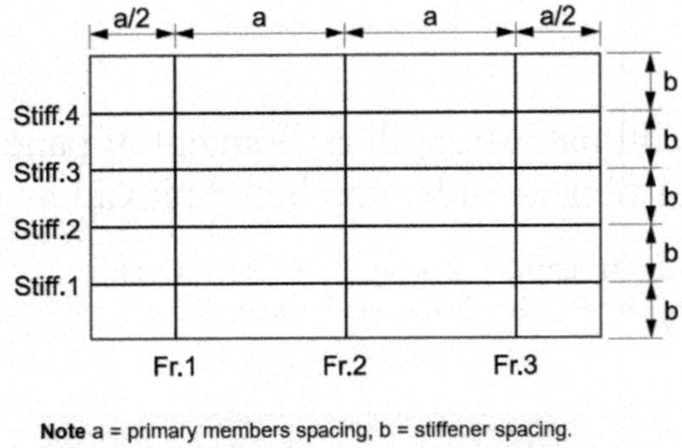

Figure 1. Modelling extent of stiffened panels according to the nonlinear FEA guidelines of LR (2016).

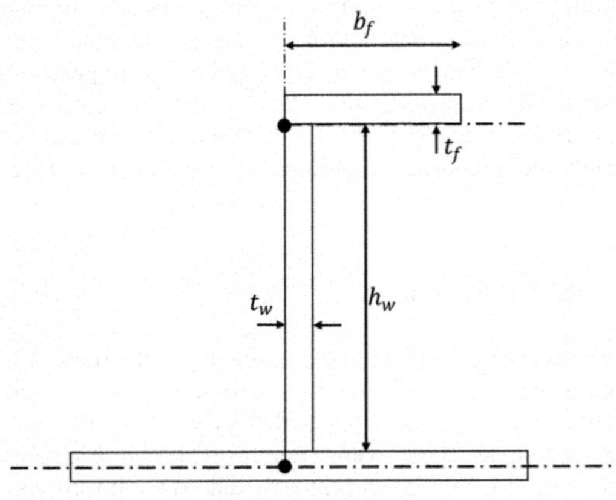

Figure 2. Cross-section of the considered stiffener profile.

2.1 *Boundary conditions*

The BCS are among the most influential parameters in the nonlinear collapse analysis of stiffened panels. Accurate implementation of the BCs is crucial to ensure that all possible buckling failure modes are captured by the numerical model. Even if the modelling extent in the longitudinal direction is sufficiently long, and the potential impact of BCs on failure behaviour is minimized, the enforced displacements and loading conditions must still reflect the actual conditions as much as possible.

The guidelines by LR (2016) for the nonlinear collapse analysis of stiffened panels had been developed to represent all possible buckling failure collapse modes under combined loading, including biaxial thrust, shear, and lateral pressure. The BCs for nonlinear collapse analyses are explained with reference to Figure 3 as follows.

- B1 except corner nodes:
  - $U_X = 0$
  - $R_Y = R_Z = 0$
- B2 except corner nodes:
  - U_X is constrained to follow C2.
  - $R_Y = R_Z = 0$

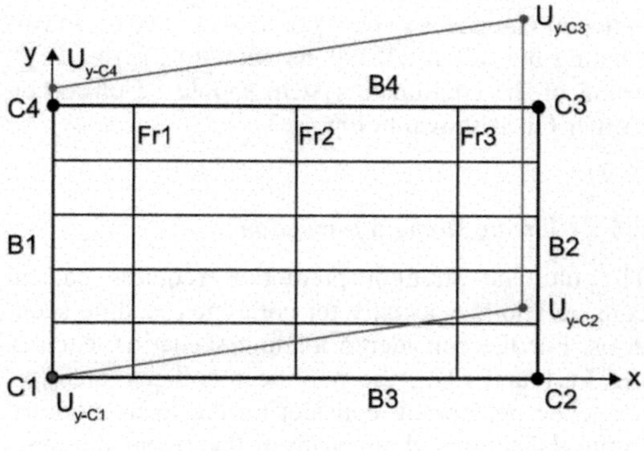

Figure 3. Representation of BCs on a schematic model.

- B3 except corner nodes:
  - U_Y is constrained to move on a straight line between C1 and C2.
  - $U_Z = 0$
  - $R_X = 0$
- B4 except corner nodes:
  - U_Y is constrained to move on a straight line between C3 and C4.
  - $U_Z = 0$
  - $R_X = 0$
- Corner - C1:
  - $U_X = U_Y = U_Z = 0$
- Corner - C2:
  - $U_Z = 0$
  - U_Y is constrained to ensure edges remain parallel.
- Corner - C3:
  - $U_Z = 0$
  - U_Y is constrained to ensure edges remain parallel.
- Corner - C4:
  - $U_X = U_Z = 0$
- Fr1, Fr2, Fr3:
  - $U_Z = 0$
  - U_Y of the stiffener webs is to follow the nodes at the intersections between webs and plate.

These BCs will be called as "Generic BCs" hereafter.

2.1.1 *Boundary conditions for eigenvalue analyses*

Local mode of initial deflections for the plating and stiffeners is to be applied from the eigenvalue analyses by selecting an appropriate eigenmode. The eigenmodes will depend on the BCs, and this effect will be more pronounced for panels with thin plates and small-sized stiffeners. This is because buckling mode interactions, including combined global buckling and local plate buckling, might occur within closely spaced eigenvalues. The aim of this study is to investigate the impact of the BCs in the eigenvalue analysis and, consequently, the shape of the initial imperfections on the ultimate strength. Accordingly, two BC cases will be considered here.

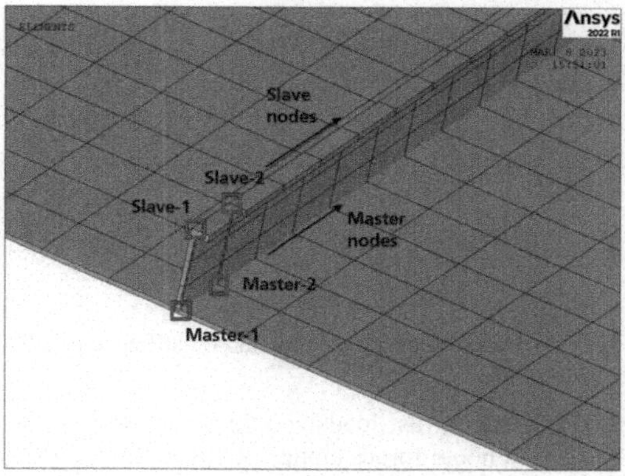

Figure 4. MPC condition along the stiffener web connecting the bottom and top nodes.

(i) LR (2016) recommends enforcing following constraints in addition to the "Generic BCs" in case of eigenvalue analysis. This case will be so called "Eigen BCs", which will be applied in case of eigenvalue analysis but will be removed later in the nonlinear collapse analysis stage.

- All BCs for the nonlinear collapse analysis will be implemented for this case,
- Additionally, $U_Z = 0$ on the stiffener bottoms.
- U_Y: *MPC* for the nodes on the stiffener bottom and web-flange intersections. Master nodes are to be located on the stiffener bottom as illustrated in Figure 4.

(ii) In the second case, the BCs for the eigenvalue and the nonlinear collapse analyses are the same, i.e., "Generic BCs".

2.2 *Loading conditions and implementation*

Only biaxial thrust loading conditions were considered, see Figure 5. The load ratios are given in Table 1.

Table 1. Biaxial thrust loading conditions.

LCs	1	2	3	4	5	6
Sx:Sy	1:0	1:0.11	1:0.54	1:1	0.43:1	0:1

In the FEA framework, the load application is to be performed according to the procedure described as follows (LR 2016).

- Longitudinal compressive load is to be applied to Corner – C2 as nodal force:

$$F_x^{C2} = S_x \cdot A_{B2} \tag{1}$$

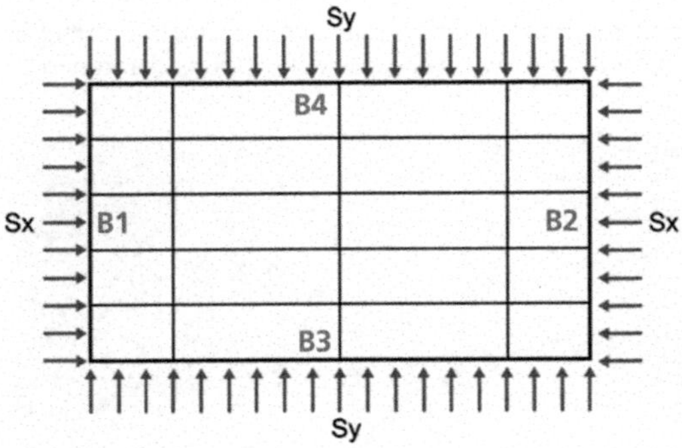

Figure 5. Biaxial load implementation for stiffened panels.

- Transverse thrust loads should be applied as distributed nodal forces along the edges B3 and B4:

$$F_y = \frac{S_y \cdot A_{B3}}{N_{B3}} \tag{2}$$

where, N_{B3} is the number of shell elements along B3 or B4, and the plate cross-sectional area of B3 is A_{B3}. The nodal forces on the corner nodes are to be half of the value calculated above.

2.3 *Initial imperfections*

The initial imperfections were considered in terms of geometric initial deflections. The aim is to predict the most conservative ultimate strength, taking into account uncertainties in the possible initial deflection modes of the panels after fabrication. In this regard, total initial deflections were applied as the superposition of local buckling and global buckling modes. The local buckling modes for the plate and stiffener webs are to be derived from eigenvalue analyses with the "Generic BCs" and "Eigen BCs" as described in Section 2.1. Then, the local mode of initial deflections is to be applied by scaling the maximum value to $b/200$.

The global mode of initial deflections is to be implemented to both plate and stiffeners according to the guidance note of LR (2016) as follows.

- Global mode initial deflections for the plate and stiffeners:

$$\delta_z = A_0 \cos\left(\frac{\pi x}{a}\right) \sin\left(\frac{\pi y}{B}\right) \tag{3}$$

- Sideway initial deflections for stiffeners:

$$\delta_y = A_0 \left(\frac{z}{h_w}\right) \cos\left(\frac{\pi x}{a}\right) \cos\left(\frac{\pi y}{B}\right) \tag{4}$$

where z is the vertical coordinate of the nodes on the stiffener web. Total breadth of the panel is $B = b \cdot (n_s + 1)$. The magnitude for the global initial deflection modes is to be taken as $A_0 = 0.001a$. In calculating the initial deflections with Eqs. (3) and (4), origin of the coordinate system should be placed on Corner-C1 as shown in Figure 3.

2.4 *Ultimate strength prediction*

The ultimate strength prediction requires careful consideration, especially for combined loading scenarios. For the considered loading scenarios, excluding LC1 and LC6, the nonlinear collapse analyses are to be performed, considering the local mode of initial deflections obtained from the eigenvalue analyses under the following conditions:

- Only longitudinal thrust,
- Only transverse thrust,
- Combined biaxial thrust with the given loading ratios.

The ultimate strength of the panels is then determined as the minimum peak load obtained from the nonlinear collapse analyses conducted with the described initial deflections. As for the LC1 and LC6 cases, obtaining the local mode of initial deflections under the considered actual loading will be sufficient for the nonlinear collapse analysis.

Further details on the FE modelling and collapse analysis of stiffened panels can be found in LR (2016).

3 NUMERICAL STUDIES

The eigenvalue analyses and ultimate strength assessments for the stiffened panels were conducted following the procedures outlined in the previous section. In this section, we will initially discuss the influence of the BCs on the eigenvalues and eigenmodes of the stiffened panels for each loading scenario. The second stage involves examining the effects of initial deflections on the collapse mode and the ultimate strength of the panels. Finally, we will analyse the impact of the number of stiffeners on the buckling and ultimate strength of the panels.

3.1 *Influence of BCs on the buckling assessment*

The impact of two different BCs, namely, "Generic BCs" and "Eigen BCs," on the buckling strength and buckling modes was thoroughly investigated. The goal is to ascertain the potential occurrence of global buckling when small-sized stiffeners are employed.

Buckling strength interaction curves were derived and are presented in Figure 6. It is evident that the BCs have a significant impact on the buckling strength of the panels, particularly in cases dominated by transverse thrust.

The results in Figure 6 indicate that the influence of the number of stiffeners on the buckling strength of these panels is insignificant. This can be explained as follows: the global buckling mode is not the

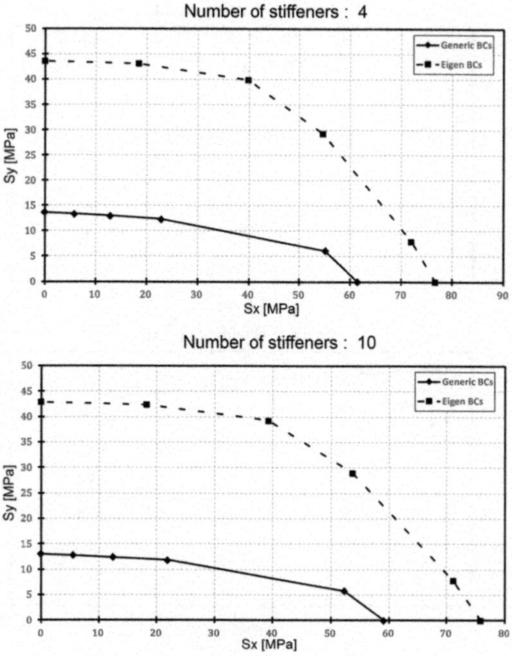

Figure 6. Influence of BCs on the buckling strength estimations.

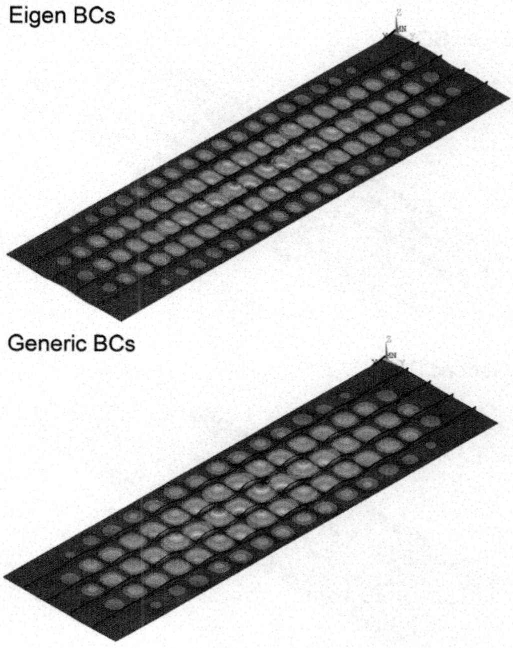

Figure 7. Buckling modes shapes for $n_s = 4$ under longitudinal thrust load.

critical mode shape for the considered panels, even with the "Generic BCs," which allows for column-type vertical displacement of the stiffeners. The buckling mode shapes for the case of $n_s = 4$ and 10 are respectively given in Figures 7 and 8.

The "Eigen BCs" prevent vertical and sideways stiffener deflections while allowing local web buckling when the web plate is slender; on the other hand, the "Generic BCs" enable the stiffeners to deform freely. The buckling mode shapes in Figures 7 and 8 show that the number of buckling half-waves in the plating is greater in the case of "Eigen BCs" than those of the "Generic BCs". Furthermore, buckling deformations occur only on the plating when the "Eigen BCs" are applied.

As stated previously, the influence of the considered BCs is more significant under transverse thrust conditions. The buckling mode shapes for the transverse load case (LC6) are given in Figures 9 and 10. The presented figures indicate that the stiffeners deform both horizontally and vertically when "Generic BCs" are employed, resulting in comparably small edge restraints for the plating between the stiffeners. This deformation condition leads to very small buckling strength, which is also smaller than the transverse buckling strength of a simply supported counterpart of the considered plating.

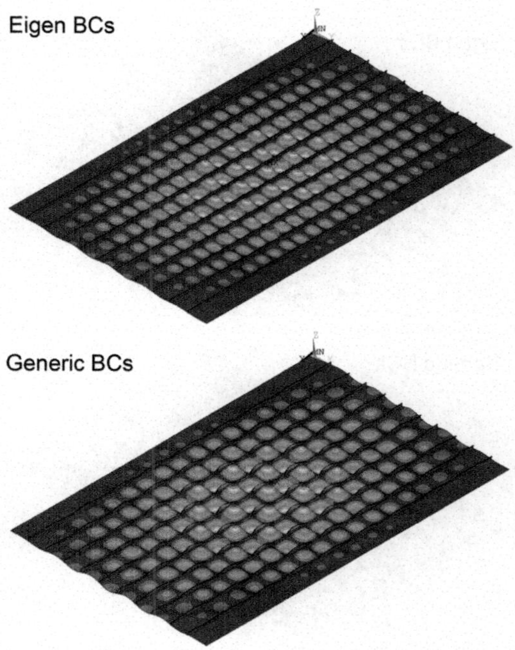

Figure 8. Buckling modes shapes for $n_s = 10$ under longitudinal thrust load.

3.2 *Impact of initial deflections on the ultimate strength*

The local mode of initial deflections derived by "Eigen BCs" will be referred to as "Tight Inits", and the initial deflections by the "Generic BCs" will be called "Loose Inits". Application of the global initial deflections has been kept identical in both cases.

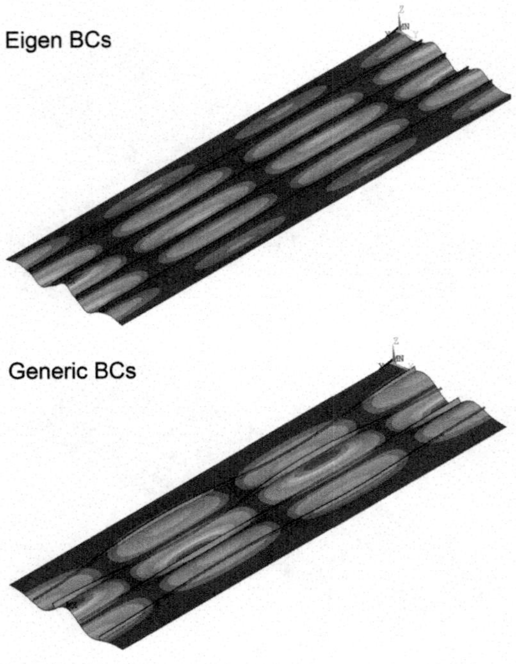

Figure 9. Buckling modes shapes for n_s = 4 under transverse thrust load.

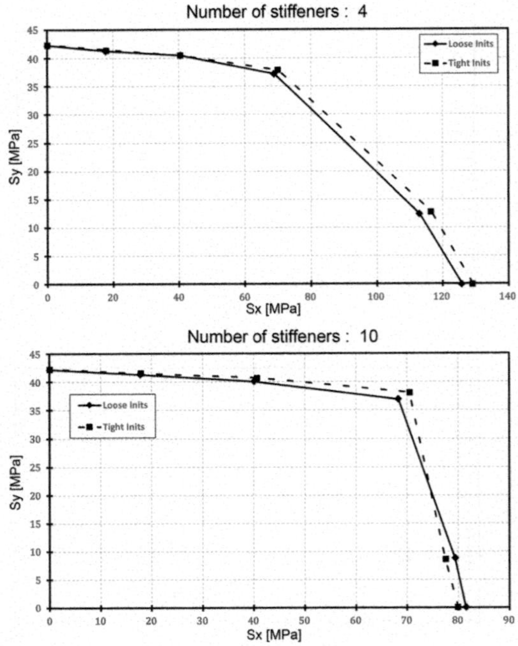

Figure 11. Impact of initial deflections on the ultimate strength.

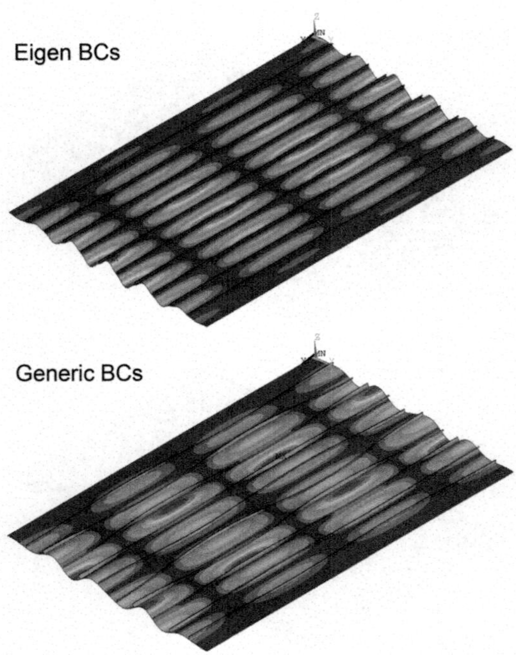

Figure 10. Buckling modes shapes for n_s = 10 under transverse thrust load.

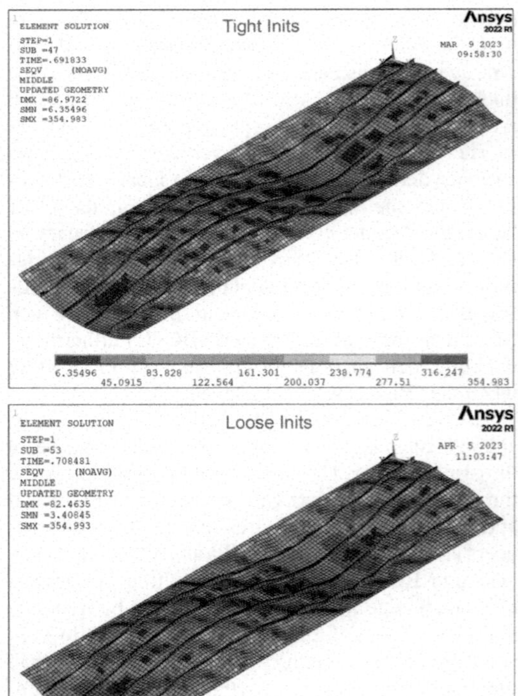

Figure 12. Collapse modes for n_s = 4 under longitudinal thrust load (LC1).

The influence of the local mode of initial deflections was examined and compared in Figure 11 for n_s = 4 and 10 cases. The given figure suggests that the influence of local mode of initial deflections is only marginal on the ultimate strength in the dominant longitudinal thrust cases. On the other hand, the impact of the initial deflections is insignificant under dominant transverse thrust loads.

The failure of the panels under pure longitudinal compression load can be explained as follows: local plate buckling occurs first; however, the panel can sustain additional load as the deflection of the stiffeners is still slight. With an increase in the compressive load, effective breadth of the plating reduces, and eventually, global buckling collapse takes place.

Figure 11 also indicates a change in the ultimate strength with respect to the number of stiffeners under longitudinal thrust loads. This is an outcome of panel collapse due to the global buckling. Under the transverse thrust conditions, however, the ultimate strength of panels is almost the same for $n_s = 4$ and 10.

The failure modes of the panel with $n_s = 4$ are given in Figure 12 for "Tight Inits" and "Loose Inits" cases. At first glance, the collapse modes of the panels with different initial deflections are similar to each other; however, a detailed examination of the deformations reveals that the local buckling mode deflections between the stiffeners are slightly different. This behaviour can be observed at the load shortening curves as well.

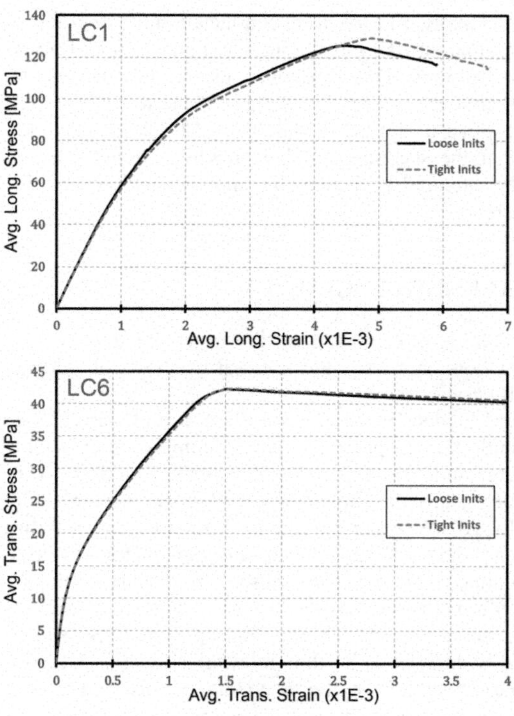

Figure 13. Load-shortening curves for $n_s = 4$ under longitudinal thrust (LC1) and transverse thrust (LC6).

The load-shortening curves of the panel with $n_s = 4$ are given in Figure 13 under longitudinal (LC1) and transverse (LC6) thrust loads. The slopes of the load-shortening curves for LC1 become slightly different beyond the average stress level of $\approx$ 60 MPa. This is mainly due to the different local mode of initial deflections imposed from the eigenvalue analyses with "Generic BCs" and "Eigen BCs". On the other hand, the load-shortening curves for LC6 exactly match each other despite the different local buckling modes of initial imperfections.

3.3 *Interaction between the buckling and ultimate strength*

The interaction curves for the eigenvalues and the ultimate strength of the panels with ten stiffeners have been derived and are presented in Figure 14. In this figure, the eigenvalues were obtained using "Generic BCs", and the corresponding eigenmodes were utilized as the local mode of initial deflections in the nonlinear collapse analysis, i.e., "Loose Inits".

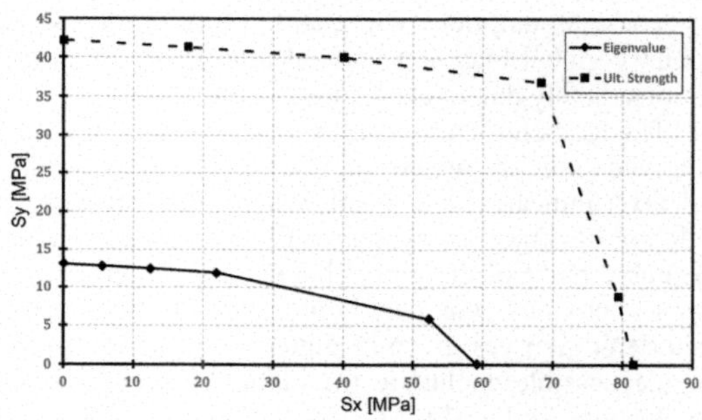

Figure 14. Buckling - ultimate strength interaction curve for $n_s = 10$ case.

The most notable observation from the interaction curve in Figure 14 is the reserve capacity of the panels beyond local plate buckling. Local plate buckling occurs rapidly once the loading is applied to the panels, and then the secondary global buckling leads to collapse when longitudinal thrust loads dominate. However, the plate can withstand additional load beyond local plate buckling due to plastic deformations along the transverse frame locations when the transverse loads are dominant.

3.4 *Elastic Large Deflection Analyses (ELDAs)*

The secondary buckling of panels under longitudinal thrust loads was examined in this section. The aim of this section is to investigate if the secondary global buckling takes place due to sole elastic instabilities. The analysis procedure has been slightly modified for ELDA as follows.

- Only local mode of initial deflections has been applied.

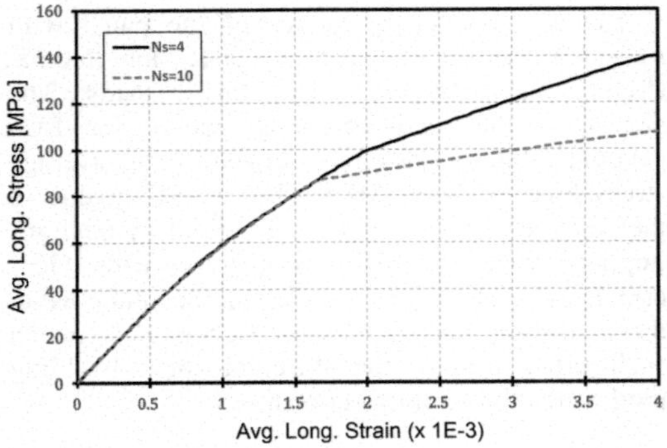

Figure 15. Load-shortening curves for $n_s = 4$ and 10 under longitudinal thrust from ELDAs.

- The local mode of initial deflections has been obtained only from eigenvalue analyses employing "Generic BCs."
- The yield stress of the material has been set to a very high value, e.g., 10 times the actual yield stress.

These assumptions will enable capturing the secondary global buckling without any plastic deformations (solely due to elastic instabilities).

The load-shortening curves for $n_s = 4$ and 10 were derived and are given in Figure 15. This figure clearly indicates the point where the secondary global buckling takes place for both cases. The knuckle point of the global buckling in Figure 15 is more obvious compared to that at the load-shortening curves given in Figure 13. This has been made possible excluding the global mode of initial deflections in the ELDAs.

The load-shortening curves show a perfect match with each other until the applied stress reaches ≈ 85 MPa, where the global buckling occurs for the panel with $n_s = 10$. The secondary global buckling for the $n_s = 4$ case takes place around 100 MPa. These two values are below the predicted ultimate strength values in Section 3.2.

At this stage, the key question is whether the global buckling modes can be captured by eigenvalue analyses when "Generic BCs" are employed. A quantitative comparison of eigenvalues with the secondary global buckling stresses obtained from ELDAs would provide valuable insights. The comparison is given in Table 2.

Table 2 reveals two crucial insights from the present calculations. Firstly, the closely spaced eigenvalues suggest that the buckling modes may be sensitive to mesh density. Secondly, the eigenvalue analysis could not capture the global buckling mode within practical limits. This finding implies that "Generic BCs" can be confidently applied for generating local buckling modes of initial deflections for the considered panels, without the risk of unintended global buckling deformations.

Table 2. Comparison of eigenvalues with the global buckling stresses from ELDAs.

	Eigenvalues [MPa]	
Modes	$n_s = 4$	$n_s = 10$
1st mode	61.44	59.14
2nd mode	61.93	59.54
3rd mode	62.52	59.91
50th mode	84.23	71.04
ELDA	≈ 100	≈ 85

4 CONCLUDING REMARKS

A series of eigenvalue analyses and nonlinear FEAs were conducted to assess the impact of BCs and initial deflections on the buckling and ultimate strength of panels with thin plates and small-sized stiffeners. The results indicate that the BCs for the stiffeners have a substantial effect on the buckling mode shapes and loads, especially in cases dominated by transverse thrust conditions.

Nonlinear FEAs were performed, incorporating the local mode of initial deflections obtained from eigenvalue analyses with two different BCs on the stiffeners. The impact of initial deflections was found to be marginal under dominant longitudinal thrust, while being deemed insignificant in cases dominated by transverse thrust loads.

Finally, ELDAs were conducted under only longitudinal thrust by excluding the global mode of initial deflections. This procedure accurately captured the secondary global buckling. "Generic BCs" were applied to the FE model. The ELDAs demonstrated that the secondary global buckling strength cannot be captured by eigenvalue analyses within practical limits.

ACKNOWLEDGEMENT

The views expressed in this paper are those of the authors and are not necessarily those of Lloyd's Register.

Lloyd's Register, its affiliates and subsidiaries and their respective officers, employees or agents are, individually and collectively, referred to in this clause as the 'Lloyd's Register Group'. The Lloyd's Register Group assumes no responsibility and shall not be liable to any person for any loss, damage or expense caused by reliance on the information or advice in this document or howsoever provided, unless that person has signed a contract with the relevant Lloyd's Register Group entity for the provision of this information or advice and in that case any responsibility or liability is exclusively on the terms and conditions set out in that contract.

REFERENCES

Gordo, J. M. (2008). Ultimate strength of imperfect plates under axial compression. In *Proceedings of XIV Congresso SOBENA 2008*, Rio de Jenerio: Brasil.

ISSC (2009). Committee 3.1 ultimate strength. In *Proceedings of the 17th International Ship and Offshore Structures Congress (ISSC 2009)*, Seoul: Korea, pp. 375–474.

ISSC (2022). Committee 3.1 ultimate strength. In *Proceedings of the 21st International Ship and Offshore Structures Congress (ISSC 2022)*, USA - Canada, pp. 395–500.

LR (2016). *ShipRight ADP: Non-linear Structural Collapse Analysis for Plates and Stiffened Panels*. UK: Lloyd's Register.

Tekgoz, M., Y. Garbatov, & C. G. Soares (2013a). Advanced guideline to predict the ultimate strength of perforated plate with edge reinforcement. In *Proceedings of the PRADS2013.*, Changwon City: Korea.

Tekgoz, M., Y. Garbatov, & C. G. Soares (2013b). Finite element modelling of the ultimate strength of stiffened plates with residual stresses. analysis and design of marine structures. In *Proceedings of the Analysis and Design of Marine Structures (Marstruct2013).*, Espoo: Finland.

Xu, M. C., D. Yanagihara, M. Fujikubo, & C. G. Soares (2013). Influence of boundary conditions on the collapse behaviour of stiffened panels under combined loads. *Mar. Struct. 34*, 205–225.

Yao, T. & M. Fujikubo (2016). *Buckling and Ultimate Strength of Ship and Ship-like Floating Structures*. Oxford: Butterworth-Heinemann.

Yao, T. & Y. Ueda (1985). The influence of complex initial deflection modes on the behaviour and ultimate strength of rectangular plate in compression. *J. Constr. Steel Res. 5*, 265–302.

Advances in Maritime Technology and Engineering – Guedes Soares & Santos (Eds)
© 2024 The Author(s), ISBN 978-1-032-83099-5

Effects of mesh density on stress intensity factor of a pitting-induced crack

M. Pang
School of Civil Engineering, Tianjin University, Tianjin, R.P. China

N.Z. Chen
School of Civil Engineering, Tianjin University, Tianjin, R.P. China
State Key Laboratory of Hydraulic Engineering Intelligent Construction and Operation, Tianjin University, Tianjin, R.P. China

ABSTRACT: A comprehensive investigation is conducted to determine the influence of mesh density on the numerical accuracy of the solutions of Stress Intensity Factor (SIF) obtained through the eXtended Finite Element Method (XFEM). Specimen models with pitting-induced cracks of diverse sizes are constructed, and both short cracks and long cracks are considered, providing a representative set of scenarios for analysis. A range of mesh densities is applied to these XFEM models. The computed SIF values at the crack tip are meticulously compared with those derived from Finite Element (FE) models employing finely detailed meshes. Based on the results of this study, recommendations regarding the optimal mesh density for XFEM models dealing with pitting-induced cracks are given.

1 INTRODUCTION

During long-term service, offshore structures are subject to the coupling effect of corrosive environments and alternating stresses. Therefore, corrosion-fatigue fracture is one of the primary failure modes of offshore structures. Pitting causes roughness on the surface of offshore structures because of non-uniform corrosion (Jakubowski, 2015). Pitting provides potential spots for fatigue crack initiation and propagation due to the localized stress concentration and the large stress amplitude caused by environmental loads. There are several stages of pitting corrosion fatigue: pit nucleation, pit-to-crack transition, short crack growth, short to long crack transition, and long crack growth (Larrosa et al., 2018). The existence of pitting-induced cracks significantly weakens the fatigue performance of structures, so a deeper understanding of the behavior of pitting-induced cracks is urgently required.

According to linear elastic fracture mechanics, the propagation rate of a fatigue crack is determined by the SIFs at the crack tip by Paris law (Paris and Erdogan, 1963). For a reliable prediction of pitting-induced fatigue crack propagation rate as well as fracture strength, accurate SIF solutions along the crack tip are needed for cracked structures. The methods commonly used for SIF calculation are displacement and stress extrapolation methods (Chan et al., 1970), the node displacement method (Chan and Kuang, 1992; Lim et al., 1992), and the J-integral method (Rice, 1968). The basic idea of all these methods is to relate the crack tip SIF to physical quantities around the crack tip, such as displacement, stress, or energy. The reliable SIF solutions for relatively simple geometries can be found in numbers of handbooks, such as BS7910 (BS7910, 2013). When dealing with complex or irregular structures, it will be difficult to find SIF solutions by purely analytical methods. Due to the development of finite element technology, SIF calculations have been widely performed using the FE method. A focused spider-type crack-tip mesh is generally required when the FE method is applied to obtain SIFs (Anderson, 2017). However, it is difficult to model a detailed crack-tip mesh for complex crack geometries, such as pitting-induced cracks. Furthermore, to calculate SIFs for a propagating crack, a remeshing technique is required to match the crack plane when using the FE method, which requires a lot of complex meshing work. This means that the FE method is not an efficient tool for predicting pitting-induced crack initiation and propagation.

The XFEM method (Moghaddam et al., 2020) can be an effective method to calculate SIFs for complex crack geometries as well as for propagating cracks. In the XFEM method, discontinuities are realized by local enrichment and partition of unity functions so that a crack can be located anywhere in a model (Moës et al., 1999). Thus, XFEM can be applied to simulate the propagation of cracks without remeshing. However, in practical applications, there are still a number of technical issues to be resolved, such as mesh density (Han et al., 2015). Because of the

DOI: 10.1201/9781003508762-56

singularity at the crack tip, the accuracy of SIF calculations based on XFEM method is highly dependent on mesh density. Besides, the computing time of the XFEM model also depends on mesh density. The computing time of the XFEM model with a sufficiently dense mesh is comparable to the FE model with a detailed mesh (Kim et al., 2019). Without remeshing required, modeling complexity and effort can be considerably reduced by using the XFEM method, but an appropriate mesh density is still needed to ensure the computational accuracy and further improve the computational efficiency. For the application of the XFEM method to investigate the behavior of pitting-induced cracks, it is imperative to analyze the mesh density effect on SIF calculation.

In this study, both FE and XFEM models of a specimen with a pitting-induced crack are established. In the XFEM models, a series of mesh densities are applied to the region of the crack tip. The SIF solutions for short and long pitting-induced cracks based on XFEM are compared with FE solutions. The influences of mesh density on the accuracy of SIF calculation for XFEM are investigated, and the recommendation of mesh density for XFEM models with pitting-induced cracks is given.

2 ANALYSIS CASES

2.1 *Model introduction*

In this study, the pre-pitted specimen model is established based on the experiment of Fatoba (Fatoba and Akid, 2022), the geometry of the specimen is shown in Figure 1. To reduce the number of meshes for calculation, half symmetric specimen model is applied for FE and XFEM models. The pit in the middle of the specimen is semi-elliptical, with depth a = 140 μm, and width $2c$ = 548 μm, as shown in Figure 2.

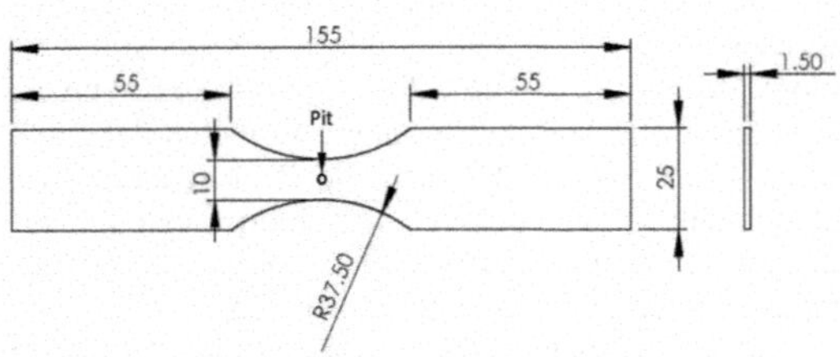

Figure 1. Specimen geometry (mm).

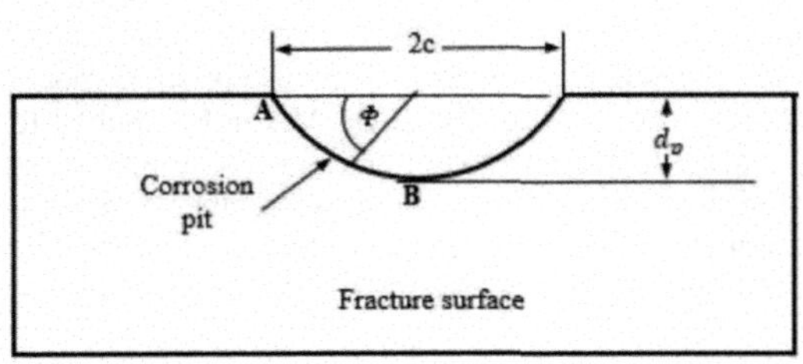

Figure 2. Pit geometry.

Table 1. Crack dimensions.

	Depth, a (μm)	Length, $2c$ (μm)
Crack 1	160	400
Crack 2	320	1348

The cracks are assumed to be semi-elliptical. The analyzed crack dimensions are summarized in Table 1. According to the experimental results (Fatoba and Akid, 2022), the cracks nucleate on the left and right edges of the pit mouth. Two cracks initiated from the pit mouth propagate individually and then coalesce with each other, which grows into one crack. Thus, two cracks are set at the edge of the pit in the case 1. In the case 2, a single crack is the coalescence of two individual cracks, so the crack is set at the middle of the pit, as shown in Figure 3.

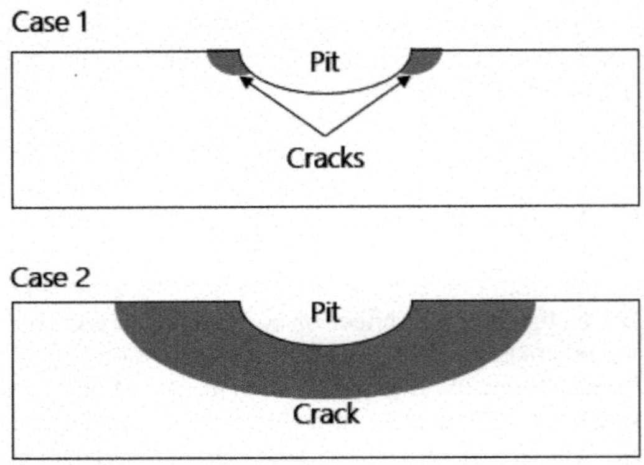

Figure 3. Location of cracks.

The material of the specimen is API X65 steel, and the mechanical properties at room temperature are presented in Table 2. Uniform tensile stress of σ = 200 MPa is applied to one end of the specimen, and another end is fixed.

Table 2. Mechanical and physical properties of API 5L X65 steel at room temperature.

Young's Modulus (GPa)	Lower Yield Strength (MPa)	Upper Yield Strength (MPa)	Tensile Strength (MPa)	Elong. (%)	Reduction of area (%)
211	516	530	614	15.7	81.5

2.2 *FE model*

To obtain sufficiently accurate solutions, the SIF is calculated using a FE model with a focused spider-type crack tip mesh. Due to the symmetry conditions, a half model is applied for SIF calculation. The short and long crack FE models with the detailed crack-tip mesh are shown in Figure 4. The 20-node brick element with reduced integrations (C3D20R) and 15-node quadratic triangular prism element (C3D15) are used to mesh the crack-tip region, the contour integral option within ABAQUS is used for SIF calculation. The minimum element size in the crack-tip region is 2 μm and the global element size is 1 mm. The line integral method is applied for the FEM SIF calculation, which the 1 ~ 10 contours are considered.

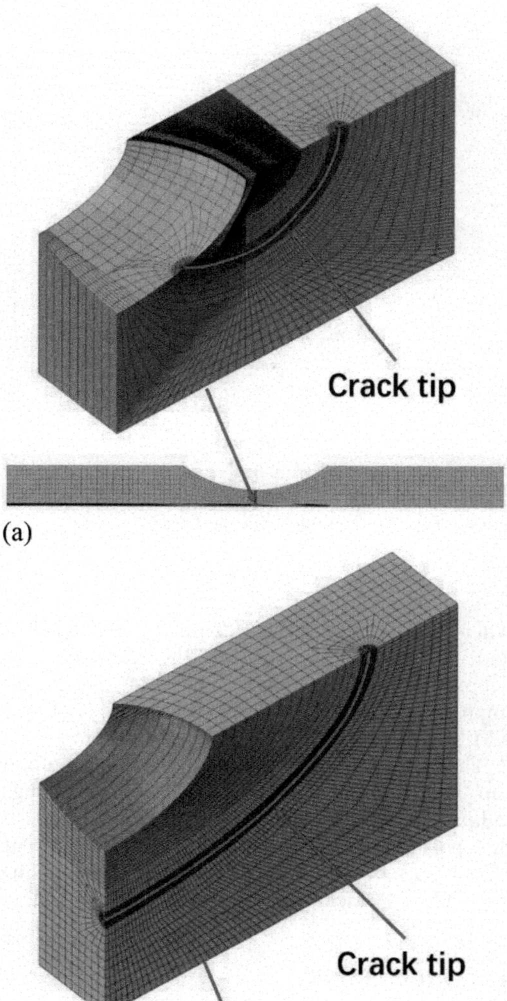

Figure 4. Detailed crack tip mesh for FE model: (a) crack 1; (b) crack 2.

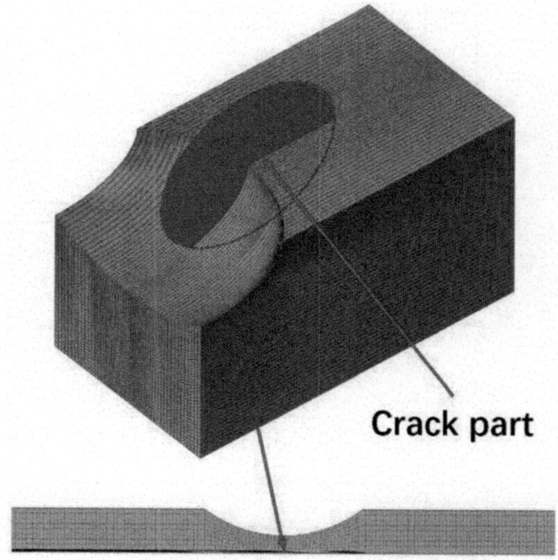

Figure 5. Mesh For Xfem Model.

2.3 *XFEM model*

The mesh density is a key factor in the accuracy and convergence of the calculations when using the XFEM to calculate the SIF. Therefore, a series of meshes are used to explore their effect on the calculations of the XFEB method. The sizes of meshes near the crack tip range from 5 to 50 μm. A half model is used with 8-node solid elements (C3D8), due to 20-node elements are not available for the XFEM analysis in ABAQUS. Since no special optimization of the mesh is required, the typical XFEM meshes in the crack-tip region are simple, as show in Figure 5. Except the crack-tip region mesh, the XFEM model and FE model have the same loading condition and material properties. The line integral method is applied for the XFEM SIF calculation, which the 1 ~ 10 contours are considered.

3 RESULTS

3.1 *Contour number effect on SIF results*

The SIFs calculated from the FEM show path-independence. However, XFEM SIF solutions show path-dependence. As shown in Figures 6 and 7, after the second contour, the FEM SIF values essentially converge. Regardless of mesh sizes, the SIF values calculated by XFEM oscillate with increasing contour number. Therefore, the average of the FEM SIF values of 2~10 contours is used as the reference value for SIF, and the average SIF value of 1~10 contours is used for XFEM SIF results.

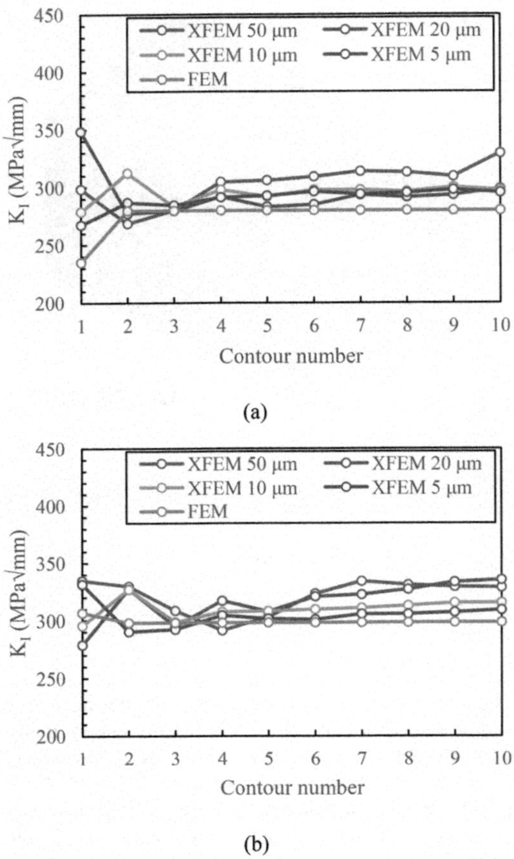

Figure 6. The SIF variations with contour number for crack 1: (a) surface point; (b) deepest point.

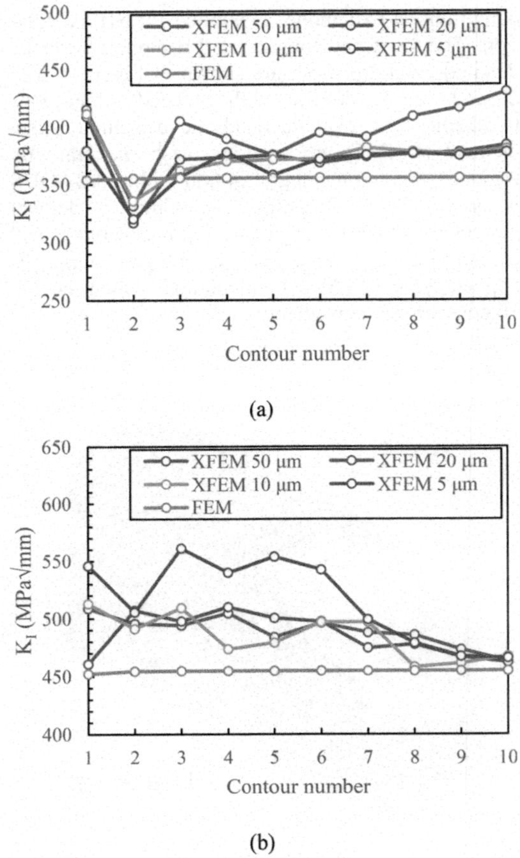

Figure 7. The SIF variations with contour number for crack 2: (a) surface point; (b) deepest point.

3.2 *Mesh density effect on SIF results*

Figures 8 and 9 show the SIF results calculated by FEM and XFEM. The FEM SIFs are used as reference values to investigate the accuracy of the XFEM SIF solutions. Compared to the FEM SIFs, the XFEM SIF values fluctuate significantly along the crack tip. To sooth the XFEM SIF results, post-processing is performed with 5th order polynomial function.

Overall, the XFEM SIF results exhibit a tendency towards higher values compared to FEM results. The errors between the reference and SIF values at crack tips calculated by XFEM models with different mesh densities are shown in Tables 3 and 4. It can be seen from the tables that the mesh density has a significant effect on the SIF calculated by the XFEM method. In most cases, the error decreases continuously with the increase of the mesh density. The accuracy of the SIF calculated by the XFEM method improves as the mesh density increases. It can be seen that the errors are around 5% when the element size is 20 μm for the crack 1 and crack 2. However, as the element size decreases further, there is no significant improvement observed in the accuracy of the XFEM SIF results.

The results indicate that the XFEM method can be used to determine the SIF of pitting-induced crack. Although the cases analyzed in this paper may be limited, the mesh density of 20 μm is recommended for reliable SIFs of pitting-induced crack using the ABAQUS XFEM.

4 CONCLUSIONS

In this study, the effects of mesh density on the reliability and accuracy of SIF solutions using XFEM method were investigated by comparing it with the FE solutions. Models of a specimen with a pitting-induced crack were analyzed, and both short crack and long crack were considered.

SIF solutions using the XFEM show oscillating behavior along the crack tip. To smooth the

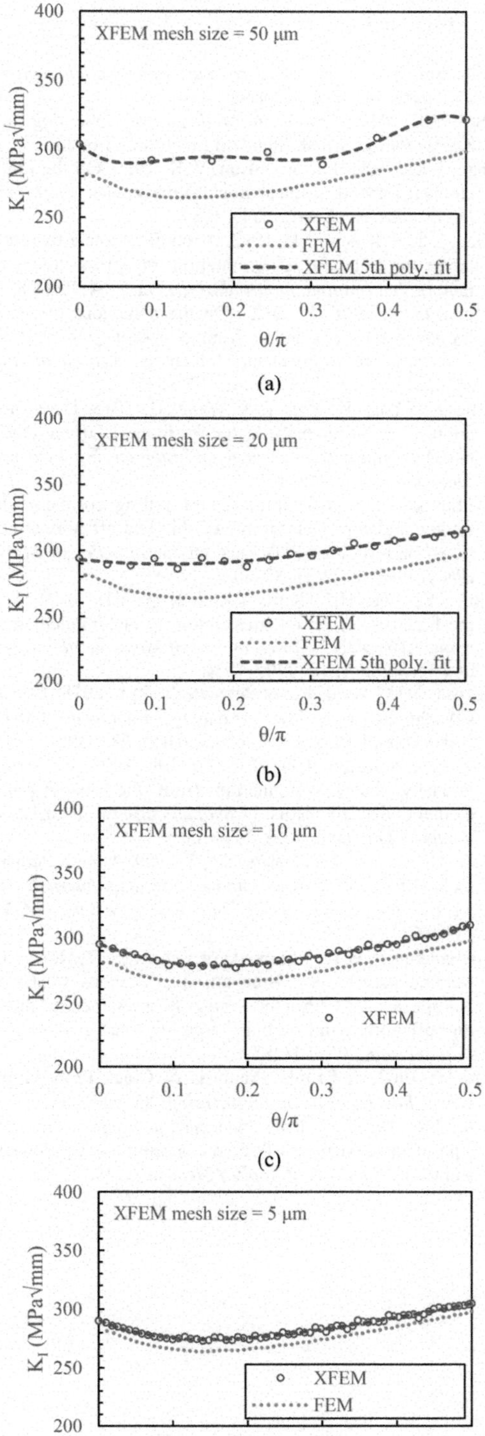

Figure 8. The mesh density effect on SIF results for crack 1: (a) XFEM mesh size = 50 μm; (b) XFEM mesh size = 20 μm; (c) XFEM mesh size = 10 μm; (d) XFEM mesh size = 5 μm.

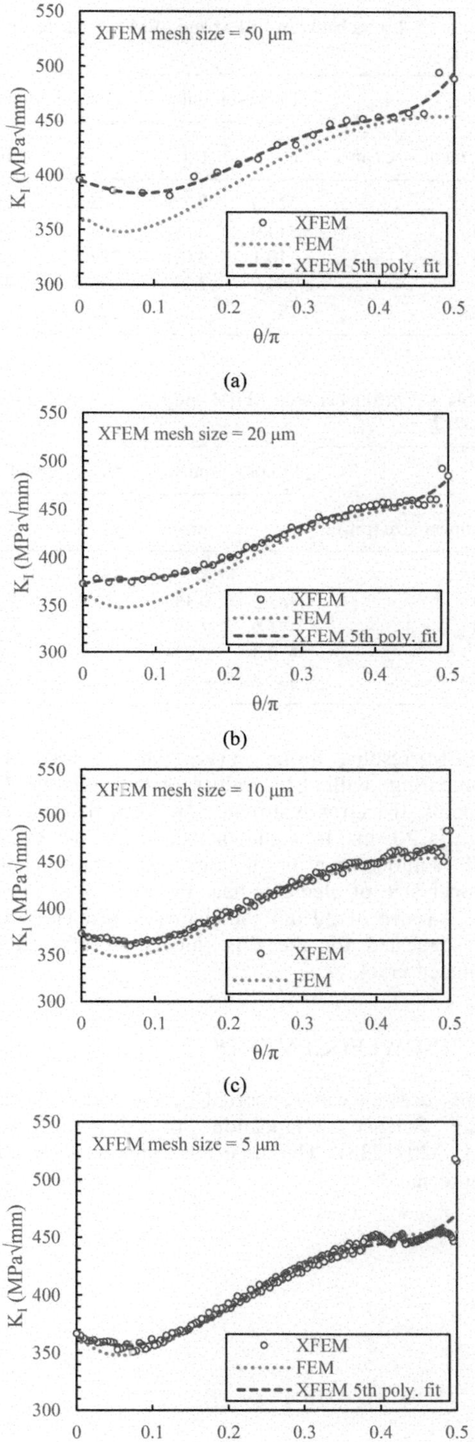

Figure 9. The mesh density effect on SIF results for crack 2: (a) XFEM mesh size = 50 μm; (b) XFEM mesh size = 20 μm; (c) XFEM mesh size = 10 μm; (d) XFEM mesh size = 5 μm.

Table 3. Errors between XFEM and FEM SIF results of crack 1.

	Deepest point		Surface point	
Element size (μm)	K_I	Error	K_I	Error
50	321.3	7.7%	303.3	8.4%
20	314.9	5.6%	293.1	4.7%
10	310.1	4.0%	295.5	5.6%
5	304.9	2.2%	290.8	3.9%
FEM	298.2		279.9	

Table 4. Errors between XFEM and FEM SIF results of crack 2.

	Deepest point		Surface point	
Element size (μm)	K_I	Error	K_I	Error
50	494.5	8.7%	395.5	11.2%
20	482.5	6.1%	372.1	4.7%
10	471.0	3.6%	374.1	5.2%
5	470.4	3.4%	363.8	2.3%
FEM	454.8		355.5	

XFEM results, fitting curves can be used. By comparing with SIF values calculated by FE models, the error is around 5% when the element size is 20 μm. It is shown that a reliable XFEM SIF solution can be obtained by choosing the proper size of elements near the crack tip. Therefore, a size of 20 μm for elements near crack tip is suggested for the SIF calculation of pitting-induced cracks.

ACKNOWLEDGEMENTS

This research was supported by the National Natural Science Foundation of China (Grant No. 52071235). The financial support is greatly appreciated.

REFERENCES

Anderson, T.L. 2017. *Fracture mechanics: fundamentals and applications*. CRC press.

BS 7910. 2013. *Guide on Methods for Assessing the Acceptability.* British Standard Institution, London.

Chan, S.K., Tuba, I.S. & Wilson, W.K. 1970. On the finite element method in linear fracture mechanics. *Engineering fracture mechanics*, 2(1), 1–17.

Chen, L.S. & Kuang, J.H. 1992. A modified linear extrapolation formula for determination of stress intensity factors. *International Journal of Fracture*, 54, R3–R8.

Fatoba, O. & Akid, R. 2022. On the behaviour of small fatigue cracks emanating from corrosion pits: Part I–The influence of mechanical factors. *Theoretical and Applied Fracture Mechanics*, 117, 103154.

Han, Q., Wang, Y., Yin, Y. & Wang, D. 2015. Determination of stress intensity factor for mode I fatigue crack based on finite element analysis. *Engineering Fracture Mechanics*, 138, 118–126.

Jakubowski, M. 2015. Influence of pitting corrosion on fatigue and corrosion fatigue of ship and offshore structures, part II: load-PIT-crack interaction. *Polish Maritime Research*, 22(3), 57–66.

Kim, J.S., Lee, H.J., Kim, Y.J. & Kim, Y.B. 2019. The mesh density effect on stress intensity factor calculation using ABAQUS XFEM. *Journal of Mechanical Science and Technology*, 33, 4909–4916.

Larrosa, N.O., Akid, R., & Ainsworth, R.A. 2018. Corrosion-fatigue: a review of damage tolerance models. *International Materials Reviews*, 63(5), 283–308.

Lim, I.L., Johnston, I.W. & Choi, S.K. 1992. On stress intensity factor computation from the quarter-point element displacements. *Communications in applied numerical methods*, 8(5), 291–300.

Moës, N., Dolbow, J. & Belytschko, T. 1999. A finite element method for crack growth without remeshing. *International journal for numerical methods in engineering*, 46(1), 131–150.

Moghaddam, B.T., Hamedany, A.M., Taylor, J., Mehmanparast, A., Brennan, F., Davies, C.M. & Nikbin, K. 2020. Structural integrity assessment of floating offshore wind turbine support structures. *Ocean Engineering*, 208, 107487.

Paris, P. 1963. A Critical Analysis of Crack Propagation Laws. *Journal of Basic Engineering*, 85, 528–534.

Rice, J.R. 1968. A Path Independent Integral and the Approximate Analysis of Strain Concentration by Notches and Cracks. *Journal of Applied Mechanics*, 35(2), 379.

Advances in Maritime Technology and Engineering – Guedes Soares & Santos (Eds)
© 2024 The Author(s), ISBN 978-1-032-83099-5

Reconstructing non-uniform strain distributions for Bragg grating sensor using memetic algorithm

H. Qiu
School of Civil Engineering, Tianjin University, Tianjin, P.R. China

N.Z. Chen
School of Civil Engineering, Tianjin University, Tianjin, P.R. China
State Key Laboratory of Hydraulic Engineering Intelligent Construction and Operation, Tianjin University, Tianjin, P.R. China

ABSTRACT: A new method based on memetic algorithm (MA) is presented to demodulate the strain profile along a fiber Bragg grating (FBG) sensor from its reflection spectrum, which is calculated by transfer matrix method. MA is a combination of population-based optimization algorithm and local search operator, which has been considered more efficient than traditional swarm intelligence optimization algorithm in many applications. In this paper, gradient descent (GD) combined with particle swarm optimization (PSO) and genetic algorithm (GA), respectively, to search for a strain distribution that produces spectrum closest to the target spectrum. Numerical simulation shows that MA is more efficient in strain distribution reconstruction than individual swarm intelligence optimization algorithm.

1 INTRODUCTION

Composite material structures are widely used in the field of marine engineering. They are frequently subjected to external perturbations and varying environmental conditions which may cause the structures to suffer from fatigue damage and/or failures (Ramakrishnan et al. 2016). Fiber Bragg grating (FBG) sensors have excellent characteristics such as anti-electromagnetic interference, reusability and small size, so they have become a reliable option for health monitoring of composite structures. FBG can be embedded into composite materials, acting in many ways as the equivalent of a human nervous system (Gros et al. 2000).

FBG sensors are used to measure various physical parameters such as strain and temperature. When measuring strain, the average strain on the grating length can be easily obtained by measuring the wavelength shift at maximum reflectivity due to the good linear relationship between the wavelength shift and the strain. However, when the strain distribution on the grating length is nonuniform, the strain cannot be identified by measuring wavelength shift at maximum reflectivity, because the reflection spectrum will be distorted (Kang et al. 2005). This situation most commonly occurs when the sensor is located near the damaged area. Therefore, it has become a widespread concern to reconstruct the non-uniform strain distribution along the FBG sensor (Peral et al. 1996, Feced et al. 1999, Skaar et al. 2001, Azaña et al. 2001). However, the above-mentioned methods need both the intensity and phase information of the reflection spectrum, but it is complex and expensive to measure the phase information in practice.

The reconstruction of strain distribution from reflection spectrum is an inverse problem. In recent years, some heuristic intelligent algorithms have been used to solve this problem, such as genetic algorithm (Gill et al. 2004), simulated annealing (Yun et al. 2005), genetic programming (Zheng et al. 2014), and particle swarm (Wang et al. 2017). The aforementioned methods belong to global optimization algorithms, requiring only the intensity information of the reflection spectrum. However, they suffer from drawbacks such as prolonged search time and susceptibility to local optima.

The Memetic Algorithm draws inspiration from cultural diffusion in human societies. Unlike the random mutations frequently observed in biological evolution, individuals involved in cultural diffusion often engage in lifelong learning, ensuring the development rather than the descent into chaos of culture.

Memetic algorithm (MA) is a combination of population-based optimization algorithm and local search operator, also known as hybrid algorithm. The aim of hybridization is to benefit from unique search capabilities of several algorithms to find high-quality solutions to problems, which are better than the solutions obtained by the individual algorithms in isolation (Omidvar et al. 2021).

In recent years, memetic algorithms have been used to solve complex real-world problems, and have shown excellent performance in many cases.

DOI: 10.1201/9781003508762-57

In this paper, we combine the gradient descent method with population-based intelligent optimization algorithms to create the Memetic Algorithm for solving the strain reconstruction problem. In Section 2, we introduce the transfer matrix method for calculating the reflection spectrum of the grating and outline the main steps of genetic algorithm, particle swarm algorithm, and the Memetic Algorithm. Section 3 discusses the performance of the proposed algorithms. Section 4 summarizes the conclusions of this paper.

2 BACKGROUND

2.1 *Theory of fiber Bragg grating*

As shown in Figure 1, the fiber Bragg grating is a product of periodic modulation of the effective refractive index (n_{eff}) of the core on the basis of ordinary optical fiber. It can reflect light waves in specific wavelength bands. The reflected spectrum exhibits a single-peak morphology, carrying information about the measured physical quantity.

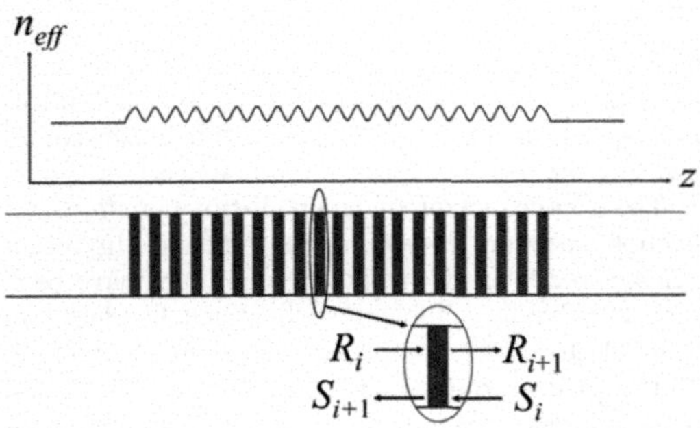

Figure 1. The refractive index distribution of the grating core.

Coupled-mode theory (Erdogan 1997) is a commonly used tool for analyzing fiber Bragg gratings. If we define R and S as the amplitudes of the forward and backward propagating modes, respectively, the following differential equation can be obtained through coupled-mode theory:

$$\begin{cases} \frac{dR(z)}{dz} = i\sigma R(z) + i\kappa S(z) \\ \frac{dS(z)}{dz} = -i\sigma S(z) - i\kappa R(z) \end{cases} \tag{1}$$

The reflection spectrum can be obtained by directly integrating the above first-order differential equation, but it is complex and time-consuming. In practice, the transfer matrix method (Yamada et al. 1987) is widely used to model the grating's spectrum. This method uniformly divides the grating into M segments, assuming each segment is subjected to uniform strain.

Define R_i and S_i to be the field amplitudes after traversing the i-th section, the propagation through this section is described as

$$\begin{bmatrix} R_i \\ S_i \end{bmatrix} = F_i \begin{bmatrix} R_{i-1} \\ S_{i-1} \end{bmatrix} \tag{2}$$

$$F_i = \begin{bmatrix} cosh(\gamma\Delta z) - i\frac{\sigma}{\gamma}sinh(\gamma\Delta z) & -i\frac{\kappa}{\gamma}sinh(\gamma\Delta z) \\ i\frac{\kappa}{\gamma}sinh(\gamma\Delta z) & cosh(\gamma\Delta z) + i\frac{\sigma}{\gamma}sinh(\gamma\Delta z) \end{bmatrix} \tag{3}$$

where $\Delta z = L/M$, signifies the length of each segment; L is the length of the FBG; and $\gamma = \sqrt{\kappa^2 - \sigma^2}$.

The 'dc' self-coupling coefficient σ, and the 'ac' coupling coefficient κ were expressed as:

$$\kappa = \frac{\pi}{\lambda}\nu\delta\overline{n_{eff}} \tag{4}$$

$$\sigma = \frac{2\pi}{\lambda}\left(n_{eff} + \delta\overline{n_{eff}}\right) - \frac{\pi}{\Lambda(z)} \tag{5}$$

Where $\delta\overline{n_{eff}}$ is the mean index change of the FBG, ν is the fringe visibility of the index variation, and $\Lambda(z)$ is calculated by:

$$\Lambda(z) = \Lambda_0[1 + (1 - P_e)\varepsilon(z)] \tag{6}$$

P_e is the effective strain-optic coefficient of the grating, and $\varepsilon(z)$ is the strain at the axial position z of the grating. Thus, the entire grating propagation mode can be expressed by

$$\begin{bmatrix} R_M \\ S_M \end{bmatrix} = F \begin{bmatrix} R_0 \\ S_0 \end{bmatrix} \tag{7}$$

$$F = \begin{bmatrix} f_{11} & f_{12} \\ f_{21} & f_{22} \end{bmatrix} = F_1 F_2 \ldots F_M \tag{8}$$

After completing the calculations for all M transfer matrices, the reflected spectrum can be computed by:

$$r(\lambda) = \left|\frac{f_{21}}{f_{11}}\right| \tag{9}$$

2.2 *Theory of strain reconstruction and memetic algorithm*

The transfer matrix method requires us to simplify the continuous strain distribution into a set of discrete strain values. In the population-based optimization algorithms, we consider such a set of strain values as an "individual". In genetic algorithms, it refers to a gene, while in particle swarm optimization, it refers to a particle.

After establishing the concept of 'individual', we need to define the fitness function to measure the quality of each individual:

$$f(P) = \sum_{N} [R_{calc}(\lambda_m, P) - R_{tar}(\lambda_m)]^2 \quad (10)$$

Where λ_m represents one of the wavelengths when the spectrum is discretized into N wavelengths and $P = [\varepsilon_1, \ldots, \varepsilon_M]$ is a strain distribution (individual). $R_{calc}(\lambda_m, P)$ and $R_{tar}(\lambda_m)$ represent the reflectivity at wavelength λ_m in the reconstructed spectrum and the target spectrum respectively. The fitness function is used to measure the proximity between the reconstructed spectrum and the target spectrum.

Given the upper and lower bounds of each strain, the task of the population-based algorithm is to search the solution space for an individual that minimizes the fitness function so that the strain distribution represented by that individual matches the applied strain distribution well.

Memetic algorithms are population-based metaheuristics composed of an evolutionary frame-work and a set of local search algorithms which are activated within the generation cycle of the external framework (Neri et al. 2012).

To create memetic algorithms, we combined a commonly used local search operator, gradient descent (GD), with genetic algorithms and particle swarm algorithms, respectively.

The gradient descent method finds local optima by moving in the direction opposite to the gradient direction. For each strain value ε_i, its movement at the t-th iteration can be represented as:

$$\varepsilon_i^t = \varepsilon_i^{t-1} - \eta \frac{\partial f}{\partial \varepsilon_i} \quad (11)$$

Where η determines the step size of each movement of ε_i. In each iteration, the gradient values of the fitness function with respect to each strain value are first calculated, and then all strain values are simultaneously updated according to Formula (11).

Genetic algorithms (GA) draw inspiration from biological evolution. They transform potential solutions to a problem into the form of genes. In each iteration, selection, reproduction, and mutation operators act on the gene population, generating a new generation. This iterative process continues until the best gene in the population reaches the convergence tolerance.

The initial concept of Particle Swarm Optimization (PSO) was to simulate the foraging behavior of bird flocks. In PSO, a swarm consists of a group of particles, each representing a solution. In each iteration, the performance of each particle is evaluated using the fitness function. The position of each particle changes based on its own experience and the collective experience of the entire population. The process is repeated iteratively until the best particle in the population reaches the convergence tolerance. PSO, with its advantages of fewer parameters and easier implementation, has become one of the most popular swarm intelligence algorithms.

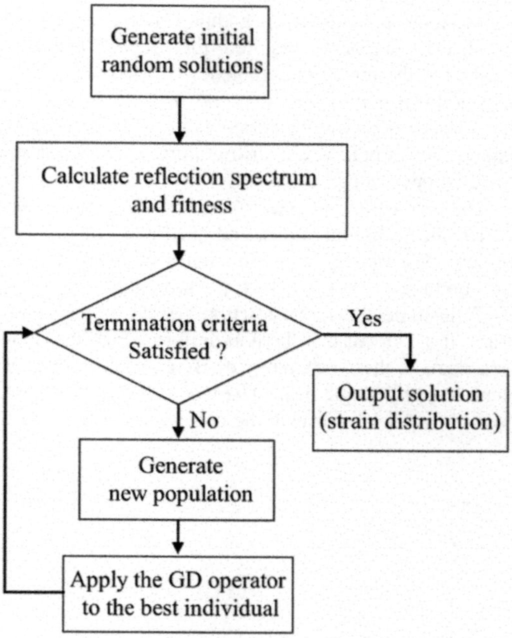

Figure 2. Flowchart of the memetic algorithm.

The principles of genetic algorithms and particle swarm optimization are detailed in references (Gill et al. 2004, Wang et al. 2017). Our innovation lies in applying the gradient descent operator to the best individual in each newly born population. Figure 2 illustrates the key steps of our approach.

The main key feature of the memetic algorithms is to combine population-based algorithm with its capability of performing wide exploration, with the gradient descent method with its capability of performing deep exploitation on the best obtained solution at each iteration.

3 RESULTS AND DISSCUSION

3.1 *Reconstruction results of MA*

Table 1. Parameters of Bragg grating sensor.

Parameter	Value
Length (L)	4mm
Mode effective index of refraction (n_{eff})	1.46
Mean index variation ($\overline{\delta n_{eff}}$)	2.5×10^{-4}
Initial Period (Λ_0)	533.22nm
Effective strain-optic coefficient (P_e)	0.26
Fringe visibility (ν)	1

The parameters of the grating used in the simulation are shown in Table 1. One of the advantages of group-based algorithms in optimization is that no gradient information is required. In addition to increasing the number of segments required for spectral response modeling, the numerical efficiency of MAs for highly non-uniform strain fields is no less than that of smooth, weakly varying strain fields. Therefore, this paper uses simple strain distributions to demonstrate the efficiency of the MAs.

The reconstruction performance of MAs was tested under linear, discontinuous, and quadratic strain distributions. For brevity, only the results of MAs formed by combining PSO and GD are presented. Figure 3, 5, 7 illustrate the reconstruction results of the spectrum. It is evident that the non-uniform strain distribution disrupts the morphology of Bragg peaks so that it is no longer a single peak. The reconstructed spectrum by MA matches well with the original spectrum.

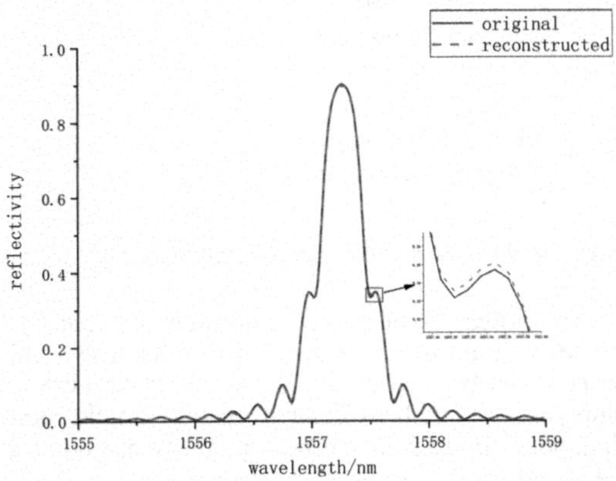

Figure 3. Spectral reconstruction result under linear strain.

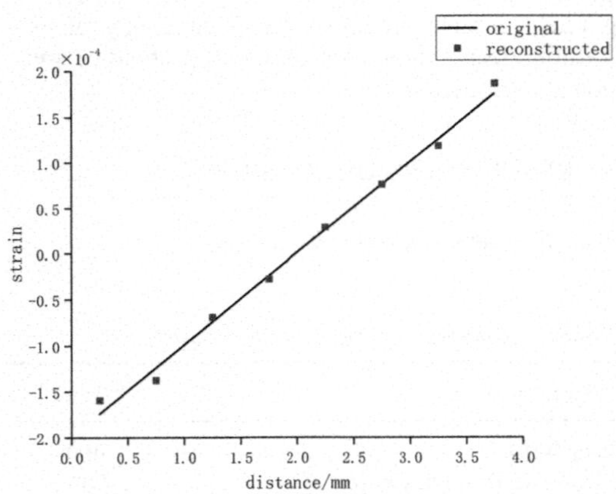

Figure 4. Reconstruction results for linear strain.

Figures 4, 6, 8 demonstrate the reconstruction results of the strain distributions. Although MA did not precisely reproduce the target strain distribution, it captured the general trend of the target strain distribution. During the simulation, it was observed that the performance of the memetic algorithm based on GA was consistently inferior to that based on PSO, possibly due to the dependence of genetic algorithms on parameter optimization, while this study directly utilized the parameter values provided in reference (Gill et al. 2004).

3.2 *Convergence performance of MA*

Figures 9-10 depict the convergence performance of the MA compared to GA/PSO. It is evident that, with the inclusion of the local search operator, MA shows an improvement in the required convergence iterations and the quality of solutions. It is also important to note that the runtime of the memetic algorithm depends on the type of local search operator used and the number of individuals influenced by the local search operator.

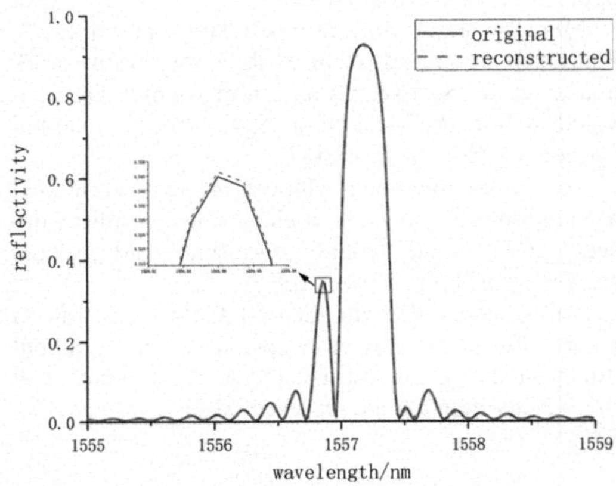

Figure 5. Spectral reconstruction result under discontinuous strain.

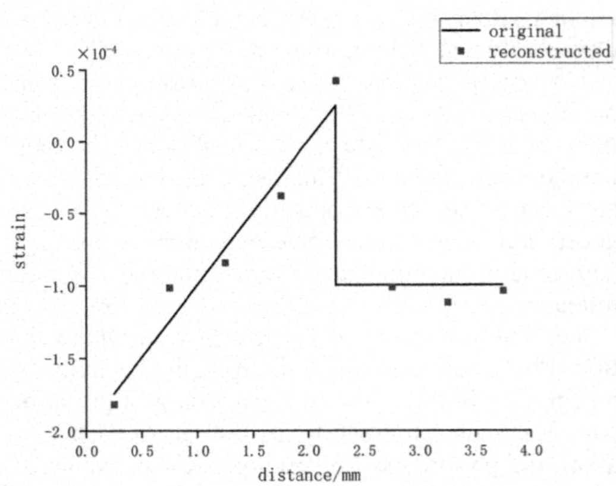

Figure 6. Reconstruction results for discontinuous strain.

Expanding the scope of the local search operator may lead to the discovery of higher-quality solutions, but it comes at the cost of longer computation time. Therefore, striking a balance between the computational accuracy and speed of the memetic algorithm is an important consideration

To eliminate randomness, 50 simulation experiments were conducted separately on linear, discontinuous, and quadratic strain distributions. The results are presented in Table 2. The average convergence rate (ACR) is defined as the percentage of experiments that converged within the convergence tolerance out of 50 experiments. The average convergence iteration (ACI) is defined as the average number of iterations used in all successful convergence experiments. Due to the differences in the performance of PSO and GA, the convergence tolerance for PSO and GA are defined as 0.002 and 0.001, respectively.

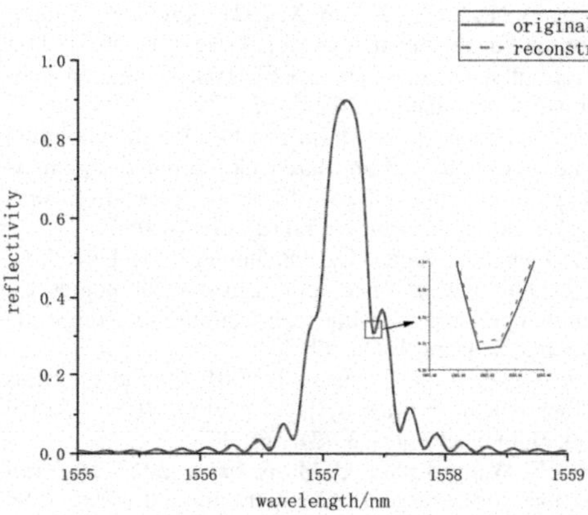

Figure 7. Spectral reconstruction result under quadratic strain.

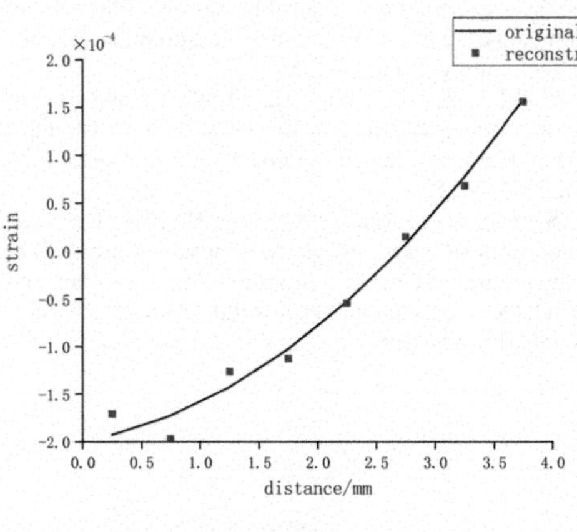

Figure 8. Reconstruction results for quadratic strain.

For fairness, the population initialization methods for GA/PSO and its memetic algorithm variants are the same. For example, the initial genes used for the *i*-th experiment in the genetic algorithm and the *i*-th experiment in the memetic algorithm are the same.

From the results, it can be observed that MA has a higher convergence success rate and requires fewer iterations compared to original individual algorithms.

The memetic algorithm is a flexible framework composed of a global search algorithm and local search operators. In this paper, we utilized a simple form of it. In practice, the search depth, frequency, and domain size of the local search operators can be dynamically adjusted based on the specific characteristics of the problem for optimal performance. We have grounds to believe that by further optimizing the search strategy of the local search operator, the memetic algorithm can continue to enhance its performance, achieving better results in strain reconstruction.

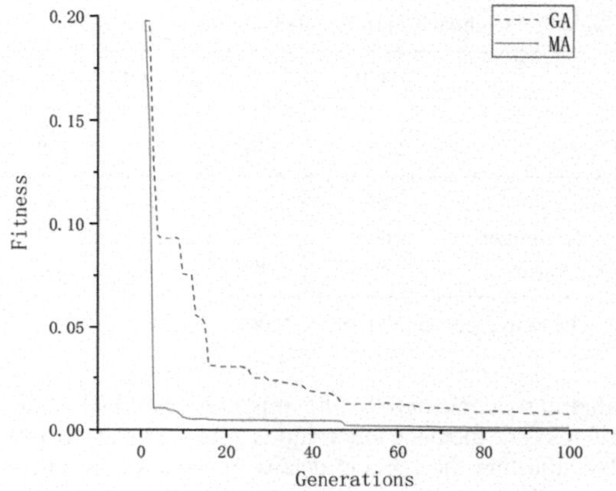

Figure 9. Convergence performance of MA compared to GA.

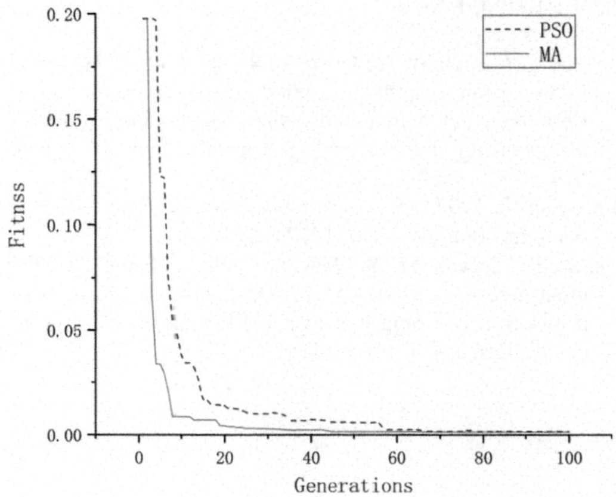

Figure 10. Convergence performance of MA compared to PSO.

4 CONCLUSIONS

A memetic algorithm, combining with population-based optimization algorithms with local search methods (using gradient descent as an example in this paper), was proposed for the reconstruction of non-uniform strain along the fiber Bragg grating sensor. Simulation results indicate an improvement in the reconstruction performance of the memetic

Table 2. Comparison of results between MA and GA.

	ACR		ACI	
Case	MA	GA	MA	GA
liner	54%	16%	57	66
Discontinuous	96%	48%	35	63
quadratic	34%	12%	63	64

* The convergence tolerance is 0.002

Table 3. Comparison of results between MA and PSO.

	ACR		ACI	
Case	MA	PSO	MA	PSO
liner	78%	34%	49	59
Discontinuous	100%	92%	18	31
Quadratic	76%	18%	50	62

* The convergence tolerance is 0.001

algorithm compared to the original individual algorithms. The memetic algorithm is a flexible framework. By adjusting the specific details of local search operators, the memetic algorithm is expected to perform better in the problem of FBG strain reconstruction.

REFERENCES

Azaña, J., Muriel, M. A., Chen, L. R., & Smith, P. W. 2001. Fiber Bragg grating period reconstruction using time-frequency signal analysis and application to distributed sensing. Journal of Lightwave Technology, 19(5), 646.

Erdogan, T. 1997. Fiber grating spectra. Journal of lightwave technology, 15(8), 1277–1294.

Feced, R., Zervas, M. N., & Muriel, M. A. 1999. An efficient inverse scattering algorithm for the design of nonuniform fiber Bragg gratings. IEEE journal of quantum electronics, 35(8), 1105–1115.

Gill, A., Peters, K., & Studer, M. 2004. Genetic algorithm for the reconstruction of Bragg grating sensor strain profiles. Measurement Science and Technology, 15(9), 1877.

Gros, X. E. 2000, November. Current and future trends in non-destructive testing of composite materials. In Annales de Chimie Science des Matériaux (Vol. 25, No. 7, pp. 539–544).

Kang, D. H., Park, S. O., Hong, C. S., & Kim, C. G. 2005. The signal characteristics of reflected spectra of fiber Bragg grating sensors with strain gradients and grating lengths. NDT & E International, 38(8), 712–718.

Neri, F., & Cotta, C. 2012. Memetic algorithms and memetic computing optimization: A literature review. Swarm and Evolutionary Computation, 2, 1–14.

Omidvar, M. N., Li, X., & Yao, X. 2021. A review of population-based metaheuristics for large-scale black-box global optimization—Part I. IEEE Transactions on Evolutionary Computation, 26(5), 802–822.

Peral, E., Capmany, J., & Marti, J. 1996. Iterative solution to the Gel'Fand-Levitan-Marchenko coupled equations and application to synthesis of fiber gratings. IEEE journal of quantum electronics, 32(12), 2078–2084.

Ramakrishnan, M., Rajan, G., Semenova, Y., & Farrell, G. 2016. Overview of fiber optic sensor technologies for strain/temperature sensing applications in composite materials. Sensors, 16(1), 99.

Skaar, J., Wang, L., & Erdogan, T. 2001. On the synthesis of fiber Bragg gratings by layer peeling. IEEE Journal of quantum electronics, 37(2), 165–173.

Wang, Z. F., Wang, J., Sui, Q. M., & Jia, L. 2017. The simultaneous measurement of temperature and mean strain based on the distorted spectra of half-encapsulated fiber Bragg gratings using improved particle swarm optimization. Optics Communications, 392, 153–161.

Yun, B., Wang, Y., Li, A., & Cui, Y. 2005. Simulated annealing evolutionary algorithm for the fibre Bragg grating distributed strain sensor. Measurement Science and Technology, 16(12), 2425.

Yamada, M., & Sakuda, K. 1987. Analysis of almost-periodic distributed feedback slab waveguides via a fundamental matrix approach. Applied optics, 26 (16), 3474–3478.

Zheng, S., Zhang, N., Xia, Y., & Wang, H. 2014. Research on non-uniform strain profile reconstruction along fiber Bragg grating via genetic programming algorithm and interrelated experimental verification. Optics Communications, 315, 338–346.

Advances in Maritime Technology and Engineering – Guedes Soares & Santos (Eds)
© 2024 The Author(s), ISBN 978-1-032-83099-5

Comparative numerical analysis of vibratory pile extraction with experiment

S. Salahshour & M.C. Ong
Department of Mechanical and Structural Engineering and Materials Science, University of Stavanger, Stavanger, Norway

N. Hinzmann
Federal Waterways Engineering and Research Institute, Hamburg, Germany

ABSTRACT: Offshore wind turbines from the early generation were initially designed with a lifespan of 20 to 25 years, leading an increased attention on decommissioning as many of these installations approach their end of operational lifetime. This research focuses on numerical investigation of pile extraction using vibratory hammer based on the large-scale tests. The primary objective is to develop a numerical tool that allows operators and industry professionals to efficiently estimate pile extraction forces with high accuracy. Accurate force estimation is crucial for operators in selecting appropriate vibratory hammer and vessel (or crane barge) combinations, subsequently leading to cost reduction in the decommissioning process. A Coupled Eulerian-Lagrangian (CEL) approach within ABAQUS/Explicit is employed in this study, adopting a modified Mohr-Coulomb (MMC) to model saturated dense sand. This study also investigates the advantages of using the vibratory hammer for pile extraction from the saturated dense sand.

1 INTRODUCTION

Offshore Wind Turbines (OWTs) typically last 20 to 25 years (Kerkvliet & Polatidis, 2016; Topham et al., 2019). When they reach the end of their lifespan, operators must decide whether to extend their life, replace them with newer models, or decommission them (Martinez Luengo & Kolios, 2015). Only a few offshore wind farms have been decommissioned so far (Salahshour et al., 2022). Removing the foundations of bottom-fixed OWTs, especially monopile foundations that make up 81.2% of OWTs in Europe (Komusanac et al., 2022), is a significant challenge during decommissioning. The current method involves partial removal, which requires time-consuming cutting processes, excavation, and dredging. Partial removal raises concerns like stubs remaining in the seabed, which can pose risks to fishing equipment and makes the seabed unsuitable for future applications. Planning future wind farms and cable routes must carefully consider these areas, ensuring safe distances from old foundations, managing potential tilting issues with other foundation types, redirecting cable routes, and considering limitations on operation vessels such as jack-ups (Hinzmann et al., 2018).

An alternative approach for monopile removal is complete removal, which is more sustainable and environmentally friendly. Among available methods for complete monopile removal, using vibratory hammer is more reliable. This method has been applied successfully in decommissioning of the Lely offshore wind farm (seen in Figure 1), and offers advantages in reducing carbon emissions, saving time, and being cost-effective.

Figure 1. Monopile removal of Lely wind park using vibratory hammer (DiesekoGroup, 2017).

Numerical methods offer substantial advantages in geotechnical problem-solving by saving time and costs compared to in situ tests or experiments. The Finite Element Method (FEM) is a powerful tool for

DOI: 10.1201/9781003508762-58

modeling soil-pile interaction. There are different FEM mesh approaches: Lagrangian, Eulerian, and Coupled Eulerian-Lagrangian (CEL). Lagrangian method follows material particles, suitable for small deformations but demanding computationally. Eulerian methods keep the grid fixed, suitable for models with significant deformations but time-consuming for large problems. CEL aims to combine both approaches. Studies have been done on Lagrangian and CEL methods for pile-soil interactions under various conditions (Staubach & Machaček, 2019; Staubach et al., 2020; Yang et al., 2020; Staubach et al., 2021).

ABAQUS, a FEM software, has been utilized for pile-soil interaction analysis. However, the classical Mohr-Coulomb (MC) soil model within ABAQUS library results in unrealistic deformations in sandy seabed due to its limitations in representing non-linear stress-strain responses in dense sand. Advanced soil models like hypoplastic constitutive models have been developed but are complex to use due to calibration needs and computational efforts.

This study applies the Modified Mohr-Coulomb (MMC) model to incorporate nonlinear effects. It conducts large deformation analyses in ABAQUS/Explicit using the CEL approach and a user-defined subroutine (VUSDFLD) to update soil parameters based on factors like accumulated plastic shear strain. The results are compared with large-scale test results on saturated dense sand conducted by Hinzmann et al. (2021) with the goal of improving the modeling of pile-soil interactions in numerical simulations.

2 MODEL FRAMEWORK

2.1 *Experimental setup*

The study's numerical model stems from tests conducted at Technical University of Braunschweig's testing facilities, exploring methods for retrieving foundation piles (Hinzmann et al., 2021). Figure 2 illustrates the experimental setup, where vibratory extraction tests were conducted within a 5 m deep and 4 m wide cylindrical bucket. The test pile, 3 m long and 0.61 m in outer diameter, is utilized. Typically, the ground level during tests remained 1 meter below the bucket's top edge. The testing process comprises five key phases including "sand installation", "probing", "pile installation", "extraction", and "probing after extraction".

The layer of sand was added in layers with thickness of 25 cm and compacted using a 350 Kilogram vibrating plate with a 50 kN centrifugal force until each layer compressed to 20 cm. Pressure sensors and pore water transducers were installed for each sand layer, repeating the process until reaching a depth of 4 m. Since the main purpose of this study is estimation of pile extraction force, the recordings of soil stress and pore water pressure are not presented. Subsequently, water filled the bucket gradually from the bottom to saturate the sand. This process was done slowly to prevent excessive pore water pressure and maintain minimal stress.

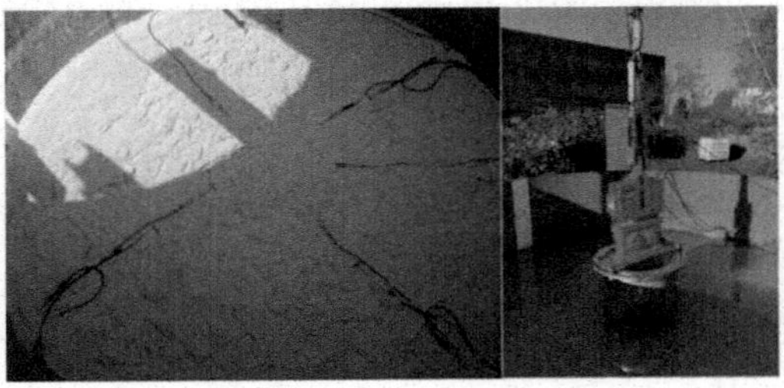

Figure 2. Test set-up, (left) layer wise installation of sand, (right) installed pile.

The open-ended test pile was driven to its intended 2.7 m depth using a diesel impact hammer, subjected to 600 blows within 60 minutes to reach the final penetration depth. However, in the numerical simulation, the time between blows is removed to reduce simulation time, and the pile is pushed into the sand at a steady velocity of 4.5 mm/s, resembling pile jacking. Comparing the spatial distribution of effective stress and density shows minor distinctions between jacking and impact driving (Staubach et al., 2020). Hence, in this numerical modeling, pile jacking represents the method for driving the pile in saturated dense sand.

The measurements of earth pressure cells installed in horizontal and vertical orientations showed that the horizontal earth pressure was k_0 ≈ 1. The high horizontal earth pressure originates from the compaction procedure. Soil conditions in offshore applications are similar due to over consolidation. The sand parameters of the test are available in Table 1.

Table 1. Saturated dense sand parameters.

Soil Parameter	symbol	value	unit
Angle of shear friction	φ'	31	[°]
Angle of friction at failure	φ_f'	37.8	[°]
Angle of friction installed	φ_{in}	35.4	[°]
Cohesion	c	0	[kPa]
Relative density	D	0.63	[-]
Max. bulk density	ρ_{max}	1.79	[g/cm^3]
Effective specific weight	γ'	10.3	[kN/m^3]
Median particle size	d_{50}	0.39	[mm]

2.2 *Numerical model*

In Lagrangian analysis, the movement of the continuum relies on material coordinates and time, commonly used in solid mechanics (Qiu et al., 2011). In contrast, Eulerian analysis involves fixed spatial nodes, enabling material flow through non-deforming elements, often applied in fluid mechanics (Qiu et al., 2011). When employing numerical analyses using the CEL method, Eulerian material is tracked within the mesh using Eulerian Volume Fraction (EVF) to determine the filled portion of each Eulerian element (Qiu et al., 2011). Lagrangian elements move freely within the Eulerian mesh until they encounter an Eulerian element filled with material (EVF $\neq$ 0). Contact between these materials is established using a general contact algorithm based on the penalty contact method (ABAQUS documentation, 2008).

Figure 3 illustrates the numerical model used for the CEL simulations in this study. The blue elements depict initially empty voids, while the red elements represent regions filled with material. The height of the void region is conservatively selected to capture soil heave or soil plugging during pile extraction. It's important to note that only material-filled elements contribute to the global force equilibrium. The pile experiences minimal deformations compared to the soil and is therefore modeled as a rigid body using Lagrangian elements. The bottom boundary of the Eulerian region is vertically constrained, while the side walls are horizontally constrained. Interaction between the pile and soil is modeled using a kinematic contact formulation in accordance with Coulomb's friction law. The friction coefficient μ is selected as tanφ'.

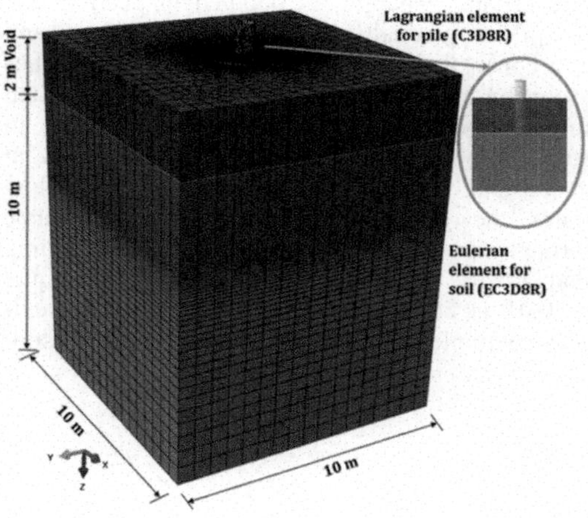

Figure 3. Finite element model to simulate pile extraction.

Similar to the experimental setup, a penetration depth of 2.7 m is considered for pile extraction. A vibratory hammer, whose characteristics are detailed in Table 2, is used for pile extraction. The high-frequency motion induced by the vibratory hammer, typically within the range of 10 to 40 Hz, causes vertical oscillation of surrounding soil particles, leading to alterations in the soil structure. This alteration significantly reduces soil resistance, causing a state of liquefaction in saturated sand. However, it is not still possible to capture changes of pore water pressure or occurrence of liquefaction when running ABAQUS/Explicit. Liquefaction is prone to occur when the acceleration of soil particles in saturated sand reaches approximately 1.5 g (Viking, 2002).

Table 2. Properties of vibratory hammer J&M Model 23.

Hammer feature	symbol	value	unit
Eccentric moment	M_e	2.65	[Kg.m]
Dynamic mass	$m_{dyn,1}$	557	[Kg]
Mass pile	$m_{dyn,2}$	132	[Kg]
Theoretical displacement	s_0	3.85	[mm]
Max. frequency	$f_{d,\,max}$	27	[Hz]
Max. centrifugal force	$F_{c,\,max}$	76	[kN]

A displacement-controlled analysis is adopted to determine the necessary force for pile extraction. The maximum centrifugal force F_c generated by the vibratory hammer depends on the eccentric moment of contra-rotating eccentric masses (typically two or another even number) and the vibration frequency f_d. It is calculated using the following equation (Holeyman, 2002):

$$F_c = M_e.\omega^2 \tag{1}$$

where

$$M_e = \sum m_{e,i}.r_{e,i} \tag{2}$$

$$\omega = 2\pi.f_d \tag{3}$$

The theoretical peak displacement amplitude of pile and vibratory hammer can be calculated as follows:

$$s_0 = \frac{M_e}{m_{dyn}} \tag{4}$$

where m_{dyn} is the dynamics mass, i.e., the mass of all vibrating parts consisting of the pile and the vibratory hammer.

The weight of the pile and vibratory hammer are not included in the simulations. Therefore, the reaction force obtained from ABAQUS postprocessing represents the soil resistance. During pile driving process, the soil resistance consists of pile shaft resistance and pile tip resistance, while during the pile extraction process, pile tip resistance is not of concern anymore. During extraction, an uplift force is applied to the pile by a crane. Thus, during vibratory extraction, pile displacement comprises two parts including cyclic oscillations induced by vibratory hammer and vertical movement which originates from the crane lift force (see Figure 4).

In this study, a nonlinear model is employed to address the dense sand nonlinear stress-strain behavior not covered by the default Mohr-Coulomb (MC) model in ABAQUS. This model encompasses crucial aspects including nonlinear pre-peak hardening and post-peak softening of internal friction angle, φ^', and dilation angle, ψ, in response to plastic shear strain, loading conditions, density, and mean effective stress (Bolton, 1986; Roy et al., 2015; Vermeer & De Borst, 1984), which is shown in Figure 5. The large deformation CEL analysis in ABAQUS/Explicit lacks the capability to simulate pile driving and extraction using the MC material model. However, the adopted model allows for effective stress analysis while incorporating the nonlinear characteristics of dense sand stress-strain responses. For further details, interested readers are directed to Nematzadeh & Shiri (2020) and Salahshour et al. (2023).

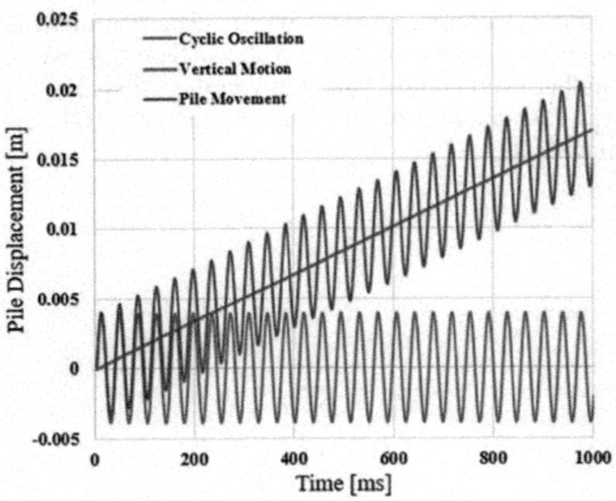

Figure 4. Pile movement during extraction.

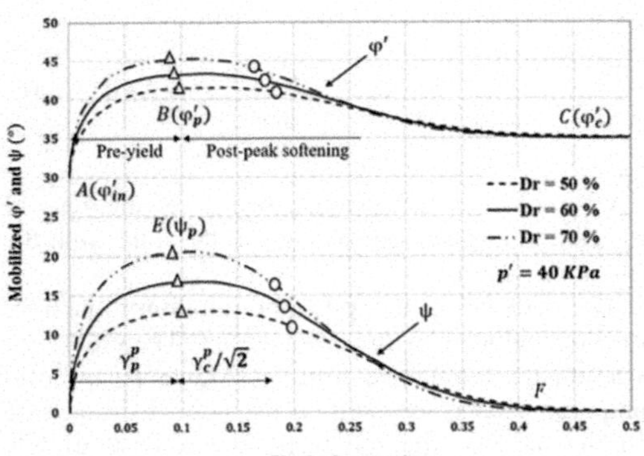

Figure 5. The mobilized φ' and ψ obtained from Nematzadeh & Shiri (2020).

A user-defined subroutine (VUSDFLD) is developed and integrated into explicit CEL analysis. At each time increment, stress and strain components are supplied to the subroutine. The subroutine initially computes the mean effective stress, p', based on the stress components. Subsequently, the principal strain components are obtained using the VSPRINC utility. The gradient of the principal plastic strain, $\Delta\gamma_p$, is then determined as the difference between its minor and major components. The sum of $\Delta\gamma_p$ over time yields the total γ_p. The parameters γ_p and p' are defined as field variables FV1 and FV2, respectively.

Utilizing the MMC model, mobilized φ' and ψ are defined as functions of FV1 and FV2 in a tabular format within the model input file. Subsequently, the ABAQUS software employs information from the subroutine to update the values of φ' and ψ based on the field variables. This process occurs concurrently during the analysis, encompassing the nonlinear stress-strain behavior of dense sand material.

2.3 *Validation of the soil model*

To verify the soil model, a comparison is conducted with a test where the only force used to extract the pile from the soil was the crane uplift force. The numerical simulation for this scenario encompasses four distinct steps. In step one, both the gravity force and geostatic stress are applied to the soil to establish equilibrium within it. The subsequent step, step two, involves the process of pile driving, wherein, as detailed in Section 2.1, the pile is gradually inserted into the dense sand at a constant velocity of 4.5 mm/s until it reaches a penetration depth of 2.7 m. Following this, step three involves a temporary stop to the pile. In step four, the pile is extracted by 2 meters at a constant velocity of 17 mm/s. Figure 6 illustrates the graph depicting the soil resistance throughout these steps. It is worth mentioning that a very small friction coefficient (FC) equal to 0.01 is considered in this case which resembles frictionless contact between the pile and soil.

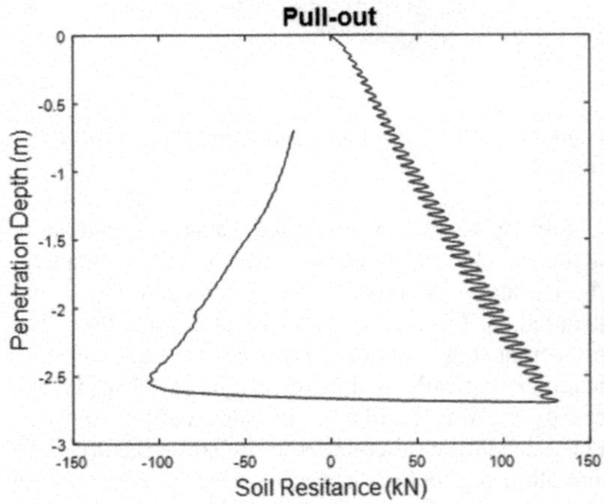

Figure 6. Soil resistance during pile driving and pulling out.

The experimental test recordings indicate that crane exerted an uplift force of 120 kN to pull out the pile. According to the numerical results, the maximum soil resistance reaches to 104.5 kN, not factoring in the pile and hammer weights. Accounting these weights, the calculated numerical uplift force required to remove the pile stands at 111 kN. This value is closely aligned with the experimental result.

The soil parameters employed for verification purpose are utilized in simulating the cyclic extraction of the pile. Normally, Young's modulus varies with the soil depth, influenced by the initial confining pressure and typically described by a power function for dense sand (Janbu, 1963). However, due to the limited depth of the soil in this study, a simplified linear variation of the Young's modulus throughout the soil depth is adopted. The soil parameters of the numerical study are provided in Table 3.

Table 3. Soil parameters in numerical modelling.

Soil Parameter	symbol	value	unit
Angle of friction	φ'	35	[°]
Peak of angle of friction	φ_p'	38	[°]
Angle of dilation	ψ	6	[°]
Peak of angle of dilation	ψ_p	17	[°]
Cohesion	c	2	[kPa]
Relative density	D	0.6	[-]
Density	ρ	1.79	[g/cm^3]
Young's modulus- top	E	45	[kN/m^2]
Young's modulus- bottom	E	7.5	[MN/m^2]
Poisson's ratio	ν	0.2	[-]

In the numerical model, a low value is allocated to soil cohesion, even though cohesionless soil is being simulated. This decision is justified by the influence of water viscosity in saturated sand, which affects soil behavior by introducing a minor degree of cohesion.

3 RESULTS AND DISCUSSION

In this numerical study, the focus is on vibratory extraction test VIB3, which was part of the DeCoMP project carried out by the Institute of Geomechanics and Geotechnics at the Technische Universität Braunschweig (IGGTUBS). Throughout this test, various parameters such as extraction frequency, f_d, pile velocity, v_{ext}, acceleration, a, and the crane's uplift force, F_{line}, were monitored and recorded for pile embedded length of 2.7 m. During the vibratory pile extraction, the pile was displaced 2.01 m and its embedded length changed from 2.7 m to 0.7 m. Figure 7 illustrates the recordings from the extraction process, while additional specifics are outlined in Table 4.

Table 4. Measured values of VIB3.

Parameter	symbol	value	unit
Free vibration	t_{free}	60	[s]
Pile extraction duration	t_{ext}	118	[s]
Pile extraction length	L_{ext}	2.01	[m]
Pile extraction speed	v_{ext}	17	[mm/s]
Frequency	f_d	27	[Hz]
Crane force	F_{line}	10	[kN]

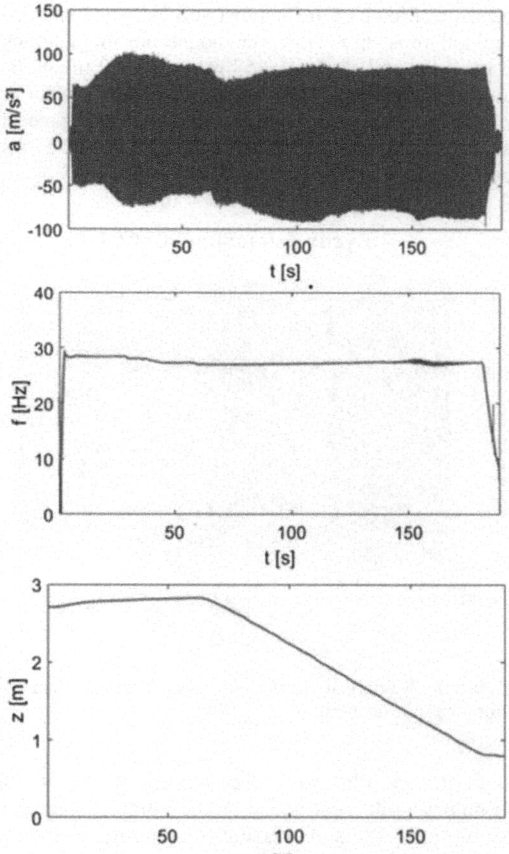

Figure 7. Recordings of pile vibratory extraction (Hinzmann et al., 2021).

As depicted in Figure 7, the extraction maintained a consistent frequency of 27 Hz throughout the entire process. In practice during vibratory pile extraction operation, the pile undergoes high-frequency vibrations until downward movement is detected (visible in third graph of Figure 7). Subsequently, a dedicated crane lifts both the pile and the vibrating hammer out of the soil. Notably, the hammer continues its vibrations during the extraction process. This technique has shown a remarkable 90% reduction in soil resistance

(Meijer, 2022). During the test, the pile experiences consistent vibrations for a duration of 60 seconds, after which the crane uplift force is applied. Throughout the initial phase of unrestricted vibration during extraction, the crane uplift force gradually rose to 10 kN. Following 60 seconds of free vibration, the exerted force by the crane was kept constant. Subsequently, adjustments were made to maintain a consistent extraction rate. However, in some moments of the test period, it became necessary to increase the force to sustain the desired extraction rate. A maximum crane uplift force of 14.5 kN was registered during the test. The extraction speed showed a steady pattern, suggesting that the centrifugal force of the hammer successfully surpassed the soil resistance. However, noticeable irregularities occurred at soil levels of 1.58 m and near 2 m, attributed to the halting of the hammer and adjustments in frequency. This pause aimed to achieve more precise control over the vibrator and to optimize its performance.

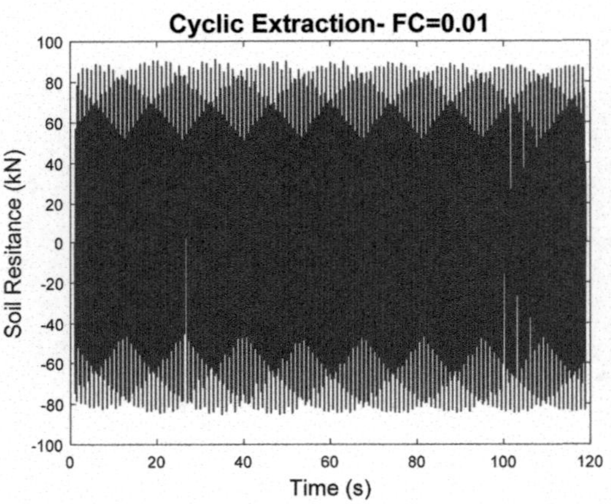

Figure 8. Numerical result of soil resistance during vibratory pile extraction.

Figure 8 illustrates the numerical result of vibratory pile extraction. The graph displays oscillations in soil resistance ranging between +90 kN and -85 kN. In order to be able to interpret the results better, the Butterworth filter is used. It is a tool for signal processing and data analysis. It's designed to smooth or filter time-dependent data, particularly useful for reducing noise or fluctuations in simulation results. The Butterworth filter helps remove high-frequency noise from a signal while preserving the overall shape and characteristics of the data. It operates based on a defined cutoff frequency. This cutoff determines which frequencies are allowed to pass through the filter. Frequencies higher than the cutoff are attenuated or removed, while lower frequencies are retained. The filtration is carried out by applying three different cutoff frequencies, illustrated in Figure 9.

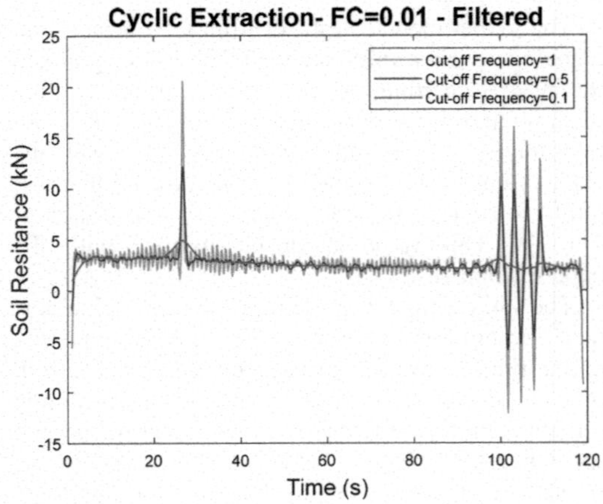

Figure 9. Soil resistance during vibratory pile extraction after filtration.

Following the filtration with a cutoff frequency of 0.1, the soil resistance gradually diminishes from 3.35 kN to 2 kN by the conclusion of the extraction process. Notably, the pile tip rests 0.7 meters below the soil surface, as detailed in Section 2.2.

Throughout the numerical modeling, the weights of the pile and hammer are excluded. When factoring in these weights, the required uplift force for pile extraction is 10.1 kN, which gradually decreases to 8.8 kN. This aligns quite well with the consistent experimental force of 10 kN applied to the vibratory hammer and pile during test.

ABAQUS/Explicit is not able to capture the changes of pore water pressure. Otherwise, it could be possible to investigate the occurrence of liquefaction through the vibratory extraction process. However, it is quite clear that soil degradation due to exposure to high cyclic loading has happened in the saturated dense sand, which has resulted in a significant reduction in soil resistance.

As described in Section 2.3, the numerical modeling initially adopts an exceptionally low friction coefficient to emulate a smooth pile surface. However, to explore the impact of friction on vibratory pile extraction, a friction coefficient of 1 (i.e., FC=1) representing a rough pile is considered. This decision is rooted in two primary reasons: first, prior research by Dutta et al. (2015) indicates that the CEL method yields improved outcomes for both smooth (μ=0) and rough (μ=1) cases; secondly, the anticipated increase in pile friction over time due to phenomena like corrosion and sand cementation.

The resulting soil resistance during vibratory extraction with FC=1 is depicted in Figure 10, oscillating between 546 kN and -622 kN. Subsequently, applying Butterworth filtration with a cutoff frequency of 0.1 to this data produces the graph shown in Figure 11. After filtration, it becomes evident that the soil resistance reaches a maximum value of 65.5 kN, approximately six times higher than observed in the frictionless

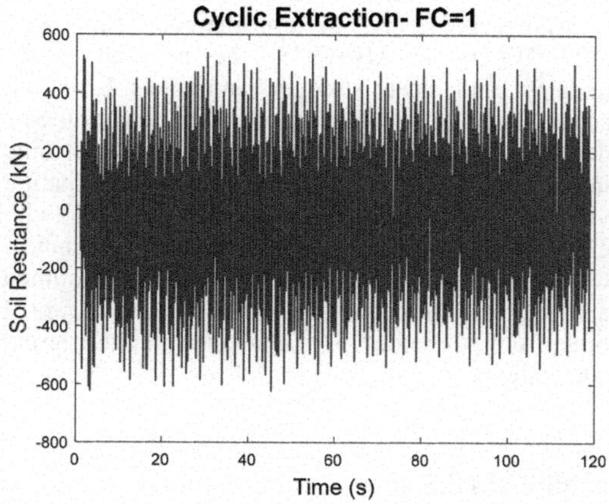

Figure 10. Numerical result of soil resistance during vibratory pile extraction for FC=1.

scenario. This conservative approach highlights the importance of considering the effect of time for pile extraction operations, which requires further detailed analysis and investigation.

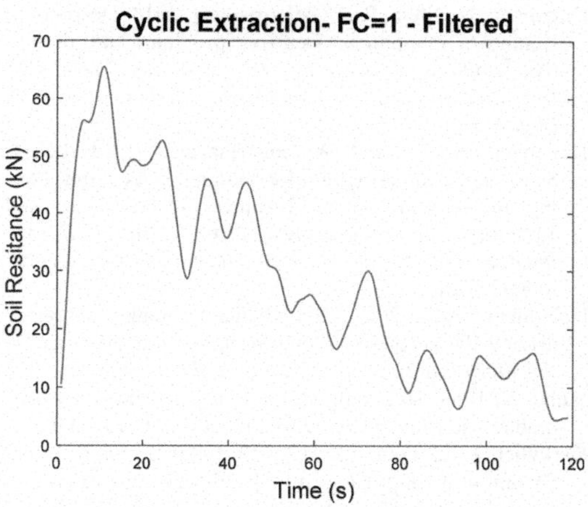

Figure 11. Soil resistance during vibratory pile extraction (FC=1) after filtration with cutoff frequency of 0.1.

The resistance to deformation in cohesionless soils is notably influenced by the shear resistance at the points where individual soil particles make contact and by the interlocking among these particles (Viking, 2002). Vibration technology introduces inertial forces ranging between 10-20g into the soil, leading to dynamic movement of individual grains. When these forces surpass the initial vertical stress within a soil volume at a particular depth, the grains essentially undergo "free fall," resulting in no force transfer between them. Consequently, the ground's resistance drops close to zero, primarily because the soil grains lack chemical or physical connections, unable to withstand tensile forces. This disconnection among soil grains leads to minimal interaction and almost zero resistance to deformation.

This condition induced by vibration is termed as soil fluidization or liquefaction, an alteration of soil from its original state to one where quasi-fluid properties prevail. Viking (2002) suggests a threshold of about 1.5g for attaining this state, which increases with higher vertical stress and relative density of the soil. It's important to note that induced shear strength reaction is not solely linked to soil liquefaction mechanisms. Viking (2002) emphasizes that soil fluidization can occur even in dried sand without pore water. However, its effect intensifies in water-saturated soils and significantly reduces shear strength.

During the experimental test, the acceleration of the pile was monitored and recorded (Figure 7), but the acceleration of soil particles was not measured. Numerical method provides possibility to measure soil particle accelerations. To examine whether free fall occurred for soil particles, accelerations at three distinct locations within the soil are extracted: Point 1, positioned near the pile tip; Point 2, situated at the middle of the pile's embedded length (1.35 meters below the soil surface); and Point 3, located at the soil surface (as depicted in Figure 12).

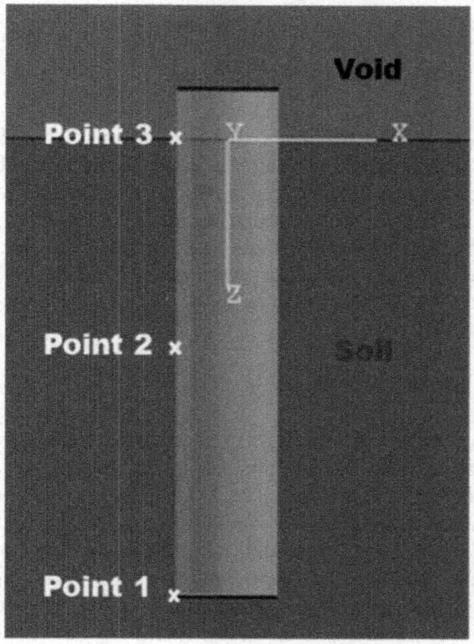

Figure 12. Locations of measuring points for extraction soil particle acceleration.

Figure 13 illustrates the accelerations obtained at these points. When the vibratory pile extraction initiates, the soil particle near the pile tip (Point 1) begins vibrating, reaching an acceleration of approximately 7g after a few seconds. However, as the pile is extracted, this point's acceleration decreases, settling at around 1 m/s until the end of the extraction process. At Point 2, the soil particle's acceleration

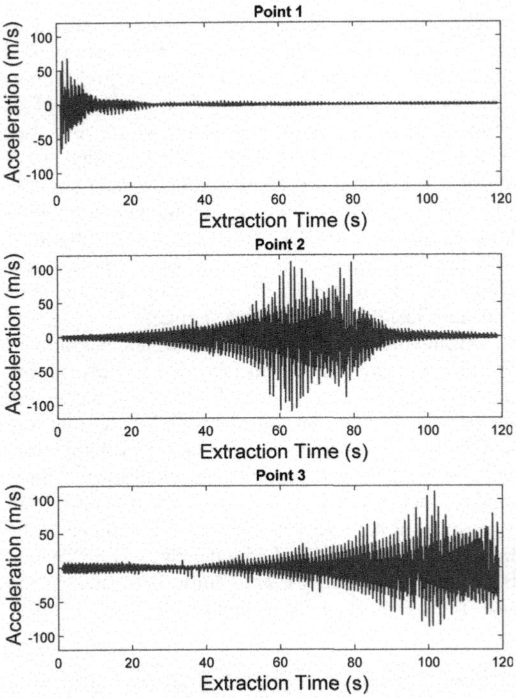

Figure 13. Soil Particles acceleration.

reaches 10g, matching the experimental recorded value for the pile acceleration during extraction. This occurs when the pile tip approaches and surpasses the measuring point, mirroring the behavior observed at Point 3. These findings suggest that accelerations in saturated dense sand are predominantly influenced by the pile's tip rather than its shaft.

4 CONCLUSION

The decommissioning of offshore wind foundations is set to grow in significance in the coming years. There is a notable interest in methods enabling quicker and complete foundation removal, driven by their eco-friendliness and cost efficiency. The complete extraction of monopiles, the most commonly used foundation type, holds substantial importance. Among available methods, vibratory extraction stands out for its potential to reduce operation time and minimize disruptions to the marine environment.

This study introduces a numerical method aimed at exploring vibratory pile extraction with the goal to offer engineers and industry professionals deeper insights. Conducted in ABAQUS/Explicit using the Coupled Eulerian-Lagrangian (CEL) approach, the analysis employs the Modified Mohr-Coulomb (MMC) model. This model accounts for dense sand's behavior, encompassing pre-peak hardening and post-peak softening, which make it capable to capture large deformations.

Implemented as a user-defined subroutine (VUSDFLD) in ABAQUS, the material model updates sand shear strength parameters based on accumulated plastic shear strain, loading conditions, density index, and confining pressure in each time increment. The numerical results for estimating vibratory pile extraction are aligned well with experimental test results, conducted at the Technical University of Braunschweig. This streamlined approach has the potential to decrease computational time, and therefore is suitable for engineering applications.

REFERENCES

ABAQUS, 2008. Documentation, Version 6.8EF.

Bolton, M.D. 1986. The strength and dilatancy of sands. *Geotechnique*. 36 (1): 65–78.

Dieseko Group BV, P. W, 2017. PVE 500M Decommissioning Windpark Lely. Youtube.

Dutta, S., Hawlader, B. & Phillips, R. 2015. Finite element modeling of partially embedded pipelines in clay seabed using Coupled Eulerian–Lagrangian method. *Can. Geotech. J.* 52(1): 58–72.

Hinzmann, N., Stein, P. & Gattermann, J. 2018. Decommissioning of monopiles, occurring problems and alternative solutions, *Proceedings of the ASME 2018- 37th International Conference on Ocean, Offshore and Arctic Engineering-OMAE2018*.

Hinzmann, N., Lehn, P. & Gattermann, J. 2021. Large-Scale Model Investigation for Monopile Decommissioning of Offshore Wind Turbines- Overpressure and Vibratory Pile. *Proceedings of the ASME 2021. 3rd International Offshore Wind Technical Conference-IOWTC2021*.

Holeyman, A.E. 2002. Soil behavior under vibratory driving. *Proceedings of the International Conference on Vibratory Pile Driving and Deep Soil Compaction*.

Janbu, N. 1963. Soil compressibility as determined by oedometer and triaxial tests. *Proc. European Conf. SMFE*.

Kerkvliet, H. & Polatidis, H. 2016. Offshore wind farms' decommissioning: a semi quantitative multi-criteria decision aid framework. *Sustainable Energy Technologies and Assessments*, 18: 69–79.

Komusanac, I., Brindley, G., Fraile, D. & Ramirez, L. 2022. Wind Energy in Europe. 2021 Statistics and the Outlook for 2022-2026.

Martinez Luengo, M. & Kolios, A. 2015. Failure mode identification and end of life scenarios of offshore wind turbines: a review. *Energies* 8(8): 8339–8354.

Meijer, D.J.M., 2022. Dynamic Modelling of Offshore Monopile Decommissioning. Master thesis, TUDelft. The Netherlands.

Nematzadeh, A. & Shiri, H. 2020. The influence of non-linear stress-strain behavior of dense sand on seabed response to ice gouging. *Cold Reg. Sci. Technol.* 170: 102929.

Qiu, G., Henke, S. & Grabe, J. 2011. Application of a Coupled Eulerian–Lagrangian approach on geomechanical problems involving large deformations. *Computers and Geotechnics* 38(1): 30–39.

Roy, K., Hawlader, B., Kenny, S. & Moore, I. 2015. Effects of post-peak softening behavior of dense sand on lateral and upward displacement of buried pipelines.

International Conference on Offshore Mechanics and Arctic Engineering.

Salahshour, S., Ong, M.C., Skaare, B. & Jiang, Z. 2022. A Perspective of Decommissioning Methods for Bottom-Fixed Offshore Wind Turbines. *International Conference on Offshore Mechanics and Arctic Engineering.*

Salahshour, S., Ong, M.C. & Nematzadeh. A. 2023. Numerical study on vibratory extraction of offshore wind turbine monopile foundations under sandy seabed condition. *Ocean Engineering*. 289–116229.

Staubach, P. & Machacek, J. 2019. Influence of relative acceleration in saturated sand: analytical approach and simulation of vibratory pile driving tests. *Comput. Geotech.* 112: 173–184.

Staubach, P., Machacek, J., Moscoso, M.C. & Wichtmann, T. 2020. Impact of the installation on the long-term cyclic behaviour of piles in sand: a numerical study. *Soil Dynam. Earthq. Eng.* 138: 106223

Staubach, P., Machacek, J., Skowronek, J. & Wichtmann, T. 2021. Vibratory pile driving in water-saturated sand: Back-analysis of model tests using a hydro-mechanically coupled CEL method. *Soils and foundations* 61(1): 144–159.

Topham, E., Gonzalez, E., McMillan, D., & João, E. 2019. Challenges of decommissioning offshore wind farms: Overview of the European experience. *Journal of Physics: Conference Series- IOP Publishing.*

Vermeer, P.A. & De Borst, R. 1984. Non-associated plasticity for soils, concrete and rock. *Heron* 29 (3).

Viking, K. 2002. Vibro-driveability - a field study of vibratory driven sheet piles in non-cohesive soils. PhD Dissertation.

Yang, Z., Gao, YY., Jardine, RJ., Guo WB. & Wang, D. 2020. Large deformation finite-element simulation of displacement-pile installation experiments in sand. *Journal of Geotechnical and Geoenvironmental Engineering* 146(6): 04020044.

Advances in Maritime Technology and Engineering – Guedes Soares & Santos (Eds)
© 2024 The Author(s), ISBN 978-1-032-83099-5

Probability model of flexural capacity of corroded steel beam

T. Shi
School of Civil Engineering, Xi'an University of Architecture, Xi'an, P.R. China
Shenzhen Technology Institute of Urban Public Safety, Shenzhen, P.R. China

Y. Wang
School of Civil Engineering, Xi'an University of Architecture, Xi'an, P.R. China

ABSTRACT: Steel components of ships exposed to offshore environments are highly susceptible to corrosion, leading to deterioration of their bearing capacity and affecting the service life and reliability of ship structures. Corrosion is significantly random due to the influence of materials and environment. This article conducts statistical analysis on the corrosion loss thickness, non-uniform corrosion degree, and corrosion morphology of corroded specimens, and introduces a random field model to characterize the corroded surface. Based on numerical simulation results of monotonic tensile tests, a probability model of the yield strength of corroded steel that follows a normal distribution is proposed. Finally, a probability model of the flexural capacity of corroded steel beams is established, considering the effects of corrosion thickness loss and strength degradation. The accuracy of the probability model is verified by comparing the experimental results with the calculated results.

1 INTRODUCTION

Corrosion is one of the main causes of structural failure in aging ships. Steel components exposed to corrosive environments are highly susceptible to corrosion, leading to deterioration of component performance. Excessive corrosion compromising the strength of steel components can lead to a catastrophic impact on structural safety. Therefore, it is necessary to study the bearing capacity of corroded steel components.

Corrosion can simultaneously change the geometric dimensions and material mechanical properties of steel components (Garbatov et al. 2014, Wang et al. 2017). Researchers (Schumacher 1979, Garbatov et al. 2007, Melchers 1999) have investigated the corrosion behavior of steel under marine environment and proposed various corrosion rate models to describe the corrosion evolution laws. Some researchers also studied the effects of corrosion on the mechanical properties of steel plate. Paik et al. (2004) employed the cross-sectional loss ratio to characterize the corrosion degree of steel plate and established a quantitative relationship between the strength of corroded steel plates and the cross-sectional loss ratio. Nakai et al. (2004, 2006) utilized the manual drilling method to simulate the steel pits and found that steel strength was affected by the depth and density of pits, and the effect of non-uniform corrosion was more significant.

Corrosion is essentially a complex random phenomenon that is influenced by various factors, such as material properties and environmental conditions (Kong et al. 2021, Rajput & Paik 2020). Therefore, the mechanical properties of corroded steel are bound to be uncertain. Wang et al. (2018) proposed a distribution model of corrosion pits based on the measurement data of ship structures in marine environments, but lacking parameter statistical laws. Silva et al. (2013) investigated the effect of corrosion thickness on the ultimate strength of corroded structural steel in marine environments. They reconstructed the corrosion surface but did not consider the correlations of corrosion variables. The probabilistic strength models for randomly pitted steel plates currently rely on assumed characteristics of random pit distribution, which may diverge substantially from real conditions.

Many researchers also have studied the bearing capacity of corroded steel beams. Vu et al. (2020) and Georgiadis & Samuelides (2019) simplified corrosion degradation to uniform thinning of structural components and studied the effect of corrosion on the bearing capacity of ship beams. Rahgoza (2009) conducted bending performance tests on corroded I-beams and theoretical analysis by reducing the cross-sectional thickness of flanges and web plates. It was found that the calculated results based on the thickness reduction method were more conservative. Wang et al. (2020) used non-contact three-dimensional (3 D) scanning system to capture the corrosion features of three naturally corroded steel beams and established a theoretical formula and a detailed calculation process based on considering the thickness reduction and strength

DOI: 10.1201/9781003508762-59

degradation to evaluate the bending resistance of corroded beams. Wang et al. (2015) proposed an approach to simulate the behaviour of steel stiffened plates subjected to weld-induced grooving corrosion using nonlinear finite element modelling. The study showed that the depth of grooving corrosion had a greater impact on structural performance than the width of grooving corrosion. Saad Elden et al. (2014) established two corrosion models based on reducing average corrosion thickness and true thickness of the corroded plates. They employed a finite element method to evaluate the ultimate strength of severely corroded box girders, accounting for the natural distribution of the pits and general corrosion degradation.

In summary, there is a lack of research on the degradation of bearing capacity of corroded steel beams considering the random corrosion morphology. Therefore, this study analyzes the stochastic characteristics of corrosion morphologies, and then propose a random field model to characterize the corroded surface. Based on monotonic tensile tests and numerical simulation results, a probability model for the yield strength of corroded steel is proposed. Comprehensively considering the effects of corrosion thickness loss and strength degradation, a probability model for the flexural bearing capacity of corroded steel beams considering different corrosion ages, degrees, and morphologies is proposed.

2 STOCHASTIC MODEL OF CORROSION FEATURES OF CORRODED STEEL

2.1 *Data sources*

The data of corrosion features of steel plate come from our previous experiments in (Wang et al. 2021). The Q235 steel plate specimens (280×50×7.2mm) are employed for accelerated corrosion testing. The surface morphologies of the corroded specimens are measured by using a PS50 profiler produced by NANOVEA. The Surfer software is employed to perform three-dimensional reconstruction of the surface morphology of the corroded specimen. Figure 1 shows the measured results of the three-dimensional morphology of some corroded steel plate specimens.

The randomness of corrosion is manifested in several aspects: the randomness of the proportion of non-uniform corrosion, the randomness of corrosion surface features, and the spatial randomness of corrosion quantity under the same corrosion ratio. The following will analyze these three aspects of randomness separately and establish the stochastic model.

2.2 *Stochastic model of non-uniform corrosion loss*

Normally, the corrosion thickness includes two parts: uniform corrosion thickness and non-uniform corrosion thickness, as shown in Figure 2. Uniform corrosion thickness does not cause degradation of material properties, while non-uniform corrosion thickness can cause degradation of material strength

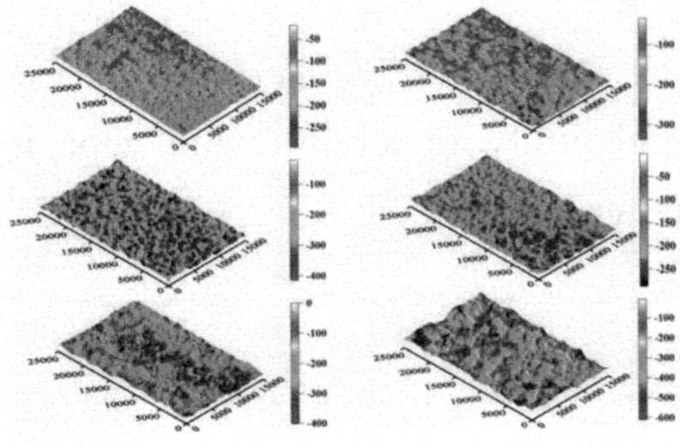

Figure 1. Three-dimensional surfaces of corroded specimens.

and ductility due to surface stress/strain concentration. Therefore, it is necessary to define the two types of corrosion as

$$\rho = \frac{M_0 - M}{M_0} = \frac{V_0 - V_c}{V_0} \times 100\% \tag{1}$$

$$\rho_u = \frac{T_0 - T_{\max,r}}{T_0} \times 100\%, \ \rho_n = \rho - \rho_u \tag{2}$$

where ρ is corrosion loss ratio, ρ_u is the uniform corrosion loss ratio, ρ_n is the non-uniform corrosion ratio; V_0 and V_c are the initial and residual volumes of the plate specimen, respectively; M_0 and M_c are the initial and residual mass of the plate specimen, respectively; T_0 and $T_{\max,r}$ are the initial and maximum resting corroded plate thickness of the specimen, respectively.

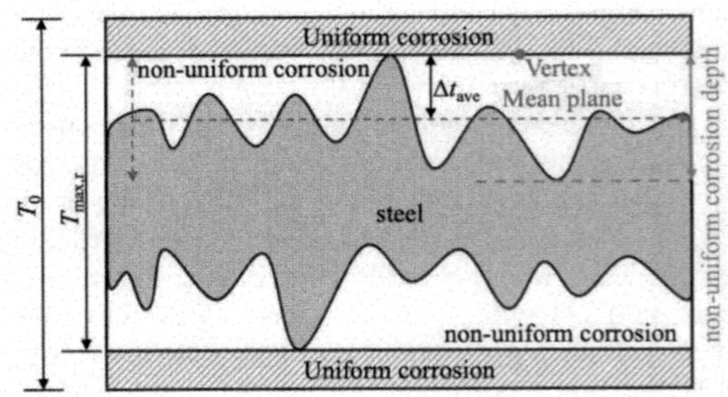

Figure 2. Description of uniform corrosion and non-uniform corrosion.

Non-uniform corrosion loss is strongly correlated with the strength of corroded steel. Therefore, the statistical characteristics of non-uniform corrosion loss to some extent affect the distribution features of the strength of corroded steel. Introducing the proportion of non-uniform corrosion loss thickness to characterize the degree of non-uniform corrosion in specimens:

$$\lambda = \rho_n / \rho \tag{3}$$

where, λ represents the proportion of non-uniform corrosion loss thickness.

By analyzing experimental data, Figure 3 shows the evolution of λ along with the corrosion equivalent thickness loss (Δt_e, i.e., ρT_0). The data indicates that when Δt_e<1.500mm, λ continues to decrease, and the corroded steel plates are in the initial stage of corrosion, with numerous shallow pits appearing on the specimen surface, exhibiting non-uniform corrosion. When 1.500<Δt_e<2.500mm, λ stabilizes within two intervals, (0.2, 0.4) and (0.6, 0.8), and the degree of corrosion further increases. With the expansion and deepening of pits, where both uniform and non-uniform corrosion occur simultaneously. Even under the same corrosion equivalent thickness loss, there is a significant difference in the values of λ due to the different corrosion morphologies, which demonstrates the stochastic nature of corrosion. For example, when Δt_e =2.0mm, the minimum and maximum difference of λ is about three times.

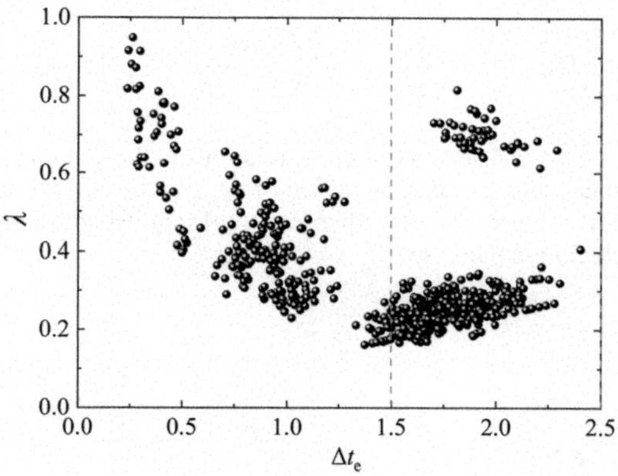

Figure 3. The relationship between λ and Δt_e.

Statistical analysis is conducted on the proportion of non-uniform corrosion loss, and the density distribution histogram of λ is shown in Figure 4. After conducting a hypothesis test on λ, it was found that the λ follows the normal distribution, as shown in Table 1. The probability density function of λ is as follows:

$$f(\lambda) = \frac{1}{\sigma_\lambda \sqrt{2\pi}} \exp\left[-\frac{1}{2}\left(\frac{\lambda - \mu_\lambda}{\sigma \lambda}\right)^2\right] \tag{4}$$

Table 1. Results of normal distribution test of λ.

Δt_e/mm	0.000~1.500	1.500~2.500	1.500~2.500
μ_λ	0.363	0.253	0.686
σ_λ	0.236	0.090	0.060

2.3 *Random field model of corrosion depth*

The corroded surfaces of steel plate specimens may also be different under the same degree of non-uniform corrosion. There, the mean and standard

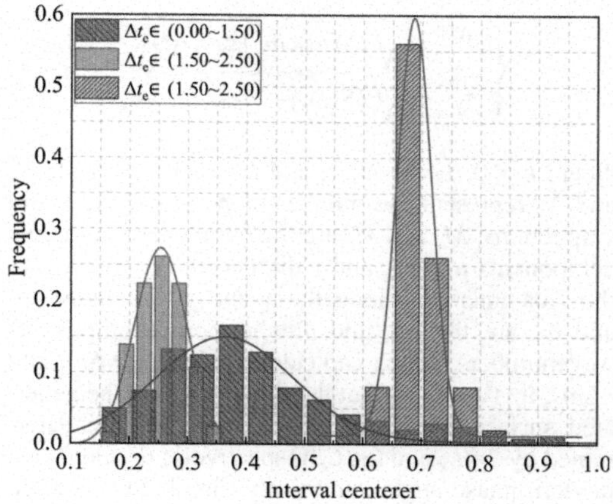

Figure 4. Density distribution histogram of λ.

deviation of non-uniform corrosion depth are utilized as the basic parameters for characterizing the surface features of corroded steel, which can be calculated by

$$\Delta t_{\text{ave}} = \left|\bar{z}(x_i, y_j)\right| = \frac{1}{MN}\sum_{i=1}^{M}\sum_{j=1}^{N}\left|z(x_i, y_j)\right| \tag{5}$$

$$t_{\text{sd}} = \sqrt{\frac{1}{MN}\left(\sum_{i=1}^{M}\sum_{j=1}^{N}\left(z(x_i, y_j) - \bar{z}(x_i, y_j)\right)^2\right)} \tag{6}$$

where Δt_{ave} is the distance between the vertex of the corroded surface and the mean plane, as shown in Figure 2. Δt_{ave} can be calculate d through the proportion of non-uniform corrosion loss (λ), i.e., $\Delta t_{ave} = \lambda \Delta t_e$; t_{sd} is the standard deviation of c non-uniform corrosion thickness. M and N are the number of scanning points in the x and y directions; $z(x_i,\ y_j)$ is the vertical coordinate value of the scanning point, i.e, non-uniform corrosion depth as shown in Figure 2; $\bar{z}(x_i,\ y_j)$ is the average value of $z(x_i,\ y_j)$.

The corroded surface morphology can be regarded as a random field of corrosion depth, which can be effectively characterized by the second type of random harmonic components. This approach accurately represents the stochastic process of the target power spectral density using a minimal number of random harmonic components (Chen & Li 2011). Due to the profiler collecting finite data in a grid-based manner, it cannot obtain continuous sample functions. Therefore, (x, y) is discretized into $(k\Delta x, l\Delta y)$. The representation of the stochastic model of corrosion depth is expressed as:

$$\begin{aligned} \eta(x,y) &= \eta(k\Delta x, l\Delta y) \\ &= \sqrt{2}\sum_{\mu=0}^{M_x-1}\sum_{\nu=0}^{M_y-1}\left[A_1^{(1)}\cos\left(\omega_\mu x + \omega_\nu y + \theta_1\right)\right. \\ &\quad \left. + A_2^{(2)}\cos\left(\omega_\mu x - \omega_\nu y + \theta_2\right)\right] \end{aligned} \tag{7}$$

$$\begin{cases} A_1^{(1)} = \sqrt{2S(\omega_\mu, \omega_\nu)\Delta\omega_1\Delta\omega_2} \\ A_2^{(2)} = \sqrt{2S(\omega_\mu, -\omega_\nu)\Delta\omega_1\Delta\omega_2} \end{cases} \quad (8)$$

where $k = 0, 1, \ldots, N_x$-1, and $l = 0, 1, \ldots, N_y$-1; N_x and N_y represent the number of points in the x and y directions; M_x and M_y are the numbers of harmonic components in the x and y directions; Δx and Δy are the data intervals along the x and y directions; ω_μ and ω_ν are the pth and qth harmonic components wavenumbers, which can calculated using $p\Delta\omega_1$ and $q\Delta\omega_2$. It is assumed that the dimensions of the generated surface are (L_x, L_y). $\Delta\omega_1$ and $\Delta\omega_1$ are determined by $2\pi/L_x$ and $2\pi/L_y$, respectively; θ_1 and θ_2 are random phase angles uniformly distributed between $[0, 2\pi]$; ω_{m1} and ω_{m2} are the upper cutoff wave numbers of the random field in the x and y directions, which can be determined by the following equation (Shinozuka & Deodatis, 1996).

$$\int_0^{\omega_1}\int_{-\omega_{m2}}^{\omega_{m2}} S(\omega_1,\omega_2)\mathrm{d}\omega_1\mathrm{d}\omega_2 = (1-\phi)\int_0^{\infty}\int_{-\infty}^{\infty} S(\omega_1,\omega_2)\mathrm{d}\omega_1\mathrm{d}\omega_2 \quad (9)$$

where $\phi \ll 1$, for example, 0.0001. To prevent of spectral aliasing, the number of generated points (N_x, N_y) should exceed twice the number of harmonic components (Shinozuka & Deodatis, 1996). The intervals $(\Delta x, \Delta y)$ of the generated points are expressed by

$$\begin{cases} \Delta x = \dfrac{L_x}{N_x - 1} = \dfrac{\pi L_x}{\omega_{m1}L_x - \pi} \\ \Delta y = \dfrac{L_y}{N_y - 1} = \dfrac{\pi L_y}{\omega_{m2}L_y - \pi} \end{cases} \quad (10)$$

In equation (9), $S(\omega_{1p}, \omega_{2q})$ is the target power spectral density (PSD) of the non-uniform corrosion depth. Based on the assumption that the distribution characteristics of the corroded surface in all directions are similar, the expression of the non-uniform corrosion depth PSD was derived by reference (Wang et al. 2019, 2021).

$$S(\omega_1, \omega_2) = t_{sd}{}^2\frac{a^2}{4\pi}\exp\left(-(b\omega_1)^2 - (b\omega_2)^2\right) \quad (11)$$

where a and b are the fitting parameters related to the peak value of the PSD and the upper limit cut-off wavenumbers (Shinozuka & Deodatis, 1996). By controlling t_{sd}, a, b, ω_{m1} and ω_{m2} of the corrosion morphology, the function $\eta(x, y)$ can be employed to randomly generate a corrosion surface with certain statistical characteristics. The relationship between t_{sd}, a and b with Δt_{ave} is given in reference (Wang et al. 2021).

$$\Delta t_{ave} = 3.48\Delta t_e{}^{0.64} \quad (12)$$

$$t_{sd} = 7.16\Delta t_{ave}{}^{0.48} \quad (13)$$

$$a = 2.33[1 - \exp(-0.022\Delta t_{ave})] \quad (14)$$

$$b = 0.62a \quad (15)$$

2.4 *Stochastic model of total corrosion loss*

The total corrosion loss (or the equivalent corrosion thickness loss Δt_e) of 360 corroded areas is obtained from previous experimental data. The mean (t_a) of Δt_e under the same age was calculated. There are ten values for t_a, i.e., t_a=0.3, 0.4, 0.5, 0.8, 1.0, 1.6, 1.7, 1.9, and 2.0. All Δt_e values were grouped and merged values by intervals of [t_a -0.05, t_a +0.05). The statistical data are divided into 10 groups. The density distribution histogram of Δt_e is presented in Figure 5. After conducting a hypothesis test, it was found that Δt_e follows the normal distribution, as shown in Table 2. The probability density function of Δt_e is as follows:

$$f(\Delta t_e) = \frac{1}{\sqrt{2\pi}\sigma_{\Delta te}}\exp\left[\left(-\frac{\Delta t_e - \mu_{\Delta te}}{\sqrt{2}\sigma_{\Delta te}}\right)^2\right] \quad (16)$$

Figure 6 shows the linear relationship between the mean and standard deviation of Δt_e. As the corrosion level increases, the dispersion of equivalent loss thickness increases.

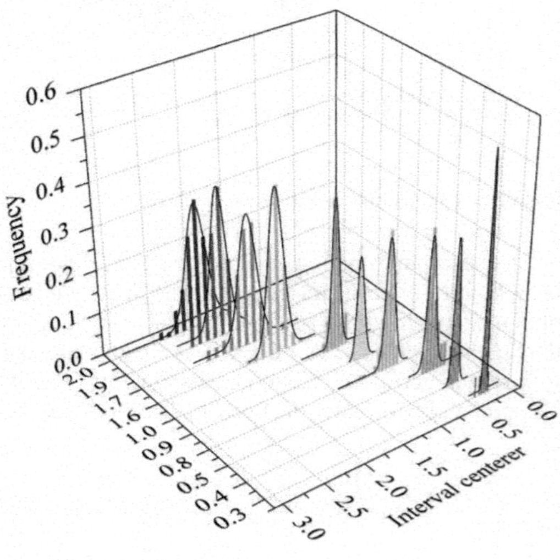

Figure 5. Density distribution histogram of Δt_e.

3 PROBABILITY MODEL OF YIELD STRENGTH OF CORRODED STEEL

3.1 *Monte Carlo simulations for monotonic tension*

Monte Carlo simulation method is the most widely used technique in probability analysis and simulation. Precise prediction and probabilistic characteristics of uncertain mechanical properties of corroded steel plates are achieved by combining Monte Carlo simulation methods with nonlinear finite element analysis.

Table 2. Results of normal distribution test of Dte.

Groups	0.3	0.4	0.5	0.8	0.9	1.0	1.6	1.7	1.9	2.0
$\mu_{\Delta te}$	0.285	0.416	0.461	0.793	0.915	1.012	1.569	1.740	1.923	1.980
$\sigma_{\Delta te}$	0.033	0.039	0.075	0.078	0.055	0.075	0.122	0.187	0.155	0.156

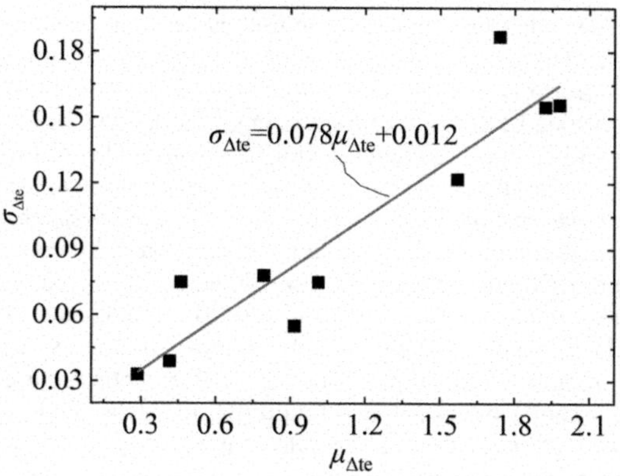

Figure 6. Relationship between $\mu_{\Delta te}$ and $\sigma_{\Delta te}$.

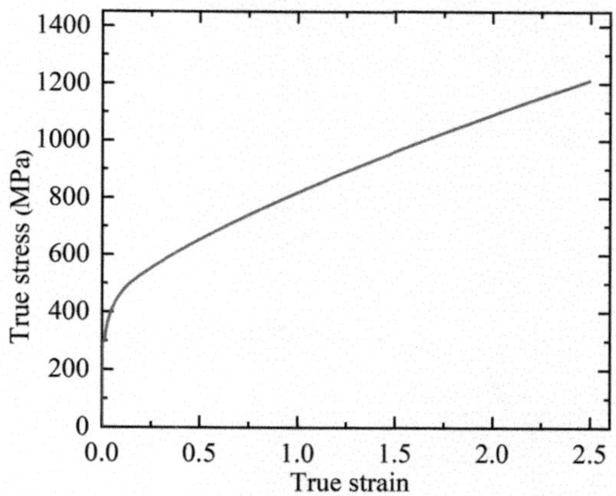

Figure 7. Constitutive relationship.

Based on a large number of random samples obtained by stochastic modelling of corrosion features of corroded steel in Section 2, the three-dimensional reconstructions of corrosion morphologies are accomplished within the finite element software ABAQUS.

The constitutive relationship is shown in Figure 7. The mesh size for the finite element model of corroded steel plate is 750μm; The initial thickness is 7.2mm; Elastic modulus is 2.4×10^5 MPa; The Poisson's ratio is 0.3. The geometry models of corroded steel plate with equivalent corrosion loss thickness (Δt_e) of 1.540mm (ρ=21.4%), 2.190mm (ρ=30.4%), 2.680mm (ρ=37.2%), and 3.090mm (ρ=42.9%) were established, and each had a simulation quantity of 100. Under the same corrosion loss ratio, the proportion of non-uniform corrosion loss (λ) is uniformly sampled at intervals of 0.05 within the range of 0 to 1. The boundary conditions of the numerical model were such that one end was fixed and the other end was constrained in the x- and z-directions, and a longitudinal tensile displacement was applied in the y-direction as shown in Figure 8(a).

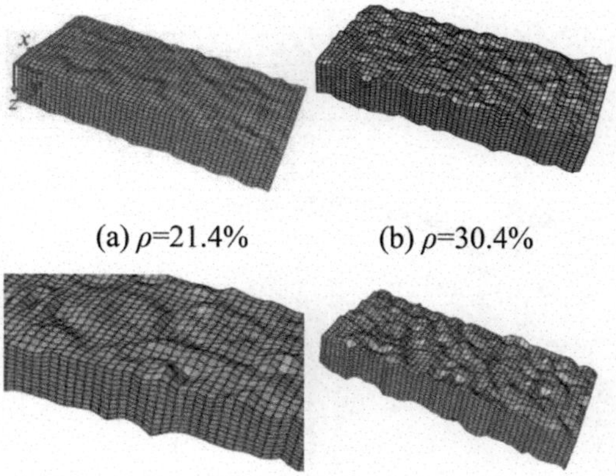

(a) ρ=21.4% (b) ρ=30.4%

(c) ρ=37.2% (d) ρ=42.9%

Figure 8. Finite element model of corroded steel plate (λ=0.6).

Figure 8 shows the finite element models of corroded steel plates with different degrees of corrosion. Obvious surface fluctuations and complex morphologies tend to appear in steel plates with greater corrosion degrees.

The nominal stress and nominal strain of all corroded steel plate are calculated according to the computed load-displacement curves by:

$$\sigma = F/BT_e,\ \varepsilon = \Delta L/L_0 \quad (17)$$

where σ is the nominal stress, is the nominal strain, B is the width of the specimen, T_e is the equivalent thickness of corrosion steel, T_e=T_0-Δt_e. and L_0 is the initial gauge length, ΔL is the elongation of the gauge length.

Figure 9 shows the calculated results of the stress-strain curve of models under monotonic tension tests. The curve results indicate an increase in curve dispersion with the rise in the corrosion ratio. Figure 10 shows relationship between λ and f_y. Under the same corrosion ratio, the yield platform gradually disappears with the increase in λ. The results indicate that when λ<0.5, the degradation of yield strength is not significant. As λ approaches 0.5, the yield strength begins to degrade, with

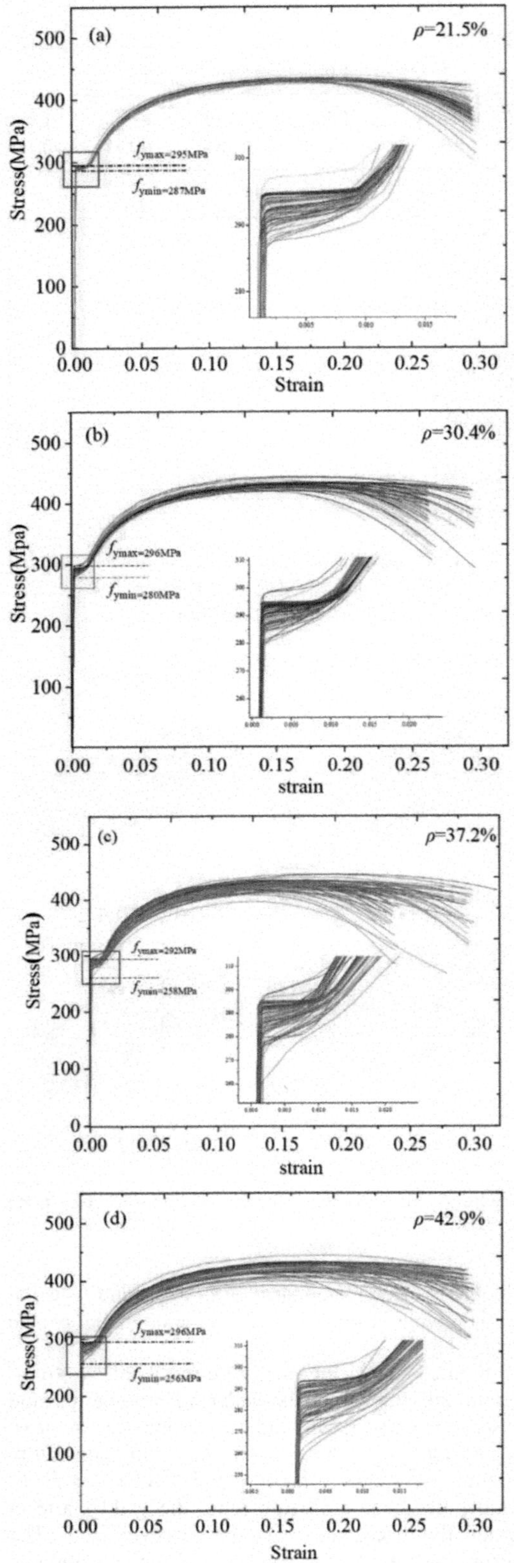

Figure 9. Simulation results of monotonic tensile stress-strain curve of corroded steel plate.

a maximum decrease of 7.8% in f_y, and a clear yield platform is still visible. However, when $\lambda \geq 0.5$, the yield strength degraded significantly, with a maximum decrease of 14.28% in f_y, but the yield platform disappeared.

3.2 *Probability model of yield strength*

The magnitude and distribution pattern of strength degradation is affected with the degree of non-uniform corrosion. The ratio of the yield strength of corroded steel to its initial strength is defined as the yield strength reduction factor ($\varphi = f_y/f_{y0}$). Due to the fact that the corrosion ratio varies with the thickness of the specimen under the same loss thickness, the yield strength reduction factor is calculated by comprehensively considering the influence of corrosion loss thickness and corrosion ratio. Figure 11 shows the density distribution histogram of the yield strength reduction factor. After conducting a hypothesis test, it was found that the data of yield strength reduction factor follows the normal distribution, as shown in Table 3.

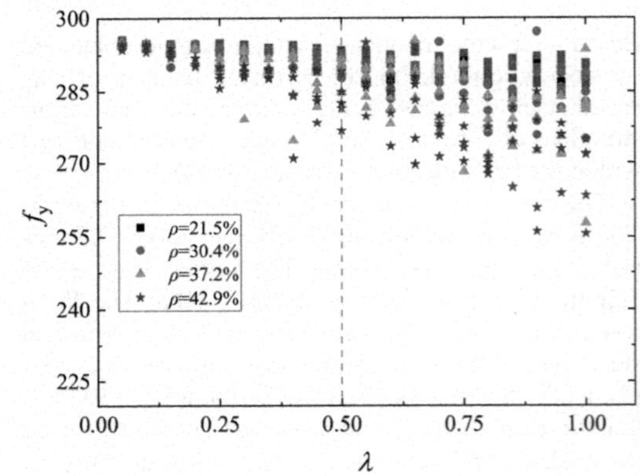

Figure 10. The relationship between λ and f_y.

The relationship between the mean and standard deviation of the yield strength reduction factor (φ) and the corrosion ratio is depicted in Figure 12. The corrosion ratio exhibits a linear relationship with both the mean and standard deviation.

$$\mu_\varphi = -0.036\rho + 1 \; \Delta t_e \in (0.000, 1.500) \tag{18}$$

$$\mu_\varphi = -0.113\rho + 1 \; \Delta t_e \in (1.500, 2.500) \tag{19}$$

$$\sigma_\varphi = 0.040\rho \; \Delta t_e \in (0.000, 1.500) \tag{20}$$

$$\sigma_\varphi = 0.103\rho \; \Delta t_e \in (1.500, 2.500) \tag{21}$$

Based on the above analysis, substitute equation (18) - (21) into the normal distribution probability density function. The probability model of yield strength is:

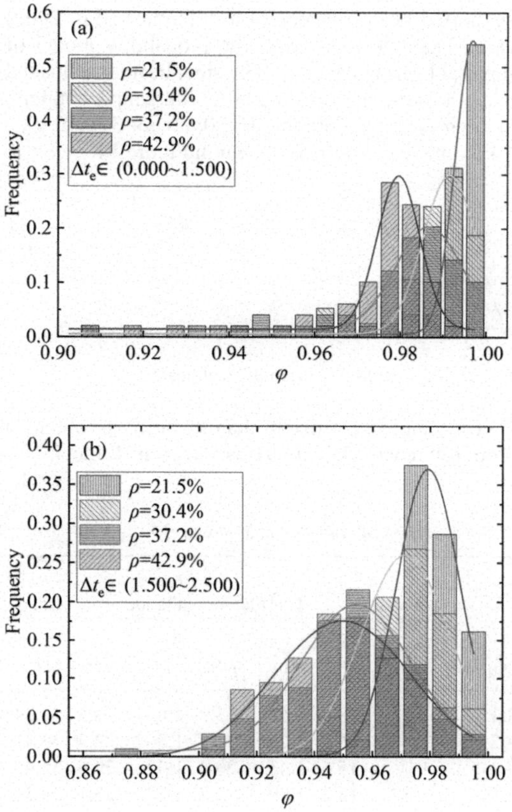

Figure 11. Density distribution histogram of φ.

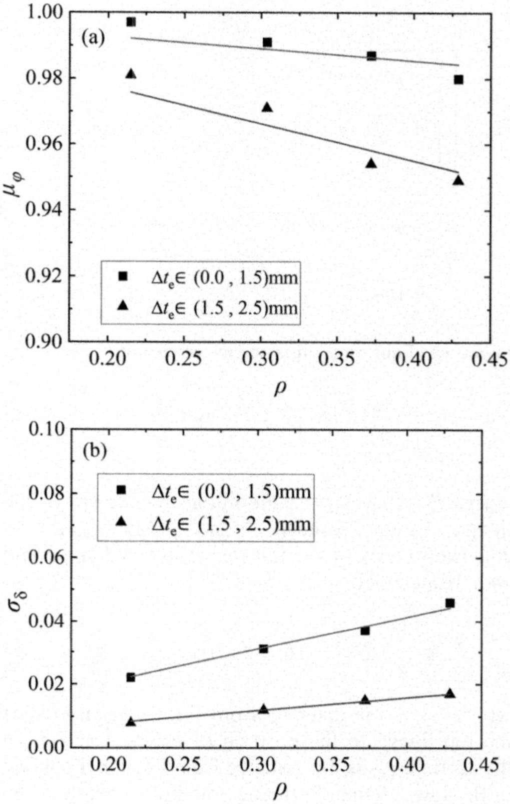

Figure 12. Relationship between ρ and μ_{φ} (a), σ_{φ} (b).

$$f(\varphi)=\begin{cases}\frac{1}{\sqrt{2\pi}0.04\rho}\exp\left(-\frac{1}{2}\left(\frac{\varphi+0.036\rho-1}{0.04\rho}\right)^2\right), \Delta t_e \in (0.0, 1.5)\\ \frac{1}{\sqrt{2\pi}0.103\rho}\exp\left(-\frac{1}{2}\left(\frac{\varphi+0.113\rho-1}{0.103\rho}\right)^2\right), \Delta t_e \in (1.5, 2.5)\end{cases} \tag{22}$$

4 PROBABILITY MODEL OF BEARING CAPACITY OF CORRODED STEEL BEAM

4.1 *Model establishment*

Before establishing the model of bearing capacity of corroded steel beam (taking channel steel as an example), the following assumptions are made:

(1) The cross section of corroded channel steel is considered equivalent to a regular cross-section, as shown in Figure 13a.
(2) The distribution of stress on the cross-section satisfies the plane section assumption, as shown in Figure 13b.
(3) The section width and height loss thickness are the same.
(4) The eccentricity of the section is ignored.
(5) The degree of elastic modulus degradation caused by corrosion is the same, i.e. $E=E_{tf}=E_{lf}=E_w$(E, E_{tf}, E_{lf}, E_w represent the elastic modulus of the whole section, top flange, lower flange, and web, respectively.)

The derivation process of the calculation formula for the flexural capacity of corroded channel steel is as follows:

$$\begin{cases}H=H_0-T_{0,\mathrm{f}}+\frac{T_{\mathrm{tf}}+T_{\mathrm{lf}}}{2}\\ B=B_0-T_{0,\mathrm{w}}+T_{\mathrm{w}}+\frac{T_{0,\mathrm{f}}}{2}-\frac{T_{\mathrm{tf}}+T_{\mathrm{lf}}}{4}\end{cases} \tag{23}$$

Table 3. Results of normal distribution test.

ρ/%	21.5	30.4	37.2	42.9	21.5	30.4	37.2	42.9
Corrosion depth/mm	0.000~1.500				1.500~2.500			
μ_{φ}	0.997	0.991	0.987	0.980	0.981	0.971	0.954	0.949
σ_{φ}	0.008	0.012	0.015	0.017	0.022	0.031	0.037	0.046

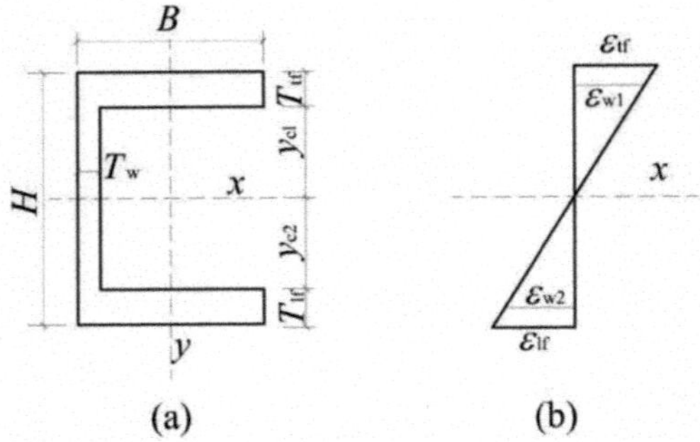

Figure 13. Equivalent cross-section schematic.

$$T_{tf}=(1-\rho_{tf})T_{0,f},T_{lf}=(1-\rho_{lf})T_{0,f},T_w=(1-\rho_w)T_{0,w} \quad (24)$$

where $T_{0,f}$ and $T_{0,w}$ are the initial thicknesses of the flange and web, respectively. ρ_{tf}, ρ_{lf}, and ρ_w are the corrosion ratios of the top flange, lower flange, and web, respectively.

$$I_x = \frac{1}{12}\left[BH^3 - (B - T_w)(y_{c1} + y_{c2})^3\right] \quad (25)$$

where y_{c1} is the distance from the lower surface of the top flange to the position of the neutral axis in the section, y_{c2} is the distance from the upper surface of the lower flange to the neutral axis position in the section. I_x is the moment of inertia of the section.

The calculation of the ultimate flexural capacity of steel beams is as follows:

$$M_x = \frac{I_x f_y}{y_{max}} \quad (26)$$

where y_{max} is the max distance of from the section to the neutral axis, i.e. $y_{max}=y_{c1}+(T_{0,f} - T_{tf})/2$; f_y is the yield stress at the weak point.

The position of the weak point under bending, the position of the neutral axis and the moment of inertia have a mapping relationship with the equivalent corrosion loss thickness, and y_{max} and I_x are related to the equivalent corrosion loss thickness.

By substituting Δt_e into equation (23) ~ (26), we can calculate I_x and y_{max}. The degree of corrosion of steel beams has a significant impact on the yield strength, and the yield stress can be calculated based on the probability model of yield strength reduction factor, i.e. $f_y = \varphi f_{y0}$.

In summary, the mean of M_x can be calculated by equation (28).

$$\mu_{Mx} = \frac{\mu_{Ix}\mu_{fy}}{\mu_{ymax}} \quad (27)$$

4.2 *Case studies*

This section takes the calculation of the flexural bearing capacity of corroded channel steel simply supported beam (uncoated) in reference (Bai. 2009) as an example to validate the probability model of flexural bearing capacity. The steel beam was loaded using a four-point bending test, with simple supports at both ends, L=600mm, H=50mm, B=37mm, $T_{0,w}$ =4.5mm, $T_{0,f}$=7mm, as shown in Figure 14.

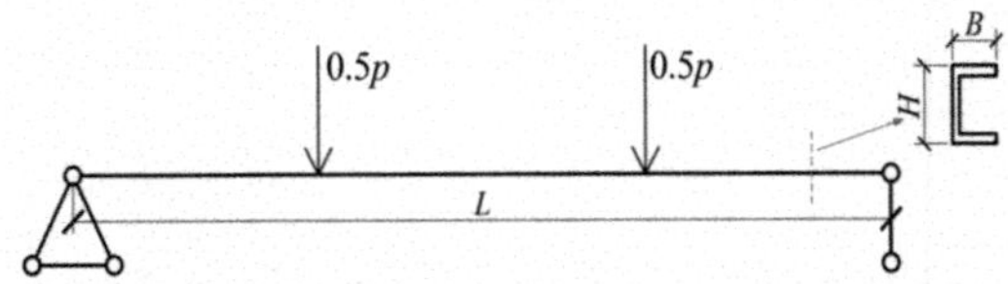

Figure 14. Loading diagram of steel beam.

The data of corrosion channel steels is obtained from reference (Bai. 2009), as shown in Table 4.

Table 4. Data of corroded channel steel.

	Web		Top Flange		Lower Flange		Load
Number	Δt_e/ mm	ρ/%	Δt_e/ mm	ρ/%	Δt_e/ mm	ρ/%	f_{max}/ kN
L1	0.83	18.4	1.67	23.9	1.63	23.3	33.34
L2	0.68	15.1	1.75	25.0	1.56	22.3	33.41
L3	0.7	15.6	1.43	20.4	1.44	20.6	35.41
L5	0.82	18.2	1.59	22.7	1.62	23.1	34.36
L6	0.79	17.6	1.68	24	1.72	24.6	33.12
L7	0.99	22.0	1.55	22.1	1.71	24.4	34.01
μ	0.80	17.8	1.61	23.0	1.61	23.0	33.9

The yield strength of uncorroded channel steel plate is f_{y0}=372MPa. The yield strength reduction factor of the top flange, lower flange, and web, as well as the yield strength after corrosion are calculated by equation (18) ~ (21), and the results are shown in Table 5

Table 5. Yield strength reduction factor and yield strength.

Location	Top Flange	Lower Flange	Web
φ	0.974	0.974	0.994
f_y	362.33	362.33	369.77

The mean of the geometric parameters of the equivalent section of corroded channel steel can be calculated based on the data in Table 4. The mean of the inertia moment of the equivalent section can be calculated by equation (23) ~ (25), and the results are shown in Table 6.

The weak point of the section is either the outer side of the top flange or the lower flange. The mean of M_x of corroded channel steel can be calculated by equation (27). The maximum bending moment

Table 6. Calculation results of geometric parameters of equivalent cross section of channel steel.

Parameter	H	B	y_{c1}/y_{c2}	B-T_w	I_x
	48.39	35.795	18.81	32.095	195705

section is located at one-third of the beam (where the load is applied) from Figure14. The ultimate bending load (P) of the beam can be calculated by

$$M_x = 0.5P \times \frac{L}{3} \tag{28}$$

The ultimate bending load (P) is 29.3kN. The mean of f_{max} in Table 4 is 33.9kN. The error between the calculated value and the experimental value is 13.5%. There is a difference between the theoretical calculation results and the experimental results, which is because the channel steel in the experiment has stiffeners, resulting in the actual ultimate load being greater than the theoretical load.

5 CONCLUSIONS

(1) A stochastic model of corrosion characteristics of corroded steel was established based on the corrosion depth random field. Through parameter control, the model can generate multiple corrosion morphology and reflect the randomness of the corrosion surface.

(2) The proportion of non-uniform corrosion loss thickness (λ) can be used to characterize the degree of non-uniform corrosion. In the initial corrosion phase of the corroded steel plate, the value of λ is relatively large, indicating that corrosion is primarily characterized by non-uniform corrosion. With the progression of corrosion, the value of λ gradually decreases, indicating the onset of uniform corrosion.

(3) A probability model for the yield strength of corroded steel plates was established by introducing parameter φ to quantify the degree of mechanical property damage caused by corrosion. The corrosion ratio exhibits a linear relationship with both the mean and standard deviation of φ.

(4) A probability model for the flexural bearing capacity of corroded channel steel beams was established based on the stochastic model for corrosion features and the yield strength of corroded steel beams. The probability model of flexural bearing capacity was verified using the experimental results from a real simply supported corroded channel steel beam as an example.

ACKNOWLEDGMENT

The authors would like to thank the National Natural Science Foundation of China (Grants No. 52278316), Shaanxi Key Research and Development Program (2022SF-473), China Postdoctoral Science Foundation (Grant No. 2021M692511), Independent Research and Development Project of State Key Laboratory of Green Building in Western China (LSZZ202220), and the Youth Innovation Team of Shaanxi Universities (2019-2022) for the financial support.

REFERENCES

Bai, Y. 2009. Experimental study and theoretical analysis on the bending behavior of the corroded channel steel, Xi'an University of Architecture and Technology, Xi'an, China.

Chen, J. B. & Li, J. 2011. Stochastic harmonic function and spectral representations. *Chin. J. Theor. Appl. Mech.*, 43 (3): 505–513.

Garbatov, Y. et al. 2007. Non-linear time-dependent corrosion wastage of deck plates of ballast and cargo tanks of tankers. *J. Offshore Mech. Arct. Eng.*, 129(1): 48.

Garbatov, Y. et al. 2014. Tensile strength assessment of corroded small-scale specimens. *Corros. Sci.*, 85: 296–303.

Georgiadis, D. & Samuelides, M. 2019. A methodology for the reassessment of hull-girder ultimate strength of a VLCC tanker based on corrosion model updating. *Ships Offshore Struct.*, 14: 270–280.

Kong, D. L. et al. 2021. Random Field Model of Corroded Steel Plate Surface in Neutral Salt Spray Environment. *KSCE J. Civ. Eng.*, 25(8): 2651–2661.

Melchers, R. E. 1999. Corrosion uncertainty modeling for steel structures. *J. Constr. Steel Res.*, 52(1): 3.

Nakai, T. et al. 2004. Effect of pitting corrosion on local strength of hold frames of bulk carriers (2nd Report)—Lateral-distortional buckling and local face buckling. *Marine Structures*, 17(8): 612–641.

Nakai, T. et al. 2006. Effect of pitting corrosion on the strength of web plates subjected to patch loading. *Thin-Walled Structures*, 44(1): 10–19.

Paik, J. K. et al. 2004. Ultimate shear strength of plate elements with pit corrosion wastage. *Thin-Walled Structures*, 42(8): 1161–1176.

Rahgozar, R. 2009. Remaining capacity assessment of corrosion damaged beams using minimum curves. *Journal of Constructional Steel Research*, 65(2): 299–307.

Rajput, A. & Paik, J. K. 2020. Effects of naturally-progressed corrosion on the chemical and mechanical properties of structural steels. *Structures*, 29: 2120–2138.

Saad-Eldeen, S. et al. 2014. Strength assessment of a severely corroded box girder subjected to bending moment. *Journal of Constructional Steel Research*, 92 (1): 90–102.

Schumacher, M. M. 1979. *Seawater Corrosion Handbook*. New Jersey: Noyes Data Corporation, 87.

Shinozuka, M. & Deodatis, G. 1996. Simulation of Multi-Dimensional Gaussian Stochastic Fields by Spectral Representation. *ASME. Appl. Mech. Rev.*, 49(1): 29–53.

Silva, J. E. et al. 2013. Ultimate strength assessment of rectangular steel plates subjected to random localized corrosion degradation. *Eng. Struct.*, 52: 295.

Vu, V. T. & Dong, D. T. 2020. Hull girder ultimate strength assessment considering local corrosion. *J Mar Sci Appl.*, 19: 693–704.

Wang, R. et al. 2018. Ultimate strength assessment of plated steel structures with random pitting corrosion damage. *J. Constr. Steel Res.*, 143: 331.

Wang, Y. et al. 2015. Ultimate strength assessment of steel stiffened plate structures with grooving corrosion damage. *Engineering Structures*, 94: 29–42.
Wang, Y. et al. 2017. Predicting the residual strength and deformability of corroded steel plate based on the corrosion morphology. *Constr. Build. Mater.*, 152: 777–793.
Wang, Y. et al. 2019. Surface Characteristics and Stochastic Model of Corroded Structural Steel Under General Atmospheric Environment. *ACTA Metall. Sin.*, 56: 148–160.
Wang, Y. et al. 2020. Flexural performance evaluation of corroded steel beam based on 3D corrosion morphology. *Structure and Infrastructure Engineering*, 16(11): 1562–1577.
Wang, Y. et al. 2021. Stochastic Model for Surface Characterization of Structural Steel Corroded in Simulated Offshore Atmosphere. *ACTA Metall. Sin.*, 57: 811–821.

Advances in Maritime Technology and Engineering – Guedes Soares & Santos (Eds)

Vibration dose value assessment from full scale measurements

G. Storhaug & G. Jagite
DNV, Oslo, Norway

ABSTRACT: Noise and vibration onboard ships may lead to discomfort mainly, excited by the engine or propeller. These vibrations are often above 5-10 Hz. In serious cases, mitigation actions must be taken, but acceptable solutions are often available. However, wave-induced vibrations may also contribute with lower frequencies below 5-10 Hz, which is difficult to avoid. While machinery vibration may be more continuous, wave-induced vibration may be more impulsive. The concept of vibration dose value (VDV) is formulated in an old ISO standard but rarely assessed. In this case, full-scale measurements from a container ship are assessed to define an upper limit of realistic hourly VDV levels as a reference. It turns out to be at a VDV level of about 10 times the comfort threshold.

1 INTRODUCTION

1.1 *General*

Noise and vibration are frequent issues onboard ships and may lead to consequences such as discomfort and fatigue cracking. The sources are mainly excitation from the engine and propeller, and the noise and vibration are often assessed by confirming acceptable levels through spot check measurements during sea trials. These vibrations are often in the range of 5 Hz and above. In case of serious noise and vibration levels, mitigation actions must be taken, but in almost all cases, acceptable solutions are possible.

All ships vibrate due to waves, and these wave-induced vibrations may amplify the stress levels, both for fatigue and extreme loading. They may, however, also contribute to discomfort, but this has rarely been assessed. The vibration frequency of the hull girder may be in the range of 0.4 Hz to 5 Hz, so less than from the propeller and engine excitation. There is, however, an old ISO standard (ISO, 1997) that formulates vibration dose values (VDVs) not only related to root mean square values useful for continuous vibrations but also to whipping types of accelerations with high peaks now and then. This approach is especially relevant for cruise ships, ferries, and super yachts but is also a known issue for container ships and some other vessel types. Recently, the acceptance level of this VDV has been re-confirmed on an icebreaker research vessel. This reference level corresponds to an hourly VDV value in the order of 0.4 $m/s^{1.75}$ (Engelbrecht and Bekker, 2023). It resembles the reference level from Griffin (2012), so this is reducing the uncertainty to the hourly comfort threshold even though there may be different perception by crew and passengers. An issue related to discomfort from such wave-induced vibrations is that the mitigation action may be more dramatic. Changing the bow shape may not be a feasible option when the vessel is already delivered. And if the bow shape should be improved, it is not straight forward to define how much change is necessary.

There are a few other studies on VDV levels. Bekker et al. (2017) stated that VDV level for the polar research vessel was higher in open waters than in ice. Matsumoto & Kunimatsu (2022) suggested that VDV thresholds of 0.2 to 0.4 over 16 hours for buildings resulted in low risk of complaints. Duarte et al. (2018) suggested that the ISO (1997) standard might be more useful for health, and Omer & Bekker (2018) had another evaluation of VDV based on daily values based on summarizing VDV values from 5-minute periods with identified slamming as a measure for complaints and sleeping disturbance. Lorenzino et al. (2022) emphasized that mood affects the perception. There are also other studies related to especially sea sickness at frequencies below 1 Hz but that is another topic.

Due to limited VDV assessment in the literature for maritime applications, measurements from a large container ship have been assessed during a storm. The intention is to illustrate how high the VDV levels may be onboard ships, even though the crew may stay away from the worst areas. These values may serve as reference values in extreme cases but also suggest that high VDVs may also be relevant for container ships. It cannot be excluded that the perceived VDVs on the bridge are correlated with voluntary speed reductions.

1.2 *The vessel*

The measurements are taken from an accelerometer placed at the forward perpendicular at the center line, measuring vertical acceleration. The vessel is an 8600 TEU vessel with a pronounced bow flare, which is

DOI: 10.1201/9781003508762-60

more severe than typically seen on many large container ships. The maximum local bow flare angle exceeds 60° (90° is flat "bottom"), which may give rise to severe slamming. The container ship has been previously assessed in full scale by Heggelund et al. (2011) and in model test by Storhaug et al. (2010). Those studies confirmed severe whipping events in moderate sea conditions and in some cases exceeding the IACS (International Association of Classification Societies) design moments, so this vessel is a good candidate to define an upper likelihood limit of realistic VDV levels for ships. The size of the vessel is otherwise rather standardized for this TEU capacity, with a length between perpendiculars of just above 320 m. For more details of the vessel, see Heggelund et al. (2011).

2 VIBRATION DOSE VALUE CONCEPT

2.1 *Preparing the new file with the correct template*

The concept of VDV is explained by ISO (1997), but for completeness, the relevant part is included in this paper. The formula for VDV, in m/s1.75, is expressed by Equation 1:

$$VDV_i = \left\{ \int_0^T \left(a_{w,i}(t)\right)^4 dt \right\}^{1/4} \quad (1)$$

where T is time equal to typically 3600 seconds (one-hour VDV), $a_{w,i}$ is the weighted acceleration in a specific direction i, where i is the longitudinal (x-), transverse (y-), or vertical (z-) direction. The weighted acceleration is taken to the power of 4 which makes the VDV concept more sensitive to high impulsive responses from especially whipping events from bow flare, bottom or stem flare slamming but occasionally also from stern slamming induced whipping (which one author also experienced even in port on a container ship from choppy waves from a large passing vessel and the whipping starting with a loud sound from the stern part and with subsequent decaying vibrations and noise from paintings on the wall). In this paper, only the vertical z acceleration is considered. If several directions are included, the combined VDV, VDV_{tot}, can be calculated based on the individual components VDV_i as in Equation 2:

$$VDV_{tot} = \left\{ \sum_i VDV_i^4 \right\}^{1/4} \quad (2)$$

The VDV is normally calculated for each hour or each half hour. The time period, T, in Equation 1 is essential for the integration, but assuming stationary conditions, the VDV may be corrected to another time period by Equation 3 based on the two time periods $T1$ or $T2$:

$$VDV_{T1} = VDV_{T2} \left(\frac{T1}{T2} \right)^{1/4} \quad (3)$$

For instance, going from half-hour VDV to hourly VDV implies an increase of 19%.

In ISO (1997), the VDV is defined from 1 to 80 Hz. It is unclear why these limits are chosen, but it is not only relevant for the marine industry. As mentioned before, it may be possible by different measures to reduce the propeller or engine excited vibrations at high frequencies, but it may be difficult to change the natural frequencies of the hull girder excited by wave loads. These natural frequencies may be below 1 Hz for nowadays large vessels, and for a 300 m beam like ship in cargo condition, the rule of thumb is 0.5 Hz and the lowest natural frequency of the vertical 2-node mode is almost linear with the length ratio (so a 600 m ship should have about 0.25 Hz and a 150 m ship about 1 Hz; these are rough estimates and there are significant difference also between loading conditions in the range of 10-20% for cargo ships and even shallow water can change the frequency significantly due to added mass effects). The lower threshold in the ISO (1997) standard is therefore regarded as too high for many large vessels for VDV levels related to wave-induced vibrations. The practice is then to identify the lowest frequency and use a frequency band filter that includes the wave induced hull girder vibrations from about 0.5 Hz to about 5 to 10 Hz but to avoid the propeller or engine excitations. For instance, a 4-blade propeller in 90 RPM gives a blade passing frequency of 4·90/60 = 6 Hz, which should be excluded, but this vibration frequency is typically more pronounced close to the propeller and not necessarily in the bow. In principle, the data should be inspected to include only the wave-induced vibrations, and this may also be automated. However, if unacceptable VDV levels are caused by the propeller or engine, then it is justified to include it, and by using different upper levels on the frequency band, the relative importance may be revealed. It may then turn out that in parts of the vessel, it could be the wave-induced vibrations that are annoying, while in other parts, it is propeller- or engine-excited vibrations that are annoying.

The ISO (1997) standard tabulates the weighting factor for the acceleration in the vertical and horizontal direction. These are displayed in Figure 1 which shows the vertical weighting function to be maximum 1 at about 6 Hz, while the horizontal weighting function is 1 already at 1.0 Hz. Intuitively this may be understood as the horizontal motion is related to keeping balance and movement of the whole-body relative to the seat or feet, while the vertical may be more related to the head movements and eyesight, but this may not be very scientific explanation as this is based on perceptions from experience by the authors onboard different vessels from small high speed vessels to large vessels. At 1 Hz horizontal acceleration is basically felt more uncomfortable than vertical acceleration, which is the dominating vibration mode considered in this paper.

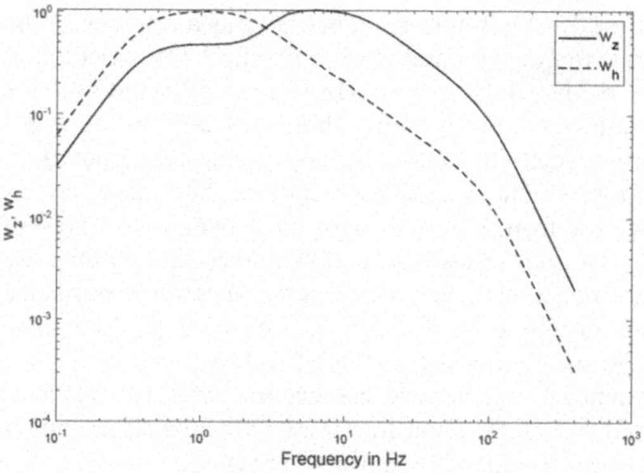

Figure 1. Weighting functions for vertical, w_z, and horizontal, w_h, directions.

Assuming that the structural response is linear, the global hull girder vibrations can be seen as the superposition of many individual modes, for example, vertical vibration modes, horizontal, torsional, and coupled hull girder modes between torsional and horizontal modes. For large container ships it is well recognized that the coupled horizontal-torsional modes have a low natural frequency, close to the vertical 2-node vibration mode and could be even lower than the vertical 2-node on large container ships. Nevertheless, considering that the accelerometer is located at the center line, only the vertical vibration modes are captured by the vertical accelerometer. For illustration purposes, the first three vertical vibration modes, together with the rigid-body motions in the vertical plane, are depicted in Figure 2. The vibration modes illustrated are analytical, but the main point is that the accelerometer is located at the bow, where all the modes tend to have close to maximum amplitudes when the excitation occur there as an impulsive load, even the rigid ones (which may not really contribute to any acceleration due to short waves at high frequencies and the cancelling effect of the forces along the hull giving small acceleration amplitudes; the vessel is basically acting like a filter). If one moves to another location than the bow, the relative amplitude may be less and therefore, the *VDVz* component is expected to be less. Selecting the bow accelerometer is then regarded as a good reference to define an expected maximum VDV level. If it was a cruise vessel or a large container ship with a tall and slender superstructure located forward, the longitudinal x-component would also be relevant, but for this vessel, the superstructure is aft. In any case the rotation of the modes makes the forward motion relevant when you are high up in the superstructure everywhere except where the vibration mode only give vertical motion like at midship for the vertical 2-node mode. For higher modes, midship can also give rotation and forward motion, but these higher modes may not be dominating at midship. For a cruise ship and container ship, the transverse y-component may be the main concern and especially towards the end of the vessels and provided that the container ship has a forward superstructure. In short, the ends of the ships are of most interest as well as deck area high above the sea surface. The specific ship design however needs to be considered and there could be local effects due to amplification of soft local structures.

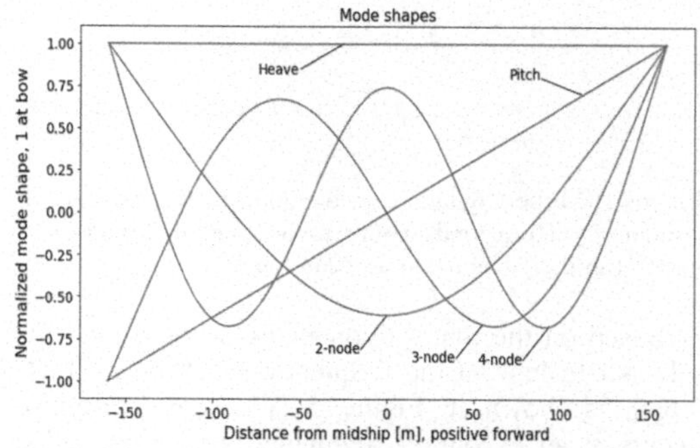

Figure 2. Illustration of analytical vertical vibration modes for a 300 m ship with constant mass and stiffness along the ship.

The sampling frequency is 42.37325 Hz, which is adequate to capture amplitudes at 4 Hz accurately but becomes questionable at higher frequencies like 8 Hz. The time series of the dynamic acceleration for the whole day is presented in Figure 3. The mean is removed for the whole day, but there are some non-zero mean in the beginning which may be difficult to explain but may be related to temperature effects. It has some maximum dynamic acceleration after about 7 hours with amplitudes up to 7-8 m/s2, which is substantial. It may look like instability peaks, however when zooming in there is no instabilities but it illustrates that several frequencies are present, but it is evident that the sampling frequency, at least for the higher frequencies, is not adequate, see Figure 4. This represents a 10 second zoom-in on the extreme negative amplitude in Figure 3.

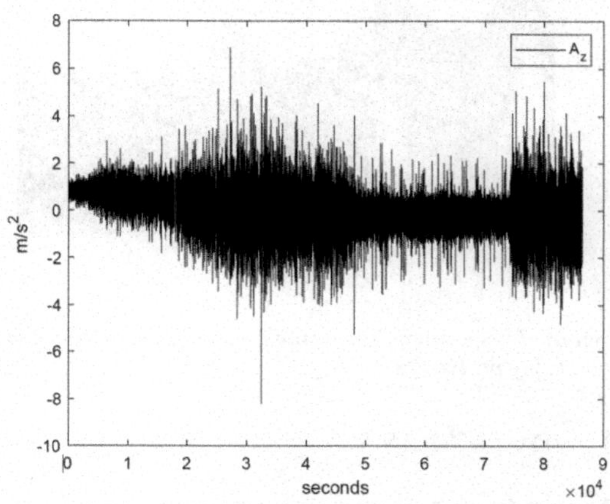

Figure 3. Vertical dynamic acceleration for the bow accelerometer.

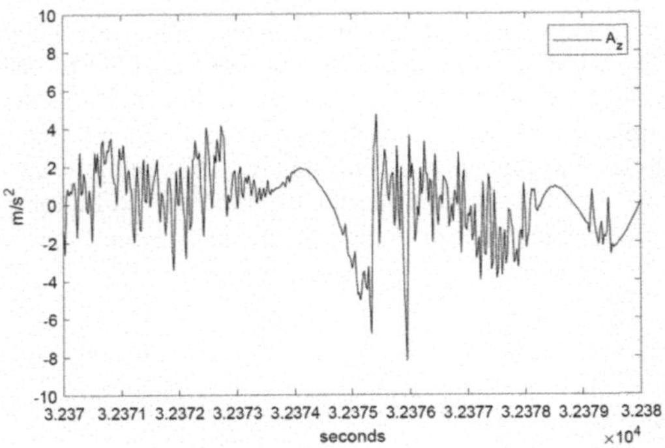

Figure 4. Vertical dynamic acceleration for the bow accelerometer – 10 seconds zoom in on one of the extreme events (unfiltered signal with mean removed).

Based on the many frequencies in Figure 4, it is relevant to look at the frequency spectrum by FFT, which is shown in Figure 5. The wave response occurs at an encounter frequency around 0.6 rad/s, i.e., 10.5 seconds. It is close to head seas due to the whipping level, and the vessel has forward speed so the peak wave period may be somewhat larger. There is however a clear 2-node vibration mode at about 3.2 rad/s, i.e., 0.51Hz and a second pronounced peak around 6.2 rad/s, i.e., 0.99 Hz, which is likely the 3-node vibration mode. Higher frequencies are more at a noise level, but with some goodwill there may be a third frequency at 9 rad/s, i.e., 1.43 Hz, which could be a 4-node, but from this, we may say that 4-node vibration is not significant overall, but from Figure 4 there are clearly peaks occurring more frequently than each second, but they occur apparently seldom based on the spectrum from the whole day. This suggest that some higher modes may not be visible from the spectrum.

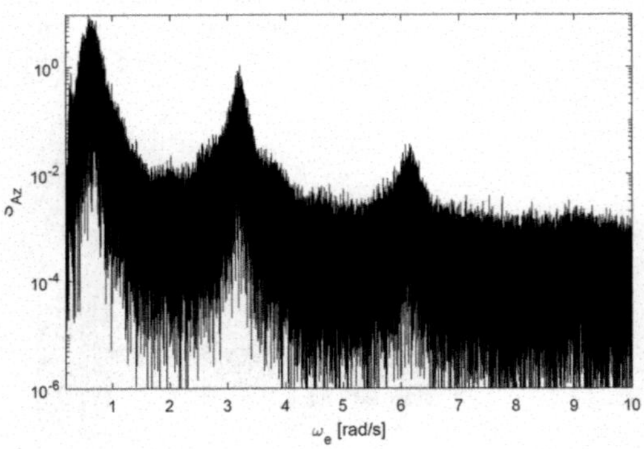

Figure 5. Spectrum of vertical acceleration for the whole day time series.

3 VDV RESULTS

As seen from the day of acceleration in Figure 3, the hourly VDV levels will vary, which is the nature of the non-stationary sea conditions. We are, however, after the maximum level per hour as a reference. The first frequency band chosen for the VDV calculation is from 0.4 Hz to 1.6 Hz (2.5 to 10.1 rad/s). This captures the two main vibration peaks in the spectrum seen in Figure 5. The results are shown in Table 1. The maximum VDVz is 2.06, however, as we see from Figure 4, higher frequencies are present in the time series now and then even though they are not revealed clearly through the spectrum, and since the acceleration is taken to the power of four, it is relevant to investigate the VDV level for a wider frequency band. Several choices are shown in Table 1, and the VDV level increases with increasing upper cut-off frequency. In these measurement data, the propeller and engine excitation are not evident from the spectrum, so it is OK to increase the upper cut-off frequency all the way to 20 Hz, even though the amplitudes at such high frequencies are questionable due to the sampling frequency of 42Hz. Based on the results, it is difficult to judge on a maximum expected VDV level as this is dependent on the upper cut-off frequency.

Considering a few more choices of the upper cut-off frequency, the expectation of maximum VDV value may be chosen as dependent on the upper cut-off frequency. For the worst hour of these 24 hours, the VDVz level is presented as a function of the upper cut-off frequency in Figure 6. The first jump of the upper cut-off frequency occur between 0.8 and 1.6 Hz is because 0.8 Hz omits the second vibration peak in the spectrum of Figure 5 related to the vertical 3-node mode. This second peak, off course, needs to be included; however, it was not that crucial. Then there is an increasing increase from 1.6 to 7.5 Hz and then a decreasing increase (second derivative is negative) from 7.5 Hz to something that looks converged at 15 Hz. Reducing the lower cut-off frequency below the lowest vibration mode from 0.4 to 0.3 Hz has limited effect, and with the upper cut-of frequency of 10 Hz, the maximum value is increased from 4.030 to 4.069, which is insignificant. The lower cut-off frequency is therefore not regarded as an issue provided that the lower vibration mode is captured.

Remembering that the propeller and engine excitation may occur at higher frequencies and considering some experience from past consultancy projects and frequencies from segmented model tests with a limited number of modes as well as the questionable quality of amplitudes because of the sampling frequency of 42 Hz an upper cut-off frequency of 10 Hz may be relevant and regarded as a maximum. The rule of thumb value for the maximum expected VDV at any location for any ship may then be about four (hourly VDV_{max}=4.0). (starting the hour a few minutes before or later would give a slightly different results so no point of adding more digits). This is ten times the threshold for discomfort ($VDV_{threshold}$ =0.4), according to Griffin (2012) and Engelbrecht and Becker (2023). Higher estimated VDV levels should be inspected for the presence of propeller or

Table 1. *VDVz* levels in m/s$^{1.75}$ for the 24 hours for several selected filtering frequency bands in Hz.

Hour	VDV	VDV	VDV	VDV	VDV
	0.4-1.6	0.4-3.2	0.4-5.0	0.4-10	0.4-20
1	0.2316	0.2369	0.2503	0.2812	0.2866
2	0.6177	0.6252	0.6490	0.7798	0.8995
3	0.7036	0.7048	0.7080	0.7208	0.8167
4	0.7153	0.7157	0.7169	0.7204	0.7234
5	0.7665	0.7763	0.7947	0.9662	1.0084
6	1.1156	1.1747	1.3306	1.6879	1.9758
7	1.6079	1.6853	1.9668	2.4644	2.7615
8	1.7415	1.8647	1.9943	2.7698	3.2380
9*	2.0631	2.2269	2.6716	4.0297	4.5276
10	1.9233	2.1242	2.5472	3.5676	3.9353
11	1.4064	1.6292	1.9386	2.7044	3.0386
12	1.0837	1.3375	1.7857	2.7163	3.2838
13	1.3415	1.5796	2.0922	2.9806	3.4521
14	0.6302	0.8840	1.2848	2.1366	2.8118
15	0.6342	0.8700	1.2486	2.0393	2.4829
16	0.3888	0.6359	0.9199	1.4985	1.8332
17	0.4336	0.5387	0.7842	1.4576	1.8223
18	0.4126	0.6373	0.9523	1.6349	2.0621
19	0.3716	0.5355	0.8172	1.2965	1.6416
20	0.3350	0.4963	0.8388	1.4441	1.8110
21	1.2432	1.4221	1.7728	2.5581	3.0095
22	1.5101	1.7540	2.2525	3.1158	3.5378
23	1.6333	1.8509	2.1813	3.1386	3.6524
24	1.3672	1.5722	2.0082	3.0761	3.4435

* Worst hour

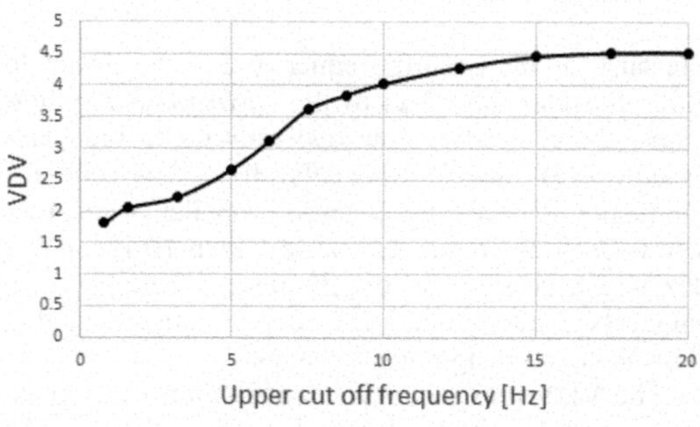

Figure 6. Maximum expected VDV in m/s1.75 as a function of upper cut-off frequency for the worst hour 9 in Table 1 of the considered measurements.

engine excitation, which may appear as very narrow peaks in the spectrum, and if higher values are achieved without additional propeller/engine excitation, there may be errors in the calculation.

It should be mentioned that the code for the calculation of the VDV has been checked against analytical calculations to verify that the weighting is correctly implemented. It is easy to make the weighting for a single frequency, but it is easy to make mistakes when weighting the spectrum, which is necessary before inverse FFT is used to reproduce the time series with weighted frequencies.

An example of the same 10 second time series from Figure 4 is given as a filtered time series between 0.4 and 8.75 Hz and the corresponding weighted time series (dashed line) in Figure 7. It may be seen that the different frequencies have different weighting and for the higher frequencies the difference is smaller according to Figure 1. Comparing Figure 7 and Figure 4 some false frequencies are introduced now and then in Figure 7. This is because of the abrupt cut-off frequency of 8.75 Hz in Figure 7 which happens to be close to the real frequency in the raw signal. This could be improved by less abrupt cut-off techniques and filters and by selecting the cut-off frequency away from physical frequencies. These false frequencies are gone if 20 Hz cut-off frequency is chosen, but this example is included to illustrate some of the traps, and that care needs to be taken. Possible inspection may be necessary not to remove too much physical frequencies.

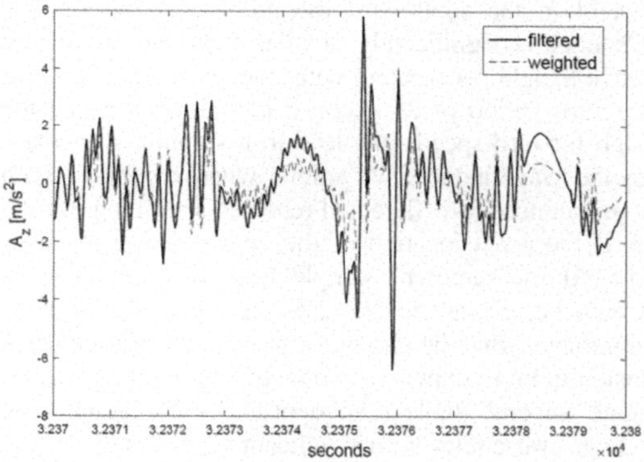

Figure 7. Vertical filtered (0.4 – 8.75 Hz) acceleration for the bow accelerometer and the corresponding weighted time series – 10 seconds zoom in on one of the extreme events.

The sea conditions during this event may be questioned. From Storhaug et al. (2024), it is already evident that this day belongs to extreme response levels even though the sea state may not be extreme. This is because the container ships may maintain a high forward speed in rough seas which in combination with pronounced bow flare, may give rise to high impacts and whipping events, which affect the maximum possible hull girder loading. This is known from other studies like Oberhagemann et al. (2015). From the vessel positions, the sea states could be retrieved, but unfortunately, the month around this date the specific GPS was not functioning (the vessel has several GPS devices but only one connected to the hull monitoring system), so accurate sea states estimates are not possible. However, when the vessel, based on the working GPS appeared in the Suez Channel on the 26th of November on the way to Asia, it suggests that 8-9 days earlier the ship was going through the English Channel passing Bay of Biscay and towards Gibraltar Straight in the southwest direction. It is known that coming out of the English Channel may imply close to head seas.

For illustration purposes, the hindcast weather data extracted from the IOWAGA1 dataset for the date of November 17th, 2010, at 21h00 is depicted in Figure 8 and Figure 9. The IOWAGA (integrated Ocean Waves for Geophysical and other Applications) dataset is developed by IFREMER (Institute Francaise de Recherche pour l'Exploitation de la Mer). More details of this data source can be found in Rascle and Ardhuin (2013), for example.

From Figure 8, it can be inferred that the maximum significant wave height on the evening of November 17th, over the entire globe, is recorded on the west coast of Ireland, at about 250 km from the shore, reaching a value of about 10.6 meters, see Figure 9. Then, Figure 9 shows that on the west border of the Bay of Biscay being the likely track of the specific vessel, the mean significant wave height is around 5.5 meters. Although the observed significant wave height in the south-west direction from the English Channel is significantly smaller than the significant wave height associated with the extreme loads for a return period of 25 years, it is well-known that the high forward speeds can lead to a significant increase in the wave-induced vibrations when combined with a pronounced bow flare and repeated slamming events.

If the exact location of the vessel during the considered measurement sample had been available, it would be relevant to correlate the VDV level against significant wave height, wave period and relative wave heading. From experience, bow quartering seas may be quite uncomfortable (experienced by the authors on a large cruise ferry with significant sea states exceeding 4 meters and with the captain taking appropriate measures) but also head seas may be severe. Even stern seas and beam seas may be significantly inducing stern slamming, but this is more relevant when the vessel is drifting, while stern slamming may become insignificant when the vessel has forward speed (above 8 knots reported by the crew in one case as below 8 knots was reported to increase the vibration level on a smaller container ship with tall and flexible superstructure due to stern slamming).

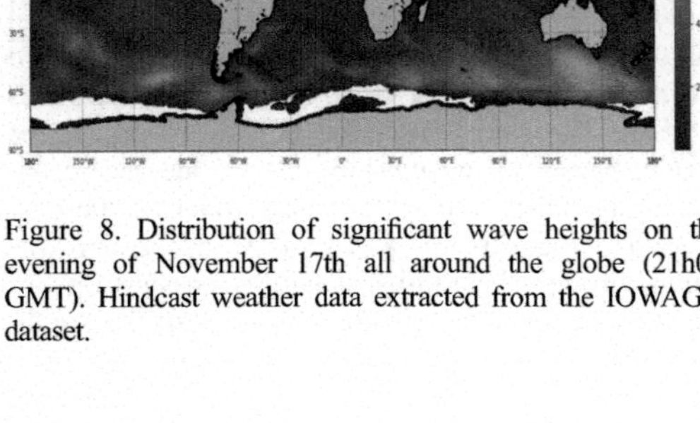

Figure 8. Distribution of significant wave heights on the evening of November 17th all around the globe (21h00 GMT). Hindcast weather data extracted from the IOWAGA dataset.

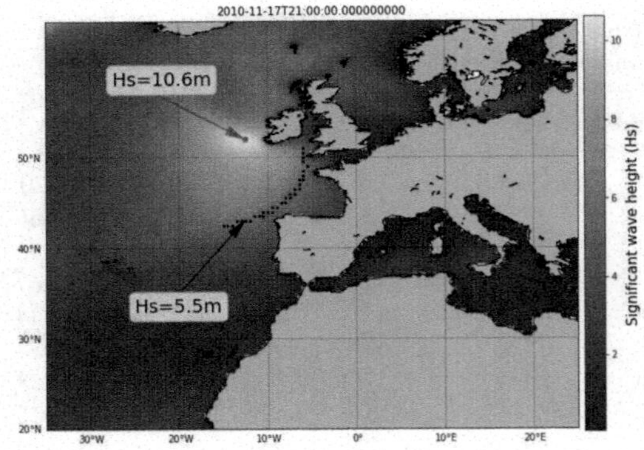

Figure 9. Local wave conditions around the English Channel on November 17th, 2010, at 21h00 GMT. Black dots indicate the isoline with a significant wave height of about 5.5 meters.

In any case it is evident that extreme VDV levels for this container ship was achieved at about 5-6 meters significant wave height. Higher sea states but a bit lower speed may give slightly higher VDV levels, but for this vessel higher dynamic acceleration levels has not been found based on measurements between 2009 and 2014, so hourly VDV_{max} of 4.0 is still considered as maximum possible, but it cannot be totally excluded that a similar vessel design with slightly worse conditions and a captain that disregards seamanship may achieve a slightly higher value.

4 DISCUSSION

Because of the cut off frequency and the desire to only consider the effect of the wave induced vibrations, there will be some uncertainties to the maximum VDV level especially if a low cut off frequency is necessary because propeller or machinery is exciting frequencies at less than 10 Hz. If 7.5 Hz is a necessary cut off frequency, then Figure 6 suggests a maximum VDV of 3.5 rather than 4.0. This is not an insignificant change.

The VDV may also have contribution from transverse and longitudinal direction. For the specific location at the bow the transverse and longitudinal component may be relatively minor especially since Equation 2 show that the three components should be combined with the power of four on each component. Let us assume that the vertical VDV is 4.0 and the horizontal components are 1.0 then the combined would be less than 1% higher. Hence, it is only when another component is of equivalent magnitude that there would be a real change in the total. If all three components would have a VDV of 4 then the combined would be only 31% larger, so basically the VDV concept is relatively insensitive to the other VDV components then the dominating one.

The location of the measurements matters however a lot. If the vertical sensor would be located at the two nodes of the vertical 2-node vibration mode, the VDV would only be affected by higher modes and the VDV level would be rather small and insignificant. If we look at Figure 2, the vertical 2-node mode suggest a vibration amplitude being half of the amplitude at the bow and at the stern. The frequency is the same, so the acceleration would then be half. Due to the acceleration being taken to the power of four in Equation 1 but then taken to the power of ¼ for the whole integration, an acceleration of half would reduce the VDV value to roughly half at the middle of the ship disregarding the higher vibration modes. The VDV level would then increase to the same level at the aft end again. The location of the measurements should therefore expected to be essential, and the location of the crew is essential in relation to what they are exposed to. Having low VDV levels where the crew and passengers are located is essential versus probability of complaints. Considering an oil tanker with a low superstructure relative to a container ship with a tall superstructure would imply a significant difference in the longitudinal VDV component not only because of different excitation and amplitudes of the hull girder vibration but also because of the distance from the center of rotation to the bridge. This makes longitudinal VDV component relatively more important high above the sea surface but also away from midship. From some other ship design and locations, it has been observed that the longitudinal VDV component have been important and dominating. On cruise ships with tall superstructures far forward the transverse VDV component in bow quarterings seas has been dominating. No other VDV levels as high as 4.0 have been experienced from other measurements so far.

5 CONCLUSIONS

Measurements of vertical acceleration from an 8600TEU container ship with pronounced bow flare have been retrieved. The bow flare is relatively extreme, implying relatively high whipping vibrations compared to other vessels, and the accelerometer is in the bow where all the vibration modes may have significant amplitudes. The measurement is taken from a day with severe weather and high response levels, so this is the foundation for establishing a rule of thumb value for extreme maximum hourly VDV possible onboard ships caused by wave-induced vibrations.

The VDV concept is explained, and the time series of acceleration is filtered and weighted depending on the frequencies (and direction), and the VDV related to impulsive loading from typical whipping events are calculated for each hour. It has been shown that the VDV level depends on the upper cut-off frequency, even though propeller and engine excitation cannot be seen from these measurements. Considering the sampling frequency and the possible contributions from engines and propellers, it is wise to consider VDV calculations based on a frequency band that covers the lowest vibration mode but excludes the propeller and engine excitation. An upper cut-off frequency in the range of 5 to 10 Hz may be relevant, and based on these measurements, a hourly VDV level of 4.0 m/s1.75 may be used as an expected upper maximum VDV level onboard ship for an upper cut-off frequency of 10 Hz. This is ten times the threshold of 0.4 which is regarded to represent the discomfort level for passengers according to Griffin (2012) and Engelbrech and Bekker (2023). Hourly VDV values above this are either caused by additional engine/propeller/equipment excitation or possibly a result of false calculations or sensor output, and for higher VDV levels proper inspection of time series and spectrum is needed as well as analytical check of the VDV calculation.

It should be emphasized that combined VDV levels for horizontal and vertical acceleration are relevant and not only the vertical VDV level, since the weight factor is high for horizontal acceleration at lower frequencies relevant for several of the vibration modes and that the vertical vibration mode also can give forward acceleration high above deck for tall and slender superstructures.

The frequency band suggested for the VDV calculation from wave-induced vibration ranging from about 0.3 to 10 Hz is way less than recommended by the ISO (1997) standard, but then again, getting rid of wave-induced vibration have drastic measures like changing the bow shape, while excessive vibrations from propeller, engine, and equipment at higher frequencies may be more common and convenient to mitigate. The effect of the latter may be included just by increasing the upper cut-off frequency to see their relative importance. For lower frequencies below the lowest vibration mode, sea sickness may become a topic. The proposed frequency ranges for VDV may be more relevant for the maritime industry while the ISO (1997) range is specified for any industry.

The comfort concept of VDV is regarded especially important for passenger vessels, cruise vessels, mega yachts, and ferries related to passengers but also for crew in relation to safety and human fatigue during execution of critical work tasks. These measurements confirm that the VDV level may be extreme on container ships, but further work needs to be done to conclude if the vibration level on the bridge and superstructure of such vessels are adequate. Even more interestingly, if the VDV level on the bridge is correlated with seamanship, so for container ships, it may simply be mitigated in practice by reducing the speed a few knots when feeling uncomfortable. This may then also affect the safety of the hull girder by reducing the whipping response as well. So, comfort may be related to safety, even for structural strength.

ACKNOWLEDGEMENTS

The authors would like to thank DNV for funding the measurement campaign on container ships, including this specific container ship. Thanks also to the master, operator, owner, and yard for their cooperation and assistance in facilitating the instrumentation and collecting the measurement data to shore. Light Structures AS also did a good job of instrumenting the vessel on behalf of DNV.

The opinions of the authors may not be the formal opinion of DNV.

REFERENCES

Bekker, A., Soal, K.I. & McMahon, K. J. 2017. Whole-body vibration exposure on board a Polar Supply and Research Vessel in open water and in ice, *Cold Regions Science and Technology*, 141, 188–200, September 2017.

Duarte, M.L.M., Araújo, P.A. & Carvalho, L.A.P. 2018. Correlation between weighted acceleration, vibration dose value and exposure time on whole body vibration comfort levels evaluation, *Safety Science*, 103, 218–224, March, 2018.

Engelbrecht, M. & Bekker, A. 2023. A discomfort threshold for impulsive whole-body vibration on a slamming prone vessel, *journal of Applied Ergonomics*, Volume, 109, 103992, 12 pages

Griffin, M.J. 2012, *Handbook of human vibrations*, Academic Press, ISBN 9780080984407, (first edition in 1996), 1008 pages

ISO 1997. Mechanical vibration and shock – evaluation of human exposure to whole-body vibration – Part 1: General requirements, International standard ISO 2631-1 Second edition 1997-05-01.

Heggelund, S.E., Storhaug, G. & Choi, B-K. 2011. Full scale measurements of fatigue and extreme loading including whipping on an 8600TEU Post Panamax container vessel in the Asia to Europe trade, *Proc. of the ASME 30th Int. Conf. on Ocean, Offshore and Artic Engineering*, OMAE2011–49378, 19-24 June, 2011, Rotterdam, Holland.

Lorenzino, M., D'Agostin, F., Rigutti, S., Bovenzi, M., Fantoni, C. & Bregant, L. 2022. Mood regulates the physiological response to whole-body vibration at low intensity, *Applied Ergonomics*, 108, 103956, 26 December, 2022.

Matsumoto, Y. & Kunimatsu, S. (2022). Evaluation of human perception thresholds of transient vibrations for the assessment of building vibration, *Applied Acoustics*, 197, 22 July, 2022, 108906.

Oberhagemann, J., Shigunov, V., Radon, M., Mumm, H. & Won, S.I. 2015, Hydrodynamic load analusis and ultimate strength check of an 18000 TEU containership, Hyel 2015, *7th international conference on hydroelasticity in marine technology*, Split, Croatia, Sept 16-19, 2015.

Omer, H. & Bekker, A. 2018. Human responses to wave slamming vibration on a polar supply and research vessel, *Applied Ergonomics*, 67, 71–82, February, 2018.

Storhaug, G., Choi, B-K., Moan, T., & Hermundstad, O.A. 2010, Consequence of whipping and springing on fatigue for a 8600TEU container vessel in different trades based on model tests, *Proc. of the 11th Intern. Symp. on Practical Design of Ships and Other Floating Structures*, PRADS2010–2048, Rio de Janeiro, Brazil.

Storhaug, G., Holtmann, M., Jagite, G., Wang, S., Dessi, D., Pais, T., Takahashi, H., Dhavalikar, S., Sagvolden, G., Diego, V. & Kim, S.J. 2024. Uncertainties in predicted quantities based on measured time series of dynamic response, ISOPE-2024, *The 34th International Ocean and Polar Engineering Conference*, 2024-TPC-0619, Rhodes, Greece, 16-21 June 2024.

Rascle, N. & Ardhuin, F. 2013. A global wave parameter database for geophysical applications. Part 2: Model validation with improved source term parameterization. *Ocean Modelling*, 70, 174–188.

Advances in Maritime Technology and Engineering – Guedes Soares & Santos (Eds)
 ISBN 978-1-032-83099-5

An approach to damage identification of ship hull structures in irregular waves

H.Y. Tang
Navigation College, Dalian Maritime University, Dalian, P.R. China
Centre for Marine Technology and Ocean Engineering (CENTEC), Instituto Superior Técnico, Universidade de Lisboa, Lisbon, Portugal

Z.C. He & D.Y. Ren
Navigation College, Dalian Maritime University, Dalian, P.R. China

C. Guedes Soares
Centre for Marine Technology and Ocean Engineering (CENTEC), Instituto Superior Técnico, Universidade de Lisboa, Lisbon, Portugal

ABSTRACT: In order to improve the safety of ship navigation, intelligent monitoring systems are installed on ships and provide early warning of potential hazards in hull structures. A hull structural damage model is established based on the finite element method, and the vertical bending deformation of ship structure during navigation is simulated in time domain due to the wave loads in irregular waves. A ship structure damage identification method is then studied by combining the particle swarm optimization and machine learning technology, and automatic damage identification is finally achieved in the monitoring system. This research can effectively improve the monitoring reliability of ship structure and provide technical support for ship intelligent navigation.

1 INTRODUCTION

With the establishment of global economic integration, international trade has achieved rapid development, which greatly drives the technological progress of the shipping industry. Intelligent ships based on computer and communication technology have become one of the important directions for the development of ship safety technology today. To observe and evaluate the hull structural condition in long-term navigation, some ship structure monitoring systems have been developed. Intelligent monitoring systems can monitor hull structures in real-time and evaluate structural damage in complex sea cases. Through the implementation of these monitoring systems, captains can get a reliable decision-making basis (Slaughter et al. 1997), which enhances the risk prevention ability of ships.

To improve ship survival ability in harsh ocean environments, Lindemann & Odland (1978) developed a simple monitoring system to measure the condition of ship motion and structure stress. Their research is a meaningful attempt. The ship and marine data system company (S&MDS) collaborated with Lloyd's register of shipping (LR) in crude carriers under the application of structural monitoring systems, and the long-term structural monitoring on actual ships was proven to be feasible (Cuffe, 1993). The Bureau Veritas (BV) also collaborated with some companies in British and Greece to monitor and evaluate the structure status in a ship comprehensive prediction system (Li, 1994). The extraction technology on wave characteristics from ship motions was achieved by Pascoal & Guedes Soares (2009), and a decision support system was developed by Perera et al. (2012) to provide advices on ship handling in bad weather, which had a monitoring system based on long strain gauges as one of its components. Some regulations and standards on monitoring systems were further proposed by American Bureau of Shipping (ABS, 2015) and Det Norske Veritas (DNV, 2011).

With the widespread application of fibre optic sensors in the field of monitoring, Hjelme (1997) designed a dynamic monitoring system for observing the hull hydroelasticity fluctuation in waves. A ship structure monitoring system for catamarans by combining the wireless communication technology with some small-scale sensors is also achieved by Lynch & Loh (2006).

According to these successful hull health monitoring cases, the ship structural monitoring and evaluation techniques had received a lot of attention. Liu & Guan (2002) designed a monitoring and evaluation system for an oil tanker. The hull vertical deformation of tankers along the ship length longitudinal is observed by some simplified beam sensors.

DOI: 10.1201/9781003508762-61

For bulk carriers, Zhang (2008) developed a structural condition monitoring system using some optical fibre sensors. Jin (2008) also used containers as monitoring objects to develop an intelligence system. Through the multi-signal parallel computing and database technology, Ren (2013) expanded the ship structure monitoring range, which incorporates both the hull longitudinal strength assessment and local yield assessment into the monitoring system.

To solve the system stability of long-term monitoring, Zhang et al. (2019) conducted a comprehensive analysis of the fluctuation characteristics of monitoring stress signals and provided corresponding denoising methods. Tang et al. (2020) optimized the sampling frequency to improve the monitoring efficiency of floating production storage and offloading (FPSO) units.

At present, the hull structure monitoring system, as a foundation for intelligent ships, has been applied on many ships (Lou, 2021). However, the identification of structural damages based on monitoring signals is still relatively difficult, which reduces the effectiveness of monitoring systems in hull structural evaluation. Therefore, it is necessary to establish a feasible method to identify potential hull structural damage.

In the present research, a structural damage identification method is studied using the finite element method and machine learning. A hull structural model with different damage degrees and positions is established, and the vertical deformation of this structural model in irregular waves is simulated further to obtain the fluctuation of structural response signals at different positions. The hull curvature difference is then calculated by the response signal, and it is used to train an artificial neural network. Finally, the structural damage in the structural model can be rapidly recognized and forecasted using a back propagation (BP) neural network. In addition, the particle swarm optimization (PSO) algorithm has also been applied to determine the relevant parameters of the BP neural network, thereby further improving the stability of neural network prediction.

This research on the autonomous identification of hull structure damage is one of the preliminary stages to make the monitoring system intelligent, and it offers a technical solution to quickly detect changes in monitoring signals caused by structural damage. The structural damage identification method will also make the diagnosis of hull structural conditions more reliable in unmanned ships.

2 FRACTAL DIMENSION CALCULATION

The fractal dimension originated from the fractal theory of geometry and is an important branch of modern mathematics. In the past two decades, numerous damage identification methods based on fractal theory have obtained some good results in fields such as civil engineering and aerospace. The fractal box dimension is one of the common fractal dimensions, it is widely used in the analysis of time-domain hull structural vibration.

2.1 *Calculation of fractal box dimension*

In a numerical space (*F*, *d*), *R* is a nonempty set of *F*. *B*(*f*, *ε*) is assumed as a closed sphere with the centre *f* and radius*ε*, *A* is another nonempty subset of *F*, and *N*(*A*, *ε*) is the number of smallest closed spheres in *A*:

$$N(A,\varepsilon) = \left\{ M : A \subset \bigcup_{i}^{M} B(f_1,\varepsilon) \right\} \tag{1}$$

where f_i is the *i*-th point of the numerical set *F*; *g*(*x*) is defined as a continuous function of the closed set *T*, which is included in the numerical set *R*. Then, *G* is regarded as a set of R^2:

$$G = \{(x,y) : x \in T \subset R, y = g(x) \subset R\} \subset R^2 \tag{2}$$

D(*g*) is defined as the fractal dimension of *g*(*x*), and is shown as follows.

$$D(g) = \lim_{\varepsilon \to 0} \left\{ \sup \frac{\log N(G,\bar{\varepsilon})}{-\log \bar{\varepsilon}} : \bar{\varepsilon} \in (0,\varepsilon) \right\} \tag{3}$$

In Equation (3), *sup* represents the set supremum, which means that all elements in the set are not higher than the value of this supremum. Wang (2005) proposed a computational expression for a discrete spatial signal set from a time-domain vibration response. In the sampling period *T*, there is a series of time-domain vibration signals (x_1, x_2,···, x_n), and the fractal box dimension of the signal is expressed as follows.

$$D = -\frac{\ln\left(\sum_{i=1}^{n-1} |x_{i+1} - x_i| / \Delta t \right)}{\ln(\Delta t)} \tag{4}$$

where *n* is the total sampling number of signal data and *Δt* is the sampling interval. The value of the box dimension is related to the total sampling time and its interval. When the sampling time is longer, the box dimension will increase. When the time interval shortens, the calculated value will be closer to the actual fractal dimension. So, the collection time should include at least three load application cycles, and the collection frequency should also be increased appropriately.

2.2 *Curvature difference based on box dimension*

When the hull structure is damaged, the total structure stiffness will change, and the vibration response signals at each monitoring position will also fluctuate. According to the definition of fractal dimension, it is known that the fractal dimension value of the monitoring point near the damage location will appear with different fluctuation characteristics. The central difference method is applied to calculate the fractal dimension curvature before and after structural damage, and the curvature difference in fractal dimension is also calculated:

$$C_i = \frac{D_{i-1} - 2D_i + D_{i+1}}{l^2} \tag{5}$$

$$\Delta C_i = C_i^b - C_i^a \tag{6}$$

where D_i is the box dimension at monitoring point i, l is the distance between adjacent monitoring points, and C_i represents the curvature value at monitoring point i. ΔC_i is the curvature difference at the monitoring point i. C_i^b is the fractal dimension curvature value before the damage happened, and C_i^a is the fractal dimension curvature value after the damage formed.

3 MACHINE LEARNING TECHNOLOGY

Various intelligent algorithms emerged in the 20th century, and these algorithms are attempted to be used in structural damage identification. The application of artificial neural networks (ANN) often gets relatively satisfactory results among intelligent algorithms. Artificial neural networks are inspired by the structure of human brains and include some powerful memory functions, which make the networks be able to handle complex nonlinear problems and have the characteristics of high adaptability and self-organization. They have been used in various applications related to structural reliability (Chojaczyk et al. 2014), and their good prediction ability also makes the ANNs an excellent tool for structural damage identification.

3.1 *Back propagation neural network*

Due to different computational frameworks and transfer functions, artificial neural networks are divided into a series of subclasses, such as self-organizing competitive neural networks, radial basis function neural network method and BP neural networks. The BP neural network has some advantages in the identification of complicated signal characteristics because it has excellent numerical stability and strong nonlinear mapping ability (Tang et al. 2023). Nowadays, neural networks have already been used to model the wave-induced loads in ships (Moreira & Guedes Soares, 2020), in connection with onboard decision support systems (Perera et al. 2012).Therefore, the BP neural network is attempted to be further applied in the identification of structural damages.

The BP neural network includes three layers which are the input layer, hidden layer and output layer. In each layer, there are a series of numerical neurons. X, K, Y denote the neurons in the input layer, hidden layer, and hidden layer, respectively. Their subscripts (M, I, J) indicate specific locations of the BP neural network. The weight function is a numerical tool to connect these neurons in different layers. The BP specific framework is shown in Figure 1. Generally, the Sigmoid function is used as the connection function in a hidden layer, while the linear transfer function is taken in the output layer:

$$f(x) = \frac{1}{1 + e^{-x}} \tag{7}$$

$$g_1(x) = \omega_{mi}x + a_m \tag{8}$$

$$g_2(x) = \omega_{ij}x + b_m \tag{9}$$

where ω_{mi}, a_m are the connection weight parameters of neurons between the input layer and hidden layer; ω_{ij}, b_m are the connection weight parameters of neurons between the hidden layer and output layer. When a time-domain vector $X(t)$ inputs into the BP neural network, a result vector $Y(t)$ will be calculated, according to the transfer function and the connection weight parameters. So, the value of connection weight parameters is critical for the effectiveness of the BP neural network.

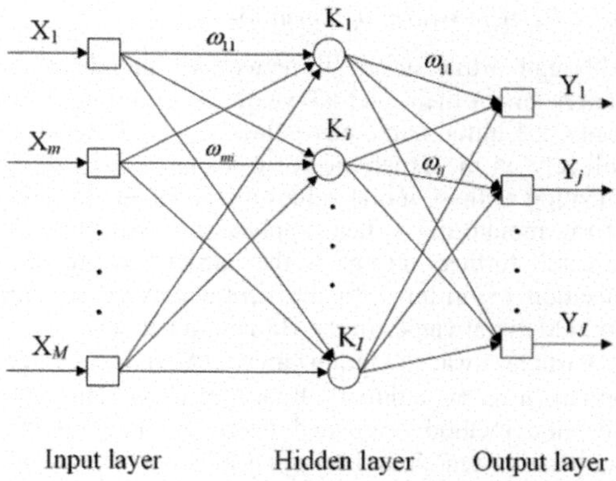

Figure 1. Framework of BP neural network.

The determination of weight parameters mainly relies on BP neural network training, which is a conventional method of supervised learning. The neural network training is an iterative calculation aimed at approaching real data. When the signal forward propagation is taken in the neural network, the total error of the BP neural network is recorded:

$$e(n) = \frac{1}{2}\sum_{j=1}^{J} e_j^2(n) \tag{10}$$

where $e_j(n)$ denotes the j-th neuron error of the output layer after the n-th iteration. To find the error minimum of the BP neural network, the connection weight of the neural network is further justified in the error back-propagation. The connection weight in the BP neural network is optimized by using the steepest descent method. The changes of weight parameters at the n-th iteration are then obtained:

$$\Delta\omega_{ij} = -\eta\frac{\partial e(n)}{\partial\omega_{ij}(n)} \tag{11}$$

$$\Delta b_i(n) = -\eta\frac{\partial e(n)}{\partial b_i(n)} \tag{12}$$

$$\Delta\omega_{mi}(n) = -\eta\frac{\partial e(n)}{\partial x_i(n)}X_m(n) \tag{13}$$

$$\Delta a_m = \eta K_i(1-K_i)X_m\sum_{m=1}^{M}\omega_{mi}(n)e(n) \tag{14}$$

where η is the learning rate; X_m is the m-th input vector in the input layer, and K_i is the i-th input vector in the hidden layer. Under the optimization of weight parameters, the calculated output in the BP neural network will be closer to the expectations in the training. The neural network training is an automatic optimization process. It finds the minimum error between the predicted values in BP and actual signal values, and the weight parameters is ultimately determined under this minimum error condition.

3.2 *Particle swarm optimization*

Although artificial neural networks can adjust the parameters of their internal weight functions automatically, the initial parameter values of neural networks still rely on the experience of researchers. The determination of these initial values increases the difficulty of the monitoring system application. Therefore, the research further introduces the particle swarm optimization algorithm into the neural network to find optimal initial values of the BP neural network.

Particle Swarm Optimization (PSO) is a global optimization evolutionary algorithm (EA). This optimization method originated from the study of bird cooperative search and foraging behaviour (Bonyadi, 2017). The particle swarm optimization algorithm is also based on iterative calculation and has the characteristics of a simple structure framework and easy implementation. In the particle swarm optimization, the optimization problem of artificial neural network parameters is transformed into a particle position problem. Based on the global and local particle optimal solutions, the algorithm will iteratively find the most suitable position for the designed particles.

The theoretical community is assumed to have h particles, which are distributed in the L-dimensional space for search behavior. The vectors of the i-th particle position and velocity are expressed as:

$$Z_i = \{z_{i1}, z_{i2}, \cdots, z_{ih}\} \qquad (i = 1, 2, \cdots, L) \tag{15}$$

$$V_i = \{v_{i1}, v_{i2}, \cdots, v_{ih}\} \qquad (i = 1, 2, \cdots, L) \tag{16}$$

The optimal position found by the i-th particle:

$$p_{best} = (p_{i1}, p_{i2}, \cdots, p_{ih}) \qquad (i = 1, 2, \cdots, L) \tag{17}$$

The global optimal position searched by the entire particle swarm is also calculated:

$$g_{best} = (g_1, g_2, \cdots, g_L) \tag{18}$$

In t-th iterative calculation, the update equation for particle velocity and position is shown as follows.

$$\begin{aligned} v_{ij}(t+1) = \gamma v_{ij}(t) + c_1 r_1(t)\left[p_{ij}(t) - x_{ij}(t)\right] \\ + c_2 r_2(t)\left[p_{gi}(t) - x_{ij}(t)\right] \end{aligned} \tag{19}$$

$$z_{ij}(t+1) = z_{ij}(t) + v_{ij}(t+1) \tag{20}$$

where γ is the inertia weight, it denotes the degree to which particles retain their search speed at the last iteration. c_1 and c_2 are the learning factors, used to constrain the maximum step size during update calculation. r_1 and r_2 are uniform random numbers in the range [0,1] to enhance the randomness of particle search. Finally, the initial weights and thresholds of the BP neural network can be determined using a particle swarm optimization algorithm.

4 HULL DAMAGE IDENTIFICATION TEST

4.1 *Simulation of hull structure deformation*

In the field of ship and ocean engineering, the finite element method is an effective way for static and dynamic analysis of hull structures, which has been widely used. The finite element software ABAQUS is applied to build the hull structural model and calculate the strain changes under designed wave loads. These stress fluctuations are assumed to be signal data received by the ship structural monitoring system.

The finite element model of the hull structure midships for a typical bulk carrier is established. The bulk carrier's length is 300m, and its width is 32.3m. The depth and design draft of the hull is 24.9m and 8.4m, respectively. The construction of the midship model mainly includes some critical structure components such as decks, longitudinal frames, girders, and hull shell, as shown in Figure 2. This model adopts shell element (element types of S3R and S4R) in ABAQUS. The total number of the hull model is 23436. The element material is set to the common steel with a yield limit of 235MPa. In the finite element model, the nodes at two ends are constrained to move as two whole sections. These terminal sections are further constrained separately in degrees of freedom. One terminal section of the hull model is constrained by the rigid fixed condition, so this section cannot be moved on the six degrees of freedom. Another terminal model section only allows vertical rotation around the hull neutral axis.

The wave load simulation is calculated using potential flow theory (Tang et al. 2023). In potential flow theory, the fluid around the target ship is considered as the ideal fluid, which can be represented by some velocity potential functions. The information of hull shape, weight distribution, stiffness distribution towards this bulk carrier is collected and the boundary

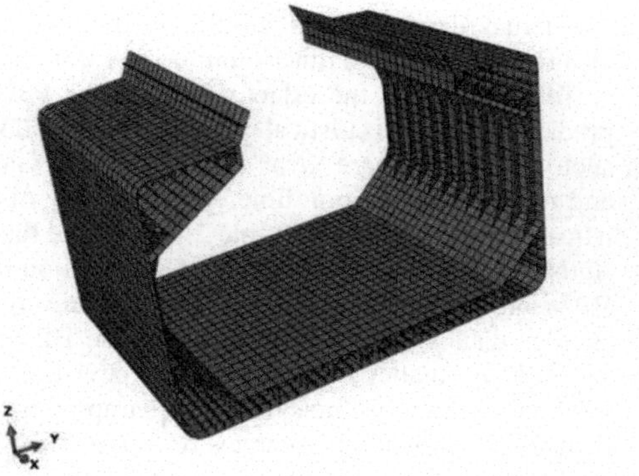

Figure 2. Finite element model of hull structure.

condition of flow field is defined. The velocity potential functions and the hull motion differential equation are solved by using the boundary element algorithm. The ship motion and hydrodynamic forces are finally simulated. According to D'Alembert principle, the bending moment amidships can be obtained in regular waves. The bending moment in irregular waves are the superposition of bending moment in different regular waves by using the ISSC wave spectrum. The wave spectrum is evenly divided into multiple parts in the frequency domain to determine the relevant parameters of regular waves. The bending moment in these designed regular waves is calculated, and the bending moment in irregular waves is synthesized finally. The time histories of midship vertical bending moment in irregular waves is shown in Figure 3.

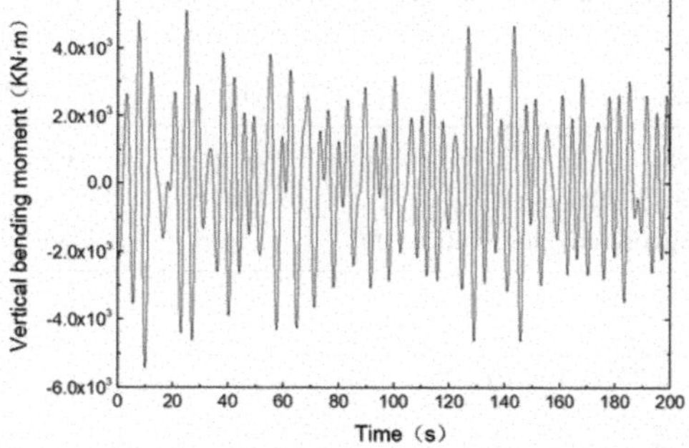

Figure 3. Time histories of vertical bending moment.

According to the hull section modulus and the calculated vertical bending moment in waves, the rotation angle on the terminal section is obtained and imposed on the finite element model. With the model terminal section rotating frequently, the finite element model will appear hogging and sagging deformation, as shown in Figure 4.

In addition, there are 49 uniformly distributed monitoring points to record the time-domain deformation fluctuation. Although these points are uniformly distributed, it is also recognized that a smaller number of points can be chosen in the future by optimizing their location as done for example by Liu et al. (2018) in an offshore platform case.

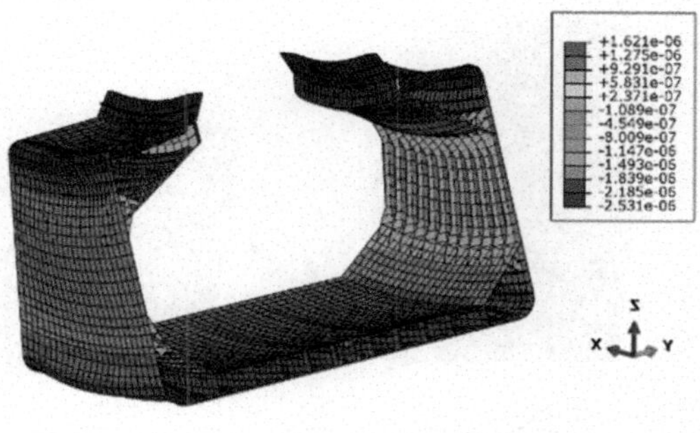

Figure 4. Hull structural sagging deformation.

Due to the randomness of wave load in irregular waves, the simulated bending moments in whole simulated time (200s) were divided into four parts, and the BP neural network was trained and tested at each part (50s). The curvature difference of structural deformation was calculated. These recorded data of curvature difference were inputs to BP neural network, and the hull structural response rule was learned by the BP neural network. The final evaluation of the BP neural network is the average results on the four time periods.

4.2 *Establishment of structural damage*

The hull structural damage, which is caused by corrosion, ageing, and fatigue, is mainly focused on the research. In some applications, the damaged members were removed from the model as they were no longer effective (Hussein, & Guedes Soares, 2009; Luis et al. 2009). In the present work, the damage is assumed to be less intense than in the previous situations and thus, the weakening of structural resistance ability after structural damage is assumed to be represented by a decrease in elastic modulus. Because the structural damage has already formed and fixed, The acceleration effect of nonlinear loads on structural damage is ignored at present. Noise interference is an important challenge in structural monitoring. It is also assumed that the singal noise has been eliminated through an excellent filter currently.

In structural damage simulation, the proportion of structural elastic modulus loss is 20%, 40%, 50% and 60%, respectively. The structural damage ratios are also regarded as the same value as the proportion of elastic modulus loss. The structural damage was considered to occur separately at these twelve locations (P1~P12), as shown in Figure 5. The vertical bending moment in irregular waves is considered not to be affected by the existence of the damage and it is applied to these structural damage models, and the dynamic curvature difference of the structural deformation was also recorded. Table1 shows the case numbers at different positions and damage ratios.

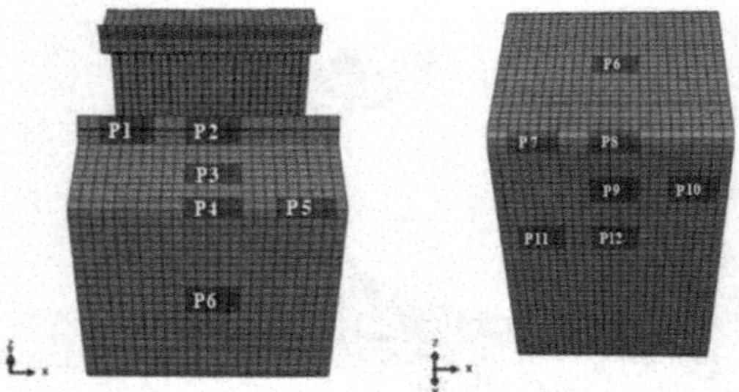

(*a*) top view (*b*) bottom view

Figure 5. Hull structural deformation.

Table 1. Statistics of case numbers at different positions and damage ratios.

	Damage ratio				
Position	0%	20%	40%	50%	60%
P1	No.1	No.2	No.3	No.4	No.5
P2		No.6	No.7	No.8	No.9
P3		No.10	No.11	No.12	No.13
P4		No.14	No.15	No.16	No.17
P5		No.18	No.19	No.20	No.21
P6		No.22	No.23	No.24	No.25
P7		No.26	No.27	No.28	No.29
P8		No.30	No.31	No.32	No.33
P9		No.34	No.35	No.36	No.37
P10		No.38	No.39	No.40	No.41
P11		No.42	No.43	No.44	No.45
P12		No.46	No.47	No.48	No.49

4.3 *Training and testing of neural network model*

There are 49 hull structural response cases, which include the deformation simulation from the hull intact model (No.1 case) and damage model (No.2~ No.49 case). The training data is obtained from the simulation of the hull intact model and random 43 damage models. The simulation data of the 5 unselected damage models are used as the testing data.

In the BP neural network, the number of neurons in both the input and output layers is 49, and the number of neurons in the hidden layer is 14. The initial value of the weight function in the BP neural network is determined using the PSO algorithm. The population size of particles is 50, and the dimension of particles is 3150. The maximum number of iterations for particle evolution is 100 in the PSO method.

After the training and optimization of the BP neural network, the BP network model can output the structural damage proportion at the corresponding position. Figure 6 shows the comparison results of damage identification at the positions of P3, P6, P9 and P12. The damage identifications by these two methods are both lower than the actual damage ratio (50%), but the damage identification by combing the BP neural network and PSO algorithm is closer to the actual damage. This similar phenomenon has also been observed in other monitoring locations.

To demonstrate the effectiveness of the PSO-BP prediction, some statistical parameters are adopted such as mean square error (MSE), max deviation, and network recognition time. The summary of statistical results is shown in Table 2. It is found that the statistical error values of damage identification using BP neural network and PSO algorithm are always smaller than the error values just by the BP neural network, no matter on MSE or max deviation. The PSO algorithm is beneficial for improving the damage identification accuracy of neural networks. In addition, the optimized neural network also requires less running time.

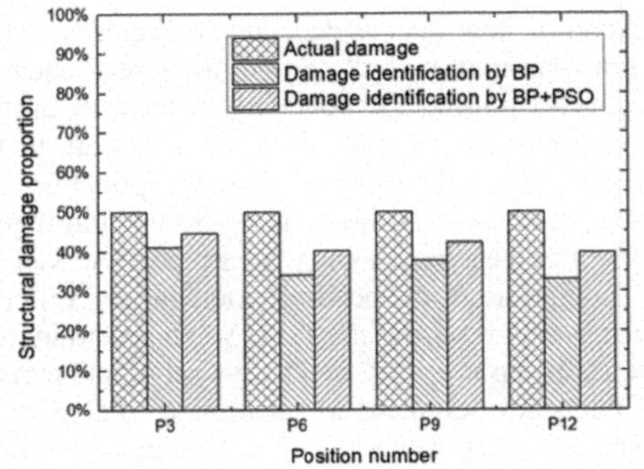

Figure 6. Comparison of damage identification proportions with different methods.

Table 2. Statistics of structural damage identification.

	Method	
Indicator	BP	PSO+BP
MSE	0.017	0.004
Max deviation	0.492	0.189
Network recognition time (s)	8.736	7.461

5 CONCLUSIONS

In the paper, the dynamic response on the vertical deformation of hull structure amidships is simulated by using the finite element method and potential flow theory. Combining particle swarm optimization with BP neural network technology, a damage identification method is finally established. The simulation of hull structural damages at different monitoring positions can be identified rapidly by using this intelligent method. In the damage identification analysis, it is found that the particle swarm optimization is useful. Its application can improve the accuracy of artificial neural networks towards hull damage identification. In addition, the efficiency of the monitoring system based on particle swarm optimization can be also improved. This research will provide preliminary technical support for subsequent hull structural warning and auxiliary decision-making in unmanned ships.

ACKNOWLEDGEMENTS

This research was supported by the National Natural Science Foundation of China (No.52201353), the Fundamental Research Funds for the Central Universities (No.3132023125), and the Innovation and Entrepreneurship Special Item from Dalian Maritime University Teaching Reform Project in 2022 (No.125). The first author has benefitted from this support to visit the Centre for Marine Technology and Ocean Engineering (CENTEC). This work contributes to the Strategic Research Plan of the Centre for Marine Technology and Ocean Engineering (CENTEC), which is financed by the Portuguese Foundation for Science and Technology (Fundação para a Ciência e Tecnologia) under contract UIDB/UIDP/00134/2020.

REFERENCES

American Bureau of Shipping. 2015, Guide for Hull Condition Monitoring Systems. Houston, USA.

Bonyadi, M. R., & Michalewicz, Z. 2017. Particle swarm optimization for single objective continuous space problems: a review. *Evolutionary Computation*. 25 (1): 1–54.

Chojaczyk, A. A., Teixeira, A. P., Neves, L. C., Cardoso, J. B., & Guedes Soares. C. 2014. Review and application of artificial neural networks models in reliability analysis of steel structures. *Structural Safety*, 52(3), 78–89.

Cuffe R.W. 1993. Transducers for Ship's Hull Stress Monitoring. *The Institution of Electrical Engineers*, 1–5.

Det Norske Veritas. 2011, Hull Monitoring Systems. DNV Rules for Classification of Ships/High Speed, Light Craft and Naval Surface Craft, Part 6, Chapter 11.

Hjelme, D.R. 1997. Application of Bragg grating sensors in the characterization of scaled marine vehicle models, *Applied Optics*, 36: 328–336.

Hussein, A. W. & Guedes Soares, C. 2009. Reliability and Residual Strength of Double Hull Tankers Designed According to the new IACS Common Structural Rules. *Ocean Engineering*. 36(17-18)1446-1459.

Jin, Y.X., Hu, X, & Shi, C.J. 2008. Container ship structural condition monitoring and evaluation system, *Journal of Shanghai Maritime University*, 29 (03): 1–4.

Li J.X. 1994. Europe Jointly Developing Ship Monitoring System, *Technology and Economy Information of Ship Bulldings Industry*. 11:40–41.

Lindemann, K., & Odland, J. 1978. On the application of hull surveillance systems for increased safety and improved structural utilization in rough weather, *Instrumentation*, 2–36.

Liu, C.M., & Guan, Y.H. 2002. Introduction of VLCC cargo oil integrated control system, *China Ship Repair*, 30–33.

Liu, K.; Yan, RJ., & Guedes Soares, C. 2018. Damage identification in offshore jacket structures based on modal flexibility. *Ocean Engineering*. 170171–185.

Lou, S.T. 2021. Development of an online fatigue assessment and strength monitoring system for ship structure, *Ship Science and Technology*, 43(3), 40–41.

Luís, R. M.; Teixeira, A. P., & Guedes Soares, C. 2009. Longitudinal Strength Reliability of a Tanker Hull Accidentally Grounded. *Structural Safety*. 31(3)224-233.

Lynch, J.P., & Loh, K.J. 2006. A summary review of wireless sensors and sensor networks for structural health monitoring, *Shock and Vibration*, 91–128.

Moreira, L.& Guedes Soares, C. 2020. Neural Network Model for Estimation of Hull Bending Moment and Shear Force on Ships in Waves. *Ocean Engineering*. 206; 107347.

Pascoal, R., Guedes Soares, C. 2009. Kalman Filtering of Vessel Motions for Ocean Wave Directional Spectrum Estimation. *Ocean Engineering*, 36(6-7): 477–488.

Perera, L. P.; Rodrigues, J. M.; Pascoal, R., & Guedes Soares, C. 2012. Development of an onboard decision support system for ship navigation under rough weather conditions. In: Rizzuto, E. & Guedes Soares, C., (Eds). *Sustainable Maritime Transportation and Exploitation of Sea Resources*. Taylor and Francis Group, UK; pp. 837–844.

Ren, H.L., Jia, L.H., Zhen C.B., Liu N., & Li, J.N. 2013. Design of Ship Structural Stress Monitoring and Strength Assessment System, *Journal of Wuhan University of Technology*, 35(8):60–65.

Slaughter, S.B., Cheung, M.C., Sucharski, D. & Cowper, B. 1997. State of the art in hull response monitoring systems. Ship Structure Committee, *Report SSC-401*, Washington, DC.

Tang, D., Chen, J., Wu, W., Jin, L., & Feng, J. 2020. Research on sampling rate selection of sensors in offshore platform SHM based on vibration. *Applied Ocean Research*, 101, 102192.

Tang, H.Y., He, Z.C., Zhang, X.K., & Wan, Q. 2023. Autonomous Recognition of a Ship Structure Characteristic Signal Based on Machine Learning, *International Journal of Offshore and Polar Engineering*, 33(2): 174–183.

Tang, H.Y., Wan, Q., & Ren, H.L., 2023. Numerical study of trimaran wave load based on time-domain Rankine method, *Brodogradnja*, 74(3):107–129.

Wang, B.Y. 2005, Detection of structural damage using fractal dimension technique. *Journal of Vibration and Shock*, 24(2):87–88.

Zhang, L. 2008. A Ship Structural Health Monitoring System Based on Fiber Bragg Grating Sensing Technology, *China Water Transport*, 8(4),260-263.

Zhang Z.L., Zhu X.J., Peng F., & Mu J.L. 2019. Characteristic analysis and de-noising method of stress monitoring data of hull structures based on HHT. *Chinese Journal of Ship Research*, 14(1): 158–164.

Advances in Maritime Technology and Engineering – Guedes Soares & Santos (Eds)

Initial structural analysis of a Portuguese 'Muleta' lateen mainsail yard

S.D. Viegas & L.S. Sutherland
Centre for Marine Technology and Ocean Engineering (CENTEC), Instituto Superior Técnico, Universidade de Lisboa, Lisbon, Portugal

P.H. Miller
Department of Naval Architecture and Marine Engineering, U.S. Coast Guard Academy, USA

ABSTRACT: The 'Muleta' was a traditional Portuguese fishing boat. The last remaining example is the replica 'Álvaro Velho' whose 16.5 m wooden lateen mainsail yard failed. This paper concerns the ongoing work to design a composite material replacement. A simple bending only analysis design gave a 40% in weight reduction. To include the effects of shear, a finite element analysis approach has also been undertaken. A baseline numerical model of the strength and stiffness of the original replica yard (conservatively of Douglas Fir) has first been developed, and then very successfully corroborated with the previous simplified analytical model, giving confidence in both approaches. Development of the numerical model to consider shear effects and to give a new, refined composite materials design is underway. The aim is to facilitate economically viable and longer lifecycle composite yard replacements for both the Muleta and the numerous other traditional sailing craft of the Tejo.

1 INTRODUCTION

The 'Muleta' was a traditional fishing boat from the Tagus (Lisbon River, the Tejo) estuary, emblematic of the towns of Seixal and Barreiro and readily identifiable by its unique 'cloud of sails' and characteristically rounded bow (Figure 1).

The name 'Muleta' is not thought to come from the obvious literal English translation of 'crutch', but perhaps from the wooden stick used to extend a small red cape used in bullfighting, which together are reminiscent of the small sails extended along the bowsprit and long bumpkin at bow and stern, respectively. Another theory is that the origins are from the Latin 'motilus' in the sense of 'sluggish' describing the motion whilst fishing, as described below (Leitão et al., 2009).

In fact, the name 'Muleta' was also used generically for different types of craft using the same fishing method, such as the 'Botes da Tartaranha', and 'Batéis do Seixal', often referred to as 'Muletas Modernas' ('modern' as opposed to the 'old' round-bowed 'Muletas Antigas') (Carrasco and Peres, 1997) as shown in Figure 2 with a more traditional bow profile.

'Tartaranha' refers to the type of fishing method and net used by the Muletas (Figure 2), which is a form of pelagic (midwater) trawling performed with the boat 'hove to' and drifting slowly abeam (beware erroneous translations to 'beam trawling' which is not the case here) to leeward with the net extending to windward and pulled through the water via lines from both sides to the ends of the bowsprit and bumpkin (which are also visible in Figure 1).

Figure 1. A 'Muleta' (Laird Clowes, 1936).

This fishing method explains the spectacular array of small sails extended fore and aft of the boat - they allow fine-tuning of the fore and aft position of the sail plan's centre of effort whilst trawling, thus enabling control of the vessel's sideways movement through the water against the variable currents acting on the large fishing net (as well as the windage and current forces acting on the vessel itself). Any sailor will immediately

DOI: 10.1201/9781003508762-62

recognise that this would have still required an extremely expert level of seamanship. When sailing to and from the river mouth fishing grounds, however, only the lateen (also 'bastardo' in Portuguese) rigged mainsail was used, and the main yard supporting this mainsail is the subject of the present paper.

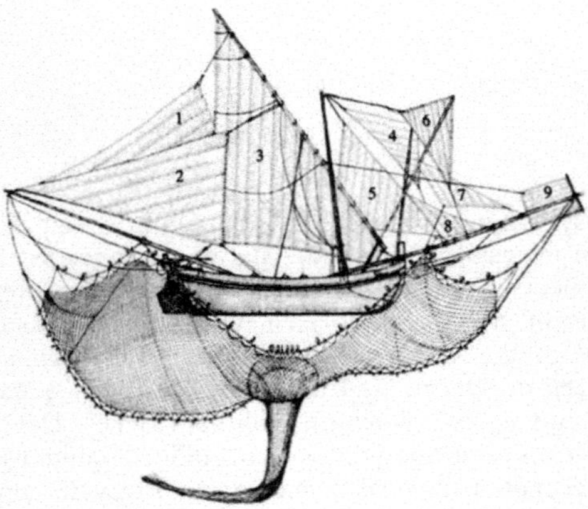

Figure 2. A 'Modern' Muleta or 'Bote da Tartaranha' (Curtinhal, 2007).

More on the fascinating history of these extraordinary boats may be found in (Baldaque da Silva, 1895; Gomes Pedrosa, 2016; Leitão et al., 2009) but sadly by 1901 there was only one working 'Antiga' Muleta and by 1928 no 'Botes de Tartaranha' of any form were left sailing (Câmara Municipal Barreiro, 2018).

However, in 2018 Jaime Ferreira da Costa & Irmão, Lda Construções e Reparações Navais constructed a replica Muleta (the 'Velho Álvaro', ref) for the City of Barreiro. The design was based on the plans of the Portuguese Naval Museum (Museu de Marinha) and the amazing depth of information and plans in (Leitão et al., 2009), and mostly based on contemporary measurements, plans and drawings published in (Pâris, 1882) from those made by Raphael Monléon (curator of the Naval Museum of Madrid) and provided by Dr Fonseca Vaz (Portuguese Marine Minister) (Leitão et al., 2009).

This boat is run as a tourist vessel by Nosso Tejo Lda (Figure 3), and whilst sightseeing trips are only undertaken under motor, the wooden main yard failed and broke as the crew were moving the vessel under sail. The Authors were commissioned to design a replacement yard in fibre reinforced polymer (FRP) composites, as Dr Sutherland (with Dr Nuno Fonseca & Eng° Mário Figueiredo) had previously done for the replica caravel Vera Cruz (Sutherland, 2023). The progress made to date is outlined in the present paper.

Table 1. Principle dimensions of replica Muleta 'Álvaro Velho'.

Dimension	(m)
LOA (including rudder)	14.6 (15.4)
Beam	4.1
Depth (to cap rail)	1.7 (2.0)
Draft	0.8
Main Yard	16.5

Figure 3. The replica Muleta 'Álvaro Velho'.

2 DESIGN METHODOLOGY

In general, the same successful approach as used for the design of the 27 m long carbon yards for the Vera Cruz was applied here; since loading and geometry were both non-standard and variable (the yard is in bending around the mast, and is then free to rotate about all three axes), and since these craft 'evolved' in a 'survival of the fittest' sense over the years, the characteristics of the 'original' Vera Cruz replica wooden yards were used as a baseline. However, this was not as straightforward as it may seem; what do we mean by the 'original' yard with respect to the Muletas?

The 'original' wooden yard supplied with the 'Álvaro Velho' was hand-crafted to dimensions according to the passed-down experience of Mestre Jaime Costa (an at least fourth-generation master boat builder with over half a century of experience (Moreira, 2018)). Measurements of this (broken) wooden yard gave length and maximum diameter of 16.5 m and 187 mm (between 10 and 13 m from the tip, the 'mast contact zone'), respectively - which are just 5% less than the diameters from the historical Museu de Marinha plans (which had to be scaled for length differences and since the yard was not dimensioned explicitly on the drawings calibrated measurements had to be used, and hence the diameters from the plans are themselves estimates). Hence the measurements taken from the 'original' replica wooden yard were assumed to be typical of a 'historic' Muleta yard.

However, unlike the historical wooden spars (which would have been fabricated from single pieces of timber since bonding using contemporary adhesives was not an option) the original replica wooden yard was shaped from spruce planks glued together using modern epoxy adhesives.

Further, would the original spars have been made from spruce, which became popular in the early part of the 19th century in yachting circles due to its lighter weight, but suffers from accelerated rotting? Eucalyptus, although available in Portugal from around 1860 (Cabral, 2019) and stiff and strong, is extremely heavy - plausibly too heavy to have been historically used. 'Pinheiro-bravo' (Maritime Pine), the native Portuguese pine species, seems to be the most likely to have been used historically.

Taking all these previous points into account, the design decision was made to use Douglas Fir as a conservatively stiff and strong pine species for the timber assumed, and the dimensions of the wooden replica spar as originally supplied by Jaime Costa when calculating the strength and stiffness behaviour of the baseline 'original' wooden yard. This design decision was thought to be appropriate even though this original (spruce) yard failed, understandably, at the mast (12 m from the tip), and, according to the crew, not under testing conditions.

Further investigations revealed that the yard had been stored unprotected for a number of years open to the hot Portuguese climate and hence environmental degradation became a prime suspect for this failure rather than under dimensioning of the spar. Also, although epoxy was used for joining the planks used together, finger joints within the planks made at the mill could not be assumed to be of such high performance resins and were hence easily degraded.

Additionally, the illustrations in Figures 22 and 23 of (Leitão et al., 2009) show very thick twisted traditional rope padding on both yard and mast at their intersection, yet Figure 3 indicates that only a thin layer of rope protection was used on the Álvaro Velho yard, and none on the mast. Further, on inspection of the numerous traditional craft in Seixal, the use of a single layer of neatly coiled modern synthetic braided rope (much 'harder' than traditional hemp twisted rope) was seen to be the norm. Further, impacts between yard and mast is now thought to be the main cause of the (repeated) failure of the original eucalyptus yards of the Vera Cruz. Finally, Figure 3 clearly shows the yard not hauled to the top of the Álvaro Velho's mast - as it is in Figures 22 and 23 of (Leitão et al., 2009) - but suspended from around 1 m of the halyard, allowing significant impacts between yard and mast, perhaps more so when sailing in lighter winds with any seaway running – the conditions when the yard broke. Hence, impact with the mast was thought to be another contributing, if not major, factor leading to the premature failure of the wooden replica Muleta yard.

Hence, the wooden replica yard, if not environmentally degraded, but well-padded and hauled right up to the hounds, would likely not have failed, and provides a suitable design baseline.

The Muleta is a smaller and far less stiff vessel than the Vera Cruz and the tip sections of the Muleta's yard significantly finer, meaning both that the maximum possible (in normal use at least) loadings will be lower and the ability to 'spill the wind' by twisting off of the head of the sail will be greater for the Muleta.

Also, it must be remembered that the use of the current vessel, that of tourist trips in light conditions at or above Lisbon, contrasts starkly with that of the original working boats that spent most of their time in the 'Mar da Muleta' ('Muleta Sea', the area between Capes Espichel and Roca) close to the edge of the continental shelf and open to the full force of the Atlantic Ocean, all year round, day in day out.

Hence, the new FRP design will be based on the strength and stiffness of a solid Douglas Fir yard with the dimensions of the original yard of the 'Álvaro Velho', an approach thought to be sufficiently conservative taking into account all of the points discussed above.

Previous experience with the 'Vera Cruz' (Sutherland, 2023) has shown that the use of FRP will both effectively eliminate rot and degradation (leading to a far longer lifecycle and hence significantly reduced environmental impact) and especially reduce weight. This latter point is of great advantage, both for the stabilising effect of less weight aloft, but also, very importantly for this tourist application (and as already confirmed for the youth sail training Vera Cruz), in the ease in which this enables the sail to be raised. This is acutely important for a tourist vessel where a short-handed crew is normal to maximise space for paying visitors, and the ability to raise the sails provides a clear advantage over the commercial competition.

Due to the ongoing (15 years) success of the 'Vera Cruz' FRP spars, the same simplified analytical design method is initially used here, as outlined in Section 3. However, since this method is known not to include the potentially significant shear stiffness and strength provided by the off axis plies, this will be refined via a Finite Element Analysis (FEA) to try to make the design less conservative (Section 4). This is thought to be justified due to the predicted light use and careful approach of the crew in such a small boat full of paying tourists.

Further, since the yard will be constructed (in-house with Eng. Mário Figueiredo), previous experience will be used to give a more 'designed for production' approach to reduce material and manufacturing costs - essential for the small company operating this historic vessel. One of the main differences from the Vera Cruz project is that to avoid the huge amount of costs associated with manufacturing bespoke moulds, the original (repaired) wooden yard will be used here as a mandrel (a cylindrical male mould).

In this paper, this ongoing work is detailed, including the simplified analytical design and the FEA model of the 'original' wooden spar to be used as a baseline for the composite spar design.

3 SIMPLIFIED ANALYTICAL DESIGN

The approach taken here is to ensure that the FRP yard is as stiff and at least as strong as the original wooden yard at metre-spaced 'stations' along the length of the spar assuming classical beam theory (Gere and Timoshenko, 1991). These calculations and the final results (after the five design cycles which were needed to obtain the required behaviour) are only outlined below in the limited space available here.

'Stiffness' is evaluated as beam Flexural Rigidity, EI, where E is the Young's Modulus and I the local cross-section second moment of area. 'Strength is evaluated as the bending moment to rupture:

$$M_r = \frac{I\sigma_r}{c} \tag{1}$$

where, σ_r is the failure strength and c the (maximum for minimum M_r) distance between the outer surface of the circular beam and the neutral axis (in this case the maximum radius, r).

Second moments of area for these circular sections were calculated using:

$$I = \frac{\pi r^4}{4} \tag{2}$$

The original wooden yard was of length 16.5 m and maximum diameter 187 mm. The measurements taken are compared with those from the Museu de Marinha Muleta plans and measurements of the mizzen yard of the Vera Cruz caravel in Figure 4.

Spruce (lighter) and also Douglas Fir (stiffer and stronger) material properties were used to give relevant bounds for the wooden spar behaviour as shown in Table 2.

Table 2. Wood material properties.

	σ_r (MN/m^2)	E (MN/m^2)
Spruce[1]	75	11000
Douglas Fir[2]	85	13400

[1](Swedish Forest Industries Federation, 2023) [2](Ross, 2010).

The calculated wooden yard bending stiffness and strength behaviours are presented in Figure 5 and Figure 6, respectively, but first, however, a necessarily brief outline of the design of the composite yard is described below.

Throughout the raw materials selection process the local availability of specific materials was often a deciding factor. Since the original yard will be used as a mandrel the dimensions of the composite yard are fixed. To reduce weight as far as possible, carbon fibre was selected as the main structural fibre reinforcement, which makes the choice of Epoxy resin as the matrix an obvious one to ensure the benefits of the carbon are not compromised. As a suitable cost to properties solution and facilitating layup of the fibres around the yard, 200 g/m^2 unidirectional (UD) standard modulus carbon was specified, and to save labour costs a 600 g/m^2 stitched multi-directional (0° /-45/ 90/ +45) architecture was used where possible.

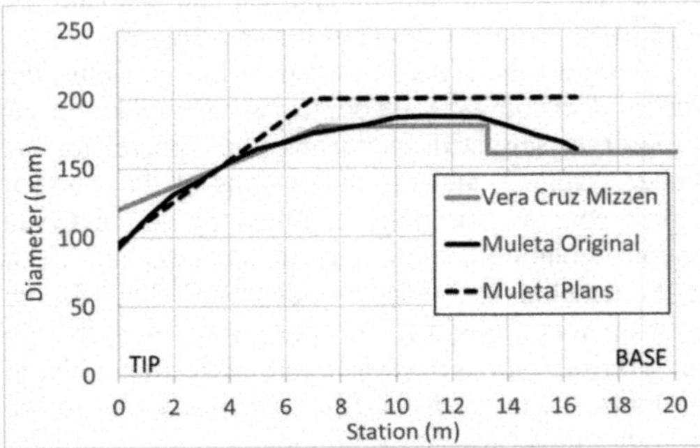

Figure 4. Yard dimensions.

Since the yard will undoubtedly often hit against the mast and elsewhere, two outer, non-structural 300 g/m^2 plain weave E-glass fibre plies were incorporated to absorb impact damage and protect the brittle carbon fibres. A single interior glass ply (the first to be laminated against the male mould) gave a degree of 'fairing' to the hand-made wooden yard surface, avoiding 'crimping' of the following UD carbon plies.

Since the yard will be infused, Sicomin SR 8100/ SD 882X Epoxy Resin Infusion (low viscocity) System was used which is DNV-GL (TAK00001FU) and LR (MATS/4612/1) approved. Previous experience has shown that a Fibre Weight Fraction (FWF) of 0.7 is attainable, equating to Fibre Volume Fractions (FVF) of approximately 0.6 for the UD carbon and 0.5 for the woven glass plies; values which are 'typical'. FVF is calculated from FWF via:

$$FVF = \frac{1}{\left[1 + \frac{\rho_f}{\rho_m}\left(\frac{1}{FWF}\right) - 1\right]} \tag{3}$$

where ρ_f and ρ_m are density of fibre and matrix, respectively (Shenoi et al., 1993).

Identifying the FVF values allowed selection of appropriate sources for the material property data required for calculations. Manufacturer's data sheets generally provided some properties for suitable FVF example laminates, and missing values were taken from those of 'AS4/3501-6 UD Carbon/Epoxy' and '120/3501-6 Woven Glass/Epoxy' from Tables A.4 & A.5 in (Daniel and Ishai, 2006) and from (USA Department of Defense, 2002, p. 2). FVF also allows calculation of the ply cured thickness, d, from:

$$d = \frac{A_W}{\rho_f \, FVF} \tag{4}$$

where A_W is the reinforcement areal weight (weight per unit area) (Shenoi et al., 1993).

The laminate stiffness and strength could now be calculated, and 'The Laminator' software ("The Laminator: Classical Analysis of Composite Laminates," 2023) was used for this (Miller, 2006). In terms of the detailed design process this cannot be detailed here due to space restrictions and commercial sensitivities, but the main design constraints were taken from the previous experiences of the authors and GL 'Design and Construction of Large Modern Yacht Rigs' (Germanischer Lloyd, 2009):

- Use a well-balanced laminate
- Groupings of the same plies < 1.5 mm thick
- Avoid grouping 90° plies
- Compressive Young's modulus E: Transverse > 20% Longitudinal
- In plane shear stiffness > 11% Longitudinal compressive Young's modulus
- Avoid steps or bonded structures with significant different stiffness (peel stresses)
- Fibre-dominated (0°/±45°/90°) Layup recommended with minimum 10% in each direction
- Quasi-isotropic laminate in the zone around the mast contact.

The 'base' yard laminate was reinforced around the mast 'zone' to achieve a quasi-isotropic layup, with this extended to the yard base to minimise 'pumping' of the spar in a seaway and to keep the centre of gravity low for easier manhandling. The

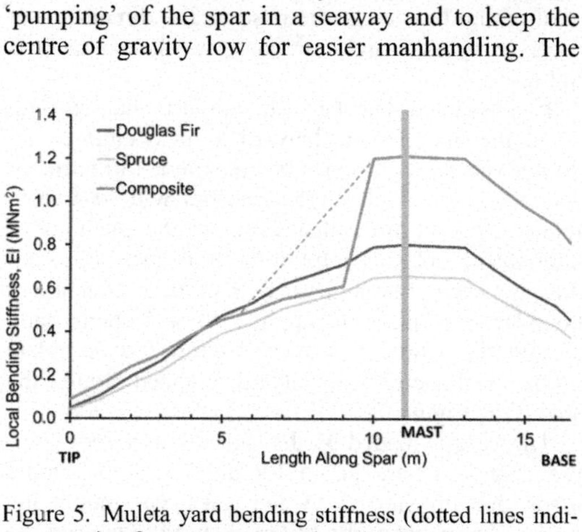

Figure 5. Muleta yard bending stiffness (dotted lines indicate tapering from mast zone to base laminate).

mast zone reinforcing laminate was tapered into the base laminate to avoid sudden stiffness changes.

Calculation of the laminate modulus and strength values enabled the same Classical Beam Theory approach as for the wooden spar to be carried out here, and compared with the latter results to complete an iterative design process. The results of this process are summarised in Figure 5 and Figure 6.

Figure 5 Shows that the stiffness of the composite yard is at least as high as those of the wooden ones in its upper part, meaning that sailing characteristics should be agreeable. As designed, the lower part of the composite yard is much stiffer than those of the wooden ones, indicating better motions and manoeuvrability.

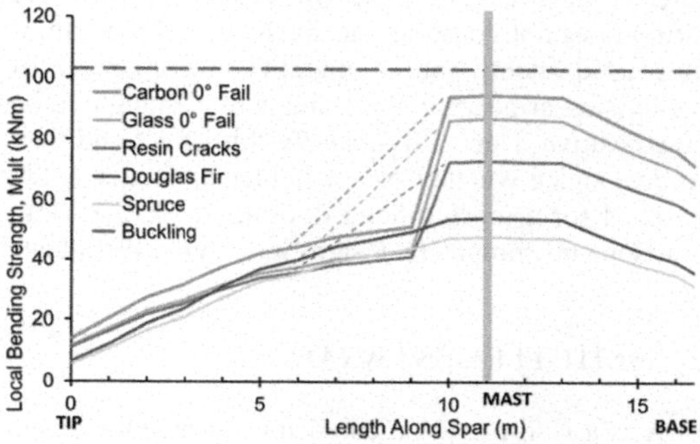

Figure 6. Muleta yard bending strength (dotted lines indicate tapering from mast zone to base laminate).

The different failure lines in Figure 6 deserve some explanation. Resin cracking, between 90° fibres, although to be avoided due to possible adverse effects on durability, does not indicate failure of the yard. The fact that this is never significantly lower than the spruce final failure values shows that this should not become a problem.

The first failure of 0° plies is the first 'real' failure, but this does not correspond to the final failure of the yard but just the very beginning of the progressive failure process as more and more fires and plies fail – the yard will still have significant load bearing capacity in reserve.

Further, since the glass fibres are non-structural, their failure is not of concern, and in fact could be beneficial in identifying if the yard has been overstressed before catastrophic failure occurs (c.f. wooden spars, which crews are accustomed to, which 'complain' before they fail, allowing time to back off and/or carry out repairs).

Hence, the main failure of interest is that of the first 0° Carbon ply, which Figure 6 shows to be consistently stronger than even the Douglas fir yard in its upper part, and significantly higher than, almost double even, in its lower part and especially at the critical intersection with the mast – and it must be remembered that this is only the very first stages of failure, and so there is also an additional large 'built-in' Safety Factor. Also relevant is that the wooden yards already have a substantial built-in, 'process of evolution', safety factor to account for non-prescribed use and events, which are of course part and parcel of sailing.

The design also passes a conservative (i.e. not considering the internal bulkheads which will be included at the mast zone) check for buckling failure of a thin walled circular tube under transverse bending (Roark et al., 2002) as also indicated in Figure 6.

Finally, the weight of this composite solution yard is only around 60% of that of a wooden spar, and hence the design is considered to be sufficiently strong as well as having the desired stiffness profile.

However, this simple approach (whilst proving its effectiveness via 15 years of service for the Vera Cruz) does not consider shear effects and the laminates used incorporate a significant proportion of off-axis shear bearing fibres and is thus thought to be conservative. Hence, to quantify this effect and then to investigate whether or not lighter laminates could be used for a more refined design, an FEA of both wooden and composite designs has been undertaken.

Table 3. Material properties of Douglas Fir used in FEA.

Property	Value (Ross, 2010)
Young's Modulus X, E_x	1.34×10^{10} (Pa)
Young's Modulus Y, E_y	9.11×10^{8} (Pa)
Young's Modulus Z, E_z	6.70×10^{8} (Pa)
Shear Modulus XY, G_{xy}	8.576×10^{8} (Pa)
Shear Modulus YZ, G_{xz}	1.045×10^{9} (Pa)
Shear Modulus XZ, G_{yz}	9.380×10^{8} (Pa)
Poisson's ratio XY, ν_{xy}	0.292
Poisson's ratio YZ, ν_{yz}	0.390
Poisson's ratio XZ, ν_{xz},	0.449
Density	438 (kg/m^3)
Tensile Strength X, σ_{Tx}	1.08×10^{8} (Pa)
Tensile Strength Y, σ_{Ty}	2.30×10^{6} (Pa)
Tensile Strength Z, σ_{Tz}	2.30×10^{6} (Pa)
Compressive Strength X, σ_{Cx}	4.99×10^{8} (Pa)
Compressive Strength Y, σ_{Cy}	5.50×10^{6} (Pa)
Compressive Strength Z, σ_{Cz}	5.50×10^{6} (Pa)
Shear Strength XY, τ_{xy}	1.40×10^{6} (Pa)
Shear Strength YZ, τ_{yz}	7.80×10^{6} (Pa)
Shear Strength XZ, τ_{xz}	1.40×10^{6} (Pa)

4 FINITE ELEMENT ANALYSIS

FEA is a well-proven method for advanced analysis of marine structures (Miller, 2006). For this project the ANSYS software package was used to numerically model the Muleta's lateen spar. At the current stage of this study a solid model of the original wooden spar based on the dimensions in Figure 4 has been developed.

Given the slow response rate to loads, linear material properties and moderate deflections, a linear static analysis was performed with large deflection (membrane stress effects) included. As through-the-thickness shear stress results were desired for the failure analysis of the solid wood spars, solid elements were used. The solid models consisted of 39971 nodes and 24002 "Solid187" elements. This solid element is a higher-order pyramidal element with 10-nodes. It can model anisotropic, orthotropic or isotropic materials.

The first step in the FEA analyses was to check the basic approach and implementation against a simplified isotropic un-tapered (with average diameter of the actual tapered spar) closed-form solution for a fixed ended, 3-point beam bend loaded at the centre using a representative force. Nominal material properties (Aluminium 6061-T6) were used. The FEA predicted deflected shape was as expected, and the predicted deflections, stress results, volume and mass closely matched the closed-form solution (Roarke et al., 2002), giving confidence that the FEA model was correctly constructed.

As discussed in section 2, the wood species considered for design purposes here was Douglas Fir, whose properties are shown in Table 2 and were taken from (Ross, 2010).

The boundary conditions applied in the FEA were based on observations from the yard rigged for sailing in photographs, e.g. Figure 3, drawings and plans, e.g. (Leitão et al., 2009), and measurements taken from the 'Velho Álvaro'. The yard has two main attachment points; at its base and at the mast.

For analysis purposes, here the yard is modelled horizontally, and x defined as along the yard (longitudinal), y as up and down and z as port and starboard (Figure 7). At the base of the yard (in Figure 3 shown forward of the mast on the deck, but on the left in the following FEA analysis figures), two lines attached to the base are made fast to the deck to port and to starboard preventing translations in the y and z direction and rotation about the x-axis. This boundary condition is applied to a 2.5 cm band around the yard circumference at the base as the dimension of the existing notch in where these two lines attach (B in Figure 7).

It is assumed that the yard should be hauled right up to the mast by the halyard as in (Leitão et al., 2009) where it will hence be constrained in terms of displacements in all three axes, but will be free to rotate about all three planes (due to the elasticity of the halyard and especially of the protective rope padding between mast and yard). This mast attachment boundary condition is applied to a 5.08cm band around the whole yard circumference to simulate a typical diameter rope halyard wrapped around the spar (C in Figure 7).

The yard is loaded by the sail, but in a very complex manner where the sail lift and drag aerodynamic loads are transferred to the yard via membrane forces. The sheet at the sail clew, used to control the sail also gives a reaction force to these aero-dynamic forces, which are transferred to the yard. It is well beyond the scope of this study to try to evaluate the loading on the spar, so a worst case where all the load is applied to the tip of the yard via the sail from the sheet (measured to be at a 40-degree angle to the yard – to the horizontal in the FEA axis system) was used (A in Figure 7). This not only a conservative approach but also a valid one since the objective of this analysis is not to predict the actual breaking loads but to ensure that future composite material solutions are at least as stiff and strong as the original wooden yard. Also, since this was a comparison between analysis methods, for simplicity inertial effects were ignored.

The FEA model was then used to validate (or otherwise) the bending failure predicted by the simplified analytical method of section 3. Applying a nominal 1.0 kN force gave a stress of 13.9 MPa and a factor of safety (FOS) for the tensile stresses in the x-direction (longitudinally) of 7.72, at 5.83 m from the base, very close to the mast (at 5.76 m from the base). A further check was that the same stress was reported at the same distance from the base (but on the other side of the yard), but due to the much higher compressive strength of Douglas Fir, the corresponding FOS was 35.80. Although the large displacement analysis used leads to nonlinearities, a simple linear calculation was used to give a predicted loading of 7.72 kN for an $FOS_{tens,x}$ of 1. As expected this gave an FOS of close to but not equal to 1, and this value was then used to make another linear prediction of the failure load in longitudinal tension, and so on until this iterative approach converged.

The result of this FEA, for an $FOS_{tens,x}$ of 1.00, was 6.70 kN at 5.83 m from the base. Figure 8 (where tensile stress is defined as positive) shows that as expected, this longitudinal stress occurs on the upper surface of the yard and that the maximum compressive stress occurs on the lower surface at the same longitudinal position, close to the mast.

The simple analytical approach of section 3 resulted in a maximum ultimate bending moment at the mast of 54 kN.m which was constant for 1.5 m either side of the mast. Using this failure bending moment value and the (maximum) bending lever at the mast of a load applied at the yard tip (16.5 m from the base), and simple trigonometry to give the force required at the same 40° to the yard as in the FEA analysis resulted in a failure load of 6.78 kN. This is a difference of little over 1% compared to the FEA result, thus delivering confidence in both FEA and the simple bending only analysis of section 3.

Figure 8 also shows the real deflected shape of the yard at bending failure, and the maximum deflection, at the tip, is 6.09 m. This appears to be a large deflection, but the deformed shape shown is certainly not excessive for a wooden spar.

However, so far the FEA model has only been used to predict the bending failure of the yard, but it was developed to investigate the possibility of shear and other effects on its behaviour. Now that the FE and simplified bending analyses have been corroborated, work is currently underway to include these aspects into the design.

5 CONCLUSIONS AND FURTHER WORK

A simple bending only analysis based design of a replacement lateen yard for a replica 'Muleta' traditional sail fishing boat has been completed. This design used the strength and stiffness of the original replica wooden yard as a baseline. In this, the conservative assumption of a Douglas Fir spar was also used alongside that of the actual Spruce of the wooden spar.

A design of a carbon and glass fibre reinforced epoxy composite material was then developed to give the same bending characteristics in the upper part (for good sailing characteristics) and increased strength (and hence stiffness) in the lower part, and especially at the mast where maximum bending moments occur, and the original wooden replica yard broke. This breakage was thought to be due to a combination of environmental degradation and/or unintended impact with the mast. This design resulted in a composite spar of just 60% the weight of the wooden one.

However, this simple analytical pure bending approach obviously does not consider the effects of shear, which are thought to be significant due to the off-axis fibre plies required for the composite solution.

Hence, a Finite Element analysis (FEA) has been undertaken to include shear, and other stresses, which again uses the original replica wooden yard as a baseline, but in this case only with the conservative assumption of Douglas Fir. This FEA has been corroborated with the simple analytical approach, giving a strength error between the two of just 1%.

Future work is underway to include the effects of shear and other stresses via FEA modelling of a composite material yard. This will then allow refinement of the current composite design, with the aim of facilitating economically viable replacement of the wooden Muleta yard, and providing a model for the numerous tourism traditional boats of the Tejo (Tagus) basin.

ACKNOWLEDGEMENTS

L.S. Sutherland was funded by the Portuguese Foundation for Science and Technology (Fundação para a Ciência e Tecnologia - FCT), under the Transitional Standard with contract IST-ID/164/2018. This work contributes to the Strategic Research Plan of the Centre for Marine Technology and Ocean Engineering (CENTEC), which is financed by FCT under contract UIDB/UIDP/00134/2020.

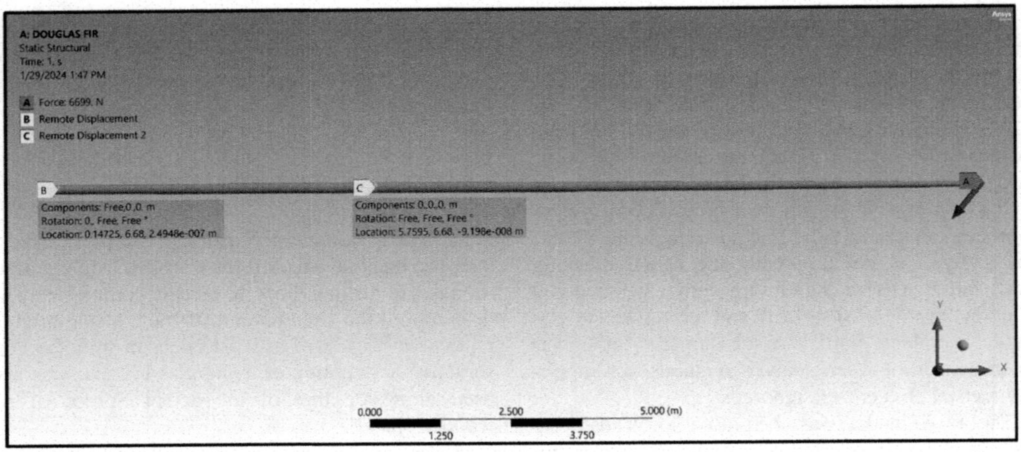

Figure 7. Wooden yard FEA boundary conditions and loading (6.7 kN).

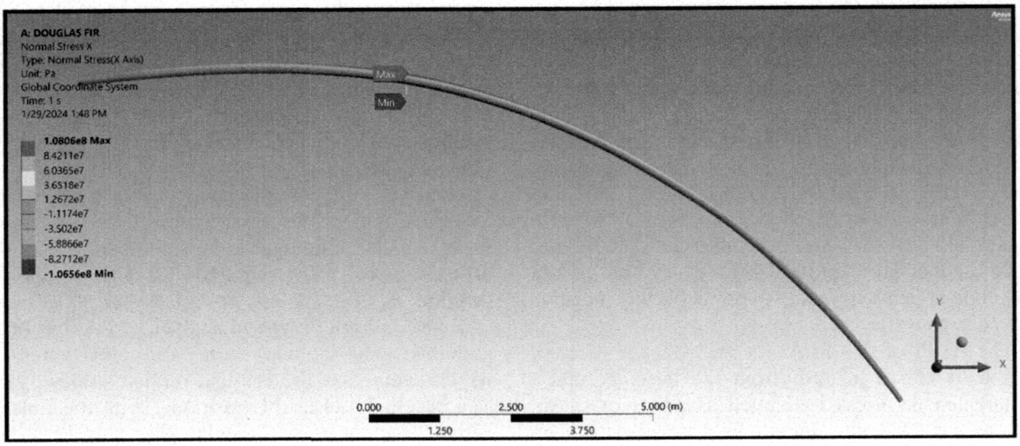

Figure 8. Wooden yard FEA real scale deflection and x-axis normal stress results at 6.7 kN.

REFERENCES

Baldaque da Silva, A.A., 1895. A Muleta. Arte Port.

Cabral, J.P., 2019. A entrada na Europa e a expansão inicial do eucalipto em Portugal Continental. Hist. Ciencia Ens. 20, 8–27. https://doi.org/10.23925/2178-2911.2019v20espp18-27.

Câmara Municipal Barreiro, 2018. Muleta Álvaro Velho [WWW Document]. URL https://www.cm-barreiro.pt/conhecer/turismo/embarcacao-tradicional-muleta/ (accessed 12.13.23).

Carrasco, E., Peres, A., 1997. Barcos do Tejo. Edições INAPA.

Curtinhal, E., 2007. Barcos, Memórias do Tejo. Ecomuseu Municipal do Seixal.

Daniel, I.M., Ishai, O., 2006. Engineering mechanics of composite materials, 2nd ed. ed. Oxford University Press, New York.

Gere, J.M., Timoshenko, S., 1991. Mechanics of materials, Third edition. ed. Chapman and Hall, London.

Germanischer Lloyd, 2009. Rules for Classification and Construction, I - Ship Technology, Part 4, Section 2: Guidelines for Design and Construction of Large Modern Yacht Rigs (Rules No. 1-4– 2 (2009)). Germanischer Lloyd, Hamburg.

Gomes Pedrosa, F., 2016. A Muleta e a Tartaranha SecXV-XX.

Laird Clowes, G.S., 1936. The Story of Sail. Eyre and Spottiswoode, London.

Leitão, M., Oliveira Simões, F., Marques da Silva, A., 2009. A Muleta. Museu de Marinha & Ecomuseu Minicipal do seixal, Seixal, Portgal.

Miller, P.H., 2006. Marine Composites Design, Part 2. Prof. Boatbuild.

Moreira, D.R., Cristiana Faria, 2018. Não é só o barco que é do rio, também Jaime Costa é do Tejo [WWW Document]. PÚBLICO. URL https://www.publico.pt/2018/08/22/local/noticia/jaime-fezse-mestre-para-por-a-navegar-sonhos-embalados-por-rios-e-mares-1841620 (accessed 12.14.23).

Pâris, F.-E., 1882. Souvenirs de marine. Collection de plans de navires et de bateaux anciens ou modernes existants ou disparus avec les éléments numériques nécessaires à leur construction par le VA Pâris. Gauthier Villars, Paris.

Roark, R.J., Young, W.C., Budynas, R.G., 2002. Roark's formulas for stress and strain, 7th ed. ed. McGraw-Hill, New York.

Ross, R. (Ed.), 2010. Wood Handbook, Wood as an Engineering Material. Forest Products Laboratory, United States Department of Agriculture Forest Service, Madison, Wisconsin, USA.

Shenoi, R.A., Wellicome, J.F., West European Graduate Education Marine Technology (Eds.), 1993. Composite Materials in Maritime Structures, Volume 1 Fundamental Aspects, Cambridge ocean technology series. Cambridge University Press, Cambridge; New York.

Sutherland, L.S., 2023. Composite Sailing Ship Spars [WWW Document]. URL https://leighsutherland.com/ship-spars/ (accessed 12.13.23).

Swedish Forest Industries Federation, 2023. Properties of softwood [WWW Document]. Swed. Wood. URL https://www.swedishwood.com/wood-facts/about-wood/from-log-to-plank/properties-of-softwood/ https://www.swedishwood.com/wood-facts/about-wood/from-log-to-plank/properties-of-softwood/ (accessed 12.18.23).

The Laminator: Classical Analysis of Composite Laminates [WWW Document], 2023. URL http://www.thelaminator.net/" (accessed 12.19.23).

USA Department of Defense, 2002. Composite Materials Handbook - Volume 2. Polymer Matrix Composites Materials Properties, Department of Defense Handbook. Department of Defense, USA.

Advances in Maritime Technology and Engineering – Guedes Soares & Santos (Eds)

Numerical study on energy dissipation of ship plates under repeated impacts

X.G. Wang & L. Zhu
Key Laboratory of High Performance Ship Technology (Wuhan University of Technology), Ministry of Education, Wuhan, P.R. China
School of Naval Architecture, Ocean and Energy Power Engineering, Wuhan University of Technology, Wuhan, P.R. China

X. He & C. Guedes Soares
Centre for Marine Technology and Ocean Engineering (CENTEC), Instituto Superior Técnico, Universidade de Lisboa, Lisbon, Portugal

ABSTRACT: Repeated impact is a common issue in naval architecture and ocean engineering, and experimental tests and numerical simulations are the most widely used tools to study this kind of problem, which can provide a clear insight into the dynamic response characteristics of structures. Experimental tests have been conducted to investigate the elastic energy and pseudo-shakedown of rectangular plates in the lab. In this paper, a corresponding numerical model of rectangular plate under repeated impacts is established, and shows good consistency with experimental tests, when friction and damping are considered. With the numerical model, the energy conversion among internal energy, kinetic energy, friction and damping dissipated energy is analyzed, and their variations during repeated impacts are explored. The internal energy consists of plastic energy and elastic energy of plate deformation, and the elastic energy increases with the plate deflection, regardless of the magnitude of impact energy. Based on the energy dissipation analysis, a general energy criterion is suggested for the study of pseudo-shakedown phenomenon.

1 INTRODUCTION

Engineering structures inevitably suffer from repeated mass impacts during their lifespan, such as supply vessels collision, dropped objects, ice floe, among others. Although the structures may undergo only small deformation and not need repair under a single impact, these impacts are prone to happen repeatedly. As a consequence, the accumulation of deformation would finally threaten the safety of life and property. Till to now, many researchers have paid attention to engineering structures suffering from repeated impacts.

Study on repeated impacts of ship structures dates back to 1970s when Jones (1973, 1977) proposed an analytical solution to estimate the damage on bow plates under repeated slamming pressures. For the repeated mass impact behavior, experimental, numerical and theoretical methods were performed on ship plates impacted by a wedge indenter (Zhu 1990). Later, Zhu & Faulkner (1996) used a rigid-perfectly plastic method to estimate the deformation of plates from a semi-submersible platform repeatedly impacted from a supply vessel. The permanent deflections and rebound velocities increased under repeated impacts. Huang et al. (2000) conducted drop tests on fully clamped square and circular plates under repeated mass impacts with spherical indenters, and found that the elastic energies increased with the increase of plastic deformation. An energy criterion was postulated to judge the occurrence of pseudo-shakedown. Jones (2014) reviewed the research works of Zhu & Faulkner (1996) and Huang et al. (2000), and considered the accumulation of mass from earlier impacts on the plate, such as loading of cargo. An empirical equation of elastic energy versus deflection of plate was proposed.

In recent years, Zhu et al. (2018a) carried out the repeated mass impact tests on fully clamped stiffened plates and the energy dissipation of stiffener was considered. He et al. (2021a, b, 2022a, b) studied the repeated impact behaviors of beams and plates by experimental tests and numerical simulations. The energy dissipated by damping was considered in the FE model, and contributed to 3%-4% of the impact energy. Cai et al. (2022) performed horizontal repeated impact tests between steel plates and wedged ice, and the energies dissipated by ice were included. Cho et al. (2014) and Truong et al. (2017) analyzed the repeated impacts of beams and grillage structures, considering the effect of sub-zero temperature. Duan et al. (2018) studied the dynamic behavior of aluminum alloy plates with surface cracks subjected to repeated impacts. Zhu et al. (2018b) and Guo et al. (2018, 2019) studied the repeated impact behavior of aluminum foam sandwich

DOI: 10.1201/9781003508762-63

plate, and analyzed the effect of sandwich structure on energy dissipation and low temperature. Zhang et al. (2021) studied the energy absorption of honeycomb sandwich plate under repeated impacts. Rezasefat et al. (2023) studied repeated impact behavior of a hybrid composite structure. Zhu et al. (2024) investigated the effects of indenter geometry on Pseudo-Shakedown of plates using experimental and a modified rigid-plastic method is proposed to predict the number of repeated impacts for a plate necessary to reach a pseudo-shakedown state.

Many researches have shown the importance of energy dissipation on engineering structures subjected to repeated impacts. However, few of them concentrated on the general analysis of energy dissipated and conversion in a repeated impact system. The energy analysis provides valuable insights into the dynamic elastoplastic behaviors of structures, and contributes to the understanding on the nature of repeated mass impacts, especially for the energy criterion.

In this paper, the energy analysis of rectangular plate under repeated impacts is carried out through numerical simulation. The elastic energy, friction dissipated energy and damping dissipated energy are analyzed in detail. The effect of these energy on energy criterion is discussed.

2 FINITE-ELEMEMT INVESTIGATION

2.1 *FE modeling*

The non-linear finite element (FE) software ABAQUS/Explicit is employed to establish the numerical model of rectangular plate under repeated impacts. The dimensions of the rectangular steel plate and rigid indenter corresponding to laboratory tests are shown in Figure 1. The edges of 34 bolt holes are completely fixed, while the boundary area (clamped in the tests) is limited not to produce any movement in Z-direction. The plate is impacted at the center by a rectangular indenter with a mass of 82.8 kg. The movement of indenter is only allowed in the Z-direction. The interaction between the indenter and plate is set as surface to surface contact.

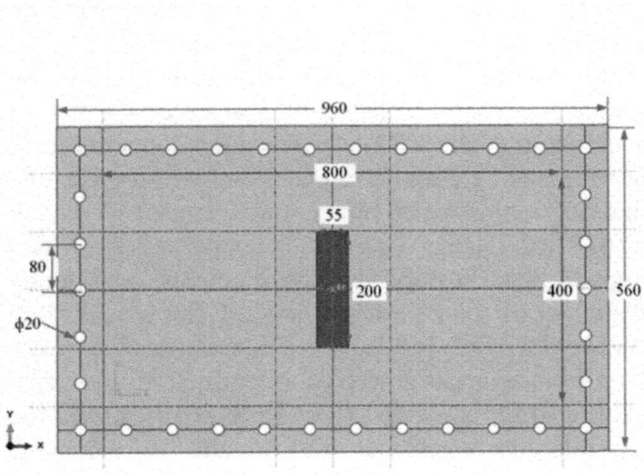

Figure 1. Finite element model of rectangular plate under mass impact.

The focus is put on the dynamic elastoplastic behavior of steel plate, and the indenter is set to be discrete rigid in the FE model. The material property of the steel plate is obtained through uniaxial quasi-static tensile test. The engineering and true stress-strain curves for plate material are shown in Figure 2, while other specifications in FE model are listed in Table 1. The coefficients of friction and damping are measured as 0.417 and 218.58 in this analysis, which will be further discussed in Sec. 3.

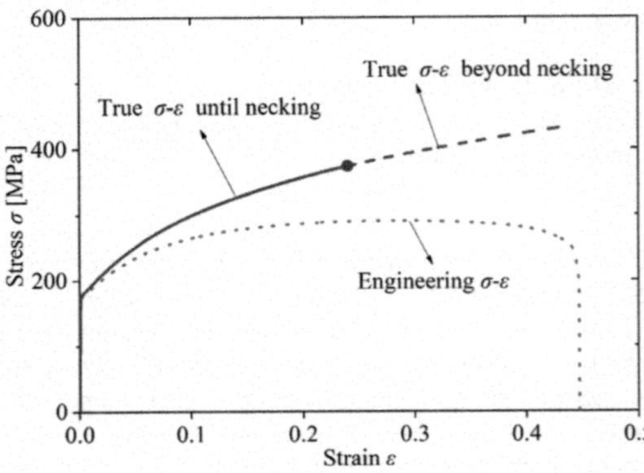

Figure 2. Engineering and true $\sigma - \varepsilon$ curves of plate material.

Table 1. The specifications in FE model.

Specification	Rectangular plate	Unit
Density	7850	kg/m^3
Elastic modulus	210	GPa
Poisson's ratio	0.3	/
Yield stress	175	MPa
Ultimate tensile stress	290	MPa
Fracture strain	0.45	/
Plate thickness	1	mm
Coefficient of friction μ	0.417	/
Coefficient of damping α	218.57	/

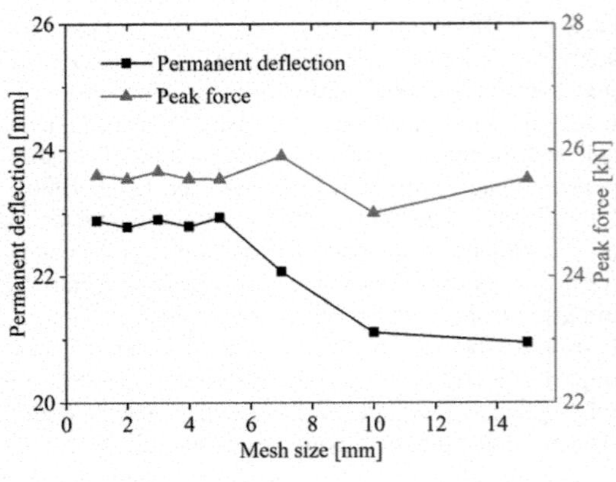

Figure 3. Convergence analysis of the mesh size on permanent deflection and peak force.

The indenter is meshed with bilinear rigid quadrilateral element (R3D4), while the plate is meshed with linear reduced integration shell element (S4R). In order to determine available size of the mesh, the analysis of mesh convergence is performed. For various mesh sizes from 1 mm to 15 mm, the permanent deflection and peak force of the plate subjected to 2.7 m/s impact of indenter is shown in Figure 3. It indicates that as the mesh size decreases to 5 mm, the permanent deflection and peak force are fairly stable. Therefore, mesh size of 5 mm is applied for the plate and indenter for the control of time spent on calculation.

2.2 *Model validation*

The corresponding repeated impact tests were conducted using the horizontal impact testing machine in Impact and Fluid-Structure Interaction Laboratory (IFSI) of Wuhan University of Technology, as shown in Figure 4. The effect of gravity can be eliminated in the design of horizontal impact. The geometry dimension of plate and indenter are the same with FE model. The impact mass is 82.8 kg, and the impact velocity is 2.7 m/s. Hence the impact energy is about 302 J.

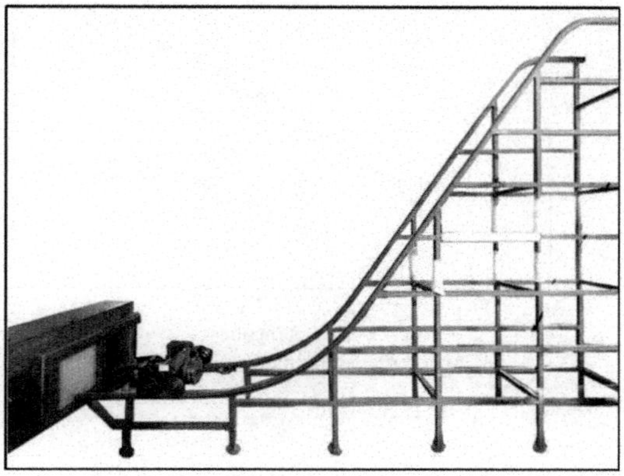

Figure 4. Experimental set-up of mass impact test.

Two groups of numerical simulation are conducted and compared to highlight the effects of friction and damping. One considers the energy dissipation of friction and damping, while the other does not. During the 1st impact, the time history curves of plate displacement are shown in Figure 5a. When the friction and damping are not considered, the plate has a maximum deflection of 26.04 mm, and has an average permanent deflection of 22.64 mm with continuous elastic vibration. The consideration of friction and damping reduces the maximum and permanent deflections to 25.61 mm and 21.86 mm, respectively. The later simulation is beneficial to obtain a steady value of permanent deflection, which is more consistent with the experimental situation. Regarding the permanent deflection, the results of the two numerical simulations are 5.3% and 1.7 % larger than the experimental result of 21.50 mm, respectively. Although the errors will accumulate with the increase of repeated impacts, the employed FE model considering friction and damping can achieve good consistency with experimental tests, as shown in Figure 5b.

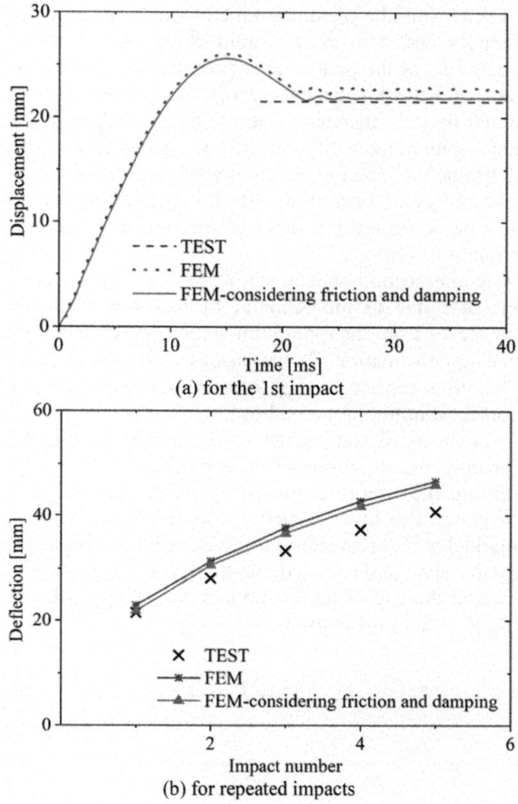

Figure 5. Validation of finite element model on plate deflections.

3 NUMERICAL STUDY

3.1 *Energy dissipation and conversion in an impact system*

In the analysis of ABAQUS/Explicit, the energy balance equation of Eq. (1) for the whole FE model is available (ABAQUS 6.14. User's manual 2014),

$$E_{total} = E_i + E_k + E_v + E_f + E_{other} \tag{1}$$

where E_i, E_k, E_v, E_f are the internal (strain) energy, kinetic energy, energy dissipated by viscous (damping) effect and energy dissipated through friction effect for the whole model, respectively. In current FE model, $E_{other} = 0$ represents that other energies are not considered, such as electrical and thermal energy. Eq. (1) reflects that the sum of these energy should be a consistent during each impact. During the 1st impact, these energies are output, as shown in Figure 6.

Obviously, the energy dissipation and conversion can be divided into three states. In the first state when $t \le t_1$, it can be observed that, at the beginning of the impact, E_{total} is equal to E_k which is comprised entirely of kinetic energy of indenter. When the indenter hits the plate, the kinetic energy of indenter reduces and converts to strain energy E_i, damping energy E_v of the plate, and friction energy E_f by contact, and the kinetic energy of plate vibration. It is noted that, E_i increases due to the increase of plate deformation, including elastic deformation and plastic deformation. The kinetic energy of indenter and kinetic energy of plate make up the renewed E_k which still has a decreasing trend till the indenter velocity reduces to zero.

In the second state when $t_1 < t \le t_2$, E_k starts to increase due to the rebound of indenter, while E_i decreases from the maximum value for the release of elastic deformation. E_v increases till the vibration of plate total ceases. E_f slightly increases due to shorter contact duration in unloading.

In the third state when $t > t_2$, i.e. the 1st impact finishes, the strain energy E_i is marked as E_p, representing the energy absorbed by plastic deformation of plate. The kinetic energy of indenter rebound is marked as E_e, representing the energy not dissipated by the plate and released due to elastic effect. Therefore, at the end of the 1st impact, the energy balance has the relationship as shown in Eq. (2).

$$E_{total} = E_p + E_e + E_v + E_f + E_{\text{other}} \tag{2}$$

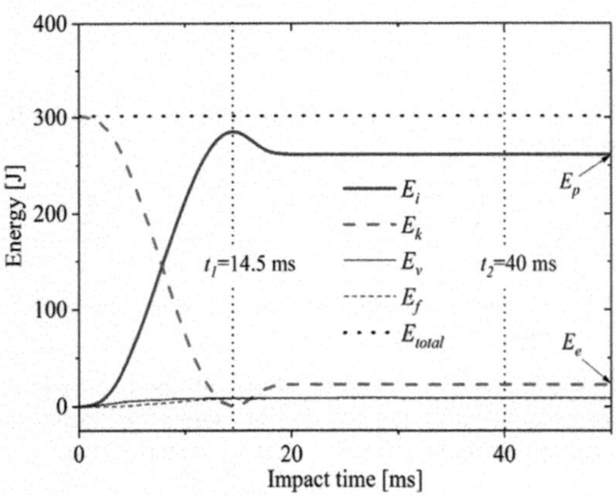

Figure 6. The energy dissipation and conversion of the system in the 1st impact.

3.2 *Energy stored due to elastic effect*

For the repeated impacts, the energy released due to elastic effects E_e is calculated according to the kinetic energy of indenter rebounding. Under different impact energies of 125 J, 250 J and 375 J (impact velocity is 2.5 m/s, and the indenter mass is 40, 80 and 120 kg, respectively), the rebound velocity increases with impact number before plate ruptures, as shown in Figure 7. It indicates the elastic energy of a plate also increases with impact number.

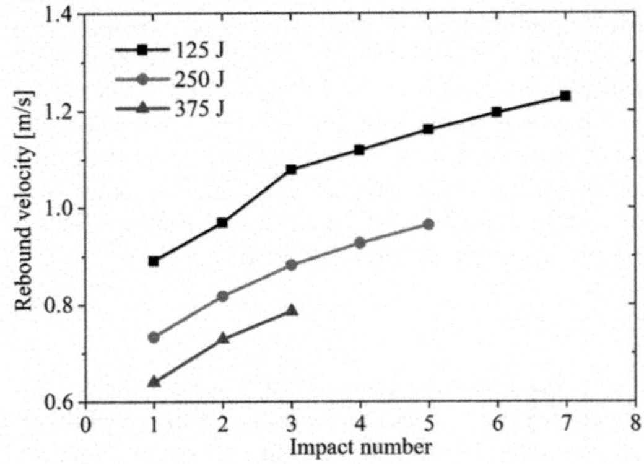

Figure 7. The rebound velocities for repeated impacts with different energies.

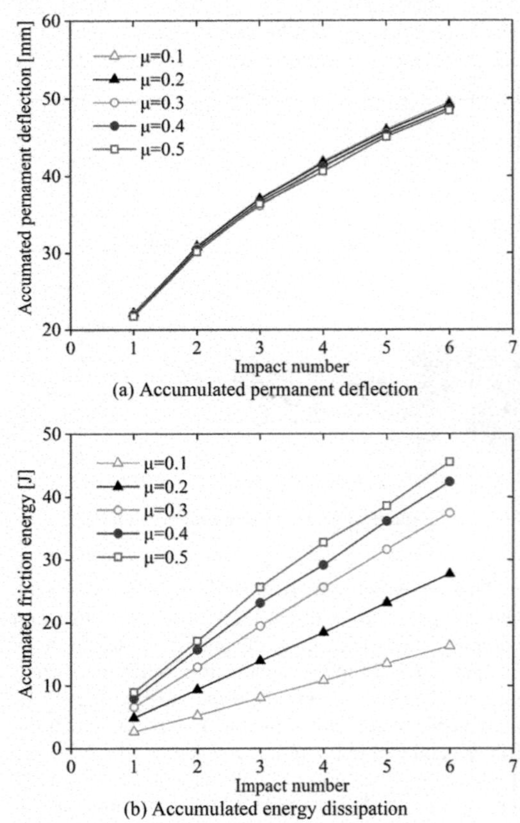

Figure 8. The effects of friction coefficient on accumulated deflection and energy dissipation under repeated impacts.

3.3 *Energy dissipated due to friction effect*

Inevitably, the dynamic friction coefficient μ has influence on the plate accumulation and energy dissipation. Figure 8a shows that the permanent deflections decrease as μ increases from 0.1 to 0.5 under repeated impacts with energy of 302 J. In terms of the accumulated energies dissipated due to friction effect, they linearly increase with impact number as shown in Figure 8b, even for different μ. As μ increases from 0.1 to 0.5, the average increments of E_f increase from 2.7 J to 7.6 J. It is noted that, this relationship provides good simplification approach for the analysis of energies dissipated due to friction effect under repeated impacts.

3.4 *Energy dissipated by damping*

In order to determine the damping coefficient of the plate, a modal analysis has been carried out through FE method. Results show that, the participant factor of the plate damping in Z-direction is the largest for the vibration of the 1st mode, followed by the 3th, 8th and 12th modes. The effective mass of the 12 vibration modes contributes to 92% mass of the plate, indicating that the analysis is relatively reliable. Accordingly, the damping coefficient α is determined as $0.4\pi/T$ (Bathe & Wilson 1976), and T is the period of mode for damping, and the reciprocal of the natural frequency. Therefore, different values of α are tested, including 38.09, 99.53, 182.69 and 237.51, according to the natural frequencies of 27.56, 72.00, 132.16 and 171.82 Hz corresponding to these vibration modes.

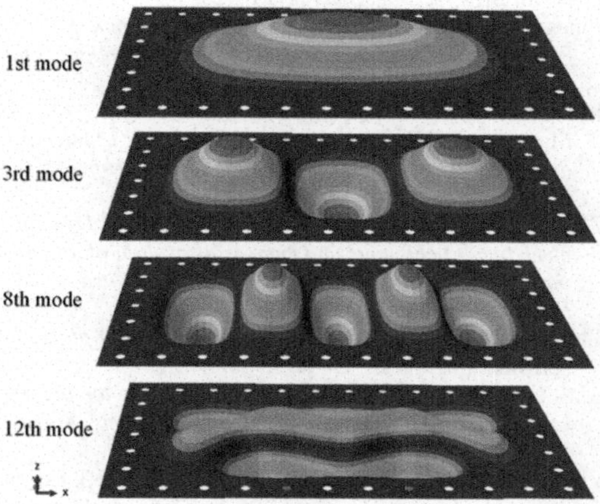

Figure 9. Modal analysis of rectangular plate in FE method.

Figure 10a shows that as the damping coefficient α becomes larger, the plate displacement stabilizes more quickly, and a smaller permanent deflection is obtained. E_v for the 1st impact is the largest, while E_v for the subsequent impacts are nearly stable, as shown in Figure 10b. As α increases from 38.09 to 237.51, E_v increase from 2.2 J to 3.8 J for the repeated impacts after the 1st impact.

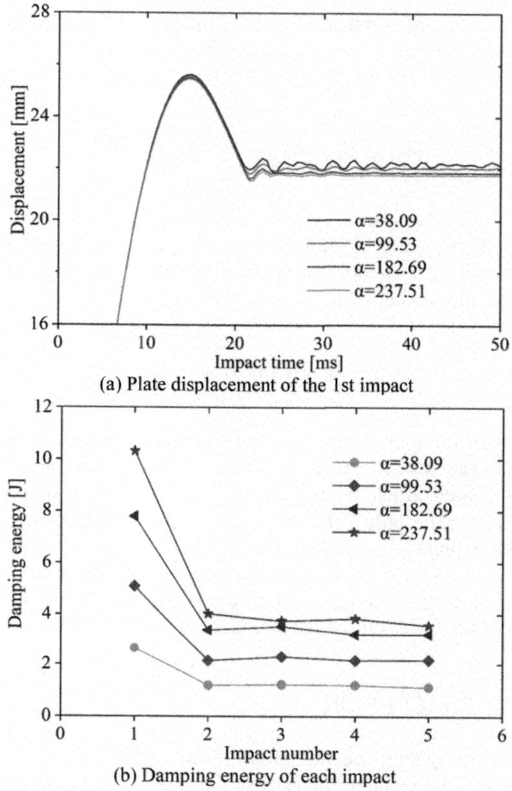

Figure 10. The effects of damping coefficient on deflection and energy dissipation under repeated impacts.

4 DISCUSSION ON ENERGY CRITERION

Huang et al. (2000) proposed an energy criterion to study pseudo-shakedown in repeated mass impacts. It indicates that the structure will achieve pseudo-shakedown under repeated impacts, if the impact energy is smaller than the maximum elastic energy of plate, i.e. $U_0 \leq E_{em}$. U_0 is the initial impact energy, and $U_0 = E_{\text{total}}$ in each independent impact. Otherwise, the plastic deformation will keep increasing and the plate may undergo rupture failure. Based on hundreds of repeated impacts simulation considering the damping energy, He & Guedes Soares (2022a) also made a conclusion that, pseudo-shakedown will happen when the internal energy is less than the elastic energy that a specimen can store.

In this study, it shows the energies dissipated by friction effect E_f and damping effect E_v contribute to almost stable proportion of energy analysis, which is not considered in the energy criterion. Based on Eq. (2), the plastic energy absorbed by the plate can be expressed as,

$$E_p = U_0 - E_e - E_v - E_f - E_{\text{other}} \quad (3)$$

In pseudo-shakedown state, the plate produces no more plastic energy under subsequent impact, i.e. $E_p \leq 0$. It means,

$$U_0 - E_e - E_v - E_f - E_{\text{other}} \leq 0 \quad (4)$$

The elastic energy increases with plate deflection, and has a maximum value E_{em}. Eq. (5) gives a general energy criterion for the judgement of pseudo-shakedown.

$$U_0 - E_v - E_f - E_{other} \leq E_{em} \quad (5)$$

In this paper, the FE model gives a prediction on the ratios of E_f and E_v to the total impact energy the repeated impact tests, i.e. 2.4 % and 1.2 %, respectively. Therefore, the modified energy criterion can be expressed as, $0.964 \cdot U_0 \leq E_{em}$.

The above analysis quantifies the internal energy of a certain impact system and proposed a general energy criterion for pseudo-shakedown.

5 CONCLUSION

In this paper, the numerical simulations are carried out to investigate the energy dissipation and conversion of rectangular plate under repeated impacts, and some conclusions are drawn.

The FE model considering friction and damping effects has been established, and validated against experimental results of both the single and repeated impacts. Numerical energy analysis shows that, the elastic energy increases with the increase of plate deflection, regardless of the magnitude of impact energy. The accumulated energy dissipated through friction effect increases linearly with impact number. After considering of energies dissipated by friction, damping and other effects, a general energy criterion is suggested, i.e. $U_0 - E_v - E_f - E_{other} \leq E_{em}$, for further study of pseudo-shakedown phenomenon.

REFERENCES

ABAQUS 6.14. 2014. *User's manual. Providence, R.I. Dassault Systèmes.*

Bathe, K.J. & Wilson, E.L. 1976. *Numerical Methods in Finite Element Analysis. PrenticeHall, Englewood Cliffs, New Jersey, USA.*

Cai, W. Zhu, L. & Qian, X.D. 2022. *Dynamic responses of steel plates under repeated ice impacts. Int J Impact Eng* 162:104129.

Cho, SR. Truong, D.D. & Shin, H.K. 2014. *Repeated lateral impacts on steel beams at room and sub-zero temperatures. Int J Impact Eng* 72:75–84.

Duan, F. Liu, J. Wang, G. and Yu. Z.L. 2018. *Dynamic behaviour of aluminium alloy plates with surface cracks subjected to repeated impacts. Ships Offshore Struct* 14 (5):478–491.

He, X. & Soares, C. G. 2021a. *Numerical study on the pseudo-shakedown of beams under repeated impacts. Ocean Eng* 242:110137.

He, X. & Soares, C. G. 2021b. *Experimental study on the dynamic behavior of beams under repeated impacts. Int J Impact Eng* 47:103724.

He, X. & Soares, C. G. 2022a. *Pseudo-shakedown of rectangular plates under repeated impacts. Marine Struct* 85:103258.

He, X. Garbatov, Y. & Soares, C. G. 2022b. *Analysis of pseudo-shakedown of rectangular plates under repeated impacts. Ocean Eng* 265:112609.

Huang, Z. Chen, Q. & Zhang, W. 2000. *Pseudo-shakedown in the collision mechanics of ship. Int J Impact Eng* 24 (1):19–31.

Guo, K.L. Zhu, L. Li, Y.G. Yu, T.X. Shenoi, A. & Zhou, Q. W. 2018. *Experimental investigation on the dynamic behavior of aluminum foam sandwich plate under repeated impacts. Compos Struct* 200, 298–305.

Guo, K.L. Zhu, L. Li, Y.G. & Yu, T.X. 2019. *Numerical study on mechanical behavior of foam core sandwich plates under repeated impact loadings. Compos Struct* 224:111030.

Truong, D.D., Shin H.K. & Cho, S.R. 2017. *Repeated lateral impacts on steel grillage structures at room and sub-zero temperatures. Int J Impact Eng* 113:40–53.

Jones, N. 1973. *Slamming damage. J Ship Res* 17 (2):80–86.

Jones, N. 1977. *Damage estimate for plating of ships and marine vehicles. In: International Symposium Practical Design in Shipbuilding (PRADS). Tokyo: Society of Naval Architects of Japan* p.121–128.

Jones, N. 2014. *Pseudo-shakedown phenomenon for the mass impact loading of plating. Int J Impact Eng* 65:33–39.

Rezasefat, M. Silva, A. Amico, S.C. Giglio, M. & Manes, A. 2023. *Repeated impact behaviour of inter-ply hybrid aramid/S2-glass epoxy laminates. Thin-Walled Struct* 186:110680.

Zhang, Y. Li, Y.G. Guo, K.L. & Zhu, L. 2021. *Dynamic mechanical behaviour and energy absorption of aluminium honeycomb sandwich panels under repeated impact loads. Ocean Eng* 219:108344.

Zhu, L. 1990. *Dynamic inelastic behavior of ship plates in collision. PhD thesis. University of Glasgow.*

Zhu, L. & Faulkner, D. 1996. *Damage estimate for plating of ships and platforms under repeated impacts. Marine Struct* 9(7):697–720.

Zhu, L. Shi, S. & Jones, N. 2018a. *Dynamic response of stiffened plates under repeated impacts. Int J Impact Eng* 117:113–122.

Zhu, L. Guo, K.L. Li, Y.G. Yu, T.X. & Zhou, Q.W. 2018b. *Experimental study on the dynamic behavior of aluminum foam sandwich plates under single and repeated impacts at low temperature. Int J Impact Eng* 114:123–132.

Zhu, L. Wang, X.G. Guo, K.L. Cai, W. Jones, N. & Dai, S. L. 2024. *Effect of indenter geometry on pseudo-shakedown of steel plate under repeated mass impacts. Int J Impact Eng* 185:104865.

Advances in Maritime Technology and Engineering – Guedes Soares & Santos (Eds)

A Graph Neural Network (GNN) based method of acoustic emission source localization for a composite panel

Z. Zhao
School of Civil Engineering, Tianjin University, Tianjin, P.R. China

N.Z. Chen
State Key Laboratory of Hydraulic Engineering Intelligent Construction Operation, Tianjin University, Tianjin, P.R. China
School of Civil Engineering, Tianjin University, Tianjin, P.R. China

C. Jiang
School of Civil Engineering, Tianjin University, Tianjin, P.R. China

ABSTRACT: An innovative deep learning method that leverages graph neural network (GNN) for the regression task of localizing acoustic emission (AE) source on a composite panel is proposed in this paper. In the proposed method, each sensor within the sensor network is treated as a node of the graph where the received AE signals serve as node features. The physical distance between each pair of sensors is measured to establish graph connectivity. First, the node features are preprocessed through normalization operation and one dimensional convolutional neural network (1D-CNN) layers. Subsequently, the GNN layers are employed to extract graph-level features that represent AE source locations. Finally, a regression layer is utilized to derive the AE source coordinates. The proposed method is validated using the experimental dataset collected from a glass fiber-reinforced plastic (GFRP) panel. The results illustrate the superior performance of the proposed method in AE source localization.

1 INTRODUCTION

Composite materials are extensively utilized in various fields of offshore engineering, such as wind turbine blades. Composite structures offer a variety of advantages such as being lightweight, corrosion-resistant, flexible in design, and cost-effective (Kalanchiam & Chinnasamy 2012). However, these structures are vulnerable to various types of damage due to the complex operating environments (Hu et al. 2016). Various structural health monitoring (SHM) methods have been developed to monitor these damages at an early stage (Zhao & Chen 2022; Chaupal & Rajendran 2023). Among them, acoustic emission (AE) is a passive monitoring method and it is sensitive to various forms of damage in composite materials. AE has garnered considerable attention as a SHM technique.

With the development of artificial intelligence technology and the advancement of data processing techniques, data-driven methods represented by deep learning have been widely applied to SHM of composite structures (Moradi et al. 2023; Bandara et al. 2023). A series of deep learning algorithms have been developed for damage source localization, which are highly robust and can overcome the influence of the complex propagation characteristics of AE waves (Zhao & Chen 2023a; Zhao & Chen 2023b). Deep learning methods can automatically extract features from monitoring data, effectively avoiding human error and feature loss. However, although conventional deep learning algorithms can capture hidden features from regular data, most methods ignore the interdependence between multiple sensors (Li et al. 2022).

Emerging graph neural network (GNN) (Scarselli et al. 2008) is introduced to address such challenges, where the graph encompasses nodes and edges, with the relationships between nodes reflected in the edges. GNNs have found success in various applications, such as numerical acceleration (Jiang & Chen 2023), natural language processing (Wu wt al. 2023), traffic prediction (Jiang et al. 2023), etc. GNNs have also been used in the field of SHM. Yu et al. (2021) proposed a graph neural network (GCN) based method for fault classification of wind turbine gearboxes. Li et al. (2020) performed fault classification of rolling bearings using vibration signals and GNN. Zhao & Chen (2023b) first introduced GNN to AE monitoring systems, and presented a spatial-temporal graph convolutional networks (STGCN) method to successfully localize

DOI: 10.1201/9781003508762-64

AE sources on composite structures. The results demonstrate that the STGCN method outperforms other methods.

In this paper, a method based on GNN is proposed for damage coordinate localization for a composite panel. Each sensor in the AE sensor network is treated as a node in the graph, with the obtained signals serving as features of the nodes. The distances between sensors are utilized to establish the connectivity of the graph. The performance of this proposed method is assessed using an experimental dataset obtained from tests conducted on a glass fiber reinforced plastic (GFRP) panel.

2 METHODOLOGY

Graphs have powerful data representation and consist of nodes and edges. Graph can be defined as $G = (V, E)$, where V denotes the nodes of the graph and E represents the edges of the graph. The number of nodes is $N = |V|$. The connectivity of the graph is determined by the adjacency matrix A. If $A_{ij} = 1$, it means that there is an edge between the two nodes $i, j \in V$.

2.1 *AE signals preprocessing*

In this paper, a method based on GNN is proposed to localize the AE source for a composite panel. Each AE sensor is considered as a node.

As the first step of the method, the AE signals are preprocessed to obtain the features of the nodes. First, the AE signal is normalized by Min-Max normalization

$$x_{nor} = \frac{x(t) - \min(x)}{\max(x) - \min(x)} \tag{1}$$

where $x(t)$ is the value of the AE signal at moment t.

The normalized AE signals are then preprocessed using several one-dimensional convolutional neural network (1D-CNN) layers with the Rectified Linear Unit (ReLU) activation function. 1D-CNN have demonstrated their efficacy as effective extractors of features from time series data (Kim et al. 2023). They exhibit adaptability to various types of time series problems and possess the capability to autonomously learn features. Furthermore, convolutional neural network (CNN) proves adept at capturing local features, thereby facilitating the extraction of crucial time-related features.

Furthermore, the AE signals are affected by the distance between the sensor and the AE source. Therefore, the features of the nodes are composed of the AE signals processed by the Min-Max normalization and 1D-CNN layers together with the position information of the AE sensors.

The next step is the construction of the graph. It is reasonable to construct the connectivity of the graph based on the distance between sensors. In this paper, the adjacency matrix is expressed as

$$A_{ij} = \frac{d_{ij} - \min(d)}{\max(d) - \min(d)} \tag{2}$$

where d_{ij} is the distance between node $i,j \in V$. d is the set of distances between nodes.

2.2 *Graph neural network (GNN)*

The graphs are then fed into the GNN layer for further extraction of graph features. GNN employs two distinct methods to execute the convolution operation on graph data: a spectral-based approach and a spatial-based approach. The spectral method stands out as a widely adopted approach for graph convolution. The spectral method captures graph-level features and is suitable for our task of graph-level regression.

The eigenvalue decomposition of the graph yields a spectral representation of the graph. For the AE signals x, the Fourier transformation and inverse Fourier transformation of x can be calculated by

$$\begin{cases} F(x) = U^T x \\ F^{-1}(x) = Ux \end{cases} \tag{3}$$

where U is the eigenvector matrix of the Laplacian matrix L

$$L = I_N - \Lambda^{-\frac{1}{2}} A \Lambda^{-\frac{1}{2}} \tag{4}$$

where $\Lambda = diag\{\lambda_0, \ldots, \lambda_{N-1}\}$ is the degree matrix of the adjacency matrix A and I_N is the identity matrix.

Chebyshev Graph Convolutional Networks (ChebNet) (Defferrard et al. 2016) is a spectral-based GNN, which has the advantages of efficient feature capture and good generalization ability. In this paper, the ChebNet is used to extract the graph features. The graph filtering function $h(\Lambda)$ of K-order Chebyshev polynomials can be expressed as

$$h(\Lambda) = \sum_{k=0}^{K-1} \theta_k T_k\left(\frac{2\Lambda}{\lambda_{\max}} - I_N\right) \tag{5}$$

The output of for a graph data input X can be calculated by

$$y = Uh(\Lambda)U^T X = \sum_{k=0}^{K-1} \theta_k T_k\left(\frac{2L}{\lambda_{\max}} - I_N\right) \tag{6}$$

2.3 *The proposed AE source localization method*

The proposed AE source localization method consists of 19 layers, as shown in Figure 1, including an input layer, three 1D-CNN layers, a reshape layer, eight ChebNet layers, a flatten layer and four fully connected layers. The AE signal is used as an input to the input layer. Three 1D-CNN layers are used for pre-processing of the AE signal, and then a reshape layer reorganizes the features of the nodes and fuses the position information of the sensors. The constructed graph will be input into the ChebNet layers for graph-level feature extraction. Then it goes through a flatten layer and a dropout layer with the dropout rate set to 0.4. Subsequently, the extracted graph features are regressed to the coordinates of the AE source by four fully connected layers.

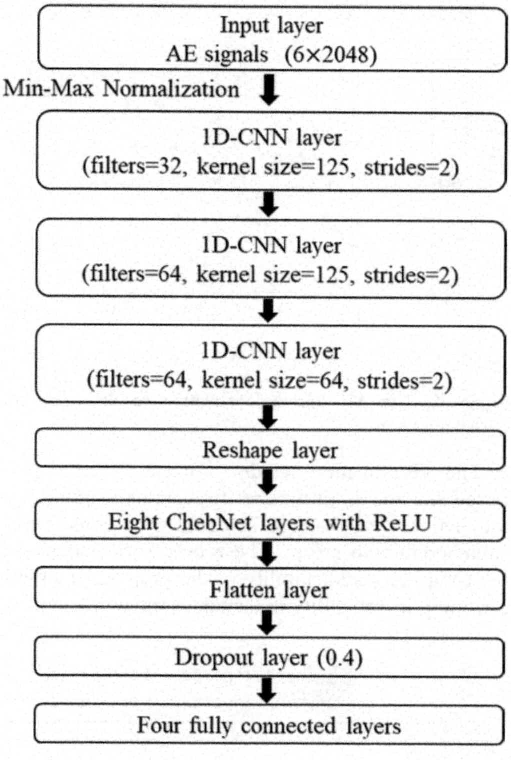

Figure 1. The structure of the proposed AE source localization method.

The performance of the model was evaluated using the mean absolute error (MAE)

$$MAE = \frac{1}{n}\sum_{nn=1}^{n} |\hat{y}_{nn} - y_{nn}| \tag{7}$$

where, y_{nn} and $\hat{y}_{nn}$ denote the actual and prediction coordinates, respectively. n is the number of samples.

The mean squared error (MSE) is used as the loss function for training

$$MSE = \frac{1}{n}\sum_{nn=1}^{n} (\hat{y}_{nn} - y_{nn})^2 \tag{8}$$

RMSprop is used as the optimizer. Learning rate wis set to 0.0001, and the maximum epoch is 150. In addition, a patience-based early stopping strategy is used to improve the generalization ability of the model and conserve computational resources, and the patience is set to 15.

The proposed method is built in the python TensorFlow framework and runs on an Intel(R) Xeon(R) Gold 6258R CPU (2.6 GHz), 256G of RAM, and NVIDIA GeForce GTX 256 hardware.

3 EXPERIMENTS

The primary objective of this experiment is to validate the proposed damage localization method outlined in this paper. The experimental was carried out on a glass fiber-reinforced plastic (GFRP) panel. GFRP is a composite material that is widely used in infrastructures, e.g. as the main fabrication material for wind turbine blades. The pencil lead break (PLB) test has conventionally been used as the reproducible artificial AE source, which was chosen as the simulated AE source for this study. To conduct the experiment, six AE sensors were strategically positioned on the GFRP panel measuring 400mm in length and width, as illustrated in Figure 2. The sensor coordinates were configured as follows: S1 (300mm, 260mm), S2 (200mm, 260mm), S3 (100mm, 260mm), S4 (100mm, 130mm), S5 (200mm, 130mm), and S6 (300mm, 130mm). To ensure signal accuracy, the AE sensors were securely affixed to the panel using the grease-high vacuum, and wooden supports were employed at the four corners of the GFRP panel. This meticulous setup was designed to guarantee the reliability of the acquired AE signals throughout the experimental process.

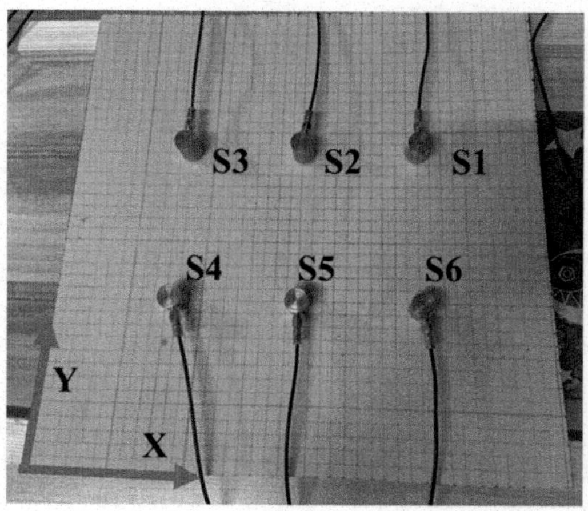

Figure 2. GFRP panel and AE sensor array.

The AE signal acquisition system is a PCI-Express-based 32-channel AE system, as shown in Figure 3. The model of the AE sensor is $R15\alpha$. The sampling frequency is 1 MHz, and the pre-trigger time is 512 μs. On the GFRP panel, the interval of each location of PLB test is 10 mm, and a total of 1,467 AE events are acquired, and the waveform length of each AE event is 2048.

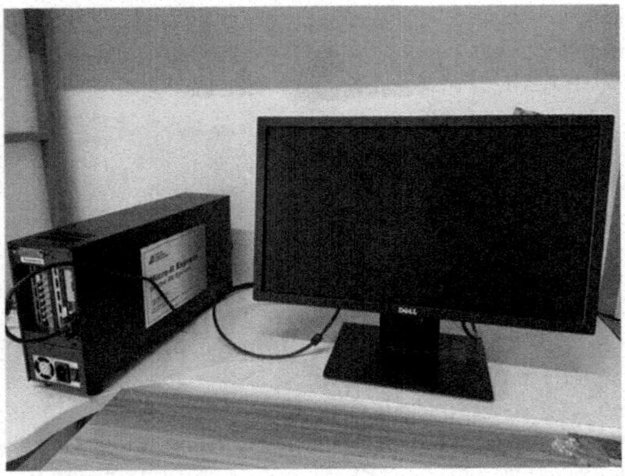

Figure 3. The AE signal acquisition system.

4 RESULTS AND DISCUSSION

This study aims to validate the proposed AE source localization method. To ensure a thorough and reliable assessment of the performance of the proposed method, the k-fold cross-validation is employed. The k-fold cross-validation maximizes the utilization of available data for deep learning model evaluation, safeguarding against the potential impact of specific dataset divisions on the performance assessment. In this paper, the proposed method undergoes validation using a 5-fold cross-validation approach. The MAE values obtained from five folds are presented in the Figure 4. The Figure offers insight into the robustness and consistency of the proposed method across different folds.

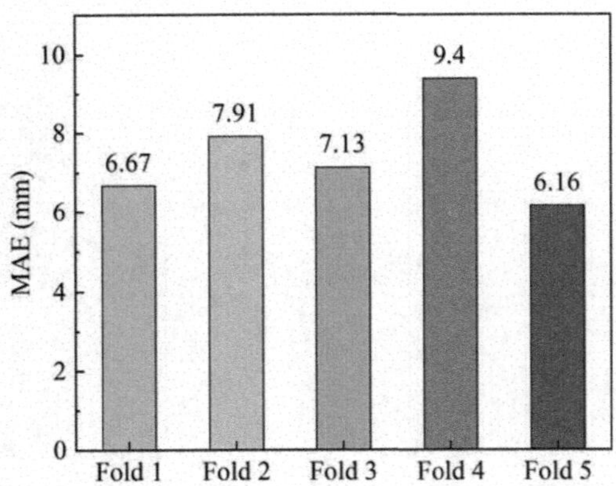

Figure 4. The AE source localization results of 5-fold cross validation.

The utilization of 5-fold cross-validation ensures a more robust application of the dataset, minimizing the likelihood of chance inaccuracies in the results. The calculated MAE values for each iteration demonstrate consistency, with none exceeding 10 mm. The average MAE across the five iterations is recorded as 7.45 mm. The results show that the proposed method effectively and accurately locates the AE sources within the GFRP panel.

Furthermore, for a more comprehensive visualization of the localization results, Figure 5 presents a mapping of the damage source localization. In this figure, the actual and predicted AE sources are linked with a red line, providing a clear representation of the comparison between the predicted and real sources.

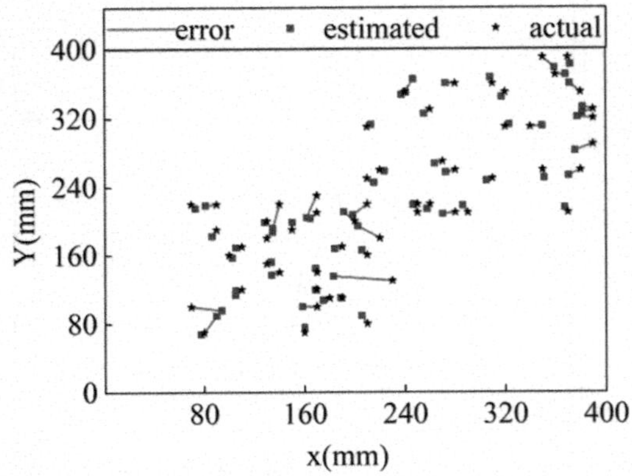

Figure 5. The AE source localization results of 5-fold cross validation.

The visualization results indicate that the proposed method is proficient in accurately predicting the majority of damage sources, with consistently small prediction errors. These outcomes underscore the robustness and stability of the proposed method, affirming its reliability in damage source prediction.

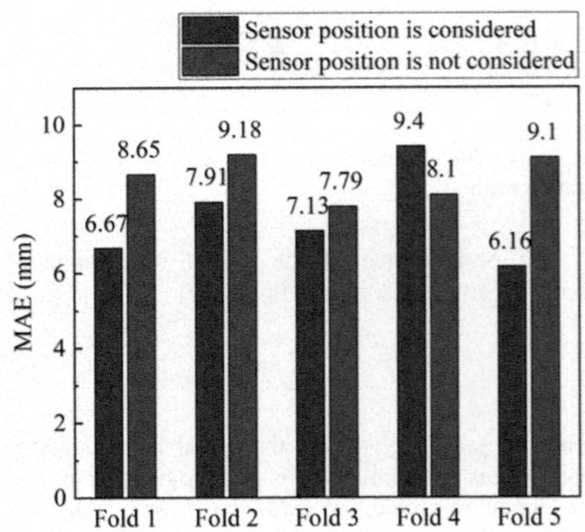

Figure 6. Comparison of the localization results for AE sources considering and not considering the position information of AE sensors.

Moreover, this study incorporates the spatial location information of the AE sensors as the key features of the node. To assess the influence of the location information of the AE sensors on the proposed method, the outcomes are juxtaposed with those obtained when excluding the sensor's spatial location information as the node features. The comparative results are visually presented in Figure 6, employing a five-fold cross-validation approach for consistency.

It can be seen from Figure 6 that there has a noticeable impact on the prediction accuracy of the AE source when incorporating the location information of the AE sensors. The prediction accuracy experiences a significant improvement with the inclusion of the location information of the AE sensors. This observation underscores the effectiveness of utilizing the spatial information of the AE sensors in enhancing the feature representation of the graph, consequently leading to an enhanced performance of the proposed method.

Furthermore, the impact of the number and placement of sensors on the positioning accuracy performance of the proposed method is further investigated. To examine the influence of the number of sensors on the positioning accuracy of the proposed method, the localization performance is studied for sensor quantities of 3, 4, 5, and 6. The sensor arrangement diagrams can be found in Figure 2, where 3, 4, 5, or 6 sensors were randomly selected from a total of 6 sensors for each scenario.

Specifically, when the number of sensors is 3, the sensors with the identifiers S1, S3, and S5 were chosen. For 4 sensors, the identifiers were S3, S4, S5, and S6. In the case of 5 sensors, the identifiers were S2, S3, S4, S5, and S6. These scenarios are compared to the positioning results obtained using 6 sensors. The 5-fold cross-validation method is employed, and the MAE is utilized to measure the accuracy of the results. The comparative results of the study are illustrated in Figure 7.

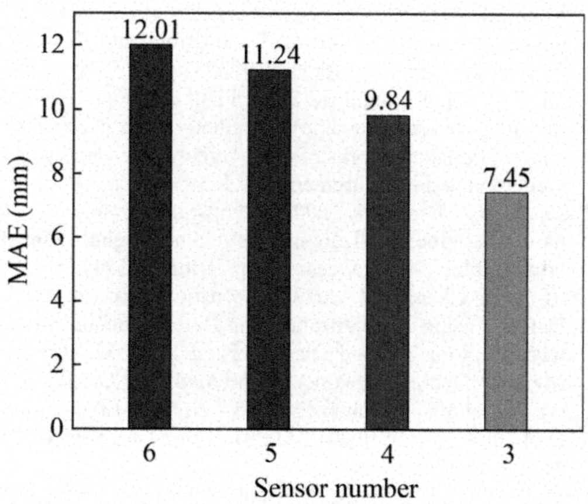

Figure 7. Comparison of the localization performance of the proposed method with different AE sensor quantities.

As illustrated in Figure 7, the comparative study results of varying sensor quantities indicate that the accuracy of the proposed method decreases as the number of sensors is reduced. In comparison to using all 6 sensors, the localization MAE increases by 32.08% when utilizing 5 sensors, 50.87% with 4 sensors, and 61.21% with 3 sensors. Despite the increase in localization error with a decrease in the number of sensors, the maximum increment in error is only 4.56mm. As the number of sensors increases, the localization error rapidly diminishes. This suggests that, while the accuracy of the proposed method decreases with a reduction in the number of sensors, it still effectively determines the coordinates of structural damage sources.

The results demonstrate the capability of the proposed method to accurately locate the coordinates of structural damage sources using only a small number of sensors. Even with a limited sensor configuration, the method exhibits the potential for precise localization, emphasizing its effectiveness in identifying the source of structural damage.

Moreover, to investigate the impact of sensor placement on the localization accuracy of the proposed method, a study was conducted focusing on the case with 4 sensors. Eight different sensor arrangement scenarios were considered, labeled as case1 through case8. The specific sensor layouts for each case are detailed in Table 1.

Table 1. Sensor layouts for each case with different sensor placement.

	Sensor number
case 1	S3, S4, S5, S6
case 2	S2, S4, S5, S6
case 3	S2, S3, S5, S6
case 4	S2, S3, S4, S6
case 5	S2, S3, S4, S5
case 6	S1, S3, S5, S6
case 7	S1, S3, S4, S6
case 8	S1, S2, S5, S6

A comparative study of the localization results for different cases is carried out. The 5-fold cross-validation method is employed, and the MAE is used to measure the accuracy of the results. The comparative study results are presented in Figure 8.

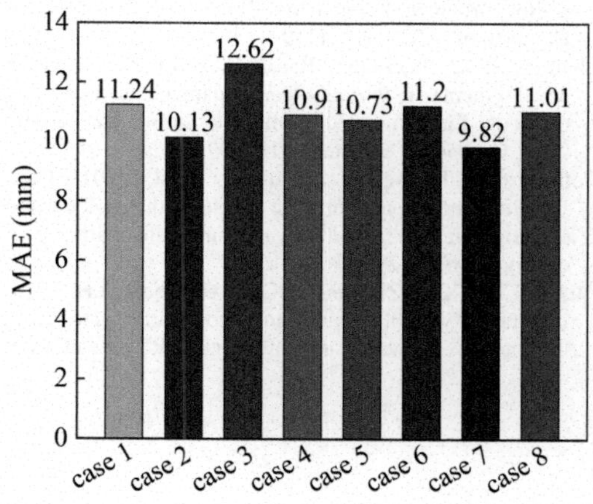

Figure 8. Comparison of the localization performance of the proposed method with different AE sensor placement.

The results indicate that the location and distribution of sensors have a certain impact on the localization accuracy of the proposed method, although the magnitude of this influence is relatively small. The maximum difference among these various cases is only 2.8 units, indicating that, when the number of sensors is fixed, the proposed method is not highly sensitive to variations in sensor position and distribution. This effectively reduces the complexity associated with sensor deployment.

5 CONCLUSIONS

An innovative GNN-based method for localizing AE sources for a composite panel was proposed in this paper. In this method, the AE sensor network was modelled as graph structure data and the connectivity of the graph was determined by the distance between AE sensors. Experimental study on a GFRP plate panel shows that the proposed method can achieve effective localization of AE sources, and the results of 5-fold cross validation reflect the robustness and effectiveness of the proposed method. Furthermore, the consideration of sensor position information as node features effectively enhances the graphical feature representation and significantly improves the performance of the proposed method for AE source localization. Moreover, the proposed method demonstrates promise in achieving accurate structural damage localization even with a reduced number of sensors. Additionally, its insensitivity to changes in sensor position enhances its practical applicability by simplifying the challenges associated with sensor arrangement.

REFERENCES

Bandara, S. Herath, M. & Epaarachchi, J. 2023. Sensory methods and machine learning based damage identification of fibre-reinforced composite structures: An introductory review. Journal of Reinforced Plastics and Composites, 42(21-22), 1119–1146.

Chaupal, P. & Rajendran, P. 2023. A review on recent developments in vibration-based damage identification methods for laminated composite structures: 2010–2022. Composite Structures, 116809.

Defferrard, M. Bresson, X. & Vandergheynst, P. 2016. Convolutional neural networks on graphs with fast localized spectral filtering. Advances in neural information processing systems, 29.

Hu, W. Choi, K.K. Zhupanska, O. & Buchholz, J.H. 2016. Integrating variable wind load, aerodynamic, and structural analyses towards accurate fatigue life prediction in composite wind turbine blades. Structural and multidisciplinary optimization, 53, 375–394.

Jiang, C. & Chen, N.Z. 2023. Graph Neural Networks (GNNs) based accelerated numerical simulation. Engineering Applications of Artificial Intelligence, 123, 106370.

Jiang, W. Luo, J. He, M. & Gu, W. 2023. Graph Neural Network for Traffic Forecasting: The Research Progress. ISPRS International Journal of Geo-Information, 12(3), 100.

Kalanchiam, M. & Chinnasamy, M. 2012. Advantages of composite materials in aircraft structures. *International Journal of Aerospace and Mechanical Engineering*, 6 (11), 2428–2432.

Kim, J. Kang, H. & Kang, P. 2023. Time-series anomaly detection with stacked Transformer representations and 1D convolutional network. Engineering Applications of Artificial Intelligence, 120, 105964.

Li, C. Mo, L. Yan, R. 2020. Rolling Bearing Fault Diagnosis Based on Horizontal Visibility Graph and Graph Neural Networks, proceedings of the 2020 International Conference on Sensing, Measurement & Data Analytics in the era of Artificial Intelligence (ICSMD). 2020. IEEE, 275–9.

Li, T. Zhou, Z. Li, S. Sun, C. Yan, R. & Chen, X. 2022. The emerging graph neural networks for intelligent fault diagnostics and prognostics: A guideline and a benchmark study. Mechanical Systems and Signal Processing, 168, 108653.

Moradi, M. Broer, A. Chiachío, J. Benedictus, R. Loutas, T.H. & Zarouchas, D. 2023. Intelligent health indicator construction for prognostics of composite structures utilizing a semi-supervised deep neural network and SHM data. Engineering Applications of Artificial Intelligence, 117, 105502.

Scarselli, F. Gori, M. Tsoi, A.C. Hagenbuchner, M. & Monfardini, G. 2008. The graph neural network model. IEEE transactions on neural networks, 20(1), 61–80.

Wu, L. Chen, Y. Shen, K. Guo, X. Gao, H. Li, S. ... & Long, B. 2023. Graph neural networks for natural language processing: A survey. Foundations and Trends® in Machine Learning, 16(2), 119–328.

Yu, X. Tang, B. & Zhang, K. 2021. Fault diagnosis of wind turbine gearbox using a novel method of fast deep graph convolutional networks. IEEE Transactions on Instrumentation and Measurement, 70, 1–14.

Zhao, Z. & Chen, N.Z. 2022. Acoustic emission based damage source localization for structural digital twin of wind turbine blades. Ocean Engineering, 265, 112552.

Zhao, Z. & Chen, N.Z. 2023a. Acoustic Emission Based Damage Source Localization for Heterogeneous Structure of Wind Turbine Blades Using Long Short-Term Memory Neural Networks. In International Conference on Offshore Mechanics and Arctic Engineering (Vol. 86847, p. V002T02A009). American Society of Mechanical Engineers.

Zhao, Z. & Chen, N.Z. 2023b. Spatial-temporal graph convolutional networks (STGCN) based method for localizing acoustic emission sources in composite panels. Composite Structures, 323, 117496.

Maintenance

Advances in Maritime Technology and Engineering – Guedes Soares & Santos (Eds)

Model predictive control framework for optimizing offshore wind O&M

M. Borsotti, R.R. Negenborn & X. Jiang
Department of Marine and Transport Technology, Technical University of Delft, Delft, The Netherlands

ABSTRACT: Offshore wind farms are a promising source of renewable energy, but they face significant challenges in terms of operation and maintenance (O&M). Traditional scheduling models often overlook the potential of condition-based maintenance (CBM). Addressing this gap, this paper introduces a novel framework, incorporating principles of Model Predictive Control (MPC), to optimize the O&M scheduling of offshore wind farms using prognostic-driven maintenance. The framework integrates probabilistic remaining useful life (RUL) prognosis in a mixed-integer linear programming (MILP) optimization model with a rolling horizon approach, in alignment with MPC's predictive and adaptive decision-making approach. The optimization model determines the optimal time to replace each component by minimizing the expected cost over the expected lifetime. This approach seeks to achieve the lowest expense while guaranteeing the highest utilization rate of each component. For the case study presented, the total O&M costs are reduced by up to 15% with respect to corrective maintenance strategies.

1 INTRODUCTION

In recent years, the global focus on renewable energy sources has intensified, partly driven by the urgent need to address climate change. Offshore wind farms have emerged as a promising and significant contributor in this domain. In 2021, a record number of offshore wind farm projects were commissioned, indicating a robust growth trajectory for this sector. The cumulative deployment of offshore wind energy is estimated to reach approximately 117 GW by 2027, and potentially 370 GW by 2031, according to the US Department of Energy (2022) (US Department of Energy 2022). This expected growth positions offshore wind as a pivotal component in achieving global electricity generation targets and in assisting countries to meet their climate and renewable energy objectives.

However, the high cost of offshore wind, compared to other renewable technologies, remains a significant challenge (NREL 2023). A critical factor influencing this cost is Operation and Maintenance (O&M), which accounts for a substantial portion of the total lifecycle costs of offshore wind projects, estimated to be between 25-30%, as opposed to 10-15% for onshore wind farms (National Renewable Energy Laboratory 2022), (van Bussel & Schöntag 1997).

Efficient and effective O&M strategies are essential to minimize downtime, enhance the performance and availability of the wind farm, and thereby reduce the overall Levelized Cost of Electricity (LCoE). These strategies include reactive and proactive maintenance (Ren et al. 2021).

Proactive strategies such as predictive maintenance are based on Condition Based Monitoring (CBM) and can substantially reduce the cost of O&M (Van Horenbeek & Pintelon 2012b), CBM involves monitoring the health and performance of wind turbine components in real-time or periodically, using sensors and other monitoring systems (Kou et al. 2022). By analyzing the collected data, trends and patterns, anomalies can be detected, allowing for timely maintenance actions (Le & Andrews 2016).

Nevertheless, traditional scheduling models in offshore wind farm maintenance primarily rely on fixed age thresholds for the planning of maintenance tasks, while overlooking the potential of CBM (May et al. 2015). This approach can often lead to inefficiencies since it does not account for the actual condition of the components, leading to scenarios where maintenance is either performed too early, resulting in unnecessary costs, or too late, leading to unexpected failures and extended downtimes (Fox et al. 2022).

In recent years, the advent of advanced control strategies has significantly impacted the operational efficiency of complex systems. Among these, Model Predictive Control (MPC) stands out as a form of control strategy that utilizes a model of the process to predict future outcomes and make decisions that optimize a set of predefined objectives over a future time horizon. It is characterized by its ability to anticipate future events and take control actions accordingly, making it an ideal framework for managing systems where the cost of operational decisions is high and the need for reliability is critical (Mayne et al. 2000).

DOI: 10.1201/9781003508762-65

Table 1. Acronyms and nomenclature.

Symbol	Definition
OFW	Offshore Wind Farm
O&M	Operation and Maintenance
CBM	Condition-Based Maintenance
MPC	Model Predictive Control
RUL	Remaining Useful Life
MILP	Mixed-Integer Linear Programming
CTV	Crew Transfer Vessel
SOV	Service Operation Vessel
LCoE	Levelized Cost of Electricity
JUV	Jack-Up Vessel
HLV	Heavy Lift Vessel
t	Time in days since the beginning of the observation period
k	The current day or time step since the beginning of component operation
t_k	The time in days since the component V has been in operation, considering the current day k
L	Length of the maintenance planning horizon
d_p	Present day in the planning horizon
τ	Time step in the planning horizon
S	Set of all possible maintenance schedules
V	Set of components in the wind turbine
j^V	Vessel used for the replacement of component V
$c^V_{preventive}$	Cost of preventive replacement of component V
$c^V_{corrective}$	Cost of corrective replacement of component V
$T^V_{preventive}$	Preventive replacement time for component V
$T^V_{corrective}$	Corrective replacement time for component V
N_{tech}	Number of technicians necessary for replacement
R_{hour}	Hourly rate of a technician
C_{el}	Cost of electricity
$P(V)$	Power curve of a turbine
v_{mean}	Average wind speed
$\phi^V_k(i)$	Probability that the RUL of component V is exactly i days, after being used for k days
$c^V_{dt}(t_k)$	Expected cost of potential downtime losses for component V over time
x^V_s	Binary decision variable indicating if component V is replaced at time s
W_s	Binary variable indicating if weather conditions are suitable at time s
$Y^{j^V}_s$	Binary variable indicating if vessel j^V is available at time s
T^V_s	Binary variable indicating if technicians with the required skills to replace component V are available at time s
$E_R[C_V(k,t_k)]$	Expected cost of replacement after t_k days
$E_{DN}[C_V(k, L)]$	Expected cost of doing nothing over the planning horizon L
$E_R[L_V(k,t_k)]$	Expected lifetime after replacement at time t_k
$E_{DN}[L_V(k,L)]$	Expected lifetime without replacement over the planning horizon L
C^V_R	Cost over expected lifetime ratio for replacement
C^V_{DN}	Cost over expected lifetime ratio for doing nothing

The integration of MPC principles into these maintenance strategies offers a promising approach to further address these challenges. By adopting a framework that includes both predictive and adaptive elements, we can more effectively balance maintenance needs with operational efficiency.

The methodological innovation in O&M scheduling proposed in this paper involves the development of a novel framework that incorporates probabilistic RUL prognosis into a mixed-integer linear programming (MILP) model. Such an approach not only aims to minimize O&M costs but also to maximize the utilization rate of each component, a critical factor in the profitability, viability, and sustainability of offshore wind farms.

1.1 *Paper structure*

This paper will explore the concept of prognostic-driven maintenance in depth, elaborating on its methodology, implementation, and benefits. It aims to contribute to the existing body of knowledge by addressing the gap in current O&M scheduling practices and demonstrating the potential advantages of a prognostic-driven approach.

An extensive review of current methodologies in O&M for offshore wind farms is outlined in Section 2, showcasing how current strategies incorporate fault detection, but often overlook the integration of fault prognosis. This section identifies gaps in the literature and positions the paper within the context of existing research. The methodology is explained in Section 3, detailing the proposed prognostic-driven maintenance model. It explains how the model is constructed, the synthetic model used for RUL prediction in the absence of real-time data, and the mathematical formulations used for the optimization of maintenance scheduling. A hypothetical case scenario is outlined in Section 4, detailing the specific components, operational parameters, and maintenance requirements of a single offshore wind turbine. This scenario serves as a practical example to illustrate the application of the prognostic-driven scheduling model. The findings from applying the model to the case scenario are finally presented in Section 5. The results section includes graphs and figures to visualize the cost implications of different maintenance strategies over time, offering insights into the model's effectiveness. Lastly, in Section 6, the reader will find a critical analysis of the model's outcomes, its implications for the O&M of offshore wind farms, and a discussion on potential areas for further research and model refinement.

2 LITERATURE REVIEW

Short-term scheduling in offshore wind farm O&M is a critical factor for ensuring operational efficiency and reliability. This horizon focuses on the daily or weekly planning of maintenance tasks, influenced by real-time factors like weather conditions, component health, and logistical considerations (Irawan et al. 2017).

Recent advancements in technology, particularly in data analytics and predictive modelling, have enabled new, sophisticated approaches to short-term scheduling. The integration of real-time data from sensors, coupled with advanced prognostic models, allows for more accurate predictions of component health and maintenance needs (Le & Andrews 2016).

Several studies have highlighted the effectiveness of prognostic-driven approaches in short-term scheduling. These case studies provide valuable insights into the practical applications and benefits of adopting a prognostic-driven approach in short-term scheduling scenarios (Van Horenbeek & Pintelon 2012a).

The ongoing research in short-term scheduling is oriented towards further refining prognostic models and integrating them with dynamic scheduling systems. The focus is on enhancing the accuracy of predictions and developing more agile scheduling tools that can respond in real-time to changes in component conditions and environmental factors.

2.1 *Short-term scheduling in offshore wind farms*

Short-term scheduling within offshore wind farms O&M remains an intricate process that demands meticulous analysis and planning. The state of the art in short-term scheduling predominantly revolves around optimizing the routing of vessels and timing of maintenance activities.

In the literature, various strategies have been proposed to tackle these challenges. The routing and scheduling problem is often addressed by optimizing the paths and schedules for vessels to service wind turbines (Irawan et al. 2017). These models aim to minimize costs by reducing travel times, optimizing resource allocation, and ensuring safe and efficient transportation (Ade Irawan et al. 2023).

However, a significant aspect of these strategies is the reliance on fixed age thresholds for maintenance actions. These thresholds are identified using methodologies that optimize the maintenance schedule based on the age of the components (Sarker & Ibn Faiz 2016). This approach has been the traditional method for maintenance planning, where tasks are scheduled at predetermined intervals, regardless of the actual condition of the components.

Age threshold optimization techniques in the literature underscore the influence of age groups and thresholds on the overall maintenance costs (Sarker & Ibn Faiz 2016),(Li et al. 2022). These techniques utilize various decision rules and stochastic models to determine the most cost-effective moments for maintenance actions (Safaei et al. 2020). However, they do not fully exploit the potential benefits of prognostic information that could further optimize scheduling.

In contrast to this traditional approach, recent advancements have seen a shift towards prognostic-driven scheduling models. These models deviate from fixed age thresholds by employing probabilistic RUL predictions to inform maintenance decisions (Li et al. 2020). Prognostic-driven scheduling considers the actual condition and predicted future state of health of the components, potentially offering a more dynamic and cost-effective maintenance strategy.

The integration of prognostics into maintenance planning for OWFs represents a paradigm shift from a purely reactive to a more predictive and proactive approach. The impact that advanced monitoring strategies can have on lifetime O&M costs for offshore wind turbines is evaluated in (Turnbull & Carroll 2021), here the authors showed a potential cost reduction of up to 8% in direct O&M costs (transport, staff and repair costs) and up to 11% reduction in lost production, where the major source of savings are obtained through early intervention to avoid failure and major component replacement.

In conclusion, while fixed age threshold optimization has served as a foundational approach for short-term scheduling in OWFs, the advent of prognostic-driven models is suggesting the beginning of a new era in maintenance strategy. This emerging approach, grounded in probabilistic RUL predictions, aims to redefine the maintenance optimization landscape, promising increased efficiency, reduced costs, and improved system reliability for OWFs.

2.2 *Short-term scheduling in other industries*

A review of contemporary literature in the domain of maintenance scheduling for aircrafts, reveals a variety of methodological approaches, each contributing uniquely to the development of more refined and predictive O&M scheduling models. Prognostic models are central to forecasting the RUL of components. The methodologies employed across recent studies vary, with some common approaches being highlighted.

Convolutional Neural Networks (CNN) and Monte-Carlo dropout have been used by (Mitici et al. 2023), (Lee & Mitici 2023), (de Pater et al. 2022), (de Pater & Mitici 2021), employing a probabilistic Remaining Useful Life approach that estimates the likelihood of component failure within a given time frame, contributing to a more precise and cost-effective scheduling process.

Similarly, Bayesian Deep Learning (BDL), Long Short-Term Memory networks (LSTM) and Feed-forward Neural Networks (FNN) are used by (Zhuang et al. 2023). Kalman Filters are employed by (Vianna & Yoneyama 2018), where degradation trends and future wear values are estimated considering an implementation of a multiple model approach of the extended Kalman filter technique. The RUL prognosis methodologies presented are summarized in Table 2

Table 2. RUL Prognosis methodologies.

Authors	CNN	BDL	LSTM	FNN	Kalman Filter	NA
(Mitici et al. 2023)	✓					
(Zhuang et al. 2023)		✓	✓	✓		
(Lee & Mitici 2023)	✓					
(de Pater et al. 2022)	✓					
(de Pater & Mitici 2021)	✓					
(Chen et al. 2021)			✓			
(Camci et al. 2019)						✓
(Vianna & Yoneyama 2018)					✓	
(Li et al. 2016)						✓

The optimization of maintenance scheduling is influenced by the predicted RUL, where various methods are employed to ensure cost-effectiveness and high component utilization.

Mixed-Integer Linear Programming (MILP) is a widely adopted method, seen in most of the reviewed papers such as (Mitici et al. 2023), (Zhuang et al. 2023), (de Pater et al. 2022),(de Pater and Mitici 2021), (Chen et al. 2021), (Camci et al. 2019), providing a structured approach to optimizing cost while maintaining high component utilization rates; in (Mitici et al. 2023) the authors claim that through their novel approach, 95.6% of unscheduled maintenance can be prevented.

The predictive maintenance planning can also be formulated as a Deep Reinforcement Learning (DRL) problem, as seen in (Lee & Mitici 2023), this approach can be particularly useful in dealing with the dynamic and complex nature of maintenance scheduling. Exhaustive search methods, although computationally intensive, provide a thorough exploration of all possible solutions and are used in (Vianna & Yoneyama 2018).

Rolling-Horizon strategies, as seen in (Mitici et al. 2023) and (Zhuang et al. 2023), are employed for their adaptability and the capacity to update decisions with new information over time. The main scheduling optimization methods discussed are summarized in Table 3.

Table 3. Scheduling optimization methods.

Authors	MILP	DRL	Exhaustive search	Rolling-Horizon
(Mitici et al. 2023)	✓			✓
(Zhuang et al. 2023)	✓			✓
(Lee & Mitici 2023)		✓		
(de Pater et al. 2022)	✓			✓
(de Pater & Mitici 2021)	✓			✓
(Chen et al. 2021)	✓			
(Camci et al. 2019)	✓			
(Vianna & Yoneyama 2018)			✓	
(Li et al. 2016)	✓			

The objective functions in these models often include maintenance costs and the utilization rate of the system. For instance, (Chen et al. 2021) minimizes the expected costs over the component's expected lifespan (cost rate) while others only minimize maintenance cost (Li, Guo, & Zhou 2016). An overview of the objective functions used by each author is presented in Table 4

Table 4. Objective functions in scheduling models.

Authors	Maintenance Cost Rate	Maintenance Cost
(Mitici et al. 2023)	✓	
(Zhuang et al. 2023)	✓	
(Lee & Mitici 2023)	✓	
(de Pater et al. 2022)	✓	
(de Pater & Mitici 2021)	✓	
(Chen et al. 2021)	✓	
(Camci et al. 2019)		✓
(Vianna & Yoneyama 2018)		✓
(Li et al. 2016)		✓

In summary, the literature presents a rich tapestry of methodologies aimed at optimizing short-term scheduling. The trend towards integrating predictive analytics and probabilistic models is clear, with a strong focus on optimizing maintenance activities based on real-time data and advanced diagnostic techniques. These studies form the bedrock upon which future operational frameworks can be developed for offshore wind farms.

2.3 *Research gap*

As we have seen, the state of the art in scheduling for offshore wind farms (OWFs) primarily focuses on optimizing the allocation of resources, minimizing travel time and costs, and ensuring the timely completion of maintenance tasks. The literature reveals that, while fault detection is commonly accounted for in current methodologies, there is a lack of integration of fault prognosis and future component health predictions into these scheduling models.

Task sequencing and scheduling play crucial roles in minimizing downtime and maximizing resource utilization, however, current approaches typically do not integrate prognostic-driven strategies that can predict the future state of health of components. This represents a significant gap in the literature, as the incorporation of such strategies could potentially lead to more effective and cost-efficient scheduling by anticipating future maintenance needs and avoiding the pitfalls of reactive maintenance approaches.

Table 5. Maintenance scheduling in literature.

Authors	Consider fault detection	Consider fault prediction
(Irawan et al. 2017)	✓	
(Dai et al. 2015)	✓	
(Stock-Williams & Swamy 2019)	✓	
(Pattison et al. 2016)	✓	
(Raknes et al. 2017)	✓	
(Ade Irawan et al. 2023)	✓	
(Nachimuthu et al. 2019)	✓	
(Lazakis & Khan 2021)	✓	
(Mazidi et al. 2017)	✓	
(Sinha et al. 2013)	✓	
(Dawid et al. 2018)	✓	
(Liu et al. 2019)	✓	
(Mitici et al. 2023)[1]	✓	✓
(Zhuang et al. 2023)[1]	✓	✓
(Tseremoglou et al. 2023)[1]	✓	✓
(Lee & Mitici 2023)[1]	✓	✓
(de Pater et al. 2022)[1]	✓	✓
(de Pater & Mitici 2021)[1]	✓	✓
(Chen et al. 2021)[1]	✓	✓
(Camci et al. 2019)[1]	✓	✓
(Vianna & Yoneyama 2018)[1]	✓	✓
(Gogu 2018)[1]	✓	✓
(Li, Guo, & Zhou 2016)[1]	✓	✓
(Rodrigues et al. 2015)[1]	✓	✓

Note: Papers marked with[1] are related to the maintenance of aircrafts, the rest is related to the maintenance of OWFs.

On the other hand, as illustrated in Table 5, while research in the field of offshore wind maintenance is currently considering only fault detection, the aircraft maintenance industry has made significant strides in incorporating prognostic information and fault predictions into its scheduling models. The advanced adoption of CBM in aircraft maintenance provides a framework from which OWF maintenance can draw inspiration. The similarities between the two industries, such as the criticality of maintenance for safety and efficiency, the high costs associated with downtime, and the technical complexity of the systems involved, justify the adoption of similar methodologies in O&M scheduling for OWFs.

Addressing these challenges, the primary objective of this research is to propose a novel framework that not only integrates a prognostic-driven maintenance approach into the O&M scheduling process of offshore wind farms but also makes use of the predictive and adaptive capabilities of MPC. This framework aims to optimize maintenance activities based on probabilistic RUL predictions, thereby reducing costs and maximizing operational efficiency and energy production. By incorporating MPC, the research seeks to bridge the gap in current O&M practices, offering a dynamic scheduling model that can anticipate and adapt to the operational state of the wind farm components.

3 METHODOLOGY

The methodology applied in this research constructs a prognostic-driven optimization model for the short-term scheduling of maintenance activities in offshore wind farms. The proposed model is designed to integrate the probabilistic estimation of Remaining Useful Life (RUL) for components within a cost optimization framework, enriched by the principles of Model Predictive Control (MPC). MPC's predictive capabilities enable the model to consider not just current component states but also forecast future conditions, thereby allowing for proactive maintenance decisions that can adapt to changing operational needs.

This chapter describes the foundational framework of the model.

3.1 *Synthetic model for probabilistic RUL prediction*

To address the inherent uncertainties in predicting the Remaining Useful Life (RUL) of offshore wind turbine components, our study employs a synthetic model. This model is instrumental in generating probabilistic RUL predictions in the absence of real-time condition-based monitoring data.

The true RUL (RUL_{true}) of the components is considered a latent variable that follows a Weibull distribution, a common choice for modeling the life of mechanical systems. The probability density function of the Weibull distribution is defined as:

$$f(t;\lambda,\alpha) = \frac{\alpha}{\lambda}\left(\frac{t}{\lambda}\right)^{\alpha-1} e^{-\left(\frac{t}{\lambda}\right)^{\alpha}} \quad (1)$$

where t is the ime in days since the beginning of the observation period, λ is the scale parameter and α is the shape parameter. To simulate the true RUL, we use inverse transform sampling, where:

$$RUL_{true} = \lambda(-\ln(U))^{1/\alpha}, \quad U \sim \text{Unif}(0,1) \quad (2)$$

where U is a random variable uniformly distributed between 0 and 1, used in the inverse transform sampling method.

Our model generates point predictions of RUL and characterizes the uncertainty around these predictions, which decreases over time as more information becomes available. The range for the standard deviation of the predictions is set by:

$$\text{std}_{max} = RUL_{true} \cdot (\text{large uncertainty}) \quad (3)$$

$$\text{std}_{min} = RUL_{true} \cdot (\text{small uncertainty}) \quad (4)$$

where std_{max} is the maximum standard deviation representing the initial uncertainty in the RUL prediction and std_{min} is the minimum standard deviation representing the reduced uncertainty as more information becomes available over time.

The confidence in the prediction increases with time, converging towards the lower bound of uncertainty. This is modeled as:

$$\text{std}(t) = \text{std}_{max} - (\text{std}_{max} - \text{std}_{min}) \cdot \left(\frac{1}{1+\exp\left(-\frac{t-a\cdot RUL_{true}}{b\cdot RUL_{true}}\right)}\right) \quad (5)$$

where std(t) is the confidence in the RUL prediction at time t, which is a function of time reflecting decreasing uncertainty. The maximum and minimum standard deviation, as well as the values of parameters a and b used in Equation 5, can be modified to represent different behaviours of the RUL predictions in time. Subsequently, the RUL prediction adjusted for error at time t is calculated as:

$$RUL_{pred}(t) = RUL_{true} - t + \text{error}(t) \quad (6)$$

with the error term modeled as a normally distributed random variable with mean 0 and standard deviation corresponding to the confidence at time t is calculated as:

$$\text{error}(t) \sim N(0, \text{std}(t)) \quad (7)$$

The synthetic model allows us to construct a probability distribution for the RUL at any given day d, denoted by ϕ^d. This distribution is a normal distribution centered around the point prediction $RUL_{pred}(t)$ with a standard deviation std(t):

$$\phi^d(t) = N\big(RUL_{pred}(t), \text{std}(t)\big) \quad (8)$$

The synthetic model serves as a stand-in for actual condition-based monitoring data, which is often scarce or unavailable. By creating a simulated environment, we can test the robustness of our prognostic framework and its underlying assumptions. Furthermore, this model provides a controlled setup to assess the performance of maintenance scheduling algorithms and to investigate the effects of different parameters on the maintenance optimization process. Despite the benefits of using real sensor data for condition monitoring and prognostics, the synthetic model offers a valuable alternative for conducting preliminary analyses.

Ultimately, while the synthetic model provides valuable insights, it is designed to be supplanted by real-life condition-based monitoring systems as they become more prevalent and integrated into wind farm O&M practices.

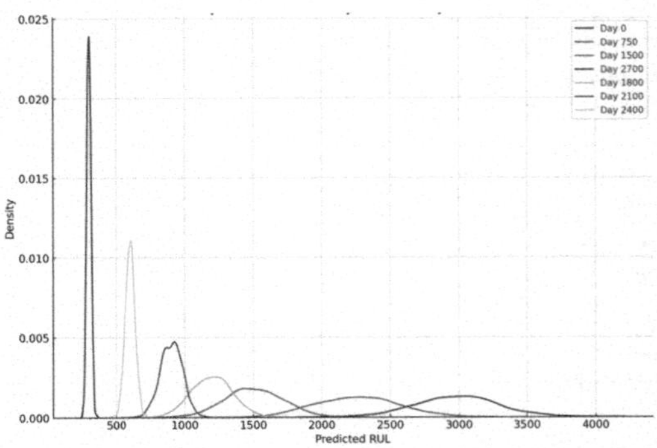

Figure 1. Probability Distributions of Synthetic Daily RUL Predictions.

In Figure 1, we observe the evolution of the probability distributions of the predicted Remaining Useful Life (RUL) at different moments of a component's operational span. This visualization is critical for understanding how the prognostic model's certainty improves as the component ages.

The graph illustrates several Gaussian distributions corresponding to different days in the component's lifecycle. Each curve represents the probability density function $\phi^d(t)$ for the RUL prediction on given days. The values of the days visualized in the graph have been chosen arbitrarily at intervals that give a comprehensive overview of the behavior of the curves during the lifetime of a component, while also highlighting the transition from early uncertainty to later confidence in RUL predictions. As time progresses, these distributions evolve in the following manner:

Early Life (Day 0, Blue Curve): Initially, the RUL distribution is wide with a peak far from zero, indicating that the component has a long expected lifespan ahead. However, the broad spread of the curve reflects significant uncertainty in this early prediction. This uncertainty is due to the lack of operational data and the inherent unpredictability at the beginning of the lifecycle.

Mid Life (Days 750 and 1500, Green and Red Curves): As the component transitions into its mid-life, the distributions start to narrow, indicating an increase in confidence regarding the RUL. The peak of the distributions starts to shift left, towards lower RUL values, as the component naturally ages and accumulates wear and tear.

Approaching End of Life (Days 1800 to 2400, Yellow to Purple Curves): As the component nears its end of life, the distributions become increasingly sharper and more skewed towards the left, indicating that the remaining lifespan is diminishing. The peak of these distributions gets closer to zero, and the narrowing of the curves signifies a higher confidence in the RUL prediction. This increased precision is likely due to the accumulation of more significant operational data and wear patterns, allowing the model to make more accurate forecasts.

The sharpening of the distributions towards the end of the component's life is a crucial aspect of the prognostic model. It represents a condition-monitoring framework where, as more operational data becomes available over time, the RUL can be predicted with greater accuracy.

3.2 *Computation of maintenance costs*

Maintenance activities for offshore wind farms, for the purpose of this study, consist of corrective and preventive replacements. The costs associated with these activities are computed by considering various factors such as transportation, labor, materials, and potential downtimes. We disregard major and minor repairs, focusing solely on replacements. Each cost component is detailed below.

The transportation cost is associated with the usage of vessels to carry out maintenance activities. It is computed based on the type of vessel used, the duration of the activity, and the distance to the wind farm. The transportation cost C_{trans} is given by:

$$C_{trans} = C_{mobil} + (F_{consumption} \times D_{travel} \times 2) + (R_{day} \times T_{duration}) \tag{9}$$

Equation (9) includes three elements: C_{mobil} is the mobilization cost i.e. the initial cost to deploy the vessel, $F_{consumption}$ is the fuel consumption, calculated by multiplying the vessel's fuel consumption rate by the travel distance D_{travel} multiplied by two to take into account both travels, the first one to reach the wind turbine and the second one to get back to the onshore base, R_{day} is the operational cost of the vessel per day, multiplied by the duration of the activity $T_{duration}$ expressed in days.

The cost of labor is determined by the number of technicians needed, the duration of the activity, and the hourly rate. The technician cost formula is:

$$C_{tech} = N_{tech} \times R_{hour} \times T_{duration} \tag{10}$$

where C_{tech} is the technician cost, N_{tech} is the number of technicians, R_{hour} is the hourly rate, and $T_{duration}$ is the duration of the maintenance activity in hours.

Downtime costs refer to the loss of revenue due to the turbine being non-operational during maintenance. It is calculated by considering the electricity cost rate, the duration of downtime, and the rated power output of the turbine, the formula is:

$$C_{dt} = C_{el} \times P(v_{mean}) \times T_{duration} \tag{11}$$

where C_{dt} is the downtime cost, C_{el} is the cost of electricity per kWh, T_{dt} is the downtime duration, and $P(v)$ represents the power curve of the turbine i.e. its power output as a function of wind speed, therefore $P(v_{mean})$ is the power output at the average wind speed at the selected location.

The total cost of maintenance is the sum of the replacement material cost which is given, transportation cost, technician cost, and downtime cost. The formula is as follows:

$$C_{maintenance} = C_{material} + C_{trans} + C_{tech} + C_{dt} \tag{12}$$

These computations ensure a comprehensive understanding of the costs involved in the maintenance of offshore wind farms.

3.3 *Expected costs*

Two key cost considerations form the basis of the model: the expected cost of replacement and the expected cost of doing nothing, denoted by

$E_R[C_V(k,t_k)]$ and $E_{DN}[C_V(k,L)]$, respectively. The expected cost of replacement after t_k days is calculated using equation (13) which computes the expected total cost by considering both the risks of failure and the costs associated with preventive actions.

Furthermore, in the offshore wind farm O&M scheduling, downtime can significantly affect the total cost due to lost production. Thus, it is necessary to incorporate the expected cost of downtime losses due to potential unexpected failures, c_{dt}^V, into the optimization model. This cost is associated with the expected production losses when the turbine is unavailable due to the failure of component V, in contrast to the downtime cost computed in Equation 11 which is used to evaluate only the downtime and power losses during the maintenance activity itself.

The expected cost of potential downtime losses is calculated over the time horizon t_k and is a function of the cost of electricity C_{el}, the probability of component V failing at each time i, denoted as $\phi_k^V(i)$, the average power production of the turbine $P(v_{mean})$, and the average electricity market price. The cost is accumulated over all possible failure moments within the planning horizon, weighted by the duration of downtime that would result from a failure at each moment.

$$E_R[C_V(k,t_k)] = c_{corrective}^V \sum_{i=0}^{t_k-1} \phi_k^V(i) + c_{preventive}^V \left(1 - \sum_{i=0}^{t_k-1} \phi_k^V(i)\right) + c_{dt}^V \sum_{i=0}^{t_k-1} \phi_k^V(i)(t_k - i - 1) \tag{13}$$

8where $E_R[C_V(k,t_k)]$ is the expected cost of replacement after the component V has been used for $k + t_k$ days. $\phi_k^V(i)$ is the probability that the RUL of component V is exactly i days, after being used for k days. $c_{preventive}$ is the cost of preventive maintenance. $c_{corrective}$ is the cost of corrective maintenance, including downtime and potential penalties and the daily cost of downtime losses due to delayed maintenance, c_{dt}^V, is given by the equation:

$$c_{dt}^V = C_{el} \cdot P(v_{mean}) \cdot 24 \tag{14}$$

The expected cost of doing nothing denotes the expected cost if no maintenance action is taken within the planning horizon L. Here, the cost is purely the cost of corrective actions, as no preventive maintenance is performed. It is calculated by summing the probabilities of failure for each day within the planning horizon and multiplying by the cost of corrective maintenance. The expected cost of doing nothing if the component is not replaced within the period $[k,k + L]$ is given by:

$$E_{DN}[C_V(k,L)] = c_{corrective}^V \sum_{i=0}^{L-1} \phi_k^V(i) + c_{dt}^V \sum_{i=0}^{L-1} \phi_k^V(i)(L - i - 1) \tag{15}$$

3.4 *Expected cost/lifetime ratio*

The model also estimates the expected lifetime for both replacement and doing nothing scenarios.

$$E_R[L_V(k,t_k)] = k + \sum_{i=0}^{t_k-1} i \cdot \phi_k^V(i) + t_k \cdot \left(1 - \sum_{i=0}^{t_k-1} \phi_k^V(i)\right) \tag{16}$$

$E_R[L_V(k,t_k)]$ is the sum of the days the component has been in service k, the remaining useful life weighted against its probability distribution $\phi_k^V(i)$, and the additional lifetime t_k if no failure occurs. This gives a comprehensive view of the expected operational life of a component considering both scenarios, failure and no failure within the time to replacement. Furthermore, it accounts for the risk of early interventions and the potential reduction in component life due to the accumulated probability of failure.

$$E_{DN}[L_V(k,L)] = k + \sum_{i=0}^{L-1} i \cdot \phi_k^V(i) + L \cdot \left(1 - \sum_{i=0}^{L-1} \phi_k^V(i)\right) \tag{17}$$

$E_{DN}[L_V(k,L)]$ takes into account the component's current service life k and the weighted sum of the remaining useful life throughout the planning horizon L in both scenarios, failure and no failure within the planning horizon.

The decision-making process involves comparing the cost over the expected lifetime for replacement and doing nothing, for this reason we defined C_R^V as the ratio of the expected replacement cost over the expected lifetime and C_{DN}^V as the ratio of the expected cost of doing nothing over the expected lifetime without intervention, offering a perspective on the long-term cost implications of deferring maintenance.

$$C_R^V = \frac{E_R[C_V(k,t_k)]}{E_R[L_V(k,t_k)]} \tag{18}$$

$$C_{DN}^{V} = \frac{E_{DN}[C_V(k,L)]}{E_{DN}[L_V(k,L)]} \quad (19)$$

3.5 *Optimization model*

The optimization model is formulated to determine the optimal time to maintain a component to minimize the total expected cost of maintenance over the expected lifetimes of the components.

The decision variable is a binary variable that indicates whether component V is to be replaced at a specific time s.

$$x_s^V = \begin{cases} 1 & \text{if component V is replaced at time s} \\ 0 & \text{if component V is not replaced at time s} \end{cases} \quad (20)$$

The model aims to minimize the sum of the cost rates over the expected lifetime for all components, providing a decision framework that balances cost with component health and operational effectiveness. The objective function to be minimized is:

$$\min \left[\sum_{V \in V} (C_R^V \cdot x_s^V + C_{DN}^V \cdot (1 - x_s^V)) \right] \quad (21)$$

The model is subject to logical constraints to ensure feasibility:

$$\sum_{s \in S} x_s^V \leq 1 \quad \forall V \in V \quad (22)$$

$$x_s^V \in \{0, 1\} \quad \forall V \in V, \forall s \in S \quad (23)$$

Equation 22 ensures that for any given component *V*, at most one maintenance activity can be scheduled within the planning horizon.

Equation 23 defines the decision variable x_s^V as binary, where $x_s^V = 1$ indicates that maintenance is scheduled for component *V* at time *s*, and $x_s^V = 0$ indicates that no maintenance is scheduled.

Although the model currently focuses on these logical constraints, it is designed to be extensible. Future iterations should include operational constraints to account for the following factors:

Weather Window: Scheduling must consider the availability of suitable weather conditions for safe maintenance operations. The constraint ensures that maintenance is only scheduled when weather conditions are favorable.

$$x_s^V \leq W_s \quad \forall V \in V, \forall s \in S \quad (24)$$

Where W_s is a binary variable indicating if weather conditions are suitable at time *s*, depending on the available weather forecast and the operational limits of the vessels.

Vessel Capacity and Availability: Maintenance scheduling depends on the availability of vessels with the necessary capacity to transport technicians and equipment to the turbines. This constraint ensures that maintenance is only scheduled when a suitable vessel is available.

$$x_s^V \leq \sum_{j \in J} Y_s^{j^V} \quad \forall V \in V, \forall s \in S \quad (25)$$

Where $Y_s^{j^V}$ is a binary variable representing the availability of vessel j^V at time *s*, j^V being a vessel with a suitable capacity for carrying out the replacement of component *V*.

Spare Parts Availability: The availability of necessary spare parts must be assured for scheduled maintenance activities. This constraint ensures that maintenance is only scheduled when the required spare parts are available.

$$x_s^V \leq Z_s^V \quad \forall V \in V, \forall s \in S \quad (26)$$

Where Z_s^V is a binary variable indicating the availability of a spare component *V* at time *s*.

Technician Availability: Scheduling must align with the availability of technicians, both in terms of numbers required and their specific skill sets. This constraint ensures that maintenance is only scheduled when the necessary technicians are available.

$$x_s^V \leq T_s^V \quad \forall V \in V, \forall s \in S \quad (27)$$

Where T_s^V is a binary variable representing the availability of technicians with the required skills to replace component *V* at time *s*.

By incorporating these additional constraints, the model will more accurately reflect the complexities of real-world offshore wind farm maintenance operations. The integration of these factors would enhance the model's capability to generate schedules that are not only cost-effective but also operationally viable.

The rolling horizon approach updates the scheduling decisions at regular intervals to reflect the current state of the farm and maintenance needs. As new information becomes available, the model recalibrates the schedule, ensuring that it remains responsive to the actual conditions and performance of the wind farm components. The maintenance planning window moves forward with each time step τ, allowing for the reevaluation of maintenance decisions based on the latest data and predictions.

The current day is updated according to:

$$k = k + \tau \tag{28}$$

where d_p is the present day and τ is the time step within the planning horizon L.

Maintenance activities planned until $k + \tau, \tau < L$ are executed according to the updated schedule.

A comprehensive overview of the methodology proposed is presented in 2.

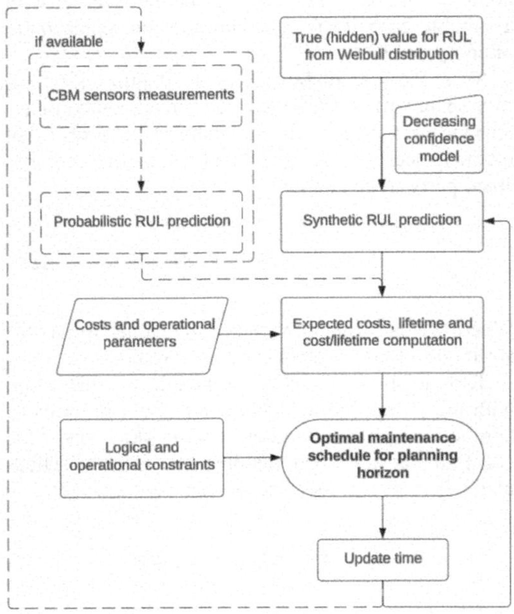

Figure 2. Flowchart of the proposed methodology.

4 CASE SCENARIO DESCRIPTION

In constructing the prognostic-driven optimization model, several assumptions were considered to reduce the complexity of the model.

The first assumption is that only replacement and preventive replacement are considered.

It is then assumed that the wind farm is always accessible and there are no delays or restrictions due to weather or other external factors, reflecting the ideal operational conditions where weather or other environmental factors do not impede maintenance activities. This could be applicable in regions with very stable and predictable weather conditions or where advanced forecasting allows for precise planning around weather constraints.

All necessary resources, including vessels, spare parts, and maintenance crews, are assumed to be available whenever required, aligning with situations where operations are well-funded, and resource management is efficient enough to maintain a surplus or quick availability of resources, possibly in smaller wind farms or farms close to supply bases.

Finally, only one maintenance activity can be performed at a time, precluding the possibility of simultaneous maintenance actions, which applies to a strategy that does not include opportunistic grouping of maintenance activities.

These assumptions could potentially lead to an underestimation of costs and scheduling times, as they do not account for delays and unavailability that often occur in real-world scenarios. While they allow for the development of an optimized maintenance schedule under ideal conditions, the actual implementation may require adjustments to account for the unpredictability of real-life operations and are left for further development of the model.

The proposed framework is applied to a single wind turbine located in an offshore wind farm characterized by specific environmental and operational conditions. Having assumed a strategy that does not capitalize on grouping maintenance tasks, addressing turbines individually becomes not just a simplification for modeling purposes, but a practical approach.

Cost data and failure rates for each component, as well as the duration of corrective and preventive replacements and the number of technicians required are presented in Table 6-8 and are taken or adjusted from (Li et al. 2022), (Golestani et al. 2023) and (Carroll et al. 2017), whereas other values are reasonably assumed and presented for the replicability of the model.

4.1 *Operational parameters*

A wind turbine consists of several critical components such as blades, bearings, gearbox, generator, and shaft. A detailed breakdown of these components includes their failure distribution parameters (shape and scale), material costs for preventive and corrective replacements, and the time required for each task type as presented in Table 6.

Table 6. Component maintenance and vessel assignments.

Component	Shape	Scale	Material Cost (*$*)	Time (*hrs*)	Assigned Vessel
Blades	3	3000	90,000	288	HLV
Bearings	2	3750	10,000	36	HLV
Gearbox	3	2400	230,000	231	HLV
Generator	2	3300	60,000	81	HLV
Shaft	1.5	7300	13,000	57	HLV

The key specifications of the wind turbine are included in Table 7.

Table 7. Turbine specifications.

Specification	Value
Rated Power Output (MW)	3
Cut-in wind speed (m/s)	3
Cut-out Wind Speed (m/s)	25
Rated Wind Speed (m/s)	12
Hub Height (m)	90

These specifications are used for determining the turbine's performance and for the computation of downtime losses. For the O&M environment, the operational parameters presented in Table 8 are considered:

The case scenario presented provides a comprehensive view of the operational, maintenance, and logistical aspects of O&M for an offshore wind turbine.

Table 8. Operational parameters.

Parameter	Value
Mean Wind Speed (m/s)	8.5
Cost of Electricity ($\$/kWh$)	0.15
Distance from Shore (km)	25
Planning horizon L ($days$)	180
Time step τ (τ)	1
Technicians for Replacement	8
Hourly Rate for Technicians Hourly Rate for Technicians ($\$/h$)	60
Vessel Day rate ($)	75000
Vessel mobilisation cost ($)	200000
Fuel consumption (L/km)	200
Vessel mean speed (km/h)	12

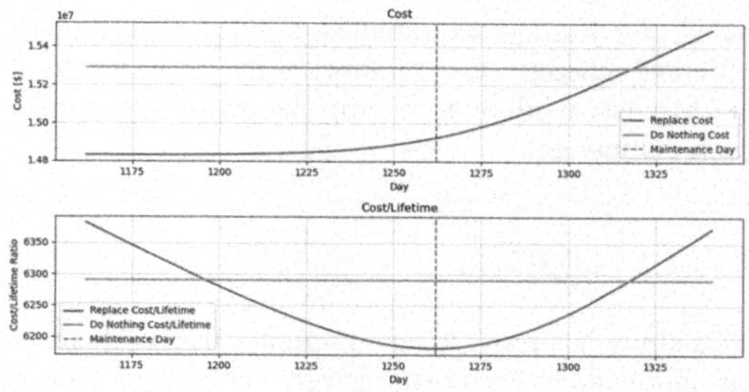

Figure 3. Optimization result on week 166, proposed replacement day: 1262.

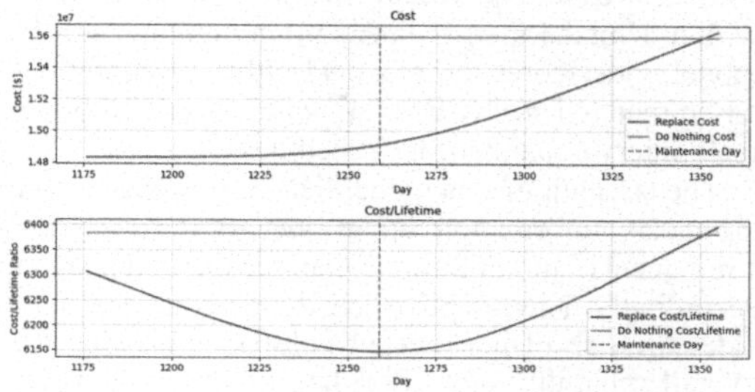

Figure 4. Optimization result on week 168, proposed replacement day: 1259.

5 RESULTS

The implementation of MPC principles within our O&M scheduling optimization framework is illustrated in Figure 3 and 4, which display the updated results over successive weeks as the optimization model is rerun with new data inputs, simulating a rolling horizon approach characteristic of MPC strategies. These figures reveal the evolving decision-making process, where maintenance schedules are adaptively adjusted based on updated probabilistic RUL forecasts, reflecting MPC's inherent adaptivity and reactivity to system feedback.

On week 166, as seen in 3, the maintenance decision aligns with the initial prognosis, suggesting a conservative approach with an earlier maintenance day proposed relative to the MPC's rolling horizon forecast. As the weeks progress, on week 168 a shift in the proposed maintenance day is noticed 4, illustrating the model's responsiveness to updated component health information. This dynamic adjustment exemplifies MPC's predictive control feature, which dictates that actions are based not only on the current state but also on predicted future states of the system.

The analysis of the expected costs and cost to lifetime ratios provides a clear indication that timely, preventive maintenance is financially prudent in the long run. The results of this model highlight the importance of scheduling maintenance before the probability of failure and the associated costs increase.

When compared to the cost of corrective maintenance, represented by the Expected cost of doing nothing, planning the replacement based on RUL prognosis can lead to a potential decrease in maintenance costs that ranges between 1 and 15%, depending on the component that requires maintenance and other operational parameters.

When focusing only on potential downtime losses the cost decrease becomes more significant, which is obvious given that other cost items in this study are assumed to be fixed and do not depend on the scheduling of maintenance activities, such as the cost of materials for replacements, the cost of technicians and the high mobilisation costs of the vessels.

The optimization results, therefore, confirm the suitability of incorporating MPC strategies within the O&M scheduling for offshore wind farms. By leveraging the predictive and adaptive capabilities of

MPC, our framework achieves a dynamic and cost-efficient schedule that minimizes the expected total O&M costs while maximizing the utilization rate of each component.

6 CONCLUSIONS

This study has provided a comprehensive analysis of an innovative prognostic-driven scheduling model for offshore wind farm operation and maintenance (O&M). The results demonstrate the model's potential to minimize O&M costs while maximizing equipment utilization rate. The integration of probabilistic remaining useful life (RUL) predictions within a mixed-integer linear programming (MILP) model underscores the practical utility of condition-based maintenance (CBM) strategies over traditional fixed interval maintenance schedules.

The strength of this model lies in its ability to use prognosis information about the RUL of critical components to optimize maintenance schedules accordingly. This adaptive scheduling approach mitigates the risk of unexpected failures, enhancing the overall reliability and efficiency of the system. By doing so, the model not only ensures the highest utilization rate of each component but also minimizes downtime, which is a significant cost driver in offshore wind farm operations.

The use of probabilistic methods to forecast component failure provides a more dynamic and responsive approach to maintenance scheduling, allowing for adjustments based on real-time data and predictions. The integration of MPC principles into the prognostic-driven scheduling model has been instrumental in demonstrating the model's capacity for adaptive and dynamic decision-making, critical for optimizing O&M. The rolling horizon approach, characteristic of MPC, has enabled the model to adjust maintenance schedules in response to continually updated RUL forecasts.

However, the model is not without its limitations. The assumption of constant accessibility and the availability of resources does not reflect the complex reality of offshore wind farm environments, where conditions are highly dynamic and often unpredictable. Weather conditions, logistical constraints, and limited resources can lead to significant deviations from the model's recommendations. Furthermore, the assumption of performing only one maintenance activity at the time does not reflect the most recent advancements in opportunistic maintenance strategies, whose advantages have been highlighted in studies such as (Li et al. 2021) and (Li et al. 2020). In the future, the integration of opportunistic maintenance strategies in the model might reveal additional benefits of a prognostic-driven strategy for O&M scheduling.

Another limitation is the model's reliance on synthetic data for RUL predictions, which, while useful in the absence of actual sensor data, may not fully capture the nuances of real-world operational conditions. The adaptation of the model to utilize real-life data from monitoring systems could enhance its accuracy and reliability. For future developments, research could focus on relaxing some of the model's simplifying assumptions to better reflect the operational challenges faced by offshore wind farms. Incorporating weather prediction models and stochastic resource availability could provide a more realistic maintenance scheduling framework. Additionally, exploring the potential for opportunistic maintenance where multiple maintenance activities are performed simultaneously could further optimize resource utilization and reduce operational costs.

In conclusion, the proposed prognostic-driven scheduling model serves as a starting point for further research on O&M scheduling for offshore wind farms. The model's integration of MPC principles, CBM strategies and probabilistic RUL predictions, make it a promising solution to the complex challenge of maintaining offshore wind farms. Future research should aim to build upon this foundation, refining the model to account for the full spectrum of real-world operational variables and constraints.

ACKNOWLEDGEMENTS

This research is financially supported by NWO as part of the Holi-DOCTOR project.

REFERENCES

Ade Irawan, C., S. Starita, H. K. Chan, M. Eskandarpour, & M. Reihaneh (2023). Routing in offshore wind farms: A multi-period location and maintenance problem with joint use of a service operation vessel and a safe transfer boat. *European Journal of Operational Research 307*(1), 328–350.

Camci, F., K. Medjaher, V. Atamuradov, & A. Berdinyazov (2019, OCT 3). Integrated maintenance and mission planning using remaining useful life information. *Engineering Optimization 51*(10), 1794–1809.

Carroll, J., A. McDonald, I. Dinwoodie, D. McMillan, M. Revie, & I. Lazakis (2017, FEB). Availability, operation and maintenance costs of offshore wind turbines with different drive train configurations. *Wind Energy 20*(2), 361–378.

Chen, C., C. Wang, N. Lu, B. Jiang, & Y. Xing (2021). A data-driven predictive maintenance strategy based on accurate failure prognostics. *Eksploatacja i Niezawodnosc-Maintenance and Reliability 23*(2), 387–394.

Dai, L., M. Stålhane, & I. B. Utne (2015). Routing and scheduling of maintenance fleet for offshore wind farms. *Wind Engineering 39*(1), 15–30.

Dawid, R., D. McMillan, & M. Revie (2018, SEP). Decision support tool for offshore wind farm vessel routing under uncertainty. *Energies 11*(9).

de Pater, I. & M. Mitici (2021, OCT). Predictive maintenance for multi-component systems of repairables with remaining-useful-life prognostics and a limited stock of spare components. *Reliability Engineering & System Safety* 214.

de Pater, I., A. Reijns, & M. Mitici (2022, MAY). Alarm-based predictive maintenance scheduling for aircraft engines with imperfect remaining useful life prognostics. *Reliability Engineering & System Safety* 221.

Fox, H., A. C. Pillai, D. Friedrich, M. Collu, T. Dawood, & L. Johanning (2022, JAN). A review of predictive and prescriptive offshore wind farm operation and maintenance. *Energies 15*(2).

Gogu, C. (2018). Prognostics and health management: current state-of-the-art for optimizing aircraft structural maintenance. In *2018 19th International conference on thermal, mechanical and multi-physics simulation and experiments in microelectronics and Microsystems (EUROSIME)*. 19th International Conference on Thermal, Mechanical and Multi-Physics Simulation and Experiments in Microelectronics and Microsystems (EuroSimE), Toulouse, FRANCE, APR 15–18, 2018.

Golestani, N., E. Arzaghi, R. Abbassi, V. Garaniya, & H. Meng (2023). A system dynamics model of offshore wind farm degradation: Enabling operation and maintenance planning under foreseen asset management impacts. *Applied Ocean Research 138*, 103685.

Irawan, C. A., D. Ouelhadj, D. Jones, M. Stalhane, & I. B. Sperstad (2017, JAN 1). Optimisation of maintenance routing and scheduling for offshore wind farms. *European Journal of Operational Research 256*(1), 76–89.

Kou, L., Y. Li, F. Zhang, X. Gong, Y. Hu, Q. Yuan, & W. Ke (2022, APR). Review on monitoring, operation and maintenance of smart offshore wind farms. *Sensors 22*(8).

Lazakis, I. & S. Khan (2021, APR 1). An optimization framework for daily route planning and scheduling of maintenance vessel activities in offshore wind farms. *Ocean engineering* 225.

Le, B. & J. Andrews (2016, APR). Modelling wind turbine degradation and maintenance. *Wind Energy 19*(4), 571–591.

Lee, J. & M. Mitici (2023, FEB). Deep reinforcement learning for predictive aircraft maintenance using probabilistic remaining-useful-life prognostics. *Reliability Engineering & System Safety* 230.

Li, M., X. Jiang, J. Carroll, & R. R. Negenborn (2022, SEP 1). A multi-objective maintenance strategy optimization framework for offshore wind farms considering uncertainty. *Applied Energy* 321.

Li, M., X. Jiang, & R. R. Negenborn (2021, JUL 1). Opportunistic maintenance for offshore wind farms with multiple-component age-based preventive dispatch. *Ocean Engineering* 231.

Li, M., M. Wang, J. Kang, L. Sun, & P. Jin (2020, NOV 15). An opportunistic maintenance strategy for offshore wind turbine system considering optimal maintenance intervals of subsystems. *Ocean Engineering* 216.

Li, Z., J. Guo, & R. Zhou (2016). Maintenance scheduling optimization based on reliability and prognostics information. In *Annual reliability and maintainability symposium 2016 proceedings*, Reliability and Maintainability Symposium. Amer Inst Aeronaut & Astronaut; Amer Soc Qual Control Elect Div; IEEE; Ints Environm Sci & Technol; Inst Ind Engineers; SAE Int; SOLE; Soc Reliabil Engineers; Syst Safety Soc; American Society of Quality Control Reliabil Div. Annual Reliability and Maintainability Symposium, Tucson, AZ, Jan 25-28, 2016.

Liu, L., Y. Fu, S. Ma, L. Huang, S. Wei, & L. Pang (2019, OCT 28). Optimal scheduling strategy of o&m task for owf. *IET Renewable Power Generation 13*(14), 2580–2586.

May, A., D. McMillan, & S. Thons (2015, NOV). Economic analysis of condition monitoring systems for offshore wind turbine sub-systems. *IET Renewable Power Generation 9*(8, SI), 900–907. Annual Conference of the European-Wind-Energy-Association (EWEA), Barcelona, SPAIN, MAR, 2014.

Mayne, D., J. Rawlings, C. Rao, & P. Scokaert (2000). Constrained model predictive control: Stability and optimality. *Automatica 36*(6), 789–814.

Mazidi, P., Y. Tohidi, & M. A. Sanz-Bobi (2017, MAR). Strategic maintenance scheduling of an offshore wind farm in a deregulated power system. *Energies 10*(3).

Mitici, M., I. de Pater, A. Barros, & Z. Zeng (2023, JUN). Dynamic predictive maintenance for multiple components using data-driven probabilistic rul prognostics: The case of turbofan engines. *Reliability Engineering & System Safety* 234.

Nachimuthu, S., M. J. Zuo, & Y. Ding (2019, APR 2). A decision-making model for corrective maintenance of offshore wind turbines considering uncertainties. *Energies 12*(8).

National Renewable Energy Laboratory (2022). Offshore wind energy market assessment 2022. National Renewable Energy Laboratory.

NREL (2023). Annual technology baseline.

Pattison, D., M. S. Garcia, W. Xie, F. Quail, M. Revie, R. I. Whitfield, & I. Irvine (2016, MAR). Intelligent integrated maintenance for wind power generation. *Wind Energy 19*(3), 547–562.

Raknes, N. T., K. Odeskaug, M. Stalhane, & L. M. Hvattum (2017, MAR). Scheduling of maintenance tasks and routing of a joint vessel fleet for multiple offshore wind farms. *Journal of Marine Science and Engineering 5*(1).

Ren, Z., A. S. Verma, Y. Li, J. J. E. Teuwen, & Z. Jiang (2021, JUL). Offshore wind turbine operations and maintenance: A state-of-the-art review. *Renewable & Sustainable Energy Reviews* 144.

Rodrigues, L. R., J. P. P. Gomes, F. A. S. Ferri, I. P. Medeiros, R. K. H. Galvao, & C. L. Nascimento Junior (2015, DEC). Use of phm information and system architecture for optimized aircraft maintenance planning. *IEEE Systems Journal 9*(4), 1197–1207.

Safaei, F., E. Chatelet, & J. Ahmadi (2020, MAY). Optimal age replacement policy for parallel and series systems with dependent components. *Reliability Engineering & System Safety* 197.

Sarker, B. R. & T. Ibn Faiz (2016, JAN). Minimizing maintenance cost for offshore wind turbines following multi-level opportunistic preventive strategy. *Renewable Energy 85*, 104–113.

Sinha, Y., J. A. Steel, J. A. Andrawus, & K. Gibson (2013, DEC). A smart software package for maintenance optimisation of offshore wind turbines. *Wind Engineering 37*(6), 569–577.

Stock-Williams, C. & S. K. Swamy (2019, APR). Automated daily maintenance planning for offshore wind farms. *Renewable Energy 133*, 1393–1403.

Tseremoglou, I., P. J. van Kessel, & B. F. Santos (2023, FEB). A comparative study of optimization models for condition-based maintenance scheduling of an aircraft fleet. *Aerospace 10*(2).

Turnbull, A. & J. Carroll (2021, AUG). Cost benefit of implementing advanced monitoring and predictive maintenance strategies for offshore wind farms. *energies 14*(16).

US Department of Energy (2022). Offshore wind market report: 2022 edition.

van Bussel, G. & C. Schöntag (1997). Operation and maintenance aspects of large offshore windfarms. In *Proceedings of the 1997 European Wind Energy Conference*, pp. 272–275.

Van Horenbeek, A. & L. Pintelon (2012a). A dynamic prognostic maintenance policy for multi-component systems. *IFAC proceedings* Volumes *45*(31), 115–120. 2nd IFAC Workshop on Advanced Maintenance Engineering, Services and Technology.

Van Horenbeek, A. & L. Pintelon (2012b). Optimal prognostic maintenance planning for multi-component systems. In C. Berenguer, A. Grall, and C. Soares (Eds.), *Advances in Safety, reliability and risk management*, pp. 910–917. Univ Technologie Troyes; European Safety & Reliabil Assoc; Reg Champagne Ardenne; Ville Troyes; Ctr Natl Rech Sci; Electricite France; GIS 3SGS; Det Norske Veritas; Grand Troyes; GDR MACS CNRS; MAIF. European Safety And Reliability Conference (ESREL), FRANCE, SEP 18-22, 2011.

Vianna, W. O. L. & T. Yoneyama (2018, JUN). Predictive maintenance optimization for aircraft redundant systems subjected to multiple wear profiles. *IEEE Systems Journal 12*(2), 1170–1181.

Zhuang, L., A. Xu, & X.-L. Wang (2023, JUN). A prognostic driven predictive maintenance framework based on bayesian deep learning. *Reliability Engineering & System Safety* 234.

Advances in Maritime Technology and Engineering – Guedes Soares & Santos (Eds)

A transformer-based fault detection framework for offshore wind turbines based on SCADA data

J.Y. Guo
Centre for Marine Technology and Ocean Engineering (CENTEC), Instituto Superior Técnico, Universidade de Lisboa, Lisbon, Portugal
School of Mechatronic Engineering, Southwest Petroleum University, Chengdu, Sichuan, P. R. China

Z.Y. Wang
School of Mechatronic Engineering, Southwest Petroleum University, Chengdu, Sichuan, P. R. China

H. Li & C. Guedes Soares
Centre for Marine Technology and Ocean Engineering (CENTEC), Instituto Superior Técnico, Universidade de Lisboa, Lisbon, Portugal

ABSTRACT: This paper presents a novel Transformer model for fault detection of offshore wind turbines based on Supervisory Control and Data Acquisition (SCADA) data. Fault detection contributes to safety and effectiveness of the energy production of offshore wind turbines. Initially, the preprocessing of SCADA data is carried out aiming at enhancing the quality and accessibility of data. Subsequently, the Pearson correlation analysis with the assistance of a Transformer model is complete to select pivotal input signals. SCADA data from multiple wind turbines recorded in the LGS-Offshore dataset validated the effectiveness and the superior performance of the proposed methodology. The outcomes of this paper contribute to failure prevention and the maintenance resource management of wind farms.

1 INTRODUCTION

As one of the most promising pollution-free and renewable energy sources, wind energy has become the centre of research both in theory and practice (Liu et al. 2023). During the past few years, the installed capacity of wind turbines (WTs) has been continuously growing and widely distributed (Diaz, & Guedes Soares, 2020). However, due to the harsh and varied operating conditions, the components of WTs are prone to failure (e.g., fans, gears, generators, bearings) (Li et al. 2022a). Therefore, it is of vital importance to perform continuous condition monitoring of WTs to improve their reliability and safety. Specifically, continuous condition monitoring and fault detection can detect potential failures at the early stage and alert the engineers to perform preventive maintenance on these WTs.

Nowadays, the fault detection of WTs can be categorized into three categories, namely, signal-trending analysis, model-based methods, and data-driven methods (Wang et al. 2023). The signal-trending analysis mainly includes vibration analysis, acoustic emission analysis, and strain analysis. However, obtaining these signals requires the installation of additional sensors, which escalates the cost (Salameh et al. 2018). Moreover, the model-based methods typically utilize the prior knowledge of the failure mechanism to construct the model and then depict the operating state of WTs or the components. When facing massive data or variable operating conditions, the capacity and applicability of these methods are severely limited (Lei et al. 2018).

On the contrary, due to the advanced feature extraction capability and mapping capability, the data-driven methods have received considerable attention (Guo et al. 2023a, Guo et al. 2023b, Wang et al. 2023). Since most WTs have the supervisory control and data acquisition (SCADA) system installed for system control and data acquisition, the condition monitoring data of critical subassemblies in WTs could be conveniently obtained and then the operating status of these critical subassemblies could be monitored (Khan et al. 2023). Therefore, it is of vital importance to perform continuous condition monitoring and fault detection according to SCADA data and reduce the costs of operation and maintenance.

Among the data-driven methods, deep learning (DL)-based methods have gained much more attention and are extensively employed. The commonly utilized DL-based methods mainly include convolutional neural network (CNN), long short-term memory (LSTM), gated recurrent unit (GRU), and auto-encoder (AE) (Rathore et al.,

DOI: 10.1201/9781003508762-66

2022; Huang et al. 2021a). Xiang et al. (2021) employed the CNN-LSTM with an attention mechanism to realize the fault detection of WTs and deduced the faulted component by prediction residuals. Wang et al. (2016) proposed a deep AE to identify the blade damage by using the reconstruction error as an indicator from SCADA data. Rahimilarki et al. (2022) proposed a time-series analysis method to convert the raw SCADA data into 2-D greyscale images and utilized CNN to classify the small anomalies in WTs. Kong et al. (2020) realized the spatio-temporal features fusion of SCADA data based on CNN-GRU, and the residual between the actual value and predicted value was calculated and utilized as the indicator to monitor the state of WTs. Xiang et al. (2022) utilized the CNN-BiGRU with an attention mechanism to extract multi-dimensional spatiotemporal features of SCADA data and detect the early fault of WTs.

The present applications of DL-based methods have improved the fault detection accuracy and reliability of WTs. However, several problems remain in the DL-based fault detection method as follows: (1) The recurrent neural network (RNN)-based methods are limited in extracting the features in SCADA data and are easily prone to gradient explosion and disappearance when processing the long-term time series, especially since the components of WTs can typically operate for months (Guo et al. 2024, Ding et al. 2022). (2) The methods listed above typically monitor the indicator of components of WTs and set the threshold of the indicator using the health stage of components (Xiang et al. 2022, Kong et al. 2020).

To overcome the constraints of RNN-based methods in processing the long-term time series, the Transformer model (Wu et al. 2023, Vaswani et al. 2017) is proposed as there is no recurrent structure in this network. The Transformer-based method can avoid the complexity of using convolutional layers or recurrent layers, thus extracting the important features embedded in raw data. Moreover, the condition monitoring interval of the health state is constructed and employed to identify the early fault of offshore WTs. The main contribution of this paper can be summarized as follows:

(1) The Transformer model is introduced to process the multi-dimensional signal of offshore WTs.
(2) The 3σ criterion is employed to construct the condition monitoring interval of the health state, and the threshold is defined based on the health state of offshore WTs.

The remainder of this paper is organized as follows: Section 2 comprehensively illustrates the proposed method. The experimental validation is performed on the real offshore WTs dataset (LGS-Offshore dataset) in Section 3. Finally, the conclusions are summarized in Section 4.

2 THE PROPOSED METHOD

2.1 *Data pre-processing and feature selection*

Data pre-processing is essential in removing noise and cleaning abnormal data, which improves the reliability of fault detection. In this paper, the data pre-processing is mainly abnormal data cleaning. After that, the feature selection is employed on the cleaned data.

Firstly, abnormal data cleaning is performed on the raw SCADA data to improve the effectiveness and reliability of the model. The common abnormal data include wind speed equal to 0, power equal to 0, missing values, and duplicate data. Therefore, such abnormal data needs to be cleaned for further study.

Finally, the feature selection using the Pearson correlation coefficient is performed on the cleaned offshore WTs SCADA data. The Pearson correlation coefficient is diffusely employed in analysing the relationship between different types of features. Specifically, the features with a coefficient larger than 0.6 are selected for fault detection and coefficients lower than 0.6 are removed. The Pearson correlation coefficient is defined by:

$$p = \frac{\sum_{i=1}^{n}(x_i - \bar{x})(y_i - \bar{y})}{\sqrt{\sum_{i=1}^{n}(x_i - \bar{x})^2}\sqrt{\sum_{i=1}^{n}(y_i - \bar{y})^2}} \tag{1}$$

where p represents the Pearson correlation coefficient, $\bar{x}$ and $\bar{y}$ denote the mean of x and y, respectively. n denotes the sample points.

2.2 *Transformer model*

Without any convolutional layers or recurrent layers, the Transformer-based method can adequately extract the features embedded in raw data with the assistance of a multi-head self-attention mechanism. The Transformer consists of three parts, namely the encoder-decoder, the positional encoding, and the multi-head self-attention (Guo et al. 2023), see Figure 1.

As shown in Figure 1, the Transformer contains three parts, namely, encoder-decoder, the positional encoding, and the multi-head self-attention. The Transformer-based methods generally contain three categories: (1) encoder-only (e.g., for classification and regression) (2) decoder-only (e.g., for language modelling) (3) encoder-decoder (e.g. for machine translation) (Ding et al. 2022). The Transformer network is first proposed by Vaswani and utilized the encoder-decoder structure for machine translation, which is just Seq2Seq problem. Nevertheless, the fault detection of offshore WTs is part of the problem that dealing with the time series, which is just a branch of the regression problem. Therefore, in this paper, the encoder-only Transformer-based model (Guo et al. 2024, Ding et al. 2022) is utilized to realize the fault detection of offshore WTs. The following subsection details the pivotal parts in the encoder part of the Transformer.

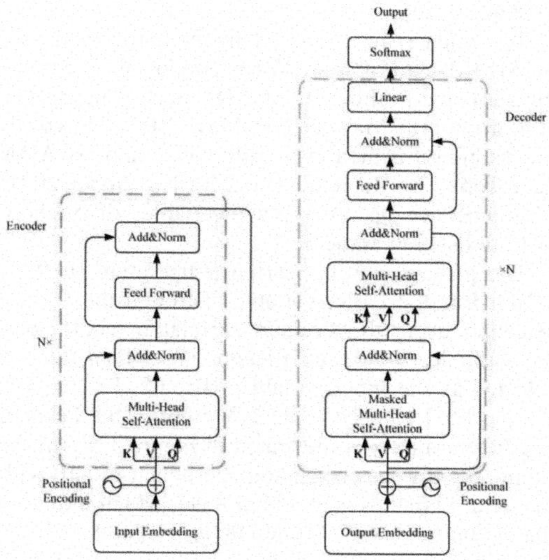

Figure 1. The structure of the Transformer.

(a) Multi-head self-attention

The multi-head self-attention mechanism allows the model to acquire crucial information from the input sequence by allotting different weights on each location. The mathematical formulation of multi-head self-attention mechanism is listed as follows:

$$Attention(Q,K,V) = softmax(\frac{QK^T}{\sqrt{d}})V \quad (2)$$

$$Multihead(Q,K,V) = Concat(h_1,h_2,\cdots,h_n)W^O \quad (3)$$

$$h_i = Attention(Q_i,K_i,V_i) \quad (4)$$

where Q, K and V denote the query, key, and value, respectively. h_i denotes the head, n represents the number of heads.

(b) The positional encoding

Traditional RNN-based methods are constrained by the gradient explosion and vanishment when processing the long-term time series. However, the Transformer network efficiently solves the constrains using the special structure. As the attention mechanism could not record the relationship between the input sequence and its sequence order. The positional encoding is employed to add the positional information that represents the sequence order. The positional encoding algorithm can be expressed as:

$$\begin{cases} pos_{(a,2i)} = \sin(\frac{a}{10000^{2i/d}}) \\ pos_{(a,2i+1)} = \cos(\frac{a}{10000^{2i/d}}) \end{cases} \quad (5)$$

where a represents the position of element in the input sequence, $2i$ and $2i+1$ represent the even and odd position of the sequence, d denotes the dimension.

2.3 *Residual calculation and threshold definition*

The residual between the actual value and the predicted value is calculated and the RMSE of residual for each day as the indicator to monitor the state of offshore WTs, denoted as follows:

$$res = y_p - y_r \quad (6)$$

$$Ind_j = \sqrt{\frac{1}{N}\sum_{i=1}^{N}(res_i)^2} \quad (7)$$

where res denotes the residual, y_p and y_r represent the predicted value and actual value, respectively, Ind_j denotes the j th indicator, N represents the sample points of each day, and in this paper, $N = 1440$.

Once the RMSE of residual for each day is obtained, the 3σ criterion (Li et al. 2015, Huang et al. 2022) is performed on this indicator to construct the condition monitoring interval of offshore WTs, expressed by:

$$U_b = m_{ind} + 3 * s_{ind} \quad (8)$$

where U_b represents the upper bound of the threshold, m_{ind} and s_{ind} denote the mean and standard deviation of the indicator, respectively.

2.4 *Flowchart of the proposed method*

The flowchart of the proposed method is illustrated in Figure 2. During the training stage, the historical SCADA data is sent into the Transformer model for training after the data pre-processing and the feature selection using the Pearson correlation coefficient. Once the Transformer model is trained, the residual and the threshold of the upper bound are calculated. Then, the same data pre-processing procedure is performed on the online SCADA data and the RMSE of residual for each testing day is obtained. Once the two continuous values exceed the condition monitoring interval, the components of offshore WTs are considered to have failures and need preventive maintenance.

3 CASE STUDY

3.1 *LGS-offshore dataset*

The offshore WTs SCADA data are from the LGS-Offshore dataset. The LGS-Offshore dataset encompasses

Table 1. The generator operation state variables of offshore WT.

No.	Variable	Units
F1	Environmental temperature	°C
F2	Cabin temperature	°C
F3	Generator winding u1 temperature	°C
F4	Generator winding u2 temperature	°C
F5	Generator winding v1 temperature	°C
F6	Generator winding v2 temperature	°C
F7	Generator winding w1 temperature	°C
F8	Generator winding w2 temperature	°C
F9	Generator cooling water temperature	°C
F10	Generator front bearing temperature	°C
F11	Generator rear bearing temperature	°C
F12	Generator torque	N·m

Table 2. The correlation coefficients for each feature.

No.	Variable	Correlation coefficient
F1	Environmental temperature	0.855
F2	Cabin temperature	0.793
F3	Generator winding u1 temperature	0.602
F4	Generator winding u2 temperature	0.626
F5	Generator winding v1 temperature	0.612
F6	Generator winding v2 temperature	0.609
F7	Generator winding w1 temperature	0.615
F8	Generator winding w2 temperature	0.619
F9	Generator cooling water temperature	0.560
F11	Generator rear bearing temperature	0.732
F12	Generator torque	0.504

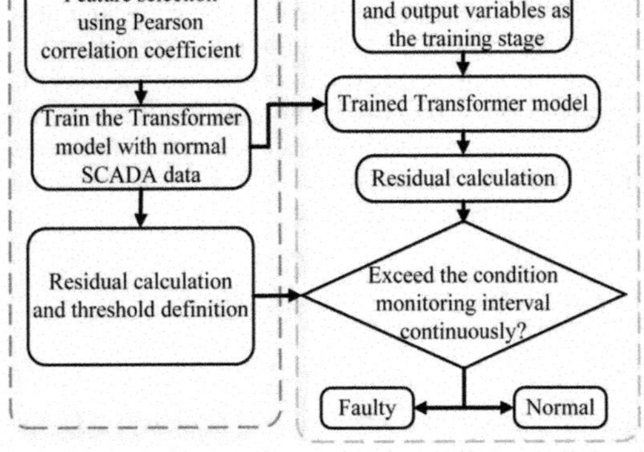

Figure 2. The flowchart of the proposed method.

313 wind turbines operating in six wind farms and 1742 failures. Moreover, the distance of the six offshore wind farms from the coast varies from 23 to 65 kilometres. The LGS-Offshore dataset is just a complement to the LGS-Onshore dataset, detailed information is shown in (Li & Guedes Soares, 2022a; Li et al. 2022b).

To validate the proposed method, the generator SCADA data of offshore WT is employed. The training set is the health state SCADA data of the generator from 11th May to 31st May, 2021. Moreover, the testing set is the run-to-failure generator SCADA data from 12th September to 25th November 2021. The generator operation state variables of offshore WT are listed in Table 1.

The generator front-bearing temperature (F10) is selected as the target variable. Through the feature selection using the Pearson correlation coefficient, the variables with a coefficient larger than 0.6 are selected as the input variables (F1, F2, F3, F4, F5, F6, F7, F8, F11), see Table 2. Moreover, the Pearson coefficient matrix is illustrated in Figure 3. From the Figure 3, only the correlation coefficient of F10 with the other variables is at a high level, which indicates that it contributes to accurate fault detection.

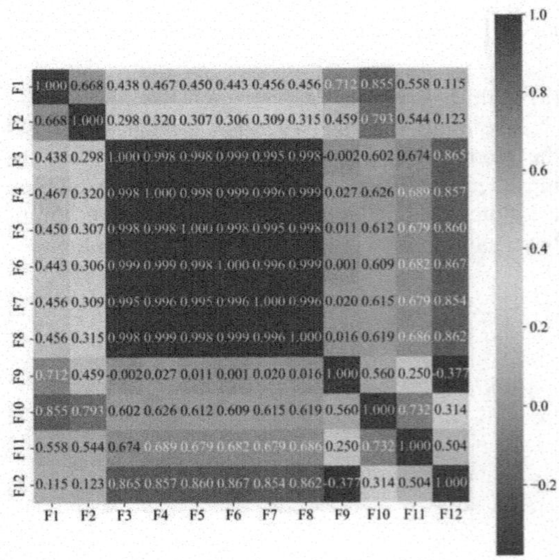

Figure 3. The Pearson coefficient matrix of the variables.

3.2 *Hyperparameters*

The selection of hyperparameters is of particularly great significance and affects the accuracy of fault detection. Therefore, the hyperparameters of the proposed method should be tuned according to the prediction performance. Figure 4 presents the comparisons of prediction performance with different hyperparameters. Specifically, the RMSE is employed as the performance metric to evaluate the prediction performance. Moreover, only one hyperparameter is changed at a time and the model is run ten times to reduce errors due to initialisation of model parameters. From Figure 4, the proposed method with the hyperparameters listed in Table 3 exhibits the superior prediction performance than with the other hyperparameters. From Table 3, the proposed method contains three encoder layers and three heads are utilized in the multi-head self-attention mechanism.

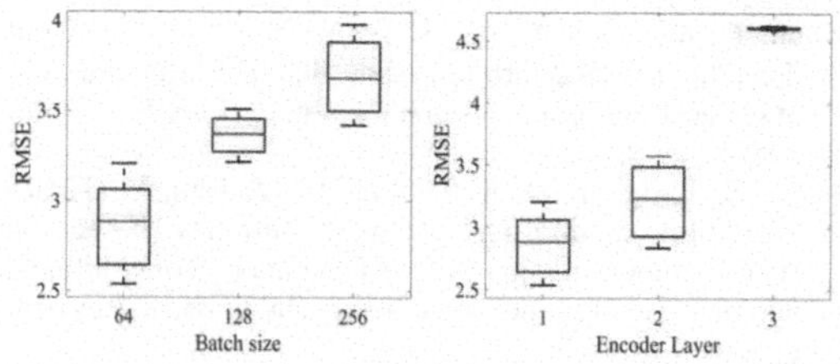

Figure 4. The performance evaluation with different hyperparameters.

Table 3. The hyperparameters of the proposed method.

Name	Parameters
d_model	9
Heads	3
Encoder layer	1
Epoch	200
Batch_size	64
Dropout rate	0.3
Optimizer	Adam
Learning rate	0.001

3.3 *Experimental results*

Once the Transformer model is trained, the residual and the threshold of the upper bound are calculated.

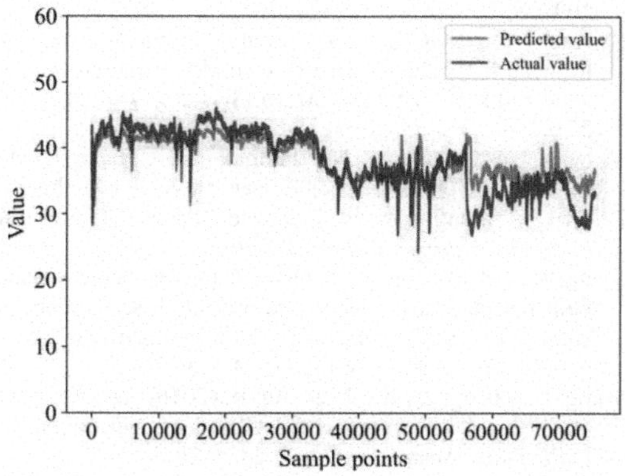

Figure 5. The predicted and actual value.

Figures 5 and 6 illustrate the predicted value and the actual value and its corresponding residual. From Figures 5 and 6, it can be seen that the difference between the predicted and actual values is small at the

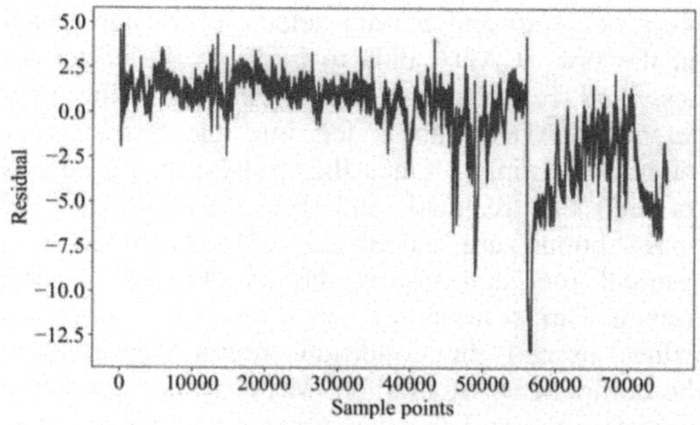

Figure 6. The residual between the actual and predicted value.

beginning stage, and as the time passes, the components of the generator begin to deteriorate, resulting in a gradual increase in the difference between the predicted and true values. After that, the RMSE of residual for each testing day can be obtained by sending the online testing SCADA data into the trained Transformer model. It can be seen from Figure 7 that the RMSE of the residual fluctuate below the upper bound of the threshold in the first thirty-five days. However, on the 40th day, the RMSE of the residual exceeds the upper bound of the threshold. This indicates that the generator is considered to have failures and needs preventive maintenance.

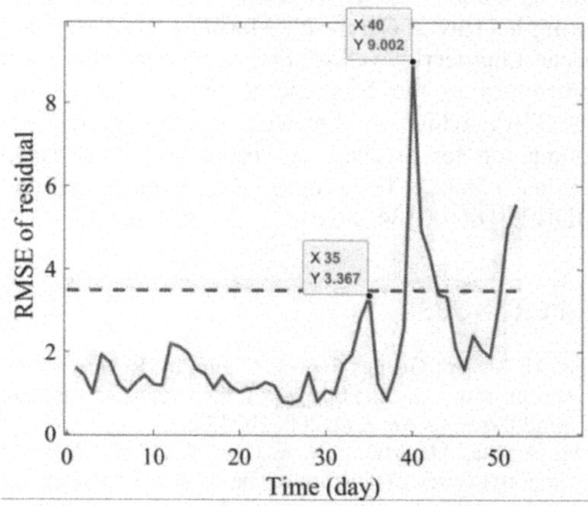

Figure 7. The RMSE of residual for each testing day.

4 CONCLUSION

This paper proposes a Transformer-based fault detection framework for offshore wind turbines based on SCADA data. The Transformer-based method can avoid the complexity of using convolutional layers or recurrent layers, thus extracting the important features embedded in raw data. The procedures of the proposed fault detection framework can be summarized as follows: (1) The data pre-processing and feature selection are performed on the raw SCADA data to improve the effectiveness and reliability of the model. (2) The processed SCADA data is fed into the Transformer model for training. Once the Transformer model is trained, the residual and the threshold of the upper bound are calculated. (3) The RMSE of residual for each testing day is obtained by the trained Transformer and once the two continuous values exceed the condition monitoring interval, the components of offshore WTs are considered to have failures and to need preventive maintenance. Finally, the superiority and effectiveness of the proposed method are validated on the LGS-Offshore dataset. The experimental results show that the proposed method can effectively perform condition monitoring and fault detection for the components of offshore wind turbines.

ACKNOWLEDGEMENTS

This work was conducted within the ARCWIND project–Adaptation and implementation of floating wind energy conversion technology for the Atlantic region (http://www.arcwind.eu/), which is co-financed by the European Regional Development Fund through the Interreg Atlantic Area Programme under contract EAPA 344/2016. The first author has been funded by the scholarship from China Scholarship Council (CSC) (Grant No. 202108515059) during his stay at Centre for Marine Technology and Ocean Engineering (CENTEC) in Lisbon. This work contributes to the Strategic Research Plan of the CENTEC, which is financed by the Portuguese Foundation for Science and Technology (Fundação para a Ciência e Tecnologia - FCT) under contract UIDB/UIDP/ 00134/2020.

REFERENCES

Diaz, H. M. and Guedes Soares, C. (2020). Review of the current status, technology and future trends of offshore wind farms. *Ocean Eng*. 209, 107381.

Ding, Y., Jia, M., Miao, Q., & Cao, Y. (2022). A novel time–frequency Transformer based on self–attention mechanism and its application in fault diagnosis of rolling bearings. *Mech. Syst. Sig. Process*. 168, 108616.

Guo, J., Wan, J. L., Yang, Y., Dai, L., Tang, A., Huang, B., … & Li, H. (2023a). A deep feature learning method for remaining useful life prediction of drilling pumps. *Energy*. 282, 128442.

Guo, J., Wang, J., Wang, Z., Gong, Y., Qi, J., Wang, G., & Tang, C. (2023b). A CNN-BiLSTM-Bootstrap integrated method for remaining useful life prediction of rolling bearings. *Qual. Reliab. Eng. Int*. 39, 1796–1813.

Guo, J., Wang, Z., Li, H., Yang, Y., Huang, C. G., Yazdi, M., & Kang, H. S. (2024). A Hybrid Prognosis Scheme for Rolling Bearings Based on a Novel Health Indicator and Nonlinear Wiener Process. *Reliab. Eng. Syst. Saf*., 110014.

Huang, C. G., Huang, H. Z., Li, Y. F., & Peng, W. (2021a). A novel deep convolutional neural network-bootstrap integrated method for RUL prediction of rolling bearing. *J. Manuf. Syst*. 61, 757–772.

Huang, C. G., Zhu, J., Han, Y., & Peng, W. (2021b). A novel Bayesian deep dual network with unsupervised domain adaptation for transfer fault prognosis across different machines. IEEE Sens. J. 22(8), 7855–7867.

Khan, P. W., Yeun, C. Y., & Byun, Y. C. (2023). Fault detection of wind turbines using SCADA data and genetic algorithm-based ensemble learning. *Eng. Fail. Anal*. 148, 107209.

Kong, Z., Tang, B., Deng, L., Liu, W., & Han, Y. (2020). Condition monitoring of wind turbines based on spatio-temporal fusion of SCADA data by convolutional neural networks and gated recurrent units. *Renewable Energy*. 146, 760–768.

Lei, Y., Li, N., Guo, L., Li, N., Yan, T., & Lin, J. (2018). Machinery health prognostics: A systematic review from data acquisition to RUL prediction. *Mech. Syst. Sig. Process*. 104, 799–834.

Li, H., & Guedes Soares, C. (2022a). Assessment of failure rates and reliability of floating offshore wind turbines. *Reliab. Eng. Syst. Saf*. 228, 108777.

Li, H., Peng, W., Huang, C. G., & Guedes Soares, C. (2022b). Failure rate assessment for onshore and floating offshore wind turbines. J. Mar. Sci. Eng. 10(12), 1965.

Li, N., Lei, Y., Lin, J., & Ding, S. X. (2015). An improved exponential model for predicting remaining useful life of rolling element bearings. IEEE Trans. Ind. Electron. 62(12), 7762–7773.

Liu, J., Yang, G., Li, X., Wang, Q., He, Y., & Yang, X. (2023). Wind turbine anomaly detection based on SCADA: A deep autoencoder enhanced by fault instances. *ISA Trans*. 139, 586–605.

Rahimilarki, R., Gao, Z., Jin, N., & Zhang, A. (2022). Convolutional neural network fault classification based on time-series analysis for benchmark wind turbine machine. *Renewable Energy*. 185, 916–931.

Rathore, M. S., & Harsha, S. P. (2022). An attention-based stacked BiLSTM framework for predicting remaining useful life of rolling bearings. Appl. Soft Comput. 131, 109765.

Salameh, J. P., Cauet, S., Etien, E., Sakout, A., & Rambault, L. (2018). Gearbox condition monitoring in wind turbines: A review. *Mech. Syst. Sig. Process*. 111, 251–264.

Vaswani, A., Shazeer, N., Parmar, N., Uszkoreit, J., Jones, L., Gomez, A. N., Kaiser, Ł. & Polosukhin, I. (2017). Attention is all you need. *Advances in neural information processing systems*, 30.

Wang, A., Pei, Y., Zhu, Y., & Qian, Z. (2023). Wind turbine fault detection and identification through self-attention-based mechanism embedded with a multivariable query pattern. *Renewable Energy*. 211, 918–937.

Wang, L., Zhang, Z., Xu, J., & Liu, R. (2016). Wind turbine blade breakage monitoring with deep autoencoders. *IEEE Trans. Smart Grid*. 9(4), 2824–2833.

Wang, Z., Guo, J., Wang, J., Yang, Y., Dai, L., Huang, C. G., & Wan, J. L. (2023). A deep learning based health indicator construction and fault prognosis with uncertainty quantification for rolling bearings. *Meas. Sci. Technol*. 34, 105105.

Wu, H., Triebe, M. J., & Sutherland, J. W. (2023). A transformer-based approach for novel fault detection

and fault classification/diagnosis in manufacturing: A rotary system application. *J. Manuf. Syst*. 67, 439–452.
Xiang, L., Li, J., Hu, A., & Zhang, Y. (2020). Deterministic and probabilistic multi-step forecasting for short-term wind speed based on secondary decomposition and a deep learning method. *Energy Convers. Manage*. 220, 113098.
Xiang, L., Wang, P., Yang, X., Hu, A., & Su, H. (2021). Fault detection of wind turbine based on SCADA data analysis using CNN and LSTM with attention mechanism. *Measurement*. 175, 109094.
Xiang, L., Yang, X., Hu, A., Su, H., & Wang, P. (2022). Condition monitoring and anomaly detection of wind turbine based on cascaded and bidirectional deep learning networks. *Appl. Energy*, 305, 117925.

Advances in Maritime Technology and Engineering – Guedes Soares & Santos (Eds)
© 2024 The Author(s), ISBN 978-1-032-83099-5

Advancing hull monitoring through physics-informed machine learning: Towards a real-time approach

S. Haberl, S.A. Eid & F. von Bock und Polach
Institute for Ship Structural Design and Analysis, Hamburg University of Technology, Germany

S. Ehlers
DLR Institute for Maritime Energy Systems, Germany

ABSTRACT: This paper introduces a novel approach to virtual hull monitoring for ships, focusing on time-domain simulations and integrating Machine Learning (ML). Conventional frequency domain methods in hull monitoring lack phase information of the waves, limiting their ability to address dynamic effects and non-linearities like steep waves. To overcome these constraints, this study combines hydrodynamic simulations in the frequency domain with Finite Element (FE) simulations, producing quasi time-series data for a comprehensive understanding of structural responses. However, real-time application of these simulations is impractical. To address this, a Physics-Informed ML concept is employed, training an ML model with simulation data to predict the ship's structural response in quasi real-time. The first findings demonstrate the feasibility of adapting this methodology to the time domain, offering a cost-effective solution for holistic ship structure monitoring. This innovative approach leverages the advantages of time-domain simulations and ML efficiency, addressing current limitations and paving the way for real-time hull monitoring.

1 INTRODUCTION

In the intricate tapestry of modern society, technological systems form the backbone, with maritime transportation as the driving force behind global trade. Ships, as complex amalgamations of various mechanical systems seamlessly interacting, play a pivotal role in this interconnected web. When considering the design of new structures, a focus on cost and environmental efficiency becomes imperative. Achieving this may involve implementing less conservative safety margins, emphasizing the need for a nuanced balance between the reductions of structural weight and safety requirements. Continuous awareness of a structure's condition throughout its operational life is crucial for successfully navigating this balance.

In addition to ensuring the safety, knowledge about the condition of ships has significant economic benefits. Information about the fatigue behavior of the ship allows for targeted fleet management and forms the basis for specific ship maintenance. All IACS (International Association of Classification Societies) members demand regular comprehensive surveys of ships. Insights into the condition of ships have the potential to reduce inspection efforts. This might extend to the point where maintenance intervals can be specifically tailored to the ship's condition, ultimately reducing costs for the ship operator.

A holistic structural health monitoring, involving the continuous surveillance of every crucial detail of a ship, has not been fully achieved to date due to the inherent complexities. The concept of structrual health monitoring is defined in Farrar and Worden (2012): Structural Health Monitoring (SHM) is a crucial process in implementing a damage detection strategy for infrastructure. It involves continuously observing a structure or mechanical system over time through periodically spaced dynamic response measurements. Extracting damage-sensitive features from these measurements, along with statistical analysis, enables the determination of the current state of the system's health.

This process can rely on various input data sources. One approach involves the use of sensors explicitly measuring structural strain at predefined locations, which are then analytically processed to formulate damage assessments. A practical application of strain measurement systems is described by Drummen et al. (2019): One system is set up with focus on global load effects, while another utilizes a local acoustic system to detect fatigue fractures. Another method, based on optimization and machine learning algorithms, is developed in Anyfantis (2021). This method identifies damaged structural elements based on anomalies in strain sensor measurement signals.

While the installation of sensors is intricate and demands the expertise of professionals, the operation phase also presents vulnerabilities. As described by Kaminski (2007), this involves continuous verification of the proper functioning of the measurement system. Despite protection, classic measurement systems such as strain gauges might suffer from moist creeping in compromising

DOI: 10.1201/9781003508762-67

measurements. Additionally, raw measurements must be filtered and condensed into meaningful statistical data, and the information needs to be transmitted to a central location for interpretation and comparison with design expectations.

Alternative methods exist that rely on implicit measurements. In this approach, the structural response is not explicitly measured in the form of strains but is instead determined implicitly. To achieve this, a readily measurable parameter, such as the movements and accelerations of the ship, is gauged on board. Statistical methods are then employed to infer the structural response based on these measurements. Lang et al. (2023) propose a machine learning method to determine fatigue rates of a ship along its route. The foundation of the machine learning model is based on a three-year measurement campaign of a 2800 TEU container ship in the Atlantic, encompassing measurements of the ship's movements and corresponding strains. Neither the methods based on explicit measurements nor those based on implicit measurements directly consider the exciting wave environment of the ship, rendering the prediction of the structural response to future wave events impossible.

As these methods rely on historical datasets including strain measuremens, they do not address the issue of the necessity for expensive and complex measurement systems. To completely eliminate the need for these, hull monitoring must transition to a fully virtual approach, entirely devoid of sensors. Methods of this nature are referred to in the literature as virtual hull monitoring. Schirmann et al. (2018) propose a hull monitoring approach for assessing fatigue damage on various routes. This involves integrating AIS (Automatic Identification System) position data with weather reports to capture the wave environment of the ship. A component stochastic analysis in the frequency domain is then employed to assess fatigue damage using a beam formulation of a simplified ship cross-section to determine the stress transfer functions. Thompson (2020) propose a similar method based on numerical simulations. The innovation of this work involves the utilization of a full-stochastic approach in the frequency domain and the incorporation of a detailed global Finite Element model for stress determination. The stress spectra obtained with the method are statistically compared with measured signals from conducted measurement runs. In the context of virtual hull monitoring, a literature review (e.g Aarsnes et al. (2021), Hulkkonen et al. (2019), among others), reveals that the majority of publications center around simulations in the frequency domain. In all of these applications, the emphasis is placed on determining fatigue damage along a predefined route.

The examination of wave conditions in the frequency domain can be conducted at various levels of accuracy. However, even the most precise representation using reconstructed 2D wave spectra (Kumar et al. (2017)) does not capture the full complexity of real sea conditions. While these studies demonstrate that the statistical analyses in the frequency domain provide sufficient accuracy for determining fatigue damage, they are unable to capture dynamic effects and non-linear load impacts from high steep waves due to the absence of phase information in the waves. Since these effects can trigger high stresses, they have an impact on fatigue damage, but can also have drastic consequences for the ultimate limit state. This limitation poses a challenge on the path towards comprehensive hull monitoring.

One approach to overcome these limitations is virtual hull monitoring in the time domain. A possible method to achieve this could look like the following: The near-field environment of the ship (wave environment) is recorded using wave radar. A machine learning method predicts the wave elevation at the time of impact with the ship (see Wedler et al. (2023)). With this information, a hydrodynamic simulation in the time domain is to be conducted using a non linear Boundary Element Solver. For each time step, the hydrodynamic pressures of the waves are transferred to a Finite Element Simulation. A detailed strength analysis could provide structural responses such as hull girder strength, buckling utilization factor, and fatigue damage (rainflow counting) within a defined time period. Given the computational intensity of these simulations, they cannot be performed in real-time. Therefore, using the described process, a dataset is created, featuring the wave elevation time series as features and the structural response as targets. This dataset trains a machine learning model, which later uses the wave radar ML model output as input to predict structural responses in real-time. As achieving this goal comprehensively is extensive, this paper presents preliminary results that aim to pave the way for achieving the goal of hull monitoring in the time domain. It delves into the practicality of the aforementioned approach. It introduces simplified simulation environments designed for dataset creation. Additionally, an initial machine learning model is trained to represent the maximum stress and its occurrence location for a given long-crested wave packet.

2 METHODOLOGY

The process commences with the creation of numerical simulation environments to represent a bulk carrier in waves. Initially, regular long-crested waves with varying wave directions and frequencies are generated, and the hydrodynamic pressures are determined using a linear boundary element solver. Subsequently, for each combination of wave direction and frequency the the hydrodynamic pressure distribution along the hull is transferred to a linear structural-mechanical model. The concatenation of the simulation results for a sequence of phase angles, yields a stress profile in the time domain. For each load case, i.e., a combination of wave frequency and direction, the wave parameters, maximum stress, and its location are determined and stored in a dataset. The dataset undergoes processing and serves as the

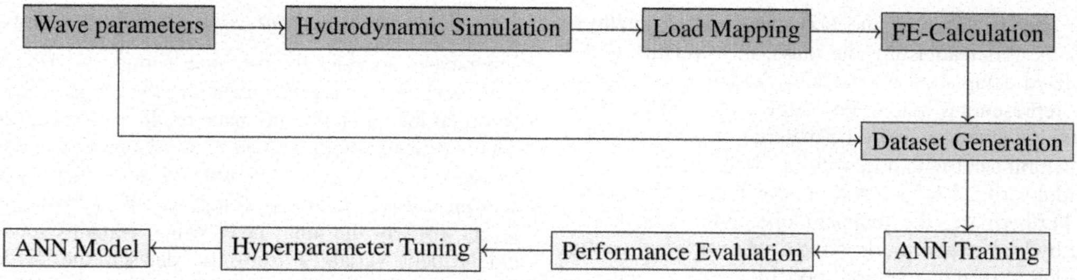

Figure 1. Workflow to create a dataset and train the ANN Model.

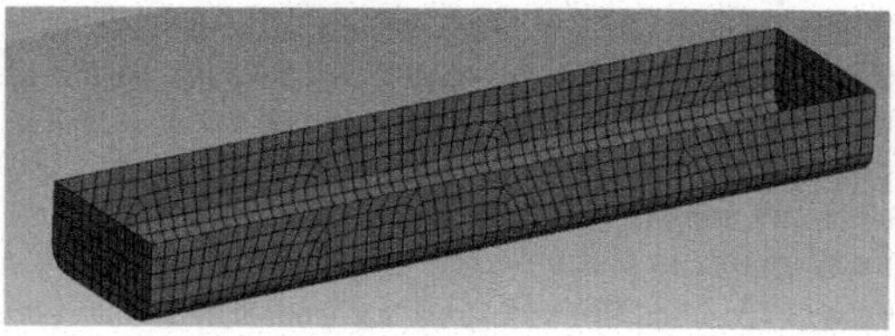

Figure 2. Discretized Aqwa model.

basis for training the neural network, capable of predicting structural responses from given wave parameters. The sequence of the process is illustrated in Figure 1, where numerical components are highlighted in blue, and machine learning elements are depicted in yellow.

2.1 *Numerical simulation and dataset generation*

For the hydrodynamic diffraction analysis, a simplified geometry of the bulk carrier consisting of the outer surface of the hull is adapted. AQWA imposes some restrictions on the elements used to discretize the model regarding the elements' size with relation to neighboring elements and wave parameters. Following these restrictions, a mesh is generated with a maximum element size of $3.5m$, resulting in about 1500 elements. The discretized model is presented in Figure 2. Furthermore, the element size governs the maximum allowable wave frequency where the element size could be chosen based on the desired value of maximum wave frequency. The maximum element size should not exceed $\frac{1}{7}$ of the wavelength (ANSYS 2015). Moreover, the minimum allowable angular frequency can be determined by the relationship as in Equation 1

$$\omega_{min} = 0.001\sqrt{\frac{g}{d}}, \tag{1}$$

where ω_{min} is the minimum allowable angular frequency in rad/sec, g is the gravitational acceleration in m/s^2 and d is the water depth in m. In addition to the maximum and minimum calculated frequencies, a number of frequencies with equal intervals can be inserted manually. To represent an important loading condition where the wavelength corresponds to the structure's length resulting in maximum bending moments, an additional frequency is inserted to represent the hogging and sagging load conditions resulting in a total of 15 frequencies. Furthermore, the wave directions can be specified based on user defined intervals from $-180°$ to $180°$, where a range of directions is chosen at $45°$ intervals to represent head, following and oblique wave conditions.

The resulting pressures acting on the hull's outer surface are stored in the user files directory containing every combination of the selected frequencies, directions, and phase angles, along with information about the element coordinates, and the generated mesh.

The result files generated from the hydrodynamic diffraction analysis are to be used by AQWA Wave to map the pressures resulting from the hydrodynamic analysis to the structural analysis environment.

For the hydrodynamic panel elements, the wave data consisting of the magnitude and phase angle of the hydrodynamic pressures is stored at the centroids of the panels for the specified frequencies and directions. The position of each integration point of the structural model is mapped onto the hydrodynamic model and the hydrodynamic panel element corresponding to the point is specified. The wave pressure at the determined location is then linearly interpolated from the nodal values obtained from the hydrodynamic diffraction analysis. Since the pressure is stored at the centroids of the panels, the pressures on the nodes are calculated from the centroids of the surrounding panels (ANSYS 2018).

This process will generate load information files that contain the pressure information applied on each finite element mesh element for all of the predefined load cases consisting of the applied wave frequencies, directions, phase angles and heights. Consequently, the pressures resulting from the hydrodynamic diffraction analysis are mapped onto the finite element model and the structural simulation environment is initialized.

The geometry used in the structural analysis is composed of a simplified model of the hull including first and second order elements like frames and stiffeners. Once the geometry is imported to ANSYS Mechanical, an artificial mass is inserted to maintain the balance between the buoyancy and gravitational effects. After adjusting the mass, the structure is discretized using shell elements, which are most suitable for representing thin walled structures, with an element edge length of $1m$. The material of the structure is structural steel with a density of $7850kg/m3$, Young's modulus of $2.10e^{11}Pa$ and Poisson's ratio of 0.3.

Furthermore, the mapping operation assumes that the hydrodynamic mesh is fixed in its original configuration, thus the structural model has to have small or no displacements for the mapping process to work. For this reason, appropriate boundary conditions have to be imposed on the FE model to result in a solvable system of finite element equations. The boundary conditions specified by the American Bureau of Shipping (ABS) for a typical bulk carrier are shown in Figure 3.

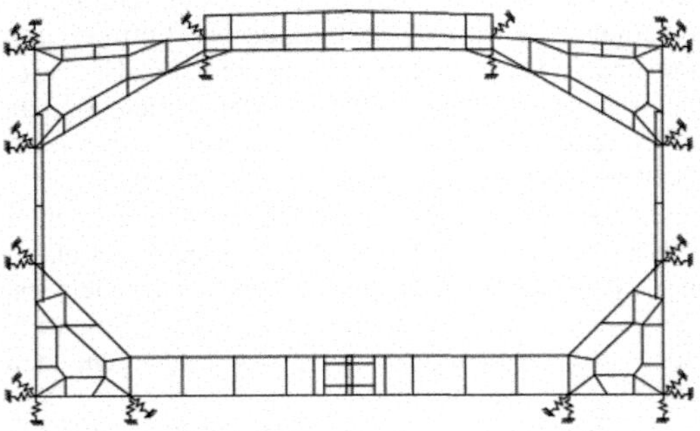

Figure 3. ABS boundary conditions specification.

As per the ABS guidelines, spring constraints have to be applied on both ends of the bulk carrier due to the significant shear and warping stresses experienced by bulk carriers in oblique sea conditions (ABS 2014). The spring system constrains the longitudinal, vertical and transversal motions of the structure. Furthermore, the spring supports should be totally fixed, having their motions set to zero in all six degrees of freedom.

Once the simulation setup is complete and the hydrodynamic pressures are imported, the structural analysis is initiated to generate the stress results. For the purpose of this work, a total of 675 test cases are implemented consisting of 15 frequencies, 9 directions and 5 phase angles assuming regular, long crested and stationary waves. Both the hydrodynamic and structural mechanics solvers are linear, so the structural response depends linearly on the wave height. Therefore, for the purposes of this study, the wave height is kept constant and serves not an input feature for the machine learning model. The results obtained from the simulations contain the element numbers along with their coordinates, the wave parameters and the equivalent von-Mises stresses on each element in a comma separated format (.csv file). The results of all the test cases are then saved to separate files ahead of the creation of the dataset for the machine learning model as will be discussed in the next section.

2.2 *Response prediction using machine learning*

The dataset provides the machine learning model with input and output variables, enabling the model to learn based on the variables and make predictions on unseen data. Artificial neural networks use advanced and iterative approaches to predict continuous or distinct results. A typical ANN model is composed of at least three layers, namely the input layer which contains sets of independent variables from the dataset, the hidden layer where the sum of all inputs and their weights is multiplied by an activation function, and the output layer where the prediction is compared to the corresponding actual output (Eremenko and de Ponteves 2018). In fact, such models would usually have multiple hidden layers to represent the complex nature of the problem. Moreover, the resulting error is back-propagated and the weights applied on the input variables are adjusted accordingly. Such a cycle forms one iteration that is performed on one observation from the dataset. A cycle through all observations in the dataset is referred to as an epoch. The epochs are then repeated till the predictions are in line with the actual results and convergence is achieved. To evaluate the model's performance, various measures can be deployed such as the mean square error, R-square coefficient, root mean square error, etc. This section sheds light on the generation of the dataset and the ANN model.

After generating the result files from the structural analysis, the data has to be adjusted into a useful form in the data preprocessing phase. The raw data exported from ANSYS Mechanical contains the results from each test case in separate files. Consequently, a Python script is implemented to extract the maximum stress from each test case with the corresponding wave parameters and appended to a new dataset. Once the dataset is generated containing the maximum stress values and their corresponding input values, the dataset is imported in Python to train the ANN model.

The Dataset has three input variables, the wave frequency, direction and phase angle, and four output variables which are the corresponding maximum equivalent von-Mises stresses, x, y, and z coordinates. Nevertheless, since there is a significant difference in the range of values of the maximum stress and location coordinates, two separate models are implemented to achieve unbiased predictions. The data is split into a training set to train the model, a validation set to give an estimation of the model's performance after each iteration, and a test set to evaluate the modelâ€™s performance on unseen data after fully training the model. 10% of the dataset will be used as the validation set, and from the remaining data, the training and test sets will be split into 80% and 20%, respectively. Furthermore, the data has to be normalized to ensure the convergence, stability and performance of the model by scaling the features and outputs to a common scale. This is done to ensure that the weights applied to the input variables in the hidden layers are unbiased due to the numerical gap among the observations.

After normalizing the data, a sequential model is used to initialize the ANN model. Sequential models are linear stacks of layers where each layer is added one after the other (Tensorflow 2023c). Initially, both, the model for predicting maximum stress and the model for predicting the location of it are composed of an input layer with three neurons corresponding to the three input variables, two hidden layers with 64 neurons in each layer and an output layer with one neuron for the stress prediction model and three neurons for the location prediction model. Moreover, the Adam (Adaptive Moment Estimation) optimizer will be used as an optimization algorithm for training the ANN model by estimating the first and second order moments, accelerating the optimization process by accumulating a weighted average of past gradients. Finally, the model is trained over 500 iterations and a batch size of 32. The specified data of the ANN models are to be understood as the initial architecture. An improved structure of the neural networks is achieved through hyperparameter tuning, which will be explained later in this chapter.

To evaluate the performance of the first run, two metrics are used, mainly the mean square error and the R-square value. The mean square error is a performance measure that calculates the sum of all differences between the predicted and actual results for all observations in the dataset. The smaller the mean square error value, the better the performance. In addition, the R-squared value measures the proportion of the variance of the dependent variable that is explained by the independent variable (Tensorflow 2023b). Since the R-square value is a ratio of the sum squared regression to the total sum of squares, the closer it is to one, the better the predictions. However, it is rare for a model to deliver accurate results on the first run. Consequently, the parameters used in the first run are tuned to achieve satisfying results in the process of hyperparameter tuning.

Hyperparameter tuning is an iterative process that includes adjusting the percentage of the training and testing sets, adding or removing hidden layers, increasing or decreasing the number of neurons in each layer, varying the activation function (Tensorflow 2023a). Mainly, a parameter is varied while other parameters are held constant until a satisfactory performance is achieved. Then, that parameter is held constant while another parameter is varied. This process can be automated to go through a predefined number of variations and obtain the set that achieves the overall best performance. Once the hyperparameter tuning process is completed, the set of parameters with the best performance is chosen.

After processing the data to generate a dataset, the machine learning model is initialized using the optimal parameters obtained from the hyperparameter tuning process. Consequently, the stress prediction model is trained with 500 iterations, four hidden layers with 512 neurons in each layer, and a learning rate of 0.001 resulting in a loss value of 0.0046. In addition, the position prediction model is trained with 2500 iterations, four hidden layers with 612 neurons in each layer, and a learning rate of 0.001 resulting in a loss value of 0.0098.

3 RESULTS

This chapter focuses on presenting the results of the machine learning model. After identifying the optimal parameters for the ANN architecture, the model is tested using the test dataset. The performance of predictions is evaluated separately for both, stress model and stress location model. The performance of

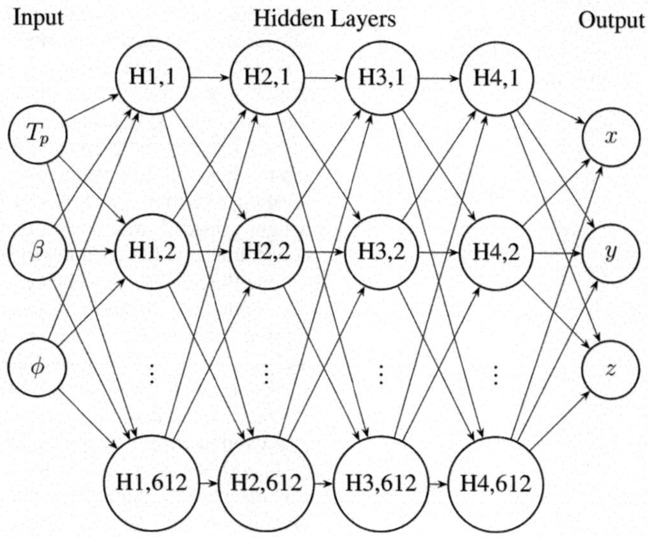

Figure 4. Architecture of the neural network for predicting the location of maximum stress.

both models is measured by finding the mean square error and the R Square value. An overview of the overall metrics can be found in Table 1. For the stress prediction model, the MSE is 1.07% and the R Square value is 0.71, whereas for the position prediction model, the mean square error is 4.63% and the R Square value is 0.65.

Table 1. Model performance for both maximum stress and location prediction.

	R squared	MSE
Stress prediction	0.7124	0.0107
Location prediction	0.6457	0.0463

A representative depiction of all data points in the test set can be seen in Figure 5. It shows a scatter plot for predicting the maximum stress values. The x-axis represents the numerically determined value, while the y-axis represents the values predicted by the machine learning model. The majority of points align along the bisector (x=y), indicating a good agreement. This observation is consistent with the examination of the residual plots. Here, each data point is represented on the abscissa, while its corresponding residual, i.e., the deviation of the predicted value from the simulated value, is depicted on the ordinate. Figure 6(a) illustrates the residuals of the maximum stress prediction, with most points clustering around the zero line of the residual. However, the results are mixed for predicting the location of the maximum stress in Figure 6(b) to (d). On one hand, the residuals of the predictions for the y and z coordinates align along the zero line. However, for the y coordinate, there are occasional outliers reaching up to 15 m. In contrast, the prediction for the z coordinate is notably more accurate, with the maximum residual being 1 m. Overall the residuals for predicting the y and z components are considered satisfactory, but this is not the case for predicting the x component, as

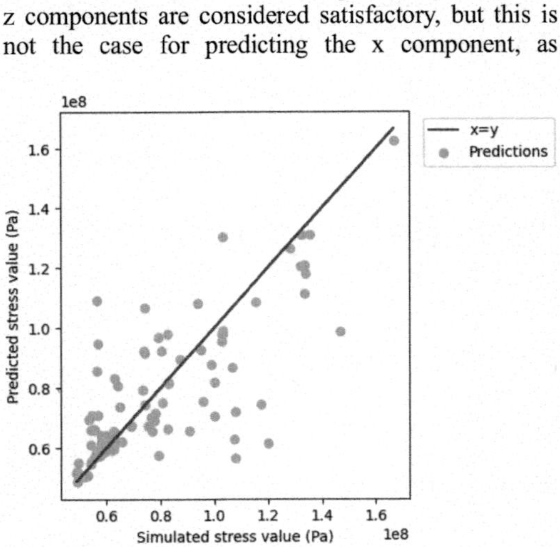

Figure 5. Stress predictions vs. simulations.

evident in Figure 6(b), where a significant dispersion around the zero line is noticeable, with outlines with up to 150m scattering. The small residual values for the prediction of the y and z components result in a recognizable pattern in their predictions. Most predicted points are located in the hatch cover coamings, aligning with the simulation results. However, the erroneous prediction of the x component prevents the assignment of this pattern to individual wave parameters. In the context of this study, the representation of the relationship between the location of maximum stress and wave parameters is therefore omitted.

Finally, the robustness of the models to unseen data is tested by simulating two new test cases in *ANSYS* and using the specified wave parameters as input to the machine learning models to compare the predeictions to the results of the simulations. It should be noted that the applied wave parameters are not included in the original dataset and are therefore completely unseen. Table 2 demonstrates the wave parameters for both test cases along with the stresses generated by *ANSYS* and the machine learning model. The predictions of the ML model show a good agreement with the results of the FE simulation. The significance of the prediction quality is addressed in the discussion in the next chapter.

4 DISCUSSION

In the previous chapter, it is demonstrated that the machine learning models provide predictions of varying quality. The interpretation of these predictions is discussed in the following section. Metrics for stress prediction are deemed acceptable, demonstrating good accuracy for all data points. This is justified by the fact that stress is a scalar quantity. A small percentage error only results in a minor deviation from the actual value. In contrast, the results for predicting the location of maximum stress are more error-prone. Additionally, the location of maximum stress can be described as a vector. The interpretation of errors in predicting this vector is less robust than predicting scalar quantities like the stress. This becomes evident when examining Figure 7. While the prediction for the location of maximum stress is only minimally offset from the actual location determined with ANSYS, the vector points to a lightly loaded area on the deck, whereas the stress concentrates on the coaming of the hatches. Therefore, despite the minimal percentage difference, the information about the location of maximum stress is considered unreliable. The methodology and the architecture of the employed artificial neural network (ANN) needs to be enhanced. Especially the prediction of the location of the maximum stress must be improved to enable its application.

In addition to evaluating the accuracy of the results, it is essential to illuminate numerical limitations in the creation of the dataset. The outlined process commences with hydrodynamic simulations of regular long-crested waves in the frequency domain. The

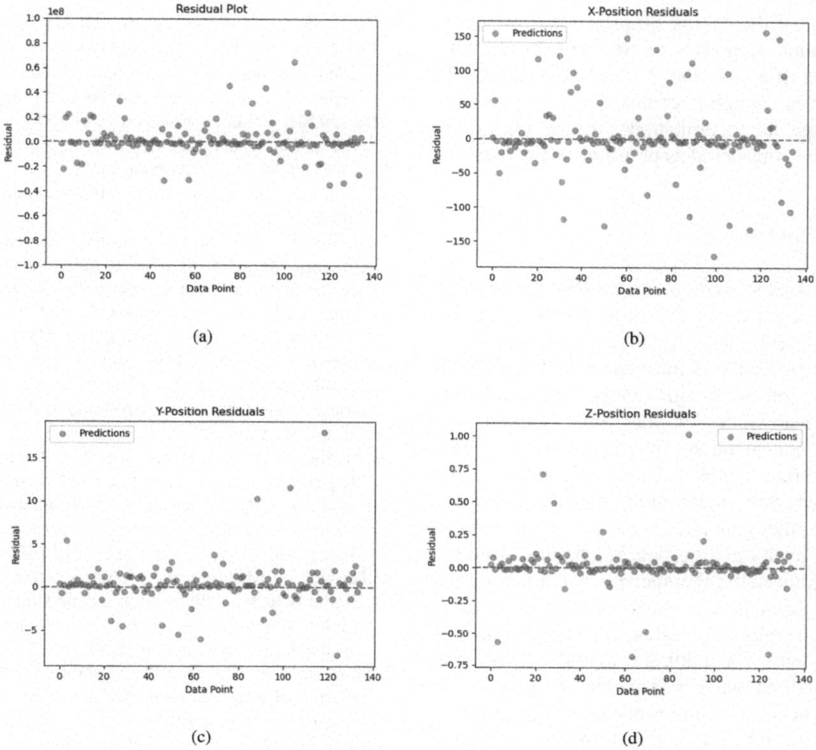

Figure 6. ANN models residual Plots of (a) Stress predictions vs. simulations, (b) x-location predictions vs. simulations, (c) y-location predictions vs. simulations and (d) z-location predictions vs. simulations.

Table 2. Ansys results vs. ANN predictions.

	Wave Parameters			Maximum Stress (MPa)	
	Frequency	Direction	Phase	ANSYS	ML Model
Case 1	0.08hz	0°	90°	64.4	64.77
Case 2	0.11hz	−180°	0°	69.3	64.74

consecutive simulation runs with successive phase angles to represent hydrodynamic loads in the time domain. However, due to limitations, the employed hydrodynamic simulation environment does not fully capture reality. Firstly, it represents regular long-crested wave packets, and time-discretized simulation in this case offers no advantages. The introduction of irregular waves would provide benefits over the frequency domain. Additionally, the simulation environment used can only depict linear wave effects, restricting the study to the global ship response as local effects of slamming waves at the bow cannot be investigated.

Moreover, the simulations in this paper assume no forward speed of the ship. In the case of frequency domain simulations, this is not a significant limitation, as ship speed only alters the encounter frequency. However, for nonlinear simulations in the time domain, this would have a significant impact. The hydrodynamic loads determined in this way are individually transferred to the mechanical model for each phase angle (load step). The load transfer between the hydrodynamic model and the mechanical model represents a one-way fluid interaction. It is assumed that the application in hull monitoring does not require a fully coupled simulation environment, making the method applicable for future studies.

Following the force transfer, a series of static linear finite element analyses, each with a load step, is conducted. This results in a temporal evolution of stresses in the entire ship hull. It should be noted that the finite

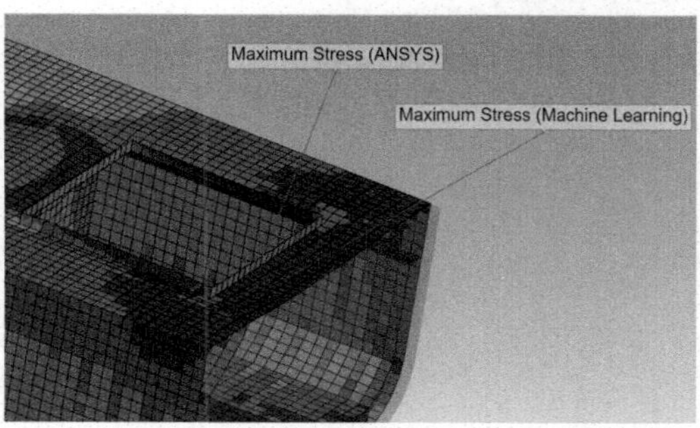

Figure 7. First test case.

element model of the ship is simplified and, in consequence, structural responses of the ship do not fully resemble those of a real ship. Nevertheless, the model is finely resolved enough to capture local stress effects. Due to the mentioned limitations, the results of this study should be considered as preliminary findings.

5 CONCLUSIONS

The aim of this paper was to demonstrate the feasibility of hull monitoring in the time domain. For this purpose, a preliminary method was presented. A simulation method was introduced to calculate the hydrodynamic pressures in regular stationary sea states. The realization of the time domain was achieved by concatenating phases of waves, each shifted by a certain angle. Subsequently, the loads for each time step were transferred to a finite element model. From the generated results, the maximum stress and its location were determined and stored in a dataset. Additionally, the spectral wave parameters of the regular sea state were also included in the dataset as input variables. The dataset created in this way serves as the basis for training a neural network that uses the described wave parameters to predict the structural response. The interpretation of the results is twofold. On the one hand, the proposed method is capable of predicting the maximum stress in the structure adequately compared to the simulations. However, the prediction of the location of the maximum stress is not satisfactory. A pattern of the location in relation to the wave parameters has not been identified yet. This should be addressed specifically in future methods. The overall functionality of this workflow was demonstrated; however, the limitations mentioned in the discussion hinder a meaningful application. To achieve the goal described in the introduction, these limitations must be eliminated in future studies. This includes implementing irregular sea states in a coupled time-domain simulation. The dataset thus created will no longer contain spectral parameters as input but rather the wave elevation time signal. The requirements for the ML prediction methodology will fundamentally change accordingly.

REFERENCES

Aarsnes, L. H., G. Storhaug, & M. Radon (2021). Utilization of structural design models in operation to monitor fatigue strength performance. In T. Okada, K. Suzuki, and Y. Kawamura (Eds.), *Practical Design of Ships and Other Floating Structures*, Volume 64 of *Lecture Notes in Civil Engineering*, pp. 622–636. Singapore: Springer Singapore.

ABS (2014). Guidance notes on safehull finite element analysis of hull structures.

ANSYS (2018). Hydrodynamic loads.

ANSYS, I. (2015). Aqwa theory manual. *ANSYS Inc.: Canonsburg, PA, USA*.

Anyfantis, K. N. (2021). An abstract approach toward the structural digital twin of ship hulls: A numerical study applied to a box girder geometry. *Proceedings of the Institution of Mechanical Engineers, Part M: Journal of Engineering for the Maritime Environment 235*(3), 718–736.

Drummen, I., A. Benhamou, R. Hageman, & K. Stambaugh (2019). Hull structure monitoring of a new class of us coast guard cutters. *ASNE Technology, Systems & Ships*.

Eremenko, K. & H. de Ponteves (2018, August). The ultimate guide to artificial neural networks (ann).

Farrar, C. R. & K. Worden (2012). *Structural health monitoring: A machine learning perspective*. Chichester, West Sussex, U.K and Hoboken, N.J: Wiley.

Hulkkonen, T., Manderbacka Teemu, & Sugimoto Kei (2019). Digital twin for monitoring remaining fatigue life of critical hull structures. *Digital Twin for Monitoring Remaining Fatigue Life of Critical Hull Structures*, 415–427.

Kaminski, M. L. (2007). Sensing and understanding fatigue lifetime of new and converted fpsos. In *Proceedings of the Offshore Technology*.

Kumar, N., D. L. Cahl, S. C. Crosby, & G. Voulgaris (2017). Bulk versus spectral wave parameters: Implications on stokes drift estimates, regional wave modeling, and hf radars applications. *Journal of Physical Oceanography 47*(6), 1413–1431.

Lang, X., J. W. Ringsberg, W. Mao, D. Wu, & C. Zhang (2023). Data-driven ship fatigue assessment based on pitch and heave motions. In J. W. Ringsberg and C. Guedes Soares (Eds.), *Advances in the Analysis and Design of Marine Structures*, pp. 95–103. London: CRC Press.

Schirmann, M., M. Collette, & J. Gose (2018). Ship motion and fatigue damage estimation via a digital twin. *Life-Cycle Analysis and Assessment in Civil Engineering: Towards an Integrated Vision*, 2075–2082.

Tensorflow (2023a). Introduction to the keras tuner.

Tensorflow (2023b). Metrics documentation.

Tensorflow (2023c). The sequential model.

Thompson, I. (2020). Virtual hull monitoring of a naval vessel using hindcast data and reconstructed 2-d wave spectra. *Marine Structures* 71.

Wedler, M., M. Stender, M. Klein, & N. Hoffmann (2023). Machine learning simulation of one-dimensional deterministic water wave propagation. *Ocean Engineering 284*, 115222.

Advances in Maritime Technology and Engineering – Guedes Soares & Santos (Eds)
© 2024 The Author(s), ISBN 978-1-032-83099-5

Predictive maintenance scheduling framework for offshore wind turbines based on condition monitoring: A review

J.B. Hes & X. Jiang
Maritime and Transport Technology, Delft University of Technology, Delft, The Netherlands

ABSTRACT: This study investigates the optimization of the operation and maintenance of offshore wind turbines based on condition monitoring data. Due to their increasingly remote and challenging location, a decision framework is proposed that optimizes the cost and risk of maintenance scheduling based on, dynamic Bayesian network based, iterative estimation of turbine lifetime. This allows for the combining of predictive and opportunistic maintenance strategies, scheduling preventative component replacements to minimize lost production, while maximizing lifetime and optimizing use of resources. Assessment of related literature and applications suggests the approach could lead to a reduction of maintenance costs that exceeds 30%. The proposed framework relies on effective fault detection and prognosis of wind turbine components, realised through the implementation of machine learning techniques on the turbine's own SCADA system. The installing of additional sensors can potentially increase the capability of this system for more advanced diagnosis and localization of a fault.

1 INTRODUCTION

The rapidly increasing demand for renewable energy sources has made wind energy one of the most promising alternatives to fossil fuels (Office of Energy Efficiency and Renewable Energy nd). With offshore wind energy projected to triple its capacity by 2026 (International Energy Agency 2021), it offers an efficient energy generation solution at a reduced environmental impact (National Grid nd).

However, the working conditions imposed by the marine environment present significant challenges for the construction, infrastructure, but mostly maintenance of offshore wind turbines (Duffy 2020). Increased loads from waves and strong winds, along with increased humidity, contribute to heightened failure rates among various turbine components, leading to costlier repairs and an increased need for maintenance (Zhu & Li 2018). Offshore working conditions also increase the wind turbine downtime required for repairs, resulting in further loss of revenue (Faulstich et al. 2011). Notably, the costs associated with offshore Operations and Maintenance (O&M) make up a considerable 23-35% of the overall investment, exceeding the 5-25% range typical for onshore wind (Dinwoodie & Mcmillan 2014, Blanco 2009, Sinha & Steel 2015b). As a consequence, 25% to 50% of the total energy generation cost of an offshore wind farm originates from maintenance expenses.

While research on onshore wind has traditionally emphasized cost-effective design and maximizing performance to minimize the initial capital cost of wind turbines (Molenaar 2003), the increased cost of O&M has brought forward new research priorities to ensure the long-term viability of the energy source (Barthelmie & Pryor 2001, Li et al. 2020).

This study investigates the implementation of a predictive maintenance decision framework that enables the application of strategic measures to enhance the critical aspects of wind turbine reliability, minimize the frequency of costly on-site visits, and aim to reduce the failure rates of high-risk components (Nikitas et al. 2020, Le & Andrews 2016, Carroll et al. 2017). Notably, the electric and control systems, gearbox, generator, and blades emerge as key contributors, accounting for 90% of the repair and replacement costs over a turbine's lifetime (Carroll et al. 2016, Artigao et al. 2018).

Recent studies into wind turbine condition monitoring are discussed to assess the prerequisites and potential capabilities of a fault prognosis system. Based on this predictive assessment of the turbine's health, it is discussed how well-informed maintenance decisions can be implemented to optimize O&M expenses, while mitigating risk and maximizing component lifetime.

2 INTELLIGENT CONDITION MONITORING

Predictive and condition-based maintenance decisions rely on accurate determination of the health of the system, which can be achieved by condition monitoring (CM) (Ren et al. 2021). A Condition Monitoring System (CMS) allows for the initiation of maintenance decisions by providing information on the condition of turbine subsystems.

DOI: 10.1201/9781003508762-68

While CM methods are initially ported from other machinery types, often based on vibration monitoring, the condition monitoring of wind turbines proved to be more complex due to the increase of stress and fatigue that originates from the inherently random load spectrum (Barszcz 2019). Combined with the increased remoteness of offshore turbines, a wide variety of online monitoring methods is developed that varies per type of component. Analysis of the operational data provided by SCADA systems as well as additionally installed sensors that perform measurements and report on the condition of the turbine allow for appropriate maintenance decisions that improve availability, reliability and lifespan, to provide the best economic maintenance solution (Walgern et al. 2017).

2.1 *SCADA data*

Signals and alarms from monitoring instrumentation, together with real-time performance and operational parameters, are collected by the SCADA system (Turbines 2005).

Thorough analysis of this comprehensive data set can potentially result in a cost-effective and easy-to-implement tool for evaluation of the wind turbine's health condition. Component specific CM approaches based on SCADA data often employ a wide array of parameters to evaluate anomalies in the turbine state, as well as accompanying minima, maxima and standard deviations. These methods generally estimate expected parameters, such as a component its temperature, and consider deviations from the simulated state as abnormal. An approach that considers the turbine as a whole, generates an expected power curve based on the set of SCADA parameters to compare with the actual power curve in the localization of anomalies (Papatheou et al. 2015, Pandit et al. 2019).

As wind turbine components deteriorate, the efficiency of the energy conversion process decreases, reducing the performance of the wind turbine. Therefore, deviations in the relationship between parameters such as power output, blade pitch angle, generator torque and rotor speed can be used as an assessment of the turbine condition and the early detection of faults (Sørensen et al. 2002).

The major difficulty of fault detection based on power output or other parameters is the variation in operating point that influences the measured data. Information on the operating point is not often available, so in abnormal events, data interpretation systems are required to distinguish between the existence of a fault or a harmless internal or external influence.

A common method to improve accuracy and robustness of the fault detection system is the deployment of machine learning methods (Zaher et al. 2009). These algorithms are well suited to process large amounts of SCADA data, while being able to deal with changes in operating point and complex system interactions (Barszcz 2019).

Supervised methods learn signal patterns for different technical states using data from both malfunctioning as well as healthy turbines. The data is labeled with information about the occurrence or type of faults, to train a machine learning model to predict the type of fault based on new input data. Unsupervised learning can also be used for fault diagnosis when there is no labeled data available. In this case, the goal is to identify patterns or anomalies in the data that may be indicative of a fault (El Bouchefry & de Souza 2020, Schneider & Xhafa 2022).

A common limitation originates from the low resolution of SCADA data, which is predominantly only stored every 10 minutes. Therefore, most studies into SCADA-based CM consider this sampling frequency (Maldonado-Correa et al. 2020). This constraint introduces the risk of overlooking spikes in temperatures, power curve anomalies or oddities in voltages and currents, causing SCADA-based CM applications difficulty to match the capability of traditional sensor-based CMSs. Recent studies investigate the possibility of using 1-s high frequency sampling of SCADA data to overcome this issue, with promising results (Bi et al. 2016, Gonzalez et al. 2017, Gonzalez et al. 2019, Lin et al. 2020).

2.2 *CM sensor data*

A common approach to CM is to equip the system with additional hardware, such as separate sensors dedicated to monitoring particular conditions like oil quality, vibrations, and sound emissions. Unlike SCADA systems, these devices can provide more precise readings at faster intervals, giving a more detailed understanding of the wind turbine's condition. However, this introduces additional costs, which are influenced by factors like measurement accuracy, sampling frequency, system capabilities, and the operating environment. Table 1 summarizes the sensing methods and their applications on turbine components according to existing research and reviews (e.g. (Pérez & Márquez 2015, Qiao & Lu 2015, Tchakoua et al. 2014, Hameed et al. 2011, GarcíaMárquez et al. 2012, Du et al. 2020)), with cost indications of implementations based on a study by Yang et al. (2014).

Analysis of a survey into commercially available CM systems performed by Crabtree et al. (2014) identifies vibration analysis applied to drive-train components as the most dominant CM method for wind turbines. In fact, 27 of the 36 total reviewed systems rely on vibration monitoring through accelerometers. After the drivetrain components, the blade is the most popular component, often associated with acoustic emission (AE) analysis (Verbruggen 2003). As most of the reviewed methods rely on time-domain, Fast Fourier Transform (FFT) or envelope analysis, there remains a notable absence in the adoption of more advanced machine learning methods.

2.3 *Wind turbine CM applications*

Based on a thorough analysis of recent studies by Hes in (Hes 2023), the most promising techniques

Table 1. Condition monitoring sensor techniques for turbine components and their costs as adapted from Yang et al, each € = ± 2000 euro.

	Blades	Rotor	Gearbox	Generator	Bearings	Tower
Vibration €	✓	✓	✓	✓	✓	✓
A.E. €€€€	✓	✓	✓		✓	✓
Ultrasonic €€	✓	✓				✓
Oil debris €€€			✓			
Strain €€€€€	✓					✓
Shock Pulse €€			✓	✓	✓	
Thermography €€€€€	✓	✓	✓	✓	✓	

for offshore wind CM, are established. Building on the comprehensive study performed by Badihi et al. (2022), the analysis discusses the capability and accuracy of CM methods from a practical perspective. The following sections brings forth recommendations based on his findings, which are summarized at the end of the section in Table 2.

2.3.1 *Blade condition monitoring*

Deterioration of the blade is quickly noticeable through a decreased performance of the turbine's power output. Combined with other SCADA parameters such as wind- and rotor speeds, studies suggest a fault warning can be produced hours (Wang et al. 2018), to days (Chen et al. 2013, Chen et al. 2015) and months (Chen et al. 2017) in advance. However, at this point, further localization and estimation of the fault severity of the fault solely based on analyzing SCADA parameters is difficult.

The most effective method to classify blade damage applies acoustic emission sensors. By clustering acoustic emissions picked up by two spaced sensors, the developed machine learning model localizes the fault. Xu et al. (2020) demonstrate accurate diagnosis of structural defects and delamination. Much cheaper vibration sensors potentially demonstrate the same diagnostic capability as AE sensors (Pacheco-Chérrez & Probst 2022), but need further research as they are not yet fully sensitive to damage and fail to obtain high accuracies (Oliveira et al. 2018). These sensors should also allow for the detection of icing and mud levels when supported by analysis of the power curve (Skrimpas et al. 2016).

Alternatively, a promising method to monitor structural defects, as well as erosion and the accumulation of mud and ice, is based on UAV-based thermography (Wang & Zhang 2017, Hwang et al. 2019, Sousa et al. 2020). However, this technique is still in an experimental stage and requires additional development.

2.3.2 *Generator condition monitoring*

Zhang & Lang (2020) demonstrate that nonlinear system frequency analysis of the power curve, as well as the generator temperature referenced with other SCADA parameters, is capable of producing a fault warning one year ahead of time. By adding analysis of generator currents, Brigham et al. (2020) demonstrate potential SCADA-based severity estimation and fault localization.

Even though some additional development of the technique is required for implementation, both methods display high accuracies. Therefore generator maintenance based only on SCADA data CM is very promising. This fact is supported by Gangsar & Tiwari (2017), who state that current-based fault detection of electrical generator faults is the most effective method for detecting and predicting generator faults.

2.3.3 *Gearbox condition monitoring*

Even though vibration analysis is still a very popular gearbox CM technique, recent literature shows that SCADA data can be used for the diagnosis and prognosis of many complex gearbox faults. By deploying a machine learning model, Rashid et al. (2020) demonstrate the detection of faults 2 months before critical failure based on analysis of the interaction between gearbox temperature and the produced power. The method is able to assess severity, but to provide a cheap preventative solution to gearbox fault detection, it is required that the model can effectively recognize the health state of the turbine and characterize the fault. Therefore, the model should be combined with models that allow for diagnosis, such as proposed by Bravo-Imaz et al. (2017) and He et al. (2020). They propose deviating between wear, crack, and pitting through the application of high-level feature extraction methods on the current signal.

If a longer prediction horizon is preferred, while also obtaining additional diagnostic information, additional vibration sensors should be installed. For instance, Teng et al. demonstrate a potential fault warning two years before critical failure by performing wavelet analysis on the vibration data (Teng et al. 2019). Other studies show that, based on vibration data, faults can be diagnosed very accurately in multiple gearbox stages (Carroll et al. 2019, Vamsi et al. 2019).

2.3.4 *Bearing condition monitoring*

Xiang et al. (2021) utilized convolutional, as well as long-, and short-term memory neural networks to analyze both temporal and spatial aspects of input data. Combined with an attention mechanism, the model predicts a bearing temperature based on SCADA data trends.

In this model, discrepancies between the predicted and actual bearing temperature trend serve as early indicators of an impending defect, signaling its emergence approximately two months before the eventual failure. Leveraging the local measurement of SCADA component temperatures, the model allows for diagnosis, facilitating the identification of the specific bearing responsible for the temperature irregularity and the evaluation of fault severity. This result is strengthened by Hu et al. (2018) and Carroll et al. (2019) who demonstrate accurate temperature-based bearing diagnosis and prognosis.

These studies suggest that temperature deviations could be detectable 4 to 6 months in advance. Although this fault prediction may not be highly accurate, about 70%, it can still provide valuable insights for strategic maintenance decisions.

While SCADA data analysis is the cheapest and least complex implementation, it is also the most accurate and effective. In the current state of bearing CM research, the implementation of SCADA-based is able to replace traditional vibration analysis. While the integration of vibration analysis alongside temperature analysis does not improve diagnostic accuracy, it does contribute to the accuracy of the RUL estimation (Carroll et al. 2019). This option could be considered if increased prognostic accuracy is required for effective maintenance decisions.

2.3.5 *Condition monitoring of other faults*

Faults in the electrical- and auxiliary system are not studied as extensively as the blade and drive-train. Literature does suggest however that most of these faults can be detected or even predicted based on the analysis of SCADA parameters such as temperatures and error codes. While some studies are still in an experimental phase, after additional development, proposed models should be able to accurately monitor converter-, controller-, transformer-, hydraulic-, yaw-, pitch- and anemometer defects (Schlechtingen & Santos 2014, Teimourzadeh Baboli et al. 2021, Chen et al. 2017, Astolfi et al. 2019, Guo & Infield 2020), respectively.

The current state of research in turbine tower and foundation CM is insufficient establish the advantages of either AE or vibration implementation. However, installing vibration sensors to allow implementation of an ANN, as proposed by Nguyen et al. (2018), while also running regression models as proposed by Oliveira et al. (2018), can provide a cost effective and accurate CM solution to both foundation cracks, tower damage, and scour. This can provide an inclusive CM solution for structure defects, in contrast to AE appliances, which are currently limited to foundation cracks (Tziavos et al. 2020).

Machine vision methods that assess component condition with cameras, can also present possible solutions to the maintenance of wind turbine structures (Badihi et al. 2022).

2.3.6 *Concluding remarks on CM application*

Study of recent literature demonstrates that SCADA data can provide an easy-to-implement and cost-effective CM solution for every component except for the tower and the foundation. While some of these methods are still in an experimental or untested shape, most component-specific cases demonstrate an accurate indication of developing faults at least 2 months before a failure turns critical. In the meanwhile, continuing analysis of deviations in the wind turbine's power curve can serve as a turbine-wide indicator for damage. Currently, modelling of the power curve based on Gaussian processes is demonstrated to be most effective method (Pandit et al. 2019, Morrison et al. 2022).

Installing additional sensors, such as vibration or acoustic emission (AE) sensors, can potentially increase the capability of a SCADA-based system, for more advanced diagnosis and localization of a fault. However, as this is not a given fact for every component, this decision should be evaluated on a component-specific basis, considering the operator's requirements.

Table 2 summarizes these findings, showing if developing defects are noticeable through power curve analysis and what the suggested methods are for online prognosis and diagnosis of faults.

Table 2. CM Recommendations summarized in a table.

Component	Power Curve	Diagnosis	Prognosis
Blade	✓	Vibration or AE	SCADA
Generator	✓	SCADA	SCADA
Gearbox	✓	SCADA or Vibration	SCADA or Vibration
Bearing	✓	SCADA or Vibration	SCADA or Vibration
Auxiliary		SCADA	SCADA
Structures		SCADA	-

3 MAINTENANCE ARCHITECTURE

Pattison et al. (2016) present an architecture to allow for structured and systematic handling of preventive or predictive maintenance decisions based on the data provided by CM techniques.

In contrast to other studies into maintenance scheduling (e.g. (Rademakers et al., Hofmann & Sperstad 2013, Stålhane et al. 2016, Ioannou et al.

2019)), this maintenance architecture supports an automated and intelligent approach to identify failure patterns or modes without the need for expert knowledge. Short-term decisions can be made concerning maintenance activities based on the current state of external influences such as environmental- and logistical conditions. In addition, the modular structure allows for individual development of each module and constant re-evaluation of the proposed solution through feedback of the state of the other modules.

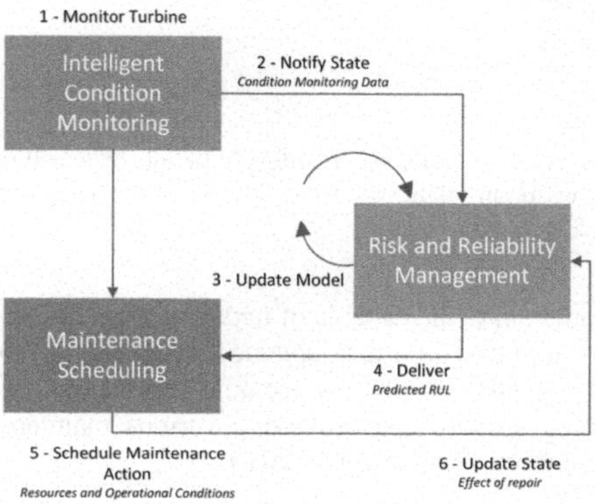

Figure 1. High-level information flow of maintenance system.

However, the architecture proposed by Pattison et al. (2016) is limited to the use of fault detection and diagnosis. As a result, the model can only directly incorporate data inputs related to the faulty component, fault severity, and the confidence level of the diagnosis. In the principles of Industry 4.0, Cachada et al. (2018) propose an intelligent and predictive maintenance architecture for early fault detection of machine failures, raising the question of how both real-time monitoring as well as prognosis can be applied to schedule maintenance actions in offshore wind maintenance. Consequently, this paper proposes an adapted architecture.

Figure 1 illustrates three modules, that collectively generate a solution in six high-level steps, from an information flow perspective.

Here, the Intelligent condition monitoring (ICM) module includes data analysis methods based on SCADA or deployed sensors to identify anomalous behavior (step 1). If any such deviation/anomaly is found, the system identifies this state and informs the Risk and Reliability Management (RRM) module (step 2). Based on the ICM input, this module evaluates a component's probability of survival across the duration of its intended service life and adapts a generic lifetime model to the observed state of the component (step 3).

This expected remaining useful life (RUL) is delivered to the Maintenance Scheduling (MS) module (step 4). Based on this, as well as ICM module fault descriptive data, the module is able to propose the best type and timing of maintenance actions (step 5). The output will subsequently impact the turbine survivability within the RRM (step 6).

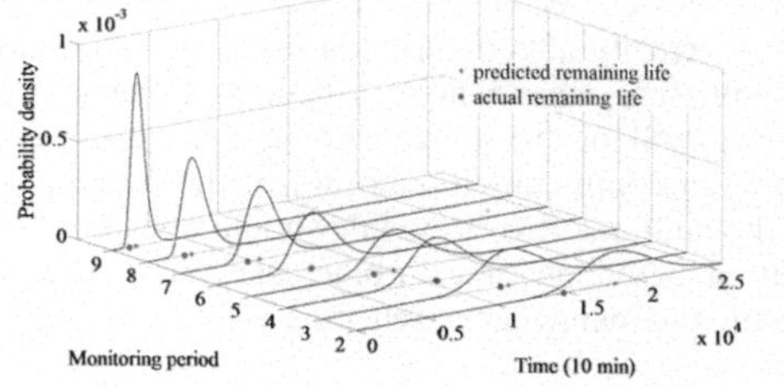

Figure 2. Comparisons of Gaussian probability distributions of RUL prediction results for turbine generator rear bearing, where each monitoring period consists of 2 weeks, up to a total of about 18 weeks at period 9 (Hu et al. 2018).

4 RISK AND RELIABILITY MANAGEMENT

Under the effect of a wide variety of external influences and a complex load pattern, a wind turbine's behaviour is difficult to predict. Combined with uncertainty concerning the effect of a maintenance action, the turbine state is under constant change.

For this reason, maintenance schedules are usually only prepared for a short period of three to seven days. Then, the schedule is updated every morning, as usually only the tasks of the first day are executed as planned (Kovacs et al. 2011).

To address this dynamic and random behaviour, the risk and reliability management module should be able to react to varying conditions and uncertainty and re-inform the scheduling process accordingly. This uncertainty can arise from the unpredictability in performance and power output due to for instance: (i) Variation in wind direction and velocity, (ii) Reliability of the turbine and grid connection, (iii) Regulatory changes in the allowable operating hours (iv) Inaccuracy of data (Dinwoodie & McMillan 2014, Ioannou et al. 2019, Dao et al. 2020).

The integration of prognosis allows for decisions to be made further into the future, requiring the increased alleviation of uncertainties due to inherently random behavior in the development of the fault. This uncertainty in degradation is influenced by factors such as: (i) Environmental conditions, including saltwater, winds, waves and temperature variations that influence the rate at which components degrade and corrode, (ii) Variation in operational factors influencing its rate of degradation, such as high loads or extended operation at rated capacity, (iii) Effectiveness and frequency of maintenance, (iv) Age of the turbine, (v) Uncertainties in the data and models.

In the estimation of remaining useful life, these uncertainties generally causes a decrease in prediction accuracy for a longer prediction horizon. This behavior can be seen in Figure 2, where the certainty of the prediction converges to the actual remaining life when the prediction horizon decreases.

As an example, Figure 2 is considered at 18 weeks lead time. Here, the model estimates RUL as a probability function between 12.5 and 25 weeks.

While this can be considered an inaccurate estimate, it can actually already provide sufficient information to initialize maintenance scheduling, because the failure will occur somewhere in that time period. Through continuous monitoring of the development of the fault, the RMM module should update the estimation based on new turbine inspection data, and update the maintenance schedule.

4.1 *Machine learning methods*

Due to the required compatibility with CM data, data-driven machine learning methods are considered (Zonta et al. 2020). Ferreira & Gonçalves (2022) reviewed machine learning methods for RUL prediction, pointing out difficulties related to traditional ML methods such as Kalman Filters and Support Vector Machines. These methods do not consider the relevance of time-series signals reflecting changes in the health condition and often rely on manual feature extraction from raw sensor data (Wu et al. 2020).

The application discussed in this paper introduces additional stochastic requirements, that can be met with approaches such as Bayesian Networks (BNs) or fuzzy logic. Introduction of temporal dependencies into these models, provide more advanced versions such as Dynamic Bayesian Networks (DBNs), which includes the Hidden Markov Model, or fuzzified long short-term memory networks. The Dynamic Bayesian network then emerges as the most promising method for the continuous refinement of estimations in the face of external disturbances that influence the lifetime prediction, such as differences in operating conditions and failure modes or unexpected correlation between faults (Straub & Kiureghian 2010, Adedipe et al. 2020).

4.2 *Dynamic Bayesian Network*

The versatility of BNs is widely acknowledged across industries, particularly because of their ability to cope with uncertainties present in organizational operations and decision-making. In structural reliability, Straub & Kiureghian (2010) implemented a DBN, which facilitates Bayesian updating of the model when new information becomes available, including both continuous and discrete random variables. Similarly, a study focusing on battery health estimation demonstrates the effectiveness of statistical forecasting (Richardson et al. 2017). While this study uses a more straightforward method based on Gaussian regression, their results as displayed in Figure 3, serve as an insightful demonstration of a continuously updated lifetime estimation at different points in time.

The applicability of DBNs for wind turbines is demonstrated by Nielsen & Sørensen (2017). To analyze the degradation process and estimate the RUL of wind turbine blades, the model produces an initial prediction based on detected damage, which is updated using CM and inspections.

Implementation of a DBN allows for the capturing of the dynamic nature of current- and future

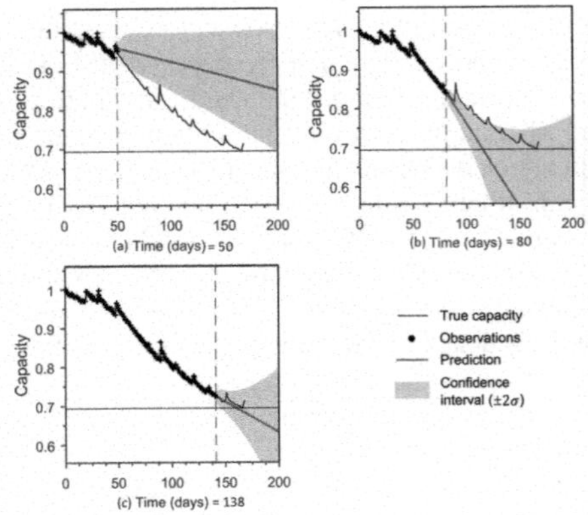

Figure 3. Remaining useful life prediction of a battery at three different points in time. Edited from: (Richardson et al. 2017).

degradation, while capable of updating the reliability of the system based on data obtained through CM. Moreover, the DBN facilitates accurate estimations of the resulting system state following various maintenance interventions (Pattison et al. 2016).

5 MAINTENANCE SCHEDULING

The ability to identify faults preemptively, preventing them from progressing into critical failures, gives time to optimize maintenance activities and costs (Sinha & Steel 2015a, Kandukuri et al. 2016). Based on SCADA or sensor data, maintenance tasks can be initiated while considering factors such as downtime, repair costs, risk of failure and the cost of lacking performance.

During maintenance scheduling, it may be necessary to temporarily halt turbine operation to extend the available time for maintenance actions. Alternatively, Fault-tolerant and Condition-based control can be applied to accommodate the effects of noncritical faults in wind turbines, or modify the operational state of components depending on their health-condition to avoid unnecessary shutdowns and missed production, and increase the time window for maintenance (Badihi et al. 2022).

By replacing components in anticipation of their failure, while also assessing the most convenient and efficient time and method, the inclusion of opportunistic scheduling strategies into predictive maintenance can provide a promising step in the minimization of maintenance costs de Pater & Mitici (2021). Research that combines predictions with maritime logistics optimisation has yet to be performed, however (Halvorsen-Weare & Nonås 2022).

To consider the application of this hybrid solution to offshore wind, first, the separate aspects of predictive and opportunistic maintenance scheduling are discussed.

5.1 *Predictive maintenance scheduling*

To determine if maintenance should be carried out preventively, Zhu et al. (2017) assess the costs originating from a preventative or, future corrective maintenance action based on stochastic lifetime predictions. Evaluation of these costs produces a control limit used to notify the need for preventative replacement. This ensures that component lifetime is maximized without unnecessary early replacements. A similar threshold is applied by Zhou & Yin (2019), while also allowing for preventative replacement of other turbine components in the same action, if the cost of this preventative action is lower than a condition-based maintenance action later on.

Instead of a cost-based threshold to maximize the effective RUL, Schenkelberg et al. (2020) propose that the balancing of a failure risk in a maintenance optimization problem can be done by estimating the impact of failure on profitability. Through the application of a DBN, the model predicts costs associated with a possible failure escalation and compares that with the cost of preventative replacement.

5.2 *Opportunistic maintenance scheduling*

Opportunistic maintenance aims to find the most ideal moment to perform maintenance based on resources and costs related to O&M. Each individual task poses its own requirements on these resources, encompassing the management of personnel and spare parts as well as the arranging and routing of vessels (Karyotakis 2011). Importantly, transportation costs constitute to 30% of a turbine's maintenance expenses (Irawan et al. 2017).

Maintenance costs can also arise due to lost production. This lost revenue is a direct consequence of maintenance activities, as turbines may need to be temporarily halted for maintenance purposes. The servicing of one turbine can result in the interruption of other turbines as well, because multiple turbines are connected in series to the power grid (Pattison et al. 2016).

Lost production can be minimized by the incorporation of forecasted wind speed, or analysis of energy prices, aligning turbine shutdown with periods of low yield. Proposed by Petros Papadopoulos & Ezzat (2023), this method demonstrates promising results in the optimization of total cost, downtime, resource utilization, and maintenance interruptions.

Nguyen et al. (2022) propose a maintenance scheduling approach, designed to maximize cost-effectiveness while minimizing lost production and environmental and safety risks. This is obtained by the intelligent optimization of resources, minimization of production losses and efficient vessel routing. The proposed model utilizes fuzzy probabilities to assess environmental- and safety risks associated with marine environments.

Incorporation of these fuzzified risks into scheduling decisions enables stochastic inputs into the model. In a similar way, a probability density function associated with a prognosis, as shown in Figure 2, can be incorporated into the scheduling decision. Increased risk of failure could be balanced with maintenance opportunities, such periods of low yield and lower safety risks due to calm weather, or increased consolidation of maintenance tasks. Careful evaluation of the risk of component failure is required when seeking maximization of the RUL, as the preventative replacement of a component is usually beneficial over the risk of having to perform a more severe repair (Turnbull & Carroll 2021).

5.3 *Optimization solution methods*

Based on the discussed literature, solving the predictive maintenance scheduling optimization problem describes the scheduling of tasks and allocation of resources such as vessels and spare parts, while optimizing risk and costs.

When reviewing maintenance optimization studies, the mathematical handling of these tasks can often be found to be based on Mixed Integer Linear Programming (MILP). Given the magnitude of transportation costs of offshore O&M, proper management and routing of a suitable fleet of vessels is required. Stålhane et al. (2016) demonstrate a MILP model for optimal fleet deployment and long-term decision-making on the fleet size and composition. This includes the acquisition and routing of crew transfer vehicles, service operation vessels, jack-up vessels and helicopters. In this decision, which is made one year ahead of time, environmental conditions as well as failures are taken as a-priori information.

When the model is confronted with unexpected weather conditions and turbine failures, unplanned corrective maintenance actions drive up the O&M cost. This also means that maintenance costs are often underestimated. To achieve a better MILP estimate of costs, an heuristic is presented by Gutierrez-Alcoba et al. (2019). While this solution increases the initial investment costs, the capability to react to uncertain events reduces the risk of unexpected cost increases.

This prompts Magnus Stålhane & Hvattum (2021) to include a broader field of strategic uncertainties as well, such as economic trends, gradual development of wind farms, and vessel technology development. These approaches show that a more realistic description of the system, as well as associated costs allows for more accurate predictions of decision outcomes, and a overall more optimal scheduling solution.

Addressing maintenance task management from a different point-of-view, Schrotenboer et al. solve the optimization problem in the context of a single maintenance provider responsible for multiple wind farms (Schrotenboer et al. 2020). In the Netherlands, for instance, offshore wind practices demonstrate that maintenance is performed by a service provider that does not own the farms and, therefore, does not risk uncertain production revenues due to alternating energy prices or the risk of production losses due to downtime.

Ge et al. (2020) as well as Sang et al. (2021) demonstrate that proper evaluation of downtime related wind costs should include power generation losses due to wake effects. The proposed approach consists of two stages, where the first stage determines the required amount of ships and technicians, and the second stage determines which turbines should be maintained while minimizing cost or travel distance per day. This allows for the short term updating of maintenance tasks if unexpected environmental conditions or faults are revealed after the initial schedule made in stage one.

The mathematical model of Irawan et al. (2017) is considered to include spare part allocation. The model includes constraints related to spare parts and describes the not only the availability of spare parts, but also limitations related to the weight and required storage space when considering transportation.

Due to the uncertain nature of the marine environment, the discussed MILP-based studies consider heuristics and complex multi-layer solution frameworks (Zhong et al. 2018). Other non-deterministic methods can also be considered, including fuzzy-based programming (Zhong et al. 2019), integration of a rolling-horizon (Petros Papadopoulos & Ezzat 2023, Papadopoulos et al. 2022), discrete heuristic optimization machine learning (Fan et al. 2019), and meta-heuristic Genetic Algorithms (Stock-Williams & Swamy 2019).

The model should be able to meet the opportunistic demands of the maintenance scheduling module within a given time-frame. When integrated into the proposed architecture, the RUL probability function from the RRM module replaces fixed predictions or heuristic approaches to address unexpected failures and degradation uncertainty. This integration allows the optimization model to make advanced decisions, leading to a more optimal solution.

5.4 *Integrated architecture*

After evaluation of the proposed architecture's required functions, a more detailed information flow visualization of the maintenance decision model can be made. Figure 4 depicts the three discussed major modules, as well as their interaction with each other and external data-sets.

Novel CM systems monitor the health of the turbine. After detection of a fault, its location and severity are used to initiate maintenance decisions. Based on the evaluation of component or turbine health, the RRM module estimates the RUL utilizing a DBN. In an iterative process, the RRM module converges step by step to refine its estimation as the turbine condition approaches its point of failure.

This estimated point of failure defines the amount of time that is available to solve the maintenance scheduling optimization problem. The required maintenance action is derived from the ICM module its diagnosis. This maintenance action includes task characteristics, such as time and cost, as well as conditional constraints such as equipment, stock or weather requirements.

Meeting the scheduling demands is done in multiple stages. First, the mid-term comprises the availability of stock and, if applicable to the service provider, the management of the fleet. This is followed by the assigning of vessels as well as personnel within any imposed environmental limitations. Finally, the short-term stage, covers the optimal routing of tasks. Based on the proposed schedule, the model verifies if a more optimal solution can be found at a higher risk of critical failure. Even when a maintenance schedule is produced the proposed solution might require updating in the face of changing environmental conditions. In turn, the performed maintenance action influences the predicted RUL.

6 COST

To estimate the potential financial impact of the implementation of the proposed maintenance strategy, several studies into the effect of predictive maintenance or optimization of a maintenance problem are discussed.

6.1 *Predictive maintenance*

Through time-based simulation of wind farm operations, Turnbull & Carroll (2021) quantify the cost

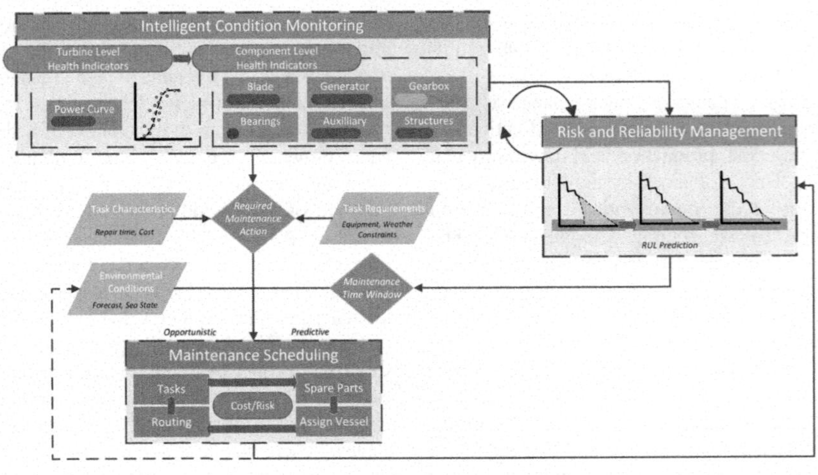

Figure 4. Integrated high-level information flow of opportunistic and predictive maintenance system.

benefits associated with different maintenance strategies, taking into consideration both direct maintenance costs and lost production.

Their results indicate that O&M costs can be reduced up to 8% along with a reduction of lost production up to 11%. However in obtaining these values, Turnbull & Carroll assume that, after fault detection of monitoring systems, only 25% of major failures of the generator and gearbox can be repaired before major replacement is required.

6.2 *Opportunistic maintenance*

A comprehensive review performed by Tusar & Sarker (2022) Maintenance reviews a vast amount of maintenance models to assess the cost benefits associated with different strategic approaches. Certain risk and reliability-based models reduce annual O&M costs by 23%, whereas opportunistic maintenance strategies suggest the ability to minimize 32% of production losses and transportation costs.

The most cost-efficient model in the review performed by Tusar & Sarker (2022) reduced the cost of maintenance by 48.2% compared to corrective maintenance. This method applied opportunistic maintenance for wind farms, proposing the efficient scheduling of tasks by grouping tasks of multiple components of a single turbine or combining tasks of multiple proximate turbines in a farm (Ding & Tian 2012). To provide enough room for these decisions, the model allows for performing incomplete or imperfect maintenance tasks, not always returning the component to a good-as-new state.

In a different approach, Besnard et al. (2011) propose stochastic scheduling at the lowest cost based on a wind forecast. The schedule is updated daily based on changes in the production and weather forecast. After evaluation of real wind data, the proposed approach demonstrated a 32% reduction in maintenance cost due to reduced production losses and saved transportation costs.

6.3 *Hybrid maintenance approach*

To assess the cost benefit of the hybrid opportunistic and predictive maintenance strategy as proposed in this paper, the model proposed by Tian et al. (2011) is considered. It uses predictive sensor data to determine a cost-effective solution that incorporates both opportunistic as well as predictive maintenance aspects. After comparing this method with traditional methods, Tian et al. suggest a cost saving of 44.42%.

In these examples, it is crucial to ensure that the benefits derived from implementing more advanced methods surpass the associated increased costs. This evaluation is contingent on the reliability and conditions of each unique component. As a consequence, the optimal maintenance solution must be assessed independently for each component (Zhong et al. 2019).

7 CONCLUSION

Optimization of the operation and maintenance costs of offshore wind turbines is realized through informed and strategic predictive decisions grounded on condition monitoring data. Based on lifetime predictions made by a dynamic Bayesian neural network, preventative component replacements are scheduled in a way that minimizes downtime and lost production, while maximizing lifetime and optimizing the use of maintenance related resources.

Studied applications consistently show a potential predicted lead time to failure of at least 2 months. Applying machine learning techniques on the turbine's own SCADA system provides an effective fault detection and prognosis system for many turbine components, while the installing of additional sensors can potentially increase the capability of this system, for more advanced diagnosis and localization of a fault.

Analysis of similar studies in offshore wind and other fields suggests the combination of predictive and opportunistic maintenance approaches could lead to significant reductions in O&M costs.

ACKNOWLEDGMENTS

This research is financially supported by NWO as part of the Holi-DOCTOR project.

REFERENCES

Adedipe, T., M. Shafiee, & E. Zio (2020). Bayesian network modelling for the wind energy industry: An overview. *Reliability Engineering and System Safety 202*, 107053.

Artigao, E., S. Martín-Martínez, A. Honrubia-Escribano, & E. Gómez-Lázaro (2018). Wind turbine reliability: A comprehensive review towards effective condition monitoring development. *Applied Energy 228*, 1569–1583.

Astolfi, D., F. Castellani, & F. Natili (2019). Wind turbine generator slip ring damage detection through temperature data analysis. *Diagnostyka Vol. 20, No. 3*, 3–9.

Badihi, H., Y. Zhang, B. Jiang, P. Pillay, & S. Rakheja (2022). A comprehensive review on signal-based and model-based condition monitoring of wind turbines: Fault diagnosis and lifetime prognosis. *Proceedings of the IEEE 110*(6), 754–806.

Barszcz, T. (2019). *Vibration-based condition monitoring of wind turbines*. Springer.

Barthelmie, R. J. & S. Pryor (2001). A review of the economics of offshore wind farms. *Wind engineering 25*(4), 203–213.

Besnard, F., M. Patriksson, A.-B. Strömberg, A. Wojciechowski, K. Fischer, & L. Bertling (2011). A stochastic model for opportunistic maintenance planning of offshore wind farms. In *2011 IEEE Trondheim PowerTech*, pp. 1–8.

Bi, R., C. Zhou, & D. M. Hepburn (2016). Applying instantaneous scada data to artificial intelligence based power curve monitoring and wtg fault forecasting. In *2016 International Conference on Smart Grid and Clean Energy Technologies (ICSGCE)*, pp. 176–181. IEEE.

Blanco, M. I. (2009). The economics of wind energy. *Renewable and sustainable energy reviews 13*(6-7), 1372–1382.

Bravo-Imaz, I., H. D. Ardakani, Z. Liu, A. Garca-Arribas, A. Arnaiz, & J. Lee (2017). Motor current signature analysis for gearbox condition monitoring under transient speeds using wavelet analysis and dual-level time synchronous averaging. *Mechanical Systems and Signal Processing 94*, 73–84.

Brigham, K., D. Zappalá, C. J. Crabtree, & C. Donaghy-Spargo (2020). Simplified automatic fault detection in wind turbine induction generators. *Wind Energy 23*(4), 1135–1144.

Cachada, A., J. Barbosa, P. Leitño, C. A. Gcraldcs, L. Deusdado, J. Costa, C. Teixeira, J. Teixeira, A. H. Moreira, P. M. Moreira, & L. Romero (2018). Maintenance 4.0: Intelligent and predictive maintenance system architecture. In *2018 IEEE 23rd International Conference on Emerging Technologies and Factory Automation (ETFA)*, Volume 1, pp. 139–146.

Carroll, J., S. Koukoura, A. McDonald, A. Charalambous, S. Weiss, & S. McArthur (2019). Wind turbine gearbox failure and remaining useful life prediction using machine learning techniques. *Wind Energy 22*(3), 360–375.

Carroll, J., A. McDonald, I. Dinwoodie, D. McMillan, M. Revie, & I. Lazakis (2017). Availability, operation and maintenance costs of offshore wind turbines with different drive train configurations. *Wind Energy 20*(2), 361–378.

Carroll, J., A. McDonald, & D. McMillan (2016). Failure rate, repair time and unscheduled o&m cost analysis of offshore wind turbines. *Wind Energy 19*(6), 1107–1119.

Chen, B., P. C. Matthews, & P. J. Tavner (2013). Wind turbine pitch faults prognosis using a-priori knowledge-based anfis. *Expert Systems with Applications 40*(17), 6863–6876.

Chen, B., P. C. Matthews, & P. J. Tavner (2015). Automated on-line fault prognosis for wind turbine pitch systems using supervisory control and data acquisition. *IET Renewable Power Generation 9*(5), 503–513.

Chen, N., R. Yu, Y. Chen, & H. Xie (2017). Hierarchical method for wind turbine prognosis using scada data. *IET Renewable Power Generation 11*(4), 403–410.

Crabtree, C., D. Zappala, & P. Tavner (2014, May). Survey of commercially available condition monitoring systems for wind turbines. Technical report, DU.

Dao, C. D., B. Kazemtabrizi, & C. J. Crabtree (2020). Offshore wind turbine reliability and operational simulation under uncertainties. *Wind Energy 23*(10), 1919–1938.

de Pater, I. & M. Mitici (2021). Predictive maintenance for multi-component systems of repairables with remaining-useful-life prognostics and a limited stock of spare components. *Reliability Engineering and System Safety 214*, 107761.

Ding, F. & Z. Tian (2012). Opportunistic maintenance for wind farms considering multi-level imperfect maintenance thresholds. *Renewable Energy 45*, 175–182.

Dinwoodie, I. & D. Mcmillan (2014). Operation and maintenance of offshore wind farms. *Eng. Technol. Ref 1*(1).

Dinwoodie, I. A. & D. McMillan (2014). Operational strategies for offshore wind turbines to mitigate failure rate uncertainty on operational costs and revenue. *IET Renewable Power Generation 8*(4), 359–366.

Du, Y., S. Zhou, X. Jing, Y. Peng, H. Wu, & N. Kwok (2020). Damage detection techniques for wind turbine blades: A review. *Mechanical Systems and Signal Processing 141*, 106445.

Duffy, T. S. P. B. P. (2020). 2019 cost of wind energy review. *NREL* (NREL/TP-5000-78471).

El Bouchefry, K. & R. S. de Souza (2020). Chapter 12 - learning in big data: Introduction to machine learning. In P. Škoda & F. Adam (Eds.), *Knowledge Discovery in Big Data from Astronomy and Earth Observation*, pp. 225–249. Elsevier.

Fan, D., Y. Ren, Q. Feng, B. Zhu, Y. Liu, & Z. Wang (2019). A hybrid heuristic optimization of maintenance routing and scheduling for offshore wind farms. *Journal of Loss Prevention in the Process Industries 62*, 103949.

Faulstich, S., B. Hahn, & P. Tavner (2011, 04). Wind turbine downtime and its importance for offshore deployment. *Wind Energy* 14.

Ferreira, C. & G. Gonçalves (2022). Remaining useful life prediction and challenges: A literature review on the use of machine learning methods. *Journal of Manufacturing Systems 63*, 550–562.

Gangsar, P. & R. Tiwari (2017). Comparative investigation of vibration and current monitoring for prediction of mechanical and electrical faults in induction motor based on multiclass-support vector machine algorithms. *Mechanical Systems and Signal Processing 94*, 464–481.

García Márquez, F. P., A. M. Tobias, J. M. Pinar Pérez, & M. Papaelias (2012). Condition monitoring of wind turbines: Techniques and methods. *Renewable Energy 46*, 169–178.

Ge, X., Q. Chen, Y. Fu, C. Chung, & Y. Mi (2020). Optimization of maintenance scheduling for offshore wind turbines considering the wake effect of arbitrary wind direction. *Electric Power Systems Research 184*, 106298.

Gonzalez, E., B. Stephen, D. Infield, & J. Melero (2017). On the use of high-frequency scada data for improved wind turbine performance monitoring. In *Journal of Physics: Conference Series*, Volume 926, pp. 012009. IOP Publishing.

Gonzalez, E., B. Stephen, D. Infield, & J. J. Melero (2019). Using high-frequency scada data for wind turbine performance monitoring: A sensitivity study. *Renewable Energy 131*, 841–853.

Guo, P. & D. Infield (2020). Wind turbine power curve modeling and monitoring with gaussian process and sprt. *IEEE Transactions on Sustainable Energy 11*(1), 107–115.

Gutierrez-Alcoba, A., E. Hendrix, G. Ortega, E. Halvorsen-Weare, & D. Haugland (2019). On offshore wind farm maintenance scheduling for decision support on vessel fleet composition. *European Journal of Operational Research 279*(1), 124–131.

Halvorsen-Weare, E. E. & L. M. Nonås (2022). Predictive maintenance logistics for offshore wind farms.

Hameed, Z., J. Vatn, & J. Heggset (2011). Challenges in the reliability and maintainability data collection for offshore wind turbines. *Renewable Energy 36*(8), 2154–2165.

He, Q., J. Zhao, G. Jiang, & P. Xie (2020). An unsupervised multiview sparse filtering approach for current-based wind turbine gearbox fault diagnosis. *IEEE Transactions on Instrumentation and Measurement 69*(8), 5569–5578.

Hes, J. (2023). Predictive maintenance of offshore wind turbines: A comprehensive review of condition monitoring technologies and maintenance solutions, unpublished manuscript. *Delft University of Technology.*

Hofmann, M. & I. B. Sperstad (2013). Nowicob–a tool for reducing the maintenance costs of offshore wind farms. *Energy Procedia 35*, 177–186.

Hu, Y., H. Li, P. Shi, Z. Chai, K. Wang, X. Xie, & Z. Chen (2018). A prediction method for the real-time remaining useful life of wind turbine bearings based on the wiener process. *Renewable Energy 127*, 452–460.

Hwang, S., Y.-K. An, & H. Sohn (2019). Continuous-wave line laser thermography for monitoring of rotating wind turbine blades. *Structural Health Monitoring 18*(4), 1010–1021.

International Energy Agency (2021). Executive summary.

Ioannou, A., A. Angus, & F. Brennan (2019). Informing parametric risk control policies for operational uncertainties of offshore wind energy assets. *Ocean Engineering 177*, 1–11.

Irawan, C. A., D. Ouelhadj, D. Jones, M. Stålhane, & I. B. Sperstad (2017). Optimisation of maintenance routing and scheduling for offshore wind farms. *European Journal of Operational Research 256*(1), 76–89.

Kandukuri, S. T., A. Klausen, H. R. Karimi, & K. G. Robbersmyr (2016). A review of diagnostics and prognostics of low-speed machinery towards wind turbine farm-level health management. *Renewable and Sustainable Energy Reviews 53*, 697–708.

Karyotakis, A. (2011). On the optimisation of operation and maintenance strategies for offshore wind farms.

Kovacs, A., G. Erdös, Z. J. Viharos, & L. Monostori (2011). A system for the detailed scheduling of wind farm maintenance. *CIRP annals 60*(1), 497–501.

Le, B. & J. Andrews (2016). Modelling wind turbine degradation and maintenance. *Wind Energy 19*(4), 571–591.

Li, J., G. Wang, Z. Li, S. Yang, W. T. Chong, & X. Xiang (2020). A review on development of offshore wind energy conversion system. *International Journal of Energy Research 44*(12), 9283–9297.

Lin, Z., X. Liu, & M. Collu (2020). Wind power prediction based on high-frequency scada data along with isolation forest and deep learning neural networks. *International Journal of Electrical Power & Energy Systems 118*, 105835.

Magnus Stålhane, Kamilla Hamre Bolstad, M. J. & L. M. Hvattum (2021). A dual-level stochastic fleet size and mix problem for offshore wind farm maintenance operations. *INFOR: Information Systems and Operational Research 59*(2), 257–289.

Maldonado-Correa, J., S. Martín-Martínez, E. Artigao, & E. Gómez-Lázaro (2020). Using scada data for wind turbine condition monitoring: A systematic literature review. *Energies 13*(12).

Molenaar, D. (2003). Cost-effective design and operation of variable speed wind turbines. *Institutional Repository.*

Morrison, R., X. Liu, & Z. Lin (2022). Anomaly detection in wind turbine scada data for power curve cleaning. *Renewable Energy 184*, 473–486.

National Grid (n.d.). Onshore vs offshore wind energy: what's the difference?

Nguyen, C.-U., T.-C. Huynh, & J.-T. Kim (2018). Vibration-based damage detection in wind turbine towers using artificial neural networks. *Structural Monitoring and Maintenance 5*(4), 507.

Nguyen, T.-A.-T., S.-Y. Chou, & T. H.-K. Yu (2022). Developing an exhaustive optimal maintenance schedule for offshore wind turbines based on risk-assessment, technical factors and cost-effective evaluation. *Energy 249*, 123613.

Nielsen, J. S. & J. D. Sørensen (2017). Bayesian estimation of remaining useful life for wind turbine blades. *Energies 10*(5).

Nikitas, G., S. Bhattacharya, & N. Vimalan (2020). 16 - wind energy. In T. M. Letcher (Ed.), *Future Energy (Third Edition)* (Third Edition ed.)., pp. 331–355. Elsevier.

Office of Energy Efficiency and Renewable Energy (n.d.). Advantages and challenges of wind energy.

Oliveira, G., F. Magalhães, î Cunha, & E. Caetano (2018). Vibration-based damage detection in a wind turbine using 1 year of data. *Structural Control and Health Monitoring 25*(11), e2238. e2238 STC-17-0154.R1.

Pacheco-Chérrez, J. & O. Probst (2022). Vibration-based damage detection in a wind turbine blade through operational modal analysis under wind excitation. *Materials Today: Proceedings 56*, 291–297. International Conference on Materials, Machines and Information Technology-2022.

Pandit, R. K., D. Infield, & A. Kolios (2019). Comparison of advanced non-parametric models for wind turbine power curves. *IET Renewable Power Generation 13*(9), 1503–1510.

Papadopoulos, P., D. W. Coit, & A. A. Ezzat (2022). Seizing opportunity: Maintenance optimization in offshore wind farms considering accessibility, production, and crew dispatch. *IEEE Transactions on Sustainable Energy 13*(1), 111–121.

Papatheou, E., N. Dervilis, A. E. Maguire, I. Antoniadou, & K. Worden (2015). A performance monitoring approach for the novel lillgrund offshore wind farm. *IEEE Transactions on Industrial Electronics 62*(10), 6636–6644.

Pattison, D., M. Segovia Garcia, W. Xie, F. Quail, M. Revie, R. I. Whitfield, & I. Irvine (2016). Intelligent integrated maintenance for wind power generation. *Wind Energy 19*(3), 547–562.

Petros Papadopoulos, D. W. C. & A. A. Ezzat (2023). Stochos: Stochastic opportunistic maintenance scheduling for offshore wind farms. *IISE Transactions 0*(0), 1–15.

Pérez, J. & F. Márquez (2015). 11 - condition monitoring and fault diagnosis in wind energy systems. In J.-L. Bessède (Ed.), *Eco-Friendly Innovation in Electricity Transmission and Distribution Networks*, pp. 221–241. Oxford: Woodhead Publishing.

Qiao, W. & D. Lu (2015). A survey on wind turbine condition monitoring and fault diagnosis—part i: Components and subsystems. *IEEE Transactions on Industrial Electronics 62*(10), 6536–6545.

Rademakers, L., H. Braam, T. Obdam, & R. vd Pieterman. Operation and maintenance cost estimator (omce) to estimate the future o&m costs of offshore wind farms.

Rashid, H., E. Khalaji, J. Rasheed, & C. Batunlu (2020). Fault prediction of wind turbine gearbox based on scada data and machine learning. In *2020 10th International Conference on Advanced Computer Information Technologies (ACIT)*, pp. 391–395.

Ren, Z., A. S. Verma, Y. Li, J. J. Teuwen, & Z. Jiang (2021). Offshore wind turbine operations and maintenance: A state-of-the-art review. *Renewable and Sustainable Energy Reviews 144*, 110886.

Richardson, R. R., M. A. Osborne, & D. A. Howey (2017). Gaussian process regression for forecasting battery state of health. *Journal of Power Sources 357*, 209–219.

Sang, M.-G., N.-K. Lee, Y.-H. Shin, C. Lee, & Y.-J. Oh (2021). An optimization model for o&m planning of floating offshore wind farm using mixed integer linear programming. *Journal of the Korea Society of Computer and Information 26*(12), 255–264.

Schenkelberg, K., U. Seidenberg, & F. Ansari (2020). Analyzing the impact of maintenance on profitability using dynamic bayesian networks. *Procedia CIRP 88*, 42–47. 13th CIRP Conference on Intelligent Computation in Manufacturing Engineering, 17-19 July 2019, Gulf of Naples, Italy.

Schlechtingen, M. & I. F. Santos (2014). Wind turbine condition monitoring based on scada data using normal behavior models. part 2: Application examples. *Applied Soft Computing 14*, 447–460.

Schneider, P. & F. Xhafa (2022). Chapter 3 - anomaly detection: Concepts and methods. In P. Schneider & F. Xhafa (Eds.), *Anomaly Detection and Complex Event Processing over IoT Data Streams*, pp. 49–66. Academic Press.

Schrotenboer, A. H., E. Ursavas, & I. F. Vis (2020). Mixed integer programming models for planning maintenance at offshore wind farms under uncertainty. *Transportation Research Part C: Emerging Technologies 112*, 180–202.

Sinha, Y. & J. A. Steel (2015a). Failure prognostic schemes and database design of a software tool for efficient management of wind turbine maintenance. *Wind Engineering 39*(4), 453–477.

Sinha, Y. & J. A. Steel (2015b). A progressive study into offshore wind farm maintenance optimisation using risk based failure analysis. *Renewable and Sustainable Energy Reviews 42*, 735–742.

Skrimpas, G. A., K. Kleani, N. Mijatovic, C. W. Sweeney, B. B. Jensen, & J. Holboell (2016). Detection of icing on wind turbine blades by means of vibration and power curve analysis. *Wind Energy 19*(10), 1819–1832.

Sørensen, B. F., L. Lading, & P. Sendrup (2002). Fundamentals for remote structural health monitoring of wind turbine blades-a pre-project.

Sousa, P. J., F. Barros, N. V. Ramos, P. J. Tavares, & P. M. Moreira (2020). Damage detection in rotating objects using position-triggered thermography. *Engineering Failure Analysis 115*, 104642.

Stock-Williams, C. & S. K. Swamy (2019). Automated daily maintenance planning for offshore wind farms. *Renewable Energy 133*, 1393–1403.

Straub, D. & A. D. Kiureghian (2010). Bayesian network enhanced with structural reliability methods: Methodology. *Journal of Engineering Mechanics 136* (10), 1248–1258.

Stålhane, M., H. Vefsnmo, E. E. Halvorsen-Weare, L. M. Hvattum, & L. M. Nonås (2016). Vessel fleet optimization for maintenance operations at offshore wind farms under uncertainty. *Energy Procedia 94*, 357–366. 13th Deep Sea Offshore Wind R&D Conference, EERA DeepWind'2016.

Tchakoua, P., R. Wamkeue, M. Ouhrouche, F. Slaoui-Hasnaoui, T. A. Tameghe, & G. Ekemb (2014). Wind turbine condition monitoring: State-of-the-art review, new trends, and future challenges. *Energies 7*(4), 2595–2630.

Teimourzadeh Baboli, P., D. Babazadeh, A. Raeiszadeh, S. Horodyvskyy, & I. Koprek (2021). Optimal temperature-based condition monitoring system for wind turbines. *Infrastructures 6*(4).

Teng, W., X. Ding, H. Cheng, C. Han, Y. Liu, & H. Mu (2019). Compound faults diagnosis and analysis for a wind turbine gearbox via a novel vibration model and empirical wavelet transform. *Renewable energy 136*, 393–402.

Tian, Z., T. Jin, B. Wu, & F. Ding (2011). Condition based maintenance optimization for wind power generation systems under continuous monitoring. *Renewable Energy 36*(5), 1502–1509.

Turbines, W. (2005). Part 1: Design requirements, iec 61400-1. *International Electrotechnical Commission: Geneva, Switzerland.*

Turnbull, A. & J. Carroll (2021). Cost benefit of implementing advanced monitoring and predictive maintenance strategies for offshore wind farms. *Energies 14*(16).

Tusar, M. I. H. & B. R. Sarker (2022). Maintenance cost minimization models for offshore wind farms: A systematic and critical review. *International Journal of Energy Research 46*(4), 3739–3765.

Tziavos, N. I., H. Hemida, S. Dirar, M. Papaelias, N. Metje, & C. Baniotopoulos (2020). Structural health monitoring of grouted connections for offshore wind turbines by means of acoustic emission: An experimental study. *Renewable energy 147*, 130–140.

Vamsi, I., G. Sabareesh, & P. Penumakala (2019). Comparison of condition monitoring techniques in assessing fault severity for a wind turbine gearbox under non-stationary loading. *Mechanical Systems and Signal Processing 124*, 1–20.

Verbruggen, T. (2003). Wind turbine operation and maintenance based on condition monitoring. *Final Report, April.*

Walgern, J., L. Peters, & R. Madlener (2017). Economic evaluation of maintenance strategies for offshore wind turbines based on condition monitoring systems.

Wang, L. & Z. Zhang (2017). Automatic detection of wind turbine blade surface cracks based on uav-taken images. *IEEE Transactions on Industrial Electronics 64*(9), 7293–7303.

Wang, L., Z. Zhang, J. Xu, & R. Liu (2018). Wind turbine blade breakage monitoring with deep autoencoders. *IEEE Transactions on Smart Grid 9*(4), 2824–2833.

Wu, J., K. Hu, Y. Cheng, H. Zhu, X. Shao, & Y. Wang (2020). Data-driven remaining useful life prediction via multiple sensor signals and deep long short-term memory neural network. *ISA transactions 97*, 241–250.

Xiang, L., P. Wang, X. Yang, A. Hu, & H. Su (2021). Fault detection of wind turbine based on scada data analysis using cnn and lstm with attention mechanism. *Measurement 175*, 109094.

Xu, D., P. Liu, Z. Chen, J. Leng, & L. Jiao (2020). Achieving robust damage mode identification of adhesive composite joints for wind turbine blade using acoustic emission and machine learning. *Composite Structures 236*, 111840.

Yang, W., P. J. Tavner, C. J. Crabtree, Y. Feng, & Y. Qiu (2014). Wind turbine condition monitoring: technical and commercial challenges. *Wind Energy 17*(5), 673–693.

Zaher, A., S. McArthur, D. Infield, & Y. Patel (2009). Online wind turbine fault detection through automated scada data analysis. *Wind Energy 12*(6), 574–593.

Zhang, S. & Z.-Q. Lang (2020). Scada-data-based wind turbine fault detection: A dynamic model sensor method. *Control Engineering Practice 102*, 104546.

Zhong, S., A. A. Pantelous, M. Beer, & J. Zhou (2018). Constrained non-linear multi-objective optimisation of preventive maintenance scheduling for offshore wind farms. *Mechanical Systems and Signal Processing 104*, 347–369.

Zhong, S., A. A. Pantelous, M. Goh, & J. Zhou (2019). A reliability-and-cost-based fuzzy approach to optimize preventive maintenance scheduling for offshore wind farms. *Mechanical Systems and Signal Processing 124*, 643–663.

Zhou, P. & P. Yin (2019). An opportunistic condition-based maintenance strategy for offshore wind farm based on predictive analytics. *Renewable and Sustainable Energy Reviews 109*, 1–9.

Zhu, C. & Y. Li (2018, 10). *Reliability Analysis of Wind Turbines.*

Zhu, Q., H. Peng, B. Timmermans, & G.-J. van Houtum (2017). A condition-based maintenance model for a single component in a system with scheduled and unscheduled downs. *International Journal of Production Economics 193*, 365–380.

Zonta, T., C. A. da Costa, R. da Rosa Righi, M. J. de Lima, E. S. da Trindade, & G. P. Li (2020). Predictive maintenance in the industry 4.0: A systematic literature review. *Computers & Industrial Engineering 150*, 106889.

Advances in Maritime Technology and Engineering – Guedes Soares & Santos (Eds)
 ISBN 978-1-032-83099-5

Gear fault data acquisition and diagnosis for ship rotating machinery

S.H. Kim, D.M. Kim, J.W. Jung, K.P. Park & S.J. Kim
Department of Autonomous Vehicle System Engineering, Chungnam National University, Daejeon, Korea

ABSTRACT: The malfunction of the re-liquefaction system to recover the boil-off gas (BOG) generated from the LNG tank of the LNG carrier can lead to economic losses and fire risks. The data-based fault diagnosis study for the LNG re-liquefaction system and the rotating machine, mostly has focused on the failure of bearings and shafts. Moreover, data gathering by intentional failure of the system is not appropriate. This study implemented a test bed to simulate gear failure which is one of the main fault modes. An algorithm was developed to classify three gear faults by selecting the system characteristic values from the time series vibration data. The results of the useful features and machine learning algorithms in the gear failure diagnosis obtained through this study will help develop a fault diagnosis algorithm for ship rotating machines, including a gear mesh, and establish a data acquisition strategy.

1 INTRODUCTION

There are various rotating devices in a ship, such as motors, generators, fans, pumps and compressors. Rotating equipment plays an important role in the operation of a ship, such as fuel supply, power generation, and propulsion, and is exposed to a continuous operating environment (Madhukar et al. 2023, Moon et al. 2012). If a failure occurs in a rotating machine, it can lead to serious human casualties as well as financial damage caused by the suspension of operation. In order to diagnose defects and predict failures in advance to proceed with efficient maintenance, studies on defect diagnosis and health monitoring of rotating equipment are being actively conducted. As rotating equipment becomes more complex, interest in Machine Condition Monitoring (MCM) technology, which detects potential defects and function degradation in the early stages, increases, which is evaluated to be connected to PHM (Prognostic and Health Management), CBM (Condition Based Maintenance), etc. to provide and efficient device management solution (Chen et al. 2018).

Researches have been focused on PHM and CBM that can explore and separate the cause of the defect after determining whether it has occurred, and even predict the condition afterwards (Tian et al. 2017). In the PHM progress stage, after measuring physical phenomena as signals, it goes through a monitoring process that extracts meaningful information from measurement data. It consists of diagnosing the defect condition of the machine and prediction future soundness and lifespan using this.

Recently, machine learning has been applied to defect diagnosis research in order to increase the accuracy of technology for diagnosing and predicting machine defects in advance. Machine defect diagnosis using machine learning is carried out by extracting features from data and classifying the types of defects (Bhadra et al. 2018). A large amount of normal and defect data is required to utilize the machine learning method in PHM technology. However, there are cost and time limitations in securing the data by generating defect is actual machine devices. Therefore, the technology for acquiring defect data necessary for machine learning is considered an important research field. In general, defect data used to develop algorithms for diagnosing and predicting defect conditions are obtained from experiments using test beds that simulate actual mechanical devices. However, the test beds for defect simulation that have been developed so far often focus on bearings, which are major components of rotating machines, rather than simulating the entire system. For example, Case Western Reverse University (CWRU) constructs a test bed that simulates bearing defects, acquires defect data, and shares it with an open source. The data provides data measured by the rotational speed and load conditions of the motor given when a defect exists in each of the inner, outer, and motor bodies (Wang et al. 2020). The PRONOSTIA system used in the IEEE PHM 2012 Data Challenge consists of three main elements: a rotating unit, a deterioration generating unit, and a signal collecting unit, and data is sampled at a high frequency of 25,600 Hz (Chaudhari et al. 2023), NASA also provides data sets related to rotors (Hasan 2022), and CWRU also presents data for defect diagnosis (Rebiai et al. 2022, Kammam et al. 2022).

DOI: 10.1201/9781003508762-69

There are various types of major failure modes of rotating equipment, such as shaft mis-alignment, mass imbalance, and gear damage, as well as bearing defects, but previous test data tend to focus mainly on defect diagnosis studies related to bearings, which are support parts. However, in the case of a rotating device containing a gear mesh that continuously makes counterformal physical contact, such as the LNG carrier re-liquefaction system targeted in this study, gear defects can act as a root cause of power transmission problems, noise, vibration, and damage to the support part.

Referring to the case of gear defect studies so far, Han et al. (2019) extracted features using empirical mode decomposition (EMD) method and injected them into particle swarm optimization support vector machine (PSO-SVM). Liang et al. (2016) decomposed the vibration signal of the planetary gear into the teeth level, showing that the RMS (Root mean square) value of the defective gear was higher than the normal RMS value. Handikherkar et al. (2021) calculates time varying mesh stiffness (TVMS) for the normal gear and the defective gear. This is reduced when the defective gear engages, resulting in a change in the vibration response. Afterwards, the acquired vibration signal was processed and input into a machine learning algorithm, and training was conducted on three algorithms: support vector machine, (SVM) random forest, and multi-layer perceptron (MLP). This study reduced the features by using a principal component analysis (PCA).

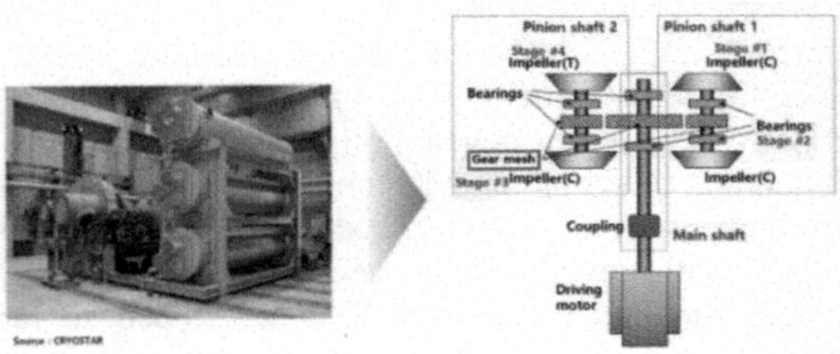

Figure 1. LNG re-liquefaction system and its rotor system (Left figure source: CRYOSTAR).

Previous studies learned a machine learning model by extracting features that were thought to have the greatest influence on the classification of defect states. This can reduce the efficiency of the model by including features with relatively small weights in learning. Therefore, in this study, in order to achieve the two goals of improving the performance of the machine learning classification model and reducing the calculation cost, the number of features learned through the process of selecting features through statistical t-test was reduced, and the performance improvement of the model according to the sampling window size was compared and analyzed.

In this study, the algorithm was studied using the data acquired by implementing a test bed that simulates the structure of the rotating body of the re-liquefaction system inside the LNG carrier.

It is a major rotating device of a ship that prevents the efficiency and economic loss of the ship by performing the function of recovering BOG generated from the LNG tank. Figure 1 shows an example of an LNG re-liquefaction system. Large-scale rotating devices such as LNG re-liquefaction systems have limitations in terms of time and cost to artificially acquire defect data for research, so the test bed was manufactured by selecting the device as a target of simulation. The purpose of this study was to focus on the study that can determine and classify the gear defects that can cause the most serious damage in the rotating structure of the LNG re-liquefaction system, and the optimal classification model and classification data generation method were analyzed through machine learning by acquiring acceleration vibration data from the accelerometer attached to the test bed.

2 FAULT SIMULATION TEST-BED CONFIGURATION

In order to secure rotational body defect data, an experimental device that simulates the rotational body structure of the LNG re-liquefaction system was developed. Vibration data was acquired by implementing three different gear defect situations among rotational body defects occurring in the actual rotational body.

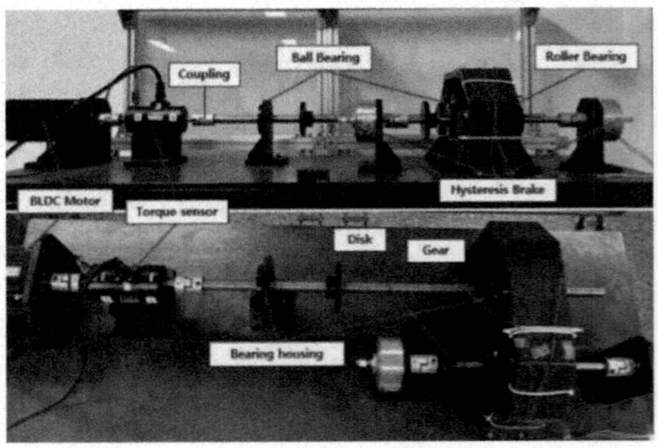

Figure 2. Layout of the designed test bed.

2.1 *Test bed design procedure*

Since the pinion shaft of the target LNG re-liquefaction system rotates at a maximum of 20,000 RPM, scaling was performed and some components were changed to operate at the laboratory level. In the process of scaling, the operating speed was lowered to maintain the dynamic characteristics of the system, and the dimensionless number related to the critical speed was matched. The following shows the non-dimensional number for scaling.

$$\pi_1 = a_i\left(\frac{d}{D}, \frac{l}{L}, etc\right) \tag{1}$$

$$\pi_2 = \frac{\omega}{\omega_{cr}} \approx \frac{\omega l^2}{Cd} \tag{2}$$

$$\pi_3 = \frac{\omega R^3}{Q} \approx \frac{\omega R}{V} \tag{3}$$

where π_1 is Geometry Scale Factor; π_2 is Critical Speed Margin; π_3 is Same Velocity Profiles.

The scale was equally lowered for the dimensionless number related to the critical speed, and as a result of selecting the gear module as 2.5, the number of teeth of the fire gear and the pinion gear was designed to be 50 and 34, respectively. In addition, the design was carried out by changing the helical gear and the fluid journal bearing to a spur gear and a rolling bearing, respectively. In addition, the design was carried out by changing the helical gear and the fluid journal bearing to a spur gear and a rolling bearing, respectively. The main specifications of the developed test bed are shown in the following Table 1.

The test bed consists of gear, bearing, electric motor, and coupling. The rotation structure is largely composed of a main shaft and an auxiliary shaft, which are connected through a gear mesh. At the end of the main shaft, the BLDC motor (3000 RPM, 2KW) is connected by coupling to rotate the main shaft. The situation in which an aerodynamic force is applied to the compressor and expansion turbine at both ends of the auxiliary shaft of the re-liquefaction system was simulated in the test bed by applying a load through the hysteresis brake. The appearance of the designed test bed is shown in Figure 2.

The driver for controlling the BLDC motor is MDROBOT's MDA2K model and communicates with the control PC through RS485 communication.

Through this, a motor control command is received from the PC and an operation notification and a status signal are transmitted to the PC.

The NI-USB 6001 board controls the hysteresis brake through the current supplier (ICS-500), and the control voltage signal are transmitted from the control PC.

Table 1. Specifications of the test-bed.

Part	Value
Main Shaft Length(m)	0.8
Pinion Shaft Length(m)	0.39
Gear Ratio	0.68
Bull Gear teeth number	50
Pinion Gear teeth number	34
Main Shaft max.speed(rpm)	3,000
Pinion Shaft max.speed(rpm)	4,400

The signal processing of the accelerometer was carried out by DT9837 (DATA Translation) which is a dynamic signal DAQ equipment. Acceleration signals measured by the accelerometer can be monitored in real time by the PC and FFT conversion can be performed.

2.2 *Implementing gear defects*

Research on the defects in various rotating bodies has been actively conducted, but the topics for gear defects have been relatively less studied. Three gear defect situations are considered: a case where half of the gear tooth is worn, a case where one of the gear teeth are completely removed, and a case where three gear teeth are removed sequentially, as shown in Figure 3.

2.3 *Vibration signal acquisition and control system*

A series of systems for acquiring data and processing signals were constructed as shown in Figure 4. Through a series of processes, data about the motor rotation speed and torque, accelerometer signal, and axial rotation speed was stored in the PC.

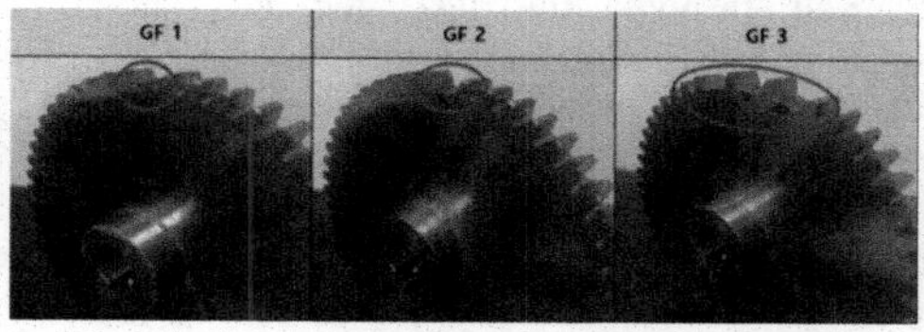

Figure 3. Three implemented gear defect cases.

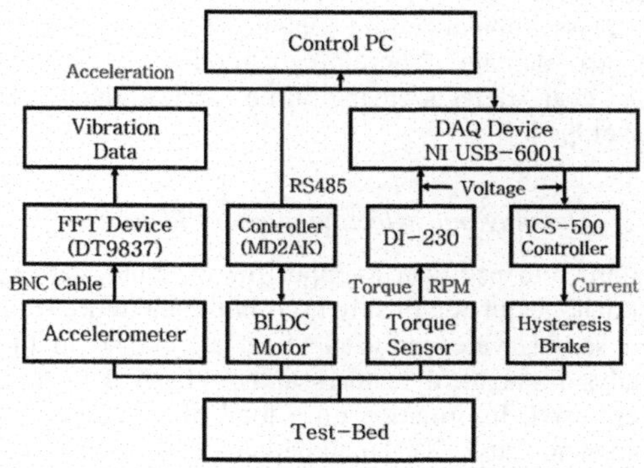

Figure 4. Vibration data acquisition flow chart.

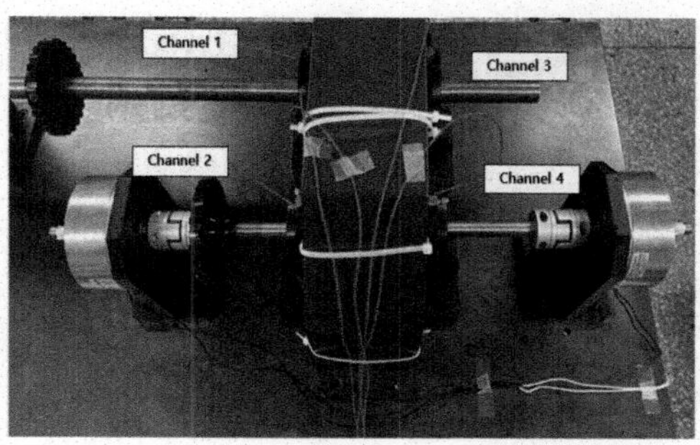

Figure 5. The location of accelerometer.

In order to acquire vibration data, signals were measured by mounting accelerometers on a total of 4 bearing housings, and its location is shown in Figure 5. The accelerometer used PCB Piezotronics' 333B30 model which

Data were acquired for 1 minute at a rotational speed of 2,300 RPM and a sampling rate of 6,000 Hz. In the normal state, 10 samples were acquired, and 6 data samples were acquired for each of the 3 gear defect states. Each sample was divided by 1.04 sec, which is the reciprocal of the rotational speed, to increase the data. In addition, in all cases, data were divided in a random section within one data sample to secure up to 1,000 divided data in each state as initial training data.

3 SELECTION OF FEATURES USING T-TEST

3.1 *Experimental results*

The time series and frequency domain signals acquired through the test bed are shown in Figure 6 and Figure 7. The vibration signal in frequency domain shows a peak in the 1X frequency component of the rotating shaft, shows no difference in normal and GF1 situations, and shows a difference in the absolute peak value in other defect conditions. However, in the case of a 1X frequency component, there is a limit to determining that it is a gear defect only with a frequency series vibration signal because it can be caused by mass imbalance and mis-alignment in addition to gear damage. Therefore, a model for classifying gear defects using time series data was developed.

3.2 *Defining and selecting effective features*

In order to use data as input for machine learning models, features typically used in the literature studied so far were analyzed (Ji et al. 2022). In this study, 4 statistical features and 8 physical features were used. Information on a total of 12 features is shown in Table 2 (Handikherkar et al. 2021, Altaf et al. 2022, Sanchez et al. 2018).

The t-test is a parametric statistical test method that compares the means between two groups, and can test the homogeneity, independence, and association between data (Rauber et al. 2021). The t-value is used as the test statistic and refers to the ratio of

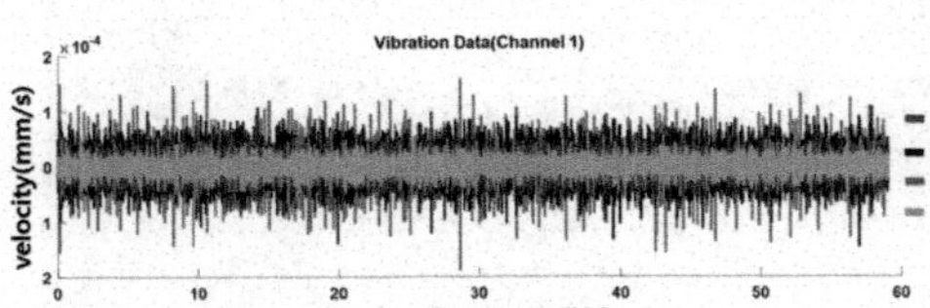

Figure 6. Time series vibration signal.

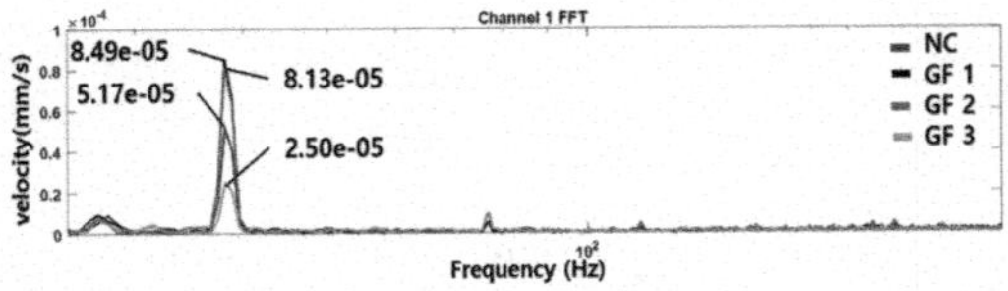

Figure 7. Frequency series vibration signal.

Table 2. Selected features.

Feature	Expression	Feature	Expression
Mean abs	$\frac{1}{N}\sum_{k=1}^{N}\lvert x_k\rvert$	Peak to Peak	$\max(x_k) - \min(x_k)$
Skewness	$\frac{N\sum_{k=1}^{N}(x_k-u)^4}{\sigma^3}$	Standard deviation	$\sqrt{\frac{1}{N}\sum_{k=1}^{N}(x_k-u)^2}$
Kurtosis	$\frac{N\sum_{k=1}^{N}(x_k-u)^4}{\left[\sum_{k=1}^{N}(x_k-u)^2\right]^2}$	Shape factor	$\frac{RMS}{\frac{1}{N}\sum_{k=1}^{N}\lvert x_k\rvert}$
Variance	$\frac{1}{N}\sum_{k=1}^{N}(x_k-u)^2$	Impulse factor	$\frac{\max(x_k)}{\frac{1}{N}\sum_{k=1}^{N}\lvert x_k\rvert}$
RMS	$\sqrt{\frac{1}{N}\sum_{k=1}^{N}x_k^2}$	Crest factor	$\left(\frac{\max(x_k)}{RMS}\right)$
Peak	$\max(x_k)$	Clearance factor	$\frac{\max(x_k)}{\frac{1}{N}\sum_{k=1}^{N}(x_k)^2}$

the difference between the standard error and the sample mean.

$$t = \frac{\bar{X} - \mu}{\frac{S}{\sqrt{n}}} \tag{4}$$

where $\bar{X}$ = mean of the difference between the two groups; μ = mean of the population; and S = standard deviation between the two groups, so the molecular term becomes the standard error. Usually, the significance level (area of the rejection area under the curve, Figure 8) is set to 0.05, and the fact that the lower area (p-value) of the probability density function is less than the significance level is a proof of the hypothesis (null hypothesis) that the average values of the two groups are the same, so the null hypothesis is rejected. In other words, it can be judged that there is a difference between the two groups. Based on this, the p-value for a combination of all two possible groups in a total of 4 classes (normal and 3 gear defects) was derived. When the null hypothesis was rejected based on the p-value, the significance (h) was set to 1, and if at least one of all possible combinations of two groups for 4 classes was not rejected, h was set to 0, and the features was excluded from the data to be used in the machine learning model. The results for this are shown in the Figure 9. As a result, the remaining 9 of the 12 features, excluding three features (Skewness, Curtis, and Clearance factor), were determined as the input of the classifier.

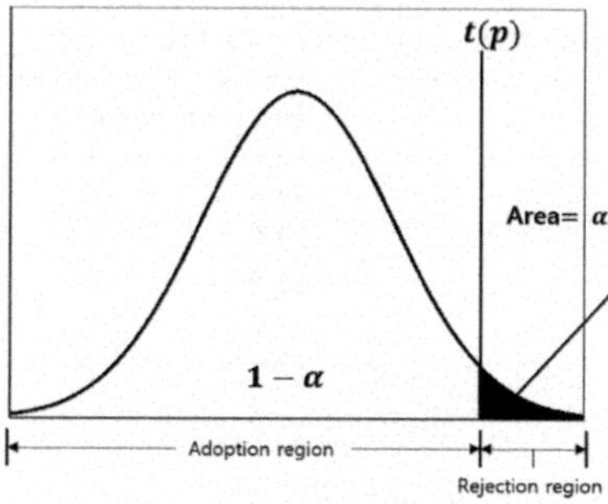

Figure 8. T-test method.

time	N-GF1	N-GF2	N-GF3	GF1-GF2	GF1-GF3	GF2-GF3
Mean abs	1	1	1	1	1	1
Variation	1	1	1	1	1	1
Standard Deviation	1	1	1	1	1	1
Skewness	0	1	1	1	1	1
Kurtosis	1	1	0	1	1	1
Peak	1	1	1	1	1	1
Peak to Peak	1	1	1	1	1	1
RMS	1	1	1	1	1	1
Shape factor	1	1	1	1	1	1
Clearance factor	1	1	1	0	1	1
Crest factor	1	1	1	1	1	1
Impulse factor	1	1	1	1	1	1

Figure 9. H results at window size 1.04sec.

4 CLASSIFICATION OF GEAR DEFECTS USING MACHINE LEARNING

The classification performance of each algorithm was compared and analyzed by applying the previously selected features to the kNN, SVM, Naive Bayes, and ANN classification models.

		Prediction	
		Positive	Negative
Actual	Positive	TP (True Positive)	FN (False Negative)
	Negative	FP (False Positive)	TN (True Negative)

Figure 10. Confusion matrix & metrics.

4.1 *Performance evaluation metrics for machine learning models*

The performance of the gear defect determination model was evaluated based on the Figure 10. The performance of the basic classifier model in the machine learning model can be quantitatively confirmed through a table called the confusion matrix. This is a matrix that organizes the prediction categories of classification and the classification categories of actual data in an intersecting form (Marom et al. 2010). Each element of the matrix can be understood as one of TP (True Positive), TN (True Negative), FP (False Positive), and FN (False Negative), and the performance of the classifier is confirmed through the indicators expressed as the relationship between them. Representative evaluation indicators include accuracy. Precision, Recall (Sensitivity), and F1- score (Hossin et al. 2015). Accuracy represents the percentage of all judgments that the model determines accurately. However, accuracy cannot be a correct indicator if the positive and negative data are imbalanced, so Precision and Recall (or Sensitivity) are used. Precision represents the percentage of actually positive among the data that the model judged to be positive. Recall represents the percentage of actual positive data that the model correctly judged as positive. Both Precision and Recall focus on positive data that are correctly judged, but the former focuses on lowering FP and the latter focuses on lowering FN, so a trade-off relationship is established between the two indicators. All indicators have a size between 0 and 1, and the closer to 1, the better the classification performance.

4.2 *Basic classification algorithms*

There are four types of classification algorithms that can be used in this study. First, the k-Nearest Neighbor (kNN) algorithm calculates the physical distance between the input data and the existing data, determines the closest data as many as k specified, and then determines where the input data belongs

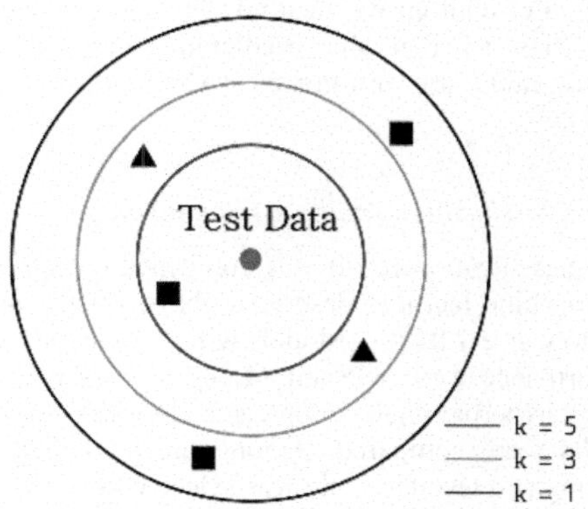

Figure 11. K- Nearest Neighbor algorithm.

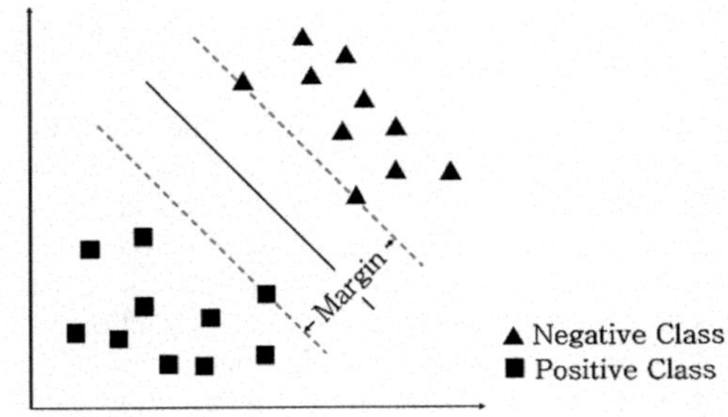

Figure 12. Support vector machine algorithm.

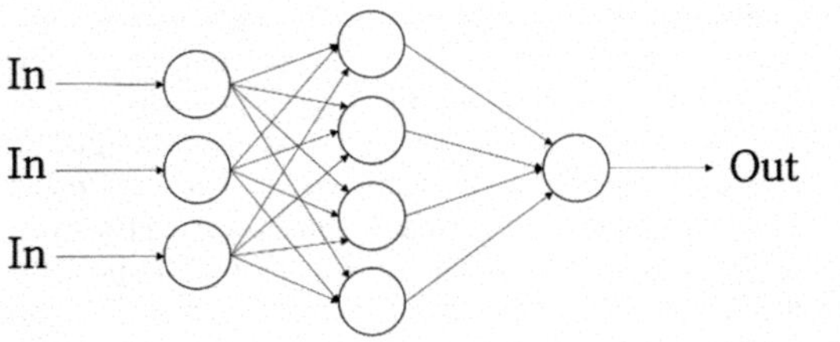

Figure 13. Artificial Neural Network algorithm.

(Tayyab et al. 2020). In addition to the commonly used Euclidean distance for calculating the distance, several other methods of distance calculation formula can be used.

SVM is an algorithm that divides a data group into a hyperplane and classifies it to maximize margin, the distance between the data closest to it (Liu et al. 2018). If the dataset consists of two linearly distinguishable groups, as shown in the Figure 12, the hyperplane can be classified as a straight line.

The Naive Bayes algorithm is based on the Bayesian theory, and 'Naive' means that all classes are mutually independent (Liu et al. 2018).

An artificial neural network (ANN) is a computational model that forms a pattern modeled after human neuron action. Hidden layers existing between the input layer and the output layer are configured to find an optimal weight and Bias so that the predicted value and the actual value are minimized (Almounajjed et al. 2021).

4.3 *Classification results using fault data*

The classification result was confirmed by applying the machine learning described above for the initial window size 1.04 sec. Models with consistently high performance were selected among models with different hyperparameters for each algorithm and the results were compared. Among the total data, the training and test data ratio was selected as 9:1.

The selected variations are weighted kNN, dense Gaussian SVM, kernel Naive bayes, and wide ANN.

Table 3. Test performance of model for 1.04 sec.

Algorithm	Test performance (%)
kNN	84.5
SVM	83.0
Naive Bayes	81.8
ANN	81.2

Table 4. Features excluded for window size.

Window Size	Excluded Feature
2.08 sec	Kurtosis Clearance Factor
3.12 sec	Skewness Kurtosis Crest Factor
4.16 sec	Clearance Factor Crest Factor

The information on the hyperparameters of the mentioned algorithms is as follows. In a weighted kNN, k is 10 and the distance is measured by the Euclidean method, and the weight of the distance is the reciprocal of the square. In a dense Gaussian SVM, the kernel function was set as a Gaussian function. In the kernel Naive Bayes algorithm, the type of kernel was set to Gauss. Finally, in the wide ANN algorithm, one complete connection layer was set and its size was set to 100, and a rectified linear unit (ReLU) was used to overcome the gradient loss problem as the activation function to convert the input. All algorithms performed cross-validation by dividing the remaining data into five equal parts, excluding the data left for testing, to prevent overfitting. Cross-validation is a method of evaluating a model by dividing data by a specified number and using each part alternately for training and testing purposes. Therefore, the generalization performance of the model can be measured more accurately.

4.4 *Performance according to the change in window size*

In the previous classification results, it shows less than 85% as shown in Table 3, and sufficient performance was not guaranteed in determining gear

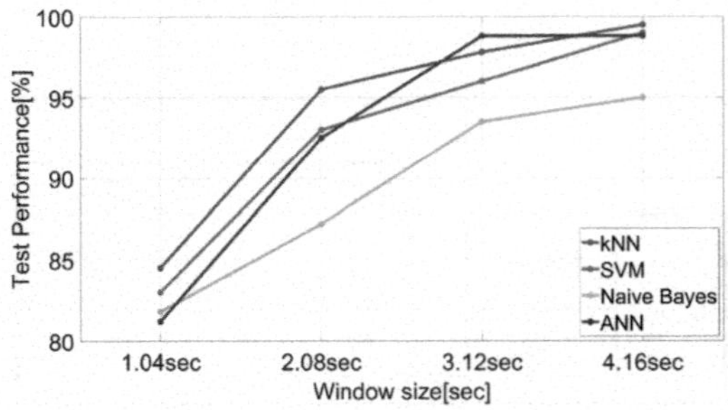

Figure 14. Model performance depending on window size.

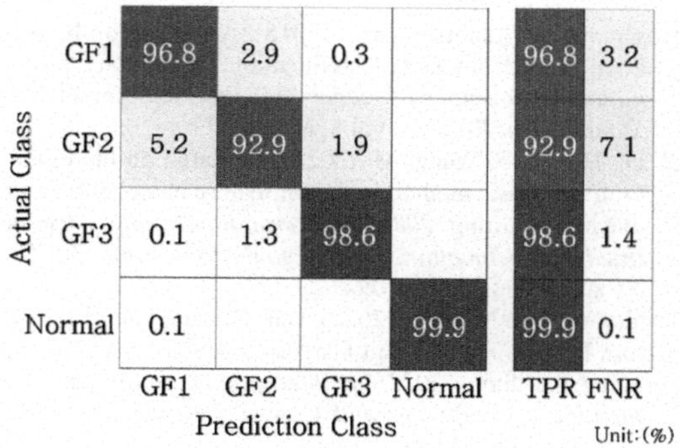

Figure 15. Confusion matrix of SVM model: window size 4.16sec, selected feature using t-test.

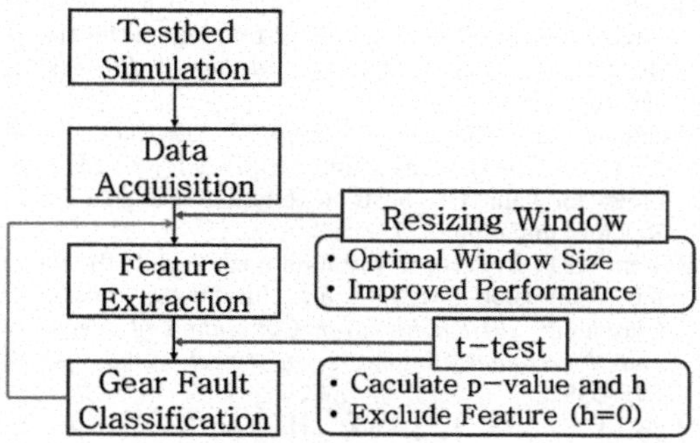

Figure 16. Flow chart of the gear fault diagnosis process.

defects. Since feature contains many statistical values, the larger the window size, the more statistical representativeness can be improved, so we increased the window size and searched for a window size to secure appropriate performance.

With 1.04 sec, the reciprocal of the rotation speed as the initial value, the window size was adjusted to four times by integer multiples of the corresponding value. Table 4 shows the features excluding each case by deriving significance in the same way as when the data was divided at 1.04 sec.

As shown in Figure 14, the performance of the classification model remarkably has been improved as the size of dividing the data increased. It can be analyzed that the wider sample better represents the characteristics of the case. Figure 15 shows the confusion matrix of the dense Gaussian SVM model (test performance 99.0%) learned by selecting the features from the segmentation size 4.16 s data that showed the highest performance in the overall model. It showed very high performance in distinguishing between the normal state and the gear defect state, and it also showed high performance in distinguishing different degrees of gear defects. Detailed consideration will be dealt with in the conclusion section.

To summarize the procedure of the gear defect diagnosis process performed in this study, features are extracted by sufficiently increasing the vibration data obtained from the test bed simulating the gear defect for use as an input for machine learning. Afterwards, the classification model is trained and tested by excluding features that do not affect the classification using the t-test technique to improve the performance of the model. This reduces the calculation cost of the classification model and improves the classification performance of the model by correlating the characteristics with higher relevance. Returning to the process of increasing the data again and adjusting the window size to divide the data, the process is repeated again. In this paper, the reciprocal of the rotational speed was taken as the initial value, and the size of the window was set by performing an integer multiple of the corresponding value. Gear defects can be classified with high accuracy through the model with the highest performance among several models. This procedure is shown in Figure 16.

5 CONCLUSION

In this study, a test bed was developed that can simulate the rotational structure of the LNG reliquefaction system and reproduce gear defects. As a result of monitoring the data acquired from the test bed, it was determined that the analysis of the time series signal would be more significant than the frequency series signal where no distinct characteristics could be found. After acquiring the data for the normal state and three gear defect cases, the work of increasing the number of data was performed in order to learn sufficient data for machine learning. After that, a total of 12 features (4 statistical features and 8 physical features) were extracted from the time series graph. After excluding features whose correlation is not clear through t-test, defect classification through machine learning was conducted. The contents of analyzing the results are as follows.

- In terms of gear defects, features that did not show a clear difference between the normal state and each defect state could be selected according to the window size.
- As the size of dividing one sample data increases, the accuracy of the classification model remarkably improves. This is judged to be due to the improved characteristics of statistically calculated features representing the system because more data is included within one sampling window.
- Classification accuracy does not increase significantly above a certain window size. Therefore, it is necessary to select an optimal window size in consideration of the classification period.
- In terms of gear defect analysis, the SVM model shows the best performance. Because the simulated gear defect situations show a high degree of clustering characteristics based on the features SVM can classify data as a boundary called Hyperplane.

The classification algorithm applied in this paper is a basic algorithm that is the basis for various defect diagnosis algorithms that have been developed a lot recently, and it is believed that a more accurate classification model can be developed by incorporating the latest deep learning technology in the future.

The test bed developed in this paper can be a useful tool to cope with the difficulty of securing defect data in the rotating machine, and if the reliability is verified by identifying the correlation between the additional defect data to be secured and the measurement data in the actual system, the proposed defect diagnosis model can be applied to the actual system.

ACKNOWLEDGEMENTS

This work was partly supported by the National Research Foundation of Korea (NRF) grant funded by the Korea government (MSIT) (No. NRF-2022R1F1A 1070449) and Korea Institute for Advancement of Technology (KIAT) grant funded by the Korea Government (MOTIE) (P0017006, HRD Program for Industrial Innovation)."

REFERENCES

Almounajjed, A., Sahoo, A.K., Kumar, M.K. & Alsebai, M.D. (2021). Investigation Techniques for Rolling Bearing Fault Diagnosis Using Machine Learning Algorithms. *2021 5th International Conference on Intelligent Computing and Control Systems (ICICCS)*, Madurai, India: 1290–1294.

Altaf, M., Akram, T., Khan, M., Iqbal, M. & Hsu, C. (2022). A New Statistical Features Based Approach for Mearing Fault Daignosis Using Vibration Signals. *Sensor*, Vol.22: 1–15.

Bhadra, R., Dutta, S., Kedia, A., Gupta, S., Panigrahy, P.S. & Chattopadhyay, P. (2018). Applied Machine Learning for Bearing Fault Prognostics. *IEEE ASPCON*, Kolkata, India: 158–162.

Chaudhari, P.A., Wadhvani, R., Rasool, A. & Gupta, M. (2023). A Review on Condition Assessment Methods of Ball Bearing. *Third International Conference on Secure Cyber Computing and Communication (ICSCCC)*, Jalandhar, India: 554–559.

Chen, G., Lu, G. & Yan, P. (2018). An Integrated Method for Condition Monitoring of Rotating Machines. *2018 Prognostics and System Health Management Conference (PHM-Chongqing)*, Chongqing, China: 144–149.

Han, D., Zhao, N. & Shi, P. (2019). Gear fault feature extraction and diagnosis method under different load excitation based on EMD, PSO-SVM and fractal box dimension. *J Mech Sci Technol 33*: 487–494.

Handikherkar, V.C. & Phalle, V.M. (2021), Gear Fault Detection using Machine Learning Techniques-A Simulation-driven Approach. *IJE TRANSACTION A: Basic* Vol.34, No.01: 212–223.

Hasan, A. & Singh, J. (2022). Fault Detection in Ball Bearing through Machine Learning Models. *2022 6th International Conference on Electronics, Communication and Aerospace Technology*, Coimbatore, India: 1–6.

Hossin, M. & Sulaiman, M.N. (2015). A Review on Evaluation Metrics for Data Classification Evaluations. *International Journal of Data Mining & Knowledge Management Process*, Vol.5, No.2: 1–11.

Ji, D., Hu, M. & Wang, M. (2022). Research on the motor fault diagnosis method based on feature engineering and machine learning. *2022 7th International Conference on Intelligent Computing and Signal Processing (ICSP)*, Xi'an, China: 1704–1709.

Kammam, P. & Neha, S. (2022). Ball Bearing Fault by Feature Extraction and Fault Diagnosis method based on AI ML Algorithms. *2022 6th International Conference on Intelligent Computing and Control Systems (ICICCS)*, Madurai, India: 1310–1314.

Liang, X., Zuo, M.J. & Liu, L. (2016). A windowing and mapping strategy for gear tooth fault detection of a planetary gearbox. *Mech. Syst. Signal Process.* vol80: 445–459.

Liu, R., Yang, B., Zio, E. & Chen, X. (2018). Artificial intelligence for fault diagnosis of rotating machinery: A review. *Mechanical Systems and Signal Processing*, Vol.108: 33–47.

Madhukar, G., Boosi, S., Adhoni, Z., Bhaskar, A. & Naik, A. (2023). A Machine Learning Based Methodology for Fault Diagnosis in Rotating Machine. *IEEE*, Raichur, India: pp. 1–5.

Marom, N.D., Rokach, L. & Shmilovici, A. (2010). Using the Confusion Matrix for Improving Ensemble Classifiers. *2010 IEEE 26-th Convention of Electrical and Electronics Engineers in Israel*, Eilat, Israel: 555–559.

Moon, D.K., Bae J.Y., Park J.H., Lee K., & Kim, H. (2012). A Development of Remote Ship Maintenance System Based on Ship Area Network. *Journal of Society of Naval Architects of Korea*, Vol 47: 751–756.

Rauber, T.W., Loca, A.L., Boldt, F.A., Rodrigues, A.L. & Varejao, F.M. (2021). An experimental methodology to evaluate machine learning methods for fault diagnosis based on vibration signals. *Expert Systems with Applications*, Vol.167 114022: 1–18.

Rebiai, M., Bengherbia, B., Zmirli, M.O. & Lachenani, S. (2022). An Interactive Educational Tool for Studying the Diagnosis of Bearing Faults Using Vibration Analysis. *2022 2nd International Conference on Advanced Electrical Engineering (ICAEE)*, Constantine, Algeria: 1–6.

Sanchez, R.V., Lucero, P., Vasques, R.E. & Cerrada, M. (2018). Feature ranking for multi-fault diagnosis of rotating machinery by using random forest and KNN. *Journal of intelligent & Fuzzy Systems*, Vol 34.: 3463–3473.

Tayyab, S., Asghar, E., Pennacchi, P. & Chatterton, S. (2020). Intelligent fault diagnosis of rotating machine elements using machine learning through optimal features extraction and selection. *Procedia Manufacturing*, Vol.51: 266–273.

Tian, J., Azarian, M.H., Pecht, M., Niu, G. & Li, C. (2017). An ensemble learning-based fault diagnosis method for rotating machinery. *2017 Prognostics and System Health Management Conference (PHM-Harbin)*, Harbin, China: 1–6.

Wang, H., Yu, Z. & Guo, L. (2020). Real-Time Online Fault Diagnosis of Rolling Bearings Based on KNN Algorithm. *2019 4th International Seminar on Computer Technology, Mechanical and Electrical Engineering (ISCME 2019)*, Chengdu, China: 1–7.

Advances in Maritime Technology and Engineering – Guedes Soares & Santos (Eds)

Determining the frequency of failure-finding tasks for ship critical systems based on the required availability

J. Sobral
Centre for Marine Technology and Ocean Engineering (CENTEC), Instituto Superior Técnico, Universidade de Lisboa, Lisbon, Portugal
Mechanical Engineering Department, ISEL – Instituto Superior de Engenharia de Lisboa, Lisbon, Portugal
CIMOSM, ISEL – Centro de Investigação em Modelação e Optimização de Sistemas Multifuncionais, Instituto Politécnico de Lisboa, Lisbon, Portugal

C. Guedes Soares
Centre for Marine Technology and Ocean Engineering (CENTEC), Instituto Superior Técnico, Universidade de Lisboa, Lisbon, Portugal

ABSTRACT: Several systems that are installed in ships can be considered critical, once a failure on a critical system may correspond sometimes to huge costs, a lack of safety for the crew, a severe environmental impact or even consequences on ship integrity. Thus, it is important to avoid such failures based on preventive or predictive maintenance, assuring a high reliability and availability of these critical systems. For failures that occur upon a degradation process and where it is possible to monitor some parameter related to that degradation mechanism, it is possible to predict the occurrence and act proactively. However, for hidden failures, mainly related to protective systems, it is difficult to predict it. For these situations, most of time the best way to know if a protective system in a dormant state is in an operational state is to perform failure-finding tasks upon a scheduled maintenance plan. A low frequency plan for these tasks will increase risk and availability, while a high frequency will increase maintenance costs. This paper refers to how to determine the frequency of such failure-finding tasks based on a desired availability, giving an example of a protective system, demonstrating the applicability and impact of the methodology proposed.

1 INTRODUCTION

Protective devices may be installed to reduce the consequences and risks associated with failures of critical systems. To be successful in this topic, the risk associated with these devices must be addressed proactively as part of a primary risk management strategy.

If a common routine maintenance task cannot be established to reduce the risk of a failure of a critical system to a tolerable level, and the referred system is protected by one or more protective devices, the question is what the protection function will do when it happens. It is expected that the protection function will take over from a function that has failed.

Based on this, when implementing a Reliability-Centered Maintenance (RCM) process to optimize the maintenance plan of an asset regarding the assessment of the consequence of the failure, one option is to identify failure-finding tasks that are suitable to be performed (Moubray, 1997).

If the optimization of the protective device will reduce the risk to a tolerable level by reducing the consequence severity, a periodic failure-finding task for the protective device will be recommended.

If a suitable failure-finding task cannot be found, then a one-time change will have to be considered for failures posing an intolerable risk to safety and the environment (the item may have to be redesigned).

The objective of this paper is to show how to define the failure-finding interval or frequency for maintenance activities when dealing with critical systems.

This paper is structured into five sections. Section 2 gives short information about critical systems and section 3 refers to failure-finding tasks. Section 4 describes the methodology proposed to identify the frequency of failure-finding tasks and section 5 gives a demonstrative example of the application of the methodology proposed. Finally, section 6 states some conclusions and possible future works on this issue.

DOI: 10.1201/9781003508762-70

2 CRITICAL SYSTEMS

A critical system is a term used to describe a system or component within a larger installation that is deemed vital to the overall functionality, performance, and safety of that installation. The criticality of a system is often determined by its role in maintaining the integrity of operations, ensuring safety, or preventing catastrophic failures. In various domains, such as aviation, healthcare, finance, and manufacturing, critical systems play a central role in the success and safety of operations.

To be considered a critical system it must fulfil some key characteristics as:

- Vital functionality: Critical systems often perform functions that are essential to the core purpose of the overall system or installation. Failure of a critical system may lead to significant consequences, including installation shutdown, safety hazards, or loss of essential services.
- Safety Implications: Critical systems are frequently associated with safety-critical functions. This means that their failure or malfunction could pose a threat to human life, the environment, or property.
- High reliability required: Critical systems typically have stringent reliability requirements to ensure continuous and dependable operation. This may involve the use of redundancy, fault tolerance, and rigorous testing to minimize the risk of failures.
- Regulatory compliance: Many critical systems operate in areas with strict regulations and standards. Compliance with these regulations is crucial to ensure the system's safety and reliability.

The identification, maintenance, and proper functioning of critical systems are of paramount importance in ensuring the overall success and safety of operations in several areas. Organizations typically implement specialized design, testing, and maintenance practices for critical systems to meet the high standards of reliability and safety demanded by their critical roles.

Related to energy critical systems, Bae & Thorp (1999) refer to a study on power systems protection mechanisms regarding false trips of line protection relays or generator protection equipment. In this study, the resulting sequence of rare events and their corresponding probability are used to detect weak links in the power system. The hidden failures of protective relays were also studied by Zeynivand and Shahrtash (2023), as one of the major causes of outages and eventually cascading failures in power systems. The analytical model developed gets the information of the recorded outputs of starters, trip signals, and relay resets as inputs, and by comparing these inputs with the reference data for the outputs of the main distance relays, the presence of hidden failures in distance relays is directly revealed.

Chen et al (2005) also refer to hidden failures in power transmission systems presenting a study about the power law distributions observed in North American blackout data, including as main parameters the system loading level, hidden failure probability, spinning reserve capacity and control strategy. The goal is to guide when and how mitigation methods might be effective.

In ships, critical systems are those that are essential for the vessel's safe navigation, operation, and the well-being of its crew. These systems play a pivotal role in ensuring the ship's functionality, safety, and compliance with maritime regulations. Follows some examples of critical systems usually present in ships:

- Navigation Systems – For example, the GPS (Global Positioning System), which provides accurate position information for navigation, the ECDIS (Electronic Chart Display and Information System), which displays electronic navigational charts and integrates with other navigation sensors, or radar systems, used for detecting other vessels, navigation hazards, and land.
- Propulsion Systems – For example, the main engine, where the primary engine is responsible for propelling the ship, and the propulsion control systems, that manage and control the ship's propulsion, including throttle control, speed control, and reversing.
- Propeller and Shafting Systems – This includes the propeller shafting, corresponding to the mechanical system connecting the main engine to the propeller, and the thruster system, which provides additional manoeuvrability, especially in confined spaces.
- Steering Systems – For example, the steering gear, which is responsible for controlling the ship's direction, and the autopilot systems, which assist in maintaining a steady course automatically.
- Communication Systems – For example, the GMDSS (Global Maritime Distress and Safety System), which ensures reliable communication during emergencies, and the VHF and HF radios, used for ship-to-ship and ship-to-shore communication.
- Safety Systems – For example, the fire detection and suppression systems, to detect and control fires on board, the lifeboat release systems, that ensure proper functioning during emergency evacuations, and the emergency alarm systems, that alert the crew in case of emergencies.
- Power Generation and Distribution Systems – Including the generators and electrical systems, that provide electrical power for various ship systems, and the power distribution systems, which ensure proper distribution of electrical power throughout the vessel.
- Ballast and Stability Systems – For example, the ballast systems, that manage the distribution of

ballast water for stability, or the stability control systems, that monitor and control the ship's stability.
- Cargo Handling Systems (for cargo vessels) – For example, the cargo control room systems, which monitor and control loading and unloading operations, and the tank gauging systems, which monitor the levels of liquids in cargo tanks.
- Weather and Environmental Monitoring Systems – For example, the anemometers and weather stations, that measure wind speed and other weather parameters, or the bow thrusters, used to aid in maneuvering during adverse weather conditions.

These systems collectively contribute to the ship's operational capabilities, safety, and compliance with maritime regulations. Regular maintenance, testing, and adherence to international standards are essential for ensuring the reliability and effectiveness of these critical systems on ships. Additionally, the International Maritime Organization (IMO) sets guidelines and regulations to ensure the safety and security of ships and their crews, namely the SOLAS Convention (Safety of Life at Sea) specifying minimum standards for the construction, equipment and operation of ships, compatible with their safety. The current SOLAS Convention includes articles setting out general obligations, amendment procedures and so on, followed by an Annex divided into 14 Chapters (IMO, n.d.).

Performing failure-finding tasks for a ship's critical system involves a systematic and thorough approach to identifying potential issues or failures that may impact the ship's safety, reliability, and performance. For a better establishment of the failure-finding tasks and definition of the failure-finding intervals, a deep comprehension of the critical systems must be developed. This comprehension involves the following activities that must be accomplished:

- Define objectives and the scope of the study - Clearly define the objectives of the failure-finding tasks. Identify specific aspects of the ship to be assessed. Determine the scope of the assessment, including the subsystems or components of the critical system.
- Review design and specifications for the critical system - Examine the ship's design documentation and specifications to understand the intended functionality and performance criteria of the critical system. Verify that the ship design complies with relevant regulations, industry standards, and best practices.
- Develop functional tests - Develop functional tests to ensure that each component of the ship system performs its intended function. Simulate various operating conditions to assess the system's response under different scenarios.
- Develop performance tests - Evaluate the performance of critical components under normal and extreme conditions. Assess the ship's response to challenging environmental factors, including rough seas, adverse weather, and emergency situations.
- Develop inspections and audits - Perform regular inspections and audits of the ship's critical system. Check for compliance with international maritime regulations, classification society standards, and the ship's own maintenance and operational procedures.
- Monitoring and surveillance - Implement continuous monitoring and surveillance systems to track the real-time performance of the critical system under study. It should be used sensors, alarms, and monitoring tools to detect anomalies or deviations from normal operating parameters.
- Root Cause Analysis - Investigate any incidents or failures that occurred to determine their root causes. Use techniques like root cause analysis to identify underlying issues. Implement corrective actions to address root causes and prevent similar failures in the future.
- Establish feedback mechanisms - Establish mechanisms for obtaining feedback from ship crew members regarding their experiences with the ship's critical system. Encourage reporting of any issues or concerns and use this feedback to continuously improve the ship's reliability and safety.
- Training and awareness - Ensure that the ship's crew is adequately trained on the operation and maintenance of the ship's critical system. Promote a culture of safety and awareness to empower crew members to identify and report potential issues.
- Documentation and reporting - Maintain comprehensive documentation of failure-finding tasks, testing results, inspections, and corrective actions taken. Generate regular reports summarizing the findings and actions taken and use this information to inform future maintenance and improvement plans.

By following these steps, ship operators can systematically identify and address potential failures, contributing to the overall safety, reliability, and efficiency of the ship system. Regular assessments and a proactive approach to maintenance and testing are key to preventing issues and ensuring the ship's continued safe operation.

All the referred systems usually have protections that are implemented to avoid the occurrence of failures, give alerts, or minimize the consequences of such failures. The way to ensure the operationality of these protective devices is to perform regular tests, called failure-finding tasks.

3 FAILURE-FINDING TASKS

Failure-finding tasks typically refer to activities or processes designed to identify and uncover potential failures or issues within a system, product, or process. The goal is to proactively discover problems before they can lead to significant consequences, such as system failures, product defects, or process breakdowns. Failure-finding tasks are an essential part of risk management.

Organizations can actively seek out potential failures through these tasks to enhance their products and process's reliability, safety, and quality. This proactive approach helps prevent problems before they reach the end-users or customers, ultimately improving overall performance and customer satisfaction, but also preventing the failure of systems, sometimes catastrophic.

With failure-finding tasks, one is checking whether the required protection function or protective device is still working as it should. It needs to ensure that the function is physically possible to be checked which sometimes is impossible to carry out because the protective device is not accessible or when it cannot be checked without its destruction. It is also needed to ensure that the failure-finding task will not increase the risk.

While predictive maintenance is usually based on the estimated P-F interval (time interval between the detection of a Potential failure and Failure occurrence), and preventive maintenance (time-based maintenance) is performed upon the knowledge of the item's useful life, the frequency of the failure-finding tasks relies on the assumption of a pre-determined or desired availability.

Usually, the main question refers to the establishment of the frequency of those tasks because a low-frequency plan for them will increase risk and potential unavailability, while a high-frequency plan will increase maintenance costs.

Several works can be referred to concerning this theme, having different approaches and objectives.

Jiang et al (2014) state the importance of predicting hidden failures in complex engineering systems and propose a model for hidden failure prognosis based on a belief rule base (BRB), where an algorithm operates together in an integrated manner to improve the precision of failure prognosis by using the qualitative knowledge and quantitative information effectively.

On the other side, Zhang et al (2021) present a study of a two-stage system with hidden failures, introducing a delay-time model to describe the evolution of the system behaviour, and presenting a preventive maintenance policy in response to the observed system state, considering the minimal expected cost rate by determining the optimal inspection interval.

Tang et al (2013) also refer to the existence of hidden or unrevealed failures and present a study about the availability of a critical system subject to hidden failures considering the downtime due to repair/replacement and the downtime due to inspection. The study considers periodic inspections performed at a constant interval, and the age of the system.

Recently, Zhang et al (2024) reported that most unexpected failures may cause huge economic losses and bring about uncontrolled risks and damage to humans and the environment. It is stated that inspection quality is an important while factor affecting maintenance decisions. Based on this, the authors studied the factor of imperfect inspection regarding multi-state systems subject to competing and hidden failures and developed a model based on the probabilities of different renewal scenarios and achieving the optimal inspection interval, which is greatly influenced by the factor of imperfect inspection.

Levitin et al (2019) also studied the importance of imperfect inspections modeling non-repairable systems subject to a delay-time failure process involving hidden and fatal failures in two stages during their missions. The authors state that the system undergoes scheduled inspections to detect hidden failures if imperfect inspections may produce false positive and negative failures. Based on this, the authors proposed probabilistic models for evaluating the performance metrics of the system considered, including mission success probability, system survival probability, expected number of inspections during the mission, and total expected losses.

Brissaud et al (2022) proposed reliability and availability models for ageing systems subject to failures detected online and hidden failures, such as safety-related systems. For the hidden failures, preventive maintenance consists of periodic proof tests that reveal the failures and, if applicable, followed by corrective maintenance. The authors state that for safety-related systems the results show that the proposed models may prevent underestimating the unavailability and inherent risks of ageing systems and enable the definition of more appropriate maintenance policies.

All these references and other studies reveal the impact of the occurrence of hidden failures when dealing with critical systems, and the importance of performing failure-finding tasks in a way that will increase the availability of such systems, reducing costs and lowering the risk for people, business, and environment.

4 METHODOLOGY PROPOSED

As stated before, critical systems are usually protected in a way to avoid undesirable consequences in case of failure.

If a protected function fails while a protective device is in a failed state it is called a multiple failure (Basson, 2019).

The probability of a multiple failure (PMF) can be determined by the expression:

$$P_{MF} = PF_{CS} \times U_{PD} \quad (1)$$

Where:

PF_{CS} = Probability of failure of the critical system (protected function)

U_{PD} = Unavailability of the protective device

Based on this, it is concluded that the probability of a multiple failure can be reduced by increasing the availability of the protective device and by establishing a good maintenance plan for the critical system, promoting a low probability of failure of the protected function.

Once protective devices are usually in a dormant state with a hidden function where preventive maintenance is not applicable or it will not increase the availability of the protective device (on the contrary, sometimes it has an adverse impact on it), the solution is to check the hidden function periodically using failure-finding tasks that are performed at regular intervals to ensure that at that moment the protective device is not in a failed state.

The frequency of a failure-finding task must be practical and based on the desired availability of the protective device given by its MTBF (Mean Time Between Failures).

The MTBF of the protected device ($MTBF_{PD}$) can be obtained by observing the total operational experience time (t_{OE}) and the number of times that a hidden failure was found on a failure-finding task (NF_{OE}) for that protective device during the operational time.

$$MTBF_{PD}=t_{OE}/NF_{OE} \quad (2)$$

Analyzing the failures, it is assumed that on average each failure occurs at half of the period between two consecutive failure-finding tasks, corresponding to an average unavailability on each period between failure-finding tasks (U_{AV}).

Based on this, it is possible to determine the time in which the protective device was in a failed state during the operational time (FS_{OE}).

$$FS_{OE}= NF_{OE} \times U_{AV} \quad (3)$$

Thus, the average unavailability of a protective device (U_{PD}) during the time of operational experience (t_{OE}) subject to hidden failures can be determined based on the total time in a failed state during the operational experience by the following expression:

$$U_{PD}= FS_{OE}/t_{OE} \quad (4)$$

Regarding the sequence of the logic presented above, there seems to exist a linear correlation between the average unavailability of the protective device (U_{PD}) and the failure-finding interval (FFI), leading to:

$$U_{PD}= NF_{OE} \times U_{AV}/t_{OE} \quad (5)$$

Considering equation (2):

$$U_{PD}=(1/MTBF_{PD}) \times U_{AV} \quad (6)$$

$$U_{PD}=U_{AV}/MTBF_{PD} \quad (7)$$

Resulting:

$$U_{PD}= 0.5 \times FFI/MTBF_{PD} \quad (8)$$

Or, to determine the failure-finding interval (FFI):

$$FFI = 2\times U_{PD}\times MTBF_{PD} \quad (9)$$

This availability does not include the unavailability of the protective device when the failure-finding task is performed or the repair time (if necessary) when a failure is observed. This can be important if a failure of the protected function occurs while the protective device is in repair, leading to an undesired multiple failure.

In the traditional RCM process, this is asked when it is assessed if the failure-finding task is worth doing, answering if the task reduces the probability of a multiple failure to a tolerable level (Basson, 2019).

Considering the expression (9), by establishing a desired availability for the protective device and assuming the existing failure data, it is possible to define the failure-finding interval for the referred device.

The desired availability is decided upon an assumed tolerance to the existing multiple failure (or a $MTBF_{MF}$), that is based on the probability that the protected function will fail during the interval (FFI) (or $MTBF_{CS}$) and on the availability of the protective device (defined in the above paragraphs).

Thus, equation (1) can be rewritten as:

$$1/MTBF_{MF}= 1/MTBF_{CS}\times U_{PD} \quad (10)$$

Or:

$$U_{PD}= MTBF_{CS}/MTBF_{MF} \quad (11)$$

Substituting Equations (11) in (9), it becomes:

$$FFI = 2 \times MTBF_{CS}\times MTBF_{PD}/MTBF_{MF} \quad (12)$$

Note that all of this logic assumes a constant failure rate (random failures) for the protected function as for the protective device, following an exponential survival distribution.

Economic analysis is also not considered in this study, including the cost of the failure-finding tasks or the cost of a multiple failure.

5 DEMONSTRATIVE EXAMPLE

This section presents an example applied to a critical ship system. The objective is to show the applicability of the proposed methodology and observe the influence of several parameters on the establishment of the failure-finding interval.

The critical system under analysis was selected based on the SOLAS Convention, namely on

Chapter II-2 (Fire protection, fire detection and fire extinction), which includes detailed fire safety provisions for all ships and specific measures for passenger ships, cargo ships and tankers.

According to IMO (n.d.), it is included in this Chapter the division of the ship into main and vertical zones by thermal and structural boundaries, the separation of accommodation spaces from the remainder of the ship by thermal and structural boundaries, the restricted use of combustible materials, the detection of any fire in the zone of origin, the containment and extinction of any fire in the space of origin, the protection of the means of escape or access for fire-fighting purposes, the ready availability of fire-extinguishing appliances, and the minimization of the possibility of ignition of flammable cargo vapour.

The present demonstrative example deals with safety systems, specifically a suppression system based on water. When working onboard a ship out to sea, the occurrence of a fire is one of the last things that the crew wants to deal with. If it occurs in the main engine room it can completely cripple the ship's mobility, as well as the risk of injury or death for crew members.

A multiple failure is related to the occurrence of a fire and the non-actuation of the sprinklers system (high-pressure system) inside de main engine room. If this happens it will lead to potential severe consequences for the main engine room or even for the ship itself.

Statistics show that between 2019 and 2021, there was a sharp increase in engine room fires on board container ships. Preliminary data gathered from public sources indicate that this trend applies to the global fleet and that for 2022 to 2023, the number of engine room fires remains at the same level as in 2021 (DNV, 2023).

Figure 1 shows the frequency of fires by accident which is notorious for the higher occurrence in the ship's engine room (Cefor, 2021).

By considering the wide range of both sources of fuel and sources of ignition within the engine room, it should come as little surprise that a large proportion of fires onboard ships start in this space. There are all sorts of flammable liquids inside the main engine room, such as tanks of heated fuel oils, lubricating oils, diesel oil, greases, and chemicals, and a fire may result when flammable oil hits hot surfaces or hot spots.

Heated fuel oils range in temperature as high as 120 to 150 degrees Celsius. It runs in kilometres of piping and often at a pressure as high as 1200 bars (in high-pressure pipes during injection for a short length). Electrical fires can also occur in engine rooms when wires short out, insulation breaks down, or wires get too hot.

When a fire occurs, the first action is to seal off the engine room to make sure it is preventing the entrance of air (oxygen) that feeds the fire. It is common to have a fixed protective CO_2 fire extinguishing system for the protection of the space, and in other situations, there are fixed extinguishing systems based on clean agents such as the HFC-227ea and the FK-5-1-12.

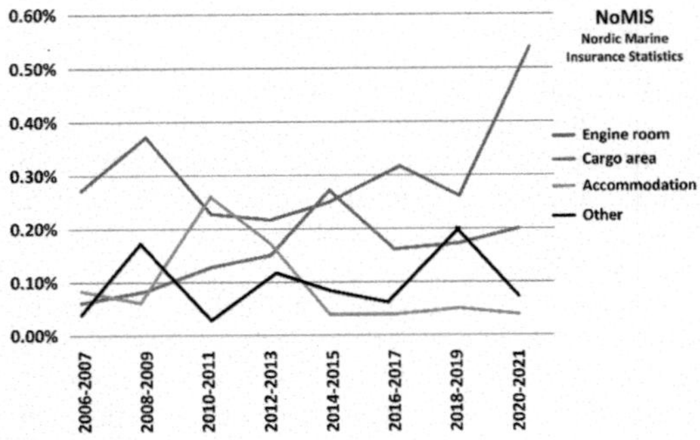

Figure 1. Frequency of fires on container vessels, by type and year (adapted from (Cefor, 2021)).

However, the water is the chief firefighting medium on a ship. This system includes a fixed pipeline running throughout the ship with hydrants at suitable points to cover all the areas known as the fire main which is fed by big fire pumps sized by regulations. There must be at least two fire pumps and an emergency fire pump in a cargo ship. The emergency fire pump must be at a remote location from the engine room on the ship. Isolating valves to isolate sections of the pipeline in the event of pipe failure are inaccessible places and clearly marked. Fire hoses and nozzles are kept coiled and ready near the hydrants.

Consider the fire extinguishing system based on water as the protective device for the main engine room in the occurrence of a fire in this compartment. Assume a water-mist system (high-pressure water system), that creates a water fog that puts out the fire while also providing a quenching or cooling effect. This system is automatically activated for two different types of detectors located in the fire zone, making also possible to operate the system remotely.

Once there is no data available regarding the probability of a fire in the main engine room to sustain the present example or even information about the efficiency of the protective system onboard ships, the information presented in this study is assumed to result from the experience of the authors.

The failure of the protective system (extinguishing system) after the occurrence of a fire in the main engine room (critical system) can be the result, for example, of a closed valve, pump failure, no water in the tanks, pipe rupture, pipe or nozzles clogged, energy failure, and other potential failure mode situations.

In this case, the failure-finding tasks incorporate inspections and checks of all valves and pipes, while performing operational tests to pumps and nozzles.

In a way to determine the frequency of the referred failure-finding tasks, if it is assumed a desired probability of a multiple failure of 1 in 1 hundred in any one year (probability of a fire occurrence and the extinguishing system failure), it means an $MTBF_{MF}$ of 100 years. Assuming also a demand rate for the protection function of 1 in 4 years, the probability of failure corresponds to an $MTBF_{CS}$ of 4 years for the critical system.

To be possible to achieve the probability for multiple failures to occur (10^{-2}), given the demand rate for the protection function, the unavailability of the protective device should be:

$$1/100 = 1/4 \times U_{PD}$$

Then,

$$U_{PD} = 1/25 = 0.04$$

The required availability for the protection system (extinguishing water mist system) is then established at 96%. Considering Equation (2):

$$FFI = 2 \times 0.04 \times MTBF_{PD}$$

$$FFI = 0.08 \times MTBF_{PD}$$

Having the information that the Mean Time Between Failures of the protective device is about 5 years, then the interval of the failure-finding task (FFI) will be defined as 0.4 years (about 5 months).

On the other side, if equation (12) is used, the same value for the failure-finding task frequency (FFI) for the protective device is also achieved:

$$FFI = 2 \times MTBF_{CS} \times MTBF_{PD}/MTBF_{MF}$$

$$FFI = 2 \times 4 \times 5/100$$

$$FFI = 0.4 years (approximately the same 5 months)$$

If the Mean Time Between Failures of the protective device is reduced to 4 years, then the FFI will be 0.32 years, corresponding to about 4 months. This has logic once less reliable the protective device, the more frequently the failure-finding tasks should be performed upon it.

If the demand for the protective device is higher, say it is called once every 2 years instead of every 4 years, and the Mean Time Between Failures of the protective device of about 2 years), then the failure-finding tasks interval (FFI) will be 0.08 years, corresponding to a monthly failure-finding task, what is coherent with the required necessities.

6 CONCLUSIONS

The present paper refers to the establishment of a frequency for failure-finding tasks based on a required availability, proposing a methodology to determine such frequency for a specific critical system of a ship.

Based on the simulation performed using assumed values according to the experience of the authors it is possible to observe that there are three factors influencing the determination of the frequency of the failure-finding tasks. These factors are the probability of failure of the critical system protected, the unavailability of the protective device and the desired availability.

It is also important to have reliable data related to the behaviour of the critical systems and protective devices to produce an accurate and efficient result for the frequency of the failure-finding tasks, avoiding the so-called multiple failures.

Following this methodology allows for the systematic development of failure-finding tasks for protective systems to achieve and maintain the desired availability level and the probability of the occurrence of multiple failures. Regular monitoring, analysis of performance data, and a commitment to continuous improvement are critical elements of this approach. The methodology proposed in this paper is suitable to be applied to other protective systems installed in ships regardless of their type, showing the potential advantages of this work and the impact on the safety of a ship if a critical system fails and the protective devices are unavailable.

ACKNOWLEDGEMENTS

This work contributes to the Strategic Research Plan of the Center for Marine Technology and Ocean Engineering (CENTEC), which is financed by FCT under contract UIDB/UIDP/00134/2020.

REFERENCES

Bae, K. & Thorp, J.S. 1999. A stochastic study of hidden failures in power system protection. *Decision Support Systems* 24: 259–268.

Basson, M. 2019. *Risk-Based Reliability Centered Maintenance.* Aladon. ISBN 978-0-8311-3632-1. Connecticut, USA. Industrial Press, Inc.

Brissaud, F. Folleau, C. & Cournuaud, B. 2022. Reliability and availability models for ageing safety-related systems. *Journal of Loss Prevention in the Process Industries* 75: 104712.

Cefor, 2021. Cefor Fire Trend Analysis. *NoMIS - The Nordic Association of Marine Insurance Statistics.* https://cefor.no/globalassets/documents/statistics/nomis/2021/cefor-fire-trend-analysis-per-december-2021.pdf, accessed on 18 december 2023.

Chen, J., Thorp, J.S. & Dobson, I. 2005. Cascading dynamics and mitigation assessment in power system disturbances via a hidden failure model. *Electrical Power and Energy Systems* 27: 318–326.

DNV. 2023. *Engine room fires - causes, contributors and preventive measures.* Technical and Regulatory News, No. 29/2023. https://www.dnv.com/news/engine-room-fires-causes-contributors-and-preventive-measures-250621, accessed on 28 december 2023.

IMO, n.d. *International Convention for the Safety of Life at Sea (SOLAS).* 1974. https://www.imo.org/en/About/Con

ventions/Pages/International-Convention-for-the-Safety-of-Life-at-Sea-(SOLAS),-1974.aspx, accessed on 03 january 2024.

Jiang J., Zhou, Z., Han, X., Zhang, B. & Ling, X. 2014. A new BRB based method to establish hidden failure prognosis model by using life data and monitoring observation. *Knowledge-Based Systems* 67: 270–277.

Levitin, G., Xing, L. & Huang, H. 2019. Cost effective scheduling of imperfect inspections in systems with hidden failures and rescue possibility. *Applied Mathematical Modelling* 68: 662–674.

Moubray, J. 1997. *RCM II – Reliability-Centered Maintenance*, Second Edition. ISBN 978-0-8311-3164-3. New-York, USA. Industrial Press Inc.

Tang, T., Lin, D., Banjevic, D. & Jardine, A.K.S. 2013. Availability of a system subject to hidden failure inspected at constant intervals with non-negligible downtime due to inspection and downtime due to repair/replacement. *Journal of Statistical Planning and Inference* 143: 176–185.

Zeynivand, M. and Shahrtash, S.M. 2023. Direct identification of Multi-Hidden failures in main distance relays of transmission systems. *Electrical Power and Energy Systems* 146: 108722.

Zhang, Y., Shen, J. & Ma, Y. 2021. An optimal preventive maintenance policy for a two-stage competing-risk system with hidden failures. *Computers & Industrial Engineering* 154: 107135.

Zhang, Y., Ouyang, L., Meng, X. & Zhu, X. 2024. Condition-based maintenance considering imperfect inspection for a multi-state system subject to competing and hidden failures. *Computers & Industrial Engineering* 188: 109856.

Advances in Maritime Technology and Engineering – Guedes Soares & Santos (Eds)
© 2024 The Author(s), ISBN 978-1-032-83099-5

Author Index

ADVANCES IN MARITIME TECHNOLOGY AND ENGINEERING

This book comprises a collection of the papers presented at the 7th International Conference on Maritime Technology and Engineering (MARTECH 2024) held in Lisbon, Portugal, on 14-16 May 2024. This Conference has evolved from the series of biannual national conferences in Portugal, which have become an international event, reflecting the internationalization of the maritime sector and its activities. MARTECH 2024 is the seventh of this new series of biannual conferences. This book comprises 142 contributions that were reviewed by an International Scientific Committee.

This book is dedicated to ship design, including ship machinery and to different aspects of hydrodynamics including wave loads, autonomous ships, renewable energies offshore, aquaculture offshore, as well as environmental modelling.

The "Proceedings in Marine Technology and Ocean Engineering" series is dedicated to the publication of proceedings of peer-reviewed international conferences dealing with various aspects of "Marine Technology and Ocean Engineering". The series includes the proceedings of the following conferences: the International Maritime Association of the Mediterranean (IMAM) conferences, the Marine Structures (MARSTRUCT) conferences, the Renewable Energies Offshore (RENEW) conferences and the Maritime Technology (MARTECH) conferences. The "Marine Technology and Ocean Engineering" series is also open to new conferences that cover topics on the sustainable exploration of marine resources in various fields, such as maritime transport and ports, usage of the ocean including coastal areas, nautical activities, the exploration and exploitation of mineral resources, the protection of the marine environment and is resources, and risk analysis, safety and reliability. The aim of the series is to stimulate advanced education and training through the wide dissemination of the results of scientific research.

Proceedings in Marine Technology and Ocean Engineering

ABOUT THE SERIES

The 'Proceedings in Marine Technology and Ocean Engineering' series is devoted to the publication of proceedings of peer-reviewed international conferences dealing with various aspects of 'Marine Technology and Ocean Engineering'. The Series includes the proceedings of the following conferences: the International Maritime Association of the Mediterranean (IMAM) Conferences, the Marine Structures (MARSTRUCT) Conferences, the Renewable Energies Offshore (RENEW) Conferences and the Maritime Technology and Engineering (MARTECH) Conferences. The 'Marine Technology and Ocean Engineering' series is also open to new conferences that cover topics on the sustainable exploration and exploitation of marine resources in various fields, such as maritime transport and ports, usage of the ocean including coastal areas, nautical activities, the exploration and exploitation of mineral resources, the protection of the marine environment and its resources, and risk analysis, safety and reliability. The aim of the series is to stimulate advanced education and training through the wide dissemination of the results of scientific research.

BOOKS IN THE SERIES

Volume 1: Advances in Renewable Energies Offshore, 2019, C. Guedes Soares (Ed.).
Volume 2: Trends in the Analysis and Design of Marine Structures, 2019, J. Parunov and C. Guedes Soares (Eds.).
Volume 3: Sustainable Development and Innovations in Marine Technologies, 2020, P. Georgiev and C. Guedes Soares (Eds.).
Volume 4: Developments in the Collision and Grounding of Ships and Offshore Structures, 2020, C. Guedes Soares (Ed.).
Volume 5: Developments in Renewable Energies Offshore, 2021, C. Guedes Soares (Ed.).
Volume 6: Developments in Maritime Technology and Engineering, 2021, C. Guedes Soares and T.A. Santos (Eds.)
Volume 7: Developments in the Analysis and Design of Marine Structures, 2021, J. Amdahl and C. Guedes Soares (Eds.)
Volume 8: Trends in Maritime Technology and Engineering, 2022, C. Guedes Soares and T.A. Santos (Eds.)
Volume 9: Sustainable Development and Innovations in Marine Technologies, 2023, Selma Ergin and C. Guedes Soares (Eds.)
Volume 10: Trends in Renewable Energies Offshore, 2023, C. Guedes Soares (Ed.)
Volume 11: Advances in the Analysis and Design of Marine Structures, 2023, J.W. Ringsberg and C. Guedes Soares (Eds.)
Volume 12: Advances in the Collision and Grounding of Ships and Offshore Structures, 2023, H. Le Sourne and C. Guedes Soares (Eds.)
Volume 13: Advances in Maritime Technology and Engineering, 2024, C. Guedes Soares and T.A. Santos (Eds.)

Proceedings in Marine Technology and Ocean Engineering (Print): ISSN: 2638-647X
Proceedings in Marine Technology and Ocean Engineering (Online): eISSN: 2638-6461

PROCEEDINGS OF THE 7TH INTERNATIONAL CONFERENCE ON MARITIME TECHNOLOGY AND ENGINEERING (MARTECH 2024), 14–16 MAY 2023, LISBON, PORTUGAL

Advances in Maritime Technology and Engineering

Celebrating 30 years of the Centre for Marine Technology and Ocean Engineering (CENTEC)

Volume 2

Edited by

C. Guedes Soares

Centre for Marine Technology and Ocean Engineering (CENTEC), Técnico Lisboa, Universidade de Lisboa, Portugal

Tiago A. Santos

Ordem dos Engenheiros, Portugal

CRC Press
Taylor & Francis Group
Boca Raton London New York

CRC Press is an imprint of the
Taylor & Francis Group, an **informa** business
A BALKEMA BOOK

Designed cover image: Authors

First published 2024
by CRC Press/Balkema
4 Park Square, Milton Park, Abingdon, Oxon, OX14 4RN

and by CRC Press/Balkema
2385 NW Executive Center Drive, Suite 320, Boca Raton FL 33431

CRC Press/Balkema is an imprint of the Taylor & Francis Group, an informa business

Typeset by Integra Software Services Pvt. Ltd., Pondicherry, India

British Library Cataloguing-in-Publication Data
A catalogue record for this book is available from the British Library

Library of Congress Cataloging-in-Publication Data
A catalog record has been requested for this book

Set:
ISBN: 978-1-032-83327-9 (hbk)
ISBN: 978-1-032-83328-6 (pbk)

Volume 1:
ISBN: 978-1-032-83099-5 (hbk)
ISBN: 978-1-032-83310-1 (pbk)
ISBN: 978-1-003-50876-2 (ebk)
DOI: 10.1201/9781003508762

Volume 2:
ISBN: 978-1-032-83107-7 (hbk)
ISBN: 978-1-032-83316-3 (pbk)
ISBN: 978-1-003-50877-9 (ebk)
DOI: 10.1201/ 9781003508779

Advances in Maritime Technology and Engineering – Guedes Soares & Santos (Eds)
© 2024 The Editor(s), ISBN 978-1-032-83107-7

Table of contents

Aquaculture offshore

Environmental modelling

Preface

Since 1987, the Naval Architecture and Marine Engineering branch of the Portuguese Association of Engineers (Ordem dos Engenheiros) and the Centre for Marine Technology and Ocean Engineering (CENTEC) of the Instituto Superior Técnico (IST), University of Lisbon, (formerly Technical University of Lisbon) have been organizing national conferences on Naval Architecture and Marine Engineering. Initially, they were organised annually and later became biannual events.

These conferences had the objective of bringing together Portuguese professionals giving them an opportunity to present and discuss the ongoing technical activities. The conferences have been typically attended by 150 to 200 participants and the number of papers presented at each one was in the order of 30 in the beginning and 140 in the latest events.

At the same time as the conferences have matured, the international contacts have also increased and the industry became more international in such a way that the fact that the conference was in Portuguese started to hinder its further development with wider participation. Therefore, a decision was made to experiment with having also papers in English, mixed with the usual papers in Portuguese. This was first implemented in the First International Conference of Maritime Technology and Engineering (MARTECH 2011), but, subsequently, five more MARTECH conferences have been organized in 2014, 2016, 2018, 2020 and 2022.

This year, the MARTECH 2024 Conference also celebrates the 30th anniversary of the foundation of the Centre for Marine Technology and Ocean Engineering (CENTEC) in 1994, which has been one of the main contributors to the organization of this Conference representing the contribution of Técnico Lisboa (IST).

For the Seventh International Conference of Maritime Technology and Engineering (MARTECH 2024), a total of around 200 abstracts have been received and 142 full papers were finally accepted. Also included in the book are abstracts of the four Keynote Lectures.

The Scientific Committee had a major role in the review process of the papers although several other anonymous reviewers have also contributed and are deserving of our thanks for the detailed comments provided to the authors allowing them to improve their papers. The participation is coming from research and industry from almost every continent, which is also a demonstration of the wide geographical reach of the conference.

The contents of the present books are organized in the main subject areas corresponding to the sessions in the Conference and within each group the papers are listed by the alphabetic order of the authors.

We want to thank all contributors for their efforts and the sponsoring Institutions for their support, hoping that this Conference will be continued and improved in the future.

C. Guedes Soares & T.A. Santos

Advances in Maritime Technology and Engineering – Guedes Soares & Santos (Eds)
© 2024 The Editor(s), ISBN 978-1-032-83107-7

Organizing Committee

Conference Chairmen
Carlos Guedes Soares, Técnico Lisboa, Universidade de Lisboa, Portugal
Dina Dimas, Ordem dos Engenheiros, Portugal

Organizing Committee
Tiago A. Santos, Ordem dos Engenheiros & Técnico Lisboa, Portugal
Ângelo Teixeira, Técnico Lisboa, Universidade de Lisboa, Portugal
José Manuel Cruz, Ordem dos Engenheiros, Portugal
Manuel Ventura, Técnico Lisboa, Universidade de Lisboa, Portugal
Francisco C. Salvado, Ordem dos Engenheiros, Portugal
José Gordo, Técnico Lisboa, Universidade de Lisboa, Portugal
Raúl Caria, Ordem dos Engenheiros, Portugal
José Gaspar, Técnico Lisboa, Universidade de Lisboa, Portugal
Pedro Pereira da Silva, Ordem dos Engenheiros, Portugal

Scientific Programme Committee
Ahmed Dursun Alkan, Yildiz Techn. University, Turkey
Ole Andreas Alsos, NTNU, Norway
Felice Arena, Mediterranea Univ of Reggio Calabria, Italy
Ermina Begovic, University of Naples, Italy
Kostas Belibassakis, NTUA, Greece
Franz von Bock und Polach, TUHH, Germany
Maria Boile, University of Piraeus, Greece
Evangelos Boulougouris, Univ. of Strathclyde, UK
Nian Zhong Chen, Tianjin University, China
Ranadev Datta, IIT Kharagpur, India
Nastia Degiuli, University of Zagreb, Croatia
Vicente Díaz-Casás, Universidad A Coruña, Spain
Leonard Domnisoru, Univ. "Dunarea de Jos, Galati, Romania
Bettar Ould el Moctar, Univ. of Duisburg-Essen, Germany
Segen F. Estefen, UFRJ, Brazil
Massimo Figari, University of Genova, Italy
C.J. Fortes, LNEC, Portugal
Yordan Garbatov, Técnico Lisboa, Portugal
Spyros Hirdaris, Aalto University, Finland
Zhiqiang Hu, Newcastle University, UK

Sponsors

Sponsored by

GRUPO ETE

DNV

Portline Bulk International S.A.

BUREAU VERITAS

LISNAVE ESTALEIROS NAVAIS, S.A.

GRUPO SOUSA

WestSEA VIANA SHIPYARD

PORTO DE SINES

Supported by

TÉCNICO LISBOA

Ship design

Advances in Maritime Technology and Engineering – Guedes Soares & Santos (Eds)

An analysis of the momentary sulfur emission peak of open-loop scrubbers applying the environmental index

E. Altarriba, T. Tanhuanpää & S. Rahiala
South-Eastern Finland University of Applied Sciences, Kotka, Finland

ABSTRACT: Sulfur scrubbers have become increasingly common in shipping due to the introduction of sulfur emission control (SECA) areas. The commonly used scrubber type, open-loop scrubbers, discharges wash water directly into the sea. In some ports, wash water discharge is forbidden, the open-loop scrubbers must be turned off, and the fuel must be switched to low-sulfur oil. When the scrubber is turned off, a momentary sulfur emission peak has regularly been observed in our measurements. In this paper, we analyze the harmfulness of this emission peak by applying an environmental index, focusing on the impacts on human health and acidification. Our results show that in the vicinity of a settlement, even a momentary peak in sulfur emissions is harmful to human health. It is important to consider this issue in the future product development of sulfur scrubbers, especially when SECA areas are expanded.

1 INTRODUCTION

1.1 *Background*

To reduce global emissions from shipping, the IMO has significantly tightened its aims, setting the goal of net zero GHG emissions by 2050 (IMO, 2023). Restricted areas for toxic emissions such as NO_X and SO_X are expanding, e.g., to the Mediterranean (IMO, 2022). A holistic change is necessary. According to Tadros et al. (2023), a new approach to ship design, fuels, numerous technical solutions, and the optimal use of ships must be considered. Different emissions require specific solutions. The formation of sulfur emissions closely follows the fuel sulfur content (Tadros et al., 2022) when reduction requires low-sulfur fuels or solutions that reduce sulfur dioxides from flue gases.

Sulfur emissions from shipping have been a typical emission component of marine diesel engines for a long time. Globally, shipping contributes to approximately 13% of anthropogenic sulfur oxide (SO_X) emissions (Aakko-Saksa et al. 2023). Sulfur dioxide and other sulfur compounds are toxic to humans and the environment, causing, among other things, acidification of the seas (Bengtsson et al. 2014). For a long time, the maximum sulfur content of fuel in maritime transport was allowed to be relatively high: 3.5% in 2012–2019 and 4.5% in 2005–2011 (IMO 2008). However, since 2020, the global sulfur content limit for fuels has been 0.5% (IMO 2016). Exceeding this limit is allowed if the ship is equipped with sulfur scrubbers and they are used during voyages.

Stricter sulfur restrictions have been used regionally. For example, the Baltic Sea has been a SECA (sulfur emission control area) area defined by the IMO, where the maximum sulfur concentration without using sulfur scrubbers has been 0.1% since 2015 (IMO 2008). The Mediterranean Sea will become a corresponding SECA area in 2025 (IMO 2022).

Experiences have been accumulated in the Baltic Sea, and shipowners have found different approaches to fulfill sulfur regulations. Some of them have switched to using low-sulfur fuel (low-sulfur MGO or residual fuel types containing less than 0.1% sulfur), some shipowners have invested in LNG engines, while some have invested in sulfur scrubbers and are able to benefit from cheaper fuel oil. There are two main types of sulfur scrubbers. In open-loop scrubbers, sulfur is washed out of flue gases with seawater, which after the washing process is pumped back into the sea. In closed-loop scrubbers, the washing process is based on the NaOH cycle, and significantly less wash water is diverted to the sea. Additionally, there are hybrid scrubber models that enable choosing between the open-loop and closed-loop operating modes (Andreasen & Mayer 2007; Lee et al. 2022; Tadros et al., 2023; Wilewska-Bien et al. 2016).

Open-loop scrubbers have raised concerns about the environmental effects of the discharged wash waters (Bengtsson et al. 2014; Karatuğ et al., 2022; Osipova et al. 2021). The wash waters contain heavy metals and PAH compounds, which can have

DOI: 10.1201/9781003508779-1

unprecedented impacts on aquatic biota. Information about this issue is scarce. Therefore, some countries, such as Germany, have prohibited the use of open-loop scrubbers in some port areas or the discharge of wash waters to the inland waterways (Jalkanen et al. 2021a). Open-loop scrubbers must be turned off before entering the areas with prohibitions, and the fuel must be changed to low-sulfur fuel to fulfill the sulfur oxide limits. In on-board emission measurements, we observed a momentary peak in sulfur emissions after the sulfur scrubbers were turned off (Altarriba et al. 2023a). Since the emission peak occurs in the vicinity of populated areas, it raises concerns about its impact on human health. This issue has been overlooked, while the wash waters have gained intensive attention, even though this has been noticed before, for example, in Teinilä et al. (2018) and Aakko-Saksa et al. (2023) state that the emissions from shipping are estimated to cause annually around 250,000 premature deaths and 6.4 million childhood asthma cases even with the global marine fuel sulfur limit of 0.5%. The authors also highlight that when the emissions occur in the vicinity of urban areas, the adverse effects are aggravated (Aakko-Saksa et al. 2023).

1.2 *Aim and scope*

In this study, we demonstrate the consequences of the sulfur emission peak, in coastal areas, with an environmental index. Particularly, the index illustrates the effects on air quality (HHPA) and acidification (AP). The aim is to increase a holistic understanding of the potential impacts of the ban on discharging wash waters in restricted areas and using open-loop scrubbers in other areas. This is an exceptionally topical issue when the SECA areas expand into new areas.

Sulfur scrubbers enable cheaper, high-sulfur fuel to be used on board. However, in some areas, the use of open-loop sulfur scrubbers is limited due to wash water discharges into the marine environment. The objective of this paper is to deal with the sulfur emissions of a ship arriving at the port in a situation where the sulfur scrubbers have been turned off and the fuel has been changed to low-sulfur fuel in accordance with the regulations in force. In on-board emission measurements, a momentary peak in sulfur emissions has been observed after the sulfur scrubbers have been turned off, which is discussed in detail by Altarriba et al. (2023). This paper studies this emission peak by utilizing an environmental index.

1.3 *Study structure*

This paper consists of the following chapters: In Section 1, the background and scope of the study are introduced. Section 2 includes applied methods and the criteria to apply them in this study. The selected methods are on-board emission measurements and analysis of obtained emission data applying the environmental index method, which is part of the sustainability index approach developed by Iannaccone et al. (2020). Section 3 contains the main results; Section 4 includes a discussion of the results and Section 5 involve conclusions.

2 METHODS

2.1 *Sustainability index*

The sustainability index is an approach that can be applied to analyze multi-criteria set of problems that sustainability includes. The sustainability index for comparing alternative fuel systems was developed by Iannaccone et al. (2020), and it includes environmental, economic, and safety indexes. The indexes are weighted with different weighting factors to create the total sustainability index, and the sum of the weighting factors is always 1. More information on the application of the sustainability index approach to the analysis of emission measurement data can be found in Altarriba et al. (2023a).

2.2 *Environmental index*

The environmental index includes four potentials (GWP100, AP, EP, and HHPA) that model the environmental effects of the most significant emission components of shipping (Bengtsson et al. 2011). The global warming potential over a 100-year timescale (GWP100) is affected by CO_2, N_2O, CH_4, and black carbon (BC), and the acidification potential (AP) is affected by NH_3, NO_2, and SO_2. Correspondingly, the eutrophication potential (EP) influencing emission components are NO_2, N_2O, and NH_3, and the human health particulate air (HHPA) is affected by SO_2, $PM_{2.5}$, PM_{10}, NH_3, and NO_2. The environmental index includes air emissions from both fuel production and the operational phase. When comparing different fuel options—and emission reduction techniques—it has been found that most emissions are generated during the use phase with conventional fuels such as HFO and MGO (Bengtsson et al. 2011).

In this study, attention is paid to the AP and HHPA potentials that are affected directly by the amount of sulfur emissions. Since the aim of this study is not to examine fuel prices or price fluctuations nor the safety aspects of related fuel systems, the limitation to the application of the environmental index can be performed. By applying the environmental index, the emission data measured on board can be analyzed in relation to the assumed adverse effects of sulfur compound emissions.

2.3 *Emission measurements*

On-board emission measurements have been carried out for each engine from the stacks by using the following emission measurement equipment. The exhaust gas samples were obtained with a PSP4000-H heated filter probe, and a processing dryer Permocure Minigas 2812T was used to dry the fuel gas sample. The analysis was performed with a portable Horiba PG350 analyzer that can analyze the amount of sulfur compounds, carbon dioxide and monoxide, nitrogen oxides, and residual oxygen fulfilling the following measurement standards (Stationary source emissions 2017a; Stationary source emissions 2019; Stationary source emissions 2017b; Stationary source emissions 2017c; Stationary source emissions 2017d; Stationary source emissions 2013). TVOC (total volatile organic compound) emissions have been measured with the propane calibrated ERSATEC SmartFID analyzer, which is based on gas chromatography. Table 1 lists the measurement standards and measuring methods that were followed.

Table 1. Emission measurement methods and followed standards.

Emission	Method	Standard
SO2	IR absorption	CEN/TS 17021/2017
NOx	Chemiluminescence	SFS-EN 14792/2017
CO	IR absorption	SFS-EN 15058/2017
CO2	IR absorption	ISO 12039/2019
O2	Paramagnetic	SFS-EN 14789/2017
TVOC	FID	SFS-EN 12619/2013

The ship's voyage data has been obtained as recordings from the ship's systems. Engine data was downloaded from the L3 Valmarine system and scrubber control data from the Ecospray control system. The following variables have been received: Engine load, shaft power, fuel consumption, scrubber status (on/off), exhaust gas temperature, and the ship's position and running speed. A more detailed description of the measurement session is presented in Altarriba et al. (2023a). The ship specifications are presented in Table 2.

Table 2. Ship specifications.

Symbol	Factor	Value
Los	Hydrodynamic length	218 m
B	Beam	30 m
T	Draught	7.1 m (max.)
DWT	Deadweight tonnage	9,653 (summer max.)
PWR	Main engine power	41,580 kW

When the ship arrives at the port, the scrubbers are turned off because it is prohibited to discharge the wash water of open-loop scrubbers into the environment. Before this, the fuel was changed to low-sulfur fuel in accordance with the terms of the SECA area. The fuel change was already performed about 2 hours beforehand to be sure that the fuel supply lines and equipment were completely emptied of sulfurous heavy fuel oil before arriving at the restricted area. The fuel change operation takes about 30-40 minutes. The measurement process and arrangements are described in more detail in Altarriba et al. (2023a).

3 RESULTS

In this paper, observations performed by onboard sulfur dioxide emission measurements, described in detail in Altarriba et al. (2023a), are presented in Figure 1. The sulfur scrubber has been turned off at the timestamp of 0:00, and a low-sulfur marine gas oil that complies with fuel regulations (under 0.1% sulfur level) has been used during the whole period. Two analyses have been performed, including the environmental potential weighting values given in Table 3.

The weighting factors of analysis 1 are based directly on the study of Iannaccone et al. (2020). According to them, the environmental index method enables the selection of weightings flexibly based on expert assessments but in all cases, the sum of the weighting factors must be 1. The weightings of analysis 2 have been chosen based on the following: The GWP100 and EP potentials are minimized (not affected by SO_2 emissions), and HHPA potential (0.5 is chosen) is emphasized due to the settlement's proximity. The AP potential is given a weight of 0.3 because shallow water areas are prone to acidification.

Table 3. Weighting values.

Option	GWP100	AP	EP	HHPA
Analysis 1	0.3	0.3	0.2	0.2
Analysis 2	0.1	0.3	0.1	0.5

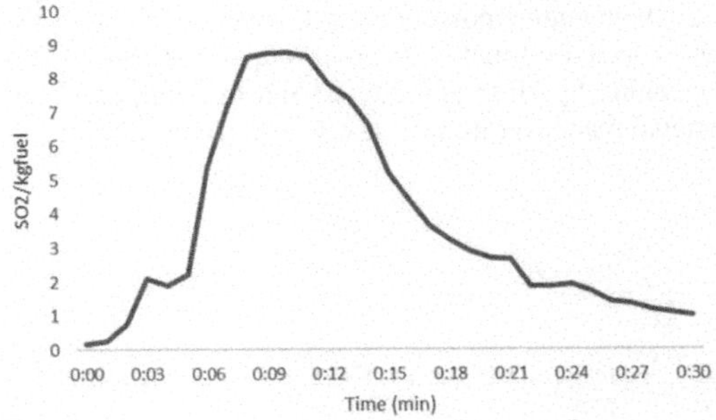

Figure 1. SO_2 emission peak.

When the scrubber is turned off, the sulfur-containing substance left in the scrubber vaporizes, gets heated by the hot exhaust gases, and causes a short but sharp sulfur emission peak. As can be seen in Figure 1, the sulfur emissions are at their highest level for only about 10 minutes.

When comparing the sulfur emission curve in Figure 1 to a curve describing the engine output power in Figure 2 (data is downloaded from the L3 Valmarine control system), it can be found that the amount of sulfur emissions does not correlate with the engine power at all. This indicates that the sulfur emissions do not originate from the current operation of the marine diesel engine and cannot be caused by sulfur-rich fuel left in the booster, fuel lines, or other components of the fuel system. This is also supported by the strength of sulfur emissions. On average, during the period presented in Figure 1, the calculated amount of SO_2 emissions would correspond to a fuel sulfur content of 3.71%, which is clearly more than the sulfur content (2.5%) of heavy fuel oil used when a ship is running with scrubbers.

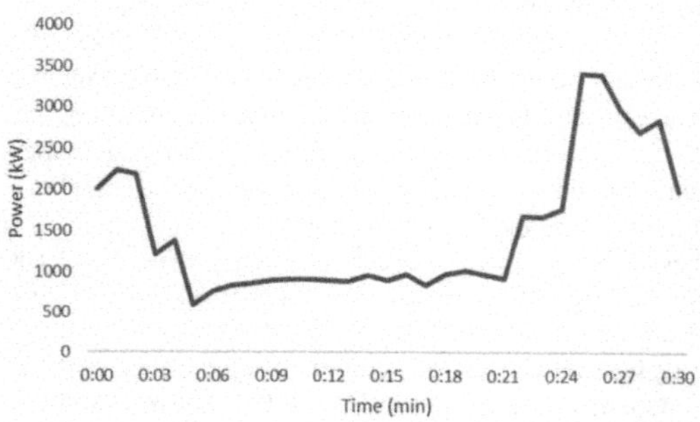

Figure 2. Main engine power curve (measured).

This study is focused only on the main engines. Exhaust gases from auxiliary engines do not pass through the scrubbers. Figure 3 shows the total power produced by the main engines, which is considered the engine power of the environmental index analysis. Based on the measurement results, the following values have been selected as initial values: The average main engine power during this period is 3300 kW; the selected estimated fuel consumption is 195 g/kWh operating by HFO and 200 g/kWh by MGO; and the annual maneuvering time is estimated to be 120 h.

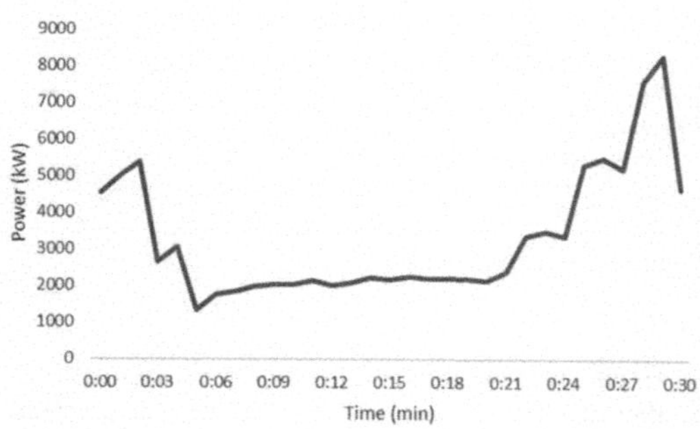

Figure 3. Engine power curves of all main engines.

The environmental index gives an insight into the adverse effects of a sulfur emission peak. The options shown in Figure 4 are the following: Option HFO 1 presents a case were arriving at the port is operated using high-sulfur heavy fuel oil and open-loop scrubbers (not allowed because of regional restrictions of wash water discharge of open-loop scrubbers). The second HFO case (HFO 2) simulates an operation without scrubbers but using heavy fuel oil containing up to 0.5% sulfur (not allowed in the Baltic Sea but fulfills global sulfur restrictions). The MGO 1 case describes a real situation where the scrubbers are turned off and the fuel used is MGO or the equivalent with a maximum sulfur content of 0.1% (without any sulfur emission peaks from scrubbers). The second MGO case (MGO 2) simulates an actual situation where the sulfur content of the fuel is assumed to be 3.71% to estimate the sulfur emission peak by utilizing the environmental index, which corresponds to the average value of the sulfur concentration (Figure 1) formed after the sulfur scrubbers were turned off.

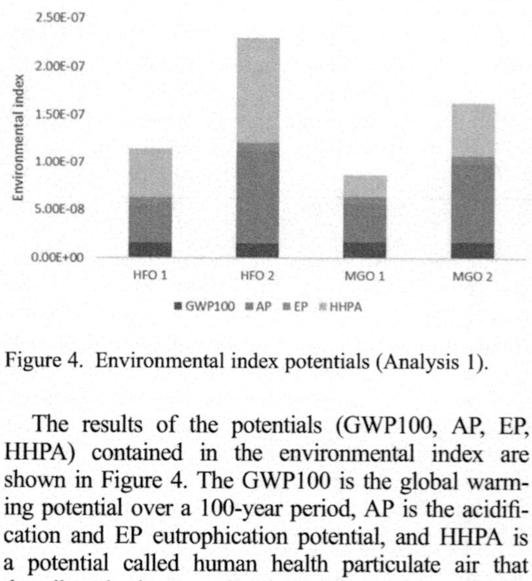

Figure 4. Environmental index potentials (Analysis 1).

The results of the potentials (GWP100, AP, EP, HHPA) contained in the environmental index are shown in Figure 4. The GWP100 is the global warming potential over a 100-year period, AP is the acidification and EP eutrophication potential, and HHPA is a potential called human health particulate air that describes the impact of emissions on humans. Higher values of each potential represent adverse effects in each potential category.

Between the alternatives HFO 1 and HFO 2, when considering air emissions, the advantages and efficiency of sulfur scrubbers come to the fore when considering air emissions. This approach does not consider emissions carried into the sea by wash water. Sulfur dioxide, as a chemical compound that acidifies the environment, raises the acidification potential high as well as the HHPA index. However, the reduction of the latter is also affected by the assumed reduction in particulate emissions, while scrubbers practically reduce them as well. The MGO 1 is the cleanest of the alternatives, which supports the validity of the current environmental regulations. Additionally, wash water emissions are not produced, unlike in the HFO

1 alternative. The MGO 2 option shows the adverse effects caused by the sulfur emission peak. The increase in the HHPA index is especially noteworthy because during this period, the ship is operating in the immediate vicinity of the settlement. The increase in the acidification potential is significant, too.

The numerical values of the potentials from Figure 4 are listed in Table 4. With these initial values, changes in the eutrophication potential do not occur. Likewise, the values of the GWP100 potential remain at almost the same level, which is mainly influenced by the different greenhouse gas emission coefficients of the fuels and the additional energy required by the scrubbers.

Table 4. Environmental index potentials (Analysis 1).

Potential	HFO 1	HFO 2	MGO 1	MGO 2
GWP100	1.59E-08	1.56E-08	1.66E-08	1.66E-08
AP	3.92E-08	9.73E-08	3.95E-08	8.26E-08
EP	7.92E-09	7.92E-09	7.93E-09	7.93E-09
HHPA	5.18E-08	1.09E-07	2.29E-08	5.51E-08

The changes in sulfur dioxide emission levels mainly concern the acidification (AP) and human health particulate air (HHPA) potentials. In Analysis 2, the weighting factors of the potentials have been selected as shown in Table 3. Global warming (GWP100) and eutrophication (EP) potentials are estimated to need less emphasis, as GWP100 is a global variable and the share of EP is marginal, and it reaches almost constant values with applied initial values. These potentials are given a 10% weightage for both. Due to the proximity of the settlement, the HHPA is considered more interesting, so it is given a 50% weightage, leaving 30% for AP. The results are shown in Figure 5, and the numerical values are presented in Table 5.

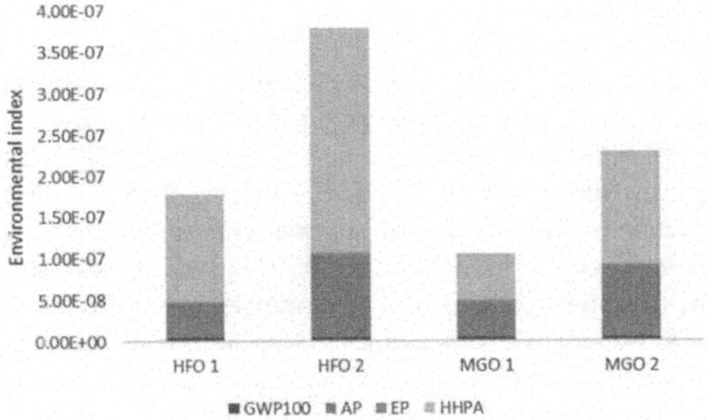

Figure 5. Environmental index potentials (Analysis 2).

Table 5. Environmental index potentials (Analysis 2).

Potential	HFO 1	HFO 2	MGO 1	MGO 2
GWP100	5.31E-09	5.21E-09	5.54E-09	5.54E-09
AP	3.92E-08	9.73E-08	3.95E-08	8.26E-08
EP	3.96E-09	3.96E-09	3.96E-09	3.96E-09
HHPA	1.29E-07	2.72E-07	5.73E-08	1.38E-07

4 DISCUSSION

The wash water emissions of open-loop scrubbers have been a topic of political debate (Karatuğ et al., 2022; Tadros et al., 2023), and many ports have restricted their use. In the Baltic Sea region, this has been the case in Germany for years (Jalkanen et al. 2021). Tightening restrictions are being planned in Sweden (Safety4sea 2023a, b), and the issue has also been included in the Finnish government program in the summer of 2023 (Valtioneuvosto 2023). In the Mediterranean region, the discharge of wash water has also been restricted in several ports, for example in Spain, France, Croatia, Israel, Turkey, and the Suez Canal (Clean Shipping International 2021). In addition, there are also regional restrictions or special permit requirements elsewhere. The wash water discharge problem does not apply to closed-loop scrubbers because produced wash water can be led to a wastewater tank in the restricted area.

Justified concern about wash water emissions has overshadowed the peak in sulfur emissions caused by turning off the open-loop scrubbers (Altarriba et al. 2023a; Teinilä et al. 2018). Even if the peak in sulfur emissions seems to be short in time, the problem is that it often occurs in the immediate vicinity of a settlement when a ship is arriving at the port. The environmental index, which we apply here, gives a perspective on the harmfulness of the issue: The HHPA index in the MGO 2 option is 240% higher than the MGO 1 option (Table 4). Accordingly, HFO 1 gives 226% and HFO 2 gives 476% higher values. The corresponding percentages for the AP potential are 209% when MGO 2 is compared to MGO 1 and 246% when comparing HFO 2 to MGO 1. In the case of HFO 1 to MGO 1, HFO 1 reaches 1% better values. The percent numbers are large, but due to the momentary peak, the traffic density level has a decisive effect on the overall level of harmfulness.

As the environmental index adds together the impacts of the fuel production phase and the operational phase (ship), the analysis considers emissions both inside and outside the studied area. As mentioned in Section 2.2, the majority of life-cycle emissions occur during the operational phase. Greenhouse gases have mainly global effects, but the effects of particulate emissions depend on the location and distribution of the emission source. These cannot be fully captured by an environmental index.

In a literature review, Toscano (2023) reviews the contribution of port emissions to sulfur, nitrogen, and particulate matter levels in urban areas in different Mediterranean countries. On average, the contribution of shipping to sulfur emissions was higher than to nitrogen and particulate matter, and the differences between countries were significant. For instance, in Croatia, shipping caused 2–4% of the sulfur emissions, while in the urban area of Naples, the share was as high as 92%. In Bandirma, the minimum contribution was 36%. Due to the high sulfur emission levels in some Mediterranean ports, Toscano (2023) concludes that the inclusion of the Mediterranean in SECA areas is of paramount importance. In many ports, SECA restrictions have efficiently reduced sulfur dioxide concentrations (Toscano 2023). Our paper questions how well the emission peak from sulfur scrubbers, demonstrated here, is addressed when calculating the efficiency of SECA areas in reducing sulfur emissions in ports. This topic might be of further research interest in ports where the use of open-loop scrubbers is banned.

After the tightening of marine fuel sulfur content restrictions, the number of ships equipped with scrubber systems has increased considerably. According to Osipova et al. (2021), the number of ships fitted with scrubbers increased from 243 in 2015 to more than 4,300 in 2020. As the prices of marine fuels rise, shipowners must carefully calculate whether it is more profitable to invest in scrubber systems or run on more expensive, low-sulfur fuel oil. In the case presented in this paper, the ship had accumulated about 60 engine hours before arriving in the restriction area of open-loop scrubbers. In terms of the intensity of the sulfur emission peak, an important research question is what will happen if engine hours accumulate significantly more than this (for example, ships arriving in Europe from the Pacific Ocean). Does the sulfur-containing substance accumulated in the scrubber increase in relation to the engine hours or do the structural solutions of the scrubber prevent a linear cumulative accumulation?

The future of marine transport fuel solutions directly affects the prevalence of open-loop scrubbers and sulfur emission peaks such as presented in this paper. Gas-powered ships (LNG or LBG) do not produce sulfur emissions, and their number is constantly increasing, although the majority of new ships are still equipped with conventional fuel oil engines (DNV 2023). The proliferation of low-carbon fuels (e.g., due to the FuelEU Maritime regulation) also reduces the sulfur emission problem because these fuels are basically low-sulfur or sulfur-free products (Aakko-Saksa et al. 2023). However, the majority of the global fleet still operates on conventional fuels, and their number is large. According to Solakivi et al. (2022), heavy fuel oil can turn out to be a surprisingly competitive alternative in terms of price, when it may prove to be an attractive option for as long as regulations allow. The ship pooling possibilities also contribute to ensuring that ships based on conventional technology can still voyage our seas for a remarkably long time.

However, the sulfur peak problem from scrubbers is clearly technical, so solutions can also be found in the development of equipment. Of course, the essential question is whether open-loop scrubbers will be accepted at all in the future. If their use is approved and the number of restricted areas for wash water discharging increases, it is worth paying attention to sulfur peaks when a scrubber is turned off. In practice, this requires that scrubber manufacturers be required to observe sulfur emission levels and prevent the mentioned peaks in the scrubber turning-off phase. In this study, only open-loop scrubbers have been studied, but more measurement-based knowledge is needed from hybrid scrubbers in relation to the transition phase from the open-loop to the closed-loop washing process, too. This is particularly interesting because the open-loop option is probably preferred as much as possible as a more economical washing process for the shipowner.

5 CONCLUSION

During the process of conducting on-board emission measurements (Altarriba et al., 2023a), it was discovered that turning off the open-loop scrubber causes a short but noticeable peak in sulfur emissions when the substance left in the scrubber vaporizes due to the hot exhaust gases. The issue has received substantial attention as the problem with the peak is that it occurs in the immediate vicinity of a settlement and affects the people directly. This study deals with the effects of the emission peak from the point of view of the environmental index. The index provides a numerical perspective to the estimated environmental effects, such as the effect on human health (HHPA potential). Based on the results, even short peaks in sulfur emissions are harmful. Thus, when considering air emissions only, with this configuration, a better result for the period of port maneuvering can be obtained if the scrubbers are left on and high-sulfur HFO is used (if regulations permit). In practice, during the product development process of sulfur scrubbers, it would be important to consider this sulfur peak and attempt to eliminate it to reduce the harmful effect on human life and the environment near ports and settlements.

ACKNOWLEDGEMENTS

The research discussed in this paper has been carried out as part of the research projects "Renewable fuels – support for the sustainable and carbon-neutral growth of maritime transport" (MERPOL, 401807) and "Benchmarking study of marine emission reduction technologies" (MEPTEK, A75791), both funded by the European Regional Development Fund (ERDF). Acknowledgments are due to the regional councils (Regional Council of Kymenlaakso and Helsinki-Uusimaa Regional Council) for organizing the funding. We thank the accredited emission measurement laboratory of Kymilabs (T197/SFS-EN ISO/IEC 17025, laboratory NB2450) for performing the emission measurements. Special thanks to the shipowners, partners, and other entities that collaborated on the project.

REFERENCES

Aakko-Saksa, P.T., Lehtoranta, K., Kuittinen, N., Järvinen, A., Jalkanen, J-P., Johnson, K., Jung, H., Ntziachristos, L., Gagné, S., Takahashi, C.,

Karjalainen, P., Rönkkö, T., Timonen, H. 2023. Reduction in greenhouse gas and other emissions from ship engines: Current trends and future options. *Progress in Energy and Combustion Science* 94: 101055.

Altarriba, E., Rahiala, S., Tanhuanpää, T. 2023a. Open-loop scrubbers and restricted waterways: a case study investigation of Travemünde port and increased sulphur emissions immediately after the scrubbers are turned off. *International Journal on Marine Navigation and Safety of Sea Transportation* 17(2): 465–471.

Altarriba, E., Rahiala, S., Tanhuanpää, T., Piispa, M. 2023b. Developing sustainable shipping and maritime transport: multi-criteria analysis between emission abatement methods. *Journal of Mechanical Engineers, Part M: Journal of Engineering for the Maritime Environment*. Published online.

Andreasen, A., Mayer, S. 2007. Use of seawater scrubbing for SO_2 removal from marine engine exhaust gas. *Energy Fuels* 21: 3274–3279.

Bengtsson, S.K., Andersson, K.E., Fridell, E. 2011. *Life cycle assessment of marine fuels. A comparative study of four fossil fuels for marine propulsion*. Gothenburg: Chalmers University of Technology.

Bengtsson, S.K., Fridell, E., Andersson, K.E. 2014. Fuels for short sea shipping: a comparative assessment with focus on environmental impact. *Proceedings of the Institution of Mechanical Engineers, Part M: Journal of Engineering for the Maritime Environment* 228(1): 44–54.

Clean shipping international. 2021. Scrubber-prohibiting countries (online). Available: www.cleanshippinginternational.com/scrubber-prohibiting-countries/

DNV. 2023. Maritime forecast to 2050. Energy transition outlook 2023. Hamburg: DNV AS.

Iannaccone, T., Landucci, G., Tugnoli, A., Salzano, E., Cozzani, V. 2020. Sustainability of cruise ship fuel systems: comparison among LNG and diesel technologies. *Journal of Cleaner Production* 260: 121069.

IMO. 2008. Report of the marine environment protection committee on its fifty-eight session. IMO resolution MEPC. 58/23 (adopted on 16 October 2008).

IMO. 2016. Effective date of implementation of the fuel oil standard in regulation 14.1.3 of MARPOL Annex VI. IMO resolution MEPC. 280(70) (adopted on 28 October 2016).

IMO. 2022. Amendments to the annex of the protocol of 1997 to amend the international convention for the prevention of pollution from ships, 1973, as modified by the protocol of 1978 relating thereto (Mediterranean Sea Emission Control Area for Sulphur Oxides and Particulate Matter). IMO resolution MEPC. 361(79) (adopted on 16 December 2022).

IMO. 2023. Revised IMO strategy on reduction of GHG emissions from ships. IMO resolution MEPC.377(80) Annex I (adopted on 7 July 2023).

Jalkanen, J-P., Johansson, L., Wilewska-Bien, M., Granhag, L., Ytreberg, E., Eriksson, K.M., Yngsell, D., Hassellöv, I-M., Magnusson, K., Raudsepp, U., Maljutenko, I., Winnes, H., Moldanova, J. 2021a. Modelling of discharges from Baltic Sea shipping. *Ocean Science* 17(3): 699–728.

Jalkanen, J-P., Majamäki, E., Johansson, L. 2021b. Emissions from Baltic Sea shipping in 2006-2020. Maritime working group. MARITIME 21-2021.

Lee, I., Chang, T., Chang, C., Truong, V.D.H., Ward, J.D. 2022. Comparison of open- and closed-loop operating strategies for exhaust gas scrubbing in marine applications: Modeling and sea-trial data. *Proceedings of the Institution of Mechanical Engineers, Part M: Journal of Engineering for the Maritime Environment*: published online.

Karatuğ, Ç., Arslanoğlu, Y., Guedes Soares, C. 2022. Feasibility analysis of the effects of scrubber installations on ships. *Journal of Marine Science and Engineering* 10: 1838.

Osipova, L., Goergeff, E., Comer, B. 2021. *Global scrubber washwater discharges under IMO's 2020 fuel sulfur limit*. Washington DC: The international council on clean transportation.

Safety4sea. 2023a. Update: Scrubber discharges bans in ports (online). Available: safety4sea.com/update-scrubber-discharges-bans-in-ports

Safety4sea. 2023b. Research: Scrubbers responsible for up to 9% of harmful emissions in Baltic Sea (online). Available: https://safety4sea.com/research-scrubbers-responsible-for-up-to-9-of-harmful-emissions-in-baltic-sea/

Solakivi, T., Paimander, A., Ojala, L. 2022. Cost competitiveness of alternative maritime fuels in the new regulatory framework. *Transportation Research Part D: Transport and Environment* 113: 103500.

Stationary source emissions. 2017a. Determination of the mass concentration of sulphur dioxide by instrumental techniques, CEN/TS 17021/2017.

Stationary source emissions. 2019. Determination of the mass concentration of carbon monoxide, carbon dioxide and oxygen in flue gas. Performance characteristics of automated measuring systems, ISO 12039/2019.

Stationary source emissions. 2017b. Determination of the mass concentration of carbon monoxide. Standard reference method: non-dispersive infrared spectrometry, SFS-EN 15058/2017.

Stationary source emissions. 2017c. Determination of mass concentration of nitrogen oxides. Standard reference method: chemiluminescence, SFS-EN 14792/2017.

Stationary source emissions. 2017d. Determination of volume concentration of oxygen. Standard reference method: paramagnetism, SFS-EN 14789/2017.

Stationary source emissions. 2013. Determination of the mass concentration of total gaseous organic carbon - continuous flame ionisation detector method, SFS-EN 12619/2013.

Tadros, M., Ventura, M., Guedes Soares C. 2020. Optimization of the performance of marine diesel engines to minimize the formation of SOx emissions. *Journal of Marine Science and Application* 19: 473–484.

Tadros, M., Ventura, M., Guedes Soares C. 2023. Review of current regulations, available technologies, and future trends in the green shipping industry. *Ocean Engineering* 280: 114670.

Teinilä, K., Aakko-Saksa, P., Jalkanen, J-P., Karjalainen, P., Bloss, M., Laakia, J., Saarikoski, S., Vesala, H., Pettinen, R., Koponen, P., Kuittinen, N., Piimäkorpi, P., Timonen, H. 2018. *Effect of aftertreatment on ship particulate and gaseous components at ship exhaust*. Helsinki: Finnish Meteorological Institute, Envisum WP2 report.

Toscano, D. 2023. The impact of shipping on air quality in the port cities of the Mediterranean area: a review. *Atmosphere* 14(7): 1180.

Valtioneuvosto. 2023. *Vahva ja välittävä Suomi. Pääministeri Petteri Orpon hallituksen ohjelma*. Helsinki: Valtioneuvoston julkaisuja 2023:58.

Wilewska-Bien, M., Lindgren, J.F., Magnusson, M., Zetterdahl, M., Salo, K., Gabrielii, C., Granhag, L., Brynolf, S. 2016. Measures to reduce discharges and emissions. In Andersson, K., Brynolf, S., Lindgren, F., Wilewska-Bien M. (eds.), *Shipping and the environment*. Gothenburg: SpringerNature.

Advances in Maritime Technology and Engineering – Guedes Soares & Santos (Eds)
© 2024 The Author(s), ISBN 978-1-032-83107-7

Design of an ammonia bunker installation for sea-going vessels - A best worst method approach

P.C. Hofste, B.N. van Veldhuizen, M.B. Duinkerken & D.L. Schott
Department of Marine Technology and Transportation, Delft University of Technology, Delft, The Netherlands

ABSTRACT: Ammonia is considered a possible fuel to reduce ship emissions, but bunker systems are not yet available. Therefore, this research conceptualizes an ammonia bunker installation suitable for sea-going vessels. According to the system engineering approach, requirements and functions are defined. For every system function, different components are considered, called alternatives. The alternatives are compared through an ordinal analysis and quantitatively evaluated using the Best Worst Method (BWM). Three stakeholders of the Port of Rotterdam are surveyed to determine the criteria weights. All assign safety the highest rating. The results of the ordinal analysis, combined with the weights given by the stakeholders, result in final configurations that differ only for the pumping and connecting functions. It is concluded that the final configuration is applicable to multiple ports due to the general prioritization of safety and similarities in the final system configuration for different stakeholder inputs.

1 INTRODUCTION

Shipping contributes about 11% to the total CO_2 emissions of the transport sector, which totals around 2 – 3% of global CO2 emissions (Olmer et al. 2018). Petroleum-based fuels, currently make up 95% of the fuel mixture (Maersk Mc-Kinney Moller Center for Zero Carbon Shipping 2021). The production and conversion of these fuels produce large amounts of sulphur oxides (SOX), nitrogen oxides (NOX), as well as carbon dioxide CO_2) and particulate matter. According to the climate goals for 2050, these emissions are dangerous and need to be reduced drastically (IMO 2022).

To accomplish this reduction in emissions, several future fuels are under consideration by the marine industry. It becomes clear that a quick-win is possible through the implementation of LNG, having the potential to cut on board SOX, NOX and particulate matter emissions by 90%, 80% and 100%, respectively, as well as reducing CO_2 emissions by around 20% (Burel et al. 2013). However, due to LNG still being a fossil fuel, it is not viable in the long run. Furthermore, the fourth Greenhouse Gas (GHG) study of the International Maritime Organisation (IMO) pointed out that the methane slip in combustion engines is larger than expected. Since methane is a strong GHG, the decrease in anthropogenic emissions is minimal (IMO 2020). Instead, hydrogen and ammonia are carbon-free hydrogen carriers that can potentially almost eliminate onboard emissions (Sveistrup et al. 2021, Afif et al. 2016, Atilhan et al. 2021, Ship&Offshore 2021). Furthermore, they can be produced through electrolysis, powered by renewables (DNV-GL 2019). Between these two future fuels, hydrogen is deemed infeasible for deep-sea shipping due to storage size and cost (Mandra 2022). Maersk Mc-Kinney Moller Center for Zero Carbon Shipping (2021) expects ammonia to make up 50% of the fuel mixture for the shipping industry by 2050. Moreover, Karatug et al. (2023) also concluded that ammonia is the best alternative between LNG, methanol and ammonia based on a comparative analysis including technical, economical, safety and environmental criteria. The use of ammonia as a fuel for deep-sea shipping is more appropriate than as an inland shipping application due to toxic challenges and associated risk (European Maritime Safety Agency 2022), therefore this research is focused on the deep-sea shipping application.

Ammonia is not yet implemented as bunker fuel. There exist significant challenges that limit its implementation. These challenges are centred around:

1. Fuel storage, logistics and bunkering
2. Onboard fuel conversion
3. Onboard safety and fuel management
4. Regulations

The four challenges listed before need to be addressed for ammonia to become a viable fuel for the shipping industry. Research into several of these topics is being done. For instance, Perčić et al. (2022) performed a Life-Cycle Analysis (LCA) and Life-Cycle Cost (LCCA) Analysis for ships with several conversion technologies and fuels. The implementation of blue ammonia in a Solid Oxide

DOI: 10.1201/9781003508779-2

Fuel Cell (SOFC) leads to significant reductions in greenhouse gasses (65% - 72%) at an acceptable cost, which is 37% - 43% higher than that of a diesel power system. Cheliotis et al. (2021) reviewed the safe use of ammonia in fuel cells in the marine industry. They conclude that it is imperative to develop large-scale production infrastructures for green or blue ammonia. Furthermore, there is a clear need for the identification of the hazards and consequences of ammonia release. Yadav & Jeong (2022) used computational fluid dynamics (CFD) to look at ammonia dispersion onboard ship engine rooms to evaluate the safety of using ammonia as a marine fuel. Duong et al. (2023) performed a safety review regarding ammonia bunkering, there is little known about the configuration of an ammonia bunker installation. For example, it is unclear what components should be used and what mitigation strategies are effective for the different hazardous properties of ammonia.

Table 1. Main characteristics of ammonia, based on LHV (Chatterjee et al. 2021).

Property	Value	Unit
Formula	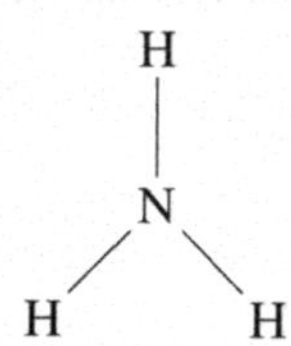	
Molar mass	17.03	g/mol
Density as gas (STP)	0.77	kg/m^3
Density as liquid	610a /680b	kg/m^3
Volumetric energy density	3.59a /3.93b	kWh/L
Gravimetric energy density	5.88	kWh/kg
Autoignition temperature	651	C
Explosive range in air	15-28	vol %
Short-term exposure limit	35	ppm
Fatality in 30 minutes	2500-4500	ppm

[a] At 20°C and 10 bar.
[b] At -33°C Atm.

The purpose of this research is to conceptualize an ammonia bunker installation that is viable for ports that handle sea-going vessels. The scope of this research includes part of an ammonia supply chain, as graphically represented in Figure 1. The boundaries of the scope are defined at the connection to an ammonia-fueled ship, as well as at the connection from an arbitrary supply modality to port-side storage.

In order to fulfil the purpose of this research, the following structure is proposed. First, a system description is given in Section 2. Then, a decision-making model is proposed in Section 3. The results of the research are shown in Section 4, and are discussed in Section 5. Finally, the conclusions are drawn in Section 6.

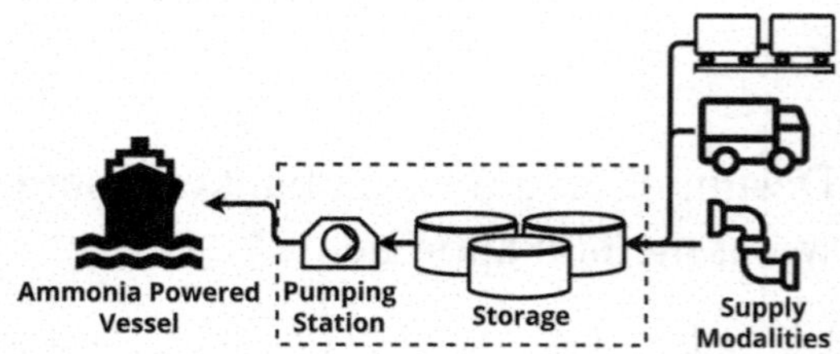

Figure 1. Bunker supply chain with the scope of this research in the dashed box.

2 SYSTEM DEFINITION

This section explains what requirements a bunkering system needs to fulfil, and of which possible components a bunkering station consists. The requirements are translated into the functions of the bunkering system. Every function can be satisfied by different system components, which are called alternatives. To define the requirements, a basic bunker installation is analysed in Section 2.1. Subsequently, the operational requirements are translated into system functions and corresponding alternatives are proposed in Section 2.2. The same is done for the safety requirements in Section 2.3.

2.1 *Requirements*

A distinction between requirements is made: operational requirements and safety requirements. Operational requirements relate to enabling the operation of the system, whereas safety requirements are imposed by the intended use of ammonia and its associated characteristics.

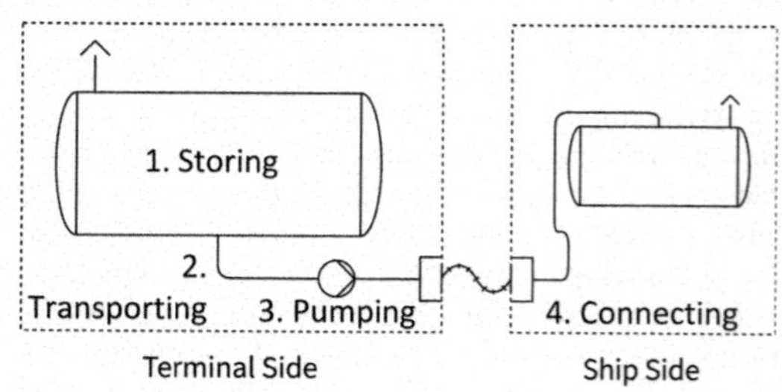

Figure 2. Basic Bunker System. The location of each operational function is labeled with its corresponding number.

2.1.1 *Operational requirements*

Figure 2 shows a schematic version of a basic bunkering system. Bunkering can be done in multiple ways, Ship to Ship (STS), Port to Ship (PTS) and Truck to Ship (TTS). These bunker methods can be generally defined as a storage vessel connected to another storage vessel via an impermanent connection. The only bunker method excluded from this definition is Portable Tank Transfer (PTT), which is considered inviable for the volumes of fuel needed for deep sea shipping and is therefore not further considered in this research (Hofste 2022). From the shown typology, the operational requirements are derived:

1. The system is required to store a medium at the terminal side
2. The system is required to have piping in place to form a connection from the terminal to the ship side through which the medium can flow
3. The system is required to have a pump to actuate the medium in a controlled manner from the terminal side storage to the ship side storage
4. The system is required to have a detachable coupling between the terminal and ship side of the installation.

2.1.2 *Safety requirements*

From the characteristics of ammonia, safety requirements can be inferred. Ammonia is toxic and thus life-threatening in concentrations above 2700 *ppm* when exposed for 10 minutes or more, can form explosive mixtures with air between 15 – 28%, and is corrosive (Centers for Disease Control 2021). The energy density of ammonia is about 50% of LNG, which in turn has about 65% of the energy density of traditional Heavy Fuel Oil (HFO) (DNV-GL 2019, Aronietis et al. 2016). Therefore, it needs to be liquefied. This is done by lowering the temperature to $-33°C$ instead of liquefaction through pressure because of better safety and higher energy density (Bartels 2008, Ekinci & Akman 2022). Furthermore, this difference in temperature means liquid ammonia will start to boil, and thus evaporate, under the influence of heat that enters the system. This Boil Off Gas (BOG) increases pressure in the system because gaseous ammonia expands to 850 times its liquid volume (Minnesota Department Of Agriculture 2022). This effect needs to be handled in the installation. Because of the aforementioned intended deep-sea application of ammonia, large vessels need to be able to use the bunker installation. Thus, the largest vessel currently in use, the MSC Irina with a deadweight tonnage (DWT) of around 240,000 (Marine Traffic 2023) needs to be able to use the installation. These characteristics associated with ammonia and the resulting derivation of requirements are graphically represented in Figure 3. The ammonia safety requirements are defined as follows:

1. It is required to have exposure mitigation systems in place with the goal of mitigating the toxic hazard associated with ammonia, in case of an unintended release.
2. It is required to have a location that is suitable such that, in case of an unintended release, there are acceptable risks of exposure to surrounding structures, people and other vulnerable objects.
3. It is required that the bunker system is able to bunker the largest vessels in service.
4. It is required that the materials used in the bunker system are able to withstand the corrosivity of ammonia.
5. It is required that there be systems and protocols in place that have the absolute goal of preventing an explosive mixture from forming.
6. It is required to have systems and protocols in place that have the absolute goal of preventing ammonia leakage into the atmosphere.
7. It is required that the bunker system is able to minimize the heat influx at all storage and transfer components in order to keep ammonia in its liquid form.
8. It is required that BOG and its consequent pressure-increasing effect is handled in the bunker system.

2.2 *Operational functions & alternatives*

The operational requirements from Section 2.1.1 are translated into four operational functions. Their corresponding locations can be found in Table 2. These operational functions are:

1. Storing
2. Transporting
3. Pumping
4. Connecting

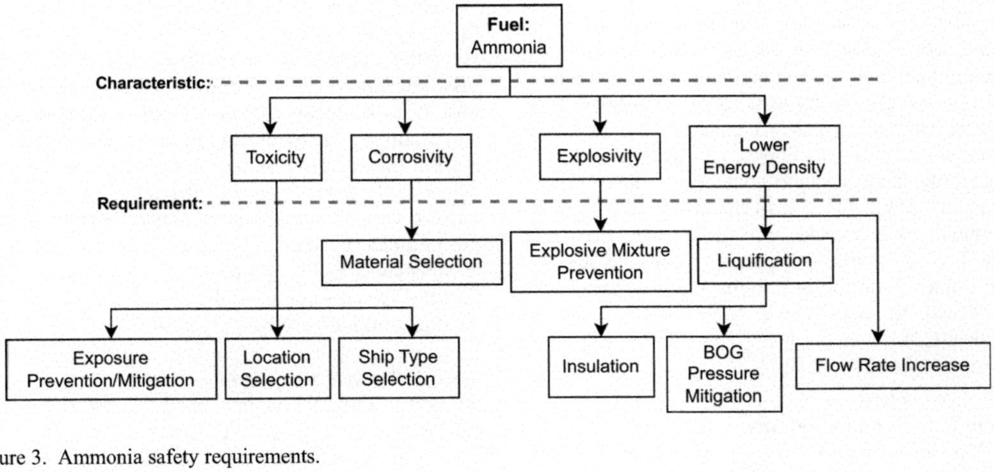

Figure 3. Ammonia safety requirements.

The solutions, or alternatives, that fulfil these functions, while complying with the requirements are described in the following sections. An overview of the alternatives per function can be found in Table 5 in the Appendix.

2.2.1 *Storing*

There are three categories relevant for ammonia storage tanks: single containment, double containment and full containment (Venkat Pattabathula et al. 2023). The difference between these storage tanks is the ability to handle failures of the primary tank. For this research, pressure-ready bullet tanks and atmospheric flat-bottom tanks are considered. Bullet tanks do not have the same definition by codes, and the containment philosophies cannot be directly copied from atmospheric storage (Rath & Krol 2013). Therefore, basic containment safety requirements are assumed for the bullet tanks for comparison purposes. Single containment storage offers no further containment than the primary tank. Double containment storage offers liquid spillage containment but does release ammonia gas. Full containment stores liquid and gas in case of a primary tank failure, thus being able to remain operational during tank failure (Venkat Pattabathula et al. 2023).

2.2.2 *Transporting*

The medium needs to be transported within the system through pipes. Possible solutions for pipes are flexible or rigid piping. Furthermore, floating flexible piping can be used. Rigid piping is preferred for on-land applications due to the restricted movement of the pipes, but a flexible floating pipe could make a floating terminal much more practical.

2.2.3 *Pumping*

There are several pumps commercially available that can handle the characteristics of ammonia, as well as offer high enough flow rates. High enough is considered to be 1000 m^3/h, which is the LNG bunker flow rate at the Gate Terminal (2022). Three pump types that are suitable for ammonia are: mag-drive pumps, canned motor pumps, and double mechanically sealed pumps, due to their hermetically sealed nature. These pumps can be dry-installed or submersed in the medium (Hermetic 2022). Canned motor pumps offer double containment, as opposed to mag-drive pumps which only have a single containment layer (Hayes Pump, inc 2023). Double mechanically sealed pumps are initially less expensive than the mag drive and canned counterparts (Magnatex Pumps 2020), but from canned motor designs to mag-drive, to mechanically sealed, the service life goes down respectively (Feldle 2022).

2.2.4 *Connecting*

Establishing a connection between the terminal side to the ship side of the system is a critical safety system because the impermanent connection is prone to human error and wear. The equipment used to realise this function consists of two parts, the connecting hose, and the coupling. Two options for the connecting hose are possible, a floating hose or a suspended hose. Concerning the coupling, three options are considered: the dry disconnect coupling, the break-away coupling and the Powered Emergency Release Coupling (PERC) (Marine Service Noord 2023). These couplings all feature a system that prevents the ammonia from leaking when the coupling is released. The latter two feature systems that automatically disconnect in case of excess tensile stress, and the PERC is able to disconnect when an Emergency Shut-Down (ESD) button is pressed.

2.3 *Safety functions & alternatives*

The safety requirements from Section 2.1.2 are translated into the following seven safety functions:

1. Bunker location
2. Substance slip prevention
3. Toxic cloud mitigation
4. Spill containment
5. BOG vapour pressure equalisation
6. Insulation
7. Purging of fuel lines

The alternatives that fulfil these functions, while complying with the requirements are described in the following sections. An overview of the discovered alternatives can be found in Table 5 in the appendix.

2.3.1 *Bunker location*

The toxic risk associated with ammonia makes it more suitable for deep-sea applications. Thus, deep-sea vessels need to be able to reach the bunker installation. Furthermore, the bunker location can have implications on the feasibility of certain mitigation strategies. The bunker locations that are handled in this research are dictated by the intended end user: Ships. Specifically, where ships can sail in order to receive the bunker fuel. Namely, at the quay, near the quay but not fully docked, or not near a quay. Hereafter referred to as onshore-, nearshore-, and offshore bunkering. Onshore bunkering offers solid ground to build on providing more available space and stability. Nearshore bunkering prevents relative motion between the bunker vessel and the jetty and offers little restriction on the orientation of the bunker vessel. Furthermore, a nearshore system is not bounded by the quay depth. Offshore bunkering is independent of port-side facilities and away from port traffic. This comes at the cost of vulnerability to weather and increased supply complexity.

2.3.2 *Substance slip prevention*

Substance slip can be prevented by the use of gas detectors coupled to an ESD, or a passive water spray system over the bunker manifold and other leak-prone objects. The gas detectors can prevent large amounts of substance slip by shutting down the installation or closing specific valves when ammonia is detected. A water spray system can leverage the solubility of ammonia in water to prevent vapours in the air around leak-prone objects. Furthermore, leaks could be detected by monitoring wastewater from this system.

2.3.3 *Toxic cloud mitigation*

Mitigating toxic clouds in case of an unintended release is very important to ensure safety. Three strategies are proposed: overpressurisation, a mesh barrier and a water curtain. Ksenia (2021) showed that through the internal overpressurisation of the insides of a ship, sufficient protection against potential ammonia hazards could be offered, and this could be utilized to protect buildings near an ammonia bunker installation. By maintaining a slightly higher than ambient pressure inside the buildings, ammonia gas in the atmosphere can not diffuse inside. The working principle of a mesh barrier is through the reduction of vortices downstream, thus increasing the concentration of ammonia on the ground surface (Barzegar Gerdroodbary et al. 2016), thus preventing long-distance migration of a toxic cloud. The operating principle of a water curtain is to leverage the reactivity of ammonia with water. This leads to the cloud being absorbed by the water and falling to the ground. This system is reported to have a possible reduction percentage of ammonia concentration as high as 90% at 10*m* behind the release location (Dandrieux et al. 2001).

2.3.4 *Spill containment*

Spill containment is incorporated in double- and full containment levels inside storage tanks. Though, it is possible to have additional spill containment. This comes in the form of a bund dyke that is able to contain the storage tank volume. The addition of modified carbon nano-particles to an ammonia pool in the dyke area around compromised storage can reduce the area of the threat zone by 50% (Namboothiri & Soman 2018). This is due to the carbon nano-particles being able to increase the absorption rate of the ammonia, which prevents additional off-gassing of an ammonia pool, thus preventing a larger ammonia cloud.

2.3.5 *BOG vapour pressure equalisation*

BOG pressure can be handled by the use of three systems: the BOG return hose, a vent with a scrubber to relieve excess pressure, or pressure can be relieved by rerouting it to a buffer tank. A BOG return hose is able to form a closed system, where excess pressure is relieved without the need for pumps back to the storage. The vent with a scrubber can relieve pressure without the need to add storage tanks or extra connections. Since ammonia needs to be gaseous to be used as a fuel (Thomas, George and Parks, George), the buffer tank can serve as an expansion tank, as well as a pressure relief system.

2.3.6 *Insulation*

Two considered options to insulate the system are mechanical and vacuum insulation. Vacuum seals can generally create a better insulating layer for a smaller diameter of piping, even up to 20 times as much as fibre or foam solutions (Va-Q-tec 2023), but are generally much more expensive than mechanical insulation. However, mechanical insulation is prone to degradation (Curtis 2007).

2.3.7 *Purging of fuel lines*

To prevent explosive mixtures, as well as preventing ammonia from spilling from fuel lines, a purging sequence is required for ammonia bunker installations (Green Shipping Programme 2021). A solution is needed that handles the vented gases because they cannot be vented into the atmosphere. This can be done with a scrubber on the vent, or by redirecting the gas mixture to a buffer tank that is connected to a gas separator. The scrubbers are only around 90% effective (Jamaludin et al. 2018), and require refilling of the medium. Gas separation units, such as distillation columns, can be costly to install and use. The effectiveness of these gas separators depend on the method used.

3 METHODOLOGY

In the previous section, system requirements, system functions, and solutions adhering to those requirements are proposed. In this section, the method for decision making between the alternatives is proposed in Section 3.1, the design criteria used to evaluate the alternatives are proposed in Section 3.2, and an example of the ordinal analysis used to rank the alternatives is given in Section 3.3. Then, the data from this analysis is combined with stakeholder input from the Port of Rotterdam, as discussed in Section 3.4, to perform a Best Worst Method (BWM) analysis. In this analysis, stakeholders are asked to evaluate the relative importance of the design criteria. Subsequently, the data from the BWM undergoes a sensitivity analysis in Section 3.5. Finally, the results are checked for infeasible combinations, and testing of the requirements is done. Afterwards, the results are unified in a final configuration.

3.1 *Decision making method*

Several Multi-Criteria Decision Making (MCDM) methods exist that use subjective weighing of design criteria by experts, such as the analytical hierarchical process (AHP), and the analytical network process (ANP) (Singh & Pant 2022).This use of subjective weighing is preferred because it circumvents researcher bias. Because it is desired to perform decision-making through stakeholder input, a straightforward and consistent survey is required. The simplicity of the survey for the decision maker is desirable to ensure more reliable results and ensure a higher response rate. With BWM it is possible to ensure survey results of sufficient quality with a low number of comparisons. Moreover, the consistency of the stakeholder input can be quickly evaluated. Additionally, the low computational time required to perform the BWM and the lack of additionally required software led to the BWM being chosen for the purposes of this research.

3.1.1 *The best worst method*

The BWM, proposed by Rezaei (2015), can solve discrete MCDM problems, which can be generally formulated as follows:

$$A = \begin{array}{c} \\ a_1 \\ a_2 \\ \vdots \\ a_m \end{array} \begin{array}{c} \begin{array}{cccc} c_1 & c_2 & \cdots & c_n \end{array} \\ \begin{pmatrix} p_{11} & p_{12} & \cdots & p_{1n} \\ p_{21} & p_{22} & \cdots & p_{2n} \\ \vdots & \vdots & \ddots & \vdots \\ p_{m1} & p_{m2} & \cdots & p_{mn} \end{pmatrix} \end{array} \tag{1}$$

where $\{a_1, a_2, \ldots, a_m\}$ is a set of feasible alternatives and $\{c_1, c_2, \ldots, c_n\}$ is a set of decision-making criteria. The values for p_{ij} correspond to the relative value for the criterion j with respect to the alternative i. The p_{ij} values are determined with an ordinal analysis as described in section 3.3. With these values, the best alternative i with the highest value V_i for a specific function can be found. This is done with Equation (2)

$$V_i = \sum_{j=1}^{n} w_j p_{ij} \tag{2}$$

Where:

$$w_j \geq 0, \sum w_j = 1 \tag{3}$$

In Equations (2) and (3), the values for $w = \{w_1, w_2, \ldots, w_n\}$ correspond to the weights associated with the importance of each criterion.

Instead of determining all relative values of all criteria, the BWM starts by determining the best (most important) criterion and the worst (least important) criterion. Subsequently, the preference of the best criterion over all the other criteria, and the preference of the other criteria over the worst criterion is determined. These preferences are allocated a number between 1 and 9. Resulting in the Best-to-Others and Others-to-Worst vectors, respectively:

$$B_B = (b_{B1}, b_{B2}, \cdots, b_{Bn}) \tag{4}$$

$$B_W = (b_{1W}, b_{2W}, \cdots, b_{nW})^T \tag{5}$$

where $b_{BB} = 1$, and $b_{WW} = 1$.

Finally, to calculate the optimal weights w_j^* and ξ^*, with '*' denoting the optimal value, the following problem can be solved:

$$min\, \xi$$

s.t.

$$\begin{aligned} &\left|\frac{w_B}{w_j} - b_{Bj}\right| \leq \xi, \forall j \\ &\left|\frac{w_j}{w_W} - b_{jW}\right| \leq \xi, \forall j \\ &\sum_j w_j = 1 \\ &w_j \geq 0, \forall j \end{aligned} \tag{6}$$

Finally, to ensure the quality of the assessment of the weights w_j, a high consistency is desired. To determine the consistency of a BWM analysis, a Consistency Ratio (CR) is proposed, as follows:

$$Consistency\,Ratio = \frac{\xi^*}{Consistency\,Index} \tag{7}$$

where *Consistency Ratio* $\in [0, 1]$, values close to 0 are more consistent, while values close to 1 are less consistent. Thus, the CR safeguards the validity of the comparisons. To calculate the consistency index the the consistency ratio needs to be determined, this is found through the following equation:

$$\xi^2 - (1 + 2b_{BW})\xi + (b_{BW}^2 - b_{BW}) = 0 \tag{8}$$

Solving Equation (8) for different values for b_{BW}, the maximum value for ξ can be found. This value is used as the consistency index. With the solutions for (2) known for each function, and the solution of (7) as a measure of the quality of the pairwise comparison, a decision for the best alternative can be made.

3.2 *Design criteria*

The design criteria to compare the bunker installations are based on a study by Satta et al. (2021), who did a SWOT analysis for an LNG bunkering station. This leads to an emphasis on reducing cost, having a reliable installation and minimizing bunker time. Additionally, energy efficiency is important from a sustainability point of view. Finally, safety must be guaranteed. Thus, the design criteria are:

- Cost
- Energy Efficiency
- Safety
- Reliability
- Bunker Time

3.3 *Ordinal analysis*

In order to find the values for p_{ij} in Equation (1), an ordinal analysis of the alternatives is performed. For each function, the alternatives are evaluated for each of the design criteria, and a ranking with respect to these design criteria is allocated. Below, as an example, the analysis for the function connecting (coupling) is shown. The resulting data is shown in Table 2, which corresponds with the matrix in equation 1. The same analysis is done for all functions and alternatives and the full analysis can be found in Hofste (2023). The full results of the analysis can be found in the appendix in Table 5.

Table 2. P_{ij} values of means that fulfil the function connecting (coupling). EE denotes energy efficiency, BT denotes Bunker time.

	Cost	EE	Safety	Reliability	BT
Dry disconnect	3	2	1	1	1
Breakaway	2	2	2	3	1
PERC	1	1	3	2	1

3.3.1 *Cost*

Since the dry disconnect coupling is a simpler version of the breakaway coupling, which in turn is a simpler version of the PERC, capital expenditure (CapEx) also goes down with increasing simplicity. CapEx, as opposed to operational expenditure (OpEx), is assumed to be the dominant cost factor due to the similarity of the use case, and thus requirements for maintenance. However, there is a distinction when comparing breakaway couplings and PERCs to dry disconnect couplings. This distinction is that there is no protocol in place for the dry disconnect coupling in case of excessive motion. This means that damage to the installation will follow such a scenario. In the case of the other two options, there is only a connector bolt replacement required.

3.3.2 *Energy efficiency*

The dry disconnect coupling, as well as the breakaway coupling, do not use any energy. The PERC, however, does use energy to force the coupling to disconnect in the case of an emergency.

3.3.3 *Safety*

The safety of the couplings increases with increased complexity. All three alternatives are designed to be leak free when the coupling is disconnected. Only the breakaway coupling and PERC are designed to release in case of excessive tensile stress on the hose. Finally, only the PERC is designed to disconnect in case of an ESD button being pressed, and can therefore be considered to be safer than the breakaway coupling.

3.3.4 *Reliability*

The fact that a dry disconnect coupling can potentially damage the installation in case of excessive tensile stress on the hose means it can cause serious calamity to the terminal. Because the PERC is in connection to the ESD and manual override systems, it will always disconnect in case of an emergency. This can potentially cause downtime in case of a human error or false alarm. Resetting this coupling can cause extra unwanted downtime, a passive breakaway coupling does not have this problem.

3.3.5 *Bunker time*

There is no discernible distinction in the complexity, possible flow or procedure when comparing these three couplings.

3.4 *Stakeholder survey*

Input from three stakeholders in the Port of Rotterdam is used to create the final input to execute the BWM. Input from these stakeholders is used to generate weights which can be used in Equation (2) to generate alternative scores. The considered stakeholders are: The Gate Terminal, the ACE Terminal and OCI N.V. The Gate Terminal is the LNG hub of Europe and LNG bunkering shares some similarities with ammonia bunkering. The ACE terminal aims to offer ammonia-cracking facilities to deliver hydrogen to their customers. This terminal will receive and store ammonia and thus needs ammonia bunkering facilities. OCI N.V. will expand to The Port of Rotterdam by creating an ammonia import terminal. They aim to create scalable ammonia jetty infrastructure that is very similar to the goal of this study, making them an interesting candidate. These three stakeholders represent the major advancements in the port of Rotterdam concerning the bunkering of ammonia. The survey that is used is based on Rezaei (2015) and can be found in Hofste (2023). The final derivation of the weights is done through the linear model provided by Rezaei (2016) and its results are presented in Table 3.3.5.

3.5 *Sensitivity analysis*

By looking at the influence of small variations in the Best-to-Others and Others-to-Worst vectors, see Equations (4) and (5), it is possible to gain insight into the robustness of the individual configurations (Mazurek et al. 2021). This is done by systematically increasing and decreasing each of the values in the

Table 3. Optimal weights based on the survey results of each respondent.

Stakeholder	Cost	Energy efficiency	Safety	Reliability	Bunker time	Consistency ratio	Accept (y/n)
Gate Terminal	0.182	0.091	0.409	0.273	0.046	0.232	y
ACE Terminal	0.228	0.152	0.400	0.069	0.152	0.200	y
OCI NV	0.052	0.141	0.587	0.141	0.078	0.222	y

survey. Here, attention is also paid if an order violation occurs, meaning a different alternative becomes the best alternative. For this research, the occurrence of an order violation following a small change in the vectors is interpreted as the new best alternative representing the interests of the stakeholder closely.

Table 4. The final configuration for the most viable bunker installation for ammonia.

Function	Final Configuration
Storing	FC flat-bottom
Transporting	Dry Rigid Pipe
Pumping	Submersed Canned Motor Pump
Connecting (hose)	Suspended Hose
Connecting (coupling	Beak Away Coupling
Bunker Location	Onshore
Substance Slip Prevention	Gas Detectors with ESD + Water Spray System over Leak-prone Object
Toxic Cloud Mitigation	Water Curtain
Additional Spill Containment	Bund Dyke + Carbon Nanoparticles
BOG Vapor Pressure Equalisation	BOG Return Hose
Insulation	Vacuum Insulation
Purging of Fuel Lines	Inerting Sequence with Gas Separator

4 RESULTS

From the analysis, it became clear that there are many similarities between the configurations of stakeholders. All three stakeholders selected safety as their most important criterion. However, they chose different least important criteria: bunker time, reliability and cost, for the Gate Terminal, ACE Terminal and OCI NV respectively. The results of the ordinal analysis, combined with the weights given by the stakeholders, resulted in final configurations that differed only in a submerged or dry-installed version of the canned motor pump for the pumping function, and the breakaway coupling or the PERC for the connecting (coupling) function. The configuration for the Gate and ACE terminals includes a breakaway coupling, as opposed to the PERC for OCI. Additionally, the configuration of both the Gate Terminal and OCI resulted in a submersed canned motor pump, whereas the ACE terminal has a dry canned motor pump.

The results of the sensitivity analysis showed that the resulting configurations for each of the stakeholders are robust, only showing sensitivity in the functions pumping and purging of fuel lines. Sensitivity in the pumping function was observed for the Gate Terminal as well as the ACE Terminal. Whereas, sensitivity in the purging of fuel lines function was only observed for the ACE Terminal.

Finally, the results are unified into the final configuration as seen in Table 4. The sensitivity analysis is used to make sure the final configuration optimally represents the overall interest of the stakeholders.

5 DISCUSSION

5.1 *Feasibility of final configuration*

Several system alternatives are infeasible to combine. With the infeasible combinations known, the final configuration can be adapted if necessary. To find the infeasible combinations, the component overview in Table 5 is analysed. The selection of a flat-bottom tank, for instance, does not combine with an offshore design. Due to the size and weight of the storage technology, it is highly impractical to install this on any offshore installation. Therefore, in an offshore design, bullet tanks are used instead of flat-bottom tanks. Secondly, the selection of an onshore design will restrict the usage of floating pipes. It could still be possible for the connecting pipes from the terminal side to the ship side to be floating pipes, but it is not possible for the pipes of a land-based installation to be floating. Thirdly, the selection of an inerting sequence with a gas separator is considered infeasible for an offshore bunker installation due to its size. None of these combinations occur in any of the stakeholder configurations.

The final configuration is influenced by the method that is used to analyse the alternatives. Although a feasible configuration is found, a notable shortcoming of the ordinal analysis is the inability to express quantitative differences between alternatives; the ordinal analysis is only able to express what alternative is better than another, and not able to express how much better that alternative is.

5.2 *Requirement testing*

In this section, it is tested if all requirements, as presented in Section 2.1.1 and 2.1.2, are respected. There are several requirements that are inherently adhered to because the selected components fulfil these requirements. This includes all operational requirements, as well as safety requirements 1, 4, 5, 6, 7, and 8. Requirements 2 and 3 are location-specific requirements that will require further investigation. Due to the location-specific nature of requirements 2 and 3, they will be tested by means of a case study in the Port of Rotterdam. This port is selected because the same port is used for the stakeholder survey.

Both the Gate Terminal and the ACE Terminal, are located on the Maasvlakte. The Maasvlakte II, which is still partially under construction, will offer more surface area for the development of extra port facilities, such as future ammonia terminals (Gasunie 2023). This location offers a large distance from neighbouring cities and dwellings and direct access to the North Sea. It is desirable to have as much

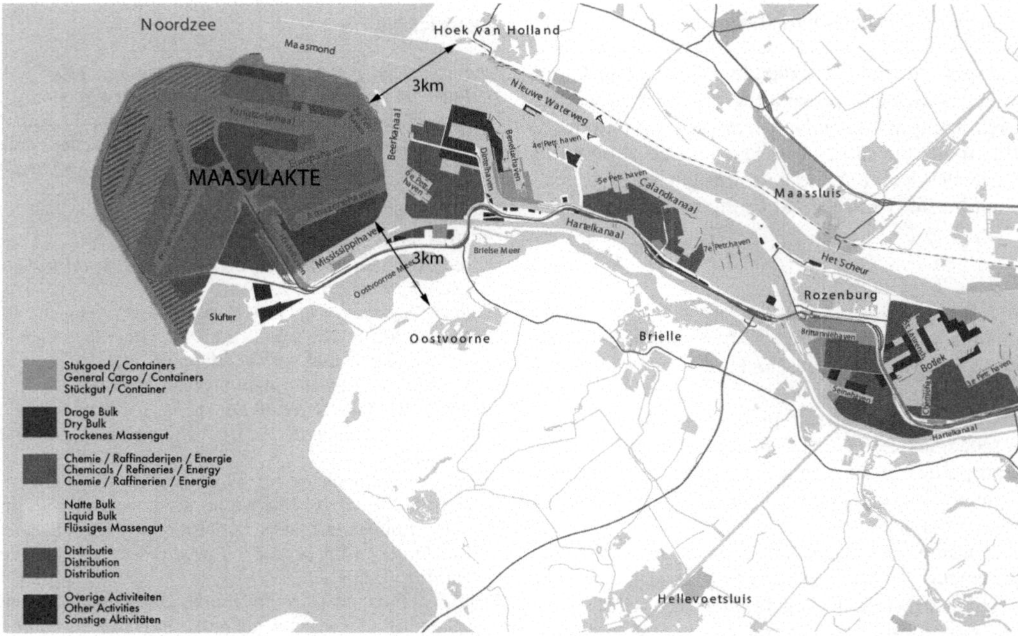

Figure 4. A map of the port of rotterdam, including different designated areas. The striped areas on the left side denote the unfinished Maasvlakte II.

distance as possible between vulnerable objects. Furthermore, the direct access to the North Sea means that larger ships can bunker directly at the proposed bunker installation, thus not restricting the use of the installation to just bunker barges.

The closest inhabited areas to the Maasvlakte II are Hoek van Holland and Oostvoorne. Both can be seen in Figure 4. The nearest dwellings are at a distance of 3 km, measured from the area of the Maasvlakte II. In a study, The Port of Amsterdam (2021) determined that the risk contour for harm to individuals at a specific frequency of 10^{-6}/*year* is 427 meters for a bunker operation with refrigerated ammonia (Wessels 2021). In this contour, there can be no vulnerable objects such as dwellings and residential areas. Furthermore, the focus area as defined by this same study, provides an area with a radius of 2624*m* in which people are insufficiently protected indoors against the consequences of accidents involving hazardous substances, without additional measures. The focus area is not restrictive on development, but indicative for relevant authorities to state what measures are sufficient to protect persons. Since both of these distances are smaller than the closest populated area, the Maasvlakte II is deemed as being sufficient for requirement 2.

The depth of the waterways to the Maasvlakte II from the Noordzee, as can be seen in Figure 4, is between 23*m* NAP to 24*m* NAP (Port of Rotterdam 2023). This means that classes of ships that have a DWT of 300,000 can safely pass into the port since their average draught is around 20m (Haider 2023). Since the DWT of one of the largest ships currently in service, the MSC Irina, is around 240,000 (Marine Insight 2023), it should be able to reach the proposed installation. If this is undesirable due to other influences, it can still be serviced by bunker barges. Subsequently, requirement 3 is satisfied.

5.3 *Applicability to other ports*

A viable configuration for an ammonia bunker installation in the Port of Rotterdam is found, see Table 4. This configuration is based on research presented in this paper combined with input from stakeholders from The Port of Rotterdam. However, broader applicability is suggested due to several factors. The different areas of expertise of the stakeholders yield similar results. Other stakeholders will most likely prioritise safety as well, which means final configurations will not be vastly different. This is further supported by the robustness of the final configuration. Furthermore, due to the expected effectiveness of mitigation strategies, risk contours will shrink substantially. Therefore, the final configuration, and its associated risk, are expected to be viable for most ports that handle sea-going vessels that are not located directly in inhabited areas.

5.4 *Suggestions for future research*

For future work, the research can be continued by replacing the ordinal analysis with a quantitative analysis because the ordinal analysis lacks nuance between the rankings. The second recommendation is to increase the amount of stakeholders that are surveyed. As of now, the only party that is surveyed are the terminal operators. Expanding this to legislators or other bodies can remove the bias that surveying a single group with similar interests might introduce. Furthermore, the survey can be expanded to multiple ports, which could create more insight into the general applicability of the final configuration.

6 CONCLUSION

Ammonia as a shipping fuel is not readily implementable as of now. Little research has been published on the bunkering of ammonia and it is unclear what system components should be used and what mitigation strategies are effective for the different hazardous properties of ammonia. Therefore, this research compares different system layouts of an ammonia bunker installation.

According to the system engineering approach, requirements and functions are proposed for this system, categorised for operational and safety purposes. For all functions, a wide range of technologies are considered as possible alternatives. These alternatives are compared quantitatively using an ordinal analysis. For every function, its alternatives are compared on cost, energy efficiency, safety, reliability and bunker time. Stakeholder input from the Port of Rotterdam is used to quantify the weights of the different criteria, which are combined with the results from the ordinal analysis to find a suitable system configuration using the Best Worst Method.

The stakeholders collectively prioritize safety. Consequently, the optimal configurations for the different stakeholders varied only in a submerged or dry-installed version of the canned motor pump for the pumping function, and the breakaway coupling or the PERC for the connecting (coupling) function. This made it possible to unify these differences into a final configuration (Tab. 4) through a sensitivity analysis. Although only stakeholders in the port of Rotterdam are interviewed, it can be concluded that the configuration is applicable to multiple ports due to the general prioritization of safety and similarities in the final system configuration for different stakeholder input.

This study proposes a framework to evaluate current and upcoming technologies for bunker installations. Since ammonia is anticipated as a future marine fuel, knowledge and system designs are needed to bunker this fuel. Terminals and port authorities can use this framework to select and engineer suitable bunkering systems.

ACKNOWLEDGEMENTS

The research presented in this paper is part of the EU-funded Horizon 2020 project MAGPIE (sMArt Green Ports as Integrated Efficient multimodal hubs, contract No.: 101036594).

REFERENCES

Thomas, George and Parks, George, year = 2006, n..U. Potential Roles of Ammonia in a Hydrogen Economy.

Afif, A., Radenahmad, N., Cheok, Q., Shams, S., Kim, J. H., & Azad, A.K. 2016. Ammonia-fed fuel cells: a comprehensive review. *Renewable and Sustainable Energy Reviews 60*, 822–835.

Aronietis, R., Sys, C., van Hassel, E., & Vanelslander, T. 2016. Forecasting port-level demand for LNG as a ship fuel: the case of the port of Antwerp. *Journal of Shipping and Trade 1*(1), 2.

Atilhan, S., Park, S., El-Halwagi, M.M., Atilhan, M., Moore, M., & Nielsen, R.B. 2021. Green hydrogen as an alternative fuel for the shipping industry. *Current Opinion in Chemical Engineering 31*, 100668.

Bartels, J.R. 2008. *A feasibility study of implementing an Ammonia Economy*. Master of Science, Iowa State University.

Barzegar Gerdroodbary, M., Mokhtari, M., Bishehsari, S., & Fallah, K. 2016. Mitigation of Ammonia Dispersion with Mesh Barrier under Various Atmospheric Stability Conditions. *Asian Journal of Atmospheric Environment 10*(3), 125–136.

Burel, F., Taccani, R., & Zuliani, N. 2013. Improving sustainability of maritime transport through utilization of Liquefied Natural Gas (LNG) for propulsion. *Energy 57*, 412–420.

Centers for Disease Control 2021. Ammonia Solution, Ammonia, Anhydrous: Lung Damaging Agent.

Chatterjee, S., Parsapur, R.K., & Huang, K.W. 2021, 12). Limitations of ammonia as a hydrogen energy carrier for the transportation sector. *ACS Energy Letters 6*, 4390–4394.

Cheliotis, M., Boulougouris, E., Trivyza, N.L., Theotokatos, G., Livanos, G., Mantalos, G., Stubos, A., Stamatakis, E., & Venetsanos, A. 2021. Review on the Safe Use of Ammonia Fuel Cells in the Maritime Industry. *Energies 14*(11), 3023.

Curtis, J.J. 2007. Vacuum-Insulated Pipe vs. Conventional Foam-Insulated Pipe. In *Offshore Technology Conference*, Houston, Texas, pp. OTC 18701.

Dandrieux, A., Dusserre, G., Ollivier, J., & Fournet, H. 2001. Effectiveness of water curtains to protect firemen in case of an accidental release of ammonia: comparison of the effectiveness for two different release rates of ammonia. *Journal of Loss Prevention in the Process Industries 14*(5), 349–355.

DNV-GL 2019. Comparison of Alternative Marine Fuels. Technical report, Norway.

Duong, P.A., Ryu, B.R., Song, M.K., Nguyen, H.V., Nam, D., & Kang, H. 2023. Safety Assessment of the Ammonia Bunkering Process in the Maritime Sector: A Review. *Energies 16*(10), 4019.

Ekinci, M.S. & Akman, A. 2022. The effect of storage temperature on threat zone caused by an ammonia release from a storage tank. *Hittite Journal of Science and Engineering 9*(2), 125–132.

European Maritime Safety Agency 2022. Potential of Ammonia as Fuel in Shipping. Technical report.

Feldle, D.G. 2022. Vertical submersible pumps in the chemical and petrochemical industry. *TECHNICAL REPORT.*

Gasunie 2023. ACE Terminal. URL: https://www.gasunie.nl/en/projects/ace-terminal.

Gate Terminal 2022. LNG Carrier Master's Marine Services Manual. Technical report.

Green Shipping Programme 2021. Ammonia as a Marine Fuel. Safety Handbook.

Haider, S. 2023. Various bulk carrier sizes and employment guide.

Hayes Pump, inc 2023. Canned Motor vs. Magnetic Drive Pumps. URL: https://blog.hayespump.com/blog/canned-motor-vs-magnetic-drive.

Hermetic 2022. Product-information TCN TCAM E-Line.

Hofste, P.C. 2022. Bunkering of Future fuels. pp. 27. Literature review.

Hofste, P.C. 2023. Design of an Ammonia Bunker Installation for Sea-going Vessels - A Best Worst Method approach. Graduation thesis.

IMO 2020. Fourth Greenhouse Gas Study 2020. Technical report.

IMO 2022. IMO's work to cut GHG emissions from ships. URL: https://www.imo.org/en/MediaCentre/HotTopics/Pages/Cutting-GHG-emissions.aspx.

Jamaludin, Z., Rollings-Scattergood, S., Lutes, K., & Vaneeckhaute, C. 2018. Evaluation of sustainable scrubbing agents for ammonia recovery from anaerobic digestate. *Bioresource Technology 270*, 596–602.

Karatu˘g, C., Ceylan, B.O., Ejder, E., & Arslano˘glu, Y. 2023. *Investigation and Examination of LNG, Methanol, and Ammonia Usage on Marine Vessels*, pp. 65–85. Springer.

Ksenia, Z. 2021. Ammonia Bunkering Of Passenger Vessel Concept Quantitative Risk Assessment. Technical report, DNV.

Maersk Mc-Kinney Moller Center for Zero Carbon Shipping 2021. Industry transition strategy. Technical report.

Magnatex Pumps 2020. Mechanical Seal vs Sealless Centrifugal Pumps. URL: https://magnatexpumps.com/blog/mechanical-seal-vs-sealless-centrifugal-pumps/.

Mandra, J.O. 2022. DNV: Hydrogen likely to have limited uptake in deep-sea shipping.

Marine Insight 2023. Ship CMA CGM JACQUES SAADE - Vessel details - AIS Marine Traffic. URL: https://www.marinetraffic.com/.

Marine Service Noord 2023. Cryogenic bunker connection. URL: https://marine-service-noord.com/en/products/alternative-fuels-and-technologies/lng/cryogenic-bunker-connection/.

Marine Traffic 2023. Ship MSC IRINA Vessel details.

Mazurek, J., Perzina, R., Ram´ık, J., & Bartl, D. 2021. A Numerical Comparison of the Sensitivity of the Geometric Mean Method, Eigenvalue Method, and Best–Worst Method. *Mathematics 9*(5), 554.

Minnesota Department Of Agriculture 2022. Physical & Chemical Properties of Anhydrous Ammonia. URL: https://www.mda.state.mn.us/physical-chemical-properties-anhydrous-ammonia.

Namboothiri, N.V. & Soman, A. 2018. Consequence assessment of anhydrous ammonia release using CFD-probit analysis. *Process Safety Progress 37*(4), 525–534.

Olmer, N., Comer, B., Roy, B., Mao, X., & Rutherford, D. 2018. Greenhouse gas emissions from global shipping, 2013-2015. Technical report. International Council on Clean Transportation.

Per˘ci´c, M., Vladimir, N., Jovanovi´c, I., & Kori˘can, M. 2022. Application of fuel cells with zero-carbon fuels in short-sea shipping. *Applied Energy 309*, 118463.

Port of Rotterdam 2023. Maritime Chart Server.

Rath, S. & Krol, M. 2013. *Comparative Risk Assessment for Different LNG-Storage Tank Concepts*. Number 31 in Chemical engineering transactions. Milano.

Rezaei, J. 2015. Best-worst multi-criteria decision-making method. *Omega 53*, 49–57.

Rezaei, J. 2016. Best-worst multi-criteria decision-making method: Some properties and a linear model. *Omega 64*, 126–130.

Satta, G., Parola, F., Vitellaro, F., & Morchio, G. 2021. LNG Bunkering Technologies in Ports: An Empirical Application of the SWOT Analysis. *KMI International Journal of Maritime Affairs and Fisheries 13*(1), 1–21.

Ship&Offshore 2021. Ammonia fuel cells for deep-sea shipping. *FUTURE FUELS & ENERGY SOLUTIONS.*

Singh, M. & Pant, M. 2022. A review of selected weighing methods in MCDM with a case study. *International Journal of Systems Assurance Engineering and Management 12*(1), 126–144.

Sveistrup, D.M., Krantz, R., Mouftier, L., & Christiansen, E.S. 2021. Ammonia as a shipping fuel. Technical report. Global Maritime Forum.

Thomas, G. & Parks, G. 2006. Potential Roles of Ammonia in a Hydrogen Economy. USA DOE positional paper.

Va-Q-tec 2023. Vacuum insulation panels as a sustainable super insulating material. URL: https://va-q-tec.com/en/news-en/vacuum-insulation-panels-as-a-sustainable-super-insulating-material/.

Venkat Pattabathula, Raghava Nayak, & Don Timbres 2023. Ammonia Storage Tanks - AmmoniaKnowHow. URL: https://ammoniaknowhow.com/ammonia-storage-tanks/.

Wessels, N. 2021. External safety study - bunkering of alternative marine fuel for seagoing vessels. Technical report. DNV-GL.

Yadav, A. & Jeong, B. 2022. Safety evaluation of using ammonia as marine fuel by analysing gas dispersion in a ship engine room using CFD. *Journal of International Maritime Safety, Environmental Affairs, and Shipping 6* (2-3), 99–116.

APPENDIX

Table 5 shows the complete overview of all discovered solutions to the proposed functions, as well as the complete results of the ordinal analysis

Table 5. Overview of the functions and their alternative and the results of the ordinal analysis.

Function	Alternative	Cost	Energy Efficiency	Safety	Reliability	Bunker Time
Storing	Single Containment Flat bottom	6	2	1	1	1
	Single Containment Bullet	3	1	2	2	1
	Double Containment Flat Bottom	5	2	4	1	1
	Double Containment Bullet	2	1	3	2	1
	Full Containment Flat Bottom	4	4	6	3	1
	Full Containment Bullet	1	3	5	3	1
Transporting	Dry Rigid Pipe	3	3	3	3	3
	Dry Flexible Pipe	2	2	2	2	2
	Floating Flexible Pipe	1	1	1	1	1
Pumping	Dry Mechanically Sealed Pump	4	1	1	1	4
	Submersed Mechanically Sealed Pump	1	2	2	2	3
	Dry Mag-drive Pump	6	3	3	3	4
	Submersed Mag-drive Pump	3	5	4	4	3
	Dry Canned Motor Pump	5	4	5	5	2
	Submersed Canned Motor Pump	2	6	6	6	1
Connecting (hose)	Suspended Hose	2	2	2	2	2
	Floating Hose	1	1	1	1	1
Connecting (coupling)	Dry Disconnect Coupling	3	2	1	1	1
	Beak Away Coupling	2	2	2	3	1
	Powered Emergency Release Coupling (PERC)	1	1	3	2	1
Bunker Location	Onshore	3	3	3	2	1
	Nearshore	2	2	2	3	3
	Offshore	1	1	1	1	2
Substance Slip Prevention	Gas Detectors with ESD	2	2	2	1	1
	Water Spray System over Leak-prone Objects	3	1	1	2	1
	Detectors + Water Spray	1	1	3	2	1
Toxic Cloud Mitigation	Mesh Barrier	5	5	1	5	1
	Water Curtain	4	4	4	4	1
	Overpressurisation	3	2	2	1	1
	Mesh Barrier + Overpressurisation	2	3	3	2	1
	Water Curtain + Overpressurisation	1	1	5	3	1
Additional Spill Containment	None	3	0	1	1	1
	Bund Dyke	2	0	2	2	1
	Bund Dyke + Carbon Nano-particles	1	0	3	3	1
BOG Vapor Pressure Equalisation	BOG Return Hose	3	3	3	3	3
	Relieve Gas in Buffer Tank	2	1	2	2	1
	Vent with Scrubber	1	2	1	1	2
Insulation	Mechanical Insulation	2	1	1	1	1
	Vacuum Insulation	1	2	2	2	2
Purging of Fuel Lines	Inerting Sequence with Scrubber	2	2	1	0	1
	Inerting Sequence with Gas Separator	1	1	2	1	1

Advances in Maritime Technology and Engineering – Guedes Soares & Santos (Eds)
© 2024 The Author(s), ISBN 978-1-032-83107-7

Modelling and simulation of a hydrogen-powered passenger catamaran ferry

L. Micoli, R. Russo, T. Coppola, V. Sorrentino, F. De Luca & C. Pensa
Department of Industrial Engineering, University of Naples Federico II, Naples, Italy

ABSTRACT: The work focuses on the modelling of a zero-emission, high-speed catamaran ferry employing a full-electric propulsion system. It addresses the global emphasis on full-electric vessels to align with IMO regulations regarding ship emissions and energy efficiency improvement. Using the AVL Cruise-M software, the research verifies the implementation of an onboard fuel cell power generating system integrated with the propulsion plant, aiming to assess dynamic performance under load variation. The catamaran is 30 meters long and 10 meters wide with a cruise speed of 20 knots. The model includes a 1600 kWe PEM system, fuelled by hydrogen stored in type IV cylinder vessels at 700 bars, and a 2-kWh capacity battery pack. Simulation results show that the battery effectively supports the PEM during the manoeuvring phase, enhancing overall performance and energy economy. The study emphasises that new hydrogen storage systems can make easier the onboard installation and increase the ship's autonomy.

1 INTRODUCTION

In the contemporary landscape of maritime transportation, sustainability has become a paramount concern, prompting the exploration of alternative fuels to mitigate environmental impacts. The imperative to reduce carbon emissions and dependence on traditional fuels has driven a profound shift towards greener solutions. Several alternative fuels, including hydrogen, biofuels, ammonia, and methanol, have been proposed (IMO's Work to Cut GHG Emissions from Ships, IMO.).

Hydrogen (H_2) emerges as a promising alternative, offering the potential to revolutionize the maritime industry. Studies have been ongoing since the beginning of the 21st century, since before the IMO strategy was formulated. Many works can be found in the literature demonstrating the feasibility of the application of H_2 on board ships. As summarized by Klebanoff et al.(2021), examinations of the applicability of H_2 in Fuel Cells (FC) and Internal Combustion Engines (ICEs) have been conducted for shore power, propulsion, and auxiliary power. In 2016, van Biert et al. (2016)reviewed different types of FC for their applicability to vessels and assessed various methods of storing or generating H_2 onboard. Pratt and Klebanoff examined the feasibility of a high-speed ferry called the SF-BREEZE(2017).

As an environmentally friendly fuel, hydrogen holds significant promise for addressing the challenges associated with conventional marine propulsion systems. However, its effective integration into transportation applications demands a comprehensive understanding of the associated issues and complexities.

Firstly, challenges are related to the storage on board ships. Unlike traditional fuels, H_2 has a low volumetric energy density, necessitating advanced storage solutions to accommodate sufficient energy for extended maritime journeys. Storage issues arise from the need to maintain H_2 in a compressed or liquefied state, introducing complexities related to space, weight, and safety considerations, particularly on board passenger vessels. Moreover, the permeability of H_2 through materials poses additional challenges for the design of robust storage systems (Bosu & Rajamohan, 2023).

The storage plays a pivotal role in energy systems, particularly in scenarios involving widespread and large-scale utilization of H_2. Several techniques have been developed to store H_2, including compression in gas cylinders, storage in metal hydride alloys, liquefaction in cryogenic tanks, adsorption onto large specific surface area materials, and using hydrogen carriers (i.e. ammonia and methanol) (Balsamo et al., 2023; Russo et al., 2022; Ahluwalia et al., 2010; Eberle et al., 2009; Schlapbach & Züttel, 2001)). Within the framework of the hydrogen economy, applications for H_2 storage can be broadly categorized into two groups which are mainly related to the gravimetric and volumetric energy densities (Moradi & Groth, 2019, Lai et al., 2015): stationary and mobile applications. Various methods have been compared, considering factors such as density, pressure, temperature, and cost. Among these methods, the compressed gas approach is employed in onboard H_2 storage for passenger cars utilizing H_2 storage tanks with pressures ranging from 350 to 700 bar. Whiston et al.

DOI: 10.1201/9781003508779-3

(2021) anticipate that compressed H_2 storage methods will dominate, constituting 67% of all H_2 storage methods by 2035 and increasing to 56% by 2050.

The onboard compressed gas H_2 system designed for FC electric vehicles (FCEVs) is divided into two main subsystems: the hydrogen storage tank and the BoP (Balance of Plant). The BoP encompasses essential elements such as a fill port, regulator, valves, sensors, etc. H_2 storage tanks are categorized into five types (David, 2005; *Research Priority Workshop on Hydrogen Safety*, n.d.):

- Type I: constructed entirely of metal.
- Type II: metal structure with a composite material wrapped around the cylindrical section of the tank. The internal pressure load is distributed between the liner and the composite layers.
- Type III: metal liner fully wrapped with carbon or glass fiber, with the composite bearing the entire load.
- Type IV: polymer liner fully wrapped with carbon or glass fiber.
- Type V: Linerless composite vessels with the entire tank composed of carbon or glass fiber (Shin & Ha, 2023).

Furthermore, the adoption of hybrid powertrains, combining H_2 technologies with other energy sources and storage systems, presents a strategic approach to enhance efficiency and overcome certain limitations of standalone H_2 systems. This integration fosters a more versatile and reliable power solution for maritime applications. Investigating these hybrid configurations contributes to a holistic exploration of sustainable propulsion strategies for ships.

In this context, the research outlined in this paper aims to advance the understanding of hydrogen's role in maritime sustainability, addressing the critical issues that impede its widespread adoption. Specifically, the study focuses on modeling a fully electric hybrid propulsion system for a passenger catamaran ferry based on the integration of a Proton Exchange Membrane FC (PEM) and a battery system, using hydrogen as a power source. Major attention is given to the utilization of innovative type IV tanks for hydrogen storage at 700 bars on board and how this affects the ship's autonomy.

2 CASE STUDY

2.1 *The ship*

The ship platform considered in the present work is the high-speed passenger ferry proposed in the feasibility study of Coppola et al., (2023). The passenger ferry is a full-electric catamaran equipped with a battery assisted FC, with an overall length of 30 m, a beam of 10 m, and is designed to carry up to 220 passengers at the cruise speed of 20 kn. Table 1 recalls the main characteristics of the considered passenger ferry.

Table 1. Main characteristics of the catamaran ferry.

L_{OA}	m	30
B_{OA}	m	10
L_{WL}	m	29.60
B	m	10
T	m	1.41
D	m	3.90
L_{CG}	m	12.63
V_{CG}	m	3.94
Displacement	t	125

The ferry provides commuter service along the Amalfi Coast, conducting daily round trips from Salerno to Capri Island with intermediate stops at the ports of Cetara, Maiori, Amalfi, and Positano. The duration of a single trip is approximately 3.28 hours, considering a laytime of 15 minutes at ports and an average maneuvering time of 3 minutes. An illustration of the sailing route is shown in Figure 1.

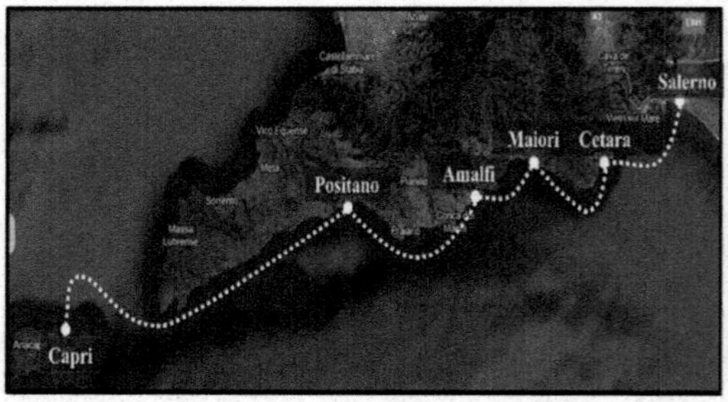

Figure 1. Sailing route of the catamaran ferry.

2.2 *The propulsion system*

The propulsion plant is a fully electric system consisting of a couple of electric propulsion motors (EMs), one for each demi hull, which move the propellers through the gearboxes and shafts. Such a plant is powered by a FC generating system integrated with a battery system (BS).

The FC system is made of PEM modules, whose specifics are similar to those of Ballard products for maritime applications (Ballard, 2024); these are reported in Table 2.

Table 2. PEM system main specifics.

Rated power	kW	200
Min. power	kW	55
Peak efficiency	%	60
Operating voltage	V	350 – 720 DC
Weight	kg	1000
Dimensions	mm	1209 x 747 x 2195

A PEM basically converts the chemical energy of pure H_2 directly into electricity and heat, according to well-known reactions described in the literature (Yuan & Wang, 2008). Major advantages of a PEM include a high electrical efficiency (up to 65 %), a low operating temperature (about 80 °C), a fast start-up (compared to other FC technologies) and a high-power density.

We assumed to store H_2 in type IV cylindric vessels at 700 bars, considering the specifics of the product from Plastic Omnium company (France), which are reported in Table 3 (Plastic Omnium, 2024).

Table 3. Main characteristics of the H_2 storage system.

Vessel type	-	IV
Operating pressure	bar	700
Material	-	Carbon fibre
Approx. weight	kg	400
Dimensions	mm	2700 x 750
H_2 Capacity	kg	32

The BS is intended to assist the PEM in providing energy during transient operations such as load demand variations, acceleration, and maneuvering phases (Truong et al., 2020). In this study, we considered the performances of the Li-ion based battery pack provided by Corvus Energy company (Corvus Energy, 2024), which is designed for lightweight applications having high power density (168 Wh/kg and 218 kWh/m^3).

3 SIMULATION MODEL

The power plant propulsion system has been modelled using the AVL Cruise-M software. AVL Cruise-M is a multidisciplinary vehicle system simulation tool for mobility concept analysis, subsystem design and layout, and virtual component integration (AVL, 2024). The implemented models use an object-oriented approach, capturing the components of each subsystem and their connections. The software allows the development of customized models with user-defined functions programmed in C.

The subsequent section provides a short description of each model subsystems and units which are shown in Figure 2. Blocks of diagram in Figure 2 can include various units which could be structured as subsystems in a hierarchy.

Table 4. Ship performances data and efficiencies.

Speed	kn	20
Total resistance	kN	74
Effective power	kW	761
1-t	-	0.985
1-w	-	0.965
ηH	-	1.021
$\eta 0$	-	0.68
ηR	-	1.00
Delivered power per axle	kW	545

The *propeller subsystem* (PS) contains the ship's resistance and the kinematics mathematical models. Details and hypothesis adopted for these models have been described in detail in Sorrentino et al., (in press.). The performances in terms of total resistance and effective power at the required speed have been assessed by means of full-scale URANS simulations, taking into account the full load condition of 125 t with the weight distribution calculated in Coppola et al., (2023). In Table 4, the relevant performance data and efficiencies calculated at the design speed of 20 kn are reported; further details are provided in Coppola et al., (2023). Based on the required effective power, two 5-bladed Wageningen B propellers have been selected (Figure 3).

The *electric motor subsystem* (*E-Motor*, EMS) is characterized by torque and efficiency curves as

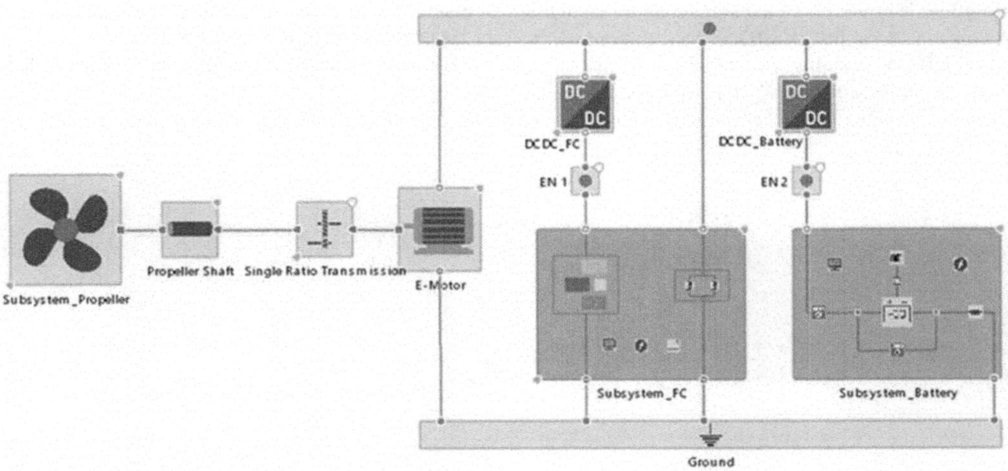

Figure 2. Block diagram of the full-electric powertrain for the catamaran demi-hull.

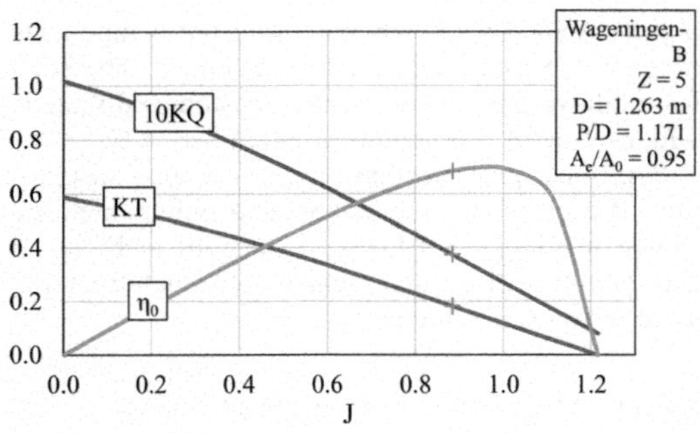

Figure 3. Propeller open water diagram of the selected propeller.

a function of voltage and speed, having a maximum efficiency value of 96% at 1600 rpm and 500 V, which corresponds to a torque of 4639 Nm and the mechanical power of 777 kW. The EMS drives the PS through a *transmission* and a *shaft* unit and is connected to the main switchboard of the ship at 500 V.

The FC model separately accounts for the components of the PEM as well as all auxiliary devices and equipment (BoP), such as cathode air supply, water and thermal management, power conditioning, etc. The model includes the following subsystems:

The *FC stacks subsystem* model features a 1D resolution along the gas channel flow and a reduced dimensionality electrochemical model. The model comprises multiple cells with variable geometry, along with other physical and chemical specifics that users can customize or select from the software's library. Users can also implement a single-cell performance (polarization curve). The FC stack model simulates the relevant thermodynamic, electrochemical, and transport processes occurring within the FC as a function of the components' material chosen.

The *anode subsystem* (AS) models the H_2 flow at the anode FC section including a recirculation branch of unused humid hydrogen that is fed back through a Venturi injector. Excess water from anode gas outflow is removed by a water separator to maintain constant humidity. A PID is used to control the Venturi injector opening and regulate the pressure. The AS contains also the H_2 tank subsystem which is made of a tank plenum supplemented with a charging valve restriction, a discharging valve restriction, a relief device valve restriction, and a relief device control function (Figure 4).

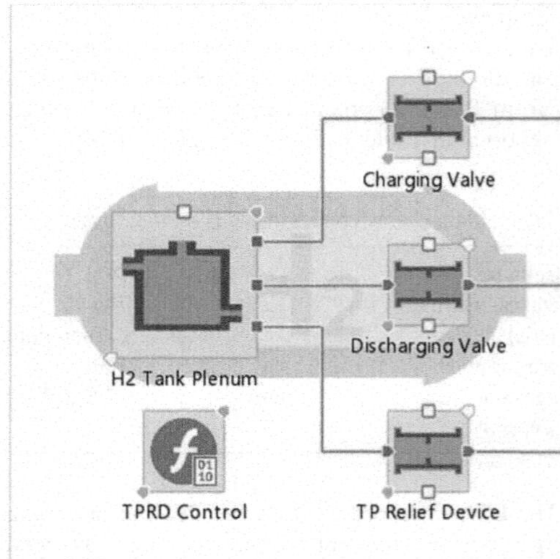

Figure 4. Block diagram of the H_2 tank subsystem.

The *cathode subsystem* consists of a compressor, which can provide up to ~0,5 kg/s of external air pumped up to 4 bar. The compressed air is cooled down and then circulated through the humidifier component before reaching the FC stack to provide the desired humidity to the membrane. Pressure and airflow are regulated via two PID controllers that control the compressor and the backpressure valve.

The *thermal management subsystem* (TMS) simulates a cooling circuit that uses environmental water to exchange heat. Circulation in the TMS is driven by a pump that is regulated considering the heat flow generated by the FC system with a correction provided by a PID to maintain the FC temperature at the constant value of 80 °C. We assumed that the FC has already reached the operating temperature, allowing a faster response to transient loads.

A scheme of the FC model's subsystems is shown in Figure 5.

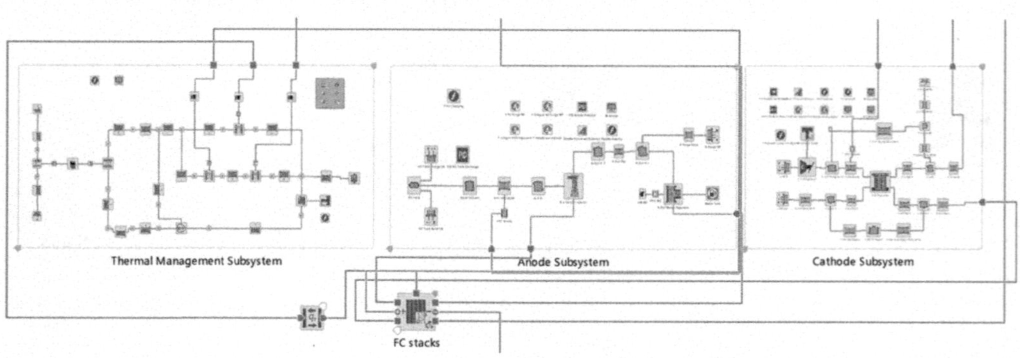

Figure 5. Block diagram of the FC subsystem.

The *battery subsystem* (BS) consists of a controlled voltage source and an Ohmic resistance used to describe the instantaneous voltage response to a current input. It includes an advanced model to predict the transient voltage response to a dynamic current load, eventually coupled with a thermal model to predict the transient thermal behavior of the battery. The BS is made of battery packs consisting of multiple cells connected in series and parallel with a battery management system (BMS) for peak and continuous power. It serves as recuperation storage, as a buffer when the FC system is active and as main storage when the power request is below the activation threshold of the FC system.

The battery system has to be specifically designed according to the case study ship and operating routing profile, for instance, varying the initial State of Charge (SoC), the operating voltage, and the single-cell configuration (series-parallel). Open circuit voltage and ohmic resistance variations as a function of the SoC are implemented, assuming common trends for LFP ($LiFePO_4$) batteries (Corvus Energy, 2024). During normal operation, the battery is charged towards 90% of SoC. If power demand rises too quickly or goes above the maximum rated power of the FC subsystem, the battery covers for the remainder.

As the EM, FC stacks, and BS operate within different voltage ranges, two *DC-DC converter* units are considered to boost the voltages up to the EM side level. Both DC-DC converters have an efficiency of 93%.

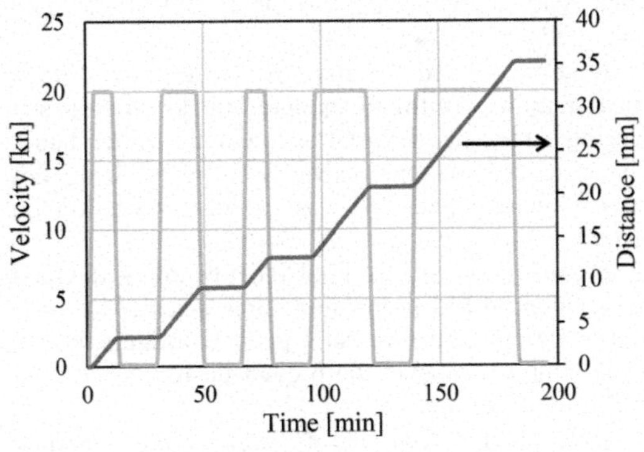

Figure 6. Ship velocity and travelled distance for the route from Salerno to Capri Island.

The performance of the powertrain system will be analyzed, considering the operating routing profile shown in Figure 6. This profile assumes a constant cruise speed of 20 knots and includes a maneuvering phase (as acceleration/deceleration) with an almost linear trend. Figure 6 also displays the predicted distances of the route.

The power during simulation is determined by the ship's velocity, which serves as the model's input.

It should be emphasized that in the catamaran, there are two separate and identically powered generators, each accounting for half of the overall power needed by the ship. These generators are located in each demi-hull and equipped with a propeller powertrain system. Considering this, simulation tests will be conducted for a single powertrain, and the results will be presented accordingly.

Moreover, since every element in the software's models has been validated and verified by experts and other authors in the literature (Liu et al., 2023; Salek et al., 2023; Wancura et al., 2023), experimental validation of the model in the present work may not be strictly necessary. The primary objective of this investigation is to determine the power requirements of each component and to predict the preliminary performance of the powertrain system and fuel consumption.

4 RESULTS

Preliminary tests were conducted to optimize the power size of each component in the powertrain system, including the configuration of both the FC and BS (number of stacks and arrangement). Emphasis was placed on optimizing H_2 consumption and related storage system, as well as the installed battery capacity, since these factors could strongly affect the ship's design due to excessive embarked weights and volumes.

The required power generating system consisted of a FC made of 8 stacks with a maximum power of 1600 kW, having a weight of 8 t and a volume of about 16 m^3, while the BS had an installed capacity of about 2 kWh, a weight of 12 kg, and volume of 10 litres.

Figure 7 (A) shows the FC and BS power variation responses to the EM power request, according to the operating routing profile of Figure 6. Essentially, the electric system operates as follows: when the electrical power demand applied to the electrical network in the powertrain by the EM exceeds the electric energy supplied by the FC (or its activation threshold value), the power demand is met by the BS. Conversely, when the power demand is lower than the FC supply, the surplus energy will charge the BS up to a SoC of 90%. Subsequently, the FC will reduce power production. According to Figure 7 (B), the BS is active during the maneuvering phase, especially during the acceleration from 0 to 20 knots. Once the EM surpasses the peak power demand, the FC charges the BS to restore the battery's SoC (negative power values in Figure 7 (B)). Consequently, the FC generates approximately 750 kW for about 1.6 minutes after each acceleration to 20 knots, for each powertrain. This charging persists even if the EM power demand decreases, eventually reaching a steady value of 714.5 kW.

Figure 8 shows the FC system H_2 consumption and the accumulated H_2 per route (one way). The average H_2 consumption at the cruise speed (20 kn) is estimated to be about 577 g/min, corresponding to

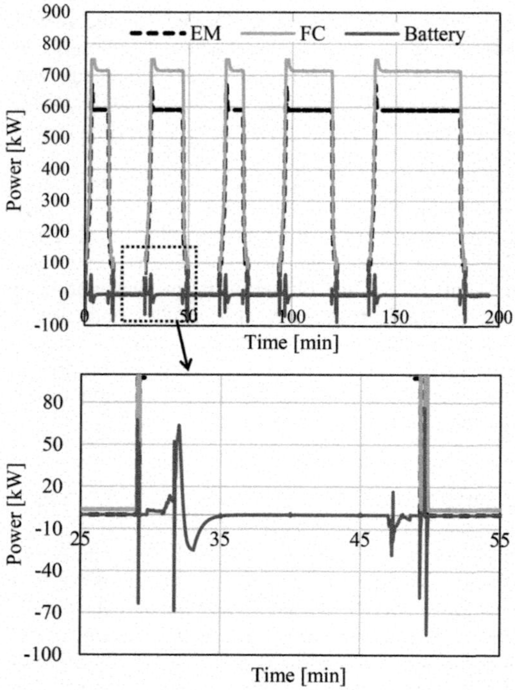

Figure 7. (A) Power variation over the routing profile for the EM, FC, and battery; (B) A detail of the battery behavior.

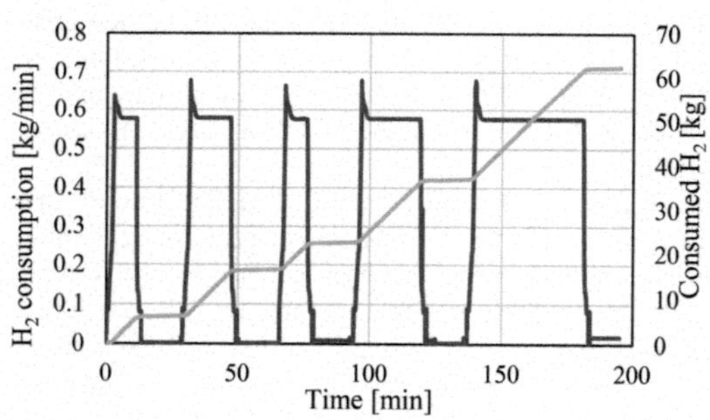

Figure 8. H_2 consumption and consumed H_2 amount over the routing profile.

the FC's efficiency of 59.5%. Despite this, during load variations and very low power supply, the FC efficiency could briefly reach values as low as about 20%.

The overall H_2 required for a one-way route is about 124.8 kg (3.5 kg_{H2}/nm). Consequently, assuming a round trip, the total amount of H_2 to be stored on board is about 250 kg, necessitating the installation of 8 cylindrical tanks. These tanks occupy a net volume of 9.5 m^3 and weigh 3.2 t. It must be noted that an additional cylindric tank should be considered as a precautional margin. Figure 9 provides a view of the catamaran ferry with a focus on the on-board location of the H_2 storage tanks, in accordance with recent regulations (Coppola et al., 2023).

Figure 9. View of the passenger ferry.

5 CONCLUSION

In this study, a fully electric powertrain was designed and analyzed for a high-speed passenger catamaran ferry. For this purpose, the powertrain was mathematically modeled using AVL Cruise-M software for a specific route (Salerno-Capri Island) and a cruise speed of 20 knots. Simulation tests were conducted to determine the power size of the main components of the powertrain, which essentially include the Fuel Cell (FC) system (1600 kW), the Battery System (BS) (2 kWh), the Electric Motor (EM) (2 x 777 kW), and to estimate hydrogen consumption (3.5 kg_{H2}/nm) and the amount of hydrogen to be stored on board for a round trip (250 kg). It was assumed that the hydrogen would be stored in nine type IV cylindrical vessels at 700 bar, located externally on the upper deck.

Results showed the need for an energy storage system alongside the FC generating system to assist during strong load variations. On the other hand, the required battery capacity is very low, not involving any particular issues for installation on board.

Future work will be finalized towards a detailed assessment of EM propulsion efficiencies and weight optimization (structures and general arrangements). This aims to enhance the project design process by reducing uncertainties.

Hydrogen-powered ferries offer green shipping solutions, promising industry revolution. Continued innovation is crucial for environmental impact reduction. Achieving maritime carbon neutrality requires ongoing efforts, but progress fuels hope for a better future.

REFERENCES

AVL. (2024). *AVL*. https://www.avl.com/en

Ballard. (2024). *Ballard*. https://www.ballard.com/

Bosu, S., & Rajamohan, N. (2023). Recent advancements in hydrogen storage - Comparative review on methods, operating conditions and challenges. *International Journal of Hydrogen Energy*. https://doi.org/10.1016/j.ijhydene.2023.01.344

Coppola, T., De Luca, F., Gambardella, S., Mancini, S., Micoli, L., Pensa, C., & Sorrentino, V. (2023). Feasibility Study of a Zero-Emission Passenger Catamaran Ferry Operating in Italian Coastal Island. *Progress in Marine Science and Technology*, *7*, 172–180. https://doi.org/10.3233/PMST230023

Corvus Energy. (2024). *Corvus Energy*. https://corvusenergy.com/?gclid=CjwKCAiAtt2tBhBDEiwALZuhALAEzJ3wrY2NFg1fa7W10uIuTfV82PnQnec8p72KFgzh-IgB_dXMfBoCKOEQAvD_BwE

David, E. (2005). An overview of advanced materials for hydrogen storage. *Journal of Materials Processing Technology, 162–163*(SPEC. ISS.), 169–177. https://doi.org/10.1016/j.jmatprotec.2005.02.027

IMO's work to cut GHG emissions from ships. (n.d.). Retrieved February 14, 2023, from https://www.imo.org/en/MediaCentre/HotTopics/Pages/Cutting-GHG-emissions.aspx

Klebanoff, L. E., Caughlan, S. A. M., Madsen, R. T., Conard, C. J., Leach, T. S., & Appelgate, T. B. (2021). Comparative study of a hybrid research vessel utilizing batteries or hydrogen fuel cells. *International Journal of Hydrogen Energy*, *46*(76), 38051–38072. https://doi.org/10.1016/j.ijhydene.2021.09.047

Liu, M., Li, Y., Xu, L., Wang, Y., & Zhao, J. (2023). General modeling and energy management optimization for the fuel cell electric tractor with mechanical shunt type. *Computers and Electronics in Agriculture*, *213*. https://doi.org/10.1016/j.compag.2023.108178

Plastic Omnium. (2024). *Plastic Omnium*. https://www.plasticomnium.com/en/

Pratt, J. (2017). *Feasibility of the SF-BREEZE: a Zero-Emission, Hydrogen Fuel Cell, High-Speed Passenger Ferry.*

Research Priority Workshop on Hydrogen Safety. (n.d.). https://doi.org/10.2760/77730

Salek, F., Abouelkhair, E., Babaie, M., Cunliffe, F., & Nock, W. (2023, April 11). Multi-Objective Optimization of the Fuel Cell Hybrid Electric Powertrain for a Class 8 Heavy-Duty Truck. *SAE Technical Papers*. https://doi.org/10.4271/2023-01-0473

Shin, H. K., & Ha, S. K. (2023). A Review on the Cost Analysis of Hydrogen Gas Storage Tanks for Fuel Cell Vehicles. In *Energies* (Vol. 16, Issue 13). Multidisciplinary Digital Publishing Institute (MDPI). https://doi.org/10.3390/en16135233

Sorrentino, V., Altosole, M., De Luca, F., Micoli, L., Mocerino, L., & Russo, R. (2024). Preliminary modeling of a ferry methanol fuel cell power plant by using AVL Cruise M software. In Soares C.G. & Santos T.A. (Eds.), *Developments in Maritime Technology and Engineering - Proceedings of the 7th International Conference on Maritime Technology and Engineering, MARTECH 2024*.

Truong, H. V. A., Dao, H. V., Do, T. C., Ho, C. M., To, X. D., Dang, T. D., & Ahn, K. K. (2020). Mapping fuzzy energy management strategy for PEM fuel cell-battery-supercapacitor hybrid excavator. *Energies*, *13*(13). https://doi.org/10.3390/en13133387

van Biert, L., Godjevac, M., Visser, K., & Aravind, P. V. (2016). A review of fuel cell systems for maritime applications. In *Journal of Power Sources* (Vol. 327, pp. 345–364). Elsevier B.V. https://doi.org/10.1016/j.jpowsour.2016.07.007

Wancura, H., Kühberger, G., & Schutting, E. (2023). Concept evaluation of a P2 MHEV SUV: application for possible EU7 boundaries. *Automotive and Engine Technology*, *8*(1), 1–16. https://doi.org/10.1007/s41104-022-00123-7

Whiston, M. M., Lima Azevedo, I. M., Litster, S., Samaras, C., Whitefoot, K. S., & Whitacre, J. F. (2021). Hydrogen Storage for Fuel Cell Electric Vehicles: Expert Elicitation and a Levelized Cost of Driving Model. *Environmental Science and Technology*, *55*(1), 553–562. https://doi.org/10.1021/acs.est.0c04145

Yuan, XZ., & Wang, H. (2008). *PEM Fuel Cell Electrocatalysts and Catalyst Layers* (Jiujun Zhang, Ed.). Springer London.

Advances in Maritime Technology and Engineering – Guedes Soares & Santos (Eds)
© 2024 The Author(s), ISBN 978-1-032-83107-7

Conceptual design of a drone carrier

R. Pinto da Costa
Damen Naval, Vlissingen, The Netherlands

M. Ventura
Centre for Marine Technology and Ocean Engineering (CENTEC), Instituto Superior Técnico, Universidade de Lisboa, Lisbon, Portugal

ABSTRACT: In this paper, the conceptual design process of a drone carrier warship will be developed, with a multi-role operational concept and requirements comparable to modern European NATO (North Atlantic Treaty Organization) navies. This conceptual design process will include the overview of the state of the art of surface warship preliminary design methodologies as well as of the drone carrier ships themselves, the development of an operational concept for the ship along with the definition of its operational requirements, a system selection process in which the systems that the ship will make use of in order to undertake its wide range of missions will be defined, the development of a ship synthesis model and, finally, the development of two ship concepts for different NATO end-users.

1 INTRODUCTION

In modern conflicts a surge of the use of naval unmanned vehicles has been seen – with significant effectiveness in both strike and reconnaissance operations as well as largely reduced acquisition and operational costs compared to other market alternatives (Kahn, 2022). The state of the art of warships used to carry and deploy these systems has consequently also seen an increase in development (RAND Corporation, 2022), with recent high profile projects to design and build drone carrier ships. Despite this recent surge in development, the state of the art of drone carriers is still limited to a small range of functional examples of built ships – most of which are either reconverted ships or vessels aimed at scientific research purposes. This makes it profoundly advantageous to study how these ships can be further used within military applications, what systems are required for these platforms to operate unmanned systems and exactly what unmanned vehicles will be needed to carry out these ship's missions, how they compare to current market solutions and what design practices best fit the design process of a naval drone carrier ship. The main objective of this work will thus be to better understand the design process of a drone carrier ship through the process of the concept design of a naval drone carrier. The paper is structured as follows:

Chapter 2 and 3 – Literature and State of the art review: An overview of ship design procedures will be performed, the differences between commercial and military ship design will be elaborate upon and a study of select process methodologies and toolsets used by institutions for warship concept design will also be conducted. Additionally, current operational drone carriers and concept designs will also be reviewed.

Chapter 4 – Operational Requirements: The ship's operational concept will be defined and it will be elaborated what operational requirements and benchmarks it will have to adhere to. In this stage, modern NATO warship benchmarks are considered as well as OPFOR (Opposing Force) benchmarks from current NATO threats. This process will end with a compilation of all the operational requirements in a User Requirements Document (URD).

Chapter 5 – System Definition: The various systems the ship will make use of to ensure that it complies with its previously defined operational requirements will be detailed. These will include all of the unmanned vehicles, platform systems, naval payload and other relevant systems for all of the drone carrier's mission configurations.

Chapter 6 - Ship Synthesis Model: A tailor-made concept design toolset was developed in the form of a ship synthesis model through which non-linear optimization algorithms can be used to find optimal concept solutions depending on various design objectives. These objectives will themselves depend on the context of the end-user for which the specific concept solution will be aimed at. The development process and all of the individual modules of the synthesis model will be detailed. The synthesis model will also undergo a validation process in order to ensure the validity of the obtained results – using

DOI: 10.1201/9781003508779-4

modern NATO warship as references for the validation stage.

Chapter 7 – Optimization Process: The optimization process performed for both concept solutions will be detailed in this stage – including all project variables, model restrictions, the context behind each optimization scenario that lead to the final two concept solutions as well as the selected non-linear optimization algorithm.

Chapter 8 – Conclusions: The work's results will be discussed in this final chapter. Both concept solution results will be analyzed and evaluated accordingly. Further future work for the project will also be defined within the context of the development of the drone carrier concept design process as well as future developments for the ship synthesis model.

2 LITERATURE REVIEW

The design of a ship is an inherently complex and challenging process. It requires a large spectrum of knowledge in several scientific areas combined with an inherently iterative process of conceptualizing, analyzing and improving until a satisfactory final solution is found. In most cases, inexact or analytical-empirical methods are used which are extrapolated from previously successful design processes or methods (Costa, 2019). One of the earliest descriptions of the modern ship-design process is the Ship Design Spiral (Andrews et al., 2008), first seen in (Evans, 1959). This design spiral diagram showcases the various iterative stages that make up the design process of a conventional ship. This diagram shows the design process starting from the owner's requirements and then following a sequence of different steps which trace back to these, repeating this cycle until a final design solution is found. However, several researchers have addressed this method's limitations and proposed other solutions that make up the state of the art of modern ship design (Bruinessen et al., 2013). Currently, the three most influential branches of ship design methodologies include: System-based design, Requirements Elucidation and System Engineering focused processes.

The most recent of the previous methodologies, are the system engineering focused processes. These methodologies provide a different perspective on the design process of complex engineering projects. This approach orients itself on the realization of successful systems by focusing on the definition of customer needs and required functionality in the early design stages, documenting requirements and then proceeding with design synthesis and validation (INCOSE, 2006; Bruinessen et al., 2013). The basis of system engineering is the decomposition of larger complex systems into their sub-subsystems which will then be further decomposed into components. Design at these levels is independent and based on the initial project requirements. A discussion of maritime applications of this methodology can be seen in (Griethuysen, 2000) where it's recognized that system engineering places engineering at the core of management processes but that the methodologies should be tailored to specific projects. For this work, this design process methodology shows particular merit, as it's capable of handling complex systems which are frequent in ship design. However, the design of innovative solutions appears to occur between different levels of decomposition: this communication is not solely done through requirements and design feedback as described in this methodology but rather contains continuously changing designs, requirements and functionalities. Constraining design to separate levels in decomposition could then, as described in (Bruinessen et al, 2013) lead to stove-piping where the independent development of systems without offering the opportunity to propose innovative solutions across these levels is mentioned as one of the issues in the system engineering process.

Naval ship design comes with its own unique set of challenges when compared to commercial ship design – particularly due to a larger focus on design for operational capability. Due to their operational concept and requirements, military ships generally have more robust structural designs when compared to commercial vessels due to their requirements to withstand combat scenarios as well as having significantly more system redundancy to ensure survivability. Naval vessels generally also have higher speed and maneuverability requirements when compared to conventional commercial ships. Military ships require a much more advanced sensor suite and a dedicated naval payload in order to support their own combat roles. Additional modern requirements related to electronic warfare, cybersecurity and information technology management also make related systems on board warships profoundly more complex and expensive than on commercial vessels. Other special considerations also have to be kept in mind in more advanced military ship designs regarding stealth technologies designed to reduce radar cross-section, acoustic signature and infrared emissions. Regarding personnel and crew, the complement on military ships tends to be much larger as well when compared to commercial ships due to the additional crew needed to operate the complex naval systems and ensure damage control onboard. Unique considerations also have to be kept in mind regarding military security and safety in spaces for weapons, ammunition and explosives' storage. Finally, while commercial ships are subject to classification society rules and international maritime regulations, military ships are often exempt from these while having their own specific military standards and regulations to adhere to which vary from nation to nation (although often standardized in NATO countries).

To face these challenges, modern navies often used a traditional point-based serial engineering process beginning with an initial concept design, moving onto a basic design and finally an execution design (Sobek et al., 1999). This method led to frequent redesigns during the design process due to the lack of expertise and information transfer between stages. This process later evolved to integrate the traditional design spiral (particularly within the US Navy) which went through adapted contract design, preliminary design and concept design stages for surface warships. This has since been further adapted in consideration that naval ship design is actually a "system of systems" type design challenge, in which the sum of various complex systems and their connections give rise to new needs in concurrent design methodologies (Kwak, 2016).

A key part of this process that also deserves a detailed analysis is the design toolset development (including ship synthesis models) as well as the preliminary concept design process (particularly the initial sizing stage). Although the initial stages of operational requirements definition and system selection all proceed with very small variations in the process between different institutions, the same is not true for these last two processes. Various research, industry and academic institutions that work with military surface warship designs go through the initial sizing of vessels in varying ways. For the particular case of drone carrier design, which are unconventional naval surface combatants, it becomes increasingly important to perform an overview of the current process and ship synthesis model state of the art. One of the most widely researched and extensively developed toolsets from the state of the art is ASSET. This ship synthesis model, developed by NAVSEA in the United States, is tailor-made to be used on naval ship design with added modules directly related to these unique design challenges (Kerns et al, 2011). ASSET allows for multi-objective optimization. However, history with the software has shown that it has various problems related to its robustness, crashes and computational time. ASSET data model and input configurations have also shown to be limited in applications, particularly in weight space and electric power estimations as well as in design flexibility for non-conventional ship designs. Despite these limitations, the model has shown to be fully representative of a complete concept-development process of a naval ship.

One of the most significant challenges associated with this ship design work is that the current literature regarding the early-stage design of drone carriers is highly limited due to the still emerging nature of these ships (Bauer et al., 2006; Kouriampalis, 2019). Additionally, although some drone carriers already exist in the current state of the art, the research and development documentation and most of the scientific work associated with these is confidential due to these including sensitive information on naval ships. Historically however, military ship design has always presented difficulties when undertaken by civilian universities. This makes the work detailed in this project exceedingly novel in nature as well as a first for Instituto Superior Técnico. One of the few available works for drone carrier concept design can be seen in (Kouriampalis, 2019) in which a vessel designed to carry a fleet of unmanned vehicles was conceptualized through the use of architecturally oriented early-stage ship design processes. Within this thesis, decision-making and ship concept design tools were developed – intending on providing a holistic means of integrating aspects of a fleet of unmanned vehicles into early stage mothership design.

Additional considerations should also be made regarding how to implement and manage innovation within the concept design of the drone carrier warship. Considering the still emerging and novel nature of naval drone carriers, the conceptual and innovative space will need to be defined for the project. To do so, it's possible to resort to fundamental design theories, namely C-K (Concept-Knowledge) Theory as seen in (Hatchuel & Weil, 2009). Within C-K Theory, designed systems are determined by a form (design synthesis), function (functional analysis and allocation) and performance (requirements and analysis). Each of these parameters provide insight into what elements are wholly conceptual or based on established knowledge. This theory is generally based on the distinction between two spaces: the concept space (C-Space) and the knowledge space (K-Space). The C-Space includes all concepts that do not yet have knowledge established in the K-Space. Concepts are generally regarded as necessary departure points within the conceptual design process, since without these the design is reduced to an optimization or small incremental improvement problem.

With the above considerations in mind, the methodology to be used was defined. This process would merge key insights from the current state of the art into a single methodology. The work will thus start with an operational concept and requirements development process as per (Kwak, 2016) and using NATO benchmarks and requirements. Following this, and considering the ship as a "system of systems", each individual system will be defined for the ship in a system definition stage. From this, a preliminary design tool will be developed in the form of a ship synthesis model, similar to the US Navy's ASSET software, that'll further elucidate the concept design and perform relevant calculations for the ship's initial sizing using (UCL, 2016) guidelines. This model will use selected modules relevant to surface warship design, and specifically altered to best fit this work. Regarding the innovation management in this work, it's possible to observe that in the ship concept design, in general, the performance of the concept vessel and form will largely be in the concept space as the drone carrier will be relatively novel in form and with generally unknown performance metrics from which to compare the vessel to.

Meanwhile, the function of the ship will be developed considering modern naval assets (particularly its operational requirements) with part of this

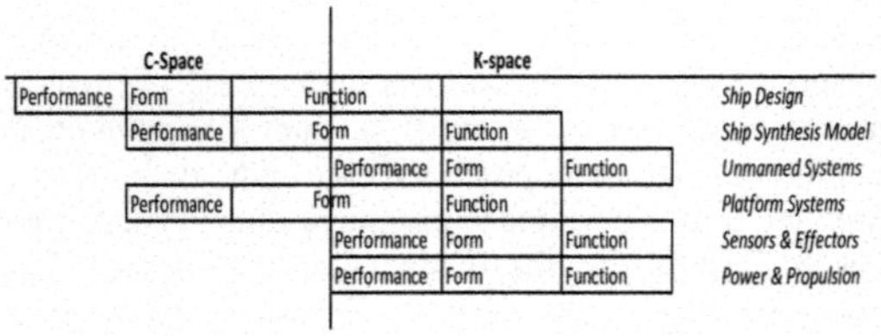

Figure 1. K and C-space distribution for the drone carrier concept.

function still lying in the concept space as the use of unmanned technologies will be leveraged in novel ways to increase the operational versatility of the warship. Regarding the ship synthesis model, the function of it is in the knowledge space as many ship synthesis models already exist in the state of the art but the form of the model will be developed both within the knowledge and concept spaces as a tailor-made tool will be programmed for the innovative design challenge. Additionally, the performance of the model will thus also reside in the concept space and will need to undergo a validation process to ensure its adequacy. A summary of the K and C-space distribution for this work can be seen bellow in Figure 1.

3 DRONE CARRIER STATE OF THE ART

Within the state of the art of drone carriers, various examples were considered and compiled in a common database. These include examples from NATO member states, such as Portugal and England, as well as other modern navies which have demonstrated a particular interest in unmanned systems development and technology such as the Chinese PLA Navy. The following ships where considered in the state of the art database: Zhu Hai Yun (PLA Navy), TCG Anadolu (Turkish Navy), Drone carrier concept ship (Portuguese Navy), UXV Combatant (BAE Systems) and Octopoda 500 (Mauric). The current state of the art of this type of naval ships is very limited when compared to other warship typologies. Despite this, some design trends can clearly be observed, particularly regarding the ships' arrangements and dimensions. The main of which, is a clear distinction between drone carriers designed to accommodate a larger amount of unmanned aerial vehicles (characterized by large hangar spaces and flight decks and being generally larger than their counterparts) and drone carriers designed to mainly operate with surface and underwater vehicles (smaller ships, with hangar spaces and flight decks designed to generally only accommodate VTOL UAVs).

4 OPERATIONAL REQUIREMENTS

Within this chapter the drone carrier's operational requirements will be defined. These will later guide the concept development process as well as serve as key-concept success indicators. This is a multi-stage process and one of the most important ones in the concept design process. Generally, this starts by defining an operational concept for the warship. From this concept, mission requirements will be defined for all of the ship's operational roles. All of these requirements are then compiled into a User Requirements Document (URD). In the case of multi-role warships, only the most stringent requirements from all of the ship's mission configurations are considered and are thus compiled into the URD. A summary of this process can be seen below in Figure 2.

4.1 *Operational concept*

Considering the current state of the art of drone carriers, the recent usage of unmanned vehicles in modern conflicts and the operational context of a modern Western European NATO nation (i.e.: Portugal, the UK or France), a set of initial design considerations were defined for the ship. These will include:

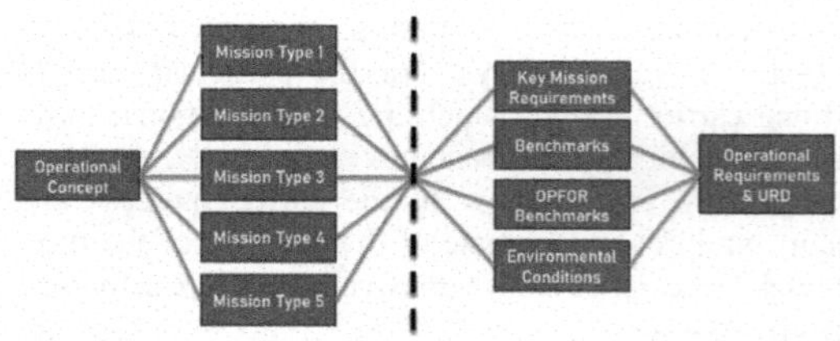

Figure 2. Operational requirements definition process.

The end-users of the warship: these will include NATO member states with a need to rapidly integrate large scale unmanned capabilities into their fleets in both wartime and peacetime scenarios.

What areas the ship sail in: In peacetime, the ship will operate principally in the end-user's EEZ. Additionally, it'll also be considered that the ship will need to be able to sail in the Baltic, Norwegian and Barents Sea as well as the navigational passageways (such as the Northwest Passage) in the Arctic Ocean in close-to-shore operations.

When the ship will operate: The ship will be designed to operate year-round in patrolling missions within the end user's EEZ as well as in NATO peacetime operations in various maritime operational areas – in missions such as disaster relief or search and rescue. Combat deployments will occur when in crisis or transition to war, these will include littoral strike and raiding operations targeting hostile country's infrastructure, naval assets or military personnel or task force support operations.

How will the concept design's success be measured: In peacetime, the ship should be able to endure long mission times characterized by high system utilization, it should be able to provide reconnaissance in as large of an area as possible with the use of its unmanned systems while guaranteeing high versatility and low operational costs when compared to current market solutions. Additional capabilities regarding disaster response operations should also be considered. As a combat vessel, the ship should be able to traverse high-risk areas while remaining mostly undetected/inconspicuous, quickly deploy a large amount of unmanned systems and then successfully extract while guaranteeing high mission success rates combined with low system loss rates for unmanned systems designed to be retrieved. Considering the high-risk nature of these raiding missions, usually with the ship not being accompanied by a substantial naval force in order to further difficult ship recognition/detection, the ship should be designed to have the lowest acquisition costs possible combined with reduced ship crews to minimize damages incurred from losing the vessel itself while allowing countries to mass produce simple, low-cost platforms for sustained mass deployment of unmanned systems - a strategy seen as increasingly effective in modern conflicts (Al Jazeera, 2022). In task force support operations, the ship should be able to be fitted with various configurations allowing it to provide differentiated support to naval task forces in the submarine, air, mine-countermeasures and surface domains.

4.2 *Platform mission range*

With the previous operational concept for the vessel in mind, a set of missions were detailed and their subsequent requirements were defined for the ship. These missions would include peacetime and wartime operations. Each of these were detailed separately, using benchmarks and mission specific details to define operational requirements which would then be compiled into a single user requirement's document.

An example of one of these missions will be littoral raiding and strike operations. In these missions, the ship will sail from friendly waters into hostile zones while remaining undetected or inconspicuous throughout. When in hostile controlled or disputed territory, the ship will sail along maritime traffic routes – exhibiting navigational behaviors, lights and AIS/Radar/EM emissions similar to a smaller vessel (i.e.: a fishing vessel returning to its home port). The ship's own hull and superstructure should be designed in order to reduce its radar signature. The drone carrier will conduct these operations after dusk in order to minimize the chance of visual detection – further viable when operating in Arctic and North Atlantic waters where daily sunlight hours during wintertime are scarce. When nearing designated targets, the ship will sail along the coastline while deploying strike assets such as UAVs, USVs or manned craft along with special operations units. This will be done underway, in the shortest amount of time possible, retrieving any designated systems the ship has deployed before returning to friendly controlled waters. If detected, the ship should be able to elude opposing forces through the use of its own self-defense weapon systems, combat UAVs and USVs as well as its own maximum speed.

In order to define operational requirements for this mission, both NATO and non-NATO ships designed for this exact purpose as well as modern NATO naval threats will be used as references. The reference surface combatant ships considered for benchmarking purposes where the Tuo Chiang-class and Visby class corvettes. Regarding the main OPFOR benchmarks and threats the ship will face, these will generally include patrol vessels, surface-to -surface land missile systems, destroyers, frigates and corvettes as well as conventional attack submarines. In addition to conventional naval and ground forces, the ship will also have to deal with mines, asymmetric threats and land based aircraft - with encounters with conventional task forces being possible but less likely. With these in mind, it's now possible to define operational requirements that will ensure the ship is able to successfully complete these types of missions.

Considering the likely threats for the vessel, the ship should be able to sail at speeds higher than the aforementioned destroyers, frigates, corvettes and patrol ships (a minimum of 35 knots, matching Udaloy Class Destroyers' top speed). It should be able to deter small naval forces made up of coastal patrol boats through the use of unmanned systems and weaponry (i.e.: using combat UAV's or VLS (Vertical Launch System) with ranges above 5,500 meters - equivalent to the range of a 9M133 Kornet portable guided missile system). The ship's unmanned systems and manned craft should be able to be deployed outside of coastal defense systems' range (at a minimum distance of 22 km to avoid the range of A-222 Bereg systems, and using long-range unmanned systems with ranges of over 150 km in areas where K-300P Bastion-P systems are present). The ship should be equipped with USVs capable of performing submarine deterrence and detection in order to handle submarine threats effectively. Defensive weapon systems should be equipped on the ship in order to ensure its own defense against missile threats from both enemy ships, coastal defenses and airborne threats. The vessel should be equipped with radar systems able to detect threats before the ship enters their effective weapon's range (a minimum range of 300 km to handle the bulk of naval air and surface threats). The ship's hull or superstructure should ideally be constructed of non-magnetic composite materials or include a payload of MCM focused USVs in order to minimize the threat of mine warfare – despite the ship most likely operating outside common high-risk mine warfare areas (such as inland waterways, canals and naval port entrances).

This process was repeated for the entirety of the ship's mission range, which would include: littoral and coastal strike operations, task force support (ASW, ASuW, AAW and MCM), search and rescue and disaster response operations.

4.3 *Area of operations and environmental conditions*

The area in which the drone carrier is designed to operate in will heavily influence the ship's requirements. Considering the developed operational concept, the drone carrier will mostly sail in the North Atlantic Ocean in peacetime and in the Norwegian, Barents, Baltic, Bering, Kara and North Seas in transition to crisis or wartime (with rare but possible crossings into the Laptev and East Siberian Seas).

In the North Atlantic, the ship will face its overall less demanding environmental conditions. The ship will mainly sail along well travelled maritime routes. It will face water temperatures with annual means varying between 24 and 16 °C - while below 60ºN latitudes - and between 16 and 2ºC - above 60ºN latitudes. The water salinity in the North Atlantic is also around 35.5 parts per a thousand. The mean significant wave heights the ship will face in this area will vary between 1.2 and 3.5 meters (sea state conditions between sea state 2 and 3). With ratios between the significant wave height and 100-year maximum wave heights ranging from 4 to 6 (equivalent to sea state 6 and 7). The drone carrier will be designed to dock in most NATO countries' main military ports - with their respective dimension restrictions to be applied in the final ship design. For example, when considering the port of Den Helder, the following restrictions should be applied: length under 200-metres and draught under 8 meters.

In the North Sea, Baltic and Artic Water Regions, the ship will mostly sail while conducting combat operations. It will also be in these waters where it'll face the bulk of the challenges related to the local environmental conditions. The ship will face surface water temperatures ranging between -5 to 2ºC, with lower significant wave heights with mean values of 1.5-metres. A significant challenge faced in these waters is also the presence of arctic sea ice throughout the year. The presence of ice varies greatly between seasons, with a current trend of sea ice reduction through the years due to global warming. Currently, the entirety of the Russian northern coastline can be traversed through the use of the Northern Sea Route. Parts of this route are only free of arctic ice for 2 months per year. However, the Dutch Bureau for Economic Policy Analysis projected in 2015 that the Northern Sea Route may be ice-free by 2030, earlier than the Northwest Passage or Transpolar Sea Route (Stoetman, 2020). With this objective in mind, the ship should comply with weather/climate specific class guidelines (DNV, 2016) – obtaining at least an ICE-C classification for floating and light ice conditions.

5 SYSTEM DEFINITION

With the vessel's operational requirements established, it's now possible to make initial design decisions regarding what systems the drone carrier will be equipped with. These will be divided onto two types: the variable payload (which will change depending on the ship's operational role and subsequent configuration) and the platform systems (which will be common between all mission configurations). This process begins by using the previously defined operational concept and requirements and, through the analysis of modern warship systems, benchmarks from modern naval vessels, NATO suppliers and considering each equipment's functionality, the various systems used on board the warship are selected. These include systems such as the hull, propulsion, sensors and effectors, energy production, unmanned vehicles and all the auxiliary platform systems. A system selection process diagram can be seen below in Figure 3.

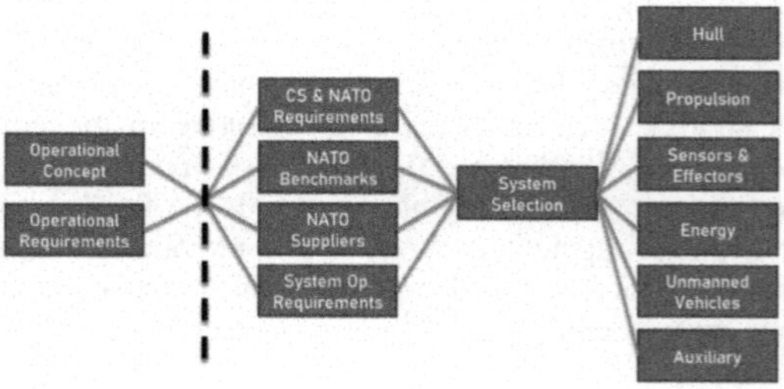

Figure 3. System selection process diagram.

The variable payload will include systems which vary between mission configurations. These will include: manned and unmanned vehicles as well as containerized mission modules. The platform systems include the fixed platform equipment such as: defensive weaponry, propulsion, sensors and all platform auxiliary equipment. Each of these systems was selected considering the ship's operational concept. A parametric analysis was performed for the most important of these while others were selected using a holistic approach to evaluate how adequate each alternative is considering the operational requirements of the drone carrier. For example, to carry out the hull selection, the following hull types were considered: monohull, small waterplane area, catamaran, trimaran and hydrofoil hulls. Each of these will be evaluated according to the following characteristics with a relevance scale of 1 (least important) to 5 (most important) being attributed to each parameters depending on their relevance to the ship's operational concept: available well dock space (5), viability for use in arctic waters (5), speed (4), acquisition costs (3), deck space (3), stability (3), maintainability/reliability (2) and maneuverability (1). This led to the choice of a monohull type hull. This process was repeated for the entirety of the ship's systems, leading to the definition of: hull type, propulsion and energy generation systems, containerized modules, sensors and effectors, decoy systems, platform auxiliary systems, aerial, surface and subsurface vehicles, among other systems.

After the conclusion of the system selection process, various mission configurations were defined considering each individual role the ship can be reconfigured to undertake. In total, 10 mission configurations were developed for the drone carrier, including: littoral strike (low to high-risk, amphibious and non-amphibious), task force support (anti-submarine, anti-aircraft, anti-surface and mine countermeasures), search and rescue and disaster response configurations. Benchmarks and individual systems were defined for each configuration. For example, for the task force support operations, a possible mission configuration can be seen in Table 1, with variable payload benchmarks detailed. These configurations take into consideration two of the three types of modularity defined within the Royal Navy's maritime modularity concept (Royal Navy, 2023). The first of these is integral modularity. This modularity type relies on locations within the platform designed to accommodate specific systems that can be removed, replaced and upgraded with a small refit period. These form an integral part of the ship structure and integrity so do not adversely affect stability of structural support (e.g. the Mark 41 VLS). The second type of modularity (seen within the task force support configurations below) is installed modularity. This modularity includes payload methods that can be replicated across multiple ships without the need for a refit period. In this case, these will include the various vehicles compatible with the ship as well as the containerized mission modules.

Table 1. Task force support configurations' payloads.

	Task Force Support			
Unmanned System	MCM	AAW	ASW	ASuW
HERO-400EC	25+	50+	25+	50+
HERO-900	20+	40+	20+	40+
HERO-1250	15+	30+	15+	30+
Kratos XQ-58	4+	4+	4+	4+
Schiebel Camcopter	4+	0	0	4+
Tekever AR3	10+	0	0	10+
Loitering USV	0	0	0	4+
Protector USV	4+	4+	2+	3+
Zodiac MACH II MM	2+	2+	2+	2+
Swarmdiver USV	10+	0	5+	0
Echo Voyager	0	0	1+	0
Loitering UUV	0	0	5+	5+
SDV Mk9	0	0	0	2+
SDV Mk8	0	0	0	2+
Modular TEU	4+	0	4+	0

6 SHIP SYNTHESIS MODEL

In this design stage, the ship synthesis model will be developed which will be used to determine the ship's main dimensions and characteristics. This model was generally based on adapted methods and calculations from (Ventura, 2014; D'Almeida, 2009; NATO, 2006) as well as procedures commonly used within the Royal Navy and standardized within (UCL, 2016). This model will be made up of various individual modules that will perform concurrent and interconnected calculations, feeding off project variables that will iteratively be varied until an optimal concept solution is found. Tailor-built for this work, the ship synthesis model will include the following modules:

Database: Compiled database of warships (20) from the past 25 years, with similar mission profiles to the designed drone carrier concept. This module is used as a reference point for a later comparison process of the obtained concept solutions;

Ship Details & User Interface: Within this module, the project variables for the ship synthesis model are inputted. These will include: the ship's length, beam, depth, draft and block coefficient (Cb). In this module, the project restrictions, operational requirements checking, comparison with the main database among other important concept solution information will also be included.

Hull Form Coefficients: Performs initial calculations and estimates regarding the ship's hull form coefficients. These will include: block coefficient (Cb), used only as a reference value, waterplane coefficient (Cwl), midship coefficient (Cm), and prismatic coefficient (Cp);

Volume Calculations: Performs initial volume estimates for the vessel, which will include both the hull and superstructure volumes. These calculations were performed based on (UCL, 2016) formulas;

Resistance Estimation: In this module a resistance estimation will take place, based on Holtrop & Mennen's method extension of 1982 (Holtrop et al., 1982). These calculations will be performed for both the ship's service and maximum speeds – leading to two separate resistance estimation modules;

Propulsion, Fuel & Range: This module will make use of the previous modules' results regarding the resistance estimation and will be used to dimension the main propulsion installation. It will also perform calculations to obtain how much fuel the ship will need to comply with its operational requirements;

Initial Stability Analysis: In this module, an initial intact stability analysis will be performed as per (UCL, 2016) procedures and considering modern NATO warship requirements;

Mission Payload Profiles: This module will serve as a database for the vessel's naval payload and configurations. It includes information regarding all of the ship systems including: weight, volume, deck area, overall dimensions, cost, required minimum amount and other relevant technical details;

Crew: In this module an initial estimate of the ship's crew will be performed. The crew will be divided into functional departments; whose personnel requirements vary with the ship data. This module's calculations are obtained regressively from Royal Navy (UK) data as well as technical specifications from installed systems;

System Groups: Within this module, all of the ship's systems are divided into functional groups that make up the full ship. For each system group a set of individual systems are detailed and calculations are performed to obtain the weight, volume and acquisition cost of each individual equipment. These results vary with inputs from other modules that run concurrently within the optimization process. It is also in this module that estimates are obtained for operational capability details – e.g. the amount of unmanned vehicles the ship is able to carry. The formulas used to estimate the above details are extracted from UCL's warship design guideline documents (UCL, 2016) or from market data. This large module is further divided into system group modules: hull, personnel, ship systems, propulsion, electrical power, naval payload, variables;

Costing: This module uses results from the entirety of the synthesis model to estimate the ship's acquisition costs in US$. This acquisition cost will be an important design driver, with its relative importance varying between end-users;

The optimal concept solution can be evaluated in different ways depend on the drone carrier's end-user. It's possible to optimize the warship concept to have the smallest initial acquisition cost, largest number of unmanned vehicles possible within the design restrictions, smallest superstructure volume (directly related to the ship's radar cross section) or any other chosen scenario. For the purposes of this work, two scenarios for different end users were developed. In the optimization process, the ship synthesis model will make use of non-linear optimization algorithms which will be detailed in a subsequent chapter. An overview diagram of the synthesis model can be seen in Figure 4:

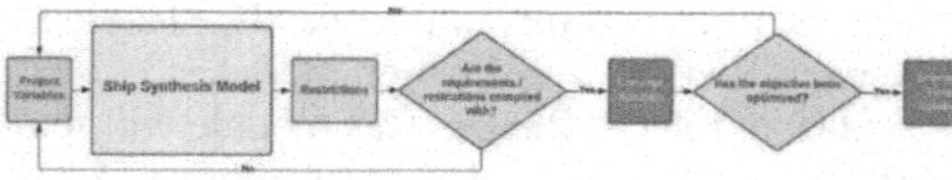

Figure 4. Ship synthesis model simplified process diagram.

Considering that the synthesis model was a custom built tool for this work, it will need to undergo a validation process. The main objective of this will be to ensure that all of the individual modules are working correctly and that the model itself is providing realistic concept design results. In order to do so, available warship data will be used to verify each individual module.

This initial stage of the validation process will be a relatively challenging one due to the fact that most warships' technical data are either classified or not publicly available through conventional sources. Following this, the concept designs obtained from the model will be compared with the warship database to ensure that the drone carrier concepts fit the expectations for modern surface warships. An example of part of this validation process can be seen below where a German Baden-Württemberg-class warship was used as a study case scenario. This ship's data was input into the model which then performed calculations for this warship. These model results were then compared with the real ship's data (Table 2).

Table 2. Comparison from the model calculations and real case.

Variable	Ship	Model	Units	Variation
LOA	149.5	149.5	m	INPUT
B	18.5	18.8	m	INPUT
D	12	12	m	INPUT
T	5.4	5.4	m	INPUT
Cb	0,495	0,495	-	INPUT
Displacement	7200	7228	t	0.38%
Resistance (Max)	32000	33123	kW	3.51%
Resistance (Svc)	21000	20605	kW	-1.88%
Complement	110	108	PAX	-1.82%
Cost	573	708	[M] GBP	5.20%
Weight Diff	-	115	t	1,60%
Volume Diff	-	-828	m^3	-3.35%
GM	-	1.15	m	Compliant

From this process, it's possible to conclude that the displacement calculations of the model for this warship class are accurate, along with the resistance and power prediction, complement, cost and weight group and volume estimates. Additionally, the stability module also considers the ship compliant with intact stability standards and places it, according to its stability characteristics, between naval ships similar to large destroyers/cruisers and frigates – which is particularly adequate considering that this ship class currently has the largest displacement of any frigate type warship worldwide. Considering the interconnectivity between each individual module of the ship synthesis model, as well as the small variations between the reference ship and the model calculations, it's possible to safely assume that the model is working correctly. This process was later repeated for three ship examples, where it was observed that the model was working correctly throughout, leading to very small variations between the model estimates and the real world data.

7 OPTIMIZATION PROCESS

Within this chapter, the optimization process which will lead to the project's concept solutions through the use of the previously developed ship synthesis model will be detailed. In general, using the ship synthesis model, selected project variables will be varied iteratively through the use of a non-linear optimization algorithm until the model converges to a final solution which best suits the project objectives. This variation process is limited by the project's restrictions and will output the final ship's main details which will then be used to further develop the drone carrier concept design. Two scenarios were developed for distinct end-users – leading to two distinct concept solutions.

The selected project variables were: length overall, beam, depth, draught and Cb. These where chosen considering that they will be the ones that have the largest impact on the final concept design and will be used by each module to determine the final ship's characteristics. The process restrictions will be related to the operational concept and requirements defined in the operational requirements and system definition chapter – which will be set for all mission configurations and operational scenarios.

The final design concept is optimized considering different types of goals or concept objectives for the end-user. Some navies might wish to have their concept design proposal optimized to comply with the operational requirements while reducing acquisition/ life cycle costs, other navies might want to set a budget and obtain the ship with the highest capability within a pre-determined acquisition cost while others might want a trade-off between capability, costs, range, complement, among other factors.

For the first scenario, a large NATO navy will be considered (e.g.: United States Navy) whose main goal will be to acquire a platform to provide its task forces with diverse unmanned vehicle support capabilities in response to the ever growing threat of aggressive foreign states. With this in mind, the design goal will be to optimize the various capability indicators of the final vessel (as well as its cost) which will be defined by the amount of unmanned vehicles the ship is able to carry in each mission configuration per millions of US$ acquisition cost – with an initial maximum acquisition cost comparable to modern multi-mission frigates. This will allow the end-user to build a fleet of drone carriers that will be able to provide it with differentiated capability in a wide range of missions.

For the second scenario, a smaller NATO navy will be considered (e.g.: Portuguese Navy) whose main goal will be to inexpensively acquire a warship that fulfils the baseline operational requirements. For this scenario, the ship's cost/capability trade-off will be disregarded, only ensuring that the vessel complies with the operational requirements while minimizing its acquisition cost. This will allow the end-user to quickly finance and acquire an initial ship prototype. This will ensure a rapid integration of these capabilities onto the end-user's armada and reduce the financial risk of the ship acquisition process – at the trade-off of reduced ship capability.

The chosen optimization algorithm for the project was the Generalized Reduced Gradient (GRG) nonlinear algorithm with multi-start enabled. This is a favorable compromise between the robustness of an evolutionary algorithm with the performance of the GRG. This algorithm creates a randomly distributed population of initial values that are independently evaluated by the traditional GRG algorithm (Arora, 2012). By increasing the initial population size, the speed of the algorithm is decreased but the chance of the found solution being a global optimum is increased (Lasdon et al., 1974). A population size of 200 was used for the GRG non-linear multi-start process along with varying algorithm seeds.

It's important to also consider that due to the programming architecture of the ship synthesis model, only single-objective optimization can be performed at this stage of the tool's development. This is adequate for the second scenario, where only a single objective function is considered (the ship's acquisition cost). However, the first scenario presents some challenges due to the fact that various objective functions are being optimized: the cost of the vessel is being minimized while its capability in each mission configuration is being maximized. In a multi-objective optimization problem, there is not typically a solution where all objective functions are optimized. Rather, Pareto optimal solutions (solutions which cannot be improved in any of the objectives without degrading at least one of these, are found – making up a Pareto front when all solutions are plotted. A feasible solution, in a minimization-minimization problem is said to (Pareto) dominate another one when:

$$x_1 \in X \; and \; x_2 \in X \tag{1}$$

$$\forall i \in \{1,\ldots,k\}, f_i \leq f_i(x_2) \; and \; \exists i \in \{1,\ldots,k\}, f_i < f_i(x_2) \tag{2}$$

The set of these optimal solutions is denoted the Pareto frontier (X*). In the case of the scenario 1 problem, it's possible to find a way to select a valid ship solution by defining a utility function that'll convert the problem into single-objective optimization. To do so, each objective will be normalized (from 0 to 1) depending on the minimum and maximum expected values. Objective function parameters will be set for all mission configuration sizes (10 in total, expressed in weight and maximized) and acquisition cost (minimized). For example, for the ASW capability (to be maximized) and acquisition cost (to be minimized) objective function parameters:

$$ASW\ Capability\ Metric_{0-1} = \frac{ASW\ Config.\ Weight}{ASW\ Config.\ W._{Max.} - ASW\ Config.\ W._{Min.}} \tag{3}$$

$$Acq.\ Cost\ Metric_{0-1} = \frac{Acq.\ Cost_{Max.} - Acq.\ Cost}{Acq.\ Cost_{Max.} - Acq.\ Cost_{Min.}} \tag{4}$$

A weight will be assigned to each of these variables, depending on their relevance to the end-user. For scenario 1, a significant weight will be given to the capability (0.7 in total) due to its importance in naval ships. Within this, the most important capability for the ship will be littoral strike (0.4, equally distributed between the 4 configurations), followed by task force support (0.2, equally distributed between the 4 configurations) and peacetime missions (0.1, equally distributed between the 2 configurations). The acquisition cost, not as important for this scenario's end user, will be assigned a weight of 0.3 to translate its importance in facilitating the acquisition and rapid integration of the first set of ships. The utility function, to be maximized, will thus be:

$$\begin{aligned} \textit{Utility Function}_{0-1} \\ &= \textit{Normalized Acq. Cost}_{0-1} \\ &\cdot 0.3 + \textit{LS Capability}_{0-1} \cdot 0.4 \quad (5) \\ &+ \textit{ASW Capability}_{0-1} \cdot 0.05 \\ &+ \ldots \end{aligned}$$

8 DISCUSSION

Using the presented methodology, both concept solutions were obtained. These results were compared with the ship database module, where it was verified that the drone carrier ships were in the bounds of the main relations set in the database – meaning these would lay in the expectation of modern warship design and supporting the validity of the model.

Comparing the scenarios' results with warships to which the UCL methodology has previously been applied, it's possible to see that the drone carriers' payload volume fraction sits on the lower end range of aircraft carriers, while the enclosed total volume is closer to a small to medium-sized frigate with payload densities lower than most warship examples but being closest to aircraft carriers. These findings are shared with both scenarios, with both fitting into the expectations for the concept design derived from the initial operational concept.

The main ship dimensions can be seen below for both iterations, in Tables 3 and 4:

Table 3. Scenario 1 main results.

SSM Module Results: Scenario 1		
Details	Results	Units
LOA	182.4	m
Beam	18.6	m
Depth	18.6	m
Draft	4.7	m
Freeboard	5	m
Cb	0.406	
Max. Speed	35	knots
Service Speed	12	knots
Propulsive Power	76	MW
LBP/B	9.83	
LOA/T	38.63	
B/T	3.93	
Displacement	6 245	t
Cost	0.374	[B] US$

Analyzing the above results, it's possible to observe a set of interesting evidences regarding the behavior of the algorithm. In both scenarios the obtained beam is very similar between concepts. This is driven by the payload selection for this class of ship – as the beam is influenced by the length of the modular containers as well as the necessary clearance between them for effective operation. The same is true for the draft and depth which are influence by the necessary minimum draft and depth for the unmanned assets used within the well dock of the ship. None of these variables are restricted by model. Rather, the ship synthesis model reacts to the selected payloads and their respective operational and technical constraints on the main platform to obtain the shown results. The selection of the length varies greatly between both scenarios. In the first iteration, the model outputs the minimum ship length in order to ensure it's able to carry the baseline payload as well as fulfill all the requirements associated with the selected systems (i.e.: flight deck length). Meanwhile, in the second iteration, the final ship's operational capability is heavily reliant on the ship length as it directly influences both the amount of unmanned vessels and aircraft the warship is able to carry in its well-dock and hangar, as well as the amount of modular containers that can be accommodated in the mission bay – the breadth and depth also influence these but the length has a larger comparable influence. This leads to the obtained results as well as an interesting notion on how system selection affects the synthesis model's behavior.

Table 4. Scenario 2 main results.

SSM Module Results: Scenario 2		
Details	Results	Units
LOA	110.9	m
Beam	18.5	m
Depth	18.5	m
Draft	4.6	m
Freeboard	6.1	m
Cb	0.579	
Max. Speed	35	knots
Service Speed	12	knots
Propulsive Power	70	MW
LBP/B	6.01	
LOA/T	24.13	
B/T	4.02	
Displacement	5 241	t
Cost	0.331	[B] US$

The proposed concept solutions comply with the initially established requirements, as well as the restrictions associated with acquisition costs and the ships' multi-role capability. It's thus possible to consider that the concept proposals are successful solutions – in the context of a concept design. This however, requires some analysis. The results of the synthesis model provide the user with a reference for successful solutions. These need elucidation and refinement as the concept transitions into the basic and detailed design. In practice, the information obtained from the model will vary slightly as designers transition into more exact methodologies and processes used in later stages. Project requirements may also change as some requirements defined in initial stages may need to be readjusted. This is inherent to the ship design process which is iterative by nature. The goal of the concept design is to provide informed decisions that'll make up the basis from which concepts will be elaborated upon. Variations can also be mitigated by the use of design margins (such as the ones used within this work).

9 CONCLUSIONS

Within this project, the initial design stages of a drone carrier ship were performed. This process began with a revision of current literature of naval ship design, concept design toolsets and innovation management. A process methodology was defined and developed throughout the entirety of the project work. A review of the current state of the art of drone carrier warships was also performed. Concluding these studies, operational concept and subsequent requirements were defined – which would be used as key metrics to gauge the success of the concept solutions. To do so, a mission range was defined for the ship, the environmental conditions of the ship's operational areas were studied and the final operational requirements were defined. From here, a system definition process was undertaken where key warship systems were selected. With this process finished, a ship synthesis model was developed. This toolset would be used to obtain the first concept solutions, and would use key insights, data and calculations obtained from UCL warship design guidelines (UCL, 2016). This model also underwent a validation process. The final results were then analysed (as seen in the previous chapter) where the concept solutions were both determined to be compliant with the design requirements and could thus be considered successful concept solutions and ready for refinement. Despite this, the outputs from a concept design process are merely indicative of possible solutions which will need to be refined in later stages. This carries with it inherent risk, which in successful concept designs is mitigated through a series of measures. This was achieved through an adequate innovation process management, the selection of proven methodologies, the use of modern warships as benchmarks for the operational concept and system definition process, the validation of developed tools and a robust and critical analysis of the final results. In general, drone carriers (although still having a significantly limited state of the art), show advantages in operational versatility, acquisition costs and future-proofing in comparison to modern surface combatant ships.

ACKNOWLEDGEMENTS

This work contributes to the Strategic Research Plan of the Centre for Marine Technology and Ocean Engineering (CENTEC), which is financed by the Portuguese Foundation for Science and Technology under contract UIDB/UIDP/00134/2020.

REFERENCES

Al Jazeera (2022). Mass drone attacks in Ukraine foreshadow the 'future of warfare'. https://www.aljazeera.com/news/2022/10/20/mass-drones-are-a-worry-for-the-future-of-warfare. Retrieved 2 February 2023.

Andrews, D.J. & Pawling, R. (2008), A Case Study in Preliminary Ship Design, IJME RINA Trans. Vol. 150, Part A1, 2008. Discussion and Authors' reply IJME, Part A3, 2008.

Arora, Jasbir S. (2012). Introduction to Optimum Design (Third Edition). Chapter 13 – More on Numerical Methods for Constrained optimum design. Pages 533–573. Academic Press.

Bauer, J.; Boudreau, B.; Bryant, C.; Casias, J.; DeMeyer, A.; DuBois, H.; Gasperetti, F.; Mack, T.; Nowlin, B.; Zane, T. (2006). TSUNAMI Maritime Threat Response Ship. Naval Postgraduate School. Monterey, California.

Bruinessen, T. van; Smulders, F.; Hopman, H. (2013). Towards a Different View on Ship Design: The development of ships observed through a social-technological perspective. TU Delft. Delft, Netherlands.

Costa, R. (2019). Challenges in the Design of a Multi-Mission Surface MUS (Project VENT-Sup EN). SNAME Maritime Convention '19. Tacoma, United States of America.

DNV (2016). DNV-RU-SHIP Pt.6 Ch.6: Cold Climate.

D'Almeida, J. (2009). Arquitectura Naval – O Dimensionamento do Navio. 1st Edition. Lisbon, Portugal: Prime books.

Evans, J.H. (1959). Basic Design Concepts. Journal of the American Society for Naval Engineers, 71: 671–678.

Griethuysen, W., (2000). Marine Design, can systems engineering cope?. Naval Engineers Journal, 39(2): 233–245.

Hatchuel, A. & Weil, B., (2009). C-K design theory: An advanced formulation. Research in Engineering Design, Volume 19: 181–192.

Holtrop, J. & Mennen G. (1982). An approximate power prediction method. International Shipbuilding Progress: 166–170.

INCOSE, (2006). Systems Engineering Handbook. International Council on Systems Engineering. New York: John Wiley and Sons.

Kahn, L. (2022). All-out drone war in Ukraine points to future. https://www.channelnewsasia.com/world/ukraine-invasion-all-out-drone-war-points-future-3203156. Retrieved 2 February 2023.

Kerns, C.; Brown, A.; Neu, W.; Xiros, N.s (2011). Naval Ship Design and Synthesis Model Architecture Using

a Model-Based Systems Engineering Approach. Virginia Polytechnic Institute and State University. Virginia, United States of America.
Kouriampalis, N. (2019). Applying Queueing Theory and Architecturally-Oriented Early Stage Ship Design to the Concept of a Vessel Deploying a Fleet of Uninhabited Vehicles. University College London. London, United Kingdom.
Kwak, H-M. (2016). Early-Phase Design of Naval Ship Using Systems Engineering and Interface Management. Master's thesis. Seoul National University. Seoul, South Korea.
Lasdon, L.; Fox, R.; Ratner, M. (1971). Nonlinear Optimization using the generalized reduced gradient method. Revue française d'automatique, informatique, recherche opérationnelle. Recherche opérationnelle, Vol. 8: 73–103.
NATO (2006). ANEP-41 Ship Costing.
RAND Corporation (2022). A New Direction for China's Defence Industry. https://www.rand.org/content/dam/rand/pubs/monographs/2005/RAND_MG334.pdf. Retrieved 10 August, 2023.
Royal Navy (2023). Maritime Operating Concept (MarOpC): Maritime Modularity Concept.
Sobek, D..; Ward, A.; Liker, J. (1999). Toyota's Principles of Set-Based Concurrent Engineering. MIT Sloan Management Review 40 (2): 67–83.
Stoetman, A. (2020). The future of Arctic security: The geopolitical pressure cooker and the consequences for the Netherlands. Clingendael Institute.
UCL (2016). Ship Design Procedure & Data Book. UCL Department of Mechanical Engineering, Naval Architecture and Marine Engineering. University College London. London, United Kingdom.
Ventura, M. (2014). Ship dimensioning in the initial design. Developments in Maritime Transportation and Exploitation of Sea Resources.

Ship machinery

Advances in Maritime Technology and Engineering – Guedes Soares & Santos (Eds)

Evaluating alternative fuels and power systems for marine hybrid propulsion

L. Maloberti, G. Adami, M. Figari & R. Zaccone
Department of Electrical, Electronics, Telecommunication Engineering and Naval Architecture (DITEN) Università di Genova, Genova, Italy

ABSTRACT: The transportation sector significantly contributes to greenhouse gas (*GHG*) emissions; thus, its decarbonization has become one of the most pressing topics for the maritime community. This work proposes an optimization-based algorithm for minimizing a vessel's greenhouse gas-polluting impact. The study aims to compare different propulsion plant layouts that combine alternative technologies in a hybrid propulsion system. The technologies considered are electric propulsion, hydrogen-fueled fuel cells, and methanol as an alternative marine fuel. The optimization aims to minimize the ship exhaust emissions of equivalent CO_2 using a Well-to-Wake approach. The algorithm has been applied to a small ferry operating in a marine-protected area. The passage through an environmentally sensitive area has led to assessing local emissions of other pollutants such as SO_x, NO_x, and particulates. Dealing with innovative solutions, it proved appropriate to make a preliminary economic calculation estimating the annual costs of fuel based on current market prices.

1 INTRODUCTION

The transportation sector is an essential element of modern society and economies, and its changes through the years reflect the growing need for commodities and interconnections. Over the past 40 years, it has undergone an increase of 250% (Bouman et al. 2017), and the forecasts are continuously increasing, seeing consistently higher demand. In parallel, transport emissions grew at an annual average rate of 1.7% from 1990 to 2022 (International Energy Agency). It is responsible for about a quarter of the European Union's (EU) GHG emissions from human activities (European Environment Agency 2023).

The maritime industry is one of the pillars of the transport sector, carrying 90% of all the world's trade (European Maritime Safety Agency). The 4th IMO *GHG* Study shows that shipping accounted for nearly 3% of global *GHG* emissions; these emissions are projected to increase up to 130% of 2008 emissions by 2050 (IMO 2021).

In 2015, the Paris Agreement set ambitious and challenging goals to limit *GHG* emissions and global average temperature rise. The decarbonization of international shipping became one of the most pressing topics in a sector strongly linked to tradition and fossil fuels. In 2018, the International Maritime Organization (IMO) developed and adopted an Initial Strategy that envisages for the first time objectives, tools, and principles for reducing *GHG* emissions from shipping (IMO 2018). In 2023, the strategy has been revised with enhanced targets and ambitions for decarbonization and net-zero *GHG* shipping.

European Commision in FuelEU Maritime initiative presents a Well-to-Wake approach to decarbonization. This part of the Fit for 55 package focuses on reducing the greenhouse gas intensity of fuels, imposing increasing restrictive limits and penalties. This regulation supports the uptake of renewable fuels and energy, especially of non-biological origin, and the use of on-shore power supply (Council of European Union 2023). Tadros et al. 2023b proposes a review of regulations and innovative technologies in the marine sector.

Regulations mainly focus on cargo and cruise vessels of more than 5000 GT, excluding part of the global fleet such as offshore, yachts, and fishing vessels. Transport & Environment 2022 highlights that, at the European Level, the total emissions from exempted vessels are significant: 25.8 Mt, including 19.7 Mt from vessels under 5000 GT.

The increasing attention to environmental impact and the challenging limits on *GHG* emissions lead to the need for a profound transformation. The maritime industry and research are collaborating to find sustainable and ecological solutions (Tadros et al. 2023).

There are several viable options to reduce the *GHG* footprint of shipping, among which low-emission fuels and alternative propulsion systems stand out (Karatug et al. 2022). Focusing on the potential substitute fuels for shipping, Liquefied Natural Gas (LNG), Liquefied Petroleum Gas (LPG), methanol, bio-fuels, and hydrogen are identified as the most promising solutions (DNV-GL 2018). Methanol and hydrogen can be obtained from different feedstocks and technologies, in distinction with

DOI: 10.1201/9781003508779-5

LNG and LPG, which are fossil products. The various production paths affect the *GHG* footprint, market availability, and costs.

From 2015, with the first retrofit to methanol, it gained interest as marine fuel for the future, arriving at 230 vessels in operation and on order (DNV-GL 2023). The main advantages of using this alcohol as fuel lie in its liquid state at atmospheric conditions and the possibility of adapting standard marine engines to run on methanol with minimum changes and a small amount of diesel as pilot oil. Methanol also offers the possibility of *GHG* reduction, especially on a WtW basis, and abatement of other pollutants such as Sulfur Oxides (SO_x), Nitrous Oxides (NO_x), and Particulate Matter (*PM*) (Ellis, J. and Tanneberger, K. for EMSA 2016). This chemical also presents some drawbacks when used as fuel onboard vessels, especially safety-, technical-, and cost-related.

In the maritime energy transition, hydrogen has great potential, particularly as a reactant in fuel cells (Benet et al. 2022). The possibility of an environmentally sustainable propulsion system powered by hydrogen has to deal with several barriers. The main challenges are the availability and cost-competitiveness of hydrogen from renewable sources and the need for cryogenic or pressurized storage tanks (MAN Energy Solution).

Emission reduction can also be achieved with innovative propulsion system configurations combining various environmentally friendly technologies. Hybrid propulsion enables greater flexibility in vessel design. It allows ship operators to accommodate environmental requirements with attention to financial issues, exploiting the benefits of the installed technologies in various operating scenarios (Belvisi et al. 2022). The use of Battery Energy Storage Systems (*BESS*) to hybridize ship power plants has been acknowledged as a technology that can reduce the carbon footprint of shipping (Karvounis et al. 2022). *BESS* can enhance the safety and reliability of the propulsion system by preventing blackouts and could represent a power reserve (Maloberti & Zaccone).

Various researchers have evaluated the environmental impact of different technologies on board ships, and even for operational scenarios such as dynamic positioning, the issue of greenhouse gas emissions has become a topic of great interest (Fruzzetti et al. 2023). Vasilikis, Geertsma, & Coraddu propose a novel digital twin for predicting fuel consumption and carbon impact for various propulsion systems, considering the influence of operational and environmental conditions. Adami & Figari 2023 studied a methodology to evaluate the feasibility of methanol as fuel onboard bulk carriers with significant attention to its environmental and economic impact. Karatuğ et al. 2023 compared the use of methanol as a marine fuel with other alternatives, i.e. ammonia and LNG. The environmental benefit of lithium-ion batteries recharged via an on-shore power grid installed on small vessels working in inland waterways is demonstrated in Maloberti et al. 2022.

Several optimization methods are proposed in the literature to minimize the consumption and emissions produced (Tadros et al. 2019, Tadros et al. 2021, Tadros et al. 2023a). This paper proposes an optimization-based approach to evaluate and minimize the carbon footprint of a hybrid propulsion system, finding the optimal power split. The system features a generator set powered by methanol, a *BESS*, and a hydrogen fuel cell. The procedure follows the Well-to-Wake emissions concept, aiming to reduce the equivalent CO_2 (CO_{2eq}) over the entire fuel chain. The methodology is applied to a small passenger ferry used as a case study. The vessel operates on a route that crosses a ``Cinque Terre'' Marine Protected Area with a speed limit. The passage through a protected area leads to the need to assess local emissions. The SO_x, NO_x, and *PM* emissions are estimated from a TtW perspective. The solutions with hybrid power plants are compared with those equipped with generator sets powered by fossil Marine Diesel Oil (MDO) used as the reference standard. The comparison metric includes technical, environmental, and economic aspects.

2 PERFORMANCE INDICATORS

The performance indicators presented in this section aim to quantify the environmental impact of a power system and are used to evaluate the cost function of the optimization problem and to compare the different propulsion configurations. For this purpose, the performance indicators include Well-to-Wake equivalent CO_2 production, also considering the production chain of fuels used and the emissions of other locally impacting pollutants. To evaluate the economic aspects, the annual fuel cost is considered.

2.1 *Greenhouse gas Well-to-Wake emissions*

To assess a ship's global warming impact, regulations require the calculation of two indices: the Energy Efficiency Design Index (EEDI) and the Carbon Intensity Indicator (CII). As the name implies, the former is calculated only at the design stage, while the latter is calculated annually and is used to rank the ship using a scale ranging from A to E. These indices are aimed only at quantifying carbon dioxide emissions using a Tank-to-Wake approach, considering only the emissions produced by the fuel used onboard. In this article, however, to compare solutions from an environmental impact perspective, the *GHG* emissions are calculated on a Well-to-Wake basis following the actual trend of primary regulators. The greenhouse gas emissions are computed considering the entire fuel chain, including the production phase.

The equivalent CO_2 emissions concept is introduced to incorporate the contribution of various GHG. The environmental impact of CO_2, Methane (CH_4) and Nitrous Oxide (N_2O) are converted to CO_{2eq} through the respective Global Warming Potential (GWP), resulting in a uniform polluting metric. GWP indicates the amount of infrared thermal radiation that a greenhouse gas retains in the atmosphere over a given time, related to the amount of radiation retained by carbon dioxide. Table 1 shows the GWP values of the leading greenhouse gases, referring to an observation period of 100 years.

The mass of equivalent CO_2 m_{CO2eq} is the sum of the emissions of all greenhouse gases, weighted by their Global Warming Potential according to the following equation:

$$m_{CO_{2eq}} = \sum_i m_i \, GWP_i \tag{1}$$

Where m_i is the mass of the GWP GHG, and GWP_i is its global warming potential.

Methanol is a chemical widely traded worldwide as a commodity (Tammy 2021) and can be produced from different feedstocks. Brown and Grey methanol from coal and natural gas represent most of the production. Approximately 0.2 Mt of green methanol is produced annually from renewable sources (Methanol Institute and IRENA 2021), resulting in small availability. The various production paths lead to different emission footprints on a WtW basis (Methanol Institute 2022). The combustion onboard of this chemical produces lower GHG emissions compared with traditional fossil fuels.

Hydrogen is considered a clean fuel with no locally-impactful emissions. Thus, the critical phases are production, storage, and transport onshore. The production technologies range from Methane Steam Reforming (SMR) without or with Carbon Capture and Storage (CCS), Methane Pyrolysis (MP) and Polymer Membrane Electrolysis (PEMEL). This work considers the Italian hydrogen production for consistency with the proposed case study. The ``green'' H_2 produced in Italy relies mainly on nonrenewable sources and, thus, is the worst solution in terms of environmental impact. The Norwegian emission factor is reported to represent the best scenario currently available.

Given the variety of methanol and hydrogen production paths and their impacts, overall Well-to-Wake emissions are assessed, and their emission factors are reported in Table 2.

Table 1. Global warming potential.

GHG	GWP
Carbon dioxide (CO_2)	1
Methane (CH_4)	25
Nitrous oxide (N_2O)	298

Table 2. WtW emission factors of diesel oil, hydrogen (Hermesmann & Müller 2022) and methanol (methanol institute 2022) from different production path.

Marine Diesel Oil	Short Name	Technology	$WtW-EF$
			$[kg_{CO_{2eq}}/kg_{fuel}]$
	MDO	-	3.87
Hydrogen	Short Name	Technology	$WtW-EF$
			$[kg_{CO_{2eq}}/kg_{fuel}]$
	Gray	SMR	11.7
	Blue	SMR-CCS (56%)	7.4
	Blue	SMR-CCS (90%)	4.6
	Turquoise	MP	7.5
	Green	PEMEL	20.4
	Green (Norway)	PEMEL	0.8
Methanol		Feedstock	
	Brown	Coal	5.97
	Gray	SMR	2.12
	Green	SMR (bio-methane)	0.2
	Green	Wood	0.3
	Green	Municipal Solid Waste	0.65
	E-methanol	Renewable Energy&Feedstock	0.09

2.2 SO$_x$, NO$_x$, *and* PM *emissions*

Annex VI of Marpol indicates that Sulphur Oxides, Nitrous Oxides, and Particulate Matter are pollutants incredibly impactful on the environment and human health. Their evaluation has to be assessed in addition to GHG emissions.

Methanol as fuel in combustion engines almost eliminates SO_x; NO_x emissions are still present but compliant with Tier II regulation. To comply with Tier III regulation, it is possible to blend methanol with 25-45 % of water (MAN Energy Solution 2021). The particulate matter is cut down by up to 95% (Ellis, J. and Tanneberger, K. for EMSA 2016). The non-GHG onboard emissions are independent of the production path. Analyzing the fuel life-cycle, there are differences in polluting impact depending on the production feedstock. SO_x total emissions are mainly attributable to the production phase and reduced by approximately 92% compared to fossil products (IMO 2016). By contrast, NO_x emissions are produced predominantly during combustion and are more than halved compared to traditional fuels (IMO 2016). Using hydrogen in fuel cells eliminates the emissions of NO_x, SO_x, and PM (European Maritime Safety Agency 2023). On a WtW basis, those pollutants are present; their intensity depends on the feedstock used during production (E4tech & BEIS 2018).

This work focuses on local non-GHG emissions, and the MDO, hydrogen and methanol emission factors are summarized in Table 3. Methanol NO_x emissions are subdivided into Tier II and Tier III compliant.

Table 3. Non-*GHG* tank-to-wake emission (Methanol Institute 2023).

	NO_x [g/kWh]	SO_x [g/kWh]	PM_{10} [g/kWh]
MDO	9	0.36	0.23
Hydrogen	0	0	0
Methanol(Tier II)	5	0.007	0.034
Methanol(Tier III)	2.2	0.007	0.034

2.3 *Cost and availability*

Installing innovative technologies powered by alternative fuels leads to the need to analyze economic and availability-related aspects. In this work, the economic assessment is centralized on fuel costs, representing most of the Operational Expenditure (OpEx). The production technology, feedstock and maturity level are factors that influence costs.

Methanol is mainly produced from natural gas via steam reformation; this represents the lowest cost production method (Ellis, J. and Tanneberger, K. for EMSA 2016). Methanol from renewable sources and energy is a small reality with only a few plants worldwide, and its cost is inevitably high compared to fossil-based methanol. Gray methanol is currently available at over 120 ports worldwide (Methanol Institute 2023). The 2023 average bunker price of gray methanol in Rotterdam port was about 320 \$/t (Ship & Bunker Website 2023). Bio- and e-methanol prices are available in the literature; there are optimistic forecasts with decreasing costs (Methanol Institute and IRENA 2021). Shipping companies and producers are engaging in strategic partnerships to scale up green methanol production. Hydrogen from fossil feedstocks is available on a large scale today, but its *GHG* footprint is not competitive with traditional marine fuels. Green hydrogen from renewable sources is the most environmentally friendly but is a nascent reality, which results in high costs and low availability (European Maritime Safety Agency 2023). Blue hydrogen is considered a low-carbon fuel compared to fossil products. It is a cost-effective option for hydrogen production as the technology is mature and infrastructures are well-established (Metwally, Ashraf). The literature provides cost in €/*GJ* for different types of hydrogen Table 4 summarizes the fuel costs used for this work.

The annual fuel cost is estimated with the following equation:

$$Fuel\,Cost = \sum_{i=1}^{n} FP_i \cdot FCons_i \qquad (2)$$

Where *FP* represents the fuel price express in *\$/t*, *FCons* is the annual fuel consumption express in tons, and the subscript *i* indicates the ith fuel type used in the various technologies installed in the power plant.

Table 4. Fuel costs.

Fuel Type		Cost [$\$/ton$]	Ref.
	MDO	900	(Ship & Bunker Website 2023)
	Gray Methanol	320	(Ship & Bunker Website 2023)
	Green Methanol	676	(Methanol Institute and IRENA 2021)
	e-Methanol	1300	(Methanol Institute and IRENA 2021)
Fuel Type		Cost [€/*GJ*(\$/*ton*)]1	Ref.
	Blue Hydrogen	35 (4586.4)	(European Maritime Safety Agency 2023)
	Green Hydrogen	75 (9828)	(European Maritime Safety Agency 2023)

[1] Conversion -1€ = 1.09 \$ (18.12.23) and $LHV\ H_2$ = 120000 *kJ/kg*

3 POWER SYSTEM MODELING AND OPTIMIZATION PROBLEM FORMULATION

The components modelled in this work are combustion engines, batteries, and fuel cells. The proposed modelling, extensively described in (Maloberti & Zaccone), use a static approach, neglecting the components' dynamics and obtaining a uniform level of approximation for the different models. It is worth noting that the models used are simplified and known from the literature, integrated to build a model of system performance for use in optimization, i.e., to compare solutions. The ability of the model to provide accurate performance estimates under specific conditions is therefore secondary to its ability to model the physical behavior of the system within the computational domain. Combustion engine modelling involves using Brake Specific Energy Consumption (*BSEC*) to represent fuel mass flow rate as a function of power output. The model can also be used with dual-fuel engines powered by any fuel. In this case, a pilot fuel is used. To estimate the specific consumption of the generic fuel and the pilot fuel, the following equation must be satisfied:

$$BSEC = SXC\,LHV_X + SPOC\,LHV_{pilot} \qquad (3)$$

Where SXC, and LHV_x are the specific consumption and the lower heating value of the generic fuel, $SPOC$, and LHV_{pilot} are the specific consumption and the lower heating value of the pilot fuel.

The State-of-Charge (SoC) of the battery is modeled according to the following equation:

$$\begin{cases} SoC(t) = SoC(0) - \frac{\beta}{E_{BESS}} \int_0^T P_{BESS}(t)dt \\ \beta = \begin{cases} & \text{if } P_{BESS} \leq 0 \\ \eta_{BESS} & \text{if } P_{BESS} > 0 \end{cases} \end{cases} \quad (4)$$

Where $SoC(0)$ is SoC at time $t = 0$, η_{BESS} is the battery efficiency, P_{BESS} is the power delivered by the $BESS$, which can be positive or negative, depending on whether it is in the charging or discharging phase, and E_{BESS} is the nominal energy stored in the battery. For technological reasons, the State-of-Charge should be kept within its central range, usually between 30% and 85%. Fuel cells are electrochemical devices that provide direct current through the hydrogen oxidation reaction. In this work, fuel cells are modeled by a purely resistive equivalent circuit, and the hydrogen mass flow rate is calculated as a function of the cell current according to the reaction stoichiometry.

The presented optimization strategy aims to calculate the optimal power split to reduce the environmental impact of the considered propulsion solution through the minimization of the equivalent CO_2 emissions for a given mission profile. The inputs to the problem are the required mission power profile and the propulsive system configuration. The mission power profile is defined in discrete time. Each instant $t_j, j = 0 \ldots T$ corresponds to changes in the required power. The variables identifying the optimal solution are the powers delivered by the generators and fuel cells, expressed as a fraction of the respective maximum power through an appropriate load coefficient α. Battery power is calculated as the difference between the load and the power delivered by the other components. The equivalent CO_2 emissions are calculated summing the contribution of the fuel cell and the contribution of the generators. The amount of equivalent CO_2 associated with the use of fuel cell is evaluated according to the following equation:

$$CO_{2eq}^{FC} = EF_{H_2} \sum_{j=0}^{T} \dot{m}_{H_2}(\alpha_{FC}^{j}) \Delta t_j \quad (5)$$

Where EF_{H_2} is the equivalent CO_2 emission factor, depending on the hydrogen production technology, and $\dot{m}_{H_2}$ is the hydrogen mass flow rate for each time instant j. The equivalent CO_2 emitted by the generators is calculated summing the emissions due to the pilot fuel:

$$CO_{2eq}^{pilot} = EF_p \sum_{j=0}^{T} \sum_{i=1}^{n} \alpha_{gen_i}^{j} P_i^{max} \, SPOC \, \Delta t_j \quad (6)$$

And the emissions due to the generic fuel:

$$CO_{2eq}^{methanol} = EF_m \sum_{j=0}^{T} \sum_{i=1}^{n} \alpha_{gen_i}^{j} P_i^{max} \, SXC \, \Delta t_j \quad (7)$$

Where subscript p and m refer to pilot fuel and methanol, respectively.

The optimization problem subject of this paper is a constrained minimum problem. The power generated by each generator and fuel cell can never be lower than zero and must always be less than the maximum power each component can deliver. The power that charges or discharges the $BESS$ cannot exceed its maximum power limits. The power balance constraint, therefore, takes the following form:

$$-P_{BESS}^{max,c} \leq P_{FC_j} + \sum_{i=1}^{n} P_{i,j} - P_{load_j} \leq P_{BESS}^{max,d} \quad (8)$$

Where P_i is the power delivered by the ith generator, P_{FC} is the power delivered by the Fuel Cell, P_{BESS} is the $BESS$ power, P_{load} is the power demand, and $P_{BESS}^{max,d}$, $P_{BESS}^{max,c}$ are the maximum power in discharge and charge, respectively.

The SoC of the $BESS$ is limited between a minimum and a maximum value, to preserve the battery state of health. The initial and final state of charge of the battery must be equal. Otherwise, the battery would have to be recharged with the onshore grid, which would impact the overall emissions calculation. Eventually, the consumption of hydrogen, methanol and pilot fuel must not exceed the respective amount available on board.

The presented constrained minimum problem has been solved using MATLAB's nonlinear programming solver "fmincon". A multi-start approach has been adopted to avoid local minima. The solver leveraged the interior point algorithm Byrd et al. 2000 with default settings. Calculations were performed using MATLAB R2021b and parallel computing toolbox. The average computational times ranged from around 500 s for configurations G and G/B to around 1300 s for configuration G/B/FC because of the larger domain dimensionality. The analysis has been performed on an AMD Ryzen 9 6900HS 4.935GHz CPU, 8-core/16-thread, with 16Gb DDR5-SDRAM.

4 CASE STUDY

The proposed methodology is applied to a ferry boat used as a case study; its main characteristics are presented in Table 5. The case-study vessel is applied in northern Italy, specifically in the Ligurian Riviera, and connected La Spezia and Levanto. The route covered by this touristic boat crosses the "Cinque Terre" marine protected area. Different configurations are presented to provide power to two electric propulsion motors (EPMs) that drive two fixed-pitch propellers, and meet the hotel load. The technologies used for the various power system layouts include 3 × 400 kW Generator Sets, a 1 MWh battery pack, and 10 × 34 kW fuel cell modules. Three configurations were evaluated; the first is composed of generator sets only (G), the second is a hybrid solution with generators and *BESS* (G/B), and the other one uses all the technologies generators, *BESS*, and fuel cells (G/B/F). The various solutions lead to displacement variations and, consequently, different required power. Specifically, it increases as the weight on board increases. To consider this variation, the required power curve for the reference configuration is scaled according to the following equation:

$$P_B^{hyb,i}(V) = \left(\frac{\Delta_{hyb,i}}{\Delta_{ref}}\right)^{\frac{2}{3}} P_B^{ref}(V) \tag{9}$$

Where hyb,i subscript represent the ith hybrid configuration, i.e. G/B or G/B/F, and *ref* subscript represent the reference configuration G. Figure 1 shows a schematic representation of the complete hybrid propulsion plant, corresponing to the G/B/F configuration.

The ship's mission profile is designed to conduct a one-day mini-cruise through the "Cinque Terre" marine protected area (Figure 2). The ship travels from La Spezia to Levanto and back, making short stops at Portovenere, Riomaggiore, Manarola, Vernazza, and Monterosso. When cruising outside the marine protected area, the ship proceeds at a cruising speed of 18 Kn, while inside the marine protected area, the cruising speed is reduced to 13 Kn. In intermediate ports, there is a 20-minute stopover, while in Levanto, there is a longer 80-minutes stopover. Figure 3 shows the power required by the mission profile of the reference configuration. The hotel load increased the power demand for the propulsion, assumed constant and equal to 50 kW. For hybrid configurations, the required power has been adapted according to the displacement variations. The algorithm is used in the three plant configurations by also considering variations in the type of fuel used to power the generators and fuel cells:

- G: the reference configuration. Only generators are considered on-board. This configuration is divided into three sub-cases that differ in the fuel used to power generator:
  - G0: Marine Diesel Oil.
  - G1: "gray" methanol, produced through natural gas reforming.
  - G2: "green" methanol produced from wood.
- G/B: In this configuration, batteries are used in addition to generators. Two sub-cases also occur in this case, distinguished by the fuel used:
  - G1/B: "gray" methanol, produced through natural gas reforming.
  - G2/B: "green" methanol produced from wood.
- G/B/F: The complete hybrid system is considered in this configuration, with even the fuel cell on board. It is split into five sub-cases, which differ in the fuel burned to power the generators and the hydrogen used in the fuel cells:
  - G1/B/F1: "gray" methanol, and "blue" hydrogen with 56% carbon capture and storage;
  - G2/B/F1: "green" methanol produced from wood, and "blue" hydrogen;
  - G1/B/F2: "gray" methanol and ``green" hydrogen from the Norwegian production toolchain;
  - G2/B/F2: "green" methanol from wood and "green" norwegian hydrogen;
  - G3/B/F2: "green" e-methanol produced using renewable electricity and feedstock, and "green" norwegian hydrogen. This sub-case was analyzed to consider a particularly virtuous case, although impractical in the current state of the technology.

Table 5. General and propulsion system specification of the case study vessel.

Length between perpendiculars	28.73 m
Breadth	7.02 m
Depth	2.60 m
Draft	1.22 m
Displacement Base	97.5 t
Displacement Hybrid 1	101 t
Displacement Hybrid 2	105 t
Cruise Speed	18 Kn
Low Speed	13 Kn
Passenger	516
Generators power	400 kW
Number of generators	3
Battery capacity	1 MWh
η_{BESS}	0.97
Fuel Cell power	340 kW
Hydrogen on-board	180 kg
Methanol on-board	1.2 t
MDO on-board	200 kg

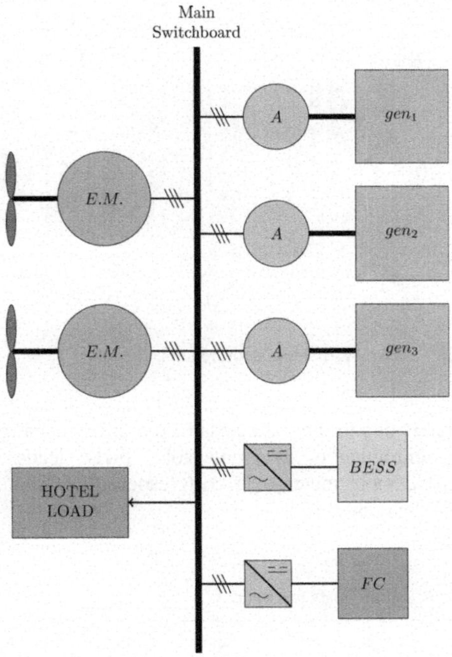

Figure 1. Propulsion system configuration scheme.

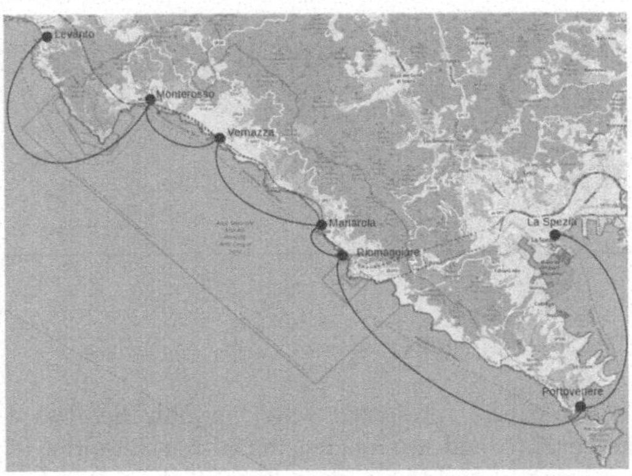

Figure 2. The ``Cinque Terre'' marine protected area.

5 RESULTS

This section presents the results of applying the proposed methodology in the case study under analysis. These results are based on the assumption of 310 working days per year, with one round trip per day. The required power profile for the vessel's mission is shown in Figure 3.

The optimization algorithm has been applied in different propulsion layouts and with several fuel types. For each case, the main result is a Power Split diagram that allows the minimization of *GHG* emissions. Figure 4 shows the graphical

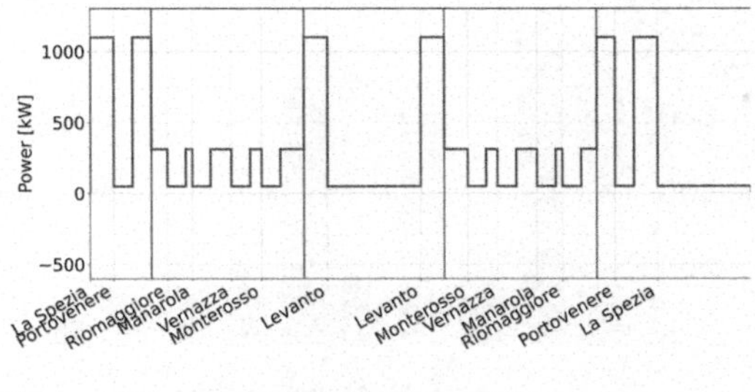

Figure 3. Required power profile.

solution for configuration G2/B/F2. In this configuration, all plant components can generate power. The emission factor of the hydrogen used in this configuration is the lowest available, corresponding to green hydrogen produced through electrolysis, with energy coming mainly from renewable sources. Meanwhile, the methanol considered is green methanol produced from wood. Figure 4 shows that the generators are used only in areas outside the marine protected area, except the stops in Manarola on the outward route-leg and the stops in Vernazza and Riomaggiore on the return route-leg.

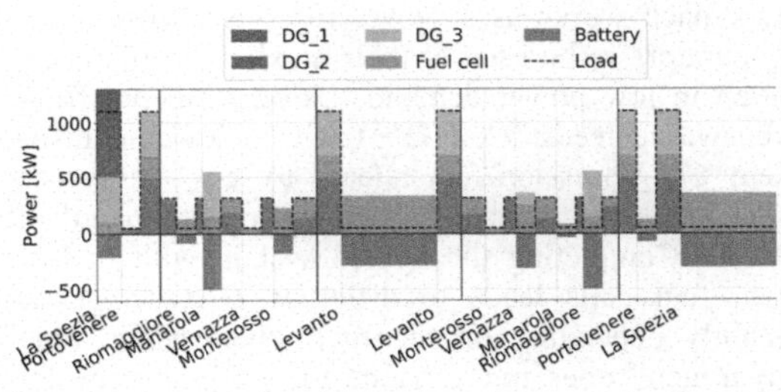

Figure 4. Configuration G2/B/F2.

Moreover, the consumption of various fuels can be obtained in every configuration and used to calculate the performance indicators (Figure 5). Figure 6 shows the results obtained for the equivalent CO_2 minimization.

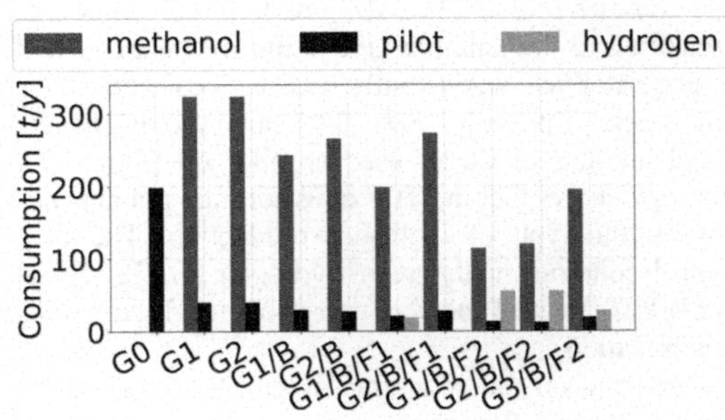

Figure 5. Annual-based consumption comparison.

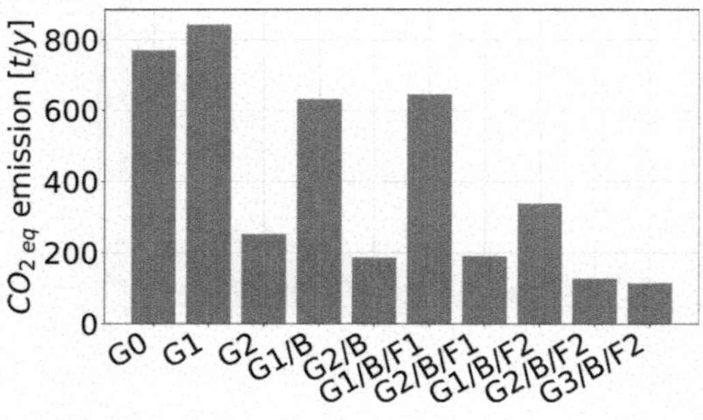

Figure 6. Equivalent CO_2 (WtW) emitted per year.

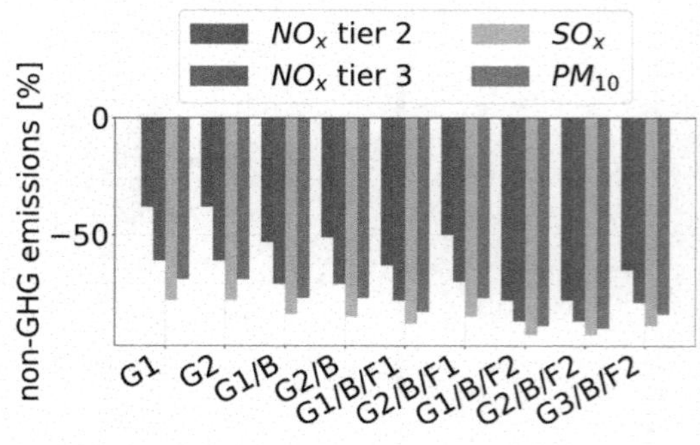

Figure 7. Non-*GHG* emission results.

The reference configuration, G0, with a traditional diesel-fuelled generator set, with almost 700 t/year of CO_{2eq} emissions, is one of the most polluting solutions. Using methanol in non-hybrid propulsion layouts is advantageous from an environmental point of view only when using renewable options. Differently, grey methanol worsens the environmental impact of the ship according to the WtW approach. With the integration of the *BESS* into the propulsion system (G1/B and G2/B), the optimization algorithm provides a decreasing in methanol consumption (Figure 5) and, consequently, of *GHG* emissions. The battery pack works as a power reserve, allowing the generators to be used at their maximum efficiency, even in low power demand. Using methanol from renewable feedstock (G2/B) with a lower equivalent CO_2 emission factor leads to less battery use and increased methanol consumption compared to G1/B. Completing the hybrid system with hydrogen fuel cells leads to different *GHG* impacts, mainly depending on the fuel type. Using "blue hydrogen" does not enhance the polluting impact of this plant solution. The optimization shows that G1/B/F1 is slightly more pollutant than G1/B, while in configuration G2/B/F1, the fuel cell are not used from the algorithm (Figure 5) due to the high emission factor of blue hydrogen. Considering renewable hydrogen allows a significant reduction in CO_{2eq} footprint and G2/B/F2 is one of the best solutions on an equivalent CO_2 basis.

The results of local non-*GHG* emissions are summarized in Figure 7. The figure represents the percentage variation of NO_x, SO_x, and PM_{10} emissions compared to the solution G0. Using methanol as fuel in generator sets significantly reduces local non-*GHG* emissions, especially SO_x attributable only to the small amount of diesel used as pilot oil (Figure 5). Figure 7 shows that the TtW emissions are not dependent on the type of methanol production. The full-hybrid solutions with green hydrogen (G1/B/F1 and G2/B/F2) are the best options for local emissions minimization.

Focusing on the economic aspect, the cost assessment for actual fuel prices is summarized in Table 6. The gray methanol solutions are the most economic solutions due to the high costs of diesel oil nowadays. The introduction of renewable fuels leads to a considerable increase in costs especially with full-hybrid layouts.

Table 6. Fuel costs per year.

Configuration	Cost [*k$/year*]	Variation
G0	179	–
G1	140	− 22%
G2	255	+ 43%
G1/B	105	− 41%
G2/B	205	+ 14%
G1/B/F1	170	− 5%
G2/B/F1	210	+ 17%
G1/B/F2	597	+ 234%
G2/B/F2	640	+ 258%
G3/B/F2	555	+ 210%

6 CONCLUSION

This paper presented an optimization-based approach aimed at reducing the carbon footprint of a marine propulsion system through the minimization of its equivalent CO_2 emissions. The use of innovative technologies and alternative fuels led to the need for a preliminary economic investigation. The considered case study is a small ferry operating in marine protected areas; thus a local non-*GHG* emissions estimation was assessed.

The optimization algorithm was applied to several propulsive configurations proposing the use of methanol and hydrogen as fuel. The optimization provides the power split for each propulsive solution that allows the minimization of *GHG* emissions on a Well-to-Wake basis. The results show that adopting a hybrid configuration reduces fuel consumption and the carbon footprint of the vessels. When renewable fuels are used, emissions can be reduced to one-eighth compared to the diesel configuration. The local impact of other pollutants is highly reduced in every solution proposed. The full-hybrid is the best option, as fuel cell technology is inherently locally zero-emission.

The preliminary analysis of fuel costs highlights the significant cost increase when considering renewable fuels from non-fossil sources. Introducing hydrogen fuel cells is noncompetitive, but a hybrid power plant gives flexibility and the possibility of adapting the system, considering both ecological and financial issues.

REFERENCES

Adami, G. & M. Figari (2023). Feasibility analysis of a methanol fuelled bulk carrier. *Modelling and Optimisation of Ship Energy Systems 2023*.

Belvisi, D., R. Zaccone, M. Figari, S. Simone, & B. Spanghero (2022). Bess-based hybrid propulsion: An application to a front line naval vessel preliminary design. In *Technology and Science for the Ships of the Future*, pp. 154–161. IOS Press.

Benet, Á., A. Villalba-Herreros, R. d'Amore Domenech, & T. J. Leo (2022). Knowledge gaps in fuel cell-based maritime hybrid power plants and alternative fuels. *Journal of Power Sources* 548.

Bouman, E., E. Lindstad, A. Rialland, & A. Strømman (2017). Stateof-the-art technologies, measures, and potential for reducing ghg emissions from shipping–a review. *Transportation Research Part D: Transport and Environment*, 52, 408–421.

Byrd, R. H., J. C. Gilbert, & J. Nocedal (2000). A trust region method based on interior point techniques for nonlinear programming. *Mathematical programming 89*, 149–185.

Council of European Union (2023). Fueleu maritime initiative: Council adopts new law to decarbonise the maritime sector.

DNV-GL (2018). Alternative fuels: the options.

DNV-GL (2023). Methanol vessel orderbook gains momentum, ammonia sees a 'breakthrough'–dnv.

E4tech & BEIS (2018). H2 emission potential literature review. Technical report.

Ellis, J. and Tanneberger, K. for EMSA (2016). Study on the use of ethyl and methyl alcohol as alternative fuels in shipping. Technical report.

European Environment Agency (2023). Transport and mobility.

European Maritime Safety Agency. Discover the eu maritime profile.

European Maritime Safety Agency (2023). Potential of hydrogen as fuel for shipping. Technical report.

Fruzzetti, C., S. Donnarumma, F. Maggiani, & M. Martelli (2023). Probabilistic operability and greenhouse gas assessment during dynamic positioning operations. *Proceedings of the Institution of Mechanical Engineers, Part M: Journal of Engineering for the Maritime Environment*, 14750902231203441.

Hermesmann, M. & T. Müller (2022). Green, turquoise, blue, or grey? environmentally friendly hydrogen production in transforming energy systems. *Progress in Energy and Combustion Science 90*, 100996.

IMO (2016). Methanol as marine fuel: Environmental benefits, technology readiness, and economic feasibility. Technical report.

IMO (2018). 2018 initial imo strategy.

IMO (2021). Fourth imo ghg study 2020. Technical report.

International Energy Agency. Tracking transport.

Karatug, C., Y. Arslanoglu, & C. G. Soares (2022). Evaluation of decarbonization strategies for existing ships. *Trends in Maritime Technology and Engineering*, 45–54.

Karatüg,, Ç.., B. O. Ceylan, E. Ejder, & Y. Arslanog lu (2023). Investigation and examination of lng, methanol, and ammonia usage on marine vessels. In *Decarbonization of Maritime Transport*, pp. 65–85. Springer.

Karvounis, P., J. L. Dantas, C. Tsoumpris, & G. Theotokatos (2022). Ship power plant decarbonisation using hybrid systems and ammonia fuel - a techno-economic-environmental analysis. *Journal of Marine Science and Engineering* 10.

Maloberti, L. & R. Zaccone. An environmentally sustainable energy management strategy for marine hybrid propulsion. *Available at SSRN 4621257*.

Maloberti, L., R. Zaccone, P. Gualeni, & P. Mazzucchelli (2022). A zeroemission ferry for inland waterways. In *Technology and Science for the Ships of the Future*, pp. 162–169. IOS Press.

MAN Energy Solution. H2 – key player in the maritime energy transition.

MAN Energy Solution (2021). The methanol fuelled man b&w lgim engine. Technical report.

Methanol Institute (2022, January). Carbon footprint of methanol. Technical report.

Methanol Institute (2023, May). Marine methanol: Future-proof shipping fuel. Technical report.

Methanol Institute and IRENA (2021). Innovation outlook: Renewable methanol. Technical report.

Metwally, Ashraf. Blue hydrogen vs green hydrogen: A closer look at production, usage, and environmental cost.

Ship & Bunker Website (2023).

Tadros, M., M. Ventura, & C. Guedes Soares (2021). Design of propeller series optimizing fuel consumption and propeller efficiency. *Journal of Marine Science and Engineering 9*(11), 1226.

Tadros, M., M. Ventura, & C. Guedes Soares (2023a). Fuel consumption analysis of single and twin-screw propulsion systems of a bulk carrier. *Journal of Marine Science and Application 22*(4), 741–750.

Tadros, M., M. Ventura, & C. Guedes Soares (2023b). Review of the imo initiatives for ship energy efficiency and their implications. *Journal of Marine Science and Application 22*(4), 662–680.

Tadros, M., M. Ventura, & C. G. Soares (2019). Optimization procedure to minimize fuel consumption of a four-stroke marine turbocharged diesel engine. *Energy 168*, 897–908.

Tadros, M., M. Ventura, & C. G. Soares (2023). Review of current regulations, available technologies, and future trends in the green shipping industry. *Ocean Engineering 280*, 114670.

Tammy, K. f. M. I. (2021, February). Methanol: a future-proof fuel. Technical report.

Transport & Environment (2022). Climate impacts of exemptions to eu's shipping proposals. Technical report.

Vasilikis, N., R. Geertsma, & A. Coraddu. A digital twin approach for maritime carbon intensity evaluation accounting for operational and environmental uncertainty. *Ocean Engineering*.

Advances in Maritime Technology and Engineering – Guedes Soares & Santos (Eds)

Analysis of ships' performance in port: AIS, experimental campaign, and simulation

L. Mocerino & F. Quaranta
University of Naples Federico II, Naples, Italy

ABSTRACT: The behavior of a passenger ship in a port, concerning power, speed, consumption, and emissions, can be investigated using various methods. To illustrate these different approaches, the outcomes of these three approaches have been compared on a case study of a passenger ship, for which the results of an experimental campaign, a bottom-up method based on AIS data, and the engine simulation results were available. The experimental campaigns were conducted along the vessel's service routes. For the bottom-up approach, AIS data were processed using a custom MATLAB code to generate a detailed schedule of all in-port operations, and emissions were estimated using a globally recognized methodology based on emission factors. The engine simulation was calibrated using test bench results and validated during sea trials. Comparisons were made across various parameters, including speed, consumption, power, and emissions.

1 INTRODUCTION

Maritime transportation plays a pivotal role in global trade and passenger transport, contributing significantly to economic development and connectivity. However, the environmental consequences of shipping activities, particularly in port areas, have raised concerns regarding air quality, human health, and overall ecosystem well-being (Adami & Figari 2023; Ay et al. 2022; Mocerino et al. 2023; Palomba et al. 2022). The top ten of the environmental priorities of ESPO (European Sea Port Organization) reported in the Environmental Report 2023 highlights some encouraging trends (ESPO, 2023). Climate change and CO_2 emissions, along with other pollutants that have global impacts, take precedence in port priorities. Following closely is air quality, addressing emissions of pollutants on a local scale such as sulfur oxides, nitrogen oxides, unburnt hydrocarbons, and particulate matter. Additional focal points include energy efficiency, noise reduction, water quality, management of ship waste, community engagement, port development, and waste management associated with port activities, especially those related to water (ESPO, 2023). Monitoring ship emissions in ports is of paramount importance to assess and mitigate the environmental impact of maritime activities, promoting sustainable practices and safeguarding air quality in coastal areas (Bojić et al. 2023; Mocerino et al. 2023). The methods employed for estimating maritime emissions can be broadly categorized into three main approaches: experimental measurements conducted on board or in port, the utilization of emission factors, and the application of numerical simulation models for engine performance (from power to emissions). Recently, researchers have developed an alternative method of emissions estimation method based on the Automatic Identification System (AIS). This type of method adopts the data automatically collected from AIS to estimate the emissions from operational data (navigation, port, anchored, etc.) (Kivekäs et al. 2014; Mocerino et al. 2023; Winther et al. 2014).

Among the various vessels navigating through ports, ferry and high-speed vessels stand out as crucial contributors to emissions due to their frequent and often prolonged stays at terminals. Ferry vessels, designed to transport passengers and cargo across water bodies, are essential components of many coastal and inland transportation networks. While they offer efficient and convenient services, the combustion of fossil fuels in ferry engines releases a variety of pollutants into the atmosphere, including particulate matter (PM), nitrogen oxides (NOx), sulfur oxides (SOx), and carbon dioxide (CO_2). The proximity of ferry terminals to urban centers intensifies the impact of these emissions, posing potential risks to public health and the environment.

This paper aims to comprehensively investigate and evaluate the emissions generated by ferry vessels during their stay and navigation in port, shedding light on the extent of their environmental footprint. By focusing on emissions profiles, the study seeks to quantify the levels of pollutants released and assess their impact on air quality in and around port areas.

Understanding the specific challenges posed by ferry emissions in port environments is crucial for the development of targeted policies and strategies aimed at minimizing their adverse effects. The objectives of this study are outlined to achieve a comprehensive

DOI: 10.1201/9781003508779-6

understanding of marine engine emissions during their port operations. This will be accomplished through a methodology that integrates measured data, estimates based on the AIS along with emission factors, and simulation results from engine modeling.

The primary objective of the experimental campaign has been to acquire reliable and accurate data on marine engine emissions during their port and cruising activities. These measurements will serve as a fundamental basis for understanding the composition and extent of pollutants released into the surrounding environment. Measured emissions will be compared with estimates derived from monitoring vessel activities through AIS data. This approach allows for an assessment of the consistency between real data and predictions based on vessel position and speed, providing a broader perspective on the environmental impact of port operations. Then, known emission coefficients for different engine types and fuels will be applied to estimate emissions, facilitating a detailed comparison between measured data and those anticipated based on standardized models. Finally, engine simulations will be conducted to predict emissions under specific operating conditions. This theoretical approach offers insights into expected emissions and will be compared with empirically measured data, contributing to the validation and enhancement of engine simulation models. Through this integrated approach, we aim to obtain a comprehensive assessment of marine engine emissions in ports, enabling cross-validation between empirical data, AIS-based estimates, and engine simulations. This work will help delineate a more accurate picture of pollution sources in port environments and aid in the development of targeted strategies for mitigating the environmental impact of maritime operations.

2 CASE STUDY

The ship chosen as the case study is *Achernar* an high-speed catamaran operating in the Mediterranean Sea in particularly in the Gulf of Naples (connecting Naples to the island of Capri, Ischia e Procida).

Figure 1. Achernar.

The main characteristics of the ships and of the engine installed on board are presented in Tables 1 and 2.

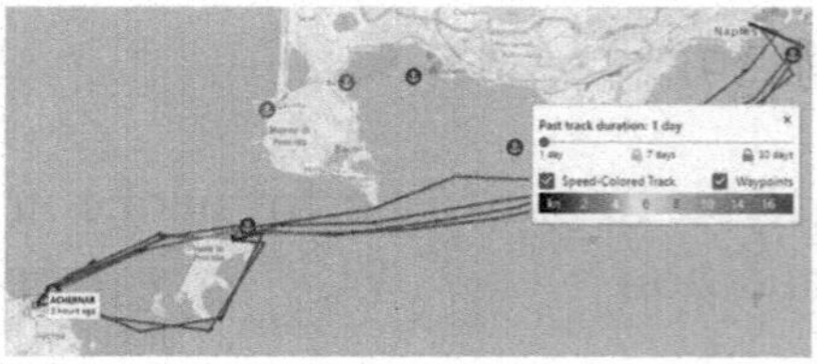

Figure 2. Typical routes (from https://www.marinetraffic.com).

Table 1. Principal data.

Ship: Achernar	Data
Hull	Catamaran
Hull material	Light alloy
Launched	1993
GT	624 (t)
DWT	120 (t)
Loa x Breadth x Draught	43.7 x 10.9 x 1.64 (m)
Main Engine	2–MTU16V396TE74L
Engine Power (kW)	4000 (kW)
Propulsion system	S71 Kamewa waterjets
Max speed (kn)	34
Passengers	354

Table 2. Engine parmeters and performance.

Engine	Data
Number of cylinders	16
Power	2000 (kW)
Bore	165 (mm)
Stroke	185 (mm)
Displacement	63.3 (L)
Compression ratio	12.3
Number of valves per cylinders	4
Mean piston speed	12.33 (m/s)
Reference air temperature and pressure	25 (°C); 1×10^5 (Pa)
IVO/IVC Intake Valve Opening/Closing angle (°)	36 BTDC/ 68 ABDC
EVO/ EVC - Exhaust valve Opening/Closing angle (°)	75 BTDC/28 ATDC
SFC (g/kWh)	220

3 EXPERIMENTAL CAMPAIGN/EMISSIONS ACHERNAR

The measurement campaign held over 2 days followed several daily trips to the small islands in the Gulf. In particular, three trips were selected for the purpose of these analyses (see Figures 3,4 and 5).

The data logging system (National Instruments) was capable of registering (2 samples/second) the main parameters of the propulsion and navigation. In some cases, an exhaust gas analyzer was used to survey the concentration of the main pollutants in the exhausts. The monitored parameters were torque on the propeller shaft (kNm) (Extensimetric torque meter, Binsfeld Eng), power (kW), rotational speed (r/min) (Magnetic pick-up), fuel consumption (cm3/s), and exhaust gas temperature (°C). The pollutants measured, i.e., HC (ppm), CO (%vol), CO_2 (%vol), and NOx (ppm), were measured together with the concentration of O_2 in the air (Gas Analyser TecnoTest). The ship's position and speed were surveyed by a global positioning system (GPS). For each day, an average behavior with the mean consumption of both engines, maximum-measured speeds, and average engine speeds and loads have been recorded. Subsequently, we computed the same parameters for the approach and port stay phases, resulting in the findings presented in Table 4. As evident from the comparison between the two tables, fuel consumption during maneuvering phases at reduced engine speeds and RPM notably increases, while both maximum and average speeds decrease. Regarding emissions focusing on NO_X, CO_2, and HC the average values found during the measurements (all: the whole route and port: during port approach maneuvers) are shown in Table 5. The biggest difference is found for unburned Hydrocarbon emissions which, as we expected, increase at low engine speeds.

Table 3. Voyage behavior.

Voyage	Mean Sfc engine left (g/kWh)	mean Sfc engine right (g/kWh)	mean speed (kn)	max speed (kn)	mean engine power (kW)
1	223.7	227.6	21.5	28.2	2540
2	227.4	252.4	17.4	28.4	2213.7
3	236.7	248.3	17.4	28.2	2068

Table 4. In port behavior.

Port	Sfc engine left (g/kWh)	Sfc engine right (g/kWh)	mean speed (kn)	max speed (kn)	mean engine power (kW)
1	249.2	251.6	7.7	20.2	592.5
2	247.08	304.9	4.01	18.9	396.1
3	267.9	292.3	4.5	17.5	403.7

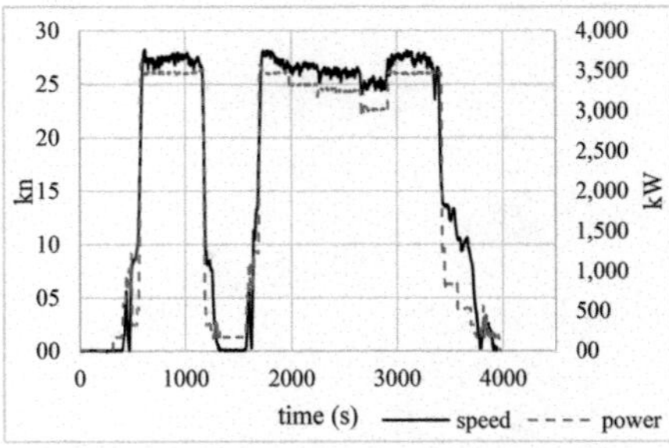

Figure 3. Voyage 1.

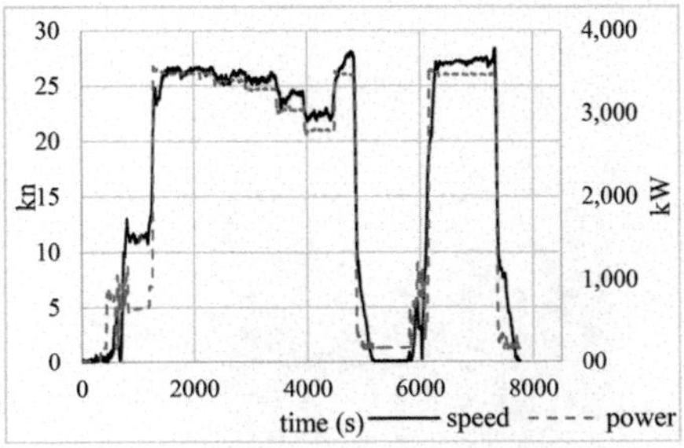

Figure 4. Voyage 2.

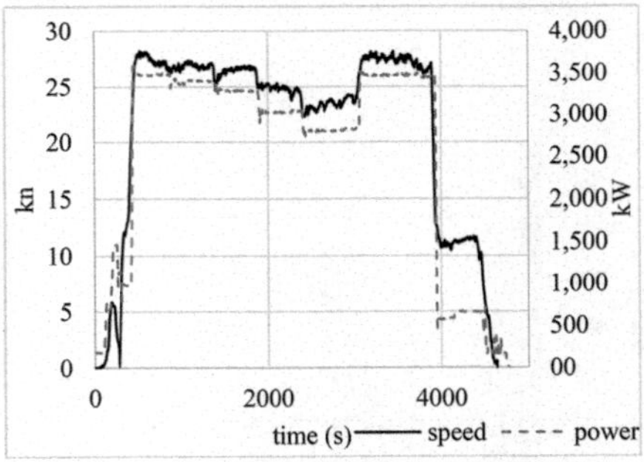

Figure 5. Voyage 3.

Table 5. Emissions measurements.

NO_X				CO_2		HC	
all	port	all	port	all	port	all	low
Ppm		g/kWh		% vol		ppm	
727.1	798.2	8.7	9.3	3.02	6.08	0.2	0.80
659.6	750.0	8.1	7.8	4.33	2.58	0.8	1.50
650.6	507.0	8.1	8.2	4.39	3.51	0.03	0.10

4 AIS DATA AND EMPIRICAL CRITERIA

The AIS data used refer to 2018 and concern all accesses to the port of Naples of the boat; in particular, the data characterise entry, manoeuvre, stopover and exit from the port of Naples. The main useful information is position (latitude, longitude), course angles, speed, and time. Figure 6 shows the position records acquired and Figure 7 the speed records.

On the speeds, a frequency analysis has been carried out for speeds greater than 0.5 kn (to exclude the numerous dwell data); the results obtained are shown in Figure 8.

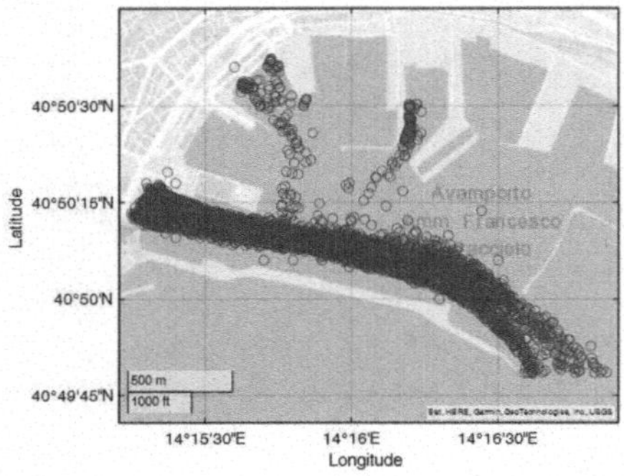

Figure 6. AIS data longitude and latitude signal.

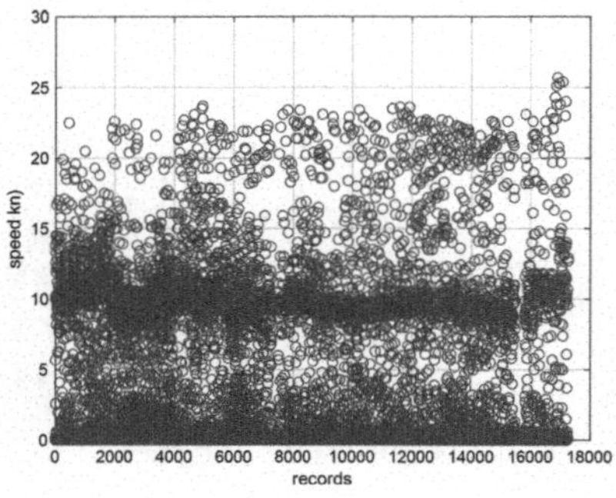

Figure 7. Speed.

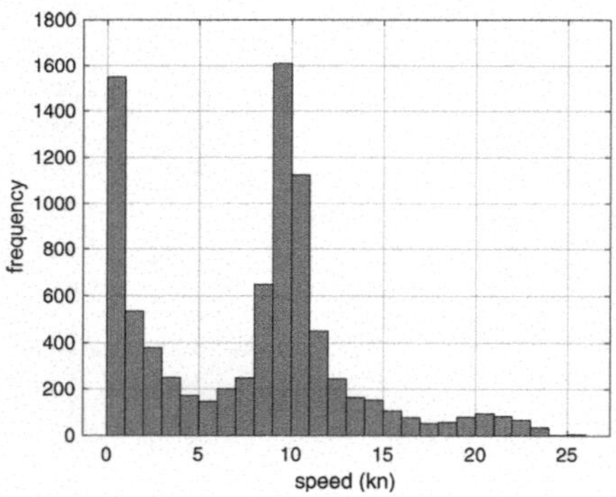

Figure 8. Frequency distribution of speed >0.5 kn.

The calculation of fuel consumption was carried out in accordance with the estimates available in Schwarzkopf et al. (2021) and Mocerino et al. (2023) for engine load factors (obtained by cubing the ratio between the recorded instantaneous speed and the vessel's operating speed) less than 0.25.

The results obtained are shown in Figure 9. Consumption is estimated by means of a third-level polynomial with the load factor as input variable and the coefficients shown below:

$$SOFC(EL) = \cdot\ 102.25 \cdot EL^2 - 170.47 \cdot EL + 274.63$$

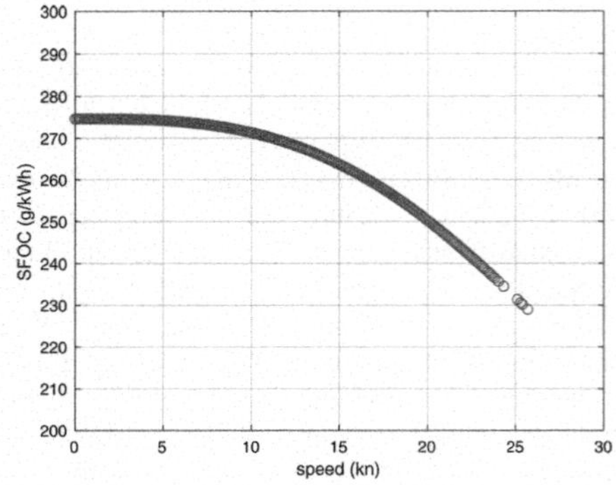

Figure 9. Specific fuel consumption.

As far as NO_X emissions are concerned, the expression used for obtaining the specific emission is as follows:

$$EF_{NOX}(EL) = 4.15 \cdot EL^2 + 5.39 \cdot EL + 7.91$$

For HC, on the other hand, an emission factor, in g/kWh, of 0.03 was used, with a correction for low loads as reported in Schwarzkopf et al. (2021).

Finally, for CO_2 emissions, an emission factor of 3.206 g/kWh was used for distillate fuel in accordance with the IMO.

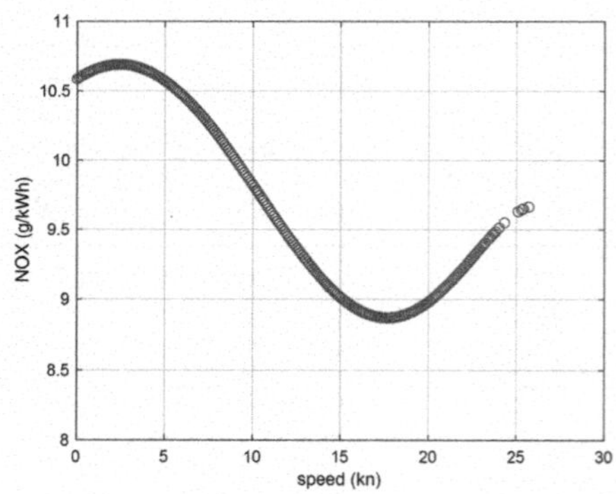

Figure 10. NO_X emissions based on AIS data + Emission factor.

On average, below 8 kn speed, the estimated specific emissions of NO_X are 10 g/kWh as regards CO_2 emissions at speeds below 8 kn range between 875 and 880 g/kWh.

5 SIMULATION

The last approach has been a simulation approach; the model has been realized in RICARDO Wave environment and the overall of the scheme is reported in Figure 11.

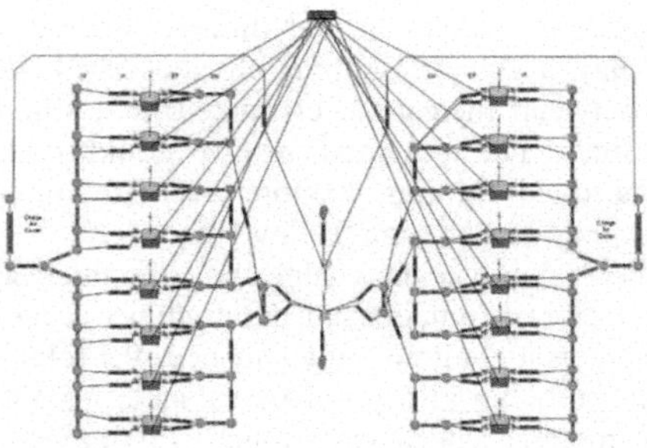

Figure 11. Engine model.

In the top center area is the engine block connected to all the cylinders, which are distributed on two banks, left and right. Each cylinder has four valves and one direct injector on the top. The turbocharger group is linked to the ambient the compressor is linked to the intercooler while the turbine is connected to the exhaust gas manifold and the ambient. For the initial conditions of a fluid used within a cylinder, the unique parameter set in a different way, than the defaults values, is the temperature, which has been set according to an average value measured during the experimental campaign equal to 305 K.

For friction, a modified form of the Chen-Flynn correlation is used to model it. In the absence of details about the turbocharger the developed procedure used turbine and compressor scaled maps.

The NOx emissions sub-model predicts emission production during combustion and exhaust in an engine cylinder element. The formation of NOx emissions is calculated using the extended Zeldovich mechanism, taking into account the two zones of combustion (burned and unburned zone). The Arrhenius multipliers of the chemical reactions in the NOx emissions model are calibrated depending on the data of NOx emissions provided from the datasheet of the manufacturer (Arrhenius pre-exponent multiplier: 1.5, Arrhenius relationship rate constant multiplier: 1.0). For HC, he recommended value, also used in this model, is "0.2"

The model has been calibrated on the results of bench tests in terms of consumption and power in accordance with Altosole et al. (2021) and Mocerino et al. (2022). Comparisons with sea trials in stationary conditions are shown in Figures 12 and 13.

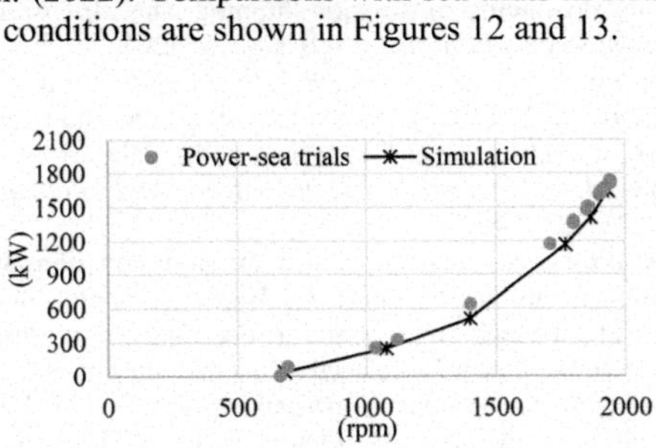

Figure 12. Validation on sea trials in stationary conditions (specific consumption).

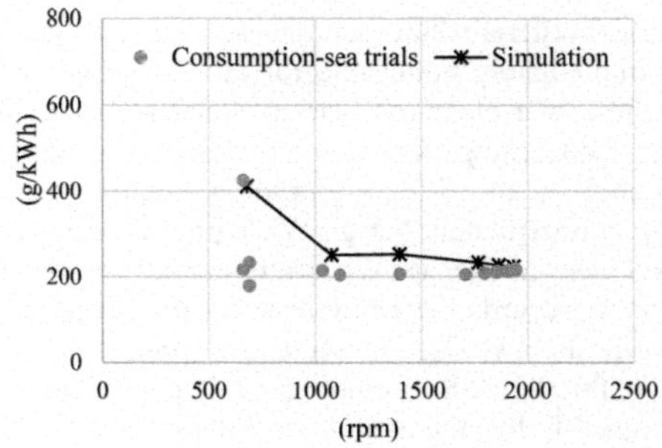

Figure 13. Validation on sea trials in stationary conditions (specific consumption).

The estimated emissions as regard NOX and HC are reported in Figures 14 and 15.

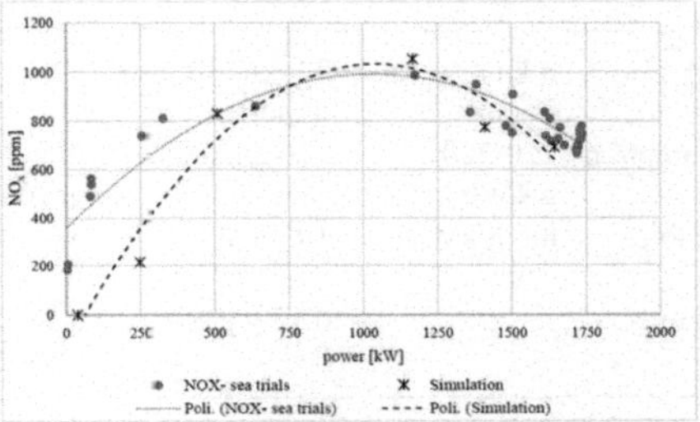

Figure 14. NOx emissions based on simulation results.

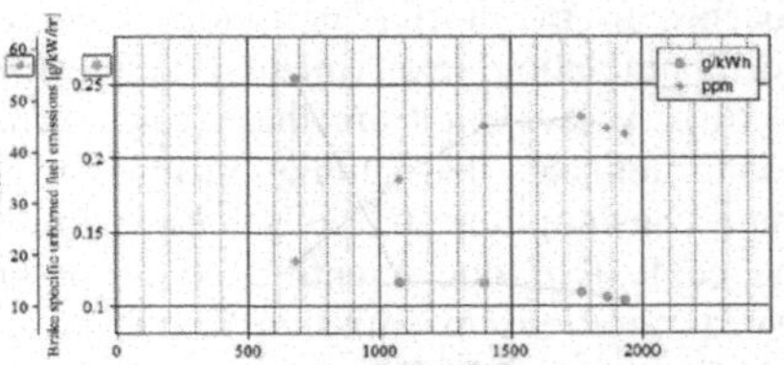

Figure 15. HC emissions based on simulation results.

6 COMPARISONS

Possible comparisons between two or all three of the proposed methods include speed (Measurements vs AIS+empirical criteria), consumption (Measurements, AIS + empirical criteria, and simulation), and emissions of NO_X, HC, and CO_2 (Measurements, AIS + empirical criteria, and simulation) (see Table 6). Comparing the speeds measured during the tests and those acquired through AIS data (considering the time elapsed between the two analyses) yields excellent results. On average, we measured 5 knots during port operations, while AIS data provides an average of 3.9 knots for the port of Naples. Regarding specific consumption, simulation at low engine loads yields values ranging from 255 to 242 g/kWh, measurements range from 250 to 300 g/kWh, while AIS +

empirical criteria results estimates around 270 g/kWh. It is immediately noticeable for all three approaches that there is a clear increase in consumption at low loads. Concerning Nitrogen Oxides, 10 g/kWh are estimated by the AIS + empirical criteria methodology, approximately 5.5 g/kWh with simulation, and an average of 8.5 g/kWh from experimental campaign. As regards the HC emissions the comparisons between the two last methods are promising: 0.15 g/kWh for the AIS + empirical criteria method and 0.18 g/kWh for the engine simulation model. The high variability of the HC measured during the experimental campaign suggests measurement spike problems at low loads.

Table 6. Comparisons.

	Experimental	AIS+EF	Simulation
Speed	5 kn as mean in port	3.9 kn as mean in port	
Consumption	250-300 g/kWh	270 g/kWh	255-242 g/kWh
Emissions NO_X	8.5 g/kWh	10 g/kWh	5.5 g/kWh
Emissions HC	-	0.15 g/kWh	0.18 g/kWh

7 CONCLUSIONS

This work seeks to compare three different approaches to monitoring emissions from a ship in port: simulation, experimental and estimation (AIS+EF). As evident from these values, the current state indicates that simulation tends to underestimate consumption and emissions at low loads. In the context of our scientific investigation, the observed variations in emission results arise from several intrinsic and extrinsic factors that merit careful consideration. Firstly, the inherent variability in measurement techniques introduces a level of uncertainty, as different instruments may have subtle differences in sensitivity and calibration. Additionally, the complex nature of the emission sources under study, coupled with the dynamic environmental conditions, contributes to the discrepancies between measured values and estimated emissions.

Moreover, the reliance on simulation models introduces a layer of approximation, where the fidelity of the model is contingent on the accuracy of underlying assumptions and input parameters. Small discrepancies may emerge because of simplifications made to facilitate computational efficiency or due to the evolving nature of emission sources over time.

Furthermore, temporal, and spatial variability in emissions, which may not be fully captured by the measurement campaign or adequately represented in the simulation model, can contribute to the observed differences. The intricacies of atmospheric dispersion and mixing processes can also play a role in shaping the observed disparities between measured, estimated, and simulated emissions. The obtained results could certainly be improved through a more accurate modeling of the engine behavior at low loads.

It is crucial to acknowledge and critically evaluate these factors to understand our findings accurately. This understanding not only enhances the robustness of our study but also provides valuable insights for future research endeavors aiming to refine emission quantification methodologies.

REFERENCES

Adami, G., & Figari, M. (2023). Feasibility Analysis of a Methanol Fuelled Bulk Carrier. *Modelling and Optimisation of Ship Energy Systems 2023*.

Altosole, M., Benvenuto, G., Campora, U., Silvestro, F., & Terlizzi, G. (2018). Efficiency improvement of a natural gas marine engine using a hybrid turbocharger. *Energies*, *11*(8), 1924.

Altosole, M., Balsamo, F., Mocerino, L., Quaranta, F., Campora, U., & Rizzuto, E. (2021). Analysis of ship performance data for the evaluation of marine engines emissions in ports. In *Maritime Technology and Engineering 5 Volume 2* (pp. 449–456). CRC Press.

Altosole, M., Balsamo, F., Campora, U., & Mocerino, L. (2021). Marine dual-fuel engines power smart management by hybrid turbocharging systems. *Journal of Marine Science and Engineering*, *9*(6), 663.

Ay, C., Seyhan, A., & Beşikçi, E. B. (2022). Quantifying ship-borne emissions in Istanbul Strait with bottom-up and machine-learning approaches. *Ocean Engineering*, *258*, 111864.

Bojić, F., Gudelj, A., & Bošnjak, R. (2023). An Analytical Model for Estimating Ship-Related Emissions in Port Areas. *Journal of Marine Science and Engineering*, *11* (12), 2377.

Ergin, S., Mocerino, L., & Quaranta, F. (2022). Possible approaches to the study of the emissions from ships during their operations in ports. In *Sustainable Development and Innovations in Marine Technologies* (pp. 371–378). CRC Press.

ESPO, 2023. https://www.espo.be/media/Top%2010%20environmental%20priorities%202023.pdf=.

Kivekäs, N., Massling, A., Grythe, H., Lange, R., Rusnak, V., Carreno, S., ... & Kristensson, A. (2014). Contribution of ship traffic to aerosol particle concentrations downwind of a major shipping lane. *Atmospheric Chemistry and Physics*, *14*(16), 8255–8267.

Mocerino, L., Soares, C. G., Rizzuto, E., Balsamo, F., & Quaranta, F. (2021). Validation of an emission model for a marine diesel engine with data from sea operations. *Journal of marine science and application*, *20*(3), 534–545.

Mocerino, L., & Quaranta, F. (2022). How emissions from cruise ships in the port of Naples changed in the COVID-19 lock down period. *Proceedings of the Institution of Mechanical Engineers, Part M: Journal of Engineering for the Maritime Environment*, 236(1), 125–130.

Mocerino, L., Murena, F., Quaranta, F., & Toscano, D. (2023). Validation of the estimated ships' emissions through an experimental campaign in port. *Ocean Engineering*, 288, 115957.

Palomba, G., Scattareggia Marchese, S., Crupi, V., & Garbatov, Y. (2022). Cost, energy efficiency and carbon footprint analysis of hybrid light-weight bulk carrier. *Journal of Marine Science and Engineering*, *10*(7), 957.

Schwarzkopf, D. A., Petrik, R., Matthias, V., Quante, M., Majamäki, E., & Jalkanen, J. P. (2021). A ship emission modeling system with scenario capabilities. *Atmospheric Environment: X*, *12*, 100132.

Tadros, M., Ventura, M., & Soares, C. G. (2023). Review of current regulations, available technologies, and future trends in the green shipping industry. *Ocean Engineering*, *280*, 114670.

Winther, M., Christensen, J. H., Plejdrup, M. S., Ravn, E. S., Eriksson, Ó. F., & Kristensen, H. O. (2014). Emission inventories for ships in the arctic based on satellite sampled AIS data. *Atmospheric Environment*, *91*, 1–14.

Advances in Maritime Technology and Engineering – Guedes Soares & Santos (Eds)

Preliminary modeling of a ferry methanol fuel cell power plant by using AVL Cruise M software

V. Sorrentino, M. Altosole, F. De Luca, L. Micoli, L. Mocerino & R. Russo
Industrial Engineering Department, University of Naples Federico II, Naples, Italy

ABSTRACT: This paper focuses on modelling and simulation of a full-electric power system for a - 25 m passenger transport vessel, equipped with an integrated Fuel Cell (FC)-battery system and using methanol as an alternative marine fuel. Each power plant component is simulated by AVL Cruise-M software to analyze the system dynamic performance under load variation. The study provides resistance and power data from the development of a new Hull Systematic Series, aimed at expanding the low-resistance speed range. Simulation results reveal that the selected battery modules effectively support the FC in meeting power demands during strong load variation. The FC system is intended to operate in steady state condition, due to the presence of methanol processing system. The average hydrogen consumption of 4.8 kg/h is estimated, and the amount of methanol to be stored on board to guarantee 6 voyages is approximately 359 kg.

1 INTRODUCTION

In recent decades, the maritime industry has faced escalating pressure to address environmental concerns and pursue sustainable alternatives (Mocerino et al. 2018). The constraint for sustainable practices within the maritime sector has become increasingly apparent, driven by both regulatory mandates and a growing global awareness of the environmental impact associated with traditional marine propulsion systems. Shipowners will be forced to reduce the environmental impact of their operations to comply with regulations that become increasingly stringent over the years. This change is expected to lead to an increase in the number of all-electric ferries along with ships adopting various configurations of hybrid or all-electric propulsion systems, often equipped with energy storage devices, such as batteries. In response to this necessity, full-electric ships and the use of Alternative Fuels (AF) have emerged as transformative solutions, promising a paradigm shift in maritime transportation towards a more environmentally conscious future (Altosole et al. 2022, 2023; Karatuğ 2023, 2024). Full-electric ships, powered by electric propulsion systems and energy storage technologies, have garnered attention as a viable means to reduce greenhouse gas emissions and enhance overall environmental stewardship (Bucci et al. 2016; Capasso et al. 2016; Karatuğ & Durmuşoğlu 2020). Alternative and renewable fuels such as LNG, hydrogen, biodiesel, biogas, and methanol, along with innovative technologies like Fuel Cells (FCs), and hybrid renewable energy systems (solar and wind), are considered possible options (Altosole et al. 2023; Bucci et al. 2016; Campora et al. 2023). All these technologies could be integrated into existing ships and new ships. The use of an all-electric power plant, with Electric Motors (EM) as prime movers, offers several benefits: space utilization, low vibration and noise, reduced fuel consumption, and increased automation and reliability (Del Pizzo 2010; Markowski & Pielecha 2019; Zhan 2022). Furthermore, by powering the EM via FCs, emissions reduction can be obtained by using AF, which drastically reduces polluting emissions (DNV 2017, Han 2012 and 2014).

Proton Exchange Membrane Fuel Cells (PEMFCs) are primarily developed for transport applications, benefiting from their high-power density (0.2-1.0 kW/kg and 200-800 kW/m^3) and very low operating temperature (approximately 80 °C). This low temperature allows for flexible operations and imposes less stringent material requirements (Balsamo et al. 2023; Xing et al. 2021). PEMFCs require pure hydrogen as fuel, achieving an efficiency in the range of 50-65%, emitting only water and electricity (Gadducci et al. 2022). However, the use of PEMFCs must consider challenges related to hydrogen production and storage, which could be particularly demanding for applications on board ships, as reported in the literature (Coppola et al. 2022).

A way to bypass hydrogen's challenges is to use molecular vectors, such as methanol (CH_3OH). There are several major advantages to using methanol, particularly for maritime applications: it is liquid at room conditions and can be stored and bunkered as traditional fuels, with only minor modification to the current infrastructure, particularly if

DOI: 10.1201/9781003508779-7

compared to the construction of terminals managing cryogenic fuels (e.g., LNG) or high-pressure gases (e.g., hydrogen) (DNV 2017, Han 2012 and 2014).

Using methanol as an energy source onboard ships allows for complying with stringent emissions regulations and favors the decarbonization of the marine sector, particularly if it is produced from renewable feedstocks or energies (green methanol) (Pluijlaar & van Biert, L. 2022).

Some FC technologies can be fuelled by methanol both directly and indirectly; in the latter case, methanol is processed (mostly through a reforming system) to produce a pure or highly concentrated hydrogen mixture (Kappis et al. 2021).

Methanol reforming systems integrated with High-Temperature PEMFC (HT-PEMFC) can be very promising power generators to be installed on board ships (Lee et al. 2020).

Compared with a PEMFC, an HT-PEMFC operates in the range of 180-220 °C, which reduces the sensitivity towards CO (and other impurities) and can be directly integrated with simpler, lighter, and cheaper methanol steam reformers to produce hydrogen (Coppola et al. 2022; Lee et al. 2020). Furthermore, it is also suitable for heat recovery increasing the overall efficiency. On the other hand, an HT-PEMFC has a lower power density, and it is not possible to cold start it.

It must be noted that the HT-PEMFC system is in the early stages of research and development, and very few commercial products with a power lower than 5 kW are available. Despite this, it shows high-potential applications, suggesting that modeling and simulation studies could promote and encourage its application in various fields (DNV 2017, Han 2012).

Simulation models dealing with FC technologies, mostly integrated with Energy Storage Systems (ESS), play a pivotal role in assessing the feasibility and optimization of ships equipped with full electric and hybrid power plants in several operational scenarios (Cavo et al., 2021; Choi et al., 2016; Gadducci et al., 2022; Haxhiu et al., 2022; Wu et al., 2018). For instance, Letafat et al. (2020) combined FC technology with the cold ironing solution to obtain a zero-emission ferry boat solution. Balestra & Schjølberg (2021) present a simulation tool for marine hybrid power plants equipped with PEMFC and batteries. In 2021, Bassam et al. developed a Matlab model of a 123 m domestic ferry, for which the effectiveness of a suitably sized hybrid HT-PEMFC/ESS propulsion system is compared to the conventional diesel propulsion system; results have been validated using real vessel working data and show good agreement with the real data.

According to this, the present work aims to model a fully electric power plant propulsion system composed of an integrated HT-PEMFC system assisted by a battery pack, using methanol as the primary power source. The study preliminarily investigates the performance of a small ferry implementing these technologies. To study the system and its performance, a ferry service, speed, and routes have been fixed to obtain data on autonomy and weight on-board based on fuel cell and battery load consumption and management. The paper is structured as follows: after this introduction, the case study and the simulation model will be presented respectively in Sections 2 and 3; finally results in full acceleration and operational routing with power management, consumption, Battery State of Charge, autonomy and weight on-board will be presented in Section 4.

2 CASE STUDY

The proposed vessel is a small-passenger ferry intended to perform short voyages in inland waterways and coastal waters, at a cruise speed of up to 12 knots. The ferry has an overall length of 25 m, a beam of 5.5 m, and a maximum capacity of 100 passengers. Figure 1 shows the ferry's external side view and the passenger deck arrangement. The power plant is composed of a fixed-pitch propeller driven by an Electric Motor (EM) coupled with a gearbox. The electric power required for the EM is supposed to be provided by an FC and an ESS-generating system.

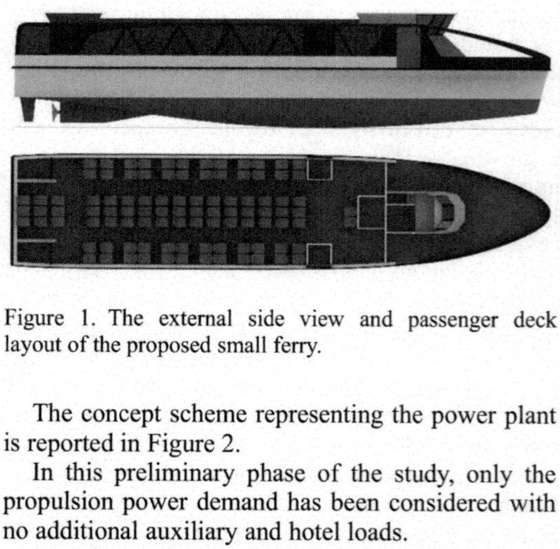

Figure 1. The external side view and passenger deck layout of the proposed small ferry.

The concept scheme representing the power plant is reported in Figure 2.

In this preliminary phase of the study, only the propulsion power demand has been considered with no additional auxiliary and hotel loads.

The ferry will operate on the Amalfi-Positano route along the Amalfi Coast at a speed of 12 knots. The distance between the ports is approximately 6.36 nm, with an estimated crossing time (from Amalfi to Positano) of 32 minutes (excluding maneuvering phases). The time spent in ports for embarking and disembarking routines is 15 minutes. Figure 3 shows a typical conceptual speed profile for a voyage consisting in a sailing from Positano to Amalfi and back.

2.1 *Hull and propeller*

The adopted hull form is particularly suitable for this type of service and it is part of the new Naples

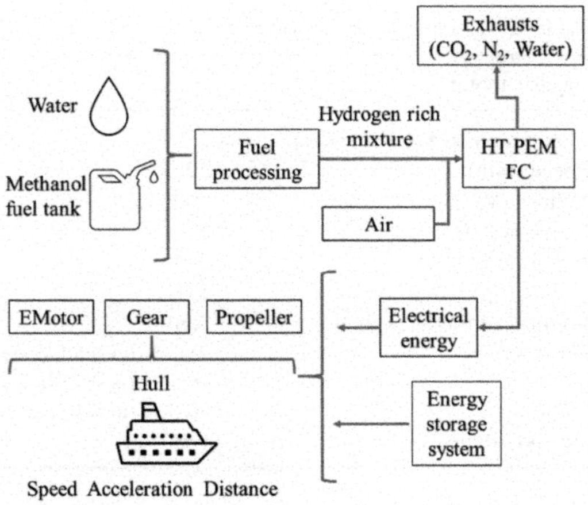

Figure 2. Concept scheme of the fully electric power plant.

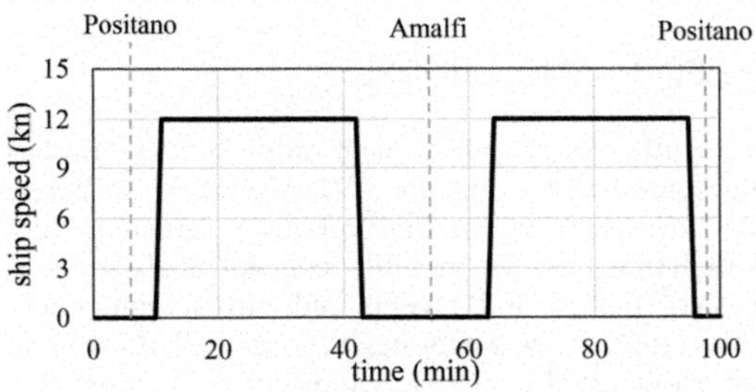

Figure 3. Conceptual ship speed profile over time for one voyage (Positano-Amalfi and back).

Displacement Systematic Series (NDSS), currently under study at the Department of Industrial Engineering (DII) of Naples, within the funded research program *PNRR Spoke 3 Waterways*. The main characteristics of the hull are reported in Table 1.

Table 1. Hull form principal dimensions.

Hull	A 57 NDSS
LWL (m)	24.79
BWL (m)	4.52
T (m)	141
Displacement (t)	57

As regards the hull performance, the total resistance (R_T) and the effective power (P_E) have been evaluated at 12 kn (Fr = 0.4) through full-scale bare hull CFD simulations and provided in Table 2. The full load displacement has been estimated referring to a database of similar ships.

The propulsive factors, thrust deduction *t* and wake fraction *w* (Miranda, 2022) and the appendages resistance (skeg, rudder, and shaft) have been evaluated according to Holtrop & Mannen (1982) procedure.

Table 2. Ship total resistance R_T and effective power P_E at the maximum speed.

v (kn)	Fr	R_T (kN)	P_E (kW)
12.0	0.41	12.67	78

Based on the estimated values of R_T and P_E of Table 2, a 4-bladed Wageningen propeller has been selected, with a diameter (D) of 0.9 m, a Pitch to Diameter (P/D) ratio of 1.039, and an expanded area-to-disc area ratio A_e/A_0 of 0.90. The propeller K_T and K_Q coefficient curves, the efficiency η_0 and the propeller working point have been shown in the screw propeller open water diagram reported in Figure 4.

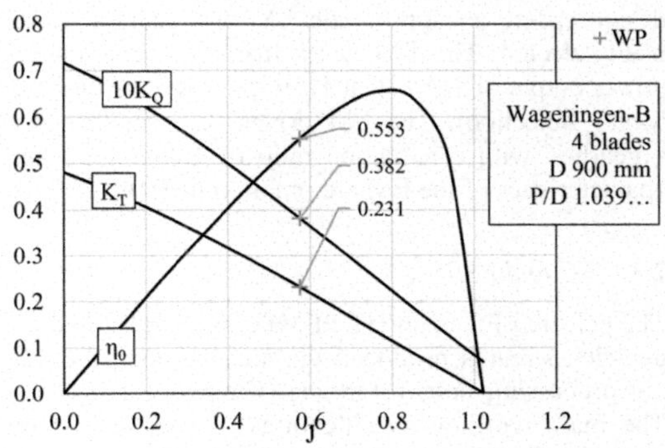

Figure 4. Selected Wageningen-B, 4-bladed propeller open water diagram, with highlighted propeller working point (cross dots).

Considering the efficiencies and the propulsive factors reported in Table 3, the total delivered power (P_D) is 147 kW, which is calculated as:

$$P_D = P_E / \eta_D \tag{1}$$

where the propulsive efficiency (η_D) can be expressed as the product of hull efficiency (η_H), propeller open water efficiency (η_0), and relative rotative efficiency (η_R):

$$\eta_D = \eta_H \eta_0 \eta_R \tag{2}$$

where,

$$\eta_H = (1 - t)/(1 - w) \tag{3}$$

The relative rotative efficiency has been assumed as η_R=1. Details are provided in Table 3 which shows the thrust deduction fraction t and the wake fraction w, and the respective thrust efficiency, *1-t*, and wake coefficient, *1-w*, calculated according to Holtrop & Mannen (H&M) as a preliminary estimation. It should be noted that the values obtained and reported in Table 3 are precautionary for this type of hull form, and a more accurate analysis through towing tank tests is planned at the DII.

Table 3. Calculated hull and propeller efficiencies.

Parameter	Value
Thrust deduction, t	0.189
Wake fraction, w	0.159
1-t	0.811
1-w	0.841
η_H	0.965
η_O	0.553
η_R	1.0
η_D	0.534

2.2 *EM and gearbox*

Based on the amount of power required to sail at the design speed, an appropriate EM and gearbox have been selected. The EM has a maximum continuous power output of 185 kW at 1119 rpm (WEG - Electric Motor Selection). The EM drives the propeller via a gearbox with a reduction ratio of 1.86. The main characteristics of the EM are reported in Table 4.

2.3 *FC and ESS*

The selected FC is an HT-PEMFC fuelled by hydrogen-rich streams produced via an integrated methanol processing unit (reformer) (Kappis et al. 2021). The main working specifications of such a system are taken from those of Advent Technologies's products (https://advent.energy/). It is presented as an integrated system fuelled by a 60/40 vol% methanol/water solution.

A battery pack assists the HT-PEMFC integrated system during the start-up phase and load variations. It is assumed to use Li-ion-based batteries, that will be recharged by the FC. It is assumed that the ESS is made of several battery single cells in series and parallel which will be adapted to the main electric circuit according to the load demand. These have a maximum capacity of 5 Ah and an operating voltage range between 2.5-3.5 V.

Table 4. EM and gearbox characteristics.

Electric motor	
Maximum power output (kW)	185
Speed (rpm)	2984
Voltage (V)	500
Frequency (Hz)	50
Rated current (A)	249
Weight (kg)	1315
Moment of Inertia (kgm^2)	3.50
Efficiency	95.3 % (at 0.50 load) 96.1 % (at 0.75 load) 96.5 % (at full load)
Gearbox Ratio	1.86

3 SIMULATION MODEL

The fully electric power plant propulsion system has been modelled using the AVL Cruise M software. AVL Cruise M is a multidisciplinary vehicle system simulation tool for mobility concept analysis, subsystem design and layout, and virtual component integration. The system has been modelled using an object-oriented approach, capturing the components of each subsystem and their connections. The software allows the development of customized models with user-defined functions programmed in C. Figure 5 shows a screen of the built-up model.

Two modes of operation have been simulated:

- *Full acceleration*: an operative mode consisting of full acceleration (from zero to design speed) and constant sailing at a given speed.

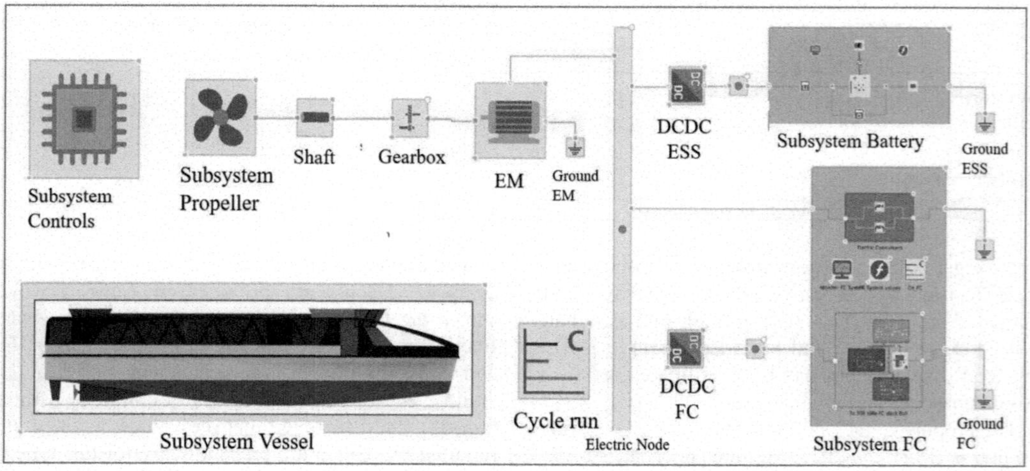

Figure 5. Overall scheme of the power plant model in AVL Cruise-M.

- *Operational route profile*: the overall system works during one voyage, following the speed profile given in Figure 3.

Simulation results will be presented in terms of system power outputs, battery State of Charge (SoC), ship's autonomy, and fuel consumption.

In the following, the description of each model sub-component has been provided, specifying the initial data required for each element the typology and the size of each component have been selected after an initial assessment of the ship's performances at the required speed. The power plant model is composed of the following sub-models end blocks:

The *vessel* and *propeller subsystems*: the ship block contains the resistance and the kinematics models. The ship resistance is modelled by a user-defined function, expressed as a resistance factor (R_T/v^2) and calculated as follows:

$$k_R = \frac{R_T}{v_{ds}^2} = \frac{(1-t)T_{ds}}{v_{ds}^2} = \frac{(1-t)K_{T_{ds}}\rho D^4 {n_{ds}}^2}{v_{ds}^2} \quad (4)$$

where the subscript *ds* stands for *design speed*, T = the delivered thrust, K_T = thrust coefficient, ρ= water density, t = thrust deduction, and n = propeller rotational speed. These are calculated at the design operational point of v_{ds}=12 kn. The advance coefficient at the design operational point is calculated as:

$$J_a = \frac{(1-w)v_{ds}^2}{n_{ds}D} \quad (5)$$

where, w = wake fraction, v = ship speed, n = propeller rotational speed, D = propeller diameter.

With the hypothesis of constant wake fraction and thrust deduction for the entire speed range, the coefficient k_R becomes constant. Therefore, the model assumes a square resistance relationship between the resistance and the ship speed. This simplification is acceptable for this preliminary work since the ferry follows a simple operative profile at a constant speed with fast accelerations. In cases of complex operative profiles at multiple speeds, a more accurate model should be considered by implementing performance data (resistance and propulsive factors) at medium and low speeds.

The ship kinematic model adopts a one-degree-of-freedom approach (1 DoF) based on the equation:

$$(M_a + \Delta)\dot{v}_s = (1-t)T - R_T(V_s) \quad (6)$$

where M_a = the added mass assumed here equal to 0.25Δ as a precautionary first attempt value. A more realistic evaluation of the added mass is going to be performed by means of towing tank tests.

The propeller thrust force (T) and torque (Q) are calculated based on the given K_T and K_Q curves obtained from the selected propeller open water test results.

$$T = K_T \rho D^4 n^2 \quad (7)$$

$$Q = K_Q \rho D^5 n^2 \quad (8)$$

The electric motor block (*EM*) is characterized by torque and efficiency curves as a function of voltage and speed.

The *shaft* and *gearbox* blocks connect the propeller to the electric motor.

The *battery subsystem* consists of a controlled voltage source and an Ohmic resistance used to describe the instantaneous voltage response to a current input. It includes an advanced model to predict the transient voltage response to a dynamic current load, eventually coupled with a thermal model to predict the transient thermal behaviour of the battery. The battery's initial SoC can be varied, while the voltage variation as a function of the SoC assumed in this study is shown in Figure 6. The main characteristics of the chosen battery are:

- Number of cells per row = 90;
- Number of cell rows = 90;
- Operating voltage = 2.5 V- 3.5 V;
- Maximum charge = 5 Ah.

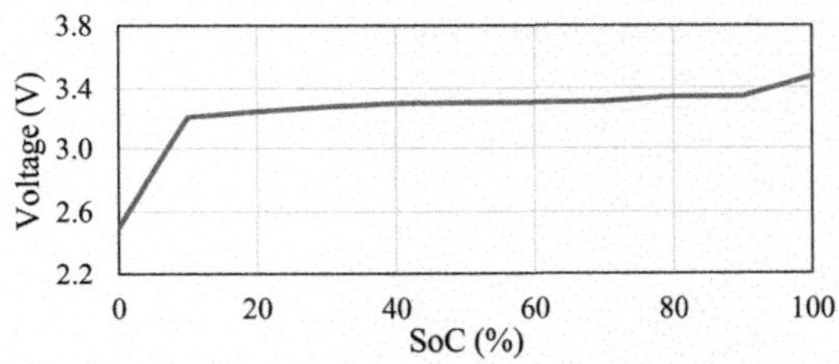

Figure 6. Voltage variation as a function of the battery stack's SoC.

Hybrid control *system* includes a section for regulation, monitoring, and real-time measurements.

There are two *DC/DC* converters with an efficiency of 93 %.

The *FC subsystem* consists of four main subsystems representing the anode, cathode, and the entire stack model, integrated with a thermal model. The Reduced Dimensionality (1D+RD) FC stack model simulates the relevant thermodynamic and electrochemical processes in the FC comprising the following domains:

- Material balance: gas flows in the anode (fuel) and cathode (air);
- Electric balance: links the component to the electric network;
- Thermal balance: internal thermal model or linked to the coolant circuit;
- Membrane-Electrode Assembly: electrochemical and transport processes in the membrane, catalyst layers and gas diffusion layers.

The single-cell performance (polarization curve) and geometry, as well as those of the stack, the components' materials, specifics, etc., can be chosen from the software library or introduced by the user and varied.

We assumed the following main specifics:

- Number of cells in a stack = 910;
- Stack voltage = from 500 V to 600 V;
- Max stack power from 90 kW.

Once the individual blocks are characterized with all the required information, the simulation takes the desired number of revolutions for the electric motor as input and provides the complete performance of each element as output.

4 RESULTS

4.1 *Full acceleration case*

The full acceleration test intends to investigate the FC and battery systems' responses to a strong acceleration from 0 to design speed (12 kn).

It is assumed to analyze the power produced as a function of time in the case of the FC system in standby condition, with the battery SoC at 90%, (no additional external forces have been considered). Results are presented in Figure 7.

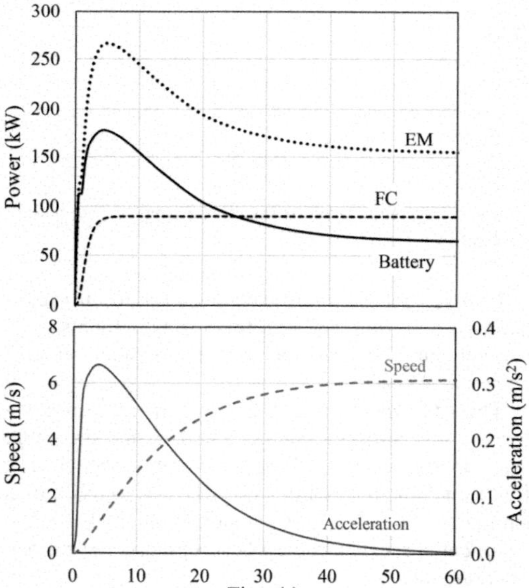

Figure 7. Time plot of Battery, FC, and EM power output responses to a ship acceleration from 0 to 12 kn.

The vessel takes approximately 60 seconds to reach its maximum speed, with a peak acceleration at 4s. As expected, the battery exhibits a faster response compared to the FC. The battery reaches a peak power at 4.5 s, promptly supplying the electric power needed by the EM. The EM power output reaches a peak of 266 kW at 5s, after which the power demand stabilizes, remaining nearly constant beyond 40 seconds.

The FC system achieves a steady-state condition after approximately 6-8 seconds. During the initial 0-25 seconds, it provides power to both its Balance of Plant (BoP) and the EM. Subsequently, a portion of the generated energy recharges the battery.

An interesting result derived from the full acceleration test is the battery discharge time. Starting at a speed of 12 knots with a SoC of 90%, it takes 94 minutes for the SoC to decrease to 35%. These values represent the capability to sail a distance of 19 nm at the specified speed without the battery depleting below a SoC of 35%.

The full acceleration test was used also to determine the ship's range in the case the FC is deactivated, relying solely on the battery to supply energy to the EM. Assuming a SoC reduction from 90% to 35% and a speed of 12 knots, the installed battery capacity allows an operational range of 40 minutes, covering 8 nautical miles.

4.2 *Operational routing case*

The performance of the full-electric propulsion system on the operational route depicted in Figure 3 has been investigated, matching the ship's speed requirements over time.

As illustrated in Figure 8, the FC system maintains a steady-state condition, generating 90 kWe, with an exception for the initial phase lasting less than 1 minute, necessary to transition from standby. Throughout the route, the FC system provides a portion of the propulsive power required by the EM. When the EM is inactive, the generated power is directed towards recharging the battery. It is important to note that this operational configuration is essential because the methanol processing system, responsible for hydrogen generation within the FC system, cannot endure dynamic conditions as they may significantly impact fuel conversion and hydrogen yield (Balsamo et al. 2023).

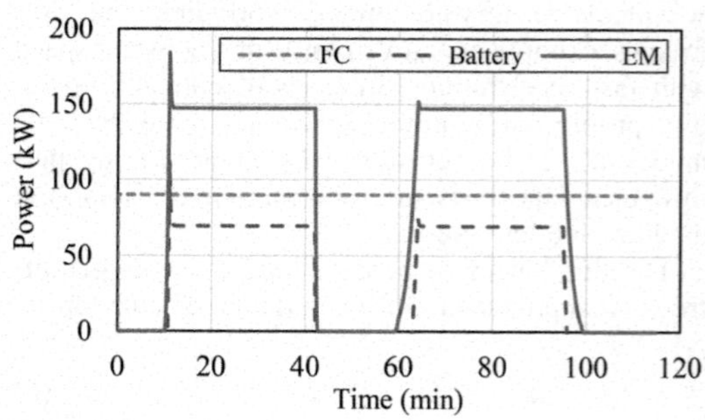

Figure 8. Power output time plot for one voyage.

The presence of the battery enables the fulfilment of the electric load for the EM, characterized by a power peak of approximately 147 kW at 12 knots. Conversely, the initial SoC, its variation within the range of 35-90%, and the total installed battery capacity are tailored to align with the presented route profile. The SoC at the end of the first route can be easily restored during the time between one daily route and another to start the cycle again. The SoC trend during the operational route (one voyage) is shown in Figure 9.

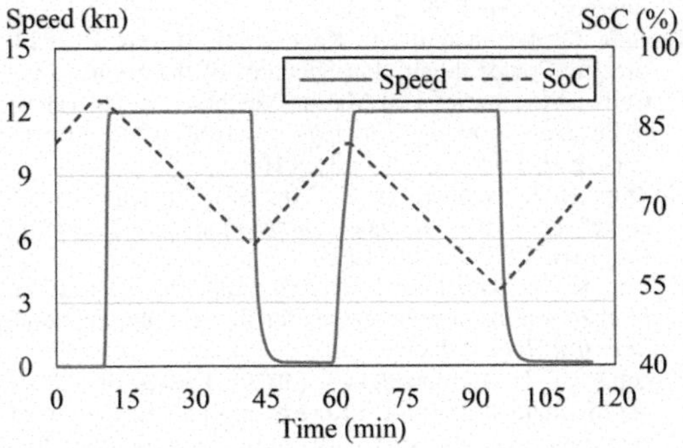

Figure 9. Speed profile and SoC trend per voyage.

It is important to note that hotel loads are currently neglected in this analysis.

The AVL Cruise M software calculates the hydrogen quantity required by the FC, which is about 9.2 kg per voyage. Considering the efficiency of the HT-PEM integrated system equal to 40%, the methanol to be stored on board can be estimated. For example, Table 5 outlines the weights and volume of methanol to perform 6 voyages per day (Positano to Amalfi and back); these values are 360 kg and 0.45 m3, respectively. In full acceleration condition, the battery capacity has been calculated according to the ship's route, which is 145 kWh (Table 5).

Table 5. Main specifics of the ship's route and required energy.

Analysis	Value
Distance (Positano-Amalfi) (nm)	6,4
Numbers of voyages/day	6
Max speed (kn)	12
Crossing time (min)	32
Required H_2 for one voyage (kg)	9,2
Required methanol for one voyage (kg)	59,8
Required methanol for 6 voyages (kg)	359
Methanol tank volume (m^3)	0.45
H_2 consumption at 12 kn (kg/h)	4.8
Continuous range at 12 kn (min)	94
Continuous range at 12 kn (nm)	19
Continuous range in full-battery mode (nm)	10
Battery capacity (kWh)	145

5 CONCLUSION

In this study, a preliminary simulation model of a power plant for a small ferry fueled by methanol was developed using *AVL Cruise M* software. The software models each component of the power plant, including the FC, battery, EM, and their auxiliaries. Simulation tests were carried out to analyze material, thermal, and electric flows. Specifically, a full acceleration operating mode (from zero to design speed) was considered to investigate dynamic responses, along with an operational route profile consisting of a ship's speed demand over time for the Positano-Amalfi route, used to estimate the fuel consumption and ship's autonomy.

The simulation shows an average hydrogen consumption of 4.8 kg/h, and the amount of methanol to be stored on board to guarantee 6 voyages is approximately 359 kg. Furthermore, the results demonstrate how the system can manage a range of over 20 nm at a constant speed of 12 knots and over 10 nm in full battery mode.

The upcoming developments will focus on enhancing various aspects of the model and its experimental validation. Three specific paths will be pursued:

- *Hull Hydrodynamic Model*: a more accurate model will be developed to account for ship performances, in terms of total resistance and propulsive factors. This will involve towing tank test data and outputs from Computational Fluid Dynamics (CFD) simulations. This improvement is particularly critical for vessels operating at multiple speeds, such as hybrid systems with zero-emission modes at low speeds. It will also involve hulls optimized for various speeds, like the currently studied Displacement Systematic Series.
- *Additional Loads and Maneuvering Phases:* this path involves simulating permanent loads in port, power requirements of maneuvering thrusters during departure and arrival phases, and optimizing battery capacity.
- *Balance of Plan Optimization:* the focus is on the system optimization strategy, considering both sizing (reducing onboard weight) and management (lowering consumption).

ACKNOWLEDGEMENTS

Acknowledgements for financial support from the EU-funded project PNRR Spoke 3, Waterways, WP Emission reduction, and Task 2.6 are sincerely expressed.

REFERENCES

Altosole, M., Campora, U., Mocerino, L., & Scamardella, A. (2022, June). Comparison between high-efficiency propulsion systems in electric ship applications. In 2022 *International Symposium on Power Electronics*, Electrical Drives, Automation and Motion (SPEEDAM) (pp. 634–639). IEEE.

Altosole, M., Balsamo, F., Mocerino, L., & Scamardella, F. (2023, June). Methanol fuelled hybrid propulsion system for a charter yacht. In 2023 *International*

Conference on Clean Electrical Power (ICCEP) (pp. 810–817). IEEE.

Balestra, L., Schjølberg, I., 2021. Modelling and simulation of a zero-emission hybrid power plant for a domestic ferry. *International Journal of Hydrogen Energy* 46, 10924–10938. https://doi.org/10.1016/j.ijhydene.2020.12.187

Balsamo F., Capasso C., Coppola T., Micoli L., Russo R., Veneri O. (2023) "A case study on High-Temperature Fuel Cells for Hybrid Electric Ship Propulsion." *Electrical Systems for Aircraft, Railway, Ship Propulsion and Road Vehicles (ESARS)*.

Bassam, A.M., Phillips, A.B., Turnock, S.R., Wilson, P.A., n.d. Design, modelling and simulation of a hybrid fuel cell propulsion system for a domestic ferry.

Bucci, V., Mauro, F., Marinò, A., Bosich, D., & Sulligoi, G. (2016, June). An innovative hybrid-electric small passenger craft for sustainable mobility in the Venice Lagoon. In 2016 *International Symposium on Power Electronics, Electrical Drives, Automation and Motion* (SPEEDAM) (pp. 1388–1395). IEEE.

Campora, U., Coppola, T., Micoli, L., Mocerino, L., & Ruggiero, V. (2023). Techno-Economic Comparison of Dual-fuel Marine Engine Waste Energy Recovery Systems. *Journal of Marine Science and Application*, 1–14.

Capasso, C., Veneri, O., Notti, E., Sala, A., Figari, M., & Martelli, M. (2016, November). Preliminary design of the hybrid propulsion architecture for the research vessel "G. Dallaporta". In 2016 *International Conference on Electrical Systems for Aircraft, Railway, Ship Propulsion and Road Vehicles & International Transportation Electrification Conference (ESARS-ITEC)* (pp. 1–6). IEEE.

Cavo, M., E. Gadducci, D. Rattazzi, M. Rivarolo, e L. Magistri. «Dynamic Analysis of PEM Fuel Cells and Metal Hydrides on a Zero-Emission Ship: A Model-Based Approach». *International Journal of Hydrogen Energy* 46, fasc. 64 (settembre 2021): 32630–44. https://doi.org/10.1016/j.ijhydene.2021.07.104.

Choi, Choeng Hoon, Sungju Yu, In-Su Han, Back-Kyun Kho, Dong-Gug Kang, Hyun Young Lee, Myung-Soo Seo, et al. «Development and Demonstration of PEM Fuel-Cell-Battery Hybrid System for Propulsion of Tourist Boat». *International Journal of Hydrogen Energy* 41, fasc. 5 (febbraio 2016): 3591–99. https://doi.org/10.1016/j.ijhydene.2015.12.186.

Coppola, T., Micoli, L., & Russo, R. (2022, June). Concept design and feasibility study of propulsion system for yacht: innovative hybrid propulsion system fueled by methanol. In 2022 *International Symposium on Power Electronics, Electrical Drives, Automation and Motion* (SPEEDAM) (pp. 683–688). IEEE.

Del Pizzo, A., Polito, R. M., Rizzo, R., & Tricoli, P. (2010, September). Design criteria of on-board propulsion for hybrid electric boats. In The *XIX International Conference on Electrical Machines-ICEM* 2010 (pp. 1–6). IEEE.

Dnv, G. (2017). *Study on the use of fuel cells in shipping. Retrieved from European Maritime Safety Agency* (EMSA): http://www.emsa.europa.eu/news-a-press-centre/external-news/item/2921-emsa-study-on-the-use-offuel-cells-in-shipping.html.

Gadducci, E., Lamberti, T., Rivarolo, M., & Magistri, L. (2022). Experimental campaign and assessment of a complete 240-kW Proton Exchange Membrane Fuel Cell power system for maritime applications. *International Journal of Hydrogen Energy*, 47(53), 22545–22558.

Han, J., Charpentier, J. F., & Tang, T. (2012, May). State of the art of fuel cells for ship applications. In 2012 *IEEE international symposium on industrial electronics* (pp. 1456–1461). IEEE.

Han, J., Charpentier, J. F., & Tang, T. (2014). An energy management system of a fuel cell/battery hybrid boat. *Energies*, *7*(5), 2799–2820.

Haxhiu, A., Abdelhakim, A., Kanerva, S., Bogen, J., 2022. Electric Power Integration Schemes of the Hybrid Fuel Cells and Batteries-Fed Marine Vessels—An Overview. *IEEE Trans. Transp. Electrific*. 8, 1885–1905. https://doi.org/10.1109/TTE.2021.3126100

Holtrop, J., & Mennen, G. G. J. (1982). An approximate power prediction method. *International Shipbuilding Progress*, 29(335), 166–170.

Kappis, K., Papavasiliou, J., & Avgouropoulos, G. (2021). Methanol reforming processes for fuel cell applications. *Energies*, *14*(24), 8442.

Karatuğ, Ç., & Durmuşoğlu, Y. (2020). Design of a solar photovoltaic system for a Ro-Ro ship and estimation of performance analysis: A case study. *Solar Energy*, *207*, 1259–1268.

Karatuğ, Ç., Tadros, M., Ventura, M., & Soares, C. G. (2023). Strategy for ship energy efficiency based on optimization model and data-driven approach. *Ocean Engineering*, *279*, 114397.

Karatuğ, Ç., Tadros, M., Ventura, M., & Soares, C. G. (2024). Decision support system for ship energy efficiency management based on an optimization model. *Energy*, 130318.

Lee, H., Jung, I., Roh, G., Na, Y., & Kang, H. (2020). Comparative analysis of on-board methane and methanol reforming systems combined with HT-PEM fuel cell and CO2 capture/liquefaction system for hydrogen-fueled ship application. *Energies*, 13(1), 224.

Letafat, A., Rafiei, M., Sheikh, M., Afshari-Igder, M., Banaei, M., Boudjadar, J., Khooban, M.H., 2020. Simultaneous energy management and optimal components sizing of a zero-emission ferry boat. *Journal of Energy Storage* 28, 101215. https://doi.org/10.1016/j.est.2020.101215

Markowski, J., & Pielecha, I. (2019). The potential of fuel cells as a drive source of maritime transport. In *IOP Conference Series: Earth and Environmental Science* (Vol. 214, No. 1, p. 012019). IOP Publishing.

Miranda, S. (2022). *Architettura Navale-Elementi di Dinamica della Nave* (Vol. 1, pp. 1–520). Attuale-LIGUORI Editore srl Via Posillipo 394 80123 NAPOLI (Italy).

Mocerino L., Quaranta, F., & Rizzuto, E. (2018). Climate changes and maritime transportation: A state of the art. *Technology and Science for the Ships of the Future*, 1005–1013.

Pluijlaar, D., & van Biert, L. (2022). Using renewable methanol, PEM fuel cells, and on-board carbon capture to reduce well-to-propeller ship emissions. In *Proceedings of the International Naval Engineering Conference* (Vol. 16, p. 27).

Xing, H., Stuart, C., Spence, S., & Chen, H. (2021). Fuel cell power systems for maritime applications: Progress and perspectives. *Sustainability*, 13(3), 1213.

Zhan, Z. (2022). Current Status and Future Development of Fuel Cell Ships in China. In Journal of Physics: Conference Series (Vol. 2160, No. 1, p. Wu012061). IOP Publishing.

Wu, P., & Bucknall, R. W. G. (2018). On the design of plug-in hybrid fuel cell and lithium battery propulsion systems for coastal ships. In *Marine Design XIII, Volume 2* (pp. 941–951). CRC Press.

Advances in Maritime Technology and Engineering – Guedes Soares & Santos (Eds)

GreenH2CM-LiA7-PT1. Fuel-cell hybrid powertrains laboratory for maritime

A. Villalba-Herreros, R. d'Amore-Domenech, V.L. Meca & T.J. Leo
Departamento de Arquitectura, Construcción y Sistemas Oceánicos y Navales (DACSON), Escuela Técnica Superior de Ingenieros Navales, Universidad Politécnica de Madrid, Madrid, Spain
Grupo de Investigación Pilas de Combustible, tecnologías del Hidrógeno y Motores Alternativos (PiCoHiMA), Universidad Politécnica de Madrid, Madrid, Spain

E. Posada
Grupo de Investigación Pilas de Combustible, tecnologías del Hidrógeno y Motores Alternativos (PiCoHiMA), Universidad Politécnica de Madrid, Madrid, Spain

ABSTRACT: Fuel cell-based power systems, when combined with secondary sources, match the effectiveness of internal combustion engines but offering higher efficiency and the potential for emission-free operation. Despite being in the early stages of development, increased research is essential. The GreenH2CM project Activity Line "Uses of hydrogen in heavy transport, aviation, and maritime sectors" includes the establishment of a specialized laboratory by PiCoHiMA at Universidad Politécnica de Madrid. This facility focuses on testing fuel cell integrations for marine and aeronautical applications, accommodating power rates up to 270 kW and simulating secondary power sources like batteries, among others. The laboratory, equipped to monitor critical parameters and replicate platform movements, enhances research capabilities for interested developers and researchers. This paper introduces the facility, providing insight into its potential contributions to fuel cell research and product development in the field.

1 INTRODUCTION

The use of fuel cell technologies onboard ships and aircrafts is getting increasing interest because of the politics adopted by different countries, as the member states of the European Union (EU), China, or the United States of America, among others, that try to reduce the Greenhouse (GHG) emissions from the marine and aviation transport sectors. In the case of the EU, the emissions produced by ships with a gross tonnage over 5000 have been included in the European Emissions Trading System (ETS) as from January 1st 2024 (Rao 2023). The ETS obliges ship operators to submit yearly a number of CO_2 emissions allowances corresponding to GHG emissions produced during the previous year. These allowances are bought at public auction and the EU progressively reduces the total quantity of allowances available to force the deployment of new and more efficient technologies with the ultimate goal of reducing GHG emissions. In this sense, the adoption of fuel cells onboard ships combined with other technologies to increase the global efficiency and reduce the GHG emissions is the key. In addition to the ETS, the EU launched in 2023 the "Fit for 55" initiative. This initiative aims to increase the use of greener fuels in the maritime and aviation sectors through the corresponding regulations, FuelEU (European Council 2023) for the maritime sector and ReFuelEU for the aviation sector, respectively. In the case of FuelEU, fuel cells are considered "zero emissions technologies", i.e., technologies that do not emit carbon dioxide (CO_2), methane (CH_4), nitrous oxide (N_2O), sulphur oxides (SO_x) or particulate matter (PM) and, from January 1st 2030, container and passenger ships should use these kind of technologies or make use of Onshore Power Supply (OPS) to generate their energy while at berth.

Power systems utilizing fuel cells combined with secondary power sources exhibit comparable effectiveness to current internal combustion engine-based powertrains. However, these fuel cell hybrid systems boast enhanced efficiency and the potential for emission-free operation. Despite these promising prospects, the development of such powertrains is in its early stages, necessitating strengthened research efforts, particularly in the fields of marine and aeronautical applications. This situation makes necessary facilities where fuel cell stacks intended for maritime and aviation sectors as well as their integration with other power sources to form hybrid power plants can be tested in a variety of conditions that reproduce the those that these solutions will face onboard.

DOI: 10.1201/9781003508779-8

Nowadays there are different laboratories spread along the word dedicated to the study and development of fuel cells stacks and its integration in hybrid powertrains. For example, in the USA can be cited the Southwest Research Institute that offers a complete list of services for testing fuel cell powertrains for the automotive industry (Southwest Research Institute 2024). Also in the USA, the National Aeronautics and Space Administration (NASA) has a fuel cell testing laboratory able to test fuel cells with power ratings up to 125 kW (National Aeronautics and Space Administration 2024). In Europe, the Spanish National Hydrogen Centre has a fuel cell testing facility equipped with a walk-in climatic chamber and a vibration test system that allows PEM fuel cell stacks up to 30 kW to be tested under extreme working conditions (Centro Nacional del Hidrógeno 2024). In The Netherlands the JRC Fuel Cell and Electrolyser Testing Laboratory has a testing station for PEM fuels cells up to 84 kW and is also equipped with a walk-in climatic chamber and a vibration test system. This facility is especially dedicated to supporting the development of regulations, codes and standards by validating test procedures and measurement methodologies for the assessment of fuel cell performance (Central European Research Infrastructure Consortium 2024). In China, it is noticeable the foundation in 2023 of the National Hydrogen Power Quality Inspection and Testing Centre conceived as a national test centre for hydrogen-powered vehicles (Leigh Collins 2023). Besides, the Asian country has other facilities such as the Institute of Fuel Cells belonging to the Shanghai Jiao Tong University, which is equipped with several fuel cell test stations (Shanghai Jiao Tong University 2024).

A review of the objectives and capabilities of the above facilities shows that, regarding mobility, most of them focus on the development of fuel cells for the road transport sector. However, the specific environmental conditions that can be found onboard a ship or an aircraft is not well studied. For example, the National Centre of Hydrogen and the JRC Fuel Cell and Electrolyser Testing Laboratory have vibration test systems to study the impact of vibrations on fuel cells but the effect of low frequency periodic movements as those given on ships (pitch, roll and yaw) suffers a lack of information. In this sense, it is crucial an installation designed to test fuel cell stacks under these conditions.

As part of the project GreenH2CM (Ciemat 2023), a valuable tool will be provided to support research on the use of fuel cells in marine and aeronautical sectors. Under the GreenH2CM Activity Line focusing on "Uses of hydrogen in heavy transport, aviation, and maritime sectors", the research group PiCoHiMA from Universidad Politécnica de Madrid (UPM) in Spain is building a laboratory specially designed to to test fuel cell integrations for marine and aeronautical applications, see Figure 1. This facility will be installed at the Escuela Técnica Superior de Ingenieros Navales (Universidad Politécnica de Madrid, 2023) in Madrid, an UPM educational centre for naval architects and marine engineers.

Among other capabilities, this installation will facilitate the testing of Proton Exchange Membrane (PEM) fuel cell stacks with power rates up to 90 kW (expandable to 270 kW), along with the simulation of secondary power sources like batteries, capacitors, or diesel generator sets.

In addition to monitoring and controlling critical stack parameters under simulated environmental conditions, the laboratory will feature a 6-degree-of-freedom hexapod to replicate platform movements to study their impact on the fuel cells performance.

This article introduces this new facility so that interested researchers and technology developers can learn about its capabilities, which can help them both in their research and in the development of new products.

The work is structured as follows. Section 2 presents the project GreenH2CM. Section 3 describes the laboratory itself including its general layout and key equipment. The capabilities of the facility are detailed in Section 4. Finally, Section 5 presents the main conclusions drawn from the construction and putting into service of this facility.

2 THE PROJECT GREENH2CM

The project GreenH2CM "Strategic positioning of the Community of Madrid in R&D&I in green hydrogen and fuel cells" is an initiative of the Regional Government of the Community of Madrid in Spain, partially founded by Community of Madrid with its own funds and the Spanish Ministry of Science and Innovation with funds from the NextGenerationEU (European Commission 2023a) instrument of the European Union trough the Recovery and Resilience Facility (European Commission 2023b).

This project pursues four main goals: (i) to develop innovative technologies for hydrogen production by low and high temperature electrolysis and other innovative routes, (ii) to increase the efficiency of hydrogen production by optimising the renewable electricity-electrolyser connection, (iii) to develop technologies that enable the optimisation of fuel cell integration in various applications and (iv) to develop technologies for safe combustion of hydrogen, pure or blended, in end-use devices (combustors, engines, turbines).

The facility presented here is covered by work package 1 (PT1) within the Activity Line 7 (LiA7) "Uses of hydrogen in heavy transport, aviation, and maritime sectors" of GreenH2CM and attends the goal iii.

This installation has been designed from scratch despite the existence of commercial solutions such as test benches manufactured by Scitech Korea (Scitech Korea 2024) or by the Canadian firm GreenLight Innovation (Greenlight Innovation 2024). This approach responds to the need for specific capabilities that these products do not address. On the one hand, the simulation of movements on board ships or aircrafts that is achieved by means of a Stewart platform. On the other hand, the ability to hybridize the fuel cell with other power

Figure 1. General view of the fuel cell and powertrains testing laboratory.

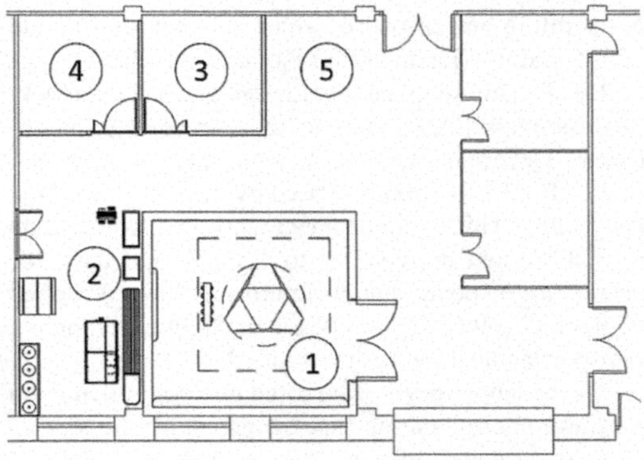

Figure 2. Distribution of the spaces of the laboratory. Fuel Cell Chamber, ①. Auxiliary Area, ②. Electronics Room, ③. Control Room, ④. Multipurpose Area, ⑤. Hydrogen Storage Area ⑥ (not represented).

sources as batteries or supercapacitors that is carried out with the use of a Power Management System. These capabilities make this installation unique.

3 LABORATORY DESCRIPTION

As can be seen in Figure 2, the laboratory is distributed into six main spaces. The Fuel Cell Chamber ① allocates the fuel cell under study that is placed onto a Stewart platform with 6 degrees of freedom together with the Hydrogen Regulation System (HRS). The Fuel Cell Cooling System (FCCS), the Fuel Cell Air System (FCAS), the Fuel Cell Chamber Ventilation System, and other ancillary equipment, are in the Auxiliary Area ②. The Electronics Room ③ houses electronic equipment as the Electrical Power Drainage System (EPDS), Power Management Systems (PMS), and communication equipment. The control and monitoring of the installation is carried out remotely from the Control Room ④ and a Multipurpose Area ⑤ will be used to place secondary power systems as battery banks. Finally, the hydrogen used as fuel will be stored in an open area outside the laboratory ⑥ (not represented in Figure 2).

The Fuel Cell Chamber is an enclosed space equipped with independent ventilation from the building that runs continuously when the installation is in operation. This ventilation minimises any risk in case of hydrogen leakage. Besides, strategically placed hydrogen sensors allows any leaks to be detected quickly cutting-off the supply of hydrogen instantly. For a higher safety, the HRS placed inside the Fuel Cell Chamber is enclosed in a sealed container with independent ventilation and hydrogen sensors.

The HRS is capable to supply a hydrogen mass flow up to 18 kg/h with a maximum pressure of 1.5 MPa. Besides, it allows the temperature of the hydrogen stream to be regulated to values between 325 K and 338 K.

The Stewart platform, see Figure 3, is supplied by the French manufacturer Symétrie (Symétrie 2023), model Mistral-XS. It allows a maximum payload of 500 kg, maximum roll, and pitch angles equal to 28º, maximum roll and pitch speeds equal to 40º/s and maximum heave acceleration equal to 9.81 m/s^2 as key values. These figures make it possible to reproduce movement and position conditions like those onboard a ship or an aircraft.

The FCCS works with deionized water and can regulate the temperature of the cooling fluid at stable temperatures between 333 K and 363 K. This range allows to take the fuel cell under study to temperatures that are usually far from the typical values to assess its performance.

From its part, the FCAS can supply and excess of 200 % of air with respect to the stoichiometric proportion in a PEM and can regulate the inlet

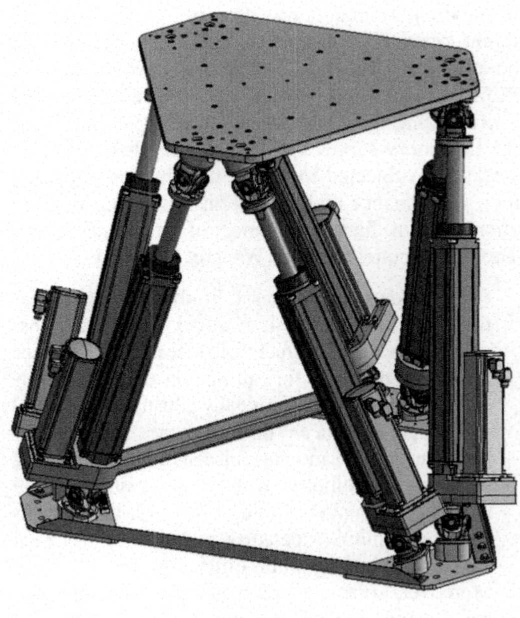

Figure 3. Stewart platform model Mistral-XS from Symétrie (reproduced with permission of Symétrie).

temperature and pressure values between 308 K and 373 K and 94 kPa and 600 kPa, respectively.

The EPDS, supplied by Elektro-Automatik (Elektro-Automatik 2023), consists of a series of programmable electronic DC-loads with energy recovery, model EA-ELR 10000, especially indicated for fuel cell testing. This equipment can work with voltages up to 2000 V and currents up to 1000 A and have the peculiarity of being able to return the power supplied by the fuel cell to the grid of the building recycling the energy generated by the fuel cell.

All the laboratory is monitored and controlled from a central control station that integrates all the signals coming from the different sensors and equipment of the installation. This integrated architecture allows to automatize most of the tasks and coordinate the different systems to avoid errors or unsafe operation.

The data generated by the different sensors distributed throughout the installation are stored by a data acquisition and storage system at appropriate frequencies up to 10 kHz in the case of the voltage and current signals of the fuel cell under study.

The hydrogen storage system consists of two racks of sixteen 50 L cylinders at 20 MPa each. The mass of hydrogen in each rack is equal to 12 kg, enough to feed a 90 kW PEM fuel cell with an average efficiency of 45 % for 4 hours. In addition, the two-rack configuration allows for continuous operation of the installation, as the out-of-service rack can be replaced while the in-service rack is in operation.

4 CAPABILITIES

The new research and testing laboratory for fuel cell-based hybrid powertrains for maritime and aeronautical environments was designed to carry out different types of experiments. These experiments can be classified into two blocks: characterization of PEM fuel cells, and study of powertrain integrations.

Regarding the characterization of PEM fuel cells, the laboratory serves as a test bench where fuel cell stacks are subjected to different techniques to study their performance under different conditions, degradation, design flaws, influence of movements, etc. These techniques include (Wang et al. 2011):

- Polarization curve analysis. Studying the polarization curve, see Figure 4, of a fuel cell is indispensable in the field of fuel cell research, serving as a fundamental tool for optimizing energy conversion efficiency. By graphically illustrating the fuel cell's performance through the relationship between voltage and current density under diverse operating conditions, researchers gain valuable insights. This understanding allows for the identification of optimal operating parameters, a critical step in enhancing both efficiency and lifespan. Moreover, precise polarization curve analysis plays a pivotal role in diagnosing and mitigating issues such as electrode degradation and mass transport limitations. As a diagnostic tool, it aids to develop targeted solutions to address these challenges, and to understand the influence of different parameters on the fuel cell performance.
- Electrochemical Impedance Spectroscopy (EIS) is a powerful diagnostic technique widely applied in the evaluation of fuel cell performance. It involves subjecting the fuel cell system to a small amplitude sinusoidal signal across a range of frequencies while measuring the resulting variations in current and voltage. By analyzing the impedance response, EIS provides valuable insights into the electrochemical processes occurring within the fuel cell and the impedance of the system. This technique enables researchers to examine phenomena such as charge transfer resistance, mass transport limitations, and electrode kinetics, crucial factors influencing the overall efficiency of the fuel cell. EIS is particularly adept at detecting subtle changes in the system over time, making it an invaluable tool for monitoring and diagnosing issues related to fuel cell degradation or malfunctions. The detailed information derived from EIS aids in improving fuel cell designs, identifying performance-limiting factors, and developing strategies to enhance durability and efficiency. Overall, Electrochemical Impedance Spectroscopy serves as a non-invasive and versatile analytical tool, contributing significantly to the advancement of fuel cell technology by providing a comprehensive understanding of the complex electrochemical processes within these energy conversion devices.
- Voltage interruption. The voltage interruption test is a method employed in fuel cell research to assess the electrochemical behavior and durability of the cell under dynamic operating conditions. This test involves intentionally interrupting the voltage supply to the fuel cell for specified durations while monitoring its response. The primary objective is to evaluate the cell's ability to withstand abrupt changes in electrical load and its capacity to recover stability upon voltage restoration. Researchers analyze the transient behavior of voltage, current, and other relevant parameters during and after the interruption to glean insights into the cell's dynamic performance and potential vulnerabilities. This test provides valuable data for optimizing control strategies, assessing the impact on overall system reliability, and informing the development of robust fuel cell technologies for diverse applications. It is a crucial tool in advancing the understanding of fuel cell behavior under realistic operating scenarios and improving their overall operational resilience.
- Durability tests. These tests are integral components of fuel cell assessment, in order to measure the long-term stability and reliability of these energy conversion devices. These tests subject fuel cells to extended operational conditions, simulating real-world usage scenarios to assess their ability to withstand stress and maintain performance over time. Factors such as temperature

variations, continuous cycling, and varying load conditions are systematically applied to emulate the challenges encountered in practical applications. Monitoring parameters like voltage, current, and impedance throughout the test duration allows researchers to identify degradation mechanisms, assess material durability and anticipate potential failure modes. Durability testing plays a pivotal role in validating the viability of fuel cell technologies for commercial and industrial applications, providing crucial data for refining designs, enhancing materials, and ultimately advancing the overall longevity and efficiency of fuel cell systems in pursuit of sustainable and resilient energy solutions.

Additionally, the utilization of the gas chromatograph Agilent 990 MicroGC in tandem with these techniques provides valuable information. This analytical tool facilitates a comprehensive examination of fuel cell exhaust gas streams. The data acquired allows determining the degradation status of the membranes within the fuel cell by analyzing the composition of both anode and cathode output streams and enables proactive decisions to address degradation issues, contributing to the optimization of fuel cell performance and longevity.

From its side, the Stewart platform allows both testing the impact of platform-induced motions on stack performance, a critical condition aboard vessels sailing in harsh conditions, and testing the fuel cell under study in different orientations. These tests will provide, for example, valuable information on the effectiveness of the design of the cathode channels to evacuate the produced water avoiding flooding of the fuel cell and their influence on the fuel cell performance.

In the case of powertrain integrations, the EPDS allows the fuel cell to be tested under any power cycle of interest. On the other hand, the inclusion between the fuel cell and the EPDS of a PMS and secondary energy sources, e.g. batteries, will allow the PMS to be tested and/or hybridised architectures to be studied.

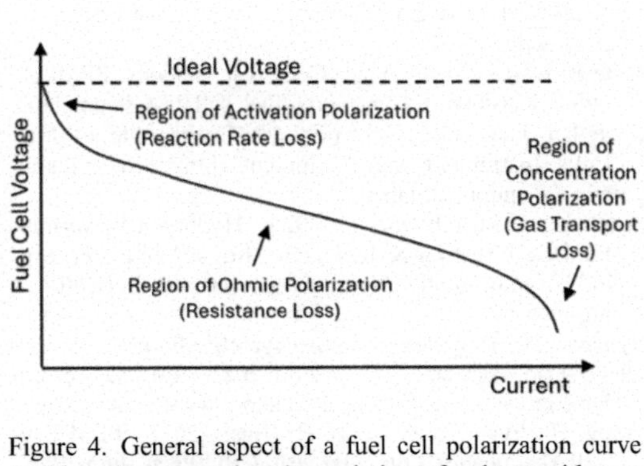

Figure 4. General aspect of a fuel cell polarization curve (solid line) representing the variation of voltage with current. The dotted line represents the ideal voltage or open circuit voltage of the fuel cell and corresponds to the electrical potential generated by the electrochemical reactions within the cell. The gap between the polarization curve and the ideal voltage is due to the losses produced by different phenomena inside the fuel cell.

The flexibility in the control of hydrogen, air, and cooling parameters allows the stack to be tested under extreme conditions to evaluate changes on its performance. Table 1 shows a list of the most relevant parameters that can be controlled by the test bench and their corresponding ranges.

Table 1. Controllable parameters of the test bench and their corresponding range of variation.

Parameter (units)	Min.	Max.
Hydrogen mass flow* (%)	-	50
Hydrogen pressure (kPa)	100	1600
Hydrogen temperature (K)	325	338
Hydrogen humidity (%)	0	100
Air mass flow* (%)	-	200
Air pressure (kPa)	94	600
Air temperature (K)	308	373
Air humidity (%)	40	100
Cooling water temperature (K)	333	363
Roll amplitude (°)	-29	29
Pitch amplitude (°)	-28	28
Heave amplitude (mm)	-215	215
Roll speed (°/s)	-40	40
Pitch speed (°/s)	-40	40
Yaw speed (°/s)	-50	50
Heave speed (m/s)	-0.625	0.625
Roll acceleration (°/s^2)	-700	700
Pitch acceleration (°/s^2)	-700	700
Heave acceleration (m/s^2)	-9.81	9.81

* Mass flow rate in excess of the stoichiometric ratio of the reaction $H_2 + \frac{1}{2}O_2 \rightarrow H_2O$.

These capabilities represent a powerful set of tools for the research and development of fuel cell technologies and hybridized fuel cell-based power plants onboard ships and aircrafts.

Regarding future developments, the facility is expected to be expanded in the medium term to cover PEM fuel cell capacities up to 270 kW, while maintaining its flexibility. This expansion will open the door to larger cells, which will most likely be in high demand for marine applications. In addition, the adaptation to support fuel cell types other than PEM and integration with the already existing diesel engine test bench next to this facility is also on the long-term roadmap.

5 CONCLUSIONS

The laboratory presented here serves as a pivotal resource for advancing both scientific and industrial endeavours in the development of innovative and efficient powertrains, particularly those integrating fuel cells with secondary power sources for maritime and aeronautical applications.

A global overview of fuel cell research facilities reveals a predominant focus on road transport, leaving a notable gap for dedicated testing environments tailored to the unique conditions of aircraft and ships.

The laboratory adaptability and advanced technical capabilities offer a platform for comprehensive testing of nearly market-ready PEM fuel cell stacks and power management systems. This extends to the evaluation of additional power sources, such as batteries and supercapacitors, under conditions reflective of real-world operational settings. This approach allows for the identification and rectification of design flaws that may elude detection in smaller-scale prototypes. Moreover, it facilitates the exploration of novel power management strategies to enhance the longevity and efficiency of power plant components.

Key to the efficacy of the laboratory is the integration of a gas chromatograph and a 6-DOF (Six Degrees of Freedom) Stewart platform, providing unparalleled capabilities for studying fuel cell-based powertrains for ships and aircraft. The gas chromatograph plays a crucial role in analysing exhaust gas streams, offering insights into membrane degradation within fuel cells. Simultaneously, the 6-DOF Stewart platform enables dynamic testing scenarios, mimicking the intricate operational conditions encountered in the targeted sectors.

Finally, it is important to note that this facility is open to international public collaboration in both public and private projects. Researchers, engineers, and interested individuals are encouraged to contact the Research Group PiCoHiMA at the Universidad Politécnica de Madrid to explore collaborative opportunities and glean further insights into the laboratory offerings. This collaborative approach underscores the commitment to fostering global partnerships in advancing research and innovation in the dynamic field of fuel cell-based power systems for maritime and aeronautical applications.

Comunidad de Madrid | Plan de Recuperación, Transformación y Resiliencia | España Puede | Financiado por la Unión Europea NextGenerationEU

ACKNOWLEDGEMENTS

This work has been conducted in the framework of the Project GreenH2CM funded by the Regional Government of Madrid and the Government of the Kingdom of Spain through the Recovery, Transformation and Resilience Plan (PRTR) funded by the European Union – NextGenerationEU. Furthermore, it was supported by Departamento de Arquitectura, Construcción y Sistemas Oceánicos y Navales (DACSON) from Universidad Politécnica de Madrid.

REFERENCES

Central European Research Infrastructure Consortium. 2024. 'JRC Fuel Cell and Electrolyser Testing Laboratory'. https://www.ceric-eric.eu/lab-instrument/jrc-fuel-cell-and-electrolyser-testing-facility/

Centro Nacional del Hidrógeno 2024. 'PEM Technology Testing Laboratory'. https://www.cnh2.es/en/cnh2/laboratorio-de-testeo-de-tecnologia-pem/

Ciemat 2023. 'Arranca el proyecto estratégico de la Comunidad de Madrid en I+D+i del hidrógeno verde y pilas de combustible liderado por el CIEMAT'. https://www.ciemat.es/portal.do?IDM=61&NM=2&identificador=2658

Daniele Rao 2023. 'FAQ: The EU ETS for shipping explained'. https://carbonmarketwatch.org/2023/12/06/faq-the-eu-ets-for-shipping-explained/#:~:text=The EU ETS for shipping will be phased in starting,the emissions reported in 2025

Elektro-Automatik 2023. ‚Elektro-Automatik'. https://elektroautomatik.com/gb/

European Commission 2023a. 'NextGenerationEU'. https://commission.europa.eu/strategy-and-policy/eu-budget/eu-borrower-investor-relations/nextgenerationeu_en

European Commission 2023b. 'The Recovery and Resilience Facility'. https://commission.europa.eu/business-economy-euro/economic-recovery/recovery-and-resilience-facility_en

European Council 2023. 'Infographic - Fit for 55: increasing the uptake of greener fuels in the aviation and maritime sectors'. https://www.consilium.europa.eu/en/infographics/fit-for-55-refueleu-and-fueleu/

Greenlight Innovation 2024. 'Hydrogen Fuel Cell Testing Equipment for PEMFC applications'. https://www.greenlightinnovation.com/products/fuel-cell

Leigh Collins 2023. 'China starts operations at new $74m national hydrogen fuel-cell vehicle testing facility in Chongqing'. Hydrogeninsight. https://www.hydrogeninsight.com/transport/china-starts-operations-at-new-74m-national-hydrogen-fuel-cell-vehicle-testing-facility-in-chongqing/2-1-1385779

National Aeronautics and Space Administration 2024. 'Fuel Cell Testing Laboratory'. https://www1.grc.nasa.gov/facilities/fctl/

Scitech Korea 2024. 'Fuel Cell Test Stations PEMFC'. https://scitechkorea.com/energy/fuel_cell_test_station.html

Shanghai Jiao Tong University 2024. 'Institute of Fuel Cells. Instruments and Equipment'. https://fuelcell.sjtu.edu.cn/equipment/label

Southwest Research Institute 2024. 'Hydrogen Powertrain Fuel Cell Testing & Research'. https://www.swri.org/industry/automotive/hydrogen-powertrain-fuel-cell-testing-research

Symétrie 2023.'Symétrie'. https://symetrie.fr/en/

Universidad Politécnica de Madrid 2023. 'Escuela Técnica Superior de Ingenieros Navales'. https://www.etsin.upm.es/

Wang, H., Yuan, X.-Z., & Li, H. (Eds.). 2011. 'PEM Fuel Cell Diagnostic Tools' (1st ed.). CRC Press. https://doi.org/10.1201/b11100

Hydrodynamics

Advances in Maritime Technology and Engineering – Guedes Soares & Santos (Eds)
© 2024 The Author(s), ISBN 978-1-032-83107-7

On the computational costs of potential models for floating bodies in waves

A. Abbasnia & C. Guedes Soares
Centre for Marine Technology and Ocean Engineering (CENTEC), Instituto Superior Técnico, Universidade de Lisboa, Lisbon, Portugal

ABSTRACT: The potential flow theory has been used to determine the response of floating bodies in waves and the boundary integral equation was implemented to solve the Laplace equation. In the linear potential models, the complexity of the free surface boundary conditions leads to the boundary conditions linearized and it makes the validity range of the solution shortened for large amplitude response for non-wall sided bodies. To treat the exact free surface boundary conditions, fully nonlinear potential models have been developed based on the mixed Eulerian-Lagrangian methods. The burden of computation of the fully nonlinear model is extremely larger than linear models which draws a rationale of this paper for delving into the results of linear and fully nonlinear models to measure the cost of computation. Hence non-wall-sided bodies are taken into comparison of the responses obtained from the in-house linear and fully nonlinear models which were developed previously.

1 INTRODUCTION

The potential flow theory has been widely used to solve the flow field over floating bodies such as marine vehicles and offshore floating platforms (Abbasnia *et al.* 2022, Harris *et al.* 2022, Liang *et al.* 2022, Tang *et al.* 2023). Regardless of the negligence of the viscous effects in the potential flow, the Laplace equation has been widely used for the hydrodynamic simulation of wave-body interaction problems rather than the Reynolds Average Navier-Stocks (RANS) equation notably due to the lower cost of computation. Since the Laplace equation is an elliptical differential equation, enormous analytical and numerical models have been developed to solve the boundary value problem for the various boundary conditions (Gao & Zou 2008, Watai *et al.* 2015, Dombre *et al.* 2019, Pan 2022). The robustness of the boundary conditions triggers different orders of solutions. Among the various boundary conditions of the wave-body interaction problems, treating free surface boundary conditions and the body boundary condition leads to two different solvers which are identified by the frequency domain and the time domain solvers.

In the frequency domain solver, the variation of the wetted surface of oscillating bodies is neglected, which limits the range of validity of the solution for the large relative motion of bodies to the free surface level. Moreover, the kinematics of water particles over the exact free surface is governed by the kinematic and dynamic free surface boundary conditions which contain the nonlinear and non-homogenous differential terms. Literarily, the free surface boundary conditions would be tracked out of the wave theories and the solution of the equation motion.

Wave theories have been derived from the different orders of Taylor's series of free surface boundary conditions. For instance, Airy wave theory as the linear wave theory originates from the coupled linearized free surface boundary condition and it is derived from coupling first-order terms of Taylor's expansions which approximate both free surface boundary conditions (Dean & Dalrymple 1991). Conversely, the fifth-order Stokes wave theories have been widely used as the robust wave theories in fully nonlinear wave-body interaction problems which was derived out of Taylor's expansion of the free surface boundary conditions up to fifth-order terms (Fenton 1985).

In the context of the motion equation, the role of the free surface boundary conditions is inevitable to obtain the hydrodynamic coefficients for the frequency domain solution. Moreover, the approximation of the kinematics of water particles at the corner points (free surface-body intersection) depends on the treatment of the free surface boundary conditions drastically (Grilli & Svendsen 1990, Abbasnia & Guedes Soares 2022). Corner points are the source of numerical instability for the time domain during the updating of the moving boundaries at each time step. As long as the determined kinematics of the water particles on the exact free surface preserve the momentum conservation during the time domain simulation, the numerical stability of the simulation remains. The kinematic of the free surface water particles at the corner points is the function of the fully nonlinear free surface boundary condition and the body boundary conditions simultaneously.

DOI: 10.1201/9781003508779-9

Boundary integral equations such as second Green's identity have been used to compute the potential flow field of the wave interaction with ship-shaped bodies in both the frequency domain and time domain. The prerequisite of the boundary integral equation is the solution of a pulsating singularity which has been derived for various lateral boundary conditions such as sea bottom and waterways walls. To treat the confined water conditions and the linearized free surface boundary conditions in the frequency domain, the image method has been implemented into the solution of the singularity $(1/r)$ and the resultant solution comes with the Principal Value (PV) integral which stands for the dynamic free surface boundary condition. The PV integral triggers the hydrodynamic coefficients which are identified by the radiation potential function and the diffraction potential function (Wehausen & Laitone 1960, Yang 2000, Abbasnia & Ghiasi 2013). Eventually, the Neumann boundary integral equation would be solved upon the linear body boundary condition.

In the time domain solution, the moving boundaries would be updated while time marching of the exact free surface. The Mixed Eulerian-Lagrangian (MEL) has been vastly used to deal with the fully nonlinear free surface boundary conditions. Correspondingly, the boundary integral equation for the time domain solution would be Dirichlet-Neuman in which the free surface is responsible for the Dirichlet boundary, and the rest of the boundaries are the Neuman boundaries at every time step. The time integration of the dynamic free surface boundary condition updates the Dirichlet condition on the exact free surface and the kinematic free surface boundary condition is used to re-allocate the free surface computational nodes in new positions. Analogously, the oscillating body boundary condition would be updated at each time step using the solution of the motion equation and the kinematics of the corner points on the body. Re-gridding of the moving boundaries is undoubtful at each time step and smoothing schemes might be applied to the free surface occasionally to avoid the spike instability (Tanizawa 2000, Kim *et al.* 2021).

An in-house code was developed by Abbasnia, *et al.* (2022) to compute the response of advancing ship-shaped bodies in which the Rankine source panel method was programmed in C++. Furthermore, a fully nonlinear model was programmed in Fortran by Abbasnia & Guedes Soares (2020) which has been used to calculate the time domain response of freely floating bodies. The fully nonlinear model was enhanced by the Non-Uniform B-Spline basis function for updating the exact free surface. The material node approach was implemented for tracking the moving boundaries and the fourth-order Runge-Kutta methodology was used for time integration.

An opportunity is taken in the present paper to demonstrate the range of validity of both linear and fully nonlinear models for freely floating ship-shaped bodies over a broad range of frequencies. Since the burden of computations of the fully nonlinear model is excessively heavier than the linear model, the costs of computations for each of the models are revealed on the same computer machine which may be interesting for ship hydrodynamic model developers, particularly for those who are going to augment the hydrodynamic solver into decision-making design algorithms.

2 MATHEMATICAL THEORIES

The Laplace equation governs the potential flow as

$$\nabla^2\phi(x,y,z,t) = 0 \text{ in } \Omega(t) \tag{1}$$

within the computational domain (Ω) and the velocity potential (ϕ) is responsible for the irrotational flow of an inviscid, incompressible, and homogenous fluid. $Oxyz$ is the right-hand coordinate system in which the z-axis is directed upward. The origin of the coordinate system is allocated to the still water level as shown in Figure 1. t is the time and θ is the progressing angle of the incident wave with wavelength λ.

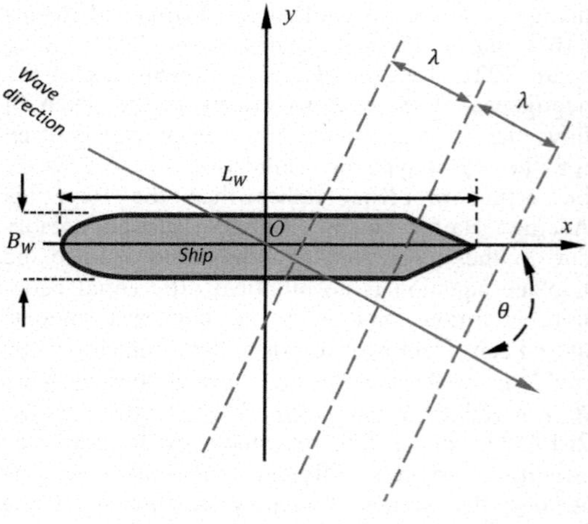

Figure 1. Definition of the coordinate system and the wave direction.

To solve Equation 1, the boundary conditions need to be specified which are given as

- Free surface boundary conditions

$$\left.\begin{array}{l}\frac{\partial\eta}{\partial t} = \frac{\partial\phi}{\partial z} - \frac{\partial\phi}{\partial x}\frac{\partial\eta}{\partial x} - \frac{\partial\phi}{\partial y}\frac{\partial\eta}{\partial y}\,(\text{KFSBC}) \\ \frac{\partial\phi}{\partial t} = -gz - \frac{1}{2}|\nabla\phi|^2(\text{DFSBC})\end{array}\right\} \text{on } \Gamma_f \tag{2}$$

where $\eta(x,y,t)$ is the wave elevation, g is the gravitational acceleration and Γ_f stands for the exact free surface.

- The body boundary

$$\frac{\partial\phi}{\partial n} = \vec{n}\cdot\left(\vec{V}+\vec{\omega}\times\vec{R}\right) on\ \Gamma_b \quad (3)$$

where $\vec{V}$ and $\vec{\omega}$ are responsible for the translational and rotational velocity of the rigid body, respectively, $\vec{R}$ is the distance vector between the centre of rotation and the body surface (Γ_b). $\vec{n}$ is the normal vector oriented outward of the fluid.

- Impermeable boundary condition

The zero normal derivative of velocity potential is specified on the sea bottom (Γ_d) as

$$\frac{\partial\phi}{\partial n} = 0 \text{ on } \Gamma_d \quad (4)$$

The rest of the boundary conditions are functions of the methodologies which are taken to discretize the Laplace equation.

2.1 *Fully nonlinear solution of the potential flow*

The second Green's identity is applied to solve the boundary value problem in the Eulerian frame which is given as:

$$c(q)\phi(q) = \int_\Gamma\left(G(q,p)\frac{\partial\phi(p)}{\partial n} - \phi(p)\frac{\partial G(q,p)}{\partial n}\right)ds \quad (5)$$

The field point (p) is distributed on the boundary surface ($\Gamma \in \Gamma_f \cup \Gamma_b \cup \Gamma_d \cup \Gamma_i \cup \Gamma_o$) and the source point ($q$) sweeps over the computational domain and $c(q)$ equals 4π,2π and zero when the source point is located inside, upon and outside of the smooth computational domain, respectively. Γ is the close boundary where Γ_i is the wave generator boundary which is defined as

$$\frac{\partial\phi}{\partial n} = -\frac{\partial\phi_w}{\partial x} on\ \Gamma_i \quad (6)$$

ϕ_w is the velocity potential function of the wave theories. Γ_o stands for the outflow boundary and lateral walls which are assumed impermeable (zero-flux of potential). $G(q,p)$ are three-dimensional problems given as

$$G(q,p) = \frac{1}{r(q,p)} \quad (7)$$

The definition of the initial conditions is necessary to obtain the unique solution from the boundary value problem which is defined as

$$\begin{matrix}\eta(x,y,t\le 0) = 0 \\ \phi(x,y,z,t\le 0) = 0\end{matrix} on\ \Gamma_f \quad (8)$$

The mixed Eulerian-Lagrangian method is exploited to implement the fully nonlinear free surface boundary condition and the instantaneous body boundary condition into the solution. Therefore, Equation 2 is reformulated in the Lagrangian manner as

$$\begin{matrix}\frac{\delta\phi}{\delta t} = -g\eta + \frac{1}{2}|\nabla\phi|^2 \\ \frac{\delta\vec{X}}{\delta t} = \nabla\phi\end{matrix} \text{ on } \Gamma_f \quad (9)$$

where $\delta/\delta t = \partial/\partial t + \upsilon^{\rightarrow}.\nabla\phi$ and $\vec{\upsilon} = \nabla\phi$ for the material node approach and $X^{\rightarrow} = (x,y,z)$ is the location of computational nodes on the exact free surface. To avoid the wave reflection from the outflow boundary and any lateral walls, the sponge layer is adopted adjacent to the walls over the free surface which is articulated as

$$\begin{matrix}\frac{\delta\vec{X}}{\delta t} = \nabla\phi - \vartheta(\tau)\left(\vec{X}-\vec{X}_e\right) \\ \frac{\delta\phi}{\delta t} = -g\eta + \frac{1}{2}(\nabla\phi.\nabla\phi) - \vartheta(\tau)(\phi-\phi_e)\end{matrix} \text{ on } S_f \quad (10)$$

in which

$$\vartheta(\tau) = \alpha\sigma\left[\frac{(\tau-\tau_0)}{\lambda}\right]^2, \tau_0 \le \tau \le \tau_1 = \tau_0 + \beta\lambda \quad (11)$$

σ is incident wave angular frequency. The length and the strength of the sponge layer are adjusted by the parameters β and α, respectively τ_0 and τ_1 represent the longitudinal positions of the sponge layer interval. The simple absorber is made when $\phi_e = 0$ and $\eta = 0$. The fourth-order Runge-Kutta time integration method is used to update the moving boundaries. The corner points need to be treated to avoid numerical instability based on the methodology which is elaborated on by Abbasnia & Guedes Soares (2022).

2.2 *Linear solution of the potential flow*

The frequency domain solution takes advantage of the feature of the velocity potential decomposition and the total velocity potential is defined as

$$\begin{aligned}\phi(x,y,z,t) = \text{Re}\Big[&a_0\phi_D(x,y,z)e^{i\sigma t} \\ &+\textstyle\sum_{j=1}^{6}\{\xi_j\phi_j(x,y,z)\}e^{i\sigma t}\Big]\end{aligned} \quad (12)$$

in which a_0 is the wave amplitude and ϕ_D is the diffraction potential which is stimulated by the scattered incident wave due to the presence of the ship hull in the progressive wave regime. The response amplitude of the ship in six degrees of freedom is written by ξ_j, $j = 1,..6$. The translational oscillations are numbered $j = 1, 2, 3$ for the surge, sway and heave motions respectively and $j = 4, 5, 6$ for the rotational oscillations of roll, pitch and yaw. ϕ_j, $j = 1,..6$ is the radiation potential for each mode of motion. The diffraction potential function can be decomposed as

$$\phi_D = \phi_w + \phi_s \tag{13}$$

in which ϕ_s is responsible for the scattering potential. Based upon the potential flow theory, the Laplace equation governs each of the velocity potential components and the primary Fredholm equation (Sutulo *et al.* 2012, Abbasnia *et al.* 2021) is employed to calculate the strength of pulsating singular points (μ), given as

$$2\pi\mu(q) + \int_\Gamma \mu(p)\frac{\partial H(q,p)}{\partial n}ds = \phi(q) \tag{14}$$

and $\Gamma \equiv \Gamma_b$ and the pulsating singular solution (H) is defined as

$$H(q,p,\sigma) = \frac{1}{r(q,p)} + \frac{1}{r'(q,p')} + \tilde{G}(q,p,\sigma) \tag{15}$$

p' is the mirrored layer of the computational nods about the undisturbed free surface ($z = 0$). The first two terms on the right-hand side of the Green function is the double-body formulation and the $\tilde{G}$ function originates from the solution of a pulsing singularity with the frequency σ upon the linearized free surface boundary conditions (Wehausen & Laitone 1960) which contains PV integral and it was regularized by Newman (1985). Analogously, to obtain the unique solution out of the boundary value problem, the boundary condition on the body needs to be designated as

$$\frac{\partial \phi}{\partial n}(q) = \begin{cases} i\sigma N^j, j = 1,...,6 & \text{for } \phi_j \\ -\frac{\partial \phi_w}{\partial n} & \text{for } \phi_s \end{cases} \tag{16}$$

where $(N^1, N^2, N^3) = \vec{n}$ and $(N^4, N^5, N^6) = \vec{R} \times \vec{n}$. When the q point coincides with the distributed nodal points (p) over the wetted surface, the strength of the pulsating Green function at the p points is computed for each mode of radiation potential and the scattering potential function due to sweeping the q point throughout the locations of nodal points. Eventually, the radiation potential for each mode of motion is obtained as

$$\phi^j(q) = \int_\Gamma \mu^j(p)H(q,p)ds, j = 1,...,6 \tag{17}$$

Similarly, the scattering potential can be computed. The solutions of radiation potential are used to determine the coupled added mass (A^{kj}) and coupled damping coefficients (D^{kj}) for each mode of motion as

$$f^{kj} = \omega^2 A^{kj} - i\omega D^{kj},\ k,j = 1,2,...,6 \tag{18}$$

where

$$f^{kj} = -\rho \int_\Gamma \frac{\partial \phi^k}{\partial n}\phi^j ds, k,j = 1,2,...,6 \tag{19}$$

and ρ is the fluid density.

2.3 *Equation of motion in the time domain*

The motion equations of the fluid and the floating body are solved simultaneously in the time domain to find the response of the rigid. To couple the motions of the potential flow and the body, the acceleration potential (Φ) is defined as

$$\vec{a} = \nabla\Phi(x,y,z,t) \tag{20}$$

in which

$$\Phi = \frac{\partial \phi}{\partial t} + \frac{1}{2}(\nabla\phi)^2 \tag{21}$$

The second term on the right-hand side of the acceleration potential does not satisfy the Laplace equation whereas it is derived from the solution of the velocity potential field. Conversely, the first term satisfies the Laplace equation, and the second Green's identity is used to compute it as given as

$$c(q)\frac{\partial \phi(q)}{\partial t} = \int_\Gamma \left(G(q,p)\frac{\partial^2 \phi(p)}{\partial t \partial n} - \frac{\partial \phi(p)}{\partial t}\frac{\partial G(q,p)}{\partial n}\right)ds \tag{22}$$

The boundary conditions of the time derivative of the velocity potential need to be assigned all over the computational domain as

- Inflow boundary

$$\frac{\partial^2 \phi}{\partial n \partial t} = -\frac{\partial^2 \phi_w}{\partial x \partial t}\ on\ \Gamma_i \tag{23}$$

- Rigid walls and bottom boundaries

$$\frac{\partial^2\phi}{\partial n\partial t} = 0 \, on \, \Gamma_d \, , \Gamma_o \tag{24}$$

- Fully nonlinear free surface boundary

$$\frac{\partial\phi}{\partial t} = -gz - \frac{1}{2}|\nabla\phi|^2 on \, \Gamma_f \tag{25}$$

- Floating body boundary

$$\frac{\partial^2\phi}{\partial n\partial t} = \vec{n}\cdot\left(\vec{V} + \vec{\omega}\times\vec{r}\right) + \tilde{Q} \tag{26}$$

where

$$\tilde{Q} = -k_n\left(\nabla\phi - \vec{V} - \vec{\omega}\times\vec{r}\right)^2 + \vec{n}\cdot\vec{\omega}\times(\vec{\omega}\times\vec{r}) + \vec{n}\cdot 2\vec{\omega}\times\left(\nabla\phi - \vec{V} - \vec{\omega}\times\vec{r}\right) - \frac{\partial}{\partial n}\left(\frac{1}{2}\nabla\phi\cdot\nabla\phi\right) \tag{27}$$

and k_n is the body surface's normal curvature along the fluid path line. The $\tilde{Q}$ term is found explicitly using the velocity potential solution. The first term on the right-hand side of the floating body boundary condition can be achieved through the motion equation as

$$\mathrm{M}^{kj}\gamma_j + \mathrm{B}^k = F_f^k + F_g^k, \; k,j = 1,..,6 \tag{28}$$

in which $\vec{V} = (\gamma_1, \gamma_2, \gamma_3), \vec{\omega} = (\gamma_4, \gamma_5, \gamma_6)$,

$$[\mathrm{M}^{kj}] = \begin{bmatrix} m & 0 & 0 & 0 & 0 & 0 \\ 0 & m & 0 & 0 & 0 & 0 \\ 0 & 0 & m & 0 & 0 & 0 \\ 0 & 0 & 0 & I_{xx} & I_{xy} & I_{xz} \\ 0 & 0 & 0 & I_{yx} & I_{yy} & I_{yz} \\ 0 & 0 & 0 & I_{zx} & I_{yz} & I_{zz} \end{bmatrix} \tag{29}$$

and

$$\{\mathrm{B}^k\} = \begin{bmatrix} 0 \\ 0 \\ 0 \\ (I_{zz} - I_{yy})\omega_y\omega_z - I_{xy}\omega_x\omega_z - I_{zx}\omega_x\omega_y + I_{yz}\left(\omega_y^2 - \omega_z^2\right) \\ (I_{xx} - I_{zz})\omega_x\omega_z - I_{yz}\omega_x\omega_y - I_{xy}\omega_z\omega_y + I_{zx}\left(\omega_z^2 - \omega_x^2\right) \\ (I_{yy} - I_{xx})\omega_x\omega_y - I_{zx}\omega_y\omega_z - I_{yz}\omega_x\omega_z + I_{xy}\left(\omega_x^2 - \omega_y^2\right) \end{bmatrix} \tag{30}$$

F_f^k is the general hydromechanics load vector given as

$$F_f^k = \int_{\Gamma_b}\left(-\frac{\partial\phi}{\partial t} - \frac{1}{2}\nabla\phi\cdot\nabla\phi - gz\right)N^k ds \tag{31}$$

and F_g^k is the gravitational load tensor. Eventually, the body boundary condition can be written as

$$\frac{\partial^2\phi}{\partial n\partial t} = \{N^k\}.[\mathrm{M}^{kj}]^{-1}\left\{\int_{\Gamma_b} -\frac{\partial\phi}{\partial t}\{N^k\}ds\right\} + \tilde{q} \tag{32}$$

where

$$\tilde{q} = \{N^k\}.[\mathrm{M}^{kj}]^{-1}\left[\int_{\Gamma_b}\left(-gz - \frac{1}{2}\nabla\phi\cdot\nabla\phi\right)\{N^k\}ds + \left\{F_g^k\right\} - \{\mathrm{B}^k\}] + \tilde{Q} + \left\{F_g^k\right\} - \{\mathrm{B}^k\}\right] + \tilde{Q} \tag{33}$$

The resultant body boundary condition demonstrates the coupling of time derivatives of velocity potential and its normal flux. Eventually, by assigning the initial condition for the time derivative of the velocity potential given as

$$\begin{matrix} \eta(x,y,t\leq 0) = 0 \\ \phi_t(x,y,z,t\leq 0) = 0 \end{matrix} on \; \Gamma_f \tag{34}$$

the instantaneous pressure hydrodynamic on the exact body boundary condition is achieved and correspondingly the kinematics of the floating body implicitly is driven in the time domain. At each time step, the response of the floating body along each mode of motion is determined by the time integration of the kinematics of the body employing the fourth-order Runge-Kutta methodology.

2.4 *Equation of motion in the frequency domain*

The form of the motion of the equation including the hydrodynamic coefficients is written as

$$\begin{aligned} \textstyle\sum_{j=1}^{6}\xi_j\left[-\sigma^2(M^{kj} + A^{kj}) + i\sigma D^{kj} + C^{kj}\right] \\ = a_0F^k, k,j = 1,2,...,6 \end{aligned} \tag{35}$$

C^{kj} is responsible for the restoring matrix which is achieved through the hydrostatic calculation in each mode of motion. The hydrodynamic exciting loads (F^k) due to the Froude-Krylov forces and the diffraction potential are derived through the Haskind relation which is written as

$$F^k = -\rho\int_{\Gamma}\left(\phi_w\frac{\partial\phi^k}{\partial n} - \phi^k\frac{\partial\phi_w}{\partial n}\right)ds, k = 1,2,...,6 \tag{36}$$

Eventually, the amplitude of the ship response in six degrees of freedom would be approximated by solving the resultant system of equations.

3 NUMERICAL IMPLEMENTATION

A theoretical Wigley hull is considered as a non-wall side ship-shaped body in the present study which is formulated as

$$y = \pm\frac{B_W}{2}\left[1-\left(\frac{z}{D_W}\right)^2\right]\left[1-\left(\frac{2x}{L_W}\right)^2\right] \quad (37)$$

where D_W represents the ship draught and the quadrilateral element is used to describe the form as shown in Figure 1.

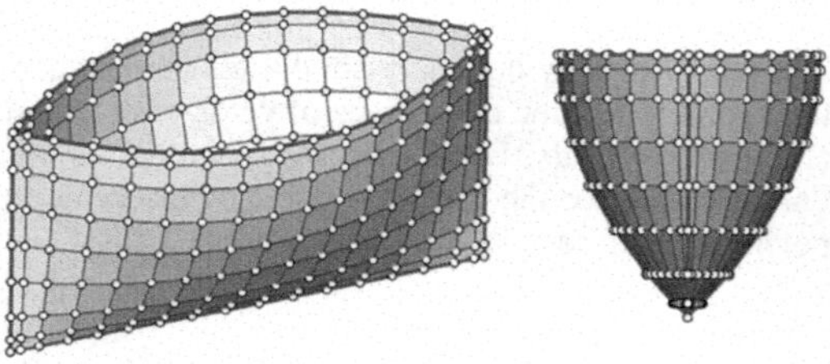

Figure 2. Discretized form of the Wigley hull by the quadrilateral element. (White point: Nodal point).

The same configuration of nodal points is used for both the frequency domain and the time domain solvers. In the frequency domain solver, the centroids of panels represent features of each element including the normal vector, element and the p point location and it is labelled by the element area. The constant distribution of the velocity potential is assumed for spatial integration in the primary Fredholm equation, and the strength of the pulsating point would be computed as the result of the boundary value problem in the frequency domain fluid solver.

In the time domain solver, the nodal points by themselves are considered. The nodal points over the verge of the water plane are the corner points which are coupled with the free surface nodal points as shown in Figure 3.

The resolution of the nodal points along the curvature of the body is 2.5 times higher than the rest of the free surface patches. The free surface encompasses three patches which are defined as the inflow patch, the body patch and the outflow patch. The inflow patch is extended from the wave generator to the body and the outflow is prolonged from the body to the outflow wall. The body patch is the refined mesh zone.

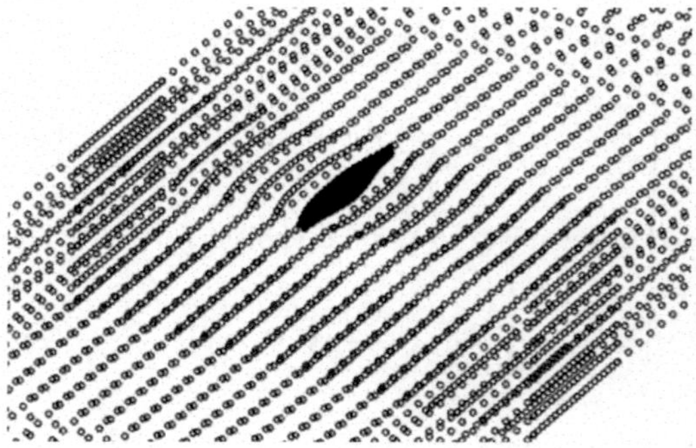

Figure 3. Configuration of the nodal points for the time domain solver.

The finer mesh zone is extruded to the lateral walls. To keep the continuity of the patches, the conservation momentum of nodal points at the intersection of patches is kept up during the simulation. For the narrow distance of the lateral walls as well as shallow water, the fluctuation of the velocity potential is elevated particularly at the tips of the bow and the stern on the exact free surface. Therefore, the accurate calculation of the corner points' kinematics may preserve the numerical stability for the large amplitude motion of the body.

Both numerical algorithms have been run on a laptop PC (Intel Core i7 CPU, 2.50 GHz, and 8 GB of RAM). The full nonlinear solver was programmed using Fortran whereas a C++ program was provided for the frequency domain solver.

4 NUMERICAL RESULTS

The geometrical characteristics of the Wigley hull are listed in Table 1.

Table 1. Geometrical characteristics of the Wigley hull.

Specification	Value
Length (L_W)	1.025 (m)
Width (B_W)	0.51225 (m)
Draft (D_W)	0.1 (m)
Displacement	0.023 (m^3)
Vertical centre of mass	0.1 (m)

Both solvers were already verified by (Abbasnia & Guedes Soares 2017, Abbasnia *et al.* 2022). To do a benchmark study on the range of validity of both solvers, the available experimental data by Journee (1992) is used which demonstrates the response of a Wigley hull under a broad range of waves in Table 2. To keep the deep waves condition the water depth is assumed 2λ.

The Wigley hull form is defined by 15 sections and 23 nodal points are responsible for the half-beam at each section. The number of elements is kept the same for both solvers and the heave and pitch motions over the range of frequencies are illustrated in Figure 4 for the head seas condition.

Table 2. Properties of the incident waves.

No	Wave steepness ($2\pi a_0/\lambda$)
1	0.03974
2	0.09869
3	0.19732
4	0.30156
5	0.35182
6	0.40208

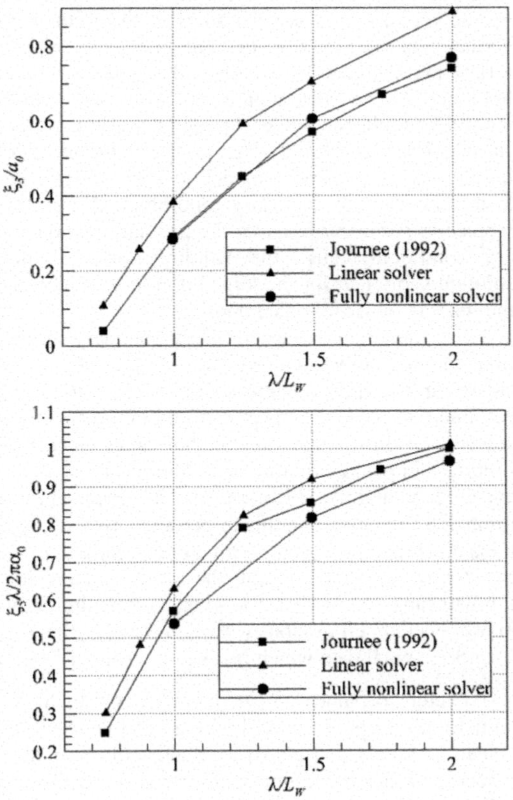

Figure 4. Heave and pitch motions of the Wigley hull for the head sea condition.

To avoid the reflection of the lateral walls into the computational domain, the artificial wave absorber is also embedded adjacent to the lateral walls and the width of the computational taken is taken enough wide (3λ). The time step size for the time domain solver is $\Delta t = T/40$ and $T = 2\pi/\sigma$ is responsible for the wave period. Five replications of either the heave motion record or the pitch response are assumed for the periodic steady state. The accuracy of the fully nonlinear solver for the larger waves ($\lambda/L_W > 1$) is outstanding whereas the discrepancy between the linear solution and the fully nonlinear results is decreased while the wavelength decreases for heave motion.

For the pitch motion, the numerical results are closer to the experimental data than the heave motion and interestingly, the difference between the numerical results is smaller than the results of heave motion for the whole range of the frequencies. The computation load for each of the solvers is not adequate to be compared since the run time of the fully nonlinear solver is excessively higher than the linear solver due to the time marching of the simulation. Table 3 demonstrates the cost of both solvers' computations for $\lambda/L_W = 1$ per iteration. The linear solution needs a single iteration to accomplish the simulation whereas the number of iterations for the fully nonlinear solver is the function of the mesh size, the time step size, and the size of the computational domain.

Table 3. CPU time for the solver.

Item	Fully nonlinear solver	Linear solver
Heave Error*	6.8%	61%
Pitch Error	3.6%	40%
CPU time per iteration	6.72 (*s*)	3.22 (*s*)

* Error=100 × $|Num. - Exp.|/|Exp.|$

A total of 735 iterations passed to reach the steady state of the time domain simulation and 1.4 hours is the actual time simulation for the fully nonlinear simulation. Despite the accuracy of the fully nonlinear simulation is remarkable, the cost of computation looks intolerable for a range of frequencies. The bottleneck of the burden of computations in the fully nonlinear models is triggered by the time integration function and the re-gridding of the moving boundaries during the simulation. The velocity potential solver and the acceleration potential solver are recursed four times per each time step and correspondingly, the regridding of the entire computational is replicated at each sub-time step of the fourth-order Runge-Kutta approach. Corner points particularly the ones that lie down on the intersection of the moving body and the exact free surface need to be updated precisely to avoid numerical instability. Therefore, applying iterative three-dimensional high-order interpolation function to real ship forms would be inevitable to define the exact wetted surface using the ship offset table at each sub-time step.

To demonstrate the robustness of both solvers for high-flare ship form, an analytical form of the Wigley hull given as

$$y = \pm \frac{B_W}{2} \left[\left(1 - \frac{|z|}{D_W} \right)^3 \right] \left[1 - \left(\frac{2x}{L_W} \right)^2 \right] \quad (38)$$

is subjected to the same simulation. The discretized geometrical model is illustrated in Figure 5. The third-order polynomial basis function is deployed to define the flare of the ship section which is identified as the extreme curvature of the sections. At the design draft, 0.0083 m^3 of the displacement is reserved by the hull.

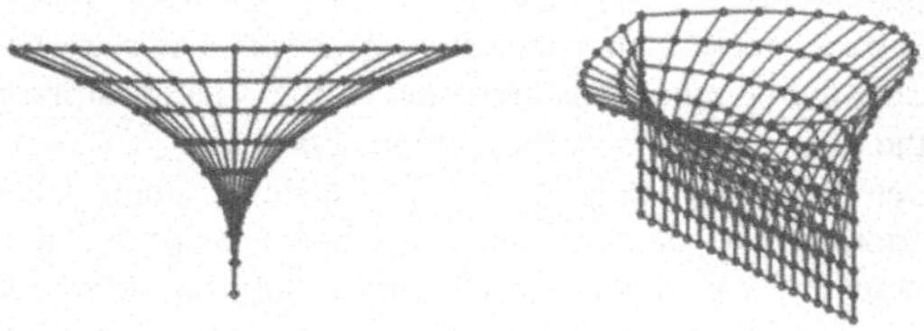

Figure 5. Discretized high-flare form of Wigley hull.

The amplitudes of heave and pitch motions are shown in Figure 6 and the comparisons highlight the differences between two solvers remarkably. For the heave motion, the 30% overestimation of the linear solution is observed at $\lambda/L_W = 2$ whereas the difference is reduced while the wavelength decreases. In contrast, the trend of the discrepancy between the linear solution and the fully nonlinear solution reverses over the range of wavelengths.

For the pitch motion, the fully nonlinear solution is larger than the linear solution across the range of wavelengths and the smallest difference between the solutions is 3% at $\lambda/L_W = 2$. While the wavelength decreases the difference grows up to 15% for $\lambda/L_W = 1$. The substantial difference for the smaller wavelengths implies that the remarkable impact of high-flare ship forms on the pitch response and the range of validity of the linear solution may not extend enough to capture it.

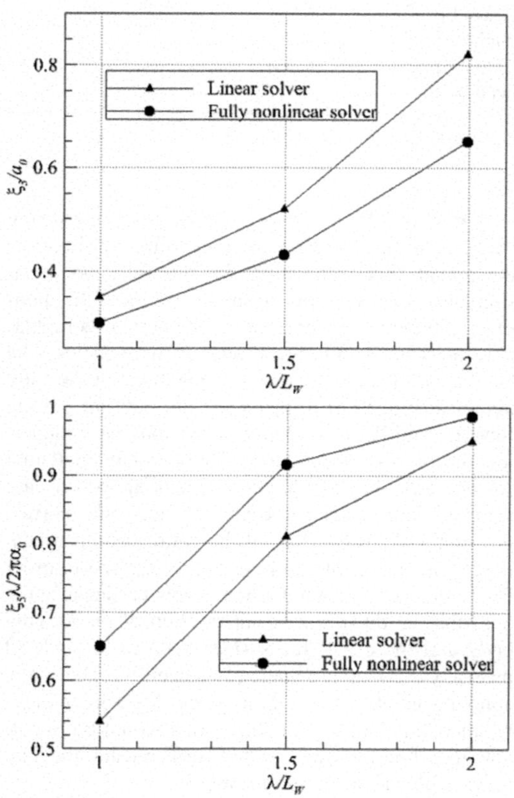

Figure 6. Heave and pitch motions of the high-flare Wigley hull for the head sea condition.

5 CONCLUSIONS

Two in-house programs were used to determine the response of freely floating bodies in the frequency domain and the time domain. The linear solver uses diffraction theories to calculate the potential flow field whereas the MEL approach is implemented into the fully nonlinear solver to apply the exact free surface boundary conditions to the solution. The limitations of the linear model to approximate the instantaneous body boundary condition and the negligence of nonlinear terms of the free surface boundary conditions cause the validity range of the solution shortened for the large amplitude motion of the floating non-wall-sided body.

In contrast, both the instant body boundary condition and the fully nonlinear free surface are considered in the fully nonlinear solver. The burden of computation and the ranges of validities of two solvers are drawn attention in this paper since the runtime simulation of the fully nonlinear solver is excessively high, which makes its applicability limited. To demonstrate this importance, two Wigley hulls were adopted to perform the computations in the frequency domain and the time domain for a broad range of waves. The responses of the body in heave and pitch motions were compared with a set of experimental data which leads to the conclusions:

- Although the linear solver predicted the trend of the body response over the range of frequency, it can be barely considered a reliable solver for the decision-making design due to the large incompatibility with the experimental results (Error > 10%).
- The influence of the fully nonlinear free surface boundary conditions and the instantaneous body boundary condition on the heave motion is larger than on the pitch motion.
- The non-wall-sided forms of the Wigley hulls amplify the inconsistency of the linear solution with the fully nonlinear solution and the experimental results.
- Despite the acceptable accuracy of the fully nonlinear solver, the huge cost of computation makes

its application narrow. The viscous flow solver is found more reasonable to be used in the design process than the fully nonlinear solver in terms of reliability.

- Preserving the numerical stability of the fully nonlinear solver is difficult while the wave steepness grows. Indeed, calibration of the fully nonlinear solver for each case study is challenging which causes the cost of computation heavy for a range of frequencies.
- For real ship shapes and their complex curvature near the free surface, re-gridding the moving boundaries magnifies the simulation run-time and the burden of computations due to iterative interpolating of the corner points at the intersection of the free surface and the floating body.

ACKNOWLEDGEMENTS

This work contributes to the Strategic Research Plan of the Centre for Marine Technology and Ocean Engineering (CENTEC), which is financed by FCT under contract UIDB/ UIDP/00134/2020.

REFERENCES

Abbasnia, A. & Ghiasi, M. 2013. Evaluation of the modified 3D free-surface Green's function for potential flow in a numerical towing tank. *Scientia Iranica* 20 (5): 1538–1549.

Abbasnia, A. & Guedes Soares, C. 2017. Exact evaluation of hydrodynamic loads on ships using NURBS surfaces and acceleration potential. *Engineering Analysis with Boundary Elements* 85: 1–12.

Abbasnia, A. & Guedes Soares, C. 2020. OpenMP parallelism in computations of three-dimensional potential numerical wave tank for fully nonlinear simulation of wave-body interaction using NURBS. *Engineering Analysis with Boundary Elements* 117: 321–331.

Abbasnia, A. & Guedes Soares, C. 2022. Spatial derivatives of boundary element solutions on the exact free surface of fully nonlinear numerical wave tanks. *Engineering Analysis with Boundary Elements* 134 (February 2021): 532–538.

Abbasnia, A., Karimirad, M., Friel, D., & Whittaker, T. 2022. Fully nonlinear dynamics of floating solar platform with twin hull by tubular floaters in ocean waves. *Ocean Engineering*, 257: 111320.

Abbasnia, A., Sutulo, S., Callewaert, B., & Guedes Soares, C. 2021. Development of a three-dimensional frequency domain seakeeping code. *In*: *Developments in Maritime Technology and Engineering*. London, UK: Taylor and Francis Group;247–253.

Abbasnia, A., Sutulo, S., & Guedes Soares, C. 2022. Three-dimensional potential seakeeping code in frequency domain for advancing ships. *In*: C. Guedes Soares and T.A. Santos, eds. *Trends in Maritime Technology and Engineering*. London, UK: Taylor and Francis;, 259–268.

Dean, R.G. & Dalrymple, R.A. 1991. *Water wave mechanics for engineers and scientists*. World Scientific, Singapore.

Dombre, E., Harris, J.C., Benoit, M., Violeau, D., & Peyrard, C. 2019. A 3D parallel boundary element method on unstructured triangular grids for fully nonlinear wave-body interactions. *Ocean Engineering* 171: 505–518.

Fenton, J.D. 1985. A Fifth-Order Stokes Theory for Steady Waves. *Journal of Waterway, Port, Coastal, and Ocean Engineering* 111 (2): 216–234.

Gao, Z. & Zou, Z. 2008. A NURBS-based high-order panel method for three-dimensional radiation and diffraction problems with forward speed. *Ocean Engineering* 35: 1271–1282.

Grilli, S.T.T. & Svendsen, I.A.A. 1990. Corner problems and global accuracy in the boundary element solution of nonlinear wave flows. *Engineering Analysis with Boundary Elements* 7 (4): 178–195.

Harris, J.C., Dombre, E., Benoit, M., Grilli, S.T., & Kuznetsov, K.I. 2022. Nonlinear time-domain wave-structure interaction: A parallel fast integral equation approach. *International Journal for Numerical Methods in Fluids* 94 (2): 188–222.

Journee, J.M.J. 1992. *Experiments and calculations on four Wigley hull forms*. Delft.

Kim, S.J., Kim, M.H., & Koo, W. 2021. Nonlinear hydrodynamics of freely floating symmetric bodies in waves by three-dimensional fully nonlinear potential-flow numerical wave tank. *Applied Ocean Research* 113 (June): 102727.

Liang, H., Shao, Y., Chua, K.H., & Magee, A. 2022. An implicit time-domain Rankine panel method for ship motions in a Non-inertial body-fixed frame of reference. *In*: *Proceedings of the ASME 2022 41st International Conference & on Ocean, Offshore and Arctic Engineering*, Hamburg: *OMAE2022-79054, V05AT06A039.*

Newman, J.N. 1985. Algorithms for the free-surface Green function. *Journal of Engineering Mathematics* 19 (1): 57–67.

Pan, Z. 2022. Application of a BEM model with dipole elements. *Ocean Engineering* 249 (August 2021): 110938.

Sutulo, S., Guedes Soares, C., & Otzen, J. 2012. Validation of potential-flow estimation of interaction forces acting upon ship hulls in parallel motion. *Journal of Ship Research* 56 (3): 129–145.

Tang, Y., Sun, S., Abbasnia, A., Guedes Soares, C., & Ren, H.-L. 2023. A fully nonlinear BEM-beam coupled solver for fluid–structure interactions of flexible ships in waves. *Journal of Fluids and Structures*, 121: 103922.

Tanizawa, K. 2000. The state of the art on numerical wave tank. *Proceeding of 4th Osaka Colloquium on Seakeeping Performance of Ships* 95–114, 17-21 October 2000, Osaka.

Watai, R.A., Dinoi, P., Ruggeri, F., Souto-Iglesias, A., & Simos, A.N. 2015. Rankine time-domain method with application to side-by-side gap flow modeling. *Applied Ocean Research* 50: 69–90.

Wehausen, J.V. & Laitone, E.V. 1960. *Surface waves*. Berlin: Springer.

Yang, S.A. 2000. On the singularities of Green's formula and its normal derivative, with an application to surface-wave-body interaction problems. *International Journal for Numerical Methods in Engineering* 47 (11): 1841–1864.

Advances in Maritime Technology and Engineering – Guedes Soares & Santos (Eds)
© 2024 The Author(s), ISBN 978-1-032-83107-7

Multivariable operability and downtime assessment of a tanker moored at an upgraded oil terminal in Leixões port

H.S. Abdelwahab & C. Guedes Soares
Centre for Marine Technology and Ocean Engineering (CENTEC), Instituto Superior Técnico, Universidade de Lisboa, Lisbon, Portugal

ABSTRACT: This paper examines the operability and downtime of a moored tanker at an upgraded berth within the port of Leixões, Portugal, employing multivariable experimental measurements. A physical model replicating the modified port layout, new bathymetry, and a planned 300 m north breakwater extension, is constructed. The tanker undergoes testing under various outer-port sea states and two mooring configurations. Two methodologies are utilized to accurately characterise and evaluate the operability and downtime of the berthed tanker: mono-parametric thresholding and conditioned thresholding at various predefined minimal operational windows. The study proposes a new comprehensive clustered operability scheme that integrates in-port and outer-port wave records, motions, and mooring loads, where the exceedance of standard operational and safety criteria indicates the potential mooring hazards, operability issues, and delays. The results highlight the effectiveness of the conditioned thresholding approach in addressing a broader range of variables and their interrelationships compared to the mono-parametric thresholding approach.

1 INTRODUCTION

The recent surge in trade demand has prompted advancements in ship design, favouring the construction of larger vessels capable of effectively reducing transportation costs. This emphasis on the scale has created a demand for robust, larger port facilities that can efficiently handle the increased cargo volumes of these sizable ships. However, challenges arise as larger ships may struggle to manoeuvre within ports or dock at existing terminals (Kim et al. 2019). This prompts the need for upgrades to entrance channels and berthing areas (APDL 2016, 2022, Van der Molen et al. 2016), potentially exposing terminals to complex wave conditions.

While the option of constructing new, larger berths with sheltering structures exists (Ahuja et al. 2010, Baker et al. 2016), this may be a concern in areas prone to severe sea conditions. Consequently, there is a critical need to develop optimized sheltering structures capable of withstanding environmental challenges without compromising the operability and safety of moored ships in ports.

The ship mooring system is typically composed of various elements that must be arranged appropriately to facilitate bunkering, embarkation, disembarkation, and cargo handling operations, while concurrently adhering to operational and safety criteria (PIANC 1995, Van der Molen 2006, Puertos del Estado 2011, OCIMF 2018, OCDI 2020).

During mooring operations, moored ships at berth may face met-ocean forces such as waves propagating with varying heights, periods, and directions from outside the port to the berthing site (Rusu & Guedes Soares 2011, 2013, Abdelwahab et al. 2021). Additionally, other environmental factors such as wind and gusts at different speeds and angles come into play (Weiler et al. 2009), along with swell and long waves (McComb et al. 2009, Troch et al. 2020, Costas et al. 2022), seiches (Pinkster 2004), tidal forces, current (Esferra et al. 2018), sea level rise (Sierra et al. 2017, Gracia et al. 2019), and passing ship effects (Zhou et al. 2021).

In extreme weather, excessive ship movements can disrupt moored ship operations, leading to downtime, reduced operational performance or operability, and potential economic losses. In other situations, it may cause a failure in the mooring system, potentially resulting in serious accidents for the crew (MAIB 2017) or damage to the environment and infrastructures in the port (NTSB 2018, DMAIB 2020) or in nearby structures, raising socio-economic concerns (Figuero et al. 2018). Hence, it is crucial to consider potential impacts on both performance and safety in mooring operations when estimating operability, the number of stops, and downtime duration.

The duration of time that ships spend moored at cargo-handling terminals plays a pivotal role in influencing overall transport costs, analogous to the significance of the duration of sea voyages. The accurate characterization and assessment of the efficiency of operation and downtime at a specific berth is

DOI: 10.1201/9781003508779-10

a challenging task (Molina-Sanchez et al. 2020). This is primarily influenced by the dynamic response of the moored ship system, considering its six degrees of freedom (6-DOF) and mooring forces. Minimizing these responses, especially in surge, sway and yaw, is crucial for reducing the cost of delay in operations and mitigating operational hazards (Rosa-Santos et al. 2014, Abdelwahab et al. 2024).

The assessment of moored ship behaviour at berth is a complex and site-specific challenge influenced by various factors (Taveira-Pinto et al. 2008, Molina-Sanchez et al. 2020). Ship-related elements, including geometry, inertia, type, cargo, and windage area, play a crucial role. Additionally, the mooring system, comprising the layout, number, and stiffness of mooring ropes, wires, and fenders, as well as the winch type, also contributes significantly.

Consideration of port infrastructure, including layout, location, bathymetry, sheltering structures, berthing facilities, and cargo handling means, is another key aspect. The interaction of multiple metocean forces with the ship and port further complicates the scenario. The proficiency of the berth operator may also affect operability (Molina-Sanchez et al. 2020). The combined effect of the abovementioned factors at a specific berth poses a sophisticated problem. Therefore, achieving a comprehensive understanding of the dynamic behaviour of moored ships requires thorough analyses of all included processes to accurately evaluate operability and downtime.

Numerical models play a crucial role in predicting the dynamic response of moored ships in ports. To simulate the evolution of waves as they propagate from deep to shallow waters and their local interactions with ships, researchers frequently integrate a wave model such as the Boussinesq model (Bingham 2000), an elliptic mild-slope model (Kwak & Pyun 2013), or a non-hydrostatic model (Dobrochinski et al. 2023) with a 3D panel model.

The numerical modelling approach allows the simulation of various scenarios, but the need for extensive information in configuration requires calibration and validation with full-scale field observations or physical model measurements (Abdelwahab et al. 2021). Full-scale data for moored ships at berths are often limited, primarily sourced from port operators (Sakakibara & Kubo 2008, Kwak & Pyun 2013) or field monitoring campaigns (Uzaki et al. 2010, López & Iglesias 2014, Figuero et al. 2019).

Physical model testing remains among the most reliable approaches for studying moored ship behaviour in ports, accurately reproducing phenomena and their nonlinear effects resulting from environmental loads. However, selecting the right scale and basin size is crucial for considering both port features and a moored ship in a test. The opportunity to test various scenarios aids in optimising the operational performance at the berth (Rosa-Santos & Taveira-Pinto 2013, Esferra et al. 2018, Santos et al. 2019, Dong et al. 2022, Abdelwahab & Guedes Soares 2023, Abdelwahab et al. 2024), port layout (Cornett et al. 2012, Baker et al. 2016, Demenet et al. 2018) and sheltering structures (Fortes et al. 2017).

Operability studies often focus on vessel performance during navigation (Guedes Soares et al. 1995, Fonseca & Guedes Soares 2002, Tello et al. 2011, Tezdogan et al. 2014) or open-sea operations (Wu 2014, Guachamin Acero et al. 2016), with limited research on moored ships in ports (Prpic-Orsic et al. 2014, Figuero et al. 2019, Molina-Sanchez et al. 2020, Costas et al. 2022, Abdelwahab et al. 2024).

Various methodologies exist for characterizing and calculating operability, with the application depending on the problem's complexity. Methods range from mono-parametric (Moes & Terblanche 2010) and conditioned thresholding, which assess variables affecting operability independently, to n-dimensional thresholding, considering multiple variables without assuming independence. Inference models offer an alternative given the numerous contributing variables (Gomez et al. 2014). Molina-Sanchez et al. (2020) proposed a new approach, recommending decision-making tools with adjustable thresholds to empower operators to manage challenges at each dock based on their experience and needs, rather than relying on standard thresholds.

In this paper, the assessment of operational efficiency and downtime of a moored tanker at a modified berth within the port of Leixões, Portugal, is conducted using multivariable experimental measurements. The modifications to the berth, aimed at enhancing operational capabilities, include a future planned 300 m extension of the north breakwater and deeper entrance channel. For this purpose, a physical model replicating the modified port layout is constructed, incorporating new bathymetry. Tests are conducted using the Esso Osaka tanker model at oil terminal A, considering various sea states and two generic mooring configurations.

The experiments focus on collecting wave measurements both outside and inside the port, monitoring ship motions, and assessing loads on lines and fenders. Analysis is conducted in both time and frequency domains to explore the interplay between measurements. Employing the zero-crossing technique and statistical tools, the entire time series of each signal is examined, representing them with various statistical measures. Additionally, frequency domain spectrum analysis yields multiple spectral parameters associated with the area below the curve.

Two methodologies, mono-parametric thresholding, and conditioned thresholding at multiple predefined minimal operational windows are employed to accurately characterize and evaluate the operability and downtime for cargo handling operations of the moored tankers at the upgraded oil terminal. Results are utilized to construct a new comprehensive clustered operability plot that integrates statistical, spectral, operational and safety parameters for the outer-port and local in-port waves, motions, and mooring loads under different tested conditions. Exceedance of operational and safety criteria is considered inoperable/unsafe, indicating potential delays, operability issues, and mooring hazards.

2 STUDY AREA: LEIXÕES OIL TERMINAL

The port of Leixões is a major Portuguese port, situated 4 km north of the Douro River near the city of Porto. It is protected by two curved breakwaters, facilitating the docking of various types of ships. The port includes an oil terminal with three berths, as shown in Figure 1. Berth (A), positioned near the port entrance to accommodate large tankers, is equipped with two breasting dolphins and a loading platform. These facilities are protected by the north breakwater, which is 700 m long and 15 m above the seawater level. However, Berth (A) is susceptible to severe waves between west-northwest directions along the coast, causing unfavourable conditions and operational disruptions (Veloso-Gomes et al. 2005).

Recently, ships that demand Leixões port increased by 77% (APDL 2022). Therefore, the Leixões port authority, APDL, has approved extensive upgrades to the port layout aimed at improving shelter conditions, maritime accessibility, manoeuvrability, operability and accommodating larger ships. The outer north breakwater will be extended by 300 m with a 20º opening angle relative to the existing breakwater. Furthermore, the entrance channel and turning basin will be dredged to deeper depths (-16.85 m from the zero hydrographic level ZHL and -15.5 m ZHL, respectively) as shown in Figure 1. For this purpose, a physical model is constructed to evaluate future modifications and enhance the operational capabilities of moored ships at berth (A).

Figure 1. Leixões port with berth location and upgrade.

3 EXPERIMENTAL STUDY

3.1 *Experimental facility and model layout*

The physical model tests were carried out at the Portuguese National Laboratory for Civil Engineering LNEC. A 3D model of berth (A) and surrounding geometry was built at a scale of 1:80. The basin is 23 m long and 22 m wide and is equipped with a mobile piston wave generator that can produce unidirectional regular/ irregular waves at a water depth up to 0.5 m. Portable water barriers were used to direct the generated waves towards the port entrance.

In this study, the Froude scaling was applied in compliance with the tank size to construct a 3D model of the entire upgraded basin of the port of Leixões as specified by (APDL 2016, 2022) and up to the entrance of Leça River at a scale of 1:80 as shown in Figure 2. The model incorporates a mortar surface to replicate the new seabed bathymetry up to -20 m ZHL at which the wave generator is settled in the tank. Furthermore, it includes the existing extended breakwaters, the adjacent beach of Matosinhos and a dissipative absorbing wave beach along the basin's boundary to reduce the reflections.

3.2 *Experimental equipment*

3.2.1 *Ship model*

The selected ship model for the study is the Esso Osaka oil tanker shown in Figure 3. The model resembles the large tanker ships that dock at berth (A) in the port of Leixões. The model was constructed at a scale of 1:80 in fibreglass-reinforced plastic and covered with a wooden deck to facilitate the installation of mooring equipment and motion detector markers.

Before wave testing, the free ship model was calibrated in calm water to obtain the main characteristics listed in Table 1. Inclination tests are conducted to estimate the metacentric height (GM_T), while decay tests assess the natural periods and damping ratios of the free-floating model.

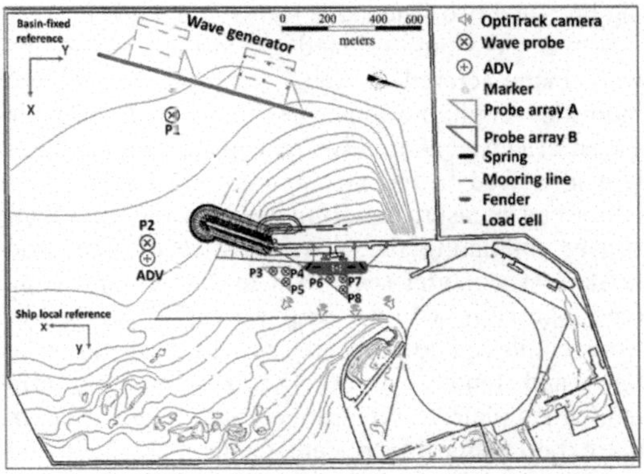

Figure 2. Physical model layout in LNEC's wave basin.

3.2.2 *Wave measurement system*

Wave conditions at four distinct locations in the basin were assessed using 8 resistive wave gauges and one Acoustic Doppler Velocimetry (ADV) with a sampling frequency of 25 Hz. The wave gauges were categorized into four groups as shown in Figure 2. Firstly, wave gauge (1) was placed near the

Table 1. Characteristics of the Osaka tanker model.

Specification	Osaka (model scale)
Length, L_{pp} (m)	3.250
Breadth, B (m)	0.530
Draft, T (m)	0.132
Displacement volume, Δ (kg)	186.8
LCG aft of mid-ship (m)	0.235
Metacentric height, GM_T (m)	0.078
Pitch radius of gyration, K_{yy} (m)	0.812
Roll radius of gyration, K_{xx} (m)	0.185
Roll natural period, (sec)	1.532
Roll damping factor, (-)	0.022

wave generator to evaluate the incident wave conditions. A second wave gauge (2) was paired with the ADV at the port entrance to characterize the wave height and direction. Group three incorporated three wave gauges (3, 4, and 5) in array A positioned in front of the moored ship. Finally, group four comprised gauges (6, 7 and 8) in Array B, situated alongside the centre of gravity of the moored model.

3.2.3 *Custom-designed mooring simulators*

The Osaka model is moored at berth (A), adjacent to the loading platform and dolphins with a generic asymmetric starboard mooring configuration. To guide the mooring lines to their connection points, frictionless fairleads were installed on both the deck and the quay close to the real locations of the mooring bitts. The nonlinear characteristics of the mooring system were derived from a similar case presented by Rosa-Santos and Taveira-Pinto (2013). Due to equipment limitations, our model tests utilized four mooring lines and two fenders, as shown in Figure 4.

Novel custom-designed mooring simulators were manufactured at LNEC to replicate targeted mooring configurations in tests and provide a cost-effective alternative to expensive equipment typically used in testing laboratories. These simulators can measure mooring loads while simultaneously simulating the characteristics of lines and fenders at a model scale. Six tension-compression type load cells with a maximum capacity of 2 kgf were employed to measure the loads on mooring lines and fenders at 50 Hz. Two rocking arm fender simulators with rubber pads were used in the experiments to replicate fender behaviour and ship-fender friction interface. Four mooring line simulators were used to model the real stiffness of lines. Additional details about the design, calibration and measurement uncertainties of the mooring simulators are explained in (Abdelwahab & Guedes Soares 2023, Abdelwahab et al. 2024).

3.2.4 *Motion capture system*

The response of the moored Osaka model was evaluated in the 6-DOF using the OptiTrack™ motion capture system. This system is composed of 4 digital

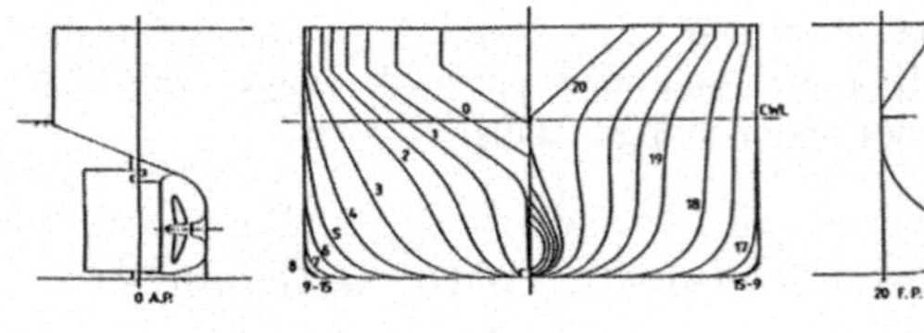

Figure 3. The body plan of the Esso Osaka tanker.

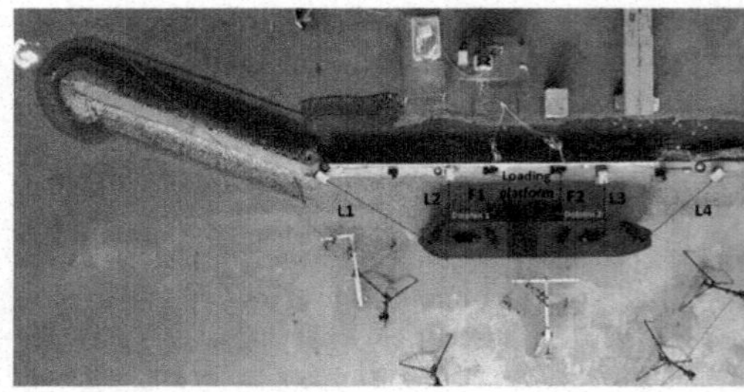

Figure 4. Mooring configuration in the physical model.

cameras that can capture the motions at the centre of gravity of the moored ship using multiple target markers mounted onboard the model, as shown in Figure 4. The time series of ship motions were measured at a sampling frequency of 120 Hz.

3.3 *Testing procedure*

The experiments were conducted under combined testing conditions. Two pretension configurations were used in this study: a low pretension rated at 110-250 kN, and a high pretension rated at 300-600 kN (full scale). Pretensions were applied to mooring elements by adjusting screws linked to the load cells. Decay tests were performed with the moored model in calm waters to estimate natural periods shown in Table 2. The analysis of the results shows that the natural periods of surge, sway and yaw motions are relatively long and vary with pretensions. However, the roll natural period is less affected by pretensions.

Before moored ship testing in waves, the tank was confirmed in calm water to avoid disturbances affecting data. The wavemaker was utilized to create long-crested irregular waves that follow a JONSWAP spectrum. The waves were generated with a direction of $\theta = 270°$ to propagate nearly perpendicular to the new breakwater, as shown in Figure 2. In this study, wave direction is referenced to the ship's local axis (90° for starboard waves, 180° for bow waves). Despite possible mixed swell and wind waves at port entrances (Guedes Soares 1984), tests focused on long-crested waves due to facility limitations. Targeted sea states at the wavemaker feature H_s from 3 to 6 m and T_p from 10 to 16 s (real scale), which were calibrated before starting tests.

The wave test procedure involves data acquisition systems collecting wave and mooring load data while tracking target markers on the moored model using the motion capture system, ensuring synchronization.

Table 2. Natural periods of the moored system.

	Natural period (s)	
Motions	Low pretension	High pretension
Surge	19.3	17.9
Sway	9.13	7.43
Roll	1.48	1.46
Pitch	8.43	6.84

Experiments were conducted at both low and high tide levels. This paper presents results specifically at the high tidal level of +4 m (ZHL), resulting in a total water depth of 20.85 m (real scale) at berthing site (A) in the entrance channel. Each 20-minute test represents about 3 hours of vessel exposure in a real-world storm, with an interval of 25 minutes between runs to minimize tank disturbances. The study focuses solely on wave-induced inoperability, excluding other factors (wind and currents), which are beyond the scope of this work.

3.4 *Data analysis*

This section analyses the multivariable measurements for waves, motions, and loads in both the time domain and frequency domain. Two approaches are employed to accurately evaluate the operability and downtime of the moored ship at the berth. The analysis of the experimental measurements yields multiple representative parameters, contributing to the creation of a new comprehensive clustered operability scheme.

3.4.1 *Time domain analysis*

The entire time series of each recorded signal, including free-surface elevation at 8 different probes, motions in 6 DOF, and 6 mooring loads, is analysed using the zero-crossing technique and other statistical tools. The statistical results encompass the mean (A_m), skewness, kurtosis, and zero-peak amplitudes such as the maximum (A_{max}) and absolute minimum amplitude ($A_{\min}$) of each measurement. Additionally, peak-to-peak analysis is applied to derive average peak-to-peak amplitudes (H_{avg}) and maximum peak-to-peak amplitudes (H_{max}) for the entire time series of each signal. Significant peak-to-peak amplitudes ($H_{1/3}$) are determined as the average of the highest third peaks of each signal.

Subsequently, key statistical parameters essential for operability and downtime assessment are selected. Waves, moored ship responses, and mooring loads for each tested case are characterized by 80 statistical parameters ([A_{max}, A_{min}, H_{max}, $H_{1/3}$] × [8 wave probes + 6 DOF + 6 Load]) calculated from the corresponding measured data.

3.4.2 *Frequency domain spectrum analysis*

The frequency domain spectrum analysis uses the Welch method, segmenting the signal into sub-records with a 50% overlap and applying a Hanning window. Fast Fourier Transform (FFT) estimates power spectra for waves, 6-DOF motions, and mooring loads. The overall energy power spectrum is obtained by averaging these spectra.

The power spectrum for each recorded signal is characterized by spectral significant amplitude (H_s), linked to the energy power spectrum $H_s = 4\sqrt{m_0}$, where m_0 is the area under the entire energy spectrum curve. The period associated with the highest spectral ordinate is defined as the peak period T_p. A low-frequency zone ranging (from 0 Hz to 0.04 Hz) is characterized by the peak period (T_{pL}) and spectral significant amplitude $H_{sL} = 4\sqrt{m_{0L}}$. The ratio of the energy in the low-frequency band to the total energy $R_L = H_{sL}^2/H_s^2$ provides insight into the most energetic zone in the spectrum, aiding in establishing the relationship between frequency bands in waves, motions, and load spectra.

Wave records from probes 3, 4, and 5 in array A were used to estimate directional spectra near the berth under various sea states. Calculations were performed using the DIWASP directional wave spectra toolbox based on the extended maximum likelihood method.

The key spectral parameters for subsequent operability and downtime assessment are selected. Waves, moored ship responses, and mooring loads are characterized by 84 spectral parameters ([H_s, R_L, T_P, T_{PL}] × 21 spectra [8 wave probes + Array A + 6 DOF + 6 Load]) calculated for each tested case.

3.4.3 *Mono-parametric operability analysis*

The operability of a moored ship at Leixões oil terminal is measured using experimental data under various testing conditions. Operational efficiency is assessed by comparing the likelihood of an abrupt system stoppage when exceeding a specific mono-parametric operational limit to the expected stoppage during normal operation. The recorded time series for motions and mooring loads are compared against zero-peak operational (PIANC 1995, Van der Molen 2006) and safety (OCIMF 2018) limits as depicted in Figure 5 (a) and (b), respectively.

The loading/unloading operation stops when the recorded value surpasses operational criteria. If the recorded value exceeds safety criteria, the operation is considered unsafe due to a mooring hazard. The mono-parametric operative threshold outlines the number of stoppages (N) during the entire operation (S_T) and the duration of each stoppage (S_i). Therefore, operational efficiency is defined through independent analysis of motion and load data as follows:

$$r_i = \left(1 - \frac{\sum_{i=1}^{N} S_i}{S_T}\right) \times 100 \tag{1}$$

3.4.4 *Conditioned threshold operability analysis*

Conditioned thresholding allows the introduction of multiple variables in operability assessment. Each

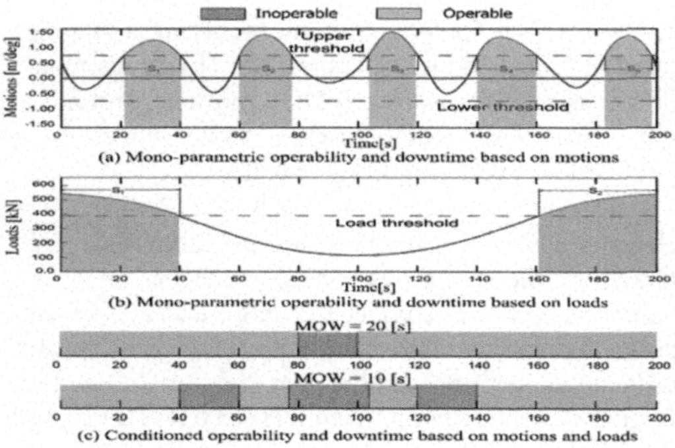

Figure 5. Operability definition based on different approaches.

variable undergoes an independent mono-parametric analysis, as depicted in Figure 5(a) and (b).

Specific limits are set for each variable, and individual operational times are determined through downtime analysis in the time domain, based on the minimal operational window (MOW), illustrated in Figure 5(c). The combined operational time is then calculated as the intersection of the operational time from various variables. In other words, the operation is assumed feasible only if all variables remain below their limits for at least the minimal operational window. In this study, the conditioned threshold operability analysis is assessed at four different values of the minimal operational window (MOW=10, 20, 30 minutes) as well as an unlimited duration.

4 RESULTS AND DISCUSSION

In this section, the results obtained from experiments are presented for the study case of the Osaka tanker moored at berth A in the port of Leixões. The study experimentally examines the relationship between outer-port sea conditions and local waves at the berth, as well as the measured motion and loads on the moored tanker, following the upgraded berth layout. Subsequently, the study evaluates the operability of the moored tanker based on clustered operability plots that consider waves, motions, and loads, at Test 3 and Test 6 and two pretensions.

4.1 *Relation between outer-port and in-port waves*

The experiments, conducted at nine sea states, assess the impact of outer-port sea conditions on local waves at the upgraded berthing site. Table 3 illustrates an example of the full-scale wave characteristics at wave probe number 3 placed in front of the ship under all tested outer-port sea conditions. Results show that the time domain estimates of the significant wave heights and peak periods outside the port align with targeted conditions and spectrum analysis.

Results reveal improved berth tranquillity after upgrades, with waves near the oil terminal's berthing site reaching a maximum elevation of 1.1 m and a height of 1.68 m, even when the highest outer-port wave elevation is 6.3 m, and the largest wave height is 9.2. The measured significant wave height reaches up to 1.1 m around the berthing site.

The wave spectra in Figures 6(a) and 7(a) for Test 3 near the berth reveal significant power at low frequencies (up to 0.04 Hz) inside the port, possibly indicating long-period waves near the moored oil tanker's bow. Reflections from nearby structures or bathymetric effects may explain these long waves. Results in Table 3 show that the energy ratio (R_L) increases with wave severity, reaching up to 12% in the berthing area. In Figures 6-9(a), long waves outside the port exhibit minimal energy with $R_L < 2\%$.

The nine wave conditions propagated within the port, forming short-crested waves around the oil terminal's berthing site. Results in Figures 6(b) and 8 (b) depict directional wave spectra obtained by Array A at the ship's bow, showing that increased wave period and height result in less scattering at the berth. Analysis of results obtained from Array A indicates that waves with heading angles θ_i between 216° and 221° may reach the berthing site under tested sea states with T_p = 10 s while heading varies between 173° and 187° in the other tested sea states as shown in Table 3.

Table 3. Full-scale characteristics of waves measured at probe 3 under different sea conditions.

Analysis type	Properties	Units	T1	T2	T3	T4	T5	T6	T7	T8	T9
Time domain statistical parameters	A_{max}	m	0.36	0.49	0.63	0.40	0.61	1.08	0.48	0.61	1.08
	A_{min}	m	0.33	0.51	0.69	0.44	0.60	0.83	0.50	0.64	0.97
	H_{max}	m	0.61	0.80	1.11	0.82	1.11	1.67	0.96	1.17	1.68
	$H_{1/3}$	m	0.37	0.49	0.61	0.45	0.60	0.89	0.53	0.69	1.03
Frequency domain spectral parameters	H_s	m	0.40	0.53	0.68	0.49	0.66	0.98	0.56	0.74	1.10
	T_P	s	9.27	9.79	10.1	13.9	14.3	14.1	15.1	15.3	15.4
	T_{PL}	s	42.5	43.6	43.1	43.4	45.8	43.6	43.2	44.2	42.2
	R_L	-	3.25	5.10	12.18	5.30	5.86	9.78	5.05	6.30	9.80
Directional spectrum	H_{sd}	m	0.38	0.49	0.62	0.44	0.6	0.89	0.52	0.66	0.98
	θ_i	o	216	217	221	178	175	187	173	174	177
	θ_r	o	14	15	14	11	10	21	337	334	351

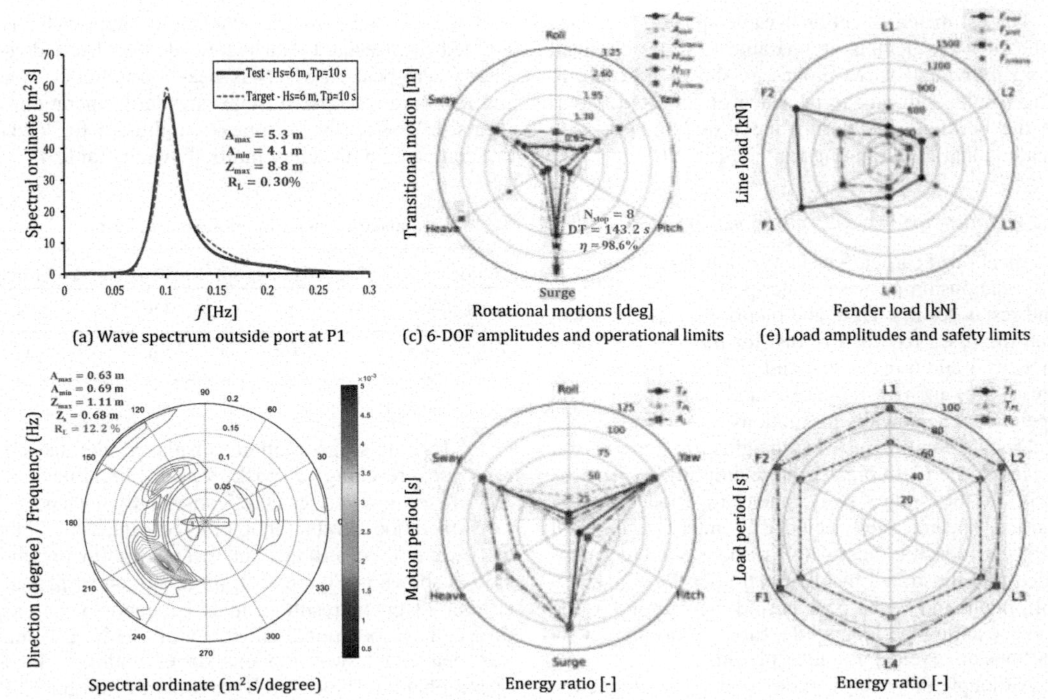

Figure 6. Clustered operability plot of the measured experimental data in Test 3 at Hs = 6 m and Tp = 10 s under low pretensions.

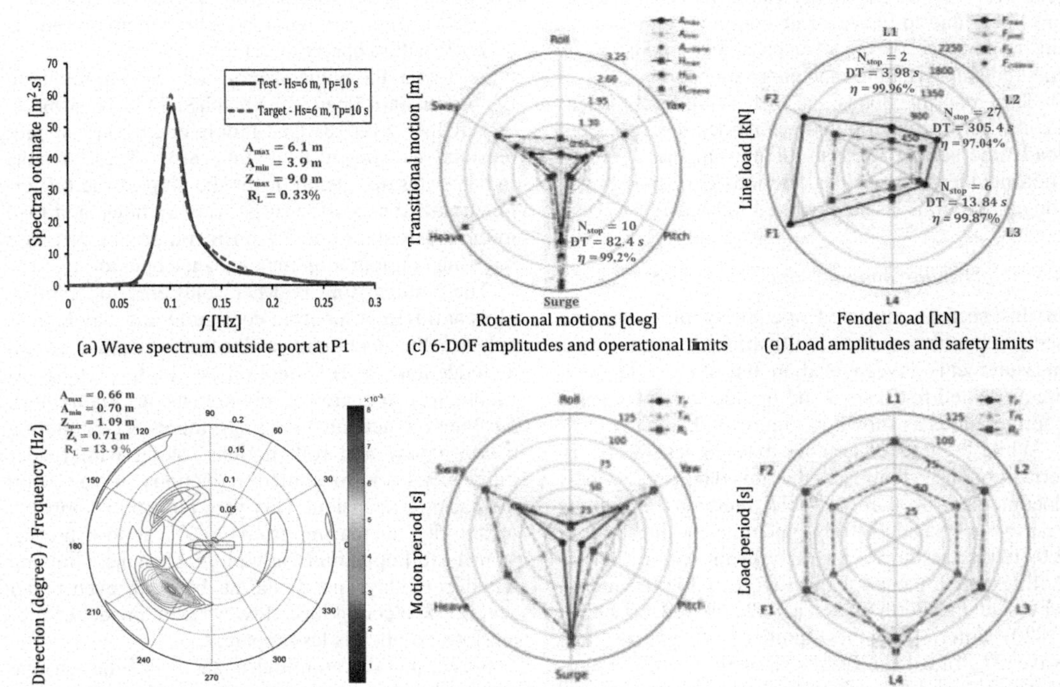

Figure 7. Clustered operability plot of the measured experimental data in Test 3 at Hs = 6 m and Tp = 10 s under high pretensions.

The estimated directional wave spectra yield significant wave heights comparable to those obtained from individual wave probes as shown in Table 3. The results suggest the presence of reflected waves at the berthing site during the tested storms, with heading angles θ_r ranging from 334° to 10°.

4.2 *Outer-port wave effects on motions and loads*

As mentioned earlier, 9 tests were conducted to assess the relationship between outer-port wave conditions and responses in 6 DOF and mooring loads. This section discusses recorded results for motions and loads in Tests 3 and 6 under low and high pretensions. Figures 6-9(c) and (d) summarize statistical, spectral and operational parameters for motions.

Despite small local wave heights, the long-period waves at the berth created large amplitudes, particularly in the surge, sway, and yaw modes. Using motion spectra (not included), energy ratios and peak periods for the 6-DOF responses were calculated and illustrated in Figures 6-9(d). Results show roll, pitch, and heave peak periods correspond to the wave exciting period, while surge, sway, and yaw motions are excited at longer periods.

Although the energy ratio in the low-frequency band of the wave spectra R_L^{wave} stays below 0.6% outside the port and rises to 14% around the berth, Tests 3 and 6 show a high energy ratio R_L^{motion} in the low-frequency band of surge, sway, heave, and yaw (Figures 6-9). High energy ratios in heave motion are likely due to heave-surge coupling, emphasizing the importance of long waves at berthing sites for moored ship dynamics and cargo operations.

The mooring loads, shown in Figures 6-9(e), exhibit zero-peak amplitudes obtained from recorded load time series. Analysis of mooring load spectra (not included) reveals peak periods corresponding to longer durations, as summarized in Figures 6-9(f).

4.3 *Operability and downtime assessment*

In this section, clustered operability plots are presented based on the multivariable experimental measurements investigated in the study. The plots are presented for Tests 3 and 6 under two pretension configurations as shown in Figures 6-9.

While the Port of Leixões may possess wave criteria at berth A, it appears that this data is not widely documented in the available literature. Adopting Puertos del Estado (2011) guidelines, a threshold of 2.0 m is assumed for significant wave height. A threshold of energy ratio (R_L) of low-frequency waves at the berth is set at 0.25, based on OCDI (2020) criteria for the significant height of long waves H_{sL} of 0.1 m. Results show that, while significant wave height criteria at the berth aren't surpassed in any test, the long-period wave energy ratio exceeds criteria in all tested conditions, as depicted in Figures 6-9. This result emphasises the significance of adapting site-specific criteria at berth (A).

The mono-parametric threshold approach is applied to measured data independently. The calculated zero-peak and peak-peak amplitudes for motions are compared with standard operational limits. Likewise, the zero-peak amplitudes for loads are compared with safety limits shown in Table 4.

Table 4. Standard criteria for motions and loads.

Motion	Surge	Sway	Heave	Roll	Yaw	Fender	Line
Units	m	m	m	deg	deg	kN	kN
A_{limit}	1.5	2.0	1.5	1.0	1.0	2450	650

In Test 6, surge motions displayed the largest response, reaching a peak-to-peak amplitude of 3.68 m at low pretension, consistently exceeding safe operation limits for a moored tanker in both tests 3 and 6. For instance, in Test 6 with low pretension as shown in Figure 8(c), surge motion surpassed criteria 23 times, resulting in 354 seconds of downtime and an operational efficiency of 96.6%. Despite this, heave and sway motions stayed within defined operational thresholds, with maximum amplitudes of 0.59 m and 2.37 m, respectively. The pitch motion exhibited the lowest angular motion at a maximum peak-to-peak amplitude of 0.62°. Yaw, however, surpassed operational 10 times at limits at 2.26°, resulting in 56 s downtime and an operational efficiency of 99.4%, while roll recorded a maximum amplitude of 1.79° within operational limits.

At Test 6, the highest zero-peak load amplitude in the breast stern line (F_{max}^{L3}) reached 743 kN, surpassing OCIMF's (2018) load failure criteria at low pretensions as shown in Figure 8(e). These results indicate unsafe operation for the ship at the oil terminal due to excessive surge, yaw motions, and load in the breast stern line 3. Consideration of a different mooring configuration may enhance operability.

The comparison between clustered operability plots at different mooring configurations may help to enhance the operability at the berth. Results reveal a reduction in roll motion at high pretensions Figure 9(c) compared to low pretensions Figure 8(c), attributed to activated damping forces. The coupling between yaw and roll motions, along with fender influence, leads to similar reductions in yaw with increasing pretension. This yaw reduction improves operability, increasing from 99.5% at low pretensions to full operability at high pretensions, allowing the ship to stay operational at the berth even under severe wave conditions. However, the impact of pretensions on pitch is less apparent.

Surge motion amplitudes showed little change despite higher pretensions, attributed to the ship's inability to establish close contact with fenders generating effective frictional forces. The increase in pretension led to high reactive fender loads on the opposite side, resulting in heightened sway motions.

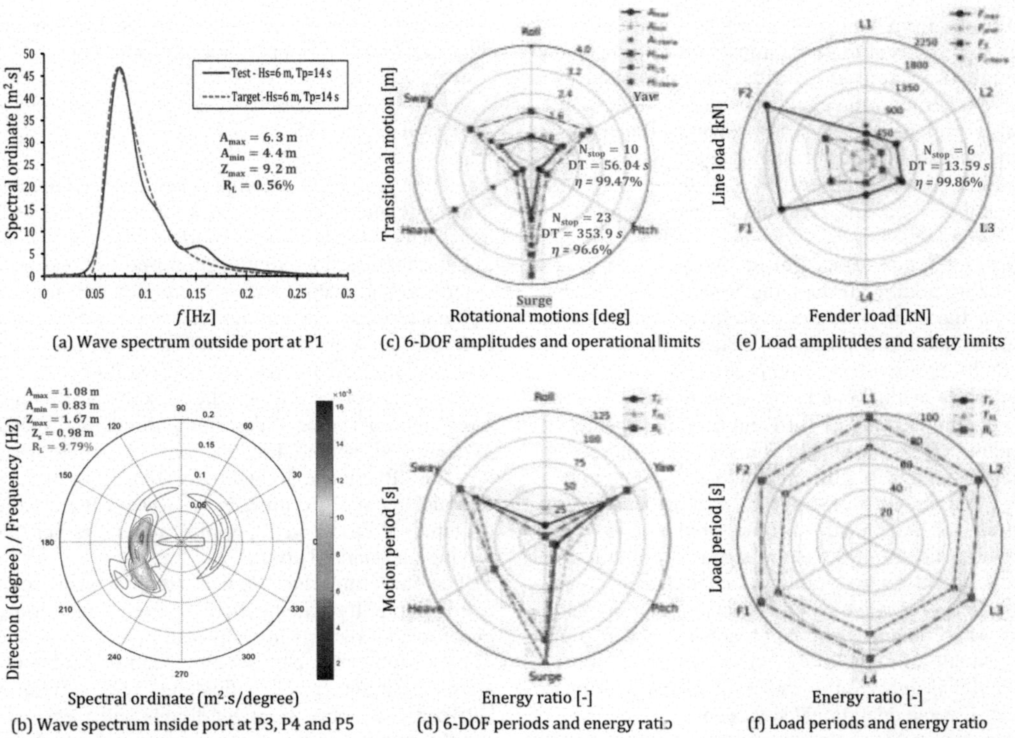

Figure 8. Clustered operability plot of the measured experimental data in Test 6 at Hs = 6 m and Tp = 14 s under low pretensions.

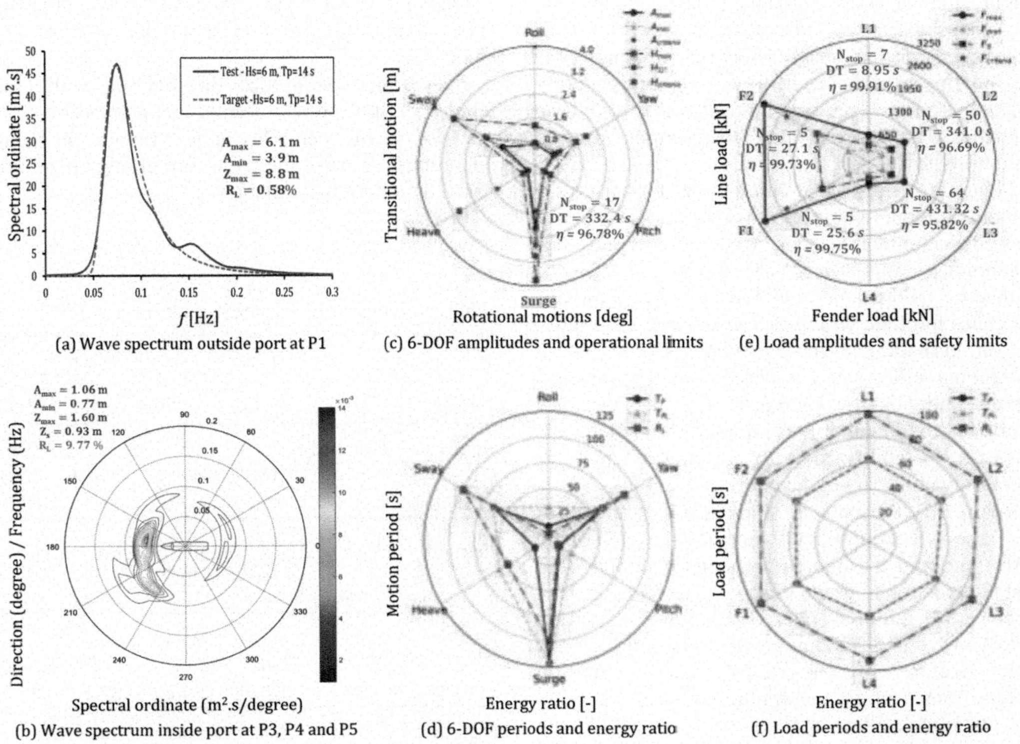

Figure 9. Clustered operability plot of the measured experimental data in Test 6 at Hs = 6 m and Tp = 14 s under high pretensions.

Results from Test 6 as shown in Figure 8(e) and Figure 9(e) reveal minimal impact on zero-peak load amplitudes in stern line 4 with changing pretensions, unlike other mooring elements significantly affected. Higher pretensions caused substantial deflection in both fenders. Stoppages due to safety issues in lines 2 and 3, reaching 50 and 64, led to downtimes of 341 and 431 seconds, resulting in operability rates of 96.7% and 95.8%.

Despite the assistance of increased pretensions in activating additional damping forces from the fenders on the ship side, the overall operability of the moored tanker did not improve significantly.

Table 5, provides a comprehensive overview of operability evaluations using the conditioned thresholding approach across different minimal operational windows. It is crucial to emphasize that these computations hinge on the meticulous examination of either maximum peak-to-peak or zero peak-to-peak amplitudes, obtained through exhaustive time-domain analyses of the entire signals of motions and loads. Notably, wave results are excluded, as prevailing criteria for waves predominantly rely on significant wave heights. This approach ensures a robust assessment of operability and downtime, offering valuable insights into the system's performance across diverse operational scenarios.

The results highlight a noteworthy challenge, with 37 disruptions during operation at low pretension in Test 6, attributed to surpassing operational limits and/or safe load thresholds. In an attempt to mitigate this operational issue, an exploration into increased pretensions was conducted. However, it did not yield the anticipated enhancement in operability conditions. Instead, it exacerbated the situation, leading to 123 stops and a consequential reduction in combined operability to 89.5%.

Figure 9(e) shows that cumulative stops from the individual mono-parametric assessment (148) surpass the combined stops from the conditioned thresholding approach (123). Consequently, the application of the conditioned thresholding approach integrates multiple variables, considering their intersection, providing a more realistic assessment of operational efficiency compared to reporting each variable individually. Therefore, it is recommended to present conditioned thresholding results alongside clustered operability plots to accurately depict the operability condition at the berth.

Figure 10 Presents an illustrative example of the assessed conditioned operability, combined downtime, and the number of stops obtained from Test 3 under a high pretension. The data is provided at 3 minimal operational window values (MOW=10, 20, 30 minutes). Additionally, the figure features the shortest operable duration (SOD), representing the minimum time for feasible operation. When SOD is less than MOW, the efficiency of operation gradually decreases. Another key variable is the longest operable duration (LOD), indicating the maximum time for maintaining feasible operation. If the LOD is less than the MOW, the operation entirely stops.

5 CONCLUSIONS

The study conducts a comprehensive assessment of downtime and operability for a moored tanker at an upgraded oil terminal in the port of Leixões. The terminal upgrades, including a future north breakwater extension and deeper channel, are aimed at improving operational capabilities. To replicate the new port layout, a physical model is constructed, and tests are conducted under various conditions.

Studying waves through the Leixões port emphasizes the importance of describing directional spectra to capture all potential interactions around the berthing site effectively. It highlights the importance of establishing specific limiting criteria for the Leixões oil terminal based on experimental data, rather than relying on standard criteria.

The study introduces a new comprehensive clustered operability scheme, derived from multivariable experimental records in a physical model study. This scheme integrates statistical, spectral, and operational parameters to illustrate the correlation between outer-port wave conditions and local wave conditions, motions, and mooring loads at the berth. By accurately assessing operability, this approach aids in evaluating and selecting the optimal mooring configuration, ensuring maximum operability and safety during moored ship operations at a berthing site.

Integrating conditioned thresholding with clustered operability plots offers a comprehensive overview of berth operability and captures the actual operational scenario instead of presenting operability based on individual variables.

Table 5. Conditioned operability at various MOWs.

		Test 3		Test 6	
Variable	MOW	Low	High	Low	High
N_{stop}	Unlimited	8	42	37	123
DT		143.2	400.7	410.2	1085
η		98.6	96.12	96.1	89.5
N_{stop}	10	5	4	3	1
DT		1231	6498	5566	10400
η		88.0	37.2	47.3	0
N_{stop}		4	2	2	1
DT	20	3085	8870	7110	10400
η		70.05	14.2	32.6	0
N_{stop}	30	1	1	2	1
DT		7237	10400	8472	10400
η		29.8	0	17.76	0

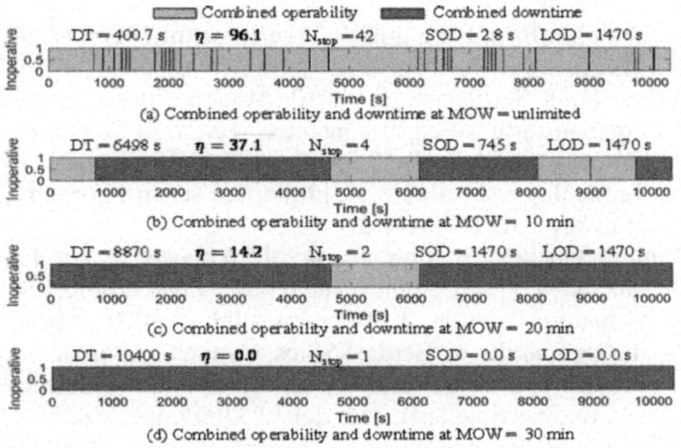

Figure 10. Conditioned operability and downtime in Test 3.

The operability analysis identifies the frequency and duration of expected stoppages while evaluating the operational efficiency of the moored Osaka tanker in the Leixoes port. These results can be translated into technical and financial data, facilitating a thorough cost-benefit assessment that considers both downtime costs and the expenses associated with modifying or constructing new sheltering structures.

Although the study is particularized for the Leixões oil terminal, the applied methodology for operability and downtime assessment through the presented clustered operability scheme can be applied to any berthing site. In conclusion, this research contributes valuable insights into enhancing the operational efficiency and safety of moored tankers in upgraded port facilities.

ACKNOWLEDGEMENTS

This work contributes to the strategic research plan of the Centre for Marine Technology and Ocean Engineering (CENTEC), which is funded by the Portuguese Foundation for Science and Technology (Fundação para a Ciência e Tecnologia - FCT) under contract UIDB/UIDP/00134/2020. The first author is funded by FCT for a PhD grant under contract 2020.06969.BD. The experimental work in this paper was conducted within the scope of the M&M SHIPS project "Manoeuvring & Moored Ships in Ports" PTDC/EMSTRA/5628/2014) funded by FCT.

REFERENCES

Abdelwahab, H.S. & Guedes Soares, C. 2023. Experimental uncertainty of a physical model of a tanker moored to a terminal in a port, Marine Structures, 87, 103331. doi:10.1016/j.marstruc.2022.103331.

Abdelwahab, H.S., Guedes Soares, C., Pinheiro, L.V., Fortes, C.J.E.M. & Santos, J.A. 2021. Experimental and numerical study of wave-induced ship motions and mooring loads of a tanker moored in Leixões port, in Guedes Soares, C. and Santos, T.A. (Eds.), Developments in Maritime Technology and Engineering. London, UK: Taylor & Francis Group, 341–350. doi:10.1201/9781003216599.

Abdelwahab, H.S., Pinheiro, L., Santos, J.A., Fortes, C.J.E. M. & Guedes Soares, C. 2024. Experimental investigation of wave severity and mooring pretension on the operability of a moored tanker in a port terminal, Ocean Engineering, 291, 116243. doi:10.1016/J.OCEANENG.2023.116243.

Ahuja, M.M., Mani, C.S. & Joshi, A.K. 2010. LNG berths for open seas: Part 2 Dahej LNG terminal, a case study, Port technology international, 46, 125–129. Available at: https://www.porttechnology.org/technical-papers/lng_berths_for_open_seas_dahej_lng_terminal_a_case_study_part_2/.

APDL 2016. Request for submission of proposal. Physical and numerical model studies of the extension of the outer breakwater and the new container terminal at the Port of Leixões (in Portuguese). Portugal. Available at: https://www.apdl.pt/pt/descricao.

APDL 2022. Extension of the outer breakwater of the port of leixões, deepening of the access channel and the rotation basin. Available at: https://www.apdl.pt/projetos-investimentos/.

Baker, S., Frank, G., Cornett, A., Williamson, D. & Kingery, D. 2016. Physical Modelling and Design Optimizations for a New Container Terminal at the Port of Moin, Costa Rica, in Ports 2016. Reston, VA: American Society of Civil Engineers, 560–569. doi:10.1061/9780784479919.057.

Bingham, H.B. 2000. A hybrid Boussinesq-panel method for predicting the motion of a moored ship, Coastal Engineering, 40(1), 21–38. doi:10.1016/S0378-3839(00)00002-8.

Cornett, A., Wijdeven, B. & Boeijinga, O.O. 2012. 3-D physical model studies of wave agitation and moored ship motions at Ashdod port, in the 8th International Conference on Coastal and Port Engineering in Developing Countries COPEDEC. Chennai, India, 1–13.

Costas, R., Figuero, A., Sande, J., Peña, E. & Guerra, A. 2022. The influence of infragravity waves, wind, and basin resonance on vessel movements and related downtime at the Outer Port of Punta Langosteira, Spain, Applied Ocean Research, 129, 103370. doi:10.1016/J.APOR.2022.103370.

Demenet, P.-F., Guisier, L. & Marcol, C. 2018. Physical and numerical modelling of ships moored in ports, in 34th PIANC World Congress. Panama: Curran Associates, Inc., 214–227.

DMAIB 2020. Marine accident report on loss of control on Stone I. Available at: https://dmaib.com/media/10212/stone-i-loss-of-control-of-on-4-january-2020.pdf.

Dobrochinski, J.P.H., Van Deyzen, A., Zijlema, M. & Van der Hout, A. 2023. Combining numerical tools to determine wave forces on moored ships, Coastal Engineering, 179, 104224. doi:10.1016/j.coastaleng.2022.104224.

Dong, G., Yan, M., Zheng, Z., Ma, X., Sun, Z. & Gao, J. 2022. Experimental investigation on the hydrodynamic response of a moored ship to low-frequency harbor oscillations, Ocean Engineering, 262, 112261. doi:10.1016/J.OCEANENG.2022.112261.

Esferra, R., Bernardino, J. & Alfredini, P. 2018. Physical modeling applied in evaluation of the safety and efficiency of vessel mooring plans, Brazilian Journal of Water Resources, 23. doi:10.1590/2318-0331.231820170192.

Figuero, A., Rodríguez, Á.R., Sande, J., Peña, E. & Rabuñal, J.R. 2018. Field measurements of angular motions of a vessel at berth : Inertial device application, Control Engineering and Applied Informatics, 20, 79–88.

Figuero, A., Sande, J., Peña, E., Alvarellos, A., Rabuñal, J. R. & Maciñeira, E. 2019. Operational thresholds of moored ships at the oil terminal of inner port of A Coruña (Spain), Ocean Engineering, 172, 599–613. doi:10.1016/j.oceaneng.2018.12.031.
Fonseca, N. & Guedes Soares, C. 2002. Sensitivity of the Expected Ships Availability to Different Seakeeping Criteria, in ASME (Ed.), Proceedings of 21st International Conference on Offshore Mechanics and Artie Engineering (OMAE). Oslo, Norway: ASME, 595–603. doi:10.1115/OMAE2002-28542.
Fortes, C.J.E.M., Lemos, R., Fortunato, A., Capitão, R., Pinheiro, L. V, Neves, M.G., Freire, P., Oliveira, F., Azevedo, A., Lopes, H., Tato, P., Brogueira Dias, E. & Luís, L. 2017. Studies carried out to improve safety conditions maritime access to the port of leixões, in Atas das 9as Jornadas Portuguesas de Engenharia Costeira e Portuária - PIANC Portugal. Lisbon: APRH, 25. doi: https://pianc.pt/wp-content/uploads/2021/03/9as-jornadas/22.pdf.
Gomez, R., Molina, R., Camarero, A. & De los Santos, F. 2014. Development of a terminal operability forecasting system: analysis of the effects that wind generates over vessel performance, Coastal Engineering Proceedings, 1 (34), 16. doi:10.9753/icce.v34.management.16.
Gracia, V., Sierra, J.P., Gómez, M., Pedrol, M., Sampé, S., García-León, M. & Gironella, X. 2019. Assessing the impact of sea level rise on port operability using LiDAR-derived digital elevation models, Remote Sensing of Environment, 232, 111318. doi:10.1016/J.RSE.2019.111318.
Guachamin Acero, W., Li, L., Gao, Z. & Moan, T. 2016. Methodology for assessment of the operational limits and operability of marine operations, Ocean Engineering, 125, 308–327. doi:10.1016/J.OCEANENG.2016.08.015.
Guedes Soares, C. 1984. Representation of double-peaked sea wave spectra, Ocean Engineering, 11(2), 185–207. doi:10.1016/0029-8018(84)90019-2.
Guedes Soares, C., Fonseca, N. & de Marco Centeno, R. 1995. Seakeeping performance of fishing vessels in the Portuguese economic zone, in Proceedings of International Conference on Seakeeping and Weather. London, England, 1–10. Available at: https://api.semanticscholar.org/CorpusID:106897671.
Kim, S., Yu, Y. & Lee, Y. 2019. A Study on the Allowable Range of Overhanging Berthing at the Port of Ulsan, Journal of the Korean Society of Marine Environment and Safety, 25(3), 313–319. doi:https://doi.org/10.7837/kosomes.2019.25.3.313.
Kwak, M. & Pyun, C. 2013. Computer Simulation of Moored Ship Motion Considering Harbor Resonance in Pohang New Harbor, in Ports 2013. Reston, VA: American Society of Civil Engineers (Proceedings), 1415–1424.~ doi:10.1061/9780784413067.145.
López, M. & Iglesias, G. 2014. Long wave effects on a vessel at berth, Applied Ocean Research, 47, 63–72. doi:10.1016/j.apor.2014.03.008.
MAIB 2017. Report on the investigation of the failure of a mooring line on board the LNG carrier Zarga. Southampton. Available at: https://assets.publishing.service.gov.uk/media/5940014e5274a5e4e000239/MAIBInvReport13_2017.pdf.
McComb, P., Johnson, D.L. & Beamsley, B. 2009. Numerical study of options to reduce swell and long wave penetration at port Geraldton, in Coasts and Ports 2009: In a Dynamic Environment. Wellington, New Zealand: Engineers Australia, 490–496. Available at: https://search.informit.org/doi/10.3316/informit.863683725499435.
Moes, H. & Terblanche, L. 2010. Motion criteria for the efficient (un)loading of container vessels, in Port Infrastructure Seminar. Delft, Holland: Delft University of Technology. Available at: https://api.semanticscholar.org/CorpusID:107763962.
Molina-Sanchez, R., Campos, Á., de Alfonso, M., de los Santos, F.J., Rodríguez-Rubio, P., Pérez-Rubio, S., Camarero-Orive, A. & Álvarez-Fanjul, E. 2020. Assessing Operability on Berthed Ships. Common Approaches, Present and Future Lines, Journal of Marine Science and Engineering, 8(4), 255. doi:10.3390/jmse8040255.
NTSB 2018. Breakaway of Containership Helsinki Bridge and Subsequent Allision with Black Falcon Cruise Terminal. Available at: https://safety4sea.com/wp-content/uploads/2018/12/NTSB-Breakaway-of-containership-Helsinki-Bridge-and-subsequent-allision-with-black-falcon-cruise-terminal-2018_12.pdf.
OCDI 2020. Technical standards and commentaries for port and harbour facilities in Japan. Tokyo, Japan: The Overseas Coastal Area Development Institute of Japan. Available at: https://ocdi.or.jp/tec_st/tec_pdf/tech_00H.pdf.
OCIMF 2018. Mooring equipment guidelines: (MEG4). London, UK: Witherby Seamanship International Ltd. Available at: https://www.ocimf.org/publications/books/mooring-equipment-guidelines-meg4.
PIANC 1995. Criteria for movements of moored ships in harbours: a practical guide - A Practical Guide, Permanent Technical Committee II. Working Group No. 2. Brussels, Belgium.
Pinkster, J.A. 2004. The influence of a free surface on passing ship effects, International shipbuilding progress, 51 (4), 313–338. Available at: https://api.semanticscholar.org/CorpusID:108681520.
Prpic-Orsic, J., Slapnicar, V. & Turk, A. 2014. Berth operability estimation related to ship motion, Transactions of Famena, 38(1), 13–24. Available at: https://hrcak.srce.hr/120153%0A%0A.
Puertos del Estado 2011. Recommendations for the project and execution in Docking and Mooring Works ROM 2.0-11 (Spanish). Ministry of Public Works, Madrid, Spain. Available at: https://www.puertos.es/es-es/BibliotecaV2/ROM2.0-11.pdf.
Rosa-Santos, P., Taveira-Pinto, F. & Veloso-Gomes, F. 2014. Experimental evaluation of the tension mooring effect on the response of moored ships, Coastal Engineering, 85, 60–71. doi:10.1016/j.coastaleng.2013.11.012.
Rosa-Santos, P.J. & Taveira-Pinto, F. 2013. Experimental study of solutions to reduce downtime problems in ocean facing ports:the Port of Leixões, Portugal, case study, Journal of Applied Water Engineering and Research, 1(1), 80–90. doi:10.1080/23249676.2013.831590.
Rusu, E. & Guedes Soares, C. 2011. Wave modelling at the entrance of ports, Ocean Engineering, 38(17–18), 2089–2109. doi:10.1016/j.oceaneng.2011.09.002.
Rusu, L. & Guedes Soares, C. 2013. Evaluation of a high-resolution wave forecasting system for the approaches to ports, Ocean Engineering, 58, 224–238. doi:10.1016/j.oceaneng.2012.11.008.
Sakakibara, S. & Kubo, M. 2008. Characteristics of low-frequency motions of ships moored inside ports and harbors on the basis of field observations, Marine Structures, 21(2–3), 196–223. doi:10.1016/j.marstruc.2007.11.002.

Santos, J.A., Pinheiro, L.V., Abdelwahab, H.S., Fortes, C.J. E.M., Pedro, F.G.L., Capitão, R.P., Hinostroza, M.A. & Guedes Soares, C. 2019. Physical modelling of motions and forces on a moored ship at the Leixões port, Defect and Diffusion Forum, 396, 60–69. doi:10.4028/www.scientific.net/DDF.396.60.

Sierra, J.P., Genius, A., Lionello, P., Mestres, M., Mösso, C. & Marzo, L. 2017. Modelling the impact of climate change on harbour operability: The Barcelona port case study, Ocean Engineering, 141, 64–78. doi:10.1016/J.OCEANENG.2017.06.002.

Taveira-Pinto, F., Veloso-Gomes, F., Rosa-Santos, P., Guedes Soares, C., Fonseca, N., Santos, J.A., Paulo-Moreira, A., Costa, P. & Brógueira-Dias, E. 2008. Analysis of the behavior of moored tankers, in Proceedings of the International Conference on Offshore Mechanics and Arctic Engineering - OMAE. Estoril, Portugal, 755–764. doi:10.1115/OMAE2008-58013.

Tello, M., Ribeiro E., Silva, S. & Guedes Soares, C. 2011. Seakeeping performance of fishing vessels in irregular waves, Ocean Engineering, 38(5–6), 763–773. doi:10.1016/J.OCEANENG.2010.12.020.

Tezdogan, T., Incecik, A. & Turan, O. 2014. Operability assessment of high speed passenger ships based on human comfort criteria, Ocean Engineering, 89, 32–52. doi:10.1016/J.OCEANENG.2014.07.009.

Troch, C., Terblanche, L., Rossouw, M., Ntantala, Z. & Plessis, G. du 2020. Long-wave mitigation study for the Port of Ngqura, Coastal Engineering Proceedings, (36v), 28. doi:10.9753/icce.v36v.papers.28.

Uzaki, K. ichi, Matsunaga, N., Nishii, Y. & Ikehata, Y. 2010. Cause and countermeasure of long-period oscillations of moored ships and the quantification of surge and heave amplitudes, Ocean Engineering, 37(2–3), 155–163. doi:10.1016/j.oceaneng.2009.12.004.

Van der Molen, W. 2006. Behaviour of moored ships in harbours. PhD, Delft University of Technology. Available at: http://resolver.tudelft.nl/uuid:1c20fa4a-ee42-4236-a69b-b895a81413af.

Van der Molen, W., Scott, D., Taylor, D. & Elliott, T. 2016. Improvement of Mooring Configurations in Geraldton Harbour, Journal of Marine Science and Engineering, 4 (1). doi:10.3390/jmse4010003.

Veloso-Gomes, F., Taveira-Pinto, F., Rosa-Santos, P., Brógueira-Dias, E. & Guedes-Lopes, H. 2005. Berthing characteristics and the behaviour of the oil terminal of Leixões harbour, in Maritime Heritage and Modern Ports. Portugal: WIT Press, 481–492.

Weiler, O., Cozijn, H., Wijdeven, B., Le-Guennec, S. & Fontaliran, F. 2009. Motions and mooring loads of an LNG-carrier moored at a jetty in a complex bathymetry, in Proceedings of the ASME 28th International Conference on Ocean, Offshore and Arctic Engineering OMAE. Honolulu, Hawaii, 425–435. doi:10.1115/OMAE2009-79420.

Wu, M. 2014. Numerical analysis of docking operation between service vessels and offshore wind turbines, Ocean Engineering, 91, 379–388. doi:10.1016/J.OCEANENG.2014.09.027.

Zhou, L., Abdelwahab, H.S. & Guedes Soares, C. 2021. Experimental and CFD investigation of the effects of a high-speed passing ship on a moored container ship, Ocean Engineering, 228, 108914. doi:10.1016/J.OCEANENG.2021.108914.

Advances in Maritime Technology and Engineering – Guedes Soares & Santos (Eds)

Computational fluid dynamics of vortex induced vibration on cylinders

V. Galvão, S. Wang & C. Guedes Soares
Centre for Marine Technology and Ocean Engineering (CENTEC), Instituto Superior Técnico, Universidade de Lisboa, Lisbon, Portugal

G. Assi
Department of Naval and Ocean Engineering, University of São Paulo, Brazil

ABSTRACT: This study investigates Flow-induced Vibration (FIV) of cylinders free-to-move in the direction perpendicular to the flow, employing experimental and numerical approaches. Vortex-Induced Vibration (VIV) and Wake-Induced Vibration are explored, the latter involving two aligned cylinders exhibiting amplitude buildup with increasing flow velocity. Experimental tests in the NDF lab at the University of São Paulo utilized channels for high and low Reynolds numbers, estimating natural frequencies and damping through decay tests. Numerical analysis with Ansys Fluent employed the Finite Volume Method, necessitating Dynamic Mesh and 6 DOF settings. Both methods yielded similar results, indicating that the lift oscillation frequency matched the vortex shedding frequency (f_s) and drag oscillation at $2f_s$. The study enhances understanding of FIV mechanisms, offering valuable insights into dynamic interactions between cylinders and fluid flow.

1 INTRODUCTION

As offshore exploration has reached ultra-deep waters, structures began to inhabit regions with sea currents of unknown direction. Thus, it became inevitable that some pipes, known as risers, would interfere with other submerged structures. Therefore, in the context of offshore engineering, it is highly important to understand this interaction between fluid and structure in order to avoid severe vibrations and collisions between risers. According to Assi (2005), some solutions to mitigate the vibrations can reach up to a cost of $ 4 million per riser installed.

The main Flow-Induced Vibrations (FIV) for cylindrical-section structures are:

1. Vortex-Induced Vibrations: Considering flow around a blunt body, two shear layers arise from two separation points resulting in the formation and shedding of vortices. Thus, the Von Kárman vortex-street appears, as shown in Figure 1. The shedding of vortices has an effect on the pressure field on the cylinder wall, making it periodic, which generates cyclic excitation forces in the body. One of the types of FIV phenomena originates the vortex-induced vibrations (VIV). This vortex emission phenomenon was better understood by Gerrard (1966), showing that the periodicity of the excitation force was associated with the change of the lift force with time. This periodicity also applies to vortex emission, thus defining the vortex emission frequency f_s. Therefore, there is a possibility of resonance if this frequency approaches the natural frequency of the system, f_N.

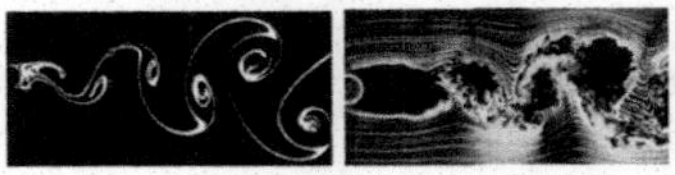

Figure 1. Von Kármán Vortex-street for the laminar and turbulent regime (Assi 2005).

2. Wake-Induced Vibrations (WIV): The effects of FIV intensify when more than one structure is exposed to the flow. When a blunt system is immersed in the wake of another blunt body positioned upstream, the phenomenon of vibration induced by the wake known as WIV which was described by Assi (2009). A curious fact demonstrated by the cited study is that the dynamic response of WIV is not resonant despite the indefinite increase in amplitude as the reduced velocity of the flow increases. Figure 2 compares the response of a cylinder under the effect of VIV and WIV.

1.1 *Objective*

The main goal of this work is to simulate vortex-induced vibrations and wake-induced vibrations on the CFD software Ansys Fluent. The parameters based for the modelling in the CFD are the same of physical experiments carried out in the NDF laboratory of the

DOI: 10.1201/9781003508779-11

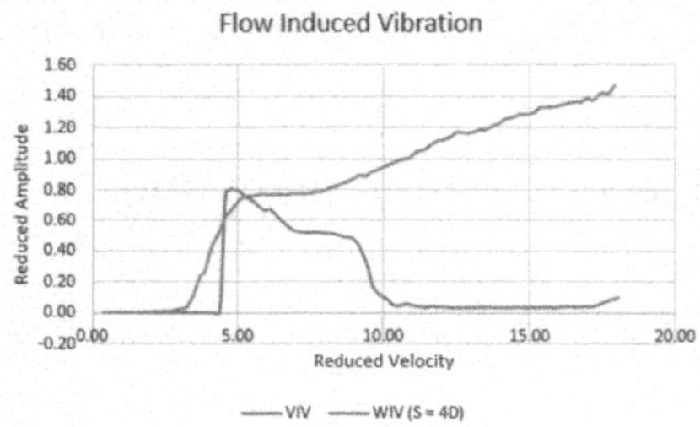

Figure 2. Dynamic response of the cylinder for VIV case and the WIV case.

University of São Paulo with two different channels, one for high Reynolds number and other for low Reynolds number. The idea is to analyze the y-motion of the cylinder produced by the lift force generated by the perpendicular water flow in two different arrangements in low and high Reynolds numbers, the computational results will be compared to these physical ones.

Several simulations were carried out for each arrangement for both ranges of reynolds number. Throughout these simulations, the reduced velocity was changed by testing other flow velocities (U) or different natural frequencies (f_N) which are obtained by using another spring constant (k) value. In the end, the relationship between the reduced velocity and reduced amplitude will be plotted.

The expected outcome for the single cylinder, according to the references, is plotted in Figure 3 starting at zero for the amplitudes and reaching a resonance peak and later going to zero. For the other arrangement, the tandem cylinders have the spacing of four times the diameter between the center of the cylinders. As seen in Figure 4, the upstream cylinder is fixed and the downstream ine is free to move in the y-direction coupled witha a spring and damper. The expected outcome at the beginning is similar to the isolated one until the resonance peak, then the amplitude starts growing as the reduced velocity increases.

The first phenomenon is the VIV response. The vortices produced by the single cylinder generate the

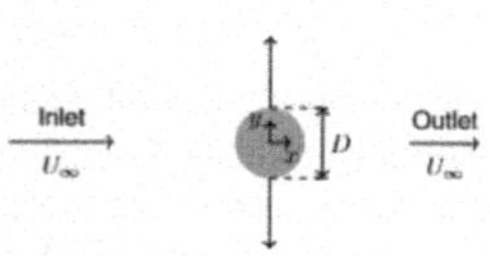

Figure 3. Single cylinder in the flow free to move in perpendicular direction.

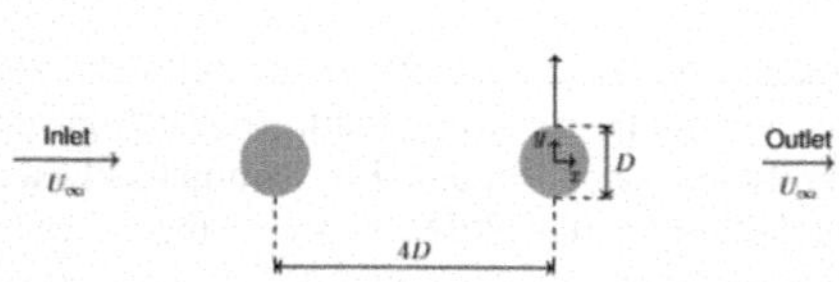

Figure 4. Two tandem cylinders with the downstream free-to-oscillate in the y-direction and the upstream one fixed.

lift variation force. In the other arrangement the lift variation is caused by the vortices produced by the upstream cylinder and also by the downstream one, the frequency of the lift force is related to aforementioned vortex shedding frequency.

2 LITERATURE REVIEW

2.1 *Flow around blunt bodies*

In an affiliated body, as can be seen in the upper part of Figure 5, the streamlines are always connected to the contour surface and do not separate from the body, which does not happen in the so-called bluff bodies forming a separate region of the flow. This phenomenon occurs when there is a discontinuity in the geometry of the body or when the transverse dimension to the flow is larger than the longitudinal dimension, as in the case of a cylinder. Figure 5 illustrates the formation of the separation for the two different types of bodies.

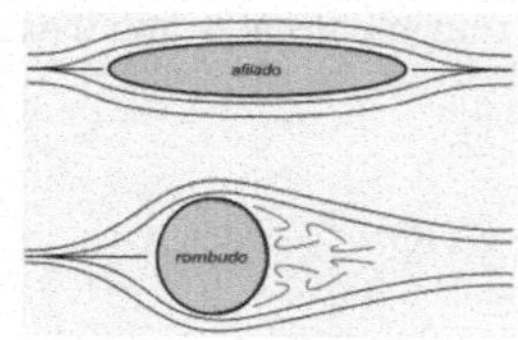

Figure 5. Flow around affiliated and blunt bodies. (Assi 2005).

This phenomenon occurs due to an adverse gradient in the flow direction, $\frac{\partial P}{\partial x} > 0$, generating lower velocity, higher pressure and consequent flow separation. It is in this region that the dynamic effects studied by this work occur.

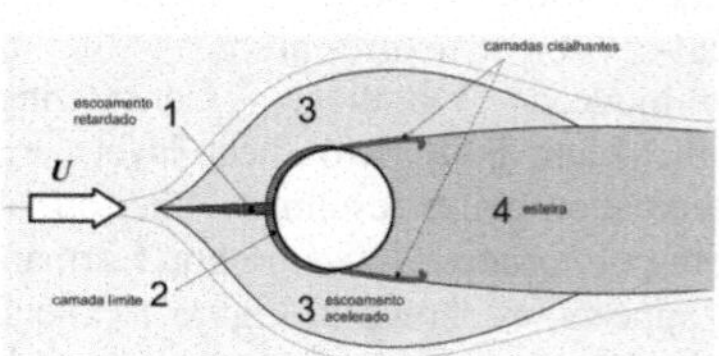

Figure 6. Map Flow around a blunt body. (Zdravkovich 1977).

According to Figure 6, the flow around the bluff bodies has four different regions: 1. Delayed Flow: It presents reduced velocity thanks to the frontal stagnation point of the body. 2. Boundary Layer: This is adhered to the body until the unavoidable pressure gradient occurs and the layer separates and forms free shear layers. 3. Accelerated flow: The presence of the body displaces and increases the speed of particles in the region. 4. Vortex street:

Completely separate flow and with reduced velocity. Region where the phenomena responsible for the vibrations induced by the flow are originated.

This phenomenon is best described by the dimensionless Reynolds number $Re = UD/v$. This parameter demonstrates the ratio of inertial and viscous forces in the flow, where D is the cylinder diameter, U is the flow velocity and v is the kinematic viscosity of the fluid.

The value of this dimensionless parameter is directly associated with the separation of the boundary layer and the formation of vortices as shown in Figure 7. In this way, low Reynolds number values indicate high viscosity and therefore there is no separation of the boundary layer, but as the Reynolds number increases, vortices appear close to the body wall due to the interaction between the shear layers. Increasing even more Re, the appearance of vortices becomes unstable giving rise to the phenomenon called Von Kárman vortex-street.

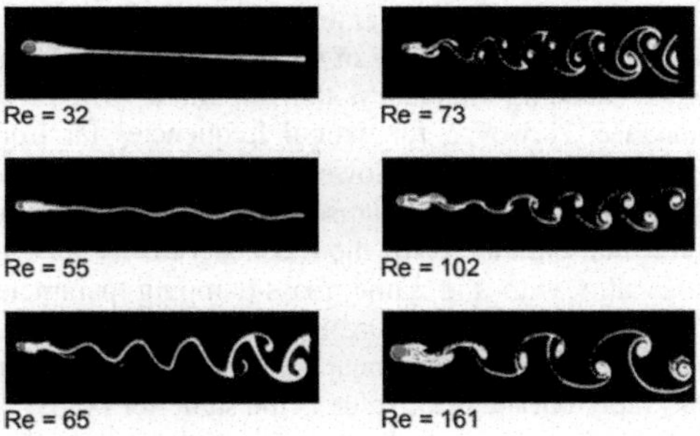

Figure 7. Vortex formation for different Reynolds Numbers. (Batchelor 2000).

Bearman (1984) wrote a review of this vortex mechanism in which he stated that the vortex-street formation is caused by the mutual interaction between the shear layer shed at the separation point. Moreover, from the Figure 7 the vortex continues growing downstream. According to Gerrard (1966), this is caused by the oppositely signed of the vorticity which ceases circulation to the benefit of the growing vortex which is then shed and, hence, it moves downstream.

2.2 *Vortex formation: Strouhal number*

While a vortex is released in the wake, another is formed on the cylinder wall in the opposite region. Bearman (1984) concluded that the shedding is caused mainly by the presence of the two shear layers not the bluff body itself. The body's presence only modifies the process by allowing interaction between the wake and the shedding flow at the separation points.

In this way, the vortex emission frequency is defined by f_s, which is dependent on the flow velocity, U. In addition, the interaction between shear layers which depends on the distance between them, in this case it's the cylinder's diameter, D. Therefore, the Strouhal Number is defined as $St = f_sD/U$ which was related with the Reynolds number for several experiments (Blevins, 1977) for the case of an isolated cylinder. As the result from this historical data it is known that for subcritical Reynolds number, $St \approx 0.2$.

2.3 *Vortex-induced vibration*

The vortex-induced vibrations are caused by the cyclic loads generated by vortices formed from the body at the separation points. This is explained by the asymmetric pressure distribution around the bluff body causing a drag in the streamwise direction and a fluid-dynamic lift force on the cross-flow direction.

A resultant force, F, is computed from the integration of this pressure field around the cylinder's surface, which is projected into drag component, F_x and lift component, F_y. According to Assi (2009), during the cycle of vortex shedding, the F_x (drag) and F_y (lift) drift according to the shedding frequency of $2f_s$ and f_s, respectively.

2.4 *System model*

Applying Newton's second law, the equation representing the motion of a cylinder oscillating under one degree of freedom can be generalized by:

$$m\ddot{y} + c\dot{y} + ky = F(t) \tag{1}$$

Where m represents the mass of the cylinder; the component $F(t)$ is the lift force, which is influenced by the vortices, c the damping factor and k the stiffness.

In addition, the analysis will occur using dimensionless parameters in order to facilitate the comparison between the experiments produced by this work and those already performed in the literature. They are described below:

1. Reduced Amplitude: Ratio between the maximum amplitude, A, and the diameter of the cylinder, D.

$$A^* = \frac{A}{D} \tag{2}$$

2. Reduced Velocity or Dimensionless velocity: Flow velocity divided by the diameter and the natural frequency, f_N.

$$U^* = \frac{U}{Df_N} \tag{3}$$

3. Reduced frequency: Ratio between the frequency of oscillation, f, and the natural frequency, f_N.

$$f^* = \frac{f}{f_N} \tag{4}$$

4. Mass parameter: Ratio between the cylinder's mass, m, and the mass of the displaced water.

$$m^* = \frac{m}{\rho \frac{\pi D^2}{4} L} \tag{5}$$

5. Damping Parameter: Ratio between damping, c, and the critical damping $(2\sqrt{km})$.

$$\zeta = \frac{c}{2\sqrt{km}} \tag{6}$$

The final response for each case tested will be two plots the reduced frequency and reduced amplitude versus reduced velocity. Another important parameter used by the literature on the subject is the combined mass-damping parameter $m^*\zeta$. According to Khalak & Williamson (1999), the higher this value, the lower the reduced amplitude value in the resonance peak.

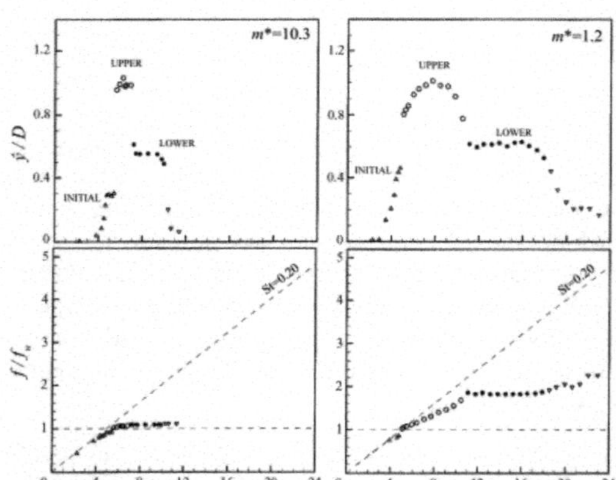

Figure 8. VIV response of experimental studies from Govardhan & Williamson (2000) for low mass and damping.

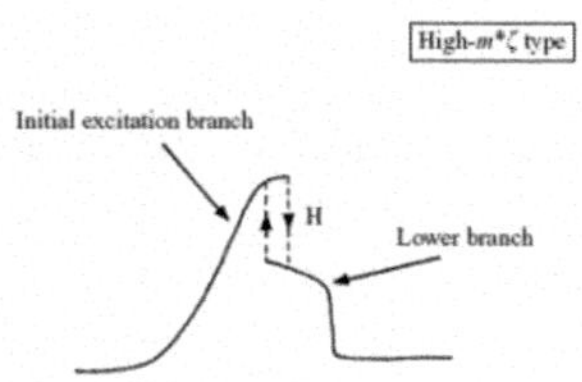

Figure 9. Typical response for high $m^*\zeta$. (Assi 2005).

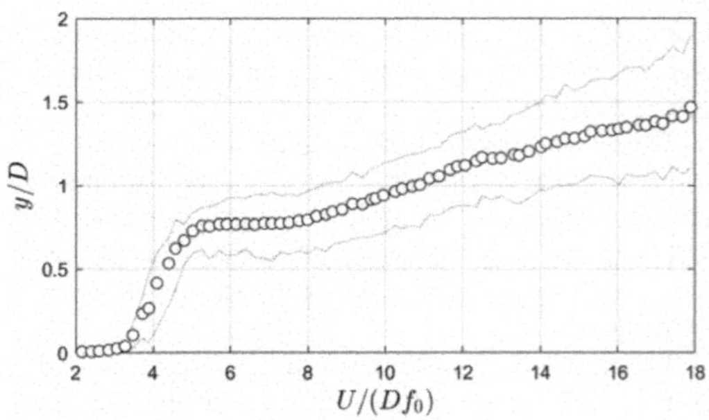

Figure 10. Flow induced vibration for a downstream cylinder, experiment produced in University of São Paulo for *Re* = 500 - 25,000 and for *S*/*D* = 4.

2.5 *Typical VIV responses*

Several studies have been carried out in order to analyze the outcome of the VIV. The experiments produced by Assi (2005), Assi (2009) and Khalak & Williamson (1999) could produce different plots for different Reynolds number and parameters.

If the natural frequency of the system is close to the vortex shedding frequency, the amplitude will be at resonance. However, for natural frequencies far from f_s, the amplitudes will be lower. An experimental result from Govardhan & Williamson (2000) could plot the reduced amplitude versus the reduced velocity for two mass ratio with the same mass-damping parameter of $m^*\zeta \approx 0.01$. From Figure 8, it can be seen that the range of U^* for the resonance is wider for smaller mass ratio, but the peak value is the same for both.

From the results of Batchelor (2000), Govardhan & Williamson (2000), Assi (2005) and Assi (2009) it is possible to state that there's a pattern for the outcome for the VIV phenomena for low mass ratio and damping for high Re. Therefore, Williamson & Govardhan (2004) identified three distinct phases of response for cylinders vibration: 1. An initial branch where the oscillations begin to develop. 2. An upper branch characterized by the maximum amplitude, typically occurring near the resonance point. 3. An lower branch where the oscillations gradually diminish and they may cease completely for high mass ratio.

Nonetheless, for high $m^*\zeta$ there are two branches. An initial excitation branch where's the amplitude starts from zero and continuously goes to the peak and then a discontinuity appears and the amplitudes reduces to the lower branch. Figure 9 shows the outcome for $m^*\zeta \geq 1$.

2.6 *Flow Interference around a pair of cylinders.*

For the single cylinder case, the origin of the vibrations is the alternate shedding of vortices, however for the case of a pair of cylinders in which the upstream is fixed and the downstream free to move in the y-direction, the vibrations are also caused by this along with the influence of the wake from the upstream cylinder.

This interference causes different results in the downstream one from the VIV case. As the reduced velocity increases, the amplitude builds-up. Figure 10 shows an example, investigations from Bokaian & Geoola (1984) classified four types of vibration for the downstream cylinder, pure vortex-induced vibration, pure wake-galloping, separated VIV and wake-galloping and combined VIV wake galloping. Hence the researchers called the phenomena wake-induced galloping. Assi (2005) stated that for the pure wake-galloping the phase angle between the displacement and the lift force should be 90°, but in his experiments the phase angle varied from 0 to 180°. Therefore, Assi stated that the the phenomena is explained by the velocity fluctuations in the cylinder's gap and vortices formed by the downstream cylinder, hence he called Wake-induced Vibrations or the acronym WIV.

In this type of experiment, another parameter that shows up is the distance between both cylinders, *S*, attributed to the parameter $\frac{S}{D}$. It influences the size of the wake from the upstream cylinder hence changing the type of vibration for the downstream cylinder. The *Re* for the downstream cylinder is considered the same as in the upstream.

2.7 *Interference from the gap between the cylinders*

For low *S/D*, the shear layers from the upstream might reattach to the downstream cylinder generating Gap-flow-switching. According to Assi (2009), the WIV only occurs with a well developed wake inside the gap. As the distancing increases, the WIV phenomena becomes more clear as can be seen in Figure 10. Nonetheless, if the *S/D* increases significantly the wake from the upstream will have low effect and the outcome will be very similar to the VIV. As stated by Assi (2005).

There's a critical value for *S/D* for the beginning of a well developed vortex street which highly depends on the Reynolds Number (Assi 2009). For Re between 1000 and 100,000, this critical value decreases when Re increases according to Lin et al. (1995) and it is inside the range 3<*S/D*<4. Nonetheless, this is not valid for laminar flow which Zdravkovich (1972) couldn't find a wake developed for *S/D* < 4 for Reynolds between 40 and 250.

2.8 *Wake-Induced vibrations models*

Several attempts have been carried out in the past decades in order to model analytically the WIV phenomena like the ones produced by Parkinson (1989) and Granger & Paidoussis (1996), but most of these authors tried to model it by using the quasi-steady models, which only works for small U^* according to Assi (2005). Therefore, WIV should be modelled based on unsteady flow theory.

The study from Yang et al. (2022) proposed a nonlinear mathematical model for the fluid force on the WIV, combining the self-excited force plus the forces with the frequencies of the vortex-shedding from the upstream and the downstream cylinder. The self-excited force modelling was based on an energy equivalent method from Gao & Zhu (2017) to identify the aerodynamic parameters. This model was validated by the data from a numerical simulation carried out in Ansys Fluent for $\Delta t = 10^{-4} s$ with *Re* from 1640 to 8240, *S/D* = 4, $m^*\zeta = 0.2\epsilon 7$.

3 METHODOLOGY

3.1 *Finite volume method*

The numerical technique applied is the Finite Volume Control method, which a small part of the domain is divided into a collection of control volumes where its faces, center and vertexes are stored by the software for the mass, momentum and energy flux computation inside of each control volume. Along with the boundary conditions with a spatial and temporal refinement the solution might be found accurately, despite being different from the continuous formulae.

$$\frac{\partial}{\partial t}\int_V \rho\phi dV + \oint_A \rho\phi\vec{v}\cdot d\vec{A} = \int_V S_\Phi dV + \oint_A \Gamma_\Phi \vec{\nabla}\phi \cdot d\vec{A} \tag{7}$$

$$\frac{\partial\rho\phi}{\partial t}V + \sum \rho\vec{V}_f\phi_f\vec{A}_f = \sum^{N\,faces} \Gamma\nabla\phi_f\vec{A}_f + S_\phi V \tag{8}$$

The transport equation (7) of an scalar variable ϕ is transformed into a discrete form equation (8).

where ρ is the density; $\vec{V}_f$ is the velocity vector; S^ϕ is the source of ϕ per unit volume; $\vec{A}$ is the surface of area Vector; Γ_ϕ is the diffusion coefficient for ϕ; $\rho\vec{V}_f\phi_f\vec{A}_f$ is the mass flux through the face and $\vec{\nabla}\phi = \frac{\partial\phi}{\partial x}\vec{i} + \frac{\partial\phi}{\partial y}\vec{j} + \frac{\partial\phi}{\partial z}\vec{k}$ is the gradient of ϕ.

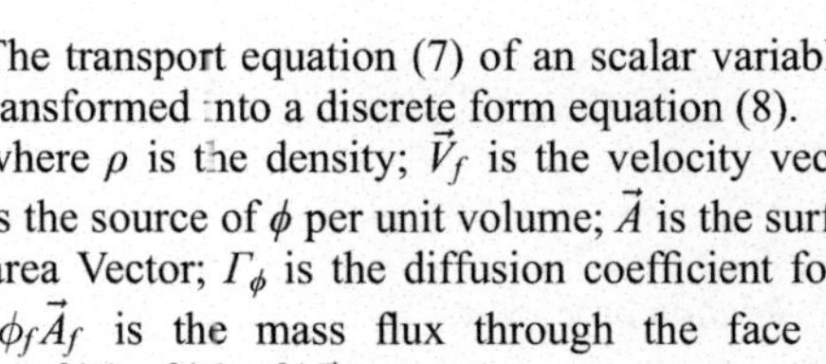

Figure 11. Elements system.

3.2 *Interpolation function*

Ansys Fluent uses the cell center approach in which the elements created in the mesh are taken as the control volumes and the unknown is at the center of the element. Therefore, the ϕ and $\nabla \phi$ computed in Equation (8) must be interpolated to the cell center, however this may lead to truncation errors like numerical oscillation for second order functions and diffusion for first order. Some methods in Fluent are listed below:

3.3 *Central differecing scheme*

This method considers the ϕ of the elements with the same interface and its distance to the centers, as shown in Figure 11, and also the gradient.

$$\phi_{CD} = \frac{1}{2}(\phi_0 + \nabla\phi_0 \cdot \vec{r}_0 + \phi_1 + \nabla\phi_1 \cdot \vec{r}_1) \quad (9)$$

3.4 *Second order upwind scheme*

In this function the ϕ at the face is computed from the ϕ at the center.

$$\phi_f = \phi + \nabla\phi \cdot \vec{r} \quad (10)$$

3.5 *Gradient*

For the computation of the Gradient, $\Delta\ \phi$, Fluent has three different methods: Least-Square cell based, Green-Gauss cell based and Green Gauss node-based. The Least-Square Cell was used. In this method the gradient is reconstructed based on the ϕ of the neighboors cells and its distance, $r - r_p$, from the cell the gradient is being computed. Thus, Fluent finds $\Delta\ \phi_P$ from the solution of the equation .

$$\phi = \phi_p + (\nabla\phi)_p \cdot (\vec{r} - \vec{r}_p) \quad (11)$$

3.6 *Solution procedure*

The velocity components (u, v, w) and pressure (P) can be found simultaneously, using coupled method in Fluent, with a linear system for all the variables or in a segregated way, the SIMPLE, SIMPLEC and PISO, schemes in several linear systems are used, one for each variable.

There are two types of solvers in Fluent: density-based which uses the mass conservation in its original form and only used the COUPLED method and the pressure-based, which uses either COUPLED or segregated methods. The mass conservation is used for finding the pressure. For this work the pressure based solver with the SIMPLE method was used. Their solution procedure is shown in Figure 12.

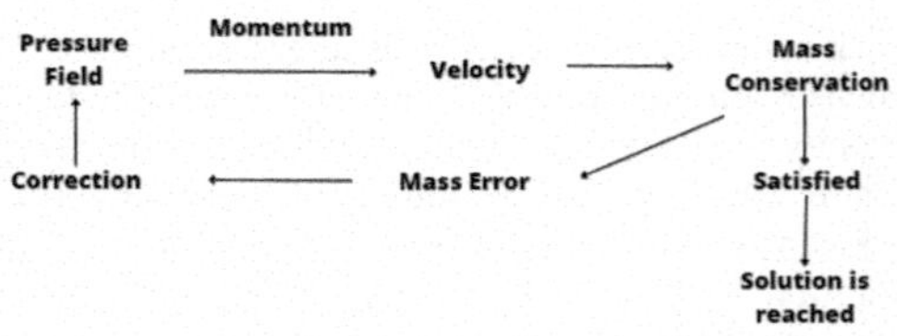

Figure 12. Pressure-Based solver used in the segregated manner.

3.7 *Dynamic mesh*

As the cylinder moves, the mesh needs to be updated, therefore dynamic mesh in Fluent must be selected. There are three types of dynamic mesh: Layering, Smoothing and Remeshing. This part demands attention because the time step for the simulation must be lower than the smallest mesh size, Δs divided by the velocity of the boundary motion, v, as shown in equation (12). Otherwise, cell with negative volume will appear and then the solution will fail.

$$\Delta t \leq \frac{\Delta s}{v} \quad (12)$$

The Layering Method creates and deletes cells as the zone grows or decreases with the boundary motion set by profile or an UDF file. It's a very useful method for in-cylinder application. The remeshing method works within an specific range and a remeshing interval, SRI, both set by the user. If some cells are out of this range after SRI iterations the domain is remeshed. This method allow great motion of the mesh.

The Smoothing method does not change the amount of cells or it displaces the nodes for absorb the motion. It has two main methods inside this one for CFD-only application, Spring-based and Diffusion.

In the spring based method, the displacement in one node, j, generates a spring force on another. In a way that the spring forces on the node i from all n connected nodes are sum, as Equation (13) shows. This force has a stiffness shown by Equation (14) which depends on the node position and also a parameter, k_{fac}, defined between 0 and 1 by the user. When $k_{fac} = 1$ the most distant elements will move more.

$$\vec{F}_i = \sum_{j}^{n} k_{ij}(\Delta\vec{x}_i - \Delta\vec{x}_j) \quad (13)$$

Where

$$k_{ij} = \frac{k_{fac}}{\sqrt{|\vec{x}_i - \vec{x}_j|}} \quad 0 \leq k_{fac} \leq 1 \quad (14)$$

The sum of the forces is set as zero, then the displacement is solved iterative as equation below shows.

$$\Delta \vec{x}_i^{m+1} = \frac{\sum_j^n k_{ij} \Delta \vec{x}_j^m}{\sum_j^n k_{ij}} \tag{15}$$

The solution of the iterative equation is controlled by tolerance convergence and by the maximum number of iterations set by the user. Thus, after the convergence the positions are updated as equation 16 shows. Furthermore, as the convergence is reached, the difference between the outcome of k_{fac} 0 or 1 is nothing.

$$\vec{x}_i^{n+1} = \vec{x}_i^n + \Delta \vec{x}_i^{Converged} \tag{16}$$

The Spring-based method is an efficient and fast method to smooth any cell. However, it's limited for unidirectional and small deformation. For higher deformation and rotational boundary motion the Diffusion is more appropriate despite being computationally more expensive. In the diffusion method the motion is ruled by equation:

$$\nabla \cdot (\gamma \nabla \vec{u}) = 0 \tag{17}$$

where $\vec{u}$ is the Displacement's velocity; γ is the Diffusion coefficient. It can be computed by two different methods based on the cell's distance, equation $\gamma = \frac{1}{d^a}$, or cell's volume, equation $\gamma = \frac{1}{V^a}$.

The diffusion parameter a is defined by the user, It shall be equal or greater than zero and less than 2. When It's zero, the smoothing in the mesh is uniform as $\gamma = 1$. When a increases, the elements deformed less close to the moving limit.

3.8 *6 DOF*

In order to analyze the motion of bodies influenced by the fluid forces, the option Six DOF on Dynamic Ansys Fluent has to be enabled. Fluent will compute forces and moments and then the acceleration is calculated. With the given initial position and velocity of the center of the gravity, the new velocity and position are computed from the integration of the acceleration. In other words, this option allows solution of the Equation (1).

Therefore, the inertia properties, m, the restoring, ky, and damping forces $c\dot{y}$ needs to be set besides the degrees of freedom. This is done by an UDF code, as shown by Figure 13.

4 EXPERIMENTAL STUDY

4.1 *Experimental setup*

The experiments were carried out in two different channels. The low Reynolds number recirculating channel has a rectangular section and has the following dimensions: length 310 cm; width 50 cm and height 30 cm. The flow rate is 0.00425 m^3/s. The high Reynolds number circulating channel have dimensions of 750 x 90 x 70 (cm) and its maximum flow rate is 0.31 m^3/s which is 40 % of the feeder bomb's maximum power. Figure 14 shows the channels and Table 1 shows the experimental data used.

The cylinder does not oscillate by itself. The structural part moves together so the cylinder and structural mass must be taken into a account. For the low Reynolds the mass parameter. In Equation (5), is $m^* = 2.29$ while for the high Reynolds channel is $m^* = 2.63$.

```
DEFINE_SDOF_PROPERTIES(VIVLf100, prop, dt, time, dtime)
{
prop[SDOF_MASS] = 0.15;
prop[SDOF_IZZ] = 0.0;
prop[SDOF_IXX] = 0.0;
prop[SDOF_IYY] = 0.0;
prop[SDOF_IXY] = 0.0;
prop[SDOF_IYZ] = 0.0;
prop[SDOF_IXZ] = 0.0;
prop[SDOF_ZERO_TRANS_X] = TRUE;
prop[SDOF_ZERO_TRANS_Y] = FALSE;
prop[SDOF_ZERO_TRANS_Z] = TRUE;
prop[SDOF_ZERO_ROT_X] = TRUE;
prop[SDOF_ZERO_ROT_Y] = TRUE;
prop[SDOF_ZERO_ROT_Z] = TRUE;
prop[SDOF_LOAD_F_Y] = -1.48*DT_CG(dt)[1] - 0.1*DT_VEL_CG(dt)[1];
}
```

Figure 13. User Defined Function for the 6 DOF properties.

Figure 14. Channels used for the physical experiments, high (right) and low (left) Reynolds number.

Table 1. Cylinder measurements used in both Chanel's.

Dimension	Variable	Low *Re*	High *Re*
Length	L	333 mm	700 mm
Diameter	D	19 mm	50 mm
Water Line height	L_w	225 and 250 mm	700 mm
Cylinder Mass	m	0.12 kg	0.6 kg
Mass of the structure	M_e	0.03 kg	3.01 kg
Oscillating mass	M_t	0.15 kg	3.61 kg
Maximum velocity	U	0.0378 m/s	0.5 m/s
Minimum velocity	U	0.0392 m/s	0.01 m/s
Mass density	ρ	1000 kg/m3	1000 kg/m3
Cinematic viscosity	ν	1E-6 m2/s	1E-6 m2/s
Minimum Reynolds	Re	644.1	500
Maximum Reynolds	Re	718.2	25,000

For each channel, the U^* was obtained by different methods. In the low Reynolds channel the natural frequency is modified by changing the pendulum height while the water flux in the channel is constant. For the other channel the natural frequency is fixed while the U^* is changed by increasing the flow velocity. The values for f_N and ζ are obtained from a decay test. For the results of the low Reynolds test several values for the natural frequencies were obtained and the range was from 0.4 Hz to 0.9 Hz considering an average value of 0.57 % for ζ. For the high Reynolds, the results were f_N = 0.54 Hz and ζ = 0.45 %.

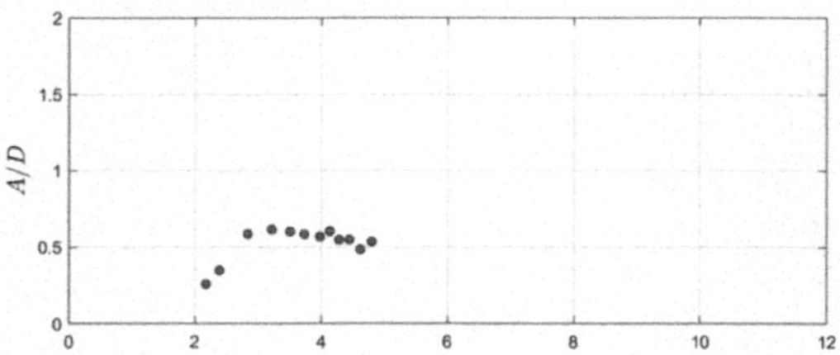

Figure 15. VIV Experiment in Low Re $m^*\zeta$=0,013.

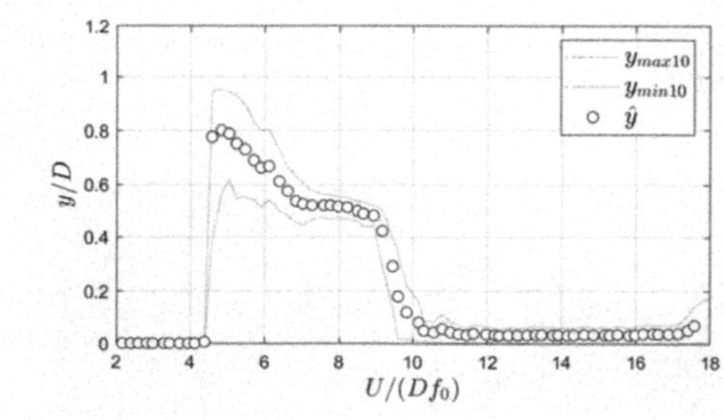

Figure 16. VIV physical experiment in high Re $m^*\zeta$ = 0,0118 - Reduced Amplitude.

For the small channel with $m^*\zeta = 0.013$, the spring constant is computed as $k = M_t(2\pi f_N)^2$ so it varies from 1.02 to 4.96 N/m and $c = 2\zeta\sqrt{km}$ so it varies from 0.0045 to 0.0098. For the high Reynolds case, $m^*\zeta = 0.0118$, k = 41.82 N/m and c = 0.11 N.s/m.

4.2 *Experimental results*

The data from the motion, Y of the cylinder was gathered by the software Labview and later processed in Matlab in order to compute the data and save the plots. The maximum amplitude is computed by the root mean square of the Y data multiplied by square root of 2, since it's not statistically accurate to use the maximum value of Y.

Figure 15 Shows the physical experiment result for the single cylinder for low Reynolds case. Figures 16 and 17 show the physical experiments results for the single cylinder and tandem arrengement for high Reynolds case, respectively.

The resonance is reached for U^* = 3.2 and A^* = 0.61. The profile is quite similar to low $m^*\zeta$ produced by Khalak & Williamson (1999) with a low decreasing in the amplitude after the resonance. The resonance is reached for U^* = 5 and A^* = 0.8. The profile is very similar to low $m^*\zeta$ produced by Khalak & Williamson (1999).

For this case the WIV phenomena is observed which the amplitude grows for higher U^* even after the resonance.

5 ANSYS FLUENT APPLICATION

The application in Ansys Fluent involves several steps:

1. The geometry was created in the Rhinoceros CAD a rectangular surface with 20D length and 500 mm width with a hole centered in (0,0) and another surface with two aligned holes spaced in 4D were created in order to reproduce WIV simulations. 2. Discretization of the geometry into many tri cells in the domain and quad cells near the walls with body size of 0.3D and edge size of 0.1D. 3. Setting of the boundary conditions: Inlet u = U and v = 0, U = 0.0378 m/s for the single cylinder and U = 0.0339 m/s for the pair of cylinder simulation; Outlet the pressure is equal to the atmospheric; Top wall and bottom wall are free-slip and the gradient of the u is zero and v = 0; The cylinder is a non-slip wall u = 0 and v = 0. 4. The system is transient, pressure-based solver, viscous model as laminar, material is the compressible liquid water; 5. Enable dynamic mesh with smoothing spring-based with k_{fac} = 0.1 because the motion is simple and it's computationaly cheper than diffusion and along with remeshing in order to improve mesh quality with a minimum mesh size of 1 mm and maximum 9 mm. 6. Enable Six DOF with an UDF as shown in Figure 13 for each natural frequency. 7. Initialization and calculation with time step of 0.01 seconds and 20 iterations per time step by using SIMPLE method.

5.1 *Vortex-induced vibrations*

Thus, nine different natural frequencies were simulated in Fluent, which has produced different amplitudes, therefore a plot A^* vs U^* could be done as depicted by Figure 19.

From the reduced amplitudes, it's possible to conclude existence of three branches. An initial where the amplitudes increases from very low values to the resonance peak. A region of resonance with a peak of A^* = 0.63 and the third branch where the reduced amplitudes stabilizes to a value of 0.51. This result shows a similar result from the simulation carried out by Assi (2005) with a very low mass parameter which the amplitudes don't go to zero after the resonance region.

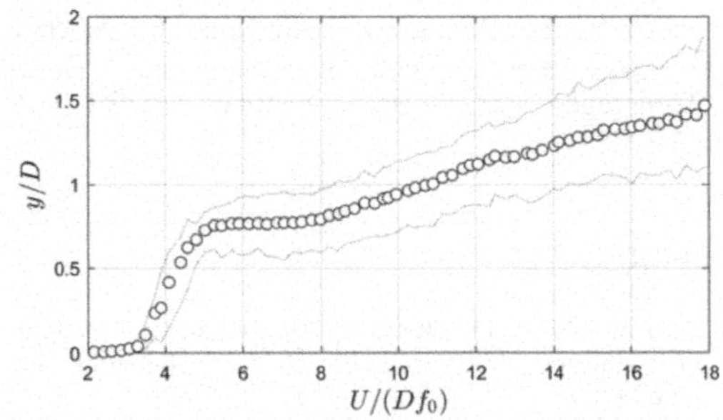

Figure 17. WIV physical experiment in High Re $m^{*}\zeta$ = 0,0118 - Reduced Amplitude.

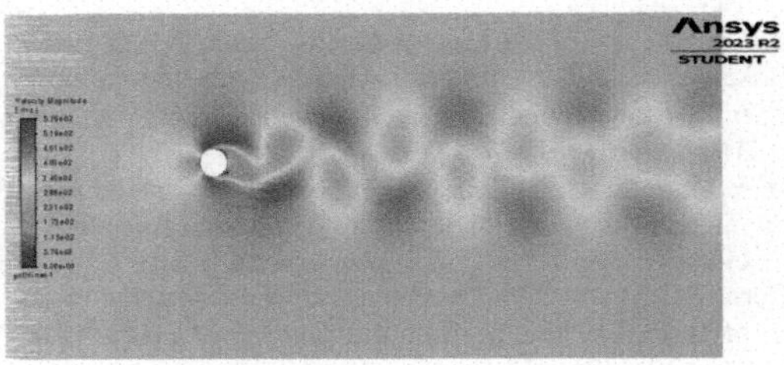

Figure 18. Velocity Contour around the cylinder for f_N = 1.4 Hz.

Lift - The time series for the lift oscillates differently for the three different branches. The first one varies from -0.02 N to 0.02 N at beginning but it doubles at f_N = 1 Hz, taking a great time for amplitude stabilization. The frequency is 0.4 Hz in this branch. For the second branch where the motion reached resonance for f_N = 0.7 Hz, the amplitude of lift goes back to 0.02 N and the lift's frequency is 0.49 Hz. However, for f_N = 0.5 Hz the lift magnitude is reduced in seven times. Besides the lift's frequency, it is reduced to 0.44 Hz. For the third branch, the lift amplitude doubles and it increases as the natural frequencies decrease until 0.0081 N. Moreover, the lift's frequency starts with 0.46 Hz for f_N = 0.3 Hz and falls to 0.37 Hz for f_N = 0.03 Hz.

5.2 *Wake-induced vibration*

Thus, with all these data gathered, the reduced amplitude can be plotted for each natural frequency simulated along with the reduced velocity, equation 3, as shown in Figure 21.

From the reduced amplitudes it's possible to conclude existence of three branches, a initial where the amplitudes increases from very low values to the resonance peak, a region of resonance and the third branch where the reduced amplitudes stabilizes to a value of 0.51. This

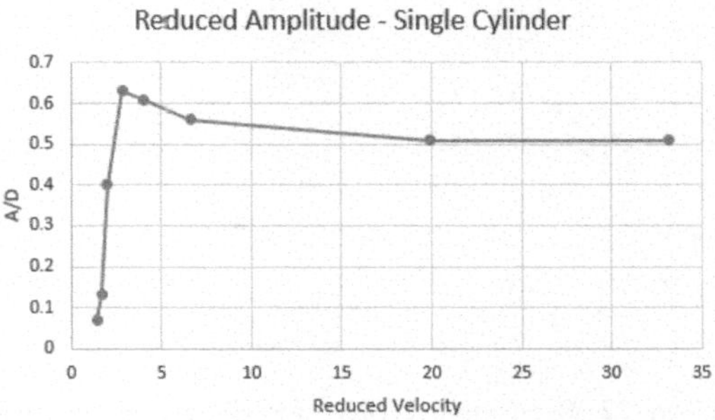

Figure 19. Cylinder's Vibration for each natural frequency simulated on Fluent for *Re* = 718.2.

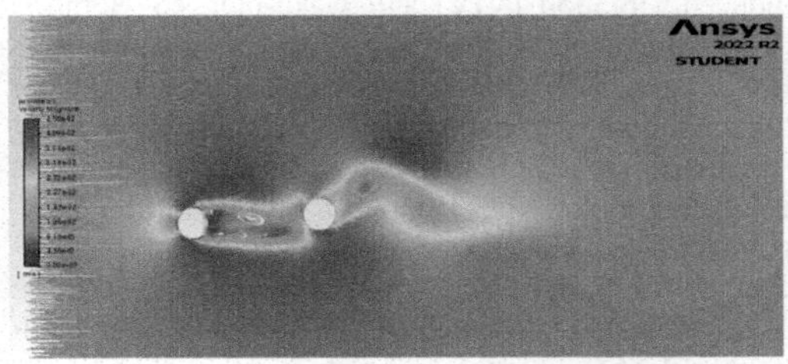

Figure 20. Velocity Countour around a pair of cylinders for f_N = 1.0 Hz.

outcome shows a similar result from the simulation carried out by Assi (2005) with a low mass parameter of 1.82 for *S*/*D* equals to 2, 3 and 4, which the amplitudes increases a little after the resonance.

Lift - The time series of the lift is different for the three branches. Initially it oscillates around -0.02 N to 0.02 N and the frequency is 0.32 Hz. At the resonance, the oscillation of the lift has very different profile and its amplitude is lower than the other cases, with 0.0007 N and the frequency is 0.31 Hz. For the third branch, the amplitude increases ten times the case from the resonance and the frequency sticks the same.

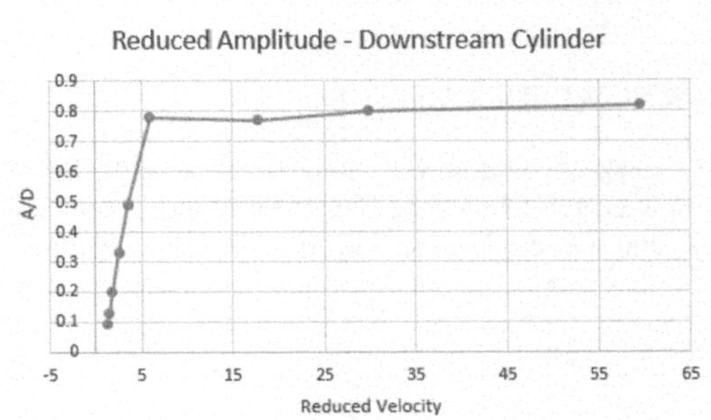

Figure 21. Downstream cylinder's vibration for each natural frequency simulated on Fluent for *Re* = 644.1.

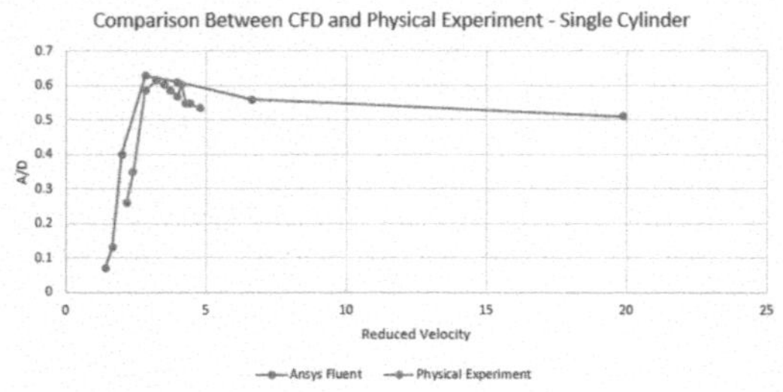

Figure 22. Comparison between results from the experimental test and CFD test for the single cylinder case.

6 CONCLUSIONS

The study concludes that the phenomena of vortex-induced vibration (VIV) and wake-induced vibration (WIV) could be successfully observed in the Ansys Fluent application for low Reynolds number simulations. A comparison between the results of the VIV simulation and the outcomes of physical experiments conducted in the NDF laboratory revealed differences, particularly in the first branch for $f_N > 0.7$ Hz. However, the difference was small at resonance, with the peak computed experimentally being only $0.01D$ smaller than the Fluent value.

Regarding the WIV results, analyzing the disparity between the laboratory results and the CFD-produced results was challenging due to limitations in the physical experiment's ability to produce lower natural frequencies. The results exhibited a common profile for low mass parameters, a gradually reducing amplitude after resonance, with peaks higher than those observed in the VIV experiment at higher reduced velocities. However, for high reynolds simulations, there was a significant deviation between the physical experiment and the one produced in Fluent.

The complexity of flow induced vibration is explained by several limitations of the CFD, including truncation errors arising from the use of discrete equations and simplifications in the modeling of restoring and damping forces. Future work could be performed, including testing higher reynolds numbers in CFD, introducing two degrees of freedom for the cylinders and adding more cylinders to the domain to gain a deeper understanding of VIV and WIV phenomena.

ACKNOWLEDGEMENTS

This work was performed within the scope of the Strategic Research Plan of the Centre for Marine Technology and Ocean Engineering (CENTEC), which is financed by the Portuguese Foundation for Science and Technology (Fundação para a Ciência e Tecnologia - FCT) under contract UIDB/UIDP/00134/2020.

REFERENCES

Assi, G. R. (2009). *Mechanisms for flow-induced vibration of interfering bluff bodies*. Ph. D. thesis, Imperial College London.

Assi, G. R. d. S. (2005). Estudo experimental do efeito de interferência no escoamento ao redor de cilindros alinhados. Master's thesis, Universidade de São Paulo.

Batchelor, C. K. (2000). *An introduction to fluid dynamics*. Cambridge university press.

Bearman, P. W. (1984). Vortex shedding from oscillating bluff bodies. *Annual review of fluid mechanics 16*(1), 195–222.

Blevins, R. D. (1977). Flow-induced vibration. *New York*.

Bokaian, A. & F. Geoola (1984). Wake-induced galloping of two interfering circular cylinders. *Journal of Fluid Mechanics 146*, 383–415.

Gao, G.-z. & L.-d. Zhu (2017). Nonlinear mathematical model of unsteady galloping force on a rectangular 2: 1 cylinder. *Journal of Fluids and Structures 70*, 47–71.

Gerrard, J. (1966). The mechanics of the formation region of vortices behind bluff bodies. *Journal of fluid mechanics 25*(2), 401–413.

Govardhan, R. & C. Williamson (2000). Modes of vortex formation and frequency response of a freely vibrating cylinder. *Journal of Fluid Mechanics 420*, 85–130.

Granger, S. & M. Paidoussis (1996). An improvement to the quasi-steady model with application to cross-flow-induced vibration of tube arrays. *Journal of Fluid Mechanics 320*, 163–184.

Khalak, A. & C. H. Williamson (1999). Motions, forces and mode transitions in vortex-induced vibrations at low mass-damping. *Journal of fluids and Structures 13*(7-8), 813–851.

Lin, J.-C., J. Towfighi, & D. Rockwell (1995). Instantaneous structure of the near-wake of a circular cylinder: on the effect of reynolds number. *Journal of Fluids and Structures 9*(4), 409–418.

Parkinson, G. (1989). Phenomena and modelling of flow-induced vibrations of bluff bodies. *Progress in Aerospace Sciences 26*(2), 169–224.

Williamson, C. H. & R. Govardhan (2004). Vortex-induced vibrations. *Annu. Rev. Fluid Mech. 36*, 413–455.

Yang, X., Z. Yan, H. Fu, L. Wang, & E. Savory (2022). Nonlinear modelling of unsteady wake galloping forces on a circular cylinder in the wake of a fixed cylinder. *Ocean Engineering 250*, 111016.

Zdravkovich, M. (1972). Smoke observations of wakes of tandem cylinders at low reynolds numbers. *The Aeronautical Journal 76*(734), 108–114.

Zdravkovich, M. (1977). Review of flow interference between two circular cylinders in various arrangements.

Advances in Maritime Technology and Engineering – Guedes Soares & Santos (Eds)
© 2024 The Author(s), ISBN 978-1-032-83107-7

Accuracy of the SPH-based solver DualSPHysics to reproduce hydrodynamic forces on bodies in a steady flow

J. González-Cao, J.M. Domínguez, I. Martínez-Estévez, A.J.C. Crespo & M. Gómez-Gesteira
Environmental Physics Laboratory, CIM-UVIGO, Universidade de Vigo, Ourense, Spain

S. Wang & C. Guedes Soares
Centre for Marine Technology and Ocean Engineering (CENTEC), Instituto Superior Técnico, University of Lisboa, Portugal

ABSTRACT: In this work, the hydrodynamics forces on semi-submerged and submerged cylinders under a steady flow are analysed using the numerical code DualSPHysics, which implements the Smoothed Particle Hydrodynamics (SPH) method. The published experimental tests which were conducted in the Shanghai Ship and Shipping Research Institute, for submerged and semi-submerged cylinders under steady flow with several velocities are reproduced. Horizontal (drag) and vertical (lift) hydrodynamics forces for each case are obtained from the numerical code DualSPHysics. The accuracy of the code is obtained by the comparison of the numerical drag and lift forces against the experimental ones. Once the accuracy of the code is assessed, the analysis of more complex problems as, for example, the interaction of waves and currents with floating bodies can be analysed.

1 INTRODUCTION

Coastal and marine engineering studies often require an in-depth analysis of the physical processes that characterize wave formation, propagation, and wave-structure interaction. Moreover, the effects of the interaction of currents alone or combined with waves with structures should also be taken into account for an appropriate analysis of these marine processes. Physical and numerical modelling are emerging as valuable tools to achieve these goals. Both tools have their advantages and drawbacks. On the one hand, physical model testing constitute is the most widely used approach for analysis purposes, although it is costly, time-consuming and strongly dependent on the accuracy of the measurement equipment. On the other hand, numerical models are a useful alternative to be used in the analysis of coastal and marine engineering. One of the main advantages of numerical modelling is its capability to simulate any scenario with reduced costs. Moreover, numerical models can provide information on physical quantities that could be difficult or almost impossible to measure in scaled models or prototypes.

Generally speaking, there are two main approaches to numerical modelling of flow fields involved in coastal and marine engineering: Eulerian (mesh-based) and Lagrangian (mesh-free). Traditional Computational Fluid Dynamics (CFD) techniques use the Eulerian approach. These mesh-based models analyse the flow field at a fixed position of the domain, hence they might require the definition of fine meshes to discretise the domain to study, for example, the wave-structure interaction problems (Kleefsman et al. 2005, Higuera et al. 2013, 2014). The Lagrangian methods, also called mesh-free methods, analyse the flow by following the movement of the fluid particles. In this case, the fluid is previously discretised in particles. These methods allow for overcoming part of the drawbacks of the mesh-based schemes, although they usually have a higher computational cost.

The Smoothed Particle Hydrodynamics (SPH) (Violeau, 2012) or the Particle Finite Element Method (PFEM) (Oñate et al. 2011) are examples of mesh-free schemes. In SPH no special tracking is used to detect the free surface and the domain is multiply connected due to the Lagrangian nature of the method. Consequently, large deformations can be efficiently treated since there is no mesh distortion, making SPH an ideal technique to study highly non-linear phenomena. The reader can find a very interesting comparison of these two approaches, both mesh-based and mesh-free methods, in González-Cao et al. (2019), where an analysis of the violent impacts of waves on marine structures is carried out.

Usually, marine structures such as floating tunnels, pipelines or even ships are affected by violent impacts derived not only by the effect of waves but from the combined effect of waves and currents. Therefore, it is essential to analyse such problems from the safety point of view of relevant marine structures. Numerical modelling

DOI: 10.1201/9781003508779-12

arises as a very valuable tool to carry out this task. To do this, the study of hydrodynamic forces onto submerged and semi-submerged cylinders reveals a very good option to face this task.

Usually, the study of flow affecting cylinders is focused on three categories: i) cylinders fully submerged in infinite depth water, ii) cylinders near the free surface, and iii) cylinders partially submerged. Some works focused on fully submerged cylinders are Morison et al. (1950) and Morison et al. (1953), who developed the Morison equation for the calculation of the horizontal forces onto the cylinders. Also, Keulegan (1958), Mercier (1973), Williamson (1985) and Tatsumo and Bearman (1990) studied the effect of oscillatory flows in fully submerged cylinders in experimental campaigns. Regarding the studies of cylinders near the free surface, fully or partially submerged, and focused on oscillatory flows, the reader can find very interesting information in the original works of Roshko (1975), Bearman and Zdravkovich (1978). Not only oscillatory flows with partially submerged cylinders were studied. For example, Lei et al., (1999) carried out some experiments focused on steady flows. More recently Kristiansen and Faltinsen, (2008) and Kristiansen (2010) analysed the interaction of wave and semi-submerged cylinders. Combined flows (steady and oscillatory) were experimentally studied by Summer et al., (1992). Another interesting work where cylinders installed in fish farming are analysed is Fu et al., (2013). Recently, Ren et al., (2019) studied the steady flow conditions in partially submerged cylinders and Ren et al. (2020) analysed similar problems but focused on oscillatory flows.

Previous examples are focused on experimental tests. The alternative CFD methodologies were also applied to such problems. Examples of these are the works conducted by Reich et al. (1997), Triantafyllou and Dimas (1989) and Sheridan et al., (1995).

This work presents a study of the numerical analysis of the hydrodynamic forces of fully and semi-submerged cylinders in steady flow utilizing the numerical model DualSPHysics. The reference data used to validate the numerical results were obtained from a set of experiments carried out at the Shanghai Ship and Shipping Research Institute (Ren et al., 2019). This work is structured as follows: Section 2 includes a brief description of the numerical model; Section 3 describes the physical model and the numerical setup Section 4 shows the numerical results of hydraulic forces exerted onto the cylinders and the drag coefficients; finally, and Section 5 concludes this work.

2 NUMERICAL MODEL: DUALSPHYSICS

DualSPHysics (Domínguez et al., 2022) is a numerical model based on the Smoothed Particle Hydrodynamics (SPH) method that was developed to apply the SPH methodology to real engineering problems. This model is open source and can be freely downloaded from http://www.dual.sphysics.org. The software can be run on either CPUs (Central Processing Units) or GPUs (Graphics Processing units). GPUs offer greater computing power than CPUs due to their highly powerful parallel computing, and, therefore, they are a good choice to accelerate SPH computations. The code also includes pre-processing and post-processing tools to make easier the tasks of generating the computational domain, defining model parameters and also the analysis of the numerical results.

SPH is a Lagrangian mesh-less method where the fluid (continuum) is discretised into a set of particles. For each particle, physical quantities (such as position, velocity, density, and pressure) are computed as an interpolation of the values of the neighbour particles. The contribution of these neighbours is weighted using a kernel function (W) that measures the contribution from the rest of the particles in terms of the particle spacing. This distance between particles is normalized using the smoothing length (h), which is the characteristic length that defines the area of influence of the kernel (Ω). The kernel presents compact support so that the contribution of particles beyond a cut-off distance (here $2h$) is not considered.

The mathematical fundamentals of SPH are based on integral interpolants. Any function F can be computed by the integral approximation.

$$F(\boldsymbol{r}) = F(\boldsymbol{r}')W(\boldsymbol{r} - \boldsymbol{r}', h)d\boldsymbol{r}' \tag{1}$$

where $\boldsymbol{r}$ and $\boldsymbol{r}$*'* are the positions where the function is computed and a temporary position variable, respectively. This function F can be expressed in a discrete form based on particles. Thus, the approximation of the function is interpolated at particle a and the summation is performed over all the particles within the region of compact support of the kernel:

$$F(\boldsymbol{r}_a) = \sum\nolimits_b F(\boldsymbol{r}_b)W(r_a - r_b, h)\frac{m_b}{\rho_b} \tag{2}$$

where the volume associated with the neighbouring particle b is m_b/ρ_b, being m_b and ρ_b mass and density respectively. The kernel function employed in this work is the Cubic Spline.

The discrete (Lagrangian form) SPH system of equations of a weakly compressible fluid (Navier-Stokes) is, according to Monaghan (1992):

$$\begin{cases} \frac{d\rho_a}{dt} = \sum_b m_b \boldsymbol{v}_{ab} \nabla_a W_{ab} + D_a \\ \frac{d\boldsymbol{v}_a}{dt} = -\sum_b m_b \left(\frac{P_b + P_a}{\rho_b\,\rho_a}\right)\nabla_a W_{ab} + \boldsymbol{g} + \Gamma_{\boldsymbol{a}} \\ \frac{d\boldsymbol{r}_a}{dt} = \boldsymbol{v}_{\boldsymbol{a}} \end{cases} \tag{3}$$

where t is time, r position, v velocity, P pressure, ρ density, m mass, g gravitational acceleration and W_{ab} the kernel function that depends on the distance between particles a and b. Γ_a describes the dissipation:

$$\Gamma_a = \sum_b m_b \frac{4\upsilon_0 \boldsymbol{r_{ab}} \nabla_a W_{ab}}{(\rho_a+\rho_b)(r_{ab}^2+0.01h^2)} \boldsymbol{v}_{ab} + \\ + \sum_b m_b \left(\frac{\tilde{\tau}_a^{ij}+\tilde{\tau}_b^{ij}}{\rho_a \rho_b}\right) \nabla^i W_{ab} \quad (4)$$

In Equation (4) υ_0 is the so-called laminar viscosity introduced by Lo & Shao, (2002). The second term of the right-hand side of Equation (4) describes the sub-particle scale model (SPS). The term D_a included in Equation (4) is the density diffusion term and is included to reduce fluctuations in the density field according to Fourtakas et al. (2019).

$$D_a = 2\delta h c_s \sum_b \left(\rho_{ba}^T - \rho_{ba}^H\right) \frac{r_{ab} \nabla_a W_{ab}}{r_{ab}^2} \frac{m_b}{\rho_b} \quad (5)$$

where c_s is the speed of sound, δ is equal to 0.1 (diffusive control coefficient) and superscripts T and H stand for the total and hydrostatic components of the density. Some inconsistencies arise from this approach but a correction was introduced by Fourtakas et al. (2019) obtaining consistent pressure fields near boundaries.

The system of equations is closed using Tait's equation of state to describe the implemented weakly compressible SPH formulation. This allows computing pressure from density values

$$P = B\left[\left(\frac{\rho}{\rho_0}\right)^\gamma - 1\right] \quad (6)$$

where γ=7 is the polytropic constant and B=$c_s^2\rho_0/\gamma$, ρ_0 the reference density of the water and c_s is the numerical speed of sound.

The modified dynamic boundary conditions (mDBC) are implemented according to English et al., (2022). This implementation improves previous ones (DBC) since it allows more accurate pressure computations and also reduces the gaps between boundary and fluid particles. The reader can find a detailed description of this implementation in English et al. (2022), Capasso et al. (2021), Altomare et al. (2021) and Suzuki et al. (2022).

The new functionality to define inlet and outlet conditions implemented in DualSPHysics was presented by Tafuni et al. (2018). This implementation allows dealing with cases where an inlet (or outlet) flow past the limits of the domain. In these cases, an inflow velocity usually needs to be prescribed at the inlet while other conditions (velocity, pressure, etc.) can be obtained from the fluid in the simulation at the outlet boundary. Figure 1 shows a scheme of the open boundary algorithm implemented in DualSPHysics to deal with such kinds of problems.

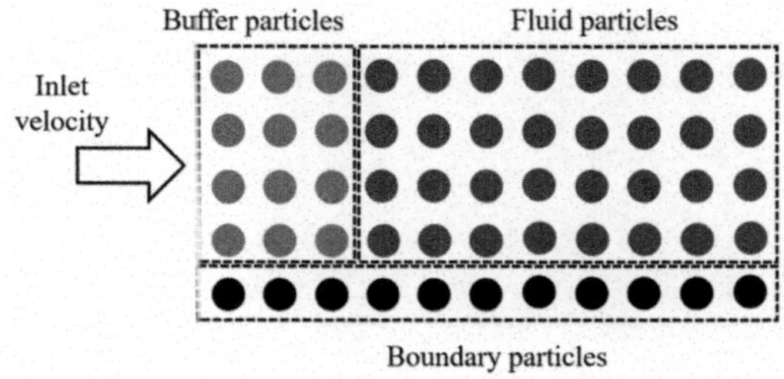

Figure 1. Scheme of the implemented open boundary algorithm (Tafuni et al., 2018).

The buffer width is chosen to be equal to or exceed the radius of the SPH kernel function to ensure that the full kernel support is only filled with the fluid particles in the proximity of an inlet or outlet. A buffer layer can be both an inlet and an outlet depending on the direction of the flow velocity. This feature allows to solve complex cases where the flow direction changes and also allows the reduction of the extension of the domain.

3 CASE OF STUDY

3.1 *Physical model*

The physical model consists of a 192 m x 10 m (length x width) water tank with a water depth equal to 4.2 m. A towing carriage with vertical and horizontal displacement is installed on the top of the water tank. This carriage moves a test model consisting of a 2 m length rigid cylinder model of radius (R_e) equal to 0.125 m. Therefore, this cylinder can move both in vertical and horizontal directions. Two groups of tests were carried out. Firstly, the cylinder is fully submerged and then it is moved with constant horizontal velocity. The range of velocities generates different regimes of flow from sub-critical to super-critical. This is called a validation group. For the second group of tests, the cylinder is semi-submerged and then it is moved again with constant horizontal velocity. Figure 2 shows a sketch of the experimental water tank along with the conditions of each group test.

From these tests, horizontal (drag) and vertical (lift) hydrodynamic forces are obtained. A detailed description of both the device and the experimental data obtained from these tests is shown in Ren et al., (2019), Ren et al., (2020) and Hu et al., (2023).

3.2 *Numerical setup*

First, the lattice nodes around the object are defined and then, particles are created only at the nodes needed to draw the desired geometry. Only the initial size of the lattice needs to be defined, which will be the initial inter-particle spacing, *dp*.

Figure 3 shows an example of the discretisation of a generic square box with a water depth equal to

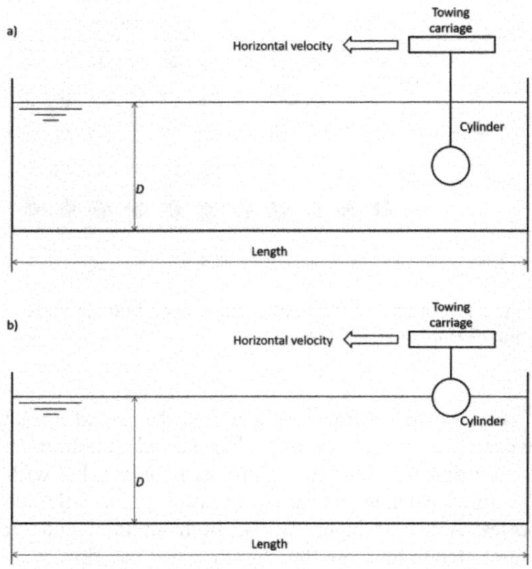

Figure 2. Sketch of the physical water tank. A) stands for fully submerged tests and b) stands for semi-submerged tests.

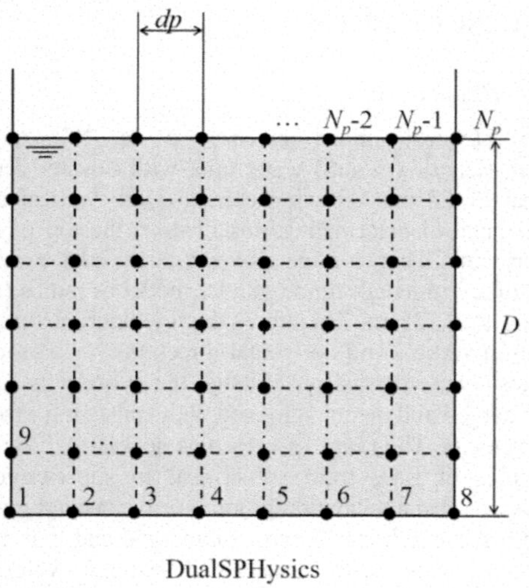

Figure 3. Discretization of a square domain with a water height of *D* using DualSPHysics.

D using DualSPHysics. The initial inter-particle spacing defined with DualSPHysics is equal to *dp* leading to *Np* particles of water.

The numerical domain defined to reproduce the experimental tests carried out in the Shanghai Ship and Shipping Research Institute is shown in Figure 4. In this case, it is a 19 m length and 4 m depth 2D water tank. The radius of the cylinder is equal to 0.125 m. In the numerical simulations, the cylinder is fixed, and a constant flow horizontal velocity is defined in the inlet boundary of the domain. This velocity is also defined as an initial condition in the fluid domain and in the outlet boundary. This configuration allows defining a smaller numerical domain than the experimental one since in the latter the cylinder moves with constant velocity in a very long domain. This configuration reduces dramatically the runtimes since the number of fluid particles defined with a value of *dp* is reduced (the lesser length, the lesser number of particles). Figure 4 shows the numerical configuration defined for the fully submerged cases. In these cases, the centre of the cylinder is located at a depth equal to 2 m. For the semi-submerged cases, the centre of the cylinder is located at a depth equal to 0 m. Two values of initial interparticle distances (*dp*) are defined in these simulations: 0.050 m and 0.025 m. The total number of initial fluid particles is 33 and 127 k, respectively. The values of the main parameters of DualSPHysics are shown in Table 1. Laminar viscosity is defined for the simulations and, for the fully submerged cases a calibration step is carried out by defining different values of the viscosity.

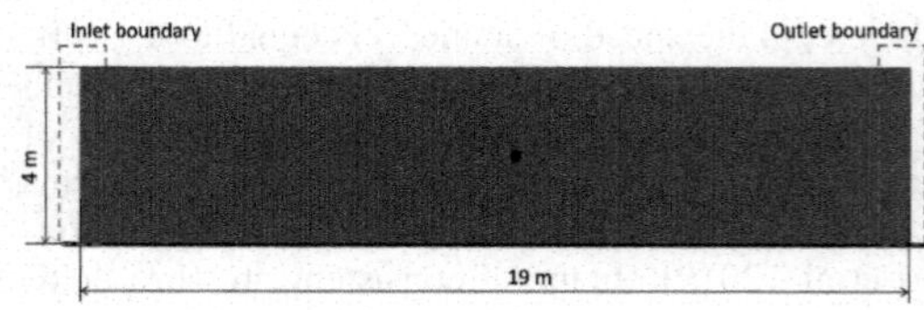

Figure 4. Sketch of the initial configuration of the numerical domain for the fully-submerged case. The blue area stands for the fluid domain and the black areas represent the fixed boundaries of the domain.

Table 1. Main Parameters of DualSPHyiscs.

Name	Value
Coefsound	20
ViscoBoundFactor	1
coefh	1.2
cflnumber	0.2
shifting	3
ShiftCoef	-2
Coarse	3.5

The physical time simulated in the numerical simulations is equal to 30 s and the sampling frequency is equal to 10 Hz. For the fully-submerged case, the inlet velocities (*V*) analysed in this work are 0.6, 1.0 1.3, 3.0 and 3.6 m s^{-1}.

4 RESULTS AND DISCUSSION

The numerical results of the draft coefficient, C_d (Equation (7)), of the fully-submerged cases obtained with both values of *dp*, along with experimental data obtained from Ren et al. (2019) as a function of the inlet velocity (*V*) are depicted in Figure 5.

$$C_d = \frac{F_x}{\rho \, R_e \, L \, V^2} \tag{7}$$

where F_x is the horizontal force, ρ (1000 kg m^{-3}) is the density of water, R_e (0.125 m) is the radius of the cylinder, L is the transversal length of the cylinder, which is set to 1m in the 2-D case, and V is the inlet velocity.

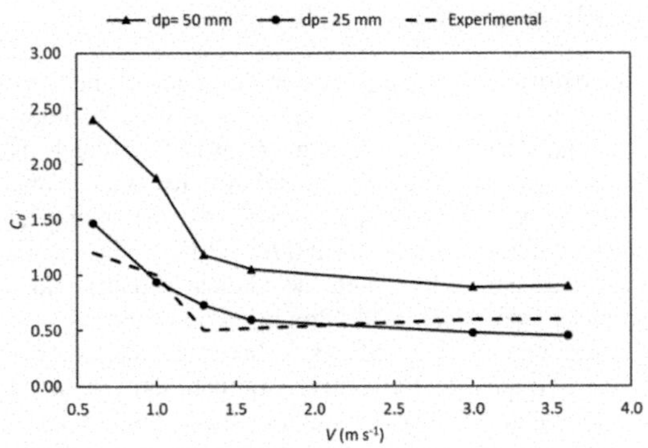

Figure 5. Numerical values of Cd obtained with dp equal to 50 mm (solid line with triangles) and 25 mm (solid line with circles) along with the experimental results obtained from Ren et al. (2019) as a function of the inlet velocity.

The numerical results of C_d presented in Figure 5, show a reasonable agreement between the values obtained with *dp* equal to 25 mm with the experimental ones. The values of C_d obtained with *dp* of 50 mm are far away from experimental data. The values of *rmsd* for each *dp* computed following Equation (8) are 0.72 (50 mm) and 0.17 (25 mm). Therefore, according to these values, the methodology can be considered validated.

$$rmsd^{dp} = \sqrt{\frac{\sum_{i=1}^{N} \left[C_d^{dp}(i) - C_d^{ref}(i)\right]^2}{N}} \tag{8}$$

In Equation (8) N is the number of tests, $C_d^{dp}(i)$ is the numerical value of C_d for *dp* and $C_d^{ref}(i)$ is the reference (experimental) value of C_d.

The mean values of the laminar viscosity for these two sets of results (*dp* equal to 50 mm and 25 mm) are 3.4x10^{-6} and 5.4x10^{-6}, respectively. Therefore, the value of *dp* equal to 25 mm along with a laminar viscosity of 5.4x10^{-6} can be considered as the "reference configuration" to analyse the semi-submerged cases. Figure 6 shows a snapshot of the numerical simulations of the fully-submerged case considering an inlet velocity of 1.3 ms^{-1}. The numerical results show that the velocity field is homogeneous in the entire domain. Near the cylinder, the velocity increases above and below the cylinder due to the reduction of the section. There is a small non-symmetric variation of the velocity field above and below the cylinder. This is due to the effect of the bottom of the numerical tank. In the upper part, there is the free surface of the domain, and the fluid particles do not interact with other particles. On the other hand, in the opposite part, it is located at the bottom of the domain and the fluid particle interacts with solid ones. Furthermore, the velocity decreases in front and behind the cylinder. A small instability is shown at the beginning of the domain near the inlet boundary. This is mainly due to numerical aspects, but it does not affect the structure of the cylinder, so it can be neglected. Near the cylinder surface, the velocity field is very homogenous and shows the areas of zero velocity above the cylinder.

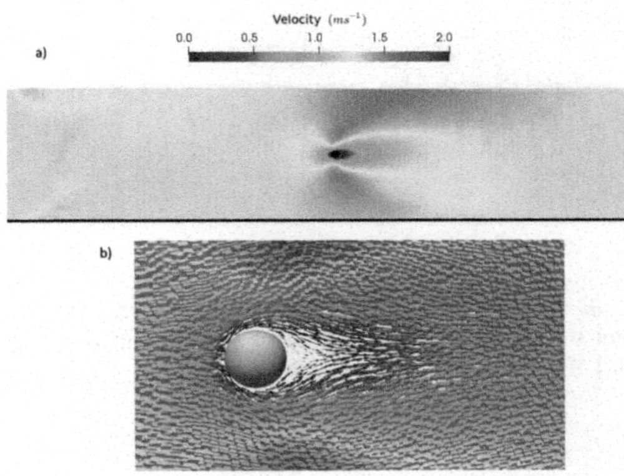

Figure 6. Snapshot of the numerical simulation of a fully-submerged cylinder with V equal to 1.3 ms-1. a) General view; b) vector velocity field near the cylinder surface.

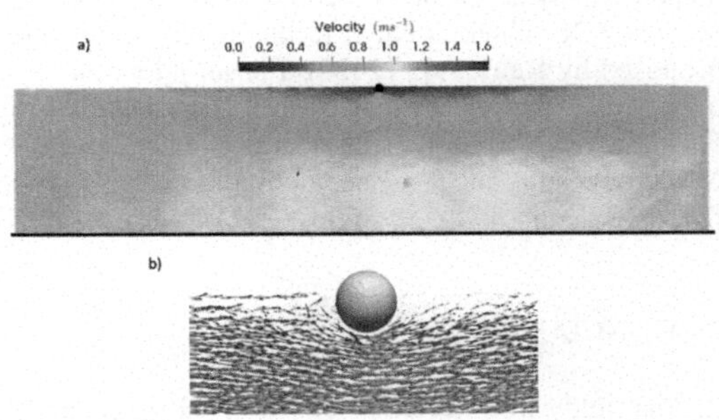

Figure 7. Snapshot of the numerical simulation of a semi-submerged cylinder with inlet velocity equal to 0.8 ms-1. a) General view; b) vector velocity field near the cylinder surface.

Once the numerical results in terms of the drag coefficient of the fully submerged case are in good agreement with the reference data, a case of a semi-submerged cylinder will be explored to analyse the capability of DualSPHysics when dealing with these problems. Figure 7 shows a snapshot of the numerical results obtained with the code for the semi-submerged case with an inlet velocity of 0.8 ms^{-1}.

Again, the general velocity field obtained for the semi-submerged case is homogeneous in the numerical domain. Close to the cylinder, the velocity reduces to near zero, and beneath the cylinder, the flow accelerates. This is due to the reduction of the section. Near the inlet boundary the velocity remains equal to 0.8 ms^{-1} and some variations can be observed near the outlet boundary. Macroscopically, the dynamics of the numerical flow are similar to the experimental observations. The time series of the hydraulic horizontal force is depicted in Figure 8.

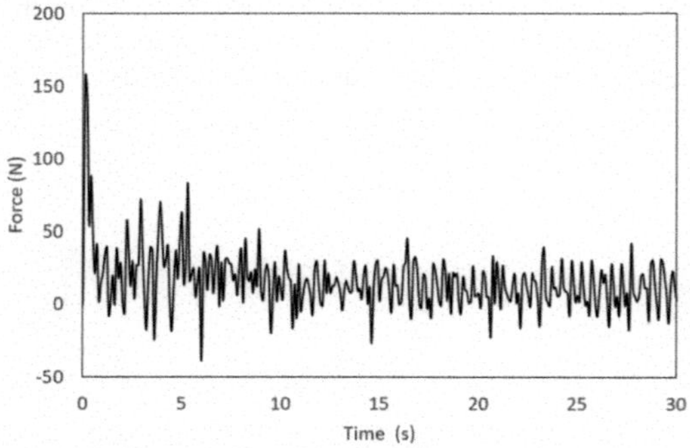

Figure 8. Time series of the horizontal force for the semi-submerged cylinder with inlet velocity equal to 0.8 ms-1.

At the beginning of the simulation, from 0 to 10 s, there is a period where the flow is irregular since the movement of the flow needs a stabilisation period. From time 10 s to the end of the simulation (30 s) the force exerted onto the cylinder is very homogeneous. The mean of the numerical horizontal force is near 13.8 N, which is equal to the experimental one proposed by Ren et al., (2020). For an inlet velocity of 1.2 ms^{-1}, the experimental mean force is equal to 53 N, while the numerical result is equal to 51.4 N. These results show the accuracy of DualSPHysics for reproducing this kind of flow.

5 CONCLUSIONS

In this study, the accuracy of the SPH-based code DualSPHysics has been studied for several cases involving fully and semi-submerged cylinders under steady flow conditions. Experimental results of drag forces for several velocities were used as reference data.

The main conclusions of the study are:

- The code is able to reproduce the C_d (draft coefficient) vs. velocity curve of fully-submerged cylinders under steady flow conditions using laminar viscosity and with an appropriate value of initial interparticle distance (*dp*), that is, a good spatial resolution.
- Using the configuration defined in the fully-submerged case the drag forces of two cases of semi-submerged cylinders were reproduced satisfactorily.
- The dynamics of the flow associated with both the fully and the semi-submerged cases are in accordance with experimental observations.

Therefore, DualSPHysics arises as a good option to reproduce such phenomena and show the ability to analyse problems such as the interaction of waves/currents with floating bodies.

ACKNOWLEDGEMENTS

This work was performed within the scope of the Strategic Research Plan of the Centre for Marine Technology and Ocean Engineering (CENTEC), which is financed by the Portuguese Foundation for Science and Technology (Fundação para a Ciência e Tecnologia) under contract UIDB/UIDP/00134/2020. This work was also funded by Xunta de Galicia (Spain) under project ED431C 2021/44 "Programa de Consolidación e Estructuración de Unidades de Investigación Competitivas (Grupos de Referencia Competitiva)" financed by Xunta de Galicia, Consellería de Cultura, Educación e Universidade, Spain. I. Martínez-Estévez acknowledges funding from Xunta de Galicia, Spain under "Programa de axudas á etapa predoutoral da Consellería de Cultura, Educación e Universidades da Xunta de Galicia" (ED481A-2021/337).

REFERENCES

Altomare, C., Suzuki, T., Domínguez, J., Barreiro, A., Crespo, A., Gómez-Gesteira, M. 2015. Numerical wave dynamics using Lagrangian approach: wave generation and passive & active wave absorption. Proceedings of the 10th SPHERIC International Workshop, Parma, Italy.

Altomare, C., Gironella, X., Crespo, A.J., 2021. Simulation of random wave overtopping by a WCSPH model. Appl. Ocean Res. 116, http://dx.doi.org/10.1016/j.apor.2021.102888.

Bearman, P.W.; Zdravkovich, M.M. Flow around a circular cylinder near a plane boundary. 1978. J. Fluid Mech, 89, 33–47.

Capasso, S., Tagliafierro, B., Güzel, H., Yilmaz, A., Dal, K., Kocaman, S., Viccione, G., Evangelista, S., 2021. A numerical validation of 3D experimental dam-break wave interaction with a sharp obstacle using DualSPHysics. Water 13 (15), http://dx.doi.org/10.3390/w13152133.

Canelas, R.B., Domínguez, J.M., Crespo, A.J.C., Gómez-Gesteira, M., Ferreira, R.M.L. 2015. A Smooth Particle Hydrodynamics discretization for the modelling of free surface flows and rigid body dynamics. International Journal for Numerical Methods in Fluids, 78(9): 581–593. DOI: 10.1002/fld.4031

Crespo, A.J.C., Gómez-Gesteira, M., Dalrymple, R.A. 2007. Boundary conditions generated by dynamic particles in SPH methods. Computers, Materials and Continua, 5 (3): 173–184.

Crespo, A.J.C., Domínguez, J.M., Rogers, B.D., Gómez-Gesteira, M., Longshaw, S., Canelas, R., Vacondio, R., Barreiro, A., García-Feal, O. 2015. DualSPHysics: Open-source parallel CFD solver based on Smoothed Particle Hydrodynamics (SPH). Computer Physics

Communications, 187: 204–216. DOI: 10.1016/j.cpc.2014.10.004.
Domínguez, J.M., Fourtakas, G., Altomare, C., Canelas, R. B., Tafuni, A., García-Feal, O., Martínez-Estévez, I., Mokos, A., Vacondio, R., Crespo, A.J.C., Rogers, B.D., Stansby, P.K., Gómez-Gesteira, M., 2022. DualSPHysics: From fluid dynamics to multiphysics problems. Comput. Part. Mech. 9 (5), 867–895. http://dx.doi.org/10.1007/s40571-021-00404-2.
English, A., Domínguez, J., Vacondio, R., Crespo, A., Stansby, P., Lind, S., Chiapponi, L., Gómez-Gesteira, M., 2022. Modified dynamic boundary conditions (mDBC) for general-purpose smoothed particle hydrodynamics (SPH): application to tank sloshing, dam break and fish pass problems. Comput. Part. Mech. 9 (5), 1–15. http://dx.doi.org/10.1007/s40571-021-00403-3, URL: https://www.scopus.com/inward/record.uri?eid=2-s2.0-85103385297&doi=10.1007%2fs40571-021-00403-3&partnerID=40&md5=9c42490a240326320537466675df4c27.
Fourtakas, G., Dominguez, J.M., Vacondio, R., Rogers, B. D., 2019. Local uniform stencil (LUST) boundary condition for arbitrary 3-D boundaries in parallel smoothed particle hydrodynamics (SPH) models. Comput. & Fluids 190, 346–361. http://dx.doi.org/10.1016/j.compfluid.2019.06.009.
Fu, S.; Xu, Y.; Hu, K.; Zhang, Y. Experimental investigation on hydrodynamics of floating cylinder in oscillatory and steady flows by forced oscillation test. 2013. Mar. Struct., 34, 41–55.
González-Cao, J., Altomare, C., Crespo, A.J.C., Domínguez, J.M., Gómez-Gesteira, M., Kisacik D., 2019. On the accuracy of DualSPHysics to assess violent collisions with coastal structures. Computers and Fluids, 179: 604–612. DOI: 10.1016/j.compfluid.2018.11.021.
Hu T., Ren H., Shen J., Niu Z., Zhang M., Xu Y., Song B., Sun T. 2023. Experimental investigation on hydrodynamic forces of semi-submerged cylinders in combined steady flow and oscillatory flow. Ocean Engineering, 268, art. no. 113612. DOI: 10.1016/j.oceaneng.2022.113612.
Jasak, H., 1996. Error Analysis and Estimation for the Finite Volume Method with Applications to Fluid Flows, Ph.D. thesis, Imperial College of Science, Technology & Medicine, London, 1996.
Jasak, H. 2009. OpenFOAM: Open source CFD in research and industry. International Journal of Naval Architecture and Ocean Engineering, 1 (2): 89–94. DOI: 10.3744/JNAOE.2009.1.2.089.
Higuera, P., Lara, J.L., Losada, I.J. 2013. Realistic wave generation and active wave absorption for Navier-Stokes models. Application to OpenFOAM®. Coastal Engineering, 71: 102–118. DOI: 10.1016/j.coastaleng.2012.07.002.
Higuera, P., Lara, J.L., Losada, I.J. 2014. Three-dimensional interaction of waves and porous coastal structures using OpenFOAM®. Part I: Formulation and validation. Coastal Engineering, 83: 243–258. DOI: 10.1016/j.coastaleng.2013.08.010.
Keulegan, G.H. Forces on cylinders and plates in an oscillating fluid. 1958. J. Res. Natl. Bur. Stand. Res. Pap, 2857, 423–440.
Kleefsman, K.M.T., Fekken, G., Veldman, A.E.P., Iwanowski, B., Buchner, B. 2005. A Volume-of-Fluid based simulation method for wave impact problems. Journal of Computational Physics, 206 (1): 363–393. DOI: 10.1016/j.jcp.2004.12.007.
Kristiansen, D.; Faltinsen, O.M. A study of wave loads on fixed horizontal cylinders in the free surface. In Proceedings of the 8th International Conference on Hydrodynamics, Nates, France, 30 September–3 October, 2008.
Kristiansen, D. Wave Induced Effects on Floaters of Aquaculture Plants; IMT: Trondheim, Norway, 2010.
Lo, E.Y.M., Shao, S., 2002. Simulation of near-shore solitary wave mechanics by an incompressible SPH method. Appl. Ocean Res. 24 (5), 275–286. http://dx.doi.org/10.1016/S0141-1187(03)00002-6, URL: https://www.sciencedirect.com/science/article/pii/S0141118703000026.
Lei, C.; Cheng, L.; Kavanagh, K. Re-examination of the effect of a plane boundary on force and vortex shedding of a circular cylinder. 1999. J. Wind Eng. Ind. Aerodyn., 80, 263–286.
Mercier, J.A. Large amplitude oscillations of a circular cylinder in low speed stream. Ph.D. Thesis, Stevens Institute of Technology, Hoboken, NJ, USA, 1973.
Molteni, D., Colagrossi, A. 2009. A simple procedure to improve the pressure evaluation in hydrodynamic context using the SPH. Computer Physics Communications, 180 (6): 861–872. DOI: 10.1016/j.cpc.2008.12.004.
Monaghan, J.J. 1992. Smoothed particle hydrodynamics. Annual Review of Astronomy and Astrophysics, 30 (1): 543–574. DOI: 10.1146/annurev.aa.30.090192.002551.
Monaghan, J.J., Kos, A., Issa, N. 2003. Fluid motion generated by impact. Journal of Waterway, Port, Coastal and Ocean Engineering, 129 (6): 250–259. DOI: 10.1061/(ASCE)0733-950X(2003)129:6(250).
Monaghan, J.J. 2005. Smoothed particle hydrodynamics. Reports on Progress in Physics, 68 (8): 1703–1759. DOI: 10.1088/0034-4885/68/8/R01.
Morison, J.R.; Johnson, J.W.; Schaaf, S.A. The Force Exerted by Surface Waves on Piles. 1950. J. Pet. Technol, 2, 149–154, doi:10.2118/950149-G.
Morison, J.R.; Johnson, J.W.; O'Brien, M.P. Experimental studies of forces on piles. 1953. Coast. Eng. Proc, 1, 25, doi:10.9753/icce.v4.25.
Oñate, E., Celigueta, M.A., Idelsohn, S.R., Salazar, F., Suárez, B. 2011. Possibilities of the particle finite element method for fluid-soil-structure interaction problems. Computational Mechanics, 48 (3): 307–318. DOI: 10.1007/s00466-011-0617-2.
Reichl, P.; Hourigan, K.; Thompson, M.C. Flow past a cylinder close to a free surface. 1997. J. Fluid Mech., 533, 269–296.
Ren, B., He, M., Dong, P., Wen, H. 2015. Nonlinear simulations of wave-induced motions of a freely floating body using WCSPH method. Applied Ocean Research, 50: 1–12. DOI: 10.1016/j.apor.2014.12.003.
Ren, H., Xu, Y., Zhang, M., Deng, S., Li, S., Fu, S., Sun, H. 2019. Hydrodynamic forces on a partially submerged cylinder at high Reynolds number in a steady flow. Applied Ocean Research, 88, 160–169. DOI: 10.1016/j.apor.2019.04.025
Ren, H., Fu, S., Liu, C., Zhang, M., Xu, Y., Deng, S. 2020. Hydrodynamic forces of a semi-submerged cylinder in an oscillatory flow. Applied Sciences (Switzerland), 10 (18), art. no. 6404. DOI: 10.3390/APP10186404.
Roshko, A. Flow Forces on a Cylinder Near a Wall or Near another Cylinder. In Proceedings of the 2nd US Nation Conference on Wind Engineering Research, Fort Collins, CO, USA, 1–7 January 1975.
Sheridan, J., Lin, J.C., Rockwell, D., 1995. Metastable states of a cylinder wake adjacent to a free surface. Phys. Fluids 7 (9), 2099–2101.
Sumer, B.M., Jensen, B.L., Freds0e, J., 1992. Pressure measurements around a pipeline exposed to combined waves and current. In: 11th Offshore Mechanics and

Arctic Engineering Conf. Canada, Calgary, pp. 113–121. June 7-11, 1992, V-A.
Suzuki, T., García-Feal, O., Domínguez, J.M., Altomare, C., 2022. Simulation of 3D overtopping flow–object–structure interaction with a calibration-based wave generation method with DualSPHysics and SWASH. Comput. Part. Mech. http://dx.doi.org/10.1007/s40571-022-00468-8,.
Tafuni A., Domínguez J.M., Vacondio R., Crespo A.J.C. 2018. A versatile algorithm for the treatment of open boundary conditions in Smoothed particle hydrodynamics GPU models. Computer Methods in Applied Mechanics and Engineering, 342, pp. 604–624 DOI: 10.1016/j.cma.2018.08.004
Tatsuno, M.; Bearman, P.W. A visual study of the flow around an oscillating circular cylinder at low Keulegan–Carpenter numbers and low Stokes numbers. 1990. J. Fluid Mech., 211, 157.
Triantafyllou, G.S.; Dimas, A.A. Interaction of two-dimensional separated flows with a free surface at low Froude numbers. 1989. Phys. Fluids, 1, 1813–1821
Taylor, K.E. 2001. Summarizing multiple aspects of model performance in a single diagram. Journal of Geophysical Research Atmospheres, 106 (D7), art. no. 2000JD900719: 7183–7192. DOI: 10.1029/2000JD90071.
Violeau, D. 2015. Fluid Mechanics and the SPH Method: Theory and Applications. Oxford University Press.
Wendland, H. 1995. Piecewise polynomial, positive definite and compactly supported radial functions of minimal degree. Advances in Computational Mathematics, 4 (1): 389–396. DOI: 10.1007/BF02123482.
Williamson, C.H.K. Sinusoidal flow relative to circular cylinders. J. Fluid Mech. 1985, 155, 141–174.
Xiang, G., Wang, S., Guedes Soares, C. 2020. Study on the motion of a freely falling horizontal cylinder into water using OpenFOAM. Ocean Engineering, 196:106811.

Advances in Maritime Technology and Engineering – Guedes Soares & Santos (Eds)
© 2024 The Author(s), ISBN 978-1-032-83107-7

Analysis of dual submerged horizontal membrane breakwater using an analytical method

Y.C. Guo & C. Guedes Soares
Centre for Marine Technology and Ocean Engineering (CENTEC), Instituto Superior Técnico, Universidade de Lisboa, Lisbon, Portugal

ABSTRACT: The performance of a dual submerged horizontal flexible membrane breakwater is analysed analytically. The mathematical model for oblique wave interaction with the membrane breakwater is established based on the linearized potential flow theory and the assumption of small amplitude structural response. The analytical solution is obtained using the matched eigenfunction expansion method, and its convergence is checked by analysing the values of reflection and transmission coefficients, as well as the non-dimensional wave forces acting on the two horizontal membranes. Furthermore, the present results are validated against published experimental data, showing good agreement. Several numerical results, including reflection and transmission coefficients, dimensionless wave forces, and membrane deflections obtained from different design parameters, are analysed to study the wave-blocking efficiency of the breakwater. It is observed that the wave reflection and transmission coefficients exhibit non-standard periodically sinusoidal displacements from the zero location.

1 INTRODUCTION

Flexible membranes have a wide range of applications in ocean engineering, of which breakwaters are an important one (Liu & Huang 2019, Li et al. 2020, Guo et al. 2020, 2022a, Wu et al. 2022). Additionally, submerged breakwaters with multi-layer horizontal membranes have been proven effective in preventing incident waves, creating a peaceful environment for various ocean structures to operate safely (Guo et al. 2023b).

However, most investigations on multi-layer breakwaters focus on horizontal rigid plates. For such breakwaters, numerous analytical studies have been conducted using the matched eigenfunction expansion method. Wang & Shen (1999) evaluated the hydrodynamic performance of a group of submerged horizontal plate wave barriers. Usha & Gayathri (2005) analytically studied incident wave reflection and transmission over a horizontal twin-plate breakwater. Liu et al. (2008) examined the hydrodynamic performance of a two-layer horizontal-plate breakwater consisting of an upper submerged horizontal porous plate and a lower submerged horizontal solid plate. Subsequently, Liu et al. (2009) considered the thickness of submerged two-layer horizontal plates.

Additionally, Cho et al. (2013) investigated the interaction between oblique incident waves and dual submerged horizontal porous plates. They found that the breakwater system becomes more effective for oblique waves due to the reduction of the effective wavelength in the direction of propagation. Moreover, when the gap is greater than 10% of the water depth, adding the lower porous plate does not significantly improve wave-blocking efficiency. Furthermore, Fang et al. (2017) derived a generalized analytical solution for wave interaction with multi-layer submerged horizontal porous plate breakwaters. They found that the submerged depth of the upper plate has the most significant influence on overall performance, and the lower plate can reinforce the obstruction of vertical fluid motion.

In addition to analytical studies, experimental investigation provides another effective method for studying and observing the interaction between waves and ocean structures (Guo et al. 2022b, 2023a). Numerous experimental studies related to multi-layer rigid plate breakwaters have been conducted. Wang et al. (2006) studied the performance of a new type of multiple-layer breakwater, consisting of several horizontal plates, through model tests. Neelamani & Gayathri (2006), as well as Guo et al. (2011), further investigated experimentally the hydrodynamic performance of twin-plate breakwaters in regular and random waves. It was observed that the transmission coefficient of the twin-plate barrier exhibits oscillating behaviour when the relative plate width is increased, attributed to the blocking and pumping effect.

Liu & Li (2014) provided additional experimental data for wave scattering by two-layer horizontal

DOI: 10.1201/9781003508779-13

porous plates, as studied by Cho et al. (2013). Through extensive model tests, Gu et al. (2017) explored the wave-dissipating performance of a twin-plate breakwater under oblique random waves. Fang et al. (2018) conducted a series of model tests to study the performance of a four-layer submerged horizontal porous plate breakwater. It was concluded that a moderate alteration of the submerged depth of each layer shifts the proportion of energy reflected or transmitted.

However, there are few existing studies on multi-layer flexible structure breakwaters compared to rigid ones. Mohapatra & Guedes Soares (2016) studied the effect of submerged flexible membranes on moored finite elastic plates in finite water depth by analysing the reflection and transmission coefficients. Subsequently, they investigated analytically the three-dimensional problem of wave interaction with a floating and a submerged flexible plate in Mohapatra & Guedes Soares (2019). The observation revealed that the reflection coefficient decreases as the length and width of the flexible plate increase, suggesting that more energy is transmitted and absorbed due to the submerged membrane beneath the floating structure.

Furthermore, Mohapatra & Guedes Soares (2021) also studied wave interaction with a horizontal flexible floating plate and a submerged plate in two dimensions, incorporating the porosity of plates. More recently, Guo et al. (2023b) conducted a series of laboratory experiments to investigate the performance of a newly proposed multi-layered flexible membrane breakwater under regular and irregular waves. The findings indicate that in regular waves, the reflection coefficient fluctuates periodically with respect to the wave period.

It is clear that progress in multi-layer horizontal flexible structure breakwaters is far from sufficient, despite their proven eco-friendliness, deployability, lightweight nature, easy installation, and cost-effectiveness. In contrast to traditional multi-layer horizontal rigid plate breakwaters, a dual horizontal membrane breakwater endures reduced wave forces and allows for the unrestricted exchange of seawater, making it durable and environmentally friendly. Therefore, in the present study, a dual submerged horizontal membrane breakwater is proposed. The mathematical model associated with oblique wave interaction with the breakwater is formulated, and an analytical solution is presented to analyze the effect of various design parameters on the membrane breakwater.

The paper is organized as follows: Section 2 demonstrates the mathematical formulation of the boundary value problem for the interaction of oblique waves with the proposed membrane breakwater. Section 3 presents the analytical solution. In Section 4, numerical results and corresponding discussions are provided. Finally, Section 5 concludes the main findings from the present analysis, clarifies the limitations of the research approach and breakwater model, and discusses the future scope of the work.

2 MATHEMATICAL FORMULATION

The physical problem of oblique waves interacting with the proposed dual submerged membrane breakwater is shown in Figure 1. Assuming that the fluid is ideal and irrotational, it satisfies potential flow theory. A three-dimensional Cartesian coordinate system (x, y, z) is used to formulate the boundary value problem, where the mean free surface coincides with the x-z horizontal plane, and the y -axis points downward positively. A progressive wave of angular frequency ω strikes the breakwater obliquely, forming an angle θ with the positive x-axis.

The two horizontal flexible membrane of length $2a$ is infinitely wide in z-direction and submerged at a depth d_1 and d_2, respectively, in the fluid region with constant water depth h. Therefore, the fluid domain can be divided into five regions: region 1 ($-\infty < x < -a$, $0 < y < h$), region 2 ($-a < x < a$, $0 < y < d_1$), region 3 ($-a < x < a$, $d_1 < y < d_2$), region 4 ($-a < x < a$, $d_2 < y < h$), and region 5 ($a < x < \infty$, $0 < y < h$). The fluid motion in each region can be described by velocity potentials $\Phi_j\ (x,y,z;\ t)=\mathrm{Re}\{\phi_j(x,\ y)\exp(ilz-i\omega t)\}$, where j=1, 2, 3, 4 and 5 refer to region 1, region 2, region 3, region 4, and region 5 respectively, with $l=k_0\sin\theta$ is the wave number in z-direction.

Further, the deflections of two submerged flexible membranes $\zeta_k\ (x,\ z;\ t)$ are of the form $\zeta_k(x, z;t)=\mathrm{Re}\{\xi_k(x)\exp(ilz-i\omega t)\}$, where k=1 and 2 correspond to the upper and lower horizontal membranes, respectively.

The incident wave potential of linear waves propagating to the positive x- and z-directions is given by

$$\phi_I = \frac{igH}{\omega}\frac{\cosh k_0(h-y)}{\cosh k_0 h}e^{i\alpha_0(x+a)} \tag{1}$$

where $\alpha_0=\sqrt{(k_0^2-l^2)}$, $i=\sqrt{-1}$, g is acceleration due to gravity, and H and ω represent the amplitude and circular frequency of the incident wave, respectively. Further, the wave number k_0 satisfies the dispersion relation $\omega^2=gk_0\tanh(k_0h)$.

The spatial velocity potentials in each region satisfy the reduced wave equation as

$$\frac{\partial^2\phi_j}{\partial x^2}+\frac{\partial^2\phi_j}{\partial y^2}-l^2\phi_j=0 \text{ for } j=1,\ 2,\ 3,\ 4,\ 5. \tag{2}$$

These velocity potentials also satisfy the free surface and the bottom boundary conditions as

$$\frac{\partial\phi_j}{\partial y}+K\phi_j=0,\ y=0,\ j=1,2,5 \tag{3}$$

$$\frac{\partial\phi_j}{\partial y}=0,\ y=h,\ j=1,4,5 \tag{4}$$

where $K=\omega^2/g$.

The kinematic boundary condition on two submerged horizontal membranes is (Guo et al. 2020a)

$$\frac{\partial \phi_{k+1}}{\partial y}=\frac{\partial \phi_{k+2}}{\partial y}=-i\omega\xi_k,\ |x|<a,\ y=d_k,\ k=1,2 \quad (5)$$

The dynamic boundary condition on two submerged horizontal flexible membrane is given as follows:

$$T_{fk}\frac{\partial^3 \phi_{k+2}}{\partial y^3}-m_{pk}\frac{\partial \phi_{k+2}}{\partial y}+K(\phi_{k+2}-\phi_{k+1})=0, y=d_k \quad (6)$$

where $T_{fk}=\sigma_k/\rho g$, $m_{pk}=m_k\omega^2/\rho g$, with σ_k and m_k are the tension and mass per unit length of the two membranes.

The far field radiated condition is assumed to take of the form

$$\phi(x,y)=\begin{cases}\phi_I(x,y)+R_0\phi_I(-x,y), & x\to-\infty\\ T_0\phi_I(x,y), & x\to+\infty\end{cases} \quad (7)$$

where R_0 and T_0 are the reflection and transmission wave amplitudes, respectively.

It is assumed that both ends of the two horizontal flexible membranes are fixed at $x=-a$ and $x=a$ and cannot move. Hence, the boundary conditions at the edges of the two submerged horizontal membranes are given by

$$\xi_k=0,\ x=\pm a,\ y=d_k \quad (8)$$

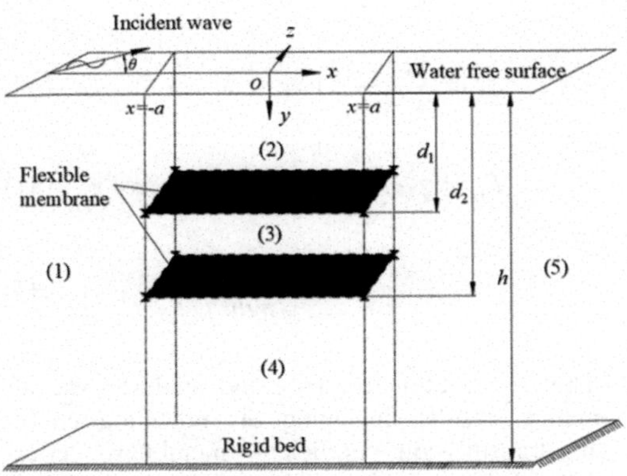

Figure 1. Schematic diagram of the interaction between an oblique wave and a dual submerged horizontal membrane breakwater.

The continuity of pressure and velocity across the vertical interfaces at $x=\pm a$ is given by

$$\phi_1=\begin{cases}\phi_2, & x=-a,\ 0\le y\le d_1\\ \phi_3, & x=-a,\ d_1<y<d_2\ ,\\ \phi_4, & x=-a,\ d_2\le y\le h\end{cases} \quad (9)$$

$$\phi_5=\begin{cases}\phi_2, & x=a,\ 0\le y\le d_1\\ \phi_3, & x=a,\ d_1<y<d_2\ ,\\ \phi_4, & x=a,\ d_2\le y\le h\end{cases} \quad (10)$$

and

$$\frac{\partial \phi_1}{\partial x}=\begin{cases}\frac{\partial \phi_2}{\partial x}, & x=-a,\ 0\le y\le d_1\\ \frac{\partial \phi_3}{\partial x}, & x=-a,\ d_1<y<d_2\ ,\\ \frac{\partial \phi_4}{\partial x}, & x=-a,\ d_2\le y\le h\end{cases} \quad (11)$$

$$\frac{\partial \phi_5}{\partial x}=\begin{cases}\frac{\partial \phi_2}{\partial x}, & x=a,\ 0\le y\le d_1\\ \frac{\partial \phi_3}{\partial x}, & x=a,\ d_1<y<d_2\\ \frac{\partial \phi_4}{\partial x}, & x=a,\ d_2\le y\le h\end{cases} \quad (12)$$

Based on the above mathematical formulas, the analytical solution procedure for determining velocity potentials in each region associated with the interaction of oblique waves with the dual submerged horizontal membrane breakwater will be discussed in detail in the next section.

3 ANALYTICAL SOLUTION

The analytical expressions for velocity potentials and associated eigenfunctions in both horizontal and vertical directions can be derived using the method of separation of variables. By applying the reduced wave equation (2) and the corresponding boundary conditions (3-7), velocity potentials for each region are obtained.

$$\phi_1=I_0f_0(y)e^{i\alpha_0(x+a)}+\sum_{n=0}^{\infty}R_nf_n(y)e^{-i\alpha_n(x+a)}, \quad (13)$$

$$\phi_2=\sum_{n=0}^{\infty}\left\{A_ne^{-i\gamma_n(x+a)}+B_ne^{i\gamma_n(x-a)}\right\}U_n(y), \quad (14)$$

$$\phi_3=\sum_{n=0}^{\infty}\left\{A_ne^{-i\gamma_n(x+a)}+B_ne^{i\gamma_n(x-a)}\right\}M_n(y), \quad (15)$$

$$\phi_4=\sum_{n=0}^{\infty}\left\{A_ne^{-i\gamma_n(x+a)}+B_ne^{i\gamma_n(x-a)}\right\}D_n(y), \quad (16)$$

$$\phi_5=\sum_{n=0}^{\infty}T_nf_n(y)e^{i\alpha_n(x-a)}, \quad (17)$$

where

$$I_0=\frac{igH}{\omega}\frac{\sqrt{\varepsilon_0}}{\cosh(k_0h)}, \quad (18)$$

$$\omega^2 = gk_n \tanh(k_n h),\ n = 0, 1, 2, \ldots, \quad (19)$$

$$\alpha_n = \sqrt{k_n^2 - l^2},\ n = 0, 1, 2, \ldots, \quad (20)$$

$$\gamma_n = \sqrt{p_n^2 - l^2},\ n = 0, 1, 2, \ldots, \quad (21)$$

with the wave number p_n, n = 0, 1, 2, ... satisfy the dispersion relation:

$$K - p_n \tanh(p_n h) = F_n\{K \tanh(p_n h) - p_n\}, \quad (22)$$

where

$$S_{1n} = T_{f1} p_n^2 - m_{p1},\ n = 0, 1, 2, \ldots, \quad (23)$$

$$S_{2n} = T_{f2} p_n^2 - m_{p2},\ n = 0, 1, 2, \ldots, \quad (24)$$

$$L_{1n} = \tanh p_n (h - d_1),\ n = 0, 1, 2, \ldots, \quad (25)$$

$$L_{2n} = \tanh p_n (h - d_2),\ n = 0, 1, 2, \ldots \quad (26)$$

$$F_n = -\frac{p_n^2 S_{1n} S_{2n} L_{1n} L_{2n} (L_{1n} - L_{2n}) - K p_n \{S_{1n} L_{1n}^2 (1 - L_{2n}^2) + S_{1n} L_{1n}^2 (1 - L_{1n}^2)\}}{p_n^2 S_{1n} S_{2n} L_{2n} (L_{1n} - L_{2n}) - K p_n \{S_{1n} L_{1n} (1 - L_{2n}^2) + S_{2n} L_{2n} (1 - L_{1n}^2)\} + K^2 (1 - L_{1n}^2)(1 - L_{2n}^2)},\ n = 0, 1, 2, 3 \ldots \quad (27)$$

The roots of dispersion relation (22) are determined using Newton's method.

The eigenfunctions $U_n(y)$'s, $M_n(y)$'s and $D_n(y)$'s are given by

$$U_n(y) = \cosh p_n (h - y) + F_n \sinh p_n (h - y), \quad (28)$$

$$M_n(y) = C_n\{\cosh p_n (h - y) - F_{2n} \sinh p_n (h - y)\}, \quad (29)$$

$$D_n(y) = C_n (1 - T_{2n}) \cosh p_n (h - y), \quad (30)$$

with

$$C_n = \frac{L_{1n} + F_n}{L_{1n} - F_{2n}}, \quad (31)$$

$$F_{2n} = \frac{S_{2n} p_n L_{2n}^2}{S_{2n} p_n L_{2n} - K(1 - L_{2n}^2)}, \quad (32)$$

$$T_{2n} = F_{2n} / L_{2n} \quad (33)$$

The eigenfunctions $f_n(y)$'s are given by

$$f_n(y) = \frac{\cosh k_n (h - y)}{\sqrt{\varepsilon_n}},\ n = 0, 1, 2, \ldots, \quad (34)$$

with

$$\varepsilon_n = \frac{2k_n h + \sinh(2k_n h)}{4k_n},\ n = 0, 1, 2, \ldots \quad (35)$$

The eigenfunctions $f_n(y)$'s are orthogonal and is defined as

$$F_{mn} = \langle f_m(y), f_n(y) \rangle = \int_0^h f_m(y) f_n(y) dy = \begin{cases} 0, & m \neq n, \\ 1, & m = n. \end{cases} \quad (36)$$

By taking into account the two continuity conditions (9-12) at vertical interfaces of different fluid domains, the fixed edge conditions (8), and the orthogonal relation (36), a linear system comprising 4(N+1) algebraic equations can be established. This system is used to determine the unknowns A_n, B_n, R_n, T_n associated with the velocity potentials (13-17). The infinite series is truncated to a finite number of terms, specifically N+1 for A_n and B_n, and N-1 for R_n and T_n.

After obtaining all the unknowns, the reflection, transmission and dissipation coefficients K_r, K_t, K_d, the deflection of each membrane ξ_k, and the non-dimensional vertical wave forces K_{fk} acting on the two submerged horizontal membranes can be computed by using the following formulas:

$$K_r = |R_0 / I_0|, \quad (37)$$

$$K_t = |T_0 / I_0|, \quad (38)$$

$$K_d = \sqrt{1 - K_r^2 - K_t^2}, \quad (39)$$

$$\xi_k = \frac{i}{\omega} \frac{\partial \phi_{k+2}}{\partial y},\ y = d_k,\ k = 1, 2, \quad (40)$$

$$F_{vk} = \rho i \omega \int_{-a}^{a} (\phi_{k+2} - \phi_{k+1}) dx, \quad (41)$$

$$K_{fk} = 1000 \frac{F_{vk}}{\rho g h^2} \quad (42)$$

The next section presents and analyses various numerical results, including the reflection coefficient, transmission coefficient, membrane deflections, and dimensionless wave forces. This analysis aims to study the effects of non-dimensional membrane tension, membrane half-length, membrane submergence depth, oblique angle, and wave period.

Matlab codes are developed to compute the numerical results. In the present computations, MATLAB-R2017b, 64-bit (win64) is used to perform calculation and all numerical computations were performed in a desktop machine with Intel ® core i7-4790 CPU with 3.60GHz processor and 32 GB of ram memory. On an average, each case takes roughly 5-7 minutes to finish.

4 NUMERICAL RESULTS AND DISCUSSION

In the context of the present work, the convergence of the method is demonstrated in Table 1 by computing the reflection and transmission coefficients, as well as non-dimensional wave forces acting on two membranes for different wave periods, considering varying numbers of terms N in the series solution. It is evident that N=20 is sufficient for all calculated results to converge up to three decimals.

Table 1. Convergence of K_r, K_t, K_{f1} and K_{f2} with increasing values of N at d_1/h=0.1, d_2/h=0.2, a/h=1.5, $\sigma_1=\sigma_2$ =15ρgh^2, θ=0.1°, $G_1=G_2$=0, and $m_1= m_2$=3 kg/m^2.

	N	K_r	K_t	K_{f1}	K_{f2}
T = 6s	1	0.2450	0.9695	0.1151	0.1896
	5	0.3992	0.9169	0.8070	1.2198
	10	0.4097	0.9122	0.8650	1.1883
	15	0.3885	0.9215	0.8843	1.2083
	20	0.3868	0.9222	0.8930	1.2153
	21	0.3865	0.9223	0.8932	1.2155
	22	0.3865	0.9223	0.8933	1.2153
T = 8s	1	0.5332	0.8460	0.6385	0.3850
	5	0.5244	0.8515	0.3522	1.5661
	10	0.4816	0.8764	0.2449	1.4003
	15	0.5028	0.8644	0.2103	1.4267
	20	0.5081	0.8613	0.1967	1.4272
	21	0.5083	0.8612	0.1965	1.4273
	22	0.5081	0.8613	0.1962	1.4275

For computational accuracy, the number of terms N in the series solution is restricted to 20 throughout the computation. In addition, all numerical computations are executed by considering fluid density $\rho = 1025\text{kgm}^{-3}$ and acceleration due to gravity $g = 9.8\text{ms}^{-2}$ unless mentioned otherwise.

Figure 2 illustrates a comparison between the current analytical results and the experimental data from Guo et al. (2023b) by presenting the reflection coefficient and transmission coefficient of the dual horizontal flexible membrane breakwater. Among the 12 experimental data points, it is observed that 9 points for K_r and K_t are situated precisely on or very close to the corresponding curves, indicating approximately 75% agreement between the analytical results and experimental data. The good agreement between the present results and the experimental data shows that the current analytical model, methods, and solutions can effectively investigate the interaction between oblique waves and the proposed membrane breakwater.

On the other hand, based on the definition of the wave energy dissipation coefficient, K_d is found to be very close to zero for all wave periods, indicating that almost no incident waves are dissipated by the breakwater. This is reasonable, given that the viscosity of the fluid and the resistance characteristics of the membrane breakwater, factors that mainly contribute to wave dissipation, are ignored in this study. As a result, the dissipation coefficient is not discussed in the following sections.

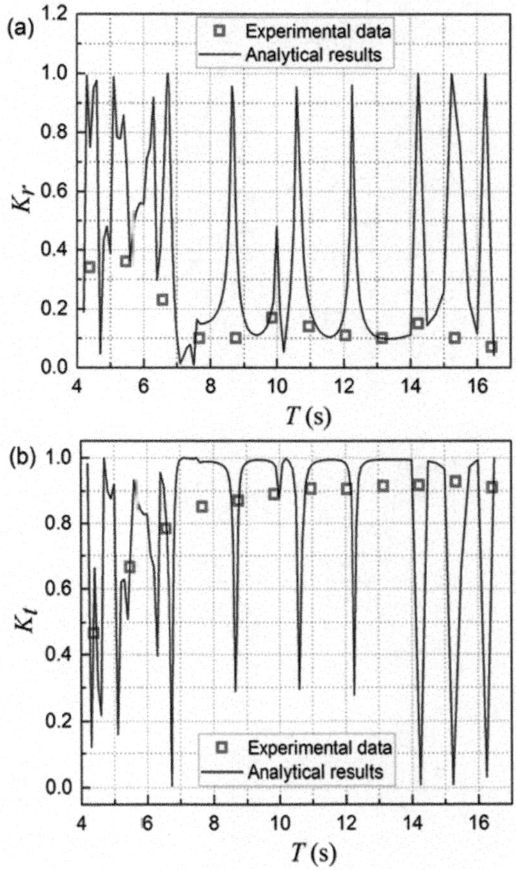

Figure 2. Comparison between the present analytical results and experimental data from Guo et al. (2023) for (a) the K_r and (b) the K_t, with a=1.5h, d_1=0.01h, d_2=0.3h, $m_1=m_2$=1, $\sigma_1=\sigma_2$=0.03ρgh^2, and θ=0.1°.

Figure 3 shows the effect of the non-dimensional half-length of the dual submerged horizontal membrane breakwater on (a) the reflection coefficient K_r, (b) the transmission coefficient K_t, (c) the non-dimensional wave force K_{f1} acting on the upper membrane, and (d) the non-dimensional wave force K_{f2} acting on the lower membrane against the non-dimensional wave number k_0h, with d_1 =0.05h, d_2=0.2h, $m_1=m_2$=1, $\sigma_1=\sigma_2$=0.25ρgh^2, and θ=0.1°. It is observed that with the increase of k_0h, both the reflection and transmission coefficients exhibit alternating peaks and troughs. The values of 0 or 1 for the wave reflection and transmission coefficients refer to zero or full reflection and transmission of incident waves. Additionally, the values of 0 or 1 for K_r and K_t at various locations may be due to the constructive or destructive interference of the incident and reflected waves by the dual submerged horizontal membranes for different non-dimensional wave numbers. This phenomenon was also observed by Behera & Sahoo (2015) when dealing with the interaction of surface gravity waves with a submerged horizontal flexible porous plate.

The number of peaks and troughs increases for a breakwater with a larger half-length. Specifically, for the longest half-length (a=3h), the number of peaks in K_r and the number of troughs in K_t are much higher than those for the shortest half-length ($a = h$). Additionally, the gaps between two adjacent peaks in K_r and between two adjacent troughs in K_t in the long-wave region ($k_0h < 2.5$) are larger compared to the gaps in the short-wave region ($k_0h >$ 2.5). Therefore, the present breakwater is more effective in reflecting and preventing short incident waves, and the longer the breakwater, the more efficient it is.

In Figures 3c, d, peaks and troughs are also observed in the dimensionless wave forces acting on two membranes, K_{f1} and K_{f2}. It is found that, although longer breakwaters lead to greater peak values of K_{f1} and K_{f2}, they do not necessarily result in greater wave forces acting on the two membranes.

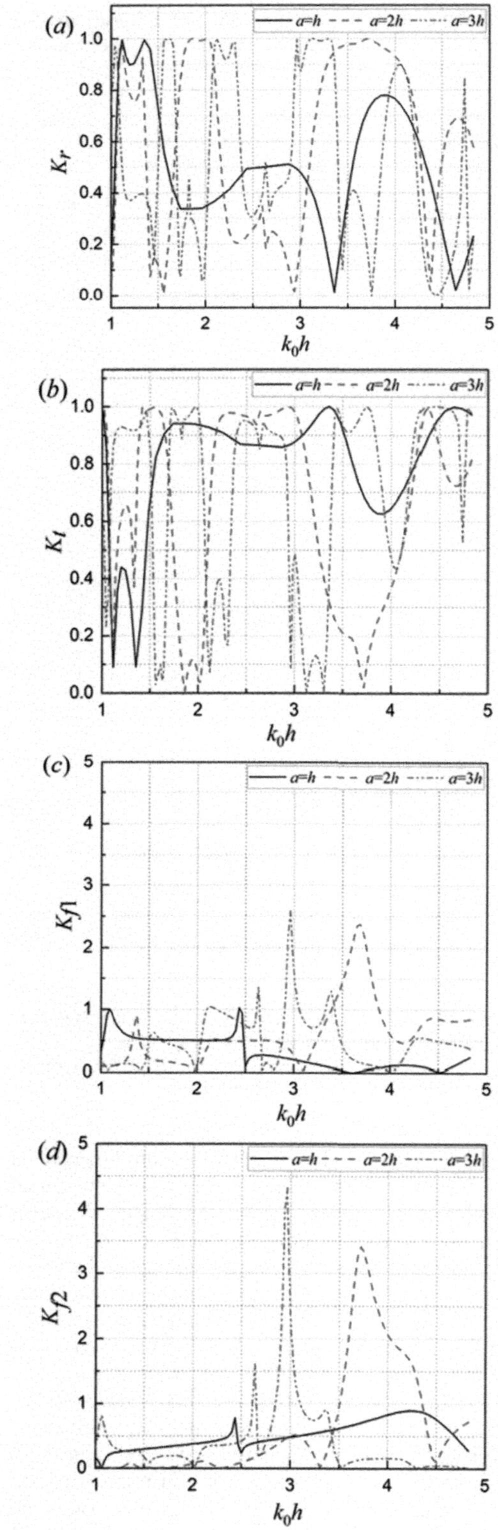

Figure 3. Effects of half-length of the breakwater on (a) the K_r, (b) the K_t, (c) the K_{f1}, and (d) the K_{f2}, with d_1=0.05h, d_2=0.2h, m_1=m_2=1 kg/m^2, σ_1=σ_2=0.25ρgh^2, and θ=0.1°.

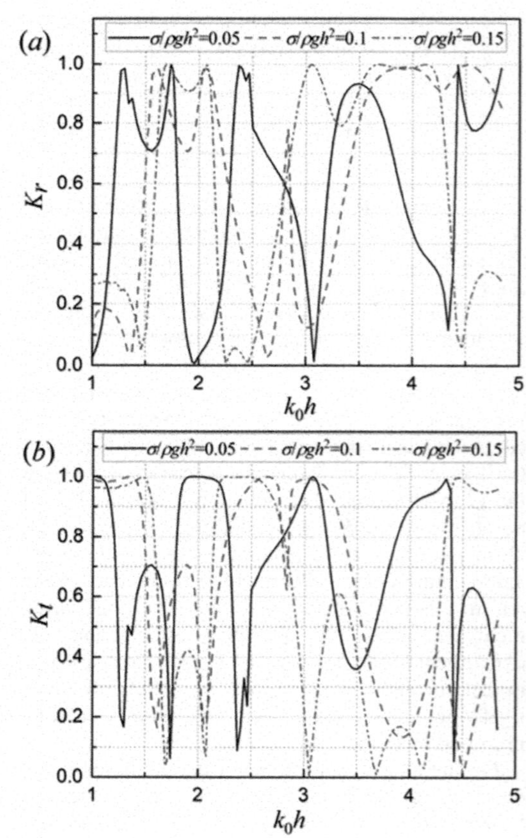

Figure 4. Effects of non-dimensional membrane tension $\sigma/\rho gh^2$ on (a) the K_r and (b) the K_t, with a=1.5h, d_1=0.05h, d_2=0.2h, m_1=m_2=1 kg/m^2, σ=σ_1=σ_2, and θ=0.1°.

Figure 4 plots the effect of the non-dimensional tension of membranes $\sigma/\rho gh^2$ on (a) the reflection coefficient K_r and (b) the transmission coefficient K_t as a function of the non-dimensional wave number k_0h, with a=1.5h, d_1=0.05h, d_2=0.2h, m_1=m_2 =1 kg/m^2, σ=σ_1=σ_2, and θ=0.1°. It is observed that different values of membrane tension lead to an obvious phase lag in the reflection and transmission coefficients K_r and K_t. This is because membrane tension represents the stiffness of the membrane, and a stiffer membrane exhibits smaller deflection and fewer variations along the direction of the breakwaters. This also implies a more intense interaction between structures and waves.

In the short-wave region ($k_0h > 3$), a higher membrane tension results in more incident wave reflection and less wave transmission. However, in the long-wave region ($k_0h < 2.5$), when the non-dimensional membrane tension $\sigma/\rho gh^2$ increases from 0.05 to 0.1, the membrane tension has a negligible effect on the values of the reflection and transmission coefficients, except for the phase lag. Conversely, greater wave reflection and smaller wave transmission are observed without phase lag when the non-dimensional membrane tension $\sigma/\rho gh^2$ increases from 0.1 to 0.15.

Figure 5 reveals the deflection of the two horizontal membranes along the length direction, with a=1.5h, d_1=0.05h, d_2=0.2h, m_1=m_2=1 kg/m^2, σ=σ_1 =σ_2, θ=0.1°, and k_0h=1.373. It is evident that both membranes with higher tensions exhibit smaller maximum deflections and fewer peaks and troughs along the length direction. Additionally, the upper membrane displays a more distinct and complex shape variation compared to the lower one, regardless of the tension. This phenomenon arises from the concentration of incident wave energy near the free surface, causing the membrane close to the free surface to experience more significant wave interaction.

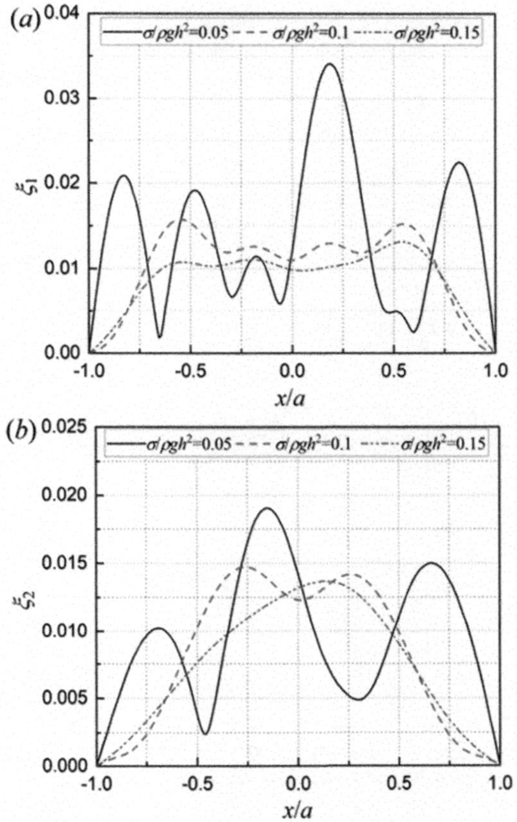

Figure 5. Effects of non-dimensional membrane tension $\sigma/\rho gh^2$ on the deflection of two membranes.

Figure 6 reveals the variation of (a) the reflection coefficient K_r, (b) the transmission coefficient K_t, (c) the non-dimensional wave force K_{f1} acting on the upper membrane, and (d) the non-dimensional wave force K_{f2} acting on the lower membrane with respect to the non-dimensional wave number k_0h, with a=1.5h, d_1=0.05h, d_2=0.2h, m_1=m_2=1 kg/m^2, σ_1=σ_2 =0.15ρgh^2, and k_0h=1.962.

It can be observed that for a very short dual membrane breakwater (a/h=0.1 and 0.5), its reflection coefficients increase, while the transmission coefficient decreases as the incident wave angle θ increases from 0° to 90°. However, for a breakwater with a longer length (a/h=1 and 1.5), the reflection coefficient shows a decreasing trend followed by an increasing trend, which is opposite to the variation of the transmission coefficient.

The two troughs in the reflection coefficient curves and the two peaks in the transmission coefficient curves result in almost zero incident wave reflection and full transmission. The maximum wave forces acting on the upper horizontal membrane are also observed at corresponding incident wave angles θ=30° for a/h =1 and θ=36° for a/h =1.5. Such a specific phenomenon needs to be avoided when designing such breakwaters. However, Figures 6c, d show that the wave forces acting on the lower membrane are quite different from those acting on the upper one. It is evident that K_{f2} is not necessarily smaller than K_{f1}, despite both approximating zero as the angle θ approaches 90°.

The variation of the deflections of the two horizontal membranes in a dual membrane breakwater is depicted in Figure 7 as a function of the dimensionless breakwater half-length x/a, with a=1.5h, d_1=0.05h, d_2=0.2h, m_1=m_2=1 kg/m^2, σ_1=σ_2=0.15ρgh^2, and k_0h=1.962. It can be observed that, for all incident wave angles except when θ is close to 90°, the two horizontal membranes exhibit non-standard periodically sinusoidal displacements from the zero location. Moreover, as θ increases from 0.01° to 30°, the displacements of the two membranes increase and then rapidly decrease, nearing zero as θ changes from 30° to 90°. It can be concluded that the two horizontal membranes reach their maximum deflection when the incident wave angle θ is between 30° and 45°,

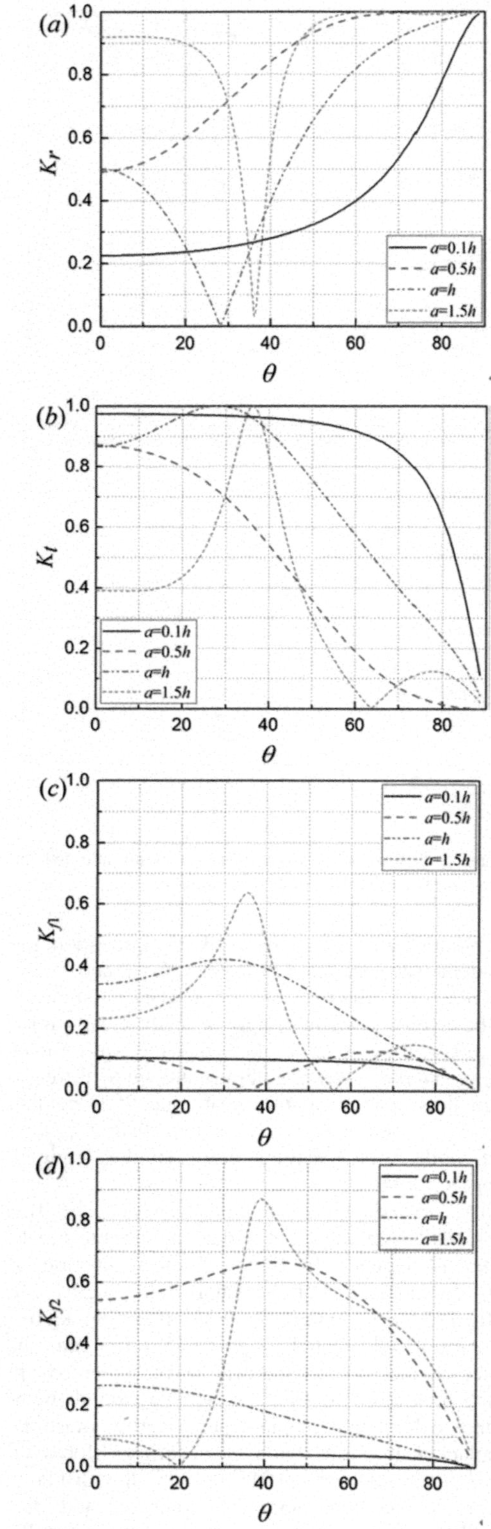

Figure 6. Variation of (a) K_r, (b) K_t, (c) K_{f1}, and (d) K_{f2} of the breakwater against incident wave angle θ, with d_1=0.05h, d_2 =0.2h, m_1=m_2=1 kg/m^2, σ_1=σ_2= 0.15ρgh^2, and k_0h=1.962.

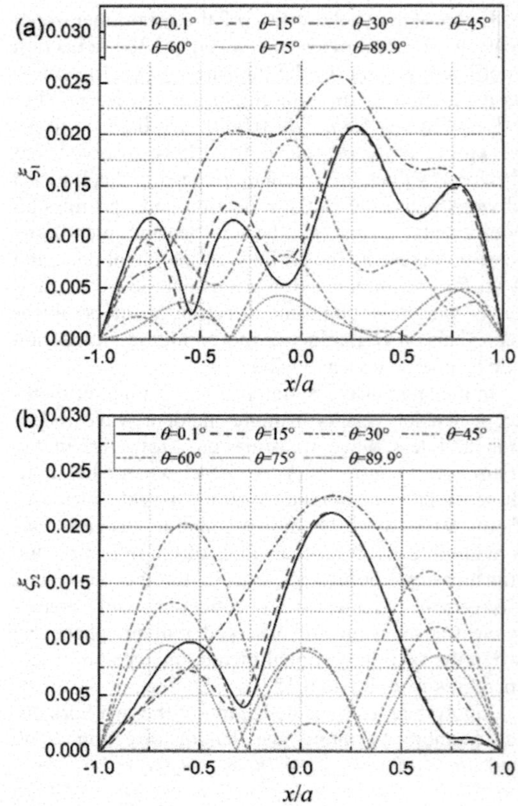

Figure 7. Variation of the deflection of two horizontal membranes, with a=1.5h, d_1=0.05h, d_2=0.2h, m_1=m_2=1 kg/m^2, σ_1=σ_2=0.15ρgh^2, and k_0h=1.962.

which is consistent with the value of θ causing the least wave reflection and the most wave transmission.

Figure 8 shows the effect of the submergence depth d_2 of the lower horizontal membrane on the reflection and transmission coefficients of the dual membrane breakwaters against the non-dimensional wave number k_0h, with a=1.5h, d_1=0.05h, m_1=m_2 =1 kg/m^2, σ_1=σ_2=0.1ρgh^2, and θ=0.1°. It is observed from Figures 8a, b that, compared with a dual membrane breakwater with a small height (d_2=0.1h), a larger submergence depth (d_2=0.2h) causes the breakwater to reflect more and transmit fewer incident waves in the short-wave region (k_0h < 1), the middle-wave region (1.8 < k_0h < 2.6), and the long-wave region (k_0h > 3.5). Therefore, it is more efficient in blocking waves.

Figures 8c, d reveal that increasing the submergence depth d_2 from 0.2h to 0.3h undermines the efficiency of the breakwater for all incident waves except in a small wave region (1.75 < k_0h < 2). Moreover, further increasing d_2 from 0.3h to 0.4h has a negligible effect on the wave prevention performance of the dual horizontal membrane breakwater. This finding is consistent with the

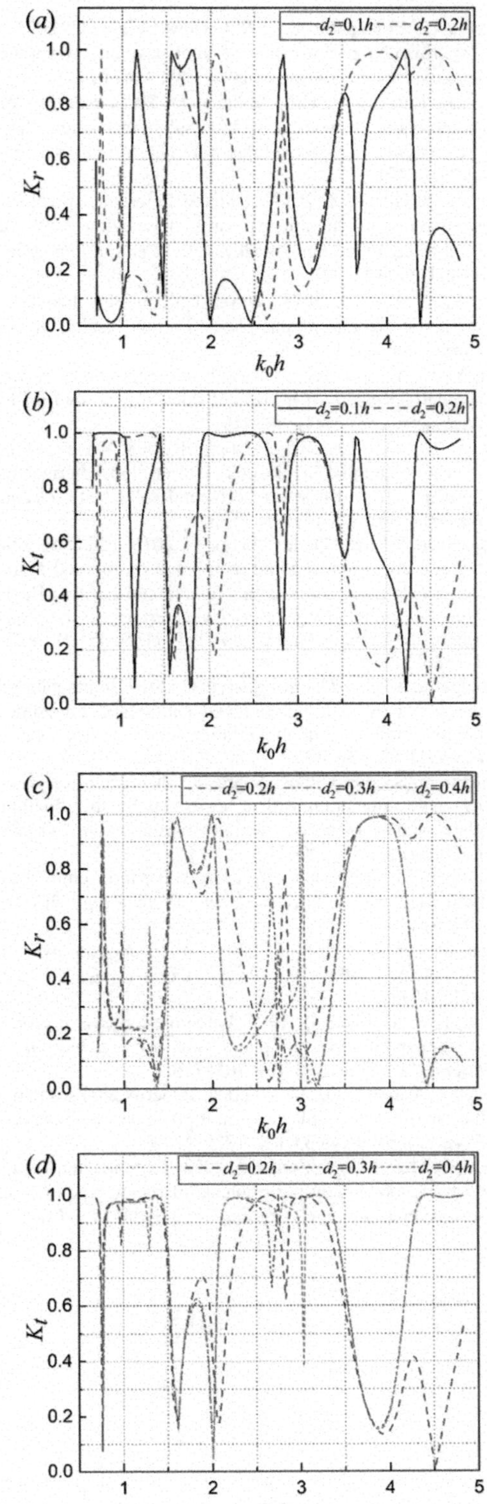

Figure 8. Effects of submergence depth d_2 on (a, c) the K_r and (b, d) the K_t, with a=1.5h, d_1=0.05h, m_1=m_2=1 kg/m^2, σ_1=σ_2=0.1ρgh^2, and θ=0.1°.

conclusion obtained by Cho et al. (2013). Therefore, it can be concluded that a submergence depth of d_2=0.2h makes such breakwaters most efficient, and the effect of greater submergence depths (d_2>0.3h) on K_r and K_t can be ignored.

5 CONCLUSIONS

In this study, the interaction between oblique wave and a dual submerged horizontal membrane breakwater is explored using the matched eigenfunction expansion method. The convergence of the present analytical solution is checked, and the accuracy of the analytical solutions is validated with experimental data published in the literature. The study investigates the impact of various modeling parameters, including the length and tension of the membrane, the submergence depth of the lower membrane, and the incident wave angle, on the performance of the membrane breakwater.

It is observed that the present breakwater is more effective in reflecting and blocking short incident waves, and a longer half-length makes the breakwater more efficient. Furthermore, the tension of membranes causes a phase lag in the reflection and transmission coefficient curves K_r and K_t. The two horizontal membranes exhibit different variations in deflection, and the upper one has a more complex pattern as it is closer to the free surface. The incident wave angle also greatly affects the performance of the breakwater. $\theta\approx 90^\circ$ leads to nearly full incident wave reflection and zero transmission, as the incident wave propagates perpendicular to the length direction of the breakwater. Moreover, it is found that increasing the height of the breakwater does not essentially result in better performance in preventing waves, and d_2=0.2h makes the breakwater most effective.

However, the analytical approach and breakwater model employed in this study did not consider the effect of membrane porosity, which undoubtedly plays a significant role in dissipating incident wave energy and thereby improving the efficiency of the proposed membrane breakwater.

Therefore, future work will concentrate on examining the influence of membrane porosities. Analytical methods suitable for handling complex dispersion relations, incorporating the effects of various porous-effect parameters of the membrane, will be explored.

ACKNOWLEDGEMENTS

The paper contributes to the project Hydroelastic behaviour of horizontal flexible floating structures for applications to Floating Breakwaters and Wave Energy Converters (HYDROELASTWEB), which is co-funded by the European Regional Development Fund (Fundo Europeu de Desenvolvimento Regional

- FEDER) and by the Portuguese Foundation for Science and Technology (Fundação para a Ciência e a Tecnologia - FCT) under contract 031488_770 (PTDC/ECI-EGC/31488/2017). The work of the first author has been supported in part by the scholarship from China Scholarship Council (CSC) under Grant No. 201707920004.

REFERENCES

Behera, H., & Sahoo, T. 2015. Hydroelastic analysis of gravity wave interaction with submerged horizontal flexible porous plate. *Journal of Fluids and Structures* 54: 643–660.

Cho, I.H., Koh, H.J., Kim, J.R. & Kim, M.H. 2013. Wave scattering by dual submerged horizontal porous plates. *Ocean engineering* 73: 149–158.

Fang, Z., Xiao, L., Kou, Y., & Li, J. 2018. Experimental study of the wave-dissipating performance of a four-layer horizontal porous-plate breakwater. *Ocean Engineering* 151: 222–233.

Fang, Z., Xiao, L. & Peng, T. 2017. Generalized analytical solution to wave interaction with submerged multi-layer horizontal porous plate breakwaters. *Journal of Engineering Mathematics* 105: 117–135.

Gu, Q., Huang, G., Zhang, N., & Hu, C. 2017. Wave-dissipating performance of twin-plate breakwater under oblique random waves. *International Journal of Offshore and Polar Engineering* 27(02): 184–192.

Guo, C.S., Zhang, N.C., Li, Y.Y., Li, J.B., Fang, Z., & Cui, C. 2011. Experimental study on the performance of twin plate breakwater. *China Ocean Engineering* 25: 645–656.

Guo, Y.C., Mohapatra, S.C., & Guedes Soares, C. 2020. Composite breakwater of a submerged horizontal flexible porous membrane with a lower rubble mound. *Applied Ocean Research* 104: 102371.

Guo, Y.C., Mohapatra, S.C., & Guedes Soares, C. 2022a. Submerged breakwater of a flexible porous membrane with a vertical flexible porous wall over variable bottom topography. *Ocean Engineering* 243: 109989.

Guo, Y.C., Mohapatra, S.C., & Guedes Soares, C. 2022b. Experimental study on the performance of an array of vertical flexible porous membrane type breakwater under regular waves. *Ocean Engineering* 264: 112328.

Guo, Y.C., Mohapatra, S.C., & Guedes Soares, C. 2023a. An experimental study on the performance of combined membrane breakwaters against incident waves. *Ocean Engineering* 280: 114793.

Guo, Y.C., Mohapatra, S.C. & Guedes Soares, C. 2023b. Experimental performance of multi-layered membrane breakwaters. *Ocean Engineering* 281: 114716.

Li, A.J., Liu, Y., Li, H.J., & Fang, H. 2020. Analysis of water wave interaction with a submerged fluid-filled semi-circular membrane breakwater. *Ocean Engineering* 197: 106901.

Liu, C., & Huang, Z. 2019. A mixed Eulerian-Lagrangian simulation of nonlinear wave interaction with a fluid-filled membrane breakwater. *Ocean Engineering* 178: 423–434.

Liu, Y., & Li, H.J. 2014. Wave scattering by dual submerged horizontal porous plates: Further results. *Ocean engineering* 81: 158–163.

Liu, Y., Li, Y.C., & Teng, B. 2009. Wave motion over two submerged layers of horizontal thick plates. *Journal of Hydrodynamics* 21(4): 453–462.

Liu, Y., Li, Y.C., Teng, B., & Dong, S. 2008. Wave motion over a submerged breakwater with an upper horizontal porous plate and a lower horizontal solid plate. *Ocean Engineering* 35(16): 1588–1596.

Mohapatra, S.C. & Guedes Soares, C. 2016. Effect of submerged horizontal flexible membrane on moored floating elastic plate. *In*: Guedes Soares and Santos (Eds), *Maritime Technology and Engineering* 3. London: Taylor & Francis Group, pp.1181–1188, ISBN: 978-1-138-03000-8.

Mohapatra, S.C., & Guedes Soares, C. 2019. Interaction of ocean waves with floating and submerged horizontal flexible structures in three-dimensions. *Applied Ocean Research* 83: 136–154.

Mohapatra, S.C., & Guedes Soares, C. 2021. Surface gravity wave interaction with a horizontal flexible floating plate and submerged flexible porous plate. *Ocean Engineering* 237: 109621.

Neelamani, S., & Gayathri, T. 2006. Wave interaction with twin plate wave barrier. *Ocean Engineering* 33(3-4): 495–516.

Usha, R., & Gayathri, T. 2005. Wave motion over a twin-plate breakwater. *Ocean Engineering* 32(8-9): 1054–1072.

Wang, K.H., & Shen, Q. 1999. Wave motion over a group of submerged horizontal plates. *International Journal of Engineering Science* 37(6): 703–715.

Wang, Y., Wang, G., & Li, G. 2006. Experimental study on the performance of the multiple-layer breakwater. *Ocean Engineering* 33(13): 1829–1839.

Wu, J.P., Mei, T.L., & Zou, Z.J. 2022. Experimental study on wave attenuation performance of a new type of free surface breakwater. *Ocean Engineering* 244: 110447.

Advances in Maritime Technology and Engineering – Guedes Soares & Santos (Eds)

Comparative analysis between numerical and analytical methods to calculate added mass

A. Jahanbakhsh & L. Moro
Memorial University of Newfoundland, St. John's, Canada

M. Islam
Ocean, Coastal and River Engineering Research Centre, National Research Council, Canada

ABSTRACT: Considering the effect of the water surrounding the hull on the ship structural dynamics is crucial in vibration analyses. This paper presents the results from different methods used to calculate added mass and compare the simulations results against experimental data. We used a scaled Post-Panamax bulk carrier model as a case study and calculated the wet natural frequencies using numerical and analytical methods. In particular, we performed structural dynamic simulations implementing two numerical methods, i.e. coupled Boundary Element (BE) and Finite Element (FE) analysis in MSC NASTRAN and vibroacoustic analysis in MSC ACTRAN. In addition, we calculated the wet natural frequencies of the model implementing the analytical formulation by Lewis. Resistance tests were performed at the National Research Council of Canada's towing tank, and during the tests we performed an experimental Operational Deflection Shape (ODS) analysis. The data from the ODS were used to validate the results from the numerical simulations. The results highlight the accuracy and viability of the tested methods.

1 INTRODUCTION

In the context of ship structural dynamics, particularly when studying ship vibrations, it is imperative to consider the influence of the surrounding water. The inertia of the displaced water impacts the inertia, dynamic characteristics, and behavior of the structure.

Various methods have been proposed in the literature to determine the added mass effect, including analytical formulations, experimental measurements, and numerical simulations.

Korotkin (2009) summarized analytical formulations to calculate the components of the added mass matrix for different geometrical cases, such as planar and 3-D bodies like ships. Pais (2018) validated the numerical simulation of added mass by implementing the analytical Lewis formulation presented in (Korotkin, 2009).

Numerical approaches integrate hydrodynamic and structural—typically finite element (FE)—approaches. A strong coupling of these methods is necessary to accurately incorporate the added mass effect. The following paragraphs explore fluid-structure interaction (FSI) simulations in this context.

Zhang et al. (2021) combined the Detached Eddy Simulation method and the finite volume solid stress method to study the vibration of a plate excited by propeller-induced pressure pulsations. The plate represented the ship's aft surface. They coupled the fluid and structural solvers using hexahedral cells, which take into account the boundary layer effect for the fluid domain and solid elements for the plate. Due to the resulting high computational cost, they considered a simple plate geometry to represent the ship's aft surface.

Zambon et al. (2021) investigated the validation of the Holden Method (HM) in predicting fluctuating forces caused by ship propellers on the hull. The study integrated experimental measurements, empirical approaches, and numerical analyses using coupled FSI. The numerical analyses took into account the added mass effect by using a fully coupled FE-BE method. They analyzed acceleration data from a 54-meter superyacht with dual propellers to assess the response of the hull structure. They accounted for frequency-dependent added mass. In order to validate HM, they revealed its tendency to overestimate propeller-induced pressures.

Pais et al. (2017) introduced a methodology based on linear dynamic finite element analysis (FEA) to predict the vibration response of a 60-meter superyacht. Their methodology thoroughly compared numerical simulations with experimental investigations conducted during the vessel's sea trial. Their research aimed to identify fundamental design parameters, including the ship's global damping ratio and added mass. The study also evaluated the ship's noise and vibration levels to assess onboard comfort.

DOI: 10.1201/9781003508779-14

Annicchiarico et al. (2008) examined propeller-induced excitations and the corresponding structural response, considering both rigid and flexible structures. The study was performed on an ellipsoid body first and then on a cruise ship. They used a coupled FE-BE approach to analyze hydroelasticity.

Tian et al. (2017) improved the precision of higher frequency FE-BE simulations by implementing smooth FE-BE coupling, known as SFEM-BEM. They focused on analyzing a cavity enclosed by shell elements in vibroacoustic scenarios, choosing an alternative to the conventional FEM approach. The introduced method showed significant enhancements in solution accuracy.

Experimental investigations were also conducted to determine the dynamic characteristics of the structures considering the added mass effect and validate numerical calculations. For instance, Eslaminajad et al. (2019) studied the vibrational behavior of a hemispherical shell welded to a covering plate equipped with a pressure control mechanism. This study explored two distinct scenarios: one involving the hemisphere filled with an inviscid compressible fluid and another one where the hemisphere was empty. Employing the Experimental Modal Analysis (EMA), the researchers captured dynamic characteristics for both configurations, including natural frequencies, damping ratios, and normal mode shapes. FEM was employed to simulate fluid-structure interaction (FSI), and the accuracy of the FE model was validated by comparing the outcomes from the simulations against EMA data derived from a roving hammer test. The researchers introduced Rayleigh proportional damping, incorporating mass and stiffness components, to conduct modal analysis for the damped system.

In addition, Rajendran et al. (2016) investigated the vertical response of an ultra-large container carrier model through a combination of numerical simulations and experimental analyses. The study focused on observing the whipping and springing behaviors of the model under severe irregular waves and regular waves with moderate amplitudes conducted in a wave tank. They employed strip theory to compute the numerical response of the model and integrated it with FEM.

FSI studies have not been limited to the analyses in the frequency domain. Datta and Guedes Soares (2020) investigated the elastic behavior of a container ship by analyzing the vertical bending moment and shear force in the time domain. The study integrated hydrodynamic forces computed using a transient panel method with a structural FE model. The outcomes emphasized the significance of considering ship structural elasticity in such analyses, as demonstrated by comparisons between rigid and flexible conditions.

This study was conducted to calculate the natural frequencies and modes of the scaled model, considering the added mass effect. It is part of a large project aimed at defining accurate procedures to test non-cavitating propeller-induced hull pressures.

Fast and cost-efficient numerical analyses were performed to investigate the added mass effect on the natural frequencies and normal modes of a scaled ship model used as a case study. The scaled ship represented a Post-Panamax bulk carrier. In the project, resistance, bollard pull, and self-propulsion tests were conducted in the atmospheric towing tank at the National Research Council of Canada's Ocean, Coastal and River Engineering Research Centre (NRC-OCRE) in St. John's, NL, Canada. During these tests, propeller-induced pressure pulsations and structural accelerations were measured, while deck-mounted accelerometers measured the global response of the model.

Differently from Zhang et al. (2021), we used a fully coupled FE-BE method to assess the added mass effect. Then, the water domain was explicitly modeled and coupled with the FE model, and vibroacoustic analyses were carried out in MSC ACTRAN. In addition, we calculated the wet natural frequencies of the model using the approximate analytical formulation proposed by Lewis.

MATLAB® codes were generated to conduct Operational Deflection Shape (ODS) analyses and validate the simulation results using the acceleration data acquired during the hydrodynamic tests.

2 METHODOLOGY

2.1 *Scaled model specification*

Figure 1 shows the Post-Panamax bulk career model that was used as a case study in this work. The model dimensions are listed in Table 1. During the bollard and self-propulsion tests, the model was equipped with a fixed-pitch four-blade propeller with a modified NACA section profile to work in non-cavitating conditions.

The ship model is made of Fiber Reinforced Plastic (FRP) (hull plating), plywood (inner box and transverse bulkheads), wooden beams (stiffeners), filling foam between the inner box and outer plating, and a synthetic polyurethane plate was used to install pressure sensors on the hull plating above the propeller (Figure 2).

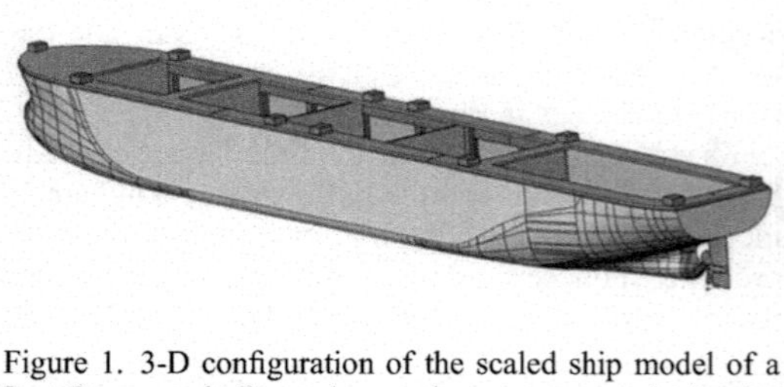

Figure 1. 3-D configuration of the scaled ship model of a Post-Panamax bulk carrier used as a case study in this research.

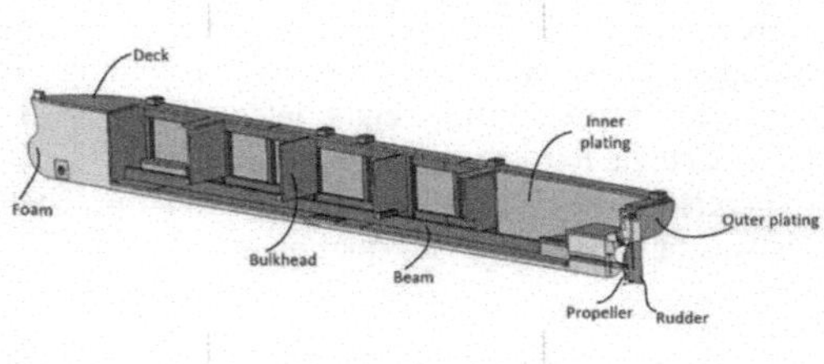

Figure 2. Scaled ship model's structural details.

Table 1. Hull dimensions of the Post-Panamax bulk carrier and corresponding ship model.

Dimensions [m]	Model	Ship
Length Overall [m]	7.142	226.00
Breadth [m]	1.020	32.25
Depth [m]	0.623	19.75
Draft [m]	0.430	13.50
Displacement [t]	2.685	85637

2.2 *Finite element model*

The model's geometry was created in Rhinoceros® and imported in HYPERMESH®, which was used to generate the FE mesh presented in Figure 3. The inner box and the outer hull were modeled using quadrilateral reduced-integration shell elements (4-noded shell elements); beam elements simulated the wooden framing inside the model. Table 2 lists the thickness of FRP and plywood surfaces and the cross-sectional dimensions of the beam elements.

The model's materials were considered linear elastic. We performed a convergence analysis to find the optimal mesh size (Moro et al., 2013). The final model comprised 22513 shell and beam elements.

Table 2. Properties of structural elements of the ship model.

Structural Element	Parameter	value
FRP layer hull	Thickness [mm]	6
FRP layer deck	Thickness [mm]	7
Plywood layer	Thickness [mm]	10
Beam element I	Cross Section [mm × mm]	95×45
Beam element II	Cross Section [mm × mm]	50×45

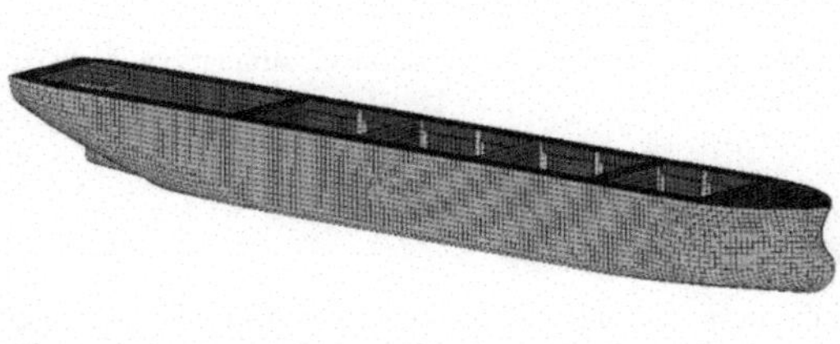

Figure 3. FE model of the scaled-ship model used for modal analysis (dry condition).

2.2.1 *FE model validation*

The FE model (dry condition) was validated, without the ballast weights, against results from an EMA test conducted by the authors on the ship model in the same configuration. Then, the weights were distributed over the FE model, simulating the weight distribution applied during the hydrodynamic test. We followed this approach since EMA could not be conducted, including the weights, as the model's structure was not able to withstand this load in dry condition.

Figure 4 shows the arrangement of nine PCB piezoelectric accelerometers, model 352C33, mounted on the model's deck and used for EMA.

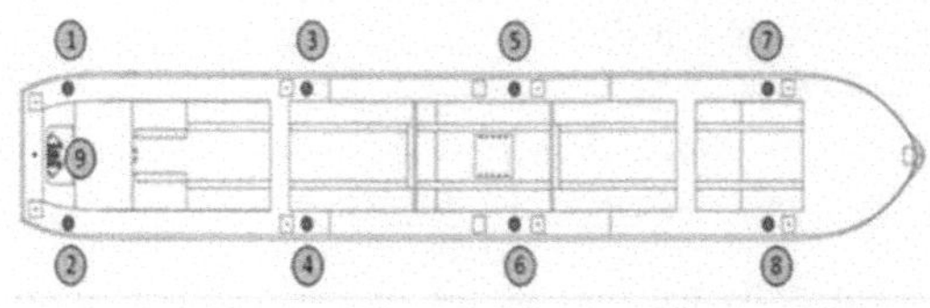

Figure 4. Arrangement of the accelerometers for EMA and ODS tests.

Figure 5. Experimental test set up for free-free EMA of the ship model in dry condition.

Vibration data were collected using NI DAQ modules, model 9234. The data were converted into digital signals for post-processing. MATLAB® codes were developed to post-process the acquired signal and calculate frequency response functions (FRFs).

Figure 5 shows the model suspended from an OEB crane using a strong back and two lifting slings to establish free-free boundary conditions.

2.3 *Numerical FSI*

2.3.1 *Coupled FE-BE method*

Several researchers (e.g., Egusquiza et al., 2009; Jaiman et al., 2014; Jaiman and Joshi, 2022; Rahman and Bhatta, 1993) have explored the added mass effect on scaled ship models. In these studies, a strongly coupled fluid-structure interaction model was used to incorporate this effect into numerical simulations. In this study, we used MSC NASTRAN to implement the FE-BE method. Point sources were distributed on the shell elements of the wetted hull. The solver employs the Helmholtz method to solve the Laplace governing equation (MSC NASTRAN, 2022). Source strength values (σ) were determined by applying relevant boundary conditions to the governing equation. A linear matrix equation based on these strength values allowed us to compute the pressure distribution and the resulting inertia force. The relation between the velocity vector$\dot{u}_i$ at the point r_i and the fluid point source strength σ_j (volume flow rate per area) acting over the area A_j, is calculated according to Equation 1:

$$\dot{u}_i = \sum_j \int_{A_j} \frac{\sigma_j e_{ij}}{\left|r_i - r_j\right|^2} dA_j \tag{1}$$

where e_{ij} is the unit vector of the direction from point j to i. The potential function, whose gradient is $\dot{u}_i$, satisfies the Laplace's equation. Thus, the relevant pressures at each point p_i, are calculated as per Equation 2:

$$p_i = \sum_j \int_{A_j} \frac{\rho \dot{\sigma}_j e_{ij}}{\left|r_i - r_j\right|^2} dA_j \tag{2}$$

Through integrating over the domain, the matrix formulations of the velocity and force vectors are calculated as Equations 3 and 4, with X and Λ as the coefficient matrices of source strengths and their time derivations:

$$\{\dot{u}\} = [\mathrm{X}][\sigma] \tag{3}$$

$$\{F\} = [\Lambda][\dot{\sigma}] \tag{4}$$

Performing the time derivation of Equation 3 and replacing $[\dot{\sigma}]$ into Equation 4 result in the force vector F as Equation 5 based on the virtual mass matrix $\left[M^f\right]$ presented in Equation 6:

$$\{F\} = \left[M^f\right][\ddot{u}] \tag{5}$$

$$\left[M^f\right] = [\Lambda][\mathrm{X}]^{-1} \tag{6}$$

2.3.2 *Vibroacoustic simulation*

The FE model was then modified to perform vibracoustic simulations in MSC ACTRAN. The acoustic elements, representing the surrounding water, were generated based on the model's draught and the outer boundary of the water volume. Figure 6 shows the FE-acoustic model. The water was modeled using a finite fluid component, and the outer surfaces of the water domain, except for the free surface, were simulated as acoustic absorbers to simulate the infinite fluid domain. Accordingly, the free surface was modeled through the pressure boundary condition (MSC, 2021).

The resulting acoustic model had 1133570 four-node tetrahedral linear acoustic elements. According to (Yucel and Arpaci, 2013), eight internodal intervals lead to accurate results.

Therefore, the maximum frequency in this analysis was obtained through Equation 7:

$$L_{\max} \le \frac{c_f}{n_{\min} \cdot f_{\max}} \tag{7}$$

where $L_{\max}$, c_f, and f_{max} are the max internodal interval, i.e., the element size in the linear elements used in this study, the speed of sound in water (1500 m/s), and the frequency limit, respectively. The number of elements per wavelength is represented by $n_{\min}$, which is set as eight in this study based on the work (Yucel and Arpaci, 2013). The maximum edge length of the acoustic elements was 0.15 m, leading to the allowable frequency limit of 1250 Hz, which is largely beyond the frequency range of interest in this problem ($\le$ 100 Hz).

Elasto-acoustic approximations through FE simulation were selected. Considering linear-elastic structural material properties, small displacement, and acoustic approximations of the fluid—homogeneous, inviscid, irrotational—the governing system of equations becomes Equation 8 (Van Antwerpen et al., 2009):

$$\begin{bmatrix} K_s - \omega^2 M_s & i\omega C_{sa} \\ i\omega {C_{sa}}^T & K_a - \omega^2 M_a \end{bmatrix} \begin{Bmatrix} U \\ \Phi \end{Bmatrix} = \begin{Bmatrix} F_s \\ F_a \end{Bmatrix} \tag{8}$$

where U and Φ are the structural nodal displacement and fluid's velocity potential, K and M are stiffness and mass matrices, respectively, C is the coupling matrix, and F is the load vector (including structural and acoustic loads). The subscripts s and a stand for structural and acoustic, respectively.

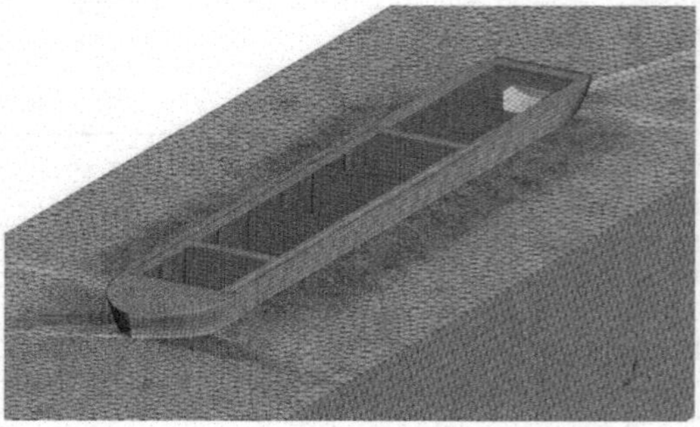

Figure 6. FE model for vibroacoustic simulations in MSC Actran.

2.4 *Analytical model*

The authors employed the analytical formula proposed by Lewis (Equation 9) to assess the effect of added mass on the hull structural response (Korotkin, 2009).

Considering$B/2T \cong 1.2$, the authors determine the added mass per length using Equation 10. The block coefficient of the ship, $C_B = 0.85$, was utilized to modify the effective length for the added mass. Subsequently, by considering the calculated added mass, the authors derived the wet natural frequencies of the model using Equation 11 (Pais, 2018):

$$k_{33} = \frac{\lambda_{33}}{(\pi/2)\,\rho\,(B/2)^2} \quad (9)$$

$$\lambda_{33} = 494\; l\; C_B = 2998\text{kg} \quad (10)$$

$$\lambda_{33} = \Delta\left[\left(\frac{\omega_{n,dry}}{\omega_{n,wet}}\right)^2 - 1\right] \quad (11)$$

where λ_{33} is the added mass of the ship in the vertical direction, k_{33} is its coefficient, Δ is the ship displacement, C_B is the block coefficient, and $\omega_{n,\ dry}$ and $\omega_{n,\ wet}$ are the natural frequencies in dry and wet conditions, respectively. $\omega_{n,\ dry}$ were calculated using the FE model presented in Section 2.2.1.

2.5 *Validation of the numerical simulations*

2.5.1 *Towing tank tests*

Towing tests were conducted at the towing tank of NRC-OCRE in St. John's, NL, Canada.

The towing tank is 200 m long, 12 m wide, and 7 m deep. The propeller was mounted on the model during the self-propulsion and bollard pull tests. A conical weighted hub was installed during resistance tests to compensate for the propeller's weight. This approach ensured the inclusion of propeller and shafting inertial effects in both test scenarios, with and without the propeller.

2.5.2 *Measurements and signal processing for ODS analysis*

Figure 4 shows the arrangement of nine PCB piezoelectric accelerometers, model 352C33, mounted on the model's deck.

Vibration data were collected using the same instrumentation used for EMA (Section 2.2.1), and FRFs were calculated in MATLAB®. ABRAVIBE MATLAB® toolbox was used to visualize the mode shapes. In the analysis, accelerometer No. 9 was used as the reference.

3 RESULTS

3.1 *Numerical results*

Figure 7 shows the model's normal modes from the FE-BE analysis performed in MSC Nastran, and Figure 8 shows the model's normal modes obtained from the simulations in MSC ACTRAN, respectively. Table 3 lists the relevant natural frequencies for both methods.

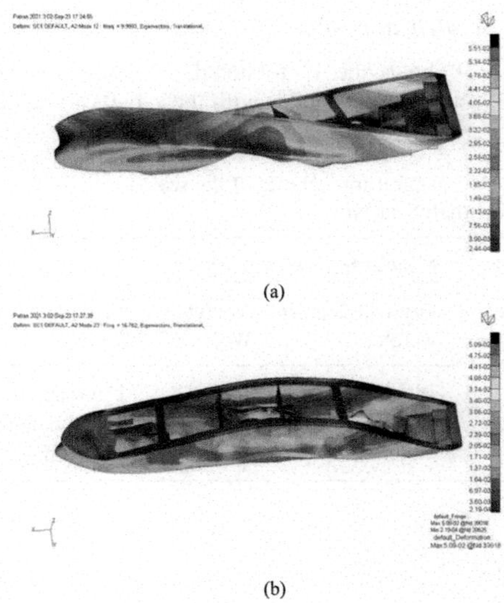

Figure 7. Normal modes from FE-BE simulations - (a) First Torsional mode, (b) first lateral bending mode.

Table 3. Natural frequencies of the scaled ship model-wet condition.

Mode number	Wet Natural frequency [Hz] FE-BE	Vibro-acoustic	Modes
1	10.0	13.0	1st torsional
2	16.8	16.3	Lateral bending
3	24.1	24.3	1st bending

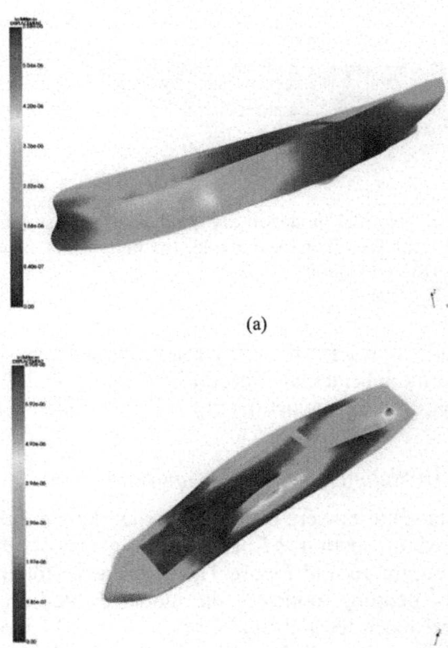

Figure 8. Normal modes from vibroacoustic simulations - (a) First Torsional mode, (b) first lateral bending mode.

3.2 *Analytical results*

Figure 9 shows the 1st torsional, 1st lateral bending, and 1st bending dry modes with these ballast weights.

Table 4. Natural frequencies of the scaled ship model, Dry vs Analytical Wet.

	Natural frequency [Hz]		
Mode number	Dry with ballast weights	Analytical Wet	Modes
1	14.64	10.06	1st torsional
2	22.67	15.58	lateral bending
3	36.18	24.85	1st bending

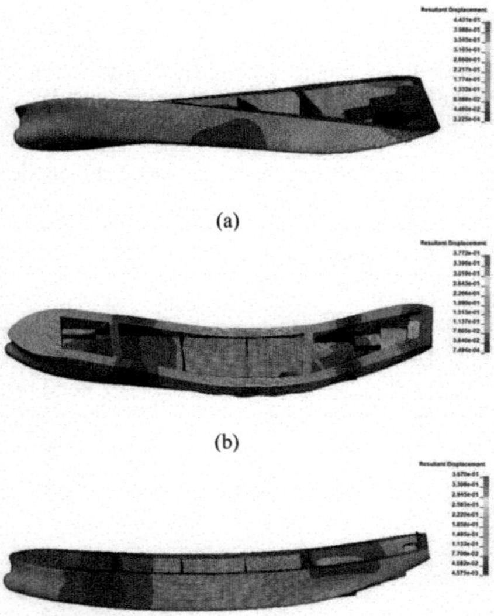

Figure 9. Normal modes in dry conditions, ballast weights added - (a) First Torsional mode, (b) first lateral bending mode, (b) first bending mode.

Table 4 presents the wet natural frequencies of the scaled model calculated according to Section 2.4 and compares them against the dry natural frequencies.

3.3 *Validation through ODS analysis*

Resistance tests were conducted under the conditions outlined in Section 2.5, i.e., utilizing the weighted hub. Figure 10 and Figure 11 show the 1st torsional and 1st bending modes of the model excited during the resistance tests at model advance speeds of 1.15 and 1.95 m/s, and shaft rotation speeds equal to 6 and 11 rps. Table 5 and Table 6 compare the 1st torsional and 1st bending modes obtained through ODS and analytical calculations.

Table 5. Natural frequencies obtained from experimental tests (ODS)-mode 1.

Natural frequency - 1st torsional mode	
Res., Vm=1.15 m/s, 6 rps	11.8 Hz
Res., Vm=1.15 m/s, 11 rps	11.0 Hz
Res., Vm=1.95 m/s, 6 rps	11.8 Hz
Res., Vm=1.95 m/s, 11 rps	11.0 Hz

Table 6. Natural frequencies obtained from experimental tests (ODS)-mode 2.

Natural frequency - 1st bending mode	
Res., Vm=1.15 m/s, 6 RPS	24.1 Hz
Res., Vm=1.15 m/s, 11 RPS	22.1 Hz
Res., Vm=1.95 m/s, 6 RPS	24.1 Hz
Res., Vm=1.95 m/s, 11 RPS	22.1 Hz

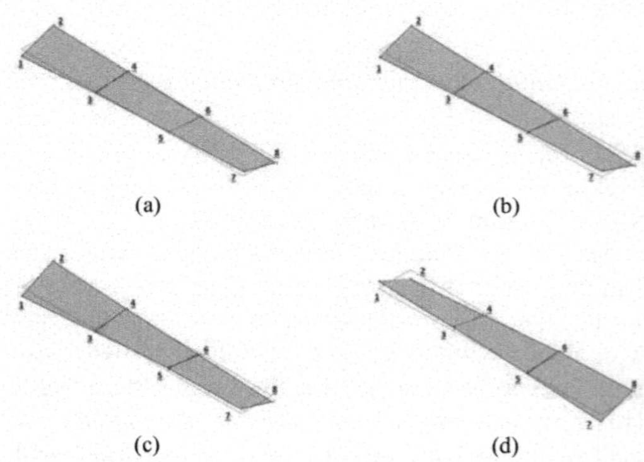

Figure 10. 1st torsional mode excited through Resistance tests, (a) Vm=1.15 (m/s), 6 RPS, (b) Vm=1.15 (m/s), 11 RPS, (c) Vm=1.95 (m/s), 6 RPS, and (d) Vm=1.95 (m/s), 11 RPS.

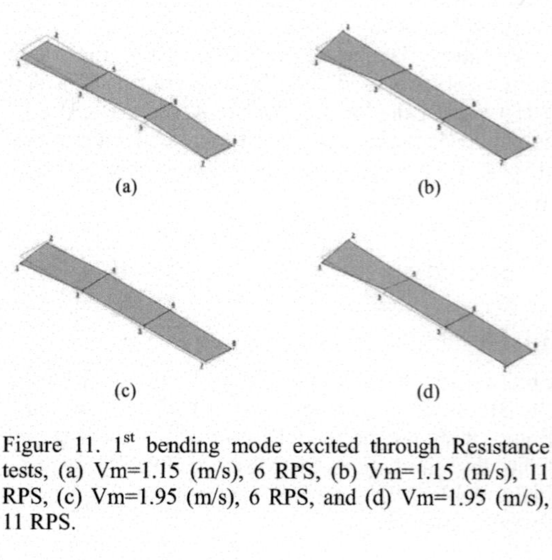

Figure 11. 1st bending mode excited through Resistance tests, (a) Vm=1.15 (m/s), 6 RPS, (b) Vm=1.15 (m/s), 11 RPS, (c) Vm=1.95 (m/s), 6 RPS, and (d) Vm=1.95 (m/s), 11 RPS.

4 DISCUSSION

4.1 *Operational deflection shape analysis*

The results from the operational deflection-shape analysis show some discrepancies for different rotational speeds. In particular, the first torsional mode's frequency ranges between 11.0 Hz and 11.8 Hz (Table 5), while the first bending mode's frequency ranges between 22.1 Hz and 24. 1 Hz (Table 6). This discrepancy is due to the low level of excitation of the model, which is mainly generated by the hydrodynamic flow surrounding the model, and the interference of the shaft, which always presents some eccentricities.

Overall, the ODS performed during the hydrodynamic tests proved to be an effective way to evaluate the dynamics of the ship model's structure and estimate its natural frequencies. While this approach is not as accurate as the EMA, it presents several advantages: it informs on the dynamic behavior of the model's structures at each instant of time during the tests, and it overcomes typical practical issues of EMA when performed on the floating model, e.g., higher energy dissipation and related low signal coherence, and model's rigid body motions.

Through the ODS, we were not able to identify the lateral bending mode of the structure since the ODS was performed using monoaxial accelerometers set up to measure the vertical response.

4.2 *FE-BE versus vibroacoustic method*

By comparing the results from the FE-BE model against those from the vibroacoustic model, we can notice that the results differ by 3 Hz from each other for the first torsional mode, with the vibroacoustic model calculating a higher frequency, while the FE-BE simulation provides the same frequency as the analytical model. When compared against the ODS results and considering an average value of the frequency obtained in the first torsional mode, the FE-BE results underestimate the natural frequency by 12%, while the vibroacoustic model overestimates the natural frequency by 13%.

The first lateral bending mode's frequency differs only 0.5 Hz between the two simulations, with the FE-BE model providing a higher frequency. The analytical results provide the lowest value of the first lateral bending, being 0.72 Hz lower than the vibroacoustic simulation. This lateral mode could not be validated through ODS for the limitations of the experimental set-up explained in Section 4.1.

From the modeling point of view, the FE-BE method is a more straightforward approach. The fluid domain does not need to be modeled explicitly, while it needs to be modeled in the vibroacoustic model. The BE method only requires the definition of the wetted hull.

The vibroacoustic method is beneficial when there is a requirement to observe acoustic pressure distribution within the fluid domain.

Finally, the analytical formulation provides accurate results based on accurate dry natural frequency data.

5 CONCLUSIONS

This paper compared different methods to predict the wet natural frequencies of a ship model. Predicting the effect of the added mass on the structural dynamics of a ship model is important when the structural dynamics may impact the quality of the data from hydrodynamic tests, for instance, the measurement of propeller-induced hull pressures. The findings indicate that FE-BE, the vibroacoustic method, and the analytical models present acceptable results compared to the natural frequencies obtained experimentally through ODS analysis. When comparing the two numerical methods, i.e., FE-BE and vibroacoustic methods, the FE-BE method does not require explicit modeling of the fluid domain, and the calculated frequency of the first torsional mode is equal to the value calculated analytically. On the other hand, the vibroacoustic method allows for calculating vibration-induced acoustic pressures in the fluid domain.

The ODS analysis has proven reliable in monitoring the dynamic behavior of the ship model structures during the hydrodynamic test and identifying the natural frequencies of the model during the resistant tests.

ACKNOWLEDGEMENTS

This study is part of the project "Propeller Induced Noise and Vibration (PINOV)", funded by Transport Canada, National Research of Council Canada, and the Department of Industry, Energy and Technology (IET) of the Government of Newfoundland and Labrador, Canada.

REFERENCES

Annicchiarico, C., Salio, M.P., Viviani, M., 2008. A coupled DBE/FE analysis for the prediction of propeller induced pressure and vibration on the ship hull. *23rd International Conference on Noise and Vibration Engineering 2008*, ISMA 2008 7, 3985–3999.

Datta, R., Guedes Soares, C., 2020. Analysis of the hydroelastic effect on a container vessel using coupled BEM–FEM method in the time domain. *Ships and Offshore Structures* 15, 393–402.

Egusquiza, E., Valero, C., Liang, Q., Coussirat, M., Seidel, U., 2009. Fluid Added Mass Effect in the Modal Response of a Pump-Turbine Impeller. In *International Design Engineering Technical Conferences and Computers and Information in Engineering Conference*, 48982:715–24.

Eslaminejad, A., Ziejewski, M., Karami, G., 2019. An experimental–numerical modal analysis for the study of shell-fluid interactions in a clamped hemispherical shell. *Applied Acoustics* 152, 110–117.

Jaiman, Rajeev Kumar, and Vaibhav Joshi. 2022. *Computational Mechanics of Fluid-Structure Interaction*. Springer.

Jaiman, R.K., Parmar, M.K., Gurugubelli, P.S., 2014. Added Mass and Aeroelastic Stability of a Flexible Plate Interacting With Mean Flow in a Confined Channel. *Journal of Applied Mechanics* 81, 041006.

Korotkin, A.I., 2009. *Added Masses of Ship Structures, Fluid Mechanics and Its Applications*. Springer Netherlands, Dordrecht.

Moro, L., Biot, M., Brocco, E., De Lorenzo, F., Nicolás, P., Vassallo, M., 2013. Hull vibration analysis of river boats. *EUT Edizioni Università di Trieste*.

MSC, 2021. *Actran 2021.1 User's Guide*.

MSC NASTRAN, 2022. *Dynamic Analysis User's Guide*.

Pais, T., 2018. Analytical and numerical computation of added mass in vibration analysis for a superyacht. *Ships and Offshore Structures* 13, 443–450.

Pais, T., Moro, L., Boote, D., Biot, M., 2017. Vibration analysis for the comfort assessment of superyachts. *Journal of Marine Science and Application* 16.

Rahman, M., Bhatta, D.D., 1993. Evaluation of added mass and damping coefficient of an oscillating circular cylinder. *Applied Mathematical Modelling* 17, 70–79.

Rajendran, S., Fonseca, N., Guedes Soares, C., 2016. A numerical investigation of the flexible vertical response of an ultra large containership in high seas compared with experiments. *Ocean Engineering* 122, 293–310.

Tian, W., Yao, L., Li, L., 2017. A Coupled Smoothed Finite Element-Boundary Element Method for Structural-Acoustic Analysis of Shell. *Archives of Acoustics* 42, 49–59.

Van Antwerpen, B., d'Udekem, D., Weißenborn, C., 2009. Vibro-Acoustic Simulation of Structure-borne Induced Radiation of Ship Windows. *Proceedings of the International Conference on Acoustics NAG/DAGA*.

Yucel, A., Arpaci, A., 2013. Free and forced vibration analyses of ship structures using the finite element method. *Journal of Marine Science and Technology (Japan)* 18, 324–338.

Zambon, A., Moro, L., Biot, M., 2021. Vibration analysis of superyachts: Validation of the Holden Method and estimation of the structural damping. *Marine Structures* 75, 102802.

Zhang, W., Ning, X., Guo, H., Wang, X., Li, F., Sun, S., 2021. Numerical analysis of propeller-excited plate vibrations. *AIP Advances* 11, 065115.

Advances in Maritime Technology and Engineering – Guedes Soares & Santos (Eds)

Uncertainty assessment of the scale effects on a submerged cylinder

H.S. Li, S. Wang & C. Guedes Soares
Centre for Marine Technology and Ocean Engineering (CENTEC), Instituto Superior Técnico, Universidade de Lisboa, Lisbon, Portugal

ABSTRACT: This study investigates scale effects on hydrodynamics of a 2D submerged circular cylinder subjected to incoming flow in Computational Fluid Dynamics simulation. The primary focus is the influence of varying scale ratios on drag coefficient, lift coefficient, and vortex shedding frequency in high Reynolds number conditions. The methodology involves *k-kl-w* turbulence model and PimpleFoam solver in OpenFOAM. Initial simulations establish baseline values for drag, lift and frequency in deep open water. Subsequent simulations examine these parameters for cylinders at different scale ratios. Numerical discretization uncertainty for each case is quantified by using a Correction Factor based Approach to verify the simulation results, allowing for an assessment of scale effect. Results indicate that three values show significant variations with scale changes, even at the same Reynolds number. This variation is more pronounced in large scale and low flow velocity conditions. The study highlights the importance of considering scale effects in Computational Fluid Dynamics of underwater objects.

1 INTRODUCTION

The cruising performance of Autonomous Underwater Vehicle (AUV) serves as a crucial indicator in pre-development, design, and evaluation phases (Xia et al., 2016). As a representative case with a high Reynolds number in fluid dynamics research, the analysis of drag coefficient, lift coefficient, and tail vortex shedding frequency holds paramount significance for the comprehensive investigation of cruising performance (Mitra et al., 2019).

For parameterized studies on the design of AUV, it is essential to perform optimization based on reliable parent hydrodynamic models such as SUBOFF (Huang & Liu, 1994) or a simplified cylinder with varying scales (Ostafichuk, 2004). Seeking the most optimal scale parameters is crucial to optimize parameters related to payload, endurance, speed, and manoeuvrability (Bidoki et al., 2018). In particular, different design schemes with varying scale ratios typically require initial numerical simulations using Computational Fluid Dynamics (CFD) tools (Vasudev et al., 2014). Leong et al. (2013) validated hydrodynamic disturbances among AUVs of varying scales using CFD based on the SUBOFF model. Kahramanoglu (2023) conducted a viscous CFD study of the scale effects on horizontal plane manoeuvring derivatives for AUVs based on the SUBOFF. On the other hand, conducting water tank experiments with a designed and scaled-down fully functional AUV can significantly reduce the substantial costs associated with directly manufacturing a full-scale AUV (de Moraes & Santiago, 2018). Fan et al. (2023) conducted water entrance simulation of scaled-down AUV using CFD tools in various gravity environments to validate their similarity and predictive accuracy compared to full-scale prototypes.

Although most CFD solvers utilize the same governing equations, the use of different time discretization methods, grid topologies, turbulence assumptions, and boundary conditions for models lead to numerical simulations that do not precisely capture scale effects observed in experiments, resulting in numerical uncertainties and errors even for different scales of a same geometry. Islam & Guedes Soares (2021) quantified the grid uncertainty in 3D circular cylinder simulations of wave-induced motion using OpenFOAM. Wang et al. (2022) studied the numerical uncertainty in mesh-based and mesh-free numerical simulations of the motion of wave floaters.

Zhang & Fu (2019) proposed an efficient Bayesian uncertainty quantification approach to assess the parameter uncertainty of the SST turbulence model in DNS methods. Cheung et al. (2011) established three competitive categories based on the Spalart-Allmaras turbulence model and used Bayesian uncertainty quantification to predict Quantities of Interest (QoI). Oliver & Moser (2011) captured the uncertainty in RANS methods arising from insufficient parameters and models using Bayesian uncertainty quantification.

The main objective of the present study is to discuss the scale effects on the drag coefficients, lift and vortex shedding for 2D cylinders in flow and quantify the numerical discretization uncertainties for each scale in the predictions from OpenFOAM. Section 2 introduces the numerical solvers used in CFD and the uncertainty estimation method. Section 3 describes the parameter setup for both the

DOI: 10.1201/9781003508779-15

benchmark case and other scaled cases. Section 4, discuss the computational results of all cases, focusing on the scale effect and its impact on uncertainty and error. Finally, Section 5 concludes the study, summarizing the key findings and their implications.

2 METHODOLOGY

2.1 *Numerical solver*

OpenFOAM is an open-source CFD solver tool (Jasak, 2009). In OpenFOAM, the viscous solver PimpleFoam combines the advantages of the simple algorithm (transient) and the PISO algorithm (steady-state) (Penttinen et al., 2011). This makes it suitable for large time-step transient simulations of incompressible flow, which is relevant to this paper. It allows vortex shedding to develop sufficiently over time and enables frequency domain analysis of the dynamic shedding process (King, 1977).

As PimpleFoam is based on SimpleFoam, it initially utilizes the steady-state Navier-Stokes equations:

$$\nabla \cdot U = 0 \tag{1}$$

$$\nabla \cdot (UU) - \nabla \cdot (\upsilon \nabla U) = -\nabla p \tag{2}$$

Omitting the intermediate derivation steps, Equation (1) & (2) ultimately arrive at the following equation:

$$A_P^n U_P^* + \Sigma A_N^n U_N^* = -\frac{1}{V_P} \Sigma p_f^n S_f \tag{3}$$

Equation (3) represents the velocity equation in OpenFOAM, also known as the momentum prediction step. By solving Equation (3), the predicted velocity U_P^* can be obtained.

The construction of the pressure Poisson equation is formally derived from the continuity equation. It is obtained by further applying the divergence operator to the momentum equation and summing the results. Considering the case of final convergence, the discretized form of the continuity equation, denoted as $\nabla \cdot U=0$, is as follows:

$$\Sigma U_{P,f}^{n+1} \cdot S_f = 0 \tag{4}$$

After derivation, the pressure Poisson equation is obtained as follows:

$$\nabla \cdot HbyA^{n+1} = \nabla \cdot (\frac{1}{A_P^n} \nabla p^{n+1}) \tag{5}$$

The final solved velocity U_P^{n+1} and pressure p^{n+1} should satisfy Equations (4) and (5), corresponding to the momentum equation and the continuity equation, respectively. However, currently, only U_P^* is obtained from the momentum prediction step, with both $HbyA_P^{n+1}$ and p^{n+1} remaining unknown. Therefore, Equation (5) cannot be solved directly. The PIMPLE iterative algorithm is designed to address this problem.

In terms of turbulence models, the PimpleFoam solver in OpenFOAM supports various improved turbulence models based on the k-ω framework, including *k-kl-ω* and *k-ω* SST (Shear-Stress Transport) (Devolder et al., 2018). Compared to the $k-\varepsilon$ model, the $k-\omega$ SST model generally performs better in predicting flow separation and adverse pressure gradient flows (Menter, 1992). Furthermore, *k-kl-ω* turbulence model takes into account the unsteadiness of turbulence and provides more accurate predictions, especially in turbulent shear layers and near-wall regions (Salimipour, 2019). It is particularly useful for capturing the unsteady processes of vortex generation at initial time steps and simulating anisotropy (Dikbaş and Baran, 2023).

Walters & Cokljat (2008) presented a three-equation turbulent viscosity model for transitional flow simulation. This model includes transport equations for the turbulent kinetic energy per unit mass *(kT)*, the laminar kinetic energy per unit mass *(kL)*, and the specific dissipation rate *(ω)*. These differential equations in an incompressible form can be written as follows:

$$\frac{Dk_T}{Dt} = P_{kT} + R_{BP} + R_{NAT} - \omega k_T - D_T + \frac{\partial}{\partial \chi_j}\left[\left(\upsilon + \frac{\alpha_T}{\alpha_k}\right)\frac{\partial k_T}{\partial \chi_j}\right] \tag{6}$$

$$\frac{Dk_L}{Dt} = P_{kL} - R_{BP} - R_{NAT} - D_L + \frac{\partial}{\partial \chi_j}\left[\nu \frac{\partial k_L}{\partial \chi_j}\right] \tag{7}$$

$$\frac{D\omega}{Dt} = C_{\omega 1}\frac{\omega}{k_T}P_{kT} + \left(\frac{C_{\omega R}}{f_W} - 1\right)\frac{\omega}{k_T}(R_{BP} + R_{NAT}) - f_W{}^2 C_{\omega 2}\omega^2 + C_{\omega 3} f_\omega \alpha_T f_W{}^2 \frac{\sqrt{k_T}}{d^3} + \frac{\partial}{\partial \chi_j}\left[\left(\nu + \frac{\alpha_T}{\sigma_\omega}\right)\frac{\partial \omega}{\partial \chi_j}\right] \tag{8}$$

where *P*, R_{BP}, R_{NAT} and *D* denote the turbulent production, bypass transition, natural transition, and near-wall dissipation term, respectively. The turbulent production terms are expressed as follows:

$$Pk_T = \nu_{T,S} S^2 \tag{9}$$

$$Pk_L = \nu_{T,l} S^2 \tag{10}$$

where S is the strain rate, and $\nu_{T,s}$ and $\nu_{T,l}$ denotes small and large-scale eddy viscosities, respectively. The small-scale eddy viscosity is defined as follows:

$$\nu_{T,S} = f_\upsilon f_{INT} C_\mu \lambda_{eff} \sqrt{k_{T,S}} \tag{11}$$

where f_υ, f_{INT}, λ_{eff} and C_μ represent the viscous damping function, intermittency damping function, wall-limited turbulence length scale, and turbulent viscosity coefficient, respectively.

In this study, the Reynolds-Averaged Navier-Stokes (RANS) approach is employed for the time-averaging of turbulent flows. This method is instrumental in solving for the mean velocities and turbulent characteristics in fluid dynamics. It implies that velocity, pressure, and other flow properties are decomposed into their mean and fluctuating components (Holloway et al., 2004). Although Large Eddy Simulation (LES) shows improved agreement with experimental data at finer mesh resolutions (Celik, 2005), the accuracy of the RANS is considered adequate for addressing flow around a cylinder (Stringer et al., 2014), as well as getting a higher efficiency than DNS (Komen et al., 2014). Consequently, this paper does not extend to irrelevant analysis of different approaches to solve turbulent flows.

In this paper, the numerical procedure used to compute the unsteady, viscous, and incompressible flow is used. The integral formulations of RANS equations for the mass and momentum conservation can be expressed by:

$$\oint_{\partial\Omega} \rho V dS = 0 \tag{12}$$

$$\frac{\partial}{\partial t}\int_{\Omega} \vec{W} d\Omega + \oint_{\partial\Omega} \vec{J} dS = 0 \tag{13}$$

where ρ is the fluid density, Ω is the control volume bounded by the closed surface $\partial\Omega$, V denotes the contravariant velocity defined as the velocity normal to the surface element dS (Teymourtash & Salimipour, 2017), W is the vector of the so-called conservative variables, and J denotes the sum of convection and diffusion fluxes expressed as follows:

$$\vec{W} = \begin{bmatrix} \rho u \\ \rho v \end{bmatrix}, \quad \vec{J} = \begin{bmatrix} \rho u V + n_x p - (\mu + \mu_t)(n_x \frac{\partial u}{\partial x} + n_y \frac{\partial u}{\partial y}) \\ \rho v V + n_y p - (\mu + \mu_t)(n_x \frac{\partial v}{\partial x} + n_y \frac{\partial v}{\partial y}) \end{bmatrix}, \tag{14}$$

$$V \equiv \vec{v}.\vec{n} = n_x u + n_y v \tag{15}$$

where n_x and n_y are the components of the outward-facing unit normal vector of the surface $\partial\Omega$ and μ and μ_t are the molecular and turbulent viscosities, respectively.

Supported by those numerical solvers, detailed settings for turbulence and grid parameters will be elaborated in the following section.

2.2 *Uncertainty estimation method*

Uncertainty analysis is a systematic study aimed at evaluating the consistency and accuracy of CFD solvers in addressing intended problems. It encompasses two key processes: verification and validation. Verification is the process of ensuring that the model implementation accurately represents the algorithms and solutions of the model being implemented. Validation, on the other hand, determines the extent to which a model's predictions adequately represent the physical phenomena being measured (Islam & Guedes Soares, 2021).

Following the ITTC (2017) guidelines for Uncertainty Analysis in CFD, the simulation error δ_S can be defined as the discrepancy between the actual value T, and the simulated value S. This error can be further decomposed into the sum of modelling error δ_{SM} and numerical error δ_{SN} in CFD (δ_S=S-T=δ_{SM}+δ_{SN}). The modelling error pertains to the inaccuracies or assumptions inherent in the numerical model itself, whereas the numerical error relates to the computation-related errors (Islam & Guedes Soares, 2019).

Verification involves assessing simulation numerical uncertainty U_{SN} and determining the error sign and magnitude when conditions permit. It also involves assessing the uncertainty in the error estimation U_{ScN}. In general, numerical errors are decomposed into iterative error δ_I, gird size error δ_G, time step error δ_T, and other parameters related errors δ_p, assuming as δ_{SN} =δ_I +δ_G,+δ_T +δ_p. In this case, the simulation numerical uncertainty is provided as Equation 16 (Islam & Guedes Soares, 2019).

$$U_{SN}^2 = U_I^2 + U_G^2 + U_T^2 + U_P^2 \tag{16}$$

For the corrected simulation approach, the solution is adjusted to generate a numerical benchmark S_C, and the estimated numerical simulation error, represented by δ^*_{SN} =δ^*_I +δ^*_G,+δ^*_T +δ^*_p. The corrected uncertainty δ_{SM} is calculated as follows (Islam & Guedes Soares, 2019):

$$U_{ScN}^2 = U_{Ic}^2 + U_{Gc}^2 + U_{Tc}^2 + U_{Pc}^2 \tag{17}$$

Validation is a procedure to evaluate the uncertainty of simulation modelling U_{SM} by employing benchmark experimental data and, when possible, estimating the sign and magnitude of the modelling error δ_{SM}. The comparison error E is determined as the disparity between the data D and the simulation value S, E=D-S=δ_D-(δ_{SM} +δ_N) (Islam & Guedes Soares, 2019).

Considering that quantifying modelling errors is very difficult (Islam & Guedes Soares, 2019), in this study, the geometric model is introduced with accuracy settings smaller than the grid resolution to disregard model error. Furthermore, the main focus of the research will involve comparisons related to scale, grid and time step.

This paper employs the uncertainty analysis method based on correction factors (Ci) as recommended by the ITTC (2017) guidelines. This method uses the Richardson extrapolation technique to estimate uncertainties in simulation results. The specific details of this approach will be presented in Section 4.

3 CASE SETUP

3.1 *Preprocessing for benchmark*

3.1.1 *Geometry and computational domain*

The base case considers a cylindrical geometry with the following dimensions: diameter $D=0.12m$, which is a typical diameter of a small AUV. According to previous work by other researchers (Stringer et al., 2014, Vijay & Kumar, 2019, Ong et al., 2009, Vakil &Green, 2009), the water tank region is defined with a length $L=40D$, height $H=20D$ and center coordinates of cylinder at $G=(2.5m, 0m)$. The initial mesh size is set to $\Delta x=0.01m$ and is isotropic. A hexahedral mesh model was selected as the initial mesh.

A rectangular region with a finer mesh is selected as the refinement area, with a length $L_{fine}=13D$, a height $H_{fine}=6D$, and a thickness of $B=0.01m$. The refinement level is set to 1 (0.5 times the initial mesh size) for the rectangular region and 3 (0.25 times the initial mesh size) near the cylindrical wall. Original mesh length set as $\Delta x_1= 0.01m$. The refinement mesh size was determined by dividing the cylinder diameter D by approximately 1/100. For computational convenience, the mesh was further refined using $\sqrt{2}$, which resulted in a final mesh size of Δx_{1fine}=0.00125m. Details of the parameters are shown in Table 1 and Figure 1.

Table 1. Size of the three main regions.

Region	Mesh length M	L m	B m	H m
Cylinder	Δx_{1fine} =0.00125	0.12	0.01	0.12
Refine mesh	Δx_{1fine} =0.00125	1.56	0.01	0.72
Water tank	Δx_1 = 0.01	4.8	0.01	2.4

A boundary prism layer mesh is introduced with 20 layers, and its thickness will be discussed in the next section. In the computational region, the physical property of the fluid is specified as seawater at a temperature of 20°C. That is, the density of the fluid is denoted as $\rho=1025kg/m^3$; The dynamic viscosity coefficient is represented by $\mu=1.006e\text{-}3\ kg/(m{\cdot}s)$; The incoming flow velocity is given as $V_1=0.5m/s$; The calculation method for determining the Reynolds number at the open water region of the flow field is as follows:

$$Re_1=\frac{\rho V_1 D_1}{\mu}=6.1e+4 \tag{18}$$

In this study, which focuses on high Reynolds number motion in infinite water depth, the *k-kl-ω* turbulence model is chosen to close the governing equations (Ong et al., 2009). In addition, the physical property in all the cases is set as incompressible, viscous water with that subcritical Reynolds number.

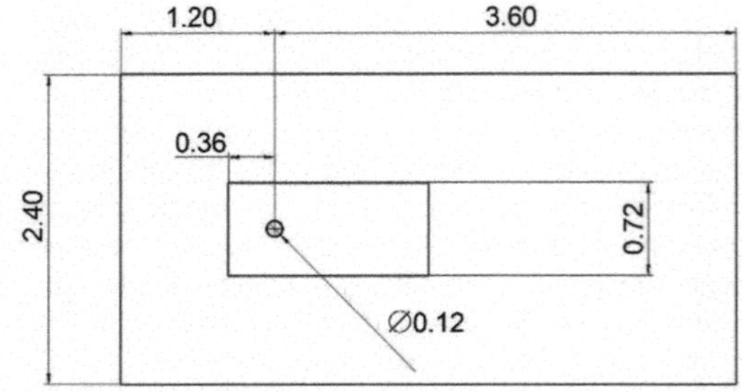

Figure 1. Size of the three main regions.

3.1.2 *Low Y+ wall treatment boundary layer mesh*

Different turbulence models have different requirements for the *y+* value, which refers to the thickness of the first mesh layer. In this study, the *k-w* SST model is selected to solve the viscous sublayer near the boundary. Therefore, a *y+* value of 1 is chosen to ensure that the first mesh layer is located within the viscous sublayer.

In the case of low *y+* values, the near-wall region is typically handled using a direct solution method. This means that in the generation of near-wall mesh, the thickness of the first layer of the boundary prism layer can be obtained based on the desired *y+* value and the formula derived from the selected wall treatment method. Therefore, assuming a target *y+* value, the first layer thickness of the boundary prism layer *y* can be calculated using the formula (Stringer et al., 2014):

$$y=\frac{\sqrt{80Ly^+}}{\mathrm{Re}^{-13/14}} \tag{19}$$

where *L* represents the characteristic length, which is the thickness of the first mesh layer in the boundary layer, and *Re* refers to the Reynolds number. Furthermore, the method for determining *y+* can be derived from Equation 19 as follows:

$$y^+=\frac{yu_\tau}{v} \tag{20}$$

where *y* represents the distance from the wall, u_z refers to the shear velocity, and *v* denotes the kinematic viscosity. Based on the calculations, the first layer thickness of the boundary prism layer is assumed that $\Delta y_1=3.4e\text{-}5m$. Given that the mesh size in the refinement region is $\Delta x_{1fine}=0.00125m$, the thickness of the outermost layer relative to the fine mesh can be calculated as *finalLayerThickness* $=1\Delta x_{fine}$.

The thickness of the first layer away from the wall can be determined as *minThickness*=$0.0272\ \Delta x_{fine}$. Based on engineering experience, the number of boundary prism layers is set as *nSurfaceLayers* =*20*, with an expansion ratio of *expansionRatio*=*1.2* for each layer (Travin et al., 2000). Overall, the boundary prism layer mesh structure near the surface of the cylinder is illustrated in Figure 2. The total number of mesh cells in the computational region is 327,546.

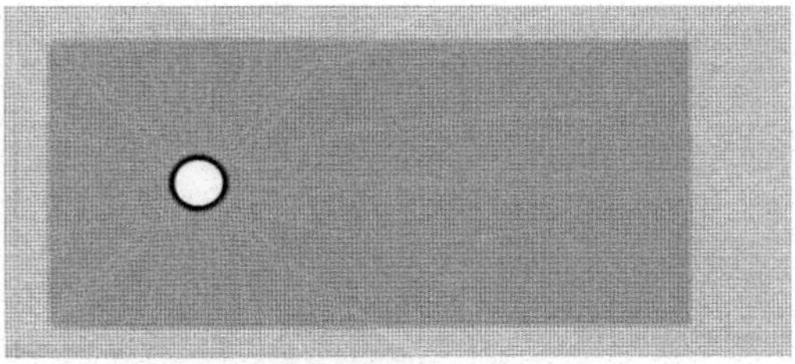

(a) Global mesh with refined region

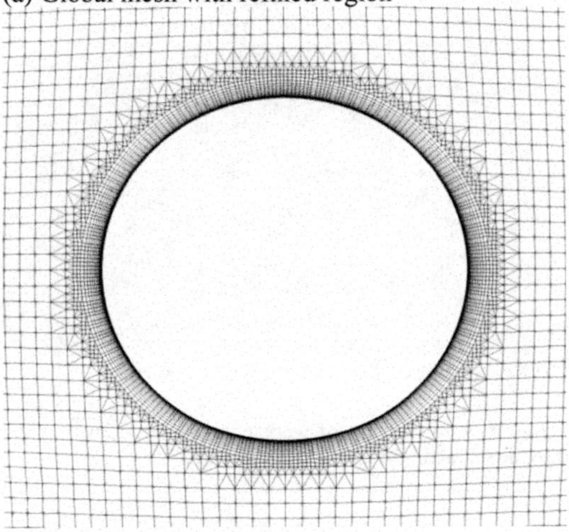

(b) Local mesh with boundary layer

Figure 2. Boundary prism layer mesh of cylinder.

3.1.3 *Solver settings*

The *k-kl-ω* turbulence model is applied in OpenFOAM, where the initial parameters for various faces of the computational domain are defined as Table 2, where *ω* represents the specific dissipation rate, *k* is the turbulence kinetic energy, *U* denotes flow velocity, *p* symbolizes pressure, and *nut* stands for turbulent viscosity.

The time step size is set as $\Delta t_1=0.01s$ and the total simulation time is $T_1=50s$. The number of internal iterations is set to 5. The flow velocity *U* is equal to the ship's speed V_1. Results are recorded at every time step of $\Delta t_{1r}=0.1s$; The maximum local Courant-Friedrichs-Lewy (CFL) number is computed as follows:

$$Co_{\max} = V_1\Delta t/\Delta x_{fine} = 1$$
$$Co_{mean} = V_1\Delta t/\Delta x = 0.125 \quad (21)$$

Regarding the numerical methods of flow around a cylinder is not overly complex, to achieve higher computational efficiency, relaxation factors for velocity, pressure, and turbulence parameters are not set (Widiatmono et al., 2022). Those values satisfy the convergence criteria for the calculations

3.2 *Scaled cases*

This section covers control cases with consistent relative mesh size to investigate the scale effect.

Table 2. The initial set of parameters.

Quantity	Face	Boundary Condition	Value
ω	Inlet	fixedValue	1e-5
	Outlet	zeroGradient	-
	Top and Bottom	symmetryPlane	-
	Cylinder	zeroGradient	-
k	Inlet	fixedValue	1e-10
	Outlet	zeroGradient	-
	Top and Bottom	symmetryPlane	-
	Cylinder	fixedValue	0
U	Inlet	fixedValue	0.5
	Outlet	zeroGradient	-
	Top and Bottom	symmetryPlane	-
	Cylinder	fixedValue	0
P	Inlet	zeroGradient	-
	Outlet	fixedValue	0
	Top and Bottom	symmetryPlane	-
	Cylinder	zeroGradient	-
nut	All	calculated	0

Based on extensive experimental data from others (Assi, 2009), it is known that the vortex field state and drag coefficient Cd of flow around a circular cylinder are solely related to the Reynolds number Re. In Section 3.1.1, it is mentioned that Re=*6.1e+4*. To analyze the impact of scale effect on drag and vortex shedding of the cylinder, a control case with large scale and low flow velocity, while maintaining an equal Reynolds number is used based on the following relationship:

$$Re_2 = \frac{\rho V_2 D_2}{\mu} = Re_1 = \frac{\rho V_1 D_1}{\mu} \quad (22)$$

Considering the geometry, mesh, and flow conditions of the base case, scaled ratios $\lambda_2=2$, $\lambda_3=4$, have been defined.

For example, when $\lambda_2=2$, then D_2 =0.24m. By solving formula (22), it is found that when *Re* is the same as the base case, the incoming flow velocity V_2 =0.05m/s. To maintain the mesh uncertainty, both the computational region and the refined region are scaled up by a factor of λ_2, while keeping the relative grid size unchanged. Therefore, the grid spacing is adjusted as $\Delta x_2=\lambda_2\Delta x_1=0.02m$. However, the boundary layer grid is recalculated based on the new parameters, ensuring that the first layer thickness in the prism layer is maintained at *min Thickness*$=\Delta y=3.4e\text{-}4m=0.0272\Delta x_{2fine}$, thus, to ensure *y*+ still 1. Due to the change in flow velocity, the development time of the vortex shedding is different, the simulation duration is set as $T_2=\lambda_2^2\ T_1=200s$ as well as set $\Delta t_2=\lambda_2^2\Delta t_1=0.04$ to ensure that the maximum Courant Number remains at 1 following:

$$Co_{\max} = V_1 \Delta t_1 / \Delta x_{1fine} = V_2 \Delta t_2 / \Delta x_{2fine} \quad (23)$$

Consequently, the numerical parameters for the scale effect control group can be summarized as follows in Table 3.

Table 3. Parameters of scaled cases.

Parameters	Quantity	Value			
Scale	λ	1	2	4	8
Cylinder Diameter	*D(m)*	0.12	0.24	0.48	0.96
Velocity	*V(m/s)*	0.5	0.25	0.125	0.0625
Cell Length	*Δx(m)*	0.01	0.02	0.04	0.08
Calculation Time	*T(s)*	50	200	800	3200
Time step	*Δt(s)*	0.01	0.04	0.16	0.64
Boundary Layer Thickness	*Δy(m)*	3.4e-4	*6.8e-4*	*1.36e-3*	*2.72e-3*
Reynold Number	*Re*	6.1e +4			

4 RESULTS AND DISCUSSION

4.1 *Results of the base case*

4.1.1 *Drag coefficient*

The transient cylinder surface of each time step is surface integrated through ParaView post-processing. The drag coefficient, F_d is calculated as:

$$F_d = 1/2 C_d \rho_w v^2 S = p \cdot Normals\_X \quad (24)$$

where p is surface pressure; *Normals_X* is a component of the surface normal vector along the *X*-axis, which can be obtained through post-processing visualization via each written time step; V is flow velocity, the cloud map of which at t*=50s* is shown in Figure 3; S is the projected area of the object to the incoming flow direction. ρ_w is the density of seawater in 20°C. Based on the derivation of Equation (5), the expression for solving the drag coefficient can be obtained as follows:

$$C_d = 2p \cdot Normals\_X / v^2 S \rho_w \quad (25)$$

In Figure 3, it can be observed from the local streamline plot along the surface of the cylinder that the vortex separation points occur in the front hemisphere of the oncoming flow (Williamson, 1996). This observation aligns with the theoretical vortex separation points expected for a given *Re=6.1e+4*.

Given V_1*=0.5m/s* and ρ_w*=1025kg/m*^{3}, for a single-layer mesh on a 2D circular cross-section, the frontal flow projected area $S = \Delta y_D = 0.012m^2$. The convergence of the drag coefficient C_D with time steps is shown in Figure 4.

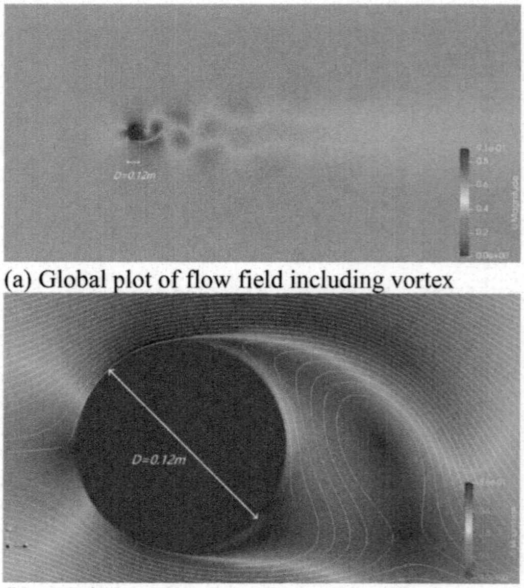

(a) Global plot of flow field including vortex

(b) Local plot of boundary layer flow filed

Figure 3. Contour plot of velocity at t =50s depicting the fully developed vortex.

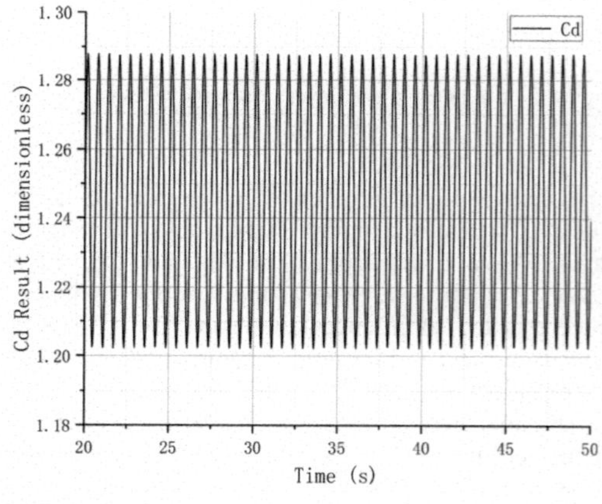

Figure 4. Time history of the drag coefficient.

Considering the time interval of t*=20s~50s* when the vortex shedding effect is fully developed, the average drag coefficient is C_D*=1.2496*. Comparing this value with the experimental drag coefficient C_{D0} =1.25 at *Re=6.1e+4* as shown in (Assi, 2009), the total numerical error is δ_{SN}*=0.032%.*

The time series of the drag force, obtained during the fully developed vortex shedding period can be approximated by a sine function as follows:

$$F_x(t) = F_{x0} + \hat{F}_x \sin(2\pi f_x t + \phi_x) \quad (26)$$

where F_{x0}=1.91N, $F^\wedge_x$=0.0659N, the frequency of drag force fluctuation is f_x= 1.60, and the phase angle is Φ_x=0.184.

4.1.2 *Lift and vortex shedding*

The lift coefficient C_L is calculated using the same method as the drag coefficient C_D, which involves integrating the surface pressure of the object and projecting it along the *Y-axis*. As a result, the lift coefficient C_L converges over time steps, as shown in Figure 5.

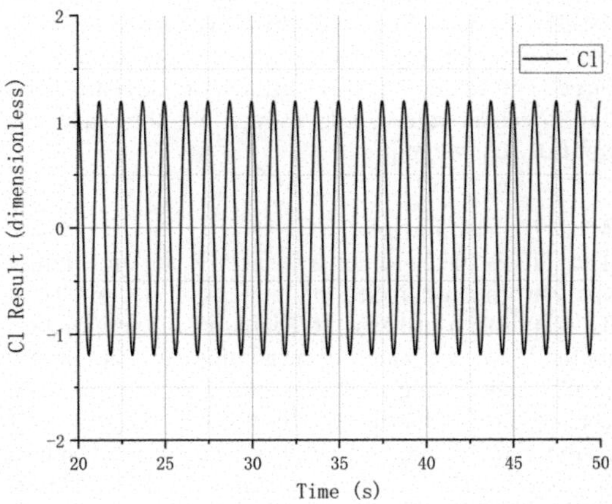

Figure 5. Time history of the lift coefficient C_L over time.

The periodic oscillations of C_L in the range of $t = 20s \sim 50s$, which is the fully developed vortex shedding period, can be approximated by the following Sine function:

$$C_y(t) = \hat{C}_y \sin(2\pi f_y t + \phi_y) \tag{27}$$

where the peak value of the lift coefficient $\hat{C'}_L$=1.197, and the dimensionless root mean square (R.M.S.) value of the lift coefficient C'_L=0.5385. Compared with the experimental value C'_{L0}=0.5, the error is δ=7.7%. The lift fluctuation frequency f_y =0.799, and the phase angle Φ_y=*-4.52*. Hereby introduce dimensionless Strouhal number (Williamson, 1996) for the frequency of lift fluctuations given by:

$$St = \frac{f_s D}{V} \tag{28}$$

When comparing to the experimental value St_0 =0.187 obtained from Norberg (2003) at *Re=6.1e+4*, the experimental vortex shedding frequency would be $f_{y0}=0.779$. Given that the lift fluctuation frequency coincides with vortex shedding frequency, numerical error $\delta_{SN}=|f_y-f_{y0}|/\ f_{y0}=2.6\%$. The resistance fluctuation frequency is typically twice as δ_1 $=|f_x-2f_y|/2f_y$=0.125%. The iterative error could be obtained like $\delta_1=|f_x-2f_y|/2f_y$=0.125%.

4.2 *Results of scaled cases*

4.2.1 *Comparison with experimental data*

To investigate the impact of grid size uncertainty for different scaled cases, this study sets three different grid sizes for each of the 4 scaling ratios $\lambda_1=1$, $\lambda_2=2$, $\lambda_3=4$, *and* $\lambda_4=8$. The impact of these variations is assessed by testing C_D, root mean square C_L and vortex shedding frequency f in the flow field.

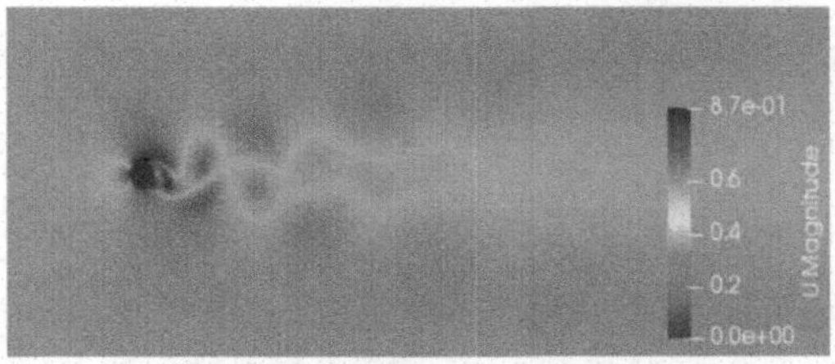

(a) λ_1=1 with V_1=0.5m/s.

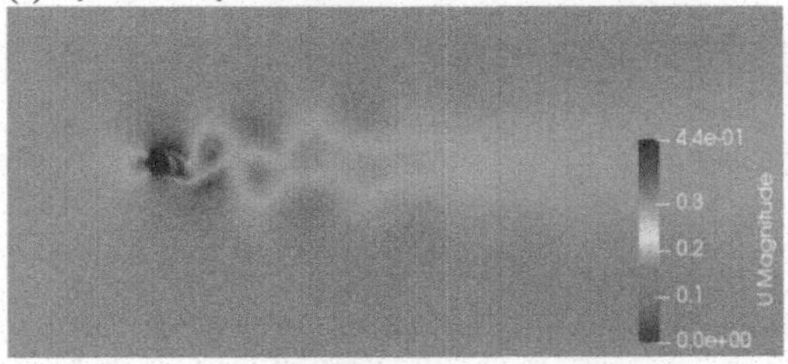

(b) λ_2=2 with V_2=0.25m/s.

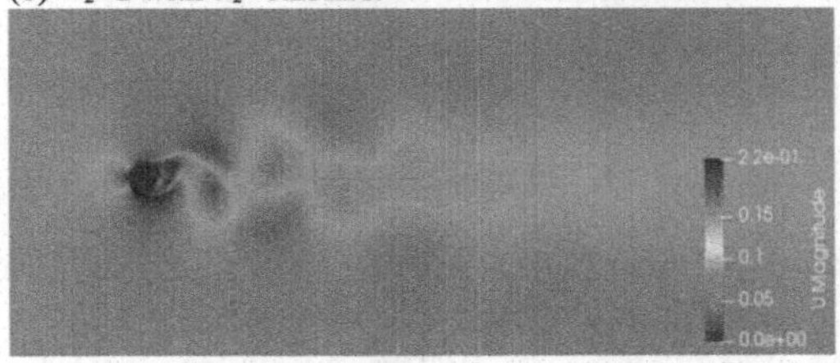

(c) λ_3=4 with V_3=0.125m/s.

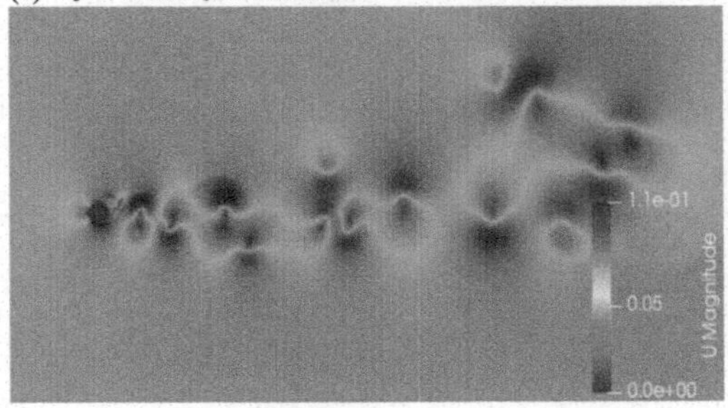

(d) λ_4=8 with V_4=0.0625m/s.

Figure 6. Contour plot of velocity at latest time step of 4 different cylinder scales.

In the previous section, the refinement mesh size of the benchmark case is $\Delta x_1=0.01m$. According to the ITTC uncertainty verification method on CFD, two additional control mesh configurations are selected for re-computation, with their edge lengths being $\Delta x/\sqrt{2}$ and $\Delta x/2$ (Procedurest, 2017). All the calculation results and errors compared with the experiment are shown in Table 4. And comparation of flow field for all 4 cylinder diameters is shown in Figure 6.

The results in Table 4 show that for both C_D and C_L, the error relative to experimental values gradually decreases as the grid is refined. However, for

Table 4. Summary of the uncertainty analysis of C_D using correction factors Ci.

λ	D	experiment	origin	medium	fine
		C_D			
1	0.12	1.25	1.2496	1.2350	1.2281
			0.035%	1.197%	1.753%
2	0.24		1.2588	1.2411	1.2327
			0.701%	0.710%	1.385%
4	0.48		1.2590	1.2409	1.2245
			0.722%	0.730%	2.040%
8	0.96		1.6692	1.7381	1.7290
			33.538%	39.047%	38.321%
		C_L			
1	0.12	0.5	0.5483	0.5254	0.5105
			9.656%	5.078%	2.100%
2	0.24		0.5549	0.5542	0.5087
			10.986%	10.846%	1.736%
4	0.48		0.5553	0.5538	0.4938
			11.060%	10.760%	1.240%
8	0.96		0.4089	0.5486	0.3535
			18.220%	9.720%	29.300%
		f			
1	0.12	0.779	0.7980	0.7990	0.7768
			2.439%	2.567%	0.282%
2	0.24	0.195	0.1994	0.1995	0.1977
			2.256%	2.308%	1.385%
4	0.48	0.0487	0.04985	0.04987	0.04937
			2.361%	2.402%	1.376%
8	0.96	0.0122	0.00997	0.009348	0.00997
			18.279%	23.374%	18.279%

the same relative grid scale, as the scale is increased (and the flow velocity is reduced), the error relative to experimental values gradually increases. As for the vortex shedding frequency, its error does not show a clear correlation with the grid scale or the physical scale of the model.

In the case with $\lambda_4=8$, due to irregular vortex shedding (Norberg, 1994, Schaefer & Eskinazi, 1959) shown in Figure 6, the discrepancy between the simulated forces and the experimental data is too large. The frequency also significantly deviates from experimental values, suggesting that this case has not fully converged.

4.2.2 *Uncertainty analysis*

The ITTC-2017 guidelines specify an uncertainty analysis method based on correction factors (Ci). This approach utilizes the Richardson extrapolation method to estimate uncertainties in simulation results, calculating multiples based on the observed convergence order and the limit order of solver precision. The Ci method employs dimensional values (Celik et al., 2008) as well as dimensionless and normalized values for computation.

In this section, for cases of monotonic convergence, uncertainties are calculated according to the Richardson extrapolation method. For cases of harmonic oscillation, time-averaged values are used for drag, and peak values in the frequency domain are taken for vortex-shedding periods. In the case of divergence, the situation is categorized as Not Estimated (NE). When the *CFL* and *Re* remain constant, the influence of the relative mesh scale on computational uncertainty interacts synergistically with the time step.

Table 5. Summary of the uncertainty analysis of C_D using correction factors Ci.

Scale	λ	1	2	4	8
Diameter	D	0.12	0.24	0.48	0.96
Output values	$Ø_{1fine}$	1.2281	1.2327	1.2245	1.7290
	$Ø_{2middle}$	1.2350	1.2411	1.2309	1.7381
	$Ø_{3origin}$	1.2496	1.2588	1.2590	1.6692
Convergence	$\epsilon_{21}/\epsilon_{32}$	0.4787	0.47789	0.2266	-0.1319
Grid convergence index	GCI^{21}_{fine}	-0.0066	-0.00783	-0.0019	-0.00100
(GCI)	GCI^{32}_{fine}	0.0137	0.01627	0.0084	-0.00752
Uncertainty	U_1	0.0066	0.00783	0.0019	0.00100
	U_2	0.0137	0.01627	0.0084	0.00752
Corrected Uncertainty	U_{1C}	0.1321%	0.1565%	0.0382%	0.0199%
	U_{2C}	0.2744%	0.3253%	0.1678%	0.1505%
Correction Factor based Approach					
Uncertainty	U_1	0.0074	0.00915	0.01089	0.01679
	U_2	0.0155	0.01915	0.04807	0.12734
Corrected Uncertainty	U_{1C}	0.0467%	0.0717%	0.4511%	0.7706%
	U_{2C}	0.0975%	0.1500%	1.9903%	5.8441%
Normalized Correction Factor based Approach					
Uncertainty	U_1	0.0060	0.00742	0.00890	0.00971
	U_2	0.0126	0.01543	0.03905	0.07327
Corrected Uncertainty	U_{1C}	0.0380%	0.0582%	0.3684%	0.4457%
	U_{2C}	0.0790%	0.1209%	1.6170%	3.3624%

In this scenario, the uncertainty analysis of the drag coefficient C_D is shown in Table 5. Similarly, the analysis for vortex shedding frequency f is detailed in Table 6. As indicated in Table 4, the lift and drag in each case group are proportionally related (Bearman, 2011). Therefore, a separate uncertainty analysis for C_L is not conducted, and the results are assumed to be the same as those for C_D.

From Table 5, it can be observed that in the scale cases with $\lambda_1=1$, $\lambda_2=2$, *and* $\lambda_3=4$, the ratios of $\epsilon_{21}/\epsilon_{32}$ are all between 0 and 1. This indicates that the drag coefficient C_D monotonically converges as the grid is refined. However, in the scale case with $\lambda_4=8$, the ratio of $\epsilon_{21}/\epsilon_{32}$ falls between 0 and -1, indicating an oscillatory convergence for C_D. In all four scale cases, the Correction Factor Ci included uncertainty U_{1C} is greater than U_{2C}, suggesting that the uncertainty in C_D increases with finer grid refinement. Under the same *Re* and *CFL* conditions, as the scale increases (and the flow velocity decreases), the uncertainty shows a divergent increase. Overall, the uncertainty in each group remains below 5.8%.

Table 6. Summary of the uncertainty analysis of f using correction factors Ci.

Scale	λ	1	2	4
Diameter	D	0.12	0.24	0.48
Output values	$Ø_{1fine}$	0.7768	0.1977	0.0494
	$Ø_{2middle}$	0.7990	0.1995	0.0499
	$Ø_{3origin}$	0.7980	0.1994	0.0499
Convergence	$\epsilon_{21}/\epsilon_{32}$	-22.200	-18.000	-25.000
Grid convergence index (GCI)	GCI_{21fine}	-0.0017	-0.00067	-0.0005
	GCI_{32fine}	-0.0001	-0.00004	0.0000
Uncertainty	U_1	0.0017	0.00067	0.0005
	U_2	0.0001	0.00004	0.0000
Corrected Uncertainty	U_{1C}	0.0337%	0.0134%	0.0105%
	U_{2C}	0.0015%	0.0007%	0.0004%
Correction Factor based Approach				
Uncertainty	U_1	0.0434	0.00350	0.00098
	U_2	0.0020	0.00019	0.00004
Corrected Uncertainty	U_{1C}	2.1166%	0.1695%	0.0479%
	U_{2C}	0.0953%	0.0094%	0.0019%
Normalized Correction Factor based Approach				
Uncertainty	U_1	0.0558	0.01768	0.01985
	U_2	0.0024	0.00097	0.00079
Corrected Uncertainty	U_{1C}	2.7248%	0.8575%	0.9712%
	U_{2C}	0.1193%	0.0472%	0.0385%

Due to the irregular vortex shedding periods in the case with $\lambda_4=8$, Table 6 only discusses the other three scales. It can be concluded that for these three scales, the $\epsilon_{21}/\epsilon_{32}$ ratios are all negative and absolute value is much greater than 1, demonstrating that there is no significant convergence trend in the vortex shedding period towards convergence as the grid is refined. Under the same *Re* and *CFL* conditions, the uncertainty decreases as the scale increases (and the flow velocity decreases). Overall, the uncertainty in each group is below 2.7%.

5 CONCLUSIONS

This paper presents a study on the scale effects on CFD results for a 2D submerged circular cylinder within the OpenFOAM environment. The research focuses on the numerical errors and uncertainties of drag, lift, and vortex shedding frequency under high Reynolds number conditions, in relation to scale ratios and grid refinement. The simulations are performed using the Pimple FOAM solver and *k-kl-w* turbulence model, while the numerical uncertainty is quantified with a Correction Factor-based approach.

The findings highlight a significant sensitivity of the drag coefficient C_D and lift coefficient C_L to the scaling of the model. This sensitivity is evident in two ways. Firstly, there is a reduction of errors compared to experimental data when the grid is refined. Secondly, as the scale of the model is enlarged, there is an increase in both computational errors and uncertainties. However, the behaviour of vortex shedding frequency appears less directly correlated to scale changes. A detailed analysis indicates a pattern of monotonic convergence in drag coefficients at smaller scales and an oscillatory convergence at larger scales.

In summary, this work highlights the importance of using same size in CFD and experimental validations to reduce errors and uncertainties.

ACKNOWLEDGEMENTS

This work was performed within the Strategic Research Plan of the Centre for Marine Technology and Ocean Engineering (CENTEC), which is financed by Portuguese Foundation for Science and Technology (Fundação para a Ciência e Tecnologia-FCT) under contract UIDB/UIDP/00134/2020.

REFERENCES

Assi, G. R., 2009, *Mechanisms for flow-induced vibration of interfering bluff bodies*. Doctoral dissertation, Department of Aeronautics, Imperial College London.

Bearman, P. W., 2011, Circular cylinder wakes and vortex-induced vibrations. *Journal of Fluids and Structures*, 27, 648–658.

Bidoki, M., Mortazavi, M. & Sabzehparvar, M., 2018, A new approach in system and tactic design optimization of an autonomous underwater vehicle by using Multidisciplinary Design Optimization. *Ocean Engineering*, 147, 517–530.

Celik, I. (2005, RANS/LES/DES/DNS: The future prospects of turbulence modeling. *Journal of Fluids Engineering*, 127(5), 829–830.

Celik, I. B., Ghia, U., Roache, P. J. & Freitas, C. J., 2008, Procedure for estimation and reporting of uncertainty due to discretization in CFD applications. *Journal of fluids Engineering-Transactions of the ASME*, 130(7), 078001.

Cheung, S. H., Oliver, T. A., Prudencio, E. E., Prudhomme, S. & Moser, R. D., 2011, Bayesian uncertainty analysis with applications to turbulence modeling. *Reliability Engineering & System Safety*, 96, 1137–1149.

De Moraes, C. C. & Santiago, R. C., 2018, AUV Scaled Model Prototyping using 3D Printing Techniques. *2018 IEEE/OES Autonomous Underwater Vehicle Workshop (AUV)*. IEEE, 1–6.

Devolder, B., Troch, P. & Rauwoens, P., 2018, Performance of a buoyancy-modified k-ω and k-ω SST turbulence model for simulating wave breaking under regular waves using OpenFOAM®. *Coastal Engineering*, 138, 49–65.

Dikbaş, E. & Baran, Ö. U., 2023, Towards Accurate Vortex Separation Simulations with RANS Using Improved k-kL Turbulence Model. *Aerospace*, 10(4), 377.

Fan, S., Shi, D., Ma, G., Zhang, W., Chen, Y., Sun, L., Yuan, K., Yang, Z. & Zhang, M., 2023, Research on similarity of water entry load for scaled-down underwater vehicle based on different model test environments. *Ocean Engineering*, 286, 115697.

Holloway, D. S., Walters, D. K. & Leylek, J. H., 2004, Prediction of unsteady, separated boundary layer over a blunt body for laminar, turbulent, and transitional flow. *International Journal for Numerical Methods in Fluids*, 45, 1291–1315.

Huang, T. & Liu, H., 1994, Measurements of flows over an axisymmetric body with various appendages in a wind tunnel: the DARPA SUBOFF experimental program. *The 19th Symposium on Naval Hydrodynamics*. Seoul, Korea, 23–28 August.

Islam, H. & Guedes Soares, C., 2019, Uncertainty analysis in ship resistance prediction using OpenFOAM. *Ocean Engineering*, 191, 105805.

Islam, H. & Guedes Soares, C., 2021, Assessment of uncertainty in the CFD simulation of the wave-induced loads on a vertical cylinder. *Marine Structures*, 80, 103088.

Jasak, H., 2009, OpenFOAM: Open source CFD in research and industry. *International Journal of Naval Architecture and Ocean Engineering*, 1, 89–94.

Kahramanoglu, E., 2023, Numerical investigation of the scale effect on the horizontal maneuvering derivatives of an underwater vehicle. *Ocean Engineering*, 272, 113883.

King, R., 1977, A review of vortex shedding research and its application. *Ocean Engineering*, 4, 141–171.

Komen, E., Shams, A., Camilo, L. & Koren, B., 2014, Quasi-DNS capabilities of OpenFOAM for different mesh types. *Computers & Fluids*, 96, 87–104.

Leong, Z. Q., Saad, K., Ranmuthugala, S. & Duffy, J., 2013, Investigation into the hydrodynamic interaction effects on an AUV operating close to a submarine. *University Of Tasmania, Conference contribution*, https://hdl.handle.net/102.100.100/525199,1-11.

Menter, F. R., 1992, *Improved two-equation k-omega turbulence models for aerodynamic flows*. In: NASA Technical Memorandum 103975. NASA, Moffett Field, CA.

Mitra, A., Panda, J. P. & Warrior, H. V., 2019, The effects of free stream turbulence on the hydrodynamic characteristics of an AUV hull form. *Ocean Engineering*, 174, 148–158.

Norberg, C., 1994, An experimental investigation of the flow around a circular cylinder: influence of aspect ratio. *Journal of Fluid Mechanics*, 258, 287–316.

Norberg, C., 2003, Fluctuating lift on a circular cylinder: review and new measurements. *Journal of Fluids and Structures*, 17, 57–96.

Oliver, T. A. & Moser, R. D., 2011, Bayesian uncertainty quantification applied to RANS turbulence models. *Journal of Physics: Conference Series*, 318(4), 042032.

Ong, M. C., Utnes, T., Holmedal, L. E., Myrhaug, D. & Pettersen, B., 2009, Numerical simulation of flow around a smooth circular cylinder at very high Reynolds numbers. *Marine Structures*, 22, 142–153.

Ostafichuk, P. M., 2004, *AUV Hydrodynamics and modelling for improved control*. Doctoral dissertation, Department of Mechanical Engineering, University of British Columbia.

Penttinen, O., Yasari, E. & Nilsson, H., 2011, A pimplefoam tutorial for channel flow, with respect to different LES models. *Practice Periodical on Structural Design and Construction*, 23, 1–23.

Procedurest, I. R., 2017, Uncertainty Analysis in CFD Verification and Validation, Methodology and Procedures. *ITTC Recommended Procedures and Guidelines*, 7.5-03-01-01, 1–12.

Salimipour, E., 2019, A modification of the k-k-ω turbulence model for simulation of short and long separation bubbles. *Computers & Fluids*, 181, 67–76.

Schaefer, J. W. & Eskinazi, S., 1959, An analysis of the vortex street generated in a viscous fluid. *Journal of Fluid Mechanics*, 6, 241–260.

Stringer, R. M., Zang, J. & Hillis, A. J., 2014, Unsteady RANS computations of flow around a circular cylinder for a wide range of Reynolds numbers. *Ocean Engineering*, 87, 1–9.

Teymourtash, A. R. & Salimipour, S. E., 2017, Compressibility effects on the flow past a rotating cylinder. *Physics of Fluids*, 29, 016101.

Travin, A., Shur, M., Strelets, M. & Spalart, P., 2000, Detached-eddy simulations past a circular cylinder. *Flow, Turbulence and Combustion*, 63, 293–313.

Vakil, A. & Green, S. I., 2009, Drag and lift coefficients of inclined finite circular cylinders at moderate Reynolds numbers. *Computers & Fluids*, 38, 1771–1781.

Vasudev, K. L., Sharma, R. & Bhattacharyya, S. K., 2014, A multi-objective optimization design framework integrated with CFD for the design of AUVs. *Methods in Oceanography*, 10, 138–165.

Vijay, A. & Kumar, A., 2019, Numerical Simulation on Flow Over Circular Cylinder With Heated Splitter Plate. *ICTEA: International Conference on Thermal Engineering* (Vol. 2019).

Walters, D. K. & Cokljat, D., 2008, A three-equation eddy-viscosity model for Reynolds-averaged Navier–Stokes simulations of transitional flow. *Journal of Fluid Engineering*. 130(12), 121401.

Wang, S., González-Cao, J., Islam, H., Gómez-Gesteira, M. & Guedes Soares, C., 2022, Uncertainty estimation of mesh-free and mesh-based simulations of the dynamics of floaters. *Ocean Engineering*, 256; 111386.

Widiatmono, R., Mahmudah, R. S. N., Hidayat, I., Nugroho, K. A. & Apriansari, R. Y., 2022, Investigation of Under-Relaxation Factors in 2D Dam Break Simulation of OpenFOAM®. *Journal of Physics: Conference Series*, 2377(1), 012014.

Williamson, C. H., 1996, Vortex dynamics in the cylinder wake. *Annual Review of Fluid Mechanics*, 28, 477–539.

Xia, G., Liu, C. & Chen, X., 2016, Multi-objective optimization for AUV conceptual design based on NSGA-II. *OCEANS 2016-Shanghai*. IEEE, 1–6.

Zhang, J. & Fu, S., 2019, An efficient approach for quantifying parameter uncertainty in the SST turbulence model. *Computers & Fluids*, 181, 173–187.

Advances in Maritime Technology and Engineering – Guedes Soares & Santos (Eds)

Conducting tests using large-scale, self-propelled models at sea

D. Liarokapis, G. Trachanas, G. Grigoropoulos & K. Belibassakis
National Technical University of Athens, LSMH, Athens, Greece

ABSTRACT: The paper describes the instrumentation and the procedure developed for testing of a large-scale ship model at sea. Although it is more common to conduct ship model experiments in towing tanks, the accuracy is highly affected by the small scale of the models. Moreover, real sea conditions cannot be accurately simulated using the laboratory facilities. In this work, the methodology as well as the instruments used to measure a self-propelled large model at sea is described in detail. The model is propelled by two electric outboard engines of 4KW each. Wave buoys were used to record the wave heights and direction. The model 6dof motions, the sailing route, the propulsive thrust, the power consumption, the wave amplitude and period and the system for measuring model accelerations are accurately recorded. Moreover, an IMU unit together with an in-house motion detector system was used to acquire the heave motions and the pitch, roll and yaw angle of the model. Their advantages and shortcomings are analyzed in detail. Selected results are presented from recent tests using a 10m ferry ship model tested at real sea conditions illustrating the method.

1 INTRODUCTION

According to the recent IMO guidelines (IMO 2023), the reduction of fuel consumption and gas emissions of ships must be reduced by 30% till 2030 and reach zero emissions till 2050. To reduce fuel consumption, most ship designers are trying to optimize the hydrodynamic efficiency of the hull (Lee et al. 2021). However, many researchers are experimented with innovative devices trying to gain thrust either by the wind (Flettner rotors, Sails at Ships) or by taking advantage of the ship motion due to operating at waves (Belibassakis & Politis 2013, Belibassakis et al. 2021, Bower & Townsend 2022, Ntouras et al. 2022).

The most common method to investigate ships performance in calm water and in waves is to conduct scaled ship model experiments in towing tanks facilities. Due to the blockage effect, scaled models should not exceed a certain value depending from the size of the towing tank. On the other hand, large scales reduce the accuracy of the prediction of full-scale performance (Mishkevich 1995, Chunyu et al. 2023). Moreover, the realistic sea condition cannot be accurately simulated at the laboratory basin.

Another approach is to use numerical simulations. CFD predictions are gaining attention lately. However, the results are usually user-dependable and needing experimental validation to ensure their accuracy.

A different approach that increases significantly the accuracy of the experimental method is to conduct self propelled large-model experiments at sea. This procedure reduces Reynolds scale effect resulting in more accurate predictions for the full scale. Moreover, the model is self-propelled (not driven by the carriage), equipped with the bilge keels, sailing at six-degrees of freedom at real sea conditions. From 2005 ITTC commented that offshore model tests are feasible and necessary. Experimental techniques for measuring ship models at sea were highlighted at ITTC 2008 and 2011.

Serter (1993), reports tests of radio controlled high-speed military models, in a Swiss lake. Grigoropoulos and Katsaounis (2004) proposed the procedure for seakeeping tests of large-scaled models at sea. Later, LSMH by the guidance of Prof. Grigoropoulos & Loukakis conducted experiments of a man-driven, 9 meter large-scale warship model at open sea at extreme wave conditions. Quintana et al. (2007) reported multiple model scale tests of US destroyer, contracting a ¼ scaled model. At 2020, Chun-yu Guo test a 25m long bulk carrier model to study the propulsion performance. More recent results have been presented by Belibassakis (2023), which are obtained in the framework of H2020 Seatech project, using a 10m large-scale self propelled model to conduct experiments at sea with and without a thrusting foil.

Although large-scale model testing is advantageous compared to towing tank experiments, measuring accuracy is debatable. The measuring equipment and measuring procedure followed must be carefully planned to reduce uncertainty. Irregular waves produced in the tanks are accurately produced and measured in the laboratory facilities compared to the single point measurements from the wave bouy's. The same applies to the synchronized measurements of the speed, the forces, the accelerations and the heave-pitch

DOI: 10.1201/9781003508779-16

motions of the large-scale model. In the following sections, the methodology developed in the laboratory of ship and marine hydrodynamics (LSMH) as well as the instruments used to measure a self-propelled large model at sea is described in detail.

2 MODEL DESCRIPTION

A ferry ship model, with length L_{BP} of 10m, $L_{BP}/B=7.67$, $B/T=3.3$ and $Cb=0.45$ has been designed and constructed at the LSMH facilities of NTUA. To create the hull, the model's frames and planking was created from plywood. To reinforce the structure and increase waterproof protection the hull was covered by fiberglass and resins. After the completion of the hull the model was painted with antifouling paint.

Figure 1. Sketch from the model in the design stage.

Complex geometries at the stern and the bow of the model were formed by engraved wood shaped by the CNC machine of the laboratory. The model was equipped with all the necessary instrumentation (light, pumps, backup batteries, etc) to issue the legal documents and to ensure safe voyage at sea.

Extra weights were added to achieve the desired draft. The weights were distributed in such manner to ensure the stability of the vessel at waves. The models breadth is unusual small compared to vessel of the same length due to its scale from a passenger ship. The relatively small breadth compared to the wavelength existing to the real sea required lowering of the CG to increase stability. Most of the additional weights were placed at the hull lowering the CG just a few cm above the WL. At the final stage, the model was tested at the towing tank and the dock for a decay test.

3 INSTRUMENTATION

3.1 *Propulsion system*

The model is propelled by two electric outboard engines of 4kW each, which is considered to be equivalent with a petrol outboard engine of 15Hp. Selection was based on CFD and experimental calculations of a 1/3 model on the towing tank.

The engines were placed to the transom in such manner, so between the two propellers there is a space of the size of a propeller in-between (Figure 2).

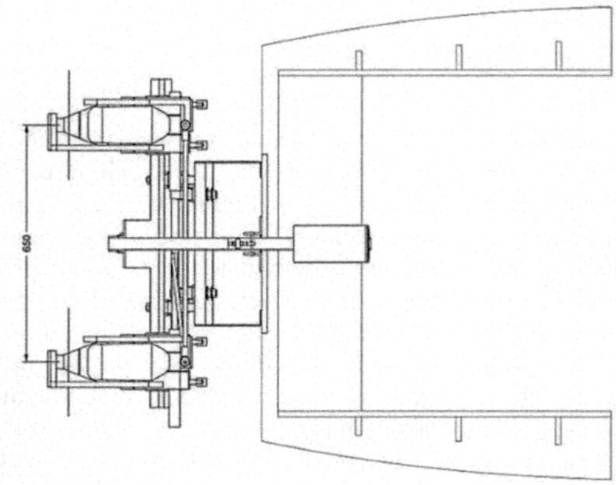

Figure 2. Outboard engine arrangement– Top view.

In the same manner, to avoid the propellers exiting the water surface due to the model motions, the engines were placed in depth were the upper tip of the propeller was 26cm bellow the water surface (Figure 3).

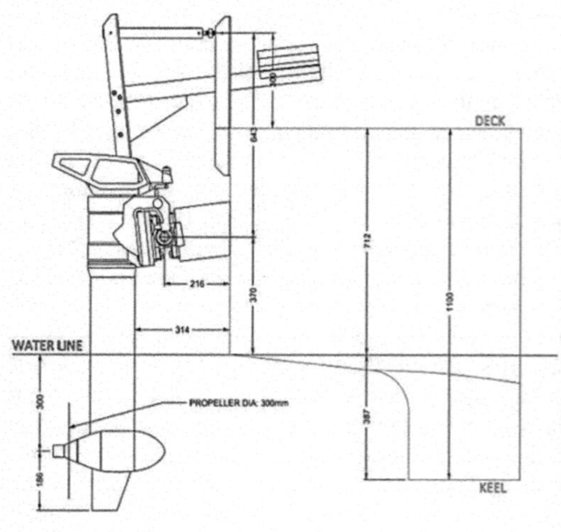

Figure 3. Engine Setting – side view.

The engines were powered from two batteries 9kW in total, providing autonomy of half day at the speed of 5kn. The control system of the motors displayed instant and total power consumption, remaining time at speed as well as the route from a GPS sensor. The motor have independent speed control and power measurement. The pilot could adjust the controller to propel at constant speed or constant consumption.

In addition, an electrical board was installed in-between the batteries and the engine, measuring instant current (I) voltage (V) and time. All the data was driven to the data acquisition board.

A back-up generator-set provided the necessary power to propel the model at reduced speed in case of a battery failure.

Engines thrust were measured by a home-made apparatus linking both engines to the model through a strain gauge and a rotational axis. The sensor is constructed by stainless steel, IP67 with non-linearity of +-0.25%, hysteresis of +-0.25% and repeatability of +-0.1%. A schematic diagram can be seen at Figure 3. Friction in the axis was assumed negligible. Added weights were added to counter balance the applied force at rest. A calibration procedure was conducted by applying vertical static forces to the propeller center.

The above apparatus was in favor compare to installing axial configuration due to simplicity reasons. The main deficit is the lack of shaft torque measurements.

3.2 *GPS data*

Two GPS units were used to capture the models route. The GPS were adjusted to acquire the models position, speed, distance and heading every one second. Typical accuracy is within couple of meters.

GARMIN. Page 2

20221003 Dat_019

Index	Elevation	Leg Distance	Leg Time	Leg Speed	Leg Course	Time	Position
1	15 ft	8 ft	0:00:01	5 kt	209.0° true	3/10/2022 1:38:55 μμ	N37° 42.773' E24° 05.282'
2	15 ft	8 ft	0:00:01	5 kt	209.0° true	3/10/2022 1:38:56 μμ	N37° 42.772' E24° 05.281'
3	13 ft	8 ft	0:00:01	5 kt	209.4° true	3/10/2022 1:38:57 μμ	N37° 42.771' E24° 05.280'
4	15 ft	8 ft	0:00:01	5 kt	209.6° true	3/10/2022 1:38:58 μμ	N37° 42.770' E24° 05.279'
5	15 ft	8 ft	0:00:01	5 kt	209.4° true	3/10/2022 1:38:59 μμ	N37° 42.768' E24° 05.279'
6	15 ft	8 ft	0:00:01	5 kt	210.0° true	3/10/2022 1:39:00 μμ	N37° 42.767' E24° 05.278'
7	15 ft	8 ft	0:00:01	5 kt	210.6° true	3/10/2022 1:39:01 μμ	N37° 42.766' E24° 05.277'
8	15 ft	8 ft	0:00:01	5 kt	211.6° true	3/10/2022 1:39:02 μμ	N37° 42.765' E24° 05.276'
9	15 ft	8 ft	0:00:01	5 kt	212.1° true	3/10/2022 1:39:03 μμ	N37° 42.764' E24° 05.275'
10	15 ft	8 ft	0:00:01	5 kt	214.3° true	3/10/2022 1:39:04 μμ	N37° 42.762' E24° 05.274'
11	15 ft	8 ft	0:00:01	5 kt	215.6° true	3/10/2022 1:39:05 μμ	N37° 42.761' E24° 05.273'
12	15 ft	8 ft	0:00:01	5 kt	215.9° true	3/10/2022 1:39:06 μμ	N37° 42.760' E24° 05.272'
13	15 ft	8 ft	0:00:01	5 kt	216.4° true	3/10/2022 1:39:07 μμ	N37° 42.759' E24° 05.271'
14	13 ft	8 ft	0:00:01	5 kt	217.4° true	3/10/2022 1:39:08 μμ	N37° 42.758' E24° 05.270'
15	13 ft	8 ft	0:00:01	5 kt	218.2° true	3/10/2022 1:39:09 μμ	N37° 42.757' E24° 05.269'

Figure 4. GPS data.

The position and speed data were cross checked with the data provided from the engine control system. In the same manner, GPS heading was valuated with the data from the magnetic compass.

At the end of each measurement, data were downloaded to Garmin Software (Figure 4). From the data, the course were the speed and heading was kept constant could be separated from the rest of the measurement for post analysis (Figure 5).

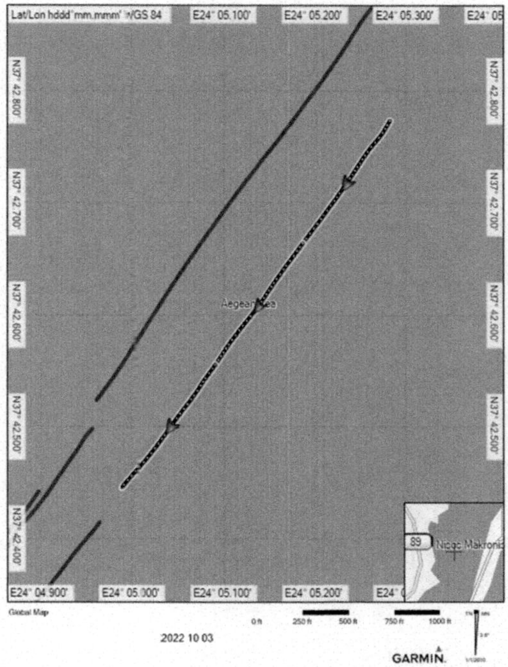

Figure 5. Single measurement from GPS data.

3.3 *Data acquisition*

Two data acquisition differential analog boards linked together were used to acquire the measurements. The DA was controlled by a PC. The first board was used to acquire the model's motions by the seven accelerometers and the total thrust of the motors. The second board was used to acquire the power consumption of the motors and the rest of the sensors. Both units used separate battery unit from the engine to increase the accuracy of the measurements. Thrust and power consumption was measured with a sampling rate of 1Hz while the rest with 50Hz. All measurements except from one GPS unit were synchronized by internal clock. The time clock of GPS was synchronized with the time of the computer at the beginning of each day. All cables routes were parallel to each other at the side of the hull to minimize interference.

3.4 *Motions measurements*

For capturing the model's six DOF motions, an IMU and a portable system using seven accelerometers (Figure 6), which was developed at the LSMH (Grigoropoulos & Politis 1999) of NTUA to record the dynamic responses of ships and models was used. The accelerometers are water tight, with housing with epoxy seal, have range of 2g, output of 0-+4V and sensitivity of 2000 mv/g.

The system had been validated against motions obtained from the dynamometer heave and pitch sensors and an optical measuring system (Grigoropoulos

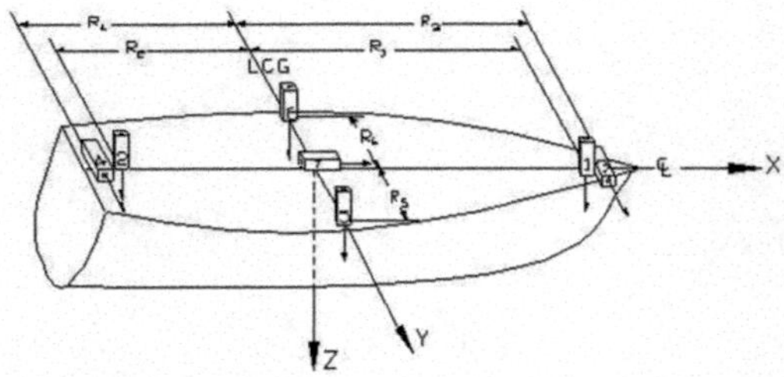

Figure 6. Sketch from the model in the design stage.

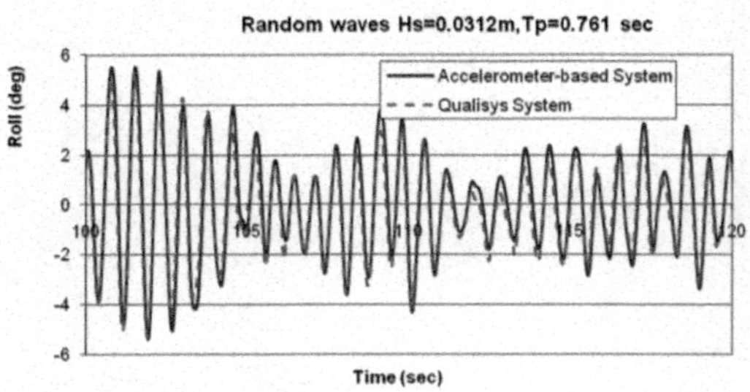

Figure 7. Roll and vertical acceleration at the stern of the FESTOS PALACE model for significant wave height H_S = 0.031 m and modal period T_P = 0.762 sec.

et al. 2011) using three cameras and markers with excellent results (Figure 7).

To ensure the accurate positioning of the accelerometers, they were placed on a wooden cross at known distance, in front of the pilot's cabin. The system was fed by a separate power supply system.

3.5 *Free surface elevation*

To record the sea condition, directional and non-directional Wave Buoys were used. Before the experiments, a support boat placed the Buoys at the predefined positions at the measuring area.

The directional bouy, developed by Neptune Science Inc (Figure 8), has a large diameter (0.75m) floating component, housing at the center the electronic hardware (electronics, battery, and antenna). The hardware is easily removable on the spot, without the need to disengage the mooring. Measurements are acquired at preset interval or on demand, with minimum period of 15 minutes. Wave data can be downloaded via a UHF radio modem, or by cable. The radio modem was placed at the support boat that was sailing at a small distance. Maximum range of the UHF signal is 8km. The buoy's outputs processed data of wave height and period wave direction as well as raw data.

The battery duration in standard mode (4Hz) was about one month. Two set of batteries were used in the experiments. For safety reasons the buoy was retrieved from the sea at the end of the day.

In addition to the directional buoy, three non-directional Buoys were placed in the measuring area.

Figure 8. Directional Wave Buoy.

The receiver of the UHF signal was placed at the coast and receives raw data. The Buoys were calibrated at the laboratory before the experiments.

The main deficit of the non-directional Buoy is the fact that they are carried out from the induced waves, consecutively changing the measuring spots.

4 MEASURING PROCEDURE

4.1 *Test cases*

Large-scale ship models at sea can be tested for calm water testing, seakeeping, propulsion characteristics, maneuverability and structural loads. In our case, the experiments referred to calm-water resistance and seakeeping at various speed with and without a flapping foil.

The model was transferred from the NTUA-LSMH workshop to the test site served by the marina of Olympic Marine S.A. in Lavrion area near Athens (Figure 9).

For the calm water resistance, repeatable tests were conducted at the exit of the port for various speeds. The duration of the measurements were about 3min, providing a good estimation of the resistance. The tests were repeated at the opposite direction taking into account sea currents at the measuring area.

For the seakeeping, a number of tests at various sea-states at various speeds were carried out. Tests were conducted at various locations near the harbor depending on the wind direction and wave height (Figure 10). The pilot was keeping the course at a specific heading trying to avoid routing alternations.

Figure 9. The self-propelled model of NTUA docked at the jetty of Olympic Marine SA premises.

Figure 10. Testing the model at moderate wave conditions.

In addition the model was tested in nearby location of the buoys to ensure the correct measurement of the sea state. Most of the tests were conducted with head or following waves.

A record was kept both by the pilot and the support boat, noting parameters that could potentially alter the measurement (wake from a big ship).

From the way points provided by the GPS the route of the model at sea was presented at Google Maps. As it can be seen from Figure 11, most of the measurements follow the same heading.

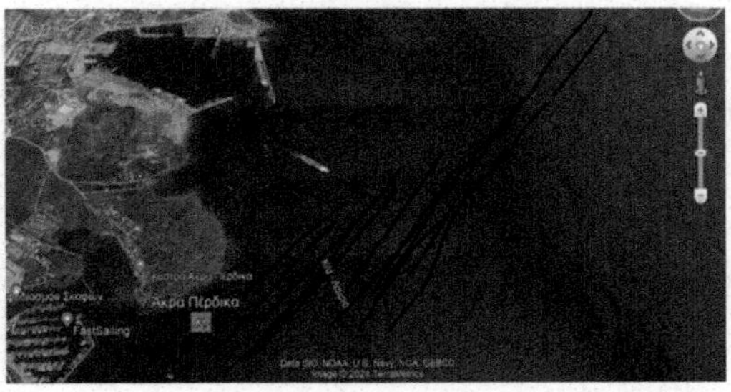

Figure 11. Day measurement from GPS Data plotted to GoogleMaps.

4.2 *Selected results*

Indicative results concerning the wave conditions during the experiments at sea model speed 5kn and the measured responses and thrust of the propellers are presented in Figures 12 for the tests without the dynamic wing at the bow, when the recorded significant wave height was Hs=0.13m and peak period Tp=1.9s. Corresponding results with the operation of the dynamic wing operating at the bow of the large NTUA model are presented in Figures 13 and 14, when the significant wave height was Hs=0.20m, and peak period Tp=1.9s. In both cases the relative direction of waves with the route(s) of the model was β=210deg, corresponding to bow-quartering waves, and the recorded speed of the model was 5.1kn.

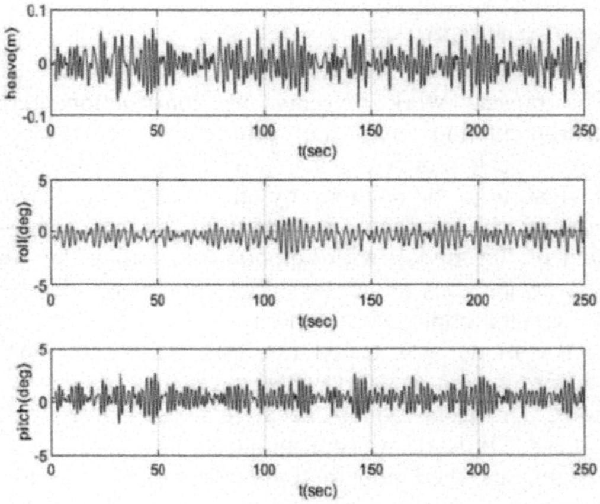

Figure 12. Measured data during tests at sea without the foil at the bow. Recorded model responses and thrust from engine propellers. Wave conditions: Hs=0.13m and Tp=1.9s.

The results clearly show that the dynamic wing technology provides a significant reduction in propulsive power required to maintain a constant speed of the ship in waves, offering also enhancement of stability due to the drop of responses and the added wave resistance.

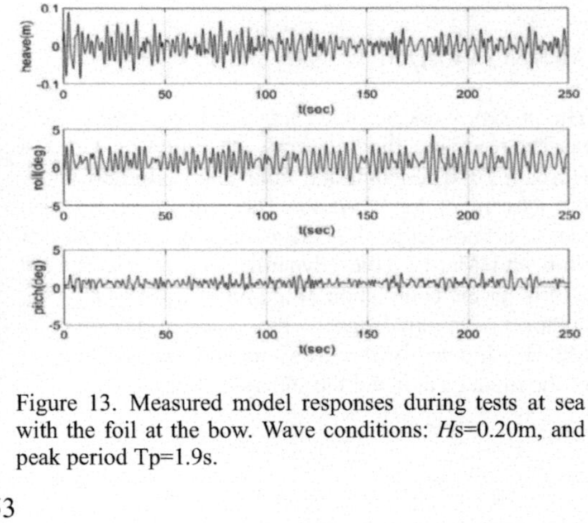

Figure 13. Measured model responses during tests at sea with the foil at the bow. Wave conditions: Hs=0.20m, and peak period Tp=1.9s.

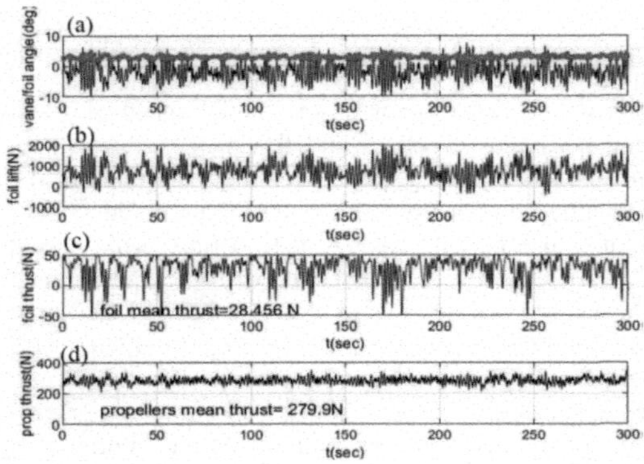

Figure 14. Dynamic foil thruster responses during tests at sea. Wave conditions as in Fig. 10.

5 CONCLUSIONS

The present work presents the construction, the instrumentation and the procedure for testing a large-scale model at sea. The 10m long model was navigating to the open sea by an onboard pilot. Two outboard electric motors were used for the propulsion of the model, with autonomy of half a day at service speed. A power generator was intergraded to the system to increase autonomy.

The model was tested for resistance seakeeping and propulsion at calm water and rough sea conditions. The tests took place at in coastal area near Athens. Thrust, power consumption, position, speed, heading and the model motions were recorded simultaneously by the data acquisition system onboard the model. The sea conditions were monitored by portable directional and non-directional buoys.

From the experiments the energy gains of a flapping foil placed at the bow of the model were evaluated. Data from a 1/3 model in the towing tank showed good agreement with the open sea tests. The model can be used to conduct a large number of experiments at sea. Future work is planned towards the study of T-foil arrangement of foils below the bow of the ship for augmenting ship propulsion in waves; see Belibassakis et al (2023).

ACKNOWLEDGEMENTS

The present work has been supported by Seatech 2020 project receiving funding from the European Union's Horizon 2020 research and innovation program under the agreement No. 857840. The authors would also like to acknowledge the personnel of the laboratory of Ship and Marine Hydrodynamics for their endlessly efforts for the completion of the project. Especially, the authors acknowledge the contribution by Mr. G. Milonas for the drawings and the construction of the model's hull and the superstructure.

REFERENCES

Belibassakis, K.A., Politis, G.K. 2013. Hydrodynamic performance of flapping wings for augmenting ship propulsion in waves. *Ocean Engineering 72, pp.* 227-240.

Belibassakis, K.A., Filippas, E. & Papadakis, G. 2021. Numerical and experimental investigation of the performance of dynamic wing for augmenting ship propulsion in head and quartering seas. *Jour. Marine Science & Engineering*. 1024.

Belibassakis K., 2023, Augmenting ship propulsion by flapping foil thrusters, *SNAME Int. Symp.on Ship Operations, Management and Economics*, Athens March 2023, SNAME-SOME-2023-018.

Belibassakis, K.A., Liarokapis, E.D., Anders, O., 2023. A novel Energy device for ferry ships in waves based on flapping thrusters. *15th Symposium in Hyper-Performance Marine Vehicles, Hiper'23, Bernried, Germany*, September 18-20.

Bower, J.A., Townsend N.C. 2022. Evaluation of bow foils on ship delivered power in waves using model tests. *Appl. Ocean Research 123*: 103148.

Chun, Y.G., Xiang, H.Z., Dagang, Z., Chao, W., Jian, F.L. & Kewei, S. 2020. Propulsion performance of large-scale ship model in real sea environment. *Ocean Engineering, Volume 210, 15 August 2020, 107440.*

Chunyu, G., Xianghai Z. & Dagang Z. 2023. Research on scale effect of resistance components for full-formed ship based on large-scale model towing test. *Journal of Marine Science abd Engineering 11*(*7*):1300.

Grigoropoulos, G.J. & Politis, C.G. 1999. A measuring system of the six degrees of motions of a moving body. *Ship Technology Research – Schiffstecgnik Journal*, VOL. 45, No. 1.

Grigoropoulos, G.J. & Katsaounis, G.M. 2004. Measuring procedures for seakeeping tests of large-scaled ship modles at sea. *13th Intl. Symposium on Measurements for Research and Industry, Athens*, September 2004.

Grigoropoulos, G.J., Damala, D. & Liarokapis, D.E. 2011. Experimental evaluation of the dynamic responses of ship models moored in scaled ports. *Sustainable Maritime Transportation and Exploitation of Sea Resources*.

Lee, H.P., Liarokapis, D.E., Thachanas, P.G. & Triantafyllou, G. 2021. Experimental investigation of the effect of superhydrophobic coating and microbubbles lubrication on the resistance of a scaled ship. *SNAME 7th International Symposium on Ship Operations, Management and Economics*, April 6-7 2021.

Mishkevich, V. 1995. Scale and roughness effects in ship performance from the designer's viewpoint. *Marine Technology, Vol. 32, No. 2, April 1995, pp.* 126-131.

Ntouras, D., Papadakis, G, Belibassakis, K.A. 2022. Ship bow wings with application to trim and resistance control in calm water and in waves. *Journal of Mar. Sci. Eng. 10(4), pp.*492.

Quantana, V., Howells, A.R. & Hettinger, L. 2007. User-centered design in a large-scale naval ship design program: Usability testing of complex military systems – DDG 1000. *Naval Engineers Journal 119(1)*: 2–33. *October 2003.*

Serter, E.H. 1993. Hydrodynamics and Naval Architecture of Depp-Vee Hull Forms. *Research – Development - Designs, Hydro Research Systems S.A., March 1993.*

Advances in Maritime Technology and Engineering – Guedes Soares & Santos (Eds)
© 2024 The Author(s), ISBN 978-1-032-83107-7

Hydrodynamic analysis and experimental verification of cutting device for propeller entanglement

J. Liu
Faculty of Information Science and Engineering, Ocean University of China, Qingdao, P.R. China
Centre for Marine Technology and Ocean Engineering (CENTEC), Instituto Superior Técnico, Universidade de Lisboa, Lisboa, Portugal

P. Qin & B. He
Faculty of Information Science and Engineering, Ocean University of China, Qingdao, P.R. China

C. Guedes Soares
Centre for Marine Technology and Ocean Engineering (CENTEC), Instituto Superior Técnico, Universidade de Lisboa, Lisboa, Portugal

ABSTRACT: A novel cutting device is proposed for solving the propeller and shaft entanglement in autonomous underwater vehicles. The rotating cutter is mounted on the propeller in an embedded manner and requires no additional power. The hydrodynamic characteristics of the propeller are predicted based on computational fluid dynamics, especially the effect of the rotating cutter on thrust and torque at different rotational speeds. Steady-state calculations are performed using the shear stress transport $k-\omega$ turbulence model and the multiple reference frames method. For open water performance, the effect of the cutter on thrust and torque is progressively apparent as rotational speed increases. However, the simulation results show that the cutter has a negligible effect on self-propulsion performance. Tank experiments demonstrate that the cutting device can effectively cut various fishing nets, cables, and seaweed, enabling the propeller to recover normal rotation. Nevertheless, there is a significant increase in propeller torque when the cutter performs cutting tasks.

1 INTRODUCTION

Propeller propulsion remains the most crucial mode of underwater propulsion and is widely used on ships and autonomous underwater vehicles (AUVs). However, once the propeller is entangled by marine debris (Pham et al. 2013, Rodríguez & Pham 2017, Chambault et al. 2018), fishing nets (Hong et al. 2017), or marine vegetation (Vanessa Anuat et al. 2022), it will cause an abnormal vibration of the shaft system, which will intensify the load on the main engine.

Propeller entanglement seriously impacts the safety of ships (Macfadyen et al. 2009) and AUVs (Chen et al. 2021), as shown in Figure 1. It even leads to substantial economic losses (Watters et al. 2010, McIlgorm et al. 2011, Kim et al. 2023).

Currently, solving entanglement problems relies on manual underwater operations, which are inefficient, costly and extremely risky. Therefore, it is essential to study propeller entanglements autonomous cutting technology.

Propeller entanglement is a common global challenge. Cho (2005) presented the challenges faced by Korea in marine debris management. The direct economic impact of marine litter on the Azores

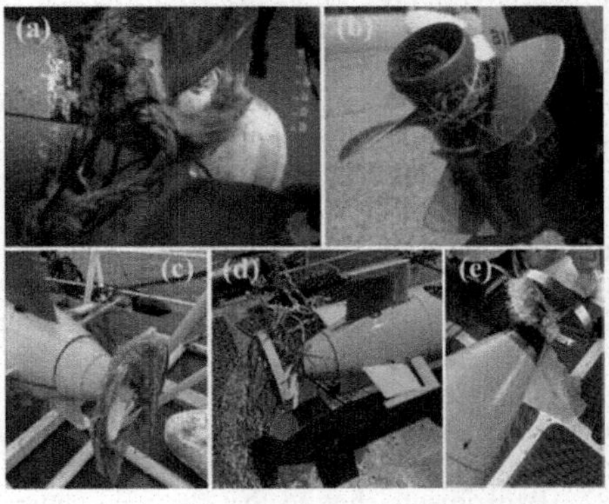

Figure 1. Propeller entanglements: (a)(b) Ship (Macfadyen et al. 2009) (c)(d)(e) AUV.

archipelago was quantified by Rodríguez et al. (2020). Marine debris jeopardises the ecological environment and causes many maritime accidents.

Propeller fault diagnosis is a prerequisite for autonomous propeller disentanglement. An underwater thruster fault diagnosis method based on load feature extraction was proposed by Gan et al.

DOI: 10.1201/9781003508779-17

(2022). The load torque model of thruster fault was established by analysing the load torque characteristics of propeller fault, such as winding, jamming, breaking and falling off. However, this method is not embedded in the underwater thruster drive system and cannot realise real-time online diagnosis.

Zhang et al. (2023) investigated an underwater thruster fault diagnosis method based on current and rotational speed signals. The analysis method based on the correlation coefficient can fully extract the fault information, which helps to improve the accuracy of underwater thruster fault diagnosis. A tight correlation exists between fault diagnosis, entanglement recognition and autonomous cutting.

Dubyoski et al. (2022) investigated the asymmetric entanglement of AUV propellers with marine vegetation using a small towing tank. The entanglement is asymmetric due to the direction of propeller rotation. In addition, sailing velocity and working depth directly impact the possibility of entanglement. However, the study only considered eelgrass entanglements.

Vanessa Anuat et al. (2022) investigated the effect of vehicle shape and propulsion on the interaction of AUV with marine vegetation. Experimental tests were conducted at different vegetation densities using either eelgrass or giant kelp. Interactions consisted primarily of disturbance, blockage, and entanglement, depending on vegetation type, density, AUV shape, and mode of propulsion. Overall, the streamlined shape is beneficial in mitigating entanglement. The effects of sailing velocity, vegetation type and density on the entanglement of torpedo-shaped AUV were systematically studied by Irgens et al. (2021). They also attempted to mitigate entanglement by moving the AUV back and forward.

A patent (Wiggins et al. 2015) provides the idea of using concentric cutting devices to break free from entanglement. Marine rope cutters (Lee et al. 2019) generally require more power, and their cutting effectiveness has been evaluated in detail by the Korea Maritime and Ocean University (Lee et al. 2019, Kim et al. 2021). An average of 78% of accidents were reduced due to propeller entanglement after installing the rope cutter.

Transient torsional vibrations of a rope cutter cutting operation were investigated by Kwon et al. (2022). Experiments showed that the rope cutter did not significantly reduce propeller power or increase structural vibration. Ding et al. (2022) adopted a fibre optic grating sensing network to monitor propeller deformation for underwater operations. Based on the inverse finite element method, the strain data collected by Fibre Bragg Grating (FBG) is converted into the deformation of the propeller to realise the deformation monitoring during the operation of the propeller. Although the above studies have achieved good results, the effect of the cutter on the propeller propulsion performance has not been fully investigated.

The effect of a rope cutter on the hydrodynamic performance of a propeller was studied with the computational fluid dynamics (CFD) method by Nyongesa et al (2022). Studies showed that the rope cutter leads to a slight reduction in water velocity behind the propeller. There was no significant difference in the pressure distribution around the propeller but a slight increase in both the thrust generated by the propeller and the torque required (Nyongesa et al. 2023). Different from ships, AUVs are generally driven by low-power motors. Therefore, cutting devices for AUVs should focus on the impact on propulsion performance. A valuable cutting device provides both effective cutting and good hydrodynamic characteristics. As the frequency of using AUVs increases, more reliable and effective cutting devices should be urgently explored to minimise the adverse effects of underwater entanglement.

In this paper, a novel cutting device is designed to solve propeller and shaft entanglement in AUVs. The effect of the cutting device on propulsive performance is investigated by numerical simulation.

2 STUDY OBJECTS

A new cutting device was designed for the Sailfish 260 AUV, which was developed independently by the underwater vehicle laboratory (UVL) of Ocean University of China. Cutting devices, propellers, and AUVs are the study objects of this paper. The parameters of the AUV and propeller are shown in Table 1, and further details can be obtained in previous study (Liu et al. 2023a).

Table 1. Parameters of the AUV and propeller.

Objects	Parameters	Values
AUV	Total length	2.75 m
	Tail length	0.42 m
	Hull diameter	0.262 m
	Tail height	0.50 m
Propeller	Blade diameter	0.21 m
	Hub diameter	0.066 m
	Number of blades	3
	Pitch ratio	0.8
	Expanded area ratio	0.35

2.1 *Cutting device*

The cutting device proposed in this paper consists of six fixed cutters and a rotating cutter. The fixed cutter is designed in the shape of a hook, which is favourable for shearing and tearing the entanglements. The six fixed cutters are distributed evenly around the drive shaft and are bolted to the cabin. The rotating cutter is mounted on the propeller in an embedded manner, with the shaft driving both together without additional power. The diameter of the rotating cutter is slightly larger than the propeller hub, and the number of blades can be set flexibly according to the diameter of the hub.

Both fixed and rotating cutters have sharp edges and are interchangeable. If the cutter is damaged, it can be replaced individually. There is a gap between the propeller hub and the hull, which is an area susceptible to entanglement. This device has a special design for the bearing cap that adjusts the thickness to the gap and adds cutting teeth. This scheme can both fill the gap and improve the cutting effect. The design of the cutting device is shown in Figure 2.

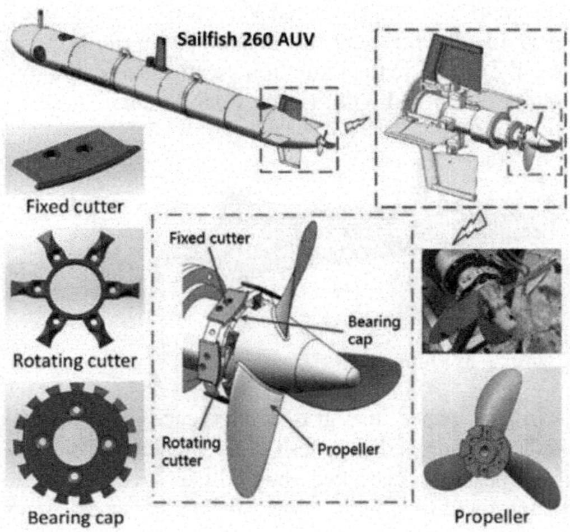

Figure 2. Model of cutting device.

2.2 *Geometric parameters*

The Wageningen B-3-35 propeller with three blades and a 35% disc ratio is used in this study. The blade diameter is 210 mm, and the hub diameter is 66 mm. The fixed cutter is of a thin, symmetrical design with a thickness of 2 mm and a curved cutting edge length of 55 mm.

The rotating cutter is hollowed out in the centre and has six blades with multi-directional cutting edges. The thickness of the rotating cutter is 5 mm, and the maximum outer diameter is 78 mm. The bearing cap is circular with a thickness of 5 mm and a maximum outer diameter of 68 mm. Fifteen cutting teeth with a triangular profile are designed outside the cap. After installation, the gap between the rotating cutter and the fixed cutter is less than 1 mm, and the gap between the propeller and the bearing cap is also less than 1 mm. The specific geometric parameters are shown in Table 2.

3 NUMERICAL SIMULATION

The numerical simulation and analysis are based on ANSYS Workbench, which allows the integration of multiple modules in a single interface with automatic data sharing. The open water and behind hull performance of the propeller are examined.

Table 2. Geometric parameters of the cutting device.

Objects	Parameters	Values
Fixed cutter	Thickness	0.002 m
	Length of cutting edge	0.055 m
	Diameter of hook circle	0.004 m
Rotating cutter	Number of cutting blades	6
	Thickness	0.005 m
	Outside diameter	0.078 m
Bearing cap	Number of cutting teeth	15
	Thickness	0.005 m
	Outside diameter	0.068 m

3.1 *Study strategy*

This paper uses the discrete propeller method, where propellers with and without cutting device are mounted behind the hull, respectively. Multiple experiments are designed to simulate the thrust and torque of the propeller by varying the inlet velocity and rotational speed to further analyse the effect of the cutting device on the propulsion performance. Based on the multiple reference frames method, the results derived from CFD calculations are used as sample points to provide data for subsequent performance prediction and comparison. The simulation procedure in the ANSYS Workbench platform is shown in Figure 3.

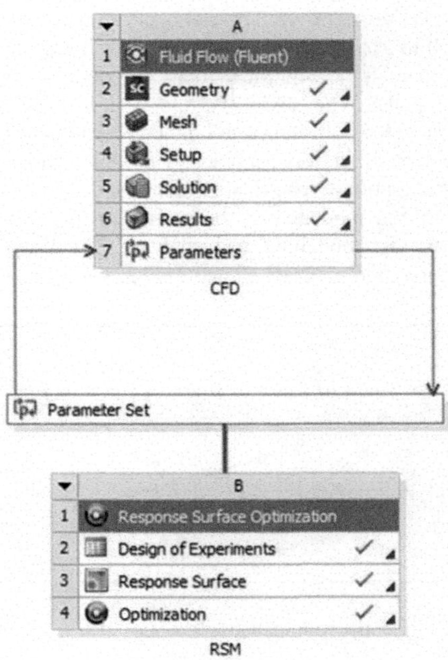

Figure 3. Numerical simulation procedure.

3.2 *Turbulence model*

In this study, the Reynolds Averaged Navier–Stokes (RANS) method is used to solve the governing equations. The y^+-insensitive wall treatment is applied, and the Shear Stress Transport (SST) turbulence model is used for steady-state calculations. The type of solver is pressure-based, and the solution method is velocity–pressure coupling based on the SIMPLE algorithm. Spatial discretisation uses a least-squares cell-based gradient with second-order pressure, momentum, turbulent kinetic energy, and specific dissipation rate in second-order upwind. The averaged continuity and momentum equations are expressed in tensor notation and Cartesian coordinates as follows:

$$\frac{\partial \rho \bar{u}_i}{\partial x_i} = 0 \quad (1)$$

$$\frac{\partial \rho \bar{u}_i}{\partial t} + \frac{\partial}{\partial x_j}\left(\rho \bar{u}_i \bar{u}_j + \rho \overline{u_i' u_j'}\right) = -\frac{\partial \bar{p}}{\partial x_i} + \frac{\partial}{\partial x_j}\left[\mu\left(\frac{\partial \bar{u}_i}{\partial x_j} + \frac{\partial \bar{u}_j}{\partial x_i}\right)\right] \quad (2)$$

where ρ is the density of the fluid, the density of fresh water is 998.2 kg/m^3. $\bar{u}_i$ is the averaged Cartesian components of the velocity vector, $\bar{p}$ is the average pressure. μ is the dynamic viscosity, and the dynamic viscosity of fresh water at 20°C is 0.001003 kg/(m·s).

3.3 *Boundary conditions*

The current study constructs a cylindrical stationary domain for a fully attached AUV and propeller. The stationary domain has a diameter of 20*D* and a length of 5*L*. *D* and *L* denote the outer diameter and length of the AUV, respectively. The stationary domain is large enough to avoid the effects of wall boundaries. In addition, there is a cylindrical rotational domain around the propeller. The stationary and rotational domains are connected with interface boundaries, which can be implemented in the geometry module employing a shared topology. The rotating cutter and propeller are coupled and modelled in the pre-processing stage to facilitate the modelling and analysis. The visualised boundary conditions are shown in Figure 4.

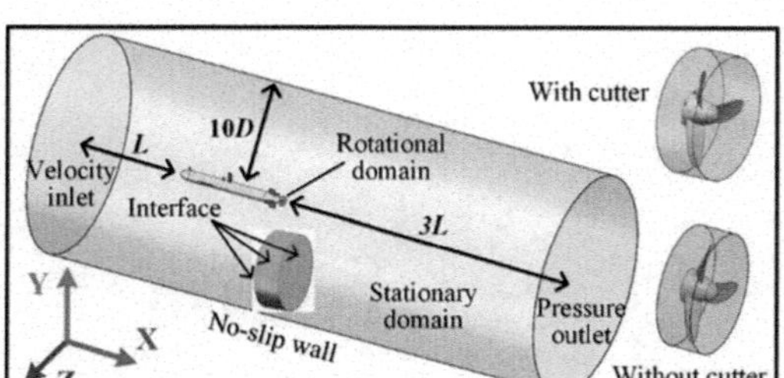

Figure 4. Boundary conditions.

The inlet and outlet of the stationary domain are defined as the velocity inlet and pressure outlet, respectively. The fluid flows in the positive direction along the x-axis. The boundary of the stationary domain, the fully attached AUV and the propeller are all no-slip walls.

3.4 *Mesh independence verification*

The unstructured mesh is autonomously meshed based on ANSYS Meshing, and the rotational domain is finer than the stationary domain. The AUV attachment, propeller and rotating cutter are locally encrypted. The hull, attachment and propeller are set up with boundary layers, and the height of the first layer and the number of layers are determined according to the desired y^+ (Liu et al. 2023b).

In this paper, the desired y^+ value for the AUV hull is 30, while the desired y^+ value for the propeller is 40. The meshing results are shown in Figure 5, and the numerical simulation results show that y^+ are all distributed within a reasonable range.

Figure 5. Meshing results.

In this paper, the grid convergence index (GCI) method is adopted for mesh convergence verification. The mesh is classified into three levels: coarse, medium and fine, with progressively smaller cell sizes in the stationary domain, the rotational domain and the propeller. Both mesh refinement ratios are close to 1.2. The boundary mesh was defined as ten layers with a growth rate of 1.2. As the mesh size is adjusted, the calculations slightly change the results, as shown in Table 3.

Table 3. Mesh refinement and calculation results.

Level	Cells	Thrust (N)	Torque (Nm)
Coarse	1,765,436	34.38	0.95
Medium	2,900,667	34.79	0.98
Fine	4,887,624	34.89	0.99

The first refinement significantly improves the calculation results and doesn't consume many computational resources. However, the second refinement does not benefit the calculations significantly but increases the preprocessing time and computational resources. Therefore, medium-sized meshes are chosen for numerical simulations in this study.

Calculations are performed using the GCI method, as shown in Table 4. The uncertainty (*U*) of both thrust and torque are within 2%, with

Table 4. Results of GCI verification.

Property	E_{21}	E_{32}	U_{21}	U_{32}	R
Thrust	0.29%	1.18%	0.12%	0.48%	0.244
Torque	1.01%	3.06%	0.63%	1.91%	0.333

discrimination ratio (R) ranging from 0 to 1. Therefore, the calculations are monotonically converged.

4 RESULTS AND DISCUSSIONS

Open water and behind hull performance are analysed for the propeller equipped with a rotating cutter and compared with the propeller without a cutter. The performance of the propeller behind the hull in the self-propulsion state is predicted by the response surface methodology and Kriging model.

4.1 *Propeller performance*

When not performing cutting tasks, the thrust, torque, and pressure distributions of the propeller at a constant high rotational speed are predicted. The data are shown in Table 5.

Table 5. Numerical simulation results at high speed.

Device	Velocity (m/s)	Speed (rpm)	Thrust (N)	Torque (Nm)
Without cutter	3	1500	149.39	4.92
With cutter	3	1500	120.48	4.17

It is evident from the calculations that the rotating cutter at high rotational speed significantly affects the thrust and torque of the propeller. At a rotational speed of 1500 rpm without entanglement, the rotating cutter results in a 19.4% decrease in propeller thrust, a 15.2% decrease in torque, and a 2.8% decrease in efficiency.

The pressure distribution around the propeller changes when the cutter is installed. The pressure difference between the blade surface and the back of the blade becomes smaller, which is the main reason for the decrease in thrust. The AUV stern pressure distribution is shown in Figure 6.

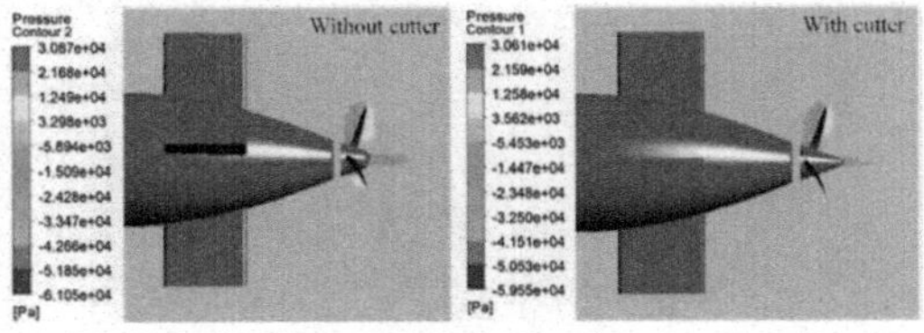

Figure 6. Pressure distribution.

To further investigate the effect of the rotating cutter, the pressure distribution of the propeller at 200 rpm, 500 rpm and 1000 rpm is simulated, as shown in Figure 7.

Whether or not the rotating cutter is fitted, when advancing forward, the propeller is positively pressurised on the surface side and negatively pressurised on the back side, and both are close to the blade guide edge area.

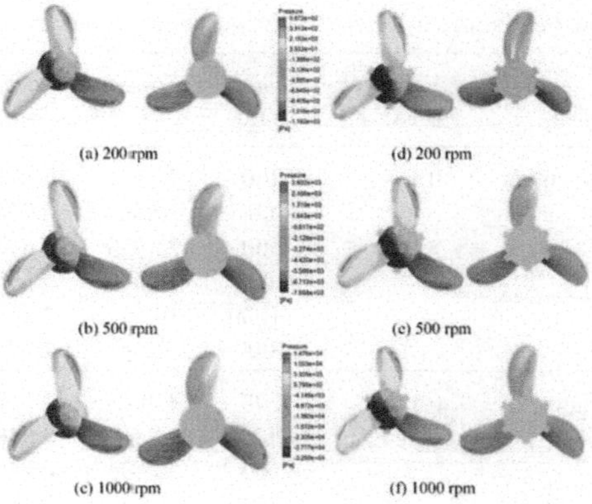

Figure 7. Pressure distribution at different rotational speeds: (a)(b)(c) Without cutter (d)(e)(f) With cutter.

As the rotational speed increases, the pressure difference between the two sides gradually becomes larger. Installing a rotating cutter leads to an increase in negative pressure on the back side and a decrease in positive pressure on the surface side, which is especially noticeable at high rotational speeds. Therefore, the rotating cutter slightly reduces the thrust, torque and efficiency of the propeller at low to medium rotational speeds, as shown in Figure 8.

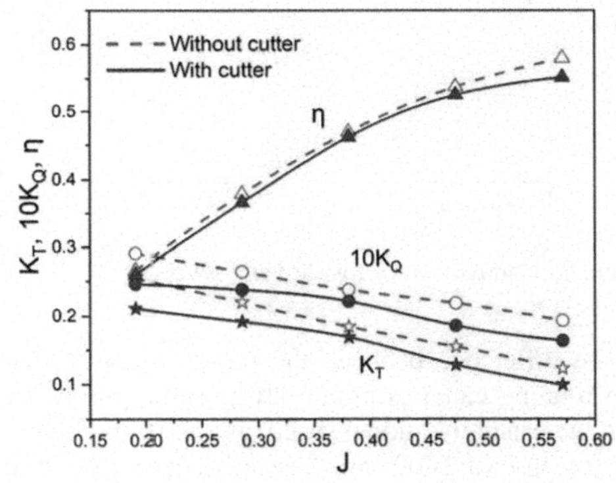

Figure 8. Propeller open water characteristics.

When performing cutting tasks, the propeller drives the rotating blade to do frequent forward and reverse rotation. In this paper, the thrust and torque under various operating conditions are obtained by changing the propeller rotational speeds and directions, as shown in Table 6. The "+" and "-" indicate forward and reverse rotation, while thrust and torque are considered only in terms of magnitude.

Whether or not the rotating cutter is fitted, forward rotation of the propeller always produces more thrust and required torque than in the reverse case. Whether in forward or reverse rotation, the propeller with a rotating cutter always produces less thrust and requires less torque than the one without a rotating cutter.

Table 6. Propeller forward and reverse performance.

Device	Velocity (m/s)	Speed (rpm)	Thrust (N)	Torque (Nm)
With cutter	0.1	+200	4.114	0.108
		-200	2.852	0.096
		+500	25.959	0.676
		-500	18.048	0.589
		+1000	105.27	2.719
		-1000	72.94	2.385
Without cutter	0.1	+200	4.491	0.111
		-200	3.131	0.098
		+500	29.496	0.709
		-500	20.131	0.596
		+1000	120.19	2.835
		-1000	80.743	2.392

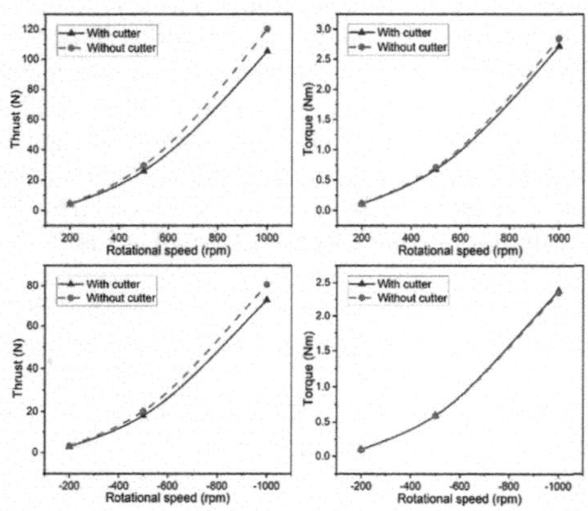

Figure 9. Comparison of forward and reverse.

The difference between the presence and absence of a rotating cutter is progressively more pronounced with increasing rotational speed, as shown in Figure 9. For the case of 1000 rpm forward rotation, the thrust and torque difference is 12.4% and 4.1%, respectively. For the case of 1000 rpm reverse rotation, the thrust and torque difference is 9.7% and 0.3%.

The stern flow field and velocity contours at different rotational speeds are shown in Figure 10. The disturbance of the rotating cutter increases gradually with the acceleration of the rotational speed. The most intuitive effect is that it disperses the centre's stern flow more.

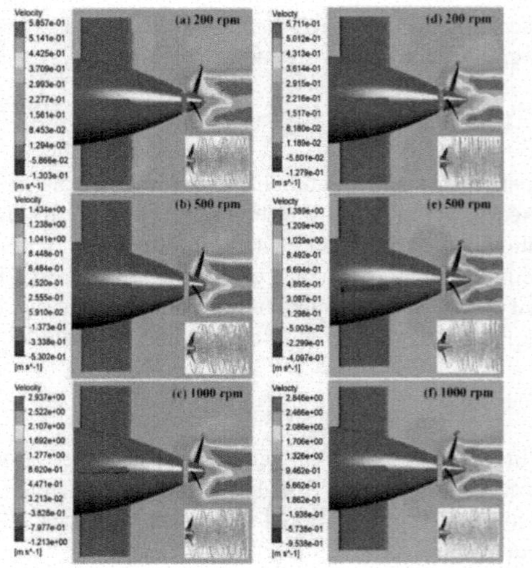

Figure 10. Stern flow field at various rotational speeds: (a)(b)(c) Without cutter (d)(e)(f) With cutter.

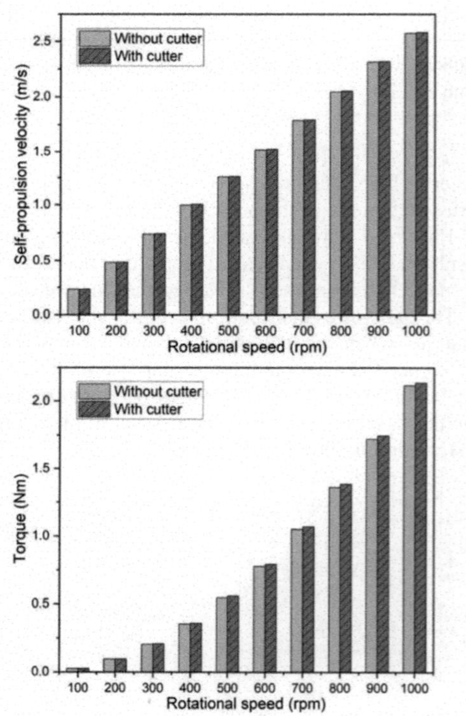

Figure 11. Self-propulsion velocity and torque at various rotational speeds.

4.2 *Self-propulsion properties*

Self-propulsion properties are reliable indicators to characterise the motion performance of AUVs, and the self-propulsion velocity and torque are usually taken as the key parameters. When the AUV drag equals the propeller thrust, it enters a self-propulsive state.

The previous section investigates the open water and behind-fixed hull performance, corresponding to the tank tests. In contrast, this section investigates how mounting a rotating cutter affects the self-propulsion velocity and torque at different rotational speeds corresponding to the actual operation. The specific findings are shown in Figure 11. Both self-propulsion velocity and torque are obtained by response surface methodology with the Kriging model. The numerical simulation process converges, and the relative prediction error is 1.84%. Therefore, the prediction results are reliable.

It can be seen that the rotating cutter does not have a significant effect on the propeller performance in the self-propulsion state. Self-propulsion velocity and torque with a rotating cutter are generally slightly higher at the same rotational speed, and the torque change is more obvious. Self-propulsion velocity remains linearly related to propeller rotational speed, and torque remains quadratically related to propeller rotational speed. In summary, the cutting device designed in this paper has little effect on the motion performance of AUV.

5 EXPERIMENTAL VERIFICATION

Tank tests are carried out using an AUV drive cabin to test the capability of the cutting device proposed in this paper. The cutting device is mounted on the drive cabin stern and propeller, and the drive cabin is fixed to the frame. Automatic control is achieved by connecting the motor via watertight cables, and data is shared to multiple terminals by a switch. A waterproof camera is placed around the cutting device to monitor and record its operation. The tests are conducted on land and in a tank, while the land test provides a more intuitive experience. Adequate tests are carried out using fishing nets, cables, plastic bags and seaweed as entanglements. The test setup is shown in Figure 12.

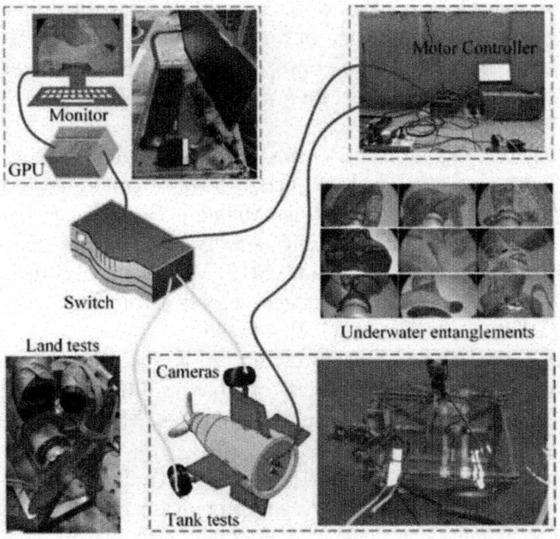

Figure 12. Land and tank tests.

The tests adopt two different strategies.

Strategy A: When entanglement occurs, the propeller rotates counter-clockwise at high speed (1000 rpm) for 3 seconds, then clockwise at high speed for 3 seconds, and the forward and reverse rotations are switched continuously and repeated in 5 groups.

Strategy B: When entanglement occurs, the propeller rotates counter-clockwise at medium speed (500 rpm) for 5 seconds, then clockwise at medium speed for 5 seconds, and the forward and reverse rotations are switched continuously to repeat five groups. Figure 13 illustrates the two cutting strategies.

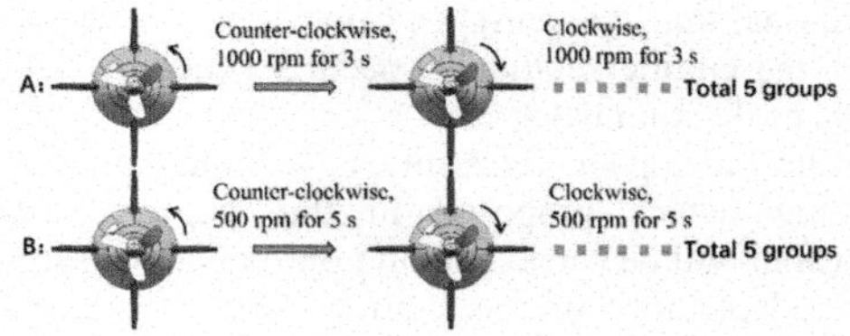

Figure 13. Different cutting strategies.

For the tests of each strategy, two types of cables of different diameters, three types of fishing nets with different weaving densities and plastic bags are selected. White nets have the largest holes, while green nets have the smallest holes. Ten experiments are repeated for the same entanglement, and the success rate is calculated. The complete test results are shown in Table 7.

Table 7. Results of the two strategy tests.

	Cables		Fishing nets			
Tests	3 mm	5 mm	White	Red	Green	Plastic bags
A	100%	90%	100%	100%	90%	100%
B	90%	70%	80%	70%	80%	90%

From the experimental results, the cutting success rate of strategy A is generally greater than that of strategy B. High-speed rotation with frequent direction switching is beneficial for cutting and getting rid of entanglements. In the case of fishing nets, the smaller the holes and the higher the density, the more challenging to cut.

6 CONCLUSIONS

In this paper, a novel cutting device with an appropriate cutting strategy is proposed to solve the propeller and shaft entanglement of AUVs effectively. The effects of the rotating cutter on the propeller's hydrodynamic characteristics and AUV's self-propulsion performance are investigated by numerical simulations. The study shows that the cutting device effectively cuts entanglements without affecting the motion performance of the AUV.

A combination of fixed and rotating cutters helps to improve cutting effectiveness. The rotating diameter of the blade should be slightly larger than the propeller hub diameter. The rotating cutter and propeller are designed separately and used in combination, so the cutter can be replaced individually when damaged. Although the rotating cutter significantly affects the pressure and stern flow distribution of a fixed-position operating propeller, it has almost no effect on the self-propulsion motion performance.

The high-speed rotation of the propeller with frequent direction switching is beneficial for cutting and getting rid of entanglements. However, when entanglement occurs, it is usually accompanied by abnormal transient currents. Considering the motor's life, the cutting rotational speed is generally below 70% of the nominal speed.

Currently, poor versatility is a limitation of the cutting device proposed in this paper. Further research will be carried out to improve the versatility and adaptability of this cutting device for various types of underwater vehicles.

ACKNOWLEDGEMENTS

This study is supported by the China Scholarship Council (Grant No. 202306330051) and the Innovative Project for Equipment Pre-study and Application (Grant No. 62602010134).

This work contributes to the Strategic Research Plan of the Centre for Marine Technology and Ocean Engineering (CENTEC), which is financed by the Portuguese Foundation for Science and Technology (Fundação para a Ciência e Tecnologia - FCT) under contract UIDB/UIDP/00134/2020.

REFERENCES

Chambault, P., F. Vandeperre, M. Machete, J. C. Lagoa, & C. K. Pham. 2018. Distribution and composition of floating macro litter off the azores archipelago and madeira (neatlantic) using opportunistic surveys. *Marine Environmental Research* 141, 225–232.

Chen, X., N. Bose, M. Brito, F. Khan, B. Thanyamanta, & T. Zou. 2021. A review of risk analysis research for the operations of autonomous underwater vehicles. *Reliability Engineering & System Safety* 216, 108011.

Cho, D. O. 2005. Challenges to marine debris management in korea. *Coastal Management* 33(4), 389–409.

Ding, G., N. Jiang, X. Gao, F. Wang, & X. Wu. 2022. Deformation monitoring of propeller underwater operation based on fiber optic grating sensing network. *Ocean Engineering* 264, 112308.

Dubyoski, J. M., J. T. Klamo, & A. G. Pollman. 2022. Asymmetric entanglement of a propeller with marine vegetation. In *OCEANS 2022, Hampton Roads*, pp. 1–10. IEEE.

Gan, W., Q. Dong, & Z. Chu. 2022. Fault diagnosis method for an underwater thruster, based on load feature extraction. *Electronics* 11(22), 3714.

Hong, S., J. Lee, & S. Lim. 2017. Navigational threats by derelict fishing gear to navy ships in the Korean seas. *Marine Pollution Bulletin* 119(2), 100–105.

Irgens, K. E., J. T. Klamo, & A. G. Pollman. 2021. Experimental assessment of entanglement for a propeller driven unmanned underwater vehicle. *Naval Engineers Journal* 133(3), 103–114.

Kim, B., H. Lee, & T. Kim. 2023. Possibility analysis of complex marine accidents caused by large marine plastic litter in the republic of korea. *Marine Policy* 152, 105591.

Kim, J.-S., Y. Seul, D.-Y. Lee, K. Park, T. H. Kim, J.-H. Choi, & W.-J. Lee. 2021. A study on the structural stability and effectiveness of rope cutter for ship's propeller. *Journal of the Korean Society of Marine Environment & Safety* 27(4), 550–556.

Kwon, W.-S., Q. D. Vuong, J.-H. Choi, J.-u. Lee, J.-w. Lee, S. H. Yoon, A. J. Nyongesa, M.-H. Park, S.-K. Yang, & W.-J. Lee. 2022. Study on the propeller rope cutter concerning transient torsional vibration due to cutting action. *Applied Sciences* 12(3), 1628.

Lee, H., J.-S. Oh, & S.-H. Choi. 2019. Development of an underwater rope-cutter device and controller for removal of propeller and shaft foreign material for small vessel. *Journal of the Korean Society of Marine Environment & Safety* 25(7), 927–935.

Lee, W., J. Kim, W. Lee, J. Kim, & J. Choi. 2019. A study oncurrent status of entanglement accident and user satisfaction survey for ship with rope cutter in domestic. *Journal of Fisheries and Marine Sciences Education* 31(6), 1536–1543.

Liu, J., M. Wang, F. Yu, S. Gao, T. Yan, & B. He. 2023a. Numerical study on the hull–propeller interaction of autonomous underwater vehicle. *Ocean Engineering* 271, 113777.

Liu, J., F. Yu, T. Yan, & B. He. 2023b. Self-propulsion performance predictions of auv based on response surface methodology. *Ocean Engineering* 287, 115923.

Macfadyen, G., T. Huntington, R. Cappell, et al. 2009. *Abandoned, lost or otherwise discarded fishing gear.* (No. 523). Food and Agriculture Organization of the United Nations (FAO).

McIlgorm, A., H. F. Campbell, & M. J. Rule. 2011. The economic cost and control of marine debris damage in the asiapacific region. *Ocean & Coastal Management* 54(9), 643–651.

Nyongesa, A. J., V. C. Pham, S. H. Yoon, W.-S. Kwon, J.-S. Kim, D. N. Ngo, J.-H. Choi, Y.-Y. Sul, & W.-J. Lee. 2022. Investigation of the effect of rope cutter on water flow behind ship propellers based on cfd analysis. *Machines* 10(5), 300.

Nyongesa, A. J., Y. Sul, & S. Yeo. 2023. Numerical investigation on the influence of rope cutter installation on water flow around a ship's propeller. *Journal of Advanced Marine Engineering and Technology (JAMET)* 47(2), 98–104.

Pham, C., J. Gomes-Pereira, E. Isidro, R. Santos, & T. Morato. 2013. Abundance of litter on condor seamount (Azores, Portugal, northeast Atlantic). *Deep Sea Research Part II: Topical Studies in Oceanography* 98, 204–208.

Rodríguez, Y. & C. K. Pham. 2017. Marine litter on the seafloor of the faial-pico passage, Azores archipelago. *Marine Pollution Bulletin* 116(1-2), 448–453.

Rodríguez, Y., A. Ressurreic¸ao, & C. K. Pham. 2020. Socio- economic impacts of marine litter for remote oceanic islands: The case of the Azores. *Marine Pollution Bulletin* 160, 111631.

Vanessa Anuat, G., J. T. Klamo, & A. G. Pollman. 2022. Effects of body geometry and propulsion type on unmanned underwater vehicle interactions with marine vegetation. *Journal of Autonomous Vehicles and Systems* 2(2), 021002.

Watters, D. L., M. M. Yoklavich, M. S. Love, & D. M. Schroeder. 2010. Assessing marine debris in deep seafloor habitats off california. *Marine Pollution Bulletin* 60(1), 131–138.

Wiggins, J., C. Zeglin, & W. Allensworth. 2015. Concentric cutting assembly, concentric cutting system, and net penetration method. US Patent 9,061,361.

Zhang, Y., J. Geng, & N. Wang. 2023. Correlation analysis and fault detection of current and speed signals for underwater thruster entanglement. *Advanced Control for Applications: Engineering and Industrial Systems* 145, 1–14.

Advances in Maritime Technology and Engineering – Guedes Soares & Santos (Eds)
© 2024 The Author(s), ISBN 978-1-032-83107-7

Development of an innovative bow shape to enhance hydrodynamic performance in a seaway

H. Orihara, I. Amaya, H. Yoshida, K. Takagishi & T. Inoue
Japan Marine United Corporation, Yokohama, Japan

ABSTRACT: An innovative bow shape, named SP-Bow (Side-Protruding Bow), has been developed to enhance hydrodynamic performance in a seaway. The concept of the developed bow shape is highly unique, representing a significant departure from existing bow shapes intended for superior performance in a seaway. Firstly, the hydrodynamic concept of the SP-Bow is described, including the hypothesis for the mechanism behind the reductions of added resistance in waves. Subsequently, the hydrodynamic performance of the SP-Bow has been evaluated through both model experiments and CFD simulations. Furthermore, the energy-saving capability of the SP-Bow for full-scale operation has been demonstrated for typical merchant ship types based on the prediction of speed-power characteristics in a seaway. Up to a 6% reduction in propulsive power in a seaway through the adaptation of the SP-Bow has been confirmed in a typical sea state.

1 INTRODUCTION

In recent years, the maritime community has increasingly recognized the importance of energy efficiency to reduce greenhouse gas emissions from ocean-going ships. Various technologies, including optimized hull forms, energy-saving devices, machinery improvements, and optimal routing, have been employed to enhance fuel efficiency in ship operations.

For large tankers and bulk carriers, which typically feature full hull forms and low service speeds, waves encountered in a seaway significantly affect the ship's fuel efficiency during operation. Thus, reducing added resistance due to waves has become an important consideration in the design of these ship types.

Given the significant influence of added resistance in waves on the operational performance of large tankers and bulk carriers, Japan Marine United Corporation (JMU) has developed a variety of technologies effective for reducing resistance in waves over the last three decades. These include Ax-Bow and LEADGE-bow shapes designed to mitigate resistance due to waves (Matsumoto et al., 2000; Hirota et al., 2005). These bow shapes have been adopted on most tankers and bulk carriers built by JMU, and their effectiveness has been verified in actual operating conditions through full-scale performance measurements (Hirota et al., 2005; Orihara et al., 2014).

More recently, in response to increasing demands for higher fuel efficiency of ocean-going ships, JMU has further developed energy-saving technologies to enhance performance, incorporating them into a series of innovative, fuel-efficient, and eco-friendly large tankers and bulk carriers.

In the present study, an innovative bow shape named SP-Bow (Side-Protruding Bow) has been developed to enhance hydrodynamic performance in a seaway. The concept of the developed bow shape is highly unique, representing a significant departure from existing bow shapes intended for superior performance in a seaway. The SP-Bow improves performance by reducing added resistance in waves. While most existing bow shapes have achieved a reduction in resistance by refining the waterline in the bow part, the SP-Bow has adopted a completely different approach, incorporating an underwater side-protrusion in the bow region.

Firstly, the hydrodynamic concept of the SP-Bow is described in the following section. Then, the description of the model experiments conducted to confirm the effectiveness of the SP-Bow is provided. After that, CFD simulation results are presented and discussed, mainly focusing on the relevant issues regarding the mechanism of reduction in added resistance by the SP-Bow. Following this, the energy-saving capability of the SP-Bow for full-scale operation is demonstrated for typical merchant ship types based on the prediction of speed-power characteristics in a seaway. Brief conclusions are presented in the final section.

2 CONCEPT OF NEW BOW SHAPE

As mentioned in the introduction, the influence of added resistance in waves on operation in a seaway

DOI: 10.1201/9781003508779-18

has been increasing in recent years, primarily due to the adaptation of fuller hull forms to increase cargo carrying capacity and the reduction of service speed to decrease fuel consumption. To address this challenge, the innovative new bow shape "SP-Bow" has been developed by JMU. This section describes the concept of the SP-Bow, focusing mainly on its hydrodynamic aspects and the hypothesis of its action in reducing added resistance in waves.

The added resistance due to waves consists of two components, as illustrated in Figure 1. One is caused by diffraction at the bow, and the other is caused by the radiation wave due to ship motion. The diffraction component primarily contributes in the shorter wavelength region, usually with a wavelength-to-ship-length ratio (λ/L) below around 0.7. For large bulk carriers and tankers exceeding 300 meters in length, most waves encountered in a seaway are irregular waves comprising mainly wave components of short wavelengths corresponding to $\lambda/L \leq 0.7$. Thus, added resistance acting on these large, full hull form ships mainly results from wave diffraction at the bow.

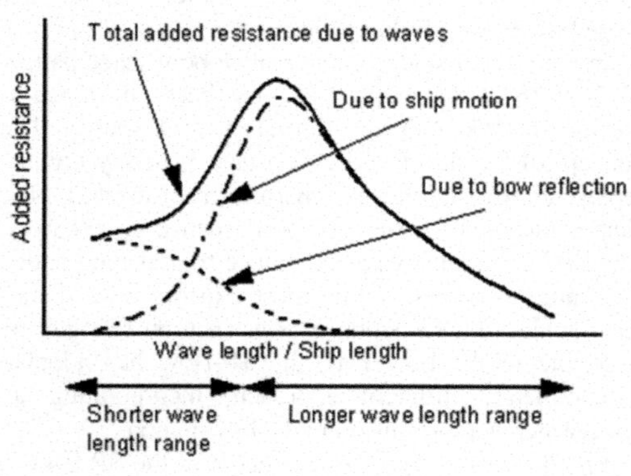

Figure 1. Characteristics of added resistance in waves.

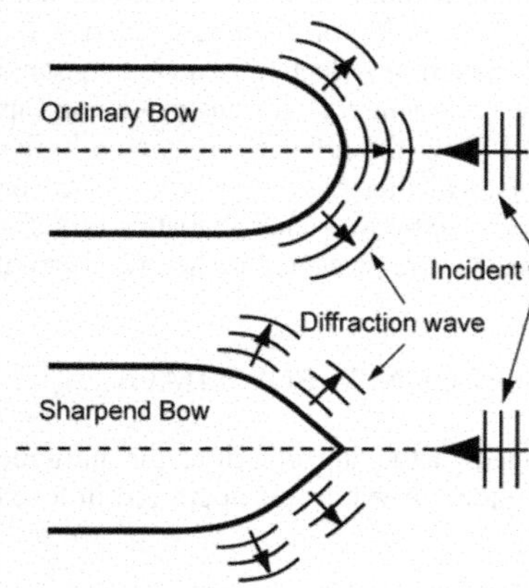

Figure 2. Wave reflection at bow.

The wave diffraction phenomenon at the bow is schematically shown in Figure 2. Wave diffraction primarily occurs due to the reflection of incident waves at the bow. This wave reflection leads to a reaction force acting on the hull in the direction opposite to the wave reflection. Since the component of the reaction force in the longitudinal (backward) direction corresponds to the added resistance due to incident waves, reducing wave reflection in the hull's longitudinal direction is effective in reducing the added resistance caused by waves.

Following this consideration, a variety of bow shapes have been developed to fundamentally reduce added resistance in shorter waves. The underlying idea is to modify the bow's waterline shape to minimize added resistance caused by waves by altering the principal direction of wave reflection from the longitudinal direction to the lateral direction (e.g.,

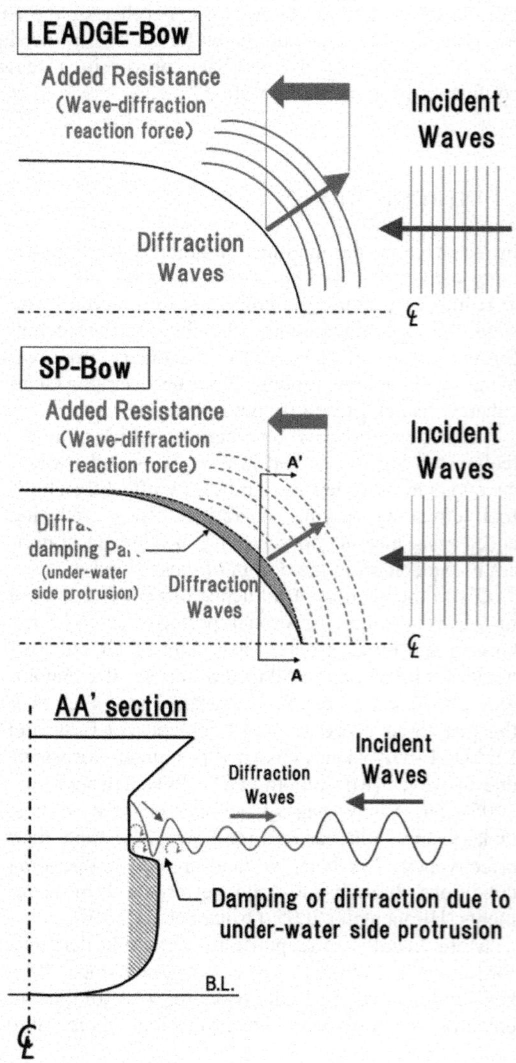

Figure 3. Mechanism of added-resistance reduction by SP-Bow shape.

Matsumoto et al., 1998; Matsumoto et al., 2000; Hirota et al., 2004; Hirota et al., 2005; Hwang et al., 2016). Among these shapes, JMU has developed the Ax-Bow (which sharpens the bow shape above the load waterline) and the LEADGE-Bow (which sharpens the entire stem line) (see Matsumoto et al., 2000; Hirota et al., 2004)

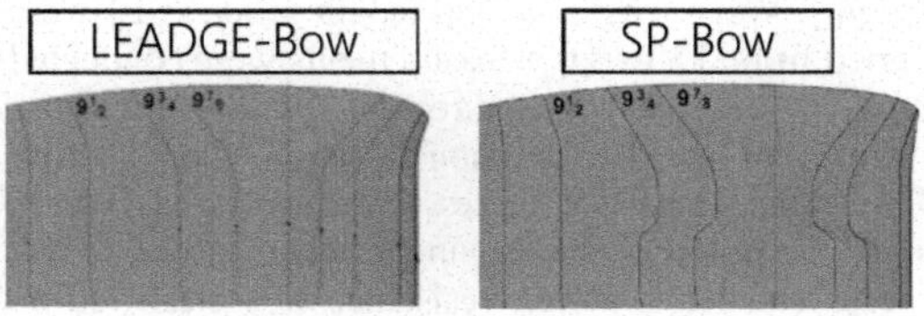

Figure 4. Comparison of bow shapes.

The SP-Bow aims to reduce the diffraction components of added resistance, similar to existing low-resistance bow shapes such as the Ax-Bow and LEADGE-Bow. However, its hydrodynamic concept in reducing resistance in waves is highly unique and represents a significant departure from existing bow shapes. While most existing bow shapes have achieved a reduction in resistance by refining the waterline in the bow part, the SP-Bow has adopted a completely different approach, incorporating an underwater side-protrusion in the bow region.

The hydrodynamic concept of the SP-Bow has been devised from the consideration of wave systems around the ship's bow in waves. Around the bow of a ship advancing in waves, two wave systems consisting of incoming waves and reflected waves (diffraction and radiation waves) exist. The momentum of reflected waves propagating in the forward direction results in the added resistance acting on the ship. The SP-Bow shape reduces this momentum component through the interaction between reflected waves from both the main hull and underwater bow protrusions. The underwater protrusion parts also dampen the diffracted waves from the main hull by reflecting wave orbital motion over the upper surface of the protruding part, further reducing the diffraction component of added resistance. Figure 3 schematically illustrates this concept compared with the existing LEADGE-Bow case. By adopting this type of underwater protrusions, hydrodynamic performance in waves can be improved without reducing hull displacement volume, as in the cases of refining bow shapes. Figure 4 shows an example of the SP-Bow shape applied to a bulk carrier compared with the LEADGE-Bow including bow cross-sectional lines.

The idea of utilizing the interaction of wave systems bears some analogy with those employed for reducing wave-making resistance in still water, such as bulbous bow shapes. However, to the best of the author's knowledge, it has not been applied to the reduction in added resistance waves of ships with forward speed. Although some studies on the wave forces acting on floating structures have employed a similar idea to reduce the steady wave forces, primarily for the reduction of mooring requirements. It seems that the utilization of the interaction of unsteady wave systems is not necessarily confined to the added resistance problem but can have wider applications in seakeeping problems, such as steady lateral force and yawing moment acting on a ship advancing in waves.

3 EXPERIMENTAL INVESTIGATION

3.1 *Model testing facilities*

The model experiments were conducted in Japan Marine United Corporation's (JMU) two model testing basins: the Tsu Ship Model Basin (TSMB: L×B×D = 240m × 18m × 8m) and the Yokohama Seakeeping and Maneuvering Basin (YSMB: L×B×D = 70m × 30m × 3m). In the TSMB, a flap-type wave maker is equipped on the west side (18m wide), allowing for the realization of head and following wave conditions. In the YSMB, flap-type wave-makers are equipped on both the east side (70m width) and the north side (30m width) of the basin, enabling the realization of unidirectional wave conditions. A tested model can be towed or freely propelled in arbitrary wave directions by adjusting the model's heading angle.

In the TSMB, experiments were conducted using a seakeeping dynamometer, allowing for the realization of four degrees of freedom of ship motion in waves by towing the model connected to the dynamometer. The dynamometer consists of a mount and a sliding platform with two heave rods. The sliding platform moves smoothly in the surge direction on rails and is balanced to a mean position using soft springs and a servo-motor. A heave rod moves smoothly in the vertical direction to allow for free heave motion. Free pitch and roll motions are achieved by connecting a ship model to the bottom of the two heave rods at the center of gravity of the model and at a point approximately 30% of the ship's length ahead of the center of gravity through gimbals. The resistance in waves is measured using two 3-component load cells placed at the lower ends of the two heave rods.

In the YSMB, experiments were conducted using a seakeeping dynamometer, allowing for the realization of six degrees of freedom of ship motion in waves by towing the model connected to the dynamometer. The dynamometer consists of a mount, a nested sliding platform, and a heave rod. The nested sliding platform moves smoothly in surge and sway directions on rails and is balanced to a mean position using soft springs and counterweights. A heave rod moves smoothly in the vertical direction to allow for free heave motion. The yawing motion is balanced with a bar-spring accommodated in the heave rod. Free pitch and roll motions are achieved by connecting a ship model to the bottom of the heave rod at the center of gravity of the model through a gimbal. The resistance and drifting force are measured using a 3-component load cell placed at the lower end of the heave rod.

3.2 *Model descriptions*

Two models were used in the experiment: a tanker (Model A) and a bulk carrier (Model B). Their dimensions are listed in Table 1. Figures 5 and 6 show Model A and Model B towed in waves, respectively. The models have hull forms typical of recent tanker and bulk carrier designs. For both Model A and B, two bow shape cases (LEADGE-Bow and SP-Bow) were tested in the experiments. Model A is of hybrid construction with paraffin and wood. Initially, the bow section of Model A is made to the LEADGE-Bow shape. After testing the LEADGE-Bow shape, the bow section of the model is reshaped to SP-Bow.

Model B is made of wood and is of a segmented type, with bow sections detachable at a location approximately 15% aft of the bow end. Two bow section models of LEADGE-Bow and SP-Bow were prepared. The models have been dynamically balanced so that the weight, the center of gravity, the longitudinal radius of gyration, and the roll natural period coincide with those of fully loaded conditions.

Table 1. Principal dimensions of the models.

Model	Length (L)	Breadth (B)	Draft (T)
Model A	7.88m	1.41m	0.48m
Model B	3.00m	0.56m	0.19m

Figure 5. Model A towed in head waves in TSMB.

Figure 6. Model B towed in head waves in YSMB.

3.3 *Condition of experiment*

In the experiment, all measurements were conducted in regular waves by towing ship models at specified advancing speeds. For Model A, the model towing speed was set at a Froude number (Fn) of 0.14, corresponding to the normal service speed of large oil tankers. The wave amplitude to ship length ratio (ζ_A/L) was set at approximately 0.5% for all wave lengths. Wave length to ship length ratios (λ/L) were varied from 0.3 to 2.0. Measurements were conducted only in the head wave direction (χ=180deg.). Four degrees of freedom motions (Surge, Heave, Roll, Pitch) and resistance forces were measured. Waves were measured at the towing carriage. Prior to the measurements in waves, resistance was measured by towing the model in still water. Response Amplitude Operators (RAOs) for added resistance and ship motions were obtained from harmonic analysis of measured time histories over a duration of more than 12 wave encounter periods (Te). Added resistance was evaluated as the difference between mean resistance in waves and resistance in still water.

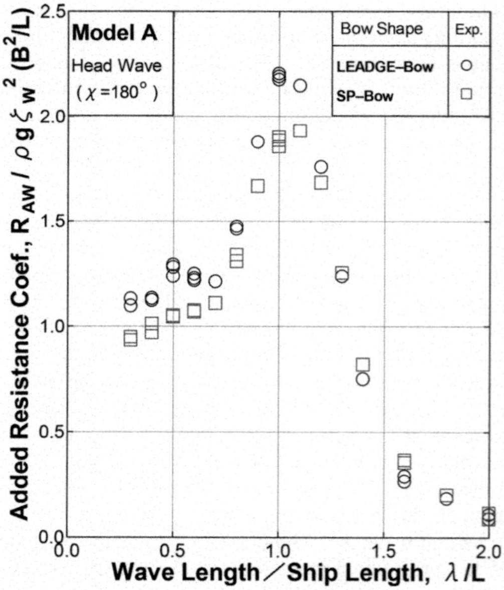

Figure 7. Comparison of normalized added resistance for a model A in head waves (χ=180deg.).

For Model B, the model towing speed was set at a Froude number (Fn) of 0.14, corresponding to the normal service speed of large bulk carriers. The wave amplitude to ship length ratio ($\zeta A/L$) was set at approximately 0.5% for all wave lengths. Measurements were conducted in four wave directions from head (χ=180deg.) to beam (χ=180deg.) directions. For head waves, wave length to ship length ratios (λ/L) were set from 0.2 to 2.0, while for other wave directions, λ/L was set from 0.2 to 1.0.

Six degrees of freedom motions and resistance forces were measured. Waves were measured at both the towing carriage and the stationary position in the basin. Prior to the measurements in waves, resistance was measured by towing the model in still water. Analysis of the measured data was carried out in a similar manner as for Model A, except for the number of ship motion components.

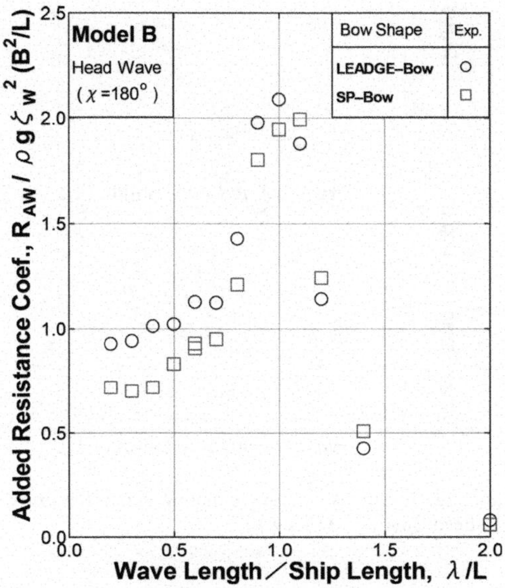

Figure 8. Comparison of normalized added resistance for a model B in head waves (χ=180deg.).

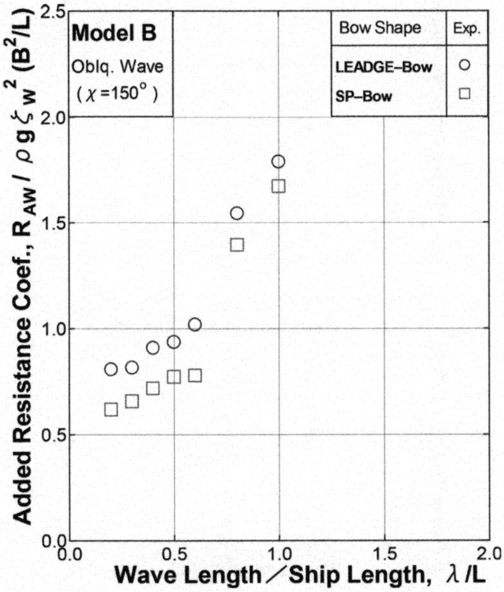

Figure 9. Comparison of normalized added resistance for a model B in oblique waves (χ=150deg.).

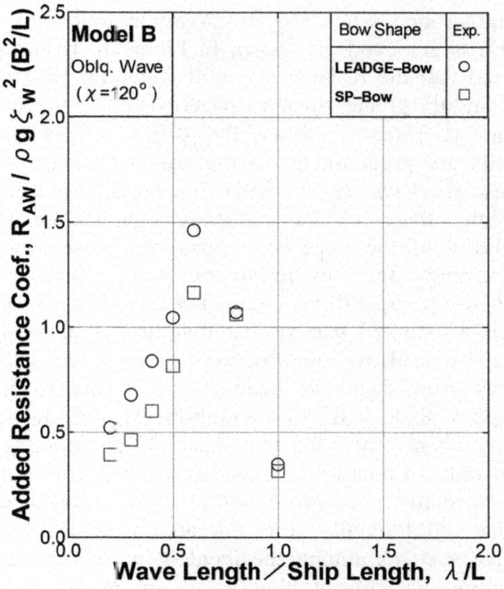

Figure 10. Comparison of normalized added resistance for a model B in oblique waves (χ=120deg.).

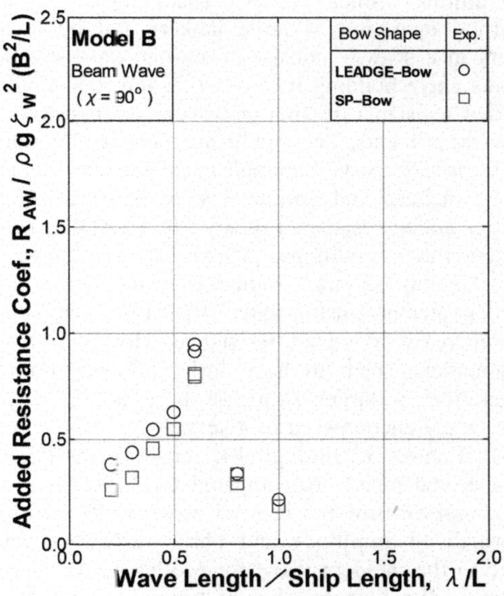

Figure 11. Comparison of normalized added resistance for a model B in oblique waves (χ=90deg.).

3.4 *Results of experiments*

Firstly, comparisons of added resistance are shown in Figures 7 through 11. As clearly seen from the Figures, SP-Bow data, indicated by open squares (□), shows a significant reduction of added resistance relative to LEADGE-Bow data, indicated by open circles (○), particularly in shorter waves with λ/L < 1.0. In the shorter wave limit around λ/L ≈ 0.3, a maximum reduction of 30% is achieved, as shown in Figure 8. It is also noted that the resistance is still water for SP-Bow is slightly greater than LEADGE-Bow in the present experiments. Since the diffraction components are predominant in the added resistance in these short waves, in which ship motions are negligible, these results explicitly demonstrate the validity of the hypothesis proposed in section 2 regarding the hydrodynamic mechanism of SP-Bow's capability in reducing added resistance. It is also noted that similar magnitudes of reduction by SP-Bow are observed in all wave directions from head to beam waves. This readily implies that diffraction components, which SP-Bow mainly intended to cope with, are dominant not only in conditions close to the head direction as normally considered but also in wider directions. Additionally, the interaction of unsteady wave systems among the incident waves and diffraction waves can play a role in reducing the added resistance. While the present SP-Bow shapes are designed mainly for the head wave condition, consideration in the other wave directions may lead to enhancing the resistance reduction capability in oblique and beam wave conditions. Since waves encountered in the normal operation of large tankers and bulk carriers in a seaway consist of shorter wave components corresponding to λ/L < 0.7, the reduction of added resistance in shorter waves, as ob-served in the experiments, has significant benefits for operation in a seaway. Although in longer wave conditions in head and oblique waves, SP-Bow shows larger added resistance relative to LEADGE-Bow. Under these conditions, relative bow motions tend to develop so that radiation waves from the underwater protrusion parts of SP-Bow may cause an increase in added resistance. However, these phenomena seem to have little influence on the operation of larger ships as these longer waves are rarely encountered in a seaway.

In Figures 12 through 14, ship motions (surge, heave, and pitch) are compared for LEADGE-Bow (○) and SP-Bow (□) in head waves in the form of normalized amplitude and phase difference relative to the incident wave for Model A. As clearly seen in the Figures, ship motions are almost the same for both bow shapes across all ship motion components. This implies that SP-Bow has little effect on ship motions and also suggests that the motion compo-nent of the added resistance is affected by local modifications of the bow shape in a similar manner as reported by previous studies (e.g., Matsumoto et al., 1998; Matsumoto et al., 2000). In this study, experimental uncertainty was judged from the multiple measurements for selected wave-length cases. Maxim differences among measured data are generally less than 5% and 2% for added resistance and ship motions respectively.

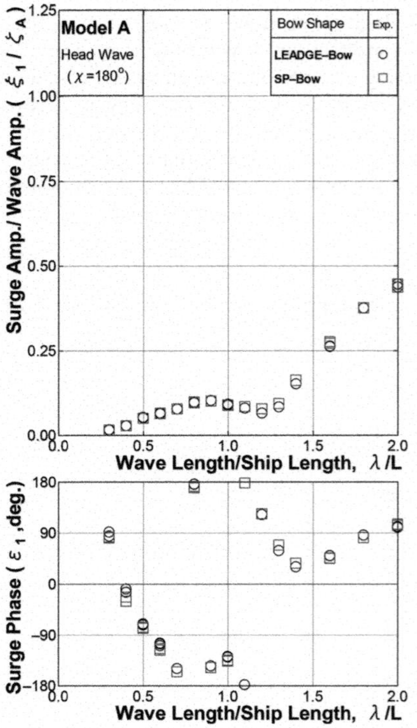

Figure 12. Comparison of surge motion RAOs for a model A in head waves (χ=180deg.).

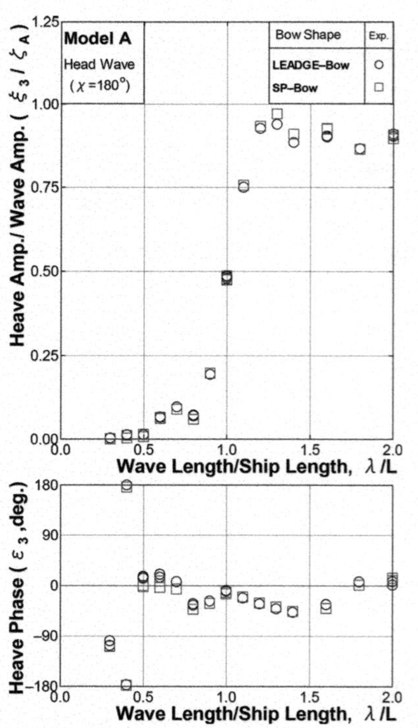

Figure 13. Comparison of heave motion RAOs for a model A in head waves (χ=180deg.).

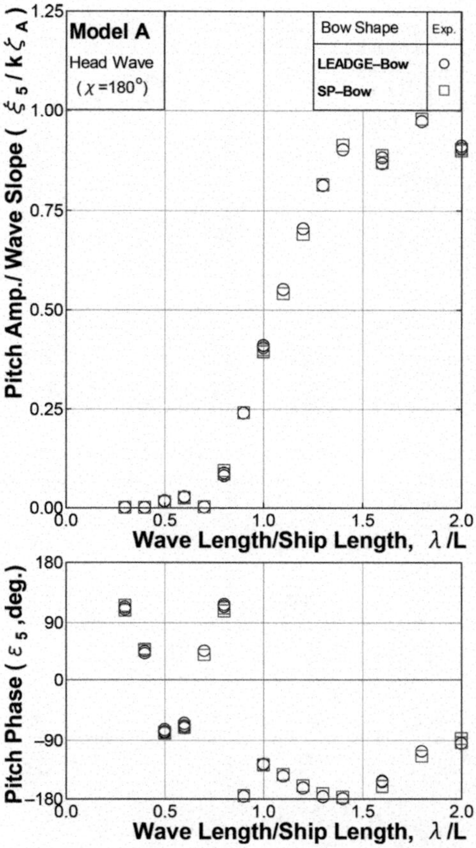

Figure 14. Comparison of pitch motion RAOs for a model A in head waves (χ=180deg.).

4 COMPUTATIONAL INVESTIGATION

4.1 *Outline of CFD code*

In this study, the Reynolds-Averaged Navier-Stokes (RANS) code called WISDAM-X was used. This code employs an overlapping grid system to implement rigorous wave generation, the interactions of ships with incident waves, and the resultant ship motions. Since the computational procedure details of WISDAM-X have been explained in Orihara and Miyata (2003), they are briefly summarized below.

The motion of the ship is solved simultaneously by combining the equations of motion of the ship body with the flow computation. A non-deforming grid is used, and the flow computation in the near field is performed on the non-inertial coordinate system fixed to the body of the ship. The effect of the ship motion on the flow solution is taken into account by adding inertia forces as the body force term in the RANS equation.

In the WISDAM-X code, a PID (Proportional, Integral, and Derivative) control is implemented to maintain a time-averaged ship advancing speed at a specified mean value during surging motion in waves. This control simulates the artificial thrust force needed to maintain the determined time-averaged speed in waves. Based on the current and determined time-averaged advancing speeds of the ship, the artificial thrust force is evaluated and added to the surging equation of motion.

Numerical uncertainties for WISDAM-X code in predicting ship motions and added resistance has thoroughly examined in the previous studies. It has shown that the magnitudes of computational size and resolutions similar to those employed in this study are sufficiently converged to grid independent state (e.g. Orihara and Miyata (2003), Orihara 2010).

4.2 *Condition of simulation*

Simulations were conducted for two bow shape cases (LEADGE-Bow and SP-Bow) for both Model A and B in head wave conditions. The simulations were performed on an overlapping grid system consisting of inner and outer grids. The numbers of grid points allocated for the grids were 183×30×89 and 356×61×101 for the inner and outer grids, respectively, for both Model calculations. Since only the head wave condition was considered, half side of the model was modeled in the simulations.

The conditions of the ship model and waves were basically the same as those of the experiment. Model speeds were set at values equivalent to Fn = 0.14 for both model cases. Regular head waves were considered, with a wave amplitude to ship length ratio (ζ_A/L) set at 0.5% for all wave lengths. For cases of wave lengths, λ/L= 0.4, 0.6, and 0.8 were selected. The ship was allowed to heave, pitch, and surge freely.

The flow was accelerated from rest to a steady advancing condition during the computational time T=0.0 to T=4.0, where T is the normalized time made dimensionless with respect to (L/U0, where U0 is the mean advancing speed). The wave computations started at T=8.0 and continued until T=24.0. Surge motions were restrained until T=8.0+8Te, where Te is the encounter wave period. Then the model was released free to surge. In addition to simulations in waves, simulations in still water conditions were conducted for both model cases for the evaluation of added resistance in waves.

4.3 *Results and discussions*

Firstly, calculated added resistances are compared with experimental data. In calculations, computed hull resistance time histories obtained from the integration of pressure and frictional stress over the instantaneous wetted surface are analyzed over a time duration of more than 12 Te to evaluate mean resistance in waves. Added resistance is calculated as the difference between the mean resistance in waves and resistance in still water.

In Figures 15 and 16, comparisons of added resistance are shown for Model A and B, respectively.

In the Figures, added resistance is normalized in the same way as in the experiments. Calculated values are shown as closed circles (●) and closed squares (■) for LEADGE-Bow and SP-Bow, respectively. As clearly shown in the Figures, the calculated re-sults agree quite well with the experimental data for both Models, including the difference between the two bow shapes for all the wave-length cases. Calcu-lated added resistances for SP-Bow are reduced by about 20% relative to LEADGE-Bow over the vari-ous wave lengths.

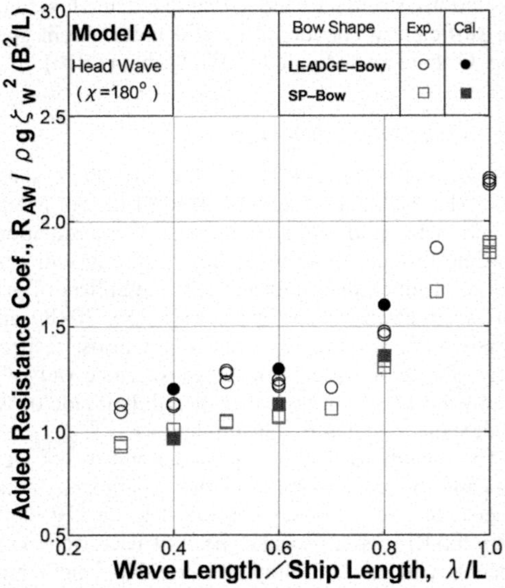

Figure 15. Comparison of calculated added resistance with experimental data for a model A in head waves (χ=180deg.).

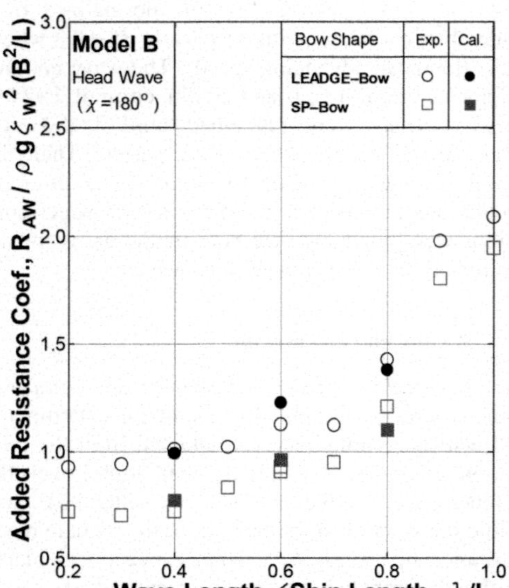

Figure 16. Comparison of calculated added resistance with experimental data for a model B in head waves (χ=180deg.).

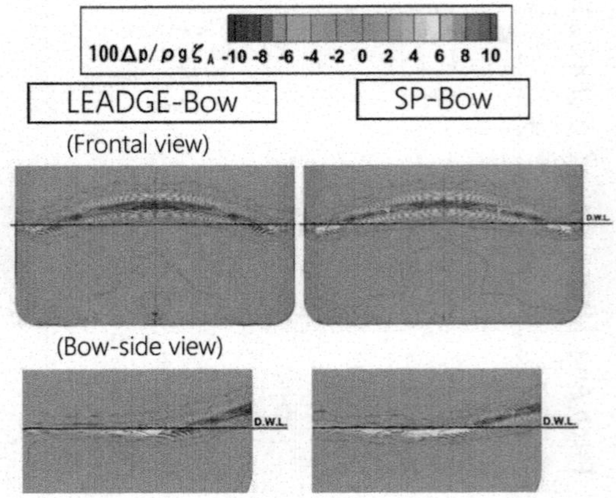

Figure 17. Comparison of computed added pressure contour maps for two bow shapes of model A in regular head waves of λ/L=0.6, 1.0 at Fn = 0.14.

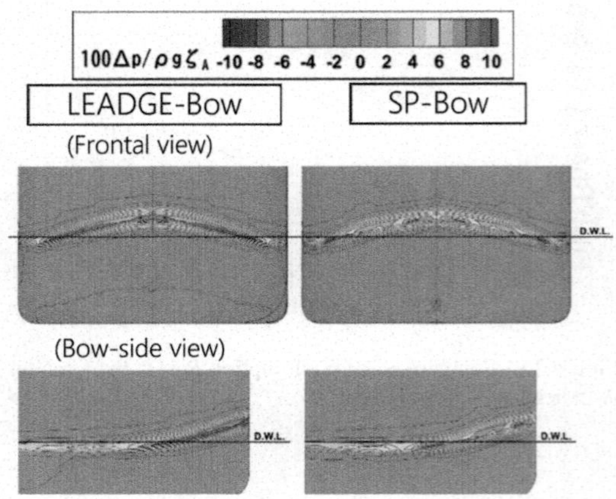

Figure 18. Comparison of computed added pressure contour maps for two bow shapes of model B in regular head waves of λ/L=0.6, 1.0 at Fn = 0.14.

To analyze the mechanism of the calculated reduction in added resistance in waves, added pressure (Δp) distributions are evaluated for the case of λ/L= 0.6. Δp are obtained as the difference in mean pressure in waves, which is evaluated over the same time duration as for the added resistance, and pressure in still water at the same forward speeds. In Figures 17 and 18, comparisons of normalized added pressure distributions are shown for Model A and B, respectively. Added pressure in waves is confined to the narrow region along the so-called swell-up lines, which is the surface free-surface profile in still water, in a similar manner as reported in previous works (e.g., Orihara & Miyata 2003, Satoh et al. 2006).

It is noted that added pressures are significantly reduced for SP-Bow in the bow region, particularly near the bow end. Since the integration of added pressure corresponds to the added resistance, these reductions in added pressure are considered to be the contributing factors to the smaller added resistance shown in Figures 15 and 16. From a wave dynamics point of view, added pressures should correspond to the magnitude of wave reflection on the hull surface. Thus, the reduction in added pressures demonstrates clearly the validity of the hypothesis proposed in section 2 describing the concept of SP-Bow. That is, SP-Bow can reduce the added resistance in waves by reducing both wave reflection on the bow and the resulting reflection forces acting as resistance to the hull.

Also of significance is the effectiveness of the CFD simulation for the examination of added resistance mechanisms. In addition to the accurate prediction of added resistance, the hydrodynamic mechanism can be thoroughly examined using detailed flow features, including surface pressure distributions.

5 EVALUATION OF FULL-SCALE PERFORMANCE IN A SEAWAY

In this section, the effectiveness of SP-Bow in energy-saving in full-scale ship operation in a seaway is examined based on the added-resistance characteristics obtained in the experiments.

Speed/Power characteristics in a seaway are evaluated based on the JTTC method (Tamura 1975), which is a kind of resistance and thrust identity method (RTIM) recommended by ITTC for the evaluation of the speed/power characteristics in waves (ITTC 2017). In the evaluation, still-water resistance, self-propulsion factors, and propeller open-water characteristics are obtained from the towing tank tests. The condition of a seaway corresponding to Beaufort (BF) scale 6 in head sea is selected, as BF 6 conditions are considered representative of sea states encountered in normal operations

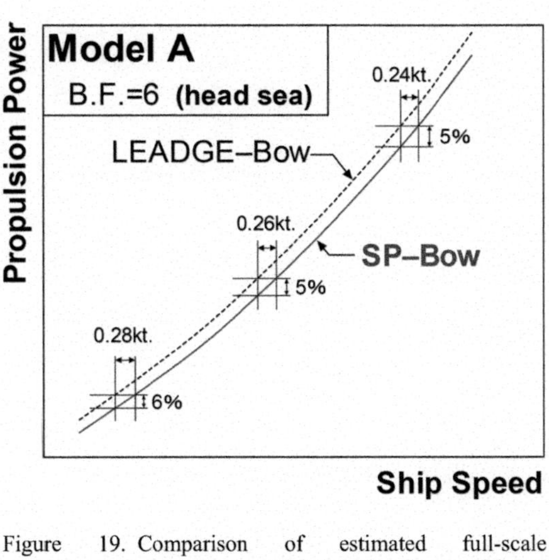

Figure 19. Comparison of estimated full-scale speed-power curves for a model A in head seas of BF=6 condition.

in a seaway. The significant wave height is set at 3m, and the mean wave period is set at 6.7 seconds. Short-crested irregular waves with the ISSC 1964 wave spectrum are employed as a representation of fully develcped wind-wave sea state. Since the evaluation is made in short-crested waves, added resistance characteristics in regular waves from head to beam directions are used.

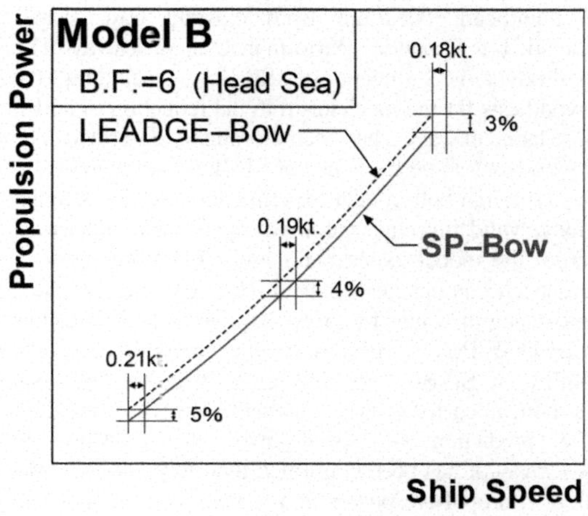

Figure 20. Comparison of estimated full-scale speed-power curves for a model A in head seas of BF=6 condition.

In Figures 19 and 20, speed/power characteristics are compared between LEADGE-Bow and SP-Bow for Model A and B at full scale, corresponding to lengths of about 300m. In the Figures, speed/power curves for LEADGE-Bow and SP-Bow over the power range in normal operations are shown in dotted and solid lines, respectively. Differences in speed are shown in the unit of knot (kt.). Up to a 6% reduction in propulsion power in a seaway is achieved by the adaptation of SP-Bow relative to LEADGE-Bow. In terms of speed, SP-Bow achieved about 0.3 knots faster than LEADGE-Bow for both Models. Since the hydrodynamic characteristics other than added resistance are almost the same for LEADGE-Bow and SP-Bow, differences in speed/power characteristics in a seaway can be considered as resulting from the added resistance in waves. For the case of low-speed full-form ships, such as tankers and bulk carriers represented by Model A and B, waves have a dominant effect on speed/power characteristics in a seaway. Thus, reduction in added resistance in waves is an effective means to improve the performance in operations normally conducted in a seaway of varying sea states.

In the present study, it is shown that added resistance reductions of around 20% as shown in section 3 result in propulsion power improvement of about 5% in BF 6 head seas. This implies the significance of the bow shape improvement achieved by SP-Bow in the reduction of hydrodynamic performance in operation.

6 CONCLUSIONS

An innovative bow shape, named SP-Bow (Side-Protruding Bow), has been developed to enhance hydrodynamic performance in a seaway. The concept of the developed bow shape is highly unique and represents a significant departure from existing bow shapes intended for superior performance in a seaway. The hydrodynamic concept of SP-Bow, including the hypothesis for the mechanism in the reduction of added resistance in waves, has been thoroughly described. The hydrodynamic performance of SP-Bow has been evaluated through both model experiments and CFD simulations, validating the proposed hypothesis effectively. From the model experiments and CFD simulations, it has been demonstrated that SP-Bow can reduce added resistance in waves by up to 30% compared to JMU's LEADGE-Bow. Furthermore, the energy-saving capability of SP-Bow for full-scale operation has been demonstrated for typical merchant ship types based on the prediction of speed-power characteristics in a seaway. It has been confirmed that up to a 6% reduction in propulsive power in a seaway can be achieved by the adaptation of SP-Bow, particularly in typical sea states. These findings highlight the significant potential of SP-Bow in enhancing the hydrodynamic performance and energy efficiency of maritime vessels operating in challenging seaway conditions.

REFERENCES

Hirota, K. et al (2004), verification of Ax-Bow effect based on full scale measurement, Journal of kansai Society of naval Architects, Japan 241: 33–40 (in Japanese).

Hirota, K. et al (2005), Development of bow shape to reduce the added resistance sue to waves and verification on full scale measurement, Proc. Int. Symp. on Marine research and Transportation, the island of Ischina, Italy.

Hwang, S. et al (2016), Experimental Study on the Bow Hull-Form Modification for Added Resistance Reduction in Waves of KVLCC2, Proc. 26th Int. Ocean and Polar Engineering Conference, Rhodes, GREECE.

ITTC (2017), Prediction of Power Increase in Irregular Wave from Model Tests, ITTC-Recommended Procedures and Guidelines 7.5-02-07-02.2.

Matsumoto, K. et al (1998), BEAK-Bow to Reduce the Wave Added Resistance at Sea, Proc. 7th Int. Symp. on Practical Design of Ships and Mobile Units, the Hauge, the Netherlands.

Matsumoto, K., Hirota, K., and Takagishi, K. (2000), Development of energy Saving Bow Shapes at Sea, Proc. 4th Osaka Colloquium on Seakeeping Performance of Ships, Osaka, JAPAN.

Orihara, H. and Miyata, H (2003). "Evaluation of added re-sistance in regular incident waves by computational fluid dynamics motion simulation using an overlapping grid system", Journal of Marine Science and Technology, 8, 47–60.

Orihara, (2010) Comparison of CFD Simulations with Experimental Data for a Tanker Model Advancing in Waves, International *Journal of Naval Architecture and Ocean engineering*, Vol. 3, No. 1 pp.1–8, Mar. 2011.

Orihara, H., Yoshida, H., and Takagishi, K. (2018), A validation study of CFD simulations of a tanker in ballast condition advancing in waves, Proc. 28th Int. Ocean and Polar Engineering Conference, Sapporo, JAPAN.

Sato, Y, Orihara, H. & Miyata, H, Practical Application of Two CFD Codes for Ship Motions in Arbitrary Waves, *Proceedings of Twenty-Sixth Symposium on Naval Hydrodynamics*, Rome, Italy, 17-22 September 2006.

Tamura, H, (1975), Speed and Power Prediction Techniques for High Block Ships applied in Nagasaki Experimental Tank. Proc. 1st Ship Technology and Research Symposium, Washigton DC, USA, 7-1 – 7-17.

Advances in Maritime Technology and Engineering – Guedes Soares & Santos (Eds)
© 2024 The Author(s), ISBN 978-1-032-83107-7

Quantifying ship impact loads on fenders: Experimental approach

L.V. Pinheiro, A.H. Gomes & C.J.E.M. Fortes
Ports and Maritime Structures Unit, Hydraulics and Environment Department, National Laboratory of Civil Engineering, Lisbon, Portugal

J. Manso & J. Marcelino
Geotechnics in Hydraulic Works Unit, Geotechnical Department, National Laboratory of Civil Engineering, Lisbon, Portugal

ABSTRACT: Docking large vessels is a delicate operation as the kinetic energy associated with the large mass of the vessel can result in high impact forces that can damage the vessel, fenders or even the quay. Berthing loads are usually quantified using design formulae based on kinetic energy and a single point of impact. Some correction factors are then used to consider the hydrodynamic mass, the ship's angle with the quay, the softness of berthing and the berth configuration. In this study, a scaled model experimental set-up was used to determine the impact forces of a ship on the fender system, including all fenders touched by the ship, during various docking maneuvers. The pattern and magnitude of the impact forces are different for each fender and are highly dependent on the approach trajectory and mass of the ship. A comparison was made of the measured values of the impacts and the design forces calculated using widely used regulations. Our findings showed some discrepancy between maximum computed forces using kinetic energy method and measured forces, suggesting it may underestimate the maximum impact force in some scenarios.

1 INTRODUCTION

1.1 *Quay and fender design*

Fenders are protective structures mounted on the sides of quays to effectively absorb collision energy and prevent damage to vessels and structures. They are typically made of a resilient material, such as rubber, plastic, or foam. Rubber fenders are the most common type and are usually made from natural or synthetic rubber. Foam fenders are made from a cellular foam material often used in high impact applications.

When properly designed and installed, fenders can significantly reduce the risk of hull damage, which can be costly and time-consuming to repair, and personal injury of crew and passengers.

It is important to correctly design berthing structures and select appropriate ship fenders. For this purpose, the impact forces of ships on the fenders during docking procedures must be well quantified.

Regulation, recommendations issued by PIANC (2002), OCIMF (1992) and other norms such as British Standard (BS 2014) or ROM2.0-11. (2012) provide a set of recommendations for the design of fendering and mooring systems for commercial vessels berthing at quays, dolphins, pontoons, and other structures. The methodology in such norms is similar.

To determine the impact energy, i.e. the maximum energy that the fendering system must absorb in the event of a collision, usually involves the following steps:

(1) Determine the maximum impact speed.
(2) Calculate the kinetic energy of the ship.
(3) Establish a desired coefficient of restitution, CR.
(4) Calculate the impact energy.
(5) Select the appropriate fendering system.

The maximum impact speed is the maximum speed at which the ship could collide with the fendering system. This can be determined from the ship's displacement and easiness of the maneuver (ranked form a, easy berthing with good conditions to e, difficult berthing with bad conditions).

Kinetic energy is the energy of motion of the ship. It can be calculated using the following equation:

$$E_c = 1/2Mv^2C_MC_EC_SC_C \quad (1)$$

where:
M is the mass of the vessel (in tonnes),
v is the maximum impact velocity or berthing velocity (in m/s),

DOI: 10.1201/9781003508779-19

C_M is the hydrodynamic mass coefficient, and is given by:

$$C_M = 1 + \frac{2D_v}{B} \quad (2)$$

D_v is the draught of vessel (in m).
B is the ship's breadth (in m).
C_E is the eccentricity coefficient, and is given by:

$$C_E = \frac{K^2 + R^2\cos^2\gamma}{K^2 + R^2} \quad (3)$$

where:
K is the radius of gyration of vessel (m), given by:

$$K = (0.19C_b + 0.11)L \quad (4)$$

R is the vector distance between the CG and point of impact
γ is the angle between R and v.
C_b is the block coefficient, given by:

$$C_b = \frac{M}{LBd\rho_w} \quad (5)$$

L is the ship's length (in m).
C_S is the softness coefficient.
C_C is the berth configuration coefficient.

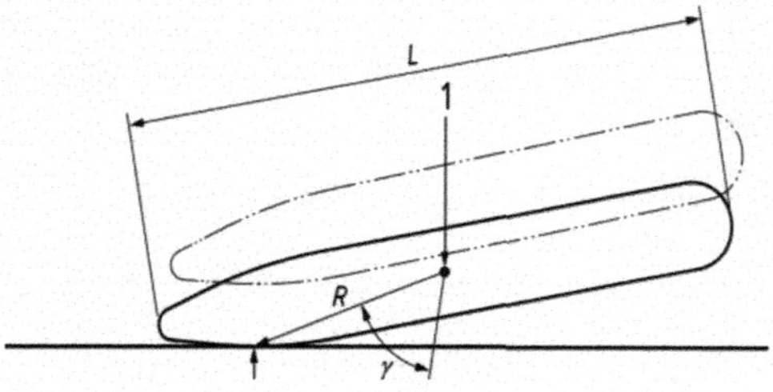

Figure 1. 1st impact of the ship at an angle with the quay.

Finally, the appropriate fendering system can be selected based on its energy absorption capacity. The fendering system must be able to absorb at least the calculated impact energy to protect the vessel and the berth from damage (Fontijn 1988).

The impact energy is the energy that is transferred to the fendering system during the collision. It can be calculated using the following equation:

$$E_I = \frac{E_c}{1 - CR^2} \quad (6)$$

CR is the coefficient of restitution, dimensionless, represents the ratio of the relative velocity after impact to the relative velocity before impact. It is typically between 0 and 1, with 0 representing a perfectly inelastic collision and 1 representing a perfectly elastic collision.

1.2 *Objectives*

The docking of a ship is usually made of a series of impacts, back and forward, on several fenders until total immobilisation of the ship, especially if the impact energy is high.

This is because the first impact is followed by an impulsive reaction force. This reaction force, together with the kinetic energy still driving the ship's CG towards the quay, produces a rotational moment around the z-axis that causes the second impact on the opposite side (lengthwise) of the ship, Figure 2.

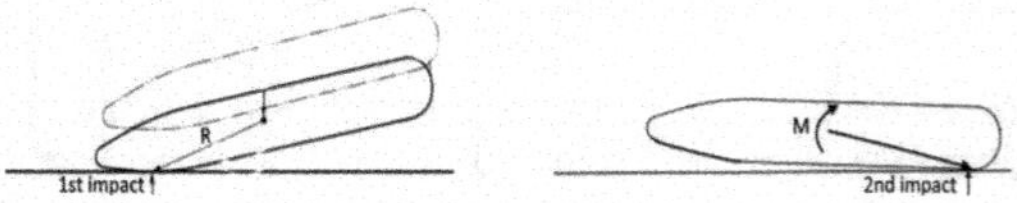

Figure 2. Berting sequence: 1st impact on the left, 2nd impact on the right.

In this study, a scaled model experimental set-up was used to determine the impact forces of a ship on the fender system, including all fenders touched by the ship, during various docking maneuvers. The pattern and magnitude of the impact forces are different for each fender and are highly dependent on the approach trajectory, velocity, and mass of the ship.

The main objective of this study is to investigate and characterize the maximum force applied to each fender and its distribution over the set of fenders. This information is important to assess the adequacy of existing design formulae if the measured values of the impacts are compliant with the design forces calculated using widely used regulations.

1.3 *Structure*

After this introduction, section 2 presents the experimental set up, the mmeasurement equipment, methods, and test conditions. Section 3 presents the results and their analysis. Section 4 states the conclusions of the study.

2 EXPERIMENTS

2.1 *Physical model tests set-up*

Physical model tests on this subject are scarce. Antolloni et al. (2017), simulated the behavior of buckling-type marine fenders using a simple physical model. Yldiz et al. (2013) performed eexperiment and finite element simulation of the effect of different strain energy functions on rubber fender. Tuleja et al. (2023) performed experimental determination of the reaction force and absorption energy values and compared them with numerically determined for cylindrical fenders. To our knowledge physical model tests with a ship's hull model and impacting a quay structure fitted with several fenders were never performed.

In this work a testing plan was devised to provide relevant information and characterise the maximum force applied to each fender and its distribution over the set of fenders.

The tests were carried out in a 4 m x 4 m section of a larger wave tank (22 m x 23 m), Figure 4. The ship model is a 1:100 scale version of an oil tanker with LOA=172.50 cm, B=24.8 cm, D=9 cm and DW=13,124 kg.

Four fenders were placed along the quay equally distanced (50cm) and carefully aligned to ensure the same distance to the vertical quay structure, Figure 3. The vertical quay structure is made of a perforated metal plate to allow some water disturbance to be absorbed. The fenders consist of the force sensor screwed to the metal vertical plate on the quay side, and a squared (10x10cm) wooden plate screwed on the seaside of the sensor, simulating the fender actual contact area with the ship's hull.

The scale of the tests is 1:100.

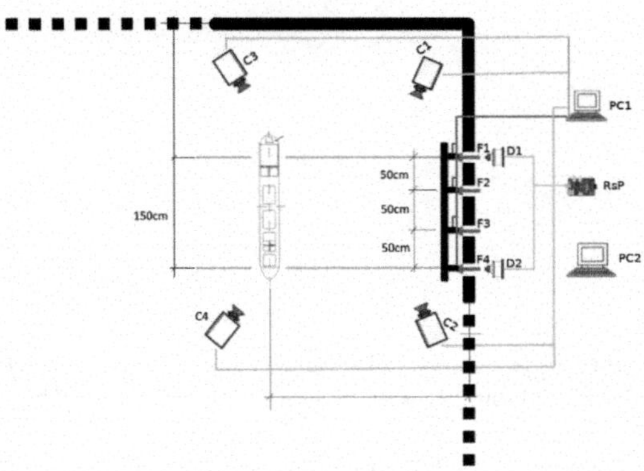

Figure 3. Experimental set-up layout.

Measurement equipment and methods

For precise measurement of both ship motion and impact forces, a comprehensive instrumentation setup was employed. An Optitrack® multi-camera system tracked the ship's speed and heading throughout the tests. Simultaneously, four force sensors, coupled with a Quantum MX data acquisition system and CatmanEasy® software, captured the magnitude and sequence of each impact on the fenders, Figures 4 and 5.

The sampling frequency used in the model test for load measurements was 50Hz and for motion measurements was 120Hz.

Tests conditions

To comprehensively characterise the impact forces of a berthing ship, a series of physical tests were carried out in controlled still water conditions. The scaled ship model was subjected to a wide range of scenarios, including ship mass varying from ballast (empty) to fully loaded, simulating different cargo capacities; approach angles ranging from strictly parallel to the quay wall to large angles (up to 70°), representing different berthing manoeuvres; approach speed ranges, smooth (normal docking) and fast (simulating accidental impacts), were investigated.

A total of 599 individual docking tests were performed, Table 1, covering a comprehensive matrix of these parameters. Each test condition was repeated between 20 and 34 times, ensuring robust statistical analysis and minimising random error.

In addition, a dedicated weight and pulley system was implemented to precisely control the approach speed, ensuring consistent and reproducible test conditions throughout the campaign.

Table 1. Testing conditions matrix.

Loading condition	Docking speed type	Docking angle type	Test series	n° tests
3/4 Loaded	Smooth docking	Lateral	T01 - T02	20
		Small angle	T03 - T04	20
		Large angle	T05 - T06	20
	Accidental docking	Lateral	T07 - T08	20
		Small angle	T09 - T010	19
		Large angle	T011 - T012	20
Fully loaded	Smooth docking	Small angle	T013 - T014	30
		Large angle	T015 - T016	32
		Stern	T017 - T018	30
	Accidental docking	Lateral	T019 - T020	34
		Small angle	T021 - T022	31
		Large angle	T023 - T024	31
Half loaded	Smooth docking	Lateral	T025 - T026	25
		Small angle	T027 - T028	26
		Large angle	T029 - T030	28
		Stern	T031 - T032	31
	Accidental docking	Lateral	T033 - T034	32
		Small angle	T035 - T036	32
		Large angle	T037 - T038	26
Ballast	Smooth docking	Small angle	T039 - T040	29
		Large angle	T041 - T042	31
		Stern	T043 - T044	32

Figure 4. Experimental setup. Ship model and pressure sensors.

Figure 5. High performance motion capture camera (left). Rigid-body target points array placed on the ship's deck (right). Force sensor.

3 RESULTS

The first instant of impact is identified from the force sensor readings and simultaneous values for horizontal velocity and angle of attack are extracted at the same time. After systematic post-processing of all tests, the maximum impact force is correlated with the docking speed. The distribution of the impact forces over time and over different fenders is also analysed. Results are presented in model scale

Figure 6 shows the results for one of the test series T010, which represents a partially (3/4) loaded ship with an approach speed of approximately 50 mm/s and an angle of 90º (parallel to the quay). This series of tests consists of 10 impact repetitions.

Figure 7 show the impact sequence of the 4 fenders: D1, D2, D3 and D4. Most of the times the most energetic impact is not the first one.

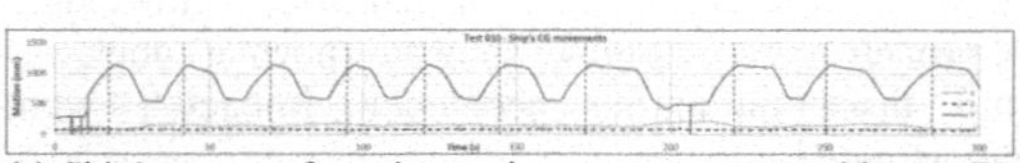

(a) Ship's centre of gravity motions: surge, sway and heave (X, Z and Y).

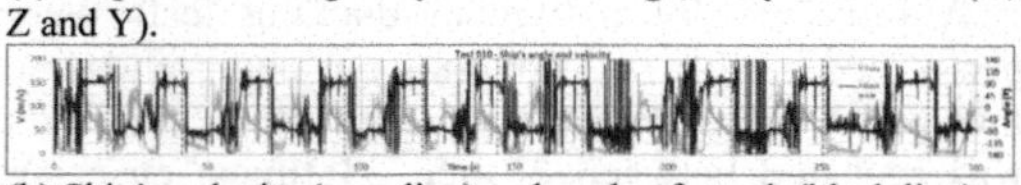

(b) Ship's velocity (grey line) and angle of attack (black line)

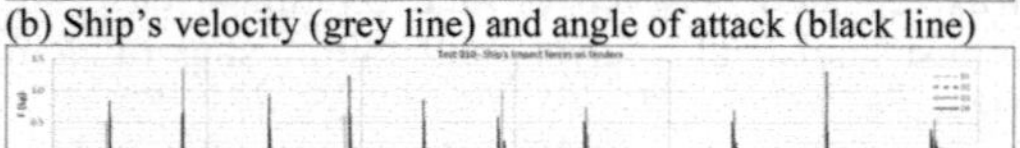

(c) Ship's impact forces on fenders: D1, D2, D3 and D4.

Figure 6. Test series T010. Raw records of the instruments.

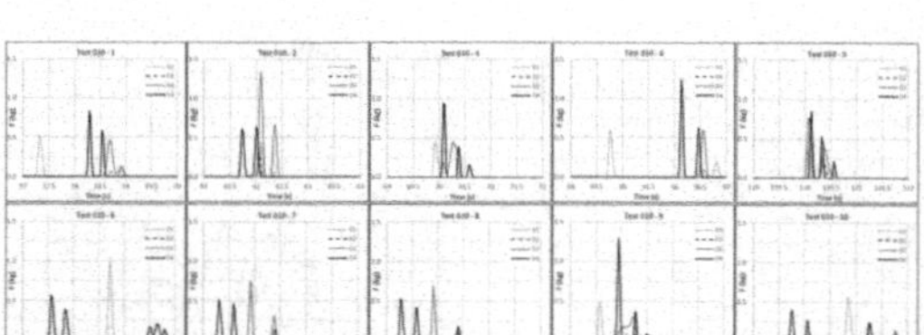

Figure 7. Test series T010. Number of impacts 1 through 10. Impact sequence of the 4 fenders: D1, D2, D3 and D4.

Figure 8 shows the results for one of the test series T033, which represents a half-loaded ship with an approach speed of approximately 200 mm/s and an angle of 90º (parallel to the quay). This series of tests consists of 15 impact repetitions.

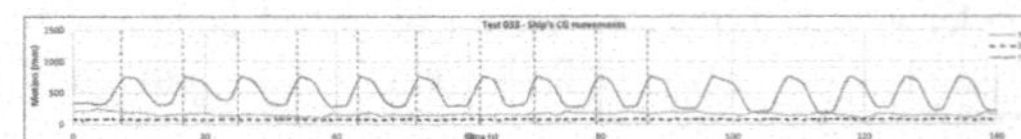

Ship's centre of gravity motions: surge, sway and heave (X, Z and Y).

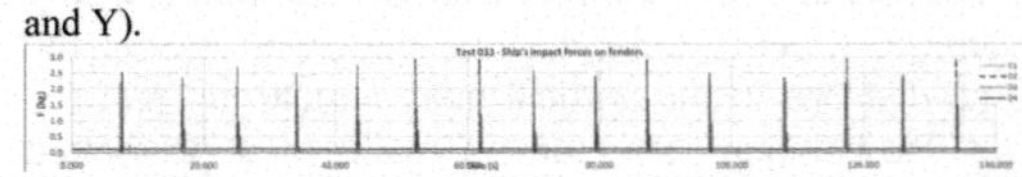

(b) Ship's velocity (grey line) and angle of attack (black line)

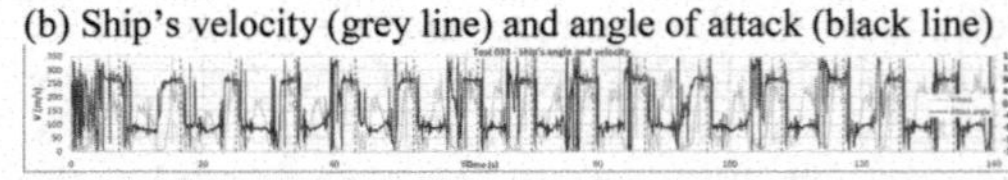

(c) Ship's impact forces on fenders: D1, D2, D3 and D4.

Figure 8. Test series T033. Raw records of the instruments.

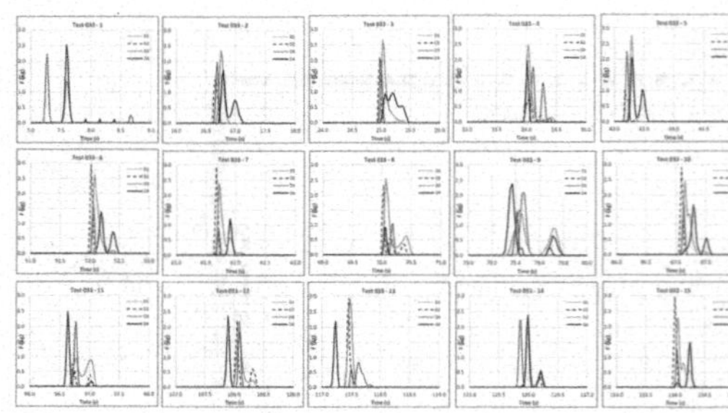

Figure 9. Test series T033. Number of impacts 1 through 15. Impact sequence of the 4 fenders: D1, D2, D3 and D4.

Figures 7 and 9 show the impact sequence of the 4 fenders: D1, D2, D3 and D4 and most of the times the most energetic impact is not the first one.

Analysing all the 598 tests, the number of tests where the 2nd impact is the most energetic is greater than, for all test conditions, Figure 10.

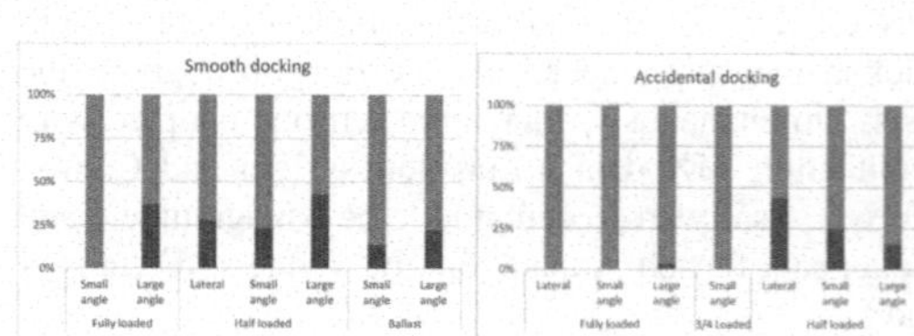

Figure 10. The number of tests where the maximum force occurs at the first impact compared with the number of tests where the 2nd impact is the most energetic.

The average percentage of Fmax exceedance over the first impact, Figure 11, is between 10% (accidental docking with large angle of a fully loaded ship) and

45% (accidental docking with small angle of a 3/4 loaded ship), averaging 27% across all test conditions.

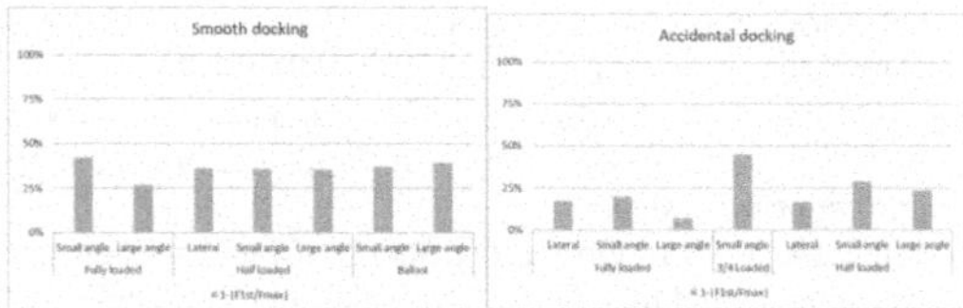

Figure 11. Average percentage of Fmax exceedance over the first impact.

Now comparing all forces on the 4 fenders with design force, F_d, obtained with British Standards (BS 6349-4:2014) formulae. Where $F_d = F_k \times 2$, which accounts for a safety factor of 2, and F_k is the characteristic force $F_k = \frac{E_C}{s}$, where E_C is the Energy of the moving vessel, obtained from equation 1, in which v is the measured berthing velocity.

In Figure 11, it can be seen that the 1st impact force is always smaller than the design Force, F_d as calculated from BS formulae. Which would be in accordance with regulations if there were not higher following impacts. However, the 2nd impact force can be, in some cases, higher than the design Force, F_d as calculated from BS formulae.

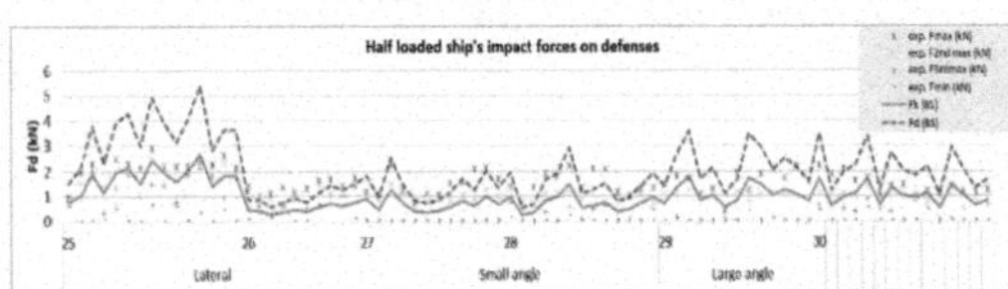

Figure 12 . Comparison of the forces on the 4 fenders, Fmax, F2nd max, F3rd max and Fmin, with Characteristic Force, Fk and design Force, Fd, as calculated from BS formulae.

Figure 13 shows the ship's impact forces on fenders for all loading and test conditions and the comparison with BS characteristic and design forces. The measured maximum impact force seldom exceeds the design force obtained using the British Standards norms.

Results show that the formulae used to design berthing structures can, in certain circumstances, underestimate the real impact force of berthing ships.

The first impact (graphs on the right) is more likely to fall below design forces from BS norms. The maximum impact falls outside regulation limits, especially for lighter vessels with large approach angles (see top left graph) and heavier vessels with small approach angles (bottom left graph). Also, smaller docking speeds appear to be most likely to fall outside of regulation limits.

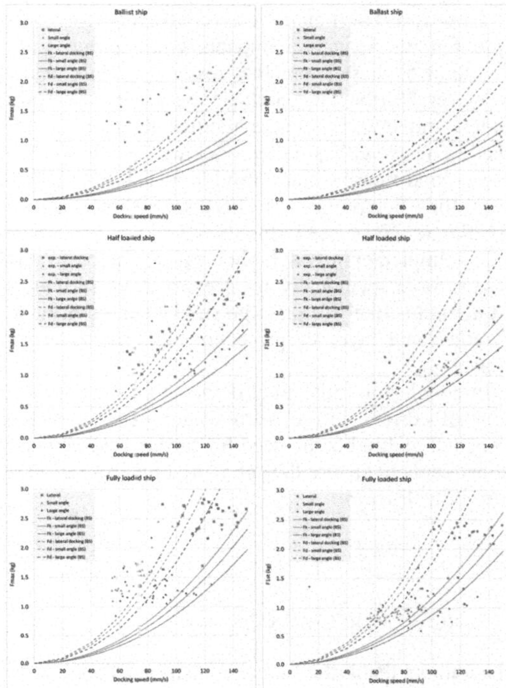

Figure 13. Ship's impact forces on fenders. Comparison of experimental results with British Standards design norms.

4 CONCLUSIONS

This study investigated the impact forces exerted by berthing ships on harbour structures, comparing measured values with design forces calculated using British Standard (BS) methods. Our findings highlight a discrepancy between predicted and actual forces, particularly for lighter vessels with large approach angles and, for heavier vessels with small approach angles. These exerted higher-than-expected forces, suggesting the BS method may underestimate the impact for such scenarios.

Furthermore, the study revealed that the maximum impact force typically occurs on the second fender hit, not the first, and is 27% higher on average. This finding emphasizes the dynamic nature of berthing events and the need to consider the possibility of more energetic subsequent impacts in design calculations.

These results call for a reconsideration of the design methodologies for berthing structures. Two key approaches are proposed:

1 - Reevaluate the existing formulae could improve the accuracy of design calculations. Revisiting the ranges of coefficients used in the energy equation within BS standards or introducing a new coefficient, C_F, to account for the observed discrepancies:

$$E_c = 1/2MV^2 C_M C_E C_S C_C \boldsymbol{C_F} \quad (7)$$

This new coefficient, C_F, coefficient for multiple fenders sequence impacts could range from 1.3 to 1.5, based on the results of this tests. However, further research is necessary to fully understand the complex dynamics involved. Conducting tests with smaller docking speeds and in wave conditions would provide valuable insights into the behaviour of berthing ships across a wider range of scenarios.

2 - Implementing a higher safety factor in the design force specifically for scenarios where underestimation is likely (e.g., large vessels) could provide an additional layer of protection against excessive impact.

In conclusion, this study demonstrates the limitations of current design methods for berthing structures and highlights the need for further research and adjustments to ensure the safety and integrity of harbour infrastructure.

ACKNOWLEDGEMENTS

The authors thank the Centre for Marine Technology and Ocean Engineering for lending the ship model. The financial support from Portuguese national blue fund "Fundo Azul", through project FA_04_2017_016 "BlueSafePort – Safety System for Manoeuvring and Moored Ships in Ports" is acknowledged.

REFERENCES

Antolloni, G.; Carbonari, S.; Gara, F.; Lorenzoni, C.; Mancinelli, A. Simple physical models to simulate the behavior of buckling-type marine fenders. *J. Waterw. Port Coast. Ocean. Eng.* 2017, 143, 04016014.

BS 6349-4:2014. (2014) British Standards. *Maritime works – Part 4: Code of practice for design of fendering and mooring systems.*

Fontijn, H.L. (1988). On the Prediction of Fender Forces at Berthing Structures Part I: Calculation Method(s). In: Bratteland, E. (eds) *Advances in Berthing and Mooring of Ships and Offshore Structures.* NATO ASI Series, vol 146. Springer, Dordrecht. https://doi.org/10.1007/978-94-009-1407-0_4

OCIMF (1992) *Mooring equipment guidelines.* Witherby e Co. Ltd.

PIANC (Permanent International Association of Navigation Congresses). *Guidelines for the design of fender system: 2002. In PIANC WG33*, Rapport du Groupe de Travail n °33 de la Commission pour La Navigation Maritime; Association Internationale Permanente des Congrès de Navigation: Brussels, Belgium, 2002.

ROM2.0-11. (2012) *Recomendaciones para obras marítimas ROM2.0-11, Tomos I y II. Recomendaciones para el proyecto y ejecución en Obras de Atraque y Amarre.* Puertos del Estado. Ministerio de Fomento.

Tuleja, J.; Kędzierska, K.; Sowa, M.; Galor, P. Evaluation of the Possibility of Increasing the Energy Absorption Efficiency of Fender Devices Using the Example of Cylindrical Fenders with Additional Structural Elements Applied. *Energies* 2023, 16, 1165.

Yildiz, F. The effect of different strain energy functions on rubber fender. Experiment and finite element simulation. *J. Elastomers Plast.* 2013, 46, 722–736.

Advances in Maritime Technology and Engineering – Guedes Soares & Santos (Eds)

Seaworthiness and resistance performance of a catamaran at preliminary design stage

F. Soardi & G. Vernengo
DITEN, Department of Electric, Electronical and Telecomunication Engineering and Naval Architecture, Polytechnic School of the University of Genoa, Genoa, Italy

ABSTRACT: In the proposed study a fast method to predict the seakeeping behavior of catamarans in waves is described and applied for the sensitivity analysis of its performance in the framework of a preliminary design process. The method is based on two-dimensional potential flow theory in which viscous forces are considered through a cross-flow drag approach. The obtained results show a good agreement with other published theoretical and experimental results. Thanks to the high computational efficiency of the approach, the sensitivity analysis is carried out in a design by optimization fashion, hence spanning over a wide design space to capture the possible existing correlations among the geometric features of the vessel and both its seaworthiness and wave resistance performance, aiming to identify the best catamaran configuration at a very early stage of the design process.

1 INTRODUCTION

Catamarans seakeeping related performance have been studied by many authors by using both Boundary Element Methods (BEM) and Reynolds Averaged Navier Stokes (RANS) (see for instance Fang & Chan (2004), Thomas et al. (2007), Castiglione et al. (2011), Broglia et al. (2011), Belga et al. (2018), Liu et al. (2020) and Vernengo et al. (2021)). Among these methods, strip-theory approaches for multi-hulls have been developed since the seventies (Ohkusu 1970, Lee et al.1971, Lee et al.1973,among others). Because of the smaller water plane area, catamarans and SWATHs do not produce large surface waves during vertical-mode oscillations. Thus, wave-making damping contribution has little influence on the overall damping effect and, in order not to overestimate the amplitude of vertical motions, it is necessary to take into account the viscous effects. For this purpose, Lee (1976) developed a cross-flow drag method for catamarans and SWATHs which was based on previous studies (Thwaites 1960). More recently the development of three-dimensional time-domain methods methods (Kring & Sclvounos 1991, Chan 1993, Fang 1996, van't Veer 1997, Holloway 1998) has introduced various non-linear effects into the seakeeping analysis for multi-hulls. However, at low Froude's numbers, strip-theories relying on two-dimensional solution seem to be more efficient than three-dimensional ones, since the hydrodynamic interactions of demi-hulls are correctly taken into account (van't Veer & Siregar (1995),Fang et al. (1996)) and, generally, the slender shape of demi-hulls is consistent with strip-theory assumptions. Therefore, even now, due to their low implementation complexity and high computational speed, two-dimensional methods are a good tool to perform the preliminary seakeeping analysis for multi-hulls.

The present method is based on two-dimensional potential flow theory in which viscous forces are considered through a cross-flow drag approach. The results obtained shown a good agreement with other published results. In addition, the simulations performed shown that, due to high computational speed, the method is able to carry out an optimization analysis concerning seaworthiness and total resistance to identify the best catamaran configuration in terms of main dimensions and distance between demi-hulls at a early stage of the design.

2 MATHEMATICAL FORMULATION

To formulate the motions problem two reference systems are introduced: a system fixed in space ($O_0 - x_0y_0z_0$) and a steady-translating one ($O - xyz$) at the mean forward speed of the ship (U). The origin (O) of the mobile inertial system ($O - xyz$) is located on the undisturbed free surface (S_0, i.e. the plane $z_0 = z = 0$). This reference system is fixed with respect to the mean position of the ship. See Figure 1 for a sketch of axis orientation and ship motions.

DOI: 10.1201/9781003508779-20

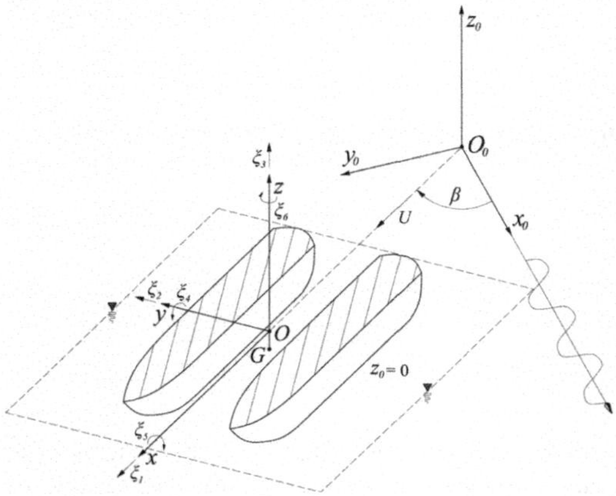

Figure 1. Reference systems and multi-hull motions.

2.1 *Equations of motions*

Due to the assumption of small amplitude motions, the dynamic of the ship in space is described by the following system of six linearized equations ($j = 1,2, \ldots,6$)

$$F_j^0 + F_j^7 + F_j^F = \sum_{k=1}^{6}[-\omega_e^2(M_{jk} + A_{jk}) + \\ +i\omega_e B_{jk} + C_{jk}]\zeta_k, \tag{1}$$

where ζ_k is the complex amplitude of k –th oscillation mode, ω_e is the encounter frequency and, i the imaginary unit. For the coefficients M_{jk}, A_{jk}, B_{jk} and C_{jk} refer to the notation of Salvesen et al. (1970). In the system (1), time dependency is omitted and total force are separated into the incident wave force (F_j^0), the diffraction force (F_j^7) and the viscous force (F_j^F). The latter is defined as

$$F_j^F = F_j^V - \sum_{k=1}^{6}[i\omega_e B_{jk}^V + C_{jk}^V]\zeta_k, \tag{2}$$

with F_j^V the j –th component of exciting viscous force and B_{jk}^V and C_{jk}^V the viscous damping and restoring coefficients, respectively. Except for M_{jk} and C_{jk}, all remaining terms in expressions (1) and (2) are unknowns.

2.3 *Cross-flow drag method for viscous effects*

To account for viscous effects, a semi-empirical cross-flow drag approach developed for aerodynamic airfoils is proposed in this study (Thwaites 1960, Lee 1976, Chan 1993). According to Thwaites's experiments, viscous forces acting on long and tapered profile can be expressed using a viscous lift coefficient (a_o) and a viscous cross-flow drag coefficient (C_D) which depend on body geometry, oscillation frequency, and the direction of motion. Assuming that the multi-hull moves in regular waves with constant forward speed (U) and oscillates in the vertical-plane modes with harmonic motion, then for each x ordinate viscous force, that acts on a hull-strip of infinitesimal area projected on the perpendicular plane to j –th direction (dA_{Pj}), results, for $j = 2,3$

$$dF_j^F(x,t) = \frac{1}{2}\rho dA_{Pj}(x)[U^2 a_o \alpha_j(x,t) + \\ + C_D \frac{8}{3\pi}\Re\{\dot{\zeta}_j^V(x)e^{i\omega_e t}\}|\dot{\zeta}_j^V(x)|], \tag{3}$$

where the cross-flow drag term has been approximated through an equi-linearization method [?]. In this expression $\dot{\zeta}_j^V$ is the complex amplitude of the relative fluid velocity with respect to the ship in the j –th direction and, α_j is the angle of attack of the flow. Note that, in addition to the section width, also other geometrical parameters of the demi-hulls sections have a considerable importance in the present cross-flow drag method as they are explicit in the calculation of the $\dot{\zeta}_j^V$ components, that is:

$$\dot{\zeta}_2^V(x) = \dot{\zeta}_2 + d_2\dot{\zeta}_4 + x\dot{\zeta}_6 - \dot{\zeta}_h(x, \pm b_{cy}, -d_2) \tag{4}$$

$$\dot{\zeta}_3^V(x) = \dot{\zeta}_3 \pm b_{cy}\dot{\zeta}_4 - x\dot{\zeta}_5 - \dot{\zeta}_v(x, \pm b_{cy}, -d_1) \tag{5}$$

here, for each ordinate x, $d_1(x)$ is the draught of the horizontally farthest point, $d_2(x)$ the half of maximum draught and, $b_{cy}(x)$ the half distance between the centers of the demi-hulls. The vertical and horizontal components of fluid velocity induced by the incident wave, which may be evaluated using potential (9), are indicated with $\dot{\zeta}_v$ and $\dot{\zeta}_h$, respectively.

Clearly the method is iterative, hence for the first cycle it requires knowledge of the amplitudes of motions obtained without the viscous effect. To obtain the viscous forces in horizontal and vertical direction, expression (3) must be integrated on length of the catamaran for each demi-hull. The viscous moments can be obtained by integrating of forces (3) multiplied by the respective rotating arm. Later the viscous coefficients and viscous exciting forces are derived from (2) by omitting real part and the term $e^{i\omega_e t}$.

2.3 *Linearized formulation of the hydrodynamic problem*

The exact formulation of the problem is derived from the flow potential theory where the velocity field is expressed as function of the total potential, $\Phi(x,t)$, which satisfies the Laplacian in all the fluid domain ($\mathcal{D}$). The uniqueness of the solution is obtained setting, on the instantaneous free surface

($\mathcal{S}$), the non linear unified condition (van't Veer 1997)

$$\frac{\partial^2\Phi}{\partial t^2}+2\nabla\Phi\cdot\nabla(\frac{\partial\Phi}{\partial t})+ \\ +\frac{1}{2}\nabla\Phi\cdot\nabla(\nabla\Phi\cdot\nabla\Phi)+g\frac{\partial\Phi}{\partial z}=0\,on\,\mathcal{S}. \tag{6}$$

In addition, on the wet hull surface ($\mathcal{S}_H$), the following body boundary condition is imposed

$$\frac{\partial\Phi}{\partial n}=n\cdot\nabla\Phi=v_n on\mathcal{S}_H, \tag{7}$$

where $v_n(x,t)$ is the normal component of velocity of the generic point on $\mathcal{S}_H$. A further boundary condition is imposed on the bottom, where infinite depth is assumed, and a suitable radiation condition at infinity is required.

Due to Laplace's operator linearity, the total velocity potential (Φ) can be expressed as

$$\Phi(\boldsymbol{x},\boldsymbol{t})=-\boldsymbol{U}\boldsymbol{x}+\phi_S(\boldsymbol{x})+ \\ +\Re\{[\phi_0(\boldsymbol{x})+\phi_7(\boldsymbol{x})+\sum_{k=1}^{6}\phi_k(\boldsymbol{x})\zeta_k]e^{i\omega_e t}\}, \tag{8}$$

where ϕ_S is the steady perturbation potential, while the unsteady complex potential has been decomposed linearly in the incident wave potential (ϕ_0), the diffraction potential (ϕ_7), and the radiation potential (ϕ_k) for the k −th oscillation mode. Based on linear wave theory, the incident wave potential results:

$$\phi_0(\boldsymbol{x})=\frac{\boldsymbol{iga}}{\boldsymbol{\omega_0}}\,e^{K_0 z}e^{-K_0(x\cos\beta-y\sin\beta)}, \tag{9}$$

with g the gravity, a the wave amplitude, β the encounter angle and ω_0 ($K_0=\frac{\omega_0^2}{g}$) the wave frequency (number).

Under the assumption that amplitudes of motions are small, flow disturbance generated by the ship is minimal and the perturbation potential (ϕ_S) is negligible compared to the term $-Ux$. As result, the boundary conditions can be linearized on undisturbed free surface ($\mathcal{S}_0$ or $z=0$), and on mean hull surface ($\mathcal{S}_{H0}$), respectively. The formulation of the linearized problem can be obtained by substituting the potential (9) in the equations of the flow potential theory and then by eliminating the negligible terms (Salvesen et al. 1970). After linearization, if the forward speed of the multi-hull is considered small and the encounter frequency very high, the term $U\frac{\partial}{\partial x}$ is negligible compared to ω_e and the unified boundary condition it's show to become:

$$-\omega_e^2\phi_k+g\frac{\partial\phi_k}{\partial z}=0\,on\,\mathcal{S}_0, k=0,1,\ldots,7. \tag{10}$$

In addition, because of linearity assumptions, unsteady problem can be separated into a problem for $U\neq 0$ and one for $U=0$ and after simplifying the m-terms (Salvesen et al. 1970), the entire hydrodynamic problem is reformulated in terms of the speed-independent potentials, $\phi_k^0(x)$, that satisfy the Laplacian, condition (10) and also the followings

$$\frac{\partial\phi_k^0}{\partial n}=i\omega_e n_k^*\,on\,\mathcal{S}_{H0},\; k=1,2,\ldots,6, \tag{11}$$

$$\frac{\partial\phi_7^0}{\partial n}=-\frac{\partial\phi_0}{\partial n}on\mathcal{S}_{H0}, \tag{12}$$

where $n_k^*(\boldsymbol{x})$ is the k −th component of the positive outward generalized normal to hull surface.

3 NUMERICAL MODEL

The demi-hull of the catamaran is compared to a long slender body (i.e. $n_1^*\simeq 0$, then all forces associated with surge are neglected), so the total hydrodynamic problem can be resolved suitably using a strip-theory method with good approximation. The complete derivation of the hydrodynamic forces and coefficients presented in this section are described in Salvesen et al. (1970).

3.1 *Hydrodynamic and exciting forces*

The radiation forces acting on the hull surface ($\mathcal{S}_{H0}$) are calculated by integrating of the respective dynamic pressure, which is achieved from the linearized Bernoulli's equation. Applying Stokes's theorem and (11), radiation forces are formulated as functions of only speed-independent hydrodynamic forces (F_j^{0k}) which, under the hypothesis of the strip-theory method, can be expressed in terms of the sectional forces $f_j^k(x)$ integrated along the length of the hull. Namely, for $j,k=2,3,\ldots,6$:

$$F_j^{0k}=-i\rho\omega_e\int_{\mathcal{S}_{H0}}\phi_k^0 n_j^*\,d\boldsymbol{S}=\int_{\mathcal{L}}f_j^k\,dx, \tag{13}$$

where:

$$f_j^k(x)=\omega_e^2 a_{jk}-i\omega_e b_{jk}=-i\rho\omega_e\int_{C_x}\psi_k n_j^*\,dl, \tag{14}$$

with C_x the contour of the section of both demi-hulls at x ordinate, ρ the fluid density, $a_{jk}(x)$ and $b_{jk}(x)$

the hydrodynamic sectional coefficients, and $\psi_k = \phi_k^0$ the radiation two-dimensional potentials.

Because the hydrodynamic forces are expressed as functions of only the components F_j^{0k}, therefore the three-dimensional hydrodynamic coefficients, A_{jk} and B_{jk}, are formulated in terms of the speed-independent coefficients A_{jk}^0 and B_{jk}^0 which are obtained by integrating the corresponding a_{jk} and b_{jk} coefficient along the hull. Actually, all speed-independent added-masses and damping coefficients are expressed in terms of only f_2^2, f_3^3, f_4^4, and f_2^4.

Again, the diffraction and wave incident forces are obtained by integrating of the relative dynamic pressure on $\mathcal{S}_{H0}$. Using the strip-theory, incident forces (F_j^0) based on potential (9) can be obtained as:

$$F_j^0 = -i\rho\omega_0 \int_{\mathcal{S}_{\mathcal{H}0}} \phi_0 n_j^* d\boldsymbol{S} = \\ = \int_{\mathcal{L}} e^{-iK_0 x \cos\beta} f_j^0\, dx, j = 2, 3, \ldots, 6, \quad (15)$$

where, for $j = 2, 3, 4$

$$f_j^0(x) = \rho g a \int_{\mathcal{C}_x} e^{K_0 z} e^{iK_0 y \sin\beta} n_j^*\, dl, \quad (16)$$

and:

$$f_5^0(x) = -x f_3^0, \quad f_6^0(x) = x f_2^0 \quad (17)$$

Finally, through the Stokes's theorem, Green's second identity, condition (12), and strip-theory assumptions, it can be shown that the diffraction force components (F_j^7) result:

$$F_j^7 = \int_{\mathcal{L}} e^{-iK_0 x \cos\beta} [f_j^7 - \tfrac{U}{i\omega_e} f_j^{U7}]\, dx + \\ + \tfrac{U}{i\omega_e} e^{-iK_0 x_A \cos\beta} f_j^{A7}, j = 2, 3, \ldots, 6, \quad (18)$$

where $f_j^{U7} = 0$ for j = 2,3,4, and:

$$f_5^{U7} = -f_3^7, \quad f_6^{U7} = f_2^7. \quad (19)$$

In Eq. (19) f_j^{A7} indicates the j −th component of the sectional diffraction force acting on the uttermost section of both demi-hulls ($\mathcal{C}_A$) and f_j^7 is the j −th component of the diffraction sectional force on $\mathcal{C}_x$, that is, for j = 2,3,4:

$$f_j^7(x) = \rho\omega_0 a \int_{\mathcal{C}_x} e^{K_0 z} e^{iK_0 y \sin\beta} \psi_j \tilde{n}\, dl, \quad (20)$$

where $\tilde{n} = i n_3^* - \sin\beta n_2^*$, and

$$f_5^7(x) = -x f_3^7, \quad f_6^7(x) = x f_2^7 \quad (21)$$

According to the assumptions made, all the forces and hydrodynamic coefficients can be evaluated only known the two-dimensional radiation potentials ψ_2, ψ_3 and ψ_4 for each ordinate of demi-hulls.

3.2 *Numerical solution of the two-dimensional boundary value problem*

To solve the two-dimensional problem in the $y-z$ plane, a source distribution method is used. The required two-dimensional potentials, $\psi_k(y,z)$, must satisfy the Laplacian, equations (10) and (11), a suitable condition on the bottom, the Sommerfield's radiation condition and, additionally, the following hydrodynamic symmetry condition on $y = 0$ plane:

$$\frac{\partial \psi_3}{\partial y}(0, z) = 0, \quad \psi_2(0, z) = \psi_4(0, z) = 0. \quad (22)$$

Due to the Green's second identity, the two-dimensional velocity potential, describing the flow perturbations generated by the demi-hulls, can be formulated as (for k = 2,3,4):

$$\psi_k(\delta) = \int_{\mathcal{C}_x^+} Q_k(\zeta) G_k^*(\delta;\zeta)\, d\zeta, \delta \in \mathcal{C}_x^+ \quad (23)$$

where $G_k^*(\delta;\zeta) = [G(\delta;\zeta) - (-1)^k G(\delta;-\bar{\zeta})]$. In particular, G is the Green's function for a two-dimensional pulsating source (Wehausen & Laitoine 1960), $\mathcal{C}_x^+$ is the contour of the section at x ordinate of the demi-hull placed in the $y > 0$ region, and Q_k is the sources complex intensity related to the k −th mode of oscillation. Here $\delta = y + iz$ and $\zeta = \xi + i\eta$ represent the field point and the source position on $\mathcal{C}_x^+$, respectively.

In this study the sources are positioned on the profile of both demi-hulls to take into account of their hydrodynamic interation. Consequently, the Green's function integrated in (23) implicitly contains the contributions of all the sources on both demi-hulls. The integration is carried out only on the contour $\mathcal{C}_x^+$ as result of geometric and hydrodynamic symmetry of the problem.

By discretization of the immersed profile ($\mathcal{C}_x^+$) into N straight segments, assuming constant source intensity on each segment and, imposing condition (11) on potential (23) at N collocation points (i.e. the midpoint of each segments), then an algebraic system of $2N$ equations is solved to determine the real and imaginary part of the unknown sources intensity Q_k. Influence coefficients and potential integrals have been evaluated through symbolic calculus as in Frank (1967) or Lee et al. (1971).

Finally, from (23), the two-dimensional potentials are evaluated numerically at each midpoint, (y_i, z_i), by using the previously determined $Q_{k,j}$ for each j −th segment. Then, the hydrodynamic sectional forces and coefficients are obtained numerically by (14) for $j,k = 2,3,4$:

$$f_j^k(x) \simeq -i\rho\,\omega_e \sum_{i=1}^{N} \psi_k(y_i, z_i) s_i n_{j,i}^* \qquad (24)$$

with s_i the length of the i −th segment and $n_{j,i}^*$ the components of the generalized normal at i −th midpoint. Numerical expressions similar to (24) can also be derived for sectional forces (16) and (20).

4 SIMULATIONS AND RESULTS

In this section a series of simulations implemented to validate the present method and, after, the results of the optimization analysis that has been carried out in order to identify the best configuration of the demi-hulls main dimensions that improves seakeeping behavior and wave resistance, are presented.

4.1 *Method validation*

To validate the method, a first set of simulations were performed for circular and triangular twin-cylinders to compare two-dimensional added mass and damping coefficients obtained from expression (24) with theoretical and experimental results published by other authors. To highlight the effect of distance between the demi-hulls, $H = 2(b_{cy} - a_{cy})$, the calculations have been repeated for different values of the ratio $\frac{b_{cy}}{a_{cy}}$, with b_{cy} the half distance between the centers of the cylinders and a_{cy} the half width. In Figure 3, the non-dimensional heave added mass and damping coefficient for circular twin-cylinders with $\frac{b_{cy}}{a_{cy}} = 1.5$ in head sea are presented.

The results obtained matches very well with the theoretical and experimental results of Lee et al. (1971) (a) and Nordenstrom et al. (1971) (b), and fairly well with experimental data. The same has been obtained from comparisons of sway and roll coefficients with numerical results by Ohkusu & Takaki (1971). In general, it has been observed that the separation distance of the cylinders has an important influence on the sectional coefficients.

For all the twin-cylinders configurations examined, the added mass and damping coefficients present a discrete number of frequencies associated to piston mode and sloshing mode phenomena. The piston mode effect is typical of heave oscillation, as Figure 3 shows. This effect depends on H and may be predict according to Molin (2001). The sloshing mode frequencies are instead characterized by heavy and also sway oscillations. Generally, an increase in the distance H results in lower first frequency of the sloshing mode and piston mode phenomena.

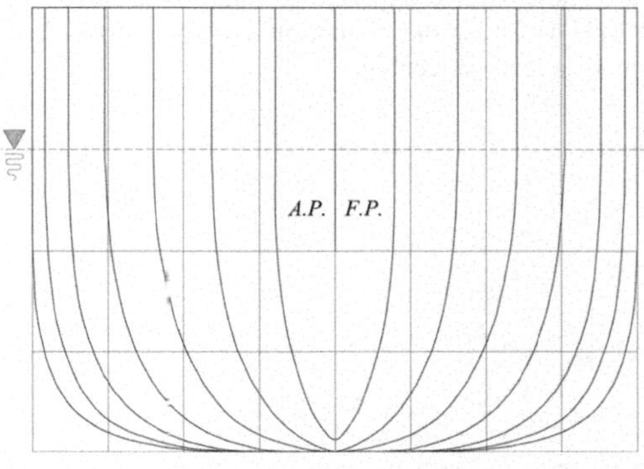

Figure 2. Ohkusu's demi-hull body plan.

Table 1. Main characteristics of the original Ohkusu's TW1 catamaran.

Characteristic	Symbol	Measurement
Length between perpendiculars	L	40.00 m
Over all beam	B_T	9.000 m
Demi-hull beam	B	3.600 m
Draught	T	1.800 m
Separation distance	H	1.800 m
Vertical center of gravity	Z_g	0.6048 m
Roll radius of gyration	ρ_{xx}	3.190 m
Pitch radius of gyration	ρ_{yy}	10.48 m
Length over beam ratio	L/B	11.11
Beam over drought ratio	B/T	2.000

Calculations have been performed for the Ohkusu's TW1 catamaran too and results are presented in terms of vertical motions responses which are compared with those of Ohkusu & Takaki (1971). The main features of the model are shown in the Table 1 and a sketch of the body plan is available in the Figure 2. In Figure 4, heave and pitch responses are showed for head sea at $Fn = \frac{U}{\sqrt{gL}} = 0.1$ with L the model length. In addiction, to confirm the accuracy and consistency of the cross-drag flow method, the corrections for viscous effects have been performed and the obtained results have been compared with those without correction. Note that $\lambda = \frac{2\pi}{K_0}$ is the wave length, and $\alpha = aK_0$ is the wave slope.

For the Ohkusu's catamaran a good agreement with the experimental and theoretical results from literature has been obtained too. Although not specific to the Ohkusu's catamaran, generally accepted values of viscous coefficients (a_o and C_D) for cross-flow drag calculations are $a_o = 0.07$ and $C_D = 0.01$

(Thwaites (1960), Lee (1976), Fang et al. (1996)). As expected, the correction using the cross-flow drag method shows a reduction in the magnitude of motion at the peak of the response. However, the Ohkusu's catamaran hull has full forms and a large water plane area compared to other catamarans or SWATH's. Thus, the reduction by cross-flow drag method is less noticeable.

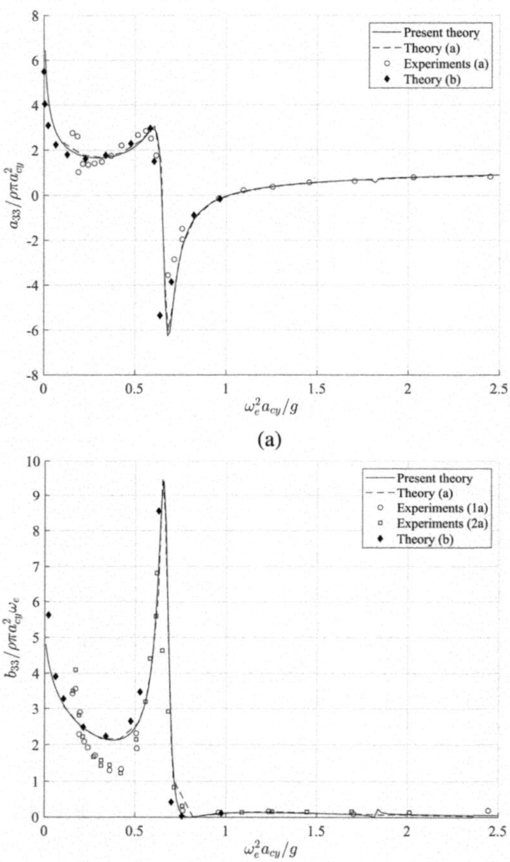

Figure 3. Non-dimensional sectional Heave added mass (a) and damping coefficient (b) for circular twin-cylinders with $\frac{b_{cy}}{a_{cy}} = 1.5$.

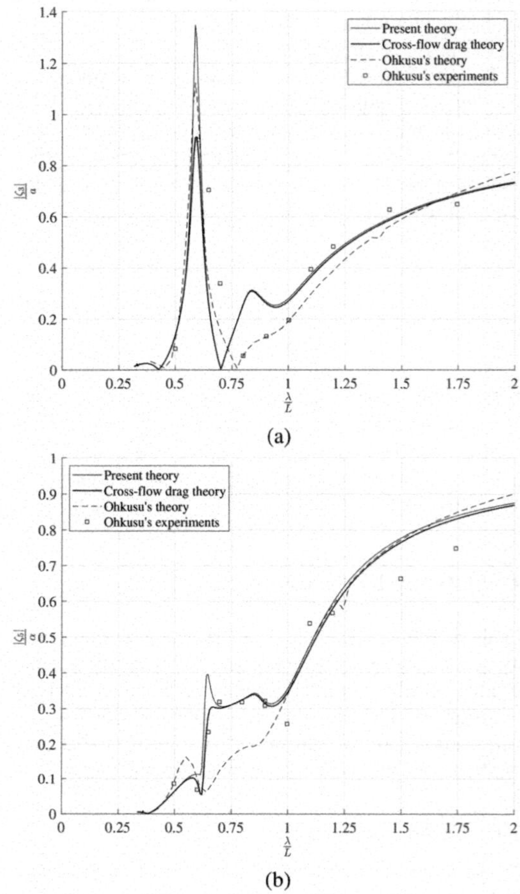

Figure 4. Heave (a) and Pitch (b) response amplitude operator with cross-flow drag correction for Ohkusu's TW1 model, in head sea at $Fn = 0.1$.

4.2 *Optimization analysis results*

Aiming at demonstrate the possibility of using the presented method in the early stages of a new design, a series of simulations have been carried out by modifying the main dimensions of demi-hull (*L*, *B* and *T*) and the separation distance (*H*) of the original Ohkusu's TW1 model. This has been achieved by using four scaling factors (λ_x, λ_y, λ_z and λ_d). The goal is to find several configurations of the multi-hull providing optimal, or increased, resistance and seakeeping performance. These have been evaluated in terms of RMS of motions, vertical acceleration, occurrence probability of certain undesirable effects, and total resistance. Since the present model does not consider non-linear effects, relatively low Froude numbers have been chosen to be consistent with the assumptions of the two-dimensional potential method.

The sensitivity study has been carried out on 1000 randomly generated configurations of the Ohkusu's TW1 model. The three scale factors λ_x, λ_y and λ_d have been chosen randomly with uniform distribution between 0.9 and 1.1, while λ_z is fixed for each configuration at 1.0. Calculations have been done for 13 angles of encounter between 0° and 180°, for three Froude numbers (0.0, 0.15 and 0.3) and for two sea states (STANAG SS3 and SS5), respectively. The calculation of the wave and total resistance coefficient (C_T) was carried out using the method proposed by Tuck (1987) and according to the ITTC'78 procedure. The form factor was evaluated according to the results of a study carried out by Armstrong (2000).

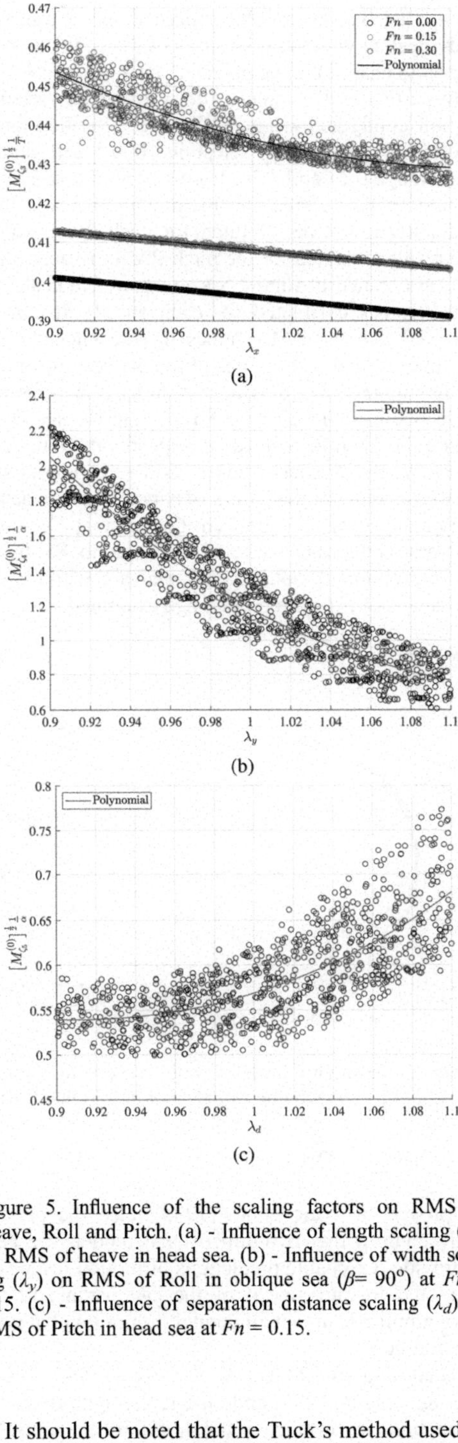

Figure 5. Influence of the scaling factors on RMS of Heave, Roll and Pitch. (a) - Influence of length scaling (λ_x) on RMS of heave in head sea. (b) - Influence of width scaling (λ_y) on RMS of Roll in oblique sea (β= 90°) at Fn = 0.15. (c) - Influence of separation distance scaling (λ_d) on RMS of Pitch in head sea at Fn = 0.15.

It should be noted that the Tuck's method used to evaluate the wave resistance does not take into account the non-linear contribution of added resistance related to second-order forces; however, according the aim of this study, also at a preliminary design stage, an approximate assessment of the resistance characteristics should be related to the seaworthiness performance analysis.

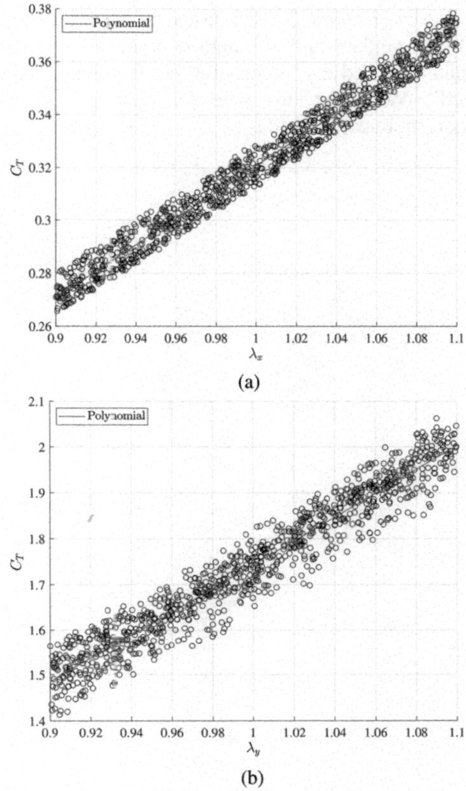

Figure 6. Influence of scaling factors on total resistance coefficient (C_T). (a) - Influence of length scaling (λ_x) on total resistance coefficient at Fn = 0.15. (b) - Influence of width scaling (λ_y) on total resistance coefficient at Fn = 0.30.

Figure 5 shows some correlations between percentage variation of L, B and H on the non-dimensional RMS of heave, roll and pitch, respectively. Instead in the Figure 6 the correlation of the variation of L and B on the total resistance coefficient (C_T) is presented. The relatively high spreading of some results corresponds to configurations in which one or more scale factors are close to the extremes. The distance H has little or no influence on vertical motions at low forward speed. For higher speed, H modestly influences vertical and roll motions. The demi-hull width B has a negligible influence on vertical motions for all tested Froude numbers, however it's present a correlation with total resistance at medium/high forward speed. Finally, the separation distance H was found to have little influence on resistance for all speeds tested. Please note that a not so evident influence of the separation distance (H) on seakeeping parameters is mainly due to the limited range chosen for the scaling factor λ_d, but also because, in general, at low Froude numbers, the hydrodynamic interaction effects between the demi-hulls occurs differently and the separation distance significantly influences the sectional hydrodynamic coefficients as well as the characteristics of the sloshing and piston mode phenomena previously mentioned. This leads to the presence of non-linear effects and to a high unpredictability of the results obtained by modifying H, therefore finding specific correlation laws with the seakeeping parameters is more complicated.

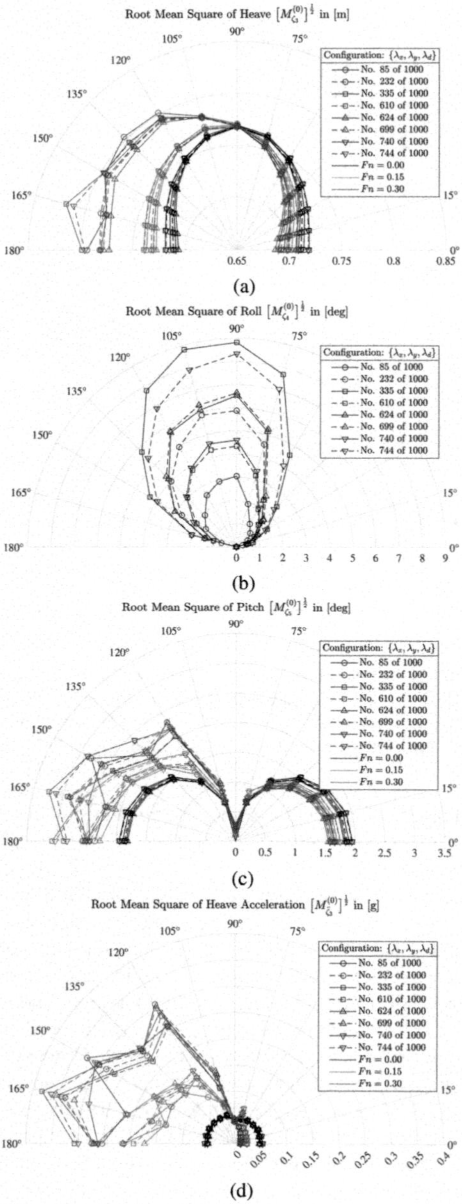

Figure 7. Comparison of Heave, Roll, Pitch and Vertical Acceleration RMS for 8 random configurations for 13 encounter angles (β). (a) - Comparison of Heave RMS. (b) - Comparison of Roll RMS at Fn = 0.15. (c) - Comparison of Pitch RMS. (d) - Comparison of Vertical Acceleration RMS.

In Figure 7, eight configurations are examined in more detail. Configuration no. 85 is the best in terms of RMS of pitch and vertical acceleration, for head sea at Fn = 0.3. The latter hull has been chose for a further study dealing with the robustness of the solutions with respect to changes in the operating parameters.

In particular, the effect of forward speed variation on hull configuration no. 85 has been verified using a sensitivity analysis by selecting 1000 speed values, in the range 0.27 and 3.30, from a normal distribution centred on the reference value at Fn = 0.3 and then evaluating the RMS of pitch for each speed value. The results obtained for the pitch RMS distribution at 180° are shown in Figure 8 (the squares indicate the actual values calculated for each speed). It can be observed that the RMS values for the pitch motion are concentrated on two bins of the discrete histogram in the range 2.2÷2.6 which includes the reference value obtained for Fn = 0.3 (i.e. 2.360°), shown in Figure 7c. Therefore, inside the speed variation range of ±10%, the prediction taken as a reference is still reliable as most of the values obtained from the sensitivity analysis are not far from the value at Fn = 0.3. This proofs the robustness of the design no. 85, found by the preliminary optimization analysis, with respect to changes of Fn, within the range examined.

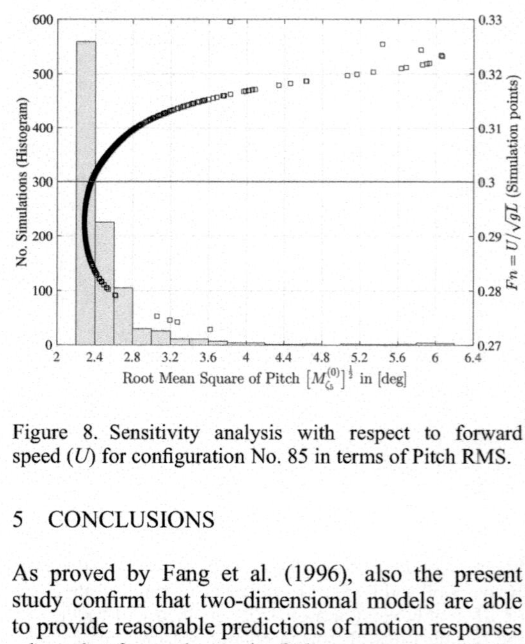

Figure 8. Sensitivity analysis with respect to forward speed (U) for configuration No. 85 in terms of Pitch RMS.

5 CONCLUSIONS

As proved by Fang et al. (1996), also the present study confirm that two-dimensional models are able to provide reasonable predictions of motion responses when the forward speed of the catamaran and the wave amplitude are small enough to not involve non-linear effects.

According to Ohkusu & Faltinsen (1990), van't Veer & Siregar (1995) and others, the hydrodynamic interactions importance of the demi-hulls depends by variables like: forward velocity (U), encounter angle (β) and distance between the demi-hulls (H). For high forward speed and elevated separation distance, hydrodynamics interactions can be ignored only if pressure field around the catamaran is not influenced by radiation waves of the demi-hulls, but for very low Froude numbers, the radiated waves move transverse from one hull to the other and interactions are present. In particular, at constant ω_e, the utmost forward speed (U) for which the hydrodynamic interactions appear can be estimated as $\frac{g}{\omega_e}\frac{L}{H}$. In general, when $\frac{L}{H} \gg \frac{U\omega_e}{g}$, demi-hulls interactions are not negligible. Although not shown in this paper, the present model proved that for very high separation distance (H) at low Froude number the demi-hulls may be considered hydrodynamically indipendent.

The sensitivity analysis revealed some correlations between the main dimensions of the multi-hull and the quantities describing seaworthiness behavior. Because of non-linear effects not considered by model, the quality of the correlations is negatively affected by higher forward speeds, nevertheless from the results, it's evident that even small variations in the main dimensions of demi-hulls can significantly affect the seaworthiness and wave resistance performance. This confirms the usefulness of a fast and reliable enough model at the early stages of the design process.

A possible optimization analysis can be refined by changing the ranges of one or all of the scale factors and using experimentally determinated viscous coefficients for the specific hull geometry. Although not reported in this article, the results of a previous sensitivity analysis carried out on Wigley's catamaran showed a correlation between the increase of these coefficients and the size of the viscous correction using the cross-flow drag method.

REFERENCES

Armstrong, N. A. (2000). *On the Viscous Resistance and Form Factor of High-Speed Catamaran-Ferry Hull Forms*. Phd thesis, University of New South Wales.

Belga, F., M. Ventura, & C. G. Soares (2018). Seakeeping optimization of a catamaran to operate as fast crew supplier at the alentejo basin. *Progress in Maritime Technology and Engineering, London, UK: Taylor & Francis*.

Broglia, R., B. Bouscasse, B. Jacob, A. Olivieri, S. Zaghi, & F. Stern (2011). Calm water and seakeeping investigation for a fast catamaran. In *Proceedings of the 11th international conference on fast sea transportation (FAST2011), Honolulu*.

Castiglione, T., F. Stern, S. Bova, & M. Kandasamy (2011). Numerical investigation of the seakeeping behavior of a catamaran advancing in regular head waves. *Ocean Engineering* 38(16), 1806–1822.

Chan, H. S. (1993). Prediction of motion and wave loads of twinhull ships. *Marine Structure* 6, 75–102.

Fang, C. & H. Chan (2004). Investigation of seakeeping characteristics of high-speed catamarans in *waves. Journal of Marine Science and Technology* 12(1), 2.

Fang, C. C. (1996). *An Investigation of Motions of Catamarans in Regular Waves*. Phd thesis, University of Glasgow.

Fang, C. C., H. S. Chan, & A. Incecik (1996). Investigation of motions of catamarans in regular waves-i. *Ocean Engineering* 23, 89–105.

Frank, W. (1967). Oscillation of cylinders in or below the freesurface of deep fluids. Internal Report 2375, Naval Ship Research and Development Center.

Holloway, D. S. (1998). *A High Froude Number Time Domain Strip Theory Applied to the Seakeeping of Semi*-SWATHs. Phd thesis, University of Tasmania.

Kring, D. & P. D. Sclavounos (1991). A new method for analyzing the seakeeping of multi-hull ships. In *Proceedings of the 1st International Conference on Fast Sea Transportation*, Trondheim, Norway, pp. 429–444.

Lee, C. M. (1976). Theoretical prediction of motion of small water-plane-area twin-hull ships in waves. Internal Report 76-0046, Naval Ship Research and Development Center.

Lee, C. M., H. D. Jones, & J. W. Bedel (1971). Added mass and damping coefficients of heaving twin cylinders in a free surface. Internal Report 3695, Naval Ship Research and Development Center.

Lee, C. M., H. D. Jones, & R. M. Curphey (1973). Prediction of motions and hydrodynamic loads of catamarans. Internal Report 3960, Naval Ship Research and Development Center.

Liu, L., X. Wang, R. He, Z. Zhang, & D. Feng (2020). Cfd prediction of stern flap effect on catamaran seakeeping behavior in long crest head wave. *Applied Ocean Research* 104, 102367.

Molin, B. (2001). On the piston and sloshing modes in moonpools. *Journal of Fluid Mechanics* 430, 27–50.

Nordenstrom, N., O. Faltinsen, & B. Pedersen (1971). Predictionof wave-induced motions and loads for catamarans. In *Proceedings of the Offshore Technology Conference*, Dallas, Texas, USA, pp. 13–58.

Ohkusu, M. (1970). On the motion of multihull ships in waves (i). Internal Report 60, Research Institute for Applied Mechanics, University of Kyushu.

Ohkusu, M. & O. Faltinsen (1990). Prediction of radiation forces on a catamaran at high froude number. In *Proceedings of 18th Symposium on Naval Hydrodynamics*, Ann Arbor, Michigan, USA, pp. 5–19.

Ohkusu, M. & M. Takaki (1971). On the motion of multi-hull ships in waves (ii). Internal Report 62, Research Institute for Applied Mechanics, University of Kyushu.

Salvesen, N., E. O. Tuck, & O. Faltinsen (1970). Ship motions and sea loads. In *Transactions of The Society of Naval Architects and Marine Engineers*, Volume 78, pp. 250–287.

Thomas, G., L. Doctors, P. Couser, & M. Hackett (2007). Catamaran motions in beam and oblique seas. In *9th Intl. Conf. on Fast Sea Transportation, Shanghai, China, September*.

Thwaites, B. (1960). *Incompressibile Aerodynamics*. Oxford, UK: Oxford University Press.

Tuck, E. O. (1987). Wave resistance of thin ship and catamarans. Internal Report T8701, Applied Mathematics Department, Univesrity of Adelaide.

van't Veer, A. P. & F. R. T. Siregar (1995). The interaction effects on a catamaran travelling with forward speed in waves. In *Proceedings of the Third International Conference on Fast Sea Transportation*, Volume 1, pp. 87–99.

van't Veer, R. (1997). Analysis of motions and loads on a catamaran vessel in waves. In *Fourth International Conference on Fast Sea Transportation*, Volume 1, pp. 439–446.

Vernengo, G., D. Villa, D. Bruzzone, & L. Bonfiglio (2021). A study on the added resistance of a catamaran advancing in waves considering variations of both operating and geometric parameters. *Ships and Offshore Structures* 16(4), 334–352.

Wehausen, J. V. & V. Laitoine (1960). *Surface Waves*. Heidelberg, Germany: Springer-Verlag.

Advances in Maritime Technology and Engineering – Guedes Soares & Santos (Eds)
© 2024 The Author(s), ISBN 978-1-032-83107-7

Hydrodynamics of ship propulsion in waves: A review

I. Sulovsky & J. Prpić-Oršić
Faculty of Engineering, University of Rijeka, Rijeka, Croatia

ABSTRACT: The focus of this paper is to review the current state of the art in estimating ship propulsion in waves, where the conventional propeller is the main source of propulsion. The hydrodynamic performance of propellers operating in waves is affected by additional disturbances introduced to the inflow of fluid into the propeller plane. Firstly, orbital wave-induced velocities affect the local fluid velocity and secondly, ship motions introduce significant spatial variations of fluid velocity in the vicinity of the propeller plane. Theoretical considerations on this matter are presented, followed by reviewing current experimental and numerical methods for estimating propulsion performance in waves.

1 INTRODUCTION

The development of ships with high energy efficiency is the backbone of the increasingly stringent rules of the regulatory bodies in the maritime industry. The efficiency of ship propulsion has a major impact on the potential reduction of harmful greenhouse gases. Therefore, the assessment of the impact of realistic environmental conditions on the ship must be rigorous and detailed. This is of particular importance within the aims of IMO (International Maritime Organization) that requires reduction of greenhouse gas emissions by 80% until 2040. and 20% by 2030.

Conventional practice in ship design implies adding sufficient power to the main engine that satisfies both calm water conditions and navigation in non-ideal weather conditions, deterioration of hull roughness due to abrasion etc. This method has proven its liability in practice; however, ships often sail under-powered which is detrimental to navigational safety. Also, ships equipped with over-powered main engines can suffer from frequent damage of the propulsion plant and vibrations due to longer periods of operation below the normal continuous rating. In this sense, detailed estimation of both the operational profile of the ship and careful selection of propulsion components (main engine, propeller) is crucial for the most economically and ecologically viable ship. In the latter, estimation of ship hydrodynamic performance in real conditions is always a challenging task due to stochastic nature of sea surface and consequently, ship motions.

The main scope of work in this paper is a comprehensive overview of literature on the matter of propulsion performance in waves and to broaden the understanding of complex hydrodynamical phenomena. This paper is organized as follows. Firstly, simplified theoretical considerations are given to give a broad understanding of the subject to the reader. Decomposition of the axial inflow velocity components occurring in wave conditions are outlined. The third chapter covers experimental methods and current guidelines for such procedures.

ITTC guidelines are briefly presented and explained. The fourth chapter covers numerical methods that are appropriate for the estimation of propulsion performance in waves. Different methods for modelling the effects of propeller rotation are explained. Starting with the low-fidelity ones, known as the body force methods and more high-fidelity approaches using AMI intefaces or overset technology. The most common simplifications of current numerical simulations are identified. Advantages and disadvantages of experimental and numerical methods are highlighted and current trends in the field are considered.

2 THEORETICAL CONSIDERATIONS

In this chapter, for the purpose of broad understanding of the problem, mathematical description of the inflow velocity in the propeller plane for wave conditions is outlined. For simplification purposes, some assumptions need to be accounted for. In the theoretical model, a ship has two degrees of freedom and a constant advancing speed. It is free to translate in vertical direction, e.g. to heave and to rotate around its center of rotation, e.g., to pitch. Also, effects of drift are omitted. The mean axial velocity of the fluid in the propeller plane behind a ship, $\overline{V_a}$, for uniform inflow conditions reads:

$$\overline{V_a}(r,\varphi) = \int_{r_h}^{R}\int_{0}^{2\pi} v_a(r,\varphi) r dr d\varphi \quad (1)$$

Where v_a is axial velocity in the propeller plane, R is the propeller radius and r_h is the hub radius, Carlton 2007. Figure 1. clarifies the 2D polar coordinate system.

DOI: 10.1201/9781003508779-21

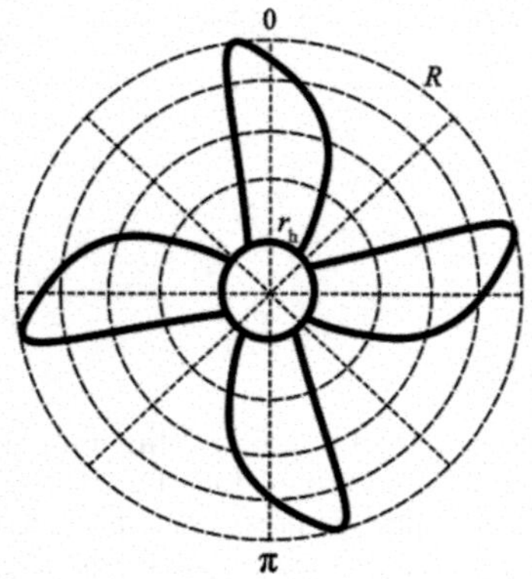

Figure 1. Polar coordinate system in the propeller plane.

However, under wave conditions, spatial variation of particle velocity introduces changes to the inflow plane of the propeller. If the distribution of orbital wave velocity, V_{orb} is taken from a linear theory, Equation (3):

$$V_{orb} = \omega\zeta\frac{\cosh k(h+z)}{\sinh kh} \tag{2}$$

where ω is the wave frequency, ζ wave amplitude, k wave number, h total depth and z being the observed depth beneath free surface. Averaging the mean value of the orbital velocities at a depth that corresponds to the lowest, -R, and highest, +R point of propeller blades, an average value of wave particle speed can be added to equation 1, which reads:

$$V_{orb} = \frac{V_{-R} + V_{+R}}{2} \tag{3}$$

Averaging of the wave particle velocities is depicted in Figure 2.

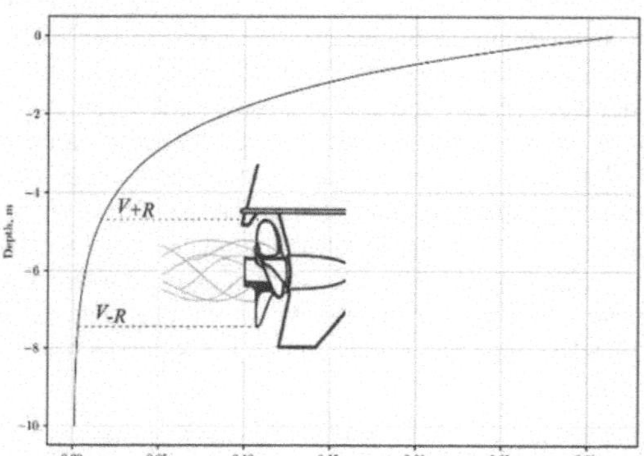

Figure 2. Averaging of the wave particle velocity.

In this manner, strictly theoretically, spatial variation of inflow conditions occurring in the propeller plane are taken into account. Until this point, equations are considered steady in time. Introducing temporal dependence in the equations yields a significant amount of complexity and uncertainty, which is attributed to the influence of ship motions on both the first and second term on the RHS of equation 3.

In still water occasions, small oscillations of ship motions influence the inflow speed into the propeller thus reducing the propulsive efficiency. Assuming the most significant effects are inherited from pitch and heave motion of the ship, fluctuating component of the axial velocity can simply be written as:

$$V_f = f(\eta_3, \eta_5) = f(t) \tag{4}$$

η_3 denotes heave amplitude while η_5 stands for pitch amplitude, by which both terms are unsteady in time. Analytical solution of the fluctuating component is exceptionally difficult to bring out since the flow field in such an environment exhibits strong turbulence effects. Therefore, numerical approximations often yield favorable results for such problems. Finally, simplified mathematical description of the inflow velocity in the propeller plane reads:

$$V_A(r, \varphi, t) = V_A(r, \varphi) + V_{orb} + V_f(t) \tag{5}$$

Such decomposition of the axial velocity component into three distinct terms should not be regarded as physically relevant. This is primarily due to complexity of adhering the ship-wave-fluid interaction in a senseful mathematical framework. Also, propeller performance in rough sea state can be affected by ventilation and out-of-water effects, especially in the non-design loading conditions. Kozlowska et al (2017) developed a predictive model for thrust losses due to the aforementioned effects.

3 EXPERIMENTAL METHODS

Experimental estimation of propulsion characteristic in waves is done exclusively with model testing in a relevant hydrodynamic facility. First experiments on propulsion performance in waves has been documented by Moor & Murdey (1970). They conducted self-propulsion model tests for three ship types in regular waves while varying propeller operational point. Their results indicate that the power increase was less if the ship model is operating below the loading of self-propulsion point. Nakamura & Naito (1975) performed resistance and self-propulsion experiments for a single-screw high speed container ship model in both regular and irregular waves.

Regarding propeller thrust and torque fluctuations, their findings indicate that the main cause is the disturbed inflow velocity into the propeller plane that can be predicted from the wave height reduction of the incident wave at the ship aft. Faltinsen (1980) points out false changes in propulsion factors if the propeller is pitching and heaving. In general, research on propulsion performance in waves is very limited for public use.

Sigmund & Moctar (2017) performed numerical investigations of propulsion in waves for a cruise ship and an ultra-large containership and compared the results with model tests conducted at the Hamburg Ship Model Basin (HSVA) that are not publicly available.

Shen et al (2014) reported thrust and torque time series of a self-propelled KCS (Kriso Container Ship) in regular waves. The experiments were conducted at FORCE Technology's towing tank in Denmark. Experimental measurements of propulsive factors in following and head waves have been performed on a Duisburg Test Case, Moctar (2012) and reported by Saettone et al. (2022). Guidelines for such experiments are given by ITTC (2014). For regular waves, the experimental procedure involves two steps. First is the estimation of self-propulsion point (SPP) in still water for a given speed and second, estimation of self-propulsion point in waves. The difference between average values of RPM, torque and thrust from still water and waves are then adopted. For estimating the propulsion in irregular waves, three different methods are proposed. Torque and revolution method (QNM) involve model tests at SPP in calm water and waves. Then, for a given wave directional spectrum, mean increase of torque and revolution rate are calculated.

This method offers an advantage in a sense that propeller characteristics are not necessary. The second method, Thrust and revolution method (TNM) is analogous to the first method but also includes estimation of wake fraction. Finally, mean power increase is calculated as a difference between needed power in still water and in waves. The last method offered by guidelines is the Resistance and Thrust identity (RTIM). Towing tests in regular waves are performed to obtain the Response Amplitude Operators (RAO) of resistance increase. Mean power increase is calculated based on the assumption that the wake fraction and propeller characteristics are the same in still water and waves.

Main advantage of the method lies in fact that only resistance towing tests are needed. Seakeeping committee of 25th ITTC concluded that these three methods offer practically the same results, as shown in Figure 3.

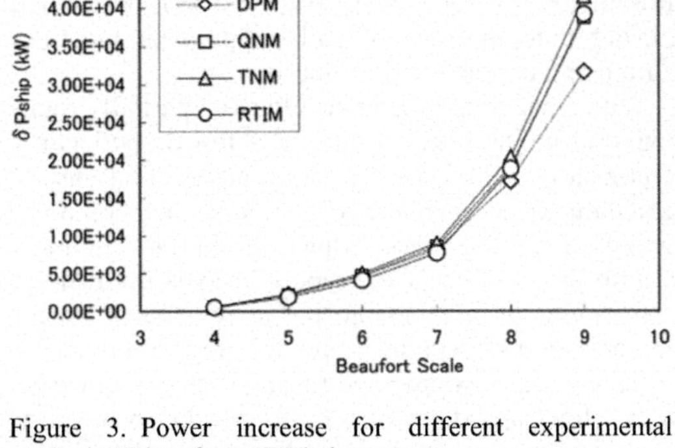

Figure 3. Power increase for different experimental methods. Taken from ITTC document.

4 NUMERICAL METHODS

Relevant numerical methods appropriate for estimation of propulsion performance in waves are almost exclusively based on Unsteady Reynolds-Averaged Navier Stokes Equations (URANS). The reason for this is twofold; (i) Computation of flow field in wave conditions in the propeller plane is hydrodynamically complex and (ii) Instantaneous disturbance of the flow due to operating propeller is highly nonlinear, turbulent and three-dimensional. Therefore, accurate predictions of propulsion in waves are computationally very demanding. The most widely used discretization of the URANS equations in computational fluid dynamics is using the finite volume method. It is an extremely versatile method based on the spatial division of control volume upon which Navier-Stokes Equations are solved. Over the years, special numerical techniques have been developed for the treatment of different terms in the Navier-Stokes equations that have been successfully implemented with the finite volume method.

It is essential to mention different techniques for modelling the effects of the propeller operating at ship aft. The most simple method, widely used both in academia and industry is the body force method, based on an actuator disk theory. In the method, pressure jump due to the propeller action is modelled as a volumetric source term on the RHS of the Navier-stokes equations that are imposed on the selected group of control volumes that correspond to the actual propeller plane.

Such approach yields favorable computational time with, however, inevitable loss of the detailed description of the flow both downstream and upstream of the propeller. Reduction of the computational time is of high interest for seakeeping calculations within viscous CFD, therefore the necessity and usability of the body force method does not need further emphasizing. A refined grid with volumetric actuator disk for propulsion simulation behind the KCS hull is depicted in Figure 4.

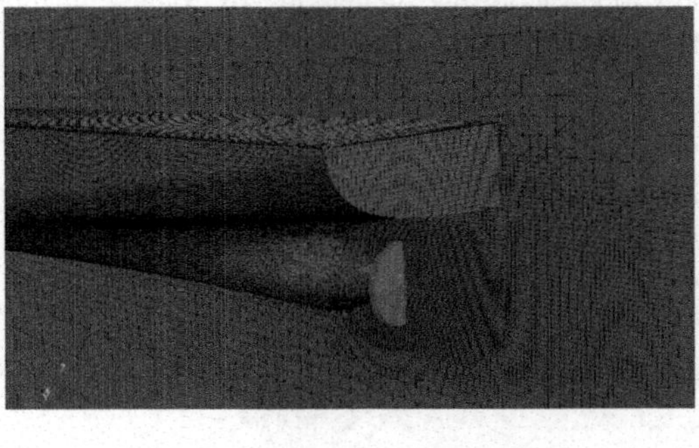

Figure 4. Actuator disk applied on the KCS hull.

There are several levels of complexity within the body force methods. Gaggero et al 2018. Computed self-propulsion characteristics along with the effective and nominal wake for KCS using four types of body force methods implemented in OpenFOAM. The simplest body force method includes thrust and torque with an even distribution in the propeller plane. More accurate body force method implies a more precise distribution of thrust and torque in the prop. plane. It most often follows the Goldstein optimum distribution of circulation along a single propeller blade, Goldstein (1929). Also, coupling of BEMt (Blade Element Momentum theory) solution and CFD can be found in several sources. Such procedure yields a more accurate results with a preservation of computational efficiency, however, in most cases the implementation is not straightforward. Feng et al (2020) presents an improved method of such coupling where corrections for induced velocities are avoided. Wu (2022) developed a body force model with rotational effects of the blades based on BEMt.

Another similar approach is reported from Bakica (2019) where the actuator disk is implemented as a boundary condition in an open-source CFD toolbox OpenFOAM®. Zhang et al (2013) used body force propeller model for evaluation of speed loss and propulsion performance evaluation in irregular head waves on a trimaran hull form. Bakica et al (2020) reports thrust losses in waves on an ESD appended ship model using actuator disk approach. Sulovsky & Prpić-Oršić investigates propulsion performance of a full-scale ferry in regular waves using the body force method.

A more complex approach of modelling the propeller is to discretize the actual three-dimensional geometry of the propeller that includes blades, hub, shaft line if present etc. The technique is based on a sliding grid methodology, better known as Arbitrary Mesh Interface (AMI) which follows the work of Farell & Madisson (2011). The computational mesh is divided into two subdomains at the boundaries of which interpolation takes place based on the Galerkin projection method. Using fully discretized propeller yields much higher computational times compared to the body force method. Increased number of cells combined with stability issues related to low Courant number contribute to a very low time-step of the simulation. An example of full propeller discretization is shown in Figure 5.

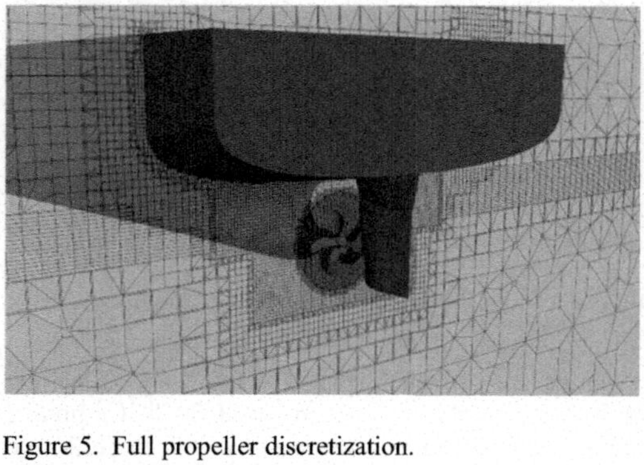

Figure 5. Full propeller discretization.

Usually, time-step is set according to the propeller revolution rate in such manner that no more than 2° of the rotation cycle is done in one time-step.

Sigmund & Moctar (2017) used sliding grid methodology on both of their numerical models. The primary advantage of this technique is the ability to capture local flow details and consequently, accurate estimation of hull-propeller interaction, however, tight similarity between disconnected mesh regions should be achieved beforehand. This is of particular interest if the ship is fitted with an energy-saving device. An example of vortical structures of the flow using sliding grid for propeller rotation is shown in Figure 6.

Figure 6. Vortical structures of the flow.

Another approach of simulating propeller rotation is using overset grid technology. It is a special numerical technique where interpolation occurs between arbitrary number of computational domains that overlapped with each other in space. It is also widely used for seakeeping computations where large ship displacements are expected. The method is even more computationally demanding than sliding grid since there are even more computational cells and the interpolation procedure between domains requires additional CPU time. Also, low-order interpolation schemes may introduce unwanted mass flux imbalance, as pointed out in Lemaire et al (2022). Shen et al (2013) used overset grids for application on KCS self-propulsion. Mofidi & Carrica (2014) performed zigzag computations of fully appended container ship using overset grid technology.

Đurašević et al (2022) developed a partially rotating grid method that combines a full propeller discretization with locally introduced non-inertial equation system, in order to reduce the computational costs. The next aspect is to be discussed within this section is the type of waves used in the estimation of propulsion in waves. Most often, researchers use regular waves due to the possibility of validating the numerical results with experimental data. However, ships very rarely or almost never in their lifetime encounter regular, periodical waves.

Therefore, realistic estimation of propulsion performance should be done with an appropriate wave spectrum, based on operational profile of the ship and the environmental conditions on the known navigational route. Three-dimensional viscous numerical simulations of spectrum waves are computationally demanding due to simulation duration that is necessary for capturing accurate wave energy that is exhibited from different wave components. In the context of propulsion estimation in irregular waves, literature is incredibly scarce. Speed loss and propulsion performance in irregular waves have been reported by Zhang et al (2021) where short-crested irregular seas are simulated with 200 distinct wave components. However, the scarcity of reported work on this matter is not a surprise since researchers are often limited by both the availability of computational resources and high license costs of appropriate software packages. Another aspect of numerical simulations worth considering is the model scale arrangement. Just as it is with regular waves, numerical simulations are conducted in model scale in order to verify the results with model test from the basin. However, scale effects play a significant role in the accurate development and design of ships, Terziev et al (2022). In essence, scale effects represent different ratios of inertial and viscous forces that occur in model and full scale. Extrapolation procedures have been developed to surpass this issue with satisfactory accuracy from an engineering perspective. Numerical environments allow us to simulate full-scale events in which special attention should be given towards resolving the wall shear stress. High computational cost of resolving turbulent boundary layers in the near wall region has proved impractical, especially for full-scale arrangement. Instead, wall functions are applied which are based on a universal distribution of near wall velocity e.g., similarity of boundary layers. In general, wall functions are applied as a boundary condition for turbulent viscosity and turbulence terms and are based on near wall cell center height y which corresponds to the distance between wall surface and cell center. In model scale events in ship hydrodynamics, the y^+ value should be in the range of $50 \approx 80$ on the wetted surface, as it is depicted in Figure 7. For KCS hull model.

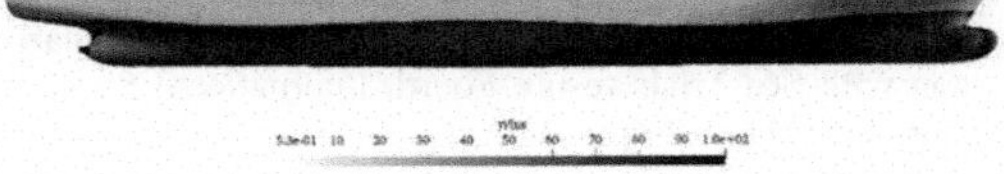

Figure 7. y^+ distribution on a KCS hull model.

Detailed description and explanation of the y^+ value can be found for example in Ferziger and Perić (2002.). Huang et al (2023) suggest that the values are within the range of $900 \approx 1000$ which also correspond to previous report from Peric (2019). Jasak et al (2018) also reports accurate self-propulsion results for two distinct ships in full-scale with y^+ ranging from $900 \approx 1100$. An example of a typical mesh topology in the vicinity of the hull for achieving the targeted y^+ is given in Figure 8.

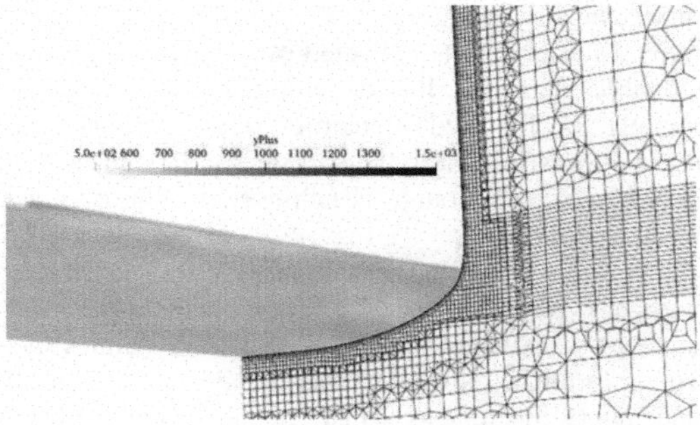

Figure 8. Mesh topology for a full-scale ferry.

In seakeeping events, the y^+ value has a mild non-uniform distribution due to orbital wave velocities and ship motions, however, this effect on the application of wall functions should be negligible but deserves future research. Perhaps the most important issue while conducting numerical simulations in full-scale is often the inability to validate the results with actual measurements. Ponkratov & Zegos (2015) did pioneering work in this field by conducting full-scale ship propulsion simulations after which the torque and thrust coefficients have been compared with sea trial measurements. On the other hand, validating full-scale propulsion simulations in wave conditions adds another significant layer of uncertainty: wave measurements. It is by far the greatest source of uncertainty due to the inherent complexity of capturing wave amplitude and wave direction in the given point in time and space.

5 CONCLUSIONS AND PERSPECTIVES

An overview of hydrodynamics of propulsion estimation in waves has been covered up to now. Theoretical considerations for the inflow velocity in the propeller plane have been derived with, however, a great deal of simplification. The main aim of such theoretical outlining is to give the reader a sense for the complexity of the given problem.

Wave particle orbital velocities, together with ship motions are mentioned as a contributing term for disturbance of the inflow velocities. Current guidelines for experimental methods are briefly addressed as they are covered in detail in the ITTC documents. Significant reports on the investigation of propulsion in waves is mentioned together with newest additions. The lack of publicly available benchmark studies on propulsion in waves is identified. Moving on to the numerical methods, key techniques for propeller discretization within finite volume method are addressed and explained. It is pointed out that the literature predominantly considers regular waves, as these do not represent the stochastic nature of the sea surface. Furthermore, the issue of scale effects and the correct treatment of near-wall velocities is addressed as the CFD community is steadily bridging over to the full-scale simulations. Measurements collected on sea trials represent valuable data for the development of numerical models in ship hydrodynamics. However, the primary purpose of sea trials is to satisfy the contractual speed, which of course is achieved in almost ideal weather conditions.

Measurements of propulsion quantities in actual seaways also contribute to valuable insights into propeller loadings at different sea states, engine dynamics etc. The unfortunate part in the data is the large amount of uncertainty inherent to actual weather conditions that the ship is experiencing. Collection of wave data is most often done by a wave radar, a device mounted on the ship's forecastle that measures geometrical characteristics of wind waves. Another approach includes purchasing the data from weather forecast providers on a daily basis. This leads us to current perspectives in the field of propulsion in waves.

The necessity to deliver high quality measurements from actual seaways becomes an insurmountable step towards achieving zero carbon shipping. Computational fluid dynamics has long-ago become a mature tool for ship design, for both model scale and full-scale assessments. Steady increase and availability of computational power holds the potential to shift the focus on the full-scale propulsion simulations in actual seaways. It will enable the development of efficient numerical towing tanks that do not degrade any of the real physical properties of the fluid. Also, propelled discretization methods are subject for further improvement, both the body force methods (increasing output quality) and full geometrical propeller discretization approaches (reducing computational costs).

ACKNOWLEDGEMENTS

This work has been fully supported by Croatian Science Foundation (HRZZ) under the project IP-2022-10-2821.

REFERENCES

Bakica A., Gatin I., Vukčević V., Jasak H. and Vladimir N. 2019. Accurate assessment of ship-propulsion characteristics using CFD, Ocean Engineering, Volume 175.

Bakica A., Vladimir N., Gatin I. and Jasak H. 2020. CFD Simulation of loadings on circular duct in calm water and waves, Ships and offshore structures, Volume 15.

Carlton. J.S., 2007. Marine propellers and propulsion. -2nd ed.

Đurasević S., Gattin I., Uroić T. and Jasak H. 2022. Partially rotating grid method for self-propulsion calculations with a double body ship model, Ocean Engineering, Volume 266, Part 15.

Faltinsen O.M. 1980. Prediction of Resistance and Propulsion of a ship in a seaway, Proceedings of the 13th symposium on naval hydrodynamics, Tokyo.

Farell P.E., Madisson J.R., 2011. Conservative interpolation between volume meshes by local Galerkin projection, Computer methods in Applied Mechanics and Engineeing, Volume 200, Iss: 1, pp 89–100.

Feng. D., Yu J., He R., Zhang Z. and Wang X. 2020. Improved body force propulsion model for ship propeller simulation, Ocean Engineering, Volume 104.

Ferziger J. and Perić M. 2002. Computational Methods for Fluid Dynamics, Springer, Berlin, Third edition.

Gaggero, S., Gaggero, T., Tani, G., Vernengo, G., Viviani, M., & Villa, D. (2018). Ship self-propulsion performance prediction by using OpenFOAM and different simplified propeller models.

Goldstein S., 1929. On the Vortex Theory of Screw Propellers, Proceedings of the Royal Society of London. Volume 123, Issue 792.

Huang L., Pena B. and Thomas G. 2023.Towards a full-scale CFD guideline for simulating a ship advancing in open water, Ship Technology Research, Volume 70.

Jasak H., Vukčević V., Gatin I. and Lalović I. 2019. CFD validation and grid sensitivity studies of full scale ship self-propulsion, International Journal of Naval Architecture and Ocean Engineeing, Volume 11, Issue 1, Pages 33–43.

Kozlowska A.M., Savio L., Steen S. 2017. Predicting Thrust Loss of Ship Propellers Due to Ventilation and Out-of-Water Effect. Journal of Ship Research 61: 198–213.

Lemaire, S., Vaz, G., van Rijswijk, M. D.., and Turnock, S. R. 2023. Influence of Interpolation Scheme on the Accuracy of Overset Method for Computing Rudder-Propeller Interaction. ASME. *J. Verif. Valid. Uncert*. March 2023; 8(1): 011002.

Moor, D.I., and Murdey D.C. 1970. Motions and Propulsion of Single Screw Models in Head Seas, Part I.Trans. RINA, Vol. 112 (1970)p.121.

Nakamura S. and Naito S. 1975. Propulsive performance of container ship in waves. Journal of the Society of Naval Architects of Japan, 15.

Ould el Moctar B., Shigunov V. and Zorn T. 2012. Duisburg Test Case: Post-Panamax Container Ship for Benchmarking., Ship Technology Research, 59 (3) (2012), pp. 50–64.

Perić M. 2019. White paper: full-scale simulation for marine design. Siemens White Pap.1-14.

Ping-Chen Wu 2022. CFD Body Force Propeller Model with Blade Rotational Effect. Applied Sciences 12, no. 21: 11273.

Ponkratov D. and Zegos C. 2015. Validation of Ship Scale CFD Self-Propulsion Simulation by the Direct Comparison with Sea Trials result, Fourth International Symposium on Marine Propulsors.

Saettone S., Taskar B., Steen S. and Andersen P. 2021. Experimental measurements of propulsive factors in following and head waves, Applied Ocean Research, Volume 111.

Seakeeping committee of the 27th ITTC, 2014. Prediction of Power Increase in Irregular Waves from Model Test.

Shen Z., Carrica P.M and Wan D., 2014 Ship Motions of KCS in Head Waves with Rotating Propeller using Overset Grid Method., Proceedings of the ASME 2014

33rd International Conference on Ocean, Offshore and Arctic Engineering, OMAE2014.
Shen Z., Wan D. and Carrica P.M. 2015. Dynamic overset grids in OpenFOAM with application to KCS self-propulsion and maneuvering. Ocean Engineering, Volume 108, Pages 287–306.
Sigmund S. and Ould el Moctar B. 2017. Numerical and experimental investigation of propulsion in waves, Ocean Engineering, 144 (2017), pp. 35–49.
Sulovsky I., Prpić-Oršić J. and Dejhalla R. 2023. CFD study on propulsion characteristics of ships subjected to waves, Proceedings of the 16th Baška GNSS Conference
Terziev M, Tezdogan T. and Incecik A. 2022. Scale effects and full-scale ship hydrodynamics: A review. Ocean Engineering, Volume 245.
Zhang L., Zhang J. and Shang Y. 2021. A practical direct URANS CFD approach for the speed loss and propulsion performance evaluation in short-crested irregular head waves, Ocean Engineering, Volume 2019.
Zhang L., Zhang J. and Shang Y. 2021. A practical direct URANS CFD approach for the speed loss and propulsion performance evaluation in short-crested irregular head waves, Ocean Engineering, Volume 2019.

Advances in Maritime Technology and Engineering – Guedes Soares & Santos (Eds)
© 2024 The Author(s), ISBN 978-1-032-83107-7

Analysis of shear flow field in ship motion using numerical simulation method

H. Wang, J. Zhang & B. Mei
Navigation College, Dalian Maritime University, Dalian, China

ABSTRACT: This study focuses on the geographical features and tidal characteristics of cape waters in major ports. Utilizing the incompressible Navier-Stokes equations and numerical simulation methods, we establish a mathematical model for tidal currents in these waters. Additionally, a ship maneuvering model is developed using the airfoil theory model within the shear flow field, simulating both dynamic and kinematic aspects of ship maneuvering. The primary objective is to analyze the impact of shear flow on ship maneuvering motions. We conclude by summarizing the associated risks of ship maneuvering in cape waters and proposing strategies to mitigate these risks through shear flow field maneuvering. This research provides valuable insights for navigators operating in similar waters, offering guidance for safe navigation in challenging cape water conditions.

1 INTRODUCTION

Cape waters pose a significant challenge to ship maneuvering due to their highly variable tidal currents, often resulting in navigational difficulties and temporary loss of control. Large seaports often have capes or headland areas with complex and variable tidal currents, making navigation challenging due to the intricate traffic flow. Even a slight error in ship maneuvering within such areas can lead to momentary loss of control, potentially resulting in hazardous situations(Sharma & Nazir, 2017). In order to adapt to the development of ports and the trend toward larger vessels, addressing the safety concerns and challenges in navigation becomes crucial. Conducting targeted and systematic studies on these specific geographical areas, identifying risk factors, and formulating preventive and control measures can significantly enhance the safety of vessel navigation within these unique terrains. The focus of this research is on large heavy-loaded vessels, specifically referring to bulk carriers and oil tankers with lengths exceeding 180 meters, which often exhibit relatively poor maneuverability (Zhao, 2014). Such vessels are particularly susceptible to the influence of non-uniform flow due to their dimensions and draft considerations (Shen, 2018). For instance, the core port area of Ningbo-Zhoushan Port includes various cape terrains such as Tunizui, Luotoujiao, and Changbingzui.

Capes refers to a pointed landform that juts into the sea. The planar contour of a cape is similar to that of a peninsula, but it is smaller in size (Li et al, 2020). Due to factors such as tidal height, tidal phase, and topography, water currents near capes can become turbulent. The cape's protrusion effect often leads to the formation of certain ranges of eddies or backflows on the "downstream" side of the cape, closer to the coastal area (Mao, 2018).

Near the confluence of strong currents and backflow areas, a clearly visible demarcation line can be observed during daylight hours (refer to Figure 1), commonly referred to by navigators as a shear line (Hu, 2003). On either side of the shear line, there are significant differences in flow velocity, flow direction, and even near-opposite directions (Wu, 2015). The shear line is not fixed and unchanging; it shifts and changes with the tidal cycle, gradually dissipating as the tidal currents weaken (Zhang, 2018).

Taking Luotoujiao in the Zhoushan Port area as an example, Luotoujiao is located at the southern tip of Damao Island. It protrudes like a pointed angle into the Luotou Waterway, forming a narrow gateway with Chuanbi Island and Liangmao Mountain on the opposite side, causing a sudden contraction of the Luotou Waterway at this point. The Luotou Waterway runs in an east-west direction with an average width of about 2.5 nautical miles, narrowing to approximately 1 nautical mile near Luotoujiao. The waters around Luotoujiao are a traffic separation scheme for ships inward and outward the Ningbo Beilun Port from the south channel. Influenced by the geography, the waters around Luotoujiao experience swift and variable currents. The western waters of Luotoujiao can have maximum flow speeds of 5 to 7 knots, while the eastern waters exhibit slower flow speeds in the inbound lane during ebb tide compared to the

DOI: 10.1201/9781003508779-22

outbound lane. As shown in Figure 2, it depicts the state of rising tidal currents around Luotoujiao.

Figure 1. Tidal shear line in cape waters during rapid rising tide.

Source: Photograph taken by the author.

Examining the mathematical model of ship motion elucidates two viable methodologies for constructing such a model. The initial approach entails formulating a straightforward linear mathematical model, albeit constrained in its applicability to specific scenarios, such as serving as a reference for controller design. Conversely, the second approach involves crafting a nonlinear mathematical model that incorporates diverse environmental factors, including sea winds. Evidently, this latter model more closely corresponds to the motion of the ship within real-world conditions.

Currently, in the research of nonlinear mathematical ship motion models, there are two main approaches. The first is the holistic model represented by Abkowitz, which generates numerous higher-order ship fluid dynamic derivatives during computations. This results in ambiguity in the physical interpretation of each derivative, hampering ship design and maneuverability studies (Tong, et al, 2018). The second approach is the separated control motion model equation proposed by the MMG group of Japan's JTTC. This model has undergone continuous development and improvement, comprehensively exploring the fluid dynamics of the hull, propeller, and rudder and their mutual influences. In 2008, Buren Xiang (Bu, 2008) established a three-degree-of-freedom control motion model for under-actuated surface ships, fully considering the model's nonlinearity, as well as disturbance factors such as wind, currents, and shallow water, along with the mechanical characteristics of the main engine and rudder. In 2012, Li (Li, 2012) chose the MMG separated ship motion equation and employed a modified shallow water low-speed model for simulation studies to account for possible shallow water effects. In 2016, Zhang & Zhang (Zhang & Zhang, 2016), inspired by the concept of MMG separated modeling, created a fluid dynamics mathematical model for the hull, propeller, and rudder in an infinite water field scenario assuming no external environmental influence. They incorporated the propeller force estimation formula and developed a mathematical model for the motion of a single propeller ship with a right-hand rotation. The MMG model has matured through practical verification and continuous refinement, becoming widely applied in ship hydrodynamics computations and significantly advancing the field of maritime simulation.

The study of tidal flow characteristics primarily involves two aspects: analysis of field measurements and numerical simulation calculations. Field observations have been effective in studying tidal flow characteristics. However, due to the high cost of observations, the observation periods for tidal flow are typically long. Additionally, limited tidal observation stations result in a relatively small dataset with limited representativeness. It's not advisable to solely infer the tidal flow characteristics of an entire sea area based solely on these limited observations. Therefore, transitioning from point observations to comprehensive insights, using computer-based tidal numerical simulations to obtain a large-scale view of tidal flow distribution in a water region is an effective and economical approach (Zhou & Xia, 2021). Utilizing numerical simulation methods can address the numerical simulation challenges presented by shear flow fields.

This paper initially analyzes the tidal flow characteristics of cape waters in major ports. Employing numerical computations, the study simulates the flow field in the Luotoujiao area of the Ningbo-Zhoushan core harbor region. A mathematical model for ship maneuvering within shear flow fields is established, and research is conducted on ship navigation within shear line flow fields.

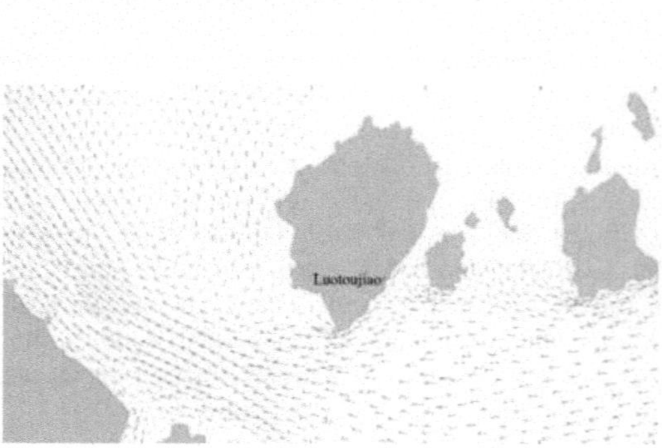

Figure 2. Distribution of rising tidal currents at Luotoujiao.

2 NUMERICAL SIMULATION OF TIDAL FLOW FIELD

Numerical simulation is a process where the equations that describe the motion of water flow are discretized using mathematical models through a method of discretization. These equations are then solved using computer programming to calculate and obtain the values of multiple variables at discrete points in both time and space. This approach quantitatively describes the flow field by providing a detailed representation of various variables.

2.1 *Step-by-step method*

The step-by-step method, also known as the time-stepping method, is used to numerically solve the unsteady Navier-Stokes equations in a separated manner. At each time step, an incomplete form of the momentum equation is integrated, yielding an approximate velocity field. In general, this velocity field does not satisfy the condition of being divergence-free. Subsequently, this velocity field is projected onto a divergence-free field without altering vorticity. This projection step is achieved by solving the Poisson equation for pressure distribution.

The Navier-Stokes equations in terms of primitive variables are provided by this equation.

$$\begin{gathered}\frac{\partial u}{\partial t}+u.\nabla u=-\nabla p+\frac{1}{Re}\nabla^2 u\\ \nabla.u=0\end{gathered} \tag{1}$$

Here, spatial discretization employs a second-order central difference scheme, while temporal discretization uses a first-order Euler scheme. When fully discretized, the aforementioned equation can be expressed.

$$\begin{gathered}\frac{u^{n+1}-u^n}{dt}=-Gp-Nu^n+\frac{1}{Re}Lu^n\\ Du^{n+1}=bc^{n+1}\end{gathered} \tag{2}$$

Alternatively, in the following matrix form

$$\begin{bmatrix} I & G \\ D & 0 \end{bmatrix}\begin{bmatrix} u^{n+1} \\ p \end{bmatrix}=\begin{bmatrix} r^n \\ 0 \end{bmatrix}+\begin{bmatrix} 0 \\ bc \end{bmatrix} \tag{3}$$

Where I is the identity matrix, N is the convection operator, G is the gradient operator, D is the divergence operator, r^n is the explicit right-hand side of the momentum equation, and bc is the boundary condition vector for the incompressibility constraint, in discretized numerical methods, n typically denotes the current time step, while $n+1$ represents the next time step.

2.2 *Spatial discretization*

Here, we have employed a staggered grid arrangement. The components of flow direction and wall-normal velocity are stored on cell faces, while the pressure values are stored at cell centers. We use explicit, second-order finite differences on the staggered grid to approximate the derivatives in the (x,y) directions. The convective term is calculated using a fully conservative divergence form, as illustrated in Figure 3.

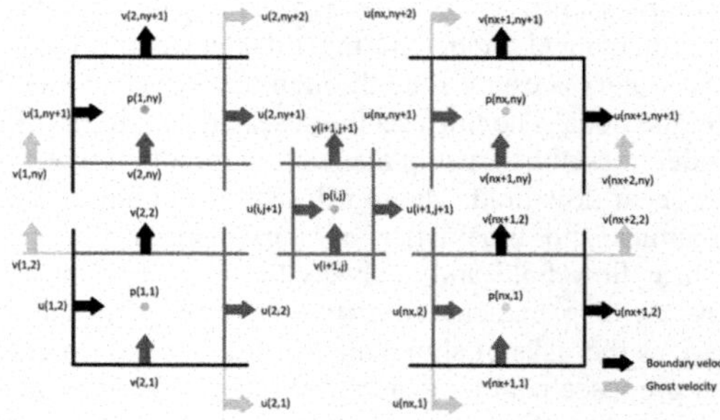

Figure 3. Staggered grid cell arrangement and its indices.

The bold black lines represent the boundaries of the computational domain. The diagram above shows the four corners and the interior of the domain. The black arrows indicate the boundary conditions, while the orange arrows represent the "ghost" velocity. The "ghost" velocity is calculated using the boundary conditions and the velocity from one step within the domain. This ensures that the interpolated velocity matches the boundary conditions. The orange velocity is referred to as the "ghost velocity," and its calculation follows the formula (4).

$$u(i,1)=2u_{bc}-u(i,2) \tag{4}$$

The actual velocity to be calculated is represented by the light blue arrow portion.

2.3 *The simulation results of tidal flow*

In this segment, a geographic structure is formulated by extracting topographic data from electronic nautical charts. The previously mentioned incompressible Navier-Stokes equations are applied, and numerical simulation methods are utilized to calculate shear flow fields in cape waters during both rising and ebbing tides. The outcomes are depicted in Figure 4.

The nearby waterway has a depth of about 100 meters. Near the western side of Luotoujiao, the water depth suddenly decreases close to the shore, with a steep slope. Isobaths such as 20, 30, and 50 meters are densely distributed, causing a sudden narrowing of the Luotou Waterway. Taking the rising tide as an example, influenced by the topography of the water area and due to the protruding effect of Luotoujiao, backflow often occurs to the northwest of Luotoujiao. As a result, an enclosed water body known as the Clamp weir water forms in the vicinity. During the rising tide, a strong shear line tends to form on the western side of Luotoujiao. When vessels sail from east to west during the rising tide, they will steer towards the western side of Luotoujiao while staying to the right and following the downstream flow. The second section of this paper focuses on the topography and tidal characteristics of cape waters in large hub ports. Based on the incompressible Navier-Stokes equations, a mathematical model of the tidal currents in the cape waters is established through numerical simulation methods. The third section uses an airfoil theory model to establish a ship maneuvering motion model in a shear flow field. The fourth section simulates the dynamics and kinematics of ship maneuvering in a shear flow field and analyzes the impact of shear flow on ship maneuvering motion. Finally, it summarizes the risks of ship maneuvering in cape waters and proposes countermeasures through maneuvering in shear flow fields, providing a reference for ship pilots navigating such waters. The specific mechanism behind this maneuver will be analyzed in the subsequent section of this article.

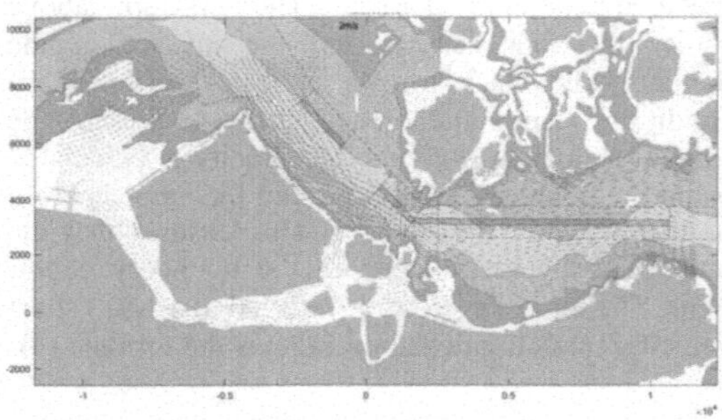

(a) Rising Tide Flow Distribution

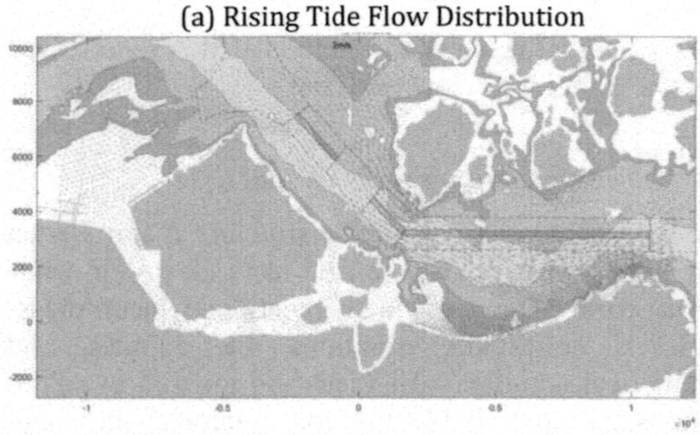

(b) The tidal current distribution of the ebb tide

Figure 4. Tidal flow simulation results.

3 THE EFFECTS OF SHEAR CURRENT ON SHIP DYNAMIC PERFORMANCE

3.1 *Characteristics of rudder force effects*

During ship maneuvering in cape waters, assuming the lateral velocity of the ship with respect to the ground is relatively small (less than the rising tide velocity), when the ship's speed decreases, the effective longitudinal velocity at the rudder location, denoted as u_R, also decreases. Simultaneously, when the angle between the ship's heading and the flow direction approaches perpendicular, the effective lateral velocity of the flow at the rudder location, v_R, increases. As depicted in Figure 5, u_R is positively correlated with the ship speed and the outflow of the propeller. After initiating a turn and experiencing a reduction in speed, u_R decreases. Moreover, as the angle between the flow direction and the ship's heading increases, v_R increases. Consequently, with the rudder angle δ remaining constant, the effective flow angle α_R at the rudder blade location decreases, leading to a decrease in rudder force and diminished rudder effectiveness.

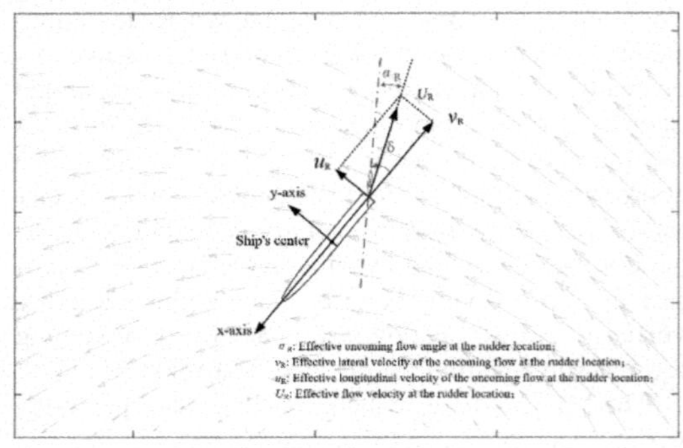

Figure 5. Schematic illustration of flow Field's impact on Rudder force.

3.2 *Characteristics of flow-induced ship body moments*

Prior to initiating a turn in cape waters, the vessel is practically within a uniform flow. As the vessel approaches and passes through the shear line while turning, it experiences non-uniform flow. The flow is stronger at the rear compared to the bow, as illustrated in Figure 6. The green arrows represent the magnitude and direction of the tidal flow, while the red area indicates the forces acting along the ship's longitudinal axis. Consequently, the forces on the ship's hull are unevenly distributed along its length, and in some cases even have opposing directions. This leads to a significant increase in flow-induced ship body moments after the turn, causing the vessel to exhibit a flow-induced deviation. This phenomenon of moments is notably different from when the vessel operates in a uniform flow environment.

3.3 *Characteristics of ship motion encountering shear lines*

When the rudder force turning moment is smaller than the flow-induced turning moment, the vessel faces difficulties in maneuvering and cannot control its deviation. If the rudder force turning moment is slightly larger than the flow-induced turning moment, and the vessel has an initial yaw rate (Gucma, et al,

2020), the suppression of deviation takes a longer time. However, if subsequent deviations lead to a significant increase in the transverse angle of the ship's stern due to flow or inconsistency between the bow and stern flow directions, causing the flow-induced turning moment to exceed the rudder force turning moment, the vessel becomes unable to control its deviation. In other words, difficulties in ship maneuvering tend to arise near shear lines, if the initial rotation velocity is superimposed, the navigation risk is further increased, making it challenging to control the vessel's heading over short periods.

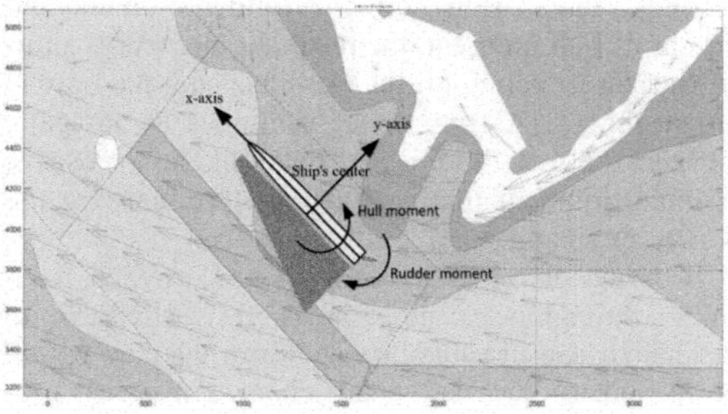

Figure 6. Illustration of Flow-Induced Forces on the Hull.

3.4 *Simulated ship maneuvering*

Cape areas often experience shear currents, which are a type of non-uniform flow. Currently, there are three types of ship maneuvering motion equations in uniform flow: the along-ship mean flow velocity (surge) model, lateral flow theory model, and wing theory model (Yang & Fang, 1998).

The ship maneuvering motion model in uniform flow is a well-established model structure. In comparison to the uniform flow model, the key distinction in non-uniform flow models lies in the lateral force on the vessel, denoted as Y_H, and the yawing moment N_H. In non-uniform flow, Y_H and N_H are functions of the longitudinal water velocity u_a, lateral water velocity v_a, and yaw angular velocity r. By computing the forces at each location along the vessel length L, the vessel's force can be obtained through integration. Similarly, integrating considering the vessel length L factor provides the vessel's moment, as shown in formula (5), where $y(x)$ represents the force on the vessel at each location along the length (Yang & Fang, 1998).

$$\begin{cases} Y_{\mathrm{H}}(u_a, v_a, r) = \int_{-\frac{L}{2}}^{\frac{L}{2}} y(x)\mathrm{d}x \\ N_{\mathrm{H}}(u_a, v_a, r) = \int_{-\frac{L}{2}}^{\frac{L}{2}} y(x)x\mathrm{d}x \end{cases} \tag{5}$$

This study employs the wing theory model to calculate $y(x)$, as depicted in formula (6). In formula (6), the calculation methods for linear lateral force $y_L(x)$, nonlinear lateral force $y_{NL}(x)$, mean longitudinal water velocity $\bar{u}_a$, and mean lateral water velocity $\bar{v}_a$ are outlined (Kashiwagi, 1986).

$$\begin{cases} y(x) = y_L(x) + y_{NL}(x) \\ y_L(x) = \frac{8}{\pi} N'_L \sqrt{\frac{0.5+x}{0.5-x}} + (Y'_L - 4N'_L)\sqrt{0.25 - x^2}] \\ y_{NL}(x) = -C_D |v'_a + xr'|(v'_a + xr') \\ \bar{u}_a = \frac{1}{L}\int_{-L/2}^{L/2} u_a dx, \bar{v}_a = \frac{1}{L}\int_{-L/2}^{L/2} v_a dx \end{cases} \tag{6}$$

The rudder force model is represented by formula (7), and the numerical values of the model parameters are the same as those in the maneuvering motion model in uniform flow (Chen, Yang, Li, Tian, Liu & Hao, 2022).

$$\begin{gathered} F_N = -\tfrac{1}{2}\rho A_R f_a U_R^2 \sin \alpha_R \\ U_R^2 = V^2(1 - w_R)^2\{1 + KG(s)\} \\ G(s) = \eta k\{2 - (2 - k)s\}s/(1 - s)^2 \\ s = 1 - (1 - w_P)u/nP \\ \eta = D_P/H_R \\ k = 0.6(1 - w_P)/(1 - w_R) \\ w_R = w_{R0} w_p / w_{p0} \\ \alpha_R = \delta - \gamma \cdot \beta_R \\ \beta_R = \beta - 2x'_R r', \quad x'_R \approx -0.5 \\ K = \begin{cases} 1.065, \delta 0 \\ 0.935, \delta < 0 \end{cases} \end{gathered} \tag{7}$$

In addressing the deviation phenomenon exhibited by vessels navigating through cape waters influenced by shear currents, the aforementioned mathematical model for ship motion is employed for computations. This entails the acquisition of data related to rudder moment and ship body moment. The alterations in moment under full rudder conditions are scrutinized, along with their consequential effects on ship deviation. The tidal flow data is obtained through the previously elucidated methodology, with the main channel flow speed approximately 3 to 5 knots. For this experiment, a large bulk carrier of 250,000 tons (Mei, 2020) is selected as the subject of study, the maximum flow velocity in the shear flow field is 5.2 knots, with its ship parameters outlined in Table 1. The numerical simulations in this study were conducted using MATLAB R2022b, a powerful programming environment for engineering and scientific computations. Calculations were performed on a computer equipped with an AMD 6800H CPU, featuring 8 cores and 16 threads, and 16GB of DDR5 RAM. This setup efficiently handles complex computational tasks. The simulation process was completed within 1 minute, indicating high efficiency in processing and the potential for repeated experiments or extensive parameter sweeps.

Table 1. Main dimensions of the experimental bulk carrier.

Main Dimensions	Actual Ship
Displacement (m^3)	312622
Distance Between Two Columns (m)	320.2
Ship Width (m)	58.0
Maximum Rudder Rate (°)	2.34
Rudder Wetted Area (m^2)	273.3
Service Speed(knot)	15.5
Froude Number	0.142
Block Coefficient	0.8098
Midship Section Coefficient	0.9980
Moulded Depth (m)	30
Draught (m)	20.8
Number of Propeller Blades	4
Propeller Diameter (m)	9.86
Propeller Pitch (0.7R)	0.721
Aspect Ratio	0.431

* Main Dimensions of the Experimental Bulk Carrier

4 SIMULATION RESULTS AND COUNTERMEASURES ANALYSE

Using the tidal flow data obtained earlier, a model of a 250,000-ton large bulk carrier is established. The simulation of ship motion in Luotou Waterway is conducted, and the results of the vessel's trajectory, moments, heading, and speed are presented in Figure 7. The initial ship speed is 9 knots, heading is 317°, and rudder angle is 35° to port. The maximum flow velocity in the shear flow field is 5.2 knots.

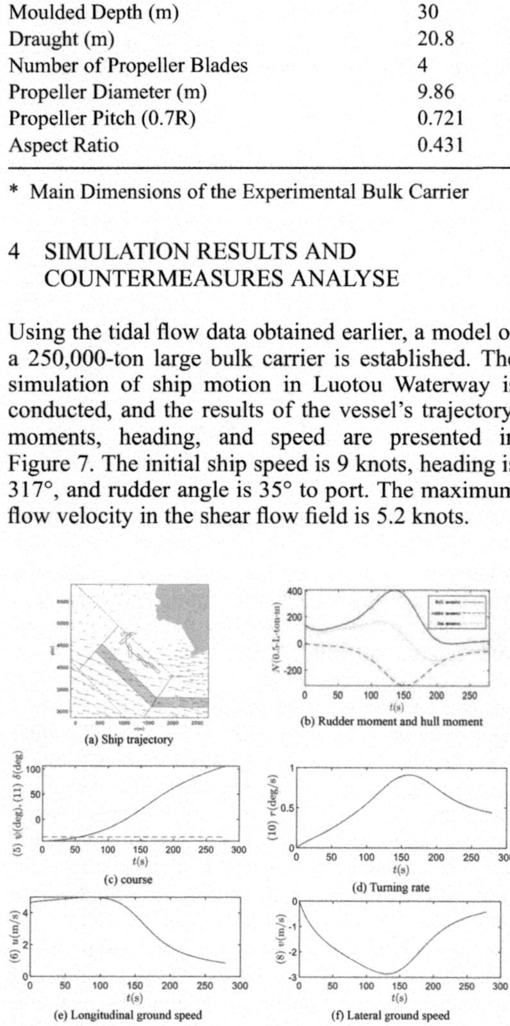

Figure 7. Simulation experiment results of ship motion in shear flow.

In Figure 7(a), it is evident that even with full left rudder, the vessel cannot suppress its tendency to turn to the right after entering the shear flow, resulting in the ship turning toward the shore side. From Figure 7(c) and 7(d), it can be observed that the vessel achieves its maximum turning rate of approximately 1° per second at around 150 seconds, after which the turning rate decreases. Figure 7(e) and 7(f) show that after 110 seconds of turning, the longitudinal speed of the vessel decreases, while the lateral speed approaches its maximum.

The analysis of the causes behind the aforementioned phenomena is as follows, based on observations from Figure 7(b): (1) At 0 seconds, the ship's hull experiences a shear flow-induced moment that acts like a thrust on the right quarter of the stern. (2) At 0 seconds, full left rudder generates a force to counteract the vessel's right deviation. (3) At 0 seconds, considering both the rudder moment and the ship's hull moment, the resultant moment is similar to a thrust acting on the stern, causing the bow to turn right. (4) As the vessel continues to navigate, the shear effect diminishes, resulting in a positive resultant moment that acts at the stern, causing the bow to turn right.

In terms of mitigation strategies: (1) When navigating through shear flow areas, it is advisable to maintain a considerable distance from the core region of the shear line and areas characterized by pronounced shear effects. (2) In regions exhibiting significant shear effects, particularly in close proximity to the shear line, it is crucial to regulate the turning rate to prevent excessively rapid or sharp maneuvers. (3) In areas featuring fast-flowing shear flow, executing a right turn poses a risk of exacerbating rightward bow deviation towards the shoreline, nearing the right boundary of the channel. During the initial deviation phase, it is recommended to promptly apply the rudder and employ a substantial rudder angle to suppress the deviation. (4) If the vessel is unable to counteract the deviation with full rudder, it should navigate out of the challenging maneuvering zone before resuming heading control, resulting in an S-shaped track during this period. (5) In severe cases, deliberate initiation of a right turn is advised, particularly in shear flow areas. Increasing the ship's speed or temporarily elevating the main engine's RPM can augment the propeller's discharge velocity, thereby assisting in mitigating the impact of shear flow and reducing the risk of maneuvering difficulties.

5 CONCLUSION AND ANALYSIS

This study is based on the incompressible Navier-Stokes equations and focuses on large bulk carriers. Utilizing numerical simulation methods, it analyzes the tidal flow characteristics in cape areas and investigates the effects of shear flow on ship's dynamic performance. The following points summarize the key findings:

(1) The results demonstrate that numerical simulation is effective in reproducing tidal flow patterns in cape areas, providing information about

the flow at various locations and highlighting the impact of terrain on tidal patterns. This forms the basis for fluid dynamics data to study ship motion in shear flow conditions.

(2) A simulation method for studying ship motion in shear flow conditions is proposed.

(3) Simulation outcomes indicate that when ships enter shear flow areas, there is a sudden increase in the ship's body moment and a decrease in rudder moment. This results in difficulties in controlling the ship's heading, causing deviations in the ship's trajectory, and making it challenging to maintain the desired course.

(4) The research findings presented in this paper provide valuable insights for ship motion in cape areas, offering ship operators an understanding of ship motion mechanics and maneuvering strategies. Furthermore, this research contributes theoretical insights for managing risks in special port waters with challenging conditions.

While current numerical simulation methods have explored the impact of shear currents on ship motion, the research has been limited to a single type of vessel. Other types of large vessels, such as ultra-large container ships, large bulk carriers, and large oil tankers, have not been studied regarding their susceptibility to shear currents. With the ongoing trend of vessel enlargement, it is imperative that future research continues to analyze the effects of tidal currents on these different types of vessels. Ship operators should be especially vigilant about the influence of estuarine circulation and shear currents. They must remain cautious throughout navigation, being mindful of the effects of tidal currents coming from different directions, both from the bow and stern of the vessel. Maintaining a prudent approach during navigation is crucial to avoid potential mishaps, urgent situations, or even maritime accidents resulting from improper handling.

REFERENCES

Bu, R. 2008. *Nonlinear feedback control of underactuated surface ships*. Dalian Maritime University.

Chen, L., Yang, P., Li, S., Tian, Y., Liu, G. & Hao, G. 2022. Grey-box identification modeling of ship maneuvering motion based on LS-SVM. *Ocean Engineering*, 266:112957.

Gucma, S., Dzwonkowski, J. & Przywarty, M. 2020. Kinematic method of determining safe fairway bend widths. *TransNav: international journal on marine navigation and safety of sea transportation*, 14(2).

Hu, Z. 2003. Concept of "Shear Line" in "Flow Field". *Marine Technology*, (6):37–37.

Kashiwagi, M. 1986. A New Method for Calculating the Maneuvering Motion of a Ship in Non-uniform Current and Its Application in Naruto Kaikyou. *The Journal of Japan Institute of Navigation*, 75:55–67.

Li, A., Ji, X., Guo, X. & Shen, K. 2020. Study on shoreline evolution caused by change of cape shape of headland bay. *Periodical of Ocean University of China*, 50 (11):81–86.

Li, R. 2012. *Study of self-berthing of the large vessel*. Dalian Maritime University.

Mao, Z. 2018. Non-uniform flow and its pilot manoeuver. *Proceedings of China Maritime Pilots 2017*: 165–169.

Mei, B. 2020. *Grey box identification modeling for ship manoeuvring motion based on free running test*. Dalian Maritime University.

Sharma, A. & Nazir, S. 2017. Distributed Situation Awareness in pilotage operations: implications and challenges. *TransNav: international journal on marine navigation and safety of sea transportation*, 11 (2):289–293.

Shen, Y. 2018 Study on VLCC berthing at Daxie Intermediate oil terminal. *Proceedings of China Maritime Pilots 2017*: 151–156.

Tong, S., Tang, X. & Zhang, H. 2018. Modeling and Simulation of Ship Maneuvering MotionBased on Non-Uniform Flow and Shallow Water Effect. *Journal of System Simulation*, 30(3):866.

Wu, Y. 2015. The Impact of Tidal Shear Lines on Vessel Maneuverability and Countermeasures. *Collection of Outstanding Papers in Chinese Nautical Science and Technology (2014)*: 408–413.

Yang, Y. & Fang, X. 1998. Simulation model of ship maneuvering motion in uneven current and it's application. *Shipbuilding of China*: 30–35.

Zhang, J. 2018. The relationship between depth contour and shearing line in current. *China Water Transport*, 18 (1):19–20.

Zhang, Q. & Zhang, X. 2016. A ship maneuvering model and application in harbor considering reversing characteristics of propeller. *Shipbuilding of China*, 57 (4):147–156.

Zhao, X. 2014. The Impact of Flow Field Shear Lines on the Maneuverability of Large Vessels and Countermeasures. *China Water Transport*, 14 (12):28–30.

Zhou, Z. & Xia, W. 2021. An analysis of tidal current characteristies cf Fangchenggang sea area. *Transactions of Oceanology and Limnology*, 43(04):62–67.

Advances in Maritime Technology and Engineering – Guedes Soares & Santos (Eds)

Rapid evaluation of pure loss of stability based on atlas method

H.M. Zhang & K. Li
School of Naval Architecture & Ocean Engineering, Dalian University of Technology, Dalian, P.R. China

ABSTRACT: This paper proposes an atlas method for rapidly evaluating pure stability loss. At first, a modeling study utilizing on fishery administration vessels as research objects. Secondly, for pure stability loss of vulnerability in different Sea Conditions, the proposed method is used to explore the parametric behavior of the significant wave height (*Hs*) and zero crossing-period (T_0). Additionally, vulnerability is computed systematically under diverse loading conditions, resulting in visual representations like 2D plots. Finally, we conduct a comparative analysis between the performance of our proposed method and the ICS-HydroSTAB, the simulations reveal that the Atlas method, as proposed, demonstrates superior efficiency in executing failure probability calculations. The main innovation of this study lies in the expedited computation time and retained precision of the calculations, potentially offering more substantial support in marine practice.

1 INTRODUCTION

Ship stability is a crucial theme in naval architecture, with considerable impact on ship design and operation. In the second round of intact stability criteria, as prescribed by the IMO, careful attention is paid to five separate modes of dynamic stability failure in wave conditions. Parametric roll, pure stability loss, surfriding, broaching, dead-ship condition, and excessive accelerations are all phenomena that merit assessment methodologies for the anticipatory prediction of pure stability loss in wave conditions. The vulnerability of pure stability loss is linked to multiple contributing factors and thus demands immediate attention. Lilienthal et al. (2007) conducted a study into the survival limit KG, which highlighted the insignificant impact of wavelength ratio on capsize behavior, according to the test results. Bassler et al. (2011) observed the phenomenon of pure loss of stability, which correlates with fluctuations in the restoring arm and is particularly pronounced in the context of astern waves. Furthermore, Liu et al. (2022) have highlighted that the velocity at which the free-running ONR Tumblehome moves forward plays a crucial role in causing the instability failure mode to occur.

The Experimental Fluid Dynamics (EFD) method is often used to evaluate the relationship between wave parameters, ship velocities, and restoring arm variation (Hashimoto 2009, Lilienthal et al. 2007, Lu et al. 2019). When assessing pure stability loss, the focus is on conditions present in astern seas (Htun et al. 2019, Lu et al. 2020, Umeda et al. 2017). The development of computing technology has facilitated the use of Computational Fluid Dynamics (CFD) to capture finer details and gain a deeper understanding of the fundamental nature of capsizing. The CFD method is widely applied in the assessment of second-generation stability (Sadat-Hosseini et al. 2010, Yao-hua et al. 2015, Liu et al. 2000, Liu et al. 2022). further investigated the application of the CFD method. An in-house unsteady ReynoldsAveraged Navier-Stokes (RANS) solver was used. Bulian (2010) proposed a probabilistic methodology to analyse a ship's vulnerability to the risk of pure loss of stability events. These events are modeled as the persistence of the metacentric height below a critical level for an extended period. The proposed methodology can be used to assess the likelihood of such events occurring. Wang et al. (2016) and Wang et al. (2017) simulated the free-running ship under course-keeping control in waves, focusing on the presentation of the hull-propeller-rudder interaction. EFD methods, with associated high costs and lengthy testing periods, face challenges simulating extreme conditions and require precise models. In contrast, CFD methods, while avoiding costly experiments, are constrained by significant computational resource demands, limiting practical engineering applications. Currently, there is a lack of a direct and effective approach to address the pressing need of mariners to evaluate their susceptibility to complete stability failure while navigating. To tackle this challenge, our investigation presents a pioneering procedure that uses graph theory to estimate complete stability loss. Thus, we are laying the foundation for the adoption of a second-generation stability vulnerability assessment.

2 METHOD

The pure stability loss vulnerability assessment criteria underwent a series of refinements before being finalized at the sub-committee on ship design and construction conference(SDC2017). To meet the criteria, two conditions must be satisfied

- The length of the vessel should be 24m or longer.

DOI: 10.1201/9781003508779-23

- For ships with higher service speeds, the Froude number should exceed 0.24.

If the speed is low, the duration over which the ship is exposed to decreased stability (reduced righting lever) is generally too short for a large heel angle to develop.

2.1 *Level 1 criteria*

The non-vulnerability conditions of the ship in the pure stability loss model are:

$$GM\,\text{min} > \left[R_{PLA}\right] \tag{1}$$

R_{PLA} indicates the benchmark value for the first denotes the minimum height of initial stability in waves after the correction of free surface. If the vessels have relatively full displacement, *GM* min can be derived from (2) based on parameters in static water. However, for other vessels, the minimum initial stability height in waves must be computed. Notably, the calculation of *GM* in waves should be based on the simplified formula for *GM* minrequiring only hydrostatic data.

$$GM\,\text{min} = KB + \frac{I_L}{\nabla} \tag{2}$$

where *KB* represents the height of the vertical center of buoyancy for the specific loading condition, measured in meters; *KG* represents the height of the vertical center of gravity for the same loading condition, and I_L represents the moment of inertia of the waterplane at the draught.

2.2 *Level 2 criteria*

If the ship does not pass the first level of vulnerability criterion, then the second level of vulnerability criterion needs to be calculated. According to the IMO proposal, (3) serves as the benchmark criterion for the second-layer pure stability loss.

$$\max(CR_1, CR_2) < [R_{PL0}] \tag{3}$$

where R_{PL0} is 0.06. If the vessel adheres to (3), it is posited that the occurrence of pure stability loss in the vessel is deemed improbable. CR_1 and CR_2 are the average weights of certain stability parameters for specific wave conditions, calculated from (4) and (5).

$$CR_1 = \sum_{N}^{i=1} W_i C1_i \tag{4}$$

$$CR_2 = \sum_{N}^{i=1} W_i C2_i \tag{5}$$

where W_i is the weighting coefficient, determined by wave conditions; $C1_i$ and $C2_i$ are the coefficients corresponding to different waves, determined by (6)-(8).

$$C1_i \begin{cases} 1\ \varphi_v < [R_{PL1}] \\ 0\ \varphi_v \leq [R_{PL1}] \end{cases} \tag{6}$$

$$C_{2i} \begin{cases} 1\ \varphi_s > [R_{PL2}] \\ 0\ \varphi_s \leq [R_{PL2}] \end{cases} \tag{7}$$

$$R_{PL3} = 8(\frac{H_i}{\lambda})dFr^2 \tag{8}$$

where φ_v is angle of vanishing stability, Unit (°); φ_s is steady-heel-angle. where R_{PL2} is 15 degrees for passenger vessels, and 25 degrees for all other types of vessels. H_i is wave height; *d* is draught.

2.3 *The impact of wave environment*

The impact of the ocean wave environment on the stable transverse inclination angle is essentially determined by two inherent characteristics of the waves: wave height and wave period. Due to the stochastic nature of ocean waves, the adoption of wave spectra is presently the most efficacious and straightforward means of character rising waves. This involves solving the assorted moments of the wave spectrum and utilising the combination of these moments to describe a range of characteristics pertaining to the waves. The technique of extracting wave spectrum features has gained widespread acceptance and universal implementation due to its effectiveness in capturing the fundamental characteristics of ocean waves. The Equations (9)-(12) expresses the correlation between the characteristic wave height and zero-crossing period of ocean waves with the spectral moments.

$$Hs = 4.0\sqrt{\mu_0} \tag{9}$$

$$T_0 = \sqrt{2\pi\frac{\mu_0}{\mu_2}} \tag{10}$$

$$\mu_0 = \int^{\omega^i} S(\omega)d\omega, i = 0, 2 \tag{11}$$

The present study adopts the North Atlantic spectrum as the reference ocean wave spectrum, with its calculation method outlined as follows: for each zero-crossing period (T_0) in the wave spectrum, there exists a corresponding significant wave height. The significant wave height ($H_{1/3,ref,i}$ ($T_{ref,j}$)) is defined as the conditionally averaged significant wave height under the condition of an average spectral period ($T_{ref,i}$):

$$H_{1/3,ref,i} = E(H_{1/3}|T_{0,i}) = \frac{1}{W_i}\sum_{j=1}^{M} P_{ij} \cdot H_{1/3,j} \quad (12)$$

where $H_{1/3,ref,i}$ is the reference meaningful wave height; At a zero-crossing period of $T_{0,i}$, P_{ij} represents the probability of occurrence for the sea condition with a significant wave height $H_{1/3,j}$.

3 RESEARCH OBJECT

Fishery administration ship (DFAS) serve as specialized maritime assets dedicated to fisheries supervision and law enforcement. Noteworthy for their compact dimensions and high operational velocity, these vessels confront challenging sea conditions, rendering them prone to stability failures throughout various operational phases. Consequently, this study has chosen fishery administration vessels as the primary research subjects. In the modeling endeavor, specific components, including the extension beyond the stern bulkhead, the representation of the stern post, and the hub, have been intentionally omitted. The three-dimensional model diagram and type line diagram of DFAS are shown in Figures 1 and 2.

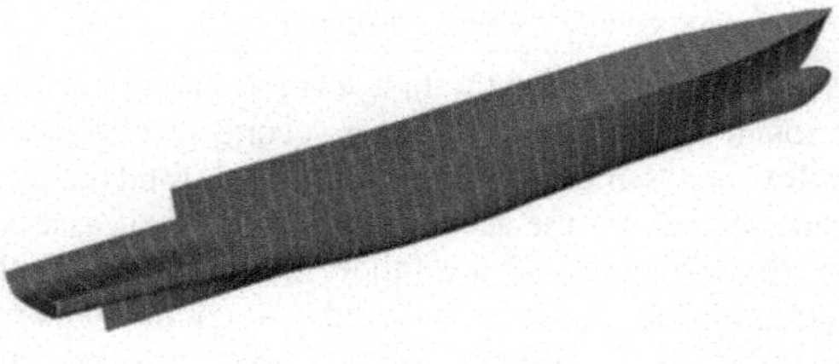

Figure 1. Three-dimensional model drawing of DFAS.

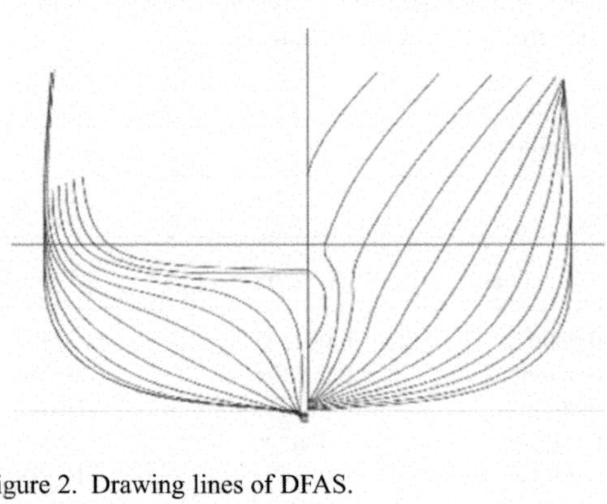

Figure 2. Drawing lines of DFAS.

Table 1. Principal dimensions of DFAS.

ship type parameters	numerical value
Length L/m	66.66
Breadth B/m	11.14
Draft T/m	4.28
Vertical center of gravity ZG/m	3.91
Displacement W/t	2107.93
rolling period Tr/s	5.72
block coefficient Cb	0.66

4 SIMULATION

Regarding pure stability loss, proportional changes in the primary dimensions of vessels in a particular series show fundamental uniformity in their impact on stability. However, the magnitude of these effects varies depending on the specific vessel series, with the following hierarchy: vertical center of gravity ZG >draft d >beam B >length L >molded depth D >other vessel parameters. In this present research, a comprehensive analysis is carried out on the multiscale conditions of fisheries enforcement vessels to examine the impact of fluctuations in both center of gravity height and draft on the susceptibility to vulnerabilities associated with pure stability loss within this particular context.

4.1 *Level 1 vulnerability diagram*

Assuming unchanged primary dimensions, this study systematically varies the loading characteristics of Dalian Fisheries Administration vessel. It is crucial to underscore that the first-layer computation for pure stability loss only necessitates static hydrodynamic calculations, without delving into stability assessments within wave dynamics. The minimum value of the initial stability height (GM min) is then calculated for 50 different loading conditions. The results are summarized in Table 2 and graphically presented in Figure 3.

Table 2 presents the calculated values for the minimum initial stability height of a vessel under different draft and center of gravity height conditions. The unit of measurement for the minimum initial stability height in this table is expressed in meters (m).

Figure 3 depicts the result of the minimum initial stability height for fisheries Administration vessel that has been exposed to diverse loading scenarios. The vertical center of gravity height has a stronger impact on the first-layer pure stability loss vulnerability compared to draught. a vessel indicates a decrease in minimum initial stability height as the vertical center of gravity height increases, highlighting a negative effect of height increase on the initial layer pure stability loss vulnerability. Further examination of the five draft spline curves in the figure shows that an increase in draft correlates with a reduction inGM min, indicating an increased detrimental effect of draft on the occurrence of pure stability loss and consequently its beneficial effect on

Table 2. Calculation results of the first layer for different loading conditions.

	Dalian Fishery Administration Ship				
	ZG=3.90m	ZG=3.95m	ZG=4.00m	ZG=4.05m	ZG=4.10m
d=4.30m	1.01251	0.96138	0.91024	0.85912	0.80796
d=4.45m	0.99002	0.93864	0.88725	0.83586	0.78447
d=4.50m	0.96643	0.91498	0.86356	0.81213	0.76071
d=4.60m	0.94398	0.89252	0.84106	0.78959	0.73814
d=4.70m	0.92261	0.87111	0.81961	0.76811	0.71662

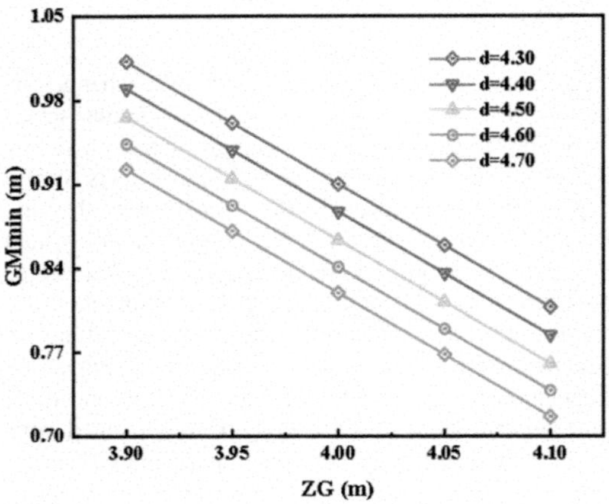

Figure 3. The first-layer diagram of pure stability loss under various loading conditions.

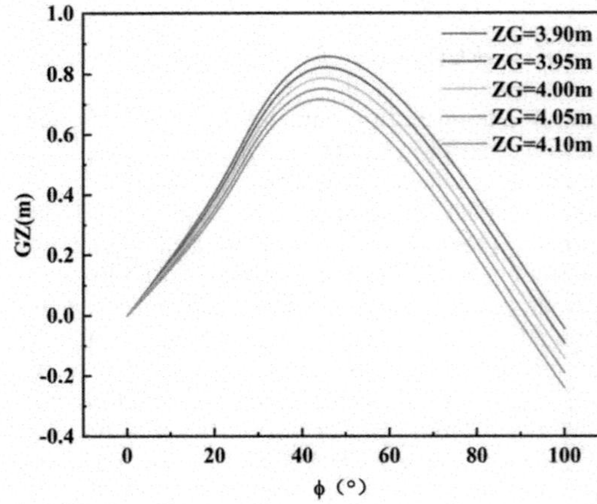

Figure 4. Curve diagram of righting lever at different center of gravity heights.

ship stability. In order to undertake a comprehensive analysis of the factors contributing to this observed trend, the restoring arm for the vessel under different vertical center of gravity heights are computed, as shown in the Figure 4.

Figure 4 shows that when parameters are kept constant(d=4.30m), raising the center of gravity leads to a consistent decrease in the curves of the vessel's righting arm. Consequently, when assessing the loading conditions of maritime vessels, the impact of different load configurations on center of gravity height and draft will be significant during operational phases. This enables the adjustment of the righting arm to affect the stability characteristics of the vessel.

4.2 *Level 2 vulnerability diagram*

If a ship fails to pass the first level vulnerability criterion for pure stability loss, a second level vulnerability assessment is required. The fundamental principle behind the second level consists of assessing the likelihood of the failure modes linked with pure stability loss. To undertake a comprehensive examination of the ships' second-layer vulnerability in distinct wave scenarios, an estimation of the probability of occurrence is essential. To conduct a comprehensive analysis of the second-layer vulnerability of vessels under distinct wave circumstances, it is vital to compute the stable transverse inclination angles in diverse wave scenarios.

Table 3. Calculation results of the first layer for different loading conditions.

	Dalian Fishery Administration Ship				
	ZG=3.90m	ZG=3.95m	ZG=4.00m	ZG=4.05m	ZG=4.10m
d=4.30m	0.000041	0.000173	0.000622	0.002143	0.003
d=4.45m	0.000045	0.000324	0.000994	0.003128	0.007494
d=4.50m	0.000121	0.000332	0.00112	0.003433	0.01022
d=4.60m	0.000135	0.000488	0.00214	0.006496	0.01278
d=4.70m	0.000139	0.000488	0.00214	0.009499	0.03538

4.2.1 *Second-layer graph of pure stability loss under average sea conditions*

This investigation leverages a sophisticated VB6.0based numerical computation program to intricately assess the maximum failure probability. It is imperative to underscore that, in this context, "average" pertains to the conversion of random waves through the utilization of the North Atlantic wave spectrum into 272 specific regular waves, subsequently undergoing a meticulous weighted averaging procedure. The utilization of a fourth-degree polynomial fitting methodology enhances the precision of the computations, capturing subtle variations with heightened accuracy. The outcomes are systematically summarized, offering a comprehensive analysis that delves into the nuanced intricacies disclosed by this precision-centric fitting procedure. The results for the maximum failure probability under various loading conditions are presented in Table 3 and Figure 5.

From Table 3, it can be observed that all scenarios pass the second-layer stability criteria, indicating that the vessel is not experiencing a pure loss of stability. but as the center of gravity height increases in Figure 5, the maximum failure probability also increases, indicating that the vessel is more susceptible to occur pure loss of stability. This observation corresponds to the conclusions drawn from the first-layer calculations. On the contrary, alterations in draft display an inverse relationship with the outcomes of the first-lever calculations. This phenomenon can be attributed to the incremental distribution of hazardous sea conditions in the North Atlantic spectrum with an increase in the vessel's draft.

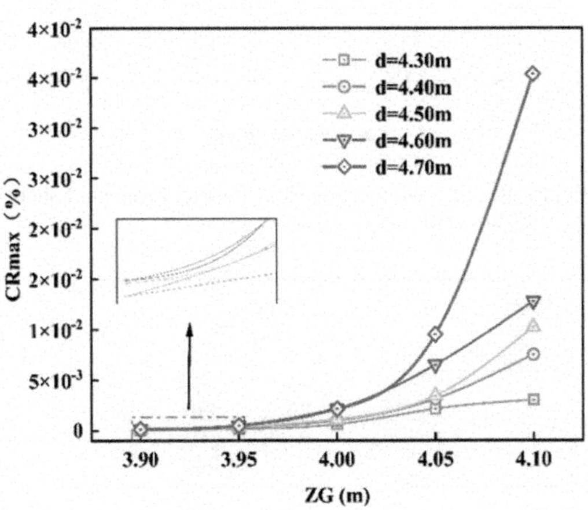

Figure 5. The second-layer diagram of pure stability loss under various loading conditions.

4.2.2 *Second-layer graph of pure stability loss under different sea conditions*

For the analysis of the susceptibility of Dalian fisheries enforcement vessels to pure stability loss under diverse maritime conditions, this study employs distinct significant wave heights (*Hs*) as wave characteristic parameters, selecting a zero-crossing period (T_0) of 7.5 seconds and keeping it constant is because during this period, the ship is most prone to happen pure stability loss (IMO,2017) .The stable transverse inclination angle serves as the computed outcome, which is then subjected to polynomial fitting. The resultant graphical representation is illustrated below.

Analysis of Figures 6-7 reveals that, as the center of gravity height increases, the steady heel angle also rises. This implies a heightened susceptibility to pure stability loss. While the change in draft is not as prominent as the shift in center of gravity height, it is still evident that an increase in draft renders the vessel more prone to experiencing pure stability loss.

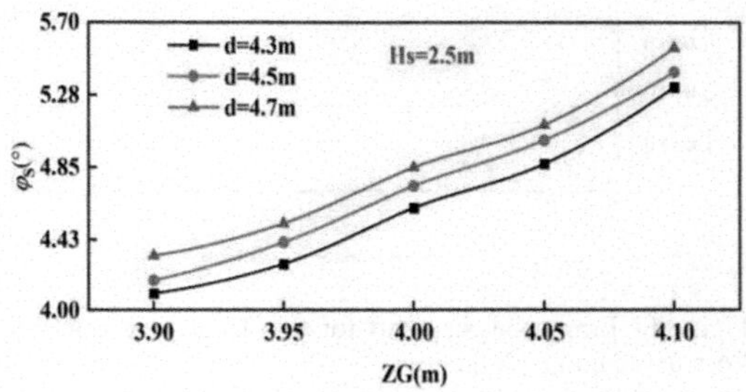

Figure 6. Steady heel angle variation with a meaningful wave height of 2.5.

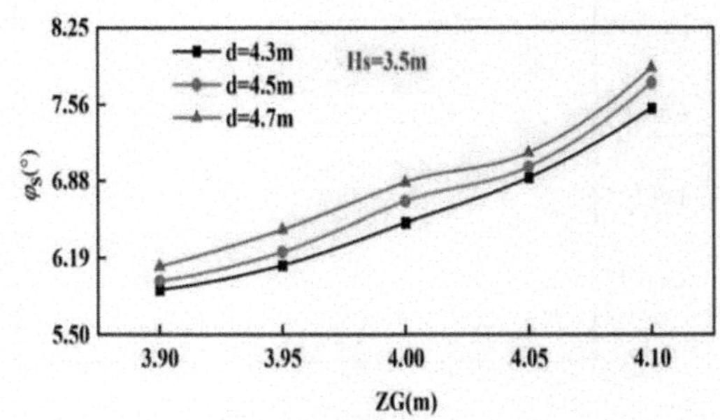

Figure 7. Steady heel angle variation with a meaningful wave height of 3.5.

5 METHOD VALIDATION FORMULATION

In the context of practical ship engineering predictions, the precise forecasting of pure stability loss, particularly the improvement of prediction accuracy in the proximity of safety critical points for failure probability, bears crucial engineering implications for the assessment of vessel safety. the vulnerability to both first-layer and second-layer pure stability loss is calculated using graph theory and the ICSHydroSTAB software under two distinct stochastic loading conditions. The stochastic loading conditions are outlined in the following Table 4: From the aforementioned Figures 8 and 9, it is ob served that, served that, based on the

Table 4. Load condition information table.

	Condition 1	Condition 2
height of center of gravity(m)	4.00	3.91
Draft(m)	4.39	4.28
Displacement(m^(3))	2199.08	2130.43

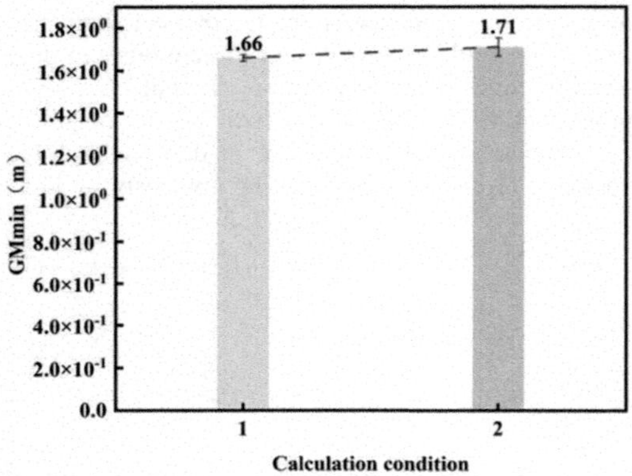

Figure 8. Error analysis chart for first-layer pure stability Loss based on graph theory.

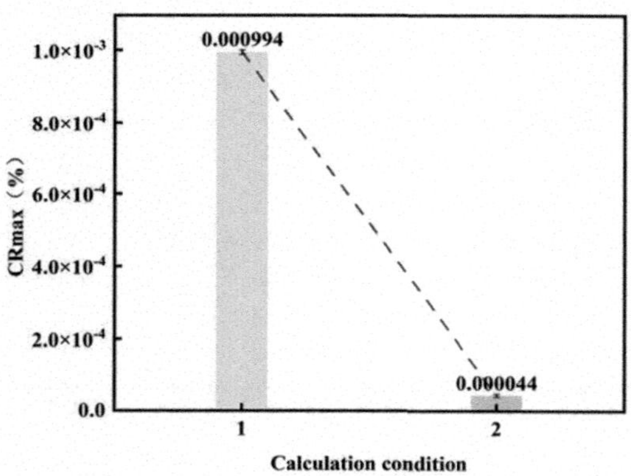

Figure 9. Error analysis chart for second-layer pure stability Loss based on graph theory.

graph theory approach, the errors in calculating the first-layer pure stability loss under stochastic loading condition 1 is 2.59%, and under stochastic loading condition 2, the error is 3.86%. The average error rate for both conditions is 3.225%, consistently below 5%, indicating that the first-layer graph theory meets the precision requirements. For the second-layer pure stability loss, the error is 3.60% under stochastic loading condition 1 and 4.23% under stochastic loading condition 2 when computed using the graph theory method. The average error for both conditions is 3.92%, demonstrating that the second-tier graph theory also meets precision requirements.

6 EFFICIENCY COMPARISON

To analyze the disparity in computational time between the graph theory approach and ICSHydro-STAB software, this study similarly opts for pure stability loss vulnerability calculations under loading conditions 1 and 2. The chart comparing time consumption is displayed in the following table: The comparative time consumption chart is presented following: In the computation of the first-layer pure stability loss, the graph theory approach exhibits computational efficiency with a runtime of 5 seconds for loading condition 1 and 4 seconds for loading condition 2. Conversely, employing the program leads to a substantial increase in computational time, necessitating data adjustments and prolonged response times, exceeding 60 seconds for both loading conditions; In the context of second-layer pure stability loss computations, the graph theory method demonstrates expeditious calculations, requiring less than 10 seconds for each loading condition. In contrast, the program demands significantly longer computational times with 362 seconds for loading condition 1 and 421 seconds for loading condition 2. In maritime scenarios, especially

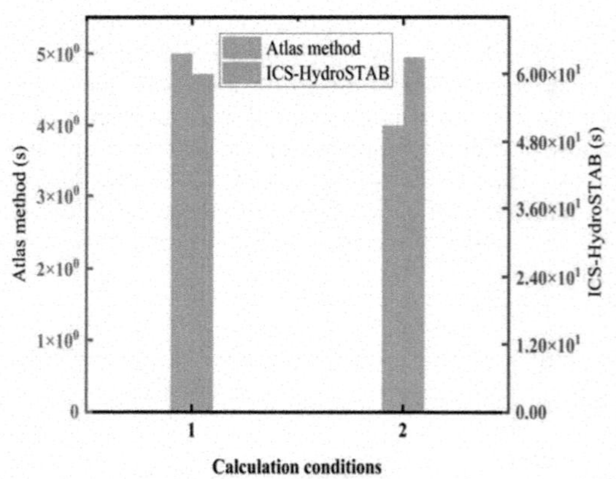

Figure 10. Computing time chart for first-layer pure stability Loss based on graph theory.

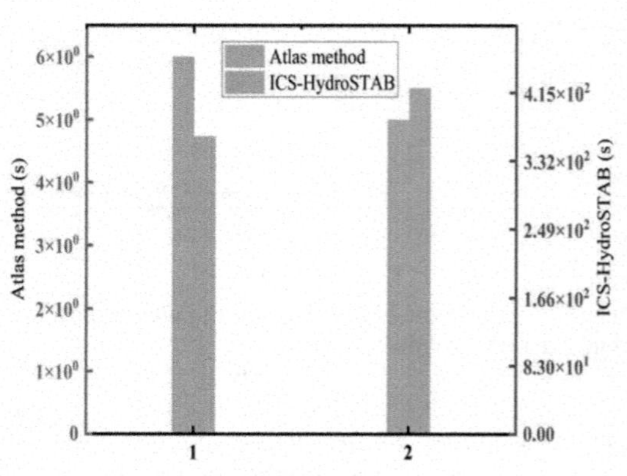

Figure 11. Computing time chart for second-layer pure stability Loss based on graph theory.

during exigent circumstances, temporal efficiency becomes a critical factor, and the elapsed time in these critical minutes significantly influences the safety outcomes for both the crew and the vessel.

7 CONCLUSION

- Predictive Efficacy of Graph Theory Approach

The proposed graph theory approach in this study proves to be an effective means of forecasting vulnerability to pure stability loss. The error rates in predicting first-tier vulnerability are consistently below 4%, while for second-layer pure stability loss, the error rates remain under 5%, meeting the IMO SDC (2017) precision requirements.

- Computational Efficiency

In terms of computational efficiency, the graph theory method evidences significant superiority over traditional software-based approaches. It is noteworthy that the time required for predicting stability loss in both the first and second layers within the graph theory method is less than 10 seconds. Conversely, program-based computations necessitate more than 60 seconds for the first layer and over 300 seconds for the second layer. This notable difference underscores the significant effectiveness of the graph theory approach, positioning it as a convincing and temporally efficient alternative in practical applications.

- Engineering Applications

This research, tailored for specific vessel types, emphasizes the potential of equipping each vessel with corresponding stability loss graphs. Such graphs empower crew members to assess the vulnerability to pure stability loss under specific loading conditions, thereby enhancing safety evaluations in practical maritime engineering applications.

- Trends and Variations

Analyzing the trends, it is evident that as the center of gravity height increases, the probability of occurring pure stability loss decreases. However, variations in draft yield distinct results for first-tier and second-tier calculations. Notably, for second-tier vulnerability, an increase in draft leads to an augmented occurrence of hazardous sea conditions, resulting in different patterns of variation.

- Unique Advantages under Various Wave Conditions

The graph theory approach has a distinct advantage in its ability to evaluate the steady heel angle in different wave conditions, which is a feature that conventional computational programs cannot assess. This exceptional capability of the graph theory approach is a crucial finding, highlighting its importance in determining maritime stability. This academic observation highlights the limitations of traditional calculation methods in addressing this specific aspect objectively, emphasizing the innovative contributions and practical significance of utilizing the graph theory methodology.

In summary, the graph theory approach not only showcases high precision in predicting pure stability loss vulnerability but also offers computational efficiency, making it a valuable tool for engineering applications. The tailored approach for specific vessels, along with insights into trends and variations, contributes to a comprehensive understanding of pure stability loss characteristics under varying loading conditions.

REFERENCES

Bassler, C. C., V. Belenky, G. Bulian, A. Francescutto, K. Spyrou, & N. Umeda (2011). Review of available methods for application to second level vulnerability criteria. *Contemporary ideas on ship stability and capsizing in waves*, 3–23.

Bulian, G. (2010). Checking vulnerability to pure loss of stability in long crested following waves: A probabilistic approach. *Ocean Engineering 37*, 1007–1026.

Hashimoto, H. (2009). Pure loss of stability of a tumblehome hull in following seas. *ISOPE International Ocean and Polar Engineering Conference*, ISOPE–I.

Htun, S. S., N. Umeda, M. Sakai, A. Matsuda, & D. Terada (2019). Water-on-deck effects on roll motions of an offshore supply vessel in regular stern quartering waves. *Ocean Engineering 188*, 106225.

Lilienthal, T., A Matsuda, & G. Thomas (2007). Dynamic stability in following seas: predictive and experimental approaches. *Journal of marine science and technology*.

Liu, L., M. Chen, X. Wang, Z. Zhang, J. Yu, & D. Feng (2000). Cfd prediction of full-scale ship parametric roll in head wave. *Ocean Engineering 2021*, 109180.

Liu, L., D. Feng, X. Wang, Z. Zhang, J. Yu, & M. Chen (2022). Study on extreme roll event with capsizing induced by pure loss of stability for the free-running onr tumblehome. *Ocean Engineering 257*, 111656.

Lu, J., M. Gu, & E. Boulougouris (2019). Model experiments and direct stability assessments on pure loss of stability of the onr tumblehome in following seas. *Ocean Engineering. 194*, 106640.

Lu, J., M. Gu, & E. Boulougouris (2020). Model experiments and direct stability assessments on pure loss of stability in stern quartering waves. *Ocean Engineering 216*, 108035.

Sadat-Hosseini, H., F. Stern, A. Olivieri, E. F. Campana, H. Hashimoto, N. Umeda, G. Bulian, & A. Francescutto (2010). Head-wave parametric rolling of a surface combatant. *Ocean Engineering 37*, 859–878.

Umeda, N., M. Osugi, M. Sakai, A. Matsuda, & D. Terada (2017). Model experiment on pure loss of stability for a ship in astern waves and its relationship with the second generation intact stability criteria. *Proceedings of the 16th International Ship Stability Workshop, Belgrade, Serbia.*, 21–26.

Wang, J., W. Zhao, & D. Wan (2016). Free maneuvering simulation of onr tumblehome using overset grid method in naoefoam-sjtu solver. *Proceedings of 31th Symposium on Naval Hydrodynamics*.

Wang, J., L. Zou, & D. Wan (2017). Cfd simulations of free running ship under course keeping control. *Ocean engineering. 141*, 450–464.

Yao-hua, Z., M. Ning, L. Jiang, & G. Xie-chong (2015). A study on applicability of cfd approach for predicting ship parametric rolling. *STAB2015.*, 464.

Advances in Maritime Technology and Engineering – Guedes Soares & Santos (Eds)
© 2024 The Author(s), ISBN 978-1-032-83107-7

CFD study on the hydrodynamic performance of a container ship under berthing speed

S. Zhang, Q. Wu & S.J. Li
State Key Laboratory of Maritime Technology and Safety, Wuhan University of Technology, Wuhan, P.R. China
School of Transportation and Logistics Engineering, Wuhan University of Technology, Wuhan, P.R. China

Y.X. Duan
State Key Laboratory of Maritime Technology and Safety, Wuhan University of Technology, Wuhan, P.R. China
School of Naval Architecture, Ocean and Energy Power Engineering, Wuhan University of Technology, Wuhan, P.R. China

J.L. Liu
State Key Laboratory of Maritime Technology and Safety, Wuhan University of Technology, Wuhan, P.R. China
Intelligent Transportation Systems Research Center, Wuhan University of Technology, Wuhan, P.R. China
National Engineering Research Center for Water Transport Safety, Wuhan, P.R. China

ABSTRACT: Ship hydrodynamic performances is the foundation of automatic berthing control. The present paper focuses on the force characteristics of ship hull under berthing speed. Computational fluid dynamics (CFD) method is adopted to perform series oblique towing tank (β=[0, 180°]) and circular motion test (r'= [-1.2, 1.2]) simulations upon an inland container ship. The hydrodynamic forces and detailed flow field characteristics are extracted and discussed. Conclusions of this work could provide reference for the hydrodynamic performance and maneuverability study on ship berthing.

1 INTRODUCTION

Automatic berthing is at the top level of ship autonomy (CCS, 2023), among which, the description of the maneuverability and hydrodynamic performances are the foundation.

In the berthing process, ships undergo larger drifting angles and greater turning angular velocity, the ship longitudinal velocity component is of the same dimension as the lateral component velocity and yaw rate (Okuda et al., 2023), or even far smaller. Thus, concerns about the hydrodynamic forces move from the friction-dominant longitudinal force to the pressure-dominant lateral force and yaw moment (Zhang et al., 2023), and the hydrodynamic forces present strong non-linearity.

The computational fluid dynamics (CFD) method can conduct captive model tests in the numerical towing tanks and has shown its potential in ship industry. Utilizing the CFD method, Benedetto (Piaggio et al., 2020a, Piaggio et al., 2020b, Piaggio et al., 2018) investigated the hydrodynamic forces of an escort tug under both slow- and high-speed conditions. Wang (Wang and Wan, 2020) discussed the stopping ability of KVLCC1. Wang (Wang et al., 2010) studied the transient flow and force characteristics of a Wigley hull with defined velocity function. Toxopeus and Zhang (Toxopeus et al., 2012, Zhang et al., 2013) studied the force tendency and flow field development of a submarine.

In the present paper, considering the actual berthing maneuver (full drifting angles and high turning rates), series oblique towing and circular motion simulations based on berthing speed are performed to investigate the hydrodynamic performance and flow filed characteristics.

2 METHODOLOGY

2.1 *Studied ship and working conditions*

The studied ship is a scaled 64TEU twin-propeller-twin-rudder inland container ship (Lu et al., 2022), with a raked shape bow and two deadwood skegs at stern section. The present paper aims to investigate the hydrodynamic performance of the bare hull, and hence, the propellers and rudders are eliminated. The principal particulars and geometry view of the ship are shown in Table 1 and Figure 1 respectively.

DOI: 10.1201/9781003508779-24

Table 1. Principal particulars of the 64TEU inland container ship.

Parameters		Full-scale	Model-scale
λ	(-)	1.0	18.000
L_{oa}	(m)	65.000	3.611
L_{wl}	(m)	64.000	3.556
B	(m)	12.550	0.697
d	(m)	2.750	0.153
A_S	(m^2)	1065.369	3.288
Δ	(m^3)	1961.088	0.336
Fr	(-)	0.0169	0.0169
U_0	(m/s)	0.225	0.101

On safety consideration, ships in berthing maneuver undergo large drifting angles and high turning rates, with rather low speed. In accordance with the actual ship berthing process, the velocity of studied ship U_0 is set as $Fr = 0.0169$. The drifting angle β ranges from 0 to 180°, and the turning rate r' covers [-1.2, 1.2]. Among which, r' is defined as $r' = L_{wl}r / U_0 = L_{wl} / R$, r the rotating angular velocity, R the turning radius. In the present paper, oblique towing tank test (OTT) and circular motion test (CMT) simulations are performed, and the CFD working conditions are collected in Table 2.

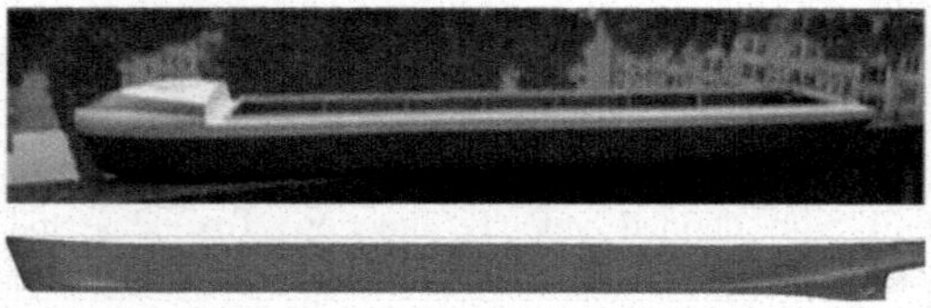

Figure 1. Geometry view of scaled 64TEU (top: model-scale ship; bottom: CAD model).

On safety consideration, ships in berthing maneuver undergo large drifting angles and high turning rates, with rather low speed. In accordance with the actual ship berthing process, the velocity of studied ship U_0 is set as $Fr = 0.0169$. The drifting angle β ranges from 0 to 180°, and the turning rate r' covers [-1.2, 1.2]. Among which, r' is defined as $r' = L_{wl} r/U_0 = L_{wl}/R$, r the rotating angular velocity, R the turning radius. In the present paper, oblique towing tank test (OTT) and circular motion test (CMT) simulations are performed, and the CFD working conditions are collected in Table 2.

Table 2. CFD working conditions.

ID	Content	r'	β (°)
1-6	Convergence analysis	0	0, 90°
7-25	OTT	0	[0, 180°]
26-139	CMT	[-1.2, 1.2]	[0, 180°]

Cases 1-6 are the validation section, mesh convergence analysis is conducted, to demonstrate the feasibility and efficiency of the CFD mesh alignment and setups. Cases 7-139 are the numerical study section, series captive model tests are simulated, to demonstrate the hydrodynamic performance and flow field characteristics of the studied ship.

2.2 *Numerical setups*

In the present paper, series low-speed simulations are performed based on CFD code Star-CCM+, the RANS solver is adopted to resolve the turbulent flow. Considering the slow speed and slight amplitude of ship in the berthing maneuver, double body (DB) method eliminating the free-surface effect is utilized. Furthermore, the moving reference frame (MRF) method (Chen et al., 2017) is used to simulate the circular motion of the studied ship.

The calculation domain and boundary conditions are shown in Figure 2. Rectangular and cylinder calculation domains are adopted for the OTT and CMT simulations, respectively. The setups follow the ITTC recommended CFD procedures (ITTC, 2014): In the rectangular domain, the inflow plane $2.5L_{wl}$ ahead of the midship point, outflow plane $3.5L_{wl}$ away from the midship point, far-field planes $2.5L_{wl}$ from the central longitudinal section, and the bottom plane extends downwards $2.0L_{wl}$ for infinite water depth. And in the cylinder domain, the radius is set as R+$2.0L_{wl}$ from the rotating center, and the water depth $1.5L_{wl}$ from the rigid surface plane.

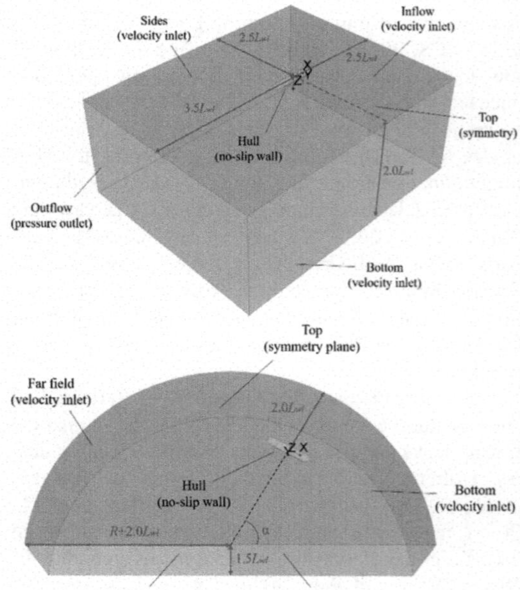

Figure 2. Calculation domain and boundary conditions (top: OTT-Rectangular domain; bottom: CMT-Cylinder domain).

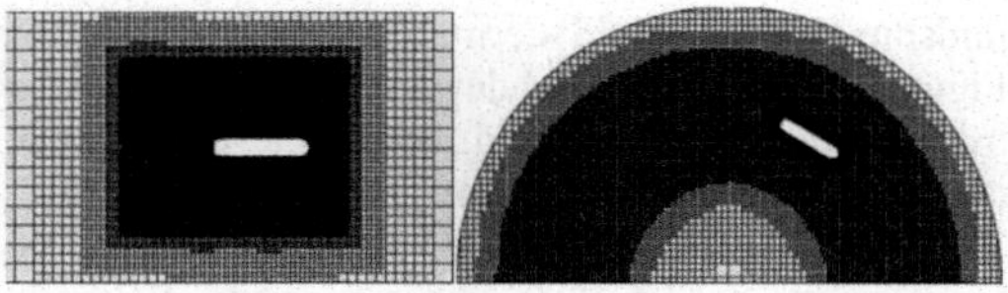

Figure 3. Overall mesh alignments (left: OTT-Rectangular domain; right: CMT-Cylinder domain).

Hexahedral mesh topology is used in the simulations, *SST* k-omega turbulent model is adopted to solve the hydrodynamic forces and capture the flow separation downstream the hull. As for the near-wall treatment, all y^+ wall treatment resolving the viscous sublayer and buffer layer is introduced (CD-adapco, 2018), and the dimensionless wall distance y^+ is set as 5. For convenience of batch calculation, large refinement boxes covering the bow, hull, skeg, the flow field development regions are used. The mesh details are shown in Figure 3.

3 RESULTS AND DISCUSSION

3.1 *Grid convergence analysis*

Grid convergence study are required to verify the feasibility and efficiency of adopted mesh strategy, before series CFD simulations. In the present paper, three sets of mesh densities upon the straight ahead ($\beta = 0°$) and cross flow ($\beta = 90°$) conditions are selected, and the growth rate of the basic grid size is defined as $r=\sqrt{2}$. The convergence conditions are concluded in Table 3. Moreover, the grid convergence index (GCI) method is adopted, to evaluate the uncertainty of the utilized numerical setups and mesh strategy, the GCI and uncertainty conditions are concluded in Table 4.

$$\begin{cases} p = \left|\ln\left|\varepsilon_{32}/\varepsilon_{21}\right|\right|/\ln r \\ \delta_{RE} = \varepsilon_{21}/(r^p - 1) \\ S_{ext}^{21} = (r^p S_1 - S_2)/(r^p - 1) \\ e_a^{21} = \left|(S_1 - S_2)/S_1\right| \\ e_{ext}^{21} = \left|(S_{ext}^{21} - S_1)/S_{ext}\right| \\ GCI = F_s\left|\delta_{RE}\right| \\ U = GCI/1.15 \end{cases} \tag{1}$$

Where p the actual accuracy, δ_{RE} the error, S_{ext}^{21} the extrapolation value based on S_1 and S_2 simulation, e_a^{21} the evaluation error, e_{ext}^{21} the extrapolation error, and F_s the safety factor, F_s=1.25.

As a result, the extracted longitudinal and lateral force indicate monotonic convergence, and the yaw moment at cross flow condition shows oscillatory convergence, where the value is small due to the small ship speed and symmetrical plan of the ship hull. In conclusion, the utilized numerical setups and mesh strategy could perform systematic simulations with satisfied accuracy and efficiency.

Table 3. Grid convergence study.

β=0	S_1	S_2	S_3	R_G	*Con.*
X (N)	-0.1251	-0.1265	-0.1328	0.22	*Mon.*
Y (N)	-	-	-	-	-
N (N·m)	-	-	-	-	-
β=90°	S_1	S_2	S_3	R_G	*Con.*
X (N)	0.2226	0.2188	0.2050	0.275	*Mon.*
Y (N)	1.3070	1.2928	1.2557	0.383	*Mon.*
N (N·m)	-0.004396	-0.003156	-0.008550	-0.230	*Osc.*

$r_G=\varepsilon_{21}/\varepsilon_{21}$, $\varepsilon_{21}=S_2$-S_1, $\varepsilon_{21}=S_3$-S_2.$0<r_G<1$: Monotonic; $r_G<0$: Oscillatory; $r_G>1$: Diverging.

Table 4. Grid convergence index and uncertainty.

β=0	p	S_{ext}^{21}	e_a^{21}(% S_1)	e_{ext}^{21}(% S_1)	GCI (% S_1)	U
X	4.34	-0.12	1.12	0.32	0.40	0.35
β=90°	p	S_{ext}^{21}	e_a^{21}(% S_1)	e_{ext}^{21}(% S_1)	GCI (% S_1)	U
X	3.72	0.22	1.71	0.64	0.81	0.71
Y	2.77	1.32	1.09	0.67	0.84	0.73
N	4.24	-0.0048	28.21	7.77	10.53	9.15

3.2 *Hydrodynamic forces and moments*

The tendencies of the hydrodynamic forces and moments due to various drifting and yaw rate conditions are shown in Figure 4. Overall, with a berthing speed, the hydrodynamic forces acting on ship hull are relatively small and the longitudinal force, lateral force, and yaw moment are in the same magnitude, the longitudinal forces are even smaller.

In the pure drifting conditions ($r' = 0$):

As for the longitudinal force *X,* as the drifting angle $\beta < 60°$, with the growth of drift angle, the force decreases, and the slope gets steeper. As the drift angle is around $\beta = 60°$, the direction of longitudinal force reverses, and points to the bow section. As $60° < \beta < 100°$, the force gradually increases, and reaches the local maximum value between 90°and 100°. As the drift angle β=110°, the longitudinal force drops to zero, which is like the stall of an airfoil. Afterwards, the longitudinal force rises with the increase of drift angle, and nearly keeps constant at the range of β = [160°, 180°].

Tendency of the lateral force Y is legible and could be indicated by 3 stages. The increment stage with $\beta < 70°$; the flat stage with $70° < \beta < 120°$, where the lateral force slightly reduces with the change of drifting angles; and the decrement stage with $\beta > 120°$, where the lateral force decreases with the increase of drift angle, and the changing rate barely differs from the increment stage.

With regard to the yaw moment *N*, 3 equal-to-zero points are found at drifting angle β = 0, 90°, 180°. In the range of β = [0°, 90°], the yaw moment gradually increases as β < 40°, within 40° < β < 60°, the moment keeps still, while the magnitude drops to zero as the drift angle increases to 90°. In the range of β = [90°, 120°], the direction of yaw moment reverses, and rapidly increases. Then gradually decreases to zero with the sustained growth of drift angle.

In the hybrid oblique-circular conditions (β, r' ≠ 0), the larger the yaw rate, the stronger the hydrodynamic forces:

With respect to the longitudinal force *X*, as with the positive yaw rate, the force experiences progressive increase- gradual decrease- reverse and increase, with the increase of drifting angle. However, as with the negative yaw rate, the force tendency indicates 5 phases. In the first phase, the longitudinal force gradually decreases, and in the second phase, direction of the force reverses and gradually climbs to the local maximum with the growth of drift angle. Still, in the third phase, the longitudinal force points towards the bow section, with a decline of force amplitude. And sequentially, with the continuous growth of drifting angle, the longitudinal force increases and decreases gradually.

Trends of the lateral force *Y* are similar to that in the pure drifting conditions. And it is worth noticing, that the positive turning rate brings stronger lateral force than the negative turning rate conditions.

Regarding the yaw moment *N*, the tendency line of negative turning rates resembles that of the pure drifting conditions. For the positive turning rates. As r' = 0.4, the amplitude of yaw moment is small, and fluctuates around the zero line with the range of β = [0, 80°]. With the drift angle increasing, the yaw moment approaches its local maximum value at β = 140°. And as β > 140°, the yaw moment gradually decreases. As r' = 0.8 and 1.2, the moment gradually increases with 0 < β < 100°, and as 100° < β < 140°, the yaw moment rapidly reaches up to the local maximum at β = 140°, and then drops with the sustained growth of drifting angle.

The hydrodynamic force tendencies could be explained by the hull geometry plan. In the straight-ahead conditions (β≤30°), the raked bow smooths and divides the incoming flow, which brings relatively small resistance. With the increase of drifting angle, the flow hits the ship board, which enhances the pressure difference between the windward side and leeward side, and such phenomenon touches the top at the cross-flow condition (β=90°). With the continuous increase of drifting angle, the ship is in the backward condition (90°<β≤180°), the stern section with complex geometry meets the flow, and as a result, leads to a relatively large magnitudes of hydrodynamic forces. Such conclusions could also be verified by the streamline distribution around ship hull.

3.3 *Velocity and streamline development*

Figure 5 shows the lateral velocity distribution on the transverse section of the bow (x/L = 0.415), midship (x/L = 0), and stern (x/L = -0.415) parts. In Figure 5(a), the flow field development characterized by velocity magnitude and streamline of the pure drifting conditions (β = 0°, 10°, 50°, 60°, 90°, 120°, 180°) are illustrated. It is shown that at the cross-flow condition (β = 90°), obvious reversed flow at the leeward section of the bow and deadwood skeg

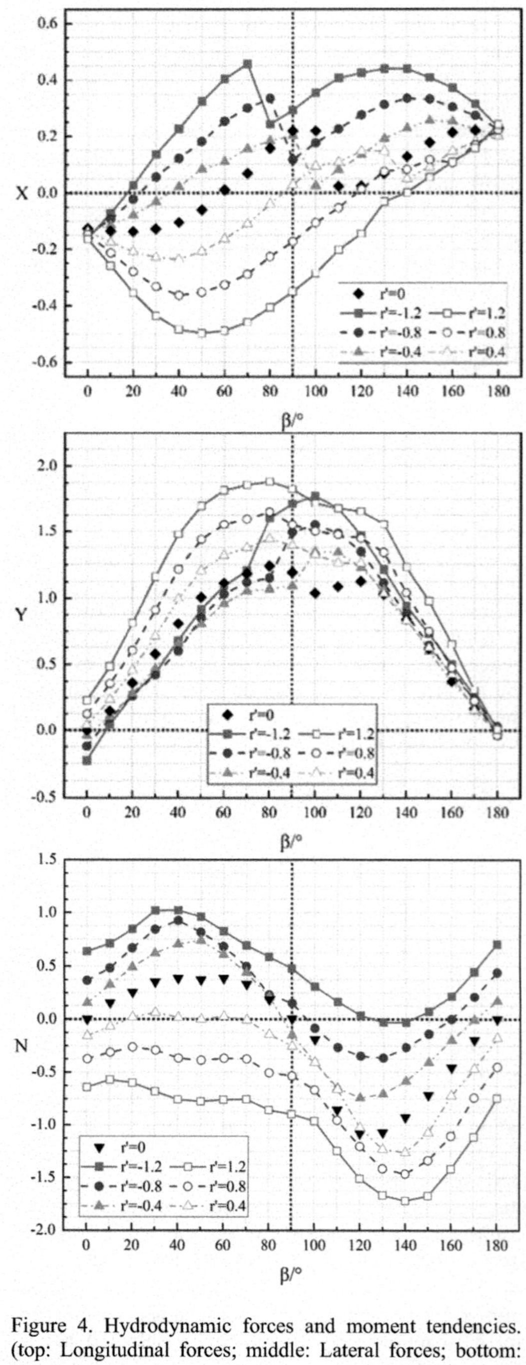

Figure 4. Hydrodynamic forces and moment tendencies. (top: Longitudinal forces; middle: Lateral forces; bottom: Yaw moment).

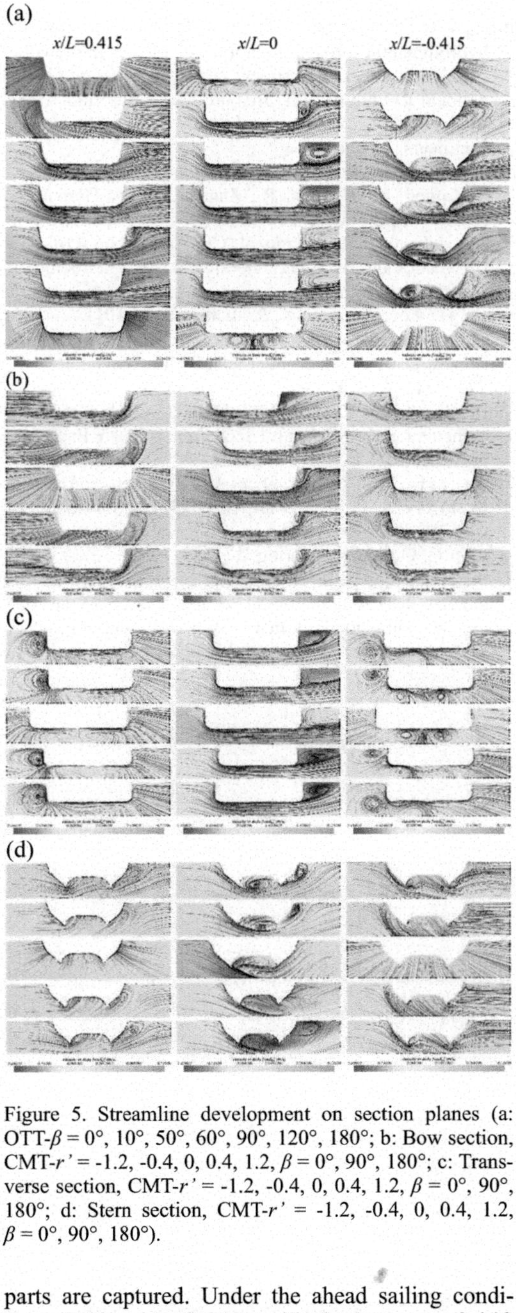

Figure 5. Streamline development on section planes (a: OTT-$\beta = 0°, 10°, 50°, 60°, 90°, 120°, 180°$; b: Bow section, CMT-$r' = -1.2, -0.4, 0, 0.4, 1.2, \beta = 0°, 90°, 180°$; c: Transverse section, CMT-$r' = -1.2, -0.4, 0, 0.4, 1.2, \beta = 0°, 90°, 180°$; d: Stern section, CMT-$r' = -1.2, -0.4, 0, 0.4, 1.2, \beta = 0°, 90°, 180°$).

parts are captured. Under the ahead sailing conditions (drift angle β<90°), with the increase of drift angle, the magnitude of lateral velocity gets higher, and stronger vorticities are found at the leeward side of the midship section and deadwood skeg. As for the astern sailing conditions (drift angle $\beta > 90°$), with the growth of drift angle, the velocity magnitude reduces, while stronger vorticities at the deadwood skeg sections are observed. Furthermore, the vortex at the windward deadwood skeg is stronger than that at the leeward deadwood skeg. In general, due to the hull profile difference, the flow field is much more complicated under the cross flow and astern sailing conditions, which also explains the simulation convergence issue.

Compared with the pure drifting tests, the flow field around the ship hull variates in the rotating arm tests. Figures 5(a)-(c) demonstrate the flow field characteristics at the bow, midship, and stern parts respectively, under coupled drift-circular conditions ($r' = -1.2, -0.4, 0, 0.4, 1.2, \beta = 0°, 90°, 180°$). Due to the circular motion, the bow section streamline develops away from the rotation center, while the streamlines at the midship and stern sections point towards the rotation center. The lateral velocity magnitude and streamline are found to become higher and more complicated with the increase of turning rate. Alike the pure drift conditions, the ridges of the deadwood skeg slows down the flow velocity, and form vortices at the leeward section, and for the twin deadwood skeg configurations, the windward skeg makes apparent velocity loss. Overall, the positive turning rate enhances the vortex, and the higher the turning rate, the stronger the vortex.

4 CONCLUSION

With reference to the actual berthing maneuver, the hydrodynamic performance and flow filed characteristics of a scaled twin-propeller-twin-rudder inland container ship are studied. Series oblique towing and circular motion simulations with full drifting and high turning rates are performed.

Generally, the higher the turning rate, the greater the hydrodynamic forces, and the stronger the vorticity. As for the longitudinal force, an obvious force drop like the stall of airfoil are observed under the $r' \leq 0$ conditions. For the lateral force, the local maximum value occurs at the cross-flow conditions. Regarding the yaw moment, as the turning rate $r' < 0$, the local maximum values were found around drift angle $\beta = 40°, 50°$, as the turning rate $r' \geq 0$, the maximum values emerge at drift angle $\beta = 130°, 140°$. Furthermore, due to the hull type and configuration of the deadwood skegs, the flow field vorticity is stronger at the astern conditions, followed by the cross-flow conditions, and the ahead conditions.

For better understand the performance of a ship in berthing maneuver, the hydrodynamic performance and flow field development of the rudder and propeller under four-quadrant conditions are put on agenda.

ACKNOWLEDGEMENTS

This research was funded by the National Key R&D Program of China, grant number 2022YFB4301402, the National Natural Science Foundation of China, grant number 52272425.

REFERENCES

CCS (2023) Rules for Intelligent Ships, Beijing, China Classification Society.

CD-ADAPCO (2018) Star-CCM+ 13.06 user guide.

Chen, J., Zou, Z., Chen, X., Xia, L. & Zou, L. (2017) CFD-based simulation of the flow around a ship in turning motion at low speed. Journal of Marine Science Technology, 22, 784–796.

ITTC (2014) Recommended Procedures and Guidelines: Practical guidelines for ship CFD applications. ITTC Report, 7.5-03-02-03.

Lu, S., Cheng, X., Liu, J., Li, S. & Yasukawa, H. (2022) Maneuvering modeling of a twin-propeller twin-rudder inland container vessel based on integrated CFD and empirical methods. Applied Ocean Research, 126, 103261.

Okuda, R., Yasukawa, H., Yamashita, T. & Matsuda, A. (2023) Maneuvering simulations at large drift angles of a ship with a flapped rudder. Applied Ocean Research, 135, 103567.

Piaggio, B., Villa, D. & Viviani, M. (2020a) Numerical analysis of escort tug manoeuvrability characteristics. Applied Ocean Research, 97, 102075.

Piaggio, B., Villa, D., Viviani, M. & Figari, M. (2020b) Numerical analysis of escort tug manoeuvrability characteristics–Part II: The skeg effect. Applied Ocean Research, 100, 102199.

Piaggio, B., Viviani, M. & Martelli, M. (2018) Escort tug hydrodynamic forces estimation in a design framework: from model test to manoeuvrability simulation. International Conference on Offshore Mechanics and Arctic Engineering. American Society of Mechanical Engineers.

Toxopeus, S., Atsavapranee, P., Wolf, E., Daum, S., Pattenden, R., Widjaja, R., Zhang, J. T. & Gerber, A. (2012) Collaborative CFD exercise for a submarine in a steady turn. International Conference on Offshore Mechanics and Arctic Engineering. American Society of Mechanical Engineers.

Wang, H., Zou, Z., Tian, X. & Huang, Y. (2010) Computation of the viscous hydrodynamic forces on a berthing ship in shallow water. Journal of Ship Mechanics, 14, 723–731.

Wang, J. & Wan, D. (2020) CFD study of ship stopping maneuver by overset grid technique. Ocean Engineering, 197, 106895.

Zhang, J., Maxwell, J. A., Gerber, A. G., Holloway, A. G. L. & Watt, G. D. (2013) Simulation of the flow over axisymmetric submarine hulls in steady turning. Ocean engineering, 57, 180–196.

Zhang, S., Wu, Q., Liu, J., He, Y. & Li, S. (2023) State-of-the-Art Review and Future Perspectives on Maneuvering Modeling for Automatic Ship Berthing. Journal of Marine Science Engineering, 11, 1824.

Hydrodynamics - seakeeping

Advances in Maritime Technology and Engineering – Guedes Soares & Santos (Eds)

Study on the behaviour of a container ship in the Mediterranean Sea area

A.-M. Chirosca & L. Rusu
"Dunarea de Jos" University of Galati, Galati, Romania

ABSTRACT: Climate change is affecting all regions of the world. They will continue to occur over many decades and represent an increasing risk factor for the shipping industry in Europe. As a reflection of climate change, Europe's regions are facing rising sea levels and more extreme weather conditions. The Mediterranean Sea provides the shortest shipping route between Europe and Asia, but it is also known to be an important transport corridor for several European countries. Unfortunately, the effects of climate change are increasingly visible in this area. The aim of this paper was to study how a container ship behaves according to the wind and wave climate identified in the Mediterranean Sea. A series of simulations were carried out to identify how mean wave resistance varies for a range of sea states according to the wind speeds identified in this area.

1 INTRODUCTION

As the economy grows, so does the demand for supplies and services. Shipping is the most popular means of transport because of its many opportunities. One of the research directions of our century is the improvement and adaptation of ships to the increasingly severe environmental conditions of recent times due to climate change.

The Mediterranean Sea connects Europe, Asia, and Africa (EMSA, 2021; Chirosca & Rusu, 2020) and represents an area of interest for maritime traffic, as can be seen from the density of maritime traffic shown in Figure 1, in terms of the transport of goods, especially oil. Maritime traffic is also intensified by passenger ships, due to the high tourism potential of this area. With the opening of the Suez Canal, the Mediterranean Sea has served as a transnational trade route that is crossed by all types of ships, ranging from cargo ships, tankers, passenger ships, ferries, fishing vessels, yachts, and even military vessels (Chirosca & Rusu, 2020; SAFEMED, 2008).

The Mediterranean Sea can be described as an oceanic system due to its complex and variable circulation. Unfortunately, the weather conditions are affecting shipping and its smooth running. A study of wind and wave parameters is therefore necessary to ensure successful and safe transport.

Investigating the effects of climate change on wind and wave conditions is a current trend in terms of simulations for the present or the near future (Bricheno & Wolf, 2018; Vettore & Guedes Soares, 2017). This paper presents the results of a study carried out on a typical container ship in irregular waves, for a range of sea states, correlated with a wind and wave climate study for the near future, particularly to navigation along the most important shipping route in the Mediterranean Sea.

Figure 1. Mediterranean Sea (Marine traffic, 2023).

2 CASE STUDY

One of the most common types of transport ships along the Mediterranean Sea are container ships, and that is why this type of ship was considered for this study. The hull under investigation is a known benchmark ship, the "Duisburg Test Case", known as DTC, and represents the hull of a container ship developed at ISMT (El. Moctar et al., 2012) and designed by the University of Duisburg-Essen. The ship for the numerical simulations was considered without appendages.The main characteristics of the hull (El. Moctar et al., 2012) are presented in Table 1 and the shape can be seen in Figure 2.

This study aims to investigate the behaviour of the container ship in irregular waves for a range of speeds from 6 knots to 14 knots specific to the Mediterranean

DOI: 10.1201/9781003508779-25

Sea area, and several sea states due to the complexity of the basin which are presented in Table 2. The sea states studied start from sea state 2, known as a light breeze when the wind speed is about 2.06 m/s to 3.09 m/s and the wave height is maximum of 0.30 m, and the worst sea state studied, sea state 8, corresponds to the strong wind when the wind speed is about 17.49 m/s to 20.58 m/s and the maximum wave height is about 7.62 m.

Table 1. Main particulars of DTC hull.

Main characteristics	Full Scale
Length between perpendiculars L_{BP} [m]	355
Waterline breadth B_{WL}[m]	51
Draught midship T_m[m]	14.5
Trim angle θ [°]	0
Volume displacement ∇ [m^3]	173467
Block coefficient c_B [-]	0.661
Wetted surface S_W[m^2]	22032

Figure 2. Duisburg Test Case hull.

Table 2. Test cases.

Speed, v [knots]	Sea state	Hs [m]	Tz [s]
	2	0.300	5.300
6	3	0.880	6.200
8	4	1.880	6.200
10	5	3.250	6.900
12	6	5.000	8.800
14	7	7.500	10.700
	8	11.500	11.600

CFD simulations, like any other method of investigating fluid flow around a body, have several advantages and disadvantages over experimental methods, and the main advantage of numerical simulations is the control over the configuration of the parameters, which makes it possible to modify them and perform several types of simulations in a short time.Numerical simulations were performed using SHIPFLOW MOTIONS, a fully nonlinear and non-stationary potential flow solver for free surface flows. The numerical simulations presented on this paper were performed based on the potential flow theory and the flow is assumed to be nonviscous, incompressible, and irrotational. The domain size for the numerical simulatios is established according to the length of the vessel. The computational domain extends 1.5 lengths forward of the hull, 3 lengths aft of the hull, 1.5 lengths to the side, 1 length above the hull and 1.5 lengths below the model keel.

The Volume of Fluid (VOF) method was applied for free surface capture and the k-ω Shear-Stress Transport (SST) turbulence model was adopted to calculate vortex viscosity which is a popular model used for near-wall treatment performance. Motions of waves are generated by means of superposition of wave components defined by ITTC-spectrum. The irregular waves were defined by introducing the wave height (Hs) and zero-crossing period in head sea conditions (Tz). The stokes 5th order wave theory was applied.

3 ANALYSIS OF THE WIND AND WAVE CLIMATE

The climate analysis consisted of studying the wave and wind conditions that directly influence the sea state and the way ships operate. In order to assess the effect of various climate change scenarios on the near future wave and wind climate in the Mediterranean Sea, wind fields from EURO-CORDEX database are analyzed, namely CNRM-ALADIN63 (Nabat et al., 2020), over the period 2041-2070. The wind fields extracted from the database were those simulated under different emissions trajectories known as Representative Concentration Pathways (RCPs), namely RCP4.5 and RCP8.5 (Moss et al., 2010). They have a spatial resolution of 0.11° and 6 hours of temporal resolution.

RCP 4.5 is considered an intermediate scenario, which considers the depletion of non-renewable fuels and for which emissions peak around 2040 and then decline. On the other hand, in the RCP 8.5 scenario, emissions continue to rise throughout the 21st century, and this scenario is used to predict mid-century emissions based on current and stated policies.

These wind fields were used to drive the SWAN (Simulating WAves Nearshore, Booij et al., 1999) model to simulate the sea state conditions in the Mediterranean Sea (see also Rusu, 2024). In wave modelling, using high-resolution wind fields is a very important issue (Rusu et al., 2008), especially in the case of semi-enclosed sees as the Mediterranean Sea. The significant wave heights (Hw) simulated by the SWAN model (3 hours temporal resolution) and interpolated along the main maritime route, together with the wind speed (V) at 10 m above sea level, are used in the next analysis.

From the data gathered a statistical analysis was performed to determine the 50th, 75th, 90th and 95th percentiles. Thus, as far as the Mediterranean Sea is concerned, the route with the highest traffic was chosen based on the data extracted and presented in Figure 1.

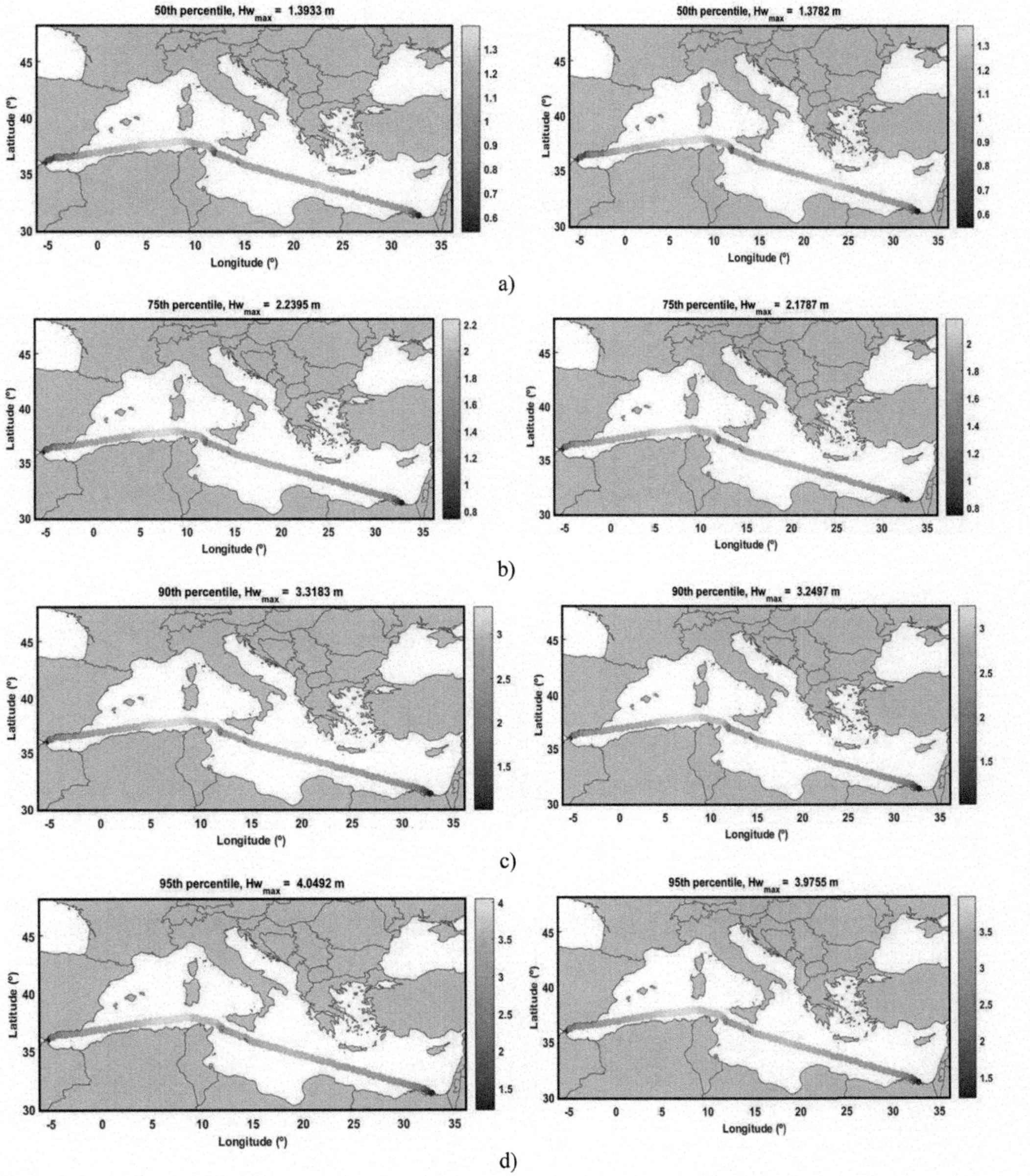

Figure 3. Comparative statistical investigation on wave height between RCP4.5 (left) and RCP8.5 (right): (a) 50th percentile; (b) 75th percentile; (c) 90th percentile; (d) 95th percentile.

As can be seen in Figure 3, in terms of the maximum values identified for the studied route and reference period, the values range from sea state 2 to sea state 8, with the same range applying to the wind speed, as can be found in Figure 4.

It can be observed that between the simulated data considering the two RCP scenarios, both regarding the wind speed and the significant wave height, there are no great differences. It is observed that for both parameters considered, the highest values along the maritime route are found in the western part of the basin as well as in the Ionian Sea.

4 NUMERICAL SIMULATION RESULTS

Prediction of ship motion in viscous flow under real conditions is of great importance for seakeeping and ship design (Nimma et al., 2018). Often ships operate in irregular waves, with rare cases where regular

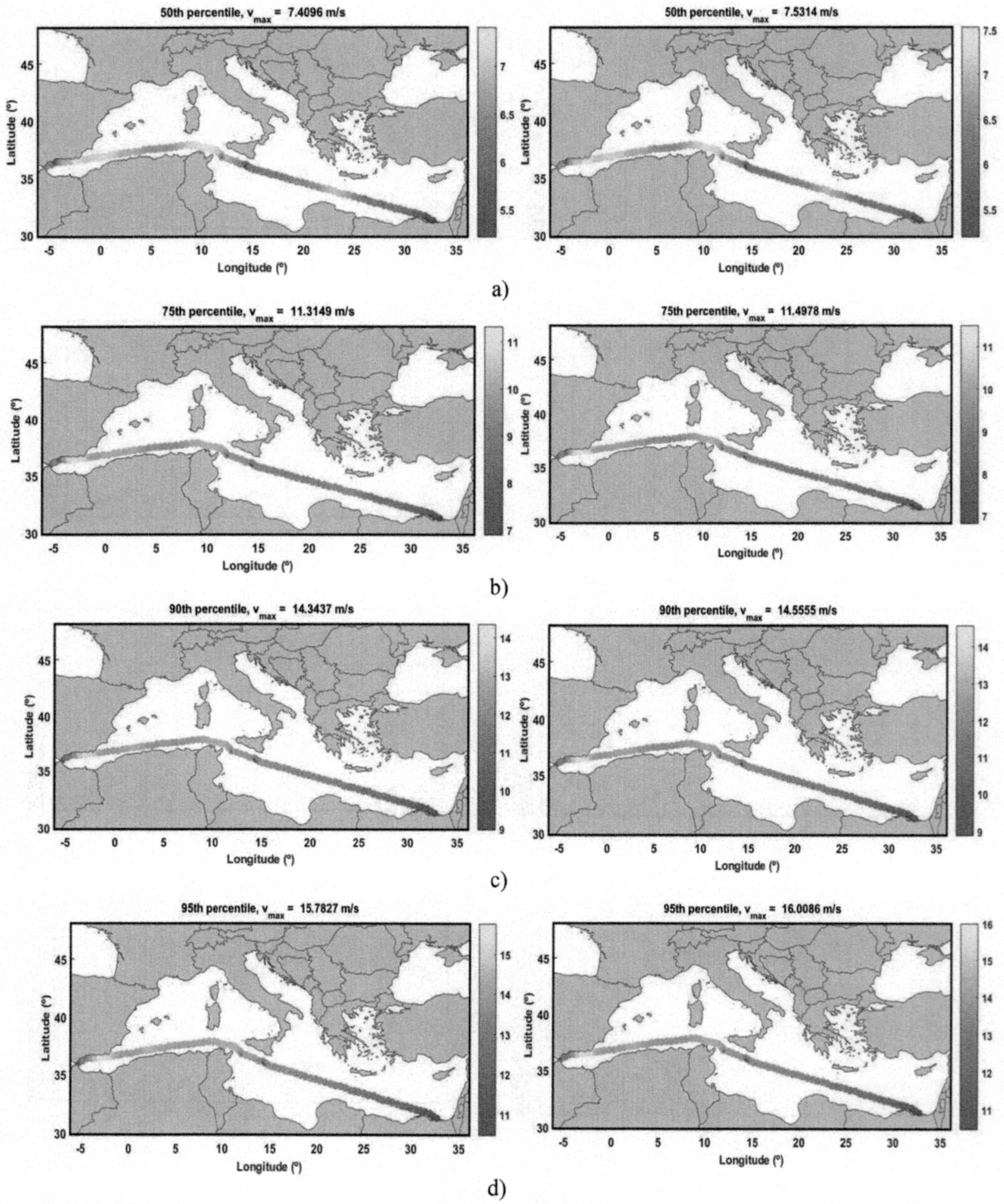

Figure 4. Comparative statistical investigation on wind speed between RCP4.5 (left) and RCP8.5 (right): (a) 50th percentile; (b) 75th percentile; (c) 90th percentile; (d) 95th percentile.

waves are encountered, and for this reason, wave resistance must be determined in order to predict the total resistance developed by the ship (Havelock, 1937; Maruo, 1957; ITCC, 2017)). For this paper, the DTC ship was studied using SHIPFLOW Motions 7.

The simulations were run on a 10-core desktop computer. A total of 40 simulations were performed and simulations ranged from 4 to 20 hours per case studied.

As the verification of the discretization grid is an important step for numerical simulations as it can affect the final results, three grids were analyzed: coarse (corresponding to a panel density factor of 0.8), medium (corresponding to a panel density factor of 0.9) and fine (corresponding to a panel density factor of 1.0) were used for numerical simulations on calm water.

Table 3 illustrates how many panels and knots were distributed on the hull and on the free surface of the grids used for the grid convergence test. These simulations were performed on the same computer used for the simulations for different sea speeds, adding a number of 3 runs to the final number of cases studied.

The grid testing was performed using Richardson extrapolation. As the numerical uncertainty obtained was within the limits of the assumptions of the method, we can conclude from the results that the solutions obtained from the SHIPFLOW runs are independent of the grid.

Table 3. Grids considered.

	Coarse	Medium	Fine
Density factor	0.8	0.9	1.0
No. of panels	25442	32446	40882
No. of nodes	26169	32266	41800

The grid with a density factor of 1.0 was selected for further simulations to better capture the flow around the hull, presented in Figure 5.

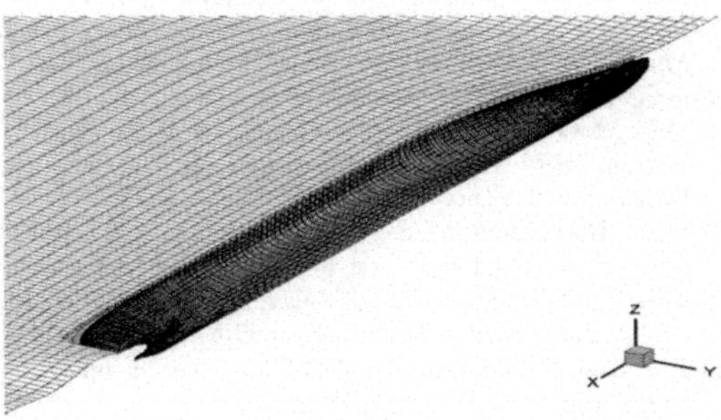

Figure 5. Grid.

Further, the results of the irregular wave simulations over the full range of speeds and sea states can be seen in Figure 6 and Table 4.

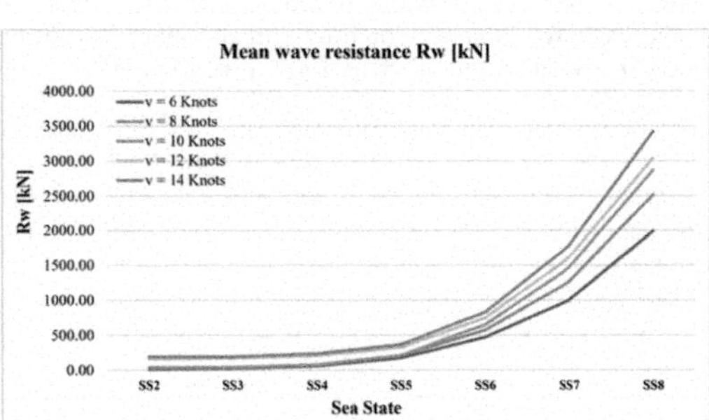

Figure 6. Mean wave resistance.

As expected, the mean wave resistance increases with speed, but also with sea state, and Figure 7 shows the ship behaviour for the lowest sea state and the highest sea state considered, at 14 knots speed. The mean wave resistance was generated as follows (Equation 1), where R_{W0} is the wave resistance in calm water and R_{AW} is the added wave resistance.

$$R_W = R_{W0} + R_{AW} \quad (1)$$

Table 4. Mean wave resistance, Rw [N].

	Speed, v [kn]				
Sea state	6	8	10	12	14
2	13	25	41	162	193
3	25	34	44	173	193
4	62	74	82	209	237
5	177	209	216	329	367
6	470	570	647	744	828
7	1001	1265	1467	1611	1782
8	1989	2516	2874	3045	3418

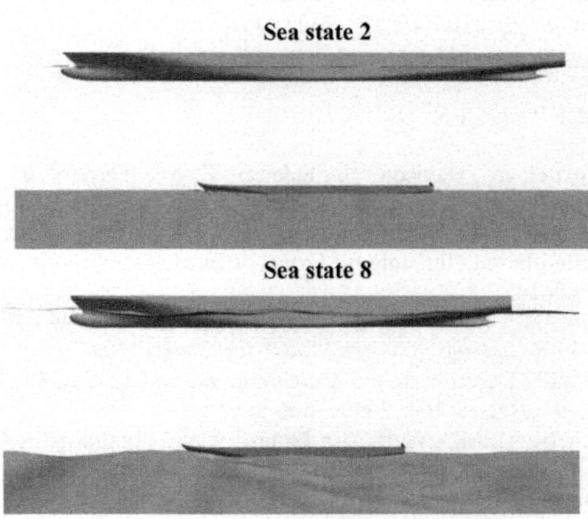

Figure 7. Wave height elevations.

5 CONCLUSIONS

From ancient times to the present day, the Mediterranean Sea has seen intense maritime traffic, both for cargo and passenger transport, but the smooth running of maritime transport is threatened by climate change. Lately, annual wind and sea averages have increased, and summers tend to be drier and hotter, and for this reason, an important step in counteracting the negative effects generated is to study climate conditions and how they will evolve in the future, which was the objective of this study.

This paper presents the behavior of a container ship for a range of speeds common for the Mediterranean Sea basin, correlated with the wind and wave climate conditions of the near future.

For the main shipping route in the Mediterranean studied in this paper, the 50th percentile showing the median values, in terms of wind conditions has the highest medium values of 7.41 m/s for RCP4.5 and 7.53 m/s for RCP8.5, and in terms of wave conditions has the highest medium values of 1.40 m for RCP4.5 and 1.38 m, corresponding to Beaufort Sea state number 4.

Regarding the 95th percentile, for the wind speed it records for RCP4.5 a value of 15.8 m/s and for RCP8.5 16.0 m/s, and for the wave height for RCP4.5 a value of 4.05 m and for RCP8.5 3.98 m, corresponding to Beaufort Sea state number 7 or 8.

ACKNOWLEDGEMENT

This work was carried out in the framework of the research project CLIMEWAR (CLimate change IMpact Evaluation on future WAve conditions at the Regional scale for the Black and Mediterranean seas marine system), supported by the Romanian Executive Agency for Higher Education, Research, Development, and Innovation Funding – UEFISCDI, grant number PN-III-P4-PCE-2021-0015.

REFERENCES

Barbariol, F., Davison, S., Falcieri, F. M., Ferretti, R., Ricchi, A., Sclavo, M. & Benetazzo, A. 2021. Wind waves in the Mediterranean Sea: An ERA5 reanalysis wind-based climatology. *Frontiers in Marine Science* 8: 760614.

Booij, N., Ris, R.C. and Holthuijsen, L.H. 1999. A Third-Generation Wave Model for Coastal Regions: 1. Model Description and Validation. *Journal of Geophysical Research* 104: 7649–7666

Bricheno, L.M., Wolf, J., Future wave conditions of Europe, in response to high-end climate change scenarios, Journal of Geophysical Research: Oceans, Vol. 123/issue 12, pp 8762–8791, 2018

Chirosca, A.M. & Rusu, L. 2020. Sea state characteristics and the maritime traffic in the European seas. In Proc. of the 20th International Multidisciplinary Scientific GeoConference SGEM2020, Albena, 18-24 August, Vol. 20, 1314–2704, https://doi.org/10.5593/sgem2020/3.1/s15.111

El. Moctar, O., Shigunov, V. & Zorn, T. 2012). Duisburg Test Case: Post-panamax container ship for benchmarking. *Ship Technology Research Schiffstechnik* 59(3): 50–64.

Havelock, T.H. The Resistance of a Ship among Waves. Proc. Roy. Soc. London A, 1937, 161, 299–308.

ITTC - International Towing Tank Conference, 2017. Uncertainty Analysis in CFD Verification and Validation Methodology and Procedures 7.5-03-01-01. 2017, pp. 1–13. Available online: https://www.ittc.info/media/8153/75-03-01-01.pdf (accessed on 10 November 2023).

EMSA - European Maritime Safety Agency, European Maritime Transport Environmental Report 2021, ISBN 978-92-9480-371-9, ISSN 1977–8449, doi:10.2800/3525

Marine Traffic: Global Ship Tracking Intelligence, Live Ships Map, https://www.marinetraffic.com, accessed in 7.12.2023.

Maruo, H. The excess resistance of a ship in rough seas. International Shipbuilding Progress 1957, 4(35), 337–345.

Moss, R.H., Edmonds, J.A., Hibbard, K.A., Manning, M. R., Rose, S.K., van Vuuren, D.P., Carter, T.R., Emori, S., Kainuma, M., Kram, T.; et al. 2010. The next Generation of Scenarios for Climate Change Research and Assessment. *Nature* 463: 747–756, doi:10.1038/nature08823.

Nabat, P., Somot, S., Cassou, C., Mallet, M., Michou, M., Bouniol, D., Decharme, B., Drugé, T., Roehrig, R. & Saint-Martin, D. 2020. Modulation of Radiative Aerosols Effects by Atmospheric Circulation over the Euro-Mediterranean Region. *Atmospheric Chemistry and Physics* 20: 8315–8349, doi:10.5194/acp-20-8315-2020.

Nimma, R.B., Arundeepan, V., Shashikala, A.P. (2018). Ship Motion in Viscous Flow under Irregular Waves. International Journal of Scientific & Engineering Research, vol. 9, no. 4, 8–13, ISSN 2229–5518

Rusu, L., Bernardino, M. & Guedes Soares, C. 2008. Influence of the wind fields on the accuracy of numerical wave modelling in offshore locations. In *Proc. of the 27th International Conference on Offshore Mechanics and Arctic Engineering* (OMAE2008), Vol. 4, pp. 637–644, Estoril, 15–20 June, ASME.

Rusu, L. 2024. An analysis of the environmental matrix in the Adriatic Sea – past and future projections. *J. Sustain. Dev. Energy Water Environ. Syst.* 12(2): 1110480. https://doi.org/10.13044/j.sdewes.d11.0480

SAFEMED- Euromed Cooperation on Maritime Safety and Prevention of Pollution from Ships, 2008. Study of Maritime Traffic Flows in the Mediterranean Sea – Final Report, July 2008. https://mmla.org.mt/wp-content/uploads/2015/07/Study-of-Maritime-Traffic-Flows-in-the-Mediterranean-Sea.pdf

Vettor, R. & Guedes Soares, C. 2017. Characterization of the expected weather conditions in the main European coastal traffic routes. *Ocean. Eng.*, 140: 244–257.

Advances in Maritime Technology and Engineering – Guedes Soares & Santos (Eds)

Uncertainty assessment for linear transfer functions from different numerical methods

M.I.P. Rodrigues, S. Wang & C. Guedes Soares
Centre for Marine Technology and Ocean Engineering (CENTEC), Instituto Superior Técnico, Universidade de Lisboa, Lisbon, Portugal

ABSTRACT: The present research focuses on assessing the uncertainties associated with seakeeping codes used to predict rigid-body responses of two different ship types, a liquified natural gas carrier and a chemical tanker. The accuracy of the calculated linear transfer functions, derived from numerical methods based on potential flow theory, is rigorously verified through comparisons with corresponding experimental benchmark studies. For the seakeeping analysis, two groups of numerical tools are employed, namely the Strip Theory Method and 3D Panel Methods. Short-term statistical models such as Frequency Independent Model Error, Coefficient of Determination, and Total Difference Approaches are used to quantify and assess the uncertainties in predictions. These measures are compared across various code-based responses for each ship and across the different ship types when experiencing the same response. The research concludes with an evaluation and discussion of the precision of all numerical solutions.

1 INTRODUCTION

Seakeeping analysis is crucial in ship design as it assesses ship performance in various sea conditions. Through this analysis, the impact of environmental factors like wind, waves, and currents on a ship's stability and navigation (Chen et al., 2013) can be understood. Seakeeping tools with various complexities have been developed according to mathematical models, but it is challenging to claim one methodology superior to the others (Hirdaris et al., 2014). The results from different seakeeping models, or the same model with different users, may not always be the same. Therefore, uncertainties do exist in the predictions of the transfer functions obtained by various numerical codes (Abdelwahab et al., 2023).

Numerous comparative studies have assessed uncertainties in different sea conditions, using 2D linear theory and complex 3D methods that account for nonlinear effects. Across all the studies surveyed, a consistent topic emerges, the importance of comparing computations in benchmark studies to validate their accuracy and evaluate the uncertainty of the assessments.

Several studies focusing on container ships have offered valuable insights. For example, Kim & Kim (2016) evaluated seakeeping codes, concentrating on wave-induced motions and loads across various wave conditions and ship speeds. The study assessed the Green's Function Boundary Element (BE) Method, the Rankine Source Boundary Element (BE) Method, and the Strip Theory Method. The findings revealed reasonable accuracy for linear responses, yet notable disparities at higher frequencies, particularly with the Strip Theory Method. Parunov et al. (2022a) raised concerns about the accuracy of linear codes, primarily due to their inability to account for nonlinear effects. However, they found the Strip Theory to be reliable across different speeds and headings, contrary to Kim & Kim (2016) conclusions. Moreover, both studies highlighted higher uncertainties in predicting wave-induced loads compared to motion predictions. Abdelwahab et al. (2022), highlighted that seakeeping codes exhibit varying degrees of accuracy, influenced by mathematical assumptions, incomplete model information and user expertise. Although certain codes may demonstrate strong agreement with experimental data at specific wave headings, none of the codes can precisely predict all motions and loads across all headings. Notably, higher uncertainties in the predictions of vertical bending moment (VBM), were observed, which aligns with prior studies.

Pacuraru et al. (2020) introduced Computational Fluid Dynamics (CFD) simulations, employing Reynolds-Averaged Navier-Stokes Equations (RANS) to address complex nonlinearities, such as those resulting from fluid viscosity, pressure, boundary conditions, and hull geometry. Their study unveiled noticeable differences when comparing the use of BE Methods to RANS Equations, especially with varying speeds and wave heights. Furthermore, Strip Theory was deemed as accurate, establishing its suitability for seakeeping analysis. In summary, theory-based quantified uncertainty analysis for wave-

DOI: 10.1201/9781003508779-26

induced motions and loads offers a more insightful accuracy assessment than code-based analysis. Expanding the research scope to encompass various hull forms is essential for a more comprehensive evaluation and the identification of uncertainty factors (Abdelwahab et al., 2023).

The research of Ley & el Moctar (2021) served as foundation, offering extensive insights across various vessel types. Their findings exposed trends into the accuracy of seakeeping prediction. Notably, the Strip Theory tended to overestimate motions, while the Rankine Source leaned toward underestimation. Both methods exhibited strong agreement with VBM outcomes, differing from prior research. Hydrodynamic nonlinearities such as green water pressures and slamming impacts challenged the accuracy of potential-based methods. Therefore, a promising but computationally intensive solution, the Field Method, was explored. However, it was found to be time-consuming and resource-intensive.

In an extension of comparative research, Lee et al. (2021) assessed both motions and added resistances. Their results unveiled significant disparities between the Strip Method and 3D Panel Methods, especially for blunt ships at high speeds. Discrepancies in the Strip Method can be attributed to limitations in accounting for high-speed effects and complex 3D hull forms. On the other hand, the accuracy of 3D Panel Methods depends on the steady flow linearization method and its ability to address speed-induced effects. As anticipated, CFD simulations showed superior accuracy, but they come with higher computational requirements, consistent with the findings of Ley & el Moctar (2021).

The state of art demonstrates the reliability of Strip Theory within acceptable limits. However, it acknowledges its limitations in addressing nonlinearities. Advanced methods like 3D Panel Methods or CFD simulations should be considered for analysis. The present study focuses on Strip Theory and 3D Panel Methods, excluding CFD simulations due to their significantly higher computational demands, making them less practical for this research.

The research employs short-term methodologies like Frequency Independent Model Error (FIME), the Coefficient of Determination (CoD), and Total Difference Approaches, to assess model uncertainties in computational simulations. FIME, introduced by Guedes Soares (1991), is applicable to all wave-induced responses and is independent of frequency. It helps determine whether computational results tend to overestimate or underestimate the experimental measurements. The CoD assesses how effectively discrete data points approximate to the regression line (Parunov et al., 2022a).

Additionally, the Total Difference quantifies the degree to which an individual numerical code deviates from the average of all codes employed (Kim & Kim, 2016). Abdelwahab et al. (2023) introduced the Modified Total Difference, a new statistical measure, which enhances the original Total Difference approach by focusing on evaluating the accuracy of computational results concerning experimental data. Seakeeping codes have varying uncertainties, even when sharing a common theoretical foundation. Therefore, several authors, including Parunov et al. (2022a) and Abdelwahab et al. (2023), recommend conducting benchmark studies with various ship types to establish a comprehensive foundation for quantifying uncertainties.

2 METHODOLOGY

2.1 *Strip theory method*

The Strip Theory Method, a well-established approach in seakeeping analysis, has been introduced by Salvesen et al. (1970). This method divides the hull into cross-sectional strips, also known as Lewis Sections (Lewis, 1929), along the longitudinal direction, as shown in Figure 1 sourced from Andersson (2018). The linearised equation of motion for the six degrees of freedom (6 DOF) of a ship is represented in the following expression, based on Newton's second law:

$$\sum_{i=1}^{6} [M_{ji} + A_{ji}]\, \ddot{x}_i + [C_{ji}]\, \dot{x}_i + [K_{ji}]\, x_i = F_j(t) \quad (1)$$

where: $i = 1 - 6$, represents the motion, $j = 1 - 6$, represents the force direction, $[M]$ the mass matrix, $[A]$ the added mass matrix. $[C]$ the damping matrix, $[K]$ the stiffness matrix and $F(t)$ is the harmonic wave force or moment applied. $\ddot{x}_i$, $\dot{x}_i$, x_i are the acceleration, velocity, and displacement in the i-th degree of freedom, respectively. For example, C_{35} represents the damping in heave due to a change in the pitch direction.

The hydrodynamic properties for each strip contribute to the total properties of the ship. The velocity potential ϕ needs to satisfy the 2D boundary problem in fluid domain. The Laplace, free surface, hull, and deep-water boundary conditions are denoted by the following expressions:

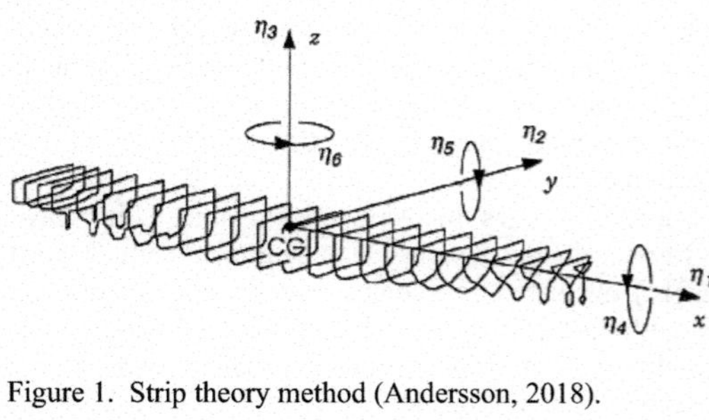

Figure 1. Strip theory method (Andersson, 2018).

$$\nabla^2 \phi_i = 0 \quad (2)$$

$$-\omega_e{}^2 \phi_i + g\frac{\partial \phi_i}{\partial z} = 0, \; z = 0 \quad (3)$$

$$\frac{\partial \phi_i}{\partial n} = V_n \quad (4)$$

$$\lim_{z \to \infty} \nabla \phi_i = 0 \quad (5)$$

where: n is the normal direction of the body and ω_e is the encounter wave frequency. The basic assumptions for the linear Strip Theory are as follows:

- The ship is a rigid body and structural deformation is not considered.
- The ship has a slender body design.
- The transversal sections are wall sided.
- The viscous damping is neglected.
- Deep-water conditions are assumed.
- Froude-Krylov hypothesis assumes that the hull has no effect on the encountered waves.

2.2 *3D panel methods*

3D Panel Methods use surfaces composed of quadrilateral panels to analyse velocity potentials and hydrodynamic coefficients. The Green's Function Method employs an integral equation that combines Green's Theorem and Source potentials to calculate the radiation and diffraction problems along the mean wet surface of the body. The equation satisfied by the radiation potentials ϕ_j is shown in:

$$2\pi\phi_j(x) + \iint_{S_b} \phi_j(\xi)\frac{\partial G(\xi;x)}{\partial n_\xi}d\xi = \iint_{S_b} n_j G(\xi;x)d\xi \quad (6)$$

where: S_b is the body mean wet surface, the source point ξ is defined by (ξ, η, ζ) and the field point is represented by $x = (x, y, z)$ The equation for the diffraction problem ϕ_D and the general form of the equation for source are given by the following expressions:

$$2\pi\phi_D(x) + \iint_{S_b} \phi_D(\xi)\frac{\partial G(\xi;x)}{\partial n_\xi}d\xi = 4\pi\phi_0(x) \quad (7)$$

$$2\pi\sigma(x) + \iint_{S_b} \sigma(\xi)\frac{\partial G(\xi;x)}{\partial n_\xi}d\xi = \frac{\partial \phi(x)}{\partial n_x} \quad (8)$$

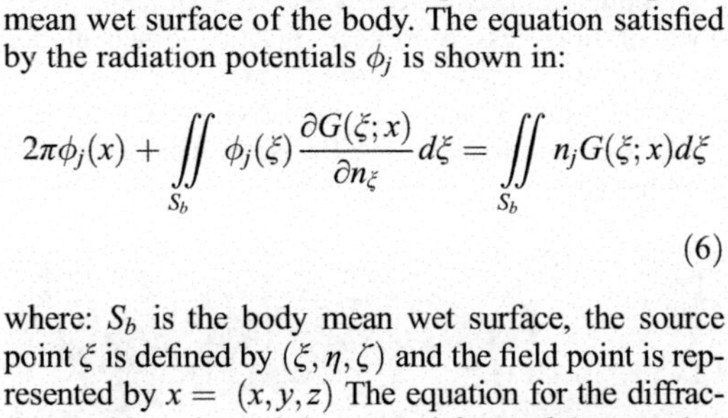

Figure 2. Rankine source BE method.

The Rankine Source Method excels at solving seakeeping analysis within potential theory by efficiently addressing complex hull and free surface boundary conditions. It entails discretising the hull and free surface into panels, as shown in Figure 2.

The ship-fixed coordinates move along with the ship as it advances with forward speed U in regular sea conditions. The coordinate system is located on the undisturbed free surface. The total velocity potential Φ is given by:

$$\Phi(x,y,z,t) = \phi_s(x,y,z) + \phi_I(x,y,z,t) + \phi_D(x,y,z,t) \quad (9)$$

where: ϕ_s is the steady velocity potential, ϕ_I is the incident velocity potential and ϕ_D is the unsteady disturbance velocity potential. In deep waters, the incident wave velocity potential is represented in:

$$\phi_I(x,y,z,t) = \frac{\zeta_A g}{\omega_0} e^{kz} \sin[k(x\cos\beta + y\,sen\beta) - \omega_e t] \quad (10)$$

where: ζ_A is the wave elevation, ω_0 is the incident wave frequency, g is the acceleration of gravity, k is the wave number and β is the incident wave angle. The function $\phi_D(x, t)$ can be obtained through the source-sink distribution method. It can be represented as the distribution of sources on the wetted surface in:

$$\frac{\partial \phi_D(x,t)}{\partial n} = \iint_{S_f+S_b} \sigma(\xi,t)\frac{\partial G(x,\xi)}{\partial n} dS \quad (11)$$

where: S_f is the mean free surface, S_b is the mean wetted hull surface and σ is the unknown source strength distributed. $G(x, \xi)$ represents the Green's Function, which satisfies the Laplace equation and denotes the Rankine source distribution.

$$G(x,\xi) = \frac{1}{r} = \frac{1}{\sqrt{(x-\xi)^2 + (y-\eta)^2 + (z-\zeta)^2}} \quad (12)$$

The solution must be obtained by distributing a constant-strength source across the integration surfaces S_f and S_b. The following expression can be obtained by using the linear free-surface condition and the body surface condition (Kara et al., 2007).

$$\iint_s \sigma(\xi,t)\frac{\partial G(x,\xi)}{\partial r_p} dS_p = \begin{cases} -\frac{1}{g}\left(\frac{\partial}{\partial t} - U\frac{\partial}{\partial x}\right)\phi_D(x,t)^2 & p \in S_f \\ \begin{cases} \dot{x}_j \vec{n}_j(t) + x_j \vec{m}_j(t) \\ -\frac{\partial \phi_I}{\partial n} \end{cases} & p \in S_b \end{cases} \quad (13)$$

2.3 *Uncertainty measures*

FIME, features a coefficient that remains invariant across wave frequencies. In addition, it is essential to account for the random dispersion around the regression line. This dispersion is quantified by CoD which effectively captures the variance of individual data points around the regression line. The measures are expressed as follows:

$$\varphi = \frac{\sum_j |\hat{H}_j||H_j|}{\sum_j |H_j|^2} \tag{14}$$

$$R^2 = 1 - \frac{\sum_j (\varphi H_j - \hat{H}_j)^2}{\sum_j (\varphi H_j - \bar{H})^2} \tag{15}$$

where: φ is FIME coefficient, H_j is the calculated transfer function, $\hat{H}_j$ is the measured transfer function and $\bar{H}$ is the average measured transfer function across all frequencies. The index j corresponds to the measurement frequency.

When $\varphi > 1$, the numerical code tends to underestimate the experimental data. Conversely, $\varphi < 1$ shows that the numerical code tends to overestimate the experimental results. If $R^2 > 0.9$, it indicates a remarkable fit to the data points, suggesting that FIME is an effective measure of uncertainty. When $R^2 = 1$, it signifies that the data points precisely align with the regression. The baseline model always predicts the average of the measurements $\bar{H}$ and has $R^2 = 0$. Codes performing worse than this baseline will yield $R^2 < 0$.

$$Mean_j = \frac{\sum_{i=1}^{N} H_{ij}}{N} \tag{16}$$

$$TD_i[\%] = \frac{\sum_{j=1} |H_{ij} - Mean_j|}{\sum_{j=1} Mean_j} \times 100 \tag{17}$$

$$MTD_i[\%] = \frac{\sum_{j=1} |H_j - \hat{H}_j|}{\sum_{j=1} \hat{H}_j} \times 100 \tag{18}$$

Kim & Kim (2016) introduced the Total Difference (TD) to evaluate how an individual numerical model deviates from the mean values of the results generated by all numerical codes. Abdelwahab et al. (2023) introduces a new approach (MTD), to quantify the uncertainty of individual codes against the experimental measurements instead. Here the index i corresponds to the outcomes derived from to the i-th numerical model utilized in the study.

3 COMPARATIVE STUDY OF SEAKEEPING METHODS

Within the EU project EXTREME SEAS, an extensive series of experiments was conducted to analyse wave-induced motions and loads on two distinct vessel types: a liquefied natural gas (LNG) carrier and a chemical tanker (CT). These experiments encompassed a wide range of sea conditions, including regular, irregular, and extreme seas. The objective of this research is to quantitatively assess the uncertainty associated with seakeeping tools based on potential-flow theory. It focuses on five seakeeping codes: Maxsurf Motions, PDStrip, HydroD/WADAM, HydroD/WASIM, and ANSYS AQWA. These analyses require data within the EXTREME SEAS project. The aim is to validate these tools against experimental measurements and quantify the uncertainties using the short-term models: FIME, CoD, and Total Difference Approaches.

The conducted seakeeping analysis concentrates solely on regular sea waves, particularly head waves (wave heading of 180º), within a frequency range of 0.1 to 1.5 (rad/s). The wave-induced responses for assessment will be heave (ξ_3), pitch (ξ_5), and the vertical bending moment (M_5).

3.1 *Description of the models*

The benchmark studies for the Liquified Natural Gas (LNG) carrier and the chemical tanker (CT) were conducted at Technische Universität Berlin (TUB), using scaled models tested without forward speed ($Fr = 0$). The seakeeping basin had 90 meters in length, 8 meters in width, and a 1-meter water depth. The models consisted of two rigid segmented sections of fiberglass reinforced plastic, joined at the midship. Both vessels were scaled down to a 1:70 model. Figure 3 displays the LNG carrier model, while Table 1 displays the full-scale characteristics of both vessels.

Figure 3. LNG carrier model test in technische universität berlin (Klein et al., 2016).

Table 1. Full scale parameters – LNG Carrier and Chemical Tanker.

	LNG Carrier	Chemical Tanker
$Lpp\ [m]$	186.90	161.00
$B\ [m]$	30.38	28.00
$D\ [m]$	18.20	13.00
$T\ [m]$	8.40	9.00
$C_b\ [-]$	0.66	0.75
$\Delta\ [kg]$	35 674 800	30 666 000
$I_{xx}\ [kg.m^2]$	4.53×10^9	2.66×10^9
$I_{yy}\ [kg.m^2]$	5.86×10^{10}	3.32×10^{10}
$I_{zz}\ [kg.m^2]$	5.76×10^{10}	3.32×10^{10}
$L_{CG}\ [m]$	94.87	82.51
$V_{CG}\ [m]$	8.40	6.15

Tables 4 and 5 present the primary parameters for each mesh, including the total number of panels in the hull surface and the average length of these panels.

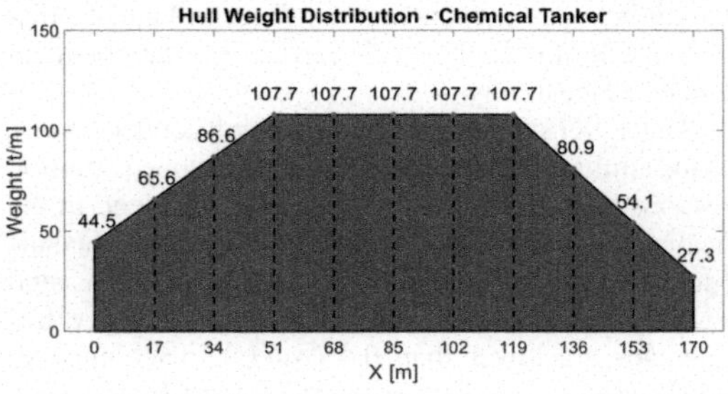

Figure 4. Hull weight distribution – Chemical tanker.

Aside from the main parameters, the mass distribution is critical in simulations, as it exerts influence on rotational motions and sectional loads. The benchmark studies supply the necessary data for establishing the mass distribution of each vessel.

The models consider not only the weight of the hull, as depicted in Figure 4, but also the added masses of essential components crucial for testing. Batteries, discs, suspensions bars, and other components been carefully considered, specifying their individual masses and precise positioning. The total mass distribution of the CT is shown in Figure 5.

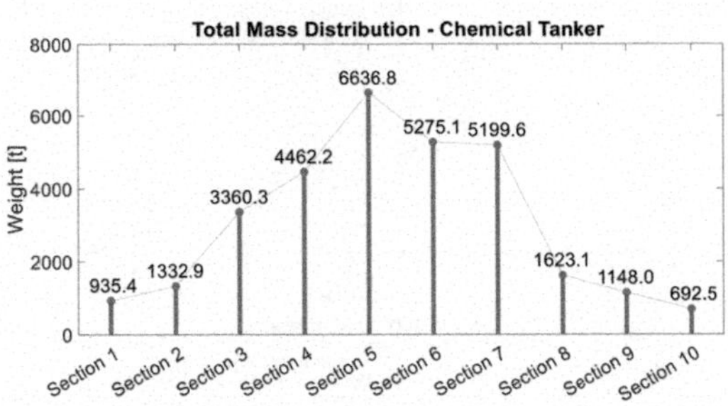

Figure 5. Total mass distribution – Chemical tanker.

For more in-depth information about the LNG carrier and the associated benchmark test, please refer to Klein et al. (2016) and Zhang et al. (2022). Similarly, for the chemical tanker, the cited references provide a wealth of information, offering extensive details. Consult Ley & el Moctar (2021) and Wang & Guedes Soares (2016) for a thorough exploration of the topic.

3.2 *Computational setup*

The accuracy of responses generated by commercial codes might be influenced by several factors such as incomplete model data, mathematical formulations, and computational setup. Table 2 summarises the selected numerical codes for each seakeeping method and outlines the potential results from each simulation.

3.2.1 *Maxsurf motion & PDStrip*

Maxsurf Motions, a software by Bentley Systems (2022), and PDStrip, an open-source hydrodynamic code (Bertram et al., 2006), both employ the Strip Theory Method. Maxsurf does not provide load-related results and does not capture nonlinear behaviour. Contrary, PDStrip does calculate sectional loads and includes some nonlinear effects. For the studied ships, around 20 and 40 sections were delineated between the perpendiculars in PDStrip and Maxsurf Motions, using 30 and 15 offset points, respectively. Figure 6 displays the cross-sections for the LNG carrier in PDStrip, while Table 3 shows the methods applied in Maxsurf Motions for the solution.

Table 2. Numerical seakeeping codes employed.

Numerical Code	Method	Results
Maxsurf Motions	Strip Theory	Motions
PDStrip	Strip Theory	Motions/ Loads
HydroD/WADAM	Green's Function	Motions/ Loads
ANSYS - AQWA	Green's Function	Motions/ Loads
HydroD/WASIM	Rankine Source	Motions/ Loads

Figure 6. LNG carrier – strip theory method.

Table 3. Methods applied – Maxsurf motions.

Ship	Transom Terms	Added Resistance
LNG carrier	No	Salvesen
Chemical Tanker	Yes	Salvesen

3.2.2 *HydroD/WADAM & ANSYS-AQWA*

HydroD/WADAM, developed by DNV-GL (2023, 2017), and ANSYS-AQWA by ANSYS Inc. (2013), both utilize a 3D radiation-diffraction theory using a panel model, known as the Green's Function BE Method. To implement these methods, a quadrilateral panel model, as shown in Figures 7 and 8, is created. Notably, ANSYS-AQWA has an advantage in introducing a forward speed correction in the body boundary condition, whereas HydroD/WADAM does not include this feature.

3.2.3 *HydroD/WASIM*

HydroD/WASIM, developed by DNV-GL (2023, 2011), is designed to compute global responses and local loadings on vessels at any forward speed, by employing Rankine Source BE Method. The simulations are executed in time domain but are also transformed to frequency domain using Fourier

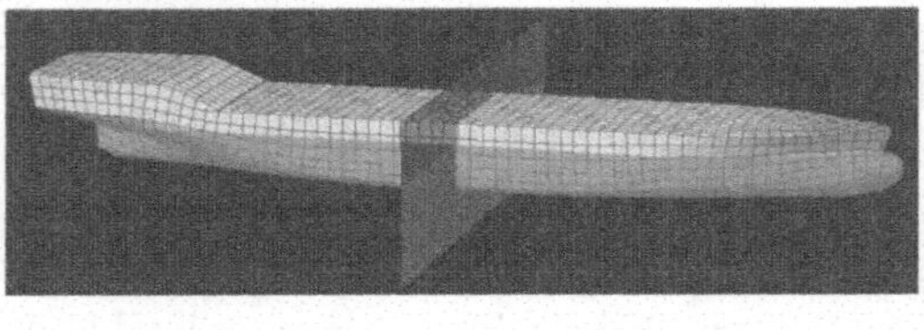

Figure 7. Chemical tanker panel model – HydroD/ WADAM.

Figure 8. Chemical tanker panel model – ANSYS-AQWA.

Table 4. Meshes details – HydroD/WADAM.

Mesh	Number of Panels	Length of Panel
LNG carrier	2501	≈ 2.4 [m]
Chemical Tanker	2211	≈ 2.5 [m]

Table 5. Meshes details – ANSYS-AQWA.

Mesh	Number of Panels	Element Size
LNG carrier	11575	1 [m]
Chemical Tanker	8477	1 [m]

transform. This approach requires the use of panels on both the structure and the free surface, as represented in Figure 2. The free surface is automatically generated and comprises a circular area with a radius of 5 vessel length, designed to cover all relevant wavelengths. Table 6 displays the number of panels for the hull and free surface of each vessel.

Table 6. Meshes details – HydroD/WASIM.

Mesh	Nº Panels Hull	Nº Panels Free Surface
LNG carrier	12000	19400
Chemical Tanker	15000	22696

4 RESULTS AND DISCUSSION

4.1 *Response Amplitude Operators (RAO)*

The transfer functions obtained in the codes are converted to nondimensional parameters to enable meaningful comparisons across the different ships and seakeeping methods. Specifically, heave RAO will be normalised against wave amplitude (ζ_a), the pitch RAO against the wave slope ($k\zeta_a$), where k is the wave number, and finally, the vertical bending moment RAO will be normalised against the product ($\rho g B L^2 \zeta_a$), where ρ is the water density, g is the acceleration of gravity, B is the breadth and L is the length between perpendiculars. Figures 9 and 10 display the RAO for the LNG carrier and the chemical tanker, respectively.

The LNG carrier and the chemical tanker, showcase similar behaviours in wave-induced motion responses. Notable discrepancies are observed in the heave RAO, particularly with PDStrip and Maxsurf Motions at $\omega \approx 0.6$ (rad/s). ANSYS-AQWA and PDStrip show inaccuracies in pitch when wave frequencies are lower than 0.4 (rad/s), reaching over 100% amplitude difference at the lowest frequencies. Notably, by default, Maxsurf Motions does not calculate motion results for frequencies below that value. Regarding the VBM, most codes demonstrate acceptable agreement, except for PDStrip, which exhibits substantial variation above 0.7 (rad/s) for the LNG carrier and 0.55 (rad/s) for the chemical tanker, which may be attributed to the mathematical formulation employed in the seakeeping code and complexities of the hull geometry.

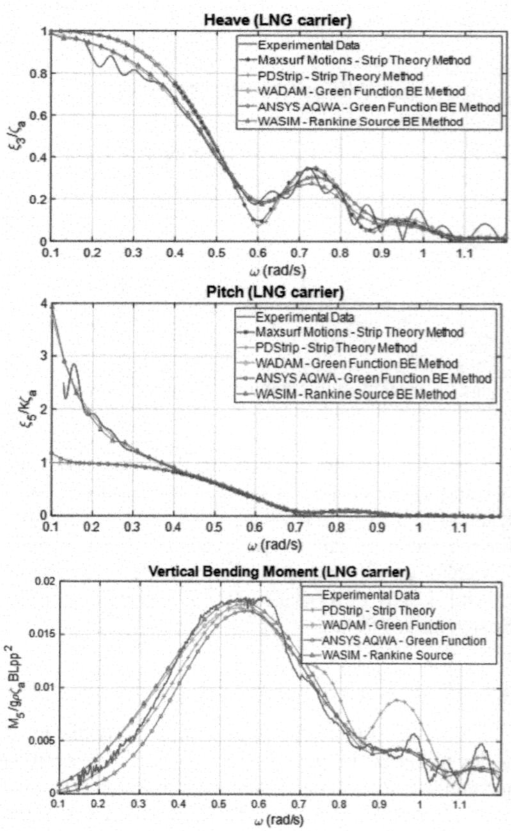

Figure 9. LNG carrier RAO.

4.2 *Uncertainty assessment*

4.2.1 *Comparison of uncertainty measures*

The comparative analysis is vital for evaluating the accuracy of each seakeeping code. Additionally, it helps us identify factors that might influence these uncertainties, given that the ships under investigation vary in size, hull form and operational conditions. The LNG carrier and the chemical tanker were tested in the same hydrodynamic laboratory, resulting in a similar dataset for analysis, allowing direct comparison between the vessels. A comparative dataset spanning from 0.25 to 1.2 (rad/s), with an increment of 0.05 (rad/s), was employed. The uncertainties from FIME are illustrated in Figure 11, while the comparative results for the Coefficient of Determination are presented in Figure 12.

1) Comparison between ships:
   The LNG carrier and the Chemical Tanker, both tested without forward speed, display comparable behaviors across the uncertainty measures for each seakeeping code. The chemical tanker exhibited slightly better heave motion results, while the LNG carrier generally performed better for VBM. These disparities are likely influenced by distinct hull geometries, particularly the blunter ship form of the chemical tanker, or possible user input inconsistencies, given that the models were constructed and tested under identical conditions.
2) Comparison between Seakeeping Codes:
   Regarding the Strip Theory codes, both codes demonstrate a comparable pattern in heave uncertainties. However, when it comes to pitch motion, Maxsurf Motions tends to outperform PDStrip. The difference arises from Maxsurf ability to automatically remove inaccurate results below a certain frequency, which PDStrip lacks. Regarding VBM, the Strip code reveals notable disparities that are consistent with the RAO observed. HydroD/WADAM and ANSYS-AQWA (Green's Function BEM codes), show comparable results for the heave motion and VBM. However, HydroD/WADAM shows slightly more consistent values. Regarding the behaviour in pitch motion, it varies depending on the accuracy for each code in certain frequency range.

As for HydroD/WASIM, which employs the Rankine Source Method, the results highlight its accuracy and reliability, especially in accounting for hydrodynamics nonlinearities. A notable comparison exists between PDStrip and ANSYS-AQWA regarding pitch motion. The observation reveals that both LNG carrier and the chemical tanker, display underestimated values. This underscores the significant impact of dataset selection on the uncertainty results.

In summary, the overall observation emphasises that even when codes share the same mathematical model, such as Maxsurf Motions and PDStrip, or

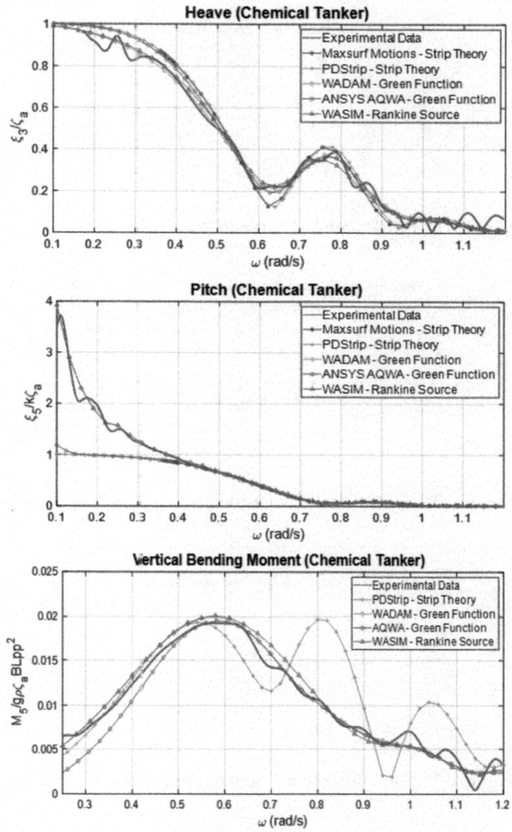

Figure 10. Chemical tanker RAO.

HydroD/WADAM and ANSYS-AQWA, they do not necessarily produce identical uncertainties. This underscores the importance of carefully considering the specific features and adjustments within each software package when interpreting results.

By examining Figures 13 and 14, a notable trend becomes apparent: there is a general variation in the differences between the two uncertainty models. This shift highlights the disparity between these approaches to quantifying model uncertainty. The Modified Total Difference emerges as the more suitable model in this context since it directly compares numerical predictions to the actual measured responses, assessing how well the numerical codes align with real-world observations. However, relying solely on this forward statistical parameter can introduce inaccuracies and may not provide a truly reliable measure of uncertainty. It is always required FIME and Coefficient of Determination (CoD).

1) Comparison between ships:
   For the LNG carrier and the chemical tanker, consistent and comparable behaviours are noted across various seakeeping codes, aligning with previous findings on uncertainty measures.

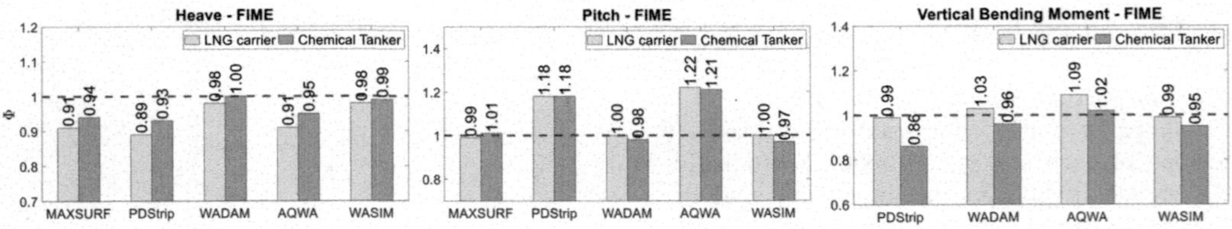

Figure 11. FIME – global comparison.

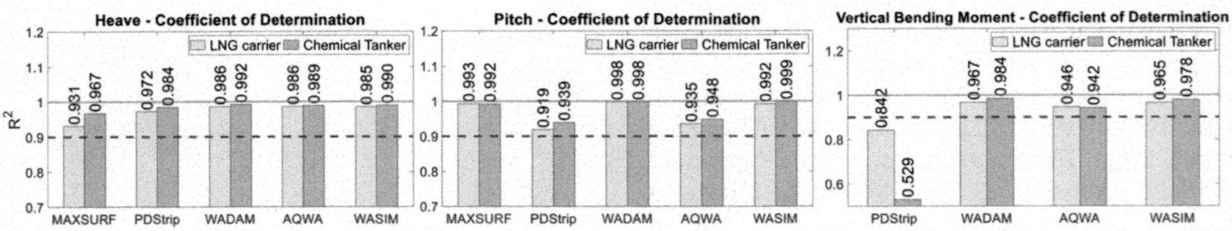

Figure 12. Coefficient of determination – global comparison.

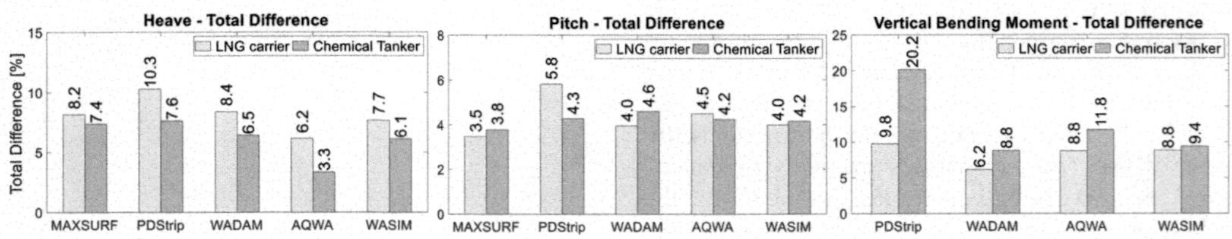

Figure 13. Total difference – global comparison.

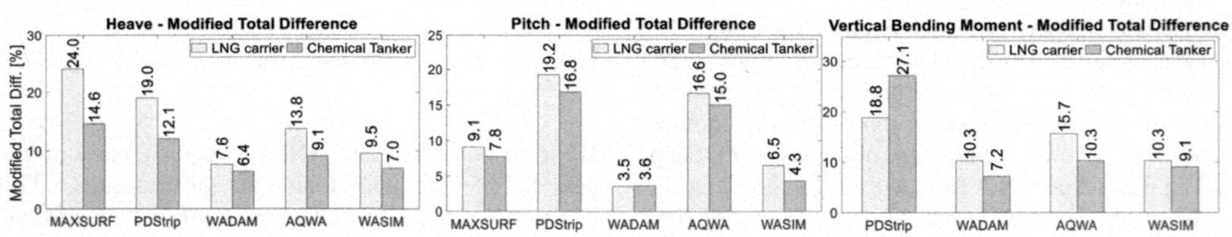

Figure 14. Modified total difference – global comparison.

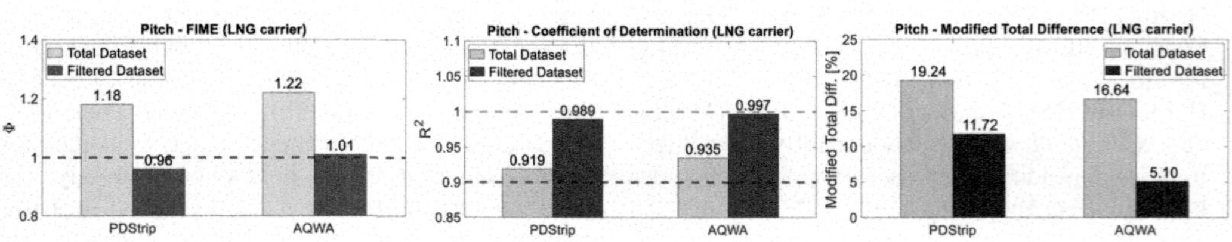

Figure 15. Pitch uncertainties for LNG carrier.

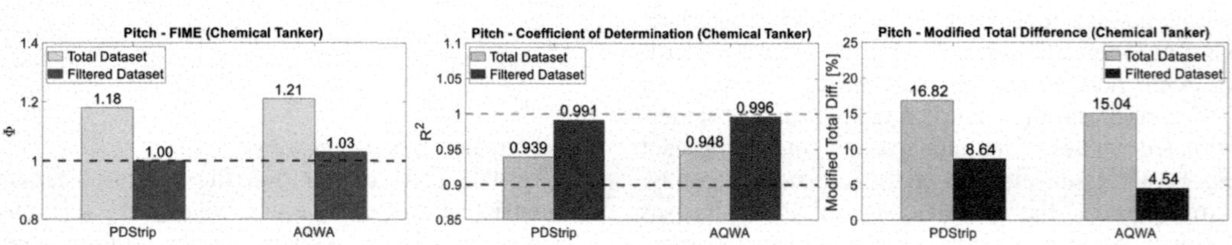

Figure 16. Pitch uncertainties for chemical tanker.

Figure 14 displays overall slightly higher differences for the LNG carrier than for the chemical tanker. An exception is evident in the VBM results for the PDStrip code, suggesting a potential influence of variations in hull geometry or possible user input inconsistencies, as previously discussed.

2) Comparison between Seakeeping Codes:Concerning the Strip Theory codes (Maxsurf Motions and PDStrip) show consistent behaviours in heave motion across the tested ships, aligning with previous uncertainty assessments. Notably, PDStrip presents slightly higher pitch differences than Maxsurf Motions since the inaccurate values are not automatically removed. Regarding VBM, the Strip code reveals significant differences when compared to the motion outcomes, consistent with findings from Kim & Kim (2016). These disparities can be attributed to the mathematical formulation employed by the code when considering the intricacies of a 3D hull and the mass distribution properties.
Concerning the Green's Function codes, HydroD/WADAM shows consistent performance. ANSYS-AQWA yields least favorable results for both vessels. Specially, for pitch motion, tends to exhibit consistently higher differences than HydroD/WADAM, which can also be attributed to disparities occurring below a wave frequency of $0.4\ (rad/s)$, similar to what was observed with PDStrip.
Overall, the uncertainties derived through the Modified Total Difference Method align with other measures, emphasizing the method's robustness in assessing uncertainties in seakeeping methods. This consistency underscores the robustness of the Modified Total Difference Method as a tool for assessing the uncertainties of seakeeping methods.

4.2.2 *Influence of dataset selection*

The influence of the dataset selection on the uncertainty models was also explored. This analysis was prompted by PDStrip and ANSYS-AQWA, which yielded highly inaccurate values for pitch motion in wave frequencies below 0.4 (rad/s). The variations of the uncertainty measures for the LNG carrier are illustrated in Figure 15, while those for the Chemical Tanker are presented in Figure 16. The Total Difference was the only uncertainty measure excluded from this evaluation, as it was already necessary to exclude that dataset range for the comparison of seakeeping methods, since it can only effectively compare methods when they provide the same number of measurement frequencies.

Upon a thorough analysis of the bar charts, a clear trend emerges: the exclusion of data below a specific wave frequency range led to improvements in all uncertainty measures. These improvements can be attributed to discrepancies in low-frequency wave responses from PDStrip (Strip Theory) and ANSYS-AQWA (3D Panel Method), which had a notable impact on overall uncertainties.

Furthermore, the assessment of PDStrip VBM responses indicates a reduced accuracy beyond a specific frequency, particularly for higher frequencies. Conversely, for rotational motion (pitch), the accuracy diminishes for lower frequencies. This evaluation underscores the substantial impact of wave frequency range selection on seakeeping code accuracy. It emphasizes that both low and high-frequency ranges can introduce inaccuracies in certain seakeeping responses, highlighting the significance of a meticulous frequency range selection for obtaining meaningful results.

5 CONCLUSIONS

This research is an uncertainty assessment that centers on the wave-induced responses for a LNG carrier and a chemical tanker, operating under regular head waves conditions, without forward speed. The expanded assessment seeks to pinpoint the factors that might influence uncertainties in seakeeping analysis.

The key findings resulting from comparison between the ships are outlined as follows:

- Both ships exhibit comparable behaviours for the wave-induced responses across all seakeeping uncertainties.
- PDStrip and ANSYS-AQWA show higher inaccuracies in pitch motion, underestimating results.
- The Modified Total Difference for the LNG carrier has slightly more pronounced disparities (1% to 10%) compared to the CT, possibly due to hull geometry differences or inconsistencies in user inputs.

The primary findings resulting from the uncertainty measurements are summarised as follows:

- The Modified Total Difference is preferred over the Total Difference, aligning with Abdelwahab et al. (2023).
- Total Difference and Modified Total Difference methods lack depth of information, requiring additional uncertainties like FIME and CoD for comprehensive evaluation. This conclusion aligns with Kim & Kim (2016).
- Tools sharing the same mathematical model do not yield identical uncertainties.
- Dataset choice significantly influences pitch uncertainties. The removal of inaccurate data improves the measures, aligning with Parunov et al. (2022b).

The principal findings regarding the seakeeping codes are as follows:

- Maxsurf Motions and PDStrip (Strip Theory) display comparable heave uncertainties but Maxsurf

Motions offers more accurate pitch results due to its frequency filtering capability.
- Both Strip Theory codes exhibit acceptable accuracy, consistent with prior studies.
- PDStrip code exhibits VBM inconsistencies for the CT, attributed to mass distribution properties and the tool's interpretation of the complex 3D hull form.
- HydroD/WADAM has proven to be reliable across all the wave-induced responses.
- ANSYS-AQWA shows more inaccuracies in sectional loads and rotational motions compared to HydroD/WADAM, potentially related to mass distribution interpretations, inaccuracies in certain frequencies or the formulation implementation.
- HydroD/WASIM demonstrates high reliability across all frequency ranges, outperforming Green's Function codes, aligning with Söding (2010). However, it requires additional time and computational effort in comparison with other seakeeping codes.

ACKNOWLEDGEMENTS

This work was performed within the Strategic Research Plan of the Centre for Marine Technology and Ocean Engineering (CENTEC), which is financed by Portuguese Foundation for Science and Technology (Fundação para a Ciência e Tecnologia-FCT) under contract UIDB/UIDP/00134/2020. The first author is indebted to Hossam Abdelwahab for his help to deal with the intricacies of HydroD/WADAM and HydroD/WASIM.

REFERENCES

Abdelwahab, H.S., Wang, S., Parunov, J. and Guedes Soares, C., 2023. A New Model Uncertainty Measure of Wave-Induced Motions and Loads on a Container Ship with Forward Speed. In Journal of Marine Science and Engineering, 11, 1042. https://doi.org/10.3390/jmse11051042

Abdelwahab, H.S., Wang, S. and Guedes Soares, C., 2022. Uncertainty assessment of wave-induced motions and loads on a container ship with a forward speed. In Trends in Maritime Technology and Engineering, 269–281. http://dx.doi.org/10.1201/9781003320272-30

Andersson, E., 2018. Application of the open source code Nemoh for modelling of added mass and damping in ship motion simulations. [Master Thesis, KTH Royal Institute of Technology, Stockholm, Sweeden]. https://www.diva-portal.org/smash/get/diva2:1263450/FULLTEXT01.pdf (accessed on 10 October 2023).

ANSYS, Inc., 2013. AQWA User's Manual Release v.15.0. (Version 15.0). Canonsburg, PA, USA. https://vdocument.in/aqwa-users-manual.html (accessed on 14Oct2023).

Bentley Systems, I., 2022. Maxsurf Motions, Windows Version 21, User Manual. https://communities.bentley.com/products/offshore/m/mediagallery/276121 (accessed on 15 Oct 2023).

Bertram, V., Veelo, B., Söding, H. and Graf, K., 2006, January. Development of a freely available strip method for seakeeping. In Proc. 5th International Conference on Computer and IT Applications in the Maritime Industries. 6, 356–368. https://www.researchgate.net/publication/262413711_Development_of_a_Freely_Available_Strip_Method_for_Seakeeping

Chen, C., Shiotani, S. and Sasa, K., 2013. Numerical ship navigation based on weather and ocean simulation. In Ocean Engineering, 69, 44–53. https://doi.org/10.1016/j.oceaneng.2013.05.019

DNV, G., 2023. HydroD - Wave load & stability analysis of fixed and floating structures v.6.2-01 (Version 6.2). SESAM User Manual. Det Norske Veritas Software. Høvik, Norway. https://sesam.dnv.com/status/hydrod/status/hydrod_releasenotes_6.2.1.pdf (accessed on 16 Oct 2023).

DNV, G., 2017. WADAM - Wave Analysis by Diffraction and Morison theory v.9.3 (Version 9.3). SESAM User Manual. Det Norske Veritas Software. Høvik, Norway. https://home.hvl.no/ansatte/tct/FTP/H2023%20Marinteknisk%20Analyse/SESAM/SESAM%20UM%20Brukermanualer/Wadam_UM.pdf (accessed on 16 Oct 2023).

DNV, G., 2011. WASIM - Wave Loads on Vessels with Forward Speed v.5.1-01 (Version 5.1). SESAM User Manual. Det Norske Veritas Software. Høvik, Norway. https://manualzz.com/doc/7270848/wasim-user-manual (accessed on 16 Oct 2023).

Guedes Soares, C., 1991. Effect of transfer function uncertainty on short-term ship responses. In Ocean Engineering, 18, 329–362. https://doi.org/10.1016/0029-8018(91)90018-L

Hirdaris, S., Bai, W., Dessi, D., Ergin, A., Gu, X., Hermundstad, O.A., Huijsmans, R., Iijima, K., Nielsen, U.D., Parunov, J., Fonseca, N., Papanikolaou, A., Argyriadis, K. & Incecik, A., 2014. Loads for use in the design of ships and offshore structures. In Ocean Engineering, 78, 131–174. https://doi.org/10.1016/j.oceaneng.2013.09.012

Kara, F., Tang, C.Q. and Vassalos, D., 2007. Time domain three-dimensional fully nonlinear computations of steady body–wave interaction problem. In Ocean Engineering, 34, 776–789. https://doi.org/10.1016/j.oceaneng.2006.04.015

Kim, Y. and Kim, J.H., 2016. Benchmark study on motions and loads of a 6750-TEU containership. In Ocean Engineering, 119, 262–273. https://doi.org/10.1016/j.oceaneng.2016.04.015

Klein, M., Clauss, G.F., Rajendran, S., Guedes Soares, C. and Onorato, M., 2016. Peregrine breathers as design waves for wave-structure interaction. In Ocean Engineering, 128, 199–212. https://doi.org/10.1016/j.oceaneng.2016.09.042

Ley, J. and el Moctar, O., 2021. A comparative study of computational methods for wave-induced motions and loads. In Journal of Marine Science and Engineering, 9, 83. https://doi.org/10.3390/jmse9010083

Lee, J.H., Kim, Y., Kim, B.S. and Gerhardt, F., 2021. Comparative study on analysis methods for added resistance of four ships in head and oblique waves. Ocean Engineering, 236, 109552. https://doi.org/10.1016/j.oceaneng.2021.109552

Lewis, F.M., 1929. The inertia of the water surrounding a vibrating ship. In 37th Meeting of The Society of Naval Architects and Marine Engineers, SNAME. Institute of Naval Architecture, New York, USA. http://resolver.tudelft.nl/uuid:56768cfa-57ca-463c-8dc1-19fb64049d22 (accessed on 14 Oct 2023).

Pacuraru, F., Domnisoru, L. and Pacuraru, S., 2020. On the comparative seakeeping analysis of the full scale KCS by several hydrodynamic approaches. In Journal of Marine Science and Engineering, 8, 962. https://doi.org/10.3390/jmse8120962

Parunov, J., Guedes Soares, C., Hirdaris, S., Iijima, K., Wang, X., Brizzolara, S., Qiu, W., Mikulić, A., Wang, S. and Abdelwahab, H.S., 2022a. Benchmark study of global linear wave loads on a container ship with forward speed. Marine Structures, 84, 103162. https://doi.org/10.1016/j.marstruc.2022.103162

Parunov, J., Guedes Soares, C., Hirdaris, S. and Wang, X., 2022b. Uncertainties in modelling the low-frequency wave-induced global loads in ships. In Marine Structures, 86, 103307. https://doi.org/10.1016/j.marstruc.2022.103307

Salvesen, N., Tuck, E.O. and Faltinsen, O., 1970. Ship motions and sea loads. Transactions SNAME. 78, 250–287. https://www.boatdesign.net/attachments/ship-motions-and-sea-loads-salvesen-tuck-faltinsen-1970-pdf.53097/

Söding, H., 2010. Boundary Element Methods for Seakeeping. Technical University Hamburg-Harburg, Duisburg, Contribution to the Lectures on Seakeeping and Sea Loads. https://www.researchgate.net/publication/260433070_Boundary_Element_Methods_for_Seakeeping

Wang, S. and Guedes Soares, C., 2016. Experimental and numerical study of the slamming load on the bow of a chemical tanker in irregular waves. In Ocean Engineering. 111, 369–383. https://doi.org/10.1016/j.oceaneng.2015.11.012

Zhang, H., Cui, J., Liao, X., Shi, H. and Guedes Soares, C., 2022. Numerical study on the vertical response of LNG carrier in abnormal waves generated with different mechanisms. In Ocean Engineering. 262, 112090. https://doi.org/10.1016/j.oceaneng.2022.112090

Advances in Maritime Technology and Engineering – Guedes Soares & Santos (Eds)
© 2024 The Author(s), ISBN 978-1-032-83107-7

Numerical study on seakeeping behavior of unmanned sailing boats under extreme wave conditions

J.J. Wang & H.T. Wang
School of Naval Architecture and Ocean Engineering, Huazhong University of Science and Technology, Wuhan, P.R. China

G. Xiang & X.B. Xiang
School of Naval Architecture and Ocean Engineering, Huazhong University of Science and Technology, Wuhan, China
Hubei Key Laboratory of Naval Architecture and Ocean Engineering Hydrodynamics (HUST), Wuhan, P.R. China
Collaborative Innovation Center for Advanced Ship and Deep-Sea Exploration, Shanghai, P.R. China

C.Guedes Soares
Centre for Marine Technology and Ocean Engineering (CENTEC), Instituto Superior Tecnico, Universidade de Lisboa, Lisbon, Portugal

ABSTRACT: The seakeeping performance of unmanned sailboat directly affects its motion response characteristics. In this study, a series of CFD simulation experiments were designed and carried out to simulate the motion response of unmanned sailing boats under different wave conditions. In the simulation process, the Reynolds Averaged Navier-Stokes equations, which are unsteady and incompressible, are used, and the turbulence model and the coupling effect in the fluid-structure interaction are taken into account. By analyzing the simulation results, we study the motion characteristics and stability of the unmanned sailboat under various wave conditions. The results show that in extreme wave environment, the stability of unmanned sailboat is obviously affected, and the navigation Angle and acceleration are significantly changed, and corresponding control strategies need to be taken to ensure its safety and reliability. Finally, the research results are discussed and summarized in this paper, which can provide reference and guidance for engineering design and Marine mission planning in related fields.

1 INTRODUCTION

Seakeeping refers to the sailing performance of a sailing boat under the condition of waves. The sea wave will affect the sailing of the sailing boat, and it will have an adverse effect on the safe sailing of the sailing boat under the condition of large waves, large waves usually mean greater wave height and greater wave power. In order to ensure the safety of sailing boat, it is very significant to study the seakeeping of sailing boat.

An et al. (2023) introduced a variety of environmental adaptive mechanisms that can provide dynamic anti-capsizing capability, and analyzed their adaptability to unmanned sailboats. Combined with hydrodynamic simulation, the stability, drag and growth efficiency of passive hydrofoils with different aspect ratios are investigated. Li et al. (2018) used the numerical simulation software STAR CCM+ to simulate the resistance characteristics of sails, and found that the ship's resistance increased with the increase of drift Angle, and the increase amplitude was larger and larger. Wang et al. (2023) studied the hydrodynamic characteristics of a six-degree-of-freedom laser Rideer-class sailing ship by means of experimental observation and numerical calculation. A set of six element force test method for sailing boat is designed, and the hydrodynamic characteristics of sailing boat are obtained through towing tank test. Shen et al. (2023) studied the path planning optimization of unmanned sailing ships in complex Marine environments, considering the maneuvering characteristics of unmanned sailing ships and the impact of sudden obstacles. Niu et al. (2023) studied the influencing factors of the root stress of the wing sail mast and its influencing rules. In view of the stress concentration at the root of the mast, a method of setting reinforcing bars to optimize the structure is proposed. Fang et al. (2023) investigated the deformation results of flexible sails under different wind conditions and the influence of deformation on aerodynamic performance. Kristian et al (2016) proposed a method to convert a sailing ship heading controller into a heading controller by adding minor corrections. Simulation results show the effectiveness of the proposed method.

DOI: 10.1201/9781003508779-27

Chen et al. (2023) obtained a four-degree-of-freedom dynamic model of an unmanned sailing ship by using Bayes regression method, and obtained the velocity polar curve of the ship by static velocity prediction algorithm. By comparing the system identification results with the constrained model pool test results, it is verified that the system identification method is accurate enough, and it is a reasonable method to construct the dynamics model of small unmanned sailboat. Liu et al (2022) explored the effect of sail Angle of attack control strategy on sail turning diameter in order to make sail have better turning performance. The effects of different sail Angle of attack direction and sail turning direction were analyzed by integrating target sail force coefficient, and the range of available sail Angle of attack under different wind conditions was obtained. Based on the responsive three-degree of freedom ship maneuvering analysis (MMG) method, the simulation model of sailing motion was established, and the sailing diameter clusters corresponding to different wind conditions and sail Angle of attack were generated respectively, and the sail Angle of attack strategy was obtained under the control of the minimum sailing diameter. The research results provide a theoretical reference for the design of sail Angle of attack maneuvering strategy to optimize the turning performance.

To sum up, the current research on unmanned sailboats mostly focuses on sail, mast stress and path planning, while there are few studies on the seakeeping and rapidity of the hull of the sailboat. In this paper, the seakeeping of unmanned sailboats is discussed.

2 MATHEMATICAL MODEL

2.1 *Governing equation*

In this paper, the compressibility of seawater and the influence of water temperature are ignored. The fluid studied is a viscous incompressible fluid, and the energy change is not considered when the submarine moves underwater, so the fluid movement must meet the mass conservation equation and the energy conservation equation.

The mass conservation equation, also known as the continuity equation, is expressed as follows:

$$\frac{\partial \rho}{\partial t}+u_x\frac{\partial(\rho u_x)}{\partial x}+u_y\frac{\partial(\rho u_y)}{\partial y}+u_z\frac{\partial(\rho u_z)}{\partial z}=0 \quad (1)$$

The momentum conservation equation, also known as the N-S equation, is expressed as follows:

$$div(u)=0 \quad (2)$$

$$\frac{\partial u_x}{\partial t}+u_x\frac{\partial u_x}{\partial x}+u_y\frac{\partial u_x}{\partial y}+u_z\frac{\partial u_x}{\partial z}=X-\frac{1}{\rho}\frac{\partial p}{\partial x}+\frac{\mu}{\rho}\nabla^2 u_x \quad (3)$$

$$\frac{\partial u_y}{\partial t}+u_x\frac{\partial u_y}{\partial x}+u_y\frac{\partial u_y}{\partial y}+u_z\frac{\partial u_y}{\partial z}=Y-\frac{1}{\rho}\frac{\partial p}{\partial y}+\frac{\mu}{\rho}\nabla^2 u_y \quad (4)$$

$$\frac{\partial u_z}{\partial t}+u_x\frac{\partial u_z}{\partial x}+u_y\frac{\partial u_z}{\partial y}+u_z\frac{\partial u_z}{\partial z}=Z-\frac{1}{\rho}\frac{\partial p}{\partial z}+\frac{\mu}{\rho}\nabla^2 u_z \quad (5)$$

Where, x, y and z represent the Cartesian coordinate system, p is the pressure, μ is the dynamic viscosity, ρ is the fluid density, and u_x, u_y and u_z are the velocity components along the x, y and z axes, respectively.

2.2 *Turbulence model*

Methods for studying turbulent motion include direct numerical simulation (DNS), Reynolds Mean Navier-Stokes method (RANS), and large eddy simulation method (LES). Because RANS method has the requirement of high precision and low requirement of computer equipment, this paper chooses to use it to simulate submarine near surface navigation.

RANS method uses a special model to represent the instantaneous turbulence pulsation as the time mean, and establishes the relationship between the turbulence pulsation value and the mean value, thus establishing a turbulence model. Turbulence models widely used in the field of ocean engineering include k-ε models (including standard k-ε, RNG k-ε) and k-ω models (including standard k-ω and shear pressure transfer k-ω). Among them, the SST k-ω model considers the influence of free shear layer on turbulent viscosity, so in the simulation solution, the SST k-ω turbulence model is selected for the simulation calculation of submarine near-surface navigation. The model can accurately predict the fluid motion state in the near wall region and the far field. The standard k-ω model is used in the near wall region, and the boundary layer edge and free shear layer are automatically switched to the SST k-ω model, and the mixed function is used for transition. The corresponding k equation and ω equation are shown in equations 6 and 7:

$$\frac{\partial}{\partial t}(\rho k)+\frac{\partial}{\partial x_i}(\rho k u_i)=\frac{\partial}{\partial x_j}\left[\Gamma_k\frac{\partial k}{\partial x_j}\right]+\widetilde{G_k}-Y_k+S_k \quad (6)$$

$$\frac{\partial}{\partial t}(\rho\omega)+\frac{\partial}{\partial x_i}(\rho\omega u_i)=\frac{\partial}{\partial x_j}\left[\Gamma_\omega\frac{\partial \omega}{\partial x_j}\right]+\widetilde{G_\omega}-Y_\omega+D_\omega+S_\omega \quad (7)$$

Where, k represents the turbulent kinetic energy, ω represents the special dissipation rate, Γ_k and Γ_ω represent the effective diffusion of k and ω respectively, Y_k and Y_ω represent the dissipation of k and ω respectively, G_k is the production term of k, G_ω~ is the production term of ω, D_ω represents the cross-diffusion term, S_k and S_ω represent the custom source term.

2.3 *VOF model*

The Volume of Fluid (VOF) method was proposed by Hirt and Nichols in 1981, in which the volume function C is defined, which represents the volume fraction of a certain fluid in a cell. When the C value is between 0 and 1, the unit contains both part of the phase and one or more other phases. The VOF method can track the interface between different phases by solving the volume fraction of the phase. The equation is as follows:

$$\frac{\partial C_q}{\partial t} + \vec{u} \cdot \nabla C_q = 0 \tag{8}$$

In the formula, the volume fraction is satisfied

$$\sum_{q=1}^{n} C_q = 1 \tag{9}$$

In N-phase fluids, the physical properties of each fluid are averaged linearly according to the volume fraction they occupy, as follows:

$$\rho = \sum_{q=1}^{n} C_q \rho_q \tag{10}$$

$$\mu = \sum_{q=1}^{n} C_q \mu_q \tag{11}$$

3 INTRODUCTION OF UNMANNED SAILING SHIPS AND NUMERICAL PREDICTION OF SEAKEEPINNG

3.1 *Introduction to the ship type used in calculation*

This paper takes the 3.5-meter-long unmanned sailboat as the research object, and Table 1 shows the main scale of the unmanned sailboat. Since this paper mainly explores the seakeeping of unmanned sailboats, the wind resistance of unmanned sailboats has not been studied at present. Figure 1 shows the schematic diagram of the unmanned sailboats.

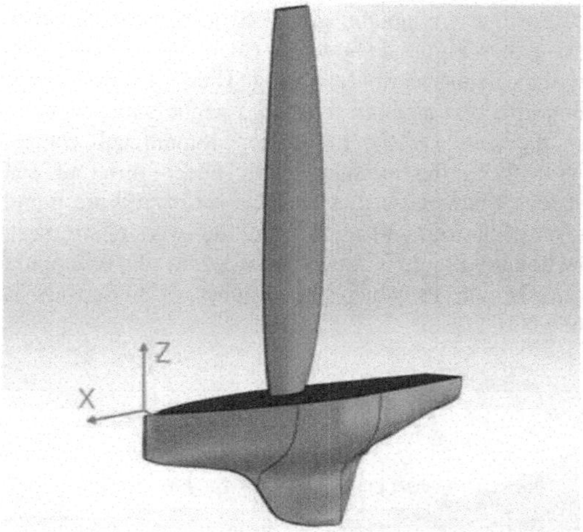

Figure 1. 3D model of sailing boat.

3.2 *Introduction to the ship type used in calculation*

The evaluation criteria of seakeeping of unmanned sailboats are mainly concerned with the motion response, stability and ship performance of the hull in the wave environment. Evaluate the motion response of unmanned sailboats in waves, including yaw and pitch, the magnitude and frequency of which are critical to the stability and controllability of the ship in waves, and have a direct impact on mission performance and equipment performance. This paper will analyze the response of sailing ships under different sea conditions, so as to verify the design standards of unmanned sailing ships that can survive under five sea conditions and can operate under four sea conditions, and explore the response of unmanned sailing ships under extreme sea conditions.

Table 1. Main scale of unmanned sailing vessel.

Argument	Numerical value
Vertical line length /m	3.78
Molded width /m	0.92
Freeboard /m	0.3
Sailing weight /kg	372.26
Center of gravity	(-1.99, 0.08, -0.38)
Moment of inertia/Kg.m^2	(211.33, 292.12, 402.11)

For the overlapping grid used in the calculation in this paper, Figure 2 shows the calculation domain used in the calculation of seakeeping. The length of the captain direction is about five times of the wavelength L, so the length of the calculation domain will change with the different wavelength under different sea states. It is necessary to divide a set of hull mesh and a set of background mesh using the overlapping mesh technique. Figure 3 shows the diagram of overlapping grid layout, in which the number of hull grids is 680,000.

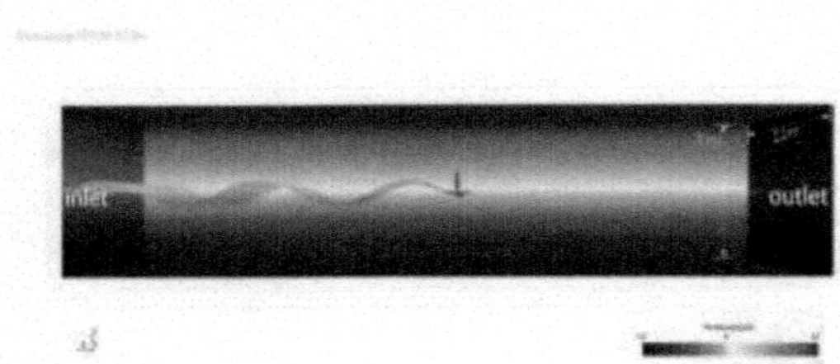

Figure 2. Schematic diagram of seakeeping calculation domain.

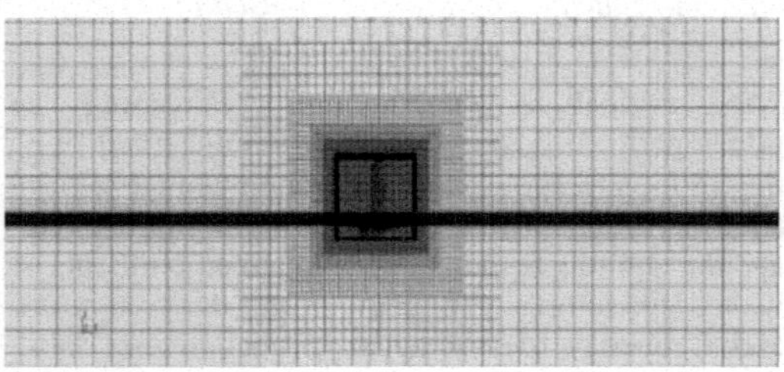

Figure 3. Schematic diagram of overlapping grid for seakeeping calculation.

The wave calculated in this paper is a regular wave, and the ratio of wave height to wavelength is 0.14. In order to explore the motion response of sailing ships in regular waves, the Settings of different calculated sea states are shown in Table 2.

Table 2. Different calculated sea state settings.

Class of sea state	Wave height /m	Wave length /m	Initial wave direction Angle /°	Real ship speed /kn
3	1	7.14	90	0
4	1.5	10.71	90	0
4	2.2	15.71	90	0
5	4	21.43	90	0

3.3 *Numerical simulation of seakeeping of different waves*

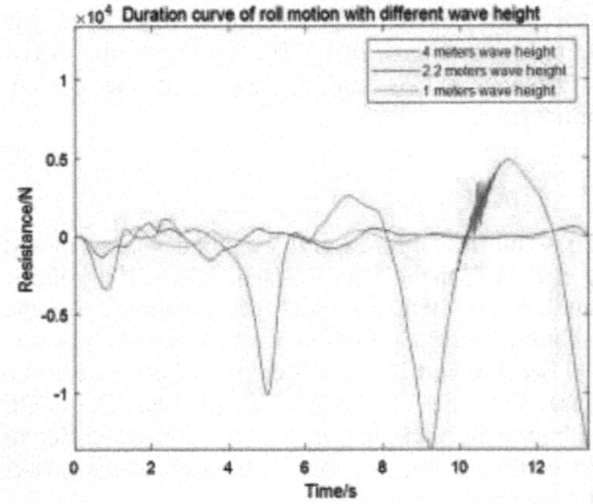

Figure 4 (a). Resistance.

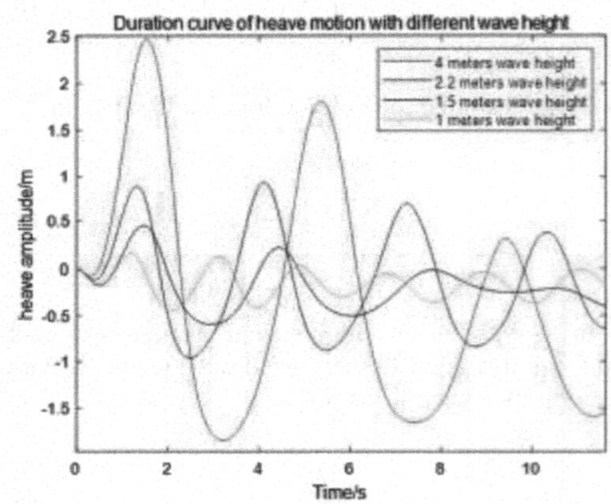

Figure 4 (b). Heave.

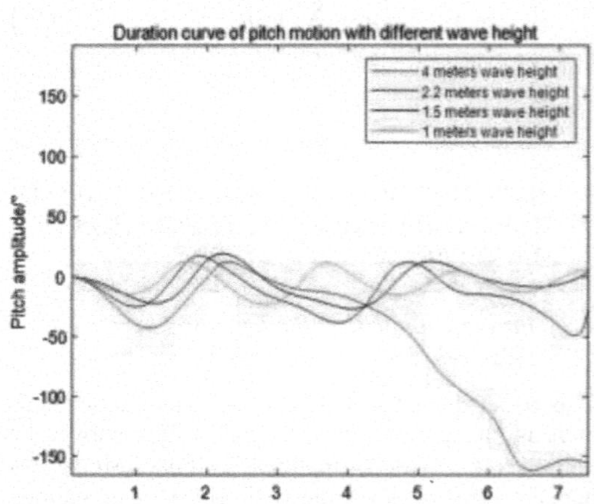

Figure 4 (c). Pitch.

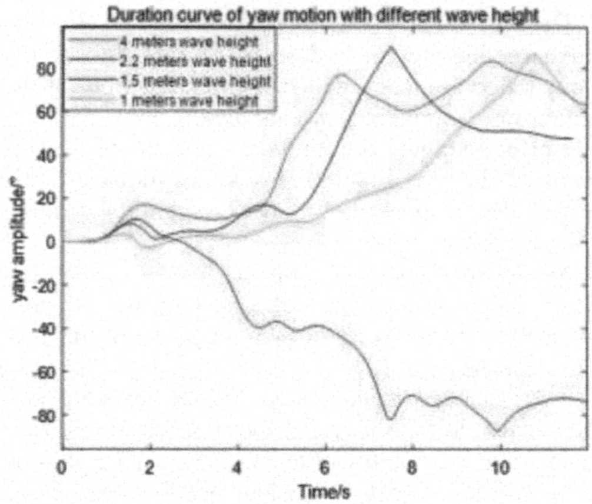

Figure 4. (d) Yaw.

The following is a numerical simulation analysis of the seakeeping performance of unmanned sailboats in different waves. Figure 4 shows the response curve of unmanned sailing vessels at 1m, 1.5 m, 2.2m and 4 m, and the resistance curve of sailing vessels at 1m, 2.2m and 4 m. As can be seen from Figure 4 (a), the resistance of unmanned sailboat shows a large fluctuation with the increase of wave height, and the lowest value of the resistance is negative. The reason is that the wave Angle of the sailboat changes at this time, and the direction of the bow is not perpendicular to the direction of the wave, resulting in negative resistance. It can be seen from Figure 4 (b) and Figure (c) that the heave and pitch motion of unmanned sailing ships will produce pulsating changes with time, that is, the value of the motion presents a certain sinusoidal characteristic, and the average amplitude of the motion increases with the increase of the wave height. It is worth mentioning that at the height of four meters, the sailboat will be airborne after passing the crest, so the pitch Angle reaches 150 degrees. Figure 4 (d) shows that the rolling motion with different wave heights has no obvious rule, and its amplitude does not change significantly with the increase of wave height.

3.4 *Numerical simulation of seakeeping of different waves*

Under the condition that the wave height is 1.5m and the design speed is 2kn, the attitude changes and wave generation of the unmanned sailboat during a wave period are observed, as shown in Figure 5.

It can be seen that unmanned sailboats undergo a series of attitude changes in a wave cycle, which are related to the wave cycle and wave height, mainly including the following stages:

Rising stage: When the boat touches the rising part of the wave, the boat will rise with the rising wave. In this stage, the front of the boat rises with the wave, while the stern follows the rising wave.

Peak stage: When the boat reaches the highest point of the wave, the hull may remain relatively stable. At this moment, the boat will lean slightly forward or backward, depending on the shape of the waves and the design of the hull.

Descent stage: As the wave goes down, the hull of the sailboat also goes down. At this stage, the front of the hull follows the wave down, while the stern lags slightly behind.

Bottom stage: When the boat reaches the lowest point of the wave, the boat experiences a moment of relative stability. This stage is accompanied by the forward tilt of the hull.

It can be seen that in the peak stage, the unmanned sailboat will slightly lift the stern, and in the descent stage, the unmanned sailboat will appear to the left. The simulated unmanned sailing boat has gone through three iterations, and the optimized design of the large aspect ratio sailing boat provides sufficient buoyancy reserve and initial stability for the current medium aspect ratio sailing boat and the small aspect ratio sailing boat, and the buoyancy reserve provides the basis for the realization of large inclination Angle and even reverse stability. And long keel can produce large motion damping, dissipate motion energy and reduce motion response. In addition, the ship adopts the inclined type near the stern waterline, which can reduce the stern wave attack and avoid large stern lifting. There is also a larger buried depth of the rudder, to a certain extent to avoid the stall problem, improve the maneuverability in the wind and waves. It is in line with the design concept of long endurance and low response of the unmanned sailboat.

In a wave cycle, the unmanned sailboat will also experience different wave changes within a wave cycle. The changes are as follows:

Initial stage: The sailboat first encounters the first half of the wave. At this stage, the front of the hull rises with the waves, forming a wave bulge. At the same time, the stern may still be under the influence of the previous wave, gradually sinking.

Figure 5 (a). Attitude of sailing boat at t=0.26s.

Figure 5 (b). Attitude of sailing boat at t=0.94s.

Figure 5 (c). Attitude of sailing boat at t=1.4s.

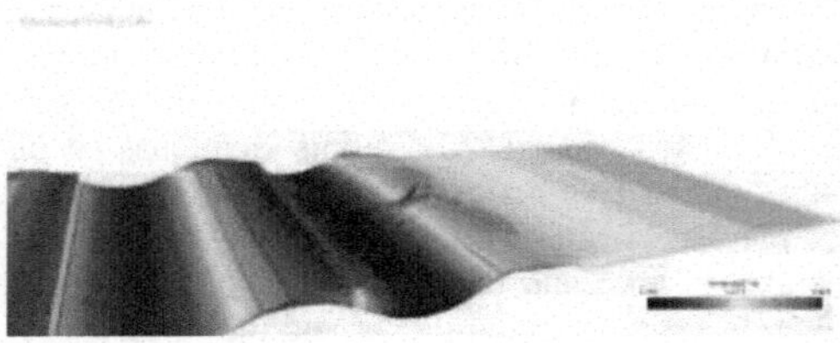

Figure 5 (d). Attitude of sailing boat at t=1.9s.

Figure 5 (e). Attitude of sailing boat at t=2.68s.

Figure 5 (f). Attitude of sailing boat at t=4.16s.

Figure 5 (g). Attitude of sailing boat at t=5.34s.

Rising stage: As the wave rises, the bulge in the front of the hull gradually increases, forming a wave peak. The ship will gradually lift up to the highest point of the wave, at which point the hull will lean forward slightly.

Descent stage: As the boat passes the crest of the wave, the hull begins to descend. In this process, the wave bulge gradually decreases, and the front of the hull drops, forming a wave flat area.

Bottom stage: The sailboat reaches the lowest point of the wave, creating a valley floor. At this stage, the ship will tilt back slightly, creating a flatter, sunken wave.

Wave advance: The sailboat leaves the current wave and begins to meet the next crest. The ship gradually returns to its horizontal position and prepares for the next wave.

These changes are continuous and the boat's hull attitude adjusts to the shape and speed of the waves. Factors such as the stability of the boat design, the position of the center of gravity of the hull, and the shape of the hull all play a key role in this process. For unmanned sailboats, the autonomous control system may need to constantly adjust sail position, rudder position and other parameters to maintain the stability and control of the hull in waves. The wave changes of a sailing vessel have important implications for both autonomous navigation and mission execution, so these aspects need to be fully considered in the design and control.

3.5 *Numerical simulation of seakeeping of different waves*

Under the condition that the wave height is 1.5m and the design speed is 2kn, the surface pressure of the unmanned sailing boat in one wave period is observed as shown in Figure 6.

The pressure on the hull surface of the unmanned sailboat will undergo complex changes as the waves change. This process is influenced by a number of factors, including the shape of the waves, their frequency, and the speed and attitude of the boat. Figure 6 shows the surface pressure variation of a sailing boat during one cycle, the following is a description of changes.

Figure 6 (a). The surface pressure of the sailing boat at t=0.26s.

Figure 6 (b). The surface pressure of the sailing boat at t=0.94s.

Figure 6 (c). The surface pressure of the sailing boat at t=1.4s.

Figure 6 (d). The surface pressure of the sailing boat at t=1.9s.

Figure 6 (e). The surface pressure of the sailing boat at t=2.68s.

Figure 6 (f). The surface pressure of the sailing boat at t=4.16s.

Figure 6 (g). The surface pressure of the sailing boat at t=5.34s.

Initial stage: When the unmanned sailboat first encounters a wave, the front of the hull is hit by the wave, resulting in higher pressure. The tail of the ship may be in a relatively low-pressure zone because the tail of the ship is still sinking under the influence of the previous wave.

Rising stage: As the wave rises, the front of the hull experiences a gradual increase in pressure as the rising wave exerts an upward force on the hull. The boat will lift, and the pressure on the surface of the hull will rise accordingly.

Peak stage: When the boat reaches the highest point of the wave, the hull experiences the highest pressure. At this moment, the energy of the wave is transferred to the hull, which leans forward slightly, resulting in higher pressure on the front of the hull.

Descent stage: As the wave descends, the pressure at the front of the hull gradually decreases because the force of the wave gradually decreases. At the same time, the hull began to sink and the stern experienced lower pressure.

Bottom stage: When the sailboat reaches the lowest point of the wave, the hull undergoes a relatively smooth state with relatively low pressure on the hull surface.

Wave advance: The sailboat leaves the current wave and the hull refaces the next crest. In this process, the hull gradually returns to the horizontal attitude, and the pressure on the hull surface gradually returns to the initial state.

The surface strength of unmanned sailboat will not only affect the sailing attitude of the sailboat, but also be a great test to the strength of the hull. Many factors are involved, including wave height, frequency, wind strength, the speed of the sailboat, hull design and structural strength. The following is a detailed description of the effect of surface pressure on hull strength.

When sailing in waves, the hull surface will be affected by the dynamic load caused by waves and wind. This dynamic load will cause alternating stress in the hull structure, which may cause fatigue damage. Under the action of dynamic load for a long time, cracks and deformation of the hull structure may occur, which will affect the strength of the hull. When sailing in waves, the front of the hull may suffer wave impact, resulting in higher surface pressure. This impact load may cause stress concentration in local structures, especially at wave peaks. Prolonged wave impact can cause structural damage, especially in rough sea conditions. Structural deformation: The pressure change of the wave may cause the structural bending, distortion or deformation of the hull. This deformation may exceed the design limit of the hull structure, cause structural fatigue, and thus affect the strength of the hull. The design of the hull and the materials used directly affect its compressive strength. Designers need to consider the overall stiffness of the hull, bending strength, torsional strength and other parameters. Reasonable structural design and material selection can reduce the stress on the hull and improve its overall strength. Some sailboat structures may require additional reinforcement to handle the extra load brought on by waves and wind. This may include the addition of reinforced structures at key points, the thickening of hull surfaces or the use of high-strength materials.

4 CONCLUSIONS

This paper conducts numerical simulation of seakeeping for the designed unmanned sailboat, and mainly draws the following conclusions:

1. When the wave height is 1m, 1.5m, 3m and 4m, the resistance of the unmanned sailboat is periodic, and the average amplitude of its trim and heave motion increases with the increase of the wave height. Under these four conditions, when the wave height is 4m, that is, level 5 sea conditions, the roll of the unmanned sailboat will be greater than 90°. Therefore, if it is necessary to continue to evaluate its survivability under level 5 sea conditions, it is necessary to continue to evaluate whether the stress at the mast is less than the allowable stress of the material. In the other three working conditions, the roll of the unmanned sailboat can be quickly righted, ensuring the normal operation of the measuring instruments on the unmanned sailboat.

2. The results of the numerical simulation of seakeeping for unmanned sailboat can be used to guide the development of unmanned sailboat model test, improve the efficiency of the experiment, and can also be mutually verified with the experiment. The next step will be to investigate the wave drag increase of unmanned sailboats in various sea states and the seakeeping of sailboats under fixed wave direction angles.

ACKNOWLEDGEMENTS

This work is partially supported by National Natural Science Foundation of China (No. 52131101), Hubei Provincial Natural Science Foundation, China for Innovation Groups (No. 2021CFA026), and by Hubei Provincial Natural Science Foundation (No. ZRMS2023001640).

This work has also been performed within the Strategic Research Plan of the Centre for Marine Technology and Ocean Engineering, which is financed by the Portuguese Foundation for Science and Technology (Fundação para a Ciência e Tecnologia - FCT) under contract UIDB/UIDP/00134 /2020.

REFERENCES

An, Yang. (2023). Research on Design and Optimiz tion Methods of unmanned sailing ships for optimal speed. Doctoral dissertation.

An, G., Xiang, G., Xiang, X., Datla, R. Resistance prediction and optimization of Tri-SWACH using hybrid surrogate model with particular consideration of outrigger layout. Ocean Engineering, 285, 115239, 2023.

Chen Yuhang, Deng DeHeng, Xu Jinsong.(2023). Dynamic model identification of unmanned sailing ships based on free self-propelled data. Marine Engineering (02),85-89 +122.

Deng Haoyun, Guan Dianzhu, Li Senmao, Wang Xian-Hui. (2023). Influence of Magnus cylinder and variable Angle flap on aerodynamic performance of unmanned sailing ships. Chinese Journal of Ship Research, (01), 170–180. The doi: 10.19693/j.i SSN. 1673-3185.02693.

Fang Zejiang, Tan Junzhe, Ji Guangying, Yuan Peng, Sun Ze and Wang Shujie. (2023). Fluid-structure coupling simulation analysis of flexible sail performance of unmanned sailing vessel. Journal of Ocean University of China (Natural Science Edition) (05), 136–143.

JIA Jinheng, Huang Yan, Zhao Wentao, Sun Zhaoyang and Yan Minxiu. Unmanned sailing boat trajectory tracking method based on model predictive control. 1-13 Chinese Journal of Ship Research. Doi: 10.19693/j.i SSN. 1673-3185.03162.

Kristian L. Wille, Vahid Hassani, Florian Sprenger (2016), Modeling and Course Control of Sailboats, IFAC-Papers Online, Volume 49.

Li Dongqin, Li Ji, Zheng Xin, Li Guo-Huan and Dai Jing-Jing. (2018). Research on resistance characteristics of sail-aided ships based on CFD. Journal of Jiangsu University of Science and Technology (Natural Science Edition) (01), 13–20.

Liu Rulei, Huang Chaoming, An Yunsheng, Liu Xulu & Lin Ye Jin. (2022). Study on sail Angle of attack under minimum turning diameter of sailing vessel. Computer Applications and Software (08), 76–81.

Niu Wenchao, Sun Zhaoyang, Sheng Zhongqi, Yu Jiancheng, Zhao Baode & Zhao Wentao.(2023). Research on the influencing factors and laws of the root stress of the wing sail mast of unmanned sailboat. China Mechanical Engineering (22), 2647–2658.

Sun Ye, Si Xiancai, Ma Jingfeng, Tan Junzhe, Zhang Xuefei and Wang Shujie. (2019). Prediction and analysis of Z-shaped maneuvering performance of small unmanned sailboat. Chinese Journal of Ship Research (02), 15 to 20. The doi: 10.19693/j.i SSN. 1673-3185.01380.

Wang Guoyu, Xu Minghui, He Zhonghua, Li Xinglong, Zhang Xiguang, and Chen Tairan(2023). Study on hydrodynamic characteristics test and numerical calculation of laser Radyr class sailing ship. Journal of Beijing Institute of Technology 1–11.

Wang H T, Xiang G, Yuan Z S, et al. Numerical study on hydrodynamic performance of ducted propeller based on improved body force model[J]. Chinese Journal of Ship Research, 2023, 18(4): 186–196. doi: 10.19693/j.issn.1673-3185.02920.

Xiang, G., Xiang, X., Datla, R., Numerical Study of a Novel Small Waterplane Area USV Advancing in Calm Water and in Waves using Higher Order Rankine Source Method, Engineering Applications of Computational Fluid Mechanics, 2023, 17 (1): 2241892.

Xiang, G., Xiang, X.B. "3D trajectory optimization of the slender body freely falling through water using cuckoo search algorithm". Ocean Engineering, 235:109354. 2021.

Yuan Peng, Xu Hong-shen, Zhang Xue-Fei, Tan Jun-Zhe, Wang Shu-Jie, Hui Wen-hai. (2020). Unmanned sailing vessel tracking based on speed selection switching line of sight method. Chinese Journal of Ship Research (01), 31 to 37. Doi: 10.19693/j.i SSN. 1673-3185.01664.

Zhipeng Shen, Wenna Ding, Yuchen Liu, Haomiao Yu (2023), Path planning optimization for unmanned sailboat in complex marine environment, Ocean Engineering, Volume 269.

Zhao Wentao, Jing Longyue, Sun Zhaoyang, Huang Yan, Xie Hualong. Dynamic collision avoidance method of unmanned sailboat based on improved fast leveling method. 1-15. Chinese Journal of Ship Research doi: 10.19693/j.i SSN. 1673-3185.03241.

Zhang Xuefei, Yuan Peng, Tan Junzhe, Wang Shujie, Xu Hongshen and Sun Ye. (2019). Heading control method of unmanned sailing boats based on fuzzy PID. Chinese Journal of Ship Research. 14 (06), 15 to 21. Doi: 10.19693/j.i SSN. 1673-3185.01518.

Advances in Maritime Technology and Engineering – Guedes Soares & Santos (Eds)

Seakeeping analysis in regular waves using a Smoothed Particle Hydrodynamics solver

S. Wang & C. Guedes Soares
Centre for Marine Technology and Ocean Engineering (CENTEC), Instituto Superior Técnico, Universidade de Lisboa, Lisbon, Portugal

J. González-Cao, J.M. Domínguez-Alonso, I. Martínez-Estévez, A.J.C. Crespo & M. Gómez-Gesteira
Environmental Physics Laboratory (EPhysLab), Centro de Investigación Mariñas (CIM-UVIGO), Universidade de Vigo, Ourense, Spain

ABSTRACT: The study investigates the behaviour of a liquified natural gas carrier in regular head waves using the Smoothed Particle Hydrodynamics meshfree method implemented in DualSPHysics. A numerical wave tank in three-dimensional domain is generated using the modified Dynamic Boundary Conditions and compared with the theoretic wave surface using Stokes equation. Heave and pitch motions of the vessel without forward speed are then predicted in time domain considering various wave heights and frequencies. Comparisons between the model tests measurement and numerical simulations are made on the time series of the motions. Considering the overall good agreement with the experiments and the low computational cost, the Smoothed Particle Hydrodynamics method can be considered a good alternative tool for seakeeping analysis for marine structures.

1 INTRODUCTION

Seakeeping analysis plays a crucial role in marine structures' design and operation, addressing the dynamic interaction between a vessel and the surrounding waves. It helps to ensure the safety of maritime operations and understand the vessel's stability and manoeuvrability in challenging sea states. It is also critical for offshore structures, cruise ships, and naval vessels operating in unpredictable and adverse weather conditions. The accuracy of seakeeping analysis also plays a significant role in the investigation of hydroelasticity of ships considering whipping/springing, and in nonlinear effects, such as slamming and green water impacts.

Model tests or full-scale tests are always an effective way of understanding the motions and loads of vessels in waves. However, the scale effects and limitations of the wave tank and facilities are the concerns in model testing, whereas the high cost and complexity are the drawbacks in full-scale trials. In addition, the performance of ships in real-world conditions is still challenging to replicate accurately in a scaled wave tank. Therefore, researchers often validate the numerical simulations using experimental data to obtain a comprehensive understanding of a ship's performance in various sea conditions.

Numerical methods with different levels of complexity, varying from linear strip theory solutions to complex computational fluid dynamics (CFD) codes, have been developed for seakeeping analysis. In the review of Hirdaris et al. (2014), linear methods (Guedes Soares & Schellin, 1998; Clauss et al., 2010; Datta et al., 2011; Shao & Faltinsen, 2011) are considered as the Level 1 approaches for which the wetted body surface is defined by the mean position of the hull under the corresponding position of the free surface where the free surface boundary conditions are applied. The hydrodynamic problem is solved in the frequency domain by either 2D, 2.5D or 3D idealisations. Fonseca & Guedes Soares (2004a, 2004b) indicated that the linear methods from Guedes Soares & Schellin (1998) and Clauss et al. (2010) could not provide accurate predictions for responses of ships with large bow flare angles.

The Level 2 Froude–Krylov nonlinear methods (Lin et al., 2007; Weems et al., 2007; Liu & Papanikolaou, 2010) were considered the most popular because they may capture many important nonlinear effects with reasonable computational time. Level 3 comprises the body-nonlinear methods where the disturbance potential is calculated for the wetted hull surface defined by the instantaneous position of the hull under the mean position of the free surface. Level 4 includes the body-exact Methods (Mortola et al., 2011; Rajendran et al., 2015; Sclavounos, 2012) in which the wetted hull surface is defined by the instantaneous position of the hull under the incident wave surface.

CFD-based tools that solve RANS equations were classified as fully nonlinear methods that include viscous effects and may address the breaking waves,

DOI: 10.1201/9781003508779-28

slamming and green water on decks. The unsteady RANS code CFD Ship-Iowa V.4 applied in Castiglione et al. (2009) discretised the equations using a finite difference approach. The code RIAM-CMEN applied by Hu et al. (2010) to a containership at Froude number 0.179 in head and bow regular waves was based on the Constrained Interpolation Profile (CIP) method. Tezdogan et al. (2015) and Jiao et al. (2021) applied the commercial CFD software Star-CCM+ which uses a finite volume method for discretising the integral formulation. The open-source code, OpenFOAM based on the Finite Volume method (FVM) has been widely applied for seakeeping analysis (Ok et al., 2017; Wei & Tezdogan, 2021).

The Smoothed Particle Hydrodynamics (SPH) method is one of the most popular Lagrangian approaches, widely used in many engineering fields and inherently able to address complex fluid-structure interactions. Capasso et al. (2023) investigated the behaviour of a planning hull in regular head waves was investigated using the open-source code DualSPHysics based on the SPH meshfree method. The study concluded that the SPH method is effective in addressing planning craft response for the tested wave conditions in both phase coupling and the motion amplitude, though the peaks of response show a range of 10% divergence. Future work regarding the validation of various cases and numerical uncertainties should be performed to confirm the accuracy and reliability of the method.

The objective of the study is to predict the heave and pitch motions of a liquified natural gas carrier (LNG) in regular head waves using the open-source mesh-free DualSPHysics code and validate the numerical simulation using experimental data from model tests.

2 NUMERICAL METHOD

DualSPHysics (Domínguez et al., 2022) is a numerical model based on the Smoothed Particle Hydrodynamics (SPH) method. This model was conceived to apply the SPH methodology to real engineering problems. This model is open source and can be freely downloaded from http://www.dual.sphysics.org. The software of DualSPHysics is highly parallelised and can be run on either multi-core Central Processing Units (CPUs) or Graphics Processing Units (GPUs). GPUs are graphics cards with powerful parallel computing that offer greater computing power than CPUs. The use of GPUs is an affordable option to accelerate SPH computations. This software also includes pre-processing and post-processing tools to make easier the tasks of generating the computational domain and defining parameters and also the analysis of the numerical results.

SPH is a Lagrangian mesh-less method where the fluid (continuum) is discretised into a set of particles. For each particle, physical quantities (such as position, velocity, density, and pressure) are computed as an interpolation of the values of the neighbour particles. The contribution of these neighbours is weighted using a kernel function (W) that measures the contribution from the rest of the particles in terms of the particle spacing. This distance between particles is normalised using the smoothing length (h), which is the characteristic length that defines the area of influence of the kernel (Ω). The kernel presents compact support so that the contribution of particles beyond a cut-off distance (here $2h$) is not considered.

The mathematical fundamentals of SPH are based on integral interpolants. Any function F can be computed by the integral approximation.

$$F(r) = \int F(r')W(r - r', h)dr' \tag{1}$$

being $\boldsymbol{r}$ the point position where the function is computed and $\boldsymbol{r'}$ is a temporary position variable. This function F can be expressed in a discrete form based on particles. Thus, the approximation of the function is interpolated at particle a and the summation is performed over all the particles within the region of compact support of the kernel:

$$F(r_a) = \sum\nolimits_b F(r_b)W(r_a - r_b, h)\frac{m_b}{\rho_b} \tag{2}$$

where the volume associated with the neighbouring particle b is m_b/ρ_b, being m_b and ρ_b mass and density, respectively. The kernel function employed in this work is the Cubic Spline.

The discrete (Lagrangian form) SPH system of equations of a weakly compressible fluid (Navier-Stokes) is, according to Monaghan (1992):

$$\begin{cases} \frac{d\rho_a}{dt} = \sum_b m_b v_{ab} \cdot \nabla_a W_{ab} + D_a \\ \frac{dv_a}{dt} = -\sum_b m_b \left(\frac{P_b + P_a}{\rho_b \rho_a}\right) \nabla_a W_{ab} + g + \Gamma_a \\ \frac{dr_a}{dt} = v_a \end{cases} \tag{3}$$

where t is time, $\boldsymbol{r}$ position, $\boldsymbol{v}$ velocity, P pressure, ρ density, m mass, $\boldsymbol{g}$ gravitational acceleration and W_{ab} the kernel function that depends on the distance between particles a and b. $\boldsymbol{\Gamma}_a$ describes the dissipation:

$$\Gamma_a = \sum\nolimits_b m_b \frac{4\upsilon_0 r_{ab} \cdot \nabla_a W_{ab}}{(\rho_a + \rho_a)(r_{qb}^2 + 0.01h^2)} v_{ab} + \\ + \sum\nolimits_b m_b \left(\frac{\vec{\tau}_a^{ij} + \vec{\tau}_b^{ij}}{\rho_a \rho_a}\right) \nabla^i W_{ab} \tag{4}$$

In Equation (4), υ_0 is the so-called laminar viscosity introduced by Lo & Shao, (2002). The second term of the right-hand side of Equation (4) describes the sub-particle scale model (SPS). The term D_a included in Equation (3) is the density diffusion term and is included to reduce fluctuations in the density field according to Fourtakas et al., (2019):

$$D_a = 2\delta h c_s \sum_b \left(\rho_{ba}^T - \rho_{ba}^H\right) \frac{r_{ab} \nabla_a \cdot W_{ab}}{r_{ab}^2} \frac{m_b}{\rho_b} \tag{5}$$

where c_s is the speed of sound, δ is equal to 0.1 (diffusive control coefficient) and superscripts T and H stand for the total and hydrostatic components of the density. Some inconsistencies arise from this approach, but a correction was introduced by Fourtakas et al. (2019) obtaining consistent pressure fields near boundaries.

The system of equations is closed using Tait's equation of state to describe the weakly compressible SPH formulation implemented in DualSPHysics. This allows computing pressure from density values as:

$$P = B\left[\left(\frac{\rho}{\rho_0}\right)^{\gamma} - 1\right] \tag{6}$$

where $\gamma = 7$ is the polytropic constant and $B = c_s^2 \rho_0 / \gamma$, $\rho_0 = 1000$ kg/m^3 the reference density.

The modified dynamic boundary conditions (mDBC) are implemented according to English et al., (2022). This implementation improves the previous DBC (Crespo et al., 2007) since it allows more accurate pressure computations and reduces the gaps between boundary and fluid particles. The reader can find a detailed description of this implementation in English et al. (2022), Capasso et al. (2021), Altomare et al. (2021) and Suzuki et al. (2022).

To generate waves DualSPHysics implements different techniques. In this work, waves are generated using the moving boundary particles method. The physical flumes are reproduced by using the displacement of a piston or flap to generate the waves. The movement of the piston (flap) is computed according to 1st or 2nd order wave theory-generated solution for regular or random waves (Altomare et al., 2017). This implementation allows the definition of absorption areas to avoid reflection (damping areas) or re-reflection of waves using an Active Wave Absorption System (AWAS).

DualSPHysics also implements the equation of the rigid body dynamics to reproduce the movement of fluid-driven objects. For this DualSPHysics discretizes the rigid body as a set of boundary particles. Each boundary particle k experiences a force $\boldsymbol{f}_k$ equal to:

$$f_k = \sum_b \frac{dv_{kb}}{dt} + g \tag{7}$$

Then the equation of the rigid body dynamics is solved following:

$$M\frac{dv}{dt} = \sum_k m_k f_k \tag{8}$$

$$I\frac{d\Omega}{dt} = \sum_k m_k \left(r_k - R_0\right) \times f_k \tag{9}$$

The term $\boldsymbol{M}$ is the total mass of the object, $\boldsymbol{I}$ the moment of inertia matrix, $\boldsymbol{V}$ the velocity, $\boldsymbol{\Omega}$ the angular velocity and $\boldsymbol{R_0}$ the centre of mass. Then, Equations (8) and (9) are integrated in time to obtain the values of $\boldsymbol{V}$ and $\boldsymbol{\Omega}$ at the beginning of the next time step. Each boundary particle belonging to the body has the following velocity:

$$v_k = V + \Omega \times (r_k - R_0) \tag{10}$$

A detailed description of this implementation and validation cases is shown in Monaghan et al. (2003), Canelas et al. (2015) and Domínguez et al. (2019b).

3 CASE OF STUDY

3.1 *Description of the model test*

As part of the EU project EXTREME SEAS, the Ocean Engineering Division of the Technical University Berlin conducted an experimental study on the ship motions and wave-induced loads of an LNG carrier in extreme seas. The seakeeping basin, with dimensions of 110 m in length, 8 m in width, and a water depth of 1m, was utilized for these experiments. The model used had a scale of 1:70 (Clauss et al., 2011).

Table 1. Main particulars of the LNG carrier.

Items	Full scale	Model scale
Loa (m)	197.13	2.816
Lpp (m)	186.9	2.67
Breath (m)	30.38	0.434
Depth (m)	18.2	0.268
Draft (m)	8.4	0.12
Disp.(t)	35675	0.104
xcg (m)	94.88	1.355
Zcg (m)	8.23	0.118
Ixx (tm^2)	4505225	0.00268
Iyy (tm^2)	58507468	0.03481
Ixx (tm^2)	57450042	0.03418

Measurements of the ship's motions in all six degrees of freedom at the centre of gravity were carried out. Various scenarios, including regular waves, irregular waves, and real-world extreme sea states, were considered. The model was fixed using an elastic suspension system with a triangular towing arrangement, ensuring fixation without inducing a moment during the tests. To restrict longitudinal motions, a spring at the front and a counterweight at the back of the model were employed. This arrangement allowed for unrestrained heave and pitch motions, along with accurate measurements of forces and moments.

The main particulars of the investigated models are listed in Table 1. The sketch of the investigated LNGC with the coordinate system and the manufactured model is shown in Figure 1, while the ship body lines are presented in Figure 2. The ship model

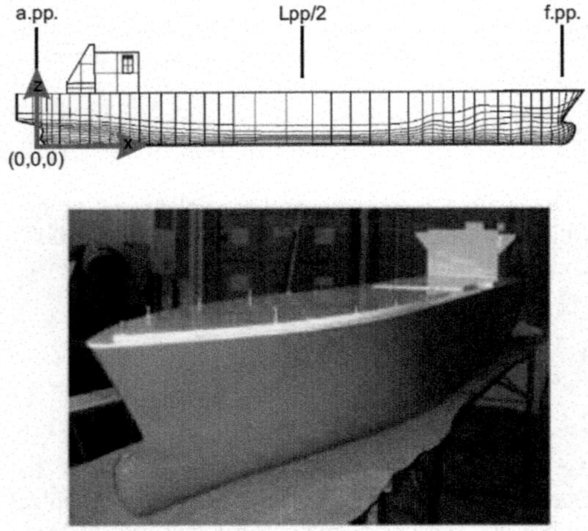

Figure 1. Sketch of the investigated LNGC with the coordinate origin and the manufactured model.

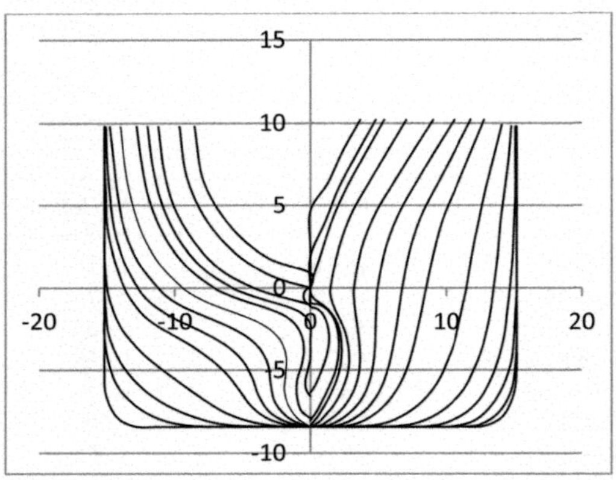

Figure 2. Cross-sections of the LNG carrier in full scale (unit:m).

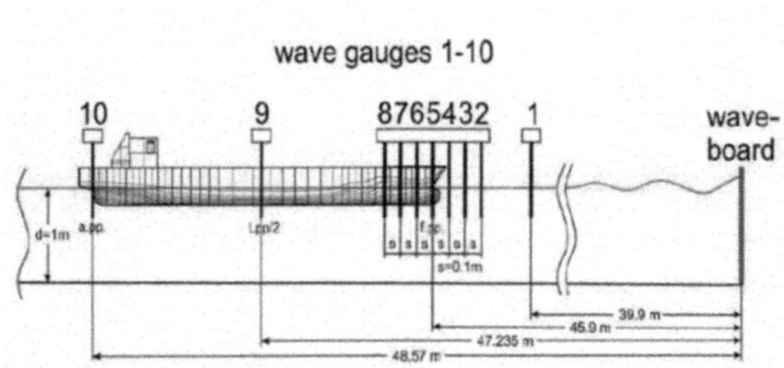

Figure 3. Sketch of the arrangement of the wave gauges.

was constructed from fiberglass-reinforced plastic (GRP) and segmented at the mid-section (Lpp/2). The two segments were linked by three force transducers, responsible for recording longitudinal forces throughout the model tests. To measure the water elevation accurately, ten wave gauges were strategically installed at predefined locations on the undisturbed water surface. Figure 3 illustrates the longitudinal arrangement of wave gauges for the tests conducted without forward speed. Considering the model's width, the gauges were positioned laterally, running parallel to the ship's orientation.

The ship's motions were recorded by an optical tracking system. The tracking system consists of a ceiling frame that carries five infrared cameras, which were shifted parallel to the moving ship model. The system enables high precision, contact-free motion tracking over large distances by following the trajectories of infrared light-emitting diodes mounted on the ship model. More details on the test facilities and test setups can be found in Klein et al. (2023). Previous studies regarding the numerical simulations of vertical response (Zhang et al., 2022), slamming loads on the hull (Wang and Guedes Soares, 2023), and the random experimental uncertainty (Wang and Guedes Soares, 2022) for the LNG carrier, show the reliability of the measurements.

In the present study, the regular wave case with λ/Lpp = 0.8, ω = 0.6421 rad/s and a wave height of 2 m in full scale is analysed. The numerical prediction of the wave surface elevation, heave and pitch motions will be compared to the measurements.

3.2 *Numerical modelling*

In this section, the numerical configuration is described. The numerical water tank, shown in Figure 4, consists of an 8 m length and 1.7 m wide 3D rectangular wave tank. The water depth was defined as equal to 1 m. The value of the initial inter-particle distance (*dp*) was set to 0.015 m. These dimensions, along with the value of *dp*, were defined by running several numerical simulations to obtain the minimum sizes required to obtain accurate results to reduce the runtimes of the simulations.

Table 2. Main parameters of DualSPHysics.

Name	Value
Coefsound	20
Viscosity	Artificial
ViscoBoundFactor	0
Visco	0.01
coefh	1.2
cflnumber	0.2
shifting	0
ShiftCoef	-2
Simulation time	40 s
Time out	0.1

Specifically, the value of *dp* was obtained after running several simulations to analyse the wave generation and propagation in the numerical tank. Figure 4 also shows the location of the wave piston generator, where the Active Wave Absorption System (AWAS)

was included and the damping zone was defined to avoid the reflection of waves at the end of the numerical water tank. The values of the main parameters of DualSPHysics are shown in Table 2. The number of fluid particles generated in this simulation was approximately 3.8 M. The simulations were performed using GPU in NVIDIA GeForce GTX 1660 SUPER. For a simulation with 40s duration, it took 70179s (around 20 hours).

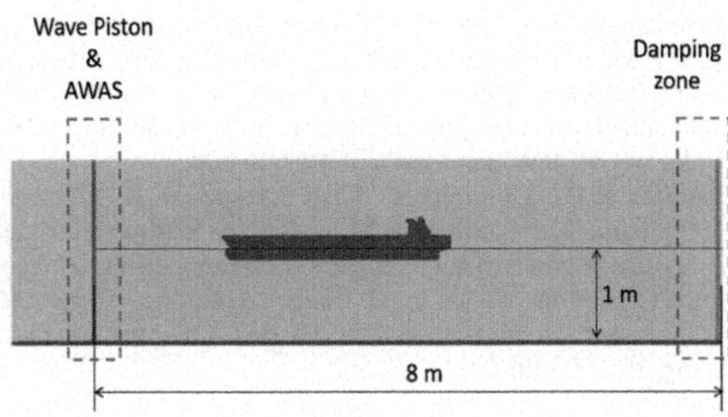

Figure 4. Computational modelling setup within DualSPHysics.

4 RESULTS AND DISCUSSIONS

First, the accuracy of DualSPHysics to generate the specific wave is analysed in this work without the hull as shown in Figure 5. The solid red line stands for the numerical results obtained with *dp* equal to 0 0.015 m and the dashed black line represents the theoretical wave. This is a second-order wave according to Le Méhauté, (2013) and DualSPHysics can automatically generate it by defining its height and period. It is seen that the numerical wave height and phase were in satisfactory agreement with the theoretical ones, though the peaks from DualSPHysics are slightly higher in the later stage. The time histories of the heave and pitch motions in the generated numerical wave tank are predicted using DualSPHysics and compared with experimental data in Figure 6.

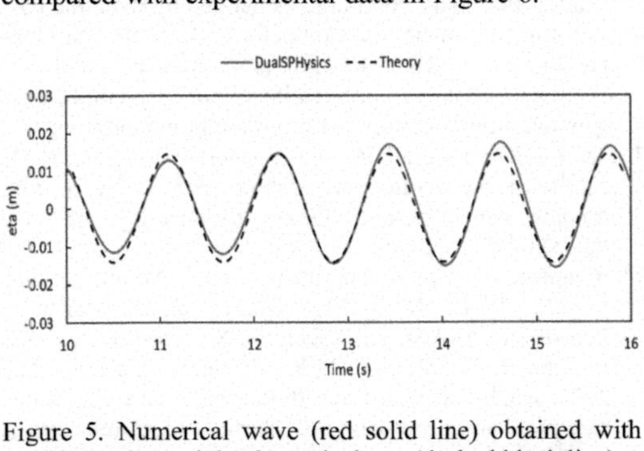

Figure 5. Numerical wave (red solid line) obtained with DualSPHysics and the theoretical one (dashed black line).

For both heave and pitch motions in regular waves, the predictions from the mesh-free method

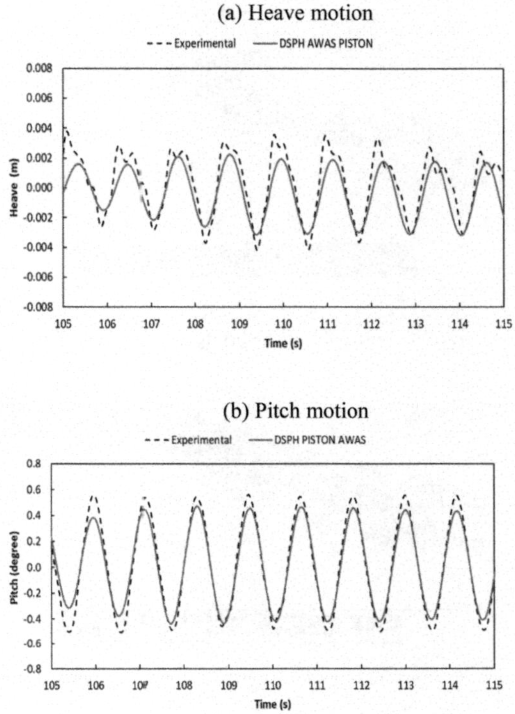

Figure 6. Comparison of the heave and pitch motion between the numerical prediction and model test. (a) heave motion; (b) pitch motion.

DualSPHysics are quite well in phase, compared to the measurements, however the peaks are slightly lower. It is probably due to the discretisation of the numerical setup. A further verification analysis is required.

Figure 7 shows the contour of the velocity field around the hull at three different time instants. The pattern of the velocity field clearly shows the propagation of the waves during the simulation.

5 CONCLUSIONS

The heave and pitch motions of an LNG carrier subjected to a head regular wave are predicted using an open-source mesh-free method DualSPHysics in the study. A 3D wave tank was generated first and validated with theoretical values on the wave surface elevation, and then the time series of heave and pitch motion were compared with the measurements from model tests. The comparisons show satisfactory predictions from the approach which could be a good alternative tool for seakeeping analysis. However, only one case and a single discretisation are considered in the present study. Further validation and verification analysis are required to confirm the accuracy and reliability of the method.

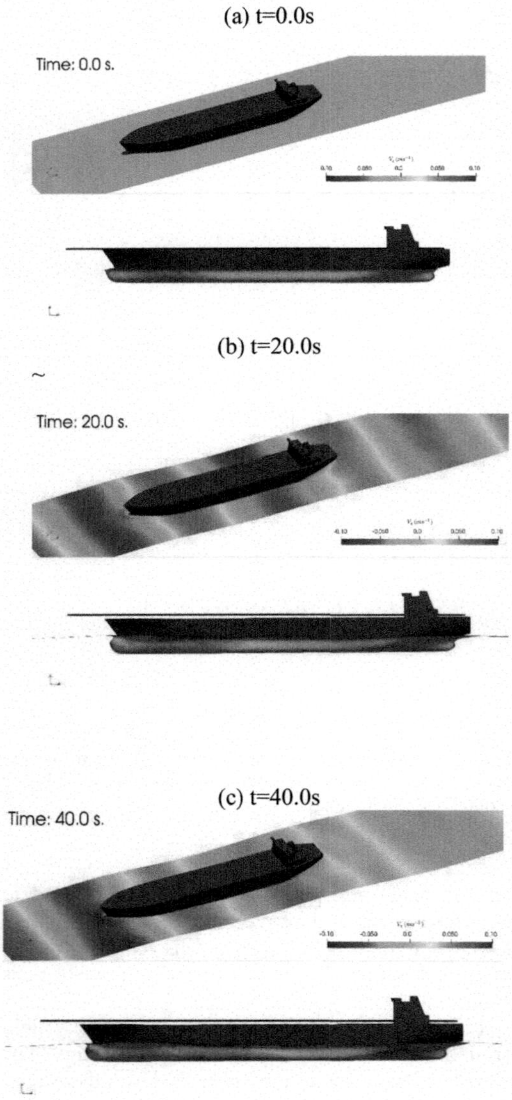

Figure 7. Velocity field contour of the simulation. (a) t=0s; (b) t=20s; (c) t=40s.

ACKNOWLEDGEMENTS

This work was performed within the scope of the Strategic Research Plan of the Centre for Marine Technology and Ocean Engineering (CENTEC), which is financed by the Portuguese Foundation for Science and Technology (Fundação para a Ciência e Tecnologia) under contract UIDB/UIDP/00134/2020. This work was also funded by Xunta de Galicia (Spain) under project ED431C 2021/44 "Programa de Consolidación e Estructuración de Unidades de Investigación Competitivas (Grupos de Referencia Competitiva)" financed by Xunta de Galicia, Consellería de Cultura, Educación e Universidade, Spain. I. Martínez-Estévez acknowledges funding from Xunta de Galicia, Spain under "Programa de axudas á etapa predoutoral da Consellería de Cultura, Educación e Universidades da Xunta de Galicia" (ED481A-2021/337).

REFERENCES

Altomare, C., Gironella, X., Crespo, A.J., 2021. Simulation of random wave overtopping by a WCSPH model. Appl. Ocean Res. 116.

Altomare, C., Domínguez, J., Crespo, A., González-Cao, J., Suzuki, T., Gómez-Gesteira, M., Troch, P., 2017. Long-crested wave generation and absorption for SPH-based DualSPHysics model. Coast. Eng. 127, 37–54. http://dx.doi.org/10.1016/j.coastaleng.2017.06.004.

Canelas, R.B., Domínguez, J.M., Crespo, A.J., Gómez-Gesteira, M., Ferreira, R.M., 2015. A Smooth Particle Hydrodynamics discretization for the modelling of free surface flows and rigid body dynamics. Internat. J. Numer. Methods Fluids 78 (9), 581–593.http://dx.doi.org/10.1002/fld.4031.

Capasso S, Tagliafierro B, Mancini S, Martínez-Estévez I, Altomare C, Domínguez JM, Viccione G. Regular Wave Seakeeping Analysis of a Planing Hull by Smoothed Particle Hydrodynamics: A Comprehensive Validation. Journal of Marine Science and Engineering. 2023; 11(4):700.

Capasso, S., Tagliafierro, B., Güzel, H., Yilmaz, A., Dal, K., Kocaman, S., Viccione, G., Evangelista, S., 2021. A numerical validation of 3D experimental dam-break wave interaction with a sharp obstacle using DualSPHysics. Water 13 (15).

Castiglione, T., Stem, F., Kandasamy, M., Bova, S., 2009. Unsteady RANS simulations for a catamaran advancing in regular waves. In: Proceedings of the 10th International Conference on Fast Sea Transportation, Athens, Greece.

Clauss, G.F., Klein, M., Dudek, M., 2011. Model test report-Test Campaign I (TCI)-LNG Carrier (TUB), EXTREME SEAS project report.

Clauss, GF, Klein, M and Dudek, M. (2010) Influence of the bow shape on loads in high and steep waves. In International Conference on Offshore Mechanics and Arctic Engineering (Vol. 49101, pp. 159–170).

Crespo, A. J., Gómez-Gesteira, M., & Dalrymple, R. A. (2007). Boundary conditions generated by dynamic particles in SPH methods. Computers, Materials and Continua, 5(3), 173–184. https://www.scholars.northwestern.edu/en/publications/boundary-conditions-generated-by-dynamic-particles-in-sph-methods

Datta, R., Rodrigues, J.M. and Guedes Soares, C., 2011. Study of the motions of fishing vessels by a time domain panel method. Ocean Engineering, 38(5-6), pp.782–792.

Domínguez, J. M., Fourtakas, G., Altomare, C., Canelas, R. B., Tafuni, A., García-Feal, O., Martínez-Estévez, I., Mokos, A., Vacondio, R., Crespo, A. J. C., Rogers, B. D., Stansby, P. K., & Gómez-Gesteira, M., 2022. DualSPHysics: From fluid dynamics to multiphysics problems. Computational Particle Mechanics, 9(5), 867–895.

Domínguez, J., Crespo, A., Hall, M., Altomare, C., Wu, M., Stratigaki, V., Troch, P., Cappietti, L., Gómez-Gesteira, M., 2019b. SPH simulation of floating structures with moorings. Coast. Eng. 153, 103560. http://dx.doi.org/10.1016/j.coastaleng.2019.103560.

English, A., Domínguez, J., Vacondio, R., Crespo, A., Stans-by, P., Lind, S., Chiapponi, L., Gómez-Gesteira, M., 2022. Modified dynamic boundary conditions (mDBC) for general-purpose smoothed particle hydrodynamics (SPH): application to tank sloshing, dam break and fish pass problems. Comput. Part. Mech. 9 (5), 1–15.

Fonseca N, Guedes Soares C. (2004a) Experimental investigation of the nonlinear effects on the vertical motions and loads of a containership in regular waves. Journal of Ship Research, 48(02), pp.118–147.

Fonseca N, Guedes Soares C. (2004b) Experimental investigation of the nonlinear effects on the statistics of vertical motions and loads of a containership in irregular waves. Journal of Ship Research, 48(02), pp.148–167.

Fourtakas, G., Dominguez, J.M., Vacondio, R., Rogers, B. D., 2019. Local uniform stencil (LUST) boundary condition for arbitrary 3-D boundaries in parallel smoothed parti-cle hydrodynamics (SPH) models. Comput. & Fluids 190, 346–361. http://dx.doi.org/10.1016/j.compfluid.2019.06.009.

Guedes Soares C, Schellin TE (1998) Nonlinear effects on long-term distributions of wave-induced loads for tankers. Journal of Offshore Mechanics and Arctic Engineering, 120(2).

Hirdaris, S.E., Bai, W., Dessi, D., Ergin, A., Gu, X., Hermundstad, O.A., Huijsmans, R., Iijima, K., Nielsen, U.D., Parunov, J. and Fonseca, N., 2014. Loads for use in the design of ships and offshore structures. Ocean Engineering, 78, pp.131–174

Hu, C., Sueyoshi, M., Miyake, R., Zhu, T., 2010. Computation of fully nonlinear wave loads on a large container ship by CIP based Cartesian grid method. In: Proceedings of the 29th International Conference on Ocean. Offshore and Arctic Engineering, Shanghai, China

Jiao, J., Huang, S., Wang, S. and Guedes Soares, C., 2021. A CFD–FEA two-way coupling method for predicting ship wave loads and hydroelastic responses. Applied Ocean Research, 117, p.102919.

Klein, M., Wang, S., Clauss, G. and Guedes Soares, C., 2023. Experimental Study on the Effect of Extreme Waves on a LNG Carrier. Journal of Marine Science and Application, 22(1), pp.52–74. https://doi.org/10.1007/s11804-023-00321-1

Le Méhauté, B. An Introduction to Hydrodynamics and Water Waves; Springer Science & Business Media: Berlin/Heidelberg, Germany, 2013.

Lin, W.M., Collette, M., Lavis, D., Jessup, S. and Kuhn, J., 2007. Recent hydrodynamic tool development and validation for motions and slam loads on ocean-going high-speed vessels. In 10th International Symposium on Practical Design of Ships and Other Floating Structures (Vol. 10).

Liu, S.K., Papanikolaou A., 2010. Time-domain hybrid method for simulating large amplitude motions of ships advancing in waves. In: Proceedings of ITTC Benchmark Study and Workshop on Seakeeping, Seoul, Korea.

Lo, E.Y.M., Shao, S., 2002. Simulation of near-shore soli-tary wave mechanics by an incompressible SPH method. Appl. Ocean Res. 24 (5), 275–286. http://dx.doi.org/10.1016/S0141-1187(03)00002-6.

Monaghan, J.J. 1992. Smoothed particle hydrodynamics. Annual Review of Astronomy and Astrophysics, 30 (1): 543–574. DOI: 10.1146/annurev.aa.30.090192.002551.

Monaghan, J., Kos, A., Issa, N., 2003. Fluid motion generated by impact. J. Waterw. Port Coast. Ocean Eng. 129 (6), 250–259. http://dx.doi.org/10.1061/(ASCE)0733-950X(2003)129:6(250).

Mortola, G., Incecik, A., Turan, O. and Hirdaris, S.E., 2011. Non linear analysis of ship motions and loads in large amplitude waves. International Journal of Maritime Engineering, 153(A2).

Ok, H.T., Lee, S.J. and Choi, J.H., 2017. Numerical simulation of motion of single and side-by-side vessels in regular waves using OpenFOAM. Ships and Offshore Structures, 12(6), 793–803.

Rajendran, S., Fonseca, N. and Guedes Soares, C., 2015. Simplified body nonlinear time domain calculation of vertical ship motions and wave loads in large amplitude waves. Ocean Engineering, 107, 157–177.

Sclavounos, P.D., 2012. Nonlinear impulse of ocean waves on floating bodies. Journal of Fluid Mechanics, 697, 316–335.

Shao, Y. and Faltinsen, O.M., 2011. Numerical study of the second-order wave loads on a ship with forward speed. In 26th International Workshop on Water Waves and Floating Bodies.

Suzuki, T., García-Feal, O., Domínguez, J.M., Altomare, C., 2022. Simulation of 3D overtopping flow–object–structure interaction with a calibration-based wave generation method with DualSPHysics and SWASH. Comput. Part. Mech. http://dx.doi.org/10.1007/s40571-022-00468-8. URL: https://www.scopus.com/inward/record.uri?eid=2-s2.0-85126228694&doi=10.1007%2fs40571-022-00468-8&partnerID=40&md5=628-cef1294546869b94814a40627f928

Tezdogan, T., Demirel, Y.K., Kellett, P., Khorasanchi, M., Incecik, A. and Turan, O., 2015. Full-scale unsteady RANS CFD simulations of ship behaviour and performance in head seas due to slow steaming. Ocean Engineering, 97, pp.186–206.

Wang, S. and Guedes Soares, C., 2023. Analysis of the experimental data of slamming loads on a liquefied natural gas carrier in abnormal waves. Journal of Offshore Mechanics and Arctic Engineering, 145(4), 041702.

Wang, S. and Guedes Soares, C., 2022. Random Experimental Uncertainty Analysis on the Model Tests of an LNG Carrier in Extreme Seas. In International Conference on Offshore Mechanics and Arctic Engineering (Vol. 85901, p. V05BT12A010). American Society of Mechanical Engineers.

Weems, K.M., Zhang, S., Lin, W.M., Kery, S.M., 2007. 3D potential flow simulation of multiple ships in waves for ship-ta-ship transfer at sea. In:Proceedings of the 10th International Symposium on Practical Design of Ships and Other Floating Structures PRADS, Houston, USA.

Wei, Y. and Tezdogan, T., 2021. A CFD study of the behavior of a crew transfer vessel in head seas using OpenFOAM. In 2nd International Conference on Ship and Marine Technology.

Zhang, H., Cui, J., Liao, X., Shi, H. and Guedes Soares, C., 2022. Numerical study on the vertical response of LNG carrier in abnormal waves generated with different mechanisms. Ocean Engineering, 262, 112090

Wave loads

Advances in Maritime Technology and Engineering – Guedes Soares & Santos (Eds)

On the estimation of rogue wave loads on passenger ship superstructures

M. Acanfora, F. De Luca & R. Pasqua
Department of Industrial Engineering, University of Naples Federico II, Naples, Italy

ABSTRACT: This research concerns the analysis of the impact loads by abnormal waves, on ship structures of passenger vessels. In particular, the study aims to point out the vulnerability of some structural elements of ship superstructures, demonstrating that the standard design loads can be relevantly exceeded. At the scope, we exploit the so-called "New Wave" technique, for simulating rogue waves by mono-directional focalization of the components of the sea spectrum, together with a numerical method for the estimation of the wave impact pressures. In this research framework, we will generate several rogue waves and we will compare the height of the wave crest with the vertical position of the superstructures. The impact pressure of the rogue wave is finally estimated and compared with the design pressure that is the value prescribed by the Classification Society Rules.

1 INTRODUCTION

Rogue waves are a serious treat for the safety of passenger ships. Starting from the New Year's wave, one of the first ever recorded rogue waves, the observation of these abnormal phenomena at sea became more frequent and dangerous. Rogue waves – named also freak, abnormal, or giant waves – are exceptionally steep and high waves compared to the surrounding waves.

In (Toffoli et al., 2003) five years ship accidents due to bad weather were analyzed; although only a few accidents were categorized as being caused by freak waves (Gunson & Lehner, 2001), it does not exclude that other ship accidents under severe weather were caused by rogue waves. A more recent study reports on the geographical distribution of the rogue waves accidents in 2006-2010, both in shallow and deep waters (Nikolkina & Didenkulova, 2011), showing that there is a serious risk of ship accidents involving rogue waves.

Several research projects were established to study the development of extreme wave events and their effects on ship structures such as: MAXWAVE, SEAMOCS, EXTREME SEAS and others, as reported and described in (Bitner-Gregersen & Gramstad, 2016). Most of these projects focused on the modeling approaches for rogue waves, on the possibility of predicting them and on the design problems of the hull girder. In particular, in EXTREME SEA project greater attention was given to the ship responses and loads in abnormal waves: effect on vertical bending moments(Rajendran et al., 2016) (Zhang & Guedes Soares, 2016) (Guo et al., 2016), slamming occurrence (Wang et al., 2016), green water events (Hu et al., 2015; Huijsmans et al., 2004; Qin et al., 2017a, 2017b). Minor importance was given to the consequences of rogue wave effects on superstructure design. Moreover, these researches were mainly conducted on cargo vessels, in particular on chemical tankers, LNG and container carriers.

In 2010, the Louis Majesty cruise vessel was hit by rogue waves (the so called three sisters) that caused the breaking of some windows at a high superstructure deck (i.e. deck number five) and two fatalities (Cavaleri et al., 2012). This type of damage scenario, with similar consequences, repeated also for the cruise ship Viking Polaris, where the rogue wave broke some windows at superstructure deck number two. These accidents, pose the attention on the design of high superstructures, in event of rogue wave impact, especially for passenger ships, such as cruise ships and ferries.

The current research study aims at disclosing the vulnerability of the structures of higher decks, showing that: the rogue wave can reach the superstructure height and the rogue wave impact loads can exceed the design loads.

The most consolidated approach in the rogue wave definition is available in (Bitner-Gregersen & DNV GL, 2017) that define it as a wave whose height or crest height exceeds some thresholds related to the significant wave height (Haver, 2000). This definition does not imply any peculiar modeling of rogue wave i.e. it is applicable both for linear and non-linear wave models. Actually linear and non-linear mechanisms can be responsible for the generation of rogue waves (Kharif & Pelinovsky, 2003): these waves can be obtained from linear models by some peculiar tuning of the wave components

DOI: 10.1201/9781003508779-29

(Varkey, 1989), or obtained by non-linear models where non-linear Schrodinger equations are the most popular and effective ones (Vitanov et al., 2013).

Linear mechanism (superposition principles of regular wave components) is suited for the generation of rogue waves in wave tank facilities (Brown & Jensen, 2001); however, some non-linear corrections are needed, in numerical simulation, accounting for very steep and high waves (Fochesato et al., 2007).

In this paper the so-called New Wave technique (Gao et al., 2016; Hann et al., 2018; Tromans et al., 1991), based on linear mechanisms, is used for the generation of deterministic rogue wave from a given sea spectrum. Such approach exploits the focalization of wave energy components, disregarding the random determination of the wave train in the time domain. The input spectra, for the rogue wave simulations, were generated for all observed sea states in the North Atlantic scatter diagram (IACS, 2001). The obtained rogue waves were analyzed in terms of height and steepness. Moreover, a preliminary analysis of the impact pressure on ship structures is carried out following the simplified approach proposed in (Stansberg, 2020). This method is based on the instantaneous steepness of the wave profile and was developed and validated for off-shore platforms hit by large waves.

The ship chosen for the case study is a ropax ferry, already used for a preliminary analysis on inertial and impact loads from rogue waves in moderate-heavy seas (Acanfora et al., 2021). In particular, for this vessel, the design pressure for the side plating was calculated as prescribed by the Classification society Rules (RINA, 2012), showing that there is a threshold height where a minimum design load applies. Among all generated rogue waves, it is possible identifying which of them can reach, i.e. can hit, the minimum design load area of the ferry under investigation.

The comparison of the minimum design load with the estimated rogue wave impact pressures will disclose the risk of a potential structural failure in the superstructure elements.

2 GENERATION OF ROGUE WAVES BY NEW WAVE TECHNIQUE

In this research study, we define the rogue wave following the approach proposed by Haver (Haver, 2000) and acknowledged by DNV-GL (Bitner-gregersen & DNV GL, 2017). A rogue wave is characterized by a height H_{max} (crest-trough) that is two times greater than the significant wave height H_s or by a crest height C_{max} that is 1.25 times greater than H_s. A wave that fulfills both the requirements is called double rogue.

$$H_{max}/H_s = 2 \text{ and/or } C_{max}/H_s = 1.25 \quad (1)$$

In order to model a rogue waves, the so-called New Wave Technique (NWT) is used. Originally, (Tromans et al., 1991) proposed New Wave methodology as an efficient approach to simulate extreme wave, using the summation of N component waves to create an isolated wave peak. The limitation of this technique stays in the possibility of simulating isolated local wave peaks leaving the rest of the surface almost calm before and after it. In (Cassidy et al., 2001), the NWT was modified in order to account for irregular sea surface before and after the wave peak, while in (Acanfora et al., 2021) rogue waves were obtained from random superposition of regular wave components. Actually in (Acanfora et al., 2021) it was necessary to generate several wave trains from the same spectra prior to identify a possible rogue wave; on the other hand, the description of the irregular sea state before and after the rogue wave event was kept.

In this paper, the use of the standard NWT is employed at the scope of generating rogue waves, focusing on their induced loads, thus disregarding the rest of the sea surface:

$$\eta(x,t) = \sum_{n=0}^{N} a_n \cos\left(k_n\left(x - x_f\right) - \omega_n\left(t - t_f\right) + \varepsilon_n\right) \quad (2)$$

Where a_n is the wave amplitude, k_n is the wave number, ω_n the wave frequency and ε_n is the phase angle related to the n wave components. While t_f and x_f are the time and the position of the generated rogue wave. Usually phase angles are all set to zero, while a sufficiently large number of wave components N is required. Wave components are obtained from discretization of the wave spectrum by:

$$a_n = \sqrt{2\ S_\zeta(\omega_n)\delta\omega_n} \quad (3)$$

For all sea states described in the scatter diagram of the North Atlantic (IACS, 2001) with non-zero occurrence, a JONSWAP wave spectrum is generated and, following, a rogue wave is obtained.

3 ESTIMATION OF WAVE PRESSURES AT SIDE PLATES OF UPPER DECKS

The estimation of the design loads for wave pressures on ship sides above the waterline, as prescribed by RINa Rules in (RINA, 2012), assumes that above the design draft *T*, the wave pressure on ship sides reduces linearly up to a minimum design load (that is then kept constant). This is done accounting for all load cases. According to RINa rules for wave pressure on ship side, the minimum design value of pressure $p_{w,min}$ acting on ship sides depends only on ship length:

$$p_{w,min} = 0,15\ C_1 L \quad (4)$$

where C_1=1 and L is the design length.

The height D*, from which $p_{w,min}$ applies, depends on the wave pressure distribution on ship sides above the design draft, accounting for all load combination cases. The height of the minimum design load h, to be compared to the rogue wave crest C, is defined as (see Figure 1):

$$h = D^* - T \quad (5)$$

The rogue wave loads on ship sides can be seen as a particular case of green water on deck. Thus, the corresponding impact pressure could be considered: quasi-static, dynamic or impulsive pressure depending on the duration of the application of the load compared to the natural period of the structure. Usually, when the duration of the impact pressure is short, its effective value is approximated to an equivalent pressure based on the approach proposed by (Paik et al., 2005).

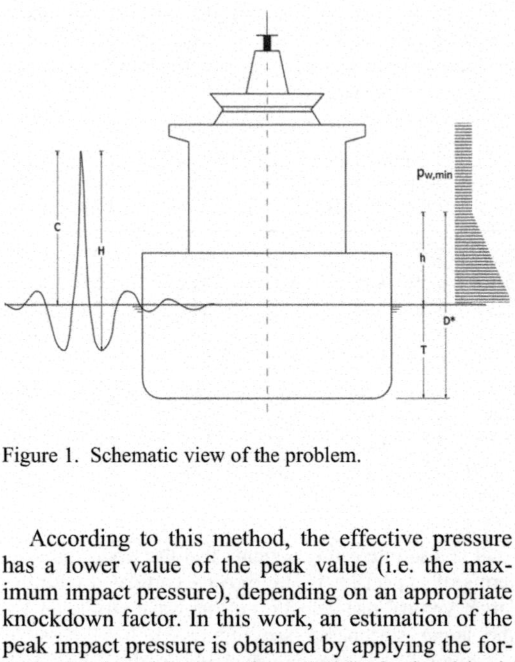

Figure 1. Schematic view of the problem.

According to this method, the effective pressure has a lower value of the peak value (i.e. the maximum impact pressure), depending on an appropriate knockdown factor. In this work, an estimation of the peak impact pressure is obtained by applying the formula developed in (Stansberg, 2020), for local horizontal slamming pressure from a long-crested head sea wave impact on a finite vertical wall. This reference method is based on the assumption that the magnitude on the impact depends on local wave steepness, and in particular, it depends on the derivative of the wave profile dζ/dt, that identifies the vertical rise velocity:

$$p_I = \rho g h + \frac{0.5\rho C_{s0}\,(d\zeta/dt)^{3.1}}{C_P^{1.1}} \quad (6)$$

Where $C_{s0'}$ is a modified slamming coefficient derived in (Stansberg, 2020), that is constant and equal to 2.5, and wave celerity C_P is based on the peak period of the discretized spectrum. This model was validated also in presence of very steep waves close to breaking, thus it is judged applicable for rogue wave cases.

In this formula, the hydrostatic contribution can be neglected. Since this formulation was developed for off-shore platforms, aiming at a straightforward application to a ship problem, hull dynamics effects (i.e. ship motions) are disregarded. Moreover, there is no mention about knockdown factors for effective pressure analysis. Therefore, in case the rogue wave hits the ship sides, we compare directly the peak impact pressure p_I against the $p_{w,min}$ value.

Although it could be questioned that this analysis is not conservative, it still provides an overall sensitivity of the ship side structures to rogue wave load problems.

4 CASE STUDY

Several rogue waves were generated from Jonswap Spectra; wave periods and significant wave height were taken from North Atlantic scatter diagram. Since a rogue wave is a rare event, it is yet more unlikely to happen within a rare sea state: therefore, sea states characterized by small occurrence (i.e. less than 0.01%) are excluded from the calculation. Following, the analysis covers 101 sea states over the total 306 ones available in the scatter diagram (see Figure 2). It is worth underlying that this assumption somewhat limits the generation of rogue waves in moderate-high sea states rather than in extreme seas.

For the application of the NWT, N=130 wave components are used. The hull chosen for the case study is a ro-pax ferry whose main geometrical characteristics and body lines are shown in Table 1 and Figure 3 respectively. Following rules requirements (RINA, 2012), it is found for this ship that the minimum design pressure by (4) is equal to 23,7 kN/m^2 from an height D*=10 m (see Figure 1 for a qualitative behavior of design pressures on ship side above the waterline).

The Figure 4 shows the rogue wave obtained by the application of the NWT from the discretized JONSWAP spectrum for the sample case H_S =5.5 m and T_Z =9.5 s.

5 RESULTS

This section summarizes the outcomes of the analysis carried out for the 101 rogue waves obtained by the NWT:

- All modeled rogue waves can be classified as a double rogue;
- In 95 cases, the wave can reach the ship side in the area of the minimum design pressure;
- Among the above cases, 9 times the condition $p_I > p_{w,min}$ is satisfied, that is the impact pressure exceeds the minimum design pressure.

Hs/Tz	1,5	2,5	3,5	4,5	5,5	6,5	7,5	8,5	9,5	10,5	11,5	12,5	13,5	14,5	15,5	16,5	17,5	18,5	SUM
0,5	0,0	0,0	1,3	133,7	865,6	1186,0	634,2	186,3	36,9	5,6	0,7	0,1	0,0	0,0	0,0	0,0	0,0	0,0	3050
1,5	0,0	0,0	0,0	29,3	986,0	4976,0	7738,0	5669,7	2375,7	703,5	160,7	30,5	5,1	0,8	0,1	0,0	0,0	0,0	22675
2,5	0,0	0,0	0,0	22,0	197,5	2158,8	6230,0	7449,5	4860,4	2066,0	644,5	160,2	33,7	6,3	1,1	0,2	0,0	0,0	23830
3,5	0,0	0,0	0,0	0,2	34,9	696,5	3226,5	5675,0	5099,1	2838,0	1114,1	337,7	84,3	18,2	3,5	0,6	0,1	0,0	19129
4,5	0,0	0,0	0,0	0,0	6,0	196,1	1354,3	3288,5	3857,5	2685,5	1275,2	455,1	130,9	31,9	6,9	1,3	0,2	0,0	13289
5,5	0,0	0,0	0,0	0,0	1,0	51,0	498,4	1602,9	2372,7	2008,3	1126,0	463,6	150,9	41,0	9,7	2,1	0,4	0,1	8328
6,5	0,0	0,0	0,0	0,0	0,2	12,6	167,0	609,3	1257,9	1268,6	825,9	386,8	140,8	42,2	10,9	2,5	0,5	0,1	4725
7,5	0,0	0,0	0,0	0,0	0,0	3,0	52,1	270,1	594,4	703,2	524,9	276,7	111,7	36,7	10,2	2,5	0,6	0,1	2586
8,5	0,0	0,0	0,0	0,0	0,0	0,7	15,4	97,9	255,9	350,6	296,9	174,6	77,6	27,7	8,4	2,2	0,5	0,1	1309
9,5	0,0	0,0	0,0	0,0	0,0	0,2	4,3	33,2	101,9	159,9	152,2	99,2	48,3	18,7	6,1	1,7	0,4	0,1	626
10,5	0,0	0,0	0,0	0,0	0,0	0,0	1,2	10,7	37,9	67,5	71,7	51,5	27,3	11,4	4,0	1,2	0,3	0,1	285
11,5	0,0	0,0	0,0	0,0	0,0	0,0	0,3	3,3	13,3	26,6	31,4	24,7	14,2	6,4	2,4	0,7	0,2	0,1	124
12,5	0,0	0,0	0,0	0,0	0,0	0,0	0,1	1,0	4,4	9,9	12,8	11,0	6,8	3,3	1,3	0,4	0,1	0,0	51
13,5	0,0	0,0	0,0	0,0	0,0	0,0	0,0	0,3	1,4	3,5	5,0	4,6	3,1	1,6	0,7	0,2	0,1	0,0	21
14,5	0,0	0,0	0,0	0,0	0,0	0,0	0,0	0,1	0,4	1,2	1,8	1,8	1,3	0,7	0,3	0,1	0,0	0,0	8
15,5	0,0	0,0	0,0	0,0	0,0	0,0	0,0	0,0	0,1	0,4	0,6	0,7	0,5	0,3	0,1	0,1	0,0	0,0	3
16,5	0,0	0,0	0,0	0,0	0,0	0,0	0,0	0,0	0,0	0,2	0,2	0,2	0,2	0,1	0,1	0,0	0,0	0,0	1
SUM	0	0	1	185	2091	9281	19922	24898	20870	12899	6245	2479	837	247	66	16	3	1	100000

Figure 2. Scatter diagram North Atlantic. In light blue, the selected sea states for the rogue wave generation.

Table 1. Main particulars of the vessel SeatechD.

Main Particular	Value
Length between perpendicular, L (m)	158.0
Breadth, B (m)	25.0
Draft, T (m)	6.1
Depth, D (m)	15.0
Immersed Volume, ∇ (m^3)	13766
Wetted Surface, S (m^2)	4356
Center of gravity above keel. KG (m)	11.5

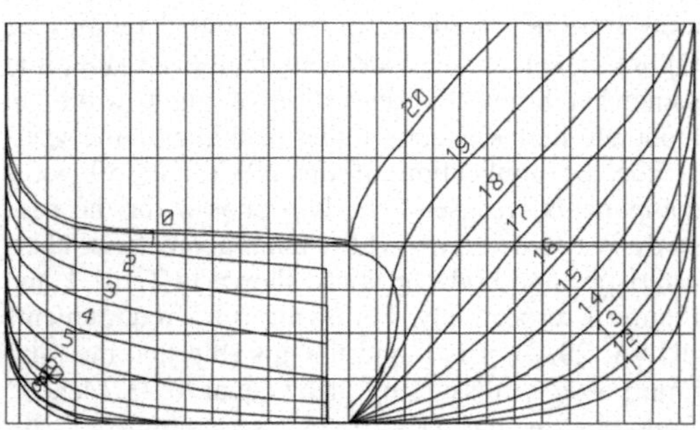

Figure 3. Hull lines of Seatech-D.

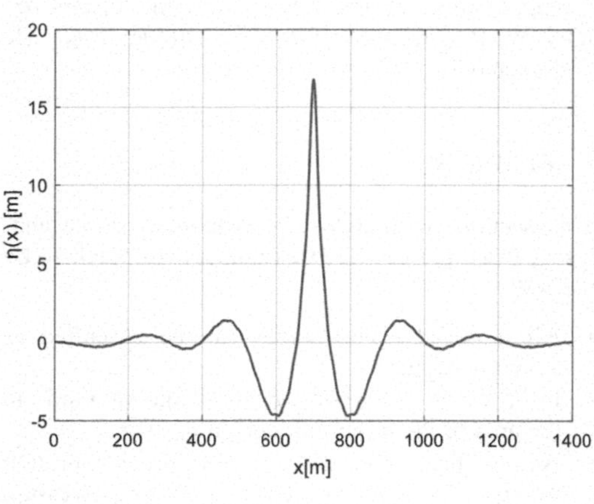

Figure 4. Rogue wave sample.

The nine cases showing the worst conditions are listed in Table 2. In particular, it is possible noticing that a dangerous rogue wave could happen even in moderate sea states (H_S=2.5 -3.5 m). The larger impact pressure, instead, was estimated for the sea state characterized by H_S=10.5 m and T_Z =8.5 s. The most alarming outcomes are the p_I values that can exceed by more than two times $p_{w,min}$, up to almost four times in the worse scenario.

Table 2. Results of the cases where $p_I > p_{w,min}$.

Hs	TZ	H_{max}/H_s	C_{max}/H_s	Above h	dζ/dt	p_i
2,50	5,50	4,22	3,11	Yes	6,02	23,82
5,50	6,50	4,26	3,18	Yes	6,53	25,51
7,50	8,50	3,99	3,12	Yes	7,76	32,46
6,50	6,50	4,26	3,18	Yes	7,72	42,82
8,50	8,50	3,99	3,12	Yes	8,80	47,85
2,50	4,50	4,32	2,85	Yes	7,37	55,65
9,50	8,50	3,99	3,12	Yes	9,83	67,55
3,50	5,50	4,22	3,11	Yes	8,43	67,60
10,50	8,50	3,99	3,12	Yes	10,87	92,12

It is also interesting noticing that the worst cases, in terms of impact loads, are directly correlated with the larger vertical rise velocities of the wave profile i.e. by the larger steepness, (see Figure 5). It is possible observing that the rogue waves obtained by the NWT exceed by large amount the thresholds defined in (1).

It is possible observing that the rogue waves obtained by the NWT exceed by large amount the thresholds defined in (1). If we compare these outcomes with the ones presented in (Acanfora et al., 2021), the technique applied in the reference work brought out lower rogue wave heights than NWT.

However, also in that research work, some estimated impact loads exceeded significantly the minimum design pressure. Apart from the wave elevation, the relevant parameter in the impact load problem is the wave steepness. Although very steep waves are rare and tend to break, they could still impact on ship side structures. The minimum design pressure in general depends on the ship length (see Figure 6, solid blue line with

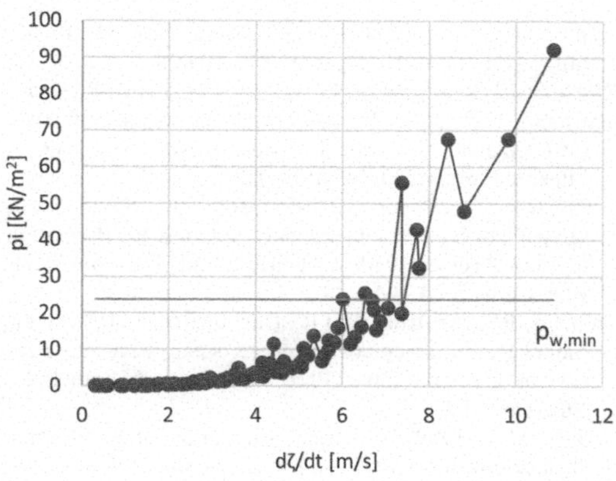

Figure 5. Peak impact pressure results.

no marker). If we neglect impact area of the rogue waves, we could generalize the obtained impact pressure outcomes (see Figure 5), related to vessel lengths.

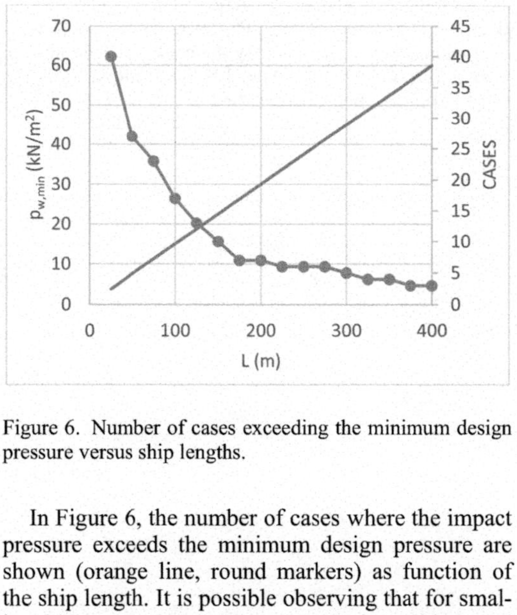

Figure 6. Number of cases exceeding the minimum design pressure versus ship lengths.

In Figure 6, the number of cases where the impact pressure exceeds the minimum design pressure are shown (orange line, round markers) as function of the ship length. It is possible observing that for smaller ships, more cases (up to 40 cases) would exceed the corresponding minimum design pressure value. However, very long ships (up to 400 m) could be still vulnerable (in 3 cases), to impact loads from rogue waves, if the superstructures in the minimum design pressure area are hit by the rogue waves.

6 CONCLUSIONS

This work regarded the study of the impact loads from rogue waves on ships superstructures. The analysis was carried out for a ropax vessel, assuming the possibility of encountering a rogue wave generated by the linear mechanisms of energy focalization. The obtained peak pressures from the waves impacting the side structures in the area of minimum design pressure, showed that there is a serious risk of structural failure in presence of very steep rogue waves, even in moderate sea states.

For the ropax under investigation, we found 9 cases where the impact pressure exceeded the minimum design pressure. It is worth underlying that minimum design pressure depends on ship length only. Therefore, lowering the ship length, more cases would exceed the corresponding minimum design pressure value.

There is certainly a room for further investigations on the topic supported by experimental research. However, the presented approach seemed adequate in underlying possible structural failures on high decks based on the current design standards.

REFERENCES

Acanfora, M., Coppola, T., Fasano, E., 2021. Estimation of rogue wave loads on ship structures by exploiting linear wave theory. Dev. Anal. Des. Mar. Struct. 3–9. https://doi.org/10.1201/9781003230373-1

Bitner-gregersen, E., Dnv GL, 2017. Rethinking rogue waves.

Bitner-Gregersen, E.M., Gramstad, O., 2016. Rogue waves. Impact on ship and offshore structures.

Brown, M.G., Jensen, A., 2001. Experiments on focusing unidirectional water waves. J. Geophys. Res. Ocean. 106, 16917–16928. https://doi.org/10.1029/2000jc000584

Cassidy, M.J., Taylor, R.E., Houlsby, G.T., 2001. Analysis of jack-up units using a Constrained NewWave methodology. Appl. Ocean Res. 23, 221–234. https://doi.org/10.1016/S0141-1187(01)00005-0

Cavaleri, L., Bertotti, L., Torrisi, L., Bitner-Gregersen, E., Serio, M., Onorato, M., 2012. Rogue waves in crossing seas: The Louis Majesty accident. J. Geophys. Res. Ocean. 117, 1–7. https://doi.org/10.1029/2012JC007923

Fochesato, C., Grilli, S., Dias, F., 2007. Numerical modeling of extreme rogue waves generated by directional energy focusing. Wave Motion 44, 395–416. https://doi.org/10.1016/j.wavemoti.2007.01.003

Gao, N., Yang, J., Zhao, W., Li, X., 2016. Numerical simulation of deterministic freak wave sequences and wave-structure interaction. Ships Offshore Struct. 11, 802–817. https://doi.org/10.1080/17445302.2015.1073864

Gunson, J., Lehner, S., 2001. Extreme Wave Conditions from Wave Model Hindcasts and from Synthetic Aperture Radar Images, in: "Design and Operation for Abnormal Conditions II", Proc. Royal Institution of Naval Architects, London, United Kingdom.

Guo, B., Bitner-Gregersen, E.M., Sun, H., Helmers, J.B., 2016. Statistics analysis of ship response in extreme seas. Ocean Eng. 119, 154–164. https://doi.org/10.1016/j.oceaneng.2016.03.060

Hann, M., Greaves, D., Raby, A., Howey, B., 2018. Use of constrained focused waves to measure extreme loading of a taut moored floating wave energy converter. Ocean Eng. 148, 33–42. https://doi.org/10.1016/j.oceaneng.2017.10.024

Haver, S., 2000. Evidences of the Existence of Freak Waves. Rogues Waves 2000 129–140.

Hu, Z., Xue, H., Tang, W., Zhang, X., 2015. A combined wave-dam-breaking model for rogue wave overtopping.

Ocean Eng. 104, 77–88. https://doi.org/10.1016/J.OCEANENG.2015.05.009
Huijsmans, R.H.M., van Groesen, E., Chung, J.S., Izumiyama, K., Sayed, M., Hong, S.W., 2004. Coupling freak wave events with green water simulations, in: Proceedings of the Fourteenth (2004) International Offshore and Polar Engineering Conference, Vol 1. pp. 366–373.
IACS, 2001. Standard Wave Data 1–4.
Kharif, C., Pelinovsky, E., 2003. Physical mechanisms of the rogue wave phenomenon. Eur. J. Mech. B/Fluids 22, 603–634. https://doi.org/10.1016/j.euromechflu.2003.09.002
Nikolkina, I., Didenkulova, I., 2011. Rogue waves in 2006–2010. Nat. Hazards Earth Syst. Sci. 11, 2913–2924. https://doi.org/10.5194/nhess-11-2913-2011
Paik, J.K., Lee, J.-M., Shin, Y.S., Wang, G., 2005. Design principles and criteria for ship structures under impact pressure loads arising from sloshing, slamming and green seas. Trans. - Soc. Nav. Archit. Mar. Eng. 112, 292–313.
Qin, H., Tang, W., Hu, Z., Guo, J., 2017a. Structural response of deck structures on the green water event caused by freak waves. J. Fluids Struct. 68, 322–338. https://doi.org/10.1016/j.jfluidstructs.2016.11.009
Qin, H., Tang, W., Xue, H., Hu, Z., 2017b. Numerical study of nonlinear freak wave impact underneath a fixed horizontal deck in 2-D space. Appl. Ocean Res. 64, 155–168. https://doi.org/10.1016/j.apor.2017.02.008
Rajendran, S., Fonseca, N., Guedes Soares, C., 2016. Body nonlinear time domain calculation of vertical ship responses in extreme seas accounting for 2nd order Froude-Krylov pressure. Appl. Ocean Res. 54, 39–52. https://doi.org/10.1016/j.apor.2015.10.008
RINA, 2012. RINA Rules, Pt B, Ch 5, Sec 5 : Sea Pressure.
Stansberg, C.T., 2020. Wave front steepness and influence on horizontal deck impact loads. J. Mar. Sci. Eng. 8. https://doi.org/10.3390/JMSE8050314
Toffoli, A., Lefevre, J., Monbaliu, J., Savina, H., Gregersen, E., 2003. Freak Waves: Clues for Prediction in Ship Accidents? Proc. Thirteen. Int. Offshore Polar Eng. Conf. 5, 23–29.
Tromans, P.S., Anatruk, A.H.R., Hagemeijer, P., 1991. New model for the kinematics of large ocean waves application as a design wave. Proc. First Int. Offshore Polar Eng. Conf. 64–71.
Varkey, M.J., 1989. Numerical simulation of wind wave surface profiles with tuned phase spectra. Indian J. Mar. Sci. 18, 177–183.
Vitanov, N.K., Chabchoub, A., Hoffmann, N., 2013. Deep-water waves: On the nonlinear Schrödinger equation and its solutions 1–11. https://doi.org/10.2478/jtam-2013-0013
Wang, S., Zhang, H.D., Guedes Soares, C., 2016. Slamming occurrence for a chemical tanker advancing in extreme waves modelled with the nonlinear Schrödinger equation. Ocean Eng. 119, 135–142, https://doi.org/10.1016/J.OCEANENG.2016.04.017
Zhang, H.D., Guedes Soares, C., 2016. Ship responses to abnormal waves simulated by the nonlinear Schrödinger equation. Ocean Eng. 119, 143–153. https://doi.org/10.1016/j.oceaneng.2016.03.064

Advances in Maritime Technology and Engineering – Guedes Soares & Santos (Eds)

Ship structural load modelling considering the temporal correlation of wave-induced load cycles

A. Kakaie & C. Guedes Soares
Centre for Marine Technology and Ocean Engineering (CENTEC), Instituto Superior Técnico, Universidade de Lisboa, Lisbon, Portugal

ABSTRACT: This study proposes explicit approaches for modelling ship structural loads including the temporal correlation of wave-induced load cycles. The available methods for ship structural load modelling are critically reviewed. The correlated wave-induced loads in each sea state are modeled using the stochastic Markov process and a closed-form formulations are derived to estimate the extreme value distribution of wave-induced effects and load combination. Finally, the applicability of the proposed framework is investigated for an Aframax tanker in the full load condition and the obtained results based on the proposed approach are compared with the existing approach.

1 INTRODUCTION

Ship structures are exposed to different stochastic load processes. The ultimate collapse failure of a hull girder subjected to still water and wave-induced loads as two main actions on ship structures is recognized as the most important failure mode with crucial economic and environmental consequences (Gaspar et al., 2016). Considering the uncertainties involved in the prediction of still water and wave-induced loads and also the stochastic description of their extreme value distribution in a proper way are required for risk and reliability assessment of ship structures (Guedes Soares, 1990a, 1990b).

Although different types of engineering models are proposed for the prediction of wave loads, they represent just simplifications of the complex physical reality and the accurate definition of the wave loads acting on the ship structures is still challenging (Parunov et al., 2022). It has been indicated that the uncertainty in the long-term wave-induced loads is larger than the uncertainty in the resistance capacity (Corak et al., 2018; Parunov & Guedes Soares, 2008; Guedes Soares & Teixeira, 2000).

The vertical wave-induced bending moments result from the dynamic forces generated by the ocean waves and their interaction with the ship structure. Several factors such as the ship's hull shape, size, speed, and the characteristics of the encountered waves all contribute to the complexity of the ship analysis and design against the wave-induced load effects (Guedes Soares, 1999; Guedes Soares & Moan, 1991). Proper design considerations and structural integrity assessments are essential to ensure the safety and reliability of the ship in various sea conditions.

An accurate prediction of extreme vertical wave-induced bending moment and its combination with still water bending moments is important for estimating the required ultimate bending capacity and therefore the sip hull girder structural design. Several factors contribute to the estimation of extreme wave-induced load effects including significant wave heights, characteristic wave period, wave spectrum and directionality and also ship characteristics and environmental conditions (Bitner-Gregersen et al., 2022; Gaspar et al., 2016).

The probabilistic modelling of wave-induced load effects has been investigated in several studies. Mao & Rychlik (2012) presented a simple approach for the prediction of extreme response using Rice's method. Lee et al. (2012) estimated the wave-induced bending moments, considering the effects o the whipping. Teixeira et al. (2013) estimated the probabilistic characteristics of the load combination factors for global still water and wave-induced of double-hull tankers. Andersen & Jensen (2014) presented full-scale measurements of the wave-induced vertical bending moment midships of a container carrier and focuses on the extreme hogging wave bending moment.

The available approaches estimate the extreme distribution based on the long-term peak values which are assumed to be independent and identical distributed pulse. However, assuming independent peak values is not realistic. The obtained autocorrelation function based on observational data and theoretical spectrum reveal temporal correlations in wave peaks (Nielsen et al., 2018; Takami et al., 2021; Yang et al., 2020). In addition, several factors such as modulation of individual waves within the group (Cherneva &

DOI: 10.1201/9781003508779-30

Guedes Soaers, 2012), wave propagation over long distances and common environmental and climatological factors may result in the degree of correlation between wave peaks, particularly in a short period of time such as one sea state.

The main aim of this study is to propose a closed-form method to estimate extreme wave-induced load effect distribution and its combined process with still water load effects, considering the temporal correlation associated with the wave cycle process. With this regard, the stochastic Markov process is applied to model the wave-induced sequences and explicit formulations are derived to estimate the extreme value distribution of wave-induced bending moment and load combination. The remainder of this paper is organized as follows: section 2 reviews the available approaches for ship structural load modelling. In section 3 a new approach is developed for ship structural load modelling regarding correlated wave loads. The applicability of the proposed approach is investigated in section 4 and the concluding remark is presented in section 5.

2 AVAILABLE APPROACHES

In this section, the available approaches for modelling the extreme value distributions and combinations of ship hull girder loads are described. The dynamic responses of the ship due to the wave-induced ship motions can be idealized as a non-stationary zero-mean Gaussian stochastic process. For an appropriate short period of time, this process is simplified as a stationary random process and the response is estimated based on linear spectrum random method.

Regarding the narrow-band stochastic process, the amplitudes of wave-induced load effects follow a conditional Rayleigh distribution in each sea state with the probability of exceeding as (Longuet-Higgins, 1984):

$$Q_S(m|R) = \exp\left(\frac{-m^2}{2R}\right) \tag{1}$$

where R is the variance of the response related to the sea state characteristics and operational condition of the ship. The variance R is calculated based on the response spectrum which is a product of a linear transfer function for a specified relative heading and significant wave height and the wave spectrum.

The long-term probability distribution of wave-induced load effects is estimated based on the weighting of the conditional probability in Equation (1) by the probability of the response variance in various sea states (Fukuda, 1967; Guedes Soares, 1991). In many practical applications, the long-term distribution of vertical wave-induced bending moment can be approximated based on the short-term ones (Campanile et al., 2017; Teixeira et al., 2013) and modelled with a Weibull distribution.

$$F_w(m) = 1 - \exp\left(-\left(\frac{m}{w}\right)^k\right) \tag{2}$$

where w and k are relevant scale and shape parameters, respectively. The shape parameter k, which depends on the ship type and environmental condition is close to 1 (Guedes Soares & Moan, 1991; Teixeira & Guedes Soares, 1998), while it is 2 in the short term. The scale parameter w is derived based on the $M_{w,rule}$, which is the characteristic wave-induced bending value with an exceedance probability level of 10^{-8}. The Weibull distribution models are the distribution of the wave load peaks at a random point in time. However, for risk and reliability assessment of ship structures, the extreme wave-induced load effects in a large number of load cycles are required (Teixeira et al., 2013).

The long-term extreme value distribution of the wave-induced bending moment has been estimated based on the distribution of the peak values which are assumed to follow the Weibull distribution. The long-term peaks are considered as an independent and identical distributed pulse and the extreme value distribution is estimated based on the Poisson square wave (PSW) model (Huang & Moan, 2008; Teixeira et al., 2013; Wen, 1990) or Ferry Borges and Castanheta (FBC) method (Borges Ferry & Castanheta, 1971) as:

$$\begin{aligned} F_{w,\max}(m) &= F_w(m)\exp\{-n_{w,T}[1 - F_w(m)]\} \\ &\approx \exp\{-n_{w,T}[1 - F_w(m)]\} \end{aligned}$$

PSW (3)

$$F_{w,\max}(m) = [F_w(m)]^{n_{w,T}}$$

(PSW 3)

where $n_{w,T}$ is the number of peak values of the wave-induced bending moment during the interval [0, T].

In contrast to the FBC model, the Poisson square model considers the uncertainty related to the number of wave cycles. However, the regarded uncertainty has an insignificant effect on the extreme value distribution of long-term wave-induced load effects (Teixeira et al., 2013).

For the design and reliability assessment of ship structures, it is required to estimate the extreme values of the combined process of the involved loads. Several stochastic methods such as the Point-Crossing (PC) (Larrabee & Cornell, 1981) method, Wen's load coincidence (WLC) method (Wen, 1977) and the FBC method (Borges Ferry & Castanheta, 1971) and deterministic methods including the Turkstra rule (Turkstra, 1972) and peak coincidence method have been applied to estimate the ship structural load combination.

Most of the applied approaches to ship structural loads assume independent and identical wave cycles. However, this assumption may not be realistic in many cases.

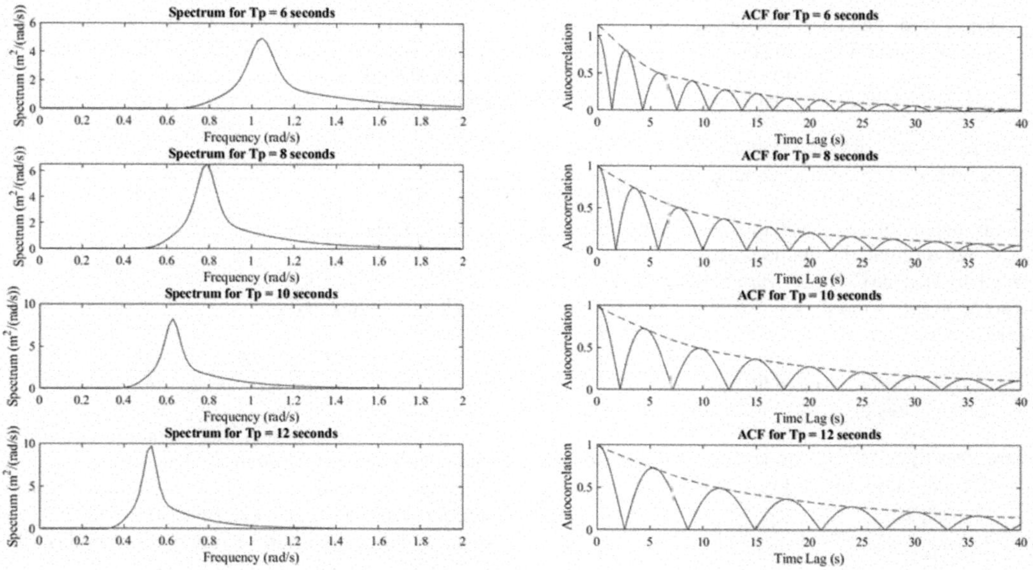

Figure 1. The JONSWAP spectrum and the absolute value of associated ACF.

Here, for clarifying the possible correlation between the wave cycles the autocorrelation function (ACF) for the JONSWAP spectrum is depicted in Figure 1. This spectrum is widely used for fully developed wind-generated waves. The JONSWAP spectrum is a modification of the Pierson-Moskovitz spectrum that considers the effect of the fetch length and is mathematically expressed as (Hasselmann et al., 1976):

$$S(\omega) = \beta H_s^2 \omega_p^4 \omega^{-5} \exp\left[-1.25\left(\frac{\omega}{\omega_p}\right)^{-4}\right]\gamma^r \quad (5)$$

which can be parameterized as (Goda, 2010):

$$\begin{aligned} \beta &= \frac{0.0624}{0.230+0.0336\gamma-0.185(1.9+\gamma)^{-1}} \times \cdots \\ &\quad [1.094 - 0.01915\ln(\gamma)] \\ r &= \exp\left[-\frac{(\omega-\omega_p)^2}{2\sigma^2\omega_p^2}\right] \\ \sigma &= \begin{cases} 0.07 & \omega \le \omega_p \\ 0.09 & \omega > \omega_p \end{cases} \end{aligned} \quad (6)$$

where ω_p represents the peak frequency (at the maximum of the spectrum), γ and σ define the shape of the spectrum. The autocorrelation function which indicates the relation of the random process with itself in different time intervals is calculated for the JONSWAP spectrum.

The spectrum and absolute value of ACF are indicated in Figure 1 for different values of peak period and assuming the significant wave height H_s of 5m. It can be observed that there is a notable autocorrelation coefficient for the short-time lags. The autocorrelation values are more than 0.5 for the time lag equal to T_p and 0.25 for the time lag equal to $2T_p$. Since the response spectrum is affected by the wave spectrum, there is also a possible correlation in the wave-induced load effects that need to be considered in the estimation of extreme value distribution.

3 EXTREME VALUE DISTRIBUTION OF CORRELATED WAVE CYCLES

The available approaches for the probabilistic modelling of wave-induced load effects assumes independent wave cycles. In this section, closed-form formulations for extreme wave-induced load effect and its combination with the still water load effect regarding the correlated wave cycles are presented. For this purpose, the Markov process is used. In the Markov process, each sequence is only dependent on the previous sequence on a temporal scale. In this study, the wave pulses in each sea state are assumed to be correlated, since they are affected by the same wave generation condition. First, the extreme value distribution of correlated wave-induce load effects is formulated for each sea state. Then the extreme wave-induced load effect is calculated for one voyage considering the sea states as independent. In the final step, the combined process between one voyage's extreme wave-induced load effect and still water load effect is derived.

Assuming k wave cycles, modelled as pulses $\{w_1, w_2, \ldots, w_k\}$ for each sea state with a reference period $[0, T_s]$, the extreme wave-induce load effect is:

$$\begin{aligned} F^s_{w,\max}(m) &= P(w_1 < m \cap w_2 < m \ldots \cap w_k < m) \\ &= P(w_1 < m)\prod_{i=1}^{k-1} P(w_{i+1} \le m | w_i \le m) \end{aligned} \quad (7)$$

Taking into account the dependency between w_{k+1}, w_k, one has (Wang, 2021):

$$w_{k+1} = \rho_w w_k + w* \tag{8}$$

where ρ_w is the correlation coefficient between w_{k+1} and w_k and w* is a normal variable with the mean of $\mu_w(1-\rho_w)$ and standard deviation of $\sigma_w(1-\rho_w^2)^{0.5}$ which is independent of w_k. since the wave cycles usually follow non-normal distributions, first they are transformed to standard normal variables using the Nataf transformation method. w_{k+1} and w_k can be transformed into normal variables U_{k+1} and U_k as:

$$\begin{aligned} F_{W_k}(w_k) &= \Phi(U_k) \\ F_{Wk+1}(w_{k+1}) &= \Phi(U_{k+1}) \end{aligned} \tag{9}$$

where Φ is the CDF of standard normal distribution. Therefore, the Equation (8) can be rewritten as:

$$U_{k+1} = \rho_U U_k + U^* \tag{10}$$

where ρ_U is the correlation coefficient in the normal space and U* is a standard normal variable with a mean of 0 and a standard deviation of $(1-\rho_U^2)^{0.5}$. it should be mentioned that ρ_u is a function of ρ_w their relationship can be found in the literature for different distributions (Der Kiureghian, 2005; Melchers & Beck, 2018).

$$\begin{aligned} &P(w_{i+1} \le m | w_i \le m) \\ &= P\left[U_{i+1} \le \Phi^{-1}\left(F_{W_{i+1}}(m)\right) | U_i \le \Phi^{-1}\left(F_{W_i}(m)\right)\right] \end{aligned} \tag{11}$$

Substituting Equation (10) into Equation (11) and assuming $\Phi^{-1}(F_{wk}(m))=m'$:

$$\begin{aligned} P(w_{i+1} \le m | w_i) &= P[\rho_U U_i + U^* \le m' | U_i \le m'] \\ &= \frac{P\left[U^* \le m' - \rho_U U_i \cap U_i \le m'\right]}{P[U_i \le m']} \\ &= \frac{\int_{-\infty}^{m'} \Phi^*(m' - \rho_U x)\,\phi(x)dx}{\Phi(m')} \end{aligned} \tag{12}$$

where Φ^* is the CDF of U*. Finally, the extreme value distribution of wave-induced load effect during each sea state is obtained by substituting Equation (12) into Equation (7) as follows:

$$\begin{aligned} F^s_{w,\max}(m) &= \Phi(m') \prod_{i=1}^{k-1} \frac{\int_{-\infty}^{m'} \Phi^*(m' - \rho_U x)\,\phi(x)dx}{\Phi(m')} \\ &= \Phi(m')\left[\frac{\int_{-\infty}^{m'} \Phi^*(m' - \rho_U x)\,\phi(x)dx}{\Phi(m')}\right]^{k-1} \end{aligned} \tag{13}$$

Equation (13) is a closed-form solution for extreme value distribution of correlated wave-induced load effect in each sea state and k is the number of waves in each sea state, which can be obtained as:

$$k = \frac{sea\,state\ duration\ (t_s)}{wave\ period\ (t_w)} \tag{14}$$

Assuming n_s number of independent sea states in each voyage, the extreme value distribution of wave-induced load effect during one voyage is obtained as:

$$\begin{aligned} F^v_{w,\max}(m) &= \left[F^s_{w,\max}\right]^{n_s} \\ &= \left[\Phi(m')\left\{\frac{\int_{-\infty}^{m'} \Phi^*(m' - \rho_U x)\,\phi(x)dx}{\Phi(m')}\right\}^{k-1}\right]^{n_s} \end{aligned} \tag{15}$$

which n_s is calculated as:

$$n_s = \frac{one\ voyage\ duration\ (t_v)}{sea\,state\ duration\ (t_s)} \tag{16}$$

It should be mentioned that Equation (3) and (4) are specific cases of Equation (15). Regarding ρ_u=0, Equation (15) leads to the FBC method if the wave number assumed to be deterministic, and the PSW method if the wave number uncertainty is described by Poisson process.

For the design and reliability assessment of ship structures, the extreme value of the combined process of wave-induced load effect and still water load effect is required. Since both of these actions involve uncertainties resulting from different sources, their extreme value does not occur simultaneously. The probability distribution of still water load effects such as bending moment has been investigated and quantified for different types of ships (Guedes Soares and Moan, 1988; Guedes Soares, 1992; Hørte, et al. 2007; Teixeira *et al.*, 2013). Therefore, the extreme value distribution of the combined process during one voyage can be calculated by the combination of one voyage still water load effect distribution $F_s(m)$ and the extreme wave-induced load effect during on voyage $F^v_{w,max}(m)$ as:

$$\begin{aligned} F^v_C(m) &= P\left[s + w^v_{\max} \le m\right] \\ &= \int_{-\infty}^{m} f_s(x) F^v_{w,\max}(m - x)dx \end{aligned} \tag{17}$$

Substituting Equation (15) into Equation (17) and considering $m'=\Phi^{-1}(F_{wk}(m))$:

$$\begin{aligned} &F^v_c(m) = \\ &\int_{-\infty}^{m} \int_{-\infty}^{\Phi^{-1}(F_w(m-x))} f_s(x) \times \ldots \\ &\left[F_w(m-x)\left\{\frac{\Phi^*\left[\Phi^{-1}(F_w(m-x)) - \rho_U y\right]\phi(y)dy}{F_w(m-x)}\right\}^{k-1}\right]^{n_s} dx \end{aligned} \tag{18}$$

Equation (18) can be used for calculating the extreme value distribution of the combined process of wave-induced load effect and still water load effects regarding the correlated wave load cycles in each sea state. having the number of voyages, the long-term distribution of the combined process can also be obtained from Equation (18).

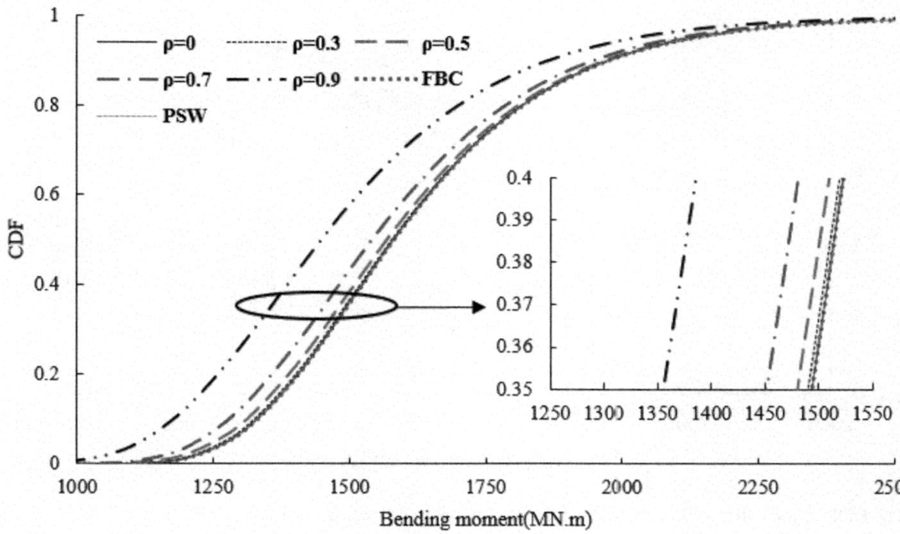

Figure 2. The CDF of the extreme wave-induced bending moments for one sea state.

For instance, assuming the deterministic n_v number of voyages the extreme value distribution of the combined process during [0, T] can be estimated as:

$$F_c^T(m) = \left[F_c^v(m)\right]^{n_v} \quad (19)$$

4 ILLUSTRATIVE EXAMPLE

In this section, a numerical example is presented to demonstrate the applicability of the proposed framework. The extreme wave-induced bending moment and its combination with the still water bending moment regarding the correlated wave loads is investigated for an Aframax tanker in the full load condition. Stillwater bending moment has a normal distribution with a mean of 1004.8 MN.m and a standard deviation of 287.1 (Hørte et al., 2007; Guedes Soares & Moan, 1988). In addition, the voyage duration is assumed to be 23.5 days as in Guedes Soares (1990a). The wave-induced bending moment follows a Weibull distribution with a scale parameter of *208.7* and a shape parameter of *1* for the mentioned tanker (Guedes Soares, 1998). The duration of a sea state is assumed to be 3 hours and the mean wave period equals to *8 s*.

The result of the extreme wave-induced bending moment for one sea state is calculated based on Equation (13) regarding ρ=0, 0.3, 0.5, 0.7 and 0.9. The obtained results are depicted in Figure 2. Regarding one sea state, the results indicate that increasing of correlation in wave-induced load cycles shifts the CDF of the extreme value distribution leftwards and predicts lower values for extreme wave-induced bending moment.

For instance, for the CDF values of 0.37 which equals the probability of exceedance (POE) of 0.63, the extreme wave-induced bending moments are 1510, 1505, 1490, 1465 and 1370 MN.m for ρ=0, 0.3, 0.5, 0.7 and 0.9 respectively. The POE of 0.63 is usually of interest for determining the characteristic values. In addition, the obtained results from Equation (13) are compared with the FBC and PSW results. It can be seen that ρ=0 in the proposed equation leads to the obtained results from FBC and PSW in which the wave loads cycles are assumed to be independent.

The results in Figure 2 imply that disregarding the correlation of wave loads leads to the overestimation of extreme value distribution for a short-term period such as one sea state.

The result of the extreme wave-induced bending moment for one voyage is calculated based on Equation (15) and depicted in Figure 3 regarding different correlation coefficients of wave loads. It can be observed that for one voyage duration, the extreme wave-induced bending moment changes only with the higher values of correlation coefficient (i.e. ρ>0.7). The medium and lower correlation does not affect the extreme wave-induced bending moment for the longer time period. This can be explained by the fact that the large number of sea states, which are assumed to be independent, diminish the effect of correlation between individual waves on the final result of extreme value distribution.

Figure 4 presents the obtained results from Equation (18) for the combination of wave-induced bending moment and still water bending moment in one voyage. The results indicate the higher values of wave correlation affect the extreme values distribution of the combined process. The lower values of wave correlation lead to approximately the same results as the FBC and PSW methods. Overall, FBC and Poisson-based methods predict the upper bound of extreme value distribution compared to the proposed approach which considers the correlated wave loads.

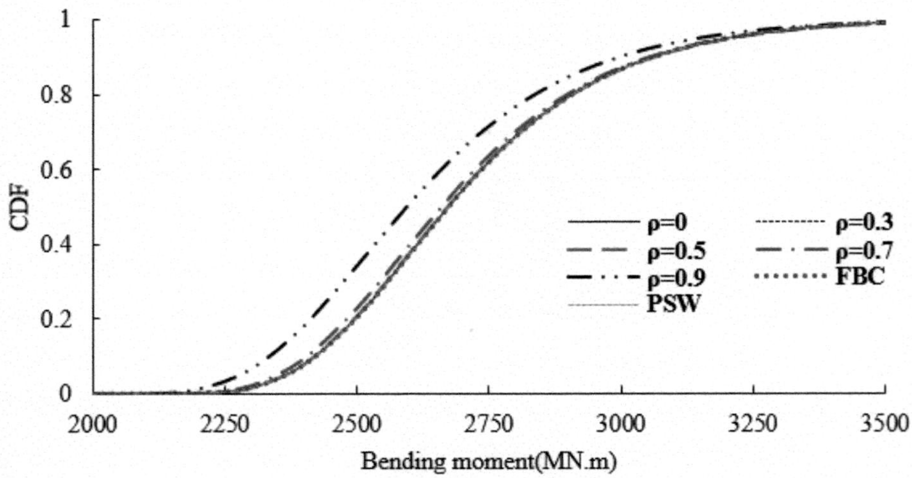

Figure 3. The CDF of the extreme wave-induced bending moments for one voyage.

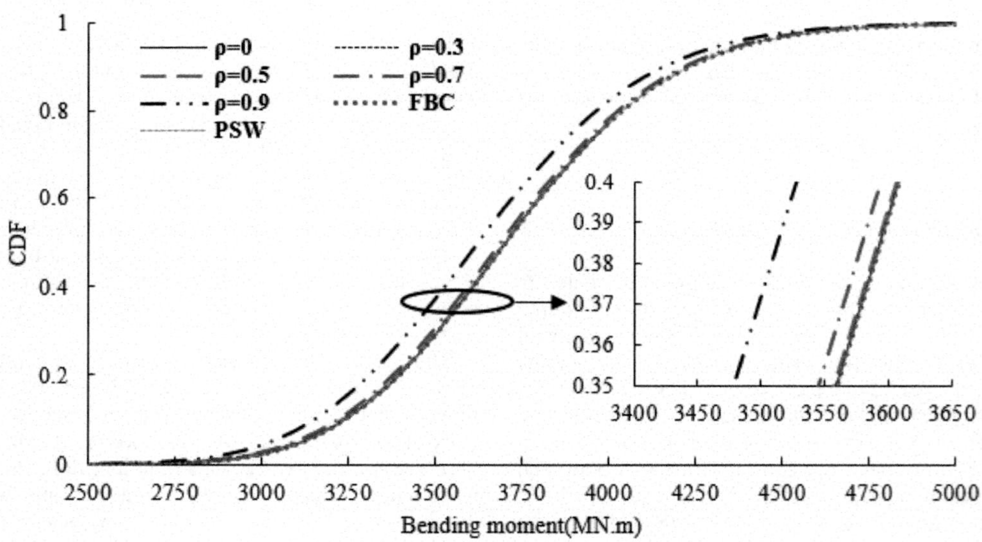

Figure 4. The CDF of the combined process for one voyage.

5 CONCLUSIONS

This paper develops a closed-form method to estimate extreme wave-induced load effect distribution and its combined process with still water load effects, considering the temporal correlation associated with the wave cycle process. Wave-induced load effect peaks in each sea state are modelled by the stochastic Markov process explicit formulations are derived and their applicability is investigated through a case study. The results indicate that disregarding the correlation of wave loads leads to overestimation of extreme value distribution for a short-term period such as one sea state. For the long-term period, the lower and medium values of wave correlation do not affect the extreme value distribution of wave-induced bending moment and the combined process. However, the higher values of the correlation coefficient lead to a lower estimation of the extreme bending moment in the investigated ship. The results and application of this study can be helpful in reconsidering and modifying the characteristics values and load factors of wave-induced load effects on ship structures. The proposed approach can be used for the other types of structures which are affected by the wave loads.

ACKNOWLEDGEMENTS

The first author has been funded by the Portuguese Foundation for Science and Technology (Fundação para a Ciência e a Tecnologia-FCT), through a doctoral fellowship under Contract no. 2023.03056. BD. This work contributes to the Strategic Research Plan of the Centre for Marine Technology and Ocean Engineering, which is financed by the Portuguese Foundation for Science and Technology (Fundação para a Ciência e Tecnologia - FCT) under contract UIDB/UIDP/00134/2020.

REFERENCES

Andersen, I. M. V., & Jensen, J. J. 2014. Measurements in a container ship of wave-induced hull girder stresses in excess of design values. *Marine Structures*, *37*: 54–85.

Bitner-Gregersen, E. M., Waseda, T., Parunov, J., Yim, S., Hirdaris, S., Ma, N., & Guedes Soares, C. 2022. Uncertainties in long-term wave modelling. *Marine Structures*, *84*: 103217.

Borges Ferry, J., & Castanheta, M. 1971. Structural safety. *Laboratorio Nacional de Engenharia Civil, Lisbon, Portugal.*

Campanile, A., Piscopo, V., & Scamardella, A. 2017. Incidence of load combination methods on time-variant oil tanker reliability in intact conditions. *Ocean Engineering*, *130*:371–384.

Cherneva, Z. and Guedes Soares, C. 2012. Non-Gaussian Wave Groups Generated in an Offshore Wave Basin. Journal of Offshore Mechanics and Arctic Engineering. 134:041602

Corak, M., Parunov, J., & Guedes Soares, C. 2018. Structural reliability analysis of container ships under combined wave and whipping loads. *Journal of Ship Research*, *62*(03): 115–133.

Der Kiureghian, A. 2005. First-and second-order reliability methods. *Engineering Design Reliability Handbook*, 14.

Fukuda, J. 1967. Theoretical determination of design wave bending moments. *Japan Shipbuilding & Marine Engineering, Volume 2, No. 2*,

Gaspar, B., Teixeira, A. P., & Guedes Soares, C. 2016. Effect of the nonlinear vertical wave-induced bending moments on the ship hull girder reliability. *Ocean Engineering*, *119*: 193–207.

Goda, Y. 2010. *Random seas and design of maritime structures* (Vol. 33). World Scientific Publishing Company.

Guedes Soares, C. 1990a. Stochastic modelling of maximum still-water load effects in ship structures. *Journal of Ship Research*, *34*(03): 199–205.

Guedes Soares, C. 1990b. Stochastic models of load effects for the primary ship structure. *Structural Safety*, *8*(1–4): 353–368.

Guedes Soares, C. 1991. Effect of transfer function uncertainty on short-term ship responses. *Ocean Engineering*, *18*(4): 329–362.

Guedes Soares, C. 1992. Combination of primary load effects in ship structures. *Probabilistic Engineering Mechanics*, *7*(2): 103–111.

Guedes Soares, C. 1998. Stochastic modelling of waves and wave induced loads. *Risk and Reliability in Marine Technology*, Guedes Soares, C. (Ed.), A.A. Balkema, The Netherlands, 197–211.

Guedes Soares, C. 1999. On the uncertainty in long-term predictions of wave induced loads on ships. *Marine Structures*, 12(3): 171–182.

Guedes Soares, C., & Moan, T. 1988. Statistical analysis of stillwater load effects in ship structures. *Transactions of the Society of Naval Architects and Marine Engineers*, 96(4), 129–156.

Guedes Soares, C., & Moan, T. 1991. Model uncertainty in the long-term distribution of wave-induced bending moments for fatigue design of ship structures. *Marine Structures*, 4(4): 295–315.

Guedes Soares, C., & Teixeira, A. P. 2000. Structural reliability of two bulk carrier designs. *Marine Structures*, *13*(2): 107–128.

Hasselmann, K., Sell, W., Ross, D. B., & Müller, P. 1976. A parametric wave prediction model. *Journal of Physical Oceanography*, *6*(2): 200–228.

Hørte, T., Wang, G., & White, N. 2007. Calibration of the hull girder ultimate capacity criterion for double hull tankers. *Proc. PRADS 2007 Conference.*

Huang, W., & Moan, T. 2008. Analytical method of combining global longitudinal loads for ocean-going ships. *Probabilistic Engineering Mechanics*, *23*(1): 64–75.

Larrabee, R. D., & Cornell, C. A. 1981. Combination of various load processes. *Journal of the Structural Division*, *107*(1): 223–239.

Lee, Y., White, N., Wang, Z., & Park, J. B. 2012. Whipping responses and whipping effects on design bending moments of a large container ship. *Proc. 6th Int. Conf. on Hydroelasticity in Marine Technology, Tokyo, Japan.*

Longuet-Higgins, M. S. 1984. Statistical properties of wave groups in a random sea state. *Philosophical Transactions of the Royal Society of London. Series A, Mathematical and Physical Sciences*, *312*(1521): 219–250.

Mao, W., & Rychlik, I. 2012. Estimation of extreme ship response. *Journal of Ship Research*, *56*(01): 23–34.

Melchers, R. E., & Beck, A. T. 2018. *Structural reliability analysis and prediction*. John Wiley & Sons.

Nielsen, U. D., Brodtkorb, A. H., & Jensen, J. J. 2018. Response predictions using the observed autocorrelation function. *Marine Structures*, *58*: 31–52.

Parunov, J., & Guedes Soares, C. 2008. Effects of Common Structural Rules on hull-girder reliability of an Aframax oil tanker. *Reliability Engineering & System Safety*, *93* (9): 1317–1327.

Parunov, J., Guedes Soares, C., Hirdaris, S., & Wang, X. 2022. Uncertainties in modelling the low-frequency wave-induced global loads in ships. *Marine Structures*, *86*, 103307.

Takami, T., Nielsen, U. D., & Jensen, J. J. 2021. Estimation of autocorrelation function and spectrum density of wave-induced responses using prolate spheroidal wave functions. *Journal of Marine Science and Technology*, *26*: 772–791.

Teixeira, A. P., & Guedes Soares, C. 1998. On the reliability of ship structures in different coastal waters. *Structural Safety and Reliability*, Shiraishi, N., Shinozuka, M. & Wen, Y.K. (Eds.), A.A. Balkema, The Netherlands, 2073–2076.

Teixeira, A. P., Guedes Soares, C., Chen, N.-Z., & Wang, G. 2013. Uncertainty analysis of load combination factors for global longitudinal bending moments of double-hull tankers. *Journal of Ship Research*, *57* (01): 42–58.

Turkstra, C. J. 1972. Theory and Structural Design Decision. Solid Mechanics Study, No. 2. *Waterloo, Ontario, Canada: University of Waterloo.*

Wang, C. 2021. *Structural Reliability and Time-Dependent Reliability*. Springer.
Wen, Y.-K. 1977. Statistical combination of extreme loads. *Journal of the Structural Division*, *103*(5): 1079–1093.
Wen, Y.-K. 1990. Structural load modelling and combination for performance and safety evaluation. *31st Edition of Developments in Civil Engineering, Vol. 31 of Tribology Series*, Elsevier, ISBN: 0444881484
Yang, W., Liang, Y., Leng, J., & Li, M. 2020. The autocorrelation function obtained from the pierson-moskowitz spectrum. *Global Oceans 2020: Singapore–US Gulf Coast*, 1–4.

Advances in Maritime Technology and Engineering – Guedes Soares & Santos (Eds)

Environmental contours and extreme vertical wave bending moments of ships in the North Atlantic

A. Mikulić & J. Parunov
Faculty of Mechanical Engineering and Naval Architecture, University of Zagreb, Zagreb, Croatia

ABSTRACT: The aim of the paper is to investigate which sea states result in the extreme vertical wave bending moments (VWBM) of ships sailing in the North Atlantic. The sea states are defined based on the wave scatter diagram provided by the International Association of Classification Societies (IACS) and encompassed by the environmental contours for the return period of 25 years. The short-term extreme VWBM is determined for head seas using calculation assumptions (wave energy spreading, wave spectrum type and ship speed) proposed by IACS, for a sea state duration of 3 hours. The short-term extreme VWBM is expressed in terms of the percentage of the lifetime extreme VWBM, calculated according to the standard long-term prediction procedure. Transfer functions of VWBM at midship are determined using closed-form expressions. Combined environmental contours – extreme VWBM diagrams are presented for four oil tankers and four container ships. The result of the study is the identification of combinations of significant wave heights and zero up-crossing periods representing the most severe sea conditions for ocean-going ships regarding VWBM. It is found that design significant wave height for 25-year return period is in the range of 11 m – 13.5 m, depending on the ship size and type.

1 INTRODUCTION

The design of novel ships is important nowadays when the alternative ship propulsion systems, new fuels and autonomous ships are considered (DNV, 2023). For innovative ships, deviating from the past designs and lacking service experience, structural design should be based on the first principle including direct calculation methods for evaluation of operational and extreme wave loads. The most important wave load component for the ship structural design is the vertical wave bending moment (VWBM) at midship, having a dominant influence on the dimensioning of the midship section modulus (IACS, 2010, 2015). The long-term probability distribution of VWBM and other wave load components is calculated according to the recommendations of the International Association of Classification Societies (IACS, 2001, 2022).

The standard procedure for the long-term wave load prediction was proposed by IACS in Recommendation No.34 "Standard Wave Data" Rev. 1 (IACS, 2001). The recommendation was based on several assumptions as the adoption of the North Atlantic scatter diagram using wave data from Global Wave Statistics (GWS) (BMT, 1986), zero ship speed, two-parameter Pierson-Moskowitz spectrum with the wave energy spreading function proportional to $\cos^2$ and equal probability of occurrence of all wave heading angles. The extreme value with the exceeding probability of 10^{-8} is prescribed as the design value of the linear wave loads, which should then be corrected for non-linear effects.

Though IACS Recommendation No. 34 Rev. 1 (IACS, 2001) has extensively been used in the past two decades, it has faced criticism because of the underlying assumptions (Parunov et al. 2004). Some of the assumptions were unconservative, as the assumptions on the zero-speed tended to reduce extreme wave loads. On the other hand, the severe assumption was employed that the ship would spend a whole lifetime sailing in the North Atlantic, which is the most severe wave environment. The GWS was criticised as the source of the statistical wave data, as their origin is in the visual wave observations from merchant ships, which are inherently inaccurate because of the human estimates (Austefjord et al. 2023). Another drawback of GWS is the difficulty in quantifying the effect of heavy weather avoidance, which is implicitly included in the visual observations from merchant ships (Guedes Soares, 1990). Furthermore, GWS contains data gathered in the period between 1949 and 1986, missing the last three decades, which may impact extreme wave statistics (Guedes Soares, 1996). A closely related question is the impact of climate change, which certainly influences the operational environment of ships, although a firm conclusion regarding the influence of climate change is still missing (Bitner-Gregersen et al. 2022). The overall effect of different assumptions was such that the IACS Rec. No. 34 Rev. 1

DOI: 10.1201/9781003508779-31

(IACS, 2001) resulted in VWBM exceeding the design values prescribed in IACS Unified Requirement S-11 and S-11a (Mikulić et al. 2023). The severity of the IACS NA scatter diagram was identified as the main reason for these excessively large VWBMs (Parunov et al. 2004).

As a response to this criticism of IACS Rec. No. 34 Rev. 1 (IACS, 2001), IACS issued the revised IACS Rec. No. 34 Rev. 2 (IACS, 2022), where some modifications are introduced. Thus, the JONSWAP wave energy spectrum is proposed instead of the Pierson-Moskowitz spectrum, an energy spreading function proportional to $\cos^3$ instead of $\cos^2$, while a minimum ship speed of 5 knots is used instead of zero speed. Another difference is that instead of extracting design VWBM for the probability level 10^{-8}, a return period of 25 years is proposed. The most important modification is the new IACS NA scatter diagram, based on the hindcast wave database, filtered for the actual shipping route using Automatic Identification System (AIS), a satellite ship positioning system available for all merchant ships (Austefjord et al. 2023).

Preliminary study of the consequences of implementing IACS Rec. No. 34 Rev. 2 (IACS, 2022), presented by Austefjord et al. (2023), concluded that the design values of wave loads are reduced compared to Rev. 1. It was also found that this reduction is not uniform, ranging between 10% and 30% depending on the type and size of the vessels, where higher reductions are achieved for shorter vessels. The comparative analysis of two revisions of IAC Rec. No. 34, focusing on the extreme VWBM amidship for oil tankers and container ships is performed by Mikulić et al. (2023). They concluded that the design values of VWBM amidship calculated by Rev. 2 are reduced by 13-19% for tankers and 16-23% for container ships, where a larger reduction is obtained for smaller ships. Although quantitative results by Mikulić et al (2023) should be taken with cautiousness because of the simplifications involved in the analysis, it is to be noted that direct analysis results were lower compared to IACS Rule VWBM (IACS 2010, IACS 2015).

As the wave scatter diagram is the most important reason for differences in the long-term prediction of extreme wave loads, a more detailed analysis of scatter diagrams from two revisions is conducted by de Hauteclocque et al. (2023). They compared visually environmental contours for 25-year return periods for relevant wave zones and found that GWS data exhibit an unrealistically broad range of periods for high significant wave heights. A similar conclusion is obtained by Mikulić et al. (2023), as shown in Figure 1, representing environmental contours for IACS scatter diagrams from two revisions. Haselsteiner et al. (2021) presented different approaches for the construction of environmental contours, while in the present study, the Inverse First Order Reliability Method (IFORM) is employed, initially proposed by Winterstein et al. (1993). It is obvious from Figure 2 that the Rev. 1 scatter diagram encompasses higher sea states and larger range of wave periods.

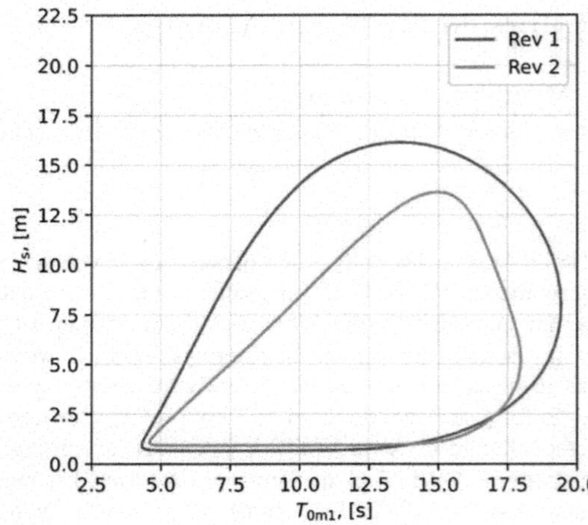

Figure 1. The comparison of 25-year IFORM contours based on joint models representing scatter diagrams from IACS Rec. No. 34 Rev. 1 and Rev. 2.

Furthermore, Mikulić et al. (2023) determined the most probable individual wave amplitude from the long-term distribution and found 16.2 m and 14 m for Rev. 1 and Rev. 2, respectively. This is another evidence of the reduced severity of the wave data proposed by IACS in Rev. 2.

The aim of the present paper is to further research the consequences of the application of IACS Rec No. 34 Rev. 2 (IACS, 2022) on extreme wave loads on ships. The research question investigated in the study reads: which sea states provide the same most probable short-term VWBM amidship as full long-term analysis performed according to IACS Rec No. 34 Rev. 2? The short-term analysis of VWBM is determined for head seas with a 3-hour duration, using the wave energy spreading, wave spectrum type, and ship speed according to Rev. 2, while the long-term analysis is performed completely in accordance with Rev. 2 (IACS, 2022). The most probable short-term VWBM for each sea state is expressed as the percentage of the most probable VWBM from the full long-term analysis. Contours of such non-dimensional VWBM are then presented in the same diagram as the IFORM contours. Such representation enables visual identification of extreme sea states where the VWBM exceeds the design value calculated by the long-term analysis. A similar analysis was presented previously for the case of the Adriatic Sea (Mikulić & Parunov, 2023).

The present study is based on the simplified calculation of transfer functions of VWBM amidship using closed-form expressions (CFEs) proposed by Jensen et al. (2004). CFEs are useful for conceptual studies where relative changes of VWBM using different assumptions are important, e.g., Jensen et al. (2008).

The methodology used for the calculation of short- and long-term distributions of VWBM is presented in Section 2. Eight ships are analysed (four oil tankers and four container ships), described in Section 3. The results of the analysis, i.e. joint diagrams of IFORM contours and VWBM are presented in Section 4. In the discussion Section 5, a comparison of the results with some published measurements of extreme sea states and ship operational limits, as well as with results obtained using assumptions from Rev. 1 (IACS, 2001), is presented. The last section of the paper provides conclusions of the study.

2 METHODOLOGY

Transfer functions of VWBM are calculated using CFE proposed by Jensen et al. (2004) as:

$$\frac{\Phi_M}{\rho g B L^2} = \kappa \frac{1-kT}{(k_e L)^2}\left[1-\cos\left(\frac{k_e L}{2}\right)-\sin\left(\frac{k_e L}{2}\right)\right] F_V\ F_C\ \sqrt[3]{|\cos\beta|} \tag{1}$$

where L, B, and T are ship waterline length, breadth, and draught respectively, κ the Smith correction factor, k_e is the effective wave number, while β is the heading angle between a ship and waves (180° represents head seas). The shape correction factor F_C and ship speed correction factor F_V are given in the original reference by Jensen et al. (2004). The Frequency Independent Model Error (FIME) of the VWBM transfer function given by Equation 1 is determined by extensive comparison to the model-scale experiments by Petranović et al. (2023).

Response spectra of VWBM in irregular seas are obtained in the domain of encounter frequencies ω_e by:

$$S_R(\omega_e|H_S,T_Z,\beta) = |\ \Phi_M\ (\omega_e|\ \beta)|^2 S_\eta(\omega_e|H_S,T_Z) \tag{2}$$

where $S_\eta(\omega_e|H_S,T_Z)$ is the encounter wave spectrum dependent on significant wave height H_S and zero up-crossing wave period T_Z, while $S_R(\omega_e|H_S,T_Z,\beta)$ is the response spectrum, which also depends on the β.

The spectral moments of the n-th order for a given heading are given as:

$$m_n = \int_0^\infty \sum\nolimits_{\theta_0-90^\circ}^{\theta_0+90^\circ} f_s(\theta)\omega_e^n \cdot S_R(\omega_e|H_S,T_Z,\theta)d\omega \tag{3}$$

where $f_s(\theta)$ is the energy spreading function for relative spreading θ around the main wave heading θ_0. More details on this expression may be found in IACS Rec. No. 34 (IACS 2001, 2022).

The average zero up-crossing response period T_z^* is given as:

$$T_z^* = 2\pi\sqrt{\frac{m_0}{m_2}} \tag{4}$$

The number of response cycles in the short-term sea state with duration T sec is calculated as:

$$n = \frac{T}{T_Z^*} \tag{5}$$

The most probable extreme value (MPEV) of VWBM in the short-term sea state with n response cycles is given as:

$$M_e^* = \sqrt{2m_0 \ln n} \tag{6}$$

The following relations are used to convert mean wave period T_{0m1} of JONSWAP and P-M spectrum to mean zero up-crossing period T_z, respectively (de Hauteclocque et al. 2023):

$$T_z = T_{0m1}/\ 1.194 \tag{7}$$

$$T_z = T_{0m1}/\ 1.21 \tag{8}$$

The average wave steepness of random sea states, calculated in the Discussion section is given as (Bitner-Gregersen et al. 1995):

$$S_{Tz} = 2\pi H_S/\ g\ T_z^{\ 2} \tag{9}$$

The long-term distribution of wave-induced responses is obtained by the standard procedure where short-term responses are weighted by the probability of occurrence of short-term sea conditions and by assuming that all heading angles between ship and waves are equally probable (e.g. Guedes Soares, 1998).

For IACS Rec. No 34 Rev. 1, the most probable extreme value is calculated for probability level 10^{-8}, while for Rev. 2 average zero up-crossing response period is calculated across all sea states enabling calculation of a total number of response cycles n_T in 25 years. The probability level of the most probable extreme response is then calculated as 10^{-n_T}, which is different than 10^{-8}.

3 SHIPS

To cover different ship types, the short- and long-term analysis is performed for oil tankers and containerships. Former ships represent full-form ships with a large block coefficient of about 0.8, while the latter ships are fine-form ships with a block coefficient less than 0.7.

The main particulars of oil tankers are selected from Michel & Osborne (2004) as average values for each of the four classes of oil tankers. Ship classes and main particulars are given in Table 1.

Container ships are those studied previously by Parunov and Ćorak (2010) and Ćorak et al. (2018). The main particulars of container ships are given in Table 2. The naming used for container ships is similar to the previous studies of those ships.

Table 1. The main particulars of oil tankers.

	L	*B*	*T*	
Ship class		m		C_b
Panamax	174.4	31.4	11.3	0.8
Aframax	229.7	41.9	13.1	0.8
Suezmax	260.8	45.8	15.9	0.8
VLCC	318.6	58.4	21.1	0.8

Table 2. The main particulars of container ships.

	L	*B*	*T*	
Ship		m		C_b
S175	175	25.4	9.5	0.57
4400TEU	261	37.1	12.6	0.61
Flokstra	270	32.2	10.85	0.60
9400TEU	333	42.8	13.17	0.67

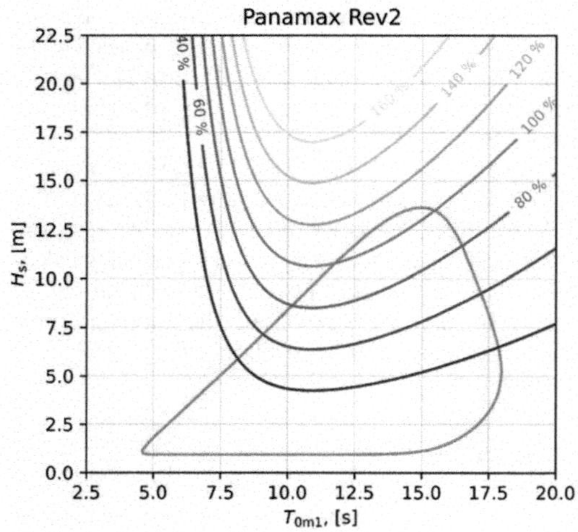

Figure 2. The IFORM 25-year contours in comparison to VWBM contours of the Panamax oil tanker.

4 RESULTS

Results for oil tankers are presented in Figure 2-5. Closed orange curves represent IFORM contours for the scatter diagram from IACS Rec. No.34 Rev. 2 (IACS, 2022), while open curves represent MPEV of the VWBM in each short-term sea state with a duration of 3 hours, expressed as a percentage of the long-term MPEV of VWBM calculated in accordance with the IACS Rec. No. 34 Rev. 2. Thus, the curve denoted as 100% represents sea states in which the short-term MPEV of VWBM is the same as the long-term MPEV of VWBM.

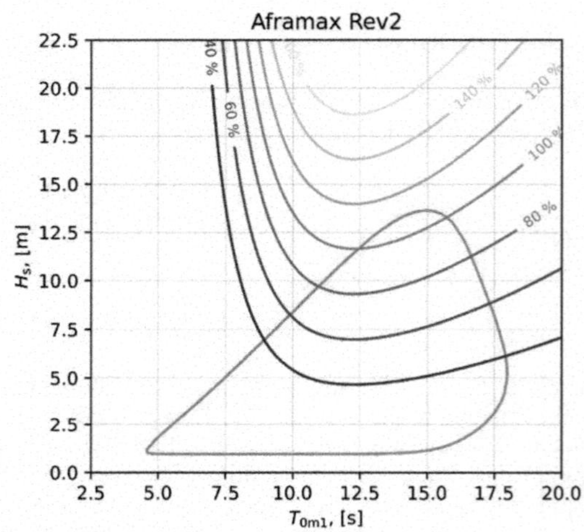

Figure 3. The IFORM 25-year contours in comparison to VWBM contours of the Aframax oil tanker.

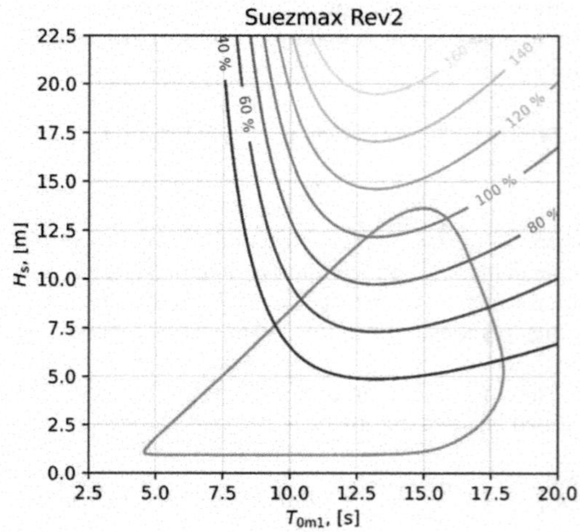

Figure 4. The IFORM 25-year contours in comparison to VWBM contours of the Suezmax oil tanker.

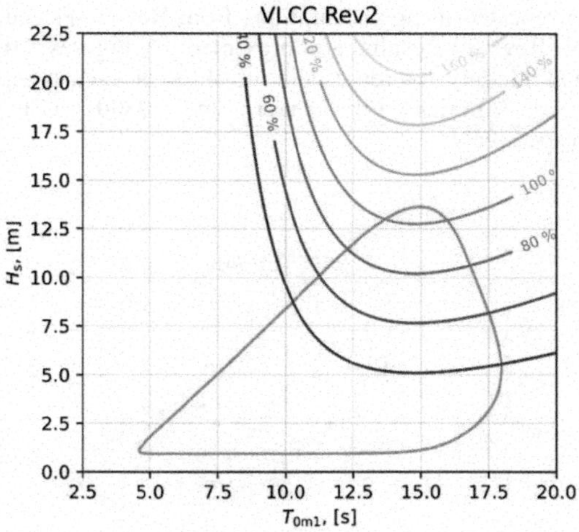

Figure 5. The IFORM 25-year contours in comparison to VWBM contours of the VLCC.

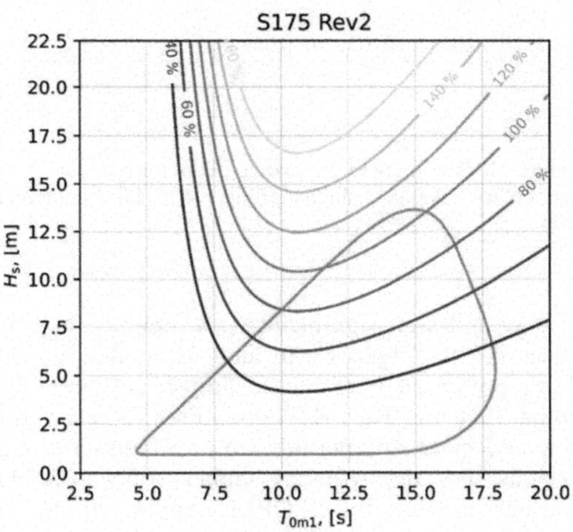

Figure 6. The IFORM 25-year contours in comparison to VWBM contours of the S-175 container ship.

Table 3. Extreme sea states for oil tankers.

	Lower		Upper	
	H_s	T_z	H_s	T_z
Ship class	m	sec	m	sec
Panamax	11	9.6	13.5	13.4
Aframax	11.5	10.5	13.5	13.4
Suezmax	12.3	10.9	13.5	13.4
VLCC	13	11.3	13.2	13.4

The extreme sea states for oil tankers are summarised in Table 3, showing sea states for which 100% VWBM contour crosses the IFORM contour. Mean periods are converted to the zero up-crossing periods using Equation 7.

Results for container ships are presented in Figures 6-9, while the extreme sea states for container ships, for which 100% VWBM contour crosses IFORM contour are shown in Table 4.

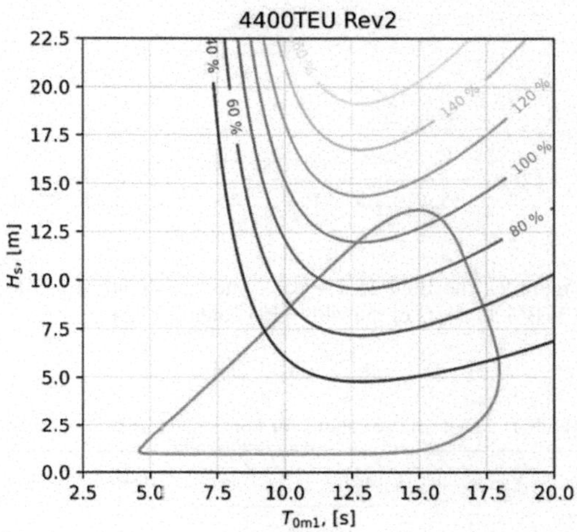

Figure 7. The IFORM 25-year contours in comparison to VWBM contours of the 4400 TEU container ship.

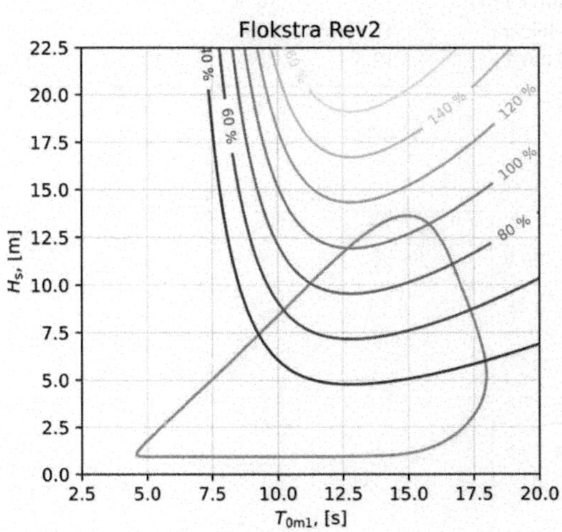

Figure 8. The IFORM 25-year contours in comparison to VWBM contours of the Flokstra container ship.

5 DISCUSSION

It is interesting to compare presented results with those according to IACS Rec. No. 34 Rev. 1 (IACS, 2001). For that purpose, the same analysis

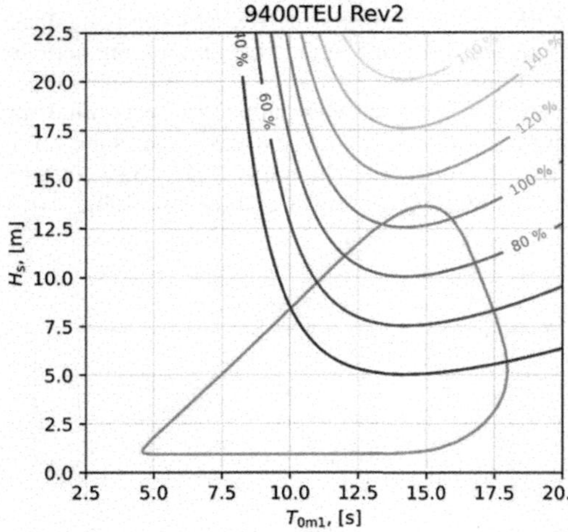

Figure 9. The IFORM 25-year contours in comparison to VWBM contours of the 9400 TEU container ship.

Table 4. Extreme sea states for container ships.

	Lower		Upper	
	H_s	T_z	H_s	T_z
Ship class	m	sec	m	sec
S-175	11	9.6	13.5	13.4
4400 TEU	12	10.5	13	13.4
Flokstra	12	10.6	13	13.4
9400 TEU	12.5	11	13	13.4

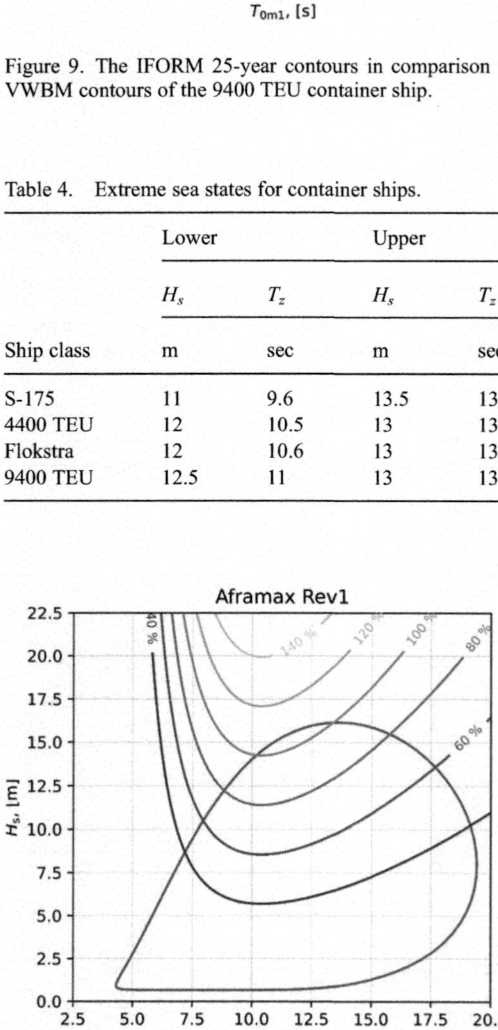

Figure 10. The IFORM 25-year contours for rev. 1 in comparison to VWBM contours of the Aframax oil tanker.

is repeated using assumptions from Rev. 1 instead of Rev. 2. Results are presented in Figures 10 and 11 only for representative ships of each type, i.e., Aframax oil tanker and 9400 TEU, respectively.

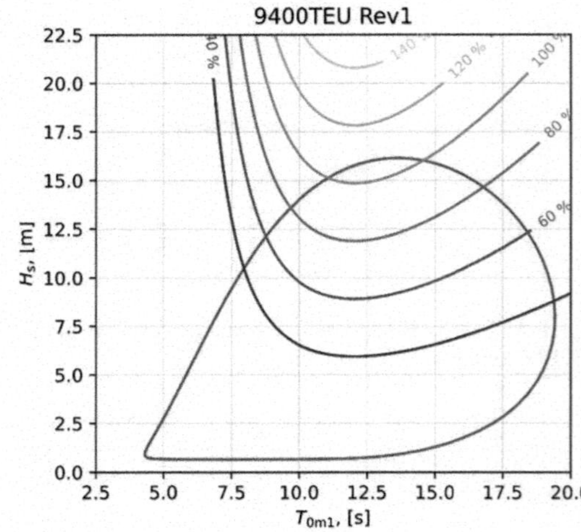

Figure 11. The IFORM 25-year contours for rev. 1 in comparison to VWBM contours of the 9400 TEU container ship.

Larger design significant wave heights may be observed in Figures 10 and 11 compared to those presented in Tables 3 and 4. It should be noted that for Rev. 1, Equation 8 is used for converting mean periods to zero up-crossing periods. For the Aframax tanker 100% VWBM contour crosses the IFORM contour at H_S = 14 m and 16 m with corresponding zero up-crossing periods T_Z of 8.4 s and 11.2 s, respectively. The former sea state corresponds to the average wave steepness of 1/8, which is an unrealistically large value. Namely, the highest wave steepness used in the design of marine structures reads 1/10 (e.g., Mitchel, 1999). Therefore, this sea state can be considered as unrealistic as such steep waves cannot persist. Latter sea state (H_S = 16 m and T_Z = 11.2 s) corresponds to the average wave steepness of 1/12, which is still higher than the average steepness of 1/15, recommended for the design of ship structures (Bittner-Gregersen et al. 1995). Corresponding average wave steepness for IACS Rec. No. 34 Rev. 2 (IACS, 2022), based on H_S and T_Z in Table 3, read 1/15 and 1/21, which are values close to the maximum recommended average wave steepness and average wave steepness of the fully developed seas, respectively.

For the 9400 TEU container ship, 100% VWBM contour crosses the IFORM contour at H_S = 15 m and 16 m with corresponding zero up-crossing periods T_Z of 9 s and 11.5 s, respectively. These values correspond to the average wave steepness of 1/8.5 and 1/13 respectively. Although this steepness is lower compared to the Aframax oil tanker, they are still unrealistically large. Corresponding values for IACS Rec. No. 34 Rev. 2 (IACS, 2022), according to the sea states given in Table 4 read 1/15 and 1/21.

It is interesting to compare presented results with wave measurements in the North Atlantic. One of the pioneering references is Ochi (1978), reporting three measured significant wave heights: 16.7 m and 15.3m for a return period of 8 years and 16.8m for a return period of 14 years. Bittner-Gregersen et al. (1995) provided measurements for the return period of 20 years at three locations around the Norwegian coast, i.e. 12.4 m, 14.9 m, and 13.5 m. Hogben (1998) reported that the most severe measured sea state in the North Sea may be described with H_S=13.6m and T_Z=10.5s. These values are larger compared to the values determined in Tables 2 and 3 from IACS Rec. No 34 Rev. 2 (IACS, 2022) since fixed measurements do not include the effect of the weather routing. Thus, de Hauteclocque et al. (2023) found that extreme significant wave heights encountered by seagoing ships in the North Atlantic are for 2 m lower compared to the extremes occurring in the same region.

Operational limits for different ship types are provided by Moan et al. (2006), reading slightly above 12 m for oil tankers and bulk carriers and up to 11 m for container ships. These figures account for heavy weather avoidance, and they are close to the extreme significant wave heights found from IACS Rec. No.34 Rev. 2 (IACS, 2022).

It should be beard in mind that there are additional phenomena that could make existing statistical wave data unconservative, namely climate changes and rogue waves. Bitner-Gregersen et al. (2022) stated that a firm conclusion about the effects of climate change is difficult to reach, but the range of changes in the operational environment is sufficiently large to influence the safety of ships.

Rogue waves are very steep and much larger than the surrounding waves, commonly defined by the criterion that the wave height exceeds twice, while the wave crest exceeds 1.25 times the significant wave height. Although responsible for several accidents of ships, rogue waves are still not explicitly included in either revision of IACS Rec. No 34 or other ship classification standards (Bitner-Gregersen & Gramstad 2015).

6 CONCLUSION

The paper investigates extreme sea states according to the recently published IACS Rec. No 34 Rev. 2 (2022), which is based on the wave hindcast database. The aim was to find design sea states, where the short-term analysis of linear VWBM would provide the same results as the long-term analysis. The results are presented as combined diagrams of IFORM environmental contours and VWBM for oil tankers and container ships of varied sizes. The results are compared to the previous revision of the same document, where the scatter diagram was obtained from GWS. The following conclusions are drawn:

- Extreme significant wave heights for a 25-year return period following the IACS Rec. No 34 Rev. 2 (IACS, 2022) are in the range of 11 m – 13.5 m, depending on the ship size and type.
- Extreme significant wave heights from IACS Rec. No 34 Rev. 1 (IACS, 2001) read from 14 m to 16 m, i.e., about 3 m higher compared to Rev. 2.
- Short-term responses of smaller ships depend on the zero up-crossing wave periods. i.e., on average wave steepness.
- Expectedly, average wave steepnesses of extreme sea states from Rev. 2 are much lower and more realistic compared to Rev 1.
- The extreme sea states from Rev. 2 seem to be closer to the operational limits of merchant ships. However, it is to be noted that ships could encounter higher sea states than their operational limits. Faulkner (1998) reported that the loss of the bulk carrier "Derbyshire" in 1980 could have occurred in a sea state with a significant wave height of 14 m.

ACKNOWLEDGEMENTS

This work has been fully supported by the Croatian Science Foundation under the project IP-2019-04-2085.

REFERENCES

Austefjord, H. N., de Hauteclocque,G., Johnson, M. C. & Zhu, T. 2023. Update of wave statistics standards for classification rules. In: *Advances in the Analysis and Design of Marine Structures* ed Ringsberg, J. W. & Guedes Soares, C. London, UK: Taylor and Francis. pp 95–205.

Bitner-Gregersen, E.M., Cramer, E.H.& LØseth, R. 1995. Uncertainties of Load Characteristics and Fatigue Damage of Ship Structures, *Marine Structures* 8:97–117.

Bitner-Gregersen E.M., Gramstad, O. 2015. Rogue waves – impact on ships and offshore structures. *DNV-GL Strategic Research & Innovation Position Paper* 05.

Bitner-Gregersen, E., Waseda, T., Parunov, J., Yim, S., Hirdaris, S., Ma, N. & Guedes Soares, C. 2022 Uncertainties in long-term wave modelling. *Marine Structures*, 84: 103217.

British Marine Technology (BMT) (Primary contributors Hogben N, Da Cunha L F & Oliver H N) *Global Wave Statistics*. Unwin Brothers Limited. 1986. London.UK.

Ćorak, M., Parunov, J. & Guedes Soares, C. 2018. Structural Reliability Analysis of Container Ships under Combined Wave and Whipping Loads. *J. Ship Res.* 62, 3: pp 115–33.

de Hauteclocque, G., Vitali Maretic, N. & Derbanne, Q. 2023. Hindcast based global wave statistics. *Appl. Ocean Res.* 130: 103438.

Det Norske Veritas (DNV) 2023., *Maritime Forecast to 2050*, available at www.dnv.com/maritime/publications/maritime-forecast-2023/index.html.

Faulkner, D., An Independent Assessment of the Sinking of the MV DERBYSHIRE. 1998. *Trans. SNAME*, 106: 59–103.

Guedes Soares, C., 1990. Effect of Heavy Weather Maneuvering on the Wave-Induced Vertical Bending Moments in Ship Structures, *Journal of Ship Research*, Vol. 34, No. 1: 60–68.

Guedes Soares, C. 1996. On the Definition of Rule Requirements for Wave Induced Vertical Bending Moments, *Marine Structures* 9, 409–425.

Guedes Soares, C. 1998. Stochastic Modelling of Waves and Wave Induced Loads, *Risk and Reliability in Marine Technology*, 197–211, Balkema, Rotterdam.

Haselsteiner, A.F., Coe, R.G., Manuel, L., Chai, W., Leira, B.J., Clarindo, G., Guedes Soares, C., Hannesdóttir, I., Dimitrov, N, Sander, A., Ohlendorf, J-H., Thoben, K. D., de Hauteclocque, G., Mackay, E.; Jonathan, P., Qiao, C.; Myers, A., Rode, A., Hildebrandt, A., Schmidt, B., Vanem, E. & Huseby, A. B. 2021. A benchmarking exercise for environmental contours. *Ocean Engineering*, Vol. 236, 109504.

Hogben, N. 1998. Discussion on: An Independent Assessment of the Sinking of the MV DERBYSHIRE. *Trans. SNAME*, 106: 59–103.

International Association of Classification Societies (IACS) 2001. *Recommendation No. 34: standard wave data.* Rev 1. London, UK.

International Association of Classification Societies (IACS) 2010. *UR S11 longitudinal strength standard.* Rev 6. London, UK.

International Association of Classification Societies (IACS). 2015. *UR S11A longitudinal strength standard for container ships*. London, UK.

International Association of Classification Societies (IACS) 2022. *Recommendation No. 34: standard wave data.* Rev 2. London, UK.

Jensen, J.J., Mansour, A.E. & Olsen, A.S. 2004. Estimation of ship motions using closed-form expressions. Ocean Engineering 31: 61–85.

Jensen, J.J., Pedersen, P.T., Shi, B., Wang, S., Petricic, M. & Mansour, A.E., 2008. Wave induced extreme hull girder loads on containerships. *Trans. SNAME*, 116: 128–151.

Michel, R. K. & Osborne, M. 2004. Oil Tankers, In: *Ship Design and Construction* ed Lamb, T., Jersey City: SNAME pp 29.1–29.41.

Mikulić, A., Ćorak, M. & Parunov, J.2023. Comparative analysis of the long-term extreme VWBM calculated according to IACS Rec. No 34 rev 1 and rev 2. *IOP Conf. Ser.: Mater. Sci. Eng.* 1288. 012030.

Mikulić, A. & Parunov, J. 2023. Environmental Contours in the Adriatic Sea for Design and Analysis of Marine Structures. *Journal of Marine Science and Engineering.* 11(5):899.

Mitchel, W. 1999. Sea Spectra Revisited, *Marine Technology*, 36, 4: 211–227.

Moan, T., Shu, Z., Drummen, I. & Almashi, H. 2006. Comparative reliability analysis of ships considering different ship types and the effect of ship operations on loads. *Trans. SNAME*, 114: 16–54.

Ochi, M.K. 1978. Wave Statistics for the Design of Ships and Ocean Structures, *Trans. SNAME*, 86: 47–76.

Parunov, J., Senjanović, I. & Pavičević, M. 2004. Use of vertical wave bending moment from hydrodynamic analysis in design of oil tankers *Transactions of the Royal Institution of Naval Architects Part A: International Journal of Maritime Engineering.* 146 a4.

Parunov, J. & Ćorak, M. 2010. Influence of environmental and operational uncertainties on wave bending moments of containerships. RINA William Froude Conference: *Advances in Theoretical and Applied Hydrodynamics – Past and Future.*

Petranović, T. Mikulić, A. Gledić, I. & Parunov, J. 2023. Comparative analysis of closed-form expressions and other commonly used seakeeping methods. *Ocean engineering*, 281; 114977.

Winterstein, S. R., Ude, T. C., Cornell, C. A., Bjerager, P. & Haver, S. 1993. Environmental parameters for extreme response: Inverse FORM with omission factors. In *Proceedings of the 6th International Conference on Structural Safety and Reliability* ICOSSAR. August 9–13, Innsbruck, Austria: 551–557.

Advances in Maritime Technology and Engineering – Guedes Soares & Santos (Eds)
 ISBN 978-1-032-83107-7

A novel segmented model of container ship for wave-induced loads test

H.L. Si & X.L. Wang
China Ship Scientific Research Center, Wuxi, China

ABSTRACT: Model experiments stand out as the most reliable method for investigating wave-induced loads on ships navigating in rough seas. To enhance precision in investigating wave-induced loads on large open ships, such as ultra-large container ships, a new type segmented ship model with a U-shaped backbone of a 20000 TEU container ship was designed and fabricated in this paper. The U-shaped backbone could simulate simultaneously the vertical, horizontal, torsional stiffness as well as the shear center of a full scale container ship. Measurement of the ship model's vertical and horizontal bending moments was facilitated by assessing the corresponding bending strains of the U-shaped backbone. A novel method, termed shear flow integration, was introduced for the direct measurement of torsional moments and shear forces. Validation of the reasonability and reliability of this method were achieved through finite element calculation and calibration of the U-shaped backbone. Based on the new type segmented model, the nonlinear vertical, horizontal, torsional springing tests were conducted for the 20000 TEU container ship. This work established a robust foundation for the accurate investigation on the wave-induced loads of ultra-large container ships.

1 INTRODUCTION

The precise estimation of wave-induced loads of ship sailing in waves is a prerequisite for designing ship with sufficient structural strength. Researchers globally had conducted extensive model tests on various ships to deal with this issue. In general, in order to conduct wave-induced loads, it is necessary to fabricate segmented ship model, which mainly consist of three parts, refer to as ship shell, the backbone and the weights. According to the requirements of test, the ship model is divided into several segments, which are connected to each other with the backbone. Each segment is equipped with a certain weight, so that the mass distribution, center of gravity and moment of inertia of ship model can satisfy the requirements. The gaps between adjacent segments are sealed by rubber membrane. The rubber membrane is generally concave, so that the wave-induced loads could transmit entirely to the backbone when the ship model is sailing in waves. So, the wave-induced loads could be obtained based on the backbone's deformation (Geng et al, 2014; Ding et al, 2015; Wang et al., 2016, Chen et al, 2012).

The ship model's backbone has three functions: (1) connecting each segment of the ship model, (2) simulating the stiffness of the ship, (3) actting as wave-induced loads transducer. The ship will vibrate with high frequency under wave excitation force when sailing in waves. In order to simulate accurately the vibration characteristics of the full-scale ship, the vertical, horizontal, torsional stiffness of the backbone have to satisfy the similarity principle. However, due to the difficulty of the design and fabricating of the backbone that can simultaneously simulate vertical, horizontal and torsional stiffness of the full-scale ship, and the vertical bending moment is the most important wave-induced load for most ships, so far, the backbone of most segmented ship model had been designed and fabricated only satisfy the similarity of vertical stiffness. For large opening ships, such as ultra-large container ships, structural strength under torsional moment has always been a focal point for designers. The shear center of ultra-large container ship locates below the baseline for a considerable distance. To accurately measure the torsional moment experienced by ultra-large container ship, not only the vertical stiffness of backbone must satisfy similarity principle, but also the horizontal, torsional stiffness and shear center have to satisfy the similarity principle.

In order to conduct wave-induced loads test more precisely, some researchers designed new type backbone to simulate container ship's stiffness. A square cross-section aluminum backbone was used to simulate a 290m long container ship's stiffness. To reduce torsional stiffness, five square holes were cut out above the aluminum backbone (Zhu et al., 2011). For better simulating the stiffness of a container ship, an upper-opened aluminum backbone was designed and fabricated (Maron & Kapsenberg, 2014). Subsequently, the backbone was placed as low as possible to allow its shear center to be below the base line of the ship model, with the goal of simulating the coupling of the horizontal bending and torsion of the hull girder.

DOI: 10.1201/9781003508779-32

Furthermore, a U-shaped and I-shaped backbone had been applied to simulate stiffness of full-scale container ship (Kim & Kim, 2014; Hong & Kim, 2014; Kim et al., 2016). The vertical and torsional springing of a river-to-sea ship had been experimentally examined using a segmented ship model equipped with a U-shaped backbone (Wang et al., 2018).

In this paper, a U-shaped backbone was designed and fabricated to simultaneously simulate the vertical, horizontal, torsional stiffness, and shear center of ultra-large container ships. Subsequently, segmented ship models with U-shaped backbone were fabricated to conduct wave-induced loads test. The shear flow integration method suitable for U-shaped backbone was proposed to measure the torsional moment on the ship. Based on the new type segmented ship model, the nonlinear vertical, horizontal, torsional springing tests were conducted for the 20000 TEU container ship in seakeeping basin.

2 SEGMENTED SHIP MODEL

2.1 *Principal of similarity*

In order to conduct wave-induced loads test, it is necessary to fabricate segmented ship model. Ship model's design and fabrication should satisfy principle of similarity. Equations (1)-(4) describe the similarities of the geometric, the vertical, horizontal, torsional stiffness and shear center, respectively.

$$\frac{L_s}{L_m} = \lambda \tag{1}$$

$$\frac{E_s I_s}{E_m I_m} = \lambda^5 \tag{2}$$

$$\frac{G_s J_s}{G_m J_m} = \lambda^5 \tag{3}$$

$$\frac{C_s}{C_m} = \lambda \tag{4}$$

In these four equations, the subscript s denotes the full-scale ship, the subscript m denotes the ship model. L is the ship line, E is the Young's modulus, I is the cross-sectional vertical or horizontal moment of inertia, G is the shear modulus, J is the cross-sectional torsional moment of inertia, C is the shear center, λ is the model scaling ratio.

2.2 *Principal dimensions of ship model*

This paper takes a 20000TEU ultra-large container ship as investigation objective. The scaling ratio of ship model is 77. The ship model was divided into 8 segments and the sections located at 4, 6, 8, $10^{+45\text{mm}}$, 12, 14, 16 station, respectively. The ship model's principal dimensions are listed in Table 1. The mass distribution of ship model is listed in Table 2. Figures 1 and 2 is the segmented ship model.

Figure 1. Segmented ship model.

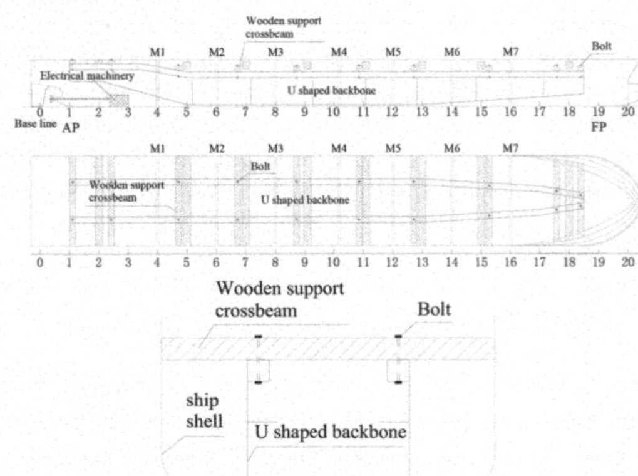

Figure 2. Schematic diagram of the composition of ship model.

Table 1. Principal dimensions of ship model.

	Full scale ship	Ship model
L_{OA} (m)	399.67	5.191
L_{PP} (m)	383	4.974
B (m)	58.6	0.761
D (m)	30.5	0.396
d_F (m)	15.2	0.174
d_A (m)	16.7	0.199
Δ (T)	260602.9	0.488
H_{CG} (m)	27.574	0.339
L_{CG} (m)	182.714	2.383

A U-shaped backbone was used to simulate container ship's stiffness and location of shear center and to connect each segment of ship model. A wooden support crossbeam in each segment is positioned above the U-shaped backbone and is fixed to U-shaped backbone by bolts.

2.3 *U-shaped backbone*

In order to conduct wave-induced loads test more precisely, based on thin-walled beam

Table 2. Mass distribution.

Station	Mass	Station	Mass
0-1	15.15	10-11	38.22
1-2	20.48	11-12	28.40
2-3	26.47	12-13	50.86
3-4	26.91	13-14	25.41
4-5	27.72	14-15	33.56
5-6	37.20	15-16	22.03
6-7	38.43	16-17	28.15
7-8	28.86	17-18	17.93
8-9	38.17	18-19	12.23
9-10	33.84	19-20	6.87

theory, a U-shaped backbone was designed and fabricated to simultaneously simulate vertical, horizontal, torsional stiffness and shear center of the full-scale ship. The U-shaped backbone and the scantling are shown in Figures 3 and 4.

Figure 3. U-shaped backbone.

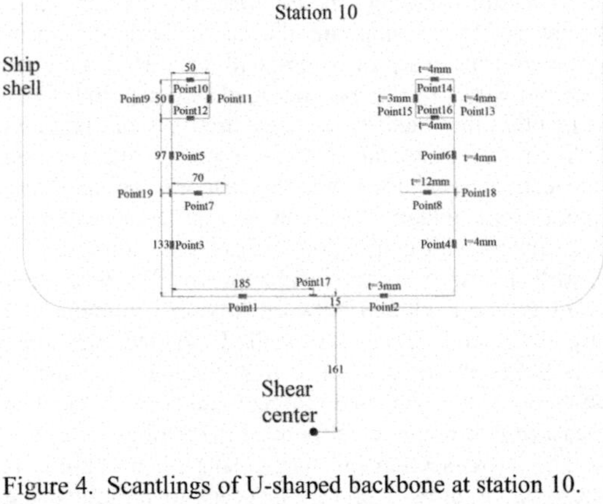

Figure 4. Scantlings of U-shaped backbone at station 10.

The material of U-shaped backbone is ABSPC. The Young's modulus and shear modulus of ABSPC material are 2.195GPa and 978.7MPa, respectively. Table 3 is the vertical, horizontal, torsional moment of inertia and shear center of U-shaped backbone at different station.

Table 3. Parameters of U-shaped backbone.

		station 6	Station 10	Station 14
Required value	Vertical moment of inertia (m^4)	4.99E-05	4.99E-05	4.10E-05
	horizontal moment of inertia (m^4)	1.54E-04	1.85E-04	1.46E-04
	torsional moment of inertia (m^4)	1.01E-06	9.31E-07	9.75E-07
	Shear center (below the baseline, m)	1.79E-01	2.16E-01	1.60E-01
Actual value	Vertical moment of inertia (m^4)	5.13E-05	5.13E-05	4.24E-05
	horizontal moment of inertia (m^4)	1.54E-04	1.54E-04	1.21E-04
	torsional moment of inertia (m^4)	1.02E-06	1.02E-06	9.56E-07
	Shear center (below the baseline, m)	1.61E-01	1.61E-01	1.37E-01
error	Vertical moment of inertia	2.81%	2.81%	3.36%
	horizontal moment of inertia	0.27%	-16.71%	-17.12%
	torsional moment of inertia	1.33%	9.96%	-1.96%
	Shear center	-10.01%	-25.26%	-14.55%

3 PRINCIPLE OF MEASUREMENT

3.1 *Vertical and horizontal bending moment*

Before the backbone was assembled with the model shell, it was necessary to calibrate the backbone to determine the relationship between the external load and the strain of the backbone. Namely, the calibration coefficient, also known as the strain coefficient. The measurement method of the calibration coefficient is shown in Figures 5 and 6. The backbone was simply supported at both ends, as shown in Figure 6, and the external load was applied at the same position from the fulcrum. The calibration coefficient was determined by gradual loading and unloading.

At the measuring section, a unidirectional strain gauge was affixed on the bottom plate at the center longitudinal cross-section of the U-shaped backbone along longitudinal direction. Simultaneously, a compensation plate was placed near measurement point, and strain gauge was affixed on the compensation plate. The two strain gauges were connected in a half-bridge configuration to form a channel for measuring vertical bending strain. Unidirectional strain gauges were affixed on both out sides of the U-shaped backbone at the height of neutral axis. The two strain gauges on each side were connected in a half-bridge configuration to create a channel for measuring horizontal bending strain. The distribution of strain gauge and half bridge configuration are shown in Figures 7 and 8. Equation (5) and (6) are the vertical and horizontal bending strain of U-shaped backbone. Equation (7) describe the calibration coefficient. M and ε is the variation of vertical or horizontal bending moment and corresponding strain at each loading and unloading, respectively. η is the calibration coefficient. The actual load acting on the ship model $M(t)$ in Equation (8) is determined based on the strain signals $\varepsilon(t)$ produced by the strain gauges during the wave-induced load test and the calibration coefficient.

Figure 5. Calibration of U-shaped backbone.

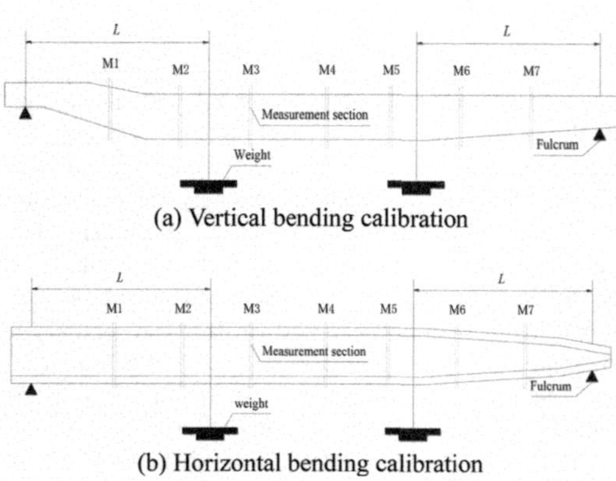

Figure 6. Schematic diagram of U-shaped backbone's calibration.

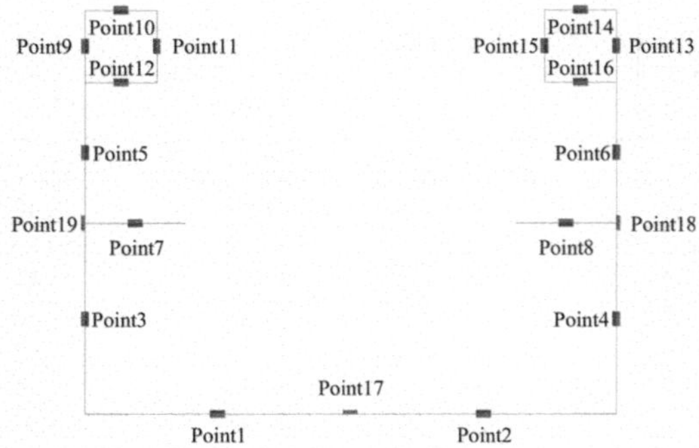

Figure 7. Distribution of strain gauge.

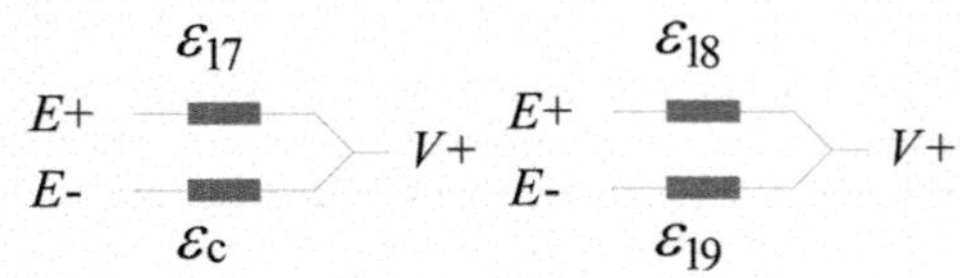

Figure 8. Half bridge configuration.

$$\varepsilon = \varepsilon_{17} - \varepsilon_c \tag{5}$$

$$\varepsilon = (\varepsilon_{18} - \varepsilon_{19})/2 \tag{6}$$

$$\eta = M/\varepsilon \tag{7}$$

$$M(t) = \eta \cdot \varepsilon(t) \tag{8}$$

3.2 *Torsional moment and shear force*

3.2.1 *Theoretical analysis*

The container ships are in a restrained torsion condition when they sail in oblique waves. The shear stress of ship structure could be resolved into two components, free torsional shear stress and secondary shear stress. The total torsional moment is the sum of free torsional moment and the secondary torsional moment. Free torsional moment is induced by free torsional shear stress while secondary torsional moment is induced by secondary shear stress. For a certain cross-section, although the torsional moment and constraint acting on ship are the same, both the shear stress and the warping stress will be varied with the location varies of the external load or constraint. Consequently, the method to measure vertical bending moment or horizontal bending moment with relation between external load and the response of unidirectional strain gauge affixed at certain point which is determined with calibration of backbone must not be used to measure the torsional moment. Because there is no constant relation between torsional moment and any other single quantities for the U-shaped backbone, it is necessary to develop a new method to measure container ship's torsional moment and shear force. The measuring principle is introduced in detail as follows.

The cross-section of the U-shaped backbone is shown in Figures 4 and 7. In order to introduce the method conveniently, the plates which make up the U-shaped backbone are numbered in the way as shown in Figure 9. Each plate has two faces which are called

Z1 and Z2. τ_{z1} is the shear stress of face Z1 and τ_{z2} is the shear stress of face Z2. For plates 1~ 8 which make up the open section, the shear stress is resolved into two components as shown in Figure 10 one is the free shear stress which follows linear distribution in thickness direction, the other one is the secondary shear stress which follows uniform distribution in thickness direction. So,

$$\tau_\omega = (\tau_{z1} + \tau_{z2})/2 \quad (9)$$

$$\tau_f = (\tau_{z2} - \tau_{z1})/2 \quad (10)$$

$$M_{\omega 1} = \int_0^{s1} \tau_\omega htds \quad (11)$$

$$M_{f1} = GJ_1\varphi' \quad (12)$$

$$M_{X1} = M_{f1} + M_{\omega 1} \quad (13)$$

$$J_1 = \frac{1}{3}\sum_{i=1}^{8} H_i t_i^3 \quad (14)$$

Where, τ_f is free torsional shear stress, τ_ω is secondary shear stress, J_1 is sum torsional inertia moment of plates 1~8, φ' is twist rate, t is the thickness of plate, and h is the force arm of secondary shear stress. The thickness and the arm relative shear center for a plate are constant. $M_{\omega 1}$ is secondary torsional moment, M_{f1} is free torsional moment, and M_{X1} is total torsional moment of the open section which consists of plates 1~8.

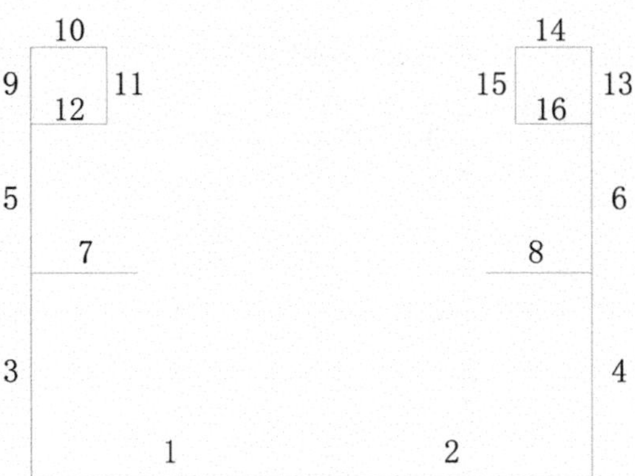

Figure 9. Plate number of U-shaped backbone.

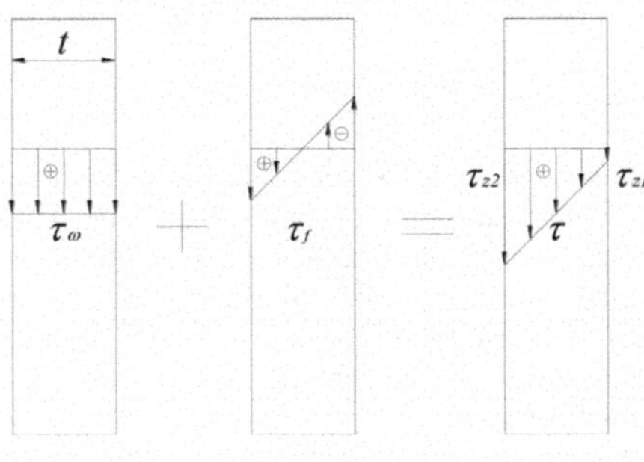

Figure 10. Distribution of stress (No.1~8).

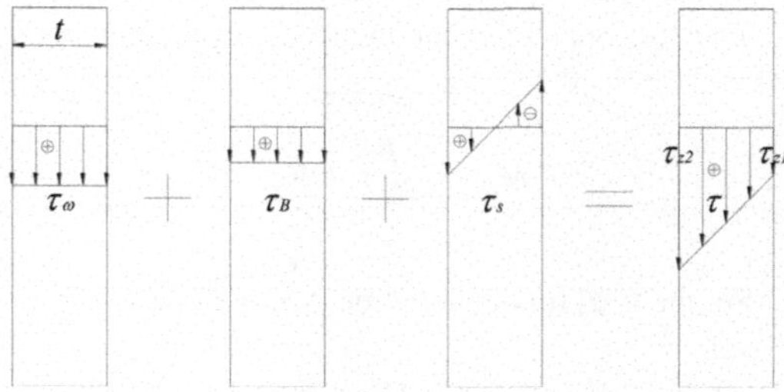

Figure 11. Distribution of stress (No.9~16).

In the case of torsion box which has a closed section including plates 9~16, the shear stress of each plate is resolved into three components as shown in Figure 11: one is the secondary shear stress which follows uniform distribution in thickness direction; the second one is Bredt shear stress which follows uniform distribution in thickness direction, the last one is Saint Venant shear stress which follows linear distribution in thickness direction. Bredt shear stress and Saint Venant shear stress are free torsional shear stress. So,

$$\tau_\omega + \tau_B = (\tau_{z1} + \tau_{z2})/2 \quad (15)$$

$$\tau_s = (\tau_{z2} - \tau_{z1})/2 \quad (16)$$

where, τ_ω is secondary shear stress. τ_B is Bredt shear stress. τ_s is Saint Venant shear stress. Torsion box's free torsional moment M_{f2} could be divided into two parts and are shown as follows:

$$M_{f2} = M_S + M_B = GJ_2\varphi' \quad (17)$$

where, M_S is the torsional moment induced by Saint Venant shear stress. M_B is the torsional moment induced by Bredt shear stress. For a cross-section, the twist rate of each plate is the same. Therefore,

$$\varphi' = \frac{M_S}{GJ_S} = \frac{M_B}{GJ_B} = \frac{M_{f2}}{GJ_2} = \frac{M_{f1}}{GJ_1} \quad (18)$$

Thus,

$$M_S = GJ_S\varphi' \quad (19)$$

$$M_B = GJ_B\varphi' = \oint_{s2} \tau_B htds \quad (20)$$

$$J_2 = J_S + J_B \quad (21)$$

$$J_S = \frac{1}{3}\sum_{i=9}^{16} H_i t_i^3 \quad (22)$$

$$J_B = \frac{8A^2}{\oint \frac{ds}{t}} \quad (23)$$

where, J_S is the sum of torsional inertia moment of plates 9~16 corresponding with the Saint Venant shear stress. J_B is the torsional inertia moment corresponding with the Bredt shear stress of the two torsion box. A is the enclosed area of each torsion box's section.

The torsional moment induced by torsion box's secondary shear stress is:

$$M_{\omega 2} = \oint_{s2} \tau_\omega htds \qquad (24)$$

So, the total torsional moment of torsion box is:

$$\begin{aligned} M_{X2} &= M_S + M_B + M_{\omega 2} \\ &= GJ_S\varphi' + GJ_B\varphi' + M_{\omega 2} \end{aligned} \qquad (25)$$

And the total torsional moment of the section is:

$$\begin{aligned} M_X &= M_{X1} + M_{X2} \\ &= M_{\omega 1} + M_{f1} + M_S + M_B + M_{\omega 2} \\ &= G(J_1 + J_S + J_B)\varphi' + M_{\omega 1} + M_{\omega 2} \\ &= G(J + J_B)\varphi' + M_{\omega 1} + M_{\omega 2} \end{aligned} \qquad (26)$$

$$J = \frac{1}{3}\sum_{i=1}^{16} H_i t_i^3 \qquad (27)$$

$$J_B = \frac{8A^2}{\oint \frac{ds}{t}} \qquad (28)$$

where, J is sum of torsional inertia moment of plates 1~16. Compared with J_B, J is very small. Compared with M_B, M_{f1} and M_S are very small and could be ignored.

It can be concluded that the total torsional moment is mostly induced by secondary shear stress and Bredt shear stress.

$$\begin{aligned} M_X &= M_{\omega 1} + M_B + M_{\omega 2} \\ &= \int_0^{s1} \tau_\omega htds + \oint_{s2} \tau_B htds + \oint_{s2} \tau_\omega htds \\ &= \int_0^{s1} \tau_\omega htds + \oint_{s2} (\tau_B + \tau_\omega) htds \\ &= \frac{1}{2}\int_0^{s} (\tau_{z1} + \tau_{z2}) htds \end{aligned} \qquad (29)$$

The vertical shear force and horizontal shear force could also be expressed as:

$$F_Z = \frac{1}{2}\int_0^{s_V} (\tau_{z1} + \tau_{z2}) tds \qquad (30)$$

$$F_H = \frac{1}{2}\int_0^{s_H} (\tau_{z1} + \tau_{z2}) tds \qquad (31)$$

Based on the theoretical analysis, if the distribution of shear stress of two faces of each plate which makes up the U-shaped backbone is known, the distribution of secondary shear stress and Bredt shear stress will be obtained. So we could obtain the total torsional moment as well as shear force acting on ship body with equations (29) ~ (31). We could call $\tau_{z1} + \tau_{z2}$ as the sum of two face's shear stress and call $\tau_{z1} - \tau_{z2}$ as the difference of two face's shear stress. In fact, τ_B and τ_ω are the shear flow of each plate.

3.2.2 *Numerical calculation*

A calculation is conducted with finite element method for a U-shaped backbone with the same cross-section along the length to validate whether the external loads obtained with Equations (29)-(31) are equal to the loads acting on the U-shaped backbone.

The model of the U-shaped backbone and its cross-section are shown in Figures 12 and 13. The length of the U-shaped backbone is 4.56 m, the width is 0.37 m and the height is 0.3 m. The left side of the U-shaped backbone is fixed rigidly and a steel frame is installed 1.1 m away from right side of the U-shaped backbone.

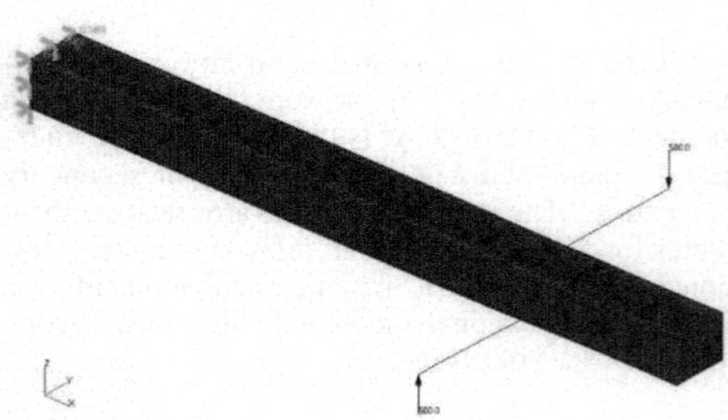

Figure 12. Finite element model.

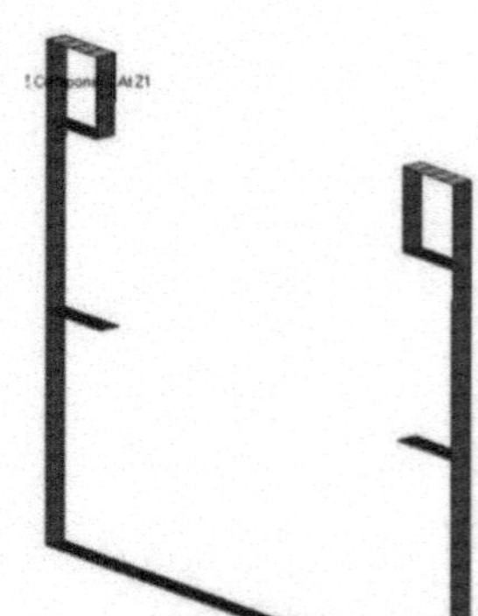

Figure 13. Section of finite element model.

Forces with opposite direction or same direction could be applied on the U-shaped backbone. Vertical and horizontal shear forces will be applied on the U-shaped backbone. The calculation consists of 6 conditions as shown in Table 4 ~ Table 6. The investigated section is located at the middle of the U-shaped backbone.

The shear stresses of each element's two faces are extracted, the free torsional moment and secondary

Table 4. Torsional moment.

	T (N·m)	F_V (N)	F_H (N)	*T_Total* (N·m)	*T_C_A* (N·m)	*T_C_M* (N·m)	*Error* (*T_C_A*)	*Error* (*T_C_M*)
1	1000	0	0	1000	994.1	1001.4	-0.59%	0.14%
2	0	2900	0	0	1.5E-04	-19.7	-	-
3	0	0	2300	753.2	749.1	761.0	-0.54%	1.03%
4	1000	2900	0	1000	994.1	981.6	-0.59%	-1.84%
5	1000	0	2300	1753.2	1743.3	1762.3	-0.57%	0.52%
6	1000	2900	2300	1753.2	1743.3	1742.6	-0.57%	-0.60%

Table 5. Vertical shear force.

	T (N·m)	F_V (N)	F_H (N)	F_V*_Total* (N)	F_V*_C_A* (N)	F_V*_C_M* (N)	*Error* (F_V*_C_A*)	*Error* (F_V*_C_*M)
1	1000	0	0	0	-3.2E-03	-205.6	-	-
2	0	2900	0	2900	2893.7	2956.0	-0.22%	1.93%
3	0	0	2300	0	8.0E-04	-49.5	-	-
4	1000	2900	0	2900	2893.7	2750.5	-0.22%	-5.16%
5	1000	0	2300	0	-2.4E-03	-255.1	-	-
6	1000	2900	2300	2900	2893.7	2701.0	-0.22%	-6.86%

Table 6. Horizontal shear force.

	T (N·m)	F_V (N)	F_H (N)	F_H*_Total* (N)	F_H*_C_A* (N)	F_H*_C_M* (N)	*Error* (F_H*_C_A*)	*Error* (F_H*_C_M*)
1	1000	0	0	0	-5.5	56.9	-	-
2	0	2900	0	0	-8.3E-04	-125.9	-	-
3	0	0	2300	2300	2293.3	2387.8	-0.29%	3.82%
4	1000	2900	0	0	-5.5	-69.0	-	-
5	1000	0	2300	2300	2287.8	2444.6	-0.53%	6.29%
6	1000	2900	2300	2300	2287.8	2318.8	-0.53%	0.82%

torsional moment of each element are calculated based on equations (9) ~ (31). The calculated results are shown in Tables 4-6. Where, T, F_V and F_H are torsional moment, vertical shear force and horizontal shear force acting directly on U-shaped backbone, respectively. *T_Total*, F_V*_Total*, F_H*_Total* are total torsional moment, total vertical force and total horizontal force acting on U-shaped backbone actually. F_V is equal to F_V*_Total* and F_H is equal to F_H*_Total.* However, *T_Total* include two components, one is the torsional moment T acting on U-shaped backbone directly, the other one is the torsional moment induced by horizontal force. *T_C_A*, F_V*_C_A* and F_H*_C_A* are the torsional moment, vertical force and horizontal force calculated with finite element method taking account of all elements of the cross-section based on equations (29) ~ (31). Error is the difference between calculated results and total external loads acting on U-shaped backbone actually. For both torsional moment and shear force, the calculated results and actual results agree with each other very well and the errors are all smaller than 0.6%. So the theoretical analysis introduced above is validated to be reasonable.

However, as it is impossible to measure the shear stress of all elements in practice, it is necessary to develop an alternative method to measure the torsional moment and shear force. The distribution of the sum of two face's shear stress are shown in Figure 14.

Based on the calculated results, it can be found that $\tau_{z1} + \tau_{z2}$ follow linear distribution in height and breadth direction. Thus, the total torsional moment and shear force could be obtained with $\tau_{z1} + \tau_{z2}$ of each plate's middle point. So, equations (29) ~ (31) could be replaced by equations (32) ~ (34).

$$M_X = M_{\omega 1} + M_B + M_{\omega 2} = \frac{1}{2}\int_0^s (\tau_{z1} + \tau_{z2}) htds = \frac{1}{2}\sum_{i=1}^{16} \tau_i h_i t_i H_i \quad (32)$$

$$F_Z = \frac{1}{2}\sum_{i=3}^{6} \tau_i t_i H_i + \frac{1}{2}\tau_9 t_9 H_9 + \frac{1}{2}\tau_{11} t_{11} H_{11} + \frac{1}{2}\tau_{13} t_{13} H_{13} + \frac{1}{2}\tau_{15} t_{15} H_{15} \quad (33)$$

$$F_H = \frac{1}{2}\tau_1 t_1 H_1 + \frac{1}{2}\tau_2 t_2 H_2 + \frac{1}{2}\tau_7 t_7 H_7 + \frac{1}{2}\tau_8 t_8 H_8 + \frac{1}{2}\tau_{10} t_{10} H_{10} + \frac{1}{2}\tau_{12} t_{12} H_{12} + \frac{1}{2}\tau_{14} t_{14} H_{14} + \frac{1}{2}\tau_{16} t_{16} H_{16} \quad (34)$$

where, τ is $\tau_{z1} + \tau_{z2}$ of each plate's middle point, F_Z is the vertical shear force, F_Y is the horizontal shear force, h is the force arm of shear stress relative shear center, t is the thickness of plate, and H is breadth or height of plate.

In practice, shear stress can be measured by strain gauges affixed at middle of each plate. However, in order to obtain $\tau_{z1} + \tau_{z2}$, it is not necessary to measure the actual shear stress τ_{z1} or τ_{z2}. The simplified method to obtain $\tau_{z1} + \tau_{z2}$ is introduced as follows. As we all know that τ_{z1} and τ_{z2} could be represented as Equations (35)-(36) in local coordinate shown in Figure 15. X axis is consistent with ship's longitudinal direction, from stern to stem. Y axis is the direction indicated by the arrow as shown in Figure 15. Compared with the height and breadth, the thickness of plate is very small. So the strain difference of two faces in X direction and Y direction is very small and can be ignored. So, Equation (37) can be simplified as Equation (38):

$$\tau_{z2} = G\gamma_{xy\_z2} = G(2\gamma_{45^o\_z2} - \gamma_{0^o\_z2} - \gamma_{90^o\_z2}) \quad (35)$$

$$-\tau_{z1} = -G\gamma_{xy\_z1} = G(2\gamma_{135^o\_z1} - \gamma_{0^o\_z1} - \gamma_{90^o\_z1}) \quad (36)$$

$$(\tau_{z2} + \tau_{z1})/2 = G(\gamma_{45^o\_z2} - \gamma_{135^o\_z1}) - G(\gamma_{0^o\_z2} - \gamma_{0^o\_z1})/2 - G(\gamma_{90^o\_z2} - \gamma_{90^o\_z1})/2 \quad (37)$$

$$(\tau_{z2} + \tau_{z1})/2 = G(\gamma_{45^o\_z2} - \gamma_{135^o\_z1}) \quad (38)$$

And the torsional moment and shear force can be represented as:

$$M_X = \int_0^s G(\gamma_{45^o\_z2} - \gamma_{135^o\_z1})htds = G\sum_{i=1}^{16} \gamma_{di} h_i t_i H_i \quad (39)$$

$$F_Z = G\sum_{i=3}^{6} \gamma_{di} t_i H_i + G\gamma_{d9} t_9 H_9 + G\gamma_{d11} t_{11} H_{11} + G\gamma_{d13} t_{13} H_{13} + G\gamma_{d15} t_{15} H_{15} \quad (40)$$

$$F_H = G\gamma_{d1} t_1 H_1 + G\gamma_{d2} t_2 H_2 + G\gamma_{d7} t_7 H_7 + G\gamma_{d8} t_8 H_8 + G\gamma_{d10} t_{10} H_{10} + G\gamma_{d12} t_{12} H_{12} + G\gamma_{d14} t_{14} H_{14} + G\gamma_{d16} t_{16} H_{16} \quad (41)$$

$$\gamma_d = \gamma_{45^o\_z2} - \gamma_{135^o\_z1} \quad (42)$$

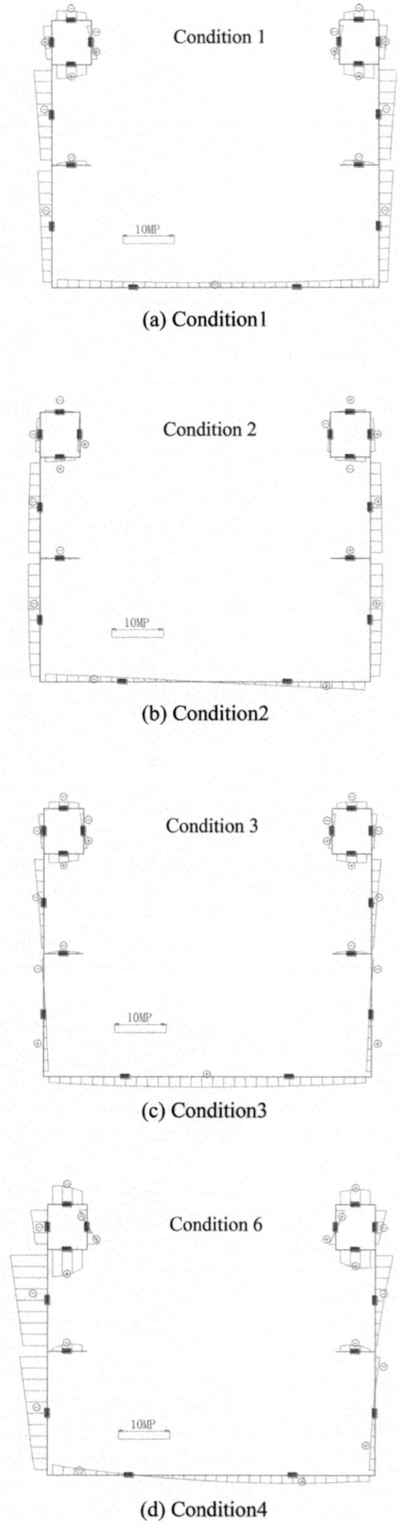

Figure 14. Distribution of the sum of the shear stress.

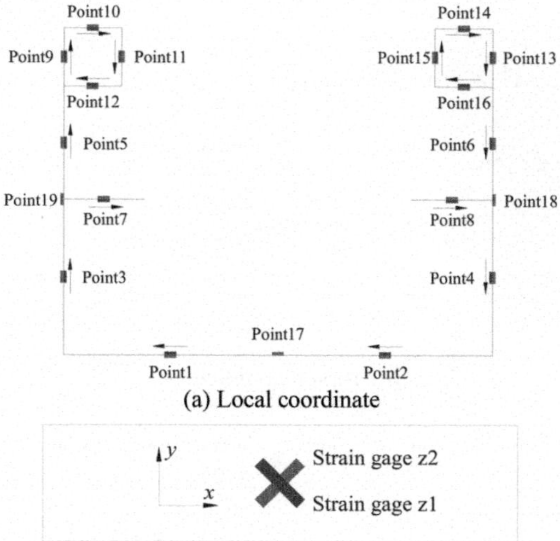

Figure 15. Distribution of strain gauge.

The method to affix strain gauges is shown in Figure 15. The faces of each plate are called *Z*2 and *Z*1. *Z*1 is inner face and *Z*2 is outer face. In local coordinate, the angle between strain gauge of *Z*2 and *X* axial is 45° and the angle between strain gauge of *Z*1 and *X* axial is 135°. The two strain gauges are vertical to each other.

In Tables 4-6, *T_C_M*, *FV_C_M* and *FH_C_M* are the torsional moment, vertical force and horizontal force calculated with finite element method taking account of only middle element's $\tau_{z1}+\tau_{z2}$ and ignoring the strain difference of two faces in *X* direction and *Y* direction. The errors of torsional moment between calculated results and actual results are all smaller than 2%. The errors of vertical shear force between calculated results and actual results are all smaller than 5% except for Condition 6 of which is 6.86%. The errors of horizontal shear force between calculated results and actual results are all smaller than 5% except for Condition 5 of which is 6.29%. Based on the results, it is reasonable to measure torsional moment, vertical shear force and horizontal shear force with this method.

3.2.3 *Calibration*

In order to validate the reasonability of the simplified method introduced above, a U-shaped backbone was designed and fabricated based on similarity theory and thin-walled beam theory. Strain gauges were affixed to the U-shaped backbone as shown Figure 15 and the U-shaped backbone was calibrated at the ship structural strength laboratory of CSSRC.

The U-shaped backbone and its calibration are shown in Figures 16-17. Figure 17 shows the way to calibrate the U-shaped backbone. Figure 17(a), (b), (c) show the calibration of vertical shear force, horizontal shear force and torsional moment, respectively. L_1 is the distance between fixed location and measured section, L_2 is distance between external load and measured section, and

Figure 16. Calibration of U-shaped backbone at laboratory.

L_3 is the distance between external load and shear center of the measured section. When the torsional moment is measured, the two forces in Figure 17(c) are in opposite direction. When the vertical force is measured, the two forces in Figure 17 (c) are in the same direction. The external force was subsequently step wisely increased, and then decreased again. The signals of the torsional moment, vertical shear force and horizontal shear force are measured and shown in Figure 18. The calibration results are shown in Tables 7-9. F_H_E, F_V_E and T_E are external horizontal shear force, vertical shear force and torsional moment respectively. F_H_M, F_V_M and T_M are measured horizontal shear force, vertical shear force and torsional moment respectively. R_F_H, R_F_V and R_T are the ratios between the external loads acting on U-shaped backbone directly and measured results. Based on the results, T_M agree very well with T_E and the ratios between them are all smaller than 1.1 for all the conditions. Although the differences between F_H_E, F_V_E and F_H_M, F_V_M are slightly higher, but the ratios between them are constant with variation of L_1, L_2 and L_3. So, we could conclude that the method introduced above is reasonable and the actual load acting on the ship model, torsional moment *T(t)*, vertical shear force *FV(t)*, horizontal shear force *FH(t)* can be obtained based on R_T, R_F_V, R_F_H and the corresponding signals $M_X(t)$, $F_Z(t)$, $F_Y(t)$ derived from wave-induced loads test.

$$T(t) = R\_T \cdot M_X(t) \tag{43}$$

$$FV(t) = R\_F_V \cdot F_Z(t) \tag{44}$$

$$FH(t) = R\_F_H \cdot F_Y(t) \tag{45}$$

4 WAVE-INDUCED LOADS TEST

The wave-induced loads test was conducted at China Ship Scientific Research Center's seakeeping basin. The result of typical condition was selected to investigate the characteristic of wave-induced loads of ultra-large container ship.

When a ship is sailing in regular waves, the incident wave frequency and wave encounter frequency

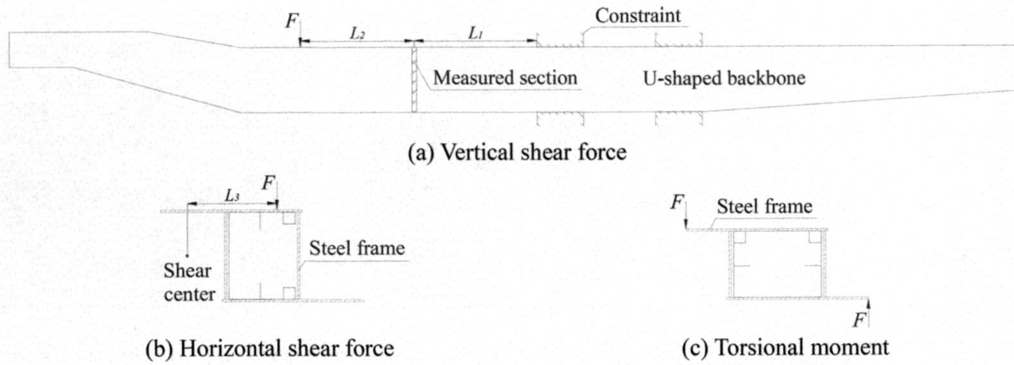

Figure 17. Schematic diagram of U-shaped backbone's calibration.

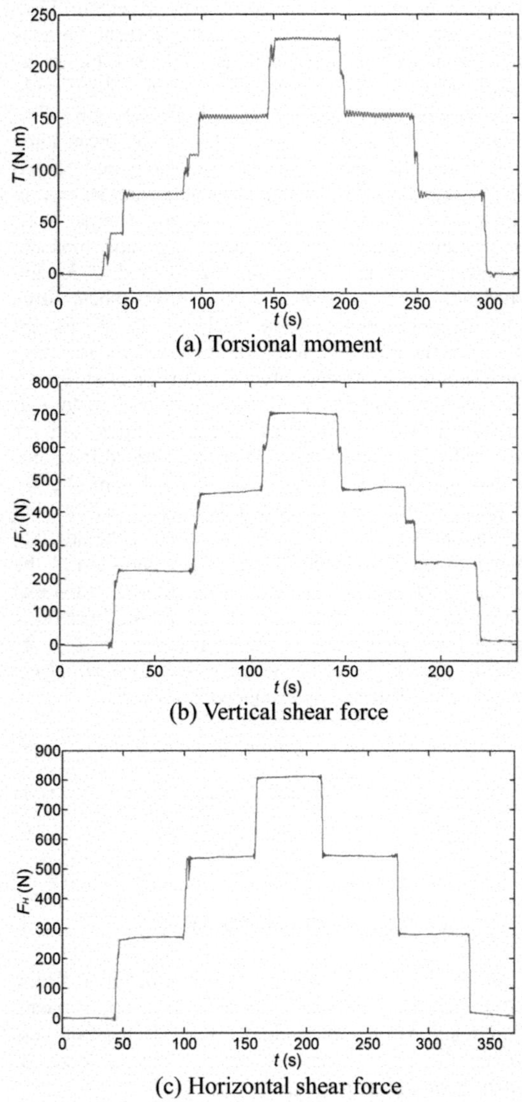

Figure 18. Signals of calibration.

Table 7. Calibration of torsional moment.

Condition	L_1(cm)	L_2(cm)	T_E(N·m)	T_M(N·m)	R_T
1	20	16	68.7	73.9	0.93
2	20	45	68.7	71.7	0.96
3	20	73	68.7	75.3	0.91
4	66	16	68.7	71.5	0.96

Table 8. Calibration of vertical shear force.

Condition	L_1(cm)	L_2(cm)	F_V_E(N)	F_V_M(N)	R_F_V
1	20	16	196.2	229.6	0.85
2	20	45	196.2	226.6	0.87
3	20	73	196.2	234.1	0.84
4	66	16	196.2	228.4	0.86

Table 9. Calibration of horizontal shear force.

Condition	L_1 (cm)	L_2 (cm)	L_3 (cm)	F_H_E (N)	F_H_M (N)	R_F_H
1	20	16	30.9	196.2	265.6	0.74
2	20	45	30.9	196.2	266.6	0.74
3	20	73	30.9	196.2	264.4	0.74
4	66	16	30.9	196.2	271.7	0.72
5	20	16	61.6	98.1	132.2	0.74
6	66	16	77.6	98.1	131.0	0.75
7	20	16	0	98.1	134.9	0.73

satisfy equation (46), where, ω_e is wave encounter frequency, ω_0 is incident wave frequency, V is ship speed, g is gravity acceleration, μ is wave heading, and $\mu = 180°$ represents heading wave. When ω_e satisfy equation (47) and $n = 1$, the ship undergoes linear springing. When $n = 2, 3, \ldots$, the ship undergoes nonlinear springing. ω_{2-node} is the first order vertical, horizontal or torsional wet natural frequency. Figure 19 is the seakeeping basin.

Figure 19. Seakeeping basin.

$$\omega_e = \omega_0 - \omega_0^2/g \cdot V\cos\mu \qquad (46)$$

$$\omega_{2-node} = n\omega_e \qquad n = 1, 2, \ldots \qquad (47)$$

The incident wave that can induce nonlinear vertical springing was determined based on Equations (46)-(47) and the first order vertical wet natural frequency which was measured by hammer test. The typical condition was listed in Table 10. The ship model undergoes double-frequency, triple-frequency, quadruple-frequency nonlinear vertical springing in heading wave condition B0503, B0507, B0511, respectively. The ship model undergoes double-frequency nonlinear vertical springing in condition B0605 of which the wave heading is 150°. The ship model undergoes double-frequency nonlinear vertical springing in condition B0704 of which the wave heading is 120°. The wet first order natural frequency is listed in Table 11.

In this investigation, the actual wave loads acting on ship are referred to as the total wave loads, referred to as *TVBM*. Fourier transform is applied to the total wave loads to extract the wave frequency component corresponding to the encountered wave

Table 10. Typical condition of test.

	Wave height		Wave length		Speed			
Condition	Full-scale (m)	Model (mm)	Full-scale (m)	Model (m)	Full-scale (knot)	Model (m/s)	Wave Heading (°)	Wave encounter frequency(Hz)
B01	5	64.9			0	0	180	
B05	5	64.9			23	1.348	180	
B06	5	64.9			23	1.348	150	
B07	5	64.9			23	1.348	120	
B0503	5	64.9	129.1	1.677	23	1.348	180	1.77
B0507	5	64.9	228.6	2.969	23	1.348	180	1.18
B0511	5	64.9	349.0	4.532	23	1.348	180	0.88
B0605	5	64.9	118.2	1.535	23	1.348	150	1.77
B0704	5	64.9	87.2	1.133	23	1.348	120	1.77
B0702	5	64.9	83.8	1.088	23	1.348	120	1.82

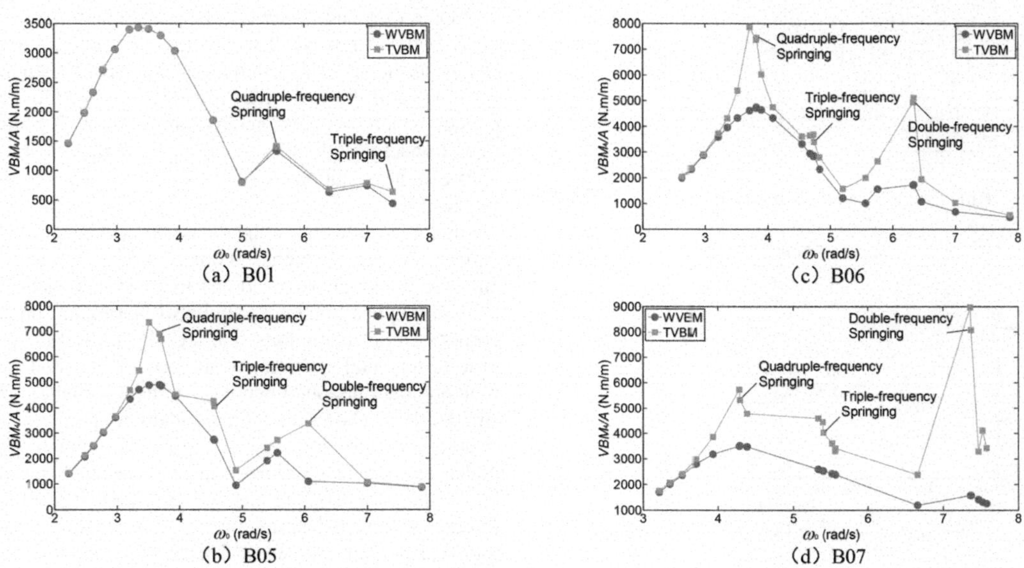

Figure 20. Vertical bending moment of midship.

Table 11. Wet first order natural frequency.

Type	Frequency(Hz)	Type	Frequency(Hz)
Vertical	3.54	Horizontal	4.64
Torsional	3.53		

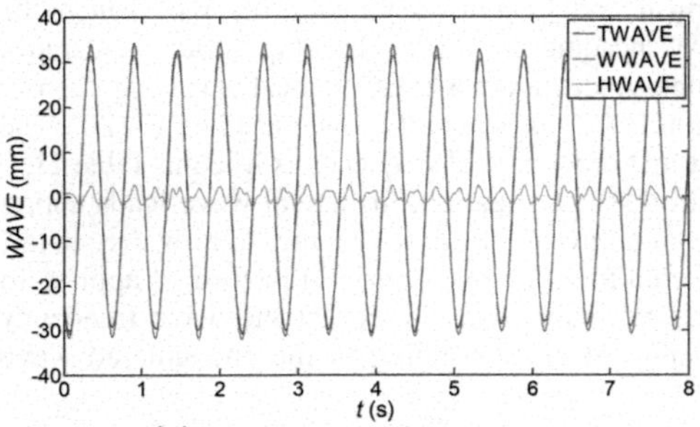

(a) Time history of incident wave

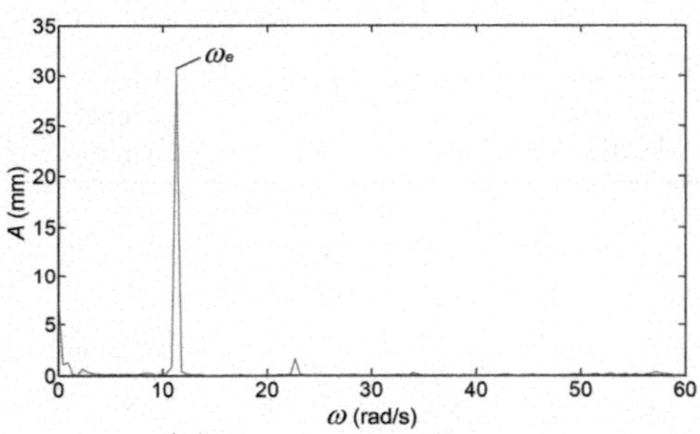

(b) Spectrum of incident wave

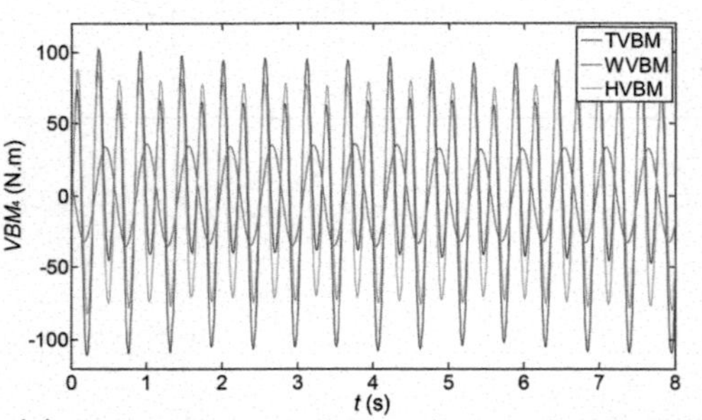

(c) Time history of vertical bending moment of midship

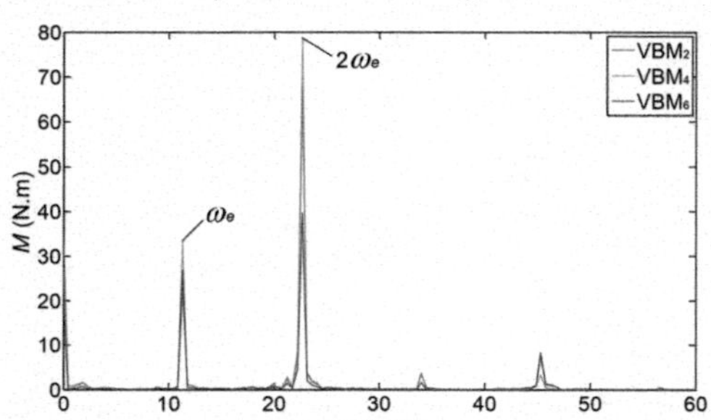

(d) Spectrum of vertical bending moment

Figure 21. Incident wave and vertical bending moment of B0503.

frequency ω_e, referred to as *WVBM*. The difference between total wave loads and wave frequency loads is high frequency loads which often includes several harmonic frequency components, referred to as *HVBM*. Figure 20 shows the vertical bending moments of midship for B01, B05, B06, B07. Figure 21 illustrates the time history and spectrum of incident wave and vertical bending moment of midship for B0503. The test results indicate that the differences between total wave loads and wave frequency component are not highly pronounced at zero speed. This implies that springing are not particularly apparent at zero speed. However, a significant difference between total wave loads and wave frequency component emerges when the ship is sailing in waves at full speed. The variation of total wave loads is typically greater than that of the wave frequency components and exhibits more pronounced discontinuities. Discontinuity points are primarily located near the nonlinear springing points.

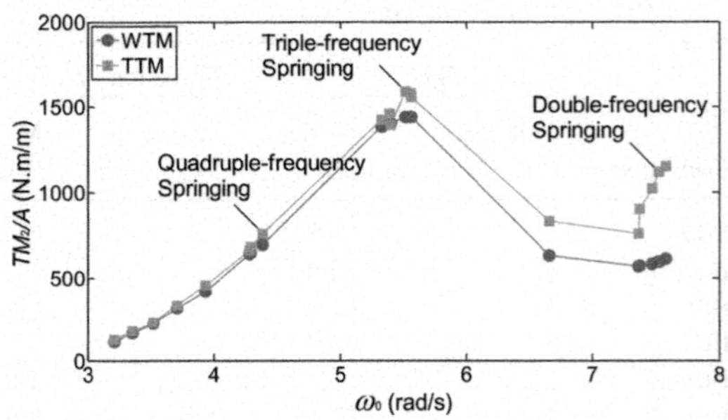

Figure 22. Torsional moment of M2 for B07.

According to the vertical bending moment spectrum in Figure 21, when the harmonic frequency exceeds the first order wet natural frequency, the corresponding harmonic frequency components can be ignored.

The incident wave that can induce nonlinear torsional springing was determined based on Equations (46)-(47) and the first order torsional wet natural frequency which was measured by hammer test. The ship model undergoes double-frequency nonlinear torsional springing in condition B0702 of which the wave heading is 120°.

Figure 22 illustrates the wave frequency components and total torsional moment under conditions B07. A comprehensive comparison reveals that the ship undergoes torsional nonlinear springing under corresponding conditions. Figure 23 illustrates the time history and spectrum of incident wave, torsional moment and horizontal bending moment for condition B0702. It is shows that, in addition to pronounced double-frequency torsional springing, the ship model undergoes noticeable triple-frequency torsional springing. The harmonic frequency corresponding triple-frequency torsional springing exceeds the first order torsional natural frequency. This appears inconsistent with the previously mentioned notion that "when the harmonic frequency exceeds the first order wet natural

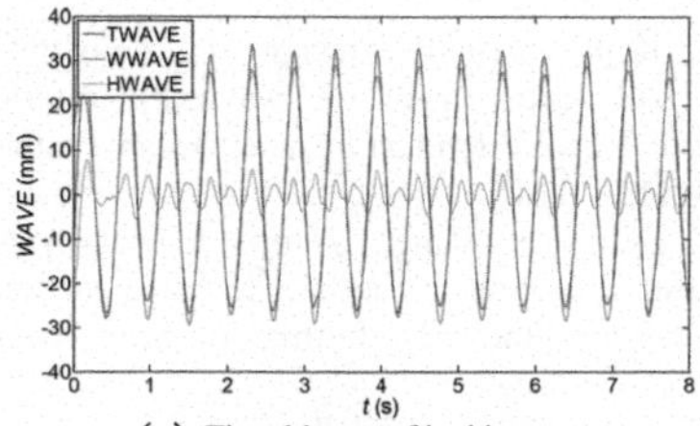

(a) Time history of incident wave

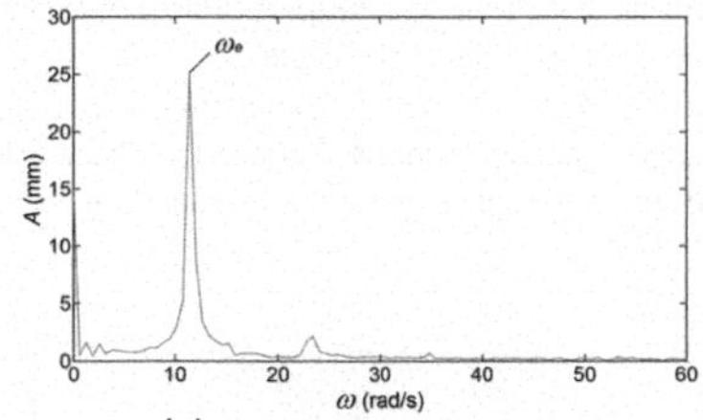

(b) Spectrum of incident wave

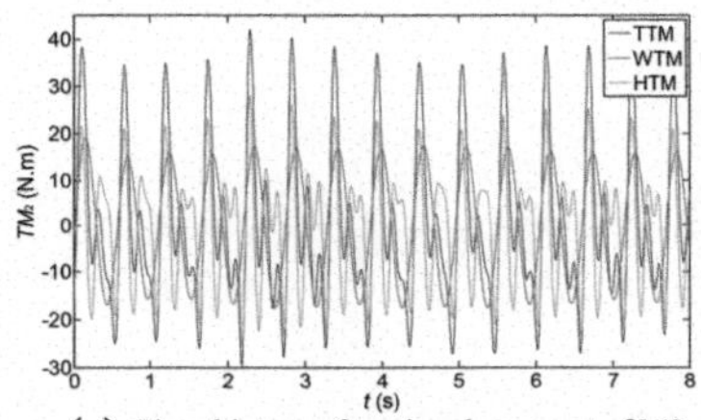

(c) Time history of torsional moment of M2

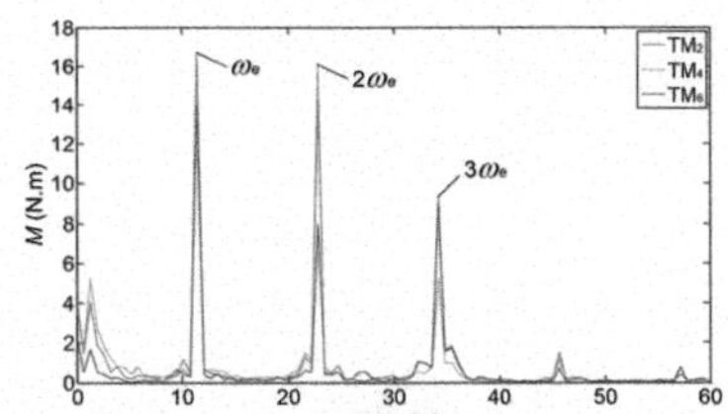

(d) Spectrum of torsional moment

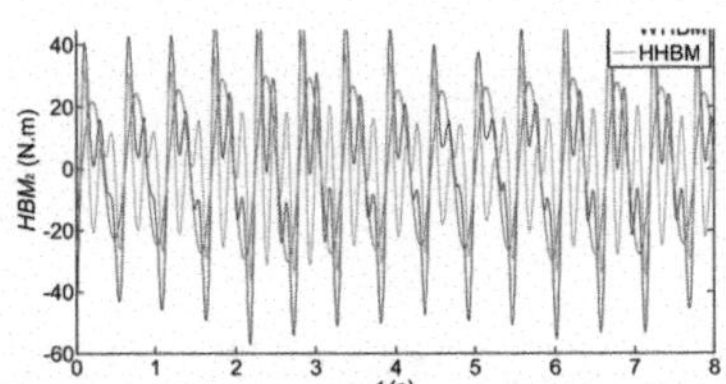

(e) Time history of horizontal bending moment of M2

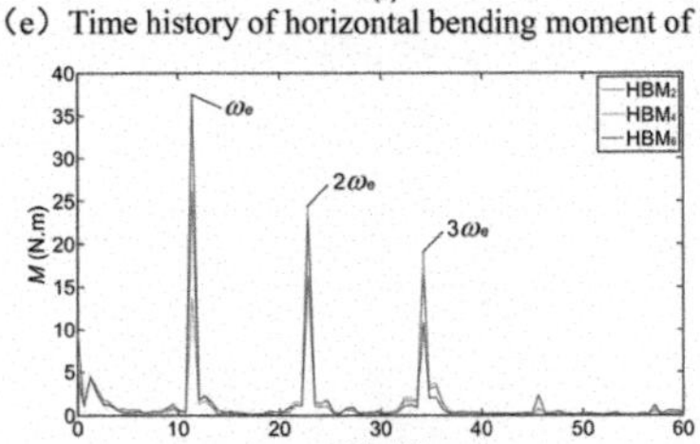

(f) Spectrum of horizontal bending moment

Figure 23. Incident wave, torsional moment and horizontal bending moment of B0702.

frequency, the corresponding harmonic frequency components can be ignored". This inconsistency primarily arises due to the ship's relatively high first order horizontal wet natural frequency. Consequently, the vessel undergoes high frequency horizontal vibration with frequencies not exceeding the first order horizontal wet natural frequency. Thus, under these conditions, the ship undergoes triple-frequency horizontal vibrations, as depicted in Figure 23(f). Due to the open structure of container ships, horizontal bending of the ship will induce torsional deformation. The triple-frequency horizontal vibration will induce corresponding triple-frequency torsional vibration shown as in Figure 23(d).

5 CONCLUSIONS

Based on the thin-walled beam theory, a U-shaped backbone was designed and fabricated to simulate stiffness and shear center of a 20000 TEU ultra-large container ship. A segmented ship model was fabricated to conduct wave-induced loads test. The investigation results are as follows:

(1) The U-shaped backbone could simultaneously simulate the vertical, horizontal, torsional stiffness and shear center of a 20000 TEU ultra-large container ship. Shear flow integration method suitable for U-shaped backbone could be effectively used to measure the torsional moment acting on container ship.
(2) The differences between total wave loads and wave frequency component are not highly pronounced at zero speed. However, a significant difference between total wave loads and wave frequency component emerges when the ship is sailing in waves at full speed. The variation of total wave loads is typically greater than that of the wave frequency components and exhibits more pronounced discontinuities which are primarily located near the nonlinear springing points.
(3) Due to the open structure of container ships, horizontal bending of the ship will induce torsional deformation. The ship model undergoes double-frequency nonlinear springing and triple-frequency vibration in the condition that the wave encounter frequency is half the first order torsional wet natural frequency.

REFERENCES

Chen, Z. Y., Ren, H. L., Li, H., Zhang, K. H. Experimental and numerical analysis of bow slamming and whipping in different sea states[J]. Journal of Ship Mechanics, 2012,16(3):246–253.

Ding, J., Wang, X. L., Hu, J. J., Liu, R. M. Experimental investigations of springing and slamming responses of a ultra-VLCC[J]. Journal of Ship Mechanics, 2015, 19 (1-2): 144–151.

Geng, Y. C., Yu, H., Hu, J. J., Gu, X. K., Si, H. L., Cheng X. P. Investigation of springing and slamming loads for river-sea ships by model test[J]. Journal of Ship Mechanics, 2014, 18(6): 635–643.

Hong, S. Y., Kim, B. W. Experimental investigations of higher-order springing and whipping-WILS project[J]. International Journal of Naval Architecture and Ocean Engineering, 2014, 6: 1160–1181.

Kim, J. H., Kim, Y. H., Korobkin, A. Comparison of fully coupled hydroelastic computation and segmented model test results for slamming and whipping loads[J]. International Journal of Naval Architecture and Ocean Engineering, 2014, 6: 1064–1081.

Kim, Y. I., Park, S. G., Kim, B. H., Ahn, I. G. Operational modal analysis on the hydroelastic response of a segmented container carrier model under oblique waves[J]. Ocean Engineering, 2016, 127: 357–367.

Maron, A., Kapsenberg, G. Design of a ship model for hydro-elastic experiments in waves[J]. International Journal of Naval Architecture and Ocean Engineering, 2014, 6: 1130–1147.

Wang, X. L., Zhao, N., Ding, J., Hu, J. J. Study of hydroelastic ship responses of a VLCC by theoretical and experimental methods[J]. Journal of Ship Mechanics, 2016, 20(1-2): 127–136.

Wang, Y. W., Wu, W. G., Liu, Z. G., Lin, X., Guo G. H. Experimental investigation on torsion and vertical springing of a river-to-sea ship[J]. Journal of Vibration and Shock, 2018, 12(37): 193–200.

Zhu, S. J., Wu, M. K., Moan, T. Experimental investigation of hull girder vibrations of a flexible backbone model in bending and torsion[J]. Applied Ocean Research, 2011, 33: 252–274.

Hydrodynamics - manoeuvring

Advances in Maritime Technology and Engineering – Guedes Soares & Santos (Eds)

System identification method of ship manoeuvring motion driven by knowledge and data

G.S. An
School of Naval Architecture and Ocean Engineering, Huazhong University of Science and Technology, Wuhan, P.R. China

G. Xiang & X.B. Xiang
School of Naval Architecture and Ocean Engineering, Huazhong University of Science and Technology, Wuhan, P.R. China
Hubei Key Laboratory of Naval Architecture and Ocean Engineering Hydrodynamics (HUST), Wuhan, P.R. China
Collaborative Innovation Center for Advanced Ship and Deep-Sea Exploration, Shanghai, P.R. China

C. Guedes Soares
Center for Marine Technology and Ocean Engineering (CENTEC), Instituto Superior Tecnico, Universidade de Lisboa, Lisboa, Portugal

ABSTRACT: This paper proposes a machine learning ship motion system identification method using a Physics-Informed Neural Network (PINN) driven by knowledge and data. The ship motion model with coefficients is incorporated into the loss function as a regularization term, enabling the PINN to conform to training data distribution and adhere to differential equations governing physical laws. Research shows that PINN can learn models with improved generalization and prediction accuracy using fewer data samples, furthermore, they exhibit high robustness. The techniques and discoveries of this study can furnish theoretical direction and methodological backing for the precise modeling of ship manoeuvring motion, thereby pioneering a novel approach in system identification theory that merges knowledge and data-driven approaches.

1 INTRODUCTION

In order to achieve intelligent control and auxiliary decision-making of ships, it is necessary to establish a mathematical model that accurately describes the ship's motion characteristics, provides corresponding dynamic constraint information for ship navigation and control systems, etc., and accurately forecasts the ship's manoeuvring trajectory, so as to achieve the best control effect (Ren et al. 2022, Hu et al. 2021, Niu et al. 2020, Wu et al. 2021). Therefore, the establishment of an accurate mathematical model of ship manoeuvring motion is a very important basic work in the research of ship manoeuvring motion prediction methods (Chen et al. 2019, Xu & Guedes Soares 2016, Zhu et al. 2019a, Sutulo & Guedes Soares 2011).

Currently, empirical formula methods (Kawahara et al. 2011, Sutulo & Guedes Soares 2011), constrained model test methods (Rhee et al. 2000), computational fluid dynamics methods (Gokce et al. 2018, Bonci et al. 2015) and system identification methods (Sajedi et al. 2019, Zhu et al. 2019b) can be used to determine the hydrodynamic coefficients of ships. The empirical formula method has a short computation time but is not very reliable and cannot provide accurate results for newly designed ships. In the constrained model test method, since the Reynolds number of the ship model and the real ship are not the same, the method is affected by the "scale effect", which affects the accuracy of the mathematical model prediction of the real ship, and the method is very costly in terms of manpower and time. The Computational Fluid Dynamics (CFD) method is highly accurate but time-consuming, and is only suitable for manoeuvring predictions at the design stage of a ship, and is not suitable for rapid updating of the mathematical model when the dynamics of the ship change.

The system identification method is a data-driven approach to the field of ship manoeuvring motion modelling, where motion data from free self-propelled model tests or real-scale ship manoeuvrability tests are generally applied to build the model. The method can either be based on the estimation of unknown parameters on the basis of a derived mechanistic model, or it can adopt a black-box modelling paradigm, where the model is constructed from the data in the absence of an a priori model.

DOI: 10.1201/9781003508779-33

The system identification method has low time and economic costs and can eliminate scale effects by modelling real ship data. More importantly, it is the only one of the ships' manoeuvring motion modelling methods with online capability. Based on these characteristics, the method can achieve rapid modelling and real-time motion prediction of ships, and if the stability and robustness of the modelling of the method can be guaranteed, it can provide solid support for the enhancement of the intelligence level of ships.

In recent years, the increasing maturity of artificial intelligence technology has provided a new method for recognising ship manoeuvring motion systems. Neural networks can theoretically approximate any nonlinear input-output system due to their very strong nonlinear fitting ability, which is very suitable for use as a discriminative modelling tool for ship motion models. Hess (2000, 2002, 2006a, 2006b) conducted a series of neural network-based simulation studies of surface ship motions early this century, adding the effects of environmental disturbances such as wind and waves to the simulation process.

Moreira and Guedes Soares (2003, 2012) carried out a modelling study of the standard manoeuvring motion of a ship using a recurrent neural network and validated it using simulation and real ship data. The results show that the model has a good prediction accuracy for the turning motion and a large prediction error for the zigzag motion. Rajesh et al. (2008, 2010) modelled and predicted the manoeuvring motion of a large tanker using a forward neural network and compared the effect of exposing different numbers of linear hydrodynamic derivatives to the neural network on the results. The results show that exposing some of the linear hydrodynamic derivatives to the neural network can improve the prediction accuracy, but the neural network does not need to know all of the linear hydrodynamic derivatives to achieve good prediction accuracy.

Yin et al. (2013, 2014, 2015, 2018) carried out several online and offline modelling studies of the ship's manoeuvring and transversal rocking motions using radial basis neural networks.

Woo et al. (2013, 2018) applied the LSTM recurrent neural network model to develop a high-order nonlinear hydrodynamic model of the equations of motion of an unmanned surface ship, which can be used in conjunction with a simplified linear hydrodynamic model to predict the manoeuvring motions of an unmanned surface ship. Peng et al. (2019) added the particle swarm optimization algorithm into the training of the LSTM network model and optimized the accuracy of the model in predicting ship motions such as transverse rocking, longitudinal rocking and longitudinal swaying. Wang et al. (2021) proposed a Bi-TPA hybrid LSTM-TPA model by incorporating the temporal pattern attention mechanism into the bidirectional LSTM neural network, and the results show that the model improves the prediction accuracy of short-term transverse rocking motion significantly than the traditional LSTM model. LSTM-TPA hybrid model by adding the temporal pattern attention mechanism to the bidirectional LSTM neural network, and the results show that this model improves the prediction accuracy of short-term transverse rocking motion significantly than the traditional LSTM model.

The cost of data acquisition can be prohibitive for many applications in the physical and engineering sciences, and as a result there has been a quest to design special network architectures that embed known a priori knowledge or laws of physics, allowing supervised learning results to satisfy limited training data and also respect the a priori knowledge, and considering that the aforementioned neural networks have the ability to approximate arbitrary nonlinear mappings, which can be combined with the a priori physical knowledge to mitigate the unexplainability introduced by the black box model, and with the continuous development of deep learning theories and techniques, this combination has led to the formation of a new model of neural networks.

In 2017, Raissi et al. (2019) from Brown University, USA, synthesized and improved the previous technical approaches in neural networks solving differential equations, formally proposed the Physics-informed Neural Networks (PINN) algorithmic framework, and for the first time proposed that the PINN method can be used not only to solve the positive problem of differential equations, i.e. the modelling problem, but also to solve the inverse problem of differential equations, i.e. the parameter identification problem. PINN is a knowledge and data driven modelling framework that allows physical constraints to be added as regular terms to the traditional neural network modelling process, reducing the need for modelling data. At the same time, PINNs also have higher discrimination accuracy compared with traditional neural networks.

After PINNs were formally proposed, it triggered a large number of related application studies in the fields of fluid mechanics, thermodynamics, biomedicine, and mechanics of materials, (Lu et al. 2021, Vadyala et al. 2022), and the more typical network structures are fractional order fPINNs network applied to unpresurizable N-S equations (Pang et al. 2019), distributional DPINNs network applied to the diagnosis of atrial fibrillation network (Dwivedi et al. 2019), Bayesian B-PINNs network applied to the design of new materials (Yang et al. 2021), and so on. However, the current applied research on system identification methods based on PINN networks is still rare.

In order to build an accurate ship manoeuvring motion model while reducing the data size and quality requirements, this paper develops a system identification method of ship manoeuvring motion using PINN driven by knowledge and data. This paper firstly introduces the basic structure and working principle of the general PINN, and then introduces the system identification modelling method of ship manoevring motion based on PINN. The robustness and generalization performance of the PINN system identification algorithm are investigated based on numerical simulation data of ship manoeuvring motion with different noise levels, and the correctness and effectiveness of the method are verified.

2 PRINCIPLE OF PINN

Consider the generalized form of a partial differential equation with parameters:

$$f(x,t,u,\partial_x u,\partial_t u,\cdots;\lambda)=0, x\in\Omega, t\in[0,T] \quad (1)$$

The initial and boundary conditions for the equations are defined as:

$$u(x,t_0)=g(x), x\in\Omega \quad (2)$$

$$u(x,t)=h(t), x\in\partial\Omega, t\in[0,T] \quad (3)$$

where the spatial variable $x\in R^d$ and the temporal variable is t, Ω is a subset of the Euclidean space R^d, $\partial\Omega$ is the boundary of Ω. The differential equation $f$contains the differential operators $\partial_x u$, $\partial_t u$, $\lambda=[\lambda 1,\lambda 2,\cdots]$ are parameters of the differential equation f, $u(x,t)$ is a theoretical solution of the partial differential equation f, $g(x)$ is the initial condition, and $h(x)$ is the Delicacy boundary condition, the Riemann boundary condition, or the hybrid boundary condition.

PINN uses a neural network to approximate the solution of a partial differential equation. It is useful to define the forecast output $\hat{u}(x,t)$ of the PINN network to approximate the solution $u(x,t)$ of the differential equation f. The loss function $L$of this PINN network is defined as:

$$L=L_{PDE}+L_{Data}+L_{IC}+L_{BC} \quad (4)$$

where L_{PDE} denotes the knowledge-driven part of the loss function, L_{Data} is the data-driven part of the loss function, L_{IC} is the initial condition-driven part of the loss function, and L_{BC} is the boundary condition-driven part of the loss function. Each part of the loss function is defined as follows:

$$L_{PDE}=f(\hat{u},\partial_t^{(k)}\hat{u},\partial_x^{(k)}\hat{u};\lambda) \quad (5)$$

$$L_{Data}=\hat{u}|_{\Omega}-u|_{Data} \quad (6)$$

$$L_{IC}=\hat{u}|_{\Omega,t_0}-g|_{\Omega,t_0} \quad (7)$$

$$L_{BC}=(\partial_n\hat{u}|_{\partial\Omega}-\partial_n h|_{\partial\Omega})+(\hat{u}|_{\partial\Omega}-h|_{\partial\Omega}) \quad (8)$$

Under the constraint of the loss function, the PINN network can not only make the error of the training data as small as possible, but also satisfy the partial differential equation as much as possible. The structure of the PINN network corresponding to the parameter-containing partial differential equation is shown in Figure 1. After inputting the data in time t and space x, the theoretical solution $u(x,t)$ of the partial differential equation is firstly approximated by the forecast output $\hat{u}(x,t)$ of the fully connected neural network, and then the residuals of the partial differential equation and the residuals of the initial and edge values are derived using the neural network auto-differentiation technique and added to the loss function as a regular term, and then the gradient descent is used finally to update the neural network weight parameters and partial differential equation physical parameters.

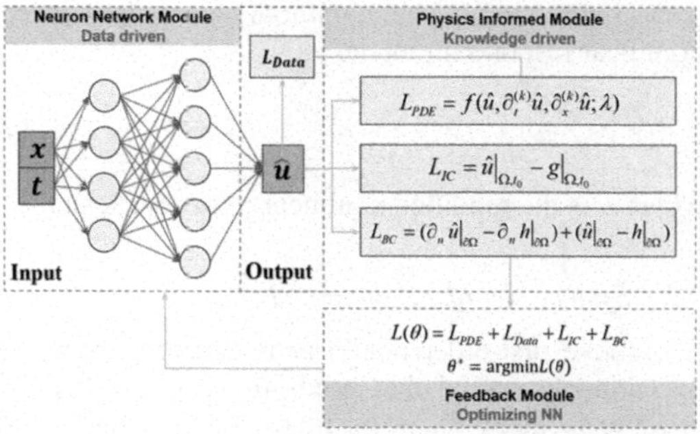

Figure 1. Schematic diagram of PINN network structure.

3 SYSTEM IDENTIFICATION METHODS AND DATA PROCESSING

3.1 *Ship manoeuvring motion responding model*

Ship motion equations (mathematical models) describe the quantitative relationship between ship control inputs and dynamic response, reflecting the kinematic and dynamic characteristics of the ship, and are in a sense a simplification and abstraction of the prototype ship manoeuvring motion system. The ship manoeuvring motion responding model is currently the most widely used ship motion model. Because of its simple expression, which reflects the response relationship between the ship's heading angle and the rudder angle during sailing, it is usually preferred for the study of ship manoeuvrability and is mainly used in the research of automatic rudder system and ship heading control. The second-order linear responding model (KT equations) of the ship can be expressed as follows:

$$T_1T_2\ddot{r}-(T_1+T_2)\dot{r}+r=K\delta+KT_3\dot{\delta} \quad (9)$$

where T_1, T_2, T_3 is the time constant, r is the ship's yaw angular velocity, K is the gain coefficient, and δ is the ship's rudder angle.

In 1956, Professor Kensaku Nomoto of Osaka University, Japan, derived the first-order KT equations based on the principle of auto-regulation for use in place of the second-order linear responding model when steering is not very frequent:

$$T\dot{r}+r=K\delta \quad (10)$$

where $T=T_1+T_2+T_3$, the equation is also called Nomoto model, and the parameters K and T are called the slewing index and stability index, respectively, which can be used to evaluate the ship's slewing and inherent stability. Since the Nomoto model is derived based on the low speed horizontal plane linear motion model, the above equations are more applicable to the ship's motion at small rudder angles. For large rudder angle motions, Nomoto and other scholars subsequently proposed a nonlinear responding model, which is corrected by adding a nonlinear term. The commonly used first-order nonlinear responding model is:

$$T\dot{r} + r + ar^3 = K\delta \tag{11}$$

where α is the modified nonlinear term coefficient.

3.2 *System identification model*

The above first-order nonlinear responding model is a continuous model that needs to be discretized in order to facilitate the acquisition of data required for system identification through numerical simulation methods, and the Eulerian difference method will be used to discretize the equations in this study. The specific form of the discretized first order nonlinear responding model is:

$$r(t+1) = (1 - \frac{\Delta t}{T})r(t) - \frac{\alpha \Delta t}{T}r^3(t) + \frac{K\Delta t}{T}\delta(t) \tag{12}$$

where Δt is the time interval of the numerical simulation, and let:

$$a = 1 - \frac{\Delta t}{T}, b = \frac{\alpha \Delta t}{T}, c = \frac{K\Delta t}{T} \tag{13}$$

Then equation 12 becomes:

$$r(t+1) = ar(t) + br^3(t) + c\delta(t) \tag{14}$$

The above equation is the final system model to be identified. The PINN is constructed through this model structure with network inputs $r(t)$, $r^3(t)$ and $\delta(t)$, network output $r(t+1)$, and a,b,c is the parameter to be identified.

The loss function of this system identification model is derived from the PINN principle and consists of a data-driven loss function and a knowledge-driven loss function, where the mean square error is used for the data-driven loss function and the knowledge-driven loss function is related to the structure of the system identification model, which is expressed as follows:

$$L_{Data} = [\hat{r}(t+1) - r(t+1)]^2 \tag{15}$$

$$L_{ODE} = r(t+1) - ar(t) - br^3(t) - c\delta(t) \tag{16}$$

$$L = L_{Data} + L_{ODE} \tag{17}$$

The system identification model is derived through the PINN principle as shown in Figure 2.

In this paper, the responding parameters of a ship are selected for numerical simulation, and the parameters are shown in Table 1, and the values of the parameters to be identified in the system identification model calculated by the above parameters are shown in Table 2. Numerical simulation of 5°/5°, 10°/10°, 15°/15°, 20°/20°, 25°/25°, 30°/30° zigzag tests is carried out respectively, the total length of the numerical simulation is 200s, the time interval $\Delta t = 0.1s$; the length of the data for each voyage was 2000, and the data of yaw angular velocity and rudder angle are collected, and the numerical simulation data of the zigzag tests of different conditions are spliced together to form the data set required for PINN training during the training of the network.

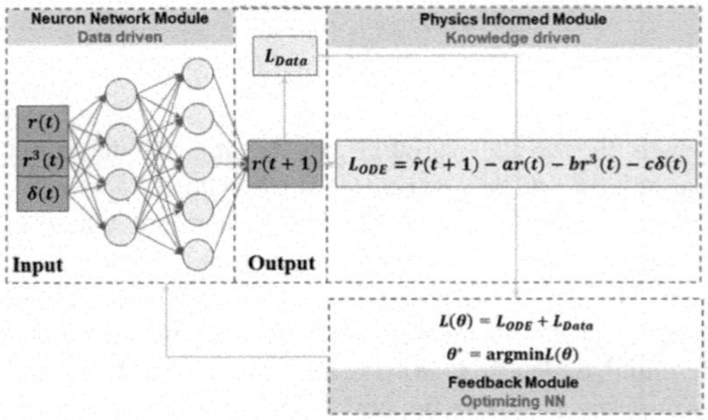

Figure 2. System identification model structure.

Table 1. KT model parameter.

Parameter	T	K	α
Value	13.514	0.216	258.405

Table 2. Coefficient to be identified

Parameter	a	b	c
Value	0.9926	-1.9121	0.0016

3.3 *Data processing*

In order to simulate the influence of measurement noise and other disturbances and to verify the robustness of the system identification algorithm, different degrees of white noise disturbances are added to the pure simulation data, and the noise disturbance data are generated as follows:

$$\zeta_i = \zeta_{i0} + \zeta^{\max} k_0 \xi_i \tag{18}$$

where ζ_i is the noise pollution data, ζ_{i0} is the pure numerical simulation data, $\zeta^{\max} k_0 \xi_i$ is the interfering signal, $\zeta^{\max}$ is the absolute value of the amplitude of the pure numerical simulation data, ξ_i is the Gaussian white noise and k_0 is used to describe the noise intensity. In this paper, three different noise levels are chosen, namely 1%, 3% and 5%. The pure numerical simulation data and the data of different noise levels under the 10°/10°, 20°/20° zigzag tests condition are shown in Figure 3.

4 SYSTEM IDENTIFICATION AND VERIFICATION

The PINN for system identification in this study consists of a 6-layer neural network with 1 input layer, 4 hidden layers and 1 output layer. Among them, the input layer contains 3 neurons, each hidden layer contains 20 neurons, and the output layer contains 1 neuron, and a fully connected network is used between the layers. The network weights and biases of each layer of the PINN are initialized using Xavier's method, and the activation functions of the neurons in each layer are chosen to be hyperbolic tangent functions, and it is chosen to optimize all the loss functions using the L-BFGS, which is a quasi-Newton, full batch gradient-based optimization algorithm. The learning rate is 0.1 and the PINN is trained for a total of 10,000 iterations.

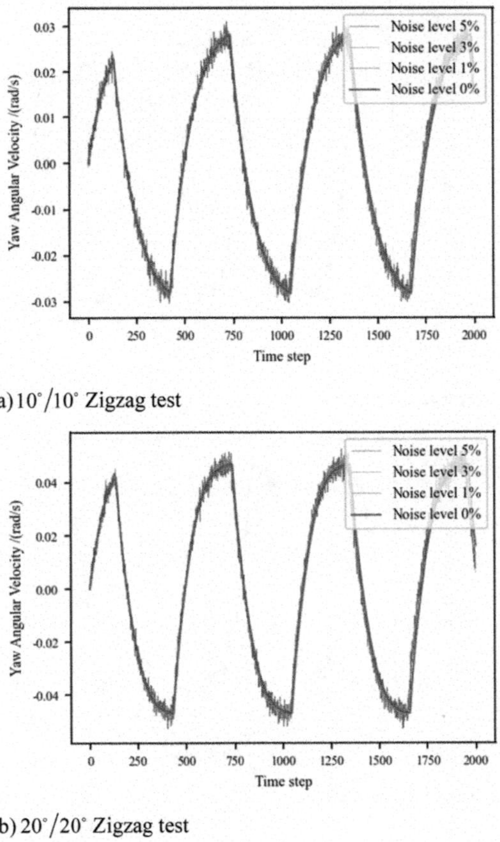

(a) 10°/10° Zigzag test

(b) 20°/20° Zigzag test

Figure 3. Zigzag test data.

4.1 *Robustness verification*

The robustness of the algorithm, when confronted with noise when identifying the system is to verify whether the PINN algorithm can maintain high performance and stability. In this study, the PINN is trained using pure data with no noise and data with different levels of noise, and system identification is performed. A total of 12000 sets of numerical simulation data of six zigzag tests sailing conditions are used as the training data. The errors between the system identification results and the true values of the three coefficients to be identified are shown in the Table 3. According to the system identification results, the ship manoeuvring motion prediction for the 20°/20° zigzag test is carried out, and the comparison between the PINN prediction results of the bow angular velocity and the numerical simulation results is shown in Figure 4.

From the errors of the system identification results, it can be seen that the accuracy of the system identification decreases for all three parameters as the noise level increases. For parameter a, the errors of the system identification results are all within 1%. For parameter c, the errors of the system identification results are all within 1% of the 1% noise level, and the errors of the system identification results increase rapidly as the noise level increases, which is due to the fact that c is two to three orders of magnitude smaller than a and b, and it is relatively easy to generate cumulative errors during the error back propagation and gradient descent of the PINN training and the parameter updating process. For parameter b, the error in the results is 10% even when the system identification is performed using pure data without noise, probably due to the fact that the hydrodynamic coefficients contained in this parameter to be identified are not sufficiently stimulated by the zigzag tests, leading to a larger error in the system identification results.

Table 3. System identification robustness verification results.

		Noise level			
Error/%		0%	1%	3%	5%
Parameter	a	0.04	0.01	0.24	0.82
	b	10.82	11.57	15.79	22.02
	c	0.63	1.25	12.50	35.00

The yaw angular velocity prediction from the system identification results show that the pure data and 1% noise data have better prediction results, and the PINN prediction results are in perfect agreement with the numerical simulation results. At 3% noise level, PINN is more accurate in predicting the extreme value of the yaw angular velocity, and there is a small error in the period, which gradually accumulates with time. For every 1000-time steps (100s of physical time), PINN predicts the yaw angular velocity with a delay of about 40-time steps (4s of physical time) and the prediction accuracy is high. At a noise level of 5%, the extreme prediction error of PINN for the yaw angular velocity is about 15%, and for every 1000-time steps (100s of physical time), PINN predicts the yaw angular velocity with a lag of about 60-time steps (6s of physical time).

From the above results, it can be concluded that the system identification method of ship manoeuvring motion using physics-informed neural network (PINN) driven by knowledge and data is feasible and effective. For pure data without noise interference and data with low noise interference, the results of ship manoeuvring motion prediction based on PINN system identification are very satisfactory. The ship manoeuvring motion prediction based on the identification results of the PINN system has a low latency for the data with large noise disturbances with a certain degree of robustness.

4.2 *Generalization performance verification*

The generalization performance of the algorithm, when confronted with smaller data sizes during the system identification process is to verify whether the PINN algorithm can maintain high performance and stability. In this study, work is carried out to validate the generalization performance of the PINN system identification algorithm based on six sets of zigzag numerical simulation data with a noise level of 3%. By using all the numerical simulation data, 4, 3, 2 and 1 groups of data are randomly selected for PINN system identification, corresponding to data sizes of 12000, 10000, 8000, 6000, 4000 and 2000, respectively. The errors of the system identification results of the three coefficients to be identified with respect to the true values are shown in the Table 4. The PINN prediction results of the yaw angular velocity are compared with the numerical simulation results of the ship manoeuvring motion prediction based on the system identification results of 12000 and 2000 data sizes for the 10°/10°, 20°/20°, 30°/30° zigzag tests, respectively, as shown in Figures 5-7.

As can be seen from the error of the system identification results, overall the data size has little effect on the error of the system identification results. In most cases, as the data scale decreases, the error of the system identification results also decreases because the PINN driven by knowledge and data not only learns the laws of data distribution, but also It can follow the physical laws governed by differential equations as much as possible. When the data influenced by noise is small, PINN can rely more on the knowledge-driven physical laws, so as to reduce the influence of noisy data on the accuracy of system recognition.

From the results of the system identification for the prediction of the yaw angular velocity, it can be seen that when the noise level is 3%, the results of training the PINN to forecast the ship manoeuvring motion with data sizes of 12,000 and 2,000 are

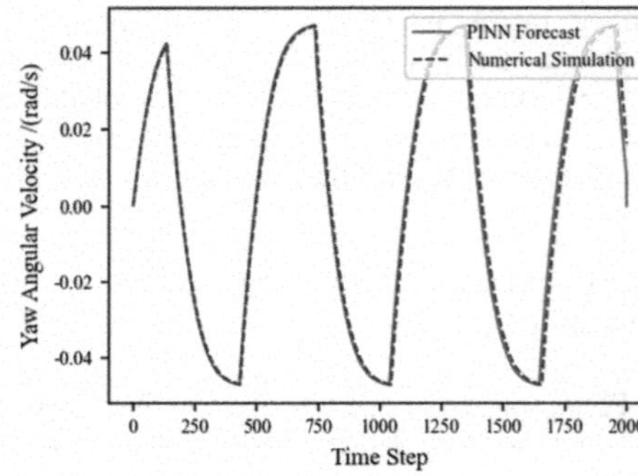

(a)Noise level 0%

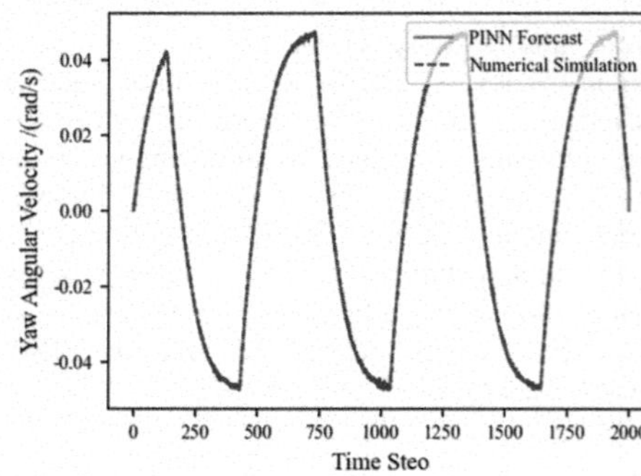

(b)Noise level 1%

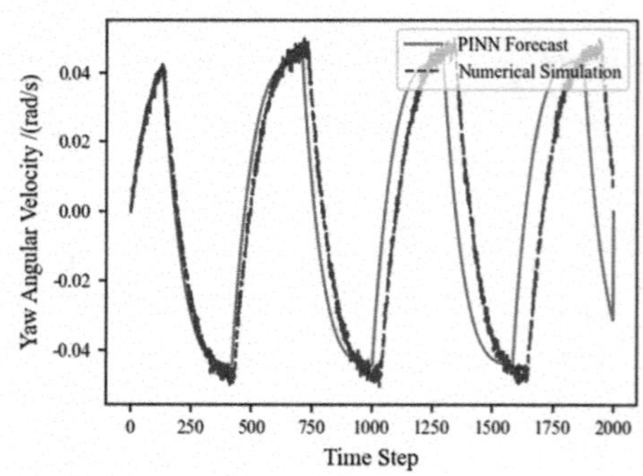

(c)Noise level 3%

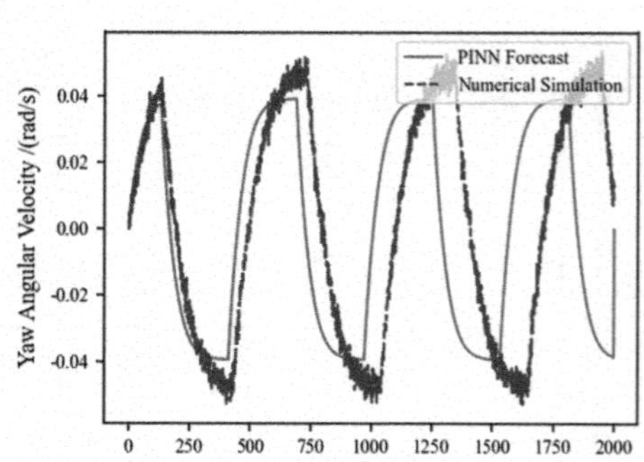

(d)Noise level 5%

Figure 4. PINN forecast results.

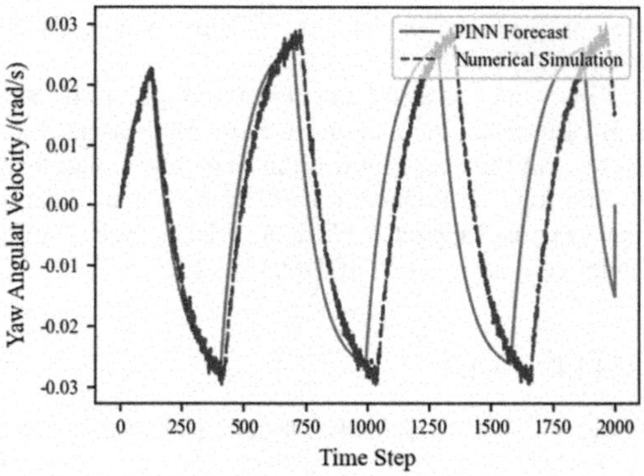

(a)Data scale = 2000

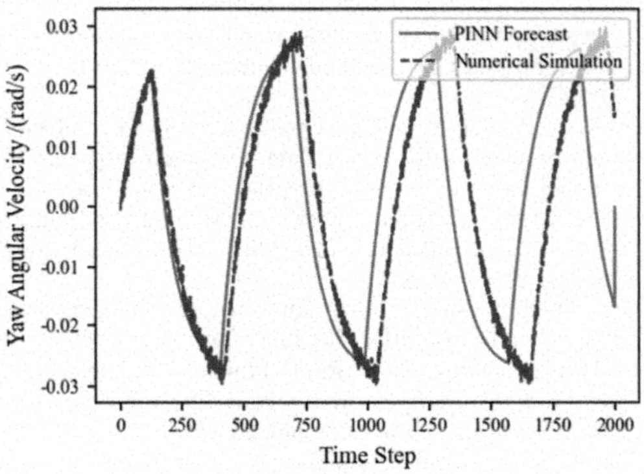

(b)Data scale = 12000

Figure 5. 10°/10° zigzag test.

basically the same, and the PINN is more accurate in predicting the extreme value of the yaw angular velocity, with small errors in the period, which gradually accumulate over time, and the prediction accuracy is higher overall. The overall forecasting accuracy is high.

Table 4. System identification generalization performance verification results.

Error/%		Parameter *a*	*b*	*c*
Data Scale	12000	0.24	15.79	12.50
	10000	0.23	15.53	11.25
	8000	0.23	15.13	11.25
	6000	0.24	14.62	10.63
	4000	0.24	13.69	11.25
	2000	0.24	12.79	11.25

From the above results, it can be concluded that the system identification method of ship manoeuvring motion using physics-informed neural network (PINN) driven by knowledge and data proposed in this study is feasible and effective. Since PINN

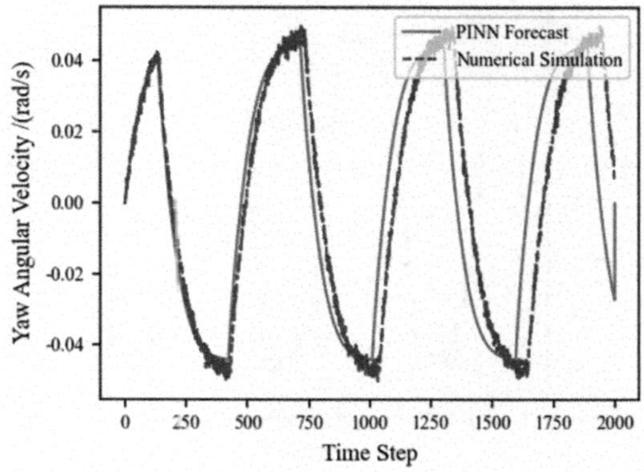

(a)Data scale = 2000

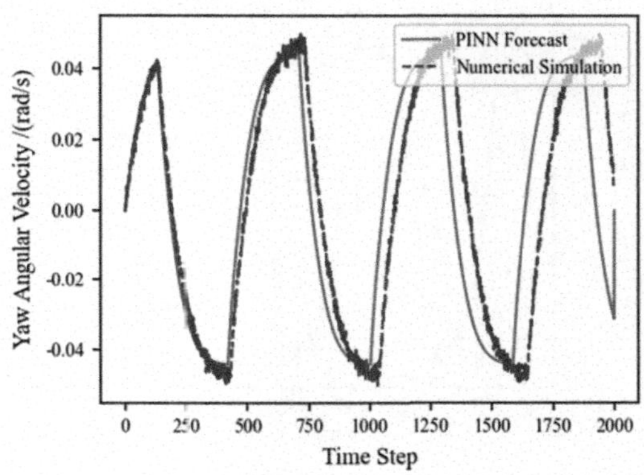

(b)Data scale = 12000

Figure 6. 20°/20° zigzag test.

incorporates the ship dynamics model containing the coefficients to be identified into the loss function as a regular term, it integrates the data-driven and knowledge-driven approaches, which can significantly reduce the data scale requirement for network training and system identification, has strong generalization performance, and can also improve the system identification accuracy.

5 CONCLUSION

This paper proposes a system identification method of ship manoeuvring motion using physics-informed neural network (PINN) driven by knowledge and data is proposed. The ship dynamics model, which contains the coefficients to be identified, is added to the loss function as a regular term. The parameters in the ship dynamics model can be identified directly by the gradient descent and parameter update algorithm during the PINN training. To construct the PINN training dataset, zigzag test data with different levels of white noise interference are obtained through numerical simulation methods. The robustness of the algorithm is studied, and the ship

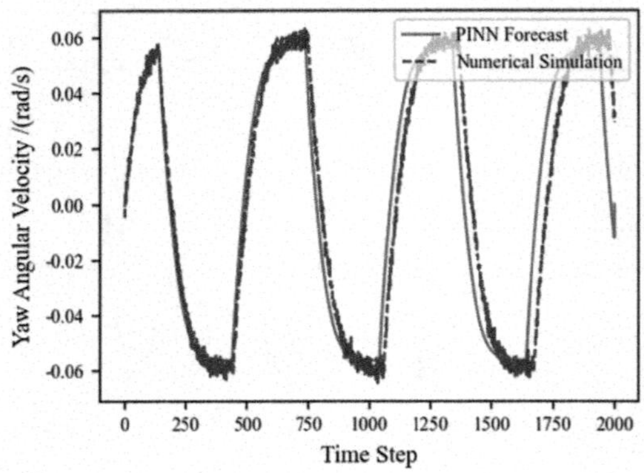

(a)Data scale = 2000

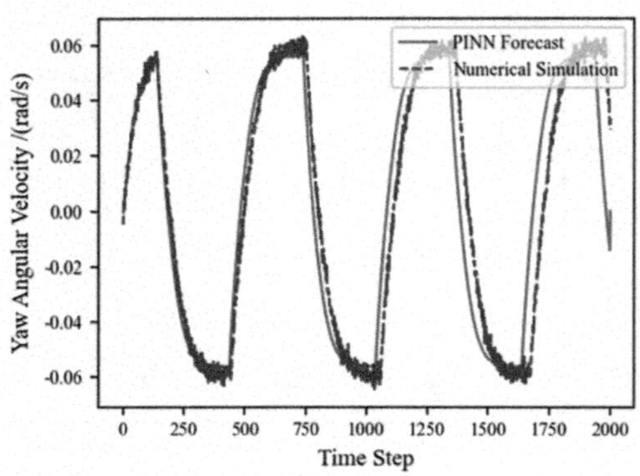

(b)Data scale = 12000

Figure 7. 30°/30° zigzag test.

manoeuvring motion prediction results based on the recognition results of the PINN system are very satisfactory for pure data without noise interference and data with small noise interference. For data with large noise disturbance, the ship manoeuvring motion prediction delay based on the recognition results of the PINN system is low, indicating certain robustness. By conducting an algorithm generalization performance study, it is found that the PINN algorithm can significantly reduce the data scale requirement for network training and system identification, and has strong generalization performance. The techniques and discoveries of this study can furnish theoretical direction and methodological backing for the precise modeling of ship manoeuvring motion, thereby pioneering a novel approach in system identification theory that merges knowledge and data-driven approaches.

ACKNOWLEDGEMENTS

This work is partially supported by Hubei Provincial Natural Science Foundation (No. ZRMS2023001640), National Natural Science Foundation of China (No. 52131101), and by the Fundamental Research Funds for the Central Universities, HUST: No.50031140038.

This work has also been performed within the Strategic Research Plan of the Centre for Marine Technology and Ocean Engineering, which is financed by the Portuguese Foundation for Science and Technology (Fundação para a Ciência e Tecnologia - FCT) under contract UIDB/UIDP/00134 /2020.

REFERENCES

Bonci, M., et al. 2015. Method for estimating parameters of practical ship manoeuvring models based on the combination of RANSE computations and System Identification. Applied ocean research, 52, 274–294.

Chen, C., et al. 2019. Parameter estimation for a ship's roll response model in shallow water using an intelligent machine learning method. Ocean Engineering, 191, 106479.

Dwivedi, V., et al. 2019. Distributed physics informed neural network for data-efficient solution to partial differential equations. arXiv preprint arXiv:1907.08967.

Gokce, M. K., & Kinaci, O. K. 2018. Numerical simulations of free roll decay of DTMB 5415. Ocean Engineering, 159, 539–551.

Hess, D. 2000. Simulation of ship manoeuvres using recursive neural networks. In 23rd Symposium on Naval Hydrodynamics. Val de Reuil, France, September 2000.

Hess, D. 2002. Using recursive neural networks for blind predictions of submarine manoeuvres. In 24th Symposium on Naval Hydrodynamics. Fukuoka, Japan, July 2002.

Hess, D. 2006. Ship maneuvering simulation in wind and waves: a nonlinear time-domain approach using recursive neural networks. In 26th Symposium on Naval Hydrodynamics. Rome, Italy, September 2006.

Hess, D., et al. 2006. Improved simulation of ship maneuvers using recursive neural networks. In 44th AIAA Aerospace Sciences Meeting and Exhibit (p. 1481).

Hu, W., et al. 2021. Digital twin: A state-of-the-art review of its enabling technologies, applications and challenges. Journal of Intelligent Manufacturing and Special Equipment, 2(1), 1–34.

Yin J.C., et al. 2015. Parametric identification of Abkowitz model for ship maneuvering motion by using partial least squares regression. Journal of Offshore Mechanics and Arctic Engineering, 137(3), 031301.

Kawahara, Y., et al 2011. A simple prediction formula of roll damping of conventional cargo ships on the basis of Ikeda's method and its limitation. Contemporary Ideas on Ship Stability and Capsizing in Waves, 465–486.

Lu, L., et al 2021. Physics-informed neural networks with hard constraints for inverse design. SIAM Journal on Scientific Computing, 43(6), B1105–B1132.

Moreira, L. & Guedes Soares, C. 2003. Comparison between manoeuvring trials and simulations with recursive neural networks. Ship Technology Research, 50(2), 77–84.

Moreira, L., & Guedes Soares, C. 2012. Recursive neural network model of catamaran manoeuvring. International Journal of Maritime Engineering, 154(A3).

Niu, H., et al. 2020. Energy efficient path planning for unmanned surface vehicle in spatially-temporally variant environment. Ocean Engineering, 196, 106766.

Pang, G., et al. 2019. fPINNs: Fractional physics-informed neural networks. SIAM Journal on Scientific Computing, 41(4), A2603–A2626.

Peng, X., et al. 2019. An improved particle swarm optimization algorithm applied to long short-term memory neural network for ship motion attitude prediction. Transactions of the Institute of Measurement and Control, 41(15), 4462–4471.

Raissi, M., et al. 2019. Physics-informed neural networks: A deep learning framework for solving forward and inverse problems involving nonlinear partial differential equations. Journal of Computational physics, 378, 686–707.

Rajesh, G., et al. 2010. System identification for nonlinear maneuvering of ships using neural network. Journal of ship Research, 54(01), 1–14.

Rajesh, G., & Bhattacharyya, S. K. 2008. System identification for nonlinear maneuvering of large tankers using artificial neural network. Applied Ocean Research, 30 (4), 256–263.

Ren, Y., et al. 2022. Dynamic analysis of a multi-column TLP floating offshore wind turbine with tendon failure scenarios. Ocean Engineering, 245, 110472.

Rhee, K., et al. 2000. An experimental study on hydrodynamic coefficients of submerged body using planar motion mechanism and coning motion device. In International Workshop on Ship Manoeuvrability at the Hamburg Ship Model Basin (pp. 1–20).

Sajedi, Y., & Bozorg, M. 2019. Robust estimation of hydrodynamic coefficients of an AUV using Kalman and H∞ filters. Ocean Engineering, 182, 386–394.

Sutulo, S., & Guedes Soares, C. 2011. Mathematical models for simulation of manoeuvring performance of ships. Marine technology and engineering, 661–698.

Vadyala, S. R., et al. 2022. A review of physics-based machine learning in civil engineering. Results in Engineering, 13, 100316.

Wang, Y., et al. 2021. Ship roll prediction algorithm based on Bi-LSTM-TPA combined model. Journal of Marine Science and Engineering, 9(4), 387.

Woo, J., & Kim, N. 2013. System Identification of an Autonomous Surface Vehicle Using Neural Networks. In ISOPE International Ocean and Polar Engineering Conference (pp. ISOPE–I). ISOPE.

Woo, J., et al. 2018. Dynamic model identification of unmanned surface vehicles using deep learning network. Applied Ocean Research, 78, 123–133.

Wu, B., et al. 2021. Navigating patterns analysis for onboard guidance support in crossing collision-avoidance operations. IEEE Intelligent Transportation Systems Magazine, 14(3), 62–77.

Xu, H., & Guedes Soares, C. 2016. Vector field path following for surface marine vessel and parameter identification based on LS-SVM. Ocean Engineering, 113, 151–161.

Yang, L., et al. 2021. B-PINNs: Bayesian physics-informed neural networks for forward and inverse PDE problems with noisy data. Journal of Computational Physics, 425, 109913.

Yin, J. C., et al. 2013. Sequential learning radial basis function network for real-time tidal level predictions. Ocean engineering, 57, 49–55.

Yin, J. C., et al. 2014. Online ship roll motion prediction based on grey sequential extreme learning machine. Neurocomputing, 129, 168–174.

Yin, J. C., et al. 2018. A real-time ship roll motion prediction using wavelet transform and variable RBF network. Ocean Engineering, 160, 10–19.

Zhu, M., et al. 2019a. A novel adaptive weighted least square support vector regression algorithm-based identification of the ship dynamic model. IEEE Access, 7, 128910–128924.

Zhu, M., et al. 2019b. Optimized support vector regression algorithm-based modeling of ship dynamics. Applied ocean research, 90, 101842.

Advances in Maritime Technology and Engineering – Guedes Soares & Santos (Eds)

Ship manoeuvrability with heel motions

V. Ferrari, R. Tonelli, A.S. Kisjes, R. Hallmann & T. Gornicz
Maritime Research Institute Netherlands (MARIN), Wageningen, The Netherlands

ABSTRACT: Depending on many parameters such as speed and steering angle, it can happen that a vessel reaches large heeling angles during a turn, which can lead to injuries on board or even capsize. Most manoeuvring simulations used to study manoeuvrability in early design stage are often based on 3 degrees of freedom, thus ignoring the heel motions. Moreover, existing international rules focus mainly on horizontal manoeuvring characteristics disregarding heeling motions due to turning. Therefore, there are few guidelines available to the designer on how to take heeling into account in manoeuvring studies. To overcome these issues, a detailed investigation was performed on heeling motions in turn and their influence on ship manoeuvrability. In particular, model tests and CFD calculations were performed for a test-case ferry. In this paper these new experimental and numerical results are presented and the impact of heel angles on the ship manoeuvrability is evaluated.

1 INTRODUCTION

The Resolution MSC 137(76) of the International Maritime Organization (IMO) guarantees minimum standard of manoeuvrability for all ships which however does not mandate requirements on the heel angles in turn. There is only the Intact stability Code formula for passenger ships (IMO 2008), known to greatly underestimate the actual heel angles (Ferrari et al. 2020, Piaggio et al. 2022). Some proposals have even been submitted to the IMO in the past in order to solve this issue (RINA 2011a, RINA 2011b, Poland 2014). Nonetheless, verifying that the heel angles do not reach excessively high values should be part of any ship design. Many types of vessels are potentially concerned by this issue: passenger ships must ensure the safety of passengers, tugs must avoid excessive heeling while operating, container ships might lose cargo due to large roll motions and other merchant vessels might experience cargo shift or sloshing due to the heel motions. Apart from the heel angles themselves, the heeling dynamics have also a noticeable impact on the hydrodynamic forces and the manoeuvring characteristics of the ship. Therefore, these must be taken into account when studying the manoeuvrability of the ship. Free running test with a self-propelled model or sea trials give the measurement of the heel angles and manoeuvring characteristics, as long as the roll motions are properly measured and recorded. However, these might not always be available in early design stage, or they might be conducted at a draught different than the one required by the IMO manoeuvring standards, as it often happens for sea trials. Most numerical models include heel motions with rather simplified formulations, if any, and usually only as an extension of the horizontal manoeuvring framework, that is, as a 3+1 degrees of freedom (DOF) model instead of a true 4 DOF model The present paper addresses this discrepancy between manoeuvrability on the horizontal plane on one side and including heel dynamics on the other. Use is made of a test-case called MARIN Ferry as generic example of passenger ships. The results of captive and free running model tests are presented and the impact of the heel is highlighted. Successively considerations on simulations are presented and finally conclusions are drawn.

2 THE TEST-CASE

2.1 *Main particulars and propulsive arrangement*

Throughout this paper reference is made to a dedicated test-case to apply the formulations investigated in the course of the study. The test-case ship used is a 190 m twin-screw twin-rudder ship ferry called MARIN Ferry. It has already been introduced in several manoeuvring studies, such as maximum heel on account of turning (Ferrari et al. 2020), heel modelling in time-domain simulations (Piaggio et al. 2022), quantification of the uncertainty of the tests (Kisjes, Quadvlieg, & Ferrari 2019), influence on manoeuvrability of the skeg sizes (Ferrari et al. 2019), and sensitivity analysis of manoeuvring models (Ferrari et al. 2022). This new ferry was designed for the purpose of assessing the manoeuvrability of passenger ships. The hull lines and appendages are representative of modern ferries for passengers and ro-ro cargo. The vessel has a twin-screw twin-spade rudder

DOI: 10.1201/9781003508779-34

configuration. The model of the MARIN Ferry was manufactured of wood to a linear scale ratio of 1:35.986. Appendages such as rudders, bilge keels and tunnel thrusters openings were modelled as well. Turbulence was stimulated using sand strips at the bow and at the leading edge of the rudders. The model is shown in Figure 1 and the main particulars are given in Table 1

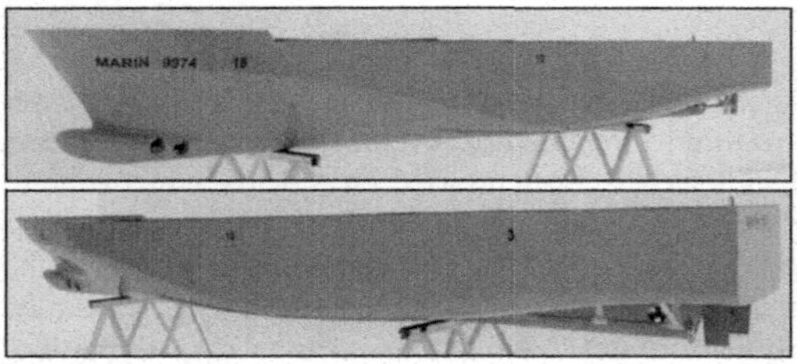

Figure 1. Fore and aft views of the MARIN Ferry.

Table 1. Main particulars of the MARIN Ferry with long skeg configuration.

Designation	Symbol	Value	Unit
Length between perpendiculars	L	190.0	m
Breadth	B	30.0	m
Draught at midship	T	7.0	m
Trim	τ	0.0	deg
Displacement in salt water	Δ	25118.3	t
LCG w.r.t. station 10	x_G	-4.02	m

2.2 *Captive model tests*

Captive manoeuvring experiments were carried out in order to measure the hydrodynamic forces of the MARIN Ferry, both for horizontal and heeled conditions. The tests were conducted in MARIN's Seakeeping and Manoeuvring Basin, which measures 170 m x 40 m x 5 m in length, width and depth respectively. The water depth during the model tests was 5.0 m on model scale, which when compared to the model scale draught of 0.19 m indicates that the results are valid for deep water. The model experiments were based on Froude's law of similitude with a linear scale ratio of 1:35.986. Full scale values are obtained by multiplying the measured model values by the appropriate scaling factor. No further corrections are applied. The manoeuvring test results are presented as full scale values. The Seakeeping and Manoeuvring Basin is equipped with an x-y carriage that can be ordered to follow and hover above a free-running model, but can also be used to execute predefined motions. This carriage can be extended with a turn table. The combination of an x-y-yaw control and the dimensions of this basin make it possible to perform standard Planar Motion Mechanism (PMM) manoeuvres as well as Computerised Planar Motion Carriage (CPMC) (Grim et al. 1976, Hallmann and Quadvlieg 2015), tests. The model was attached to the carriage such that it can freely heave and pitch. The freedom to heave and pitch is obtained by a set-up consisting of two cardanic joints. Both joints have a pitch and heave freedom and one of the joints has an additional surge freedom. This way the model is free to pitch and heave but restricted in all other directions. Both cardanic joints are mounted on top of two force measurement frames. The forces are measured in 6 degrees of freedom. To achieve the 0, -10 and -20 degrees of static heel, wedges were made to be fitted inside the model below the 6-component frame. The point of rotation was at the waterline. The model was still free to heave and pitch in this new orientation due to the freedom in heave and pitch of the captive set-up. The rudder forces are measured (rudder-fixed) with a 3-component transducer. In addition, two potentiometers are used to measure the heave and pitch. An overview of the captive test set-up is shown in Figure 2 and a detail of the model in captive

Figure 2. The MARIN Ferry model in captive set-up.

Figure 3. Close view of the heeled set-up.

heeled condition is in Figure 3. Prior to the model tests, static loads have been applied to verify the correct functioning of the model transducers, the measurement system as well as the data analysis. The forces were applied on the waterline and centreline for the surge force and at midship for the sway force. It can been concluded that the transducers and measurement system were measuring correctly and the data-analysis was combining them appropriately. The transducers used to measure the forces were regularly zeroed throughout the test campaign. To enable corrections for small misalignments, this zeroing was always performed with the model at a yaw angle of zero (aligned with the longitudinal axis of the basin) and a zero lateral position (sub carriage in the middle of the main carriage). The roll moment due to the static heel was measured before these transducers were zeroed. The following analysis was performed:

1. Combination of the 12 force transducers to determine the surge force, sway force, roll moment and yaw moment.
2. Correction of the measured forces for centrifugal forces, sideways accelerations and inertia in case of rotation around the vertical axis.
3. Correction for misalignment of the turn table.
4. Calculation of the heave and pitch based on the location of the two potentiometers.
5. For the quasi-steady tests, separation the samples into bins and determination of the average for each bin.
6. Determination of the harmonic components for the oscillation tests.

The overall forces are presented in a yawed earth-fixed or non-rolled ship-fixed reference frame around midship and waterline. The longitudinal axis is positive to the bow, the transverse axis is positive to port side and the vertical axis is positive up, perpendicular to the free surface. It must be noted that the latter (non-rolled ship-fixed) is more accurate because the six-component frames follow the pitch of the model. Nonetheless, the pitch is small and is not further taken into account. The rudder forces are presented in a rudder-fixed reference frame. The longitudinal force X, transverse force Y, heel moment K and yaw moment N are made non-dimensional with the following formulae:

$$X' = \frac{X}{\frac{1}{2}\rho LTU^2}, \quad Y' = \frac{Y}{\frac{1}{2}\rho LTU^2},$$
$$K' = \frac{K}{\frac{1}{2}\rho LT^2U^2}, \quad N' = \frac{N}{\frac{1}{2}\rho L^2TU^2}, \tag{1}$$

where $U = \sqrt{u^2 + v^2 + (rL)^2}$ is the generalised ship speed (Sutulo 1994) (Ferrari & Quadvlieg 2016). These forces are expressed as function of the heel angle ϕ, drift angle β and non-dimensional rate of turn r', defined as:

$$\beta = arctan\frac{v}{u}, \quad r' = \frac{rL}{v}. \tag{2}$$

The captive test programme is shown in Table 2 Result of the captive tests, in particular for the heel moment, are shown in the following chapters.

Table 2. Programme of the captive model tests.

Series	Description
Configuration 0	*Static heel angle of 0 deg*
Series 0-1	Load-variation tests
Series 0-2	Drift angle-variation tests
	Static drift angle tests
Series 0-3	Turn on the spot
	Yaw rate-variation tests
	Static yaw rate tests
Series 0-4	Yaw rate-variation tests with a drift angle
Series 0-5	Rudder variation tests
Series 0-6	Rudder variation tests with a drift angle
Series 0-7	Rudder variation tests with a yaw rate
Series 0-8	Oscillation tests
Configuration 1	*Static heel angle of -10 deg*
Series 1-2	Drift angle-variation tests
	Static drift angle tests
Series 1-3	Yaw rate-variation tests
	Static yaw rate tests
Series 1-4	Yaw rate-variation tests with a drift angle
Configuration 2	*Static heel angle of -20 deg*
Series 2-2	Drift angle-variation tests
	Static drift angle tests
Series 2-3	Turn on the spot
	Yaw rate-variation tests
	Static yaw rate tests
Series 2-4	Yaw rate-variation tests with a drift angle
Series 2-5	Rudder variation tests
Series 2-6	Rudder variation tests with a drift angle
Series 2-7	Rudder variation tests with a yaw rate
Series 2-8	Oscillation tests

2.3 *Captive CFD simulations*

A series of captive unsteady CFD simulations have been conducted to better understand the relationship between ship motions and hydrodynamic forces. The sailing conditions used in the simulations were similar to the captive model test experiments. All of the simulations were conducted on full scale. The simulations were carried out in RANS solver ReFRESCO. It is a MARIN in-house CFD code able to solve multiphase turbulent flow. The solver has been extensively validated for hydrodynamic applications, with recent results related to manoeuvring published in (Hoydonck et al. 2019, Bhushan et al. 2019). The $k - \omega$ SST 2003 (Menter, Kuntz, & Langtry 2003) turbulence model was used in the simulations. Although in ReFRESCO, several more advanced turbulence models are available, it is believed that the $k - \omega$ model is a good compromise between accuracy, robustness and computational complexity for typical manoeuvring simulations. The simulations resolve the ship's calm water wave system by means of the volume of fluid method. Like in the captive model tests, the heel angle is imposed and kept constant during simulations. However, the vessel is free to heave and pitch. Due to the complexity of the hull, with shafts, rudder and headbox, an unstructured numerical grid is used, with hexahedral cells and clustering towards the hull surface to capture the flow gradients in the boundary layer. Special refinement zones with highly regular cells are used around the aft body to capture complex flow bilge keels and rudders better. Additional refinement zones were applied around the free surface level to better capture the ship-generated wave system. A box shape domain is used. The size of the computational domain is coupled with the ship's

Figure 4. Aft view of the grid used for the MARIN Ferry CFD calculations.

length between perpendiculars (*L*), and the following dimensions are used (length, width, height): 9*L* x 6*L* x 6*L*. The numerical grids were generated using the commercial software Hexpress. The typical size of the grid is about 26 million cells. The grid is shown in Figure 4.

2.4 *Free running model tests*

The free running tests were performed in the same Seakeeping and Manoeuvring Basin as the captive tests with the same model. All tests were conducted with the model in self-propelled and free-running mode. The only connections between the model and the carriage consisted of free-hanging wires for measurement signals and power relay. These cables did not restrict the motions of the model. Before each manoeuvre, the model was accelerated by the carriage until the desired approach speed was reached. The rate of revolution of the propeller corresponded to the self-propulsion point of the model for the approach speed. During this phase an autopilot ensures that the model sailed on a straight course and that the accelerations became zero. When a steady state was reached, the experiment begun and the free-running model performed the manoeuvre followed by the carriage. A snapshot of the

Table 3. Inputs of the free running tests.

Input type	Value
Skeg configurations	Short, Long
GMt	1.6 m, 2.0 m, 3.0 m
Manoeuvres	Zig-zags, Turning circles, Turn-on-spot
Rudder angles (ZZ)	10/10, 20/20
Rudder angles (TC)	15 deg, 25 deg, 35 deg
Approach speed (ZZ)	25 kn
Approach speed (TC)	15 kn, 20 kn, 25 kn

Figure 5. The MARIN ferry model during a zig-zag run.

model during a 20/20 zig-zag manoeuvre is shown in Figure 5. The test programme is shown on Table 3. The tests consisted mainly in zig-zags, turning circles and turn-on-spot manoeuvre using the propellers in push-pull mode. Tests were performed with 2 skeg configurations and 3 loading conditions. Results of the free running tests are presented in the following chapters as well.

3 HYDROSTATICS AND HYDRODYNAMIC FORCES

3.1 *Hydrostatics*

The first obvious impact of the metacentric height is the change in the restoring arm GZ. The curves of the MARIN Ferry for the 3 GMt tested during the free

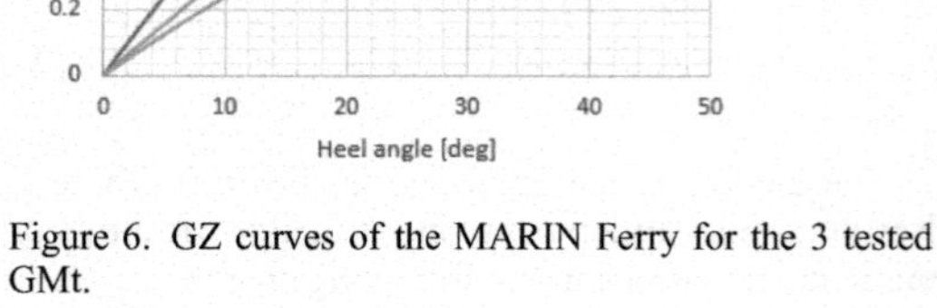

Figure 6. GZ curves of the MARIN Ferry for the 3 tested GMt.

running experiments are shown in Figure 6 An important consequence of the lower GZ and higher heel angles is that the range of validity of the assumption GZ ≈ GMt decreases. When higher heel angles are reached, the difference between the actual GZ curve and its linearization with GMt increases. This is particularly noticeable for the lower GMt since the ship will more easily and quickly reach high heel angles.

3.2 *Hydrodynamic heel moment*

The captive model tests provide insight into the hydrodynamic hull and rudder forces acting on the vessel. When looking at roll motions two aspects are of particular importance: the hydrodynamic heel moment *K* and the influence of heel angle on the other forces and moments.

3.2.1 *Arm of the heel moment*

An approach widely used in literature (Hirano and Takashina 1980, Hirano 1980, Kim et al. 2007, Fukui et al. 2015) is to express the heeling moment as the hydrodynamic sway force *Y* times a suitable arm *h*: $K \approx -Yh$. However, the arm is actually not constant and is itself a function of sway velocity, rate of turn and heel angle. This can be seen for instance in Figure 7. It which shows the non-dimensional arm of the heel moment, calculated as

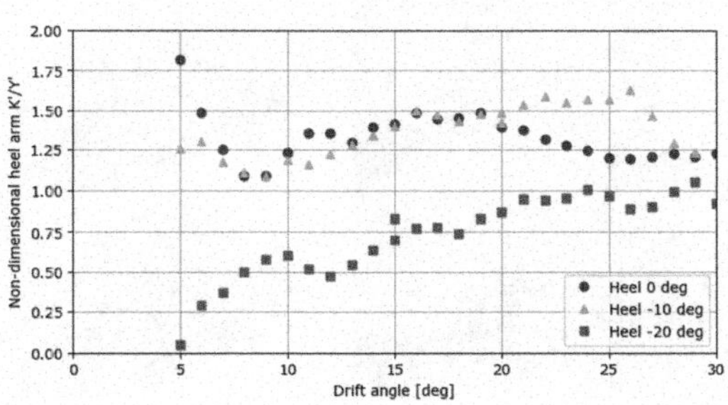

Figure 7. Non-dimensial heel arm from MARIN Ferry captive tests.

the ratio of the non-dimensional hull heel moment K' by the non-dimensional transverse hull force Y'. Therefore, an accurate modelling with the sway times arm approach would require a dedicated formulation not only for the sway force but also for the arm and moreover the arm has to be calculated as the ratio of two measured quantities, which adds to its uncertainty. For these reasons it is suggested to look at the *K* moment itself rather than the arm.

3.2.2 *Heel moment as function of drift, rotation and heel*

The influence of the drift angle on *K* is shown in Figure 8 It can be seen that the drift angle has a large impact on the heel moment, even at small drift angles. There is a peak around 45° and also the value at 90 ° is relatively high. On the other hand, the influence of rotation on heeling is much smaller, as shown in

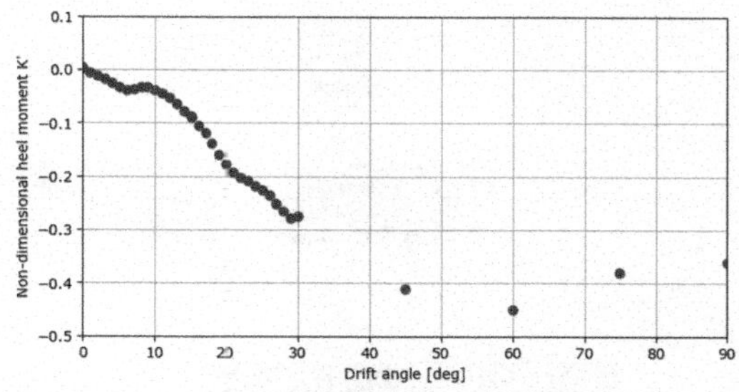

Figure 8. Pure drift non-dimensional heel moment up to 90 deg drift angle.

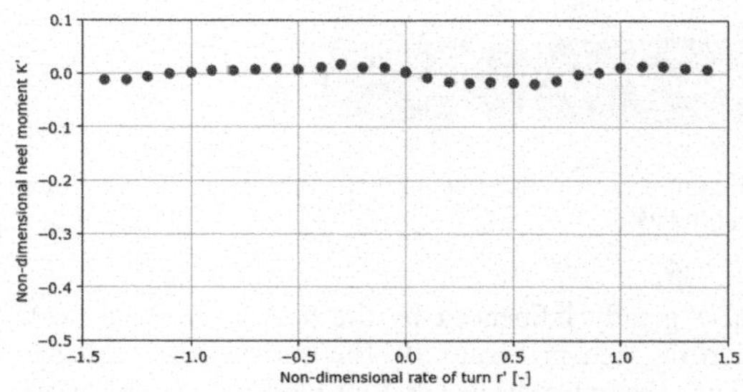

Figure 9. Pure rotation non-dimensional heel moment.

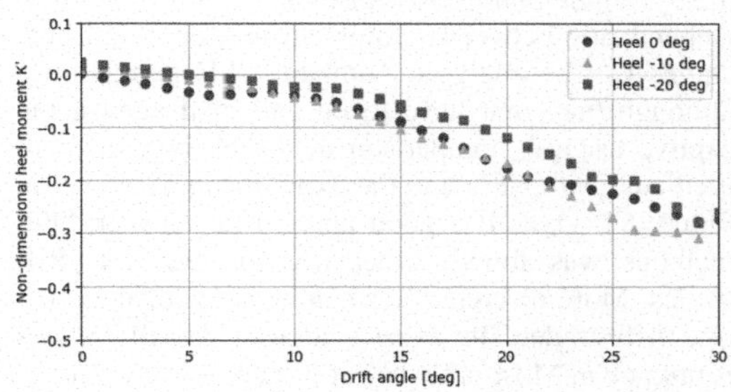

Figure 10. Combined drift and heel non-dimensional heel moment.

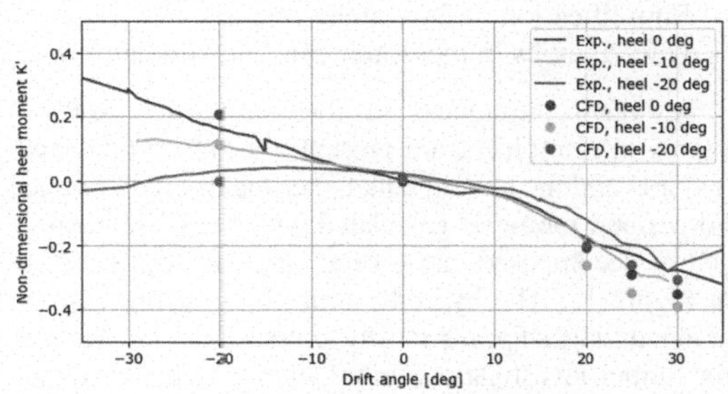

Figure 11. Comparison between CFD and model test results for the combined drift-and-heel heel moment.

Figure 9. The y-axis has the same scale as the previous plot, to emphasise that the rotation-induced heel moment is approximately one order of magnitude smaller than the drift-induced one. In addition to the influence of horizontal motions, the heel moment is

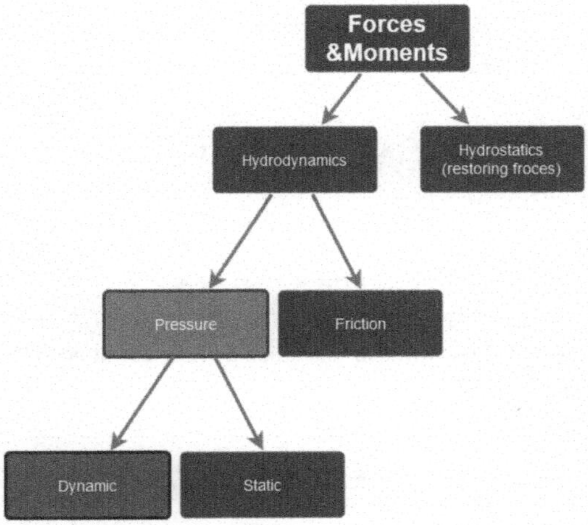

Figure 12. Decomposition of the CFD forces and moments.

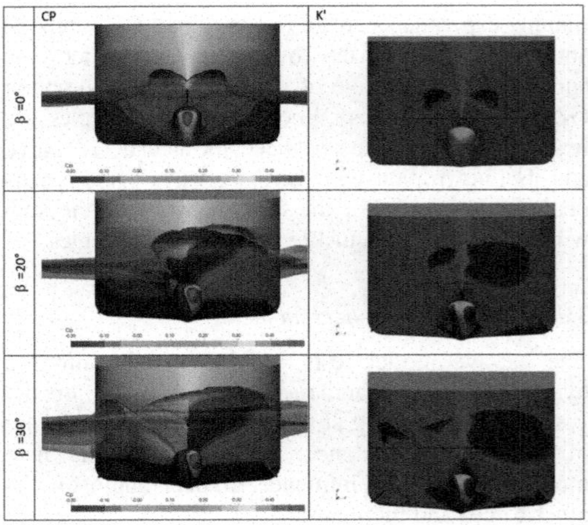

Figure 13. Dynamic pressure (left column) and heel moment contribution (right column) fields at the forebody as function of drift angle.

also greatly influenced by the heel angle itself. This can be seen in Figure 10 The curves greatly change when the hull is heeled and for moderate combinations of drift and heel even the sign changes.

3.2.3 *Detailed analysis with CFD*

Dependency between *K* moment and drift and heel angles can be analysed with the CFD simulations, although they first have to be validated against the captive tests. A comparison of the heeling moment predicted with CFD and the experiments is shown in Figure 11. Overall a good agreement between both methods was found, especially for negative drift angles. More discrepancies can be seen for the positive drift angle. The exact source of the differences is unknown. Most probably, it is a combination of:

- Scale effects ship scale was used in simulations while towing tank perform mode scale experiments;
- Simplification of flow modelling in CFD;
- Uncertainties in measurements and simulations.

Although differences in forecasting the absolute values in *K* moment are present, the trends predicted by CFD and the towing tank experiments are the same. Forces and moments calculated with the CFD method can be decomposed into several components, as shown in Figure 12. The hydrodynamic forces acting on the manoeuvring ship are mostly pressure driven. Hence, for further investigation, only dynamic pressure results were analysed, unless otherwise stated. Figure 16 presents the build-up of non-dimensional *K* moment, integrated from bow to stern, as function of drift and heel angles. Regarding the pure drift case, it can be concluded that the *K* moment is mostly generated at the midship and aft body. The fore body seems to give the opposite contribution, reducing the final value of the rolling moment. In Figure 13, the right-hand side column presents the contribution to the *K* moment in

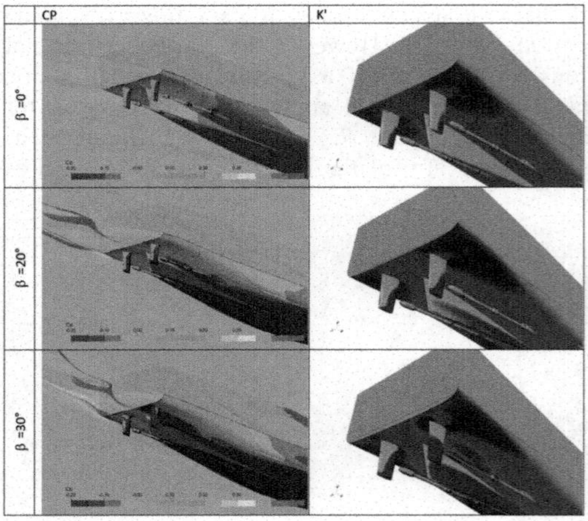

Figure 14. Dynamic pressure (left column) and heel moment contribution (right column) fields at the aft body as function of drift angle.

the form of distribution on the hull surface. The red colour indicates strong positive contribution to the *K* moment values, while the blue implies strong negative *K* values. The left-hand side presents a corresponding dynamic pressure field and bow wave shape. The bow seems to have a significant impact on the rolling moment. For example, in the straight-ahead motion, the total *K* moment is close to zero due to the symmetrical bow wave shape. However, due to asymmetry in flow caused by the drift angle, the bow wave on the outward side is more dominant. Additionally, the flow around the bulbous bow and the under pressure zone on the bottom of the vessel play essential roles in generating *K* moment during drift motion. The *K* moment contribution and the corresponding dynamic pressure field in the aft body are shown in Figure 14

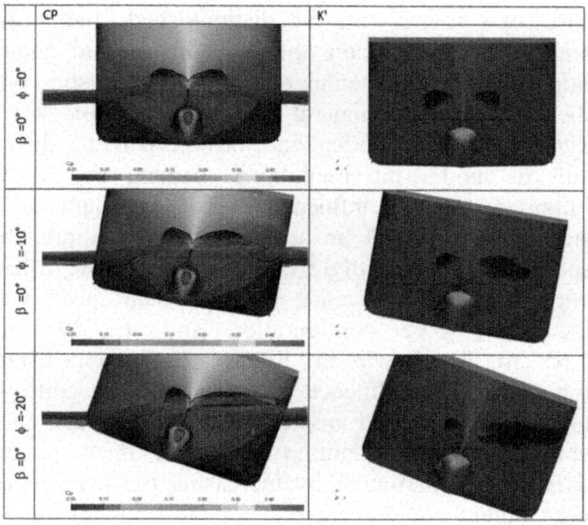

Figure 15. Dynamic pressure (left column) and heel moment contribution (right column) fields at the fore body in the function of heel angle.

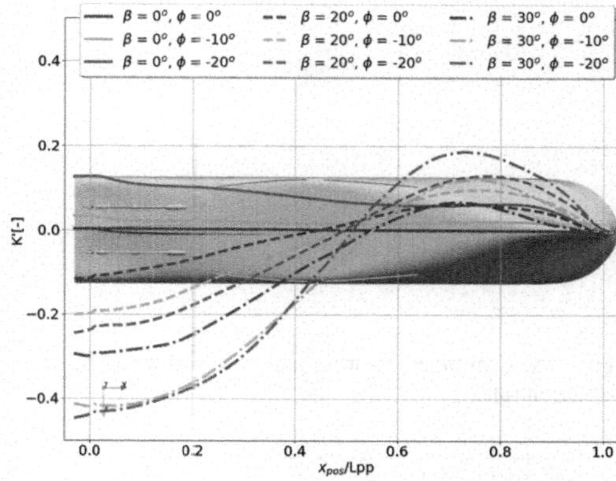

Figure 16. Build-up of the heel moment along the ship in the function of both drift and heel angles.

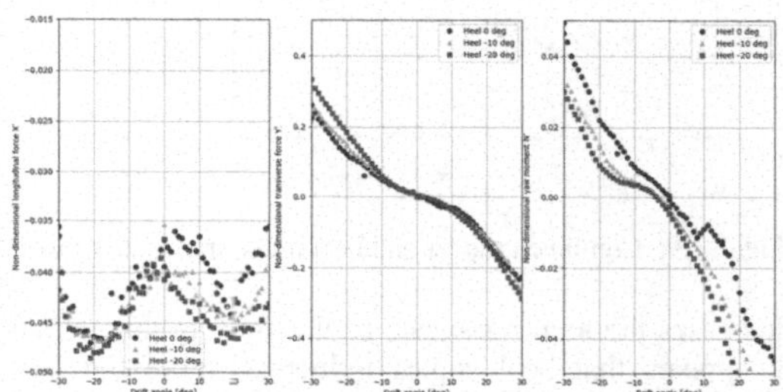

Figure 17. Combined drift and heel results as function of drift angle.

When the drift angle increases, the flow around the bilges and skeg reduces the local dynamic pressure field. As an effect, it gives a negative rolling moment visible on the skeg and aft part of the bilges. The relation between K moment and heel angle, for the straight-ahead motion, is presented as well in Figure 16. It can be seen that the rolling moment is mainly produced on the fore body. For the larger heel angles ($|\phi| \geq 20$ °), a minor but still clear contribution comes from the aft body. Finally, a small increase in the K moment is also visible in the location of head boxes and rudders. Figure 15 shows the K moment contribution and the corresponding dynamic pressure field as function of the heel angle. Similar to pure drift motion, the bow wave seems to have a dominant contribution to the rolling moment. With an increase in the heel angle, the flow asymmetry gets more substantial and the bow wave on the port side gets bigger. The asymmetry of the flow around the bulbous bow is less crucial. The coupling effect of the heel and drift angles on the build-up of K moment is shown again in Figure 16 It can be seen that when the ship is sailing straight ahead, the rolling moment is positive. However, when the vessel starts to drift, the K moment is strongly negative. Similar to previous cases, the K moment is mostly The build-up on the midship and stern. And the most substantial gradients are observed around midship. The final values of the K moment strongly depend on both drift and heel moment. The large heel angles ($|\phi| \geq 20°$) are more critical in the fore body region. The small and moderate ($|\phi| \geq 10°$) appears to be essential both for the bow and stern area. Figures 25 and 26 in the Appendix illustrate the source of K moment on the hull sailing with multiple drift and heel angles. It seems that the flow around the bilges has a dominant role in generating K moment. It can be seen that with an increase in heel angle, the contribution of bilges gets stronger. On the other hand, the flow around the skeg is altered, so from a strong negative K moment (at $\phi = 0°$), it changes to strong positive values (at ϕ= -20°).

3.3 *Influence of heel on other hull forces*

The other aspect is how the heel influences the other forces and moment. Due to the heeled hull the longitudinal force X, transverse force Y and yaw moment N will have a different shape as function of drift and rotation when compared to the horizontal case. This aspect is less studied in literature, with some investigations presented in (Kim et al. 2007, Yeo et al. 2015, Yasukawa et al. 2016). For the MARIN Ferry an example of this phenomena is shown in Figures 17 and 18 for the pure drift and pure rotation cases, respectively. It can be seen that the Y force and even more so the N moment in heeled conditions have large difference with the horizontal cases. Instead, the trend for the X force is still noticeable but less clear. Also in this case the CFD results allow to analyse the distribution of the forces and moments along the ship length. Results from the CFD regarding the influence of drift and heel on the horizontal forces and moment are shown in the Appendix, from Figure 28 to 31. The figures in that section present the sum of pressure and friction contributions. It can be seen that the heel angle

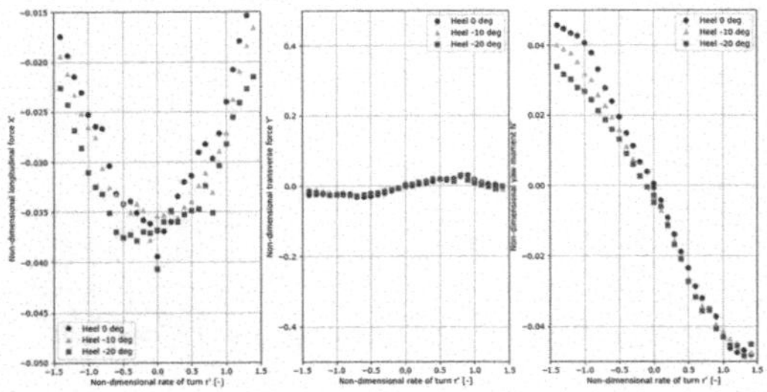

Figure 18. Combined rotation and heel results as function of rotation rate.

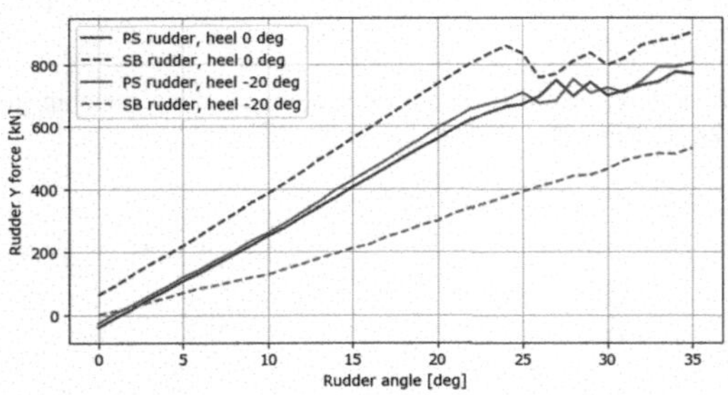

Figure 19. Combined rudder and heel transverse rudder force.

increases the total resistance of the ship. In this regard it appears that heel angles influences more the fore body, whereas the trends in the aft body are less clear. It can also be noticed that in general the total *Y* force increases with heel angle for the high drift angles, although this trend is not so clear for a moderate drift angle. In the *Y* force distribution pictures, the heel angle affects the contributions of skeg, rudders, bilges and midship. Finally the yawing moment in the pure drift manoeuvre is difficult to analyse. The total *N* moment is a difference between two large values (from fore and aft body), therefore it seems that the heel angle changes the sign of total moment.

3.4 *Influence of heel on rudder forces*

The rudder forces are also influenced by the heel angles. The measurement of the transverse rudder forces are shown for the port side and starboard rudder, for the horizontal and -20 ° heel cases in Figure 19. The port side rudder is not much influenced by the heeling, but the transverse force generated by the starboard one is much decreased, due to its partial emergence out of water as can be seen in previously shown in Figure 3.

4 MANOEUVRING CHARACTERISTICS

4.1 *Heel angles in turn*

Also the free running model tests provide insightful information on the heel dynamics and their impact on other manoeuvring aspects. Also in this case two topics can be distinguished: the heel angles themselves on the one hand, and their impact on the other manoeuvring characteristics on the other. The general phenomenon of heel dynamics is well understood, despite it being difficult to model in simulations. Among the most important factors influencing the maximum and steady heel angles in a turn are the approach speed, the rudder angle and the GMt. The influence of the rudder angle and approach speed for the short skeg configuration and a GMt of 1.65 m is shown in Figure 20, taken from bib_Mio1. The influence of GMt and skeg configuration is presented instead in Figure 21 All the above-mentioned factors play a role in the heel dynamics in particular by increasing the maximum heel angle.

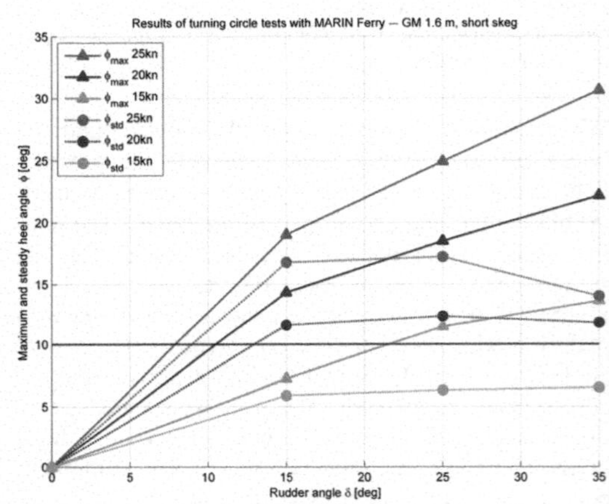

Figure 20. Maximum and steady heel angles in turn for short skeg configuration and GMt 1.65 m.

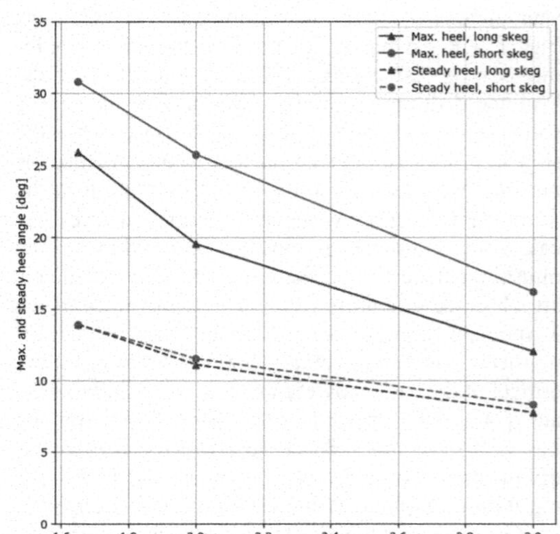

Figure 21. Maximum and steady heel angles in turn at 25 kn approach speed and 35° rudder angle.

4.2 *Influence of heel on manoeuvring characteristics*

Just as important is the impact of the GMt and the heel dynamics on the conventional manoeuvring characteristics. Indeed, a lower GMt does not only increase the heel angles, but also affects the horizontal manoeuvrability of the ship. This is shown in Figures 22 to 24. The impact of heel angles on overshoot angles at an approach speed of 25 kn is shown in Figure 22. It can be seen that the influence is particularly high, especially for the 2nd overshoot of the zig-zag 10/10 and the 2nd overshoot of the zig-zag 20/20. The latter case is also the most critical: it is the manoeuvring characteristics that gets the closest to the IMO limit. Therefore, if caution is not exerted in the design phase, there might be a risk of a ship not fulfilling the IMO manoeuvring standards just because of too low GMt. More generally it can be said that a ship with low GMt has worse yaw checking ability and therefore is more difficult to control, which makes the vessel even more unsafe. The differences due to the skeg configuration are analysed more into detail in Ferrari et al. (2019). Another characteristic derived from the zig-zag 10/10 is the initial turning distance, which is plotted in Figure 23. There appears to be practically no influence of the heel angle on this characteristic. Therefore, the worsening of the yaw checking ability due to lower GMt is not even compensated by an increase of the initial turning ability. Finally, the characteristics of the turning circle, performed at 25 kn and with a 35° rudder angle, are presented in Figure 24. The dimension of the turning circle, in particular the tactical dimeter, appear to be slightly decreasing, which indicates an improved turning ability mainly for the steady part of the turn. An

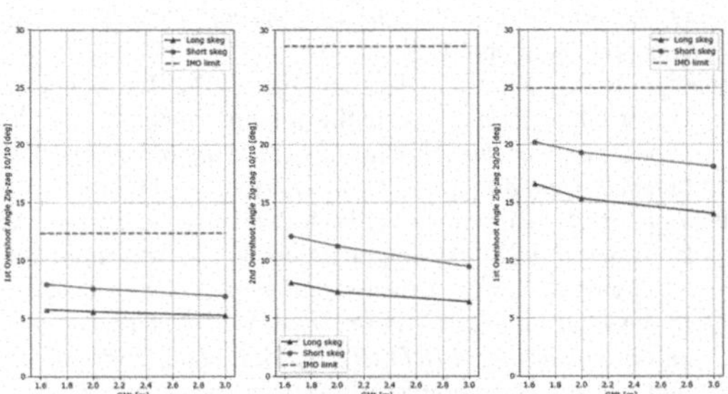

Figure 22. Overshoot angles at 25 kn approach speed.

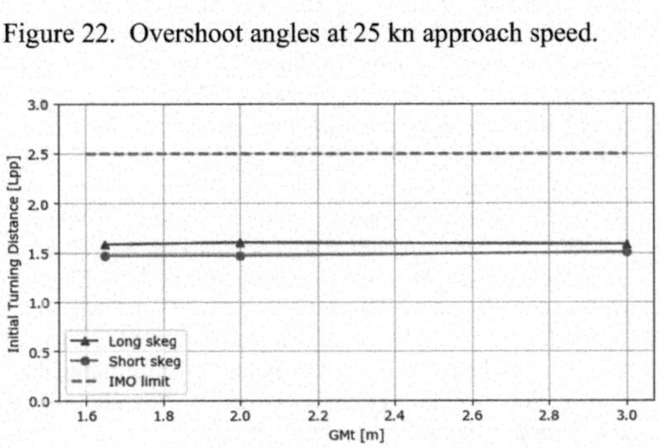

Figure 23. Initial turning distance at 25 kn approach speed.

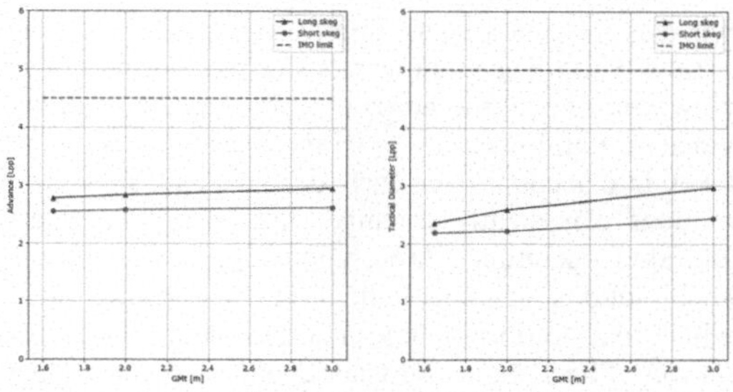

Figure 24. Turning circle characteristics at 25 kn approach speed and 35° rudder angle.

overview of how the manoeuvring characteristics change as function of GMt is presented in Table 4. As explained previously, the worsening of the yaw checking ability is only partly compensated by an improvement of the turning ability and in general it can be said that there is a net worsening of manoeuvrability. This is especially true if heel angles are taken into account as well, since they almost double when switching from 3.00 m to 1.65 m GMt.

5 CONCLUSIONS

This paper showed the influence of heel on ship manoeuvrability, based on the results of captive model tests, free running model tests and CFD calculations. It has been shown that the hydrodynamic heel moment should be investigated as a moment on its own rather than being expressed as the transverse force times an arm. The influence of drift and rotation on this moment was shown based on the captive test results. It has also been discussed how the other forces and moment are affected by the heel angle. These horizontal forces and moments can vary significantly when in presence of heel and it is important to take these changes into account for an accurate modelling of all hull forces. This is also applicable to rudder forces which are also affected by the heel angles, especially in case of emergence of one of the rudders out of the water. CFD calculations showed that taking heel angles into account is essential for accurately determining roll moment. Moreover free surface effects, especially the bow wave, seem to strongly contribute to the heel moment build-up. The skeg predictably plays a critical role in generating the heel moment, however, more unexpected is the contributions of the bilge. Another important aspect is the heel dynamic in turn. The well-known relationship between rudder angles, transverse metacentric height, approach speed and heel angles is repeated here. However, what is less studied, and shown in the present paper, is how the metacentric height and heel angles influence the conventional manoeuvring characteristics. The yaw checking ability clearly worsens. Especially the 1st overshoot angle of the zig-zag 20/20 may become critically close to the limit of the corresponding IMO criterion. The initial turning ability seems to be unaffected, whereas there is a small improvement in the turning ability, with reduced dimensions of the turning circle. However, overall the horizontal ship manoeuvrability worsens with lower GMt. All these aspects should be taken into account from early design stage, since heel motions can greatly endanger the crew, passengers and ship operations. Most approaches either ignore heel motions or model them in a simplified way, which may lead to greatly underestimating the actual impact of heel angles. Instead, a numerical model that incorporates all these heel dynamics will better predict the heel motions, thus preventing unsafe ship designs or future dangerous operations.

Table 4. Influence of heel on manoeuvring characteristics (OS: overshoot angle, ITD: initial turning distance).

Characteristic	Value	Value	Diff. [%]
GMt [m]	3.0	1.65	
Long skeg configuration			
1st OS 10/10 [deg]	5.24	5.75	+9.8
2nd OS 10/10 [deg]	6.39	8.07	+26.4
1st OS 20/20 [deg]	14.02	16.56	+18.2
ITD [Lpp]	1.59	1.58	-0.1
Advance [Lpp]	2.94	2.77	-5.8
Tactical Diamter [Lpp]	2.96	2.36	-20.4
Max. heel [deg]	12.02	25.88	+115.3
Steady heel [deg]	7.72	13.92	+80.3
Short skeg configuration			
1st OS 10/10 [deg]	6.89	7.94	+15.2
2nd OS 10/10 [deg]	9.43	12.12	+28.5
1st OS 20/20 [deg]	18.08	20.2	+11.7
ITD [Lpp]	1.51	1.47	-2.4
Advance [Lpp]	2.62	2.55	-2.9
Tactical Diamter [Lpp]	2.44	2.19	-10.2
Max. heel [deg]	16.19	30.78	+90.1
Steady heel [deg]	8.26	13.91	+68.4

REFERENCES

Bhushan, S., H. Yoon, F. Stern, E. Guilmineau, M. Visonneau, S. Toxopeus, C. Simonsen, S. Aram, S.-E. Kim, & G. Grigoropoulos (2019). Assessment of computational fluid dynamic for surface combatant 5415 at straight ahead and static drift 20 deg. *Journal of Fluids Engineering doi10.1115/1.4041229*.

Ferrari, V., T. Gornicz, A. Kisjes, & F. Quadvlieg (2019). Influence of skeg on ship manoeuvrability at high and low speeds. In *Proceedings of the 14th International Symposium on Practical Design of Ships and other Floating Structures (PRADS 2019)*, Yokohama, Japan.

Ferrari, V. & F.H.H.A. Quadvlieg (2016). Preliminary considerations on a unified model for hydrodynamic forces. In C.Guedes Soares and T. Santos (Eds.), *Maritime Technology and Engineering 3*, pp. 251–258. London, UK: Taylor & Francis Group.

Ferrari, V., R. Tonelli, A. Kisjes, & R. Hallmann (2022). Manoeuvring experiments, mathematical model and sensitivity analysis for test-case ferry. In *6th International Conference on Maritime Technology and Engineering (MARTECH 2022)*, Lisbon, Portugal.

Ferrari, V., H. van den Boom, A. Kisjes, & F. Quadvlieg (2020). Heel angles in turn and passenger safety. In *Proceedings of the RINA Conference on Sustainable and Safe Passenger Ships*, Athens, Greece.

Fukui, Y., Y. Yoshimura, H. Yano, & M. Kondo (2015). Investigation of effect of roll motion on manoeuvrability using 4-DOF mathematical model. In *Proceedings of the 13th International Conference on Marine Simulation and Ship Manoeuvrability (MARSIM 2015)*, Newcastle upon Tyne, U.K.

Grim, O., P. Oltmann, S. Sharma, & K. Wolff (1976). Cpmc - a novel facility for planar motion testing of ship models. In *Proceedings of the 11th ONR Symposium on Naval Hydrodynamics*, London, U.K.

Hallmann, R. & F. Quadvlieg (2015). Instationary captive model tests. In *Proceedings of the 13th International Conference on Marine Simulation and Ship Manoeuvrability (MARSIM 2015)*, Newcastle upon Tyne, U.K.

Hirano, M. (1980). A practical calculation method for ship manoeuvring motion at initial design stage (in Japanese). *Journal of the Society of Naval Architects of Japan* 147.

Hirano, M. & J. Takashina (1980). A calculation of ship turning motion taking into account coupling effect due to heel into consideration (in Japanese). *Transactions of the West-Japan Society of Naval Architects* 59.

Hoydonck, W., S. Toxopeus, K. Eloot, K. Bhawsinka, P. Queutey, & M. Visonneau (2019). Bank effects for KVLCC2. *Journal of Marine Science and Technology doi10.1007/s00773-018-0545-3*.

IMO (2008). *International Code on Intact Stability Section 3.1.2*. London, U.K.: International Maritime Organization.

Kim, Y. G., S. Y. Kim, H. T. Kim, S. W. Lee, & B. S. Yu (2007). Prediction of the manoeuvrability of a large containership with twin propellers and twin rudders. *Journal of Marine Science Technology 12*:130-138.

Kisjes, A., F. Quadvlieg, & V. Ferrari (2019). Uncertainty analysis of free running manoeuvring model tests on a modern ferry with emphasis on heel angles. In *Proceedings of the 38th International Conference on Ocean, Offshore and Arctic Engineering (OMAE 2019)*, Glasgow, Scotland, U.K.

Menter, F., M. Kuntz, & R. Langtry (2003). Ten years of industrial experience with the sst turbulence mode. In *Proceedings of the 4th International Symposium on Turbulence, Heat and Mass Transfer*, Antalya, Turkey.

Piaggio, B., B. Rimini, D. Villa, M. Viviani, V. Ferrari, & R. Tonelli (2022). An investigation on the heel influence on manoeuvrability: a twin-screw roro vessel case study. In *6th International Conference on Maritime Technology and Engineering (MARTECH 2022)*, Lisbon, Portugal.

Poland (2014). *Development of Amendments to the Criterion for Maximum Angle of Heel in Turns of the 2008 IS Code Proposal for the Structure of the Criterion for Maximum Angle of Heel in Turns of the 2008 IS Code.* International Maritime Organization: IMO SDC 1/14/1.

RINA (2011a). *Development of Amendments to the Criterion for Maximum Angle of Heel in Turns of the 2008 IS Code Proposed Amendment to the 2008 IS Code.* The Royal Institution of Naval Architects: IMO SLF 54/12.

RINA (2011b). *Development of Amendments to the Criterion for Maximum Angle of Heel in Turns of the 2008 IS Code Revised Proposal for Amending the 2008 IS Code.* The Royal Institution of Naval Architects: IMO SLF 55/12.

Sutulo, S. (1994). Extension of polynomial ship mathematical models to arbitrary manoeuvres. In *Proceedings of the International Shipbuilding Conference in commemoration of the centenary of the Krylov Ship Research Institute (ISC 1994)*, Saint Petersburg, Russia.

Yasukawa, H., N. Hirata, & Y. Yamazaki (2016). Effect of bilge keels on manoeuvrability of a fine ship. *Journal of Marine Science Technology 23(1)*.

Yeo, D., Y. G. Kim, & K. Yun (2015). A study on captive maneouvring model tests of a ship at large heel condition. In *Proceedings of the 13th International Conference on Marine Simulation and Ship Manoeuvrability (MARSIM 2015)*, Newcastle upon Tyne, U.K.

APPENDIX ADDITIONAL CFD RESULTS

This appendix presents additional results from the CFD calculations for the combined drift-and-heel cases.

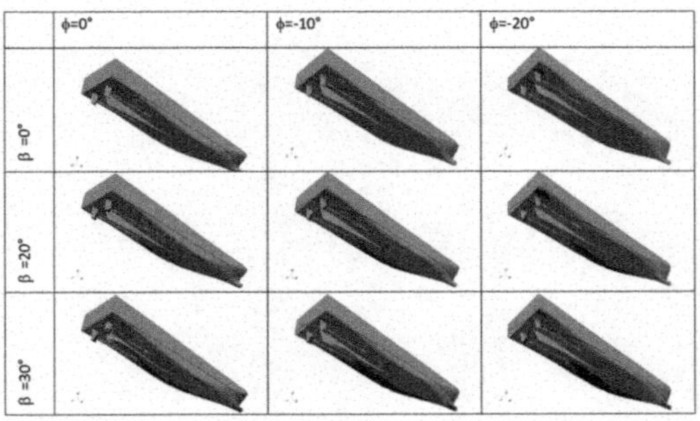

Figure 25. Heel moment contribution field as function of drift and heel angle starboard side view.

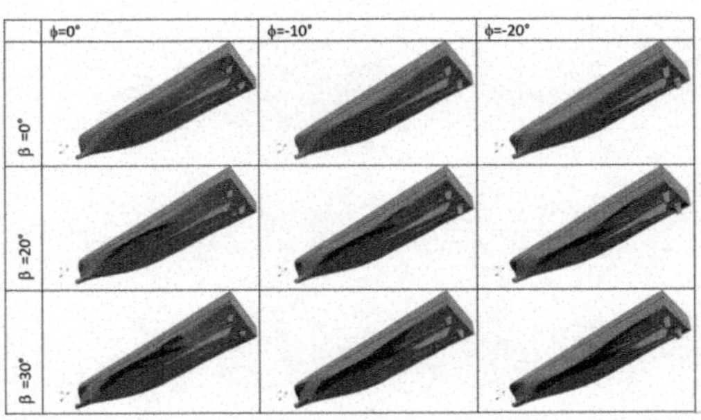

Figure 26. Heel moment contribution field as function of drift and heel angle portside view.

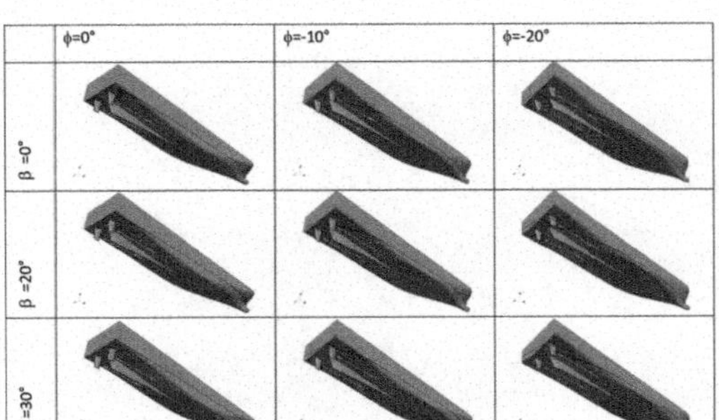

Figure 27. Longitudinal force contribution field as function of drift and heel angle starboard view.

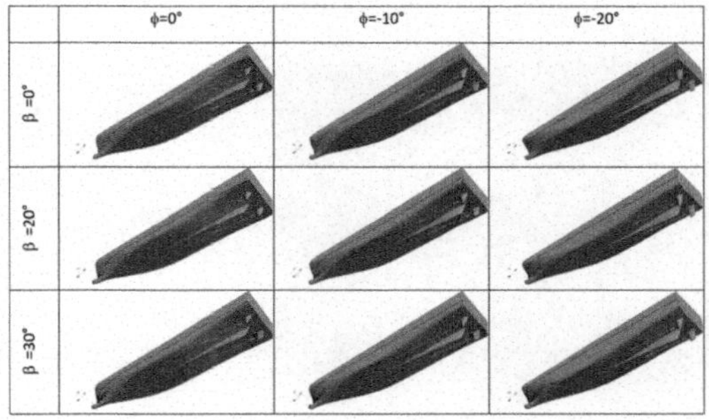

Figure 28. Longitudinal force contribution field as function of drift and heel angle portside view.

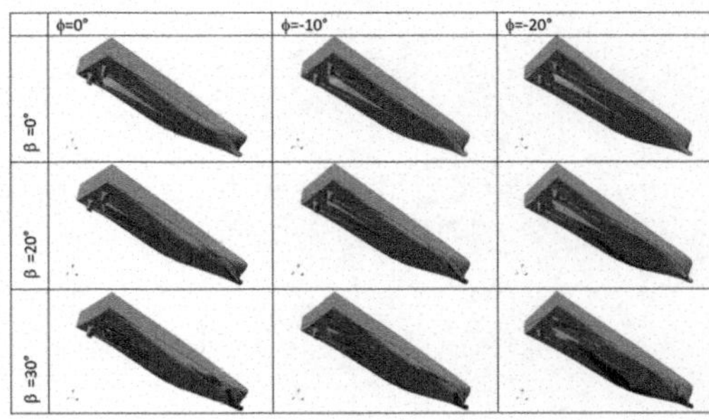

Figure 29. Transverse force contribution field as function of drift and heel angle starboard view.

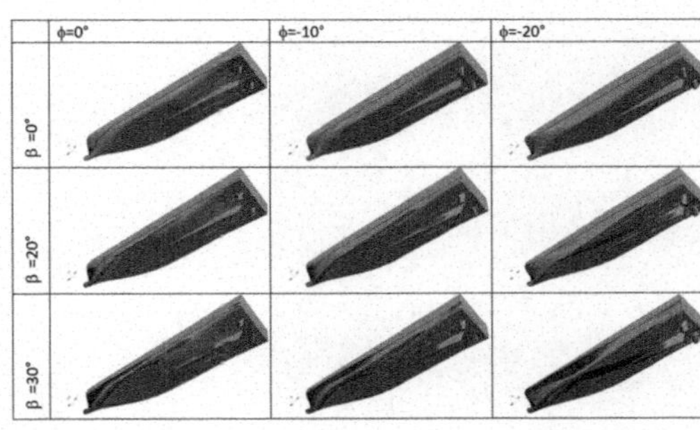

Figure 30. Transverse force contribution field as function of drift and heel angle portside view.

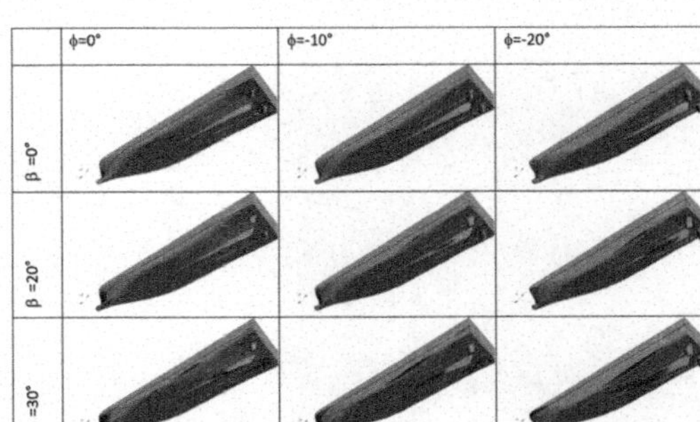

Figure 31. Yaw moment contribution field as e function of drift and heel angle starboard view.

Advances in Maritime Technology and Engineering – Guedes Soares & Santos (Eds)
© 2024 The Author(s), ISBN 978-1-032-83107-7

Spherical asymptotic approximation of hull manoeuvring model

V. Ferrari
Maritime Research Institute Netherlands (MARIN), Wageningen, The Netherlands

S. Sutulo & C. Guedes Soares
Centre for Marine Technology and Ocean Engineering (CENTEC), Instituto Superior Técnico, Universidade de Lisboa, Lisbon, Portugal

ABSTRACT: For performing time-domain simulations, a manoeuvring model of the hull forces is required. In some cases it is convenient to define an asymptotic approximation of such model. This results in a simpler model, easier to fit, even if the validity range is reduced. In this study, the asymptotic approximation problem is investigated through two different methods based on spherical harmonic expansion. The first method consists in applying a Taylor series to the spherical harmonics, the second one in linking the linear hydrodynamic coefficients to the harmonic ones. Both methods are applied to experimental results and it is shown that both approximated models fit well the experimental data. The approximation works best when closer to the condition sailing straight, nonetheless the error is still acceptable even in a larger range. In addition, the mathematical formulations becomes simpler, since it is no longer required to evaluate the complete spherical harmonic functions.

1 INTRODUCTION

For most manoeuvring simulations, it is necessary to derive a manoeuvring model of the hull forces, which expresses the hydrodynamic forces as mathematical functions depending on the kinematic parameters. There are several methods for mathematically describing these forces (Sutulo & Guedes Soares 2011), which range from simple linear models to more sophisticated ones involving several polynomials of varying degree. The definition and complexity of a model depend on several factors, among which the manoeuvre types to be simulated is of fundamental importance.

When dealing with manoeuvres restricted to a specific range of motions, such as small zig-zags not deviating too much from the sailing straight condition, it is often convenient to define an asymptotic approximation of the mathematical model. This ensures a simpler model, easier to fit, although at cost of the validity range. In the present study, the asymptotic approximation problem is investigated in case of a model based on a spherical harmonic expansion. This model, introduced in Ferrari, Sutulo, & Guedes Soares (2022), consists in expressing the hull forces and moments in spherical coordinates and apply the spherical harmonic expansion technique. It is robust and can cover any type of manoeuvre that the ship may perform. The fitting procedure is applied to the whole range of forces and moments at once, instead of having to separate the cases pure drift, pure rotation and combined drift and rotation. The coefficients deriving from the harmonic expansion provide much valuable information and insight on the hydrodynamic forces and moments of the hull.

Based on this spherical harmonic expansion, two methods are proposed in this study for obtaining an asymptotic expansion of the model. The first one consists in applying a Taylor series expansion of the spherical harmonics. The functions composing these special functions, that is, the cosine, sine and associated Legendre polynomials, are expanded as Taylor series of arbitrary degree. This has the advantage of simplifying the expression of the spherical harmonics. The second method consists in deriving explicit formulae for calculating the conventional linear coefficients based on the harmonic coefficients. In this way there is an explicit relationship between harmonic and linear coefficients, which can in turn be used as a stand-alone linear model or as part of a more complex polynomial model.

Both methods are applied to the captive hull force measurements of the KCS test-case and compared with a full harmonic expansion. It is shown how well the approximated methods fit the data, both for small pure drift and for small pure rotations. In the end conclusions are drawn on the asymptotic approximation of spherical harmonic expansions.

DOI: 10.1201/9781003508779-35

2 BACKGROUND ON MODELLING HYDRODYNAMIC HULL FORCES

2.1 *Basic definitions*

It is assumed that the ship is a rigid body sailing in unrestricted, calm and deep water. The reference system is centred at the point O, defined as the intersection of the waterline plane, the symmetry plane and station 10. The longitudinal x axis is aligned with the ship's centreline and positive forward, the transverse y axis is positive to starboard and the vertical z axis positive downwards. It is assumed that the surge, sway and yaw motions are dominant and that the others can be discarded, in other words only 3 degrees of freedom are considered in this study. Extension to higher degrees is also possible but will not be covered in the present study.

Ship motions are represented by three variables: the surge velocity u, the sway velocity v and the rotation r around the vertical axis called rate of turn, also known as rotation rate, rate of yaw or yaw rate. The resulting ship speed is:

$$V = \sqrt{u^2 + v^2}. \tag{1}$$

The three kinematic variables u,v,r define the ship motions in 3 degrees of freedom. Two other non-dimensional values are usually defined in manoeuvring studies for better differentiating between types of manoeuvres. The first one is the drift angle β, which represents the angle between V and the x axis, the second one is the non-dimensional rate of turn r':

$$\beta = arctan\tfrac{v}{u} \quad r' = \tfrac{rL}{V}, \tag{2}$$

where L is the length between perpendiculars. It must be noted that some authors define the drift angle with a minus sign. Moreover the above expression is not applicable when $u < 0$. In this regard it is possible to change the definition of the drift angle, so that the cases $u \geq 0$ and $u < 0$ are evaluated separately and explicitly (Sutulo & Kim 1997).In any case, with the two variables β and r' it is possible to much better define the type of manoeuvre.

2.2 *The hydrodynamic hull forces*

The total force acting on the hull is modelled as a longitudinal force X, a transverse force Y and a moment acting on the yaw rotation N, referred to collectively as hydrodynamic hull forces, and all dependent on the velocities and rotation u,v,r.

It is important that the hull forces reflect the port side/starboard symmetry of the hull, in other words that the mathematical functions have the correct parity, such as $X(-\beta, -r') = +X(\beta,r')$, $Y(-\beta, -r') = -Y(\beta,r')$ and $N(-\beta, -r') = -N(\beta,r')$.

2.3 *Mathematical models*

Many mathematical models have been developed to represent hydrodynamic hull force. They are usually based on multivariate Taylor expansions Abkowitz (1964), with some minor variations (Norrbin 1970, Ogawa & Kasai 1978). Most of them make use among others of the non-dimensional rate of turn r'. However, the latter has the disadvantage that at very low speeds for $V \rightarrow 0$ the value r' tends to infinity, as deriving directly from its definition shown in Equation 2.

In order to avoid this issue, some alternatives definitions of the non-dimensional rate of turn have been proposed in the past. A typical approach (Sutulo 1994) is to introduce a *generalised ship speed*:

$$U = \sqrt{u^2 + v^2 + (rL)^2}, \tag{3}$$

which can be used instead of V and is always defined even in case of turn-on-spot manoeuvre. The generalised ship speed also allows to introduce a robust form of the non-dimensional values that is well defined also in case of turn-on-spot:

$$F'' = \frac{F}{\frac{1}{2}\rho L^a T^b U^2}. \tag{4}$$

where ρ is the water density, L the length between perpendiculars, T the draught, a and b are appropriate coefficients depending on the force or moment, typically $a = b = 1$ for X,Y and $a = 2, b = 1$ for N.

A more detailed literature review on formulations for the non-dimensional rate of turn is presented in Ferrari & Quadvlieg (2016).

The relationship between the velocities and drift angle shown in Equation 2, which involves trigonometric functions, suggests another method for fitting the hydrodynamic forces: a Fourier series expansion in β instead of a Taylor series in u and v, for instance to use a multivariate Fourier series as function of both the drift angle β and the non-dimensional rate of turn r'' (Sutulo 1994). This model uses a trigonometric expansion both in β and r'', thus ensuring the robustness of the method and making it particularly well suited for any manoeuvre. The method presented in this study, based on initial research by the authors (Ferrari, Sutulo, & Guedes Soares 2022), further expands this approach.

3 HYDRODYNAMIC HULL FORCES IN SPHERICAL COORDINATES

In the previous section it has been shown that it is possible to change the coordinate base of the hydrodynamic force or moment F from the Cartesian coordinates u,v,rL to any valid combination of the variables $u, v, rL, V, \beta, u', v', r', r''$, using the relationships presented in Sections 2.1 and 2.3. As an example, switching from u,v,r to V,β,r'' can be interpreted as equivalent to a mapping from Cartesian to cylindrical coordinates.

It is of course possible to use alternative mappings and in this study the *spherical coordinate system* is applied. It is defined as follows:

$$U = \sqrt{u^2 + v^2 + (rL)^2},$$
$$\beta = arctan\frac{v}{u}, \quad \gamma = arccos\frac{rL}{U}. \tag{5}$$

The inverse transformations are:

$$u = U\cos\beta\sin\gamma, \quad v = U\sin\beta\sin\gamma,$$
$$rL = U\cos\gamma, \tag{6}$$

from which the following auxiliary variables can also be defined:

$$u'' = \frac{u}{U} = \cos\beta\sin\gamma, \quad v'' = \frac{v}{U} = \sin\beta\sin\gamma,$$
$$r'' = \cos\gamma. \tag{7}$$

The drift angle β has the same definition as in Equation 2 and retains the same physical meaning. In mathematical terms it represents the azimuth angle. Also in this case attention must be paid to its correct implementation for the case $u < 0$. The variable U is the same generalised ship speed already introduced in Equation 3 and represents the radius of the sphere. Finally, the angle γ is a *polar non-dimensional rate of turn*, defined between 0 and π. It represents the polar angle between the generalised speed U and the vertical rL axis.

The main advantage of the spherical coordinate system is that, first of all it, is robust since it is well defined for any type of manoeuvre and, secondly, it enables the application of the spherical harmonic expansion. This technique, as will be seen later, is well studied in literature and applied in many fields.

With Equation 5 the hydrodynamic force or moment F is defined on the (U,β,γ)-sphere, which covers all arbitrary manoeuvres. The case $\gamma = \frac{\pi}{2}$ represents a pure drift motion with no rotation. Similarly, the cases $\gamma = 0$ and $\gamma = \pi$ represent the turn-on-spot manoeuvre, corresponding to the poles of the (U,β,γ)-sphere. All other cases represent a combination of drift and rotation.

Moreover, as explained in Section 2.2, it is reasonable to assume that the force is proportional to the square of the speed, in this case U^2. This means that the function F'' defined in Equation 4 does not depend on U but instead depends only the two angles: $F'' = F''(\beta,\gamma)$. From a mathematical point of view, this is equivalent to saying that F'' is defined on the unit sphere, which allows to disregard the radius in all subsequent calculations presented in this chapter.

4 THE KCS TEST-CASE

In this study use is made of experimental results from the well-known KCS test-case. The main particulars are presented in Table 1.

Table 1. Main particulars of KCS test-case.

Designation	Symbol	Value	Unit
Length between perpendiculars	L	229.4	m
Breadth	B	32.2	m
Draught midship	T	10.8	m
Volume	∇	52086.6	m3
Block coefficient	C_B	0.653	-
Froude number	Fr	0.0976	-
Model scale	λ	37.89	-

The test campaign is described into detail in Tonelli & Hallmann (2020) but more generally studies on the KCS are often found in literature (Quadvlieg et al. 2015). The results of the captive test campaign provide a total of 1749 experimental points, covering the whole range of β and γ as shown in Figure 1. At each point corresponds a measurement of the bare hull X force, Y force and N moment. Also the heeling moment K has been measured, but it will not be considered in this study.

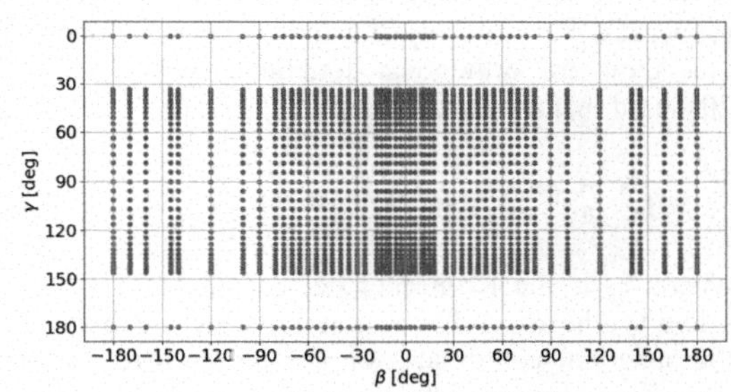

Figure 1. Experimental points of the horizontal motions available for the KCS.

The measured forces do satisfy the parity relationships, that is, the X force is an even function and both Y and N are odd This is an important point, since correct experimental points will lead to correct regression coefficients, which in turn will ensure parity also for the mathematical model as will be discussed later.

5 THE SPHERICAL HARMONICS

It can be proven that any real or complex function f defined on the sphere, which is well-behaved and bounded, can be expressed as a series (Hobson 1955):

$$f(\beta,\gamma) = \sum_{\ell=0}^{\infty}\sum_{m=-\ell}^{\ell} c_\ell^m \mathcal{Y}_\ell^m(\beta,\gamma), \tag{8}$$

where $\mathcal{Y}_\ell^m$ is the *spherical harmonic* of *degree* ℓ and *order m*, and c_ℓ^m are the *harmonic coefficients*. Since it is reasonable to assume that hydrodynamic forces are bounded and well-behaved for practical purposes, then the above can be applied to those forces.

Spherical harmonics can be interpreted as a generalisation of the Fourier series on the sphere. They form a complete set of orthogonal functions and thus form a basis for functions defined on the sphere as defined above. They are used in a wide range of applications.

The formal definition of spherical harmonic is:

$$\mathcal{Y}_\ell^m(\beta,\gamma) = \sqrt{\frac{2\ell+1}{4\pi}\frac{(\ell-m)!}{(\ell+m)!}}e^{im\beta}P_\ell^m(\cos\gamma), \quad (9)$$

where $P_\ell^m(\cdot)$ is the associated Legendre polynomial of degree ℓ and order m. For clarity the ($cos\gamma$) part is omitted for the rest of the document, so that implicitly $P_\ell^m = P_\ell^m(\cos\gamma)$.

It must be noted that some authors use different conventions regarding the normalisation factor represented by the square root term. Similarly, some authors write explicitly in the definition of spherical harmonic a factor $^{(-1)}m$, called Condon-Shortley phase, which in this study is assumed to be already included in the definition of the associated Legendre polynomial.

The spherical harmonics have several special properties, but arguably the most important one is that they are orthogonal functions (Olver et al. 2010). In mathematical terms:

$$\int_0^{2\pi}\int_0^{\pi}\mathcal{Y}_{\ell_0}^{m_0}\mathcal{Y}_{\ell_1}^{m_1}\sin\gamma d\gamma d\beta = \delta_{\ell_0\ell_1}\delta_{m_0m_1}, \quad (10)$$

where δ_{jk} is the Kronecker delta, equal to 1 if $j = k$ and 0 otherwise. The $sin\gamma$ term is the Jacobian of the spherical coordinates on the unit sphere. Later two consequences of the orthogonal property will be discussed.

The spherical harmonics are complex-valued functions, however in practice it is more convenient to work with real-valued ones. This real function, indicated with $\mathcal{Y}_{\ell m}$, is confusingly called in literature the *real spherical harmonic* but is not the same as just the real part of the complex-valued harmonic. In fact, the mapping between complex-valued and real-valued spherical harmonics is defined as follows (Dahlen & Tromp 1998):

$$\mathcal{Y}_{\ell m} = \begin{cases} \sqrt{2}\Re e\{\mathcal{Y}_\ell^m\} & \text{if } m>0, \\ \mathcal{Y}_\ell^0 & \text{if } m=0, \\ \sqrt{2}\Im m\{\mathcal{Y}_\ell^{|m|}\} & \text{if } m<0, \end{cases} \quad (11)$$

where $\Re e\{\cdot\}$ and $\Im m\{\cdot\}$ indicate the real part and imaginary part of a complex-valued function, respectively, and $\sqrt{2}$ is a normalizing factor. Aslo the real spherical harmonics satisfy the orthogonality property shown in Equation 10 (Dahlen & Tromp 1998).

The real spherical harmonics up to degree 2 are shown in Figure 2.

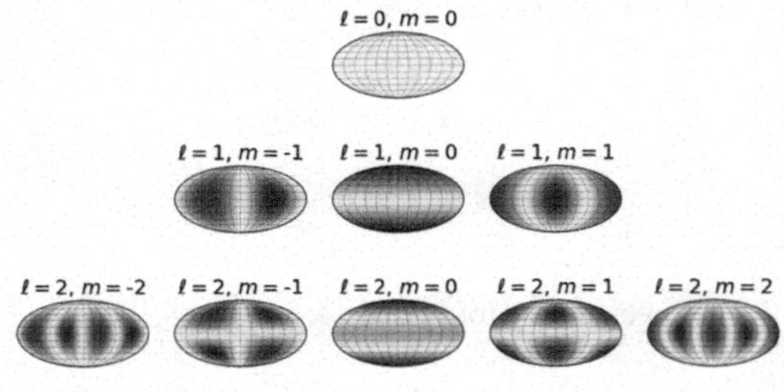

Figure 2. Qualitative view of the real spherical harmonics up to degree 2.

One practical aspect is that the series in Equation 8 cannot be computed to infinity. Therefore, a *maximum degree* or *approximation degree L* has to be chosen, up to which the harmonic series is calculated. This results in an approximation of the non-dimensional force or moment F'', indicated with $\hat{F}''$. This approximation is expressed as a sum of real spherical harmonics $\mathcal{Y}_{\ell m}$ up to degree L multiplied by real harmonic coefficients $c_{\ell m}$:

$$F'' \approx \hat{F}'' = \sum_{\ell=0}^{L}\sum_{m=-\ell}^{\ell} c_{\ell m}\mathcal{Y}_{\ell m}. \quad (12)$$

It can be proven that such an expansion is unique (Ferrers 1877) and that increasing the degree of approximation and adding more harmonics is guaranteed to improve the accuracy of the modelling. Both these properties are particularly useful when performing a fit.

6 FITTING OF THE HYDRODYNAMIC FORCES

The core problem of any harmonic analysis is deriving the harmonic coefficients and this applies also to the present case. A practical way is to use the least-squares method (Shen & Chung 2006). Let $f_p = f(\beta_p, \gamma_p)$ be the p-th measurement out of a total of P measurements, with $0 \le p < P$. These are the elements of the vector **f** of experimental values. Successively, a maximum degree L has to be chosen and the following index q is defined for convenience, depending on the total number $Q = (L + 1)^2$ of harmonics:

$$q(\ell,m) = \ell^2 + \ell + m, \quad 0 \le q < Q, \quad (13)$$

This index allows to express all harmonics as a vector with components $\mathcal{Y}_{q(\ell,m)} = \mathcal{Y}_{\ell m}$, instead of having to use separately the two indices ℓ and m. In this way the regressor matrix **Y** can be defined, with size $P \times Q$:

$$\mathbf{Y}=\begin{bmatrix} \mathcal{Y}_0(\beta_0,\gamma_0) & \cdots & \mathcal{Y}_{Q-1}(\beta_0,\gamma_0) \\ \vdots & \ddots & \vdots \\ \mathcal{Y}_0(\beta_{P-1},\gamma_{P-1}) & \cdots & \mathcal{Y}_{Q-1}(\beta_{P-1},\gamma_{P-1}) \end{bmatrix}. \quad (14)$$

With the matrix of known harmonics $\mathbf{Y}$ and the vector of experimental values $\mathbf{f}$, the linear system of equations $\mathbf{Yc} = \mathbf{f}$ is set-up, where $\mathbf{c}$ is the vector of unknown harmonic coefficients, with components $c_{q(\ell,m)} = c_{\ell m}$. The system can be solved with the well-known least-squares technique: $\mathbf{c} = \left(\mathbf{Y}^T\mathbf{Y}\right)^{-1}\mathbf{Y}^T\mathbf{f}$, thus obtaining all coefficients required for constructing the approximate function $\hat{F}''$ using Equation 12.

The spherical harmonics comprise even and odd functions, as shown in Figure 2, and therefore the parity conditions of the hydrodynamic forces presented previously can be satisfied by correctly setting the harmonic coefficients. This is done implicitly by the least-squares method as long as the experimental data points are reliable and satisfy the parity conditions themselves.

More information on fitting the hydrodynamic forces with spherical harmonics can be found in Ferrari, Sutulo, & Guedes Soares (2022).

7 ASYMPTOTIC APPROXIMATION

7.1 *Introduction*

For application of soft manoeuvres, such as the standard zig-zags and turning circles, it is useful to define the asymptotic approximation of the manoeuvring mathematical model to small drift angles and rotation rates. In the present study this is done by applying a Taylor series expansion to the real spherical harmonics. The expansion point is $(0, \frac{\pi}{2})$, corresponding to the ship sailing straight condition.

Firstly Equations 9 and 11 are combined:

$$\mathcal{Y}_{\ell m} = \begin{cases} \sqrt{2}\Re e\{k_{\ell m} e^{im\beta} P_\ell^m\} & \text{if } m>0, \\ Y_\ell^0 & \text{if } m=0, \\ \sqrt{2}\Im m\{k_{\ell |m|} e^{i|m|\beta} P_\ell^{|m|}\} & \text{if } m<0, \end{cases} \quad (15)$$

where

$$k_{\ell m} = \sqrt{\frac{2\ell+1}{4\pi}\frac{(\ell-m)!}{(\ell+m)!}}, \quad (16)$$

defined for convenience. It can be seen that this mapping involves distinguishing the real and imaginary part of only the complex exponential $e^{im\beta}$, because the coefficient $k_{\ell m}$ and the associated Legendre polynomials P_ℓ^m are all real-valued. Thus, the complex exponential becomes: $\Re e\{e^{im\beta}\} = \beta$ and $\Im m\{e^{i|m|\beta}\} = \sin|m|\beta$. Combining all the previous equations, the following relationship is obtained:

$$\mathcal{Y}_{\ell m} = \begin{cases} \sqrt{2}k_{\ell m}\cos m\beta P_\ell^m & \text{if } m>0, \\ k_{\ell\ell} P_\ell^0(\gamma) & \text{if } m=0, \\ \sqrt{2}k_{\ell|m|}\sin|m|\beta P_\ell^{|m|} & \text{if } m<0, \end{cases} \quad (17)$$

Therefore, the functions to be expanded are the cosine $cosm\beta$, the sine $sin \mid m \mid \beta$ and the associated Legendre polynomials $P_\ell^m(\cos\gamma)$. Each function is expanded as a separate Taylor series, rather than expanding one single multivariate function. On the one hand this might reduce the accuracy of the expansion, on the other hand it offers a much simpler approach since cross-derivatives need not to be calculated.

7.2 *Expansion of sine and cosine functions*

The cosine and sine are function of the drift angle only and since the expansion point for the drift is $\beta = 0$, the expansion corresponds to a simple Maclaurin series using the well-known formulae:

$$\begin{aligned} \beta &= \sum_{j=0}^{\infty}\frac{(-1)^j}{(2j)!}(m\beta)^{2j}, \\ \sin|m|\beta &= \sum_{j=0}^{\infty}\frac{(-1)^j}{(2j+1)!}(|m|\beta)^{2j+1}. \end{aligned} \quad (18)$$

7.3 *Expansion of associated Legendre polynomials*

The associated Legendre polynomials $P_\ell^m(\cdot)$ are function of the polar rate of turn γ. As first step, the general definition of the Taylor series is applied to the associated Legendre polynomial around the point $\gamma = \frac{\pi}{2}$, which represents the no-rotation condition:

$$P_\ell^m = \sum_{j=0}^{\infty}\frac{1}{j!}\frac{d^j P_\ell^m}{d\gamma^j}\bigg|_{\gamma=\frac{\pi}{2}}\left(\gamma-\frac{\pi}{2}\right)^j. \quad (19)$$

For calculating the derivatives there are recursive formulae available (Bosch 2000, cited in Slevinsky & Safouhi 2009, Fukushima 2018):

$$\frac{d^j P_\ell^m}{d\gamma^j} = \begin{cases} \frac{1}{2}\left[\frac{d^{j-1}P_\ell^{m+1}}{d\gamma^{j-1}} - a_{\ell m}\frac{d^{j-1}P_\ell^{m-1}}{d\gamma^{j-1}}\right] & \text{if } m>0, \\ \frac{d^{j-1}P_\ell^1}{d\gamma^{j-1}} & \text{if } m=0, \end{cases} \quad (20)$$

where $a_{\ell m} = (\ell+m)(\ell-m+1)$ and using the convention that $P_\ell^m = 0$ if $m > \ell$. The case $m < 0$ is not relevant since, according to Equation 11, only associated Legendre polynomials with positive order need eventually to be computed. The possibility to find an explicit formula of the Taylor expansion of the associated Legendre polynomials is left for further investigation.

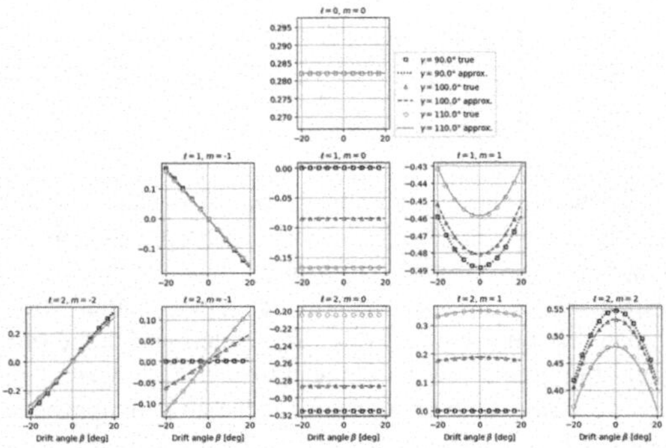

Figure 3. Comparison between explicit formula and numerical differentiation applied to the 3rd derivative of the real spherical harmonics.

7.4 *General formula for approximated spherical harmonics*

Substituting the Taylor series of the sine, cosine and associated Legendre functions into Equations 9 and 11, it is possible to have an approximation $\hat{\mathcal{Y}}_{\ell m}$ of the real spherical harmonics $\mathcal{Y}_{\ell m}$ up to a certain degree J:

$$\hat{\mathcal{Y}}_{\ell m} = \begin{cases} \sqrt{2}k_{\ell m}\left[\sum_{j=0}^{J}\frac{(-1)^j}{(2j)!}(m\beta)^{2j}\right] \\ \cdot\left[\sum_{j=0}^{J}\frac{1}{j!}\frac{d^j\mathcal{P}_\ell^m}{d\gamma^j}\bigg|_{\frac{\pi}{2}}\left(\gamma-\frac{\pi}{2}\right)^j\right] \text{ if } m>0, \\ k_{\ell 0}\left[\sum_{j=0}^{J}\frac{1}{j!}\frac{d^j\mathcal{P}_\ell^1}{d\gamma^j}\bigg|_{\frac{\pi}{2}}\left(\gamma-\frac{\pi}{2}\right)^j\right] \text{ if } m=0, \\ \sqrt{2}k_{\ell|m|}\left[\sum_{j=0}^{J}\frac{(-1)^j}{(2j+1)!}(m\beta)^{2j+1}\right] \\ \cdot\left[\sum_{j=0}^{J}\frac{1}{j!}\frac{d^j\mathcal{P}_\ell^{|m|}}{d\gamma^j}\bigg|_{\frac{\pi}{2}}\left(\gamma-\frac{\pi}{2}\right)^j\right] \text{ if } m<0. \end{cases} \quad (21)$$

An application of the above formulae for a degree $J = 3$ is shown on Figure 3. It can be seen that also in this case there is a good overall match between explicit formulae and numerical integration, as long as the drift and rotation angles are not too far from the expansion point $\left(0, \frac{\pi}{2}\right)$.

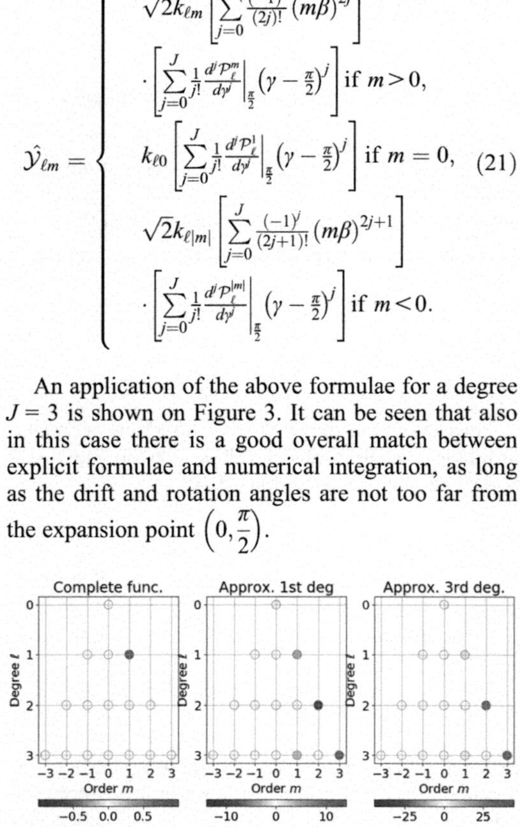

Figure 4. Verification of orthogonality of the 1st and 3rd degree asymptotic approximations for the real spherical harmonic of degree 1 and order 1.

7.5 *Partial orthogonality of the approximation*

It must be noted that the asymptotic approximations do not fully satisfy the orthogonality property. This is verified numerically in the following example. Taking the harmonic $\mathcal{Y}_{1,1}$ as example: in case of complete spherical functions, Equation 10 for the case $\ell_0 = 1$, $m_0 = 1$. The question is if the same holds even when using an asymptotic approximation based on $\hat{\mathcal{Y}}_{\ell m}$.

The integral is evaluated numerically using the method presented in Atkinson & Han (2012), based on a Delaunay triangulation and averaged evaluation of the function at each triangle. In this case the sphere is discretised with a 15° spacing grid. The approximation is performed using first the complete harmonic, then a 1st degree asymptotic approximation and finally a 3rd degree one. The values of the integral are reported as function of harmonic degree and order in Figure 4 for the three cases considered. It can be seen that for the complete harmonic function the integral is indeed equal to 1 only when $\ell = m = 1$. For the approximation cases, the orthogonality is only partially applicable. For most cases the integral is 0 and close to 1 when $\ell = m = 1$, however there are some other combinations of degree ℓ and order m that result in a non-zero integral.

7.6 *Fit with approximated spherical harmonics*

Despite not being fully orthogonal polynomials, the approximated real spherical harmornics can still be used for fitting the hydrodynamic forces, just like the full harmonics. This is done by applying the same procedure described in Section 6, except that use is made of the approximated spherical harmonics $\hat{\mathcal{Y}}_{\ell m}$ rather than the real spherical harmonics $\mathcal{Y}_{\ell m}$. Results of such a fit for some sections of the sphere are shown in Figures 5 and 6 for the Y force and N moment, respectively. Each figure shows the corresponding non-dimensional force as function of the drift angle for three fixed polar non-dimensional rates of turn (top plot) and as function of the polar non-dimensional rate of turn for three fixed drift angles (bottom plot). The asymptotic expansions are performed considering approximation degrees $J = 1$ and $J = 3$. These two fits are compared to the experimental data and to the original full harmonic expansion. Both the asymptotic and full expansions are computed up to a harmonic degree $L = 3$. It can be seen that the two asymptotic expansions converge to the full one for $\beta \to 0$ and $\gamma \to \frac{\pi}{2}$. At higher values of drift and rotation the scatter among the different sort of approximation increases, nonetheless all fits still remain reasonably good in the neighbourhood of the expansion point $\left(0, \frac{\pi}{2}\right)$, proving the validity of the asymptotic approximation.

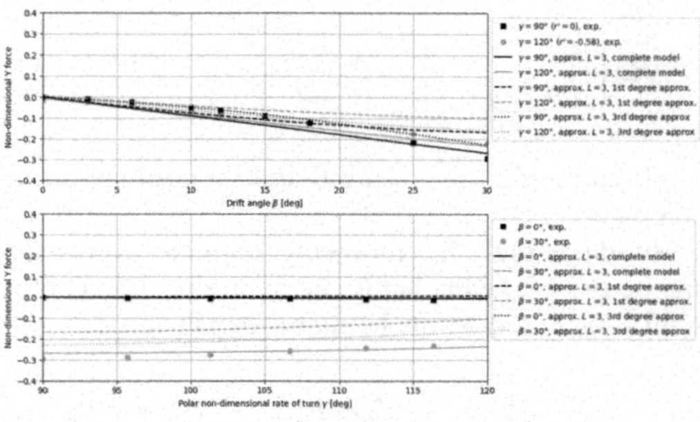

Figure 5. Asymptotic approximation of the non-dimensional Y force for the KCS.

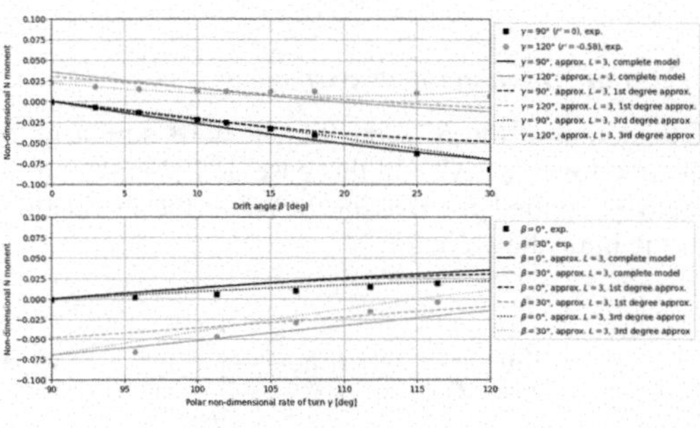

Figure 6. Asymptotic approximation of the non-dimensional N moment for the KCS.

8 CALCULATIONS OF THE LINEAR HYDRODYNAMIC COEFFICIENTS

8.1 *Definitions of the hydrodynamic coefficients*

The harmonic expansion expresses the hydrodynamic forces in terms of harmonic coefficients rather than hydrodynamic derivatives, however in some cases it might be required or convenient to have also the values of the hydrodynamic coefficients. This could happen for instance if there is a need to convert the spherical harmonic model to a different manoeuvring formulation. This conversion can be performed easily since the spherical harmonic model is based on explicit mathematical expressions, which can be used to calculate rigorously any coefficient without the need of fitting techniques. As explained in Section 2.3, there are many possible coefficients that can be defined, but the general approach for deriving them from the harmonic expansion is the same. In this section explicit conversion formulae are derived for a few particularly relevant coefficients.

In particular, a simple model for very small drift angles and rotation rates, is as follows, adapted from Norrbin (1970):

$$F'' \approx F''_{uv}u''v'' + F''_{ur}u''r'' \tag{22}$$

The above equations are meant mainly for the Y force and N moment but can be applied also to the X force, as long as the force is taken net of the ship resistance. These uv and ur coefficients represent the derivative of the hydrodynamic function at the origin. Thus, these coefficients are the most relevant for the present study as they represent the manoeuvring conditions close to sailing straight. Explicit formulae for calculating them based on a spherical harmonic expansions are presented in the following sections.

Based on this physical interpretation, the following equations are used for defining the coefficients:

$$F''_{uv} = \lim_{\beta\to 0}\frac{F''}{u''v''}\Big|_{\gamma=\frac{\pi}{2}},\quad F''_{ur} = \lim_{\gamma\to\frac{\pi}{2}}\frac{F''}{u''r''}\Big|_{\beta=0}, \tag{23}$$

where $k\pi$ with k equal to 0 or 1 represents the fact the the turn-on-spot condition is defined by the coordinates $\gamma = 0$ or $\gamma = \pi$, depending on the direction of rotation.

8.2 *Pure drift linear coefficient*

The limit of the F''_{uv} coefficient is not defined since it tends to 0/0. Nonetheless, it is possible to apply de l'Hôpital's rule and calculate the limit of the first derivatives instead:

$$\lim_{\beta\to 0}\frac{F''}{u''v''}\Bigg|_{\gamma=\frac{\pi}{2}} = \lim_{\beta\to 0}\frac{\frac{\partial F''}{\partial\beta}}{\frac{\partial(u''v'')}{\partial\beta}}\Bigg|_{\gamma=\frac{\pi}{2}}. \tag{24}$$

This is allowed since the limit of the derivatives exists and the functions are well-behaved. Recalling the following property of the limits:

$$\lim_{\beta\to 0}\frac{\frac{\partial F''}{\partial\beta}}{\frac{\partial(u''v'')}{\partial\beta}}\Bigg|_{\gamma=\frac{\pi}{2}} = \frac{\lim_{\beta\to 0}\frac{\partial F''}{\partial\beta}\Big|_{\gamma=\frac{\pi}{2}}}{\lim_{\beta\to 0}\frac{\partial(u''v'')}{\partial\beta}\Big|_{\gamma=\frac{\pi}{2}}}, \tag{25}$$

and noticing that, based on Equation 7:

$$\begin{aligned}\lim_{\beta\to 0}\frac{\partial(u''v'')}{\partial\beta}\Big|_{\gamma=\frac{\pi}{2}} &= \lim_{\beta\to 0}\frac{\partial\left(\cos\beta\sin\beta\sin^2\gamma\right)}{\partial\beta}\Bigg|_{\gamma=\frac{\pi}{2}}\\ &= \lim_{\beta\to 0}\frac{d(\cos\beta\sin\beta)}{d\beta}\\ &= \lim_{\beta\to 0}\left(\cos^2\beta - \sin^2\beta\right) = 1,\end{aligned} \tag{26}$$

then the expression for calculating the linear coefficient F''_{uv} can be reduced to:

$$F''_{uv} = \lim_{\beta\to 0}\frac{\partial F''}{\partial\beta}\Big|_{\gamma=\frac{\pi}{2}}. \tag{27}$$

Using the harmonic expansion $\hat{F}''$ it is possible to derive the explicit mathematical expression for the above coefficient. Substituting the definition of the harmonic expansion shown in Equation 12 into Equation 27:

$$\lim_{\beta\to 0}\frac{\partial \hat{F}''}{\partial \beta}\bigg|_{\gamma=\frac{\pi}{2}} = \lim_{\beta\to 0}\frac{\partial \sum_{\ell=0}^{L}\sum_{m=-\ell}^{\ell} c_{\ell m}\mathcal{Y}_{\ell m}}{\partial \beta}\Bigg|_{\gamma=\frac{\pi}{2}} = \lim_{\beta\to 0}\sum_{\ell=0}^{L}\sum_{m=-\ell}^{\ell} c_{\ell m}\frac{\partial \mathcal{Y}_{\ell m}}{\partial \beta}\bigg|_{\gamma=\frac{\pi}{2}} \tag{28}$$

In case of pure drift, when $\gamma = \frac{\pi}{2}$, the expression for complex-valued spherical harmonics can be simplified as (Olver et al. 2010):

$$\mathcal{Y}_\ell^m\left(\beta,\frac{\pi}{2}\right) = \begin{cases} b_{\ell m}e^{im\beta} & \text{if } \ell+m \text{ even,}\\ 0 & \text{if } \ell+m \text{ odd.}\end{cases} \tag{29}$$

where for simplicity the following coefficient is introduced:

$$b_{\ell m} = \frac{(-1)^{\frac{\ell+m}{2}}}{2^\ell\left(\frac{1}{2}\ell-\frac{1}{2}m\right)!\left(\frac{1}{2}\ell+\frac{1}{2}m\right)!} \cdot\sqrt{\frac{2\ell+1}{4\pi}(\ell-m)!(\ell+m)!} \tag{30}$$

The derivative of the above equation with respect to β is simply:

$$\frac{\partial Y_\ell^m}{\partial \beta}\bigg|_{\gamma=\frac{\pi}{2}} = \begin{cases} imb_{\ell m}e^{im\beta} & \text{if } \ell+m \text{ even,}\\ 0 & \text{if } \ell+m \text{ odd.}\end{cases} \tag{31}$$

However, this expression still refers to the complex-valued version of the spherical harmonic. In order to obtain the derivative of the real spherical harmonic, use is made of the mapping shown in Equation 11, except that it is applied to the derivatives with respect to β:

$$\frac{\partial Y_{\ell m}}{\partial \beta}\bigg|_{\gamma=\frac{\pi}{2}} = \begin{cases} \sqrt{2}\,\Re e\left\{\frac{\partial Y_\ell^m}{\partial \beta}\Big|_{\gamma=\frac{\pi}{2}}\right\} & \text{if } m>0,\\ \frac{\partial Y_\ell^0}{\partial \beta}\Big|_{\gamma=\frac{\pi}{2}} & \text{if } m=0,\\ \sqrt{2}\,\Im m\left\{\frac{\partial Y_\ell^{|m|}}{\partial \beta}\Big|_{\gamma=\frac{\pi}{2}}\right\} & \text{if } m<0.\end{cases} \tag{32}$$

The spherical harmonics are continuous everywhere, so in order to evaluate the limit $\beta \to 0$ one simply has to substitute $\beta = 0$ into Equations 32 and 31. In the latter, the exponential $e^{im\beta}$ simply becomes 1, leaving only the $imb_{\ell m}$ term. In this case, from the expression shown in Equation 31 it follows immediately that:

$$\Re e\left\{\frac{\partial Y_\ell^m}{\partial \beta}\bigg|_{\beta=0,\gamma=\frac{\pi}{2}}\right\} = 0,\quad \frac{\partial Y_\ell^0}{\partial \beta}\bigg|_{\beta=0,\gamma=\frac{\pi}{2}} = 0, \tag{33}$$

so all harmonics $m \geq 0$ are 0 and do not need to be computed.

Summarizing, the following explicit expression can be used for calculating the hydrodynamic coefficient $\hat{F}''_{uv}$ based on the harmonic expansion:

$$\hat{F}''_{uv} = \sum_{\ell=0}^{L}\sum_{m=-\ell}^{-1} c_{\ell m}\begin{cases}\sqrt{2}|m|b_{\ell m} & \text{if } \ell+|m| \text{ even,}\\ 0 & \text{if } \ell+|m| \text{ odd.}\end{cases} \tag{34}$$

8.3 *Pure rotation linear coefficient*

Calculation of the F''_{ur} coefficient starts similarly to the one for F''_{uv}. Also in this case de l'Hôpital's theorem is applied for making possible the evaluation of the limit shown in 23:

$$\lim_{\gamma\to\frac{\pi}{2}}\frac{F''}{u''r''}\bigg|_{\beta=0} = \lim_{\gamma\to\frac{\pi}{2}}\frac{\frac{\partial F''}{\partial\gamma}}{\frac{\partial(u''r'')}{\partial\gamma}}\Bigg|_{\beta=0}, \tag{35}$$

and the denominator can be similarly simplified as:

$$\lim_{\gamma\to\frac{\pi}{2}}\frac{\partial(u''r'')}{\partial\gamma}\bigg|_{\beta=0} = \lim_{\gamma\to\frac{\pi}{2}}\frac{\partial(\cos\beta\sin\gamma\cos\gamma)}{\partial\gamma}\bigg|_{\beta=0} = \lim_{\gamma\to\frac{\pi}{2}}\frac{d(\sin\gamma\cos\gamma)}{d\gamma} = -1, \tag{36}$$

which means that only the numerator of Equation 35 needs to be evaluated, as long as the -1 factor is not forgotten. The function F'' is replaced by its approximation $\hat{F}''$ which in turn is expressed through its definition shown Equation 12, obtaining the following equation:

$$\lim_{\gamma\to\frac{\pi}{2}}\frac{1}{-1}\frac{\partial\hat{F}''}{\partial\gamma}\bigg|_{\beta=0} = \lim_{\gamma\to\frac{\pi}{2}}\frac{1}{-1}\sum_{\ell=0}^{L}\sum_{m=-\ell}^{\ell} c_{\ell m}\frac{\partial\mathcal{Y}_{\ell m}}{\partial\gamma}\bigg|_{\beta=0} \tag{37}$$

where the -1 term is written explicitly for clarity. Unlike the pure drift case, there does not seem to be a special formula for spherical harmonics in case of pure rotation. Nonetheless, it is still possible to find a closed expression for the $\hat{F}''_{ur}$ coefficient by taking advantage of the definition shown in Equation 9. First, the limit of the derivative of the complex-valued spherical harmonic $\mathcal{Y}_\ell^m$ is analysed:

$$\lim_{\gamma\to\frac{\pi}{2}}\frac{\partial\mathcal{Y}_\ell^m}{\partial\gamma}\bigg|_{\beta=0} = \lim_{\gamma\to\frac{\pi}{2}}\frac{\partial}{\partial\gamma}\left[k_{\ell m}e^{im\beta}P_\ell^m\right]\Big|_{\beta=0} = \lim_{\gamma\to\frac{\pi}{2}} k_{\ell m}\frac{dP_\ell^m}{d\gamma}. \tag{38}$$

The derivative can be further decomposed using the chain rule of derivation:

$$\frac{dP_\ell^m}{d\gamma} = \frac{dP_\ell^m(\cos\gamma)}{d\cos\gamma}\frac{d\cos\gamma}{d\gamma} = \frac{dP_\ell^m(\cos\gamma)}{d\cos\gamma}(-\sin\gamma) \tag{39}$$

By taking the limit $\gamma = \frac{\pi}{2}$ the term $-\, sin\gamma$ reduces to -1 and $cos\gamma = 0$, but the derivative itself still remains to be evaluated. Fortunately there is an expression for the derivative of the associated Legendre polynomial calculated at the origin (Olver et al. 2010):

$$\left.\frac{dP_\ell^m}{d\cos\gamma}\right|_{\gamma=\frac{\pi}{2}} = \frac{-2^{m+1}\sqrt{\pi}}{\Gamma\left(\frac{1}{2}\ell - \frac{1}{2}m + \frac{1}{2}\right)\Gamma\left(-\frac{1}{2}\ell - \frac{1}{2}m\right)}. \tag{40}$$

There are other possible formulae for the derivative (Bosch 2000, cited in Slevinsky & Safouhi 2009, Fukushima 2018), which however are expressed as function of associated Legendre polynomials of lower degrees and therefore are less interesting when looking for an explicit expression. Based on the previous formulae the following coefficient is introduced for simplicity:

$$g_{\ell m} = -k_{\ell m}\frac{2^{m+1}\sqrt{\pi}}{\Gamma(\frac{1}{2}\ell - \frac{1}{2}m + \frac{1}{2})\Gamma(-\frac{1}{2}\ell - \frac{1}{2}m)}. \tag{41}$$

The -1 coming from the derivative of $cos\gamma$ cancels the -1 factor coming from the denominator. Moreover, since there are no imaginary numbers involved, the real spherical harmonic is equal to the real part of the complex-valued spherical harmonic multiplied by a factor $\sqrt{2}$, following the usual Equation 11. The imaginary part is 0 so all harmonics with order $m < 0$ are 0. Finally the computation of the summation can be further simplified by taking advantage of the following property of the Gamma function: the reciprocal of the Gamma function is zero when the argument is a negative integer, that is, $\frac{1}{\Gamma(-k)} = 0$ if k is a positive integer. This means that $\frac{1}{\Gamma(-\frac{1}{2}\ell - \frac{1}{2}m)} = 0$ if $\ell + m$ is even. Therefore, combining everything together, the following expression for the $\hat{F}''_{ur}$ coefficient is obtained:

$$\hat{F}''_{ur} = \sum_{\ell=0}^{L}\sum_{m=0}^{\ell} c_{\ell m}\begin{cases}\sqrt{2}g_{\ell m} & \text{if } \ell + m \text{ odd, } m > 0,\\ g_{\ell 0} & \text{if } \ell + m \text{ odd, } m = 0,\\ 0 & \text{otherwise.}\end{cases} \tag{42}$$

8.4 *Results*

Figures 7 and 8 show the results of the calculations applied to the KCS Y force and N moment, respectively. It can be seen that the linear coefficients $\hat{F}''_{uv}$ and $\hat{F}''_{ur}$ do correspond to the derivative at the origin of the forces for the pure drift and pure rotation cases. The curves approximate well the hydrodynamic forces close to the origin, for small values of β and $\frac{\pi}{2} - \gamma$.

Similar expression can be derived for other coefficients as needed. In this way it is possible to easily convert from the spherical harmonic model to a conventional polynomial model and in particular using uv and ur coefficients allows to model manoeuvres close to the saling straight condition.

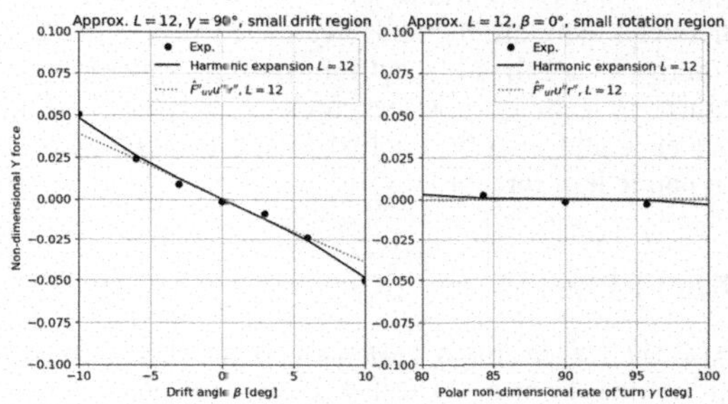

Figure 7. Approximation with hydrodynamic coefficients of the non-dimensional Y force for the KCS.

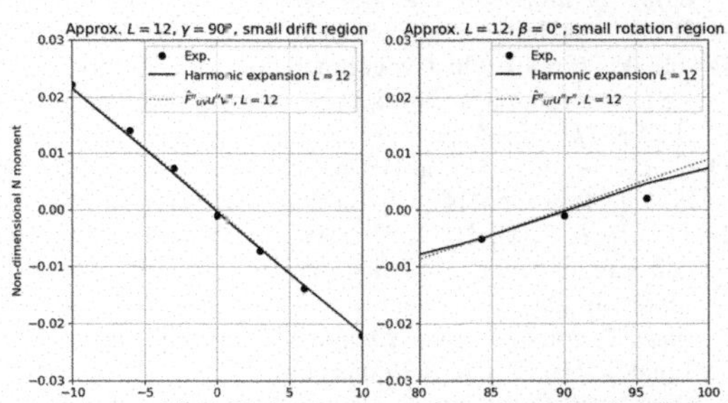

Figure 8. Approximation with hydrodynamic coefficients of the non-dimensional N moment for the KCS.

9 CONCLUSIONS

This study investigated the asymptotic approximations of a manoeuvring model for hydrodynamic forces based on spherical harmonics. By applying this approximation, the mathematical formulations becomes simpler, since it is no longer required to evaluate the complete spherical harmonic functions but only an approximated version thereof.

Two methods were presented for deriving this asymptotic approximation. The first one consists in expanding in Taylor series each term of the spherical harmonic functions. Despite not being fully orthogonal polynomials, it has been shown that the approximated real spherical harmonics can still be used for fitting the hydrodynamic forces, just like the full harmonics. The second one presents an explicit relationship between harmonic coefficients and linear hydrodynamic coefficients. In both cases, the approximated models fit well the experimental data taken from captive tests of the KCS test-case. The closer to the condition sailing straight the better the approximation, nonetheless even in a larger range the error is still acceptable. In addition,

As further possibilities, explicit formulae for the approximation of associated Legendre polynomials, which are part of the spherical harmonic functions, could be found. A general formula for finding any classic hydrodynamic coefficient could also be of interest, for freely converting between spherical harmonics and conventional polynomial models. Alternatively, there are methods available in literature for finding a localized spectrum on the sphere with tapering techniques (Wieczorek & Simons 2005), which could also be applied to the case of hydrodynamic hull forces.

REFERENCES

Abkowitz, M. A. (1964). Lectures on ship hydrodynamics - steering and manoeuvrability. Technical Report No. Hy-5, Hydro- and Aerodynamics Laboratory, Lyngby, Denmark.

Atkinson, K. & W. Han (2012). *Spherical harmonics and approximations on the unit sphere: an introduction*. Berlin and Heidelberg, Germany: Springer-Verlag.

Bosch, W. (2000). On the computation of derivatives of Legendre functions. *Physics and Chemistry of the Earth A-25(9-11)*, 655–659.

Dahlen, F. A. & J. Tromp (1998). *Theoretical Global Seismology*. Princeton, NJ, USA: Princeton University Press.

Ferrari, V. & F. H. H. A. Quadvlieg (2016). Preliminary considerations on a unified model for hydrodynamic forces. In C. Guedes Soares and T. Santos (Eds.), *Maritime Technology and Engineering 3*, pp. 251–258. London, UK: Taylor & Francis Group.

Ferrari, V., S. Sutulo, & C. Guedes Soares (2022). Spherical harmonic expansion of hydrodynamic hull forces. In C. Guedes Soares and T. Santos (Eds.), *Trends in Maritime Technology and Engineering*, pp. 337–345. London, UK: Taylor & Francis Group.

Ferrers, N. M. (1877). *An elementary treatise on spherical harmonics and subjects connected with them*. London, UK: McMillan and Co.

Fukushima, T. (2018). Fast computation of sine/cosine series coefficients of associated Legendre function of arbitrary high degree and order. *Journal of Geodetic Science 8(1)*, 162–173.

Hobson, E. W. (1955). *The theory of spherical and ellipsoidal harmonics*. New York, NY, USA: Chelsea Publishing Company.

Norrbin, N. H. (1970). Theory and observations on the use of a mathematical model for ship manoeuvring in deep and confined waters. In *Proceedings of the 8th Symposium on Naval Hydrodynamics*, Pasadena, CA, USA.

Ogawa, A. & H. Kasai (1978). On the mathematical model of manoeuvring motion of ships. *International Shipbuilding Progress 25(292)*, 306–319.

Olver, F. W. J., D. W. Lozier, R. F. Boisvert, & C. W. Clark (2010). *NIST Handbook of mathematical functions*. London, UK: Cambridge University Press.

Quadvlieg, F. H. H. A., C. Simonsen, J. Otzen, & F. Stern (2015). Review of the SIMMAN 2014 workshop on the state of the art of prediction techniques for ship manoeuvrability. In *Proceedings of the 13th International Conference on Marine Simulation and Ship Manoeuvrability (MARSIM 2015)*, Newcastle upon Tyne, UK.

Shen, L. & M. K. Chung (2006). Large-scale modeling of parametric surfaces using spherical harmonics. In *Proceedings of the 3rd International Symposium on 3D Data Processing, Visualization and Transmission (3DPVT 2006)*, Chapel Hill, NC, USA.

Slevinsky, R. M. & H. Safouhi (2009). New formulae for higher order derivatives and applications. *Journal of Computational and Applied Mathematics 233*, 405–419.

Sutulo, S. & C. Guedes Soares (2011). Mathematical models for simulation of manoeuvring performance of ships. In C. Guedes Soares and et al. (Eds.), *Marine Technology and Engineering*, pp. 661–698. London: Taylor & Francis Group.

Sutulo, S. & Y.-G. Kim (1997). Structure of mathematical model for description of hydrodynamic forces and moments acting upon the hull of a submersible in arbitrary 3d manoeuvring motion. In *Proceedings of the Annual Spring Meeting, Society of Naval Architects of Korea, 25–26 April*, Chinhae, Republic of Korea, pp. 164–167.

Sutulo, S. V. (1994). Extension of polynomial ship mathematical models to arbitrary manoeuvres. In *Proceedings of the International Shipbuilding Conference in commemoration of the centenary of the Krylov Ship Research Institute (ISC 1994)*, Saint Petersburg, Russia.

Tonelli, R. & R. Hallmann (2020). Captive manoeuvring model tests of KCS. Technical Report 30406-1-SMB, MARIN, Wageningen, the Netherlands.

Wieczorek, M. A. & F. J. Simons (2005). Localized spectral analysis on the sphere. *Geophysical Journal International 162(3)*, 655–675.

Advances in Maritime Technology and Engineering – Guedes Soares & Santos (Eds)

Dynamic evolution of ship state prediction using adaptive unscented Kalman filtering in zigzag manoeuvring tests

A. Ghassemzadeh, H.T. Xu & C. Guedes Soares
Centre for Marine Technology and Ocean Engineering (CENTEC), Instituto Superior Técnico, Universidade de Lisboa, Lisboa, Portugal

ABSTRACT: This paper explores ship state prediction during zigzag manoeuvres using the Unscented Kalman Filter. To address uncertainties, it introduces the Adaptive Unscented Kalman Filter, emphasizing covariance matching. The evaluation compares conventional Unscented Kalman Filter and Adaptive Unscented Kalman Filter, highlighting Adaptive Unscented Kalman Filter's superior accuracy in state estimation.The analysis extends to the adaptive algorithm's dynamic behavior, revealing a gradual evolution of error covariance over time.The covariance converges to a constant value after around 200 iterations, revealing a critical nuance in the system's uncertainty dynamic. Moreover, the limitations of relying solely on prior estimations based on sensor specifications are underscored. The study suggests adapting filtering algorithms for evolving uncertainty in ship dynamic, emphasizing continuous refinement for accurate state estimations in dynamic maritime environments, especially through the use of the Unscented Kalman Filter.

1 INTRODUCTION

In the domain of autonomous maritime craft, the navigation system stands as a fundamental pillar. A crucial element in evaluating the system's effectiveness lies in the vigilant monitoring of its operational condition. Concerning autonomous ships, these conditions primarily involve the vessel's location and speed (Sayyaadi & Ghassemzadeh, 2018; Xu & Guedes Soares, 2016). This scrutiny ensures a comprehensive understanding of the vessels' real-time state, allowing for informed decision-making and enhancing the safety and efficiency of autonomous operations (Xu et al., 2021a,b).

It is crucial to recognize that the reliability and accuracy of the data feeding into these parameters are a potential Achilles' heel in the control of autonomous ships (Xu & Guedes Soares, 2018). Any compromise or inaccuracy in the data sources can jeopardize the system's integrity, emphasizing the need for robust data security measures to fortify the Achilles' heel and ensure the reliable functionality of autonomous ships (Hinostroza et al., 2018). However, accurately quantifying these conditions poses a significant challenge. The unpredictable and complex nature of the maritime environment subjects all physical systems to erratic disturbances of a random nature. Consequently, determining the conditions of such systems is inherently unpredictable. To tackle this issue, essential tools such as Global Positioning System (GPS) receivers come into play, enabling the measurement and observation of conditions (Hinostroza et al., 2021). Yet, these measurements are rarely perfect, they invariably harbour noise arising from the electronic and mechanical constituents of the measuring apparatus (Jazwinski, 1970).

Navigating this conundrum constitutes what is commonly known as filtering or estimation the intricate art of deducing the accurate system state while contending with the inevitability of uncertainty and noise (Jazwinski, 1970).

Numerous suboptimal methodologies have been devised to address the intricacies of filtering and estimating navigation data. Notably, the realm includes a range of Kalman Filters (KF), a lineage pioneered by Rudolf Emil Kalman (Dove, 1977; Fossen & Perez, 2009; Fu et al., 2019; Kalman, 1960). Further enhancements to this lineage encompass the Extended Kalman Filter (EKF) (Assaf et al., 2018; Gobbo et al., 2001; Saeedmanesh et al., 2021), revered as the prevalent algorithm for navigating navigation state estimations (Chen et al., 2023). The panorama also encompasses the Unscented Kalman Filter (Julier et al., 1995; Julier & Uhlmann, 2004; Peng et al., 2007), as well as Particle Filter (Lian et al., 2019; Wang et al., 2023; Wang et al., 2021). It is worth noting that among this array of filtering methodologies, the Extended Kalman Filter asserts its dominance as the most extensively employed strategy (Xiong et al., 2006).

Fossen & Perez (2009) suggested that the forces that significantly influence ship motion control, namely wind, current, and nonlinear wave forces, fall within the ambit of low-frequency forces. The authors employed the Kalman Filter to surgically isolate high-frequency motions, a tactical manoeuvre

DOI: 10.1201/9781003508779-36

aimed at bolstering dynamic positioning and course-keeping. Subsequently, exclusively low-frequency states were harnessed as feedback signals for the controller. Their study, focused on monitoring the rudder angle, elucidated that the Kalman Filter efficaciously attenuates low-frequency motions.

Similarly, Joshi & Saha (2021) delved into the intricate interplay of dynamic wave and wind forces, which exert substantial influence over both the transverse and longitudinal motion of ships. With the paramount objective of averting untoward events such as capsizing and instability, the authors ingeniously infused noise into measurements to simulate pragmatic observations.

Expanding the purview to other domains, an abundance of research studies on state and parameter estimation can be found (Feng et al., 2023; Isermann & Münchhof, 2011; Awasthi & Raj, 2021; Tamas et al., 2008; Xu et al., 2018; Perera et al., 2015; Sutulo, and Guedes Soares, 2014, 2023; Ghassemzadeh et al., 2023). All the previously mentioned research efforts converge toward a common conclusion: achieving enhanced accuracy requires embracing nonlinear models and sophisticated techniques, prominently illustrated by the introduction of nonlinear observers such as the Extended Kalman Filter and Unscented Kalman Filter.

Nonlinear Kalman filter adaptations, explored by researchers for enhanced navigation data precision and extended applications, offer cost-effective solutions. Fossen (2022) applied the Extended Kalman Filter for a budget-friendly ship path-following controller, while (Bipin & Sunil Nag, 2023) used the Unscented Kalman Filter for a reliable, low-cost self-driving car navigation system. Challenges include understanding noise statistics and assessing algorithm performance during measurement data outages (Jwo & Cho, 2007; Matisko & Havlena, 2010; Song & Han, 2008; Bistrovs & Kluga, 2013; Zhang et al., 2015; El Sabbagh et al., 2023). Empirical estimation techniques (Mayers and Tapley, 1976) and adapting noise statistics in time-varying conditions (Deng et al., 2019) highlight the importance of these considerations for effective Kalman Filter application.

Based on previous research, there is a pressing need for adaptive filter techniques in exploring the performance of Kalman Filter variations for autonomous ship navigation. The heightened vulnerability of ships to environmental disturbances presents a unique challenge, necessitating a thorough examination of algorithm performance in maritime conditions.

Motivated by this predicament, the present research endeavours to meticulously scrutinize the performance of the Unscented Kalman filter and Adaptive Unscented Kalman filter through the prism of experimental test data. The study is underpinned by measurements derived from GPS sensors, encompassing longitude, latitude, and altitude data. Simultaneously, heading, surge, and sway accelerations will be sourced from an Inertial Measurement Unit (IMU).

The forthcoming sections of this paper will comprehensively elucidate the principal objectives undertaken, which are as follows:

- Implementation and Adaptation of Adaptive UKF: Addressing the challenge of comprehending noise statistics, this paper introduces an AUKF algorithm. The development of this AUKF is grounded in a thorough analysis of the residual and estimation errors, ensuring a robust approach to enhancing navigation system performance.
- Comparative Performance Analysis: The study will undertake an in-depth assessment of UKF, and the proposed AUKF. Employing experimental test data, the investigation will meticulously scrutinize the degradation of position error associated with each algorithm.

The subsequent sections of this paper are structured as follows: In Section 2, a dynamic model pertinent to ship navigation will be presented. Moving forward, the UKF and AUKF algorithms will be expounded upon in Section 3. Section 4 will encompass experimental testing and performance assessment, followed by a comprehensive discussion of algorithmic efficacy in Section 5. Finally, Section 6 will encapsulate the conclusion drawn from the study's outcomes.

2 DYNAMIC MODEL

The dynamic model utilized in this research builds upon a framework previously employed in (Fossen & Fossen, 2021), itself grounded in the 2D target-tracking models put forth by Li & Jilkov, (2003). Central to this model is the key assumption that the target's accelerations along the surge and sway axes are stochastic white-noise processes that are independent of each other. This premise is substantiated by the inherent nature of autonomous ships, which generally lack the capability to generate aggressive linear accelerations. As a result, the assumption of a constant linear acceleration remains applicable to this specific ship category. With this premise in place, the ship's position can be elegantly represented as follows:

$$\dot{X} = u\cos(\psi) - v\sin(\psi) \tag{1}$$

$$\dot{X} = u\cos(\psi) - v\sin(\psi) \tag{2}$$

$$\dot{u} = a_{surge} \tag{3}$$

$$\dot{v} = a_{sway} \tag{4}$$

$$\dot{a}_{surge} = w_1 \tag{5}$$

$$\dot{a}_{surge} = w_2 \tag{6}$$

$$\dot{\psi} = w_3 \tag{7}$$

where, in Equations 1-7, (X,Y) correspond to the position coordinates of the ship, while u and v denote the surge and sway velocities of the ship, respectively and ψ represents the ship's heading angle. The accelerations of the ship are denoted by a_{surge} and a_{sway}, and the terms $w_1$to w_3 encompass white Gaussian noise components. Therefore, the state of the ship, Figure 1, can be represented as:

$$x = [X,\ Y\ ]^T \tag{8}$$

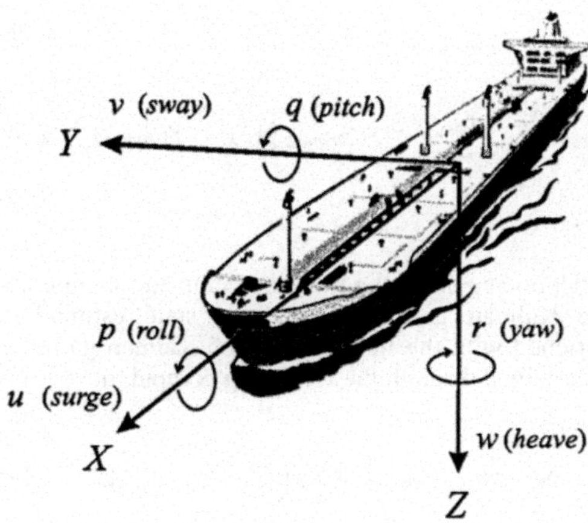

Figure 1. Ship coordinate system (Ghassemzadeh et al., 2023).

As previously highlighted, the assumption of constant accelerations and a fixed rate of change in heading holds true. It's noteworthy that these terms can be directly measured utilizing IMU, thereby attributing the white Gaussian noise in Equation 5 through Equation 7 to the inherent variability of sensor measurements.

3 THE UNSCENTED KALMAN FILTER

The primary objective of discrete-time system filtering is the estimation of $p(x_k|y_{1:k})$, the marginal distribution of the current state,x_k, based on previous measurements within the measurement sequence, $y_{1:k} = \{y_1, \ldots, y_k\}$.

The sequential Bayesian filter presents an optimal solution to the discrete-time nonlinear filtering problem by employing a cyclical process of prediction and correction steps. During the prediction step, the algorithm estimates $p(x_k|y_{1:k-1})$, representing the conditional probability distribution of the state at time instant k, considering both the measurements available at time k−1 and the transition probability$p(x_k|x_{k-1})$, determined by the dynamic model. The correction step incorporates the measurement y_k and the distribution $p(y_k|x_k)$, as specified by the measurement model, to generate $p(x_k|y_{1:k})$, the corrected conditional probability distribution of x_k.

The mathematical representation of the sequential Bayesian filter is articulated through the following equations:

$$p(x_k|y_{1:k-1}) = \int p(x_k|x_{k-1})p(x_k|y_{1:k-1})dx_{k-1} \tag{9}$$

$$p(y_k|y_{1:k-1}) = \int p(y_k|x_k)p(x_k|y_{1:k-1})dx_k \tag{10}$$

$$p(x_k|y_{1:k}) = \frac{p(y_k|x_k)p(x_k|y_{1:k-1})}{p(y_k|y_{1:k-1})} \tag{11}$$

Under specific assumptions, the solution of equations 9 to 11 is achievable. In the context of linear dynamic systems or those linearized and under the assumption of Gaussian noise, a deterministic solution for equations 9 to 11 can be derived. However, when confronted with a nonlinear dynamic system, the necessity arises to employ approximate solutions like UKF.

3.1 *Unscented Kalman Filter*

A pragmatic approach to address the inherent nonlinearity inherent in solving Equation 9 to 11 involves the utilization of the UKF. Illustrating the procedural intricacies associated with this algorithm, the system can be described by the following discrete-time equations:

$$x_{k+1} = f(x_k, u_k, t_k) + W_k \tag{12}$$

$$y_k = Hx_k + V_k \tag{13}$$

where, x, and y are states and measurements, f is the nonlinear dynamic function and H is the measurement matrices. The variables W_k and V_k represent the process and measurement noise matrices, respectively, and are assumed to follow Gaussian distributions as follows:

$$W_k \sim \mathcal{N}(0, Q) \tag{14}$$

$$V_k \sim \mathcal{N}(0, R) \tag{15}$$

Given an initial posterior estimation(x_0, P_0) the UKF proceeds as follows:

$$\hat{x}_0 = \Xi(x_0) \tag{16}$$

$$P_0 = \Xi\left[(x_0 - \hat{x}_0)(x_0 - \hat{x}_0)^T\right] \tag{17}$$

It is important to highlight that the notation "^" signifies the optimal estimation of the state. The core principle underpinning UKF is the utilization of the unscented transform. This mathematical framework plays a pivotal role in estimating the outcome resulting from the application of a nonlinear transformation to a probability distribution. This distribution is characterized exclusively by a finite set of statistics.

In the process of executing the unscented transform, the preliminary step involves the choice of a designated set of sigma points. The derivation of these sigma points and their corresponding weights is contingent upon the subsequent calculations:

$$\chi_{k-1} = \left[\hat{x}_{k-1}, \hat{x}_{k-1} \pm \sqrt{(n+\lambda)P_{k-1}}\right] \tag{18}$$

$$\Theta_0{}^m = \frac{\lambda}{n+\lambda}, \ \Theta_0{}^c = \frac{\lambda}{n+\lambda}\left(1 - \alpha^2 + \beta\right) \tag{19}$$

$$\Theta_i{}^m = \Theta_i{}^c = \frac{1}{2(n+\lambda)} \tag{20}$$

where, χ_{k-1} represents the set of sigma points computed from the previous time step, where n stands for the number of states. Additionally, λ signifies a constant value, while Θ signifies the corresponding weights associated with the sigma points. Once the sigma points and their respective weights are determined, they are advanced through the nonlinear model to predict the subsequent state and estimate the error covariance. This process can be expressed as follows:

$$\chi_{k|k-1}{}^i = f\left(\chi_{k-1}^i\right) \tag{21}$$

$$\hat{x}_{k|k-1,sampels} = \sum_{i-0}^{2n} \Theta_i \chi_{k|k-1}{}^i \tag{22}$$

$$P_{k|k-1,sampels} = \sum_{i=0}^{2n} \Theta_i\left(\chi_{k|k-1}{}^i - \hat{x}_{k|k-1,sampels}\right)\left(\chi_{k|k-1}{}^i - \hat{x}_{k|k-1,sampels}\right)^T + Q_{k-1} \tag{23}$$

Referring to the nonlinear measurement, the propagated sigma points are transformed to generate an output estimate, and the estimation output is computed as follows:

$$\hat{y}_k^i = H\chi_{k|k-1}^i \tag{24}$$

$$\hat{y}_{k,sampels} = \sum_{i=0}^{2n} \Theta_i \, \hat{y}_k{}^i \tag{25}$$

The Kalman gains, derived from the propagated sigma points, are calculated as follows:

$$K_k = P_{xy}P_y^{-1} \tag{26}$$

$$P_{xy} = \sum_{i=0}^{2n} \Theta_i\left(\chi_{k|k-1}{}^i - \hat{x}_{k|k-1,sampels}\right)\left(\tilde{y}_k{}^i - \hat{y}_{k,sampels}\right)^T \tag{27}$$

$$P_y = \sum_{i=0}^{2n} \Theta_i\left(\tilde{y}_k{}^i - \hat{y}_{k,Sampels}\right)\left(\tilde{y}_k{}^i - \hat{y}_{k,sampels}\right)^T + R \tag{28}$$

Upon acquiring the measurements and computing the Kalman gains, the updated state estimation, enriched with the influence of the Kalman gains, is ascertained through the following computation:

$$\hat{x}_k = \hat{x}_{k|k-1,sampels} + K_k\left(y_k - \hat{y}_{k,sampels}\right) \tag{29}$$

$$P_K = P_{k|k-1,sampels} - K_k P_y K_k^T \tag{30}$$

Within the framework of Equations 20 through 27, the matrices Q and R assume notable significance, exerting a direct influence on the computation of Kalman gains. This underscores the pivotal role played by these matrices in the operational dynamic of the algorithm. Following this, a comprehensive algorithm for adjusting error covariance is presented in detail.

3.2 *Adaptive Unscented Kalman filter*

A comprehensive exploration, rooted in seminal works like Myers & Tapley (1976), underscores the imperative to systematically calibrate statistical data, encompassing both the process and measurements. This adjustment becomes crucial for the precise accommodation of inherent noise and uncertainties inherent in both the data and the estimation process.

Precision in the characterization of error covariance matrices associated with individual sensor measurements stands as a crucial yet formidable undertaking. In this investigation, a methodology following the approach outlined by Deng et al. (2019) and Myers & Tapley (1976) is adopted, incorporating a dynamic process for real-time refinement of the process noise covariance, denoted as Q. This refinement is realized through the integration of an adaptive algorithm-based covariance matching estimation technique. Within this framework, real-time computation of scaling factors is undertaken to enhance both estimation accuracy and the efficacy of the filtering process. The algorithm is delineated as follows:

$$\hat{Q}_k = \eta_k \hat{Q}_{k-1} \tag{31}$$

where η_k represents an adjustment factor that modifies the process error covariance. To determine this adjustment factor, an evaluation is conducted involving two key components: the actual filter error, which signifies the disparity between the filter's estimation and the true state, and the theoretical prediction error inherent in a UKF.

Initially, compute the residual sequence for the preceding m epochs using the following procedure:

$$\tilde{r}_k^m = y_k^m - H\hat{x}_k^m \tag{32}$$

$$\hat{\tilde{r}} = \frac{1}{m}\sum_{i=k-m+1}^{k} \tilde{r}_k^i \tag{33}$$

$$\epsilon_{\tilde{y}} = \frac{1}{m}\sum_{i=k-m+1}^{k} \left(\tilde{r}_k^i - \hat{\tilde{r}}\right)\left(\tilde{r}_k^i - \hat{\tilde{r}}\right)^T \tag{34}$$

The algorithm employed to derive the empirical formula for calculating adaptive process error covariance is structured in a batch form. Notably, within this context, the parameter m signifies the width of the moving window. If the value of k is smaller than m then m is assigned the value of k. This computation revolves around the application of $\epsilon_{\tilde{y}}$, representing:

$$\check{R}_k = \epsilon_{\tilde{y}} + P_y \tag{35}$$

The real filter error, w_k, and the theoretical prediction error, U'_K, are calculated as follows:

$$W_k = \sum_{i=0}^{2n} \Theta_i(\hat{y}_k{}^i - \hat{y}_{k,sampels}) \left(\hat{y}_k{}^i - \hat{y}_{k,sampels}\right)^T \tag{36}$$

$$U'_K = \check{R}_k - (P_y + \left(\hat{y}_{k,sampels} - \hat{y}_k\right) * \left(\hat{y}_{k,sampels} - \hat{y}_k\right)^T) \tag{37}$$

Then η_k is calculated as follows:

$$d_k = \frac{tr\left(\left(\hat{y}_{k,sampels} - \tilde{y}_k\right) * \left(\hat{y}_{k,sampels} - \tilde{y}_k\right)^T\right)}{tr\left(\check{R}_k - HP_K H^T\right)} \tag{38}$$

$$\eta_k = \begin{cases} 1 & d_k \le 1 \\ \frac{|tr(U'_K)|}{tr(w_k)} & d_k > 1 \end{cases} \tag{39}$$

where, $tr(.)$ is trace of matrix.

In summation, the algorithm dynamically adapts the process noise covariance $\hat{Q}_k$ contingent upon an adaptive scaling factor. This factor is modulated by the interplay between the theoretically postulated and empirically observed innovation covariances.

4 EXPERIMENTAL DATA

To validate the performance of the proposed algorithm, the free-running ship model manoeuvring tests are used (Costa et al., 2021). These trials served as a crucial effort to gather precise and dynamic data for meticulous algorithm assessment. The model ship's detailed specifications are presented in Figure 2 and Table 1. By subjecting the model to real-world conditions, the trials aimed to validate algorithm capabilities and offer insights into their adaptability, robustness, and potential integration into genuine maritime scenarios.

Figure 2. Model equipped with multi-sensors (Costa et al., 2021).

In manoeuvring tests, a crucial examination utilized the 20-20 ZigZag manoeuvre, employing IMU and GPS sensors to capture detailed data on position, acceleration, and heading. The GPS ensured precise vessel positioning, while the IMU provided valuable insights into acceleration along surge and sway axes, as well as heading orientation. Empirical data obtained from experimental tests employing both IMU and GPS sensors serve as the foundational basis for evaluating the efficacy of the proposed algorithm. The raw data collected from these sensors are harnessed to scrutinize the algorithm's performance.

Table 1. Model Specifications (Costa et al., 2021).

Main parameters	
Length(m)	3.24
Breadth(m)	0.43
Draft(m)	0.14
Mass(kg)	108.58

5 SIMULATION

In this section, the simulation unfolds in two distinct stages. Initially, the UKF is appraised under the assumption of a constant error covariance, denoted as Q. Subsequently, the adaptive algorithm is implemented, entailing the dynamic adjustment of parameters to optimize performance. Comparative analyses of the outcomes for both algorithmic scenarios are meticulously presented, offering insights into the algorithm's adaptability and efficiency. This systematic approach provides a robust foundation for assessing the algorithm's usefulness in real-world applications, particularly in contexts where IMU and GPS data integration is critical for accurate navigation and tracking.

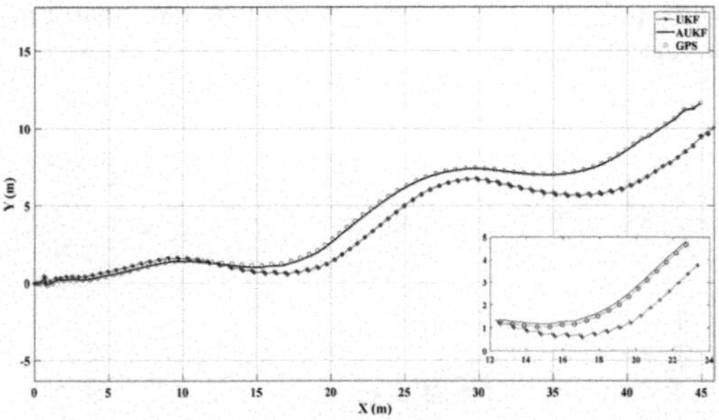

Figure 3. Analysis of Vessel Trajectory: Comparative Study of Ship Trajectory Data Refinement Utilizing UKF, and AUKF, under Homogeneous Sampling Rates for IMU and GPS during Zigzag Manoeuvres.

The analytical scrutiny of Figure 3 reveals a discernible disparity in the comparative performance between the UKF algorithm and the Adaptive UKF algorithm. Specifically, under the adaptive scenario, it becomes evident that the estimation errors are reduced. Nevertheless, a nuanced examination of the figure exposes an anomaly during course alterations at specific turning points. Within these instances, the error magnitudes surpass those observed during the remainder of the manoeuvre. This phenomenon can be elucidated by considering a primary contributing factor. The error escalation coincides with alterations in course attributed to zigzag manoeuvres.

During these manoeuvres, the ship's acceleration undergoes fluctuations. However, the algorithm assumes a constant acceleration during such manoeuvres, leading to a perceptible error amplification. This nuanced assessment sheds light on the intricacies of the algorithmic response to dynamic navigational scenarios, thereby facilitating a more comprehensive understanding of its limitations and strengths in real-world applications.

Examination of Figure 4 shows that the error dynamic associated with tracking the designated trajectory remains within acceptable bounds for both the X and Y axes. However, a discernible distinction emerges when scrutinizing the dynamic errors in both directions, X and Y. Notably, the error magnitudes for the AUKF algorithm converge more closely to zero.

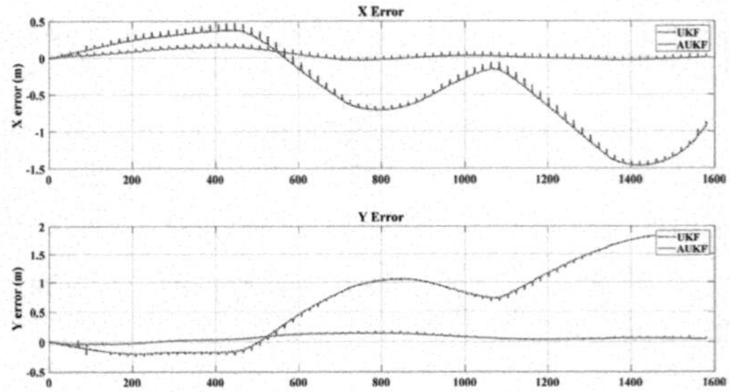

Figure 4. Error dynamics of X and Y Tracking: A comparative illustration of error evolution over time for both the UKF and the AUKF estimation algorithms.

Consistent with prior observations, a detailed examination of the error dynamics reveals that, during the ship's course adjustments, error magnitudes reach maximum points at approximately (0.4749, 1.8492) for UKF and (0.1939, 0.1715) for AUKF.

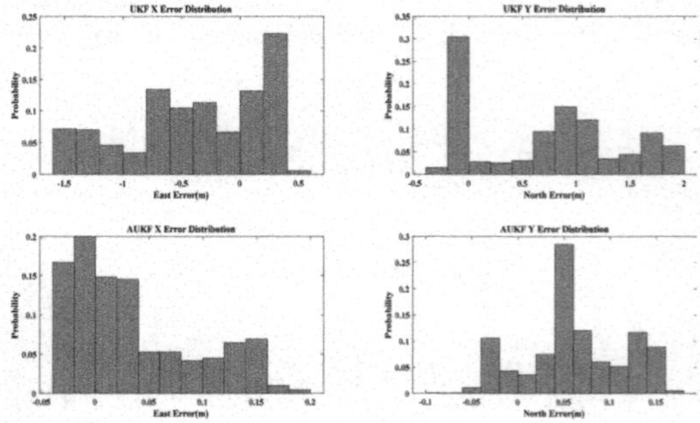

Figure 5. Error Distribution Histograms for X and Y Directions: Comparative Analysis of Tracking Accuracy employing the UKF and the AUKF Estimation Algorithms.

Even within these critical points, the advantages of AUKF over UKF are evident, underscoring the superior performance of the adaptive algorithm in mitigating dynamic errors and maintaining trajectory precision. This superiority of AUKF over UKF is further evident when analysing the distribution of error dynamics in both X and Y directions, as illustrated in Figures 5 and 6. According to the data, it is evident that the error is more centred around zero for AUKF compared to UKF, reinforcing the robustness and efficiency of the adaptive algorithm in minimizing tracking errors.

The comparative box plot, Figure 6, reveals that the error distribution for AUKF exhibits smaller interquartile ranges (IQRs) and narrower boxes compared to UKF. This suggests that AUKF demonstrates reduced variability and a more concentrated error distribution, indicating potentially improved performance in error reduction when compared to UKF.

As explicated in Section 3, the algorithmic adaptation process hinges upon a singular adjusting factor. To elucidate the efficacy of the adaptation algorithm, the temporal evolution of this adjusting factor across all iterations is graphically portrayed in Figure 7.

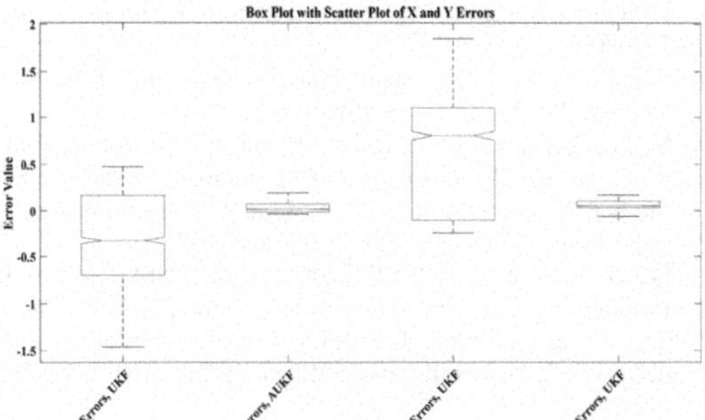

Figure 6. Comparative box plot illustrating the distribution of errors in X and Y.

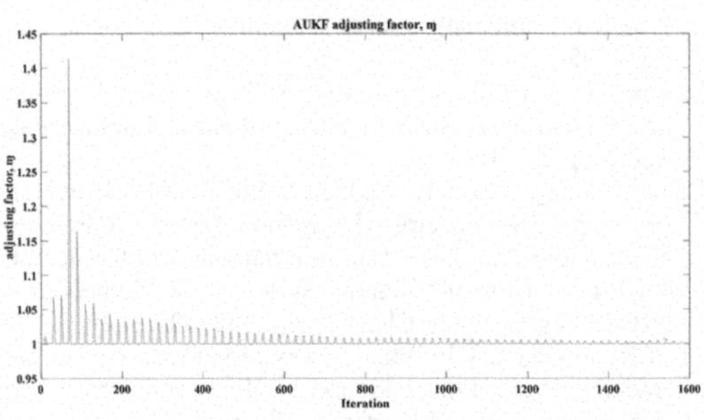

Figure 7. Dynamic Evolution of the Adaptive Algorithm's Adjusting Parameter, η_k.

Notably, the figure underscores that, at the initiation of manoeuvres, the adjusting factor attains its zenith, reaching a maximum value of 1.4142. Subsequently, as time progresses, this parameter predominantly remains proximate to unity. The intrinsic nature of this parameter, contingent upon residual and estimation errors, becomes particularly conspicuous during these course-changing intervals. In the intricate dance of dynamic systems, the adjusting factor not only responds to heightened errors but also exhibits a dynamic interplay with external environmental factors, subtly adapting its influence on the estimation process. Consequently, as the algorithm navigates through evolving scenarios, the adjusting factor's ability to augment the error covariance proves pivotal in ensuring not only the mitigation of estimation errors but also the optimization of overall algorithmic performance over extended durations.

6 CONCLUSIONS

The development of the adaptive Unscented Kalman Filter marks a significant milestone in addressing the challenges posed by uncertainty in error covariance and sensitivity of estimation algorithms. Through the integration of the actual filter error and theoretical prediction error, the AUKF has demonstrated substantial advancements, particularly evident in its enhanced error dynamic during rigorous testing, and zigzag manoeuvring.

Comparative analyses highlight the AUKF's ability to concentrate errors around zero, underscoring its superiority over the conventional UKF with constant error covariance. This concentrated error distribution attests to the AUKF's capacity to provide more accurate and reliable estimations.

Furthermore, the research findings underscore the crucial role of adaptive techniques in dealing with variable environmental conditions, emphasizing the AUKF's dynamic adjustment of error covariance in response to changing manoeuvring conditions. Its adaptability is particularly crucial in real-world applications where uncertainties abound, making it a robust solution. The dual emphasis on adaptability and stability, manifested in the AUKF's boundedness of estimation errors, contributes valuable insights to the field of estimation algorithms.

This research reinforces the significance of these attributes in ensuring reliable and precise information in diverse operational scenarios, marking the AUKF as a promising solution for real-world applications requiring stable and adaptable estimation algorithms.

ACKNOWLEDGEMENTS

The first author has been funded by the Portuguese Foundation for Science and Technology (Fundação para a Ciência e a Tecnologia-FCT), through a doctoral fellowship under Contract no. 2023.03496. BD. This work contributes to the Strategic Research

Plan of the Centre for Marine Technology and Ocean Engineering, which is financed by the Portuguese Foundation for Science and Technology (Fundação para a Ciência e Tecnologia - FCT) under contract UIDB/UIDP/00134/2020.

REFERENCES

Assaf, M. H., Petriu, E. M., & Groza, V. 2018. Ship track estimation using GPS data and Kalman Filter. 2018 IEEE International Instrumentation and Measurement Technology Conference (I2MTC), Houston, TX, USA, pp. 1–6.

Awasthi, V., & Raj, K. 2021. A Survey of Kalman Filter Algorithms and Variants in State Estimation. Current Approaches in Science and Technology Research Vol. 15, pp. 1–14.

Bipin, K., & Sunil Nag, P. V. 2023. UKF/H-Infinity Filter for Low-Cost Localization in Self-driving Cars. In: Chokkadi, S., Bandyopadhyay, R. (eds) Smart Sensors Measurement and Instrumentation. Lecture Notes in Electrical Engineering, vol 957. Springer, Singapore, pp. 379–392.

Bistrovs, V., & Kluga, A. 2013. The Analysis of the UKF-Based Navigation Algorithm during GPS Outage. Electronics and Electrical Engineering, 19(10), pp. 13–16.

Chen, S., Xiong, X., Wen, Y., Jian, J., & Huang, Y. 2023. State Compensation for Maritime Autonomous Surface Ships' Remote Control. Journal of Marine Science and Engineering, 11(2), 450.

Costa, A. C., Xu, H., & Guedes Soares, C. 2021. Robust Parameter Estimation of an Empirical Manoeuvring Model Using Free-Running Model Tests. Journal of Marine Science and Engineering, 9(11), 1302.

Deng, F., Yang, H.L., & Wang, L.J. 2019. Adaptive Unscented Kalman Filter Based Estimation and Filtering for Dynamic Positioning with Model Uncertainties. International Journal of Control, Automation and Systems, 17(3), pp. 667–678.

Dove, M.J. 1977. Kalman Filter Techniques in Marine Integrated Navigation Systems. The Journal of Navigation, 30(1), pp. 135–145.

El Sabbagh, M.S., Maher, A., Abozied, M.A.H., & Kamel, A.M. 2023. Promoting navigation system efficiency during GPS outage via cascaded neural networks: A novel AI based approach. Mechatronics, 94, 103026.

Feng, S., Li, X., Zhang, S., Jian, Z., Duan, H., & Wang, Z. 2023. A review: state estimation based on hybrid models of Kalman filter and neural network. Systems Science & Control Engineering, 11(1), 2173682.

Fossen, S., & Fossen, T.I. 2021. Five-State Extended Kalman Filter for Estimation of Speed over Ground (SOG), Course over Ground (COG) and Course Rate of Unmanned Surface Vehicles (USVs): Experimental Results. Sensors, 21(23):7910.

Fossen, T.I. 2022. Line-of-sight path-following control utilizing an extended Kalman filter for estimation of speed and course over ground from GNSS positions. Journal of Marine Science and Technology, 27(1), pp. 806–813.

Fossen, T.I., & Perez, T. 2009. Kalman filtering for positioning and heading control of ships and offshorerigs. IEEE Control Systems Magazine, 29(6), pp. 32–46.

Fu, B., Liu, J., & Wang, Q. 2019. Multi-Sensor Integrated Navigation System for Ships Based on Adaptive Kalman Filter. 2019 IEEE International Conference on Mechatronics and Automation (ICMA), Tianjin, China, pp. 186–191.

Ghassemzadeh, A., Xu, H., & Guedes Soares, C. 2023. Path following control using robust sliding mode for an autonomous surface vessel based on ε- support vector regression identified steering model. Ocean Engineering, 288, 116085.

Gobbo, D.D, Napolitano, M., Famouri, P., & Innocenti, M. 2001. Experimental application of extended Kalman filtering for sensor validation. IEEE Transactions on Control Systems Technology, 9(2), pp. 376–380.

Hinostroza, M. A., Xu, H. T., and Guedes Soares, C., 2018, Path-planning and path-following control system for autonomous surface vessel, in Maritime Transportation and Harvesting of Sea Resources, Guedes Soares, C. and A. P. Teixeira, Eds. London, UK: Taylor & Francis Group, pp. 991–998.

Hinostroza M. A., Xu, H. T., and Guedes Soares C., 2021. Experimental results of the cooperative operation of autonomous surface vehicles navigating in complex marine environment, Ocean Engineering, 219, 108256. DOI: 10.1016/j.oceaneng.2020.108256.

Isermann, R., & Münchhof, M. 2011. State and Parameter Estimation by Kalman Filtering. State and Parameter Estimation by Kalman Filtering. In: Identification of Dynamic Systems. Springer, Berlin, Heidelberg, pp. 539–551.

Jazwinski, A.H., 1970. Stochastic Processes and Filtering Theory, Academic Press, Pittsburgh.

Joshi, K., & Saha, N. 2021. Estimation of Ship Heave and Pitch Under Wave Loads Using Kalman Filtering. In: Sundar, V., Sannasiraj, S.A., Sriram, V., Nowbuth, M.D. (eds) Proceedings of the Fifth International Conference in Ocean Engineering (ICOE2019). Lecture Notes in Civil Engineering, vol 106, Springer, Singapore, pp. 371–384.

Julier, S.J., & Uhlmann, J.K. 2004. Unscented filtering and nonlinear estimation. Proceedings of the IEEE, 92(3), pp. 401–422.

Julier, S.J., Uhlmann, J.K., & Durrant-Whyte, H.F. 1995. A new approach for filtering nonlinear systems. Proceedings of 1995 American Control Conference - ACC'95, Seattle, WA, USA, vol. 3, pp. 1628–1632.

Jwo, D.J., & Cho, T.S. 2007. A practical note on evaluating Kalman filter performance optimality and degradation. Applied Mathematics and Computation, 193(2), pp. 482–505.

Kalman, R.E. 1960. A New Approach to Linear Filtering and Prediction Problems. Journal of Basic Engineering, 82(1), pp. 35–45.

Lian, Y., Yang, L., Lu, L., Sun, J., & Lu, Y. 2019. Research on Ship AIS Trajectory Estimation Based on Particle Filter Algorithm. 2019 11th International Conference on Intelligent Human-Machine Systems and Cybernetics (IHMSC) Hangzhou, China, 1, pp. 305–308.

Li, X.R., & Jilkov, V.P. 2003. Survey of maneuvering target tracking. Part I. Dynamic models. IEEE Transactions on Aerospace and Electronic Systems, 39(4), pp. 1333–1364.

Matisko, P., & Havlena, V. 2010. Noise covariances estimation for Kalman filter tuning. IFAC Proceedings Volumes, 43(10), pp. 31–36.

Myers, K., & Tapley, B. 1976. Adaptive sequential estimation with unknown noise statistics. IEEE Transactions on Automatic Control, 21(4), pp. 520–523.

Peng, Y., Han, J., & Song, Q. 2007. Tracking Control of Underactuated Surface Ships: Using Unscented Kalman Filter to Estimate the Uncertain Parameters. 2007 International Conference on Mechatronics and Automation, pp. 1884–1889.

Perera, L. P., Oliveira, P., and Guedes Soares, C. 2015. System Identification of Nonlinear Vessel Steering. Journal of Offshore Mechanics and Arctic Engineering, 137(3):031302

Saeedmanesh, M., Kouvelas, A., & Geroliminis, N. 2021. An extended Kalman filter approach for real-time state estimation in multi-region MFD urban networks. Transportation Research Part C: Emerging Technologies, 132, 103384.

Sayyaadi, H., & Ghassemzadeh, A. 2018. Control of Multiple Underwater Vessels to Converge to a Desired Pattern. International Journal of Maritime Technology, 9 (1), pp. 51–57.

Song, Q., & Han, J.D. 2008. An Adaptive UKF Algorithm for the State and Parameter Estimations of a Mobile Robot. Acta Automatica Sinica, 34(1), pp. 72–79.

Sutulo, S. and Guedes Soares, C. 2014. An Algorithm for Offline Identification of Ship Manoeuvring Mathematical Models after Free-Running Tests. Ocean Engineering. 79, pp.10–25.

Sutulo, S. and Guedes Soares, C. 2023. Application of an offline identification algorithm for adjusting parameters of a modular manoeuvring mathematical model. Ocean Engineering. 279:114328.

Tamas, L., Lazea, G., Robotin, R., Marcu, C., Herle, S., & Szekely, Z. 2008. State estimation based on Kalman filtering techniques in navigation. 2008 IEEE International Conference on Automation, Quality and Testing, Robotics, 2, pp. 147–152.

Wang, X., Wang, H., Wang, H., & Liang, H. 2023. Control of Ship' Course in Waves Based on Particle Filter Algorithm. In R. Kountchev, K. Nakamatsu, W. Wang, & R. Kountcheva (Eds.), Proceedings of the World Conference on Intelligent and 3-D Technologies (WCI3DT 2022). Smart Innovation, Systems and Technologies, vol 323. Springer, Singapore, pp. 509–517.

Wang, Y., Perera, L.P., & Bjørn-morten Batalden. 2021. Particle Filter Based Ship State and Parameter Estimation for Vessel Maneuvers. The 31st International Ocean and Polar Engineering Conference, Rhodes, Greece, ISOPE-I-21-4176.

Xiong, K., Zhang, H.Y., & Chan, C.W. 2006. Performance evaluation of UKF-based nonlinear filtering. Automatica, 42(2), pp. 261–270.

Xu, H. T. and Guedes Soares, C., 2016. Waypoint-following for a marine surface ship model based on vector field guidance law, in Maritime Technology and Engineering 3, vol. 1, Guedes Soares, C. and T. A. Santos, Eds. London, UK: Taylor & Francis Group, pp. 409–418.

Xu, H. T., and Guedes Soares C., 2018. An optimized energy-efficient path following algorithm for underactuated marine surface ship model, International Journal of Maritime Engineering, 160(A4), pp. A413–A423. doi: 10.3940/rina.ijme.2018.a4.505.

Xu, H, Hassani, V, Hinostroza, MA, & Guedes Soares, C. 2018. Real-Time Parameter Estimation of Nonlinear Vessel Steering Model Using Support Vector Machine. Proceedings of the ASME 2018 37th International Conference on Ocean, Offshore and Arctic Engineering. Vol. 11B: Honoring Symposium for Professor Carlos Guedes Soares on Marine Technology and Ocean Engineering. Madrid, Spain. June 17–22, 2018. V11BT12A009. ASME.

Xu, H. T., Oliveira P. and Guedes Soares C., 2021a. L1 adaptive backstepping control for path-following of underactuated marine surface ships, European Journal of Control. 58, pp. 357–372.

Xu, H. T., Hinostroza M. A., and Guedes Soares C., 2021b. Modified vector field path-following control system for an underactuated autonomous surface ship model in the presence of static obstacles, Journal of Marine Science and Engineering. 9(6), 652.

Zhang, B., Chu, H., Sun, T., Jia, H., Guo, L., & Zhang, Y. 2015. Error Prediction for SINS/GPS after GPS Outage Based on Hybrid KF-UKF. Mathematical Problems in Engineering, 2015, pp. 1–9.

Advances in Maritime Technology and Engineering – Guedes Soares & Santos (Eds)

Assessment of the twin-ASD vessel maneuverability using a semi-empirical method

Y.Y. He & L.Y. Chen
State Key Laboratory of Maritime Technology and Safety, Wuhan University of Technology, Wuhan, P.R. China
School of Navigation, Wuhan University of Technology, Wuhan, P.R. China

Q.S. Zeng
State Key Laboratory of Maritime Technology and Safety, Wuhan University of Technology, Wuhan, P.R. China
Green & Smart River-Sea-going Ship, Cruise, and Yacht Research Center, Wuhan University of Technology, Wuhan, P.R. China

S. Zhang
State Key Laboratory of Maritime Technology and Safety, Wuhan University of Technology, Wuhan, P.R. China
School of Transportation and Logistics Engineering, Wuhan University of Technology, Wuhan, P.R. China

ABSTRACT: To establish the 3-DOF MMG maneuvering model of the twin-ASD vessel "WillLead I", a modified semi-empirical method is introduced. Firstly, towing tank tests and propeller open-water tests are implemented to validate the numerical method. Then, the Oblique Towing Tests (OTT) and circular motion tests (CMT) are carried out to obtain the hydrodynamic coefficients of the bare hull. Moreover, the thrust identification method calculates the self-propulsion factors in straight-moving conditions, and the wake coefficient in maneuvering motions is estimated with an empirical formula. Standard turning and Zigzag maneuvers are performed and compared with free running model tests data to verify the feasibility of the adopted method.

1 INTRODUCTION

The better flexibility makes the azimuth stern drive (ASD) ships a hot spot in the studies of autonomous unmanned surface vessels, among which the precise maneuvering model is the solid foundation. There are three main approaches to ship maneuvering prediction: data-based or empirical formula, ship model tests, and numerical simulations based on CFD (Ren et al., 2022). As a powerful and efficient tool, the CFD technique has gained growing popularity.

Maciej (2017) and Wu. et al. (2023) conduct a set of PMM (planar motion mechanism) and open water tests to obtain hydrodynamic derivatives and azimuth thruster hydrodynamic coefficients of the 3-DOF MMG model. The comparison of experimental and numerical results demonstrates that the MMG approach could describe the motion of the twin-ASD ship. Meanwhile, CFD is a time- and cost-saving approach in maneuvering predictions. Piaggio. et al. (2020; 2022) implement pure drift, pure yaw, drift + yaw, and drift + heel and establish the 4-DOF MMG maneuvering model of the escort tug. The computed results and maneuvering indexes agree well with experimental data, which illustrates that the RANS has the potential to build a mathematical maneuvering model. However, in these investigations, the pod effects are neglected.

Zhang. et al (2023) utilize Stat CCM+ software to carry out virtual oblique towing tank tests, circular motion tests, open water tests, and self-propulsion tests and establish the 3-DOF MMG maneuvering model of the twin-screw ASD tug "QiuXin V". Comparisons are made with the free-running tests, the results indicate that the pod could produce additional steering force, and this force should be taken into account under low-speed conditions.

In the present work, the CFD method is utilized, and the pod effects are introduced to build a modified 3-DOF MMG maneuvering equation to express the motion of the twin-ASD vessel.

2 METHODOLOGY

2.1 *Studied ship and working conditions*

The research object "WillLead I" is a self-designed autonomous surface vessel equipped with twin ducted

DOI: 10.1201/9781003508779-37

azimuth propulsion units at the stern. The geometry of the physical model and CAD model are presented in Figure 1. The principal particulars of the ship hull and thrusters are listed in Table 1 (He et al., 2022).

Aim to obtain the hydrodynamic derivatives of the 3-DOF MMG maneuvering model, the computational conditions are listed in Table 2. The oblique towing tests (OTT) are carried out for drift angles β of ±20°, ±16°, ±12°, ±8°, ±4°, and 0° to obtain the sway-velocity-dependent hydrodynamic derivatives. In the circular motion tests (CMT), the r' varies from 0.2 to 0.8 at 0.2 intervals, and the β varies from -20° to 20° with an interval of 4° to get the yaw-rate-dependent and yaw-drift cross-coupled hydrodynamic derivatives.

Figure 1. Geometry view of "WillLead I" (left: physical model; right: CAD model).

Table 1. Principal particulars of the "WillLead I".

Parameters	Symbol	Value
Ship length	L_{oa} (m)	2.000
Length between perpendiculars	L_{wl} (m)	1.830
Breadth	B (m)	0.674
Draft	d (m)	0.211
Displacement	∇(kg)	140.4
Block coefficient	C_b(-)	0.542
Wetted surface area	A_S(m^2)	1.576
Number of propeller blades	Z(-)	4
Propeller diameter	D_p(mm)	160.0
Pitch ratio	P(-)	1.0
Hub ratio	(-)	0.2
Span ratio	A_E/A_O(-)	0.7
Pitch at 0.7 cord length	$0.7c_R$(-)	0.0875
Pod surface area	A_p(m^2)	0.0207
Pod aspect ratio	Λ(-)	1.2

The non-dimensional rotating rate r', which is defined by the following equation:

$$r' = \frac{L_{wl}r}{U} = \frac{L_{wl}}{R}, \tag{1}$$

where, L_{wl} is the length of the waterline, r is the rotating angular velocity, and R is the turning radius.

The open water tests are performed to get the open water performance curve and determine the propeller rotation speed. Then, the self-propulsion tests are carried out to calculate the self-propulsion factors in straight-moving conditions.

Table 2. The computational matrix.

	r'	β(°)	J^*
OTT	/	-20 ~ 20	/
CMT	0.2,-0.8	-20 ~ 20	/
Open water tests	/	/	0 ~ 0.8
Self-propulsion	/	/	0

* J is the advance coefficient, which is defined as Equation 2

$$J = \frac{V_A}{nD_p}, \tag{2}$$

where, V_A is the advance speed in open water; n is the rotational speed.

2.2 *Numerical setups*

RANS code Star-CCM+ enclosed by Shear Stress Transport (SST) k-ω turbulence model is adopted to resolve the flow characteristics. Due to the low design speed (U=0.424m/s, Fr=0.0479), the double body method that eliminates the free-surface effect is used.

As depicted in Figure 2a, a cuboid computational domain is adopted for the ship hull. The dimension of the domain is $1.0L_{oa}$ from the bow to the inlet boundary, 2.0 L_{oa} from the stern to the outlet boundary, 1.0 L_{oa} from the center-line to the side boundary, and 1.0 L_{oa} from the keel to the bottom boundary. The inlet, side, and bottom of the domain are set as velocity inlet, the outlet is treated as pressure outlet, the ship hull is defined as a no-slip wall, and the top plane is replaced by a plane of symmetry in the double body calculations.

For the thruster, as shown in Figure 2b, a cylindrical computational domain is chosen. The diameter of the domain is $20D_p$, the inlet boundary is $10D_p$ ahead of the propeller, and the distance between the propeller and the outlet boundary is $20D_p$. Given the fixed flow velocity, the inlet is set as a velocity inlet, the outlet is treated as a pressure outlet, the far field is defined as symmetric, and the no-slip wall boundary condition is applied on the propeller.

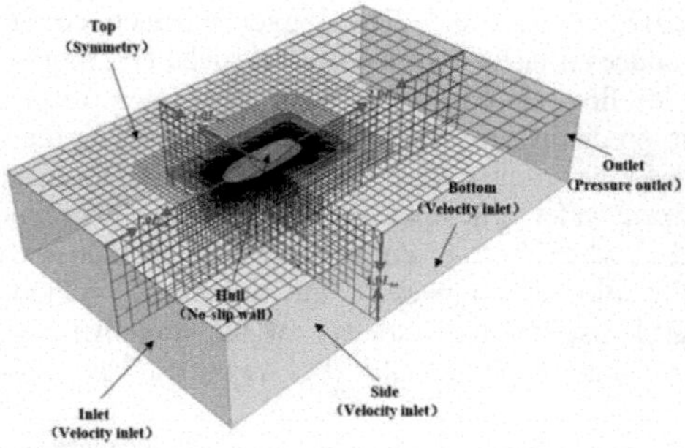

(a) Ship hull;

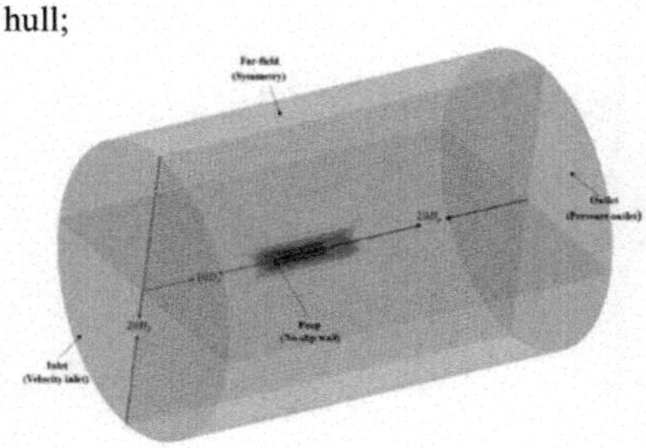

(b) Thruster;

Figure 2. Calculation domain and boundary conditions.

The grids are discretized by the automatic meshing facility in Star-CCM+, and the hexahedral mesh is selected.

The mesh details of the ship hull are shown in Figure 3a. The basic mesh size is set as 0.04m (L_{oa}/50). The prism layer mesher is adopted to capture the boundary layers, and 6 cell layers with a growth rate of 1.2 are utilized to create orthogonal prismatic cells normal to the walls. The averaged $y^+$is around 5. All y^+wall treatment is used to treat the flow near the ship hull, and a surface remesher tool is employed to enhance the overall quality of the surface. Besides, the local refinement boxes covering the hull and the skeg are implemented to further increase the calculation accuracy.

The generated grids of the thruster are given in Figure 3b. The sliding mesh method is applied in propeller modeling, and the computational domain divides two sub-regions: static and rotating. The static domain contains the duct and pod, and the rotating domain contains the blades and hub. Set the contacting surface of two sub-domains as the interface to enable the information exchange (Gong et al., 2020; 2018). The cylindrical volume refinement (size = 1.25%D_p) is implemented in the rotating region. Moreover, the denser meshes are applied to the tight clearances between the propeller blades and duct (size = 0.3125%D_p), and the downstream (size = 1.25%D_p). The averaged $y^+$of propeller is about 2.

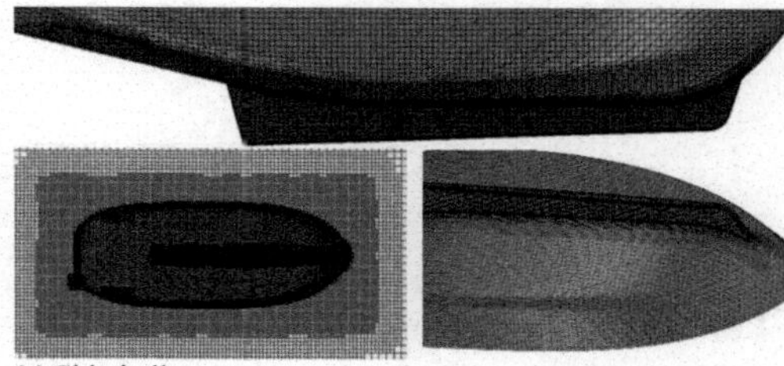

(a) Ship hull;

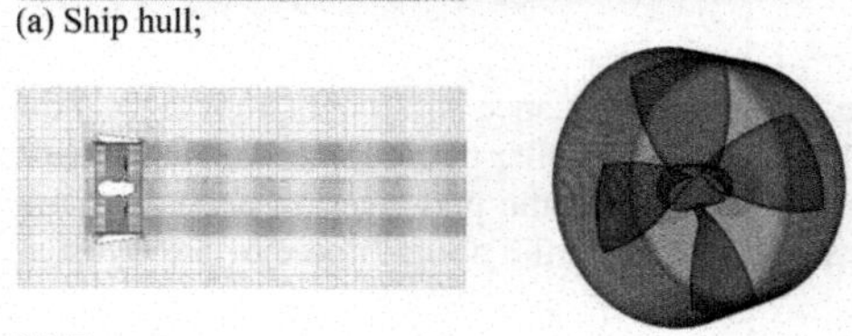

(b) Thruster;

Figure 3. Mesh details.

2.3 *MMG model*

The 3-DOF MMG ship model in still water can be expressed as (Yasukawa and Yoshimura, 2015):

$$\begin{cases} X = X_I + X_H + X_P + X_R \\ Y = Y_I + Y_H + Y_P + Y_R \\ N = N_I + N_H + N_P + N_R \end{cases}, \tag{3}$$

where, the subscript *I*, *H*, *P*, *N* mean inertial item hull, propeller and rudder, respectively.

For the ASD vessel, the pod in front of the thruster acts as a rudder and induces additional steering force (Zhang et al., 2023). The modified 3-DOF MMG model is written as Equation 4. According to the potential flow theory, the inertial items are calculated by Equation 5.

$$\begin{cases} X = X_I + X_H + X_P + X_{Pod} \\ Y = Y_I + Y_H + Y_P + Y_{Pod} \\ N = N_I + N_H + N_P + N_{Pod} \end{cases}. \tag{4}$$

$$\begin{cases} X_I = -(m_x\dot{u} - m_y vr) \\ Y_I = -(m_y\dot{v} + m_x ur) \\ N_I = -J_{zz}\dot{r} \end{cases}, \tag{5}$$

where, m_x, m_y are the added masses along longitudinal and transverse directions; J_{zz} is the added moment of inertia.

The viscous items of hull, propeller and pod could be expressed as Equations 6, 7, and 8, respectively.

$$\begin{cases} X'_H = -R_0 + X_{vv}v^2 + X_{vr}vr + X_{rr}r^2 + X_{vvvv}v^4 \\ Y'_H = Y_v v + Y_r r + Y_{vvv}v^3 + Y_{vvr}v^2 r + Y_{vrr}vr^2 + Y_{rrr}r^3 \\ N'_H = N_v v + N_r r + N_{vvv}v^3 + N_{vvr}v^2 r + N_{vrr}vr^2 + N_{rrr}r^3 \end{cases}, \tag{6}$$

$$\begin{cases} T_p = \rho n_{p\{p\}}^2 D_p^4 K_{T\{p\}} \\ T_s = \rho n_{p\{s\}}^2 D_p^4 K_{T\{s\}} \end{cases}, \quad (7)$$

$$\begin{cases} X_{Pod} = (1 - t_R)\left(F_{N\{p\}} \sin \delta_p + F_{N\{s\}} \sin \delta_s\right) \\ Y_{Pod} = (1 + \alpha_H)\left(F_{N\{p\}} \cos \delta_p + F_{N\{s\}} \cos \delta_s\right) \\ N_{Pod} = (x_P + \alpha_H x_H)\left(F_{N\{p\}} \cos \delta_p + F_{N\{s\}} \cos \delta_s\right) \end{cases}, \quad (8)$$

where, R_0 is the resistance in straight moving conditions; $T_{\{p,s\}}$ is the propeller thrust force; K_T is the thrust coefficient; t_R, α_H are the pod steering deduction and increase factors; F_N is the normal force of the thruster.

3 RESULTS AND DISCUSSION

3.1 *Validation*

A series of straight-moving captive model tests under different velocities are conducted in the 132m long, 0.8m wide, and 2m deep towing tank at Wuhan University of Technology in China. To eliminate the influence of the scale effect, the longitudinal force X is dimensionless transformed into:

$$X' = \frac{X}{0.5\rho U^2 L_{PP} T}, \quad (9)$$

where, ρ is the fluid density at 17.8°C; U is the ship speed.

Figure 4 shows the comparison of non-dimension longitudinal force X' between the experimental results and the calculation results with different velocities, the blue regions are the 5% error bands. When the $Fr = 0.0957$ and 0.1118, the error is 5.40% and 5.22%, respectively, slightly exceeding the 5% threshold. Overall, the calculated results are in good agreement with the test data in most conditions, which indicates that the numerical method adopted in this paper could be used to compute the bare hull resistance accurately.

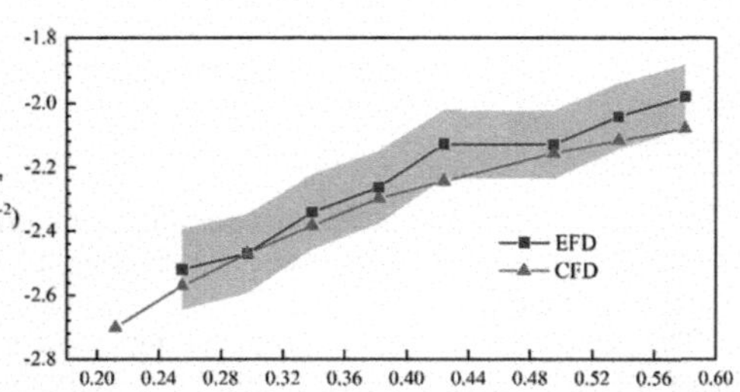

Figure 4. Comparison of X' with different velocities (He et al., 2023).

Figure 5 compares thrust K_T, torque K_Q, and open water efficiency η_0 when J is 0 ~ 0.8. The computed results agree with the test data of MARIN (Carlton, 2018) at $J < 0.8$. The bigger advance coefficient induces a faster ship navigation speed and more complex flow disturbances, which cause more difficulties in predicting propeller performance by using the CFD strategy. Hence, the calculated thrust, torque, and open water efficiency are smaller than experimental data when $J = 0.8$. For most cases, the error is within the allowable range, so the numerical method and grid discretization strategy of the propeller adopted in this paper could be regarded as reliable.

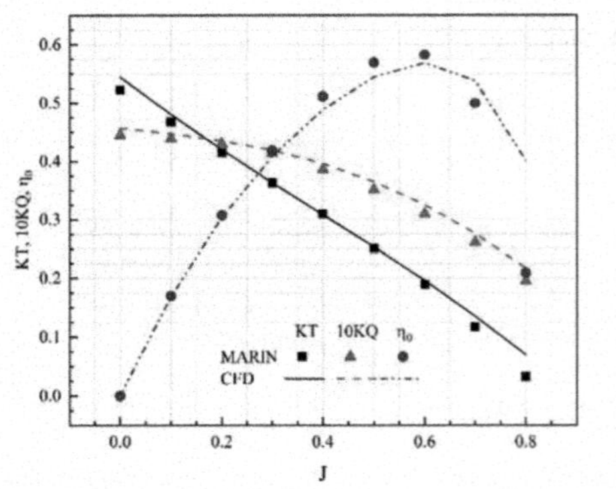

Figure 5. Comparison of the open water curve.

3.2 *Hydrodynamic forces and moment*

Figure 6 shows the computed longitudinal force X, lateral force Y, and yaw moment N, all the results are defined as dimensionless quantities ($Y' = Y/0.5\rho U^2 L_{pp} T, N' = N/0.5\rho U^2 L^2{}_{PP} T$).

During the OTT tests ($r' = 0$), the longitudinal velocity u, the lateral velocity v, and the yaw rate r are expressed as Equation 10, and the X, Y, N acting on the ship are defined as Equation 11. Due to sin 5°≈0.0872, when the drift angle is small ($|\beta| < 5°$), the lateral velocity v is small enough to be ignored. Therefore, the longitudinal force shows a steady trend, while the lateral force and yaw moment are close to zero at the small drift angle conditions. Overall, with the increase of drift angle, $|X'|$ decreases, while $|Y'|$ and $|N'|$ rise.

$$\begin{cases} u = U \cos \beta \\ v = U \sin \beta \\ r = 0 rad/s \end{cases}. \quad (10)$$

$$\begin{cases} X_H \approx X_0 + X_{vv} v^2 \\ Y_H \approx Y_v v + Y_{vvv} v^3 \\ N_H \approx N_v v + N_{vvv} v^3 \end{cases}. \quad (11)$$

During the CMT tests ($r' \neq 0$), the changes of forces and moment are monotone, X', Y', and N' are growing with the increases of drift angle. The yaw rate has different effects on the longitudinal force, lateral force, and yaw moment. When $\beta > 0$, X', Y', and N' decrease as the r' increases; when $\beta < 0$, with the rises of r', the X' increases, while the Y' and N' drop.

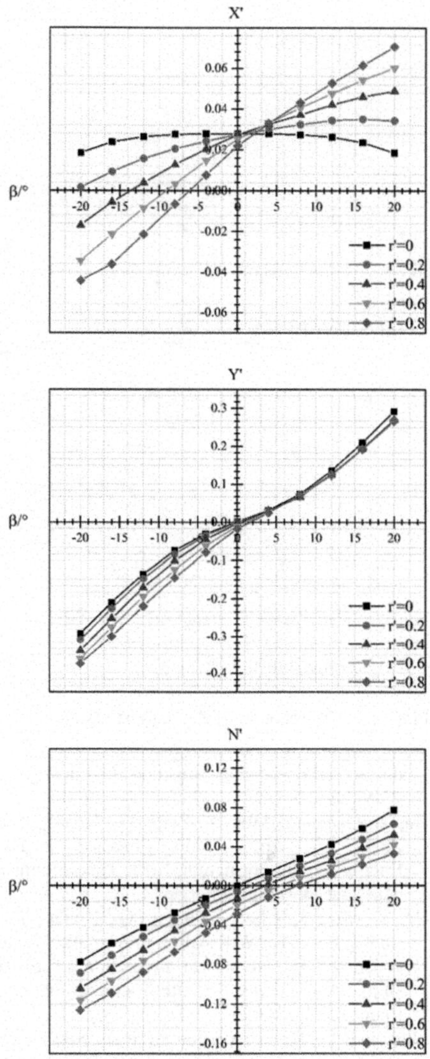

(a) X';

(b) Y';

(c) N';

Figure 6. Numerical results of OTT and CMT.

3.3 *Turning circle and Zigzag maneuvers*

The simulations of starboard turning maneuvering motion at 5°, 10°, and 15°, and the Zigzag maneuvering motion at 10°/ 10°, 20°/ 20° are performed and compared with free running model tests data to verify the feasibility of the adopted method.

The comparison of trajectories and maneuvering indexes in turning motions are shown in Figure 7 and Table 3. The comparisons of Zigzag test results are presented in Figure 8 and Table 4. Generally speaking, the computed advance of turning circle $A_D^{'}$, the tactical diameter $T_D^{'}$ and the steady turning radius $R_S^{'}$ agree with those of experimental measurement. However, there is a relatively large estimation error for the transfer of the turning circle $T_R^{'}$, and as the turning angle increases, the error is growing. In addition, the large discrepancies could also be seen in the first and second overshoot angles. The reasons which lead to the errors may include:

1) The working conditions mainly focus on small drift angles. With the increase of rudder angle, the flow separation phenomena occur, the prediction accuracy of CFD reduces, and there exists a significant difference in big turning angle condition;
2) The added mass, the added moment of inertia, and the wake coefficient in maneuvering motions are estimated with empirical formulas. The existing formulas are not suitable for the new type of ship;
3) During the free-running tests, the external disturbance is unavoidable.

4 CONCLUSION

This paper uses a modified semi-empirical method to establish a 3-DOF MMG maneuvering model for the twin-ASD vessel. The results indicate that the adopted method is more suitable for the small rudder angle maneuvers. The method should be further improved in future work to enhance its applicability.

Table 3. Comparison of the turning indexes.

	Turning criteria	Sim.	Exp.	Error
$\delta = 5°$	$A_D^{'}$	7.1898	6.2581	14.89%
	$T_R^{'}$	5.1165	3.8164	34.07%
	$T_D^{'}$	11.7735	10.2014	15.41%
	$R_S^{'}$	5.5	5.0029	9.94%
$\delta = 10°$	$A_D^{'}$	4.2830	4.2757	0.17%
	$T_R^{'}$	2.3527	1.5708	49.78%
	$T_D^{'}$	5.8989	5.3910	9.42%
	$R_S^{'}$	2.75	2.7373	0.46%
$\delta = 15°$	$A_D^{'}$	2.8148	3.2501	13.39%
	$T_R^{'}$	1.2943	0.8280	56.33%
	$T_D^{'}$	3.1767	3.5450	10.39%
	$R_S^{'}$	1.25	1.9071	34.46%

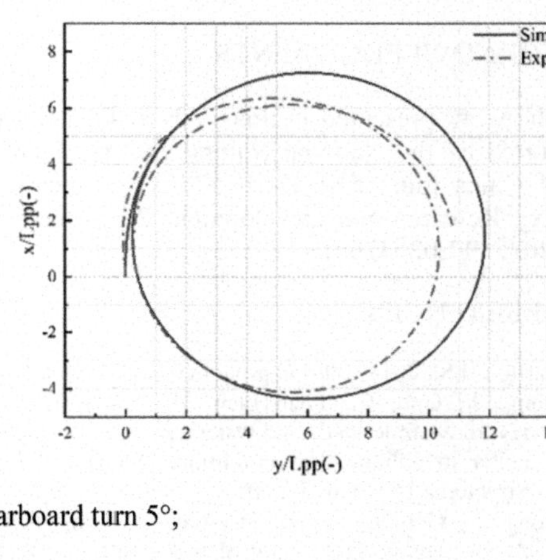

(a) Starboard turn 5°;

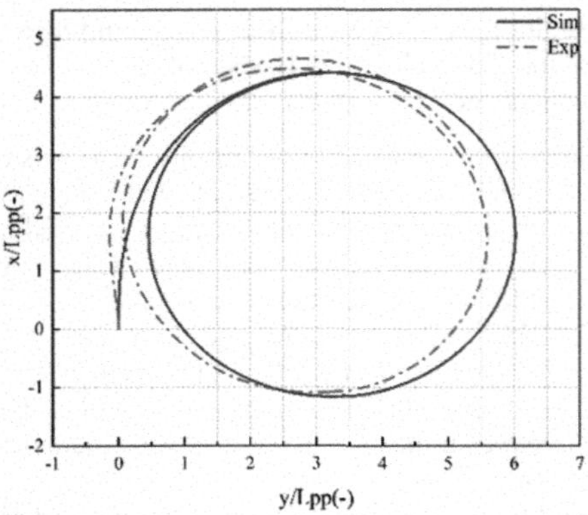

(b) Starboard turn 10°;

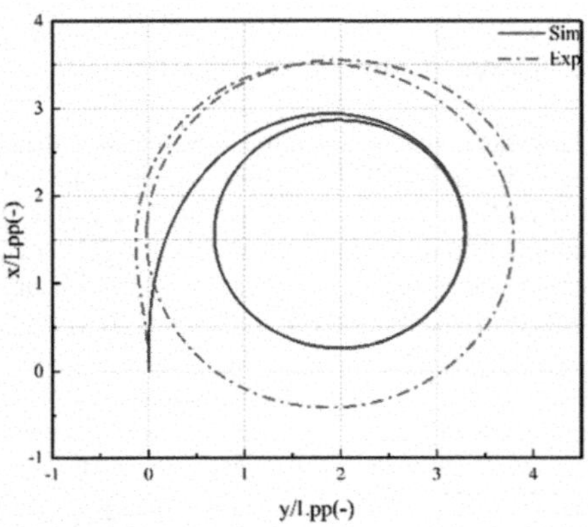

(c) Starboard turn 15°;

Figure 7. Comparison of turning test results.

Table 4. Comparison of the overshoot angles.

Zigzag criteria	Sim.	Exp.	Error
$\psi_{O1}(10^\circ/10^\circ)$	4.4041	8.9800	50.96%
$\psi_{O2}(10^\circ/10^\circ)$	4.7465	9.4900	49.98%
$\psi_{O1}(20^\circ/20^\circ)$	14.5924	29.8400	51.10%
$\psi_{O2}(20^\circ/20^\circ)$	15.6640	26.3500	40.55%

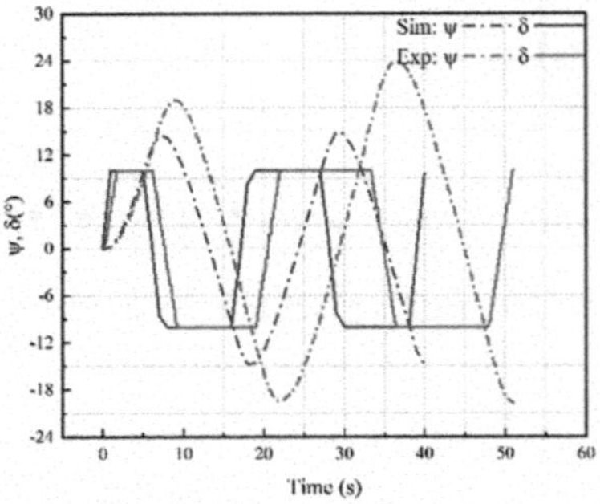

(a) Starboard turn 5° ;

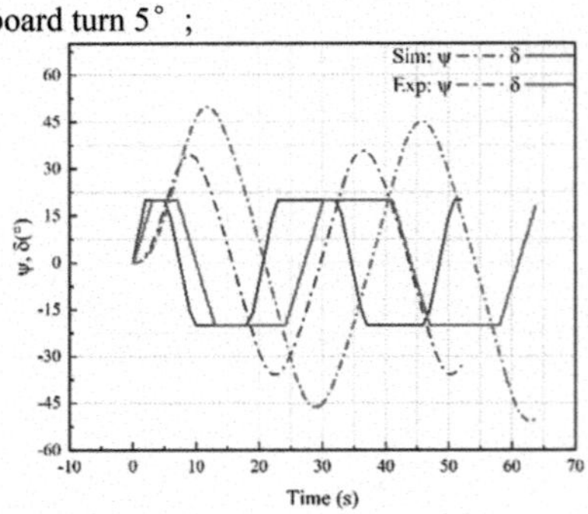

(a) Starboard turn 5° ;

Figure 8. Comparison of Zigzag test results.

ACKNOWLEDGEMENTS

The work presented in this study is financially supported by the National Natural Science Foundation of China grant 52271364, 52271367 and National Key Research and Development Program of China (2023YFE0203700).

REFERENCES

Carlton, J.S., 2018. Marine propellers and propulsion.

Gong, J., Guo, C., PhanThien, N., Khoo, B.C., 2020. Hydrodynamic loads and wake dynamics of ducted propeller in oblique flow conditions. Ships and Offshore Structures 15 (6), 645–660.

Gong, J., Guo, C., Zhao, D., Wu, T., Song, K., 2018. A comparative DES study of wake vortex evolution for ducted and non-ducted propellers. Ocean engineering 160, 78–93.

He, Y., Chen, L., Mou, J., Zeng, Q., Huang, Y., Chen, P., Zhang, S., 2023. Ship Emission Reduction via Energy-Saving Formation. IEEE Transactions on Intelligent Transportation Systems, 1–16.

He, Y., Mou, J., Chen, L., Zeng, Q., Huang, Y., Chen, P., Zhang, S., 2022. Will sailing in formation reduce energy consumption? Numerical prediction of resistance for ships in different formation configurations. Applied Energy 312, 118695.

Maciej, R., 2017. Prediction of manoeuvring abilities of 10000 DWT pod-driven coastal tanker. Ocean engineering 136, 201–208.

Piaggio, B., Villa, D., Viviani, M., 2020. Numerical analysis of escort tug manoeuvrability characteristics. Applied Ocean Research 97, 102075.

Piaggio, B., Viviani, M., Martelli, M., Figari, M., 2022. Z-Drive Escort Tug manoeuvrability model and simulation, Part II: A full-scale validation. Ocean engineering 259, 111881.

Ren, J.-G., Zhan, C.-S., Zhang, Y.-C., Liu, Z.-Y., Chang, H.-C., Feng, B.-W., 2022. Research on the influence of trimaran side hull layout on its manoeuvrability. Ocean engineering 266, 112944.

Wu, T., Li, R., Chen, Q., Pi, G., Wan, S., Liu, Q., 2023. A Numerical Study on Modeling Ship Maneuvering Performance Using Twin Azimuth Thrusters. Journal of Marine Science Engineering 11 (11), 2167s.

Yasukawa, H., Yoshimura, Y., 2015. Introduction of MMG standard method for ship maneuvering predictions. Journal of marine science technology 20, 37–52.

Zhang, S., Wu, Q., Liu, J., He, Y., Li, S., 2023. Twin-screw ASD tug maneuvering prediction based on integrated CFD and empirical methods. Ocean engineering 269, 113489.

Advances in Maritime Technology and Engineering – Guedes Soares & Santos (Eds)
© 2024 The Author(s), ISBN 978-1-032-83107-7

Setup, collection, and processing of manoeuvring full-scale sea trials data of navy ships

P. Pires da Silva
CINAV - Portuguese Navy Research Center, Almada, Portugal
Centre for Marine Technology and Ocean Engineering (CENTEC), Instituto Superior Técnico, Universidade de Lisboa, Lisbon, Portugal

S. Sutulo & C. Guedes Soares
Centre of Marine Technology and Ocean Engineering (CENTEC), Instituto Superior Técnico, Universidade de Lisboa, Lisbon, Portugal

ABSTRACT: A data acquisition system used in manoeuvrability tests on naval ships is described. Data are recorded using a portable setup of sensors and also the ships' own sensors. The parameters recorded are: the ship's instantaneous geographic coordinates; the speed relative to the ground; the heading; the heading, yaw and roll angles, the corresponding speeds and accelerations; forward and drift accelerations; the wind, the speed relative to the water, the instantaneous propellers'pitch, the shafts' rpm, power and torque. Sea trials are carried out in waves and calm seas. Given the lack of an own ship data fusion network, the data fusion is produced digitally in post-processing through data selection, data synchronization and data filtering algorithms. This work discusses the challenges associated with the organization and execution of the full-scale sea trials, the data post processing and the results obtained, namely the kinematics for the turning, zigzag and spiral manoeuvres.

1 INTRODUCTION

Suitable and reliable mathematical models of ships are very important for improving ship design and even more for the adequate and proficient training of the ships' crews. That is so when considering these models being implemented in bridge simulators combined with interactive controls and virtual reality, contributing to improving the practice of operators, who will be able to control ships more effectively and efficiently and with a higher level of safety.

This work aims to address the development of the full-scale trials during the NAVAD project, in its aspects of preparation, execution and processing of data from manoeuvrability full-scale sea trials carried out both in calm waters and in rough seas. These trials aim to produce data that can validate mathematical models for simulating manoeuvering motions of surface naval ships in adverse weather conditions, i.e. under the action of wind and waves.

1.1 *Review of origin and relevance of full-scale sea trials*

Manoeuvrability simulation and associated research have become very important as they serve as the basis for ship simulators of varying complexity for training purposes.

The software running in these simulators comprises two relatively independent and equally important components:

- the mathematical model, which describes the dynamics of the ship, together with the necessary environmental factors,
- the graphical interface necessary for realistic interaction between the operator and the simulated ship.

The graphical interface is the visible component, the dynamic mathematical model is hidden and is only felt through the simulator's reactions to control actions and simulated exogenous factors, such as wind, current and waves. Generally, the central model must correspond to a given real ship and its responses are, in the ideal case, identical to those of the real ship, although complete similarity cannot yet be achieved (Sutulo and Guedes Soares, 2011). The study and validation of manoeuvrability models as the core of the system is a demanding, complex, and non-trivial task.

In the 1990s, the exponential growth in computing power and its availability at an affordable cost eliminated simulator hardware as a major limiting factor in the application of simulation in ship design. Instead, the cost of obtaining data for mathematical models, processing that data, and developing visual scenarios has emerged as the dominant costs in simulating maritime environments (Webster, 1992).

DOI: 10.1201/9781003508779-38

In the second decade of the 21st century, according to Zhu et al. (2017), the main methods for estimating a manoeuvrability model include tank tests with captive models (Sutulo and Guedes Soares, 2006), (Sutulo and Kim, 1998), system identification combined with manoeuvrability tests on ships or unmanned model tests (Sutulo and Guedes Soares, 2023), (Xu and Guedes Soares, 2019) and computational fluid dynamics (CFD) (Islam and Guedes Soares, 2018). Among the methods mentioned above, system identification combined with testing on real ships or an unmanned scale model has become an attractive and sufficient combination of techniques to estimate the ship's manoeuvrability model due to its relatively high ratio cost-benefit (Zhu et al., 2017).

Uncertainty derived from different sources is underlying and inherent to all manoeuvrability model estimation methods, as discussed by (Pires da Silva and Guedes Soares, 2016), both in the ship design phase and during its operational life, once the variables that influence manoeuvrability performance (e.g. displacement, trim and draft) estimated during the design phase will only be confirmed during sea trials and vary throughout the operational life cycle. In real tests, these uncertainties must be added to those derived from the weather and waves and those from each of the sensors and data acquisition equipment.

1.2 *Full-scale sea trials: Execution and data acquisition challenges*

Although preferable, full-scale manoeuvrability sea trials are time-consuming and expensive. Their preparation involves the selection of ships, study and analysis of the ship's own sensor capabilities and data network technology and architecture. Also involves:

- the analysis and evaluation of the potential need, configuration and technological level of external data acquisition,
- detailed planning and budgeting of the external data acquisition setup,
- detailed planning and budgeting of the ship at sea for trials.

Furthermore, carrying out full-scale sea trials under planned weather and wave conditions is and will always be a major challenge, for obvious reasons. It is therefore not surprising that good data available from large-scale sea trials is scarce. However, some research using full-scale tests on military platforms in the open sea has been carried out, (Guedes Soares et al., 2002, 2004), where results from full-scale manoeuvrability trials carried out on two speedboats are presented. and a corvette, operated by the Portuguese Navy, to identify their manoeuvring characteristics. The test plan was established taking into account both the recommendations of the code for manoeuvrability tests proposed by IMO (1993, 1994), now replaced by IMO, (2002) and an investigation previously carried out by Alves Francisco & Guedes Soares (2000) which indicated the manoeuvres potentially most likely to efficiently identify a ship's hydrodynamic characteristics.

Sutulo et al. (2010) using previous lessons learnt from full-scale trials and technological advancement in data acquisition equipment describe a set of equipment and software for measuring and recording kinematic parameters of surface vessels. The set included a fibre-optic gyro (FOG) compass, a high-precision global positioning system (GPS) unit, a yaw rate sensor and an anemometer. A special application, developed in the LabView® G language, performed synchronized data recording. This data acquisition system was applied to full-scale passive tests, i.e., instrumented observations of the normal operation of ships in March 2008. The authors conclude that the data acquisition system did not appear to solve all the problems of the full-scale trials. The most significant difficulties in assembling the set of sensors on the ship came from the need to route the very heavy and long coaxial cables for the GPS antenna and anemometer. The main result of the study was a relatively high-quality full-scale data set, obtained as a result of instrumented observations of the berthing and unberthing manoeuvres.

2 PREPARATION OF DATA ACQUISITION SYSTEM FOR SEA TRIALS

The success of full-scale sea trials, and the quality of the data collected, depend on many factors, among which the quality of sensors and their setup is of primary importance. However, full-scale trials, especially those for scientific purposes, are generally quite expensive and as a result are rarely executed. There is no standard, off-shelf equipment sets on the market, and each hydrodynamic centre develops its own system, using, where possible, available sensors developed for other applications, connected to a controller with port modules and to a laptop PC where all measured parameters are recorded during trials.

The NAVAD project and the coordination of full-scale sea trials involved researchers from CENTEC and CINAV, as well as interactions with the Portuguese Navy Operational Command, the Portuguese Navy Directorate of ships and the crews of the designated ships.

Initially, the trials were planned to be carried out on three vessels representing three hull types: an Auxiliary Oiler Replenisher (AOR), an Oceanic Patrol Vessel (OPV) and a Coastal Patrol Vessel (CPV), operated by the Portuguese Navy, and representing different operational profiles and different levels of sensing and integration technology, ranging from fully stand-alone analogue sensors to fully integrated digital networks.

Due to the COVID19 pandemic, the operational planning, difficulties arising from competing Navy missions, suitable weather conditions and deadlines for carrying out the two series of tests per ship, one in calm seas and the other in rough seas, it was only possible to execute full-scale trials on two types of ships, the CPV Tejo class and the OPV Viana do Castelo class. Th complete set of planned trials was only possible for the OPVs.

2.1 *Relevant parameters and data acquisition system*

During the trials, the following parameters were recorded: the ship's WGS-84 instantaneous geographic coordinates; the ship's speed over ground (SOG); the speed of the ship relative to the water; the heading, the course over ground (COG); the heading; the rudder angle; the true wind vector; the relative wind vector; engine rpm; the rpm of the shafts; the torque of diesel engines; the power of diesel engines; the instantaneous pitch of the propeller; the surge, the roll and pitch angles, the corresponding accelerations; surge and sway accelerations.

Full-scale manoeuvring trials will be performed with three ships: a Replenishment Ship (RS), an Offshore Patrol Vessel (OPV), and a Coastal Patrol Vessel (CPV), operated by the Portuguese Navy, and representing different operational profiles and different levels of the sensing technology and integration, varying from fully standalone analogic sensors to fully integrated digital networks.

For tests in waves, the wave radar sensor WM-2 providing relative displacements of the system with respect to water surface will be installed. The collected data together with those related to the ship motions will be sufficient for estimating the two-dimensional wave spectrum.

Offshore Patrol Vessel (OPV)

Coastal Patrol Vessel (CPV)

Figure 1. Designated ships. Courtesy of Portuguese Navy under NAVAD project.

Due to the different levels of sensors' integration, in some ships is possible to have the same signal from two systems, allowing redundancy as shown on the "source" column of Table 1. Preventing possible difficulties in collecting some signals, they are categorized as "critical", "important", "if possible" and "important for rough weather", according to their criticality for proper system identification as presented in Table 1.

Depending on the type of ship, part of the variables can be obtained by on-board sensors (e.g. GPS position and time, rudder angle, anemometer, SOG, COG) as redundancy to portable external sensors provided by CENTEC. Transversal to all designated ships is the critical need to obtain the rudder angle in real time by digital signal (non-redundant signal).

2.2 *Data collection setup*

The setup for these trials used raw data from the ship's own sensors as redundant (except for rudder angle) to the CENTEC's parallel sensor array. Sensors installed on the ships included a fiber-optic gyrocompass (FOG): IXSea Octans Mk III, a Crossbow NAV440 IMU inertial unit, a 10 Hz high-precision GPS unit, "Davis Vantage" wireless weather station, National data collector Instruments cRIO 9037 with C# modules and portable PC. Table 2 in Pires da Silva, et al. (2021) list in detail these external CENTEC equipment and sensors.

It is possible to obtain part of the signals in Table 1 from the ECDIS system, which can provide data (GPS, anemometer, odometer, gyrocompass) from the integrated navigation system (INS) in NMEA protocol. However, this method requires a separate and sequential start of data recording for each sensor. Therefore, data synchronization for all sensors is required to be performed digitally during the post-processing process. Figure 2 presents an overview of the data collection setup.

Besides the images on board of the CENTEC sensors and equipment presented in Figure 2, connections to the ship's own sensors and control circuits in order to have a full data collection setup were performed and are presented in Figure 3. A diagram is presented with the layout of the data connection setup in ships with INS, PMS and ECDIS and respective data network.

In all designated ships it is possible to have part of the signals of Table 1 in ASCII files, concerning sensors that provide data to the ECDIS (GPS, anemometer, odometer, gyrocompass) from navigation network (INS) in NMEA protocol. This is possible, since there is a "*device master*" equipment on the bridge, where signals can be recorded to the laptop using an Ethernet port (RJ45). The "*device master*" has a tool (*PortVision* software) where using the "*Wcom2 - Test terminal*", the data in NMEA protocol can be viewed and recorded by COM emulated ports. This method requires separate and sequential start of data recording, by each COM port, that is, by each sensor. In data analysis, the date/hour/minute/second

Table 1. List of Parameters to be collected during full-scale sea trials, respective source and sampling rate.

ID	Parameter	Name	Signal Source	Sampling rate (Hz)
Critical Parameters				
1	U	speed relative to water	Ship Odometer - ECDIS	2
2	V	Speed Over Ground (SOG)	Ship GPS – ECDIS* and Portable GPS unit	2/10
3	ψ	heading angle	Ship ECDIS* and Portable IMU**	2/10
4	R	rate of yaw	Ship PMS*** and Portable IMU**	< 1 /10
5	ξ, η, χ :	Cartesian coordinates relative to the ground and Course Over Ground – (COG)	Ship GPS – ECDIS* and Portable GPS unit	2/10
6	δ	rudder angle	Ship PMS***, INS****, or rudder angle indicator on the bridge	< 1 /10/analogue
7	WGS-84	instant geographic coordinates	Ship GPS – ECDIS* and Portable GPS unit.	2/10
8	$\dot{u}$,$\dot{v}$, w,	ship accelerations in 3DOF	Ship gyro e and portable IMU**	2/10
Important				
9	V_w, χ_w	Relative wind speed and direction	Ship ECDIS* or PMS*** and portable Wi-Fi weather station	2/< 1
If Possible				
10	ϕ	roll angle	Ship gyro and Portable IMU**	2/10
11	θ	pitch angle	Ship gyro and Portable IMU**	2/10
12	w:	heave speed	Ship gyro and Portable IMU**	2/10
13	$\dot{\phi}$, $\dot{\theta}$, $\dot{\psi}$:	rotation rates in 3 DOF	Ship gyro and Portable IMU**	2/10
14	N	shafts rotation frequencies (rpm)	Ship PMS***	< 1
15	P	Propeller pitch (if aplicável)	Ship PMS***	< 1
		Rudder order	Not available	
		Engine order	Not available	
Important for rough weather				
	η	Wave height	wave radar sensor TSK WM-2	> 10 GHz

* Electronic Chart Display and Information System; ** Inertial Measurement Unit; *** Platform Management System; **** Integrated Navigation System

Figure 2. Sensors external to the ship (CENTEC) and customized workstation configuration for data collection.

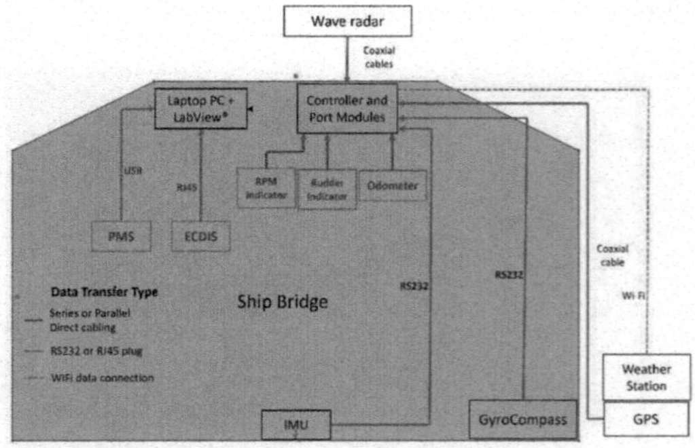

Figure 3. Scheme of on-board setup for Data Acquisition. Ships with Digital networks.

recording start of each sensor are different, although no greater difficulties are predicted given the clock of the signals being unique, that of ship GPS.

In ships with Platform Management System (PMS) it is possible to collect digital propulsion data in *"csv"* files, including the rudder angle. However, INS and PMS signals are not synchronized in time, which always makes it necessary to use the independent set of sensors and CENTEC data collection system, with connection to some on-board sensors.

3 FULL-SCALE SEA TRIALS CAMPAIGNS

3.1 *Planning*

In addition to the technical details of the data acquisition equipment, its configuration and connection to the ship, there are management challenges to be faced so that full-scale sea trials can take place. Thus, in these sea trials campaigns, CENTEC collaborated with the Portuguese Navy through the Marine Research Center (CINAV) for planning and managing sea trials works with: information on available ships and dates from the Navy Operational Command (COMNAV), technical information and authorization for assemblies and tests on board, from the Ships Directorate. Finally, there is direct coordination with ships for assembly, testing and execution of trials. These relationships for the operational management of sea trials are outlined in Figure 4.

For the NAVAD project, careful consideration was given to the fact that the trials conducted were intended for research purposes rather than those regularly performed for the approval of a new or post-overhauled ship. Additionally, factors such as risk assessment, resource constraints, and the impracticality of expecting ideal environmental conditions aligned with Navy operational planning were taken into account. Consequently, a two-fold approach was devised for each ship, comprising Series 1 trials for maneuvers conducted under optimal conditions, ideally devoid of wind and waves or up to sea/wind Douglas 3/Beaufort 3 conditions, and Series 2 tests for maneuvers in sea/wind wave conditions surpassing Douglas 3/Beaufort 3 thresholds.

A total of 7 hours required for series 1 and 8 hours for series 2 were estimated to complete the full-scale sea trials on each ship. The execution effectively corresponded to plan, except for the sea trials in waves, which took around 10 hours, due to the need to pre-position the ship for the necessary heading-swell orientation.

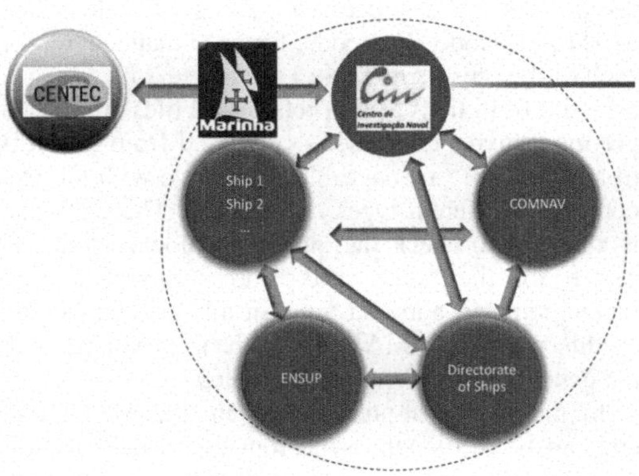

Figure 4. Entities and relations involved in coordinating and planning sea trials.

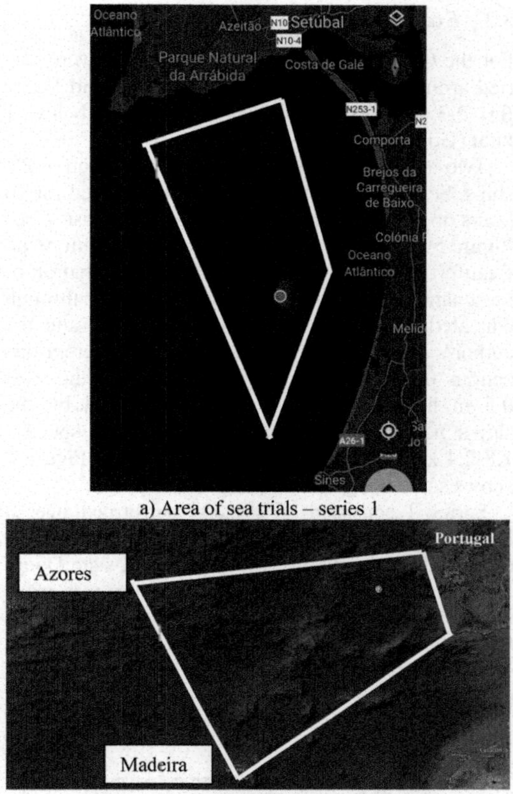

a) Area of sea trials – series 1

b) Area of sea trials – series 2

Figure 5. Areas of sea trials campaigns.

Overall, the planning and execution cycle of each test campaign corresponds to at least 3 days of preparation and execution per ship and per series. Post-processing and data analysis was carried out after the trials and lasted 3 to 4 months. It included the development of all code for converting data from each sensor to standard MATLAB® data types, time alignment of start/end files, data synchronization, debugging and filtering. At the end of this process, it is possible to verify that some test data are useless for various reasons, most of them related to the partial lack of some specific data of a "critical" signal as classified in Table 1.

As part of preparation of the sea trials and while alongside before departure is necessary to have:

- actual draught measurement,
- actual vertical and longitudinal position of the centre of mass of the ship.
- list of tanks with filling liquid characteristics and level (before and after the trials).

Series 1 took place in the area between Sesimbra and Sines as can be seen in Figure 5. Series 2 took place in the open Atlantic Ocean in the area between the archipelagos of Mainland Portugal, Azores and Madeira (see Figure 5).

3.2 *Execution*

For the CPV ships, two sets of trials trials were carried around the island of Madeira, with wind Beaufort 2-3 and waves Douglas 0-1, current below 1 knot (Goulão, 2018).

Two series of tests were carried out on OPV ships. Series 1 consisted of 31 tests performed for 10 hours in a maritime area close to the coast - see Figure 5 a) - with a depth of around 150-200m, wind Beaufort 1-4, sea state Douglas 2. The estimation of the characteristics of the waves was made through educated guesses between the ship's crew and the author's experience at sea: waves from different directions 0-0.5 m high over a swell around the west 0.5 m high height. The current indicated by the ship's navigation system was variable in speed - 1.8-2.4 kts and direction - SW, NNW, SE. Figure 6 shows some images from this test campaign.

Series 2 consisted of 25 tests performed over 8 hours in a sea area shown in Figure 5 b) with a depth of around 5000m, wind Beaufort 5-6, sea state Douglas 4-5. Estimation of wave characteristics was made through educated guesses between the ship's crew and the author's maritime experience at sea: 1.5 m high NE waves over a 2.5 m high NW swell. Figure 7 shows some images from this trials campaign.

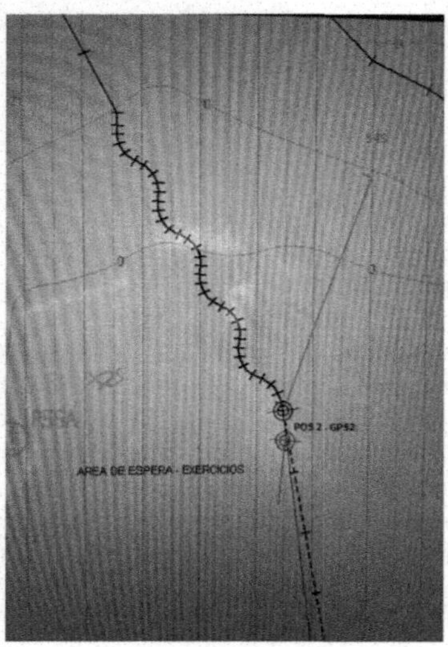

Figure 6. Images from sea trials – series 1.

3.3 *Data post-processing*

After the sea trials, the collected data had to be analyzed and post-processed using a dedicated MATLAB® code. This code, for each manoeuvre, for each of the sensors in Table 1 loads a text file and a ". csv" file from the PMS. A total of 12 files are loaded per manoeuvre. Each file is converted from raw text to a MATLAB® manageable ".mat" file with all data content as different types of variables. Then, all data is analysed to check the quality of time vectors and search for file "start" and "end" times to ensure a common start and end time for all data. At the end of this process, a MATLAB® file is produced with the general name "*Trial_matched.mat*".

In the next post-processing step, the MATLAB® code analyses, digitally synchronizes the data and filters the first stage when necessary. A Gaussian weighted moving average window, appropriately tuned for each signal, was applied to filter noisy signals. At

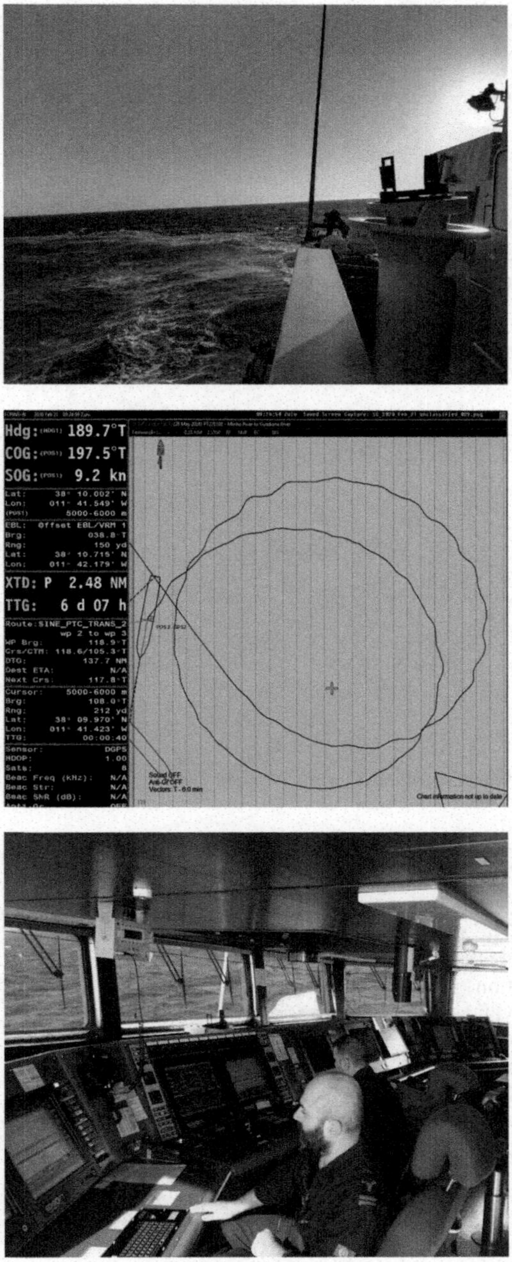

Figure 7. Images from sea trials – series 2.

the end of this process, a "*Trial_Synchronized.mat*" file is saved.

Finally, a script, called "*Trial Analysis.m*", performs the detailed comparison of variables from different sources in a time window, within a specific trial, *e.g.* zigzag, selected by the user, and can perform a second filtering stage. The output of "*Trial Analysis.m*" is the trial data file that can be used for system identification purposes, with the generic name "*Trial_ID_Data.mat*". Figure 8 shows the flowchart of the MATLAB® data post-processing code.

Within the data analysis process, one of the main tasks is to estimate the ship's trajectory from GPS data. The primary coordinates obtained from the GPS receiver via the NMEA protocol are traditional geographic coordinates according to the WGS-84 reference, that is, latitude φ and longitude λ. They are written to the data file as character vectors whose first two (in latitude, φ) or three (in longitude, λ) digits represent degrees and the remaining six digits are minutes including their fractional parts (four digits).

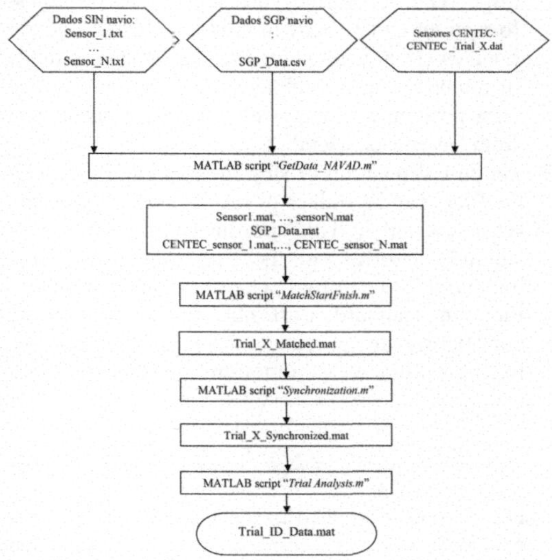

Figure 8. Flowchart for full-scale sea trial data post-processing.

To trace the trajectories, these Earth geographic coordinates are transformed into Cartesian coordinates and then transformed into the movement of the ship's centroid as follows:

1. An arc length in metres, corresponding to an angle φ on a circumference over a sphere of radius R is given as:

$$arc = \varphi R \text{ or } arc = \frac{\varphi R \pi}{180} \text{ if } \varphi \text{ is in degr} \quad (1)$$

2. Considering the International Earth Rotation and Reference Systems Service (IERS), the WGS-84 ellipsoid, mean radius of semi-axes (R1) equals 6371008:7714 metres, the WGS-84 ellipsoid, Radius of Sphere of Equal Area (R2) equals 6371007:1809 metres and the WGS-84 ellipsoid, Radius of Sphere of Equal Volume (R3) equals 6371000:7900 metres. It is usually assumed that the Earth WGS-84 ellipsoid mean radius is 6371000 metres. Therefore, for the IERS WGS-84 any conversion of an arc length to metres for φ is in minutes in minutes is given by:

$$arc = \frac{R\pi}{180}\frac{\varphi}{60} \quad (2)$$

However, it is generally considered that the conversion is given by a factor $k = 1852m/min.arc$. Therefore equation (2) maybe written as:

$$arc = k\varphi \quad (3)$$

3. The recorded data is pre-processed from the raw NMEA data and then the values latitude, φ and longitude, λ are read which results in some values not having direct physical meaning. Then the digits corresponding to degrees are separated and transformed into minutes by multiplying them by 60 arc minutes. They are summed to the remaining digits of minutes, and it is obtained a vector of absolute GPS position in minutes of arc, from Equator and from Greenwich Meridian.
4. The initial position of ship in the trial is given by the first φ and λ values of the obtained vector and set as φ_0 and λ_0 This coordinate (φ_0, λ_0) is considered the origin of the ship trajectory in the beginning of each trial. Therefore, each ship position in standard Cartesian Earth coordinates considering an equirectangular projection, relative to (φ_0, λ_0) in meters of motion is given as:

$$\eta_E = -k(\lambda - \lambda_0)\cos\varphi$$
$$\xi_N = k(\varphi - \varphi_0) \quad (4)$$

where η_E axis is directed to the East and the ξ_N axis to the North, $\cos\varphi$ is a correction of the arc travelled in longitude due to latitude since the arc length in longitude diminishes from Equator to the Poles.

As these full-scale trials areas are located entirely in the North-Western hemisphere (2nd quadrant of the Earth grid whose origin is at the intersection of the Equator and the Greenwich meridian), there is the need for the first minus sign in the first equation. Thus all longitude values are negative relatively to the referred origin. If a ship navigates through deferent quadrants a dedicated algorithm must considered in order to have correct representations of the ship's trajectories relatively to its first absolute position.

Equation (4) is valid if the antenna of the sensor, e.g. GPS is coincident with the ship's origin C. When that is not the case, a transformation of coordinates is necessary to the bring the centre of motion of the antenna to the origin of the ship reference frame, *C*. Then a more generalized trajectory equation is:

$$\eta_E = -k(\lambda - \lambda_0)\cos\varphi - (x_{GPS}\cos\psi + y_{GPS}\sin\psi)$$

$$\xi_N = k(\varphi - \varphi_0) - (x_{GPS}\sin\psi - y_{GPS}\cos\psi) \quad (5)$$

where x_{GPS} and y_{GPS} are the coordinates of the GPS position on board and ψ is the actual true heading.

A transformation of ψ in Earth polar coordinates to mathematical polar coordinates - ψ_m - is usually needed, where the transformation equation is given as $\psi_m = \pm 90$. Then equation (5) may be solved as:

$$\eta_E = -k(\lambda - \lambda_0)\cos\varphi - (x_{GPS}\cos\psi_m + y_{GPS}\sin\psi_m)$$

$$\xi_N = k(\varphi - \varphi_0) - (x_{GPS}\sin\psi_m - y_{GPS}\cos\psi_m) \quad (6)$$

Figure 9 Shows the coordinate frames and the relative position of the GPS on board. Figure 10 shows examples of two turning manoeuvres post-processed trial trajectories.

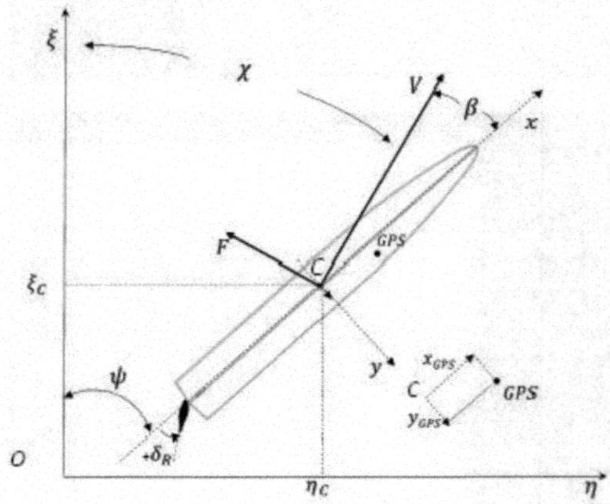

Figure 9. Coordinate frames. GPS position and coordinates.

4 TRIALS RESULTS

The results of the post-processing process are kinematic data used for further research, e.g. identification of the ship's mathematical model, whose validation and verification goes through several processes, one of which is the adaptation of the trajectory to the real trajectory recorded in sea trials. Figure 10 shows examples of this verification.

5 CONCLUSIONS

This work addresses the development of the NAVAD project, in its aspects of preparation, execution and processing of data from full-scale maneuverability sea trials, necessary not only to characterize the manoeuvrability properties of ships but also to validate mathematical models for simulation of surface naval ships.

During this complex process challenges of planning and executing full-scale trials arise due to the inexistence of standardised data collection setups. Extensive on-site research is performed in order to develop and costumize data collection setups adapted to each ship technological layout.

Other practical challenges are overcome related to operational planning, suitable weather conditions

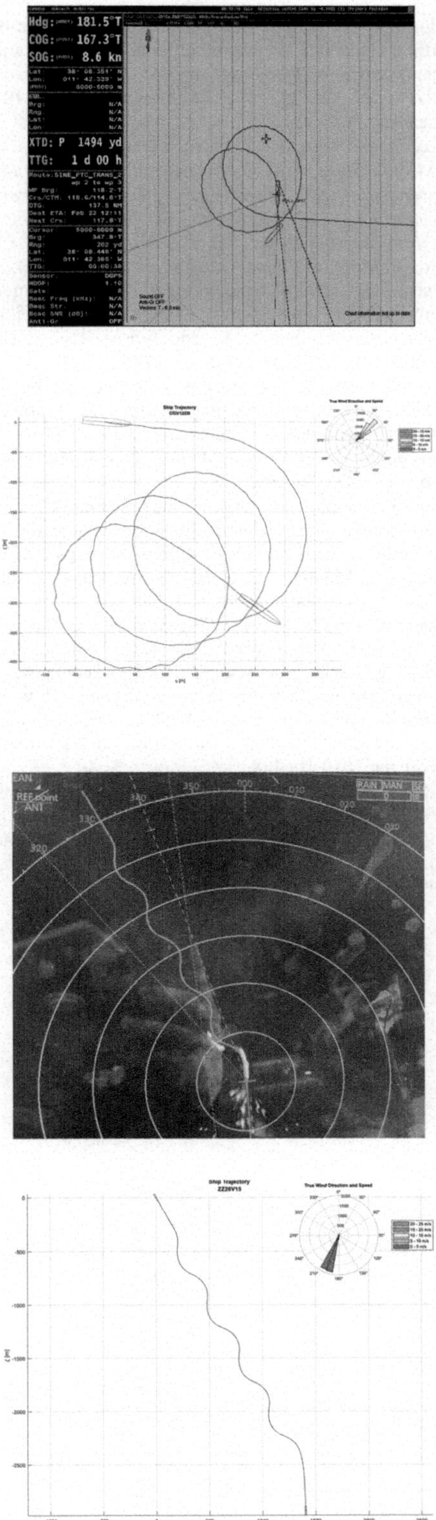

Figure 10. Example of trial trajectories after post-processing and their equivalent images on ECDIS or radar screen.

and deadlines for executing the two series of tests per ship, one in calm seas and another one in rough seas. As result, only one complete test plan in a Viana do Castelo class OPV is achieved from the six initially planned.

A partially redundant system for monitoring and acquiring data in full-scaletrials, is developed by CINAV and CENTEC, proved to be sufficiently suitable for the planned purposes. However, it is found that nowadays with the quantity and quality of sensors and networks existent on board, it is possible to adequately customize the ship's network and data outputs to obtain the necessary data for any manoeuvrability trial without the need for the difficult edification of portable setups.

The validation of the results is done during and after the post-processing of the raw data and show that the adaptation of the trajectory to the real trajectory recorded in sea trials has a good fit.

ACKNOWLEDGEMENTS

This work was performed within the NAVAD project "Simulation of manoeuvrability of ships in adverse weather conditions", which is co-funded by the European Regional Development Fund (Fundo Europeu de Desenvolvimento Regional - FEDER) funds through the Lisbon Regional Operational Program - COMPETE 2020 (PORLISBOA-01-0145-FEDER -32037) and by the Portuguese Foundation for Science and Technology (Fundação para a Ciência e a Tecnologia - FCT) under contract 02/SAICT/032037/2017. This work contributes to the Strategic Research Plan of the Centre for Marine Technology and Ocean Engineering (CENTEC), which is financed by the Portuguese Foundation for Science and Technology under contract UIDB/UIDP/00134/2020.

REFERENCES

IMO (International Maritime Organization). (1993). *IMO Resolution A.751(18): Interim Standards for Ship Manoeuverability.* International Maritime Organization. London: IMO.

Crane, C., Eda, H., and Landsburg, A. (1989). Controllability. In E. Lewis (Ed.), *Principles of Naval Architecture* (pp. 191–365). Jersey City, N.J.: SNAME.

Francisco, R., and Guedes Soares, C. (2000). Identification and Simulation of Ship Maneuvering Characteristics (in Portuguese). in: . *O Mar e os Desafios do Futuro.* Lisboa: Edições Salamandra, 185–209.

Goulão, J. (2018). *Identification of a ship's manoeuvrability parameters (in portuguese).* Portuguese Navy. Lisbon: Portuguese Naval Academy. Retrieved from http://hdl.handle.net/10400.26/30276

Guedes Soares, C., Sutulo, S., Francisco, A., Santos, F.M., and Moreira, L. (1999). Full-scale measurements of the manoeuvering capabilities of a catamaran. *RINA International Conference on Hydrodynamics of High Speed Craft* (pp. 1–12). London: RINA.

Guedes Soares, C., Sutulo, S., Francisco, R., Moreira, L., and Laranjinha, M. (2002). Maneuverability tests on fast crafts and corvettes (in Portuguese). in: Guedes

Soares, C Beirão Reis, J. Martins Guerreiro, M. b, (Eds.).. *O Mar Fonte de Desenvolvimento Sustentado* (pp.). Lisboa: Salamandra, 251–270.

IMO (International Maritime Organization). (1994). Explanatory notes to the interim standards for ship manoeuvrability. *Circular MSC/Circ. 644*. London: IMO.

Islam, H., and Guedes Soares, C. (2018). Estimation of hydrodynamic derivatives of a container ship using PMM simulation in OpenFOAM. *Ocean Engineering, 164*, 414–425.

Mandel, P. (1965). Ship maneuvering and control. In J. Comstock (Ed.), *Principles of Naval Architecture* (pp. 463–606). New York: SNAME.

Pires da Silva, P., and Guedes Soares, C. (2016). Uncertainty modeling in ship maneuvering models. in: C. Guedes Soares, and Santos T.A., (Eds.) *Marine Technology and Engineering 3 UK: London*: Taylor and Francis Group, 361–369.

Pires da Silva, P., Sutulo, S., Hinostroza, M., & Soares, C. G. (2021). Instrumentation and data acquisition system for full-scale maneuverability tests on board of naval surface ships. in: C. Guedes Soares, & T. Santos (Eds.), *Maritime Technology and Engineering, UK, CRC Press*, 5, 227-234

Schoenherr, K. (1939). Steering. In E. Henry, L. Chapman, and Rossell (Ed.), *Principles of Naval Architecture* (pp. 197–233). New York: SNAME.

Sutulo, S., and Guedes Soares, C. (2006). Development of a multi-factor regression model of ship maneuvering forces based on optimized captive-model tests. *Journal of Ship Research, 50*, 311–333.

Sutulo, S., and Guedes Soares, C. (2011). Mathematical models for simulation of manoeuvring performance of ships. in: Guedes Soares, C. Garbatov Y. Fonseca N. and Teixeira A. P., (Eds.), *Maritime Technology and Engineering London, UK*: Taylor and Francis Group, 661–698.

Sutulo, S., and Guedes Soares, C. (2023). Aplication of an offline algorithm for adjusting parameters of a modular manoeuvring mathematical model. *Ocean Engineering, 279*, 114328. doi:https://doi.org/10.1016/j.oceaneng.2023.114328

Sutulo, S., and Kim, S. Y. (1998). Systematic Approach to PMM/Rotating Arm Experiment Planning, Parameter Estimation, and Uncertainty Analysis. *International Symposium and workshop on Force Acting on a Manoeuvering Vessel*. Val de Reuil, France.

Sutulo, S., Moreira, L., and Guedes Soares, C. (2000). Analysis of the effect of wind and current in ship manoeuvring trials (in Portuguese), *O Mar e os Desafios do Futuro*, Guedes Soares, C. and Beirão Reis J., (Eds). Lisboa: Edições Salamandra, 211–230

Sutulo, S., Paço, A., and Guedes Soares, C. (2009). Full-scale observations of berthing and unberthing processes of fast displacement catamarans. *FAST 2009 - 10th International Conference on Fast Sea Transportation*, (pp. 491–503). Athens.

Sutulo, S., Pascoal, R., and Guedes Soares, C. (2010). Instrumentation and Data Acquisition System for Maneuverability Tests and Observations on Ships (in Portuguese). in: Monteiro, C. Guedes Soares and C. Costa, C. (Eds.). Lisboa: Salamandra, 525–538.

Webster, W. (1992). *Shiphandling Simulation, Application to Waterway Design*. Washington D.C.: National Press.

Xu, H., and Guedes Soares, C. (2019). Hydrodynamic coefficient estimation for ship manoeuvring in shallow water using an optimal truncated LS-SVM. *Ocean Engineering, 191*, 106488. doi:https://doi.org/10.1016/j.oceaneng.2019.106488

Zhu, M., Hahn, A., Wen, Y., and Bolles, A. (2017, September 7). Identification-based simplified model of large container ships using support vector machines and artificial bee colony algorithm. *Applied Ocean Research, 68*, 249–261.

Advances in Maritime Technology and Engineering – Guedes Soares & Santos (Eds)

Autonomous ship maneuvering in instantiated environments using deep reinforcement learning

R. Zhang, X. Qin & M. Pan
Navigation College, Dalian Maritime University, Dalian City, P.R. China

S. Loughney
School of Engineering, Liverpool John Moores University, Liverpool, UK

ABSTRACT: Deep Reinforcement Learning (DRL) excels in complex decision-making, notably in autonomous ship navigation, by enabling effective handling in intricate maritime settings. However, most studies simplify agents to points on basic maps or grids, which doesn't reflect real-world complexity. To counter this, we've created a detailed maritime navigation simulation with agents as specific ship models, incorporating dynamic and static obstacles to mimic real navigational challenges. Our approach utilizes an enhanced deep Q network (DQN) with a novel experience replay method to boost training efficiency and a reward function aligned with the International Regulations for Preventing Collisions at Sea (COLREGs), ensuring compliance during navigation. Testing in this realistic environment demonstrated our method's ability to safely and efficiently navigate, avoiding collisions in complex, busy maritime zones. The improved DQN algorithm further enhances our model's robustness and practicality, showing promise for safer autonomous maritime navigation.

1 INTRODUCTION

Intelligent ships will play a crucial role in the modern shipping industry, representing not only the forefront of technological advancement but also the key to innovation in efficiency, cost, and safety in the shipping industry. (Munim 2019) provides a comprehensive analysis of the development, benefits, and applications of autonomous ships (AS). By adopting advanced automation and artificial intelligence technologies, autonomous ships can optimize navigation routes, reduce fuel consumption and emissions, thereby significantly enhancing shipping efficiency. At the same time, the introduction of automated systems reduces dependence on human resources, lowers operational costs, and mitigates risks associated with human errors to some extent. Furthermore, the potential of autonomous ships in enhancing maritime safety cannot be overlooked. They are capable of real-time monitoring and responding to complex marine environments and emergencies, effectively reducing the incidence of accidents. Therefore, in the pursuit of more efficient, economical, and safer shipping services, the study and application of autonomous ship control is particularly important, representing a significant trend in the future development of the shipping industry.

Despite the great potential of intelligent ship technology, its practical research faces a series of complex and critical challenges. Firstly, current algorithms have limited generalization capabilities in real marine navigation environments, meaning they may not respond optimally to previously unencountered or abnormal situations. For instance, in complex intersecting environments with multiple types of ships, algorithms need to adapt flexibly while maintaining navigational efficiency and safety. (Wu et al. 2023) demonstrates that environmental information includes data about obstacles and the status of incoming ships, this is a huge challenge for adaptability in different scenarios. Additionally, (Zhou et al. 2020) notes that further research is needed, especially on safe navigation behavior of ships under the semantics of COLREGs, requiring algorithms not only to accurately identify these obstacles but also to formulate convention-compliant response strategies in a very short time.

(Wang et al. 2023) presents a novel algorithm for maritime autonomous surface ships (MASS) collision avoidance in a continuous state space, using an approximate representation model to optimize collision avoidance strategies in dynamic target encounter situations, thus improving learning efficiency in large-scale continuous state spaces. These research indicates that autonomous ship systems must have a high tolerance for long-tail effects, and decision-making neural networks require a sufficiently large state space to represent the marine navigation environment. At the level of ship maneuvering, the impact of wind, waves, and currents under micro conditions, as well as the influence of ocean currents,

DOI: 10.1201/9781003508779-39

sea conditions, and meteorological factors under macro conditions, all have a significant impact on the navigation of ships, (Peng et al. 2017) constructed a dynamic control law based on the echo state network at the dynamical level, The stability of the closed-loop system is analyzed through input-state stability and cascading theory.

Instantiated environments effectively balance the complexity of the environment with the simplicity of model scenarios. Instantiated environments refer to virtual scenes of ship navigation environments, simulated in a minimalist and real-time manner using visualization technology. These environments encompass not only common aspects like ship appearance, performance, and navigational behavior but also simulate various emergencies and dynamic conditions such as different weather conditions, traffic density in waterways, and even maritime emergencies. The core value of this simulation technology lies in its ability to provide unpredictable traffic scenarios, enabling researchers to thoroughly test and optimize intelligent ship algorithms and systems without any physical risk. By conducting extensive simulations in these instantaneous environments, the capability of intelligent ships to handle the complexity and unpredictability of the real world can be significantly enhanced. Furthermore, these environments allow researchers to replicate specific maritime conditions, thereby gaining a deeper understanding of intelligent ship behavior in various extreme situations.

Instantiated environments, combined with deep reinforcement learning, can solve most ship maneuvering decision problems. Deep reinforcement learning integrates the powerful feature extraction capability of deep learning with the decision-making mechanism of reinforcement learning. Its basic principle is to learn how to make the best decisions in a given task through interaction with the environment. In this process, an agent (in the context of ship maneuvering, the ship itself) learns by trying different actions and observing the results, thus identifying which actions lead to more favorable outcomes. Deep learning networks are used to interpret complex inputs (such as sensor data) and assist the agent in understanding its environment. (Sivaraj et al. 2023) explore various deep reinforcement learning algorithms for controlling a ship's path-following in different conditions, with the state space including factors like yaw rate, heading error, and cross-track error. This representation aids in accurately assessing the algorithms' effectiveness in real-world-like maritime conditions. (Yang & Han 2023) focuses on advancing reinforcement learning techniques for autonomous navigation of ships to avoid collisions in dynamic environments. The state space in the experimental environment is detailed and considers multiple factors influencing ship navigation, such as environmental conditions, ship dynamics, and obstacle characteristics.

The application of deep reinforcement learning in the field of autonomous ship control demonstrates its immense potential in handling uncertainty and making real-time decisions. During navigation, ships frequently encounter changing environments and unpredictable situations, such as sudden severe weather and busy maritime traffic. Deep reinforcement learning algorithms, by learning from past experiences, can gradually understand and predict potential patterns and dynamics in these complex environments, enabling autonomous ships to make more effective and rational decisions in the face of uncertainty. (Teitgen et al. 2023) presents a novel approach to autonomous ship navigation in dense maritime environments, introduces a unique approach to trajectory planning using collision grids, the state space also defined by a radar-like collision grid surrounding the ship. (Wei & Kuo 2022) addressing the problem of multi-ship collision avoidance in congested waterways, the state space is defined as the set of environmental information received by the ship at each time step.

(Xie et al. 2020) explores a novel approach for enhancing ship collision avoidance using a combination of different learning methods, the state space for collision avoidance is defined in terms of relative motion states, such as Distance to Closest Point of Approach (DCPA), Time to Closest Point of Approach (TCPA), relative distance (RT), and relative position direction (hT). These states represent various original motion states and are suitable for policy learning in the context of collision avoidance.

There are still many studies on applying deep reinforcement learning to the field of navigation, such as: (Wang et al., 2022) presents an innovative approach to control the path-following of underactuated ships, this paper developing a data-driven controller that integrates various components like the object environment, action strategies, and reward systems. (Chun et al. 2021) introduces a method for autonomous ship navigation that assesses collision risk and then generates an avoidance path, the state space is designed to be comprehensive and detailed to effectively guide the DRL process, with certain elements normalized for consistency and effectiveness in learning. (Sawada et al. 2021) proposes a novel approach for ship collision avoidance using deep reinforcement learning (DRL) in continuous action spaces, the Obstacle Zone by Target (OZT) method is utilized to compute areas where collisions might occur in the future, based on dynamic information of ships.

However, some studies do not specifically specify the state space, and the space complexity will have a direct impact on the decision-making algorithm. for example: (Chen et al. 2021) presents a cooperative collision avoidance method for multiple ships using a Multi-Agent Deep Reinforcement Learning (MADRL) algorithm, this paper validated in three typical scenarios: head-on, overtaking, and crossing situations. (Zheng et al. 2020) proposes a novel approach using inverse reinforcement learning (IRL) for autonomous collision avoidance in merchant ships, the state space in the experiments is modeled to represent various aspects crucial for ship collision avoidance. (Jiang et al. 2022) integrates attention-based deep reinforcement learning (ADRL) with a focus on risk assessment and motion planning, but the state representation in this method wasn't detailed in these paper.

This study focuses on exploring strategies and techniques for autonomous ship control using Deep Reinforcement Learning (DRL) in instantiated environments. By creating a simplified yet representative maritime navigation simulation environment that reflects the ship control action space, we can simulate various real-world navigation scenarios, including dynamic and static obstacles, different weather conditions, and multi-ship interactions. This provides an ideal platform for in-depth study and optimization of autonomous ship navigation strategies. This paper primarily utilizes an enhanced Deep Q-Network (DQN) model for training. The model is specially optimized to suit complex ship navigation needs, including an efficient experience replay mechanism and a meticulously designed reward function. The experience replay mechanism enables the model to learn key decisions from past experiences more effectively, while the reward function aims to guide the model in making safe and efficient navigational decisions, in compliance with the International Regulations for Preventing Collisions at Sea (COLREGs).

2 EXPERIMENTAL ENVIRONMENT

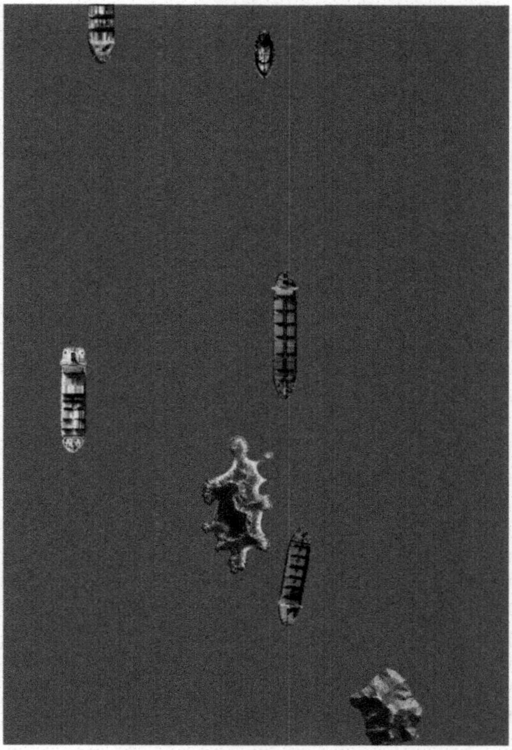

Figure 1. Instantiated ship navigation environment.

As shown in Figure 1, we have constructed a simulation environment for ship navigation, implemented using the Python language and the pygame library. This environment simulates real-world maritime scenarios, including obstacles such as islands and other ships, providing a practical training and testing platform for Deep Reinforcement Learning (DRL) models. The experimental environment is designed to simulate the navigation of ships in the ocean, particularly in the presence of obstacles such as islands and other vessels. This provides an experimental basis for studying the autonomous navigation capabilities and decision-making processes of ships.

By conducting experiments in this environment, our goal is to enhance the performance of DRL algorithms in complex maritime environments, particularly in aspects such as path planning and collision avoidance. The design of the experimental environment not only supports decision-making for ships in dynamic scenarios but also facilitates the study of how ships respond and adapt under different conditions. This provides an effective experimental platform for in-depth exploration and enhancement of the capabilities of autonomous ship navigation systems.

Instantiated environments play a crucial role in the study of autonomous decision-making and maneuvering of ships, especially in testing and validating autonomous decision-making algorithms, such as deep reinforcement learning algorithms. These environments provide a highly realistic and controllable simulation platform, enabling researchers to conduct comprehensive and in-depth testing of algorithms without the need for actual sea trials. For example, in instantaneous environments, researchers can simulate various complex marine conditions and emergency scenarios to test the responsiveness and adaptability of autonomous ship decision-making systems under extreme and non-standard conditions.

Deep reinforcement learning algorithms, especially in the field of ship maneuvering, rely on continuous interaction with the environment to learn optimal decision-making. Instantaneous environments provide such an interactive platform, allowing algorithms to self-improve and adapt under various simulated conditions. This approach not only accelerates the learning process but also significantly enhances the robustness and generalization capabilities of algorithms through extensive simulated trials. Furthermore, instantaneous environments allow for repeat experiments and scenario reproduction, which is crucial for algorithm evaluation and comparison, as it provides a standardized and consistent testing environment.

3 METHODOLOGY

3.1 *Deep reinforcement learning*

Reinforcement learning is the process where an agent learns by interacting with the environment and receiving rewards from it. It differs from traditional supervised and unsupervised learning in that it does not rely on a predefined dataset, but instead learns through rewards or feedback from the environment in response to its actions. Deep learning offers powerful perceptual abilities to handle complex, high-dimensional environmental features, such as in image recognition, speech recognition, and machine translation. DRL leverages these perceptual capabilities to extract features from high-dimensional data and interact with the environment to make decisions. Since DeepMind's introduction of Deep Q Networks (DQN) in 2015, DRL has achieved significant accomplishments in various application scenarios, such as gaming, navigation, and multi-agent collaboration, sometimes even surpassing human performance.

DRL demonstrates the capability to reach or surpass human decision-making levels in closed, static, and deterministic environments, as evidenced in areas like gaming and robotic control. DRL algorithms are primarily divided into value function algorithms and policy gradient algorithms. Value function algorithms indirectly obtain policies through iteratively updating value functions, while policy gradient algorithms directly select actions through policy networks. In ship maneuvering, DRL can effectively handle high-dimensional environmental features and make complex navigational decisions. However, DRL faces challenges including the exploration of large action spaces, handling sparse and delayed reward signals, and robustness against environmental noise. In autonomous ship navigation, these challenges can lead to inefficient learning and unstable decision-making, thereby affecting the safety and efficiency of the vessel.

In autonomous ships navigation tasks, DRL can be used for multiple aspects, including controller optimization, path and trajectory planning, motion planning, and dynamic route planning. These tasks involve the development of advanced driving strategies, scenario-based strategy learning, and prediction of intentions of traffic participants such as pedestrians and vehicles. DRL is capable of handling these complex navigation tasks, providing strategies for safety and risk assessment execution. Traditional methods in autonomous driving often include rule-based decision-making, PID control, etc. In contrast, DRL methods are better suited to handling uncertainties and complex environmental dynamics. For example, DRL can learn optimal driving strategies in dynamic environments, whereas traditional methods might require predefined rules and parameters. However, DRL faces numerous challenges in practical applications, such as dealing with high-dimensional state spaces, designing reward functions, and the disparity between simulators and the real world.

3.2 *Deep Q networks*

Deep Q Networks (DQN), as an advanced method combining deep learning with reinforcement learning, have been proven highly effective in handling high-dimensional perceptual inputs and making complex decisions. The DQN model algorithm for navigation environments is as shown in Algorithm 1. We will discuss the network architecture, particularly improvements to the experience replay method and the design of the reward function, and how the DQN model can optimize the path planning and maneuvering strategies of ships, including how to handle dynamic obstacles, uncertain environmental conditions, and effective decision-making. We will also discuss potential improvements to the DQN model to enhance its performance and robustness in applications of autonomous ship navigation.

Algorithm 1. Improved Deep Q-Network Algorithm.

1: Initialize replay memory D to capacity N
2: Initialize action-value function Q with random weights
3: Initialize target action-value function $\hat{Q}$ with weights $\theta^{-} = \theta$
4: **for** episode = 1, M **do**
5: Observe initial state S_1
6: **for** t = 1, T **do**
7: With probability ε select a random action a_1
8: otherwise select $a_t = \max_a Q(s_t, a: \theta)$
9: Execute action a_t and observe reward r_t and new state s_{t+1}
10: Store transition (s_t, a_t, r_t, s_{t+1}) in D
11: Sample random mini-batch of transitions from D
12: Set $y_j = r_j + \gamma \max_{a'} \hat{Q}(s_{j+1}, a'; \theta^{-})$ for non-terminal states
13: Perform a gradient descent step on $(y_j - Q(s_j, a_j; \theta))^2$
14: Every C steps reset $\hat{Q} = Q$
15: **end for**
16: **end for**

The DQN algorithm used in this study is closely integrated with the specific requirements of ship navigation in its design. By incorporating an experience replay mechanism and prioritized replay, the model can effectively learn important decisions from past experiences. Concurrently, considering the safety and efficiency in ship navigation, the reward function is designed to reward actions that avoid collisions, maintain stable conditions, and adhere to COLREGs rules for navigation. This DQN model structure, which integrates safety, efficiency, and rule compliance, provides robust decision support for autonomous ship navigation, ensuring effectiveness and safety in complex and dynamic maritime environments.

(1) Initialization D
First, initialize a replay memory named D with a capacity of N. This memory is used to store the experiences of the agent and for replay and learning during training. Then initialize the action-value function Q, which is used to estimate the value of executing a specific action in

a given state. Initially, this function uses random weights. Next, initialize the target action-value function $\hat{Q}$, which is used to stabilize the learning process. Initially, the weights θ^{-} of $\hat{Q}$ are set to be the same as those of Q.

(2) Processing each episode
The algorithm iterates through each episode, from the 1st to the M th episode. At the start of each episode, observe the initial state s_1.

(3) Processing each time step
At each time step t, the agent decides whether to randomly select an action a_1 (with probability ε, ε-greedy strategy), or choose the optimal action according to the current Q function (i.e., maximizing $Q(s_t, a;\theta)$). After executing the chosen action a_t, the agent observes a reward r_t and a new state s_{t+1}. Store the transition (state, action, reward, new state) (s_t, a_t, r_t, s_{t+1}) in the replay memory D. Randomly sample a batch of transitions from D for training. For non-terminal states, set $y_j = r_j + \gamma \max_{a'} \hat{Q}(s_{j+1}, a'; \theta^{-})$, where γ is the discount factor, used to calculate the present value of future rewards. Use these sampled data to perform gradient descent optimization on the Q function, to reduce the difference between predicted value and actual obtained value. Every C steps, update the weights of the target network $\hat{Q}$ to be equal to the current weights of the Q function. This helps improve the stability of learning.

In this study, the action space is considered to be discrete values such as acceleration, deceleration, turning left, turning right, and their combinations, making it relatively simple. The focus is on how the characterization of the state space and the reward function directly relates to the model's generalization capability. Therefore, this section will primarily introduce how to construct the model's state space and reward function.

The state space refers to the collection of all states that can be perceived by the agent. In autonomous ship navigation, the state space includes information such as the ship's position, speed, heading, as well as surrounding environmental information like the position, size, and shape of obstacles. In the instantiated maritime navigation environment, the dimension of the state space is N, where N is the number of state variables. The dimension of the state space depends on the chosen state variables and can be adjusted according to specific circumstances. For the specific requirements of ship navigation, we evaluated various uncertain state spaces. A common approach is using a fixed-size state representation, selecting the top N objects most relevant to the ship on the screen to ensure consistency in the state space. Although this simplifies the problem, it may lose important information beyond this scope. The feature engineering approach aggregates information from multiple objects into high-level features, such as average position or quantity, thereby reducing input dimensions but potentially sacrificing some detailed information. In the instantiated maritime navigation environment, we chose a fixed-size state representation to maintain consistency in the state space while retaining sufficient detailed information.

The reward function is the reward or punishment an agent receives based on environmental feedback after executing an action. The design of the reward function directly affects the learning efficiency and performance of the model. In autonomous ship navigation, the design of the reward function should consider the safety and efficiency of the ship. For example, the reward function can encourage a ship to maintain a safe distance to avoid collisions with other ships; it can also encourage staying in the center of the navigation channel to avoid deviating from the course. Additionally, the reward function can encourage the ship to make correct decisions in emergencies, such as avoiding collisions with obstacles. In the instantiated maritime navigation environment, we designed a series of reward functions to encourage the ship to make correct decisions in various situations. These reward functions are based on the International Regulations for Preventing Collisions at Sea (COLREGs), to ensure the safety of the ship and the necessity of adhering to COLREGs. Furthermore, we designed a series of punishment functions to penalize the ship for making incorrect decisions in different scenarios.

3.3 *Algorithm improvements*

3.3.1 *Long-short term experience replay*

Experience replay is a key technique in deep reinforcement learning that allows the model to store and reuse past learning experiences. In this process, each step of the agent's operation, including state, action, reward, and new state, is recorded. These records are then stored in a data structure known as the experience replay buffer. During subsequent training, these experiences are randomly drawn and reused, helping to break the correlation between data samples, thereby enhancing the model's learning efficiency and generalization ability.

Common optimization directions in experience replay include prioritized replay and the separation of short-term and long-term experience replay. Prioritized replay adjusts the probability of experiences being reused based on their importance (for example, based on the size of the reward), allowing more important experiences to participate more frequently in training. Short-term and long-term experience replay differentiate between recent and historical experiences, employing different strategies for recent and long-term learning processes to meet different learning requirements.

To address the specific needs of ship navigation, we will adopt a strategy that combines long-term and short-term experience replay, with an optimization of reward weight priorities for long-term experience replay. Specifically, for crucial actions like turning, we will set higher priority weights to ensure that these important experiences are reviewed and learned more frequently. The purpose of this optimization is to enhance the robustness of the model, particularly in complex and dynamic maritime environments, ensuring that the model effectively learns and remembers decisions and actions critical to successful navigation.

$$Q(s_t,a_t)\leftarrow Q(s_t,a_t)+\alpha\cdot\omega(s_t,a_t)\times\left[r_t+\gamma\max_a Q(s_{t+1},a)-Q(s_t,a_t)\right] \quad (1)$$

In the given formula, $Q(s_t, a_t)$ represents the Q-value associated with a specific state s_t and action a_t. The update operator $\leftarrow$ signifies the process of updating this Q-value. The learning rate α plays a crucial role in determining how much new information influences the update, essentially controlling the learning speed. Additionally, $\omega(s_t,a_t)$ denotes the priority weight of the experience, assigning higher weights to actions that yield higher rewards. The term r_t is the immediate reward received after taking action a_t. The discount factor γ is crucial in balancing the importance of immediate versus future rewards, with higher values placing more emphasis on future outcomes. Finally, $\max_a Q(s_{t+1},a)$. calculates the maximum Q-value among all possible actions in the subsequent state s_{t+1}, guiding the selection of the next action by forecasting its potential reward outcome.

As shown in equation (2), when long-term memory storage is incorporated into the DQN algorithm, it not only utilizes current learning experiences but can also learn from past experiences. This is achieved through the use of replay memory, which stores previous states, actions, and rewards. During training, random sampling from this memory is used to reduce data correlation and prevent overfitting. This approach helps to improve the stability and performance of the algorithm.

$$Q(s_t,a_t)\leftarrow Q(s_t,a_t)+\alpha[r_t+\gamma\max_a Q(s_{t+1},a)-Q(s_t,a_t)] \quad (2)$$

In this formula, $Q(s_t, a_t)$ represents the Q-value for the current state s_t and action a_t. This value is adjusted by the learning rate α, incorporating the immediate reward r_t and an estimate of future rewards, calculated using the discount factor γ and the maximum Q-value for all possible actions in the next state s_{t+1}, $\max_a Q(s_{t+1},a)$. This update rule helps the agent to learn the strategy of taking optimal actions in various states.

In the augmented DQN algorithm with priority weighted long-term memory, the formula integrates an additional factor, the priority weight, into the standard update mechanism. This approach emphasizes learning from more significant experiences. The formula, split for a two-column LaTeX layout, is as follows:

$$Q(s_t,a_t)\leftarrow Q(s_t,a_t)+\alpha\cdot\omega(s_t,a_t)[r_t+\gamma\max_a Q(s_{t+1},a)\quad -Q(s_t,a_t)] \quad (3)$$

In the updated Deep Q-Network algorithm with priority weighting, the formula $Q(s_t,a_t)\leftarrow Q(s_t,a_t)+\alpha\cdot\omega(s_t,a_t)[r_t+\gamma\max_a Q(s_{t+1},a)-Q(s_t,a_t)]$ encapsulates several key elements. $Q(s_t,a_t)$ is the Q-value, representing the expected utility of taking action a_t in state s_t. The update operator $\leftarrow$ denotes the process of revising this Q-value. The learning rate α indicates how much the new information will override the old, balancing between exploration and exploitation. Significantly, $\omega(s_t,a_t)$ introduces a priority weight for the experience at this state-action pair, emphasizing learning from more rewarding experiences. r_t is the immediate reward received after the action, while γ serves as the discount factor, reducing the emphasis on distant future rewards. Lastly, $\max_a Q(s_{t+1},a)$ reflects the maximum expected reward for the next state, guiding the choice towards actions that promise higher future returns.

In this optimization process, by introducing priority weights $\omega(s_t,a_t)$, the model is ensured to focus more on crucial and highly rewarded actions during learning, thereby increasing the efficiency and depth of memory for important decisions. This approach enables the model to respond more accurately and stably when facing critical navigation decisions.

3.3.2 *Reward function design*

In reinforcement learning, the reward function plays a crucial role, providing necessary guidance for the agent, helping it learn how to complete tasks and maximize its total rewards. An effective reward function should accurately reflect the objectives and requirements of the task, enabling the agent to continuously optimize its behavior through interaction with the environment. When designing the reward function, it should be ensured that it is concise, avoiding being overly complex or ambiguous, so that the agent can clearly understand its objectives.

In the training of decision neural networks for ship navigation, a variety of factors should be comprehensively considered to ensure that the model can effectively learn and make reasonable decisions. First and foremost, safety is the most important consideration, including avoiding collisions with other ships or obstacles. Efficiency is also crucial, requiring optimization of navigation routes to reduce travel time or energy consumption. Additionally, adherence to maritime traffic rules (such as COLREGs) is equally important for ensuring navigational safety. Moreover, considering environmental factors, such as changes in weather and sea conditions, places higher demands on the adaptability of the model.

The reward function in this study is designed as follows:

$$\text{Reward}=R_{\text{collision}}+R_{\text{maintain\_status}}+R_{\text{game\_over}}+R_{\text{level\_complete}}+R_{\text{specific\_objective}},$$

where each component is defined as:

- $R_{\text{collision}}$: A positive reward of +5 is given if an action successfully avoids a collision.
- $R_{\text{maintain\_status}}$: A positive reward of +1 is given if the ship maintains a stable status for a certain period (10 seconds).

- $R_{\text{game\_over}}$: A negative reward of -500 is given if a collision occurs, leading to the end of the game.
- $R_{\text{level\_complete}}$: A positive reward of +5 is given if other ships disappear from the screen, and an additional +2 is awarded if islands disappear from the screen as the own ship progresses.
- $R_{\text{specific\_objective}}$: An extra positive reward of +3 is given if a specific collision avoidance action is executed correctly, such as a proper right turn to give way.

The reward function designed in this study integrates multiple factors, including safety, efficiency, and rule compliance, aiming to guide the ship's agent in learning how to complete navigational tasks safely and effectively. By setting clear rewards or punishments for specific actions in various scenarios, the reward function enables the model to not only optimize the navigation route but also adhere to COLREGs rules, ensuring safety during navigation. Such a design helps to improve the adaptability and accuracy of the model in actual navigational environments.

4 EXPERIMENT AND RESULT ANALYSIS

4.1 *Experimental settings*

In our study, the DQN algorithm was augmented with priority-weighted long-short term memory and a custom reward function to effectively train a model for maritime navigation. The algorithm configuration was carefully designed to optimize performance and learning efficiency. The learning rate was set to 0.001, balancing the need for rapid learning while avoiding the instability associated with higher rates. The discount factor, crucial for calculating the future reward, was set at 0.99, ensuring that future rewards have substantial influence on the current decision-making process. The capacity of the priority-weighted replay memory was set to 50,000 experiences, allowing the model to learn from a wide range of past experiences while prioritizing more significant ones.

Our model employed a deep neural network with three hidden layers. The first layer consisted of 64 neurons, the second 128, and the third 64, with ReLU (Rectified Linear Unit) activation functions to introduce non-linearity. The input layer's size corresponded to the dimensionality of the state space, and the output layer had a neuron for each possible action, providing the estimated Q-values. This architecture was chosen for its ability to efficiently process the complex input data while providing accurate action predictions.

4.2 *Training process*

Considering the complexity of the experimental scenario, we have divided the experiment into two phases.

(1) Initial Learning Phase
The training commenced in a static environment, where the agent (a single ship) learned the basic maneuvers without the interference of external factors. This phase focused on understanding the fundamental dynamics of boat control and response to various actions, using the custom reward function to guide the learning process.

(2) Advanced Learning Phase
Once the agent achieved proficiency in basic navigation, the training environment was enriched with dynamic elements, including other moving vessels. This phase aimed to teach the agent advanced navigation skills, like collision avoidance and path optimization, in a more realistic and challenging setting. The difficulty of tasks was gradually increased. We started with simple objectives, such as reaching a designated location, and progressively introduced more complex scenarios involving navigation through crowded waters and reacting to unpredictable movements of other vessels.

Throughout the training process, the agent's performance was continuously monitored. Criteria such as the success rate of completing routes without collision, time taken to reach destinations, and the ability to adapt to dynamic environments were used to evaluate and iteratively improve the model. The training utilized a combination of real-time data and stored experiences from the priority-weighted replay memory. This approach allowed the agent to learn from both immediate interactions with the environment and past experiences, ensuring a comprehensive learning process.

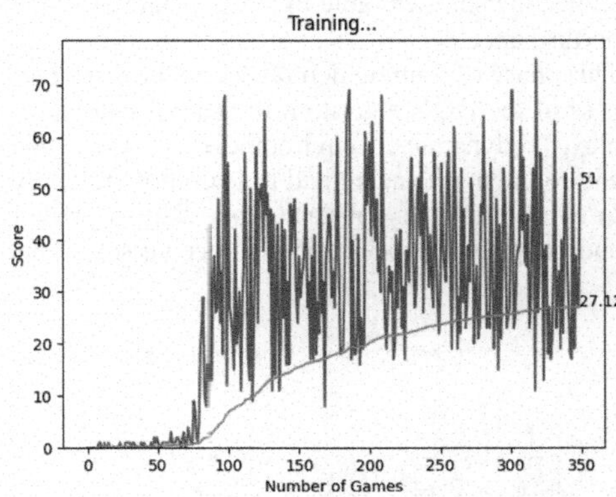

Figure 2. Training results in static environment.

4.3 *Result analysis*

Through extensive experiments, we have verified the effectiveness and robustness of the DQN algorithm in the field of ship maneuvering. Our experimental results indicate that the DQN algorithm can effectively learn and make decisions in complex, dynamic, and uncertain maritime environments, thereby achieving efficient autonomous navigation of ships.

(1) Learning Process in a Static Environment
As shown in Figure 2, in the first phase of the experiment, the objective was to teach the ship to avoid collisions in a static environment. The experimental results showed that initially, the ship hardly gained any rewards, meaning it frequently collided with islands. This phenomenon is normal in the learning process of the DQN algorithm, as initially the agent lacks understanding of the environment, and its actions are more based on random exploration rather than strategic guidance. However, after about 50 explorations, the ship learned to avoid collisions with islands, indicating that the agent began to understand and apply knowledge learned from the environment.
This progress reflects a key advantage of the DQN algorithm: improving strategies through continuous trial and error and feedback learning. With more and more exploration, the agent gradually developed effective action strategies by continuously updating its Q-value network. The success of this phase is attributed to the ability of DQN to learn from experience and adjust actions to maximize the expectation of future rewards.

(2) Advanced Training in a Instantiated Environment
Figure 3 demonstrate the experiment placed the ship in a instantiated environment with moving vessels. In this more challenging environment, the agent needed to learn how to avoid collisions while dealing with the uncertainty and dynamics of other ships. After about 2000 training iterations, the ship was able to reach a state of stable performance.
This phase of training demonstrated the effectiveness of the DQN algorithm in dealing with complex, dynamic environments. In this environment, the agent had to learn not only how to avoid stationary obstacles but also to predict and adapt to the behavior of other vessels. This required the agent to quickly learn and make adaptive adjustments from the constantly changing environmental feedback. The experimental results showed that, after thousands of training iterations, the agent gradually developed an understanding of the dynamic environment, effectively avoiding collisions while maintaining navigational efficiency.
The reason the DQN algorithm requires repeated practice to maintain robustness in instantiated environments is mainly that its learning process depends on interaction with the environment. In the initial stages, the agent's decisions are based on limited knowledge, and with an increasing number of interactions, it gradually accumulates experience and optimizes its decision-making strategies. Furthermore, the uncertainty and complexity of dynamic environments require the agent to adapt flexibly to new situations and changes. Therefore, through extensive training, the agent can exhibit higher adaptability and robustness when facing the unknown and changes.
The replay memory and target networks in the DQN algorithm play a significant role in enhancing robustness. Replay memory allows the agent to learn from past experiences, not just recent ones, helping to form a more comprehensive strategy. The target network provides a stable goal to guide the agent's learning process. These mechanisms work together to make the DQN algorithm exhibit good robustness and adaptability in complex environments.

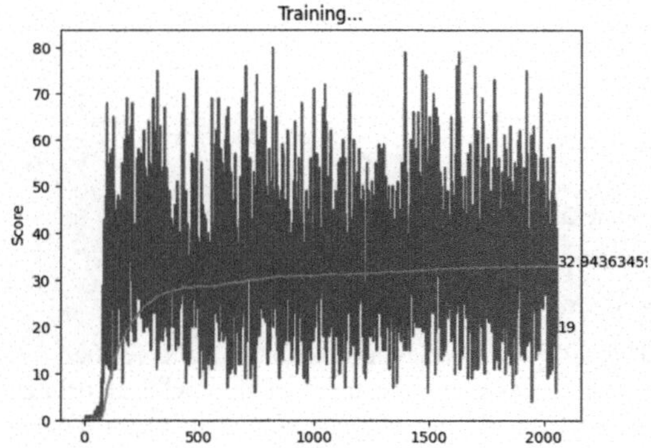

Figure 3. Training results in instantiated environment.

(3) Limitations
Due to the limitations of the pygame library, our experiments did not fully consider the impact of marine environmental factors such as wind, waves, and currents on ship navigation. The impact of these factors on ship navigation is complex and requires more detailed modeling and simulation. Furthermore, our experiments did not consider the impact of the ship's hull design and maneuvering characteristics on navigation. The impact of these factors is also complex and requires more detailed modeling and simulation. From the perspective of reward function design, we did not consider enough COLREGs rules. For example, in further research, the logic of ship navigation lights at night could be incorporated, which is an area that requires further improvement.

5 CONCLUSION

We developed an instantiated marine navigation environment, where each agent is represented by a specific ship model with particular maneuvering characteristics. Our training environment uses a bird's-eye view, including dynamic and static obstacles, closely simulating real-world navigation conditions. In this environment, we trained using an improved DQN algorithm. To enhance training efficiency, we employed a new strategy, combines long-term and short-term experience replay method that avoids repetitive experiences during training. We tested this approach in the instantiated environment, and the results demonstrate that it can achieve autonomous navigation within this environment and avoid collisions with other ships during training. The enhanced DQN algorithm further improved the robustness and practicality of our model. Through training in both static and dynamic environments, the agents learned how to navigate effectively while avoiding collisions. Although the initial learning process may take longer, as experience accumulates, the agents can adapt to more complex and unpredictable environments, demonstrating the powerful potential and adaptability of the DQN algorithm.

ACKNOWLEDGMENTS

This research was partially sponsored by 2023 DMU navigation college first-class interdisciplinary research project (2023JXA(15)), and supported by International cooperation training program for innovative talents of Chinese Scholarships Council (Grant No. CSC [2022] 2260).

REFERENCES

Chen, Chen, Feng Ma, Xiaobin Xu, Yuwang Chen & Jin Wang (2021). "A novel ship collision avoidance awareness approach for cooperating ships using multi-agent deep reinforcement learning". *Journal of Marine Science and Engineering* 9.10, p. 1056.

Chun, Do-Hyun, Myung-Il Roh, Hye-Won Lee, Jisang Ha & Donghun Yu (2021). "Deep reinforcement learning-based collision avoidance for an autonomous ship". *Ocean Engineering* 234, p. 109216.

Jiang, Lingling, Lanxuan An, Xinyu Zhang, Chengbo Wang & Xinjian Wang (2022). "A human-like collision avoidance method for autonomous ship with attention-based deep reinforcement learning". *Ocean Engineering* 264, p. 112378.

Munim, Ziaul Haque (2019). "Autonomous ships: a review, innovative applications and future maritime business models". *Supply Chain Forum: An International Journal*. Vol. 20. 4. Taylor & Francis, pp. 266–279.

Peng, Zhouhua, Jun Wang & Dan Wang (2017). "Distributed containment maneuvering of multiple marine vessels via neurodynamics-based output feedback". *IEEE Transactions on Industrial Electronics* 64.5, pp. 3831–3839.

Sawada, Ryohei, Keiji Sato & Takahiro Majima (2021). "Automatic ship collision avoidance using deep reinforcement learning with LSTM in continuous action spaces". *Journal of Marine Science and Technology* 26, pp. 509–524.

Sivaraj, Sivaraman, Awanish Dubey & Suresh Rajendran (2023). "On the performance of different Deep Reinforcement Learning based controllers for the path-following of a ship". *Ocean Engineering* 286, p. 115607.

Teitgen, Raphael, Bruno Monsuez, Romain Kukla, R Pasquier & G Foinet (2023). "Dynamic trajectory planning for ships in dense environment using collision grid with deep reinforcement learning". *Ocean Engineering* 281, p. 114807.

Wang, Xinyu Zhang, Zaili Yang, Musa Bashir & Kwangil Lee (2023). "Collision avoidance for autonomous ship using deep reinforcement learning and prior-knowledge-based approximate representation". *Frontiers in Marine Science* 9, p. 1084763.

Wei, Guan & Wang Kuo (2022). "COLREGs-compliant multiship collision avoidance based on multi-agent reinforcement learning technique". *Journal of Marine Science and Engineering* 10.10, p. 1431.

Wu, Chuanbo, Wangneng Yu, Guangze Li & Weiqiang Liao (2023). "Deep reinforcement learning with dynamic window approach based collision avoidance path planning for maritime autonomous surface ships". *Ocean Engineering* 284, p. 115208.

Xie, Shuo, Xiumin Chu, Mao Zheng & Chenguang Liu (2020). "A composite learning method for multi-ship collision avoidance based on reinforcement learning and inverse control". *Neurocomputing* 411, pp. 375–392.

Yang, Xiao & Qilong Han (2023). "Improved reinforcement learning for collision-free local path planning of dynamic obstacle". *Ocean Engineering* 283, p. 115040.

Zheng, Mao, Shuo Xie, Xiumin Chu, Tianquan Zhu & Guohao Tian (2020). "Research on autonomous collision avoidance of merchant ship based on inverse reinforcement learning". *International Journal of Advanced Robotic Systems* 17.6, p. 1729881420969081.

Zhou, Xiang-Yu, Jin-Jing Huang, Feng-WuWang, Zhao-LinWu & Zheng-Jiang Liu (2020). "A study of the application barriers to the use of autonomous ships posed by the good seamanship requirement of COLREGs". *The Journal of Navigation* 73.3, pp. 710–725.

Autonomous ships

Advances in Maritime Technology and Engineering – Guedes Soares & Santos (Eds)

Operational support framework for maritime autonomous surface ships under onshore operation centers

M. Adnan
Department of Technology and Safety, UiT The Arctic University of Norway, Norway

L.P. Perera
Department of Technology and Safety, UiT The Arctic University of Norway, Norway
SINTEF Digital, SINTEF AS, Oslo, Norway

ABSTRACT: In recent years, the concept of Maritime Autonomous Surface Ships (MASSs) has gotten attention both from the research community and shipping industry. The recent development in advanced sensing, communication, and control technologies makes it possible to operate such vessels remotely through Onshore Operation Centers (OOCs). However, there are still several challenges that need to be addressed before such ships can be used safely and effectively in ocean environments. To handle these challenges faced by MASSs, a large-scale acceptance procedure from commercial industries and regulatory authorities, an on OOC framework is proposed in this research study. However, this study is limited to an OOC operational support framework for operational monitoring, guidance, and support. It shows how the proposed framework will support online MASS operations in ocean environments. This opens the ways to develop and operate the MASS concept through OOCs to control future vessels remotely by the crew sitting in onshore centers. It will enhance the MASSs technology acceptance both from the technological and regular regulatory perspectives. The study also identifies such challenges and their viable solutions in the aspect of operational support framework as well.

1 INTRODUCTION

Autonomous ships can introduce new and exciting technology concepts that could revolutionize the shipping industry. It is expected that there will be self-driving vessels that can navigate themselves without human input in the future. Autonomous ships can be operated by remotely or independently, depending on the availability of resources and infrastructure with respect to the desired level of autonomy. Autonomous ships can be programmed with many different goals for their missions, such as simply carrying cargo from port A to B, delivering medical supplies across dangerous territories, monitoring purposes, etc. Once such vessels have been programmed with mission goals, that can create internal maps in bridge systems, i.e. which ship routes such vessels should take (based on available information), as well as which operational conditions should execute (based on previous experiences). This allows future vessels to operate more efficiently than human operators because such vessels do not need pilot guidance through every route; instead, such vessels can rely on their autonomous navigation system (Li & Yuen (2022)).

Autonomous ships also got attention from multiple users ranging from commercial shipping companies, search and rescue operators, military units, for monitoring purposes, surveying, etc., with various reasons such as less operational cost in harsh sea weather, increase in safety requirements from policymakers. The shortage of human labor force in the shipping industry, etc. (Ziajka-Poznańska & Montewka (2021)) can also be considered as another reason. Advanced automation technologies such as IoT sensors, AI, data science, etc., enable autonomous marine system implementations much faster. Recent research and development activities show that autonomous ships will soon be available for commercial use on a large scale (Porathe (2014)), specifically for short sea shipping.

Different research and international organizations, such as the International Maritime Organization (IMO), Lloyd's Register, Bureau Veritas, Norwegian Forum for Autonomous Ships (NFAS), Rolls-Royce, and UK Marine Industries Alliance, define the level of autonomy based on different schemes in autonomous shipping according to their understanding (Kim et al. 2022). The International Maritime Organization (IMO) defines autonomy into four degrees, which are usually considered standards in the maritime domain, as described below (IMO (2018)).

DOI: 10.1201/9781003508779-40

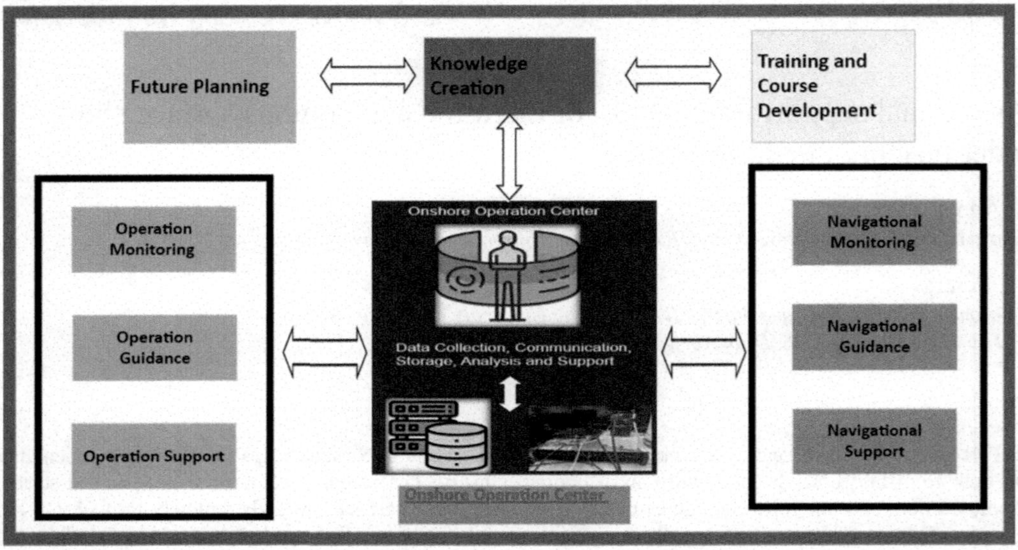

Figure 1. Onshore operation center concept.

- Level 1: Ships have advanced automation and decision support systems with onboard crew.
- Level 2: Remotely controlled ship with an onboard crew.
- Level 3: Remotely controlled ship without an onboard crew.
- Level 4: Fully Autonomous ship

Other organizations, such as Rolls-Royce, Bureau Veritas, NFAS, and UK Marine Industries Alliance, also included the supplementary detailed classification levels of ship autonomy (Kim et al. 2022).

Adopting autonomous ship technologies will follow the steady shift step-by-step based on available technologies and their maturity level, starting from the lowest (level 1) towards the highest level (level 4). It is impossible to shift directly to a higher autonomy level (level 4: Fully Autonomous) without adequately evaluating the available technologies and systems, where various lessons to be learned. To ensure navigation safety, it is recommended to operate autonomous ships under the control or supervision of onshore operations centers (OOCs) at all levels of ship routes (Manno (2014)). The OOCs will play a key role in ensuring safe operations in autonomous shipping. Managing autonomous ships through OOCs will enhance the safety and trust of society and policymakers. It will shift the seafaring crew members' responsibilities from onboard ships to onshore office environments.

The rest of the paper is organized as Section 2, which will give an overview of the proposed OOC and its operational support framework concept; Section 3 will provide details about the operational support framework's major functions in terms of monitoring, guidance, and support, and Section 4 will conclude the overall study.

2 ONSHORE OPERATION CENTER DESIGN

The Arctic University of Norway (UiT) team is currently working on a OOC development project (uit.no/autoship), which will help the shipping industry's upcoming requirements for online operational support. The OOC is responsible for data collection, communication, storage, analysis, and support for operational and navigation aspects. The OOC also takes part and supports knowledge creation, and that knowledge can be used to support future ship navigators with course curricula. The course development will be utilized to train the maritime operation workforce to fulfill future market needs as per future technology. This center will also provide advanced training and knowledge to the shipping industry professionals for better resource management. The future industry will need these advanced OOCs for online operational support for shipping logistic operations based on autonomous navigation principles.

The current study introduces monitoring, guidance, and support functions for the operational aspect of future autonomous shipping. It discusses how the proposed framework for operation is linked with OOCs for data collection, communication, storage, analysis, and support, as shown in Figure 1. The main goal of this study is to introduce a high-level overview of the OOCs in terms of operational perspective, specifically for autonomous shipping, as shown in Figure 2. Operators in OOCs are divided into three groups based on their expertise and

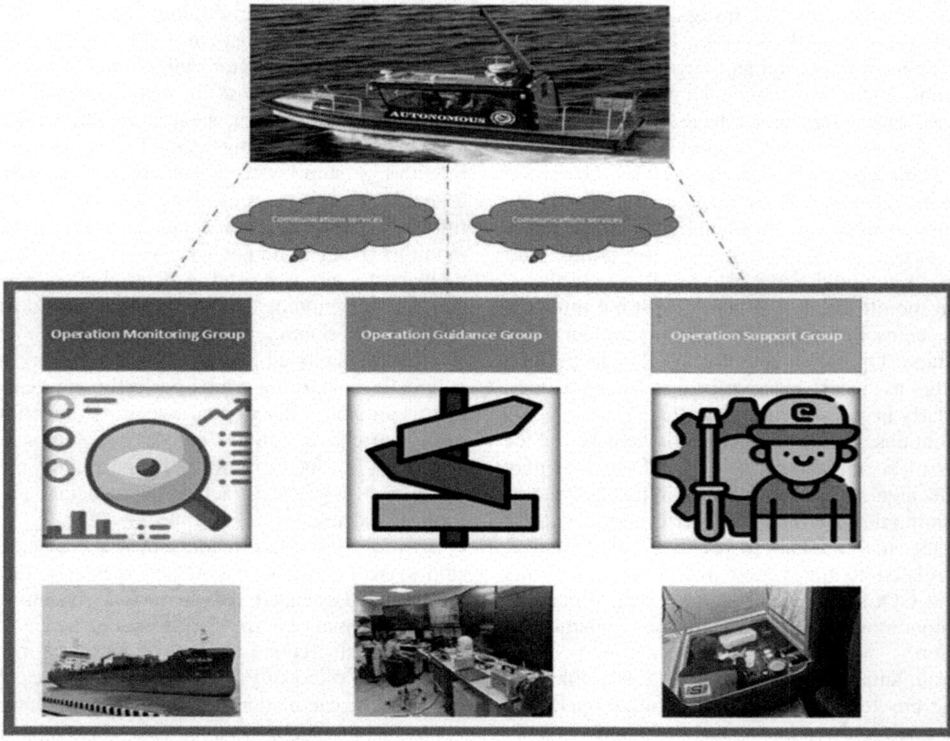

Figure 2. Operational support framework for autonomous shipping.

training: operation monitoring, guidance, and support groups.These operator divisions will help to manage the task more optimally than among operators who would overlook all diverse functions (like monitoring, guidance, and support) in OOCs. This study will address the following points in detail.

1) Operational support framework
2) Why are these functions necessary for the shipping industry
3) The respective challenges
4) Some viable solutions

These autonomous ships are not only self-deriving vessels but also need other support functions from OOCs for safe operation guarantees.

3 OOC OPERATION FRAMEWORK

3.1 *Operation monitoring*

Due to the absence of human crew onboard autonomous ships, vessel monitoring is crucial for safe operation in the sea environment. Operation monitoring is essential not only for ship navigation but also to create a safe sea environment. The operation monitoring process involves observing ship system operations, overlooking the respective environment, identifying navigation problems, and developing solutions. The purpose of monitoring is to provide services and collect information about ship system performance so that it can be used to make appropriate decisions, determine actions, and improve system performance. The OOC operators will observe autonomous ship operation remotely. This will help to identify navigation or operation issues and take corrective actions remotely as needed, just like the human crew onboard that performs the monitoring tasks.

In autonomous ships, monitoring can be performed automatically by using an advanced AI-based system. For some critical problems during ship operations, sometimes OOC operators need to watch the specific events or behaviors that occur over time and then record them for in-depth analysis. This type of monitoring usually provides more insight into what happened during a particular event or period of the problem. The automated monitoring systems make the OOC operator's job easier to manage because an operator only needs to look at warning messages generated by the systems, not into all historical data collected from different sensors to make conclusions. Automated monitoring systems use advanced technology such as IoT sensors, devices, and AI algorithms to understand the respective situations in systems automatically with greater precision without direct human involvement (Jacq et al. 2018).

If one is going to answer the question of 'Why is operation monitoring important?': The maritime industry is responsible for transporting roughly 80 % of all manufactured goods worldwide, making it one of the most important and large-scale industries (Kim et al. 2022). Unfortunately, some high-profile accidents with cargo ships have occurred due to a lack of monitoring and supervision, even from onboard ship operators (Cao et al. 2023). Operation monitoring is essential because it lets you know when any operational problems (like engine part failure, device failure, system response failure, etc) are happening and the root cause of those problems. Without monitoring the system downtime information, it seems impossible to fix the problem at an early stage. Operation monitoring also helps ship navigators avoid expensive mistakes by identifying issues early in stages (Yang et al. 2019).

Continuous ship monitoring will help to reduce the risk of accidents and system failures that may cause accidents. The operation monitoring systems will communicate online ship performance data to OOCs for analysis and to identify any technical issues related to the system that may arise. This way, the OOCs can gain the maximum efficiency of autonomous ships regarding operational utilization.

Overall, ship operation monitoring is essential to ensuring the vessel's safety and effective operation. This can help reduce the overall operation costs for ships (Veitch et al., 2022). Some of the operation monitoring challenges and solutions can be classified as: i) Safety and security issues, ii) System cost, iii) Monitoring the maintenance process, iv) Data availability issues, v) Limited visibility, and vi) Capacity limitation and communication delay.

3.1.1 *Operation monitoring solutions*

Safety and Security Issues: The ship needs a monitoring system that can identify any problems in its operation, such as fire, flooding, or leaking oil. It also requires the installation of alarm systems to ensure the safety of the vessel, cargo, crew, other vessels, and onboard crew in the surrounding area of the ship. To manage the monitoring challenges, one solution to this problem is developing alarm systems based on intelligent sensors for monitoring each critical ship system (ABS (2016)). Ship structural monitoring is an essential aspect of ship safety. Autonomous ships should be equipped with structural monitoring systems such as hull health status, engine, generators, etc., which will help ensure the vessel, cargo, and crew safety.

Safety protocols and procedures are essential to protect the respective ship, cargo, and crew. High-resolution onboard cameras enhance the remote OOC operator capabilities to overview the issues in operating environments and ship systems. Such intelligent remote monitoring can help to identify and capture issues at the initial stages, allowing the OOC operators to control the situations in case of possible accidents

System Cost: Developing and maintaining a monitoring system that can work intelligently and automatically to monitor autonomous ship operations is costly because it demands much more sophisticated equipment, such as intelligent sensors and AI-based computing technologies. The cost of monitoring ship systems, manually can also be higher because it requires many people to work continuously. However, the accident ratio in manual monitoring can be higher and riskier, so it will eventually cost more in the end. Advanced manufacturing and digital technologies now significantly reduce the costs of developing advanced monitoring systems (Oikos-International(2017)). On the other hand, automatic monitoring works perfectly and generates alarms automatically to gain operators' attention for some corrective actions before possible accident situations. So, such systems will significantly eliminate human negligence factors by reducing possible future accidents.

Monitoring the Maintenance Process: Since every ship system must be maintained regularly, there is always an associated risk that some systems may have abnormal events during maintenance operations, which may cause a severe critical incident like loss of price cargo or loss of life at sea. Such maintenance can be done at ports with ship docks for vessel repairs. So, monitoring maintenance processes for better safety is essential. Computer vision techniques allow OOC operators to watch maintenance processes remotely and ensure it follows the standards and guidelines. Developing such systems can also be challenging in the maritime domain. Researchers have already applied IoT-based methods combined with a selected algorithm (i.e. YOLOv3) for vessel maintenance, design, and deployment monitoring to fulfill the demands of the IoTs-based ship managing scheme (Chen (2022)).

Data Availability Issue: Online data availability may also be challenging for these monitoring systems for better performance. While there are some sources of data available online (such as weather information), many other types of information (such as pollution levels and locally changing weather situations) can be accessed through third parties who have their proprietary systems. The solutions can be investing in advanced technologies and intelligent sensors to optimize the efficiency and performance of autonomous ships, including using IoTs,predictive analytics, and machine learning to minimize downtime and improve the efficiency of the ship's operation.

Limited Visibility: One of the significant challenges in operation monitoring is that vessels may be operating in ocean environments with limited visibility. For example, ships may be sailing in fog or other conditions that make it difficult for OOC operators to have adequate situation awareness. This can make difficult ship navigational situations due to the absence of physical crew onboard in autonomous ships. The OOC operators rely only on sensor information, and sensors may not provide correct or all data due to limited visibility of ship operating areas or communication challenges. A solution to this problem is investing in research and development activities to utilize advanced sensors and technology capable of overseeing such situations. It will also include developing intelligent solutions for autonomous ships, including new control and decision-making/supporting systems based on multiple sensors to overcome these limited visibility issues.

Capacity Limitation and Communication Delay: The other significant challenges in autonomous ships are limited data transfer capacity and communication delays with OOCs. Based on satellite communication for online data transfer from high-speed sensors (big data), another concern is the development of remote ship control systems. One solution to this big data problem is to send only features or essential information to OOCs for data processing and decision-making. Most initial data preprocessing can be done at local computing infrastructure to oversee the limited data transfer issue. As an example, OOCs can receive ship motion data as a substitute for camera images and replicate the operational condition at sea in virtual reality environments (Ota & Okazaki (2022)) based on the respective mathematical models. Communication delay can be managed by 5G and 6G technologies that allow faster communication with a much larger data transfer capacity. Developing and implementing advanced communication technology systems, such as satellite and wireless communication systems, will enable more efficient online monitoring and remote intervention in autonomous ships.

3.2 *Operation guidance*

Operation guidance is an integral part of OOC functions, as it allows OOC operators to remotely intervene with autonomous ships and provide oversight to ensure protected and efficient operations. The OOC operators can provide guidance in different scenarios, such as for maintenance of ship systems, technicians may need guidelines related to installed systems for maintenance purposes. Operational guidance is a crucial element in filling the gaps of the onboard human crew through OOCs. It will support autonomous systems in distinct phases of ship operations, like speed selection, direction, orientation, etc., for safe operation assurance (Gupta et al. 2022).

One option for guidance is to use the Automatic Identification System (AIS) in future ships. The AIS is a system that broadcasts information about ship positions and speeds at sea. It can inform other vessels about vessel locations, although it does not provide any control or command over vessels. The OOCs operators must know current vessel operating locations and sea environment conditions to keep safe autonomous ship operations. It can guide ships during the voyage planning phase for safe operations.

If one is going to answer the question of 'Why is operation guidance important?': Operation guidance in autonomous ships will enhance efficiency and safety and significantly reduce operating costs. Operation guidance will help optimize the ship's performance by providing feedback and recommendations to ship control systems to function safely and efficiently. Operation guidance can also provide additional flexibility and adaptability to the ship's operations. For example, if a vessel encounters unexpected obstacles or challenges, the OOC operator can provide additional guidance and assistance to help the vessel operate safely. Even some factors, such as route blockage or political situations, etc., cannot be modeled in autonomous navigation with enough accuracy, and these events must require guidance from OOCs to maintain safe and secure operations.

The AUVs can be used with GPS coordinates to help and guide them through treacherous waters without relying on more expensive equipment like radar systems or standard radio transmitters installed onboard vessels (Fan et al. 2022). Overall, operation guidance is an essential function of OOCs for autonomous shipping operations, as it grants remote guidance to ensure safe and efficient ship maneuvers.

Some of the operation guidance challenges and viable solutions can be categorized under the following sections i) Lack of trained human resources, ii) Cyber security issue, iii) System failure issue and iv) Human-machine friendly interface issue.

3.2.1 *Operation guidance solutions*

Lack Of Trained Human Resources: Operation guidance faces the challenge of a lack of trained human resources, such as OOC operators, and how OOC operators will provide necessary supervision in case of system failure situations. No standard core curriculum exists to train OOC operators for autonomous shipping. There is a lack of research and development activities in this area to find the operational and functional requirements of autonomous ship systems. One solution is to invest in advanced training programs and find the required modification in existing teaching and research infrastructures. To continue the operation of the autonomous vessels and gain maximum performance efficiency, there is a need to dedicate the emergency maintenance staff ready to provide necessary and immediate support wherever required. In such events of system failures, the OOC operators may utilize AR or VR technology for guidance (Ota & Okazaki (2022)). Even during cargo loading and unloading stages, the OOC operators can provide guidelines to port operators or authorities if required.

Cyber Security Issue: How OOCs will provide guidance is another challenge in the event of cyber-attacks on autonomous ships. The challenge is that autonomous vessels face fast-growing exposure to cyber threats, uniquely due to fast-growing digitalization applications of the maritime sector, which places vessels and their onboard systems exposed to cyberattacks (Kanwal et al.2022). All system components must go through relevant cyber security standard tests before installation onboard. There is a need for robust and reliable communication systems to enable online monitoring and remote intervention and guidance in the case of cyber-attacks detected in ship systems. Such ship systems need continuous guidance from OOCs to avoid any fatal incidents due to cyber-attacks.

System Failure Issue: Another challenge is that some systems may not work well when encountering unexpected situations, such as malfunctioning equipment or obstacles in their path if a ship encounters a problem while operating autonomously. In that case, ship systems need enough guidance so that can decide how to handle it (e.g., by altering course or slowing down) without being overwhelmed by the situation itself—which could lead to near-miss or collision situation, i.e, If things go wrong too quickly or if there isn't enough time available for decision making (Zhao 2021). Developing and implementing practical algorithms and control systems to enable proper guidance are needed for autonomous ships to operate safely and efficiently.

Human-machine Friendly Interface Issue: To provide on-demand online guidance to autonomous systems with higher efficiency, human-machine user-friendly interfaces are needed between OOC operators and vessel systems. Without a friendly user interface, controlling multiple ships simultaneously by a single operator through an OOC seems impossible. This requires much more computing power than currently onboard ships, so developers must look for efficient solutions to improve their system performance and reduce costs by using existing hardware(Komianos (2018)). However, proper guidance from OOC operators is still required for optimized resource utilization and early recovery from system failures (Kurt & Aymelek (2022)).

3.3 *Operation support*

Operation Support is the critical part of the OOC operation support framework for autonomous ships that coordinates and supports vessel operations. Operation support refers to the various systems and processes in place to ensure the smooth and efficient functioning of ship operations. This may include system maintenance, navigation and communication systems, i.e. system technologies, and processes essential for vessels to operate safely and effectively. The operation support framework can provide all mandatory support needed to operate autonomous ships. It will also ensure the optimality of the operational planning phase or actions suggested by autonomous systems both economically and safely (Abaei et al., 2022).

If one is going to answer the question of 'Why is operation support important?': Operation support is the most crucial element in operating autonomous ships through the proposed OOC framework for multiple reasons, such as keeping ships in appropriate operational modes, improved security, system failure handling, equipment functions failure identification and replacement, and other supports. Operation support can help with remote intervention, if necessary through OCCs.

Furthermore, support is needed and valuable to maintain the efficiency and effectiveness of autonomous systems during ship operations. Through the operation support functions in OOCs, ship operational efficiency can be gained by providing immediate support to vessel systems, if required. The absence of a trained human crew highlights the importance of operation support functions. Operational support is essential to deal with advanced systems handling in case of system failure occurs (Fan et al. 2022). It is one of the critical issues of autonomous vessel development programs to improve safety and security standards, which will also help in terms of society's acceptance of autonomous technology.

Without proper operation support functions, it seems impossible to coordinate autonomous system actions. If a ship is fully autonomous, it will perform its operations and does not need a crew. This means that no trained personnel are on board to perform these functions, which requires ship systems to assign qualified people who can provide adequate support for operating ships autonomously from OOCs (Utne et al. 2020).

Some of the operation support challenges and solutions can be categorized as: i) Maintenance support, ii) Lack of training facility, iii) Expensive supervision functions, iv) Maintenance challenge, v) Robust control system, vi) Required resource management challenge and vii) Integration of support system functions.

3.3.1 *Operation support solutions*

Maintenance Support: One of the significant challenges of operation support in autonomous ships is ensuring safe vessel system operations as per defined criteria. Autonomous ships rely heavily on automation technology, which can be vulnerable to system failures or malfunctions. As a result, operation support and processes must be designed and implemented to detect and address potential system issues before they become severe situations, including backup systems and procedures to ensure continuous ship operations during system failure or other emergency circumstances.

One way to address such challenges is to design a robust Human-computer interaction system to provide continuous operational support. Advanced technologies, including AI, data science, intelligent sensors, and hardware devices, make implementing operation support functions more robust and accessible. The OOC operational support functions will ensure continuous ship operation and support during system response failures or other emergency situations (Murray & Perera (2020)).

Lack of Training Facility: There is a lack of training institutions and resources in terms of OOCs and autonomous vessel training requirements. This will create the required skill gaps for the maritime industry. The absence of crew members onboard may develop a lack of communication between OOC operators and autonomous vessels, leading to confusion or even mistakes during ship operations. Operators trained in traditional ship navigation must learn to communicate via OOC communication interfaces instead of being told what to do (Kooij et al. 2018). Updating the training course curriculum according to international standards is necessary to train OOC operators concerning operational support in future vessels.

Expensive Supervision Functions: Autonomous ships require continuous supervision by human operators through OOCs. This requires costly equipment, such as radar technology, that detects objects up to 10 kilometers from vessels. Vision and AI-based solutions can handle this challenge optimally rather than rely on expensive hardware solutions such as radar in some situations.

Maintenance Challenge: Maintenance in autonomous vessels is more challenging due to the absence of humans on board. Because autonomous ship systems are built using advanced technologies, they may need more up-to-date trained personnel for maintenance tasks than conventional ships. Another potential solution is to provide operation support teams with the necessary training and resources to maintain and repair such ships, effectively. This may include access to specialized tools and equipment, continuous training, and support to ensure that operation support teams are equipped and capable of successfully managing the maintenance and repairing tasks on autonomous ships. With in a selected time period, all operators of the operation support team need to refresh hands-on training courses based on modern technology to provide better operation supports.

Robust Control System: A robust controlling system interface is needed to make the operation support team's tasks more efficient and responsive. The control interface will provide online information such as vessel detection, tracking, state estimation, navigational trajectory prediction, and available control action functions on a single screen (Perera et al. 2012). Future ship systems should be equipped with enough robustness and trustworthy properties. OOC personnel should know and be trained to provide support in all critical situations that usually occur in harsh ocean environments.

Required Resource Management Challenge: Maintaining the required resources, such as power and spare parts, is challenging in autonomous ships. To keep autonomous vessel fleets operational, robust planning is required for arranging the power resources, such as refiling bunker oil and recharging batteries. Operating vessel fleets can be a challenging task without a proper operational support framework with OOCs. By utilizing automation and sensor technologies, it is possible to design robust human-machine interfaces by considering the human element in autonomous systems (Mallam et al. 2020). Therefore, considering autonomous vessels, a fully functional operational support frame is required.

Integration of Support System functions: The OOC functional support team must provide online support and the resources required to maintain continuous operations. The OOC system integrations need more effort to offer competent solutions for supervising and managing operation support teams. By implementing these solutions, OOCs can ensure vessel maneuvering is unharmed and efficient in future ships. Also, they can support continuous development and deployment of autonomous ships at a large scale.

4 CONCLUSION

To implement the autonomous ship concept into reality, OOC will play a crucial role in fulfilling the acceptance criteria from society and policymakers. The OOC will provide the monitoring capability to ship operations from land-based locations, where the domain experts analyze ship operations in online and take corrective actions for better performance management. Through the OOC systems, it is possible to detect or forecast the potential problems in ship operation. The OOC operators can interfere or take control of autonomous ships at any time if any problem forecast may lead to collision situations or dangerous situations for ship navigation. This study also enlists the challenges and their solution, including future work in terms of monitoring, guidance, and support in the aspects of the proposed OOC operation support framework.

ACKNOWLEDGEMENTS

This work is supported by the Norwegian Ministry of Education and Research, Norway, under the MARKOM II project title "Onshore Operation Center for Remotely Controlled Vessels (OOC)" (with the contract number PMK-2022\10014.) that is a development project for maritime competence established by the Norwegian Ministry of Education and Research in cooperation with the Norwegian Ministry of Trade, Industry, and Fisheries.

REFERENCES

Abaei, M.M., Hekkenberg, R., BahooToroody, A., Banda, O.V. and van Gelder, P., 2022. A probabilistic model to evaluate the resilience of unattended machinery plants in autonomous ships. Reliability Engineering & System Safety, 219, p.108176.

American Bureau of Shipping (ABS) 2016. The Application of Cyber-Security Principles to Marine and

Offshore Operations. [ONLINE] Available at: https://safety4sea.com/wpcontent/uploads/2018/04/ABS-Cyber_Security-2016_02.pdf. [Accessed 31 July 2023].
Chen, J., 2022. IOT Monitoring System for Ship Operation Management Based on YOLOv3 Algorithm. Journal of Control Science and Engineering, 2022.
Cao, Y., Wang, X., Yang, Z., Wang, J., Wang, H. and Liu, Z., 2023. Research in marine accidents: A bibliometric analysis, systematic review, and future directions. Ocean Engineering, 284, p.115048.
Fan, C., Montewka, J. and Zhang, D., 2022. A risk comparison framework for autonomous ships navigation. Reliability Engineering & System Safety, 226, p.108709.
Gupta, P., Kim, Y.R., Steen, S. and Rasheed, A., 2022. Data processing framework for ship performance analysis. arXiv preprint arXiv:2202.01000.
IMO., 2018.Working group report in 100th session of IMO Maritime Safety Committee for the regulatory scoping exercise for the use of maritime autonomous surface ships (MASS). Maritime Safety Committee 100th session, MSC 100/ WP.8
Jacq, O., Boudvin, X., Brosset, D., Kermarrec, Y. and Simonin, J., 2018, October. Detecting and hunting cyberthreats in a maritime environment: Specification and experimentation of a maritime cybersecurity operations centre. In 2018 2nd Cyber Security in Networking Conference (CSNet) (pp. 1–8). IEEE.
Komianos, A., 2018. The autonomous shipping era. operational, regulatory, and quality challenges. TransNav: International Journal on Marine Navigation and Safety of Sea Transportation, 12(2).
Kooij, C., Loonstijn, M., Hekkenberg, R.G. and Visser, K., 2018. Towards autonomous shipping: Operational challenges of unmanned short sea cargo vessels. Marine Design XIII, Taylor and Francis Group Espoo, pp.871–880.
Kim, T.E., Perera, L.P., Sollid, M.P., Batalden, B.M. and Sydnes, A.K., 2022. Publisher Correction: Safety challenges related to autonomous ships in mixed navigational environments. WMU Journal of Maritime Affairs, 21(2), pp.273–273.
Kanwal, K., Shi, W., Kontovas, C., Yang, Z. and Chang, C. H., 2022. Maritime cybersecurity: are onboard systems ready?. Maritime Policy & Management, pp.1–19.
Kurt, I. and Aymelek, M., 2022. Operational and economic advantages of autonomous ships and their perceived impacts on port operations. Maritime Economics & Logistics, 24(2), pp.302–326.
Li, X. and Yuen, K.F., 2022. Autonomous ships: A study of critical success factors. Maritime Economics & Logistics, 24(2), pp.228–254.
Manno, G., 2014. A pathway towards more sustainable shipping in 2050: A possible future for automation and remote operations: Gabriele Manno. In Maritime-Port Technology and Development (pp. 51–64). CRC Press.
Mallam, S.C., Nazir, S. and Sharma, A., 2020. The human element in future Maritime Operations–perceived impact of autonomous shipping. Ergonomics, 63(3), pp.334–345.
Murray, B. and Perera, L.P., 2020. A dual linear autoencoder approach for vessel trajectory prediction using historical AIS data. Ocean engineering, 209, p.107478.
Oikos-International. 2017. Digital Economy and Sustainability. [ONLINE] Available at: https://oikos-international.org/wp-content/uploads/2015/06/oikos-Associate-Report-2017-Digital-Economy-and-Sustainability.pdf. [Accessed 31 July 2023].
Ota, H., Hirai, Y. and Okazaki, T., 2022, October. Remote ship control system using Virtual Reality. In 2022 IEEE International Conference on Systems, Man, and Cybernetics (SMC) (pp. 2223–2228). IEEE.
Perera, L.P., Oliveira, P. and Soares, C.G., 2012. Maritime traffic monitoring based on vessel detection, tracking, state estimation, and trajectory prediction. IEEE Transactions on Intelligent Transportation Systems, 13(3), pp.1188–1200.
Porathe, T., 2014. Remote monitoring and control of unmanned vessels–the MUNIN shore control centre. In Proceedings of the 13th International Conference on Computer Applications and Information Technology in the Maritime Industries (COMPIT '14) (pp. 460–467).
Utne, I.B., Rokseth, B., Sørensen, A.J. and Vinnem, J.E., 2020. Towards supervisory risk control of autonomous ships. Reliability Engineering & System Safety, 196, p.106757.
Veitch, E. and Alsos, O.A., 2022. A systematic review of human-AI interaction in autonomous ship systems. Safety science, 152, p.105778.
Yang, X., Lian, J.J. and Ren, H., 2019. Research on shore-based intelligent vessel support system based on multi-source navigation sensors simulation. International Journal of Distributed Sensor Networks, 15(7), p.1550147719864780.
Ziajka-Poznańska, E. and Montewka, J., 2021. Costs and benefits of autonomous shipping—a literature review. Applied Sciences, 11(10), p.4553.
Zhao, T., Qiu, J., Wen, S. and Zhu, M., 2021, October. Efficient onboard energy storage system sizing for all-electric ship microgrids via optimized navigation routing. In 2021 IEEE Industry Applications Society Annual Meeting (IAS) (pp. 1–8). IEEE.

Advances in Maritime Technology and Engineering – Guedes Soares & Santos (Eds)

Navigational support framework for maritime autonomous surface ships under onshore operation centers

M. Adnan
Department of Technology and Safety, UiT The Arctic University of Norway, Norway

L.P. Perera
Department of Technology and Safety, UiT The Arctic University of Norway, Norway
SINTEF Digital, SINTEF AS, Oslo, Norway

ABSTRACT: To navigate, Maritime Autonomous Surface Ships (MASSs) must be able to determine their states. i.e., the positions and velocities, etc., by utilizing onboard IoT, and ship intelligence systems. Autonomous ship navigation can heavily rely on machine learning algorithms that can predict how vessels will maneuver in the future, based on past behavior, where vessel state estimation can play a main role. Due to multiple factors such as system failures, bad weather situations, local geographical conditions, and lack of system robustness, sometimes intelligent ship navigation systems may degrade the performance, i.e., may fail to respond or fail to find the safest route. To ensure safe and optimum operations in a selected sea area, MASSs need the necessary supporting tools for safe navigation. In this study, a navigational support framework of navigation monitoring, guidance, and control for MASS is presented in the aspects of Onshore Operation Centers (OOCs), including their respective challenges and possible solutions.

1 INTRODUCTION

The concept of autonomous shipping has received attention recently due to its multiple benefits over conventional shipping. On the other hand, the crew shortage in the maritime domain is another major consideration and that can shift conventional shipping to automated shipping. The willingness of workers to perform maritime professions is decreasing, and there is an irreversible drift in the maritime domain due to long periods away from land, harsh sea environment conditions, communication difficulties, etc. The shortage of experienced sailors can compromise the safety considerations of the maritime industry. Handling the decreasing interest of humans in the maritime profession needs to be tackled through accelerated development in advanced automated and AI-based sensors and systems (Nguyen (2019)).

Autonomous ships must be able to make intelligent navigation decisions based on the information gathered from their sensors and perception systems, i.e., ship intelligence. This involves improving sophisticated, intelligent control algorithms and systems that can also be influenced by multiple factors, such as the vessel's current position & speed, distance to other vessels in its vicinity, rough weather, and sea conditions. Autonomous ships must be able to communicate with other vessels and onshore operation centers (OOC) to exchange information and coordinate vessel movements for safe navigation. This requires the development of a reliable and secure communication framework that can be operated under various weather and navigation conditions (Noel et al. 2019).

The research and development studies in autonomous ship navigation can be clustered into two major categories based on classical mechanics and soft computing. The classical mechanics methods are based on mathematical theories and models and soft computing methods belong to advanced artificial intelligence (AI) and machine learning (ML) techniques (Perera et al. 2014). System-based navigation is an emerging technology not only in shipping, especially for autonomous shipping but also in the automotive industry (Divakarla et al. 2019). It is a requirement that this technology should improve safety and trustworthiness in shipping traffic management type applications. The maritime industry adopts the standardized Automatic Identification System (AIS) to know the exact position and location of vessel operations (Ozoga & Montewka (2018)) and that should be further integrated into future shipping.

This study proposes a navigation support framework that can consist of various considerations. The cutting-edge satellite technology available through

DOI: 10.1201/9781003508779-41

GNSS technology has allowed highly accurate vessel positions for future vessels.

Since emission reduction is the major concern in future shipping operations. An important part of the fuel consumption monitoring of these vessels can be important, where respective fuel monitoring systems should be placed. Vessel monitoring systems (VMS) can be an integrated component supported by communication facilities that can observe vessels such as cargo delivery, fishing boats, and even passenger ships (Deng (2015)). The vessel traffic service (VTS) assists in areas with high traffic density, where support is required. However, the VTS will not take control over vessels under the current situation.

However, that concept can be implemented concerning autonomous shipping. Furthermore, there is a need for an organizing mechanism for the traffic when resources become insufficient (Porathe (2016)). In such cases, a navigation support framework is needed to handle such situations safely and efficiently. Therefore, appropriate technologies are crucial for vessels involved with human lives, i.e., cruise ships, because of the many human lives on board. One should note that shipping accidents can result in many losses of human lives, environmental damage, economic devastations, etc.

The concept of E-Navigation based route planning has been considered by the shipping industry previously as a challenging task. The weather conditions can always influence vessel seakeeping and maneuvering conditions. These meteorological factors can influence ocean-going vessels, especially in rough weather conditions. Therefore, an intelligent advanced predictor system is needed to forecast the required resources on each selected shipping route, such as bunker fuel expenditure, sailing time, and assessment of fulfilled ocean-sea conditions at each waypoint.

Implementing ship fuel consumption-related solutions in autonomous shipping needs human-in-the-loop systems for better planning, i.e. resource management, and optimal solutions, at least for the near future. In navigation monitoring, it is considered that there is a need for advanced condition monitoring and condition-based maintenance type applications of marine engine and propulsion systems to support such autonomous vessels (Orlandi et al. 2021). In such situations, the OOC operators can observe the voyage planning phase and make navigation decisions based on weather and vessel traffic information for safe navigation.

The incompatibility issues between the existing maritime infrastructure and autonomous vessels are considerable. To navigate safely, autonomous

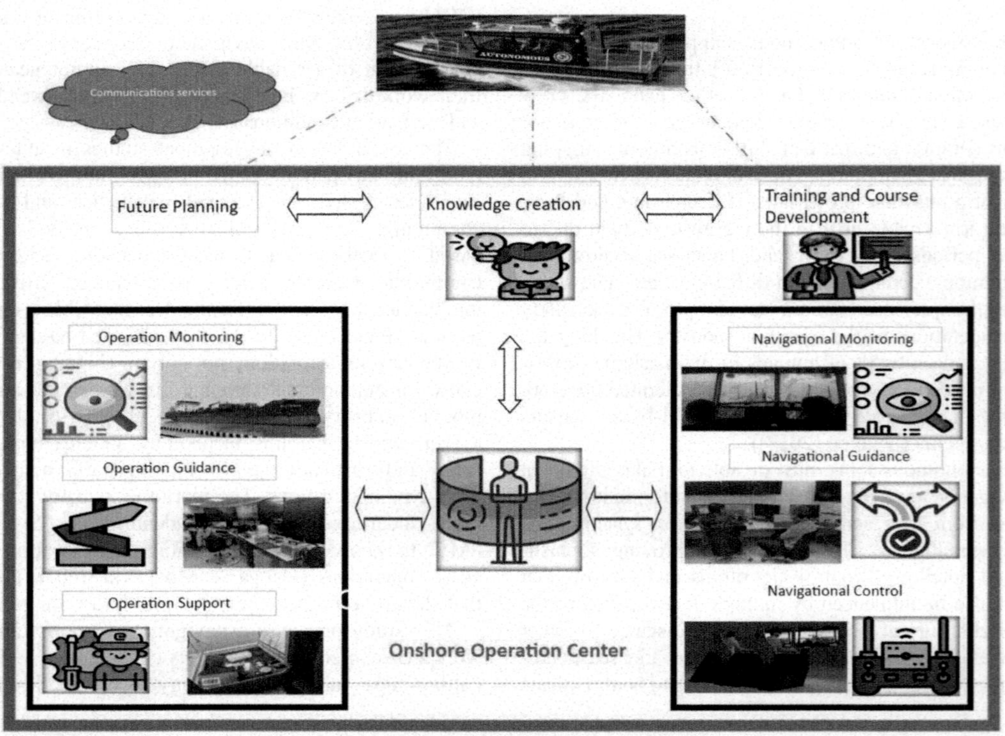

Figure 1. Onshore operation center.

navigation systems may need help from experienced human crew, especially in complex operation environments, i.e. a mixed environment where manned, autonomous, and remotely controlled vessels are interacting, with rough weather conditions. The currently available technology may not be able to reduce maritime accidents significantly in such environments. However, autonomous vessels should have the potential to handle such kinds of accidents better than humans. Such vessels also still face multiple challenges, such as system trustworthiness and cyber security issues under possible communication failure situations.

Utilizing autonomous ships in all types of environments is expected and therefore still required to solve multiple challenges concerning the safety and operational requirements of such ships, i.e., operating area local law compliance, safe environment, other near-operated vessels, cargo, etc., (Felski & Zwolak (2020)). These autonomous navigation technology failures can occur due to several reasons, i.e., low visibility of the operating area, sensor failure situations, extreme weather, wrong system prediction, etc. Such conditions create the need for OOCs that should be facilitated with a navigation support framework to facilitate autonomous vessels in such situations with higher safety considerations since maritime transport demands a high safety standard (Nguyen (2018)).

To overcome these technological gaps, the proposed navigational support framework for MASS under OOCs will play an essential role in this regard. The proposed navigational support framework is a theoretical contribution to provide an overview of the concept of OOCs for future MASS operation control. The study highlights the major functions required, challenges, and viable solutions regarding navigational monitoring, guidance, and support aspects. In the next part of this paper, a high-level overview of the developed steps in an OOC and navigational support framework is presented to support autonomous shipping in the future.

2 ONSHORE OPERATION CENTER DESIGN

Various research and industrial consortiums are on the path towards developing the concept of onshore-based operation center infrastructure to support autonomous vessels efficiently and safely. Hence, an advanced maritime remote-control structure is needed to support autonomous vessels, especially in fully autonomous mode. The concept of onshore operation centers will play a key role in accepting autonomous navigation socially and technically, by enhancing the safety conditions of autonomous vessels.

The Arctic University of Norway (UiT) is also developing the same concept known as an onshore operation center (OOC) to support the operation of a small-scale research vessel, as shown in Figure 1. It is expected that this vessel in autonomous navigation will be utilized on a small scale for short sea shipping, i.e. the main reason for selecting small-scale vessels for this research study. One should note that the OOC will help support both operational and navigation aspects of autonomous ships in the future. It has been categorized as the navigation support framework and that can also be utilized to create the required knowledge that will be utilized to train future OOC operators. That is the main contribution of this study.

The OOC will be equipped with data analysis tools that can be utilized with data science and machine learning techniques to provide information to the OOC operator for planning and making decisions while handling autonomous vessels at sea. It is expected that most of the navigation and operation activities in autonomous vessels will be done through the OOC soon.

Based on onboard sensor information, all planning and decision-making processes related to vessel navigation and operation will be done through OOCs. Critical navigation situations, i.e., complex, and narrow passage navigation, can be done through OOC human operators, i.e., humans in the loop system, to avoid any collision in critical sea areas, as required.

The continuous navigational monitoring functionality of the OOC will play a vital role in this regard.

This study focuses on introducing the navigational framework of the OOC to support the safe and secure navigation of autonomous vessels, as shown in Figure 2. Furthermore, this study will highlight an overview of autonomous navigation, challenges, and viable solutions that can be implemented in the OOC platform. The OOC navigation framework is comprised of listed three significant elements:

1. Navigation Monitoring
2. Navigation Guidance
3. Navigation Control

Various data handling capabilities should be facilitated onboard vessels to handle the respective data sets that are collected onboard vessels. Modern vessels are designed to collect and communicate such data sets, but future, these data sets should be analyzed onboard vessels to make appropriate decisions. In navigation monitoring, there is future work for developing a digital helmsman, i.e., a digital captain-like cloned from human navigator vision and actions, which will navigate future vessels.

It is expected that ocean-going vessels will collect such datasets for real-world ship navigation and utilize them to develop digital navigators. Deep learning-based neural networks, i.e., machine learning algorithms, will be utilized for developing such digital navigators (Perera & MO (2018)). One should also be noted that such data sets can also be created under a bridge simulator environment. In the next session, we will explain each element of onshore operations concerning navigation in detail, including their importance, challenges, and possible solutions.

3 OOC NAVIGATIONAL SUPPORT FRAMEWORK

3.1 *Navigation monitoring*

Navigation monitoring in autonomous ships refers to monitoring their movements and position to ensure that they follow their intended course and stay within their designated operational area, specifically under rough weather conditions. Navigation monitoring typically involves tracking the vessel's locations, speeds, and directions, i.e., course and heading, using GPS, radar, lidar, and other navigational aids such as electronic chart display systems (ECDIS) that allow the ship to locate its surroundings in a digital environment. The onboard sensor data can be analyzed and utilized by OOC operators to navigate future vessels safely and more efficiently, specifically in crises. This information will be shared with maritime authorities and local administration to issue warnings or instructions to avoid hazardous situations or major obstacles. Hence, navigation monitoring is a critical step in future shipping operations, and it will be a part of the future maritime laws that are related to navigation safety including vessel traffic situations. It will help to operate future vessels according to a set of rules and regulations, concerning vessel behavior in certain restricted operating areas.

A network of onboard sensors or IoT must be integrated in a time-synchronized manner to control future vessels, and that information should also be communicated to OOCs. In either case, autonomous vessels should be able to provide online updates on the ship's position, course, and speed and alert the operators in OOCs regarding any potential problems or deviations from the intended route. For the safety purpose of the self-directed vessel, the information system's acquisition and perception are essentially related to navigation monitoring. To improve the integrity of autonomous navigation the respective tasks can be handled by AI-based machine learning algorithms that can be adopted by various modern technologies (Ma et al. 2020). Navigation monitoring is essential in an autonomous ship for several reasons, some of which are listed below including challenges and possible solutions in detail. The respective navigation monitoring challenges can be categorized as: i) Data Collection, ii) Sensor Fault Identification, iii) Robust Communication Infrastructure, iv) System Safety Measures and Protocols v) Remote Situation Awareness, vi) Continuous Changing Weather Conditions, vii) Compliance with Maritime Traffic Laws and viii) Safety and Security.

3.2 *Navigation monitoring solutions*

Data Collection and its Applications: Navigation monitoring can also collect valuable data by utilizing

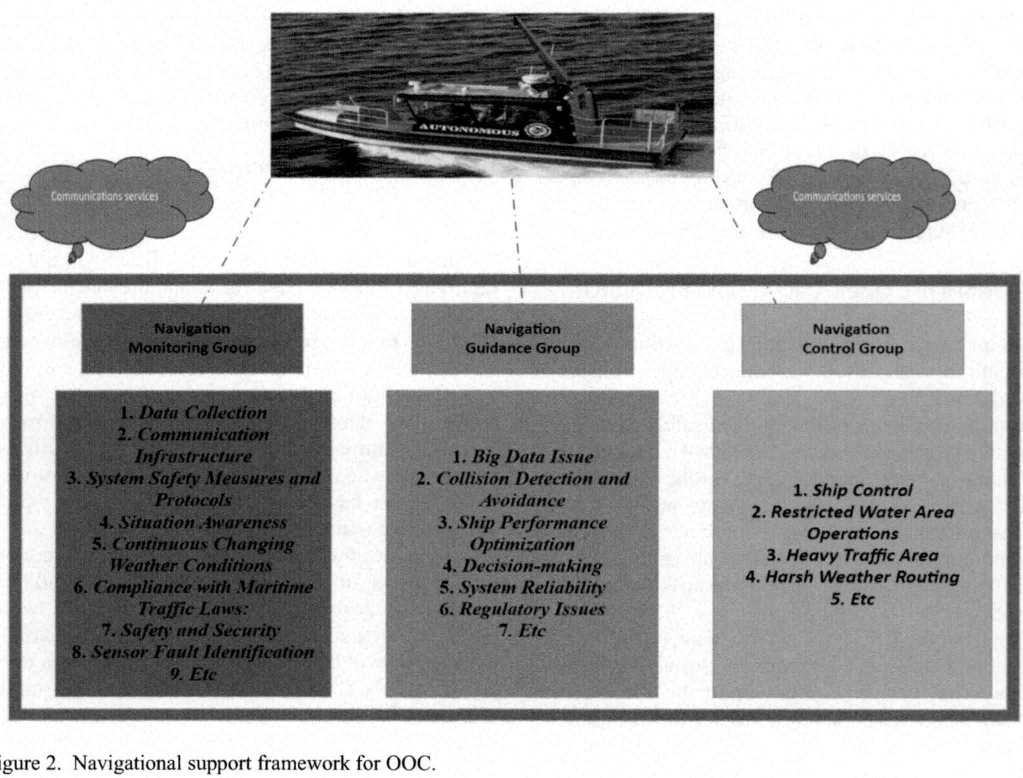

Figure 2. Navigational support framework for OOC.

different sensors, such as cameras and IoT devices in the onboard ship system. The navigation monitoring data can be used to improve its energy efficiency and identify optimal ship operation conditions. That can also be used to optimize ship's routes and vessel speeds, especially in emission control areas to improve fuel consumption and reduce emissions. An accurate position, vessel speed, and heading information are necessary to ensure the vessels reach their destination as planned under route planning type applications. The demand for exact position data is much more critical when a ship departs or arrives at a harbor at a given time interval.

Hence, a data collection system is crucial in having such information from onboard sensors or IoT. Various machine learning techniques can be applied to analyze such data sets e.g., the principal component analysis or singular values decomposition method can be one solution for detecting the large dataset's hidden structure to observe ship navigation/trajectory information (Perera (2016), Murray & Perera (2022)).

Sensor Fault Identification: In general, advanced systems that can detect possible sensor or system failures should be a part of future autonomous shipping. Due to the absence of onboard crew in autonomous ships, any mitigation actions on potential sensor or system failures can be complex. Sensor or system fault detection is critical because the automation system relies on sensor data for decision-making and systems for decision execution.

Identifying, isolating, and recovering data anomalies can improve the integrity of ship navigation from the routing perspective. A fault detection scheme can be utilized to detect the outliers of sensor data that can be outliers of the respective parameters (Perera (2016)). Such instant fault detection, isolation, and recovery applications can support future ship navigation. Immediate detection of system failures in autonomous ships based on sensor failure information can assist in preventing critical ship navigation situations (Perera (2016)).

Advanced data analysis frameworks can be developed to capture, isolate, and recover sensor faults and system abnormal events at the initial stage that can eventually support future ship navigation requirements. The sensor's generated information, i.e., an essential step in autonomous navigation systems, can also be supported through OOCs, i.e., in more critical situations, to ensure reliability for the decision-making process.

Robust Communication Infrastructure: Autonomous ships must be able to communicate with other vessels and shore-based systems to send and receive information that relates to the respective journey. This requires reliable and secure communication schemes that can functions in rough weather conditions. Communication without delay between ships and OOCs is required continuously during the ship operation phase. However, navigation systems can be developed to cope with short-time communication interruptions, specifically in non-critical ship navigation situations.

The transmission cost of large-scale data sets can be higher due to expensive satellite communication systems. That can eventually challenge autonomous ship navigation as one of the main concerns for the maritime sector. In deep learning, some appropriate solutions to reduce the size of data sets can be implemented to handle such tasks; the auto-encoder technique has shown great potential for data compression, especially with sensor data. The architecture of the autoencoding method for compressing and decompressing data sets related to vessel operation and navigation parameters must be transmitted over communication networks for the OOC operator's overview (Perera & MO (2018)). That process can also help to reduce and recover data anomalies.

System Safety Measures and Protocols: Ensuring the trustworthiness of autonomous operations in future vessels is crucial. This can be achieved by developing robust safety measures and protocol monitoring schemes for detecting failures or malfunctions at various system levels, as well as guaranteeing appropriate system behavior in crises.

The ship monitoring systems can also benefit from the AIS at a global level with the Long-Range Identification and Tracking (LRIT) scheme. The standardized support model adoption in autonomous systems makes development easier and faster by integrating advanced automation and navigation systems with onboard processes.

While various system developers integrate such automation and navigation systems, therefore having standardized modular systems for such systems can reduce system-level faults and failures. Therefore, having a proper system architecture with standardized modular systems can expedite the development of autonomous shipping (Chen et al. 2022).

Remote Situation Awareness: The probability of an operator or system losing sufficient understanding of situational awareness in autonomous ship navigation is substantially greater than in conventional shipping due to the absence of the onboard crew. Autonomous vessels operating through OOCs will introduce higher navigation safety issues because of the lack of understanding of the operating environment information. Therefore, transforming the navigation monitoring information into precise knowledge of the respective situational awareness should be done, where the respective technologies can be adopted adequately.

Autonomous ships must be able to perceive and interpret their surroundings as a part of situational awareness to navigate safely and efficiently. This involves using a variety of sensors and perception algorithms to identify and monitor other vessels, obstacles, and other relevant features of the environment. The beneficial way to navigate is to plan the passage as safely as possible to the sideways, i.e., digital ship routes, of the channel to prevent a last-minute evasive manoeuver. However, unexpected situations may lead to such maneuvers and future vessels should also have the capabilities to overcome such situations. On the one hand, autonomous vessels can increase human life's safety and security because they are unmanned or understaffed (Porathe (2016)), but that can also raise navigation safety issues, specifically in a mixed navigation environment.

These sensors can be combined with advanced perception algorithms to accurately interpret and understand the environment, including other vessels and obstacles. The vessels should have reliable system intelligence in making their conclusions and determining procedures. The vessel navigation systems should have adequate reliability to operate, i.e. the decision made by digital navigator-type technology.

Inspired by the recent adaptation of autonomous and automated systems in other transport modes, i.e., digital drivers in driverless cars, it also has influenced the development and utilization of autonomous technology in the shipping industry. Since there are many challenges in making autonomous shipping a reality, there are various discussions on these disruptive autonomous technologies (Zhou et al. 2019).

Continuous Changing Weather Conditions: The radar data can be cluttered during rough weather conditions. In such situations, autonomous navigation systems can be challenged by sensor noise conditions, therefore the OOC operator's help can be crucial for maintaining safe navigation conditions. Due to changing weather conditions, autonomous systems may not perform the respective tasks as per the defined criteria due to harsh environmental factors that can affect the system's efficiency and performance. Researchers have designed and developed different techniques, such as rain and snow suppression algorithms, to enhance the performance of radar object observation in severe weather conditions (Qin & Chen (2019)).

Compliance with Maritime Traffic Laws: Navigation monitoring can also help to ensure that ships stay within their designated ship routes and comply with relevant local and international regulations. This is especially important in situations, where vessels operating in restricted or sensitive areas, such as war zones, and in the vicinity of ports or other marine infrastructure; Aerial drones have been used by various research groups to monitor ship emissions in harbour areas to enforce the environmental laws (Deng et al. 2022).

Safety and Security: One of the primary goals of navigation monitoring is to ensure the safety of the vessel, cargo, and its crew, if any, onboard. One should note that there may be a requirement for the onboard crew in future vessels, to maintain and operate complex marine engines, specifically in deep-sea shipping.

By continuously monitoring the ship's position and movements, detecting, and preventing potential collisions with other vessels or obstacles is possible. Online navigational safety monitoring systems have been developed, such as LNG ship risk in inland waters. The global maritime distress and safety system (GMDSS) system uses maritime security data and a shared communication network to catch corrupted signals from vessels (Deng (2015)). Autonomous vessels will encounter distinct circumstances; the OOC operators should have knowledge and experience in handling such situations in a safe navigation manner (Kim et al. 2022).To improve the situation awareness of OOC operators in a mixed environment, the requirements of modern technologies and machine learning algorithms are essential (Zhou et al. 2019).

According to the Piracy Reporting Center of the International Maritime Bureau (IMB), oceanic piracy has risen for several years; maritime transport now faces increased risks and security issues (Kim et al. 2022). VMS is the most used framework in all types of vessels. Using the VMS system makes it easier to locate the ship and find abnormalities in the operation of vessels at sea. Preventing and minimizing the impact of piracy can be a difficult process, however, required technologies should be developed to support autonomous navigation in shipping.

3.3 *Navigation guidance*

Navigation guidance refers to the algorithms, systems, and protocols used to guide the vessel's movements and ensure that it navigates safely and efficiently. The OOC operators can provide navigation guidance where autonomous systems may fail, resulting in unexpected safety critical situations. The navigation system needs continuous guidance to operate safely and securely in rough weather, narrow passages, high-traffic zones, and port areas for future vessels.

The autonomous guidance and navigation (AGN) system is expected to be a key element in upcoming maritime navigation due to its more secure features. The verdict shows the worth of AGN systems in sea navigation and human observations of their abilities and requirements for future development. Automatic pilot systems are the chief development part of AGN systems, and their applications have been the dream of shipbuilders for numerous years. Recent developments in sensors, computing infrastructure, intelligent equipment or devices, and AI-based systems make the desire into reality in next-generation AGN systems (Smierzchalski (1998), Ahmed & Hasegawa (2016)). Avoiding accidents at sea is one of the core responsibilities of secure shipping that attracts considerable attention (Statheros et al. 2008). Intelligent systems make automatic maneuvering possible by developing intelligent systems for guidance, navigation, and control (Li et al. 2021).

If we are going to answer the question of 'Why is Navigation Guidance important?': Navigation guidance is essential in autonomous ships for several reasons. For example, safeguarding the secure navigation of autonomous vessels is paramount. Ship collisions and groundings are still the most significant causes of accidents. According to COLREG, collision prevention is critical when dealing with autonomous ships (Vagale et al. 2021).

Collisions at sea cause significant disturbances regarding safety, security, and damage. For example, crashes in numerous circumstances direct to irreversible environmental and economic damages. To minimize the risk of collision, some decision support and control systems based on the human-in-loop concept are recommended to be set up on vessels. It will help the vessel navigate through complicated crossings with the help of the OOC navigational operator's support. Most seaports have Vessel traffic services (VTS) systems to assess the secure paths of ships arriving and departing the port (Zhang et al. 2015).

Sea conditions and weather are highly variable and uncertain during sailing in cold climates, i.e., arctic ship navigation. This will raise the accident risk over the next couple of years, mainly at the end of the shipping season (Bolbot et al. 2023).

Navigation guidance in autonomous shipping faces multiple challenges. One of the significant functions of OOCs is to provide navigation guidance to the shipping fleet for safe operation. Some of the challenges and their possible solution are mentioned below. The navigation guidance challenges can be categorized as: i) Big Data Issue, ii) Collision Detection and Avoidance, iii) Ship Performance Optimization, iv) Decision-making, v) System Reliability and v) Regulatory issues.

3.4 *Navigation guidance solutions*

Big Data Issue: The industrial IoT collects large-scale data sets called 'Big Data' and that can also introduce various challenges. The shipping industry utilizes such datasets to understand vessel behavior, which can be used to optimize decision-making processes such as route planning, optimal engine operations, condition-based maintenance, navigation guidance, etc.

Data processing can always be associated with higher costs, which can be associated with low-quality data sets. The existing IT networks should be further improved to support the decision-making process of autonomous ships, which should be more robust and accessible based on real-time sensor data (Perera & Mo (2018)).

The big data challenge can be handled by developing advanced analytics associated with machine learning algorithms. Big data can be used for predictive analysis, such as predicting additional costs related to the ship's lifecycle. Based on the analyzed data, robust decisions can be made to control the vessel's operation in the future. Taking such actions by the OOC can improve organizational functions and decision-making capabilities Perera (2017).

Collision Detection and Avoidance: The problem of ship route planning, formulation of appropriate cost functions, and consideration of topological and kinematic constraints are still challenging tasks to avoid collision at sea. Intelligent decision-making capabilities based on navigation guidance from the OOC must be a unified part of the upcoming AGN system to advance autonomous sea navigation systems. Researchers already developed various techniques to handle such challenges (Perera et al. (2009), Zaccone & Martelli (2020)).

The goal is to significantly improve waterway safety by avoiding collisions, relieving the navigator of tedious routine work, and supporting challenging situations, especially at night or in fog conditions at sea (Zaccone (2021), Nielsen & Jensen (2011)). Based on AI techniques, several methods have been developed to overcome ship collision situations, such as fuzzy logic (Zhang et al. 2015), neural networks, feedback controllers for speed control & path tracking (Vagale et al. 2021) and dynamic predictive guidance technique (Kozynchenko, al & Kozynchenko (2018)). These techniques are introduced for guidance systems to increase the effect of ship course track retention (Kurowski et al. 2019). With the help of these available solutions, OOC operators can provide navigation guidance more robustly for safe navigation at sea. The humans in the loop situation will significantly enhance the robustness of autonomous navigation in shipping.

Ship Performance Optimization: Navigation guidance can also help to optimize ship routes and vessel speeds. Continuously updating the ship's position and movements based on current sea conditions to avoid unfavorable conditions and take necessary fuel-saving measures will increase the ship's energy efficiency. It is possible to adjust the route to avoid unfavorable weather conditions or other obstacles, which can help to reduce fuel consumption and operational costs. Ship energy efficiency is also be associated with vessel fuel consumption, where the respective emissions should be reduced. Fuel cost is a significant part of the vessel operating cost; optimizing fuel consumption can increase profit and other economic benefits while reducing the respective emissions (Chen et al. 2022).

Reliable path planning can be considered a dynamic optimization approach that can have a challenging solution. The navigation guidance functions will ensure the ship's optimization capabilities regarding operational aspects like on-time cargo delivery, fuel optimization through route planning guidance and harbor operations, and safe operation guarantees.

Decision-making: The decision-making technology in autonomous vessels should be robust enough because that can play a crucial role in oceanic autonomy. Autonomous ships must be able to make intelligent navigation decisions based on the information they have gathered from their sensors and perception systems. This requires the development of sophisticated decision-making algorithms that can consider multiple factors, such as the ship's current position and speed conditions and the weather and sea conditions; navigational charts information including water depths, average topographies of the seabed, directional risks, assistances to routing, sea current data, harbours, buildings, and bridges, etc. (Zhou et al. 2019). The OOC operators need to monitor the navigation plan continuously and will interfere with the system-generated decision-making as required, specifically in critical navigation sea areas and conditions.

System Reliability: Guaranteeing the reliability consistency of autonomous vessel system operations is a critical challenge. This includes developing strategies and protocols for detecting and responding to failures or malfunctions and developing robust safety measures to prevent accidents and injuries. Without the OOC's help, it is sometimes impossible for an autonomous system to handle such situations due to multiple factors such as a political situation or war zone, system functions failure or response failure, future weather conditions, and those factors that are usually not modeled into the system quickly. Navigation guidance is one of the essential functions OOC operators perform to enhance the security and trustworthiness of autonomous ship operations.

Regulatory issues: The development and placement of autonomous vessels, including their operations, are subject to various legal and regulatory frameworks, including liability, insurance, and safety considerations. Handling these issues will be an essential core challenge for the shipping industry in developing and deploying autonomous navigation systems. This can be supported through OOCs by careful planning and compliance with relevant laws and regulations, as well as working with regulatory bodies to make sure that autonomous vessels are safer to navigate (Ringbom & Veel (2017)).

The navigation guidance functions will solve legal and regulatory problems through the OOC operators. The OOC operators can utilize the navigation guidance functions in case of emergencies or system failure occur. These legal and regulatory issues are critical challenges in deploying autonomous ships in the future (Rodseth & Burmeister (2015)). Through the OOC human-in-the-loop-system design, many legal issues s can be solved.

3.5 *Navigation control*

There is a need to develop intelligent navigation control systems to support future ship operations through OOCs, especially when making decisions in complex navigation situations, which will directly contribute to navigation safety. The impact of waves in coarse weather conditions is one of the circumstances that impair the operational efficiency of ships. Hence, deliberate reasoning in the decision-making process for ship handling plays an essential role in navigation. The OOC operators will receive support from navigation systems with real-time sensor data and advanced analytics to ensure vessel safety concerns (Mousazadeh et al. 2018).

The ship course controllers were usually designed with the assumption of a constant vessel speed. However, sailing vessels typically need to reduce or adjust the speed to formulate berthing type maneuvers in the fairway area. The existing course tracking system tracks the ship in an open sea area by assuming the speed is constant, which cannot be used to track and navigate vessels in fairway areas because ship speed is continuously changing (Min & Zhang (2021)). Detecting and avoiding other vessel groups sailing nearby is one of the main jobs in ship navigation, as near-miss situations with them can potentially cause serious risks to ship collisions (Perera et al. 2012). In such situations, navigation control & and support are essential functions to continue safe ship operations.

In adverse weather conditions, each maneuver must also be checked to evaluate whether it is safe & secure in terms of ship steadiness point of view. In such situations, OOC navigation control is needed by a human-in-the-loop system to handle such critical circumstances. The OOC operators may utilize some advanced intelligent AI-based system to predict the future outcome based on the current situation from the analytical approaches before providing any recommendations or support for necessary navigation control. This will enhance the robustness of the navigation system significantly through the proposed OOC navigational support framework.

If we are going to answer the question of 'Why is Navigation control important?': The navigation control functions of an autonomous ship is essential for ensuring the safety and efficiency of its operations where the automatic system fails to respond. For example, it can help to prevent the ship from running aground or colliding with other vessels or obstacles, and it can also help to optimize the ship's route and speed to minimize fuel consumption and additional costs. There is a need to build a prototype system that can monitor several parameters related to the movement of ships and process this data to provide OOC operators with the consequences of different decisions about the ship's handling.

The Navigation control system is still under development, and it can utilize the AI-based decision support system to solve complex situations with sample and reliable solutions. Some existing systems use the idea of a Multi-ARPA (MARPA) system (Tsou et al. (2010), Nishizaki et al. (2018). The driving force behind the OOC's proposed build architecture is the changing international nature of shipping that needs to create standards for autonomous ship control and information exchange to increase efficiency further, reduce the possibility of accidents, decrease greenhouse gas (GHG) emissions, and enhance safety in the maritime industry (Kawaguchi et al. 2009). Some of the navigation control challenges and viable solutions can be categorized as: i) Ship Control, ii) Operations in Restricted Navigation Areas, ii) Heavy Traffic Areas and iv) Harsh Area Weather Routing

3.6 *Navigation control solutions*

Ship Control: Some navigation regions may need ship control support for safe operations, like narrow passages, severe weather conditions, low visibility areas, sensor system failure situations, etc. It is optionally possible to display only COLREGS-compliant maneuvers through smart decision support systems. The emplacement of these kinds of data allows the navigator to select effective maneuvers in situations where choices are restricted by weather conditions, where the ship stability features of ocean-going vessels can be challenged (Rødseth (2011)).

The OOC operators can use advanced sensors-based systems for all motion measurements and wave spectrum estimation schemes, approximating confined sea states in real-time to provide optimum navigation control of autonomous ships whenever needed (Mousazadeh et al. 2018). Rudder control keeps vessels on the planned track, and the vessel's speed is reduced slowly to reach the anchorage area during berthing situations. At low speeds, rudder control may not work adequately, and vessels can lose their maneuvering capabilities. Then, the vessel heading is adjusted using bow thrusters supported by auxiliary engines (Min & Zhang (2021)). The OOC team must be ready for necessary navigation control support in all these situations.

The advanced analytics system makes it possible to estimate the navigation threat, which will help for safe maneuvers. Using only AIS data for ship collision avoidance is challenging (Szlapczynski & Krata (2018)). A navigational control system capable of AIS synthesis is needed, augmented with a mark of intellect generated from a physical model for capturing group navigation dynamics of ocean-going vessels (Ozoga & Montewka (2018)).

Operations in Restricted Water Areas: The operation in restricting water needs navigation control & support for an autonomous system via OOC in terms of speed, gas emission, etc. To handle such a situation, integration of the dynamic changes into an automatic navigation handling system is complex. Electrifying the vessel and cargo will allow marine traffic to follow and build an infrastructure that can withstand the effects of climate change. Investing in recent green technologies will also reduce emissions (Porathe (2016)). The maritime community has realized the importance of new green technology, and the e-navigation (from the International Maritime Organization - IMO) and e-Maritime (from the European Union - EU) initiatives testify to this (Zhang et al. 2015).

Heavy Traffic Areas: Maritime traffic has increased exponentially in recent years expanding the possibility of ship-to-ship collisions. It could lead to one of the fatal accident's reasons in the future if it is not handled correctly. AI-based support tools could help avoid collision by providing estimations based on sensors and prediction systems, decreasing the load on navigational operators, and enhancing maritime domain safety and security (Nilsson et al. 2009). To ensure the navigation is safe and secure in a high-traffic area, the OOC operators must take control and act on time to escape from the collision based on operating area environment information, prediction, and other vessels' behavior (Li et al. 2022).

Harsh Area Weather Routing: July to October is the lowest risk session for cargo ships conducting commercial voyages throughout arctic waters. Navigation support or recommendations in critical routes, such as the optimal safe turn angle, speed, route selection, etc., are essential. As the commercial activities on arctic routes increase significantly each year, secure and safe navigation is the main concern in the multifaceted arctic environment. Operating in such complicated areas requires continuous essential navigation control & support for a safe operation guarantee. The OOC operator will utilize the AI-based system to handle such situations. Researchers developed onboard decision support systems for safe navigation in harsh weather & sea conditions using a neural network to find the safe ship course and speeds based on the current operating sea environment parameters (Perera et al. 2012). A dynamic Bayesian network can forecast arctic waters' spatial and temporal threats, creating a chart of navigational risks of these areas' seawater (Nilsson et al. 2009) OOC operators will provide necessary control & support to the autonomous system based on an advanced AI-analytical system through the proposed OOC navigational support framework.

4 CONCLUSION

This study presents the functionality of onshore operation centers and shows how these centers will help the future needs of maritime operations in terms of navigational support framework in autonomous shipping. This study also highlights the future control functions required in OOCs and autonomous ships regarding the navigational perspective. A comprehensive overview of navigational support methodology is proposed to address the challenges of autonomous shipping in the future. OOCs will increase the competitiveness of the maritime sector while providing required safe navigation features, i.e. humans in the loop. The data collected by sensors, satellite data, navigation systems, intelligent algorithms, and human operators in the loop make the system more robust and trustworthy. The proposed OOC navigational framework will help to reduce the risks of maritime accidents in the future and enhance the large-scale acceptance of autonomous technology by solving the concerns raised by regulatory authorities and the maritime industry.

ACKNOWLEDGEMENTS

This work is supported by the Norwegian Ministry of Education and Research, Norway under the MARKOM II project title "Onshore Operation Center for Remotely Controlled Vessels (OOC)" (with the contract number PMK-2022\10014.) is a development program for maritime competence. The program is established by the Norwegian Ministry of Education and Research in cooperation with the Norwegian Ministry of Trade, Industry and Fisheries.

REFERENCES

Ahmed, Y.A. and Hasegawa, K., 2016. Fuzzy reasoned waypoint controller for automatic ship guidance. IFAC-PapersOnLine, 49(23), pp.604–609.

Bolbot, V., Theotokatos, G., Wennersberg, L.A., Faivre, J., Vassalos, D., Boulougouris, E., Jan Rødseth, Ø., Andersen, P., Pauwelyn, A.S. and Van Coillie, A., 2023. A novel risk assessment process: Application to an autonomous inland waterways ship. Proceedings of the Institution of Mechanical Engineers, Part O: Journal of Risk and Reliability, 237(2), pp.436–458.

Chen, Q., Wu, W., Guo, Y., Li, J. and Wei, F., 2022. Environmental impact, treatment technology and monitoring system of ship domestic sewage: A review. Science of The Total Environment, 811, p.151410.

Deng, J., 2015. Real-time navigation monitoring system research for LNG-fuelled ships in inland water. Journal of Maritime Research, 12(2), pp.95–102

Divakarla, K.P., Emadi, A., Razavi, S., Habibi, S. and Yan, F., 2019. A review of autonomous vehicle technology landscape. International Journal of Electric and Hybrid Vehicles, 11(4), pp.320–345.

Deng, M., Peng, S., Xie, X., Jiang, Z., Hu, J. and Qi, Z., 2022. A diffused mini-sniffing sensor for monitoring SO2 emissions compliance of navigating ships. Sensors, 22(14), p.5198.

Felski, A. and Zwolak, K., 2020. The ocean-going autonomous ship—Challenges and threats. Journal of Marine Science and Engineering, 8(1), p.41.

International Maritime Bureau, [ONLINE] Available at: https://www.icc-ccs.org/ [Accessed 31 July 2023]

Kawaguchi, A., Inaishi, M., Kondo, H. and Kondo, M., 2009. Towards the development of intelligent navigation support systems for group shipping and global marine traffic control. IET Intelligent Transport Systems, 3(3), pp.257–267.

Kozynchenko, A.I. and Kozynchenko, S.A., 2018. Applying the dynamic predictive guidance to ship collision avoidance: Crossing case study simulation. Ocean Engineering, 164, pp.640–649.

Komianos, A., 2018. The autonomous shipping era. Operational, regulatory, and quality challenges. TransNav: International Journal on Marine Navigation and Safety of Sea Transportation, 12(2).

Kurowski, M., Roy, S., Gehrt, J.J., Damerius, R., Büskens, C., Abel, D. and Jeinsch, T., 2019, June. Multi-vehicle guidance, navigation, and control towards autonomous ship manoeuvring in confined waters. In 2019 18th European Control Conference (ECC) (pp. 2559–2564). IEEE.

Kim, T.E., Perera, L.P., Sollid, M.P., Batalden, B.M. and Sydnes, A.K., 2022. Safety challenges related to autonomous ships in mixed navigational environments. WMU Journal of Maritime Affairs, 21(2), pp.141–159.

Li, X., Stephenson, S.R., Lynch, A.H., Goldstein, M.A., Bailey, D.A. and Veland, S., 2021. Arctic shipping guidance from the CMIP6 ensemble on operational and infrastructural timescales. Climatic Change, 167, pp.1–19.

Li, Z., Yao, C., Zhu, X., Gao, G. and Hu, S., 2022. A decision support model for ship navigation in Arctic waters based on dynamic risk assessment. Ocean Engineering, 244, p.110427.

Mousazadeh, H., Jafarbiglu, H., Abdolmaleki, H., Omrani, E., Monhaseri, F., Abdollahzadeh, M.R., Mohammadi-Aghdam, A., Kiapei, A., Salmani-Zakaria, Y. and Makhsoos, A., 2018. Developing a navigation, guidance and obstacle avoidance algorithm for an Unmanned Surface Vehicle (USV) by algorithms fusion. Ocean Engineering, 159, pp.56–65.

Ma, X., Shen, J., Liu, Y. and Qiao, W., 2020. A methodology to evaluate the effectiveness of intelligent ship navigational information monitoring system. IEEE Access, 8, pp.193544–193559.

Min, B. and Zhang, X., 2021. Concise robust fuzzy nonlinear feedback track-keeping control for ships using multi-technique improved LOS guidance. Ocean Engineering, 224, p.108734.

Murray, B. and Perera, L.P., 2022. Ship behaviour prediction via trajectory extraction-based clustering for maritime situation awareness. Journal of Ocean Engineering and Science, 7(1), pp.1–13.

Nilsson, R., Gärling, T. and Lützhöft, M., 2009. An experimental simulation study of advanced decision support system for ship navigation. Transportation research part F: traffic psychology and behaviour, 12(3), pp.188–197.

Nielsen, U.D. and Jensen, J.J., 2011. A novel approach for navigational guidance of ships using onboard monitoring systems. Ocean Engineering, 38(2-3), pp.444–455.

Nishizaki, C., Terayama, M., Okazaki, T. and Shoji, R., 2018, June. Development of a navigation support system to predict the new course of the ship. In 2018 World Automation Congress (WAC) (pp. 1–5). IEEE.

Nguyen, V.S., 2019. Research on a support system for automatic ship navigation in fairway. Future Internet, 11(2), p.38.

Noel, A., Shreyanka, K., Gowtham, K. and Satya, K., 2019, November. Autonomous ship navigation methods: A review. In Proceedings of the Conference Proceedings of ICMET OMAN.

Orlandi, A., Cappugi, A., Mari, R., Pasi, F. and Ortolani, A., 2021. Meteorological navigation by integrating metocean forecast data and ship performance models into an ecdis-like e-navigation prototype interface. Journal of Marine Science and Engineering, 9 (5), p.502.

Ożoga, B. and Montewka, J., 2018. Towards a decision support system for maritime navigation on heavily trafficked basins. Ocean Engineering, 159, pp.88–97.

Perera, L.P., Carvalho, J.P. and Soares, C.G., 2009, November. Autonomous guidance and navigation based on the COLREGs rules and regulations of collision avoidance. In Proceedings of the international workshop advanced ship design for pollution prevention (pp. 205–216).

Perera, L.P., Rodrigues, J.M., Pascoal, R. and Soares, C.G., 2012. Development of an onboard decision support system for ship navigation under rough weather conditions. Sustainable Maritime Transportation and Exploitation of Sea Resources, 837, pp.837–844.

Perera, L.P., Ferrari, V., Santos, F.P., Hinostroza, M.A. and Soares, C.G., 2014. Experimental evaluations on ship autonomous navigation and collision avoidance by intelligent guidance. IEEE Journal of Oceanic Engineering, 40(2), pp.374–387.

Porathe, T., 2016. A navigating navigator onboard or a monitoring operator ashore? Towards safe, effective, and sustainable maritime transportation: findings from five recent EU projects. Transportation Research Procedia, 14, pp.233–242.

Perera, L.P., 2016. Statistical filter based sensor and DAQ fault detection for onboard ship performance and navigation monitoring systems. IFAC-PapersOnLine, 49 (23), pp.323–328.

Perera, L.P., 2016. Marine engine centered localized models for sensor fault detection under ship performance monitoring. IFAC-PapersOnLine, 49(28), pp.91–96.

Perera, L.P., 2017. Handling big data in ship performance and navigation monitoring. Smart Ship Technology, pp.89–97.

Perera, L.P. and Mo, B., 2018. Ship performance and navigation data compression and communication under autoencoder system architecture. Journal of Ocean Engineering and Science, 3(2), pp.133–143.

Perera, L.P. and Mo, B., 2018, June. An overview of data veracity issues in ship performance and navigation monitoring. In International Conference on Offshore Mechanics and Arctic Engineering (Vol. 51333, p. V11BT12A004). American Society of Mechanical Engineers.

Qin, Y. and Chen, Y., 2019. Signal processing algorithm of ship navigation radar based on azimuth distance monitoring. International Journal of Metrology and Quality Engineering, 10, p.12.

Rødseth, Ø.J., 2011, October. A maritime ITS architecture for e-navigation and e-maritime: Supporting environment friendly ship transport. In 2011 14th International IEEE Conference on Intelligent Transportation Systems (ITSC) (pp. 1156–1161). IEEE.

Rødseth, Ø.J. and Burmeister, H.C., 2015. Risk assessment for an unmanned merchant ship. TransNav, International Journal on Marine Navigation and Safety Od Sea Transportation, 9(3), pp.357–364.

Ringbom, H.M. and Veal, R., 2017. Unmanned ships and the international regulatory framework. Journal of International Maritime Law, 23(2), pp.100–118.

Smierzchalski, R., 1998. Evolutionary guidance system for ship in collisions situation at sea. IFAC Proceedings Volumes, 31(3), pp.129–134.

Statheros, T., Howells, G. and Maier, K.M., 2008. Autonomous ship collision avoidance navigation concepts, technologies and techniques. The journal of Navigation, 61(1), pp.129–142.

Szlapczynski, R. and Krata, P., 2018. Determining and visualizing safe motion parameters of a ship navigating in severe weather conditions. Ocean Engineering, 158, pp.263–274.

Tsou, M.C., Kao, S.L. and Su, C.M., 2010. Decision support from genetic algorithms for ship collision avoidance route planning and alerts. The Journal of Navigation, 63(1), pp.167–182.

Vagale, A., Oucheikh, R., Bye, R.T., Osen, O.L. and Fossen, T.I., 2021. Path planning and collision avoidance for autonomous surface vehicles I: a review. Journal of Marine Science and Technology, pp.1–15.

Zhang, J., Zhang, D., Yan, X., Haugen, S. and Soares, C. G., 2015. A distributed anti-collision decision support formulation in multi-ship encounter situations under COLREGs. Ocean Engineering, 105, pp.336–348.

Zhang, Y., Wang, X. and Wu, H., 2015. A distributed cooperative guidance law for salvo attack of multiple anti-ship missiles. Chinese Journal of Aeronautics, 28 (5), pp.1438–1450.

Zhou, X.Y., Liu, Z.J., Wu, Z.L. and Wang, F.W., 2019. Quantitative processing of situation awareness for autonomous ships navigation. TransNav: International Journal on Marine Navigation and Safety of Sea Transportation, 13(1).

Zaccone, R. and Martelli, M., 2020. A collision avoidance algorithm for ship guidance applications. Journal of Marine Engineering & Technology, 19(sup1), pp.62–75.

Zaccone, R., 2021. COLREG-compliant optimal path planning for real-time guidance and control of autonomous ships. Journal of Marine Science and Engineering, 9(4), p.405.

Advances in Maritime Technology and Engineering – Guedes Soares & Santos (Eds)
 ISBN 978-1-032-83107-7

Thrust and structural analysis of an unmanned sailboat subject to omnidirectional winds

K.P. Rao & M. Tan
School of Naval Architecture and Ocean Engineering, Huazhong University of Science and Technology, Wuhan, P.R. China

G. Xiang, X.B. Xiang & Y. Liu
School of Naval Architecture and Ocean Engineering, Huazhong University of Science and Technology, Wuhan, P.R. China
Hubei Key Laboratory of Naval Architecture and Ocean Engineering Hydrodynamics (HUST), Wuhan, P.R. China
Collaborative Innovation Center for Advanced Ship and Deep-Sea Exploration, Shanghai, P.R. China

ABSTRACT: Unmanned sailboats, with the advantages of long endurance and zero emissions, have demonstrated significant potential in tasks such as marine scientific research, environmental monitoring, and underwater exploration. As the power source for unmanned sailboats, the analysis and optimization of sailing thrust and the resulting structural strength are of great importance in enhancing the performance of unmanned sailboats. This paper investigates and analyzes the propulsion performance and structural strength of an unmanned sailboat used for ocean observation. Firstly, the sail thrust of the unmanned sailboat in omni-directional wind directions (0-360 degrees) is computed using Computational Fluid Dynamics (CFD) methods, and thrust coefficient curves of the sails varying with the wind direction are generated. Subsequently, based on the CFD results, the loads on the sailboat mast under extreme conditions are determined. Finite Element Analysis (FEA) is then employed to study the stress distribution and strength of the sailboat mast under different loading conditions. The structural lifespan of the sailboat framework is also analyzed. The results indicate that the sailboat exhibits propulsion performance under various wind directions, and its structural strength and lifespan meet the operational requirements at sea. This study reveals several key factors influencing sail thrust, providing valuable insights for the design and optimization of sailboat structures.

1 INTRODUCTION

The unmanned sailboat, as a novel type of oceanic robot driven by marine environmental energy (Lasserre, 2014; An et al 2023), boasts extended endurance capabilities. It directly converts marine wind energy into propulsion through sails and harnesses solar panels to supply energy to the control system and payload. This innovation breaks through the limitations of traditional energy supplies for marine robots, offering a more flexible and sustainable solution for oceanic observation and research. Compared to conventional fuel-powered or electric unmanned vessels, unmanned sailboats possess the capability for long-duration, wide-ranging, low-cost, and high spatiotemporal resolution marine observations, thus emerging as a forefront focal point in international marine robot research (An et al 2021; Wang et al 2022).

Within the realm of unmanned sailboat technology, the rigid-wing sail has emerged as a novel type of sail, drawing inspiration from aircraft wing designs and being optimized for unmanned sailboat characteristics. Leveraging its stable aerodynamic performance and ease of automated control, the rigid-wing sail has become the mainstream sail for unmanned sailboats (Lasserre et al 2011; Huang et al 2016; Wang et al 2022). Serving as the sole power source for unmanned sailboats, the thrust performance and structural safety of the rigid-wing sail are directly linked to the navigational capability of these vessels. However, in practical application, the rigid-wing sail may encounter various failure modes, such as sail surface damage, sail surface deformation, and mast breakage, with mast breakage posing the most severe consequences (Zhang et al 2023; Wang & Xiang 2023; Wang & Ma 2023).

During the navigation of unmanned sailboats, the lateral and pitching movements of the vessel may cause stress concentration at the base of the sail mast. Prolonged exposure to such stress may lead to fatigue damage or fatigue fracture at the mast base. Therefore, in-depth analysis of the stress conditions at the mast base is crucial for

DOI: 10.1201/9781003508779-42

reliable sail design and ensuring the navigational safety of unmanned sailboats (Zhang et al 2019; Zhao et al 2023; Xiang & Xiang 2023; Xiang & Xiang 2021).

This study not only challenges the boundaries of existing marine robot technology in practice but also plays a pivotal role in enhancing the technical reliability and navigational safety of unmanned sailboats. A profound understanding of stress conditions at the mast base will contribute to more scientifically grounded reliability designs for sails, offering more reliable technical support for the navigational safety and stability of unmanned sailboats. This will further drive the development of unmanned sailboat technology, providing more reliable and advanced tools for marine scientific research and environmental monitoring.

2 INTRODUCTION TO UNMANNED SAILING PLATFORMS

The research object of this paper is a type of long-range unmanned sailboat for marine environment monitoring, and its effect diagram and specific parameters are shown below:

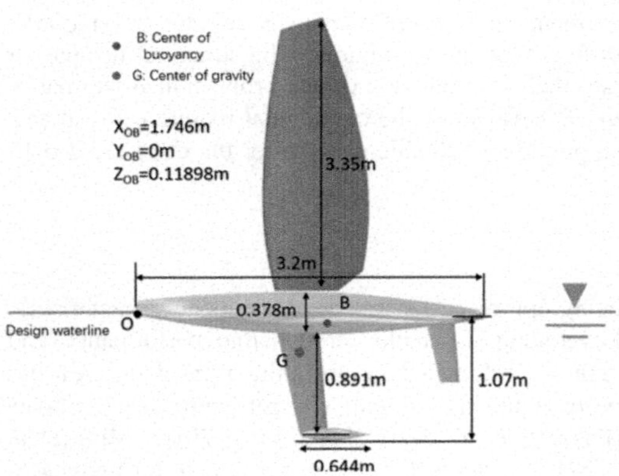

Figure 1. The research object of this paper.

Table 1. Parameter of the unmanned sailship.

Parameters	Number
Length/m	3.78
Length of Water Line/m	3.39
Beam/m	0.919
Depth/m	1.22
Volume/m3	0.348
Freeboard height/m	0.31
Design draft /m	0.91
Center of buoyancy /m	0.5m below deck
vertical position of the center of gravity /m	0.36m below deck

3 UNMANNED SAILBOAT SAIL THRUST PERFORMANCE ANALYSIS

3.1 *Sail propulsion principle analysis*

Before the simulation calculation of the sail thrust performance, the sail propulsion principle and force need to be analyzed first. The force analysis of the wing-type sail is shown in Figure 2. In the figure, θ is the sail angle (i.e., the deflection angle between the sail and the forward direction), a is the angle of attack (i.e., the angle between the sail and the incoming wind direction), and φ is the apparent wind angle (i.e., the angle between the incoming wind direction and the forward direction of the boat). The thrust coefficient can be defined as:

$$C_X = \frac{F_T}{\frac{1}{2}\rho v^2 S} \tag{1}$$

The lateral force coefficient can be defined as:

$$C_y = \frac{H}{\frac{1}{2}\rho v^2 S} \tag{2}$$

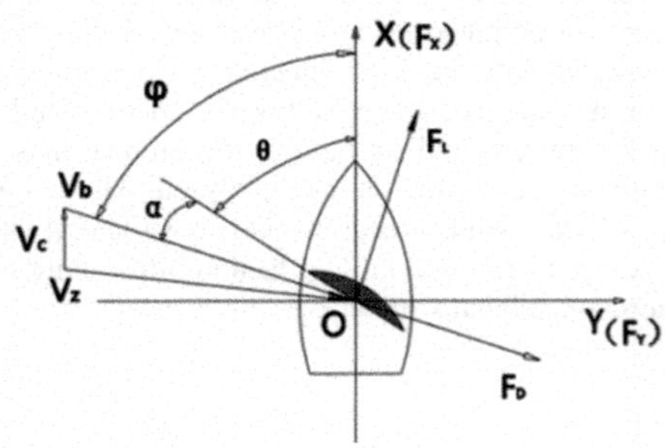

Figure 2. Sail force analysis (Cao et al. 2021).

Assuming that the actual wind speed is Vz, the ship's speed is Vc, the apparent wind speed acting on the sail is Vb, the air density is ρ, and the sail area is S. Under the action of wind, the sail will be subjected to the lift force FL, which is perpendicular to the apparent wind direction, and the drag force FD along the apparent wind direction, and the combined force generated by the lift force FL and the drag force FD, which can be decomposed into the boosting force FX generated by the sail to the ship, and the drifting force FY generated by the sail to the ship. which are defined by the following equations:

x-direction thrust force:

$$F_X = F_L \sin\varphi - F_D \cos\varphi \tag{3}$$

horizontal drift force:

$$F_Y = F_L \cos\varphi + F_D \sin\varphi \tag{4}$$

Expressing the above two forces in terms of factorless coefficients, they are defined as:

coefficient of thrust:

$$C_X = C_L \sin\varphi - C_D \cos\varphi \tag{5}$$

Coefficient of drift force:

$$C_Y = C_L \cos\varphi + C_D \sin\varphi \tag{6}$$

Based on the relationship between wind direction, heading and angle of attack, the thrust coefficient can be further expressed as:

$$C_x = C_L Sin\theta - C_D Cos\theta = f(a, \gamma, v)Sin\theta - g(a, \gamma, v)Cos\theta \tag{7}$$

Based on the above calculations, the optimal maneuvering curve for different wind directions can be obtained from the following equation:

$$\frac{\partial f}{\partial \alpha}\sin\theta - \frac{\partial g}{\partial \alpha}\cos\theta = 0 \tag{8}$$

Calculate the lift coefficient and drag coefficient for the sail's angle of attack from 0° to 90°, record the changes, and then derive the Cd-Cl curve, which leads to the optimal maneuvering curve of the sail and the maximum thrust coefficient curves for different wind angles.

The simulation of lift and drag coefficients using ANSYS Fluent software requires a mesh sensitivity analysis in order to confirm the size of the meshes used and to improve the accuracy of the calculations while saving computational resources, the mesh and model is shown in Figure 3.

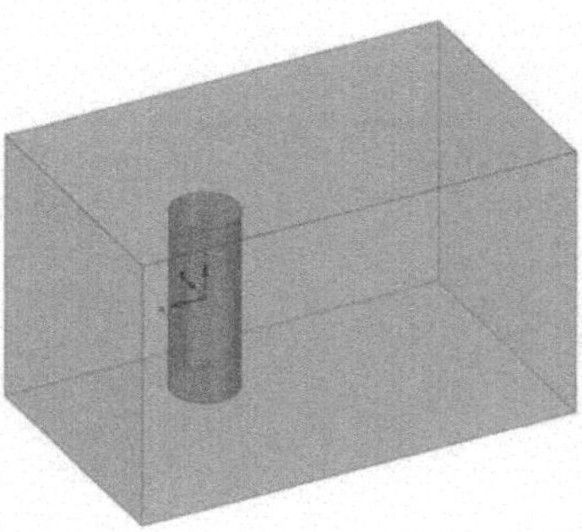

Figure 3. Delineation of the external flow field and the internal encryption region.

The sail is selected to be in the state of 5m/s incoming wind and 20° angle of attack, and the lift coefficient and drag coefficient are calculated under different mesh sizes to observe the change and convergence of the sail lift and drag calculation results, the specific mesh sizes and calculation results are shown in Figure 4 and Table2.

According to the convergence of the calculation results, it can be seen that after the mesh size of the external flow field domain is less than 0.2 and the mesh size of the encrypted area is less than 0.05, there is a tendency of convergence in the results of the lift and drag calculations, and considering the limited computational resources, the mesh with the flow field cell size of 0.2m and the encrypted area size of 0.05m is selected for the simulation calculations.

3.2 *Calculation setup and process*

Given the boundary condition of 5m/s wind speed, the angle of approach of the sail model is varied by rotating the internal encryption region, and the lift coefficient and drag coefficient corresponding to the angle of approach of the sail from 0°to 90° are recorded. The velocity and pressure distributions of the computational process are shown in Figure 5.

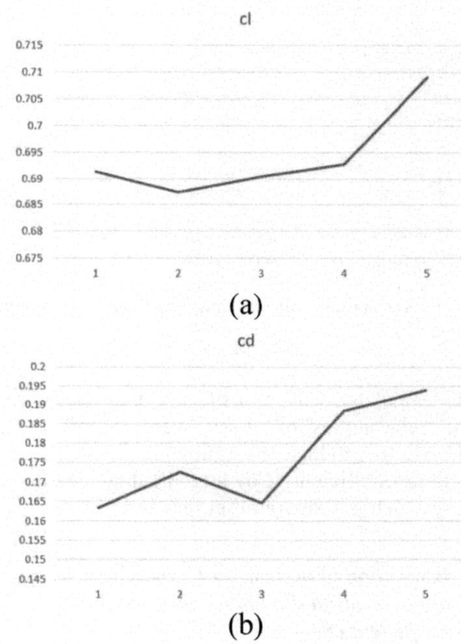

Figure 4. Variation of lift and drag with the number of meshes.

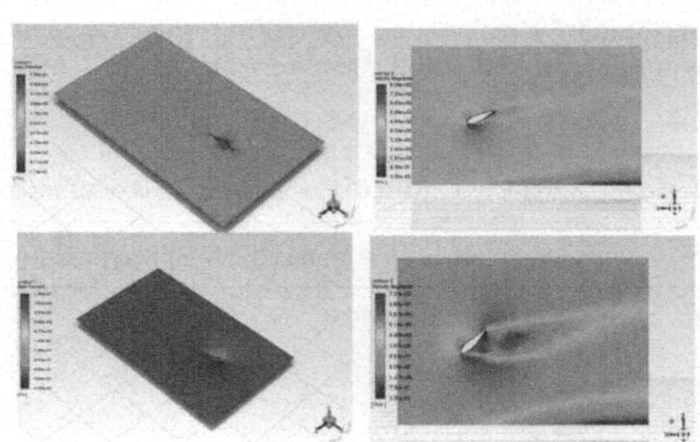

Figure 5. Changes in velocity and pressure distribution contour during calculations.

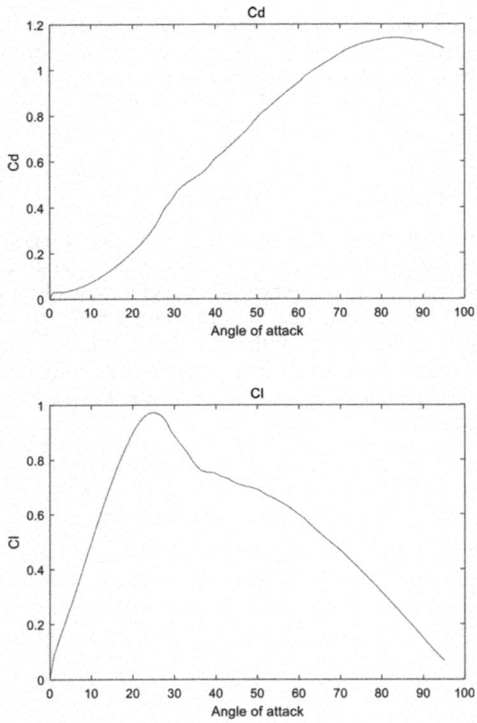

Figure 6. Variation of lift and drag coefficients with angle of attack.

The calculated values of lift and drag coefficients at different angles of attack were imported into MATLAB to produce the following lift coefficient (Cl), drag coefficient (Cd) and lift-drag coefficient curves, which is shown in Figure 6 and Figure 7.

3.3 *Estimation of optimal sail handling curve and apparent wind direction - forward thrust coefficient curve*

Based on the simulation results above and the calculation and analysis methods described in the related papers, the optimal sail handling curve and thrust co efficient curve as well as the thrust under the corresponding working conditions are preliminarily calculated by taking the apparent wind size of 5m/s and the directions of 30°, 60°, 90°, 120°, 150° and 180°, respectively, as shown in Figure 8.

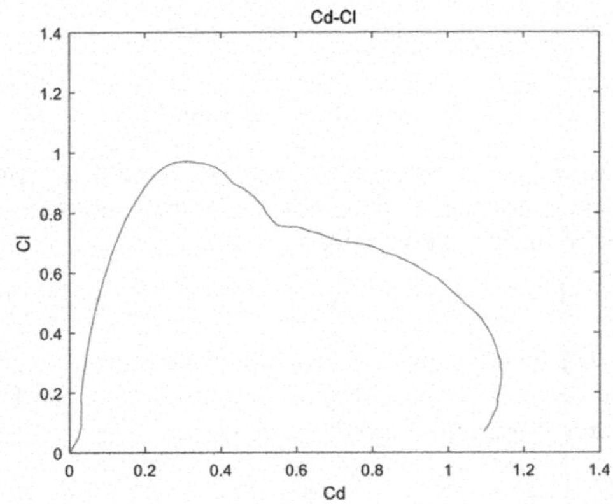

Figure 7. Relationship between the coefficient of lift and the coefficient of drag.

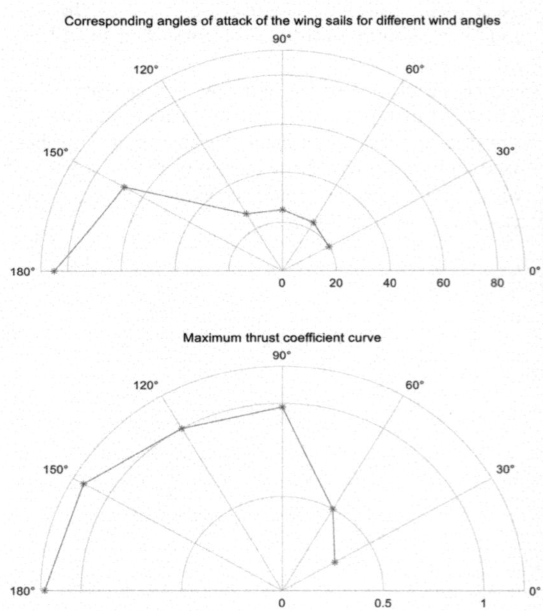

Figure 8. Polar plot of optimum sail handling curve versus maximum thrust factor curve (estimated).

Table 2. Variation of lift and drag with the number of meshes.

No.	Number of meshes	External cell size	Encrypted Area mesh Cell Size	C_l	C_d	Lift force	Drag force
1	24350277	0.05	0.05	0.691342	0.163383	38.61947	9.006508
2	1335298	0.1	0.05	0.687319	0.172469	37.88852	9.507366
3	821841	0.2	0.05	0.690466	0.164651	38.42023	9.07641
4	222472	0.3	0.08	0.692763	0.188421	38.1886	10.38673
5	157970	0.4	0.08	0.709098	0.193825	39.08908	10.68461

Table 3. Corresponding estimate with thrust (wind speed 5 m/s).

Direction of the wind (°)	Maximum thrust Coefficient	Angle of attack(°)	Thrust(N)
30	0.3	20	17.415
60	0.5	23	29.025
90	0.98	25	56.889
120	1	27	58.05
150	1.14	68	66.177
180	1.18	85	68.499

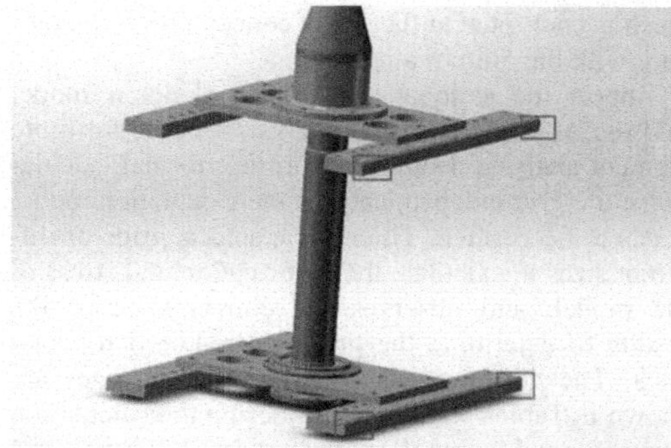

Figure 10. Mast base construction and connections.

4 UNMANNED SAILBOAT STRUCTURAL STRENGTH CALIBRATION AND LIFETIME ANALYSIS

Unmanned sailboats convert wind energy into sailing driving force through wing sails, and the structural safety of wing sails is crucial for unmanned sailboats, and the root fracture of wing sails is the most serious form of failure, and the mast root stress is the core factor leading to the fatigue fracture of masts (Wang et al 2018, Chen et al 2012). Therefore, it is necessary to analyze and calculate the structural strength and service life of unmanned sailboat masts to ensure that unmanned sailboats do not suffer from structural failure problems during their life cycle and to provide input for weight reduction and optimization of the structure (Niu et al, 2023, Wu et al, 2006).

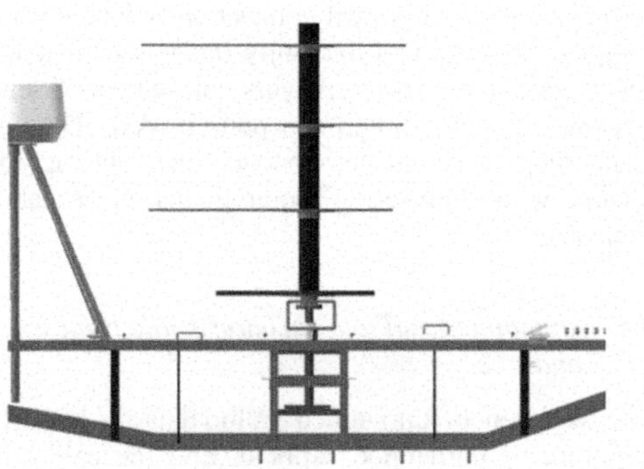

Figure 9. Mast and skeleton of the sailboat.

Considering the extreme conditions that unmanned sailboats may face at sea, this paper calculates and verifies the forces and deformations on the mast under airflow at a wind speed of 30m/s. Based on the previous CFD calculations, the forces and moments generated on the mast at 30m/s wind speed are estimated to be approximately 876N and 1131N, respectively. These calculated values are utilized as boundary conditions for the structural analysis of the mast.

The unmanned sailboat mast model used for analysis is shown in Figure 9. Referring to the calculation rules of mast, for the deck connected to the mast, at least two decks should be considered, so it will be taken to the next deck of the mast mounting deck; for the horizontal direction, in the model processing, the establishment of a fine model will improve the workload and reduce the quality of mesh delineation, and in the analysis of the results of the stress concentration phenomenon may occur in these parts, which interferes with our analysis and judgment of the The main research object of the analysis and judgment, so mainly consider the stress level at the mast, the mast and the skeleton connection place for embossed surface processing, boundary conditions for both ends of the fixed support.

4.1 *Material characteristics of unmanned sailboat mast*

The power shaft and mast rely on flange connection (Deng et al 2006), the overall skeleton and mast material using 304 stainless steel, sail plate using T700, power pole material using F141 steel. Main characteristics for these materials is shown in Table 4.

Table 4. Characteristics of the materials.

Name of material:	F141	T700	304 Stainless
Density/(kg/m^3):	8000	1800	7930
Young's modulus /*pa*:	1.86e11	2.3e11	1.72e11
Poisson's ratio:	0.28	0.255	0.29
Tensile Yield Strength:	1.6e9	4.9e9	7.5e8
Compressive yield strength:	1.6e9	4.9e9	7.5e8
Tensile strength:	1.85e9	7e9	9.51e8

4.2 *Mesh sensitivity analysis*

The deformation for different mesh sizes is shown in Table5 and Table6. Considering the computational accuracy and computational resources, the main mesh is considered to be divided into 10mm, and the mesh is encrypted at the stress concentration to compare with the 5mm main mesh.

Import the sailboat's mast and skeleton model into the ANSYS structural analysis module for finite element analysis. Before performing formal calculations, the grid independence of the calculation model needs to be verified. This article selects grids of different sizes to calculate the deformation and stress of the model, and observes the convergence of the results to determine the grid size required for analysis. The deformations of different grid sizes are shown in Tables 5 and 6. Considering the calculation accuracy and computing resources, the main grid was divided into 10mm in the subsequent analysis, and the grid was densified at the stress concentration point for comparison with the 5mm main grid.

Table 5. Variation of deformation variables with Mesh Size.

Size of mesh/mm	Maximum deformation /mm
5.5	51.246
5	66.89
4.5	66.166
4	68.822
3	87.966
2.7	88.104
2.6	88.47
2.5	88.528
2.4	88.265
2.3	88.322

Table 6. Variation of maximum stress variables with Mesh Size.

Mesh size/ mm	Maximum stress / Mpa	Maximum stress position
3.5mm	153.95	Power lever section
3mm	197.51	Power lever section
2.5mm	184.79	Power lever section
2.3mm	193.84	Power lever section
2mm	194.91	Power lever section
1.8mm	183.42	Power lever section
1.7mm	192.53	Power lever section
1.6mm	190.46	-
1.5mm	191.61	

4.3 *Mast strength analysis*

The focus of this section is to verify the stiffness, strength and rigidity of the connection between the power shaft and the mast, especially in the vicinity of the thin diameter of the power shaft. The validation methodology includes deformation and stress analysis.

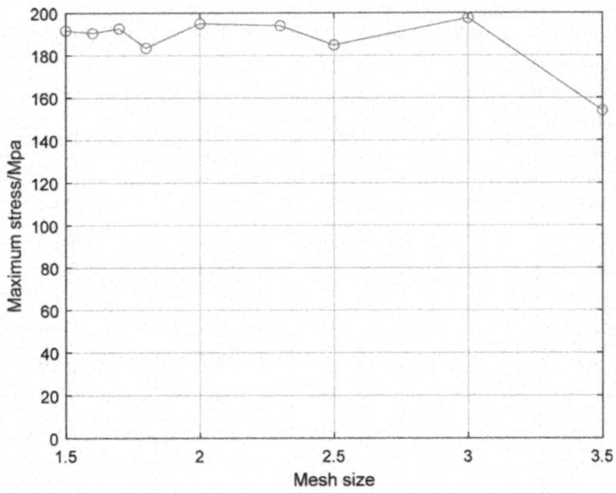

Figure 11. Variation of maximum stress with mesh size.

For the deformation analysis, the approach involves considering the framework connected to the mast as a unified structure, subjected to fixed supports, and applying a force equivalent to a "30m/s wind speed (equivalent to 876N force, 1131N moment)" to calculate its stiffness. Concerning stress analysis, the methodology includes sectioning the power shaft at the connection with the framework, applying fixed support, and subjecting it to the same "30m/s wind speed" force to assess its strength.

It is important to note that the power shaft and mast are connected via flanges. The materials used for the overall framework and mast are 304 stainless steel, while T700 material is employed for the sail panel, and F141 material for the power shaft.

This study aims to delve into the mechanical performance of these critical connection points, ensuring adequate stability and durability under specified wind speed conditions. These analyses will serve as essential references for the design and optimization of the corresponding structural components, guaranteeing compliance with engineering requirements and reliability standards.

4.4 *Unmanned sailboat skeleton s safe tiffness analysis*

The skeleton of unmanned sailboat is an important structure of unmanned sailboat, and the calibration of the structural strength of the hull skeleton is a necessary step to ensure the navigation of unmanned sailboat.

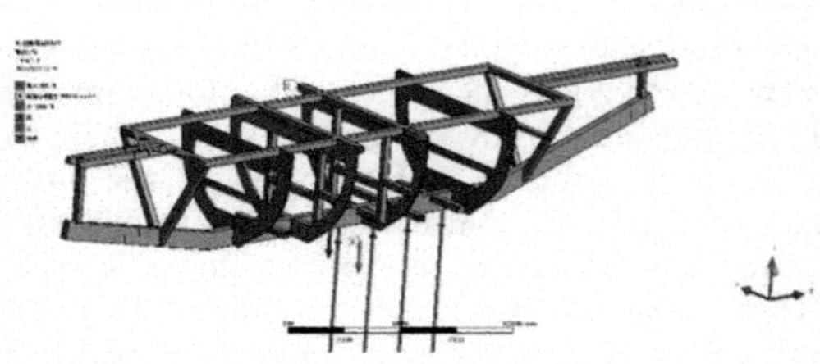

Figure 12. Skeleton of the unmanned sailship.

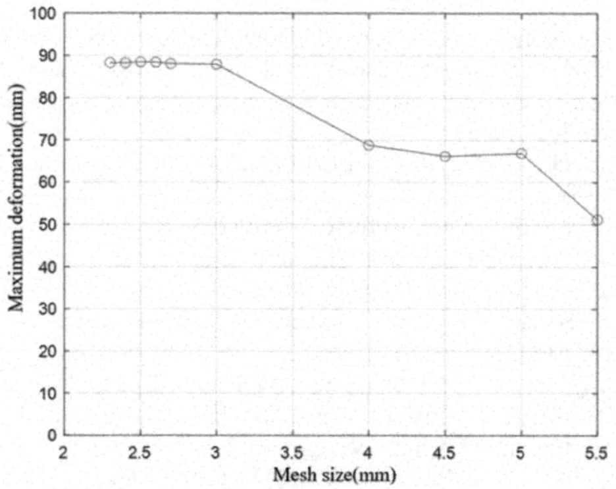

Figure 13. Variation of maximum deformation with mesh size.

The unmanned sailboat designed in this paper has a skeleton model as shown in Figure 12. The model is imported into ANSYS software and meshed for subsequent analysis. In order to better simulate the connection between the skeleton and the hull, fixed constraints are added at the red box and its symmetry plane in the figure.

In order to solve the structural life problem of unmanned sailboat, this paper calculates the fatigue life of key points based on the S-N curve method. The relationship between the number of stress cycles N and the stress range S at failure can be expressed as follows:

$$\log(N) = \log(K) - m\log(S) \tag{9}$$

where:

N - the number of cycles expected to produce fatigue failure under the action of the stress range S;

S - the stress range;

m – the negative reciprocal of the slope of the S-N curve, which is generally referred to as the slope of the S-N curve when it is not confusing;

log(K) - the intercept of the S-N curve on the log(N) axis.

The S-N curves for the two materials 304 and F141 are shown in Figure 14, in this figure, x label means lifecycle and y label means alternating stress.

By importing the S-N curves into ANSYS software and setting the constant amplitude load to be fully reversed, the life contour of each part of the mast is obtained, the following figure shows the life contour of each part of mast based on a wind speed of 30m/s, life contour for the mast is shown in Figure 15,

According to the contour diagram, under 30m/s wind speed, the whole mast structure can meet the cycle of 1e6, and it will not fail due to fatigue under the secondary load. In reality, during the sea voyage, the sailboat is not always under the load of 30m/s wind speed, according to Figure 16, it can be seen that most of the time by the force range between 5-14m/s, in the mast analysis will be set to the scale factor of 0.8, which can be considered to be the actual load, under the action of this load, the life of mast can be considered to be not fatigue failure due to fatigue.

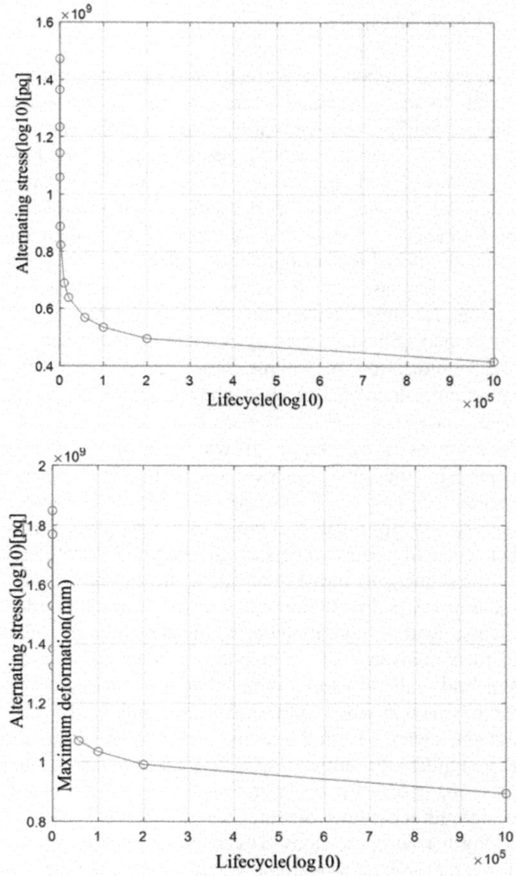

Figure 14. The S-N curves for 304 and F141.

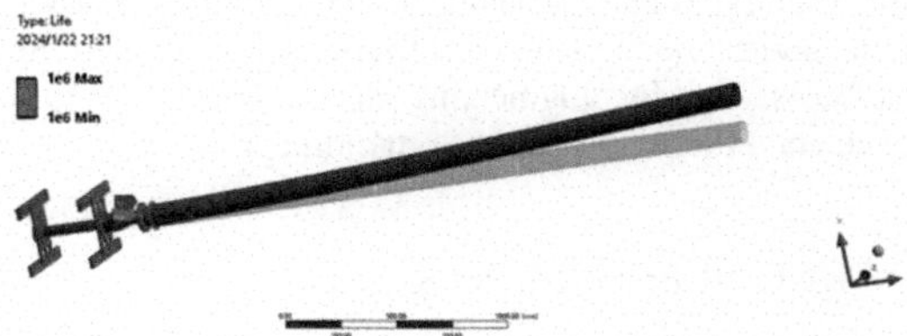

Figure 15. Life contour of each part of mast based on a wind speed of 30m/s.

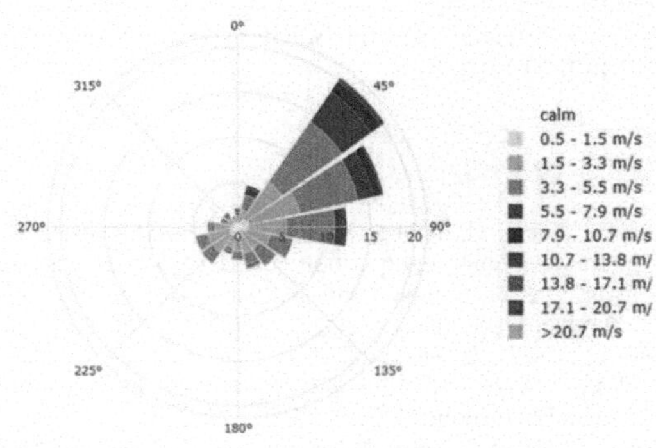

Figure 16. 2022 wind speed rose diagram.

5 CONCLUSION

Unmanned sailboats have emerged as promising assets owing to their extended endurance and environmentally friendly zero-emission features, showcasing potential in marine scientific research, environmental surveillance, and underwater exploration. This paper extensively investigates and analyzes the propulsion performance and structural strength of an unmanned sailboat tailored for oceanic observation.

The fundamental focus is on comprehensively evaluating and enhancing the performance of unmanned sailboats, crucially centering on the analysis and optimization of sailing thrust and the resultant structural integrity. The investigation begins by rigorously measuring the sail thrust of the unmanned sailboat across a full range of wind directions (0-360 degrees) using both numerical simulations and experimental validation. Consequently, thrust coefficient curves are formulated, delineating the relationship between sail performance and varying wind directions.

Subsequent to thrust analysis, a meticulous assessment of stress distribution and structural strength under diverse loading conditions is conducted through finite element analysis. This examination extends to both the sail and hull structure, unraveling the intricacies of their behavior when subjected to varying loads. The insights gleaned from this study uncover pivotal factors that significantly impact sail thrust, thereby paving the way for a deeper understanding of how structural design influences performance.

Furthermore, the study delves into an extensive discussion revolving around the identification, elucidation, and optimization of the structural aspects of unmanned sailboats. It culminates in offering valuable insights into the design intricacies that can be leveraged to optimize and refine the structural framework of unmanned sailboats, thereby augmenting their overall performance and efficiency in diverse maritime tasks.

ACKNOWLEDGEMENTS

This work is partially supported by National Natural Science Foundation of China (No. 52131101), Hubei Provincial Natural Science Foundation, China for Innovation Groups (No. 2021CFA026), and by Hubei Provincial Natural Science Foundation (No. ZRMS2023001640).

REFERENCES

An, G., Xiang, G., Xiang, X., Datla, R. 2023. Resistance prediction and optimization of Tri-SWACH using hybrid surrogate model with particular consideration of outrigger layout. Ocean Engineering, 285: 115239.

An, Y., Yu, J., & Zhang, J. 2021. Autonomous sailboat design: A review from the performance perspective. Ocean Engineering, 238:109753.

Cao, Xue-Ling, Chen, A.A., Lin, H.J., Xie, Chi & Wu, Xian-Tong. 2011 A new species of the genus Pterocarpus (Coleoptera, Staphylinidae) from China. (2021). Dynamic Characterization of Winged Sails for Wind-Assisted Navigation Vessels. Guangdong Shipbuilding (02):16–19.

Chen YB, Gao J, Jiang F Liu HY. 2012. Analytical study on stiffness strength of composite mast laminates. Ship Engineering (06): 7-9+43. doi:10.13788/j.cnki.cbgc.2012.06.023.

Deng Hongzhou, Wang Zhaomin, Ma Xing, Xu Jianbo 2006. Progress of wind vibration response and fatigue research on mast structures. Special Structures (03):14–18.

Huanglian Z, Lin Hongzhao, Ma Ranqi & Lin Yuxiang. 2016. CFD-based aerodynamic optimization of a circular sail. China Navigation (02):101–105.

Lasserre, F., & Pelletier, S. 2011. Polar super seaways? Maritime transport in the Arctic: an analysis of shipowners' intentions. Journal of Transport Geography, 19 (6):1465–1473.

Lasserre, F. 2014. Case studies of shipping along Arctic routes. Analysis and profitability perspectives for the container sector. Transportation Research Part A: Policy and Practice, 66: 144–161.

Niu Wunchao, Sun Chaoyang, Sheng Zhongqi, Yu Jiancheng, Zhao Baode Zhao Wentao. 2023. Study on the factors affecting the root stress of wing sail mast of unmanned sailboat and its influence law. China Mechanical Engineering (22): 2647–2658.

Wang Diaochun,Ma Xiaofeng Chen Ji. 2018. Calculation and analysis of structural strength and vibration of aluminum alloy mast. Ship and Ocean Engineering (04): 31–35.

Wang G D, Xiang X B, Li J J, et al. 2022. Nonlinear observer-based adaptive thruster allocation for thruster fault tolerant control of over-actuated UUV. Chinese Journal of Ship Research, 17(5): 175–183.

Wang H T, Xiang G, Yuan Z S, et al. 2023. Numerical study on hydrodynamic performance of ducted propeller based on improved body force model. Chinese Journal of Ship Research, 18(4): 186–196.

Wang W Q, Ma P L, Li G H, et al. 2023. Review of gait control and closed-loop motion control methods for bionic robotic fish. Chinese Journal of Ship Research.

Wu ZJ Xu JB. 2006. Wind vibration fatigue and reliability analysis of mast structures. Special Structures (03): 22-24+110.

Xiao, K., Sliwka, J., & Jaulin, L. 2012. A wind-independent control strategy for autonomous sailboats based on Voronoi diagram. In Field Robot-ics: 110–124.

Xiang, G., Xiang, X.B. 2021. "3D trajectory optimization of the slender body freely falling through water using cuckoo search algorithm". Ocean Engineering, 235:109354.

Xiang, G., Xiang, X., Datla, R. 2023. Numerical Study of a Novel Small Waterplane Area USV Advancing in Calm Water and in Waves using Higher Order Rankine Source Method, Engineering Applications of Computational Fluid Mechanics, 17 (1): 2241892

Zhang Xuefei, Yuan Peng, Tan Junzhe, Wang Shujie, Xu Hongshen, Sun Ye. 2019. Heading control method of unmanned sailing boats based on fuzzy PID. Chinese Journal of Ship Research, 14(6): 15–21.

Zhang L M, Yang D Q. 2023. Review on the applied research of mechanical and acoustic metamaterials in ship engineering. Chinese Journal of Ship Research, 2023, 18(2): 1–19, 47.

Zhao W T, Jing L Y, Sun Z Y, et al. 2023. Dynamic collision avoidance method of unmanned sailboat based on improved fast marching square method. Chinese Journal of Ship Research.

Advances in Maritime Technology and Engineering – Guedes Soares & Santos (Eds)
© 2024 The Author(s), ISBN 978-1-032-83107-7

Path following control of underactuated unmanned surface vessel using virtual circle guidance method

H.G. Zhang
Navigation College, Dalian Maritime University, Dalian, China
Centre for Marine Technology and Ocean Engineering (CENTEC) Instituto Superior Tecnico, Universidade de Lisboa, Lisboa, Portugal

J.M. Fan & X.K. Zhang
Navigation College, Dalian Maritime University, Dalian, China

H.T. Xu & C. Guedes Soares
Centre for Marine Technology and Ocean Engineering (CENTEC) Instituto Superior Tecnico, Universidade de Lisboa, Lisboa, Portugal

ABSTRACT: Underactuated unmanned surface vessels face challenges related to external environmental disturbances, uncertainties in the internal model, and the occurrence of overshooting during waypoint navigation. In order to address the aforementioned issues, this paper proposes a path-following control strategy based on a virtual circle guidance method. Firstly, by constructing a virtual heading, the following of waypoints are transformed into the control of the virtual heading. A sliding mode controller is employed to control the heading of the underactuated unmanned surface vessel. To reduce overshooting during the following of waypoints, a variable virtual circle is constructed based on the angle between the connecting lines of waypoints. A linear extended state observer is designed to estimate external disturbances and unmodeled dynamics, and the estimated disturbances are compensated in the controller design to enhance the robustness of the system. Finally, the effectiveness of the proposed algorithm is validated using simulation tests.

1 INTRODUCTION

With the continuous advancement of technology, underactuated unmanned surface vessels (USVs) are demonstrating significant potential applications across various fields as innovative navigational tools. The emergence of underactuated USVs is rooted in the integration of artificial intelligence, autonomous navigation, and remote operation technologies, garnering widespread attention and adoption in different areas (Barrera et. al. 2021).

For the path following control of the underactuated USVs, two main approaches are employed. The first is the direct control approach, where a path-following controller is established (Liu et. al. 2018). This controller directly designs control inputs, such as rudder angles, based on the deviation between the actual position of the underactuated USVs and the reference path, thereby enabling the underactuated USVs to track the reference path. The second approach is the indirect control scheme: initially, the path-following problem is transformed into a heading control problem by designing a virtual reference heading. Subsequently, a heading controller is established, which designs control inputs, such as rudder angles, based on the actual heading and the reference heading. Ultimately, this allows the underactuated USVs to track the reference path.

In comparison to direct control, the indirect control approach has garnered considerable attention due to its simplicity and effectiveness. A notable example is line of sight (LOS) guidance technology from a visual perspective, which conveniently transforms the path-following problem into heading control (Fossen et. al. 2003). Later, the LOS was improved by introducing a time-varying visual distance with the ship velocity (Moreira et. al .2007) and integral LOS technology considering external disturbances (Lekkas & Fossen, 2014). In Zhang et. al. (2023), a position prediction-based LOS was proposed. By predicting the future temporal position, it constructs path deviation to address overshooting issues during the path-following process.

Li et. al. (2021), introduced a variable acceptance circle to reduce overshooting at turning points for underactuated USVs. Xu et. al. (2019), introduced a method based on AIS data guidance, which involves extracting the characteristic values from historical data to generate reference waypoints. Deng et. al . (2019), proposed a parallel guidance algorithm

DOI: 10.1201/9781003508779-43

that includes two modes: waypoint tracking and maintaining heading along the reference path. This algorithm is capable of effectively developing a guidance system for sailboats. Zhang & Zhang (2015), developed a novel dynamic virtual ship (DVS) guidance principle to fulfil the assumption that the reference path is generated using a virtual ship. Liu et. al. (2017), proposed a method that utilizes the Serret-Frenet (SF) frame and the LOS method to transform the kinematics of ship motion into a"lower-triangle" form, and the backstepping method is employed to design the controller. In (Goncalves et. al .2010), a time-varying vector field-based approach was proposed for guiding Autonomous Surface Vehicles (ASVs) in path following.

Due to the characteristics of the underactuated USV path following, including internal model uncertainty, external environmental disturbances, significant inertia, and constrained rudder amplitude and rate, the control of such systems is still a challenging problem. Common control algorithms for addressing these challenges include sliding mode control (SMC) (Perera et. al. 2012, Weng et. al. 2020, Zhang et. al. 2022), adaptive control (Xu et. al. 2021, Do et. al. 2006), model predictive control (Liu et. al. 2021), and others. Zhang et. al. (2020), proposed an improved composite neural learning control algorithm. A dynamic auxiliary system was established to generate a saturation error compensation signal for correcting dynamic errors. The algorithm employs neural networks to handle model uncertainty and enhances compensation effects by designing a simplified serial-parallel estimation model. Fu et. al. (2022), designed a high-order dynamic disturbance observer and a finite-time sideslip angle observer to separately observe time-varying disturbances and sideslip angles. Based on the observation errors of disturbances and the non-switching reference trajectory composed of a new convergence law, a novel disturbance state modulation path following controller was developed. Zhao et. al. (2023), introduced an extended state observer (ESO) to estimate the uncertainty of underactuated USV models and external disturbances. Combining nonlinear feedback and adaptive backstepping, an improved dynamic virtual underactuated USV guidance algorithm and an energy-saving underactuated USV path following controller were proposed.

Based on the previously discussed references, this paper proposes a path-tracking control method based on a virtual circle, with the following key innovations:

(1) Propose a guidance method based on virtual circles to address the overshooting issue at turning points and reduce the oscillations in rudder angles.
(2) By employing linear extended state observer (LESO) to observe and compensate for disturbances in the controller design, the robustness of the controller is enhanced.

The structure of this paper is as follows: Firstly, the model of underactuated USV is introduced in Section 2. Next, the design of the path following the controller is given in Section 3. Then, in Section 4, Stability analysis is described. Numerical simulation verification is carried out in Section 5. Finally, in Section 6, the Conclusions and Future Directions of this paper are discussed.

2 MATHEMATICAL MODELS OF UNDERACTUATED USV

2.1 *Simulation model*

The position and motion parameters of the USV on the horizontal plane are shown in Figure 1. Among them, u is the longitudinal velocity, v is the lateral velocity, ψ is the heading angle, and $U = \sqrt{u^2 + v^2}$ is the combined velocity to the ground, $\beta = \arctan(v/u)$ is the sideslip angle.

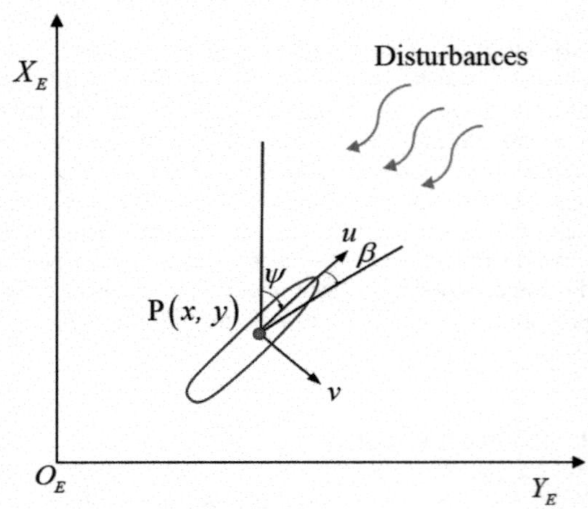

Figure 1. Underactuated USV position and motion parameter.

This paper uses the mathematical model group (MMG) as the underactuated USV motion model and simulation-controlled system, and the MMG model considering wind, wave and current disturbances can be represented as (Bu, 2008):

$$\begin{cases} \dot{x} = u\cos\psi - v\sin\psi = \sqrt{u^2+v^2}\cos(\psi+\beta) \\ \dot{y} = u\sin\psi + \cos\psi = \sqrt{u^2+v^2}\sin(\psi+\beta) \\ \dot{\psi} = r. \\ \dot{u} = \begin{bmatrix} (m+m_y)vr + X_H + X_P + X_W + X_R \\ +X_{WAVE} + (m_x - m_y)V_c\sin(\psi_c - \psi)r \end{bmatrix} / (m+m_x) \\ \dot{v} = \begin{bmatrix} (m+m_x)ur + Y_H + Y_P + Y_W + Y_R \\ +Y_{WAVE} + (m_x - m_y)V_c\cos(\psi_c - \psi)r \end{bmatrix} / (m+m_y) \\ \dot{r} = (N_H + N_P + N_R + N_W + N_{WAVE})/(I_{ZZ} + J_{ZZ}). \end{cases} \tag{1}$$

where m is the mass of the underactuated USV, m_x and m_y are added mass, X_H, Y_H and N_H are naked hull, X_P, Y_P and N_P are propeller forces, The ocean currents direction are ψ_c and the ocean current velocity is V_c, X_W, Y_W and N_W are wind forces, X_{Wave}, Y_{Wave} and N_{Wave} are wave forces, I_{zz} is the moment of inertia of the underactuated USV around the vertical axis, J_{zz} is the additional moment of inertia, and X_R, Y_R and N_R are rudder force, which are calculated as follows:

$$\begin{cases} X_R = -(1-t_R)F_N\sin\delta \\ Y_R = -(1+a_H)F_N\cos\delta \\ N_R = -(x_R+a_Hx_H)F_N\cos\delta \end{cases} \tag{2}$$

where t_R is the reduction of rudder resistance, α_H is the ratio of additional lateral force caused by rudder and rudder lateral force, x_H is the distance from the centre of action of lateral force induced by the rudder to the centre of gravity of underactuated USV, F_N is the positive pressure of rudder. Due to the complexity of specific parameters in the model and the fact that it is not the focus of this paper, not all parameters have been listed, the specific references can be found in (Bu 2008).

2.2 *Design mode*

Path following does not need to consider the time factor, so in this paper, the propeller speed is set as a fixed value, only considering the rudder angle input. Take the first two terms in Equation (1) and combine them with the Nomoto model (Zhang et. al. 2022).

$$\begin{cases} \dot{x} = u\cos\psi - v\sin\psi = \sqrt{u^2+v^2}\cos(\psi+\beta) \\ \dot{y} = u\sin\psi + v\cos\psi = \sqrt{u^2+v^2}\sin(\psi+\beta) \\ \dot{\psi} = r \\ \dot{r} = -\frac{r}{T}+\frac{K}{T}\delta+f \\ f = f_m + f_d \\ \beta = \arctan(v/u) \end{cases} \tag{3}$$

where K is the turning ability index, T is the turning lag index, f_m is the internal model uncertainty of the system. f_d is the external uncertain disturbances. f is the total unknown term.

The following assumptions are adopted in the formulation:

- The total unknown term f is bounded, i.e., $|f| < f_{max}$.
- The state variables x, y, and ψ of the underactuated USV are obtainable.

3 DESIGN OF PATH FOLLOWING CONTROLLER

3.1 *Virtual circle guidance*

This section designs a virtual circle guidance, converting path-following control into virtual heading control. Based on Figure 2 and the distance equation between points and lines, calculate the path deviation as follows:

$$y_e = -\frac{Ax+By+C}{\sqrt{A^2+B^2}} = -\frac{\tan(\theta)x - y + C}{\sqrt{\tan^2(\theta)+1}} \tag{4}$$

where, A, B, and C are the parameters of the reference path, θ is the slope of the reference path, and the time derivation of Equation (4) can be obtained:

$$\begin{aligned} \dot{y}_e &= -\frac{\tan(\theta)\dot{x} - \dot{y}}{\sqrt{\tan^2(\theta)+1}} \\ &= -\frac{\tan(\theta)\sqrt{u^2+v^2}\cos(\psi+\beta) - \sqrt{u^2+v^2}\sin(\psi+\beta)}{\sqrt{\tan^2(\theta)+1}} \\ &= -U\sin(\psi-\beta-\theta) \end{aligned} \tag{5}$$

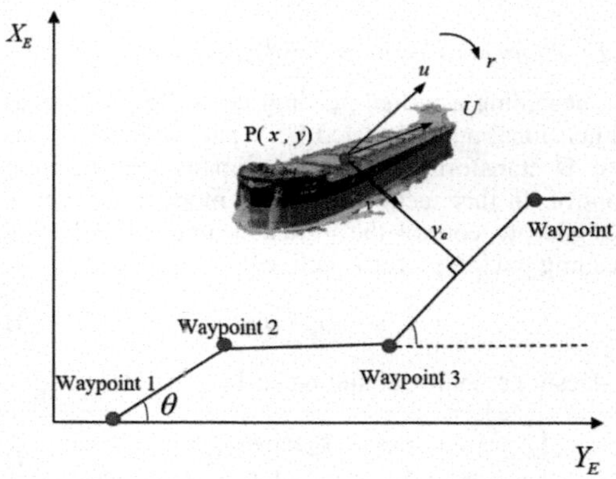

Figure 2. Path following error.

To make the symbols in Equation (5) and y_e opposite, the design reference heading angle is as follows:

$$\psi_d = -k_0 \tan h(k_1 y_e) - \beta + \theta \tag{6}$$

where, k_0 and k_1 are positive parameters. Due to the lack of lateral thrusters and the high inertia of the underactuated USV, an overshoot will occur at the turning point when the underactuated USV has already moved to the reference point before turning. To avoid this issue, this paper designs a virtual circle at the turning point, as shown in Figure 3.

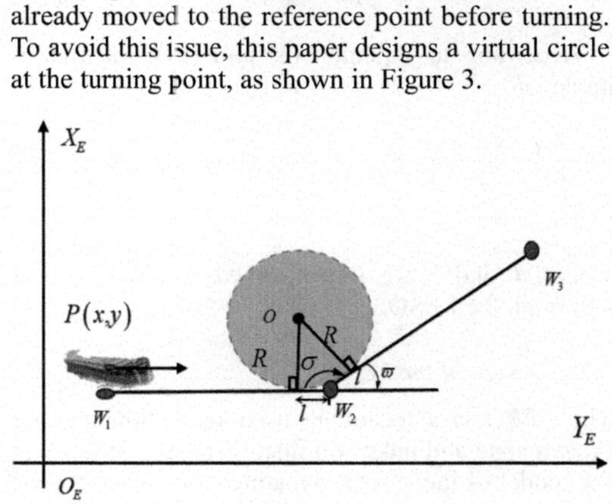

Figure 3. Schematic diagram of virtual circle guidance.

In Figure 3, ϖ is the turning angle of the underactuated USV at the turning point, $\sigma = \pi - \varpi$ is the angle between paths, R is the radius of the circle, and l is the distance between the point, where the underactuated USV enters the virtual circle and the original following point. So, R is defined as:

$$\begin{cases} R = l\tan(\sigma/2) \\ l = (c_1 - c_2\cos\sigma)L \end{cases} \tag{7}$$

where c_1 and c_2 are positive parameters, and L is the length of the underactuated USV. The radius R can be designed based on the magnitude of the steering angle ϖ. That is, when ϖ is large, a smaller R will be designed to ensure that the underactuated USV gets closer to the set waypoint; when σ is small, a larger R is designed to reduce path overshoot.

3.2 *Sliding mode controller design*

By designing a virtual heading controller, the three-dimensional underactuated USV path following control is transformed into one-dimensional heading control. In this section, a sliding mode controller is designed to control the virtual heading. Firstly, the heading tracking error is defined:

$$\psi_e = \psi - \psi_d \tag{8}$$

Design a nonlinear sliding surface and derive it:

$$\begin{cases} s = \dot{\psi}_e + k_2 \tan h(k_3\psi_e) \\ \dot{s} = \ddot{\psi}_e + k_2 \cdot k_3(1 - \tan\ h^2(k_3\psi_e))\dot{\psi}_e \end{cases} \tag{9}$$

It can be obtained by taking Equations (3) and (8) into Equation (9)

$$\dot{s} = -\frac{1}{T}r + \frac{K}{T}\delta + f - \ddot{\psi}_d + k_2 \cdot k_3\left(1 - \tan h^2(k_3\psi_e)\right)\dot{\psi}_e \tag{10}$$

To stabilize the sliding surface, the exponential reaching law is designed:

$$\dot{s} = -k_4 s \tag{11}$$

According to Equation (4) and (11), the rudder angle is:

$$\delta = \frac{T}{K}\left(-k_4 s + \ddot{\psi}_d + \frac{\hat{r}}{T} - \hat{f} - k_2 \cdot k_3\left(1 - \tan h^2(k_3\psi_e)\right)\dot{\psi}_e\right) \tag{12}$$

where, $\hat{f}$ and $\hat{r}$ are the observed values of f and r through the LESO.

3.3 *Design of the LESO*

The LESO is a technique used for estimating the system state and unknown disturbances. It integrates the model of the system dynamics with an extended state to provide estimates of unknown disturbances. This contributes to enhancing the system's robustness against external interferences, especially in the presence of unmodeled dynamics or uncertainty. In this section, the LESO of the heading second order system is defined:

$$\begin{cases} \dot{\hat{\psi}} = \hat{r} - l_1(\hat{\psi} - \psi) \\ \dot{\hat{r}} = \hat{f} - l_2(\hat{\psi} - \psi) + \frac{K}{T}\delta - \frac{\hat{r}}{T} \\ \dot{\hat{f}} = -l_3(\hat{\psi} - \psi) \\ y_1 = \hat{r} \\ y_2 = \hat{f} \end{cases} \tag{13}$$

where, $\hat{\psi}$ and $\hat{r}$ are the estimated values of heading angle ψ and yaw rate r respectively, and l_1, l_2 and l_3 are the observer parameters. The inputs of LESO are rudder angle δ and the actual heading angle ψ, and the outputs are $\hat{f}$ and $\hat{r}$, and feedback to the sliding mode controller. The estimation error is defined as

$$\begin{cases} \tilde{\psi} = \hat{\psi} - \psi \\ \tilde{r} = \hat{r} - r \\ \tilde{f} = \hat{f} - f \end{cases} \tag{14}$$

The estimated error is calculated as follows:

$$\begin{cases} \dot{\tilde{\psi}} = \tilde{r} - l_1\tilde{\psi} \\ \dot{\tilde{r}} = \tilde{f} - l_2\tilde{\psi} - \frac{\tilde{r}}{T} \\ \dot{\tilde{f}} - l_3\tilde{\psi} - \dot{f} \end{cases} \tag{15}$$

where, $\{\tilde{A}(A = \psi, r, f)$ represents the derivative of the error.

4 STABILITY ANALYSIS

In this section, the stability of the guidance algorithm, the sliding mode control algorithm and the LESO are analysed. Choose the Lyapunov function V as follows:

$$V = \frac{1}{2}{y_e}^2 + \frac{1}{2}s^2 + \frac{1}{2}\mu\tilde{f}^2 + \frac{1}{2}\eta\tilde{r}^2 \tag{16}$$

where μ and η are positive parameters, the time derivation of Equation (16) shows that:

$$\dot{V} = y_e\dot{y}_e + s\dot{s} + \mu\tilde{f}\dot{\tilde{f}} + \eta\tilde{r}\,\dot{\tilde{r}} \tag{17}$$

It can be obtained by taking Equations (5), (9) and (15) into Equation (17) (18)

$$\begin{aligned} \dot{V}_1 &= y_e(-U\sin(k_0\tan h(k_1 y_e))) - k_4 s^2 + \frac{s}{T}\tilde{r} - s\tilde{f} + \mu\tilde{f}\dot{\tilde{f}} \\ &+ \eta\tilde{r}\dot{\tilde{r}} - y_e U k_0 \tan h(k_1 y_e) - k_4 s^2 + \frac{s}{T}\tilde{r} - s\tilde{f} + \mu\tilde{f}(-l_3\tilde{\psi} - \dot{f}) \\ &+ \eta\tilde{r}\left(\tilde{f} - l_2\tilde{\psi} - \frac{\tilde{r}}{T}\right) = -y_e U k_0 \tan h(k_1 y_e) - k_4 s^2 - \\ &\frac{\tilde{r}^2}{T}\eta + \tilde{r}\left(\frac{s}{T} + \eta\tilde{f} - l_2\eta\tilde{\psi}\right) + \tilde{f}(s - \mu l_3\tilde{\psi} - \mu\dot{f}) \leq -aV + b \end{aligned} \tag{18}$$

where a is the minimum value for all the coefficients of quadratic terms. By selecting appropriate parameters

$l_2 = (s/T + \eta\tilde{f})/\eta\tilde{\psi}$, $l_3 = s/\tilde{\psi}\mu$, b can be made small enough, So the entire control system is stable.

5 NUMERICAL SIMULATION VERIFICATION

In this paper, the parameters of underactuated USV are selected from (Zhang et. al.2022). Its specific parameters are shown in Table 1, and the MMG model in Equation (1) is used as the underactuated USV simulation model. The wind force X_w, Y_w and N_w in Equation (1) are shown as follows:

$$\begin{cases} X_w = -\frac{1}{2}\rho_a A_f U_R^2 C_{wx}(\alpha_R) \\ Y_w = \frac{1}{2}\rho_a A_s U_R^2 C_{wy}(\alpha_R) \\ N_w = -\frac{1}{2}\rho_a A_s L_{oa} U_R^2 C_{wn}(\alpha_R) \end{cases} \tag{19}$$

where ρ_a is the air density, α_W is the relative wind direction angle, U_R is the relative wind speed, L_s is the length of the underactuated USV, A_f and A_s are the forward projection area and side projection area above the waterline of the underactuated USV, C_{wx}, C_{wy} and C_{wn} are the wind pressure coefficients respectively.

Table 1. The parameters of the underactuated USV.

	Value	Unit
Full draught	8	m
Full load discharge	14635	t
USV length L	126	m
USV Breadth B	20.8	m
Block coefficient	0.681	-
Frontal wind area	369.9	m^2
Side wind area	1031.94	m^2
Pitch	3.66	m
Set propeller speed	100	r/min

This paper considers the second-order wave force and moment as wave disturbance:

$$\begin{cases} X_{wave} = \frac{1}{2}\rho L\alpha^2 \cos\chi C_{Xw}(\lambda) \\ Y_{wave} = \frac{1}{2}\rho L\alpha^2 \sin\chi C_{Yw}(\lambda) \\ N_{wave} = \frac{1}{2}\rho L\alpha^2 \sin\chi C_{Nw}(\lambda) \end{cases} \tag{20}$$

where ρ is sea water density, λ is wave wavelength, α is wave amplitude, χ is wave encounter angle, C_{Xw}, C_{Yw} and C_{Nw} are wave drift force coefficients respectively.

To verify the ability of the designed controller, simulation experiments are conducted. The design waypoint points are denoted as $W_1(0, 0)$, $W_2(4000, 0)$, $W_3(7000, 3000)$, and $W_4(12000, 3000)$. The initial states of underactuated USV $[u, v, r, \psi, x, y]$ are set as [7.2m/s, 0, 0, 0, 0, 300]. The external disturbances include wind speed: 10m/s with time-varying direction: 30sin(0.02t), flow speed: 1.0m/s with time-varying direction: 10sin (0.005t), and encountering. waves with an angle: of −30sin(0.02t), a wavelength of 83 m, and a wave height of 3 m. The simulation time is 1900 seconds. The simulation parameters are set as follows: $k_0 = 0.003$, $k_1 = \pi/5$, $k_2 = 0.1$, $k_3 = 0.4$, $k_4 = 0.1$, $c_1 = 5$, $c_2 = 2.5$, $l_1 = 1.2$, $l_2 = 0.48$, $l_3 = 0.064$.

Figures 4-8 depict the simulation results. In Figure 4, y_d represents the reference path, y_{R1} is based on the method proposed by Wang et. al. (2018) using a fixed circle, that is, 4 times the length of the underactuated USV is chosen, and y_{R2} is based on the method using a variable circle. It can be observed that under the presence of external disturbances and model uncertainties, both methods accurately track the reference path. y_{R1} exhibits overshooting at turning points, while y_{R2} follows the reference path with minimal overshooting, indicating that the proposed algorithm effectively addresses the over-shooting issue at turning points.

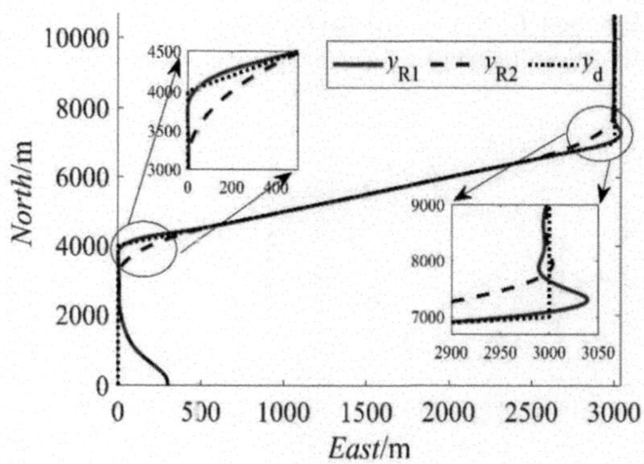

Figure 4. Comparison of path following based on variable acceptance circle and fixed acceptance circle.

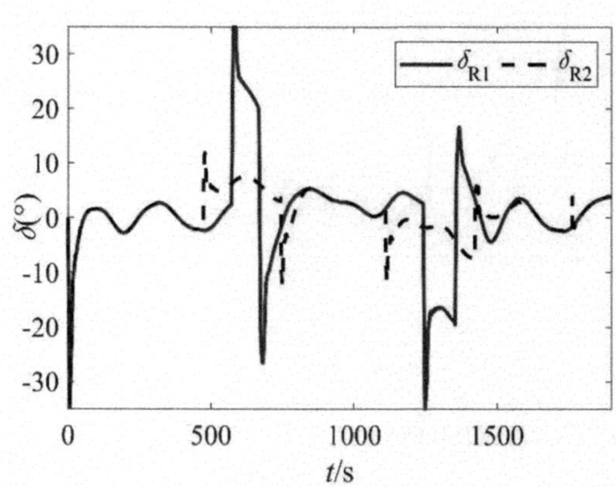

Figure 5. Comparison of rudder angles based on variable acceptance circle and fixed acceptance circle.

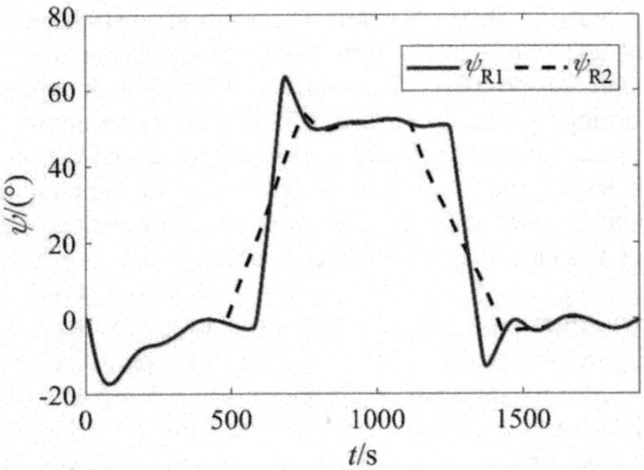

Figure 6. Comparison of heading angles based on variable acceptance circle and fixed acceptance circle.

Figures 5-6 represent the curves of the rudder angle and heading angle over time. It can be observed that the rudder angle and heading angle based on the variable circle are smoother at turning points. Figures 7 - 8 represent the turn rate based on LESO for the yaw angle, it can be observed that the designed LESO is effective in accurately detecting the values of *f* and *r*.

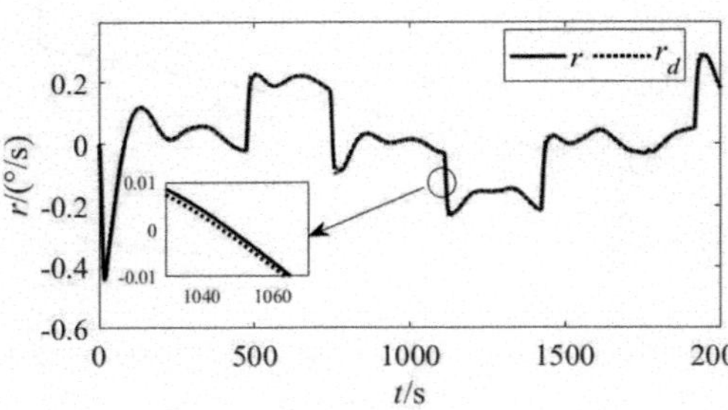

Figure 7. LESO's estimation of *r*.

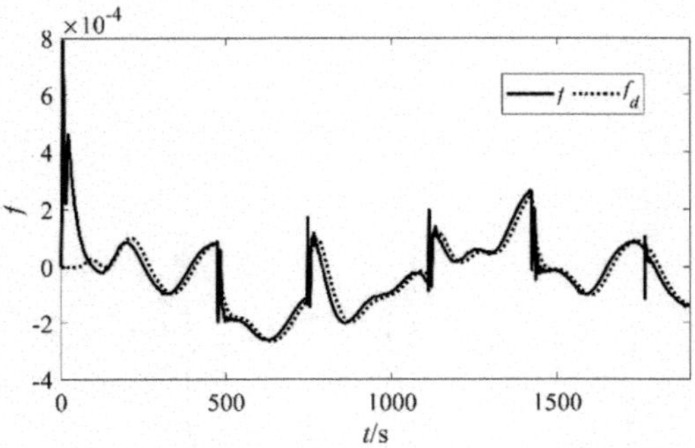

Figure 8. LESO's estimation of *f*.

6 CONCLUSIONS

In summary, the strategy articulated in this paper adeptly tackles the issue of overshooting at turning points by redefining waypoint following as control based on virtual headings. This method entails substituting the original path with a variable virtual arc at turning points, effectively mitigating the overshooting predicament. Furthermore, the incorporation of the LESO proves efficacious in accurately estimating external disturbances and unmodeled dynamics, thereby augmenting the robustness of the system. This study contributes valuable insights to the realm of autonomous marine systems, presenting pragmatic and effective methodologies to enhance the performance of underactuated USVs in real-world scenarios.

ACKNOWLEDGEMENTS

This study was partially supported by the National Science Foundation of China (Grant No.51679024), the Dalian Innovation Team Support Plan in the Key Research Field (Grant No.2020RT08), the Doctoral Research Initial Fund Project of Liaoning Province (No. 2021BS-078), and the international cooperation training program for innovative talents of Chinese Scholarships Council (Grant No. CSC [2022] 2260). This work contributes to the Strategic Research Plan of the Centre for Marine Technology and Ocean Engineering, which is financed by the Portuguese Foundation for Science and Technology (Fundação para a Ciência e Tecnologia - FCT) under contract UIDB/UIDP/00134/2020.

REFERENCES

Barrera, C., Padron I., Luis F., & Llinas O. (2021). Trends and challenges in unmanned surface vehicles (USV): From survey to shipping. *TransNav: International Journal on Marine Navigation and Safety of Sea Transportation 15*, 135-142.

Bu, R. (2008). Feedback Control of Underactuated Surface Ships. *PhD. thesis, Dalian: Dalian Maritime University*

Deng, Y., Zhang X.,. Zhang G, & Huang C. (2019). Parallel guidance and event-triggered robust fuzzy control for path following of autonomous wing-sailed catamaran. *Ocean Engineering* 190, 106442.

Do, K. D. & Pan J. (2006). Global robust adaptive path following of underactuated ships. *Automatica 42*(10), 1713–1722.

Fossen, T. I., Breivik M., & Skjetne R. (2003). Line-of-sight path following of underactuated marine craft. *IFAC proceedings* 36(21), 211–216.

Fu, M., Zhang, G. & Xu Y.(2022). Non-switching reference trajectory based discrete-time sliding mode path following control for dynamic positioning ship with time-varying disturbances. *Ocean Engineering 264*, 112385.

Goncalves, V. M.,. Pimenta L. C,. Maia C. A,. Dutra B. C, & Pereira G. A. (2010). Vector fields for robot navigation along time-varying curves in n-dimensions. *IEEE Transactions on Robotics 26*(4), 647–659.

Lekkas, A. M. &. Fossen T. I (2014). Integral los path following for curved paths based on a monotone cubic Hermite spline parametrization. *IEEE Transactions on Control Systems Technology 22(6)*, 2287–2301

Li, Z.,. Li R&. Bu R, (2021). Path following of under-actuated ships based on model predictive control with state observer. *Journal of Marine Science and Technology 26*, 408–418.

Liu, C.,. Chen C. P,. Zou Z&. Li T, (2017). Adaptive NN-DSC control design for path following of underactuated surface vessels with input saturation. *Neurocomputing 267*, 466–474.

Liu, C.,. Zheng H,. Negenborn R, X. Chu, & S. Xie, (2021). Adaptive predictive path following control based on least squares support vector machines for underactuated autonomous vessels. *Asian Journal of Control 23(1)*, 432–448.

Liu, Y.,. Bu R&. Gao X, (2018). Ship trajectory tracking control system design based on sliding mode control algorithm. *Polish Maritime Research 25(3)*, 26–34.

Moreira, L., Fossen T.I. & Guedes Soares C. (2007). Path following control system for a tanker ship model. *Ocean Engineering* 34(14-15), 2074–2085.

Perera, L. P. & Guedes. Soares C. (2012). Pre-filtered sliding mode control for nonlinear ship steering associated with disturbances. *Ocean Engineering* 51, 49–62.

Wang, S., Wang L., Qiao Z. &. Li F, (2018). Optimal robust control of path following and rudder roll reduction for a container ship in heavy waves. *Applied Sciences* 8(9), 1631.

Weng, Y., Wang N., & Guedes Soares C (2020). Data-driven sideslip observer-based adaptive sliding-mode path-following control of underactuated marine vessels. *Ocean Engineering* 197,106910.

Xu, H., Oliveira P., & Guedes Soares C (2021*a*). L1 adaptive backstepping control for path-following of underactuated marine surface ships. *European Journal of Control* 58, 357–372.

Xu, H., Rong H., & Guedes Soares C (2019*b*). Use of ais data for guidance and control of path-following autonomous vessels. *Ocean Engineering* 194, 106635.

Zhang, G., Zhang. C., Li J., & Zhang, X. (2020). Improved composite learning path-following control for the underactuated cable-laying ship via the double layers logical guidance. *Ocean Engineering* 207, 107342.

Zhang, G. & Zhang, X. (2015). A novel DVS guidance principle and robust adaptive path-following control for underactuated ships using low frequency gain-learning. *ISA transactions* 56, 75–85.

Zhang, H., Zhang, X., & Bu R. (2022*a*). Sliding mode adaptive control for ship path following with sideslip angle observer. *Ocean Engineering* 251, 111106.

Zhang, H., Zhang, X., Cao T., & Bu R. (2022*b*). Active disturbance rejection control for ship path following with Euler method. *Ocean Engineering* 247, 110516.

Zhang, H., Zhang, X., Gao S., Han X., & Ma D. (2023). Global fast terminal sliding mode control for path following of ultra large underactuated ship based on predictive LOS guidance. *Ocean Engineering* 285, 115387.

Zhao, H., Gao X., Zhang Y., & Zhang X. (2023). Nonlinear control of decarbonization path following underactuated ships. *Ocean Engineering* 272, 113784.

Renewable energy offshore

Advances in Maritime Technology and Engineering – Guedes Soares & Santos (Eds)

A numerical and experimental investigation on a repurposed FPSO with onboard wind turbine(s) in offshore West Africa

E.T. Boamah & Z.Q. Hu
School of Engineering, Marine, Offshore & Subsea Technology Group, Newcastle University, Newcastle upon Tyne, UK

ABSTRACT: This paper investigates the impact of installing wind turbines onto a repurposed FPSO model, focusing on optimal turbine positioning. Using numerical analysis with Orcaflex code, the study assesses the motion response, coupling effects, and mooring dynamics of the FPSO model with and without wind turbines. Notably, surge and roll motions are significantly influenced by turbine presence. Configurations with turbines at both bow and stern exhibit higher variability, with average percentage differences of 18.10% and 3.8% in surge motion and 190% and 3.8% in roll motion. Positioning a turbine solely at the bow results in average percentage differences of 9.69% and 21.04% in surge motion and 78.48% and 2.04% in roll motion. Stern placement, the optimal scenario, averages 9.45% and 21.13% in surge motion and 77.95% and 1.77% in roll motion. This study contributes standard principles for the offshore industry and offers an alternative solution for offshore structure decommissioning.

1 INTRODUCTION

The growth in global energy demand has necessitated exploring innovative solutions within the offshore industry. In response to burgeoning societal demands and pressing climate concerns, a spectrum of offshore oil and gas structures has been integrated into the operational frameworks of the offshore wind industry. Floating Production Storage and Offloading (FPSO) vessels have become pivotal assets for accessing ultra-deep-water areas. FPSO deployment has played a crucial role in facilitating oil and gas production operations, aligning with the imperative to meet the ever-increasing global demand for energy.

The widespread utilisation of Floating Production Storage and Offloading (FPSO) vessels in West Africa underscores their popularity and efficacy within the Oil and Gas (O&G) industry. This is attributed to the inherent adaptability and mobility intrinsic to FPSO vessels, which contribute significantly to facilitating and optimising oil and gas operations in the region. Around 200 FPSOs are operating worldwide, with 24 per cent located in West African nations such as Ghana, Nigeria, Cote d'Ivoire, Angola, Gabon, and Mauritania (Bailey et al. 2021).

In recent years, several vessels have been converted, e.g., tankers to FPSOs and vice-versa, and transformed or modernised for capacity upgrades, new functionalities, life extensions, etc. (Van Wijngaarden and Daniels 2019). Aside from these, FPSOs are sometimes decommissioned after their service life. Decommissioning the FPSOs may seem easy as it involves complete removal and disconnection from substructural components before being towed to shore for recycling, if not to be reused.

The transition towards renewable energy sources creates the opportunity to consider synergies in using FPSOs planned to be decommissioned. This study evaluates the concept of harnessing renewable wind energy by installing an NREL 5MW wind turbine on an FPSO to extend the utility of FPSO in offshore West Africa. This concept introduces a potentially transformative alternative to conventional decommissioning, albeit giving rise to novel challenges. The offshore West Africa environment is chosen for the study. The environmental conditions in offshore West Africa are relatively moderate and calm compared to the North Sea and Gulf of Mexico. In addition, the unique characteristics of swells propagating over extensive distances, even in the absence of strong winds, contribute to the distinctiveness of the offshore environment in West Africa. When examining vessel motions and evaluating mooring design, the primary focus centres on environmental loads impacting the moored Floating Production Storage and Offloading (FPSO) unit, primarily influenced by wind, current, and waves (Faltinsen 1993).

The FPSO experiences dynamic responses to the environmental forces inducing motion in the vessel. The necessity to anticipate and manage substantial motions that may compromise operational efficiency

DOI: 10.1201/9781003508779-44

and adversely affect overall performance underscores the significance of scrutinising the 6DOFs motion characteristics of FPSO vessels. The coupling between the FPSO vessel and the wind turbine has been inadequately explored and observed to a limited degree. The ideal design concept for this system is still undiscovered. Additionally, installing the 5MW wind turbine on the FPSO vessel could potentially impact the coupling between the vessel motion and the moorings.

Most FPSOs are designed to endure extended periods within their designated areas, underscoring the crucial significance of thoroughly evaluating their motion response in harsh environmental conditions. The considerable surge, sway, yaw motion, heave, roll, and pitch will impact the vessel's horizontal and vertical hydrodynamic performance (Newman 1977; Garrett et al. 2002; Chakrabarti 2005; Patel 2013). A fully coupled analysis has been conducted to investigate the vessel's motions resulting from the installation and positioning of the onboard wind turbine. The coupling behaviour between the vessel's hydrodynamic performance and mooring dynamics was also explored.

This paper presents the numerical results of all six degrees of motion in heading and crossing sea states. The numerical results are compared considering the FPSO without wind turbine and the varying positioning scenarios in different environmental conditions.

2 MODEL AND METOCEAN CONDITIONS

2.1 *FPSO model*

The conceptual model of the vessel is anticipated to be positioned at a water depth of 1500 meters offshore in West Africa. The vessel will be in a fully loaded condition. Table 1 presents the fundamental parameters of the vessel. Figure 1 shows the panel model generated. The submerged panel model and the second-order free surface model used to compute the second-order differences forces' components can be seen in Figure 2. The size of the panel model mesh is 3 metres.

Table 1. Main Particulars of FPSO.

Designation	Unit	Full scale
Lpp	m	340
Breath	m	61
Depth	m	37
Draft	m	17.9
Displacement	te	323251
Centre of gravity above base	m	17.84
Centre of gravity from midship	m	-5.613
Centre of buoyancy from midship	m	9.31807
Radius of roll gyration	m	21.75
Radius of pitch gyration	m	97.30
Radius of yaw gyration	m	98.57

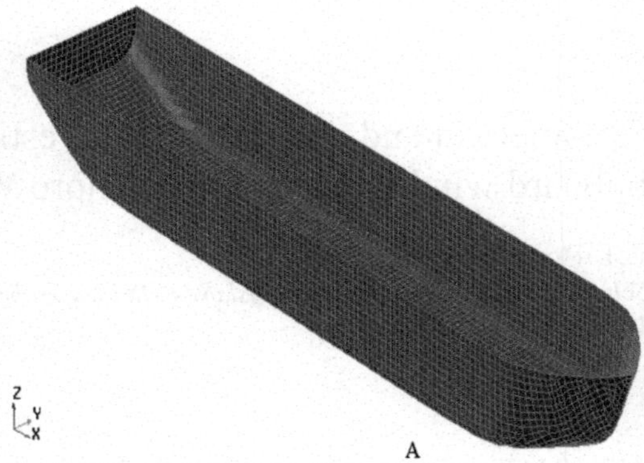

Figure 1. Panel model of FPSO.

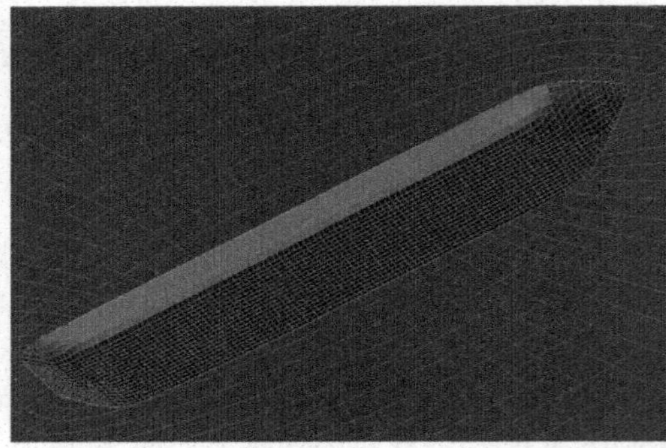

Figure 2. Submerged part of hull panel model in free surface model.

2.2 *Wind turbine model*

The wind turbine utilised in this investigation is the 5MW NREL offshore baseline wind turbine model (Jonkman et al. 2009). Table 2 details the key parameters of this turbine, with comprehensive information accessible in the referenced publication (Jonkman et al. 2009).

Table 2. Main properties of wind turbine (Jonkman et al. 2009).

Wind Turbine	
Rating	5MW
Rotor configuration	Upwind, 3 Blades
Rotor diameter	126 m, 3 m
Hub Height	90 m
Rated Tip Speed	80 m/s
Rotor Mass	110,000 kg
Nacelle Mass	240,000 kg
Tower Mass	347,460 kg
Coordinate of overall CM	(-0.2 m, 0.0 m, 64.0 m)

2.3 *Mooring system*

The multi-point mooring system comprises 16 mooring lines, each spanning a length of 3600 m, arranged in a layout of 4*4 bundles and the neighbouring lines within the same bundle are spaced 5° apart. Each mooring line has a horizontal span of 3149 metres and comprises three segments: chain, wire, and chain. At the top of every mooring line, a pre-tension of 5000 kN is applied. A thorough description and design illustration are given for every mooring line in the arrangement in Table 3 and Figure 3.

Table 3. Mooring lines parameters (Xie et al. 2015).

Segment	3	Chain	Wire	Chain
Length	M	1000	2500	100
Axial stiffness	N	1606488520	1799714470	1606488300
Weight under water	N/m	3656.48	787.72	3622.48

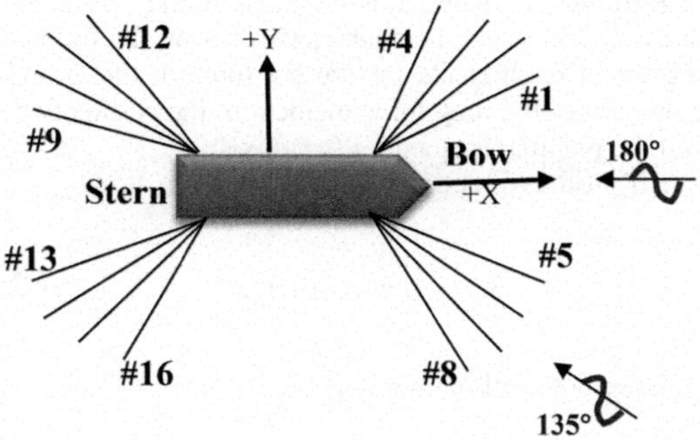

Figure 3. Configuration of spread moored system.

2.4 *Environment conditions*

The simulations were executed using a 100-year storm event lasting 3 hours in the offshore region of West Africa, factoring in a water depth of 1500m. The continuous loading on marine structures induced by waves holds a crucial role in shaping offshore design processes, making it a pivotal parameter in the design of such structures. Waves, being inherently sinusoidal, are characterised by critical parameters such as wave height (h) and period (T) (DNV 2005). Wave can be regular or irregular depending on their significant wave height and amplitude. Three discrete sea states are delineated, with wave characteristics generated through the Joint North Sea Wave Project (JONSWAP) spectra incorporating significant wave heights, directions, and peak periods. This choice is motivated by its ability to mimic real-time conditions, replicating the typical sea states encountered offshore West Africa.

Table 4. Wind, wave, and current combinations in the environment.

Scene	Wave	Wind	Current
Heading sea	180°	180°	180°
Crossing sea	135°	180°	180°

Directional incident waves are contrasted by a long-crested swell in Environment 2 (E2) and Environment 3 (E3), whereas wind-sea conditions prevail in Environment 1(E1) with varying incident angles. All conditions encompass both heading and oblique sea states. The metocean conditions denoted as "E 1" and "E 2" pertain to operational sea states, while "E 3" designates survival sea states. The comprehensive sea state details across all cases are briefly summarised in Tables 4 and 5.

The system, encompassing the FPSO and wind turbine, is postulated to undergo rigid body motion. System motions are described within a coordinate system established by a designated reference origin. The traditional degrees of freedom (DOFs) disclose these motions. Specifically, the translational motions are represented by surge, sway, and heave (3), while the rotational motions are represented by roll, pitch, and yaw.

3 METHODOLOGY

The adopted methodology for this research encompasses both numerical analysis and model testing. However, the focus of this paper is specifically directed towards the numerical study.

Table 5. Metocean data (100-year sea return).

Environment.	Hs (m)	Tp (s)	Wind (1Hr Aver.)	Surface Current Vc (m/s)	Seabed Current (5m above)	Soil Stiffness (Kn/m/m)
1	2.15	8.4	10	2.00	0 3	1067
2	2.55	12.1	12	2.00	0 3	1067
3	4.5	15	16	2.00	0 3	1067

3.1 *Numerical study*

The numerical analysis tool SESAM/Genie generates a hydrodynamic model. SESAM/HydroD is then utilised for uncoupled frequency domain hydrodynamic analysis to assess hydrodynamic coefficients within a first-order solution. The occurrence of resonant motion in large floating moored structures, particularly in rigid body modes, is typically not induced by second-order sum-frequency excitation loads. As a result, the RAO and QTFs' different data, extracted from HydroD, were seamlessly integrated into Orcaflex. A fully coupled time domain dynamic analysis is conducted in Orcaflex under the specified environmental conditions. This analysis delves into the dynamic interactions of the coupled wind turbine and vessel while examining the mooring line coupling or damping behaviour.

The assessment of the time histories of the vessel's motion responses in six degrees of freedom is significant as this will help better understand the vessel's hydrodynamics with or without the installation of wind turbines. The NREL 5MW turbine has dedicated models for the generator, gearbox, hub, and blades incorporated in Orcaflex. External Python functions and controllers facilitate blade pitch and control generator torque and motion. The aerodynamic loading is computed in OrcaFlex using a blade element momentum (BEM) method adapted from AeroDyn.

3.2 *Background of theory*

An FPSO's hydrodynamic properties are the complex interaction between fluid and structure. The diffraction problem and the radiation problem in the context of potential flow theory mainly characterise this interaction. While the latter concentrates on how the floating structure behaves on the surrounding fluid when movement happens, the former considers fluid dynamics and its possible consequences on the floating system. FPSO system can be assessed using the Morison and potential theory to estimate the hydrodynamic loads.

Aerodynamic, hydrodynamic, and mooring loads are experienced simultaneously by the integrated wind turbine and FPSO system. Both the mooring loads and the hydrodynamic loads will be present. In the temporal domain, the equation of motion (Faltinsen 1993):

$$(\mathrm{p},\mathrm{a})+\mathrm{C}(\mathrm{p},\mathrm{v})+\mathrm{K}(\mathrm{p})=\mathrm{F}(\mathrm{p},\mathrm{v},\mathrm{t}) \quad (1)$$

In the provided equation, the symbols are defined as follows: The system's inertia load is shown by M (p, a), the damping load is indicated by C (p, v), the stiffness load is indicated by K(p), and the external load is represented by F (p, v, t). Furthermore, the velocity, position, acceleration, and simulation time correlate with the vectors (v, p, a, and t). The equation of motion for every free body and every line node is as follows:

$$\mathrm{M}(\mathrm{p})\mathrm{a}=\mathrm{F}(\mathrm{p},\mathrm{v},\mathrm{t})-\mathrm{C}(\mathrm{p},\mathrm{v})-\mathrm{K}(\mathrm{p}) \quad (2)$$

The acceleration vector at the initial time step is acquired to calculate values for each time step. This process entails taking into account acceleration values at both the conclusion of the current time step and the commencement of the new time step, denoted as (t + dt) of v, p (Orcina 2023). Applying the generalized-α integration approach is suitable for formulating the system's equation of motion after the time step, incorporating parameters such as mass, forces, damping, moments, etc. (Chung and Hulbert 1993).

3.2.1 *Aerodynamic loads*

Aerodynamic loads during the wind turbine's operational phases can be calculated using the Blade Element Momentum theory (BEM). The inflow (ω) at the aerodynamic centre, occurring at an angle α, induces the aerodynamic loads. The wake effects resulting from the angle of attack and direction of inflow must, therefore, be considered by the blade frame. The aerodynamic forces can be characterised using the following set of coefficients for the lift force in the ω and z directions, the drag force incident in the w direction, and the pitching moment of the z-axis:

Lift Force (direction ω and z);

$$\mathrm{f_L}=1/2\rho \mathrm{AC_l}(\alpha)|\omega|^2 \quad (3)$$

Drag Force (direction ω);

$$\mathrm{f_D}={}^{1/2}\rho \mathrm{AC_d}(\alpha)|\omega|^2 \quad (4)$$

Pitching moment (z);

$$\mathrm{mz}={}^{1/2}\rho \mathrm{AcCm}(\alpha)|\mathrm{w}|^2 \quad (5)$$

Within the specified context, ρ represents the air density, A signifies the element area, and $C_l(\alpha)$, C_d (α), and $C_m(\alpha)$ stand for the lift, drag, and moment coefficients, respectively. These coefficients are obtained for the angle of attack α by applying linear interpolation. Applying the Blade Element Momentum (BEM) theory enables the incorporation of momentum conservation principles and the blade sectional theory. This facilitates the determination of segment aerodynamic centre v; induction factors a and a′, and the rotor's quasi-steady velocity vq. Subsequently, these parameters are instrumental in calculating the dynamic induced velocity and, consequently, the dynamic inflow.

3.2.2 *Hydrodynamic loads and mooring loads*

The calculation of hydrodynamic loads on lines can be facilitated by applying the Morrison equation, detailed in reference (Morison et al. 1950). This equation is expressed as:

$$f = Cm\Delta a_f + 1/2\rho CdA|v_f|v_f \qquad (6)$$

Where f is the body's fluid force per unit length, the parameters consist of the following: ρ is the water density; Cd is the drag coefficient; A is the drag area; vf is the fluid velocity relative to Earth; Δ is the mass of fluid displaced by the body; and Cm is the inertia coefficient.

The equation may be modified through the deduction of $ca\Delta a_b$. From the inertia force reflecting the impact of the body's motion. This results in:

$$f = (Cm\Delta a_f - Ca\Delta a_b) + 1/2\rho CdA|v_f|v_f \qquad (7)$$

Considering the fluid acceleration concerning Earth, which includes the Froude-Krylov component and the variations in fluid flow surrounding the body brought about by adding mass, a force is required to maintain this acceleration. This force, expressed by the equation, is proportionate to the body's acceleration.

$$fb = (m + Ca\Delta)\, ab \qquad (8)$$

Where fb denotes the total force acting on the body, m signifies the body's mass, and CaΔ is a constant linked to the body's shape and displacement.

4 RESULTS AND DISCUSSION

4.1 *Time domain analysis*

A numerical simulation was performed on the FPSO model, exposed to a 100-year return storm lasting 3 hours in offshore West Africa. In examining the impacts of wind turbine positioning on the FPSO, a comparative assessment was conducted, analysing the time histories of the FPSO vessel's motion response as well as the statistical data comprising the mean and standard deviation of the 6DOFs and effective tensions at the fairleads. Discrepancies identified in the numerical results led to conclusions drawn regarding the impact of wind turbine positioning.

4.1.1 *Analysis of wave, current and wind combinations*

In the preliminary phase, three cases were investigated, comprising the FPSO model devoid of wind turbines, exposed to different environmental conditions—namely, wave only (C1), wave and current (C2) and wave, current, and wind (C3). The specific environmental setup for this section involved an oblique sea state with angles set at 135 degrees for wave, 180 degrees for current, and 180 degrees for wind, employing wave conditions of 4.5m significant wave height (Hs) and 15s wave peak period (Tp).

The discrepancies in the six degrees of freedom statistical data, added mass and damping forces arising from various combinations of environmental conditions are illustrated in Table 6 and Figure 5 below. The added mass and damping force noted the relatively substantial shift of the time histories in C2 and C3. Comparing C2 and C3 with C1, notable percentage differences (Figure 4) are observed in the mean response values of surge, sway, roll, pitch, and yaw, while there is a slight difference in the heave motion. The percentage differences of the 6DOFs between C1 vs C2 and C1 vs C3 were relatively the same. This underscores the pivotal role of the interaction of environmental factors on the behaviour of a vessel.

Table 6. Statistical mean values for the six degrees of freedom (6DOFs) in the Oblique Sea States.

Mean Response Data						
Cases	Surge	Sway	Heave	Roll	Pitch	Yaw
C1	-3.506	3.745	-0.485	0.012	0.024	0.181
C2	-50.426	3.008	-0.492	0.009	0.053	0.057
C3	-50.482	3.008	-0.492	0.009	0.053	0.057

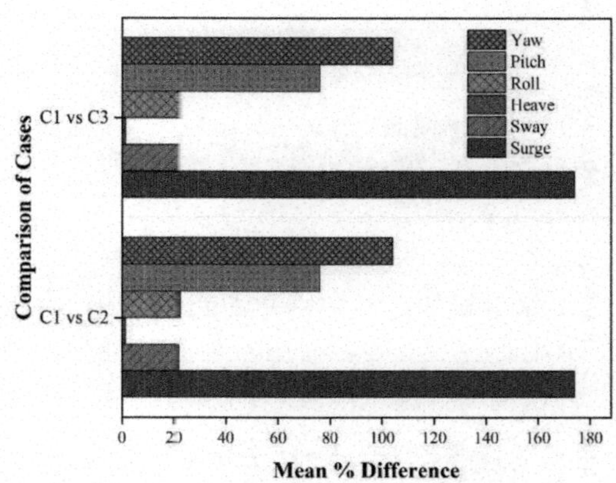

Figure 4. Comparison of mean percentage differences for the six Degrees of Freedom (6DOFs) in the Oblique Sea States considering the three cases.

4.1.2 *Analysis of 6DOFs of FPSO model in the three environments*

The time history graphs (see Figures 6 and 7) and the statistical data (Tables 7 and 8) of the vessel's motions across six degrees of freedom are assessed to obtain

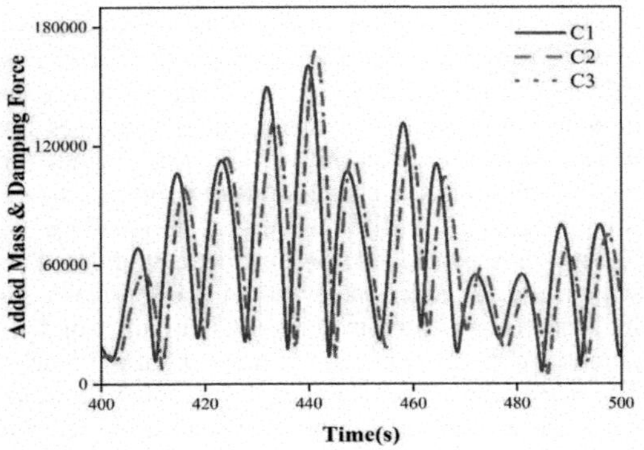

Figure 5. Comparison of added mass and damping force with three different environmental conditions: wave only,' 'wave and current,' and 'wave, current, and wind.

deeper insights into the hydrodynamic performance of the FPSO model. The mean and standard deviation are apt for analysing motion response values' motion dynamics and variance. The obtained data underscores the significant variability of the 6DOFs in the three environments. From Table 7, Environment 3 (E3) consistently exhibits the highest mean motion responses and standard deviation, followed by Environment 2 and 1. This phenomenon may be attributed to the prevalence of swells resulting from the high energy interaction between wind loads and oblique wave loads, which persist over an extended duration. Therefore, this may have caused the vessel's pronounced lateral and rotational motions in Environment . The surge, sway, and roll motion responses exhibit relatively high fluctuations around their mean values compared to heave, pitch, and yaw in all three environments. The negative values observed in the mean and maximum surge values signify the vessel's aftward movement, indicating motion towards the stern instead of the conventional forward surge motion. This phenomenon is likely attributed to the non-collinear alignment of environmental factors. This non-collinearity induces significant energy excitation, particularly in irregular wave patterns (Chakrabarti 2005).

Roll motion must be less than 3 degrees because it is essential to the onboard equipment's safety (Rezende et al. 2008; Xie et al. 2015). In Environment 3, the maximum roll motion observed was 5.97, compared to

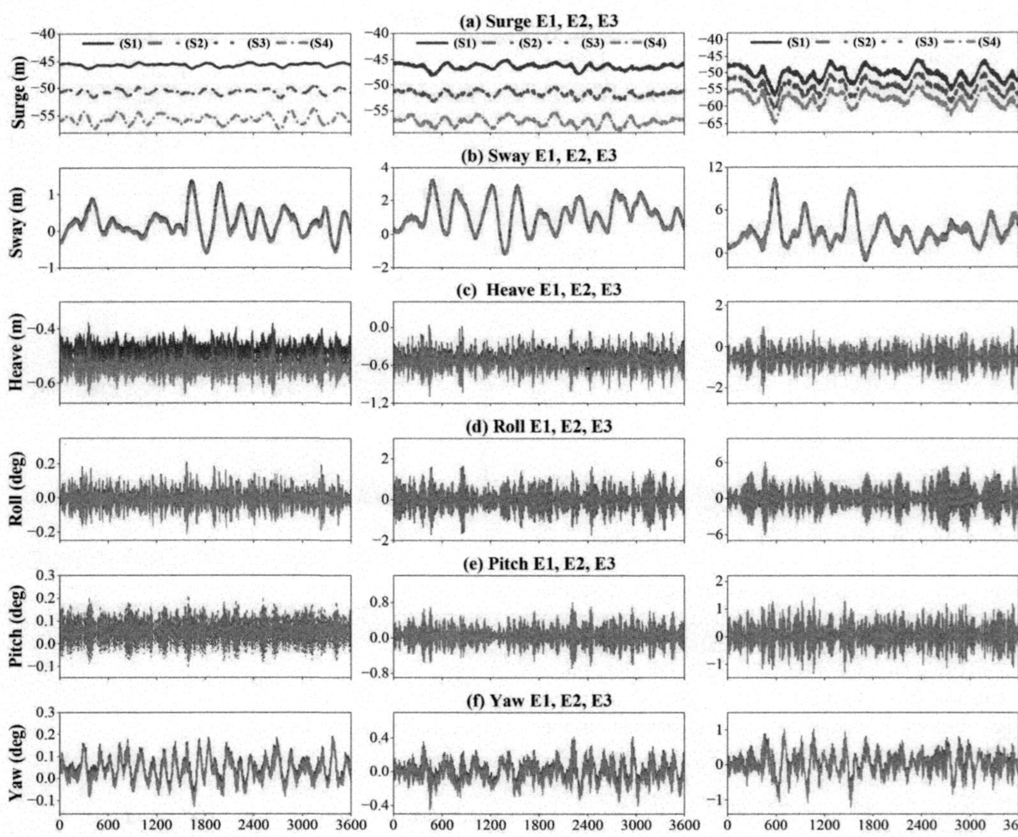

Figure 6. Time histories of motion response in 6DOFs for scenarios: (S1) – FPSO only, (S2) -FPSO + Wind turbine at Bow, (S3)- FPSO + Wind turbine at Stern. (S4)- FPSO + Wind turbines at Bow and Stern in Environments 1, 2 and 3.

1.65 in Environment 2 and 0.21 in Environment 1. This may not warrant the safety of equipment and onboard personnel. Accordingly, it is imperative to implement appropriate dampening mechanisms to mitigate the ensuing response effectively.

4.1.3 *Analysis of FPSO model + Wind Turbine(s) in 6DOFs in the three Positioning scenarios in all environments.*

The scenarios under consideration involve distinct wind turbine placements: one at the Bow (S2), one at the Stern (S3), and two turbines positioned at both the Bow and Stern (S4). The initial scenario, denoted as 'S1,' represents the baseline condition ('FPSO only'), where no wind turbine is installed on board—an aspect that has been previously discussed. The analysis focuses on assessing the variations in the time histories of motion responses in 6DOFs associated with the designated wind turbine positioning scenarios.

Upon closer examination of the distinct scenarios involving the wind turbine placements, variations in the percentage differences across the six degrees of freedom(6DOFs) are discerned. As seen in Figure 7, a consistent pattern emerges, revealing relatively lower mean percentage differences between scenarios S1 vs S2 and S1 vs S3 and higher differences between S1 vs S4. S2 and S3 observe mean surge response increases, exhibiting similar percentage differences against the base scenario (S1) across all three environments. In contrast, scenario S4 demonstrates the highest response and percentage difference (Table 7). Furthermore, surge motion displays the most significant variability in response to the mean values, with percentages of 53.07% for S1, 53.04% for S2, and 96.09% for S4.

Similar trends were noted in Environments 2 and 3, albeit with less fluctuation in values compared to Environment 1 (Tables 7 and 8).

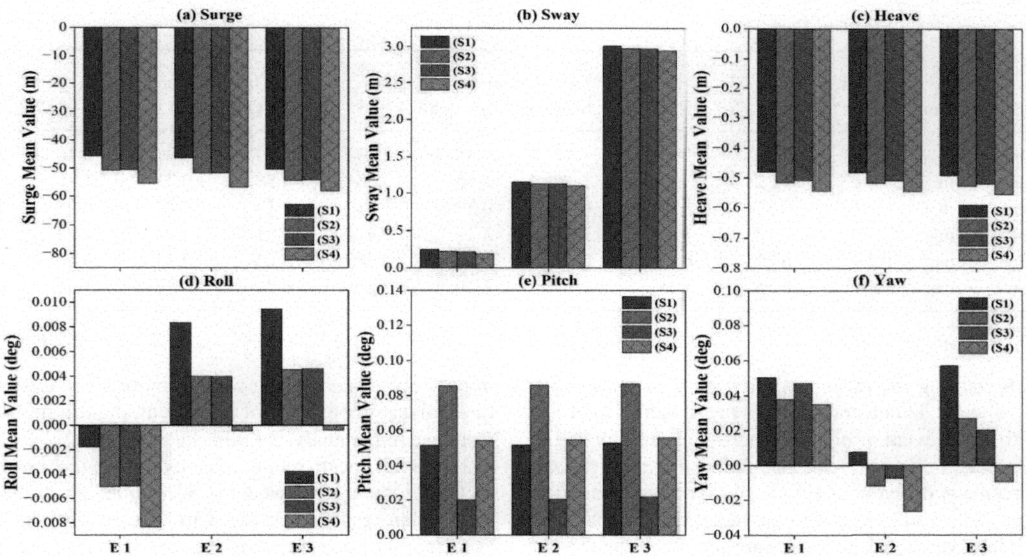

Figure 7. Mean Statistics of the six degrees of freedom of motions of the FPSO considering scenarios (S1) – FPSO only, (S2)- FPSO +Wind turbine at Bow, (S3)- FPSO +Wind turbine at Stern. (S4)- FPSO +Wind turbines at Bow and Stern in Environment 1,2,3.

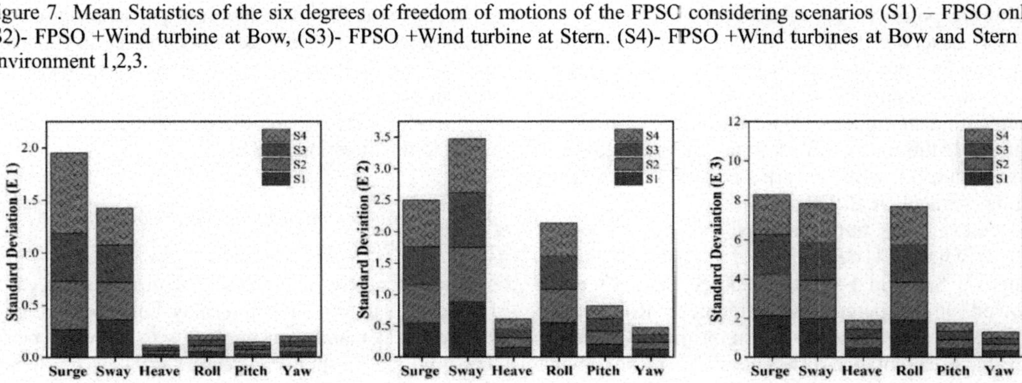

Figure 8. Standard deviation of the motion response in six degrees of freedom of motions of the FPSO considering scenarios (S1) – FPSO only, (S2)- FPSO + Wind turbine at Bow, (S3)- FPSO + Wind turbine at Stern. (S4)- FPSO + Wind turbines at Bow and Stern in Environment 1,2,3.

Table 7. Percentage Difference of Mean Motion Response in the 6DOFs considering the scenarios (S1) – FPSO only, (S2)- FPSO +Wind turbine at Bow, (S3)- FPSO +Wind turbine at Stern. (S4)- FPSO +Wind turbines at Bow and Stern S1, S2, S3, S4 in Environment 1,2,3.

		Mean Response								
		% Difference in E1			% Difference in E2			% Difference in E3		
Motion	Unit	S1 vs S2	S1 vs S3	S1 vs S4	S1 vs S2	S1 vs S3	S1 vs S4	S1 vs S2	S1 vs S3	S1 vs S4
Surge	M	10.55	10.33	19.68	10.94	10.71	20.36	7.6	7.31	14.27
Sway	M	10.48	12.80	23.88	1.93	2.42	4.19	1.05	1.22	2.1
Heave	M	7.26	5.70	12.57	7.25	5.69	12.54	7.06	5.54	12.22
Roll	Deg	94.72	94.34	128.97	70.7	70.82	224.34	70.01	68.63	218.48
Pitch	Deg	49.79	87.25	5.40	49.63	86.65	5.42	48.58	82.41	5.66
Yaw	Deg	28.18	6.74	36.32	917.19	12868.65	361.52	72.27	95.21	279.63

Table 8. Percentage Difference of Standard Deviation of Motion Response in the 6DOFs considering the scenarios (S1) – FPSO only, (S2)- FPSO +Wind turbine at Bow, (S3)- FPSO +Wind turbine at Stern. (S4)- FPSO +Wind turbines at Bow and Stern S1, S2, S3, S4 in Environment 1,2,3.

		Standard Deviation								
		% Difference in E1			% Difference in E2			% Difference in E3		
Motion	Unit	S1 vs S2	S1 vs S3	S1 vs S4	S1 vs S2	S1 vs S3	S1 vs S4	S1 vs S2	S1 vs S3	S1 vs S4
Surge	m	53.07	53.04	96.09	8.37	7.38	28.74	1.69	2.97	5.16
Sway	m	1.17	1.22	2.33	1.16	1.35	2.58	0.58	0.73	1.38
Heave	m	0.37	0.33	0.04	0.28	0.39	0.11	0.19	0.24	0.05
Roll	deg	2.28	1.58	3.77	3.02	3.1	6.4	0.81	0.64	1.23
Pitch	deg	0.24	0.21	0.42	0.14	0.14	0.28	0.24	0.32	0.56
Yaw	deg	0.64	0.86	1.69	1.5	1.39	2.86	0.24	1.24	1.56

Regarding the mean sway response, scenarios S2, S3, and S4 demonstrated lower mean values than S1, with less deviation observed across all environments than S1 (Table 7). Heave exhibited the lowest mean percentage differences among the six degrees of freedom, with values closer to the mean values of S1.

However, heave was significantly affected in scenario S4. The consistent increase in heave mean motion response values across scenarios S2, S3, and S4 was observed in all three environments, displaying similar percentage differences. Interestingly, heave was minimally influenced even with swells, maintaining its trend across the scenarios. In Environment 1, the mean roll motion values increased in S2, S3, and S4, with scenarios S2 and S3 exhibiting similar percentage differences.

However, this trend changed in Environments 2 and 3, where S1 demonstrated higher roll motion than S2, S3, and S4. Scenarios S2 and S3 again showed similar percentage differences. Roll motion emerges as a significant concern, primarily due to its elevated percentage differences in mean values observed across scenarios S2, S3, and S4 (Table 7).

Additionally, there is a notable increase in the percentage difference in standard deviation for scenarios S2, S3, and S4, further underscoring the importance of addressing the roll motion response in the analysis importance of addressing the roll motion response in the analysis (Tables 7 and 8).

For pitch mean motion, S2 exhibited the lowest mean response, followed by S4, while the highest pitch mean response occurred in S3 across all environments. The variance around the mean values in each scenario was low in all environments.

In analysing the yaw motion response, a significant mean percentage difference was observed in Environment 2, particularly with S1 vs S3, displaying the highest percentage. This discrepancy arises from the yaw motion occurring in the opposite direction compared to the direction of S1.

4.2 *Motion response spectra and coupling analysis*

Motion spectra were analysed across all degrees of freedom to explore the interplay between motions, spectral peaks, and their proximity to the wave peak. The motion response spectra in the frequency domain of the six degrees of freedom (DOF) motions were calculated using Fast Fourier Transformation (FFT) from the time series data. Figure 9 and Tables 9, 10, and 11 (Appendix) illustrate the spectra motions and highest peak percentage difference of all motions in 6DOFs within the three environments.

The observations reveal that the peak energy of the motion spectra for all six degrees of freedom (DOFs) consistently occurs at the same frequency, irrespective of the installation of wind turbines, but changes in varying environments. The following is a summary of the findings:

In environments 1 and 2, the surge motion response shows a peak energy correlated with a low frequency of 0.018 rad/s in all four scenarios. However, in Environment 3, there is a notable shift to a lower frequency of 0.013 rad/s for the peak energy across all scenarios, indicating a significant influence of swell. In Environment 1, scenario S1 displays an amplitude of 67.71 m^2s/rad. The peak response level at the resonance frequency increases in scenarios S2, S3, and S4, with percentage differences of 145.23%, 145.21%, and 181.76%, respectively, compared to scenario S1 (Tables 1, 2, and 3 in the Appendix). This trend is similarly observed in the percentage differences for the peak response level at the resonance frequency between scenarios S1 vs S2, S1 vs S3, and S1 vs S4. However, in Environment 3, there is a decrease in scenarios S2, S3, and S4 regarding the peak response level at the resonance frequency. The motion spectrum peak energy percentage differences are 7.33% for S1 vs S2, 8.97% for S1 vs S3, and 17.01% for S1 vs S4. The motion response spectra indicate that the global response is predominantly influenced by energy within the low-frequency range. In contrast, the contribution from wave frequency response is negligible in the surge direction.

In the sway motion response spectra, a peak energy level reduction was observed when wind turbines were integrated into the vessel across all three environmental conditions. Scenario S4 exhibited minimal sway motion response peak energy with higher percentage differences, followed by S3 and S2, as shown in Figure 9 and Tables 1, 2, and 3 (Appendix).

The yaw motion response was influenced by both low frequency and wave frequency, with the low frequency exerting the most significant impact. Environment 1's highest peak magnitude was observed at 2.69 deg^2s/rad with a frequency of 0.038 rad/s; in Environment 2, it was 6.5 deg^2s/rad at a frequency of 0.0502 rad/s. Environment 3's highest peak magnitude reached 81.64 deg^2s/rad at the same frequency as in Environment 2. A consistent trend was noted in Environments 1 and 3, where the peak yaw motion response spectra amplitudes decreased when wind turbines were installed. Specifically, Scenario S4 exhibited the lowest amplitude of the yaw motion spectra peak energy, followed by S2 and S3 in both Environments 1 and 3. However, in Environment 2, there was an increase in the highest peak energy of the yaw motion spectra with wind turbine installation. Specifically, S4 displayed the highest peak energy at 6.88 deg^2s/rad with a percentage difference of 5.57%, followed by S3 at 6.80 deg^2s/rad with a percentage difference of 4.35%, and finally, S2 at 6.62 deg^2s/rad with a percentage difference of 1.71%.

The roll, heave, and pitch wave frequency motions were investigated in the three scenarios. The peak energy with the highest magnitude is P1, while the second most significant peak is P2 (Figure 9). A coupled effect between heave and pitch motion responses was observed in Environment 1 at a resonance frequency of 0.502 rad/s for both pitch and heave motion responses, which affected the roll at the same frequency (Figure 9E1).

The heave motion spectra in Environment 1 reveal the presence of two significant peaks (Figure 9c). The primary peak, denoted as P1 and occurring at a frequency of 0.591 rad/s, shows that Scenario S2 exhibits the minor peak energy magnitude, followed by S4 and S3. This observation is attributed to the peak energy value of S2 closely resembling that of the reference scenario, S1. Analysing the second significant peak, labelled P2 and resulting from the coupling of heave and pitch, S2 shows the highest peak energy in the heave spectra in Environment 1, with a percentage difference of 1.05%, closely resembling the base scenario S1. S4 follows with the second-highest peak energy magnitude, displaying a percentage difference of 2.75% against scenario S1, while S3 exhibits the lowest peak energy for peak P2, with a percentage difference of 3.72%.

In Environment 2, the highest peak energy of 8.89 m^2s/rad is observed in Scenario S2, representing a percentage difference of 0.9% compared to S1 (8.81 m^2s/rad). S4 is the second highest, with a peak energy of 8.79 m^2s/rad, showing a percentage difference of 0.20% Scenario S3 exhibits the lowest peak energy at 8.71 m^2s/rad, with a percentage difference of 1.08%. A similar trend was observed with the second significant peak energy of heave motion spectra in Environment 1, which is found in Environment 3. Here, Scenario S2 surpasses the highest peak energy of the reference scenario S1 with a percentage difference of 0.2%. At the same time, Scenarios S3 and S4 demonstrate lower peak energy values compared to Scenario S1, with a higher percentage difference of 0.55% each.

For the roll motion response spectra, without wind turbine integration (S1), the highest peak energy is 0.89 deg^2s/rad (P1) at 0.65 rad/s, 0.78 deg^2s/rad (P2) at 0.502 rad/s in Environment 1, 181.28 deg^2s/rad (P1) at 0.389 rad/s in Environment 2, and 2218.23 deg^2s/rad at 0.377 rad/s in Environment 3. The analysis of the motion spectra revealed disparities in the highest peak energy (P1) when wind turbines were introduced, involving both decreases and increases compared to the base scenario S1. In Environment 1, when wind turbine(s) are installed, S4 shows the lowest peak energy for Peaks (P1) and (P2), displaying a percentage difference of 7.09% and 12.90%, respectively, followed by S2 3.80% (P1) and S3 7.38% p2). A similar trend occurred in Environments 2 and 3, where S4 displayed the lowest magnitude of peak energy in the roll motion response spectra (Figure 9d), followed by S3 and S2. But, in Environment 3, there were increases in the peak energies as S2 (2243.84 deg^{2}s/rad) showed a percentage difference of 1.15% compared to S1 and S3 (2233.24 deg^{2}s/rad) also showing a percentage difference of 0.67%.

The pitch motion spectra responses for the vessel without wind turbines (S1) exhibited magnitudes of peak energy at 1.15 deg2s/rad at a frequency of 0.502 rad/s in Environment 1, 12.7 deg2s/rad at a frequency of 0.439 rad/s in Environment 2, and 49.63 deg2s/rad at a frequency of 0.49 rad/s in Environment 3. An evident trend of increasing energy in the highest peak was consistently observed with the installation of wind turbines across all three environments. Specifically, Scenario S4 displayed the highest energy, followed by S3 and S2, compared to the base scenario S1. The percentages were, however, low across all scenarios.

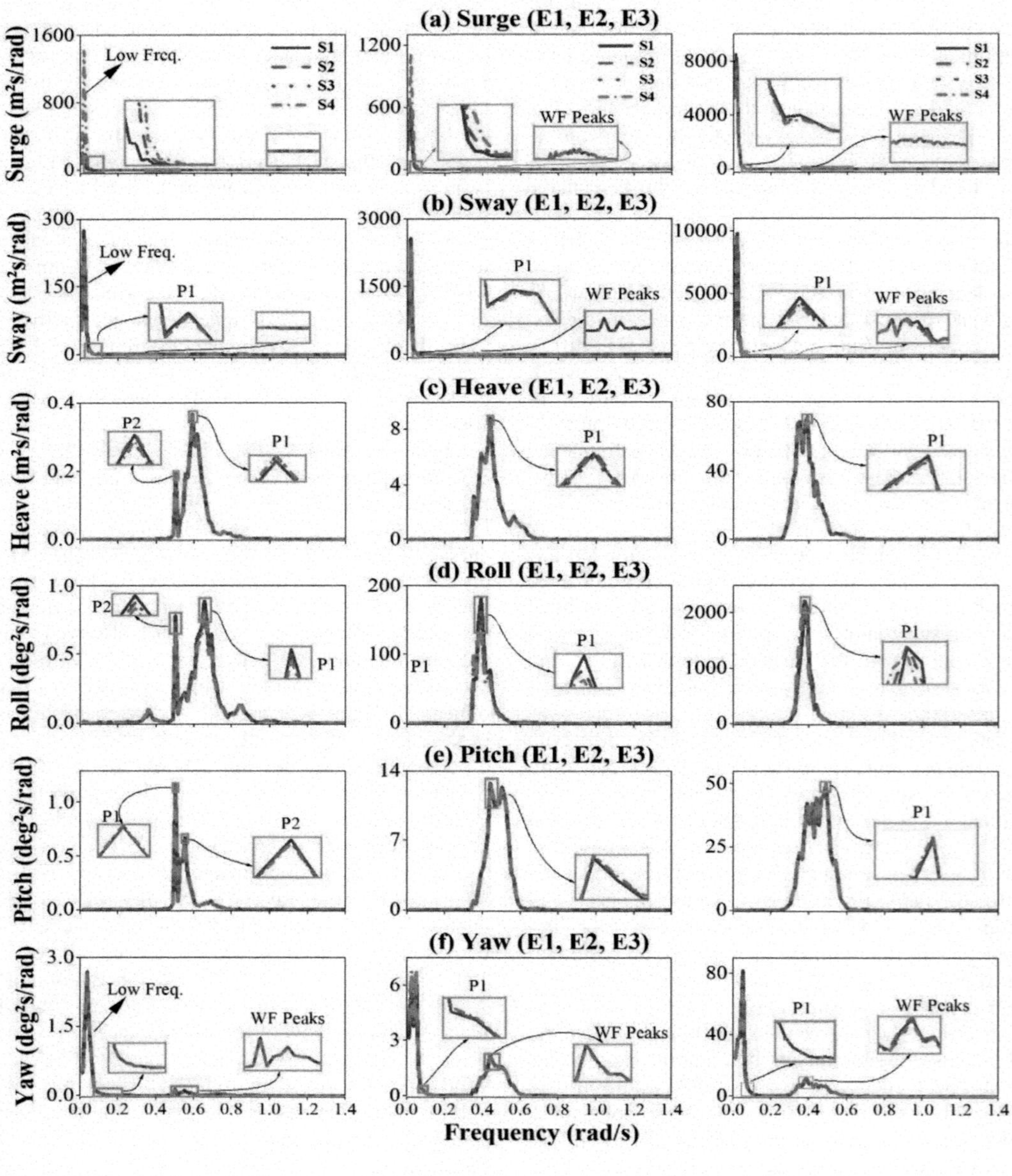

Figure 9. Motion spectral behaviour of the six degrees of freedom of motions for (S1) – FPSO only, (S2)- FPSO +Wind turbine at Bow, (S3)- FPSO + Wind turbine at Stern. (S4)- FPSO + Wind turbines at Bow and Stern (S4) in Environment 1,2,3.

The coupling effects between surge, heave, roll, and pitch are evident from their interconnected responses in the spectral motion response for line tension (see Figure 10g). This reveals the influence of both low-frequency and wave-frequency ranges on the restorative performance of moorings. Nonetheless, the energy magnitude within the wave frequency range predominantly shapes the global response of the most tensioned line 8. Conversely, the contribution of the low-frequency response to the tension experienced by the most stressed mooring line is considered negligible.

Several observations emerged after comparing the frequencies of the highest peak energy in the Jonswap wave spectrum with those of the motion response spectra for the six degrees of freedom. In Environment 3, the peak frequencies of the surge, heave, and roll motion spectra are closely aligned with the peak frequencies of the wave spectrum, suggesting that wind load may suppress the low-frequency motion range while leading to a more significant mean drift. This underscores the suppressing effect of wind loads on motions, indicating that wave load in this environment primarily influences heave motion and line tension force.

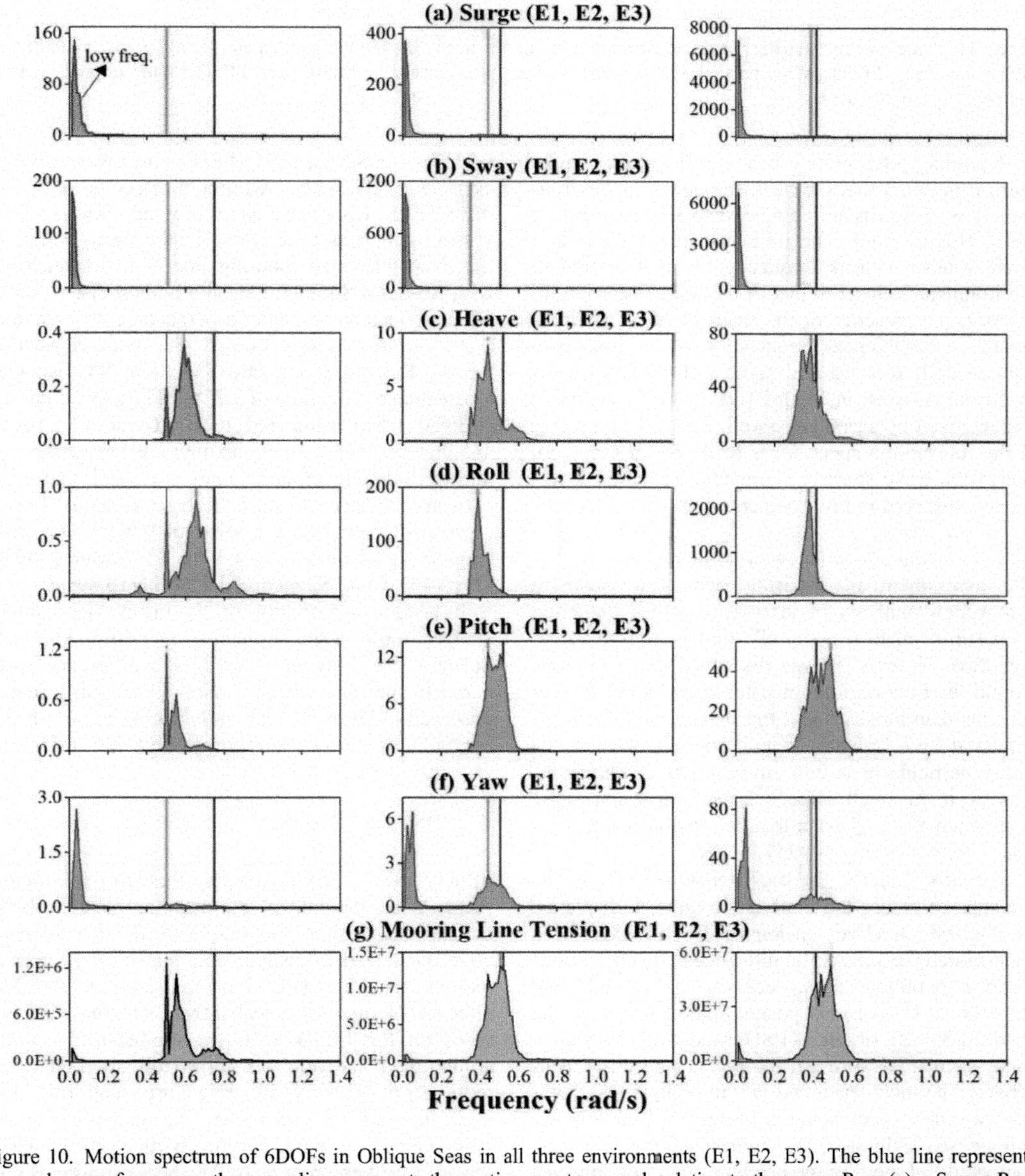

Figure 10. Motion spectrum of 6DOFs in Oblique Seas in all three environments (E1, E2, E3). The blue line represents the peak wave frequency; the green line represents the motion spectrum peak relative to the wave. Row (a) – Surge, Row (b) – Sway, Row (c) – Heave, Row (d) – Roll, Row (e) – Pitch, Row (f) – Yaw, Row (g) – Line Effective Tension.

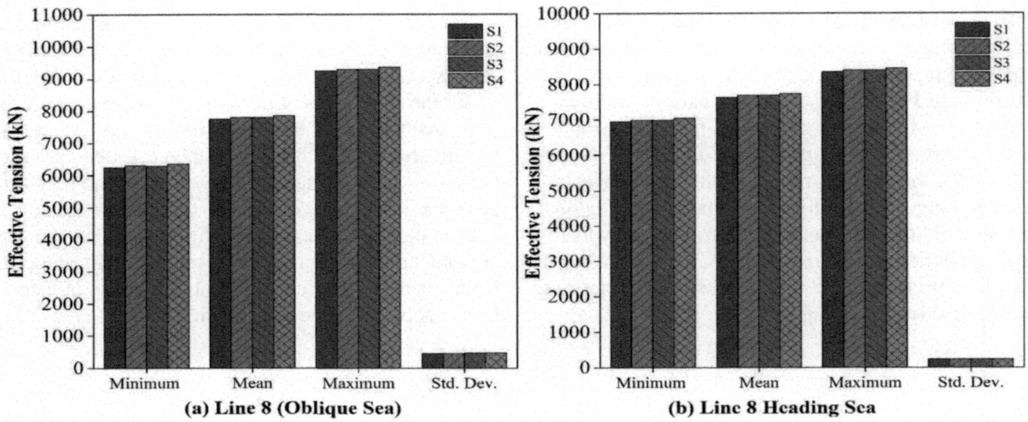

Figure 11. Time domain Statistics of the most tensioned line for Scenarios in Oblique Sea and Heading Sea states: (S1) – FPSO only, (S2) -FPSO +Wind turbine at Bow, (S3)- FPSO +Wind turbine at Stern. (S4)- FPSO +Wind turbines at Bow and Stern in Environments 3.

Regarding pitch motion peaks in the motion spectrum, it is noted that the peak frequency in Environment 1 is more distant compared to Environments 2 and 3 (Figure 10e). The influence of the secondary swell brings the peak frequency of pitch motion in Environment 2 closer to that of the wave spectrum. In contrast, the presence of the main swell in Environment 3 causes the peak frequency of the motion spectrum to shift forward and surpass that of the wave spectrum. As seen in Figure 10d, in Environment 2, characterised by secondary swell, the peak frequency of the roll motion spectrum is relatively distant from that of the wave spectrum compared to the peak frequency observed in Environments 1 and 3.

4.2.1 *Mooring dynamics and analysis*

The assessment of mooring behaviour involved a thorough analysis of statistical values extracted from time histories in all three environmental scenarios. Notably, it was discerned that moorings around the bow region consistently exhibited higher tensions than those around the stern across all environmental and sea state conditions. In oblique sea states, particularly within environment 3, characterised by main swell effects, Line 8 in scenario S1 manifested the highest effective tension at the fairlead area, registering at 9257.26 kN.

Similarly, Line 8, the most tensioned line in the heading sea across the three environments, displayed the highest effective tension of 8341.40 kN. This underscored the substantial influence of swell-induced motions in oblique seas, necessitating a heightened drag force. The various positioning scenarios of the wind turbine(s) on the FPSO model were meticulously compared to reveal the percentage differences between the most tensioned line in oblique and heading sea states. As depicted in Figure 11, Line 8 in the oblique sea exhibited the highest statistical values compared to Line 8 in the heading sea. In oblique sea conditions, the mean effective tension of line 8 at the fairleads for Scenario S1 (FPSO only) was recorded as 7752.46 kN, with a maximum effective tension of 9257.93 kN. Upon installation of wind turbines on the vessel, there was an increase in the mean values of the most tensioned mooring line 8 at the fairleads, with Scenario S4 exhibiting the highest rise in mean effective tension, showing a percentage difference of 1.29%, followed by S3 and S2, both registering 0.64%. Regarding standard deviation, S4 displayed a percentage difference of 3.37%, S3 (2.15%), and S2 (1.38%) when compared to S1 (with a value of 445.06). Assessing the maximum effective tension during designing mooring lines is critical, and comparisons are made against the base scenario. In this regard, S4 exhibited a tension of 9363.38 kN with a percentage difference of 1.19%, S3 showed 9300.84 kN (0.46%), and S2 recorded 9293 kN (0.39%).

In heading sea conditions, the mean effective tension of line 8 at the fairlead was 7636.07 kN, with Scenario S4 showing a percentage difference compared to the base scenario. Notably, the differences observed in S4 are significantly higher in both heading and oblique sea states compared to S2 and S3.

5 CONCLUSION

FPSO model in six degrees of freedom, specifically focusing on the impact of installing wind turbines onboard. Both time and frequency domain analyses were conducted to analyse data statistically, observe motion spectrum behaviour, and assess coupling effects. The findings revealed that installing wind turbines on the FPSO leads to significant deviations around the mean values of surge, roll, sway, and yaw motions, particularly affecting surge and roll. The wind turbines onboard reduce the magnitude of the sway motion response spectra, particularly in wind-sea and swell environments. The heave motion is notably influenced when the wind turbine is positioned at the bow of the FPSO. The peak magnitude of the roll motion spectrum decreases with wind turbine installation, irrespective of their placement, except in scenarios with the main swell environment, where placing turbines at the bow or stern exceeded the base scenario.

Furthermore, the peak energy of the pitch motion spectrum increases when wind turbines are installed on the FPSO, with the scenario involving wind turbines at both the bow and stern exhibiting the highest increase, followed by scenarios involving turbines at only the bow or stern. The low-frequency components associated with the surge, sway, and yaw motion spectra were found to be affected by swell conditions. Additionally, the most tensioned mooring line was identified to be located at the bow area, particularly in scenarios where wind turbines were installed at both the bow and stern. Positioning a wind turbine at the bow and stern results in relatively minor percentage differences compared to simultaneously installing two wind turbines at the bow and stern. This disparity could be attributed to wind load variations in the numerical simulations. Although all positioning scenarios of the wind turbine appeared promising, further investigations must be undertaken.

ACKNOWLEDGEMENTS

The authors gratefully acknowledge the Ghana Scholarship Secretariat's financial support for this research project.

REFERENCES

Bailey J, Clare C, David R. 2021. FPSOs: Overcoming challenges to unlock potential. White Case [Internet]. [Accessed 2022 Jun 13]. https://www.whitecase.com/publications/insight/fpsos-overcoming-challenges-unlock-potential

Chakrabarti S. 2005. Handbook of Offshore Engineering (2-volume set). Elsevier.

Chung J, Hulbert GM. 1993. A time integration algorithm for structural dynamics with improved numerical dissipation: The generalized-α method. J Appl Mech Trans ASME. 60(2).

DNV. 2005. RP-C203- Fatigue design of offshore steel structures. Recommended Practices DNVRPC203, 126.

Faltinsen O. 1993. Sea Loads on Ships and Offshore Structures (Vol. 1). Cambridge University Press.

Garrett DL, Chappell JF, Gordon RB. 2002. Global Performance of Floating Production Systems. In: Offshore Technology Conference (pp. OTC–14230).

Jonkman J, Butterfield S, Musial W, Scott G. 2009. Definition of a 5-MW Reference Wind Turbine for Offshore System Development, NREL Technical Report No. TP-500-38060. Nrel.(February).

Morison JR, Johnson JW, Schaaf SA. 1950. The Force Exerted by Surface Waves on Piles. Journal of Petroleum Technology, 2(05), pp. 149–154.

Newman JN. 1977. Marine Hydrodynamics. In: MIT Press.

Orcina. 2023. Dynamic Analysis using Orcaflex [Internet]. [accessed 2023 Jun 20]. https://www.orcina.com/

Patel MH. 2013. Dynamics of Offshore Structures. [place unknown]: Butterworth-Heinemann.

Rezende F, Chen XB, Ferreira MD. 2008. Simulation of second-order roll motions of a FPSO. In: Proc Int Conf Offshore Mech Arct Eng - OMAE. Vol. 1.

Van Wijngaarden AM, Daniels N. 2019. Upgrading and conversion opportunities for floating offshore units. In: Proc Annu Offshore Technol Conf. Vol. 2019-May.

Xie ZT, Yang JM, Hu ZQ, Zhao WH, Zhao JR. 2015. The horizontal stability of an FLNG with different turret locations. Int J Nav Archit Ocean Eng. 7(2).

APPENDIX

Table 1. Percentage differences of the motion response spectra in the six degrees of freedom in Environment 1 (E1), Significant Motion Spectrum Peak Response 1 (P1).

Scenarios	Surge P1	Sway P1	Heave P1	Roll P1	Pitch P1	Yaw P1
S1 vs S2	-145.23	-4.06	-0.14	3.80	0.29	-1.14
S1 vs S3	-145.21	-4.28	-1.96	3.52	0.39	-1.00
S1 vs S4	-181.76	-8.15	-2.05	7.09	0.72	-1.82

Table 2. Percentage differences of the motion response spectra in the six degrees of freedom in Environment 2 (E2), Significant Motion Spectrum Peak Response 1 (P1).

Scenarios	Surge P1	Sway P1	Heave P1	Roll P1	Pitch P1	Yaw P1
S1 vs S2	-12.56	2.95	-0.90	11.61	-0.55	-1.71
S1 vs S3	-10.37	3.23	1.08	12.04	-0.88	-4.35
S1 vs S4	-10.37	6.26	0.20	25.03	-0.88	-5.57

Table 3. Percentage differences of the motion response spectra in the six degrees of freedom in Environment 3 (E3), Significant Motion Spectrum Peak Response 1 (P1).

Scenarios	Surge P1	Sway P1	Heave P1	Roll P1	Pitch P1	Yaw P1
S1 vs S2	7.33	1.78	-0.19	-1.15	-0.41	4.12
S1 vs S3	8.97	2.13	0.55	-0.67	-0.52	3.73
S1 vs S4	17.01	3.84	0.55	2.98	-0.90	9.28

Advances in Maritime Technology and Engineering – Guedes Soares & Santos (Eds)

Overview of the floating offshore photovoltaic energy potential

G. Clemente, S. Ramos-Marin & C. Guedes Soares
Centre for Marine Technology and Ocean Engineering (CENTEC), Instituto Superior Técnico, Universidade de Lisboa, Lisbon, Portugal

ABSTRACT: An overview of the state of the art of the floating photovoltaic renewable energy technology is given. The main characteristics of floating photovoltaic systems are outlined, as well as the features influencing on their preformance. Moreover, existing concepts and ongoing projects around the world are compiled and briefly described. The advantages, drawbacks and potential hazards for this technology are highlighted. Offshore solar photovoltaic systems allow upscaling the installed capacity and increment the power generation, significantly reducing greenhouse gas emissions. These systems specially contribute to boost renewable energy generation in islands with minimal land availability. The environmental risks and potential hazards for this technology are also discussed, such as water level fluctuations, severe storms, earthquakes, and tsunamis. Despite some concerning hazards, the successful performance and reliability of some existing floating solar farms has raised trustability and interest for this technology among several countries committed to a prompt energy transition.

1 INTRODUCTION

The current framework of energy transition increases the interest in investing in renewable energy technologies. They represent powerful tools to move away from the use of fossil sources (coal, oil, natural gas, etc.) and reduce environmental pollution. Despite some of the advantages of fossil fuels (generally cheaper than renewable energy sources, high energy efficiency and convenient transport), several governments around the world are engaged with the implementation of new measures that stand up for the reduction of greenhouse gas emissions given off by the excessive use of these fossil fuels (Wim *et al.* 2022). Nowadays renewable energy sources constitute about 13.5% of production energy worldwide, and this percentage raised up by about one point between 2019 and 2020 (Bellavia 2022). The offshore component of reneable energy is dominated by wind energy, which has seen a large increase in recent years (Diaz & Guedes Soares 2020).

Solar, wind, tidal, geothermal, hydroelectric and biomass energy sources have the potential to mitigate the effects of ever-rising temperatures, as they allow for the generation of energy without releasing carbon dioxide and other greenhouse gases into the atmosphere. Among them, photovoltaic (PV) modules, used all over the world for the production of solar energy, absorb sunlight by converting the energy into a usable form of electric current. The solar modules have no moving parts, which significantly reduces maintenance costs ensuring highly reliable systems (Huang *et al.*, 2023). In addition, photovoltaic systems are especially cost-effective compared to other energy sources in remote places where it is difficult and economically unfeasible to reach with traditional electric lines (Remote Energy 2021).

However, an important drawback of onshore photovoltaic systems is the scarce land availability, as they compete for space with other agricultural and industrial purposes, especially in densely populated regions. A further disadvantage is the degradation of the PV array performance caused by high operating temperatures of the PV cells within onshore systems. Some of the downsides associated with onshore PV systems can be efficiently overcome with the use of the emerging floating PV technologies (FPV). Photovoltaic panels are increasingly being installed in various places, including seas, lakes, rivers, dams, and water treatment plants.

The performance of a photovoltaics system on land and at sea has been simulated and compared by Golroodbari & van Sark (2020). The results show that the relative annual average output energy is between 13% to 18% higher at sea compared with land. However, special concerns arise regarding to the mooring lines, which must be perfectly designed to survive in extreme weather conditions, avoiding damage and displacement of the system (Cazzaniga *et al.* 2018).

This work overviews the state of the art of floating photovoltaic renewable energy technology. The main characteristics of FPV systems are described in Section 2, both offshore and inland water bodies. Section 3 outlines the features influencing the performance of FPV energy systems. Section 4 reviews some of the existing FPV concepts, while Sections 5

DOI: 10.1201/9781003508779-45

and 6 put forward some ongoing projects around the world and Europe. In Section 7, the advantages, drawbacks and potential hazards of this technology have been highlighted, and Section 8 concludes with the main highlights of the overview.

2 FLOATING PV CHARACTERISTICS

The structural characteristics of floating PV panels are very similar to those of land-mounted systems, except that PV modules are mounted on floating structures. The floating photovoltaic system typically consists of floating high-density polyethylene pontoons, on which the various parts of the photovoltaic fields are positioned, including inverters, and metal frames (Figure 1).

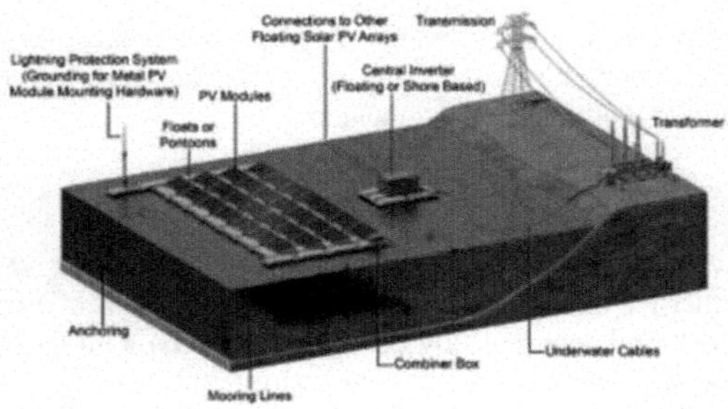

Figure 1. Typical large-scale floating PV system with its key components (Lee et al., 2020).

It is essential to correctly size a photovoltaic field to obtain high energy conversion efficiencies, also allowing to reduce the possibility of damage to the system itself (Kumar *et al.* 2021).

The inverter is a fundamental element of the photovoltaic system, as it converts the direct current derived from solar energy into alternating current destined for the public grid. The inverter can be placed both on the floating platform and on the ground and there are mainly two types: a central inverter or a multi-string inverter (Maraj *et al.* 2022).

Once produced, solar energy is taken from the panel and transferred to the ground (Figure 1). The energy produced can then be fed directly into the grid or stored in batteries. In most of the projects carried out to date, the cables are kept above the water level, guaranteeing perfectly waterproof conditions, also resisting high temperatures to avoid current leakage. High temperature resistance, water proof and robust cables are to be used to provide a long service. Other electrical components such as inverters and batteries remain on land (Sahu *et al.* 2016).

The pontoons and floats (Figure 2) represent one of the most important parts of the whole structure. They stay on the surface, floating, maintaining a constant ratio between buoyancy and own weight. They represent a fundamental part of the system as they constitute a foundation platform for the various elements. They are mainly made of high-density polyethylene (HDPE), as it has excellent characteristics, including high tensile strength, and resistance to UV rays and corrosion. Medium-density polyethene (MDPE) is also used in some applications (Sahu *et al.* 2016).

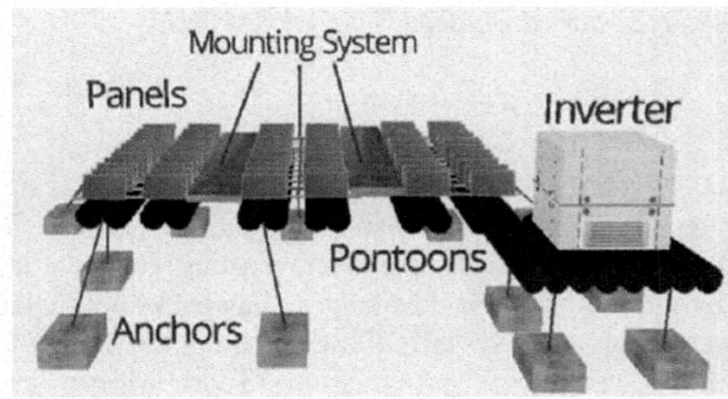

Figure 2. Pontoons and floats (Wang & Lund 2022).

The mooring and anchoring system represents a further fundamental part of the installation, as it must avoid excessive movement of the system caused by wind, changes in water level, big waves, etc. Therefore, when sizing a mooring system, it is necessary to consider the possible environmental situations that may occur during the lifecycle of the system. The mooring system is usually made with polyester nylon ropes fixed on the shore and anchored at each corner (Kumar *et al.* 2021). The approaches adopted to design mooring systems are similar to the ones used for other renewable energy devices (Xu et al. 2019).

3 PV OFFSHORE SYSTEM PERFORMANCE

The electricity production of the floating PV systems is determined by various parameters. The most important are:

- Orientation of the photovoltaic modules:

To obtain a greater performance, photovoltaic panels are generally inclined with an optimal angle of inclination relative to the horizontal surface to chase the maximum incidence of sun's rays (Figure 3). This is especially relevant for regions with a high ratio of direct to diffuse solar radiation throughout the year. In offshore parks, however, the tracking systems used in onshore conditions are not suitable, since steep angles of inclination at panels located at sea make them prone to be hit by very strong winds. Additionally, the movement of the offshore floating structure in response to incoming waves can influence the insolation on PV modules, although limited research exists quantifying this effect (Trapani *et al.* 2013, Magkouris *et al.*, 21). The effects of waves on the floating PV performance was presented in (Magkouris *et al.*, 21), indicating significant variations of the performance index ranging from 0 to 15% depending on the sea state.

Figure 4 shows the effect that the response of a floating structure to incoming waves can have on the effective slope of a PV module. The possible degrees of freedom of a moving offshore PV module motion are three, namely pitch, yaw and roll movements, as shown in Figure 5.

Figure 3. Orientation of the photovoltaic modules (Institute of Electrical and Electronics Engineers. 2012).

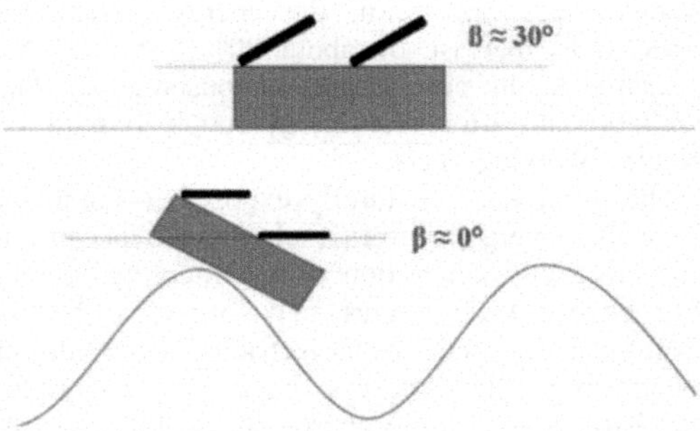

Figure 4. Floating platform at steady state (TOP) and Floating platform responding to incoming wave (BOTTOM) (Bugeja et al. 2021).

Consequently, often many offshore photovoltaic concepts have an angle of inclination equal to zero. This setup leads to a lower power production peak at noon, although a higher production in the early morning and late afternoon partially compensates for this phenomenon (Wim *et al.* 2022). Moreover, the dirt losses increase as the angle of inclination decreases, which lowers the total energy produced by the system. The dirt loss can reduce the energy produced by the PV system up to 2 Wh/Wp (El Baqqal *et al.* 2020).

- Temperature:

The output voltage, and thus the output power of PV modules decreases with increasing temperatures. However, offshore photovoltaic systems entail some advantages regarding the efficiency of the solar park, as the water acts as a natural refrigeration system, decreasing the temperature of the panels and reaching higher efficiency values (5-15% higher than

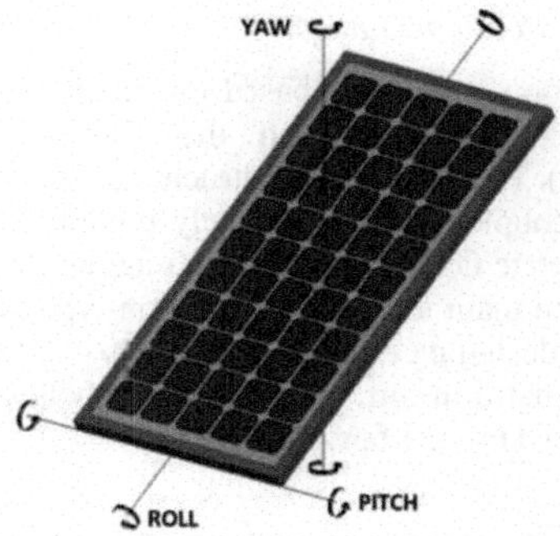

Figure 5. Solar panels (Bugeja et al. 2021).

conventional rooftop panels) (Lin 2021, Favetti 2022). The real cooling effect though is very much dependent on the construction and the materials of the floating structures on which the PV modules are mounted. Advantages also arise from an environmental point of view, if panels are installed in the right position, as it reduces the potential growth of toxic algae, as well as the evaporation effect of the water (Wim *et al.* 2022).

- Soiling and fouling:

It is essential to keep the PV panels clean as it allows the efficiency of the entire system to be kept high. Dirt accumulated on solar panels can cause a significant reduction in solar energy production. Some studies have recently confirmed that low system maintenance levels can lead to a loss of efficiency of 15-30%, thus shortening the duration of the panels' life cycle, which is usually 30 years (Wim *et al.* 2022).

4 FLOATING PV TECHNOLOGY CONCEPTS

In recent years, many increasingly sophisticated designs have been analyzed to withstand the harsh conditions present in the seas. Initially, these designs were only tested in lakes or rivers. A classification is made regarding the position of the photovoltaic modules concerning the waterline: superficial or pontoon systems. In the first case, the photovoltaic modules are installed on the water surface, in the second case, an intermediate floating platform is arranged (Claus & López 2022).

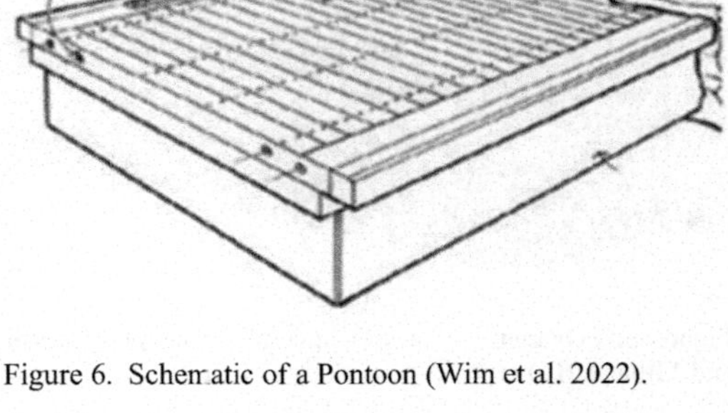

Figure 6. Schematic of a Pontoon (Wim et al. 2022).

4.1 *The Pontoon concepts*

The pontoon concept is based on floating pontoons, where the wind load on the structure is low (Figure 6). Pontoons are substructures that are mechanically coupled into a massively modular structure. The complete floating structure is moored (Wim *et al.* 2022). The main idea behind pontoon-type systems is based on designing a raft or pontoon to create a stable floating platform on which solar panels are later installed (Claus & López 2022).

4.2 *The truss concepts*

A truss structure is equipped with floats and mooring. Above the truss structure, a platform containing the solar panels is placed (Figure 7). The platform is located at a certain height above the waterline, thus avoiding direct contact with the waves and mainly allowing for the reduction of the encrustation on the solar panels due to seawater residues on the surfaces themselves. It also allows for to reduction of the mechanical load on the solar panels which is lower due to the absence of contact with the waves (Wim *et al.* 2022).

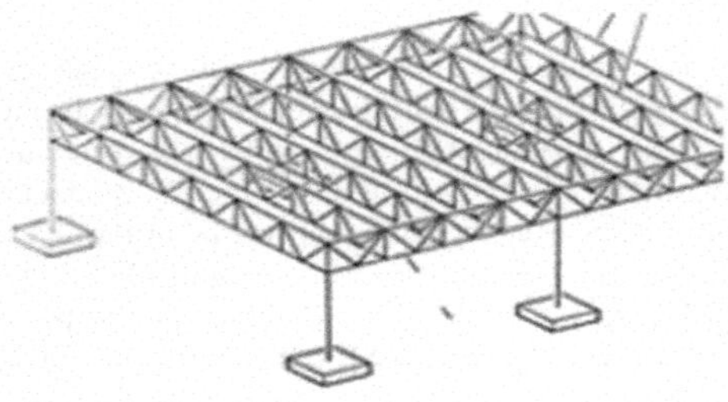

Figure 7. Schematic of a truss concept (Wim et al. 2022).

4.3 *The soft and flex concept*

The basic idea of the soft & flex concept (Figure 8) is to let the floating structure move as much as possible with the waves, to obtain a lower amount of mechanical energy on the structure, therefore reducing the mooring forces. This implies a lower cost of the structure, compared to the plant analyzed previously. Solar panels must be flexible enough as they must be able to follow the deformation of the float (Trapani *et al.* 2013).

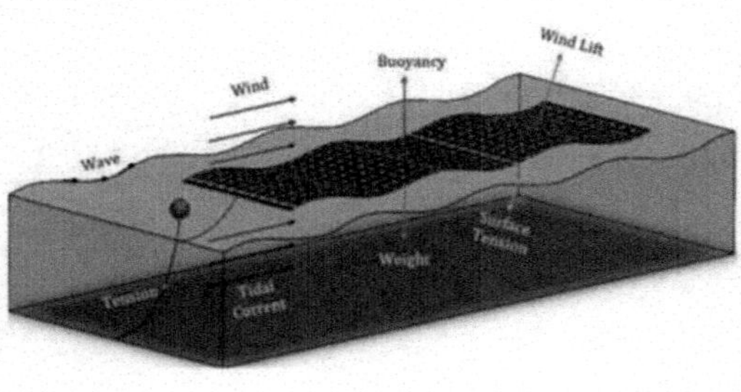

Figure 8. Schematic of a soft and flex concept (Trapani and Millar 2016).

5 OFFSHORE PV ENERGY IN THE WORLD

Although the FPV technology is increasingly getting new attention from stakeholders worldwide, not many offshore FPV systems have been installed in open water so far, while few projects are planned to be set up in the upcoming years.

5.1 *Photovoltaic systems in Asia*

- Maldives islands

The energy system in the Republic of the Maldives Islands, located southwest of Sri Lanka and India in the central Indian Ocean, relies mostly on imported fossil fuels as the primary source of energy to supply the rapidly growing economic activity (mostly boosted by the tourism industry). Nearly 80% of the petroleum product imported into the Maldives is diesel, which is used for power generation via diesel generators on the main island. Fuel transport via navigation makes the Maldives one of the regions with the highest cost of energy production in South Asia. Moreover, in the short term, the energy demand is expected to increase by about 20% (ADB 2018), compared to the current rate of consumption, thus a reliable and affordable energy supply is required for the following years.

The Maldives is naturally exposed to abundant renewable energy sources (solar, wind, ocean...), thus electricity generation using emerging renewable energy technologies may prove to be an alternative to high-cost electricity from diesel generators in the Maldives. As of the year 2020, the share of renewable energy in the total energy mix of Maldives was about 5.6 %, out of that total renewable energy generated over the country, 95 % of it is made using solar energy, 1 % using bioenergy and 4 % using wind energy (IRENA 2022).

The intensive solar radiation exposure of the region has put the Maldives in the spotlight for offshore photovoltaic energy developers. Existing research analyzed the electrical performance of floating thin-film photovoltaic systems to be installed in the Maldives Islands. Small-scale 5 MW offshore systems were designed as a compromise between energy production and environmental impact. Two PV modules were placed on the main float of the floating structure made of HDPE and connected by connecting bolts and pins, installing 6930 pontoons to support the whole PV system (SWIMSOL 2023).

As offshore solar photovoltaic modules play a fundamental role by increasing the energy supply reliability and reducing greenhouse effects, further offshore PV projects are expected to increment the solar energy penetration into the Maldivian's electric mix.

- Indonesia:

The Indonesian state-owned utility PT PLN recently started running the Indonesia's largest floating solar array up to date, a 561-kW floating solar plant built on a pond measuring 1 hectare in the Tambak Lorok coastal area of Semarang (island of Java) (Figure 9). The floating plant is part of a larger 920 kW PV project under construction, and it is expected to produce 1.4 million kWh a year, helping to reduce CO2 gas emissions by up to 1,304 tons (Meza 2023).

Figure 9. Floating solar PV plant installed on a pond in the Tambak Lorok, coastal area of Semarang (island of Java) (Meza 2023).

- Singapore

One of the world's largest floating solar PV farms, with around 60 MWp (122,000 solar panels), was installed in 2021 in the Tengeh Reservoir (Singapore) (Figure 10). It produces enough electricity to power the island's water treatment plants. The electricity generated by this plant is expected to make Singapore one of the few countries in the world to have a water treatment system powered entirely by renewable energy. Environmental impact assessments stated a reduced impact on wildlife and water quality (Lin 2021).

Figure 10. General view of one of the world's largest floating solar panel farms in Singapore (Lin 2021).

5.2 *Photovoltaic systems in Africa*

- Ghana

Following Ghana's strategy to promote access to reliable, clean, and affordable electricity, the first Hydro-Solar Hybrid power generating system in the West African subregion has been recently commissioned, which also includes a 1 MW Floating Solar PV System (Figure 11). It is of the first phase of a 250MWp solar project, which is being implemented in phases of 50MWp. The project generates renewable energy from solar resources that can operate during the day to complement existing hydropower production (Power Africa 2022).

Figure 11. One MW floating solar at Bui, Bono East Region (Ghana) (Power Africa 2022).

- The Seychelles

In the African archipelago of Seychelles, a project for the construction of a 5.8 MW floating solar power plant in the Providence lagoon, near Mahé, has been approved and is scheduled to start in the final quarter of 2023. Under the agreement, the French renewable power producer Qair will develop, build, and operate the floating solar plant to supply renewable energy to the national grid (Shetty 2023).

- Zimbabwe

The low efficiency of coal-fired power plants in Zimbabwe, which usually leads to several electricity shortages, has encouraged the government to implement a 2 GW solar energy portfolio to address the problem. An offshore photovoltaic plant is already planned to be built in the next years by the China Energy Engineering Corporation, which has proposed a project of a 1 GW floating solar plant with about 1.8 million solar modules in Zimbabwe, at the Kariba Dam on the Zambezi River. This project will be developed on the reservoir on top of the Kariba dam, which represents the largest artificial lake in the world (Norman 2023).

6 OFFSHORE PHOTOVOLTAIC IN EUROPE

6.1 *Overview of offshore floating solar systems developers in Europe*

In Europe, different developers of offshore solar panels evolved as the technology grew and spread

(Table 1). Currently, several pilot plant projects exist, which are expected to evolve towards further stages of commercialization in the upcoming years.

6.1.1 *Bluewater*

Bluewater Energy Services, a Dutch engineering consultancy, envisions building an offshore floating solar project in the North Sea with flexible, floating thin-film photovoltaic modules (Figure 12) (Santos 2022). The system consists of multiple individual floats moored to a shared mooring grid, and the floats are made of flexible, air-filled double-wall fabric. In the event of overpressure, this material combines excellent stability of the floating room and such low global stiffness that the float can follow the wave profile perfectly. Some studies state that flexible floating solar technology offers less resistance to waves, therefore facilitating the use of lighter weight floats and anchors, which greatly reduce the total cost (Wim *et al.* 2022b). If the offshore pilot project is successful, the team plans to build a commercial system ranging in size from 1 MW to 5 MW at a wind farm in the North Sea by 2024 (Santos 2022).

Table 1. Solar offshore European technology developers (Wim et al. 2022).

System Developer	Country	Concept	TRL medium exposure	TRL high exposure
Oceans of energy	NL	Pontoon	5	4
Bluewater	NL	Soft&Flex	4	4
Solar duck	NL	Truss	5	4
Tractebel-Engie	Belgium	Truss	4	4
Swimsol	Austria	Truss	8	n.a.
Moss Maritime	Norway	Pontoon	4	4
Ocean Sun	Norway	Fishfarm	8	5

Figure 12. Bluewater technology (Wim et al. 2022).

6.1.2 *Moss Maritime*

Moss Maritime, a Norwegian engineering company active in the offshore oil and renewable energy sector, has developed a concept based on floating 10 x 10 m^2 photovoltaic platforms (Figure 13). It is a classic truss concept, with square bridges and cubic floating elements. The units are flexibly connected to larger systems, allowing the system to perfectly match the surface to the longer wavelength waves (Wim *et al.* 2022).

Figure 13. Moss Maritime concept (*Wim* et al. *2022*).

6.1.3 *Tractebel-Engie*

Tractebel-Engie is a well-known developer of floating photovoltaic systems on in-land water bodies. A 1 MWp plant was recently installed on a Terhills Resort reservoir near the Hoge Kempen National Park (Belgium) (TRACTEBEL- Engie 2021). In 2020, they announced the development of three floating PV panels with 30 MW capacity within the 52.2 MW Batalha Hydroelectric Project (Bellini 2020). The photovoltaic system consists of about 7,000 solar panels with a total capacity of 3 MWh. One third of the plant is to be installed on an artificial lake and the remaining part on the ground (Wim *et al.* 2022). A test model of this technology can be seen in Figure 14.

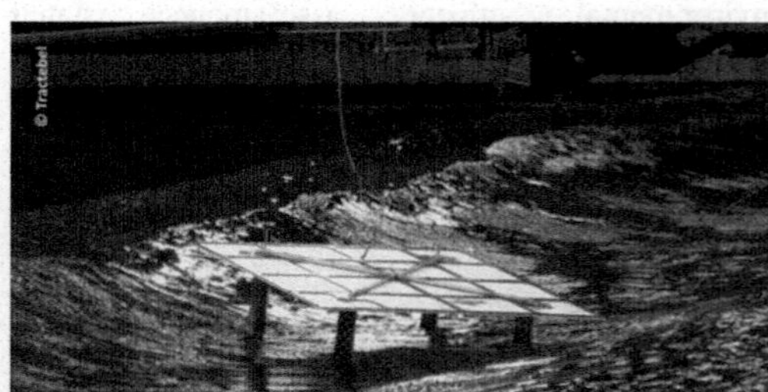

Figure 14. Model test of Tractebel system (*Wim* et al. *2022*).

6.1.4 *Solar Duck technology*

The concept of SolarDuck is based on the use of triangular platforms, mostly made of aluminum, which float a few meters above the water on cylinders (Wim *et al.* 2022). To capture a larger portion of the sun, the PV modules are mounted at an east-west angle direction. SolarDuck is developing a pilot 65 kW floating PV array to be connected to a 10-kW electrolyser to produce hydrogen. The pilot plant is being deployed on the Dutch River of Waal (Rhine), located near IJzendoorn, a village in the province of Gelderland (Solarduck 2021). The system relies on the company's own floating technology that resembles an offshore oil platform. The floating structures are triangular and measure 16 x 16 x 16 meters (Figure 15). They are connected in a flexible way to create large-scale installations. These platforms allow for the placement of photovoltaic modules more than three meters above the water surface. According to the company, this innovative configuration allows the use of conventional photovoltaic modules with high efficiency without having to obtain specially designed and excessively expensive products.

Figure 15. Solar duck system (Bellini 2023).

6.1.5 *Grafenworth technology*

The Grafenwörth floating solar power plant houses more than 45,000 PV panels and is built on the water surface of a former sand and gravel quarry. After testing the plant, the company subsequently installed 15 of them, for a total of approximately 230 MW worldwide (Figure 16) (Ajsa Habibic 2023).

Figure 16. Grafenworth System (Ajsa Habibic 2023).

6.1.6 *Protevs technology*

Protevs, developed by SolarisFloat using research and development from Portuguese universities, is a real innovation in floating solar PV (Deena 2022). Compared to fixed structures, Protevs allows dual-axis rotation of the bifacial solar panels and therefore an increase in energy production of 30 to 40%. The water also acts as a natural cooling system, lowering the temperature of the individual modules and greatly increasing efficiency.

The panel "islands" are made up of 180 panels; each platform can turn on itself and each panel individually performs a tilting movement (Figure 17). It is a very easy-to-install, modular, demountable, and reliable solution. Furthermore, Protevs states the system constitutes a reduced environmental impact.

Protevs have been successfully installed in Holland in the Oostvoornse Meer Lake. The implemented solution, with tracking on one axis, contains 130 modules of floating photovoltaic solar panels, with a capacity of 50.7 kW (Offshore Energy 2022).

Figure 17. Protevs (Favetti 2022).

7 ADVANTAGES AND DISADVANTAGES

Offshore installations entail several advantages but also some negative effects that should be assessed during any offshore PV plant design process.

- Advantages:

Space Availability: Offshore PV systems avoid the competition of land use with other conflicting users (agriculture, industry, architecture, etc. It also reduces the in-land visual impact.

Structure: The overall system features a compact structure, which is easy to move to better capture the solar energy in the different seasons of the year. Floating PV systems equipped with tracking features can increase energy by 15%-20%.

Less evaporation: FPV plants lower the percentage of evaporated water so that more water can be used for irrigation or other purposes.

Low shading: the likelihood of shading FPV is lower due to a more exposed installation environment. Furthermore, the presence of dust on the panel is significantly lower than in inland desert regions or areas with heavy traffic. This lowers the dirt losses and allows us to achieve significantly higher energy efficiencies.

Cost: Offshore photovoltaics can significantly reduce the costs of a photovoltaic plant, mainly in locations with high land management costs (Aritra Gosh 2023a).

- <u>Disadvantages:</u>

Durability: Traditional PV modules are made for terrestrial climates. FPV technologies are constantly exposed to the corrosive effects of salt, humidity, and intensive mechanical forces of wind, and waves, which can easily degrade the system, significantly reducing its performance.

Frozen ambient temperature: The fluctuations of ambient temperature can be problematic, especially if it varies between below zero and above zero degrees. The whole system may freeze when temperatures drop below zero. Anchoring systems can also be affected due to snow.

Marine and aquatic life: The presence of floating structures on the open sea often creates problems of light transmission through the water layers, which subsequently reduces the process of chlorophyll photosynthesis and therefore the growth of algae.

Conflicting users: Other activities performed in the same marine environment could be severely affected (such as fishing, navigation, recreation, etc) (Aritra Gosh 2023b).

8 CONCLUSIONS

An overview of the floating PV renewable energy has been presented in this paper, both offshore and on in-land water bodies. The main characteristics of FPV systems have been outlined, as well as the features influencing on their preformance. Moreover, existing concepts and ongoing FPV projects around the world, have been compiled and briefly described. Also, the advantages, drawbacks and potential hazards for this technology have been highlited.

The floating photovoltaic conversion technology is still very recent and innovative technology but is suffering a quick growth linked to the raising interest for an energy transition framework, as it allows high efficiency rates with low environmental impacts. Bringing these floating modules offshore at the sea is increasing in popularity as well, as it permits upscaling the installed capacity by avoiding some of the environmental impacts linked to the floating technology on in-land water bodies and making the most possibly efficient use of the marine space.

Offshore floating photovoltaic energy is expected to grow sifnificantly in the following decades as as land availability keeps and global energy targets become tighter. Consequently, it is important to highlight that during any design process of an offshore floating solar energy plant, there is a need to always evaluate not only the advantages, but also the potential hazards linked to each of type of novel concept at each potential location.

ACKNOWLEDGEMENTS

This work contributes to the Strategic Research Plan of the Centre for Marine Technology and Ocean Engineering (CENTEC), which is financed by the Portuguese Foundation for Science and Technology (Fundação para a Ciência e Tecnologia - FCT) under contract UIDB/UIDP/00134/2020. The second author has been financed by FCT under the grant 2020.06618.BD.

This work has been partially funded by the project "Novel eco-cementitious materials and components for durable, competitive, and bio-inspired offshore floating PV substructures (NaturSea-PV)", funded by the Horizon-Widera-2022-Access-07-01 Programme of the European Union under contract number GA nº 101084348. However, the views and opinions expressed are those of the author(s) only and do not necessarily reflect those of the European Union or the European Climate Infrastructure and environment Executive Agency (CINEA). Neither the European Union nor the granting authority can be held responsible for them

REFERENCES

ADB, 2018. Maldives: Economic Update 2019 [online]. Available from: https://www.adb.org/sites/default/files/institutional-document/544946/maldives-economic-update-2019.pdf [Accessed 11 Jan 2024].

Ajsa Habibic, 2023. BayWa r.e. inaugurates largest floating solar plant in Central Europe [online]. Available from: https://www.offshore-energy.biz/baywa-r-e-inaugurates-largest-floating-solar-plant-in-central-europe/ [Accessed 10 Jul 2023].

Gosh, A. 2023. A comprehensive review of water based PV: Flotavoltaics, under water, offshore & canal top. *Ocean Engineering*, 281, 115044. Doi: 10.1016/j.oceaneng.2023.115044

El Baqqal, Y., Laarabi, B., Dahrouch, A., & Barhdadi, A., 2020. Assessment of soiling effect on PV module glass transmittance in Moroccan capital region. *Environmental Science and Pollution Research*, 27 (44510–44518).

Bellavia, E., 2022. Quanto costano le fonti dei energia rinnovabile in Italia e nel mondo [How important are renewable energy sources in Italy and in the world] [online]. Available from: https://www.wecanjob.it/archivio21_quanto-contano-fonti-energia-rinnovabile-italia-mondo_0_811.html#:~:text=Quanto%20contano%20le%20energie%20rinnovabili,il%202019%20e%20il%202020. [Accessed 27 Jun 2023].

Bellini, E., 2020. Brazilian hydropower dam to host 30 MW of floating solar [online]. *PV Magazine*. Available from: https://www.pv-magazine.com/2020/02/14/brazilian-hydropower-dam-to-host-30-mw-of-floating-solar/ [Accessed 19 Jul 2023].

Bellini, E., 2023. Japan's first offshore floating solar demonstrator [online]. *2022*. Available from: https://www.pv-magazine.com/2022/12/12/japans-first-offshore-floating-solar-demonstrator/ [Accessed 14 Jul 2023].

Bugeja, R., Mule' Stagno, L., & Branche, N., 2021. The effect of wave response motion on the insolation on offshore photovoltaic installations. *Solar Energy Advances*, 1, 100008.

Cazzaniga, R., Cicu, M., Rosa-Clot, M., Rosa-Clot, P., Tina, G.M., & Ventura, C., 2018. Floating photovoltaic plants: Performance analysis and design solutions. *Renewable and Sustainable Energy Reviews*, 81, 1730–1741.

Claus, R. & López, M., 2022. Key issues in the design of floating photovoltaic structures for the marine environment. *Renewable and Sustainable Energy Reviews*, 164, 112502. Doi: 10.1016/j.rser.2022.112502

Deena, T., 2022. A Portuguese company's innovative floating solar panels stalk the Sun's movements [online]. *Interesting Engineering*. Available from: https://interestingengineering.com/innovation/floating-solar-follow-suns-movements [Accessed 19 Jul 2023].

Diaz, H.M. & Guedes Soares, C. 2020. Review of the current status, technology and future trends of offshore wind farms. *Ocean Engineering*. 209:107381.

Favetti, G., 2022. Un impianto solare offshore: è Protevs, un parco solare galleggiante in grado di seguire il Sole [An offshore solar plant: it is Protevs, a floating solar park able to follow the Sun] [online]. Available from: https://auto.hwupgrade.it/news/energie-rinnovabili/un-impianto-solare-offshore-e-protevs-un-parco-solare-galleggiante-in-grado-di-seguire-il-sole_112051.html [Accessed 5 Jul 2023].

Golroodbari, S.Z. & van Sark, W., 2020. Simulation of performance differences between offshore and land-based photovoltaic systems. *Prog. Photovolt*, 28, 873–886.

Hu, X.N., Fang, G.S., Ge, Y.J. & Guedes Soares, C. 2024. Uncertainty in the estimation of extreme ocean winds in Portugal. *Advances in Maritime Technology and Engineering*. Guedes Soares, C. & Santos T. A., (Eds.), Taylor and Francis Group, London, UK.

Huang, G., Tang, Y., Chen, X., Chen, M. & Jiang, Y.. 2023. A Comprehensive Review of Floating Solar Plants and Potentials for Offshore Applications. *Journal of Marine Science and Engineering*, 11, 2064. https://doi.org/10.3390/jmse11112064

Institute of Electrical and Electronics Engineers., 2012. *2012 Asia-Pacific Power and Energy Engineering Conference : proceedings : March 27-29, 2012, Shanghai, China*. IEEE.

IRENA, 2022. *Renewable energy statistics 2022 - International Renewable Agency.*

Kumar, M., Mohammed Niyaz, H., & Gupta, R., 2021. Challenges and opportunities towards the development of floating photovoltaic systems. *Solar Energy Materials and Solar Cells*, 233, 111408.

Lee, N., Grunwald, U., Rosenlieb, E., Mirletz, H., Aznar, A., Spencer, R., Cox, S., 2020. Hybrid floating solar photovoltaics-hydropower systems: Benefits and global assessment of technical potential. *Renew. Energy*, 162, 1415–1427.

Lin, C., 2021. Singapore unveils one of the world's biggest floating solar panel farms [online]. *Reuters*. Available from: https://www.reuters.com/business/energy/singapore-unveils-one-worlds-biggest-floating-solar-panel-farms-2021-07-14/ [Accessed 20 Jul 2023].

Magkouris, A., Belibassakis, K., Rusu, E., 2021. Hydrodynamic Analysis of Twin-Hull Structures Supporting Floating PV Systems in Offshore and Coastal Regions. *Energies*, 14, 5979. https://doi.org/10.3390/en14185979

Maraj, A., Kërtusha, X., & Lushnjari, A., 2022. Energy performance evaluation for a floating photovoltaic system located on the reservoir of a hydro power plant under the mediterranean climate conditions during a - sunny day and a cloudy-one. *Energy Conversion and Management: X*, 16.

Meza, E., 2023. Indonesia's largest floating solar plant goes online [online]. *2023*. Available from: https://www.pv-magazine.com/2023/04/13/indonesias-largest-floating-solar-plant-goes-online/#:~:text=Indonesia's%20PT%20PLN%20says%20it,1.4%20million%20kWh%20a%20year [Accessed 24 Jul 2023].

Norman, W., 2023. China state energy firm to build 1GW floating PV plant in Zimbabwe [online]. *PV Tech*. Available from: https://www.pv-tech.org/china-state-energy-firm-to-build-1gw-floating-pv-plant-in-zimbabwe/ [Accessed 20 Jul 2023].

Offshore Energy, 2022. EDP inaugurates 'pioneering' 5MW floating solar plant in Portugal [online]. Available from: https://www.offshore-energy.biz/edp-inaugurates-pioneering-5mw-floating-solar-plant-in-portugal/ [Accessed 26 Jun 2023].

Power Africa, 2022. West Africa's First Solar-Hydro Hybrid Plant Operational in Ghana [online]. Available from: https://powerafrica.medium.com/west-africas-first-solar-hydro-hybrid-plant-operational-in-ghana-62571b5cf54e [Accessed 24 Jul 2023].

Remote Energy, 2021. What is solar energy? [online]. Available from: https://www.remoteenergy.org/blog/what-is-solar-energy [Accessed 27 Jun 2023].

Sahu, A., Yadav, N., & Sudhakar, K., 2016. Floating photovoltaic power plant: A review. *Renewable and Sustainable Energy Reviews*, 66, 815–824.

Santos, B., 2022. Dutch developer secures funds for flexible floating solar pilot in North Sea [online]. Available from: https://www.pv-magazine.com/2022/12/02/dutch-developer-secures-funds-for-flexible-floating-solar-pilot-in-north-sea/ [Accessed 6 Jul 2023].

Shetty, S., 2023. Qair to Build Seychelles' First Floating Solar Power Plant [online]. *Solar Quarter*. Available from: https://solarquarter.com/2023/04/11/qair-to-build-seychelles-first-floating-solar-power-plant/#:~:text=The%205.8%20MWp%20future%20renewable,hydrogen%20powered%2Dvessel%20Energy%20Observer. [Accessed 24 Jul 2023].

Solarduck, 2021. Solarduck unveils design of its offshore floating solar technology [online]. Available from: https://solarduck.tech/solarduck-unveils-design-of-its-offshore-floating-solar-technology/ [Accessed 6 Jul 2023].

SWIMSOL, 2023. Recent swimsol solar energy projects [online]. Available from: https://swimsol.com/solar-projects-offshore-solarsea-and-rooftop/ [Accessed 19 Jul 2023].

TRACTEBEL- Engie, 2021. Floating solar panels at the Terhills Resort [online]. Available from: https://tractebel-engie.be/en/news/2021/floating-solar-panels-at-the-terhills-resort [Accessed 19 Jul 2023].

Trapani, K. & Millar, D.L., 2016. Hydrodynamic Overview Of Flexible Floating Thin Film PV Arrays. *In*: *Offshore Energy and Storage Symposium*, Malta, 2016.

Trapani, K., Millar, D.L., & Smith, H.C.M., 2013. Novel offshore application of photovoltaics in comparison to conventional marine renewable energy technologies. *Renewable Energy*, 50, 879–888.

Wang, J. & Lund, P.D., 2022. Review of Recent Offshore Photovoltaics Development. *Energies*, 5(20),7462.

Wim, S., Kingma, A., Hoogeland, M., vd Brink, R., Kroon, J., & Folkerts, W., 2022. *Challenges and potential for offshore solar* [online]. Available from: https://topsectorenergie.nl/sites/default/files/uploads/20220331_RAP_Challenges%20and%20potentialfor%20offshore%20solar_Final.pdf [Accessed 6 Jul 2023].

Xu, S., Wang, S. & Guedes Soares, C. 2019. Review of mooring design for floating wave energy converters. *Renewable and Sustainable Energy Reviews*. 111, pp. 595–621

Advances in Maritime Technology and Engineering – Guedes Soares & Santos (Eds)
© 2024 The Author(s), ISBN 978-1-032-83107-7

Wave energy converter power take-off with active oil-hydraulic accumulator

J.F. Gaspar, C. Zeng, H.T. Xu, T.S. Hallak & C. Guedes Soares
Centre for Marine Technology and Ocean Engineering (CENTEC), Instituto Superior Técnico, Universidade de Lisboa, Lisbon, Portugal

M.J.G.C. Mendes
Centre for Marine Technology and Ocean Engineering (CENTEC), Instituto Superior Técnico, Universidade de Lisboa, Lisboa, Portugal
CIMOSM, ISEL-Instituto Superior de Engenharia de Lisboa, Instituto Politécnico de Lisboa, Portugal

ABSTRACT: A new concept of an oil-hydraulic accumulator is proposed in this work to improve the harvesting performance of wave energy converters. The oil-hydraulic power take-off converts the reciprocating motion of the wave energy converter prime mover into the rotation of the generator shaft. Oil-hydraulic power take-offs integrated with accumulators are commonly classified as constant-pressure type. Conventional accumulators reduce the pressure fluctuation in the hydraulic systems but also impose limitations on the adaptability of the power take-offs to different sea states. This accumulator consists of two standard accumulators whose gas chambers are interconnected so that the charging characteristics of the accumulator can be adjusted by compressing or decompressing the gas in the shared chambers. Thus, compared to the conventional power take-offs, the power take-off integrated with the active accumulator system can be actively adapted to different sea states and ultimately improve the harvesting performance of the wave energy converter. The work focuses on the numerical implementation of the concept and the results indicate the accumulator concept brings significant advantages compared to the conventional in terms of better power output over a wide range of sea states.

1 INTRODUCTION

Power take-off (PTO) devices play a crucial role in converting the captured oscillating energy from wave energy converters (WECs) into usable electricity (Hansen et al., 2011; Hansen & Kramer, 2011; Jusoh et al., 2019). This energy is harnessed through the hydrodynamic interaction between waves and the WEC's prime mover (such as a rotating flap), and the resulting motion of the prime mover is then transferred to the PTO via an interface, such as an articulated arm. Various PTO technologies have been extensively studied to effectively convert the interface motions into electrical energy, including the implementation of oil-hydraulic technology (Guedes Soares et al., 2012; Li et al., 2021; Zhang et al., 2012). This technology is particularly well-suited for low-frequency, high-power density waves, as it is capable of generating controllable high forces even at low velocities. Additionally, it facilitates easy rectification of the bidirectional oil movement within compact hydraulic cylinders, further enhancing its applicability and performance.

Oil-hydraulic PTOs integrated with accumulators are commonly classified as constant-pressure type. The oil-hydraulic accumulators store energy during periods of high wave crests and release it when waves are smaller. Moreover, they reduce PTO pressure fluctuations. However, it constrains the adaptability of the PTOs to different sea states by maintaining a constant pressure (Gaspar et al., 2017; Gaspar et al., 2016a, 2016b; Hansen et al., 2013).

In this study, a novel accumulator concept named the active accumulator (AA) is introduced. It consists of two accumulators with interconnected gas chambers, and with adjustable parameters in function of different sea states.

An open-source solver, WEC-Sim (So et al., 2015), is used to carry out WEC simulations. To interface with the target PTO, an oscillating surge wave energy converter (OSWEC) is used.

This paper is organized into four sections. In Section 2, the WEC numerical models are presented, providing details on the used modelling approach. Section 3 discusses the simulation setups and presents the results. Finally, in Section 4, conclusions are drawn and future outlooks and potential areas for future research are provided.

DOI: 10.1201/9781003508779-46

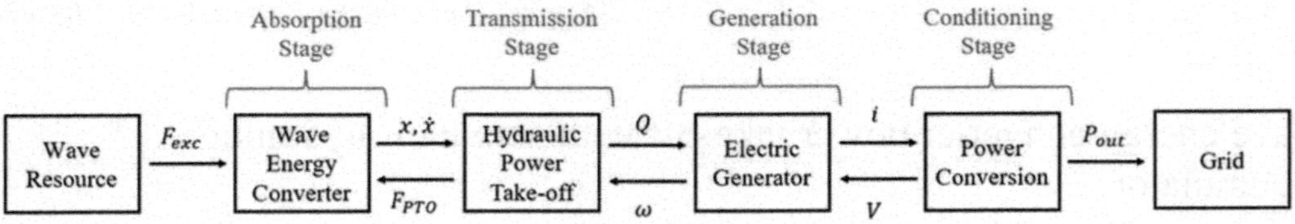

Figure 1. Wave-to-wire (W2W) generic model. The wave resource is harvested by the Prime Mover (PM) (absorption stage) and converted to electrical power by the PTO (transmission, generation, and conditioning stages). Adapted from (Coe et al., 2021).

2 METHODOLOGY

The wave-to-wire generic model, depicted in Figure 1, illustrates the overall system from wave energy capture to power generation. However, this study specifically focuses on the hydraulic PTO. Therefore, the transmission stage, which encompasses the hydraulic PTO, is the central focus of this study. To work with the transmission stage, the models of the wave resource and absorption stage are included. In this section, mathematic and numerical models of the selected parts of the WEC are presented.

2.1 *Wave resources*

The WEC is assumed to absorb energy from irregular waves in a fully developed sea, where the wind-induced waves are no longer being augmented (Coe et al., 2021). The Pierson-Moskowitz spectrum is adopted to describe the waves through parameters of significant wave height H_s and peak wave frequency ω_p (Tucker and Pitt, 2001):

$$S(f) = \frac{5}{16}\frac{\omega_p^4}{\omega^5} H_s^2 \exp\left(-\frac{5}{4}\left(\frac{\omega_p}{\omega}\right)^4\right) \quad (1)$$

where ω is the angular frequency of waves and the peak wave frequency is the reciprocal of the peak wave period ($\omega_p = \frac{2\pi}{T_p}$).

The time-average power transported by the irregular wave per unit width is represented by:

$$P_w = \rho g \int_0^\infty S(\omega) c_g(\omega) d\omega \quad (2)$$

where $c_g(f)$ is the group velocity of a planer wave of the waves.

2.2 *Absorption stage*

The absorption stage consists of a prime mover (PM) and an interface. The PM is directed and excited by forces exerted by the waves which is transmitted to the hydraulic PTO via an interface.

2.2.1 *Hydrodynamics of the prime mover*

WEC-Sim, a Matlab-based open-source program developed by Sandia National Laboratories (SNL) and the National Renewable Energy Laboratory (NREL) is used in this study to simulate the WEC dynamics (So et al., 2015). The code simulates the intricate interplay among incident waves, WEC motion and PTO mechanism in the time domain (Amini et al., 2022).

The OSWEC PM is flap-type and rotates around the hinges of its base. Thus, the PM equation of motion is expressed concerning the flap pitch angle:

$$J_{PM}\ddot{\theta} = M_{exc}(t) + M_{rad}(t) + M_{hs}(t) + M_{PTO}(t) \quad (3)$$

where, J_{PM} is the moment of inertia around hinge axis, $M_{exc}(t)$ the wave excitation moment, $M_{rad}(t)$ the PM radiation moment, $M_{hs}(t)$ the hydrostatic moment and M_{PTO} the PTO damping moment.

The excitation, radiation and hydrostatic moments are calculated according to the hydrodynamic coefficients given by the frequency-domain boundary element method solver (BEM). The linear BEM solver WAMIT is applied, which assumes the fluid is incompressible, inviscid and irrotational. In addition, the oscillating body and the waves have small amplitudes (Penalba et al., 2017). Then, the Cummins formulation is derived from the equations of motion:

$$\begin{aligned}(J_{PM} + J_{A,\infty})\ddot{\theta}(t) + \int_0^t k_r(t-\tau)\dot{\theta}(\tau)d\tau \\ + k_{hs}(t)\theta(t) = M_{exc}(t) + M_{PTO}(t)\end{aligned} \quad (4)$$

where, $J_{A,\infty}$ is the added moment of inertia at infinite frequency, k_r the radiation impulse response function (IRF) and k_{hs} the PM hydrostatic restoring coefficient.

The Froude-Krylov forces and moments act on the WEC device due to the incident wave pressure. The excitation moment (M_{exc}) is the summation of the physical part of the Froude-Krylov moments at each wave frequency (Calvário et al., 2020; Nolte and Ertekin, 2014):

$$M_{exc}(t) = \sum_i A_w(\omega_i)|\Gamma(\omega_i)|\cos(-\omega_i t + \phi_i + \angle\Gamma(\omega_i)) \quad (5)$$

where, $\Gamma(\omega_i)$, and $A_w(\omega)$ are the Froude-Krylov force transfer function and corresponding wave amplitude obtained from a specific frequency in an irregular sea state, respectively. $A_w(\omega_i)$ is then be obtained according to the sea spectrum:

$$A_w(\omega_i) = \sqrt{2S(\omega_i)\Delta\omega_i} \quad (6)$$

2.2.2 *Interface*

The wave's kinetic energy is converted into the PM rotational motion, which is subsequently converted into the linear motion of the hydraulic cylinder, before being absorbed by the hydraulic PTO.

The PM and interface with the hydraulic PTO are parameterized, as presented in Figure 2. Accordingly, the linear position $x(t)$ and speed $\dot{x}(t)$ of the cylinder joint (A) are determined with:

$$x(t) = \sqrt{(r\cos\theta - h_o)^2 + (d_o - r\sin\theta)^2} \quad (7)$$

$$\dot{x}(t) = \frac{r\cos\theta(d_o - r\sin\theta)\dot{\theta} - r\sin\theta(r\cos\theta - h_o)\dot{\theta}}{\sqrt{(r\cos\theta - h_o)^2 + (d_o - r\sin\theta)^2}} \quad (8)$$

The PTO moment (M_{PTO}) can be determined based on the linear force exerted by the hydraulic cylinder force, represented as F_{PTO}.

$$M_{PTO} = rF_{PTO}(\cos\gamma\cos\theta - \sin\gamma\sin\theta) \quad (9)$$

2.3 *Hydraulic transmission stage*

Figure 3 illustrates the schematic representation of the PTO system. The reciprocating rotational motion of the PM is converted into the linear motion of the piston in the double-acting hydraulic cylinder that pumps oil through ports "A" and "B". The bi-directional flow is converted into the unidirectional flow by the rectifier consisting of four check valves (CV). The pumped fluid flows into the PTO system via node "C", and one part of the fluid is stored in the high-pressure accumulator (HPA), and the other is delivered to the hydraulic motor. The fluid passed by the motor arrives at node "D". The low-pressure accumulator (LPA) stores part of the fluid for the pressure control on the low-pressure side. The fluid enters the rectifier and is drawn into the cylinder.

The generator is driven by a hydraulic motor to produce electricity. It is assumed that the motor axis is attached to a flying wheel (FW) for the stabilization of the motor's rotational speed. Besides, the generator damping can be adjusted by the electrical circuits, and the motor rotational speed is set as constant.

The AA system is installed in the (HPA) to store the high-pressure fluid, as presented in Figure 3.

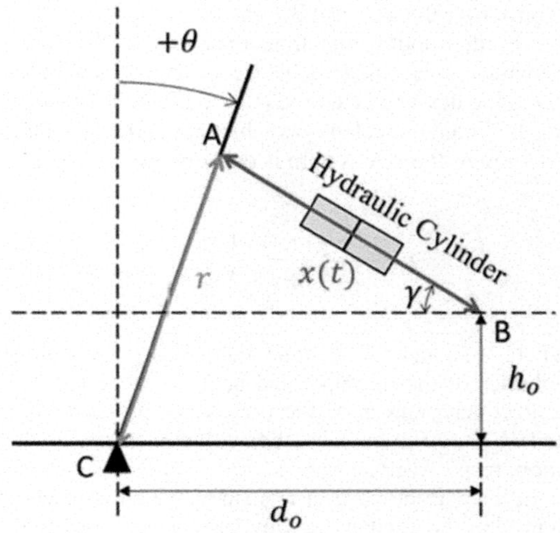

Figure 2. Schematic of the relative position between the adjustable rod attached to the hydraulic cylinder and the PM. Horizontal and vertical offset between the PM hinge C and the cylinder joint B is denoted as d_o and h_o, respectively.

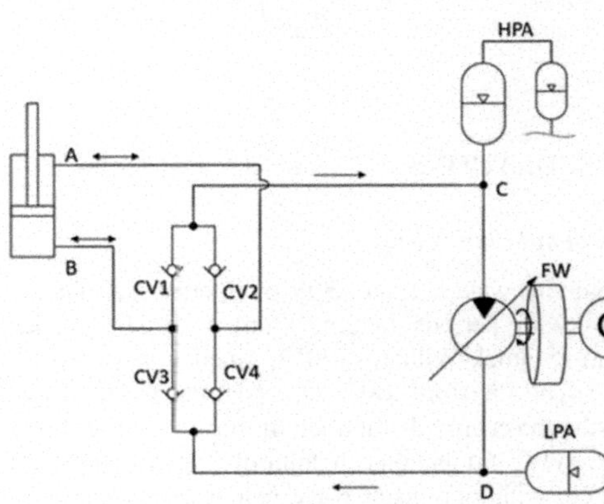

Figure 3. Schematic of the hydraulic PTO model. Arrows indicate oil flow directions (Grant contract PTDC/EME-REN/29044/2017).

2.4 *Active accumulator*

In this study, the PTO system incorporates the AA system, located on the high-pressure side, as presented in Figure 3 (HPA). Unlike conventional accumulators that maintain predetermined parameters, such as pre-charge pressure and initial gas chamber volume, throughout operation, the AA system offers controllable pneumatic properties. The AA system consists of two interconnected accumulators. The larger and principal accumulator is connected to the PTO circuit via the oil chamber and is responsible for maintaining oil pressure by delivering oil to the PTO system. The smaller auxiliary accumulator allows for modification of the characteristics of the AA system by enabling control over the gas chamber volume.

The thermodynamic transformation in the accumulators is assumed to be isentropic, which holds given the device cycle time (Cargo et al., 2014). Due to the interconnected gas chambers, the working pressure of the AA system is described as:

$$p_x = p_0 \left(\frac{V_{P,0} + V_{A,0}}{V_{P,x} + V_{A,x}} \right)^{n_0} \tag{10}$$

where, $V_{P,0}$ and $V_{A,0}$ denotes the initial gas chamber volumes of the principal and auxiliary accumulator, respectively, and p_0 is the pre-charge pressure. $V_{p,x}$ and $V_{A,x}$ are the gas chamber volumes during operation, respectively.

In the current stage of this study, the simultaneous control of the auxiliary accumulator is not considered, and $V_{A,x}$ remains constant during simulations. The maximum pressure of the AA system is denoted as P_{max}, which corresponds to the minimum gas chamber volume of the principal accumulator:

$$V_{P,\,min} = (V_0 + V_A) \left(\frac{p_0}{p_{max}} \right)^{1/n_0} - V_{A,x} \tag{11}$$

3 SIMULATIONS

3.1 *Simulation set-up*

For gravity waves, the energy of interest is from the waves with periods ranging from 1 to 30 seconds (Gunn & Stock-Williams, 2012; Guedes Soares et al. 2014). The Pierson-Moskowitz spectrum is utilized to describe the energy distribution of the sea states applied in the WEC simulations. A total of five sea states are selected, with significant wave height increasing from 0.75m to 2.75m and peak wave period from 5.45s to 7.78s. The specific parameters corresponding to these sea states are provided in Table 1 and the spectrum diagram is shown in Figure 4. In Table 1, H_s denotes the significant wave height, T_p the period of the peak wave and P_w the wave energy flux per unit of wave-crest length.

The working conditions of a conventional accumulator can be represented by a pressure-volume curve in the range between the pre-charge pressure (p_0) and the maximum pressure (p_{max}). For the AA system, another dimension of control is introduced by the adjustable gas chamber of the auxiliary accumulator and operating pressure (Figure 5). The yellow area in the contour indicates that the gas chamber pressure exceeds 350 bar, the maximum pressure. Therefore, the left bottom triangular area in yellow is forbidden during the operation. In addition, since $V_{A,x}$ is predetermined before each simulation, the corresponding pressure when $V_{p,x} = V_{P0}$ is the minimum pressure of the AA system.

Table 1. Parameters of Pierson-Moskowitz spectrum.

Sea States	H_s (m)	T_p (s)	P_w (kW)
1	0.75	5.45	1.41
2	1.25	5.98	4.36
3	1.75	6.59	9.51
4	2.25	7.22	17.25
5	2.75	7.78	27.63

The main parameters of the AA system used in the simulations are listed in Table 2.

Table 2. Active accumulator main parameters.

Symbol	Parameters	Value and Unit
$V_{P,0}$	Initial Gas Chamber Volume of the Principal Accumulator	5 m^3
$V_{A,0}$	Initial Gas Chamber Volume of the Auxiliary Accumulator	3.5 m^3
p_0	Pre-charge Pressure	350 Bar
p_{max}	Maximum Pressure	105 *Bat*
n_0	Gas polytrophic index	1.4

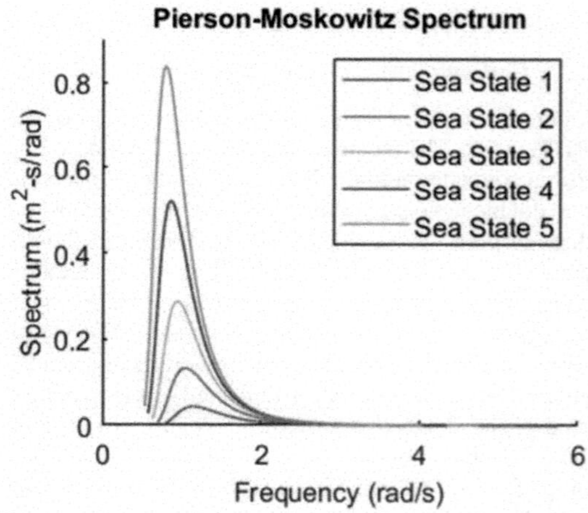

Figure 4. Spectrum diagram for selected sea states.

The AA system is designed to be able to withstand energy from the heaviest sea state 5. Thus, for this sea state simulation, the auxiliary accumulator is not pre-loaded, which means $V_{A,x} = V_{A,0}$.

3.2 *Results and discussion*

To evaluate the PTO performance, the ratio between the time average shaft power of the hydraulic motor (P_m) and the average wave energy flux along the width of the WEC device (l_{WEC}), is used:

$$C_P = \frac{P_m}{P_w l_{WEC}} \tag{12}$$

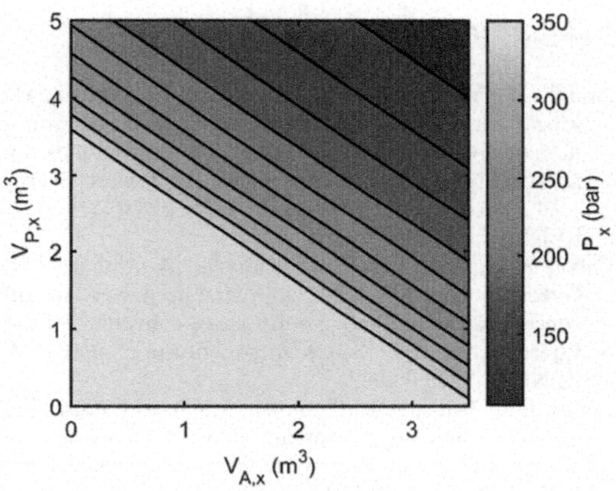

Figure. 5. Relationship between the output pressure (P_x) and the gas chamber volumes in the principal and auxiliary accumulators ($V_{P,x}$, $V_{A,x}$).

A series of simulations are performed for the WEC across five sea states, incorporating different AA setups. The power coefficients (Cp) and $V_{A,x}$, are presented in Figure 6. The color gradients and text within the zones depict variations in coefficient values.

The dark blue zones indicate that the AA system is non-functional in the simulations due to the AA pressure falling below the minimum one. This is caused by less energetic sea states and a smaller volume of the auxiliary accumulator gas chamber ($V_{A,x}$). In such cases, the small AA pressure may fall below the increased equivalent pre-charge pressure, generated by the more compressed gas chamber, and consequently, the AA system ceases its operation.

The yellow zones indicate cases in which the AA pressure exceeds the maximum pressure (p_{max}), resulting in the AA system ceasing to function. It happens in the high sea states and when $V_{A,x}$ is small. More energetic sea states can certainly increase the pressure on the high-pressure side of the PTO circuit, and the more compressed AA gas chamber may also contribute to the increase of the AA system press.

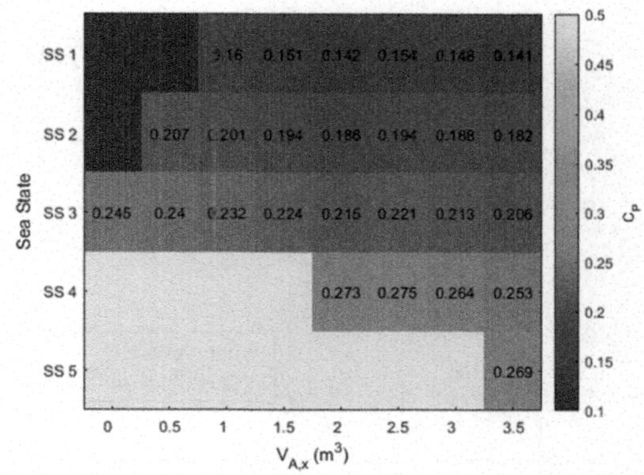

Figure 6. WEC Power coefficients for different active accumulator settings across all sea states.

It is also observed from Figure 6, that in small sea states (SS 1-3), the PTO performs better when the gas chamber of the auxiliary accumulator is compressed. To further study this case, the performance of the AA system in the same sea state (SS 3) and with three different $V_{A,x}$ values is analyzed. The results are presented in Figure 7. It can be concluded from this figure that the compressing of the auxiliary accumulator gas chamber increases the AA pressure in the same sea state. The increase of the oil pressure in the PTO high-pressure side results in a better power output.

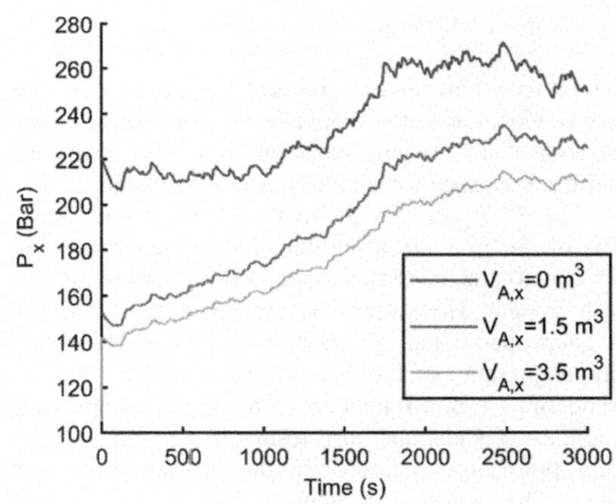

Figure 7. Active accumulator pressure variation.

Then, the overall performance of the AA system is compared to the performance of the conventional accumulator. The optimal AA system setting for each sea state, as indicated in Figure 6, is selected to represent the performance of the AA system. Since the parameters cannot be changed, the conventional accumulator should be operational for all selected sea states. Consequently, the accumulator equivalent to the AA system in the case of $V_{A,x} = 3.5\ m^3$ is chosen to represent the conventional accumulator.

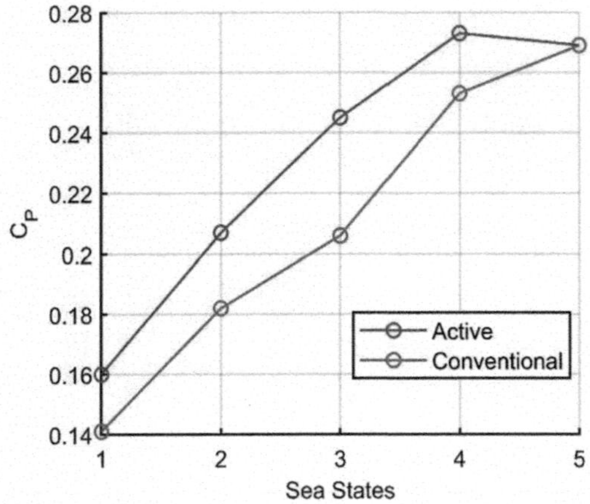

Figure 8. WEC Power coefficients for different active accumulator settings in five sea states.

The results are displayed in Figure 8, which indicates that the PTO with the AA accumulator outperforms the PTO with the conventional accumulator in all sea states except SS5 because both are sized to support the same operational conditions at this sea state. If the conventional accumulator is optimized for a more frequent sea state, such as SS3, the AA system may outperform in terms of adaptability to more sea states.

4 CONCLUSIONS

The AA system offers significant advantages over the conventional one. It is adaptable to a wide range of sea states and its charging parameters may be optimized for each sea state for a better power coefficient.

The PTO performs optimally when the gas chamber of the auxiliary accumulator is compressed. This is due to the increased operational pressure of the AA system. However, it is crucial to note that the higher operational pressure may exceed the maximum pressure of the AA system in high-sea states, rendering it non-functional. The more compressed auxiliary gas chamber also results in a higher equivalent pre-charge pressure. In the less energetic sea state, the working pressure may fall below the equivalent pre-charge pressure, leading the active accumulator system to cease its operation. Therefore, there exists a lower limit for adjusting the volume of the gas chamber in the auxiliary accumulator when the system operates in high or low-sea states.

At the current stage of this research, the AA system is not controlled in the simulations. A suitable controller will be developed to explore the potential of the active accumulator concept and enhance its performance.

In addition, the study will also focus on investigating the mechanism for controlling the auxiliary accumulator, which will provide further insights and details into the overall concept.

ACKNOWLEDGEMENTS

This work was performed within the Strategic Research Plan of the Centre for Marine Technology and Ocean Engineering, which is financed by the Portuguese Foundation for Science and Technology (Fundação para a Ciência e Tecnologia - FCT) under contract UIDB/UIDP/00134/2020, the project "Experimental simulation of oil-hydraulic Power Take-Off systems for Wave Energy Converters", funded by Portuguese Foundation for Science and Technology (FCT) under contract PTDC/EME-REN /29044/2017 and the project "Variable geometry Wave Energy Conversion system for floating platforms" funded by FCT under contract PTDC/EME-REN/0242/2020.

REFERENCES

Amini, E., Mehdipour, H., Faraggiana, E., Golbaz, D., Mozaffari, S., Bracco, G., Neshat, M., 2022. Optimization of hydraulic power take-off system settings for point absorber wave energy converter. Renew Energy 194, 938–954. https://doi.org/10.1016/J.RENENE.2022.05.164

Calvário, M., Gaspar, J.F., Kamarlouei, M., Hallak, T.S., Guedes Soares, C., 2020. Oil-hydraulic power take-off concept for an oscillating wave surge converter. Renew Energy 159, 1297–1309. https://doi.org/10.1016/J.RENENE.2020.06.002

Cargo, C.J., Hillis, A.J., Plummer, A.R., 2014. Optimisation and control of a hydraulic power take-off unit for a wave energy converter in irregular waves. Proceedings of the Institution of Mechanical Engineers, Part A: Journal of Power and Energy 228, 462–479. https://doi.org/10.1177/0957650913519619/ASSET/IMAGES/LARGE/10.1177_0957650913519619-FIG18.JPEG

Coe, R.G., Bacelli, G., Forbush, D., 2021. A practical approach to wave energy modelling and control. Renewable and Sustainable Energy Reviews 142, 110791. https://doi.org/10.1016/J.RSER.2021.110791

Gaspar, J.F., Calvário, M., Kamarlouei, M., Guedes Soares, C., 2016a. Power take-off concept for wave energy converters based on oil-hydraulic transformer units. Renew Energy 86, 1232–1246. https://doi.org/10.1016/J.RENENE.2015.09.035

Gaspar, J.F., Kamarlouei, M., Sinha, A., Xu, H., Calvário, M., Faÿ, F.X., Robles, E., Guedes Soares, C., 2017. Analysis of electrical drive speed control limitations of a power take-off system for wave energy converters. Renew Energy 113, 335–346. https://doi.org/10.1016/j.renene.2017.05.085

Gaspar, J.F., Kamarlouei, M., Sinha, A., Xu, H., Calvário, M., Faÿ, F.-X., Robles, E., Guedes Soares, C., 2016b. Speed control of oil-hydraulic power take-off system for oscillating body-type wave energy converters. Renew Energy 97, 769–783. https://doi.org/10.1016/j.renene.2016.06.015

Guedes Soares, C.; Bento, A. R.; Goncalves, M.; Silva, D., and Martinho, P. 2014. Numerical evaluation of the wave energy resource along the Atlantic European coast. Computers & Geosciences. 71:37–49.

Guedes Soares, C., Bhattacharjee, J., Tello, M., Pietra, L., 2012. Review and classification of wave energy converters, in Guedes Soares, C., Garbatov, Y., Sutulo, S., Santos, T.A. (Eds.), Maritime Engineering and Technology. Taylor and Francis Group, Lisbon, Portugal, pp. 585–594. https://doi.org/10.1201/B12726-82Gunn

K., Stock-Williams, C., 2012. Quantifying the global wave power resource. Renew Energy 44, 296–304. https://doi.org/10.1016/J.RENENE.2012.01.101

Hansen, R.H., Andersen, T.O., Pedersen, H.C., 2011. Model based design of efficient power take-off systems for wave energy converters, in: Sairiala, H., T. Koskinen, K. (Eds.), Proc. 12th Scandinavian International Conference on Fluid Power. Tampere, Finland.

Hansen, R.H., Kramer, M.M., 2011. Modelling and Control of the Wavestar Prototype, in: S. Bahaj, A. (Ed.), Proceedings of the 9th European Wave and Tidal Energy Conference, EWTEC 2011. University of Southampton, Southampton, UK.

Hansen, R.H., Kramer, M.M., Vidal, E., 2013. Discrete Displacement Hydraulic Power Take-Off System for the Wavestar Wave Energy Converter. Energies 2013, Vol. 6, Pages 4001-4044 6, 4001–4044. https://doi.org/10.3390/EN6084001

Jusoh, M.A., Ibrahim, M.Z., Daud, M.Z., Albani, A., Yusop, Z.M., 2019. Hydraulic Power Take-Off Concepts for Wave Energy Conversion System: A Review. Energies 2019, Vol. 12, Page 4510 12, 4510. https://doi.org/10.3390/EN12234510

Li, C., Zhang, D., Zhang, W., Liu, X., Tan, M., Si, Y., Qian, P., 2021. A Constant-Pressure Hydraulic PTO System for a Wave Energy Converter Based on a Hydraulic Transformer and Multi-Chamber Cylinder. Energies 2022, Vol. 15, Page 241 15, 241. https://doi.org/10.3390/EN15010241

Nolte, J.D., Ertekin, R.C., 2014. Wave power calculations for a wave energy conversion device connected to a drogue. Journal of Renewable and Sustainable Energy 6. https://doi.org/10.1063/1.4862785/286204

Penalba, M., Kelly, T., Ringwood, J., 2017. Using NEMOH for Modelling Wave Energy Converters: A Comparative Study with WAMIT, in: 12th European Wave and Tidal Energy Conference. Cork, Ireland.

So, R., Simmons, A., Brekken, T., Ruehl, K., Michelen, C., 2015. Development of PTO-Sim: A Power Performance Module for the Open-Source Wave Energy Converter Code WEC-Sim, in: ASME 2015 34th International Conference on Ocean, Offshore and Arctic Engineering. American Society of Mechanical Engineers, St. John's, Newfoundland, Canada, p. V009T09A032. https://doi.org/10.1115/OMAE2015-42074

Tucker, M.J., Pitt, E.G., 2001. Waves in ocean engineering. Elsevier Science.

Zhang, D., Li, W., Lin, Y., Bao, J., 2012. An overview of hydraulic systems in wave energy application in China. Renewable and Sustainable Energy Reviews 16, 4522–4526. https://doi.org/10.1016/J.RSER.2012.04.005.

Advances in Maritime Technology and Engineering – Guedes Soares & Santos (Eds)

Wave energy converter arrays performance in variable water depth regions

Th. Gerostathis
Department of Naval Architecture, School of Engineering, University of Western Attica, Egaleo, Athens, Greece

A. Magkouris & K. Belibassakis
School of Naval Architecture & Marine Engineering, National Technical University of Athens (NTUA), Zografos, Athens, Greece

ABSTRACT: Evaluating the performance of wave energy converter (WEC) arrays, of the type of simple floaters, operating in coastal areas, based on the local climatology and the geographical characteristics of the installation region, is important for the effective design of WEC parks. In this study, a numerical model is proposed to estimate the performance of WEC arrays in areas of varying depth, considering the interactions with the seafloor topography. The presented methodology comprises a 3D coupled-mode water waves model, combined with a boundary element method that addresses the diffraction/radiation problem and evaluates the flow details at the local scale of the WECs. A case study is performed, at a selected coastal site in a nearshore area in Greece. Using long-term climate data, seasonal and annual power output is estimated, highlighting the methodology and its usability as a supporting tool for the optimization of WEC deployments.

1 INTRODUCTION

The development of sustainable energy policies is currently aided by marine renewable energy sources (Guo & Ringwood, 2021), with one of the most widespread technologies being the wave energy converters (WECs) of the form of heaving point absorbers (see Figure 1), in offshore and coastal locations (Babarit, 2017). The optimization of WEC parks depends on the estimation of the power absorption capacity and performance of WECs. Due to hydrodynamic interactions between several floating devices, referred to as intra-array interactions (Stratigaki et al., 2014), the overall power absorbed by a WEC array of may be increased or lowered in comparison to the power provided by an equal number of individual devices. The power absorption capacity can also be impacted by the existence of irregularities concerning the local seabed topography, particularly at low frequencies when the various subfields of the flow considerably interact with the seabed (Belibassakis et al., 2018). Several methodologies have been applied to evaluate the response of various forms of WEC layouts and the resulting power output, including time and frequency domain boundary element methods (BEM), computational fluid dynamics models (Li & Yu, 2012) and reduced models (Bonovas et al. 2022). The model presented and discussed in this work is based on linear wave theory and consists of a frequency domain BEM, based on irrotational and non-viscous flow assumptions, supplemented by the 3D Consistent Coupled Mode Model - CCMM, (Belibassakis et al. 2001), to simulate wave propagation over general bathymetry. It must be noted that the influence of fluid viscosity is not considered and this may lead to significant deviations at some frequencies (Gao, et al., 2019). However, the 3D BEM approach allows for the modelling of any WEC shape, while it can still produce exploitable results in a significant range of frequencies, which could be utilized to estimate the WEC layout performance, using spectral average characteristics.

Figure 1. Point absorber wave park comprising 9 WECs.

2 MATHEMATICAL FORMULATION

An array of (one degree of freedom) heaving point absorber WECs is considered, comprising M identical floaters, operating in a coastal region, where the local seabed topography may present any variation, while the whole arrangement is subjected to incident harmonic waves. Each device response is calculated while accounting for the hydrodynamic interactions between various bodies and the seafloor. Following

DOI: 10.1201/9781003508779-47

standard hydrodynamic theory, the total field is defined by superimposing the incident, the diffracted and M radiated fields. Each of these fields is represented by a corresponding potential $\Phi_0(\mathbf{x};t)$, $\Phi_d(\mathbf{x};t)$ and $\Phi_m(\mathbf{x};t), m = 1,2,\ldots,M$, respectively, and thus the resulting velocities $\mathbf{v}_m(\mathbf{x};t)$ are given by the gradients $\nabla\Phi_m(\mathbf{x};t)$, $m = 0,\ d,\ 1,2,\ldots,M$.

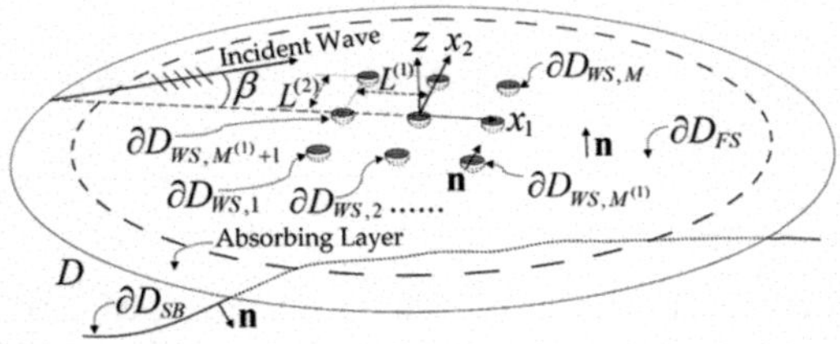

Figure 2. Sketch of a WEC array of nine ($M^{(1)} = M^{(2)} = 3$) identical devices, depicting the Domain D and the boundary ∂D components, along with several parameters.

The coordinate system $\mathbf{x} = (\mathbf{x}_h, z) = (x_1, x_2, z)$ is used, with the $(x_1, x_2, z) = (0,0,0)$ point placed at Mean Water Level (MWL) so that the WEC array is symmetric with respect to both the planes $x_1 = 0$ and $x_2 = 0$ (see Figure 2). $M^{(1)}$ and $M^{(2)}$ are the numbers of WECs in the x_1 and x_2-directions and therefore, $M = M^{(1)}M^{(2)}$. Within the framework of this study, each WEC's submerged volume includes a cylindrical part of radius a and draft T_C, and an oblate semi-spheroidal component, with semi-major axis of length a and corresponding minor axis of length T_s (see Figures 2, 3), resulting in a total draft of $T = T_C + T_s$.

The normal unit vector to the boundary ∂D, directed to the exterior of the water, is represented by $\mathbf{n} = (n_1, n_2, n_3)$ (see Figure 2). The distance between the centers of the waterplanes of two neighboring WECs' in the x_1 and x_2-directions respectively is $L^{(1)}$ and $L^{(2)}$. Finally, the WEC indexing begins from the WEC in the 3rd quadrant of the x_1x_2 plane which has the longer distance from the point (0,0) (see Figure 2) and then continues along the x_1- axis until the last WEC $M^{(1)}$. The following device is the one positioned in the next row of the layout with waterplane midpoint having the same x_1- coordinate as the first device. The indexing is similar until the last device (M).

Assuming harmonic time dependency of the form $\exp(-i\omega t)$, the interaction problem is handled in the frequency domain, where ω is the angular frequency and i the imaginary unit, by utilizing the following velocity potential representation:

$$\Phi(\mathbf{x};t) = \Phi_0(\mathbf{x};t) + \Phi_d(\mathbf{x};t) + \sum_{m=1}^{M} \tfrac{d\xi_m}{dt}\Phi_m(\mathbf{x};t) = \mathrm{Re}\{-\tfrac{igH}{2\omega}\varphi(\mathbf{x}) \cdot exp(-i\omega t)\}. \tag{1}$$

where H denotes the wave height, g is the gravitational acceleration, ξ_m is the amplitude, in the complex domain, of the m-th WEC's response in the z-direction, and

$$\varphi(\mathbf{x}) = -i\omega\left\{\varphi_0(\mathbf{x}) + \varphi_d(\mathbf{x}) + \sum_{m=1}^{M} \xi_m\,\varphi_m(\mathbf{x})\right\} \tag{2}$$

is the complex potential in the frequency domain. In Eq. (2) $[-i\omega\varphi_0(\mathbf{x})]$ and $[-i\omega\varphi_d(\mathbf{x})]$ respectively represent complex amplitudes corresponding to the incident and the diffracted fields. $\varphi_m(\mathbf{x})$, $m = 1,2,\ldots,M$ is the radiation field complex amplitude of unitary oscillations in z-direction of the m-th device.

2.1 *Formulation of the incident field*

The incident wave field $\Phi_0(\mathbf{x};t)$, propagating over the uneven bottom profile, is given by the consistent coupled mode model – CCMM, presented in (Athanassoulis & Belibassakis, 1999) and extended in three dimensions by Belibassakis et al. (2001). For the case of a parallel-contour bathymetry $h(x_1)$ free of mild bottom slope assumptions, as shown in Figure 2, the incident field is expressed as,

$$\varphi_0(\mathbf{x}) = \exp(ik_1 \sin(\beta)x_2)\varphi_I(x_1, z) \tag{3}$$

where $\varphi_I(x_1, z)$ is the solution of a one-dimensional version of the CCMM for oblique-incident waves with incident wavenumber k_1 and direction β over the one-dimensional bathymetry $h(x_1)$. The potential $\varphi_I(x_1, z)$ accepts the following representation using local modes,

$$\varphi_I(x_1, z) = \phi_{I,-1}(x_1)Z_{-1}(z;x_1) + \sum_{n=0}^{\infty} \phi_{I,n}(x_1)Z_n(z;x_1) \tag{4}$$

where $\phi_{I,0}(x_1)Z_0(z;x_1)$ is the propagating mode, while $\phi_{I,n}(x_1)Z_n(z;x_1)$, $n = 1,\ldots,\infty$ are the evanescent modes. The term $\phi_{I,-1}(x_1)Z_{-1}(z;x_1)$ is a correction term, the so-called sloping-bottom mode, that accounts for the accurate satisfaction of the bottom boundary condition at $z = -h(x_1)$. The eigenfunctions Z_n, $n \geq 0$ are solutions of vertical Sturm-Liouville problems, in the local intervals $-h(x_1) \leq z \leq 0$. Using Eqs. (3, 4), a system of partial differential equations is obtained for the complex amplitudes $\phi_{I,n}(x_1)$. The introduction of the slopping bottom mode accelerates the convergence of the local mode series (Eq. 4), leading to using only four to six terms totally, even for seabed slopes of more than 100%; see Belibassakis et al. (2001) for more details.

2.2 *BEM formulation for the problems for the diffraction and radiation fields*

The M radiation and the diffraction fields are estimated by boundary value problems (BVPs), consisting of the Laplace equation as well as appropriately defined boundary conditions (BCs) on the various boundary surfaces. The latter include the free surface of the water (∂D_{FS}), the submerged surfaces of the floaters ($\partial D_{WS,m}$, $m = 1,2,\ldots M$) and the boundary of the seabed (∂D_{SB}); see also Figure 2. Zero normal velocity BCs apply to the various solid boundaries and the standard linear free surface boundary condition (FSBC) applies to ∂D_{FS}. The radiated and diffracted fields are evaluated as solutions to the BVPs:

$$\nabla^2 \varphi_m(\mathbf{x}) = 0, \ (\mathbf{x}) \in D, \ m = d, 1, 2, \ldots, M \quad (5a)$$

$$\frac{\partial \varphi_m(\mathbf{x})}{\partial n} - \mu(\mathbf{x}_h;\ \omega)\varphi_m(\mathbf{x}) = 0, \quad \mathbf{x} \in \partial D_{FS}, \ m = d, 1, 2, \ldots, M, \quad (5b)$$

$$\frac{\partial \varphi_d(\mathbf{x})}{\partial n} = -\frac{\partial \varphi_0(\mathbf{x})}{\partial n}, \quad \mathbf{x} \in \bigcup_{m=1}^{M} \partial D_{WS,m}, \quad (5c)$$

$$\frac{\partial \varphi_m(\mathbf{x})}{\partial n} = \delta_{m\ell}\, n_3^{\ell}(\mathbf{x}), \quad \mathbf{x} \in \partial D_{WS,\ell}, \ m, \ell = 1, 2, \ldots, M, \quad (5d)$$

$$\frac{\partial \varphi_m(\mathbf{x})}{\partial n} = 0, \quad \mathbf{x} \in \partial D_{SB}, \ m = d, 1, 2, \ldots, M \quad (5e)$$

Eq. (5d), which involves the Kronecker delta, states that the m-th radiation field is computed by applying a unitary vertical oscillation of the m-th submerged surface, while the remaining devices are treated as immobile. The frequency parameter $\mu = \omega^2/g$ involved in Eq. (5b), is expressed as spatially variable to account for proper radiation conditions at infinity. More specifically, the solutions obtained by the BVPs expressed by Eqs. (5) propagate undisturbed towards all directions, indicating that the flow domain extends at infinity. The domain is truncated by adopting a perfectly matched layer (PML); see Figure 2. To this end, an absorbing layer is introduced, which surrounds the flow field and is used to artificially dissipate outgoing wave solutions far from the WEC array (Belibassakis et al., 2018). The thickness of the PML, which has to be approximately one local wavelength, defines the efficiency in the achieved artificial dissipation and the minimization of errors due to numerical reflections. In the context of this study, the layer thickness is set to one mean wavelength $\bar{\lambda} = 2\pi/\bar{k}_0$, amplified by 15% to take into account longer waves at the deeper areas of the domain. $\bar{k}_0$ is the real root of water waves dispersion equation at the mean depth $(\bar{h})$ of the rectangular region defined by the WEC park. The PML with the above parameters has been found to achieve sufficient dissipation at the deeper areas of the domain for the case of the local bathymetric data and the frequency range examined in this work, as described in the sequel. However, the PML parameters might require readjustment for lower frequencies or domains with greater depth deviations between deep and shallow regions. An imaginary part is introduced to the frequency parameter within the layer, to implement the PML. The latter parameter is defined as:

$$\mu(\mathbf{x}_h;\ \omega) = \begin{cases} \omega^2 g^{-1}, & |\mathbf{x}_h| < R_{PML} \\ \frac{\omega^2}{g}\left(1 + ic\frac{(|\mathbf{x}_h| - R_{PML})^n}{\bar{\lambda}^n}\right)^2, & |\mathbf{x}_h| R_{PML} \end{cases} \quad (6)$$

where R_{PML} is the PML activation radius (see Figure 2). The parameters *n, c* are used as optimized in previous works (Belibassakis et al., 2018) aiming toward the optimization of the layer's efficiency. The BVPs defined by Eqs (5) are numerically solved by a low-order panel method, using 4-node quadrilateral boundary elements. The complex potential functions φ_m, $m = d, 1, 2, \ldots, M$, are represented by:

$$\varphi_m(\mathbf{x}) = \int_{\partial D} \sigma_m(\mathbf{x}')\, G(\mathbf{x}'|\mathbf{x}) dS(\mathbf{x}'), \quad \mathbf{x} \in D, \ \mathbf{x}' \in \partial D, \ m = d, 1, 2, \ldots M \quad (7)$$

where $G(\mathbf{x}'|\mathbf{x}) = |\mathbf{x}' - \mathbf{x}|^{-1}/4\pi$ is the Green's function for the Laplace equation in 3D and $\sigma_m(\mathbf{x}')$, $m = d, 1, 2, \ldots, M$ are strength distributions, on the boundary, evaluated by the low order panel method, assuming that they are piecewise constant. The distributions $\sigma_m(\mathbf{x}')$, combined with a discretized form of Eq. (7), define the potentials φ_m, $m = d, 1, 2, \ldots, M$. Particularly, the potentials φ_m and the corresponding velocities are estimated by:

$$\varphi(\mathbf{x}) = \sum_{\kappa} \sigma_\kappa \Phi_\kappa(\mathbf{x}), \ \nabla\varphi(\mathbf{x}) = \sum_{\kappa} \sigma_\kappa \mathbf{U}_\kappa(\mathbf{x}) \quad (8)$$

where summation spans all the panels, indexed by κ and σ_κ is the strength of the distribution on the κ-*th* panel, which is complex, as a result of the complex values of the frequency parameter in the absorbing layer. Induced potential to the field point $\mathbf{x}$ from the κ-*th* element (with unitary distributed singularity), is denoted by Φ_κ, while the corresponding induced velocity is denoted by $\mathbf{U}_\kappa$. Numerical solutions are evaluated by satisfying corresponding BCs at the centroid of each individual panel on the separate surfaces of the domain boundary, applying a collocation method. Constant normal dipole distributions are used on each panel, and thus the induced potential matrix can be analytically calculated via the solid angle; see (Newman, 1986). Moreover, induced velocity is computed by repetitively applying the Biot–Savart law, given the equivalence of a constant dipole quadrilateral element to a vortex ring (Katz & Plotkin, 2001).

The discretization is performed by combining distinct meshes of the various boundary surfaces, with continuous junctions between the various parts of the mesh, resulting in global boundary continuity. In particular, the mesh of the wetted surface of an individual WEC comprises a part that models the lateral surface of the cylindrical part of the submerged volume, combined with a part that models the oblate semi-spheroid. Moreover, the free surface surrounding each device is discretized by a rectangular block with dimensions $L^{(1)}$ and $L^{(2)}$, in the x_1 and x_2-directions, properly centered at the midpoint of the floater's circular waterplane. The latter block is modelled by a radial mesh, using reduced element size near the device, to ensure better result quality. An example of a WEC's mesh and the corresponding mesh of the free surface block are presented in Figure 3(a). The mesh illustrated in Figure 3(a) is used M times to define the mesh of the WEC array, as illustrated in Figure 3(b) (for the case $M^{(1)} = M^{(2)} = 2$). Next, a radial mesh is defined, that surrounds the WEC array, representing the remaining free surface, up to a predetermined number of mean wavelengths $(\bar{\lambda})$, and therefore the PML activation radius as well as the external dimensions of the mesh, are frequency dependent. To retain minimum number of elements per wavelength in the radial and angular directions at the outer section of the free surface mesh, as illustrated in Figure 3(b), the nodes along the azimuthal direction

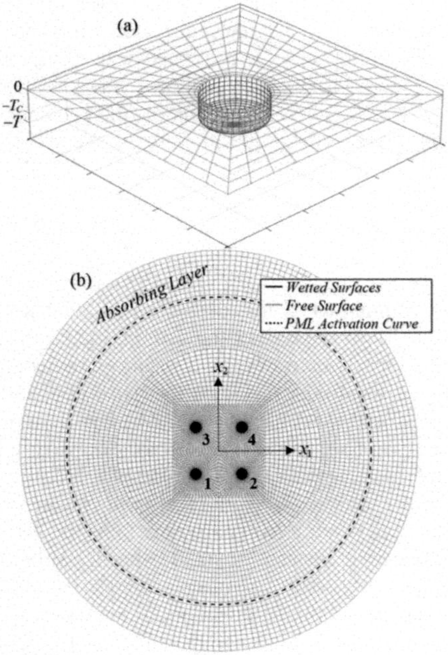

Figure 3. (a) Mesh of a WEC's submerged surface and corresponding free surface mesh. (b) WEC array mesh consisting of 4 devices (top view), demonstrating the WEC indexing and the PML activation boundary.

are doubled at a given radius. The x_1, x_2 coordinates of the seabed mesh below the WEC array are defined by a rectangular mesh. The remaining seabed mesh is structured similarly to the one of the free surface, while the vertical coordinate can be used to simulate the local seabed irregularities.

After evaluating the diffracted and radiated subfields, the complex amplitudes of the devices' heave responses $(\xi_m, m = 1, 2, \ldots, \mathrm{M})$ are obtained as solution to the system of M equations:

$$\begin{bmatrix} -\omega^2(\mathbf{M} + \mathbf{A}(\omega)) \\ -i\omega(\mathbf{B}(\omega) + \mathbf{B}_{PTO}) + (\mathbf{C} + \mathbf{C}_{PTO}) \end{bmatrix} \xi = \mathbf{F}_0 + \mathbf{F}_d. \tag{9}$$

where $\mathbf{M}$ and $\mathbf{C}$ are the diagonal matrices respectively modelling the inertia of the WECs ($\mathbf{M}_{ii} = \rho\, V_{WEC}$ where V_{WEC} is the submerged volume and ρ the water density) and the hydrostatic restoring forces $\left(\mathbf{C}_{ii} = \rho g \pi \alpha^2\right)$. $\mathbf{B}_{PTO} = B_{PTO}\mathbf{I}$ models the Power Take-Off (PTO) system energy extraction, which is achieved by introducing a damping coefficient, and $\mathbf{C}_{PTO} = C_{PTO}\mathbf{I}$ is the stiffness of the PTO system. The added mass and hydrodynamic damping corresponding matrices ($\mathbf{A}(\omega)$ and $\mathbf{B}(\omega)$) are computed using the radiation fields as:

$$\omega^2 A_{m\ell} + i\omega B_{m\ell} = \rho \int_{\partial D_{WS,m}} \varphi_\ell(\mathbf{x}) \cdot n_3(\mathbf{x}) dS(\mathbf{x}), \mathbf{x} \in \partial D_{WS,m} \tag{10}$$

for $m, \ell = 1, 2, \ldots, M$. Furthermore, the excitation forces (Froude-Krylov and diffraction) exerted on the *m-th* WEC are evaluated by integrating the induced pressure by each field multiplied by the vertical component of $\mathbf{n}$,

$$F_{m,\ell} = \omega^2 \rho \int_{\partial D_{WS,m}} \varphi_\ell(\mathbf{x}) \cdot n_3(\mathbf{x}) dS(\mathbf{x}), \quad x \in \partial D_{WS,m},\ \ell = 0, d, \tag{11}$$

for $m = 1, 2, \ldots, M$. Next, each WEC's mean output power is computed as:

$$P_{OUT,m}(\omega) = 0.5\, \omega^2 B_{PTO} |\xi_m|^2, \tag{12}$$

assuming the PTO system is 100% efficient. The normalized power $\bar{P}_m(\omega)$, taking into account the local topography and the intra-array interactions, is defined dividing the above result by the incident power flux over a cross section equal to a device's waterline diameter, at the depth $h_{\max}$ of the incidence region. This power flux is used for normalization since the wave numbers and amplitudes near each of the WECs may differ due to the depth changes. Considering an incident waveheight equal to $H = 2A$ at the area of wave incidence, $\bar{P}_m(\omega)$ is equal to,

$$\bar{P}_m(\omega) = P_{OUT,m}(\omega)\left[0.25\, \rho g H^2 C_g(h_{\max}, \omega)\ a\right]^{-1} \tag{13}$$

where C_g ($h_{\max}$, ω) is the wave group velocity, evaluated at the depth of the incidence region.

3 CASE STUDY

We examine a WEC array comprising 9 devices operating north of the coastal area of Kasos Island, in the southern Aegean Sea. For simplicity, the bathymetry profile in the domain D is defined by an one dimensional section of the local seabed topography, parallel to the x_1-axis (see Figures 1, 4), assuming the isobaths to be approximately parallel while approaching the shoreline. The wave climate in the examined area is represented by a long-term time series, spanning the time interval between the start of 2013 and the end of 2022, at the offshore point located at 35°30′00″N–27°00′00″E, as obtained from the ERA5 database (Dee, et al., 2011). The available information includes, among others, significant wave height H_S, mean energy wave period T_{-10}, mean wave direction θ_m in $6h$ time intervals. Nearshore wave data at the position of the target point, at the side of incidence of the BEM model, with geographical coordinates 35°26′22″ N–26°59′20″ E, are derived by an offshore to nearshore (OtN) transformation technique that makes use of the SWAN (Booij et al., 1999) spectral model, as shown in Figure 4. Offshore sea states, represented by appropriately defined parameters are known all along the seaward boundary and are utilized for the reconstruction

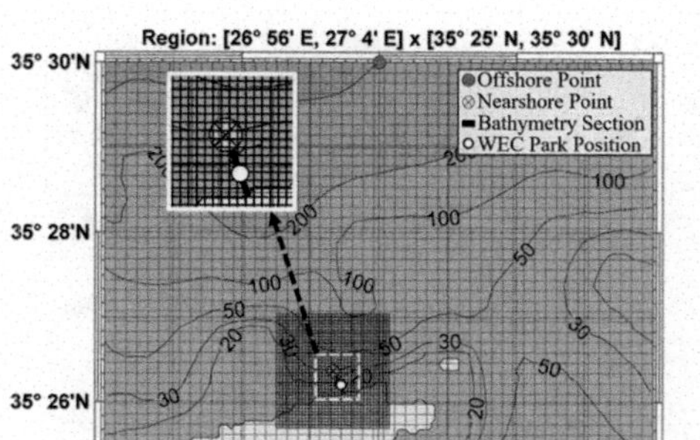

Figure 4. Numerical mesh for applying the SWAN model and seabed topography north of the coast of Kassos Island, illustrating the position of the WEC park and the section that corresponds to the bathymetric data used.

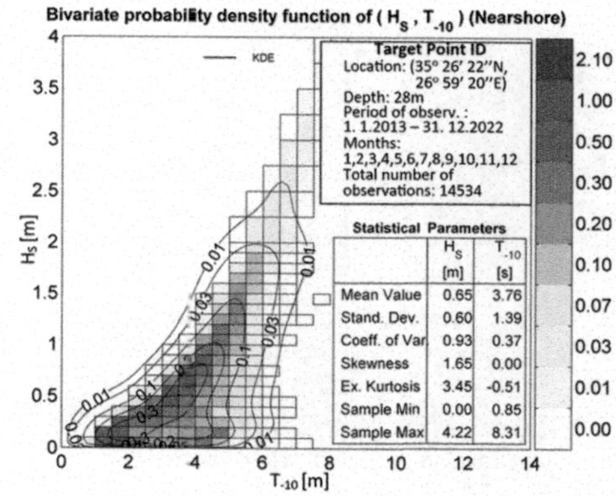

Figure 5. Nearshore wave data at the target point located at depth h = 28 m, as computed by the OtN transformation.

of the directional spectra $S(f, \theta)$, using the JONSWAP frequency spectrum. The bathymetric data are obtained from (GEBCO Compilation Group, 2023). The nearshore wave data at the target nearshore point are shown in Figure 5, at depth $h = 28m$, which is used as the incidence depth in the CCMM.

3.1 *Application of the BEM model*

Figure 6 depicts the real part of the propagating field (consisting of the incident field and the diffracted field), for the configuration of 9 WECs with diameter $2a = 3m$, draft $T = a$ and $T_C = 0.8a$ placed in a 3x3 arrangement with $L^{(1)} = L^{(2)} = 10a$. All results in the present section refer to this setup. The subplots in Figure 6 show the potential on the x_1x_2 plane, multiplied by $(i\omega/g)$ to represent the free surface elevation $\eta(x_1, x_2)$. Figure 7 shows indicative radiation fields generated by the WECs 1 and 5. It can be seen that for relatively high periods (see Figures 6a, 7a and 7b) the interaction of the fields with the seafloor is considerable, while the lower the period (see Figures 6b, 7c and 7d) the more the floating devices affect both the diffracted as well as the radiated fields. Figure 8

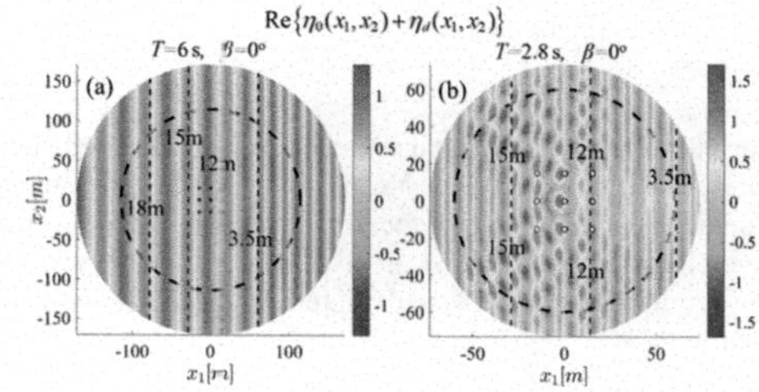

Figure 6. Superposition of the incident and the diffracted subfields. Normal incidence on the WEC park and incident wave periods (a) T=6s and (b) T=2.8s. Dashed lines represent isobaths. The PML activation curve is denoted by a bold dashed line.

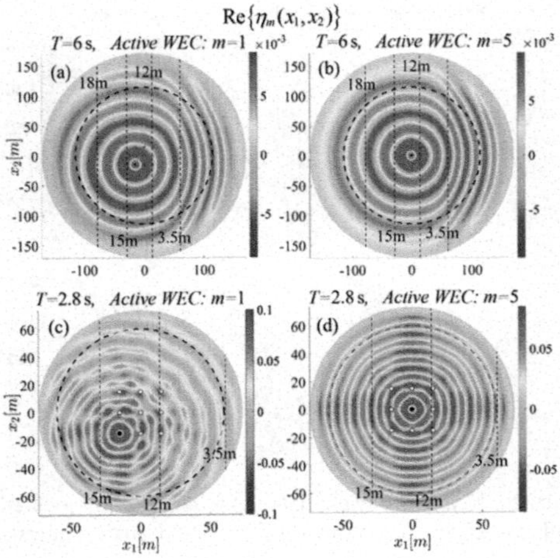

Figure 7. Radiation fields created by the WECs 1(a, b) and 5(c, d). Incident wave periods T=6s (a, b) and T=2.8s (b, d). Dashed lines represent isobaths. The PML internal boundary is denoted by a bold dashed line.

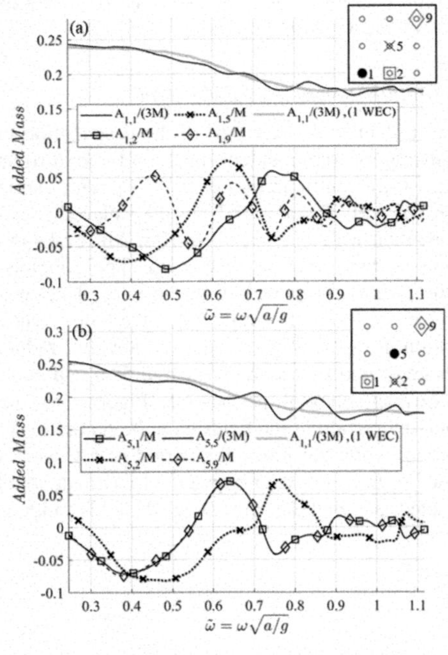

Figure 8. Normalized added mass on WEC (a) 1 and (b) 5 by the WECs 1,2,5 and 9. Normalized added mass of an isolated WEC deployed in the region and positioned at x_1 =x_2=0 (location of WEC 5), is also shown for comparison. The layout of the WECs is depicted in the corresponding legend.

depicts added mass values on WECs 1 and 5 by the WECs 1,2,5 and 9. The normalized added mass of a single WEC operating in the region and positioned at x_1=x_2=0 (location of WEC 5), is also illustrated for comparison. As it can be seen in Figure 8b, the added mass induced on WEC 5 by the WECs 1 and 9 is almost identical due to symmetry on the x_1x_2 plane, with minor differences at the low frequencies range, triggered by bathymetric variations. Corresponding results as regards the hydrodynamic damping are shown in Figure 9. The non-dimensional Froude-Krylov and diffraction forces on WECs 1 and 5, (normalized by $\rho g\pi A\alpha^2$, where A is the incident wave amplitude at the incidence region), are depicted in Figure 10. Corresponding results regarding an isolated WEC operating in the region and positioned at x_1=x_2=0 (location of WEC 5), are also shown for comparison. WECs 1 and 5 Response Amplitude operators (RAOs) are shown in Figure 11, as evaluated by Eq. (9). The stiffness of the PTO system is set to 10% and thus the hydrostatic restoring coefficient is $C_{PTO} = 0.1\rho g\pi\alpha^2$, corresponding to magnitudes used in the literature (Anderlini et al. 2018) and the PTO damping is defined as $B_{PTO} = jB_{av.}^{(1\ WEC)}$ where $B_{av.}^{(1\ WEC)}$ is the mean hydrodynamic damping of a single WEC deployed in the examined area (see Figures 9a, b - gray line) and j is a multiplying factor. Thus, the PTO damping parameter equals $j2.046\ kNs/m$. Figure 12 illustrates the normalized power output by the devices 1 and 5 for various values of the factor j. The q factor of the WEC park, is estimated by the formula,

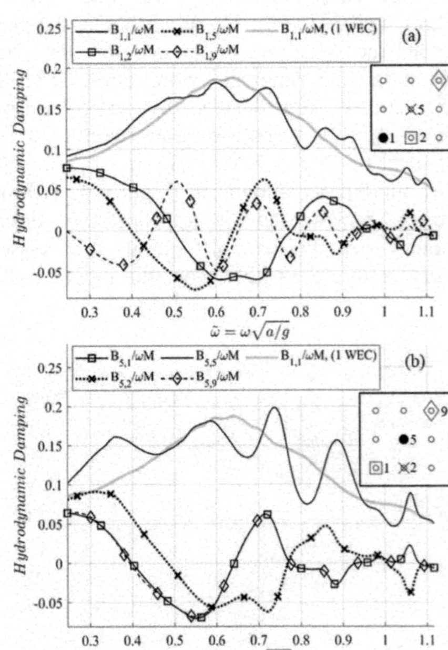

Figure 9. Normalized hydrodynamic damping contributions on WEC (a) 1 and (b) 5 by the WECs 1,2,5 and 9. The normalized damping of an isolated WEC deployed in the region and positioned at x_1=x_2=0 (location of WEC 5), is also plotted for comparison. The layout of the WECs is depicted in the auxiliary legend.

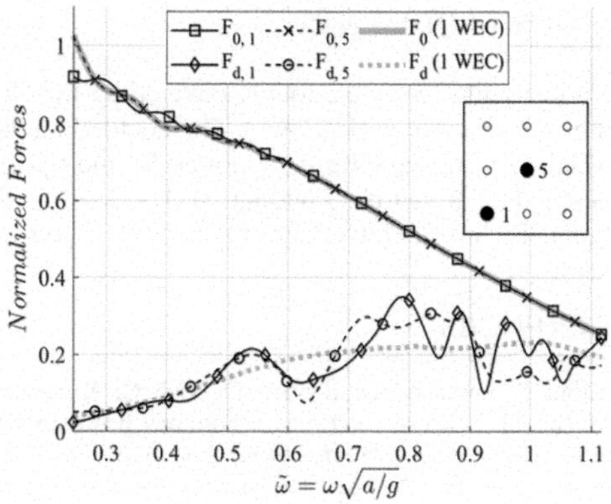

Figure 10. Normalized diffraction and Froude-Krylov forces on WEC (a) 1 and (b) 5. (Normal incidence, $\beta = 0°$) The normalized forces exerted on an isolated WEC operating in the region and positioned at x_1=x_2=0 (location of WEC 5), are also plotted for comparison. The layout of the WECs is depicted in the auxiliary legend.

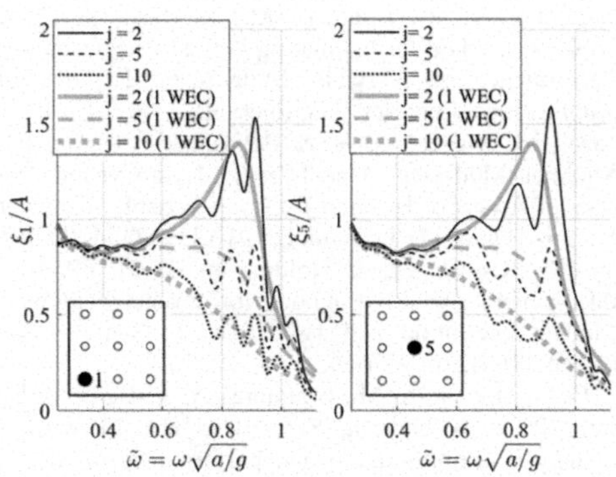

Figure 11. RAOs of WEC (a) 1 and (b) 5 (incidence normal to depth contours). The RAOs of an isolated WEC deployed in the region and positioned at x_1=x_2=0 (location of WEC 5), are also plotted for comparison. The layout of the WECs is depicted in the auxiliary legend. ($C_{PTO} = 0.1\rho g\pi\alpha^2$, $B_{PTO} = jB_{av.}^{(1\ WEC)}$).

$$q(\omega) = \left[\sum_{m=1}^{9} \bar{P}_m(\omega)\right]\left[9\bar{P}_{(1\ WEC)}(\omega)\right]^{-1} \quad (14)$$

and is shown in Figure 13 for various values of the multiplying factor j. The value j=21 was found to maximize the mean annual absorbed power. Assuming all incident energy to have direction normal to the bathymetry contours, the TMA model (Massel, 1989) is used to reconstruct wave spectra of the 10-year-long time series of sea states with 6h temporal resolution:

$$S_{TMA}(\omega; H_S, T_{-10}) = S_{JON}(\omega; H_S, T_{-10})f(\omega; h) \quad (15)$$

where $S_{JON}(\omega; H_S,\ T_{-10})$ is the JONSWAP spectrum, while $f(\omega;\ h)$ denotes the TMA filter function.

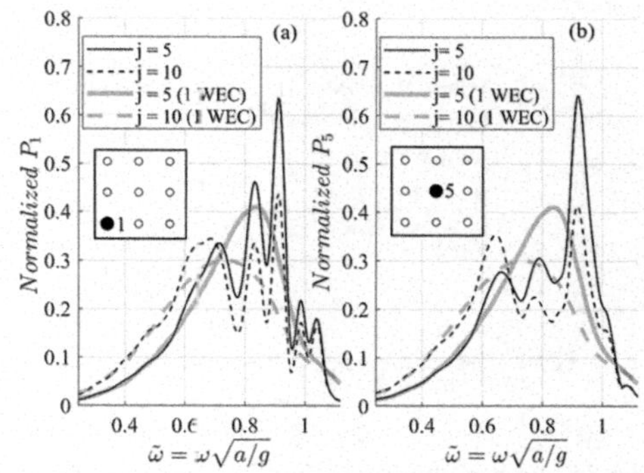

Figure 12. Normalized Power Output by the WECs (a) 1 and (b) 5 (incidence normal to depth contours). The Power Output of an isolated WEC deployed in the region and positioned at x_1=x_2=0 (location of WEC 5), are also shown for comparison. The layout of the WECs is depicted in the auxiliary legend. ($C_{PTO} = 0.1\rho g\pi\alpha^2$, $B_{PTO} = jB_{av.}^{(1\ WEC)}$).

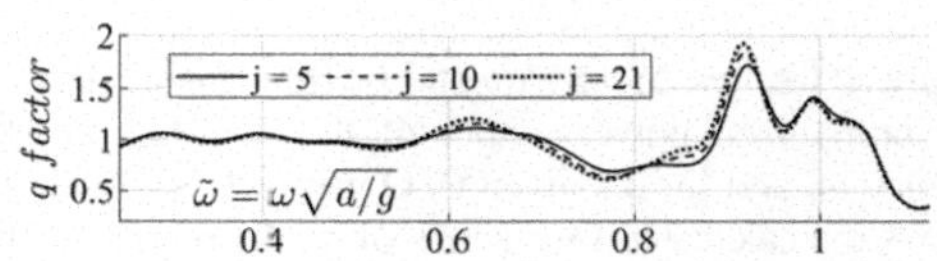

Figure 13. Q factor of the 3x3 WEC arrangement for different PTO damping values (incidence normal to depth contours, $C_{PTO} = 0.1\rho g\pi\alpha^2$, $B_{PTO} = jB_{av.}^{(1\ WEC)}$).

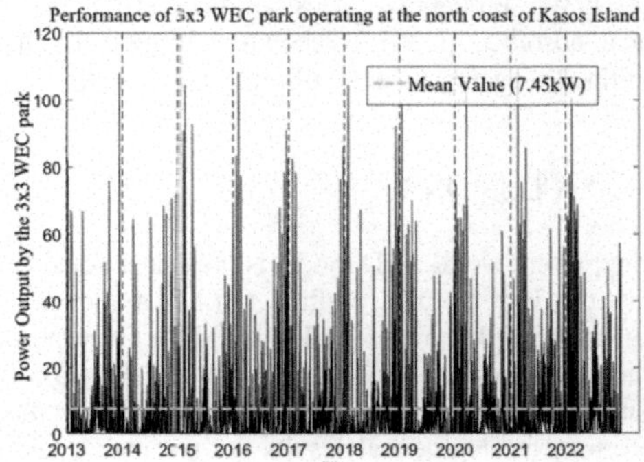

Figure 14. Time series of power output (kW) by the 3x3-WEC array in the 10-year-long period. The mean value is indicated by using a dashed line. ($C_{PTO} = 0.1\rho g\pi\alpha^2$, $B_{PTO} = 21B_{av.}^{(1\ WEC)}$).

Each reconstructed spectrum is discretized into a number of frequencies $\{\omega_i = i\Delta\omega,\ i = 1, 2, 3\ldots\}$, where $\Delta\omega$ is a constant spacing, and the output power by the 9-WEC array, for each spectrum $S(\omega_i)$, is estimated as follows:

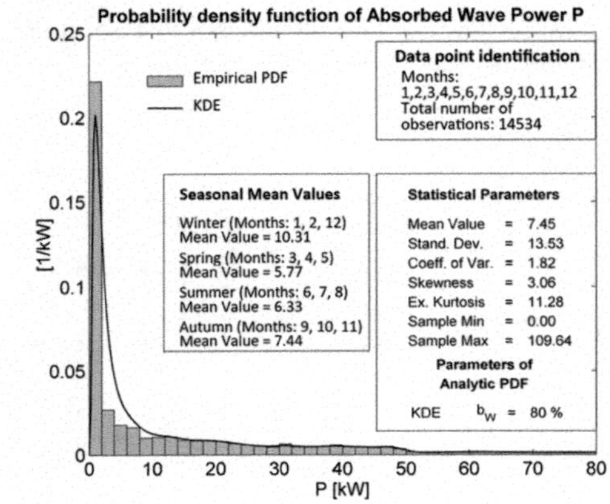

Figure 15. Probability density function of annual absorbed wave power by the 3x3-WEC array, and seasonal mean values ($C_{PTO} = 0.1\rho g\pi a^2$, $B_{PTO} = 21B_{av.}^{(1\ WEC)}$).

$$P = \sum_{m=1}^{9} \sum_{i} \rho g a\, C_g(h_{\max}, \omega_i) A^2(\omega_i) \bar{P}_m(\omega_i; B_{PTO}) \tag{16}$$

where $A^2(\omega_i) = 2S(\omega_i)\Delta\omega$ and $\bar{P}_m(\omega_i;\ B_{PTO})$ denotes the normalized power output of the m-WEC, at frequency ω_i and for the given PTO parameters. Figure 14 shows the calculated time series of power output by the examined 3x3-WEC arrangement at the north coast of Kasos Island for $j=21$, which was found to maximize the annual power absorption. The mean annual power absorbed by the arrangement is 7.45kW, leading to a total annual energy production of 65.3MWh. The power production ranges from 0kW to a maximum of 109.64kW. Probability density function (PDF) of annual absorbed power is shown in Figure 15, also reporting the seasonal values of mean power output.

4 CONCLUSIONS

In the present work, a 3D model is presented consisting of a boundary element method (BEM) model and a consistent coupled mode model (CCMM), is proposed, aiming to assess the performance of WEC arrays comprising multiple heaving devices, operating in nearshore regions. Numerical results are presented, taking into account various parameters on power performance, such as the interactions between multiple floaters, irregular seabed profiles as well as the parameters of the power take off (PTO) system. Moreover, a case study is performed that makes use of data by the proposed model combined with wave climate data at the north coast of Kasos Island, obtained by an offshore to nearshore transformation methodology. The results indicate a mean power production of 65.3 MWh for the optimum PTO damping parameter. The method can be used to optimize the WEC arrangement, based on local climatology, which will be addressed in future works.

ACKNOWLEDGMENTS

The registration fees were totally covered by the University of West Attica (UNIWA). The doctoral studies of Magkouris A. have been supported by the Special Account for Research Funding (E.L.K.E.) of the National Technical University of Athens (NTUA).

REFERENCES

Anderlini, E., Forehand, D., Bannon, E., Xiao, Q., & Abusara, M. (2018). Reactive control of a two-body point absorber using reinforcement learning. Ocean Eng, 650–658.

Athanassoulis, G. A., & Belibassakis, K. A. (1999). A coupled-mode model for the refraction – diffraction of linear waves over steep three-dimensional bathymetry. J. Fluid Mech (389), 275–301.

Babarit, A. (2017). Ocean Wave Energy Conversion, 1st ed. Amsterdam. The Netherlands: ISTE Press-Elsevier.

Belibassakis, K. A., Athanassoulis, G. A., & Gerostathis, T. P. (2001). A coupled-mode model for the refraction - diffraction of linear waves over steep three-dimensional bathymetry. Appl. Ocean Res., 319–336.

Belibassakis, K. A., Bonovas, M., & Rusu, E. (2018). A Novel Method for Estimating Wave Energy Converter Performance in Variable Bathymetry Regions and Applications. Energies (11). Energies.

Bonovas, M., Magkouris, A., & Belibassakis, K. A. (2022). Modified Mild-Slope Model for the Hydrodynamic Analysis of Arrays of Heaving WECs in Variable Bathymetry Regions. Fluids (7), https://doi.org/10.3390/fluids7060183.

Booij, N., Ris, R. C., & Holthuijsen, L. H. (1999). A third-generation wave model for coastal regions: 1. Model description and validation. J. Geophys. Res. Ocean. (104), 7649–7666.

Dee, D. P., Uppala, S. M., Simmons, A. J., Berrisford, P., Poli, P., & Kobayashi, S. (2011). The ERA-Interim reanalysis: Configuration and performance of the data assimilation system. Q. J. R. Meteorol. Soc., 553–597.

Gao, J., Zang, J., Chen, L., Chen, Q., Ding, H., & Liu, Y. (2019). On hydrodynamic characteristics of gap resonance between two fixed bodies in close proximity. Ocean Eng, 173, 28–44 https://doi.org/10.1016/j.oceaneng.2018.12.052.

GEBCO Compilation Group. (2023). GEBCO 2023. Grid. Retrieved from https://www.gebco.net/data_and_products/gridded_bathymetry_data/gebco_2023.

Guo, B., & Ringwood, J. V. (2021). A review of wave energy technology from a research and commercial perspective. 2021;15: IET Renew. Power Gener., 3065–3090 https://doi.org/10.1049/rpg2.12302.

Katz, J., & Plotkin, A. (2001). Low Speed Aerodynamics. San Francisco, CA, USA, McGraw-Hill.

Li, Y., & Yu, Y. H. (2012). A synthesis of numerical methods for modeling wave energy converter-point absorbers. Renew. Sust. Energ. Rev., 16(6), 4352–4364 https://doi.org/10.1016/j.rser.2011.11.008.

Massel, S. (1989). Hydrodynamics of Coastal Zones. Amsterdam, The Netherlands: Elsevier.

Newman, J. (1986). Distributions of sources and normal dipoles over a quadrilateral panel. J. Eng. Math., 113–126.

Stratigaki, V., Troch, P., Stallard, T., Forehand, D., Kofoed, J. P., Folley, M., … Kirkegaard, J. (2014). Wave Basin Experiments with Large Wave Energy Converter Arrays to Study Interactions between the Converters and Effects on Other Users in the Sea and the Coastal Area. Energies, 7, 701–734 doi:10.3390/en7020701.

Advances in Maritime Technology and Engineering – Guedes Soares & Santos (Eds)

Comparative study of 10MW wind turbines on different semi-submersibles

Y.J. Hong & J.X. Yue
School of Naval Architecture, Ocean and Energy Power Engineering, Wuhan University of Technology, Wuhan, Hubei, P.R. China

ABSTRACT: For Floating Offshore Wind Turbines (FOWTs), the structure layout of the platform plays an important role. In the paper, two different types of semi-submersibles for FOWTs were comparatively analyzed. Firstly, a frequency domain hydrodynamic analysis was conducted to obtain the Response Amplitude Operators (RAOs) of the FOWTs. Secondly, a fully coupled analysis in the time domain was performed under the combined wind, wave, and current loadings to investigate the influence of different water depths on hydrodynamic responses. Thirdly, wave load was determined using the Morrison equation and potential flow theory, and the structural and fatigue performance under this load was verified. The findings reveal that, among the two platforms with similar displacement, Type I exhibited better pitch performance but poorer heave performance compared to Type II. Additionally, Type I had inferior fatigue life due to stress concentration at the connection between the support and the column.

1 INTRODUCTION

In order to improve the structure energy system and reduce the proportion of fossil energy, China proposes "carbon peak" and "carbon neutrality" plan by 2030 and 2060. As a kind of clean energy, offshore wind power has the advantages of environmental friendly, cost-effectiveness and sustainability. With the gradual depletion of offshore wind resources, developing offshore wind power to the deep and distant sea is an inevitable trend. The fixed offshore wind turbine platform is usually chosen when the water depth is shallow (less than 30 m). But the cost of fixed offshore wind turbine platforms increases dramatically with water depth increasing (Chen, 2019).Therefore, FOWTs provide more suitable solution to develop the deep and distant sea wind energy.

The types of floating platforms are mainly composed of spar, tension leg platform (TLP), semi-submersible and barge (Campanile et al., 2018, L and V, 2016). Among the above platforms, semi-submersible is the most technologically advanced and commercially valuable, which can be constructed, assembled, and commissioned on land. Semi-submersible platforms are characterised by their large waterline area, which makes them self-floating and easy to tow at sea with good maneuverability. They can also maintain stability by generating a restoring moment through the waterline surface (Le et al., 2023).

Various design concepts of semi-submersible platforms for FOWTs have been proposed by domestic and foreign researchers. Roddier et al. (Roddier et al., 2010) presented a three-column semisubmersible FOWT WindFloat whose turbine is supported by a side column and braces are used to connect columns. In addition, several scientific research projects have proposed well-known concepts. One of these is the Offshore Code Comparison Collaboration Continuation (OC4) DeepCwind, which was proposed by Robertson et al (Robertson et al., 2014). It is a four-column semisubmersible FOWT whose center column bears a 5 MW wind turbine, and the columns are also connected by some braces (Bai et al., 2023). Luan et al. (2016) designed a 5 MW-CSC four-column unsupported semi-submersible FOWT. The center column and three side columns are connected by three pontoons. Song (2021) proposed a floating platform suitable for a water depth of 50 m. The platform consists of a center column that supports a 10 MW wind turbine. The center column is connected to three inclined side columns by pontoons. Wang (2014) proposed a 10 MW-CSC platform for a water depth of 200 m. The platform's basic layout is similar to the 5 MW-CSC but with modified dimensions to support a 10 MW wind turbine. Cao et al. (Cao et al., 2021; 2020) designed a four-column platform named SPIC. It was designed for 200 m water depths supporting a 10 MW wind turbine. The center column was connected to three partially tilted side columns by three pontoons and three braces.

The above newly designed platforms can support high-power turbines. However, there have been relatively few studies conducted on large turbine platforms for shallow and medium water depths. The current research on semi-submersible FOWTs mainly focuses on the conceptual design of new floating wind turbine platform, and comparative

DOI: 10.1201/9781003508779-48

performance analyses of existing semi-submersible platforms with different structural layouts.

Hydrodynamic and time-domain were analyzed by hydrodynamic software AQWA. The hydrodynamic characteristics were studied using numerical simulation in the frequency domain. A coupled numerical model of a wind turbine-tower-floating platform-mooring system was established to analyze two floating platforms in the time domain under the combined effects of wind, waves, and currents. The adaptability of the two floating platforms to different water depths was then analyzed to provide some reference for the design of floating platforms under different sea conditions. The wave loads on the platforms were calculated using Morrison's equation and potential flow theory (Yue et al., 2020). Finally, the structural strength and fatigue performance of both platforms were calculated under the wave loads by finite element software ANSYS in order to validate the applicability of the proposed platforms.

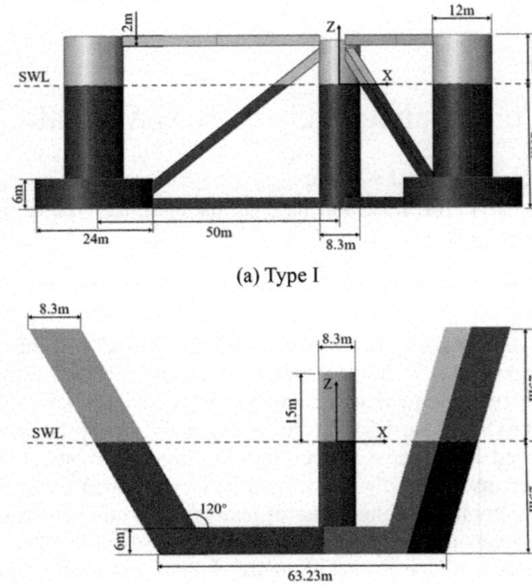

(a) Type I

(b) Type II

Figure 1. The semi-submersible platforms of the (a) Type I, (b) Type II.

2 10 MW SEMI-SUBMERSIBLE WIND TURBINE PLATFORM

2.1 *Parameters of semi-submersible wind turbine platforms*

The wind turbine and tower of the new semi-submersible floating wind turbine are chosen based on the DTU 10MW horizontal axis wind turbine design by the Danish Technical University (Bak et al., 2013) for efficient power generation under severe offshore conditions, with a rated wind speed of 11.4 m/s, a cut-in wind speed of 4.0m/s, and a cut-out wind speed of 25.0 m/s, and with variable speed pitch control. In this paper, two new kinds of semi-submersible platforms 10MW FOWT are designed based on existing platforms. The proposed platforms consist of a centre column, three side columns, and three pontoons, as shown in Figures 1 and 2. The centre column connects the upper tower and unit, and the side columns are loaded with ballast water to adjust the platform draft. The operating water depth of the FOWT is 200 m. Table 1 displays the main parameters of the two proposed semi-submersible platforms, which are determined based on the design sea state and the wind turbine parameters.

Table 1. Main parameters of the semi-submersible platforms.

Parameter	Type I	Type II
Center of gravity below SWL (m)	-8.8399	-7.9356
Depth of platform base below SWL (m)	25.0	25.0
Displacement (m^3)	13900	14400
Platform steel mass (t)	5861	6017
Columns/pontoon wall thickness (m)	0.04	0.04
Water depth (m)	200	200
Platform steel density (kg/m^3)	7850	7850
Ballast water density (kg/m^3)	1025	1025

2.2 *Mooring system*

To investigate the performance difference caused by the platforms alone, it is necessary to eliminate the influence of the mooring system. Therefore, the line and properties of the mooring systems are kept same for the two platforms. The mooring system is symmetrically arranged to the X-axis and divided into three groups. Each group has an angle of 120 deg between each two catenary lines, and each has a pretension setting of 2750 kN. Catenary with a nominal diameter of 142 mm were selected (Liu, 2018). Figure 2 shows a schematic diagram of the mooring system arrangement, with wind and wave currents in the positive direction of the X-axis. Table 2 displays the parameters of the catenary.

3 COMPARISON OF HYDRODYNAMIC PERFORMANCE

3.1 *Environmental conditions*

The amplitude-frequency response analysis is used to obtain the RAO of the two platforms under the action of regular waves, based on which the motion response and mooring tension of the two platforms coupled with the action of wind and wave currents are calculated under the time-domain course. The calculation conditions include the working condition and survival condition, with constant wind and uniform flow, and the unified selection of the

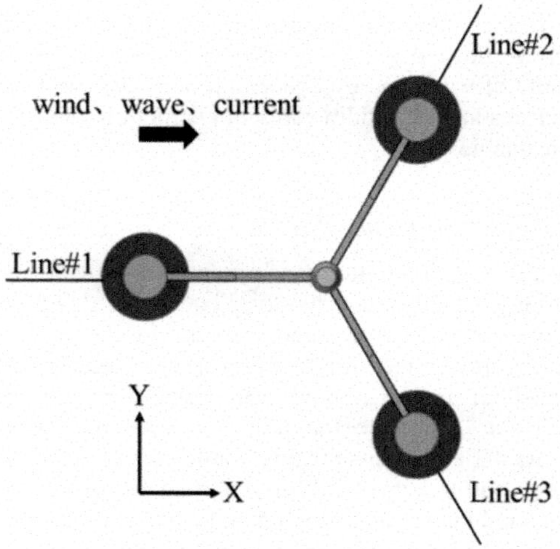

(a) Type I

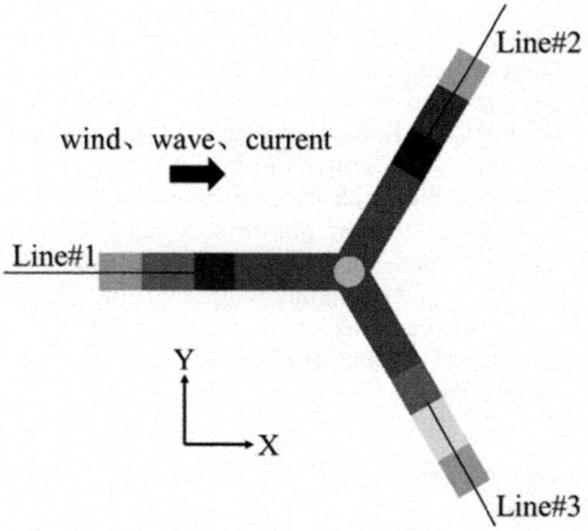

(b) Type II

Figure 2. The layout of the mooring systems of the (a) Type I, (b) Type II platforms.

Table 2. The main parameters of the mooring system.

Parameter	Value
Number of mooring lines	3
Angle between each Group of lines (deg)	120
Nominal diameter (m)	0.142
Mooring radius (m)	870
Equivalent axial stiffness (kN)	3413694.9
Break load (kN)	21061.0
Pretension (kN)	2750

JOWSWAP spectrum for the wave loads. The general sea state in the South China Sea is used for the working conditions. The sea state of one in fifty years is used as the survival condition. The marine environmental parameters are shown in Table 3.

Table 3. Marine environmental parameters.

Parameter	Working Condition	Survival Condition
H_s (m)	5.49	7.50
T_p (s)	11.3	13.0
Gamma	3.3	3.3
U_w (m/s)	11.4	40.0
U_c (m/s)	0.68	2.00

In the table, H_s is the significant wave height; T_p is the spectral peak period; *Gamma* is the spectral peak factor; U_w is the wind speed; U_c is the current speed.

3.2 *RAOs in frequency domain*

The aim of the hydrodynamic calculations in the frequency domain is to analyze the motion response of the new semi-submersible wind turbine under the influence of waves with different periods. This analysis will provide hydrodynamic parameters reflecting the turbine's water performance. In the design process, the natural frequency of the platform should avoid the periodic range of wave energy concentration, and resonance will be avoided. The wave angle calculation range for the two symmetric platforms is 0-180 deg with an interval step of 30 deg, and the wave period calculation range is 2-60 s. The frequency domain calculation has no mooring catenary constraints, and the viscous damping of the structure is set to 10% of the critical damping, following the recommendations of Roddier et al. (2011). This correction is made to the calculation results.

Response amplitude operators (RAOs) are functions that describe the relationship between the incident wave excitation and the motion response of the floating platform. The RAO curves for the floating platform in the three degrees of freedom (surge, heave, and pitch) are shown in Figure 3. The figure shows that the longitudinal surge RAO curve of the floating platform generally decreases as the incident wave frequency increases. Platform I has a relatively large surge RAO at the same frequency, as the catenary constraint is not considered. The surge RAO is at its maximum when the wave direction angle is 0 deg, as it is co-linear with the direction of the surge. Conversely, when waves are incident at 90 deg, the surge RAO is close to zero. Waves incident in the vertical direction do not contribute to the surge response. The heave RAO curves for platforms I and II peaked at 26 s and 22 s, respectively. The peak heave RAO for platform I is 1.24 m, approximately 1.13 times that of platform II. The heave RAOs of the floating platform are essentially the same for different wave incidence angles. In terms of pitch direction, both platform I and platform II exhibit a clear bimodal phenomenon, with peaks at 25 s and 17 s, respectively. The two peaks of platform I similar, while the second peak of platform II is approximately 60% of the first peak. The pitch RAO of platform II is greater than that of platform I. At the first peak, the peak of platform II is approximately twice as prominent as that of platform I.

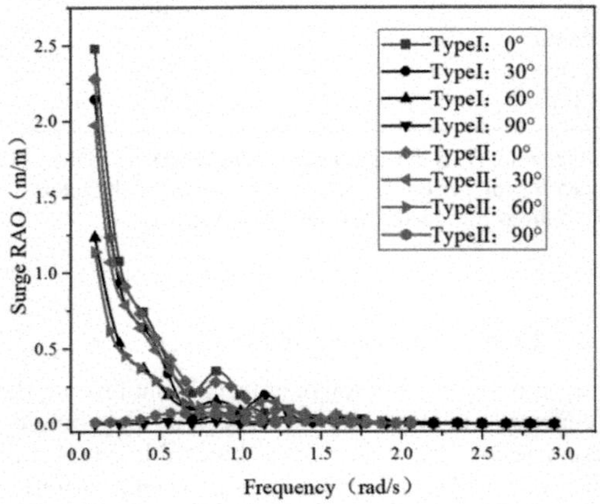

(a) Surge

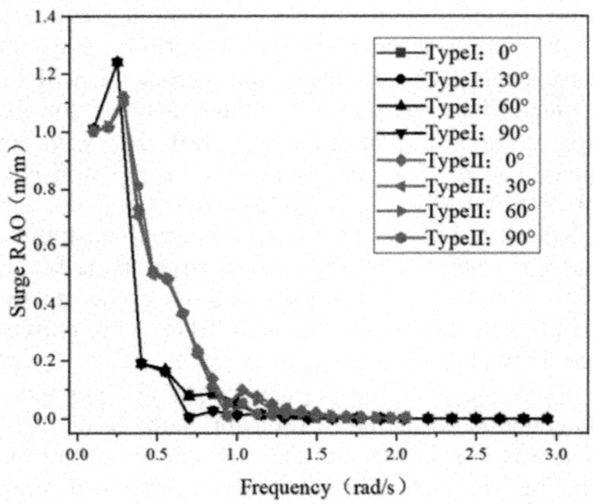

(b) Heave

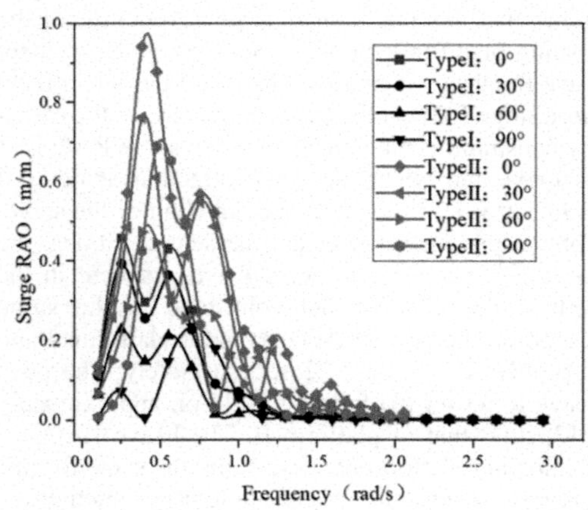

(c) Pitch

Figure 3. Motion response amplitude operators of the two platforms.

In summary, the intrinsic period of the two platforms' three degrees of freedom avoids the period range of wave energy concentration, which can prevent resonance. Additionally, the structural design is reasonable.

3.3 *Motions in time domain*

To analyze the motion performance of the two platforms under different working condition, this paper applies wind, waves, and currents in the positive X direction for a simulation duration of 2000 s. In statistical analysis, remove the data from the first 200 s to eliminate the impact caused by startup. Table 4 shows the statistical values of the two platforms' surge, pitch, and surge motion response under the combined action of wind and wave currents. Figures 4 and 5 show the time-range curves of each response parameter of the two platforms under different working condition.

It can be seen that:

(1) Working Condition. In terms of surge time domain response, the peak values of the two platforms are similar, but the time history curve of platform II fluctuates greatly and has poor stability. In terms of heave time domain response, the peak value of platform I is about 1.86 times that of platform II, and the mean value of platform I is much larger than that of platform II. In terms of pitch time domain response, the peak value of platform II is 3.16 times that of platform I, but their mean values are similar.

Table 4. Statistical results of the platform motions in different cases.

		Working Condition		Survival Condition	
Objective	Item	Maximum	Minimum	Maximum	Minimum
Type I	Surge (m)	5.421	0.664	7.089	3.373
	Heave (m)	-3.935	-5.454	-4.157	-6.242
	Pitch (deg)	0.898	-1.094	0.784	-3.861
Type II	Surge (m)	5.331	0.047	22.510	15.783
	Heave (m)	2.545	-2.934	1.995	-2.663
	Pitch (deg)	3.456	-3.031	4.857	-3.862

(2) Survival Condition. Due to the spectral peak frequency being closer to the natural frequency of the platform's surge compared to the working condition, the surge response of the two

platforms significantly increases. For FOWTs, it is usually required that the peak pitch motion during extreme operating conditions does not exceed 15 deg, while the peak pitch motion of platform I and platform II are 3.861 deg and 4.857 deg, respectively. Therefore, under extreme operating conditions, the pitch motion of the two platforms meets the requirements of the specifications (Dnv, 2018).

3.4 *Comparison mooring system tensions*

The mooring system of floating offshore wind power can limit the motion response of the wind turbine platform, ensure the normal operation of the wind turbine. Therefore, studying the dynamic response of mooring systems is of great significance. This paper aims to study the dynamic response of mooring systems on two different platforms under different sea conditions, and verify the safety of catenary according to standards (Ccs, 2020a, 2020b, 2021). Figure 6 shows the time

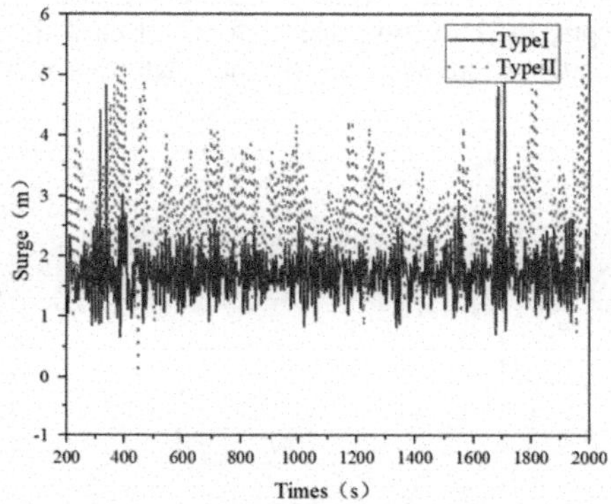

(a) Time history curve of surge under working condition

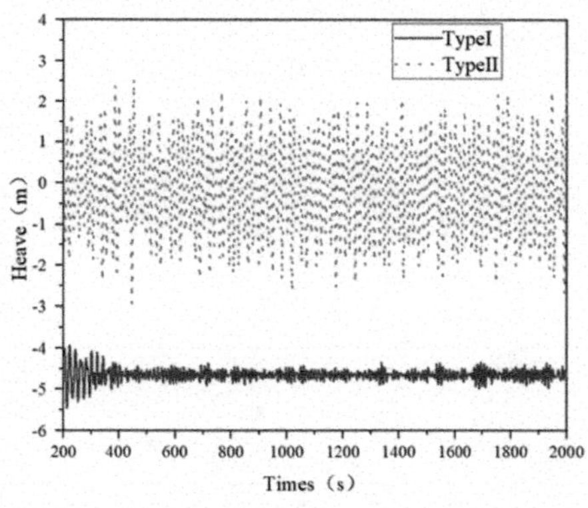

(b) Time history curve of heave under working condition

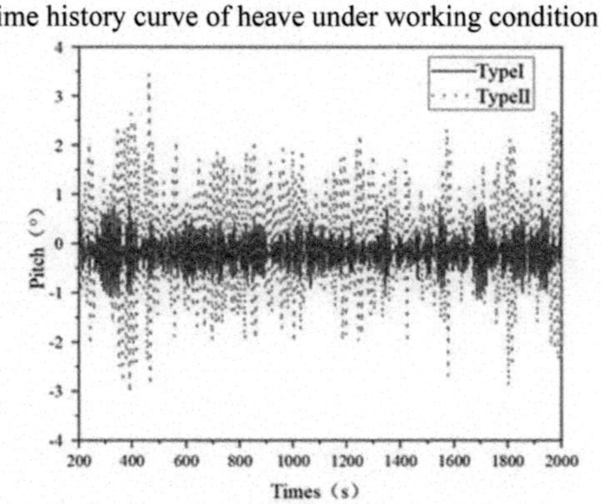

(c) Time history curve of pitch under working condition

Figure 4. Time history curves of dynamic response parameters for two platforms under working condition.

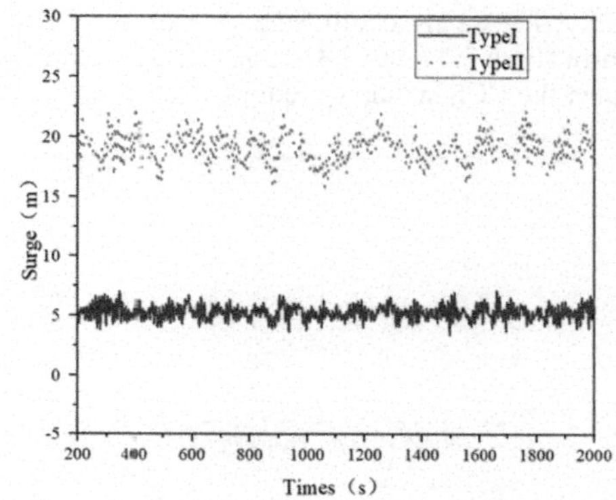

(a) Time history curve of surge under survival condition

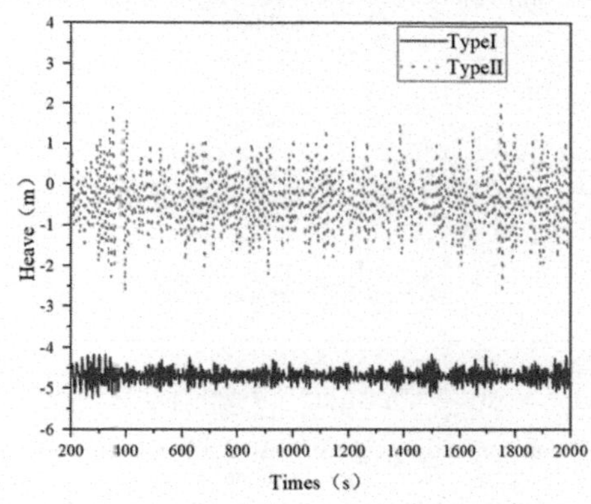

(b) Time history curve of heave under survival condition

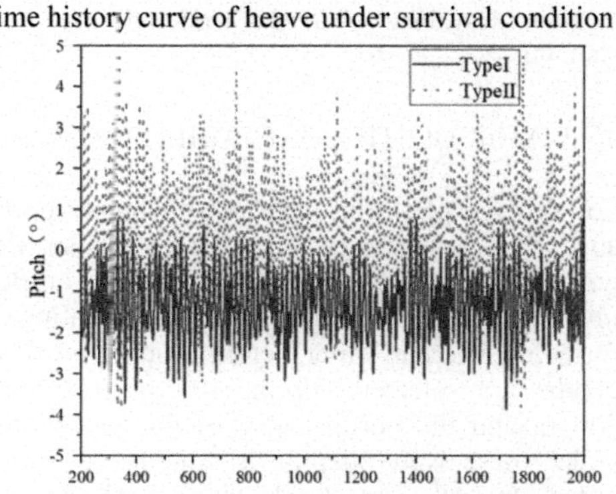

(c) Time history curve of pitch under survival condition

Figure 5. Time history curves of dynamic response parameters for two platforms under survival condition.

history curves of the No.1 catenary with the highest stress on both platforms under working and survival conditions.

Under working condition, the pulling force of the No. 1 catenary of platform II is 4684 kN, which is about 1.4 times that of platform I. Platform I and II's safety coefficients are 6.33 and 4.50, respectively, which meet the CCS standards of 2.25.

Under the survival condition, the pulling force of the No. 1 catenary of platform II is 9127 kN, which is about 1.53 times that of platform I. The safety coefficients of platform II are 3.52 and 2.31, respectively. The safety coefficients of platform I and platform II are 3.52 and 2.31 respectively, both of which meet the CCS standards requirement of 1.67.

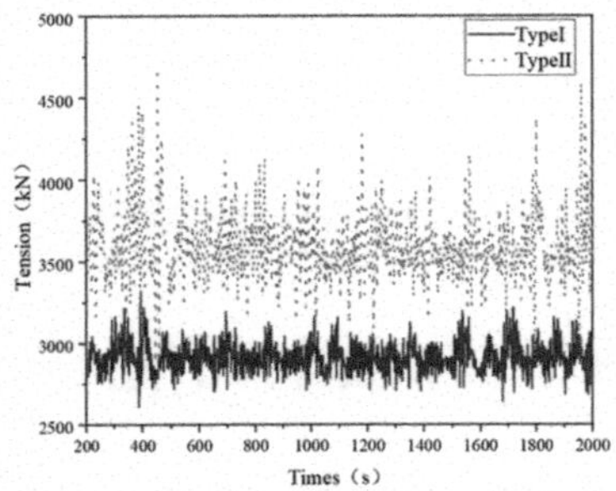

(a) Time history curve of tension of line#1 under working condition

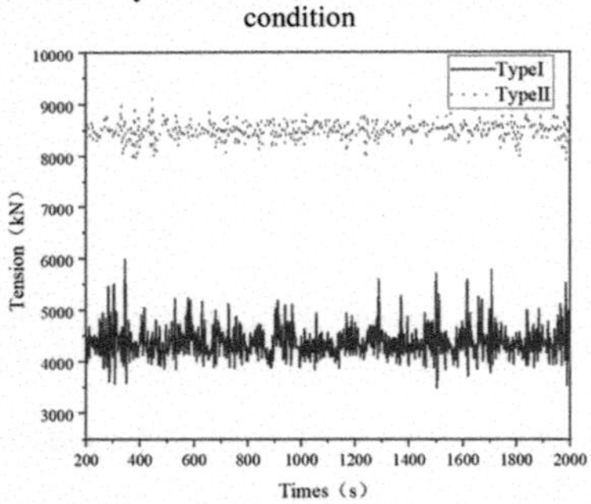

(b) Time history curve of tension of line#1 under survival condition

Figure 6. Time history curves of tension for Line 1 on two platforms.

4 WATER DEPTH ADAPTABILITY ANALYSIS

For floating platforms, the hydrodynamic performance will be different at different water depths with different wet surface shapes. In order to investigate the adaptability of the floating platform to different water depths, four typical water depths are selected: 50 m, 100 m, 150 m, and 200 m and the performance of the two platforms under these water depths is calculated and compared. In order to simplify the analysis method and draw effective conclusions, the length of the catenary is adjusted accordingly to ensure that the top preload is the same, which is 2750 kN, and the rest of the mooring parameters and load parameters are kept the same as those of the time domain course analysis. The water depth adaptability analysis only calculates the working condition to find out the law of response of the two platforms with the change of water depth. Then, it derives the different water depth adaptability of the two platforms.

The changes in RAO for different water depths of platform I and platform II are shown in Figures 7 and 8, respectively. In shallow water, the draft-to-depth ratio is large, and all three degrees of freedom show obvious amplification and are easily affected by the change of water depth. With the increase of water depth, the reduction rate of RAO of each degree of freedom decreases gradually, and stabilizes at 200 m.

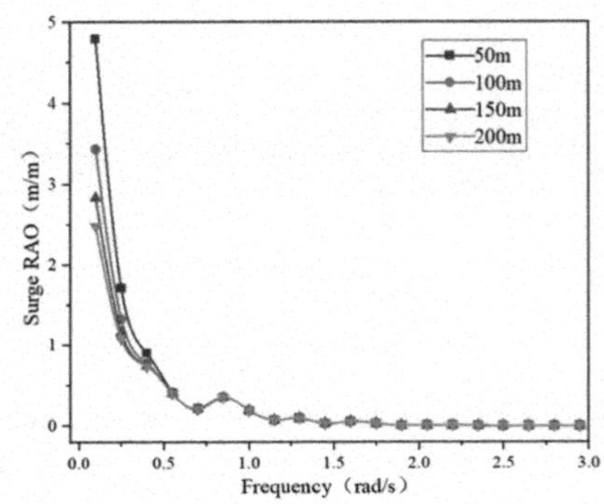

(a) Surge

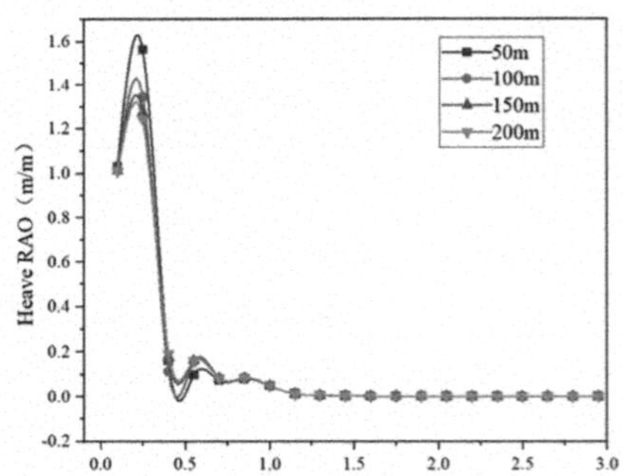

(b) Heave

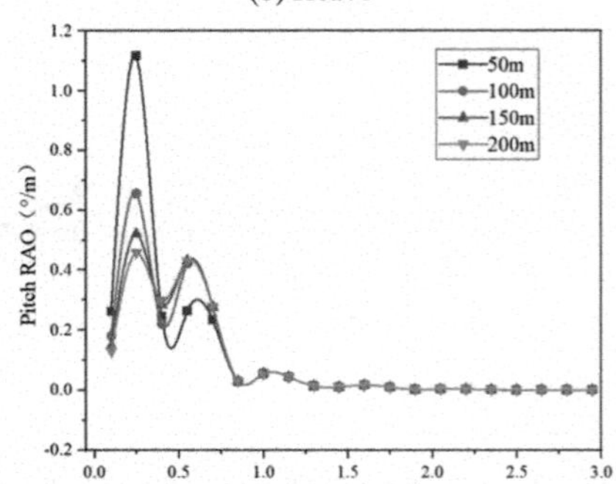

(c) Pitch

Figure 7. Motion response amplitude operator of Type I platform under different water depths.

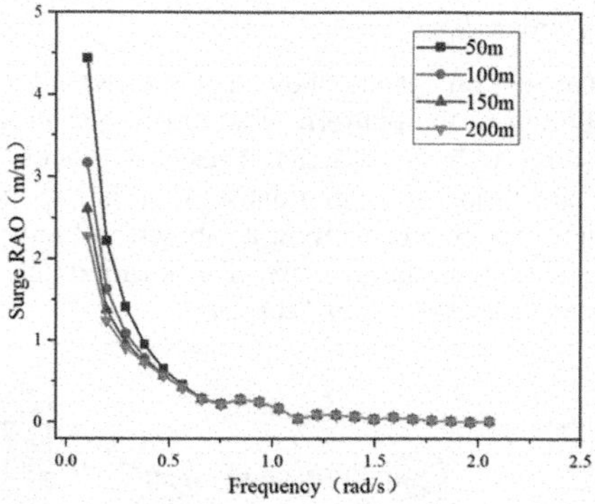

(a) Surge

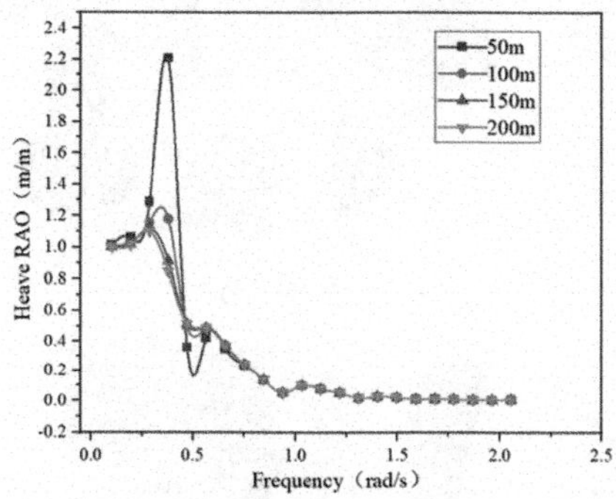

(b) Heave

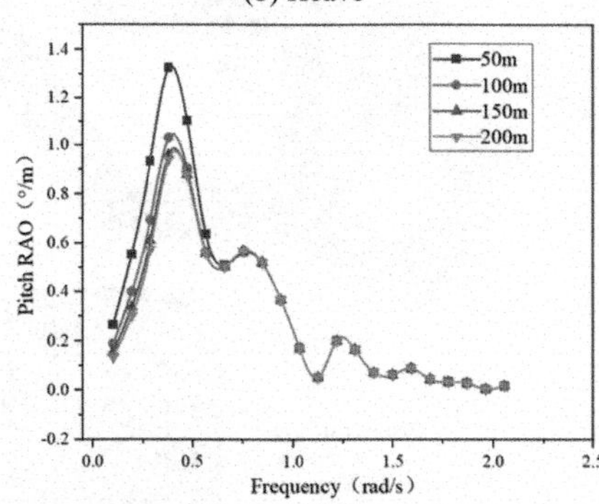

(c) Pitch

Figure 8. Motion response amplitude operator of Type II platform under different water depths.

The RAO change rules of the two platforms at different water depths are calculated and used to compare the time domain behavior at different water depths. Only the results of the 3 degrees of freedom responses, such as surge, heave, and pitch, are listed and shown in bar charts in Figure 9. As shown in Figure 9, with the increase of water depth, the recovery stiffness of the catenary decreases, leading to a decrease in the platform's restoring force and an increase in horizontal displacement. The surge response of both platforms is similar from 50 m to 200 m. In terms of heave response, the heave response of both platforms increases with the increase of water depth, but the rate of increase is relatively small. The heave response of platform I is about 1.3 times that of platform II for the same water depth. In terms of pitch response, platform I does not change much with increasing water depth, and platform II decreases with increasing water depth. The pitch response of platform II is much larger than that of platform I. In summary, platform I has good adaptability in water depths of 50 m-100 m, but its heave performance is poor. Platform II has good adaptability at depths of 100 m-150 m, but its disadvantage is poor pitch stability and susceptibility to environmental loads.

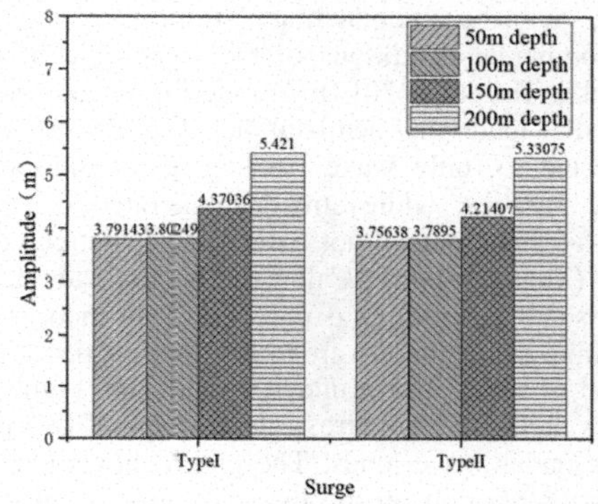

(a) Surge

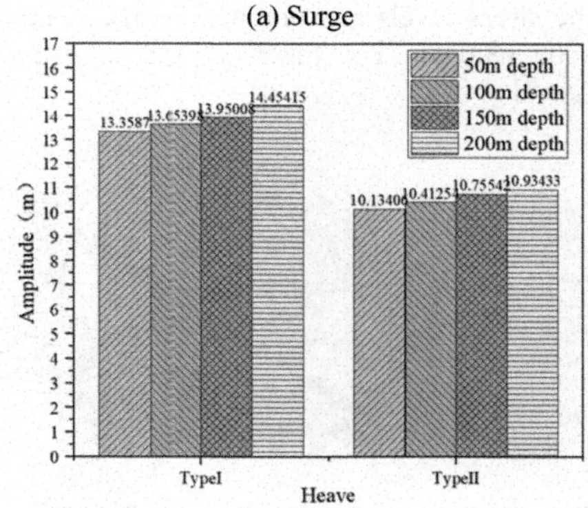

(b) Heave

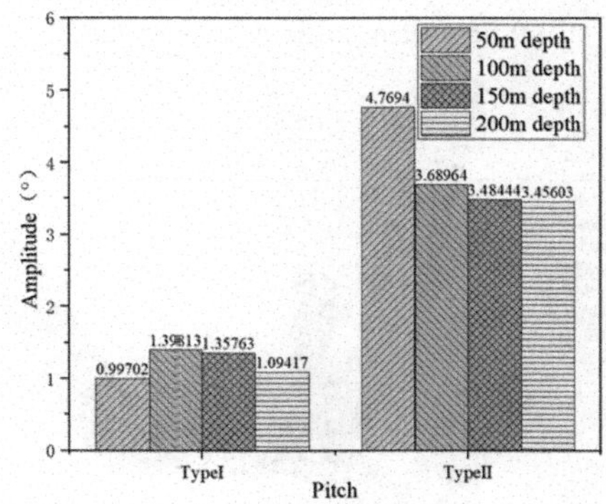

(c) Pitch

Figure 9. Time domain statistical analysis of two platforms with different water depths.

5 COMPARISON OF STRUCTURE STRENGTH

5.1 *Structural stress*

The overall stress response of the two platforms using the finite element method. Using shell 181 elements for grid partitioning, the number of elements for platform I and platform II are 336409 and 388351, respectively. In addition, when conducting strength analysis of floating platform structures, at least six degrees of freedom of the platform need to be constrained to avoid the rigid motion of the floating platform. In the lifecycle of the platform, the survival time of the working condition is extremely short, so when conducting fatigue analysis, the environmental conditions of the working condition are taken (Li et al., 2012). Among the various environmental loads that semi-submersible platforms are subjected to, only wave loads generate significant cyclic stresses, while stresses generated by other loads are temporarily not considered in fatigue analysis (Zhang, 2006). Calculate wave loads using Morrison's equation and potential flow theory, and use finite element software to obtain the stress distribution of the platform under wave loads. Figure 10 shows the equivalent stress cloud maps of two platforms under wave loads. The graph shows that the stress distribution on the two platforms is uniform, and the stress levels are similar. But the maximum stress on platform I is much greater than that on platform II.

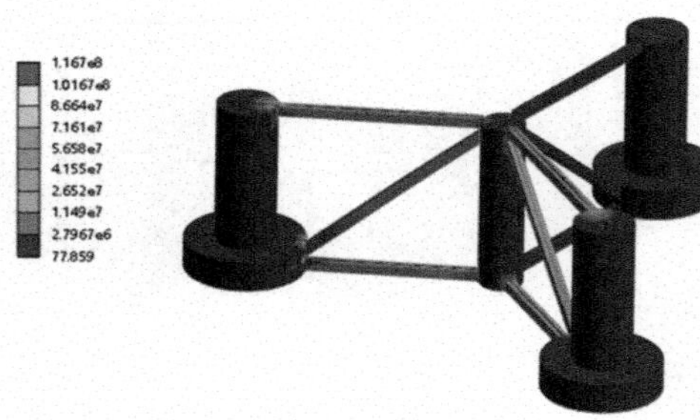

(a) Equivalent stress cloud of Type I

(b) Equivalent stress cloud of Type II

Figure 10. Equivalent stress cloud of two platforms.

5.2 *Fatigue life*

The calculation results show that the overall stress distribution of the platform is uniform, but there are also some high stress areas. These areas with high stress are prone to fatigue damage, so hot spots are selected in high stress areas, as shown in Figure 11, to calculate their fatigue life. The locations of high stress areas are shown in Table 5.

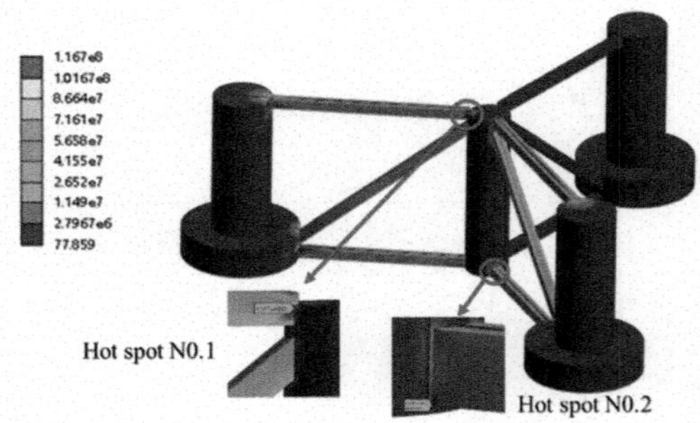

(a) Hot spots of type I

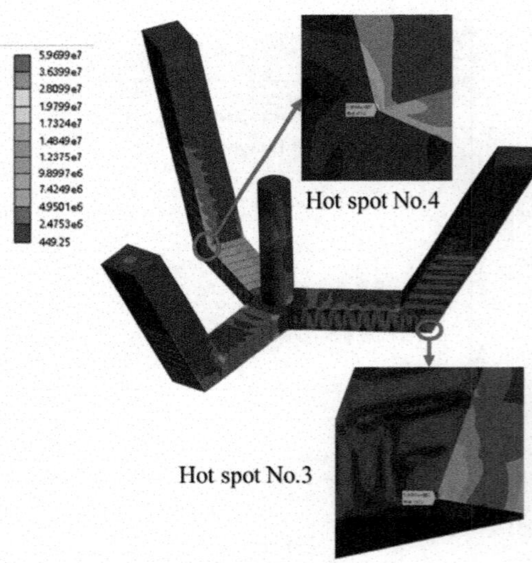

(b) Hot spots of type II

Figure 11. The location of hot spots on the two platforms.

Table 5. Location of high stress areas.

	Hot spot No.	Location
Type I	1	Connection between the column and the horizontal upper brace
	2	Connection between the column and the horizontal lower brace
Type II	3	Connection between the side column and the pontoon
	4	Connection between the side column and the pontoon

The hot spot stress results in each high stress area are calculated and shown in Table 6. The hot spot stress of platform I is generally higher than that of platform II.

Table 6. Hot spot stress.

	Hot spot No.	Stress (MPa)
Type I	1	116.7
	2	66.7
Type II	3	59.7
	4	39.1

The material used for the hot spots of the platforms is EH-36 high-strength steel, with a yield strength of 360 MPa and an elastic modulus of 206 GPa. Except for hotspot 1 which is in the air, the other three hotspots are all in seawater. Therefore, according to ABS specifications, the ABS-E (CP) curves of the S-N curves of non-tubular joint under cathodic protection in air and seawater are selected for fatigue analysis of the hot spots. The parameters of the S-N curve are shown in Table 7.

Table 7. S-N curve parameters.

S-N	A	m	C	r	N_Q	S_Q (MPa)
ABS-E(A)	1.04e12	3.0	2.30e15	5.0	1.01e7	47.0
ABS-E(CP)	4.16e11	3.0	2.30e15	5.0	1.01e6	74.4

The fatigue life of each fatigue hot spot is calculated based on the S-N curve, as shown in Table 8. The table shows that the stress of both hot spots on platform I is greater than that on platform II, with hot spot N0.1 having the highest stress, resulting in the lowest fatigue life.

Table 8. Fatigue damage and fatigue life.

	Hot spot No.	Number of cycles
Type I	1	6.54E+05
	2	1.74E+06
Type II	3	3.03E+06
	4	2.52E+07

6 CONCLUSIONS

This paper Studied the hydrodynamic performance of two semi-submersible floating platforms in the frequency domain using numerical simulation methods, considering wind load, wave load, ocean current load, and the coupling between the platform and mooring system. The coupled dynamic response characteristics of the two semi-submersible platforms under different water depths were calculated in the time domain, and the hydrodynamic performance of the two platforms was compared and analyzed. The Morrison equation and potential flow theory were used to calculate wave loads and analyze the structural strength of two platforms under wave loads. The stress obtained is then substituted into the S-N curve to analyze the fatigue performance of the two platforms. The main conclusions obtained were as follows:

(1) The natural periods of the two semi-submersible platforms can effectively avoid the periodic range of wave energy density and resonance. Therefore, the structural design of the two platforms is reasonable.
(2) The time-domain motion performance of platform I is better than platform II. The heave and pitch motion peaks of platform I are smaller than those of platform II, but the heave performance of platform I is relatively poor.
(3) The mooring performance of the two platforms meets the regulatory requirements in both working and survival conditions, with platform I having relatively better mooring performance.
(4) The heave and pitch RAO of platform I don't vary much with water depth, thus having better adaptability in shallow water. Platform II is more adaptable in moderate water depths of 100 m-150 m.
(5) The stress levels and distributions of the two platforms are similar, and their strength satisfies the specification requirements. The hot spot stress of platform I is greater than that of platform II, and the fatigue life is relatively small due to different stress concentration between the columns and supports.

REFERENCES

Bai, H., Zhang, M., Yuan, W. & Xu, K. (2023) Conceptual design, parameter optimization and performance investigation of a 10MW semi-submersible floating wind turbine in shallow water. *Ocean Engineering*, 281.

Bak, C., F,Z., Bitsche, R., Kim, T., Yde, A., Henriksen, L. C., Hansen, M. H., Blasques, J. P. A. A., Gaunaa, M. & Natarajan, A. (2013) The DTU 10-MW reference wind turbine. *In*: *Danish Wind Power Research*.

Campanile, A., Piscopo, V. & Scamardella, A. (2018) Mooring design and selection for floating offshore wind turbines on intermediate and deep water depths. *Ocean Engineering*, 148, 349–360.

Cao, Q., Xiao, L., Cheng, Z. & Liu, M. (2021) Dynamic responses of a 10 MW semi-submersible wind turbine at an intermediate water depth: A comprehensive numerical and experimental comparison. *Ocean Engineering*, 232.

Cao, Q., Xiao, L., Guo, X. & Liu, M. (2020) Second-order responses of a conceptual semi-submersible 10 MW wind turbine using full quadratic transfer functions. *Renewable Energy*, 153, 653–668.

Ccs (2020a) Classification Specification for Offshore Floating Units. Beijing, China Communications Press.

Ccs (2020b) Classification Specification for Offshore Mobile Platforms. Beijing, China Communications Press.

Ccs (2021) Guidelines for offshore floating wind turbine platforms. Beijing, China Communications Press.

Chen, J. (2019) Study on the rigid-flexible coupled multibody dynamics and the characteristics of floating offshore wind turbines. Shanghai, Shanghai Jiao Tong University.

Dnv (2018) Floating wind turbine structures: DNVGL-ST-0019. Det Norsk Veritas.

L,C.-S. & V,D.-C. (2016) *Floating offshore wind farms*, Springer International Publishing.

Le, C., Kuo, L., Puyang, Z. & Hong-Yan, D. (2023) Dynamic response characteristics of a new semi-submersible floating wind turbine in different current velocity conditions. *Journal of Ship Mechanics*, 27.

Li, Z., Li, L. & Chen, M. (2012) Fatigue Life Analysis of Deep Water Semi submersible Platforms Based on S-N Curve. *OCEAN TECHNOLOGY*, 31.

Liu, Z. (2018) Study on Hydrodynamic and Tensioning Mooring of Typical Working Condition of a Deepwater Semi-submersible Plateform.

Luan, C., Gao, Z. & Moan, T. (2016) Design and analysis of a braceless steel 5-mw semisubmersible wind turbine. Int. Conf. Offshore Mech. Arctic Eng.

Robertson, A., Jonkman, J. & Masciola, M. (2014) Definition of the Semisubmersible Floating System for Phase II of OC4. National Renewable Energy Lab.

Roddier, D., Cermelli, C. & Aubault, A. (2010) WindFloat: a floating foundation for offshore wind turbines. *J. Renew. Sustain. Energy*.

Roddier, D., Peiffer, A. & Aubault, A. (2011) A generic 5MW WindFloat for numerical tool validation & comparison against a generic Spar. *Proceedings of the ASME 2011 30th International Conference on Ocean, Offshore and Arctic Engineering*. Rotterdam.

Song, Z. (2021) Design and stability analysis of semi-submersible floating platform for large offshore wind turbines. *periodical of Ocean University of China*, 51.

Wang, Q. (2014) Design of a Steel Pontoon-type Semi-submersible Floater Supporting the Dtu 10mw Reference turbine.

Yue, J., Kang, W., Mao, W., Chen, P. & Wang, X. (2020) Prediction of dynamic responses of FSRU-LNGC side-by-side mooring system. *Ocean Engineering*, 195.

Zhang, J. (2006) Fatigue reliability analysis of typical joints in a semi-submersible drilling ship. *SHIP ENGINEERING*, 28.

Advances in Maritime Technology and Engineering – Guedes Soares & Santos (Eds)

Research on mooring line design for a 12MW floating offshore wind turbine

D. Jeong, K. Kim & C. Shim
Department of Naval Architecture and Ocean Engineering, Mokpo National University, Korea

M.S. Kim
SURF R&D Center, Mokpo National University, Korea

ABSTRACT: The surge in demand for sustainable energy solutions has placed Floating Offshore Wind Turbines (FOWTs) in the spotlight. The quest for improved stability and cost effectiveness in FOWTs is paramount. The mooring system is a critical determinant of FOWTs' operational performance and safety. Essential parameters for the engineering of mooring lines encompass their weight, length, diameter, stiffness, and type of the mooring line. This research was conducted with a comprehensive evaluation of these parameters, examining their synergistic effects on mooring systems. The focus of the investigation was the mooring design for a high capacity, 12 MW FOWT positioned in the environmental conditions of the southwestern sea of South Korea, contending with wind, wave, and current forces, as well as extreme weather scenarios.

1 INTRODUCTION

1.1 *Background of study*

The Floating Offshore Wind Turbine (FOWT) is a renewable energy technology currently under development, and research and development are being conducted worldwide to harness the potential of offshore wind power generation. With the continuous growth of offshore wind power generation, there is increasing interest in mooring lines for maintaining the position of floating offshore wind turbines.

Huang and Yang (2021) presents the modeling and optimization of mooring lines for Floating Offshore Wind Turbines (FOWT) located at various water depths ranging from 50m to 100m. Environmental conditions including Ultimate Limit State (ULS), Fatigue Limit State (FLS), and Maximum Operating Sea State (MOSS) were taken into account. The research identified the parameters that determine the cost of mooring lines depending on water depth. Liu and Yu (2022) conducted a study on the dynamic behavior of FOWT considering aerodynamic loads, wave loads, current loads, and mooring loads. Their results reported that the diameter has little influence on the static shape of the mooring line, and that the stiffness of the mooring system can be effectively increased by reducing the length of the mooring line and increasing its diameter. Eghbali et al. (2018) evaluated the reliability of Single Point Mooring (SPM) under various environmental conditions. They employed the First Order Reliability method (FORM) to calculate reliability, and used the Most Probable failure Point (MPP) method to locate the target point in the function, confirming that the mooring line maintains safe performance. Niranjan and Ramisetti (2023) investigated the dynamic behavior of a large FOWT exposed to extreme loads and mooring line failures. They reported increased displacements in surge and sway directions upon mooring line failure, compared to the intact state, accompanied by changes in yaw angle and increased tension in the remaining undamaged mooring lines. Bae et al. (2017) conducted a series of numerical simulations on the performance changes of a Floating Offshore Wind Turbine (FOWT) with a broken mooring line. They reported that the occurrence leads to changes in the tension of the remaining mooring lines and yaw changes, negatively impacting power production and structural fatigue life. Benassai (2015) analyzed the number of mooring lines to minimize the total mass for the optimal design of a mooring system.

Kang and Lee (2018) conducted a study on the optimization of design parameters for mooring systems through sampling techniques. Scenarios for mooring system parameters were created using the Latin Hypercube Sampling method, a probabilistic approach.

Ferreira et al. (2017) conducted a study using Design of Experiments (DOE) to assess the varying

DOI: 10.1201/9781003508779-49

impacts of the components of a mooring system on the system itself. They analyzed the system by statically adjusting the design variables to optimally fit a taut mooring system to a full depth mooring system

1.2 *Research objective*

The objective of this study is to perform the optimal design of mooring lines for a 12MW floating offshore wind turbine (FOWT). To achieve this, the OrcaFlex program was used to create a 12MW FOWT model and input the offshore conditions in the southwestern sea. The goal is to improve the position maintenance and mooring stability of the FOWT by applying mooring lines. The study analyzes the tension in the mooring lines and the dynamic response of the FOWT when varying the number of mooring lines connected to it.

2 LOADS AND STRUCTURES

2.1 *Environmental load*

In this study, one-year environmental load and 50-year extreme environmental load were defined. The steady state annual environmental load was determined by examining the wind resources for offshore wind in Korea using the wind resource map provided by the National Institute of Meteorological Sciences. The wind speed and direction in Korea were reviewed to understand the wind flow and intensity. The analysis was based on the observations from the state-of-the-art marine, meteorological, and environmental monitoring system located in Uido, Sinan-gun, Jeollanam-do, in the southwestern sea of South Korea. Environmental data from the year 2021 were analyzed and classified into wave, current, and wind categories. For the Serviceability Limit State (SLS), the most probable extreme environmental conditions occurring in Uido, Sinan-gun over a one-year period were selected as specified in Table 1.

Table 1. Serviceability limit state

Description		SLS	Unit
Wave	Significant wave height	4.5	m
	Peak period	9.5	Sec
	Direction	180	degree
	Spectrum type	JONSWAP	-
Wind	1 Hour mean speed	18	m/s
	Direction	180	degree
	Spectrum type	Constant	-
Current	Current speed	1.27	m/s
	Direction	180	degree
Water Depth		150	m

2.2 *Ultimate Limit State (ULS)*

The ULS design loads, which consider tropical cyclone conditions as suggested in IEC 61400-3, were defined and set according to Table 2. IEC 61400-3 adopts a 50-year return period for extreme external conditions.

In this study, environmental conditions were set for three directions of mooring lines, including wave, wind, and current directions. Figure 1 illustrates the application of wind and current directions to the 12MW FOWT.

Table 2. Ultimate limit state.

Description		SLS	Unit
Wave	Significant wave height	4.5	m
	Peak period	9.5	Sec
	Direction	180	degree
	Spectrum type	JONSWAP	-
Wind	1 Hour mean speed	18	m/s
	Direction	180	degree
	Spectrum type	Constant	-
Current	Current speed	1.27	m/s
	Direction	180	degree
Water Depth		150	m

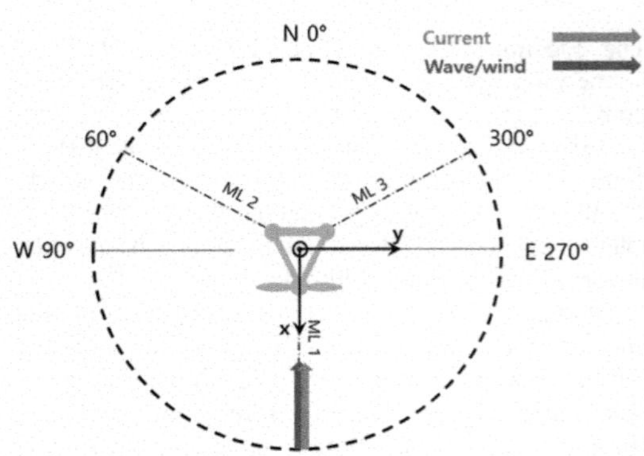

Figure 1. Load case diagram.

2.3 *Platform*

In this study, the specifications of the WINDMOOR 12MW wind turbine developed by SINTEF in Norway were considered. The detailed specifications of the applied wind turbine can be found in the reference (SINTEF, 2021), and the specifications are presented in Table 3.

Floating substructures were modeled using the SESAM program, such as Figure 2. Table 4 shows the main dimensions of the hull properties (including ballast).

Table 3. Main properties of the 12MW wind turbine.

Parameter	Properties
Turbine rating	NREL 12MW
Blade length (m)	105.4
Hub diameter (m)	6
Hub Height (m)	131.7
Draft (m)	15.5
Nacelle mass (kg)	600,000
tower mass (kg)	1,161,600
Blade mass (kg)	3 * 63024
Hub mass (kg)	60,000
Total mass (kg)	14,176,000
Cut-in/rated/cut-out wind speed (m/s)	4.0/10.6/25.0
Cut-in/ rated rotor speed (rpm)	5.5/7.8
Generator efficiency (%)	94

Figure 2. Floating substructure.

Table 4. Main properties of the full.

Property	Value
Column dimeter(m)	15.0
Column height(m)	31.0
Pontoon width(m)	10.0
Pontoon height(m)	4.0
Center-center distance(m)	61.0
Deck beam width(m)	3.5
Deck beam height(m)	3.5
Total substructure mass(t)	11,974.0

2.4 *Mooring line*

The coordinates of the fairlead and anchor points for the mooring line of the semisubmersible WIND MOOR 12MW wind turbine, applied in this study, are shown in Table 5, and the direction of the mooring line can be seen in Figure 3.

The mooring line consists of three to six segments and is designed with three types of elements, as shown in Table 6. The type, length and axial stiffness of each segment can be confirmed.

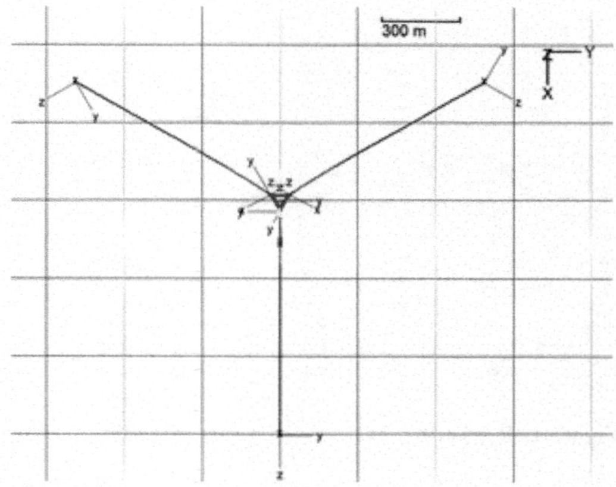

Figure 3. Plan view of the mooring system.

Table 5. Fairlead and anchor coordinates.

	Fairlead			Anchor		
	x(m)	y(m)	z(m)	x(m)	y(m)	z(m)
ML1	42.7	0.0	0.0	700	0	-150
ML2	-21.4	37.0	0.0	-350	606	-150
ML3	-21.4	-37.0	0.0	-350	-606	-150

Table 6. Mooring line segment properties.

Type	Length(m)	Axial stiff. (MN)	MBL (kN)
130mm stud. chair	32.4	1443.0	12,663
210mm steel wire rope	220	1782.0	15,400
130mm stud. chair	645	1443.0	12,663

3 12MW FOWT SIMULATION

3.1 *Response Amplitude Operator (RAO)*

In this study, environmental load conditions were set for dynamic analysis, and SESAM was used for hydrodynamic modeling. Hydrodynamic analysis was conducted using WADAM, and OrcaFlex was utilized for motion analysis by considering WADAM results, environmental loads, platform, mooring lines, turbine, and floating body modeling. Subsequently, the maximum tension and stability of the mooring lines were evaluated based on the calculation results of each analysis case. Through WADAM analysis, the RAO was obtained at 15° intervals from 0° to 180°. Figures 4-6 show the wave form directions of 0°, 90°, and 180°, respectively.

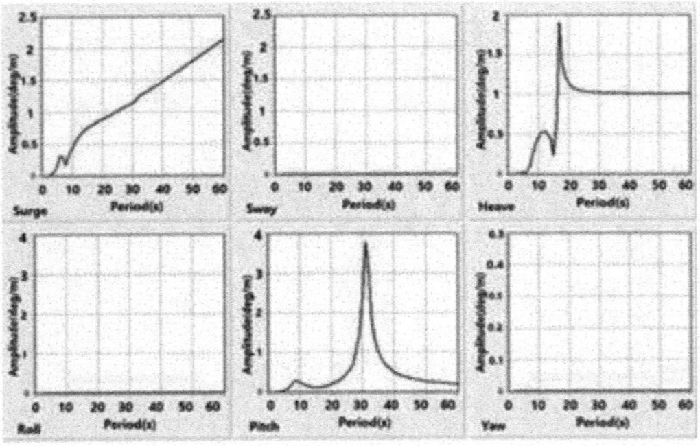

Figure 4. RAOs (0°).

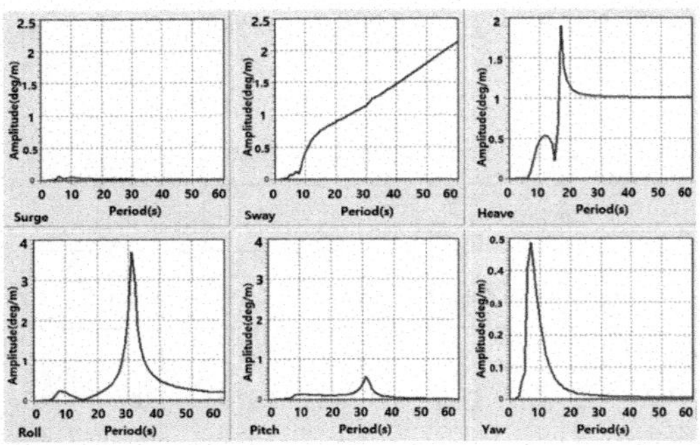

Figure 5. RAOs (90°).

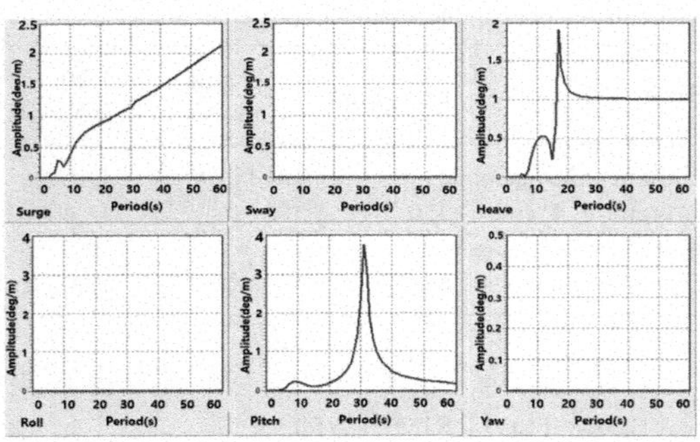

Figure 6. RAOs (180°).

3.2 *Load Case (LC)*

For the analysis cases, the number of mooring lines was varied to compare the maximum tension values. Three existing mooring lines were selected, and additional cases with four, five, and six mooring lines were analyzed. The load cases were summarized as shown in Table 7. The load cases, ranging from LC1 to LC8, consider the environmental load conditions for the SLS. For LC9 to LC16, the load cases are based on the ULS conditions. LC1 is designed with 3 mooring lines, while LC2 to LC4 incorporate an additional mooring line (ML1, ML2, ML3) individually,

Table 7. Load case.

LC	Envi. Load	Number of mooring lines
1		3
2		4 (Comp. ML1)
3		4 (Comp. ML2)
4	SLS	4 (Comp. ML3)
5		5 (Exc. ML1)
6		5 (Exc. ML2)
7		5 (Exc. ML3)
8		6
9		3
10		4 (Comp. ML1)
11		4 (Comp. ML2)
12		4 (Comp. ML3)
13	ULS	5 (Exc. ML1)
14		5 (Exc. ML2)
15		5 (Exc. ML3)
16		6

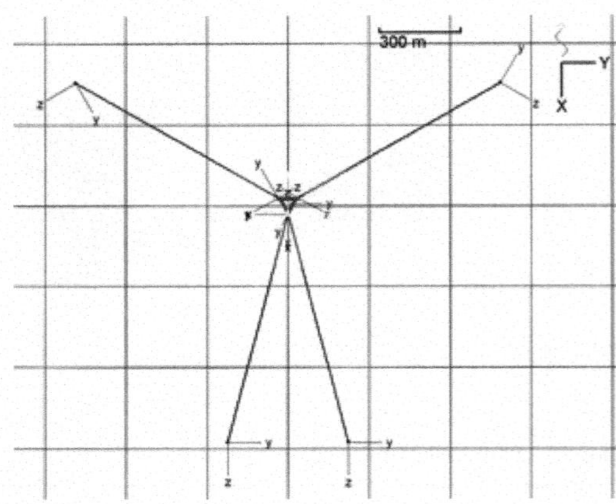

Figure 7. Plan view of the mooring system (LC2).

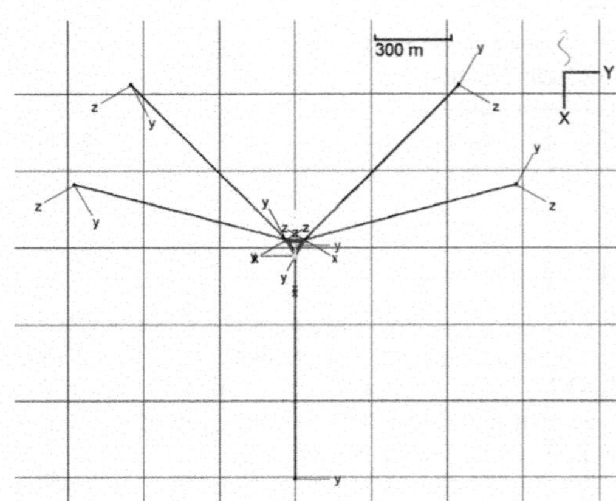

Figure 8. Plan view of the mooring system (LC5).

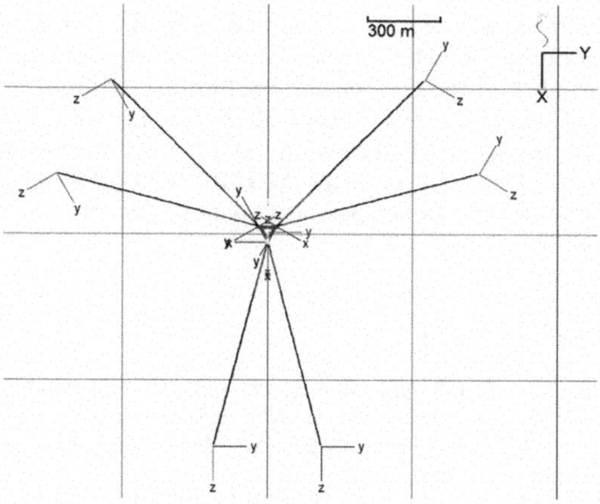

Figure 9. Plan view of the mooring system (LC8).

resulting in a total of 4 mooring lines. LC8 consists of two mooring lines for each ML1, ML2, and ML3. LC5 to LC7 follow a similar configuration as LC8, with one mooring line removed from ML1, ML2, and ML3, resulting in a total of 5 mooring lines.

For LC1, the model configuration can be observed in Figure 3, while for LC2 with 4 mooring lines, the model configuration is shown in Figure 7. LC5 has 5 mooring lines, and the corresponding model configuration is depicted in Figure 8. The model configuration with 6 mooring lines (LC8) is presented in Figure 9.

3.3 *Dynamic analysis*

To conduct the dynamic analysis, a thorough examination of the tensile force variation over time was performed for approximately 3 hours. The analysis aimed to determine the maximum tensile force in the section where a constant value was observed, as depicted in Figure 10.

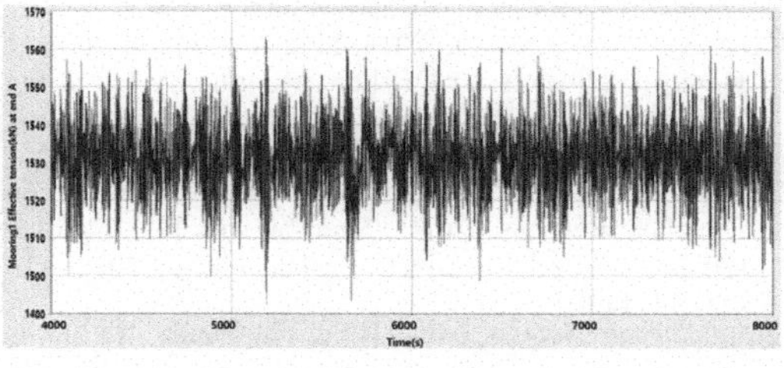

Figure 10. Time history graph (ML1).

Table 8 presents the maximum tension of ML1 for the load cases indicated in Table 7.

Table 8. Dynamic analysis result (Maximum tension).

LC	Max. tension (kN)
1	5,817.42
2	3,136.76
3	7,247.21
4	7,231.59
5	8,424.93
6	4,591.87
7	5,758.10
8	5,093.94
9	12,887.43
10	7,592.71
11	15,598.76
12	15,593.23
13	17,551.70
14	10,357.96
15	10,374.19
16	10,619.82

4 RESULTS

4.1 *Mooring system stability*

To verify compliance with the safety requirements specified in the regulations, the maximum allowable tension of the FOWT mooring lines was calculated. The maximum allowable tension of the mooring lines can be determined based on the Minimum Breaking Load (MBL). The percentage of MBL should be smaller than the calculated value, and the safety factor should be greater than the specified value.

To ensure the stability of the mooring, the applicable regulations were examined. Table 9 presents the tension limit and safety coefficient of the mooring line tensile force obtained from the relevant institutional and classification regulations during the Intact and Dynamic analyses.

Table 9. Line tension requirements.

Institute	Condition	Tension limit	Safety factor	Max. tension
API (2005)	Intact	60%	1.67	12,663kN
	Damaged	80%	1.25	

4.2 *Mooring system stability results*

Through dynamic analysis, the tension in the mooring lines was calculated based on the environmental conditions. The mooring stability was evaluated by assessing the safety coefficient and the MBL% for each condition. The MBL used for the analysis was 12,663kN, considering a chain diameter of 130 mm and a chain grade of R3(DAIHAN., 2018). The safety coefficient and MBL% were categorized into Steady and Extreme conditions. The data from each analysis case were analyzed to determine the stability of the mooring lines based on the material properties and the length of each segment.

The safety assessment results indicated that all load cases from LC1 to LC8, which represent the SLS environmental conditions, were found to be satisfied. However, for the ULS conditions, it was observed that only LC10, LC14, LC15, and LC16 met the safety requirements.

5 CONCLUSION

In this study, an investigation was conducted to examine the influence of the number of mooring lines on a 12MW floating offshore wind turbine (FOWT). Analysis was performed considering the environmental conditions of Serviceability Limit State (SLS) and Ultimate Limit State (ULS), and the tension values of the mooring lines were analyzed to evaluate the mooring stability. The impact of varying the number of mooring lines on the mooring system was analyzed, and the results indicated that the highest tension occurred in ML1.

After assessing the mooring stability based on the influence of the number of mooring lines and relevant regulations, it was found that LC2 and LC10, which have different environmental conditions but the same analysis conditions, satisfied the mooring stability requirements with the fewest number of mooring lines. According to the analysis results in this study, LC2, which has the fewest number of mooring lines and satisfies the mooring stability, was found to be the most economical option. This is because a lower number of mooring lines result in lower costs and maintenance associated with their installation. However, it should be noted that in this study, the analysis was conducted with wind, wave, and current set at 180° in the ML1 direction, making it challenging to determine the optimal number of mooring lines. In future research, it is recommended to consider various environmental conditions to make a more accurate comparison.

Key design variables necessary for the design of mooring lines include the weight, diameter, length, and stiffness of the mooring line, and each of these design variables can have different impacts on the mooring system depending on the type of mooring. This study examined the tension values and motions of mooring lines according to changes in key design variables for design of mooring lines, and aims to perform design of mooring lines using the Response Surface Methodology (RSM) with the results obtained from future calculations.

ACKNOWLEDGEMENTS

This result was supported by "Regional Innovation Strategy (RIS)" through the National Research Foundation of Korea (NRF) funded by the Ministry of Education (MOE) (Foundation Task Management Number: Gwangju Jeonnam Platform 2021RIS-002)

This work was supported Korea institute for Advancement of Technology (KIAT) grant funded by the Korea Government (MOTIE) (P0017006, The competency Development Program for Industry Specialist)

REFERENCES

American Petroleum Institute (API). (2005) Recommended Practice for Design and Analysis of Station keeping systems for Floating Structures. API Recommend Practice 2SK., 3rd Edition.

Bae, Y.H., Kim, M.H. and Kim, H.C. 2017. Per-formance changes of a floating offshore wind turbine with broken moor-ing line. Renewable Energy. (101) pp:364-375.

Benassai, G., Campanile, A., Piscopo, V. and Scamardella, A. (2015) Optimixation of Mooring Systems for Floating Off-shore Wind Turbines. Coastal Engineering Journal. Vol. 57, No. 04, 1550021

Daihan Anchor Chain Mfg. Co., Ltd, 2018. Off-shore mooring chain & Anchor chain.

Eghbali, B., Daghigh, M., Daghigh, Y. and Azarsina, F., 2018. Mooring Line Reliability Analysis of Single Point Moor-ing (SPM) Sys-tem under Extreme Wave and Current Condi-tions. International Journal of Maritime Tech-nology. (9) pp:41-49.

Ferreia, F.M.G., Lages, E.N., Alfonso, S.M.B., Lyra, P.R. M., 2017. Using Design of Experiments and Design Optimization to Determine Statically Equivalent Mooring System on Truncated Water Depth. Journal of Offshore Mechanics and Arctic Engi-neering, 139(4), 041302.

Huang, W.H. and Yang, R.Y., 2021. Water Depth Variation Influence on the Mooring Line Design for FOWT within Shal-low Water Region. Journal of Marine Science and Engi-neering, 2021.04. 9(4):409.14.

IEC (International Electrotechnical Commission)., 2020, IEC 61400-3-1. INTERNATIONAL STANDARD. Wind energy generation systems – Part 3-1: Design requirements for fixed offshore wind turbines.

Kang, S. W. and Lee, S.J., 2018. Study on Optimization of Design Parameters for Offshore Mooring System using Sam-pling Method. Journal of Ocean Engineering and Technology. 32(4); 215–221

Liu, B. and Yu, J., 2022. Effect of Mooring Parameters on Dynamic Responses of a Semi-Submersible Floating Offshore Wind Turbine. Sustainability. 2022, 14 (21):14012.

Niranjan, R., and Ramisetti, S. B. 2023. Dynamic Response of 15 MW Floating Wind Turbine with Non-Redundant and Redundant Mooring Systems Under Extreme and Accidental Conditions. Journal of Offshore Mechanics and Arctic Engi-neering. 145 (6): 062002.

Souza, C.E.S., Berthelsen, P.A., Eliassen, L., Bachynski, E. and Haslum, H. (2021) Definition of the INO WINDMOOR 12MW base case floating wind turbine, SINTEF Report., NO OC202 A-044

Advances in Maritime Technology and Engineering – Guedes Soares & Santos (Eds)

Floating offshore wind turbine motion prediction with neural network-integrated simulations

Antonio Medina-Manuel & Antonio Souto-Iglesias
CEHINAV, DACSON, ETSIN
Universidad Politécnica de Madrid (UPM), Madrid, Spain

Rafael Molina Sánchez
ETSI de Caminos, Canales y Puertos
Universidad Politécnica de Madrid (UPM), CEHINAV-UPM, Madrid, Spain

ABSTRACT: This paper details a study on simulating a Floating Offshore Wind Turbine (FOWT) Spar, specifically the OC3 reference design, using OpenFAST integrated with ANSYS AQWA. The simulations, validated against literature, accurately capture the Spar's six degrees of freedom motions. Utilizing this data, a Long-Short Term Memory (LSTM) model was developed, using wave characteristics to predict the Spar's motion time series. This neural network enhances understanding of the Spar's dynamics under various conditions and offers a tool for real-time monitoring and control in offshore wind energy systems. This work combines advanced simulations and AI to provide insights into FOWT Spar dynamics, improving design and operation in offshore wind energy, while also reducing computational time.

1 INTRODUCTION

Seakeeping simulations are pivotal in designing and analyzing offshore structures, ensuring safety and operational efficiency in the face of dynamic marine environments. The OC3 spar wind platform, a widely recognized benchmark for floating wind turbines, has been subject to extensive hydrodynamic studies to assess its response to wave-induced motions (Xu and Day 2021, Yang et al. 2021).

Traditional methodologies have predominantly utilized frequency domain analyses, with tools such as WAMIT being integral in providing potential flow solutions that form the basis for subsequent time-domain simulations in OpenFAST (Jonkman & Buhl 2005). This workflow has established a foundation for seakeeping simulations analysis, where frequency domain data capture the essence of wave-structure interactions through diffraction and radiation potential theory (Newman 1977).

Emerging from this established practice is a novel approach that harnesses the versatility of hydrodynamic analysis tools, such as ANSYS Aqwa, renowned for their robust frequency domain solutions (Wang et al. 2014). The transmutation of these solutions into a form amenable to time-domain simulation platforms as OpenFAST presents a compelling narrative in computational fluid dynamics. Such a transformation is achievable through sophisticated translation tools like BemRosetta, capable of adapting frequency domain data into suitable inputs for time-domain applications, enabling a more comprehensive hydrodynamic assessment (Zabala et al. 2021). Such hybrid approaches broaden the accuracy and efficiency horizons for marine structure simulations.

Very often, all these seakeeping simulations imply a significant computational cost (Jonkman 2007). Therefore, this study encapsulates the state-of-the-art by integrating machine learning with these advanced hydrodynamic simulation techniques. In the realm of predictive modelling, Long Short-Term Memory (LSTM) neural networks have emerged as a powerful tool, capable of learning from sequential data and capturing temporal dependencies (Hochreiter & Schmidhuber 1997, Gers et al. 2000).

Recent breakthroughs with neural networks have demonstrated significant potential in hydrodynamic applications: Bjørni et al. (2021) investigated the effect of machine learning methods on predicting the mooring line tension of a spar floating wind turbine, with average solid correlation except for severe sea states. Silva & Maki (2022) developed a methodology (LSTM) neural network to represent the 6 DoF motions of a ship operating at 20 knots in long-crested irregular seas. The accuracy of such a model was studied based on the wave elevation input and the training data quantity. Novel integrated architectures based on LSTM encoder and decoder and a convolutional neural network have been used to forecast ship motion using the encoded initial state

DOI: 10.1201/9781003508779-50

vector and foreseen information on the ocean wave field around a vessel (Lee et al. 2023).

This study presents a novel application of LSTM networks trained with diverse datasets from this hybrid simulation workflow. This approach aims to predict the 6DoF motions of the OC3 spar wind platform. The developed network promises to accelerate the prediction process and enhance the fidelity of motion forecasts, potentially transforming seakeeping analyses in marine sciences.

2 CASE OF STUDY

The OC3-Hywind model (Jonkman & Musial 2010) is a reference wind turbine that comprises a spar-type floating wind turbine (FWT) of 5 megawatts, anchored by three catenary mooring lines. This concept has been chosen in the current work as proof of concept due to its extensive validation of both experimental and numerical data, providing a robust foundation for comparative analysis. The availability of literature on the seakeeping behaviour of the OC3-Hywind model facilitates the establishment of a benchmark against the performance and accuracy of the numerical simulations carried out in the present study. Nevertheless, it is intended to extend this study to different types of FOWTs in future works.

The actual draft of the platform is 120 m and, between 4 m and 12 m below the sea water level (SWL), extends a linearly tapering conical transition region that connects two cylinder diameters of 9.4 m below the taper with a slender one of 6.5 m. Figure 1 illustrates the concept. More details on the inertia, mass and mooring properties of the platform can be found in NREL Report (Jonkman & Musial 2010).

Figure 1. Sketch of the NREL 5-MW wind turbine on the OC3-Hywind spar. Source: (Jonkman & Musial 2010).

In current simulations, the water depth was taken to be 320 m following the available literature, and the waves are propagating along the x-axis. The wave directionality will be included in future works.

3 NUMERICAL SIMULATION FRAMEWORK

OpenFAST is a time-domain open-source simulation tool analyzing wind turbines. In the numerical model, the spar FWT system is modelled by using the platform properties described above, then it was calibrated and fine tuned following the available literature. A similar methodology was adopted in (Browning et al. 2014).

The platform's hydrodynamic calculations were obtained by merging potential flow theory with Morison's equation within the HydroDyn module. Initially, hydrodynamic coefficients, namely radiation damping ($B(\omega)$), frequency-dependent added mass ($A(\omega)$) and hydrostatic restoring, ($C^{hydrostatic}$) must be computed by a frequency domain code, such as wamit. Subsequently, these coefficients are used in the time domain by replacing the integral over all frequency-dependent incident-wave-excitation forces (Cummins and Taylor 1962) with a convolution integral. However, in HydroDyn, the convolution integral is implemented by applying Fourier transforms following Oglive 1964, Duarte et al. 2013.

In the present research, although HydroDyn (OpenFAST) primarily processes data files generated by WAMIT, a different approach was employed for the frequency domain analysis. The analysis was conducted using ANSYS Aqwa, a commercial software for undertaking hydrodynamic and mooring analyses. This choice necessitated an additional step in the workflow: translating ANSYS Aqwa's output files into a format compatible with HydroDyn. To facilitate this conversion, BemRosetta, an open-source tool, was utilized (Zabala et al. 2021). BemRosetta specializes in converting various Boundary Element Method (BEM) solver formats into formats readable by HydroDyn. Figure 2 displays the numerical simulation framework employed in current study.

This conversion is critical for ensuring the seamless integration of ANSYS Aqwa's detailed frequency domain analysis into HydroDyn's time-domain simulation environment, thereby marrying the strengths of both software packages into a unified analytical framework. The advantages of using ANSYS Aqwa over WAMIT or any other frequency domain simulation tool include its user-friendly interface integrated with ANSYS Workbench, comprehensive hydrodynamic analysis capabilities, and advanced visualization and post-processing tools. The use of a translator also offers you the possibility of coupling different available

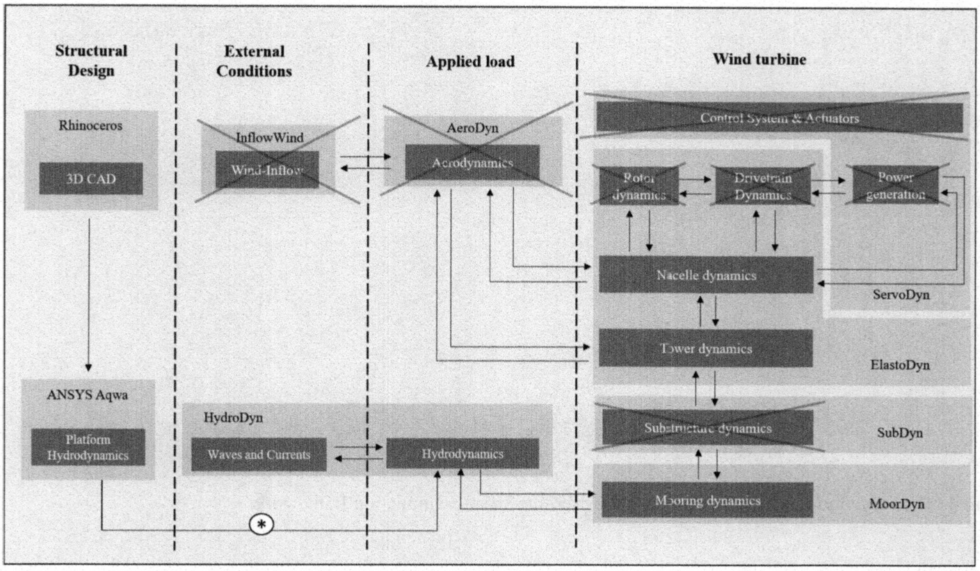

Figure 2. Numerical simulation framework was used in this study. The red crosses indicate the non-used submodules. (*) BemRosetta is used in this link to convert F.D. analysis to .1, .3, .4, .9, .12 and .hst files readable in OpenFAST.

software tools. In addition, the mooring system is modelled with MoorDyn with the data presented in Jonkman (2007).

For all DoFs, numerical decay tests were performed to calibrate the numerical model. This fine-tuning consisted of augmenting with the nonlinear viscous-drag term of $C_D = 0.6$ and an additional linear damping matrix in surge, sway, heave and yaw, which aligns with the NREL Report (Jonkman 2007). Figure 3 presents an example of a time series of platform motions for both the experimental decay test and the numerical decay test of the tuned numerical model.

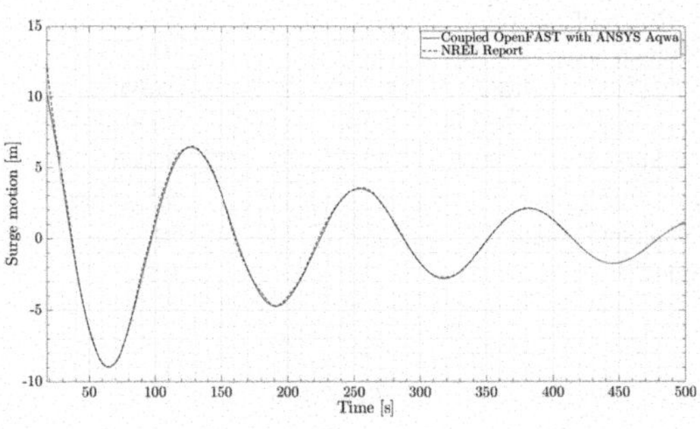

Figure 3. Surge platform motions in the numerical and NREL Report Jonkman and Musial (2010) decay tests.

Furthermore, motion RAOs were obtained by running a white noise spectrum to validate the numerical model with the available numerical and experimental literature (Jonkman & Musial 2010, Tomasicchio et al. 2018, Lopez-Olocco et al. 2023). The wave tested was propagating with 180 degrees heading, without wind. In future work, it is recommended to explore the use of several directionalities and wind time histories. In Figure 4, a comparison of the motion RAOs obtained with the developed numerical model and the ones obtained in Jonkman & Musial (2010) is displayed.

4 NEURAL NETWORK MODEL

Several data-driven and machine learning (ML) models have been designed and developed for floating bodies motion prediction; however, significant gaps still exist as many models are classical approaches. Regarding the Artificial Neural Network (ANN), feedforward Neural Networks (FNN) have been used in many time series forecasting Abdulkarim & Garko (2015) and also in FOWTs mooring lines prediction Bjørni et al. (2021). The FNN model is a network without recurrent links, which means the signals only pass in one direction. Data passes through the input layer for processing, where each element computes a weighted sum of its inputs. These values then serve as inputs for the next layer, continuing this process until the output is generated.

In the 80's, a new ANN was born for modelling time series, Recurrent Neural Networks (RNN). They are capable of storing information for a long time in the hidden states, and they get the output using feedback from their units. However, RNNs suffer from vanishing or exploding gradients during training. This means that as the sequence gets longer, RNNs tend to forget the earlier information, making it challenging to learn dependencies between distant elements in the sequence. To address this

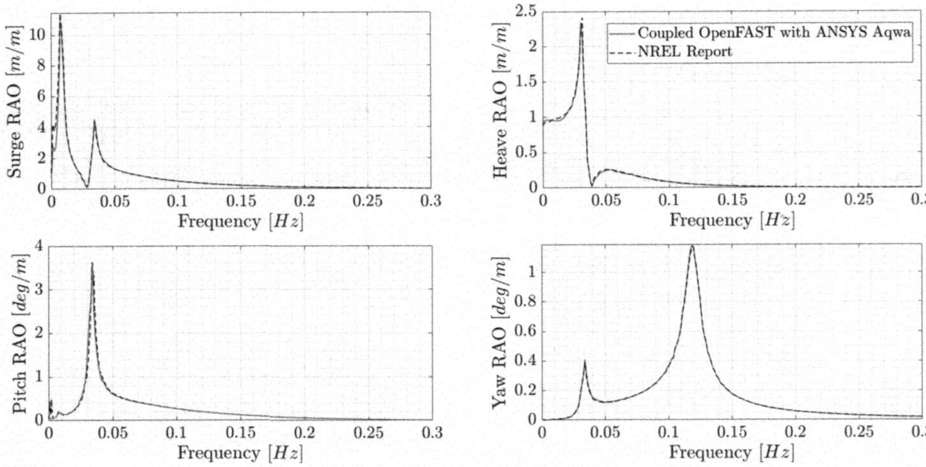

Figure 4. OC3 Hywind RAOs validation with the presented model simulation framework.

challenge, the concept of Long Short-Term Memory (LSTM) networks, a family of Recurrent Neural Networks (RNNs), was introduced by Hochreiter and Schmidhuber 1997.

In this study an LSTM-based Neural Network was employed, implemented using PyTorch (FAIR 2016), to address the sequence learning challenges in modelling FOWT platforms. These FOWTs exhibit wave-excited responses where the motion is influenced by both current and previous states (Silva & Maki 2024). The LSTM's ability to remember long-term dependencies is essential for capturing these dynamics, offering a comprehensive understanding of the motion responses influenced by wave elevations and their temporal context. This approach, distinct from traditional Feedforward Neural Networks, ensures more accurate and reliable predictions of the complex, time-dependent behaviour of FOWT platforms.

4.1 *Neural network architecture*

The developed NN features five hidden LSTM layers where the output of each previous layer becomes the input to the subsequent layer. These layers, each comprising 100 cells, employ the tanh activation function. This facilitates the model's ability to learn complex mappings from inputs to outputs and is particularly effective with the normalization used herein. The last LSTM layer is then followed by a dense layer with linear activation, which generates the temporal output of the neural network.

The LSTM receives the wave elevation η_{t_i} as the input vector for an explicit time t. The output from the developed model is each DoF motion ξ_{t_i} with $t_i \in t$ for this wave excitation. During training, the network performs a forward pass using updated weights (W_{ih} and W_{hh}) and biases (b_{ih} and b_{hh}), and computes gradients for backpropagation. Adam optimizer has been employed in the present work to optimize these parameters and minimize the loss. The LSTM also sequentially utilizes the previous timestep's hidden state (h_{t-1}). The sequence length selected herein was 50 time steps.

$$h_t = \sigma(\eta_{t_i} W_{ih}^T + b_{ih} + h_{t-1} W_{hh}^T + b_{hh}) \tag{1}$$

In this model an initial learning rate of 0.005 with a decay schedule each 1000 epochs decreased 0.5 was applied. All the hyperparameters used in the developed model are displayed in Table 1.

Table 1. Training matrix, neural network architecture, and hyper-parameters for developed model.

Properties	Value
Number of layers	6
Hidden size	100
Initial learning rate	0.005
Learning rate schedule	StepLR (γ=0.5)
Optimizer	Adam
Loss function	MSELoss
Sequence length	50
Epochs	100

4.2 *Database and test conditions*

To enable the developed NN model to predict all six DoF motions in realistic sea states, the wave scatter data used corresponds to Aberdeenshire's coast (see Figure 5). This location is significant as it is where the 30 MW Hywind Scotland pilot wind farm is situated (Burns et al. 2022).From the mentioned wave scatter, a selection of 300 irregular wave realization cases with a unique 180-degree wave heading forms the Latin Hypercube used in the NN. The selected

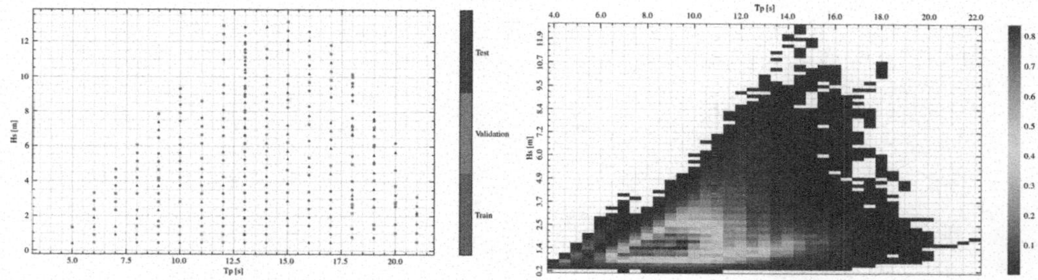

((a)) Split.

((b)) Scatter.

Figure 5. Database Scatter from Aberdeenshire's coast.

pairs of significant wave heights (H_S) and peak periods (T_P) were run during one-hour simulations in OpenFAST. Wind excitation was left for future works.

The time series output from numerical simulations of each corresponding movement is used as the training output of the network. In addition, the input provided to the model is the wave log excitation. The inputs and outputs are then standardized between -1 and 1 with the minimum and maximum signals' values. Scaling of the features avoids instances of one quantity dominating the training process. Separate datasets are withheld from the training process to validate the model and test its performance from within the above mentioned Latin Hypercube since, a priori, the model can interpolate but cannot extrapolate. Figure 5 presents the random dataset split between training, validation and test. The split has been done with 65%, 15% and 20% of the total dataset for training, validation and testing respectively.

All the datasets have been reprocessed to check if the statistical data from this split are similar. In addition, the simulated data was sampled at a frequency of 100Hz. However, when employing LSTM models, high sampling frequencies lead to redundant information, as consecutive data samples tend to be highly similar. Therefore, in the present study, data sequences were downsampled, enabling the model to focus on long-term dependencies. Nyquist theorem (Shannon 1949) was employed to select the minimum re-sampling frequency. Finally, the smoothness and continuity of the first and second derivatives of the signals have been checked to avoid noise learning.

5 RESULTS

The presented ANN model was trained to predict a 150-second time series for all DoF motions. In future work, it is planned to predict longer time series. Throughout the training process, the loss function (L), both for training and validation datasets, was monitored until it established a low value. It ensures that the developed model has no overfitting or underfitting issues. An example of the loss curves convergence for heave motion over the epochs is presented in Figure 6. As can be appreciated, both training and validation losses display similar tendencies, which means that the model is not overfitted. The loss function used was MSE averaged over all datasets (see Eq. (2)).

$$L(\xi, \hat{\xi}) = \frac{1}{N}\sum_{j=1}^{N}\left(\frac{1}{T}\sum_{t_i=1}^{T}(\xi_{t_i}^{j} - \hat{\xi}_{t_i}^{j}2)\right), \qquad (2)$$

where N is the total number of cases used in each dataset, T is the total length of each time serie, $\xi_{t_i}^{j}$ is target single DoF motion output and $\hat{\xi}_{t_i}^{j}$ is the predicted motion.

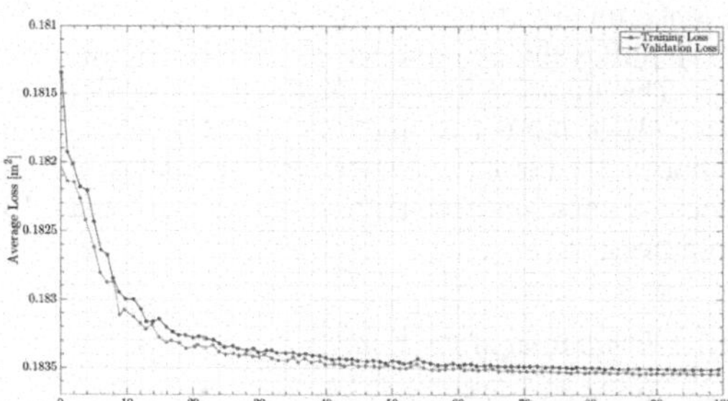

Figure 6. Loss convergence curve over the epochs while training heave motion.

A total number of 100 epochs was needed to train the model without any early stop. The computational time needed to train the ANN is over 60 hours with a CPU of 16 GB memory and 2.40 GHz. The final training loss of the model was kept around 10^{-3} in its normalized form.

Since the simulated cases only include 180 degrees wave directionality, the following results will focus on the exited DoFs, i.e. surge, heave and pitch motions. The predicted output from the trained model for some waves are presented in Figure 7. These waves belong to the test dataset and correspond to specific parts of

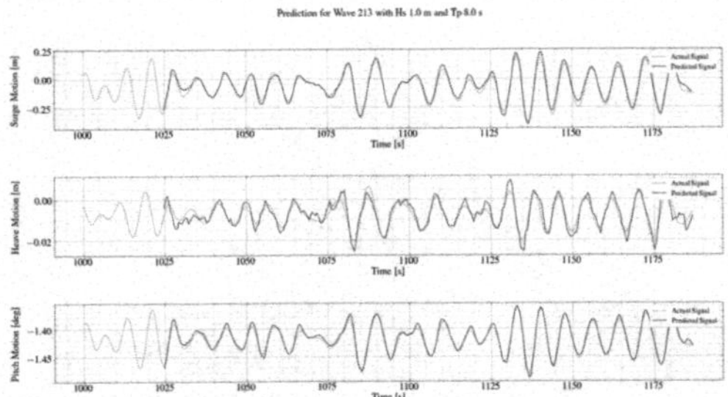

((a)) *Wave 213* with H_S 1m and T_P 8s

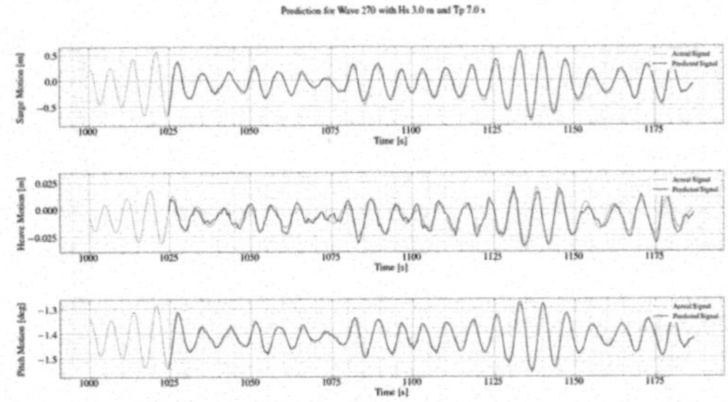

((b)) *Wave 270* with H_S 3m and T_P 7s

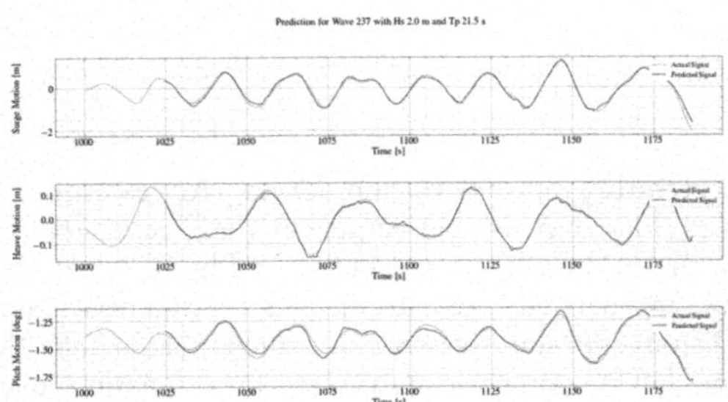

((c)) *Wave 237* with H_S 2m and T_P 21.5s

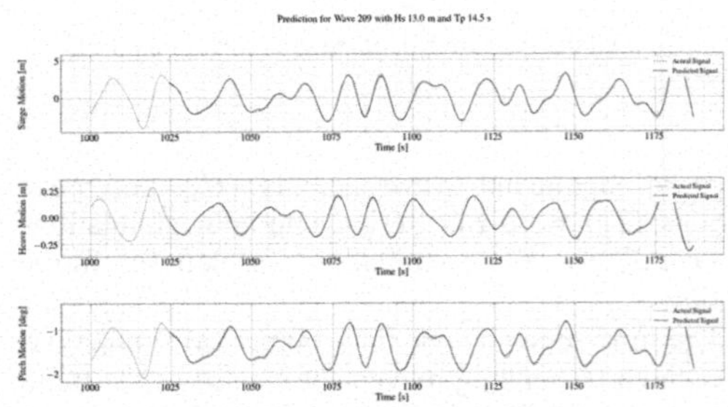

((d)) *Wave 209* with with H_S 13m and T_P 14.5s

Figure 7. Comparison of time series predictions of motions with numerical simulations for specified wave trains tested.

the Latin Hypercube and on the Convex Hull, i.e. wave scatter in Figure 5: *Wave 213* with H_S 1m and T_P 8s represents the operational wave with higher occurrence probability; on the other hand, *Wave 270* with H_S 3m and T_P 7s and *Wave 237* with H_S 2m and T_P 21.5s are in the lowest and highest peak period limits, respectively, with low observed events; finally, *Wave 209* with H_S 13m and T_P 14.5s exemplifies the extreme wave for the utilized scatter. The model takes less than 3s to predict 150 seconds of the 6 DoF motions for each wave train while the same case numerically simulated within equal computation conditions takes more than 10 minutes.

5.1 *Evaluation of the model*

Once the model achieves accurate predictions, a comprehensive model evaluation is necessary. Although the overall time series comparison in Figure 7 suggests a strong correlation, further analysis were conducted by obtaining and evaluating the heteroskedasticity (R^2 score). These metrics provide deeper insights into the model's performance, assessing its precision and the distribution characteristics of its predictions.

Figure 8 presents the residual values between the predicted and target values for the wave cases displayed previously. Each point in the scatter represents a single time step motion amplitude prediction where the input wave elevation is identical between numerical simulation and LSTM. The perfect 1:1 correlation would be in the straight black line. It can be seen that the error distribution for surge response is similar for all motion values. In *Wave 237*, the residual errors seem to increase for negative surge values, which corresponds to the end part of the time series in Figure 7 (c). The heave response presents the most significant residual errors compared with the rest of the DoFs, which could be due to its small motions. For pitch, the error distribution shows larger quantities for the mean of the pitch angle motions. Besides, the values for the R2 score are always above 0.95 for surge and pitch responses, which evince the fair performance of the developed model. However, for heave prediction in low periods (below 10s), it displays worse performance.

To assess the model's robustness, new irregular wave scenarios not included in the training, validation, or test datasets were simulated using OpenFAST using a different random phase time series generation with distinct seeds. Subsequently, the model was employed to predict the platform's response to these conditions. Notably, this involved an extremal irregular wave scenario (H_S 15 m and T_P 20 s) that fell outside the Convex Hull of the wave scatter diagram used for training, thereby testing the model's ability to generalize and perform under novel, challenging conditions. Also, regular waves were run with periods and height within the scatter utilized, ensuring that the ANN model correctly handles both simple and complex scenarios. Furthermore, it allows for a more comprehensive evaluation of the model's robustness and its ability to generalize. Time series with a comparison between the target and prediction of the 3 DoFs for some of these cases are shown in Figure 9. The axes' scale in this figure for the same motions change depending on the wave excitation.

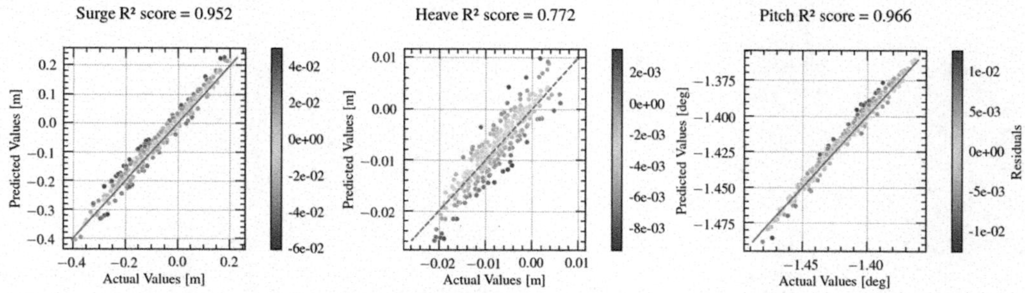

((a)) *Wave 213* with H_S 1m and T_P 8s

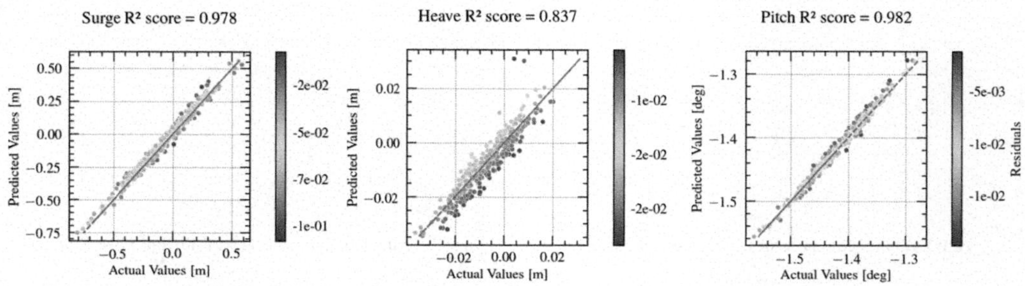

((b)) *Wave 270* with H_S 3m and T_P 7s

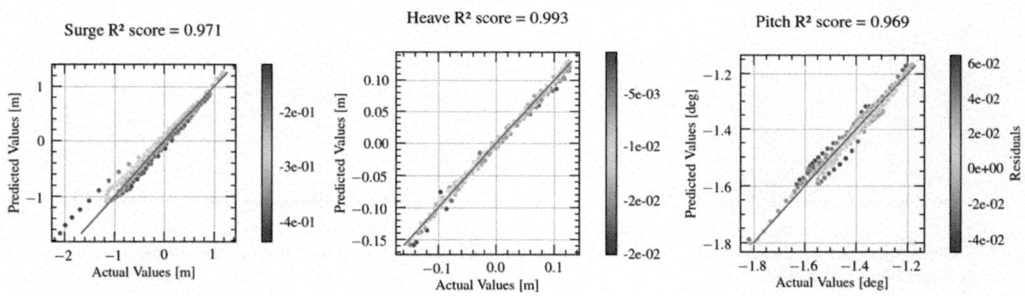

((c)) *Wave 237* with H_S 2m and T_P 21.5s

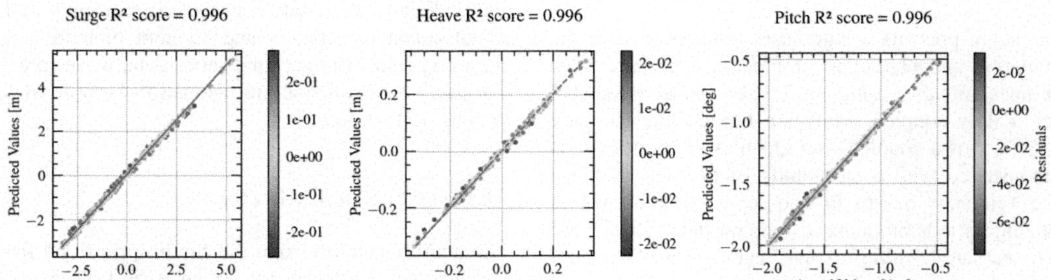

((d)) *Wave 209* with with H_S 13m and T_P 14.5s

Figure 8. R^2 scores and residual values for specified wave trains tested.

It can be noticed that the major differences are found in heave response prediction. The main concern of the authors in this regard is that the heave motion response is much lower than the others. It also affects the normalization between 1 and -1 using maximum and minimum values. Nevertheless, both regular and irregular wave excitation using different seed realizations appear to be quite accurate.

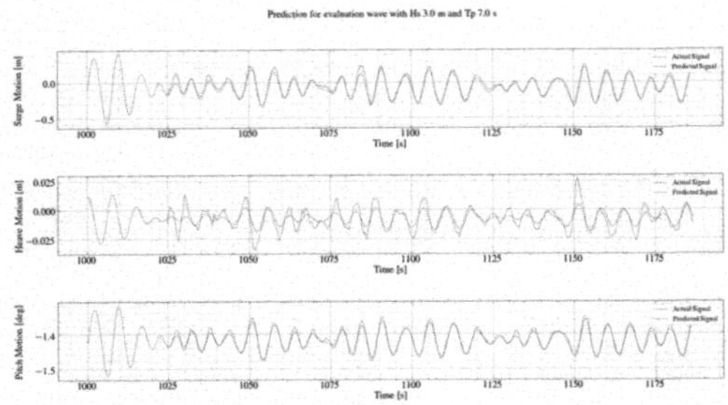

((**a**)) New seed of irregular sea state with H_S 3m and T_P 7s.

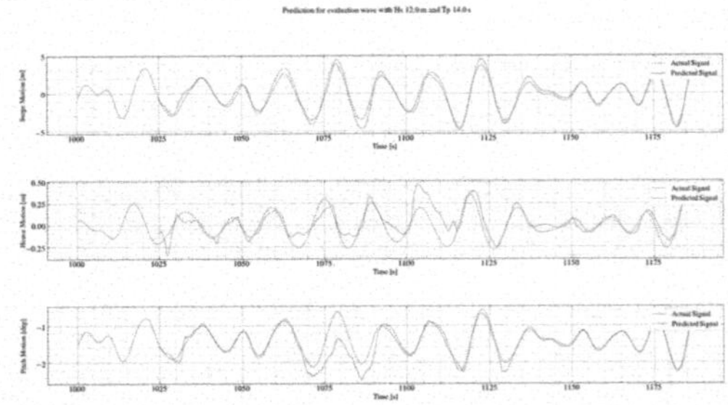

((**b**)) New seed of irregular sea state with H_S 13m and T_P 14s.

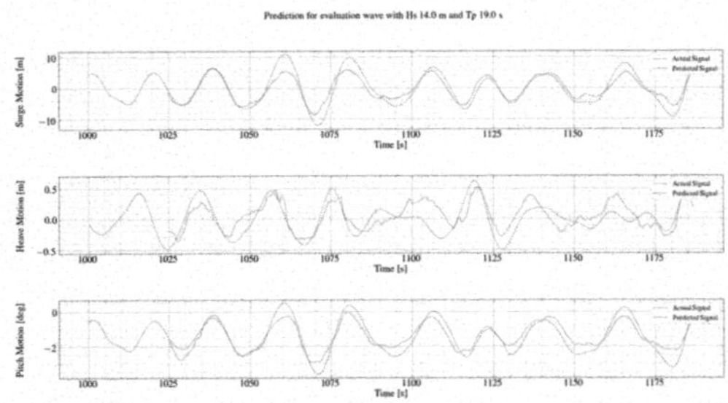

((**c**)) Extreme sea state with H_S 15m and T_P 20s.

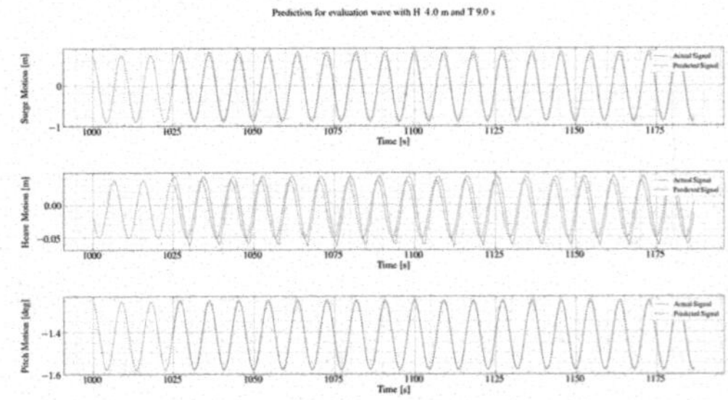

((**d**)) Regular wave with H 4m and T 9s.

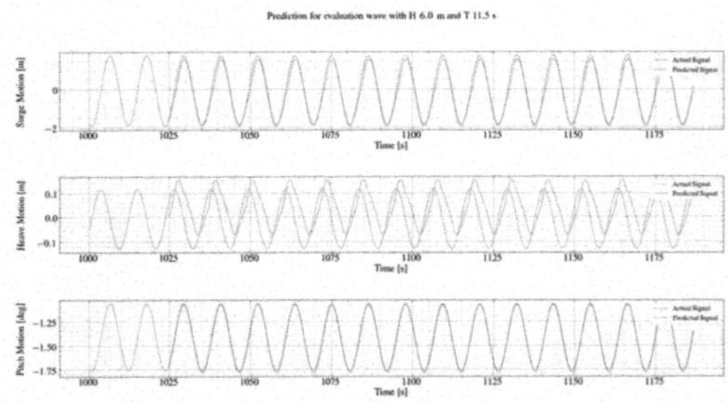

((**e**)) Regular wave with H 6m and T 11.5s.

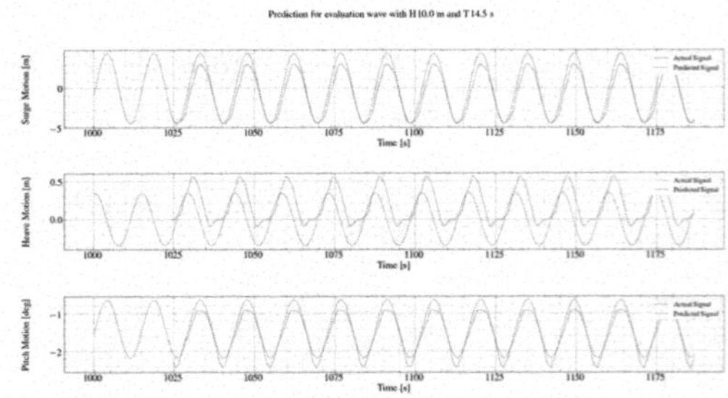

((**f**)) Regular wave with H 10m and T 14.5s.

Figure 9. Comparison of predictions with numerical simulations of platform response time series during new wave scenarios to evaluate the ANN model.

6 CONCLUSIONS

This study presents a significant advancement in the simulation and prediction of Floating Offshore Wind Turbines motions using an LSTM model integrated with a fully coupled simulation framework. The new ANSYS Aqwa coupled with OpenFAST framework is demonstrated with a case study of the spar platform OC3 reference design in 180-degree heading waves utilizing various amounts of training data. Three hundred sea states within a wave scatter from Aberdeenshire's coast were simulated to obtain the motion responses. The developed ANN model can predict complex motion time series, saving almost one order of magnitude in computational cost. The evaluation of the model's performance using R^2 scores of the data highlighted its precision, robustness and reliability when predicting platform responses under wave conditions. This research bridges the gap between advanced numerical simulations and artificial intelligence techniques, providing valuable insights into the dynamic response of FOWT Spar platforms. The methodologies and findings of this study contribute significantly to the advancement of efficient and resilient offshore wind energy systems. Longer time series data, wave spreading and wind excitation are planned to be utilized for training in the future.

ACKNOWLEDGEMENTS

The authors acknowledge the funding received from the Spanish Ministry for Science and Innovation through projects FOWT-DAMP2 (References: PID2021-123437OB-C21, FOWT-PLATE-MOOR and TED2021-130951B-100).

REFERENCES

Abdulkarim, S. & A. Garko (2015). Evaluating feed-forward and elman recurrent neural network performances in time series forecasting. *Indian Journal of Pure and Applied Mathematics 1*, 145–151.

Bjørni,F., S. Lien, T. Midtgarden, G. Kulia, A. Verma, & Z. Jiang (2021). Prediction of dynamic mooring responses of a floating wind turbine using an artificial neural network. *IOP Conference Series: Materials Science and Engineering 1201*, 012023.

Browning, J., J. Jonkman, A. Robertson, & A. Goupee (2014). Calibration and validation of a spar-type floating offshore wind turbine model using the fast dynamic simulation tool. *Journal of Physics: Conference Series 555*, 012015.

Burns, R. D., S. B. Martin, M. A. Wood, C. C. Wilson, E. Lumsden, & F. Pace (2022). Hywind Scotland Floating Offshore Wind Farm. Technical report, JASCO Applied Sciences.

Cummins, W. & D. W. Taylor (1962). *The Impulse Response Function and Ship Motions*. Report (David W. Taylor Model Basin). Department of the Navy, David Taylor Model Basin.

Duarte, T., M. Alves, J. Jonkman, & A. Sarmento (2013). State-space realization of the wave-radiation force within fast. *Proceedings of the International Conference on Offshore Mechanics and Arctic Engineering, OMAE2013-10375,* 08 - 13 Jun, 2013.

FAIR (2016). PyTorch documentation.

Gers, F., J. Schmidhuber, & F. Cummins (2000). Learning to forget: Continual prediction with lstm. *Neural computation 12*, 2451–71.

Hochreiter, S. & J. Schmidhuber (1997). Long short-term memory. *Neural computation 9*, 1735–80.

Jonkman, J. & W. Musial (2010). Offshore code comparison collaboration (oc3) for iea wind task 23 offshore wind technology and deployment. In *NREL/TP-5000-48191*, Volume 303.

Jonkman, J. M. (2007). Dynamics modeling and loads analysis of an offshore floating wind turbine. In *NREL/TP-500-41958*.

Jonkman, J. M. & M. L. Buhl Jr (2005). Fast user's guide - updated august 2005.

Lee, J.-H., J. Lee, Y. Kim, & Y. Ahn (2023). Prediction of wave-induced ship motions based on integrated neural network system and spatiotemporal wave-field data. *Physics of Fluids 35*(9), 097127.

Lopez-Olocco, T., G. Liang, A. Medina-Manuel, L. S. Ynocente, Z. Jiang, & A. Souto-Iglesias (2023). Experimental comparison of a dual-spar floating wind farm with shared mooring against a single floating wind turbine under wave conditions. *Engineering Structures 292*, 116475.

Newman, J, N. (1977). Marine hydrodynamics. *MIT Press* 90.

Oglive, T. (1964). Recent progress toward the understanding and prediction of ship motions. *Proceedings of the 5th Symposium on Naval Hydrodynamics 1*, 80.

Shannon, C. (1949). Communication in the presence of noise. *Proceedings of the IRE 37*(1), 10–21.

Silva, K. M. & K. J. Maki (2022). Data-driven system identification of 6-dof ship motion in waves with neural networks. *Applied Ocean Research 125*, 103222.

Silva, K. M. & K. J. Maki (2024). Implementation of the critical wave groups method with computational fluid dynamics and neural networks. *Ocean Engineering 292*, 116468.

Tomasicchio, G. R., F. D'Alessandro, A. M. Avossa, L. Riefolo, E. Musci, F. Ricciardelli, & D. Vicinanza (2018). Experimental modelling of the dynamic behaviour of a spar buoy wind turbine. *Renewable Energy 127*, 412–432.

Wamit, I. *WAMIT User's Guide*.

Wang, S., B. Shi, Z. Lin, & W. Zou (2014). A hydrodynamic analysis of offshore platform based on the aqwa. *Applied Mechanics and Materials 615*, 301–304.

Xu, X. & S. Day (2021). Experimental investigation on dynamic responses of a spar-type offshore floating wind turbine and its mooring system behaviour. *Ocean Engineering 236*, 109488.

Yang, J., Y.-P. He, Y.-S. Zhao, Y.-L. Shao, & Z.-L. Han (2021). Experimental and numerical studies on the low-frequency responses of a spar-type floating offshore wind turbine. *Ocean Engineering 222*, 108571.

Zabala, I., Y. Peña Sanchez, T. Kelly, J. Henriques, M. Penalba, N. Faedo, J. Ringwood, & J. Ilzarbe (2021). Bemrosetta: An open-source hydrodynamic coefficients converter and viewer integrated with nemoh and foamm. In *Proceedings of the 14th European Wave and Tidal Energy Conference.*, 05–09 Sep, Plymouth, UK.

Advances in Maritime Technology and Engineering – Guedes Soares & Santos (Eds)

Performance analysis of freely heaving U-OWC integrated with Π-shaped breakwater using Boundary Element Method

R. Muduli & D. Karmakar
Department of Water Resources and Ocean Engineering, National Institute of Technology Karnataka, Surathkal, Mangalore, India

C. Guedes Soares
Centre for Marine Technology and Ocean Engineering (CENTEC), Instituto Superior Técnico, Universidade de Lisboa, Lisbon, Portugal

ABSTRACT: In the present study, a free heave-only U-shaped OWC device integrated with a Π-shaped breakwater is analysed considering three different OWC chamber bottom profiles. The hydrodynamic performance is assessed on analysing the theoretical maximum efficiency, radiation susceptance and conductance coefficients. The influence of the draft of U-OWC and width of internal chamber on the hydrodynamic performance is analysed using Boundary Element Method (BEM). The peak efficiency is observed when resonance occurs as observed in the case with traditional OWC devices. Moreover, the radiation susceptance coefficient (indicative of energy left unutilised) is noted to be zero at peak efficiency and the radiation conductance coefficient (indicative of energy transferred into the system) is noted to have peak values. The study found that, irrespective of the chamber bottom profile, on increasing the width of the "U" channel, the energy conversion efficiency of the U-OWC gets diminished but increasing the width of the internal chamber enhances the efficiency of the device. A larger draft resulted in better efficiency of the device, which is intuitively expected, however a draft 0.75h is preferable than a draft of 0.9h as both configurations have almost similar performance.

1 INTRODUCTION

Climate change has forced to prioritize and replace the fast-depleting fossil fuel resources with renewable energy sources (Wang et al., 2018). Ocean energy is one of the cleanest sources of renewable energy having high power density (Cruz, 2007). Among different forms of ocean energy, the wave energy is the most underutilized form of energy. In order to convert wave energy into usable electrical energy, Wave Energy Converter (WEC) devices are used.

The renewable energy such as wind and solar energy systems can produce power of around 20-30% but the WECs can produce power up to 90% of power depending on the design and application (Ozkop & Atlas, 2017). Amongst the various types of WEC devices, the Oscillating Water Column (OWC) device is the most versatile and widely used due to its simplicity and uniqueness, as it has no moving parts inside the water which prolongs its life and its adaptability to be used onshore or offshore (Heath, 2012; Falcão, 2015).

The high construction and maintenance costs of WECs can be reduced by incorporating the energy devices into other marine structures such as breakwaters. Numerous studies have been conducted on the effectiveness of onshore or nearshore OWC integrated into breakwaters (Cong et al., 2022; Zheng et al., 2019). However, the coastal bottom profiles provide restrictions on the construction of shore attached OWC devices (Zhou et al., 2021; Wang et al., 2022). Furthermore, the seabed at the shoreline influences wave motion, resulting in a decrease in wave heights (Cong et al., 2022). In contrast, the deep sea has high power density, which would improve the energy capture efficiency of the OWC.

Offshore OWCs generally consists of an open bottom chamber, but other geometries such as the Backward Bent Duck Buoy (BBDB) (Masuda and McCormick, 1986), L-shaped OWC (Rezanejad et al., 2019), the double chamber OWC (Rezanejad & Guedes Soares, 2021) and U-shaped OWC (Muduli et al., 2024) are also been studied.

Luo et al. (2014) developed a fully nonlinear CFD model to analyse a heave-only OWC device reporting the significant effect of turbine damping and mooring stiffness on the power capture efficiency. A 2D CFD model of a freely heaving only OWC with a non-linear power take-off system was analysed by Connell et al. (2018) using ANSYS. Additionally, a 3D Numerical Wave Tank (NWT) was built to simulate the shape of a spar buoy OWC. A completely dynamic model of a spar buoy OWC was realized

DOI: 10.1201/9781003508779-51

with the combination of the one degree of freedom, 3D NWT, and non-linear PTO boundary condition.

Apart from the cost-sharing benefit, an integrated WEC and breakwater system can diminish the wave stresses on the breakwater, extending its lifespan (Graw, 1996). OWC's simple design makes it easy to incorporate them into breakwaters. A system of five OWCs built into a caisson breakwater at Sakata Port, Japan in 1989 was the first prototype of a WEC integrated into a breakwater to be deployed (Ohneda et al., 1991, Takahashi et al., 1992). Breakwater integrated with OWC was recently implemented during 2008-2010, at the Mutriku port in Spain (Torre-Enciso et al., 2009).

The U-shaped OWC design was developed by Boccotti (2003), and it was subsequently analytically demonstrated (Boccotti, 2007) that the U-OWC device outperforms a regular OWC device because of its higher eigen period and greater air chamber pressure. The wave power extraction of a floating box-type breakwater with dual pneumatic chambers was experimentally studied by He et al. (2017). The study found that the front chamber generated more power than the rear chamber and the draft of the front chamber required to reach its peak power extraction was shallower. Furthermore, wave period had a greater impact on power extraction than water depth. The study suggested that power extraction could be enhanced by a smaller chamber width.

An integrated system made up of a floating stationary breakwater and an offshore heaving OWC device with a specific distance between them was analysed by Wang et al. (2022). The study noted that an isolated freely-heaving OWC was able to extract wave power within a very narrow frequency range, whereas the presence of a floating fixed breakwater helped to increase the peak efficiency to a considerably larger extent.

On considering the growing interest in the research study of integrating offshore OWCs and breakwaters, the present study is carried out to analyse the hydrodynamic performance of a freely heaving U-OWC integrated with a Π-breakwater using Boundary Element Method (BEM) (Katsikadelis, 2016). Two different bottom profiles (inclined and curved) of the U-OWC internal chamber are considered in the hydrodynamic analysis. Non-dimensional radiation conductance and radiation susceptance coefficients as well as theoretical maximum efficiency are analysed for the integrated system's performance. The parameters are studied as a function of non-dimensional wavenumber while considering normal wave incidence. The effect of modifying various geometrical properties of the hybrid structure on the system performance is investigated.

2 THEORETICAL FORMULATION

The schematic diagram of the inclined and curved bottom U-OWC integrated with a Π-breakwater is presented in Figure 1(a,b). 2D coordinate system is considered for the theoretical analysis of U-OWC combined with Π-breakwater. The x-axis is considered horizontal, and z-axis, vertically downward negative. The height of the front barrier wall of U-OWC is denoted by T_2, the top lip wall by T_1 and the draft is denoted by D. The clearance between the front wall and top wall, i.e., the "U" channel width is denoted by a. The width of the internal chamber of the U-OWC is B. Therefore, the length of entire bottom base of the U-OWC is $a + B$. The Wells turbine is fitted at top of the chamber, directly over the internal free surface. The width and draft of the pi-breakwater are denoted by B_1 and D_1 respectively. Two auxiliary vertical boundaries are considered, Γ_{in} at $x = 0$ and Γ_{out} at "h" distance behind the Π-breakwater.

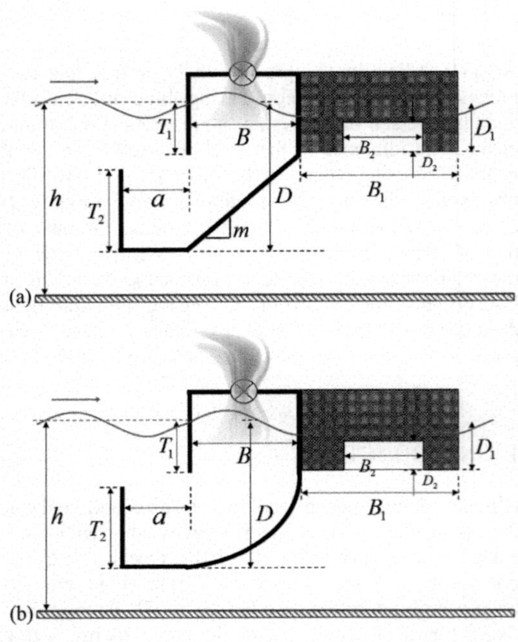

Figure 1. (a) Inclined and (b) Curved bottom U-OWC integrated with Π-breakwater.

2.1 *Governing equation and boundary conditions*

The fluid is considered inviscid, incompressible, irrotational and time-harmonic with angular frequency ω, so, the velocity potential exists, and is of the form $\Phi(x, z, t) = Re\{\phi(x, z)e^{-i\omega t}\}$. Therefore, $\phi(x, z)$ satisfies the Laplace equation.

$$\frac{\partial^2 \phi}{\partial x^2} + \frac{\partial^2 \phi}{\partial z^2} = 0. \tag{1}$$

The velocity potential $\phi(x, z)$ is decomposed into two parts (Zheng et al., 2004) as

$$\phi = \phi^S + \phi^H + \left(\frac{i\omega p}{\rho g}\right)\phi^R, \tag{2}$$

where, ϕ^S denotes the scattered velocity potential associated with the scattering of a plane incident wave in the absence of an imposed pressure on the internal free surface of the U-OWC chamber and ϕ^R denotes the radiated velocity potential, which is the solution to the radiation problem due to imposed pressure on the internal free surface. The scattered potential ϕ^S can be further decomposed into incident potential ϕ^I and diffracted potential ϕ^D. ϕ^H denotes the radiated potential considering heave motion of the body. The incident potential can be obtained analytically using linear water wave theory as

$$\phi^I = \frac{i\omega}{k_0}\frac{\cosh k_0(h-z)}{\sinh kh}e^{-ikx}, \tag{3}$$

where k_0 represents the wave number, and is the positive real root of the dispersion relation $\omega^2 = gk \tanh kh$.

The boundary condition at the mean free surfaces is given by

$$\frac{\partial \phi}{\partial n} = \frac{\omega^2}{g}\phi \text{ on } \Gamma_{FSe1} \text{ and } \Gamma_{FSe2} \tag{4}$$

$$\frac{\partial \phi}{\partial n} = \frac{\omega^2}{g}\phi + \frac{i\omega p}{\rho g} \text{ on } \Gamma_{FSi} \tag{5}$$

where, g is the gravitational acceleration and $\partial/\partial n$ represents normal derivative. The rigid sea bed boundary condition is given by

$$\frac{\partial \phi}{\partial n} = 0 \text{ on } \Gamma_{SB} \tag{6}$$

The boundary conditions on the impermeable U-OWC and Π-breakwater are given by

$$\frac{\partial \phi^H}{\partial n} = 1; \quad \frac{\partial \phi^{S,R}}{\partial n} = 0 \text{ on horizontal surfaces} \tag{7}$$

$$\frac{\partial \phi^H}{\partial n} = 0; \quad \frac{\partial \phi^{S,R}}{\partial n} = 0 \text{ on vertical surfaces} \tag{8}$$

Now, ϕ^S and ϕ^R satisfy the boundary condition Equations (4)-(6), but the Equation (4) is modified as follows

$$\frac{\delta \phi^S}{\delta n} = \frac{\omega^2}{g}\phi^S \text{ on } \Gamma_{FSe1}, \Gamma_{FSe2}, \Gamma_{FSi} \tag{9}$$

$$\frac{\delta \phi^R}{\delta n} = \begin{cases} \frac{\omega^2}{g}\phi^R \text{ on } \Gamma_{FSe1}, \Gamma_{FSe2} \\ \frac{\omega^2}{g}\phi^R + 1 \text{ on } \Gamma_{FSi} \end{cases} \tag{10}$$

Finally, the Sommerfeld condition for the far-field boundary condition is given by

$$\begin{cases} \frac{\partial(\phi^S - \phi^I)}{\partial n} = ik_0(\phi^S - \phi^I) \text{ on } \Gamma_{in} \text{ and } \Gamma_{out} \\ \frac{\partial \phi^R}{\partial n} = ik_0\phi^R \text{ on } \Gamma_{in} \text{ and } \Gamma_{out} \end{cases} \tag{11}$$

2.2 *Hydrodynamic parameters*

In the present section, the relationships and explicit expressions for key hydrodynamic parameters associated with the performance of U-OWC integrated with Π-breakwater is discussed. The time harmonic induced volume flux across the internal free surface (Evans & Porter, 1995) is given by

$$q = \int_{FSi} \frac{\partial \phi}{\partial y} dx = q^S - \frac{i\omega p}{\rho g} q^R \tag{12}$$

where q^S and q^R are the volume fluxes across the internal free surface FSi for the scattering and radiation problems, respectively. The volume flux for the radiation problem can be further decomposed into real and imaginary parts as

$$\frac{i\omega p}{\rho g} q^R = -(\bar{B} - i\bar{A})p \tag{13}$$

where A and B are called radiation susceptance and radiation conductance parameters (Evans & Porter, 1995), and are analogous to the added mass and the radiation damping coefficients in a rigid body system respectively. The expressions for these two parameters are given by

$$\bar{A} = \frac{\omega}{\rho g}\text{Re}\{q^R\} \text{ and } \bar{B} = \frac{\omega}{\rho g} Im\{q^R\} \tag{14}$$

The dimensionless forms of radiation susceptance and conductance is given by

$$\mu = \frac{\rho g}{\omega b}\bar{A} \text{ and } \nu = \frac{\rho g}{\omega b}\bar{B} \tag{15}$$

and, the dimensionless form of η_{max} is given by

$$\eta_{\max} = \frac{2\nu}{\nu + \sqrt{\nu^2 + \mu^2}} = \frac{2}{1 + \sqrt{1 + \left(\frac{\mu}{\nu}\right)^2}} \tag{16}$$

3 METHOD OF SOLUTION

The numerical solution is performed using Boundary Element Method (BEM). To obtain the solution for the boundary value problem, the weighted residue and Green-Gauss theorem is applied to the Laplace equation as given in Equation (1). The generalised form of boundary integral equation is expressed as

$$c(P)\phi(P) + \int_\Gamma \phi \frac{\partial G}{\partial n} d\Gamma = \int_\Gamma G \frac{\partial \phi}{\partial n} d\Gamma \tag{17}$$

where ϕ is the unknown flow potential, $\partial\phi/\partial n$ is the derivative of the potential relative to normal unit vector on the boundary, G is the fundamental solution to Laplace equation and Γ denotes the computational domain boundary. The free term coefficient $c(P)$ depends on the shape of the boundary and is given by

$$c(P) = \begin{cases} 1 \text{ for } P \equiv (\xi, \tau) \in \Omega, \\ \dfrac{1}{2} \text{ for } P \equiv (\xi, \tau) \in \Gamma \text{ and } \Gamma \text{ is smooth} \\ 0 \text{ for } P \equiv (\xi, \tau) \text{ outside } \Omega \end{cases} \quad (18)$$

Since the present method considers constant boundary elements with smooth boundary, the free term coefficient becomes

$$c(P) = \frac{1}{2} \quad (19)$$

The fundamental solution G, known as free-space Green's function, and its normal derivative $\partial G/\partial n$ corresponding to Laplace equation in two dimensions are given by

$$G = \frac{1}{2\pi} \ln r \text{ and } \frac{\partial G}{\partial n} = \frac{1}{2\pi r} \frac{\partial r}{\partial n} \quad (20)$$

where r is the distance between the field point $X(x, y)$ and the source point $P(\xi, \tau)$ given by

$$r = \sqrt{(\xi^2 - x^2) + (\tau^2 - y^2)} \quad (21)$$

The boundary is assumed to be divided into 'N' constant elements. The value of ϕ and G are assumed to be constant over each element and assumed to be equal to the value at the mid-element node. Further, the Equation (17) is discretised into 'N' constant boundary elements before applying the boundary conditions, and is expressed as

$$\frac{1}{2}\phi(P) + \sum_{j=1}^{N} \left(\int_{\Gamma_j} \phi_j \frac{\partial G}{\partial n} d\Gamma \right) = \sum_{j=1}^{N} \left(\int_{\Gamma_j} G_j \frac{\partial \phi}{\partial n} d\Gamma \right) \quad (22)$$

where, Γ_j is the boundary in the jth element. Now, considering $\int_{\Gamma_j} \frac{\partial G}{\partial n} d\Gamma - \frac{1}{2}\delta_{ij} = H_{ij}$, where δ_{ij} is the Kronecker delta which is defined as $\delta_{ij} = 0$ for $i \neq j$ and $\delta_{ij} = 1$ for $i = j$, and $\int_{\Gamma_j} G_j d\Gamma = Q_{ij}$, the Equation (22) can be rewritten as

$$-\frac{1}{2}\phi(P) + \sum_{j=1}^{N} (H_{ij}\phi_j) = \sum_{j=1}^{N} \left(Q_{ij} \frac{\partial \phi}{\partial n} \right) \quad (23)$$

The line integrals H_{ij} and Q_{ij} represent the influence functions between element i, at which the fundamental solution is applied, and any other element J under consideration. Now, to obtain the efficiency and radiation parameters pertaining to the OWC, the boundary conditions given in Equations (4-11) are applied on Equation (23), and integral equation for radiated potential is derived. Similarly, using boundary conditions on Equation (23), integral equation for scattered potential and for potentials in heave mode is derived. The derived integral equations are discretised and the values of H_{ij} and Q_{ij} are calculated using a standard Gaussian quadrature rule as per Katsikadelis (2016). Finally, using the method of point collocation, in which the source point $P(\xi, \tau)$ runs over each boundary element, final matrix forms of the equations are obtained which are then solved using MATLAB, and the various unknown velocity potentials and hydrodynamic parameters are calculated.

4 RESULTS AND DISCUSSIONS

The hydrodynamic performance of fixed floating U-OWCs with different bottom profiles integrated with Π-breakwater is analysed. The analysis performed using BEM approach is validated with the numerical result previously published results in the literature. The results related with changes in various geometrical and structural features of the integrated system are then examined in the following sub-sections.

4.1 *Validation of numerical model*

The BEM model used in the present study is validated by comparing the results of theoretical maximum efficiency of OWC with Evans and Porter (1995). The setup considered for validation is a bottom fixed OWC, considering the geometrical parameters as $B/h = 1$, $a/h = 1/4$, density of sea water $\rho = 1025\ Kg/m^3$. The theoretical maximum efficiency is calculated as a function of the non-dimensional wave number kh and the results are compared in Figure 2. It is observed that for $0 \leq kh \leq 1$, the η_{max} obtained using both methods show a similar steep rising limb, but there is a slight shift towards higher kh in the values obtained from the present BEM model. In the wavenumber range $1 \leq kh \leq 3$, the values from both studies matches for most of the part, differing only slightly at the extremes. In the range of $3 \leq kh \leq 4$, values obtained from the present model shift slightly towards lower kh. The apparent phase shift in the results between Evans and Porter (1995) and the present model can be explained by the different numerical methodologies utilized in the two methods. The study by Evans and Porter (1995) employed the eigenfunction expansion method, whereas the present study employs the Boundary Element Method. Velocity is estimated only for the elements on the boundary of the problem in the present BEM model, considering the effect of all other elements on one element, whereas the eigenfunction method computes the velocity potential of the entire volume of water.

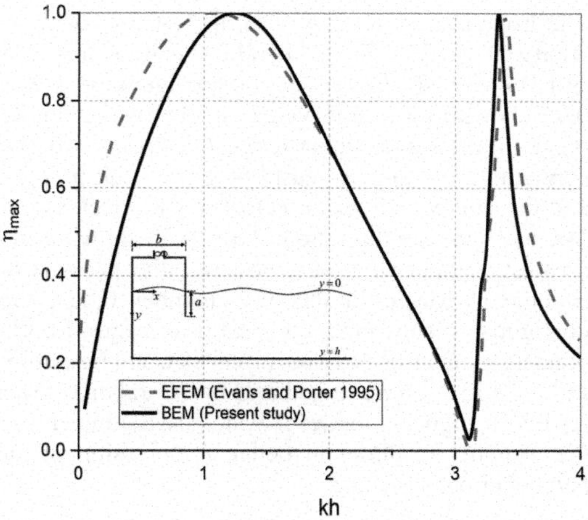

Figure 2. Validation of present BEM model with Evans and Porter (1995).

4.2 *Hydrodynamic performance of hybrid system*

The hydrodynamic performance, in terms of maximum theoretical efficiency η_{max}, non-dimensional radiation susceptance μ and radiation conductance υ of freely heaving U-OWC integrated with Π-breakwater is discussed in this section. Variations are made in the relative internal chamber width and relative draft of UWOC for both bottom profiles of U-OWC, i.e., inclined, and curved.

The volume of air getting compressed inside the internal chamber of the U-OWC is affected only by the changing volume due to heave motion, and hence, the parameters studied here take into consideration only the heave mode of motion of the system. The geometrical parameters of the base model considered are $B/h = 1$, $a/h = 0.25$, $T_1/h = 0.15$, $T_2/h = 0.60$ and $D/h = 0.75$ for the U-OWC, for the pi-breakwater as $B_1/h = 1$, $D_1/h = 0.50$, $B_2/h = 0.5$ and $D_2/h = 0.2$.

4.2.1 *Inclined bottom U-OWC with Π-breakwater*

Inclined bottom U-OWC integrated with pi- breakwater is analysed to study the effect of varying chamber width and draft of U-OWC on the energy conversion efficiency of the U-OWC. The effect of varying the relative internal chamber width *B/h* on the η_{max}, μ and υ is studied in Figure 3(a-c) for a freely heaving inclined bottom U-OWC integrated with pi-breakwater. In the short-wave regime of $0.0 \leq kh \leq 1.0$, the configuration with least *B/h* shows widest peak of efficiency (Figure 3a).

In this wavenumber range, as the *B/h* is decreased, more energy is transferred into the system denoted by higher peaks of υ (Figure 3c), however, more energy is also left unutilised denoted by increasing troughs of μ (Figure 3b). In the wavenumber range $1.0 \leq kh \leq 1.5$, all the configurations show multiple narrow peaks as well as one trough value. In this wavenumber range, configuration $B/h = 0.25$ shows an additional trough in η_{ma} which is lower than all other configurations. In the rest of the wavenumber range, $1.5 \leq kh \leq 4.0$, with increase in *B/h*, the number of occurrences of peak efficiency increases, however the peaks also become narrower. This increase in relative chamber width essentially increases the operational wavenumber range where the hybrid device gives maximum efficiency. A wider internal chamber ensures more volume of air available to operate the turbine and hence, better performance.

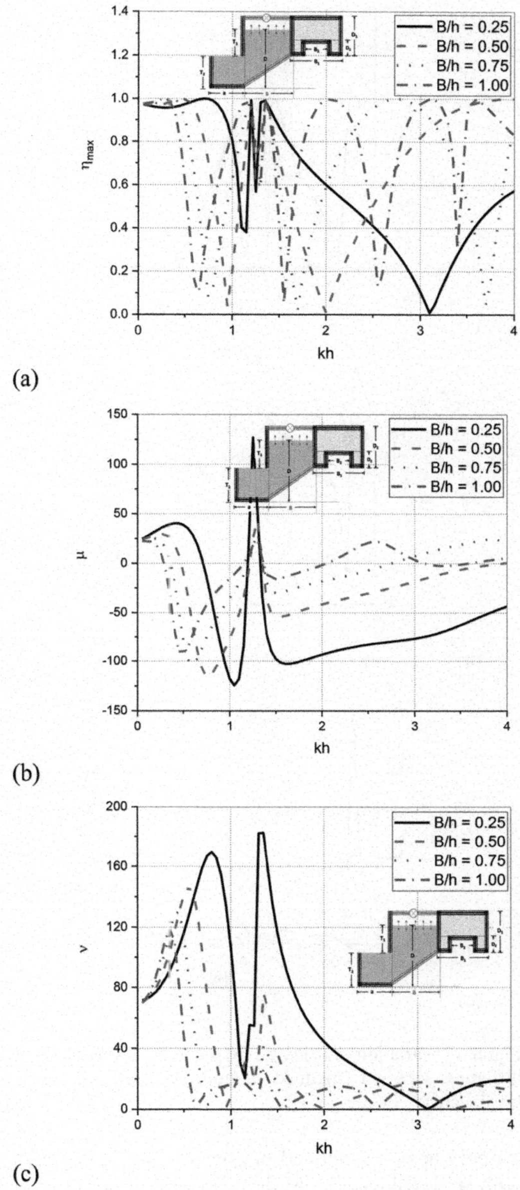

Figure 3. Variation of (a) η_{max}, (b) μ and (c) υ versus *kh* for different B/h of inclined bottom U-OWC with Π-breakwater.

The effect of varying the relative device draft *D/h* on the η_{max}, μ and υ is studied in Figure 4 (a-c) for

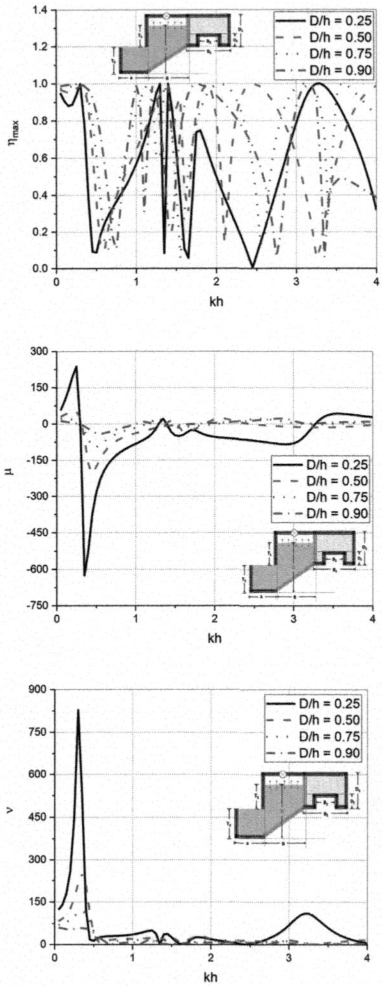

(a)

(b)

(c)

Figure 4. Variation of (a) η_{max}, (b) μ and (c) υ versus kh for different D/h of inclined bottom U-OWC with Π-breakwater.

a freely heaving inclined bottom U-OWC integrated with pi-breakwater. In the short-wave regime of wavenumbers $0.0 \leq kh \leq 0.5$, the efficiency η_{max} (Figure 4a) for all configurations reaches peak value and this peak is maintained for wider wavenumbers as the *D/h* is increased. In this wavenumber range, the behaviours of μ (Figure 4b) and υ (Figure 4c) for configuration *D/h* = 0.25 is the most exaggerated showing extremely high peaks indicating huge amount of energy being left unutilised while at the same time lot of energy being transferred into the system, however resulting in comparable efficiency as other configurations.

Within the wavenumber range $0.5 \leq kh \leq 1.5$, the various configurations show two more peaks in efficiency, with the peaks shifting towards lower *kh* as *D/h* is increased. In this wavenumber range the configurations also show a trough in value of η_{max}. The trough in value of η_{max} is highest for configuration *D/h* = 0.25 followed by *D/h* = 0.90, *D/h* = 0.75 and *D/h* = 0.50 . Moreover, the configuration *D/h* = 0.90 shows two trough values in this wavenumber range. In the wavenumber range $1.5 \leq kh \leq 4.0$, the configurations show at least one more peak in η_{max}, and the number of peaks increases with decrease in *D/h*, with the configuration *D/h* = 0.50 showing the greatest number of peaks, however the peaks also become narrower. In this wavenumber range, the configuration showing the widest peak values for the greatest number of peak occurrences is configuration *D/h* = 0.75 . Increasing the U-OWC relative draft results in higher volume of water oscillating inside the chamber resulting in better compression of air and better performance.

4.2.2 *Curved bottom U-OWC with Π-breakwater*

Curved bottom U-OWC integrated with pi-breakwater is also studied in terms of energy conversion efficiency, radiation susceptance and conductance parameters while varying the relative internal chamber width and the relative draft of U-OWC. The effect of varying the relative internal chamber width *B/h* on the η_{max}, μ and υ for a freely heaving curved bottom U-OWC integrated with pi-breakwater is studied in Figure 5(a-c).

Within the wavenumber range $0.0 \leq kh \leq 1.0$, the configurations show a peak value in η_{max} (Figure 5a) with the peak being wider as *B/h* is decreased, a behaviour observed earlier in the case of inclined bottom U-OWC (Figure 3a). Also, in this wavenumber range, the troughs in value of μ (Figure 5b) reach lower values and the peaks in value of υ (Figure 5c) reach higher values and shift towards higher *kh* as *B/h* is decreased. In the wavenumber range $1.0 \leq kh \leq 1.5$, the configurations show two more peaks in efficiency, and around $kh \approx 1.3$, the configuration $B/h = 0.25$ shows an exaggerated peak value of μ and υ as well as a deep trough in the value of η_{max}.

The widest peak in η_{max} for all configurations occurs in the intermediate and long-wave regime of $1.5 \leq kh \leq 4.0$, wherein the peaks get narrower and shift towards lower *kh* as *B/h* is increased, except for the configuration *B/h* = 1.0 showing two peaks and the configuration *B/h* = 0.25 showing only a deep trough and no peak value. From this behaviour, it can be inferred that increasing the chamber width improves the efficiency of the device which is expected as a wider chamber means more volume of air being pushed through the turbine.

The effect of varying the relative draft *D/h* on the η_{max}, μ and υ for a freely heaving curved bottom U-OWC integrated with pi-breakwater is studied in Figure 6(a-c). In the wavenumber range $0.0 \leq kh \leq 0.5$, the configurations show a peak in η_{max} (Figure 6a), with the peak being maintained for a wider range as D/h is increased. Within this wavenumber range, the configuration *D/h* = 0.25 shows exaggerated peaks in values of μ (Figure 6b) and υ

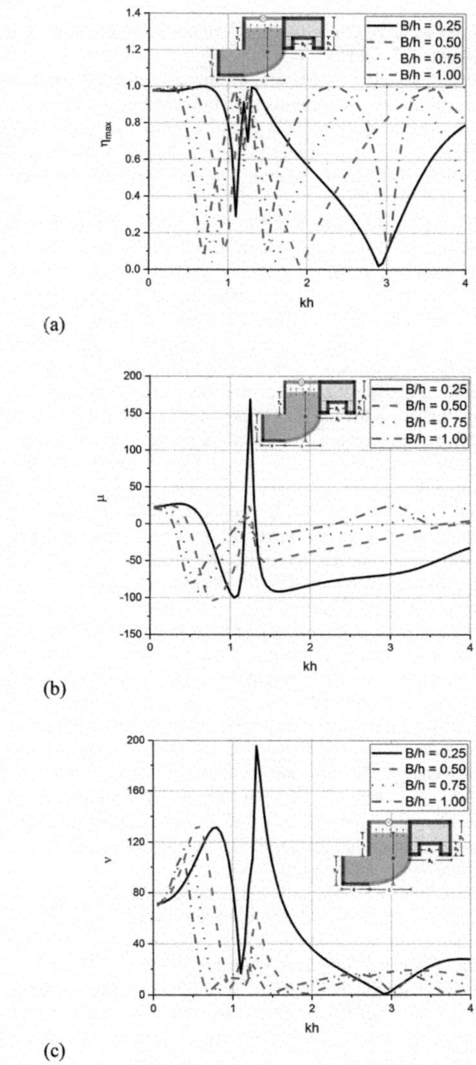

Figure 5. Variation of (a) η_{max}, (b) μ and (c) υ versus *kh* for different B/h of curved bottom U-OWC with Π-breakwater.

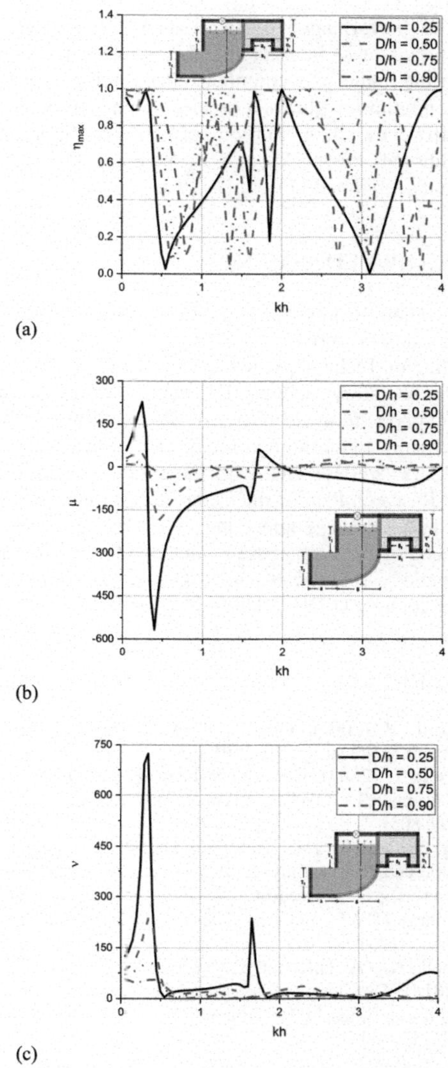

Figure 6. Variation of (a) η_{max}, (b) μ and (c) υ versus *kh* for different *D/h* of curved bottom U-OWC with Π-breakwater.

(Figure 6c) and emerges as the least efficient configuration. Within $0.5 \leq kh \leq 1.5$, the configurations show at least one more peak in η_{max}, except for the configuration *D/h* = 0.25. The configuration *D/h* = 0.50 shows two peaks, both reaching maximum efficiency. The widest peaks in efficiency for all configurations are observed in the wavenumber range of $1.5 \leq kh \leq 4.0$, with all configurations showing at least two peaks wherein configuration *D/h* = 0.75 shows the widest peaks and configuration *D/h* = 0.25 shows the narrowest peaks. The trough in value of η_{max} shown by configuration *D/h* = 0.25 at *kh* = 3.1 reaches till zero, lower than any other configuration. Increasing the draft of the device improves the energy conversion efficiency as expected since longer column of water compresses larger volume of air through the turbine.

5 CONCLUSIONS

In the present study, the hydrodynamic performance of a freely heave only U-OWC with inclined and curved bottom integrated with Π- breakwater is analysed. The conclusions drawn from the study are as follows:

- A wider relative internal chamber width *B/h* results in better U-OWC device efficiency irrespective of the chamber bottom profile.
- Irrespective of the bottom profile, a larger draft intuitively indicates towards more energy output, which is what was observed, but a draft of *D/h* = 0.75 is preferable than *D/h* = 0.90 as both configurations have almost similar performance while the shorter draft can lead to significant savings in construction costs.

- The performance of the U-OWC is similar in both cases of internal chamber profiles. However, in the higher wavenumber range, inclined bottom configuration shows a greater number of peaks in efficiency than curved bottom configuration, although the peaks of the latter are wider than that of inclined bottom U-OWC.

ACKNOWLEDGEMENTS

The authors express their gratitude to the Ministry of Education, Government of India, and the National Institute of Technology, Karnataka, Surathkal, for providing necessary facilities. DK acknowledges the partial support from Ministry of Ports, Shipping and Waterways, India through the research grant no. DW/01013(13)/2/2021. This work contributes to the Strategic Research Plan of the Centre for Marine Technology and Ocean Engineering (CENTEC), which is financed by the Portuguese Foundation for Science and Technology (Fundação para a Ciência e Tecnologia) under contract UIDB/UIDP/00134/2020.

REFERENCES

Boccotti, P. (2003). On a new wave energy absorber. *Ocean Engineering*, 30, 1191–1200.

Boccotti, P. (2007). Comparison between a U-OWC and a conventional OWC. *Ocean Engineering*, 34(5-6), 799–805.

Cong, P., Teng, B., Liu, Y. & Ning, D. (2022). A numerical approach for hydrodynamic performance evaluation of multi-degree-of-freedom floating oscillating water column (OWC) devices. *Journal of Fluids and Structures*, *114*, 103730.

Connell, K. O., Thiebaut, F., Kelly, G. & Cashman, A. (2018). Development of a free heaving OWC model with non-linear PTO interaction. *Renewable Energy*, *117*, 108–115.

Cruz, J. (2007). *Ocean wave energy: current status and future perspectives*. Springer Science & Business Media.

Dean, R.G. & Dalrymple, R.A. (1991). Water wave mechanics for engineers and scientists (Vol. 2). *World scientific publishing company*.

Evans, D.V. (1982). Wave-power absorption by systems of oscillating surface pressure distributions. *Journal of Fluid Mechanics*, 114, 481–499.

Evans, D.V. & Porter, R. (1995). Hydrodynamic characteristics of an oscillating water column device. *Applied Ocean Research*, 17, 155–164.

Falcão, A., 2015. Developments in oscillating water column wave energy converters and air turbines, in: Guedes Soares, C. (Ed.), Renewable Energies Offshore. Taylor & Francis Group, London, pp. 3–11.

Graw, K. U. (1996). Wave energy breakwaters-a device comparison. *Proc. Conference in Ocean Engineering, Madras, INDIA*.

He, F., Leng, J. & Zhao, X. (2017). An experimental investigation into the wave power extraction of a floating box-type breakwater with dual pneumatic chambers. *Applied Ocean Research*, 67, 21–30.

Heath, T.V. (2012). A review of oscillating water columns. *Philosophical Transactions of the Royal Society A: Mathematical, Physical and Engineering Sciences*, *370* (1959), 235–245.

Katsikadelis, J.T. (2016). The Boundary Element Method for Engineers and Scientists, *Academic Press, Oxford*.

Luo, Y., Wang, Z., Peng, G., Xiao, Y., Zhai, L., Liu, X. & Zhang, Q. (2014). Numerical simulation of a heave-only floating OWC (oscillating water column) device. *Energy*, *76*, 799–806.

Masuda, Y. & McCormick, M. E. (1986). Experiences in pneumatic wave energy conversion in Japan. *Coastal Engineering*, 1–33.

Muduli, R., Patil, S. B. & Karmakar, D. (2024). Hydrodynamic performance of pile restrained U-shaped OWC device using boundary element method. *Engineering Analysis with Boundary Elements*, *158*, 139–159.

Ohneda, H., Igarashi, S., Shinbo, O., Sekihara, S., Suzuki, K., Kubota, H., Ogino, H. & Morita, H. (1991). Construction procedure of a wave power extracting caisson breakwater. *Proc. 3rd Symp. Ocean Energy Utilization, Tokyo*, *1991*, 171–179.

Ozkop, E. & Altas, I. H. (2017). Control, power and electrical components in wave energy conversion systems: A review of the technologies. *Renewable and Sustainable Energy Reviews*, *67*, 106–115.

Rezanejad, K. and Guedes Soares, C. Hydrodynamic investigation of a novel concept of Oscillating Water Column type wave energy converter device. Journal of Offshore Mechanics and Arctic Engineering. 2021; 143:042003.

Rezanejad, K., Souto-Iglesias, A. & Guedes Soares, C. (2019). Experimental investigation on the hydrodynamic performance of an L-shaped duct oscillating water column wave energy converter. *Ocean Engineering*, *173*, 388–398.

Takahashi, S., Nakada, H., Ohneda, H. & Shikamori, M. (1992). Wave power conversion by a prototype wave power extracting caisson in Sakata port. *Coastal Engineering 1992*, 3440–3453.

Torre-Enciso, T., Ortubia, I., López de Aguileta, L.I. & Marqués, J. (2009). Mutriku Wave Power Plant: from the thinking out to the reality. *Proc. 8th European Wave and Tidal Energy Conference, Uppsala, Sweden*, 319–329.

Wang, C. and Zhang, Y. (2021). Hydrodynamic performance of an offshore Oscillating Water Column device mounted over an immersed horizontal plate: A numerical study. *Energy*, 222, 119964.

Wang, C., Ma, T. & Zhang, Y. (2022). Semi-analytical study on an integrated-system with separated heaving OWC and breakwater: Structure size optimization and gap resonance utilization. *Ocean Engineering*, 245, 110319.

Wang, C., Zhang, Y. & Deng, Z. (2022). Hydrodynamic performance of a heaving oscillating water column device restrained by a spring-damper system. *Renewable Energy*, *187*, 331–346.

Wang, L., Isberg, J. and Tedeschi, E. (2018). Review of control strategies for wave energy conversion systems and their validation: the wave-to-wire approach. Renewable and Sustainable Energy Reviews, 81, 366–379.

Zheng, S., Antonini, A., Zhang, Y., Greaves, D., Miles, J. & Iglesias, G. (2019). Wave power extraction from multiple oscillating water columns along a straight coast. *Journal of Fluid mechanics*, *878*, 445–480.

Zheng, Y. H., You, Y. G. & Shen, Y. M. (2004). On the radiation and diffraction of water waves by a rectangular buoy. *Ocean engineering*, *31*(8-9), 1063–1082.

Zhou, Y., Ning, D., Liang, D. & Cai, S. (2021). Nonlinear hydrodynamic analysis of an offshore oscillating water column wave energy converter. *Renewable and Sustainable Energy Reviews*, *145*, 111086.

Advances in Maritime Technology and Engineering – Guedes Soares & Santos (Eds)

Sensitivity analysis of loads on subsea power cables during installation

M. Peres
Centre for Marine Technology and Ocean Engineering (CENTEC), Instituto Superior Técnico Universidade de Lisboa, Lisbon, Portugal
Subsea7 Portugal, Lisbon, Portugal

S. Wang & C. Guedes Soares
Centre for Marine Technology and Ocean Engineering (CENTEC), Instituto Superior Técnico Universidade de Lisboa, Lisbon, Portugal

ABSTRACT: In this work, several analyses were performed to evaluate the response of different submarine power cables during normal lay under environmental loads in a number of scenarios for a fixed offshore wind farm. To do so, a numerical model was built in the commercial software OrcaFlex that enables the performance of static and dynamic analyses. Over different cable weights, layback distances, configurations and environmental conditions, the cable response was analysed. The main focus to evaluate the occurrence of top tension and compression loads in the touchdown point. Later, the vessel motion was correlated to the cable behaviour. The results indicated that the cable weight plays an important role in the final response, as well as the layback distance and water depth. Deploying the cable in J-lay method allows a higher workability when compared to standard S-lay deployment method. Finally, a strong correlation between the chute vertical downwards velocity and the cable tension and compression on the touchdown point was encountered.

1 INTRODUCTION

The race to mitigate climate change has driven the offshore wind towards protagonism based on IRENA (2022) projections. Ambitious targets have been set by governments and alongside, engineering challenges rise to ensure an efficient and safe installation. Mamatsopoulos et al. (2020) stated that the installation of subsea power cables is a complex and demanding task, mainly linked to extreme environmental conditions associated with the dynamic behaviour of the system, that leads to an increased engineering cost and detailed planning to mitigate the risks, according to Calavia et al. (2022) and Worzyk (2009). Efficient cable installation requires extensive analysis with appropriate and reliable tools to avoid uncontrolled responses and further damages Díaz and Guedes Soares (2022).

Okkerstrøm et al. (2021) investigated the key parameters related to cable failure during installation and identified a series of parameters to optimize the laying process. Nesje et al. (1999) stated that compression loads are one of the causes of failure of flexible pipes, however, many suppliers specify a limit of 0kN for axial compression, according to Reda (2020).

The influence of external parameters on the tension response was addressed by Jeng and Zhou (2013) and Yoshizawa. and Yabuta (1983). Analysis tools to estimate the cable dynamic response has been a topic of reserch over the years, such as Zajac (1957), Walton and Polachek (1960), Vaz and Patel (1995) and Vaz and Patel (2000). Research on cable modelling indicates a complexity to accurately represent the mechanical properties during analysis, mainly the bending stiffness, topic of study by Dreyer and Van Vuuren (1999), Tan et al. (2007) and Péronne et al. (2015).

Compression limits were identified as one of the main limiting criteria during the installation of power cables by Ribeiro et al. (2003) and Drumond et al. (2018), mainly due to the vessel motion, as studied by yas (2014) and Aranha et al. (2001). According to Calavia et al. (2022), there is a strong correlation between the chute velocity and the dynamic behaviour of the touchdown point. Loos (2017) also affirmed that compression is one of the main limiting criteria for cable installation.

Current industry standards provide guidelines on analysis methodologies, e.g., DNV (2014), DNV (2017), DNV (2016a) and DNV (2016b). However, according to Reda (2020), there are neither industry standards to aid the determination of compression limits in submarine cables nor guidance on the consequences of long term exposure to compression loads for the integrity of the cables.

DOI: 10.1201/9781003508779-52

In this work, a sensitivity analysis is performed to evaluate the influence of several deployment variables on the cable's response and mechanical integrity, mainly the compression and top tension during normal lay in shallow waters. To assess the integrity of the cable under different scenarios, a numerical model of the vessel and power cable is developed using the commercial software OrcaFlex to perform static and dynamic analyses. First, the static equilibrium is obtained considering only the cable self-weight for different cables, water depths, pre-tensions and configurations. Pre-tension loads are defined by different layback distances and serve as base models for the time-domain analysis. Several environmental conditions are applied, such as waves and currents, and the cable dynamic response is monitored for every case. Later, an analysis on the vessel motion influence on the cable response is also performed.

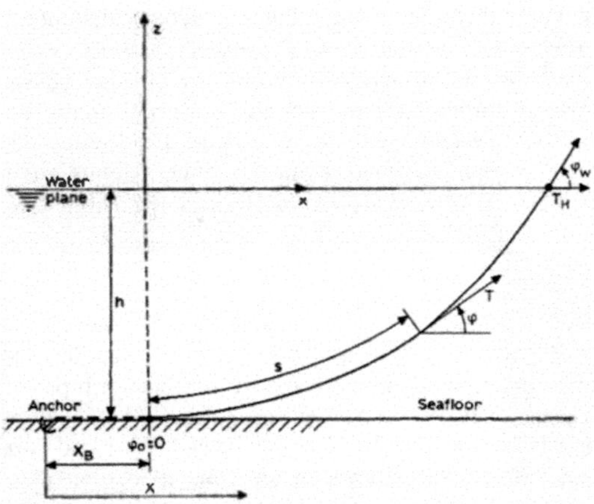

Figure 1. Catenary shape.

2 METHODOLOGY

2.1 *Catenary equations*

During cable laying, the cable assumes a catenary configuration from the departure point to the touch-down point. Simplified static analysis can be performed by using catenary equations. Analytical solutions for the catenary line are derived from the following assumptions from Worzyk (2009):

1. Bending stiffness is neglected;
2. Drag forces are neglected;
3. The line has a uniform weight per meter.

Faltinsen (1990) derived the catenary equations for the inelastic line. Considering that the seabed is at $z_0 = -h$ (Figure 1), the suspended length s of the cable is calculated by:

$$s = \frac{T_H}{w} sinh\left(\frac{w}{T_H}x\right) \tag{1}$$

where:
s: suspended length of the cable;
T_H: horizontal tension;
w: weight per unit length of the line.

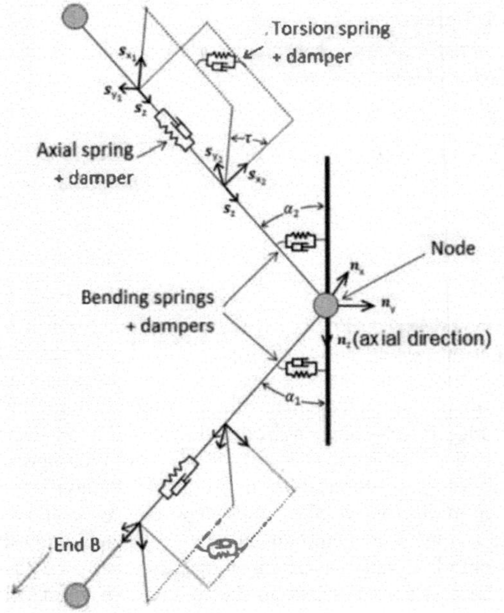

Figure 2. Lumped mass method.

The cable configuration is then given by:

$$z + h = \frac{T_H}{w}\left[cosh\left(\frac{w}{T_H}x\right) - 1\right] \tag{2}$$

The tension on the line is calculated by:

$$T = T_H + wh + (w + \rho g A)z \tag{3}$$

$$T_z = ws \tag{4}$$

where T_z is the vertical component of the tension.

2.2 *Lumped mass method*

Inter array cables are modelled as line objects with specified properties. OrcaFlex uses the lumped mass method (LMM) for line modelling as shown in Figure 2. The line is modelled as a series of nodes connected to each other by massless line segments. Axial, bending and torsional stiffness are represented by a series of spring-dampers and the remaining properties are lumped on the nodes.

2.3 *Cable mechanical properties*

The axial strain (ε) is defined as the ratio between the variation of the length by the unstreched length of the line, i.e., the length when unpressured and unstressed.

$$T_w = EA^o \tag{5}$$

Torsional stiffness relates the twist on the cable and the torsional moment required to achieve that twist angle. It is the slope of the torsional moment-twist per unit length.

$$GJ = \frac{M}{\alpha} \tag{6}$$

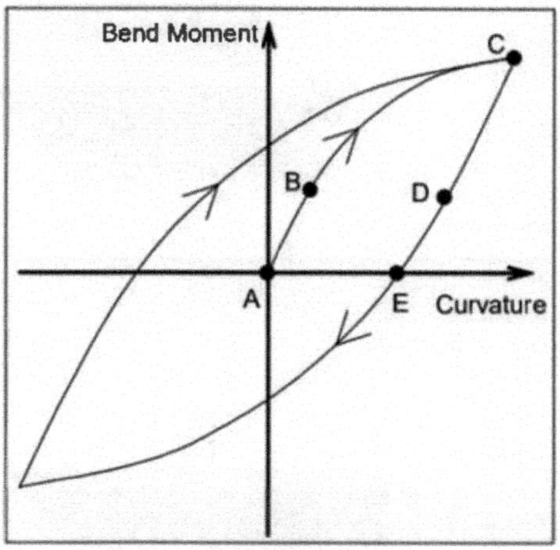

Figure 3. Bending moment hysteresis loop.

where:

GJ = torsional stiffness, $kN.m^2$

M = torsional moment, $kN.m$

α = twist angle, rad/m

Generally, unbonded flexible pipes and power cables exhibit a hysteretic nonlinear behaviour, according to Tan et al. (2007). This property is illustrated in Figure 3.

2.4 *Modelling of sea states*

Regular waves are modelled with Dean Stream 5th order wave theory, described by Equation. 7:

$$\psi(x,z) = B_0 z + \sum_{j=1}^{N} B_j \frac{sinh(jkz)}{cosh(jk)} cos(jkx) \tag{7}$$

where k is the wave number to be determined. Numerical solution is needed to determine coefficients B_j and k. The order of the stream function is the value of N.

According to DNV (2014), a realistic sea state is better described by a random wave model, either by a deterministic or by a stochastic method. Irregular waves are described by a JONSWAP wave spectrum $S_j(\omega)$ as:

$$S_j(\omega) = A_\gamma S_{PM}(\omega) \gamma^{exp\left(-0.5\left(\frac{\omega-\omega p}{\sigma \omega p}\right)^2\right)} \tag{8}$$

where:

S_{PM} is the Pierson-Moskowitz spectrum;

γ is the peak shape parameter;

σ is the spectral width parameter.

The peak shape parameter γ is calculated as a function of the significant wave height and peak period as follows:

$$\gamma = 5 \; for \; \frac{T_p}{\sqrt{H_s}} \leq 3.6 \tag{9}$$

$$\gamma = exp\left(575 - 115\frac{T_p}{\sqrt{H_s}}\right) \; for \; 3.6 < \frac{T_p}{\sqrt{H_s}} < 5 \tag{10}$$

$$\gamma = 1 \; for \; 5 \leq \frac{T_p}{\sqrt{H_s}} \tag{11}$$

Although ocean waves are random in nature, it is a common practice to perform analysis by representing the extreme wave with a regular wave model with a characteristic wave height, period and direction. From the wave height described as a Rayleigh Distribution, the following is derived:

$$H_{max} = 1.86 * H_s \tag{12}$$

The associated wave period can be calculated by (Orcina, 2023)

$$1.05 * T_z < T_{max} < 1.4 * T_z \tag{13}$$

where T_z is the zero up-crossing period. Taking the average value and considering the relationship between $Tz = 0.777 * T_p$ from Journée and Massie (2001), T_{max} can be calculated as:

$$T_{max} = 0.9 * T_p \tag{14}$$

Vertical current profile is modelled with a 7th power law, according to DNV (2014) as:

$$v_c(z) = v(0)\left(\frac{d+z}{d}\right)^{1/7} \; for \; z \leq 0 \tag{15}$$

where:

$v_c(z)$: total current velocity at level z;

z: distance from still water level, positive downwards;

$v(0)$: velocity at the still water level;

d: water depth.

2.5 *Environmental loads*

The relative motion between ocean currents and subsea power cables creates a force in the opposite direction that is proportional to the fluid density, velocity and projected area as:

$$f = \frac{1}{2}\rho C_D A U^2 \quad (16)$$

where:
ρ: fluid density;
A: projected area normal to the flow;
U: relative velocity;
C_D: drag coefficient.

The drag coefficient C_D is the non-dimensional drag force (DNV, 2014) and can be computed as a function of the Reynoulds number Re:

$$Re = \frac{U_a D}{\nu} \quad (17)$$

where U_a is the flow velocity amplitude, D is the diameter of the cable and v is the kinematic viscosity of the fluid.

Wave loads acting on slender structures can be computed by using the empirical equation derived by Morison et al. (1950). In his study, Morison et al. (1950) predicted the force exerted by waves on a vertical pile by superimposing the inertia force and the quadratic drag force as:

$$f = \rho C_M \frac{\pi D^2}{4}\dot{u} + \frac{1}{2}\rho C_D D|u|u \quad (18)$$

where:
f: horizontal force per unit length;
C_M: inertia coefficient;
D: pile diameter;
u: wave particle velocity;
$\dot{u}$: fluid acceleration.

The total load on a submerged body can be computed by means of the inertia and drag forces contributions using Morison equation. The dominance of the drag or inertia can be measured by the Keulegan Carpenter number (KC) as follows:

$$KC = 2\pi \frac{x_a}{D} \quad (19)$$

where x_a is the amplitude of the displacement of the body. In general, the following can be assumed, as per (Journee and Massie, 2001):

- For $KC < 3$, inertia force is dominant and drag can be neglected;
- For $3 \leq KC < 15$, the drag becomes significant;
- For $15 \leq KC < 45$, the full Morison equation must me applied;
- For $KC > 45$, drag force is dominant and inertia can be neglected.

3 DESCRIPTION OF THE MODEL

3.1 *Scenarios definition*

The defined scenarios for the sensitivity analysis are:

- Scenario 1: cable lay in S-lay with different bending stiffness;
- Scenario 2: cable lay in S-lay with different weights, pre-tensions and water depths;
- Scenario 3: cable lay in J-lay lay method

Three different cables and a cable laying vessel (CLV) were modelled in OrcaFlex. Dynamic analysis were performed over all models by including waves and currents. The design limits of the cable were investigated.

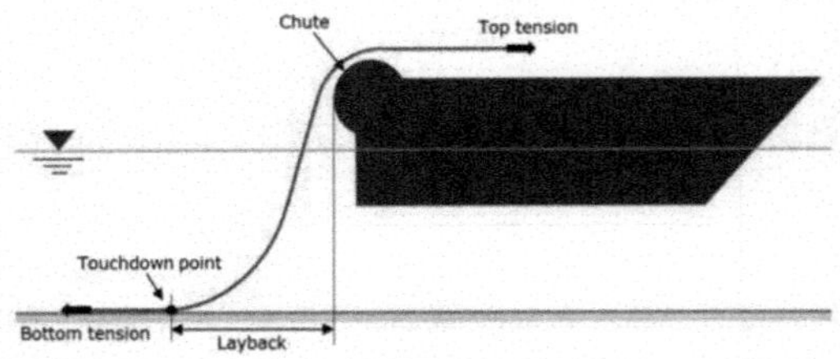

Figure 4. S-lay model.

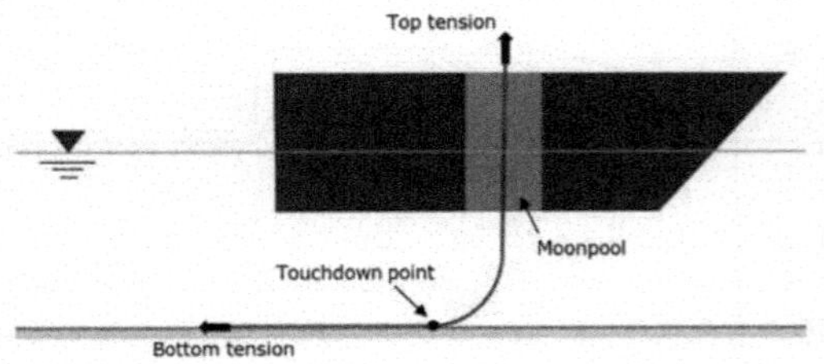

Figure 5. J-lay model.

3.2 *Laying configurations*

In the first and second scenarios the cable will be deployed in an S-lay method over the stern chute. The cable was modelled as fixed to the vessel on the deck level with nearly 6m of cable length forwards of the chute (Figure 4). The cable is positioned over the starboard side of the chute. On the other end, the cable was set as anchored to the seabed. For the J-lay model the cable is deployed through a moonpool that was modelled as an elastic solid cylinder located in the center of the vessel (Figure 5).

3.3 *Environmental conditions*

Based on a scatter table and current data of the installation site, the environmental conditions were modelled according to Table 2. Currents are modelled with a surface speed of 0.8 ms^{-1} and are set as colinear with the wave headings, based on Orcina (2018).

3.4 *Cable modelling and hydrodynamic data*

The properties of the cables are listed in Table 3. Cables were modelled with a hysteretic nonlinear bending stiffness. The drag coefficient (C_D) was modelled as a function of the Reynolds number for a smooth surface according to DNV (2014), as illustrated in Figure 6. The added mass coefficient C_A was based on DNV (2014) recommendations for a cylindrical object, taken as 1.0. The inertia coefficient was calculated by $C_M = C_A + 1$. The OrcaFlex final model is shown in 7.

Table 1. Wave scatter table.

		T_p range (s)						
	0	2	4	6	8	10	12	
H_s (m)		2	4	6	8	10	12	14
0	1	-	0%	5%	13%	11%	5%	1%
1	2	-	0%	3%	17%	14%	9%	1%
2	3	-	0%	2%	5%	9%	2%	0%
3	4	-	-	-	0%	1%	0%	0%

Table 2. Sea states for dynamic analysis.

Parameter	Min.	Max	Step
H_s (m)	0.5	3	0.5
T_p (s)	5	12	1

Table 3. Cables specifications.

Parameter	Cable A	Cable B	Cable C	Unit
Conductor material	Copper			-
Outer diameter	140.3	144.8	160.2	[m]
Weight in air	33	36	42	[kg/m]
Weight in water	18	20	25	[kg/m]
Axial stiffness	98	144	322	[MN]
Torsional stiffness	6.2	9	19.2	[kN.m^2]
Bending stiffness	Hysteretic nonlinear			[-]

3.5 *Simulation setup*

The normal lay operation occurs with the vessel with a forward speed similar to the payout rate of the power cable in order to maintain the layback distance. According to DNV (2014), however, it is recommended to model the vessel as stationary, representing the constant layback.

The setup of the simulation time must be carefully done. For the models in this study, the simulation duration is as 2 * T_p for the build-up period and 5 * T_p for Stage 1. Several combinations of parameters lead to a high number of simulations. Table 4 summarizes the simulation parameters used for the dynamic analysis.

4 RESULTS AND DISCUSSION

4.1 *Static results*

Static loads were analysed for Cables A, B and C in a range of layback distances resulting in bottom tensions from 1kN to 10kN over three water depths (20m, 35m and 50m). Figure 8 shows the variation of the top tension with the layback distance for all cables over a 35m water depth. A relationship between the cables' self-weight and the top tension can be observed. A greater layback distance also means that more cable is suspended, thus increasing the tension on the top.

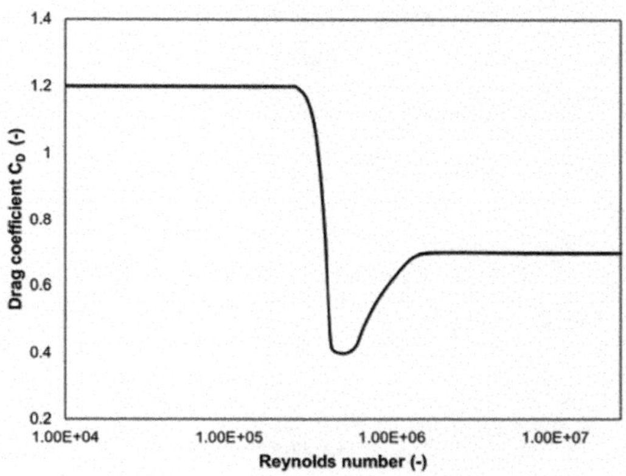

Figure 6. Modelling of drag coefficient.

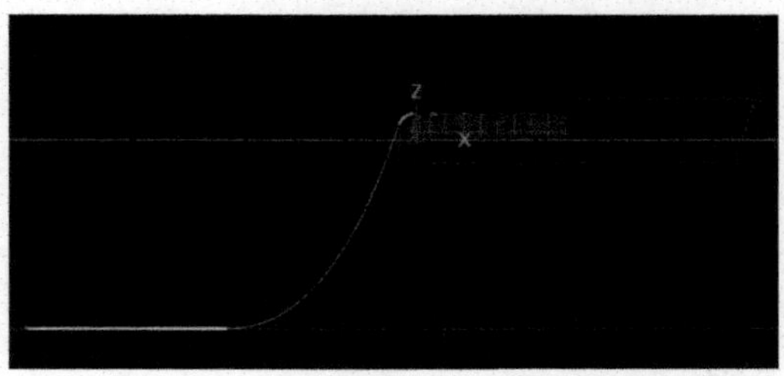

Figure 7. OrcaFlex numerical model in S-lay.

Table 4. Dynamic analysis parameters.

Parameter	Description
Wave height	0.5*m* to 3.0*m*
Wave period	5*s* to 12*s*
Wave direction	0 *deg* to 330 *deg*
Current speed (surface)	0.8 *m/s*
Current direction	Co-linear with waves
Simulation time	5 * T_p
Time step	0.01 *s*
Integration method	Implicit

4.2 *Dynamic results*

4.2.1 *Wave model*

Irregular wave results were also compared to a regular wave approach for a selected case. Comparison was made with a normalised compression in order to verify the magnitude of the tension loss in relation to the initial tension. The calculation of the dimensionless compression is done as:

$$C(-) = \frac{T_0 - T_b}{T_0} \tag{20}$$

where:
T_0 is the initial bottom tension;
T_b is the measured bottom tension.

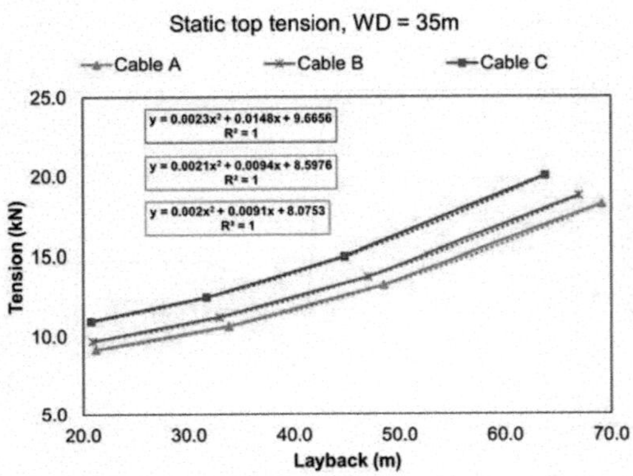

Figure 8. Variation of the static top tension with the layback distance for WD = 35m.

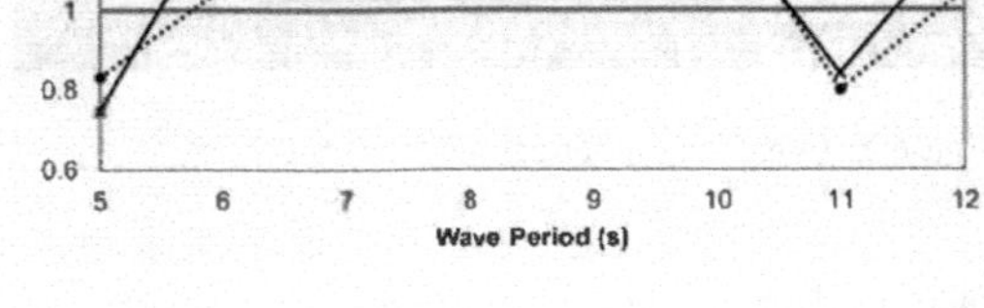

Figure 9. Normalised compression with regular and irregular waves.

The results for different wave periods ranging from 5*s* to 12*s* for a wave heading of 270 *deg* is presented in 9. Overall, the response of the system presents more conservative results with regular waves.

4.2.2 *Bending stiffness*

Figure 10 compares the minimum effective tension along the arc length for bending stiffness. Linear and

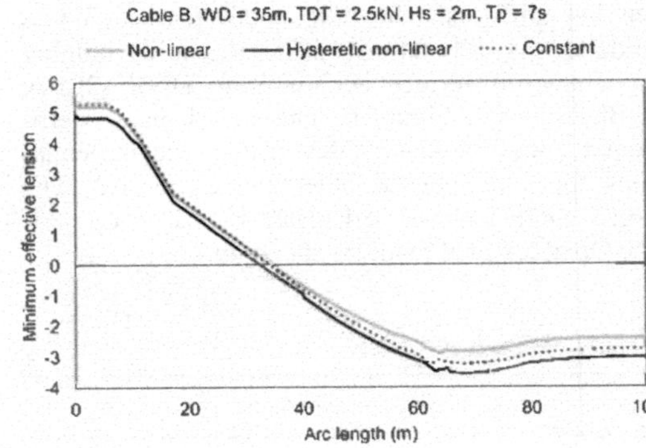

Figure 10. Minimum effective tension along arc for different bending stiffnesses.

non-linear bending stiffness tend to underestimate the compression loads on touchdown point (around 61m arc length). Even though the minimum tension along the hanging cable does not present significant differences, the tension on the bottom is higher when hysteretic non-linear bending stiffness is modelled.

4.2.3 *Effect of cable self-weight and initial bottom tension*

A verification of the compression behaviour for different cable self-weights was performed with different water depths and initial bottom tensions. In this scenario, the analysed cables are listed in 3. Layback distances were callibrated at pre-tensions of 1*kN*, 2.5*kN*, 5*kN* and 10*kN* at water depths of 20*m*, 35*m* and 50*m*.

Figures 11, 12 and 13 show the plots of the normalised compression for cables A, B and C at a water depth of 20m and initial bottom tension of 5kN as a function of the wave peak period for different wave headings. Cables A and B present very similar results, although slight differences are noticed. At 7*s* and 225*deg* heading, Cable C presented a compression peak, but overall the tension drop is smaller for this cable, even though compression takes place in all cases.

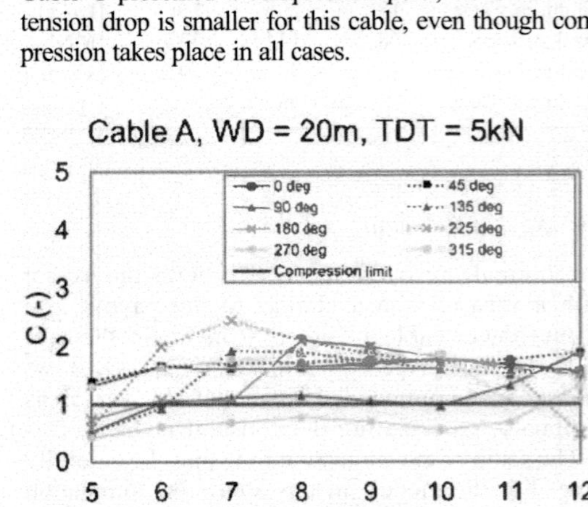

Figure 11. Normalised compression vs. wave period. Cables A, Maximum values for whole simulation, WD = 20m, TDT = 5kN.

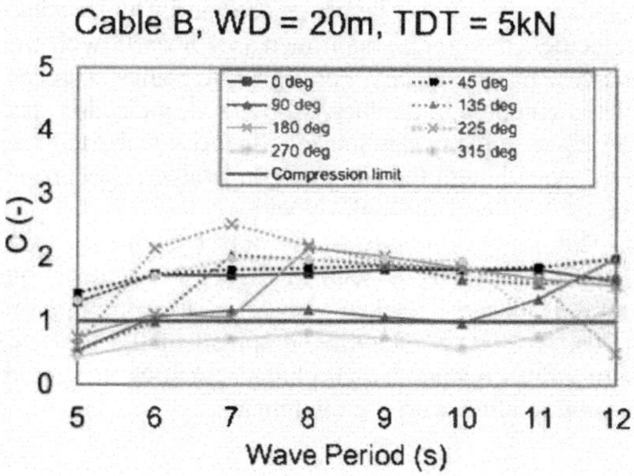

Figure 12. Normalised compression vs. wave period. Cables B, Maximum values for whole simulation, WD = 20m, TDT = 5kN.

The effect of the water depth is shown in Figures 14, 15 and 16. To maintain the initial target bottom tension for deeper waters means that the layback distance must be smaller to compensate for the weight of the longer suspended cable. The implication of shorter laybacks is that the cable tends to present a"snaking" behaviour when the chute descends. If the initial bottom tension is small, the tension loss is higher and compression is accentuated.

Lower bottom tensions present a greater tension loss on the touchdown point. For all analysed wave characteristics, the minimum normalised tension on the bottom was above 1, i.e., the cable is under compression. The effect increases with the increase of the significant wave height.

4.2.4 *Effect of deployment configuration*

In this scenario, one cable type was analysed with the J-lay method of installation in four different static bottom tensions. The cable is deployed through a moonpool located near the midship section. Tension loss on touchdown point is amplified at beam seas for

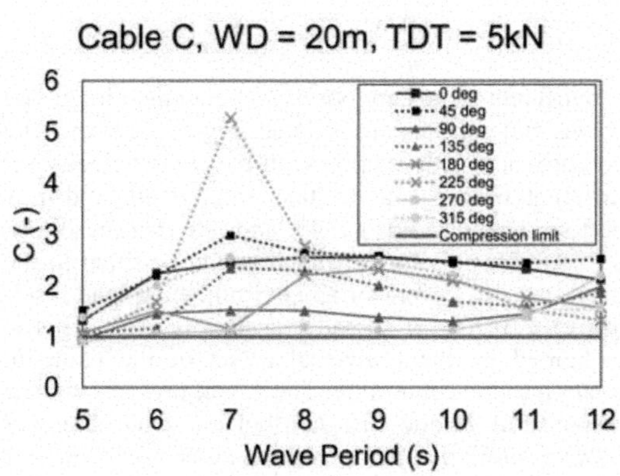

Figure 13. Normalised compression vs. wave period. Cables C, Maximum values for whole simulation, WD = 20m, TDT = 5kN.

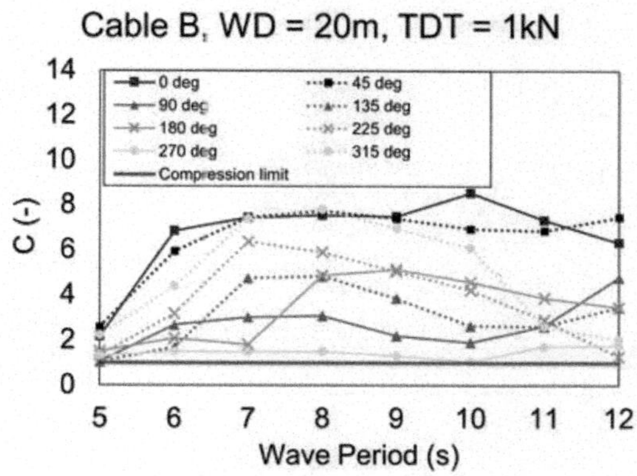

Figure 14. Normalised compression vs. wave period. Cable B, Maximum values for whole simulation, WD = 20m, TDT = 1kN.

J-lay installation, rather than quarterly waves observed in the S-lay deployment method. The compression peak periods occur mostly at 8s, decrease at 10s and increase again for higher wave periods (Figure 17).

The different response of the cable for S-lay and J-lay methods can be explained based on the motion amplitude of both points for the same incident wave. Figure 18 shows the wave elevation for $H_s = 3.0m$ and $T_p = 12s$ and the chute and moonpool vertical amplitude response. The stern chute Z amplitude is clearly amplified in relation to the wave elevation, while the moonpool motion has a smaller amplitude given its position along the vessel.

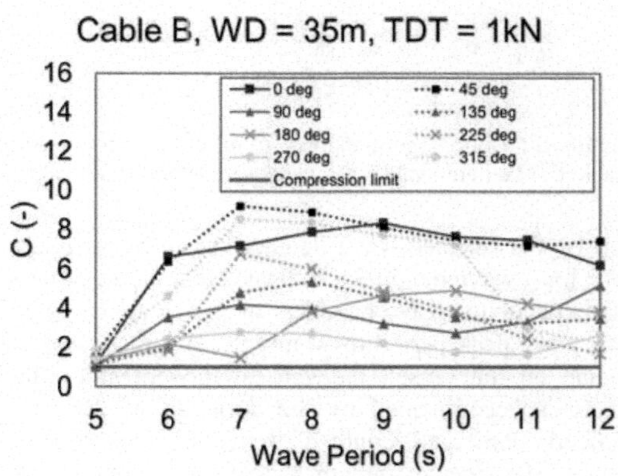

Figure 15. Normalised compression vs. wave period. Cable B, Maximum values for whole simulation, WD = 35m, TDT = 1kN.

4.2.5 *Peak tension RAOs*

Investigation on the effect of the wave period in regular waves with the top tension was performed. Figures 19 show the peak tension response amplitude operators for Cable A at a 20m water depth. The RAOs were obtained by dividing the maximum value of the tension on the

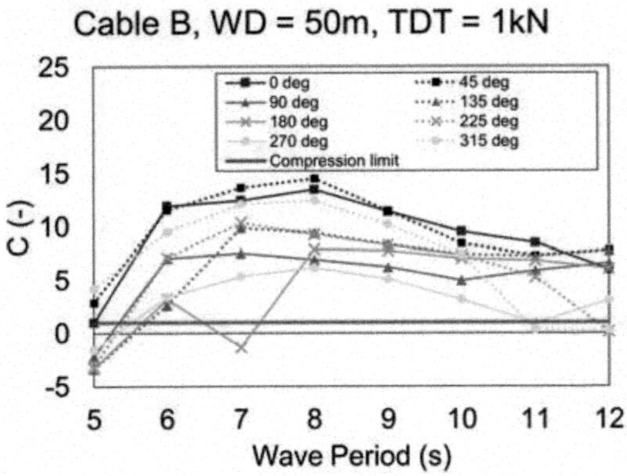

Figure 16. Normalised compression vs. wave period. Cable B, Maximum values for whole simulation, WD = 50m, TDT = 1kN.

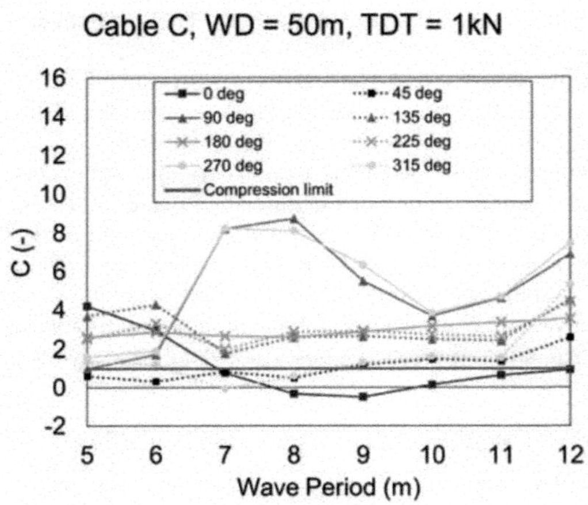

Figure 17. Normalised compression vs. wave period. J-lay, Cable C, Maximum values for whole simulation, WD = 50m.

top by the selected wave height. The response operator for a wave height $3m$ indicates that the peak tensions occur for wave headings of $45deg$ and $315deg$ at a wave peak period of $7s$ and $8s$. When the waves incide on the vessel, the vertical movement of the chute is accompanied by the variation of the suspended submerged length of the cable. This oscillation of the catenary length affects the weight of the cable that the tensioner is holding, having significant impact on the peak tensions.

4.3 *Correlation between vessel motion and cable response*

Scatter diagrams in Figure 20 show the normalised compression for cables A, B and C as a function of the vertical velocity measured on the chute for a water depth of $20m$ and different bottom tensions. Each point represent one case, i.e, a wave with a defined height, period and heading. The analyses were performed with regular waves. It can be observed that for lower chute velocities, the compression increases linearly with the increase of the velocity. For higher velocities, a spread of the compression values is observed, indicating that the chute velocity has lower influence on the tension loss, even though the compression is always occurring (no C(-) below 1 for high velocities).

The relation between the top tension and the chute GZ velocity is seen in Figure 21. A different trend is observed when compared to compression loads. At higher velocities, a spread of the top tension values occurs. Small chute velocities presented a good relation with the top tension.

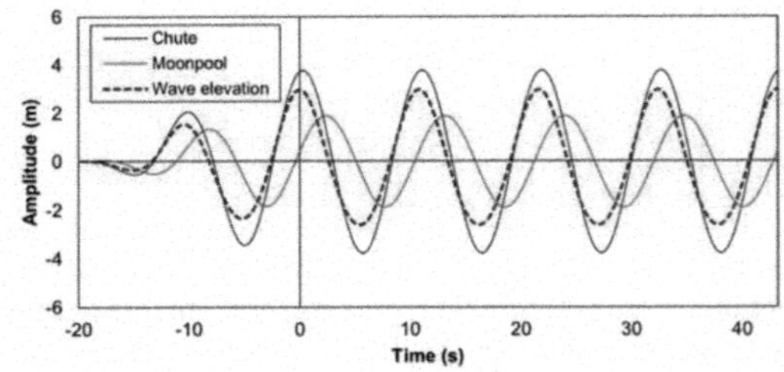

Figure 18. Wave, chute and moonpool Z amplitude. Hs = 3.0m, Tp = 12s, dir = 0deg.

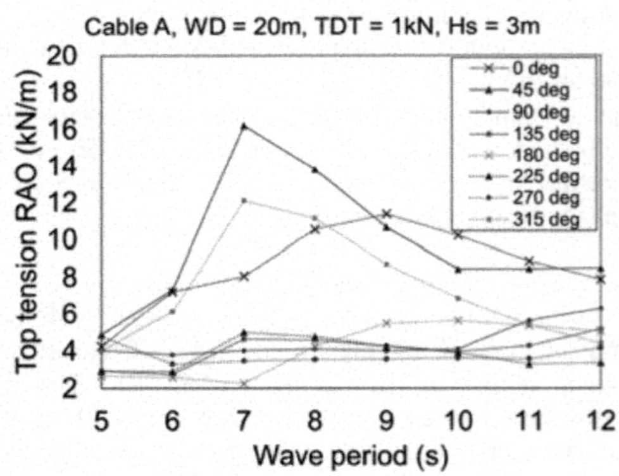

Figure 19. Tension RAOs.

Different trends are observed when the directional waves and currents are isolated. Figure 22 shows the compression with relation with the chute velocity for different headings at a $20m$ water depth and $1kN$ bottom tension. For 0, 45 and 90 *deg* headings, a 2nd order polynomial function can be fitted to the diagrams. However, for an environmental load heading of 270 *deg*, a spreader scatter is seen. This is explained by the tranversal offset of the chute in relation to the centre line. The vessel presents similar response at 90 and 270 *deg*, but the chute does not behave similarly. Same observations are made for compression loads, but with a linear relationship.

The results indicate that the system is dominated by drag forces. By computing the KC number as described in 2.5, taking that the cable displacement

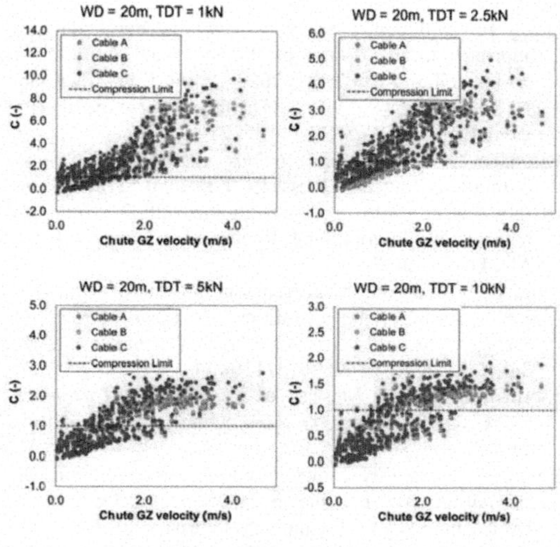

Figure 20. Normalised compression vs. chute velocity. Maximum values for whole simulation, WD = 20m.

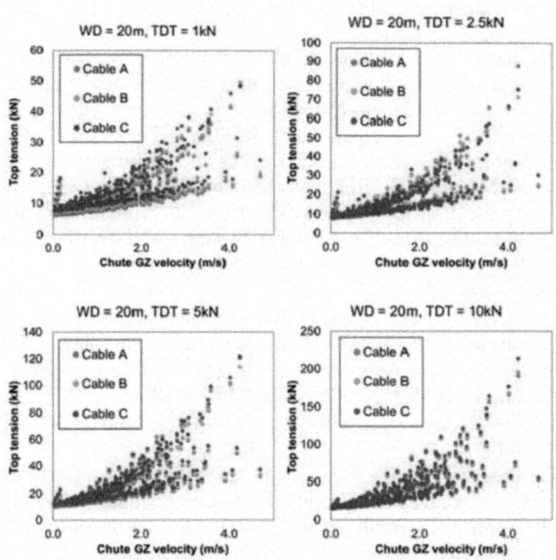

Figure 21. Top tension vs. chute velocity. Maximum values for whole simulation, WD = 20m.

is above 1.5m for most cases, $KC > 45$, meaning that the drag force is dominat and inertia can be neglected.

5 CONCLUSIONS

The following conclusions regarding the modelling of the system have been pointed out in this thesis:

- By comparing the model with different bending stiffnesses, the linear one resulted in higher top tension, but the hysteretic non-linear bending stiffness had a more conservative approach on the compression on the touchdown point.

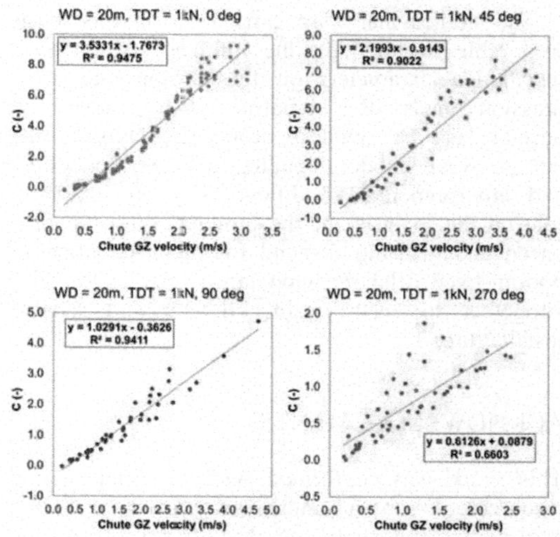

Figure 22. Norm. compression at different directions vs. chute velocity. Max. values for whole simulation, WD = 20m, all cables.

- The J-lay deployment method presented a higher operable weather window that the S-lay deployment method. Workability reached up to 3.0m for most cases in J-lay, while some sea states reduced the operable window to 1.0m wave heights. It was observed that the vertical motions of the vessel were not significantly transferred to the cable when the departure point is placed near the midship.
- A regular wave approach is more conservative than irregular waves for both compression and top tension. For compression loads, a difference of up to 30% was observed, while for the tension, the values with regular waves were more than double of irregular waves.
- Heavier cables in small bottom tensions present a wider variation of the bottom tension under different wave periods. The slacker cable tends to bend more, and consequently, the touchdown point movement is aggravated, originating compressive loads. For Cable C at a water depth of 50m and a touchdown tension of 1kN, the normalised compression reached around 12 at its peak.
- In larger laybacks, the extra length of the suspended cable also increases the actual compression load on the touchdown point. Even though the tension drop is not high, it is more likely that compression limits will be exceeded in such cases. The same observation is valid for top tension loads.
- A strong correlation between the cable compression on the touchdown point and the chute velocity was observed, with correlation coefficients of up to 0.97. It indicates that the system is dominated by drag forces.

The studies that were carried out in this thesis were able to highlight the influence of different installation parameters on the tension and compression loads of submarine power cables on normal lay. The cables' properties, layback distances, wave model, bending stiffness modelling and environmental conditions play an important role on the integrity of the product. With the fast-paced and growing demand for the installation of bottom-fixed offshore wind farms, challenges arise altogether to ensure safe, efficient and reliable installation.

ACKNOWLEDGEMENTS

This work was performed with the support and resources of Subsea7 and Seaway Offshore Cables and within the scope of the Strategic Research Plan of the Centre for Marine Technology and Ocean Engineering (CENTEC), which is financed by the Portuguese Foundation for Science and Technology (Fundação para a Ciência e Tecnologia - FCT) under contract UIDB/UIDP/00134/2020.

REFERENCES

Integrity of Subsea Control Umbilical, volume Volume 6A: Pipeline and Riser Technology of *International Conference on Offshore Mechanics and Arctic Engineering*, 06 2014. doi: 10.1115/OMAE2014-23778. URL https://doi.org/10.1115/OMAE2014-23778.

Aranha, J., Pinto, M., & da Silva, R. On the dynamic compression of risers: an analytic expression for the critical load. *Applied Ocean Research*, 23(2):83–91, 2001. ISSN 0141-1187. doi: https://doi.org/10.1016/S0141-1187(01)00009-8.

Calavia, J. R., Paalvast, M., Ledoux, B., & Kymmell, J. Derivation of Vessel Motions as Operational Limits and Motion Forecasting With 2D Wave Spectra: Case Study of Inter-Array Subsea Cable Installation in the North Sea. volume 3 of *International Conference on Offshore Mechanics and Arctic Engineering*, page V003T04A008, 06 2022. doi: 10.1115/OMAE2022-79459.

DNV. *Environmental Conditions and Environmental Loads*. Det Norske Veritas AS, April 2014. DNV-RP-C205.

DNV. *Subsea power cables for wind power plants*. DNV AS, June 2016a. DNV-ST-0359.

DNV. *Subsea power cables in shallow waters*. DNV AS, March 2016b. DNV-RP-0360.

DNV. *Modelling and analysis of marine operations*. Det Norske Veritas AS, July 2017. DNV-RP-N103.

Dreyer, T. & Van Vuuren, J. H. A comparison between continuous and discrete modelling of cables with bending stiffness. *Applied Mathematical Modelling*, 23 (7):527–541, 1999. ISSN 0307-904X. doi: https://doi.org/10.1016/S0307-904X(98)10097-5.

Drumond, G. P., Pasqualino, I. P., Pinheiro, B. C., & Estefen, S. F. Pipelines, risers and umbilicals failures: A literature review. *Ocean Engineering*, 148:412–425, 2018. ISSN 0029-8018. doi: https://doi.org/10.1016/j.oceaneng.2017.11.035.

Díaz, H. & Guedes Soares, C. Multicriteria decision approach to the design of floating wind farm export cables. *Energies*, 15(18), 2022. ISSN 1996-1073. doi: 10.3390/en15186593. URL https://www.mdpi.com/1996-1073/15/18/6593.

Faltinsen, O. M. *Sea loads on ships and offshore structures*. Cambridge University Press, 1990. ISBN 0 521 45870 6.

IRENA. Renewable energy statistics 2022. Technical report, The International Renewable Energy Agency, Abu Dhabi, 2022.

Jeng, D.-S. & Zhou, X.-L. Tension analysis of submarine cables during laying operations. *The Open Civil Engineering Journal*, 7:282–291, December 2013. doi:10.2174/1874149520131130009.

Journée, J. M. J. & Massie, W. W. *Offshore Hydromechanics*. Delft University of Technology, 1st edition, 2001.

Loos, B. Operability limits based on vessel motions for submarine power cable installation. Master's thesis, Delft University of Technology, February 2017.

Mamatsopoulos, V. A., Michailides, C., Theotokoglou, E. E., & Onoufriou, T. Critical water depth and installation curves for submarine cable deployment process. *Journal of Marine Science and Engineering*, 8(11), 2020. ISSN 2077-1312. doi: 10.3390/jmse8110838.

Morison, J. R., Johnson, J. W., & Schaaf, S. A. The force exerted by surface waves on piles. *Journal of Petroleum Technology*, 2:149–154, 1950.

Nesje, J. D., Aggarwal, R. K., Petrauskas, C., Vinnem, J. E., Keolanui, G. L., Hoffman, J., & McDonnell, R. Risk Assessment Technology and its Application to Tanker Based Floating Production Storage and Offloading (FPSO) Systems. All Days:OTC–10998–MS, 05 1999. doi: 10.4043/10998-MS.

Okkerstrøm, L., Gudmestad, O. T., & Pedersen, E. Simulation of peak tension loads in subsea power cables during installation. *IOP Conference Series: Materials Science and Engineering*, 1201(1):012010, nov 2021. doi: 10.1088/1757-899X/1201/1/012010.

Orcina. *Compressive Waves: Knowledge Base Article*, 2018.

Péronne, S., Izarn, C., Estrier, P., Caro, O., Leroy, J.-M., & Charliac, D. Flexible pipe hysteretic bending behavior: Comparison with experimental characterization and finite element method. Volume 5A: Pipeline and Riser Technology:V05AT04A032, 05 2015. doi: 10.1115/OMAE2015-41281. URL https://doi.org/10.1115/OMAE2015-41281.

Reda, A. *Numerical Modelling of Subsea Cables and Pipelines*. Phd thesis, Curtin University, August 2020.

Ribeiro, E., de Sousa, J. R., Ellwanger, G., & Lima, E. On the tension-compression behaviour of flexible risers. *Proceedings of the International Offshore and Polar Engineering Conference*, pages 652–659, 01 2003.

Tan, Z., Quiggin, P., & Sheldrake, T. Time domain simulation of the 3d bending hysteresis behaviour of an unbonded flexible riser. 26th International Conference on Offshore Mechanics and Arctic Engineering, June 2007. OMAE2007–29315.

Vaz, M. & Patel, M. Transient behaviour of towed marine cables in two dimensions. *Applied Ocean Research*, 17 (3):143–153, 1995. ISSN 0141-1187. doi: https://doi.org/10.1016/0141-1187(95)00012-7.

Vaz, M. & Patel, M. Three-dimensional behaviour of elastic marine cables in sheared currents. *Applied Ocean*

Research, 22(1):45–53, 2000. ISSN 0141-1187. doi: http://dx.doi.org/10.1016/S0141-1187(99)00023-1.

Walton, T. S. & Polachek, H. Calculation of transient motion of submerged cables. *Mathematics of Computation*, 14(69):27–46, 1960. ISSN 0025-5718. doi: 10.1090/s0025-5718-1960-0116470-5.

Worzyk, T. *Submarine Power Cables: Design, Installation, Repair, Environmental Aspects*. Springer-Verlag Berlin Heidelberg, 2009. doi:10.1007/978-3-642-01270-9.

Yoshizawa, N. & Yabuta, T. Study on submarine cable tension during laying. *IEEE Journal of Oceanic Engineering*, 8(4):293–299, 1983. doi: 10.1109/JOE.1983.1145573.

Zajac, E. E. Dynamics and kinematics of the laying and recovery of submarine cable. *Bell System Technical Journal*, 36(5):1129–1207, 1957.

Advances in Maritime Technology and Engineering – Guedes Soares & Santos (Eds)

Dynamic performance of Torus wave energy converter combined with offshore wind turbine semi-submersible platform

B. Sebastian & D. Karmakar
Department of Water Resources and Ocean Engineering, National Institute of Technology Karnataka, Surathkal, Mangalore, India

C. Guedes Soares
Centre for Marine Technology and Ocean Engineering (CENTEC), Instituto Superior Técnico, Universidade de Lisboa, Lisbon, Portugal

ABSTRACT: This article investigates the effect of incorporating a heaving wave energy converter (WEC), namely Torus on DeepCwind Semi-submersible platform (SSP) supporting a 5MW wind turbine. The hydrodynamic performance of the hybrid torus-SSP is studied using linear diffraction/radiation code WAMIT. The coupled dynamic analysis of the platform is carried out in the OpenFAST tool which takes the hydrodynamic parameters from WAMIT as input. The motion responses of the hybrid system are obtained in time domain for different irregular sea states. Statistics of motion of hybrid system is computed and compared with the initial platform to quantify the variation. The power absorption of the WEC is computed using a stand-alone MATLAB code. The maximum power is found to be absorbed at the resonance period of the semi-submersible platform. The results indicate that incorporating Torus does not cause a considerable change in the platform dynamics while rendering higher power output to the overall system.

1 INTRODUCTION

Marine renewable energy (MRE) is a promising energy source which has the potential to cater to the need of growing global energy demands. The development of MRE is seen vital for achieving a carbon neutral economy and to nexus the gap on cutting down energy from fossil fuels. Oceans possess abundant sources of energy in the form wind, waves and currents. The economical extraction of energy from offshore environment is challenging due to the harsh marine conditions and high capital investment costs (Sinha et al., 2016). Over the years the offshore wind industry has seen rapid developments in technology (Diaz & Guedes Soares, 2020) and consistently falling cost of energy (Zhang et al., 2023). The levelized cost of energy (LCOE) from offshore wind has currently become competitive with that of fossil fuels which is a milestone for the wind sector that previously had to depend on government subsidies (Jansen et al., 2020).

The wave energy sector has also been developing parallelly with the offshore wind industry; however, the wave energy conversion technology is not converged like the offshore wind turbine with many diverse concepts. Waves have high energy density, predictability, consistency and wide spread availability compared to wind, and have the potential to cater the ever-increasing demand for energy (Wan et al., 2015; Sinha et al., 2016; Kim et al., 2019). In the current technological readiness level, it is anticipated to take more than a decade for wave energy converters (WEC) to be competitive with offshore wind energy. One strategy to reduce the LCOE of energy from both wave and offshore wind is to combine both on to a single platform creating hybrid systems (Castro-Santos, et al. 2016; Ghafari et al., 2021; Lee et al., 2023).

Hybrid platforms have become a subject of research in recent times with the successful installation of several offshore wind farms. The offshore wind turbines are increasing in size and they are placed long distances of the order of 1 km apart to avoid aerodynamic wake effects (Muliawan et al., 2013). The large ocean space between the wind platforms also provides sufficient space for WECs. There are a multitude of benefits from hybrid systems on the choice of appropriate WECs. Integrating WECs not only generates additional power but also enhances the power output of the overall system by acting to mitigate excessive motion responses of the floating platform which improves the efficiency of wind turbine (Gaspar et al., 2021; Zhang et al., 2022). The total capital expenditure can be brought

DOI: 10.1201/9781003508779-53

down by sharing substructure, mooring lines, power cables and logistics (Perez-Collazo et al., 2015). The hybrid system can also reduce the down time of the stand-alone wind turbine with waves being available in the absence of wind (Hu et al., 2020). All these advantages make the option of hybrid platforms attractive. However, the concept of hybrids is still in the early developing stage and requires intensive research shedding light into various options and configurations of wind-wave systems for optimization.

The study presents the numerical investigation on a hybrid system integrating Torus WEC on DeepCwind Semi-submersible platform (SSP) (Robertson et al. 2017). DeepCwind SSP is a floating platform designed to host a 5 MW wind turbine in deep water conditions and it has been the subject of study in several research works on hybrid wind-wave energy systems (Zhang et al., 2023). The Torus is a heaving WEC which pertains to the category of point absorbers. The WEC is integrated on the central column of the floating SSP. Numerical simulations are performed using the linear diffraction/radiation code WAMIT to analyze the hydrodynamic performance of the hybrid platform. Multi-body analysis has been performed to obtain the relative motion between the platform and WEC. The hydrodynamic coefficients like added mass and damping coefficient and the motion responses of the hybrid platform is compared against that of the SSP. Using the hydrodynamic parameters and excitation forces from WAMIT as input coupled dynamic analysis is carried out using OpenFAST tool for irregular sea states. The power absorption has been computed using MATLAB code for different PTO damping parameters.

2 MODEL DESCRIPTION

The hybrid system comprises of DeepCwind Semi-submersible platform (SSP) and a heave type Torus WEC. The DeepCwind SSP has three side columns which provide the restoring moment to keep the platform stable and a central column which supports an NREL 5 MW wind turbine. Both the wind turbine and Torus are integrated to the central column of the floater. The three offset columns have a wider base at the bottom which serves to suppress excess motion in heave. All the columns are interconnected by horizontal and diagonal braces. The geometrical properties of SSP are given in Table 1. The Torus is a donut shaped WEC which is installed on the central column by mechanical connections such as a pully slide system which allow free motion in the vertical direction. However, due to the presence of interconnecting braces the displacements of the WEC is restricted to 4 m above MSL and 3 m below MSL by means of mechanical end stoppers so as to avoid contact. The Torus is neutrally buoyant and does not contribute additional buoyancy to the system. A hydraulic power take-off system is considered in the present study and power is generated by the relative motion between the Torus and the floating platform. The principal dimensions of WEC are given in Table 2. Three equally spaced catenary moorings with fairleads at the top of base columns are provided for station keeping, the details of which are summarized in Table 1. The design parameters of DeepCwind SSP and the mooring lines are kept the same as those of the original design in OC4 project.

Figure 1. Schematic representation of Hybrid Torus WEC integrated with DeepCwind SSP.

Table 1. Properties of the DeepCwind semi-submersible platform (Jonkman et al., 2009).

Property	Value
Draft of the platform below MSL	20 m
Displacement of water	14,200 ton
Mass including ballast	13,444 ton
Platform CM below MSL	14.4 m
Roll gyration	31.61 m
Pitch gyration	32.34 m
Yaw gyration	32.17 m
Number of Mooring lines	3
Depth to fairleads below MSL	14 m
Depth to Anchor below MSL	200 m
Unstretched length of Mooring line	835.5 m

Table 2. Properties of the Torus WEC.

Property	Value
Draft of the WEC below MSL	3 m
Height of Torus	7 m
Outer diameter of Torus	16 m
Inner diameter of OWC chamber	8 m
Centre of Mass below MSL	1 m

3 NUMERICAL MODELLING

The present study explores the integration of a Semi-submersible platform and a Torus WEC system. In this combined concept, the torus moves vertically along the central column to harness wave energy, while the wind turbine generates power from the wind. The numerical analysis models the system as two rigid bodies, a Semi-submersible floating wind turbine and a torus, connected through mechanical couplings at their interfaces. The hydrodynamic simulation tool WAMIT, based on the panel method and three-dimensional potential flow theory (Lee, 1995), is employed to model the hydrodynamic interaction of two body system comprising the Semi-submersible floating wind turbine combined with a heaving torus-type WEC. The Boundary Element Method (BEM) is utilized to discretize the structure's surface into panels. This approach accounts for panel interactions, imposing boundary conditions to compute surface displacement and velocity potential, subsequently used to calculate various hydrodynamic quantities. The motion responses of the platform in the frequency domain are given by the following equation.

$$\sum_{j=1}^{6}\begin{bmatrix}-\omega^2\{M+A_{ij}(\omega)\}+\\ i\omega B_{ij}(\omega)+C_{ij}(\omega)\end{bmatrix}X_j(\omega)=F_i(\omega) \quad (1)$$

$$\text{for} i=1,2,\ldots,6$$

where M is the mass of the floating body, A_{ij} depicts the added mass matrix of the platform for the incident wave frequency ω, B_{ij} is the coefficient of radiation damping of the body, C_{ij} is the hydrostatic restoring coefficient and F_i depicts the excitation force on the floating body. Here, i and j represent the mode coupling of the six degrees of freedom (DOF) of the floating body.

The added mass A_{ij} and damping coefficient B_{ij} for water density ρ and total velocity potential (Lee, 1995) are derived from the relation given by

$$A_{ij}-\frac{i}{\omega}B_{ij}=\rho\iint_{S_b}n_i\varphi_j dS \quad (2)$$

The excitation force X_i for the body structure using direct integration of hydrodynamic pressure (Lee, 1995) for diffraction velocity potential φ_j is given by

$$X_i=-i\omega\rho\iint_{S_b}n_i\varphi_j dS \quad (3)$$

To investigate the hydrodynamic interaction between two floating bodies, WAMIT takes into account the mutual interaction between the bodies through radiation and diffraction effects. The radiation problem involves the wave excitation experienced by a floating body as it moves in calm water. The hydrodynamic coupling between two floating bodies is addressed by integrating the radiation and diffraction problems of each body into a coupled system of equations.

The coupled dynamic analysis is performed using the aero-hydro-servo-elastic code OpenFAST. Blade Element Momentum Theory (BEMT), encompassing both Blade Element Theory and Momentum Theory, is employed to compute the aerodynamic forces acting on wind turbine blades. This theory is formulated for a quasi-static two-dimensional airflow around the blade, partitioned into sections of varying radius. In BEMT the blades of the wind turbine are divided into many small sections of varying radius dr, Axial force dF_A and Tangential force dF_θ acting on each section are determined as a linear sum of lift and drag forces as

$$dF_A=0.5\rho {V_R}^2 c\,[C_L sin\phi + C_D cos\phi]\,dr \quad (4)$$

$$dF_\theta=0.5\rho {V_R}^2 c\,[C_L cos\phi - C_D sin\phi]\,dr \quad (5)$$

where, ρ is the density of air, c is the chord length of each blade section, V_R is relative wind velocity acting on the blades, ϕ is the angle of relative wind velocity with respect to the plane of rotation, C_L and C_D are the coefficients of lift and drag respectively which are the non-dimensional representative of corresponding lift and drag forces. The governing equation of motion of the floating platform in time domain is the Cummins' equation (Cummins, 1962) given by

$$\sum_{i=1}^{6}\begin{matrix}\{M+A_{ij}(\infty)\}\ddot{x}_j(t)+\int_0^t K_{ij}(t-\tau)\dot{x}_j(\tau)d\tau \\ +B_{i,ext}\dot{x}_j(t)+\left(C_{ij}+C_{ext}\right)x_j(t)\end{matrix}=F_{i,ex_t}(t)$$

$$\text{for } i=1,2,\ldots,6 \quad (6)$$

$$K_{ij}(t)=\frac{2}{\pi}\int_0^\infty B_{ij}(\omega)\cos\,\omega t d\,\omega \quad (7)$$

where $K_{ij}(\tau)$ represents the kernel of convolution which incorporates the memory effects of the free surface, $B_{i,ext}$ is the nonlinear damping coefficient due to viscous drag, C_{ext} is the additional hydrostatic stiffness provided by mooring lines. $\ddot{x}_j$, $\dot{x}_j$, x_j are the acceleration, velocity and displacement of the body respectively and $F_{i,ext}$ is the total external load on the structure causing platform excitation.

The power absorption of the wave energy device determines the feasibility of the system in economic terms. A linear PTO has been assumed for computing the power absorption in the present work. The PTO generates power from relative motion between the floating platform and the Torus. The following equation is used to calculating power from Torus WEC:

$$P_{abs} = B_{pto}u_{rel}^2 \quad (8)$$

where B_{pto} is the damping coefficient due to PTO system, u_{rel} is the relative velocity between the platform and WEC which can be expressed as the relation given by,

$$u_{rel} = i\omega(x_3^{platform} - x_3^{torus}) \quad (9)$$

where $x_3^{platform}$ and x_3^{torus} are respectively the heave amplitude of floating platform and Torus WEC.

4 RESULTS AND DISCUSSIONS

The paper presents the dynamic analysis of a hybrid concept combining DeepCwind semi-submersible platform with a heaving Torus WEC. The Torus is integrated to the central column of the SSP and the relative motion between floating platform and WEC is used to generate power. The numerical analysis is performed using the linear diffraction/radiation code WAMIT to compute the hydrodynamic coefficients and relative motion responses of the hybrid system in regular waves. The simulations have been conducted in the frequency domain for wave period range of 1s to 30s which is the most frequently occurring waves in the category of gravity waves. Unidirectional waves propagating in cardinal X direction are considered for the study. Rigid body analysis has been performed to determine the added mass, damping coefficient and RAOs of the system. Multi-body analysis has been performed to obtain the relative motion between the SSP and WEC which is then used to calculate the power. The power absorbed by WEC is computed using a stand-alone code in MATLAB.

4.1 *Variation in added mass*

The presence of Torus WEC converter integrated to the floating SSP modifies the hydrodynamic coefficients of the initial platform. Figure 2(a-c) compares the variation in added mass of initial SSP and hybrid platform for the normal range of wave periods. It is observed that the added mass of the SSP and hybrid remains the same for surge and sway directions and for roll and pitch rotations owing to the symmetry of the initial SSP and hybrid system. Hence sway and roll added mass have not been included.

It can be seen that there is no considerable deviation in the added mass of surge, sway and yaw of the hybrid platform from the SSP. The peak values of added mass are exhibited in the wave period range of 5s to 12s for all motions. The frequency for which maximum added mass occurs also remains unchanged for the degrees of freedom in horizontal plane.

4.2 *Variation in damping coefficient*

The damping of the WEC is an important factor which affects the motion responses in heave, roll and pitch. Figure 3(a-c) presents the damping coefficient of both the DeepCwind SSP and combined platform in different degrees of freedom. The presence of Torus does not cause a considerable change in the damping coefficient in horizontal degrees of freedom whereas there is enhancement in peak values of

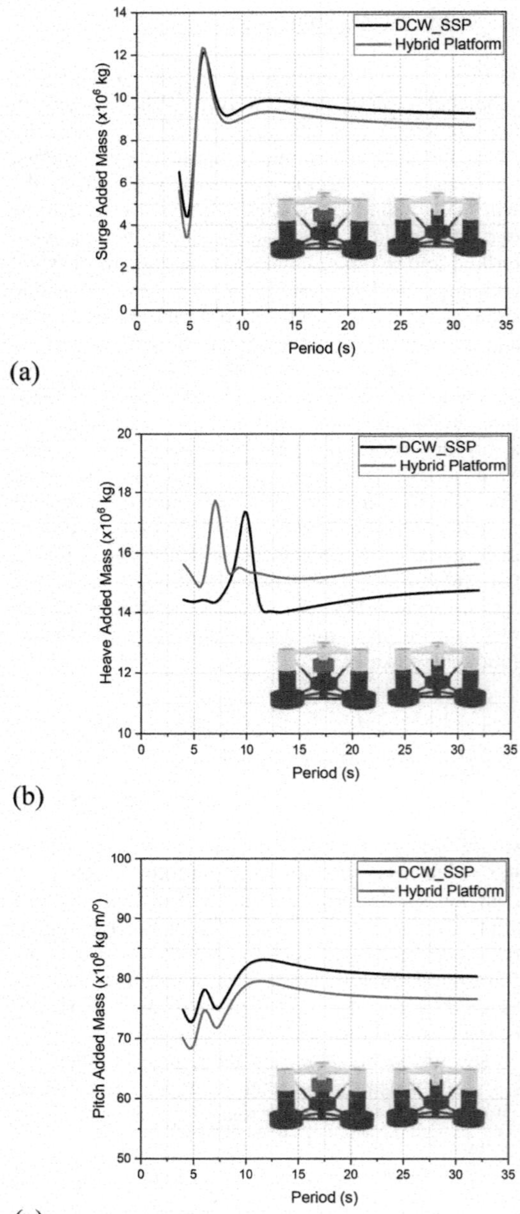

Figure 2. Comparison of added mass for (a) Surge, (b) Heave, and (c) Pitch for DeepCwind SSP and Hybrid platform.

damping for vertical motions. It is noted that the damping coefficient is significant only in period range of 5s to 15s beyond which it becomes small in magnitude. The frequency corresponding to the maximum damping coefficient is the same for the SSP and hybrid platform. The increase in the water plane area due to addition of WEC is not significant compared to the water plane area of SSP to offer substantial resistance to motion of the system.

4.3 *Response amplitude operator*

The RAOs are used to evaluate the motion responses of the platform relative to the amplitude of incident waves. Unit amplitude regular waves with varying time periods are employed for computing the RAOs of the platform in six DOF. The direction of incident waves is taken as 0° to the platform in the frame of reference. Only the motions in surge, heave and pitch are significant in magnitude as the waves are propagating in the X direction.

The significant motions of the DeepCwind SSP and Hybrid system are presented in Figure 4(a-c). The surge RAO is not enhanced by the introduction of Torus in the case of the hybrid system, however, the peak exhibited in wave excitation region has shifted to a slightly lower time period. The magnitude of surge has not increased due to the negligible variation in hydrodynamic coefficients in surge even after integrating the WEC. The heave RAO is slightly amplified by the introduction of Torus WEC as observed from Figure 4(b) with the peak shifting to lower period. The peak in heave is due to heave resonance of SSP at the wave period of 17s. The addition of Torus has modified the heave resonance frequency of the hybrid system. Similarly, the pitch resonance of the initial platform has also been brought to a lower wave period by the WEC in hybrid platform. However, the peak in pitch RAO slightly reduced for the overall system. Figure 4(d) shows the heave RAO of Torus WEC relative to the SSP. The heave RAO of Torus has been obtained from multibody analysis using WAMIT tool, which treats the SSP and torus as a two-body system. It is observed that there are two peaks for heave RAO of which the higher peak is caused by the resonance of the SSP.

4.4 *Time domain analysis*

The motion responses of the platform under four irregular sea states have been studied and presented in this subsection. The environmental conditions pertaining to that of North Sea of Atlantic Ocean have been taken from Muliawan et al. (2013). The metocean conditions of the four representative sea states defined by Significant wave height (H_s), Peak spectral period (T_p) and mean wind speed (U_{mean}) are given in Table 3. The first three sea states correspond to the operational condition of the wind turbine while the fourth sea state is that of the survival condition where the wind turbine is parked.

The time domain analysis has been carried out using the tool OpenFAST, which is capable of accounting for the aero-hydro-servo-elastic dynamics of the floating wind turbine system. The results of WAMIT viz., the hydrodynamic coefficients and excitation forces are the basic inputs for OpenFAST code. The irregular sea states are simulated using JONSWAP spectrum which is specifically developed for North Sea. An NREL 5MW wind turbine for which the DeepCwind platform was initially

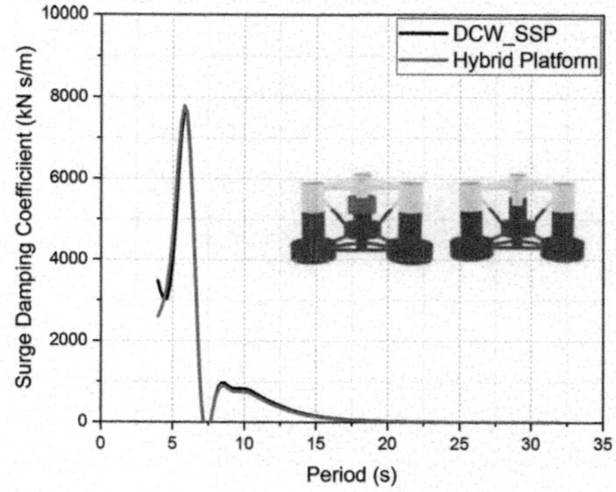

(a)

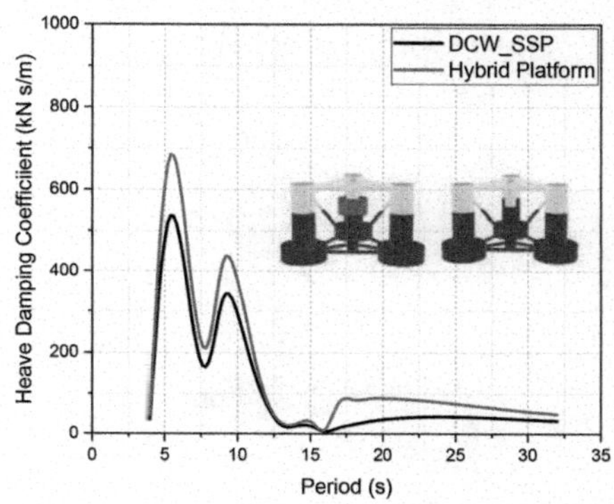

(b)

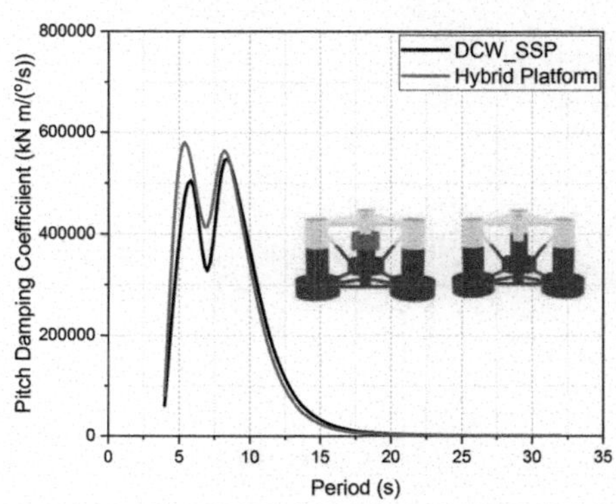

(c)

Figure 3. Comparison of damping coefficient for (a) Surge, (b) Heave, and (c) Pitch for DeepCwind SSP and Hybrid platform.

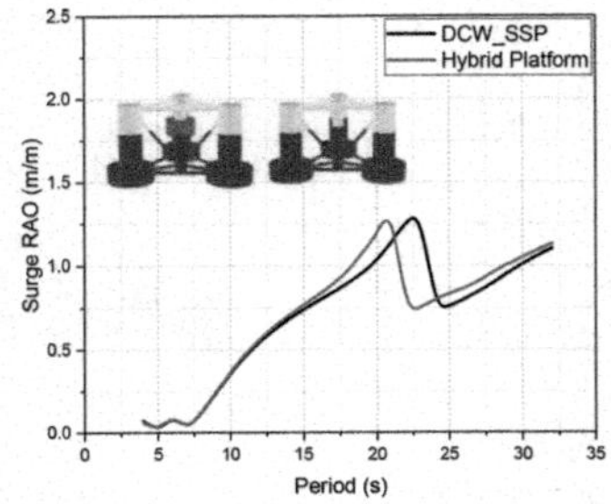

(a)

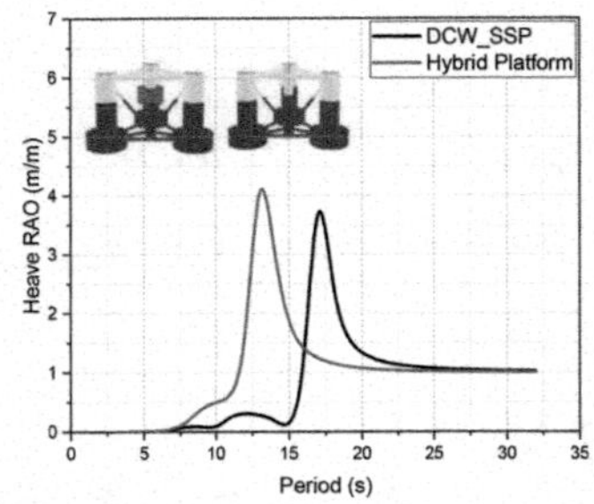

(b)

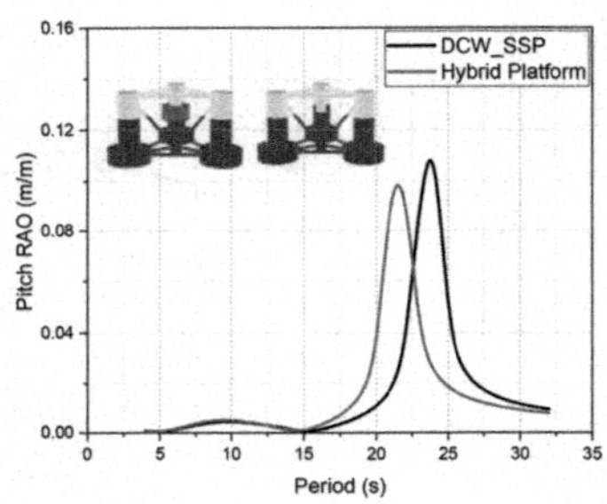

(c)

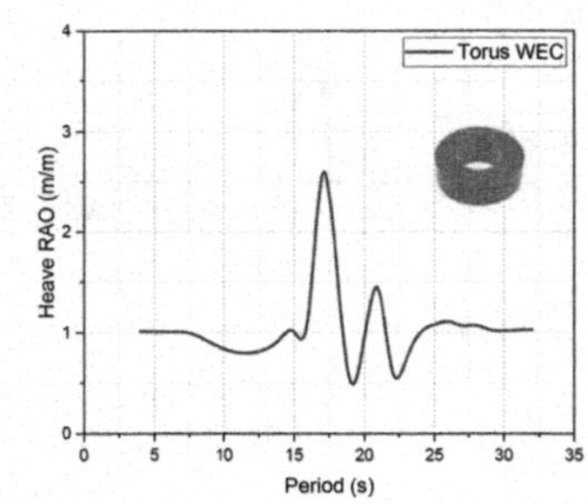

(d)

Figure 4. Comparison of RAO for (a) Surge, (b) Heave, and (c) Pitch for DeepCwind SSP and Hybrid platform and (d) Relative heave RAO of Torus WEC.

Table 3. Environmental conditions.

Sea States	H_s (m)	T_p (s)	U_{mean} (m/s)	Status
EC1	2.5	9.8	8	Operational
EC2	3.0	10	11.2	Operational
EC3	3.6	10.2	14	Operational
EC4	5.9	11.3	25	Parked

designed is employed for coupled dynamic analysis. The waves are assumed to be incident on the structure from the X-direction alone. The simulations have been performed for a duration of 4000s and the first 2000s have been omitted to avoid the influence of transient effects. The motion statistics of the initial and hybrid platforms for surge, heave and pitch are shown in Figures 5-7 as they are the dominant motion responses for the particular wave direction and positioning of the floating platform. It can be observed from Figure 5 that the mean surge has

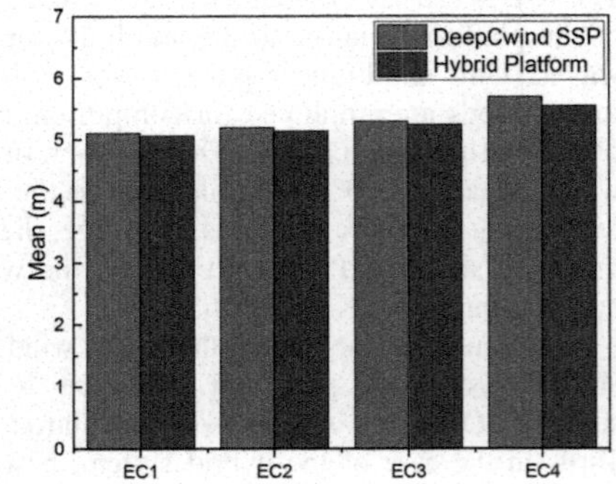

(a)

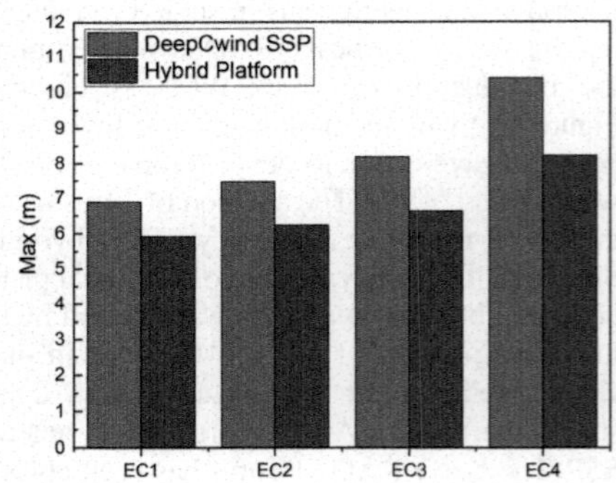

(b)

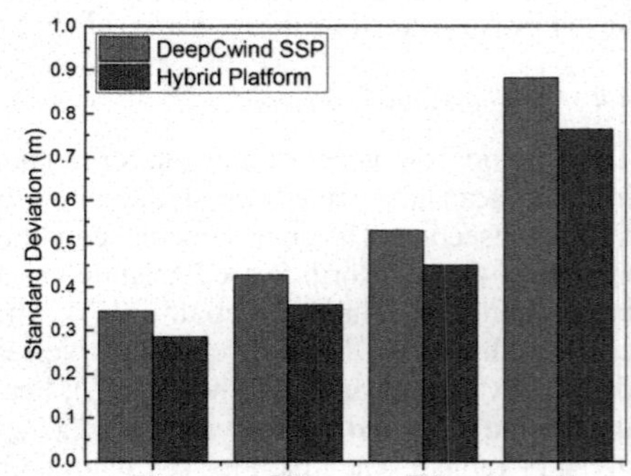

(c)

Figure 5. Torus WEC integrated with DeepCwind SSP platform surge motion statistics.

slightly decreased for the hybrid platform. of amplified resonance of the hybrid platform.

In the case of maximum surge and standard deviation (SD) there is a difference of considerable margin between the statistics of initial and hybrid platform. The presence of Torus has lowered maximum surge and SD. The WEC creates additional drag to the platform which is reflected in reduced surge. The reduction in SD implies less turbulence of the system. This improved stability is the result of increased water plane area of the hybrid platform enhancing its restoring coefficient. In Figure 6 the statistics of heave motion reveal that the net heave is in the downward direction for the operational conditions of the wind turbine while it is upward for parked wind turbine state. The integration of Torus has decreased the average and maximum heave magnitude which is due to the enhanced damping by the WEC. However, the SD is slightly larger for the hybrid system which is likely the cause

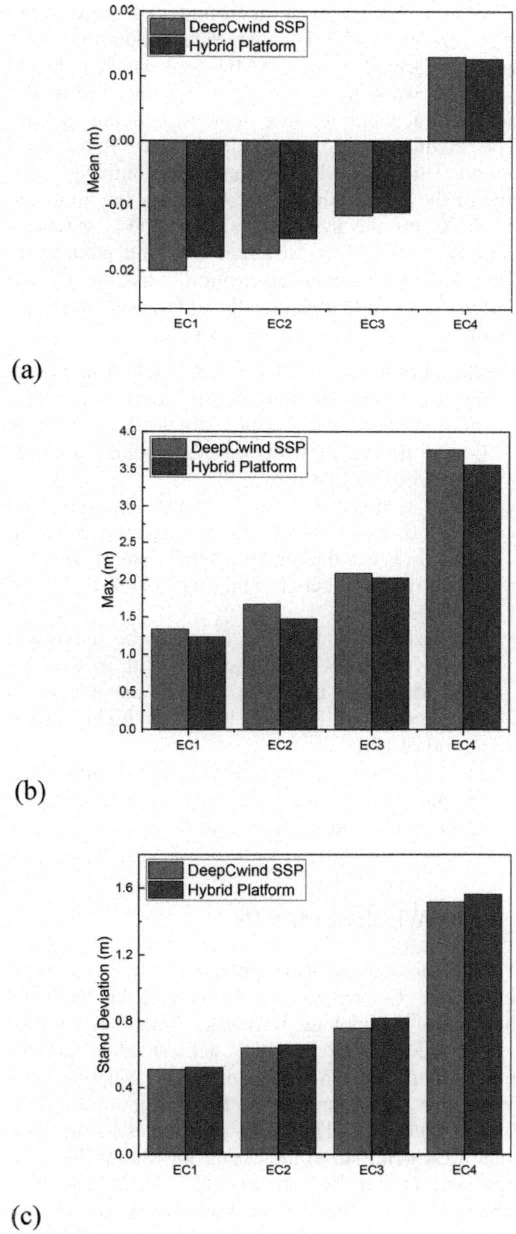

Figure 6. Torus WEC integrated with DeepCwind SSP platform heave motion statistics.

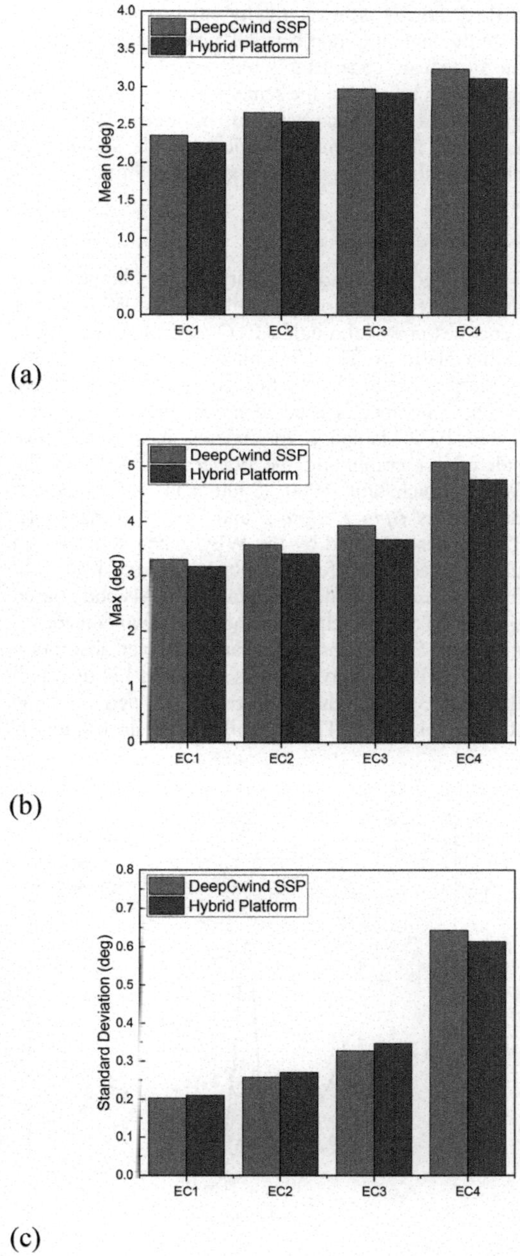

Figure 7. Torus WEC integrated with DeepCwind SSP platform pitch motion statistics.

From the pitch motion statistics, it is noted that the mean and maximum of pitch rotations are slightly lower when WEC is attached to the SSP. This may be attributed to the additional damping in pitch offered by the WEC which effectively brings down response magnitudes. The average and maximum pitch value does not vary drastically with higher sea states for both operational and parked condition of the wind turbine. The SD on the other hand is higher for the hybrid system which indicates more platform perturbations in pitch. It is to be noted that when the wind turbine is parked the SD is lower for hybrid system. In parked state the inclining moments and aerodynamic loads on the system are considerably reduced which may be the underlying reason for the same. It can be inferred from the results the incorporation of torus WEC has improved the dynamic behaviour of the system and enhanced the stability of the floating platform.

4.5 *Power absorbed*

The power absorption capacity is one factor which determines the economic viability of the WEC. Torus generates power through a PTO system connecting it to the platform. The PTO makes use of the relative motion between the bodies for power generation. A linear damper is used to model the PTO system. Three PTO damping coefficients have been considered for computing the power absorption of the WEC which are 1000, 3000 and 5000 kN-s/m denoted as *Bpto-1, Bpto-2* and *Bpto-3* respectively. The power absorbed by the WEC when subjected to unit regular waves for time period varying from 3 to 30s has been computed using a MATLAB code based on Eq. (8) and presented for the three *Bpto* considered in Figure 5. It is observed that the power absorption is higher for two wave periods of 7s and 17s of which the later corresponds to the resonance period of the floating platform. The overall power absorption is found to increase with higher PTO damping. For *Bpto-1* and *Bpto-*2 the higher power absorption occurs at lower wave period of 7s while for *Bpto-3* the power absorption is maximum at period of 17s. It is also noted that the power produced beyond the period of 20s is quite small in magnitude due to smaller relative motion between the bodies.

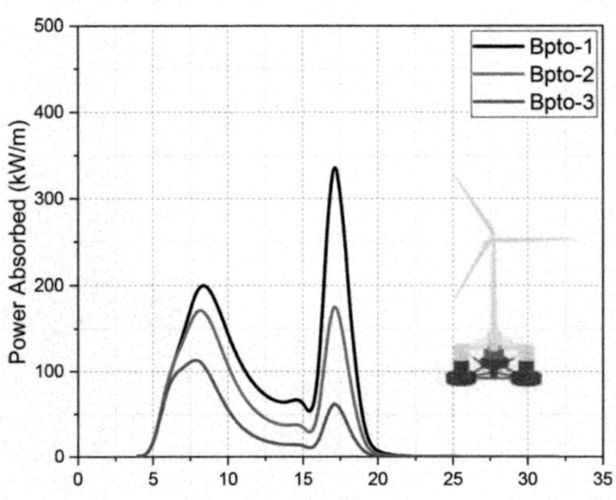

Figure 8. Power Absorption of torus WEC for different PTO damping.

5 CONCLUSIONS

The paper presents the frequency domain analysis to account for the effect of incorporating a Torus WEC on DeepCwind SSP. The numerical simulations have been performed using WAMIT tool which is based on linear wave hydrodynamics and thus the model does not account for non-linearities of the system. The hydrodynamic coefficients and RAOs are obtained for the hybrid system and compared with that of the DeepCwind SSP to quantify the influence of WEC on the system. A linear PTO system is employed with three different damping parameters to assess the power absorption capacity of the system. The following are the inferences from the study:

- The introduction of Torus has resulted in lowering of resonance periods in heave and pitch motion. The observations show that there is limited impact of WEC on the motion responses of the floating platform.
- The motion statistics reveal that the incorporation of WEC has reduced the motion responses in surge, heave and pitch by a small margin which is indicative of better stability of the hybrid platform.
- The power absorption is found to be maximum for wave periods of 7s and 17s. With increase in PTO damping the peak power absorption is observed to shift from lower to higher wave period of 17s.
- The integration of Torus WEC has limited impact on the floating platform while contributing to higher power production from the system.

ACKNOWLEDGEMENTS

The authors express their gratitude to the Ministry of Education, Government of India, and the National Institute of Technology, Karnataka, Surathkal, for providing necessary facilities. DK acknowledges the partial support from Ministry of Ports, Shipping and Waterways, India through the research grant no. DW/01013(13)/2/2021. This work contributes to the Strategic Research Plan of the Centre for Marine Technology and Ocean Engineering (CENTEC), which is financed by the Portuguese Foundation for Science and Technology (Fundação para a Ciência e Tecnologia) under contract UIDB/UIDP/00134/2020.

REFERENCES

Castro-Santos, L.; Martins, E., and Guedes Soares, C. (2016). Cost assessment methodology for combined wind and wave floating offshore renewable energy systems. *Renewable Energy*. 97; 866–880.

Cummins, W. E. (1962). The impulse response function and ship motions. *Ship Technology Research (Schiffstechnik)*, Vol. 9, pp. 101–109.

Diaz, H. M. and Guedes Soares, C. (2020). Review of the current status, technology and future trends of offshore wind farms. *Ocean Engineering*. 209; 107381.

Gaspar, J. F., Kamarlouei, M., Thiebaut, F., and Guedes Soares, C. (2021). Compensation of a hybrid platform dynamics using wave energy converters in different sea state conditions. *Renewable Energy*, 177, 871–883.

Ghafari, H. R., Neisi, A., Ghassemi, H., and Iranmanesh, M. (2021). Power production of the hybrid Wavestar point absorber mounted around the Hywind spar platform and its dynamic response. *Journal of Renewable and Sustainable Energy*, 13(3), 33308.

Giorgi, G., Gomes, R. P., Henriques, J. C., Gato, L. M., Bracco, G., and Mattiazzo, G. (2020). Detecting parametric resonance in a floating oscillating water column device for wave energy conversion: Numerical simulations and validation with physical model tests. *Applied Energy*, 276, 115421.

Hu, J., Zhou, B., Vogel, C., Liu, P., Willden, R., Sun, K. and Collu, M. (2020). Optimal design and performance analysis of a hybrid system combing a floating wind platform and wave energy converters. *Applied Energy*, 269, 114998.

Kim, S. J., Koo, W., & Shin, M. J. (2019). Numerical and experimental study on a hemispheric point-absorber-type wave energy converter with a hydraulic power take-off system. *Renewable Energy*, *135*, 1260–1269.

Lee, C. H. (1995). WAMIT theory manual. Massachusetts Institute of Technology, Department of Ocean Engineering.

Lee, C. F., Tryfonidis, C., and Ong, M. C. (2023). Power Performance and Response Analysis of a Semi-Submersible Wind Turbine Combined with Flap-Type and Torus Wave Energy Converters. *Journal of Offshore Mechanics and Arctic Engineering*, 145(4), 87–105.

Muliawan, M. J., Karimirad, M., and Moan, T. (2013). Dynamic response and power performance of a combined spar-type floating wind turbine and coaxial floating wave energy converter. *Renewable Energy*, 50, 47–57.

Robertson, A. N.; Wendt, F.; Jonkman, J. M.; Popko, W.; Dagher, H.; Gueydon, S.; Qvist, J.; Vittori, F.; Azcona, J.; Uzunoglu, E.; Guedes Soares, C.; Harries, R.; Yde, A.; Galinos, C.; Hermans, K.; de Vaal, J. B.; Bozonnet, P.; Bouy, L.; Bayati, I.; Bergua, R.; Galvan, J.; Mendikoa, I.; Sanchez, C. B.; Shin, H.; Oh, S.; Molins, C., and Debruyne, Y. (2017). OC5 Project Phase II: Validation of Global Loads of the DeepCwind Floating Semisubmersible Wind Turbine. *Energy Procedia*. 137:38–57.

Sinha, A., Karmakar, D., and Guedes Soares, C. (2016). Shallow water effects on wave energy converters with hydraulic power take-off system. *The International Journal of Ocean and Climate Systems*, 7(3), 108–117.

Wan, L., Gao, Z., and Moan, T. (2015). Experimental and numerical study of hydrodynamic responses of a combined wind and wave energy converter concept in survival modes. *Coastal Engineering*, 104, 151–169.

Zhang, D., Chen, Z., Liu, X., Sun, J., Yu, H., Zeng, W., and Si, Y. (2022). A coupled numerical framework for hybrid floating offshore wind turbine and oscillating water column wave energy converters. *Energy Conversion and Management*, 267, 115933.

Zhang, H., Zhang, N., & Cao, X. (2023). Conceptualization and dynamic response of an integrated system with a semi-submersible floating wind turbine and two types of wave energy converters. *Ocean Engineering*, 269, 113517.

Advances in Maritime Technology and Engineering – Guedes Soares & Santos (Eds)

Test bench for the study of CO2 storage using physical adsorption for AIP systems

A. Villalba-Herreros, R. d'Amore-Domenech, V.L. Meca & T.J. Leo
Departamento de Arquitectura, Construcción y Sistemas Oceánicos y Navales (DACSON), Escuela Técnica Superior de Ingenieros Navales, Universidad Politécnica de Madrid, Madrid, Spain
Grupo de Investigación Pilas de Combustible, tecnologías del Hidrógeno y Motores Alternativos (PiCoHiMA), Universidad Politécnica de Madrid, Madrid, Spain

D. Díaz-Cuenca
Grupo de Investigación Pilas de Combustible, Tecnologías del Hidrógeno y Motores Alternativos (PiCoHiMA), Universidad Politécnica de Madrid, Madrid, Spain

ABSTRACT: Air-Independent Propulsion is crucial for underwater navigation where access to oxygen is limited. Commonly, battery-based solutions have been used, but their relatively low energy content poses challenges. To overcome this, alternatives like Direct Methanol Fuel Cells and combustion engines are considered despite they produce CO2 emissions. However, expelling the CO2 produced is energetically demanding while underwater, and storing it as a compressed gas is impractical. An alternative consists of a CO2 capture system utilizing physical adsorption, offering advantages such as low energy requirements, ease of handling, and minimal noise. Employing physical adsorbents like Zeolites or Metal Organic Framework materials the challenge is to optimize CO2 adsorption capacity. The main complexity lies in managing temperature increments during CO2-adsorbent contact, affecting adsorption efficiency. With this objective in mind, a test bench has been built to study this CO2 capture mechanism and improve it to be used in AIP systems for underwater navigation.

1 INTRODUCTION

Air-Independent Propulsion (AIP) is present in applications where free access to oxygen from the atmosphere is not possible. This is the case of vehicles for underwater navigation where solutions based on batteries are common. However, these solutions are not always suitable because of their relatively low energy content. To tackle this situation, power plants based on Direct Methanol Fuel Cells (DMFC), or Combustion Engines (CI) can be considered providing enhanced energy density values. However, these solutions produce CO_2 (Martínez et al., 2012).

Since expelling the produced CO_2 is energetically expensive and disturbing to the surrounding environment while underwater and storing it onboard as a compressed gas is energy-demanding (Villalba-Herreros et al. 2021), it is necessary to look for an alternative CO_2 Capture and Storage (CO2CS) mechanism that consumes little energy and is easy to operate.

This work proposes a CO2CS system that uses physical adsorption on a solid adsorbent as the capture mechanism. Such a system would offer over other systems valuable advantages such as low energy demand, easy handling, and absence of noise. However, although physical adsorption is conceptually simple, there are many factors that affect it, making it necessary to study it from an empirical approach to enhance its CO_2 capture capacity reducing the amount of adsorbent embarked.

The key problem in that the contact of CO2 with physical adsorbent materials such as Zeolites or Metal Organic Framework materials leads to an increase in temperature due to the isosteric heat of adsorption generated, which in turn decreases their adsorption capacity (Lee et al. 2002). An adequate spread of CO2 in the adsorbent material to reduce temperature peaks together with an efficient removal of the generated heat can significantly improve the capture capacity of the adsorbent material. However, the complexity of the fluid-dynamic phenomena occurring in an adsorption bed is notably high. During CO2 adsorption, several phenomena occur simultaneously: CO2 mass transport, CO2 mass diffusion, heat conduction and heat diffusion (Dantas

DOI: 10.1201/9781003508779-54

et al. 2011, Wang et al. 2014). In addition, they occur while the CO2 mass varies. Under these conditions, the approach to this problem from an analytical point of view is very complicated and tools like Computer Fluid Dynamics need the support of experimental data.

To study physical adsorption and increase its performance, this work presents a test bench specifically built with this proposal.

2 STATE OF THE ART

Submarines use physical adsorption to capture the CO2 produced by human breathing. In this case, CO2 scrubbers typically use a solution of Monoethanolamine (MEA) that captures the CO2 present in the air stream circulating through the scrubber using fans. Such a solution has the advantage of being reusable as the MEA solution can be regenerated heating it at high temperatures. It should not be confused with classical non-regenerative CO2 scrubbers using *soda lime* (NaOH + CaO) or LiOH, where CO2 is captured by means of a reaction with these compounds, i.e., it is *absorbed* (Mazurek 2015).

However, the use of liquid solutions is not the most suitable solution in some cases, especially where available power is limited, and auxiliary systems must be simple and quiet. For these cases, the use of physical adsorption in solid state may be more appropriate. This is a common method used for post-combustion CO2 capture (Ru-Shuai et al. 2021). Nevertheless, no public information on the use of this technology for AIP systems could be found.

3 CO_2 ADSORPTION MECHANISM

Adsorption is a surface phenomenon where molecules or atoms adhere to the surface of a solid or liquid material. It occurs due to intermolecular forces between the adsorbate (the substance being adsorbed) and the adsorbent (the surface it adheres to). The mechanisms involved in adsorption can be categorized into two main types: physisorption and chemisorption.

Physical adsorption or physisorption is a common technique used, among others, for CO_2 capture. Many materials are suited for this purpose, highlighting zeolites and Metal Organic Frameworks (MOF) (Ko- ytsoumpa, et al. 2018). This technique has advantages against other CO_2 capture mechanisms, such as ab- sorption or chemical adsorption. The adsorbent is in a solid state, simplifying its handling. In addition, no chemical reactions take place as the bonding between CO_2 and the adsorbent material is based on Van der Waals forces (Encyclopædia Britannica 2013), facilitating the reuse of the adsorbent material after a re- generation phase that can consist of a temperature and/or pressure swing process (Ellison et al. 2021). Besides, adsorption can take place at relatively low pressures, ideally at ambient pressure; this is an essential feature as it is not necessary to compress the gas to high pressures, thus saving energy.

Most studies on adsorption are based on liquid ad- sorption beds. This configuration minimizes the problems caused by temperature increments when CO_2 joints the adsorbent surface due to the isosteric heat of adsorption produced. This heat is linked to the strength of the interactions between the adsorbate and the adsorbent molecules (Bao et al. 2011) and has an adverse effect in the adsorption mechanism. Figure 1 shows the uptake isotherms of CO2 in Zeolite 13X (Lee et al. 2002). As can be observed, the up- take capacity is lower the higher the adsorbent temperature is. This behaviour can be observed in many other adsorbent materials of interest for CO2 capture (Younas et al. 2016).

A liquid adsorption bed could not be practicable due to volume restrictions, then a more compact solution would be key. If the generated heat is not conveniently removed and the temperature rise is not controlled, the uptake capacity of the adsorbent can be seriously affected. This situation is critical in confined environments such as submarines, autonomous underwater vehicles, or spacecraft.

Figure 2 shows the results of a CO_2 capture experiment using beads of Zeolite 13X contained in a spherical flask equipped with a set of four K-type thermocouples evenly distributed around the gas inlet point located in the centre of the flask (Villalba-Herreros, A. 2021). This simple and compact solution could be attractive for its application in confined environments as those cited in the previous paragraph. However, it was determined that the uptake capacity of the Zeolite 13X used was only 41.83 % of its uptake capacity at 298 K and 100 kPa. It is observed that the thermocouple just in front of the gas inlet experimented an abrupt rise of temperature equal to 46 K, from 298 K to 344 K. Contrarily, the thermocouple behind the gas inlet was slightly heated experimenting an increase in temperature of only about 6 K. Considering that the experiment was carried out until an equilibrium pressure was reached and attending the uptake behaviour of zeolites, see Figure 1, it is expected that hotter zeolite beads contain a lower amount of CO2 than colder ones. To improve these figures, two improvements can be implemented: increasing the pressure of the CO2 flow or reducing the temperature of the adsorbent. In the first case, a notable amount of mechanical energy could be needed, resulting in increased energy consumption and, consequently, in CO_2 production. In the second case, if the uncaptured CO_2 is recirculated and used as a heat removal agent using the environment as a heat sink, the additional energy consumed for CO_2 capture could be maintained at an acceptable level.

Considering the potential benefits of the proposed solution based on controlling temperature by circulating CO_2 as a heat removal agent in a cartridge filled with an adsorbent material, a wide field of research arises. The phenomena that occur in such a solution are multiple and complex, including mass transportation, fluid dynamics, heat convection or heat conduction, all of them depending on many factors, such as initial conditions and geometry. The theoretical study of all these coupled phenomena is a complex task that empirical data must support.

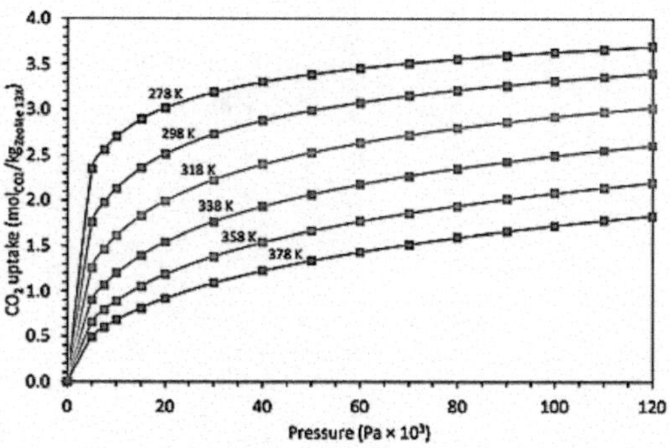

Figure 1. Zeolite 13X CO2 adsorption isotherms. (Lee et al. 2002).

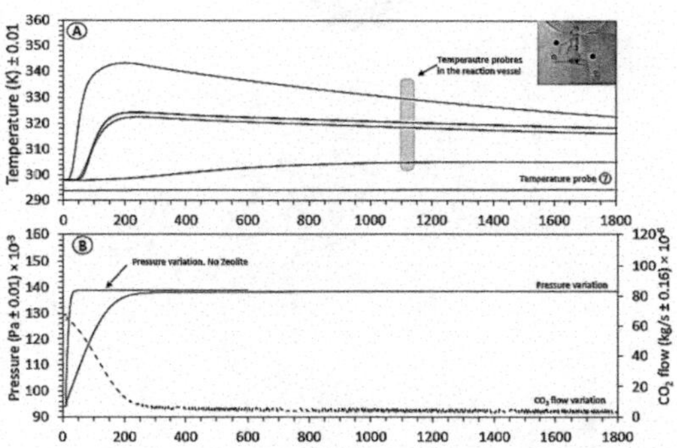

Figure 2. Temperature and pressure evolution in a spherical compact solution for Zeolite 13X adsorption. (Villalba-Herreros 2021).

An interesting matter to investigate is the geometry of the volume where the CO_2 capture occurs. In the experimental setup cited above, CO_2 was released at the centre of a sphere. However, what would have happened if it had been evenly distributed along different points of that sphere? What would happen if instead of a sphere the containing volume was a cylindrical cartridge. Different geometries and the way CO_2 come into contact with the adsorbent material can significantly impact on temperatures distribution and therefore on how much CO_2 is captured. Taking all these considerations into account, constructing a test bench that allows investigating more compact solutions with enhanced CO_2 capture capacity and low energy consumption, as the presented here, makes sense.

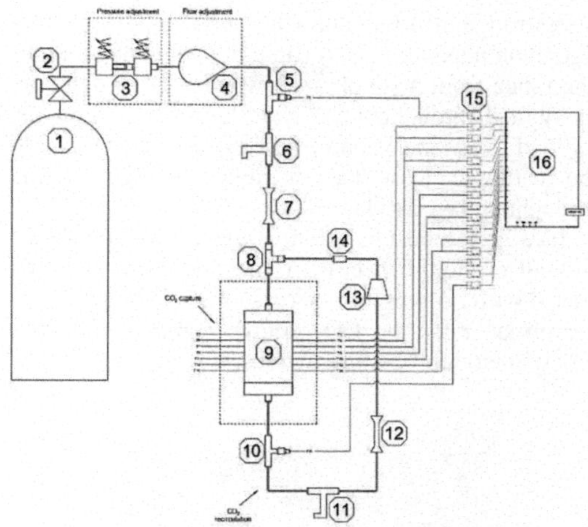

Figure 3. CO2 capture test bench and thermocouples distribution. Source: self-made. ① CO2 cylinder. ② Control valve. ③ Pres- sure adjustment. ④ Flow adjustment. ⑤ Inlet temperature sensor. ⑥ Inlet pressure sensor. ⑦ Inlet flow meter. ⑧ T-fitting. ⑨ Cartridge. ⑩ Recirculation temperature sensor. ⑪ Recirculation pressure sensor. ⑫ Recirculation flow meter. ⑬ Recirculator. ⑭ Non-return valve. ⑮ Thermocouple sensor MAX 6675. 16 ARDUINO MEGA 2560 board.

4 TEST BENCH

Figure 3 shows a sketch of the test bench used to investigate CO_2 adsorption.

First, a cylinder with high purity CO_2 (1) equipped with a shut-off valve (2) acts as the CO_2 source. Following, a pressure adjustment valve (3) and a rotameter (4) allow a fine control of pressure and volumetric flow. Next, a temperature sensor (K-type thermocouple) (5), a pressure sensor (6) and a volumetric flow meter (7) allow to measure and record temperature, pressure, and flow data, respectively. Next, a T fitting (8) allows gas recirculation. Further ahead there is the core of the test bench, the cartridge (9) where the adsorbent material is introduced, and CO_2 adsorption takes place. The non-adsorbed CO_2 is used as heat removal agent and is fed into the recirculation loop where it is tempered and its temperature, pressure and volumetric flow are measured by sensors (10), (11) and (12), respectively. A small fan (13) increases the pressure of the flow allowing its recirculation. Finally, a non-return valve (14) ensures that there is no circulation in the opposite direction. Temperature distribution inside the cartridge (9) is recorded through a series of twelve thermocouples evenly distributed radially and vertically along it, as can be seen in Figure 4. They are connected to MAX6675 sensors (15) that send the information to an Arduino Mega 2560 board (16). The interface for data recording was built in LabVIEW where data are dumped and their evolution can be viewed in real time to control the experimental setup, see Figure 5. The variation of the CO_2 parameters and the geometry of the cartridge inlet will allow to investigate which configuration maximizes CO_2 capture. Inlet configurations can range from a simple inlet where CO_2 is naturally distributed throughout the cartridge section from the top end to the inclusion of perforated tubes parallel to the cartridge axis that distribute the gas evenly throughout the cartridge volume.

Ideally, the best solution would be one where the full adsorption capacity of the adsorbent material I used with no pressure increase that could affect the performance of the CO_2 source (DMFC, CI, etc.) and without temperature increases.

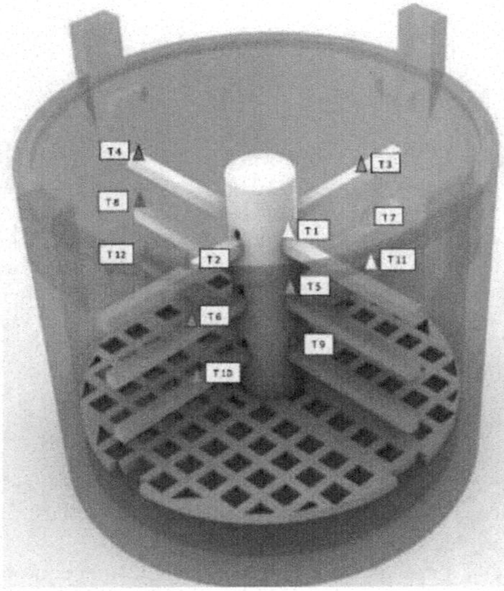

Figure 4. Thermocouple distribution inside of the cartridge (9).

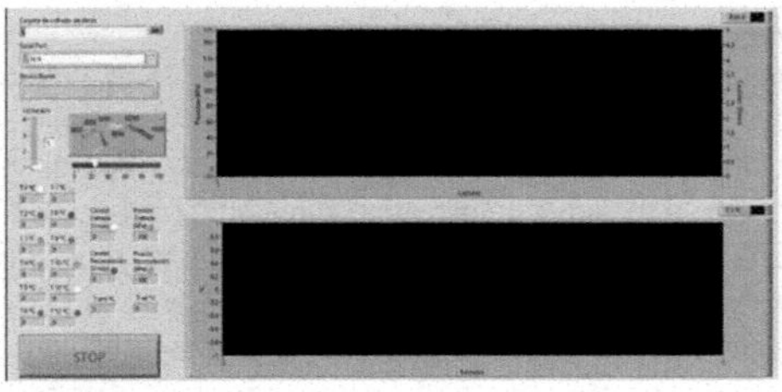

Figure 5. Data recording program interface programmed in LabVIEW.

5 METHODOLOGY

The methodology proposed to study physical adsorption consists of carrying out a series of experiments under the same conditions but varying the geometry of the CO_2 inlet in the adsorption cartridge, i.e., varying the way in which the CO_2 is distributed in the ad- sorbent material. The aim is to relate the key parameters of this geometry to the adsorption capacity used. The basic design will consist of a single inlet in which the CO_2 is naturally distributed over the entire section of the cartridge, from the top to the bottom.

This configuration shall be used as a reference. Other CO_2 inlet designs will distribute the gas evenly throughout the cartridge using different geometries.

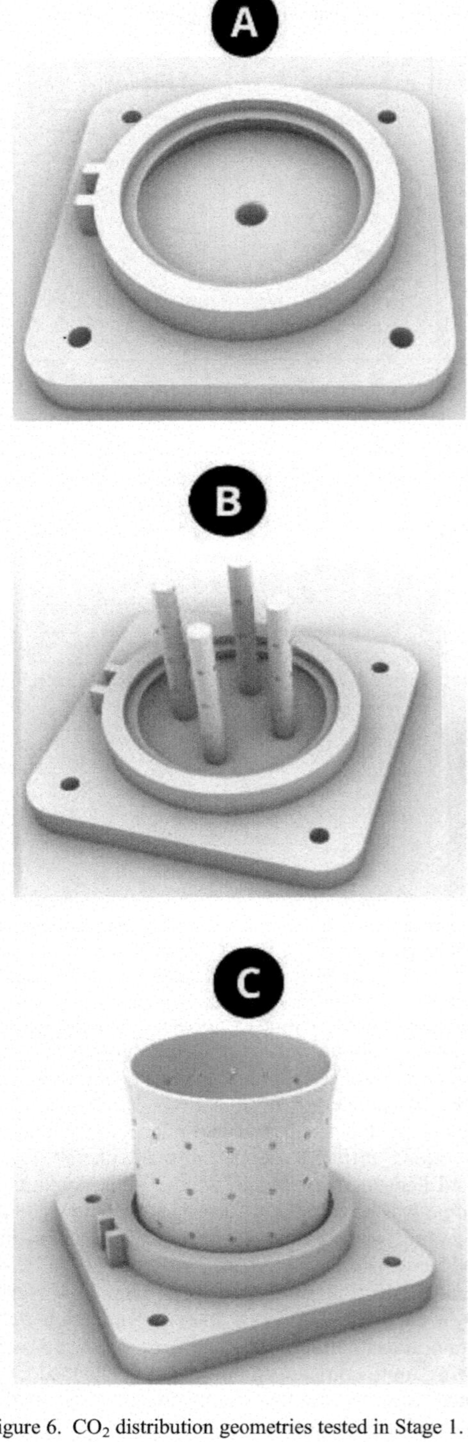

Figure 6. CO_2 distribution geometries tested in Stage 1.

The test conditions have been set at a volumetric flow rate of 3.5 l/min of CO_2 and a pressure of 0.5 bar, which will be standard conditions for a 5 W DMFC stack. A three-stages process has been designed to seek the best geometry for capturing the highest amount of CO_2. In Stage 1, the three limit geometries represented in Figure 6 will be studied to determinate which of them gives the best results regarding CO_2 capture. In Stage 2, the chosen geometry in Stage 1 will be optimized using Response Surface Methodology (RSM) (Raymond H. Myers. Et al., 2016). Finally, in Stage 3, the target geometry will be constructed to study and confirm its superior performance in CO_2 capture.

The Stage 1 will consist of the study of the limit geometries shown in Figure 6. In geometry A, the gas enters uniformly from top to bottom of the cartridge, the zeolites at the top will be the first to start adsorbing CO_2 and therefore experience a temperature increase. This will then happen in the other layers of the geometry. In geometry B the gas enters through a series of tubes in contact with the adsorbent. In geometry C the CO_2 enters through the walls of the cartridge in contact with the adsorbent. To ensure equivalence between geometries B and C, they will be constructed with equal area in contact with the zeolite, the same number of holes, and the same displaced volume. From among these seven geometries, the best one will be chosen, and a more in-depth analysis of the variables involved will be carried out.

Furthermore, in the context of geometries B and C, experimentation will be conducted to introduce CO_2 either at the bottom or top of the adsorbent mass. Consequently, a total of seven geometries will be employed in Stage 1, comprising one for A, three for B and three more for C. The same cartridge will be used to test all the geometries. This cartridge will house the thermocouple tower, which will have twelve type K thermocouples at three heights and four distances from the center. Figure 7 shows the total number of elements required to carry out the Stage 1 experiments (the cartridge, along with the thermocouple array and the seven lids).

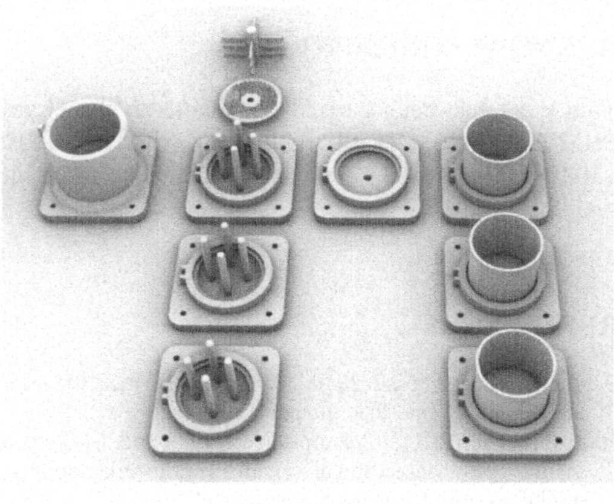

Figure 7. Elements for the initial set of experiments (Stage 1).

Stage 2 will consist of the study of the selected geometry applying RMS (Response Surface Methodology). RSM is used to model and analyze the relationship between one or more input variables (factors) and an output variable (response) in a process or system. The main objective of RSM is to optimize the desired response by adjusting the input factors within a specific range (Raymond H. Myers., et al. 2016). This is achieved by constructing mathematical models that represent the relationship between the input factors and the response, usually through polynomial equations. Experiments designed using RSM typically involve conducting a set of experiments at specific points in the design space, where the levels of the input factors are varied according to a pre-established experimental design. Then, mathematical models are fitted to the data obtained from these experiments to predict the behavior of the system within the study range and to identify the best combination of factors that maximize or minimize the desired response (André et al. 2010). In this case, an analysis of three variables will be carried out, which will involve the construction of another fourteen geometries that will be designed when the time comes with the aim of being able to analyze and reach the best solution of the different geometries.

Finally, in Stage 3, the geometry that is obtained as the best for CO_2 capture will be constructed and tested to ensure that the optimization process has been done correctly.

Each experiment starts with the initial weighing of the adsorbent sample. Then, the sample is placed in the cartridge (9), that is sealed through four steel rods at a specific torque pair, and before connecting it to the T fitting (8) closing the circuit, a uniform flow of CO_2 at a determined pressure is established. When this CO_2 flow is stable the cartridge (9) is connected to the T joint (8) closing the circuit and allowing the CO_2 to flow through the adsorbent material. Then, the process is allowed to evolve until the it reaches equilibrium. This is achieved when the adsorbent is no longer capable of capturing a greater amount of CO_2 and no CO_2 is released from the CO_2 storage cylinder (1). During this process the temperature of the different sensors is registered to analyse the variation of the temperature field. At this stage the cartridge (9) is opened, and the adsorbent sample weighted again to determine the total amount of CO_2 captured. Finally, the adsorbent sample is put into a laboratory stove at 510 K for 14 h to regenerate it by temperature swing.

6 PRELIMINARY RESULTS

Figure 8 shows the graphical output of the data recorded in a preliminary experiment for geometry A. In this experiment, CO_2 flows uniformly distributed across the cross-section of the cylindrical cartridge from its top to its bottom. The results show a progressive increase of the temperature field form the upper part of the cartridge (CO_2 inlet) towards its bottom (CO_2 outlet).

The preliminary results with this geometry indicate that zeolite 13X B can capture between 10-12% of its weight depending on the established conditions and reaches temperatures over 353 K (80 °C). It shows improvement at low volumetric flow rates and high pressures.

Taking these results as a reference, new geometries of the CO_2 inlet combined with different values of pressure and CO_2 volumetric flow will be tested. Analysis of the results will allow conclusions to be drawn as to which geometries lead to lower temperature increases and higher CO_2 capture.

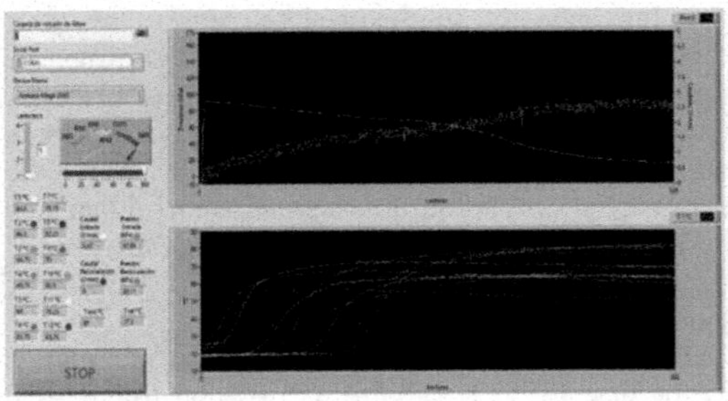

Figure 8. Evolution of temperature and pressure inside the CO2 capture cartridge when CO2 enters uniformly from top to bottom of the cartridge.

7 FUTURE WORKS

Another series of geometric designs can be tested that could be interesting in future iterations to improve the results if necessary. All these new geometries are currently being studied in industrial applications. The issue they pose for this study is that they require non-passive cooling, which entails additional energy expenditure as well as extra equipment.

The first geometry of interest would be the hollow fibre one inspired in the work of F. Rezai et al. (Rezaei et al. 2014). It utilizes hollow fibres containing fine particles of the adsorbent material. As the CO_2 circulates around the outside of the fibres, it diffuses into the fibres and is captured. The heat of adsorption produced is transferred to the water that circulates inside the fibres and is evacuated.

Another solution based on the work by M. Schaefer et al. (Schaefer et al. 2019) involves the modification of the geometry C, see Figure 6. In this solution, the cylindrical wall separating the CO_2 inlet flow from the adsorbent material is replaced by a fabric that allows a uniform distribution of CO_2 over the entire length of the cartridge.

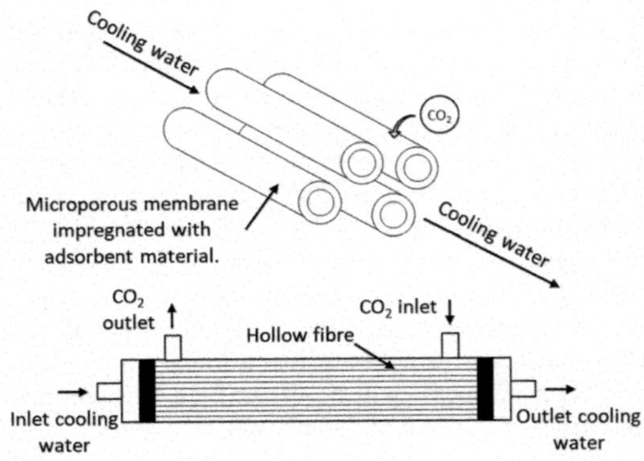

Figure 9. CO_2 capture using hollow fibres.

8 CONCLUSIONS

This work presents a test bench designed and built to study physical adsorption with the aim to develop compact CO_2 capture systems that enhance the CO_2 uptake capacity while keeping a low energy demand for their use in confined spaces along with Air Independent Propulsion systems (AIP).

The main difficulty in designing such a system is the effective removal of the isosteric heat of adsorption produced. This heat leads to an increase in temperature decreasing the capture capacity of the adsorbent material. Minimizing these temperature peaks will benefit from a greater adsorption capacity and may reduce the volume and mass of adsorbent needed. This fact is of great importance when dealing with AIP systems in confined environments like Autonomous Underwater Vehicles where weight and volume control greatly impact the design.

Physical adsorption implies multiple and complex phenomena that occur simultaneously making its theoretical study a complex task. To aid in the study and improvement of this CO_2 capture solution a test bench as the presented here is a valuable asset.

ACKNOWLEDGEMENTS

This work has been carried out thanks to the Grant PID2021-124263OB-I00 funded by MCIN/AEI/10.13039/501100011033 and by "ERDF a way of making Europe".

REFERENCES

André I. Khuri., et al. 2010. 'Response surface methodology'. Wiley Interdisciplinary Reviews.

Bao, Z., et al. 2011. 'Adsorption of CO_2 and CH_4 on a magne- sium-based metal organic framework'. Journal of Colloid and Interface Science.

Dantas T., et al. 2012. 'Modeling of the fixed-bed adsorption of carbon dioxide and a carbon dioxidenitrogen mixture on zeolite 13X'. Brazilian Journal of Chemical Engineering

Ellison, C., et al. 2021. 'Comparison of microwave and conven- tional heating for CO_2 desorption from zeolite 13X'. Interna- tional Journal of Greenhouse Gas.

Encyclopædia Britannica 2013 Inc. Adsorption. https://www.bri-tannica.com/science/adsorption. [Online; accessed 11-Octo- ber-2023]

Koytsoumpa, E.I., et al. 2018. 'The CO_2 economy: Review of CO2 capture and reuse technologies'. The Journal of Super- critical Fluids.

Lee, J.-S., et al. 2002. 'Adsorption Equilibria of CO_2 on Zeolite 13X and Zeolite X/Activated Carbon Composite'. Journal of Chemical & Engineering.

Martínez I., et al. 2012. 'Simple model for CO_2 absorption in a bubbling water column'. Brazilian Journal of Chemical Engineering

Mazurek W. 2015 "Key developments in submarine air monitoring and air purification during the past 20 years". In: SAMAP 2015. Submarine Air Monitoring & Air Purification – Inter- national Symposium. Den Helder, 2015

Raymond H. Myers., et al. 2016. 'Response Surface Methodology: Process and Product Optimization Using Designed Experiments'. John Wiley & Sons Inc; N. : 4 edition.

Rezaei F., et al. 2014. 'Modeling of rapid temperature swing adsorption using hollow fiber sorbents'. Chem Eng Science.

Ru-Shuai L., et al. 2021, 'Advances in Post-Combustion CO_2 Capture by Physical Adsorption: From Materials Innovation to Separation Practice'. ChemSusChem, 14, pp. 1428–1471

Schaefer M., et al. 2019. 'Modeling and simulation of closed low-pressure zeolite adsorbers for thermal energy storage'. Int J Heat Mass Transf.

Villalba Herreros, A. 2021. 'Carbon dioxide treatment method for autonomous underwater vehicles powered by direct methanol fuel cells: A multi-criteria decision analysis approach'. Journal of Power Sources.

Villalba Herreros, A. 2021. 'Direct Methanol Fuel Cell-based power plant for Autonomous Underwater Vehicles. Endur- ance increase and Carbon Dioxide management in a confined Environment'. Tesis Doctoral, E.T.S.I. Navales (UPM).

Wang J., et al. 2014. 'Recent advances in solid sorbents for CO_2 capture and new development trends'. Energy and Environmental Science

Younas M., et al. 2016. 'Feasibility of CO_2 adsorption by solid adsorbents: a review on low-temperature systems'. Interna- tional Journal of Environmental Science and Technology.

Advances in Maritime Technology and Engineering – Guedes Soares & Santos (Eds)

Coupling analysis of a semi-submersible platform with an array of wave energy devices

X.L. Zhao
College of Shipbuilding Engineering, Harbin Engineering University, Harbin, P.R. China
International Joint Laboratory of Naval Architecture and Offshore Technology between Harbin Engineering University and University of Lisbon, Harbin, P.R. China
Centre for Marine Technology and Ocean Engineering (CENTEC), Instituto Superior Técnico, Universidade de Lisboa, Lisbon, Portugal

R. Yan, J. Geng, L.D. Zhang & J. Wang
College of Shipbuilding Engineering, Harbin Engineering University, Harbin, P.R. China

C. Guedes Soares
Centre for Marine Technology and Ocean Engineering (CENTEC), Instituto Superior Técnico, Universidade de Lisboa, Lisbon, Portugal
International Joint Laboratory of Naval Architecture and Offshore Technology between Harbin Engineering University and University of Lisbon, Harbin, P.R. China

ABSTRACT: The hydrodynamic performance of a floating semi-submersible platform for wave power extraction is investigated. The floating system consists of a triple-column platform and an array of eagle-type oscillating bodies, of which the relative motion of the two aspect drives the Power Take-Off (PTO) system to extract wave power. The Lagrange multiplier method is employed to formulate the frequency-domain motion equation for the present multi-body system. The investigation focuses on the hydrodynamic analysis of the floating platform and the oscillating bodies and the influence of the key parameters, such as PTO damping. Results indicate that, comparing with the isolated platform, the presence of the wave energy device array significantly modifies the hydrodynamic responses of the platform.

1 INTRODUCTION

Due to its renewable and large-reserve advantages, ocean wave energy has become a promising energy source facing offshore activities. The high construction costs and lower survivability of wave energy converters (WECs) are key obstacles to commercial operation. Many authors have proposed the concept of integrating wave energy devices with conventional marine structures, enabling cost-sharing, space utilization, and synergistic effects. This provides viable solutions to reduce the levelized cost of electricity of WECs (Zhao et al., 2019b). This kind of integration significantly enhances the competitiveness of wave energy devices.

Zhao et al. (2019a) proposed an integrated system that consists of a floating breakwater and an array of heaving bodies and conducted experiments to explore the hydrodynamic performance of the system. They found that the reflected waves produced by the breakwater improved the efficiency of the wave energy devices. Zhu et al. (2020) investigated the influence of the oscillating water column (OWC) WECs on the performance of the semi-submersible floating wind turbine. They concluded that the integration of the OWCs with proper design can reduce the motion of the floating platform. Zhao et al. (2023) numerically investigate the performance of a slender floating platform coupled with an array of oscillating bodies. They derive a frequency-domain motion equation that can consider the hydroelasticity of the platform and the complex hinged connection between the platform and the oscillating bodies.

Muliawan et al. (2013) considered the hybrid concept of a floating wind turbine and an axisymmetric WEC. They performed coupled analyses in the time domain to study the motion behaviour and power generation of the hybrid concept under operational conditions. They point out that the total electricity production of the hybrid concept was 10-15% higher than that of the standalone floating wind turbine. Hu et al. (2020) numerically investigated the hydrodynamic performance of a floating-wind-platform-

DOI: 10.1201/9781003508779-55

based wave power extraction system. The study revealed that, at lower frequencies, the incorporation of WECs reduced the maximum horizontal forces and pitch moments on the platform.

Gaspar et al. (2021) experiments suggest that WECs located on the downwind and upwind sides of the floating platform played different roles in platform motion mitigation/amplification. Kamarlouei et al. (2020) carried out experimental studies on the co-location of WECs and floating offshore platforms. The results showed that the interaction between buoys and the platform increased the heave and pitch motions of the platform, which is also affected by the damping of the mooring system. Bispo et al. (2022a) presented a numerical investigation of wave interaction with a moored articulated very large floating structure. In further, Bispo et al. (2022b) numerically explored the wave power extraction performance of a moored horizontal submerged structure consisting of a set of hinged plates attached to a vertical flap.

Cheng et al. (2022) performed hydrodynamic analysis and power extraction evaluation of multi-degree-of-freedom (DoF) WECs integrated into a semi-submersible platform. They found that the integration of an oscillating buoy device with a larger radius and deeper draft may reduce wave loads on the platform. In general, both positive and negative hydrodynamic interactions of the floating platform and the wave energy devices are found for the above numerical/experimental investigations. The specific mutual interactions may be different for different kinds of platforms and WECs.

This paper aims to investigate the hydrodynamic performance of a floating semi-submersible platform with a wave energy device array. The corresponding numerical methods are developed for such kind of floating multi-body system and parametrical investigations are presented. The remainder of this paper is structured as follows: Section 2 presents the derivation of the mathematical model. Section 3 provides the results and discussion, and finally, Section 4 presents the conclusion.

2 MATHEMATICAL MODEL

The device under consideration in this study is a multi-body system, comprising an array of 15 wave energy devices attached to a semi-submersible platform. As illustrated in Figure 1, each buoy in the array is numbered from Buoy 1-15. The layout of the system is presented in Figure 1. The relative motion between the buoys and the platform drives the PTO system to extract wave power. The draft of the semi-submersible platform is 9.7 m, and the gravitational centre is detailed in Table 1.

Based on the Lagrange multiplier method, the motion equation of such kind of system can be derived as:

Table 1. The gravitational centre of the platform and the buoys.

Body number	Gravitational centre (m)
Platform	(0.00, 0.00, 0.00)
Buoy 1	(34.35, -11.49, 0.00)
Buoy 2	(23.95, -17.49, 0.00)
Buoy 3	(13.56, -23.49, 0.00)
Buoy 4	(3.17, -29.49, 0.00)
Buoy 5	(-7.22, -35.49, 0.00)
Buoy 6	(-27.12, -24.00, 0.00)
Buoy 7	(-27.12, -12.00, 0.00)
Buoy 8	(-27.12, 0.00, 0.00)

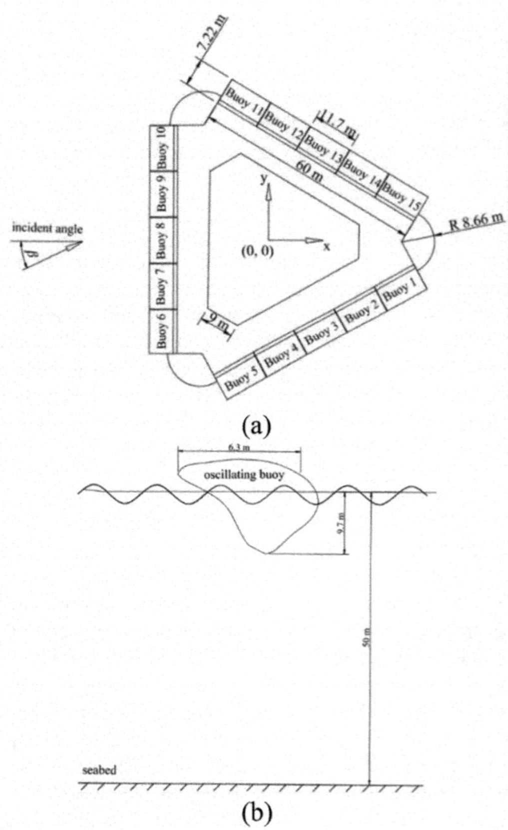

Figure 1. (a) Layout of the floating wave power system and (b) the dimensions of the oscillating buoy.

$$\begin{pmatrix} -\omega^2(\mathbf{M}' + \mu) - \mathrm{i}\omega(\lambda + \lambda_{\mathbf{pto}}) + \mathbf{C}' & \mathbf{L}^{\mathbf{T}} \\ \mathbf{L} & 0 \end{pmatrix} \begin{pmatrix} \zeta \\ \nu \end{pmatrix} = \begin{pmatrix} \mathbf{F}_{\mathrm{e}} \\ 0 \end{pmatrix} \tag{1}$$

where ω is the incident wave frequency, i is the imaginary unit, the diagonal matrix of $\boldsymbol{\lambda}_{\mathbf{pto}} =$

$C_{\text{pto}} \cdot \text{diag}(\boldsymbol{\lambda}_{\text{pto}})$ is the PTO damping matrix. $\mathbf{M}'$ and $\mathbf{C}'$ represent the mass matrix and restoring force stiffness matrix, respectively, and can be expressed as:

$$\mathrm{M}' = \begin{bmatrix} \mathrm{M}_1 & \cdots & \mathbf{0} \\ \vdots & \ddots & \vdots \\ \mathbf{0} & \cdots & \mathrm{M}_\mathrm{N} \end{bmatrix}_{6\mathrm{N}\times 6\mathrm{N}} \tag{2}$$

and

$$\mathrm{C}' = \begin{bmatrix} \mathrm{C}_1 & \cdots & \mathbf{0} \\ \vdots & \ddots & \vdots \\ \mathbf{0} & \cdots & \mathrm{C}_\mathrm{N} \end{bmatrix}_{6\mathrm{N}\times 6\mathrm{N}} \tag{3}$$

Here $\mathbf{M}_i$ and $\mathbf{C}_i$ (i = 1, 2, …, N) are the mass matrix and static water restoring force matrix for each floating body in six DoF, N (= 16) represent the body number of the present system. In this paper, the first body denotes the platform, while the second body refers to Buoy 1, and so forth.

The motion response amplitude vector $\boldsymbol{\zeta}$ and the wave excitation force vector $\mathbf{F}_\mathrm{e}$ are written as:

$$\zeta = \{\zeta_1, \zeta_2, \cdots, \zeta_N\}^T \tag{4}$$

$$\mathbf{F}_\mathrm{e} = \{\mathbf{F}_1, \mathbf{F}_2, \ldots, \mathbf{F}_N\}^T \tag{5}$$

where $\zeta_i = \{\zeta_i^{(1)}, \zeta_i^{(2)}, \zeta_i^{(3)}, \zeta_i^{(4)}, \zeta_i^{(5)}, \zeta_i^{(6)}\}$ and $\mathbf{F}_i = \{f_i^{(1)}, f_i^{(2)}, f_i^{(3)}, f_i^{(4)}, f_i^{(5)}, f_i^{(6)}\}$, $\zeta_i^{(j)}$ and $f_i^{(j)}$ represent the motion response and wave excitation force of the i-th body in the j-th DoF, where j stands for surge, heave, sway, roll, pitch, and yaw with j = 1, 2, 3, 4, 5, and 6, respectively. The hydrodynamic coefficient matrices of $\boldsymbol{\mu}$ and $\boldsymbol{\lambda}$ and the wave exciting force vector of $\mathbf{F}_\mathrm{e}$ can be calculated by using the potential-flow solver NEMOH (Babarit & Delhommeau, 2015). The boundary mesh shown in Figure 2 is used to execute the numerical calculation.

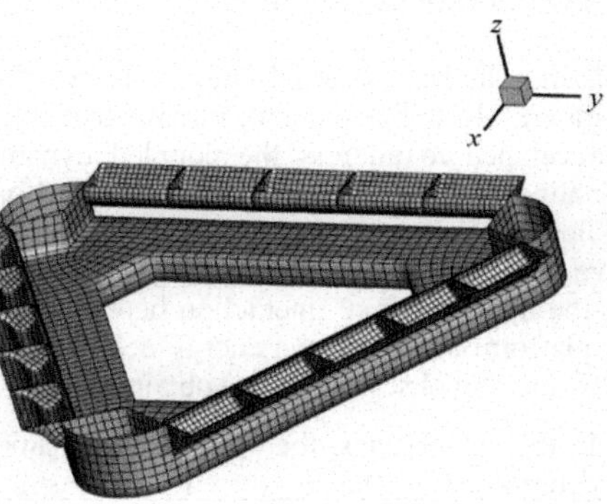

Figure 2. Boundary element mesh of the platform with an array of buoys. Mesh data is generated using Rhino software. 5000 panels for the platform and 400 panels for each buoy.

The relative pitch motion between the floating platform and the array of floating buoys is of interest in this study. The sway, surge, heave, roll, and yaw DoF at the connection point between the platform and the buoy satisfy the displacement continuity conditions. Here we introduce a local coordinate system O'-$x'y'z'$. The angle between the $O'x'$ axis and the positive x-axis is denoted as θ, the angle between the $O'x'$ axis and the O-xy plane is denoted as ψ, and the angle between the $O'y'$ axis and the O-xy plane is denoted as γ, where O-xyz is the global coordinate system. Please note that, in the present study, the yaw motion of the platform does not contribute to the relative motion of the buoy and the platform, as the angle $\gamma = 0$.

The motion constraint matrix $\mathbf{L}$ is composed of 5 $(N\text{-}1)\times 6N$ blocks:

$$\mathbf{L}_{5n\times 6N} = \begin{bmatrix} \mathbf{L}_{1,} & \mathbf{L}_{1,2} & \cdots & \mathbf{0} & \cdots & \mathbf{0} \\ \vdots & \vdots & \ddots & \vdots & \vdots & \vdots \\ \mathbf{L}_{i,} & \mathbf{0} & \vdots & \mathbf{L}_{i,i+1} & \vdots & \mathbf{0} \\ \vdots & \vdots & \vdots & \vdots & \ddots & \vdots \\ \mathbf{L}_{N-1,1} & \mathbf{0} & \cdots & \mathbf{0} & \cdots & \mathbf{L}_{N-1,N} \end{bmatrix} \tag{6}$$

Note that except for the blocks on the diagonal and in the first column, the remaining blocks are zero matrices. $\mathbf{L}_{i,1}$ and $\mathbf{L}_{i,i+1}$ (i=1, 2, …, N-1) can be derived as:

$$\mathbf{L}_{i,1} = \begin{bmatrix} 1 & 0 & 0 & 0 & -(z_1 - z_{c,i}) & (y_1 - y_{c,i}) \\ 0 & 1 & 0 & -(z_{c,i} - z_1) & 0 & -(x_1 - x_{c,i}) \\ 0 & 0 & 1 & -(y_1 - y_{c,i}) & (x_1 - x_{c,i}) & 0 \\ 0 & 0 & 0 & \cos\theta\cos\psi & \sin\theta\cos\psi & \sin\psi \\ 0 & 0 & 0 & \begin{matrix}\sin\theta\sin\gamma \\ -\cos\theta\sin\psi\cos\gamma\end{matrix} & \begin{matrix}-\sin\theta\sin\psi\cos\gamma \\ -\cos\theta\sin\gamma\end{matrix} & \cos\psi\cos\gamma \end{bmatrix}$$

$$\mathbf{L}_{i,i+1} = \begin{bmatrix} -1 & 0 & 0 & 0 & (z_i - z_{c,i}) & -(y_i - y_{c,i}) \\ 0 & -1 & 0 & (z_{c,i} - z_i) & 0 & (x_i - x_{c,i}) \\ 0 & 0 & -1 & (y_i - y_{c,i}) & -(x_i - x_{c,i}) & 0 \\ 0 & 0 & 0 & -\cos\theta\cos\psi & -\sin\theta\cos\psi & -\sin\psi \\ 0 & 0 & 0 & \begin{matrix}-(\sin\theta\sin\gamma \\ -\cos\theta\sin\psi\cos\gamma)\end{matrix} & \begin{matrix}\sin\theta\sin\psi\cos\gamma \\ +\cos\theta\sin\gamma\end{matrix} & -\cos\psi\cos\gamma \end{bmatrix}$$

where (x_1, y_1, z_1) represent the centre of gravity coordinates of the platform, (x_i, y_i, z_i) represent the centre of gravity coordinates of Buoy i, $(x_{c,i}, y_{c,i}, z_{c,i})$ represent the coordinates of hinge points between Buoy i and the platform.

The relative motion ξ_i between the Buoy i (i = 1, 2, …, N-1) and the floating platform can be calculated as calculated as:

$$\xi_i = \cos\theta \cdot \zeta_{i+1}^{(5)} - \sin\theta \cdot \zeta_{i+1}^{(4)} + \sin\theta \cdot \zeta_1^{(4)} + \cos\theta \cdot \zeta_1^{(5)} \tag{7}$$

The total power P_{total} captured by the buoy array, the hydrodynamic efficiency η_i corresponding to Buoy m and the total efficiency η_{total} can be expressed as:

$$P_{\text{total}} = \sum_{i=1}^{N-1} P_i \tag{8}$$

$$\eta_i = \frac{P_i}{P_{\text{incident}}} \tag{9}$$

$$\eta_{\text{total}} = \frac{P_{\text{total}}}{P_{\text{incident}}} \tag{10}$$

where P_i is the power generated by the Buoy i, P_{incident} is the incident wave power. P_{incident} and P_i can be calculated as:

$$P_i = \frac{1}{2}\omega^2 C_{\text{pto}} |\xi_i|^2 \tag{11}$$

and

$$P_{\text{incident}} = \frac{1}{2}\rho g \chi A^2 \frac{\omega}{k} \left\{ 1 + \frac{2kh}{\sinh(2kh)} \right\} \tag{12}$$

where χ, ρ, g, k, and h represent the projection width of the buoy that is perpendicular to the incident waves, water density, gravitational acceleration, wave number, and water depth, respectively. k and ω satisfy the dispersion relation of $\omega^2 = gk \tanh kh$. A is the wave amplitude, which is set as 1 m in this study.

3 RESULTS AND DISCUSSION

Figure 3 illustrates the total hydrodynamic efficiency for different PTO damping levels (C_{pto} varies from 1.0E3~1.0E8 N·m·s).

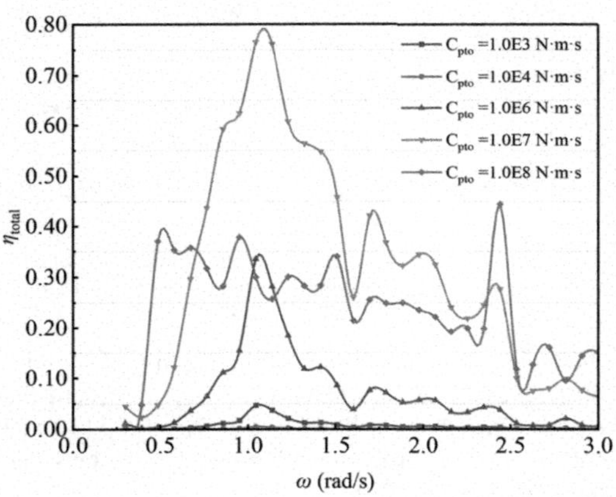

Figure 3. Results of the total hydrodynamic efficiency for different PTO damping ($\beta = 0°$).

It is observed that, within the scope of the present calculations, the PTO damping of C_{pto} = 1.0E7 N·m·s gives more excellent power extraction performance. The corresponding peaked efficiency is $\eta_{\text{total}} \approx 0.8$ at ω = 1.23 rad/s approximately. Both lighter and heavier PTO damping lead to the poor performance of wave power extraction.

The impact of the PTO damping on the motion characteristics of the floating platform is of great interest here. The platform responses in surge, heave, and pitch DoF for cases with C_{pto} = 0, 1.0E3, 1.0E4, 1.0E6, 1.0E7, and 1.0E8 N·m·s are shown in Figure 4. It is evident that PTO damping significantly modifies the motion responses of the floating platform in the frequency range near ω = 0.3 and 1.0 rad/s. Slight modifications of the platform response can be observed in the remaining frequency range.

As for the heave DoF, the responses corresponding to the PTO damping of C_{pto} = 1.0E3 N·m·s are larger than that of the remained PTO damping. There is a soaring in pitch responses at a lower frequency range for the case of 1.0E3 N·m·s. However, such kind of soaring disappears in case of C_{pto} > 1.0E4 N·m·s.

In general, the responses of the platform in surge, heave and pitch DoF corresponding to C_{pto} = 1.0E3 N·m·s is larger than that of heavier PTO damping cases. In addition, one can observe that the surge, heave, and pitch motion of the isolated platform (see Figure 5) is relatively smaller than that equipped with the wave energy device array in the low-frequency region.

Figure 6 presents the hydrodynamic efficiency corresponding to each buoy and the relative pitch motion between the platform and the buoy for the case of C_{pto} = 1.0E7 N·m·s and $\beta = 0°$. Because the buoys are symmetrical about the x-axis that aligns parallel with the incident wave direction, only the results corresponding to Buoy 1~9 are plotted. It can be observed that the pitch response corresponding to Buoy 6 ~ 9 is larger than that corresponding to Buoy 1 ~ 5, which are located on the leeward side of the platform.

A similar trend is observed for the hydrodynamic efficiency shown in Figure 6(a). The response shown in Figure 6 indicates that there exists a larger response in surge and heave DoF at the lower frequency region (ω < 1.0 rad/s). It is mainly reflected in that a peak in relative pitch motion is observed for Buoy 6-8 at ω = 0.8 rad/s. However, the relative pitch motion increases as ω approaches to 0.25 rad/s. The frequency range characterized by larger relative responses shown in Figure 6 (b) does not align with the frequency range at which the floaters achieve peaked energy conversion efficiency.

4 CONCLUSIONS

Utilizing linear potential flow theory and the Lagrange Multiplier method, a mathematical model is developed to address the coupled dynamics of a floating platform with an array of hinged oscillating bodies. The impact of PTO damping and incident wave angle is numerically investigated also exploring the hydrodynamic interaction between the floating platform and the wave energy device array. The following key observations are obtained:

1) In the calculations, the optimal PTO damping is determined to be C_{pto} = 1.0E7 N·s·m. Both lighter and heavier PTO damping result in relatively poor wave power extraction performance.
2) PTO damping significantly influences the motion response of the platform in the frequency region of ω < 1.25 rad/s. Specifically, the case of C_{pto} =

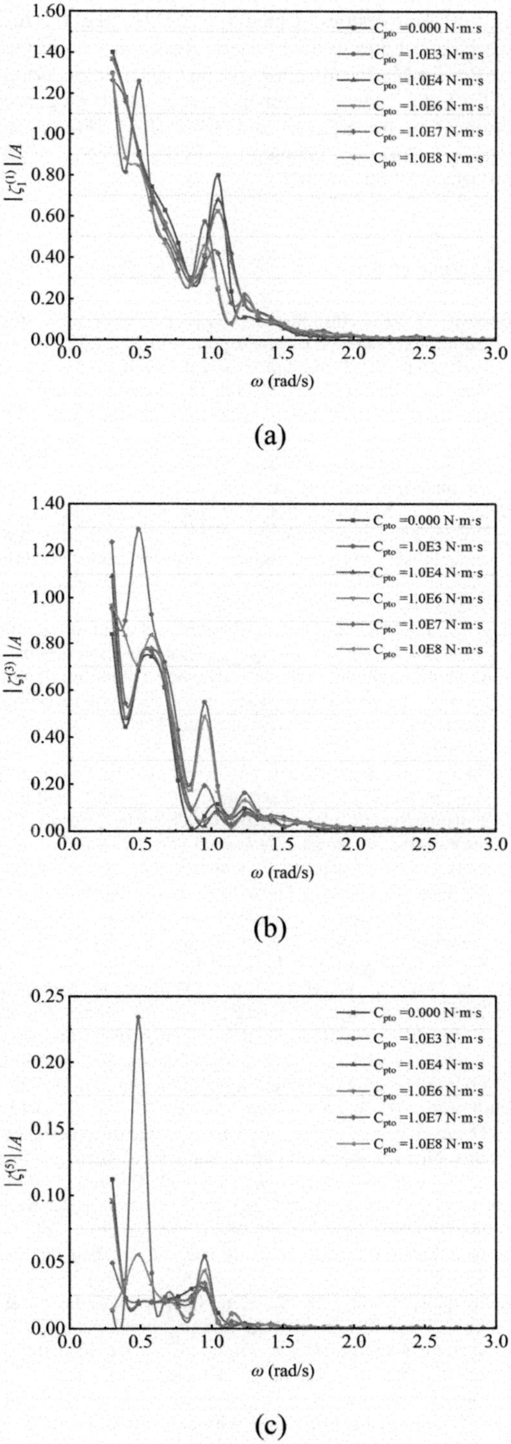

Figure 4. Results of the platform response in surge (a), heave (b) and pitch (c) DoF for different PTO damping (β = 0°).

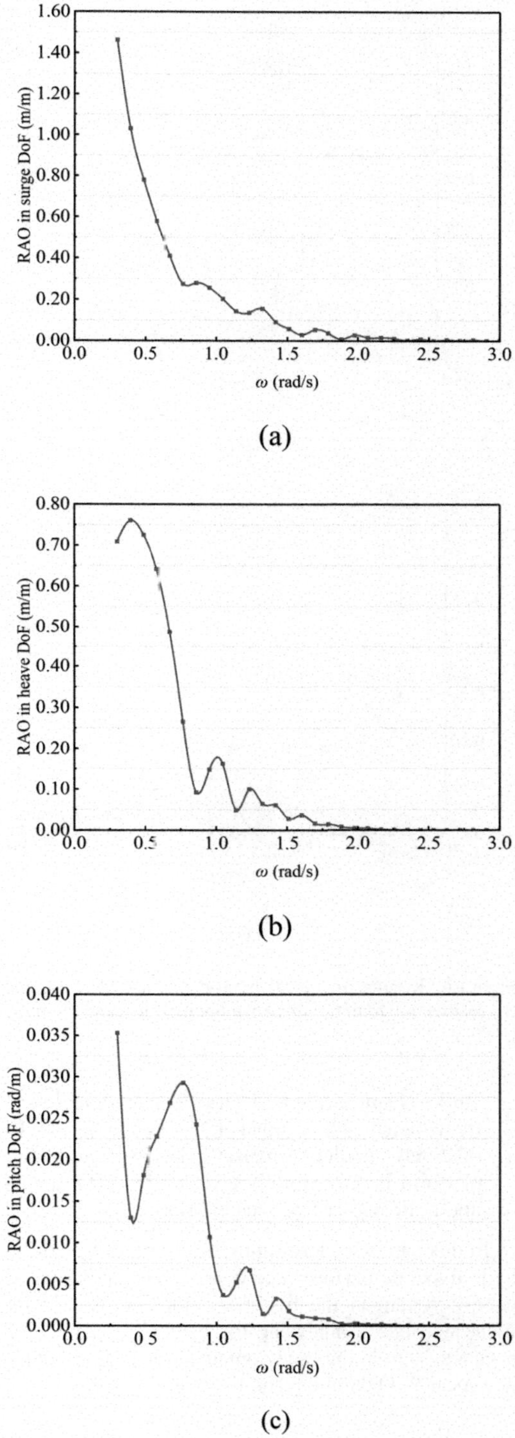

Figure 5. Response amplitude operator (RAO) of the platform in surge (a), heave (b) and pitch (c) DoF (β = 0°).

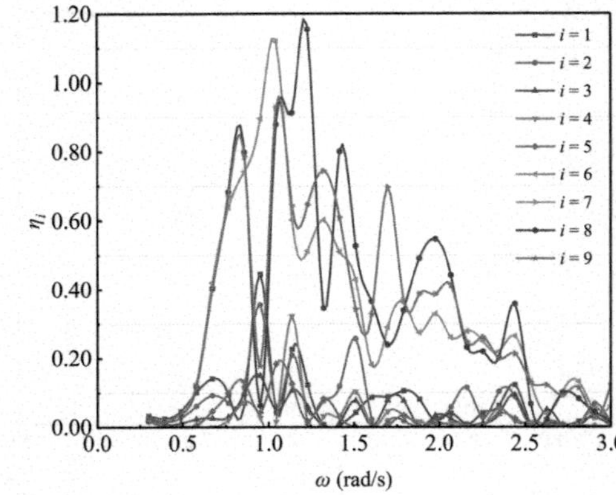

(a)

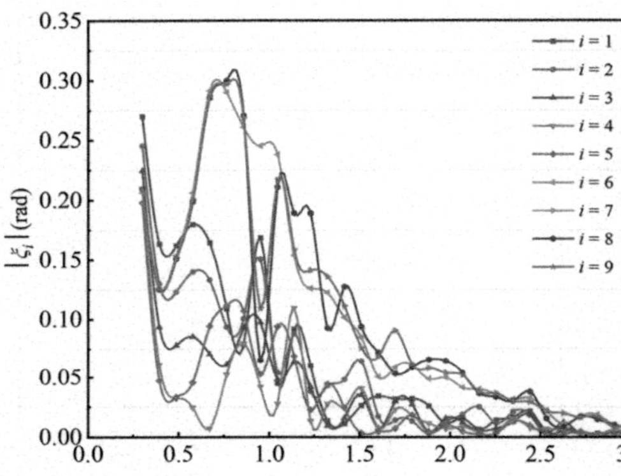

(b)

Figure 6. Results of the (a) hydrodynamic efficiency and (b) relative pitch motion between the platform and the buoy (β= 0°).

1.0E3 N·s·m exhibits a large response at lower frequencies, while other PTO damping led to relatively smaller responses. The response of the platform is only slightly affected by PTO damping in the higher frequency region.

In this paper, we have numerically investigated the hydrodynamic performance of a novel floating wave energy system in the frequency domain. The future focus will be on analysing the coupled dynamics of such a system in the time domain, taking into account nonlinear PTO damping and the mooring system

ACKNOWLEDGEMENTS

This work is supported by the National Key Research and Development Project of China (2019YFB 1504403) and Fundamental Research Funds for the Central Universities, China (3072023JC0101). This work contributes to the Strategic Research Plan of the Centre for Marine Technology and Ocean Engineering (CENTEC), which is financed by the Portuguese Foundation for Science and Technology (Fundação para a Ciência e Tecnologia - FCT) under contract UIDB/UIDP/00134/2020.

REFERENCES

Babarit, A. & Delhommeau, G. 2015. Theoretical and numerical aspects of the open source BEM solver NEMOH. *11th European Wave and Tidal Energy Conference (EWTEC 2015), Sep 2015*. France: Nantes.

Bispo, I. B. S., Mohapatra, S. C., & Guedes Soares C. 2022a. Numerical analysis of a moored very large floating structure composed by a set of hinged plates. *Ocean Engineering*, 253: 110785

Bispo, I. B. S., Mohapatra, S. C., & Guedes Soares C. 2022b. Numerical model of a WEC-type attachment of a moored submerged horizontal set of articulated plates. *Trends in Maritime Technology and Engineering*, 335–344.

Cheng, Y., Dai, S., Ji, C., Cocard, M., Yuan, Z. & Incecik A. 2022. Energy conversion and hydrodynamic analysis of multi-degree-of-freedom wave energy converters integrated into a semi-submersible platform. *Energy Conversion and Management* 252: 115075.

Gaspar, J. F., Kamarlouei, M., Thiebaut, F. & Guedes Soares C. 2021. Compensation of a hybrid platform dynamics using wave energy converters in different sea state conditions. *Renewable Energy* 177: 871–883.

Kamarlouei, M., Gaspar, J F., Calvario, M., Hallak, T. S., Mendes, M. J. G. C., Thiebaut, F. & Guedes Soares, C. 2020. Experimental analysis of wave energy converters concentrically attached on a floating offshore platform. *Renewable Energy* 152: 1171–1185.

Hu, J., Zhou, B., Vogel, C., Liu, P., Willden, R., Sun, K., Zang, J., Geng, J., Jin, P., Cui, L., Jiang, B. & Collu, M. 2020. Optimal design and performance analysis of a hybrid system combing a floating wind platform and wave energy converters. *Applied Energy* 269: 114998.

Muliawan, M. J., Karimirad, M. & Moan, T. 2013. Dynamic response and power performance of a combined Spar-type floating wind turbine and coaxial floating wave energy converter. *Renewable Energy* 50: 47–57.

Zhao, X. L., Ning, D. Z. & Liang, D. F. 2019a. Experimental investigation on hydrodynamic performance of a breakwater-integrated WEC system. *Ocean Engineering* 171: 25–32.

Zhao, X. L., Ning, D. Z., Zou, Q. P., Qiao, D. S. & Cai, S. Q. 2019b. Hybrid floating breakwater-WEC system: A review. *Ocean Engineering* 186: 106126.

Zhao, X., Xue, F., Chen, L., Göteman, M., Han, D., Geng, J. & Sun, S. 2023. Hydrodynamic analysis of a floating platform coupled with an array of oscillating bodies. *Ocean Engineering* 287: 115439.

Zhu, H., Hu, C., Sueyoshi, M. & Yoshida, S. 2020. Integration of a semisubmersible floating wind turbine and wave energy converters: an experimental study on motion reduction. *Journal of Marine Science and Technology* 25(3): 667–674.

Aquaculture offshore

Advances in Maritime Technology and Engineering – Guedes Soares & Santos (Eds)

The invisible pollutant in water: Nanoplastics in seafood

Y.Z. Ahmed, M. Li & Y. Song
Department of Marine Engineering, Dalian Maritime University, China

H. Amjad & M. Ali
School of Civil and Environmental Engineering, Institute of Environmental Sciences and Engineering, National University of Sciences and Technology, Pakistan

ABSTRACT: The prevalence of nanoplastics in seawater has raised concerns within the seafood industry due to the unknown long-term effects of these particles. Wild-caught seafood, in particular, is highly susceptible to nanoplastics contamination, as recent research indicates that these particles can induce various physiological problems in fish and accumulate in higher trophic levels. Detecting and identifying nanoplastics require sophisticated methods. Current approaches involve the use of transmission/scanning electron microscopy and spectroscopy techniques like Raman and Fourier transform infrared spectroscopy (FTIR), as well as thermal and dynamic light scattering (DLS). Recent advancements in micro/nanofluidic technology present a simpler solution for nanoplastics determination, particularly for sorting and counting. These essential steps in nanoplastics determination are crucial for understanding the long-term effects on both marine organisms and humans. This mini review will highlight some recent developments in nanoplastics research in seafood and way forward towards better understanding of these particles in fisheries sector.

1 INTRODUCTION

1.1 *Plastic particles in water*

Plastics are extremely resistant to breakdown and can withstand extreme environment for centuries. Large plastic litters will most likely decompose into microplastics (MPs) ranging from 1-5000 μm (Thompson et al., 2004) and further breaks down into nano-plastics (NPs) ranging from 0.1 μm or less (Gigault et al., 2020) before being completely decomposed. Depending on origin, micro-nano plastics (MNPs) might be classified into primary or secondary plastics. The primary MNPs are those that are directly introduced into the environment as microfibers, microbeads, micropellets, and different ways, mostly by anthropogenic means (Thompson et al., 2004). Whereas, the secondary MPs are produced when large or mesoplasticlitters break down in the environment as a result of chemical, biological, and physical factors. Among these forces include oxidation, heat, UV radiation, mechanical forces, and biological processes (Peng et al., 2020).

Recent discoveries have revealed the prevalence and magnitude of MNPs contamination, even in the world's most distant locations (Materi et al., 2021, Napper et al., 2020, Allen et al., 2019). The microplastics pollutions have become a matter of debate concerning stress to sea ecosystem (Figure 1). The properties of material plastics such as lightness, flexibility, strength, and durability make them versatile. Their favourable features lead to widespread consumption, but microplastic and nanoplastic contamination is a global concern that harms wildlife, the food chain, and public health (Cverenkárová et al., 2021). Research related to various water bodies such as freshwater (Koelmans et al., 2019, Wagner et al., 2014), groundwater (Samandra et al., 2022, Chia et al., 2021), terrestrial and aquatic biota have led towards accelerated NP research. For extended waters such as marine systems, the NP pollution is causing major concern (Barnes et al., 2009), with effects mostly affecting marine biota (Gall & Thompson, 2015). All plastic that flow and become part of oceans which act as the final sink for plastic (Thompson et al., 2004, Coleman & Wehle, 1984) causing irreversible damange. Fisheries industry is directly effected with this pollution. Direct consumption from polluted source will have its ramifications. Extensive research is required for islotion and identification of nanoparticles to achieve lasting solutions.

This mini review discusses some common methods for NPs detection, sorting, counting that can be used as a singular technique or in combination for NPs observation in seafood. For identification and quantification of NPs in environmental samples, instruments based on the principle of spectroscopy, microscopy and sensor based detection were mostly in use (Cai et al., 2021, Toussaint et al., 2019). The extraction of

DOI: 10.1201/9781003508779-56

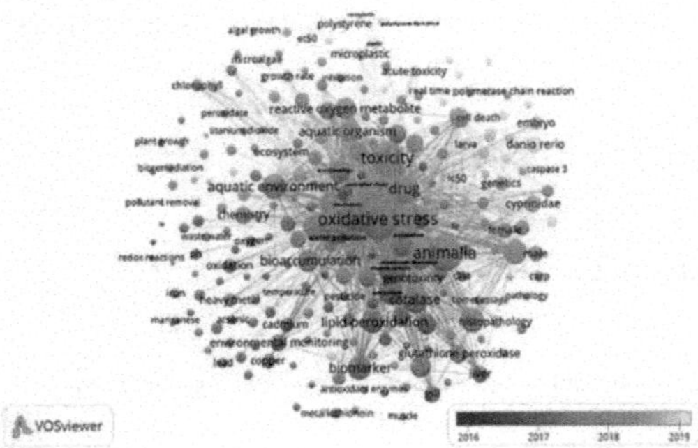

Figure 1. Trending terms related to plastic particle pollution (Geremia et al., (2023) © the authors, 2023, MDPI, CCBYND 4.0).

NPs from seafood samples involves various techniques such as filtration, chemical digestion, sieving, and evapouration process. The extremely small size of nanoplastics makes the extraction process a lot difficult (Barbosa et al., 2020). Chang et al. (2019) devised a method to remove nanoplastics from fish and oyster tissues to address the lack of knowledge about them. Nanoplastics were extracted and validated using centrifugal concentration, membrane filtration, enzymatic digestion, and purification. The Corolase-lipase method digested non-homogenized tissues well. This successful strategy, which retained nanoplastic structure, yielded high recoveries and might be utilised as a pilot programme for future nanoplastic detection methods. Overall, the study improved our understanding of nanoplastics in seafood, an important environmental topic. The basic steps from identification to quantification of NPs are illustrated in Figure 2. Further research is required to understand polymer toxicity, dynamics, and standardise NPs assessment techniques in marine environment to address seafood safety concerns.

1.2 *Nanoplastic in fisheries*

Research on the ubiquity, incidence, and transmission of NPs from the lower trophic levels to higher trophic levels and ultimately to humans is rapidly advancing. The majority of these studies are focused on the ingestion of MNPs by phytoplankton, zooplankton, oysters, crustaceans, and fishes (Benson et al., 2022) (Table 1). Plastic particles within the aquatic environment undergo fluctuations in their physical attributes, subsequently influencing their sedimentation, conveyance, retention dynamics, and bio accessibility (Krause et al., 2021). The bio accessibility and assimilation of NPs within the aquatic trophic web manifest across all hierarchical strata, as evidenced by the identification in entities spanning benthic planktonic organisms to expansive specimens in the aqueous column (Digka et al., 2018, Lima et al., 2014). The bio accessibility and assimilation processes if NPs hinge on various determinants, including density, chromaticity, size, and the prevalence of polymer particles, as well as hydrodynamic circumstances such as water flow kinetics.

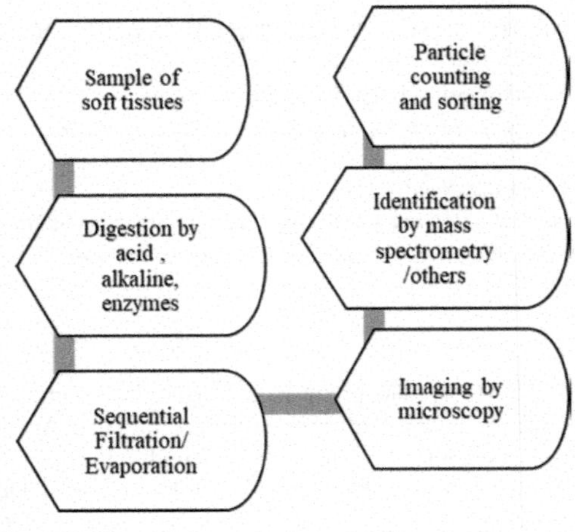

Figure 2. Work flow for analysis of nanoplastics (Vitali et al., 2022).

1.3 *Effect in marine organisms*

Numerous research have investigated the ecotoxicological ramifications of NPs across diverse aquatic organisms, revealing their tendency of cellular membrane penetration, tissue accumulation, and integration into food chains (Lee et al., 2019, Toussaint et al., 2019, Chae et al., 2018, Bouwmeester et al., 2015). Brandts et al. (2022) documented that 44 nm polystyrene nanoplastics caused bioaccumulation in muscles and livers of goldfish *Carassius auratus* and caused damage to DNA. In another study, Mattsson et al. (2017) revealed that consumption of nano plastic contaminated *Daphnia magna* by fish caused brain damage and behavioral pattern changes in fish. These discoveries have made human health at risk as humans consume nano plastics mostly through the consumption of tainted seafood especially fish (Selvam et al., 2020, Wong et al., 2020, Bradney et al 2019, Schymanski et al., 2018). Moreover, NPs could also cause substantial adversative effects on the behavioral patterns, consumption habits, growth rates, reproductive processes, and overall survival of aquatic organisms. Chen et al. (2017) documented that 50-nm NPs, when exposed at a concentration of 1 mg/L, markedly impeded locomotion in zebrafish larvae, and diminished the overall body length of the larvae. As, NPs have a considerably smaller diamer than MPs, they are easily able to move from different animal tissues and can cross biological barriers. Research revealed that 50nm fluorescent NPs in the digestive tract of the marine copepod *Paracyclopina nana* might swiftly spread throughout the body through cellular membranes of the digestive organs. In contrast, under identical conditions, microbeads measuring 500nm and 6μm were primarily restricted to the digestive organs of the copepod (Jeong et al., 2017).

Table 1. Nanoparticles in marine orgaisms.

Organism	Material	Particle Size nm	Source
Zooplankton	LDPE, PS, HDPE	<1	(Nadal et al., 2016)
Oysters	Fiber	0.01-5	(Wang et al., 2019b)
Mollusca	PET, PE, Microfibers	0.01-5	(Naji et al., 2018)
Mussels	Fragments, Fiber	<250	(Kolandhasamy et al., 2018)
Fish	PES, Fibers	-	(Su et al., 2019)
Zebra fish	PS	56	(Chen et al., 2017)
Sea Urchin	-	40	(Torre et al., 2014)

* LDPE (low density polyethylene), PS (polystyrene), HDPE (high density polyethylene),PET (polyethylene terephthalate), PE (polyethylene) and PES (polyester).

2 METHODS OF DETECTING NANO PLASTICS

2.1 *Transmission electron microscopy*

Transmission electron microscopy (TEM) is a strong imaging technique, which has been used to characterize nanoparticles since its birth in the twentieth century. It visualizes the ultrafine surfaces and has become an important technique for researching nanoparticles. The technique works on electron beam principles where transmission occurs across thin samples. The electron beam interacts with the specimens and produces high-resolution pictures through electros being spread over it. The technique detects nanoparticles excellently by generating ultra-power resolution and lets experts see elements as tiny as nanometers. TEM was used to observe and quantifiy nanoplastics in fish cells (Yang et al., 2023). TEM can be used to detect the size and shape of NPs (Ekvall et al., 2019). Woods and Stirling (2018) used TEM to demonstrate the ultrastructure of both cells and tissues. Schorb et al. (2019) also experimented with TEM to discover cell particles found in a plastic specimen.The drawback is that due to NPs low density the TEM may impede the detection in tissue samples (Triebskorn et al., 2019). Modified techniques using TEM in comibination may be tested for NPs detection in fish tissues.

2.2 *Scanning electron microscopy*

Scanning electron microscopy (SEM) has become a strong tool for characterizing nanoparticles. SEM offers comprehensive details regarding the surface image. SEM works on the principle of an electronic beam that scans the surface side of a sample and provides premium images. This technique can provide synergetic details regarding surface morphology followed by chemical properties. This cohesive technique improves knowledge about the nanoparticle aspects, with SEM providing resolution in sub-nanometer. SEM usually requires precoating of samples with conductive particles. This may cause hinderance for NPs detection as stand alone technique but in combination may provide relevant results. Sobhani et al., (2020) used Raman imaging with SEM to observe NPs of 100nm.

Vagner et al. (2022) used SEM to observe samples of european seabass where translocation of fluorescent polystyrene nanoplastics (PS-NPs) occured with a diameter of 50 nm over the intestinal barrier. PS–NPs on the luminal side increased serosal fluorescence intensity by 15–90 minutes. This confirmed that NPs penetrated the intestinal epithelium of the seabass. Guerrera et al. (2021) examined the impact of MP/NP contamination in fish species that were used as models to interpret consequences in human health. The study examined plastic particle entrance, absorption, and bioaccumulation in fish organs and the complex factors affecting plastic particle toxicity. SEM with other techniques were used and its findings revealed the extent of MP/NP distribution and effects in fish tissues. Vagner et al. (2022) also confirmed NPs translocation over the intestinal barrier in European Seabass. The work showed luminal-to-serosal migration of 50-nm NPs using additional techniques with SEM. SEM showed that nanoplastics can penetrate through marine animals' digestive epithelium, revealing fish exposure effects.

2.3 *Raman spectroscopy*

Raman Spectroscopy (RS) has also become a strong tool for characterizing nanoparticles. RS generates information regarding chemical properties that help in the identification and detailed scrutiny of nanoparticles. RS, however, scatters monochrome shade and produces information about vibrations and chemical properties found within molecules. Both SEM and RS cater to a broad variety of micro or nanoparticles along with those having metal or semi-conductors and biological elements. Ruan et al., (2024) detected NPs of 20nm in water samples using RS. Zhang et al., (2021) used RS for detection of MPs in crab and fish. Zhou et al., (2021) used surface enhanced RS to detect 50nm NPs in environmental samples. Modification in sample preparation in future may lead to successful RS- NPs detetion in marine fisheries.

2.4 *Fourier transform infrared spectroscopy*

Fourier transform infrared spectroscopy (FTIR) is a strong analytical tool that identifies and characterizes materials when they absorb infrared rays. When detecting nanoparticles, the technique provides information about the molecular composition and particle arrangement. It works on the principle of determining the absorption level of infrared radiation when it forms interactions with the given sample (Yang et al., 2023). A FTIR is used to translate the raw information into a broad range to reveal unique peaks related to particular changes within the molecules (Theakstone et al., 2021). The technique effectively detects tiny particles and captures their exceptional vibrational spectrum. It is effective when large particles are examined because its sensitivity reduces for particles below a specific size, usually falling in a range of a few nano meters. The technique can be used to detect found in polymers and organic plastics. Hazeem et al., (2020) invsetigated the toxicity of NPs in micro algae using FTIR. Peng et al., (2024) used FTIR to studied the effects of nanoplastics using fibroblast cell lines of fish and human. Garcia-Torne et al., (2023) assessed nanoplastic composition in the gastrointestinal tracts of fresh water fish using FTIR. and found most plastics particles are composed of polyethylene. Wang et al., (2023) examined NPs in digestive system of marine fish *Sebastes schlegelii* and identified polystyrene NPs using FTIR peak signals.

2.5 *Thermal and dynamic light scattering*

Thermal and dynamic light scattering (DLS) is a strong analytical tool that is used to research nanoparticles in terms of their size, delivery, shape, dimensions, and dynamics. By determining the changes in scattered light frequency, this technique generates information regarding the characteristics of nanoparticles. Thermal and dynamic scattering measures variations found in particle size distribution and thermal movements by inspecting the frequency changes in scattered light that occurred due to Brownian movement (Russo et al., 2021). Dynamic Light Scattering evaluates the complex behavior of nanoparticles in suspension by reviewing the changes in the scattered light as time passes. The technique effectively detects nanoparticles with size 1nm and it moves across to several micrometers (Gross-Rother et al., 2020). It performs well in reviewing small elements and provides analysis about diffusion co-efficient. This method provides an accurate dimension of the nanoparticle size followed by its characters. It is dynamic and can be used to detect metallic particles, polymers, and liposomes.

Brandts et al., (2018) used DLS to determine effects of NPs 40nm in marine fish *Dicentrarchus labrax* and found that the immune system is effected by presence of NPs. Bergami et al., (2017) used DLS to observe NPs in marine plankton *Dunaliella tertiolecta* and *Artemia franciscana* and found the NPs exhibits some level of toxicity towards these orgamisms. In another study by Brandts et al., (2022), DLS technique was used to determine accumulation levels of NPs in marine fish liver and muscles. It was observed that there were severe long term damanges to fish up to DNA level upon chronic exposure to NPs. Tallec et al., (2019) studied NPs bevaviour with respect to different salininties of water.

2.6 *Atomic force microscopy*

Atomic force microscopy (AFM) is a high-resolution imaging method that has roots in the discipline of nanotechnology and nanoparticle research. It visualizes and characterizes particles within their atomic and molecular surfaces. The technique operates on the principle of scanning sharply on the surface of a sample to measure the interactive forces between the tip and the sample. The tip's motion is documented to create topographical imagery so that details about surface features can be provided (Vitali et al., 2022). It can be used to detect nanoparticles with size limits from a few nanometers to micrometers and provide lateral as well as vertical resolution. The technique is useful to study nanoparticles with distinctive sizes. It is a dynamic tool and can be used to identify metal nanoparticles, polymers, nanocomposites, and biological constituents.

Fu etal., (2020) applied AFM technique to identify and characterize the NPs from marine environment samples. Zhao et al., (2023) usd AFM to observe NPs formed by rotifers in aquatic environment and observed that NPs were found in digestive system of marine samples. Mandemaker et al., (2023) discussed the significance of using AFM in organisms including fish and observed that combining AFM with other spectroscopic methods favoured the over all observation of results.

2.7 *Nuclear magnetic resonance*

Nuclear magnetic resonance (NMR) is a strong spectroscopic tool that is often used in chemistry and material sciences. Considering the research of nanoparticles, the tool is useful because it tells about the structure, composition, and dynamics of tiny particles that the human eye cannot see. The working principle relating to NMR is that it uses the magnetic properties of atoms such as protons when they are exposed to the strong magnetic field and interact with radio frequency pulses. The protons take up pulses and re-release electromagnetic rays. The resultant NMR continuum reflects information regarding the local ecosystem and interactions of the nucleus found in a sample. It can detect nanoparticles in their liquid interface. While its sensitivity declines with decreasing size of a particle, the tool is useful to characterize nanoparticles with sizes varying from nanometres to large sales. It can be used to identify organic constituents. Bashirova et al. (2023) used NMR method in Zebra fish samples and found that PET-NP causes various metabolic and oxidative stress. Afrose et al., (2024) utilized NMR technique to analyze NPs toxicity in Seabass fish tissues.

2.8 *Resistive pulse sensor technology*

Resistive pulse sensor (RPS) It is a unique technology that is prominent because of detecting and characterizing nanoparticles accurately. It measures the changes within the electric resistance field since particles move across a nanoscale pole in a microfluidic chip system (Figure 3). The technique provides comprehensive and tag-free analysis of nanoparticles. It is based on coulter counter and operates on a rule that when tiny particles move across a nanopore, they bring in temporal changes within the e-resistance field. Such changes are observed and interpreted to find out key aspects in terms of size, concentration, and surface changes within the nanoparticles. RPS detects nanoparticles from a few nanometers to micrometers and its sensitivity to fine transitions lets particles differentiate based on their sizes and compositions (June Soo Kim et al., 2023). It is suitable for detecting liposomes, polymers, and biologically associated nanoplastics. Chen et al., (2022) discussed the potential of RPS for NPs observations in water samples. Huber et al., (2023) also emphisized the use of RPS for characterization and quantification of NPs. RPS is an alternative method to detect particles between 5μm to 60nm and more with additional methods. It is frequently used in medical and chemical research of nanoparticle study of 800nm below and can be easily applied for food testing (Caputo et al., 2021, Calzolai et al., 2012) hence it shows promising potential for NPs study in seafood samples.

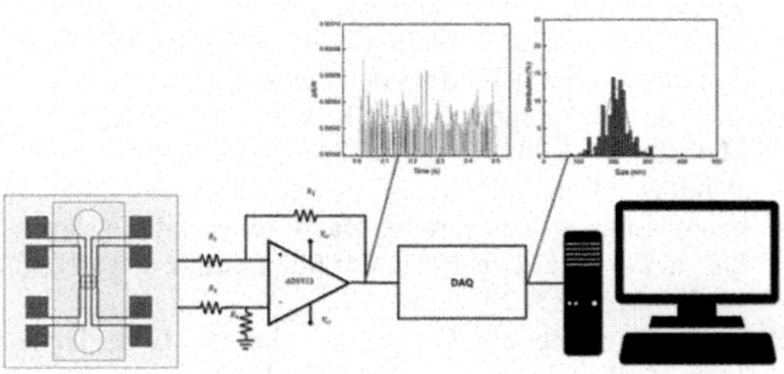

Figure 3. Microfluidic chip generating signals with hlelp of resistive pulse sensor that is observed using computer software system (Kim et al., (2023) © the authors, 2023, The Royal Society of Chemistry, CCBY3.0).

3 FUTURE OUTCOME

We reviewed some important methods to detect nanoplastics in fisheries. Transmission electron microscopy (TEM) is essential for nanoplastic research because it directly observes the shape, size, and aggregation of nanoparticles. This approach helps us understand the physical properties and behavior of nanoplastics in freshwater ecosystems. SEM and Raman spectroscopy combine morphological and chemical data for a comprehensive investigation. Researchers can use Raman spectroscopy to show the molecular composition of nanoplastics in fisheries and scanning electron microscopy (SEM) to disclose surface features to link the physical structure of these materials to their chemical properties.

Since FTIR makes it possible to identify functional groups, which helps predict the effects that nanoplastics would have on the environment in fresh water, it is crucial for characterising the chemical composition of nanoplastics. Thermal and dynamic light scattering (DLS) can be used to assess the size distribution and stability of nanoplastics. DLS can also demonstrate the mobility of the nanoplastics and their interactions with aquatic animals. The technique of atomic force microscopy (AFM) facilitates surface characterisation studies and high-resolution imaging, contributing to the understanding of the impact of nanoplastics on the environment. Nuclear magnetic resonance (NMR) spectroscopy helps determine the reactivity and bioavailability of nanoplastics by revealing molecular structure and dynamics. Resistive pulse sensor (RPS) technology is used to assess nanoplastics in freshwater environments and provides useful data. This method detects large concentrations of nanoplastics. This set of methodologies represents an all-inclusive toolkit for the examination of nanoplastics in the seafood sector and their impact on the environment.

RPS seems to be a cost-effective solution because it provides high sensitivity and detects individual particles of plastics. This aspect is critical when experts trace concentrations of minor plastics in water. Moreover, RPS has the potential to generate real-time insight and data concerning the presence of nanoplastics in the water (Pollard et al., 2020). It is beneficial for knowing about the changes found in the environment and responding immediately to the problems when they occur. RPS detects plastic particles cost-effectively and unlike other analytical methods, it is advantageous for setting initial setup and run sample-wise analysis. The technique is simple to use and requires less complicated samples to detect nanoplastics (Platt et al., 2020).

Gangadoo et al., (2020) research showed that marine animals were exposed to nanoplastics at levels previously reported in a highly controlled laboratory environment. To make tissue sample evaluation easier, another senor based method of impedance pulse technology was applied. The ability to identify and concentrate nanoplastic particles was made possible by this technology, which was very beneficial. The results of this study showed that this method allows for real-time monitoring of the absorption of nanoplastics by marine life. Leppanen et al. (2022) developed a plant-derived cellulose nanofibril (CNF) system with the ability to collect nanoplastics and microplastics in aqueous solution. Similarly, organic filters for seafood NPs studies requires integration in method development to make the testing procedure more sustainable. More research is needed to identify nanoplastics efficiently, sustainably, in real time with minimum effort and high throughput. It is important to focus on RPS and similar sensor based methods as a solution to the problem. An investigation is needed to find out the sensitivity and specificity of RPS that can help in the identification of micro and nano plastic particles specifically for seafood samples. It must involve modified designs in the microfluidic chip +sensor so that overall performance can be improved.

Moreover, methods must be caliberated that can precisely determine the size and scale of nanoplastics while using RPS. It will help in understanding the unique resistive pulse signs related to the size and compositions of different nanoplastics. In a comprehensive analysis by Chen et al. (2022) it was concluded that the current progress in the detection and removal of nanoplastics from water systems including sensor based techniques performed admirably in the detection of NPs. New research must focus on key factors including volume, depth, and contamination at the time of retrieving the sample of biological samples from aqueous medium. In addition, further investigations are required where future scholars can find out how RPS can integrate with other tools to generate a comprehensive analysis of nanoplastics. Future studies must generate insights either qualitatively, quantitatively, or both to facilitate testing, standardization, regulation, prevention and awareness of NPs in fisheries sector.

4 CONCLUSION

The growing concern of nanoplastics in fisheries was highlighted, and how these minute particles get access to marine species and cause harmful effect. It is an escalating concern of the seafood industry. The impacts of nanoplastics are long-term and need more exploration, particularly in seafood. Nanoplastics contaminate seafood, and physiological effects along with the accumulation of minor plastics in high tropical areas pose dynamic challenges for both marine species and human beings. Literature highlights the use of advanced detection methods that can potentially identify, sort and count nanoplastics. Undoubtedly, remarkable measures have been taken to understand the phenomena, but it is important to identify the limitations of existing techniques and work towards its solution. Such methods are costly and complicated because they involve laboratory work and they might cause hindrance in the general adoption.

To overcome such limitations, more studies are needed that emphasize on coseffective refinement and standardization of detection methodologies. Moreover, combined attempts could generate a more dynamic analysis of nanoplastics and their impact on the sea environment. The implications can move further than scientific investigation, which may reach the sphere of ecological reforms and practices adopted by the seafood industry. Regulations (using example of European Union for MNPs safety) must be revised to consider the risk posed by nanoplastics in seafood. This is needed to foster sustainable measures and reduce waste significantly.

REFERENCES

Afrose, S., Tran, T.K.A., O'Connor, W., Pannerselvan, L., Carbery, M., Fielder, S., Subhaschandrabose, S. and Palanisami, T., (2024). Organ-specific distribution and size-dependent toxicity of polystyrene nanoplastics in Australian bass (*Macquaria novemaculeata*). *Environmental Pollution*, 341, p.122996.

Agarwal, N., Nair, M. S., Mazumder, A., & Poluri, K. M. (2018). Characterization of nanomaterials using nuclear magnetic resonance spectroscopy. In *Characterization of nanomaterials* pp. 61–102. Woodhead Publishing. https://doi.org/10.1016/B978-0-08-101973-3.00003-1

Agarwal, U.P., Ralph, S.A., Reiner, R.S. and Baez, C. (2018). New cellulose crystallinity estimation method that differentiates between organized and crystalline phases. *Carbohydrate polymers*, 190, pp.262–270. https://doi.org/10.1016/j.carbpol.2018.03.003

Allen, S., Allen, D., Phoenix, V.R., Le Roux, G., Durántez Jiménez, P., Simonneau, A., Binet, S. and Galop, D. (2019). Atmospheric transport and deposition of microplastics in a remote mountain catchment. *Nature Geoscience*, [online] 12(5), pp.339–344. doi:https://doi.org/10.1038/s41561-019-0335-5.

Barbosa, F., Adeyemi, J.A., Bocato, M.Z., Comas, A. and Campiglia, A. (2020). A critical viewpoint on current issues, limitations, and future research needs on micro- and nanoplastic studies: From the detection to the toxicological assessment. *Environmental Research*, [online] 182, p.109089. doi:https://doi.org/10.1016/j.envres.2019.109089.

Barnes, D.K.A., Galgani, F., Thompson, R.C., Barlaz, M., (2009). Accumulation and fragmentation of plastic debris in global environments. *Philosophical Transactions of the Royal Society.* B364 1526, pp.1985–1998. doi:https://doi.org/10.1098/rstb.2008.0205.

Bashirova, N., Poppitz, D., Klüver, N., Scholz, S., Matysik, J., & Alia, A. (2023). A mechanistic understanding of the effects of polyethylene terephthalate nanoplastics in the zebrafish Danio rerio embryo. *Scientific Reports*, 131, p.1891. doi:https://doi.org/10.1038/s41598-023-28712-y

Benson, N.U., Agboola, O.D., Fred-Ahmadu, O.H., De-la-Torre, G.E., Oluwalana, A. and Williams, A.B. (2022). Micro(nano)plastics Prevalence, Food Web Interactions, and Toxicity Assessment in Aquatic Organisms: A Review. *Frontiers in Marine Science*, 9. doi:https://doi.org/10.3389/fmars.2022.851281.

Bergami, E., Pugnalini, S., Vannuccini, M.L., Manfra, L., Faleri, C., Savorelli, F., Dawson, K.A. and Corsi, I.J.A. T., (2017). Long-term toxicity of surface-charged polystyrene nanoplastics to marine planktonic species *Dunaliella tertiolecta* and *Artemia franciscana*. *Aquatic toxicology*, 189, pp.159–169.

Bouwmeester, H., Hollman, P.C.H. and Peters, R.J.B. (2015). Potential Health Impact of Environmentally Released Micro- and Nanoplastics in the Human Food Production Chain: Experiences from Nanotoxicology. *Environmental Science & Technology*, 49(15), pp.8932–8947. doi:https://doi.org/10.1021/acs.est.5b01090.

Bradney, L., Wijesekara, H., Palansooriya, K.N., Obadamudalige, N., Bolan, N.S., Ok, Y.S., Rinklebe, J., Kim, K.-H. and Kirkham, M.B. (2019). Particulate plastics as a vector for toxic trace-element uptake by aquatic and terrestrial organisms and human health risk. *Environment International*, 131, p.104937. doi:https://doi.org/10.1016/j.envint.2019.104937.

Brandts, I., Cánovas, M., Tvarijonaviciute, A., Llorca, M., Vega, A., Farrè, M., Pastor, J., Roher, N., & Teles, M. (2022). Nanoplastics are bioaccumulated in fish liver and muscle and cause DNA damage after a chronic exposure. *Environmental Research*, *212*, p.113433. https://doi.org/10.1016/j.envres.2022.113433

Brandts, I., Cánovas, M., Tvarijonaviciute, A., Llorca, M., Vega, A., Farré, M., Pastor, J., Roher, N. and Teles, M., (2022). Nanoplastics are bioaccumulated in fish liver and muscle and cause DNA damage after a chronic exposure. *Environmental research*, 212, p.113433.

Brandts, I., Teles, M., Tvarijonaviciute, A., Pereira, M.L., Martins, M.A., Tort, L. and Oliveira, M., (2018). Effects of polymethylmethacrylate nanoplastics on *Dicentrarchus labrax*. *Genomics*, 110(6), pp.435–441.

Calzolai, L., Gilliland, D. and Rossi, F., (2012). Measuring nanoparticles size distribution in food and consumer products: a review. *Food Additives & Contaminants: Part A*, 29(8), pp.1183–1193.

Caputo, F., Vogel, R., Savage, J., Vella, G., Law, A., Della Camera, G., Hannon, G., Peacock, B., Mehn, D., Ponti, J. and Geiss, O., (2021). Measuring particle size distribution and mass concentration of nanoplastics and microplastics: addressing some analytical challenges in the sub-micron size range. *Journal of Colloid and Interface Science*, 588, pp.401–417.

Chae, Y., Kim, D., Kim, S.W. and An, Y.-J. (2018). Trophic transfer and individual impact of nano-sized polystyrene in a four-species freshwater food chain. *Scientific Reports*, 8(1). doi:https://doi.org/10.1038/s41598-017-18849-y.

Chang, X., Xue, Y., Li, J., Zou, L. and Tang, M. (2019). Potential health impact of environmental micro- and nanoplastics pollution. *Journal of Applied Toxicology*, 40(1), pp.4–15. doi:https://doi.org/10.1002/jat.3915.

Chang, Y. S., Chou, S. H., Jhang, Y. J., Wu, T. S., Lin, L. X., Soo, Y. L., & Hsiao, I. L. (2022). Extraction method development for nanoplastics from oyster and fish tissues. *Science of the Total Environment*, 814, p.152675.

Chen, Q., Gundlach, M., Yang, S., Jiang, J., Velki, M., Yin, D. and Hollert, H. (2017). Quantitative investigation of the mechanisms of microplastics and nanoplastics toward zebrafish larvae locomotor activity. *Science of The Total Environment*, 584-585, pp.1022–1031. doi: https://doi.org/10.1016/j.scitotenv.2017.01.156.

Chen, Z., Wei, W., Liu, X., & Ni, B. J. 2022. Emerging electrochemical techniques for identifying and removing micro/nanoplastics in urban waters. *Water Research*, 221, p.118846. doi:https://doi.org/10.1016/j.watres.2022.118846

Chia, R.W., Lee, J.-Y., Kim, H. and Jang, J. (2021). Microplastic pollution in soil and groundwater: a review. *Environmental Chemistry Letters*, 19(6), pp.4211–4224. doi:https://doi.org/10.1007/s10311-021-01297-6.

Coleman, F.C., Wehle, D.H.S. (1984). Plastic pollution: a worldwide oceanic problem. *Parks*, 9(1), pp.9–12.

Cverenkárová, K., Valachovičová, M., Mackuľak, T., Žemlička, L., & Bírošová, L. (2021). Microplastics in the food chain. *Life*, 11(12), p.1349. doi:https://doi.org/10.3390/life11121349

Digka, N., Tsangaris, C., Torre, M., Anastasopoulou, A. and Zeri, C. (2018). Microplastics in mussels and fish from the Northern Ionian Sea. *Marine Pollution Bulletin*, 135, pp.30–40. doi:https://doi.org/10.1016/j.marpolbul.2018.06.063.

Ekvall, M.T., Lundqvist, M., Kelpsiene, E., Šileikis, E., Gunnarsson, S.B. and Cedervall, T. (2019). Nanoplastics formed during the mechanical breakdown of daily-use polystyrene products. *Nanoscale advances*, *1*(3), pp.1055–1061.

El, H., Gigault, J., Maxit, B., Grassl, B., Reynaud, S. (2020). Nanoplastic from mechanically degraded primary and secondary microplastics for environmental assessments. *NanoImpact*, 17, p.100206. doi:https://doi.org/10.1016/j.impact.2019.100206

Fu, W., Min, J., Jiang, W., Li, Y. and Zhang, W. (2020). Separation, characterization and identification of microplastics and nanoplastics in the environment. *Science of the total environment*, 721, p.137561.

Gall, S.C. and Thompson, R.C. (2015). The impact of debris on marine life. *Marine Pollution Bulletin*, [online] 92(1-2), pp.170–179. doi:https://doi.org/10.1016/j.marpolbul.2014.12.041.

Gangadoo, S., Owen, S., Rajapaksha, P., Plaisted, K., Cheeseman, S., Haddara, H., … & Chapman, J. (2020). Nano-plastics and their analytical characterisation and fate in the marine environment: From source to sea. *Science of the Total Environment*, 732, p.138792. doi: https://doi.org/10.1016/j.scitotenv.2020.138792

Garcia-Torné, M., Abad, E., Almeida, D., Llorca, M. and Farré, M., (2022). Assessment of Micro-and Nanoplastic Composition (Polymers and Additives) in the Gastrointestinal Tracts of Ebro River Fishes. *Molecules*, 28(1), p.239.

Geremia, E., Muscari Tomajoli, M.T., Murano, C., Petito, A. and Fasciolo, G. (2023). The Impact of Micro- and Nanoplastics on Aquatic Organisms: Mechanisms of Oxidative Stress and Implications for Human Health—A Review. *Environments*, [online] 10(9), p.161. doi:https://doi.org/10.3390/environments10090161.

Gross-Rother, J., Blech, M., Preis, E., Bakowsky, U., & Garidel, P. (2020). Particle detection and characterization for biopharmaceutical applications: current principles of established and alternative techniques. *Pharmaceutics*, 12(11), p.1112.

Guerrera, M. C., Aragona, M., Porcino, C., Fazio, F., Laurà, R., Levanti, M., … & Germanà, A. (2021). Micro and nano plastics distribution in fish as model organisms: histopathology, blood response and bioaccumulation in different organs. *Applied Sciences*, 11 (13), p.5768.

Hallan, S. S., Sguizzato, M., Esposito, E., & Cortesi, R. (2021). Challenges in the physical characterization of lipid nanoparticles. *Pharmaceutics*, 134, p.549. doi: https://doi.org/10.3390/pharmaceutics13040549

Hansda, C., Maiti, P., Singha, T., Pal, M., Hussain, S. A., Paul, S., & Paul, P. K. (2018). Photophysical study of the interaction between ZnO nanoparticles and globular protein bovine serum albumin in solution and in a layer-by-layer self-assembled film. *Journal of Physics and Chemistry of Solids*, 121, pp.110–120. doi:https://doi.org/10.1016/j.jpcs.2018.05.029

Hazeem, L.J., Yesilay, G., Bououdina, M., Perna, S., Cetin, D., Suludere, Z., Barras, A. and Boukherroub, R., (2020). Investigation of the toxic effects of different polystyrene micro-and nanoplastics on microalgae Chlorella vulgaris by analysis of cell viability, pigment content, oxidative stress and ultrastructural changes. *Marine Pollution Bulletin*, 156, p.111278.

Huber, M.J., Ivleva, N.P., Booth, A.M., Beer, I., Bianchi, I., Drexel, R., Geiss, O., Mehn, D., Meier, F., Molska, A. and Parot, J., (2023). Physicochemical characterization and quantification of nanoplastics: applicability, limitations and complementarity of batch and fractionation methods. *Analytical and Bioanalytical Chemistry*, pp.1–25.

Jeong, C.-B., Kang, H.-M., Lee, M.-C., Kim, D.-H., Han, J., Hwang, D.-S., Souissi, S., Lee, S.-J., Shin, K.-H.,

Park, H.G. and Lee, J.-S. (2017). Adverse effects of microplastics and oxidative stress-induced MAPK/Nrf2 pathway-mediated defense mechanisms in the marine copepod Paracyclopina nana. *Scientific Reports*, [online] 7 (1), pp.1–11. doi:https://doi.org/10.1038/srep41323.

Jin, J., Hines, W.A., Kuo, C.-H., Perry, D.M., Poyraz, A.S., Xia, Y., Zaidi, T., Nieh, M.-P. and Suib, S.L. (2015). Magnetic studies of mesoporous nanostructured iron oxide materials synthesized by one-step soft-templating. *Dalton Transactions*, [online] 4426, pp.11943–11953. doi:https://doi.org/10.1039/C5DT01388G.

June Soo Kim, Sang Hoon Kwon, Jae Yong Lee, Seung Deok Kim, Da Ye Kim, Kim, H.-J., Jang, N., Wang, J., Han, M. and Seong H K. (2023). High-throughput multi-gate microfluidic resistive pulse sensing for biological nanoparticle detection. *Lab on a Chip*, 23(7), pp.1945–1953. doi:https://doi.org/10.1039/d2lc01064j.

Jyoti, K., Baunthiyal, M., & Singh, A. (2016). Characterization of silver nanoparticles synthesized using Urtica dioica Linn. Leaves and their synergistic effects with antibiotics. *Journal of Radiation Research and Applied Sciences*, 93, pp.217–227. doi:https://doi.org/10.1016/j.jrras.2015.10.002

Kobayashi, T., Singappuli-Arachchige, D., Wang, Z., Slowing, I. I., & Pruski, M. (2017). Spatial distribution of organic functional groups supported on mesoporous silica nanoparticles: A study by conventional and DNP-enhanced 29 Si solid-state NMR. *Physical Chemistry Chemical Physics*, 193, pp.1781–1789. doi:https://doi.org/10.1039/C6CP07642D.

Koelmans, A.A., Mohamed Nor, N.H., Hermsen, E., Kooi, M., Mintenig, S.M., De France, J., (2019). Microplastics in freshwaters and drinking water: critical review and assessment of data quality. *Water Resrach*. 155, pp.410–422. doi:https://doi.org/10.1016/j.watres.2019. 02.054.

Krause, S., Baranov, V., Nel, H. A., Drummond, J. D., Kukkola, A., Hoellein, T., et al. (2021). Gathering at the top? Environmental controls of microplastic uptake and biomagnification in freshwater food webs. *Environmental Pollution*, 268, p.115750. doi: 10.1016/j.envpol.2020.115750

Lee, W.S., Cho, H.-J., Kim, E., Huh, Y.H., Kim, H.-J., Kim, B., Kang, T., Lee, J.-S., Jeong, J., (2019). Bioaccumulation of polystyrene nanoplastics and their effect on the toxicity of Au ions in Zebra fish embryos. *Nanoscale*, 11(7), pp.3173–3185. doi:https://doi.rg/10.1039/c8nr09321k.

Leppänen, I., Lappalainen, T., Lohtander, T., Jonkergouw, C., Arola, S., & Tammelin, T. (2022). Capturing colloidal nano-and microplastics with plant-based nanocellulose networks. *Nature Communications*, 131, p.1814. doi: https://doi.org/10.1038/s41467-022-29446-7

Lima, A. R. A., Costa, M. F., and Barletta, M. (2014). Distribution patterns of microplastics within the plankton of a tropical estuary. *Environmental Research*. 132, pp.146–155. doi:https://doi.org/10.1016/j.envres.2014.03.031

Mandemaker, L.D. and Meirer, F., (2023). Spectro-Microscopic Techniques for Studying Nanoplastics in the Environment and in Organisms. *Angewandte Chemie International Edition*, 62(2), p.e202210494.

Materić, D., Ludewig, E., Brunner, D., Röckmann, T. and Holzinger, R. (2021). Nanoplastics transport to the remote, high-altitude Alps. *Environmental Pollution*, 288, p.117697. doi:https://doi.org/10.1016/j.envpol.2021.117697.

Mattsson, K., Johnson, E. V., Malmendal, A., Linse, S., Hansson, L., &Cedervall, T. (2017). Brain damage and behavioural disorders in fish induced by plastic nanoparticles delivered through the food chain. *Scientific Reports*, 71. https://doi.org/10.1038/s41598-017-10813-0

Murugan, P., Sivaperumal, P., Balu, S., Arya, S., Atchudan, R. and A.K. Sundramoorthy (2023). Recent advances on the methods developed for the identification and detection of emerging contaminant microplastics: a review. *RSC Advances*, [online] 13(51), pp.36223–36241. doi:https://doi.org/10.1039/D3RA05420A.

Napper, I.E., Davies, B.F.R., Clifford, H., Elvin, S., Koldewey, H.J., Mayewski, P.A., Miner, K.R., Potocki, M., Elmore, A.C., Gajurel, A.P. and Thompson, R.C. (2020). Reaching New Heights in Plastic Pollution—Preliminary Findings of Microplastics on Mount Everest. *One Earth*, 3(5), pp.621–630. doi: https://doi.org/10.1016/j.oneear.2020.10.020.

Oliveira Jr, G. P., Porto, W. F., Palu, C. C., Pereira, L. M., Petriz, B., Almeida, J. A., ... & Pereira, R. W. (2018). Effects of acute aerobic exercise on rats serum extracellular vesicles diameter, concentration and small RNAs content. *Frontiers in physiology*, 9, pp.532. doi:https://doi.org/10.3389/fphys.2018.0053

Peng, L., Fu, D., Qi, H., Lan, C. Q., Yu, H., & Ge, C. (2020). Micro- and nano-plastics in marine environment: Source, distribution and threats — A review. *Science of the Total Environment*, 698, p.134254. doi: https://doi.org/10.1016/j.scitotenv.2019.134254

Peng, M., Félix, R.C., Canário, A.V. and Power, D.M., (2024). The physiological effect of polystyrene nanoplastic particles on fish and human fibroblasts. *Science of The Total Environment*, p.169979.

Pinilla-Torres, A. M., Sanchez-Dominguez, C. N., Basilio-Bernabe, K., Carrion-Garcia, P. Y., Roacho-Perez, J. A., Garza-Treviño, E. N., ... & Sanchez-Dominguez, M. (2023). Green Synthesis of Mesquite-Gum-Stabilized Gold Nanoparticles for Biomedical Applications: Physicochemical Properties and Biocompatibility Assessment. *Polymers*, 1517, p.3533. doi:https://doi.org/10.3390/polym15173533

Platt, M., Hunsicker, E. and Pollard, M., (2020). A Tunable 3D Printed Microfluidic Resistive Pulse Sensor for the Characterisation of Algae and Microplastics. doi https://doi.org/10.26434/chemrxiv.12249833.v1.

Pollard, M., Hunsicker, E. and Platt, M. (2020). A tunable three-dimensional printed microfluidic resistive pulse sensor for the characterization of algae and microplastics. *ACS sensors*, 58, pp.2578–2586. doi: https://doi.org/10.1021/acssensors.0c00987

Ruan, X., Xie, L., Liu, J., Ge, Q., Liu, Y., Li, K., You, W., Huang, T. and Zhang, L., (2024). Rapid detection of nanoplastics down to 20 nm in water by surface-enhanced raman spectroscopy. *Journal of Hazardous Materials*, 462, p.132702.

Russo, P. S., Streletzky, K. A., Gorman, A., Huberty, W., & Zhang, X. (2021). Characterization of polymers by dynamic light scattering. In *Molecular Characterization of Polymers* pp. 441–498. Elsevier.

Samandra, S., Johnston, J.M., Jaeger, J.E., Symons, B., Xie, S., Currell, M., Ellis, A.V., Clarke, B.O., (2022). Microplastic contamination of an unconfined groundwater aquifer in Victoria. Australia. *Science of Total Environment*. 802, p.149727. doi:https://doi.org/10.1016/j.scitotenv.2021.149727

Schorb, M., Haberbosch, I., Hagen, W.J., Schwab, Y. and Mastronarde, D.N. (2019). Software tools for automated

transmission electron microscopy. *Nature methods*, 166, pp.471–477. doi:https://doi.org/10.1038/s41592-019-0396-9
Schymanski, D., Goldbeck, C., Humpf, H.-U., and Fürst, P. (2018). Analysis of microplastics in water by micro-Raman spectroscopy: release of plastic particles from different packaging into mineral water. *Water Research*. 129, pp.154–162. doi:https://doi.org/10.1016/j.watres.2017.11.011
Smith, J. R., Olusanya, T. O., & Lamprou, D. A. (2018). Characterization of drug delivery vehicles using atomic force microscopy: current status. *Expert opinion on drug delivery*, 1512, pp.1211–1221. doi:https://doi.org/10.1080/17425247.2018.1546693
Sobhani, Z., Zhang, X., Gibson, C., Naidu, R., Megharaj, M., & Fang, C. (2020). Identification and visualisation of microplastics/nanoplastics by Raman imaging i: Down to 100 nm. *Water research*, 174, p.115658.
Tallec, K., Blard, O., González-Fernández, C., Brotons, G., Berchel, M., Soudant, P., Huvet, A. and Paul-Pont, I., (2019). Surface functionalization determines behavior of nanoplastic solutions in model aquatic environments. *Chemosphere*, 225, pp.639–646.
Tallec, K., Huvet, A., Di Poi, C., Gonzalez-Fernandez, C., Lambert, C., Petton, B., LeGoic, N., Berchel, M., Soudant, P., Paul-Pont, I., 2018. Nanoplastics impaired oyster free living stages, gametes and embryos. *Environmental Pollution*, 242, pp.1226–1235. doi:https://doi.org/10.1016/j.envpol.2018.08.020.
Theakstone, A. G., Rinaldi, C., Butler, H. J., Cameron, J. M., Confield, L. R., Rutherford, S. H., … & Baker, M. J. (2021). Fourier-transform infrared spectroscopy of biofluids: A practical approach. *Translational Biophotonics*, 3 (2), e202000025.
Thompson, R.C. (2004). Lost at Sea: Where Is All the Plastic? *Science*, [online] 304(5672), pp.838–838. doi:https://doi.org/10.1126/science.1094559.
Toussaint, B., Raffael, B., Angers-Loustau, A., Gilliland, D., Kestens, V., Petrillo, M., Rio-Echevarria, I.M. and Van den Eede, G. (2019). Review of micro- and nanoplastic contamination in the food chain. *Food Additives & Contaminants: Part A*, 36(5), pp.639–673. doi:https://doi.org/10.1080/19440049.2019.1583381.
Triebskorn, R., Braunbeck, T., Grummt, T., Hanslik, L., Huppertsberg, S., Jekel, M., Knepper, T.P., Krais, S., Müller, Y.K., Pittroff, M. and Ruhl, A.S., (2019). Relevance of nano-and microplastics for freshwater ecosystems: a critical review. *TrAC Trends in Analytical Chemistry*, 110, pp.375–392.
Vagner, M., Boudry, G., Courcot, L., Vincent, D., Dehaut, A., Duflos, G., … & Zambonino-Infante, J. L. (2022). Experimental evidence that polystyrene nanoplastics cross the intestinal barrier of European seabass. *Environment International*, 166, p.107340.
Vitali, C., Peters, R. J., Janssen, H. G., Nielen, M. W., & Ruggeri, F. S. 2022. Microplastics and nanoplastics in food, water, and beverages, part II. Methods. *TrAC Trends in Analytical Chemistry*, 157, p.116819. doi:https://doi.org/10.1016/j.trac.2022.116819
Wagner, M., Scherer, C., Alvarez-Muñoz, D., Brennholt, N., Bourrain, X., Buchinger, S., Fries, E., Grosbois, C., Klasmeier, J., Marti, T., Rodriguez-Mozaz, S., Urbatzka, R., Vethaak, A.D., Winther-Nielsen, M., Reifferscheid, G., (2014). Microplastics in freshwater ecosystems: what we know and what we need to know. *Environmental Science of Europe*, 26(1), p.12. doi:https://doi.org/10.1186/s12302-014-0012-7.
Wang, H., Yang, L., Sui, X., Karahan, H. E., Wang, X., & Chen, Y. (2018). Selective synthesis of single walled carbon nanotubes on metal iron, nickel or cobalt sulfate-based catalysts. *Carbon*, 129, pp.128–136. 10.1016/j.carbon.2017.12.002
Wang, X., Wang, H. and Liu, X., (2023). Accumulation of Pd-doped Polystyrene Nanoplastics in the Digestive Tract of Sebastes Schlegelii. In *E3S Web of Conferences* (Vol. 406, p. 01007). EDP Sciences.
Wong, S. L., Nyakuma, B. B., Wong, K. Y., Lee, C. T., Lee, T. H., and Lee, C. H. (2020). Microplastics and nanoplastics in global food webs: a bibliometric analysis 2009–2019. *Marine Pollution* Bulletin, 158, p.111432. doi:https://doi.org/10.1016/j.marpolbul.2020.111432
Woods, A.E. and Stirling, J.W. (2018). Transmission electron microscopy. *Bancroft's theeory and Practice of Histological Techniques*, pp.434–475.
Yadav, S., Senapati, S., Kulkarni, S. S., & Singh, J. P. (2023). A SERS based clinical study on HIV-1 viral load quantification and determination of disease prognosis. *Journal of Photochemistry and Photobiology B: Biology*, 239, p.112629. doi:https://doi.org/10.1016/j.jphotobiol.2022.112629
Yang, M. and Wang, W.X., (2023). Recognition and movement of polystyrene nanoplastics in fish cells. *Environmental Pollution*, 316, p.120627.
Yang, Y., Meng, L., Zhang, J., Gao, Y., Hao, Z., Liu, Y., … & Liu, S. (2023). Near-Infrared Light-Driven MXene/Liquid Crystal Elastomer Bimorph Membranes for Closed-Loop Controlled Self-Sensing Bionic Robots. *Advanced Science*, p.2307862.
Yilmaz, G., Demir, B., Timur, S., & Becer, C. R. (2016). Poly methacrylic acid-coated gold nanoparticles: functional platforms for theranostic applications. *Biomacromolecules*, 179, pp.2901–2911. doi:https://doi.org/10.1021/acs.biomac.6b00706.
Zhang, S., Sun, Y., Liu, B. and Li, R., (2021). Full size microplastics in crab and fish collected from the mangrove wetland of Beibu Gulf: Evidences from Raman Tweezers (1–20 μm) and spectroscopy (20–5000 μm). *Science of the Total Environment*, 759, p.143504.
Zhao, J., Lan, R., Wang, Z., Su, W., Song, D., Xue, R., Liu, Z., Liu, X., Dai, Y., Yue, T. and Xing, B., (2023). Microplastic fragmentation by rotifers in aquatic ecosystems contributes to global nanoplastic pollution. *Nature Nanotechnology*, pp.1–9.
Zhou, X.X., Liu, R., Hao, L.T. and Liu, J.F., (2021). Identification of polystyrene nanoplastics using surface enhanced Raman spectroscopy. *Talanta*, 221, p.121552.

Advances in Maritime Technology and Engineering – Guedes Soares & Santos (Eds)

Sustainable fish farms and optimal use of the fish food

O.T. Gudmestad
University of Stavanger, Norway
UiT The Arctic University of Norway, Tromsø, Norway

B. Heidari
Quantafuel, Oslo, Norway

E. Yitzhak
StS Group, Stavanger, Norway

ABSTRACT: The open sea fish farm industry is under considerable pressure to ensure that fish farming represents a sustainable industry. Many challenges are present related to fish health due to fish lies, considerable density of fish with limited room to move, accelerations of the cage in large waves and needs to limit the use of antibiotic in the offshore environment. Much good work is done to increase fish health, design safe cages and limit pollution from fish farms. In this paper a discussion is presented of how to ensure that the food stays longer in the cage, enabling the fish to consume a larger part of the food and limiting the food that sinks to the seafloor, thus limiting the pollution at the seafloor.

1 INTRODUCTION

1.1 *Aquaculture challenges*

Fish farming poses many challenges related to fish health, farming technology and pollution.

In modern fish farms, fish health is monitored closely to ensure that the product delivered to the market is health and free from any illness. Challenges related to limit fish lies that will feed on the living fish are met by placing fish cages in fresh water where circulation is ensured. The health is monitored by veterinaries, ensuring the use of healthy food, avoiding overcrowding in the cages and by limiting the use of antibiotics.

The technology of developing suitable fish cages is evolving to ensure that the cages and the anchor systems are designed to resist the wave and current conditions at site (Faltinsen & Shen, 2017), kept clean from marine growth, are anchored such that the net accelerations are kept within tolerable limits for the fish. Different types of open, semi-closed and closed cages are, furthermore, being developed (Shen et al., 2022). Of main concern is to ensure that fish do not escape, breeding with wild fish and potentially destroying the genes of the wild fish population.

The pollution from fish cages is considered to represent a serious concern as the pollution are due to several factors; the manure from the fish, the contamination spread to wild fish through water, due to medicine given to the farmed fish. Also, food that is not consumed by the fish in the cage will sink to the seafloor, representing another source of seafloor pollution.

1.2 *Objective of the paper*

In this paper, a method to increase the time the food stays in the fish cage will be explored. When the food sinks at a slower speed, the fish will have the possibility to consume more of the food and the pollution at the seafloor will be reduced, providing a contribution to the sustainability of fish farming.

In this research, we studied the time of sinking of food pellets in two types of water: fresh water and sea water. In addition, two cases with and without current are considered. Finally, by changing the shape of the food pellets, without changing the composition of the food, a comparison of these cases is presented. Also, analytical, and numerical models are presented to give a better background for the results of experiments and analysis.

2 LITERATURE REVIEW

2.1 *Parameters affecting the sinking time of food pellets*

The sinking time of pellets is dependent on gravity force, buoyancy and drag. The gravity force is dependent on the mass and gravity acceleration. The

DOI: 10.1201/9781003508779-57

buoyancy force is dependent on the water density and the volume of the pellets. The other force that can be manipulated for better results and can improve the sinking time is the drag force. The drag force depends on the geometry, the velocity, and the drag coefficient.

The density is dependent on the properties of the pellets. Khater et al. (2014), studied the physical and mechanical properties of fish food pellets. Those properties included: actual diameter, expansion rate, surface area, volume, weight, bulk density, durability, floatability, moisture content, water stability, repose angle and crushing load. Wang & al., have studied the effects of varying the density. In the present research, however, we have used commercial food pellets as a basis and do not vary the density, as the nutrition properties of the pellets might not be changed.

Another solution for changing the density without any change in pellets is to make some bubbles around the pellets or creating a hole in the pellets. Yang et al. (2003) presented research to form very small bubbles at the solid-water interface.

Chow & Adams, (2011), worked on prediction of drag coefficient for free falling rigid cylindrical particle at moderate Reynolds number. Jones & Knudsen, (1961), worked on drag coefficient at low Reynolds number, however, their research is for Reynolds number about 1000, while the present research has been done for Reynolds number less than 2. A somewhat relevant theme is the sinking of a net for catching fish (Zhou et al., 2015).

All in all, however, no significant experiment and modeling has been identified related to the sinking time of pellets in different water types.

3 ANALYTICAL SOLUTION

3.1 *Velocity and displacement as function of time*

The forces acting on the pellets are the gravitational force, the buoyancy, and the drag force. As the drag force increase with the velocity, the sinking velocity will reach a constant "terminal velocity" after some time. The equation for the sinking can according to Newton's second law be written as:

$$\left(\left(\rho_p - \rho_w\right)g\Delta - \frac{1}{2}\rho_w A C_D v^2\right) = m\,dv/dt \quad (1)$$

where:
ρ_p is the density of the pellets
ρ_w is the density of water
Δ is the volume of the pellet
A is the area of the pellet
C_D is the drag coefficient
g is the gravity acceleration
m is the mass of the pellet

The term$(\rho_p - \rho_w)g\Delta$ represents the gravity force minus the buoyancy force and the term $\frac{1}{2}\rho_w A C_D v^2$ represents the drag force acting oppositely to sinking.

Rearranging:

$$dt = \int \frac{m dv}{\left(\left(\rho_p - \rho_w\right)g\Delta - \frac{1}{2}\rho_w A C_D v^2\right)} \quad (2)$$

And integrating:

$$\frac{b}{2a}\ln\left(\frac{a+v}{a-v}\right) \quad (3)$$

Thus, the velocity of the sinking is given by:

$$v = a\frac{e^{\frac{2at}{b}} - 1}{e^{\frac{2at}{b}} + 1} \quad (4)$$

where:

$$a^2 = \frac{\left(\rho_p - \rho_w\right)g\Delta}{\frac{1}{2}\rho_w A C_D} \quad (5)$$

$$b = \frac{m}{\frac{1}{2}\rho_w A C_D} \quad (6)$$

Here we are assuming that v = 0 at t = 0.

Selecting input data as given in Table 1, the terminal velocity v= 0.337m/s is reached after 0.8 sec, Figure 1.

Table 1. Random sample data.

Drag Coefficient	0.8	
Food density	1500	kg/m ^{3}
Food Radius	0.005	m
Food length	0.01	m
Water density	1025	kg/m ^{3}
Gravity	9.806	m/s ^{2}

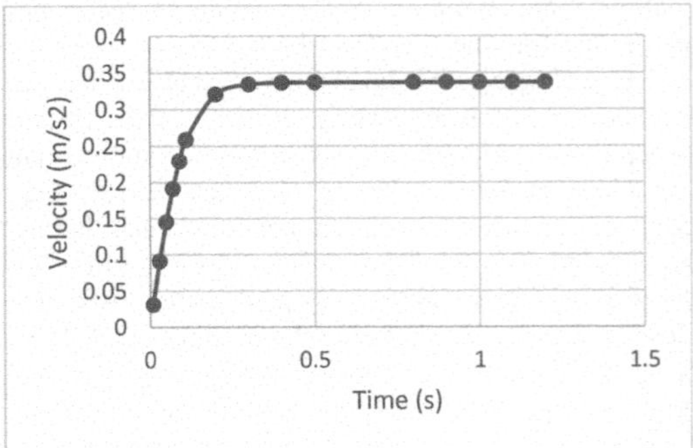

Figure 1. Velocity profile during pellet sinking.

The terminal velocity could also be found from:

$$\left(\rho_p - \rho_w\right)g\Delta = \frac{1}{2}\rho_w A C_D v^2 \quad (7)$$

Integrating equation (4), we find that the sinking after time t is given by:

$$y = a\left(\frac{b}{2a}\ln\left(e^{\frac{2at}{b}}+1\right) - t + \frac{b}{2a}\ln\left(e^{\frac{2at}{b}}+1\right)\right) - bln(2) \tag{8}$$

Figure 2 gives the displacement as function of time during sinking.

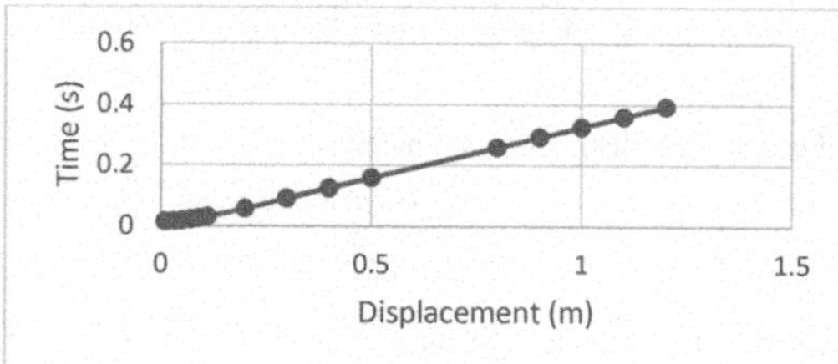

Figure 2. Displacement of pellets as function of time during sinking.

As it is predicted, due to constant velocity during the most part of sinking time, the displacement changes linearly versus time.

3.2 *Effects of varying the input parameters.*

Varying the input parameters, we find the time for the pellets to sink 0.5m. The following results are given for the effect of the drag coefficient, the density of the pellets and the water density (fresh water and salt water) as shown in Tables 2, 3 and 4, respectively:

Table 2. Drag coefficient effect; time to sink 0.5m.

Cd	Time (s)	% difference
0.8	1.52	
0.9	1.61	5.92
1	1.69	10.56
1.1	1.77	16.45
1.2	1.88	23.68
1.3	1.92	26.32

Table 3. Food density effect; time to sink 0.5m.

Density (kg/m^3)	Time (s)	% difference
1500	1.52	
1480	1.55	1.97
1460	1.59	4.52
1440	1.63	7.24
1420	1.67	9.87
1400	1.71	12.50

Table 4. Water density effect, time to sink 0.5m.

Density kg/m^3	Time (s)	% difference
1000	1.47	
1025	1.52	3.40

It should be noted that the sinking time is not dependent on whether the cage is open or closed, however, in waves, the motion of the water particles, the wave kinematics, will affect the sinking time. We have assumed no water movements in the above analysis and in the experiments reported in the next chapter.

The effect of current is not discussed above; however, a current will represent a horizontal force that will cause the pellets to float out of the fish cage.

4 EXPERIMENTS

4.1 *Experimental setup*

Measurements of sinking time for commercially available pellets were carried out at the laboratory of the University of Stavanger, using:

- A cylindrical tank ø 50cm, height 100cm.
- To simulate a current, the tank was placed on a rotating plate.
- Three types of pellets (available from a commercial company) were dropped from 10 cm above the water surface in the cylindrical tank, Table 5, and Figure 3.

Table 5. Size of pellets.

Pellets type	L (mm)	D (mm)
I	10	12
II	10	10
III	6	5

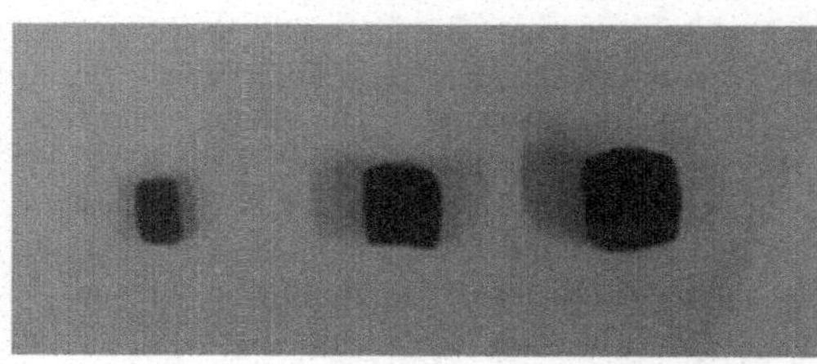

Figure 3. Types of pellets.

Water current at radius r from centerline:

$$V(r) = rpm * \frac{2\pi}{60} * r = \frac{rpm * \pi}{30} * r \tag{9}$$

4.2 *Fish pellets characteristics*

- Length of pellets: For finding the dimensions and the density of the pellets, 40 samples of each type were selected. Then average dimensions of them were used for analyzing the results from experiments. The reason for using the average of samples was that the pellets had different sizes and they were not exactly similar. In addition, for measuring the length, three different lengths were selected, and the average was used in calculations where length is needed.
- Diameter of pellets: For calculating the average diameter, both ends and center measurements of the diameter in three different directions were done, Figure 4.

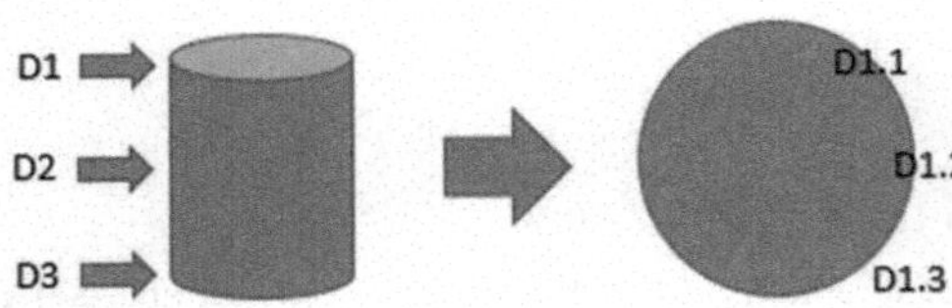

Figure 4. Average diameter of the pellets.

$$D_{avg} = \frac{D1.1 + D1.2 + \ldots.. + D3.3}{9} \tag{10}$$

4.3 *Water bubbles forming around the pellets*

From some basic experiments, it was identified that bubbles form around the pellets (See also Yang et al., 2003). These bubbles cause a significant change in the "effective density" of the pellets and consequently increase the time of sinking. In addition, bubbles change the surface shape of pellets and more eddy currents form in the flow around the pellets. Both the changes in density and drag force increase the time of sinking, hence if it is possible to have more bubbles, then the sinking time is improved.

As a new method, it is suggested to make some change on the surface of pellets to improve the bubble formation. But it should be considered that certain changes are not acceptable, due to the production method. The new suggested shapes are three channels on the surface, a hole in the centerline of food cylinder and combination of channel and hole, Figure 5. The channels trap bubbles and cause a significant change in drag force. The hole also traps the bubble, but the drag does not change drag.

4.4 *The experiments' design*

For preparing statistics, for each case (Table 6), 40 samples were selected and the time of sinking for each pellet was measured.

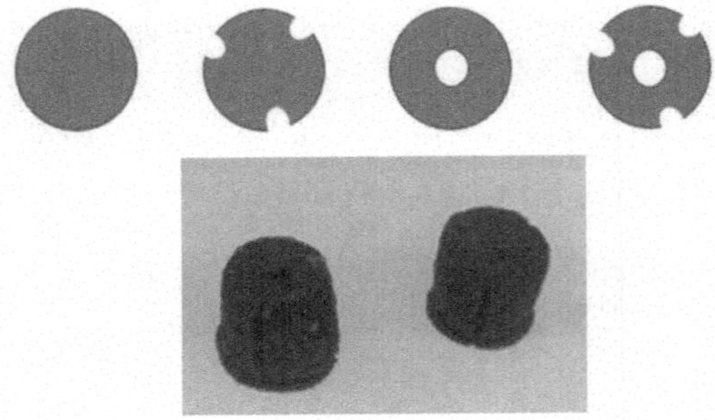

Figure 5. The shape of tested pellets.

Table 6. Experiments' design.

water type	Current type	Food Size	Food Type
1. Sea water 2. Fresh water	Still	small	Simple
			Hole
			Channel
			Hole+Channel
		Middle	Simple
			Hole
			Channel
			Hole+Channel
		Large	Simple
			Hole
			Channel
			Hole+Channel
	Current	small	Simple
			Hole
			Channel
			Hole+Channel
		Middle	Simple
			Hole
			Channel
			Hole+Channel
		Large	Simple
			Hole
			Channel
			Hole+Channel

5 RESULTS OF THE EXPERIMENTS

5.1 *The random samples used*

By using 40 random sample for each type of pellets, the average length, diameter, mass, and density were found. Then, for each type of pellets, the average values, and standard deviations (SDs) are presented in Tables 7.1-2-3.

Table 7.1. Type 1 (large size).

	L (mm)	D (mm)	m (kg)	Density (kg/m^3)
Mean	9.90	11.16	1.11	1148.60
SD	0.46	0.38	0.08	39.35

Table 7.2. Type 2 (medium size).

	L (mm)	D (mm)	m (kg)	Density (kg/m^3)
Mean	9.42	8.70	0.62	1110.87
SD	1.38	0.44	0.09	36.09

Table 7.3. Type 3 (small size).

	L (mm)	D (mm)	m (kg)	Density (kg/m^3)
Mean	6.64	4.91	0.14	1110.79
SD	1.06	0.09	0.03	76.04

The reason for the differences in density for the different types may be due to different production method. In addition, the SD is high and affects the sinking time significantly. This problem is due to a non-uniform shape and dimension that makes some errors in calculating the exact density.

5.2 *Sinking time, summary of results*

After performing 40 experiments for each case mentioned in the last chapter, the average measured times are presented in Tables 8.1-2. Also, Standard deviation and the percent of change in sinking time with respect to the simple non-manipulated pellets are calculated.

The results from the experiments are the most important part of the research.

Table 8.1. Measured time for sinking 1.0 m in salt still water.

Food Size	Food Type	Mean time	SD	% Diff
Small	Simple	8.0998	0.7472	
	Hole	10.839	2.0789	33.82
	Channel	9.8258	1.356	21.31
	Hole+Channel	13.0441	2.7973	61.04
Me-dium	Simple	6.6736	0.6765	
	Hole	7.7803	1.3333	16.58
	Channel	8.0963	2.4459	21.32
	Hole+Channel	7.5003	1.2471	12.39
Large	Simple	5.3955	0.3819	
	Hole	5.8283	0.4492	8.02
	Channel	5.8238	0.4704	7.94
	Hole+Channel	6.3408	0.8104	17.52

The time of sinking is the first parameter that should be analyzed. When the results for each size of pellets are compared, it has been observed, that the large pellets sink faster. The reason is probably the higher density for the large pellets. But for medium and small pellets densities are equal, while the sinking time for medium pellets is less than for the small pellets. The reason is due to a larger drag to gravity force ratio for small pellets and the larger relative surface where bubbles are formed.

Table 8.2. Measured time for sinking 1.0 m in fresh still water.

Food Size	Food Type	Mean time	SD	% Diff
Small	Simple	7.64475	0.65704	
	Hole	10.3738	2.46392	35.70
	Channel	9.474	1.08877	23.93
	Hole+Channel	14.429	3.52424	88.74
Me-dium	Simple	6.44975	0.85381	
	Hole	6.58675	0.65283	2.12
	Channel	7.1095	1.64862	10.23
	Hole+Channel	7.296	1.84777	13.12
Large	Simple	5.34	0.82702	
	Hole	5.6359	1.03014	5.54
	Channel	5.43	0.48753	1.69
	Hole+Channel	5.8855	0.5414	10.22

The second point is the large SD and some non-expected measured time for the sinking. There are many reasons for these effects, and there is no control on some of them. The most important reasons are:

- Large variance in the length of pellets.
- Large variance in the density due to not very exact measurements.
- Different shapes of pellets have been observed and most of them are not exact cylindrical.
- Different surfaces also cause different drag coefficient.
- Bubbles that form on the surface of pellets. There is no control on the size and number of them but the bubbles have a large effect on the "effective density" of the pellets and consequently on sinking time.

Third, the change in the shape of the pellets by making a hole or channels, significantly increases the time of sinking. Combination of hole and channel has more effect on improving sinking time. It seems the effect of this change is more effective for small pellets. The reason for this phenomenon, is due to significant changes in drag force and density at the same time. Because for small pellets the bubbles are much bigger compared to the size of pellets. But to understand which one is more effective; it is suggested to do further analysis and calculation.

5.3 *Comparison of fresh and salt water*

To observe the effect of water type, the results from experiment are compared, between fresh and sea water. Tables 8.1 and 8.2 compare the sinking time for fresh and sea water and show the percentage of change in the sinking time in sea water with respect to the fresh water. As the analytical solution showed, the effect of the change in water type was not very high. Experiments also show the same results on average. However, for some cases, sinking times become shorter for sea water. The reason is the large variance in measured time due to reasons that are mentioned above.

5.4 *Effects of current*

Based on the experiments (the results are not shown in the tables above), it was difficult to say anything about the effect of current on sinking time. The percentage variation varies irregularly. However, on average the current improves the time of sinking. This increase is not very high, about 7.12 %.

It seems there is only one reason which causes this increase. This reason is, when pellets are falling down the orientation of the pellets is important for defining the value of the drag force, because the drag coefficient and cross section can change the drag force. Therefore, when there is current, it improves the drag force by changing the orientation of the sinking from vertical to horizontal. Predicting the drag value is very difficult and needs more accurate model and study because the change is due to both drag coefficient and cross-sectional area change.

5.5 *Probability density functions for sinking time.*

The probability density functions for sinking time of the pellets in still sea water are shown in Figure 6. Note the legend: **Simple-Hole-Channel-hole and channel.**

The first important result from Figure 6 is for the simple pellets (blue line on the charts). For simple pellets the variance is less than for the were formed around the "simple pellets".

Also, the variances for shapes with hole or channels (Grey and red Line on the charts) are less than for the type with a combination of them (orange line in the charts). The reason is also due to bubble formation around them.

As the mean value showed, the bubble and the hole improve the sinking time and the combination of them has more effect on the sinking time.

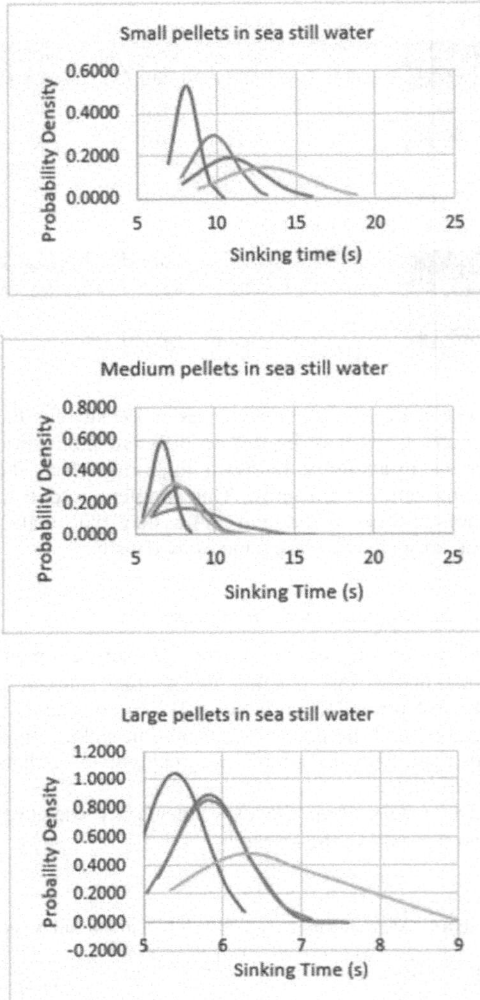

Figure 6. Probability density functions for sinking time in still salt water. The figure shows the sinking time for different pellet sizes; small, medium size and small pellets.

5.6 *The variation of the drag coefficient.*

The drag coefficient varies as the pellets change direction during the sinking. As the "effective density" also varies with the formation of the bubbles, it is very difficult to study the effects of each one separately, based on the experiments' results. Therefore, some analytical calculations performed on the results from the experiments can help to analyze each factor separately. For studying the effect of the drag coefficient, the density of food is assumed to be constant, assuming it has no effect on the results and the drag coefficient for each case of experiment is calculated. The average drag-coefficient and percentage of change compared to simple shape pellets can cause a high change, especially for small pellets. In addition, changing the shape of pellets causes significant improvement in the sinking time. (Heidari & Yitzhak, 2017).

5.7 *Effect of change in food density*

It is not proposed to change the density of the commercial food product. However, when performing the experiments, it was seen that bubbles form around pellets and cause changes in the "effective density" of pellets. So, for analysis of the bubbles' effect, three different diameters of bubbles based on observation in experiments were considered. 1 to 3 bubbles are added to pellets then the change in density is calculated, Table 9.

When the results from the analytical solution for the case with bubbles are compared to experimental results, it was seen that for small pellets the main

Table 9. Bubbles' effect on Time of sinking; percent of change in sinking time compared to pellets without bubbles.

Large		No. of bubble			
		0	1	2	3
Diameter of bubble (mm)	1	5.39	5.395(0.09)	5.41(0.37)	5.42(0.56)
	2	5.39	5.49(1.86)	5.61(4.08)	5.73(6.31)
	3	5.39	5.78(7.24)	6.27(16.33)	6.88(27.64)

Middle		No. of bubble			
		0	1	2	3
Diameter of bubble (mm)	1	6.67	6.69(0.3)	6.73(0.9)	6.78(1.65)
	2	6.67	6.99(4.8)	7.4(10.94)	7.86(17.84)
	3	6.67	8.06(20.84)	10.93(63.87)	22.83(242.28)

Small		No. of bubble			
		0	1	2	3
Diameter of bubble (mm)	1	8.1	8.29(2.35)	8.54(5.34)	8.8(8.64)
	2	8.1	10.57(30.49)	18.5(128.4)	NA
	3	8.1	NA	NA	NA

role for improving time is played by bubbles and change in density.

In addition, when the results of sinking time are compared to analytical results, it has been seen that the channel and hole helped considerably to make more bubbles. This is the reason for a larger increase in sinking time of small pellets than large pellets, when channels and a hole are made in them.

6 CONCLUSIONS AND SUGGESTIONS

6.1 *Conclusions*

- The results from measuring the sizes of the pellets show that the pellets' sizes are not the same and have a large SD, especially with respect to length. It has been seen that the large pellets have a higher density than other types. One of the most important reasons for the large variance in sinking time is the large variance in the size of pellets.
- From the analytical analysis, there should be no change in the time of sinking due to sea current as the pellets tend to float horizontally in a slight current. But there is a little change in the sinking time for the current, as the drag force may be changed. So, the sinking time improves, but not in high percent and the average percent of change is about 7%.
- The most important part of this research is the study of the sinking time of pellets by experiments with the aim of increasing the sinking time. The results of the experiments demonstrate that a hole, channels, and a combination of them, improve the sinking time.

On the other hand, there are some "strange" measurements that are caused by many reasons such as different sizes, shapes, and especially large bubbles. Bubbles, that most of the time form around the new shape of pellets, seem to be one of the main reasons for improving the time of sinking and the large variance in sinking time. The values for the variance in most cases are high. It shows that the time varies significantly between pellets in each case. There are some reasons for this; like a large difference in the sizes and shapes of pellets that causes pellets to sink at different velocities. In addition, different surface tensions cause bubbles' size and numbers to change during sinking, which also cause the time of sinking to vary.

- When the probability function for each case is drawn, it can be found that the mean sinking time improves with holes or channels. The second result is that when the shapes are changed, the variance is much higher. The other point is that in most cases, the variance is higher for pellets with both a hole and channels than for cases with only a hole or just channels.
- Drag force is one of the main forces affecting the pellets during sinking. Therefore, variation in the drag coefficient causes significant improvement in time of sinking compared with lower drag coefficient. It is observed that the drag due to changes in the shape of pellets plays the main role for larger pellets, but for small pellets other parameters have more effects.
- As it has been seen, bubbles cause variations in the "effective density" of pellets, when they are sinking. Results from an analytical solution show for small pellets, bubble formation is more effective for improving the time of sinking. In addition, analytical results demonstrate that big bubbles can cause the pellets to float on the water, due to a lower density than water. This is the reason for some measured large strange sinking time and floating pellets.
- A numerical model would show the flow around the pellets. Obviously, by making channels, there is more eddies and subsequently more drag force due to the vortices. Another interesting subject is that bubbles create more drag force by forming more eddies during the flow of water around the pellets during the sinking.

6.2 *Suggestions*

To have more control and to get better results one should prepare exact sizes of the pellets. So, the test can be performed on other material with almost the same density that can be made in different shapes, such as plastic. Also, numerical models are very helpful for more analysis but one will need exact results from the plastic pellets experiment. Bubbles, as it has been seen in experiments and the analytical solution, play a very important role in change of the sinking time. So, it is necessary to have more attention to this field. If the surface tension between the bubble and the pellets' surface is improved, then significantly higher value for sinking time will be achieved, especially for small pellets.

So, there are three main suggestions for improvement of results:

- Making some definite shapes and sizes and using design of experiments to get a very exact solution, then using optimization methods to achieve the best results.
- By using the data from exact experiments, make numerical optimization of the change in the model, if it is need.
- Search for ways to improve surface tension for fish foods, but one that should not be too high, since it may cause the pellets to float on water.

Further to these suggestions, the experiments should be carried out on a larger scale to document that the effects of the bubbles are still valid when sinking in actual aquaculture cages.

ACKNOWLEDGEMENTS

The authors will express thanks to the persons at the laboratory of the University of Stavanger; to Yaseen A. Amith and John Grønli, who were assisting during the experiments. The second and third authors; Heidari and Yitzhak, acknowledge the opportunity to study for their master of science degree at the University of Stavanger, Norway.

REFERENCES

Chow, A.C. & Adams, E.E. 2011, Prediction of Drag Coefficient and Secondary Motion of Free-Falling Rigid Cylindrical Particles with and without Curvature at Moderate Reynolds Number, *Journal of Hydraulic Engineering*, Vol. 137, No. 11, pp. 1406–1414.

Faltinsen, O. & Shen, Y. 2018. Wave and current effects on floating fish farms. *Journal of Marine Science and Applications*, Vol 17, pp. 284–296.

Heidari, B. & Yitzhak, E. 2017. Experimental Study on the Sinking Time of Fish Pellets. University of Stavanger, unpublished report.

Jones, A.M. Knudsen, J.G. 1961, Drag coefficients at low Reynolds numbers for flow past immersed bodies, *American Institute of Chemical Engineers Journal*, Vol. 7, p 20–25. https://doi.org/10.1002/aic.690070107

Khater, E., Bahnasawy, A.H. & Ali, S.A. 2014, Physical and Mechanical Properties of Fish Feed Pellets, *Journal of Food Processing and Technology*, Vol 5, Issue 10.

Shen, Y., Firoozkoohi, R., Greco, M. & Faltinsen, O.M. 2022. Comparative investigation: Closed versus semi-closed vertical cylinder-shaped fish cage in waves. *Ocean Engineering* 245,110397. https://doi.org/10.1016/j.oceaneng.2021.110397

Yang, J., Duan, J., Fornasiero, D. & Ralston, J. 2003, Very Small Bubble Formation at the Solid-Water Interface, *Journal of Physics and Chemistry*, Vol 107, pp. 6139–6147.

Wang, H., Ma, S., Yang, J., Qin, Y, Cheng, H, Xue, M.& Li, J. 2020. Optimization of the process parameters for extruded commercial sinking fish feed with mixed plant protein sources. *Journal of Food Process Engineering*. https://doi.org/10.1111/jfpe.13599

Zhou, C., Xu, L., Zhang, X, & Ye,X. 2015, Application of Numerical Simulation for Analysis of Sinking Characteristics of Purse Seine, *Journal of Ocean University of China*, Vol. 14, pp. 135–142

Advances in Maritime Technology and Engineering – Guedes Soares & Santos (Eds)
© 2024 The Author(s), ISBN 978-1-032-83107-7

Preliminary study on the hydrodynamics of a bottom-supported aquaculture platform and net cage

J. Ji
Centre for Marine Technology and Ocean Engineering (CENTEC), Instituto Superior Técnico, Universidade de Lisboa, Lisbon, Portugal

L. Zhou
School of Naval Architecture, Ocean and Energy Power Engineering, Wuhan University of Technology, Wuhan, P.R. China
Key Laboratory of High-Performance Ship Technology, Wuhan, P.R. China

B. Liu
School of Naval Architecture, Ocean and Energy Power Engineering, Wuhan University of Technology, Wuhan, P.R. China
Green & Smart River-Sea-Going Ship, Cruise and Yacht Research Centre, Wuhan University of Technology, Wuhan, P.R. China

C. Guedes Soares
Centre for Marine Technology and Ocean Engineering (CENTEC), Instituto Superior Técnico, Universidade de Lisboa, Lisbon, Portugal

ABSTRACT: In present paper, the influence of double layer nets and the net solidity on the flow field of a bottom-supported aquaculture platform in current is analyzed and discussed. A numerical method combined the realizable k-ε turbulence model and the porous media model is established based on the Reynolds-Averaged Navier-Stokes equation to describe the flow. The hydrodynamic variations of double layer nets with different solidities are evaluated. The results indicate that the reduction of flow velocity of the aquaculture cage with double layer nets is higher than that of the traditional fish cage. The increase of the net solidity leads to a decrease in current velocity, and simultaneously expands the low-speed area downstream of the platform. The vortices caused by the pipes may cause damage to the structure and should not be ignored.

1 INTRODUCTION

To meet the global population's demand for seafood, fish cages are being developed in more and more countries and have generated several research fields and hotspots (Xu et al., 2020). Harsh current and wave environments place higher demands on the safety and reliability of aquaculture platforms. The net parameters will directly affect water exchange and the flow field inside and around the aquaculture platform, and the flow field is crucial for the survival of farmed fish (Chu et al., 2019). The interaction mechanism between the net cage and wave or current is complex because the fishing net is a kind of small-scale flexible structure. Hence, the research on the fluid-net structure interactions has become a crucial process in net cage design and optimization. Research on cages needs to consider how to model and simplify fishing nets accurately.

A series of experiments and numerical analyses of different types of net panels and fish cages were conducted to investigate the hydrodynamic characteristics and the interaction between current and aquaculture platforms.

The net solidity S_n represents the ratio of the projected area to the outline area of the netting. Zhan et al. (2006) experimentally investigated the effects of net solidities, incidence angles and mesh patterns on the forces acting on nets. Lader et al. (2007) investigated the wave forces on three net panels with different solidities and discussed the blocking effect of the nets. Tang et al. (2011) studied the dynamic characteristics of dual pontoon floating structures with and without fishing nets through physical tests and numerical calculations. Guo et al. (2022, 2023) have studied the effect of different levels of solidity on the flow through net and porous panels.

DOI: 10.1201/9781003508779-58

In the field of numerical research methods, several different numerical models have been applied to simulate and simplify nets. Lee et al. (2005) established a computational model of a fish cage based on the mass-spring model. Kristiansen et al. (2012) numerically analyzed flexible fishing nets in current through a screen type of force model. Yao et al. (2016) combined the mass spring model with the finite volume method to numerically simulate a fish cage in current.

Liu et al. (2021) formulated an analytical model under the linearized wave theory and small structural response, and the influence of different uniform current speeds and time on the cage shapes were analyzed. Liu et al. (2022) used the Morison equation to calculate the hydrodynamic load, and the mooring force in a fish cage array subjected to currents and waves is investigated using the finite element method based on Ansys/APDL, and verified by experiments (Liu et al. 2023). Mohapatra et al. (2024) established an analytical model of flexible net cage systems based on the linear water wave theory and small-amplitude wave response assumption, and verified it by the published simulation and experimental results. Bi et al. (2014) proposed a numerical method that combines the lumped mass model with the porous media method. Zhao et al. (2013a) adopted a porous media model and the volume of fluid (VOF) method to study the effect of the net solidity and the attack angle of net panels on the flow field. Liu et al. (2021) investigated the blocking effect of fish nets on flow in a semi-submersible aquaculture platform. In recent years, many researchers have studied and optimized the porous media method and validated it through a series of experiments (Patursson et al., 2010; Zhao et al., 2013b; Chen et al., 2017; Guo et al., 2020; Shen et al., 2018).

In this paper, the flow field of a bottom-supported aquaculture platform with different net solidities is simulated. A numerical model combined with the realizable k-ε turbulence model and the porous media model is established to study the influence of double layer net with different net solidity on the flow field. The flow field variation and velocity attenuation of the bottom-supported aquaculture platform are compared and analyzed.

2 NUMERICAL METHOD

In this paper, the Computational Fluid Dynamics (CFD) method based on the Reynolds-Averaged Navier-Stokes (RANS) equation and the Volume of Fluid (VOF) method is adopted to analyze the effects of different parameters on the aquaculture platform flow field.

2.1 *Governing equations*

The finite volume method is used to govern the solution of the equations, and the continuity and momentum equations are obtained by averaging the Navier-Stokes equations, which can be expressed as:

Continuity equation :

$$\frac{\partial \rho}{\partial t}+\frac{\partial(\rho \bar{u}_i)}{\partial x_i}=0 \tag{1}$$

Momentum equation:

$$\frac{\partial(\rho \bar{u}_i)}{\partial t}+\frac{\partial(\rho \bar{u}_i \bar{u}_j)}{\partial x_j}=-\frac{\partial P}{\partial x_i}+\rho g_i+\frac{\partial}{\partial x_j}(\mu+\mu_t)\left(\frac{\partial \bar{u}_i}{\partial x_j}+\frac{\partial \bar{u}_j}{\partial x_i}\right)+S_i \tag{2}$$

where t is time; ρ is the density of the fluid; x_i and x_j (i, j=1,2,3) are the coordinate component; $\bar{u}_i$ and $\bar{u}_j$ are the component of the average velocity vector; g_i is the acceleration due to gravity; $P=\bar{p}+2\rho k/3$, where $\bar{p}$ is pressure and k is the turbulence kinetic energy; μ is the viscosity coefficient of the fluid; μ_t is the eddy viscosity; $-\rho\bar{u}_i\bar{u}_j$are the Reynolds stress, and S_i is a source term used to describe the added resistance of the net that can be calculated as:

$$S_i=-\left(D_{ij}\mu\bar{u}+C_{ij}\frac{1}{2}\rho|\bar{u}|\bar{u}\right) \tag{3}$$

where $\bar{u}$ is the flow velocity, D_{ij} is the viscous resistance coefficient and the C_{ij} is the inertial resistance coefficient. The material matrices consisting of the porous media resistance coefficients D_{ij} and C_{ij} are obtained experimentally, which can be written as:

$$D_{ij}=\begin{pmatrix} D_n & 0 & 0 \\ 0 & D_t & 0 \\ 0 & 0 & D_t \end{pmatrix}, C_{ij}=\begin{pmatrix} C_n & 0 & 0 \\ 0 & C_t & 0 \\ 0 & 0 & C_t \end{pmatrix} \tag{4}$$

where D_n is the normal viscous resistance coefficient, D_t is the tangential viscous resistance coefficient, C_n is the normal inertial resistance coefficient, and C_t is the tangential inertial resistance coefficient.

2.2 *Turbulence model*

The realizable *k-ε* turbulence model is chosen to describe the flow, which can effectively make the performance of fluid in the flow process conform to the physical law of turbulent flow (Shih et al.,1995). The turbulent kinetic energy *k* and the turbulent dissipation rate *ε* are:

Turbulence kinetic energy equation:

$$\frac{\partial(\rho k)}{\partial t}+\frac{\partial(\rho k u_i)}{\partial x_i}=\frac{\partial}{\partial x_j}\left[\left(\mu+\frac{\mu_t}{\sigma_k}\right)\frac{\partial k}{\partial x_j}\right]+G_k+G_b-\rho\varepsilon-Y_M+S_k \tag{5}$$

Turbulent dissipation rate equation:

$$\frac{\partial(\rho\varepsilon u_i)}{\partial x_i} = \frac{\partial}{\partial x_j}\left[\left(\mu + \frac{\mu_t}{\sigma_\varepsilon}\right)\frac{\partial\varepsilon}{\partial xj}\right] + \rho C_1 S\varepsilon$$

$$-\rho C_2 \frac{\varepsilon^2}{k + \sqrt{\nu\varepsilon}} + C_{1\varepsilon}\frac{\varepsilon}{k}C_{3\varepsilon}G_b + S_\varepsilon \quad (6)$$

$$C_1 = max\left[0.43, \frac{\eta}{\eta+5}\right], \eta = S\frac{k}{\varepsilon},$$

$$S = \sqrt{2S_{ij}S_{ij}}, S_{ij} = \frac{1}{2}\left(\frac{\partial u_i}{\partial x_j} + \frac{\partial u_j}{\partial x_i}\right) \quad (7)$$

where, σ_k, σ_ε are the turbulent Prandtl number; S_k and S_ε is the turbulent kinetic energy; ε is dissipation rate; G_b and Y_M due to buoyancy and compressibility are not necessary when modeling incompressible fluid with no temperature fluctuations, hence $G_b = 0$ and $Y_M = 0$. $C_{1\varepsilon}$, $C_{3\varepsilon}$ and C_2 are constants; G_k represents the turbulent kinetic energy generation term caused by the change of average velocity.

2.3 *Porous media model*

The momentum equation of the porous media model contains an additional source term, which considers the resistance of flow transferred by porous media. The hydrodynamic force acting on the porous media which can be written as:

$$F = S_i\lambda A \quad (8)$$

where λ is the thickness of the porous media, A is the area of the plane net, S_i is the source term for the momentum equation that can be described as follows:

$$F_d = \left(D_n\mu u + C_n\frac{1}{2}\rho|u|u\right)\lambda A \quad (9)$$

$$F_l = \left(D_t\mu u + C_t\frac{1}{2}\rho|u|u\right)\lambda \quad (10)$$

where F_d is the drag force of the plane net, and the direction is parallel to the flow direction; F_l is the lift force of the plane net, and the direction is perpendicular to the flow direction. These general equations are solved using the CFD software STAR-CCM+ 15.02.

3 NUMERICAL MODEL

3.1 *Geometry*

The bottom-supported aquaculture platform and cage model as presented in Figure 1 is simulated in this paper. The bottom-supported aquaculture platform is a regular octagon structure composed of frame structures, 8 outside nets, 8 inside nets and 8 bottom nets. The knotless nylon net (2.8 mm twine thickness, 29 mm mesh size, and the solidity $S_n = 0.2$) are chosen.

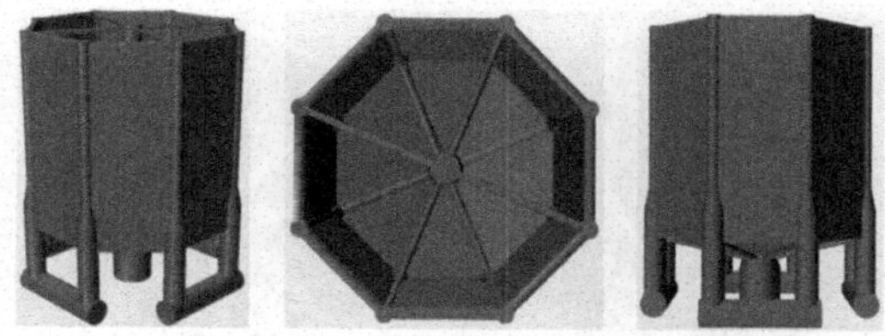

Figure 1. Geometric descriptions of the aquaculture platform and cage model.

The main dimensions of the platform are given in Table 1, and the selected scale for frame structure was 1:20.

Table 1. The structure parameters of the bottom-supported aquaculture platform.

Component	Symbol	Unit	Model value
Length	L	m	1.3
Width	B	m	1.3
Height	D	m	1.65
Draft	T	m	1.2
Center of gravity (Xg)	LCG	m	1.94 E-03
Center of gravity (Yg)	TCG	m	2.76 E-03
Center of gravity (Zg)	VCG	m	0.848
Displacement	Δ	kg	76.76

3.2 *Computational domain and mesh*

The computational domain and boundary conditions are illustrated in Figure 2. The distance between the inlet boundary and the center of the aquaculture cage model is 3 m, while the one between the outlet boundary and the center of the aquaculture platform model is 6.5 m to ensure that the wake is fully developed.

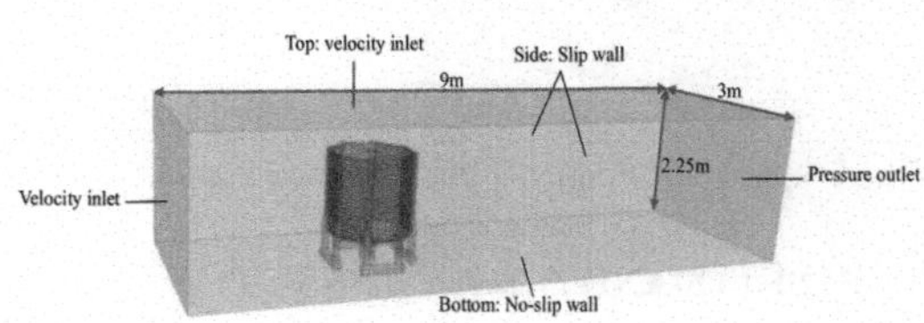

Figure 2. Sketch of the Computational domain and boundary conditions (unit: m).

As depicted in Figure 3, the plane net is replaced by porous media model and the thickness of the porous media is 50 mm. The coefficients that have been validated through experiments and simulations are chosen: D_n=75,083m^{-2}, D_t=38,307m^{-2}, C_n =4.985m^{-1}, C_t=1.660 m^{-1}.

Figure 4 shows the grid distribution of the numerical model. The computational domain is divided using the trimmed cell mesher, and the mesh density can be locally increased or decreased by using the volume mesh density factors to define the volumetric controls. The mesh is refined at the free surface, frame pipe,

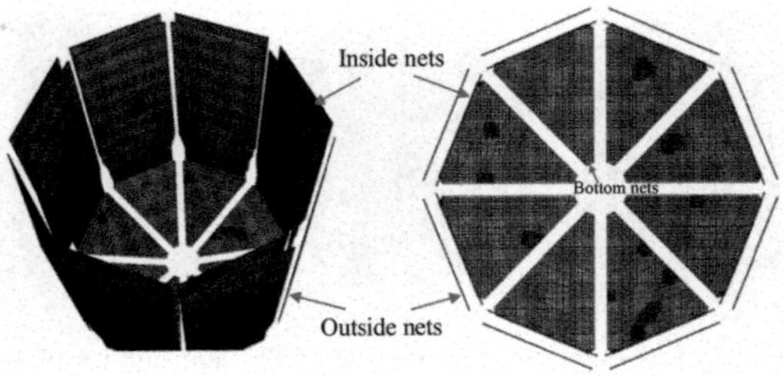

(a)the porous media model

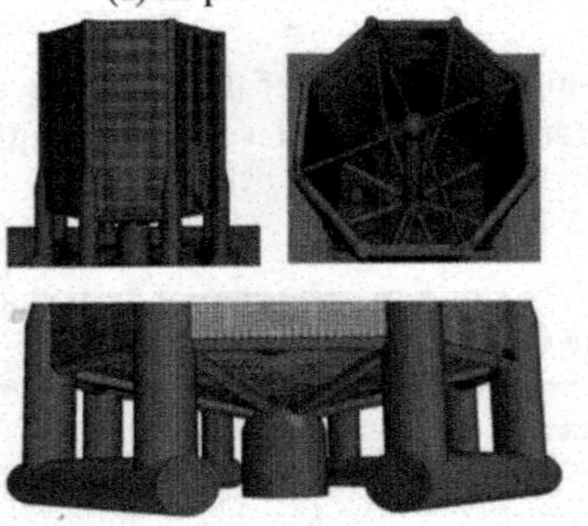

(b) the aquaculture platform and cage model

Figure 3. Meshing of the bottom-supported aquaculture platform.

flume wall, and the bottom. The simulation time is 40s to ensure the full development of the wake.

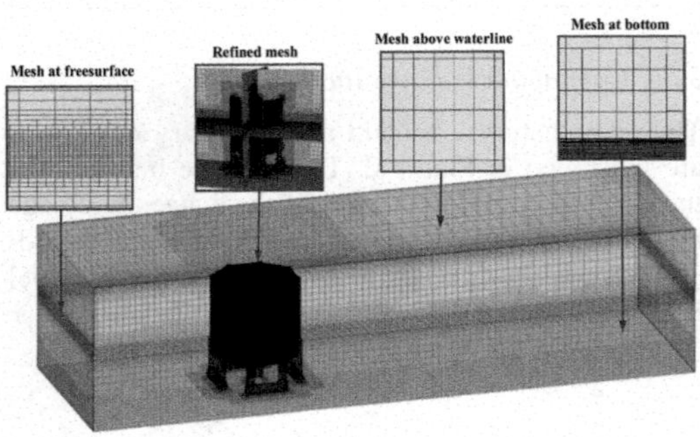

Figure 4. Grid distribution of the numerical model.

4 RESULTS AND DISCUSSIONS

4.1 *The influence of the net solidity*

The distribution of the flow velocity and the drag force acting on the aquaculture platform can be affected by the net solidity. A suitable net solidity is essential for the safety of the aquaculture platform and the stability of the flow fluctuations. In this section, three net solidities: S_n=0.13, S_n=0.20 and S_n=0.317 are adopted to calculate the flow field inside and around the bottom-supported aquaculture platform.

The incoming flow velocity is 0.28 m/s, and the x-y section of z = 1.1 m is selected and analyzed for the flow velocity.

As shown in Figure 5, an obvious flow velocity reduction exists along the centerline. The flow velocity downstream of the aquaculture platform

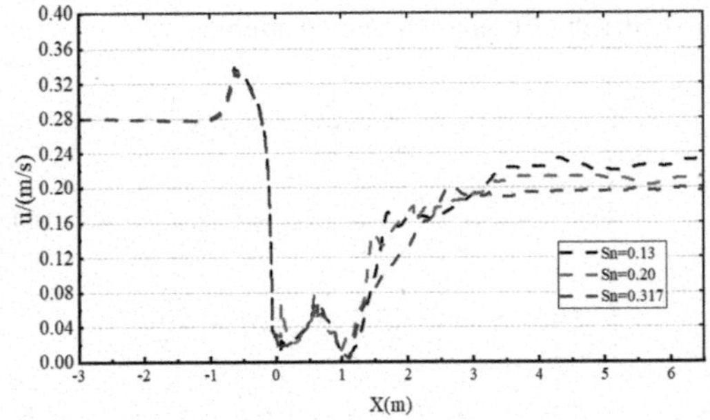

(a) The velocity magnitude along the direction of incoming flow at the centerline.

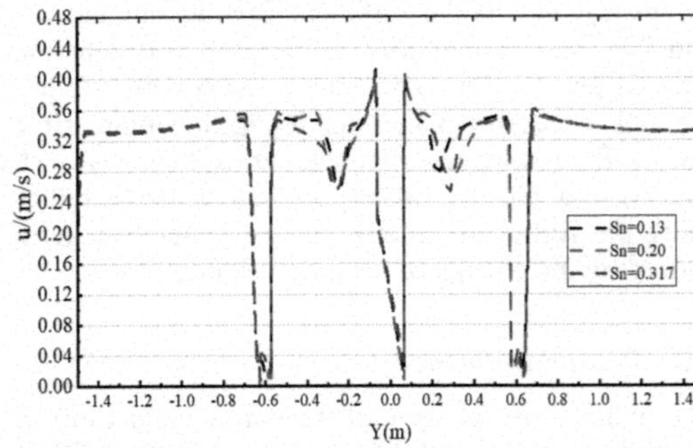

(b) The velocity magnitude along the direction perpendicular to incoming flow.

Figure 5. Flow velocity magnitude at different net solidities.

decreases significantly with the increase of the net solidities. It is indicated that the flow inside and around the platform fluctuates within a certain range between x=-0.65 m and x=0.65 m. The flow velocity in the platform fluctuates in different ranges because the aquaculture platform is a complex system with frames and nets. The flow velocity increases between the aquaculture platform and walls, probably because the pressure energy is converted into kinetic energy when the flow passes over the columns and pontoons.

It is indicated in Figures 6 and 7 that the widths of the velocity reduction regions downstream from the platform are approximately equal. The minimum flow velocity can be obtained on the downstream side of the center column, and the maximum flow velocity attenuation percentage is 75.4 %. The velocity at the flanks of the wake is approximately 3% greater than the incoming flow.

The flow velocity distribution inside the aquaculture platform is unstable, and it can be observed from the Figure 7 that there are vortices on both sides of the center column.

4.2 *The influence of double-layer nets with different solidities*

Various numerical models of the bottom-supported aquaculture platforms are simulated to evaluate the impact of double layer nets on the flow field. Figure 8 exhibits the flow velocity distribution along

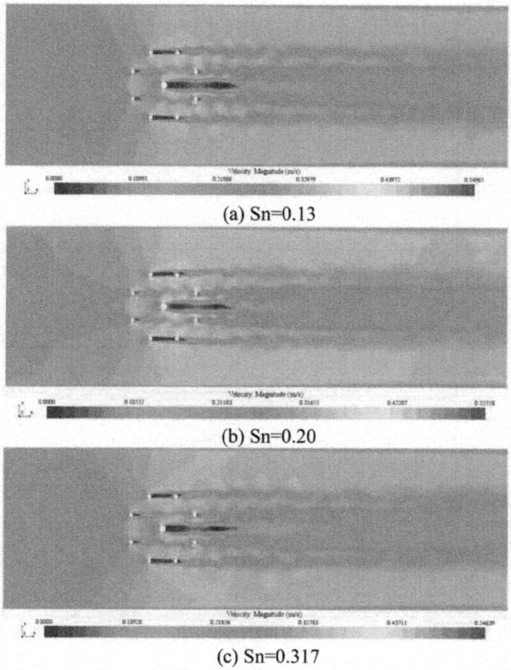

Figure 6. Flow velocity distribution of different net solidities at the x-y section of z=1.1 m.

the centerline of the aquaculture platform, which is 0.1 m below the water surface (z=1.1m).

It can be found from Figures 8 and 9 that the side nets have a great influence on the attenuation of the flow velocity. The flow velocity distribution is complicated and fluctuates, especially in front of the platform, this phenomenon may be caused by the decrease of the flow area and the combined effect of columns and nets. According to the streamlines on the horizontal plane (Z=1.1m), there is an obvious attenuation of flow velocity near the platform, the velocity increases in front of the platform and apparently decreases behind the double layer nets.

Furthermore, the flow velocity distribution inside and around the aquaculture platform tends to be uniform as the number of side nets is reduced.

The velocity attenuation downstream of double layer side nets is stronger than that of single layer side nets and without side net, and it may cause the shedding of the vortex, which is attributed to the blocking effect of the nets. The double-layer net helps alleviate the current velocity inside the aquaculture platform and avoids injury to the farmed fish caused by the high flow velocity.

It can be observed that the wake width behind the aquaculture platform is little affected by the net solidities of double layer nets. The velocity attenuates in different amplitude when the flow passes through the aquaculture platform. A wake region of low flow velocity can be observed downstream of the platform, the flow velocity gradually recovers

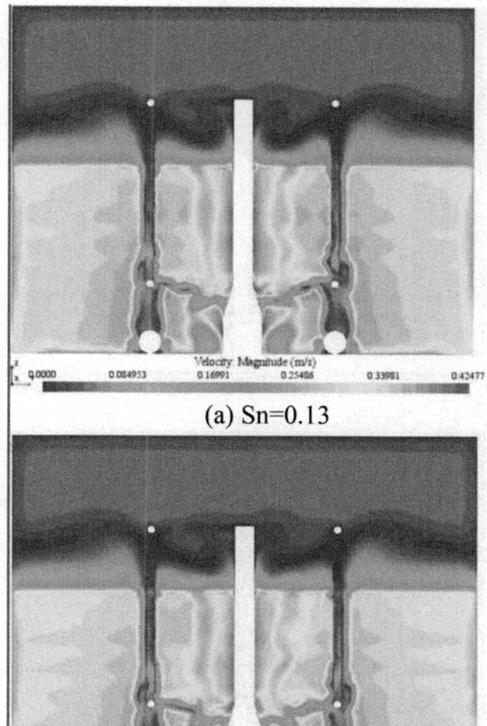

Figure 7. Flow velocity distribution of different net solidities at the y-z section of x=0 m.

and eventually stable behind the platform as the distance from the platform increases.

Figures 10 and 11 present a comparison of flow velocity distribution, it can be observed that when the platform is equipped with double layer nets, the change of the net solidity of the inner side nets has a relatively large effect on the flow velocity distribution.

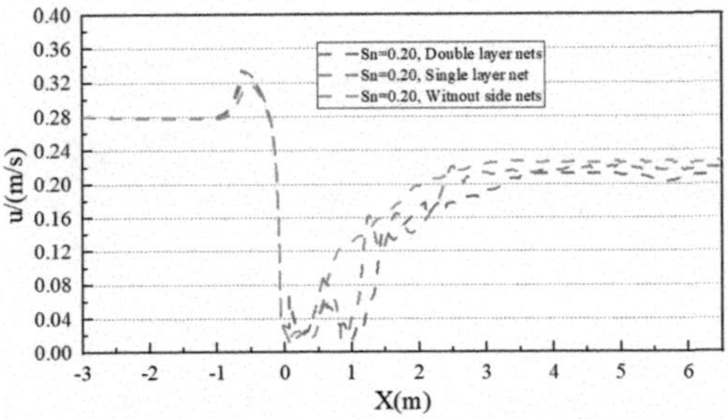

(a) Flow velocity magnitude of different models along the direction of incoming flow at the centerline.

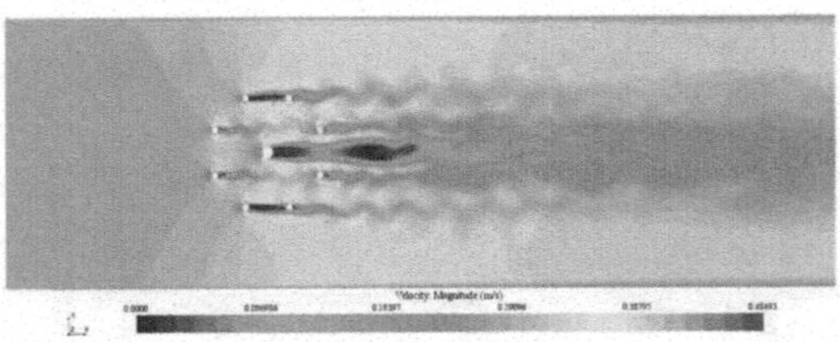

(b) Sn=0.20, Single layer net.

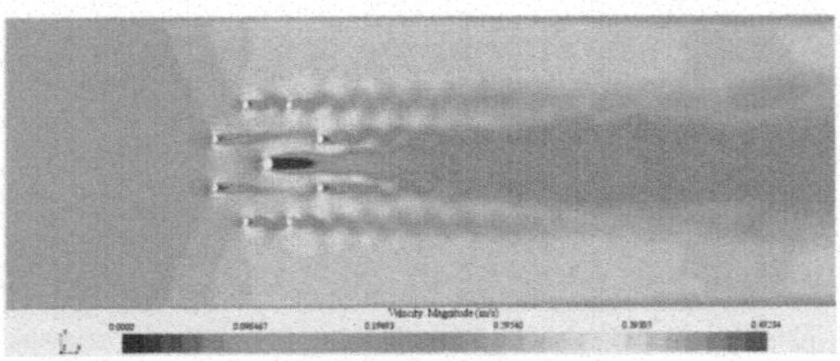

(c) Sn=0.20, Without side nets.

Figure 8. Flow velocity magnitude and distribution for different models at the x-y section of z=1.1 m.

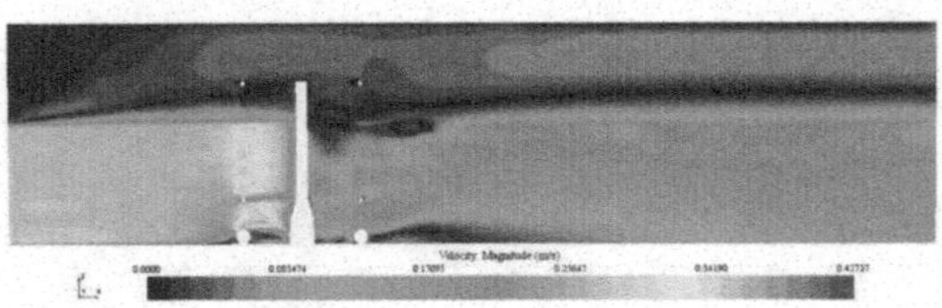

(a) Sn=0.20, Double layer nets.

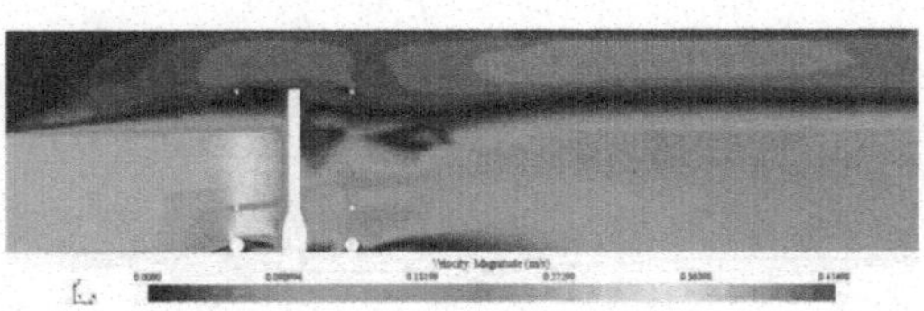

(b) Sn=0.20, Single layer net.

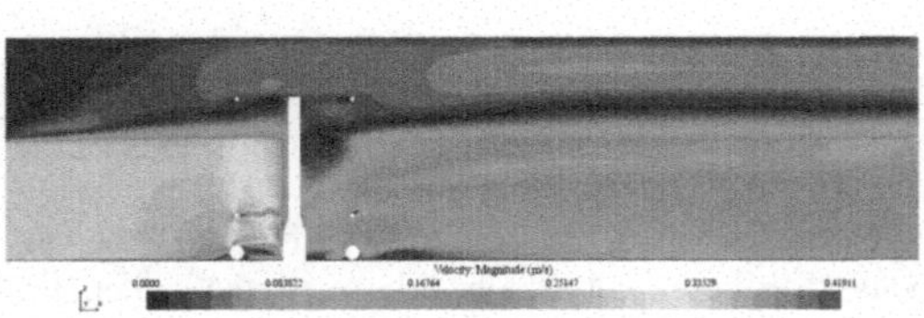

(c) Sn=0.20, Without side nets.

Figure 9. Flow velocity distribution of different models at the x-z section of y=0 m.

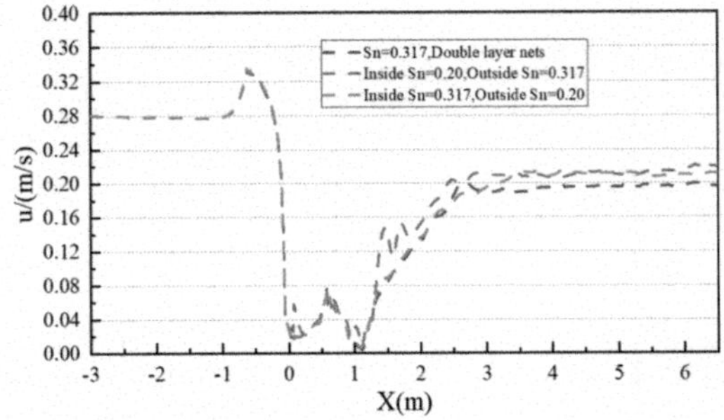

(a)

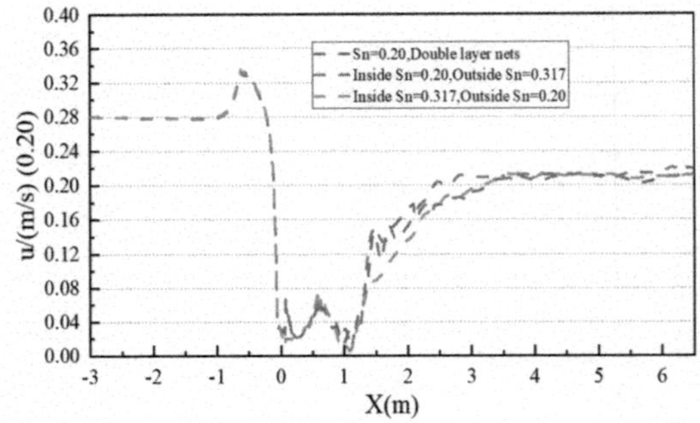

(b)

Figure 10. Comparison of flow velocity magnitude for different models along incoming flow direction.

5 CONCLUSIONS

In the present study, the flow fields inside and around the bottom-supported aquaculture platform and cage are analyzed using the porous media model and the RANS method with the realizable k-ε turbulence model. The main conclusions are as follows:

The presence of the nets changes the magnitude and affects the distribution of flow velocity. The increase of net solidity will lead to a significant attenuation of the flow velocity inside and downstream of the aquaculture platform.

In the upstream area of the platform, high flow velocity may have a strong effect on the structure. Therefore, the strength of pipes at this location should be increased to avoid damage. In the area inside the aquaculture platform, the flow velocity drops sharply and fluctuates irregularly. It should be noted that the vortices exist near the columns may lead to a dramatic fluctuation regarding the current velocity distribution and the drag force. In the downstream area of the platform, the low flow velocity may cause waste accumulation, which is worthy of attention.

ACKNOWLEDGEMENTS

This work is supported by the China Scholarship Council under Grant: 202306950024. The work is supported by the National Natural Science Foundation of China (Grant No. 52271326). This work contributes to the Strategic Research Plan of the Centre for Marine Technology and Ocean Engineering

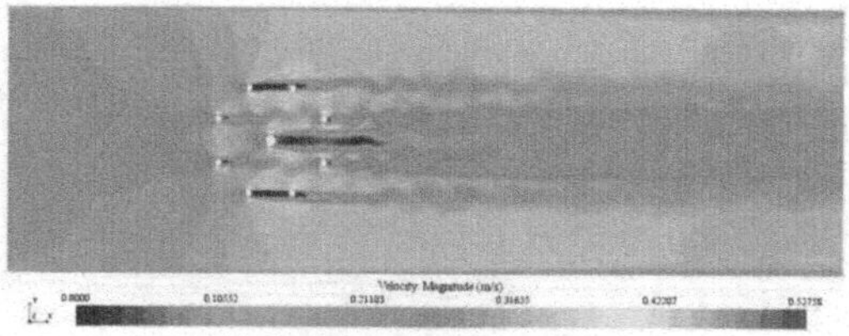

(a) Sn=0.20, Double layer nets.

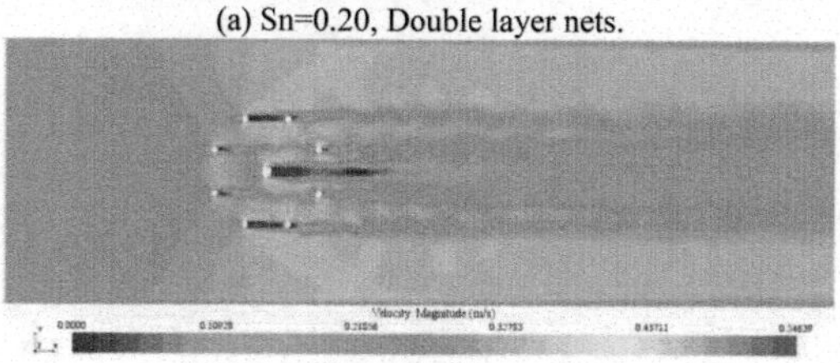

(b) Sn=0.317, Double layer nets.

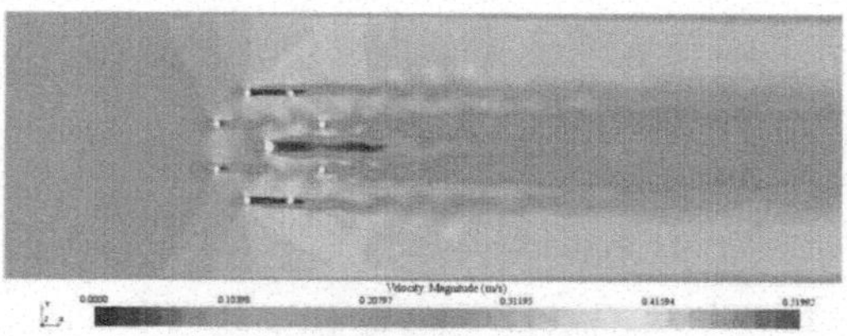

(c) Inside Sn=0.20, Outside Sn=0.317.

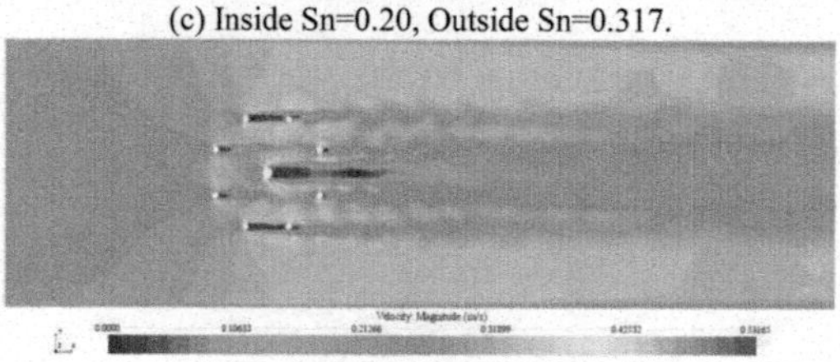

(d) Inside Sn=0.317, Outside Sn=0.20.

Figure 11. Flow velocity distribution for different models along incoming flow direction.

(CENTEC), which is financed by the Portuguese Foundation for Science and Technology (Fundação para a Ciência e Tecnologia - FCT) under contract UIDB/UIDP/00134/2020.

REFERENCES

Bi, C.W., Zhao, Y.P., Dong, G.H., Xu, T.J., Gui, F.-K., 2014. Numerical simulation of the interaction between flow and flexible nets. Journal of Fluids and Structures. 45, 180–201.

Chen H., Christensen E.D., 2017. Development of a numerical model for fluid-structure interaction analysis of flow through and around an aquaculture net cage. Ocean Engineering. 142, 597–615.

Chu, Y.I., Wang, C.M., Park, J.C., Lader, P.F., 2020. Review of cage and containment tank designs for offshore fish farming. Aquaculture. 519, 734928.

Guo, Y.C., Liu, ZC., Mohapatra, S.C. & Guedes Soares, C. 2022. Experimental investigation of an array of vertical flexible net-type structures under regular waves. *Trends in Maritime Technology and Engineering*, Guedes Soares, C. & Santos T.A. (Eds.), Taylor and Francis Group, London, UK, pp. 573–580.

Guo, Y.C., Mohapatra, S.C., Guedes Soares, C., 2020. Review of developments in porous membranes and net-type structures for breakwaters and fish cages. Ocean Engineering. 200, 107027.

Guo, Y.C., Mohapatra, S.C. & Guedes Soares, C. 2023. Experimental performance of multi-layered membrane breakwaters. *Ocean Engineering*. 281:114716.

Kristiansen, T., Faltinsen, O.M., 2012. Modelling of current loads on aquaculture net cages. Journal of Fluids and Structures. 34, 218–235.

Lader, P., Jensen, A., Sveen, J.K., Fredheim, A., Enerhaug, B., Fredriksson, D. 2007. Experimental investigation of wave forces on net structures. Applied Ocean Research 29: 112–127.

Lee, C.W., Kim, H.S., Lee, G.H., Koo, K.Y., Choe, M.Y., Cha, B.J., Jeong, S.J., 2005. Computation modeling of the moored flexible structures. In Guedes Soares, C. Garbatov Y. & Fonseca, N. (eds), Maritime Transportation and Exploitation of Ocean and Coastal Resources: Taylor & Francis Group, London. 1239–1243.

Liu, H.F., Bi, C.W., Zhao, Y.P., 2021. Hydrodynamic assessment of a semi-submersible aquaculture platform in uniform fluid environment. Ocean Engineering. 237,109656.

Liu, Z.C., Guedes Soares, C., 2023. Sensitivity analysis of a numerical model of the dynamics of gravity cages subjected to current and waves. Ocean Engineering. 287, 115715

Liu, Z.C., Mohapatra, S.C., Guedes Soares, C., 2021. Finite Element Analysis of the Effect of Currents on the Dynamics of a Moored Flexible Cylindrical Net Cage. Journal of Marine Science and Engineering. 9, 159.

Liu, Z.C., Wang, S., Guedes Soares, C.,2022. Numerical Study on the Mooring Force in an Offshore Fish Cage Array. Journal of Marine Science and Engineering. 10(3), 331.

Mohapatra, S.C., Guedes Soares, C., 2024. A semi-analytical model of an array of moored floating flexible offshore net cages under current loads. Ocean Engineering. 291, 116309.

Patursson, Ø., Swift, M.R., Tsukrov, I., Simonsen, K., Baldwin, K., Fredriksson, D.W., Celikkol, B., 2010. Development of a porous media model with application to flow through and around a plane net. Ocean Engineering. 37 (2), 314–324.

Shen, Y.G., Greco, M., Faltinsen, O.M., et al., 2018. Numerical and experimental investigations on mooring loads of a marine fish farm in waves and current. Journal of Fluids and Structures. 79, 115–136.

Shih, T.H., Liou, W.W., Shabbir, A., Yang, Z., Zhu, J., 1995. A new k-e eddy viscosity model for high reynolds number turbulent flows. Computers & Fluids. 24(3), 227–238.

Tang, H.J., Huang, C.C., Chen, W.M., 2011. Dynamics of dual pontoon floating structure for cage aquaculture in a two-dimensional numerical wave tank. Journal of Fluids and Structures. 27, 918–936.

Xu, Z., Qin, H., 2020. Fluid-structure interactions of cage based aquaculture: from structures to organisms. Ocean Engineering. 217, 107961.

Yao, Y., Chen, Y., Zhou, H. & Yang, H., 2016. Numerical modelling of current loads on a net cage considering fluid-structure interaction. J Journal of Fluids and Structures. 62,350–366.

Zhan, J.M., Jia, X.P., Li, Y.S., Sun, M.G., Guo, G.X., Hu, Y.Z., 2006. Analytical and experimental investigation of drag on nets of fish cages. Aquaculture Engineering. 35 (1),91-101.

Zhao, Y., Bi, C., Dong, G., Gui, F., Cui, Y., Guan, C., Xu, T., 2013a. Numerical simulation of the flow around fishing plane nets using the porous media model. Ocean Engineering. 62, 25–37.

Zhao, Y.P., Bi, C.W., Dong, G.H., et al., 2013b. Numerical simulation of the flow field inside and around gravity cages. Aquaculture Engineering. 52, 1–13.

Advances in Maritime Technology and Engineering – Guedes Soares & Santos (Eds)

Numerical study on the effect of cage reinforcement on the behaviour of the circular gravity cage

Z.C. Liu & C. Guedes Soares
Centre for Marine Technology and Ocean Engineering (CENTEC), Instituto Superior Técnico, Universidade de Lisboa, Lisbon, Portugal

ABSTRACT: A method of reinforcing gravity cages by connecting the bottom ring and floating collar with vertical beams is explored. The objective is to prevent the volume of the gravity cage from reducing much when subjected to currents and waves. A numerical model of a full-scale gravity cage is built. The mooring forces and the cage volume subjected to uniform flow are calculated. Four beams connected to the bottom ring and the floating collar, are evenly distributed along the circumference of the collar. The thickness and stiffness of those components are varied and the results are compared with the ones of the cages with no reinforcement. It is found that the beams can contribute to protecting the cage volume, but they increase the mooring forces. In addition, from the calculation of full-scale gravity cages, there is no tilting of the cages as observed in the towing tests.

1 INTRODUCTION

Gravity-based net cages have high cost-effectiveness due to their simple structures and large aquaculture volumes (Liu et al. 2022). The nets are expanded by the floating collar and bottom ring. However, due to the flexible nature of the structure, the volume of the gravity cage reduces when subjected to currents and waves, which impacts the living environment of the fish and brings negative effects on their growth and health. Hence, numerous studies have delved into the hydrodynamic characteristics of gravity fish cages.

Moe et al. (2005) investigated stress distribution on the net structure and implemented inclined cross ropes to mitigate stress concentration. Lee et al. (2005) established a computational model to simulate the dynamic behaviour of moored flexible structures. Berstad & Tronstad (2005) formulated a numerical model for a fish farm and explored the effects of regular and irregular waves on mooring forces. Lader et al. (2008) measured the net deformation of gravity cages at salmon farms and estimated the cage volumes in the flow.

DeCew et al. (2013) utilized acoustic sensors to measure the net deformation of gravity cages. They used the results to verify their numerical calculations. Fu et al. (2013) conducted a forced oscillation test to investigate the hydrodynamic coefficients for a section of the floating collar.

Endresen et al. (2013) took into account the influence of wake effects between adjacent twines and those upstream on downstream twines in their numerical models. Bai et al. (2016) assessed stress concentration at the junction of the collar and the mooring rope, subsequently determining the fatigue life of the collar.

Moe Føre et al. (2016) examined the deformation and hydrodynamic forces acting on cages with solidities ranging from 0.19 to 0.43. They conducted a comparative analysis between experimental outcomes and numerical model predictions. Bernardo & Guedes Soares (2021) employed SIMA to compute the motion of a fish cage under the influence of currents and waves. Føre et al. (2021) conducted experiments to determine the drag coefficients of net panels with varying solidities.

Mohapatra et al. (2021) obtained a good agreement between their numerical and analytical calculations for a gravity cage subjected to waves. Zhang et al. (2021) found the maximum mooring force in a complex mooring system for a farm with 36 cone-shaped cages. Liu et al. (2021a) used analytical results to verify their numerical model and calculated the mooring forces of a single cage system (Liu et al., 2021b) and a cage array (Liu et al., 2022). Liu & Guedes Soares (2022a and 2022b) measured the wave heights before and after small-scale cage models and assessed the impacts of the cage on the waves. In their subsequent work, they enriched the experimental content by measuring the mooring forces and the motion of net cages (Liu et al., 2023). Mohapatra & Guedes Soares (2024) built an analytical model to calculate the mooring forces and the cage motions of a cage array.

In this work, a full-scale numerical model is established to explore a method of reinforcing gravity cages. Besides the ropes in general cages, the bottom

DOI: 10.1201/9781003508779-59

ring and the collar are connected by four beams that are evenly distributed along the circumference of the collar. The cage volumes and the mooring forces of the cages with and without the reinforcements are compared. More information on the numerical model can be found in Section 2. The calculated results are in Section 3. At last, a conclusion is drawn in Section 4.

2 NUMERICAL MODEL

The finite element method is employed to construct the numerical model of a full-scale gravity cage using Ansys/APDL. The entire fish cage, including the mooring system, is composed of slender structures. Therefore, line elements are utilized to simulate all components. Depending on whether a section has bending stiffness and the cross-sectional shapes, beam, pipe, and link elements are employed for simulation. For instance, the collar and handle sections, having a hollow cross-section, are modelled using pipe elements, while the bottom ring and reinforcing beams are modelled by beam elements, and the netting and ropes are simulated using link elements. Additional details regarding the numerical model can be found in Liu & Guedes Soares (2023).

The entire numerical model is shown in Figure 1.

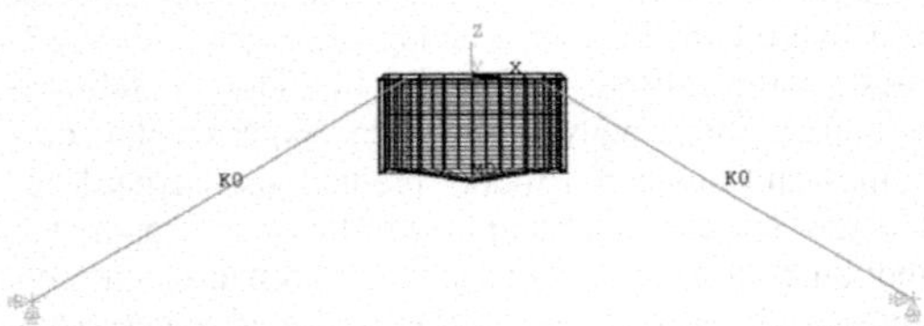

Figure 1. The model of the gravity fish cage with a mooring system.

The diameter of the inner collar is 40 m and the one of the outer collar is 42.4 m. The depth of the nets is 23 m. The bottom ring has the same diameter as the outer collar. It is connected to the bottom of the nets and the outer collar by ropes. The nets have a cone-shaped bottom with a depth of 3 m. The solidity of the nets is 0.225. Four mooring ropes are connected to the outer collar and the seabed. The water depth is assumed to be 50 m. The rest dimensions are listed in Table 1. The numerical model includes 8049 elements and for each case, it took 35-40 hours to calculate.

Furthermore, if the link elements have the same thickness as the nets, typically ranging between 1.5 mm to 2 mm, the model would involve an excessive number of elements for computation. Thus, the nets are simplified into a new structure composed of horizontal and vertical link elements intersecting. To keep the same properties as the real nets, the cross-section area, the density and Young's modulus are kept the same as the original structure. On the other hand, the hydrodynamic coefficients are increased according to the ratio of the projected area reduction. Additional details regarding the method for simplifying the net structure can be found in Liu et al. (2021a).

Two different vertical beams are utilized to reinforce the gravity cage and their effects are compared. The properties of the beams are listed in Table 2.

Table 1. The parameters and the dimensions of the FEA model.

Component	Dimension/Parameter	Value
Handle	Density(kg/m^3)	450
	Height above the collar (m)	1.2
	Cross-section diameter (m)	0.08
	Thickness (mm)	15
Collar	Density(kg/m^3)	450
	Diameter(m)	40 for the inner collar 42.4 for the outer collar
	Cross-section diameter (m)	0.5
	Thickness (mm)	50
Net	Material type	Nylon
	Young's modulus (GPa)	1.4
	Twine thickness (mm)	2.7
	Solidity	0.225
	Density (kg/m3)	1140
	Depth (m)	20 for the cylinder part 3 for the tapered bottom
Bottom ring	Weight/length (kg/m)	32
	Diameter(m)	42.4
Counter-weight at the top of the tapered bottom	Weight (kg)	150
Mooring ropes	Material type	HMPE
	Stiffness, AE (N)	1.25×108
	Breaking strength (kN)	840

Table 2. The parameters of the vertical beams.

No.	1	2
Density (kg/m^3)	7800	7800
Diameter (mm)	60	100
Young's modulus (GPa)	200	200

The Morison equation is used to calculate the hydrodynamic load on the structure, which is expressed as (Faltinsen, 1990):

$$f = \frac{1}{2}\rho C_D D(u - \dot{x})|u - \dot{x}| + \rho C_M \frac{\pi D^2}{4}\dot{u} - \rho(C_M - 1)\frac{\pi D^2}{4}\ddot{x} \quad (1)$$

where f is the hydrodynamic force per length, C_M is the mass coefficient, C_d is the drag coefficient, ρ is the density of seawater, D is the diameter of the slender structure, u and $\dot{u}$ are the velocity and acceleration of the water particle, $\dot{x}$ and $\ddot{x}$ are the velocity and acceleration of the structure element. The current velocity is 0.5 m/s or 1 m/s. The wave height and period are 1 m and 8 s, respectively.

3 RESULTS

The results of the calculations are shown in this Section. The shapes and the cage volumes of the cages with and without reinforcement are compared in Section 3.1. The drag forces are compared in Section 3.2.

3.1 *Deformations and volumes of the cages with and without vertical beams*

The deformations of the three models are plotted in Figure 2. It can be directly observed from the figures that, under the protection of the beams, the vertical displacement of the net decreases and the reduction in depth also slows down. To calculate the cage volume, the coordinates of the nodes representing the netting are recorded and then input into CAD software. The shape of the net enclosure is formed into a solid body, and the volume of this body, which is calculated by this software, represents the volume of the cage, shown in Figure 3.

The calculated cage volumes are shown in Table 3. When the cage is not protected by the beams, the average volume is 68 % of the original volume at a current velocity of 0.5 m/s. When the current velocity increases to 1 m/s, the cage volume reduces to 32.1 %.

When the net cage is protected by Beam 1 and 2, the volumes are 69.3 % and 72.2 % of the original volume at the velocity of 0.5 m/s, respectively, while the values reduce to 35.8 % and 41.5 % at a current velocity of 1 m/s. The calculated results show that the vertical beams can be used to protect the cage volume, and the effectiveness increases as the beams become thicker.

Furthermore, the effectiveness of the beams increases with the current velocity. At a velocity of 0.5 m/s, the volumes of the cage with Beam 1 and 2 are 1.9 % and 6.2 % larger than the one of the cage with no reinforcement. However, when the velocity increases to 1 m/s. the volumes of the cage with Beam 1 and 2 are 11.5 % and 29.3 % larger than the one of the cage with no reinforcement.

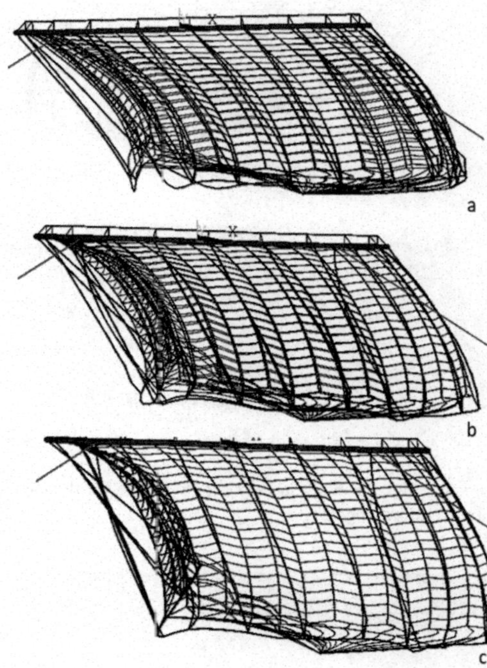

Figure 2. The deformation of the cage models (a, the model with no reinforcement, b, the model reinforced by Beam 1, c, the model reinforced by Beam 2).

Table 3. The cage volumes of the three models.

Reinforced beam	Cage volume at 0.5 m/s	Cage volume at 1 m/s
No	68 %	32.1 %
1	69.3 %	35.8 %
2	72.2 %	41.5 %

In addition, Figure 4 shows the cage shape in laboratory towing tests. Due to the vertical beams, the cage tilts when the current velocity is large. However, it does not happen for the full-scale model in this work. This is because the directions of the mooring ropes are not as horizontal as the one of the towing ropes in the towing tests. In addition, for the full-scale cage, the buoyancy of the collar is too large to be pulled into the water.

3.2 *Mooring forces of the cages with and without vertical beams*

The calculated mooring forces of the three models are listed in Table 4. The beams effectively increase

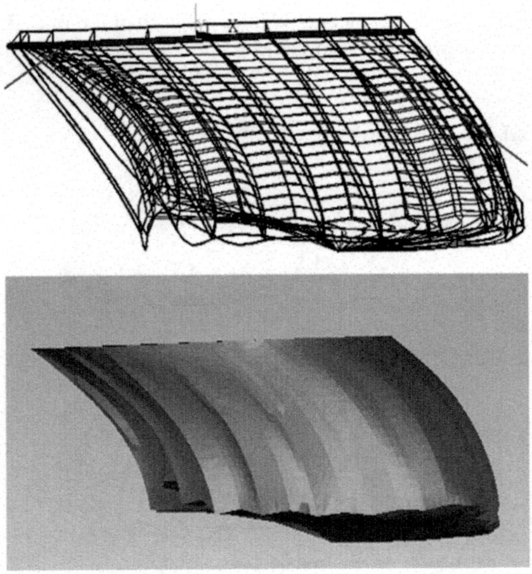

Figure 3. The solid body in the CAD software for the cage without reinforcement.

Figure 4. Tilting of the small-scale cage model in the towing tests.

the cage volumes, but they also simultaneously increase the mooring forces. When the current velocity is 0.5 m/s, the maximum mooring forces of the cage with Beam 1 and 2 are 63.7 kN and 68.7 kN, which are 3.1 kN and 8.1 kN larger than the one of the cage with no reinforcement. However, the effect from the beams reduces with the increase of the current velocity. At a velocity of 1 m/s, the maximum mooring forces of the cage with Beam 1 and 2 are only 0.6 kN and 2 kN larger than the one of the cage with no reinforcement.

The increase in mooring forces is generated by two aspects. Firstly, the hydrodynamic forces acting on the beams increase the mooring forces. Secondly, when the beams increase the cage volume, they also simultaneously increase the projected area of the cage, which results in larger mooring forces.

The numerical calculation in this work proves that the application of vertical beams can protect the cage volumes effectively. However, further research needs to be conducted before this design is applied, including the study of the connection between the collar, the bottom ring and the vertical beams, the stress distribution of the beams and the corrosion protection for the beams.

Table 4. Mooring forces of the three models.

Reinforced beam	Max value at 0.5 m/s	Max value at 1 m/s
No	60.6 kN	12.6 kN
1	63.7 kN	13.2 kN
2	68.7 kN	14.6 kN

4 CONCLUSIONS

This paper presents a numerical model of a full-scale cage. To protect the cage volumes in the currents and waves, a design connecting the bottom ring and floating collar by vertical beams is studied. The cage volume and the mooring forces are calculated and compared with the results of the cage without the reinforcement. Four beams, connected to both the bottom ring and floating collar, are evenly distributed along the circumference of the collar. Beams with different thicknesses and stiffness are applied. It has been concluded that:

1. The vertical beams can be used to protect the cage volume, and the effectiveness increases as the beams become thicker. For the beams applied in this work, the volume is increased by 6.2 % when the cross-section diameter of the beams is 100 mm. It reduces to 1.9 % when the cross-section diameter of the beams is 60 mm.
2. Unlike the results from the previous towing experiment, there is no tilting occurring for the full-scale calculation due to the directions of the mooring ropes and the large buoyancy of the collar.
3. Both the mean values and the amplitudes of the mooring forces are increased by the vertical beams. It is from the hydrodynamic loads on the beams and the increase of the projected area.
4. More studies are necessary before this design is applied, including the study of the connection between the collar, the bottom ring and the vertical beams, the stress distribution of the beams and the corrosion protection for the beams.

ACKNOWLEDGEMENTS

The first author has been funded by the Portuguese Foundation for Science and Technology (Fundação para a Ciência e a Tecnologia-FCT), through a doctoral fellowship under the Contract no. SFRH/BD/147178/2019.

This work contributes to the Strategic Research Plan of the Centre for Marine Technology and Ocean Engineering (CENTEC), which is financed by the Portuguese Foundation for Science and Technology (Fundação para a Ciência e Tecnologia) under contract UIDB/UIDP/00134/2020.

REFERENCES

Bai, X., Xu, T., Zhao, Y., Dong, G. and Bi, C., 2016. Fatigue assessment for the floating collar of a fish cage using the de-terministic method in waves. *Aquacultural Engineering*. 74: 131–142.

Bernardo, T. A. and Guedes Soares, C. 2021. Validation of tools for the analysis of offshore aquaculture installations. Guedes Soares, C. & Santos T. A., (Eds.). *Developments in Maritime Technology and Engineering*. London, UK: Taylor and Francis; 2021. Vol 2: pp. 685–692.

Berstad, A.J., & Tronstad, H., 2005. Response from current and regular/irregular waves on a typical polyethylene fish farm. In Guedes Soares, C. Garbatov Y. & Fonseca, N. (eds), *Maritime Transportation and Exploitation of Ocean and Coastal Resources*: Taylor & Francis Group, London: 1189–1196.

DeCew, J., Fredriksson, D.W., Lader, P.F., Chambers, M., Howell, W.H., Osienki, M., Celikkol, B., Frank, K. and Høy, E., 2013. Field measurements of cage deformation using acoustic sensors. *Aquacultural. Eng*, 57: 114–125.

Endresen, P. C., Fore, M., Fredheim, A., Kristiansen, D. and Enerhaug, B., 2013, Numerical Modelling of Wake Effect on Aquaculture Nets, *Proc. of 32nd International Conference on Ocean, Offshore and Arctic Engineering*, ASME Paper No. OMAE2013-11446.

Faltinsen O.M., 1990. Stationkeeping. In: *Sea Loads on Ships and Offshore Structures*. Cambridge University Press, pp.257–270.

Faltinsen. O.M., 1990. Viscous Wave Loads and Damping. In: *Sea Loads on Ships and Offshore Structures*. Cambridge University Press, pp.223–226.

Fu, S., Xu, Y., Hu, K. and Zhang, Y., 2013. Experimental investigation on hydrodynamics of floating cylinder in oscillatory and steady flows by forced oscillation test. *Marine Structures*, 34: 41–55.

Føre, H. M., Endresen, P. C., Norvik, C., and Lader, P., 2021. Hydrodynamic Loads on Net Panels With Different Solidities. "*ASME. J. Offshore Mech. Arct. Eng.*143 (5): 051901.

Føre, H. M., Lader, P., Lien, E. and Hopperstad, O., 2016. Structural response of high solidity net cage models in uniform flow. *Journal of Fluids and Structures*, 65: 180–195.

Moe, H., Fredheim, A. and Heide, M.A., 2005. New net cage designs to prevent tearing during handling. In Guedes Soares, C. Garbatov Y. & Fonseca, N. (eds), *Maritime Transportation and Exploitation of Ocean and Coastal Resources*: Taylor & Francis Group, London. 1265–1272.

Mohapatra, S. C.; Bernardo, T. A., and Guedes Soares, C. 2021. Dynamic wave induced loads on a moored flexible cylindrical net cage with analytical and numerical model simulations. *Applied Ocean Research*. 110:102591.

Mohapatra, S. C and Guedes Soares, C., 2024. A semi-analytical model of an array of moored floating flexible offshore net cages under current loads. *Ocean Engineering*. 291: 116309

Lader, P., Dempster, T., Fredheim, A. and Jensen. F., 2008. Current induced net deformations in full-scale cages for Atlantic salmon (Salmo salar). *Aquacultural Eng*. 38: 52–65.

Lee, C.W., Kim, H.S., Lee, G.H., Koo, K.Y., Choe, M.Y., Cha, B.J. and Jeong, S.J., 2005. Computation modeling of the moored flexible structures. In Guedes Soares, C. Garbatov Y. & Fonseca, N. (eds), *Maritime Transportation and Exploitation of Ocean and Coastal Resources*: Taylor & Francis Group, London. 1239–1243.

Liu, ZC., Mohapatra, S. C., and Guedes Soares, C., 2021a. Finite Element Analysis of the Effect of Currents on the Dynamics of a Moored Flexible Cylindrical Net Cage. *Journal of Marine Science and Engineering*, Vol. 9: 159.

Liu, Z., and C. Guedes Soares., 2022a. Preliminary experiments of the behaviour of circular gravity cage in linear waves. in: Guedes Soares, C. & Santos T. A., (Eds.). *Trends in Maritime Technology and Engineering*, London, UK: Taylor and Francis, Volume 2: 599–604.

Liu, Z. and Guedes Soares, C., 2022b. Experimental study of the behaviour of a circular gravity cage in linear waves. *Aquac. Eng*. 99: 102291.

Liu, Z. and Guedes Soares, C., 2023. Sensitivity analysis of a numerical model of the dynamics of gravity cages subjected to current and waves. *Ocean Eng*. Vol. 287: 115715.

Liu, Z., Garbatov, Y. and Guedes Soares, C., 2021b. Numerical modelling of full-scale aquaculture cages under uniform flow. *Developments in Maritime Technology and Engineering*, Guedes Soares, C. & Santos T.A., (Eds.), Taylor and Francis, London, UK, Vol 2: 705–712.

Liu, Z., Wang, S., Guedes Soares, C., 2022. Numerical Study on the Mooring Force in an Offshore Fish Cage Array., *Journal of Marine Science and Engineering*. 10 (3): 331.

Liu, Z., Xu, H. and Guedes Soares, C., 2023. Experimental study on the mooring forces and motions of a fish cage under regular waves. *Ocean Engineering*, 280: 114612.

Vegard, A., 2016. Modelling of Aquaculture Net Cages in SIMA. Project Report. Marintek, Norway.

Zhang, DP., Bai, Y., and Guedes Soares, C. 2021. Dynamic Analysis of an Array of Semi-rigid "Sea Station" Fish Cages subjected to waves. *Aquacultural Engineering*. 94:102172.

Advances in Maritime Technology and Engineering – Guedes Soares & Santos (Eds)

Numerical study on the application of gravity cages on the coast of Algarve

Z.C. Liu, S. Ramos-Marin & C. Guedes Soares
Centre for Marine Technology and Ocean Engineering (CENTEC), Instituto Superior Técnico, Universidade de Lisboa, Lisbon, Portugal

ABSTRACT: A numerical model is built to study the performance of gravity cages on the coast of Algarve. A review is made of the situation in Portugal concerning offshore aquaculture and the areas in Algarve that are planned for this activity are identified. From the previous study, the high-tenacity polyester multifilament fibre has excellent performance for a gravity cage system, which is used as the mooring material in this work. The mooring ropes are modelled by nonlinear spring elements. Data on the wave conditions on the coast of Algarve is used. The amplitudes of the mooring force and the distributions of mooring force amplitudes are calculated. From the results, the maximum mooring force is below the break force of the material. The amplitude distribution of the mooring forces and the maximum amplitude are shown. The results are compared with the previous studies on the mooring forces in the coasts of Madeira and Azores for the application of the same gravity cage system.

1 INTRODUCTION

Nowadays, driven by overfishing and the escalating demand for fish in human society, the conventional fishing industry has struggled to fulfil societal needs. Aquaculture has emerged as a vital source of sustenance. According to the State of World Fisheries and Aquaculture (FAO, 2020), capture production in both inland and marine waters has shown minimal growth since the 1990s. Conversely, the production from aquaculture in both inland and marine environments continues to expand.

Gravity cages and cage arrays are extensively employed in offshore aquaculture due to their substantial cultivation capacity and straightforward operational procedures. As a country with an extensive coastline and island groups, Portugal has a very promising prospect for the development of aquaculture (Madsen & Chkoniya, 2021).

There is a substantial amount of research related to gravity cages. Berstad & Tronstad (2005) developed a numerical model for a fish farm, exploring the impact of both regular and irregular waves on mooring forces. Lee et al. (2005) constructed a numerical model to simulate the dynamic behaviour of moored flexible structures. Moe et al. (2005) calculated the stress distribution of the nets during the lifting of a fish cage. Lader et al. (2008) employed pressure sensors to gauge net deformation and calculated the volumes of full-scale cages at two distinct fish farms. The flexible structure's deformation led to a reduction in cage volume, concurrently diminishing the total drag force compared to a rigid structure. DeCew et al. (2013) utilized acoustic sensors for measuring gravity cage deformation. Fu et al. (2013) conducted forced oscillation tests to determine the mass and drag coefficients of a floating cylinder structure across varying Reynolds numbers (Re) and Keulegan-Carpenter numbers (KC).

Endresen et al. (2013) incorporated the reduction in flow velocity caused by nets into their calculations of drag forces, effectively minimizing disparities between numerical and physical models. Bai et al. (2016) computed the stress concentration on a collar made of high-density polyethylene (HDPE) and assessed the fatigue life based on their numerical findings. Kristiansen & Faltinsen (2015) constructed a physical model of a small-scale cage featuring flexible nets and conducted measurements of mooring forces.

Gansel et al. (2018) conducted towing experiments on a full-scale fish cage in a fjord environment with minimal currents and waves. Measurements of both drag force and volume reduction were taken at various towing velocities. The study revealed that existing current calculation methods tend to overestimate drag forces, particularly at higher flow velocities.

Mohapatra et al. (2021) determined the hydrodynamic load on a gravity cage using both numerical and analytical approaches, achieving satisfactory agreement between the two methods. Dong et al. (2021) constructed experiments to investigate the hydrodynamic characteristics of a flexible net cage under uniform flow. This study encompassed aspects

DOI: 10.1201/9781003508779-60

such as drag force, remaining cage volume, and the flow field inside and outside the net cage. Their findings revealed a robust negative correlation between drag force and cage volumes. Additionally, the research noted that the reduction in current speed inside the net cage is influenced by both bottom weights and incoming current speed.

A numerical model was built and used to calculate the cage volumes (Liu et al., 2021a) and mooring forces (Liu et al., 2021b). Moe Føre et al. (2021) conducted measurements of drag coefficients and velocity reduction for net materials with varying solidities. Dong et al. (2021) employed a method involving the insertion of vertical pipes within a cage to simulate a fish school. Their study focused on analysing the flow field within the cage and the drag forces. Miao et al. (2021) constructed a model of a semi-submersible offshore fish farm to investigate the interaction between the fish farm and waves. Liu & Guedes Soares (2022a and 2022b) built a physical model and measured the wave heights before and after the gravity cage model in a flume.

In this paper, the Finite Element Method (FEM) is applied to build a numerical model and explore the opportunity of applying gravity cage system around the Coast of Algarve. From the previous study, the high-tenacity polyester multifilament fibre has excellent performance for a gravity cage system, which is used as the mooring material in this work. Line elements including link elements, beam elements, pipe elements and non-linear spring elements are used to model the entire structure according to the different properties of the components. Data on the wave conditions from the coast of Algarve is used (Silva et al. 2018). The JONSWAP spectrum is adopted to model the irregular waves in the numerical model. The potential locations to apply the gravity cage systems are discussed in Section 2. The properties and details of the numerical model are shown in Section 3. The results are in Section 4 and the conclusion is in Section 5.

2 POTENTIAL LOCATIONS FOR GRAVITY CAGE SYSTEMS

With a consumption of 59,9 kg/inhabitant, Portugal stands out as the major EU consumer of fishery and aquaculture products and ranks 3rd in the world (topped only by Iceland and Japan) (Madsen and Chkoniya 2021, EU Commission 2023). According to the last data available, aquaculture production in Portugal reached 17900 tons, with a value of 162.8 million euros in 2021 (INE 2023).

At the end of 2021, there were 1252 establishments licensed for aquaculture in Portugal, and the total production reached 17,900 tons, with a value of 162.8 million euros. Most of the industry is devoted to the production of mollusc and bivalves (clams, oysters, scallops, and mussels), but also fish (turbot, seabream, seabass, flounder, corvina, bream, sargo, snapper and tuna). Most of the production is developed in earthen ponds and brackish waters, while only 2.2% of the total licensed establishments in 2021 were set up floating offshore (INE 2023).

Nonetheless, there is an increasing interest in the industry for investing in new offshore and open sea fish farming opportunities, as it accounts for wider availability of space, fewer licensing limitations, and higher production rates. Thus, the expansion of offshore aquaculture production has been set up as one of the main targets towards the blue economy in Portugal within the National Ocean Strategy (DGPM 2021) and the Aquaculture Strategy Plan (2021-2030) (DGRM 2022). Besides the already ongoing licenced areas, the Maritime Spatial Planning also contemplates different offshore areas that can be destined for potential aquaculture purposes in the future.

Based on different spatial, technical, and environmental criteria, the General Direction for Marine Resources (DGRM) delimited several areas as adequate for the exploitation of offshore aquaculture farms in Portugal. Those can be found in the national Geoportal of the Portuguese Sea (DGRM 2024). The areas are divided into parcels, some of them are currently licensed with an ongoing production of mostly bivalve molluscs, but also fish (TUPEM 2024). However, some areas are still under licensing or construction process, while others have been identified as areas of interest for the potential deployment of future aquaculture infrastructure. Overall, the total area proposed as available for aquaculture production extends over 95,89 km^2 (PSOEM 2019).

Figure 1 shows the offshore locations with existing or potential aquaculture exploitation in Portugal, while Table 1 contains some of their characteristics and species grown in each of those areas.

The south coast of the Algarve (south coast of Portugal), which is generally exposed to milder sea states and extreme climate events than the rest of the country's coasts, constitutes a specially interesting region for the development of offshore mollusc and fish farming.

However, the growth rate of the sector is highly dependent on new fish farm designs and cage technologies able to adapt to the rough sea conditions of the country and other techno-economic constraints, making possible the use of previously untapped spaces and achieving efficiency the operation by increasing the productivity and lowering the installation and intermediate costs.

Thus, for this study, the spatial and climate characteristics of a point within the areas designated as the potential for fish-farm exploitation in the most occidental side of the Algarve region have been used as a reference for analysis of potential deployment of the proposed fish cage design. The coordinates of the reference point are (37°N, 8°43′W). The water depth is 56 m. The significant wave height is 0.79 m, and the spectrum peak frequency is 5.94 s. The chosen reference point is shown in Figure 2.

Table 1. Characteristics of the aquaculture areas in southwest Algarve and reference point (PSOEM 2019, DGRM 2024).

Nr	Status	Location	AREA (km^2)	Species
0	Construction	Sagres	0.37	mollusc/bivalves (mussels)
1	Active	Sagres	1.47	mollusc/bivalves (mussels)
2	Active	Sagres	0.47	mollusc/bivalves (oyster)
4	Expired	Sagres	0.36	-
5	Construction	Ponta da Piedade	0.90	mollusc/bivalves (mussel/ oyster)
	Active	Alvor	3.96	mollusc/bivalves (mussels)
7	Active	Tavira	3.67	fish (tuna)
8	Active	Faro	4.45	fish (tuna)
9	Active	Barranco-Zavial	0.74	mollusc/bivalves (oyster)
10	Potential	Sagres	7.81	mollusc/bivalves (mussels)
11	Potential	Ponta da Piedade	15.86	mollusc/bivalves (mussels)
12	Potential	Ponta da Piedade	5.94	fish
13	Potential	Sagres	23.39	fish
14	Potential	VRSA	6.76	-
15	Under Construction/ Licensing	Cancela Velha	13.73	mollusc/bivalves
16	Active/Licensing/ Construction parcels	Olhão		mollusc/bivalves/fish (scallops, mussels, oysters, flounder, corvina, bream, sea bass, sargo, snapper)
17	Licensing	Aveiro	12.89	mollusc/bivalves
18	Potential	Aveiro	13.49	-
19	Active	Peniche	0.08	mollusc/bivalves
20	Potential	Peniche	0.15	-
21	Liscencing	Matosinhos	-	-
22	Liscencing	Espinho	-	-

Figure 1. Areas delineated as available for the exploitation of offshore aquaculture in Portugal (PSOEM 2019, DGRM 2024).

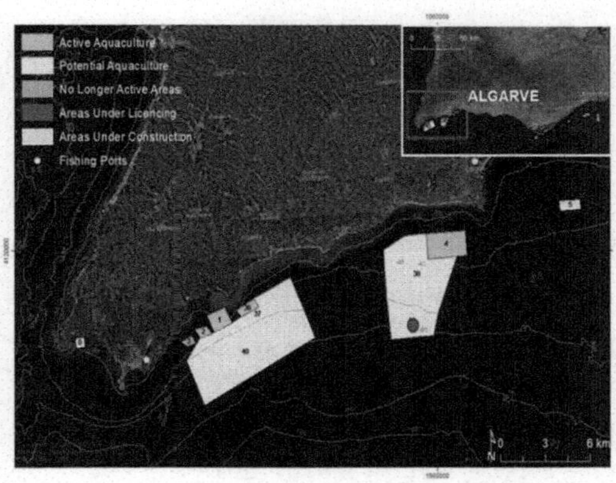

Figure 2. The reference location chosen for analysis lies within an area of interest for potential aquaculture (red dot).

3 NUMERICAL MODEL

A numerical model is constructed to compute mooring forces under the influence of currents and waves using FEM. The mooring material is high-tenacity polyester multifilament fibre, which is modelled by non-linear spring elements. The property of the material is plotted in Figure 3. Compared to traditional materials, this type of material exhibits greater elongation under the same tensile force and possesses higher break forces.

In addition, line elements are utilized to simulate all components. Depending on whether a section has bending stiffness and the cross-sectional shapes, beam, pipe, and link elements are employed for simulation.

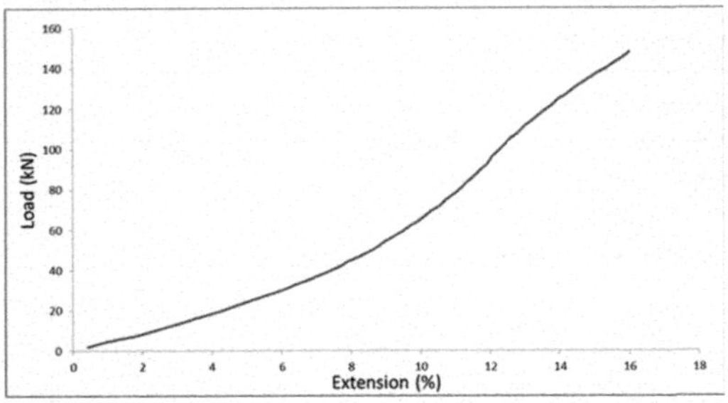

Figure 3. The load-extension curve of the high-tenacity polyester multifilament fibre.

For instance, the collar and handle sections, having a hollow cross-section, are modelled using pipe elements, while the bottom ring and reinforcing columns are modelled by beam elements, and the netting and ropes are simulated using link elements. Additional details regarding the numerical model can be found in Liu et al. (2022) and Liu and Guedes Soares (2023). The properties of the model are listed in Table 2. The numerical model is shown in Figure 4. The current velocity is 0.5 m/s in the calculation.

Furthermore, the potential location for the application of the gravity cage is shown in Figure 2. The latitude is 37° and the longitude is -8.7°. The water depth is 56 m. The significant wave height is 0.79 m and the spectrum peak frequency is 5.94 s. The irregular sea states are represented by the JONSWAP spectrum, which is given as:

$$s(f) = \beta_J H_S^2 \frac{f^{-5}}{f_p^{-4}} exp\left[-1.25\left(\frac{f_p}{f}\right)^4\right]\gamma^r \quad (1)$$

$$\beta_J = \left(\frac{0.06238}{0.23 + 0.0336\gamma - 0.185(1.9+\gamma)^{-1}}\right)(1.094 - 0.01915\ln(\gamma)) \quad (2)$$

$$r = \exp\left[-\frac{(f-f_P)^2}{2\mu^2 f_P^2}\right] \quad (3)$$

$$\mu = \begin{cases} 0.07, f \le f_P \\ 0.09, f > f_p \end{cases} \quad (4)$$

where γ is the peak enhancement factor and equal to 3.3 in this work, f is the wave frequency, f_P is the spectrum peak frequency and H_S is the significant wave height.

4 RESULTS

The calculation results are shown in this section. Section 3.1 shows the calculated mooring forces in the

Table 2. The dimensions and material parameters of the numerical model.

Component	Dimension/Parameter	Value
Handle	Density(kg/m^3)	450
	Height above the collar (m)	1.2
	Cross-section diameter (m)	0.08
	Thickness (mm)	15
Collar	Density(kg/m^3)	450
	Diameter(m)	40 for the inner collar 42.4 for the outer collar
	Cross-section diameter (m)	0.5
	Thickness (mm)	50
Net	Material type	Nylon
	Young's modulus (GPa)	1.4
	Twine thickness (mm)	2.7
	Solidity	0.225
	Density (kg/m3)	1140
	Depth (m)	12 for the cylinder part 3 for the tapered bottom
Bottom ring	Weight/length (kg/m)	32
	Diameter(m)	42.4
Counterweight at the top of the tapered bottom	Weight (kg)	150

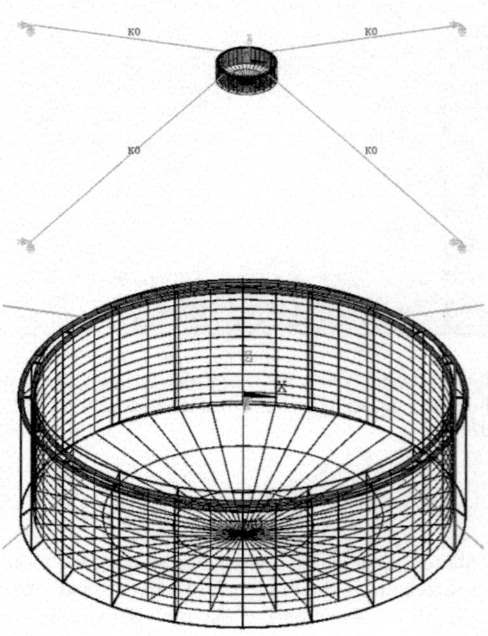

Figure 4. The numerical model of the gravity cage system.

time domain and the results of the amplitudes using the zero-down-crossing method. The comparison between the mooring forces on the coast of Algarve and the ones on the coast of Madeira and Ilha Terceira from the previous studies are shown in Section 3.2

4.1 *Mooring forces for a gravity cage system in the Algarve*

The mooring forces in the time domain are plotted in Figure 5. A total of 477 seconds of motion around the equilibrium position is calculated for the cage model. The maximum mooring force is 35.6 kN and the minimum value is 32.9 kN. The average mooring force is 34.4 kN.

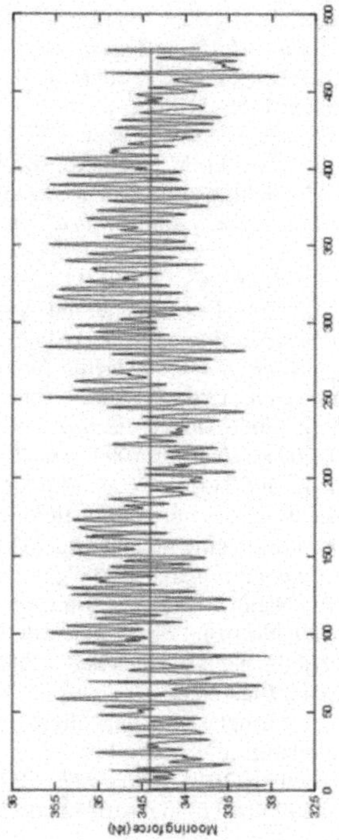

Figure 5. The mooring forces in the time domain.

In the zero-down-crossing method, a trough is defined as a period during which the force fluctuates from one average value to the next, consistently remaining below the average value throughout the duration. Conversely, a crest is defined as a period in which the mooring force consistently exceeds the average value. An entire period of mooring forces is defined as a trough and the following crest. The amplitude of the period is half of the difference between the maximum and minimum values in that period. The maximum amplitude of the mooring forces is 1.16 kN while the average amplitude is 0.56 kN.

The distribution of the amplitudes is shown in Fig. 6. 29.8% of the amplitudes fall within the range of 0.2 kN to 0.4 kN. 86.9 % of the amplitudes are less than 0.8 kN. The results can be used in further calculations for the fatigue life of the mooring ropes. However, due to the large computational load, the current calculation duration is insufficient. In addition, extreme wave conditions also need to be taken into account in future calculations.

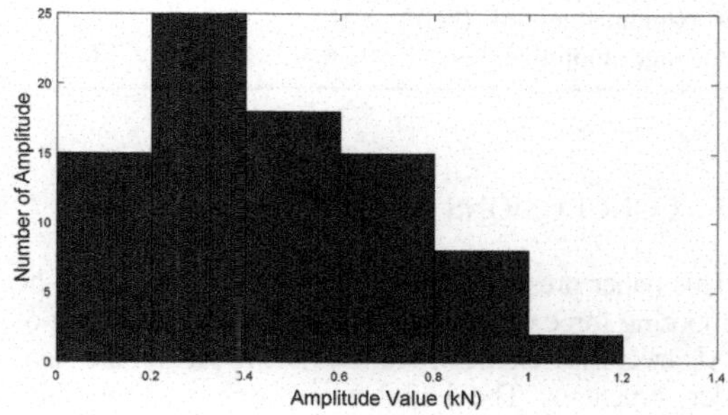

Figure 6. The distribution of the mooring forces on the coast of Algarve.

4.2 *Comparison between the mooring forces in the coast of Algarve, Madeira and Azores*

In this section, the calculated mooring forces on the coast of Algarve are compared with the results for the coast of Madeira and Azores from previous studies. The states of the three locations are listed in Table 3 and the results are in Table 4. Due to the similar water depths and mooring lengths, the average mooring forces on the coast of Algarve and Madeira are close. However, the significant wave height on the coast of Algarve is much smaller than the ones on the coast of Madeira. Therefore, the maximum amplitude of the mooring forces on the coast of Algarve is only 43 % of the one on the coast of Madeira, and the average amplitude on the coast of Algarve is 46.7 % of the one on the coast of Madeira.

On the other hand, due to the deeper seabed and the longer mooring ropes, the average mooring force on the coast of Azores is smaller than the ones on the coast of Algarve and Madeira. In addition, it also contributes to reducing the amplitudes of the mooring force. Even though the significant wave height on the coast of the Azores is approximately twice as much as the one on the coast of Algarve, the amplitudes of the two locations are close.

Table 3. States of the three locations on the coast of Algarve, Madeira and Azores.

Location	Algarve	Madeira	Azores
Latitude	37°N	32.73°N	38.65°N
Longitude	8.4°W	16.69°W	27.05°W
Water depth (h)	56 m	59.3 m	143.4 m
Significant wave height (Hs)	0.79 m	1.67 m	1.3 m
Spectrum peak frequency (Tp)	5.94 s	10.67 s	10.6 s

Table 4. Comparison of the mooring forces on the coast of Algarve, Madeira and Azores.

Location	Algarve	Madeira	Azores
Maximum value (kN)	35.6	37.5	34.1
Minimum Value (kN)	32.9	31.2	31
Average value (kN)	34.4	34.1	32.4
Maximum amplitude (kN)	1.16	2.7	1.3
Average amplitude (kN)	0.56	1.2	0.5

5 CONCLUSIONS

This paper presents a numerical model to calculate the mooring force of a gravity cage system on the coast of Algarve. Line elements are used the model all the slender structures. The high-tenacity polyester multifilament fibre is used as the mooring material and nonlinear spring elements are used to model the mooring ropes. Data on waves from January 2005 to July 2012 on the coast of Algarve is considered. The JONSWAP spectrum is used to model the irregular waves in the numerical model. The zero-down-crossing method is applied to count the amplitude of the mooring force and the distributions of mooring force amplitudes are calculated. The results are compared with previous calculations of the application of the same system on the coast of Madeira and the Azores. It has been concluded that:

1. When the current velocity is 0.5 m/s, the maximum mooring force is 35.6 kN, which is much smaller than the break force of the mooring ropes.
2. Due to the small significant wave height on the coast, the amplitudes of the mooring force are quite small. 86.9 % of the amplitudes are smaller than 0.8 kN.
3. The average mooring forces on the coast of Algarve and Madeira are close due to the similar water depth and mooring length. However, the amplitude on the coast of the Algarve is much smaller due to the low significant wave height. On the other hand, the average value and amplitudes on the coast of the Azores are smaller due to the longer mooring ropes.
4. The mooring system is simplified into four mooring lines in this work. Models that are closer to the real system need to be established. In addition, the velocity reductions inside and after the cages should be considered in future calculations.

ACKNOWLEDGEMENTS

The first and second authors have been funded by the Portuguese Foundation for Science and Technology (Fundação para a Ciência e a Tecnologia-FCT), through a doctoral fellowship under the Contracts nos. SFRH/BD/147178/2019 and SFRH/2020.06618.BD, respectively. This work contributes to the Strategic Research Plan of the Centre for Marine Technology and Ocean Engineering (CENTEC), which is financed by the Portuguese Foundation for Science and Technology (Fundação para a Ciência e Tecnologia - FCT) under contract UIDB/UIDP/00134/2020.

REFERENCES

Bai, X., Xu, T., Zhao, Y., Dong, G. and Bi, C., 2016. Fatigue assessment for the floating collar of a fish cage using the deterministic method in waves. *Aquacultural Engineering*, 74: 131–142.

Berstad, A.J. and Tronstad, H., 2005. Response from current and regular/irregular waves on a typical polyethylene fish farm. In Guedes Soares, C. Garbatov Y. & Fonseca, N. (eds), *Maritime Transportation and Exploitation of Ocean and Coastal Resources*: Taylor & Francis Group, London. 1189–1196.

DeCew, J., Fredriksson, D.W., Lader, P.F., Chambers, M., Howell, W.H., Osienki, M., Celikkol, B., Frank, K. and Høy, E., 2013. Field measurements of cage deformation using acoustic sensors. *Aquacultural Engineering*, 57: 114–125.

Dong, S, Park, S, Zhou, J, Li, Q, Yoshida, T. and Kitazawa, D. "Flow Field Inside and Around a Square Fish Cage Considering Fish School Swimming Pattern." *Proceedings of the ASME 2021 40th International Conference on Ocean, Offshore and Arctic Engineering. Volume 5: Ocean Space Utilization.* Virtual, Online. June 21–30, 2021. V005T05A003. ASME.

Dong, S., You, X. and Hu, F., 2021. Experimental investigation on the fluid–structure interaction of a flexible net cage used to farm Pacific bluefin tuna (Thunnus orientalis). *Ocean Engineering*, 226: 108872.

DGPM, 2021. National Ocean Strategy (2021-2030) [online]. Available from: https://www.dgpm.mm.gov.pt/_files/ugd/eb00d2_b2cf9034fcc84867be8d08d69435c3bc.pdf [Accessed 29 Jan 2024].

DGRM, 2022. Portuguese Aquaculture Strategy Plan (2021-2030) [online]. Available from: https://www.dgrm.pt/documents/20143/45612/PT_PEA_2021_2030.pdf/37c9c077-f248-ff56-3de9-0ffe12c89f89 [Accessed 29 Jan 2024].

DGRM, 2024. Geoportal do Mar [online]. Available from: https://www.dgrm.pt/geoportal [Accessed 1 Feb 2024].

Endresen, P. C., Fore, M., Fredheim, A., Kristiansen, D. and Enerhaug, B., 2013. Numerical Modelling of Wake Effect on Aquaculture Nets, *Proc. of 32nd International Conference on Ocean, Offshore and Arctic Engineering*, ASME Paper No. OMAE2013-11446.

EU Commission, 2023. The EU fish market – 2023 edition [online]. Available from: https://data.europa.eu/doi/10.2771/38507 [Accessed 28 Jan 2024].

FAO. 2020. *The State of World Fisheries and Aquaculture 2020.* Fisheries and Aquaculture Department, Food and Agriculture Organization of the United Nations. Publishing Policy and Support Branch, Office of Knowledge Exchange, Research and Extension, Rome, Italy.

Fu, S., Xu, Y., Hu, K. and Zhang, Y., 2013. Experimental investigation on hydrodynamics of floating cylinder in oscillatory and steady flows by forced oscillation test. *Marine Structures*, 34: 41–55.

Gansel, L., Oppedal, F., Birkevold, J. & Tuene, S., 2018. Drag forces and deformation of aquaculture cages—Full-scale towing tests in the field. *Aquacultural Engineering*, 81: 46–56.

INE, 2023. Estatistícas da Pesca 2022 [online]. Available from: https://www.ine.pt/xportal/xmain?xpid=INE&xpgid=ine_publicacoes&PUBLICACOESpub_boui=66322600&PUBLICACOESmodo=2 [Accessed 28 Jan 2024].

Kristiansen, T. and Faltinsen, O.M., 2015. Experimental and numerical study of an aquaculture net cage with floater in waves and current, *J. Fluids Struct*, 54: 1–26.

Lader, P., Dempster, T., Fredheim, A. & Jensen. F., 2008. Current induced net deformations in full-scale cages for Atlantic salmon (Salmo salar). *Aquacultural Engineering*, 38: 52–65.

Lee, C.W., Kim, H.S., Lee, G.H., Koo, K.Y., Choe, M.Y., Cha, B.J. & Jeong, S.J., 2005. Computation modeling of the moored flexible structures. In Guedes Soares, C. Garbatov Y. & Fonseca, N. (eds), *Maritime Transportation and Exploitation of Ocean and Coastal Resources*: Taylor & Francis Group, London. 1239–1243.

Liu, Z., Garbatov, Y. and Guedes Soares, C., 2021a. Numerical modelling of full-scale aquaculture cages under uniform flow. In Guedes Soares, C., Santos, T. A., Eds.; *Developments in Maritime Technology and Engineering; Taylor and Francis*: London, UK; Volume 2: 705–712.

Liu, Z., and C. Guedes Soares., 2022a. Preliminary experiments of the behaviour of circular gravity cage in linear waves. in: Guedes Soares, C. & Santos T. A., (Eds.). *Trends in Maritime Technology and Engineering*, London, UK: Taylor and Francis; Volume 2: 599–604.

Liu, Z. and Guedes Soares, C., 2022b. Experimental study of the behaviour of a circular gravity cage in linear waves. *Aquac. Eng*, 99: 102291.

Liu, Z. and Guedes Soares, C., 2023. Sensitivity analysis of a numerical model of the dynamics of gravity cages subjected to current and waves. *Ocean Eng*, Vol. 287: 115715.

Liu, Z., Mohapatra, S. and Guedes Soares, C., 2021b. Finite Element Analysis of the Effect of Currents on the Dynamics of a Moored Flexible Cylindrical Net Cage. *Journal of Marine Science and Engineering*, 9: 159.

Liu, Z., Wang, S., Guedes Soares, C., 2022. Numerical Study on the Mooring Force in an Offshore Fish Cage Array., *Journal of Marine Science and Engineering*, 10 (3): 331.

Liu, Z., Xu, H. and Guedes Soares, C., 2023. Experimental study on the mooring forces and motions of a fish cage under regular waves. *Ocean Engineering*, 280: 114612.

Madsen, A.O. and Chkoniya, V., 2021. Tendencies Regarding Fish Consumption-The Case of Portugal (Europe's Leader & 3rd in the World). *European Journal of Interdisciplinary Studies*, 7 (1): 12–35.

Miao, Y., Ding, J., Tian, C., Chen, X. & Fan, Y., 2021. Experimental and numerical study of a semi-submersible offshore fish farm under waves. *Ocean Engineering*, 225: 108794.

Moe Føre, H, Endresen, PC, & Bjelland, HV. "Load Coefficients and Dimensions of Raschel Knitted Netting Materials in Fish Farms." *Proceedings of the ASME 2021 40th International Conference on Ocean, Offshore and Arctic Engineering. Volume 6: Ocean Engineering.* Virtual, Online. June 21–30, 2021. V006T06A044. ASME.

Moe, H., Fredheim, A. & Heide, M.A., 2005. New net cage designs to prevent tearing during handling. In Guedes Soares, C. Garbatov Y. & Fonseca, N. (eds), *Maritime Transportation and Exploitation of Ocean and Coastal Resources*: Taylor & Francis Group, London. 1265–1272.

Mohapatra, S. C., Bernardo, T. A., & Guedes Soares, C. 2021. Dynamic wave induced loads on a moored flexible cylin-drical net cage with analytical and numerical model simulations. *Applied Ocean Research*, 110: 102591.

PSOEM, 2019. Espacialização de Usos e Atividades - Continente e Plataforma Continental Estendida [online]. Available from: https://webgis.dgrm.mm.gov.pt/portal/sharing/rest/content/items/5521c32624d94ab4b59403bebab1e011/data [Accessed 29 Jan 2024].

Silva, D.; Martinho, P., and Guedes Soares, C. 2018; Wave energy distribution along the Portuguese continental coast based on a thirty three years hindcast. *Renewable Energy*, 127(4):1067–1075.

TUPEM, 2024. Títulos de utilização privativa do espaço marítimo [online]. Available from: https://www.psoem.pt/tupem-2/ [Accessed 28 Jan 2024].

Advances in Maritime Technology and Engineering – Guedes Soares & Santos (Eds)
© 2024 The Author(s), ISBN 978-1-032-83107-7

Numerical modeling of a steel-framed offshore fish cage: Code development and verification

Y. Ma, L. Li & M.C. Ong
Department of Mechanical and Structural Engineering and Materials Science, University of Stavanger, Stavanger, Norway

Z. Jiang
Department of Engineering Sciences, University of Agder, Grimstad, Norway

ABSTRACT: Many existing technologies of commercial fish farming involve the utilization of large-volume offshore fish cages positioned with catenary mooring lines. This necessitates the refinement of design guidelines and analysis tools specific to these steel-frame structures, as well as fish farms arrayed with such structures. Although certain standard tools offer high-fidelity computational capabilities, their lack of flexibility poses obstacles for efficient design and optimization of complex fish cage arrays. To complement existing simulation codes, a swift modeling and simulation code in MATLAB is proposed in this work. In the code, an analytical method is used for modeling the quasi-static behavior of catenary line. Apart from this, the time-domain equations of motion for a floating body also include the Morison and the potential flow hydrodynamics for the hull, and the screen model for fish net hydrodynamics. Verification of the code involves comparisons with numerical results from commercial software for a case study on a semi-submersible fish cage. The results show the applicability of the proposed numerical approach for preliminary design of offshore fish farm systems. This efficient code can be extended to evaluate complex fish cage array configurations.

1 INTRODUCTION

Large-scale offshore fish farming has obtained traction since the trial operation of projects such as semi-submersible fish cages 'Ocean Farm 1' (Drønen, 2023) and 'Shenlan 1' (Evans, 2019), as well as vessel-shape fish farm 'Havfarm' (Lysvold, 2021). Many thousand tonnes of fish can be raised up inside these giant structures at offshore areas, where fast ocean current helps to improve water quality and reduce infections (Outram, 2021). These fish farms are featured by steel hulls designed to withstand possible rough sea conditions at offshore farming sites. The operational insights gained from such new-type aquacultural solutions are catalyzing increased interest and investment in this sector.

With growing experience gained from trial operations of such semi-submersible steel-framed fish cage structures, there is an anticipated expansion of the larger-scale remote offshore fish farming market. The arrangement of fish farms using such structures, exemplified in Figure 1, emerges as a notable site planning solution.

The design of such fish-cage systems requires both physical model experiments and appropriate numerical simulations to ensure the structural reliability. Bore, Amdahl and Kristiansen (2017) developed simulation-based approach in the finite element (FE) package USFOS for a thorough examination on structural response of the frame of the 'Ocean Farm 1' cage, focusing on its ultimate limit state (ULS) design purpose. Moreover, Yu *et al.* (2019) expanded upon this numerical model to simulate accidental limit state scenarios of ship collisions. These studies provide insights into the assessment of reliability of fish cage hull structure.

Apart from the hull structure, a safe mooring system design is also needed. Li *et al.*, (2019) devised an optimization framework tailored specifically for the mooring system of a vessel-shape fish farm, employing a comparable numerical analysis model. Additionally, a surrogate method is employed to establish a meta model, which is used within a gradient-based optimization framework. However, when analyzing the mooring system for fish cages organized in an array-like configuration, as depicted in Figure 1, establishing meta model based on a large number of FE-model-based time domain analysis can prove challenging in terms of time consumption. This necessitates a faster calculation of mooring force and accurate estimation of wave and

DOI: 10.1201/9781003508779-61

current loads to enable swift and precise preliminary designs of the mooring system. Optimization task can be thereby accelerated in the early phase before implementing more-detailed design within a numerical model with higher fidelity.

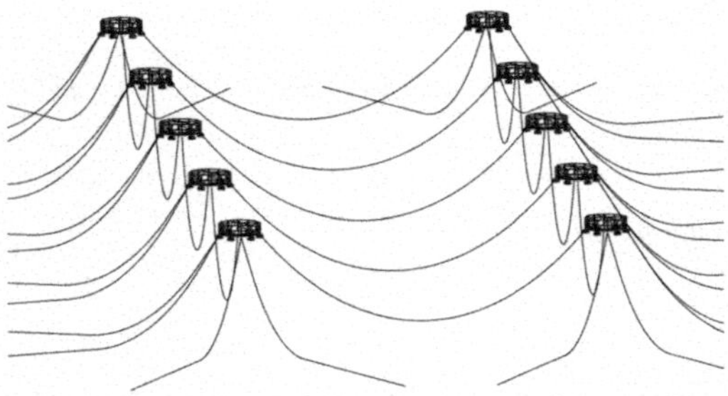

Figure 1. Example of a 10-cage shared mooring system.

Quasi-static mooring model is deemed as one efficient tool, which comes from the expressions deducted by Irvine (2022). This model describes a continuous constructive relationship and provides computational efficiency, which has been used in some analytical tools (Masciola, Jonkman and Robertson, 2013; Orcina, 2017). This model underestimates wave-frequency and other high-frequency fairlead tension response, but agrees well in the prediction of low-frequency mooring line tension response and dynamic platform motions (Hall, Buckham and Crawford, 2014). For a fish cage system, on which low-frequency tension response is important (Jin *et al.*, 2021; Ma *et al.*, 2023), quasi-static mooring model can be used for fast simulations (Hall *et al.*, 2022) or eigenvalue evaluations (Liang, Jiang and Merz, 2021; Wilson *et al.*, 2021). This method is still suitable for design tasks, where a larger safety factor is recommended (DNV, 2015; Chu *et al.*, 2023). Therefore, the numerical model for offshore fish farm system with a quasi-static mooring system can be used for fast parametric iterations in the design and optimization tasks. A customized in-house is developed based on this idea.

In this study, prior to applying the analysis to a complex fish cage array, the proposed numerical approach is verified through modeling a single 'Ocean Farm 1' structure as shown in Figure 2. This is to prove the applicability of the proposed numerical scheme. Since Jin *et al.* (2021) and Ma *et al.* (2023) developed the numerical model of 'Ocean Farm 1' fish cage using numerical simulation application SIMA (Sintef Ocean, 2022), this commercial software is used for verifying our in-house code.

In Section 2, we elaborate on the theory and methodology applied. Following this, Section 3 outlines the numerical setups utilized for code verification. Subsequently, Section 4 delves into the discussion of the verification results. Relevant conclusions and recommendations are discussed in Section 5.

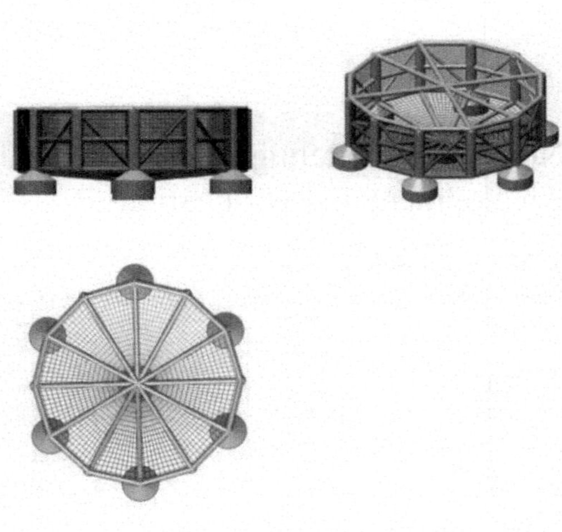

Figure 2. 3D image of the 'Ocean Farm 1' Structure.

2 THEORY & METHODOLOGY

A customized code package has been developed to deal with time-consuming tasks including mooring optimization and system layout determinations. The workflow of the proposed numerical approach is outlined in Figure 3. The fundamental objective involves assembling and solving equations of motions, specifically a set of ordinary differential equations (ODEs). These equations are currently solved using customized explicit ODE solver, such as Verlet integration (Verlet, 1967) or Runge-Kutta family (Butcher, 1963). Throughout the process, while determining the velocities and displacements of the floating unit, relevant force vectors encompassing memory effect of fluid structure interaction, drag force and mooring force are updated iteratively. Section 2.1 elaborates on the specifics of the time-domain equations of motion, followed by a discussion of the hydrodynamic force model in Section 2.2. The quasi-static mooring model is subsequently expanded in Section 2.3.

2.1 *Time-domain modeling*

The dynamic equilibrium of an arbitrary floating system in a time-domain manner can be written as the form in Equation 1. $\boldsymbol{M}$ is the mass matrix of the floating structure; $\boldsymbol{A}_\infty$ is the added mass matrix and $\boldsymbol{B}$ is the damping matrix. $\boldsymbol{C}$ is the linear stiffness matrix, which usually contains hydrostatic stiffness terms. $\overrightarrow{f_{exc}}$ is the vector of excitation force.

$$(\boldsymbol{M}+\boldsymbol{A}_\infty)\,\ddot{\vec{x}}(t)+\boldsymbol{B}\dot{\vec{x}}(t)+\int_{-\infty}^{\infty}\boldsymbol{k}(t-\tau)\,\dot{\vec{x}}(\tau)d\tau+\boldsymbol{C}\vec{x}(t)=\overrightarrow{f_{ext}}(t) \tag{1}$$

The term $\int_{-\infty}^{\infty}\boldsymbol{k}(t-\tau)\,\dot{\vec{x}}(\tau)\,d\tau$ is the impulse response function derived from Cummins's equation (Cummins, 1962), representing the convolution nature of the time-domain analysis. $\vec{x}(t)$ denotes the

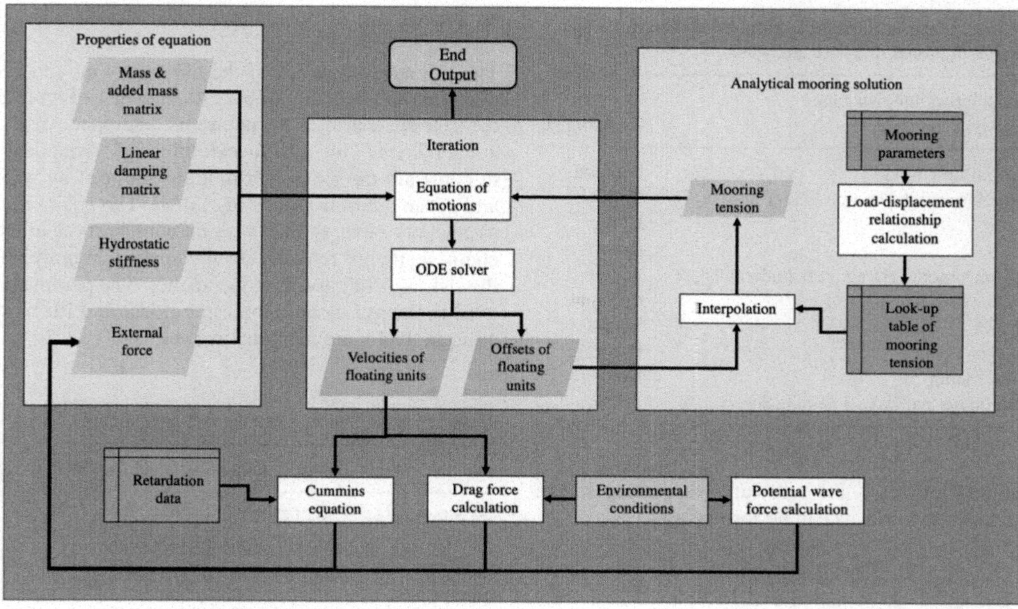

Figure 3. Workflow of proposed numerical simulation approach.

motion in time domain, and $\boldsymbol{k}(t-\tau)$ is the retardation function in time domain, which is also a matrix. Terms $\boldsymbol{A}_\infty$, $\boldsymbol{B}$ and $\boldsymbol{k}(\boldsymbol{t}-\tau)$ are estimated from frequency-dependent added mass and potential damping. The general external force includes the contribution from hydrodynamic load and restoring load from mooring lines. The equation can be reshaped into

$$(\boldsymbol{M}+\boldsymbol{A}_\infty)\,\ddot{\vec{x}}(t)+\boldsymbol{B}\,\dot{\vec{x}}(t)+\boldsymbol{C}\vec{x}(t)=\overrightarrow{f_{wave}}(t)+\overrightarrow{f_{drag}}(t)+\overrightarrow{f_{moor}}(t)-\int_{-\infty}^{\infty}\boldsymbol{k}(t-\tau)\,\dot{\vec{x}}(\tau)d\tau \tag{2}$$

2.2 *Hydrodynamic model*

The wave load acting on the main frame of the offshore fish cage is defined as a diffraction problem (Faltinsen, 1993; Newman, 2018). Given an arbitrary sine wave

$$\zeta=e^{kz}\zeta_a sin(\omega t-kx\,cos\beta-ky\,sin\beta+\in) \tag{3}$$

ζ_a is the wave amplitude; ω is the wave frequency; k is the wave number; ϵ is the phase angle in radius, and β is the wave propagation direction in XY plane. The wave load can be determined using

$$\overrightarrow{f_{wave}}(\mathrm{t})=e^{kz}\zeta_a sin(\omega t-kx\,cos\beta-ky\,sin\beta++\vec{\alpha}(\omega,\beta))\odot\overrightarrow{X_a}(\omega,\beta) \tag{4}$$

$\overrightarrow{X_a}$ and $\vec{\alpha}$ are the amplitude and phase angle of the wave excitation force, which are vectors with same dimension as $\vec{x}(t)$. These terms are dependent on $\omega$and β. Notably, transfer functions of wave excitation, frequency-dependent added mass and damping are estimated based on potential flow theory.

In addition to the potential load, viscous load effect is also important. The viscous force imposed on the cylinders and beams of the fish cage hull and can be solved using Morison's equation (Morison, Johnson and Schaaf, 1950). The drag force calculation in a discretized form for a moving body in oscillatory flow is noted as

$$\overrightarrow{f_{drag}}=\frac{1}{2}\rho C_D dA\cdot\overrightarrow{u_{rel}}\cdot\left|\overrightarrow{u_{rel}}\right| \tag{5}$$

Equation 5 can be adjusted to describe the loads on the fish net panels. The drag and lift coefficients are used in the screen model (Løland, 1991; Cheng *et al.*, 2020). The drag and lift coefficients are expressed as the functions of solidity ratio (S_n) and induction angle of fluid (θ). The expressions are listed in Equation 6, which represent the screen model proposed by Løland (1991).

$$C_D(S_n,\theta)=0.04-\left(-0.04+0.33S_n+6.54S_n^2-4.88Sn^3\right)cos(\theta)$$

$$C_L(S_n,\theta)=\left(-0.05S_n+2.3S_n^2-1.76S_n^3\right)sin(2\theta) \tag{6}$$

The hydrodynamic models discussed forehead is summarized in Table 1.

2.3 *Mooring model*

When a catenary line has two types of material as shown in Figure 4, the relationship between the 2D

Table 1. Numerical model applied to offshore fish cage structure in global response analysis.

Semi-submersible fish cage Hydrodynamic model	Components
Potential flow model (Radiation and diffraction loads)	Pontoons Columns Beams Braces
Morison's model (Drag on cylinders)	Pontoons Columns Beams Braces
Screen model (Drag on net panels)	Nets

configuration ($x(s)$ and $z(s)$) and supporting forces (H and V) at fairleads can be formed as

$$x(s) = \frac{Hs}{E_1A_1} + \frac{H}{w_1} sinh^{-1}\left(\frac{V - w_2s_2}{H}\right) + \frac{Hs_2}{E_2A_2} + \frac{H}{w_2}\left[sinh^{-1}\left(\frac{V}{H}\right) - sinh^{-1}\left(\frac{V - w_2s_2}{H}\right)\right]$$

$$z(s) = \frac{w_1s}{E_1A_1}\left[\frac{V - w_2s_2}{w_1s_1} - \frac{1}{2}\right] + \frac{H}{w_1}\left[\sqrt{1 + \left(\frac{V - w_2s_2}{H}\right)^2} - 1\right] + \frac{w_2s_2^2}{E_2A_2}\left[\frac{V}{w_2s_2} - \frac{1}{2}\right] + \frac{H}{w_2}\left[\sqrt{1 + \left(\frac{V}{H}\right)^2} - \sqrt{1 + \left(\frac{V - w_2s_2}{H}\right)^2}\right] \tag{7}$$

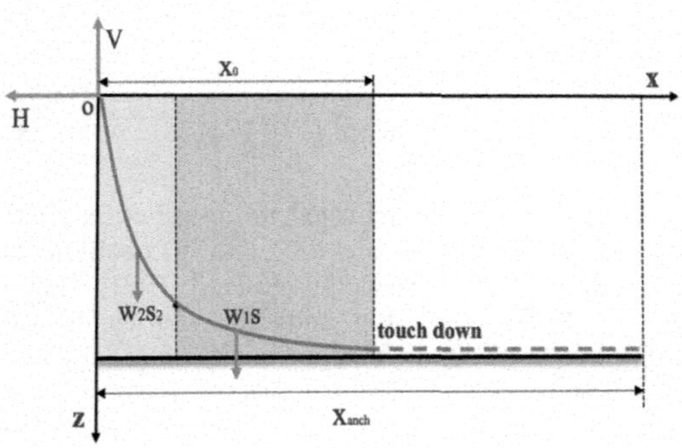

Figure 4. Illustration of analytical mooring model.

In Equation 7, w_1 and w_2 are the sectional weight of two segments with different materials. s_1 and s_2 are the length of two segments. The lifted section of the second segment can be expressed as $s = s_1 - s_{TD}$; s_{TD} denotes the sectional length of the touch-down part. E_1 and E_2 are the elastic modules of the two segments, and A_1 and A_2 are the cross-sectional areas. If removing the expressions involving w_2 and s_2, the equations can be used for modeling single-segment mooring line or other category of cable structure.

3 NUMERICAL SETUPS

The primary properties of the fish cage hull and fish nets are included in Table 2. As shown in Figure 5, the 'Ocean Farm 1' structure is equipped with 8 anchor lines. This design exhibits high redundancy to constrain the slow-drift motions induced by wave and ocean currents (Ma *et al.*, 2023). The positioning of anchors ensures that a significant portion of the chain segments remains on the seabed, minimizing the risk of lifting towards the anchor. The parameters used in the quasi-static mooring model and FE modeling in SIMA are included in Table 3.

Table 2. Information about the fish cage model.

Structure type	Semi-submersible steel-framed fish cage
Mass& Inertia	M: 12840 [Tonnes] I_{xx} & I_{yy}: $1.5x10^{10}$ [kgm^4]
Cylinder structures:	12 side vertical columns 38 horizontal beams 38 cross braces 1 central column Drag coefficient (C_D): 1.2
Fish nets:	12 side vertical net panels: 28x34 [m^2] 12 cone net panels: 0.5x28x53 [m^2] 1 central net panel: 110x34 [m^2] Solidity ratio (S_n): 0.162

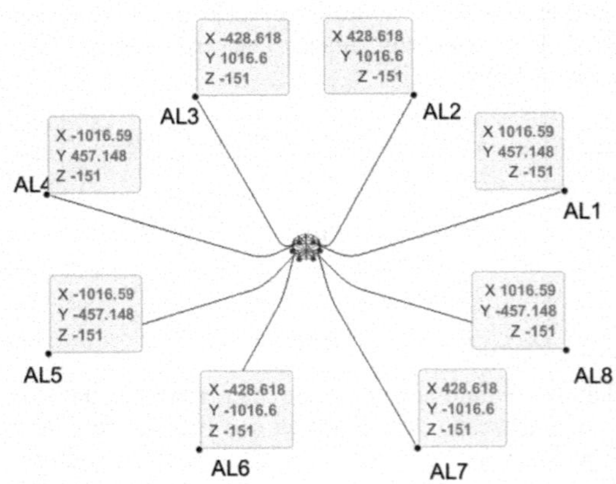

Figure 5. Fish cage with its mooring system.

Other properties including added mass matrix, linear damping matrix, hydrostatic restoring stiffness matrix as well as retardation functions are attached in the Appendix.

The numerical tests conducted in this study are detailed in Table 4. A sequence of pull-out tests is executed to demonstrate the capability of our analytical model in estimating surge and sway stiffness. Subsequently, decay tests are conducted to verify the correctness of implementing and solving Equation 1. The regular wave test is performed to present the

Table 3. Mooring system information.

Mooring system	8 anchor lines 2 segments: chain + rope
Segment 1 - Chain:	Length: 1000 [m] Submerged weight: 980 [N/m] EA: 5.08×10^9
Segment 2 - Rope:	Length: 100 [m] Submerged weight: 29.4 [N/m] EA: 1.47×10^9

wave excitation representation. Lastly, current tests are executed to show the predictive capability of drag force within our numerical framework.

Table 4. Numerical test matrix.

Test type	Setups	Purpose
Pull out test	In-house code: – Potential damping of hull only – FE mooring model SIMA: – Potential damping of hull only – Analytical mooring model	Verify the stiffness provided by mooring lines from the analytical mooring model
Decay test	Surge: 1200 [kN] Heave: 12000 [kN] Pitch: 120000 [kNm] In-house code & SIMA: – Potential damping of hull only – Linearized mooring stiffness	Verify the natural frequencies and the potential damping of the hull structure
Wave test	Wave elevation: 1 [m] Wave period: 10 [s] In-house code & SIMA: – Potential damping of hull only – Linearized mooring stiffness	Verify the wave model, wave forces and global responses under regular waves.
Current test	Uc = 0.25, 0.5, 0.75 [m/s] Uniform current profile In-house code: – Drag force of cylinders and fish nets calculated using slaved nodes – Linearized mooring stiffness SIMA: – Drag force of cylinders and fish nets calculated using FE beam elements – Linearized mooring stiffness	Verify the hydrodynamic drag force model for hull structure and fishing nets

4 RESULTS & DISCUSSIONS

4.1 *Pull-out test*

A series of Pull-out tests, as displayed in Figure 6, is performed to test the stiffness in surge and sway motions provided by the mooring system. The obtained load-displacement relationship is shown in Figure 7. The load-displacement relationship using analytical catenary equations matches the results from SIMA using FE mooring model and measured results of Jin *et al.* (2021). For small oscillations, the linearized surge and sway stiffness is approximately 3.7×10^4 N/m. This linearized stiffness is used in the following decay test and regular wave test to eliminate the numerical differences due to different numerical models.

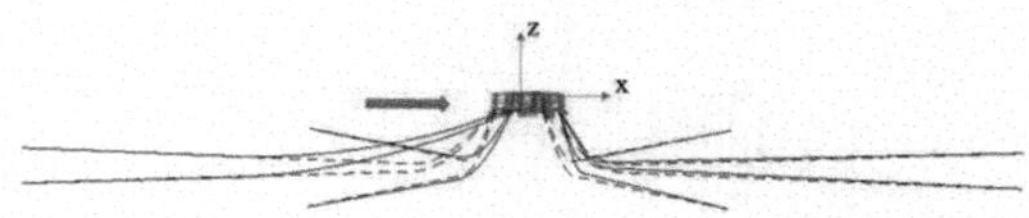

Figure 6. Illustration of pull-out test.

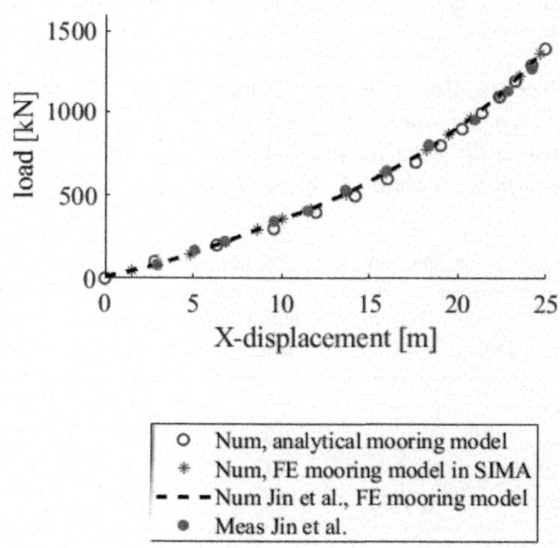

Figure 7. Results of pull-out test.

4.2 *Decay test*

The decay test is performed to the floating ofshore fish cage to validate the modeling and solving of the dynamic system as noted in Equation 1. The natural periods claculated using the in-house code and SIMA are compared with the results of Jin *et al.* (2021) in Table 5. This shows that the modeling in both our in-house code and SIMA is reasonable.

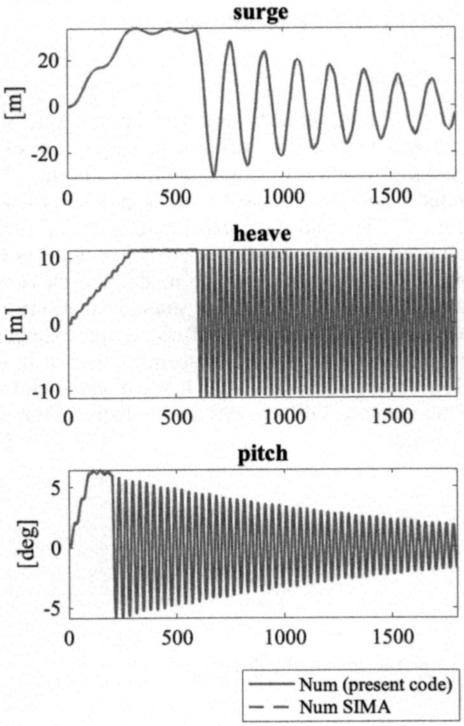

Figure 8. Results of free decay tests to the hull structure without fishing nets.

The results in Figure 7 shows good agreement between the numerical model in SIMA and our in-house code. This means that the solving of the free-decay process is correctly performed.

Table 5. Comparison of natural periods.

Model\DoF	Natural periods [s] Surge	Heave	Pitch
Num	170	29	30
Num SIMA	170	29	30
Num Jin et al. (2021)	173	27	31.5
Meas Jin et al. (2021)	171	27	31.5

4.3 *Regular wave test*

For verifying the wave excitation calculation, a fundamental regular wave test is advocated for comparing time series of solved wave excitation forces and global responses. A linear airy wave series is generated using Equation 3, as shown in

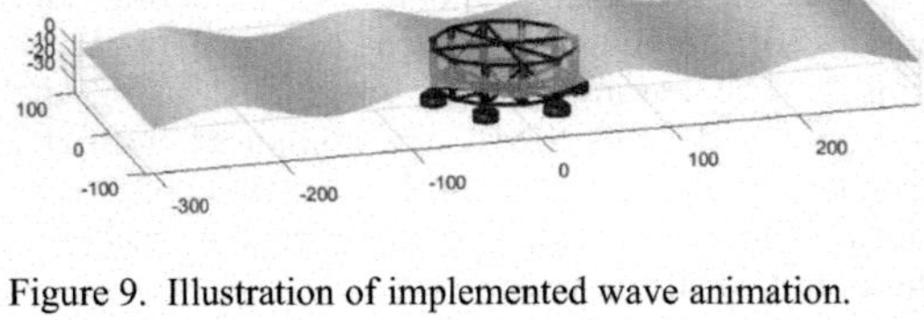

Figure 9. Illustration of implemented wave animation.

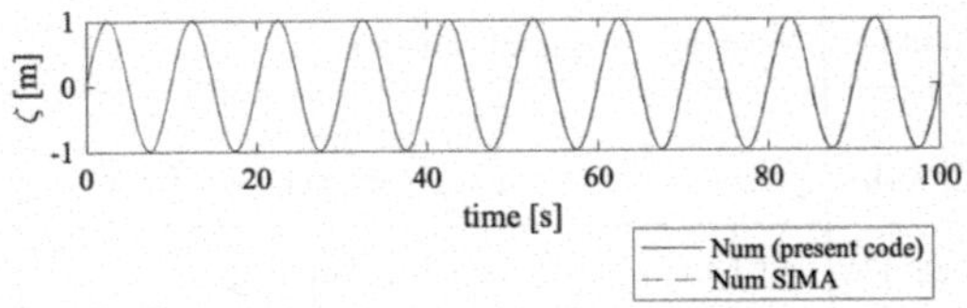

Figure 10. Wave elevation time series at initial condition.

Figure 9. The generated wave at the local coordinate origin of the fish cage is compared with that in SIMA in Figure 10.

Using Equation 4, the wave excitation force can be obtained as displayed in Figure 11. The wave excitation forces obtained from our code demonstrate strong agreement with those obtained from SIMA.

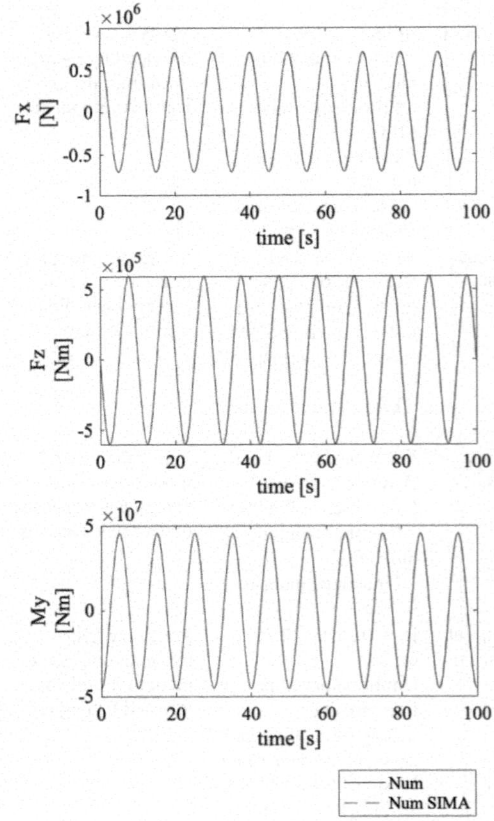

Figure 11. Wave forces in 0-degree direction.

The results of global motion time series are solved and compared in Figure 12. In general, the solved motions using the in-house code agree well with those from SIMA.

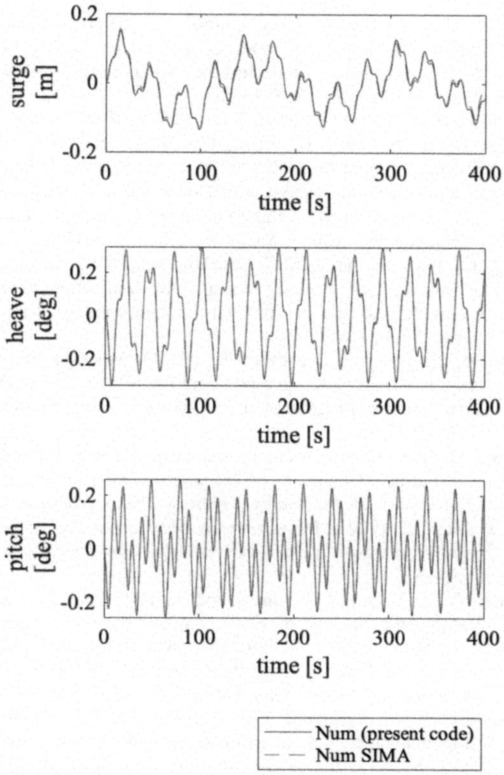

Figure 12. Dynamic motion time series under wave excitation.

4.4 *Current test*

The current test is performed to test the implementation of drag calculation for both cylinder components of the hull structure and fish nets. A uniform current profile is set up as shown in Figure 13.

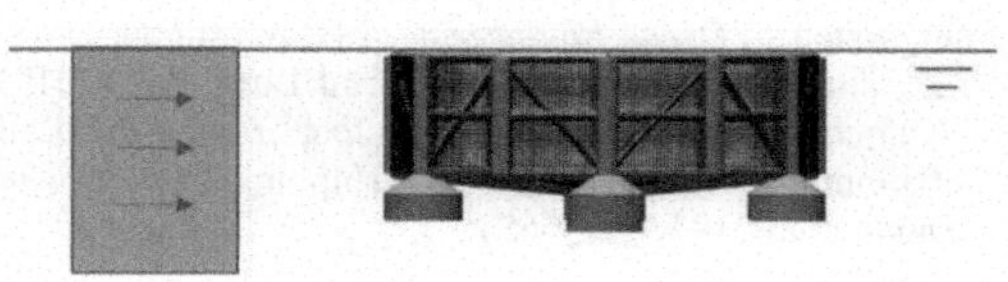

Figure 13. Uniform current profile.

In general, the results using the in-house code show good agreements with the results from SIMA toolbox, as presented in Figure 14 and Table 6. The in-house code captures the relationship between drag force and current velocity. The difference between the results of the self-developed tool and those of SIMA application is possibly due to difference in numerical discretization.

Table 6. Current test results.

Current speed [m/s]	Displacement [m] 0.25	0.5	0.75
Num (present code). drag on nets only	1.572	6.280	14.13
Num SIMA. drag on nets only	1.571	6.255	14.07
Num. (present code). drag on nets & cylinders	3.127	12.51	28.15
Num SIMA. drag on nets & cylinders	3.106	12.42	27.95

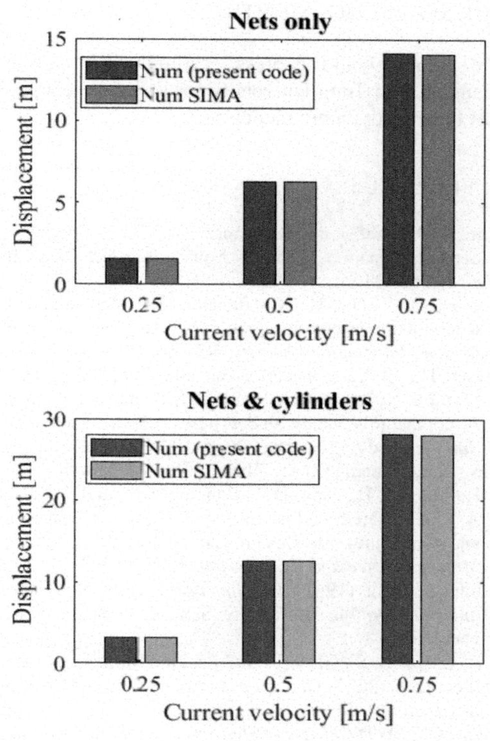

Figure 14. Comparison of current test results.

5 CONCLUSIONS & FUTURE WORKS

This study presents the development of an in-house numerical code for the purpose of having fast simulations of offshore fish farm systems. The toolbox incorporates quasi-static mooring reactions, integrating Morison's model and potential flow hydrodynamic model for the hull structure, alongside the screen model for simulating fishing net hydrodynamics.

The numerical modeling of the mooring line in a quasi-static approach aligns closely with the finite element model during pull-out tests. Additionally, decay tests demonstrate accurate numerical solutions for the equations of motions, encompassing potential-flow added mass and damping, mooring stiffness, as well as hydrostatic stiffness. Regular wave tests show reasonable solution of wave excitation loads and wave excited motions. Finally, the present tests yield an acceptable approximation of drag forces acting on both fish nets and cylindrical structures.

Acknowledging the toolbox's limitations, we recognize the need for further development to encompass complex scenarios involving drag loads from coupled wave and current, which hold significant importance in various load cases. Furthermore, refining the modeling to implement a more accurate representation of the physical current profile is crucial for enhancing the precision of drag load predictions.

ACKNOWLEDGEMENTS

The author would express gratitude to J. Jin in Shanghai Jiao Tong University for the technical support to hydrodynamic modeling.

REFERENCES

Bore, P., Amdahl, J. and Kristiansen, D. (2017) 'Modelling of Hydrodynamic Loads on Aquaculture Net Cages by a Modified Morison Model'.

Butcher, J.C. (1963) 'Coefficients for the study of Runge-Kutta integration processes', *Journal of the Australian Mathematical Society*, 3(2), pp. 185–201.

Cheng, H., Li L., Aarsæthe, KG. and Ong, MC. (2020) 'Typical hydrodynamic models for aquaculture nets: A comparative study under pure current conditions', *Aquacultural Engineering*, 90, p. 102070.

Chu, Y.I., Wang, C.M., Zhang, H., Abdussamie, N., Karampour, H., Jeng, D.S., Baumeister, J. and Aland, P. A. (2023) 'Offshore Fish Farms: A Review of Standards and Guidelines for Design and Analysis', *Journal of Marine Science and Engineering*, 11(4), p. 762.

Cummins, W.E. (1962) *The Impulse Response Function and Ship Motions*. Institut flir Schiffbau der Universitit Hamburg.

DNV (2015) 'DNVGL-OS-E301 Position Mooring.' DNV.

Drønen, O.A. (2023) *Ocean Farm 1 cost owner NOK 100m last year.*

Evans, O. (2019) *Chinese offshore mega salmon farm to be erected this year, SalmonBusiness.*

Faltinsen, O. (1993) *Sea Loads on Ships and Offshore Structures*. Cambridge University Press.

Hall, M., Housner, S., Zalkind, D., Bortolotti, P., Ogden, D. and Barter, G. (2022) 'An Open-Source Frequency-Domain Model for Floating Wind Turbine Design Optimization', *Journal of Physics: Conference Series*, 2265(4), p. 042020.

Hall, M., Buckham, B. and Crawford, C. (2014) 'Evaluating the importance of mooring line model fidelity in floating offshore wind turbine simulations', *Wind Energy*, 17(12), pp. 1835–1853.

Irvine, H.M. (1981) *Cable structures (Structural Mechanics).*

Jin, J., Su, B., Dou, R., Luan, C., Li, L., Nygaard, I., Fonseca, N. and Gao, Z. (2021) 'Numerical modelling of hydrodynamic responses of Ocean Farm 1 in waves and current and validation against model test measurements', *Marine Structures*, 78, p. 103017.

Li, L., Jiang, Z., Ong, M.C. and Hu, W. (2019) 'Design optimization of mooring system: An application to a vessel-shaped offshore fish farm', *Engineering Structures*, 197, p. 109363.

Liang, G., Jiang, Z. and Merz, K. (2021) 'Mooring Analysis of a Dual-Spar Floating Wind Farm With a Shared Line', *Journal of Offshore Mechanics and Arctic Engineering*, 143(6).

Løland, G. (1991) 'Current Forces on and Flow through Fish Farms', *Univeristy of Trondheim, Doctors Thesis, MTA Report 78, Institute for Hydrodynamics, Norway.*

Lysvold, S.S. (2021) *Havfarmen til Nordlaks skulle berge oppdrettsnæringa – i deler av anlegget dør hver fjerde laks, NRK.*

Ma, Y., Li, L., Ong, M.C., Jin, J. and Su, B. (2023) 'Design and dynamic analysis of a co-existence solution between a semi-submersible offshore fish cage and a spar-type floating offshore wind turbine', *Journal of Offshore Mechanics and Arctic Engineering*, pp. 1–30.

Masciola, M., Jonkman, J. and Robertson, A. (2013) 'Implementation of a Multisegmented, Quasi-Static Cable Model', in. *The Twenty-third International Offshore and Polar Engineering Conference*, One Petro.

Morison, J.R., Johnson, J.W. and Schaaf, S.A. (1950) 'The Force Exerted by Surface Waves on Piles', *Journal of Petroleum Technology*, 2(05), pp. 149–154.

Newman, J.N. (2018) *Marine Hydrodynamics*. The MIT Press.

Orcina (2017) *OrcaFlex 11.3a Documentation.*

Outram, R. (2021) 'SalMar says Ocean Farm has been a success', *Fish Farmer Magazine*, 2 March.

Sintef Ocean (2022) *SIMA Documentation.*

Verlet, L. (1967) 'Computer "Experiments" on Classical Fluids. I. Thermodynamical Properties of Lennard-Jones Molecules', *Physical Review*, 159(1), pp. 98–103.

Wilson, S., Hall, M., Housner, S. and Sirnivas, S. (2021) 'Linearized modeling and optimization of shared mooring systems', *Ocean Engineering*, 241, p. 110009.

Yu, Z., Amdahl, J., Kristiansen, D. and Bore, P.T. (2019) 'Numerical analysis of local and global responses of an offshore fish farm subjected to ship impacts', *Ocean Engineering*, 194, p. 106653.

APPENDIX

Table A1. Added mass matrix at infinity frequency in 6 DoFs.

A∞	1	2	3	4	5	6
1	9.3×10^6 [kg]					
2		9.3×10^6 [kg]				
3			7.3×10^6 [kg]			
4				1.49×10^{10} [kgm^2]		
5					1.49×10^{10} [kgm^2]	
6						2.08×10^{10} [kgm^2]

Table A2. Linear damping matrix in 6 DoFs.

B	1	2	3	4	5	6
1	44283 [kg/s]					
2		44283 [kg/s]				
3			2602 [kg/s]			
4				$3.6\text{x}10^6$[kgm^2/s]		
5					$3.6\text{x}10^6$[kgm^2/s]	
6						$1.18\text{x}10^9$[kgm^2/s]

Table A3. Hydrostatic restoring stiffness matrix in 6 DoFs.

C	1	2	3	4	5	6
1						
2						
3			$1.08\text{x}10^6$ [N/m]	$-2.52\text{x}10^5$ [N]		
4			$-2.52\text{x}10^5$ [N]	$1.11\text{x}10^9$ [Nm]	$-3.15\text{x}10^6$ [Nm]	$7.62\text{x}10^6$ [Nm]
5				$-3.15\text{x}10^6$ [Nm]	$1.11\text{x}10^9$ [Nm]	$4.59\text{x}10^6$ [Nm]
6				$7.62\text{x}10^6$ [Nm]	$4.59\text{x}10^6$ [Nm]	

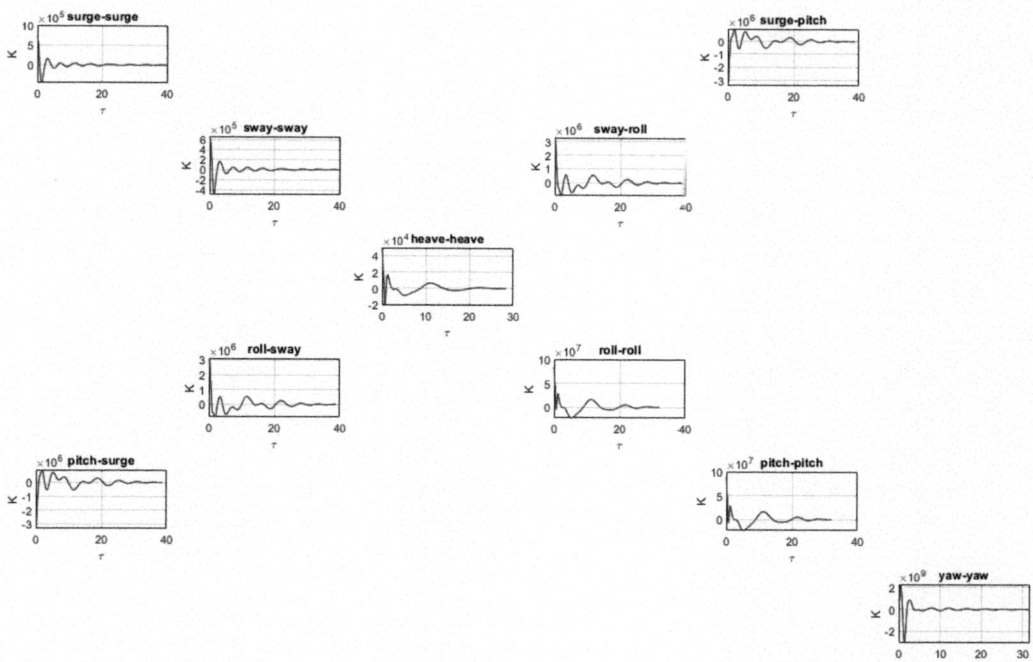

Figure A1. Retardation functions in 6 DoFs.

Advances in Maritime Technology and Engineering – Guedes Soares & Santos (Eds)

Drag analysis in auxiliary ships used in artisanal fishing ships in Perú

D.O. Sagástegui & P.A. Flores
Pontificia Universidad Católica del Perú, Lima, Perú

ABSTRACT: The Peruvian fishing sector, crucial to the national economy, depends largely on the efficiency of its ships, particularly in the configuration of the propulsion system. The lack of research in this field has led to inefficient operations, resulting in economic losses and limiting fishing days. This research focuses on the analysis of the trawling of "Pangas", essential ships in both industrial and artisanal fishing. Given the importance of artisanal fishermen, this sector becomes the main focus. The methodology uses Ansys Fluent computational methods, validating boundary conditions with experimental references. Five pangas are analyzed, revealing drag force curves, coefficients and relationships between speed and resistance, providing crucial data to improve efficiency in the artisanal fishing sector.

1 INTRODUCTION

Nowadays, the naval industry of Perú has a high impact in the national economy of the last 10 years, which has represented 1.2% of the Gross Domestic Product (IEES, 2021), reason why it is considered an influential industry and this activity goes hand in hand with the operation of fishing ships: industrial and artisanal. The industrial fishing sector has been developing using advanced methodology and international regulations. However, there is a forgotten sector in terms of engineering and research called the artisanal fishing sector. In Perú, there are around 20 000 ships (OCEANA, 2020) which are designed and built without any type hydrodynamic and drag analysis. These aspects are important in the propulsion configuration process, since through them it is possible to select the optimal engine, gearbox and propeller to avoid the effects of overloading, overheating or overrevving in the engine. Furthermore, propeller can suffer cavitation phenomenon and a degradation in its structural integrity causing downtime that translates into economic losses for fishermen.

In the artisanal fishing ships, an auxiliary ship is used to transport passengers or cargo when it is fishing season. This ship is called Panga, which is the main focus of this research as they face an ongoing problem of misconfiguration of their propulsion system. This leads to cost overruns and motor overload problems. In the present research, 5 artisanal pangas with different displacements in a range of speeds are analyzed. Computational fluid dynamics is used to perform the analysis using Ansys Fluent software which has implemented a RANS solver model (Elkafas et al., 2019). Ansys software recommends the SST k-omega turbulence model when highly accurate resolution of boundary layers is desired and where forces or heat transfer in the wall are key to the simulation (aerodynamic drag, turbomachinery blade performance, transfer of heat). Additionally, two validation cases are presented: Wigley hull and DTC hull, with which it is demonstrated that Ansys is able to predict the drag force experienced by a ship during sailing. The convergence of the solution will be comprised of the Couple method for pressure-velocity coupling and a second-order scheme for spatial discretization. The research was carried out in a range of 2000 iterations with a time step of 0.04 s. Through this research it is expected to formulate a criterion that allows the propulsion system of artisanal auxiliary vessels to be configured optimally.

Figure 1. Artisanal fishing ships.

DOI: 10.1201/9781003508779-62

2 ARTISANAL FISHING SECTOR

Peruvian artisanal fishing is an activity that is in continuous growth, because it is easier to take a place into this sector. Nowadays there is an illegal loophole that prevents different ships from being categorized accurately (Cuba et al., 2018). The major number of artisanal fishing ships is located in the North and South of Peru. Typical ships are manufactured of wood and known as "Viking ships" (Malpica, 2018).

2.1 *Auxiliary ship - Panga*

An auxiliary ship – Panga – is used for fishing activities, it features a rounded edge and a square stern with a propeller protected by grilles (CORPESCA, 2017). The panga performs 2 main functions in its operation (Villar, 2018):

- Lengthening and seizing of the net, keeping the bow end of the net in a certain position while the ship performs the purse seine maneuver.
- It serves as a rescue ship and embarkation or disembarkation of passengers

Figure 2. Panga sailing to the port of Chimbote – Perú.

To formulate a useful criterion of drag analysis for this kind of ships, 5 pangas were taken into account as statistical data. These have length over all, beam and depth dimensions, which are submerged to different drafts.

Table 1. Parameters of the Pangas study.

Ship	Length(m)	Beam (m)	Depth (m)	Cb
Panga 1	6.15	3.07	1.265	0.573
Panga 2	5.5	2.1	1.33	0.568
Panga 3	7	3.5	1.295	0.608
Panga 4	6.5	3.25	1.296	0.581
Panga 5	5	2.49	1.295	0.585

The pangas were measured in the port and at sea, due to the fact that the fishing ships are in operation. Most of the Pangas have a rustic appearance. The Pangas have been considered from the ports of Chimbote in the North of Peru and the ports of Ilo in the south of Peru. A highlight feature of these ships is that they are built with a length over all equivalent to twice the value of the beam and with a constant depth.

Figure 3. ALBATROS SHIP – Panga 1.

Figure 4. Panga MANUEL – Panga 2.

Figure 5. PIMENTEL SHIP – Panga 3.

Figure 6. MI AURELITA SHIP – Panga 4.

Figure 7. RAFAEL 2 SHIP – Panga 5.

3 METHODOLOGY

This section details the methodology used in the development of this research. It details everything from the geometric design until the configurations involved in computational fluid dynamic analysis carried out in Ansys fluent.

3.1 *Geometric design*

To design the hull of ships used for validation and study cases, 3D design software was necessary. Furthermore, for Wigley hull was through mathematical equations, since this is a mathematic model. The design of each study hull is described below.

3.1.1 *Wigley hull*

Wigley hull is a mathematical model which is described by a mathematical equation (Equation 1) (Ayub & Furukawa, 2020). This type of hull is symmetric with a very fine bow and parabolic waterlines. This is a typical model when it is necessary a numerical validation.

$$y\,(x,z) = \frac{B}{2}\left\{1 - \left(\frac{x}{l}\right)^2\right\}\left\{1 - \left(\frac{z}{d}\right)^2\right\} \quad (1)$$

Where:
B: beam of ship
l: Half ship length
d: Draft

The Wigley dimensions are based on the experimental results according to Ship Research Institute (SRI) presented in the ITTC report (Andrun et al., 2018). They are shown in Table 2.

Table 2. Wigley hull dimensions and experimental conditions according to Ship Research Institute (SRI).

Item	Value
Length over all (m)	4
Beam (m)	0.4
Draft (m)	0.25
Wetted area (m^2)	2.36
Displacement (kg)	177.2
Water temperature (°C)	10.6
Trim (m)	0
Sink and Trim	Free

Figure 8. Wigley Hull.

3.1.2 *Duisburg Test Case hull (DTC)*

DTC hull is a single-screw ship with a bulbous, developed at the Institute of Ship Technology, Ocean Engineering and Transport System (ISMT). There are only virtual CAD geometries of this kind of ship. The dimensions used in this research are shown in Table 3 (el Moctar et al., 2015). They are on a scale

Table 3. DTC hull dimensions and experimental conditions.

Item	Value	Full Scale
L_{pp} (m)	5.976	355.0
B_{wl} (m)	0.859	51.0
T_m (m)	0.244	14.5
S_w (m^2)	6.243	22032.0
C_B	0.661	0.661
V (m^3)	0.827	173467.0
Trim (°)	0.0	0.0
Vd (kn)	3.244	25
ρ (kg/m^3)	998.8	-
ν_M (m^2/s)	1.09 10^-6	-

of 1:5904, since performing a full-scale simulation requires a higher computational cost and number of elements.

Figure 9. Duisburg Test Case hull.

3.1.3 *Auxiliary ship*

Panga hull was designed based on Hull Shape Lines. The shape of the panga is with a rounded bow and a flat stern. This shape is typical of artisanal fishing boats in the ports of Peru. To carry out the 3D modelling of the hull, the Rhinoceros software was used, using the surface technique: sweep to rails. Subsequently, the hull is closed as it will be exported to Ansys Workbench where it will be locked in a domain, which is detailed in the next part of the methodology.

Figure 10. Auxiliary ship hull.

3.2 *Fluid domain and boundary conditions*

3.2.1 *Fluid domain*

An important part of numerical simulation is the size of the fluid domain (Figure 11), which must be large enough to ensure accuracy and simulation efficiency. The fluid domain was based on the International Towing Tank Conference (ITTC)'s recommendation in order to prevent flow reflections (Elkafas et al., 2019). With the aim of reducing the computational time, in this research half of the ship is considered, placing

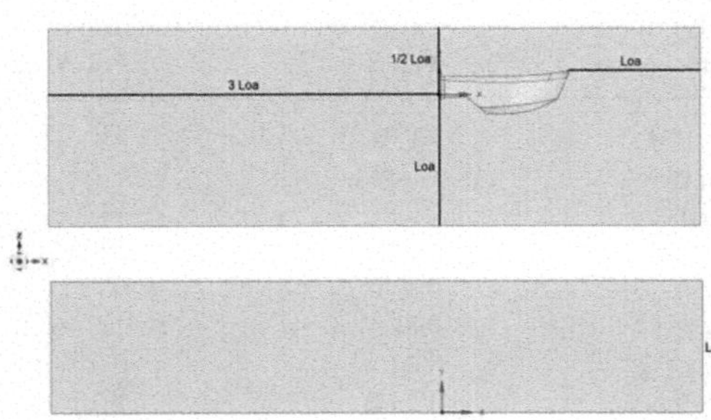

Figure 11. Auxiliary Ship Hull.

certain conditions that allow the drag force to be calculated in that half. In addition, the outlet is located at a distance of 3 times the length far enough away from the hull, as the flow in this area is unstable.

3.2.2 *Boundary conditions*

Operation conditions in ships are too much important when accurate analysis is required. The boundary conditions shown in Figure 12 are used for validation and study cases. They are composed of an inlet condition, through which the flow enters at a certain speed; a condition of symmetry, which is designated in the middle of the hull, side, top, and bottom of the domain; a wall condition for the hull, which drag force is captured, and an outlet condition, in which the flow of water continues its path without any unstable. It is considered a body of influence, as it allows the fine-shaped mesh to be captured near the hull of the panga.

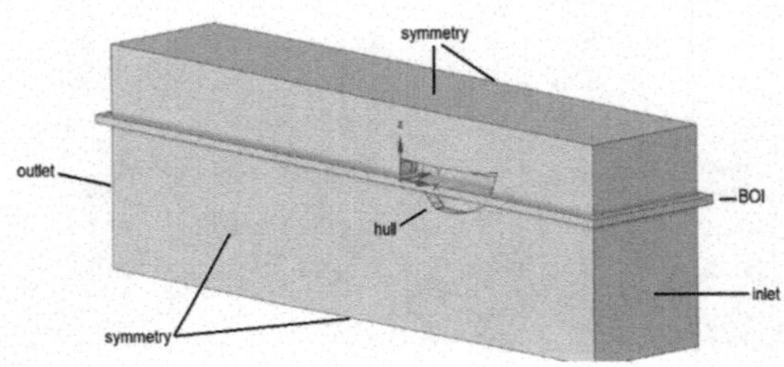

Figure 12. Boundary conditions.

3.2.3 *Turbulence model – SST k-ω*

To develop this research was chosen the Shear Stress Transport (SST), because it is a method which allows high precision in capturing the boundary layer. Furthermore, Ansys Fluent recommend to use SST k-ω model turbulence in numerical simulations where drag forces are key. This is hybrid model that combines the advantages of both k-ε and k-ω models. This model is characterized by two primary aspects: a zonal adjustment of model coefficients and a restriction on the increase of eddy viscosity in flows experiencing fast strain. Zonal modeling employs Wilcox's k-ω model in proximity to solid walls, while adopting the standard k-ε model (in a k-ω formulation) near the edges of the boundary layer and within free-shear layers (Bardina et al.,1997).

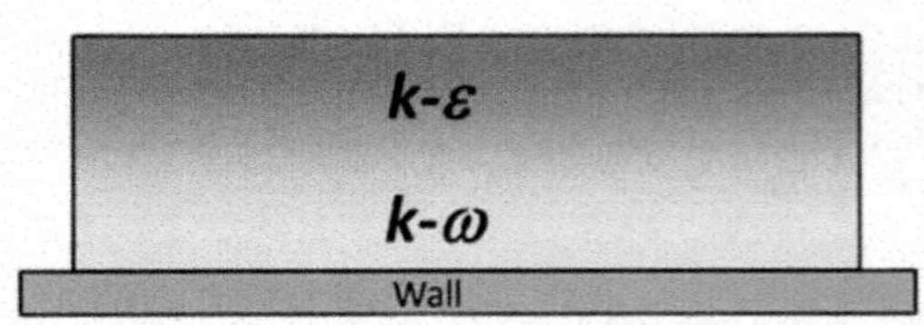

Figure 13. SST turbulence model (ANSYS, 2015).

3.2.4 *Mesh*

Meshing is a crucial step in the development of computational analysis, influencing the accuracy, efficiency and reliability of the results. In this research, a Mosaic Poly – Hexcore mesh is used, featuring an approximate 48% decrease in the overall element count, yields comparable results to the conventional Hexcore mesh, while requiring approximately 41% less computational time (Zore et al., 2019). Developers at ANSYS Fluent conducted a survey of the CFD meshing field, observing a widespread preference for hexahedral elements due to their accuracy and efficiency. Simultaneously, they recognized the advantages of polyhedral elements, which are particularly suitable for intricate geometries and demonstrate greater efficiency compared to tetrahedral elements (ANSYS, 2020).

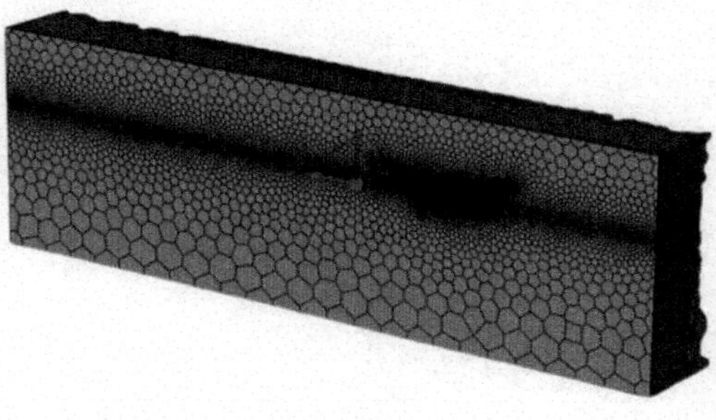

Figure 14. Mosaic Poly – Hexcore mesh.

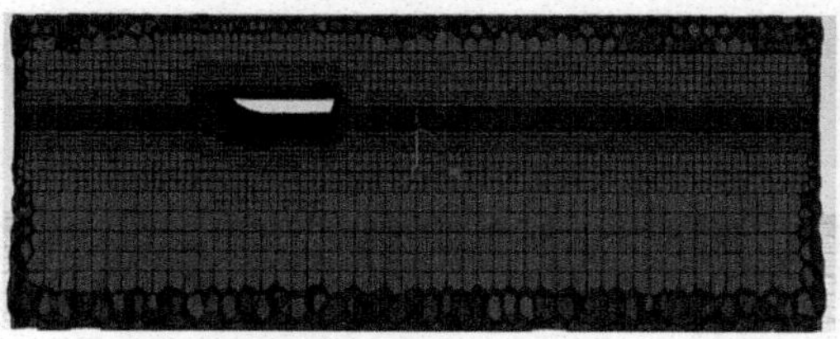

Figure 15. Mosaic Poly – Hexcore mesh of Panga.

In addition, boundary layer was considered with the goal of capturing all the viscous effects adjacent to the wall, as this is an important consideration in turbulent flow analysis. The ITTC provides guidelines for calculating the thickness of the first layer that makes up the boundary layer.

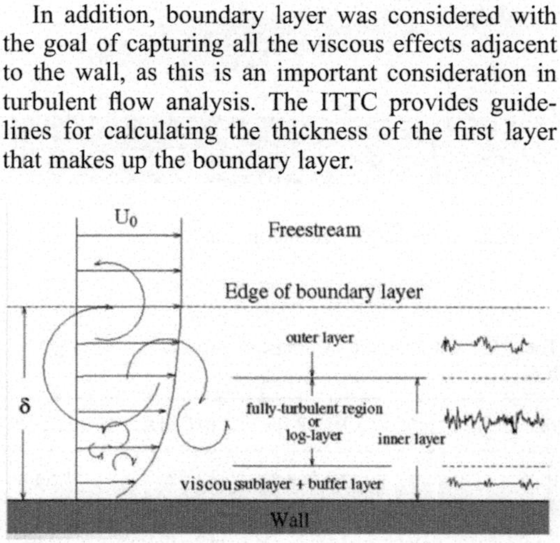

Figure 16. Turbulent boundary layers (Ansys,2015).

These guidelines are shown below (ITTC, 2014):

$$y^+ = \frac{\rho U_\tau y}{\mu} \tag{2}$$

$$U_\tau = \sqrt{\frac{\tau_w}{\rho}} \tag{3}$$

$$\tau_w = \frac{1}{2} C_f \rho U_\infty{}^2 \tag{4}$$

$$C_f = 0.058 Re_l{}^{-0.2} \tag{5}$$

$$Re_l = \frac{\rho V L_{OA}}{\mu} \tag{6}$$

Where:
ρ: Density (kg/m^3)
μ: Dynamic viscosity (kg/ms)
C_f: Skin friction coefficient
U_∞: Velocity (m/s)
L_{OA}: Length over all (m)
τ_w: Wall shear stress (kg/m.s^2)
y: First layer thickness (m)
Re: Reynolds number

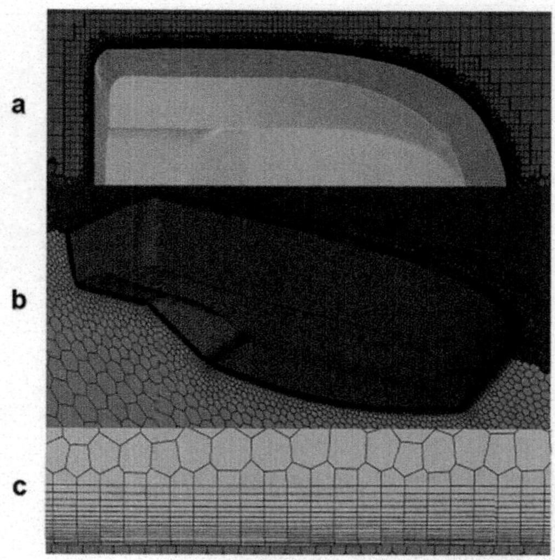

Figure 17. a) Poly Hexcore mesh – top view. b) Poly Hexcore mesh - 3D. c) Boundary layers.

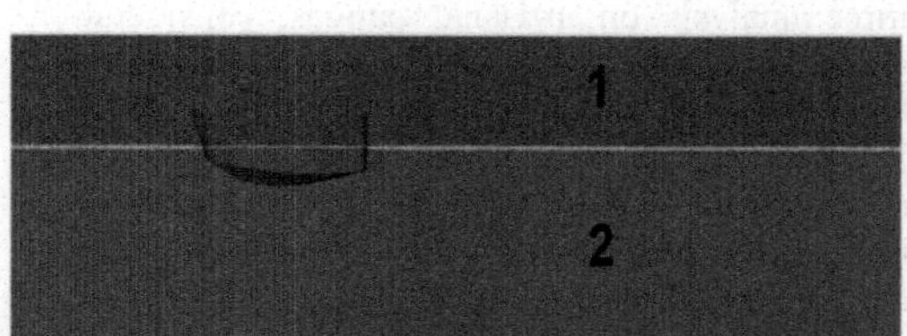

Figure 18. Flux Domain: 1) air. 2) water.

3.2.5 *Convergence*

The convergence of the solution is described by the variables presented in the residuals panel (Figure 19) and the employed tolerances. In all three cases of analysis in this research (Wigley Hull, DTC Hull, and Panga), the same tolerances as listed in Table 4. are applied. Additionally, the Couple method is used to relate velocity and pressure, along with a second-order scheme against the wind for spatial discretization in the momentum, energy, and specific dissipation rate equations.

In the aforementioned three cases, 2000 iterations are employed with a time step of 0.04 s. It is observed that the solution converges in a lower number of iterations for each analyzed velocity.

Table 4. Absolute criteria in the convergence.

Item	Absolute Criteria
continuity	0.001
x-velocity	0.001
y-velocity	0.001
z-velocity	0.001
k	0.001
omega	0.001
vf-water	0.001

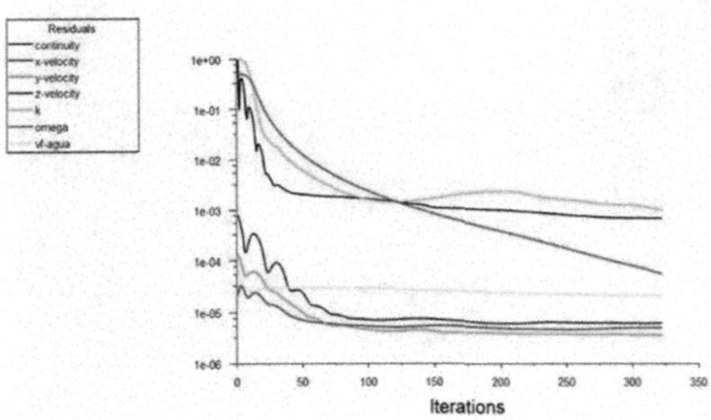

Figure 19. Residuals panel of Wigley hull.

4 RESULTS

4.1 *Validation case study*

In the absence of experimental methodology for drag force analysis on artisanal pangas, considerations ensuring the precision of the found drag force values have been inherited from validated cases. For the Wigley and DTC Hull cases, poly-hexcore-type meshes were applied, assigning an appropriate initial thickness to capture boundary layer effects based on the theory provided by Ansys.

Additionally, the analysis domain is used for both validation cases based on the lengths over all, as depicted in Figure 11, and the boundary conditions applied were the same for both validation cases (Figure 12). The results obtained in both cases under the aforementioned conditions demonstrate a low percentage error.

4.1.1 *Wigley hull*

The results obtained from the simulation are compared with the values obtained in the channel test. A minimum error of 0.29% is observed for fn 0.267, and a maximum error of 1.54% for fn of 0.25. The low % error obtained in the validation of the wigley validates the mesh size, the first thickness of the boundary layer and the boundary conditions.

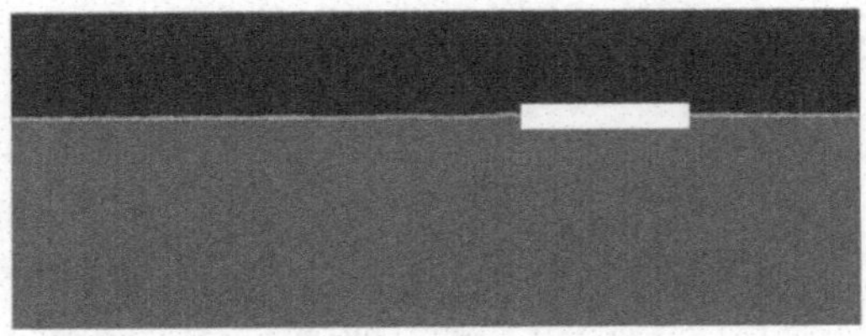

Figure 20. Flux domain – Wigley case.

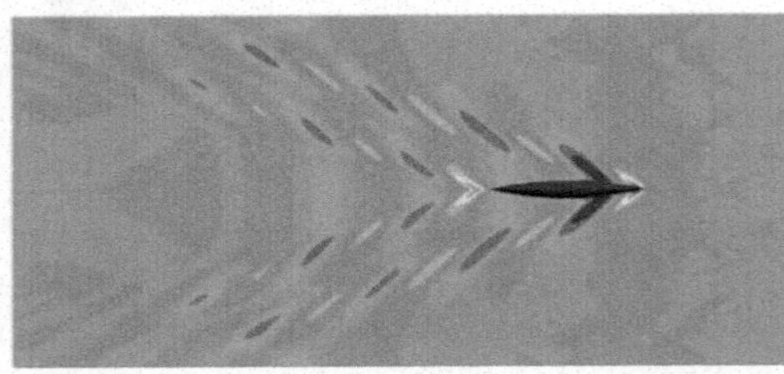

Figure 21. Wave pattern of Wigley – fn 0.316.

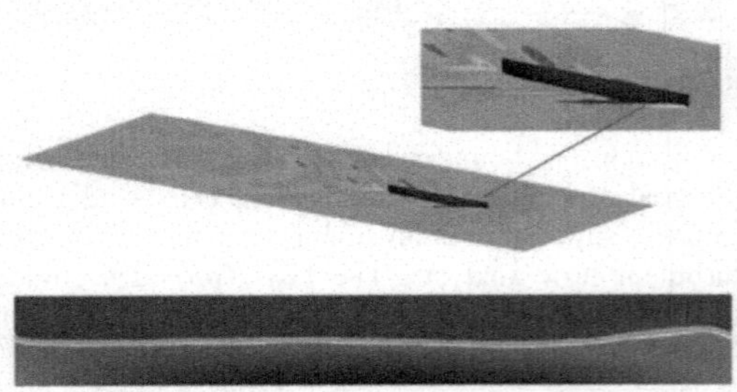

Figure 22. Free surface elevation – fn 0.316.

Table 5. Comparison of drag force results for Wigley hull.

fn	v (m/s)	R (SRI)	R(CFD)	error %
0.25	1.566	12.544 N	12.7372 N	1.5404
0.267	1.6725	13.975 N	14.0159 N	0.2928
0.289	1.8103	17.932 N	18.1320 N	1.1157
0.316	1.9794	23.538 N	23.4620 N	0.3228

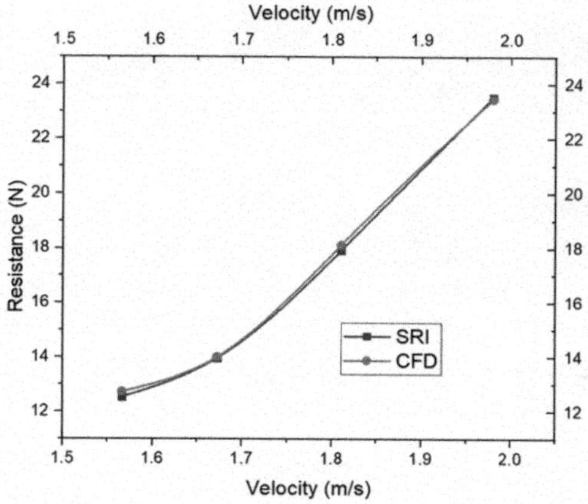

Figure 23. Comparative graph of drag force results.

4.1.2 *DTC hull*

The drag forces results obtained from model tests conducted in in the model test basins SVA Potsdam were compared with the drag force results from CFD analysis. The percentage error is shown in Table 6. It is observed a minimum error of 0.33 % for fn 1.401 and a maximum error of 1.81 % for fn 1.335. These values show an acceptable accuracy in the calculation of the drag forces.

Figure 24. Flux domain – DTC hull case.

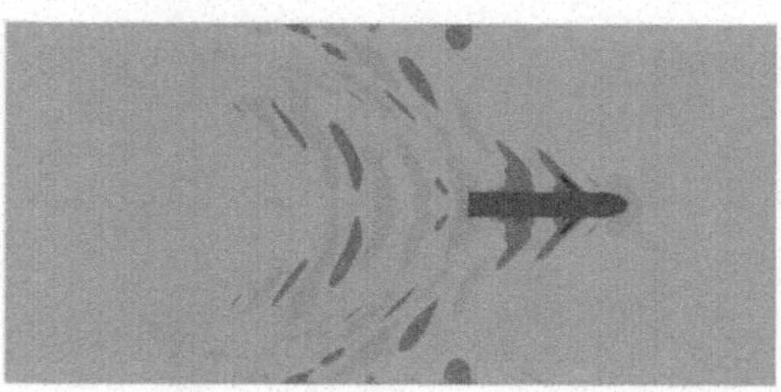

Figure 25. Wave pattern of DTC hull – 1.668 m/s.

Table 6. Comparison of drag force results for DTC hull.

v (m/s)	Rexp	Rcfd	error %
1.335	20.34 N	19.9717 N	1.81061947
1.401	22.06 N	21.9870 N	0.33091568
1.469	24.14 N	24.0208 N	0.49378625
1.535	26.46 N	25.9910 N	1.77248677
1.602	28.99 N	28.5858 N	1.39427389
1.668	31.83 N	32.0294 N	0.62645303

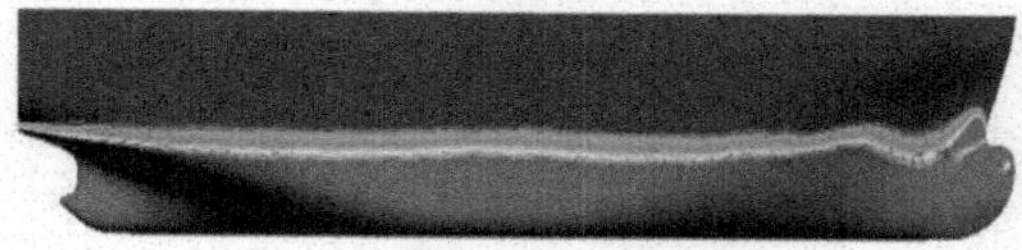

Figure 26. Free surface elevation – 1.668 m/s.

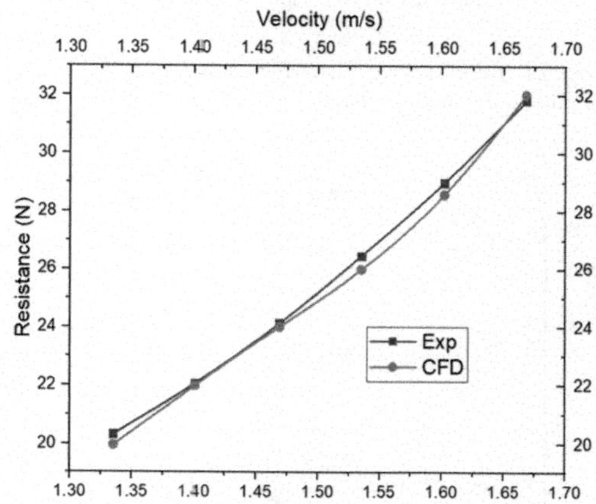

Figure 27. Comparative graph of drag force results.

4.1.3 *Auxiliary hull - Panga*

Having validated the methodology proposed in this research, we proceed to perform the drag force analysis for the 5 pangas. Seawater density (1025.86 kg/m^3) and dynamic viscosity (0.001219 kg/ms) were used for this simulation.

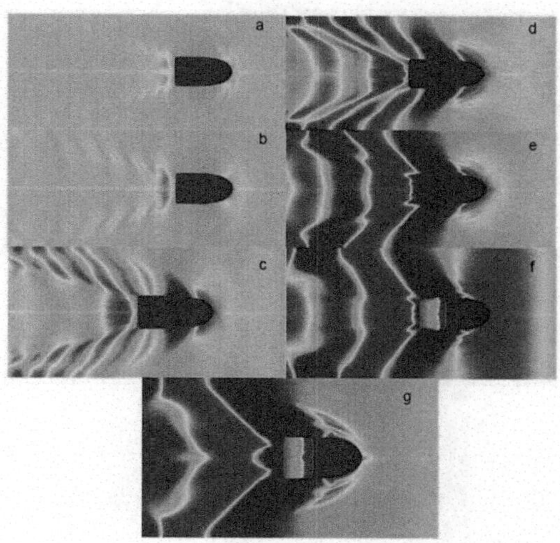

Figure 28. Wave pattern of Panga 1: a) 3 m/s. b) 4 m/s. c) 5 m/s. d) 6 m/s. e) 7 m/s. f) 8 m/s. g) 9 m/s.

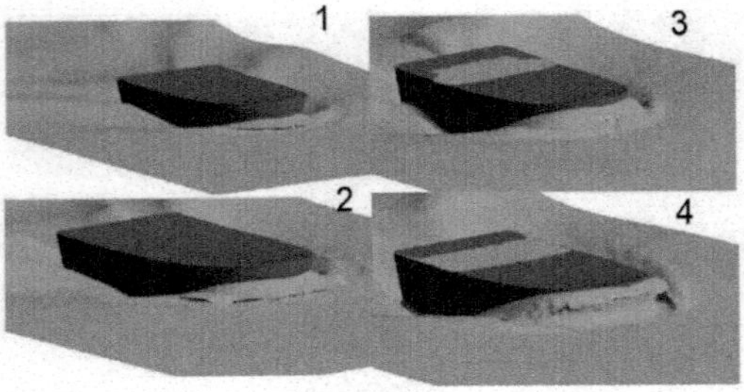

Figure 29. Free surface elevation -Panga 1: 1) 6 m/s. 2) 7 m/s. 3) 8 m/s. 4) 9 m/s.

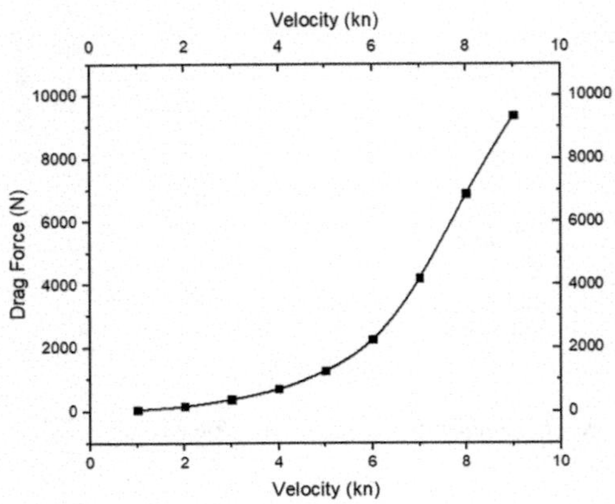

Figure 30. Graph velocity (kn) vs Drag Force (N) – Panga 1.

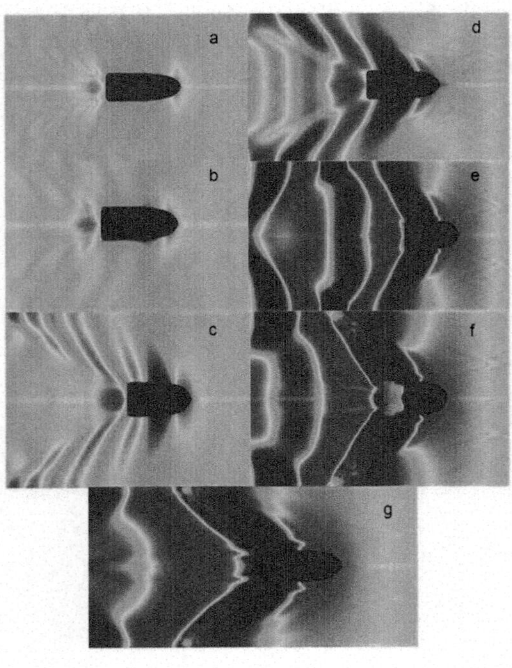

Figure 31. Wave pattern of Panga 2: a) 3 m/s. b) 4 m/s. c) 5 m/s. d) 6 m/s. e) 7 m/s. f) 8 m/s. g) 9 m/s.

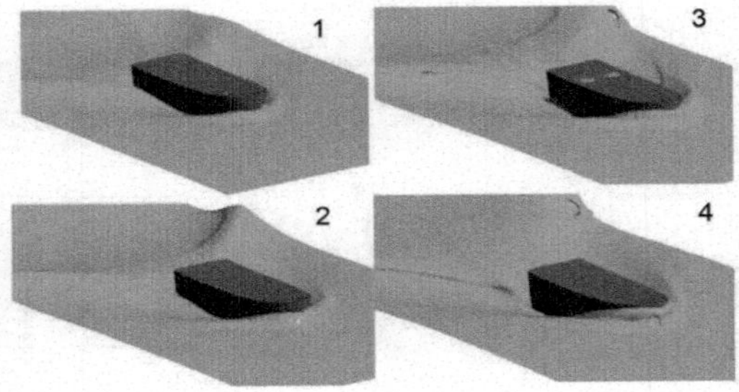

Figure 32. Free surface elevation -Panga 2: 1) 6 m/s. 2) 7 m/s. 3) 8 m/s. 4) 9 m/s.

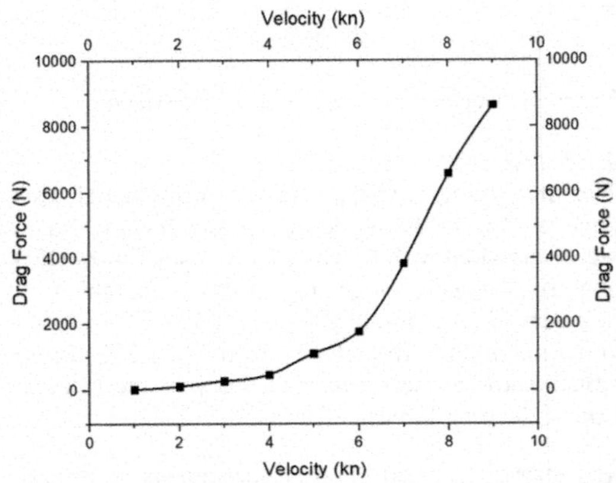

Figure 33. Graph velocity (kn) vs Drag Force (N) – Panga 2.

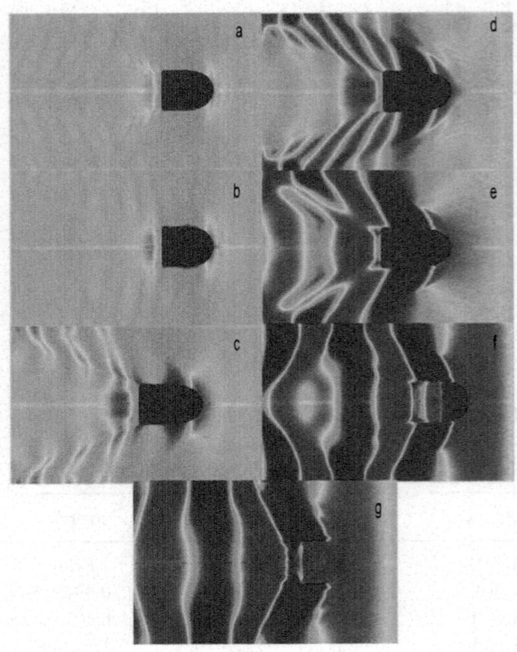

Figure 34. Wave pattern of Panga 3: a) 3 m/s. b) 4 m/s. c) 5 m/s. d) 6 m/s. e) 7 m/s. f) 8 m/s. g) 9 m/s.

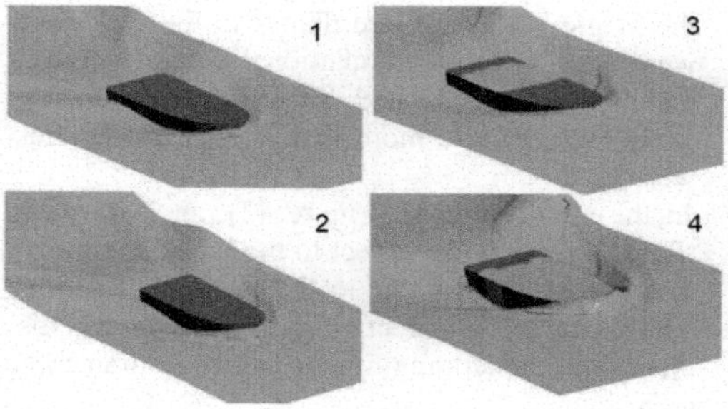

Figure 35. Free surface elevation -Panga 3: 1) 6 m/s. 2) 7 m/s. 3) 8 m/s. 4) 9 m/s.

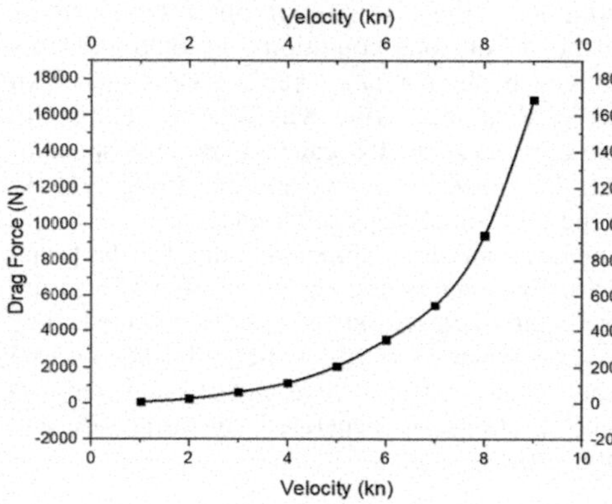

Figure 36. Graph velocity (kn) vs Drag Force (N) – Panga 3.

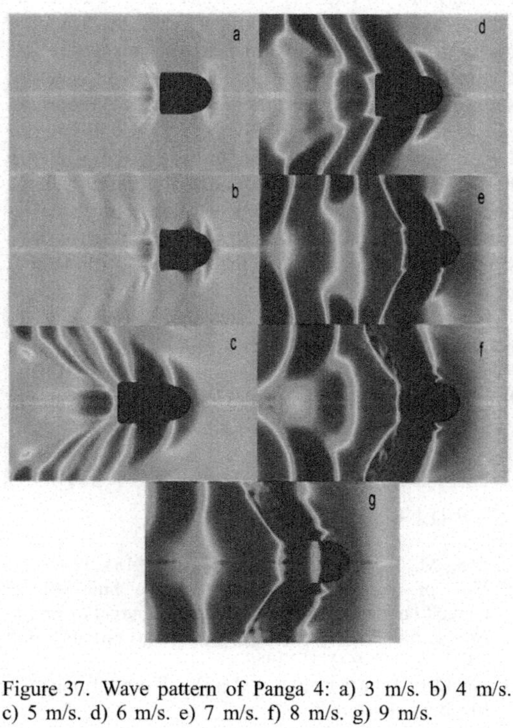

Figure 37. Wave pattern of Panga 4: a) 3 m/s. b) 4 m/s. c) 5 m/s. d) 6 m/s. e) 7 m/s. f) 8 m/s. g) 9 m/s.

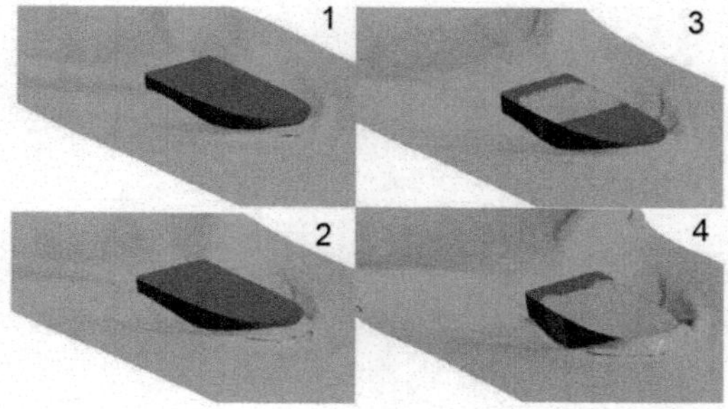

Figure 38. Free surface elevation -Panga 4: 1) 6 m/s. 2) 7 m/s. 3) 8 m/s. 4) 9 m/s.

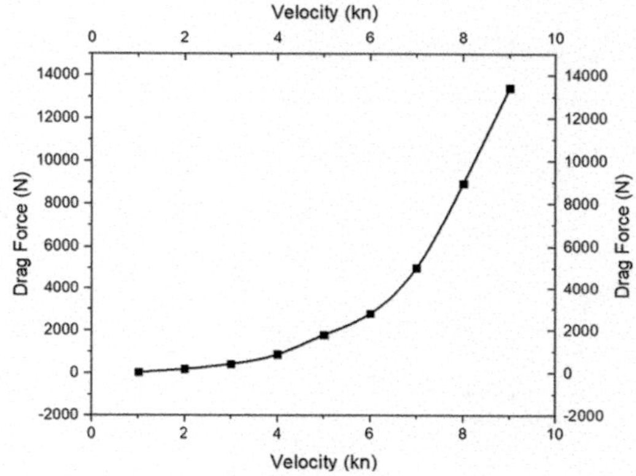

Figure 39. Graph velocity (kn) vs Drag Force (N) – Panga 4.

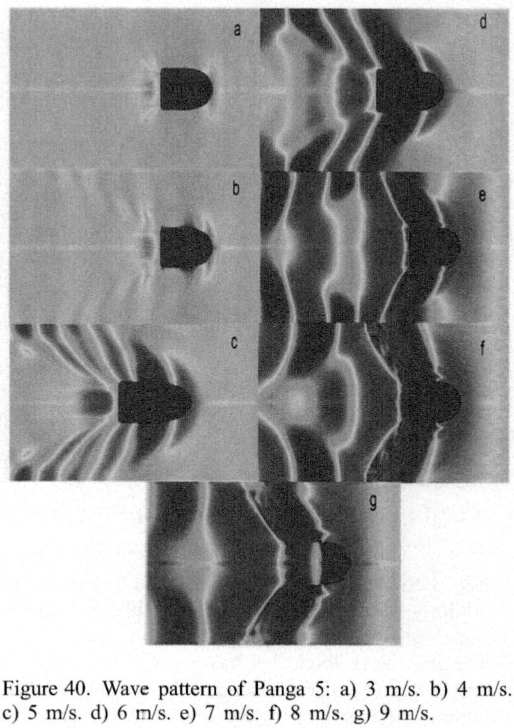

Figure 40. Wave pattern of Panga 5: a) 3 m/s. b) 4 m/s. c) 5 m/s. d) 6 m/s. e) 7 m/s. f) 8 m/s. g) 9 m/s.

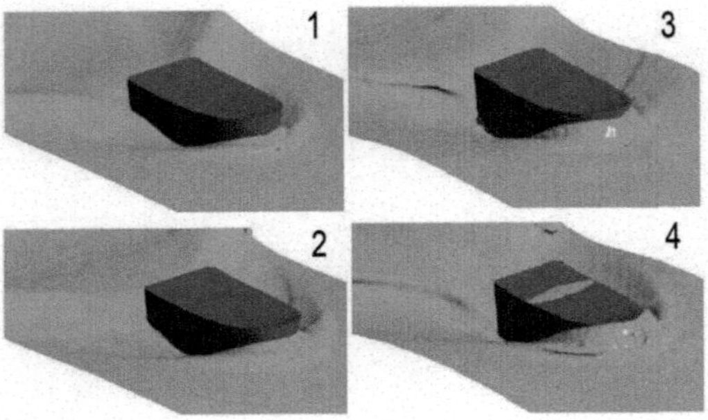

Figure 41. Free surface elevation -Panga 5: 1) 6 m/s. 2) 7 m/s. 3) 8 m/s. 4) 9 m/s.

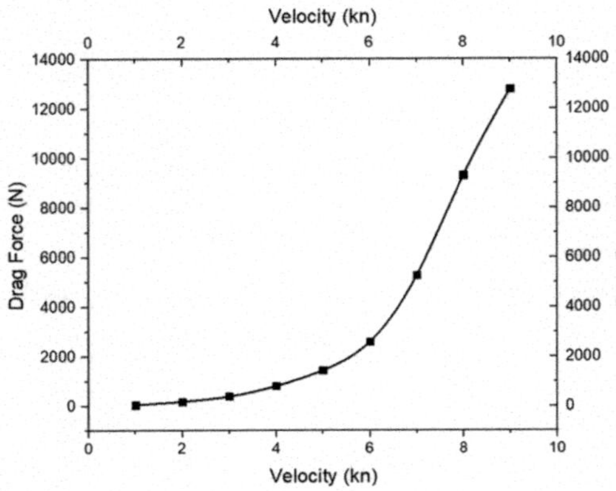

Figure 42. Graph velocity (kn) vs Drag Force (N) – Panga 5.

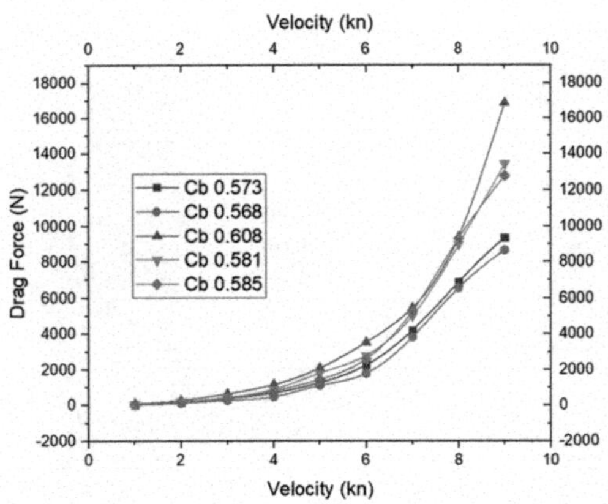

Figure 43. General graph velocity (kn) vs Drag Force (N).

5 CONCLUSIONS AND RECOMMENDATIONS

- Based on Wigley and DTC Hull results, the methodology used in this research is able to predict with high accuracy the drag forces on a ship.
- The first layer thickness has a decisive influence on the capture of the viscous sublayer and on the number of elements in the mesh. In the case of the Wigley, a little more than 2 million elements were used, with a computational cost of approximately 4 hours per speed. For DTC Hull and Auxiliary Hull, a little more than 1 million elements were used.
- In the auxiliary case (Figure 43), an increase in the drag force with respect to the block coefficient was observed. It is due to the fact that the block coefficient has a direct proportion with the displacement, generating higher values of drag force on a ship.
- The free surface elevation results show the entry of water into the deck area for panga 1, panga 3, panga 5 and panga 5, at navigation speeds between 8 and 9 knots. This can lead to instability during the sailing, so through this research limits can be set on load conditions. Furthermore, this behavior of the panga at speeds greater than 7 kn may explain the reason why artisanal pangas in Peru only occasionally reach a maximum speed of 6 kn. However, the most common sailing velocity on these artisanal ships is 4 knots.
- Pangas are auxiliary ships, intended for the transport of passengers and cargo, so high speeds are not required. Based on the results obtained, they are recommended to sail at speeds between 4 to 5 knots, since moderate wave patterns and free surface elevations are generated, allowing safe and optimal navigation.

ACKNOWLEDGEMENTS

I would like to express my sincere gratitude to the Pontifical Catholic University of Peru for providing me with the opportunity to lead this research. The commitment to academic excellence and unwavering support have been instrumental in the development of this study on the analysis of drag force in Pangas. I want to extend my gratitude to my co-author, Pedro Flores, whose dedication and collaboration have been invaluable at every stage of this project. His expertise and commitment to scientific quality have significantly contributed to the depth and robustness of our analysis.

I am deeply thankful for the opportunity to research with the Pontifical Catholic University of Peru and with Pedro Flores on this research, and I hope that this study contributes to the advancement of knowledge in the field of Pangas hydrodynamics.

REFERENCES

Andrun, M., Blagojevic, B. & Basic, J. (2018). The influence of numerical parameters in the finite-volume method on the Wigley hull resistance. Journal of Engineering for the Maritime Environment pp1–10. doi: 10.1177/1475090218812956

Ansys. (2015). Introduction to ANSYS Fluent: Turbulence Modeling. Lecture 7.

Ansys. (2020). ANSYS Fluent Mosaic Technology Automatically Combines Disparate Meshes with Polyhedral Elements for Fast, Accurate Flow Resolution. Available in: https://www.ansys.com/content/dam/resource-center/white-paper/ansys-fluent-mosaic-technology-wp.pdf

Ayub,F & Furukawa,K.(2020). Prediction of Hydrodynamic Forces Acting on Wigley Hull based on SQCM. IOP Conf. Ser.: Earth Environ. Sci. 557 012015

Bardina, J., Huang, P. & Coakley, T. (1997). Turbulence Modeling Validation, Testing, and Development. National Aeronautics and Space Administration (NASA). Moffett Field, California.

Corpesca S.A. (2017). Método y Sistema de Pesca Semiautomático y con mando a distancia para controlar una embarcación auxiliar o panga que Evita la accidentabilidad de los operarios al interior de dicha embarcación auxiliar o panga. Organización Mundial de la Propiedad Intelectual.

Cuba, R., Nina, W., Castro, E &Villegas, J. (2018). A new approach for the Categorization of Boats for Artisanal Fishing in Peru. 16th LACCEI International Multi-Conference for Engineering, Education, and Technology: "Innovation in Education and Inclusion". Lima, Perú. doi: http://dx.doi.org/10.18687/LACCEI2018.1.1.355

Elkafas, A., Elgohary, M & Zeid, A. (2019). Numerical Study on the hydrodynamic drag force of a container ship model. Alexandria Engineering Journal. Vol 58 pp 849–859. doi: https://doi.org/10.1016/j.aej.2019.07.004

El Moctar, O., Shigunov,V. & Zorn, T. (2015). Duisburg Test Case: Post-Panamax Container Ship for Benchmarking. Ship Technology Research. Vol 59 pp 50–64. doi: https://doi.org/10.1179/str.2012.59.3.004

IEES. (2021). Reporte sectorial: Industria Pesquera Consumo Humano Directo. Available: https://sni.org.pe/wp-content/uploads/2022/01/30-Pesca-CHD-y-Acuicultura.pdf

ITTC. (2014). Practical Guidelines for ship CFD Applications. ITTC – Recommended Procedures and Guidelines. pp 1–20 Available: https://www.ittc.info/media/8165/75-03-02-03.pdf

Kajitani, H., Miyata, H., Ikehata, M., Tanaka, H & Adachi, H. (1983). The Summary of the Cooperative Experiment on Wigley Parabolic Model in Japan. Defense Technical Information Center. Available in: https://apps.dtic.mil/sti/citations/ADP003037

Malpica, O. & Bardales, J. (2018). Determinación de la estabilidad de una pequeña embarcación pesquera de madera de diseño predominante en la zona norte del litoral peruano Vol 79 pp 483–495. doi: https://doi.org/10.21704/ac.v79i2.1260

OCEANA. (2020). Fortalecer la sostenibilidad de la pesquería artisanal. Available in: https://peru.oceana.org/wp-content/uploads/sites/22/agenda_azul_propuesta_01.pdf

Villar, E. (2018). Buque Atunero Congelador de 3.700 m^3. Universidade da Coruña. La Coruña, España.

Zore, K., Parkhi, G., Sasanpuri,B. & Varghese, A. (2019). Ansys Mosaic Poly- Hexcore mesh for high-lift aircraft configuration. 21 st Annual CFD Symposium, Bangalore.

Advances in Maritime Technology and Engineering – Guedes Soares & Santos (Eds)

CFD simulation of a plane net in current

M.D. Viegas, S. Wang & C. Guedes Soares
Centre for Marine Technology and Ocean Engineering (CENTEC), Instituto Superior Técnico, Universidade de Lisboa, Lisbon, Portugal

ABSTRACT: A numerical approach is proposed for simulating the hydrodynamic characteristics of a plane net in current. The fluid-structure interaction model relies on one-way coupling between a fluid model, to simulate the flow field around the undeformed structure, and on a mechanical structural model to determine the deformation caused by fluid pressure. This model is implemented for different current conditions. Validation of the numerical model is performed by comparing drag force results with published experimental data. In this study, the netting is modelled as a three-dimensional structure instead of a porous medium that has been applied commonly. Findings from the fluid model indicate that variations in current velocity have minimal impact on flow velocity behaviour but have significant effects in flow magnitude and drag force. In the model, flow reduction can be observed downstream from the net, aligning with experimental tests and prior research. The results suggest that the knotted net outperforms the knotless net and that although the fluid pressure generates deformation, it does not cause material failure. The findings demonstrate that the current numerical simulation developed in ANSYS FLUENT, is suitable for analysing hydrodynamic forces acting on aquaculture structures.

1 INTRODUCTION

Due to the growth in global population the demand and supply for food presents a constant challenge. Aquaculture, commonly known as fish farming, is an important contributor to address that demand. According to a statistic report developed by the Food and Agriculture Organization (FAO) of the United Nations, in 2020 aquaculture was responsible for 49 percent of the world sea food supply. The report also states that since 1990 aquaculture production has increased linearly, while capture fisheries production has stagnated (FAO, 2022).

As a result of the footprint of humans on the planet, environmental conditions have become more unstable, making them unreliable, which presents a challenge for maintaining an efficient production of fish. Farm sites are typically located in stagnant coastal areas, which can lead to parasite infections, oxygen shortages, and waste pollution. To address these challenges, the industry is looking to extend aquaculture operations into the open sea, which presents new challenges in terms of structural design and fish survival.

The study of fish cages and their behaviour regarding ocean environment is important to assure efficient production. Numerical models have been associated with wave theory to study the effect of design parameters and mooring forces on cage behaviour. These studies brought insight into the increase of cage displacement and mooring force when associated with cage dimensions and mooring line stiffness (Liu et al. 2021; Mohapatra et al. 2021). Further investigations into the subject were performed by Liu & Guedes Soares, (2022) that not only studied the cage displacement and mooring forces associated with current velocities but also concluded that wave parameters such as height and period have little effect on the mean value of the maximum mooring force. More recently, the wave-current interaction was studied when associated with an array of floating flexible cylindrical net cages, presenting results regarding the horizontal displacement of the cage and wave force on the cage (Mohapatra & Guedes Soares, 2024).

Experimental tests have always been considered a reliable way to predict the structure behaviour. Despite demanding a high initial investment, they provide an accurate way to study fluid structure interaction and the influence of environmental conditions. It is an approach that has been used for several investigations, Lader et al. (2007) used wave makers to study the influence of wave frequencies and net density on drag force, measuring separately the vertical and horizontal force with the use of force transducers. On the other hand, Liu et al. (2012) used a forced oscillation device to simulate ocean conditions, designing a load cell to measure forces on the frame. They performed the experiment for both the frame with and without the net and considered the net force as the difference between the two measurements. More recently, Xu et al. (2021) tested different types of nets in extreme waves at a laboratory in China and found a proportional relation between net

DOI: 10.1201/9781003508779-63

solidity and hydrodynamic loads. These results are supported by other experiments, such as Føre et al. (2021) that reached the same conclusion when developing towing tests in Norway. Further experiments were performed by Liu & Guedes Soares, (2022) who conducted an experimental study to observe the circular gravity cage response in linear waves using a wave maker, a rock breakwater and a membrane with holes to simulate biofouling. A similar experiment was performed by Liu & Guedes Soares, (2022a) who intended to observe the floating cage under different wave regimes and check the possibility of measuring the decay in wave height associated with its passage thought the cage.

In marine aquaculture facilities, the net assumes a paramount role as it stands as the singular barrier safeguarding the environment against potential fish escapes. There are two main types of netting: knotted and knotless. Initially knotted netting was the prevalent choice due to its familiarity from the fishing industry, it is considered a durable and easily repairable net, however their protruding knots can harm fish, especially when fish density is high or during net changes. Nowadays, knotless nets are the dominant choice for aquaculture structure, not only because they are safer for fish but are also 50% lighter than knotted nets, more cost-effective to produce, highly resistant to abrasion, easier to handle, and stronger (Cardia & Lovatelli, 2015).

The study of net type structures is not only crucial to the aquaculture field, but it is also being associated with coastal protection. The use of multi-membrane type structure has been studied to understand their effect on wave reflection, transmission, and wave energy dissipation (Guo et al. 2022a). Further studies introduced the possibility of membranes with different porosities, dependent on their position in relation with the incoming waves (Guo et al. 2022b). Experimental studies on these breakwater structures have been further developed with the addition of spring motions and horizontal membranes (Guo et al. 2023a, 2023b).

Recent improvements in modelling approaches have emerged. Cheng et al. (2022) introduced a coupled model using OpenFOAM for the fluid model and Code_Aster for the structural model. To simulate the net, they incorporated the porous media model in the fluid model and employed the finite element method in the structural model. Wang et al. (2022) reviewed the existing screen force model for knotless nets and proposed a combination of the Reynolds Averaged Navier-Stokes (RANS) method, applied close to the body, and Large Eddy Simulation (LES) further from the body. This approach aims to enhance the accuracy of classical RANS models while simultaneously reducing the computational cost.

A noteworthy development is the research conducted by Zou et al. (2022), where they introduced a one-way coupling method to study the flow field around knotless netting. They implemented their model in ANSYS, employing the k-ω SST turbulence model for the fluid model and assuming a non-linear finite element model with a beam element to simulate the net behaviour in the structural model. This numerical approach demonstrated that Fluid-Structure Interaction (FSI) studies could be performed effectively without relying on the porous media model.

Computational Fluid Dynamics (CFD) stands out as a highly advantageous approach for simulating Fluid Structure Interaction (FSI). Unlike many other numerical methods, CFD offers superior versatility and precision by integrating Finite Element Method (FEM) theory with fluid dynamics principles. This combination allows CFD to provide a comprehensive analysis of the complex interactions between fluid flow and structural behaviour.

The use of commercial software for CFD simulations are fast emerging in the engineering industry since they can adapt to the user specific case study and be implemented in complex geometries. Already several publications have mentioned using programs such as OpenFOAM to implement their CFD simulations choosing the VOF method and k-ω Shear Stress Transport (SST) turbulence models to solve the Navier-Stokes equations (Katsidoniotaki et al. 2023). ANSYS FLUENT proves to be a reliable software to simulate flow around structures, being already validate as a porous media fluid model with the use of the realizable k-epsilon model to study the hydrodynamic loads on a plane net (Bi et al. 2018).

In the present paper, a CFD simulation of a plane net is developed by FLUENT. Section 2 introduces the details of the numerical model, with emphasis on the turbulence model used and boundary conditions of the simulation. The validation and results of the simulation are discussed in Section 3 and a conclusion is drawn in Section 4.

2 NUMERICAL APPROACH

The numerical approach to simulate the interaction between fluid and structure includes two numerical models: the fluid model and the mechanical structural model. The fluid model can simulate the flow field around the body with no deformation, and the mechanical structural model can simulate the deformation of the body when subjected to the fluid pressure. The development of the numerical model was based on research and testing.

The employed approach is a one-way FSI simulation, so the models are semi-independent from each other. While the fluid model is totally independent in the production of its results, the mechanical structural model depends on the fluid model calculations of pressure on the structure to determine the deformation of the object. The mechanical structural model can only be developed after the fluid model is validated. The flow chart defined in Figure 1 represents the sequence of development of the coupling numerical model, where Model 1 is the fluid model developed in FLUENT and Model 2 is the mechanical structural model developed in Static Structural.

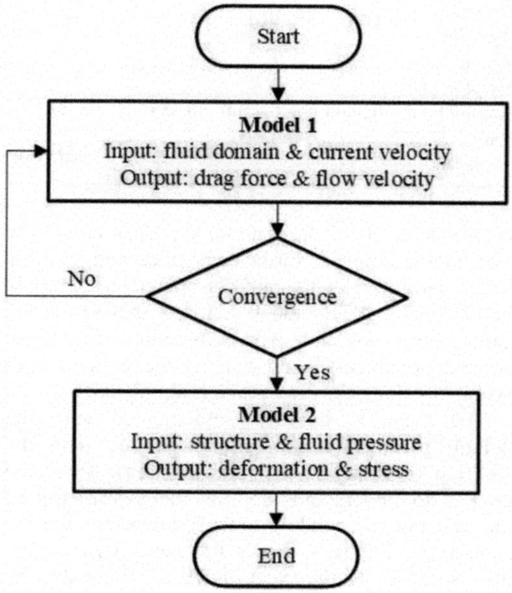

Figure 1. Overview of proposed numerical model.

2.1 *Description of the structure*

The structure characteristics are defined by the experimental study that is used to validate the model (Bi et al. 2018).

Table 1. Specifications of the plane net structure.

Component	Parameter	Dimension
Tank	Length	22 m
	width	0.45 m
	Depth	0.6 m
Frame	Height of structure	400 mm
	Diameter of cross-section	6 mm
	Material	Steel
Net	Length of mesh	20.6 mm
	Thickness of twine	1.6 mm
	Material	knotted polyamide

The structure consists of a diamond net attached to a frame, which serves as a support for the net, Figure 2. The main dimensions of the model components are listed in Table 1. To mimic the knotted net a sphere of 2.5 times the thickness of the twine is modelled.

2.2 *The fluid model*

The fluid model considered the simulation with current. It was considered that the model is in a steady

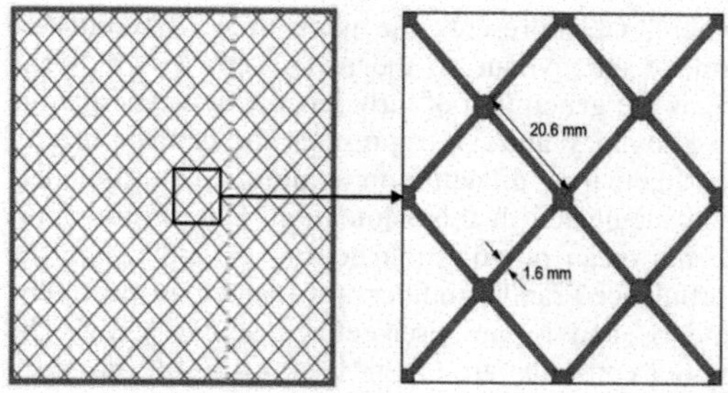

Figure 2. Configuration of plane net studied and diamond mesh.

state and used the realizable k-epsilon turbulence model with enhanced wall treatment.

2.2.1 *Governing equations and assumptions*

The realizable k-epsilon model is based on the Reynolds-averaged Navier-Stokes (RANS) equations, which governing equations can be written in tensor form as:

- Continuity equation

$$\frac{\partial \rho}{\partial t}+\frac{\partial}{\partial x_i}(\rho u_i)=0 \quad (1)$$

- Momentum equation:

$$\frac{\partial}{\partial t}(\rho u_i)+\frac{\partial}{\partial x_j}(\rho u_i u_j)=-\frac{\partial p}{\partial x_i}+\frac{\partial}{\partial x_j}\left(-\rho\overline{u_i' u_j'}\right)+\frac{\partial}{\partial x_j}\left[\mu\left(\frac{\partial u_i}{\partial x_j}+\frac{\partial u_j}{\partial x_i}-\frac{2}{3}\delta_{ij}\frac{\partial u_l}{\partial x_l}\right)\right] \quad (2)$$

These equations are derived by taking the time average of the governing equations of fluid flow and including the conservation of mass, momentum, and energy.

The model is based on model transport equations for the turbulence kinetic energy (k) and its dissipation rate (ε). The model transport equation for k is derived from the exact equation, while the model transport equation for ε is obtained using physical reasoning and bears little resemblance to its mathematically exact counterpart. In the derivation of the model, the assumption is that the flow is fully turbulent, and the effects of molecular viscosity are negligible. The transport equations are as follows:

$$\frac{\partial}{\partial t}(\rho k)+\frac{\partial}{\partial x_j}(\rho k u_j)=\frac{\partial}{\partial x_j}\left[\left(\mu+\frac{\mu_t}{\sigma_k}\right)\frac{\partial k}{\partial x_j}\right]+G_k+G_b-\rho\varepsilon-Y_M+S_k \quad (3)$$

$$\frac{\partial}{\partial t}(\rho\varepsilon)+\frac{\partial}{\partial x_j}(\rho\varepsilon u_j)=\frac{\partial}{\partial x_j}\left[\left(\mu+\frac{\mu_t}{\sigma_\varepsilon}\right)\frac{\partial\varepsilon}{\partial x_j}\right]+\rho C_1 S\varepsilon-\rho C_2\frac{\varepsilon^2}{k+\sqrt{\nu\varepsilon}}+C_{1\varepsilon}\frac{\varepsilon}{k}C_{3\varepsilon}G_b+S_\varepsilon \quad (4)$$

where, G_k represents the generation of turbulence kinetic energy due to the mean velocity gradients, G_b is the generation of turbulence kinetic energy due to buoyancy and Y_M represents the contribution of the fluctuating dilatation in compressible turbulence, both neglected in this simulation. S is the modulus of the mean rate-of-strain tensor, σ_k and σ_ε are the turbulence Prandtl numbers for k and ε, respectively, and S_k and S_ε are user-defined source terms The model constants are $C_{1\varepsilon} = 1.44$, $C_2 = 1.9$, $\sigma_k = 1.0$ and $\sigma_\varepsilon = 1.2$. (Abrahamson, 2022).

By solving these two transport equations along with the governing equations of fluid flow, the k-ε model provides information about the turbulence characteristics in the flow field. This information is then used to calculate the Reynolds stresses, which are essential for accurately predicting flow behaviour and turbulence effects.

It is important to note that while the k-ε model is widely used and provides reasonably accurate results for many engineering applications, it does have limitations.

The commercial software ANSYS Fluent is used as a tool to solve the differential equations with following hypotheses:

- The gravity effects are neglected.
- The surface tension effects are not modelled.
- The structures have no deformation and rotation motion.

2.2.2 *Mesh and boundary conditions*

A cartesian coordinate system (x,y,z) plane is introduced, and the (x,y)plane is placed in the flow of the water, while the z-axis is located in the axis of the body, as is illustrated in Figure 3. The figure also presents the conditions of the simulation. The body is fully submerged and positioned with its span almost along the entire cross section of the tank. This is designed to simulate a plane net as a part of the whole net cage and ensure that all the water travelled through the net.

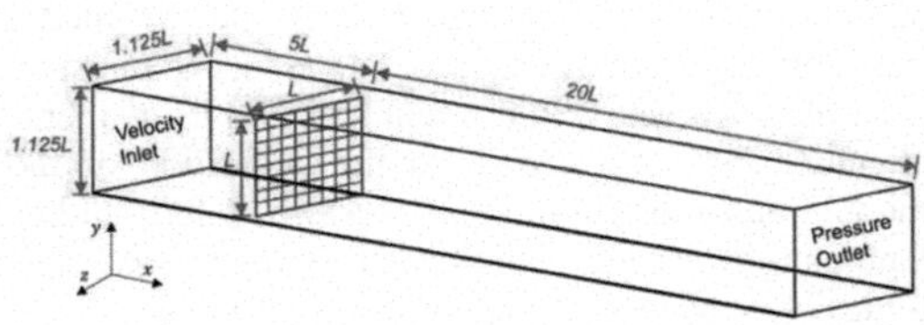

Figure 3. Computational domain of plane net.

The structure is considered as void, and the body behaves as a stationary wall. This is crucial to simulate the flow around the structure. The walls of the tank are considered with negligible frictional and drag, so the fluid velocity at the wall is zero and there is no relative motion between the fluid and the wall. To simulate these conditions, the walls motion

Table 2. Properties of net materials.

Material	PA6	PE	PET	HDPE
Density [kg/m^3]	1140	950	1339	958.5
Young Modulus [Pa]	$1.111*10^9$	$1.1*10^9$	$2.898*10^9$	$1.08*10^9$
Poisson ratio	0.3499	0.4200	0.3887	0.4183
Tensile Yield Stress [Pa]	$4.313*10^7$	$2.5*10^7$	$5.244*10^7$	$2.839*10^7$

is considered stationary with no slip shear condition. The inlet is defined as a velocity inlet, and five different current velocities are considered: $u_0 = 0.058$, 0.117, 0.171, 0.216 and 0.254 m/s that propagate along the x-axis in positive direction. To reduce computational cost, only half of the domain with symmetry along the (x,y) plane is simulated.

The mesh is initially implemented with the default parameters and later improved with the addition of sizing to the relevant areas. Since the goal is to understand the flow behaviour around the net, the surrounding area is important for the simulation. Figure 4 shows the mesh style of the fluid structure in the (x,y) plane. A higher density mesh can be observed in the area where the net is positioned. To implement this mesh, the element size is adjusted to 50 mm and the overall details of the mesh and the pinch tolerance are adjusted. In addition, element sizing is added to the net (0.8 mm) to improve skewness and orthogonality.

The mesh orientation and density at the plane net can be observed in detail in Figure 5. This refined mesh allows for an accurate study of the flow behaviour since it only creates a surface area loss of approximately 2%.

Figure 4. Mesh style along the symmetry plane (x-y).

Figure 5. Mesh style of plane net.

2.2.3 *Numerical algorithm solution*
The segregated SIMPLEC algorithm with Warped Face Gradient Correction (WFGC) is used to iterate the coupling between pressure and velocity. The discretization is implemented with a second-order scheme for pressure, second-order upwind scheme for momentum and turbulent parameters. To initialise the solution, the hybrid method is used and checked for convergency of1e – 6.

The calculation procedure is considered valid if the residuals are inferior to 0.0001 and the difference between two iterations of drag force calculations is inferior to 0.001. This parameter is added to the solution monitoring and to the convergency conditions.

2.2.4 *Drag coefficient*
The goal is to obtain the results such as drag force, drag coefficient, and flow velocity. The drag coefficient is added in FLUENT as a solution report. The software calculated the drag coefficient based on the reference values inputted in the solver.

$$C_d = \frac{2F_d}{\rho A u_0^2} \tag{5}$$

where F_d was the drag force, ρ the water density, A the effective area of the structure and u_0 the inlet flow velocity. In this study drag coefficient was obtained based on both the outline area of the structure ($A_{outline}$) and the projected area ($A_{projected}$) which were considered for the (y,z) plane.

2.3 *The mechanical structural model*

The mechanical structural model is introduced to obtain the deformation on the structure. The workbench tool Static Structural that uses the Mechanical APDL solver to solve the equations of motion was used.

2.3.1 *Governing equations and assumptions*
This simulation is a linear static analysis where the displacement is solved for the material properties matrix, (K) and the force that is being applied on the structure, (F) as follows:

$$[K]\{x\} = \{F\} \tag{6}$$

It is assumed that the structures have a flexible behaviour and that the material is linearly elastic, the deflections are small, [K] is constant. Inertial effects are also neglected, such as mass and damping, and it is assumed that the force does not change with time, {F} is statically applied.

2.3.2 *Materials*
For the analysis, the Young's modulus and Poisson ratio are crucial. These parameters are handled in the Engineering Data where an extended set of materials are characterized.

In the present study, the net is a plastic, PA type 6 commonly known as nylon and the frame is structural steel. In addition, the model was implemented for the other typical used materials in netting, PE, High Density PE (HDPE) and PET whose properties are defined in Table 2.

2.3.3 *Mesh and boundary conditions*
The cartesian system in the mechanical structural model remains the same as the fluid model. However, in these analyses the domain (water tank) is suppressed and only the net is studied. To study the deformation, the frame of the net was considered as a fixed support, which enabled the net to move freely. The fluid pressure acting on the net is imported from FLUENT and defined as a distribution of pressure applied on the structure as defined in Figure 6.

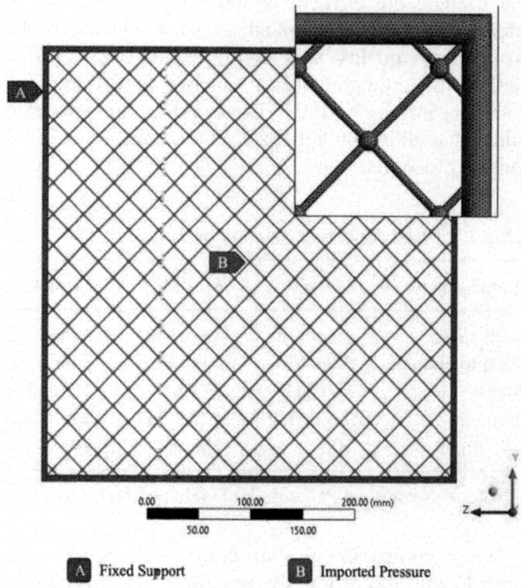

Figure 6. Plane net conditions in mechanical structural model.

2.3.4 *Von-Mises stress*
The Von Mises stress is a theoretical stress metric employed for estimating yield failure criteria in ductile materials and finds common application in fatigue strength calculations, where it is assigned a positive or negative value based on the dominant Principal stress. In contrast, principal stress represents a more tangible and directly measurable form of stress. The von Mises stress is highly valuable as it offers a single value for direct comparison with a material's yield strength or tensile strength.

If the von Mises stress is superior to the material's yield stress, then the material is inclined to plastic deform or fail.

3 VALIDATION AND RESULTS

3.1 *Convergency study*

Considering mesh size is directly associated with mesh quality, a convergence study is performed to determine the influence of mesh size on the results. Three mesh sizes, 1.5 mm, 1.0 mm, and 0.8 mm are selected for the net body. The mesh sizes are defined as surface sizing attributed to the net. It must be noted that the pinch tolerance in the advanced mesh settings is adjusted for the different sizes to prevent overlap of information (Table 3). The drag force for the different types of mesh is obtained and compared.

When defining the mesh, it is important to ensure good quality of surface mesh, since mesh quality is directly associated with convergence of results. Two parameters are used to evaluate mesh quality: skewness, which determines how close to an ideal equilateral triangle the element is and orthogonality, which refers to the angle between adjacent mesh elements. To assure mesh quality it is confirmed that for all size meshes the orthogonality is superior to 0.1 and the skewness inferior to 0.95. These values are reference values that allow the initialization to converge to 1e-06 without problems.

Table 3. Mesh convergency study parameters.

Characteristics	Mesh A	Mesh B	Mesh C
Mesh size	1.5e-3 m	1.0e-3 m	8.0e-4 m
Pinch tolerance	5.0e-4 m	5.0e-4 m	5.0e-5 m
Nodes	644.311	772.318	1.500.579
Elements	3.632.312	4.367.764	8.606.706
CPU time*	10h	15h	20h

* Note: It is run on a PC with 2.90 GHz and 16.0 GB

When comparing the structure generated by the mesh C with the originally modelled, it is possible to observe a reduction in volume of 2.9% for the net, 2.1% of the frame and overall, 2.7% reduction of volume of the whole structure. It is considered that this error is small enough.

As can be observed in Figure 7, as the incoming velocity increases so does the discrepancy between results for the different meshes. When the incoming velocity is high, there are often rapid changes in flow direction, velocity gradients, and the formation of small-scale features such as vortices or eddies. These dynamic and turbulent flow phenomena require a finer mesh resolution to accurately capture the complex flow patterns and fluid behaviour.

With a coarser mesh (larger element size), the simulation does not adequately resolve the details of the flow, which leads to inaccurate results. The coarser mesh may not capture the smaller scales of the flow phenomenon, resulting in a loss of information and potentially introducing errors into the simulation. It is found that by reducing the element size the drag force

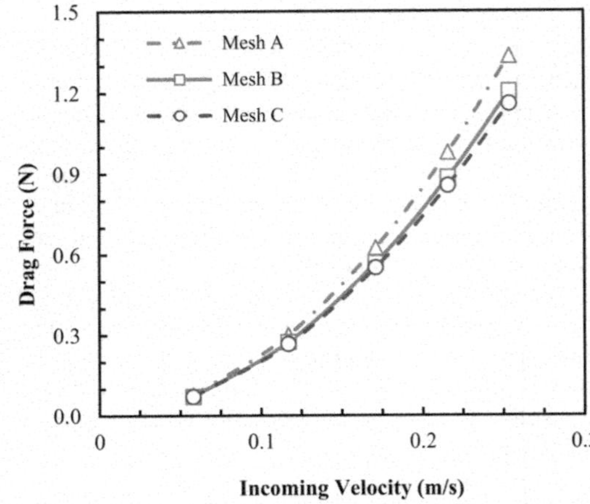

Figure 7. Comparison of drag force for different mesh sizes.

results present a significant improvement, resulting in a reduction of the average error from 19% to 5% when compared to the experimental results (Bi et al. 2018).

3.2 *Drag force*

The hydrodynamic loads that result from the fluid model are obtained for both the realizable k-epsilon and the SST k-omega models (Figure 8). The overall results of drag are in good agreement with the experimental values. For the realizable k-epsilon model the discrepancy starts at 20% for the lowest incoming velocity and improves to 1.5% at the highest incoming velocity. In contrast, the SST k-omega model starts with discrepancy of results of 9% increasing to 16% for the highest incoming velocity.

Considering that the experimental tests used a load cell with accuracy of 0.1 N, while the results for the lowest incoming velocities range around the

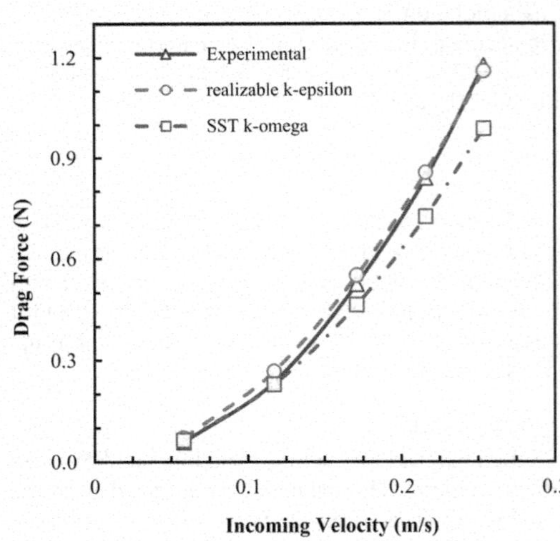

Figure 8. Validation of numerical model for plane net.

values of 0.05 N to 0.08 N, the experimental results for low incoming velocity can be deemed unreliable. As such, the increased discrepancy for low current velocity in the realizable k-epsilon model can be justified and the model considered valid.

The drag coefficient is obtained for an incoming velocity of $u_0 = 0.254$ m/s. As previously mentioned, the drag coefficient calculation depends on the user input area in the reference values (5). As such, the drag coefficient is obtained for the outline area $C_{doutline} = 0.22688$ and the projected area $C_{dprojected} = 1.1417$. When compared to Figure 7 of Bi et al. (2018) both values present good agreement of results with errors below 5%.

3.3 *Flow velocity*

Flow velocity is an important parameter since it not only affects fish behaviour and their health, but also the structure motion. Figure 9 presents the dimensionless flow velocity along the horizontal centreline of the domain for different incoming velocities. The distance to the net is nondimensionalized by the length of the plane net L.

As can be seen, the flow velocity approaches the net at a constant rate, until it encounters the net, where it drops to zero. In the close vicinity of the back of the net the flow velocity propagates in negative direction until 0,02L downstream from the net. This phenomenon, considered as wake, occurs due to fluid separation from the net, presenting an occurring peak at approximately 0,01L downstream from the net. This phenomenon is similar for the different incoming velocities, which implies that the flow maintains a constant behaviour for different current velocities. A more detailed view can be observed in Figure 11.

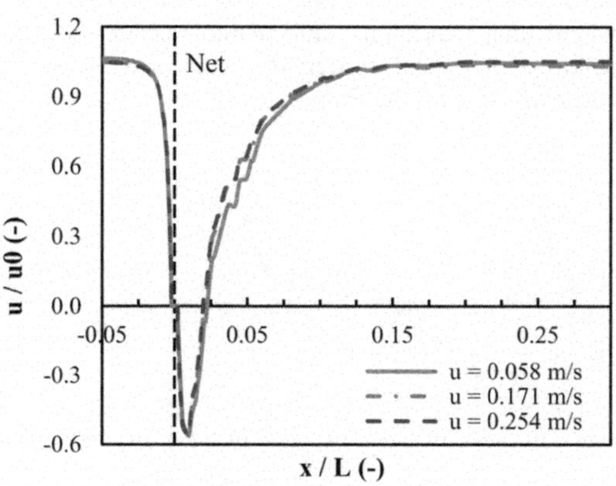

Figure 9. Detail of flow velocity along the centreline of the domain.

In Figure 10 the flow that passes at the centre of the diamond mesh ($y = 0.0145m$) is compared to the

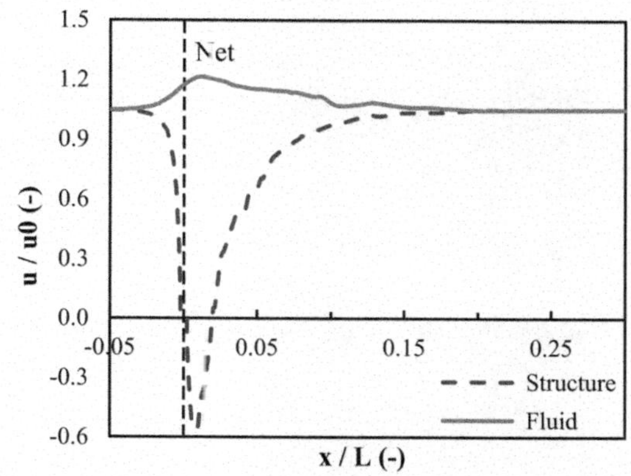

Figure 10. Flow velocity around the net.

flow that encounters the knot of the net ($y = 0.0m$). As expected, when the flow passes through the middle of the diamond mesh, the flow velocity increases. This phenomenon is generated by the reduce in fluid area, since the net creates a blocking effect in the current passage. After this passage the flow increases exponentially until stabilising at the same position, 0.20L downstream form the net.

Figure 11 demonstrates more accurately the flow velocity along the mid plane (x,y). It can be observed that the initial increase in velocity, of about 11%, occurs due do the small formation of boundary layers along the tank walls (blue and green).

The turbulence that follows the passage of fluid through the net can be associated with the passage of a fluid around a cylinder. As the flow surrounds the net, the flow initially accelerates (warm colours) but is followed by deceleration (cold colours). This creates increase in pressure which generates reverse fluid and consequently flow detachment from the net. This phenomenon is intensified by the increase in incoming current velocity as the changes in flow occur more rapidly.

The accumulation of the several individual passages of fluid through the many cylinders that generate the net create a reduction of flow velocity downstream from the net, that does not prevail, with the flow velocity stabilizing quickly.

The flow velocity at 0.8 meters downstream from the net when this is subjected to an incoming velocity $u_0 = 0.254$ m/s is 0.2611 m/s. It represents an error of 9% when compared to the experimental result. Although the error is small, the discrepancy of results can be caused by the boundary layer effect in the tank walls that produces an increase in incoming velocity of 11%.

Based on the presented results the proposed numerical model is considered valid. In addition, the simulation is performed for the knotless net since in the aquaculture market this net is considered more efficient (Cardia & Lovatelli, 2015).

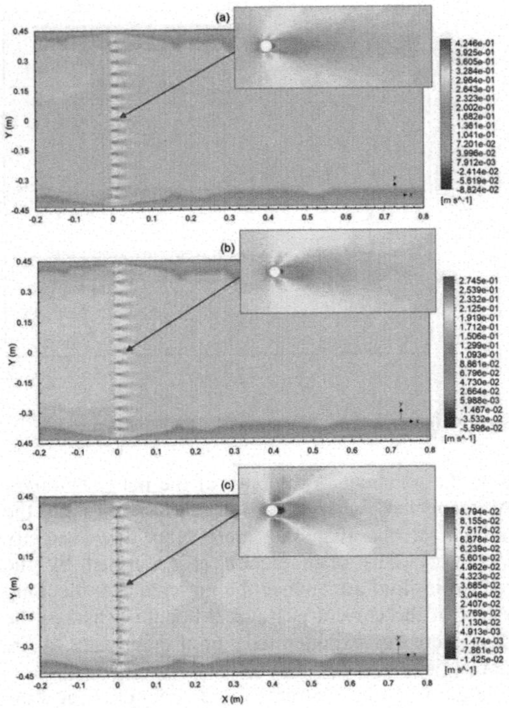

Figure 11. Flow velocity u at the mid plane (x-y) for: (a) u_0= 0.254 m/s; (b) u_0=0.171 m/s;(c) u_0=0.058 m/s.

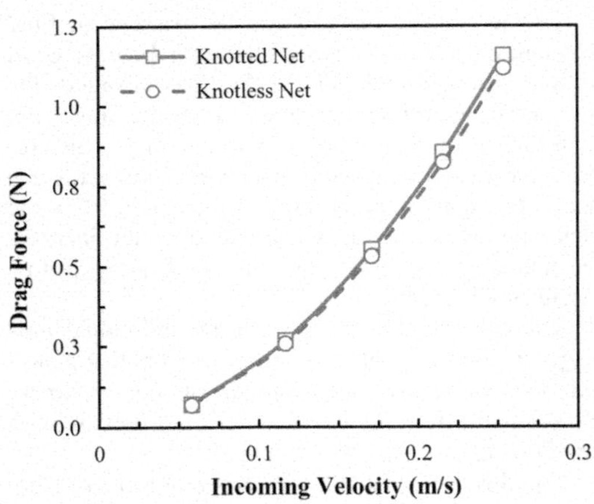

Figure 12. Drag force on knotted net vs knotless net.

Figure 12 illustrates that the knotted net experiences higher drag compared to the knotless net, which aligns with the expectations because the presence of knots leads to an increase in the surface area of the net. Considering that this increase in surface area can be interpreted as a slight elevation in net solidity, the findings are aligned with results from previous studies.

Figure 13 compares the flow behaviour near the net along the centreline of the domain between a knotted and a knotless net. It can be viewed that while the flow reaction is similar, for the knotless net the backflow

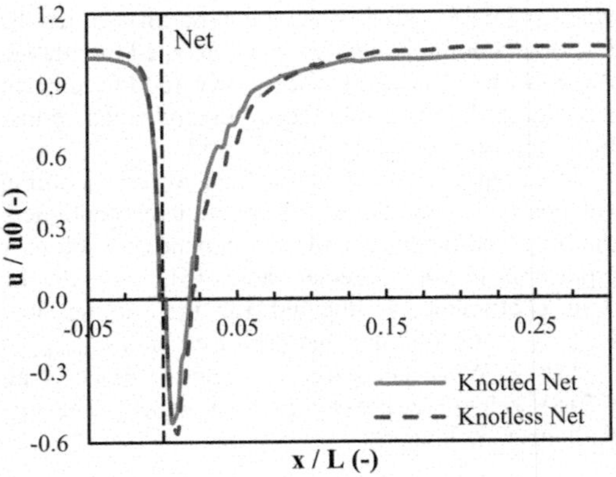

Figure 13. Flow velocity around knotted net vs knotless net.

peak presents superior magnitude, and the stable conditions are reached further away from the net. This result can be associated with the change in encounter diameter in the Reynolds number equation. For the knotless net, the encountering diameter is smaller which for the same velocity results in a higher Reynolds number. This increased parameter is associated with a bigger tendency of flow separation and formation of vortices, leading to an increase in backflow.

Although the different nets present a similar response to fluid encounter, it is noted that the plane net is a small element that is incorporated within a larger framework. The small difference in performance could significantly improve the overall operation of the system.

3.4 *Deformation and stress*

The total deformation along the net for a current velocity of u_0 = 0.254 m/s can be observed in Figure 14 where the maximum deformation occurs at the centre of the net followed by an even decrease of deformation until the frame.

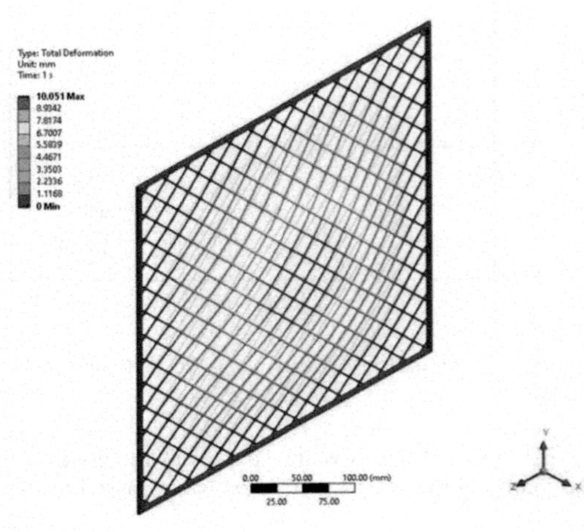

Figure 14. Plane net deformation.

It has been observed that the deformation behaviour remains consistent across various incoming velocities, with the deformation magnitude following a proportional path when associated to the encountering velocity. This proportional relation was expected since it was also viewed in the fluid model for hydrodynamic characteristics. It is found that the stronger the current the more the net deforms. It can also be observed that close to the frame the net presents little to no deformation. This information is important since it can be used in design parameters to reduce net deformation.

In addition, the deformation is studied for the different nets, knotted and knotless. The different nets present similar results for maximum deformation (Figure 15), however, the knotted net significantly reduces the average deformation of the net, especially for higher incoming velocities where this discrepancy is more significant, as shown in Figure 16. The knot not only helps with the distribution of loads (more surface) but it also represents a bigger concentration of material (denser).

The Von-Mises stress is obtained for both the knotted and knotless net to evaluate their behaviour. As described in Section 2.3.4, the Von-Mises can evaluate if the material is at risk of failure or plastic deformation.

As can be observed in Figure 17 and Figure 18 the material never exceeds the yield stress and always presents elastic behaviour. The critical stress points are positioned at the net connection to the frame for all sides. The same phenomenon is observed for the knotless net, yet the stress is more elevated. The reduction in stress can be observed at the knot. The knot provides increased surface area which results in a distribution of pressure along more structure surface.

The results are analysed more in depth for both nets and the maximum stress obtained for different incoming velocities.

In Figure 19 it can be examined that the maximum stress values are almost identical while the knotted net presents some reduction in stress, the improvement is minimal and only significant for high incoming velocities.

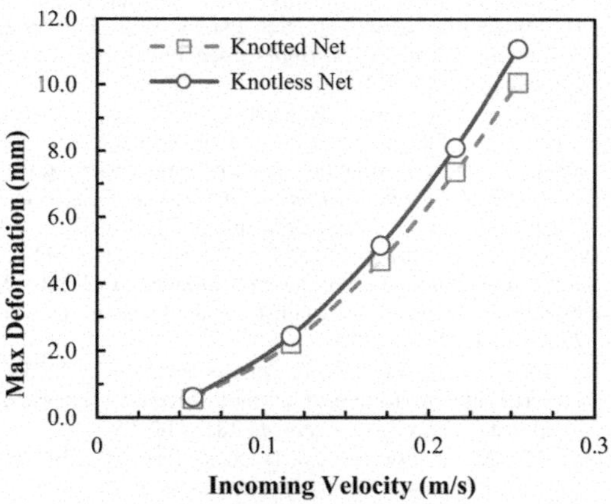

Figure 15. Maximum deformation of the knotted net vs knotless net.

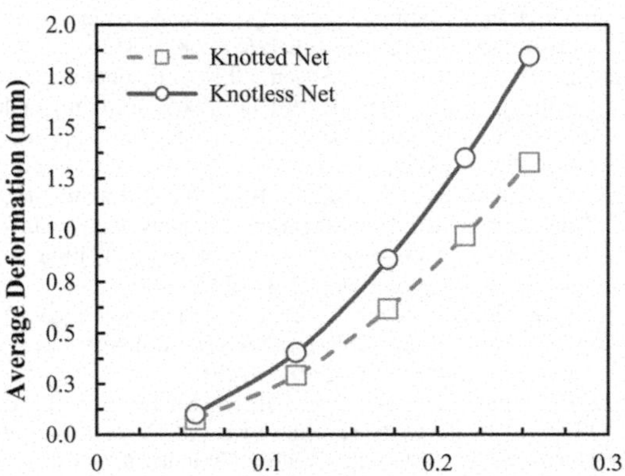

Figure 16. Average deformation of the knotted net vs knotless net.

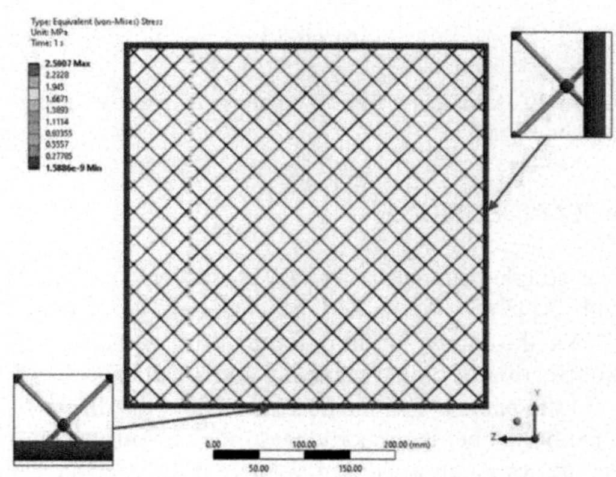

Figure 17. Von-Mises stress of knotted net for u=0.254 m/s.

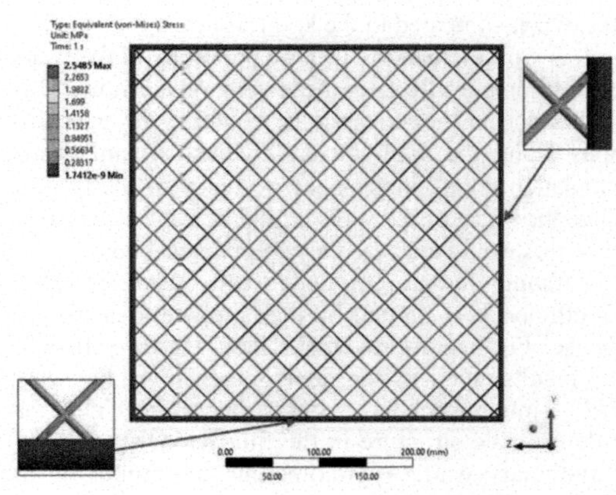

Figure 18. Von-Mises stress of knotless net for u=0.254 m/s.

The knotted net case showcases a better performance than the knotless net. Although it has an increase in load, it significantly reduces deformation. It should be indicated that the deformation of the net generates loss of volume in the fish cage, which is inefficient for the farm operation. A more detailed studied on the relation between deformation of fish cages and loss of volume has been performed by Liu et al. (2023).

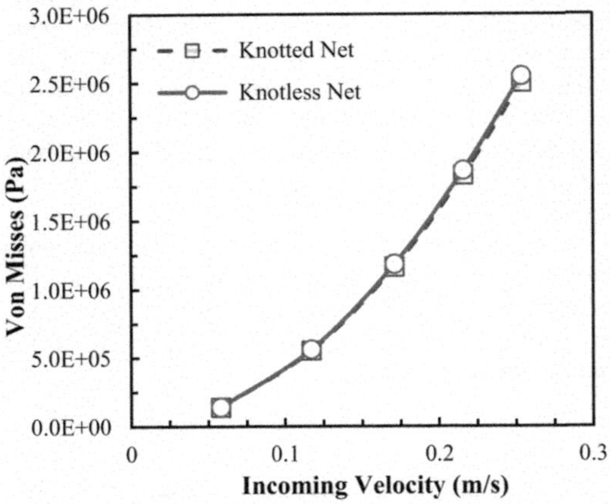

Figure 19. Maximum Von-Mises stress of the knotted net vs knotless net.

4 CONCLUSIONS

The coupled numerical model is implemented with both ANSYS Fluent and Mechanical. This model allows the study of drag force, flow velocity and deformation of both a plane net and a fish cage.

In the plane net simulation, it is found that the drag force on the net is proportional to the current velocity, that the variation in current velocity did not affect the behaviour of flow velocity and that the proximity of the tank walls to the net produces an increase in incoming velocity of about 11%. In addition, it was observed that the knotted net presents reduced deformation of the net, especially for higher incoming velocities, when compared to the knotless net.

It is worth remarking that the proximity of the walls of the domain affect the behaviour of the flow. This occurs due to the existence of boundary layer conditions along the wall. Increasing the domain of the simulation could improve the accuracy of the results, since the effect of the walls would be neglected. However, this would increase the computational cost.

Although the present model was considered valid, its efficiency is limited by the computational power in use. Further improvements could greatly benefit the results, such as the development of a two-way FSI simulation by considering the net as a deformable structure in the fluid model instead of a stationary wall. Approximating the simulation to a more realistic scenario.

ACKNOWLEDGEMENTS

The work contributes to the Strategic Research Plan of the Centre for Marine Technology and Ocean Engineering (CENTEC), which is financed by FCT under contract UIDB/UIDP/00134/2020.

REFERENCES

Abrahamson, J., 2022. Fluent Theory Guide.

Bi, C.-W., Zhao, Y.-P., Dong, G.-H., Wu, Z.-M., Zhang, Y., & Xu, T.-J., 2018. Drag on and flow through the hydroid-fouled nets in currents. *Ocean Engineering* 161, 195–204.

Cardia, F., & Lovatelli, A., 2015. Aquaculture operations in floating HDPE cages: a field handbook. *Food and Agriculture Organization of the United States*, Rome.

Cheng, H., Ong, M.C., Li, L., & Chen, H., 2022. Development of a coupling algorithm for fluid-structure interaction analysis of submerged aquaculture nets. *Ocean Engineering* 243, 110208.

FAO, 2022. *The State of World Fisheries and Aquaculture* 2022. FAO.

Føre, H.M., Endresen, P.C., Norvik, C., & Lader, P., 2021. Hydrodynamic Loads on Net Panels With Different Solidities. *Journal of Offshore Mechanics and Arctic Engineering* 143, 051901.

Guo, Y.C., Liu, Z.C., Mohapatra, S.C., & Guedes Soares, C., 2022a. Experimental investigation of an array of vertical flexible net-type structures under regular waves, in: *Trends in Maritime Technology and Engineering Volume 2*: CRC Press, London, pp. 573–580.

Guo, Y.C., Mohapatra, S.C., & Guedes Soares, C., 2023a. An experimental study on the performance of combined membrane breakwaters against incident waves. *Ocean Engineering* 280, 114793.

Guo, Y.C., Mohapatra, S.C., & Guedes Soares, C., 2023b. Experimental performance of multi-layered membrane breakwaters. *Ocean Engineering* 281, 114716.

Guo, Y.C., Mohapatra, S.C., & Guedes Soares, C., 2022b. Experimental study on the performance of an array of vertical flexible porous membrane type breakwater under regular waves. *Ocean Engineering* 264, 112328.

Katsidoniotaki, E., Shahroozi, Z., Eskilsson, C., Palm, J., Engström, J., & Göteman, M., 2023. Validation of a CFD model for wave energy system dynamics in extreme waves. *Ocean Engineering* 268, 113320.

Lader, P., Jensen, A., Sveen, J.K., Fredheim, A., Enerhaug, B., & Fredriksson, D., 2007. Experimental investigation of wave forces on net structures. *Applied Ocean Research* 29, 112–127.

Liu, L., Kinoshita, T., Wan, R., Bao, W., & Itakura, H., 2012. Experimental investigation and analysis of hydrodynamic characteristics of a net panel oscillating in water. *Ocean Engineering* 47, 19–29.

Liu, Z., & Guedes Soares, C., 2022. Experimental study of the behaviour of a circular gravity cage in linear waves. *Aquacultural Engineering* 99, 102291.

Liu, Z., Mohapatra, S.C., & Guedes Soares, C., 2021. Finite Element Analysis of the Effect of Currents on the Dynamics of a Moored Flexible Cylindrical Net Cage. *JMSE* 9, 159.

Liu, Z., Xu, H., & Guedes Soares, C., 2023. Experimental study on the mooring forces and motions of a fish cage under regular waves. *Ocean Engineering* 280, 114612.

Liu, Z.C., & Guedes Soares, C., 2022. Numerical study on the mooring force of a gravity-type fish cage under currents and waves, in: *Trends in Maritime Technology and Engineering Volume 2*: CRC Press, London, pp. 591–597.

Mohapatra, S.C., Bernardo, T.A., & Guedes Soares, C., 2021. Dynamic wave induced loads on a moored flexible cylindrical net cage with analytical and numerical model simulations. *Applied Ocean Research* 110, 102591.

Mohapatra, S.C., & Guedes Soares, C., 2024. A semi-analytical model of an array of moored floating flexible offshore net cages under current loads. *Ocean Engineering* 291, 116309.

Wang, G., Martin, T., Huang, L., & Bihs, H., 2022. An improved screen force model based on CFD simulations of the hydrodynamic loads on knotless net panels. *Applied Ocean Research* 118, 102965.

Xu, T.-J., Dong, G.-H., Tang, M.-F., Liu, J., & Guo, W.-J., 2021. Experimental analysis of hydrodynamic forces on net panel in extreme waves. *Applied Ocean Research* 107, 102495.

Zou, B., Thierry, N.N.B., Tang, H., Xu, L., Dong, S., & Hu, F., 2022. The deformation characteristics and flow field around knotless polyethylene netting based on fluid structure interaction (FSI) one-way coupling. *Aquaculture and Fisheries* 7, 89–102.

Environmental modelling

Advances in Maritime Technology and Engineering – Guedes Soares & Santos (Eds)
© 2024 The Author(s), ISBN 978-1-032-83107-7

Forecast uncertainty considerations in ship weather routing

M. Balas & J. Prpić-Oršić
Faculty of Engineering, University of Rijeka, Rijeka, Croatia

M. Valčić
Maritime Department, University of Zadar, Zadar, Croatia

ABSTRACT: The global maritime industry has undergone tremendous change due to a number of circumstances that require more environmentally friendly and effective shipping methods. With an intention of contributing to the broader discussion on sustainable shipping practices, this paper addresses the topics of weather routing and the ongoing attempts to minimize the industry's environmental impact. Weather routing determines the optimum route for a ship, subject to fuel consumption, ship characteristics, behavior and weather forecasts. Significant risks and uncertainties associated with weather routing still must be further considered. Weather forecasts are essential for ship weather routing and therefore need to be as accurate as possible. However, the initial data for weather forecasts calculation is subject to a fair amount of uncertainties. In this paper, two main objectives are discussed, i.e. how to model forecast uncertainties and how to implement that model into an optimal ship routing problem.

1 INTRODUCTION

Climate change, global warming and extreme weather events are re progressing ever faster. Energy efficient and economical shipping solutions are in demand now more than ever as the weather affects companies supply chains in new and surprising ways. Incorrect weather forecasts can lead to delays of miles or days at best, and fatalities at worst. An essential part of cost-effective shipping operations is efficient and sustainable marine transportation. Advances in voyage optimization and weather routing techniques is mainly driven by the ongoing need to improve financial stability, energy efficiency and safety while complying with emission limits. In the past, numerous different shortest path algorithms have been developed as a result of ship weather routing research. The various methods can be divided into three categories: grid-based methods, evolutionary optimization, and calculus of variations. The quality of the proposed route depends on the quality of the data as well as the optimization problem's modeling and the method used to solve it. However, only some uncertainties regarding ship routing have been studied. Over the past decade the shipping industry has experienced exponential growth and continues to do so. In light of its importance to the global economy, many unresolved problems regarding uncertainty and reliability arise. While there are a decent number of papers published addressing these topics, some questions remain unanswered. Weather forecasting is still a technologically complicated problem, and making an accurate prediction is impossible. There are three main reasons why weather forecasting is not and can never be accurate: data analysis uncertainties, model limitations, and the chaotic and unpredictable nature of the atmosphere. Prior to making a weather forecast, quantitative data about the current state of the atmosphere is collected and by using knowledge of its processes, predictions are made how the atmosphere will change in the future. To minimize the human error and offer the best support for decision making, reliable data is of crucial importance, and for that matter addressing uncertainties in weather forecasts is essential for ship weather routing. On that note, the implementation of weather uncertainties into ship routing optimization problems remains a challenge. One of the most important reasons behind this is related to weather forecast providers that use a variety of different algorithms and methods, many of which they never fully disclose for commercial reasons. As a result, not all weather and environmental forecasts are equally accurate and reliable. The main focus of this paper are the quantification and implementation considerations of weather forecasts and their uncertainties into optimization algorithms, as a basis for more reliable weather routing. By incorporating weather uncertainties in voyage optimization problems, they can be solved with stochastic programming methods. This approach, along with probability techniques, has proven successful in

DOI: 10.1201/9781003508779-64

solving various industrial optimization problems that involve uncertainty.

This paper is organized as follows. In Section 2, a general background and relevant literature overview is given for the topic of ship weather routing. In Section 3, the concept of a weather forecast is explained, along with its uncertainties and quantification challenges. In Section 4, a discussion about stochastic optimization and a general optimization framework are provided. Finally, a conclusion is drawn in Section 5.

2 SHIP ROUTING BACKGROUND

The International Maritime Organization's (IMO) greenhouse gas (GHG) strategy aims to reduce carbon emissions from international shipping by 40 % by 2030 and 70 % by 2050 (IMO, 2023). Fuel consumption minimization and gas emissions reduction have been studied in the past (Armstrong, 2013), also proposing a methodology for estimating the attainable speed in moderate and severe sea states, associated with CO2 emissions (Prpić-Oršić & Faltinsen, 2012). Some influences on CO2 emissions and fuel consumption, such as loading condition and the initial speed, have been presented by Prpić-Oršić et al. (2016). However, the comprehensive combination of environmental and economic impacts has rarely been considered. Approaches to optimal ship routing have differentiated over the years, with various methods being used, ranging from modelling the weather routing problem as a constrained graph problem, a constrained nonlinear optimization problem or as combination of both (Walther et al., 2016). Several graph theory algorithms were quite common, such as the A-star (Shin et al., 2020), or the well-known Dijkstra algorithm (Silveira et al., 2019), along with the Dempster-Shafer theory (Neumann, 2016). Dynamic programming, which decomposes routing into a multi-stage decision process, has been widely used to address problems such as deterministic minimum fuel consumption by applying optimal control theory (Avgouelas, 2008). The future of shipping is in autonomous navigation, with most of the recent research focusing on readily available ship performance models, or digital twins, as presented by Wei et al. (2023), as well as continuous dynamic optimal control of unmanned ships (Wang et al., 2022). Nature inspired optimization algorithms are also gaining popularity, with genetic algorithms being used in ship routing to enhancing the ability of finding the optimal solution (Zhou et al., 2023), while improving the robustness of the algorithm through various evolutionary strategies. Ship weather routing, as a part of voyage planning, is defined as determining the optimum ship's route, considering fuel consumption, the ship characteristics and most importantly, weather forecasts and sea states (Shao et al., 2011). An important factor in any voyage optimization and ship routing is the ship's speed loss under various environmental conditions (Sasa et al., 2019, Prpić-Oršić et al., 2020, Vitali et al., 2020). The importance of ship's motions and added resistance due to weather effects has been early recognized on as a crucial factor in ship routing (Kwon, 1981). Statistical analysis of wave effects on ships has also been studied (Chen et al., 2021), further highlighting the importance of ship motions during navigation. Wave height and period uncertainty have only recently been quantified using wave hindcasts (Kodaira et al., 2023, de Hauteclocque et al., 2023). More recently, different methodologies for the prediction of uncertainties associated with to ship fuel consumption and ensemble predictions have been investigated, with fuel consumption predictions repeated for each part of the ensemble and compared with probabilistic approaches in which wave parameters are modelled (Vettor & Guedes Soares, 2022). In current and future work, quantified uncertainties of weather forecasts and associated environmental conditions, combined with realistic ship performance models should be used in voyage planning to meet the growing maritime industry demands regarding the reliability and safety of the process.

3 WEATHER FORECASTS

A weather forecast can be defined the result of predicting the state of the atmosphere for a particular place and time using a combination of science and technology. In risk evaluation, a wide range of possible forecast scenarios is considered, not only the scenario which is deemed most likely. For risk management purposes, forecast confidence and uncertainty information are essential for optimal decision-making regarding safety of life and property. Each forecast user has unique requirements subject to their specific operations, meaning that informative decision-making requires effective and continuous communication between forecasters and users. Forecast quality and reliability are quantified using specific metrics, such as the forecast skill, which is a measure of accuracy compared to a baseline prediction. Potential meteorological events in the future are dictated by the event's size, impact and timing, with predictability being higher for larger-scale features at shorter lead times. A forecast can be considered skillful when if it informs the specific user for their actions and is used to make decisions.

There are several different types of forecasts in terms of the time period they cover, ranging from very short to long-term forecasts. In ship routing, medium-range forecasts are most commonly used, which are typically defined as those with lead time periods of up to 10 days. As with any forecast, the predictability progressively decreases over time due to the chaotic nature of the atmosphere. Today, forecasts are usually generated by a specialized computer software or a weather model. A set of meteorological data acquired by weather stations, radars, satellites, and other weather devices in order to produce a weather forecast at specific locations on Earth. These models are also frequently referred to as numerical weather forecast models and they consider a variety of data. There are over 150 external variables in models that can be entered or generated by them. The points at which a weather model solves equations to produce weather forecasts are not randomly distributed; rather, they constitute a grid that can either span the entire globe or just a portion of it. Two main types of weather models are available:

1. **Deterministic**. GFS (Global Forecast System) and ECMWF, ICON (Icosahedral Nonhydrostatic). GFS is an American model, ECMWF is a model from England, and ICON is from Germany,
2. **Ensemble** (blended or combined forecasts).

In a deterministic weather model, a single set of forecasts which are believed to be the most likely is generated. Deterministic models are considered good for the first few days, but later in the forecast their uncertainty grows with lead time. Ensemble weather models are generated by processing a number of variations of the deterministic models. It is generally accepted that ensemble models are better for forecasting over longer time periods.

The input weather data used for modelling often comes as a specific file format called GRIB (General Regularly distributed Information in Binary form) and is a WMO (World Meteorological Organization) standard format that defines space-time grids. For example, for each time stamp, and each point of the grid, the intensity and direction of the wind is indicated. The grid length or grid spacing is usually 0.1 degree which translates to 10 km distance. Nevertheless, it's crucial to keep in mind that GRIB files are computer-generated forecast files from a National Weather Service computer. They are provided as-is and passed on without review. There is little assurance that the required data will be available, or even accurate. Since the information-providing systems and the computer models are automated, they are susceptible to a number of inaccuracies or malfunctions. Such limitations must be considered before using the given data.

Local and global models are two additional types of models. The distance between two grid points, which is a measure of resolution, varies for both of these types. Since it is challenging to anticipate weather in general, the larger the area, the more complicated it is. However, all local models assume a global forecast before making their own predictions. The three primary global models include the ECMWF and ICON from Europe, as well as the GFS from the United States. There is a general consensus that the ECMWF, a European global seamless forecast model, is the most trustworthy model currently in use. It makes advantage of the 4D assimilation principle, which enables the model to be continuously updated as new satellite or other input data become available. It has different levels of resolution and its forecast lead time is 10 days with a 3-hour time step (ECMWF, 2023).

3.1 *Marine weather*

The atmospheric and wave conditions at sea, which primarily account for sea states and wind data, are referred to as marine weather. Since they usually cover larger regions of the ocean, marine forecasts are slightly more complex compared to onshore weather forecasts. Therefore, the predicted elements are often expressed as a range rather than a specific metric. Marine forecasts usually include the following parameters: wind (including direction, speed, and wind gust), sea state, large waves, horizontal visibility, fog or mist information, other types of obscuration to visibility, thunderstorms information and tropical cyclones (NOAA National Weather Service, 2023). The key aspect of marine weather forecasting is the advance notice of hazardous events.

Ranges provided in a marine forecast describe average conditions for a specific time period (often 12 hours), while the actual circumstances may be higher or lower than the anticipated range. For safety reasons, masters and chief officers should plan for conditions above and below the ranges during the voyage. The performance of large ships can be affected up to 80% by receiving inaccurate marine meteorological data (SOFAR Ocean, 2023). When navigating in adverse conditions, the resistance often increases by 50-100% of the total ship resistance in calm weather. According to MAN Energy Solutions' study of the trading circumstances for a typical 140,000 dwt bulk carrier, the increased resistance, or sea margin, may exceed extreme values of up to 220% on certain voyages (Marine Insight, 2023).

In terms of data collecting methods, marine weather forecasting continues to be affected. The integration of buoys, aircraft, satellites, and various other data collection systems makes accurate forecasting difficult. This is further limited by inadequate amounts of data, reporting discrepancies, and the ongoing challenge of collecting information from remote regions. Some commercial forecast providers have better resources for gathering weather data. Typically, mariners and shipping companies select their preferred commercial forecast providers. However, vital and urgent safety information is still transmitted through VHF (very high frequency), MF (medium frequency), NAVTEX (NAVigational TeleX), HF (high frequency), SSB Radio (Single Side Band), and Inmarsat/Iridium EGC (Enhanced Group Calling). The GMDSS's (Global Maritime Distress and Safety System) primary means for transmitting Marine Safety Information (MSI), NAVTEX, should be used by all vessels operating more than thirty nautical miles offshore. The optimization of ship routes benefits greatly from marine weather forecasts, also reducing insurance costs, preventing overtime surcharges for crew, and additional port fees. All things considered, this helps shipping corporations in establishing healthier profit margins which can be invested in fleet optimization and achieving decarbonization targets.

3.2 *Uncertainties*

Many different organizations and industries, including shipping and marine industry, need timely and precise weather forecasts, but also require quantification of their uncertainties. Due to the complexity and chaotic nature of atmospheric processes, as well as knowledge limitations, there inevitably remains a level of uncertainty in the weather model output. A reliable uncertainty quantification is essential to sustainable decision-making for adopting efficient strategies. The weather model input errors, its structure and parameters are the main causes for said uncertainties. Decision support systems are not able to aid users in making informed assessments of possible risks associated with uncertain weather data.

Initial conditions in a system and its data directly affect the uncertainty of the final result. The initial datasets that are used for calculating forecasts can never be completely accurate. They contain different uncertainties and errors since they often come from different sources and systems which have their own limitations, such as weather stations or weather satellite systems. For that reason, all weather models make somewhat inaccurate weather forecasts.

To visualize the process of collecting weather data, it is worth mentioning that the World Meteorological Organization collects them from 15 satellites, 100 stationary buoys, 600 drifting buoys, 3,000 aircraft, 7,300 ships, and some 10,000 land-based stations (WMO, 2023). When uncertainties of the weather routing process are properly quantified, it allows for the reliability of predictions to be quantified as well. The overall quantification of these uncertainties reflects on the accuracy expectations of the selected route. Two main factors in any forecast can lead to a decrease in accuracy as the lead-time period increases: uncertain initial conditions, and a numerical model of the atmospheric system with approximations used in its construction.

These two sources of inaccuracy should not be considered independently since the initial conditions themselves are constructed in part by integrating the underlying numerical model as part of the initialization process through data assimilation (ECMWF, 2023). Depending on the initial state of the atmosphere, even a relatively small amount of uncertainty in the initial conditions will eventually result in uncertainty in the forecast. In light of this, the NWP community began using probabilistic forecasting techniques, especially beyond the deterministic lead time limit of one week. Early applications of probabilistic techniques for numerical weather forecasting relied on the application of small-scale random perturbations to the atmospheric state variables (temperature, wind, and pressure) in the examined initial condition. Due to the nonlinearity of the atmosphere, the perturbations are intensified by chaotic atmospheric processes, making each forecast diverse from the other (Slingo & Palmer, 2011). The ECMWF weather prediction model, for instance, is run 51 times with slightly varying initial circumstances. The perturbed members, a total of 50 extra integrations, are initiated with somewhat modified initial conditions (ECMWF, 2023).

The degree to which the forecast's uncertainties are accurately constituted, dictates how well the probabilistic methods for weather forecasting perform. These ensemble uncertainties directly affect fuel consumption predictions in weather routing, which was recently addressed in Vettor & Guedes Soares (2022). It is crucial to make a distinction between model uncertainty which stems from inadequate representation of sub-grid scale phenomena because of the model's resolution and model uncertainty that results from imperfect knowledge of the actual system.

3.3 *Quantification challenges*

More recently, forecast uncertainty in weather data has been quantified using conditional probability and updating models such as auto-regressive (AR) model, AR exogenous input model. Adaptive neuro fuzzy inference system model proved to be an effective measure for improving the accuracy and reliability of probabilistic forecasts (Ba et al., 2019). The non-linear weather conditions have been processed and forecasted using deep learning as well, creating a powerful AI-based weather forecasting system that produces stronger deterministic forecast results than the operational Integrated Forecasting System (IFS) on all tested weather variables against reanalysis data (Bi et al., 2023). Weather generators such as CLIMARK and LARS-WG have been used to interpret future projections and assess the forecast uncertainty (Vesely et al., 2019). Hidden Markov Models (HMM) have been developed to forecast droughts and measured using different forecast skill criteria which provided a good agreement between observed and forecasted values in terms of time series on various lead time (Khadr, 2016). Monte Carlo Method (MCM) based approach for processing weather forecast data using the 24-h-ahead Support Vector Machine (SVM) was proposed by Zhao et al. (2018). A simple probability-based approach was proposed in Vettor et al. (2021), accounting for ensemble weather uncertainties.

When it comes to ship weather routing, upon choosing the reference ship and time period, its routes and schedule need to be explored and weather data along the chosen route provided. Since most shipping companies use commercial weather forecast providers, they receive weather updates directly on board. As a step toward quantifying the forecast uncertainty, the on-board received forecast data is needed for proper model calibration. Real-time weather forecast data as well as historical climate reanalysis data are necessary for modeling and optimization to achieve better routing. Monte Carlo Method is an effective method for assessing the likelihood that an event will occur when dealing with extreme values. In the case of weather forecasting, the MCM should be based on actual weather data, i.e., the same data used for the actual forecast, as an attempt to generate a scenario analysis for a range of meteorological parameters. A limitation to consider when modelling weather patterns, is that the distribution will vary based on the geographical location of the observed area (Baccon & Lunardi, 2016). In order to properly model weather simulations, it is necessary to understand the distribution of the time series to be used for MCM. The simulated forecast data and the actual forecast data obtained from the forecast providers must be compared and the predictive accuracy must be evaluated using model fit measures.

After the designated observation period has elapsed, the actual future weather data can be obtained and used as a reference system with respect to which the actual and the generated forecast data are calibrated. Comparison of the final three datasets can be done using several methods, one of which is the Triple Collocation Method (TCM), a unique method for assessing the uncertainty of three different data sets with independently occurring errors. It has found widespread application in a variety of geographical disciplines. Recent studies indicate that TCM is capable of merging datasets from different sources (Lyu et al., 2021). However, explicit assumptions about the random and systematic error distributions are required, i.e., a realistic error model must be defined and tested (Stoffelen & Vogelzang, 2012). The representation error and calculation methods for TCM are further explained and detailed in research by Stoffelen & Vogelzang (2012), but the used method depends on the data availability.

4 STOCHASTIC OPTIMIZATION

The methodology of stochastic optimization acknowledges that there is underlying uncertainty in the input model parameters, and incorporates probabilistic distributions. Multiple scenarios can be considered for informed decision-making under various potential outcomes. This approach also mitigates risks, and enhances the overall reliability of its solutions. The diverse fields of stochastic optimization that might be relevant for solving different ship routing problems include simulation optimization, optimal control, robust optimization and reinforcement learning. Simulation optimization requires the use of a simulator to generate and compare different designs or scenarios. Its application can be further extended and explored within the context of ship routing. Optimal control is typically used for deterministic problems with uncertain parameters and additive noise in the transition function. As with its implementation in the control of aircrafts and robots, it can also be applied to ships, ultimately paving the way for autonomous shipping. Robust optimization is an extension of stochastic search where the goal is to find the best design x (e.g. a route option), under the worst instance of a random variable W (e.g. wind speed, wave height). Reinforcement learning started out as a different form of approximate dynamic programming, but has since evolved. It is essentially a class of sequential decision problems, and its possible adoption into ship routing can be further explored in future studies. Detailed information about the fields mentioned above can be found in literature by Powell (2019). A general framework for utilizing stochastic optimization methods for solving ship weather routing problems is presented in Figure 1 and discussed in the following subchapters.

4.1 *General framework outline*

Implementing stochastic optimization within the framework of ship weather routing involves a comprehensive approach in order to address the inherent uncertainties in weather forecasts. After the decision variables, objective function and constraints are clearly stated, a set of scenarios for the range of possible weather outcomes can be generated.

This method integrates probabilistic distributions that account for the dynamic nature of weather patterns. Probabilities are then assigned to each ship routing scenario within the chosen area, based on historical data and weather forecasts. By considering various weather scenarios and their probabilities, this allows route adaptation to changing conditions, potentially reducing travel time and fuel consumption, which can be the optimization's objective functions. Although stochastic optimization methods offer invaluable potential advantages in assessing uncertainties within ship weather routing since they acknowledge and adapt to unpredictable weather patterns, the use of probabilistic forecasts can often be computationally complex and might not always capture the full spectrum of uncertainties accurately (Powell, 2019). Robust optimization, on the other hand, emphasizes the worst-case scenarios, but might generate conventional routes that impact the ship's operational efficiency even when faced with moder-ate weather conditions. When the optimization mod-el accurately incorporates the problem and the under-lying uncertainty, it can be validated, and its parameters adjusted.

4.2 *Discussion*

Stochastic optimization methodologies can account for the unpredictability of weather forecasts, enabling flexible routing options that can adjust to varying weather patterns. Robust

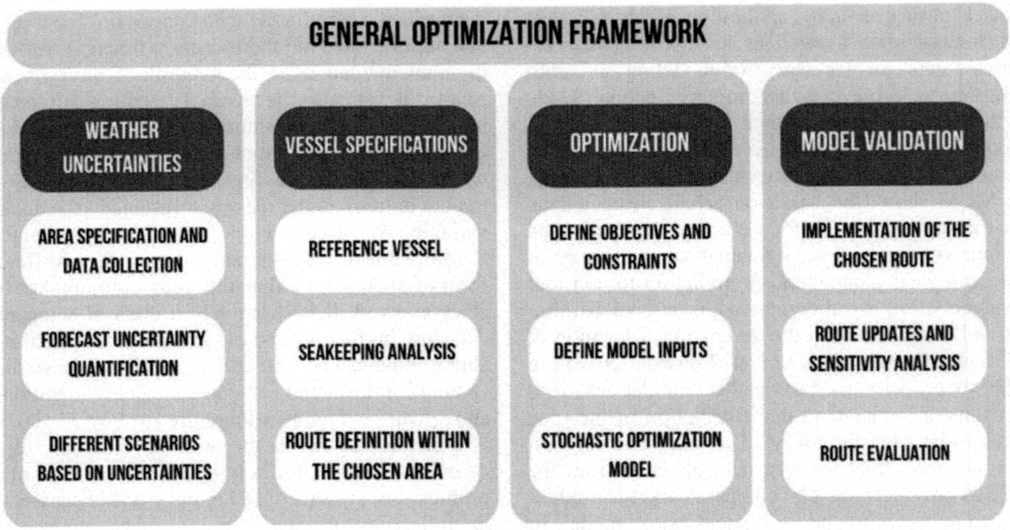

Figure 1. A general optimization framework for implementing uncertainties.

optimization mitigates the impact of forecast uncertainty so that routes can remain reliable even in the worst-case weather scenarios, enhancing the overall safety. However, these approaches have their limitations. Incorporating weather forecast uncertainty can significantly increase computational complexity, potentially affecting the real-time decision-making onboard ships. Moreover, relying on imperfect weather forecasts can impact the final reliability of the optimized routes. Balancing adaptability with decision-making under forecast uncertainty can be a problem. Most stochastic optimization methods require continuous improvement in order to compensate between adaptability and reliability, while considering the limitations of weather forecasts. Balancing these approaches while acknowledging the weather forecast limitations in real-time operations should be considered in future work.

For future research, stochastic optimization models that can account for a wider range of uncertainties need to be further investigated as well. This includes fuel cost uncertainties or vessel's speed loss uncertainties. With recent advancements in data collection and AI solutions, real-time weather data can potentially be used for simultaneously updating the optimized route. Reinforcement learning can be used to gain insight from historical data and continuously adapt to current weather patterns. Risk-aware approaches to ship routing, which consider the potential risks associated with severe weather conditions, should be explored in terms of safety. The frequent changes in environmental regulations also need to be kept up with to minimize the shipping industry's imprint and meet the sustainability goals. The future of this research field offers many possibilities for improving the efficiency, safety, and sustainability in the maritime sector.

5 CONCLUSION

Advances in satellite technology have greatly improved cyclone and adverse weather forecasting, and the international broadcasts of marine weather observations has made it possible to communicate important marine weather forecasts on a much wider degree. The number of fatalities caused by extreme weather conditions at sea has been significantly reduced thanks to two advances. However, the accuracy and uncertainty of weather forecasts continues to decline. In today's economy, where maritime shipping has a major role, shipowners are looking into sustainable solutions mainly due to the global climate change. IMO regulations targeting ships and shipping routes are constantly changed and updated. Fuel consumption and gas emissions are being regulated as well. To satisfy all the mentioned parameters, weather routing and route optimization approaches offer solution strategies complying with specified requirements. Any improvements in ship voyage planning, and weather routing respectively, directly positively affect the supply chain and logistics. Reliable data is essential for optimizing ship routes, and according to the IMO, weather routing saves at least 3% of fuel consumption.

To summarize, an optimal and satisfactory route has several aspects:

1. It saves fuel and time; it calculates the routes with the lowest costs,
2. It is safe; the risks of severe weather conditions are avoided,
3. It reduces the captain's workload; with a good routing and decision support system, the captain makes well-informed decisions,

4. It improves operational efficiency; downtime and damage to ship and cargo are reduced.

The integration of weather forecast uncertainties in stochastic optimization models within the context of ship routing is a good approach that offers considerable potential for meeting the aspects mentioned above. However, due to the large amount of data involved, stochastic optimization can sometimes be very computationally intensive.

In current and future work, quantified uncertainties of weather forecasts and associated environmental conditions should be used in combination with realistic ship performance models in voyage planning in order to meet the growing demands of the maritime industry on the reliability and safety of the shipping process. Route optimization in weather routing also has a series of limitations that should be considered when choosing the appropriate algorithm. The inaccuracy of GRIB files predictions, their space/time interpolation and unrealistic polars are just some of the limitations mentioned. For this reason, different routing algorithms with a focus on robust optimization and simulation optimization need to be further explored. In route optimization under uncertainty of weather and ship speed, no uniform solutions are investigated, but rather a distribution of solutions for each scenario. In this context, the field of stochastic optimization might be the best choice to obtain satisfactory solutions.

ACKNOWLEDGEMENTS

This work has been fully supported by the Croatian Science Foundation under the project REliaBility and Uncertainty of Ship decision support system for safe and greener navigation (REBUS), IP-2022-10-2821.

REFERENCES

Armstrong, V. N. 2013. Vessel optimization for low carbon shipping. Ocean Engineering, 73:195–207.

Avgouelas, K. 2008. Optimal Ship Routing. Thesis: Massachusetts Institute of Technology.

Ba, H., Guo, S., Zhong, Y., He, S., Wu X. 2019. Quantification of the forecast uncertainty using conditional probability and updating models. Hydrology Research. 50.

Baccon, ALP., Lunardi, JT. 2016. Testing the shape of distributions of weather data. Journal of Physics: Conference Series 738.

Bi, K., Xie, L., Zhang, H. et al. 2023. Accurate medium-range global weather forecasting with 3D neural networks. Nature 619, 533–538.

Burton, J. 1986. Robert Fitzroy and the Early History of the Meteorological Office. *The British Journal for the History of Science*, 19(2), 147–176.

Chen, C., Kenji, S., Prpić-Oršić, J., Mizojiri, T. 2021. Statistical analysis of waves' effects on ship navigation using high-resolution numerical wave simulation and shipboard measurements. Ocean engineering, 229: 1–14.

Chen, H. 1978. A dynamic program for minimum cost ship routing under uncertainty. PhD thesis: Massachusetts Institute of Technology.

de Hauteclocque, G., Vitali Maretic, N., Derbanne, Q. 2023. Hindcast based global wave statistics. Applied Ocean Research. 130.

Dickson, T., Farr, H., Sear, D., I R Blake, J. 2019. Uncertainty in marine weather routing. Applied Ocean Research, 88: 138–146.

European Centre for Medium-Range Weather Forecasts (ECMWF), Quantifying forecast uncertainty.https://www.ecmwf.int/en/research/modelling-and-prediction/quantifying-forecast-uncertainty (accessed: 20/11/2023)

F. Lyu et al. 2021. Precipitation Merging Based on the Triple Collocation Method Across Mainland China. IEEE Transactions on Geoscience and Remote Sensing, 59: 3161–3176.

International Maritime Organization (IMO). Improving the energy efficiency of ships. Istanbul Technical University, Research Project.

International Maritime Organization (IMO). Initial IMO Strategy on Reduction of GHG Emissions from Ships. https://www.imo.org/en/OurWork/Environment/Pages/2023-IMO-Strategy-on-Reduction-of-GHG-Emissions-from-Ships.aspx (accessed: 10/12/2023)

Khadr, M. 2016. Forecasting of meteorological drought using Hidden Markov Model (case study: The upper Blue Nile river basin, Ethiopia). Ain Shams Engineering Journal, 7 (1): 47–56.

Kodaira, T., Sasmal, K., Miratsu, R., Fukui, T., Zhu, T., Waseda, T. 2023. Uncertainty in wave hindcasts in the North Atlantic Ocean, Marine Structures, 89.

Kwon, Y. 1981. The effect of weather, particularly short sea waves, on ship speed performance. PhD thesis: Newcastle University.

Marine Insight, Shipping news. https://www.marineinsight.com/category/shipping-news/ (accessed: 13/12/2023)

National Centers for Environmental Prediction (NOAA). (2014). Marine Weather Forecasting in the National Weather Service (NWS). Mariners Weather Log, 58(3).

Neumann, T, 2016. Vessel route planning with uncertain data. The International Journal on Marine Navigation and Safety of Sea Transportation, 10:459–464.

Powell, W.B. 2019. Reinforcement learning and stochastic optimization: A unified framework for sequential decisions. Hoboken, New Jersey: John Wiley & Sons.

Prpić-Oršić, J. Sasa, K., Valčić, M., Faltinsen, O. M.. 2020. Uncertainties of ship speed loss evaluation under real weather conditions. Journal of offshore mechanics and arctic engineering, 142, 3.

Prpić-Oršić, J., Faltinsen, O. M. 2012. Estimation of ship speed loss and associated CO2 emissions in a seaway. Ocean engineering, 44, 1: 1–10.

Prpić-Oršić, J., Vettor, R., Faltinsen, O. M., Guedes Soares, C. 2016. The influence of route choice and operating conditions on fuel consumption and CO2 emission of ships. Journal of marine science and technology, 21:1–24.

Sasa, K., Takeuchi, K., Chen, C., Faltinsen, O. M., Prpić-Oršić, J., Valčić, M., Mrakovčić, T., Herai, N. 2019. Evaluation of Speed Loss in Bulk Carriers with Actual Data from Rough Sea Voyages. Ocean engineering, 187:1–19.

Shao, W., Zhou, P., Thong, S. K. 2011. Development of a novel forward dynamic programming method for weather routing. Journal of marine science and technology, 17:239–251.

Shin, Y. W., Abebe, M., Noh, Y., Lee, S., Lee, I., Kim, D., Bae, J., Kim, K. 2020. Near-Optimal Weather Routing by Using Improved A* Algorithm. Applied Sciences. 10.

Silveira, P., Teixeira, A. P., Guedes Soares, C. 2019. AIS Based Shipping Routes Using the Dijkstra Algorithm. The International Journal on Marine Navigation and Safety of Sea Transportation, 13:565–571.

Slingo, J., Palmer, T. 2011. Uncertainty in weather and climate prediction. Philos Trans A Math Phys Eng Sci, 369, 4751–4767.

SOFAR Ocean: What is Marine Weather and How Can We Monitor It? https://www.sofarocean.com/posts/what-is-marine-weather-and-how-can-we-monitor-it (accessed: 13/08/2023)

Stoffelen, A., Vogelzang, J. 2012. Triple collocation. EUMETSAT Satellite Application Facility on Numerical Weather Prediction (NWP SAF).

Sun, W.; Tang, S.; Liu, X.; Zhou, S.; Wei, J. 2022. An Improved Ship Weather Routing Framework for CII Reduction Accounting for Wind-Assisted Rotors. J. Mar. Sci. Eng. 10.

Vesely, F.M., Paleari, L., Movedi, E. et al. 2019. Quantifying Uncertainty Due to Stochastic Weather Generators in Climate Change Impact Studies. Sci Rep 9, 9258.

Vettor, R., Guedes Soares, C. 2016. Development of a ship weather routing system. Ocean Engineering, 123: 1–14.

Vettor, R.; Bergamini, G., Guedes Soares, C. 2021. A comprehensive approach to account for weather uncertainties in ship route optimization. Journal of Maritime Science and Engineering; 9:1434.

Vettor, R., Guedes Soares, C. 2022. Reflecting the uncertainties of ensemble weather forecasts on the predictions of ship fuel consumption. Ocean Engineering, 250.

Vitali, N., Prpić-Oršić, J., Guedes Soares, C. 2020. Coupling voyage and weather data to estimate speed loss of container ships in realistic conditions. Ocean engineering, 210.

Walther, L., Rizvanolli, A., Wendebourg, M., Jahn, C. 2016. Modelling and Optimization Algorithms in Ship Weather Routing, International Journal of e-Navigation and Maritime Economy, 4: 31–45.

Wang, X., Zhao, X., Wang, G., Wang, Q., Feng, K. 2022. Weather Route Optimization Method of Unmanned Ship Based on Continuous Dynamic Optimal Control. Sustainability, 14.

Wei, Q., Liu, Y., Dong, Y., Li, T., Li, W. 2023. A digital twin framework for real-time ship routing considering decarbonization regulatory compliance. Ocean Engineering 278.

World Meteorological Organization (WMO). (2023). WMO Bulletin: Early Warnings for All. Vol. 72 (1). Geneva. https://library.wmo.int/idurl/4/64131 (accessed: 13/12/2023)

Zhao, Jing & Duan, Yaoqi & Liu, Xiaojuan. 2018. Uncertainty Analysis of Weather Forecast Data for Cooling Load Forecasting Based on the Monte Carlo Method. Energies. 11.

Zhou, P., Zhou, Z., Wang, Y., Wang, H. 2023. Ship Weather Routing Based on Hybrid Genetic Algorithm Under Complicated Sea Conditions. Journal of Ocean University of China, 22:28–42.

Advances in Maritime Technology and Engineering – Guedes Soares & Santos (Eds)
© 2024 The Author(s), ISBN 978-1-032-83107-7

Assessing wind and wave climate in the North Atlantic using ERA20 reanalysis data

M. Bernardino & C. Guedes Soares
Centre for Marine Technology and Ocean Engineering (CENTEC), Instituto Superior Técnico, Universidade de Lisboa, Lisbon, Portugal

ABSTRACT: The present work aims to evaluate changes in marine climate in the North Atlantic using the ERA20C reanalysis data set available from the ECMWF, that extends from the beginning of the 20th century until the first decade of the 21st century. To assess the adequality of this data set, sea state variables as significant wave heigh of total sea, significant wave height of the swell component and mean wave period and also 10 meters wind magnitude (W10) are extracted for 4 locations in the North Atlantic and compared with ERA-Interim data for the overlapping period. Long-term statistical characteristics of the marine conditions, for the 110 years of ERA20 data are analysed as well as interannual variability and changes in the seasonal cycle. The relation between winter marine climate and the North Atlantic Oscillation is evaluated and a positive correlation with a correlation coefficient above 0.5, in the area to the north-east of the domain and a significant negative correlation area centred around 30° N is found.

1 INTRODUCTION

An improved understanding of marine climatology is necessary for numerous marine and coastal activities, such as offshore industries, the assessment of flooding risk, the design of marine structures and the evaluation of wave energy resources (Laugel, et al. 2014). Extreme waves, storms and storminess, in general, have a major impact upon coastal populations and ecosystems, and changes in the severity and occurrence of storm surges and ocean wave climate are some examples of how climate change affects coastal regions (Mori et al., 2013). How the marine climate changed in the past century, in terms of trends and variability can give some information of what to expect in the future. Changes in wave and wind climate are expected to have the largest impact on marine structure design in comparison with other environmental phenomena (Bitner-Gregersen, et al., 2018).

The WASA (Waves and Storms in the North Atlantic) project (WASA group, 1998) aimed to assess changes in wave climate conditions over the north-northeast Atlantic and the North Sea. Using a 40-year hindcast concluded the wave climate had gone through considerable changes during the 20th century and linked these changes to the North Atlantic Oscillation index (NAO). Later, other studies looked at the North Atlantic sub-basin wave and wind climate variability (e.g. Wolf et al., 2002, Stopa et al., 2013, Bertin et al., 2013, Lerma et al., 2015, Kumar et al., 2016 and Echevarria et al., 2019, Bernardino et al., 2021, Bernardino et al., 2023). Observations from different sources, including modelling reanalyzes (e.g. Gulev and Hasse 1998, Sterl and Caires 2005, Pilar et al. 2008, Vose et al. 2014) suggest that wave heights in the North Atlantic have increased during the last quarter of the twentieth century. Semedo et al. (2011), using the ERA-40 wave reanalysis, studied the global wind sea and swell wave climates separately, and how they are linked in the overall significant wave height climate, showing that the positive trend in the significant wave height is mostly a propagated effect due to swell waves. All these studies have linked this increase in wave height to a positive phase of the NAO, and a generalized increase in the surface wind speeds, mainly after 1975 until the end of the twentieth century.

Most of the products of ocean reanalyses, available for climate research cover just the last few decades, because of the lack of observations before the 1970s. However, for research on long-term changes and climate decadal variability, an ocean reanalysis covering a longer period, for example, the whole twentieth century, is fundamental.

A reanalysis data set, ERA-20C (Poli et al., 2016) was published in 2014 by the European Centre for Medium-Range Weather Forecast (ECMWF), which includes dynamically produced wave information. This is an improvement over previous data sets because it extends further into the past, providing a long, gap-free gridded record of many climate variables. This dataset was used by Dada et al. (2016) to study how climate variability affected waves and coastline evolution in the Niger delta coast. Patra

DOI: 10.1201/9781003508779-65

and Bhaskaran (2016) used monthly means derived from daily wave height information for the 1992 to 2010 period to validate WAM simulations for the Bay of Bengal, and Kumar et al. (2016) also used ERA-20C to perform a global study of extreme wave heights and its relation to climate variability since 1950.

The purpose of this work is to contribute to the characterization of the wave climate in the North Atlantic basin, in particular an evaluation on the long-term changes and inter-annual variability on marine characteristics such as significant wave height of the total sea, significant wave height of swell and wind magnitude using this reanalysis data-set. It aims to investigate if these changes found by previous studies extend further in the past until the beginning of the last century and if the relation between surface wave regimes and the atmosphere has been present during the last century.

2 DATA SETS

ERA-20C is an ECMWF reanalysis product for the twentieth century (1900-2010) produced by the ERA-CLIM project. It relies on a recent version of the ECMWF's Integrated Forecast System. The model has 91 levels between surface and 0.01hPa or about 80Km altitude. The horizontal resolution is spectral triangular truncation T159 or approximately 125 km. It assimilates marine surface winds from ICOADSv2.5.1 and the surface and mean sea level pressure from ISPDv3.2.6 and ICOADSv2.5.1. A quality control procedure is included in ERA-20C to limit the assimilation of bad observations so, ERA-20C is meant to reproduce the actual weather evolution and the observed occurrence of synoptic systems and it is expected to improve the low-frequency variability and mean state of the atmosphere with respect to its predecessor ERA-20CM by reducing differences between model results and real conditions (Poli et al. 2015, 2016).

Comparing with previous data sets that extend less into the past, the ERA-20C (in recent decades) is less accurate as it assimilates less data, for instance, it does not assimilate upper-air data.

Comparing ERA-20C with the NOAA 20th-Century Reanalysis (20CR), used by Compo et al. (2011) to evaluate marine climate change during the last century, ERA-20C is shorter but provides finer horizontal and vertical resolutions. There are differences in the data assimilated and in the data assimilation schemes leading to ERA-20C performing better in relatively well-observed regions such as Western Europe, while 20CR being in advantage in of sparse observations in regions such as the Southern Hemisphere middle to high latitudes.

In the present work the ERA-Interim reanalysis data set will be used as a proxy for reality as it includes more observations than ERA-20C or 20CR, during the covered period, as in this case the reanalysis assimilates all conventional and satellite observations. It is based on a 2006 release of the Integrated Forecast System (Cy31r2). The spatial resolution of the data set is approximately 80 km (T255 spectral) on 60 vertical levels from the surface up to 0.1 hPa. The data assimilation is based on a 12-hourly four-dimensional variational analysis (4D-Var) with adaptive estimation of biases in satellite radiance data. The wave model incorporated into the IFS (Integrated Forecast System) is based on the Wave Modeling Project (WAM) approach (Komen et al. 1994), and the version used in ERA-Interim includes several enhancements, both in physics and numeric, over the version that was used in the previous reanalysis ERA-40. Also, the wave model used in Era-Interim includes shallow-water physics (Janssen et al, 2005 and Janssen 2008). The updated version of WAM includes a revised formulation of ocean wave dissipation, which has reduced the root-mean-square error in the wave period against the buoy data and shows that the error in the estimate of the wave period is much smaller than that in ERA-40 (Bidlot et al., 2007). A full description of ERA-Interim can be found in Dee et al. (2011).

Although older than ERA5, ERA-Interim is still an adequate reanalysis data set to be used in this way. It is more widely validated against different types of observations and in validation of global wind wave hindcasts using ERA5, MERRA2, ERA-Interim and CFSRv2 reanalyses, performed by Sharmar & Markina (2020) and Sharmar et al., (2021), it was concluded that although ERA5 demonstrated the best agreement with both satellite altimetry and buoy measurements, all hindcasts show good correspondence with the observational data and thus can be used for further investigation of the common patterns of interannual variability in winds and wave heights in the wave climate studies.

For this study, ERA-20C significant wave height of the total sea state (Hs), significant wave height of the swell component (Sw), 10 meters wind magnitude (W10) and mean wave period (TM) with 6 hours temporal resolution were extracted for the North Atlantic Ocean (12° to 70° N; 80° to 0° W) in a 1° by 1 ° resolution grid (see Figure 1), for the period 1901 to 2010 in a total of 110 years. ERA-Interim data were also extracted, from the ECMWF website, for the period 1981-2010 in a total of 30 years, also with 6 hours temporal resolution and upscaled to 1° by 1° resolution grid to be comparable with the ERA-20C data.

A comparison between the two data sets is made in the next section using the common 30 years period, but for the long-term marine conditions 110 years of the ERA-20C data were used.

The North Atlantic Oscillation (NAO) is one of the major modes of variability of the Northern Hemisphere atmosphere. It is particularly important in winter when it exerts a strong control on the climate of the Northern Hemisphere (Hurrel, 1995).

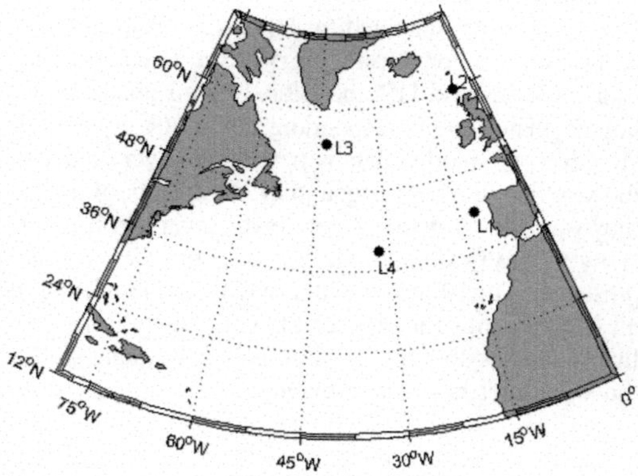

Figure 1. North Atlantic domain and location of the four points. L1 (42N, 12W), L2 (60N, 7W), L3 (55N, 43W) and L4 (39N, 32W).

Traditionally, the NAO is defined as the normalized pressure difference between a station on the Azores and one on Iceland and many authors (Pokorná and Huth, 2015) use station-based indices to study the effects of NAO on surface climate elements, both on continental and regional and country-wide scales. NAO data used here was obtained from the CRU (Climatic Research Unit) website (https://crudata.uea.ac.uk/cru/data/nao/) and it is calculated using observation data from Gibraltar and SW Iceland as it is described in Jones et al. (1997). This data set has also been used by Burningham and French (2013) to assess the relation between NAO and the wind climate and storminess, and by Osborn (2011) to explain the relation between the extreme winter temperatures in 2009/2010 and the unusually low NAO index in the same period. Time series of the mean winter (December, January, February) NAO index and the corresponding 5-year running mean, are presented in Figure 2.

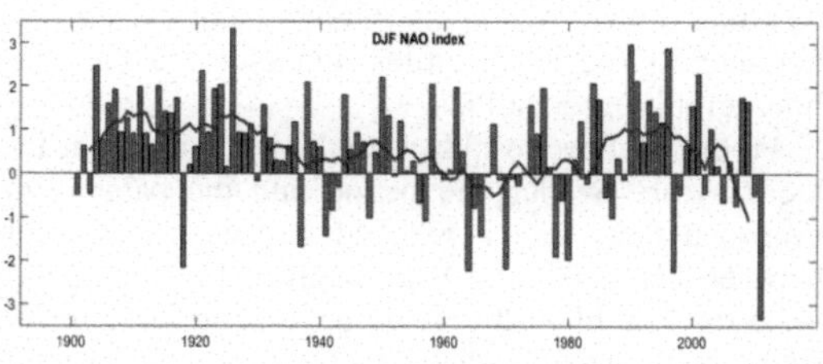

Figure 2. Mean winter (DJF) North Atlantic Oscillation time series and five-year running mean.

3 COMPARISON BETWEEN ERA-INTERIM AND ERA-20C

ERA-interim reanalysis data set has been widely used in climate impact applications and was used as a proxy for reality when observations are not available. Different parameters, from significant wave height to wave energy or wave direction, associated with marine climate may be interesting for different types of applications (e.g. Casas-Prat and Sierra, 2010, Sierra and Casas-Prat, 2014, Perez et al., 2017, Vettor and Guedes Soares, 2020). In this work the ERA-20C reanalysis will be compared with ERA-Interim in four locations in the North Atlantic, only in terms of significant wave height of total sea (H_s), significant wave height of swell (S_w), 10 meters wind magnitude (W10) and mean wave period (TM). The four sites (Figure 1) located off the Iberian Peninsula (L1: 42N, 12W), in the north of Scotland (L2: 60N, 7W), south of Greenland (L3: 55N, 43W) and in the Azores (L4: 39N, 32W).

A thirty-year overlapping period between the two reanalyzes (1981-2010) is analyzed for the four parameters, evaluating mean values, extremes, and variability. To compare the statistical characteristics of the two reanalysis data sets, marine data (H_s, S_w, W_{10} and TM) are extracted for four locations. For each location and parameter, the long-term mean, standard deviation, coefficient of variation and 95th percentile are computed and results can be seen in Table 1.

Table 1. Long term mean, standard deviation, coefficient of variation and 95th percentile for Hs, Sw, W10 and TM in the four locations.

		Mean				STD			
		Hs	Sw	W10	TM	Hs	Sw	W10	TM
L1	E-int	2.59	2.23	7.61	9.14	1.18	1.03	3.28	1.91
	E20C	2.35	2.01	7.67	8.47	1.11	0.94	3.29	1.81
L2	E-int	3.09	2.44	9.08	8.99	1.60	1.14	3.93	1.77
	E20C	2.83	2.19	9.17	8.24	1.53	1.05	3.98	1.66
L3	E-int	3.27	2.54	9.50	8.77	1.56	1.07	4.06	1.52
	E20C	3.06	2.35	9.50	8.15	1.53	0.99	3.98	1.42
L4	E-int	2.59	2.26	7.57	9.01	1.26	0.99	3.42	1.57
	E20C	2.38	2.05	7.60	8.28	1.21	0.93	3.39	1.53
		CV				P95			
		Hs	Sw	W10	TM	Hs	Sw	W10	TM
L1	E-int	0.45	0.46	0.43	0.21	4.90	4.27	13.26	12.60
	E20C	0.47	0.47	0.43	0.21	4.54	3.88	13.32	11.80
L2	E-int	0.52	0.47	0.43	0.20	6.18	4.66	16.03	12.16
	E20C	0.54	0.48	0.43	0.20	5.81	4.22	16.16	11.24
L3	E-int	0,48	0.42	0.43	0.17	6.27	4.56	16.56	11.41
	E20C	0.50	0.42	0.42	0.17	6.02	4.25	16.50	10.66
L4	E-int	0.49	0.44	0.45	0.17	5.01	4.19	13.78	11.84
	E20C	0.51	0.45	0.45	0.18	4.71	3.86	13.67	11.10

Observing the results, it can be seen that for both H_S and S_W in all four locations, ERA-Interim shows higher long term mean values, variability and extremes than ERA20C. Regarding 10m wind magnitude, the opposite behaviour can be observed, with a higher long term mean, variability and extremes being observed for ERA-20C.

A statistical comparison between values obtained from each dataset was performed computing the

Pearson's correlation coefficient (r), mean absolute error (MEA), bias (BIAS) and root mean squared error (RMSE) from equations 1 to 4.

$$r = \frac{\sum_{i=1}^{n}(M_i - \tilde{M})(S_i - \tilde{S})}{\left(\sum_{i=1}^{n}(M_i - \tilde{M})^2 \sum_{i=1}^{n}(S_i - \tilde{S})^2\right)^{\frac{1}{2}}} \quad (1)$$

$$MAE = \frac{\sum_{i=1}^{n}|S_i - M_i|}{n} \quad (2)$$

$$BIAS = \frac{\sum_{i=1}^{n}(S_i - M_i)}{n} \quad (3)$$

$$RMSE = \sqrt{\frac{\sum_{i=1}^{n}(S_i - M_i)^2}{n}} \quad (4)$$

where Mi represents ERA-Interim, Si represents ERA-20C, and are the corresponding mean values and n the number of observations.

In Table 2 the results of the statistical comparison between ERA-Interim and ERA-20C are presented for significant wave height of the total sea, significant wave height of swell 10m wind magnitude and mean wave period, in the four locations.

Table 2. Statistical comparison between ERA-Interim and ERA-20C significant wave height of total sea (Hs), significant wave height of swell (Sw), wind magnitude at 10 m height (W10) and mean wave period (TM), in the four locations.

	L1	L2	L3	L4	L1	L2	L3	L4
	H_S				Sw			
R	0.98	0.98	0.97	0.98	0.98	0.97	0.94	0.97
MAE	0.13	0.15	0.15	0.12	0.12	0.14	0.14	0.12
BIAS	0.12	0.13	0.11	0.11	0.11	0.13	0.09	0.10
RMSE	0.23	0.27	0.30	0.22	0.22	0.27	0.28	0.22
	W10				TM			
R	0.95	0.96	0.94	0.95	0.97	0.96	0.95	0.95
MAE	0.39	0.43	0.51	0.41	0.70	0.75	0.66	0.76
BIAS	-0.03	-0.04	0.00	-0.02	0.67	0.74	0.62	0.73
RMSE	0.72	0.81	0.96	0.76	0.82	0.84	0.79	0.89

It can be seen that the correlation coefficient between data series is very good, over 94%, reaching as high as 98% for S_w in location L1 and H_s in locations L1, L2 and L4. The bias takes positive values from 0.09 to 0.13, for H_s and S_w the highest value being found for location L2. These values indicate that ERA-20C underestimates ERA-Interim for Hs and Sw. Regarding wind magnitude, the bias is negative with values between 0.04 m/s and zero, showing a slight overestimation of ERA20C in comparison to ERA-Interim. RMSE shows values lower than 30 cm for all the four locations for Hs and Sw, lower than 1 m/s for wind speed and lower than 0.9 s for the mean wave period.

Q-Q plots are obtained for H_s and S_w, W_{10} and TM, at the four locations. In Figure 3 it can be seen that for four locations and HS, the ERA-Interim quantiles are higher than the corresponding ERA-20C quantiles. However, the plotted data is nevertheless very close to the x=y line. For higher quantiles, the quality of the fitting was not as good as for lower quantiles. Figures showing the Q-Q plots for S_w, W_{10} and TM, are not presented. For 10 meters wind magnitude, the fitting in all locations lies for most of the data range in the x=y line indicating that the distribution of both data set for those parameters is very similar.

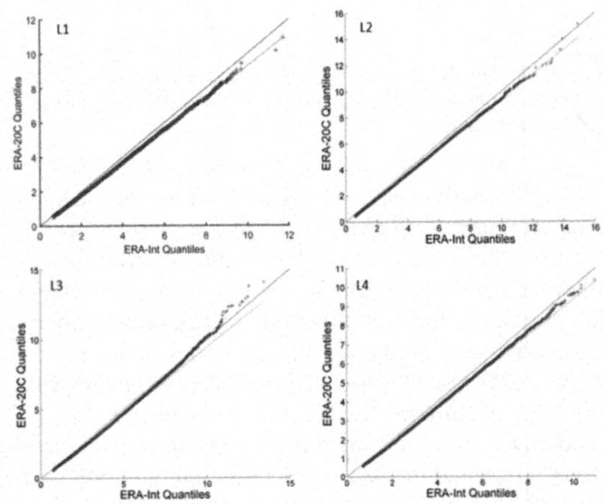

Figure 3. Quantile - quantile plot for Hs in the four selected locations. Solid line represents a perfect fit between ERA-Interim and ERA-20C quantiles and dashed line represents the best-fit slope.

No significant differences are found between the different locations although location L3 seem to be where the best agreement is found. For the mean wave period, the Q-Q plots show that ERA-Interim quantiles are also slightly higher than the corresponding ERA-20C quantiles, but close to the x=y line, for all four locations.

From the above results, it can be concluded that for the 30-years overlapping period and the parameters evaluated there is a very good statistical agreement between ERA-20C and ERA-Interim. This is a good indicator that ERA-20C is an appropriate data set to study the long-term climate change in the marine environment in the North Atlantic.

4 NORTH ATLANTIC MEAN CONDITIONS AND INTERANNUAL VARIABILITY

To evaluate the long-term characteristics of the marine conditions, for the 110 years of data available (1901-2010), the mean value, 95th and 99th percentiles are computed for H_s, S_w and W_{10}, using 3-hours temporal resolution information for the four locations and are presented in Table 3. Only these three parameters are considered in this study, but others such as direction or energy might also be interesting, in particular regarding the impacts of climate change in coastal areas.

Table 3. Long term mean of annual mean values, 95th and 99th percentiles, for Hs (m), Sw (m) and W10 (m/s).

	Location	Mean	P95	P99
Hs (m)	L1	2.20	4.35	5.77
	L2	2.62	5.40	7.08
	L3	2.87	5.74	7.62
	L4	2.24	4.53	6.09
Sw (m)	L1	1.89	3.71	4.87
	L2	2.04	3.94	5.12
	L3	2.22	4.06	5.09
	L4	1.94	3.72	4.85
W10 (m/s)	L1	7.34	12.97	15.41
	L2	8.82	15.59	18.22
	L3	9.14	16.10	18.93
	L4	7.27	13.40	16.16

It can be seen, comparing the different locations, that the mean value of HS, shows its lowest value in L1 (2.2m) and the highest in L3 (2.87m). Comparing the mean values of H_S with the ones of S_W shows that the last ones have slightly higher values. Regarding extremes, they reach 5.74 m for the 95th percentile and 7.62 m for the 99th percentile for HS in location L3. Wind magnitude at 10 meters presents mean values from 7.27 m/s (L4) and 9.14 m/s (L3) but reaches 15.41m/s (L1) and 18.93 m/s (L3) for the 99th percentile. For the three parameters, the variability between the four locations is lower for mean values and increases for extremes. Location (L3), south of Greenland is where the highest values are found for mean and extremes for the three parameters.

To study changes in the seasonal cycle, the 110 years are divided into four subsets: 1901-1920, 1921-1950; 1951-1980 and 1981-2010. The first period only includes 20 years, while the remaining three spans 30 years. For each period, and each season and each location, the mean values of Hs, Sw and W10 were computed. In Figure 4, the Hs seasonal cycle obtained for each period for location L3 is presented in the top panel. It can be observed that although the shape of the seasonal cycle remains the same, there is an increase in mean seasonal values since the beginning of the 20th century, in particular from the first half to the second half of the century. During the first half of the century, changes are small and in spring a small decrease can be observed. After 1950 there is an increase that is less marked in summer. Furthermore, from 1951-1980 to 1981- 2010, a small decrease in autumn mean value can also be observed. The same type of changes in seasonal values can be observed for S_W in the same location, in the middle panel of Figure 4. Seasonal values are in general smaller than the ones found for H_S, except for the summer values that are approximately the same. An overall increase in seasonal values since the beginning of the 20th century, more marked since 1950, can also be seen.

Observing W10 seasonal cycles for the four periods that are presented in the same figure, shows that changes in the magnitude of the season values are smaller than the ones observed for other variables and that the most noticeable feature is the difference between the periods before 1950 and after 1950.

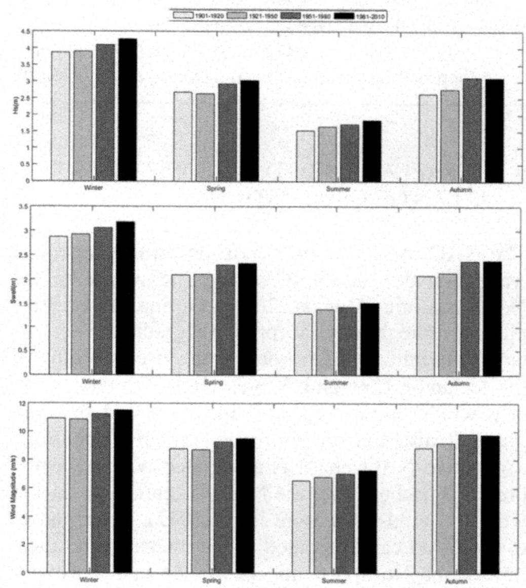

Figure 4. Seasonal cycle for mean Hs, mean Sw and mean W10 for different periods in L3.

To investigate interannual variability, the mean value, 95th and 99th percentiles are computed for each year and plotted for each location. Also, for each location and for each period, the interannual variability (IAV) is computed following Stopa et al., (2013).

$$\mathrm{IAV} = \frac{\sigma_{\overline{x_k}}}{\overline{x}} \tag{5}$$

where the indices k refers to the year.

Results obtained for P2, for mean annual value, 95th annual percentile and 99th annual percentile, are presented in Table 4.

Interannual variability in wind seems to be lower than in H_S and in S_W. In all three parameters, the interannual variability is more marked since the 1980 in the higher percentile values, but for the annual mean, is the period 1921-1950 that presents the higher variability. The same type of analysis was performed for the other three locations (not shown) and the conclusions are similar with differences found mainly in the magnitude.

Table 4. Interannual variability of annual mean values, 95th and 99th percentiles, for HS (m), SW (m) and W10 (m/s).

		1901-1920	1921-1950	1951-1980	1981-2010	1901-2010
HS	Mean	0.0597	0.0734	0.0487	0.0692	0.0883
	P95	0.0957	0.0929	0.0667	0.1022	0.1010
	P99	0.1049	0.1033	0.0783	0.1145	0.1105
SW	Mean	0.0446	0.0609	0.0428	0.0560	0.0765
	P95	0.0758	0.0792	0.0572	0.0877	0.0875
	P99	0.0864	0.0998	0.0639	0.1026	0.0955
W10	Mean	0.0394	0.0425	0.0301	0.0410	0.0493
	P95	0.0403	0.0412	0.0335	0.0433	0.0467
	P99	0.0391	0.0411	0.0384	0.0528	0.0485

5 CORRELATION WITH THE NORTH ATLANTIC OSCILLATION

The NAO index is considered the primary agent of wave climate variability during the winter in the North Atlantic (Dupuis, 2006). During the positive phase of the NAO, the pressure gradient over the North Atlantic increases due to the strengthening of the Icelandic Low and Azores High. Stronger westerly winds, associated with the increased pressure gradient, also generate larger waves. On the contrary, a negative NAO phase is associated with a weaker pressure gradient over the North Atlantic and slacker westerly winds (Gleeson et al., 2017). Both phases of the NAO can be related to the intensity and location of the North Atlantic jet stream and the extra-tropical storm track and length (Visbeck et al, 2001). In the positive phase, the storm track moves more to the north-eastward, towards the northern part of the North Atlantic. In the negative phase, the storm tracks move south-eastward, towards southwest Europe. So, generally, positive NAO index winters are associated with a north-eastward shift in the Atlantic storm activity with enhanced activity from Newfoundland into northern Europe and a modest decrease in activity to the south.

Positive NAO index winters are also typified by more intense and frequent storms in the vicinity of Iceland and the Norwegian (Hurrell & Deser, 2010). The NAO index has suffered long term changes during the last 150 years. Pinto & Raible (2012), showed that the NAO shifted phase on the 1860-2010 period, from a predominate negative before 1910 to a more negative phase in the 1920s, followed by a negative trend until the end of the 1960s. Later data and until de mid-1990's a strong positive trend is observed. In the following 15 years, a negative trend of the indices is present, with NAO reaching minimum values during the winters of 2009/2010 and 2010/2011.

The Pearson correlation coefficient was computed between the winter station-based NAO index (Jones et al., 1997), and winter means HS, SW and W10

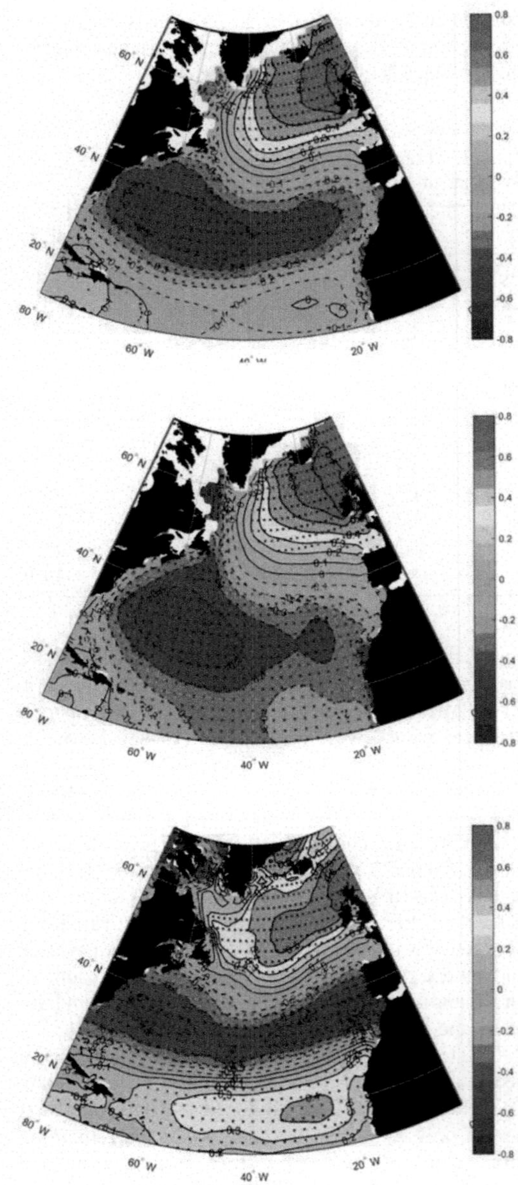

Figure 5. Correlation coefficient between winter NAO index and mean winter H_S (top), S_W (middle) and W_{10} (bottom)

(Figure 5). The areas where the correlation coefficient is significant at 95% level are marked in the Figure.

It can be observed that coherent regions are correlated with NAO during the winter. Regarding HS and SW, Figure 10 top and 10 middle show a southwest-northeast gradient with a large positive correlation (exceeding 0.6) area to the north-east of the studied domain and a significant negative correlation area centred around 30° N with values exceeding -0.5. Correlation coefficients are very low, as expected, in the subtropical region for HS but not so low for SW where correlation coefficients reach -0.3. Observing Figure 5 bottom, where the correlation coefficient between winter NAO and winter 10 m wind magnitude is presented, only weak correlation (between -0.5 and 0.6) can be observed, positive above 40 o N and below 20 o N and negative between these two latitudes.

6 DISCUSSION

The results obtained for the marine climate in the North Atlantic from the ERA-20C reanalysis, are in line with the ones found by Dada et al. (2016) for the Niger Delta coast and Ulzia et al. (2017) for the Bay of Biscay using the same reanalysis data set. Furthermore, they are coherent with Bertin et al. (2013) who conducted a 109-year numerical wind-wave hindcast over the North Atlantic, north of 40 ° N, based on the 20RC wind and ice cover and the WW3 model.

The correlation between marine variables and the NAO extends the results from Woolf et al. (2002), Dodet et al. (2010) and Martinez-Asensio et al. (2016), to a longer period that reaches the beginning of the 20th century, confirming the impact of the NAO in the winter wave marine climate of the North Atlantic Ocean. They are also coherent with results found by Semedo et al., (2015) for the North Sea, Gallagher et al., (2014) for the Northeast Atlantic and Charles et al., (2012) for the Bay of Biscay and Rangel-Buitrago. Kumar et al. (2016) analysing the second half of the 20th century, also concluded that extreme wave heights during the winter are driven by an increase in wind speed, which can to some extend be linked to the NAO.

Pinto et al. (2009) identified a strong relationship between the occurrence of extreme storms over the North Atlantic basin and strong positive NAO indices. The strong NAO trend observed from the 60's decade to the mid-'90s, may explain the increasing storminess during this period, which would reflect a strong positive trend in significant wave height, observed by different authors. During the first half of the 20th century, the observed trends in NAO are not as strong and decadal oscillations of the index are observed, so, a weaker increased in storminess may be expected. A discussion of the mechanisms associated with changes in the NAO index can be found in Pinto & Raible (2012).

7 CONCLUSIONS

Statistical comparison between ERA-20C and ERA-Interim is performed in four locations, using an overlapping period of 30-years, showing a good agreement between the two data sets and indicating that the ERA-20C is an adequate dataset to be used in the present work.

Mean and extreme H_S, S_W and W_{10} values are computed for the four locations using 110 years of 6h data from ERA-20C. It can be seen that the values are generally larger on the open ocean than on coastal areas and larger at high latitudes. Looking at the seasonal cycle it can be observed that H_S, S_W and W_{10} are seasonal through the North Atlantic, peaking in winter, particularly at higher latitudes. The shape of the seasonal cycle does not change during the 110 years but there wis an increase in seasonal values through time in particular from the first to the second half of the 20th century.

Evaluation of the correlation between winter NAO and winter mean H_S, S_W, a large positive correlation (exceeding 0.6) area can be identified to the northeast of the studied domain and a significant negative correlation area centred around 30o N with values lower than -0.5 with a southwest-northeast gradient. For W10 only weak correlations (between -0.5 and 0.6) are found. Long term variability of the NAO, in particular, a strong positive trend observed from the 1960s to the mid-1990s, associated with a positive relation between NAO and HS and SW in the northeast of the Atlantic Ocean, could be an explanation of the strong trends of H_S and S_W observed in ERA20C during the 2nd half of the 20th century. In the first half of the 20th century although the NAO takes mainly positive values its decadal variability reflects in a weaker positive trend. After the mid-1990's the downward trend in NAO is seen to be accompanied by a decrease in storminess in the North Atlantic.

ACKNOWLEDGEMENTS

The work contributes to the Strategic Research Plan of the Centre for Marine Technology and Ocean Engineering, also financed by the Portuguese Foundation for Science and Technology (Fundação para a Ciência e Tecnologia-FCT) under contract UIDB/UIDP/00134/2020.

REFERENCES

Bernardino, M., Goncalves, M., & Guedes Soares, C. (2021). Marine climate projections toward the end of the twenty-first century in the north Atlantic. *Journal of Offshore Mechanics and Arctic Engineering*, 143(6), 061201.

Bernardino, M., Gonçalves, M., Campos, R. M., & Guedes Soares, C. (2023). Extremes and variability of wind and waves across the oceans until the end of the 21st century. *Ocean Engineering*, *275*, 114081.

Bertin, X., Prouteau, E., Letetrel, C., 2013. A significant increase in wave height in the North Atlantic Ocean over the 20th century. *Global and Planetary Change*, 106, 77–83.

Bidlot, J., Janseen, P.A.E.M., Abdalla, S., 2007. A revised formulation of ocean wave dissipation and its model impact. *ECMWF Tech. Memo. 509*, 27 pp.

Bitner-Gregersen, E. M., Vanem, E., Gramstad, O., Hørte, T., Aarnes, O. J., Reistad, M., Breivik, Ø., Magnusson, A.K., Natvig, B., 2018. Climate change and safe design of ship structures. *Ocean Engineering*,149, 226–237.

Casas Prat, M., and Sierra Pedrico, J. P.,2010. Trend analysis of wave storminess: wave direction and its impact on harbour agitation. *Natural Hazards and Earth System Sciences*, 10, 2327–2340.
Charles E., Idier D., Thiebot J., Le Cozannet G., Pedreros R., Ardhuin F., Planton S., 2012. Present wave climate in the Bay of Biscay: Spatiotemporal variability and trends from 1958 to 2001. *J of Clim* 25 (6):2020–2039.
Compo, G.P., Whitaker, J.S., Sardeshmukh, P.D., Matsui, N., Allan, R.J., Yin, X., Gleason, B.E. Jr, Vose, R.S., Rutledge, G., Bessemoulin, P., Bronnimann, S., Brunet, M., Crouthamel, R.I., Grant, A. N., Groisman, P.Y., Jones, P.D., Kruk, M.C., Kruger, A. C., Marshall, G.J., Maugeri, M., Mok, H.Y., Nordli, O., Ross, T.F., Trigo, R.M., Wang, X.L., Woodruff, S.D., Worley, S.J., 2011. The twentieth century reanalysis project. Quarterly *Journal of the Royal Meteorological Society*, 137(654), 1–28.
Dada, O.A.; Li, G.; Qiao, L.; Ma, Y.; Ding, D.; Xu, J.; Li, P.; Yang, J., 2016. Response of waves and coastline evolution to climate variability off the Niger Delta coast during the past 110 years. *Journal of Marine Systems*, 160, 64–80.
Dee, D.P., Uppala, S.M., Simmons, A.J., Berrisford, P., Poli, P., Kobayashi, S., Andrae, U., Balmaseda, M.A., Balsamo, G., Bauer, P., Bechtold, P., Beljaars, A.C.M., Van de Berg, L., Bidlot, J., Bormann, N., Delsol, C., Dragani, R., Fuentes, M., Geer, A. J., Haimberger, L., Healy, S.B., Hersbach, H., Hólm, E.V., Isaksen, L., Kållberg, P., Köhler, M., Matricardi, M., McNally, A. P., Monge-Sanz, B. M., Morcrette, J.-J., Park, B.-K., Peubey, C., Rosnay, P., Tavolato, C., Thépaut F. and Vitart, J.-N., 2011. The ERA-Interim reanalysis: Configuration and performance of the data assimilation system. *Quart. J. R. Meteorol. Soc.*, 137, 553–597.
Dupuis, H., Michel, D., Sottolichio, A., 2006: Wave climate evolution in the Bay of Biscay over two decades. *Journal of Marine Systems*, 63(3–4), 105–114.
Echevarria E.R., Hemer M.A., Holbrook N.J., 2019. Seasonal Variability of the Global Spectral Wind Wave Climate. *Journal of Geophysical Research: Oceans*, 124 (4): 2924–2939.
Gallagher, S., Tiron, R., and Dias, F., 2014. A long-term nearshore wave hindcast for Ireland: Atlantic and Irish Sea coasts (1979–2012). *Ocean Dynamics*, 64(8), 1163–1180.
Gleeson, E., Gallagher, S., Clancy, C., and Dias, F., 2017. NAO and extreme ocean states in the Northeast Atlantic Ocean, *Adv. Sci. Res.*, 14, 23–33, https://doi.org/10.5194/asr-14-23-2017.
Gulev, S.K., Hasse, L., 1998. North Atlantic wind waves and wind stress fields from voluntary observing data. *J. Phys. Oceanogr.*, 28, 1107–1130.
Hurrell, J. W., and Deser, C., 2010. North Atlantic climate variability: the role of the North Atlantic Oscillation. *Journal of Marine Systems*, 79(3-4), 231–244.
Janssen, P.A.E.M., 2008. Progress in ocean wave forecasting. J. Comput. Phys., 227, 3572–3594.
Janssen, P.A.E.M., Bidlot, J., Abdalla, S., Hersbach, H., 2005. Progress in ocean wave forecasting at ECMWF. *Tech. Memo. 478*, ECMWF: Reading, UK.
Jones, P.D., Jonsson, T., Wheeler, D., 1997. Extension to the North Atlantic Oscillation using early instrumental pressure observations from Gibraltar and South-West Iceland. *International Journal of Climatology*, 17, 1433–1450.
Komen, G.J., Cavaleri, L., Donelan, M., Hasselman, K., Hasselmann, S., Janssen, P.A.E.M., 1994. *Dynamics and Modeling of Ocean Waves*. Cambridge University Press (532 pp.).
Kumar, P., Min, S. K., Weller, E., Lee, H., Wang, X. L., 2016. Influence of climate variability on extreme ocean surface wave heights assessed from ERA-interim and ERA-20C. *Journal of Climate*, 29(11), 4031–4046.
Laugel, A., Menendez, M., Benoit, M., Mattarolo, G., Méndez, F., 2014. Wave climate projections along the French coastline: Dynamical versus statistical downscaling methods. *Ocean Modelling*, 84, 35–50.
Lerma, A. N., Bulteau, T., Lecacheux, S., and Idier, D., 2015. Spatial variability of extreme wave height along the Atlantic and channel French coast. *Ocean Engineering*, 97, 175–185.
Martínez-Asensio, A., Tsimplis, M.N., Marcos, M., Feng, X., Gomis, D., Jordà, G., Josey, S.A., 2016. Response of the North Atlantic wave climate to atmospheric modes of variability. International *Journal of Climatology*, 36(3), 1210–1225.
Mori, N., Shimura, T., Yasuda, T., and Mase, H., 2013. Multimodel climate projections of ocean surface variables under different climate scenarios—Future change of waves, sea level and wind. *Ocean Engineering*, 71, 122–129.
Patra, A., Bhaskaran, P. K., 2016. Trends in wind-wave climate over the head Bay of Bengal region. International *Journal of Climatology*, 36(13), 4222–4240.
Perez, J., Menendez, M., and Losada, I. J., 2017. GOW2: A global wave hindcast for coastal applications. *Coastal Engineering*, 124, 1–11.
Pilar, P., Guedes Soares, C., Carretero J.C., 2008. 44-year wave hindcast for the North East Atlantic European coast. *Coastal Engineering*, 55(11), 861–871.
Pinto, J. G., and Raible, C. C., 2012. Past and recent changes in the North Atlantic oscillation. Wiley Interdisciplinary Reviews: *Climate Change*, 3(1), 79–90.
Pinto, J. G., Zacharias, S., Fink, A. H., Leckebusch, G. C., and Ulbrich, U., 2009. Factors contributing to the development of extreme North Atlantic cyclones and their relationship with the NAO. *Climate Dynamics*, 32(5), 711–737.
Pokorná, L., and Huth, R., 2015. Climate impacts of the NAO are sensitive to how the NAO is defined. *Theoretical and Applied Climatology*, 119(3-4), 639–652.
Poli, P., Hersbach, H., Berrisford, P., Dee, D., Simmons, A., Laloyaux, P., 2015. ERA-20C deterministic. ECMWF ERA *Rep. 20*, 48 pp. [Available online at http://www.ecmwf.int/en/elibrary/11700-era-20c-deterministic.]
Poli, P., Hersbach, H., Dee, D., Berrisford, P., Simmons, A., Vitart, F., Laloyaux, P., Tan, D., Peubey, C., Thépaut, J., Trémolet, Y., Hólm, E., Bonavita, M., Isaksen, L., Fisher, M., 2016. ERA-20C: An atmospheric reanalysis of the twentieth century. *Journal of Climate*, 29(11), 4083–4097.
Semedo, A., A. Rutgersson, A. Sterl, 2011. A Global View on Wind Sea and Swell Waves Climate and Variability from ERA-40. *J. Clim*, 24, 1461–1479.
Semedo, A., Vettor, R., Breivik, Ø., Sterl, A., Reistad, M., Guedes Soares, C., Lima D., 2015. The wind sea and swell waves climate in the Nordic seas. *Ocean Dynamics*, 65(2), 223–240.
Sharmar, V. D., Markina, M. Y., & Gulev, S. K., 2021. Global ocean wind-wave model hindcasts forced by different reanalyzes: A comparative assessment. *Journal of Geophysical Research: Oceans*, 126(1), e2020JC016710.

Sharmar, V., & Markina, M. (2020, November). Validation of global wind wave hindcasts using ERA5, MERRA2, ERA-Interim and CFSRv2 reanalyzes. In IOP Conference Series: Earth and Environmental Science (Vol. 606, No. 1, p. 012056). IOP Publishing.

Sierra, J. P., and Casas-Prat, M., 2014. Analysis of potential impacts on coastal areas due to changes in wave conditions. *Climatic change*, 124(4), 861–876.

Sterl, A. and S. Caires, 2005. Climatology, Variability and Extrema of Ocean Waves - The Web-based KNMI/ERA-40 Wave Atlas. International *Journal of Climatology*, 25(7), 963–997, doi:10.1029/joc.1175.

Stopa, J.E., Cheung, K.F., Tolman, H.L., Chawla, A., 2013. Patterns and cycles in the climate forecast system reanalysis wind and wave data. *Ocean Modelling*, 70, 207–220.

Ulazia, A., Penalba, M., Ibarra-Berastegui, G., Ringwood, J., Saénz, J. 2017. Wave energy trends over the Bay of Biscay and the consequences for wave energy converters. *Energy*, 141, 624–634.

Vettor, R., and Guedes Soares, C., 2020. A global view on bimodal wave spectra and crossing seas from ERA-interim. *Ocean Engineering*, 210, 107439.

Visbeck, M. H., Hurrell, J. W., Polvani, L., and Cullen, H. M., 2001. The North Atlantic Oscillation: past, present, and future. *Proceedings of the National Academy of Sciences*, 98(23), 12876–12877.

Vose, R., Applequist, S., Easterling, D., Karl, T., Peterson, T., Shein, K., Bourassa, M., Pryor, S., Barthelmie, R., Blanton, B., Bromirski, P., Brooks, H., Degaetano, A., Dole, R., Jensen, R., Katz, R., Klink, K., Kruk, M., Kunkel, K., Maccracken, M., Thomas, B., Walsh, J., Wang, X., Wehner, M., Wuebbles, D., Young, R., 2014. Monitoring and Understanding Changes in Extremes: Extratropical Storms, Winds, and Waves. *Bull. Amer. Meteor. Soc.*, 95, 377–386. doi: http://dx.doi.org/10.1175/BAMS-D-12-00162.1

WASA Group, 1998. Changing waves and storms in the Northeast Atlantic? *Bull. Amer. Meteor. Soc.*, 79, 741–760.

Wolf, J., and Woolf, D. K., 2006. Waves and climate change in the northeast Atlantic. *Geophysical Research Letters*, 33(6).

Woolf, D.K., Challenor, P.G. and Cotton P.D., 2002. Variability and predictability of the North Atlantic wave climate, *J. Geophys. Res.*, 107(C10), 3145, doi:10.1029/2001JC001124

Advances in Maritime Technology and Engineering – Guedes Soares & Santos (Eds)

Estimation of environmental wave contours along the Portuguese continental coast using the Burr distribution

G. Clarindo & C. Guedes Soares

Centre for Marine Technology and Ocean Engineering (CENTEC), Instituto Superior Técnico, Universidade de Lisboa, Lisbon, Portugal

ABSTRACT: The Burr distribution is adopted to model the environmental contours along the Portuguese continental coast using the Inverse-First Order Reliability Method. The resulting final contours constructed considering the Burr-Lognormal as a probabilistic model are compared with the contours constructed with Generalised Extreme Value-Lognormal distributions as baseline. The high-quality ERA5 dataset is used to demonstrate the approach. The Burr-Lognormal contours are determined along the Portuguese coast. The numerical implementation showed that the proposed scheme is a valid practical approach to accomplish step one in the contour construction. Moreover, the new scheme demonstrates high flexibility in characterizing rare and scattered events, despite the challenging complexity that consist of this task.

1 INTRODUCTION

The randomness of the environmental loads needs to be properly described for the design, operation, and overall management of marine structures. Considering the operational thresholds, extreme values of wave height and periods will be undesirable and induce unwelcome loads on the structure. In this way, the safe and unsafe domain where the structure will be submitted must be evaluated as accurately as possible, otherwise the installations can be seriously compromised. So, engineers and designers consider the loads from historical environmental observations when designing marine structures, ships, offshore platforms, or coastal defences. Mitigating these loads and ensuring the structural integrity of offshore platforms is crucial for their efficient operation and long-term viability in harnessing renewable wind energy. Therefore, comprehensive understanding, analysis, and design considerations related to marine loads play a pivotal role in the success of offshore wind energy projects.

Nowadays, the Environmental Wave Contours (EWC) are widely used to identify the critical sea states in which the safety domain of the structures should be evaluated, as a result of the joint probability of historical observations. A reliable estimation of wave variables is vital for the assessment of extreme loads. For this step, a novel distribution is implemented to model the marginal wave height observations. Burr distribution was recently proposed by Clarindo & Guedes Soares (2024) as a useful alternative for probabilistic modelling for marginal values of significant wave heights (H_s).

Nevertheless, to be more realistic and to add more usefulness, the Burr distribution is applied jointly and conditioning a secondary variable, defined by wave peak periods (T_p), which follows the traditional Lognormal distribution, thus resulting in a bivariate statistical Burr-Lognormal scheme. To finish, the distributions of the fitted parameters are employed to construct the EWC using the Inverse-First Order Reliability Method (I-FORM) for 10 and 100 years of return period. The advantage of I-FORM is related to its practical implementation, as well as the estimation of structural reliability parameters based on predominant marine conditions. In this way, it provides a compact and informative long-term approach to the foremost marine load characteristics for a given region.

The main objective is to develop a distribution of wave-induced environmental loads along the Portuguese coast. For this, the climatic reanalysis data of the ERA5 database are used (Hersbach et al. 2020), which are provided by the European Center for Medium-Range Weather Forecasts (ECMWF). The resulting database can be used in different applications for ocean engineering, being useful in navigation, estimation of energy potential, coastal defence planning, oil and gas industry in general, implementation of aquaculture farms, and others. However, the information on the distribution of wave-induced loads is beneficial in the planning and design of marine energy conversion devices, providing a point of view of structural reliability by modelling environment-structure interaction. The points studied are distributed in the South, Centre, and North regions along the coast of Portugal, regions that have been identified trough the

DOI: 10.1201/9781003508779-66

use of Regional Frequency Analysis (Lucas et al., 2017; Campos & Guedes Soares, 2018). Thus, the improved environmental contour method based on Burr-Lognormal is used to determine the characteristics of environmental wave loads in each region.

For the current work, the joint modelling of significant wave heights (H_s) and the corresponding conditional wave periods ($T_p|H_s$) are investigated. The Conditional Modelling Approach is efficient for different purposes as discussed by Guedes Soares et al. (1988), Bitner-Gregersen et al. (1991), Ferreira et al. (2002), Jonathan et al. (2010), Dong et al. (2013), Muraleedharan (2015) and Hallak et. al (2022). All those different applications showed no consensus in the community regarding multivariate modelling of met-ocean data. In addition, nowadays there is an increasing demand for an effective joint probabilistic model that can represent the joint statistical behaviour of the met-ocean parameters.

In this vein, the paper introduces an alternative distribution that is applied in the modelling of the sea conditions never used to derive wave contours on the Portuguese coast. The parameters from the novel distribution are generated by fitting the observations and implemented to construct the contours that can be directly compared with contours constructed using other parametric distributions, such as Generalized Extreme Values (GEV) and Weibull. The parameters estimated for the Burr-Lognormal model can be proposed in this paper as a baseline model to predict future conditions.

The interesting feature is that the Burr distribution also employs three parameters and consequently does not increase the model complexity. Conversely, it is necessary to test the flexibility of the parameters in fitting the distributions which can help to identify the extreme combinations of variables, even if they are considerably spread in the scatterplot. Differently from the mentioned parametric models, instead of a location parameter, Burr is defined by a second shape parameter offering more flexibility for the statistical scheme to ensure a good fit at the whole as well as the tail of the observations.

This study investigates the performance of implementing the Burr-Lognormal statistical scheme to jointly derive the wave contour of significant wave heights and peak periods. It focuses on the implementation of the new statistical scheme. Moreover, a comparison with the traditionally used GEV model is illustrated. The final contour associated with each parametric model can reveal some important characteristics such as: what model offers more conservative estimates as well as what model can dangerously underestimate extreme values. With this crucial information, a design engineer can decide which will be more appropriate for any practical case. In addition, for a deep discussion that presents statistical tests of univariate fitting processes including Chi-square (χ^2) and the Kolmogorov-Smirnov (KS) test, the estimation of Return Values (RV's) to test the extrapolation for long-periods as well as the degree of uncertainty, which can be read in the earlier implementation by Clarindo & Guedes Soares (2024).

The novel statistical schemes selected to model the sea state observations are described in section 2. The section introduces the Burr parameters applied to fit H_s distribution as well as the Lognormal parameters used to fit T_p observations. Section 3 is dedicated to describing the datasets applied in the implementation, and the map of the location for each region. Next, section 4 illustrates how the fitted parameters are used to derive wave contours based on the I-FORM approach. Section 5 exposes the final resulting wave contours for all regions. The closing section 6 expresses the remarkable conclusions as well as the foremost considerations resulting from the implementation of the novel statistical scheme to derive wave contours.

2 STATISTICAL SCHEME FOR MODELING SEA STATES

The suggested Burr-Lognormal distributions to model the sea state observations are described. Firstly, the tree-parametric Burr model is applied to characterize the marginal distribution of the primary variable, which consists of H_s observations. Following, the conditional approach scheme is applied, which consists of modelling the T_p variables following the Lognormal model, briefly described in the next subsection. The objective is to model the distribution of the dependent variable T_p as a function of the marginal distribution of H_s that are tested with the Burr model. So, the parameters resulting from the conditional process are applied in the derivations of the wave contours. Thus, it complies with the first step in the general process of derivation of environmental contours, shortly defined next.

The EWC methods generally involve three main stages: (i) statistical modelling: defining a statistical model able to jointly characterize the sea states (Burr-Lognormal), (ii) contour construction: computing the EWC, the I-FORM approach is used for this task, (iii) Design criteria: selecting extreme load conditions along the contour for design applications (Figure 1).

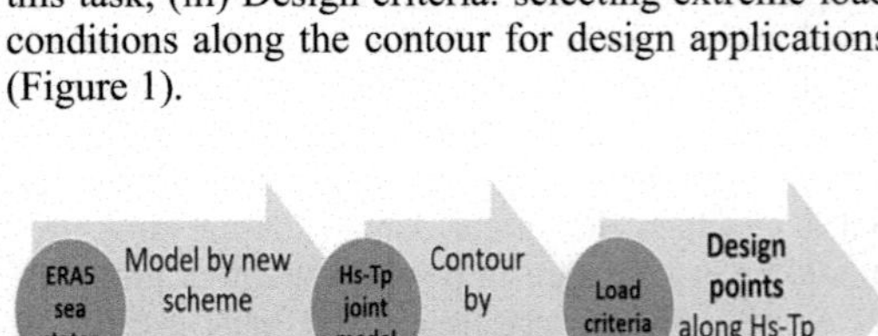

Figure 1. The general framework used for environmental contour derivation.

2.1 *Burr distribution model*

Burr (1942) introduced a two-parameter distribution based on typical scale (α) and shape (β) parameters.

Sometime later an additional shape (β_2) parameter was introduced by Tadikamalla (1980) improving its flexibility. Nowadays, the Burr type XII distribution model is defined by three parameters consisting of one scale parameter and two shape parameters respectively. This formulation can express a wide range of probability distribution shapes allowing to modeling of many types of observations. The Burr type XII distribution is a modified three-parameter family of distributions on the positive real line. They are widely used in extreme probabilistic problems that consider a parameterization of several correlated meteorological variables. Rodriguez (1977) offered a pioneer and useful guide for using the Burr distribution and Al-Husseine et al. (1991) presented a detailed characterization of the Burr model performance. The CDF of the Burr distribution is:

$$F(\alpha,\beta_1,\beta_2) = 1 - \frac{1}{\left(1 + \left(\frac{x}{\alpha}\right)^{\beta_1}\right)^{\beta_2}} \quad (1)$$

where β_1 and β_2 are the shape parameters and α is the scale parameter. The PDF is:

$$f(\alpha,\beta_1,\beta_2) = \frac{\frac{\beta_1\beta_2}{\alpha}\left(\frac{x}{\alpha}\right)^{\beta_1-1}}{\left(1 + \left(\frac{x}{\alpha}\right)^{\beta_1}\right)^{\beta_2+1}} \quad (2)$$

where $x > 0, \alpha > 0, \beta_1 > 0, \beta_2 > 0$.

The density formulations are known as the Burr type XII distribution, nonetheless, in this text, it is just named the Burr model.

2.2 *Conditional Lognormal distribution*

The Conditional approach is adopted to model the joint distribution following the Lognormal distribution. After fitting the marginal distribution of H_s by the three parameters of Burr, the conditional scheme is applied. So, the conditional Lognormal distribution of T_p that is associated with the H_s marginal is given by:

$$f_{T_p|H_s}(T_p|H_s) = \frac{1}{\sigma(H_s)T_p\sqrt{2\pi}} exp\left(-\frac{(lnT_p - \mu(H_s))^2}{2\sigma(H_s)^2}\right) \quad (3)$$

The next two equations describe the free parameters of T_p as functions of H_s observations. In this way, the expected Log (mean) and Log (standard deviation) values of T_p are described according to the defined parameters:

$$\mu(H_s) = mean[ln\ T_p|H_s] = a_1 + a_2\ H_s^{a_3} \quad (4a)$$

$$\sigma(H_s) = std[ln\ T_p|H_s] = b_1 + b_2\ exp^{\ b_3 H_s} \quad (4b)$$

Table 1 describes the baseline tested in statistical modelling for implementation to construct the sea state contours. In addition to being fitted with the Burr model, the study will also test the parameters fitted by the GEV model to construct the wave contours for all regions of the coast.

Table 1. Statistical baseline used to derive the wave contours by I-FORM approach for all regions.

Statistical scheme tested	
Marginal	Conditional
H_s (m)	T_p (s)
Burr	Lognormal
GEV	Lognormal

3 DATA SETS AND LOCATION

The ECMWF ERA5 (Fifth Generation of the European Centre for Medium-Range Weather Forecasts Reanalysis) is a state-of-the-art global atmospheric and oceanic reanalysis dataset (Hersbach et al. 2020). In this way, it is important to mention that the reanalysis refers to the process of combining observational data from various sources, such as weather, oceanographic and satellites with a numerical earth system model to create a consistent and continuous record of the climatic atmosphere and oceanic conditions over a certain period. So, ERA5 is the latest and most advanced reanalysis product released by ECMWF. It provides a comprehensive set of atmospheric, land, and oceanic variables at a high spatial and temporal resolution that include. Temporal Coverage: ERA5 covers the period from 1979 to the present, making it suitable for a wide range of climate and weather-related studies. Spatial Resolution: It has a horizontal grid resolution of approximately 31 kilometres globally, being 0.5° for marine spatial resolution.

ERA5 database is widely used in climate research, atmospheric science, and environmental modelling due to its high quality and comprehensive nature. Researchers use it to study past weather patterns, analyze climate trends, validate climate models, and conduct various other types of scientific investigations. For the present study, sea wave variables that define the intensity and frequency of sea states such as H_s and T_p are used. So, the data correspond to 41 years of hourly observations of significant wave height and peak periods, thus 368.186 observations.

A Regional Frequency Analysis was applied to wave (Lucas et al, 2017) and wind (Campos & Guedes Soares, 2018) conditions along the Portuguese coast. Using that information and assuming that the wave conditions on the coast could correspond to those areas, the coast of Portugal was regionalized and three locations were selected for the present study. The coast of Portugal was divided into the South, Center, and North regions according to the top panel from Figure 2 and Table 2.

Table 2. Information of the locations for ERA5 datasets along the Portugal offshore regions.

Regions	Location	Position (lat/log)
North	Aguçadoura	41.62/-9.00
Central	Peniche	39.37/-9.56
South	Algarve	36.84/-7.59

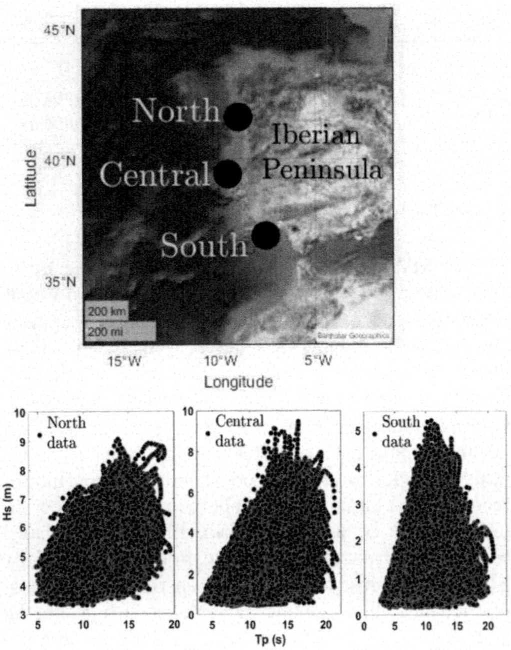

Figure 2. (Top panel) Locations of ERA5 wave data along the Portuguese coast. (Bottom) The scatter diagram for each dataset.

Due to the random nature of the variables, the historical observations include events considered rare. Generally, the maximum of the primary variable along the contour will not coincide with the maximum of the secondary variable. Moreover, characterizing these events requires that the model is flexible enough to fit the distribution of the variables. Sufficiently flexible to represent the peak as well as the tail of the distributions. This ensures a comprehensive and accurate portrayal of the entire range of possible outcomes. The term "flexible enough to fit the distribution" emphasizes the need for a model that can accommodate various shapes and patterns in the data. Different events may follow different statistical distributions, and a flexible model can adjust these variations. In a practical context, fulfilling this requirement is vital for risk mitigation, as it addresses severe events, playing a key role in enhancing the accuracy of forecasting and the effectiveness of risk management strategies.

Figure 2 (bottom panel), shows the scatter diagram of the areas of interest where rare events start to spread. This is an example of the flexibility that the statistical scheme must have, capable of identifying these zones and approximating the values. From the failure perspective, rare events contribute and can cause structural failure because they hold severe wave loads. This is the main justification for the importance of correctly representing the dispersion trend.

4 I-FORM WAVE CONTOUR CONSTRUCTION

Perhaps the most applied method to derive EWC is the Inverse-FORM (Winterstein et al. 1993) which can relate to reliability estimations. For the I-FORM operation, a circle with a radius equal to a reliability index (β_r) is computed for a given probability of failure. So, to obtain βr, the probability of failure (P_f) is firstly estimated by:

$$P_f = \frac{T_{ss}}{365 \times 24 \times T_r} \tag{5}$$

where Tss is the sea state duration in hours, Tr is the return period in years, and (365 x 24) is the factor to convert the number of years to hours. Finally, the relation between β_r and P_f is given by:

$$1 - P_f = \Phi\ (\beta_r) \tag{6}$$

Subsequently, the circle is transformed into the physical envirormental variables using the Rosenblatt transformation. The values of variables (u_1, u_2) in the standard normal space are given by:

$$u_1 = \beta_r cos\theta; u_2 = \beta_r\ sin\ \theta$$

where β_r is the radius of the circle and Θ is the angle between the radius of the circle and the u_1-axis varying between π and $-\pi$. According to the Rosenblatt transformation (Rosenblatt, 1942), the relation between the physical space-X and the standard normal space-U is given by:

$$\Phi(u_1) = F_{H_s}(H_s)$$

$$\Phi(u_2|u_1) = F_{T_p|H_s}(T_p|H_s)$$

Therefore, the marginal distribution of H_s and conditional T_p that relate the sea states variables can be defined using the Rosenblatt transformation following

$$H_s = F_{H_s}{}^{-1}(\Phi(u_1)) \tag{7}$$

$$T_p|H_s = F_{T_p|H_s}{}^{-1}(\Phi(u_2|u_1)) \tag{8}$$

In this way, environmental contours for determining extreme wave conditions are derived.

5 RESULTS AND DISCUSSIONS

5.1 *Fitting the distribution model to the datasets*

Assuming the joint conditional scheme defined previously, the EWCs are derived. Firstly, the H_s data is fitted using the Burr distribution. Figure 4 shows the histogram, highlighting the fit curve for the global observations. Figure 4 also presents the fitting for the Cumulative Density Function (CDF) as well as the quantiles computed. Following, Figure 5 shows the fitting of conditional model structures based on the Maximum Likelihood method. Table 3 describes the free statistical parameters resulting from fitting the marginal H_s by Burr univariate model. Considering that the available models are parametrized differently, the marginal H_s distributions are also fitted by GEV. Thus, Table 3 shows the fitted-free parameters for Burr, which are: one scale (α) and two shape parameters (β_1 and β_2), as well as the traditional GEV model parameters defined by one scale (α), shape (β) and one location (γ) parameter, respectively. These parameters are used to construct contours based on GEV-Lognormal for comparison. The results of fitting for conditional structure using Maximum Likelihood functions are illustrated in Table 4 and Figure 5.

As a result, the structural reliability can be expressed in terms of either the probability of failure or the reliability index. In general, all resulting information represents the failure status for various system modes. Thus, it is possible to better characterize the Limit State Function (LSF). The I-FORM method linearizes the LSF around the most probable point in the design space and estimates the probability of failure using a given distribution.

In short, $G(X) \leq 0$, where: G(X) is the Limit State Function, (X) is a vector of random variables representing the input parameters. So, the reliability analysis involves evaluating the probability of failure (P_f), which is the probability that the limit state function exceeds zero. Accordingly, the P_f represents the probability of the LSF at the negative value. Thus, considering the relationships represented in equations 6 and 7, and knowing that the ERA5 time step of measurement is hourly. The P_f for 10 years is $1.141e^{-05}$ presenting a reliability index (β_r) of 4.23. Furthermore, the value of P_f for 100 years of return period is $1.141e^{-06}$ with β_r of 4.72.

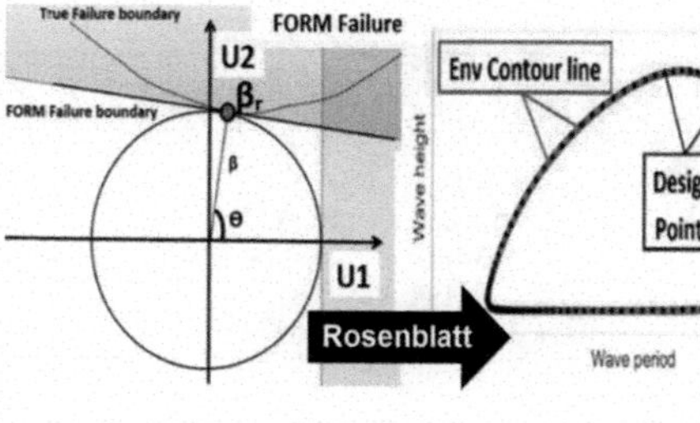

Figure 3. The Inverse-First Order Reliability Method (I-FORM) to construct the contours by the Rosenblatt performances to transform the variables.

Table 3. Estimated parameters of Burr and GEV models for H_s.

Models	Burr			GEV		
Parameters	α	β_1	β_2	α	β	γ
North	4.395	34.708	0.458	0.011	0.546	4.498
Central	3.075	4.080	2.775	0.010	0.572	1.642
South	1.568	4.841	2.142	0.020	0.295	0.756

Table 4. The conditional statistical parameters computed by Lognormal for T_p distributions.

Regions		Conditional Lognormal		
North	a_i	0.061	0.993	0.325
	b_i	0.188	-0.032	-0.120
Central	a_i	0.468	1.541	0.181
	b_i	0.001	0.241	-0.120
South	a_i	1.786	0.169	0.759
	b_i	0.341	-0.091	0.125

5.2 *Environmental wave contours*

The values of free parameters from Tables 3 and 4 are applied to construct the wave H_s-T_p environmental contours. Firstly, the EWCs resulting from the Burr-Lognormal bivariate distribution for all regions are shown. The contours are derived from 10-and -100 years of period return. Next, the EWCs are compared with those resulting from the GEV-Lognormal. Figure 6, shows the performance of the applied novel statistical schemes overlaid with the historical observations. The bivariate conditional scheme used in the current work presents good flexibility in terms of the capability to model extreme conditions, as well as the zones where the occurrence of severe conditions begins to scatter becoming rare.

In these cases, the computed H_s values for 10- and 100-years of return periods (Tr) are 8.51m and 9.73m for the North region. The Central area presents H_s = 8.25m for 10 years and 10.27m for 100 years. The remaining South region presents a maximum value of H_s = 4.71m and 5.86m for 10-and-100 years respectively.

Moreover, Figure 7, illustrates the comparison with GEV-Lognormal contours for the Algarve region. So, the maximum values computed for H_s in that area are 4.53m and 5.41m. The highest values along the EWC are summarized in Table 5.

The Table expresses the maximum values and the associated variables. In this way, it is possible to properly compute the sea states that induce the loads. The variances in the resulting values are due to the selected statistical scheme for modelling the joint distribution. These variations are evident for all regions. However, for short periods of return, the

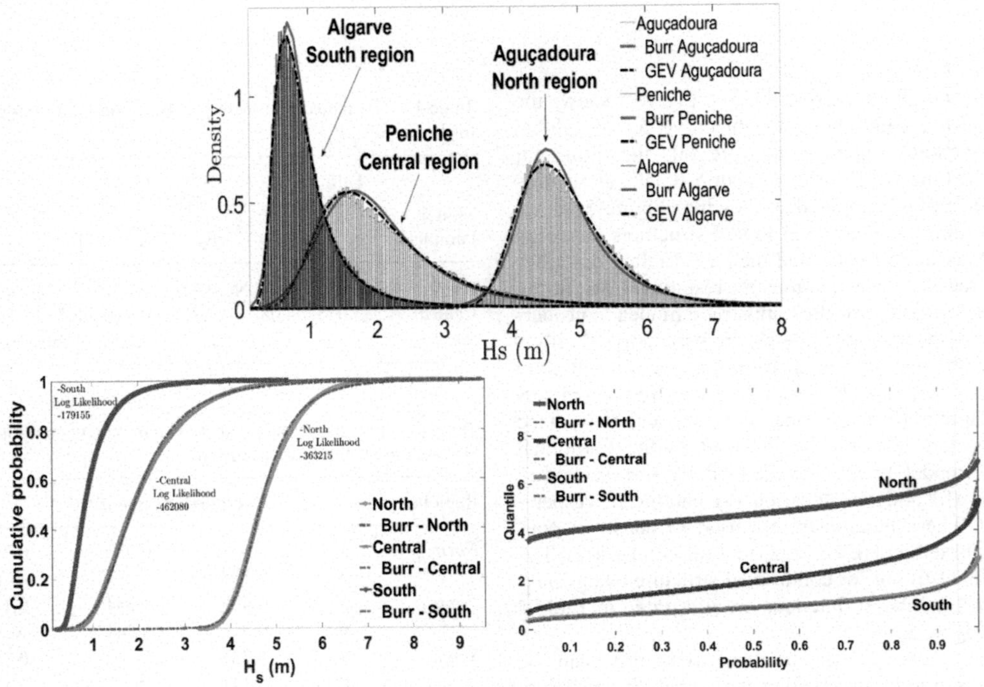

Figure 4. Histogram of H_s densities distribution of global observations fitted by Burr and GEV models. The CDF and quantile (bottom right panel) fits are also presented.

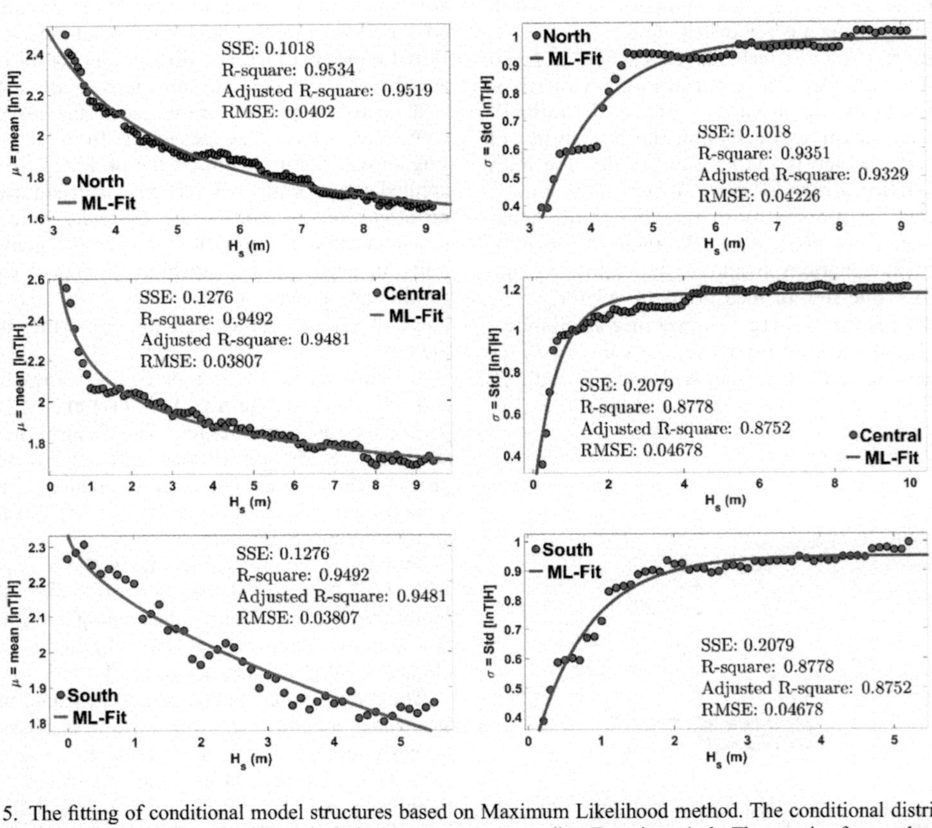

Figure 5. The fitting of conditional model structures based on Maximum Likelihood method. The conditional distributions follow the Log (mean) and Log (standard deviation) parameters, according Equations 4a-b. The metrics for goodness of fit are shown.

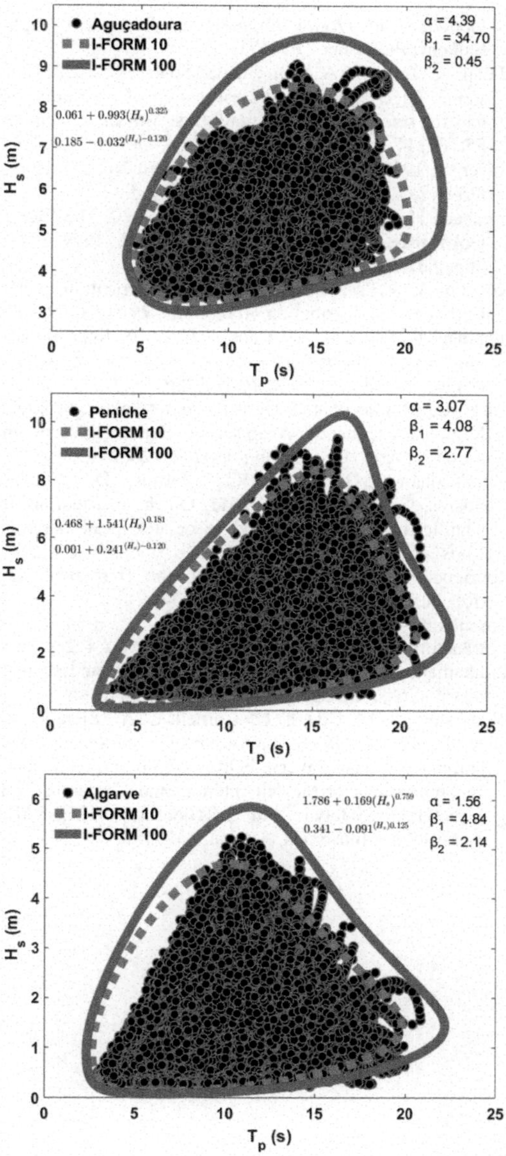

Figure 6. The I-FORM wave contour derived by Burr-Lognormal statistical scheme for each region.

Table 5. Design sea states computed along the Burr-Lognormal wave contours.

Regions	Tr (years)	Maximum H_s and associated T_p	Maximum T_p and associated H_s
North	10	8.51-13.84	20.13-5.33
	100	9.73-15.14	22.11-5.70
Central	10	8.25-15.65	20.84-2.46
	100	10.27-16.73	22.84-2.47
South	10	4.71-10.08	19.78-1.38
	100	5.86-11.43	22.19-1.39

differences can be negligible. From the point of view of wave-induced loads on the structure, the contours support the identification of the areas considered dangerous that denote the boundary of safe and unsafe domains. In this domain, the maximum of the primary variable along the contour will not coincide with the maximum of the secondary variable. So, the combination of maximum variables and the associated values as design points along the contour are priority to estimate. Thus, the values are illustrated in Table 5.

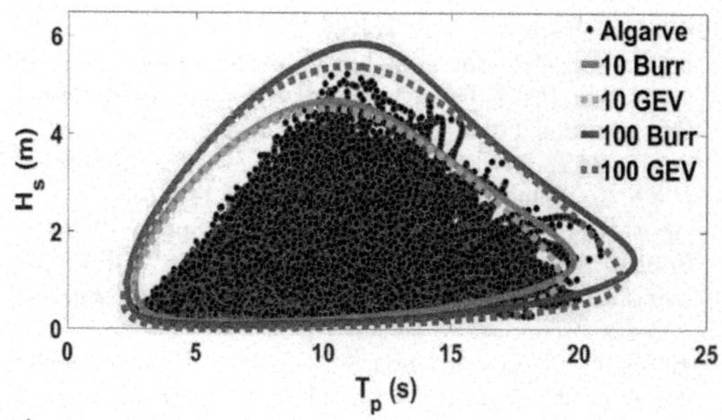

Figure 7. Comparison of wave contours derived by the Burr-Lognormal and GEV-Lognormal schemes.

6 CONCLUSION

Due to the low probability of occurrence of events that induce extreme loads, statistical models play a vital role in the construction of wave contours. Thus, an alternative bivariate scheme never used for this purpose is successfully implemented. Therefore, as shown in the figures presented in the results section, the implemented scheme was able to identify these events, despite of the complexity. Then, the discrepancies in the obtained resulting values stem from the chosen statistical model for representing the joint distribution. These variations are observable across all regions along the Portuguese continental coast. Nevertheless, for shorter return periods, the discrepancies may be insignificant. However, the contour resulting from Burr proved to be more conservationist, presenting slightly higher return period values adding a safety aspect. Moreover, the number of points outside the contours is practically negligible for the 100-years contours, which is crucial for calculating reliable design points. Otherwise, it represents a poor predictive power of future conditions.

To conclude, remembering that there is no universal consensus that defines the best scheme or metric to compare performance, it is recommended to contrast the resulting contours from other schemes. This way, the sea state that induce extreme loads used to design the marine structures gain more robustness. So, the tested and implemented scheme demonstrated to be a practical alternative valid in many situations.

ACKNOWLEDGEMENTS

The first author had the support of the Portuguese Foundation for Science and Technology (FCT), with the contract SFRH/BD/143567/2019. This work contributes to the Strategic Plan of the Lisbon Center for Naval and Ocean Engineering and Technology, which is also financed by the FCT under the contract UIDB/UIDP/00134/2020.

REFERENCES

Bitner-Gregersen, E.M. & Haver, S. 1991. Joint environmental model for reliability calculations. In: Jin Chung, S. (Ed.), Proceedings of the First International Offshore and Polar Engineering Conference. *The International Society of Offshore and Polar Engineering (ISOPE)*, August, Edinburgh, U.K., pp. 11–16.

Campos, R. M. & Guedes Soares, C. 2018. Spatial distribution of offshore wind statistics on the coast of Portugal using Regional Frequency Analysis. *Renewable Energy.* 123:806–816.

Clarindo, G. & Guedes Soares, C. 2024. Environmental contours of sea states by the I-FORM approach derived with the Burr-Lognormal statistical model. *Ocean Engineering*, 291, 116315.

Dong, S., Wang, N., Liu, W. & Guedes Soares, C. 2013. Bivariate Maximum Entropy Distribution of Significant Wave Height and Peak Period. *Ocean Engineering*. 59; 86–99.

Guedes Soares, C., Lopes, L. C. & Costa, M. 1988. Wave climate modelling for engineering purposes. Schreffler, B. A. and Zienkiewicz O. C., Eds. *Computer Modelling in Ocean Engineering*. Rotterdam: A.A. Balkema Publishers: 69–175.

Hallak, T. Teixeira, A. P. & Guedes Soares, C. 2022. Epistemic uncertainties on the estimation of minimum air gap for semi-submersible platforms. *Marine Structures* 85:103244.

Haver, S. 1985. Wave climate off northern Norway. *Appl. Ocean Res*. 7 (2), 8592.

Hersbach H, Bell B, Berrisford P, et al. 2020. The ERA5 global reanalysis. Q J R Meteorol Soc. 146: 1999–2049. https://dx.doi.org/10.1002/qj.3803

Hussaini-Al & Essam, K. 1991. A characterization of the Burr type XII distribution. *Appl. Math. Lett. 4,1*, 59–61.

Jonathan, P., Flynn, J. & Ewans, K.C. 2010. Joint modelling of wave spectral parameters for extreme sea states. *Ocean Engineering; (37)*: 1070–1080.

Lucas, C.; Muraleedharan, G., and Guedes Soares, C. 2017. Regional Frequency Analysis of Extreme Waves in a Coastal Area. *Coastal Engineering*. 12681–95.

Muraleedharan, G., Lucas, C., Martins, D., Guedes Soares, C. & Kurup, P.G. 2015. On the distribution of significant wave height and associated peak periods. *Coast. Eng*. 10342–10351.

Rodriguez, R.N. 1977. A guide to the Burr type XII distributions. Biometrika 64, 129–134.

Rosenblatt, M. 1952. Remarks on a multivariate transformation. Ann. Math. Stat. 23 (3), 470–472.

Tadikamalla, P.R. 1980. A look at the Burr and related distributions. Int. Stat. Rev. 48 (3), 337–344.

Winterstein, S. R., Ude, T. C., Cornell, C. A., Bjerager, P. & Haver, S. 1993. Environmental parameters for extreme response: Inverse FORM with omission factors. In Proceedings of the 6th International Conference on Structural Safety and Reliability ICOSSAR. August 9-13, Innsbruck, Austria: 551–557.

Advances in Maritime Technology and Engineering – Guedes Soares & Santos (Eds)
© 2024 The Author(s), ISBN 978-1-032-83107-7

Uncertainty assessment of direct Monte Carlo contours due to the sample variability

G. Clarindo & C. Guedes Soares
Centre for Marine Technology and Ocean Engineering (CENTEC), Instituto Superior Técnico, Universidade de Lisboa, Lisbon, Portugal

ABSTRACT: The uncertainty evaluation of a direct method to derive environmental contours is stated. The characterization of the environmental observations at sea is computed due to the sample variability leading to the important sources of uncertainty. The sample variability is based on 5, 10 and 15 years of sample for location A and 1, 5 and 25 for location B. Direct simulations are employed to construct environmental contour for return periods of 20 and 50 years following a fitted model. For this task, a bootstrap technique is performed. The distribution of significant wave heights is based on the Weibull three parameters model. The dependence of wave periods (location A) and near-surface wind velocity (location B) are based on the Lognormal and Weibull two parameters, respectively. The quantification and illustration of uncertainties are based on confidence intervals calculated from the several bootstrapped contours samples.

1 INTRODUCTION

For the effective design of marine structures, the wave-induced loads need to be computed and represented as accurately as possible. As the structures are exposed to different sea conditions, the estimate of the occurrence of extreme wave-induced loads as well as the combination of these and other load processes, is a crucial task (Guedes Soares, 1990, 1992). Despite the computational advances in estimating the dynamic responses of structural systems, it is still a complex problem.

It is not enough to determine the extreme value of significant wave height (H_s), but it is necessary to determine the characteristic period of the sea state, mainly associated with the extreme H_s. So, initially, the long-term climate typically models the characteristic period (T) jointly and conditioned by the H_s distributions, based on concepts of the Conditional Modelling Approach (CMA). The following references are pioneers and relevant in the application of CMA using sea states, Guedes Soares et al. (1988), Mathisen & Bitner-Gregersen (1990), Repko et al. (2004), Jonathan et al. (2010). Moreover, this situation has been recently reviewed by Bitner-Gregersen et al. (2022).

In the present work, two different sources of hindcast data are used for fitting the joint probability distribution of the long-term observations of met-ocean variables for further environmental contour construction. As the main objective is to determine the final contour as well as identify the uncertainties using statistical techniques (bootstrap) due to the different sample sizes, the choice of hindcast data seems to be the reasonable option as it provides a large time series of historical observations. From a structural response perspective, the classical full long-term analysis is considered the most accurate approach to estimate the long-term extreme response as it accounts for all sea states (see detail in Guedes Soares & Moan, 1991). Nevertheless, this process is considered inefficient due to the large number of sea states required for such analysis, whereas a huge percentage of these simulations are not contributing to the extreme response. In an effort for a more efficient way to predict the extreme response for a given return period, the environmental contour method has been adopted, based on the Inverse-First Order Reliability Method (I-FORM) suggested by Winterstein et al. (1993).

Environmental Contours (EC) are useful in the design of marine structures to identify extreme environmental conditions that induce extreme loads and responses. thus making them ideal for structural reliability tasks. Nevertheless, recently attention has been given to the fact that different methods exist for establishing such contours, and in some cases, significant differences may be obtained from the various methods. Thus, different practices of EC can be found in the literature, for example, description marine climate (Haver & Nyhus (1986), Huseby et al. 2013), for determining the response of marine structures (Leira, 2008), can be used to identify

DOI: 10.1201/9781003508779-67

extreme sea states as a cause of coastal risk (Clarindo & Guedes Soares 2022a), for modelling design mechanical structural points (Manuel et al. 2018, Haselsteiner et al. 2021), long-term loads based on short-term analyses (Haver & Winterstein 2009), estimation of the joint extreme wave and wind loads applied to offshore wind platforms in the North Atlantic (Clarindo & Guedes Soares 2022b). Moreover, it is also used to determine the potential wave loads in energy converters for island territories (Clarindo & Guedes Soares (2021a)) as well as applied for prediction of extreme environmental loads in semi-submerged structures (Karmakar et al. 2016). In short, all the previous applications commonly include an approach with three main steps: (a) statistical modelling, (b) contour construction and (c) design condition. Design engineers typically use these steps to obtain a set of reliable marine conditions for several types of projects.

Recently, Haselsteiner et al. (2019) proposed a comparative study focused on the estimation of extreme conditions aiming to quantify uncertainty. The results of the research teams that contributed to the blind comparison study can be found in Haselsteiner et al. (2021). Examination of the presented contours highlighted significant differences between contours derived using different methods. However, in addition to the methods, the variability between the contours seemed to be related to the sample size. This source of uncertainty is evaluated and demonstrated in the present work using confidence intervals. So, the main objective is to define a final contour, which is the result of a robust statistical approach. The secondary objective is to demonstrate the inverse relationship between sample size and ranges of variations. In other words, the more observations used to define the final contour, the lower the estimated statistical variability.

In an effort for an efficient way to predict the extreme response for a given return period, the EC method that is based on the Inverse-First Order Reliability Method (I-FORM) was suggested by Winterstein et al. (1993). Thus, the contour lines derived from the probability density of wave observations are habitually used in design of marine structures to identify extreme combinations. Despite the efficient performance shown by the I-FORM approach, Huseby et al. (2014, 2015) suggested an alternative method based on different mathematical principles, known as direct method. The overall process is performed using simulations of the joint observations directly in the variable space. In the case of the I-FORM method, it is necessary to accomplish a series of transformations in the space of variables. The direct simulations technique used to construct the contour is Monte Carlo, so the name used hereafter to refer to the direct method is DMC.

Several studies have identified important sources of uncertainty, some related to statistical modeling (step a), including the performance of the main and some alternatives distributions. Other studies focus only on step "b" (contour construction) comparing construction methods, as initially many of them assume different mathematical principles, although they try to solve the same problem. However, in the overall process of implementing EC, it is possible to identify a crucial source of uncertainty that can dangerously affect steps "a" and consequently the step "b", which is known as sampling variability.

Silva-González et al. (2015) studied the effect of uncertainties associated with the EC on the extreme response for offshore structure. The results showed the variability faced in step "a" of the process, which represents the importance of selecting a good conditional joint model to fit the historical observations. Clarindo et al. (2021b) compare both approaches increasing the accuracy and efficiency due to reducing the statistical variability by employing the importance sampling techniques for deriving environmental contours.

Uncertainties involved in step "b" of the procedure are evidenced by Vanem and Bitner-Gregersen (2015) when comparing EC methods for establishing structural design criteria. Additionally, the practical consequences of the adopted approach are discussed. Uncertainties involved in step (b) of the procedure are verified by Vanem (2017) when comparing the extreme structural responses from different approaches used to construct the contours. In particular, the alternative DMC method always produces convex contours, whereas the traditional I-FORM contours may be either convex or non-convex.

In short, when uncertainties are produced by the size of the time series, they can lead to inexact values in steps "a" and "b". Reducing and modeling statistical uncertainties in deriving EC due to limited sample sizes can be accomplished by replicating the contours many times. To better understand the level of uncertainty, the bootstrap sampling procedure is implemented for replication. The procedure is based on a study by Gramstad et al. (2018) and extended by Vanem et al. (2019), and also defined as a baseline by Haselsteiner et al. (2019). In this vein, after performing the bootstrap procedure, it leads to 1000 different contours. These contours are then used to calculate the confidence interval and a final contour.

It is necessary to model the joint distribution of H_s and some variables to characterize the wave period (T). A common approach to achieve the joint distribution of ocean variables is by the CMA, deeply discussed by Haver (1987) as well as Guedes Soares et al. (1988), in which a marginal distribution of H_s is combined with various conditional distributions of T, generally represented by the spectral peak period (T_p). Nonetheless, several statistical models of joint distributions have also been proposed, such as: Athanassoulis et al. (1994), Ferreira and Guedes Soares (2002), Dong et al. (2013) as well as the recently Burr-Lognormal statistical scheme suggested by Clarindo and Guedes Soares (2023) to derive wave contours. For design purposes, it is essential to consider the extreme states that can affect the survivability of the structures, so the environmental contours have become a popular method to achieve these results in a computationally economical way.

The current article discusses and assesses the uncertainties involved in deriving environmental contours using the Direct Monte Carlo (DMC) approach. Such uncertainties are evaluated according to the different sample sizes of waves and winds observations from two hindcast databases. This work is also in line with the Benchmark (BM) exercises in estimating extreme environmental conditions (Haselsteiner et al. 2021), in which the investigation is here extended to include the wave contours, considering H_s-T_p as well, since the exercise from the BM only checks the uncertainties from H_s-U_{10} contour. The paper is organized as follows: Section 2 presents a brief discussion of the main steps of EC derivation and how DMC contours can be constructed as well as the description of the conditional model applied. Section 3 is dedicated to information related to the hindcast datasets used and the main results. Section 4 presents discussions and closing with relevant conclusions in section 5.

2 ENVIRONMENTAL CONTOUR STEPS

In the Environmental Contours (EC), the time histories of environmental states are principally approximated by statistical models. In the prediction of waves and winds characteristics relating to loads, several efforts have been performed in the last decades as suggested in various reports of Committee I.1 Environment of the International Ship and Offshore Structures Congress (ISSC). A major task in the design process of an offshore structure is to evaluate the structural integrity of a candidate. This evaluation covers fatigue and extreme loads.

The widely used international standard based on the International Electrotechnical Committee (IEC 2019) describes the design process for offshore practices and formulates requirements for structural reliability and is followed in the present study for the cases of offshore wind turbines. The EC concept applied in structural design permits the consideration of extreme environmental conditions independently of a particular structure. Then, the contour lines correspond to a set of environmental states, that can be used to explore design options. Characteristically, the joint distribution of significant wave height and wave period is wanted, and other parameters, such as wind velocity, might also be relevant, which is addressed in the current work.

Today there is a wide variety of methods to establish an EC, mainly based on Inverse-FORM (see pioneers' studies such as: Haver (1987) and Moan et al. (2005)). However, the alternative Direct Monte Carlo (DMC) techniques to construct the EC, despite being a simple and direct method can offer robust and accurate estimates of extreme combinations of critical conditions. Therefore, it is the target method in this work. In this vein, the work aims to evaluate the uncertainties involved in the construction of the EC based on the DMC approach due to different sample sizes represented by years of environmental observations. It is also important to say that in this paper, the term "contour derivation" is used to describe the combination of steps (a) and (b) as shown in Figure 1.

In short, the central objective is to explore as well as illustrate that the uncertainty increases when the sample size decreases applying sea states (H_s-T_p) and wind states (H_s-U_{10}). The uncertainties can be appraised using confidence interval values applying basic statistical metrics such as the inter-quantile ranges. The final comparative results describe the performance of the bootstrap approach to replicate the contours hundreds of times in order to determine their uncertainties for different sample sizes.

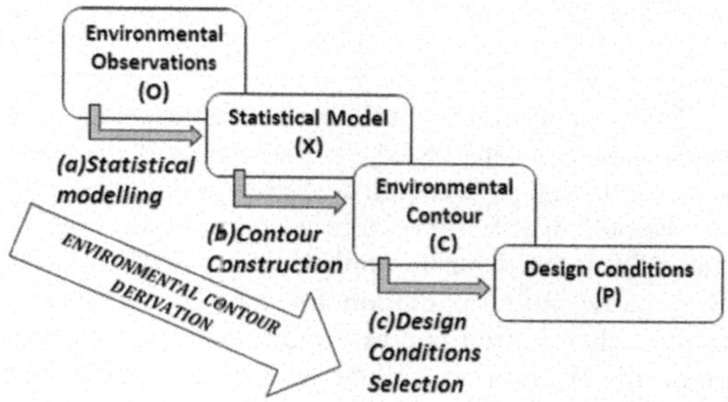

Figure 1. Description of the general EC steps: statistical modelling, contour construction and design condition selection. This paper focuses on uncertainties assessment included on contour construction.

2.1 *Statistical modelling*

Generally, the first step for any method of EC construction consists in modelling the observations based on a proper statistical model. The procedure involves fitting the observations to some probabilistic scheme able to, preferably, model the joint distributions of variables. Consequently, there are uncertainties related to this procedure, and consequently, the results will be dependent on the specific choice of parametric model for this task as well as the level of sophistication of the statistical scheme. An effective and widely used approach is to jointly model the distributions of the secondary variables (T_z and U_{10}) as a function of the distribution of the primary variable (H_s). For the considered scheme, the primary variable will be the significant height of the waves.

Many approaches attempt to characterize the distributions of these dependency schemes for metocean observations. In principle, Haver (1985) proposed a method for extreme (H_s-T_p) data modelling the tail of H_s using a 2-parameter Weibull form and then modelling the distribution of T_p for a given H_s

range using a Lognormal form, reminiscing that the pioneer application of Lognormal distribution to approximate wave periods are made by Ochi (1978). Further, the replacement of the 2-parameter Weibull distribution for H_s by the modified form 3-parameter Weibull was suggested by Mathisen and Bitner-Gregersen (1990), who showed in a comparative study that the bivariate distributions based on 3-parameters of Weibull and Lognormal provide a better fit to the Norwegian waters data than the bivariate Haver distribution (based on 2-parameter Weibull and the classical Lognormal).

For the currently work, the model structure consists in estimating the marginal Probability Density Function (PDF) for H_s and a conditional PDF for variables dependents such as T_z and U_{10}. The joint function can be written as a conditional model of $f(H_s|T_z)$ and $f(H_s|U_{10})$ as follows:

$$f_{H_s,\ T_z}(h,t) = f_{H_s}(h) f_{T_z|H_s\ (t|h)} \tag{1a}$$

$$f_{H_s,\ U_{10}}(h,u) = f_{H_s}(h) f_{U_{10}|H_s\ (u|h)} \tag{1b}$$

The conditional schemes represented in Equations 1a-b are based on the marginal distribution of H_s that is assumed to follow a 3-parameters of Weibull distribution while a conditional Lognormal PDF is used for T_z applied in point A. In point B a change in the conditional scheme is performed, the U_{10} distribution is now conditioned by 2-parameters of Weibull and their statistical conditional dependence is based on the Lognormal distribution. The parameters of the Lognormal distribution for wave period are dependent of H_s, as shown in the following equations:

$$\mu(H_s) = E[ln\ T_z|H_s\ ] = \ a_1 + a_2\ H_s^{\ a_3} \tag{2a}$$

$$\sigma(H_s) = Sd[ln\ T_z|H_s\ ] = \ b_1 + b_2 exp^{b_3 H_s} \tag{2b}$$

So, the wind speed observations are conditioned according to the parameters exposed next:

$$\alpha(H_s) = Sc[U_{10}|H_s] = \ c_1 + c_2\ H_s^{\ c_3} \tag{3a}$$

$$\beta(H_s) = Sh[U_{10}|H_s] = \ d_1 + d_2\ H_s^{\ d_3} \tag{3b}$$

where the coefficients a_i, b_i, c_i and d_i with i = 1, 2, 3 are estimated from the data. So, "E" and "Sd" are mean and the standard deviation parameters, coming from the Lognormal fitting procedure. While "Sc" and "Sh" are the scale and shape parameters from the Weibull-based conditional fit, described next.

2.1.1 *Fitting Weibull procedure*

Depending on the values of the parameters, the Weibull distribution can be used for a variety of extreme assessment applications. Note that in the H_s is assumed the most general distribution form (3-parameters) and the appropriate modifications to obtain the other used form such as, the 2-parameters form is obtained by setting γ=0.

$$f_x(X) = \frac{\beta}{\alpha}\left(\frac{x-\gamma}{\alpha}\right)^{\beta-1} e^{-\left(\frac{x-\gamma}{\alpha}\right)^{\beta}} \tag{4a}$$

$$f_x(X) = \frac{\beta}{\alpha}\left(\frac{x}{\alpha}\right)^{\beta-1} e^{-\left(\frac{x}{\alpha}\right)^{\beta}} \tag{4b}$$

where γ, α and β are parameters of location, scale and shape of the probability function, respectively.

2.1.2 *Fitting Lognormal procedure*

The Lognormal distribution is considered essential to the probabilistic modelling of environmental observations because it does not consider negative values. A random variable x has a Lognormal probability distribution when ln(x) is normal. In this case, the PDF are defined by:

$$f_x(x) = \frac{1}{\sqrt{2\pi}\ \sigma x}\ exp\left(\frac{ln(x)-\mu}{2\sigma^2}\right)^2 ;x>0 \tag{5}$$

where μ and σ are the mean and standard deviation parameters. The statistical model scheme applied in this work to derive the EC of sea and wind-sea states is currently recommended by DNV-RP-C205 (DNV 2021) in their recent 19th edition of practice that gives guidance for modelling, analysis as well as prediction of marine conditions. Moreover, offer guidance for calculating environmental loads acting on marine structures.

2.2 *Contour construction*

Based on the estimated joint distribution function of the environmental parameters, the contour lines may be constructed based on direct Monte Carlo simulations. This has been done for the contour lines corresponding to 1- and 20-years return periods for point A and 1-and-50 years for point B, respectively. Step (b) in the contour derivation process consists in constructing the EC lines based on the statistical model described previously and this step can be computed with a variety of different methods.

Probably the most popular approach to constructing EC is based on the I-FORM. proposed by Winterstein et al. (1993). However, the definition proposed by Huseby et al. (2013, 2014, 2015) is adopted in this work. This alternative approach to derive the EC does not need to perform any transformations and can be constructed directly on the real physical space. The construction of the EC based on direct Monte Carlo simulations are summarized by the following main steps:

(1) Generate an appropriate sample of the joint environmental distribution using Monte Carlo simulations. The numbers of suitable random samples simulated by Monte Carlo are the same number for each dataset evaluated.
(2) Assume an angle (Θ) situated between π and –π, and for each sample generated in (1) estimate its projection that takes the form of a straight line:

$$X_i = T_i cos(\theta) + H_i sin(\theta) \quad (6a)$$

$$Y_i = U_i cos(\theta) + H_i sin(\theta) \quad (6b)$$

(3) Sort these resulting X_i and Y_i projections in ascending order $X_1 \leq X_2 \leq X_3 \ldots X_n$.Then let (T(n), H(n) and U(n), H(n)) denote the corresponding contour samples.

(4) Estimate the number of samples that will be kept within the desired probability of failure (P_f) in a half-space so-called $\Pi(\Theta)$, such that it only contains k numbers of samples.

$$P_f = \frac{n-k}{n} \rightarrow k = n\left(1 - P_f\right) \quad (7)$$

(5) Assume another angle and repeat the steps, excluding step (1). Regarding t and u represent the T_z and U_{10} projections, practically computed with the same formulation, while h is H_s projections respectively. The next equations represent how the projections are estimated for each variable, in which the combinations generate the contours.

$$t_i = \frac{sin(\theta + \Delta\theta)\ C(\theta) - sin(\theta)C(\theta + \Delta\theta)}{sin(\theta + \Delta\theta)\ cos(\theta) - sin(\theta)cos(\theta + \Delta\theta)} \quad (8a)$$

$$u_i = \frac{sin(\theta + \Delta\theta)\ C(\theta) - sin(\theta)C(\theta + \Delta\theta)}{sin(\theta + \Delta\theta)\ cos(\theta) - sin(\theta)cos(\theta + \Delta\theta)} \quad (8b)$$

$$h_i = \frac{-cos(\theta + \Delta\theta)\ C(\theta) + cos(\theta)C(\theta + \Delta\theta)}{sin(\theta + \Delta\theta)\ cos(\theta) - sin(\theta)cos(\theta + \Delta\theta)} \quad (8c)$$

The remaining C(Θ) denotes the value of the projection onto the Θ-line and ΔΘ refers to the angular interval between two tangents. As can be seen, this methodology operates Monte Carlo directly in the real space using simulations of a fitted joint conditioned environmental model that does not require any transformation. Then the different parts of the environmental contours would then correspond to tangents in the original space. The results of previous equations represent the individual projections that are used jointly to construct the final contour. So, contours were constructed using 10º as the angular intervals, thus generating 35 points of intersection, which can be defined as variable coordinates (h35, t35 and h35, u35). This setup of angular resolution is considered reasonable by many authors, discussed and used by the pioneers of the method, as well as in the most recent benchmark study.

3 DATA AND LOCATIONS

3.1 *Hindcast data*

Two different locations of the ocean are considered to construct the contours, with differences in environmental combinations and sample sizes, as well as return periods. Location A is situated near the island of Fuerteventura (Canary Islands, Spain), and location B is positioned in the North Sea, offshore of the Germany coast. The providers of the hindcast data are the Port and Coast Authority from Spain Government for A (Simar hindcast nodes, see Pilar et al. 2008) and the organizers of the benchmark exercise on estimating extreme environmental conditions (Haselsteiner et al. 2019) for B (coastDat-2), respectively. Figure 2 and Table 1 provide additional information on the locations.

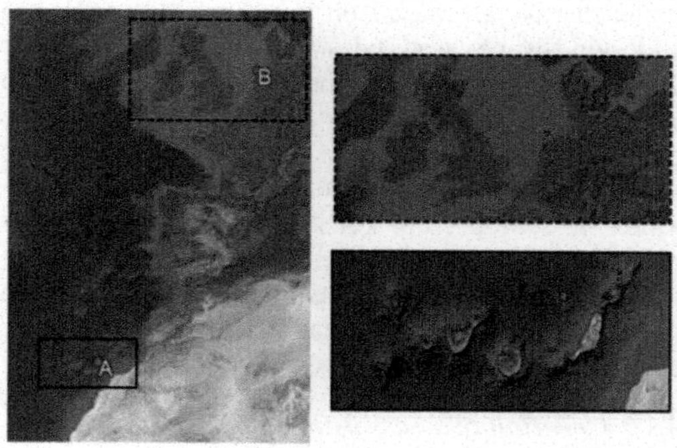

Figure 2. Map of the locations for studied hindcast points.

To assess the variability due to sample size, several random data are selected to generate sub-samples with different sizes from the original data as done by Gramstad et al. (2018), reinforced by Vanem et al. (2019) and defined as the baseline by Haselsteiner et al. (2019) for exercise two of proposed Benchmark and performed by (Haselsteiner et al. 2021). The number of data observations corresponding to each sample size of the studied area is indicated in Table 2, as well as the return period considered to construct the contours. For location A, different sample sizes corresponding to 5, 10 and 15 years of data are considered and 1, 5 and 25 years of data for location B, respectively.

Table 1. Information of study cases.

Location	Position	Source	Frequency	Global data
A	28.25N 15.25W	Simar Node	3-hourly	22 years
B	54.00N 06.57E	coastDat-2	1-hourly	25 years

Table 2. Numbers of observations included in each sample size. It is interesting to note that the number of observations in each sample increases.

Location	Sample size (years)	Number of data	Env. Contour (Return period)
A	5	14250	1 and 20 years H_s-T_z
	10	28500	
	15	42810	
B	1	8766	1 and 50 years H_s-U_{10}
	5	43830	
	25	219150	

4 RESULTS AND DISCUSSION

Environmental contours are constructed based on a direct approach computed by Monte Carlo simulations using two different databases. The observations for each case are based on wave sea states (H_s-T_z) and wind-wave sea states (H_s-U_{10}). The contour uncertainty is visually characterized due to the sample variability, and the confidence intervals are computed for each sample size established by years of observations. Using bootstrap techniques, thousands of contours are constructed, and the IQR is plotted in an overlay to perform the visual inspection. The IQR from the bootstrap examination is also plotted overlaid with the scattered data to check the consistency of fit procedures.

4.1 *Uncertainty assess using bootstrap*

For each location and each sample size of data described in Table 2, a bootstrap technique is used, and thousand contour lines are generated by drawing samples from the fitted parametric distribution model. The parameters used to construct the EC for each sample size are shown in Table 3 and 4. So, the subsamples with different sizes are randomly drawn from the total block of observations. In this vein, a resampling technique can be applied to assess the uncertainty associated in the EC derivation. It involves drawing multiple samples with replacement from the observed data to create synthetic datasets. This process provides an empirical estimate of the sampling distribution that can be used to quantify uncertainty.

Table 3. The H_s statistical parameters for locations A and B fitted by Weibull 3 parameters.

Location A (years)	*α(scale)*	*β(shape)*	*γ(Location)*
5	1.67	2.04	0.39
10	1.68	2.06	0.35
15	1.72	2.11	0.36
Location B	*α(scale)*	*β(shape)*	*γ(Location)*
1	0.97	0.56	0.14
5	1.53	1.36	0.11
25	1.33	1.77	0.10

Then, the distribution of the new block of samples are re-fitted. For this task, the EC based on 20-and 50-years return periods is considered for wave states (A) and wind-wave states (B), respectively. In many practical cases, only an individual environmental contour is assumed to be sufficient to extract design values for sea load assess of marine structures. To avoid this dangerous practice, the EC is constructed for the thousand samples.

For evidence of an effective reduction in confidence intervals, the inter-quantile values are computed for each bootstrapped sample size. Note that the range of variation decreases as the sample size increases, and as a consequence, the confidence interval becomes narrower. The scattered data present a good agreement with the EC lines based on their fitted parameters as well as a good predictive power being more robust as the data sample size is increased. The reduction of statistical uncertainties due to limited sample sizes are performed by replicating the contours many times.

To better understand the bootstrap sampling procedure and how to implement for the replication, please check the references such as Gramstad et al. (2018) that investigate the uncertainty of environmental contours for joint distribution of significant wave height and wave period (sea states) for different sample sizes (10, 20, 30 and 100 years of data), but not include the joint distribution of significant wave height and the wind velocity (wind states) as investigate in this work.

The method was extended by Vanem et al. (2019), and also defined as the baseline by Haselsteiner et al. (2019) for exercise two of BM (see Haselsteiner et al. 2021 for the final results). The mentioned studies confirm that the contours resulting from the bootstrap process can be used to estimate confidence intervals. Evaluating this information and overlaying all resulting contours show that uncertainty decreases with increasing sample size for wave as well as wind-wave environmental contours. This is roughly apparent in the bootstrap overlay plot (Figures 3 and 4 on left panels) and more evident in the confidence intervals plots overlaid with scattered data (Figures 3 and 4 on right panels). It is observed that the width of the confidence intervals decreases for larger sizes of samples.

Table 4. The parameters from conditional statistical scheme for T_z and U_{10} dependences, according Eq. 1a-b. Remembering that sea states are based on lognormal and wind states follow the weibull 2 parameters.

Conditional T_z (s) Lognormal					Conditional U_{10} (m) Weibull 2P				
		$i=1$	$i=2$	$i=3$			$i=1$	$i=2$	$i=3$
5	a_i	1.05	0.60	0.16	1	a_i	0.27	3.80	0.17
	b_i	0.00	0.89	-0.37		b_i	1.22	0.11	-0.33
10	a_i	2.02	0.42	0.41	5	a_i	0.47	6.99	0.29
	b_i	0.00	0.48	-0.16		b_i	0.00	4.49	0.16
15	a_i	1.47	0.72	0.51	25	a_i	1.37	7.93	0.32
	b_i	0.00	0.53	-0.49		b_i	0.00	3.11	0.10

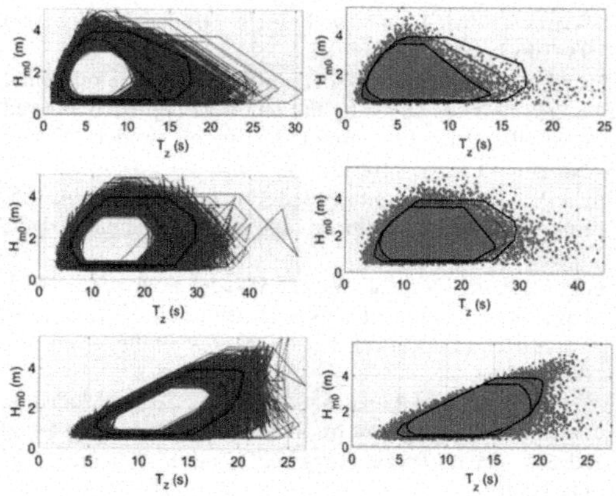

Figure 3. Environmental wave contour lines for 20-years periods of return based on DMC simulations. The left panels showed the thousands of bootstraps EC overlaid with the IQR. The right panels shown overlaid the IQR with the scatter data considered.

Regarding the construction of the environmental contour, there is no standard sample size. The sample size you need depends entirely on the nature of your research questions, what you hope to achieve with your study, and what resources are available to develop your goals. In general, a larger sample size is often preferred to provide more reliable estimates of extreme quantiles. Nevertheless, the availability of data is not always within the researcher's control, and in some cases, limited data may be the only option. In practice, consulting relevant literature, seeking guidance from statistical experts, and conducting sensitivity analyses can help researchers make informed decisions about sample size in the construction of environmental contours.

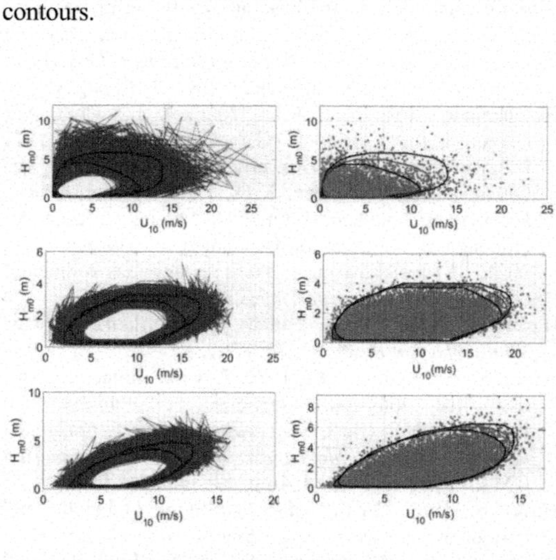

Figure 4. Environmental wind contour lines for 50-years periods of return based on DMC simulations. The left panels showed the thousands of bootstraps EC overlaid with the IQR. The right panels shown overlaid the IQR with the scatter data considered.

The method to derive EC presents some level of uncertainty concerning limited datasets. So, it is an important consideration and needs to be characterized correctly whenever possible. However, when collecting quantitative and qualitative data, the general rule of thumb is to have as many observations as possible, which will allow to reduce the ranges of variation and robustly define a representative environmental contour (see Figure 5). Hence, it can be concluded that, in general, the differences between EC for each sample size are significant even if the uncertainty due to finite data samples are considered. Bearing in mind that there is no universal metric that compares the performance of each method to construct the contour, it is recommended to evaluate its uncertainties using the bootstrap technique regardless of the sample size used. However, the results presented in the current work show that it is favorable to have as many measurements as possible since contours constructed with few state observations will present substantial uncertainty. Thus, to determine a final representative environmental contour that can be used to support projects for the design of offshore structures, it is reasonable to construct hundreds of environmental contours and then compute the average contour or, in the case of the current work, the median contour, which corresponds to the value of the quantiles 50%.

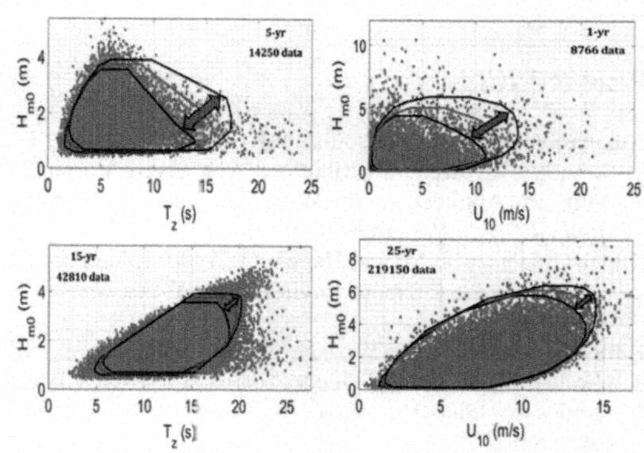

Figure 5. Demonstration of the decrease in the variation range as the sample size increases, thus the confidence interval becomes narrower. Sea states are represented in the left column and wind states in the right column of the panel.

5 CONCLUSIONS

The estimation of environmental loads is a crucial point to reliably design any marine structure. Thus, evaluating and reducing statistical uncertainties involved in the process is highly recommended. Of the various sources of uncertainty, the current work evaluated those related to the size of the samples of sea and wind-sea states. Some data sets with different sample sizes are then generated to assess the uncertainty. The degree of variability amongst the resulting contours is noticeable for all of the comparison cases. However, when the number of observations is increased, the uncertainty is reduced. Thus, it is recommended to consider as many observations as possible. Another way to robustly define a contour is to apply the bootstrap technique to derive the contour many times and then estimate the average final contour. This simple procedure proved to be efficient and robust in reducing the uncertainties involved in the construction of the contour using the direct Monte Carlo technique, even when the number of observations is limited.

Finally, the appropriate sample size depends on several factors, including the characteristics of the environmental variable, the desired level of accuracy, and the specific goals of the analysis.

ACKNOWLEDGEMENTS

The first author had the support of the Portuguese Foundation for Science and Technology (FCT), with the contract SFRH/BD/143567/2019. This work contributes to the Strategic Research Plan of the Centre for Marine Technology and Ocean Engineering (CENTEC), which is financed by the Portuguese Foundation for Science and Technology (Fundação para a Ciência e Tecnologia - FCT) under contract UIDB/UIDP/00134/2020.

REFERENCES

Athanassoulis, G.A., Skarsoulis, E.K. and Belibassakis, K. A., 1994. Bivariate Distributions with given Marginals with an Application to Wave Climate Description; *Applied Ocean Research*, 16, pp. 1–17.

Bitner-Gregersen, E.M. and Hagen, O., 1990. Uncertainties in Data for the Offshore Environment. *Struct. Safety*, 7, pp. 11–34.

Bitner-Gregersen E.M., Waseda, T., Parunov J., Yim, S., Hirdaris S., M. N., and Guedes Soares, C., 2022. Uncertainties in long-term wave modelling. *Marine Structures*; 84; 103217.

Clarindo, G. and Guedes Soares, C., 2021a. Environmental wave contours for the west coast of Fuerteventura. *Developments in Renewable Energies Offshore*, Guedes Soares (Ed.). Taylor & Francis Group, London; 3–10.

Clarindo, G., Teixeira, A.P. and Guedes Soares, C., 2021b. Environmental wave contours by Inverse FORM and Monte Carlo Simulation with variance reduction techniques. *Ocean Engineering*; 228:108916.

Clarindo, G. and Guedes Soares, C., 2022a. Long-term probabilistic identification of extreme sea-states as causes of coastal risk due to wave severity. *Trends in Maritime Technology and Engineering*. Guedes Soares (Ed.). Taylor & Francis Group, London, Vol 2, pp 299–306.

Clarindo, G. and Guedes Soares, C., 2022b. Extreme wave contours for floating wind turbine sites in the Atlantic area. *Trends in Renewable Energies Offshore*. Guedes Soares (Ed.). Taylor & Francis Group, London; pp 93–100.

Clarindo, G. and Guedes Soares, C., 2023. Environmental contours of sea states by the I-FORM approach derived with the Burr-Lognormal statistical model. *Ocean Engineering* 291, 116315.

Dnv GL., 2021. Recommended Practice DNV-RP-C205: Environmental Conditions and Environmental Loads, *Technical report*.

Dong, S., Wang, N., Liu, W., and Guedes Soares, C., 2013. Bivariate Maximum Entropy Distribution of Significant Wave Height and Peak Period. *Ocean Engineering*. 59; 86–99.

Ferreira, J. A. and Guedes Soares, C., 2002. Modelling Bivariate Distributions of Significant Wave Height and Mean Wave Period. *Applied Ocean Research*. 24 (1):31–45.

Guedes Soares, C., Lopes, L. C. and Costa, M., 1988. Wave climate modelling for engineering purposes. Schreffler, B. A. and Zienkiewicz O. C., Eds. *Computer Modelling in Ocean Engineering*. Rotterdam: A.A. Balkema Publishers: 69–175.

Guedes Soares, C., 1990. Stochastic Models of Load Effects for the Primary Ship Structure. *Structural Safety*. 81-4:353–368.

Guedes Soares, C. and Moan, T., 1991. Model uncertainty in the long-term distribution of wave-induced bending moments for fatigue design of ship structures. *Marine Structures*; 4(4):295–315.

Guedes Soares, C., 1992. Combination of Primary Load Effects in Ship Structures. *Probabilistic Engineering Mechanics*. 72:103–111.

Gramstad, O., Vanem, E. and Bitner-Gregersen, E. M., 2018. Uncertainty of environmental contours due to sampling variability. In *Proc. 37th International Conference on Ocean, Offshore and Arctic Engineering* OMAE 2018., American Society of Mechanical Engineers ASME.

Haselsteiner, A. F., Coe, R. G., Manuel, L., Nguyen, P. T., Martin, N. and Eckert-Gallup A., 2019. A benchmark exercise on estimating extreme environmental conditions: methodology and baseline results. In proceedings of the ASME – *38th International Conference on Ocean, Offshore and Arctic engineering* OMAE. – June 9-14 – Glasgow, Scotland – UK – 1–10 pages.

Haselsteiner, A.F., Coe, R.G., Manuel, L., Chai, W., Leira, B.J., Clarindo, G., Guedes Soares, C., Hannesdóttir, I., Dimitrov, N, Sander, A., Ohlendorf, J-H., Thoben, K.D., Hauteclocque, G. De, Mackay, E.; Jonathan, P., Qiao, C., Myers, A., Rode, A., Hildebrandt, A., Schmidt, B., Vanem, E., Huseby, A.B., 2021. A benchmarking exercise for environmental contours. *Ocean Engineering*, 236, 109504.

Haver, S., 1985. Wave climate off northern Norway. *Applied Ocean Research*, 72: 85–92.

Haver, S. and Nyhus, A.K., 1986. A wave climate description for long-term response calculations. In: Lunardini, V.J. (Ed.), *Proceedings of the Fifth International Offshore Mechanics and Arctic Engineering* OMAE Symposium; vol. 4-pp. 27–34. April 13- 18.

Haver, S., 1987. On the joint distribution of heights and periods of sea waves. *Ocean Engineering*, 14, 359–376.

Haver, S. and Winterstein, S., 2009. Environmental contour lines: A method for estimating long term extremes by a short-term analysis. *Transactions of the Society of Naval Architects and Marine Engineers*, 116, pp. 116–127.

Huseby, A.B., Vanem, E. and Natvig, B., 2013. A new approach to environmental contours for ocean engineering applications based on direct Monte Carlo simulations. *Ocean Engineering*. 60: 124–135.

Huseby, A.B., Vanem, E. and Natvig, B., 2014. A new Monte Carlo method for environmental contour estimation. In: *Safety and Reliability, Theory and Applications*, Nowakowski, T., Mlynczak, M., Jodejko-Pietruczuk, A. & Werbinska-Wojciechowska, S. (Eds.), CRC Press, The Netherlands.

Huseby, A.B., Vanem, E. and Natvig, B., 2015. Alternative environmental contours for structural reliability analysis. *Structural Safety*, 54, pp. 32–45.

International Electrotechnical Commission – IEC, 2019. Wind energy generation systems - Part 3-1: Design requirements for fixed offshore wind turbines. *Technical report* IEC 61400-3-1. 2019.

Jonathan, P., Flynn, J., Ewans, K.C., 2010. Joint modelling of wave spectral parameters for extreme sea states. *Ocean Engineering; (37)*: 1070–1080.

Karmakar, D., Bagbanci, H., and Guedes Soares, C., 2016. Long-term extreme load prediction of spar and semi-submersible floating wind turbines using the Environmental Contour Method. *J. Offshore Mech. Arctic Eng*. 138 (2), 021601.

Leira, B. J., 2008. A comparison of stochastic process models for definition of design contours. *Structural Safety*, 30, pp. 493–505.

Manuel, L., Nguyen, P.T., Canning J., Coe, R., Eckert-Gallup A. and Martin N., 2018. Alternative approaches to develop environmental contours from met-ocean data. *J. Ocean Eng. Mar. Energy*; 4, 293–310.

Mathisen, J., and Bitner-Gregersen, E.M., 1990. Joint distributions for significant wave height and wave zero-up-crossing period. *Appl. Ocean Res*. 12, 93–103.

Moan, T., Gao, Z., and Ayala-Uraga, E., 2005. Uncertainty of wave induced response of marine structures due to long-term variation of extra-tropical wave conditions. *Marine Structures*: 184: 359–382.

Ochi, M. K., 1978. On Long-term statistics for ocean and coastal waves. *In: Proceedings. 16th Conf. on Coastal Eng*. 1, 59–75.

Pilar, P, Guedes Soares, C., and Carretero, J.C., 2008.44-year wave hindcast for the northeast Atlantic European coast. *Coastal Engineering*, 55, 861–871

Silva-González, F.L., Vázques-Hernandez, A.O., Sagrilo, L. V.S. and Cuamatzi, R., 2015. The effect of some uncertainties associated with the environmental contour lines definition on the extreme response of an FPSO under hurricane conditions. *Appl. Ocean Res*. 53, 190–199.

Vanem, E. and Bitner-Gregersen, E.M., 2015. Alternative environmental contours for marine structural design a comparison study. *Journal of Offshore Mechanics and Arctic Engineering*, 137, 051601.

Vanem, E., 2017. A comparison study on the estimation of extreme structural response from different environmental contour methods. *Marine Structures*, 56, 137–162.

Vanem, E., Gramstad, O., and Bitner-Gregersen, E.M., 2019. A simulation study on the uncertainty of environmental contours due to sampling variability for different estimation methods. *Appl. Ocean Res*. 91, 101870.

Winterstein, S. R., Ude, T. C., Cornell, C. A., Bjerager, P. and Haver, S., 1993. Environmental parameters for extreme response: Inverse FORM with omission factors. *In Proceedings of the 6th International Conference on Structural Safety and Reliability* ICOSSAR. August 9-13, Innsbruck, Austria: 551–557.

Advances in Maritime Technology and Engineering – Guedes Soares & Santos (Eds)

On the performance of wave models in coastal areas

T. Gavazzoni, D. Silva & C. Guedes Soares
Centre for Marine Technology and Ocean Engineering (CENTEC), Instituto Superior Técnico, Universidade de Lisboa, Lisbon, Portugal

ABSTRACT: The objective of this study is to compare the accuracy in predicting waves in coastal areas of the Wavewatch3 model, which has been used mostly for deep waters, with the SWAN model, which is often used close to the coasts. The wave conditions in the target area, the coastal area of Figueira da Foz, located at the center of Portugal's continental coast, are evaluated with both models for a 3-month hindcast (December 1993 - February 1994). The wave prediction system consists of a mosaic of three nested regular grids (North Atlantic, Portugal continental coast, and Figueira da Foz). The models are validated against three buoys using statistical parameters.

1 INTRODUCTION

The measurements of wave parameters by buoys are a realistic and accurate tool to assess ocean weather conditions. They measure the wave field at one physical location, so they do not have enough information to accurately predict wave conditions in a wider geographic area.

A reliable knowledge of sea state conditions in specific sites can be obtained by characterizing the wave climate using the directional wave spectrum (Portilla-Yandúm et al. 2019, Lucas et al. 2011, Lucas & Guedes Soares 2015a, b, Lucas et al. 2019). This is of great importance for a variety of applications, such as offshore engineering, coastal management, and marine safety.

The third-generation models describe the evolution of wave spectrum by solving the spectral energy balance equation, and they are an efficient tool widely used to assess spatial wave conditions for a long time period. The WAM (WAve model) (WAMDI Group, 1988), WAVEWATCH III (WW3) (Tolman, 1991, 2009) and SWAN (Simulating WAves Nearshore) (Booij et al., 1999) are third-generation spectral wave models, that are typically used to hindcast, nowcast, and forecast waves. The WAM and WW3 models are more often applied for global and regional scale modelling, while the SWAN model is most often used in nearshore applications (Amrutha et al. 2016, Silva et al. 2018).

A global model uses relatively coarse resolution, which is generally insufficient to characterize the phenomena near the coast, where coastline details include features too small to be seen in such resolution. A scheme of nested models, one coastal model with high resolution nested to one large-scale model with a coarse grid, is the commonly accepted solution (Bertotti & Cavaleri, 2009).

Although the SWAN model is more suitable in nearshore waters, the aim of this work is the assessment of the WW3 model performance in a coastal area, since the model version (6.07) used here is capable of simulating waves at physically relevant scales as needed for coastal applications.

The application of the WW3 model to evaluate wave conditions in coastal areas has not been addressed very often. A relevant application found in the literature is from Smith et al. (2018), who used an unstructured grid to evaluate four storms, Hurricane Irene, a mixed sea/swell event, Hurricane Sandy and an event with oblique 5 m offshore waves.

However, studies using WW3 and SWAN in a nesting scheme approach have been conducted by various authors. Gonçalves et al. (2018) studied the wave energy resource available on the Atlantic coast of France. They used a 33-year wave hindcast, using the WW3 model for the Atlantic Ocean and the SWAN model for the western French coast. Bento et al. (2018) and Silva et al. (2018) conducted a similar study, assessing wave energy from a 33-year wave hindcast using the same wave prediction strategy. However, the studies were conducted in different locations, with Bento et al. (2018) focusing on the Northern Spanish coast and Silva et al. (2018) focusing on Portugal's Continental Coast. Wu et al. (2020) used WW3 and SWAN to provide accurate long-term wave hindcast in the nearshore region of the U.S. West Coast. Silva & Guedes Soares (2020), presented a hindcast system of 5 years (2012-2016) to provide the sea state conditions around the Madeira archipelago. The authors validated the

DOI: 10.1201/9781003508779-68

results of the wave prediction system against satellite data using statistical parameters.

So the question addressed in this paper is whether it is necessary to continue using the two models to obtain good predictions in the nearshore or whether the present version of WW3 can be used throughout, although refining the mesh closer to the coast.

2 WAVE MODELS THEORETICAL FORMULATIONS

The WW3 and SWAN are third-generation spectral wave models, that use the Eulerian approach for the balance equation of wave spectrum to solve the non-linear interactions of the waves. The balance wave spectrum equation is given by:

$$\frac{DN}{Dt} = \frac{S}{\sigma} \tag{1}$$

where N represents the wave action density spectrum, S the general sources and sinks and σ the relative frequency (Booij et al. 1999, WW3DG 2019).

Normally, the action density spectrum (N) is most considered in wave models, rather than the energy density spectrum, since it is conserved in the presence of currents (Witham, 1965).

The evolution of wave spectrum Equation (1) in time, geographical and spectral spaces, in the WW3 model is given by:

$$\frac{\partial N}{\partial t} + \nabla_x .(\dot{x}N) + \frac{\partial}{\partial k}(\dot{k}N) + \frac{\partial}{\partial \theta}(\dot{\theta}N) = \frac{S}{\sigma} \tag{2}$$

where:

$$\dot{x} = c_g + U \tag{3}$$

$$\dot{k} = -\frac{\partial \sigma}{\partial d}\frac{\partial d}{\partial s} - k\frac{\partial U}{\partial s} \tag{4}$$

$$\dot{\theta} = -\frac{1}{k}\left[\frac{\partial \sigma}{\partial d}\frac{\partial d}{} - k\frac{\partial U}{}\right] \tag{5}$$

where c_g is the group velocity of wave energy in x, y spatial space, U is the mean current velocity, d is the water depth, σ is the relative frequency, k is the wave number, s and m are the space coordinates of wave propagation direction θ (WW3DG, 2019).

The net source term S, in the WW3 model in deep water, consists of an atmosphere-wave interaction term S_{in}; a nonlinear wave-wave interaction term S_{nl} and a wave-ocean interaction, generally dominated by wave-breaking S_{ds}. The input term S_{in} is dominated only by the exponential wind-wave growth term, however, to provide more realistic initial wave growth, a linear input term (S_{ln}) is added. In shallow water, additional processes are considered, such as wave-bottom interactions S_{bot} and in extremely shallow water an additional breaking term (S_{db}) depth-induced breaking (WW3DG, 2019). Triad wave-wave interactions, scattering of waves by bottom features, wave-ice interactions and reflection were not considered in the WW3 model in this work.

$$S = S_{ln} + S_{in} + S_{nl} + S_{ds} + S_{bot} + S_{db} + \ldots \tag{6}$$

The SWAN model, the evolution of the wave spectrum equation (1), is given by:

$$\frac{\partial N}{\partial t} + \frac{\partial c_x N}{\partial x} + \frac{\partial c_y N}{\partial y} + \frac{\partial c_\sigma N}{\partial \sigma} + \frac{\partial c_\theta N}{\partial \theta} = \frac{S_{tot}}{\sigma} \tag{7}$$

where c_x, c_y, c_σ, c_θ represent propagation velocity in the geographical and spectral spaces x, y, σ, θ, respectively. The fourth term represents the change in relative frequency due to variations in water column depth and currents. The fifth term represents the depth-induced breaking and the net source term is represented by S_{tot}:

$$S_{tot} = S_{in} + S_{nl3} + S_{nl4} + S_{ds,w} + S_{ds,b} + S_{ds,br} + \ldots \tag{8}$$

which accounts for atmospheric input (wave generation by wind S_{in}), white capping dissipation ($S_{ds,w}$), non-linear wave-wave interactions (quadruplet S_{nl4}) and in shallow waters, the dissipation by bottom friction ($S_{ds,b}$), depth induced wave breaking ($S_{ds,br}$) and non-linear wave-wave interactions (triad S_{nl3}) (Booij et al., 1999).

3 METHODOLOGY

3.1 *Wave prediction systems*

Three wave prediction systems, using a mosaic of 3 nested regular grids (Figure 1), are implemented in this study for the same period of time, as shown in Table 1.

These grids make the connection between deep and shallow waters, with increasing resolutions towards the coast. The first grid covers the entire North Atlantic basin and accounts for wave generation induced by wind and deep water propagation, providing boundary conditions to the intermediate grid. The intermediate grid is on the scale of the Portugal continental coast and acts as a bridge between deep and shallow waters, providing the wave spectral boundary conditions to the nearshore grid, which is the coastal area of Figueira da Foz, the target area of this study.

Simulation 1 uses the WW3 (version 6.07) model in all three areas. Simulation 2 uses the WW3 model in area 1 and SWAN (version 41.31) in areas 2 and 3, while Simulation 3 uses WW3 in areas 1 and 2 and SWAN in area 3. To ensure consistency in the hindcast results, the same forcing and spatial resolutions are used, as described in the next section.

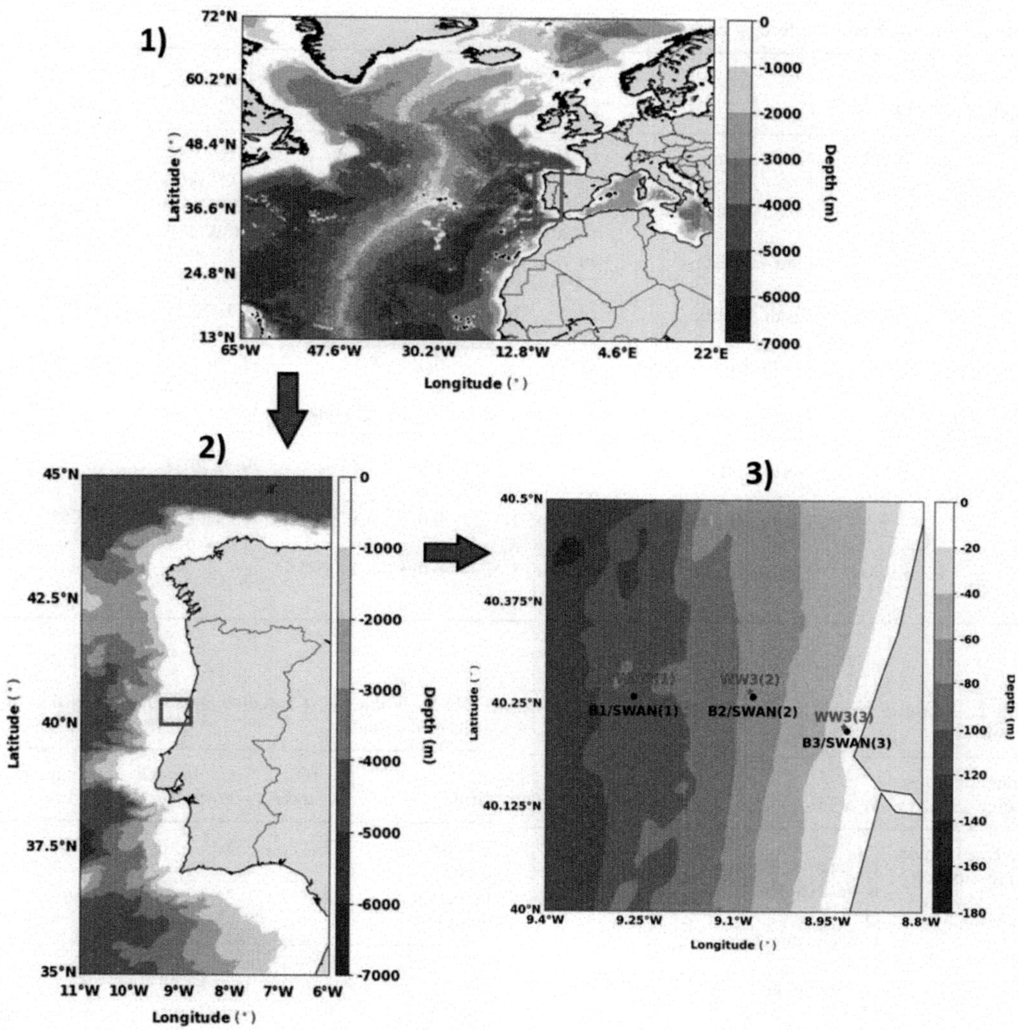

Figure 1. Wave prediction system areas. For Simulation 1, the three areas use the WW3 model. For Simulation 2, the areas 1) use the WW3 model and the areas 2) and 3) the SWAN model. For Simulation 3, areas 1) and 2) use the WW3 model and area 3) the SWAN model. Area 3) also illustrates the position of the point used for the model's validation.

3.2 *Bathymetry data*

The wave models use the bathymetry from GEBCO (General Bathymetric Chart of the Oceans) database (Weatherall et. al. 2015), which has a spatial resolution of 30 arcsecs (≅922m or 0.0083°). The same bathymetry resolution is used for all the areas, but with different computational grids, as presented in Table 1.

3.3 *Wind data*

For the wind fields input the models used the ERA5 (Hersbach & Dee, 2016) database with a time step of 3h and spatial resolutions of 0.5°in in the area of North Atlantic and 0.25° in the area of Portugal's continental coast. To save computation time, the wind was not used in the high-resolution area (target area) since it has an insignificant influence on the results, as shown by Rusu et al. (2008). The computational grids for wind input are presented in Table 1.

3.4 *Physical processes*

The formulations for the physical processes used by WW3 and SWAN are given in Table 2. For wave generation, both models use the Cavaleri & Malanotte-Rizolli (1981) formulation, which proposed VENICE model for wind wave prediction. The model is based on the ray technique and is also suitable for shallow-water applications.

The non-linear wave-wave interactions (quadruplets) are introduced in WW3 and SWAN coarse

Table 1. Configuration and forcing of the grids used in the simulations.

			Computational grids		Forcing (wind)	
Simulation	Model	Grids	LAT/LON (°)	ΔX ×ΔY (°)	LAT/LON (°)	ΔX ×ΔY (°)
1	WW3	North Atlantic	13°N/72°N 65°W/22°E	1.54° x 1.46°	9°N/75°N 70.5°W/30°E	0.5° x 0.5°
		Continental Portugal	35°N/45°N 11°W/6°W	0.05° x 0.1°	35°N/45°N 11°W/6°W	0.25° x 0.25°
		Figueira da Foz	40°N/40.5°N 9.4°W/8.8°W	0.0083° x 0.0083°	-	-
2	WW3	North Atlantic	13°N/72°N 65°W/22°E	1.54° x 1.46°	9°N/75°N 70.5°W/30°E	0.5° x 0.5°
	SWAN	Continental Portugal	35°N/45°N 11°W/6°W	0.05° x 0.1°	35°N/45°N 11°W/6°W	0.25° x 0.25°
		Figueira da Foz	40°N/40.5°N 9.4°W/8.8°W	0.0083° x 0.0083°	-	-
3	WW3	North Atlantic	13°N/72°N 65°W/22°E	1.54° x 1.46°	9°N/75°N 70.5°W/30°E	0.5° x 0.5°
		Continental Portugal	35°N/45°N 11°W/6°W	0.05° x 0.1°	35°N/45°N 11°W/6°W	0.25° x 0.25°
	SWAN	Figueira da Foz	40°N/40.5°N 9.4°W/8.8°W	0.0083° x 0.0083°	-	-

Table 2. Main source physical processes used in the models.

Parameters (source terms)	WW3 (6.07)	SWAN (41.31)
Wave generation and dissipation	Cavaleri & Malanotte-Rizolli (1981). Hasselmann et al. (1985). Ardhuin et al. (2010).	Cavaleri & Malanotte-Rizolli (1981). Komen et al. (1984). Hasselmann (1985). Janssen (1991).
Bottom friction	JONSWAP - Hasselmann et al. (1973).	JONSWAP - Hasselmann et al. (1973).
Depth induce wave breaking	Battjes-Janssen (1978).	Battjes-Janssen (1978).

grid (continental Portugal area). The WW3 model employs the discrete interaction approximation (DIA) proposed by Hasselmann et al. (1985), while SWAN uses the default formulation of fully explicit computation of the nonlinear transfer with DIA per sweep.

WW3 uses a third-order scheme for wave propagation, which was described by Leonard (1979), Davis & More (1982), and Leonard (1991). On the other hand, SWAN employs two different numerical schemes for the coarse grid and local area. For the coarse grid, the waves are propagated using a second-order upwind scheme with third-order diffusion called S&L, which was proposed by Stelling & Leedertse (1992). For the local areas, SWAN uses a first-order scheme called backward space, and backward time (BSBT).

Table 3. Frequencies and directions numbers used in the grids of the models.

Grids	North Atlantic	Continental Portugal	Figueira da Foz
Frequencies number	32	32	32
Directions number	24	36	36

The numbers of frequencies and directions used in each grid of the models are presented in Table 3. The minimum frequency used is 0.0373 Hz in both models.

4 MODEL VALIDATION

The simulation output results for the parameters *Hs* (significant wave height) and *Tp* (peak period) are validated against three buoys located at different depths in the target area (Figure 1, area 3). Testing the models for different water depths can give a perspective on how the models behave for different depths. The geographic coordinates, the depth and the wave parameters available on these buoys are specified in Table 4. The wave models are run for the same period as buoys, comprising a hindcast period of 3 months (December 1993 – February 1994), with a temporal resolution of 3 hours. In B3 there are only 26 days of the buoy data.

The points extracted from the WW3 model for validation purposes are not exactly coincident with the buoy's locations, but the closest ones, while in SWAN the points are exactly coincident (Figure 1, area 3) and their coordinates are displayed in Table 5.

Table 4. Buoy's location.

Buoys	Location	Depth (m)	Wave parameters
B1	9.24°/40.26°	92	Hs, Tp
B2	-9.07°/40.26°	72	Hs
B3	-8.92°/40.22°	19	Hs

Table 5. Location of WW3 and SWAN points.

(WW3 Points)		(SWAN Points)	
WW3(1)	-9.2417°/40.2667°	SWAN(1)	-9.24°/40.26°
WW3(2)	-9.0750°/40.2667°	SWAN(2)	-9.07°/40.26°
WW3(3)	-8.9250°/40.2250°	SWAN(3)	-8.92°/40.22°

The evaluation is done through the statistical parameters Bias, Root Mean Square Error (RMSE), Scatter Index (SI) and Pearson's Correlation Coefficient (r), which formulas are:

$$Bias = \frac{\sum_{i=1}^{n}(X_i - Y_i)}{n} \tag{9}$$

$$RMSE = \sqrt{\frac{\sum_{i=1}^{n}(X_i - Y_i)^2}{n}} \tag{10}$$

$$SI = \frac{RMSE}{\tilde{X}} \tag{11}$$

$$r = \frac{\sum_{i=1}^{n}(X_i - \tilde{X})(Y_i - \tilde{Y})}{\left(\sum_{i=1}^{n}(X_i - \tilde{X})^2 \sum_{i=1}^{n}(Y_i - \tilde{Y})^2\right)^{1/2}} \tag{12}$$

where X_i corresponds to measurement values, Y_i to simulated values and n is the number of points used.

The statistical parameter results for the three Simulations are presented in Tables 6, 7 and 8.

Table 6. Statistical parameters of Simulation 1 (WW3 × Buoys).

Simulation 1		RMSE	BIAS	SI	r
Tp (s)	B1	1.48	0.38	0.11	0.74
Hs (m)	B1	0.34	0.04	0.10	0.96
	B2	0.49	0.26	0.15	0.95
	B3	0.31	0.09	0.12	0.92

Table 7. Statistical parameters of Simulation 2 (SWAN× Buoys).

Simulation 2		RMSE	BIAS	SI	r
Tp (s)	B1	2.78	1.51	0.21	0.58
Hs (m)	B1	0.65	-0.49	0.20	0.94
	B2	0.50	-0.18	0.15	0.94
	B3	0.39	0.20	0.15	0.90

Table 8. Statistical parameters of Simulation 3 (SWAN× Buoys).

Simulation 3		RMSE	BIAS	SI	r
Tp (s)	B1	1.55	0.33	0.12	0.70
Hs (m)	B1	0.35	0.03	0.10	0.96
	B2	0.50	0.29	0.15	0.96
	B3	0.47	0.31	0.18	0.89

It was expected that the performance of SWAN would be better than WW3 since the simulations are near the coast, but instead, the outcomes show that in general the WW3 model gives the best statistical values for all buoys.

For Simulation 1, the values of SI, BIAS, and r are optimal in B1, while RMSE is best in B3. The worst performance of this Simulation is for B2, the intermediate buoy, which was not expected. The reason for this could be due to some particularities in the bathymetry not well resolved by the resolution used, and also the buoy could be more affected by phenomena (current, tides) not accounted for in the models. Simulation 3 presents similar results to Simulation 1, except for B3, which has the worst results. This Simulation only uses the SWAN model in the target area, and this could be the reason for the similarities in the results. For Simulation 2, which used SWAN in the last two areas (Portugal continental coast and Figueira da Foz), the differences are higher and the results are worse, except for B3, where Simulation 2 performs better than Simulation 3. Therefore, the usage of different models in the intermediate area will influence the results in the nearshore buoy.

The efficiency of the models can be related to the area of study, SWAN could perform better in some regions and WW3 in others. Both are spectral models, but while WW3 model negligence the diffraction effect, in SWAN model this physical effect can be included. However, the study area is a region without obstruction in the coast and on the way to it, becomes an optimal area to perform this first test. In this particular case, WW3 could be used to simulate the wave conditions in shallow waters instead of SWAN. On the other hand, it is necessary to conduct more studies of this type to reach more solid conclusions.

Figure 2 displays the scatter plots, while Figures 3 and 4 present the time series of Hs for B3, which were selected due to the highest differences in the results, and are related to Simulations 2 and 3.

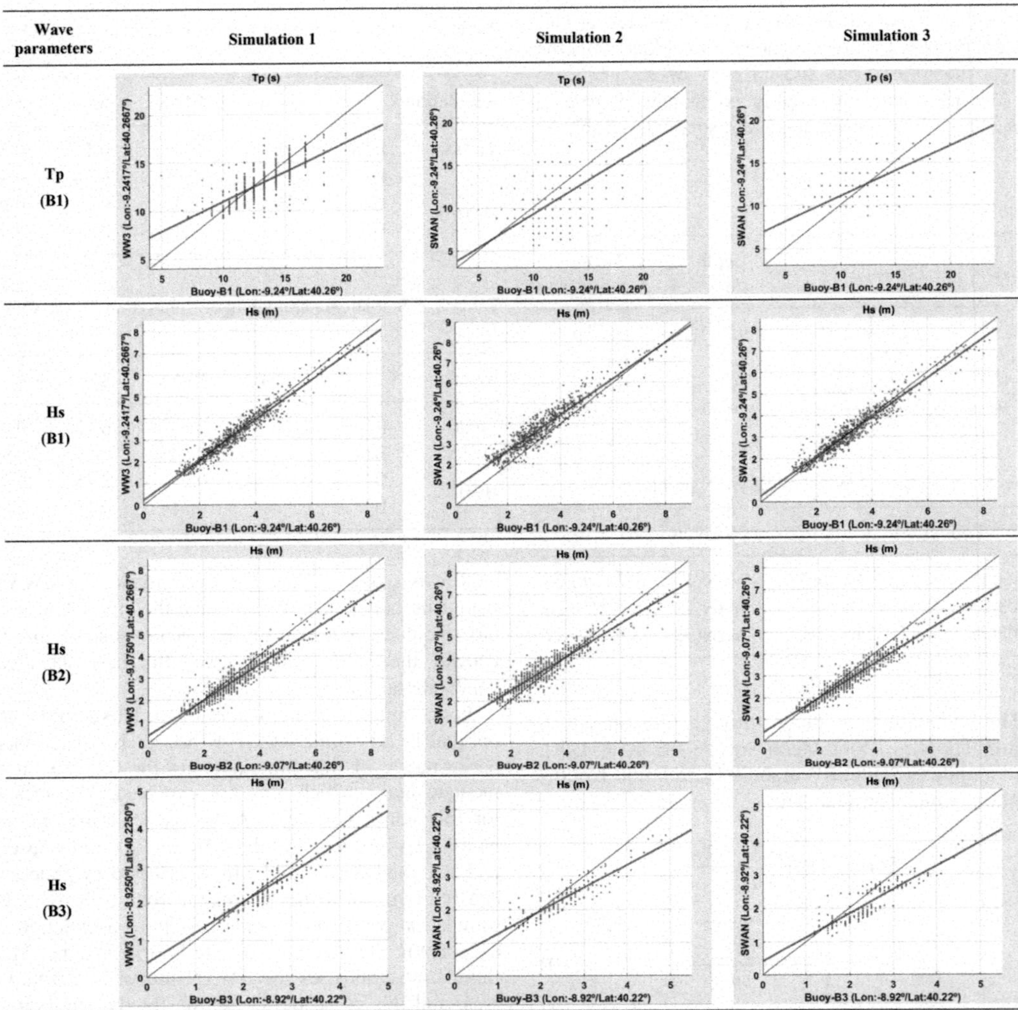

Figure 2. Scatter plots of the wave parameters Hs (significant wave height) and Tp (peak wave period) for the three simulations performed. The first row illustrates the dispersion of Tp simulated in relation to buoy B1, while the subsequent rows show the dispersion of Hs simulated for buoys B1, B2, and B3, respectively.

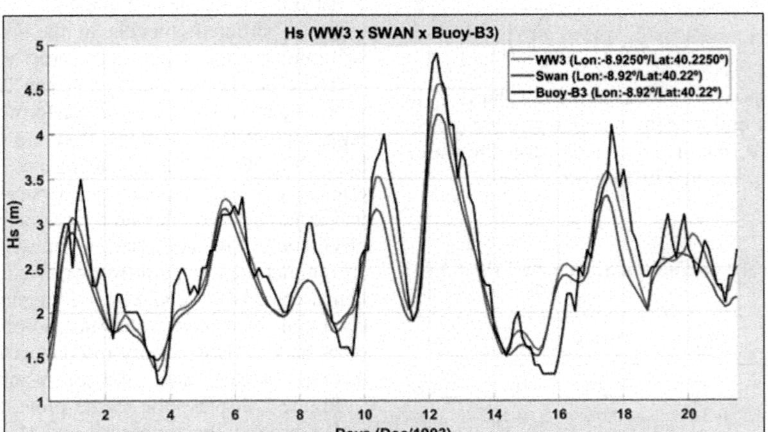

Figure 3. Hs for WW3 x SWAN (simulation 2) x B3.

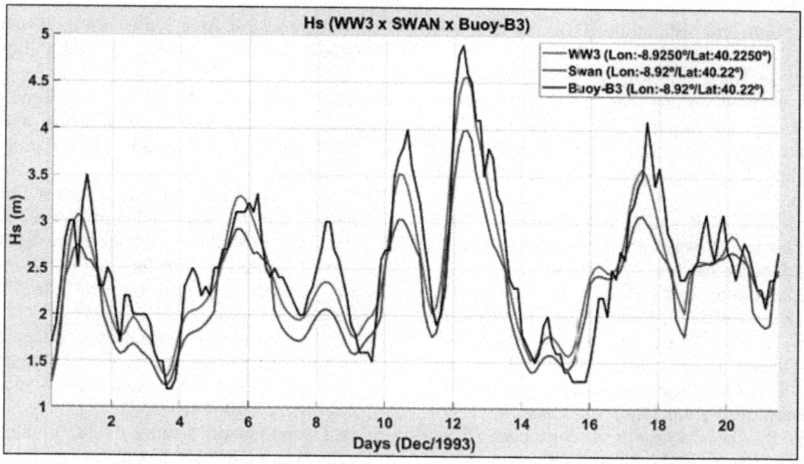

Figure 4. Hs for WW3 x SWAN (simulation 3) x B3.

The scatter plots can show visually what was said before. Simulation 1 gives better results with lower scattered data, and Simulation 3 presents similar results. For these simulations, the wave system implemented underestimates the buoys data, while for Simulation 2 the wave system overestimates the buoys measurements, except for B3 in the Tp parameter.

The time series shows that for the buoy nearest to the coast (B3), WW3 presents better results than the SWAN model, in both simulations.

5 CONCLUSIONS

This study presents a comparison between WW3 and SWAN models in predicting the wave conditions in coastal waters. The results show that WW3, in this particular area (Figueira da Foz), performs better than the SWAN model.

The usage of different models in the intermediate area influences the results in the nearshore buoy. Both models present worse results at B2, the intermediate buoy when compared to the other points. This could be due to some unique bathymetric features that were not well resolved by the resolution used. Additionally, this point could be more affected by phenomena such as currents and tides that were not properly accounted for in the models.

Based on the results of the models for this area of study, it appears that nesting with the SWAN model may not be worthwhile and it would be more beneficial to use only the WW3 model to study the nearshore target area. This conclusion still does not allow final recommendations to be made and more studies are necessary for a more definite conclusion to be reached.

ACKNOWLEDGEMENTS

This study has been performed under the project "WAVEFAI - Operational Wave Forecast using Artificial Intelligence", which is funded by the Portuguese Foundation for Science and Technology (FCT-Fundação para a Ciência e a Tecnologia) under contract CIRCNA.OCT.0300.2019_1801P.01023. This work contributes to the Strategic Research Plan of the Center for Marine Technology and Ocean Engineering (CENTEC), which is financed by FCT under contract UIDB/UIDP/00134/2020.

REFERENCES

Amrutha, M. M., Kumar, V. S., Sandhya, K. G., Nair, T. M. B. and Rathod, J. L. 2016. Wave hindcast studies using SWAN nested in WAVEWATCH III - Comparison with measured nearshore buoy data off Karwar, eastern Arabian Sea. *Ocean Engineering, 119*, 114–124.

Ardhuin, F., Rogers, W. E., Babanin, A. V., Filipot, J., R., Magne, A., Roland, A., van der Westhuysen, A., Queffeulou, P., Lefevre, J., Aouf, L. and Collard, F. 2010, Semiempirical dissipation source functions for ocean waves. Part I: Definition, calibration, and validation, *Journal of Physical Oceanography.*, 40, 1,917–1,941.

Battjes, J. A. and Janssen, J. P. F. M. 1978. Energy loss and set-up due to breaking of random waves, in *Proc. 16th Int. Conf. Coastal Eng.*, pp. 569–587, ASCE.

Bento, A. R., Martinho, P. and Guedes Soares, C. 2018. Wave energy assessment for Northern Spain from a 33-year hindcast. *Renewable Energy*, 127, 457–322-333.

Bertotti, L. and Cavaleri, L. 2009. Large and small scale wave forecast in the Mediterranean Sea. *Natural Hazards and Earth System Science*, *9*(3), 779–788.

Booij, N., Ris, R.C. and Holthuijsen, L. H. 1999. A third generation wave model for coastal regions. Part 1:

Model description and validation. *J Geoph Res* 104: 7649–7666.
Cavaleri, L. and Malanotte-Rizzoli, P. 1981. Wind-wave prediction in shallow water: Theory and applications., *Journal of Geophysical Research* 86: 10,961–10,973.
Davis, R. W. and More, E. F. 1982. A numerical study of vortex shedding from rectangles, *J Fluid Mechanics* 116: 475–506.
Gonçalves, M., Martinho, P. and Guedes Soares, C. 2018. A 33-year hindcast on wave energy assessment in the western French coast. *Energy*. 165: 790–801.
Hasselmann, K., Barnett, T. P., Bouws, E., Carlson, H., Cartwright, D. E., Enke, K., Ewing, J. A., Gienapp, H., Hasselmann, D. E., Kruseman, P., Meerburg, A., Müller, P., Olbers, D. J., Richter, K., Sell, W. and Walden, H. 1973. Measurements of wind-wave growth and swell decay during the Joint North Sea Wave Project (JONSWAP). *Ergänzungsheft zur Deutschen Hydrographischen Zeitschrift*, Reihe A(8), 12, 95 pp.
Hasselmann, S. and Hasselmann, K. 1985. Computations and parameterizations of the nonlinear energy transfer in a gravity-wave spectrum, Part I: A new method for efficient computations of the exact nonlinear transfer integral. *Journal of Physical Oceanography* 15: 1,369–1,377.
Hasselmann, S., Hasselmann, K., Allender, J. H. and Barnett, T. P. 1985. Computations and parameterizations of the nonlinear energy transfer in a gravity-wave spectrum, Part II: Parameterizations of the nonlinear energy transfer for application in wave models. *J Phys Oceanog* 15: 1,378–1,391.
Hersbach, H. and Dee, D., 2016. ERA5 reanalysis is in production. *ECMWF Newsletter*. 147, 7. https://www.ecmwf.int/en/newsletter/147/news/era5-reanalysis-production
Janssen, P. A. E. M. 1991. Quasi-linear theory of wind wave generation applied to wave forecasting, *Journal of Physical Oceanography* 21: 1,631–1,642.
Komen, G. J., Hasselmann, S. and Hasselmann, K. 1984. On the existence of a fully developed wind-sea spectrum, *Journal of Physical Oceanography* 14: 1,271–1,285.
Leonard, B. P. 1979. A stable and accurate convective modelling procedure based on quadratic upstream interpolation. *Computer Methods Applied Mechanics Engineering* 18: 59–98.
Leonard, B. P. 1991. The ULTIMATE conservative difference scheme applied to unsteady one-dimensional advection. *Computer Methods Applied Mechanics Enging* 88: 17–74.
Lucas, C., Boukhanovsky, A. and Guedes Soares, C. 2011. Modelling the climatic variability of directional wave spectra. *Ocean Engineering* 38:11–12.
Lucas, C. and Guedes Soares, C. 2015a. On the modelling of swell spectra. *Ocean Engineering* 108:749–759.
Lucas, C. and Guedes Soares, C. 2015b. Bivariate distributions of significant wave height and mean wave period of combined sea states. *Ocean Engineering* 106:341–353.
Lucas, C., Silva, D. and Guedes Soares, C. 2019. Assessment of the wave spectral characteristics in the Portuguese test zone. *Advances in Renewable Energies Offshore*. Guedes Soares, C. (Ed.), Taylor & Francis Group, London, UK, 59–66.
Portilla-Yandún, J., Barbariol, F., Benetazzo, A. and Cavaleri, L. 2019. On the statistical analysis of ocean wave directional spectra. *Ocean Engineering* 189: 106361.
Rusu, L., Bernardino, M. and Guedes Soares, C. 2008. Influence of the wind fields on the accuracy of numerical wave modelling in offshore locations, in: *Proceedings of the ASME 27th Int Conf on Offshore Mechanics and Arctic Engineering OMAE57861, (Estoril, Portugal)*.
Silva, D., Martinho, P. and Guedes Soares, C. 2018. Wave energy distribution along the Portuguese continental coast based on a thirty three years hindcast. *Renewable Energy*. 127, 1064–1075.
Silva, D. and Guedes Soares, C. 2020. Validation with satellite data of SWAN model for wave conditions at the Madeira archipelago. In Guedes Soares, C. & Santos T. A. (Eds.). *Developments in Maritime Technology and Engineering*, London, U.K: Taylor and Francis.
Smith, J., Hesser, T., Roland, A. and Bryant, M. 2018. Validation of Unstructured WAVEWATCH III for Nearshore Waves. *Coastal Engineering Proceedings*. https://doi.org/10.9753/ICCE.V36.WAVES.55.
Stelling, G.S. and Leendertse, I.I. 1992. Approximation of convective processes by cyclic AOI methods, in: *Proceedings of the 2nd International Conference on Estuarine and Coastal Modeling*, ASCE, Tampa, Florida, pp. 771e782.
Tolman, H. L. 1991. A third-generation model for wind waves on slowly varying, unsteady and inhomogeneous depths and currents, *Journal of Physical Oceanography* 21, 782–797.
Tolman, H. L. 2009. User manual and system documentation of WAVEWATCH III TM version 3.14. *Technical note*, MMAB Contribution, 276(220).
WAMDI Group. 1988. The WAM model - a third generation ocean wave prediction model. *Journal of Physical Oceanography* 18: 1755–1810.
WW3DG 2019. *The WAVEWATCH III® Development Group (WW3DG), (2019)*: *User manual and system documentation of WAVEWATCH III® version 6.07*. Tech. Note 333, NOAA/NWS/NCEP/MMAB, College Park, MD, USA, 465
Weatherall P., Marks, K. M., Jakobsson, M., Schmitt, T., Tani, S., Arndt, J. E., Rovere, M., Chayes, D., Ferrini, V. and Wigley, R. 2015. A new digital bathymetric model of the world's oceans. *Earth and Space Science*, 2, 331–345, doi: 10.1002/2015EA000107.
Whitham, G. Á. 1965. A general approach to linear and non-linear dispersive waves using a Lagrangian. *Journal of Fluid Mechanics* 22(2): 273–283.
Wu, W., Wang, T., Zhaoqing, Y. and Gacía-Medina, G.2020. Development and validation of a high-resolution regional wave hindcast model for U.S. West Coast wave resource characterization. *Renewable Energy*152:736–753.

Advances in Maritime Technology and Engineering – Guedes Soares & Santos (Eds)

Assessing climate change in the Azores area in the near future

M. Gonçalves, M. Bernardino & C. Guedes Soares
Centre for Marine Technology and Ocean Engineering (CENTEC), Instituto Superior Técnico, Universidade de Lisboa, Lisbon, Portugal

ABSTRACT: The aim of this work is to provide information on the wave conditions along the Azores Islands. This case study seeks to examine the availability of changing sea state conditions in a near future (2030-2054). For this, wind input fields and ice cover data from an RCP8.5 EC-Earth system q downscaled with the Weather Research and Forecasting model. The wave spectral boundary conditions are from a preceding North Atlantic basin WAVEWATCH III wave model run. With those wind fields and boundary conditions, the wavemodel is run to produce the wave information at the Azores Islands used in this analysis. A statistical analysis of the wave information produced from downscaled reanalysis data for present and near future climate simulations is performed. Mean values, standard deviation and 99th percentiles are evaluated for the significant wave height, mean period and 10-m wind fields. Changes between recent past (1990-2014) and near future are evaluated and maps of these empirical statics are presented.

1 INTRODUCTION

Current climate is usually studied using reanalysis data or satellite data, but future climate, usually divided into different periods, historical (current), near future, mid-20th century, and the end of the 21st century, is often studied using global or at least large-scale global waves or statistical models that are forced by atmospheric data from the most advanced climate models.

The third-generation wave models are state-of-the-art tools for modelling wave generation and transformation in offshore and nearshore areas Cavalieri et al. 2007). In recent years many studies regarding the accuracy of wave models have been made for the Iberian coast, focussing on the Portuguese coast and the Portuguese harbours (Gonçalves et al., 2009, 2015; Rusu et al., 2008b, 2011; Rusu & Guedes Soares, 2013), for the Madeira islands (Rusu et al, 2008a; Silva, & Guedes Soares, 2021) and also for the Azores islands (Rusu & Guedes Soares, 2012a, 2012b).

Information on the changes in the wind and wave system over the past few decades and how they will change in the future is of the utmost importance. The purpose of the Coupled Model Intercomparison Project (CMIP) is to better understand past, present, and future climate change. The CMIP simulations for the Earth System until the end of the 21st century take into account different types of emission scenarios (RCPs), but these simulations do not include ocean waves (IPCC, 2007).

Extensive research has been conducted on the global and regional wave climate, with a focus on trends and variation in average values and high percentages. In some global studies, wave climates have been studied in response to future climate scenarios (Semedo & Wu, 2013).

The Coordinated Ocean Wave Climate Project (COWCLIP) (Hemer et al., 2012, 2013) was established with the support of the Joint Technical Commission for Oceanography and Marine Meteorology World Meteorological Organisation (JCOMM/WMO) to address the need to pay attention to future changes in ocean climate in the 21st century.

Examples of these studies are shown in Morin et al. (2018), which evaluate changes in annual average height, the average period of the waves, and the average direction of the waves using multi-method wave forecasts (RCP4.5 and RCP8.5 emission scenarios).

Bricheno & Wolf (2018) analysed the historical and future wave conditions on the European Atlantic coast and estimated that most of the European Atlantic coast's average wave height would be reduced by 0.2m in RCP4.5 and RCP8.5 scenarios by 2100. The maximum annual wave height and the 99th percentile increase in some areas to 0.5–1 m, but with a more complex spatial pattern.

In Gonçalves & Guedes Soares (2121), a 5-year evaluation was carried out using the spectral boundary conditions from WWIII (Tolman 1991) provided by Ifremer and the SWAN model (Simulation of the NWN coastal area) (Booij et al. 1999) was adopted to evaluate the wave transformations in the Azores coast area, showing the geographical distribution of

DOI: 10.1201/9781003508779-69

significant wave height and wave energy, as well as monthly averages, annual and seasonal changes.

The purpose of this study is to provide a validation of the simulations forced with the EC-EARTH, downscaled with the WRF (Weather Research and Forecasting) model and statistically comparing them to the ECMWF Reanalysis ERA5 simulations (Hersbach et al., 2020), over 25 years, to contribute to the state-of-the-art information regarding present wave climate in the Azores Islands.

The EC-Earth (Hazeleger et al., 2010) wind fields used to force WWII were previously presented and validated in the North Atlantic by Bernardino et al. (2021). The results show that the reduction in wave energy resources is generalized in the North Atlantic. Furthermore, a statistical analysis of the wave information produced from downscaled reanalysis data for the recent past (1990-2014) and near-future climate simulations (2030-2054) is performed, and then changes in the wave climate are studied.

2 MODELS

2.1 *Wave models*

In this study, the state-of-the-art spectral wave model WAVEWATCH III, (WWIIIDG, 2016) is used to characterize the wave generation in the Azores area.

The WWIII model was developed by the Marine Modelling and Analysis Branch (MMAB), of the Environmental Centre (EMC), of the National Centers for Environmental Prediction (NCEP), and resolves an advection-type energy balance equation:

$$\frac{DN}{Dt} = \frac{S}{\sigma} \tag{1}$$

where N is the action density spectrum, σ is the relative frequency and S characterises the source terms.

The right side of the equation holds for the kinematics of the model, while the left side characterises the physical processes that generate, dissipate, and distribute the wave energy (the nonlinear effects: wind-wave interactions, quadruplet wave-wave interactions, and dissipation through white capping and bottom friction). Source terms include wind-wave interactions, quadruplet wave-wave interactions, and dissipation through white capping and bottom friction.

2.2 *Wind models*

Produced by the Swedish Meteorological and Hydrological Institute (SMHI) for a RCP8.5 scenario, EC-EARTH is an earth-system model developed by a consortium of European research institutions and researchers, based on state-of-the-art models for the atmosphere, the ocean, and sea ice and the biosphere.

The simulations of the EC-EARTH system (Hazeleger et al., 2010) followed the CMIP5 protocol for long-term simulations (Taylor et al., 2012), which was implemented for the historical period of 1850 to 2005 and for the continuing future period of 2006 to 2100.

In contrast with CMIP3, the CMIP5 experiments include more comprehensive models that are driven by the concentration or emission scenarios consistent with the RCPs that are described in Moss et al. (2010) and defined in the Fifth Assessment Report of the IPCC.

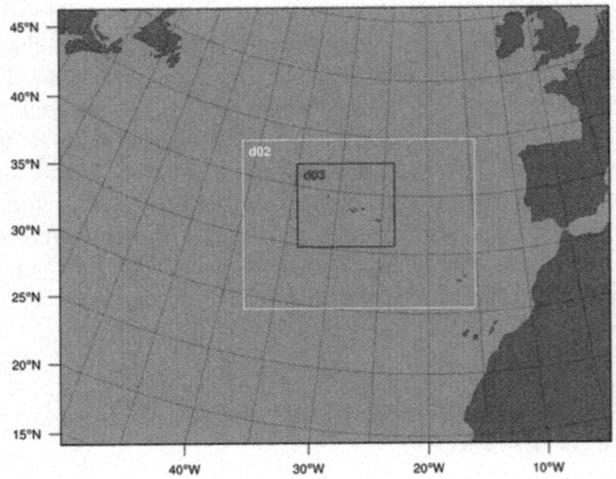

Figure 1. WRF preprocessing system domain.

The WRF model version 4.3 (Skamarock et al. 2019) was used to dynamically downscale the EC-Earth model data to the regional scale. The simulation domain includes two nested domains on a two-way nested grid, with 115 x 9, 145 x 106, and 181 x 157 grid points and a resolution of 45, 15 and 5 km, respectively (Figure 1). Spectral nudging was applied, allowing the representation of the small scales while keeping the simulation close to the driving state (Miguez-Macho et al., 2004). The main physical parametrizations used are the WRF Single-Moment 6-class scheme (Hong & Lim, 2006) for microphysics parameterization; the Community Atmosphere Model (CAM) for long wave physics parametrization (Collins et al., 2004); the Rapid Radiative Transfer (RRTMG) scheme (Iacono et al., 2008) for short wave physics parametrization; the Revised Fifth-Generation Penn State/NCAR Mesoscale Model (MM5) Monin-Obukhov scheme for surface layer parametrization (Jiménez et al., 2012); the unified Noah land-surface model (Tewari et al., 2004); the YSU scheme for the PBL scheme (Hong, et al, 2006) and the Kain-Fritsch cumulus parametrization (Kain, 2004).

3 SYSTEM SET-UP DESCRIPTION

To generate waves for the North Atlantic basin, the WWIII wave model, version 5.16 (WWIIIDG, 2016), is used, with 32 frequencies and 24 directions (Table 1), the model is implemented with wind and sea-ice cover from EC-EARTH simulations, produced by the Swedish Meteorological and Hydrological Institute (SMHI), for an RCP8.5 scenario (Van Vuuren et al., 2011).

The second level of WWIII is implemented in the Azores Islands (Figure 2) with boundary conditions from the previous level and wind fields downscaled from WRF (Table 2).

Table 1. Parameterizations of the WWIII model for the North Atlantic Basin.

Parameters	WW III
Spatial resolution	0.5°
Propagation	Spherical
Number Frequencies	32
Minimum Frequency (Hz)	0.0373
Number Directions	24
Wind input	RCP8.5 EC-EARTH
Wind input time step	3 hours
Wind field spatial resolution	55km
Output time step	3 hours

Table 2. Parameterizations of the WWIII model for Azores.

Parameters	WW III
Spatial resolution	0.05°
Propagation	Spherical
Number Frequencies	32
Minimum Frequency (Hz)	0.0373
Number Directions	24
Wind input	WRF
Wind input time step	3 hours
Wind field spatial resolution	5km
Output time step	3 hours

Both bathymetry levels are generated using the GRIDGEN software package (Chawla & Tolman, 2007), created to generate grids for WWIII using MATLAB. The resolutions used are 0.5 by 0.5 and

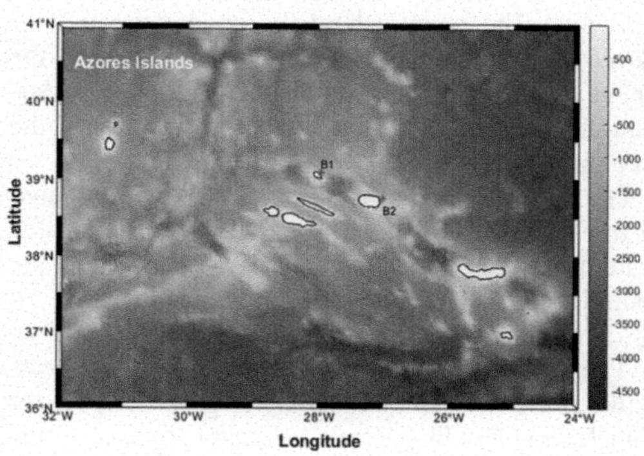

Figure 2. Area of the azores domain.

0.05x0.05, for the North Atlantic basin and the Azores Islands, respectively.

The system is implemented between 1 January 1990 to 31 December 2014, covering 25 years of the present, and between 1 January 2030 to 31 December 2054, covering 25 years of the future, suitable for climate analysis.

4 CHARACTERIZATION OF THE WAVE RESOURCE

4.1 *Statistical validation*

The accuracy of the simulations is evaluated by comparing the RCP8.5 $WWIII_{past}$ data with the $WWIII_{ERA5}$. Comparisons are made using a histogram and a quantile-quantile plot between the two simulations at the Graciosa (B1) and Praia Vitória (B2) buoy locations (Figure 2).

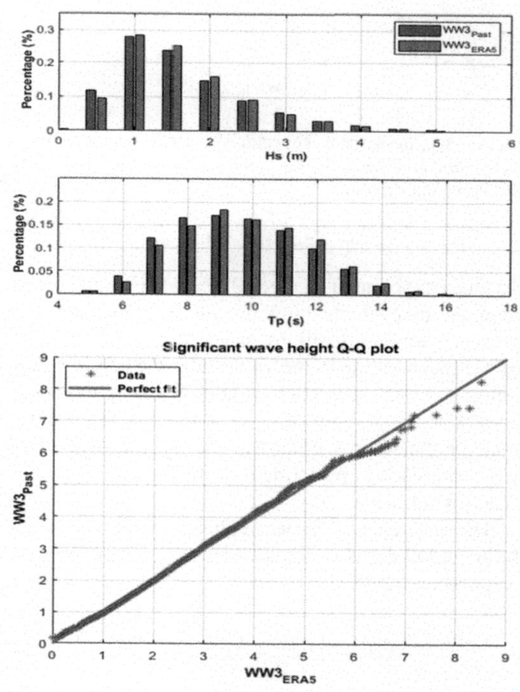

Figure 3. Comparison between the RCP8.5 $WWIII_{past}$ simulation and the $WWIII_{ERA5}$ simulation at the location of the Graciosa buoy.

As can be observed in Figure 3 and Figure 4, in general, there is a good correspondence between the two simulations, showing a good performance of the model. The Hs Q-Q plots (on the right) compare the two probability distributions by plotting their quantiles against each other, showing that the two datasets follow a similar statistic up to 4 5m.

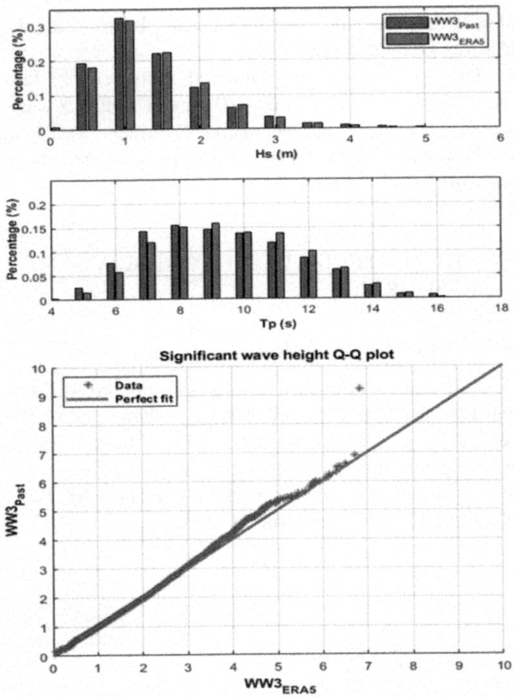

Figure 4. Comparison between the RCP 8.5 $WWIII_{past}$ simulation and the $WWIII_{ERA5}$ simulation at the location of the Praia Vitoria buoy.

4.2 *Wave climate analysis*

To analyze potential changes in the wave climate, it is required to evaluate how the model represents the present climate conditions. Therefore, a comparison of the mean and extreme values statistics for the significant wave height (*Hs*) and peak period (*Tp*) parameters is performed. The wave climate estimation for the Azores islands is obtained from the 25-year simulation performed with $Hs_{\text{Near-future}}$ and Hs_{Past} wind fields.

The discrepancy between the past and the near-future simulations is obtained by subtracting the corresponding fields:

$$\Delta Hs = Hs_{\text{Past}} - Hs_{\text{Near-future}} \qquad (2)$$

Positive values indicate that the past simulations overestimate near-future simulations and negative values that they underestimate them.

The differences between the empirical statistics of the two simulation data can be observed in Figure 5. The results show that the discrepancy between the mean values is low, varying from -0.15 and 0.15m and -0.1s and 0.1s for the *Hs* and *Tp*, respectively. The results show a general increase of the resource level in the near-future simulations. The Hs_{Past} simulation tends to underestimate (negative values) $Hs_{\text{Near-future}}$ simulation in all domains. Regarding *Tp*, the negative values show the underestimation of the Hs_{Past} mostly around the eastern side of the islands, with more emphasis around the São Miguel's Island.

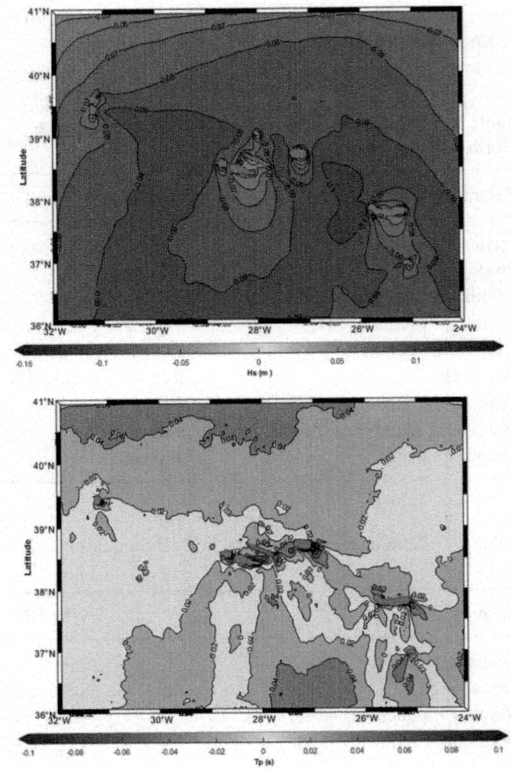

Figure 5. Difference between the mean distribution of the significant wave height (top) and the peak period (bottom).

The 99th percentile is calculated for the two parameters and is presented in Figure 6. Results show that the difference between the two simulations is also low, varying from -0.6m and 0.6m and from -0.6s and 0.6s for *Hs* and *Tp*, respectively. An increase of the resource in almost the entire domain is visible. The results show the underestimation of the Hs_{Past} simulation, with more weight around the island of São Miguel. It is also visible an overestimation of the Hs_{Past} in the northern boundary of the domain.

For *Tp*, the overestimation of the Hs_{Past} compared to the $Hs_{\text{Near-future}}$, mostly in the north area of the system domain, and the negative values in the central area, showing the underestimation of Hs_{Past} compared to the $Hs_{\text{Near-future}}$.

The results presented show that the discrepancies between the two simulations are in general small for both *Hs* and *Tp*. In general, an overestimation of the Hs_{Past} simulation for the *Hs* parameter and an underestimation of the *Tp* parameter are visible.

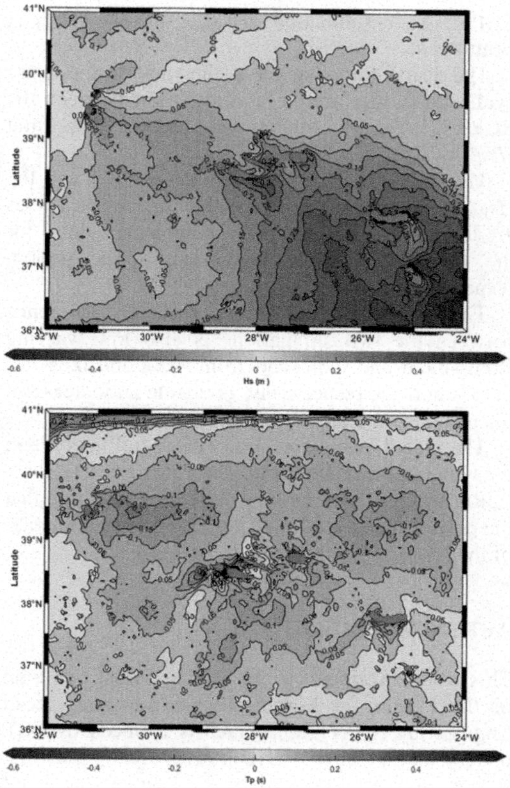

Figure 6. Difference between the 99th percentile distribution of the significant wave height (top) and the peak period (bottom).

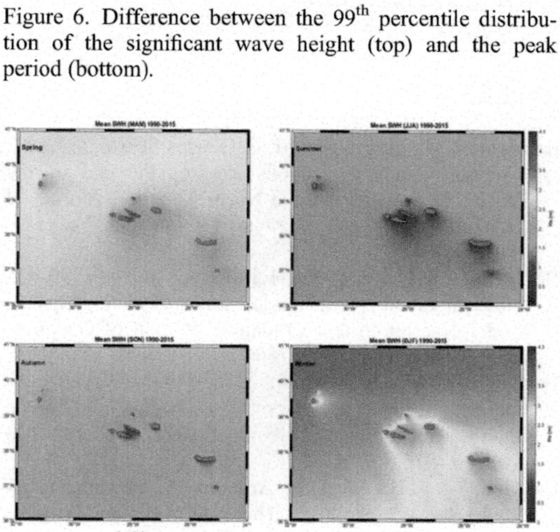

Figure 7. Seasonal distribution of the significant wave height for 1990-2014.

Figure 7 and Figure 8 show the seasonal distribution of the significant wave height. There is a small decrease in the mean wave resource in the near-future simulation compared with the past. The maximum values for the winter period are 3.67m for the past simulations and 3.65m for the near-future simulation. In the spring the values are 2.92m and 2.80m, respectively for the past and near future.

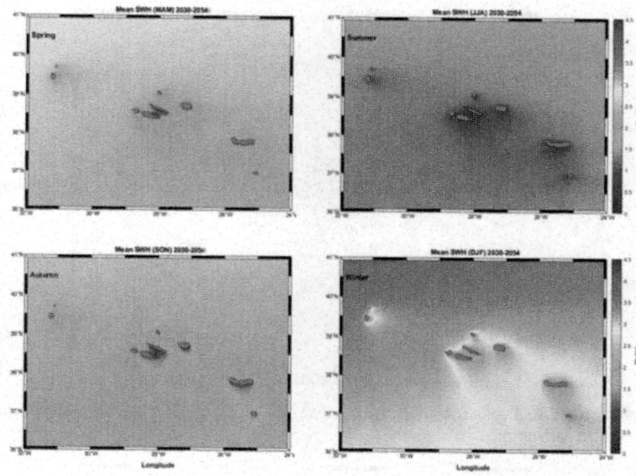

Figure 8. Seasonal distribution of the significant wave height for 2030-2054.

To study the variability of the available wave resource, three variability indices are estimated: The mean annual variability (MAV), the inter-annual variability (IAV) and the coefficient of variation (COV). The differences between the two simulations are evaluated for each index.

The mean annual variability index (MAV) stands for the average of the annual standard deviation normalised by the annual average, measuring the spread of data, also indicates the seasonal extremes, and is given by:

$$MAV = \overline{\left(\frac{\sigma_k}{\overline{x_k}}\right)} \tag{3}$$

The interannual variability index (IAV) gives the year-to-year variability throughout the period and is defined by the standard deviation of the annual means normalised by the overall mean. This relationship is given by:

$$IAV = \frac{\sigma_{\overline{x_k}}}{\overline{x}} \tag{4}$$

The coefficient of variation (COV) characterizes the amount of variability relative to the mean and is taken by dividing the standard deviation (σ) by the mean (μ).

$$\text{COV} = \frac{\sigma}{\mu} \tag{5}$$

The difference between the two WWIII periods for the MAV, IAV and COV indices can be observed in Figure 9.

The results show the MAV index discrepancies between the two simulations, the results show the overestimation (positive values) of Hs_{past} for the north coast of the domain and the underestimation (negative values) of $Hs_{\text{Near-future}}$ on the southeast coast of the islands. The low levels of variations differ from -0.5 and 0.5.

The IAV index differences results show an overestimation of the Hs_{past} in practically the entire

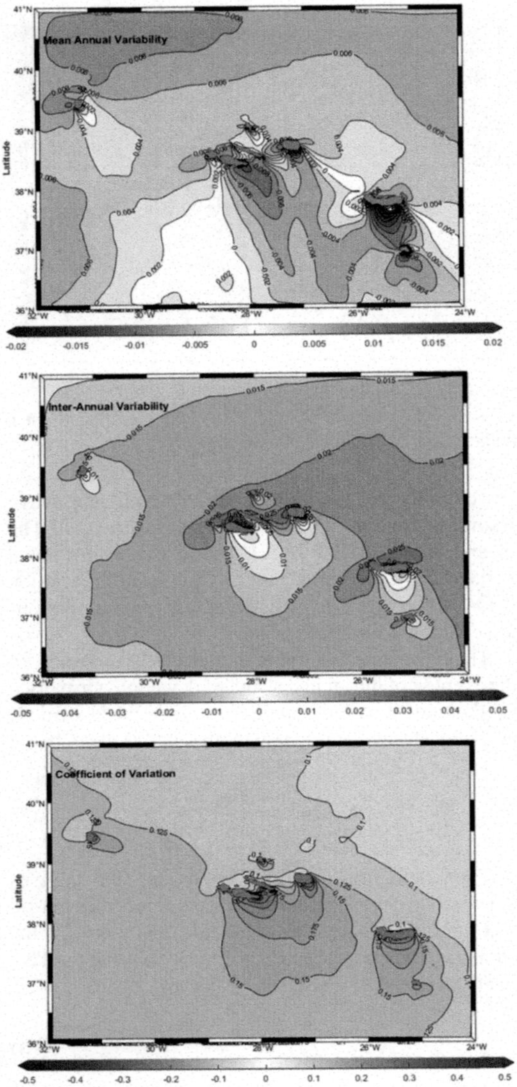

Figure 9. Difference between the MAV, IAV and COV indices for the significant wave height distribution.

domain. With higher values around the coast of the Central group (Terceira, Graciosa, São Jorge, Pico and Faial). Still, it is visible that the discrepancies are low (ranging between -0.05 and 0.05).

The results of the difference in the COV index indicate an overestimation of the Hs_{Past} simulations (positive values), indicating that this simulation has a higher coefficient of variability when compared with $Hs_{\text{Near-future}}$, for almost the entire domain.

5 CONCLUSIONS

This study is a contribution to the state-of-the-art knowledge of the present and future wave climate in the Azores Islands. The statistical validation of the system shows a good agreement between the WWIII RCP8.5 past simulation (Hs_{Past}) and $\text{WWIII}_{\text{ERA5}}$ simulation.

The near-future changes in the wave resource are evaluated by performing a comparison between the RCP8.5 Hs_{Past} simulations (1990-2014) and $Hs_{\text{Near-future}}$ (2030-2054).

The differences between the $Hs_{\text{Near-future}}$ and the Hs_{Past} are assessed, showing that the mean values are less than 0.15m and 0.1s, in absolute values, for the significant wave height and peak period, respectively.

The 99th percentile shows that the difference between the two simulations is also low, varying from -0.6m and 0.6m and from -0.6s and 0.6s, for the *Hs* and *Tp*, respectively. Is visible an increase of the resource in almost the entire domain.

The global evaluation shows a general increase in the wave resource available.

Regarding the variability indices, it is visible that the discrepancies are low. There is an overestimation of the $Hs_{\text{Near-future}}$ for most of the domain.

ACKNOWLEDGEMENTS

This work contributes to the Strategic Research Plan of the Centre for Marine Technology and Ocean Engineering (CENTEC), which is financed by FCT under contract UIDB/ UIDP/00134/2020. The first author has been financed by FCT under the grant SFRH/BD/149858/2019.

REFERENCES

Bernardino, M., Goncalves, M. & Guedes Soares, C. 2021, "Marine climate projections towards the end of the Twenty First century in the North Atlantic", Journal of Offshore Mechanics and Arctic Engineering, Vol. 143, 061201.

Booij N., Ris R.C. & Holthuijsen L.H. 1999. A third-generation wave model for coastal regions, 1, Model description and validation. Journal of Geophysical Research. 104: 7649–7666.

Bricheno, L. M., & Wolf, J., 2018, Future WaveConditions of Europe, in Response to High-End Climate Change Scenarios, J. Geophys. Res. Oceans, 123 (12), pp. 8762–8791.

Cavaleri, L., J.-H. Alves, F. Ardhuin, A. Babanin, et al. (2007). Wave modelling - The state of the art. Progress in Ocean-ography 75(4), 603–674.

Chawla, A. & Tolman, H., L. 2007. Automated grid generation for WAVEWATCH III. Technical Bulletin 254, NCEP/NOAA/NWS, National Center for Environmental Prediction, Washington, DC

Collins W.D., et al., 2004. Description of the NCAR Community Atmosphere Model (CAM 3.0), National Center for Atmospheric Research Technical Note, NCAR

Gonçalves, M. & Guedes Soares, C. 2021, Assessment of the wave resource in the Azores coastal area, Developments in Maritime Technology and Engineering, Guedes Soares, C. & Santos T.A., (Eds.), Taylor and Francis, London, UK, Vol 2, pp. 569–576.

Gonçalves, M., Bernardino, M. & Guedes Soares, C. 2021, Assessing climate change effects on the wave energy in the Canary Islands", Developments in Renewable Energies Offshore, Guedes Soares, C. (Ed.), Taylor and Francis, London, UK, pp. 19–25.

Gonçalves, M., Rusu, E. & Guedes Soares, C., 2009, Comparison of the Swan and Stwave Models on the Óbidos Coast, Proceedings of the 6th Seminar on Coastal and Port Engineering (in Portuguese), 8–9 October 2009, Funchal.

Gonçalves, M., Rusu, E., & Guedes Soares, C., 2015, Evaluation of Two Spectral Wave Models in Coastal Areas, Journal of Coastal Research, Volume 31, Issue 2: 326–339

Hazeleger, W., C. Severijns, T. Semmler, S. Stefanescu, S. Yang, X. Wang, K. Wyser, E. Dutra, R. Bintanja, B. van den Hurk, T. van Noije, F. Selten, A. Sterl & coauthors, 2010. Approach in Action Bull. Amer. Meteor. Soc., 91, 1357–1363.

Hemer, M.; Fan, Y.; Mori, N.; Semedo, A.; Wang, X.L. 2013. Projected changes in wave climate from a multi-model ensemble. Nat. Clim. Chang., 3, 471–476.

Hemer, M.; Wang, X.L.; Weisse, R.; Swail, V.R. 2012. Advancing wind-waves climate science: The COWCLIP project. Bull. Am. Meteorol. Soc, 93, 791–796.

Hersbach, H., Bell, B., Berrisford, P., Hirahara, S., Horányi, A., Muñoz-Sabater, J., Nicolas, J., Peubey, C., Radu, R., Schepers, …, Thépaut, J. N., 2020, The ERA5 Global Reanalysis, Q. J. R. Metereol. Soc., 146(730), pp. 1999–2049.

Hong, S. & Lim, J. 2006 The WRF Single-Moment 6-Class Microphysics Scheme (WSM6). Journal of the Korean Meteorological Society, 42, 129–151.

Hong, S.-Y., Noh, Y. & Dudhia, J. 2006 A New Vertical Diffusion Package with an Explicit Treatment of Entrainment Processes. Monthly Weather Review, 134, 2318–2341.

IPCC, 2007, Climate Change 2007. Impacts, Adaptation and Vulnerability, Contribution of Working Group II to the Fourth Assessment Report of the Intergovernmental Panel on Climate Change, M. L. Parry, O. F. Canziani, J. P. Palutikof, P. J. van der Linden, and C. E. Hanson, eds., Cambridge University Press, Cambridge, UK, p. 976.

Jiménez, P. A., Dudhia, J., González-Rouco, J. F., Navarro, J., Montávez, J. P., & García-Bustamante, E. 2012. A Revised Scheme for the WRF Surface Layer Formulation, Monthly Weather Review, 140(3),898–918

Kain, J.S. 2004. The Kain–Fritsch Convective Parameterization: An Update, Journal of Applied Meteorology 43, 1: 170–181.

Miguez-Macho, G., G. L. Stenchikov, & A. Robock, 2004, Spectral nudging to eliminate the effects of domain position and geometry in regional climate model simulations, J. Geophys. Res., 109, D13104

Morim, J., Hemer, M., Cartwright, N., Strauss, D., and Andutta, F., 2018, On the concordance of 21st Century Wind-Wave Climate Projections, Global Planet. Change, 167, pp. 160–171.

Moss, R. H., Edmonds, J. A., Hibbard, K. A., Manning, M. R., Rose, S. K., Van Vuuren, D. P., Carter, T. R., Emori, S., Kainuma, M., Kram, T., & Meehl, G. A., 2010, The Next Generation of Scenarios for Climate Change Research and Assessment, Nature, 463, pp. 747–756.

Rusu, E. Pilar, P., & Guedes Soares, C., 2008a. Evaluation of the wave conditions in Madeira archipelago with spectral models. Ocean Engineering, 35(13), 1357–1371.

Rusu, E., Gonçalves, M. & Guedes Soares, C. 2011, Evaluation of the Wave Transformation in an Open Bay with Two Spectral Models, Ocean Engineering, Vol. 38(16), pp. 1763–1781

Rusu, L. & Guedes Soares, C., 2012a. Wave energy assessments in the Azores islands. Renewable Energy, 45(9), 183–196.

Rusu, L., Pilar, P. & Guedes Soares, C. 2012b, Modelling the Wave Conditions in the Archipelago of Azores, Maritime Engineering and Technology, Guedes Soares et al. (Eds), Taylor & Francis Group, London, UK, pp. 533–538

Rusu, L., Pilar, P., & Guedes Soares, C., 2008b. Hindcast of the wave conditions along the west Iberian coast. Coastal Eng. 55(11), 906–919.

Semedo, A., Weisse, R., Behrens, A., Sterl, A., Bengtsson, L., and Günther, H., 2013, Projection of Global Wave Climate Change Toward the End of the Twenty-First Century, J. Clim., 26(21), pp. 8269–8288.

Silva, D. and Guedes Soares, C. 2021 Assessment of the wave power resource at Madeira archipelago with SWAN model Guedes Soares, C., (Ed.). *Developments in Renewable Energies Offshore*, London, UK: Taylor and Francis; 45–54.

Skamarock, W. C., J. B. Klemp, J. Dudhia, D. O. Gill, Z. Liu, J. Berner, W. Wang, J. G. Powers, M. G. Duda, D. M. Barker, & X.-Y. Huang, 2019: A Description of the Advanced Research WRF Version 4. NCAR Tech. Note NCAR/TN-556+STR, 145 pp. doi:10.5065/1dfh-6p97

Taylor, K. E,. Stouffer, R. J. and Meehl G. A. 2012. An Overview of CMIP5 and the Experiment Design. Bull. Amer. Meteor. Soc., 93, 485–498.

Tewari, M., Chen, F., Wang, W., Dudhia, J., Lemone, M. A., Mitchell, K. E. 2004. Implementation and verification of the unified Noah land-surface model in the WRF model [presentation]. In 20th Conf. on Weather Analysis and Forecasting/16th Conf. on Numerical Weather Prediction. American Meteorological Society: Seattle, WA, US.

Tolman, H., 1991. A third-generation model for wind waves on slowly varying, unsteady, and inhomogeneous depths and currents. Journal of Physical Oceanography 21 (6), 782–797.

Wang, C., Zhang, L., Lee, S. K., Wu, L., and Mechoso, C. R., 2014, A Global Perspective on CMIP5 Climate Model Biases, Nat. Clim. Change, 4(3), pp. 201–205.

WAVEWATCH III Development Group (WW3DG), (2016). User manual and system documentation of WAVEWATCH III version 5.16. Tech Note 329, NOAA/NWS/NCEP/MMAB, College Park, MD, USA.

Advances in Maritime Technology and Engineering – Guedes Soares & Santos (Eds)
© 2024 The Author(s), ISBN 978-1-032-83107-7

Uncertainty in the estimation of extreme winds offshore Portugal

X.N. Hu
State Key Lab of Disaster Reduction in Civil Engineering, Tongji University, Shanghai, P.R. China
Centre for Marine Technology and Ocean Engineering (CENTEC), Instituto Superior Tecnico, Universidade de Lisboa, Lisboa, Portugal

G.S. Fang & Y.J. Ge
State Key Lab of Disaster Reduction in Civil Engineering, Tongji University, Shanghai, P.R. China

C. Guedes Soares
Centre for Marine Technology and Ocean Engineering (CENTEC), Instituto Superior Tecnico, Universidade de Lisboa, Lisboa, Portugal

ABSTRACT: Extreme value estimation of ocean wind speed is typically based on frequency analysis of collected observed data and the quantile estimation with high return periods usually depends on extrapolation, inevitably introducing large uncertainties as the length of most observed dataset is finite. To better estimate extreme values, the quantification of quantile uncertainty with the standard error is implemented. Based on reanalysis winds of three Portuguese meteorological stations, this study carries out extreme value analysis using generalized extreme value and Gamma distributions. Combining the maximum likelihood method, simplified models of the standard error are utilized to quantify estimation uncertainties. Results show that extreme analysis with the standard error can provide more precise estimated quantiles, especially for high return periods. Besides, to obtain more information of the entire quantile distribution, the quadratic polynomial model is proposed to calculate the skewness and kurtosis of quantiles.

1 INTRODUCTION

Precise estimation of extreme ocean winds is important to ocean activities like ship navigation. Structural design and performance assessment of marine structures like offshore platforms, floating wind turbines and sea crossing bridges are also closely related to wind statistics (Taflanidis et al. 2013; Olivar et al. 2020). The estimation of extreme winds is generally conducted based on frequency analysis which can obtain quantiles of certain return periods by collecting observed data over several years. However, limited by finite sample size, the estimation of extreme values with high return periods usually depends on extrapolating the probability density function (PDF) or cumulative distribution function (CDF) (Drissia et al. 2019; Shen & Wei 2021). During this process, there exist great uncertainties in the selection of distribution types and parameter estimation, which inevitably leads to deviations in extreme wind speed estimation and affects its accuracy (Calenda et al. 2009).

To better estimate extreme values, some studies believe that such uncertainty can be regarded as a sampling error and quantified with the standard deviation or variance (Simiu & Scanlan, 1986; Gumbel, 1958). Gumbel (1958) and Abild et al. (1992) provided the asymptotic standard error expression of the Gumbel distribution based on the method of moments and the method of maximum likelihood estimation (MLE), respectively. Through complicated iteration and numerical calculation, Madsen et al. (1997) obtained the asymptotic covariance matrix of parameter estimators for the generalized extreme value (GEV) distribution based on the MLE. Although the matrix can be directly applied to calculate the standard error, it is non-continuous for the GEV shape parameter. Additionally, Bobee (1973) derived the standard error result for the Pearson type III distribution using the method of moments, which is typically based on the asymptotically normal assumption. Moreover, by Monte Carlo simulation without assuming an asymptotic normal distribution (Garrote et al. 2021), Hu et al. (2023) proposed simplified models for calculating the standard error of GEV, generalized Pareto and Pearson type III distributions based on MLE method.

Although the standard error provides important information for uncertainty analysis of extreme value quantiles, it is the second-order moment of quantiles. To obtain the entire extreme quantile distribution which is significant to optimal decisions, higher-order moments of quantiles are essential.

Therefore, this study aims to analyze uncertainties of extreme winds in three Portuguese coastal stations by calculating moments of quantiles. Numerical

DOI: 10.1201/9781003508779-70

predictions of the wind conditions in the Portuguese coastal and offshore areas have been produced by Salvacao & Guedes Soares, (2018) and the influence of assimilating satellite data in the predictions has been recently explored (Salvacao et al. 2022). Variability of extreme ocean winds and waves has been analyzed by Bernardino et al. (2023), considering past and future climates. An analysis of extremes of wind speed in the Portuguese offshore area was made by Campos & Guedes Soares, (2018), who identified various areas with different extreme wind regimes, by using regional frequency analysis. The three locations considered in this study are located in different areas identified in that work, having thus different extreme value characteristics.

The remainder of this paper is organized as follows. Section 2 generally describes related theories involving GEV and Gamma distributions. Section 3 introduces basic information on wind data used in this study. Detailed uncertainty analysis is presented in Section 4. Section 5 gives the conclusions of this study.

2 METHODOLOGY

2.1 *GEV distribution*

The GEV distribution is typically applied to the statistical analysis and make inferences regarding the probability of extreme events. It can be regarded as a single-family distribution which includes three fundamental distribution functions: Fisher–Tippett Type I, Type II, and Type III probability distributions. The distribution functions of the GEV distribution are given by:

$$F(x) = \exp\left\{ -[1 + k\frac{(x-\mu)}{\sigma}]^{-1/k} \right\} (k \neq 0) \quad (1a)$$

$$F(x) = \exp[-\exp(-\frac{x-\mu}{\sigma})](k = 0) \quad (1b)$$

where k, σ, and μ are the shape, scale, and location parameters, respectively. When k=0, k>0 and k<0, the GEV degenerates into Fisher–Tippett Type I (also known as Gumbel distribution), Type II, and Type III distributions, respectively. For the quantile x_T with return period T, it can be calculated as:

$$x_T = F^{-1}(P) = F^{-1}(1 - 1/T) \quad (2)$$

For the GEV distribution fitted by the MLE method, the standard error is the function of the sample size n, the return period and distribution parameters, which can be expressed as (Hu et al., 2023):

$$S(x_T) = \frac{\sigma}{\sqrt{n}}\{A(T) \cdot \exp[B(T) \cdot k]\}^{1/2} \quad (3a)$$

$$A(T) = 0.75T - 0.39 \quad (3b)$$

$$B(T) = 1.75y + 1.07 \quad (3c)$$

where y=-ln(-ln(1-1/T)) is the reduced variable.

2.2 *Gamma distribution*

The Gamma distribution is a two-parameter family of continuous probability distributions. Its probability function can be formulated as:

$$F(x) = \frac{1}{b^a \Gamma(a)} \int_0^x t^{a-1} e^{\frac{-t}{b}} dt \quad (4)$$

where a is the shape parameter, b is the scale parameter, $\Gamma(a)$ is the gamma function. When fitting the Gamma distribution using the MLE method, the standard error of quantiles can be calculated by (Hu et al., 2023):

$$S(x_T) = \frac{b}{\sqrt{n}}[A(T) \cdot a^2 + B(T) \cdot a + C(T)] \quad (5a)$$

$$A(T) = (-0.22\ln(T) - 1.61) \cdot 10^{-3} \quad (5b)$$

$$B(T) = 0.036\ln(T) + 0.18 \quad (5c)$$

$$C(T) = 1.31\ln(T) - 0.24 \quad (5d)$$

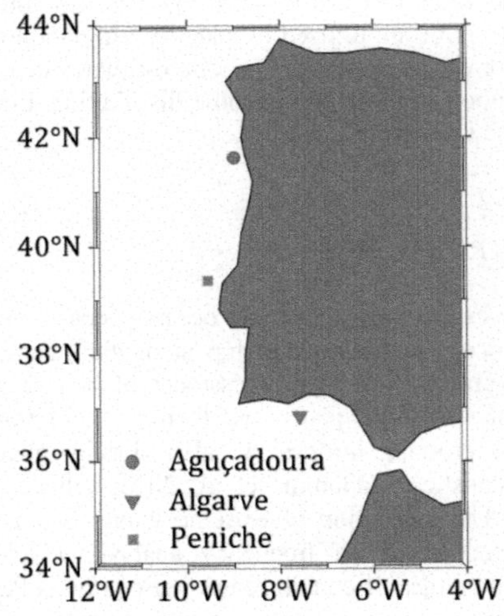

Figure 1. Location of three stations.

3 REANALYSIS DATA OF WIND SPEED

3.1 *ERA5 dataset*

ERA5 dataset is the fifth-generation atmospheric reanalysis dataset of the global climate developed by the Copernicus Climate Change Service (C3S) at the European Centre for Medium-Range Weather Forecasts (ECMWF) (Hersbach et al. 2020). Based on ERA-Interim product, ERA5 has enhanced its accuracy by integrating additional sources of high altitude and surface observations with an improved multidimensional data assimilation system. It provides data from 1940 to the present. Furthermore, the atmospheric model has a horizontal resolution of

0.25°×0.25° (latitude×longitude) and a temporal resolution of 1 hour. Reanalysis data can be downloaded on single levels (specific heights) or pressure levels according to users' requirements. Given the long duration of the records, this data set is adopted in this study instead of the more specific study of Salvacao et al (2022), which covers a shorter period.

3.2 *Analysis data of meteorological stations*

Reanalysis of wind data from three coastal meteorological stations in Portugal is utilized in this study. Their positions are shown in Figure 1. Wind speeds from 1979.01.01 to 2020.12.31 are collected every hour with a total of 368,184 records. As can be seen in Figure 2, wind conditions of the three stations are similar and most records of wind speed fall in the range from 0 to 25 m/s.

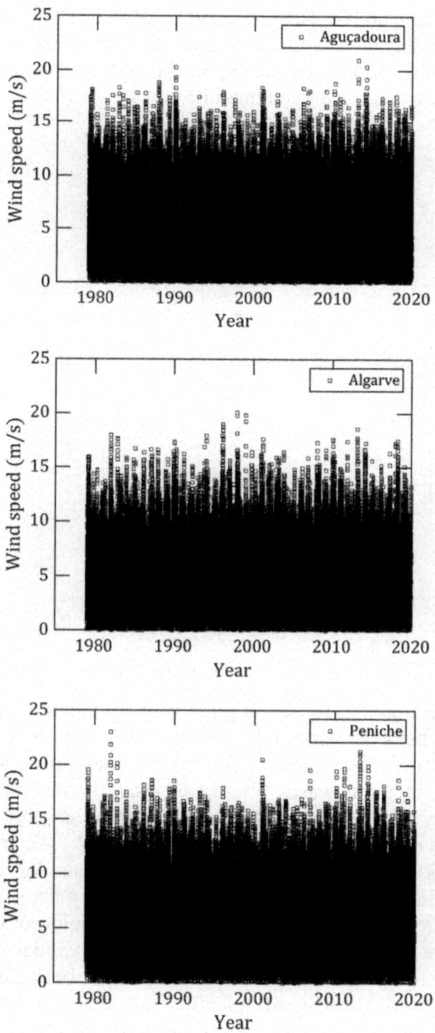

Figure 2. Wind records of three stations.

4 UNCERTAINTY ANALYSIS OF EXTREME WINDS

4.1 *Distribution fitting*

In frequency analysis of extreme values, a widely used sampling method is the annual maximum (AM) method, which is appropriate for long-term (more than 20 years) observation data (Cook, 1985). As a result, for each station, there is a sample set whose size is 42. To fit this sample set, the maximum likelihood method is employed here. Two approaches, Kolmogorov–Smirnov (K–S) test and the root mean square error (RMSE), are introduced to verify the accuracy of distribution fitting. After K–S test, the obtained p-value can reflect the probability that a sample set obeys the target distribution, and a p-value higher than 0.05 indicates that K–S test does not reject the null hypothesis at the 5% significance level. Similarly, RMSE represents fitting error. In other words, the lower the RMSE value is, the better the fitting effect is.

Table 1. Fitting results of three sample sets.

Station	Dist.	*k/a*	*σ/b*	*μ*	p-value	RMSE
Aguça-doura	GEV	-0.14	1.31	16.6	0.79	0.22
	Gam	141.5	0.12	-	0.89	0.25
Algarve	GEV	-0.24	1.47	15.9	0.85	0.20
	Gam	121.8	0.14	-	0.88	0.20
Peniche	GEV	0.04	1.48	16.7	0.85	0.24
	Gam	85.5	0.21	-	0.32	0.42

* Gam: Gamma

Fitting results are shown in Figure 3 and Table 1. As can be seen in Figure 3, for sample sets of Aguçadoura and Algarve stations, both fitting curves of GEV and Gamma distributions are close to the empirical CDF curve, while for the Peniche station, it is distinct that the GEV distribution can give a better fitting to the sample set. Table 1 presents fitting parameters and test results which provide quantified indexes. All p-values are higher than 0.05, indicating that both GEV and Gamma distributions are suitable for the three stations. RMSE values contribute to further selection of best estimation. For the Algarve station, RMSE values of GEV and Gamma distributions are equal to each other, meaning that the two models have the same accuracy. Considering that the GEV distribution presents better performance for the other two stations, the GEV distribution is selected for the Algarve station, which can eliminate the impact of different distribution models on uncertainty analysis and comparison. As a result, GEV is utilized for distribution fitting of all three sample sets.

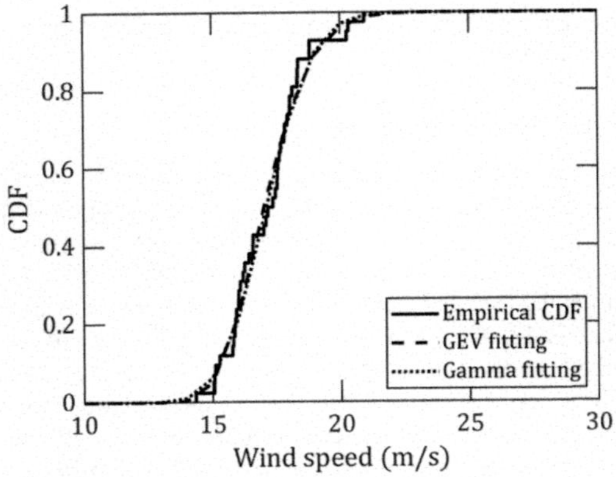

(a) Aguçadoura

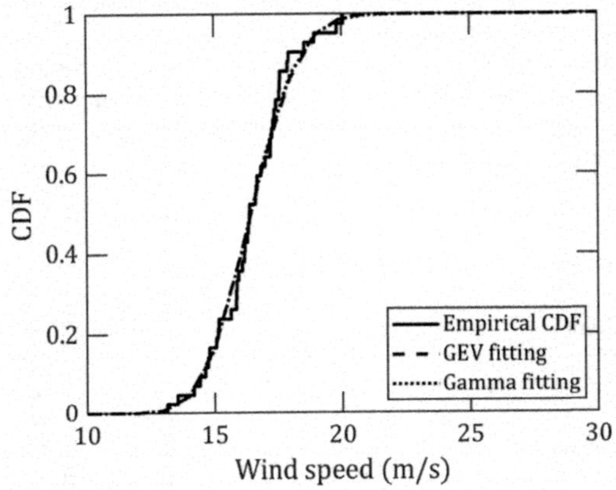

(b) Algarve

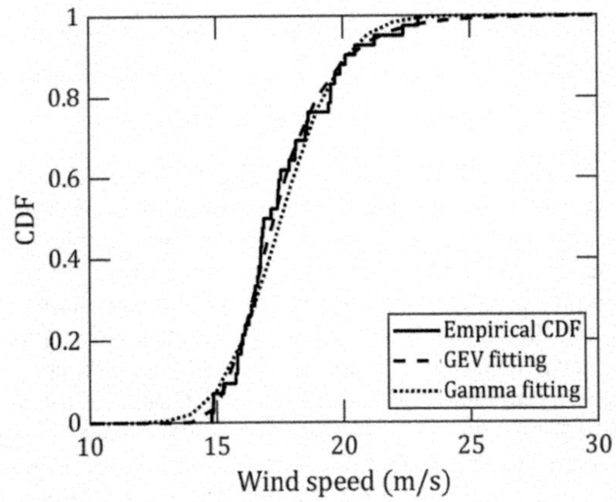

(c) Peniche

Figure 3. Fitting curves of three sample sets.

4.2 *The standard error*

With fitted probability distributions of three stations, Figure 4 represents extreme wind quantiles of certain return periods and corresponding standard errors which can be calculated as Equation 3. As can be seen, in general, the quantile increases with the increase of the return period. Among the three stations, extreme wind speeds in Peniche station are the highest, which is related to various factors like climate conditions and geographical location. Furthermore, the standard errors in Peniche station are also the largest, which mainly depends on the values of distribution parameters, especially the shape parameter k. To be more specific, as listed in Table 1, the k-value of GEV distribution in Peniche is larger than that in two other stations when there is no large difference in the σ-values in three stations. It reflects the law described in Equation 3 that the standard error will increase with the increase of the shape parameter, return period and the scale parameter.

In addition, Figure 5 further shows uncertainties of extreme wind speeds. Assuming that the extreme wind quantile obeys normal distribution, the standard error can be taken as the standard deviation parameter of normal distribution, and then the distribution of the quantile can be determined. As shown in Figure 5, for all three stations, when the return period varies form 10 years to 100 years, the peak value of PDF gradually decreases and the shape of distribution curve becomes wider and wider, meaning that the uncertainty of extreme wind quantile gets larger.

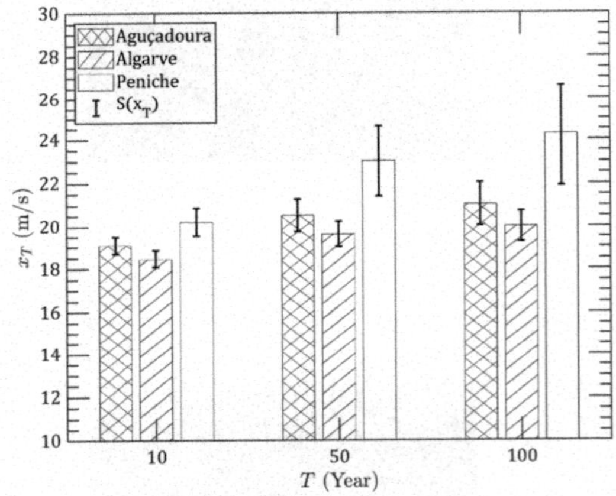

Figure 4. Wind speeds of certain return periods.

4.3 *Skewness and kurtosis of the extreme quantile*

In Section 4.2, the assumption is given that the quantile obeys the normal distribution. However, it is just an approximate assumption which does not strictly hold in many cases. To better quantify the uncertainty of quantiles, it is worthy to further explore the higher-order moments of quantiles, i.e., skewness and kurtosis. Considering that skewness and kurtosis are dimensionless variables normalized by the standard deviation of the sample set, they should not correlate with the scale parameter, which is validated as shown in Figure 6. As can be seen, for both skewness and kurtosis of quantiles of different return periods, the values merely fluctuate slightly and almost keep constant at the average level even if the scale parameter varies from 1 to 10.

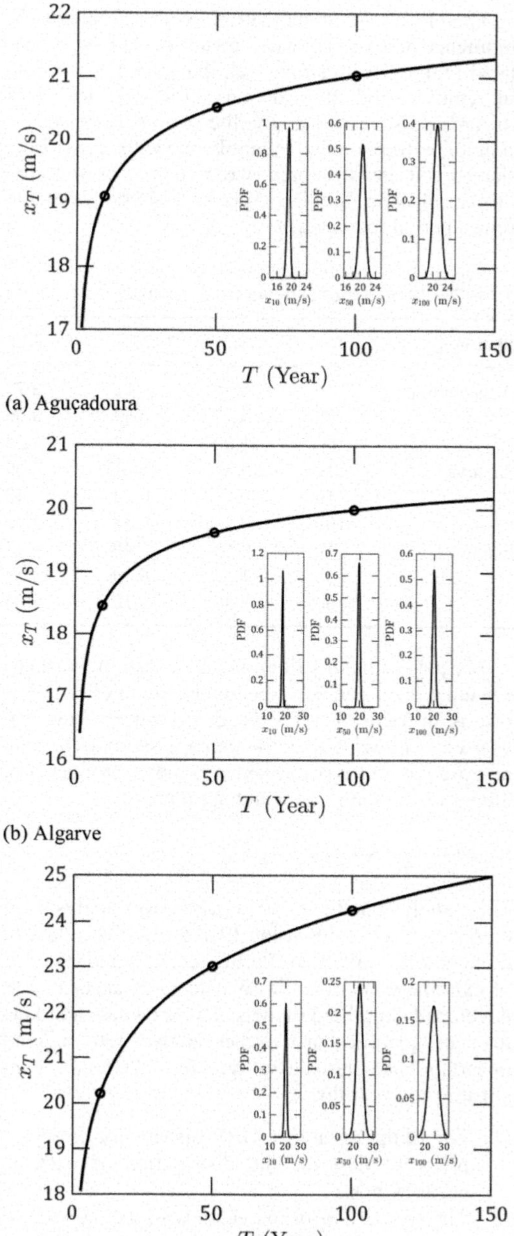

(a) Aguçadoura

(b) Algarve

(c) Peniche

Figure 5. Uncertainties of extreme wind speeds.

Therefore, the skewness and kurtosis should be the function of the sample size n, the return period T and the shape parameter k. To determine the function, the quadratic polynomial model is used:

$$\eta = h_0 + \sum_{i=1}^{3} h_i q_i + \sum_{i=1}^{3} h_{ii} q_i^2 + \sum_{i,j(i<j)}^{3} h_{ij} q_i q_j \quad (6)$$

where η is the skewness (SK) or kurtosis (KU) in this study, $\boldsymbol{h}$= [h_0, h_1, h_2, h_3, h_{11}, h_{22}, h_{33}, h_{12}, h_{23}, h_{13}] is the undetermined coefficient vector, $\boldsymbol{q}$= [n, k, T] is the vector of variables.

Skewness and kurtosis values of quantile corresponding to various values of $\boldsymbol{q}$ can be obtained by Monte Carlo simulations. The main steps are as follows:

(1) Assume that extreme samples obey the distribution with given parameter $\boldsymbol{a}_0$, and calculate corresponding quantiles x_{T0};
(2) Generate 100 sets of extreme series by random simulation based on the original parameter $\boldsymbol{a}_0$, and obtain new distribution parameter $\boldsymbol{a}_i$ and quantiles x_{Ti} (i=1,2, …, 100) for each set;
(3) Calculate skewness and kurtosis of quantiles based on x_{T0} and x_{Ti}.

It should be noted that to facilitate the regression analysis, the original parameter $\boldsymbol{a}_0$ is artificially set in this study. More details can be seen in Hu et al. (2023).

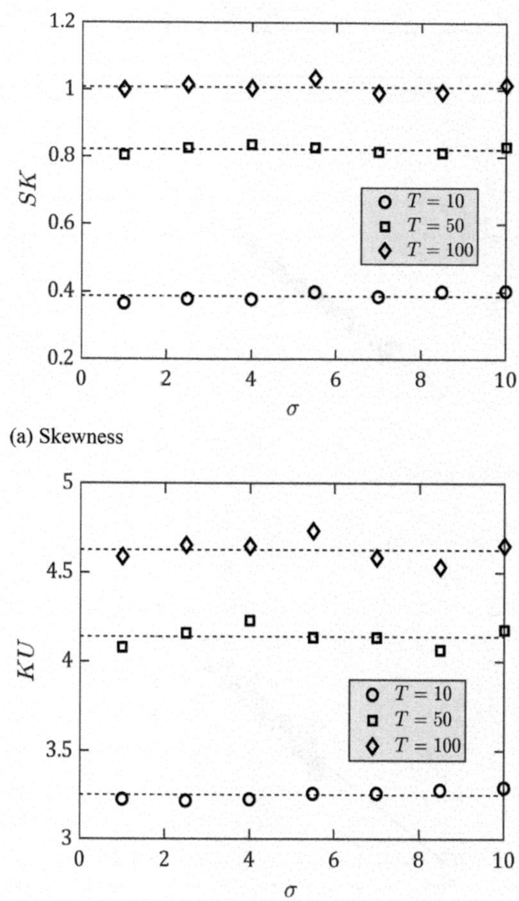

(a) Skewness

(b) Kurtosis

Figure 6. Variation of skewness and kurtosis with σ.

Using Equation 6 to fit the simulated values, as depicted in Figure 7, fitted results coincide well with simulated values. As listed in Table 2, absolute values of coefficients h_2 and h_{22} of both the skewness and kurtosis are larger compared with other coefficients, indicating that the shape parameter is very influential. By contrast, h_{11}, h_{33} and h_{13} are extremely close to zero, meaning that three items, n^2, T^2 and nT, have little impact on skewness and kurtosis.

Table 2. Fitted coefficients of Equation 6.

Coefficient	SK	KU
h_0	8.53E-01	4.80
h_1	-2.07E-02	-7.11E-02
h_2	1.88	4.35
h_3	1.54E-02	4.59E-02
h_{11}	1.41E-04	5.22E-04
h_{22}	-5.79E-01	3.61
h_{33}	-3.80E-05	-9.01E-05
h_{12}	-1.10E-02	-5.57E-02
h_{23}	1.03E-02	3.38E-02
h_{13}	-9.20E-05	-3.35E-04

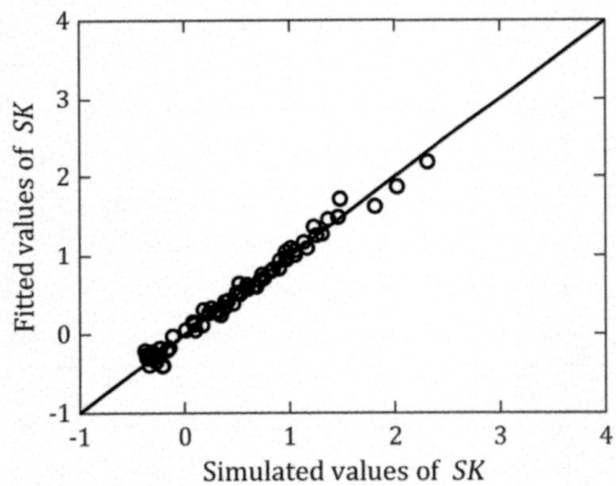

(a) Skewness

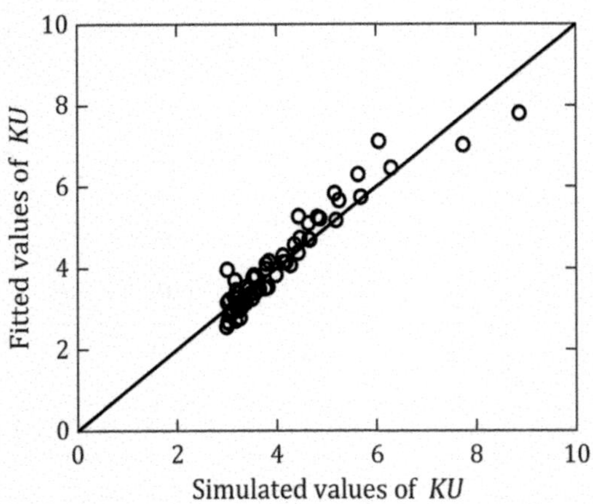

(b) Kurtosis

Figure 7. Verification of fitting results.

Therefore, in combination with Equation 3, moments of extreme wind quantiles can be calculated (Table 3). It can be seen that with the increase of return period, the skewness gradually increases towards positive values. In other words, the distribution of extreme wind quantile presents a positive skew trend and the main part is more concentrated on the left side. Similarly, the kurtosis grows as the return period increases.

Table 3. Moments of extreme wind quantiles.

Station	*T*	*S*	*SK*	*KU*
Aguçadoura	10	0.41	0.12	2.79
	50	0.77	0.43	3.66
	100	1.00	0.65	4.34
Algarve	10	0.38	-0.05	2.69
	50	0.61	0.22	3.42
	100	0.74	0.38	3.93
Peniche	10	0.66	0.40	3.14
	50	1.62	0.79	4.26
	100	2.36	1.10	5.24

For the normal distribution, skewness and kurtosis are equal to 0 and 3, respectively. According to the results in Table 3, only the distribution of the 10-year extreme wind quantile can approximate normal distribution. For quantiles with higher return periods, the normal assumption is not appropriate.

5 CONCLUSIONS

This study conducts an uncertainty analysis of extreme ocean winds using ERA5 reanalysis data of three coastal stations in Portugal. The standard error of extreme wind quantiles is calculated according to developed simplified models. The skewness and kurtosis of quantiles are further analyzed to explore uncertain characteristics of quantiles. The main conclusions are as follows:

(1) For all three stations, GEV distribution is appropriate to describe the distribution of extreme ocean winds.
(2) The standard error increases with the increase of the shape parameter, return period and scale parameter. Among the three stations, the standard error of extreme wind quantile in Peniche station is the largest.
(3) Based on the GEV distribution, the quadratic polynomial model is proposed to calculate the skewness and kurtosis of extreme wind quantiles. In this model, the skewness and kurtosis do not correlate with the scale parameter, and the shape parameter should be the dominant factor.
(4) As the return period grows, the values of skewness and kurtosis will also increase. The normal distribution is no longer appropriate to describe the distribution of extreme quantiles with high return periods.

ACKNOWLEDGEMENTS

The first author is financed by the China Scholarship Council and the National Natural Science Foundation of China under grant No. 52278520. This work contributes to the Strategic Research Plan of the Centre for Marine Technology and Ocean Engineering (CENTEC), which is financed by the Portuguese Foundation for Science and Technology (Fundação para a Ciência e Tecnologia - FCT) under contract UIDB/UIDP/00134/2020.

REFERENCES

Abild, J., Andersen, E.Y., Rosbjerg, D. 1992. The climate of extreme winds at the Great Belt, Denmark. *Journal of Wind Engineering & Industrial Aerodynamics*, 41–44: 521–532.

Bernardino, M., Gonçalves, M., Campos, R.M., et al. 2023. Extremes and variability of wind and waves across the oceans until the end of the 21st century. *Ocean Engineering*, 275: 114081.

Bobée, B. 1973. Sample error of the T-year events computed by fitting a Pearson Type III distribution. *Water Resources Research*, 9(5): 1264–1270.

Calenda, G., Mancini, C.P., Volpi, E. 2009. Selection of the probabilistic model of extreme floods: The case of the River Tiber in Rome. *Journal of Hydrology*, 371:1–11.

Campos, R. M. and Guedes Soares, C. 2018. Spatial distribution of offshore wind statistics on the coast of Portugal using Regional Frequency Analysis. *Renewable Energy*. 123806–816.

Cook, N.J. 1985. *The Designer's Guide to Wind Loading of Building Structures. Part 1: Background, Damage Survey, Wind Data and Structural Classification*. Garston: Building Research Establishment & London: Butterworths.

Drissia, T.K., Jothiprakash, V., Anitha, A.B. 2019. Flood Frequency Analysis Using L Moments: A Comparison between At-Site and Regional Approach. *Water Resources Mancgement*, 33: 1013–1037.

Garrote, J., Pena, E., Diez-Herrero, A. 2021. Probabilistic Flood Hazard Maps from Monte Carlo Derived Peak Flow Values—An Application to Flood Risk Management in Zamora City (Spain). *Applied Science*, 11:6629.

Gumbel, E.J. 1958. *Statistics of extremes*. New York: Columbia University Press.

Hersbach, H., Bell, B., Berrisford, P., et al. 2020, The ERA5 Global Reanalysis, *Q. J. R. Metereol. Soc.*, 146 (730), pp. 1999–2049.

Hu, X.N., Fang, G.S., Yang, J.Y., et al. 2023. Simplified models for uncertainty quantification of extreme events using Monte Carlo technique. *Reliability Engineering and System Safety* 230:108935.

Madsen, H., Rasmussen, P.F., Rosbjerg, D. 1997. Comparison of annual maximum series and partial duration series methods for modeling extreme hydrologic events.1. At-site modeling. *Water Resources Research*, 33(4): 747–757

Olivar, O.J.R., Mayorga, S.Z., Giraldo, F.M., et al. 2020. The effects of extreme winds on atmospheric storage tanks. *Reliability Engineering & System Safety*, 195:106686.

Salvacao, N.; Bentamy, A., and Guedes Soares, C. 2022. Developing a new wind dataset by blending satellite data and WRF model wind predictions. *Renewable Energy*. 198283–295.

Salvacao, N. and Guedes Soares, C. 2018. Wind resource assessment offshore the Atlantic Iberian coast with the WRF model. *Energy*. 145; 276–287.

Shen, Z., Wei, K. 2021. Stochastic model of tropical cyclones along China coast including the effects of spatial heterogeneity and ocean feedback. *Reliability Engineering and System Safety*, 216:108000.

Simiu, E., Scanlan, R.H. 1986. *Wind effects on structures: An introduction to wind engineering*. New York: John Wiley and Sons Ltd.

Taflanidis, A., Loukogeorgaki, E., Angelides, D.C. 2013. Offshore wind turbine risk quantification/evaluation under extreme environmental conditions. *Reliability Engineering & System Safety*, 115:19–32.

Advances in Maritime Technology and Engineering – Guedes Soares & Santos (Eds)

Improving the accuracy of significant wave height hindcast data with long short-term memory models

M. Latas, A. Ali, D. Silva & C. Guedes Soares
Centre for Marine Technology and Ocean Engineering (CENTEC), Instituto Superior Técnico, Universidade de Lisboa, Lisbon, Portugal

ABSTRACT: To determine the capabilities of long short-term memory (LSTM) neural networks to extend the skill of wave forecasts, the ERA5 hindcast database is used as a test case. Data from a buoy located in the North Atlantic Ocean near Portugal continental coast is used as the reference value. The error between the ERA5 data set and the wave buoy records is used to train a LSTM model, which is then used to correct other samples of the ERA5 database. ERA5 feature selection is performed by computing the variance inflation factor and by applying sequential feature selection with forward selection, which resulted in the selection of 5 features for training the models. After extensive testing, an optimal number of 50 neurons and a filtering window size of 10 is selected, yielding LSTM models that improve the ERA5 estimates. The accuracy of the data sets adjusted with the LSTM model is compared with the one of the initial ERA5 data using four metrics: the normalized bias, the scatter index, the normalized root mean squared error, and the correlation coefficient.

1 INTRODUCTION

Wind and sea state conditions have a significant impact on marine activities such as shipping, fisheries, offshore operations, and nautical recreation. Characterizing the wave conditions is essential to determine the responses of marine structures to wave forcing, which in turn enables their design to properly operate under those conditions (Silva & Guedes Soares 2021, Lucas et al. 2019). Accurate wind and wave forecasts are essential to alert about extreme events, allowing mitigation measures to be implemented to improve safety.

Waves can be directly measured with in situ or remote devices, namely with buoys in the first case and with radar or satellites in the second, providing accurate and realistic outcomes. However, this data is insufficient for covering all the areas and times, and it can contain gaps due to device failure.

Third-generation wave models are widely accepted as an efficient tool for assessing wave conditions, as they provide the necessary temporal and spatial range for a detailed study of wave conditions (Cavaleri et al. 2007). These models have been used to reanalyze historical datasets and to produce extensive databases of wind and wave data, aiming to improve the skill of the predictions, with increasingly more complete models. However, wave modeling is subject to uncertainties that can arise from various factors, such as the limitations of the model to simulate the area of study, suboptimal parametrization introduced by the user, limitations of the computer that performs the calculations, or inaccurate input data. The quality of the various data sets has been studied by several authors, using as reference values buoy and satellite measured data, such as Campos & Guedes Soares (2016, 2017) where other studies are referred.

If in a given ocean area the error of the predictions can be quantified and eventually modelled, it is expected that the wave predictions can then be corrected by the error model to improve the accuracy of the assessments. One way to achieve this goal is to use artificial neural networks (ANNs), one type of machine learning model inspired by the structure and function of human brains, that use the knowledge of the past data to predict future outcomes, in which neurons are aggregated into layers. ANNs have an input, output and one or more hidden, or intermediate, layers.

Deep neural networks (DNNs) are deep learning algorithms based on ANNs, in which the term "deep" is related to the number of the layers used by the network (Lazzeri 2020), a number that should be greater than three including inputs and outputs. Recurrent neural networks (RNN) (Brownlee 2017) are a type of ANN that is specifically designed to solve sequence problems. They use information from previous time steps as inputs to predict the next time step, allowing the network to propagate information through time. The vanishing gradient problem is responsible for the short-term memory of this algorithm.

During model training, the backpropagated gradient calculations shrink exponentially, approaching zero, and eventually vanishing and causing the model to stop learning (Sarang 2021). Conversely, the exploding gradient problem can occur, in which large error gradients accumulate and eventually blow up, crashing the model (Sarang 2021).

DOI: 10.1201/9781003508779-71

To deal with the vanishing gradient problem present in traditional RNNs, Hochreiter & Schmidhuber (1997) developed the long short-term memory (LSTM) model. This network is capable of learning long-term dependencies at a given time from previous time steps, which is useful for modeling time series (Brownlee 2017, Lazzeri 2020). Fan et al. (2020) were among the first to use LSTM to forecast the significant wave height at various sites.

The application of ANNs to environmental analysis and forecasting has been increasing in recent years. Recently, researchers have utilized ANN models to predict wave heights and reconstruct ocean waves. Feng et al. (2022) compared the accuracy of RNN, LSTM and gated recurrent unit (GRU) networks in forecasting the significant wave heights at various time intervals (1, 3, 6, 12, and 24 hours). Zhou et al. (2021) combined an empirical mode decomposition (EMD) with LSTM networks to improve the prediction of significant wave heights and to achieve better forecast performance for extended time horizons. Kagemoto (2020) investigated how artificial intelligence, especially RNNs and LSTM models, can be used to predict water-surface wave trains, using both simulated and observed sea wave data.

Campos & Guedes Soares (2016), Campos et al. (2019), and Costa et al. (2023) proposed a hybrid model that combines numerical wave models with ANNs as an alternative methodology to reduce bias. The present study is a follow-up of these authors' work but for the ERA5 hindcast in Portugal's continental coast. The input features are from ERA5 data, while the residue prediction (i.e., the difference between the measurement and the model) is evaluated using an LSTM algorithm trained with buoy data.

2 DATA DESCRIPTION

Various factors, such as maintenance activities, storms, and navigation incidents, can lead to buoy failures, resulting in data gaps and temporal disruptions in the buoys' time series until repairs are completed (Rao & Mandal 2005). A crucial initial step involves selecting a relatively extensive time frame with minimal gaps and missing data, as the LSTM algorithm relies on preserving valuable information from preceding data sequences to facilitate the processing of new data.

Following rigorous quality control, the chosen period spans three years, from 2014 to 2016, and involves buoy 6200084 (Cabo Sileiro, located at 42.120 N 9.430 W) provided by Puertas del Estado (PdE), situated in the North Atlantic Ocean, as depicted in Figure 1. The selection of this specific buoy is not only based on the occurrence of very few gaps in the data but also on the unique metocean conditions.

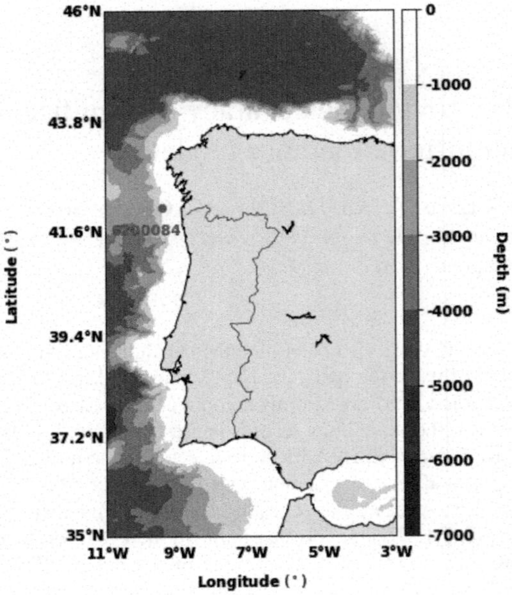

Figure 1. Buoy 6200084 (Cabo Sileiro) from the Puertos del Estado (PdE).

The ERA5 dataset is a 5th generation atmospheric reanalysis data produced by European Centre for Medium-Range Weather Forecasts (ECMWF) (Hersbach et al. 2020) and archived by the Copernicus Climate Change Service (C3S). The raw dataset comprises information related to 36 variables for a grid with a spatial resolution of 31 km and a temporal resolution of 1 hour, aligning with the time resolution of buoy measurements. The Era5 data were interpolated for the same point of the cabo Sillero buoy. To construct the input data space, a series of adjustments are implemented through the application of the following variable transformations:

$$wspd = \sqrt{u^2 + v^2} \tag{1}$$

$$ud = sin\left(\frac{2\pi}{180}d\right) \tag{2}$$

$$vd = cos\left(\frac{2\pi}{180}d\right) \tag{3}$$

These equations convert the wind components, u and v at 10 m, into wind speed (wspd) (Equation 1) and the directional variables to their corresponding sine and cosine (Equation 2 and 3), where d is the direction. This transformation is necessary to ensure that angles can be effectively utilized by the machine learning model. Furthermore, a logarithmic transformation is employed for significant wave height (swh) by using the log function of the Numpy

Python library, which computes the natural logarithm, element-wise. The rationale behind this transformation, as highlighted in Campos et al. (2019), is that it has been verified to enhance results due to the more homogeneous distribution of values, thus improving the results of the modelling process.The focus of this study is to improve the accuracy of ERA5 predictions by training an LSTM model, fed with ERA5 features, with buoy data. For comparing the accuracy of adjusted swh using LSTM with ERA5 estimates of swh, a set of four accuracy metrics are introduced. The choice of these metrics is adopted from the study of Mentaschi et al. (2013) and it aligns with the evaluation methodologies presented in Campos et al. (2019, 2020). The error metrics are the normalized bias (NBias), the scatter index (SI), the normalized root mean squared error (NRMSE), and the correlation coefficient (CC), which are defined as follows:

$$NBias = \frac{\sum_{i=1}^{n}(x_i - y_i)}{\sum_{i=1}^{n} y_i} \tag{4}$$

$$SI = \sqrt{\frac{\sum_{i=1}^{n}[(x_i - \bar{x}) - (y_i - \bar{y})]^2}{\sum_{i=1}^{n} y_i^2}} \tag{5}$$

$$NRMSE = \sqrt{\frac{\sum_{i=1}^{n}[(x_i - y_i)]^2}{\sum_{i=1}^{n} y_i^2}} \tag{6}$$

$$CC = \frac{\sum_{i=1}^{n}(y_i - \bar{y})(x_i - \bar{x})}{\sqrt{\sum_{i=1}^{n}(y_i - \bar{y})^2 \sum_{i=1}^{n}(x_i - \bar{x})^2}} \tag{7}$$

where x represents the model data either from the ERA5 estimates or the LSTM model predictions, y corresponds to the buoy measurements, and the bar symbol denotes the corresponding mean value. This comprehensive set of metrics provides a detailed evaluation of the model's performance, covering various aspects of systematic and scatter errors, as well as signal trends in comparison to observational data.

3 METHODOLOGY

3.1 *Feature selection*

The reduction of the dimensionality of the feature space is of decisive importance for the creation of efficient machine learning models. To accomplish such reduction two feature selection techniques are applied successively, the computation of the variance inflation factor (VIF) (Katrutsa & Strijov 2017) and sequential feature selection (SFS) with forward selection (Ferri et al. 1994). Feature extraction is another approach to dimensionality reduction, which can complement or be used as an alternative to feature selection techniques (Raschka & Mirjalili 2019). This work applies a single method of feature extraction, principal component analysis (PCA) (Jolliffe 2002), after the computation of the VIF and SFS.

The computation of the VIF permits the quantification of the increase of the variance of an estimated regression coefficient due to collinearity, a characteristic which results in unstable models of excessive complexity (Katrutsa & Strijov 2017). The VIF is defined for feature i as

$$\mathrm{VIF}_i = \frac{1}{1 - r_i^2} \tag{8}$$

where r_i^2 is the coefficient of determination, or multiple correlation coefficient, which gives the proportion of variance in the i-th feature associated with the remaining i-1 features (Thompson et al. 2017). The VIF can take values in $[1, +\infty]$ and a VIF superior to 10 is often taken as an indication of high collinearity, although other values are also found in the literature (Thompson et al. 2017). In this work, and following Costa et al. (2023), a cut-off value of 5 is set, meaning that starting with the feature with the highest VIF, variables are dropped iteratively until the resulting feature space includes only features with a VIF smaller than the cut-off value. The standard computation of the VIF is modified for this work, again following Costa et al. (2023), by precluding the removal of the variables Lshw and wspd during the dropout process, even when both had a VIF larger than the cut-off value. This decision is justified by observing that those two variables are known to be essential for the prediction of the target variable.

The SFS with forward selection is a search algorithm that is used to reduce the feature space to a space of lower dimensionality (Aha & Bankert 1996). This algorithm starts with zero features, then it finds a feature that minimizes a score (for instance, the mean squared error) when an estimator is trained on that feature, and finally adds that feature to the set of selected features. The algorithm repeats the above procedure by iteratively adding another feature to the set of selected features and completing the remaining steps.

The estimator selected is the random forest regressor (RFR) (Callens et al. 2020), which fits a number of decision tree regressors on various subsamples of the dataset and uses averaging to improve the predictive accuracy and control over-fitting (Pedregosa et al. 2011). Decision tree regressors start at the root of the tree and then branch into two or more directions, depending on variable outcomes, until a leaf node is reached and the result is given. The number of trees in the forest is set at 50, to reduce complexity, and the minimum number of samples required to be at a leaf node is set at 10. Each execution of the SFS method with the RFR as the estimator is repeated 10 times to account for the stochasticity of the algorithm, following Costa et al. (2023).

The Karhunen–Loeve transform (KLT) (Gerbrands 1981), also known as PCA, is an unsupervised dimensionality reduction technique that aims to reduce a set of variables to a set of linearly uncorrelated variables called principal components (Jolliffe 2002). This reduction is achieved by transforming or projecting data in a d-dimensional space onto data in a k-dimensional subspace, where k≤d, such that the greatest proportion of the variance of the original data is retained. The principal components in the subspace of dimension k correspond to mutually orthogonal axes, which can be interpreted as the directions of maximum variance. Since PCA is sensitive to data scaling, the features are standardized by removing the mean and by scaling to unit variance before the application of the algorithm. The percentage of the total variance to be kept is set at 95%.

3 1 1 *Results*

The computation of the VIF results in the elimination of 13 of the 41 features of the feature space, namely the features tmax, hmax, i10fg, shts, mwp, hgt, 10fg, mpww, mpts, shww, m0wp, atmp and swh1. All but one (Lswh) of the 28 variables that are kept have a VIF below the cut-off value of 5 that is adopted, and all have a value below the cut-off of 10, the less conservative and widely used value considered in the literature. Lswh has a VIF of 6.233, while the variable with the next largest VIF is wdw with 4.659.

To further reduce the feature space, now with 28 features, five executions of the SFS algorithm are performed with two fixed features, Lswh and wspd, and with different settings regarding the number of features to select. The SFS is performed using the SequentialFeatureSelection function, part of the mlxtend Python library (Raschka 2018). One execution of the algorithm fixes the number of features between 3 and 5, two executions fix the number between 3 and 6, and in the last two executions the string arguments 'best' and 'parsimonious' are passed to the function, the former corresponding to the computation of a feature subset with the best cross-validation performance, and the latter to smallest feature subset that is within one standard error of the cross-validation performance. Of the five executions, the ones that fix the number of features between 3 and 5 and one of those that fix the number between 3 and 6 yield the smallest feature space, with the same 5 features, namely mwp2, u850, umwd2, wspd, and Lswh. These are the features that are selected for the training of the models. The remaining executions yield feature spaces with between 6 and 17 features.

PCA does not further reduce the size of the feature space in the executions that yield 5 features, so the models are not trained with the data transformed by this technique.

3.2 *Long short-term memory neural network*

A typical LSTM unit comprises a cell, an output gate, an input gate, and a forget gate (Hochreiter & Schmidhuber 1997). The cell is capable of remembering values over arbitrary time intervals, while the input, output and forget gates intricately regulate the flow of information into and out of the cell. The structure of an LSTM cell at time step t is illustrated in Figure 2.

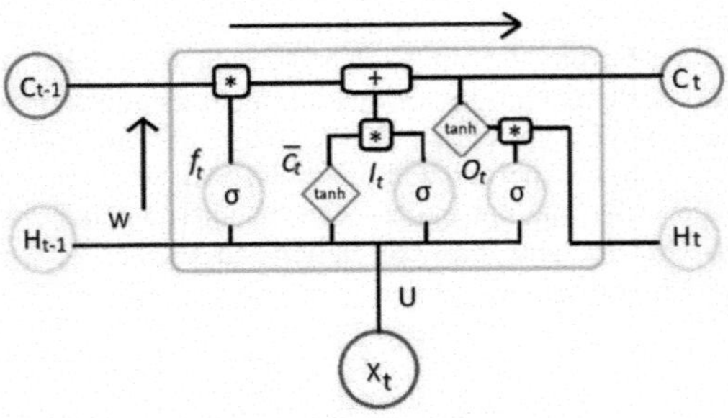

Figure 2. Network architecture of an LSTM model. The symbols + and * represent element-wise addition and multiplication, respectively.

The components of Figure 2 include the forget gate f, tasked with removing unnecessary information from the cell state, the candidate layer $\overline{C}$, which holds a repository of all conceivable candidate values that could potentially be incorporated into the cell state, the input gate I, which is responsible for introducing new information into the cell state, the output gate O, playing a role in determining the subsequent hidden state, the hidden state H, which encapsulates information pertaining to immediate previous events, and serving as a form of short-term memory, and finally the cell state C, which functions as the long-term memory repository of the model. The weight vectors (W and U) associated with each gate contribute to the regulation of information flow within the LSTM architecture. This distinctive set of components and their respective functions collectively form the intricate structure of an LSTM model, allowing for effective management of short-term and long-term dependencies in sequential data processing. The network equations are:

$$f_t = \sigma(X_t * U_f + H_{t-1} * W_f + b_f) \quad (9)$$

$$\overline{C_t} = tanh(X_t * U_c + H_{t-1} * W_c + b_c) \quad (10)$$

$$I_t = \sigma(X_t * U_i + H_{t-1} * W_i + b_i) \quad (11)$$

$$O_t = \sigma(X_t * U_o + H_{t-1} * W_o + b_o) \quad (12)$$

$$C_t = f_t * C_{t-1} + I_t + \overline{C_t} \quad (13)$$

$$H_t = O_t * tanh(C_t) \quad (14)$$

The LSTM models prove highly effective in making predictions based on time series data, particularly when there are no lags between significant events. Maintaining the temporal sequence of events is crucial for the network to discern patterns accurately, and to do so data shuffling must be avoided. Each layer's input must adhere to a three-dimensional structure, encompassing the number of samples, time steps, and features. Consequently, the dataset must undergo preparation and transformation before being fed into the network. Let X be an input space with j features and y the output space for the prediction of a single variable. If k represents the desired number of time steps, then $(X_1(t-1),\ldots,X_j(t-1),\ldots,X_1(t-k),\ldots,X_j(t\text{–}k))$ is an input sample, while the target corresponds to $y(t+p)$, where p indicates the number of time steps for the prediction.

Table 1. Hyperparameters used in the LSTM model implemented in Python's Keras library.

Hyperparameter	Range
Number of neurons	10, 50, 90, 130, 170
Window size	5, 10, 15, 20, 25
Batch size	32 (default)
Dropout	0.0, 0.2, 0.4
Learning rate	0.001 (default)
Optimizer	Adam
Epochs	100
Activation function	tanh (default)

The primary focus of the study revolves around identifying the optimal number of neurons for fixed sets of filtering window sizes. An extensive testing phase involves varying the number of neurons from 10 to 170 with increments of 40, coupled with filtering windows ranging from 5 to 25 in increments of 5, for a time step of 10 samples, which corresponds to the number of past events that the network should be learned from. The model is executed five times with random seeds to evaluate its precision. In total, the model is run 125 times for each increment of neurons, window size and random seeds. Additionally, experiments are made by setting the dropout to 0, 0.2, and 0.4. This aleads to the conclusion that the default dropout value of 0 is the most suitable for the dataset, as it facilitates a robust generalization of the model.

Model runs are conducted with five distinct features: Lswh, representing the logarithm of significant wave height, wspd, denoting wind speed, u850, indicating the U-component of wind, mwp2, representing the mean wave period of the second swell partition, and umwd2, indicating the U-component of mean wave direction of second swell partition.

To assess the performance of the LSTM model, the number of neurons and window size are varied while maintaining a fixed epoch of 100 and a batch size of 32 (the default). Throughout this evaluation process, the Adam optimizer is used and the mean squared error (MSE) loss function is adopted. The choice of the MSE is particularly suitable for the present regression problem, aligning with the objective of accurately predicting significant wave height. This systematic approach allows the thorough analysis of the model's performance under different configurations, permitting insights into the optimal setup for the predictive tasks. The years 2014–2015 and 2016 are used to train and test the model, respectively. The hyperparameters used in this work are presented in Table 1. The models are implemented in Python's Keras library.

4 RESULTS AND DISCUSSION

The findings pertaining to the search for the optimal filtering window sizes and number of neurons are detailed in Tables 2 to 5, which display the error metrics for the various combinations of number of neurons and window size.

Table 2. Nbias for the number of neurons and window size.

	Window size				
Neurons	5	10	15	20	25
10	-0.023	-0.037	-0.007	-0.021	-0.011
50	-0.008	0.000	-0.001	-0.000	-0.037
90	-0.043	-0.038	-0.021	-0.025	-0.006
130	-0.054	-0.014	-0.003	0.007	-0.013
170	-0.011	-0.025	-0.017	-0.004	-0.042

Table 3. SI for the number of neurons and window size.

	Window size				
Neurons	5	10	15	20	25
10	0.011	0.012	0.012	0.014	0.017
50	0.014	0.014	0.014	0.016	0.017
90	0.012	0.012	0.014	0.015	0.019
130	0.012	0.014	0.015	0.016	0.018
170	0.014	0.013	0.015	0.017	0.018

The results in Tables 2 to 5 are the mean of the results of the LSTM model for 5 different random seeds. It can be observed that with 10 neurons the

Table 4. NRMSE for the number of neurons and window size.

	Window size				
Neurons	5	10	15	20	25
10	0.107	0.114	0.112	0.120	0.133
50	0.126	0.120	0.118	0.129	0.138
90	0.117	0.117	0.121	0.126	0.140
130	0.125	0.123	0.126	0.133	0.151
170	0.119	0.121	0.128	0.131	0.141

Table 5. CC for the number of neurons and window size.

	Window size				
Neurons	5	10	15	20	25
10	0.976	0.974	0.973	0.968	0.961
50	0.971	0.972	0.971	0.966	0.960
90	0.971	0.971	0.970	0.967	0.959
130	0.971	0.970	0.969	0.966	0.959
170	0.970	0.969	0.968	0.965	0.957

optimal filtering window size is 5, as evidenced by the significant improvements in the corresponding values of the SI, NRMSE, and CC error metrics. On the other hand, the value of the NBias for that combination of neurons and window size is the second highest, rendering it less effective for the accurate prediction of swh using the LSTM model.

Similarly, when considering 50 neurons, higher values of the CC can be seen, exceeding 0.97 for filtering window sizes of 5, 10, and 15. Despite this, the lowest filtering window size of 5 displays substantial NBias variance (as shown in Figure 3), leading to the identification of the filtering windows of sizes 10 and 15 as optimal. After careful analysis of Table 2 to 5, it is ascertained that the LSTM model that shows the best performance for the dataset has 50 neurons and a filtering window size of 10.

Furthermore, in Table 2 it can be observed that with 10, 90, 130, and 170 neurons the NBias values are all worse than the corresponding value for 50 neurons. Additionally, a decrease in CC is observed not only with larger filtering window sizes but also with an increase in the number of neurons. Although the highest CC is observed for 10 neurons, implementing the LSTM model with such a low neuron count is impractical due to poorer results in other accuracy metrics.

The second highest CC is observed for 50 neurons, further supporting its choice as the best

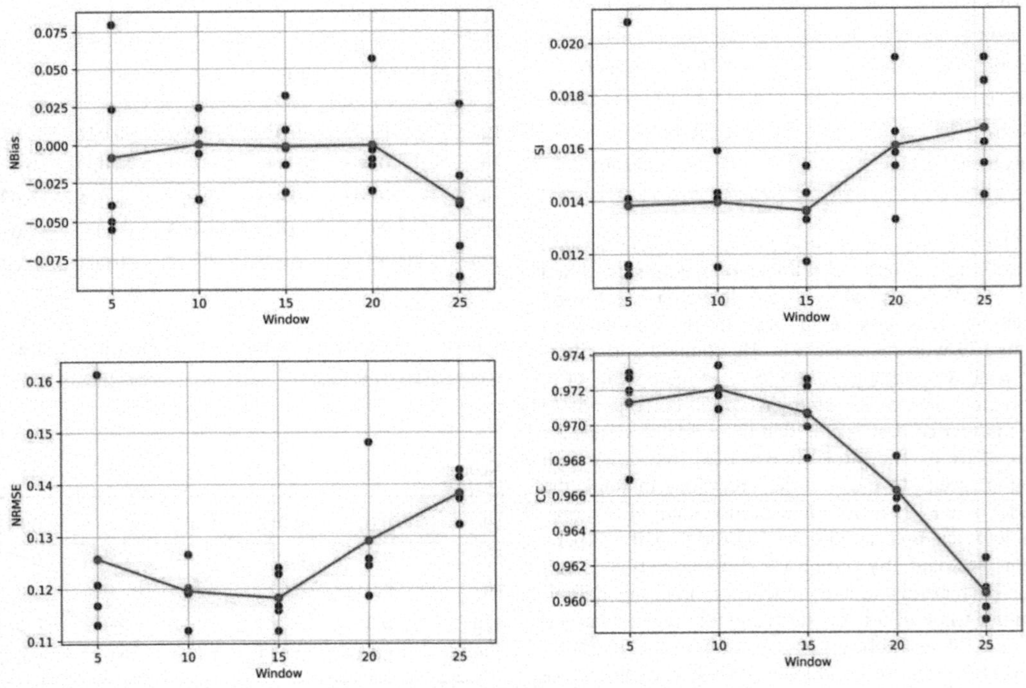

Figure 3. Results of the LSTM model with 50 neurons and different window sizes. The plots share the same x axis, which shows the filtering window sizes in a range from 5 to 25 with a step of 5. The plots exhibit the error metrics of variable swh. The blue points on a given window show the response of the LSTM models for different random seeds, while the red points represent their mean.

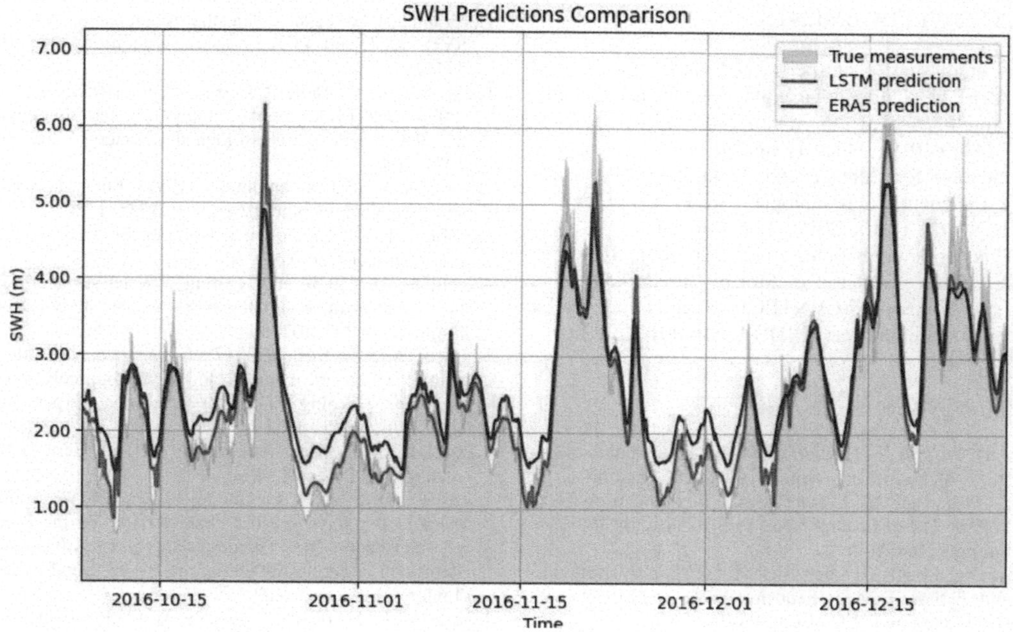

Figure 4. LSTM model estimates (in red), ERA5 estimates (in blue), and buoy measurements (in gray) from October to December 2016.

value. In addition, the NRMSE values highlight the advantages of utilizing 50 neurons and a filtering window size of 10 compared to the other configurations considered in this study.

Figure 4 shows the results of implementing an LSTM model with 50 neurons and a filtering window of 10 for the month of October, November and December 2016. These months are selected due to the higher peaks of swh in that period. The curves display an overlap of three time series including the LSTM model estimates, the ERA5 estimates, and the buoy measurements. Figure 4 shows the capability of the chosen LSTM model in predicting the residues of swh very accurately. Figure 5 is the split violin plot of the distribution of the ERA5 and LSTM model residues of the swh variable.

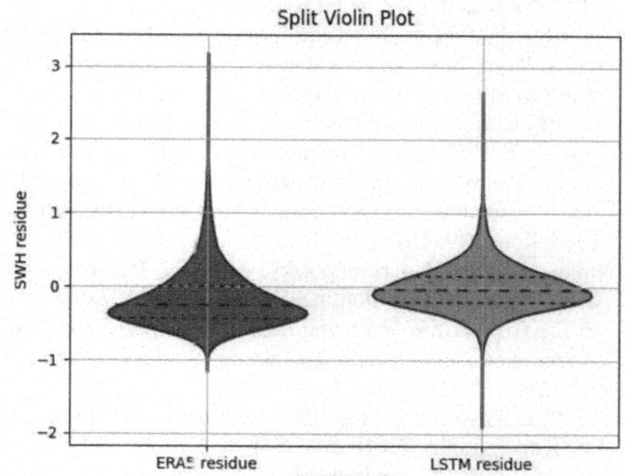

Figure 5. Distribution of the ERA5 and LSTM model residues.

5 CONCLUSIONS

This research extends the hybrid model proposed by Campos & Guedes Soares (2016), Campos et al. (2019), and Costa et al. (2023) to enhance the ERA5 predictions along the Portuguese continental coast. Leveraging ERA5 data as input features and an LSTM algorithm trained with buoy data, the aim of this work is to minimize the bias. The methodology adopted involves rigorous quality control of precise buoy measurements from the PdE. Four accuracy metrics are introduced to evaluate the model's performance.

This investigation focuses on optimizing the LSTM model's performance by varying the number of neurons and filtering window sizes. Extensive experiments are conducted with a wide range of parameters, and the model is executed 125 times for each increment of neurons, window size, and random seed. The results, detailed in Tables 2 to 5, illustrate the search for the optimal number of neurons and filtering window sizes. As optimal emerged the values of 50 neurons and filtering window size of 10, which exhibited higher CC values, exceeding 0.97. Overall, it is believed that this systematic approach offers valuable insights into optimizing LSTM models for accurate wave height predictions.

ACKNOWLEDGEMENTS

This study has been performed under the project "WAVEFAI - Operational Wave Forecast using Artificial Intelligence", (http://doi.org/10.54499/ CIRCNA/ OCT/0300/2019) which is funded by the Portuguese Foundation for Science and Technology (Fundação para a Ciência e a Tecnologia – FCT) under contract CIRCNA.OCT.0300.2019 _1801P.01023.

This work contributes to the Strategic Research Plan of the Center for Marine Technology and Ocean Engineering (CENTEC), which is financed by the FCT under contract UIDB/UIDP/00134/2020.

REFERENCES

Aha, D. W. & R. L. Bankert (1996). A comparative evaluation of sequential feature selection algorithms. In D. Fisher and H. J. Lenz (Eds.), Learning from Data, Volume 112 of Lecture Notes in Statistics, pp. 199–206. Springer New York.

Brownlee, J. (2017). Long Short-term Memory Networks with Python: Develop Sequence Prediction Models with Deep Learning (v1.0 ed.). Jason Brownlee.

Callens, A., D. Morichon, S. Abadie, M. Delpey, et al. (2020). Using Random forest and Gradient boosting trees to improve wave forecast at a specific location. Applied Ocean Research 104, 102339.

Campos, R. & C. Guedes Soares (2016). Comparison of HIPOCAS and ERA wind and wave reanalyses in the North Atlantic Ocean. Ocean Engineering 112, 320–334

Campos, R. M. & C. Guedes Soares (2017). Assessment of three wind reanalyses in the North Atlantic Ocean. Journal of Operational Oceanography 10(1), 30–44.

Campos, R. M., V. Krasnopolsky, J.-H. Alves, & S. G. Penny (2020). Improving NCEP's global-scale wave ensemble averages using neural networks. Ocean Modelling 149, 101617.

Campos, R. M., V. Krasnopolsky, J.-H. G. M. Alves, & S. G. Penny (2019). Nonlinear wave ensemble averaging in the Gulf of Mexico using neural networks. Journal of Atmospheric and Oceanic Technology 36(1), 113–127.

Cavaleri, L., J.-H. Alves, F. Ardhuin, A. Babanin, et al. (2007). Wave modelling—The state of the art. Progress in Oceanography 75(4), 603–674.

Costa, M. O., R. M. Campos, & C. Guedes Soares (2023). Enhancing the accuracy of metocean hindcasts with machine learning models. Ocean Engineering 287, 115724.

Fan, S., N. Xiao, & S. Dong (2020). A novel model to predict significant wave height based on long short-term memory network. Ocean Engineering 205, 107298.

Feng, Z., P. Hu, S. Li, & D. Mo (2022). Prediction of significant wave height in offshore China based on the machine learning method. Journal of Marine Science and Engineering 10(6), 836.

Ferri, F., P. Pudil, M. Hatef, & J. Kittler (1994). Comparative study of techniques for large-scale feature selection. In E. S. Gelsema and L. S. Kanal (Eds.), Pattern Recognition in Practice IV, Volume 16 of Machine Intelligence and Pattern Recognition, pp. 403–413. Elsevier.

Gerbrands, J. J. (1981). On the relationships between SVD, KLT and PCA. Pattern Recognition 14(1–6), 375–381.

Hersbach, H., B. Bell, P. Berrisford, S. Hirahara, et al. (2020). The ERA5 global reanalysis. Quarterly Journal of the Royal Meteorological Society 146(730), 1999–2049.

Hochreiter, S. & J. Schmidhuber (1997). Long short-term memory. Neural Computation 9(8), 1735–1780.

Jolliffe, I. T. (2002). Principal Component Analysis (2nd ed.). Springer New York, NY.

Kagemoto, H. (2020). Forecasting a water-surface wave train with artificial intelligence: A case study. Ocean Engineering 207, 107380.

Katrutsa, A. & V. Strijov (2017). Comprehensive study of feature selection methods to solve multicollinearity problem according to evaluation criteria. Expert Systems with Applications 76, 1–11.

Lazzeri, F. (2020). Machine Learning for Time Series Forecasting with Python®. Wiley.

Lucas, C., D. Silva, & C. Guedes Soares (2019). Assessment of the wave spectral characteristics in the Portuguese test zone. In C. Guedes Soares (Ed.), Advances in Renewable Energies Offshore, pp. 59–66. London, UK: Taylor & Francis Group.

Mentaschi, L., G. Besio, F. Cassola, & A. Mazzino (2013). Problems in rmse-based wave model validations. Ocean Modelling 72, 53–58.

Pedregosa, F., G. Varoquaux, A. Gramfort, V. Michel, et al. (2011). Scikit-learn: Machine learning in Python. Journal of Machine Learning Research 12 (Oct), 2825–2830.

Rao, S. & S. Mandal (2005). Hindcasting of storm waves using neural networks. Ocean Engineering 32(5–6), 667–684.

Raschka, S. (2018). MLxtend: Providing machine learning and data science utilities and extensions to Python's scientific computing stack. The Journal of Open Source Software 3(24).

Raschka, S. & V. Mirjalili (2019). Python Machine Learning: Machine Learning and Deep Learning with Python, Scikit-learn, and TensorFlow 2 (3rd ed.). Packt Publishing.

Sarang, P. (2021). Artificial Neural Networks with TensorFlow 2: ANN Architecture Machine Learning Projects. Berkeley, CA: Apress.

Silva, D. & C. Guedes Soares (2021). Validation with satellite data of SWAN model for wave conditions at the Madeira archipelago. In Maritime Technology and Engineering 5, Guedes Soares, C. & Santos T. A., (Eds.). London, UK: Taylor and Francis; Vol 2, 665–672.

Thompson, C. G., R. S. Kim, A. M. Aloe, & B. J. Becker (2017). Extracting the variance inflation factor and other multicollinearity diagnostics from typical regression results. Basic and Applied Social Psychology 39(2), 81–90.

Zhou, S., B. J.Bethel, W.Sun, Y.Zhao, W.Xie, & C.Dong (2021). Improving significant wave height forecasts using a joint empirical mode decomposition–long short-term memory network. Journal of Marine Science and Engineering9(7), 744.

Advances in Maritime Technology and Engineering – Guedes Soares & Santos (Eds)
© 2024 The Author(s), ISBN 978-1-032-83107-7

Overview of the wave energy resources in European regional seas

S. Ramos-Marin & C. Guedes Soares
Centre for Marine Technology and Ocean Engineering (CENTEC), Instituto Superior Técnico, Universidade de Lisboa, Lisbon, Portugal

ABSTRACT: An overview is provided of the different approaches used for the wave energy characterization, and of the existing wave energy resource levels along the European continental coasts. Studies are classified by basins, regional seas, and countries. Most energetic areas all over Europe are highlighted and discussed. Main outcomes indicate that a complete resource assessment should address both temporal and spatial variables and must retrieve different metrics to evaluate the resource at the different levels (theoretical, practicable, technical, and socio-economic) with sufficient resolution. The Scottish, Irish and Iberian coasts have been the regions where the highest theoretical wave power potential is found (~30 and 60 kW/m). Remarkable potential has also been found in island regions of Iceland and Macaronesia, whose average wave power potential (between 25 and 59 kW/m) could be a key contribution for the islands to reduce their fossil fuels dependency and move towards the self-sustainability.

1 INTRODUCTION

The wave energy resource is characterized by one of the highest energy densities among other marine renewable energy sources, as it has small energy losses even for long distances of propagation and is more persistent compared to offshore wind energy.

The first wave energy resource characterization at the European level, which used a standardized approach and similar attributes in the wave dataset, was implemented in the late 1990s and was named after "the Wave Energy Resource Atlas", or "WERATLAS" project (Pontes *et al.* 1996). Derived from this characterization, the computed European offshore resource stood at 320 GW (comprising 290 GW along the Atlantic coast and 30 GW in the Mediterranean). Since then, Europe has been the continent leading the research path of wave energy exploration.

To enhance uniformity in atmospheric forcing and refine the spatial and temporal resolution of the initial assessments, the HIPOCAS project (Guedes Soares, 2008) has generated hindcast data for various European seas. Within this initiative, a database encompassing climatological parameters and statistics was formulated by various authors and entities. They conducted wave resource evaluations utilizing the WAM model over a 44-year time span using finely resolved nested grids (with spatial resolution ranging from 2° offshore to 0.05° in coastal regions and temporal resolutions between 1 and 3 hours) across various European seas: the Southern North Sea (Weisse *et al.* 2002, Weisse and Günther 2007); a limited region in the German Bight (Gaslikova & Weisse 2006); the Mediterranean basin (Sotillo *et al.* 2005, Ratsimandresy *et al.* 2008); the North East Atlantic, encompassing the Azores and Canary Islands (Pilar *et al.* 2008, Gonçalves et al. 2020); the Western Iberian Coast (Rusu *et al.* 2008, Silva et al. 2015, 2018); the Black Sea (Cherneva *et al.* 2008); the Baltic Sea (Cieślikiewicz & Paplińska-Swerpel 2008); and the Irish Sea (Vijaykumar *et al.* 2004).

The impact of tidal currents on wave power was assessed by various authors (Saruwatari *et al.* 2013, Hashemi & Neill 2014, Guillou 2015, 2017, Draycott *et al.* 2019). Regions characterized by lower wave energy and shorter wave periods were found to be more sensitive to tides. Its mean impact was found almost negligible over most of the Northwest European shelf, but highly significant in some regions, such as Orkney, the west coast of Ireland, the west coast of Scotland, Pembrokeshire coast (UK), the northwest coast of France, and the north coast of Spain. In these situations, wave predictions need to account for changing tides (Silva et al. 2016).

The wave resource over some of those regions within the Western coast of the European Atlantic were assessed by Guedes Soares *et al.* (2014) and by Gleizon *et al.* (2017). The latter run and compiled different wave hindcast over a 7-year period using WW3 and SWAN models, with spatial resolutions ranging between 0.001 and 0.5° and temporal resolutions between 1 and 6 h. The result was an atlas of fine-resolution maps of wave statistics in sites of potential interest, over five distinct coastal zones: Scotland (UK), Ireland, France, Galicia (Spain), and

DOI: 10.1201/9781003508779-72

Portugal. The resulting maps (Figure 1) showed that "in coastal areas the wave power can vary significantly over short distances, especially in the presence of irregular coastlines such as in Galicia, Bretagne, and Ireland, or in the vicinity of islands and archipelagos such as those in Scotland, Cotentin, or Charente" (Gleizon *et al.* 2017). Consequently, the energy resource around complex coastlines require the use of fine-resolution grids, and operating over unstructured meshes is highly recommended.

From these outcomes, multiple sites in Europe emerged as compelling prospects for wave energy exploration showing relevant values of power density, with a pronounced emphasis on the West European Atlantic coast.

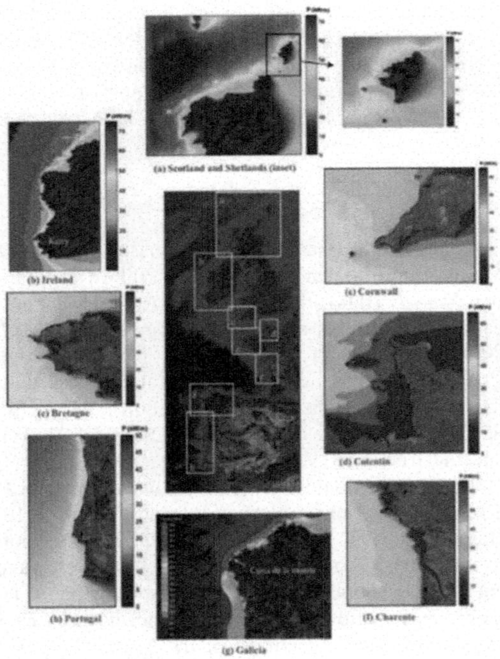

Figure 1. Mean wave power density along the European Atlantic Coast estimated by (Gleizon *et al.* 2017).

The most alluring locations identified included the coasts of Ireland, the UK, France, Galicia (Spain), and Portugal. Further studies aimed at characterizing the wave energy resource at smaller extents, either at sea, regional or local level, seeking a finer temporal and spatial resolution over special regions of interest, with a focus on spotting the most suitable areas for the deployment of wave energy converters.

As ocean wave power exhibits strong variability at time and spatial scales, one of the first steps towards wave energy exploitation consists of the assessment and mapping of the wave resource, able to capture the sea state conditions, the wave energy characteristics, and its variations in the local of interest.

In this paper, the main methods for wave energy characterization are reviewed, as well as the existing wave energy resource assessments historically performed over European open waters. The studies are classified by basins, regional seas, and countries. The most energetic areas all over Europe are highlighted and discussed.

The paper is organized as follows: Section 2 contextualizes the different approaches used for wave energy characterization. Section 3 overviews the common wave energy estimators and metrics. Section 4 presents a review on wave energy resource studies in the European regional seas and countries. Section 5 concludes with the main findings of the review.

2 METHODS FOR WAVE ENERGY CHARACTERIZATION

Historically, various data sources and methods have been utilized to describe the wave energy resource: observational measurements, numerical models, and predictive analytical models.

2.1 *Observations & measurements*

2.1.1 *In-situ*

The earliest available wave ocean data came from visual observations made by moving ships. These have been stored by meteorological organizations from the mid-1850s onwards (Gulev, 2003; Vettor & Guedes Soares, 2021). Nowadays, in-situ measurements are obtained by several devices. At the shore and near-shore area, buoys, wave staffs, echo sounders, pressure, and acoustic probes are often used. In the open sea, wave-measuring buoys are most widely used to record a time series of instantaneous buoy displacement, which allows the characterization of the sea surface elevation. Through the time series analysis of heave records, some integrated wave parameters can be directly calculated, such as H_s (significant wave height), T_z (zero-crossing period), T_s (significant wave period) or T_e (energy period). If spectral analysis is performed, data can be retrieved about the mean spectral frequency (f_m, w_m) and mean spectral period (T_m), the spectral peak frequency (f_p, w_p) and period (T_p), and the wave energy density over frequencies (m_0).

If the slope and the azimuth are also measured, directional parameters can also be obtained, such as the directional wave spectra over frequencies and directions $S(f, \theta)$, mean wave direction (θ_m), and mean direction of spreading (σ_θ) (Smith *et al.* 2013). The ideal information to fully characterize the wave conditions in a location should be given by the average spectra (Lucas et al. 2023), but in most cases this information is not readily available.

As in-situ data is generally highly accurate but scarce, it is most often used to validate extensive numerical wave models, assess uncertainties, and introduce calibration coefficients (Defne *et al.* 2009, Guillou 2020). Detailed in-situ wave statistics from different countries have been made freely available on the Internet by several organizations throughout the last decade (Table 1).

2.1.2 *Remote sensing*

Remote sensing techniques for acquiring ocean data involve the use of laser or radar systems either from land, aircraft, vessels, or satellites. Land-based radars, which can retrieve directional wave spectra S (*f*,*θ*) and currents (*U*,*θ*), are useful to observe incident waves on shoreline wave energy devices.

Table 1. Example of organizations providing wave in-situ observation data.

Institution*	Region	Device	Reference
NOAA-NDBC	Global	Buoys	www.ndbc.noaa.gov
AODN	Australia	Buoys, Radar	portal.aodn.org.au
BODC	UK	Buoys, Ships, Radar, Pressure and Motion Sensors	www.bodc.ac.uk
IH	Portugal	Buoys	www.hidrografico.pt
PE	Spain, Portugal, UK, Ireland	Buoys, Boats	www.puertos.es
RON	Italy	Buoys	eurogoos.eu/member-product/italian-national-data-buoy-network-ron-2
CANDHIS	France	Buoys	candhis.cerema.fr

* NOAA-NDBC: National Oceanic and Atmospheric Administration - National Data Buoy Center; AODN: Australian Ocean Data Network; BODC: British Oceanographic Data Center; IH: Instituto Hidrografico; PE: Puertos del Estado; RON: Rete Ondametrica Nazionale; CANDHIS: Centre d'Archivage National des Données de Houle In Situ.

Some ground radar data sources are those given by MIROS Microwave radars, WAMOS II radars or HF radars. However, the most widespread devices are radar instruments set up aboard satellites.

Two technologies of satellite radars currently provide ocean wave data: altimeter and SAR (Synthetic-aperture radar). Altimeter data provides good estimates of H_s and wind speed at 10 m height by quantifying the time spent by a radar pulse to travel from the satellite to the sea surface and return, although to compute an approximate T_z, predictive analytical models are needed. SAR satellites exploit a large bandwidth of the electromagnetic spectrum and estimate directional wave spectra.

Although less accurate than in-situ measurements, the exploitation of satellite data enables the long-term assessment of the sea state and the mapping of the wave power density where site measurements are not available. Nonetheless, their observations show important spatial and temporal limitations due to their widely spared ground tracks and orbital periods of around 10–35 days. Often, data from different satellite tracking needs to be merged to obtain as much continuous data coverage as possible.

Satellite data show remarkable uncertainties in the vicinity of the coastline, and its use is most recommended for the evaluation of sea state variability rather than extreme values. A detailed quality assessment and validation of the wave parameters measured by different satellites against buoy data in different regions can be found in (Mackay *et al.* 2008, Pontes *et al.* 2009, Campos & Guedes Soares 2016, Yaakob *et al.* 2016, Ribal & Young 2019, Kang *et al.* 2021).

Table 2. Satellite spacecraft with incorporated wave observation instruments (SAR).

Institution	Satellite	Time Period	Sensor Type
JAXA	JERS-1	1992 - 1998	SAR
	ALOS-1	2006 - 2011	SAR
	ALOS-2	2014 - Ongoing	SAR
CSA	RADARSAT-1	1995 - 2013	SAR
	RADARSAT-2	2007 - Ongoing	SAR
DLR	TerraSAR-X	2007 - Ongoing	SAR
	TanDEM-X	2010 - Ongoing	SAR
ISRO	RISAT-1, NISAR	2012 - Ongoing	SAR
CNAE	SAOCOM	2018 - Ongoing	SAR
ASI	COSMO-Skymed	2010 - Ongoing	SAR
INTA	PAZ	2018 - Ongoing	SAR
KARI	KOMPSat-5	2013 - Ongoing	SAR

Satellite data acquisition started in 1985 with the launch of Geostat. From then on, a series of other satellites given in Tables 2 and 3 have given access to almost continuous wave and wind measurements over decades.

2.2 *Numerical simulation models*

Wind-wave numerical models are widely used in areas where no in-situ or remote observations are available. They can simulate the growth, decay, and propagation of ocean waves by solving the wave energy balance equation, forced by environmental boundary condition inputs, such as wind fields produced by atmospheric models, bathymetric information and ice cover data. The energy balance equation reveals the evolution of the wave spectrum, which is determined by the rate of change of the sea state and by a physical source function, which represents the sum of associated physical processes such as wind input, dissipation by wave breaking, ice covering or nonlinear interactions.

Table 3. Satellite spacecraft with incorporated wave observation instruments (altimeter/SAR).

Institution	Satellite	Time Period	Sensor Type
NASA	GEOSAT	1985 - 1990	Altimeter
	GFO	1998 - 2008	Altimeter
ESA	SENTINEL-3A	2016 - Ongoing	Altimeter
	ENVISAT	2002 - 2012	Altimeter, SAR
	ERS-1	1991 - 2000	Altimeter, SAR
	ERS-2	1995 - 2011	Altimeter, SAR
	CRYOSAT-2	2010 - Ongoing	Altimeter
NASA, CNES	TOPEX/ Poseidon	1992 - 2005	Altimeter
	JASON-1	2001 - 2013	Altimeter
	JASON-2	2008 - 2019	Altimeter
	JASON-3	2016 - Ongoing	Altimeter
CNSA	HY-2A	2011 - 2020	Altimeter
CNES, ISRO	SARAL	2013 - Ongoing	Altimeter

Commonly, numerical simulations produce information on the directional spectra at every point of the grid, which describe the distribution of energy density in both the frequency and direction domains $S(f,\theta)$. From this spectrum, few parameters can be integrated, such as m_n, H_s, T_e, T_p, T_m, θ_m and E_w. Due to storage limitations, some models only retain those integrated parameters. As T_p is often easier to find in simulated databases than Te, a calibration coefficient (α) can be introduced to determine T_e from the available T_p, as $T_e = \alpha T_p$ (Guillou 2020).

Numerical models have increased their sophistication over the years in the way their equations represent the physics of wave dynamics. These improvements have consolidated three generations of resource assessment, whose formulation principles are explained in detail in Janssen (2008).

Most widely used today are third-generation models. The first third-generation model is the Wave Model (WAM), created in 1984, and it has been in routine operation by the ECMWF up to date (WAMDI Group 1988). Other third generation models differ in some aspects of their formulation, such as: the numerical technics used to solve the wave energy evolution equation, the number of source terms involved, the nature and parameterizations of the physical processes, or their ability to generate spectral results for shallow or deep-water locations.

Another established model is Wave Watch III (Tolman *et al.* 2014), which is in operation by NOAA and utilizes alternative equations for the physics of wind input and dissipation. The SWAN model (Holthuijsen *et al.* 2006), focuses on the physics of waves in the nearshore and shallow-water area by modelling bottom friction, depth and current-induced refraction, shoaling and triads interactions. MIKE21-SW1 (DHI, n.d.), has the ability of considering wind generation, non-linear and sediment transport interactions and TOMAWAC (Opentelemac, n.d.) incorporates shallow water mechanics and wind-generated waves. A review of the different wave models applied to resource assessment and their latest developments has been given by Lavidas & Venugopal (2018a).

2.2.1 *Forecast, nowcast and hindcast*

Wind-wave numerical models can be run following three different methods. When run forward in time using predicted wind fields as input, the resulting information is known as a *forecast*. When run using improved forecast fields with the assimilation of measured data, the output is known as *nowcast*, and it represents the best possible estimate of the situation at a given past time. *Hindcast* databases arise when models are run to estimate the wave conditions that have occurred for a given set of past weather conditions. Finally, *reanalysis* are a blend of observations with past short-range weather forecasts rerun with modern weather forecasting models. They provide the most complete picture currently possible of past weather and climate. When compared against reliable in situ and satellite measurements, reanalyses serve as a rich repository of information as input to force finer resolution regional or local hindcasts for wave energy resource characterization. Several entities, presented in Table 4, have been operating global climate simulations for years and still provide some of the most exploited databases of reanalysis products (Reistad *et al.* 2011, Guillou *et al.* 2020).

2.2.2 *Model specifications*

Relevant guidelines were established by the Marine Renewable Energy Technical Committee of the International Electrotechnical Commission (IEC 2015) to develop appropriate, accurate and homogeneous wave energy resource assessments.

The IEC-62600-101 categorizes resource assessment studies into three classes: reconnaissance, feasibility, and design, as per Table 5. Moreover, all wave energy resource assessments must cover a minimum period of 10 years with results available at a minimum temporal resolution of 3 hours.

Nonetheless, the proposed specifications have been yet applied to a limited number of resource assessments due to the required complexity, sophistication, and computational cost of running refined numerical simulations and validating them against extended field measurements.

2.2.3 *Predictive analytical models*

Most often used for wave forecasting, predictive models are statistical techniques that use historical data to estimate and predict future outcomes, sometimes also using machine learning and data mining. These models include Regression Techniques, Generalized

Table 4. Main global reanalysis products exploited for wave resource assessment.

Corporation	Product	Temporal coverage	Spatial Resolution	Temporal resolution	Reference
ECMWF	ERA-40	1957 - 2002	1.5° x 1.5°	6hr	(Caires *et al.* 2004, Uppala *et al.* 2005)
ECMWF	ERA-Interim	1979 - 2019	1° x 1°	6hr	(Dee *et al.* 2011)
ECMWF	ERA-5	1979 - Present	0.5° x 0.5°	1hr	(Hersbach *et al.* 2020)
NOAA - (NCEP/ NCAR)	NRA-40	1948 - Present	1.9°	6hr	(Kalnay *et al.* 1996, Saha *et al.* 2010)
IHCantabria	GOW	1948 - 2008	1.5° x 1°; 0.125° x 0.125° *	1hr	(Losada *et al.* 2010, Reguero *et al.* 2012)
JMA	JRA-25	1979 - 2004	1.25° x 1.25°	6hr	(ONOGI *et al.* 2007)
FNMOC	NCOM	2010 - 2013	0.125° x 0.125°	3hr	(NOAA 2021)
IFREMER	WAVERYS	1993 - 2019	0.2° x 0.2°	3 hr	(Law-Chune *et al.* 2021)

* 0.125 x 0.125 over the Mediterranean Sea

Table 5. IEC-62600-101 model setup recommendations (Ramos & Ringwood 2016, Guillou *et al.* 2020).

Category	Domain	Spatial extent	Min. Spatial Resolution	Min. Temporal Resolution	Min. Wave Frequencies	Min. Azimuthal Directions
Reconnaissance	Oceanic	≥ 300 km	5 km (~0.05°)	3 h	25	24
Feasibility	Nearshore	20- 500 km	500 m (~0.005°)	3 h	25	24
Design	Wave Farm	≤ 25 km	50 m (~0.0005°)	1 h	25	24

Linear Model, Random Forest, K-Nearest, K-Means, Neural Networks, Genetic Programming and Bayesian Networks.

Studies have found these analytical methods to work generally better for short time horizons of 3 -18 h, while numerical simulation models were found to be more accurate for longer horizons of 18–20 h and beyond (Serras *et al.* 2019). The combination of both techniques, statistical and physics-based models, are better predictors than when used individually (Serras *et al.* 2019; Costa *et al.*, 2023).

3 WAVE ESTIMATORS AND METRICS

The energy conversion process can be characterized at various levels, starting from the theoretically available resource to delivering power to the grid for the final user. This review categorizes resource characterization levels as theoretical, practicable, technical, and socio-economic.

The *theoretical potential* pertains to the raw available wave resource and its climatic conditions. It is often described through parameters such as significant wave height, peak and energy periods, mean direction, available power density, and long-term statistics. Assessing this level involves analyzing wave climate, extremes, and energy resource statistics.

Practicable potential considers the energy potential of areas where extraction is feasible and accessible, excluding constrained areas due to technical, environmental, or conflict-related reasons. Evaluation at this level requires marine spatial planning metrics and techniques.

Technical potential is linked to WEC performance and the energy transmission system to shore. Not all WEC technologies are equally suitable for every wave environment. The conversion of available power to output power depends on factors like power matrix, rated power, sea state, WEC dimensions, working principle, and PTO. Delivered Power represents the energy transferred from the WEC to the grid, accounting for losses during transport and integration, estimated ranging from 6 to 10%.

Socio-economic potential extends beyond pure energy potential, considering social, economic, and environmental criteria influencing wave farm returns. Uihlein & Magagna (2015) provide a comprehensive classification and review of criteria at the practicable and socio-economic levels for a thorough resource assessment for exploitation.

4 THE WAVE RESOURCE IN EUROPE

Through this paper, wave resource characterizations performed within the European region have been compiled and classified by basin, regional seas, and countries as per Figure 2. Regions, where test site areas exist for the support of real testing of different WEC devices and prototypes at open sea, have been generally the subject of more detailed wave resource characterizations and analysis. Those test sites have been compiled in Table 6 and are also displayed in Figure 2.

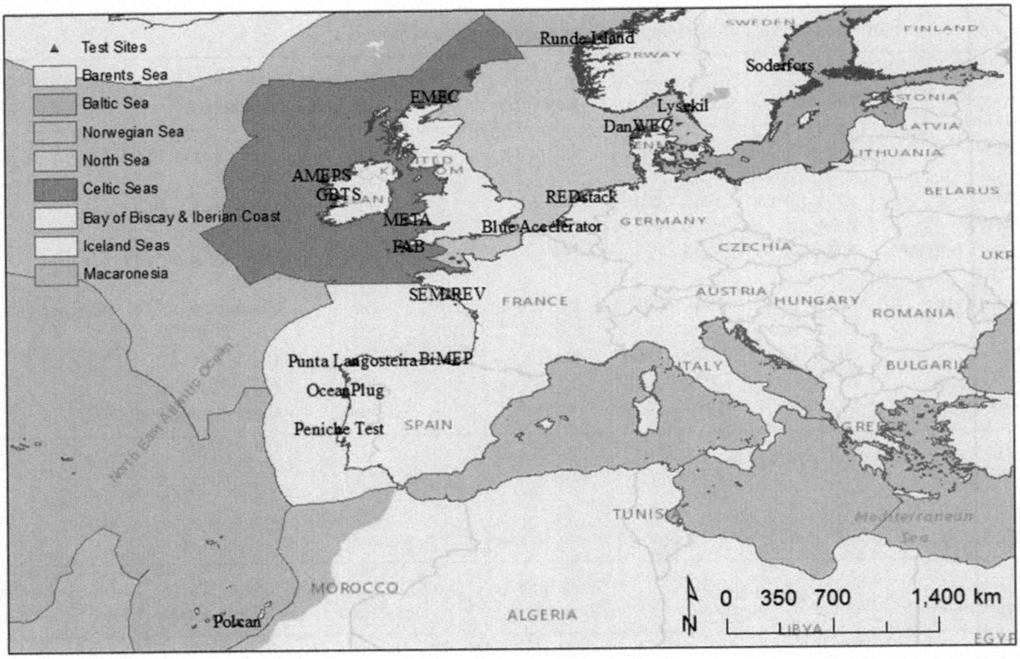

Figure 2. Wave Energy test sites existing in the NE Atlantic and the Baltic Sea.

4.1 *Celtic shelf*

Several wave energy characterization studies have been performed over the Celtic Shelf, which covers the west coast of the UK (ABP MER 2008, Carbon Trust 2012, Lavidas & Venugopal 2018b, 2018a), the full coast of Ireland (ESBI 2005, Gallagher *et al.* 2013, 2014, O'Connell *et al.* 2020) and the very north shore of France (Guillou 2015, 2017, Guillou & Chapalain 2015, Maisondieu 2015, Gonçalves et al, 2014b, 2018, Sergent *et al.* 2020).

In the UK, the potential extraction of wave energy from offshore locations was theoretically estimated to reach up to 95 TWh/yr, while the practically and economically extractable energy ranged between 32 and 42 TW/yr (Carbon Trust 2012). Research indicates that the southwest corner of Cornwall, along with the North and West Coasts of Scotland, stands out as the most appealing areas for offshore devices, specially the regions of The Hebrides, Orkney, and Pentland Firth.

Along the shores of The Hebrides and Shetlands, the spatial distribution of wave power was approximated to be between 40 and 45 kW/m, with peak values reaching up to 650 kW/m. For Orkney, the annual wave power density was assessed to be between 25 and 35 kW/m (Venugopal & Nemalidinne 2015). Various wave power assessments were also conducted, with special emphasis on the Wave Hub & FAB test sites (Smith *et al.* 2013, van Nieuwkoop *et al.* 2013), where diverse wave energy converters have already undergone testing. The west side of the Isles of Scilly exhibited an average wave power of about 30 kW/m, the northwest-facing Cornish coast ranged from approximately 10 to 25 kW/m, and the southeast-facing Cornish coast registered values in the range of 2 to 15 kW/m (Van Nieuwkoop *et al.* 2013).

In Ireland, an average theoretical power flux ranging from 50 to 60 kW/m was identified along the West Coast, situated within 25 km of the Mayo and Kerry coastlines, where the WestWave test site is situated (ESBI 2005). On a local scale, mean yearly power densities of 50 kW/m and 3 kW/m were approximated at the WestWave, Galway Bay, and Killar Point test sites, respectively (Atan *et al.* 2018). Northern Ireland showcases elevated energy values reaching up to 160 kW/m, particularly in the nearshore vicinity of Belmullet (Bento *et al.* 2011). These values experience a rapid decline when progressing eastward toward the Irish Sea, affected by the island's shadow effect, with the same parameter plummeting to as low as 10 kW/m. Despite wind waves serving as the primary contributor to the annual average wave power density in certain areas of this region (Arinaga & Cheung 2012), the wave climate off the North and West coasts of Ireland has been identified as one of the most advantageous settings for potential exploration of wave energy.

As part of France's territory, the Iroise Sea has been highlighted as among the most indicated regions for harvesting marine renewable energy in Europe, notwithstanding its substantial maritime traffic, fishing operations, and recreational pursuits. Research has indicated that the yearly mean wave power in that area has the potential to attain levels of up to 45 kW/m (Smith and Maisondieu 2014).

Table 6. Existing wave energy test sites in the NE Atlantic Ocean.

Test Site	Location	Country	Regional Sea	Start-End Date
WestWave	Killlard Point	Ireland	Celtic Sea	Planned Project
AMETS	Belmullet/Broadhaven	Ireland	Celtic Sea	Planned Project
GBTS	Galway Bay	Ireland	Celtic Sea	2006- Ongoing
WaveHub	Pembrokeshire	UK	Celtic Sea	2010-Ongoing
FAB	Falmouth Bay	UK	Celtic Sea	2011- Ongoing
META	Milford Haven Waterway in Pembrokeshire	UK	Celtic Sea	Ongoing
EMEC	PFOW	UK	Celtic Sea	2005- Ongoing
Farr Bay	PFOW	UK	Celtic Sea	Cancelled Project
SEM-REV	Le Crôisic	France	Bay of Biscay and Iberian Coastal Shelf	2009- Ongoing
BiMEP	Bilbao	Spain	Bay of Biscay and Iberian Coastal Shelf	2008-ongoing
Mutriku	Mutriku	Spain	Bay of Biscay and Iberian Coastal Shelf	2011- Ongoing
Punta Langosteira	Galicia	Spain	Bay of Biscay and Iberian Coastal Shelf	2018- Ongoing
PolCan	Gran Canaria	Spain	Macaronesia	2008- Ongoing
Pilot Zone (Ocean Plug)	Viana do Castelo	Portugal	Bay of Biscay and Iberian Coastal Shelf	Planned Project
Pilot Zone (Ocean Plug)	São Pedro de Moel	Portugal	Bay of Biscay and Iberian Coastal Shelf	Finished Project (moved to Viana do Castelo)
Pico OWC	Pico Island (Azores)	Portugal	Macaronesia	Finished Project (1999- 2018)
Aguçadoura	Póvoa do Vrazim	Portugal	Bay of Biscay and Iberian Coastal Shelf	2007- Ongoing
Peniche Test	Peniche	Portugal	Bay of Biscay and Iberian Coastal Shelf	2007- Ongoing
REDstack	Afsluitdijk	Netherlands	North Sea	Planned Project
Blue Accelerator	Port of Ostende	Belgium	North Sea	2020- Ongoing
DanWEC	Hanstholm	Denmark	North Sea	2009- Ongoing
DanWEC NB (Folkecenter)	Nissum Bredning	Denmark	North Sea	2000- Ongoing
REC	Runde Island	Norway	Norwegian Sea	2009- Ongoing
Lysekil	Lysekil	Sweeden	North Sea	2006- Ongoing
Söderfors	Dalälven	Sweeden	Baltic Sea	2013- Ongoing

4.2 *North, Norwegian and Barents Seas*

The North, Norwegian and Barents seas lie at the northern side of Europe covering coastlines from northern France up to Svalbard. The majority of wave resource information in these seas comes from coarse oceanic models applied to larger areas. These assessments are adequate for large domains but with few limitations to characterize higher resolution domains with shallow water effects and nearshore complexities (Lavidas & Venugopal 2018a). Consequently, authors aimed at increasing the accuracy of the wave resource models in this area (Christakos *et al.* 2020, Haakenstad *et al.* 2020).

The regional NORA10 and NORA10EI hindcasts (Reistad *et al.* 2011) were exploited by (Christakos *et al.* 2020), who found the highest coastal wave power density in the Norwegian Sea, as it is dominated by swell conditions instead of wind waves. The average wave power was estimated between 20 and 41 kW/m (Falnes 1993). Conversely, areas located in the North Sea and the Barents Sea presented a higher contribution of wind sea to the total wave energy flux than swell.

The North Sea, characterized by its shallow waters (average depth of 90 m), experiences minimal shading from the UK in most regions. The wave power potential was identified to surpass 15 kW/m in extremely close coastal zones, gradually diminishing as one moves southwards toward the English Channel, where the values were estimated at around 5 kW/m. Only the wave resource in the most exposed northern segment of the North Sea was found to be on par with that of the West European coastline. Nevertheless, the significance of a milder wave climate and the accessible characteristics of the North Sea offer advantageous attributes for the development and installation of WECs (Lavidas & Polinder 2019).

In the Danish Nort Sea, assessments were done over the coast of Hanstholm (Fernández-Chozas *et al.* 2013), a particularly interesting location as it hosts the grid-connected Danish Marine Test Site "Dan WEC". To date, the Wave Star, Dexawave and Exowave devices have been tested on this site.

4.3 *Macaronesia and Iceland Sea*

The wave energy potential is considerable across the European Atlantic Islands, including the Canary Islands, Madeira, Azores, and Iceland. Multiple researchers delved into the examination of wave energy potential in the Canary Islands (Chiri *et al.* 2013, Gonçalves *et al.* 2014a, 2020, Rusu 2014, Rodríguez *et al.* 2015), revealing a noteworthy spatial variability around the archipelago. The North and Northeast sides of the islands exhibit elevated average annual values of available wave power density (20 - 32 kW/m), in contrast to the Southern regions (4 - 13 kW/m) (Fernández Prieto *et al.* 2019).

Concerning the Portuguese territories, in-depth evaluations yielded annual average power densities of 30-60 kW/m in the Azores (Rusu & Guedes Soares 2012b, Gonçalves & Guedes Soares 2021) and an average of 14 kW/m in the Madeira Islands (Rusu & Guedes Soares 2012a, Silva & Guedes Soares, 2021). Assessments were also conducted to appraise the marine space availability for both nearshore and offshore wave energy exploration in the Azorean archipelago (Ramos *et al.* 2021). Investigations also focused on the techno-economic suitability of different marine areas around Madeira for the exploitation of wave energy (Ramos *et al.* 2020).

Moving further north, Iceland emerges as a particularly compelling locale for wave energy exploration, given its self-sufficiency and lack of interconnections with other countries. The southern regions of the island proved more attractive for establishing a WEC farm, with an average Pw over four decades reaching 59 kW/m—markedly higher compared to less than 10 kW/m in the northern areas (Penalba *et al.* 2020).

4.4 *Bay of Biscay and Iberian Coast*

In France, most investigations appraising the wave energy potential were conducted along the French West Coast, specifically in the Bay of Biscay (Boudière *et al.* 2013, Dufour *et al.* 2014, Gonçalves *et al.* 2014b, 2018, Maisondieu 2015, Sierra *et al.* 2017, Ulazia *et al.* 2017, Sergent *et al.* 2020). Findings indicate a substantial energy resource in the vicinity of Le Croisic, ranging between 25 and 30 kW/m (Mattarolo *et al.* 2009, Gonçalves *et al.* 2014b, 2018). The offshore waters of Le Croisic house the grid-connected SEM-REV test site, providing operational settings for demonstrations of WECs and wind turbines.

In Spain, the most energetically rich region along the coast is situated around the Galician coast, boasting an average wave power of approximately 35-40 kW/m in deep waters (Losada *et al.* 2010, Bento *et al.* 2012, 2018). The mean wave power density diminishes from west to east, with deep-water areas along the Cantabrian Coast receiving around 30 kW/m. The Basque Country presently hosts two wave energy generation test sites: BiMEP (located on the northern coast of Bilbao) and Mutriku (positioned more easterly between the cities of Bilbao and San Sebastian). BiMEP housed the initial grid-connected offshore wave energy converter in Spain, the point absorber MARMOK-A-5 (Thomaz *et al.* 2019), situated about 4 km offshore and supplying electrical energy to the grid from December 2016 to June 2019. Thomaz et al. (2019) present an Operations & Maintenance model calibrated with actual data from this wave energy device, estimating socio-economic indices like the Levelized Cost of Energy (LCOE) for various case studies. Conversely, Mutriku is a wave energy generating and testing facility with 14 Oscillating Water Column (OWC) devices located at a breakwater, operational since 2011. It stands as the only wave farm globally providing electricity to the grid approximately 74.4% of the time. Operational aspects, such as its mean capacity factor and seasonal variability, have been documented (Ibarra-Berastegi *et al.* 2018, Serras *et al.* 2019).

In Portugal, (Pontes *et al.* 2005) estimated a typical annual average wave power of 25 kW/m in 2005, with greater wave heights and power occurring at exposed sites. The observed trend indicates that wave energy is highest in the north and slightly decreases toward the south, a pattern reaffirmed in subsequent studies (Rusu & Guedes Soares 2009, Silva *et al.* 2013, 2015, 2018, Bento *et al.* 2014). Presently, a limited number of areas off the Portuguese coast are designated for supporting the testing of offshore wave and wind energy prototypes. The "Ocean Plug-Portuguese Pilot Zone" was initially established in São Pedro de Moel (Leiria) and later relocated to Viana do Castelo (northern coast), currently positioned in depths between 85 and 100m (Ocean Plug 2021). This project pioneered globally in implementing a maritime zone for installing pre-commercial and commercial phases of WECs. A small area about 900m off the coast of Peniche is presently hosting experimental tests of the WaveRoller WEC (WaveRoller 2021), and a collaborative OceanLab in Aguçadoura is in the establishment process to support future marine energy technology experiments.

4.5 *Enclosed Basins: Mediterranean and Black Sea*

Enclosed basins embody the least dynamic seas, where waves are frequently diminutive, short-crested, and heavily reliant on wind dynamics. The confined space and limited fetch lengths impede the evolution of waves into swells. Despite being characterized by subdued wave power density levels, prevailing research asserts the potential for successful exploitation of wave energy in the Mediterranean, Black, and Baltic Seas by appropriately downscaling advanced WEC technologies (Bozzi *et al.* 2018).

Evaluations in the Black Sea suggest time-average power densities ranging between 1.6 - 6 kW/m and Hs between 0.6 and 1 m (Akpınar *et al.* 2017, Rusu 2018). The Baltic Sea manifests an average annual Hs < 1.5m and Pw ≈ 0.7 -1.5 kW/m (Björkqvist *et al.* 2014, 2018). In the Mediterranean Sea, waves exhibit average Hs from 0.05-1.25 (Soukissian *et al.* 2017) and Pw ≈ 2-12 kW/m. While the region spanning from Sardinia to the Balearic Islands has been historically characterized as the most energetic in the Mediterranean (9.4 - 12 kW/m) (Besio *et al.* 2016, Soukissian *et al.* 2017, Bozzi *et al.* 2018), it also presents a relatively low efficiency for WECs due to a significant portion of the available energy being derived from extreme and infrequent events. In contrast, coastal zones in the Gulf of Lion, the Sicily channel, the Alboran Sea, the Libyan coast, Crete, and Cyprus exhibit optimal performance for scaled WECs (Bozzi *et al.* 2018)).

5 CONCLUSIONS

A theoretical background on different approaches used for the wave data acquisition and energy characterization has been provided, and existing wave energy resource assessments along the European continental coasts have been reviewed.

It was found that, for an appropriate wave energy characterization, a complete resource assessment should address both temporal and spatial variables and retrieve different metrics to evaluate the resource at the four levels of the wave resource (theoretical, practicable, technical, and socio-economic) at sufficient resolution.

Overall, the region receiving the greatest interest in wave resource characterization was the Bay of Biscay and the Iberian Coastal shelf, closely followed by the Celtic Seas. The Barents, Norwegian and Iceland Seas were the regions less covered. Within the Bay of Biscay, Spain has been the country whose wave resource has been assessed the most, while the UK has been the most studied country within the Celtic shelf.

Almost consensually, the Scottish, Irish and Iberian coasts have been the regions where the highest annual average power density was generally found (~30 and 60 kW/m). Remarkable potential has also been found in island regions of Iceland and Macaronesia, whose great mean wave power potential (between 25 and 59 kW/m) could be a key contribution for the islands to reduce their fossil fuels dependency and move towards the self-sustainability.

ACKNOWLEDGEMENTS

This work contributes to the Strategic Research Plan of the Centre for Marine Technology and Ocean Engineering (CENTEC), which is financed by the Portuguese Foundation for Science and Technology (Fundação para a Ciência e Tecnologia - FCT) under contract UIDB/UIDP/00134/2020. The first author is financed by FCT under the grant 2020.06618.BD.

REFERENCES

Abp Mer, 2008. Atlas of UK Marine Renewable Energy Resources [online]. Available from: http://www.renewables-atlas.info/ [Accessed 12 Nov 2021].

Akpınar, A., Bingölbali, B., & Van Vledder, G.P., 2017. Long-term analysis of wave power potential in the Black Sea, based on 31-year SWAN simulations. *Ocean Engineering*, 130 (December 2016), 482–497.

Arinaga, R.A. & Cheung, K.F., 2012. Atlas of global wave energy from 10 years of reanalysis and hindcast data. *Renewable Energy*, 39 (1), 49–64.

Atan, R., Goggins, J., & Nash, S., 2018. Galway Bay – The 1/4 scale wave energy test site? A detailed wave energy resource assessment and investigation of scaling factors. *Renewable Energy*, 119, 217–234.

Bento, A.R., Martinho, P., Campos, R., & Guedes Soares, C., 2011. Modelling wave energy resources in the Irish West Coast. *In*: *Proceedings of the International Conference on Offshore Mechanics and Arctic Engineering - OMAE*. 945–953.

Bento, A.R., Martinho, P., & Guedes Soares, C., 2012. Modelling wave energy for the north coast of Spain. *In*: *Maritime Engineering and Technology*. 563–570.

Bento, A.R., Martinho, P., & Guedes Soares, C., 2018. Wave energy assessment for Northern Spain from a 33-year hindcast. *Renewable Energy*, 127, 322–333.

Bento, A.R., Rusu, E., Martinho, P., & Guedes Soares, C., 2014. Assessment of the changes induced by a wave energy farm in the nearshore wave conditions. *Computers and Geosciences*, 71 (1), 50–61.

Besio, G., Mentaschi, L., & Mazzino, A., 2016. Wave energy resource assessment in the Mediterranean Sea on the basis of a 35-year hindcast. *Energy*, 94, 50–63.

Björkqvist, J.V., Lukas, I., Alari, V., van Vledder, G.P., Hulst, S., Pettersson, H., Behrens, A., & Männik, A., 2014. On the wave energy potential along the eastern Baltic Sea coast *Renewable Energy*, 71, 221–233.

Björkqvist, J.V., Lukas, I., Alari, V., van Vledder, G.P., Hulst, S., Pettersson, H., Behrens, A., & Männik, A., 2018. Comparing a 41-year model hindcast with decades of wave measurements from the Baltic Sea. *Ocean Engineering*, 152, 57–71.

Boudière, E., Maisondieu, C., Ardhuin, F., Accensi, M., Pineau-Guillou, L., & Lepesqueur, J., 2013. A suitable metocean hindcast database for the design of Marine energy converters. *International Journal of Marine Energy*, 3–4 (December).

Bozzi, S., Besio, G., & Passoni, G., 2018. Wave power technologies for the Mediterranean offshore: Scaling and performance analysis. *Coastal Engineering*, 136, 130–146.

Caires, S., Sterl, a, Komen, G., & Swail, V.R., 2004. The web-based KNMI/ERA-40 global wave climatology atlas. *Bulletin of the World Meteorological Organization*, 53 (1996), 142–146.

Campos, R. M. & Guedes Soares C. 2016. Comparison and Assessment of Three Wave Hindcasts in the North Atlantic Ocean. *Journal of Operational Oceanography*. 9(1):26–44.

Carbon Trust, 2012. *UK Wave Resource Study*. (Report No. B/C001/29603/003).
Cherneva, Z., Andreeva, N., Pilar, P., Valchev, N., Petrova, P., & Guedes Soares, C., 2008. Validation of the WAMC4 wave model for the Black Sea. *Coastal Engineering*, 55 (11), 881–893.
Chiri, H., Pacheco, M., & Rodríguez, G., 2013. Spatial variability of wave energy resources around the Canary Islands. *In*: *WIT Transactions on Ecology and the Environment*. 15–26.
Christakos, K., Varlas, G., Cheliotis, I., Spyrou, C., Aarnes, O.J., & Furevik, B.R., 2020. Characterization of wind-sea-and swell-induced wave energy along the Norwegian coast. *Atmosphere*, 11 (2), 1–20.
Cieślikiewicz, W. & Paplińska-Swerpel, B., 2008. A 44-year hindcast of wind wave fields over the Baltic Sea. *Coastal Engineering*, 55 (11), 894–905.
Costa, M. O.; Campos, R. M., & Guedes Soares, C. 2023. Enhancing the accuracy of metocean hindcasts with machine learning models. *Ocean Engineering*, 287: 115724.
Dee, D., Uppala, S.M., Simmons, A.J., Berrisford, P., et al. 2011. The ERA-Interim reanalysis: configuration and performance of the data assimilation system. *Quarterly Journal of the Royal Meteorological Society*, 137, 553–597.
Defne, Z., Haas, K.A., & Fritz, H.M., 2009. Wave power potential along the Atlantic coast of the southeastern USA. *Renewable Energy*, 34 (10), 2197–2205.
DHI, n.d. MIKE21 [online]. Available from: http://www.mikebydhi.com/products/mike-21.
Draycott, S., Sellar, B., Davey, T., Noble, D.R., Venugopal, V., & Ingram, D.M., 2019. Capture and Simulation of the Ocean Environment for Offshore Renewable Energy. *Renewable and Sustainable Energy Reviews*, 104 (September 2018), 15–29.
Dufour, G., Michard, B., Cosquer, E., & Fernagu, E., 2014. EMACOP project: Assessment of wave energy resource along France's coastlines. *In*: *Coasts, Marine Structures and Breakwaters 2013: From Sea to Shore - Meeting the Challenges of the Sea*. 306–315.
ESBI, 2005. Accessible Wave Energy Resource Atlas:Ireland:2005 (Report No. 4D404A -R2)
Falnes, J., 1993. Research and Development in Ocean-Wave Energy in Norway. *International Symposium on Ocean Energy Development*.
Fernández Prieto, L., Rodríguez Rodríguez, G., & Schallenberg Rodríguez, J., 2019. Wave energy to power a desalination plant in the north of Gran Canaria Island: Wave resource, socioeconomic and environmental assessment. *Journal of Environmental Management*, 231 546–551.
Fernández-Chozas, J., Helstrup Jensen, N.E., Sørensen, H. C., Kofoed, J.P., & Kabuth, A., 2013. Predictability of the power output of three wave energy technologies in the Danish North Sea. *International Journal of Marine Energy*, 1, 84–98.
Gallagher, S., Tiron, R., & Dias, F., 2013. A detailed investigation of the nearshore wave climate and the nearshore wave energy resource on the west coast of Ireland. *In*: *Proceedings of the International Conference on Offshore Mechanics and Arctic Engineering - OMAE*.
Gallagher, S., Tiron, R., & Dias, F., 2014. A long-term nearshore wave hindcast for Ireland: Atlantic and Irish sea coasts (1979-2012). *Ocean Dynamics*, 64 (8), 1163–1180.
Gaslikova, L. & Weisse, R., 2006. Estimating near-shore wave statistics from regional hindcasts using downscaling techniques. *Ocean Dynamics*, 56 (1), 26–35.
Gleizon, P., Campuzano, F., Carracedo, P., Martinez, A., Goggins, J., Atan, R., & Nash, S., 2017. Marine renewable energy: Resource characterization and physical effects. *In*: *Marine Renewable Energy: Resource Characterization and Physical Effects*. 1–387.
Gonçalves, M. & Guedes Soares, C., 2021. Assessment of the wave energy resource in the Azores coastal area. *In*: Guedes Soares, C., (Ed.). *Developments in Renewable Energies Offshore*, London, UK: Taylor and Francis; 26–33.
Gonçalves, M., Martinho, P., & Guedes Soares, C., 2014a. Assessment of wave energy in the Canary Islands. *Renewable Energy*, 68, 774–784.
Gonçalves, M., Martinho, P., & Guedes Soares, C., 2014b. Wave energy conditions in the western French coast. *Renewable Energy*, 62, 155–163.
Gonçalves, M., Martinho, P., & Guedes Soares, C., 2018. A 33-year hindcast on wave energy assessment in the western French coast. *Energy*, 165, 790–801.
Gonçalves, M., Martinho, P., & Guedes Soares, C., 2020. Wave energy assessment based on a 33-year hindcast for the Canary Islands. *Renewable Energy*, 152, 259–269.
Guedes Soares, C. 2008. Hindcast of Dynamic Processes of the Ocean and Coastal Areas of Europe. *Coastal Engineering*. 55(11):825–826.
Guedes Soares, C.; Bento, A. R.; Goncalves, M.; Silva, D., & Martinho, P. 2014. Numerical evaluation of the wave energy resource along the Atlantic European coast. *Computers & Geosciences*. 71:37–49.
Guillou, N., 2015. Evaluation of wave energy potential in the Sea of Iroise with two spectral models. *Ocean Engineering*, 106, 141–151.
Guillou, N., 2017. Modelling effects of tidal currents on waves at a tidal stream energy site. *Renewable Energy*, 114, 180–190.
Guillou, N., 2020. Estimating wave energy flux from significant wave height and peak period. *Renewable Energy*, 155, 1383–1393.
Guillou, N. & Chapalain, G., 2015. Numerical modelling of nearshore wave energy resource in the SeaofIroise. *Renewable Energy*, 83 (June 2016), 942–953.
Guillou, N., Lavidas, G., & Chapalain, G., 2020. Wave energy resource assessment for exploitation-A review. *Journal of Marine Science and Engineering*, 8 (9).
Gulev, S.K., Grigorieva, V., Sterl, A., Woolf D. 2003, Assessment of the reliability of wave observations from voluntary observing ships: Insights from the validation of a global wind wave climatology based on voluntary observing ship data. J. Geophys. Res. Oceans, 108, pp. 3236–3236.
Haakenstad, H., Breivik, Ø., Reistad, M., & Aarnes, O.J., 2020. NORA10EI: A revised regional atmosphere-wave hindcast for the North Sea, the Norwegian Sea and the Barents Sea. *International Journal of Climatology*, 40 (10), 4347–4373.
Hashemi, M.R. & Neill, S.P., 2014. The role of tides in shelf-scale simulations of the wave energy resource. *Renewable Energy*, 69, 300–310.
Hersbach, H., Bell, B., Berrisford, P., Hirahara, S., et al. 2020. The ERA5 global reanalysis. *Quarterly Journal of the Royal Meteorological Society*, 146 (730), 1999–2049.
Holthuijsen, L.H., Booji, N., Haagsma, J.G., Kieftenburg, a. T. M.M., Ris, R.C., van derWesthuysen, a. J., & Zijlema, M., 2006. User Manual SWAN - Version 41.20, 137.

Ibarra-Berastegi, G., Sáenz, J., Ulazia, A., Serras, P., Esnaola, G., & Garcia-Soto, C., 2018. Electricity production, capacity factor, and plant efficiency index at the Mutriku wave farm (2014–2016). *Ocean Engineering*, 147 (May 2017), 20–29.

IEC, 2015. *International Electrotechnical Commission Technical Specification 62600-101*.

Iglesias, G. & Carballo, R., 2010. Wave power for La Isla Bonita. *Energy*, 35 (12), 5013–5021.

Iglesias, G. & Carballo, R., 2011. Wave resource in El Hierro-an island towards energy self-sufficiency. *Renewable Energy*, 36 (2), 689–698.

Janssen, P.A.E.M., 2008. Progress in ocean wave forecasting. *Journal of Computational Physics*, 227 (7), 3572–3594.

Kalnay, E., Kanamitsu, M., Kistler, R., Collins, W., et al. 1996. The NCEP/NCAR 40-Year Reanalysis Project. *Bulletin of the American Meteorological Society*, 77 (3), 437–472.

Kang, J., Mao, R., Chang, Y., & Fu, H., 2021. Comparative analysis of significant wave height between a new Southern Ocean buoy and satellite altimeter. *Atmospheric and Oceanic Science Letters*, 14 (5), 100044.

Lavidas, G. & Polinder, H., 2019. North sea wave database (NSWD) and the need for reliable resource data: A 38 year database for metocean and wave energy assessments. *Atmosphere*, 10 (9), 1–27.

Lavidas, G. & Venugopal, V., 2018a. Application of numerical wave models at European coastlines: A review. *Renewable and Sustainable Energy Reviews*, 92 (May), 489–500.

Lavidas, G. & Venugopal, V., 2018b. Characterising the wave power potential of the Scottish coastal environment. *International Journal of Sustainable Energy*, 37 (7), 684–703.

Law-Chune, S., Aouf, L., Dalphinet, A., Levier, B., Drillet, Y., & Drevillon, M., 2021. WAVERYS: a CMEMS global wave reanalysis during the altimetry period. *Ocean Dynamics*, 71 (3), 357–378.

Losada, I.J., Mendez, F.J., Vidal, C., Camus, P., & Izaguirre, C., 2010. Spatial and temporal variability of nearshore wave energy resources along Spain: Methodology and results. *In*: *MTS/IEEE Seattle, OCEANS 2010*. 1–8.

Mackay, E., Bahaj, A.B., Retzler, C., & Challenor, P., 2008. Wave energy resource assessment using satellite altimeter data. *Proceedings of the International Conference on Offshore Mechanics and Arctic Engineering - OMAE*, 6 (April 2016), 861–870.

Magagna, D. & Uihlein, A., 2015. Ocean energy development in Europe: Current status and future perspectives. *International Journal of Marine Energy*, 11, 84–104.

Maisondieu, C., 2015. WEC Survivability Threshold and Extractable Wave Power WEC Survivability Threshold and Extractable Wave Power. *In*: *Proceedings of the 11th European Wave and Tidal Energy Conference 6-11th Sept 2015, Nantes, France*. 1–8.

Mattarolo, G., Lafon, F., & Benoit, M., 2009. Wave energy resource off the French coasts: the ANEMOC database applied to the energy yield evaluation of Wave Energy Converters. *In*: *8th European Wave and Tidal Energy*. 247–255.

Van Nieuwkoop, J.C.C., Smith, H.C.M., Smith, G.H., & Johanning, L., 2013. Wave resource assessment along the Cornish coast (UK) from a 23-year hindcast dataset validated against buoy measurements. *Renewable Energy*, 58, 1–14.

NOAA, 2021. Global Navy Coastal Ocean Model [online]. *n.d.* Available from: https://www.ncei.noaa.gov/products/weather-climate-models/global-navy-coastal-ocean [Accessed 31 Oct 2021].

Ocean Plug [online], 2021. *2021*. Available from: https://tethys.pnnl.gov/project-sites/ocean-plug-portuguese-pilot-zone [Accessed 24 Dec 2021].

O'Connell, R., de Montera, L., Peters, J.L., & Horion, S., 2020. An updated assessment of Ireland's wave energy resource using satellite data assimilation and a revised wave period ratio. *Renewable Energy*, 160, 1431–1444.

Onogi, K., Tsutsui, J., Koide, H., Sakamoto, M., et al. 2007. The JRA-25 Reanalysis. *Journal of the Meteorological Society of Japan. Ser. II*, 85 (3), 369–432.

Penalba, M., Ulazia, A., Saénz, J., & Ringwood, J. v., 2020. Impact of long-term resource variations on wave energy Farms: The Icelandic case. *Energy*, 192.

Pilar, P., Guedes Soares, C., & Carretero, J.C., 2008. 44-year wave hindcast for the North East Atlantic European coast. *Coastal Engineering*, 55 (11), 861–871.

Pontes, M.T., Aguiar, R., & Pires, H.O., 2005. A nearshore wave energy atlas for Portugal. *Journal of Offshore Mechanics and Arctic Engineering*, 127 (3), 249–255.

Pontes, M.T., Athanassoulis, G.A., Barstow, S., Cavaleri, L., Holmes, B., Mollison, D., & Oliveira-Pires, H., 1996. An atlas of the wave energy resource in Europe. *Journal of Offshore Mechanics and Arctic Engineering*, 118 (4), 307–309.

Pontes, M.T., Bruck, M., & Lehner, S., 2009. Assessing the Wave Energy Resource Using Remote Sensed Data. *Aperture*, 106–111.

Ramos, S., Díaz, H., Guedes Soares, C., & Lavidas, G., 2020. Identifying compatible locations for wave energy exploration with different wave energy devices in Madeira Islands. *In*: C. Guedes Soares, ed. *Developments in Renewable Energies Offshore*. London: Taylor & Francis Group, 111–122.

Ramos, S., Gonçalves, M., & Guedes Soares, C., 2021. A method for identifying compatible locations for wave energy exploration with different WECS. *In*: *Proceedings of the ASME 2021 40th International Conference on Ocean, Offshore and Arctic Engineering*. 1–10.

Ramos, V. & Ringwood, J. V., 2016. Exploring the utility and effectiveness of the IEC (International Electrotechnical Commission) wave energy resource assessment and characterisation standard: A case study. *Energy*, 107, 668–682.

Ratsimandresy, A.W., Sotillo, M.G., Carretero Albiach, J. C., Álvarez Fanjul, E., & Hajji, H., 2008. A 44-year high-resolution ocean and atmospheric hindcast for the Mediterranean Basin developed within the HIPOCAS Project. *Coastal Engineering*, 55 (11), 827–842.

Reguero, B.G., Menéndez, M., Méndez, F.J., Mínguez, R., & Losada, I.J., 2012. A Global Ocean Wave (GOW) calibrated reanalysis from 1948 onwards. *Coastal Engineering*, 65, 38–55.

Reistad, M., Breivik, Ø., Haakenstad, H., Aarnes, O.J., Furevik, B.R., & Bidlot, J.R., 2011. A high-resolution hindcast of wind and waves for the North Sea, the Norwegian Sea, and the Barents Sea. *Journal of Geophysical Research: Oceans*, 116 (5), 1–34.

Ribal, A. & Young, I.R., 2019. 33 Years of Globally Calibrated Wave Height and Wind Speed Data Based on Altimeter Observations. *Scientific Data*, 6 (1), 1–15.

Rodríguez, G.R., Cabrera, L., & Pacheco, M., 2015. Assessing and modeling annual patterns in wave energy resources off Canary archipelago. *In*: Guedes Soares, C.,

(Ed.), *Renewable Energies Offshore*. London, UK: Taylor & Francis Group
Rusu, E., 2014. Evaluation of the wave energy conversion efficiency in various coastal environments. *Energies*, 7 (6), 4002–4018.
Rusu, E., 2018. Study of the Wave Energy Propagation Patterns in the Western Black Sea. *Applied Sciences (Switzerland)*, 8 (6).
Rusu, E. & Guedes Soares, C., 2009. Numerical modelling to estimate the spatial distribution of the wave energy in the Portuguese nearshore. *Renewable Energy*, 34 (6), 1501–1516.
Rusu, E. & Guedes Soares, C., 2012a. Wave energy pattern around the Madeira Islands. *Energy*, 45 (1), 771–785.
Rusu, L. & Guedes Soares, C., 2012b. Wave energy assessments in the Azores islands. *Renewable Energy*, 45, 183–196.
Rusu, L., Pilar, P., & Guedes Soares, C., 2008. Hindcast of the wave conditions along the west Iberian coast. *Coastal Engineering*, 55 (11), 906–919.
Saha, S., Moorthi, S., Pan, H.-L., Wu, X., Wang, J., et al. 2010. The NCEP Climate Forecast System Reanalysis. *Bulletin of the American Meteorological Society*, 91 (8), 1015–1058.
Saruwatari, A., Ingram, D.M., & Cradden, L., 2013. Wave-current interaction effects on marine energy converters. *Ocean Engineering*, 73, 106–118.
Sergent, P., Baudry, V., Bonviller, A. De, Michard, B., & Dugor, J., 2020. Numerical assessment of onshore wave energy in France: wave energy, conversion and cost. *Journal of Marine Science and Engineering*, 8 (11), 1–37.
Serras, P., Ibarra-Berastegi, G., Sáenz, J., & Ulazia, A., 2019. Combining random forests and physics-based models to forecast the electricity generated by ocean waves: A case study of the Mutriku wave farm. *Ocean Engineering*, 189.
Sierra, J.P., White, A., Mösso, C., & Mestres, M., 2017. Assessment of the intra-annual and inter-annual variability of the wave energy resource in the Bay of Biscay (France). *Energy*, 141, 853–868.
Silva, D. & Guedes Soares, C. 2021. Assessment of the wave power resource at Madeira archipelago with SWAN model Guedes Soares, C., (Ed.). *Developments in Renewable Energies Offshore*, London, UK: Taylor and Francis; 45–54.
Silva, D., Bento, A.R., Martinho, P., & Guedes Soares, C., 2015. High resolution local wave energy modelling in the Iberian Peninsula. *Energy*, 91, 1099–1112.
Silva, D., Martinho, P., & Guedes Soares, C., 2018. Wave energy distribution along the Portuguese continental coast based on a thirty three years hindcast. *Renewable Energy*, 127, 1064–1075.
Silva, D., Rusu, E., & Guedes Soares, C., 2013. Evaluation of various technologies for wave energy conversion in the Portuguese nearshore. *Energies*, 6 (3), 1344–1364.
Silva, D., Rusu, E. & Guedes Soares, C. 2016. High-resolution wave energy assessment in shallow water accounting for tides. *Energies*. 9(9), pp.761–779.
Smith, H.C.M., Haverson, D., & Smith, G.H., 2013. A wave energy resource assessment case study: Review, analysis and lessons learnt. *Renewable Energy*, 60, 510–521.
Smith, H.C.M. & Maisondieu, C., 2014. *Resource Assessment for Cornwall, Isles of Scilly and PNMI.*
Sotillo, M.G., Ratsimandresy, A.W., Carretero, J.C., Bentamy, A., Valero, F., & González-Rouco, F., 2005. A high-resolution 44-year atmospheric hindcast for the Mediterranean Basin: Contribution to the regional improvement of global reanalysis. *Climate Dynamics*, 25 (2–3), 219–236.
Soukissian, T.H., Denaxa, D., Karathanasi, F., Prospathopoulos, A., Sarantakos, K., Iona, A., Georgantas, K., & Mavrakos, S., 2017. Marine renewable energy in the Mediterranean Sea: Status and perspectives. *Energies*, 10 (10), 1–56.
Thomaz, T.B., Crooks, D., Medina-Lopez, E., van Velzen, L., Jeffrey, H., Mendia, J.L., Arias, R.R., & Minguela, P.R., 2019. O&M models for ocean energy converters: Calibrating through real sea data. *Energies*, 12 (13).
Tolman, H.L., Accensi, M., Alves, H., Ardhuin, F., et al. 2014. User manual and system documentation of WAVEWATCH III R version 4.18.
Opentelemac, 2024. TOMAWAC - Wave propagation in coastal areas [online]. Available from: https://www.opentelemac.org/index.php/modules-list/20-tomawac [Accessed 11 Jan 2024].
Ulazia, A., Penalba, M., Ibarra-Berastegui, G., Ringwood, J., & Saénz, J., 2017. Wave energy trends over the Bay of Biscay and the consequences for wave energy converters. *Energy*, 141, 624–634.
Uppala, S.M., Kållberg, P.W., Simmons, A.J., et al. 2005. The ERA-40 re-analysis. *Quarterly Journal of the Royal Meteorological Society*, 131 (612), 2961–3012.
Venugopal, V. & Nemalidinne, R., 2015. Wave resource assessment for Scottish waters using a large scale North Atlantic spectral wave model. *Renewable Energy*, 76, 503–525.
Vettor, R. & Guedes Soares, C. 2021. On the accuracy of voluntary observing ship's records. Journal of Offshore Mechanics and Arctic Engineering. 143(5):054501.
Vijaykumar, N., Gault, J., Devoy, R., Dunne, D., & Mahony, C.O., 2004. Computational Modeling of Environmental Processes: A Hindcast of Wind Atlas over Irish Waters. *Computing*, 3–8.
WAMDI Group, 1988. The WAM Model: A Third Generation Ocean Wave Prediction Model. *Journal of Physical Oceanography*, 18 (12), 1775–1810.
WaveRoller [online], 2021. Available from: https://aw-energy.com/waveroller/ [Accessed 24 Dec 2021].
Weisse, R. & Günther, H., 2007. Wave climate and long-term changes for the Southern North Sea obtained from a high-resolution hindcast 1958-2002. *Ocean Dynamics*, 57 (3), 161–172.
Weisse, R., Weisse, R., Günther, H., & Feser, F., 2002. A 40-year high-resolution wind and wave hindcast for the Southern North Sea, (May), 97–104.
Yaakob, O., Hashim, F.E., Mohd Omar, K., Md Din, A.H., & Koh, K.K., 2016. Satellite-based wave data and wave energy resource assessment for South China Sea. *Renewable Energy*, 88, 359–371.

Advances in Maritime Technology and Engineering – Guedes Soares & Santos (Eds)

Accuracy improvement of wave forecasts with long short-term memory models

M. Ré Henriques & C. Guedes Soares
Centre for Marine Technology and Ocean Engineering (CENTEC), Instituto Superior Técnico, Universidade de Lisboa, Lisbon, Portugal

ABSTRACT: This study aims at improving the significant wave height and wave peak period predictions of a wave model, using buoy data located at three points offshore the Iberian coast. This is achieved by applying one type of recurring neural network, the Long Short-Term Memory, to model the discrepancies between forecast values and buoy data. After modelling the error time series, the neural network results are used to correct the forecast values. The correlation of the data corrected by the neural network model improves for both variables, when compared with the correlation of the forecast model. The correlation of the corrected data is better for the significant wave height than for the wave peak period.

1 INTRODUCTION

Currently, numerical models based on energy balance are the most adopted operational methods for wave prediction. The development in numerical models has significantly improved wind and wave data by providing better spatial and temporal resolution through reanalysis projects (Cavaleri et al., 2007). Several hindcast data sets have been produced and efforts have been made to establish their accuracy by comparing them with observations (Bromwich & Wang, 2008; Campos & Guedes Soares, 2016).

It is considered that in-situ observations from buoys offer the most precise measurements and are an essential source of information (Gemmrich et al., 2011; Menéndez et al., 2008). However, the buoy data are geographically limited, because the availability of high-quality in situ data is normally limited to nearshore locations and often has several gaps or short time series. Gaps in buoy time series can occur for various reasons such as bad weather, staff errors, equipment failure and maintenance (Wang et al., 2023). It is important to understand ocean wave characteristics, especially the significant wave heights, because this information is very useful in ocean engineering and coastal ocean engineering applications, as well as ocean resource management (Bitner-Gregersen et al., 2022).

The numerical wave modelling community faces the big challenge of simulating the complex processes that are involved in the interaction of waves and currents (Ponce de León & Guedes Soares, 2021). In the North Atlantic Ocean, storms are frequent and the study of these extreme events is important due to their intensity activity and severity (Campos et al., 2022), and the accuracy of wave models has an essential role in disaster prevention.

Computing power has become increasingly available in recent years, especially in the use of Artificial Intelligence (AI) and Machine Learning (ML), for a large range of tasks (Makarynskyy, 2004), including forecasting (Korstanje, 2021). For forecasting over a large spatial and temporal domain, the numerical models are more useful, on the other hand for point forecasts at specified locations, time series statistical models like Artificial Neural Networks (ANN) are needed. Initially, the ANN was successfully applied to hydrological predictions, it was only later that it was applied to oceanic wave prediction (Jain & Deo, 2007).

The technologies that provide an effective solution for prediction and regression problems are those based on large amounts of data, such as Recurrent Neural Network (RNN), Long Short-Term Memory (LSTM) and other algorithms (Gao et al., 2021). The RNNs are a special type of Neural Network (NN) designed for sequence problems, that take inputs from previous time steps to predict the next time steps, and the LSTMs are a type of RNN (Brownlee, 2017).

Auto-regressive ANN's have been used to obtain forecasts of significant wave height measurements at three wave rider buoy locations along the west coast of India (Jain & Deo, 2007). Fan et al., (2020) and others mentioned in Wang et al., (2023), have published work applying ANN to predict or reconstruct missing data of significant wave height (Hs) time series recorded by buoys. Londhe et al., (2016) have used the ANN technique to improve the difference between the observed

DOI: 10.1201/9781003508779-73

and forecasted wave height (error) at four-wave buoy locations along the Indian coastline.

Costa et al., (2023) have developed models based on multilayer perceptron (MLP) and LSTM algorithms to correct the bias between ERA5 reanalysis of wind and wave data and buoy measurements from NOAA's National Data Buoy Center (NDBC), for two locations of North Atlantic Ocean, where the output variables are the difference between the reanalysis and observations of significant wave height and the wind speed. They concluded that the LSTM algorithm had a better performance than the MLP and this work follows up by exploring more the capabilities of the LSTM model in this type of function.

This work follows up by using the LSTM algorithm to improve the predictions of an in-house forecast model that covers the offshore area of Portugal. It is based on the model presented by Guedes Soares et al. (2011) that is being enhanced with more recent versions of models and forcing data. Comparing the forecasts with buoy data leads to an error time series that is then modelled by the LSTM model, which is then used to improve the skill of the ocean model.

The structure of this paper is as follows: Section 2 mentions the data used as input; Section 3 describes the method to predict the filtered data; Section 4 shows a description of the results and in Section 5 the discussion and conclusions of the study are given.

2 DATA

Hourly data from 1 June 2022 to 1 June 2023, for three buoys located offshore the Iberian coast was downloaded from the Copernicus Marine Environment Monitoring Service in Situ Thematic Assembly Centre (CMEMS In Situ TAC, www.marineinsitu.eu/dashboard/). The selected buoys were: two buoys from the Portuguese Hydrographic Institute (IH) (https://www.hidrografico.pt/boias): Leixões (WMO identification numbers: 6200191) and Sines (WMO identification numbers: 6201081), and one buoy from the Spanish Ports of the State (PdE) (https://www.puertos.es/enus/oceanografia/Pages/portus.aspx): Cabo Sileiro (WMO identification numbers: 6200084). The locations of these buoy stations can be found in Figure 1 and are listed in Table 1. The parameters used in this study are significant wave height (Hs) and wave peak period (Tp).

The wave model used in this study is WAVEWATCH III (WW3) (Tolman, 1991) applied to the Portuguese continental coast and for the same period of buoy data. WW3 is a third-generation wave model developed at NOAA/NCEP and the data used has a grid resolution of 0.05° x 0.05° with a temporal resolution of one hour. Model grid points are selected to be as close as possible to the coordinates of the buoys.

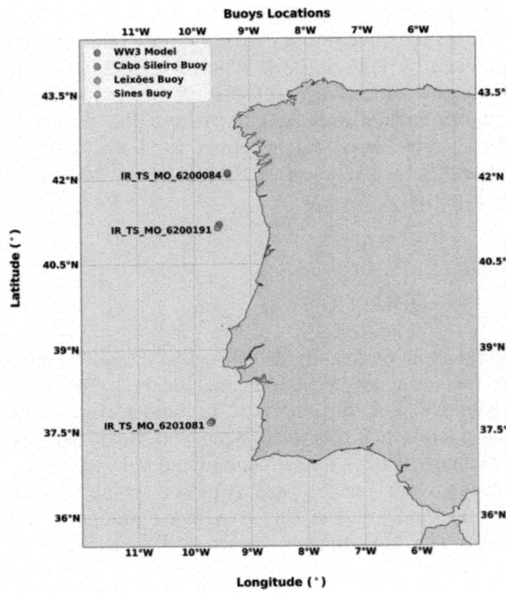

Figure 1. Buoys and WW3 points locations.

Buoy time series often have missing data due to different reasons that led to periods in which the system was not operational. This is not usual in operational forecast systems, but the present one, which is still under development, had some periods without data, due to problems of synchronization of the various components in particular the access to the forcing wind fields. Table 1 and Table 2 indicate the missing data in both data sets.

Table 1. Buoys locations and the number of missing data for the three locations.

		Latitude	Longitude	Water column depth	Distance to shore	Percentage of missing data
Buoy	WMO	degrees	degrees	m	km	%
Cabo Sileiro	6200084	42.12	-9.4	600	45	7.0
Leixões	6200191	41.15	-9.58	1600	77	36.6
Sines	6201081	37.67	-9.73	1768	80	22.1

Table 2. WW3 model locations and the number of missing data for the three locations.

	Latitude	Longitude	Percentage of missing data
WW3	degrees	degrees	%
Cabo Sileiro	42.1	-9.40	8.4
Leixões	41.2	-9.55	8.4
Sines	37.7	-9.7	8.4

Figure 2 shows the time series of Hs and Tp for the three locations, for the buoys and the WW3 data. Figure 3 shows the scatter plots of Hs and Tp between WW3 and buoy data, for the three locations, as well as the linear regression fits to the data. It can be noted that some buoy data are too underestimated by the WW3 predictions.

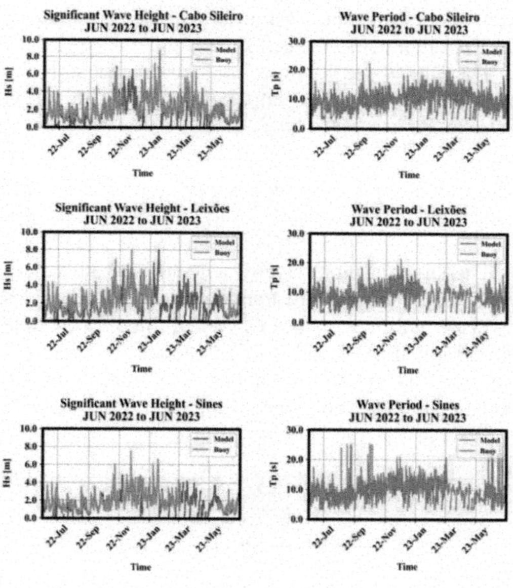

Figure 2. Observed (green and blue lines) and WW3 (red and orange lines) time series of Significant Wave Height (left) and Wave Period (right), for the three locations.

3 METHODS

3.1 *WW3 errors for Hs and Tp*

The errors of Hs and Tp between buoy data and WW3 data are calculated for the three locations. To calculate the Hs or Tp errors between the buoy data and the WW3 data the following equation was used:

$$\text{error}(t) = X_{WW3}(t) - X_{buoy}(t) \qquad (1)$$

where $X_{WW3}(t)$ and $X_{buoy}(t)$ are respectively the values predicted by the WAVEWATCH III model and observed by the buoy at instant t, for Hs and Tp.

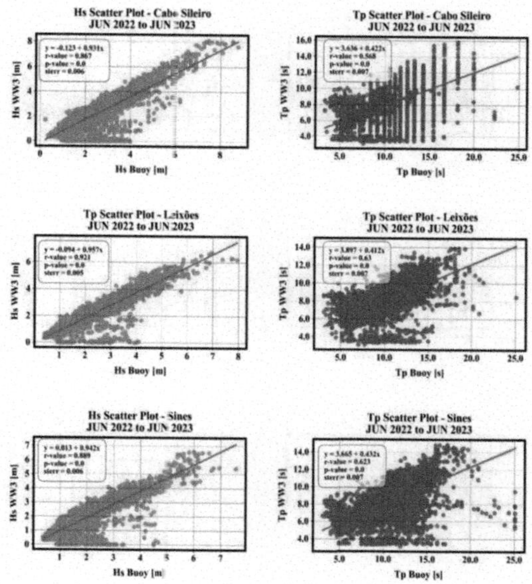

Figure 3. Scatter plots, including linear regression fit results, between WW3 and buoy data for the three locations and for Significant Wave Height (left) and Wave Peak Period (right).

Firstly, outliers are removed from the time series of Hs and Tp from the buoys and the WW3 predictions. For this purpose, the standard deviation of the error (σ_{error}) is calculated for each location. Table 3 gives the values of the standard deviation of the errors for Hs and Tp, and for each location.

Table 3. Standard Deviation of Significant Wave Height Error and Wave Period Error, for the three locations.

	$\sigma_{Hs\text{-}error}$	$\sigma_{Tp\text{-}error}$
Locations	m	s
Cabo Sileiro	0.625	2.247
Leixões	0.474	2.532
Sines	0.484	2.331

In this work, the outlier values in observed and model data of both significant wave height and wave period time series are values greater than twice the value of the standard deviation of the time series for the respective locations. After removing these outliers from buoy and WW3 data sets, the errors of these filtered model data are also computed (Figure 4).

The missing values in these plots are due to the missing data in the WW3 model data. Table 4 shows the missing data size in error data before and after the filter application.

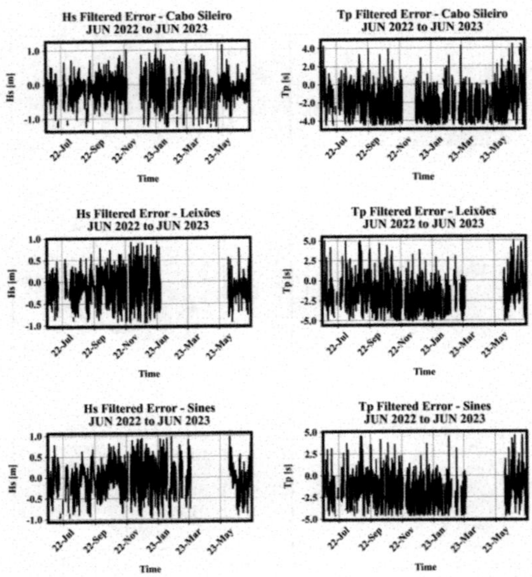

Figure 4. Filtered errors of Significant Wave Height (left) and Wave Peak Period (right) of the model with respect to buoy data, for the three locations.

Table 4. Filtered errors of Significant Wave Height (left) and Wave Period (right) of the model with respect to buoy data, for the three locations.

	Percentage of missing data	
Locations	Before filter	After filter
Cabo Sileiro	15.4 %	21.3 %
Leixões	40.4 %	43.2 %
Sines	28.8 %	32.4 %

3.2 *ANN model structure*

An artificial neural network consists of the interconnection of neurons or nodes, i.e., each neuron in the input layer is connected to all the neurons in the first hidden layer, and each neuron in the first hidden layer is connected to all neurons in the second hidden layer, and finally, each neuron in the last hidden layer is connected to all neurons in the output layer. So, the information flows from the input until it reaches the output layer (Zaki & Meira, 2020).

LSTMs can learn long-term dependencies, which is useful when needed to model time series. They have internal mechanisms called "gates" and "cell states" that can regulate the flow of information. These cells learn when to allow information to enter, leave or be deleted, through the iterative process of making guesses, backpropagation errors, and adjusting weights via gradient descent.

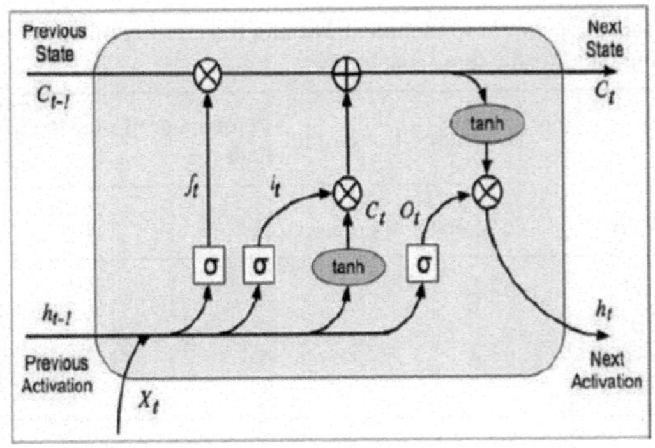

Figure 5. LSTM architecture (Sarang, 2021).

The network requires cells (Figure 5) for flow information (Sarang, 2021). The first cell is a forget gate layer and it has a sigmoid function to decide which information stays or goes. In other words, the False output tells the network to forget the information, and True outputs ask it to keep the information for each number in the cell state C_{t-1}.

The h_{t-1} is the previous hidden state, the x_t is the current input and the output is binary. The forget gate can be expressed by:

$$f_t = \sigma\left(W_f.[h_{t-1}, x_t] + b_f\right) \tag{2}$$

The sigmoid input layer (i_t) decides which values will be updated, and the *tanh* activation layer creates a vector with new values $\widetilde{C}_t$. When the old cell state is multiplied by f_t and $i_t * C_t$ is added to it, the old cell state is updated to a new cell state, and the update gate (C_t) is obtained:

$$i_t = \sigma(W_i.[h_{t-1}, x_t] + b_i) \tag{3}$$

$$\widetilde{C}_t = tanh(W_{c*}[h_{t-1}, x_t] + b_c) \tag{4}$$

$$C_t = f_t * C_t + i_t * \widetilde{C}_t \tag{5}$$

Finally, to obtain the output gate (h_t), the sigmoid layer decides which part of the cell state is going to be kept. That cell is fed by a *tanh* activation function and multiplied from the output of the sigmoid gate.

An open-source machine learning and artificial intelligence software - Tensorflow - library for multiple machine learning tasks for Python (https://www.tensorflow.org/), specifically the Keras library, is used to build an ANN model to estimate the missing values in the forecast errors. The model consists of two hidden layers called LSTMs, where the first is also the input layer, and an output layer called "Dense". The input data for the model included missing values, that are replaced with 0 because the model does not recognize the "Not a Number (NaN)" code or missing values. The model structure is illustrated in Figure 6, and the parameters that are used are represented in Table 5.

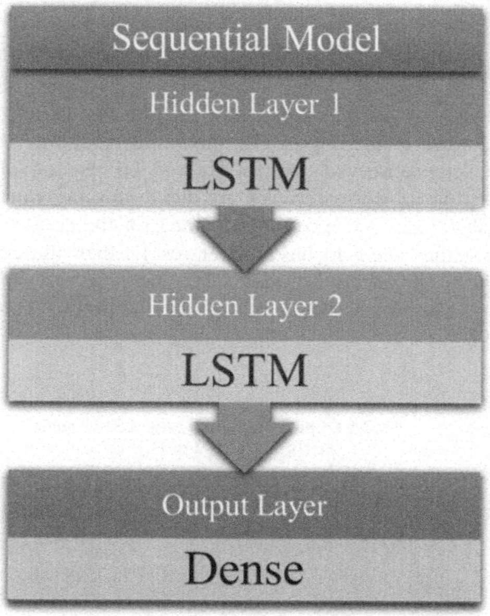

Figure 6. Model structure.

The number of steps represents the number of previous time steps that are used to estimate the next value, and the number of features is the number of variables that are used or the input dimension. In each Hidden layer, 50 neurons are used. In both Hidden Layers, the activation function used is the Rectified Linear Unit (ReLU). The ReLU function is the most widely used activation function that outputs 0 for negative inputs and 1 otherwise (Sarang, 2021). The model parameters are shown in Table 5.

Table 5. Model parameters.

Layer	Type	Parameters
Hidden 1 (Input)	LSTM	input shape (number of steps = 3, number of features = 3) number of neurons = 50 activation function = ReLU
Hidden 2	LSTM	number of neurons = 50 activation function = ReLU
Output	Dense	number of neurons = 3

3.3 *ANN model training*

The model is trained and compiled using the parameters listed in Table 6. For compiling the model, the optimization algorithm Adaptive Moment Estimation (Adam) is used to shift the model's weight parameters to minimize the loss function (Gridin, 2022). Adam computes adaptive learning rates for each parameter and keeps an exponentially declining average of previous gradients (Gridin, 2022). The loss function used to evaluate the network is the Mean Squared Error (MSE). The metric that is used is MSE and its goal is to analyse the network and find out if it is overfitting or underfitting the training data (Brownlee, 2017). The MSE takes the errors per row of data, squares those errors, and then takes the average of them (Korstanje, 2021). It can be expressed by:

$$MSE = \frac{1}{n}\sum (y_t - \hat{y}_t)^2 \tag{6}$$

The Root Mean Square Error (RMSE) and the Mean Absolute Error (MAE) are calculated to evaluate the model result, and can be expressed by (Korstanje, 2021):

$$RMSE = \sqrt{MSE} \tag{7}$$

$$MAE = \frac{1}{n}\sum |y_t - \hat{y}_t| \tag{8}$$

where MAE is calculated by taking the absolute difference between the n pairs of predicted ($\hat{y}_t$) and actual (y_t) values. The MAE is the average of those absolute errors.

For model training, the input data is a list of tensors. In this case, the errors of the filtered data of Hs and Tp for the three locations, are used. This list is divided by the amount of data, time steps for the next time step prediction and the number of variables or features. The batch size is the number of samples in the dataset that is used to train the network, i.e., the amount of data is divided by the number of batch size and therefore

Table 6. Parameters of model compile and model training.

Parameters			
Model Compile		Model Training	
Optimization Algorithm	Adam	Input data	(Amount of data = 39312, Time step = 3, Number of features = 3)
Learning rate	10^{-3}; 10^{-4}		
Loss Function	MSE	Batch size	32
Metrics	MSE	Epochs	50
		Validation Split	0.15 (15%) to 0.56 (56%)

this allows the network parameters to update the number of times the amount of data is divided. The number of epochs is the number of iterations that each batch goes through the model. The validation split is the percentage of training data that is used to train the model. The amount of data per location used as validation set during the model train is 9480, which corresponds to a total of 9480 x 3 = 28440 input values.

4 RESULTS

After the training, the model is evaluated to avoid overfitting and underfitting behaviours during the training phase. A model with a good fit occurs when the performance of the model is good on both training set and validation set, this means that train and validation loss curves decrease and stabilize at certain epoch (Brownlee, 2017).

Figure 7 shows the loss metrics plot of the ANN model for variables Hs (a) and Tp (b), in other words, it shows the behaviour of the ANN model during the training phase and helps to test new configurations to obtain better performance from the model.

Table 7. Evaluation of filtered Hs and Tp error predictions, for the three locations.

	Significant Wave Height			Wave Period		
	MSE	RMSE	MAE	MSE	RMSE	MAE
Locations	m^2	m^2	m	s^2	s	s
Cabo Sileiro	0.09	0.29	0.23	4.67	2.16	2.02
Leixões	0.04	0.21	0.17	4.85	2.20	2.05
Sines	0.05	0.23	0.17	3.87	1.97	1.83

Table 7 shows the measures of performance of the LSTM model predictions for the filtered errors of Hs and Tp respectively, for the three locations

For both variables, the training loss curves are decreasing until they begin to stabilize at a certain point. On the other hand, both validation loss curves also started to decrease, but for the training of the ANN model of the Hs variable (a) this curve shows some oscillations, which may indicate that the model has difficulties to generalize to the validation data and is being sensitive to variations. On the other hand, during the training of the ANN model of the Tp variable (b) this curve began to stabilize after ten epochs. However, although oscillations have been observed, this does not mean that the ANN model performed poorly. Figure 8 shows the predicted error for variables Hs (a) and Tp (b), and it can be observed that the ANN models predicted values in data gaps.

After the error predictions by neural networks, corrected Hs and Tp observed data are computed with the predicted filtered error with the following expression:

$$X_{corrected} = X_{WW3} - \varepsilon_{Predicted} \tag{9}$$

where the $X_{corrected}$ are the corrected values, the X_{WW3} are the original WW3 values and $\varepsilon_{Predicted}$ are the values of predicted filtered error.

Figure 9 and Figure 10 show the results of the corrected observed data for Hs and Tp respectively. The missing corrected data are due to missing values in WW3 data. The results obtained for the corrected Hs values seem to have better results than the corrected Tp values because when time series show large variations, ANN model cannot predict them well.

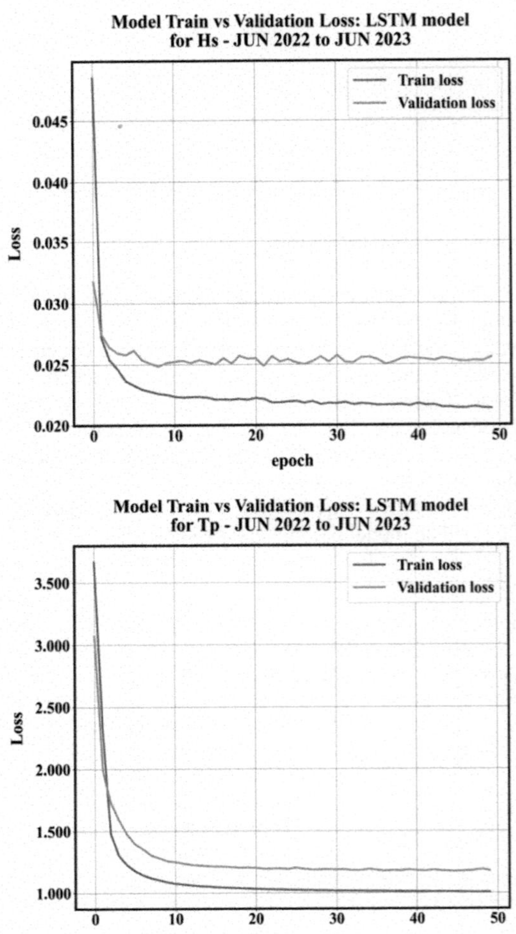

Figure 7. Loss Metrics plot of ANN model for variable Hs (a) and Tp (b).

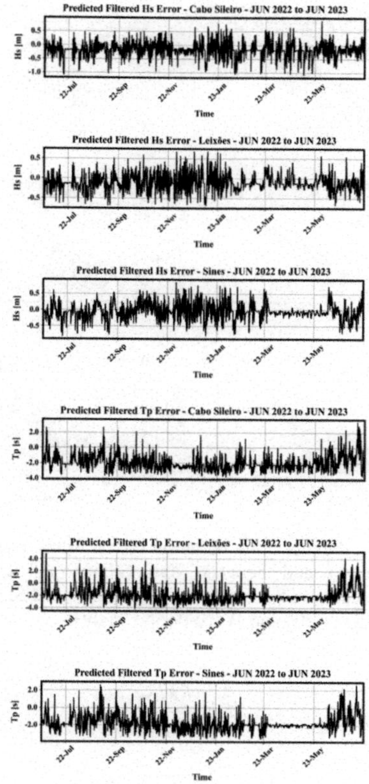

Figure 8. Predicted Filtered Error of Significant Wave Height (a) and Wave Period (b) by LSTM model.

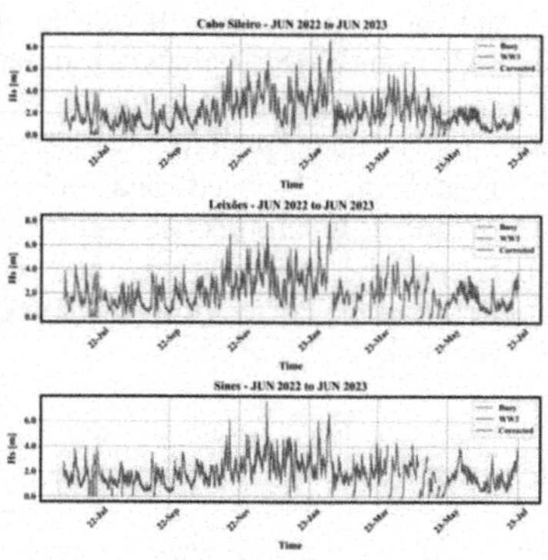

Figure 9. Corrected observed values (red lines), buoys values (green lines) and WW3 values (blue lines) for the significant wave height.

Table 8 and Table 9 show the results of the statistical analysis namely mean, standard deviation (std), maximum (max) and minimum (min) values, skewness (skew) and kurtosis of observations values by the buoy, WW3 values and corrected values by LSTM model for each location, for Hs and Tp variables respectively. It can be noted that the ANN model cannot predict large peak values, such as maxima and minima. However, in the data corrected by the ANN model, the peak values obtained are close to the observed values.

To determine the Pearson correlation coefficient (r), scatter index (SI) and Bias in these two tables the following expressions (Wang et al., 2023) are used:

$$r = \frac{\sum_i^n ((x_i - \bar{x})(y_i - \bar{y}))}{\sqrt{\sum_i^n (x_i - \bar{x})^2 \sum_i^n (y_i - \bar{y})^2}} \tag{10}$$

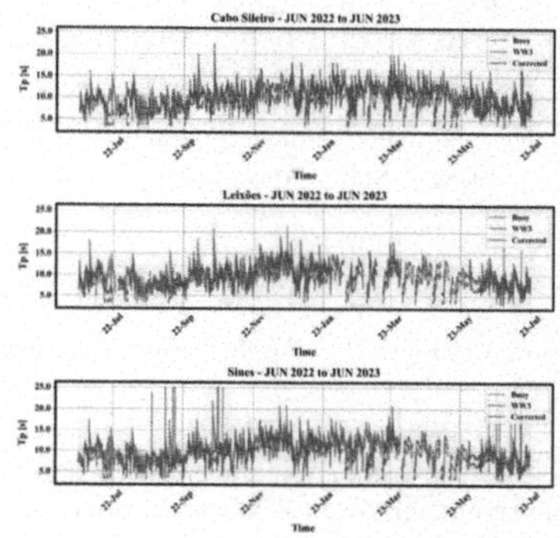

Figure 10. Corrected observed values (red lines), buoys values (green lines) and WW3 values (blue lines) for the wave period.

Table 8. Statistical analysis of observed, predicted by WW3, and corrected by LSTM model significant wave height values, for each location.

Significant Wave Height	Statistical Parameter					
Cabo Sileiro	Mean	Std	Max	Min	Skew	Kurtosis
Obs.	2.15	1.14	8.79	0.23	1.30	2.58
WW3	2.01	1.33	8.10	0.00	1.05	1.39
Correct.	2.17	1.30	8.40	0.16	1.06	1.38
Leixões	Mean	Std	Max	Min	Skew	Kurtosis
Obs.	2.12	1.15	7.94	0.41	0.90	0.70
WW3	2.02	1.26	8.04	0.00	0.91	1.32
Correct.	2.12	.23	7.96	0.10	0.93	1.39
Sines	Mean	Std	Max	Min	Skew	Kurtosis
Obs.	1.95	0.98	7.56	0.38	1.13	1.74
WW3	1.84	1.05	6.59	0.00	0.60	0.63
Correct.	1.88	1.02	6.52	-0.01	0.71	0.84

Table 9. Statistical analysis of observed, predicted by WW3, and corrected by LSTM model wave period values, for each location.

Wave Period						
	Statistical Parameter					
Cabo Sileiro	Mean	Std	Max	Min	Skew	Kurtosis
Obs.	10.70	2.65	24.80	3.32	0.10	0.22
WW3	8.30	2.05	15.81	3.11	-0.02	0.09
Correct.	10.25	2.35	18.32	3.58	-0.02	-0.55
Leixões	Mean	Std	Max	Min	Skew	Kurtosis
Obs.	9.97	2.70	25.08	3.05	0.37	0.19
WW3	8.28	2.05	15.58	3.11	0.02	0.07
Correct.	10.10	2.42	17.80	4.41	-0.04	-0.57
Sines	Mean	Std	Max	Min	Skew	Kurtosis
Obs.	10.48	2.94	25.08	3.05	0.25	0.62
WW3	8.27	2.16	15.51	3.11	0.05	-0.15
Correct.	10.03	2.50	17.88	3.43	0.03	-0.56

Table 10. Pearson correlation and error measures of the significant wave height from WW3 and from corrected by LSTM, for each location.

Significant Wave Height Error						
	Cabo Sileiro		Leixões		Sines	
Measures	WW3	Correct.	WW3	Correct.	WW3	Correct.
r	0.87	0.89	0.92	0.94	0.89	0.93
RMSE	0.68	0.56	0.51	0.41	0.49	0.39
MAE	0.27	0.12	0.19	0.08	0.10	0.07
SI	0.32	0.26	0.24	0.19	0.25	0.20
Bias	-0.27	-0.12	-0.19	-0.08	-0.10	-0.07

Table 11. Pearson correlation and error measures of the wave period from WW3 and corrected by LSTM, for each location.

Wave Period Error						
	Cabo Sileiro		Leixões		Sines	
Measures	WW3	Correct.	WW3	Correct.	WW3	Correct.
r	0.57	0.69	0.63	0.70	0.62	0.73
RMSE	3.40	2.06	2.90	2.03	3.27	2.14
MAE	2.55	0.63	1.99	0.38	2.30	0.59
SI	0.32	0.19	0.29	0.20	0.31	0.20
Bias	-2.55	-0.63	-1.99	-0.38	-2.30	-0.59

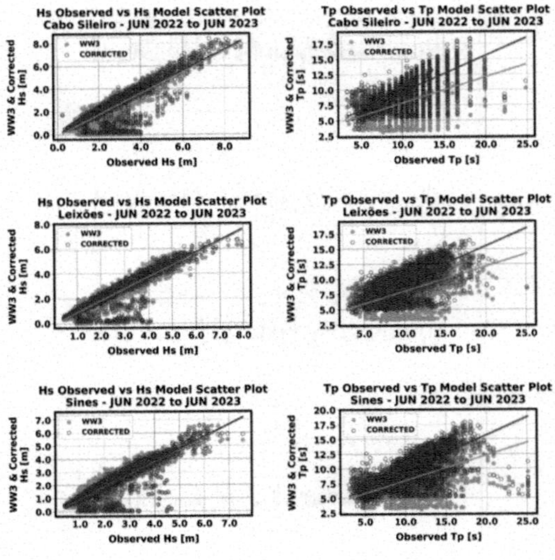

Figure 11. Scatter Plot between observations and WW3 (blue points) and between observations and corrected results (green points), for the three locations and Significant Wave Height (left) and Wave Period (right).

$$SI = \frac{RMSE}{\bar{x}} \tag{11}$$

$$Bias = \frac{1}{n}\sum^{y_t} -\hat{y_t} \tag{12}$$

Table 10 and Table 11 show the Pearson correlation and error, measures of Hs and Tp respectively, for both WW3 predictions and predictions corrected by LSTM and for each location. It can be observed that lower RMSE and MSE values are obtained for observed and corrected data. Figure 11 shows the scatter plots of Hs and Tp between observed data and WW3 model and observed data and corrected data by ANN model. It can be observed that the correlation between the observations and corrected values is better than the correlation between observations and initial values.

5 CONCLUSIONS

This work uses the Long Short-Term Memory model, which is one type of recurrent neural network, to improve the numerical wave model forecasts at three buoys along the Iberian Peninsula coast. For each buoy, the error of the wave model is calculated. The error time series are used as input to an ANN model that is fit to the data. The ANN model predictions are then used to correct the forecast values, for Hs and Tp for each location.

After getting the corrected data with the ANN model, the corrected Hs values have better agreement than the corrected Tp values. This is because the Hs data values have smaller variability than the Tp data.

For both variables, the results of the MSE and RMSE between observed data and corrected data by ANN model were lower than between observed data and WW3 model forecasts, at all locations. Despite the problematic nature of the data sets that had several missing data, the results have shown that the LSTMs algorithms can improve the results of numerical forecast models, showing that they can be valuable to operate coupled with a wave forecast model.

ACKNOWLEDGEMENTS

This work has been performed under the project "WAVEFAI - Operational Wave Forecast using Artificial Intelligence", which is funded by the Portuguese Foundation for Science and Technology (Fundação para a Ciência e a Tecnologia - FCT) under contract CIRCNA.OCT.0300.2019_1801P.01023. This work contributes to the Strategic Research Plan of the Centre for Marine Technology and Ocean Engineering (CENTEC), which is financed by FCT under contract UIDB/UIDP/00134/2020.

REFERENCES

Bitner-Gregersen, E.; Waseda, T.; Parunov, J.; Yim, S.; Hirdaris, S.; Ma, N., and Guedes Soares, C. 2022. Uncertainties in long-term wave modelling. *Marine Structures*. 84:103217.

Bromwich, D. H., & Wang, S.-H. (2008). A review of the temporal and spatial variability of Arctic and Antarctic atmospheric circulation based upon ERA-40. *Dynamics of Atmospheres and Oceans*, 44(3–4), 213–243. https://doi.org/10.1016/j.dynatmoce.2007.09.001

Brownlee, J. (2017). Long Short-Term Memory Networks With Python. *Machine Learning Mastery With Python*, *1* (1), 228.

Campos, R. M.; Bernardino, M.; Goncalves, M., and Guedes Soares, C. (2022) Assessment of metocean forecasts for Hurricane Lorenzo in the Azores Archipelago. *Ocean Engineering*. 243:110292

Campos, R. M., & Guedes Soares, C. (2016). Comparison and assessment of three wave hindcasts in the North Atlantic Ocean. *Journal of Operational Oceanography*, *9* (1), 26–44. https://doi.org/10.1080/1755876X.2016.1200249

Cavaleri, L., Alves, J.-H.G.M., Ardhuin, F., Babanin, A., Banner, M., Belibassakis, K., Benoit, M., Donelan, M., Groeneweg, J., Herbers, T.H.C., et al. 2007. Wave modelling – the state of the art. *Progress Oceanogr.* 75:603–674.

Costa, M. O., Campos, R. M., & Guedes Soares, C. (2023). Enhancing the accuracy of metocean hindcasts with machine learning models. *Ocean Engineering*, *287*, 115724. https://doi.org/10.1016/j.oceaneng.2023.115724

Fan, S., Xiao, N., & Dong, S. (2020). A novel model to predict significant wave height based on long short-term memory network. *Ocean Engineering*, *205*, 107298. https://doi.org/10.1016/j.oceaneng.2020.107298

Gao, S., Huang, J., Li, Y., Liu, G., Bi, F., & Bai, Z. (2021). A forecasting model for wave heights based on a long short-term memory neural network. *Acta Oceanologica Sinica*, *40*(1), 62–69. https://doi.org/10.1007/s13131-020-1680-3

Gemmrich, J., Thomas, B., & Bouchard, R. (2011). Observational changes and trends in northeast Pacific wave records. *Geophysical Research Letters*, *38*(22), n/a-n/a. https://doi.org/10.1029/2011GL049518

Goodfellow, I., Bengio, Y., & Courville, A. (2016). *Deep Learning: Adaptive Computation and Machine Learning. MIT Press.*

Gridin, I. (2022). *Learning, Time Series Forecasting using Deep Learning: Combining PyTorch, RNN, TCN, and Deep Neural Network Models to Provide Production-Ready Production-Ready Prediction Solutions* (Vol. 4, Issue 1). BPB Publications, India.

Guedes Soares, C. Rusu, L.; Bernardino, M., and Pilar, P. (2011) An Operational Forecasting System for the Portuguese Continental Coastal Area. *Journal of Operational Oceanography*. 4(2):17–27.

Jain, P., & Deo, M. C. (2007). Real-time wave forecasts off the western Indian coast. *Applied Ocean Research*, *29* (1–2), 72–79. https://doi.org/10.1016/j.apor.2007.05.003

Korstanje, J. (2021). Advanced Forecasting with Python. In *Advanced Forecasting with Python*. https://doi.org/10.1007/978-1-4842-7150-6.

Londhe, S. N., Shah, S., Dixit, P. R., Nair, T. M. B., Sirisha, P., & Jain, R. (2016). A Coupled Numerical and Artificial Neural Network Model for Improving Location Specific Wave Forecast. *Applied Ocean Research*, 59, 483–491. https://doi.org/10.1016/j.apor.2016.07.004

Makarynskyy, O. (2004). Improving wave predictions with artificial neural networks. *Ocean Engineering*, 31(5–6), 709–724. https://doi.org/10.1016/j.oceaneng.2003.05.003

Menéndez, M., Méndez, F. J., Losada, I. J., & Graham, N. E. (2008). Variability of extreme wave heights in the northeast Pacific Ocean based on buoy measurements. *Geophysical Research Letters*, 35(22), L22607. https://doi.org/10.1029/2008GL035394

Ponce de León, S., & Guedes Soares, C. (2021). Numerical modelling of the effects of the Gulf Stream on the wave characteristics. *Journal of Marine Science and Engineering*, 9(1), 1–15. https://doi.org/10.3390/jmse9010042

Sarang, P. (2021). Artificial Neural Networks with TensorFlow 2. In *Artificial Neural Networks with TensorFlow 2*. Apress. https://doi.org/10.1007/978-1-4842-6150-7

Tolman, H. L. (1991). A Third-Generation Model for Wind Waves on Slowly Varying, Unsteady, and Inhomogeneous Depths and Currents. *Journal of Physical Oceanography*, 21(6), 782–797.

Wang, J., Wen, K., & Deng, F. (2023). Filling gaps in significant wave height time series records using bidirectional gated recurrent unit and cressman analysis. *Dynamics of Atmospheres and Oceans*, 101, 101339. https://doi.org/10.1016/j.dynatmoce.2022.101339

Zaki, M. J., & Meira, W. J. (2020). Data Mining and Machine Learning Fundamental Concepts and Algorithms. In *Journal of Chemical Information and Modeling* (Vol. 53, Issue 9).

Advances in Maritime Technology and Engineering – Guedes Soares & Santos (Eds)

A high-resolution wave energy assessment for the Sines region

M. Sancho, M. Bernardino & C. Guedes Soares
Centre for Marine Technology and Ocean Engineering (CENTEC), Instituto Superior Técnico, Universidade de Lisboa, Lisbon, Portugal

ABSTRACT: Global warming is a present concern, and ambitious global targets for low-carbon energy sources have been set. Therefore, it is important to investigate other renewable sources like wave energy and Portugal stands as a prime location for wave energy extraction. High-resolution simulations in the vicinity of the port of Sines covered a 22-year period (1989-2010) forced with boundary conditions from the CLIMENA project and winds from the ERA-INTERIM database. To further enhance coastal resolution, a nested domain within the Iberian Peninsula area was performed. Calibration was carried out using in-situ data from an oceanographic buoy near the port of Sines. Different interest points were chosen to test seven different WECs for average electric power production. The coastal area south of Sines seems to give interesting results regrading power production and a location in this area is studied in further detail. From the different WEC's tested, the Wave Dragon is the one producing the best results in this evaluation, although with some depth limitations.

1 INTRODUCTION

Global warming and the escalating global energy demand are compelling concerns and the European Union (EU) has committed to an 80% reduction in greenhouse gas emissions by 2050 (Ruester et al. 2014) awakening a growing focus on sustainable energy sources. However, despite the importance of renewable energy, it currently constitutes less than 30% of the global primary energy supply, with only 12.8% in 2021 (Yolcan 2023). Sustainable energy development strategies effectively address global warming emphasizing three pivotal technological changes: energy savings, production efficiency enhancements, and the shift from fossil fuels to renewables (Lund 2007).

Among renewable energy sources, wave energy stands out possessing a colossal estimated global potential of approximately 2TW (Gunn & StockWilliams 2012). Accurate prediction of maximum wave energy locations is essential, given the significance of wave power for sustainability, therefore, technological innovations and numerical modelling have contributed significantly to the evolution of wave energy converters (WECs) (Aderinto & Li 2018, Babarit 2017).

Historically, European coastal studies, such as the European wave energy atlas (Pontes 1998), relied on buoy observations. However, these methods have limitations, particularly in forecasting future conditions and spatial distribution. Consequently, numerical models like SWAN (Booij et al. 1999) developed at Delft University, effectively address coastal limitations by solving the spectral action balance equation, allowing for the consideration of depth-induced wave breaking and triad wave-wave interactions in shallow waters without spectrum restrictions. These models are critical for understanding wave energy potential and for optimizing its utilization within sustainable energy strategies, therefore, studies are continually refining parameterizations to enhance their accuracy (Rusu & Guedes Soares 2013, Bento et al. 2018, Rusu et al. 2008, Rusu & Guedes Soares 2009).

With its extensive Atlantic coastline, Portugal presents an ideal location for marine renewable energy projects. Already the country has set ambitious targets of 47% renewable energy by 2030, 72% by 2040, and 88% by 2050 (International Energy Agency 2021). Portugal's strategic ports, particularly Sines, play a vital role in energy supply, positioning the nation as a hub for transmitting renewable energy (EDP 2022). Studies along the Portuguese coast emphasize substantial wave energy potential. A 33-year simulation (Silva et al. 2018) from 1979 to 2012 reveals higher wave energy during winter (38.8 kW/m) compared to summer (6.4 kW/m), with a 28% deficit at Sines. Anastas et al. (2022) performed a 40-year study on the vicinity of the port of Sines obtaining an overall mean exploitable energy flux of 35.8 kW/m on the west side of the port. Another study (Silva et al. 2015) spanning 2009-2012 identifies peak energy months as February and November, with August being the least energetic. Guedes Soares et al. (2014) reports maximum winter averages of 98.0 kW/m in Ireland, 57.1 kW/m in Portugal, and 58.5 kW/m in Northern Spain. Silva

DOI: 10.1201/9781003508779-74

et al. (2013) assesses various wave energy technologies, highlighting maximum wave power of 478 kW/m in the northern region and 467 kW/m in the Portuguese central area. Additionally, Rusu (2014) ranks the Wave Dragon as the most potent Wave Energy Converter, with an expected maximum electric power of 956 kW and the best capture width. These findings collectively underscore the substantial and diverse wave energy potential along the Portuguese coast, offering insights into sustainable energy generation.

This study aims to calibrate the SWAN model for improved parameterizations at various levels, conduct a regional assessment of wave energy potential using the calibrated model, and identify optimal locations and converters for effective wave energy extraction. Chapter 2 comprises the data and methodology, chapter 3 shows all the results obtained, chapter 4 englobes the discussion of the results and comparison with the relevant literature and lastly, chapter 5 presents the conclusions.

2 DATA AND METHODS

2.1 *Data*

The wind data was obtained from the reanalyze ERA-INTERIM database (Dee et al. 2011) created by ECMWF (European Centre for Medium-Range Weather Forecasts) (Setchell 2020). The wind conditions with a time step of six hours were retrieved from a grid located at 35°N-45°N and 12°W-2°W with a resolution of 0.1 o x 0.1 o, the same as on the level 1 domain.

Observational wave data was obtained from a wave rider type directional buoy located offshore of the port of Sines at 37.9211°N and −8.9289°W (Guedes Soares 2011) and maintained by the Hydrographic Institute (Portuguese Navy).

The bathymetry data used in the SWAN model for levels 1 and 2 encloses almost the totality of the North Atlantic Ocean and covers approximately the location of 30°N-52°N and 43°W-0°W, with a grid resolution of 0.0083333° in both axes. For the level 3 implementation, a more refined bathymetry was used. It was obtained from the "EMODnet" website (EMODnet 2023) which covers the area of the port of Sines with a grid resolution of approximately 0.001° located at 35.9349°N-40.5797°N and 11.6505°W-6.3411°W.

2.2 *Wave model*

Due to the increased complexity of shallower waters, the action balance approach is used to characterize the sea state. Equation 1 represents the Eulerian spectral action balance in coastal waters with finite water depth and accounts for the action density N(σ, θ), which is defined as the ratio of energy density E(σ, θ) to relative radian frequency σ (SWAN 2019). It describes the rate of change of action density, accounting for energy transport between currents and waves, frequency-shifting, and its propagation. It also accounts for shoaling, refraction, diffraction, bottom friction, depth-induced wave breaking and triad wave-wave interactions. Additionally, it considers the effects already present in deep water formulations, such as wave generation by wind and white-capping.

$$\frac{\partial N}{\partial t}+\frac{\partial C_x N}{\partial x}+\frac{\partial C_y N}{\partial y}+\frac{\partial C_\sigma N}{\partial \sigma}+\frac{\partial C_\theta N}{\partial \theta}=\frac{S_{tot}}{\sigma} \quad (1)$$

The left side represents the rate of change of the action density N and the right side Stot represents all effects of generation, wave-wave interactions and dissipation. $c_{g,x}$, $c_{g,y}$ are the components of the group velocity, c_θ is the refraction or diffraction-induced turning rate for an individual wave component and c_σ accounts for the frequency-shifting of the waves. The terms in S_{tot} can be further subdivided into quadruplet wave-wave interactions, triplet wave-wave interactions, white-capping, bottom friction, and depth induced wave breaking.

This equation provides a fundamental framework for understanding wave energy dynamics in coastal regions with complex depth variations and currents (Holthuijsen 2010).

2.3 *Wave energy converters*

The recent evolution of wave energy extraction technologies facilitated by advancements in numerical simulations using computational fluid dynamics are enabling the testing and prediction of the converter's behavior. However, compared to wind and solar energy, only a few WECs have achieved large-scale commercial success (Aderinto & Li 2018). This work evaluates seven converters to identify the most suitable option for the conditions near the port of Sines: Aquabuoy, Pelamis (Jahangir et al. 2023), OEbuoy (Lavelle & Kofoed 2011), Wave dragon (Kofoed et al. 2006), Oceantec (Salcedo et al. 2008), Seabased, and Wavebob (Babarit et al. 2012).

Table 1. Computational grids for the geographic areas.

Level	Lat/long	$\Delta x/\Delta y$	Boundary conditions
Level 1	35°N-45°N/ 12°W-2°W	0.1°x0.1°	WAM
Level 2.1	36.5°N-39.83°N/ 11°W-7.66°W	0.033° x0.033°	Level 1
Level 2.2	37.5°N-39°N/ 10°W-8.6°W	0.02° x0.02°	Level 1
Level 3	37.7°N-38.2°N /9.2°W-8.7°W	0.005° x0.005°	Level 2

These converters are assessed based on water depth, type (the ratio of wavelength to interacting part size) and working principle (Lopez et al. 2013). Table 2 shows the basic characteristics of the seven converters.

Table 2. Basic characteristics of chosen converters

Converter	Depth	Type	Working principle	Max rated power
AquaBuoy	>50	Point absorber	OBS	250
OEbuoy	deep	Point absorber	OWC	2800
Pelamis	50-60	Attenuator	OBS	750
WaveDragon	>25	Terminator	Overtopping	7000
Oceantec	50-100	Attenuator	OBS	500
Seabased	40-100	Point absorber	OBS	15
Wavebob	deep	Point absorber	OBS	1000

2.4 *Model implementation*

In this study, the wave model SWAN version 41.31 (Booij et al. 1999) was implemented with spherical coordinates using a set of boundary conditions for the desired domain in the specified grid.

For all levels, the SWAN model was run in nonstationary mode with a time step of 20 minutes starting on 01/01/1989 at 00:00h with the end of the implementation on 31/12/2010 at 18:00h. The wave parameters were calculated for the specific point of the buoy offshore of the port of Sines with a time step of 3 hours and for the totality of the grid points on each nested implementation with a time step of 6 hours. The output wave spectrum possesses 30 frequencies and 34 directions for each grid's point. The minimum frequency used in the model was $0.05s^{-1}$ (T=20s) and the maximum frequency used was $0.6s^{-1}$ (T=1.67s).

2.5 *Study region*

The study was conducted within the vicinity of the port of Sines. While historically the town focused on maritime activities, the rise of petrochemical plants, and coal and oil terminals in the late 20th century, made Sines a crucial economic hub. The port of Sines, Portugal's largest artificial port, plays a vital role in the nation's energy supply and trade, exemplifying the city's dynamic evolution blending tradition and modernity on Portugal's shores (Município de Sines 2023).

Initially, the simulation covered the geographic region of the Iberian Peninsula, approximately between 35°N-45°N and 12°W-2°W (level 1 domain in Figure 1). Subsequently, a series of nested runs utilized boundary conditions from the larger domains to refine the model's resolution near the coast. The last nested grid extended to approximately 37.7° N 38.2° N and 9.2 ° W-8.7 ° W (level 3 domain in Figure 1), specifically targeting the area surrounding the port of Sines. This approach not only enhances the resolution near the coast but also identifies potential wave energy conversion sites and facilitates the modelling of shallow-water physical processes (Silva et al. 2018). A summary of the implementation grids is provided in Table 1.

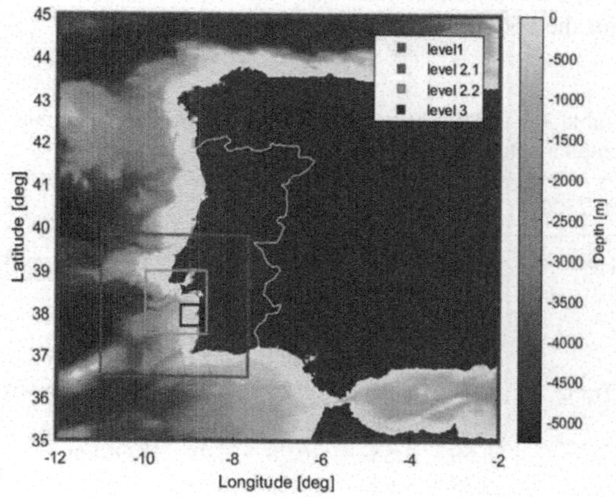

Figure 1. Implementation domains of SWAN.

3 RESULTS

3.1 *Model validation*

The physical processes of the SWAN model have default parameterizations, which is the easiest and most common way to implement them on the model. However, there is no guarantee that those are the most adequate or even the ones that provide the best results in the chosen domain. To test them, the simulated data (Yi) was compared against the observations (Xi) from the Sines' buoy using equations 2 to 5.

$$BIAS = \frac{\sum_{i=1}^{n}(X_i - Y_i)}{n} \quad (2)$$

$$RMSE = \sqrt{\frac{\sum_{i=1}^{n}(X_i - Y_i)^2}{n}} \quad (3)$$

$$SI = \frac{RMSE}{\bar{X}} \quad (4)$$

$$r = \frac{\sum_{i=1}^{n}(X_i - \bar{X})(Y_i - \bar{Y})}{\left(\sum_{i=1}^{n}(X_i - \bar{X})^2 \sum_{i=1}^{n}(Y_i - \bar{Y})^2\right)^{\frac{1}{2}}} \quad (5)$$

Initially, the parameters for the third-generation mode for wind input were tested in the level 1 domain for a period of 1 month (December 2003). The four

GEN3 command input options tested were the JANSsen (Cavaleri & Rizzoli 1981), KOMen (Komen et al. 1984), WESTHuyson (Yan 1987) and ST6 (Rogers et al. 2012). In all these tests, the wave components were propagated using a second-order upwind scheme with third-order diffusion backward space, backward time BSBT, default parameterizations for whitecapping and depth induce wave braking, deactivated triad wave-wave interactions and the JONSWAP method for the bottom friction.

Table 3. GEN3 parameters comparison in level 1 domain implementation - December 2003.

n=2720	Bias	RMSE	SI	r	Method
Hs	-0.33	0.64	0.36	0.78	WESTHuysen
	-0.47	0.74	0.42	0.77	JANSen
	-0.35	0.65	0.37	0.79	KOMen
	-0.30	0.61	0.34	0.81	ST6
Tm02	-0.44	1.52	0.23	0.56	WESTHuysen
	0.43	2.05	0.30	0.28	JANSen
	-0.50	1.85	0.28	0.49	KOMen
	-0.38	1.62	0.24	0.54	ST6

From Table 3, it is clear that the ST6 method is the one that provides closer results to the Sines' buoy wave data and the one chosen for the following simulations. Then three experiments were made where the resolutions of 100x100 bins (default), 150x150 bins and 200x200 bins were tested. The default resolution had by far the lowest running time (12 times faster than the slowest run), which is one considerable aspect to have in mind when running this type of simulation, therefore, the default resolution was used for the rest of the study. Upon completing the tests for the level 1 domain implementation, the subsequent nested runs were analyzed to determine which nested option better transports the information from one level domain to the others and provides the best statistical results in the level 3 domain. The full year of 2003 was used for comparison where three distinct nested options were tested. The first test followed the nesting approach employed by Rusu & Guedes Soares (2013), where a two-nested study in the Sines area used the level 2.2 domain and the level 3 domain. The second experiment is similar to the first only altering the size of the level 2.2 domain (now called level 2.1 domain) by reducing its perimeter to one-third smaller than the level 1 domain implementation's. The final test encompassed a direct transition from the level 1 domain to the level 3 domain, bypassing the intermediate levels (2.1 and 2.2). All three experiments maintained the same location for the level 3 domain as indicated in the referenced literature. Analyzing Table 4, it is possible to see the resemblance between all the statistical values except for the bias, which is clearly lower on the third test that directly jumps from the level 1 domain to the level 3 domain, which was used for the consequent simulations.

Table 4. Nests comparison in level 3 domain – 2003.

n=2720	Bias	RMSE	SI	r	Test
Hs	-0.304	0.515	0.292	0.913	1
	-0.299	0.511	0.290	0.912	2
	-0.269	0.500	0.283	0.909	3
Tm02	-0.128	1.418	0.222	0.791	1
	-0.084	1.420	0.222	0.787	2
	0.021	1.421	0.222	0.782	3

From this part of the validation process and beyond, the comparison between the energy transport from the model and the buoy was added. Therefore, equation 6 was used to calculate the energy transport in the buoy's location.

$$PW = \frac{\rho g^2}{64\pi} T_e H_s^2 \tag{6}$$

where ρ is the water density (1025 kg/m3), g the acceleration of gravity (9.805 m/s2), Hs the significant wave height (m), T_e the wave energy period (s) defined as $T_e = \alpha T_p$ where alpha was assumed as 1 (Cornett 2006, Boronowski et al. 2010) and Tp the peak period. Four tests were performed and then compared to the model with default physical parameters taken out of the previous test (Test 3 from Table 4). All the tests (except the default) were implemented with the new bathymetry described in section 2.1 and the experiments are as follows: the first test contains the new bathymetry and has still the wind input on the model; the second test is the same as the first but without the wind input on the model; the third test will be like the first but with the triads wave-wave interactions turned on and the fourth and final test will follow the formulation of the third (triads on) but without the wind input on the model. The results are shown in Table 5.

The choice of the best parameterizations came down to the analysis of the bias. The priority in this type of choice is always to go for the one that presents the best results for both the significant wave height and the energy transport. However, because the "no Wind + Triads" test excels in the significant wave height and the "Wind" experiment on the Energy, it was decided to give priority to the significant wave height since this test is related to the waves in coastal waters theory (Holthuijsen 2010). This means that the "no Wind + Triads" test provides the best parameterizations for the final simulation.

Following part of the methodology for validating the model proposed by Abu Zed et al. (2022), one last group of tests was performed. The goal was to test the effect of several coefficients for determining the rate of white-capping dissipation on the model, as the literature suggests.

The coefficients chosen were the same as proposed by the author (between 0.15e-4 and

0.315e-4), having values higher and lower than the default one. The results showed that the model with a white-capping dissipation rate coefficient of 0.15e4 is the one providing the best results. The parameterizations for the final run were the following: ST6 method for the third-generation mode for wind input, resolution of 100x100 bins, a jump from level 1 straight to level 3, triads turned on coupled with the deactivation of the wind input and a whitecaping dissipation rate coefficient of 0.15e-4. Finally, the comparison between the totality of the simulated data against the buoy's observations can be performed. The analysis was made for a period of 10 years, from January 1993 to December 2003, due to the lack of data available for the Sines buoy. Table 6 shows the final results.

Table 5. Wind and triads comparison in level 3 domain.

n=2720	Bias	RMSE	SI	r	Method
Hs	-0.269	0.500	0.283	0.909	Default
	-0.257	0.486	0.275	0.912	Wind
	-0.226	0.468	0.265	0.913	no Wind
	-0.280	0.500	0.283	0.912	Wind + Triads
	-0.217	0.470	0.267	0.910	no Wind + Triads
Tm02	0.021	1.421	0.222	0.782	Default
	0.034	1.429	0.223	0.779	Wind
	-0.553	1.421	0.222	0.790	no Wind
	0.008	1.437	0.225	0.779	Wind + Triads
	-0.542	1.425	0.223	0.786	no Wind + Triads
Energy	-0.254	18.741	0.743	0.889	Default
	-0.059	19.076	0.756	0.886	Wind
	-0.612	18.844	0.747	0.888	no Wind
	-0.554	18.390	0.729	0.894	Wind + Triads
	0.225	18.986	0.753	0.887	no Wind + Triads

Table 6. Final run comparison in level 3 domain- 1993 to 2003.

n=2720	Bias	RMSE	SI	r
Hs	-0.255	0.475	0.281	0.903
Tm02	-1.063	1.588	0.246	0.807
Energy	-0.830	16.475	0.729	0.883

3.2 *Energy assessment*

3.2.1 *Regional energy*

One of the many advantages of the SWAN model is the computation of the energy transport components (W/m, i.e. energy transport per unit length of wavefront), which can provide as output the energy transport both in the x and y axis, using the following equations (Silva et al. 2018):

$$p_x = \rho g \iint c_x E(\sigma,\theta) d\sigma d\theta \quad (7)$$

$$p_y = \rho g \iint c_y E(\sigma,\theta) d\sigma d\theta \quad (8)$$

where c_x and c_y are the propagation velocities of the wave energy in the geographical grid and $E(\sigma, \theta)$ the wave energy spectrum. Hence, the total wave energy transport (wave power) is:

$$P = \sqrt{P_x^2 + P_y^2} \quad (9)$$

In the totality of the 22 years of simulation, the maximum energy recorded was 432 Kw/m. The average energy obtained within the domain was 23.7 kW/m with a mean distribution maximum of 27.4 kW/m lo- cated just northwest of the port of Sines. Figure 2 shows the distribution map of the average energy.

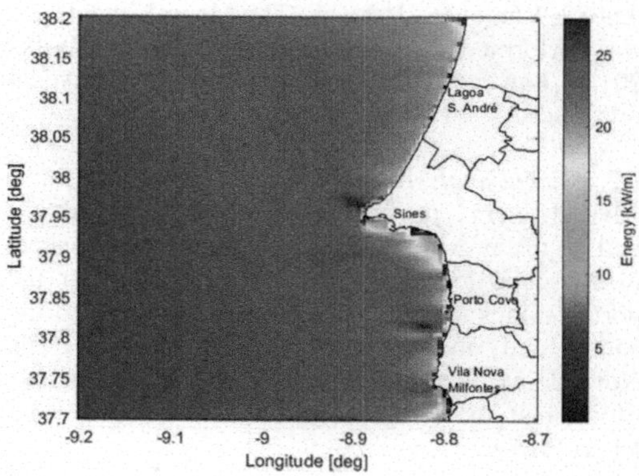

Figure 2. Energy transport over 22 years within the domain.

Seasonal averages from the totality of the 22 years of the simulation were computed. Following the trend of the significant wave height, the winter was the most energetic season within this domain, with a maximum of 48 kW/m and an average of 41 kW/m, while the summer was the least energetic season with a maximum of 10 kW/m and an average of 8 kW/m.

The seasonal variability index remained almost constant throughout the entire domain, with larger values to the north and south of the port, likely due to the shadow effect from the port and Cabo Espichel (Ponce de Leon & Guedes Soares 2005), reaching a maximum of 1.98 and an average of 1.39. These values are still lower than the ones for higher latitudes and within the expected spectrum for the location where the study is being performed, as shown by (Cornett 2008).

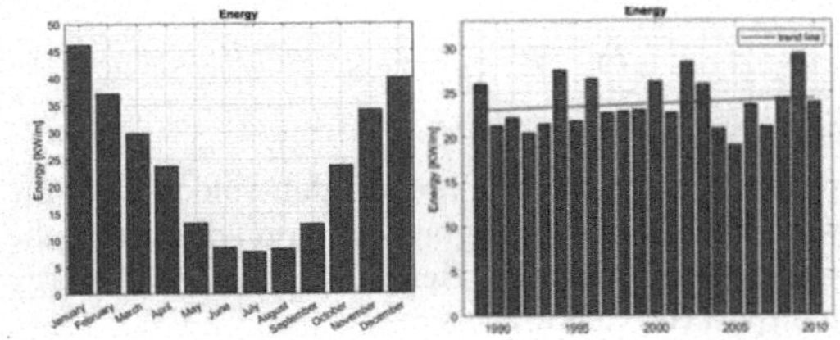

Figure 3. Monthly and yearly distribution for Energy transport.

Figure 3 shows that the most energetic month was January with an average energy of 46 kW/m, while the least energetic month was July with an average energy of 7.8 kW/m. The most energetic year was 2009, with an average energy of 29 kW/m, while the least energetic year was 2005, with an average energy of 19 kW/m. The trend line shows that the average energy increases throughout the years with a mean rate of 60.8 W/m per year or 0.608 kW/m every 10 years. The Mann-Kendall test was performed at a significance level of 5% (Wilks 2011), however, the trend presents no statistical relevance.

3.2.2 *Local analysis*

Several energy peaks near the coast were identified in the previous subsections and three clusters of energetic locations were chosen: two near the port of Sines (one to the northwest and one to the south) and one between Porto Covo and Vila Nova de MilFontes. Figure 4 shows the six points of interest that were selected and labelled from P1 to P6.

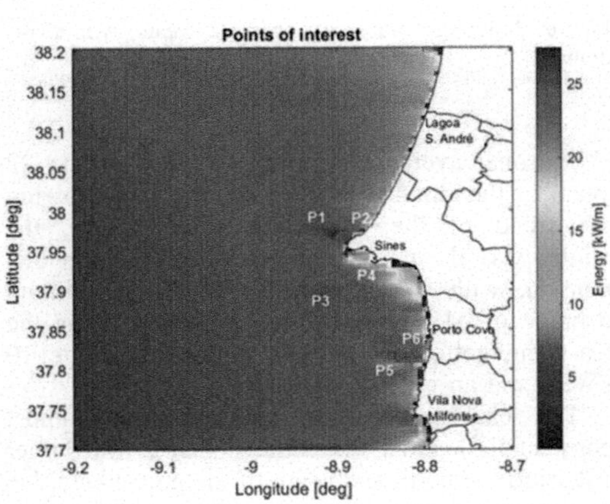

Figure 4. Selected points of interest for energy extraction.

Table 7 shows a resume of the energetic power, the significant height, the energetic period, the coefficient of variation related to the energy, the average electric production provided by the most powerful WEC (Wave Dragon) and the depth of each of the six points.

Table 7. Sines points of interest resume for 22 years of simulation.

Point	Energy [kW/m]	Hs [m]	Te [s]	Cv [-]	Pe [kW]	Depth [m]
P1	27.7	1.97	11.60	1.33	1647	24.3
P2	27.9	1.99	11.72	1.30	1667	18.0
P3	25.2	1.93	11.29	1.31	1639	55.5
P4	25.2	1.92	11.29	1.32	1630	55.0
P5	27.2	1.97	11.57	1.28	1685	20.5
P6	27.6	2.02	11.67	1.26	1743	15.4

Through the analysis of Table 7, it is clear that the most energetic site (P2) does not correspond to the most electric production. Nevertheless, the site with the lowest coefficient of variation (P6) is the one that provides the best average electric production, being only the third most energetic location. Energy power is not the only or the most crucial driver for electric power production. Consistency and reliability of the most frequent sea states is also one key aspect that should always be considered.

The sea state characterization at the point which provided the largest average electrical power production (P6) was performed in which all three wave parameters are distinguished in Figure 5.

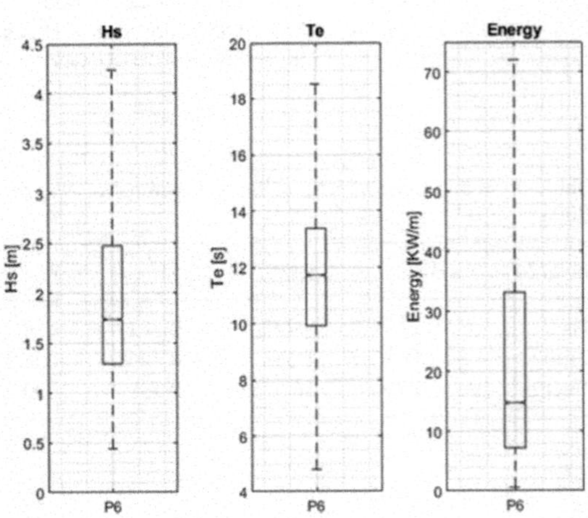

Figure 5. Box plots for Hs, Te and energy for P6.

Analyzing them separately, the box plot for the significant wave height shows that in point P6 the median is 1.73 meters with a lower adjacent of 0.44 meters and Figure 5: Box plots for Hs, Te and energy for P6 an upper adjacent of 4.24 meters. The box plot for the energetic period shows that in point P6 the median is 11.7 seconds with a lower adjacent of 4.8 seconds and an upper adjacent of 18.5. The box plot for the energetic period shows that in point P6 the median is 11.7 seconds with a lower adjacent of 4.46 seconds and an upper adjacent of 18.37. The box plot for the energy transport shows that in point P6 the median is 14.81 kW/m with a lower adjacent of 0.63 kW/m and upper adjacent of 72.02 kW/m.

To complete the characterization of point P6, the seasonal and yearly distributions are presented in Figure 6.

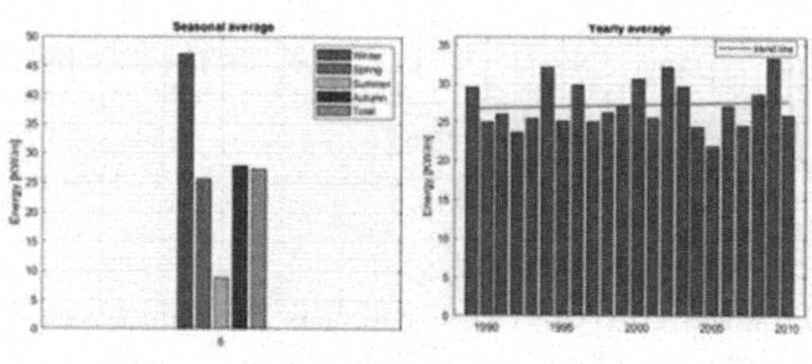

Figure 6. Seasonal and yearly distribution for energy transport.

As expected, the winter season is the most energetic with an average value over the 22 years of simulation of 47 kW/m, while the summer is the least energetic with an average of 8.9 kW/m. Spring and autumn are very alike with an average energy of 25.5 and 27.8 kW/m, respectively. The total average energy for this location is 27.3 kW/m, which is higher than the average energy in the entire domain. The most energetic year was 2009 with 33.4 kW/m and the least energetic year was 2005 with an average of 21.8 kW/m.

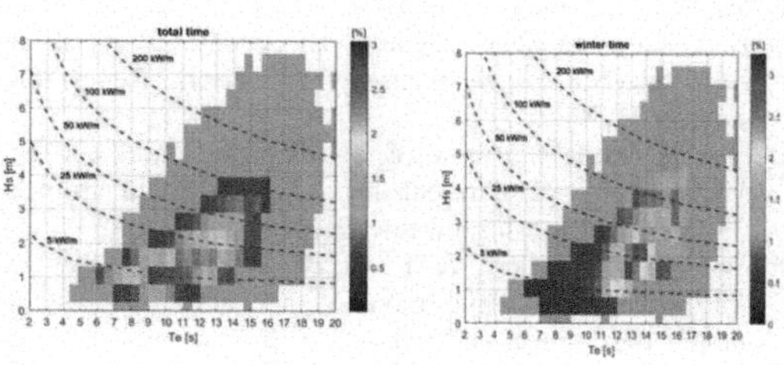

Figure 7. Scatter diagrams with isolines represented (P6).

The trend over the 22 years is positive and has an energetic increment of 46 W/m per year or 0.46 kW/m per 10 years which is lower than the average trend over the entire domain, although with no significant trend under the Mann-Kendall test. The diagrams of Figure 7 consolidate the results written above. In the total time, the majority of the sea states produce an energy output between 5 kW/m and 25 kW/m which is within the expected however, in the winter time, the sea states are less dispersed through the diagram and present larger values of both Hs and Te with an energy output between 25 kW/m and 50 kW/m, making the winter the best and most reliable season to harvest wave energy.

4 DISCUSSION

During the calibration of the model, it was observed that some physical parameterizations perform better in one domain and worse in others. For the level 1 domain, was the third-generation mode for wind input ST6 coupled with default parameterizations for the rest of the physical phenomena (quadruplet wave-wave interactions, whitecapping and bottom friction). For the level 3 domain, was the third-generation mode for wind input ST6 with no wind input (no wave generation by wind) coupled with triads wave-wave interactions activated, default bottom friction and whitecapping with a dissipation rate coefficient of 0.15e-4. Hence, for the large domain (level 1) the wave generation by wind and the quadruplet wave-wave interactions were the most relevant phenomena while for the smaller domain (level 3) the triad wave-wave interactions and the whitecapping became the most suitable. These findings are coherent with the theory presented by Holthuijsen (2010) and with several studies where nested domains were used (Rusu & Guedes Soares 2012, Rusu 2018). A resume of the physical parameters used in each level is shown in Table 8.

The singular nature of this study makes the comparison with other literature not so direct. Many studies cover the sea state and energetic conditions for the west Iberian coast, but only a few characterize the coast of the vicinity of the port of Sines. From the literature presented, the mean values for the energy transport on the Portuguese coast vary between 6 and 58 kW/m, reaching maximums of 80 kW/m. Although the methodology and data used for each study are distinct, these values clearly show that the results for the port of Sines domain shown in Table 9 are within the expected.

Another essential aspect to take from this study is the depths of each chosen point. When comparing those with the functionality depths of the WECs present in Table 2, it is possible to see that only the points P3 and P4 have depths that are within the functionality spectrum of the converter Wave Dragon, which is the most powerful. This means that deploying this type of WEC in the location that provides the largest average electric power production (P6) would not be possible. Therefore, having the depth into account, point P3 would be the most appropriate. A solution could be scaling the converters in size to fulfil the demands of the extraction location in terms of wave climate and functionality depth (Rusu 2014).

Table 8. Resume of the physical parameterizations used in levels 1 and 3 domains.

SWAN grid	Level 1	Level 3
Bathymetry	resolution 0.008°	resolution 0.001° (EMODnet 2023)
Wind forcing	ERA-INTERIM database (Setchell 2020)	Deactivated
Quadruplets	Activated (default)	Activated (default)
Triads	Deactivated	Activated (default)
Whitecapping	Default	Default with dissipation rate coefficient of 0.15e-4
Bottom friction	Default	Default

Table 9. Sines monthly energetic resume for 22 years of simulation.

kW/m	Jan	Feb	Mar	Apr	May	Jun	Jul	Aug	Sep	Oct	Nov	Dec	Total
	46.07	37.06	29.72	23.59	13.12	8.67	7.84	8.35	12.94	23.65	34.04	39.83	23.74

5 CONCLUSIONS

Three major conclusions can be taken from this work. Firstly, the most relevant physical phenomena in the level 1 domain are the wave generation by wind and the quadruplet wave-wave interactions, while in the level 3 domain are the triad wave-wave interaction and whitecapping dissipation. Secondly, the oceanic conditions surrounding the port of Sines are characterized by average significant wave heights of 1.98 meters with values between 1.3 and 2.5 meters, average mean wave periods of 7.3 seconds with values for the energetic period between 10 and 13 seconds and average energetic power of 23.7 kW/m with values between 7 and 33 kW/m. Finally, regarding only the electric production of the most powerful WEC, the best location for WEC deployment would be point P6. Taking into account the electric production, the functionality depths of the converters and the routes and mooring locations, point P3 would be the best choice. Nevertheless, the sea around the port of Sines holds significant potential for renewable energy, especially wave energy. The irregular bathymetry that concentrates energy near the harbor combined with the consistent sea state conditions makes it an attractive site for wave energy extraction. The port, a crucial liquid cargo route, offers infrastructure and high-tension connections to feed the electric grid or the port itself. While a shadowing effect may affect energy potential, it provides shelter, increasing the converter's longevity and facilitating maintenance. To optimize converter deployment, factors like depth, avoiding shipping routes and mooring points must be considered.

ACKNOWLEDGEMENTS

This work contributes to the Strategic Research Plan of the Center for Marine Technology and Ocean Engineering (CENTEC), which is financed by FCT under contract UIDB/UIDP/00134/2020.

REFERENCES

Abu Zed, A. A., R. M. Kansoh, M. M. Iskander, & M. Elkholy (2022). Wind and wave climate southeastern of the mediterranean sea based on a high-resolution SWAN model. Dynamics of Atmospheres and Oceans 99, 101311.

Aderinto, T. & H. Li (2018). Ocean wave energy converters: Status and challenges. Energies 11(5), 1250.

Anastas, G., J. A. Santos, C. Fortes, & L. V. Pinheiro (2022). Energy assessment of potential locations for owc installation at the Portuguese coast. Renewable Energy 200, 37–47.

Babarit, A. (2017). Ocean Wave Energy Conver-sion: Resource, Technologies and Performance. Elsevier.

Babarit, A., J. Hals, M. Muliawan, A. Kurniawan, T. Moan, & J. Krokstad (2012). Numerical bench-marking study of a selection of wave energy converters. Renewable Energy 41, 44–63.

Bento, A. R., P. Martinho, & C. Guedes Soares (2018). Wave energy assessment for northern Spain from a 33-year hindcast. Renewable Energy 127, 322–333.

Booij, N., R. C. Ris, & L. H. Holthuijsen (1999). A third-generation wave model for coastal regions: 1. model description and validation. Journal of Geophysical Research: Oceans 104(C4), 7649–7666.

Boronowski, S., P. Wild, A. Rowe, & G. Cornelis Van Kooten (2010). Integration of wave power in haida gwaii. Renewable Energy 35(11), 2415–2421.

Cavaleri, L. & P. M. Rizzoli (1981). Wind wave prediction in shallow water: Theory and applications. Journal of Geophysical Research: Oceans 86(C11), 10961–10973.

Cornett, A. (2006). Inventory of Canada's marine renewable energy resources. Canadian Hydraulics Centre, National Research Council of Canada: Ottawa, ON, Canada.

Cornett, A. (2008). A Global Wave Energy Resource Assessment, Volume 50. Publication Title: In: Proceedings of the eighteenth international offshore and polar conference, Vancover, Canada.

Dee, D. P., S. M. Uppala, A. J. Simmons, P. Berris-ford, P. Poli, S. Kobayashi, U. Andrae, M. Balmaseda, G. Balsamo, d. P. Bauer, et al. (2011). The era-interim reanalysis: Configuration and performance of the data assimilation system. Quarterly Journal of the royal meteorological society 137(656), 553–597.
EDP (2022). Sines' active future. Visited on 13/12/2022.
EMODnet (2023). EMODnet map viewer. Visited on 20/03/2022.
Guedes Soares, C. (2011). An operational wave forecasting system for the Portuguese continental coastal area. Journal of Operational Oceanography 4, 17–27.
Guedes Soares, C., A. R. Bento, M. Gonçalves, D. Silva, & P. Martinho (2014). Numerical evalution of the wave energy resource along the Atlantic European coast. Computers & Geosciences 71, 37–49.
Gunn, K. & C. Stock-Williams (2012). Quantifying the global wave power resource. Renewable Energy 44, 296–304.
Holthuijsen, L. H. (2010, February). Waves in Oce-anic and Coastal Waters. Cambridge University Press. International Energy Agency (2021).
Portugal 2021 Energy Policy Review. IEA Energy Policy Reviews. OECD.
Jahangir, M. H., R. Alimohamadi, & M. Montazeri (2023). Performance comparison of pelamis, wavestar, langley, oscillating water column and aqua buoy wave energy converters supplying is-lands energy demands. Energy Reports 9, 5111–5124.
Kofoed, J. P., P. Frigaard, E. Friis-Madsen, & H. C. Sørensen (2006). Prototype testing of the wave energy converter wave dragon. Renewable En-ergy 31 (2), 181–189.
Komen, G., S. Hasselmann, & K. Hasselmann (1984). On the existence of a fully developed wind-sea spectrum. Journal of physical oceanog-raphy 14(8), 1271–1285.
Lavelle, J. & J. P. Kofoed (2011). Power production analysis of the OE buoy WEC for the CORES project. Lund, H. (2007). Renewable energy strategies for sustain-able development. energy 32(6), 912–919.
Lopez, I., J. Andreu, S. Ceballos, I. Mart ´ ´ınez De Alegr´ıa & I. Kortabarria, (2013). Review of wave energy technologies and the necessary power-equipment. Renewable and Sustainable Energy Reviews 27, 413–434.
Municipio de Sines (2023). Sines. Visited on 26-10-2023.
Ponce de Leon, S. & C. Guedes Soares (2005). On the sheltering ´ effect of islands in ocean wave models. Journal of Geophysical Research: Oceans 110(C9).
Pontes, M. T. (1998). Assessing the european wave energy resource. Journal of Offshore Mechanics and Arctic Engineering 120, 226–231.
Rogers, W. E., A. V. Babanin, & D. W. Wang (2012). Observation-consistent input and whitecapping dissipation in a model for wind-generated sur-face waves: Description and simple calculations. Journal of Atmospheric and Oceanic Technology 29(9), 1329–1346.
Ruester, S., S. Schwenen, M. Finger, & J.-M. Gla-chant (2014). A post-2020 EU energy technolo-gy policy: Revisiting the strategic energy tech-nology plan. Energy Policy 66, 209–217.
Rusu, E. (2014). Evaluation of the wave energy con-version efficiency in various coastal environ-ments. Energies 7(6), 4002–4018.
Rusu, E. (2018, April). Numerical modeling of the wave energy propagation in the iberian near-shore. Energies 11(4), 980.
Rusu, E. & C. Guedes Soares (2009). Numerical modelling to estimate the spatial distribution of the wave energy in the Portuguese nearshore. Renewable Energy 34(6), 1501–1516.
Rusu, L. & C. Guedes Soares (2012). Wave energy assessments in the azores islands. Renewable Energy 45, 183–196.
Rusu, L. & C. Guedes Soares (2013). Evaluation of a high-resolution wave forecasting system for the approaches to ports. Ocean Engineering 58, 224–238.
Rusu, L., P. Pilar, & C. Guedes Soares (2008). Hindcast of the wave conditions along the west iberian coast. Coastal Engineering 55(11), 906–919.
Salcedo, F., P. Ruiz-Minguela, R. Rodriguez, P. Ric-ci, & M. Santos (2008). OCEANTEC: Sea trials of a quarter scale prototype. In Proceedings of 8th European wave tidal energy conference, pp. 460–5.
Setchell, H. (2020). ECMWF reanalysis - interim. Visited on 27- 09-2023.
Silva, D., A. R. Bento, P. Martinho, & C. Guedes Soares (2015). High resolution local wave ener-gy modelling in the Iberian Peninsula. Energy (Oxf.) 91, 1099–1112.
Silva, D., P. Martinho, & C. Guedes Soares (2018). Wave energy distribution along the Portuguese continental coast based on a thirty-three years hindcast. Renewable Energy 127, 1064–1075.
Silva, D., E. Rusu, & C. Guedes Soares (2013). Evaluation of various technologies for wave en-ergy conversion in the Portuguese nearshore. Energies 6(3), 1344–1364.
SWAN (2019). SWAN scientific and technical documentation Cycle III Version 41.20. Delft Uni-versity of Technology.
Wilks, D. S. (2011). Statistical Methods in the At-mospheric Sciences, Volume 14. Academic Press.
Yan, L. (1987). An improved wind input source term for third generation ocean wave modelling. Sci-entific report. Koninklijk Nederlands Meteorol-ogisch Instituut.
Yolcan, O. O. (2023). World energy outlook and state of renewable energy: 10-year evaluation. Innovation and Green Development 2(4), 100070.

Advances in Maritime Technology and Engineering – Guedes Soares & Santos (Eds)

Nearshore wind-wave spectrum estimation with a low-cost video system

M. Vieira & C. Guedes Soares
Centre for Marine Technology and Ocean Engineering (CENTEC), Instituto Superior Técnico, Universidade de Lisboa, Lisbon, Portugal

ABSTRACT: Two conventional, low-cost action cameras with a freeware synchronization technique are used to estimate the time-varying three-dimensional sea surface elevation map and its associated wave number frequency spectrum. Spatio-temporal measurements in very shallow water are conducted with the cameras and are validated with a moored pressure sensor as single point data reference inside the reconstructed area. The error in significant wave height is of the order of few centimeters and the treatment of the images obtained is perfectly able to identify the peak period of the in-situ measured spectrum. Although not in the ideal conditions of perfect light diffusion and camera settings, the stereo video uncertainties could be reduced to a quantified error of the order of millimeters.

1 INTRODUCTION

Understanding the nearshore wave dynamics is essential for both scientific and societal aspects. Despite the nearshore represents only 10^{-3} of the ocean's surface area, it is the part that is best known to the global population and is a critical resource for industry, ecosystem services, commerce, recreation and defence (Holman, 2013). However, direct measurements of the directional spectrum of ocean waves with in-situ instruments are logistically challenging and expensive in this part of the ocean.

The bottom shear stresses induced by waves in the surf zone can be significantly intense making it impossible to attach instruments or cables below the waves; in addition, tension in the anchor lines can interfere with measurements and produce inaccurate records during higher energy conditions. Pressure sensors rely on the assumption of the hydrostatic water column, underestimating the wave break and not representing the asymmetries developed around wave crests. The turbulence created in the water column also makes the use of acoustic sensors unfeasible. Satellites are unreliable close to shore and cannot measure the same area for a long continuous time. Radars on the X-band have been used for measurements relatively close to shore (Izquierdo & Guedes Soares 2005), but even those become problematic very close to the beach. As a result, innovative measurement techniques are still needed to better assess the complexity of fluid dynamics processes in these regions.

The advancement of optical remote sensing techniques in synchronization with low-cost platforms and new open-source algorithms provides the possibility of collecting large amounts of information at a reasonable instrumental and computational cost (Rodríguez-Padilla, 2021). In this context, stereo video techniques, which employ two or more cameras to record the same area of the sea surface, providing the spatial sea surface elevation and its evolution over time, can allow low-cost devices to capture a good estimation of the wave field and associated spectrum.

For a single-valued surface elevation $\eta(x,y,t)$, the full spectrum of the ocean surface is the three-dimensional (3D) spectrum $E(kx,ky,f)$. This spectrum is only accessible to instruments that cover the spatial behaviour of ocean waves with high acquisition rates, such as stereo video systems based on stereo reconstruction from visible spectrum (Benetazzo, 2006; Bergamasco et al., 2017; Vieira et al., 2020; Guimarães et al., 2020; Bergamasco, 2021; Pistellato, 2021), lidar cameras in which also operate visible light or close to it (Kabel 2019), infrared imagery (Sutherland & Melville, 2013), or using polarimetric slope sensing (Zappa et al., 2012). From the full 3-D spectrum, other spectrum products, such as 2D and 1D spectra, can be derived.

The 2D directional spectra of wind-generated waves $E(\theta,f)$ are broadly applied in many fields as basic air-sea interaction, upper mixed layer dynamics, wave forecasting, satellite surveillance and engineering design of marine structures and vehicles (Donelan et al., 1985). The directional spectra provide a more detailed representation of sea conditions and their broader availability has prompted the description of sea conditions based on climatic spectra (Lucas et al., 2011) instead of relying on the conventional joint distribution of significant wave height and mean period (Bitner-Gregersen et al., 2022).

DOI: 10.1201/9781003508779-75

In this context, use is made here of two conventional, low-cost, action cameras with a freeware synchronization technique to estimate the time-varying three-dimensional water surface elevation map and its associated directional wave spectrum. Spatio-temporal measurements in very-shallow water are conducted using moored pressure sensors as single-point data reference inside the reconstructed area.

This paper is organized as follows. The methodology applied to acquire sea waves data is explained in Section 2. Section 3 presents the results and analysis of the reconstructed wave fields and the uncertainties on the stereo-reconstruction are given in Section 4. Section 5 accounts for the conclusions.

2 METHODOLOGY

2.1 *Hardware – Wave data acquisition*

The data showcased in this study were gathered on May 26, 2021, near a sandy beach at Costa da Caparica in Portugal. The Low-Cost Wave Acquisition Stereo System (Vieira et al., 2020) employed here consists of a pair of synchronized GoPro Hero 9 Black action cameras. These cameras feature a 1/2.3" Complementary Metal Oxide Semiconductor (CMOS) digital sensor, operating at a rate of 24 frames per second. The spatial resolution is 1920 × 1080 pixels, with an aperture of f/2.8 and a focal length of 27 mm in a narrow field of view (FoV) mode. The recorded video data is stored in an MP4 H.265 compression format. It is important to note that the methodology introduced is not limited to a specific camera brand, and the quality of stereo video reconstruction depends on factors like camera quality, experimental settings, and environmental conditions.

The GoPro stereo system is arranged with a baseline (distance between cameras) of 0.96m, set up on tripods in a simple and cost-effective manner. Considering the chosen experimental configuration, which factors in the inclination of the cameras relative to the still water level and the chosen baseline, the measured area spanned from 10 m (closest point) to 80 m (farthest point) away from the imaged area. The ratio of the distance between cameras to the target distance can vary based on the experiment setup and camera lens selection. Recent studies utilizing stereo video techniques propose ratios between 0.025 and 0.1 (Benetazzo, 2017; Vieira, 2020; Guimarães, 2020). In the present experiments, this ratio ranges from 0.05 in the near field, close to the stereo system, to 0.03 in the far field around the wave gauge position. Within the imaged area, a pressure sensor is positioned at a depth of about 2 m, approximately 25 m away from the cameras, with a sampling rate of 1 Hz. Three samples of 28-minute videos are recorded at a 24 Hz sample rate, resulting in 20,160 frames after subsampling to 12 Hz. The findings presented in this study are based on the examination of a single stereo video record.

2.2 *Software – Wave data processing*

2.2.1 *Image synchronization and camera calibration*

In the context of geometrical camera calibration, the objective is to determine the camera parameters that define the relationship between 3D reference coordinates and 2D image coordinates (Clarke, 1998). The calibration process entails estimating intrinsic parameters, extrinsic parameters, and distortion coefficients. Intrinsic calibration provides the focal length vector, coordinates of the digital sensor's central point, and lens distortion vectors. Extrinsic calibration enables the determination of the relative position of the two cameras. The intrinsic calibration step utilizes the MATLAB toolbox — cameraCalibrator (Zhang, 2000), using 30 images with a chessboard at various distances and angles relative to the camera. The mean reprojection error, calculated from the calibration images, is 0.5 pixels.

Ensuring a robust correlation between matching points is crucial for the synchronization scheme implemented, turning its operational execution into a complex task. In this study, a method for stereo video frame synchronization based on Time Lagged Cross Correlation (TLCC) of the audio signal was employed (Vieira et al., 2020). This method synchronizes two independent video records based on background noise, accurately determining the time lag between records. As anticipated, the correct number of 3D reconstructed points, even with varying sample rates, is achieved with no lag between frame images.

2.2.2 *Three-dimensional ocean surface reconstruction*

The reconstruction of the spatio-temporal wave field involves a series of three sequential steps. The initial step is extrinsic calibration, where the relative spatial position of the two cameras is determined. This calibration, relying on known intrinsic parameters, is automatically recovered by analyzing a subset of stereo frames. In the second step, each stereo pair is processed to generate a dense 3-D point cloud of the observed sea surface. The third step includes filtering each point cloud and interpolating them onto a regular grid, resulting in a data cube containing spatial surface elevation over time (Figure 1).

The first two steps make use of the standard WASS pipeline (Wave Acquisition Stereo System) (Bergamasco, 2017), while Matlab scripts are employed for the final step.

The fundamental principle of stereo reconstruction relies on understanding the imaging model of the two cameras (intrinsic parameters) and their relative arrangement described by a rotation matrix and a translation vector (extrinsic parameters). The correspondence problem is critical in this process, involving the establishment of pixel-to-pixel correspondences between points on each camera. Once established, the exiting rays of the two cameras intersect to recover 3-D coordinates through triangulation, utilizing extrinsic camera parameters.

The WASS pipeline incorporates robust point matching and auto-calibration, particularly beneficial when the relative camera position varies during measurements compared to the extrinsic calibration. This is especially relevant in stereo video setups covering large areas, where on-field extrinsic calibration may be challenging. A least-squares plane fitting is performed for each 3-D point cloud to estimate the mean sea plane position in the right camera reference frame. To minimize uncertainty, the Gaussian nature of the sea surface elevation is leveraged to average plane parameters across each record. The average seaplane is then used to compute the transformation mapping 3-D data from the camera to a geographic reference frame. Each geographically aligned point cloud is linearly interpolated on a regular horizontal 2-D grid, resulting in a discretized surface elevation map suitable for wave analysis.

3 RESULTS

Each pair of synchronized video images produces a three-dimensional cloud of points representing elevation which is then gridded into a 10x10 m surface with a 5 cm resolution resulting in a 3-D sea surface elevation map η(x,y,t) (Figure 1). Remote sensing instruments provide the advantage of capturing directional surface elevation fields, allowing for the direct calculation of the three-dimensional frequency-wavenumber spectrum. This spectrum unveils crucial wave properties, including non-linear interactions and energy characteristics. Additionally, derived parameters such as Mean squared slope, 3-D wavenumber, 2-D directional spectrum $E(f,\theta)$ and 1-D variance spectrum $E(f)$ can be extracted from the measured data.

The spatio-temporal wave dynamics within the camera's field of view enable the computation of key sea state parameters, such as significant wave height (Hs), peak wave period (Tp), and peak wave direction (Dp). The choice of the instrument in real ocean wave measurements has an impact on acquiring frequency limits, influencing the integration of the frequency-wavenumber spectrum $E(kx,ky,f)$. Defining these limits is crucial and can affect the estimation of integrated variables based on the energy level at the frequency's limits (Guimarães, et al. 2018).

Describing the sea surface gravity wave spectrum in terms of wavelength or wavenumber introduces an added complexity with the latter being a vector, unlike the scalar nature of frequency. Both representations convey information about energy dispersion over oscillatory components. However, focusing on a single variable may provide energy information but not the wave phase. In this study, spatio-temporal statistics in the short term are addressed, considering the requirements of stationarity and homogeneity in the sea state representation (Ochi, 1998). The 3-D wavenumber and frequency spectra are obtained by applying a 3-D Fast Fourier Transform (FFT) to convert physical space into a three-dimensional spectral space.

Figure 1. 3-D sea surface elevation map $\eta(x,y,t)$ measured by the Low-Cost stereo video system in Costa da Caparica at 6:20 pm on 26 May 2021 during a super moon event.

The auto and cross spectra obtained from the measured data are estimated using the methodology outlined by Welch (1967). Fourier transforms are applied to time series comprising 21,600 samples, incorporating a 50% overlap and utilizing a Hamming window in all three dimensions for the elevation maps $\eta(x,y,t)$. The time intervals for analysis are set at 85.25 seconds (1024 frames), and the modulus squared of the three-dimensional Fourier transform is averaged over 39 overlapping time intervals. This results in 20 independent intervals, providing 40 degrees of freedom for each spectral density in the (kx,ky,f) space.

The raw sample rate of the sea surface data acquired by the Low-Cost WASS is 24 Hz. However, for wind wave analysis, it is recommended to sample at least four times the dominant frequency (Ardhuin, 2020). To streamline data processing, the surface elevation data are subsampled to 12 Hz in all records. To prevent aliasing of energy in the sea surface above the Nyquist frequency, a low-pass filter at 0.5 Hz is applied before performing the FFT. This filtering effectively removes signals that would otherwise be aliased.

Figure 2 illustrates a similar pattern in reproducing the overall spectral shape and the peak of the spectrum. However, it is noteworthy that the stereo video system tends to overestimate the energy for frequencies higher than approximately 0.15 Hz. Although subtle differences in wave spectral characteristics emerge between the stereo video and pressure sensor records, these distinctions do not significantly impact the integral wave parameters, as indicated in the box corner of Figure 2.

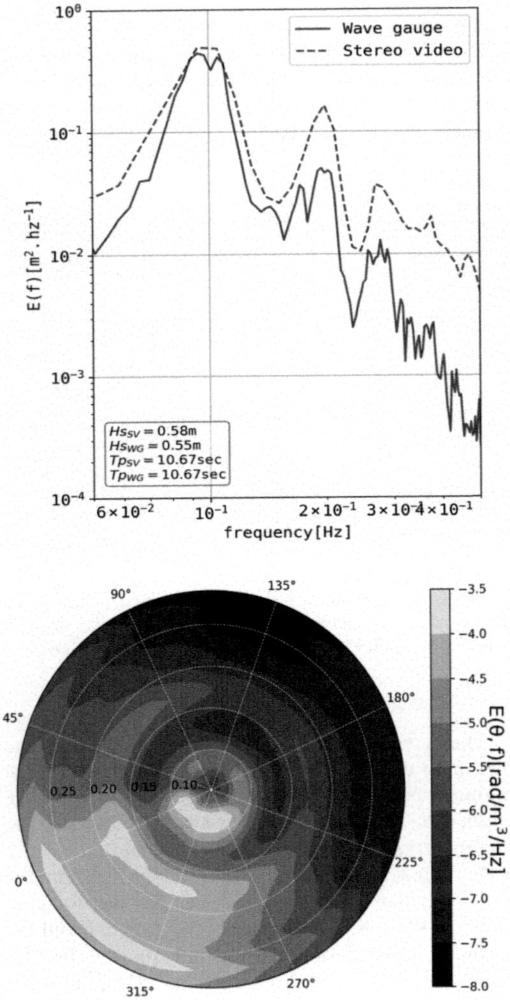

Figure 2. (top) E(*f*) 1-D frequency spectrum from a 28-minute interval: the line style differentiates the instruments used. (bottom) 2-D directional spectrum E(θ,*f*) calculated from stereo video data.

4 UNCERTAINTY ANALYSIS

During the 3D reconstruction process, identifying errors in corresponding points is a challenging task. Various factors, including image quality, environmental conditions (such as sun glint, water transparency, and rain), synchronization methods, and experiment setup, directly impact the matching process. In most cases, matching errors are negligible if there is sufficient texture on the water surfaces (Benetazzo, 2006) and are minimized when the wave steepness is significantly larger than the inclination of the stereo cameras' optical axis relative to the mean sea level (Jahne, 1993).

The uncertainty in recovering 3D coordinates can be assessed through quantization error (Rodriguez, 1990; Benetazzo et al., 2016), which depends on factors such as camera cell size, pixel numbers, focal length, baseline distance between cameras, camera orientation, and the distance between the cameras and the target. For this specific configuration, the estimated root mean square quantization error for sea elevation is approximately 2mm (z-axis). These values are sensitive to the cameras used, distance from the surface, shooting angles, and sea state conditions. A more comprehensive discussion about expected errors and precision is provided in (Rodriguez, 1990). In summary, the evaluation of low-cost stereo observations and the analysis of related errors indicate that the new methodology yields meaningful data for wave measurements.

5 CONCLUSIONS

In this study, a cost-effective and straightforward method is introduced for estimating the 3D ocean surface topography and its directional spectrum in the surf zone, particularly challenging due to gravity wave forces and hydrodynamics in the near-shore zone. The approach, utilizing open-source libraries, eliminates the need for specialized equipment, cables, or dedicated batteries. Although reliant on adequate lighting and a stable platform near the surface, the reconstruction quality depends on lighting conditions, experimental setup, and camera quality. Post-processing requires storage space and processing power but can be easily executed on a regular personal computer. The method's limitation is typical for any stereo video system. Despite these considerations, the methodology provides accurate wave measurements, with errors in significant wave height within a few centimetres. Under ideal conditions, stereo video uncertainties can be minimized to millimetres, showcasing the simplicity and practicality of using conventional, budget-friendly video devices for versatile stereo video systems.

ACKNOWLEDGEMENTS

This first has been supported by the Portuguese Foundation for Science and Technology (Fundação para a Ciência e a Tecnologia - FCT), through the grant 2022.10643.BDANA

This work contributes to the Strategic Research Plan of the Centre for Marine Technology and Ocean Engineering (CENTEC), which is financed by the Portuguese Foundation for Science and Technology (Fundação para a Ciência e Tecnologia - FCT) under contract UIDB/UIDP/00134/2020.

REFERENCES

Ardhuin, F. 2021. Ocean waves in geosciences. doi: 10.13140/RG.2.2.16019.78888/7

Benetazzo, A. 2006. Measurements of short water waves using stereo matched image sequences. Coastal Engineering, 53 (12), 1013–1032.

Benetazzo, A., Barbariol, F., Bergamasco, F., Torsello, A., Carniel, S. &Sclavo, M. 2016. Stereo wave imaging from moving vessels: Practical use and applications. Coastal Engineering, 109, 114–127.

Benetazzo, A., Ardhuin, F., Bergamasco, F., Cavaleri, L., Guimarães, P.V., Schwendeman, M., Sclavo, M., Thomson, J. & Torsello, A. 2017. On the shape and likelihood of oceanic rogue waves. Scientific Reports, 7, 8276.

Bergamasco, F., Torsello, A., Sclavo, M., Barbariol, F. & Benetazzo, A. 2017. Wass: An open-source pipeline for 3d stereo reconstruction of ocean waves. Computers & Geosciences, 107, 28–36.

Bergamasco, F., Benetazzo, A., Yoo, J., Torsello A., Barbariol, F., Jin-Yong, J., Jae-Seol, S. & Cavaleri, L. 2021. Toward real-time optical estimation of ocean waves' space-time fields. Computers & Geosciences, 147, 104666, ISSN 0098-3004, http://dx.doi.org/10.1016/j.cageo.2020.104666.

Bitner-Gregersen, E., Waseda, T., Parunov, J., Yim, S., Hirdaris, S., Ma, N. & Guedes Soares, C. 2022. Uncertainties in long-term wave modelling. Marine Structures. 84:103217.

Clarke, T. A. & Fryer, J. G.. 1998. The development of camera calibration methods and models. The Photogrammetric Record, 16, 51–66.

Donelan, M. A., Hamilton, J. & Hui, W. H. 1985. Directional spectra of wind generated waves. Philosophical Transactions of the Royal Society of London. Series A, Mathematical and Physical Sciences, 315 (1534), 509{562.

Guimarães, P.V., Ardhuin, F., Bergamasco, F., Leckler, F., Filipot, J.F., Shim, J.S., Dulov, V. & Benetazzo, A. 2020. A data set of sea surface stereo images to resolve space-time wave fields. Scientific Data, 7, 145.

Guimarães, P.V., Ardhuin, F., Sutherland, P., Accensi, M., Hamon, M., Pérignon, Y., Thomson, J., Benetazzo, A. & Ferrant, P. 2018. A surface kinematics buoy (SKIB) for wave–current interaction studies. Ocean Science, 14, 1449–1460.

Holman, R. & Haller, M. C. 2013. Remote sensing of the nearshore. Annual review of marine science, 5:95–113.

Holthuijsen L. H. 2007. Waves in Oceanic and Coastal Waters. Cambridge: Cambridge University Press. doi:10.1017/CBO9780511618536

Izquierdo, P. & Guedes Soares, C. 2005. Analysis of Sea Waves and Wind from X- Band Radar. Ocean Engineering. 32(11–12):1404–1409.

Jähne, B. 1993. Spatio-Temporal Image Processing: Theory and Scientific Applications, Springer: Berlin/Heidelberg, Germany, Volume 751.

Kabel, T., Georgakis, C.T. & Rod Zeeberg, A. 2019 Mapping ocean waves using new lidar equipment. In Proceedings of the Twenty-ninth (2019) International Ocean and Polar Engineering Conference, pp. 2558–2562. International Society of Offshore and Polar Engineers

Leckler, F., Ardhuin, F., Peureux, C., Benetazzo, A., Bergamasco, F. & Dulov, V. 2015. Analysis and interpretation of frequency-wavenumber spectra of young wind waves. Journal of Physical Oceanography, 45, 2484–2496. doi:10.1175/JPO-D-14-0237.1.

Lucas, C., Boukhanovsky, A. &Guedes Soares, C. 2011. Modelling the Climatic Variability of Directional Wave Spectra. Ocean Engineering. 38(11–12):1283–1290.

Ma, Y., Soatto, S., Kosecká, J. & Sastry, S. 2004. An Invitation to 3-D Vision. Springer.

Pistellato, M., Bergamasco, F., Torsello, A., Barbariol, F., Yoo, J., Jeong, J.-Y. & Benetazzo, A. 2021. A Physics-Driven CNN Model for Real-Time Sea Waves 3D Reconstruction. Remote Sensing. 13, 3780. http://dx.doi.org/10.3390/rs13183780

Rodríguez-Padilla. 2021. Remotely-sensed rip current dynamics and morphological control in high- energy beach environments. Geophysics [physics.geo-ph]. Université de Bordeaux. English. NNT: 2021BORD0218. tel-03436157

Rodriguez, J.J. & Aggarwal, J.K. 1990. Stochastic analysis of stereo quantization error. IEEE Trans. Pattern Anal. Mach. Intell. 12, 467–470.

Sutherland, P. & Melville, W. 2013. Field measurements and scaling of ocean surface wave-breaking statistics. Geophysical Research Letters, 40. doi:10.1002/grl.50584

Vieira, M., Guimarães, P. V., Violante-Carvalho, N., Benetazzo, A., Bergamasco, F. & Pereira, H. 2020. A low-cost stereo video system for measuring directional wind waves. Journal of Marine Science and Engineering, 8 (11). doi:10.3390/jmse8110831

Welch, P. 1967. The use of fast fourier transform for the estimation of power spectra: A method based on time averaging over short, modified periodograms. IEEE Transactions on Audio and Electroacoustics, 15–02, 70–73.

Zappa, C. J., Banner, M. L., Schultz, H., Gemmrich, J. R., P.Morison, R., LeBel, D. A. & Dickey, T. 2012. An overview of sea state conditions and air-sea fluxes during radyo. Journal of Geophysical Research, 117.

Zhang, Z.2000. A Flexible New Technique for Camera Calibration. IEEE Transactions on Pattern Analysis and Machine Intelligence, Vol. 22, pp. 1330–1334.

Author Index